Discussing the diagrams in the book and watching the BioNow videos in class helps me connect the key terms and ideas to each other and to the real world. I find it much easier to remember information if I apply the connections made in class to the overall real life themes provided at the beginning of each chapter.

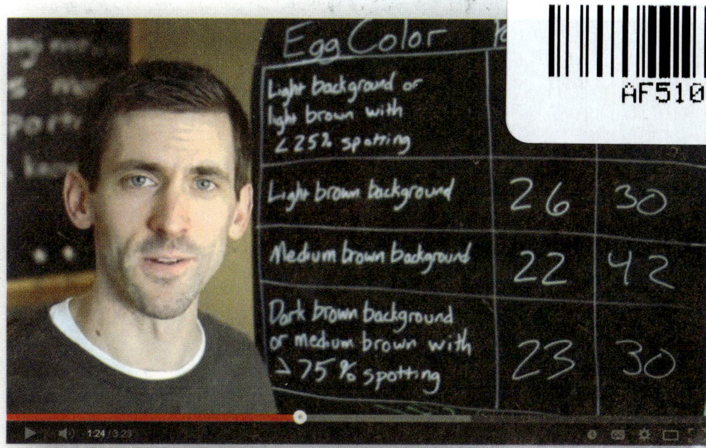

After class, I like to test myself further on concepts and processes to make sure I have grasped everything that the instructor focused on. Connect provides tutorials that allow me to watch videos and study diagrams of processes and then test myself on them.

I love how Connect keeps track of what I am struggling with so when I'm using SmartBook or LearnSmart, I'm constantly brought back to certain topics until all are understood.

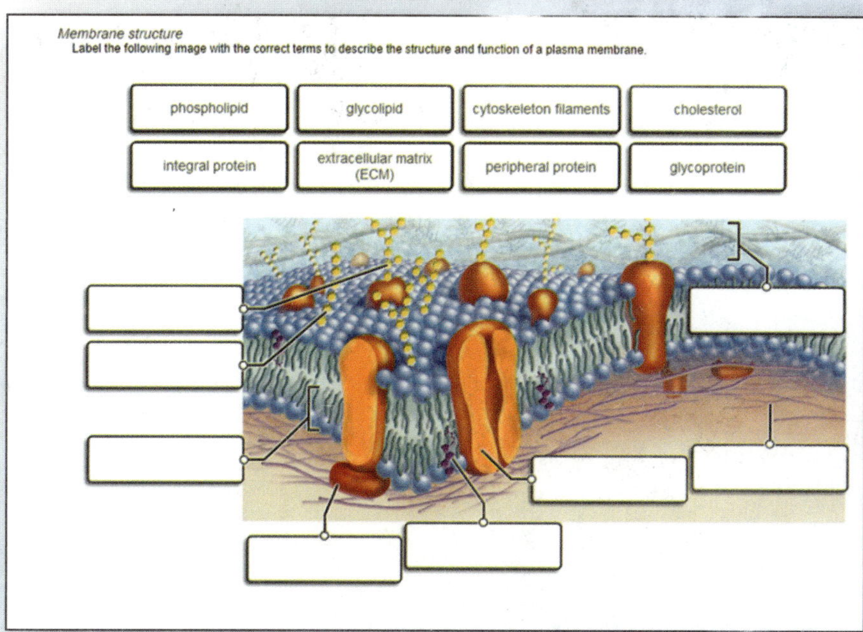

Post-Class
Connect

Apply content and build critical thinking

Before a big exam, I use all of the components of this book together to review important material. I first use LearnSmart as a refresher of underlying concepts and ideas.

Next, I move to SmartBook and reread the highlighted areas that I am really struggling with. Then I take another look at the tutorials to have a fresh image of the processes in my head.

Finally, I read through the themes at the beginning and end of the chapter to remind myself why I am learning the material. Sometimes relating a question on a test to real life helps me answer it.

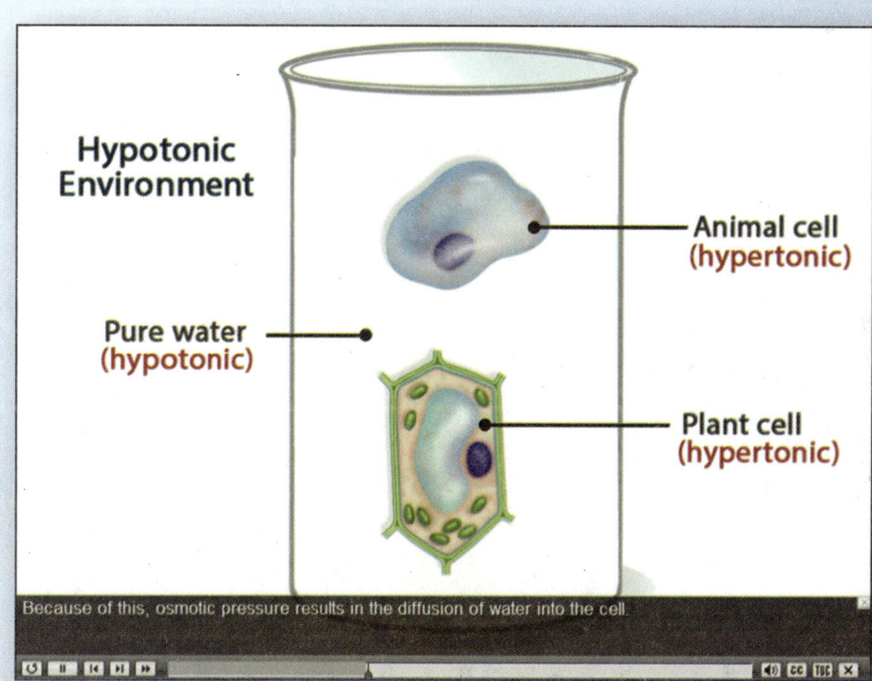

Brief Contents

1 A View of Life 1

UNIT 1 The Cell 18

2 Basic Chemistry 19
3 The Chemistry of Organic Molecules 35
4 Cell Structure and Function 57
5 Membrane Structure and Function 82
6 Metabolism: Energy and Enzymes 100
7 Photosynthesis 114
8 Cellular Respiration 129

UNIT 2 Genetic Basis of Life 146

9 The Cell Cycle and Cellular Reproduction 147
10 Meiosis and Sexual Reproduction 166
11 Mendelian Patterns of Inheritance 186
12 Molecular Biology of the Gene 207
13 Regulation of Gene Expression 228
14 Biotechnology and Genomics 244

UNIT 3 Evolution 260

15 Darwin and Evolution 261
16 How Populations Evolve 279
17 Speciation and Macroevolution 296
18 Origin and History of Life 317
19 Taxonomy, Systematics, and Phylogeny 337

UNIT 4 Microbiology and Evolution 352

20 Viruses, Bacteria, and Archaea 353
21 Protist Evolution and Diversity 373
22 Fungi Evolution and Diversity 395

UNIT 5 Plant Evolution and Biology 410

23 Plant Evolution and Diversity 411
24 Flowering Plants: Structure and Organization 435
25 Flowering Plants: Nutrition and Transport 456
26 Flowering Plants: Control of Growth Responses 476
27 Flowering Plants: Reproduction 495

UNIT 6 Animal Evolution and Diversity 512

28 Invertebrate Evolution 513
29 Vertebrate Evolution 544
30 Human Evolution 564

UNIT 7 Comparative Animal Biology 580

31 Animal Organization and Homeostasis 581
32 Circulation and Cardiovascular Systems 600
33 The Lymphatic and Immune Systems 621
34 Digestive Systems and Nutrition 641
35 Respiratory Systems 659
36 Body Fluid Regulation and Excretory
37 Neurons and Nervous Systems 691
38 Sense Organs 714
39 Locomotion and Support Systems 733
40 Hormones and Endocrine Systems 750
41 Reproductive Systems 770
42 Animal Development 793

UNIT 8 Behavior and Ecology 816

43 Behavioral Ecology 817
44 Population Ecology 836
45 Community and Ecosystem Ecology 855
46 Major Ecosystems of the Biosphere 881
47 Conservation of Biodiversity 905

TWELFTH EDITION

BIOLOGY

Sylvia S. Mader

Michael Windelspecht
Appalachian State University

With contributions by

Jason Carlson
St. Cloud Technical and Community College

David Cox
Lincoln Land Community College

Gretel Guest
Durham Technical Community College

Jeffrey Isaacson
Nebraska Wesleyan University

Mc Graw Hill Education

BIOLOGY, TWELFTH EDITION

Published by McGraw-Hill Education, 2 Penn Plaza, New York, NY 10121.

Some ancillaries, including electronic and print components, may not be available to customers outside the United States.

This book is printed on acid-free paper.

1 2 3 4 5 6 7 8 9 0 DOW/DOW 1 0 9 8 7 6 5

ISBN 978-0-07-802426-9
MHID 0-07-802426-9

Senior Vice President, Products & Markets: *Kurt L. Strand*
Vice President, General Manager, Products & Markets: *Marty Lange*
Vice President, Content Design & Delivery: *Kimberly Meriwether David*
Managing Director: *Michael S. Hackett*
Brand Manager: *Chris Loewenberg*
Director, Product Development: *Rose Koos*
Product Developer: *Anne Winch*
Marketing Manager: *Christine Ho*
Digital Product Developer: *Christine Carlson*
Director, Content Design & Delivery: *Linda Avenarius*
Program Manager: *Angela R. FitzPatrick*
Content Project Managers: *Jayne L. Klein/Sherry L. Kane*
Buyer: *Sandy Ludovissy*
Design: *David W. Hash*
Content Licensing Specialists: *Lori Hancock/Beth Thole*
Cover Image: Lionfish (*Pteropterus radiata*) and Squarespot anthias (*Pseudanthias pleurotaenia*) with soft corals in the ocean ©Panoramic Images/Getty Images
Compositor: *Electronic Publishing Services Inc., NYC*
Typeface: *10/12 Times Lt Std*
Printer: *R. R. Donnelley*

Library of Congress Cataloging-in-Publication Data

Mader, Sylvia S.
Biology / Sylvia S. Mader, Michael Windelspecht, Appalachian State University ; with contributions by April Cognato. — Twelfth edition.
pages cm
I. Windelspecht, Michael, 1963- II. Cognato, April. III. Title.
QH308.2.M23 2016
570--dc23

2014022212

www.mhhe.com

About the Authors

Sylvia S. Mader Sylvia S. Mader has authored several nationally recognized biology texts published by McGraw-Hill. Educated at Bryn Mawr College, Harvard University, Tufts University, and Nova Southeastern University, she holds degrees in both Biology and Education. Over the years she has taught at University of Massachusetts, Lowell; Massachusetts Bay Community College; Suffolk University; and Nathan Mayhew Seminars. Her ability to reach out to science-shy students led to the writing of her first text, *Inquiry into Life,* that is now in its thirteenth edition. Highly acclaimed for her crisp and entertaining writing style, her books have become models for others who write in the field of biology.

Although her writing schedule is always quite demanding, Dr. Mader enjoys taking time to visit and explore the various ecosystems of the biosphere. Her several trips to the Florida Everglades and Caribbean coral reefs resulted in talks she has given to various groups around the country. She has visited the tundra in Alaska, the taiga in the Canadian Rockies, the Sonoran Desert in Arizona, and tropical rain forests in South America and Australia. A photo safari to the Serengeti in Kenya resulted in a number of photographs for her texts. She was thrilled to think of walking in Darwin's steps when she journeyed to the Galápagos Islands with a group of biology educators. Dr. Mader was also a member of a group of biology educators who traveled to China to meet with their Chinese counterparts and exchange ideas about the teaching of modern-day biology.

Michael Windelspecht As an educator, Dr. Windelspecht has taught introductory biology, genetics, and human genetics in the online, traditional, and hybrid environments at community colleges, comprehensive universities, and military institutions. For over a decade he served as the Introductory Biology Coordinator at Appalachian State University where he directed a program that enrolled over 4,500 students annually. He currently serves as an adjunct professor of biology at ASU where he teaches nonmajors biology and human genetics in the online and hybrid formats. He was educated at Michigan State University and the University of South Florida. Dr. Windelspecht is also active in promoting the scientific literacy of secondary school educators. He has led multiple workshops on integrating water quality research into the science curriculum, and has spent several summers teaching Pakistani middle school teachers.

As an author, Dr. Windelspecht has published five reference textbooks, and multiple print and online lab manuals. He served as the series editor for a ten-volume work on the human body. For years Dr. Windelspecht has been active in the development of multimedia resources for the online and hybrid science classrooms. Along with his wife, Sandra, he owns a multimedia production company, Ricochet Creative Productions, which actively develops and assesses new technologies for the science classroom.

Contributors

Jason Carlson is a Biology Instructor at St. Cloud Technical and Community College in Minnesota where he teaches introductory biology, microbiology, nutrition, and human biology. Before entering higher education, he was a middle and high school science teacher with education from the University of Idaho, Bemidji State University, and St. Cloud State University. In the classroom, he supports a student-driven applied curriculum with relevant and hands-on research and investigation.

Dave Cox serves as Associate Professor of Biology at Lincoln Land Community College, in Springfield, Illinois. He was educated at Illinois College and Western Illinois University. As an educator, Professor Cox teaches introductory biology for nonmajors in the traditional classroom format as well as in a hybrid format. He also teaches biology for majors, and marine biology and biological field studies as study-abroad courses in Belize. He is the co-owner of Howler Publications, a company that specializes in scientific study abroad courses. Professor Cox served as a contributor to the fourteenth edition of *Inquiry* and the thirteenth edition of *Human Biology*.

Gretel Guest is a Professor of Biology at Durham Technical Community College, in Durham, North Carolina. She has been teaching nonmajors and majors Biology, Microbiology, and Genetics for more than 15 years. Dr. Guest was educated in the field of botany at the University of Florida, and received her Ph.D. in Plant Sciences from the University of Georgia. She is also a Visiting Scholar at Duke University's Graduate School. There she serves the Preparing Future Faculty program by mentoring post-doctoral and graduate students interested in teaching careers. Dr. Guest was a contributor to the fourth edition of *Essentials of Biology*.

Jeffrey Isaacson is an Associate Professor of Biology at Nebraska Wesleyan University, where he teaches courses in microbiology, immunology, pathophysiology, infectious disease, and senior research. He also serves as the Assistant Provost for Integrative and Experiential Learning. Dr. Isaacson was educated at Nebraska Wesleyan, Kansas State College of Veterinary Medicine, and Iowa State University. He worked as a small-animal veterinarian in Nevada and California, and completed a post-doctoral fellowship in the Department of Immunology at the Mayo Clinic in Minnesota. Dr. Isaacson has been a significant contributor and coauthor for three editions of *Inquiry Into Life,* for the eleventh edition of *Biology,* and is a frequent contributor to McGraw-Hill's LearnSmart adaptive learning program for several textbooks.

Preface

Goals of the Twelfth Edition

The mission of Dr. Sylvia Mader's text, *Biology,* has always been to give students an understanding of biological concepts and a working knowledge of the scientific process. However, like the world around us, the process of teaching science is changing rapidly. Increasingly, instructors are being asked to engage their students by making content more relevant, while still providing students with a firm foundation in those core principles on which biology is founded. These changes are clearly outlined in the AAAS/NSF report, *Vision and Change in Undergraduate Biology Education* (2009). The eleventh edition of *Biology* was one of the first texts to address the principles of *Vision and Change* by integrating themes within the text. In this edition we expand on that effort with the development of a number of new resources and processes.

In addition to the evolution of the introductory biology curriculum, students and instructors are increasingly requesting digital resources to utilize as learning resources. McGraw-Hill Education has long been an innovator in the development of digital resources, and the *Biology* text, and its authors, are at the forefront of the integration of these technologies into the science classroom.

In this edition, the authors focused on the following areas:

1. utilization of the data from the LearnSmart adaptive learning platforms to identify content areas within the text that students demonstrated difficulty in mastering,
2. further development of the themes that connect the content of the text across multiple chapters,
3. development of a new series of videos and websites to introduce relevancy and engage students in the content,
4. refinement of digital assets to provide a more effective assessment of learning outcomes to enable instructors in the flipped, online, and hybrid teaching environments.

Relevancy

The use of real world examples to demonstrate the importance of biology in the lives of students is a key component of *Vision and Change* and an effective teaching strategy for introductory biology. The development of relevancy-based resources is a major focus for the authors of the Mader series of texts. Some examples include:

- A series of new chapter openers to introduce relevancy to the chapter. The authors chose topics that would be of interest to a nonscience major, and represent what would typically be found on a major news source.
- The development of new relevancy-based videos, BioNow Sessions, that offer relevant, applied classroom resources to allow students to feel that they can actually do and learn biology themselves. For more on these, see page ix.

- A new website, RicochetScience.com that provides updates on news and stories that are interesting to nonscience majors. The Biology101 project links these resources to the major topics of the text. The site also features videos to assist the students in recognizing the relevancy of what they are learning in the classroom.

Ricochet Science™

The Vision and Change document clearly identifies the need to integrate core concepts throughout the curriculum. We recognize that scientific literacy is not based upon the memorization of a series of facts. Instead, learning is based on establishing associations and links between what, at first glance, appear to be diverse topics. The main themes we have chosen to emphasize include:

- Evolution
- Nature of Science
- Biological Systems

These themes are integrated into all aspects of the textbook, from the unit learning outcomes to the theme-based feature readings in the text. At the start of each chapter, "Following the Themes" introduces the relationship of the chapter's content to each of the themes. At the end of each chapter, "Connecting the Concepts with the Themes" not only reminds the student of the relationships between chapter content and the three core themes, but also acts as a prelude to topics in the next few chapters of the text. In essence, the themes act as the threads that unite the concepts throughout the text, enabling the student to see relationships from the molecular to ecosystem levels of biology.

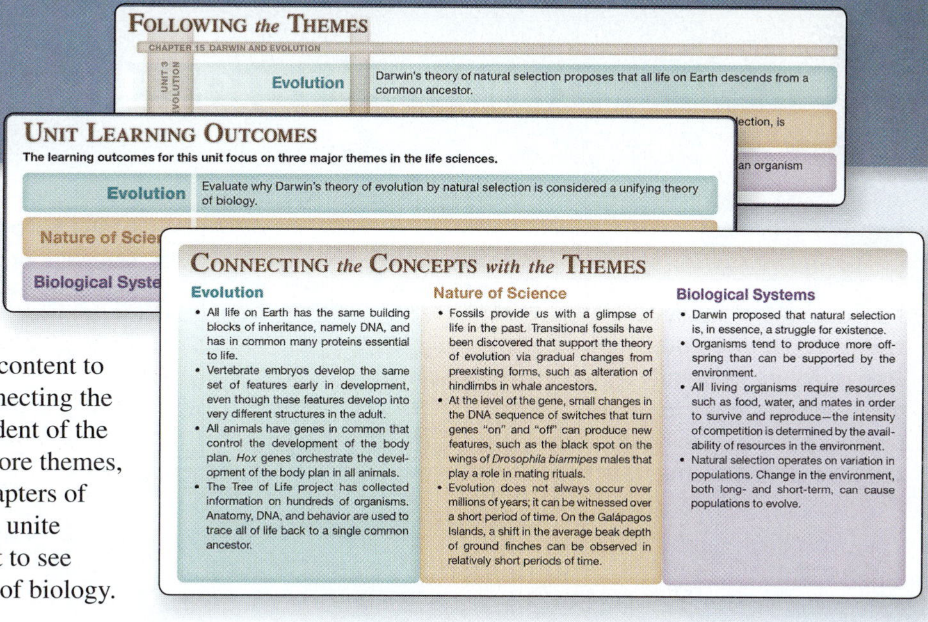
FOLLOWING the THEMES

CHAPTER 15 DARWIN AND EVOLUTION

UNIT 3 EVOLUTION

Evolution	Darwin's theory of natural selection proposes that all life on Earth descends from a common ancestor.

UNIT LEARNING OUTCOMES

The learning outcomes for this unit focus on three major themes in the life sciences.

Evolution	Evaluate why Darwin's theory of evolution by natural selection is considered a unifying theory of biology.
Nature of Scie…	
Biological Syste…	

CONNECTING the CONCEPTS with the THEMES

Evolution

- All life on Earth has the same building blocks of inheritance, namely DNA, and has in common many proteins essential to life.
- Vertebrate embryos develop the same set of features early in development, even though these features develop into very different structures in the adult.
- All animals have genes in common that control the development of the body plan. *Hox* genes orchestrate the development of the body plan in all animals.
- The Tree of Life project has collected information on hundreds of organisms. Anatomy, DNA, and behavior are used to trace all of life back to a single common ancestor.

Nature of Science

- Fossils provide us with a glimpse of life in the past. Transitional fossils have been discovered that support the theory of evolution via gradual changes from preexisting forms, such as alteration of hindlimbs in whale ancestors.
- At the level of the gene, small changes in the DNA sequence of switches that turn genes "on" and "off" can produce new features, such as the black spot on the wings of *Drosophila biarmipes* males that play a role in mating rituals.
- Evolution does not always occur over millions of years; it can be witnessed over a short period of time. On the Galápagos Islands, a shift in the average beak depth of ground finches can be observed in relatively short periods of time.

Biological Systems

- Darwin proposed that natural selection is, in essence, a struggle for existence.
- Organisms tend to produce more offspring than can be supported by the environment.
- All living organisms require resources such as food, water, and mates in order to survive and reproduce—the intensity of competition is determined by the availability of resources in the environment.
- Natural selection operates on variation in populations. Change in the environment, both long- and short-term, can cause populations to evolve.

Evolution Theme

Evolutionary change, along with the mechanism of natural selection, represents the unifying concept of the biological sciences. In essence, biological evolution is the thread that links all life together. Throughout this textbook, feature readings on this theme both demonstrate the process of evolution and illustrate how scientists study and measure evolutionary change. By following this theme through the book, students develop a better understanding of why evolution is a dynamic process, and one that has shaped, and will continue to influence, life on this planet.

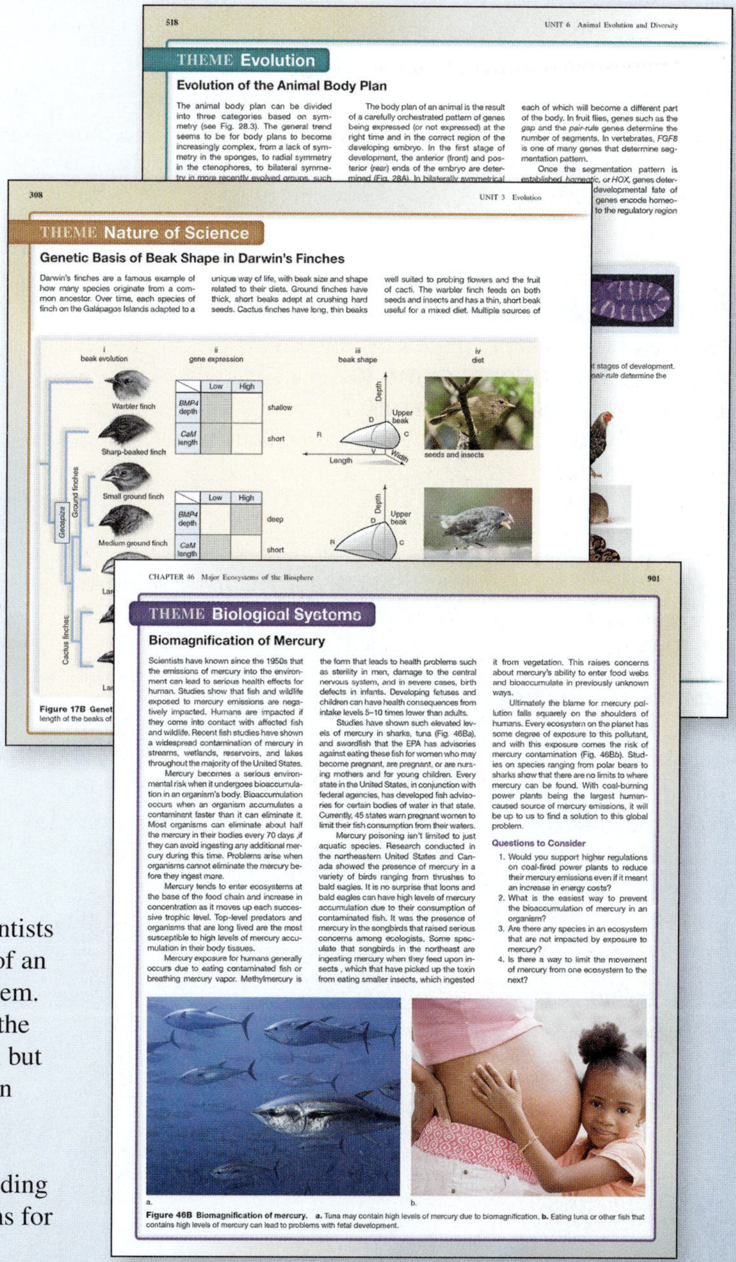
THEME Evolution

Evolution of the Animal Body Plan

THEME Nature of Science

Genetic Basis of Beak Shape in Darwin's Finches

THEME Biological Systems

Biomagnification of Mercury

Figure 46B Biomagnification of mercury. a. Tuna may contain high levels of mercury due to biomagnification. b. Eating tuna or other fish that contains high levels of mercury can lead to problems with fetal development.

Nature of Science Theme

Through the processes of observation, the application of the scientific method, and the use of both inductive and deductive reasoning, scientists study life. To develop a deeper understanding of the biological sciences, students must appreciate that the study of life is a process, and that this process has application in their everyday lives. This theme focuses not only on how biologists do science, but also on the influences scientific inquiry has on our understanding of our world.

Biological Systems Theme

From cells to ecosystems, all life is interconnected. Increasingly, scientists are becoming aware that small changes in the chemical composition of an ecosystem can have a tremendous influence on the life in that ecosystem. This theme was chosen to provide a holistic approach to the study of the life sciences, by demonstrating not only that all life is interconnected, but also that the principles regulating life at the cellular level play a role in physiology and ecosystem biology as well.

Assessment To help instructors assess their student's understanding of these core concepts, we have designed a series of Connect questions for each theme in each unit of the text .

Author's Guide to Using the Textbook

I encourage my students to use the **Before You Begin** feature to identify concepts they need to review before beginning to read the chapter content.

Before You Begin

Before beginning this chapter, take a few moments to review the following discussions.

Figure 6.1 How does energy flow in biological systems?

Section 6.3 What role do enzymes play in regulating metabolic processes?

Section 6.4 How are redox reactions and membranes used to conduct cellular work?

I use **LearnSmart Labs** to encourage critical thinking, teach scientific processes, and to integrate lab activities into the classroom environment.

ENGAGE

LEARNSMART LABS®

The following LearnSmart Labs contain exercises that are related to the content of this chapter:

- Photosynthesis

Pre-Class

Built-in Preview and Review Tools

Students come to class prepared

During Class

Discussion Questions and Activities

Flip class and engage students

Learning Outcomes at the start of each section provide a preview of the content to come, while the **Check Your Progress** feature at the end of the section helps my students assess how well they understood the material. The learning outcomes are the same ones used in **Connect,** so I can easily assign a quiz to assess which topics I need to clarify during class.

7.2 The Process of Photosynthesis

Learning Outcomes

Upon completion of this section, you should be able to

1. Describe the overall process of photosynthesis.
2. Compare ene
3. Compare carb reaction.

Check Your Progress 7.2

1. Explain how redox reactions are used in photosynthesis.
2. Describe the role of enzymes during photosynthesis.

THINKING SCIENTIFICALLY

1. In the image below, the alga is the autotroph and undergoes photosynthesis, whereas the bacteria (small, black dots) are the heterotrophs. Explain why the bacteria are clustered at the far ends after the alga was exposed to different color wavelengths of light.

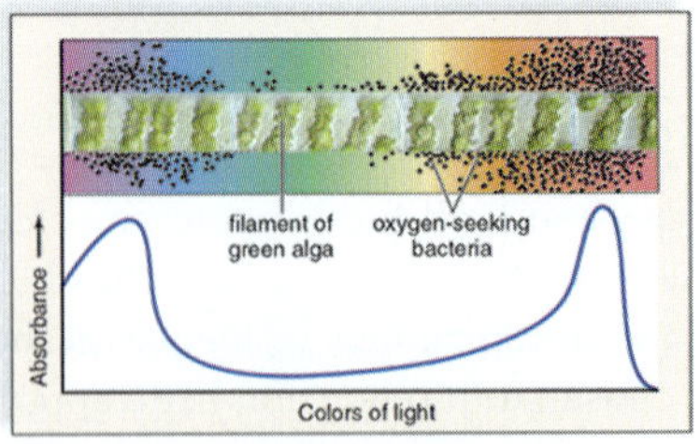

2. A Belgian doctor, Jan Baptista van Helmont (1580–1644) planted a small willow tree in a pot of soil. He weighed the tree and the soil. The tree was watered for 5 years and weighed 74.4 kg more than when he began, and the soil lost 57 g of mass. Explain what accounts for the plant's increase in biomass.

I use the **Questions to Consider** at the end of each reading and **Thinking Scientifically** questions at the end of the chapter as the basis for class discussions and active learning exercises.

MEDIA STUDY TOOLS http://connect.mheducation.com

SMARTBOOK® Study smart. SmartBook helps you maximize your study time by identifying what you know and don't know. The recharge feature in SmartBook helps you prepare for the test by identifying the material that you are most likely to forget.

connect BIOLOGY Find even more resources to learn the chapter concepts including animations, tutorial videos, and interactive practice questions.

Animations	3D Animations	Tutorials
7.3 ATP Production in the Electron Transport Chain • Proton Pump	**7.1** Photosynthesis: Structure of a Chloroplast **7.3** Photosynthesis: Properties of Light • Photosynthesis: Light-Dependent Reactions **7.4** Photosynthesis: Calvin Cycle	**7.3** Noncyclic Photosynthesis **7.4** Calvin Cycle

Media Study Tools includes a table that shows students the animations, videos, and multimedia assets that are available to further explain difficult topics. These may be used as tutorials for the students, and I may assign the accompanying Connect activities to gauge whether my students understand the content.

Post-Class

Assessment and Integration

Apply content and build critical thinking

Traditional end-of-chapter summaries and review questions provide students with an opportunity for low-stakes assessment of their comprehension of the chapter's topics.

Features like **Following the Themes** and **Connecting the Concepts with the Themes** help them understand how the main concepts of the chapter relate to each other, building a deeper understanding of the content.

CONNECTING *the* CONCEPTS *with the* THEMES

Evolution

- Plants have evolved to capture solar energy and store it in carbon-based organic nutrients. These are passed on to organisms that have evolved to feed on plants, in turn to organisms that have evolved to feed on the plant-eaters, and so on in a food web.
- Plants are called autotrophs because they make their own organic food. Heterotrophs are organisms that take in organic food made by other organisms.

Nature of Science

- The amount of carbon dioxide in the atmosphere is increasing steadily, in part because of burning fossil fuels. This buildup of carbon dioxide causes global climate change, and we can use the tools of science to understand the scope of this change.
- Scientists are learning how to manipulate photosynthesis to create commercially important products such as oils and turpentines.

Biological Systems

- Autotrophs take in carbon dioxide when they photosynthesize. Carbon dioxide is returned to the atmosphere when autotrophs and heterotrophs carry on cellular respiration. In this way, carbon atoms cycle through living organisms.
- Energy does not cycle; therefore, all life is dependent on the ability of plants to capture solar energy and produce carbohydrate molecules from CO_2.

Author's Guide to the Digital Classroom

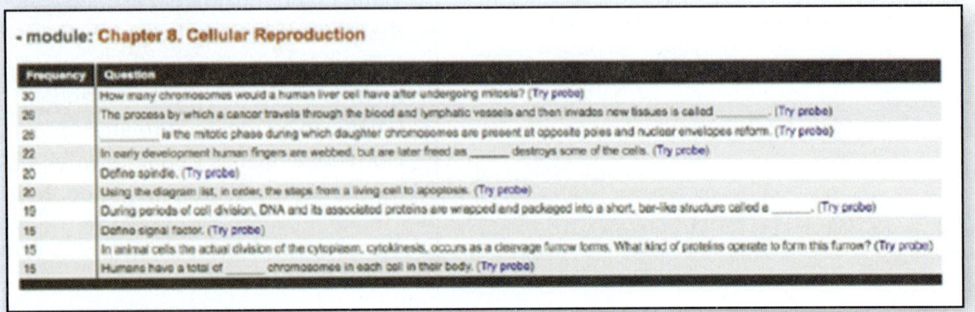

- module: Chapter 8. Cellular Reproduction

Frequency	Question
30	How many chromosomes would a human liver cell have after undergoing mitosis? (Try probe)
26	The process by which a cancer travels through the blood and lymphatic vessels and then invades new tissues is called ________. (Try probe)
26	________ is the mitotic phase during which daughter chromosomes are present at opposite poles and nuclear envelopes reform. (Try probe)
22	In early development human fingers are webbed, but are later freed as ______ destroys some of the cells. (Try probe)
20	Define spindle. (Try probe)
20	Using the diagram list, in order, the steps from a living cell to apoptosis. (Try probe)
19	During periods of cell division, DNA and its associated proteins are wrapped and packaged into a short, bar-like structure called a ______. (Try probe)
15	Define signal factor. (Try probe)
15	In animal cells the actual division of the cytoplasm, cytokinesis, occurs as a cleavage furrow forms. What kind of proteins operate to form this furrow? (Try probe)
15	Humans have a total of ______ chromosomes in each cell in their body. (Try probe)

Using reports from within the LearnSmart system, especially the Most Missed Questions report, I am able to identify areas of content that my students are struggling with before they enter the lecture.

Pre-Class

LearnSmart & SmartBook

Students come to class prepared

During Class

Instructor Resources

Flip class and engage students

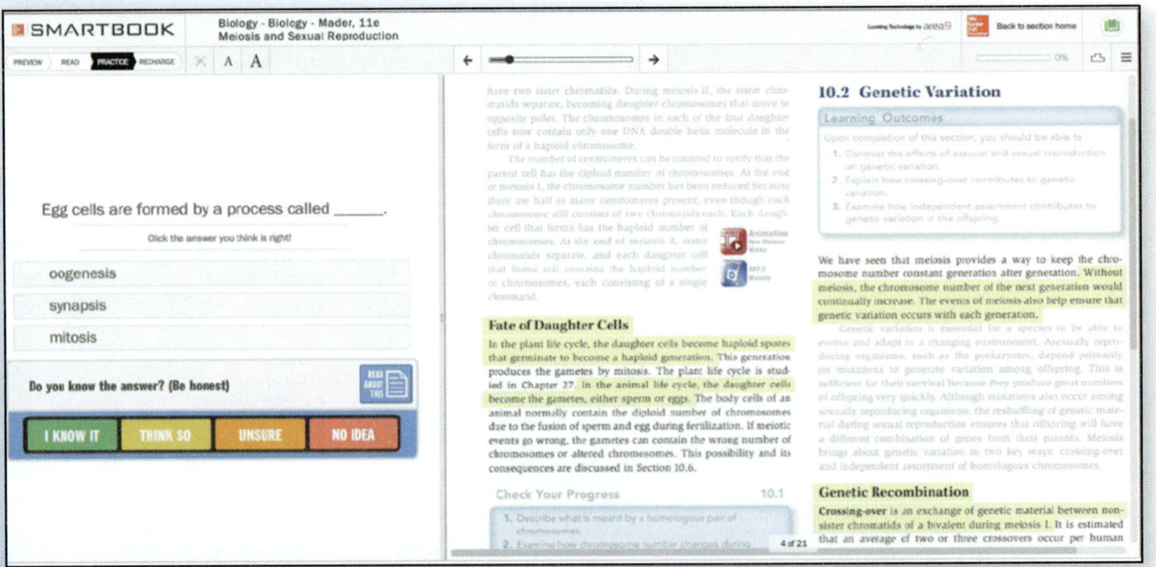

I generally assign 20-30 minutes of SmartBook 3-5 days before class. The assignments cover only the core topics for the upcoming lesson.

During class I can focus on engaging the students with the relevancy of the content using the BioNow Sessions videos, active learning exercises, and animations. Tegrity lecture capture lets my students review these concepts later.

Post-Class

Connect

Apply content and build critical thinking

Using feedback from the LearnSmart reports, I am able to design Connect assignments that act as tutorials that target the concepts my students are struggling with.

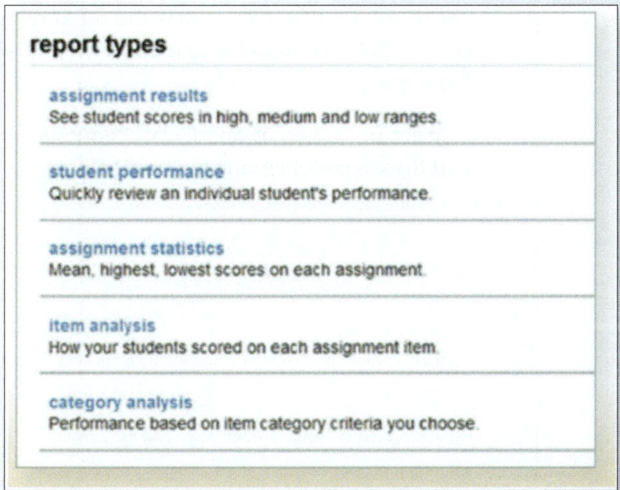

The Connect reports allow me to assess whether my students have met the learning objectives.

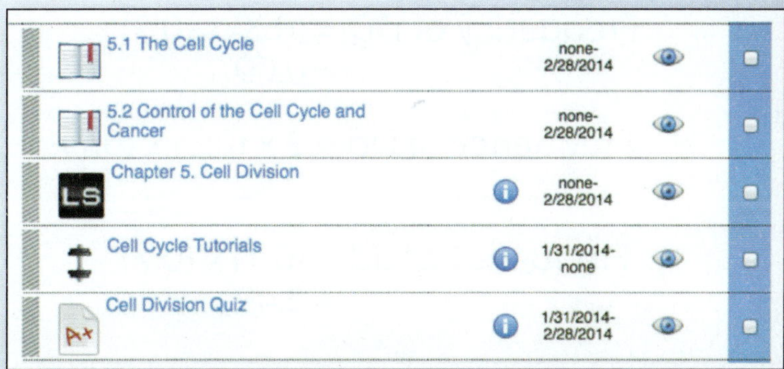

The quizzing option within Connect allows me to develop assessments for any classroom environment.

Engaging Your Students

Today's science classroom relies heavily on the use of digital assets, including animations and videos, to engage students and reinforce difficult concepts. Biology 12e includes two resources specifically designed for the introductory science class to help you achieve these goals.

BioNow Sessions Videos

A relevant, applied approach allows your students to feel they can actually do and learn biology themselves. While tying directly to the content of your course, the videos help students relate their daily lives to the biology you teach and then connect what they learn back to their lives.

Each video provides an engaging and entertaining story about applying the science of biology to a real situation or problem. Attention is taken to use tools and techniques that any regular person could perform, so your students see the science as something they could do and understand.

A 27-video series narrated and produced by author Jason Carlson

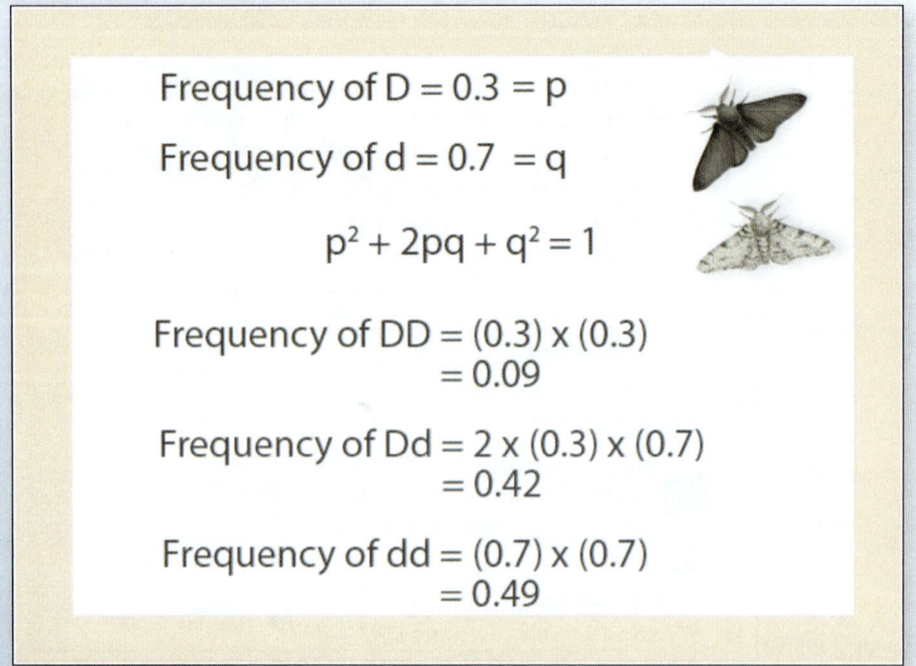

A 36-animation series narrated by Michael Windelspecht and produced by Ricochet Creative Productions, LLC

Our new tutorials were prepared to assist students in understanding some of the more difficult topics in biology. Each of the videos explores a specific figure in the text.

For students, these act as informal office hours, where they can review the most difficult concepts in the chapter at a pace which helps them learn.

Instructors of hybrid and flipped courses will find these useful as online supplements.

Readings

THEME Evolution

Metagenomics 256
The Anatomy of Speciation 299
Carboniferous Forests 421
Evolutionary History of Maize 427
Survival Mechanisms of Plants 438
The Chemical Ecology of Plants 484
Plants and Their Pollinators 500
Evolution of the Animal Body Plan 518
Sexual Selection in Male Bowerbirds 830
Interactions and Coevolution 865

THEME Biological Systems

The Impact of Acid Deposition 31
How Cells Talk to One Another 87
Tropical Rain Forest Destruction and Climate Change 122
Moving Genes Between Species: Green Fluorescent Protein and Cells 219
Same but Not the Same—the Role of Epigenetics 237
African Sleeping Sickness 388
Pathogenic Protists and Climate Change 386
Deadly Fungi 404
The Concept of Water Potential 467
Would You Eat Insects? 538
UV Rays: Too Much Exposure or Too Little? 593
AIDS and Opportunistic Infections 633
Drugs of Abuse 700
Preventing Transmission of STDs 787
Preventing and Testing for Birth Defects 805
When a Population Grows Too Large 848
Biomagnification of Mercury 901

THEME Nature of Science

Saturated and Trans Fats in Foods 44
Microscopy Today 60
Enzyme Inhibitors Can Spell Death 107
Fermentation and Food Production 135
The G1 Checkpoint 150
Reproductive and Therapeutic Cloning 157
Meiosis and the Parthenogenic Lizards 171
Hemophilia and the Royal Families of Europe 203
Testing for Genetic Disorders 254
The Tree of Life: 150 Years of Support for the Theory of Evolution by Natural Selection 272
Inbreeding in Populations 290
Genetic Basis of Beak Shape in Darwin's Finches 308
DNA Barcoding of Life 342
Flu Viruses 359
DIY Bio 362
Bryophytes—Frozen in Time 418
The Many Uses of Bamboo 446
Plants Can Be Used for Cleaning and Discovery of Minerals 462
Why So Many Scientists Work with *Arabidopsis* 488
Vertebrates and Human Medicine 555
A Genomic Comparison of *Homo sapiens* and Chimpanzees 568
Some Major Questions Remaining to Be Answered About Human Evolution 576
Regenerative Medicine 588
Recent Findings About Preventing Cardiovascular Disease 612
How Horseshoe Crabs Save Human Lives 617
Cancer Vaccines: Becoming a Reality 635
Should You Go Gluten-Free? 650
New Approaches to Treating Obesity 655
Artificial Lung Technology 671
Is "Vaping" Safer Than Smoking? 673
The Misuse of Erythropoietin in Sports 686
An Accidental Experimental Model for Parkinson Disease 706
Artificial Retinas Come into Focus 723
The Accidental Discovery of Botox 744
Identifying Insulin as a Chemical Messenger 764
Preimplantation Genetic Diagnosis 784
Do Animals Have Emotions? 825
Island Biogeography Pertains to Biodiversity 860
Global Climate Change 877
Wildlife Conservation and DNA 887
Overexploitation of Asian Turtles 915
Emiquon Floodplain Restoration 919

Overview of Content Changes to *Biology,* Twelfth Edition

Chapter 1: A View of Life has been reorganized to provide a briefer overview of biology as a science. The content on the scientific process (section 1.3) has been reworked with new examples and a new section (1.4) has been added that explores some of the major challenges facing science.

Unit 1: The Cell

Chapter 2: Basic Chemistry starts with new content on the search for life on Mars. **Chapter 3: The Chemistry of Organic Molecules** opens with a look at trans fats in common foods. **Chapter 5: Membrane Structure and Function** begins with a new opening article on chili peppers and calcium channels. **Chapter 6: Metabolism: Energy and Enzymes** includes new material on the function of ATP in cells. The content on redox reactions now focuses more on the processes of photosynthesis and cellular respiration. **Chapter 7: Photosynthesis** begins with new content on biofuels.

Unit 2: Genetic Basis of Life

Chapter 9: The Cell Cycle and Cellular Reproduction now contains information on the structure of a eukaryotic chromosome (section 9.2). **Chapter 10: Meiosis and Sexual Reproduction** starts with a new essay on the importance of meiosis and a new featured reading, "Meiosis and the Parthenogenic Lizards." The figure comparing meiosis I and II (Fig. 10.5) has been reworked to provide an easier comparison of the two processes. **Chapter 11: Mendelian Patterns of Inheritance** begins with a new essay on PKU sensitivity. A new featured reading examines hemophilia and the royal families of Europe. The content on polygenic inheritance now contains references to the genetics of skin coloring (Fig. 11.17). **Chapter 12: Molecular Biology of the Gene** has a new chapter opener that explains the genetic basis of skin, hair, and eye coloration. The content on the eukaryotic chromosome has been moved to chapter 9. There is a new illustration on semiconservative replication (Fig 12.6) and a new featured reading that examines GFP protein. **Chapter 13: Regulation of Gene Expression** starts with new content on how gene regulation relates to the physiology and behavior of primates. A new featured reading, "Same but Not the Same—the Role of Epigenetics," has been included on epigenetic inheritance. **Chapter 14: Biotechnology and Genomics** opens with a new essay on how biotechnology is being used to treat dental disease. New illustrations on the PCR reaction (Fig. 14.3) and the nature of transposons (Fig. 14.10) are included.

Unit 3: Evolution

Chapter 16: How Populations Evolve opens with an essay on MRSA evolution. **Chapter 17: Speciation and Macroevolution** contains an updated illustration on allopatric speciation (Fig. 17.8).

Unit 4: Microbiology and Evolution

Chapter 20: Viruses, Bacteria, and Archaea contains a new featured reading, "DIY Bio," that examines synthetic biology. A new illustration (Fig. 20.8) on gram staining is included. **Chapter 21: Protist Evolution and Diversity** has been restructured to give more emphasis on the supergroup classification system. The chapter begins with a new opener on *Naegleria fowleri,* and contains a new featured reading on pathogenic protists and climate change. The evolutionary relationships in **Chapter 22: Fungi Evolution and Diversity** now includes the microsporidia (Fig. 22.1).

Unit 5: Plant Evolution and Biology

Chapter 23: Plant Evolution and Diversity contains a new featured reading, "Bryophytes—Frozen in Time." **Chapter 24: Flowering Plants: Structure and Organization** begins with new content on the importance of the neem tree. The chapter has been reorganized to start with content on plant cells and tissues before exploring organ systems. **Chapter 25: Flowering Plants: Nutrition and Transport** now contains information on hydroponics (Fig. 25.2) and effects of nutrient deficiencies on plants (Fig. 25.3). Table 26.1 in **Chapter 26: Flowering Plants: Control of Growth Responses** now contains the chemical structures of the plant hormones. A new featured reading explores coevolution and the chemical ecology of plants. **Chapter 27: Flowering Plants: Reproduction** begins with a new essay on the economic importance of flowers.

Unit 6: Animal Evolution and Diversity

Chapter 28: Invertebrate Evolution begins with new content that examines the importance of colony-collapse disorder in honeybees. A new featured reading, "Would You Eat Insects?," discusses the potential benefits of using insects as a food source. **Chapter 30: Human Evolution** contains a new featured reading that explores some of the remaining questions on human evolution.

Unit 7: Comparative Animal Biology

Chapter 31: Animal Organization and Homeostasis begins with an essay that examines the importance of homeostasis for astronauts. **Chapter 32: Circulation and Cardiovascular Systems** opens with material on cardiovascular-related diseases and the NFL. **Chapter 33: The Lymphatic and Immune Systems** has a new opener on foods and anaphylactic shock. **Chapter 34: Digestive Systems and Nutrition** contains a new featured reading on gluten-free diets. **Chapter 35: Respiratory Systems** has a new figure (Fig. 35.8) explaining the relationship between air pressure and volume of a container. The chapter also contains a new featured reading on the health aspects of using e-cigarettes. **Chapter 37: Neurons and Nervous Systems** begins with new content on Parkinson disease. **Chapter 39: Locomotion and Support Systems** starts with a new essay on Olympian Gabby Douglas. **Chapter 41: Reproductive Systems** has a new opener that explores variations between the sexes in the animal kingdom.

Unit 8: Behavior and Ecology

Chapter 43: Behavioral Ecology starts with content on behavior and communication in honeybees. The featured reading on animals and emotions has been updated to include recent developments. **Chapter 44: Population Ecology** contains a new illustration on the environmental impact of developed countries. The predator-prey relationships and content on global climate change in **Chapter 45: Community and Ecosystem Ecology** has been updated to include more recent data. **Chapter 47: Conservation of Biodiversity** now begins with an essay on the impact of invasive species.

The Next Generation of Textbook Reviews: Heat Map Technology

The twelfth edition of *Biology* is the first textbook in the Mader series which utilized the data derived from the LearnSmart platform as a form of review,

The premise is very straightforward. Students don't know what they don't know—but LearnSmart does. By compiling data from all of the probes answered by all of the students, and then overlaying that data on the text, we are able to visualize areas of content where the students are having problems.

The authors were able to use this information to not only identify areas of the text that the students were having problems with, but also areas that needed additional digital resources, such as tutorials and new Connect questions.

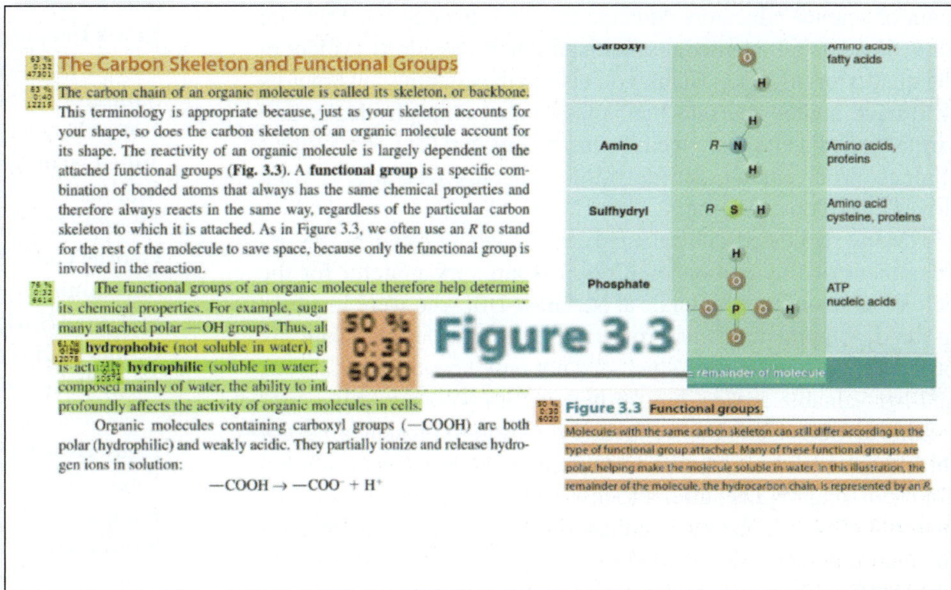

Acknowledgments

Dr. Sylvia Mader represents one of the icons of science education. Her dedication to her students, coupled to her clear, concise writing style, has benefited the education of thousands of students over the past four decades. As an educator, it is an honor to continue her legacy, and to bring her message to the next generation of students.

As always, I had the privilege to work with the phenomenal team of science educators and coauthors on this edition. They are all dedicated and talented teachers, and their passion is evident in the quality of this text. Thank you also to the countless instructors who have invited me into their classrooms, both physically and virtually, to discuss their needs as instructors and the needs of their students. Your energy, and devotion to quality teaching, is what drives a textbook revision.

Many dedicated and talented individuals assisted in the development of this edition of *Biology.* I am very grateful for the help of so many professionals at McGraw-Hill who were involved in the development of this project. In particular, let me thank my product developer, Anne Winch, for not only keeping me on track and her valuable advice, but for her endless patience. My editor for this text was Chris Loewenberg. From start to finish a project of this magnitude can take over 18 months, and Chris has the natural ability of keeping his authors focused and in reminding me of the importance we are making in education. Thanks also to my marketing manager, Chris Ho, who offers a unique insight on the needs of our students. No modern team would be complete without digital support, and for that I thank Eric Weber and Christine Carlson.

Production of this text was directed by Angela Fitzpatrick and Jayne Klein, who faithfully steered this project through the publication process. I was very lucky to have Dawnelle Krouse, Deb Debord, and Rose Kramer as proofreaders and copy editors. Today's textbooks are visual productions, and so I need to thank the creative talents of David Hash. Lori Hancock and Evelyn Jo Johnson did a superb job of finding just the right photographs and micrographs. Electronic Publishing Services produced this textbook, emphasizing pedagogy and beauty to arrive at the best presentation on the page.

Who I am, as an educator and an author, is a direct reflection of what I have learned from my students. Education is a mutualistic relationship, and it is my honest opinion that while I am a teacher, both my professional and personal life have been enriched by interactions with my students. They have encouraged me to learn more, teach better, and never stop questioning the world around me.

Last, but never least, I want to acknowledge my wife, Sandra. You have never wavered in your support of my projects. Devin and Kayla, your natural curiosity of the world we live in gives me the energy to want to make the world a better place.

Michael Windelspecht
Blowing Rock, NC

Twelfth Edition Reviewers

LaQuetta Anderson, *Grambling State University*
Isaac Barjis, *City University of New York*
Gladys Bolding, *Georgia Perimeter College*
Bertha M. Byrd, *Wayne County Community College District*
Sarah Clark, *Howard Community College*
Lewis Deaton, *University of Louisiana at Lafayette*
Angela Edwards, *Trident Technical College*
Salman Elawad, *Chattahoochee Valley Community College*
Victor Fet, *Marshall University*
Julie Fischer, *Wallace Community College*
Monica Frazier, *Columbus State University*
Melanie Glasscock, *Wallace State Community College*
George Goff, *Wayne County Community College District*
Shashuna J. Gray, *Germanna Community College*
Sylvester Hackworth, *Bishop State Community College*
Cameron Harmon, *Fayetteville Technical Community College*
Zinat Hassanpour, *Cabarrus College of Health Sciences/Rowan Cabarrus Community College*
Holly Hereau, *Macomb Community College*
Dagne Hill, *Grambling State University*
Kimberly Brown, *Mississippi Gulf Coast Community College*
Ryan Lazik, *Pacific College of Oriental Medicine*
Lynne Lohmeier, *Mississippi Gulf Coast Community College*
Geralyne Lopez-de-Victoria, *Midlands Technical College*
Tiffany McFalls-Smith, *Elizabethtown Community and Technical College*
Christian Nwamba, *Wayne County Community College District*
Tom Reeves, *Midlands Technical College*
Lyndell Robinson, *Lincoln Land Community College*
William Simcik, *Lone Star College-Tomball*
Viji Sitther, *Morgan State University*
Phillip Snider, *Gadsden State Community College*
Kimberly Sonanstine, *Wallace Community College*
Chris Sorenson, *St. Cloud Technical and Community College*
Salvatore A. Sparace, *Clemson University*
Marinko Sremac, *Mount Wachusett Community College*
Todd Tolar, *Wallace Community College*
Frances Turner, *Howard Community College*
Alanna M. Tynes, *Lone Star College-Tomball*
Amale Wardani, *Lincoln Land Community College*

Contents

Preface iv

1 A View of Life 1
- 1.1 The Characteristics of Life 2
- 1.2 Evolution and the Classification of Life 6
- 1.3 The Process of Science 9
- 1.4 Challenges Facing Science 13

UNIT 1 The Cell 18

2 Basic Chemistry 19
- 2.1 Chemical Elements 20
- 2.2 Molecules and Compounds 24
- 2.3 Chemistry of Water 26
- 2.4 Acids and Bases 30

3 The Chemistry of Organic Molecules 35
- 3.1 Organic Molecules 36
- 3.2 Carbohydrates 39
- 3.3 Lipids 42
- 3.4 Proteins 46
- 3.5 Nucleic Acids 50

4 Cell Structure and Function 57
- 4.1 Cellular Level of Organization 58
- 4.2 Prokaryotic Cells 62
- 4.3 Introduction to Eukaryotic Cells 64
- 4.4 The Nucleus and Ribosomes 67
- 4.5 The Endomembrane System 69
- 4.6 Microbodies and Vacuoles 72
- 4.7 The Energy-Related Organelles 73
- 4.8 The Cytoskeleton 75

5 Membrane Structure and Function 82
- 5.1 Plasma Membrane Structure and Function 83
- 5.2 Passive Transport Across a Membrane 88
- 5.3 Active Transport Across a Membrane 91
- 5.4 Modification of Cell Surfaces 95

6 Metabolism: Energy and Enzymes 100
- 6.1 Cells and the Flow of Energy 101
- 6.2 Metabolic Reactions and Energy Transformations 103
- 6.3 Metabolic Pathways and Enzymes 105
- 6.4 Oxidation-Reduction Reactions and Metabolism 109

7 Photosynthesis 114
- 7.1 Photosynthetic Organisms 115
- 7.2 The Process of Photosynthesis 117
- 7.3 Plants Convert Solar Energy 119
- 7.4 Plants Fix Carbon Dioxide 123
- 7.5 Other Types of Photosynthesis 125

8 Cellular Respiration 129
- 8.1 Overview of Cellular Respiration 130
- 8.2 Outside the Mitochondria: Glycolysis 132
- 8.3 Outside the Mitochondria: Fermentation 134
- 8.4 Inside the Mitochondria 136
- 8.5 Metabolism 141

UNIT 2 Genetic Basis of Life 146

9 The Cell Cycle and Cellular Reproduction 147
- 9.1 The Cell Cycle 148
- 9.2 The Eukaryotic Chromosome 151
- 9.3 Mitosis and Cytokinesis 152
- 9.4 The Cell Cycle and Cancer 158
- 9.5 Prokaryotic Cell Division 161

10 Meiosis and Sexual Reproduction 166
- 10.1 Overview of Meiosis 167
- 10.2 Genetic Variation 169
- 10.3 The Phases of Meiosis 172
- 10.4 Meiosis Compared to Mitosis 174
- 10.5 The Cycle of Life 176
- 10.6 Changes in Chromosome Number and Structure 177

11 Mendelian Patterns of Inheritance 186
- 11.1 Gregor Mendel 187
- 11.2 Mendel's Laws 188
- 11.3 Mendelian Patterns of Inheritance and Human Disease 194
- 11.4 Beyond Mendelian Inheritance 198

12 Molecular Biology of the Gene 207
- 12.1 The Genetic Material 208
- 12.2 Replication of DNA 211
- 12.3 The Genetic Code of Life 216
- 12.4 First Step: Transcription 218
- 12.5 Second Step: Translation 220

13 Regulation of Gene Expression 228
13.1 Prokaryotic Regulation 229
13.2 Eukaryotic Regulation 232
13.3 Gene Mutations 238

14 Biotechnology and Genomics 244
14.1 DNA Cloning 245
14.2 Biotechnology Products 247
14.3 Gene Therapy 250
14.4 Genomics 251

UNIT 3 Evolution 260

15 Darwin and Evolution 261
15.1 History of Evolutionary Thought 262
15.2 Darwin's Theory of Evolution 265
15.3 Evidence for Evolution 270

16 How Populations Evolve 279
16.1 Genes, Populations, and Evolution 280
16.2 Natural Selection 286
16.3 Maintenance of Diversity 290

17 Speciation and Macroevolution 296
17.1 How New Species Evolve 297
17.2 Modes of Speciation 303
17.3 Principles of Macroevolution 308

18 Origin and History of Life 317
18.1 Origin of Life 318
18.2 History of Life 323
18.3 Geological Factors That Influence Evolution 332

19 Taxonomy, Systematics, and Phylogeny 337
19.1 Systematic Biology 338
19.2 The Three-Domain System 341
19.3 Phylogeny 344

UNIT 4 Microbiology and Evolution 352

20 Viruses, Bacteria, and Archaea 353
20.1 Viruses, Viroids, and Prions 354
20.2 The Prokaryotes 360
20.3 The Bacteria 363
20.4 The Archaea 368

21 Protist Evolution and Diversity 373
21.1 General Biology of Protists 374
21.2 Supergroup Archaeplastida 377
21.3 Supergroup Chromalveolata 380
21.4 Supergroup Excavata 385
21.5 Supergroups Amoebozoa, Opisthokonta, and Rhizaria 389

22 Fungi Evolution and Diversity 395
22.1 Evolution and Characteristics of Fungi 396
22.2 Diversity of Fungi 398
22.3 Symbiotic Relationships of Fungi 405

UNIT 5 Plant Evolution and Biology 410

23 Plant Evolution and Diversity 411
23.1 Ancestry and Features of Land Plants 412
23.2 Evolution of Bryophytes: Colonization of Land 415
23.3 Evolution of Lycophytes: Vascular Tissue 417
23.4 Evolution of Pteridophytes: Megaphylls 419
23.5 Evolution of Seed Plants: Full Adaptation to Land 423

24 Flowering Plants: Structure and Organization 435
24.1 Cells and Tissues of Flowering Plants 436
24.2 Organs of Flowering Plants 440
24.3 Organization and Diversity of Roots 442
24.4 Organization and Diversity of Stems 445
24.5 Organization and Diversity of Leaves 450

25 Flowering Plants: Nutrition and Transport 456
25.1 Plant Nutrition and Soil 457
25.2 Water and Mineral Uptake 461
25.3 Transport Mechanisms in Plants 465

26 Flowering Plants: Control of Growth Responses 476
26.1 Plant Hormones 477
26.2 Plant Growth and Movement Responses 485
26.3 Plant Responses to Phytochrome 489

27 Flowering Plants: Reproduction 495
27.1 Sexual Reproductive Strategies 496
27.2 Seed Development 502
27.3 Fruit Types and Seed Dispersal 504
27.4 Asexual Reproductive Strategies 507

UNIT 6 Animal Evolution and Diversity 512

28 Invertebrate Evolution 513
28.1 Evolution of Animals 514
28.2 The Simplest Invertebrates 521
28.3 Diversity Among the Lophotrochozoans 524
28.4 Diversity of the Ecdysozoans 532
28.5 Invertebrate Deuterostomes 539

29 Vertebrate Evolution 544
29.1 The Chordates 545
29.2 The Vertebrates 547
29.3 The Fishes 548
29.4 The Amphibians 551
29.5 The Reptiles 553
29.6 The Mammals 559

30 Human Evolution 564
30.1 Evolution of Primates 565
30.2 Evolution of Humanlike Hominins 569
30.3 Evolution of Early Genus *Homo* 572
30.4 Evolution of Later Genus *Homo* 573

UNIT 7 Comparative Animal Biology 580

31 Animal Organization and Homeostasis 581
31.1 Types of Tissues 582
31.2 Organs, Organ Systems, and Body Cavities 589
31.3 The Integumentary System 591
31.4 Homeostasis 594

32 Circulation and Cardiovascular Systems 600
32.1 Transport in Invertebrates 601
32.2 Transport in Vertebrates 603
32.3 The Human Cardiovascular System 605
32.4 Blood 613

33 The Lymphatic and Immune Systems 621
33.1 Evolution of Immune Systems 622
33.2 The Lymphatic System 623
33.3 Innate Immune Defenses 625
33.4 Adaptive Immune Defenses 628
33.5 Immune System Disorders and Adverse Reactions 636

34 Digestive Systems and Nutrition 641
34.1 Digestive Tracts 642
34.2 The Human Digestive System 645
34.3 Digestive Enzymes 651
34.4 Nutrition and Human Health 652

35 Respiratory Systems 659
35.1 Gas-Exchange Surfaces 660
35.2 Breathing and Transport of Gases 665
35.3 Respiration and Human Health 669

36 Body Fluid Regulation and Excretory Systems 677
36.1 Animal Excretory Systems 678
36.2 The Human Urinary System 681

37 Neurons and Nervous Systems 691
37.1 Evolution of the Nervous System 692
37.2 Nervous Tissue 695
37.3 The Central Nervous System 699
37.4 The Peripheral Nervous System 707

38 Sense Organs 714
38.1 Sensory Receptors 715
38.2 Chemical Senses 716
38.3 Sense of Vision 718
38.4 Senses of Hearing and Balance 724
38.5 Somatic Senses 729

39 Locomotion and Support Systems 733
39.1 Diversity of Skeletons 734
39.2 The Human Skeletal System 736
39.3 The Muscular System 742

40 Hormones and Endocrine Systems 750
40.1 Animal Hormones 751
40.2 Hypothalamus and Pituitary Gland 758
40.3 Other Endocrine Glands and Hormones 758

41 Reproductive Systems 770
41.1 How Animals Reproduce 771
41.2 Human Male Reproductive System 773
41.3 Human Female Reproductive System 777
41.4 Control of Human Reproduction 781
41.5 Sexually Transmitted Diseases 785

42 Animal Development 793
42.1 Early Developmental Stages 794
42.2 Developmental Processes 798
42.3 Human Embryonic and Fetal Development 802
42.4 The Aging Process 809

UNIT 8 Behavior and Ecology 816

43 Behavioral Ecology 817
43.1 Inheritance Influences Behavior 818
43.2 The Environment Influences Behavior 820
43.3 Animal Communication 824
43.4 Behaviors That Increase Fitness 828

44 Population Ecology 836
44.1 Scope of Ecology 837
44.2 Demographics of Populations 838
44.3 Population Growth Models 841
44.4 Regulation of Population Size 844
44.5 Life History Patterns 846
44.6 Human Population Growth 849

45 Community and Ecosystem Ecology 855
45.1 Ecology of Communities 856
45.2 Community Development 866
45.3 Dynamics of an Ecosystem 868

46 Major Ecosystems of the Biosphere 881
46.1 Climate and the Biosphere 882
46.2 Terrestrial Ecosystems 885
46.3 Aquatic Ecosystems 895

47 Conservation of Biodiversity 905
47.1 Conservation Biology and Biodiversity 906
47.2 Value of Biodiversity 908
47.3 Causes of Extinction 911
47.4 Conservation Techniques 916

Appendices

A Answer Key A-1
B Tree of Life A-16

Glossary G-1
Credits C-1
Index I-1

The themes of evolution, the nature of science, and biological systems are important to understanding biology.

1

A View of Life

Chapter Outline

1.1 The Characteristics of Life 2

1.2 Evolution and the Classification of Life 6

1.3 The Process of Science 9

1.4 Challenges Facing Science 13

Our planet is home to a staggering diversity of life. It is estimated that there are over 15 million different species, including our species, *Homo sapiens,* that inhabit the globe. Furthermore, life may be found everywhere, from the deepest trenches in the oceans to the tops of the highest mountains. Biology is the area of scientific study that focuses on understanding all aspects of living organisms. To further our understanding of what it means to be alive, biologists explore life from the molecular level of the information in our genes to the large-scale ecological interactions of multiple species and their environments.

In this text, we are going to focus on three themes that define these explorations. The first is evolution—the central theme of biology and the explanation for how life adapts and changes over time. The second theme is the nature of science. Science is a process that relies on experimentation and hypothesis testing to validate its findings. The third theme is biological systems. Throughout this text you will discover that life is interconnected at many levels, from similarities in our genetic information to the cycling of nutrients in ecosystems.

As we proceed through this chapter, consider how we as humans are interconnected with other species by these three themes.

As you read through this chapter, think about the following questions:

1. Why is evolution a central theme of the biological sciences?
2. In what ways is life interconnected?
3. How do scientists use the scientific method to study life?

Following *the* Themes

CHAPTER 1 A VIEW OF LIFE

Evolution	Understanding the scientific process, the theory of evolution, and the interaction of biological systems is important in the study of biology.
Nature of Science	Scientists make observations, form hypotheses, and conduct experiments in an attempt to understand the principles of life.
Biological Systems	From communities of organisms to individual cells, all life is based on atoms and molecules.

1.1 The Characteristics of Life

Learning Outcomes

Upon completion of this section, you should be able to

1. Distinguish among the levels of biological organization.
2. Identify the basic characteristics of life.

Biology is the scientific study of life. Life on Earth takes on a staggering variety of forms, often functioning and behaving in ways strange to humans. For example, gastric-brooding frogs swallow their embryos and give birth to them later by throwing them up! Some species of puffballs, a type of fungus, are capable of producing trillions of spores when they reproduce. Fetal sand sharks kill and eat their siblings while still inside their mother. Some *Ophrys* orchids look so much like female bees that male bees try to mate with them. Octopuses and squid have remarkable problem-solving abilities despite a small brain. Some bacteria live their entire life in 15 minutes, while bristlecone pine trees outlive 10 generations of humans. Simply put, from the deepest oceanic trenches to the upper reaches of the atmosphere, life is plentiful and diverse.

Figure 1.1 illustrates the major groups of living organisms. From left to right, bacteria are widely distributed, microscopic organisms with a very simple structure. A *Paramecium* is an example of a microscopic protist. Protists are larger in size and more complex than bacteria. The other organisms in Figure 1.1 are easily seen with the naked eye. They can be distinguished by how they get their food. A morel is a fungus that digests its food externally. A sunflower is a photosynthetic plant that makes its own food, and an octopus is an aquatic animal that ingests its food.

Although life is tremendously diverse, it may be defined by several basic characteristics that are shared by all organisms. Like nonliving things, organisms are composed of chemical elements. Also, organisms obey the same laws of chemistry and physics that govern everything within the universe. The characteristics of life, however, provide insight into the unique nature of life, and help to distinguish living organisms from nonliving things.

Life Is Organized

The complex organization of life (Fig. 1.2) begins with **atoms,** the basic units of matter. Atoms combine to form small **molecules,** which join to form larger molecules within a **cell,** the smallest, most basic unit of life. Although a cell is alive, it is made from nonliving molecules. Some cells, such as single-celled *Paramecium,* live independently. In some cases, single-celled organisms clump together to form colonies, as does the alga *Volvox.*

Many living organisms are **multicellular,** meaning they contain more than one cell. In multicellular organisms, similar cells combine to form a **tissue**—for example, the nerve and muscle tissues of animals. Tissues make up **organs,** such as the brain or a leaf. Organs work together to form **organ systems;** for example, the brain works with the spinal cord and a network of nerves to form the nervous system. Organ systems are joined together to form an **organism,** such as an elephant.

The levels of biological organization extend beyond the individual organism. All the members of one species (a group of similar, interbreeding organisms) in a particular area belong to a **population.** A nearby forest may have a population of gray squirrels and a population of white oaks, for example. The populations of various animals and plants in the forest make up a **community.** The community of populations interacts with the physical environment (water, land, climate) to form an **ecosystem.** Collectively, all the Earth's ecosystems make up the **biosphere.**

You should recognize from Figure 1.2 that each level of biological organization builds upon the previous level and is more complex. Moving up the hierarchy, each level acquires new *emergent properties,* or new, unique characteristics, that are determined by the interactions between the individual parts. For example, when cells are broken down into bits of membrane and liquids, these parts themselves cannot carry out all the basic characteristics of life. However, all the levels of biological organization are interconnected and function as biological systems. For example, a change in carbon dioxide concentrations (a small molecule) may negatively influence the operation of organs, organisms, and entire ecosystems. In other words, life is interconnected at a variety of levels.

Bacteria *Paramecium* Morel Sunflower Octopus

Figure 1.1 Diversity of life. Biology is the scientific study of life. This is a sample of the many diverse forms of life that are found on planet Earth.

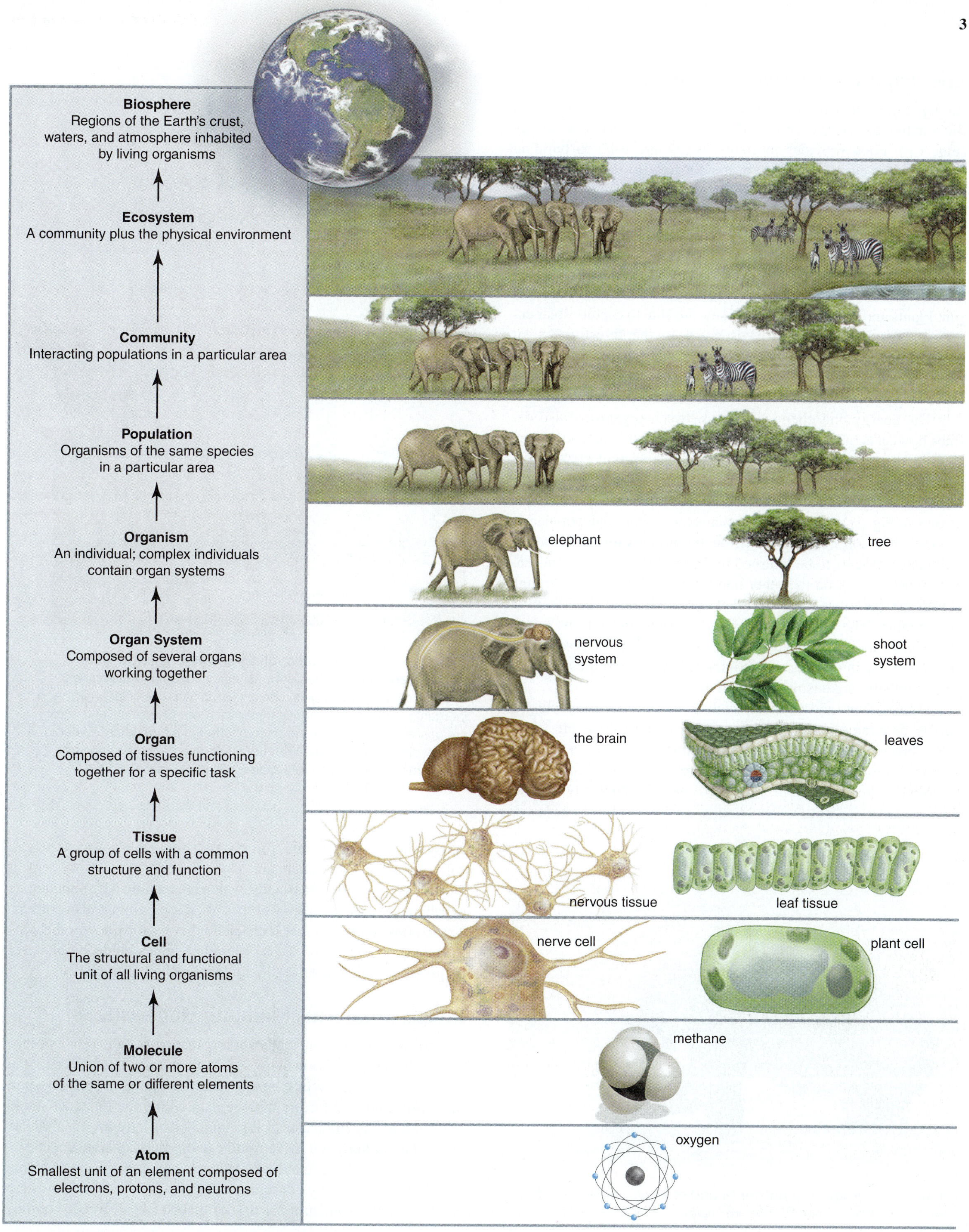

Figure 1.2 Levels of biological organization. The basic functional unit of life is the cell, which is built from nonliving molecules and atoms.

Life Requires Materials and Energy

Living organisms cannot maintain their organization or carry on life's activities without an outside source of nutrients and energy (Fig. 1.3). Food provides nutrients, which are used as building blocks or for energy. **Energy** is the capacity to do work, and it takes work to maintain the organization of the cell and the organism. When cells use nutrient molecules to make their parts and products, they carry out a sequence of chemical reactions. The term **metabolism** (Gk. *meta,* "change") encompasses all the chemical reactions that occur in a cell.

The ultimate source of energy for nearly all life on Earth is the sun. Plants and certain other organisms are able to capture solar energy and carry on **photosynthesis,** a process that transforms solar energy into the chemical energy of organic nutrient molecules. All life on Earth acquires energy by metabolizing nutrient molecules made by photosynthesizers. This applies even to plants themselves.

The energy and chemical flow between organisms also defines how an ecosystem functions (Fig. 1.4). Within an ecosystem, chemical cycling and energy flow begin when producers, such as grasses, take in solar energy and inorganic nutrients to produce food (organic nutrients) by photosynthesis. Chemical cycling (aqua arrows in Fig. 1.4) occurs as chemicals move from one population to another in a food chain, until death and decomposition allow inorganic nutrients to be returned to the producers once again. Energy (red arrows), on the other hand, flows from the sun through plants and the other members of the food chain as they feed on one another. The energy gradually dissipates and returns to the atmosphere as heat. Because energy does not cycle, ecosystems could not stay in existence without solar energy and the ability of photosynthetic organisms to absorb it.

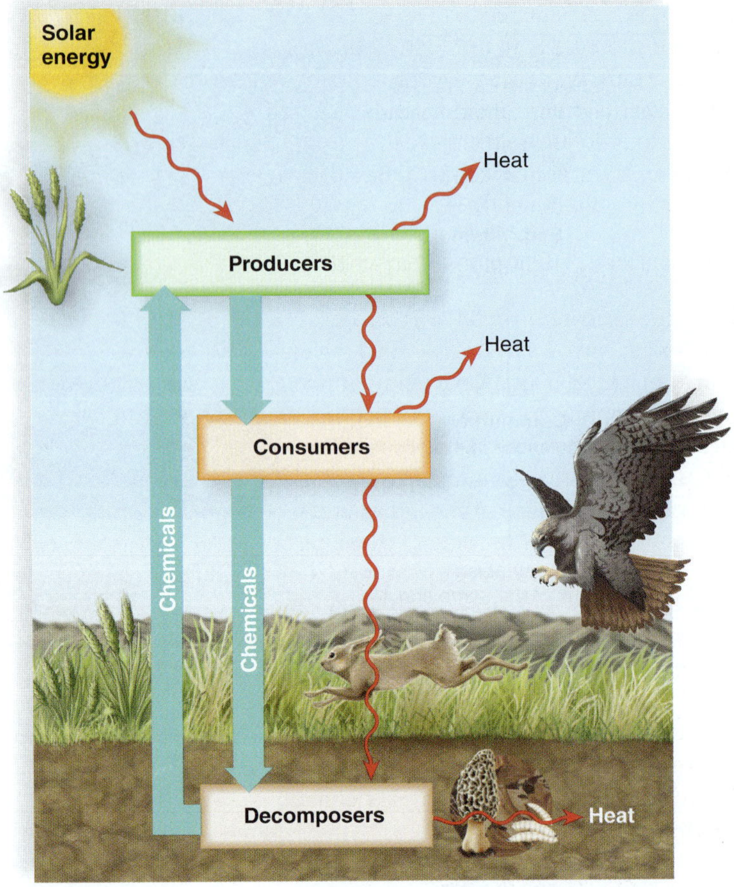

Figure 1.4 Chemical cycling and energy flow in an ecosystem. In an ecosystem, chemical cycling (aqua arrows) and energy flow (red arrows) begin when plants use solar energy and inorganic nutrients to produce their own food. Chemicals and energy are passed from one population to another in a food chain. Eventually, energy dissipates as heat. With the death and decomposition of organisms, chemicals are returned to living plants once more.

Energy flow and nutrient cycling in an ecosystem climate largely determine not only where different ecosystems are found in the biosphere but also what communities are found in the ecosystem. For example, deserts exist in areas of minimal rain, while forests require much rain. The two most biologically diverse ecosystems—tropical rain forests and coral reefs—occur where solar energy is most abundant. One example of an ecosystem in North America is the grasslands, which are inhabited by populations of rabbits, hawks, and various types of grasses, among many others. These populations interact with each other by forming food chains in which one population feeds on another. For example, rabbits feed on grasses, while hawks feed on rabbits and other organisms.

Figure 1.3 Acquiring nutrients and energy. All life, including this bear and the fish, need to acquire energy.

Living Organisms Maintain Homeostasis

To survive, it is imperative that an organism maintain a state of biological balance, or **homeostasis** (Gk. *homoios,* "like"; *stasis,* "the same"). For life to continue, temperature, moisture level, acidity, and other physiological factors must remain within the tolerance range of the organism. Homeostasis is maintained by systems that monitor internal conditions and make routine and necessary adjustments.

Organisms have intricate feedback and control mechanisms that do not require any conscious activity. These mechanisms may be controlled by one or more tissues themselves or by the nervous system. When you are studying and forget to eat lunch, your liver releases stored sugar to keep blood sugar levels within normal limits. Many organisms depend on behavior to regulate their internal environment. In animals, these behaviors are controlled by the nervous system and are usually not consciously controlled. For example, a lizard may raise its internal temperature by basking in the sun, or cool down by moving into the shade.

Living Organisms Respond

Living organisms interact with the environment as well as with other organisms. Even single-celled organisms can respond to their environment. In some, the beating of microscopic hairs or, in others, the snapping of whiplike tails moves them toward or away from light or chemicals. Multicellular organisms can manage more complex responses. A vulture can detect a carcass a kilometer away and soar toward dinner. A monarch butterfly can sense the approach of fall and begin its flight south, where resources are still abundant.

The ability to respond often results in movement: The leaves of a land plant turn toward the sun, and animals dart toward safety. Appropriate responses help ensure the survival of the organism and allow it to carry on its daily activities. All together, these activities are termed the *behavior* of the organism. Organisms display a variety of behaviors as they maintain homeostasis and search and compete for energy, nutrients, shelter, and mates. Many organisms display complex communication, hunting, and defense behaviors.

Living Organisms Reproduce and Develop

Life comes only from life. All forms of life have the ability to **reproduce,** or make another organism like itself. Bacteria, protists, and other single-celled organisms simply split in two. In most multicellular organisms, the reproductive process begins with the pairing of a sperm from one partner and an egg from the other partner. The union of sperm and egg, followed by many cell divisions, results in an immature stage, which proceeds through stages of **development,** or change, to become an adult.

When living organisms reproduce, their **genes,** or genetic instructions, are passed on to the next generation. Random combinations of sperm and egg, each of which contains a unique collection of genes, ensure that the offspring has new and different characteristics. An embryo develops into a whale, a yellow daffodil, or a human because of the specific set of genes it inherits from its parents. In all organisms, the genes are made of long **DNA** (**deoxyribonucleic acid**) molecules. DNA provides the blueprint, or instructions, for the organization and metabolism of the particular organism. All cells in a multicellular organism contain the same set of genes, but only certain genes are turned on in each type of specialized cell. You may notice that not all members of a species are exactly the same, and that there are obvious differences between species. These differences are the result of **mutations,** or inheritable changes in the genetic information. Mutation provides an important source of variation in the genetic information. However, not all mutations are bad—the observable differences in eye and hair color are examples of mutations.

Mutations help create a staggering diversity of life, even within a group of otherwise identical organisms. Sometimes, organisms inherit characteristics that allow them to be more suited to their way of life.

Living Organisms Have Adaptations

Adaptations are modifications that make organisms better able to function in a particular environment. For example, penguins are adapted to an aquatic existence in the Antarctic. An extra layer of downy feathers is covered by short, thick feathers, which form a waterproof coat. Layers of blubber also keep the birds warm in cold water. Most birds have forelimbs proportioned for flying, but penguins have stubby, flattened wings suitable for swimming. Their feet and tails serve as rudders in the water, but the flat feet also allow them to walk on land. Penguins also have many behavioral adaptations to living in the Antarctic. Penguins often slide on their bellies across the snow in order to conserve energy when moving quickly (Fig. 1.5). They carry their eggs—one or at most two—on their feet, where the eggs are protected by a pouch of skin. This also allows the birds to huddle together for warmth while standing erect and incubating the eggs.

From penguins to giant sequoia trees, life on Earth is very diverse, because over long periods of time, organisms respond to ever-changing environments by developing new adaptations.

Figure 1.5 Living organisms have adaptations. Penguins have evolved complex behaviors, such as sliding across ice to conserve energy, to adapt to their environment.

These adaptations are unintentional, but they provide the framework for evolutionary change. **Evolution** (L. *evolutio,* "an unrolling") includes the way in which populations of organisms change over the course of many generations to become more suited to their environments. All living organisms have the capacity to evolve, and the process of evolution constantly reshapes every species on the planet, potentially providing a way for organisms to persist, despite a changing environment. We will take a closer look at this process in the next section.

MP3 Life Characteristics

Check Your Progress 1.1

1. Distinguish between an ecosystem and a population in the levels of biological organization.
2. List the common characteristics of all living organisms.
3. Explain how adaptations relate to evolutionary change.

1.2 Evolution and the Classification of Life

Learning Outcomes

Upon completion of this section, you should be able to

1. Explain the relationship between the process of natural selection and evolutionary change.
2. Distinguish among the three domains of life.

Despite diversity in form, function, and lifestyle, organisms share the same basic characteristics. As mentioned, they are all composed of cells organized in a similar manner. Their genes are composed of DNA, and they carry out the same metabolic reactions to acquire energy and maintain their organization. The unity of life suggests that they are descended from a common ancestor—the first cell or cells.

Evolution—the Core Concept of Biology

The phrase "common descent with modification" sums up the process of evolution, because it means that as descent occurs from common ancestors, so do modifications that cause organisms to be adapted to their environment. Through many observations and experiments, Charles Darwin came to the conclusion that **natural selection** is the process that makes modification—that is, adaptation—possible.

Natural Selection

During the process of natural selection, some aspect of the environment selects which traits are more apt to be passed on to the next generation. The selective agent can be an abiotic agent (part of the physical environment, such as altitude), or it can be a biotic agent (part of the living environment, such as a deer). Figure 1.6 shows how the dietary habits of deer might eventually affect the characteristics of the leaves of a particular land plant.

Mutations fuel natural selection, because mutation introduces variations among the members of a population. In Figure 1.6, a plant species generally produces smooth leaves, but a mutation occurs that causes one plant to have leaves that are covered with small extensions, or "hairs." The plant with hairy leaves has an advantage, because the deer (the selective agent) prefer to eat smooth leaves, not hairy leaves. Therefore, the plant with hairy leaves survives best and produces more seeds than most of its neighbors. As a result, generations later most plants of this species produce hairy leaves.

As with this example, Darwin realized that although all individuals within a population have the potential to reproduce, not all do so with the same success. Prevention of reproduction can be the result of a number of factors, including an inability to capture resources, as when long-necked but not short-necked giraffes can reach their food source, or an inability to escape being eaten because long legs, but not short legs, can carry an animal to safety.

Whatever the example, it can be seen that organisms with advantageous traits can produce more offspring than those that lack them. In this way, living organisms change over time, and these changes are passed on from one generation to the next. Over long periods of time, the introduction of newer, more advantageous traits into a population may drastically reshape a species. Natural

Figure 1.6 Natural selection. Natural selection selects for or against new traits introduced into a population by mutations. Over many generations, selective forces such as competition, predation, and the physical environment alter the makeup of a population, favoring those more suited to the environment and lifestyle.

selection tends to sculpt a species to fit its environment and lifestyle and can create new species from existing ones. The end result is the diversity of life classified into the three domains of life (Fig. 1.7).

Organizing Diversity

An evolutionary tree is like a family tree. Just as a family tree shows how a group of people have descended from one couple, an evolutionary tree traces the ancestry of life on Earth to a common ancestor (Fig. 1.7). One couple can have diverse children, and likewise a population can be a common ancestor to several other groups, each adapted to a particular set of environmental conditions. In this way, over time, diverse life-forms have arisen. Evolution may be considered the unifying concept of biology, because it explains so many aspects of it, including how living organisms arose from a single ancestor.

Because life is so diverse, it is helpful to group organisms into categories. **Taxonomy** (Gk. *tasso,* "arrange"; *nomos,* "usage") is the discipline of identifying and grouping organisms according to certain rules. Taxonomy makes sense out of the bewildering variety of life on Earth and is meant to provide valuable insight into evolution. **Systematics** is the study of the evolutionary relationships between organisms. As systematists learn more about living organisms, the taxonomy often changes. DNA technology is now widely used by systematists to revise current information and to discover previously unknown relationships between organisms.

Several of the basic classification categories, or *taxa,* going from least inclusive to most inclusive, are **species, genus, family, order, class, phylum, kingdom,** and **domain** (Table 1.1). The least inclusive category, species (L. *species,* "model, kind"), is defined as a group of interbreeding individuals. Each successive classification category above species contains more types of organisms than the preceding one. Species placed within one genus share many specific characteristics and are the most closely related, while species placed in the same kingdom share only general characteristics with one another. For example, all species in the genus *Pisum* look pretty much the same—that is, like pea plants—but species in the plant kingdom can be quite varied, as is evident when we compare grasses to trees. Species placed in different domains are the most distantly related.

Table 1.1 Levels of Classification

Category	Human	Corn
Domain	Eukarya	Eukarya
Kingdom	Animalia	Plantae
Phylum	Chordata	Anthophyta
Class	Mammalia	Monocotyledones
Order	Primates	Commelinales
Family	Hominidae	Poaceae
Genus	*Homo*	*Zea*
Species*	*H. sapiens*	*Z. mays*

*To specify an organism, you must use the full binomial name, such as *Homo sapiens.*

Domains

Current biochemical evidence suggests that there are three domains: **domain Bacteria, domain Archaea,** and **domain Eukarya.** Figure 1.7 shows how the domains are believed to be related. Both domain Bacteria and domain Archaea may have evolved from the first common ancestor soon after life began. These two domains contain the **prokaryotes,** which lack the membrane-bound nucleus found in

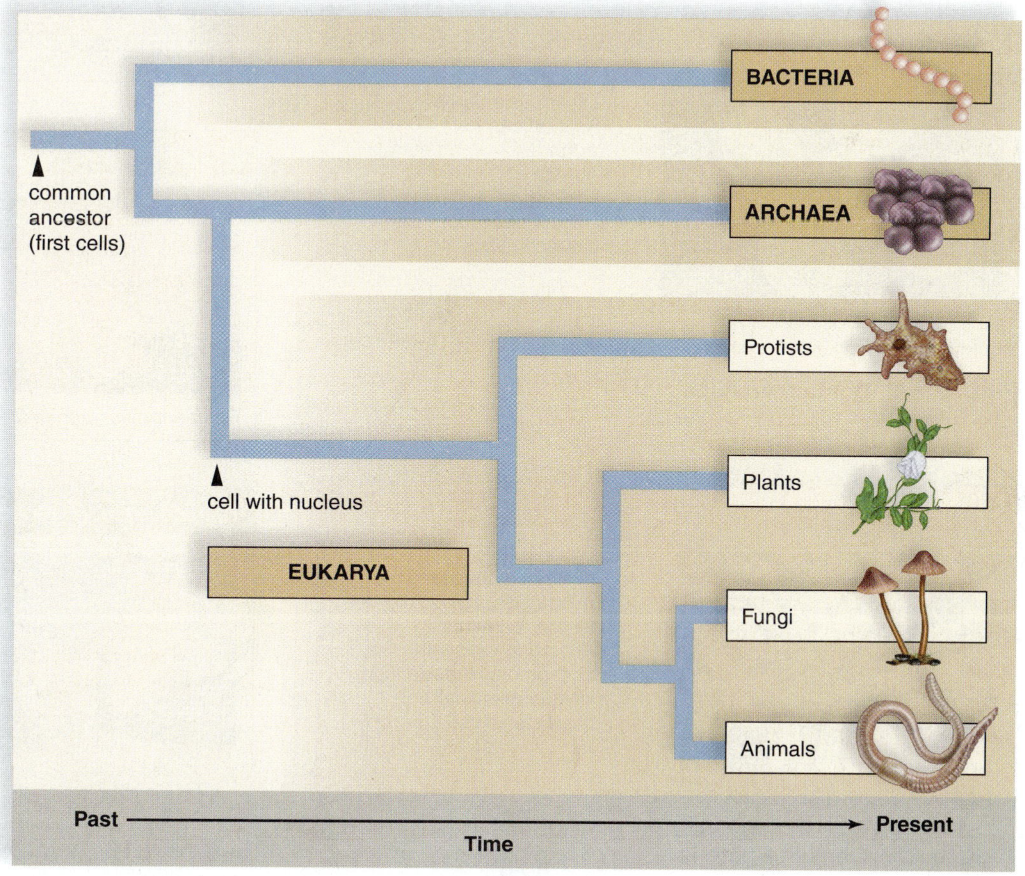

Figure 1.7 Evolutionary tree of life. As existing organisms change over time, they give rise to new species. Evolutionary studies show that all living organisms arose from a common ancestor about 4 billion years ago. Domain Archaea and domain Bacteria include the prokaryotes. Domain Eukarya includes both single-celled and multicellular organisms that possess a membrane-bound nucleus.

the **eukaryotes** of domain Eukarya. However, archaea organize their DNA differently than bacteria, and their cell walls and membranes are chemically more similar to eukaryotes than to bacteria. So, the conclusion is that eukarya split off from the archaeal line of descent.

Animation Three Domains

Prokaryotes are structurally simple but metabolically complex. Archaea (Fig. 1.8) can live in aquatic environments that lack oxygen or are too salty, too hot, or too acidic for most other organisms. Perhaps these environments are similar to those of the primitive Earth, and archaea (Gk. *archae,* "ancient") are the least evolved forms of life, as their name implies. Bacteria (Fig. 1.9) are variously adapted to living almost anywhere—in the water, soil, and atmosphere, as well as on our skin and in our mouth and large intestine.

Taxonomists are in the process of deciding how to categorize archaea and bacteria into kingdoms. Domain Eukarya, on the other hand, contains four major groups of organisms (Fig. 1.10). **Protists,** which comprise a number of kingdoms, range from single-celled forms to a few multicellular ones. Some are photosynthesizers, and some must acquire their food. Common protists include algae, the protozoans, and the water molds. Figure 1.7 shows that plants, fungi, and animals most likely evolved from protists. **Plants** (kingdom Plantae) are multicellular photosynthetic organisms. Example plants include azaleas, zinnias, and pines. Among the **fungi** (kingdom Fungi) are the familiar molds and mushrooms that, along with bacteria, help decompose dead organisms. **Animals** (kingdom Animalia) are multicellular organisms that must ingest and process their food. Aardvarks, jellyfish, and zebras are representative animals.

Scientific Name

Biologists use **binomial nomenclature** to assign each living organism a two-part name called a scientific name. For example, the scientific name for mistletoe is *Phoradendron tomentosum.* The first word is the genus, and the second word is the species designation (*specific epithet*) of each species within a genus. The genus may be abbreviated (e.g., *P. tomentosum*) and, if the species

Domain Archaea

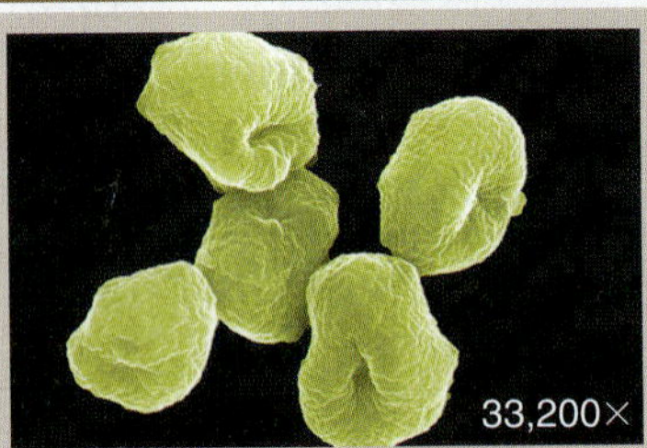

- Prokaryotic cells of various shapes
- Adaptations to extreme environments
- Absorb or chemosynthesize food
- Unique chemical characteristics

Sulfolobus, an archaean

Figure 1.8 **Domain Archaea.**

Domain Bacteria

- Prokaryotic cells of various shapes
- Adaptations to all environments
- Absorb, photosynthesize, or chemosynthesize food
- Unique chemical characteristics

Escherichia coli, a bacterium

Figure 1.9 **Domain Bacteria.**

Domain Eukarya: Protists

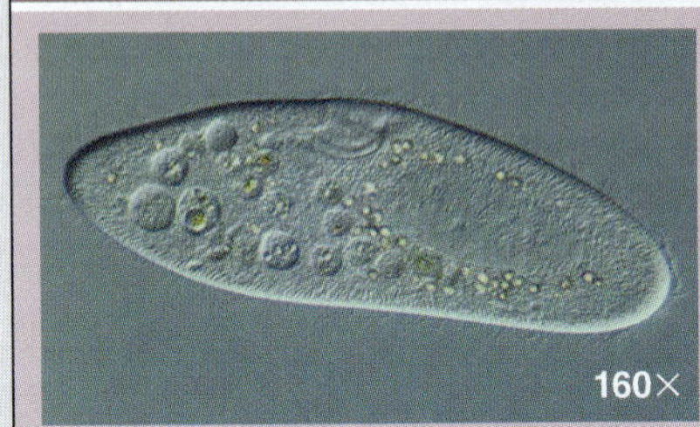

- Algae, protozoans, slime molds, and water molds
- Complex single cell (sometimes filaments, colonies, or even multicellular)
- Absorb, photosynthesize, or ingest food

Paramecium, a single-celled protozoan

Domain Eukarya: Kingdom Fungi

- Molds, mushrooms, yeasts, and ringworms
- Mostly multicellular filaments with specialized, complex cells
- Absorb food

Amanita, a mushroom

Domain Eukarya: Kingdom Plantae

- Certain algae, mosses, ferns, conifers, and flowering plants
- Multicellular, usually with specialized tissues, containing complex cells
- Photosynthesize food

Phalaenopsis, orchid, a flowering plant

Domain Eukarya: Kingdom Animalia

- Sponges, worms, insects, fishes, frogs, turtles, birds, and mammals
- Multicellular with specialized tissues containing complex cells
- Ingest food

Vulpes, a red fox

Figure 1.10 **Domain Eukarya.**

has not been determined, it may simply be indicated with a generic abbreviation (e.g., *Phoradendron* sp.). Scientific names are universally used by biologists to avoid confusion. Common names tend to overlap and often differ depending on locality and the language of a particular country. But scientific names are based on Latin, a universally used language that not too long ago was well known by most scholars.

Check Your Progress 1.2

1. Explain how natural selection results in new adaptations within a species.
2. List the levels of taxonomic classification from most inclusive to least inclusive.
3. Describe the differences that might be used to distinguish among the various kingdoms of domain Eukarya.

1.3 The Process of Science

Learning Outcomes

Upon completion of this section, you should be able to

1. Identify the components of the scientific method.
2. Distinguish between a theory and a hypothesis.
3. Analyze a scientific experiment and identify the hypothesis, experiment, control groups, and conclusions.

The process of science pertains to the study of biology. As you can see from Figure 1.2, the multiple stages of biological organization mean that life can be studied at a variety of levels. Some biological disciplines are cytology, the study of cells; anatomy, the study of structure; physiology, the study of function; botany, the study of plants; zoology, the study of animals; genetics, the study of heredity; and ecology, the study of the interrelationships between organisms and their environment.

Religion, aesthetics, ethics, and science are all ways in which human beings seek order in the natural world. The nature of scientific inquiry differs from these other ways of knowing and learning, because the scientific process uses the **scientific method,** a standard series of steps used in gaining new knowledge that is widely accepted among scientists. The scientific method (Fig. 1.11) acts as a guideline for scientific studies. Scientists often modify or adapt the process to suit their particular field of study.

Observation

Scientists believe that nature is orderly and measurable—that natural laws, such as the law of gravity, do not change with time—and that a natural event, or *phenomenon,* can be understood more fully through **observation—**a formal way of "seeing what happens."

Scientists use all of their senses in making observations. The behavior of chimpanzees can be observed through visual means, the disposition of a skunk can be observed through olfactory means, and the warning rattles of a rattlesnake provide auditory information of imminent danger. Scientists also extend the ability of their senses by using instruments; for example, the microscope enables us to see objects that could never be seen by the naked eye. Finally, scientists may expand their understanding even further by taking advantage of the knowledge and experiences of other scientists. For instance, they may look up past studies at the library or on the Internet, or they may write or speak to others who are researching similar topics.

Hypothesis

After making observations and gathering knowledge about a phenomenon, a scientist uses inductive reasoning to formulate a possible explanation. **Inductive reasoning** occurs whenever a person uses creative thinking to combine isolated facts into a cohesive whole. In some cases, chance alone may help a scientist arrive at an idea.

One famous case pertains to the antibiotic penicillin, which was discovered in 1928. While examining a petri dish of bacteria that had

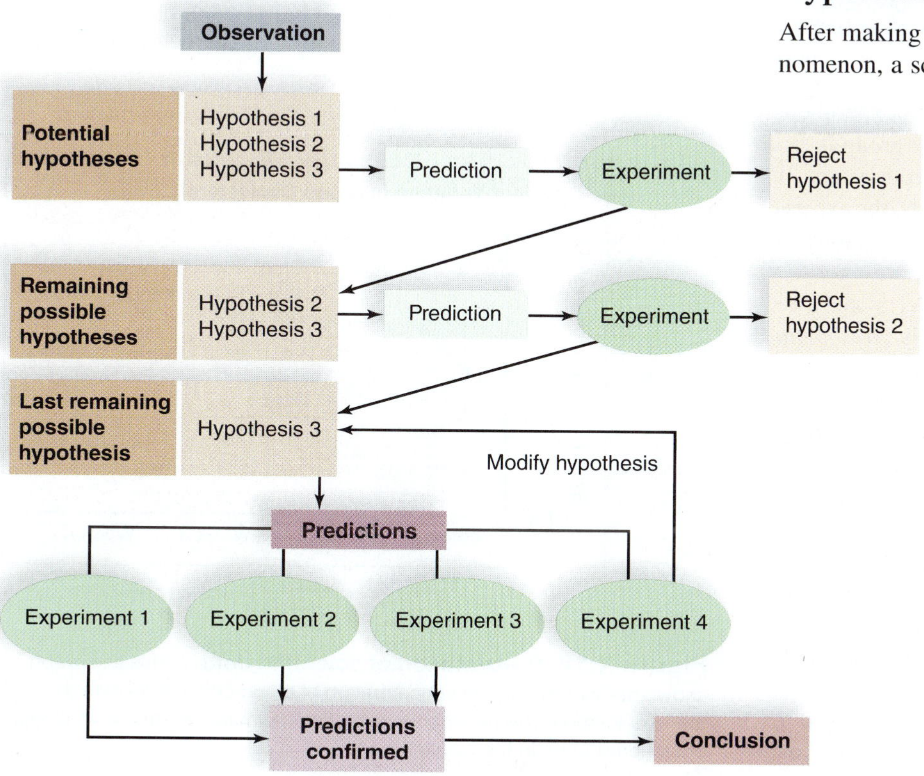

Figure 1.11 Flow diagram for the scientific method. On the basis of new and/or previous observations, a scientist formulates a hypothesis. The hypothesis is used to develop predictions to be tested by further experiments and/or observations, and new data either support or do not support the hypothesis. Following an experiment, a scientist often chooses to retest the same hypothesis or to test a related hypothesis. Conclusions from many different but related experiments may lead to the development of a scientific theory. For example, studies pertaining to development, anatomy, and fossil remains all support the theory of evolution.

become contaminated with the mold *Penicillium,* Alexander Flemming (1881–1955) observed an area that was free of bacteria. Flemming, an early expert on antibacterial substances, reasoned that the mold might have been producing an antibacterial compound.

We call such a possible explanation for a natural event a **hypothesis.** A hypothesis is not merely a guess; rather, it is an informed statement that can be tested in a manner suited to the processes of science.

All of a scientist's past experiences, no matter what they might be, have the potential to influence the formation of a hypothesis. But a scientist considers only hypotheses that can be tested. Moral and religious beliefs, while very important in the lives of many people, differ between cultures and through time and may not be scientifically testable.

Predictions and Experiments

Scientists often perform an **experiment,** which is a series of procedures, to test a hypothesis. To determine how to test a hypothesis, a scientist uses deductive reasoning. **Deductive reasoning** involves "if, then" logic. In designing the experiment, the scientist may make a **prediction,** or an expected outcome, based on knowledge of the factors in the experiment.

The manner in which a scientist intends to conduct an experiment is called the **experimental design.** A good experimental design ensures that scientists are examining the contribution of a specific variable, called the **experimental variable,** to the observation. The result is termed the **responding variable,** or dependent variable, because it is due to the experimental variable:

Experimental Variable (Independent Variable)	**Responding Variable** (Dependent Variable)
Factor of the experiment being tested	Result or change that occurs due to the experimental variable

To ensure that the results will be meaningful, an experiment contains both test groups and a **control** group. A test group is exposed to the experimental variable, but the control group is not. If the control group and test groups show the same results, the experimenter knows that the hypothesis predicting a difference between them is not supported.

Scientists often use **model** organisms and model systems to test a hypothesis. Model organisms, such as the fruit fly *Drosophila melanogaster* or the mouse *Mus musculus,* are chosen because they allow the researcher to control aspects of the experiment, such as age and genetic background. Cell biologists may use mice for modeling the effects of a new drug. Like model organisms, model systems allow the scientist to control specific variables and environmental conditions in a way that may not be possible in the natural environment. For example, ecologists may use computer programs to model how human activities will affect the climate of a specific ecosystem. While models provide useful information, they do not always answer the original question completely. For example, medicine that is effective in mice should ideally be tested in humans, and ecological experiments that are conducted using computer simulations need to be verified by field experiments. Biologists, and all other scientists, continuously design and revise their experiments to better understand how different factors may influence their original observation.

Presenting and Analyzing the Data

The **data,** or results, from scientific experiments may be presented in a variety of formats, including tables and graphs. A graph shows the relationship between two quantities. In many graphs, the experimental variable is plotted on the *x*-axis (horizontal), and the result is plotted along the *y*-axis (vertical). Graphs are useful tools to summarize data in a clear and simplified manner. For example, the line graph in Figure 1.12 shows the variation in the concentration of blood cholesterol over a four-week study. The bars above each data point represent the variation, or standard error, in the results. The title and labels can assist you in reading a graph; therefore, when looking at a graph, first check the two axes to determine what the graph pertains to. By looking at this graph, we know that the blood cholesterol levels were highest during week 2, and we can see to what degree the values varied over the course of the study.

Statistical Data

Most authors who publish research articles use statistics to help them evaluate their experimental data. In statistics, the standard error, or standard deviation, tells us how uncertain a particular value is. Suppose you predict how many hurricanes Florida will have next year by calculating the average number during the past 10 years. If the number of hurricanes per year varies widely, your standard error will be larger than if the number per year is usually about the same. In other words, the standard error tells you how far off the average could be. If the average number of hurricanes is four and the standard error is ± 2, then your prediction of four hurricanes is between two and six hurricanes. In Figure 1.12, the standard error is represented by the bars above and below each data point. This provides a visual indication of the statistical analysis of the data.

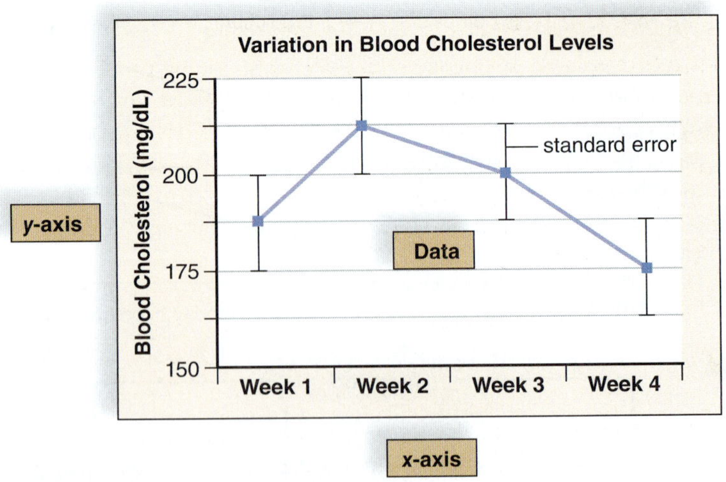

Figure 1.12 Presentation of scientific data. This line graph shows the variation in the concentration of blood cholesterol over a four-week study. The bars above each data point represent the variation, or standard error, in the results.

Statistical Significance

When scientists conduct an experiment, there is always the possibility that the results are due to chance or to some factor other than the experimental variable. Investigators take into account several factors when they calculate the probability value (p) that their results were due to chance alone. If the probability value is low, researchers describe the results as statistically significant. A probability value of less than 5% (usually written as $p < 0.05$) is acceptable; even so, keep in mind that the lower the p value, the less likely it is that the results are due to chance. Therefore, the lower the p value, the greater the confidence the investigators and you can have in the results. Depending on the type of study, most scientists like to have a p value of < 0.05, but p values of < 0.001 are common in many studies.

Scientific Publications

Scientific studies are customarily published in scientific journals (Fig. 1.13), so that all aspects of a study are available to the scientific community. Before information is published in scientific journals, it is typically reviewed by experts, who ensure that the research is credible, accurate, unbiased, and well executed. Another scientist should be able to read about an experiment in a scientific journal, repeat the experiment in a different location, and get the same (or very similar) results. Some articles are rejected for publication by reviewers when they believe there is something questionable about the design of an experiment or the manner in which it was conducted. This process of rejection is important in science since it causes researchers to critically review their hypotheses, predictions, and experimental designs, so that their next attempt will more adequately address their hypothesis. Often, it takes several rounds of revision before research is accepted for publication in a scientific journal.

Scientific magazines (Fig. 1.13), such as *Scientific American,* differ from scientific journals in that they report scientific findings to the general public. The information in these articles is usually obtained from articles first published in scientific journals.

Scientific Theory

The ultimate goal of science is to understand the natural world in terms of scientific **theories,** which are concepts that join together well-supported and related hypotheses. In ordinary speech, the word *theory* refers to a speculative idea. In contrast, a scientific theory is supported by a broad range of observations, experiments, and data, often from a variety of disciplines. Some of the basic theories of biology are:

Theory	*Concept*
Cell	All organisms are composed of cells, and new cells come only from preexisting cells.
Homeostasis	The internal environment of an organism stays relatively constant—within a range that is protective of life.
Evolution	All living organisms have a common ancestor, but each is adapted to a particular way of life.

As stated earlier, the theory of evolution is the unifying concept of biology because it pertains to many different aspects of life. For example, the theory of evolution enables scientists to understand the history of life, as well as the anatomy, physiology, and embryological development of organisms. Even behavior can be described through evolution, as we will see in a study discussed later in this chapter.

The theory of evolution has been a fruitful scientific theory, meaning that it has helped scientists generate new hypotheses. Because this theory has been supported by so many observations and experiments for over 100 years, some biologists refer to the **principle** of evolution, a term sometimes used for theories that are generally accepted by an overwhelming number of scientists. The term **law** instead of principle is preferred by some. For instance, in a subsequent chapter concerning energy relationships, we will examine the laws of thermodynamics.

An Example of the Scientific Method

We now know that most stomach and intestinal ulcers (open sores) are caused by the bacterium *Helicobacter pylori.* Let's say investigators want to determine which of two antibiotics is best for the treatment of an ulcer. When clinicians do an experiment, they try to vary just the

Figure 1.13 Scientific publications. Scientific journals, such as *Science,* are scholarly journals in which researchers share their findings with other scientists. Scientific magazines, such as *Scientific American, New Scientist,* and *Science News,* contain articles that are usually written by reporters for a broader audience.

experimental variables—in this case, the medications being tested. A control group is not given the medications, but one or more test groups receive them. If by chance the control group shows the same results as a test group, the investigators immediately know that the results of their study are invalid, because the medications may have had nothing to do with the results. The study depicted in Figure 1.14 shows how investigators may study this hypothesis:

> Hypothesis: Newly discovered antibiotic B is a better treatment for ulcers than antibiotic A, which is in current use.

Experimental Design

Next, the investigators might decide to use three experimental groups: one control group and two test groups. It is important to reduce the number of possible variables (differences), such as sex, weight, and other illnesses, among the groups. Therefore, the investigators randomly divide a very large group of volunteers equally into the three groups. The hope is that any differences will be distributed evenly among the three groups. This is possible only if the investigators have a large number of volunteers.

The three groups are to be treated like this:

> Control group: Subjects with ulcers are not treated with either antibiotic.
> Test group 1: Subjects with ulcers are treated with antibiotic A.
> Test group 2: Subjects with ulcers are treated with antibiotic B.

After the investigators have determined that all volunteers do have ulcers, they will want the subjects to think they are all receiving the same treatment. This is an additional way to protect the results from any influence other than the medication. To achieve this end, the subjects in the control group can receive a placebo, a treatment that appears to be the same as that administered to the other two groups but actually contains no medication. In this study, the use of a placebo would help ensure the same dedication by all subjects to the study.

Results and Conclusion

After two weeks of administering the same amount of medication (or placebo) in the same way, the stomach and intestinal linings of each subject are examined to determine if ulcers are still present. Endoscopy is a procedure that involves inserting an endoscope (a small, flexible tube with a tiny camera on the end) down the throat and into the stomach and the upper part of the small intestine. Then, the doctor can see the lining of these organs and can check for ulcers. Tests performed during an endoscopy can also determine if *Helicobacter pylori* is present.

Because endoscopy is somewhat subjective, it is probably best if the examiner is not aware of which group the subject is in; otherwise, the examiner's prejudice may influence the examination. When neither the patient nor the technician is aware of the specific treatment, it is called a double-blind study.

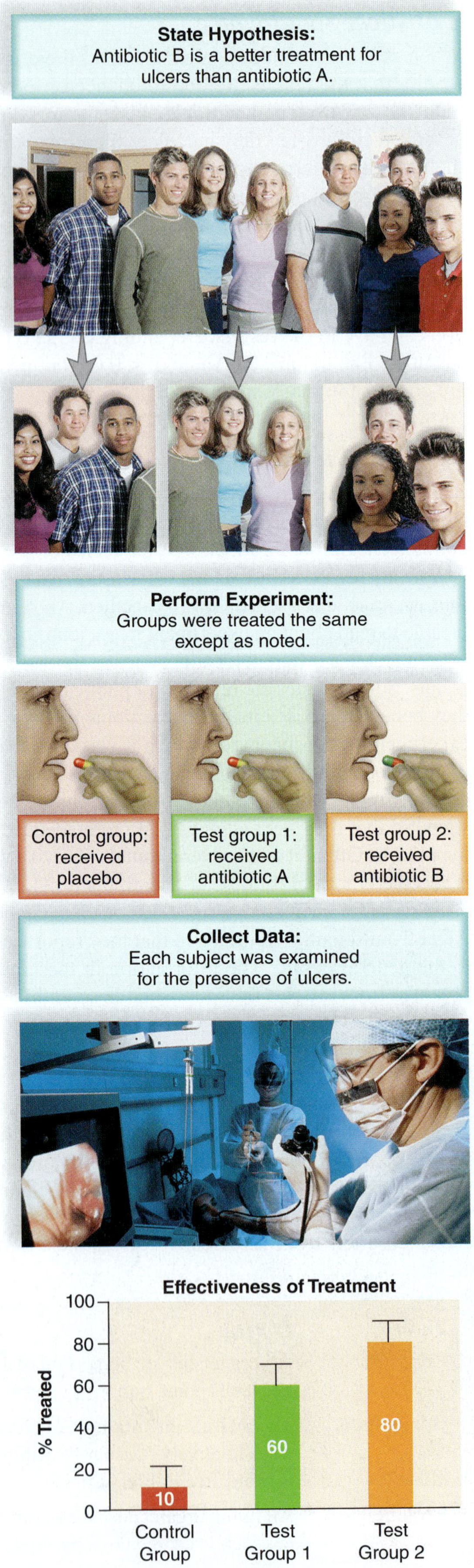

Figure 1.14 Example of a controlled study. In this study, a large number of people were divided into three groups. The control group received a placebo and no medication. One of the test groups received medication A, and the other test group received medication B. The results are depicted in a graph, and it shows that medication B was a more effective treatment than medication A for the treatment of ulcers.

In this study, the investigators may decide to determine the effectiveness of the medication by the percentage of people who no longer have ulcers. So, if 20 people out of 100 still have ulcers, the medication is 80% effective. The difference in effectiveness is easily read in the graph portion of Figure 1.14.

Conclusion: On the basis of their data, the investigators conclude that their hypothesis has been supported.

Check Your Progress 1.3

1. Identify the role of the experimental variable in an experiment.
2. Distinguish between the roles of the test group and the control group in an experiment.
3. Describe the process by which a scientist may test a hypothesis about an observation.

1.4 Challenges Facing Science

Learning Outcomes

Upon completion of this section, you should be able to

1. Distinguish between science and technology.
2. Summarize the major challenges facing science and society.

As we have learned in this chapter, science is a systematic way of acquiring knowledge about the natural world. Science is a slightly different endeavor than technology. **Technology** is the application of scientific knowledge to the interests of humans. Scientific investigations are the basis for the majority of our technological advances. As is often the case, a new technology, such as your cell phone or a new drug, is based on years of scientific investigations. However, despite our scientific and technological advances, there are many challenges facing society. In this section, we will explore a few of those critical challenges that scientists are actively investigating.

Biodiversity and Habitat Loss

Biodiversity is the total number and relative abundance of species, the variability of their genes, and the different ecosystems in which they live. The present biodiversity of our planet has been estimated to be as high as 15 million species, and so far, less than 2 million have been identified and named. **Extinction** is the death of a species or larger classification category. It is estimated that presently we are losing hundreds of species per day due to human activities and that as much as 38% of all species, including most primates, birds, and amphibians, may be in danger of extinction before the end of the century. Many biologists are alarmed about the present rate of extinction and hypothesize it may eventually rival the rates of the five mass extinctions that occurred during our planet's history. The last mass extinction, about 65 million years ago, caused many plant and animal species, including the dinosaurs, to become extinct.

The two most biologically diverse ecosystems—tropical rain forests and coral reefs—are home to many organisms. These ecosystems are also threatened by human activities. The canopy of the tropical rain forest alone supports a variety of organisms, including orchids, insects, and monkeys. Coral reefs, which are found just offshore of the continents and islands near the equator, are built up from calcium carbonate skeletons of sea animals called corals. Reefs provide a habitat for many animals, including jellyfish, sponges, snails, crabs, lobsters, sea turtles, moray eels, and some of the world's most colorful fishes (Fig. 1.15*a*). Like tropical rain

a. Healthy coral reef

1975 Minimal coral death

1985 Some coral death with no fish present

1995 Coral bleaching with limited chance of recovery

2004 Coral is black from sedimentation; bleaching still evident

b.

Figure 1.15 Coral reef, a marine ecosystem. **a.** Coral reefs, a type of ecosystem found in tropical seas, contain many diverse forms of life, a few of which are shown here. **b.** Various human activities have caused catastrophic damage to this coral reef off the coast of Florida, as shown over the course of 29 years. Preserving biodiversity is a modern-day challenge of great proportions.

forests, coral reefs are severely threatened as the human population increases in size. Some reefs are 50 million years old, yet in just a few decades, human activities have destroyed an estimated 25% of all coral reefs and seriously degraded another 30% (Fig. 1.15*b*). At this rate, nearly three-quarters could be destroyed within 40 years. Similar statistics are available for tropical rain forests.

The destruction of healthy ecosystems has many unintended effects. For example, we depend on them for food, medicines, and various raw materials. Draining of the natural wetlands of the Mississippi and Ohio Rivers and the construction of levees have worsened flooding problems, making once fertile farmland undesirable. The destruction of South American rain forests has killed many species that may have yielded the next miracle drug and has decreased the availability of many types of lumber. We are only now beginning to realize that we depend on ecosystems even more for the services they provide. Just as chemical cycling occurs within a single ecosystem, so all ecosystems keep chemicals cycling throughout the biosphere. The workings of ecosystems ensure that the environmental conditions of the biosphere are suitable for the continued existence of humans. And several studies show that ecosystems cannot function properly unless they remain biologically diverse. We will explore the concept of biodiversity in greater detail in Chapters 44 through 47 of the text.

Emerging Diseases

Over the past decade, avian influenza (H5N1 and H7N9), swine flu (H1N1), and severe acute respiratory syndrome (SARS) have been in the news. These are considered new, or emerging, diseases. Where do emerging diseases come from? Some of them may result from new and/or increased exposure to animals or insect populations that act as vectors for disease. Changes in human behavior and use of technology can also result in new diseases. SARS is thought to have arisen in Guandong, China, due to the consumption of civets, a type of exotic cat considered a delicacy. The civets were possibly infected by exposure to horseshoe bats sold in open markets. Legionnaires' disease emerged in 1976 due to bacterial contamination of a large air-conditioning system in a hotel. The bacteria thrived in the cooling tower used as the water source for the air-conditioning system. In addition, globalization results in the transport of diseases all over the world that were previously restricted to isolated communities. The first SARS cases were reported in southern China the week of November 16, 2002. By the end of February 2003, SARS had reached nine countries/provinces, mostly through airline travel. Some pathogens mutate and change hosts, jumping from birds to humans, for example. Before 1997, avian flu was thought to affect only birds. A mutated strain jumped to humans in the 1997 outbreak. To control that epidemic, officials killed 1.5 million chickens to remove the source of the virus. New forms of avian influenza (bird flu) are being discovered every few years. Each of these has the potential to cause health problems for humans across the globe. Scientists investigate not only the causes of these diseases (for example, the viruses) but also their effects on our bodies and the mechanisms by which they are transmitted. We will take a closer look at viruses in Chapter 20 of the text.

Climate Change

The term **climate change** refers to changes in the normal cycles of the Earth's climate that may be attributed to human activity. Climate change is primarily due to an imbalance in the chemical cycling of the element carbon. Normally, carbon is cycled within an ecosystem. However, due to human activities, more carbon dioxide is being released into the atmosphere than is being removed. In 1850, atmospheric CO_2 was at about 280 parts per million (ppm); today, it is over 400 ppm. This increase is largely due to the burning of fossil fuels and the destruction of forests to make way for farmland and pasture. Today, the amount of carbon dioxide released into the atmosphere is about twice the amount that remains in the atmosphere. It's believed that most of this dissolves in the ocean. The increased amount of carbon dioxide (and other gases) in the atmosphere is causing a rise in temperature called **global warming.** These gases allow the sun's rays to pass through, but they absorb and radiate heat back to Earth, a phenomenon called the greenhouse effect.

Global warming is contributing to climate change, which is causing significant changes in many of the Earth's ecosystems. We will examine climate change in more detail in Chapter 46.

Check Your Progress 1.4

1. Explain how a new technology differs from a scientific discovery.
2. Explain why the conservation of biodiversity is important to human society.
3. Summarize how emerging diseases and climate change have the potential for influencing the entire human population.

CONNECTING the CONCEPTS with the THEMES

Evolution

- Evolution is the core concept of biology; it explains how species develop adaptations to an ever-changing environment.
- Natural selection is the mechanism by which evolutionary change occurs.

Nature of Science

- Science is based on the ability to observe the natural world and then formulate hypotheses as potential explanations for these observations.
- Scientists use an ordered series of events, called the scientific method, to construct experiments that explore the structure of the natural world.

Biological Systems

- All life is based on atoms and molecules, which in turn are involved in the formation of a cell, the basic unit of all life.
- Members of a species form populations. Populations of different species in a given area are called a community. The interaction of a community with the environment is called an ecosystem.
- Ecosystems are characterized by energy flow and chemical cycling.

MEDIA STUDY TOOLS http://connect.mheducation.com

SMARTBOOK® Study smart. SmartBook helps you maximize your study time by identifying what you know and don't know. The recharge feature in SmartBook helps you prepare for the test by identifying the material that you are most likely to forget.

connect® BIOLOGY Find even more resources to learn the chapter concepts including animations, tutorial videos, and interactive practice questions.

MP3 File	Animation
1.1 Life Characteristics	**1.2** Three Domains

SUMMARIZE

1.1 The Characteristics of Life

Biology is the area of science that studies life. Although living organisms are diverse, they have certain characteristics in common. Living organisms:

- *are organized.* These levels of organization extend from the **cell** (the basic unit of life) to **multicellular tissues, organs** and **organ systems. Atoms** and **molecules** are the nonliving components of cells. Above the level of the cell, organisms are organized into **populations** and **communities. Ecosystems** and the **biosphere** represent the highest levels of biological organization.
- *require materials and energy.* All living organisms need an outside source of materials and **energy. Metabolism** is the term used to summarize these chemical reactions in the cell. Photosynthesis is an example of a metabolic process.
- *maintain homeostasis.* **Homeostasis** is the ability to maintain a stable internal environment.
- *respond to stimuli.* These stimuli help an organism react to changes in its environment, such as the presence of food.
- *reproduce and develop.* Organisms reproduce to pass on their genetic information, included in the **genes** of their **DNA,** to the next generation. **Mutations** introduce variation into the DNA. **Development** is the series of steps that an organism proceeds through to become an adult.
- *have adaptations.* **Adaptations** allow an organism to exist in a particular environment. **Evolution** is the accumulation of these changes over multiple generations.

1.2 Evolution and the Classification of Life

Life on Earth is diverse, but the theory of evolution unifies life and describes how all living organisms evolved from a common ancestor. **Natural selection** describes the process by which living organisms are descended from a common ancestor. Mutations occur within a population, creating new traits. The agents of natural selection, present in both biological and physical environments, shape species over time and may create new species from existing ones.

In **taxonomy,** organisms are assigned an italicized **binomial** nomenclature that consists of the genus and the specific epithet. From the least inclusive to the most inclusive category, each **species** belongs to a **genus, family, order, class, phylum, kingdom,** and finally **domain.** Systematics is the study of evolutionary relationships between species.

The three domains of life are Archaea, Bacteria, and Eukarya. **Domain Archaea** and **domain Bacteria** contain **prokaryotic** organisms that are structurally simple but metabolically complex. **Domain Eukarya** contains the **eukaryotic** protists, fungi, plants, and animals. **Protists** range from single-celled to multicellular organisms and include the protozoans and most algae. Among the **fungi** are the familiar molds and mushrooms. **Plants** are well known as the multicellular photosynthesizers of the world, while **animals** are multicellular and ingest their food. An evolutionary tree shows how the domains are related by way of common ancestors.

1.3 The Process of Science

When studying the natural world, scientists use a process called the **scientific method.**

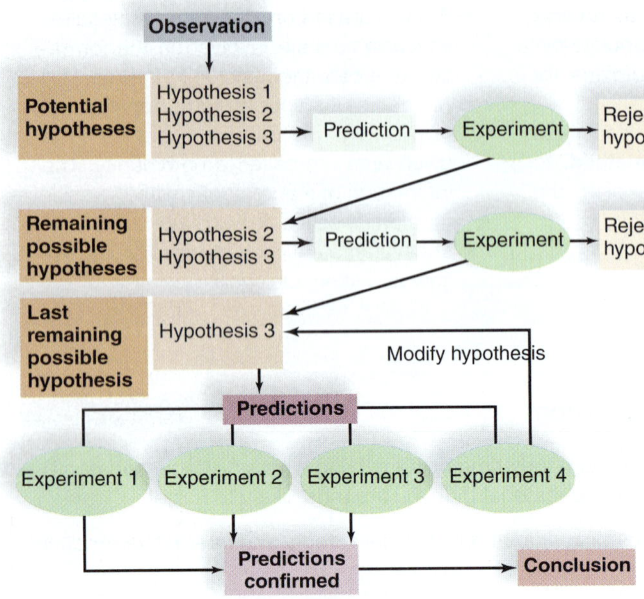

- **Observations,** along with previous data, are used to formulate a hypothesis. **Inductive reasoning** allows a scientist to combine facts into a **hypothesis.**
- New observations and/or **experiments** are carried out in order to test the hypothesis. **Deductive reasoning** allows for the development of a **prediction** of what may occur as a result of the experiment. A good **experimental design** includes an **experimental variable** and a **control** group. Scientists may use models and **model** organisms in their experimental design.
- The data from the experimental and observational results are analyzed, often using statistical methods. The results are often presented in tables or graphs for ease of interpretation.
- A conclusion is made as to whether the results support the hypothesis or do not support the hypothesis.
- The results may be submitted to a scientific publication for review by the scientific community.
- Over time multiple conclusions in a particular area may allow scientists to arrive at a **theory** (or **principle** or **law**), such as the cell theory or the theory of evolution. The theory of evolution is a unifying concept of biology.

1.4 Challenges Facing Science

While science investigates the principles of the natural world, **technology** applies this knowledge to the needs of society. Some challenges that scientists are investigating include:

- The loss of **biodiversity** and habitats such as coral reefs and rain forests. This often results in the **extinction** of species.
- Emerging diseases, such as avian influenza and SARS
- The impact of **climate change** and **global warming**

ASSESS

Choose the best answer for each question.

1.1 The Characteristics of Life

1. Which of these is not a property of all living organisms?
   a. organization
   b. acquisition of materials and energy
   c. care for their offspring
   d. reproduction
   e. responding to the environment
2. The level of organization that includes cells of similar structure and function is
   a. an organ.
   b. a tissue.
   c. an organ system.
   d. an organism.
3. The process that involves passing on genetic information between generations is called
   a. natural selection.
   b. reproduction.
   c. development.
   d. metabolism.

1.2 Evolution and the Classification of Life

4. Which of the following includes prokaryotic organisms?
   a. protists
   b. fungi
   c. archaea
   d. plants
5. The most inclusive level of classification is
   a. species.
   b. kingdom.
   c. domain.
   d. phylum.
6. The process by which evolution occurs is called
   a. natural selection.
   b. development.
   c. reproduction.
   d. taxonomy.

1.3 The Process of Science

7. After formulating a hypothesis, a scientist
   a. proves the hypothesis to be true or false.
   b. tests the hypothesis.
   c. decides how to best avoid having a control.
   d. makes sure environmental conditions are just right.
   e. formulates a scientific theory.
8. Experiments examine the contribution of the ________ to the observation.
   a. responding variable
   b. control group
   c. standard deviation
   d. experimental variable
9. Which of the following is not correctly linked?
   a. model: a representation of an object used in an experiment
   b. standard deviation: a form of statistical analysis
   c. principle: a theory that is not supported by experimental evidence
   d. data: the results of an experiment or observation

1.4 Challenges Facing Science

10. Which of the following applies scientific knowledge to the needs of society?
    a. evolution
    b. taxonomy
    c. systematics
    d. technology
11. Which of the following represents the permanent loss of a species?
    a. natural selection
    b. greenhouse effect
    c. extinction
    d. climate change
12. H5N1 and SARS are examples of
    a. extinct species.
    b. forms of greenhouse gases.
    c. endangered habitats.
    d. emerging diseases.

ENGAGE

LEARNSMART LABS®

The following LearnSmart Labs contain exercises that are related to the content of this chapter:

- Scientific Method

THINKING SCIENTIFICALLY

1. An investigator spills dye on a culture plate and notices that the bacteria live, despite their exposure to sunlight. He decides to test if the dye is protective against ultraviolet (UV) light. He exposes one group of culture plates containing bacteria and dye and another group containing only bacteria to UV light. The bacteria on all plates die. Complete the following diagram.

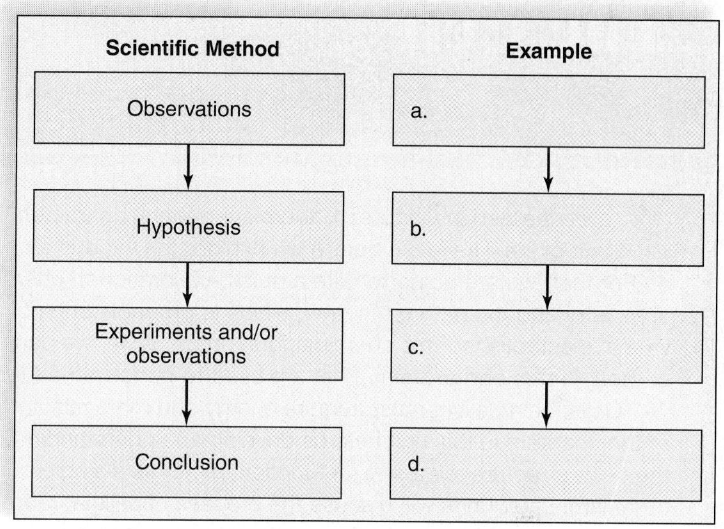

2. You want to grow large tomatoes, and you notice that a name-brand fertilizer claims to produce larger plants than a generic brand. How would you test this claim?
3. A scientist wishes to test her hypothesis that a commonly used drug causes heart attacks in some individuals. What kind of study should she initiate? What would you expect her experimental and responding variables to be?

UNIT 1

The Cell

As we learned in Chapter 1, there are general characteristics that all life shares. We also know that the cell is the basic unit of life. However, before we explore the various functions of the cell, we need to understand what cells are made of. For that, we are going to take a quick exploration of chemistry. We will start with developing an understanding of basic chemistry and the nature of water, which is probably one of the most important molecules for life as we know it. Once we have established this chemical foundation of life, we can proceed to the structure of the organic molecules, such as carbohydrates and proteins, that are used to perform the functions of the cell.

Cells, being alive, must acquire energy and materials and maintain an internal environment by homeostasis. The majority of the chapters in this unit help us develop an understanding of how cells accomplish these goals. We will also explore how the cell's structure relates to its function, either as a single-celled organism or as part of a multicellular tissue, organ, or organism. Later units will discuss the process of cellular reproduction and response to stimuli.

Since the cell forms the foundation for all life, your understanding of these concepts will serve you well as you move into the later parts of this text.

Unit Outline

Chapter 2 Basic Chemistry 19

Chapter 3 The Chemistry of Organic Molecules 35

Chapter 4 Cell Structure and Function 57

Chapter 5 Membrane Structure and Function 82

Chapter 6 Metabolism: Energy and Enzymes 100

Chapter 7 Photosynthesis 114

Chapter 8 Cellular Respiration 129

Unit Learning Outcomes

The learning outcomes for this unit focus on three major themes in the life sciences.

Evolution	Examine how inanimate elements can be combined to produce a living cell.
Nature of Science	Describe how science is used to investigate cellular phenomena.
Biological Systems	Evaluate how cellular components work together in order to function and live.

The *Curiosity* rover on Mars.

2

Basic Chemistry

On August 6, 2012, NASA's *Curiosity* rover successfully landed on the surface of Mars. Previous missions, including the long-lived *Spirit* and *Opportunity* rovers, focused on exploring the planet and detecting whether water once existed on Mars. *Curiosity* was designed to explore whether Mars at one time may have had the conditions to support life by looking for elements that we know are associated with life on Earth.

Curiosity possesses a collection of highly sophisticated instruments that can detect trace levels of specific elements and minerals in the Martian soil and rocks. For example, ChemCam uses a small laser to blast away portions of rocks. As the rocks are vaporized, another instrument records the types of elements and molecules that are released. ChemCam can determine whether the rocks were formed in the presence of water, a molecule that is essential for life as we know it. Another set of experiments is called SAM (Sample Analysis at Mars), which contains an instrument, called a spectrometer, that can be used to detect the presence of carbon, hydrogen, nitrogen, and oxygen in the Martian soil. Other spectrometers on *Curiosity* are also able to detect the presence of elements and chemical compounds that are associated with life. In its first year of operation, *Curiosity* detected water in the soil of Mars, and it is providing insights into whether the conditions on Mars may have supported life in the past. In the process, we may better understand how life evolved on our planet.

As you read through this chapter, think about the following questions:

1. Why are scientists looking for carbon, hydrogen, oxygen, and nitrogen on Mars?
2. What is an isotope, and how would a scientific instrument detect its presence?
3. Why is water considered to be so important to life?

Chapter Outline

2.1 Chemical Elements 20
2.2 Molecules and Compounds 24
2.3 Chemistry of Water 26
2.4 Acids and Bases 30

Before You Begin

Before beginning this chapter, take a few moments to review the following discussions.

Section 1.1 What are the general characteristics shared by all living organisms?

Section 1.3 How does the scientific process help us understand the natural world?

Following *the* Themes

CHAPTER 2 BASIC CHEMISTRY

UNIT 1 THE CELL

Evolution	Chemicals form the basis of living organisms, which evolve by changing their chemistry over time.
Nature of Science	Knowledge of chemicals is used to understand the scientific basis of life.
Biological Systems	Chemical elements are combined into molecular compounds, which are used to build cells, the basic units of life.

2.1 Chemical Elements

Learning Outcomes

Upon completion of this section, you should be able to

1. Describe how protons, neutrons, and electrons relate to atomic structure.
2. Use the periodic table to evaluate relationships between atomic number and mass number.
3. Describe how variations in an atomic nucleus account for its physical properties.
4. Determine how electrons are configured around a nucleus.

Throw a ball, pat your dog, rake leaves, turn a page; everything we touch—from the water we drink to the air we breathe—is composed of matter. **Matter** refers to anything that takes up space and has mass. Although matter has many diverse forms—anything from molten lava to kidney stones—it exists in only four distinct states: solid, liquid, gas, or plasma.

Elements

All matter, both nonliving and living, is composed of basic substances called **elements.** An element is a substance that cannot be broken down to simpler substances by ordinary chemical means. Each element has its own unique properties, such as density, solubility, melting point, and reactivity. It is quite remarkable that, in the known universe, there are only 92 naturally occurring elements (see Appendix C) that serve as the building blocks of matter. Other elements have been artificially constructed by physicists and are not biologically important.

Both the Earth's crust and all organisms are composed of elements, but they differ as to which ones are common. Only six elements—carbon, hydrogen, nitrogen, oxygen, phosphorus, and sulfur—are basic to life and make up about 95% of the body weight of organisms. The properties of these elements are essential to the uniqueness of cells and organisms, such as both the human and the tree in Figure 2.1. Other elements, such as potassium, calcium, iron, magnesium, and zinc, are also important to life.

Atoms

In the early 1800s, the English scientist John Dalton (1776–1844) developed the *atomic theory,* which says that elements consist of tiny particles called **atoms** (Gk. *atomos,* "uncut, indivisible"). An atom is the smallest part of an element that displays the properties of the element. An element and its atoms share the same name. One or two letters create the **atomic symbol** that stands for this name. For example, the symbol H means a hydrogen atom, the symbol Rn stands for radon, and the symbol Na (L. *natrium*) is used for a sodium atom.

Physicists have identified a number of subatomic particles that make up atoms. The three best-known subatomic particles are positively charged **protons,** uncharged **neutrons,** and negatively charged **electrons.** Protons and neutrons are located within

Figure 2.1 A comparison of the elements that make up the Earth's crust and living organisms. The graph inset shows that the Earth's crust primarily contains the elements silicon (Si), aluminum (Al), and oxygen (O). Living organisms, such as the tree and human, primarily contain the elements oxygen (O), nitrogen (N), carbon (C), and hydrogen (H). Biological molecules also often contain the elements sulfur (S) and phosphorus (P).

the nucleus of an atom, and electrons move about the nucleus. Figure 2.2 shows the arrangement of the subatomic particles in a helium atom, which has only two electrons. Since the precise location of the electrons is difficult to establish, we often indicate their probable positions using shading (Fig. 2.2*a*). When we are using a model of an atom—for example, to predict a chemical reaction—we indicate the average location of the electrons using **electron shells** (Fig. 2.2*b*).

The concept of an atom has changed greatly since Dalton's day. Today's physicists are using high-energy supercolliders, such as the Large Hadron Collider in Europe, to explore the intricate structure of the atom.

It is also important to note that the majority of an atom is empty space. If an atom could be drawn the size of a football field, the nucleus would be like a gumball in the center of the field, and the electrons would be tiny specks whirling about in the upper stands. We should also realize that both of the models in Figure 2.2 indicate only where the electrons are expected to be most of the time. In our analogy, the electrons might very well stray outside the stadium at times.

Atomic Number and Mass Number

Atoms have not only an atomic symbol but also an atomic number and a mass number. All the atoms of an element have the same number of protons housed in the nucleus. This is called the **atomic number,** which accounts for the unique properties of this type of atom. Generally, atoms are assumed to be electrically neutral, meaning that the number of electrons is the same as the number of protons in the atom. The atomic number tells you not only the number of protons but also the number of electrons.

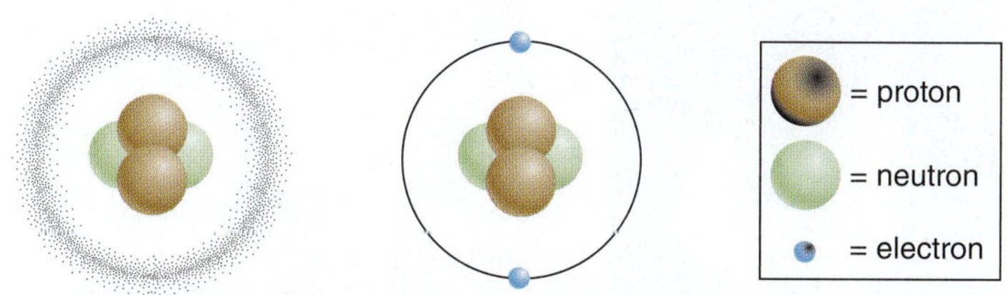

a. b.

Subatomic Particles			
Particle	Electric Charge	Atomic Mass Unit (AMU)	Location
Proton	+1	1	Nucleus
Neutron	0	1	Nucleus
Electron	–1	0	Electron shell

c.

Figure 2.2 Model of helium (He). Atoms contain subatomic particles, which are located as shown. Protons and neutrons are found within the nucleus, and electrons are outside the nucleus. **a.** The shading shows the probable location of the electrons in the helium atom. **b.** The average location of an electron is sometimes represented by an electron shell. **c.** The electric charge and the atomic mass units (AMU) of the subatomic particles vary as shown.

Each atom also has its own **mass number,** which is the sum of the protons and neutrons in the nucleus. Protons and neutrons are assigned one atomic mass unit (AMU) each. Electrons are so small that their AMU is considered zero in most calculations (Fig. 2.2*c*). By convention, when an atom stands alone (and not in the periodic table, discussed next), the atomic number is written as a subscript to the lower left of the atomic symbol. The mass number is written as a superscript to the upper left of the atomic symbol:

mass number — 12; atomic number — 6; $^{12}_{6}C$ — atomic symbol

Whereas each atom of an element has the same atomic number, the number of neutrons may vary slightly. **Isotopes** (Gk. *isos,* "equal") are atoms of the same element that differ in the number of neutrons. For example, the element carbon has three naturally occurring isotopes:

$$^{12}_{6}C \qquad ^{13}_{6}C \qquad ^{14}_{6}C$$

It is important to note that the term *mass* is used, not *weight,* because mass is constant, while weight changes according to the gravitational force of a body. The gravitational force of the Earth is greater than that of the moon; therefore, substances weigh less on the moon, even though their mass has not changed.

The term **atomic mass** refers to the average mass for all the isotopes of that atom. Since the majority of carbon is carbon 12, the atomic mass of carbon is closer to 12 than to 13 or 14. To determine the number of neutrons from the atomic mass, subtract the number of protons from the atomic mass and take the closest whole number.

6 — atomic number
C — atomic symbol
12.01 — atomic mass

The Periodic Table

Once chemists discovered a number of the elements, they began to realize that even though each element consists of a different atom, certain chemical and physical characteristics recur. The periodic table, developed by the Russian chemist Dmitri Mendeleev (1834–1907), was constructed as a way to group the elements, and therefore atoms, according to these characteristics.

Figure 2.3 is a portion of the periodic table, which is shown in total in Appendix C. In the periodic table, the horizontal rows are called periods, and the vertical columns are called groups. The atomic number of every atom in a period increases by one if you read from left to right. All the atoms in a group share similar chemical characteristics, namely in the type of chemical bonds that they form. For example, the atoms in group VIII are called the noble gases, because they are inert and rarely react with another atom. Helium, neon, argon, and krypton are all examples of noble gases.

Periods	I	II	III	IV	V	VI	VII	VIII
1	1 (atomic number) **H** (atomic symbol) 1.008							2 **He** 4.003 (atomic mass)
2	3 **Li** 6.941	4 **Be** 9.012	5 **B** 10.81	6 **C** 12.01	7 **N** 14.01	8 **O** 16.00	9 **F** 19.00	10 **Ne** 20.18
3	11 **Na** 22.99	12 **Mg** 24.31	13 **Al** 26.98	14 **Si** 28.09	15 **P** 30.97	16 **S** 32.07	17 **Cl** 35.45	18 **Ar** 39.95
4	19 **K** 39.10	20 **Ca** 40.08	31 **Ga** 69.72	32 **Ge** 72.59	33 **As** 74.92	34 **Se** 78.96	35 **Br** 79.90	36 **Kr** 83.60

Groups

Figure 2.3 A portion of the periodic table. In the periodic table, elements are listed in the order of their atomic numbers but are arranged so that each element is placed in a group (vertical column) and period (horizontal row). All the atoms in a particular group have the same number of valence electrons and therefore share common chemical characteristics. Each period shows the number of electron shells for an element. This abbreviated periodic table contains the elements most important in biology; the complete periodic table is in Appendix C.

Radioactive Isotopes

Some isotopes of an element are unstable, or radioactive. For example, unlike the other two isotopes of carbon, carbon 14 changes over time into nitrogen 14, which is a stable isotope of the element nitrogen. As carbon 14 decays, it releases various types of energy in the form of rays and subatomic particles. The radiation given off by radioactive isotopes can be detected in various ways. The Geiger counter is an instrument that is commonly used to detect radiation. In 1896, the French physicist Antoine-Henri Becquerel (1852–1908) discovered that a sample of uranium would produce a bright image on a photographic plate even in the dark, and a similar method of detecting radiation is still in use today. Marie Curie (1867–1934), who worked with Becquerel, coined the term *radioactivity* and contributed much to its study. Today, biologists use radiation to date objects from our distant past, to create images, and to trace the movement of substances in the body.

Low Levels of Radiation

The chemical behavior of a radioactive isotope is essentially the same as that of the stable isotopes of an element. This means that you can put a small amount of radioactive isotope in a sample and it becomes a *tracer* by which to detect molecular changes. Melvin Calvin and his co-workers used carbon 14 to detect all the various reactions that occur during the process of photosynthesis (see Chapter 7).

The importance of chemistry to medicine is nowhere more evident than in the many medical uses of radioactive isotopes. Specific tracers are used in imaging the body's organs and tissues. For example, after a patient drinks a solution containing a minute amount of iodine 131, it becomes concentrated in the thyroid—the only organ to take it up. A subsequent image of the thyroid indicates whether it is healthy in structure and function (Fig. 2.4*a*).

Positron-emission tomography (PET) is a way to determine the comparative activity of tissues. Radioactively labeled glucose, which emits a subatomic particle known as a positron, is injected into the body. The radiation given off is detected by sensors and analyzed by a computer. The result is a color image that shows which tissues have taken up the glucose and are therefore metabolically active. The red areas surrounded by green in Figure 2.4*b* indicate which areas of the brain are most active. PET scans of the brain are used to evaluate patients who have memory disorders of an undetermined cause or suspected brain tumors or seizure disorders that might benefit from surgery. PET scans, utilizing radioactive thallium, can detect signs of coronary artery disease and low blood flow to the heart.

High Levels of Radiation

Radioactive substances in the environment can harm cells, damage DNA, and cause cancer. When Marie Curie was studying radiation, its harmful effects were not known, and she and many of her co-workers developed cancer. The release of radioactive particles following a nuclear power plant accident, as occurred in Japan in 2011 following a tsunami, can have far-reaching and long-lasting effects on human health. The harmful effects of radiation can be put to good use, however (Fig. 2.5). Radiation from radioactive isotopes has been used for many years to sterilize medical and dental products. Radiation is now used to sterilize the U.S. mail and other packages to free them of possible pathogens, such as anthrax spores. High radiation is often used to kill cancer cells. Targeted radioisotopes can be introduced into the body, so that the subatomic particles emitted destroy only cancer cells, with little risk to the rest of the body.

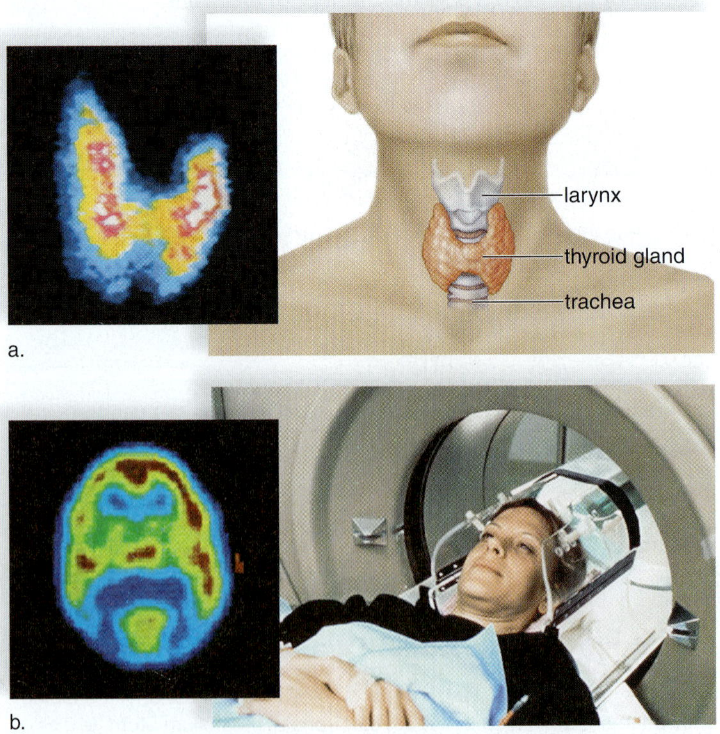

Figure 2.4 Low levels of radiation. **a.** Medical scan of the thyroid gland (colored image) indicates the presence of a tumor that does not take up radioactive iodine. **b.** A positron-emission tomography (PET) scan reveals which portions of the brain are most active (green and red colors).

a. b.

Figure 2.5 High levels of radiation. **a.** Radiation used to kill bacteria and fungi on peaches reduces spoilage and allows them to stay fresh for a longer period of time. **b.** Physicians use targeted radiation therapy to kill cancer cells.

Electrons and Energy

Various models may be used to illustrate the structure of a single atom. While the number of neutrons and protons may easily be depicted, since they are located in the nucleus, it is not possible to determine the precise location of any individual electron at any given moment. One of the more common models is the Bohr model (Fig. 2.6), developed by the physicist Niels Bohr (1885–1962).

In the Bohr model, the electron shells (also called electron orbitals) about the nucleus are used to represent the average energy levels of an electron. Because the negatively charged electrons are attracted to the positively charged nucleus, it takes energy to push them away and keep them in their own shell. The more distant the shell, the more energy it takes. Therefore, it is more accurate to speak of electrons as being at particular energy levels in relation to the nucleus. Electrons may move between energy levels. For example, when we explore the processes of photosynthesis, you will learn that when atoms absorb the energy of the sun, electrons are boosted to a higher energy level. Later, as the electrons return to their original energy level, energy is released and transformed into chemical energy. This chemical energy supports all life on Earth; therefore, our very existence is dependent on the energy of electrons.

Let's take a more detailed look at the Bohr models depicted in Figure 2.6. The first shell is closest to the nucleus and can contain two electrons; the second shell can contain eight electrons. In all atoms, the lower shells are filled with electrons before the next higher level contains any electrons.

The sulfur atom, with an atomic number of 16, has two electrons in the first shell, eight electrons in the second shell, and six electrons in the outer, third shell. Revisit the periodic table (see Fig. 2.3), and note that sulfur is in the third period. In other words, the period tells you how many shells an atom has. Also note that sulfur is in group VI. The group tells you how many electrons an atom has in its outer shell.

Regardless of how many shells an atom has, the outermost shell is called the **valence shell.** The valence shell is important, because it determines many of an atom's chemical properties. If an atom has only one shell, the valence shell is complete when it has two electrons. In atoms with more than one shell, the valence shell is most stable when it has eight electrons. This is called the **octet rule.** Each atom in a group within the periodic table has the same number of electrons in its valence shell. As mentioned previously, all the atoms in group VIII of the periodic table have eight electrons in their valence shell. These elements are also called the noble gases, because they do not ordinarily react.

The electrons in the valence shells play an important role in determining how an element undergoes chemical reactions. Atoms with fewer than eight electrons in the outer shell react with other atoms in such a way that after the reaction each has a stable outer shell. As we will see, the number of electrons in an atom's valence shell determines whether the atom gives up, accepts, or shares electrons to acquire eight electrons in the outer shell.

Check Your Progress 2.1

1. Contrast atomic number and mass number.
2. Examine the periods and groups from the periodic table to determine the electron configuration of chlorine.
3. Explain how two isotopes of an element vary with regard to their atomic structure.

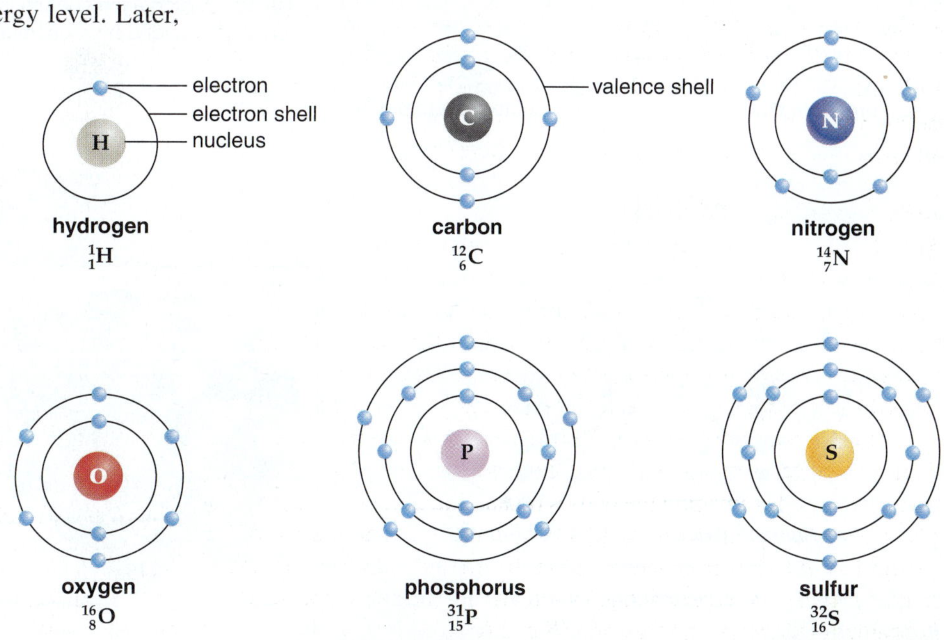

Figure 2.6 Bohr models of atoms. Electrons orbit the nucleus at particular energy levels (electron shells). The first shell contains up to two electrons, and thereafter each shell is most stable when it contains 8 electrons. Atoms with an atomic number above 20 may have more electrons in their outer shells. The outermost, or valence, shell helps determine the atom's chemical properties and how many other elements it can interact with.

2.2 Molecules and Compounds

Learning Outcomes

Upon completion of this section, you should be able to

1. Describe how elements are combined into molecules and compounds.
2. List the different types of bonds that occur between elements.
3. Explain the difference between a polar and a nonpolar covalent bond.

A **molecule** exists when two or more elements bond together; it is the smallest part of a compound that retains its chemical properties. A **compound** is a molecule containing at least two different elements. In practice, these two terms are used interchangeably, but in biology we usually speak of molecules. Water (H_2O) is a molecule that contains atoms of hydrogen and oxygen. A **formula** tells you the number of each kind of atom in a molecule. For example, the formula for glucose is:

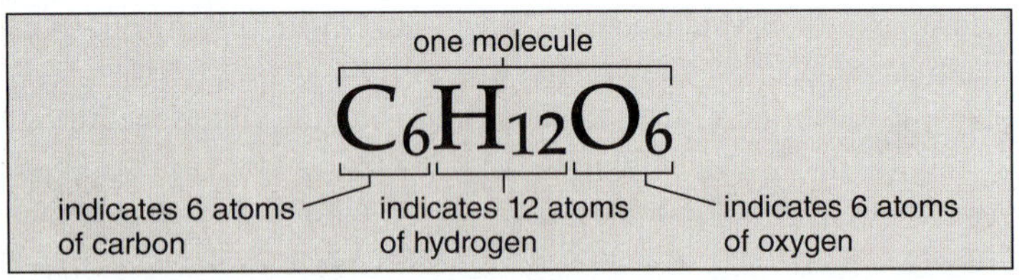

Electrons possess energy, as do the bonds between atoms. Organisms are directly dependent on chemical-bond energy to maintain their organization. As you may know, organisms routinely break down glucose, the sugar shown above, to obtain energy. When a chemical reaction occurs, as when glucose is broken down, electrons shift in their relationship to one another, and energy is released. Spontaneous reactions, which occur freely, always release energy.

MP3 Chemical Bonding

Ionic Bonding

Sodium (Na), with only one electron in its valence shell, tends to be an electron donor (Fig. 2.7*a*). Once it gives up this electron, the second shell, with its stable configuration of eight electrons, becomes its outer shell. Chlorine (Cl), on the other hand, tends to be an electron acceptor. Its valence shell has seven electrons, so if it acquires only one more electron it has a stable outer shell. When a sodium atom and a chlorine atom come together, an electron is transferred from the sodium atom to the chlorine atom. Now both atoms have eight electrons in their outer shells.

This electron transfer, however, causes a charge imbalance in each atom. After giving up an electron, the sodium atom has one more proton than it has electrons; therefore, it has a net charge of +1 (symbolized by Na^+). After accepting an electron, the chlorine atom has one more electron than it has protons; therefore, it has a net charge of −1 (symbolized by Cl^-). These charged particles are called **ions.** Sodium (Na^+) and chloride (Cl^-) are not the only biologically important ions. Some, such as potassium (K^+), are formed by the transfer of a single electron to another atom; others, such as calcium (Ca^{2+}) and magnesium (Mg^{2+}), are formed by the transfer of two electrons.

Ionic compounds are held together by an attraction between negatively and positively charged ions, called an **ionic bond.** When sodium reacts with chlorine, an ionic compound called sodium chloride (NaCl) results. Sodium chloride is an example of a salt. It is commonly called table salt, because it is used to season food (Fig. 2.7*b*). Salts are solid substances that usually separate and exist as individual ions in water, as discussed on page 28.

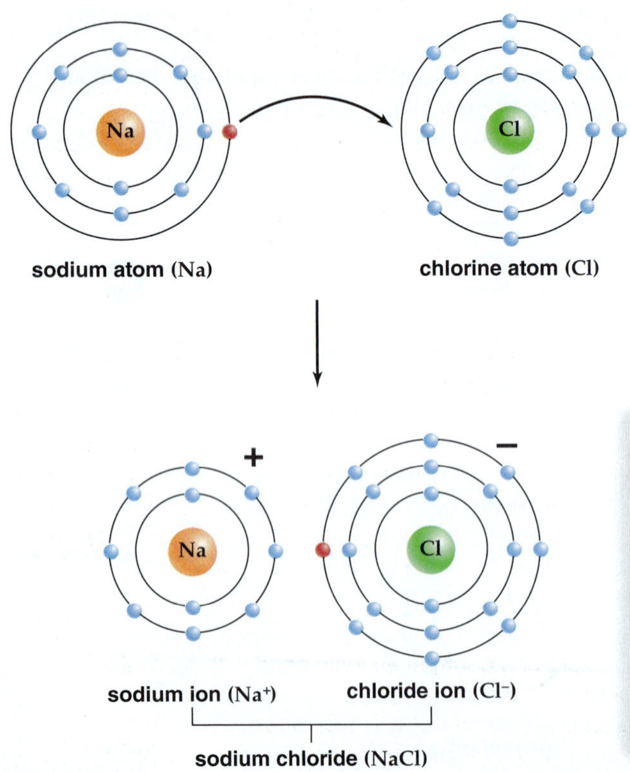

a.

b.

Figure 2.7 Formation of sodium chloride (table salt). **a.** During the formation of sodium chloride, an electron is transferred from the sodium atom to the chlorine atom. At the completion of the reaction, each atom has eight electrons in the outer shell, but each also carries a charge as shown. **b.** In a sodium chloride crystal, ionic bonding between Na^+ and Cl^- causes the atoms to assume a three-dimensional lattice in which each sodium ion is surrounded by six chloride ions, and each chloride ion is surrounded by six sodium ions. The result is crystals of salt as in table salt.

Covalent Bonding

A **covalent bond** results when two atoms share electrons in such a way that each atom has an octet of electrons in the outer shell (or two electrons, in the case of hydrogen). In a hydrogen atom, the outer shell is complete when it contains two electrons. If hydrogen is in the presence of a strong electron acceptor, it gives up its electron to become a hydrogen ion (H^+). But if this is not possible, hydrogen can share with another atom and thereby have a completed outer shell. For example, one hydrogen atom will share with another hydrogen atom. Their two electron shells overlap, and the electrons are shared between them (Fig. 2.8*a*). Because they share the electron pair, each atom has a completed outer shell.

A more common way to symbolize that atoms are sharing electrons is to draw a line between the two atoms, as in the structural formula H—H. Just as a handshake requires two hands, one from each person, a covalent bond between two atoms requires two electrons, one from each atom. In a molecular formula, the line is omitted and the molecule is simply written as H_2.

Sometimes, atoms share more than one pair of electrons to complete their octets. A double covalent bond occurs when two atoms share two pairs of electrons (Fig. 2.8*b*). To show that oxygen gas (O_2) contains a double bond, the molecule can be written as O=O. It is also possible for atoms to form triple covalent bonds, as in nitrogen gas (N_2), which can be written as N≡N. Single covalent bonds between atoms are quite strong, but double and triple bonds are even stronger.

Electron Model	Structural Formula	Molecular Formula
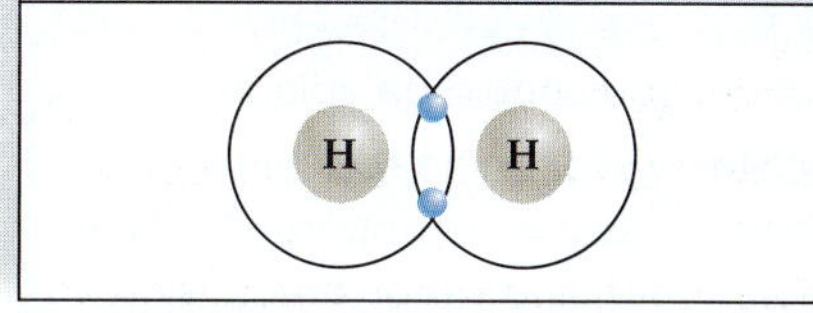	H—H	H_2

a. Hydrogen gas

Electron Model	Structural Formula	Molecular Formula
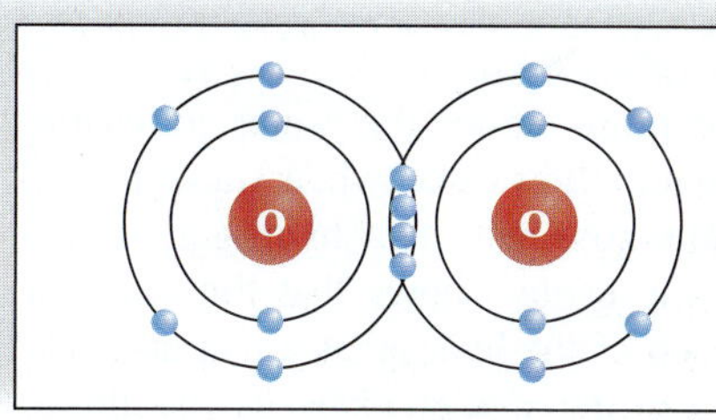	O=O	O_2

b. Oxygen gas

Electron Model	Structural Formula	Molecular Formula
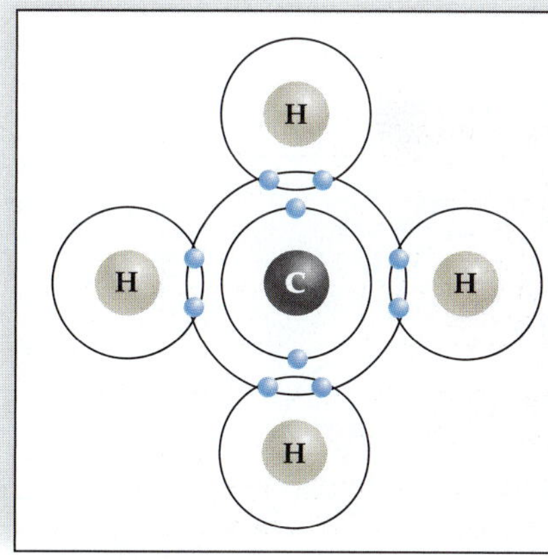	H—C(—H)(—H)—H	CH_4

c. Methane

Figure 2.8 Covalently bonded molecules. In a covalent bond, atoms share electrons, allowing each atom to have a completed outer shell. **a.** A molecule of hydrogen (H_2) contains two hydrogen atoms sharing a pair of electrons. This single covalent bond can be represented in any of the three ways shown. **b.** A molecule of oxygen (O_2) contains two oxygen atoms sharing two pairs of electrons. This results in a double covalent bond. **c.** A molecule of methane (CH_4) contains one carbon atom bonded to four hydrogen atoms.

Nonpolar and Polar Covalent Bonds

When the sharing of electrons between two atoms is equal, the covalent bond is said to be a **nonpolar covalent bond.** However, in some cases one atom is able to attract electrons to a greater degree than the other atom. In this case, we say that the atom that has a greater attraction for a shared pair of electrons has a greater **electronegativity.** When electrons are not shared equally, the covalent bond is a **polar covalent bond.**

The shape of a molecule may also influence whether it is polar or nonpolar. While carbon is larger and has more protons than a hydrogen atom, the symmetrical nature of a methane molecule cancels out any polarities; thus, methane is a nonpolar molecule. Not so in water, which has this shape:

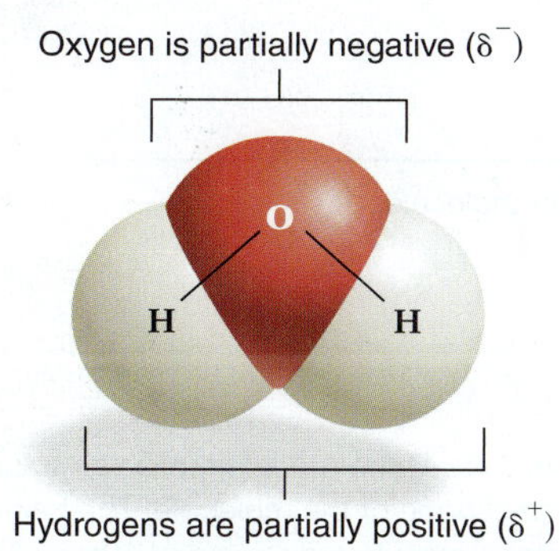

In water, the oxygen atom is more electronegative than the hydrogen atoms; as a result, water molecules are polar. Moreover, because of its nonsymmetrical shape, the polar bonds cannot cancel each other, and water is a polar molecule. The more electronegative end of the molecule is designated slightly negative (δ^-), and the hydrogens are designated slightly positive (δ^+).

Water is not the only polar molecule in living organisms. For example, the amine group (—NH_2) is polar, and this causes amino acids and nucleic acids to exhibit polarity, as we will see in the next chapter. The polarity of molecules affects how they interact with other molecules.

Animation Ionic Versus Covalent Bonding

Check Your Progress 2.2

1. Compare and contrast an ionic bond with a covalent bond.
2. Describe the process by which ions are formed.
3. Explain why methane is nonpolar but water is polar.

2.3 Chemistry of Water

Learning Outcomes

Upon completion of this section, you should be able to

1. Describe how water associates with other molecules in solution.
2. Describe why the properties of water are important to life.
3. Analyze how water's solid, liquid, and vapor states allow life to exist on Earth.

Figure 2.9*a* recaps what we know about the water molecule. The structural formula at the top shows that when water forms, an oxygen atom is sharing electrons with two hydrogen atoms. The ball-and-stick model in the center shows that the covalent bonds between oxygen and each of the hydrogens are at an angle of 104.5°. Finally, the space-filling model gives us the three-dimensional shape of the molecule and indicates its polarity.

In biology, we often state that structure relates to function. This is true at a variety of organizational levels, including molecules such as water. For example, hormones have specific shapes that allow them to be recognized by the cells in the body. We can stay well only when antibodies recognize the shapes of disease-causing agents, the way a key fits a lock, and are able to remove them.

The shape of a water molecule and its polarity result in the formation of hydrogen bonds. A **hydrogen bond** is caused by the attraction of a slightly positive hydrogen to a slightly negative atom in the vicinity. In carbon dioxide, O = C = O, a slight difference in polarity between carbon and the oxygens is present, but because carbon dioxide is symmetrical, the opposing charges cancel one another and hydrogen bonding does not occur.

Hydrogen Bonding

The dotted lines in Figure 2.9*b* indicate that the hydrogen atoms in one water molecule are attracted to the oxygen atoms in other

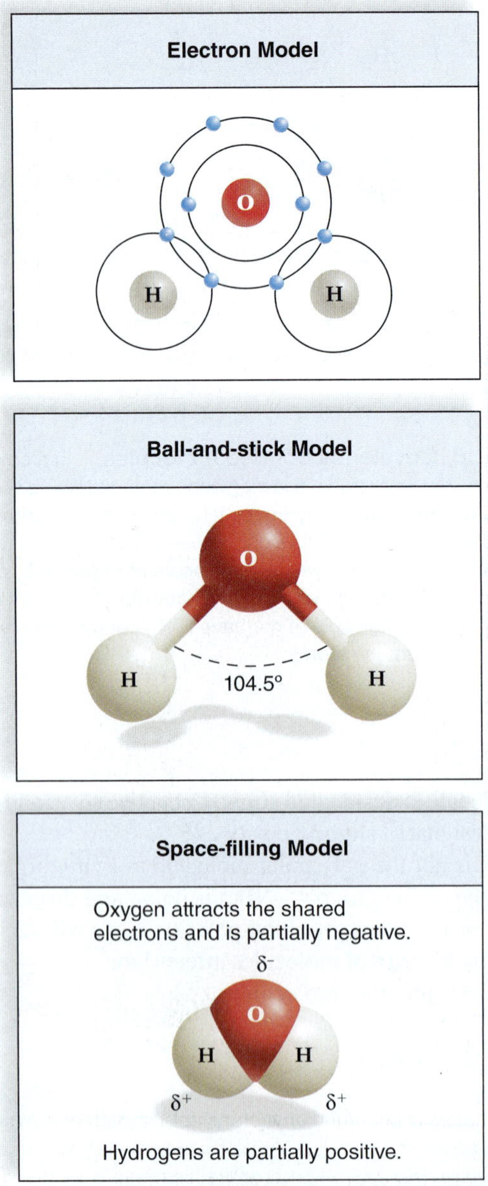

a. Water (H_2O)

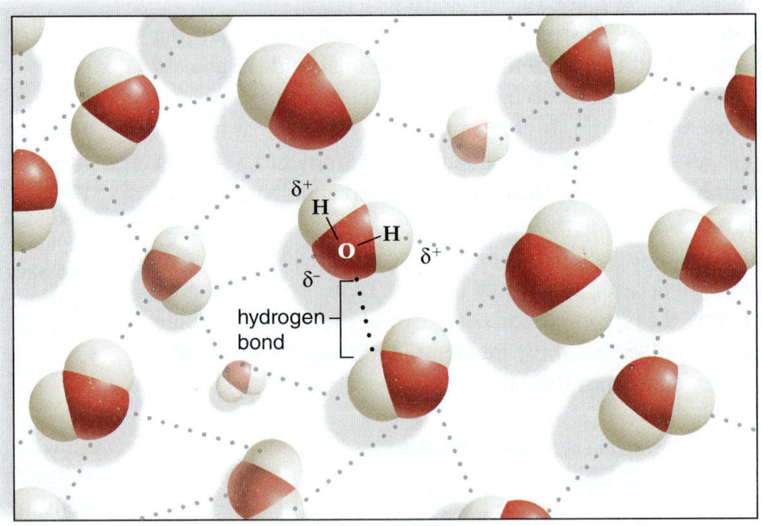

b. Hydrogen bonding between water molecules

Figure 2.9 Water molecule. **a.** Three models for the structure of water. The electron model does not indicate the shape of the molecule. The ball-and-stick model shows that the two bonds in a water molecule are angled at 104.5°. The space-filling model also shows the V shape of a water molecule. **b.** Hydrogen bonding between water molecules. Each water molecule can hydrogen bond with up to four other molecules, in three dimensions. When in a liquid state, water is constantly forming and breaking hydrogen bonds.

Tutorial Hydrogen Bonds

water molecules. Each of these hydrogen bonds is weaker than an ionic or covalent bond. The dotted lines indicate that hydrogen bonds are more easily broken than the other bonds.

Hydrogen bonding is not unique to water. Other biological molecules, such as DNA, have polar covalent bonds involving an electropositive hydrogen and usually an electronegative oxygen or nitrogen. In these instances, a hydrogen bond can occur within the same molecule or between nearby molecules.

Although a single hydrogen bond is more easily broken than a single covalent bond, multiple hydrogen bonds are collectively quite strong. Hydrogen bonds between cellular molecules help maintain their proper structure and function. For example, hydrogen bonds hold the two strands of DNA together. When DNA makes a copy of itself, hydrogen bonds easily break, allowing DNA to unzip. But normally, the hydrogen bonds add stability to the DNA molecule. Similarly, the shape of protein molecules is often maintained by hydrogen bonding between different parts of the same molecule. As we will see, many of the important properties of water are the result of hydrogen bonding.

Properties of Water

The first cell(s) evolved in water, and all living organisms are 70–90% water. Because of hydrogen bonding, water molecules cling together, and this association gives water its unique chemical properties. Without hydrogen bonding between molecules, water would freeze at –100°C and boil at –91°C, making most of the water on Earth steam, and life unlikely. Hydrogen bonding is responsible for water being a liquid at temperatures typically found on the Earth's surface. It freezes at 0°C and boils at 100°C. These and other unique properties of water make it essential to the existence of life as we know it. As noted in the chapter opener, the search for life on other planets often begins with the search for water.

Water Has a High Heat Capacity

A **calorie** is the amount of heat energy needed to raise the temperature of 1 g of water 1°C. In comparison, other covalently bonded liquids require input of only about half this amount of energy to rise in temperature 1°C. The many hydrogen bonds that link water molecules together help water absorb heat without a great change in temperature. Converting 1 g of the coldest liquid water to ice requires the loss of 80 calories of heat energy (Fig. 2.10*a*). Water holds on to its heat, and its temperature falls more slowly than that of other liquids. This property of water is important not only for aquatic organisms but for all life.

Because the temperature of water rises and falls slowly, organisms are better able to maintain their normal internal temperatures and are protected from rapid temperature changes.

Water Has a High Heat of Evaporation

When water boils, it evaporates, meaning that it vaporizes into the environment. Converting 1 g of the hottest water to a gas requires an input of 540 calories of energy. Water has a high heat of evaporation because hydrogen bonds must be broken before water boils.

Water's high heat of vaporization gives animals in a hot environment an efficient way to release excess body heat. When an animal sweats, or gets splashed, body heat is used to vaporize water, thus cooling the animal (Fig. 2.10*b*). Because of water's high heat of vaporization and ability to hold on to its heat, temperatures along the coasts are moderate. During the summer, the ocean absorbs and stores solar heat, and during the winter, the ocean releases it slowly. In contrast, the interior regions of continents experience abrupt changes in temperatures.

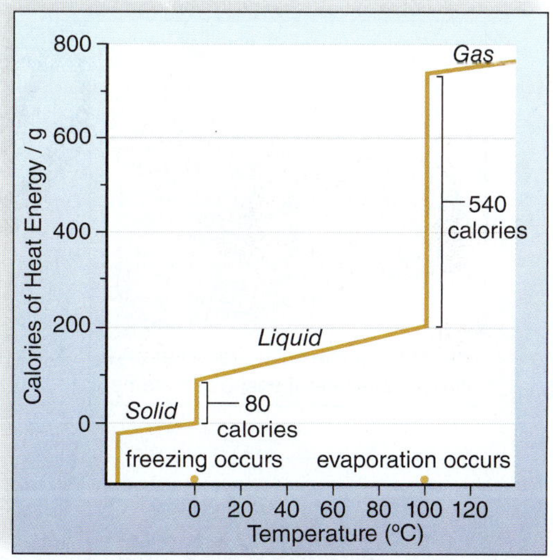

a. Calories lost when 1 g of liquid water freezes and calories required when 1 g of liquid water evaporates.

b. Bodies of organisms cool when their heat is used to evaporate water.

Figure 2.10 Temperature and water. **a.** Water can be a solid, a liquid, or a gas at naturally occurring environmental temperatures. At room temperature and pressure, water is a liquid. When water freezes and becomes a solid (ice), it gives off heat, and this heat can help keep the environmental temperature higher than expected. On the other hand, when water evaporates, it takes up a large amount of heat as it changes from a liquid to a gas. **b.** This means that splashing water on the body will help keep body temperature within a normal range.

Water Is a Solvent

Due to its polarity, water facilitates chemical reactions, both outside and within living systems. As a solvent, it dissolves a great number of substances, especially those that are also polar. A **solution** contains dissolved substances, which are then called **solutes.** When ionic salts—for example, sodium chloride (NaCl)—are put into water, the negative ends of the water molecules are attracted to the sodium ions, and the positive ends of the water molecules are attracted to the chloride ions. This attraction causes the sodium ions and the chloride ions to separate, or dissociate, in water.

Water is also a solvent for larger polar molecules, such as ammonia (NH_3).

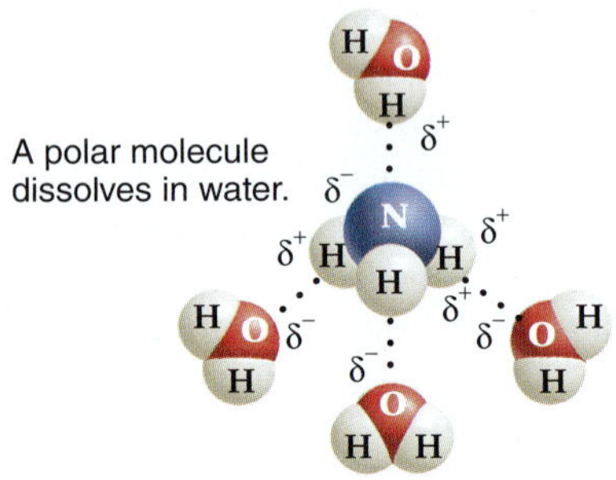

Molecules that can attract water are said to be **hydrophilic** (Gk. *hydrias,* "of water"; *phileo,* "love"). When ions and molecules disperse in water, they move about and collide, allowing reactions to occur. Nonionized and nonpolar molecules that cannot attract water are said to be **hydrophobic** (Gk. *phobos,* "fear"). Hydrophilic molecules tend to attract other polar molecules; similarly, hydrophobic substances usually associate with other nonpolar molecules. Gasoline contains nonpolar molecules; therefore, it does not mix with water and is hydrophobic.

Water Molecules Are Cohesive and Adhesive

Cohesion refers to the ability of water molecules to cling to each other due to hydrogen bonding. At any moment in time, a water molecule can form hydrogen bonds with at most four other water molecules. Because of cohesion, water exists as a liquid under the conditions of temperature and pressure present at the Earth's surface. The strong cohesion of water molecules is apparent because water flows freely, yet water molecules do not separate from each other.

Adhesion refers to the ability of water molecules to cling to other polar surfaces. This is a result of water's polarity. Multicellular animals often contain internal vessels in which water assists the transport of nutrients and wastes, because the cohesion and adhesion of water allow blood to fill the tubular vessels of the cardiovascular system. For example, the liquid portion of our blood, which transports dissolved and suspended substances about the body, is 90% water.

Cohesion and adhesion also contribute to the transport of water in plants. Plants have their roots anchored in the soil, where they absorb water, but the leaves are uplifted and exposed to solar energy. Water evaporating from the leaves is immediately replaced with water molecules from transport vessels that extend from the roots to the leaves (Fig. 2.11). Because water molecules are

Figure 2.11 Water molecules are cohesive and adhesive. Cohesion and adhesion play an important role in the movement of water in a plant. When water evaporates from the leaves, the water column is pulled upward due to the cohesion of water molecules to one another and the adhesion of water molecules to the sides of the vessels. This capillary action is critical for plants to function.

cohesive, a tension is created that pulls the water column up from the roots. Adhesion of water to the walls of the transport vessels also helps prevent the water column from breaking apart. This capillary action is essential to plant life, as will be discussed in Chapter 25.

Because water molecules are attracted to each other, they cling together where the liquid surface is exposed to air. The stronger the force between molecules in a liquid, the greater the **surface tension.** Water's high surface tension makes it possible for humans to skip rocks on water. Water striders, a common insect, can even walk on the surface of a pond without breaking the surface.

Frozen Water (Ice) Is Less Dense Than Liquid Water

As liquid water cools, the molecules come closer together. Water is most dense at 4°C, but the water molecules are still moving about (Fig. 2.12). At temperatures below 4°C, only vibrational movement occurs, and hydrogen bonding becomes more rigid but also more open. This means that water expands as it reaches 0°C and freezes, which is why cans of soda burst when placed in a freezer, or why frost heaves make northern roads bumpy in the winter. It also means that ice is less dense than liquid water, and therefore ice floats on liquid water.

If ice did not float on water, it would sink to the bottom, and ponds, lakes, and perhaps even the ocean would freeze solid, making life impossible in the water and on land. Instead, bodies of water always freeze from the top down. When a body of water freezes on the surface, the ice acts as an insulator to prevent the water below it from freezing. This allows aquatic organisms to survive the winter. As ice melts in the spring, it draws heat from the environment, helping prevent a sudden change in temperature, which might be harmful to life.

Check Your Progress 2.3

1. Explain how water's structure relates to the formation of hydrogen bonds.
2. Explain how hydrogen bonds relate to the properties of water.
3. Explain how spraying water on your body helps cool it off.

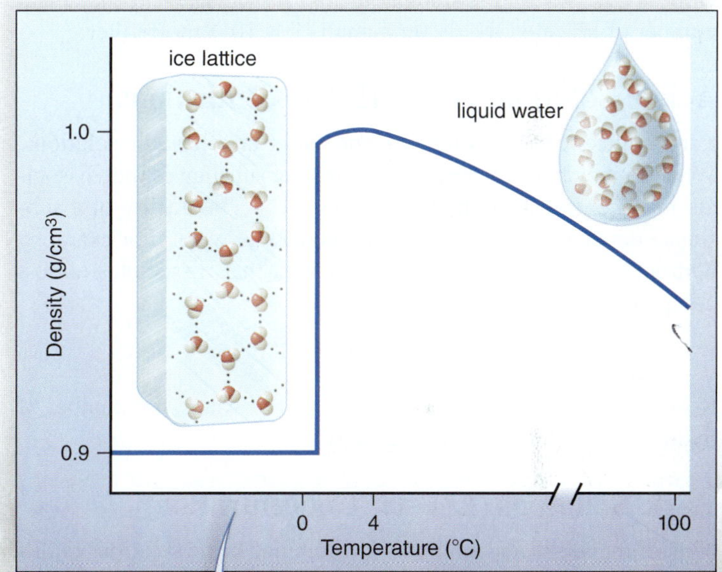

a.

b.

Figure 2.12 Ice is less dense than water. **a.** Water is more dense at 4°C than at 0°C. Most substances contract when they solidify, but water expands when it freezes, because in ice, water molecules form a lattice in which the hydrogen bonds are farther apart than in liquid water. **b.** This property of water allows ice to flow, providing habitats for some aquatic species and protecting other species that live beneath the ice.

2.4 Acids and Bases

Learning Outcomes

Upon completion of this section, you should be able to

1. Distinguish between an acid and a base.
2. Explain the relationship betwteen H^+ or OH^- concentration and pH.
3. Analyze how buffers prevent large pH changes in solutions.

When water ionizes, it releases an equal number of **hydrogen ions (H^+)** (sometimes just called protons[1]) and **hydroxide ions (OH^-)**:

$$\underset{\text{water}}{H{-}O{-}H} \rightleftharpoons \underset{\text{hydrogen ion}}{H^+} + \underset{\text{hydroxide ion}}{OH^-}$$

Only a few water molecules at a time dissociate, and the actual number of H^+ and OH^- is very small (1×10^{-7} moles/liter).[2]

Acidic Solutions (High H^+ Concentrations)

Lemon juice, vinegar, tomatoes, and coffee are all acidic solutions. What do they have in common? **Acids** are substances that dissociate in water, releasing hydrogen ions (H^+). The acidity of a substance depends on how fully it dissociates in water. For example, hydrochloric acid (HCl) is a strong acid that dissociates almost completely in this manner:

$$HCl \longrightarrow H^+ + Cl^-$$

If hydrochloric acid is added to a beaker of water, the number of hydrogen ions (H^+) increases greatly.

Basic Solutions (Low H^+ Concentration)

Milk of magnesia and ammonia are common basic solutions familiar to most people. **Bases** are substances that either take up hydrogen ions (H^+) or release hydroxide ions (OH^-). For example, sodium hydroxide (NaOH) is a strong base that dissociates almost completely in this manner:

$$NaOH \longrightarrow Na^+ + OH^-$$

If sodium hydroxide is added to a beaker of water, the number of hydroxide ions increases.

pH Scale

The **pH scale** is used to indicate the acidity or basicity (alkalinity) of a solution.[3] The pH scale (Fig. 2.13) ranges from 0 to 14. A pH of 7 represents a neutral state in which the hydrogen ion and hydroxide ion concentrations are equal. A pH below 7 is an acidic solution, because the hydrogen ion concentration is greater than the hydroxide concentration. A pH above 7 is basic, because the $[OH^-]$ is greater than the $[H^+]$. Further, as we move down the pH scale from pH 14 to pH 0, each unit is 10 times more acidic than the previous unit. As we move up the scale from 0 to 14, each unit is 10 times more basic than the previous unit. Therefore, pH 5 is 100 times more acidic than pH 7 and 100 times more basic than pH 3.

The pH scale was devised to eliminate the use of cumbersome numbers. For example, the possible hydrogen ion concentrations of a solution are on the left of this listing and the pH is on the right:

$[H^+]$ (moles per liter)		pH
0.000001	$= 1 \times 10^{-6}$	6
0.0000001	$= 1 \times 10^{-7}$	7
0.00000001	$= 1 \times 10^{-8}$	8

To further illustrate the relationship between hydrogen ion concentration and pH, consider the following question. Which of the pH values listed indicates a higher hydrogen ion concentration $[H^+]$ than pH 7, and therefore would be an acidic solution? A number with a smaller negative exponent indicates a greater quantity of hydrogen ions than one with a larger negative exponent. Therefore, pH 6 is an acidic solution.

The Biological Systems feature, "The Impact of Acid Deposition," describes detrimental environmental consequences of low pH rain and snow. In humans, pH needs to be maintained within a narrow range, or there are health consequences. The pH of blood is around 7.4, and blood is buffered in the manner described next to keep the pH within a normal range.

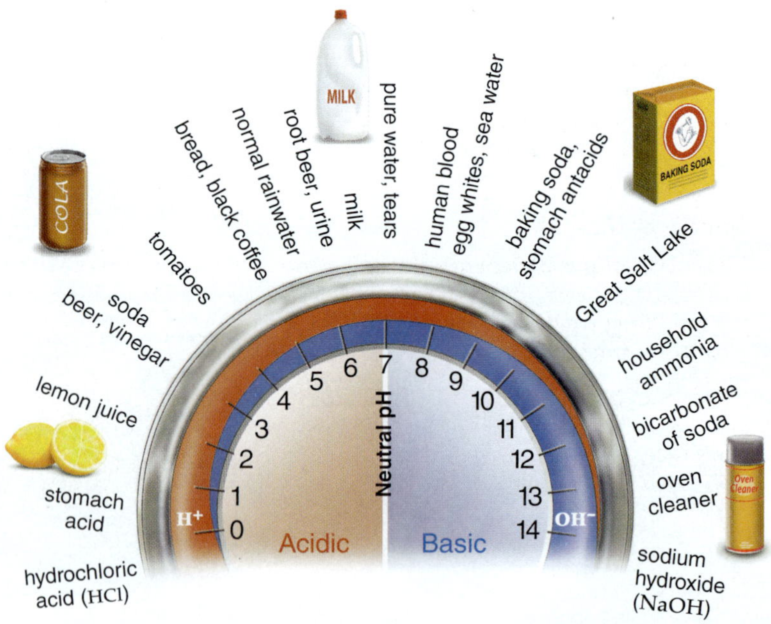

Figure 2.13 The pH scale. The pH scale ranges from 0 to 14, with 0 being the most acidic and 14 being the most basic. pH 7 (neutral pH) has equal amounts of hydrogen ions (H^+) and hydroxide ions (OH^-). An acidic pH has more H^+ than OH^- and a basic pH has more OH^- than H^+.

1 A hydrogen atom contains one electron and one proton. A hydrogen ion has only one proton, so it is often simply called a proton.

2 In chemistry, a mole is defined as 6.02×10^{23} of any atom, molecule, or ion. For example, 6.02×10^{23} atoms of 12C would have a mass of exactly 12 g. The same number of glucose molecules (1 mole) would have a mass of 180 g.

3 pH is defined as the negative log of the hydrogen ion concentration $[H^+]$. A log is the power to which 10 must be raised to produce a given number.

THEME Biological Systems

The Impact of Acid Deposition

Acid Deposition

Normally, rainwater has a pH of about 5.6 because the carbon dioxide in the air combines with water to give a weak solution of carbonic acid. Acid deposition includes rain or snow that has a pH of less than 5, as well as dry acidic particles that fall to Earth from the atmosphere.

When fossil fuels such as coal, oil, and gasoline are burned, sulfur dioxide and nitrogen oxides combine with water to produce sulfuric and nitric acids. These pollutants are generally found eastward of where they originated because of wind patterns. The use of very tall smokestacks causes them to be carried even hundreds of miles away. For example, acid rain in southeastern Canada results from the burning of fossil fuels in factories and power plants in the midwestern United States.

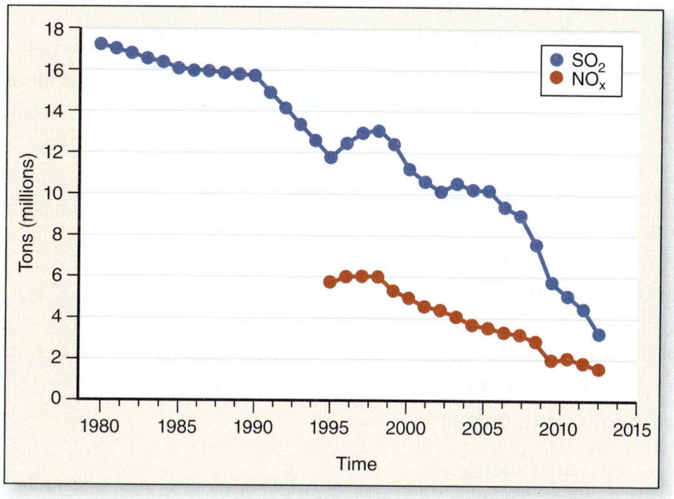

Figure 2A Trends in U.S. acid rain emissions. The burning of fossil fuels in factories, automobiles, and other industrial processes produces chemicals like SO_2 and NO_x that lead to acid deposition and destruction of the environment. Clean air legislation and stricter emission standards over the past several decades have resulted in steady decreases in SO_2 and NO_x, chemicals that lead to acid rain. Source: EPA.gov

Impact on Lakes

Acid rain adversely affects many aspects of biological systems. Aluminum may leach from the soil of lakes, particularly in areas where the soil is thin and lacks limestone (calcium carbonate, or $CaCO_3$) as a buffer. Acid rain may convert mercury in lake bottom sediments to toxic methyl mercury. Methyl mercury accumulates in fish, which wildlife and people eat. Over time, methyl mercury can accumulate in body tissues and cause serious sensory and muscular health problems. Acid rain in Canada and New England has caused hundreds of lakes to be devoid of fish, and in some cases, any life at all.

Impact on Forests

The leaves of plants damaged by acid rain can no longer carry on photosynthesis as before. When plants are under stress, they become susceptible to diseases and pests of all types. Forests on mountaintops receive more rain than those at lower levels; therefore, they are more affected by acid rain. Forests are also damaged when toxic chemicals such as aluminum are leached from the soil. These kill soil fungi that assist roots in acquiring the nutrients trees need. In New England and the southern Appalachians, millions of acres of high-elevation forests have been devastated. Sulfur dioxide and nitrogen oxides, the main precursors of acid rain, have been steadily decreasing in the United States due to clean air legislation and strict emission limits (see Fig. 2A).

Impact on Humans and Structures

Humans may be affected by acid rain. Inhaling dry sulfate and nitrate particles appears to increase the occurrence of respiratory illnesses, such as asthma. Buildings and monuments made of limestone and marble break down when exposed to acid rain. The paint on homes and automobiles is likewise degraded. However, damage to natural systems and human structures due to acid rain is likely to decrease if we continue efforts to reduce chemicals that contribute to acid rain.

Questions to Consider

1. What acid rain trends are evident from the EPA data?
2. Considering that manufacturing is essential to our national interests, how might we modify industrial processing to reduce sulfur dioxide and nitrogen oxide contamination?
3. How might we prevent methyl mercury from entering biological systems and reduce the amount already present?

Buffers and pH

A **buffer** is a chemical or a combination of chemicals that keeps pH within normal limits. Many commercial products, such as shampoos or deodorants, are buffered as an added incentive for us to buy them.

In living organisms, the pH of body fluids is maintained within a narrow range, or else molecules don't function correctly and our health suffers. The pH of our blood when we are healthy is always about 7.4—that is, just slightly basic (alkaline). If the blood pH drops to about 7, acidosis results. If the blood pH rises to about 7.8, alkalosis results. Both conditions can be life threatening, so the blood pH must be kept around 7.4. Normally, pH stability is possible because the body has built-in mechanisms to prevent pH changes. Buffers are one of these important mechanisms.

Buffers help keep the pH within normal limits because they are chemicals or combinations of chemicals that take up excess hydrogen ions (H^+) or hydroxide ions (OH^-). For example, carbonic acid (H_2CO_3) is a weak acid that minimally dissociates and then re-forms in the following manner:

$$\underset{\text{carbonic acid}}{H_2CO_3} \underset{\text{re-forms}}{\overset{\text{dissociates}}{\rightleftharpoons}} H^+ + \underset{\text{bicarbonate ion}}{HCO_3^-}$$

Blood always contains a combination of some carbonic acid and some bicarbonate ions. When hydrogen ions (H^+) are added to blood, the following reaction reduces acidity:

$$H^+ + HCO_3^- \longrightarrow H_2CO_3$$

When hydroxide ions (OH^-) are added to blood, this reaction reduces basicity:

$$OH^- + H_2CO_3 \longrightarrow HCO_3^- + H_2O$$

These reactions prevent any significant change in blood pH.

Check Your Progress 2.4

1. Explain the difference in H^+ concentration between an acid and a base.
2. Determine how much more acidic a pH of 2.0 is than a pH of 4.0.
3. Summarize how buffers play an important role in the physiology of living organisms.

CONNECTING *the* CONCEPTS *with the* THEMES

Evolution

- Six of the 92 naturally occurring elements—carbon, hydrogen, nitrogen, oxygen, phosphorus, and sulfur—make up 95% of the body mass of organisms on Earth.
- Elements may be combined by chemical bonds to form the structures associated with life.
- Our undertanding of life focuses on its interaction with water. Our search for water on other planets underscores the central role that water plays in living matter.

Nature of Science

- Scientific discoveries about chemistry, water, and carbon help us understand substances that are built from these atoms.
- Our understanding of life continues to change over time as we uncover fundamental truths about chemistry. Items we use every day, including food, plastics, and medicine, are products of ongoing research in chemistry.

Biological Systems

- All of life consists of systems, from the atomic and molecular levels to the ecosystem and planet levels.
- Like all systems, chemistry consists of inputs and outputs, with resulting changes in energy.
- Many elements exist as ions in the body, and their functions depend on their charged nature.
- Cells consist largely of water, a molecule that contains only hydrogen and oxygen atoms. Because of oxygen's strong electronegativity, water is a polar molecule, making hydrogen bonding possible.

MEDIA STUDY TOOLS http://connect.mheducation.com

SMARTBOOK® Study smart. SmartBook helps you maximize your study time by identifying what you know and don't know. The recharge feature in SmartBook helps you prepare for the test by identifying the material that you are most likely to forget.

connect BIOLOGY Find even more resources to learn the chapter concepts including animations, tutorial videos, and interactive practice questions.

MP3 File	Animation	Tutorial
2.2 Chemical Bonding	**2.2** Ionic Versus Covalent Bonding	**2.3** Hydrogen Bonds

SUMMARIZE

2.1 Chemical Elements

Both living organisms and nonliving things are composed of **matter** consisting of **elements.** The most significant elements that are found in living organisms include carbon, hydrogen, nitrogen, oxygen, phosphorus, and sulfur.

Elements are identified by a unique **atomic symbol,** such as Na for sodium. Elements contain **atoms,** and atoms contain subatomic particles. **Protons** have positive charges, **neutrons** are uncharged, and **electrons** have negative charges. Protons and neutrons in the nucleus determine the **mass number** of an atom. The **atomic number** indicates the number of protons and the number of electrons in electrically neutral atoms.

Isotopes are atoms of a single element that differ in their numbers of neutrons. The **atomic mass** of an element is the average of the naturally occurring isotopes. Radioactive isotopes have many uses, including serving as tracers in biological experiments and medical procedures.

Electrons occupy energy levels (**electron shells**) at discrete distances from the nucleus. The number of electrons in the **valence shell** determines the reactivity of an atom. The first shell is complete when it is occupied by two electrons. In atoms up through calcium, number 20, every shell beyond the first shell is stable when eight electrons are present. The **octet rule** states that atoms react with one another in order to have a stable valence shell. Most atoms do not have filled outer shells, and this causes them to react with one another to satisfy the octet rule. In the process, they form compounds and/or molecules.

2.2 Molecules and Compounds

Compounds and **molecules** are formed when elements associate with each other. A **formula** tells you the number of each kind of element in a compound or molecule. **Ions** are formed when atoms lose or gain one or more electrons to achieve a completed valence shell. The interaction of oppositely charged ions forms **ionic bonds.**

Covalent bonds occur when atoms share one or more pairs of electrons. There are single, double, and triple covalent bonds.

In **nonpolar covalent bonds** the atoms share the electrons equally. In **polar covalent bonds,** the sharing of electrons is not equal. If the molecule forms polar covalent bonds, the more **electronegative** atom carries a slightly negative charge and the other atom carries a slightly positive charge.

2.3 Chemistry of Water

Water is an essential molecule for life. The polarity of water molecules forms hydrogen bonds between water molecules. A **hydrogen bond** is a weak attraction between a slightly positive hydrogen atom and a slightly negative oxygen or nitrogen atom within the same or a different molecule. Hydrogen bonds help maintain the structure and function of cellular molecules.

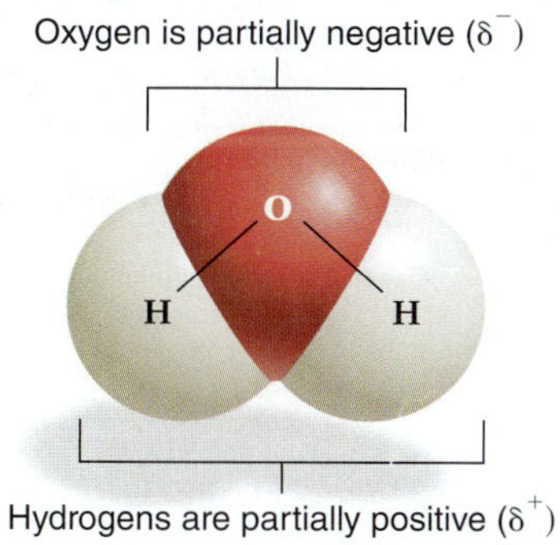

Water's polarity and hydrogen bonding account for its unique properties, which allow life to exist and carry on cellular activities. These properties include

- Water has a high heat capacity. Water can absorb heat (measure in **calories**) without a great change in temperature.
- Water has a high heat of evaporation. A large amount of heat is required to cause liquid water to change to a gas.
- Water is a solvent. Water allows the formation of **solutions** with many different **solutes.** Molecules that attract water are **hydrophilic,** whereas molecules that repel water are **hydrophobic.**
- Water is cohesive and adhesive. **Cohesion** allows water molecules to cling together and accounts for the **surface tension** of water. **Adhesion** allows water to cling to surfaces, such as internal transport vessels.
- Frozen water is less dense than liquid water. This allows ice to float on liquid water.

2.4 Acids and Bases

pH is a measure of how acidic or basic a substance is. A small fraction of water molecules dissociate to produce an equal number of **hydrogen ions** and **hydroxide ions.** The **pH scale** indicates the concentration of hydrogen ions. Solutions with equal numbers of H^+ and OH^- are termed neutral.

Acids release hydrogen ions in water; these solutions have a pH less than 7. **Bases** either release hydroxide ions or take up hydrogen ions; these solutions have a pH greater than 7. Cells are sensitive to pH changes. Biological systems often contain **buffers** that help keep the pH within a normal range.

ASSESS

Choose the best answer for each question.

2.1 Chemical Elements

1. The atomic number tells you the
   a. number of neutrons in the nucleus.
   b. number of protons in the atom.
   c. atomic mass of the atom.
   d. number of its electrons if the atom is neutral.
   e. Both b and d are correct.
2. Isotopes differ in their
   a. number of protons.
   b. atomic number.
   c. number of neutrons.
   d. number of electrons.
3. The periodic table provides us with what information?
   a. the atomic number, symbol, and mass
   b. how many shells an atom has
   c. how many electrons are in the outer shell
   d. All of these are correct.
4. Which of the subatomic particles contributes almost no weight to an atom?
   a. protons in the electron shells
   b. electrons in the nucleus
   c. neutrons in the nucleus
   d. electrons at various energy levels

2.2 Molecules and Compounds

5. An atom that has two electrons in the valence shell, such as magnesium, would most likely
   a. share to acquire a completed outer shell.
   b. lose these two electrons and become a negatively charged ion.
   c. lose these two electrons and become a positively charged ion.
   d. bind with carbon by way of hydrogen bonds.
   e. bind with another magnesium atom to satisfy its energy needs.
6. When an atom gains electrons, it
   a. forms a negatively charged ion.
   b. forms a positively charged ion.
   c. forms covalent bonds.
   d. forms ionic bonds.
   e. gains atomic mass.
7. An unequal sharing of electrons is a characteristic of a/an
   a. ionic bond.
   b. polar covalent bond.
   c. nonpolar covalent bond.
   d. All of these are correct.

2.3 Chemistry of Water

8. Hydrogen bonds are formed as a result of which of the following?
   a. ionic bonds
   b. nonpolar covalent bonds
   c. polar covalent bonds
   d. None of these are correct.
9. Which of these properties of water can be attributed to hydrogen bonding between water molecules?
   a. Water stabilizes temperature inside and outside the cell.
   b. Water molecules are cohesive.
   c. Water is a solvent for many molecules.
   d. Ice floats on liquid water.
   e. All of the above are correct.

For questions 10–13, match the statements with a property of water in the key.

KEY:

a. Water flows because it is cohesive.
b. Water holds its heat.
c. Water has neutral pH.
d. Water has a high heat of vaporization.

10. Sweating helps cool us off.
11. Our blood is composed mostly of water and cells.
12. Our blood is just about pH 7.
13. We usually maintain a normal body temperature.

2.4 Acids and Bases

14. $H_2CO_3/NaHCO_3$ is a buffer system in the body. What effect will the addition of an acid have on the pH of a solution that is buffered?
    a. The pH will rise.
    b. The pH will lower.
    c. The pH will not change.
    d. All of these are correct.
15. Which of these best describes the changes that occur when a solution goes from pH 5 to pH 7?
    a. The solution is now 100 times more acidic.
    b. The solution is now 100 times more basic.
    c. The hydrogen ion concentration decreases by only a factor of 20, as the solution goes from basic to acidic.
    d. The hydrogen ion concentration changes by only a factor of 20, as the solution goes from acidic to basic.

ENGAGE

LEARNSMART LABS®

The following LearnSmart Labs contain exercises that are related to the content of this chapter:

- pH and Cells

THINKING SCIENTIFICALLY

1. Natural phenomena often require an explanation. Based on how sodium chloride dissociates in water (see page 28) and Figure 2.12, explain why the oceans don't freeze.
2. Melvin Calvin used radioactive carbon (as a tracer) to discover a series of molecules that form during photosynthesis. Explain why carbon behaves chemically the same, even when radioactive.
3. On a hot summer day, you decide to dive into a swimming pool. Before you begin your dive, you notice that the surface of the water is smooth and continuous. After the dive, you discover that some water droplets are clinging to your skin and that your skin temperature feels cooler. Explain these observations based on the properties of water.
4. Like carbon, silicon has four electrons in its outer shell, yet life evolved to be carbon-based. What is there about silicon's structure that might prevent it from sharing with four other elements and prevent it from forming the many varied shapes of carbon molecules?
5. Antacids are a common over-the-counter remedy for heartburn, a condition caused by an overabundance of H^+ ions in the stomach. Based on what you know about pH, how do the chemicals in antacids work?

Trans fats are an organic molecule found in processed foods.

3

The Chemistry of Organic Molecules

Chapter Outline

3.1 Organic Molecules 36
3.2 Carbohydrates 39
3.3 Lipids 42
3.4 Proteins 46
3.5 Nucleic Acids 50

Before You Begin

Before beginning this chapter, take a few moments to review the following discussions.

Section 2.1 What is a covalent bond?

Section 2.2 What information about a molecule can be obtained from its structural and molecular formulas?

Section 2.3 What is the difference between hydrophobic and hydrophilic molecules?

All life is interconnected, because it uses a common set of chemicals to build larger, more complex molecules. These macromolecules—namely, carbohydrates, lipids, proteins, and nucleic acids—are combined to produce different structures, which lead them to have different functions. However, at times, some of these macromolecules can produce health issues in humans. For example, historically we have associated cardiovascular health problems with high levels of cholesterol and fat in the diet. However, we now know that trans fats are worse offenders than saturated fats when it comes to cardiovascular health. While trans fats are found naturally in small quantities in milk and meat products, the majority of trans fats are obtained from processed foods. Trans fats are formed from liquid unsaturated fats when hydrogen has been added to them. This process, called hydrogenation, makes a liquid fat resemble a solid saturated fat. Why is this done? Historically, the food industry added trans fats not only to increase the shelf life of processed food but also to enhance the flavor and texture.The Food and Drug Administration (FDA) now directs that trans fat content must be disclosed on Nutrition Facts panels of food products. In response, many food manufacturers and restaurants have lowered the trans fat content of their foods.

An understanding of the structure and function of carbohydrates, lipids, proteins, and nucleic acids is important not only for knowing the main building blocks of cells but also for understanding how to establish a healthy diet.

As you read through this chapter, think about the following questions:

1. What are the general characteristics of a lipid?
2. How does the structure of a trans fat differ from that of a saturated fat?

Following *the* Themes

CHAPTER 3 THE CHEMISTRY OF ORGANIC MOLECULES

UNIT 1 THE CELL

Evolution	The diversity of biological life is the result of changes in DNA sequences, one of the many organic molecules found in cells.
Nature of Science	An understanding of a molecule's structure provides insight into its function.
Biological Systems	Organic molecules form the functional basis for all cellular systems.

3.1 Organic Molecules

Learning Outcomes

Upon completion of this section, you should be able to

1. Explain how the properties of carbon enable it to produce diverse organic molecules.
2. Explain the relationship between a functional group and the chemical reactivity of an organic molecule.
3. Compare the role of dehydration synthesis and hydrolytic reactions in organic chemistry.

Chemists of the nineteenth century thought that the molecules of cells must contain a vital force, so they divided chemistry into **organic chemistry,** the chemistry of living organisms, and **inorganic chemistry,** the chemistry of nonliving matter. We still use this terminology, even though many types of organic molecules can now be synthesized in the laboratory. Today, we simply use the term **organic** to identify those molecules and compounds that contain both carbon and hydrogen atoms.

There are only four classes of organic molecules in any living organism: carbohydrates, lipids, proteins, and nucleic acids. Collectively, these are called the **biomolecules;** despite the limited number of types, their functions in a cell are quite diverse. A bacterial cell contains some 5,000 different organic molecules, and a plant or an animal cell has twice that number. The diversity of life (Fig. 3.1) is possible because of the diversity of organic molecules. Despite their functional differences, the variety of organic molecules is based on the unique chemical properties of the carbon atom.

MP3 Organic Molecules

The Carbon Atom

What is there about carbon that makes organic molecules the same but also different? Carbon is quite small, with only a total of six electrons: two electrons in the first shell and four electrons in the outer shell. To acquire four electrons to complete its outer shell, a carbon atom almost always forms covalent bonds. Carbon can form covalent bonds with as many as four other elements.

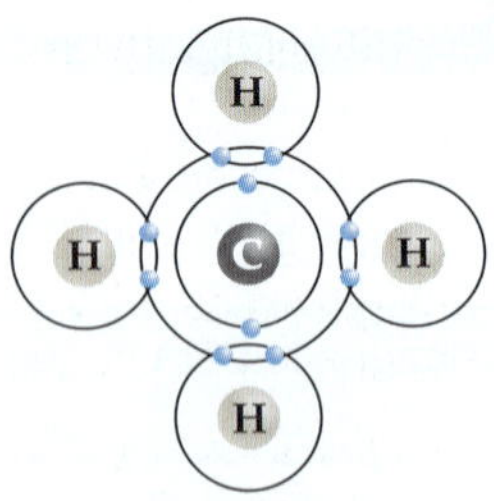

Methane

Generally, carbon forms those bonds with other atoms of carbon, plus hydrogen, nitrogen, oxygen, phosphorus, and sulfur—the same elements that make up most of the weight of living organisms (see Fig. 2.1). The ability of carbon to share electrons with other carbon atoms plays an important role in establishing the shape, and therefore the function, of the biomolecules. This is because the C—C bond is very stable and allows the formation of long carbon chains. The molecules termed hydrocarbons, such as the octane molecule below, are chains of carbon atoms that have additional bonds exclusively with hydrogen atoms.

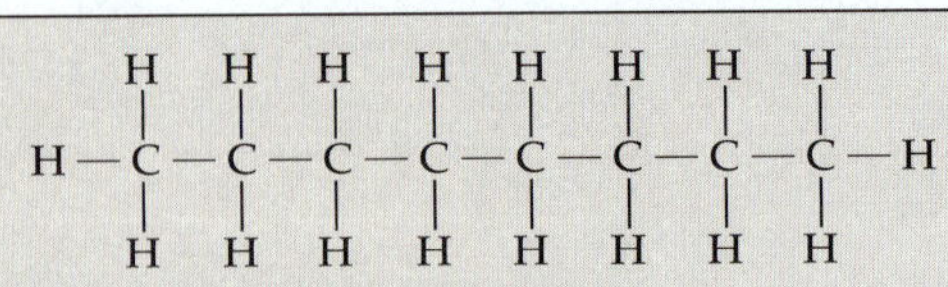

octane

Figure 3.1 Carbon and life. Carbon is the basis of life as we know it. The structure of carbon allows for the formation of (**a**) the lipids that store energy in this canola plant; (**b**) carbohydrates that provide structure for this tree; (**c**) the proteins that form the hemoglobin of red blood cells; and (**d**) the genetic material that allows this lioness to pass on information to her offspring.

In addition to long, linear molecules, hydrocarbons may also form a ring structure when placed in a water environment:

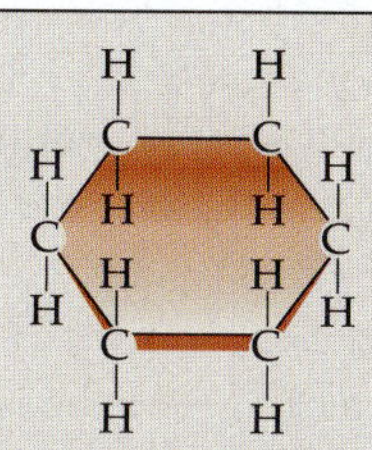

cyclohexane

In addition to forming single bonds, carbon can form double bonds with itself and other atoms. Double bonds aren't as flexible as single bonds, and they restrict the movement of bonded atoms. Double bonds affect a molecule's shape and therefore influence its function. The presence of double bonds is one way to distinguish between saturated and unsaturated fats, which are important to heart health. Carbon is also capable of forming triple bonds with itself, as in acetylene, H—C≡C—H, a gas used in industrial welding. Branches may also form at any carbon atom, allowing the formation of long, complex carbon chains. This flexibility makes carbon the ideal building block for biomolecules, and it plays an important role in establishing the diversity of organic molecules that we observe in nature.

The Carbon Skeleton and Functional Groups

The carbon chain of an organic molecule is called its skeleton, or backbone. Just as a skeleton accounts for your body's shape, so does the carbon skeleton of an organic molecule account for its shape. The diversity of vertebrates, species with a backbone, results from the overall shapes of the organisms and the types of appendages (fins, wings, limbs) they have developed. Likewise, the diversity of organic molecules comes from the attachment of different functional groups to the carbon skeleton.

A **functional group** is a specific combination of bonded atoms that always has the same chemical properties and therefore always reacts in the same way, regardless of the carbon skeleton to which it is attached. The majority of the chemical reactivity of a biomolecule can be attributed to its functional groups, rather than to the carbon skeleton to which it is attached.

Typically, the carbon skeleton acts as a framework for the positioning of the functional groups. Table 3.1 lists some of the more common funtional groups. The *R* indicates the "remainder" of the molecule. This is the place on the functional group that attaches to the carbon skeleton.

The configuration of the functional groups determines the properties of the biomolecule. For example, the addition of an —OH (hydroxyl group) to a carbon skeleton turns that molecule into an alcohol. When an —OH replaces one of the hydrogens in ethane, a 2-carbon hydrocarbon, it becomes ethanol, a type of alcohol that is familiar, because humans can consume it. Whereas ethane, like other hydrocarbons, is hydrophobic (not soluble in water), ethanol is hydrophilic (soluble in water), because the —OH functional group makes the otherwise nonpolar carbon skeleton polar. Because cells are 70–90% water, the ability to interact with and be soluble in water profoundly affects the function of organic molecules in cells.

Table 3.1 Functional Groups

Group	Structure	Compound	Significance
Hydroxyl	*R*—OH	Alcohol as in ethanol	Polar, forms hydrogen bond Present in sugars, some amino acids
Carbonyl	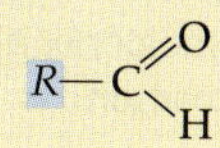	Aldehyde as in formaldehyde	Polar Present in sugars
	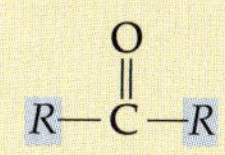	Ketone as in acetone	Polar Present in sugars
Carboxyl (acidic)	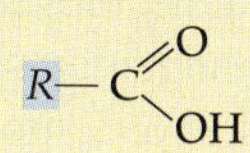	Carboxylic acid as in acetic acid	Polar, acidic Present in fatty acids, amino acids
Amino	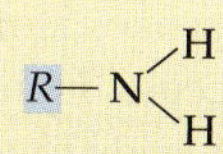	Amine as in tryptophan	Polar, basic, forms hydrogen bonds Present in amino acids
Sulfhydryl	*R*—SH	Thiol as in ethanethiol	Forms disulfide bonds Present in some amino acids
Phosphate	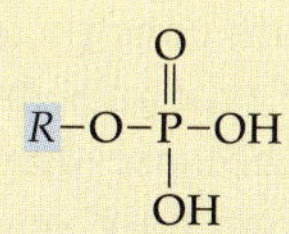	Organic phosphate as in phosphorylated molecules	Polar, acidic Present in nucleotides, phospholipids

R = remainder of molecule

Another example is organic molecules that contain carboxyl groups (—COOH). Carboxyl groups are highly polar. In a water environment, they tend to ionize and release hydrogen ions in solution, therfore acting as an acid:

$$-COOH \longrightarrow -COO^- + H^+$$

The attached functional groups determine not only the polarity of an organic molecule but also the types of reactions it will undergo. You will see that alcohols react with carboxyl groups when a fat forms, and that carboxyl groups react with amino groups during protein formation.

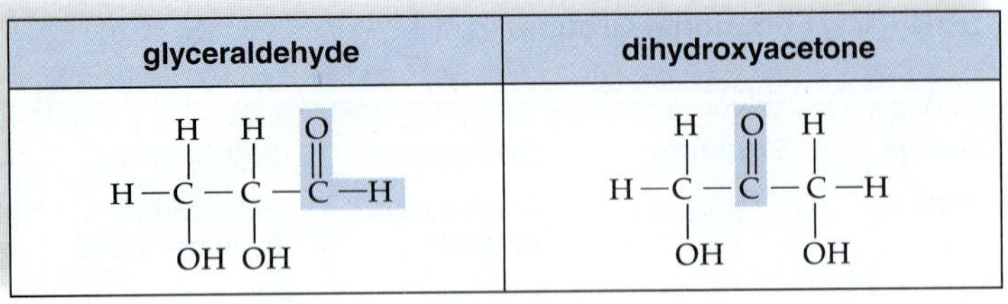

Figure 3.2 Isomers. Isomers have the same molecular formula but different atomic configurations. Both of these compounds have the formula $C_3H_6O_3$. In glyceraldehyde, a colorless crystalline solid, oxygen is double-bonded to an end carbon. In dihydroxyacetone, a white crystalline solid, oxygen is double-bonded to the middle carbon.

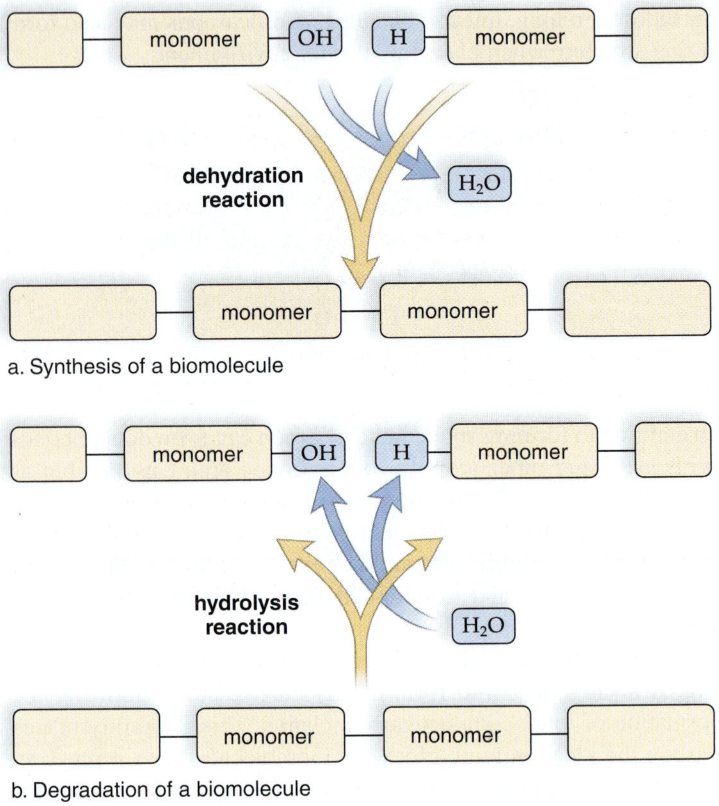

Figure 3.3 Synthesis and degradation of biomolecules. **a.** In cells, synthesis often occurs when subunits bond during a dehydration reaction (removal of H_2O). **b.** Degradation occurs when the subunits separate during a hydrolysis reaction (the addition of H_2O).

Isomers

Isomers (Gk. *isos,* "equal") are organic molecules that have identical molecular formulas but different arrangements of atoms. The two molecules in Figure 3.2 are isomers of one another; they have the same molecular formula but different functional groups. Therefore, we would expect them to have different properties and react differently in chemical reactions. In essence, isomers are variations in the molecular structure of a molecule. Isomers are another example of how the chemistry of carbon leads to variations in the structure of organic molecules.

The Biomolecules of Cells

Many of the biomolecules you are familiar with, such as carbohydrates, lipids, proteins, and nucleic acids, are macromolecules, meaning that they contain smaller subunits joined together (Table 3.2). The carbohydrates, proteins, and nucleic acids are referred to as **polymers,** since they are constructed by linking together a large number of the same type of subunit, called a **monomer.** Lipids are not polymers, because they contain two different types of subunits (glycerol and fatty acids). Polymers may vary considerably in length. Just as a train increases in length when boxcars are hitched together one by one, so a polymer gets longer as monomers bond to one another.

Synthesis and Degradation

To build, or synthesize, a macromolecule, the cell uses a condensation reaction. This is commonly called a **dehydration reaction,** because the equivalent of a water molecule—that is, an —OH (hydroxyl group) and an —H (hydrogen atom)—is removed as subunits are joined. Therefore, water molecules are formed as biomolecules are synthesized (Fig. 3.3*a*).

To break down biomolecules, a cell uses an opposite type of reaction. During a **hydrolysis reaction** (Gk. *hydro,* "water"; *lyse,* "break"), an —OH group from water attaches to one subunit, and an —H from water attaches to the other subunit. In other words, hydrolytic reactions break down biomolecules by adding water to them (Fig. 3.3*b*).

These reactions rarely occur spontaneously. Usually, special molecules called *enzymes* act as catalysts that allow the reaction to occur or speed up the rate of the reaction. We will take a closer look at enzymes in Chapter 6.

Table 3.2 Biomolecules

Category	Subunits (monomers)	Polymer
Carbohydrates*	Monosaccharide	Polysaccharide
Lipids	Glycerol and fatty acids	Fat
Proteins*	Amino acids	Polypeptide
Nucleic acids*	Nucleotide	DNA, RNA

*form polymers

Check Your Progress 3.1

1. Describe the properties of a carbon atom that make it ideally suited to produce varied carbon skeletons.
2. Explain why the substitution of a carboxyl group for a hydroxyl group in a biomolecule would change the molecule's function.
3. Explain why water is needed for the breakdown of a biomolecule.

3.2 Carbohydrates

Learning Outcomes

Upon completion of this section, you should be able to

1. Summarize the role of carbohydrates in a cell.
2. Distinguish among the forms of carbohydrates.
3. Compare the energy and structural uses of starch, glycogen, and cellulose.

Carbohydrates are almost universally used as an immediate energy source in living organisms, but in some organisms they also have a structural function (Fig. 3.4). The majority of carbohydrates have a carbon to hydrogen to oxygen ratio of 1:2:1 (CH_2O). The term *carbohydrate* (literally, carbon-water) includes single sugar molecules and chains of sugars. Chain length varies from a few sugars to hundreds of sugars. The monomer subunits, called monosaccharides, are assembled into long polymer chains called polysaccharides.

a. Cell walls contain cellulose.

b. Shell contains chitin.

c. Cell walls contain peptidoglycan.

Figure 3.4 Carbohydrates as structural materials. **a.** Plants, such as the cactus shown here, have the carbohydrate cellulose in their cell walls. **b.** The shell of a crab contains a carbohydrate called chitin. **c.** The cell walls of bacteria contain a carbohydrate known as peptidoglycan.

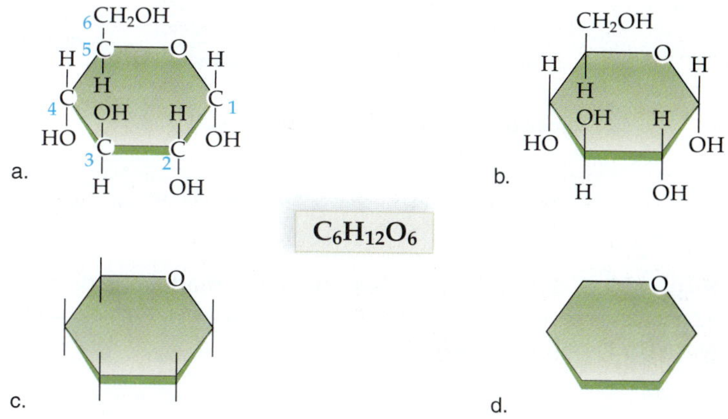

Figure 3.5 Glucose. Glucose is the common form of monosaccharide that provides energy for cells. Each of these structural formulas is glucose. **a.** The carbon skeleton and all attached groups are shown. **b.** The carbon skeleton is omitted. **c.** The carbon skeleton and attached groups are omitted. **d.** Only the ring shape, which includes one oxygen atom, remains.

Monosaccharides

Monosaccharides (Gk. *monos,* "single"; *sacchar,* "sugar") consist of only a single sugar molecule and are commonly called simple sugars. A monosaccharide can have a carbon backbone of three to seven carbons. For example, **pentoses** (Gk. *pent,* "five") are monosacharides with five carbons, and **hexoses** (Gk. *hex,* "six") are monosaccharides with six carbons. Monosaccharides have a large number of hydroxyl groups, and the presence of this polar functional group makes them soluble in water.

An example of a hexose is **glucose** (Fig. 3.5). Glucose has a molecular formula of $C_6H_{12}O_6$. Despite the fact that glucose has several isomers, such as fructose and galactose, we usually think of $C_6H_{12}O_6$ as glucose. Glucose is critical to biological function and is the major source of cellular fuel for all living organisms. Glucose is the molecule that is broken down and converted into stored chemical energy (ATP) during cellular respiration in nearly all types of organisms.

Ribose and **deoxyribose,** with five carbon atoms, are pentose sugars that are significant because they are found, respectively, in the nucleic acids RNA and DNA. RNA and DNA will be discussed in more detail when we examine nucleic acids later in the chapter.

Disaccharides

A **disaccharide** contains two monosaccharides that have joined during a dehydration reaction. Figure 3.6 shows how the disaccharide maltose (an ingredient used in brewing) is formed when two glucose molecules bond together. Note the position of the bond that results when the $-OH$ groups participating in the reaction project below the ring. When the enzymes in our digestive system break this bond, the result is two glucose molecules.

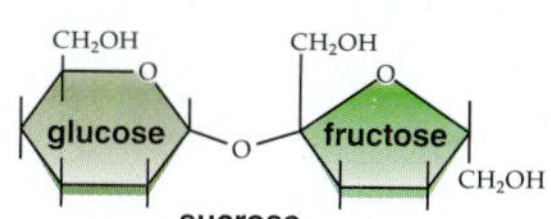

Sucrose (the structure shown above) is another disaccharide of special interest, because it is sugar we use at home to sweeten

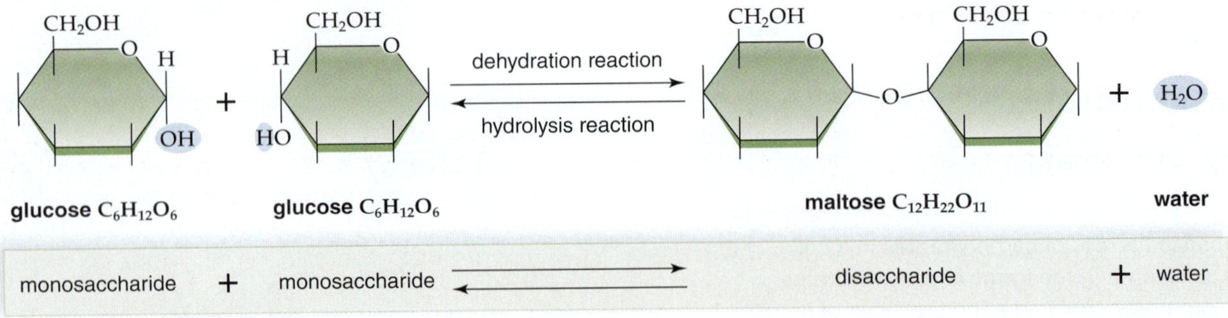

Figure 3.6 Synthesis and degradation of maltose, a disaccharide. Synthesis of maltose occurs following a dehydration reaction when a bond forms between two glucose molecules, and water is removed. Degradation of maltose occurs following a hydrolysis reaction when this bond is broken by the addition of water.

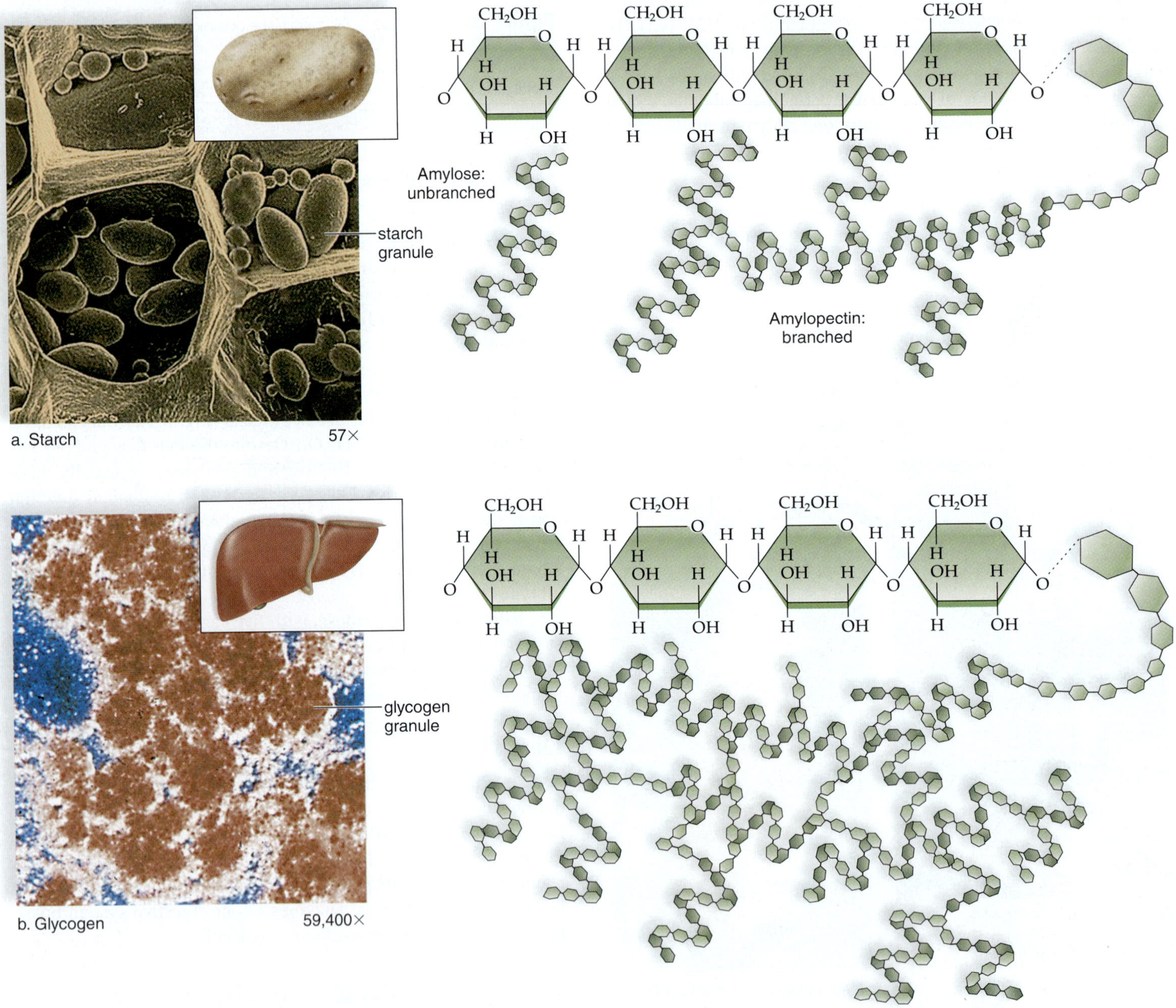

Figure 3.7 Starch and glycogen structure and function. **a.** The electron micrograph shows the location of starch in plant cells. Starch is a chain of glucose molecules that can be branched or unbranched. **b.** The electron micrograph shows glycogen deposits in a portion of a liver cell. Glycogen is a highly branched polymer of glucose molecules.

our food. Sucrose is also the form in which sugar is transported in plants. We acquire sucrose from plants such as sugarcane and sugar beets. You may also have heard of lactose, a disaccharide found in milk. Lactose is glucose combined with galactose. Individuals who are *lactose intolerant* lack the enzyme (called lactase) that breaks down lactose. To prevent gastrointestinal discomfort, they can either avoid foods that contain lactose (e.g., dairy products) or take nutritional supplements that contain the enzyme.

Polysaccharides: Energy-Storage Molecules

Polysaccharides are long polymers of monosaccharides. Due to their length, they are sometimes referred to as complex carbohydrates. Some types of polysaccharides function as short-term energy-storage molecules. When an organism requires energy, the polysaccharide is broken down to release sugar molecules. The helical shape of the polysaccharides in Figure 3.7 exposes the sugar linkages to the hydrolytic enzymes that can break them down.

Plants store glucose as **starch.** The cells of a potato contain granules, where starch resides during winter until energy is needed for growth in the spring. Notice in Figure 3.7*a* that starch exists in two forms: One form (amylose) is unbranched and the other (amylopectin) is branched. When a polysaccharide is branched, there is no main carbon chain, because new chains occur at regular intervals, always at the sixth carbon of the monomer.

Animals store glucose as **glycogen.** In our bodies and those of other vertebrates, liver cells contain granules, where glycogen is stored until needed. The storage and release of glucose from liver cells are controlled by hormones. After we eat, the release of the hormone insulin from the pancreas promotes the storage of glucose as glycogen. Notice in Figure 3.7*b* that glycogen is even more branched than starch.

Polysaccharides serve as storage molecules because they are not as soluble in water and are much larger than a simple sugar. Therefore, polysaccharides cannot easily pass through the plasma membrane, a sheetlike structure that encloses cells.

Polysaccharides: Structural Molecules

Structural polysaccharides include **cellulose** in plants, **chitin** in animals and fungi, and **peptidoglycan** in bacteria (see Fig. 3.4). In all three, monomers are joined by the type of bond shown for cellulose in Figure 3.8. The cellulose monomer is simply glucose, but

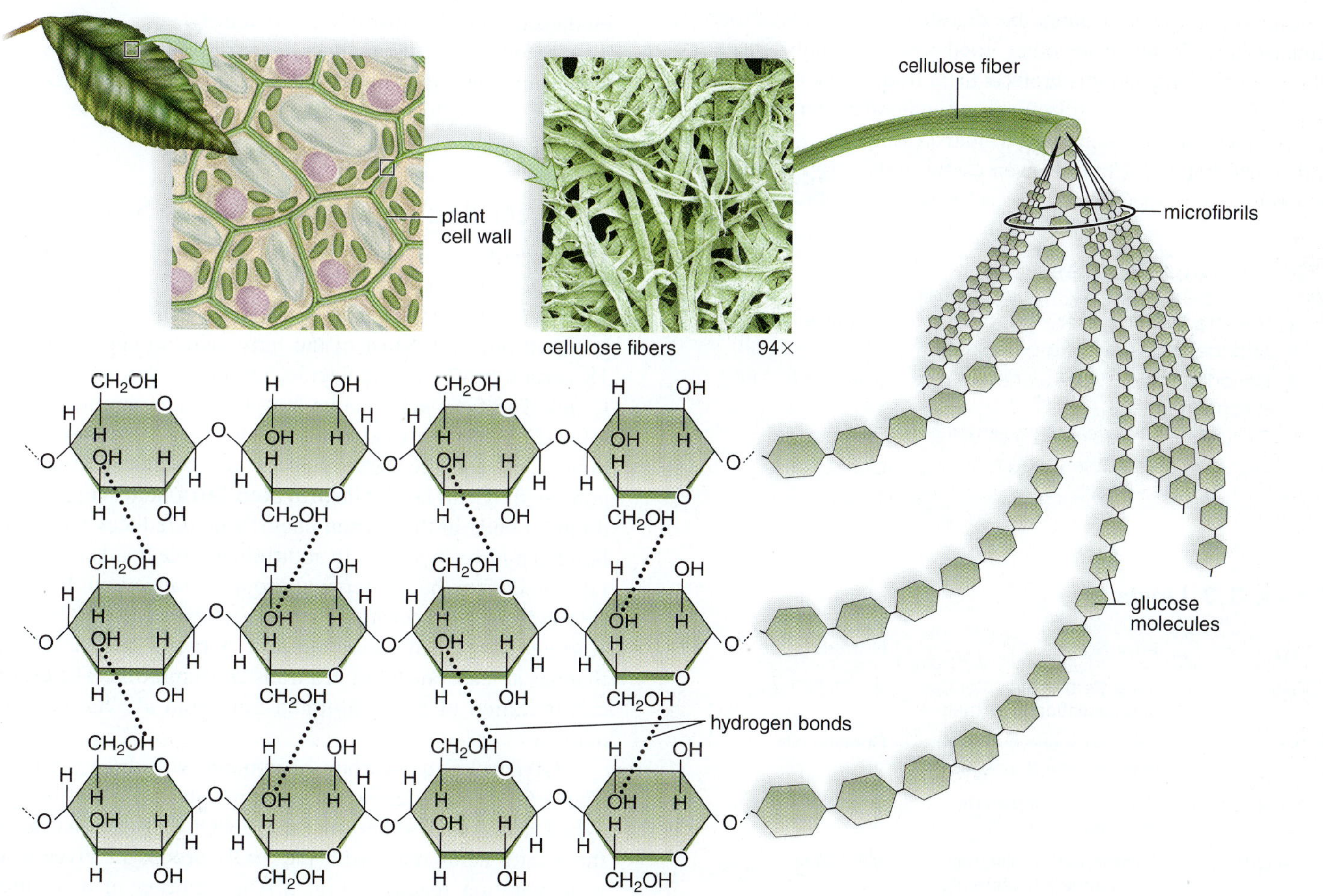

Figure 3.8 Cellulose fibrils. Cellulose fibers criss-cross in plant cell walls for added strength. A cellulose fiber contains several microfibrils, each a polymer of glucose molecules—notice that the linkage bonds differ from those of starch. Every other glucose is flipped, permitting hydrogen bonding and greater strength between the microfibrils.

in chitin, the monomer has an attached amino group. The structure of peptidoglycan is even more complex, because each monomer also has an amino acid chain. In both cases, the addition of a functional group to the glucose monomer changes its chemical properties.

Cellulose is the most abundant carbohydrate and, in fact, the most abundant organic molecule on Earth—over 100 billion tons of cellulose are produced by plants each year. Wood, a cellulose plant product, is used for construction, and cotton is used for cloth. Because of the structure of the bonds between the glucose molecules, animals are not able to digest cellulose. However, some microorganisms can. The protozoans in the gut of termites enable these insects to digest wood. In cows and other ruminants, microorganisms break down cellulose in a special digestive-tract pouch before the "cud" is returned to the mouth for more chewing and reswallowing. In rabbits, microorganisms digest cellulose in a pouch, where it is packaged into pellets. In order to make use of these nutrient pellets, rabbits have to reswallow them as soon as they pass out at the anus. For other animals, including humans, that have no means of digesting cellulose, cellulose serves as dietary fiber, which maintains regularity of fecal elimination.

Chitin is found in fungal cell walls and in the exoskeletons of crabs and related animals, such as lobsters, scorpions, and insects. Chitin, like cellulose, cannot be digested by animals; however, humans have found many other good uses for chitin. Seeds are coated with chitin, and this protects them from attack by soil fungi. Because chitin also has antibacterial and antiviral properties, it is processed and used in medicine as a wound dressing and suture material. Chitin is even useful in the production of cosmetics and various foods.

MP3 Carbohydrates

Check Your Progress 3.2

1. Summarize the general characteristics of carbohydrates and their roles in living organisms.
2. Describe how monosaccharides are combined to form disaccharides.
3. Explain why humans cannot utilize the glucose in cellulose as a nutrient source.

Table 3.3 Lipids

Type	Functions	Human Uses
Fats	Long-term energy storage and insulation in animals	Butter, lard
Oils	Long-term energy storage in plants and their seeds	Cooking oils
Phospholipids	Component of plasma membrane	Food additive
Steroids	Component of plasma membrane (cholesterol), sex hormones	Medicines
Waxes	Protection, prevention of water loss (cuticle of plant surfaces), beeswax, earwax	Candles, polishes

3.3 Lipids

Learning Outcomes

Upon completion of this section, you should be able to

1. Describe why lipids are essential to living organisms.
2. Distinguish between saturated and unsaturated fatty acids.
3. Contrast the structures of fats, phospholipids, and steroids.
4. Compare the functions of phospholipids and steroids in cells.

A variety of organic compounds are classified as **lipids** (Gk. *lipos,* "fat") (Table 3.3). These compounds are insoluble in water due to their hydrocarbon chains. Hydrogens bonded only to carbon are nonpolar and have no tendency to form hydrogen bonds with water molecules. **Fats,** more fomally called the triglycerides, are the primary lipid used by animals for both insulation and long-term energy storage. Fat is distrubuted throughout the body, but the majority is found just beneath the skin of most animals, where it helps retain body heat. Triglycerides in plants are commonly referred to as **oils.** We are familiar with fats and oils, because we use them as foods and for cooking. However, increasingly they are also being used as a form of alternative fuel source, such as biodiesel, for our industrialized societies.

In addition to the triglycerides, phospholipids, steroids, and waxes are important lipids found in living organisms. The next sections will explore the structure and function of these classes of lipids.

Triglycerides: Long-Term Energy Storage

Triglycerides contain two types of subunit molecules: fatty acids and glycerol. Each **fatty acid** consists of a long hydrocarbon chain with an even number of carbons and a $-COOH$ (carboxyl) group at one end. Most of the fatty acids in cells contain 16 or 18 carbon atoms per molecule, although smaller ones are also found. The fatty acid chains may be either saturated or unsaturated (Fig. 3.9). **Saturated fatty acids** (Fig. 3.9*a*) have no double bonds between the carbon atoms and contain as many hydrogens as they can hold. **Unsaturated fatty acids** (Fig. 3.9*b*) have double bonds in the carbon chain, which reduces the number of bonded hydrogen atoms. In addition, double bonds in unsaturated fatty acids may have chemical groups arranged on the same side (termed *cis configuration*) or on opposite sides (termed *trans configuration*) of the double bond. A **trans fat** is a triglyceride that has at least one bond in a trans configuration. The cis or trans configuration of an unsaturated fatty acid affects its biological activity.

Glycerol is a 3-carbon compound with three $-OH$ groups. The $-OH$ groups are polar, making glycerol soluble in water. When a fat or an oil forms, the $-COOH$ functional groups of three fatty acids react with the $-OH$ groups of glycerol during a dehydration reaction (Fig. 3.10), resulting in a fat molecule and three molecules of water. Fats and oils are degraded during a hydrolysis reaction. Notice that triglycerides have many nonpolar $C-H$ bonds; therefore, they do not mix with water. Even though cooking oils and water are both liquid, they do not mix, even after

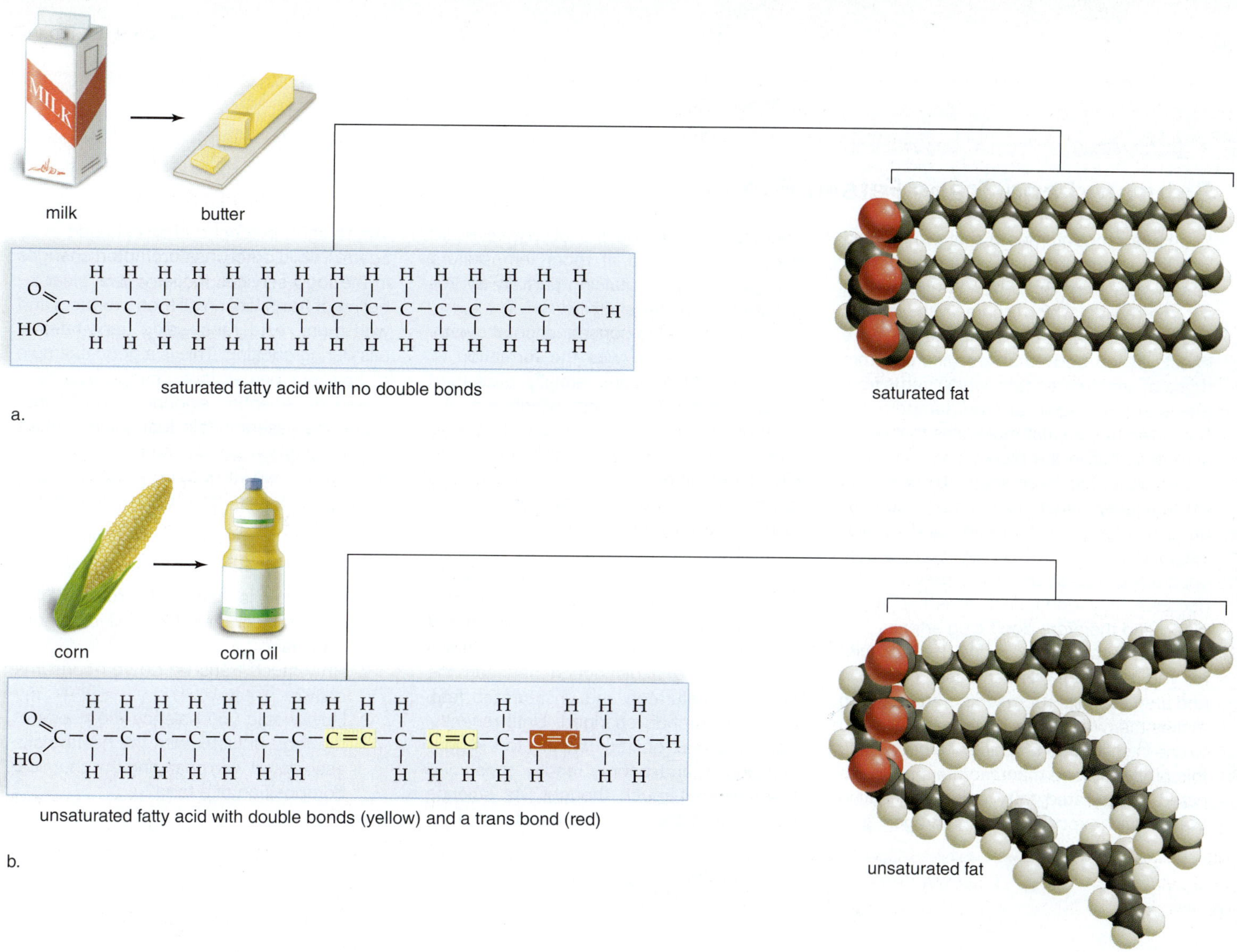

Figure 3.9 Saturated and unsaturated fatty acids. A fatty acid has a carboxyl group attached to a long hydrocarbon chain. **a.** If there are no double bonds between the carbons in the chain, the fatty acid is saturated. **b.** If there are double bonds between some of the carbons, the fatty acid is unsaturated and a kink occurs in the chain.

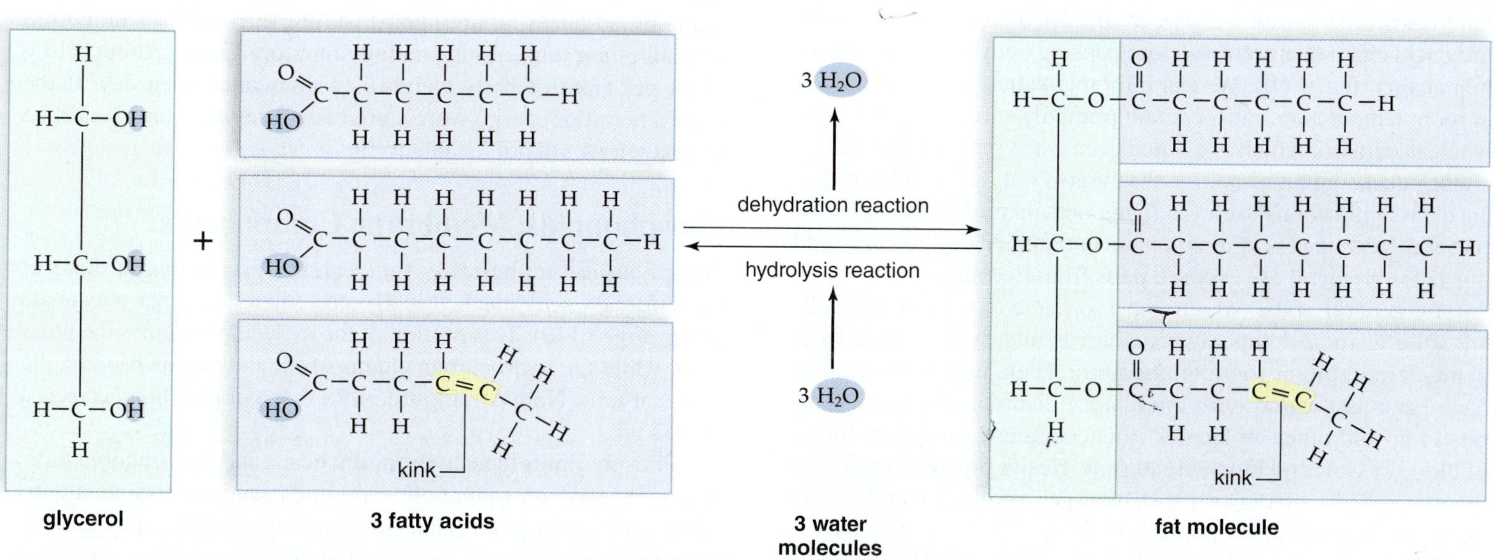

Figure 3.10 Synthesis and degradation of a triglyceride. Following a dehydration reaction, glycerol is bonded to three fatty acid molecules as fat forms and water is given off. Following a hydrolysis reaction, the bonds are broken due to the addition of water.

THEME Nature of Science

Saturated and Trans Fats in Foods

You have probably heard that you should limit the amount of saturated fats and trans fats in your diet. But why? We know that saturated fats, which come from animals and are solid at room temperature, have effects in the body that are different from those of unsaturated fats, which come from plants and are liquid at room temperature. Saturated fats are flat molecules that easily stick together in the blood, and too much saturated fat has been shown by scientists to negatively affect heart health, contributing to clogging of arteries and cardiovascular disease (CVD). By comparison, unsaturated fats seem to help prevent CVD, because they don't stick together in the blood and therefore don't clog arteries.

Unsaturated fats might be healthier for you, but plant oils can easily go rancid and aren't solid at room temperature, which makes them more difficult to cook with and to use in solid food products. To get around this problem, food manufacturers hydrogenated unsaturated fatty acids by heating the oil and exposing it to hydrogen gas. This treatment made the otherwise liquid plant oils semisolid at room temperature and gave foods containing partially hydrogenated oils better shelf life.

An unintended consequence of hydrogenation, however, was the formation of trans fats. Many commercially packaged foods contain trans fats, which recently have been shown to increase LDL (sometimes called bad cholesterol) and lower HDL (sometimes called good cholesterol) levels in the blood. Trans fat consumption also appears to increase risk of CVD and heart attack.

At one point, investigators thought that the total amount of lipid in the diet caused coronary and other heart-related diseases. As scientific evidence accumulated showing a distinction between the effects of saturated and unsaturated fats, public perception changed. Until recently, trans fats were of little concern to the general public, and people readily consumed them without much thought. As science has brought the negative effects of trans fats to light, perception has changed once again. Public outcry has prompted changes in the food services industry, with clear labeling of trans fats on all food products and with more restaurants using trans fat–free oils during cooking. This is a good example of how our perceptions change over time based on scientific evidence, and it illustrates the essential role that science plays in the common good. Science constantly refines what we know as new evidence provides greater insights into how we function and live.

Questions to Consider

1. What is the chemical structure of a trans fat compared to that of a nontrans fat?
2. How much trans fat do you consume daily?
3. How would you balance the needs of the food industry with the health risks associated with changing the chemical composition of a food?

shaking, because the nonpolar oil and polar water are chemically incompatible.

Triglycerides containing fatty acids with unsaturated bonds melt at a lower temperature than those containing only saturated fatty acids. The reason is that a double bond creates a kink in the fatty acid chain that prevents close packing between the hydrocarbon chains (Fig. 3.10). We can infer that butter, a fat that is solid at room temperature, must contain primarily saturated fatty acids, whereas corn oil, which is a liquid even when placed in the refrigerator, must contain primarily unsaturated fatty acids. This difference has applications useful to living organisms. For example, the feet of reindeer and penguins contain unsaturated triglycerides, and this helps protect those exposed parts from freezing.

In general, fats, which most often come from animals, are solid at room temperature, whereas oils, which come from plants, are liquid at room temperature. Diets high in animal fat have been associated with circulatory disorders, because saturated fats and other molecules can accumulate inside the lining of blood vessels and block blood flow. Health organizations have recommended replacing fat with oils such as olive oil and canola oil in our diet whenever possible.

Nearly all animals use fat rather than glycogen for long-term energy storage. Gram for gram, fat stores more energy than glycogen. The C—H bonds of fatty acids make them a richer source of chemical energy than glycogen, because more bonds with stored energy are present in fatty acids; in contrast, glycogen has many C—OH bonds, which are less energetic bonds. Also, fat droplets do not contain water, because they are nonpolar. Small birds, such as the broad-tailed hummingbird, store a great deal of fat before they start their long spring and fall migratory flights. About 0.15 g of fat per gram of body weight is accumulated each day. If the same amount of energy were stored as glycogen, a bird would be so heavy it would not be able to fly.

Phospholipids: Membrane Components

Phospholipids are basically triglycerides, except that in place of the third fatty acid attached to glycerol, there is a polar phosphate group (Fig. 3.11*a*). This portion of the molecule becomes the polar head, while the hydrocarbon chains of the fatty acids become the nonpolar tails. Notice in Figure 3.11*a* that a double bond causes a tail to kink.

Phospholipids have hydrophilic heads and hydrophobic tails. When exposed to water, phospholipids tend to arrange themselves so that the polar heads are oriented toward water and the nonpolar fatty acid tails are oriented away from water. In living organisms, which are made mostly of water, phospholipids tend to become a bilayer (double layer), because the polar heads

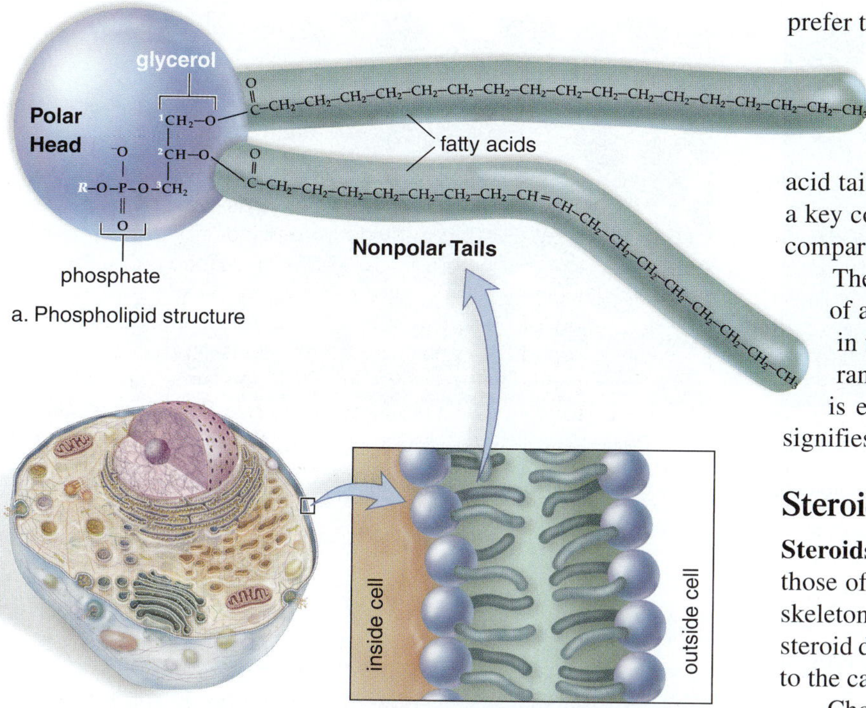

a. Phospholipid structure

b. Plasma membrane of a cell

Figure 3.11 Phospholipids form membranes. **a.** Phospholipids are constructed like fats, except that in place of the third fatty acid, they have a polar phosphate group. The hydrophilic (polar) head is soluble in water, whereas the two hydrophobic (nonpolar) tails are not. A tail has a kink wherever there is an unsaturated bond. **b.** Because of their structure, phospholipids form a bilayer that serves as the major component of a cell's plasma membrane. The fluidity of the plasma membrane is affected by kinks in the phospholipids' tails.

prefer to interact with other polar molecules, such as water. Conversely, the nonpolar tails associate together and stay away from polar water molecules. Thus, phospholipids arrange themselves like a "sandwich," with the polar heads facing the outside (the bread slices) and the fatty acid tails on the inside (the filling). This phospholipid bilayer is a key component used to keep cells separate from the biological compartments within cells.

The plasma membrane that surrounds cells consists primarily of a phospholipid bilayer (Fig. 3.11*b*). The presence of kinks in the tails causes the plasma membrane to be fluid across a range of temperatures found in nature. A plasma membrane is essential to the structure and function of a cell, and this signifies the importance of phospholipids to living organisms.

Steroids: Four Fused Rings

Steroids are lipids with structures that are entirely different from those of triglycerides and phospholipids. Steroid molecules have skeletons of four fused carbon rings (Fig. 3.12*a*). Each type of steroid differs primarily by the types of functional groups attached to the carbon skeleton.

Cholesterol is an essential component of an animal cell's plasma membrane, where it provides physical stability. Cholesterol is the precursor of several other steroids, such as the sex hormones testosterone and estrogen (Fig. 3.12*b, c*). The male sex hormone, testosterone, is formed primarily in the testes, and the female sex hormone, estrogen, is formed primarily in the ovaries. Testosterone and estrogen differ only by the functional groups attached to the same carbon skeleton, yet each has its own profound effect on the body and sexuality of an animal. Human and plant estrogens

a. Cholesterol

b. Testosterone

c. Estrogen

Figure 3.12 Steroid diversity. **a.** Built like cholesterol, (**b**) testosterone and (**c**) estrogen have different effects on the body due to different functional groups attached to the same carbon skeleton. Testosterone is the male sex hormone active in peacocks (*left*), and estrogen is the female sex hormone active in peahens (*right*). These hormones are present in many living creatures.

a. b.

Figure 3.13 Waxes. Waxes are a type of lipid. **a.** Fruits are protected by a waxy coating, which is visible on these plums. **b.** Bees secrete the wax that allows them to build a comb, where they store honey.

are similar in structure, and if estrogen therapy is recommended, some women prefer taking soy products in preference to estrogen from animals.

Cholesterol can also contribute to circulatory disorders. The presence of cholesterol encourages the accumulation of fatty material inside the lining of blood vessels, which decreases the size of the opening and thereby can result in high blood pressure. Cholesterol-lowering medications are available.

Waxes

In **waxes,** long-chain fatty acids are connected to carbon chains containing alcohol functional groups. Waxes are solid at normal temperatures, because they have a high melting point. Being hydrophobic, they are also waterproof and resistant to degradation. In many plants, waxes, along with other molecules, form a protective cuticle (covering) that prevents the loss of water from all exposed parts (Fig. 3.13*a*). In many animals, waxes are involved in skin and fur maintenance. In humans, wax is produced by glands in the outer ear canal. Earwax contains cerumin, an organic compound that, at the very least, repels insects and in some cases even kills them. It also traps dust and dirt, preventing these contaminants from reaching the eardrum. A honeybee produces beeswax in glands on the underside of its abdomen. Beeswax is used to make the six-sided cells of the comb, where honey is stored (Fig. 3.13*b*). Honey contains the sugars fructose and glucose, breakdown products of the sugar sucrose. In humans, waxes are produced by glands in the ear. These waxes waterproof the ear canal and prevent the growth of bacteria.

MP3 Lipids

Check Your Progress 3.3

1. List the functions of triglycerides, phospholipids, steroids, and waxes.
2. Contrast the structure of a saturated fatty acid with that of an unsaturated fatty acid.
3. Explain why phospholipids form a bilayer in water.

3.4 Proteins

Learning Outcomes

Upon completion of this section, you should be able to

1. Describe the functions of proteins in cells.
2. Explain how a polypeptide is constructed from amino acids.
3. Compare the four levels of protein structure.
4. Understand the factors that affect protein structure and function.

Proteins (Gk. *proteios,* "first place") are of primary importance to the structure and function of cells. As much as 50% of the dry weight of most cells consists of proteins. Several hundred thousand proteins have been identified. The following are some of their many functions in animals:

- *Metabolism* Enzyme proteins bring reactants together and thereby speed chemical reactions in cells. They are specific for one particular type of reaction and function best at specific body temperatures and pH.
- *Support* Some proteins have a structural function. For example, keratin makes up hair and nails, while collagen gives strength to ligaments, tendons, and skin.
- *Transport* Channel and carrier proteins in the plasma membrane regulate what substances enter and exit cells. Other proteins transport molecules in the blood of animals; hemoglobin is a complex protein that transports oxygen to tissues and cells.
- *Defense* Antibodies are proteins of our immune system that combine with foreign substances, called antigens. Antibodies bind and prevent antigens from destroying cells and upsetting homeostasis.
- *Regulation* Some hormones are proteins that regulate how cells behave. They serve as intercellular messengers that influence cell metabolism. The hormone insulin regulates how much glucose is in the blood and in cells; the presence of growth hormone during childhood and adolescence determines the height of an individual.
- *Motion* The contractile proteins actin and myosin allow parts of cells to move and cause muscles to contract. Muscle contraction

allows animals to travel from place to place. All cells contain proteins that move cell components to different internal locations. Without such proteins, cells would not be able to function.

Proteins are such a major part of living organisms that tissues and cells of the body can sometimes be characterized by the proteins they contain or produce. For example, muscle cells contain large amounts of actin and myosin for contraction; red blood cells are filled with hemoglobin for oxygen transport; and support tissues, such as ligaments and tendons, contain the protein collagen, which is composed of tough fibers.

Amino Acids: Protein Monomers

Proteins are polymers constructed from amino acid monomers. The name **amino acid** is used because one of the functional groups in the amino acid is $-NH_2$ (an amino group) and another is $-COOH$ (an acid group). The third group is called an *R* (variable) group. The structure of an amino acid is as follows:

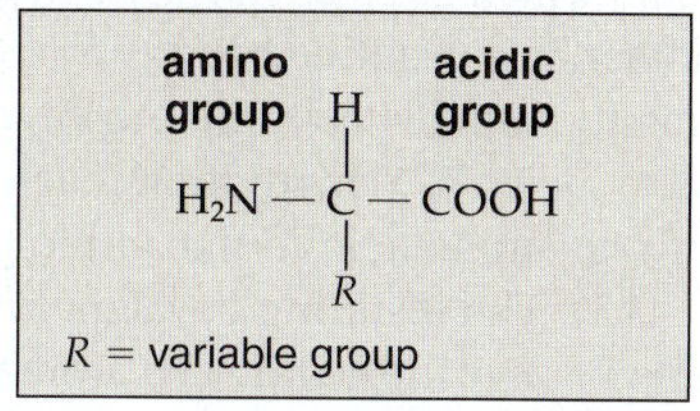

Note that the central carbon atom in an amino acid bonds to a hydrogen atom and to three other groups of atoms, one of which is the *R* group. Amino acids differ according to their particular *R* group (Fig. 3.14). The *R* groups range in complexity from a single

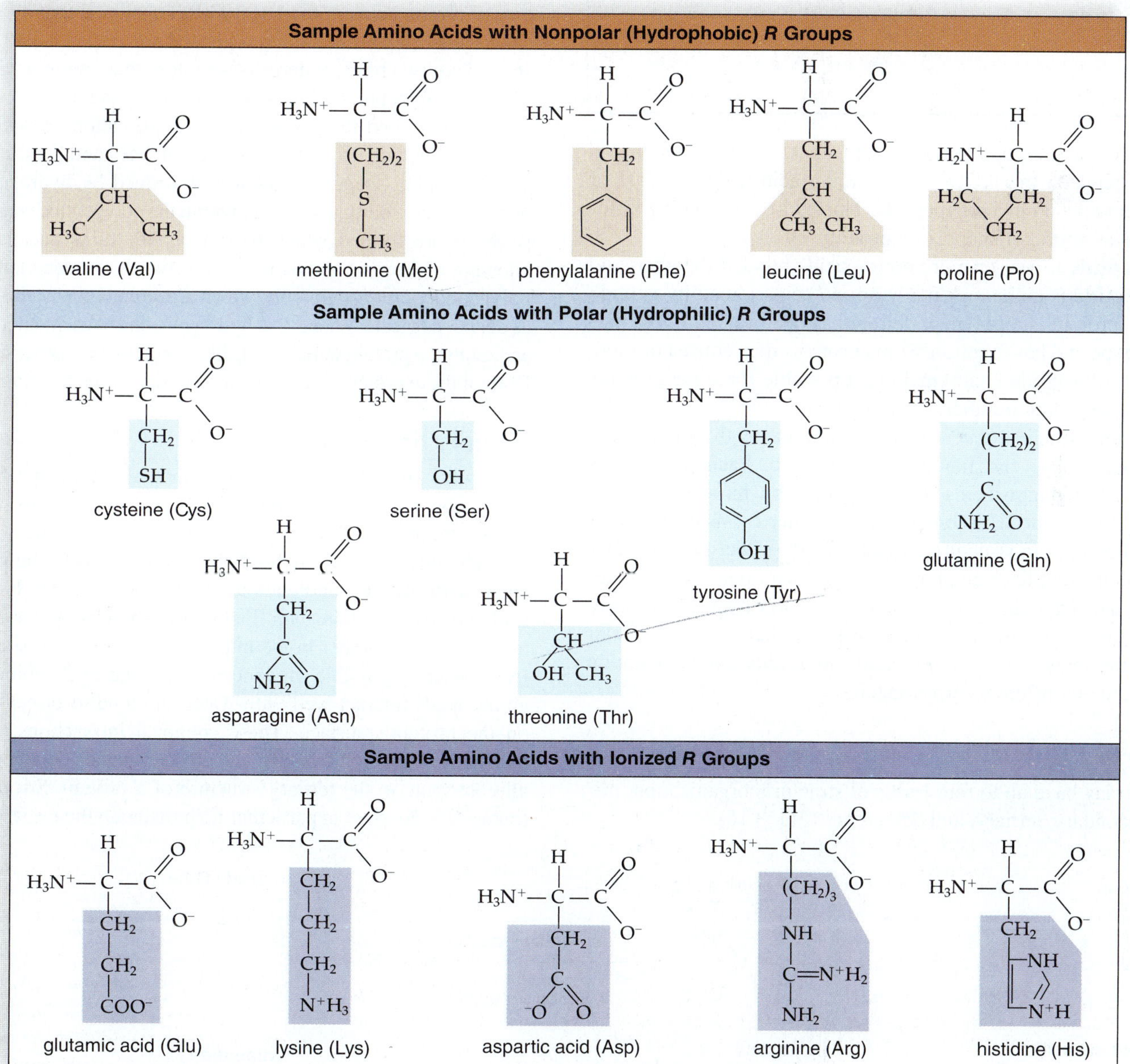

Figure 3.14 Amino acids. Polypeptides contain 20 different kinds of amino acids, some of which are shown here. Amino acids differ by the particular *R* group (shaded area of the molecule) attached to the central carbon. Some *R* groups are nonpolar and hydrophobic (top), some are polar and hydrophilic (center), and some are ionized and hydrophilic (bottom). The amino acids are shown in ionized form.

hydrogen atom to complicated ring compounds. Some *R* groups are polar and associate with water, whereas others are nonpolar and do not. Also, the amino acid cysteine has an *R* group that ends with a −SH (sulfide) group, which often covalently connects one chain of amino acids to another by a disulfide bond, −S−S−. Several other amino acids commonly found in cells are shown in Figure 3.14.

Amino acids are linked by dehydration reactions that link the carboxyl group of one amino acid to the amino group of another amino acid (Fig. 3.15). The resulting covalent bond between two amino acids is called a **peptide bond.** The atoms associated with the peptide bond share the electrons unevenly, because oxygen is more electronegative than nitrogen. Therefore, the hydrogen attached to the nitrogen has a slightly positive charge, while the oxygen has a slightly negative charge:

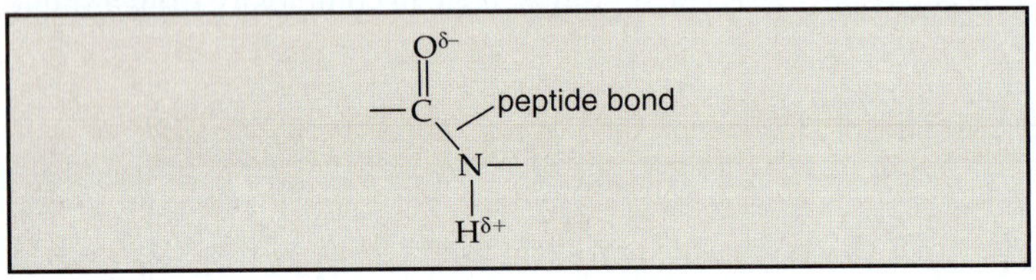

The polarity of the peptide bond means that hydrogen bonding is possible between the −CO of one amino acid and the −NH of another amino acid in a polypeptide. This hydrogen bonding influences the structure, or shape, of a protein.

A **peptide** is two or more amino acids bonded together, and a **polypeptide** is a chain of many amino acids joined by peptide bonds. A protein is a polypeptide that has been folded into a particular shape and has function. Some proteins may consist of more than one polypeptide chain, making it possible for some proteins to have a very large number of amino acids.

The amino acid sequence greatly influences the final three-dimensional shape and function of a protein. Each protein has a sequence of amino acids that is defined by information contained within a gene. This amino acid sequence forms the basis for all levels of protein structure, which directly affect protein function. Proteins that have an abnormal sequence often have a three-dimensional shape that causes them to function improperly. From an evolutionary perspective, we also know that, for a particular protein, the sequences of amino acids are highly similar within a species and are different across species.

Shape of Proteins

Proteins may have up to four levels of structural organization: primary, secondary, tertiary, and quaternary (Fig. 3.16).

Primary Structure

The *primary structure* of a protein is the linear sequence of amino acids. Just as millions of different words can be constructed from just 26 letters in the English alphabet, so can hundreds of thousands of different polypeptides be built from just 20 amino acids. To make a new word in English, all that is required is to vary the number and sequence of a few letters. Likewise, changing the sequence of 20 amino acids in a polypeptide can produce a huge array of different proteins. The sequence of the amino acids in the primary structure is determined by the information contained within genes, which are part of the DNA of the cell (see Chapter 12).

Secondary Structure

The *secondary structure* of a protein occurs when the polypeptide coils or folds in a particular way (Fig. 3.16).

Linus Pauling and Robert Corey, who began studying the structure of amino acids in the late 1930s, concluded that a coiling they called an α (alpha) helix and a pleated sheet they called a β (beta) sheet were two basic patterns of structure that amino acids assumed within a polypeptide. The names came from the fact that the α helix was the first, and the β sheet the second, pattern they discovered. Each polypeptide can have multiple α helices and β pleated sheets.

The spiral shape of α helices is formed by hydrogen bonding between every fourth amino acid within the polypeptide chain, whereas β sheets are formed when the polypeptide turns back upon itself, allowing hydrogen bonding to occur between extended lengths of the polypeptide. *Fibrous proteins,* which are structural proteins, exist only as helices or pleated sheets that hydrogen bond to each other. Examples are keratin, a protein in hair, and silk, a protein that forms spider webs. Both of these proteins have only a secondary structure (Fig. 3.17).

Tertiary Structure

A *tertiary structure* is the folding that results in the final three-dimensional shape of a polypeptide. *Globular proteins,* which tend to ball up into rounded shapes, have tertiary structure.

The interaction of hydrophobic amino acids in the polypeptide chain with the surrounding water is a major factor in how proteins fold into, and maintain, their final shape. These nonpolar amino acids tend to group together in the interior of a protein, to be as far away from water as possible. In contrast, the polar hydrophilic and ionic amino acids interact well with water and tend to orient themselves on the protein's surface. These chemical interactions, along with hydrogen bonds, ionic bonds, and covalent bonds between *R* groups, all contribute to the tertiary structure of a protein. Strong disulfide linkages (−S−S−) in particular help maintain the tertiary shape.

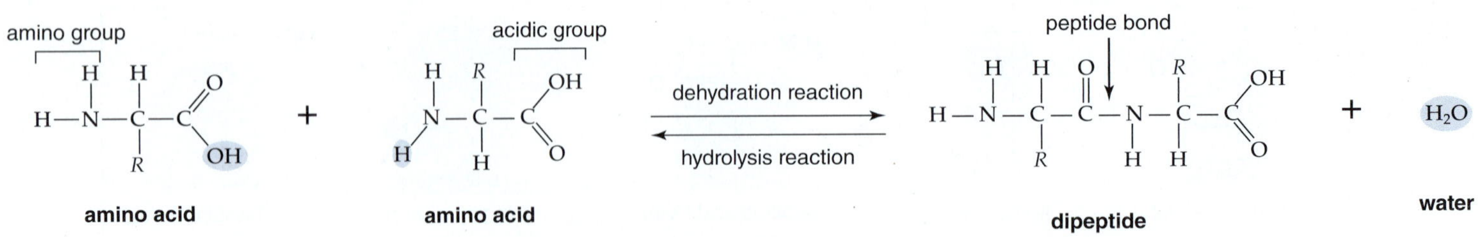

Figure 3.15 Synthesis and degradation of a peptide. Following a dehydration reaction, a peptide bond joins two amino acids and a water molecule is released. Following a hydrolysis reaction, the bond is broken due to the addition of water.

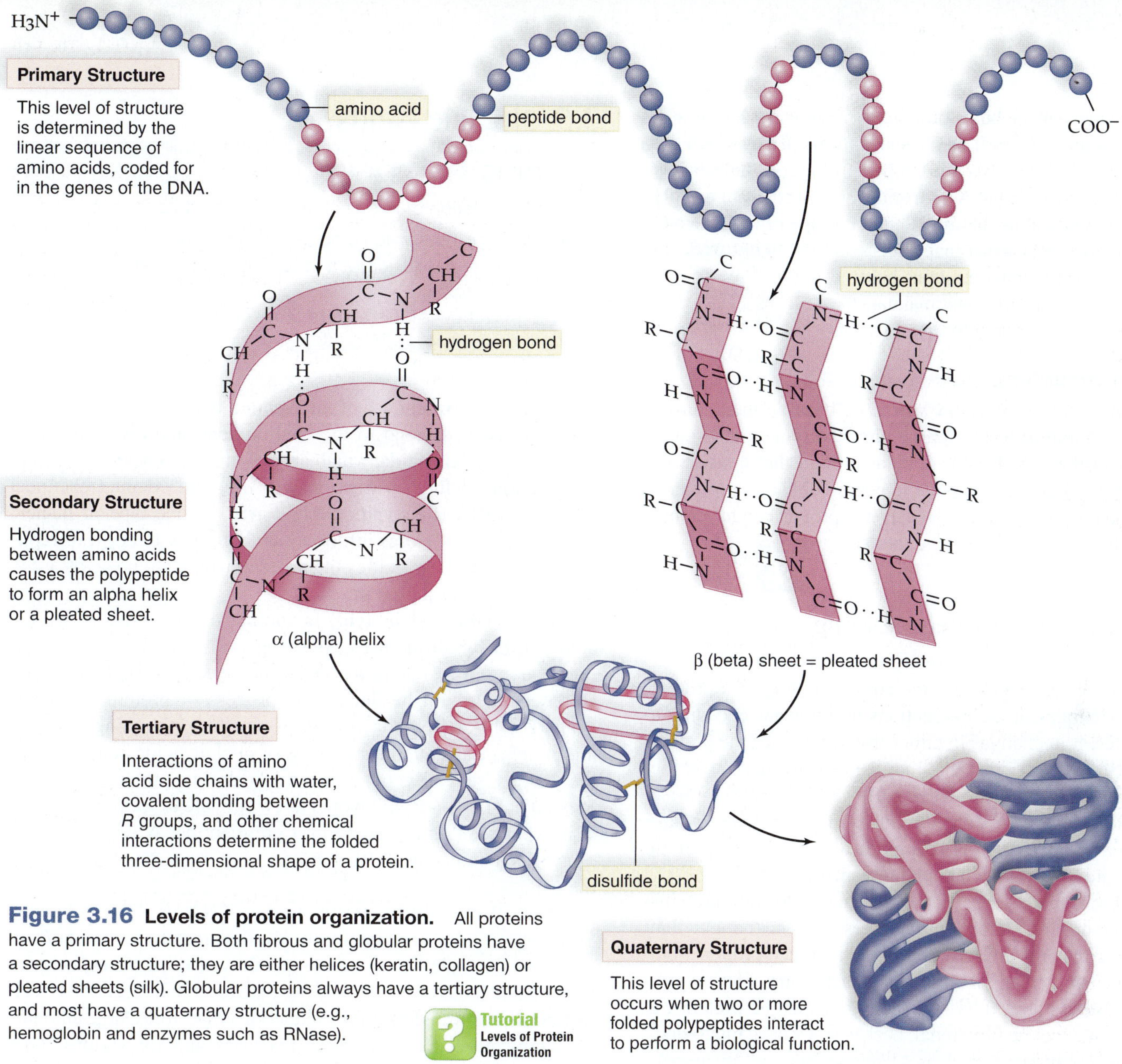

Figure 3.16 Levels of protein organization. All proteins have a primary structure. Both fibrous and globular proteins have a secondary structure; they are either helices (keratin, collagen) or pleated sheets (silk). Globular proteins always have a tertiary structure, and most have a quaternary structure (e.g., hemoglobin and enzymes such as RNase).

a. b. c.

Figure 3.17 Fibrous proteins. Fibrous proteins are structural proteins. **a.** Keratin—found, for example, in hair, horns, and hooves—exemplifies fibrous proteins that are helical for most of their length. **b.** A chemical treatment, called a perm, may be used to alter the secondary structure of the keratin proteins. **c.** Silk made by spiders is fibrous proteins that are pleated sheets for most of their length. Hydrogen bonding between parts of the molecule causes the pleated sheet to double back on itself.

Most enzymes are globular proteins. Enzymes work best at a specific temperature, and each one has an optimal pH, at which the rate of the reaction is highest. At this temperature and pH, the enzyme can maintain its normal shape. A high temperature and change in pH can disrupt the interactions that maintain the shape of the enzyme. When a protein loses its natural shape, it is said to be **denatured.** An organism can die if too many proteins become denatured, because it can no longer maintain the metabolic processes necessary for life.

Animation Protein Denaturation

Quaternary Structure

Some proteins have a *quaternary structure,* because they consist of more than one polypeptide. Hemoglobin, the protein that transports oxygen in the blood, is a globular protein that consists of four polypeptides. Each polypeptide in hemoglobin has a primary, secondary, and tertiary structure. However, a protein can have only two polypeptides and still have quaternary structure.

MP3 Proteins

The Importance of Protein Folding

The function of a protein is directly associated with its three-dimensional structure. Therefore, the correct folding of a protein is important. Changes in the instructions in the genes encoding a protein may result in changes in either the structure of the protein or the way it folds. At times, this may be deterimental, as is the case for diseases such as cystic fibrosis. However, sometimes these changes are less damaging. Minor changes in the proteins associated with hair or eye color are examples of variation in protein structure that are not detrimental to an organism.

Cells contain *chaperone proteins,* which help new proteins fold into their normal shape. Initially, researchers thought that chaperone proteins only ensured that proteins folded properly, but now it appears that they might correct any misfolding of a new protein. In any case, without fully functioning chaperone proteins, a cell's proteins may not be functional, because they have misfolded. Several diseases in humans, such as Alzheimer disease, are associated with misshapen proteins.

Other diseases in humans are due to misfolded proteins, but the cause may be different. For years, investigators have been studying fatal brain diseases, known as TSEs,[1] that have no cure because no infective agent could be found. Mad cow disease is a well-known example of a TSE disease. Now it appears that TSE diseases might be due to misfolded proteins, called **prions,** that cause other proteins of the same type to fold the wrong way, too.

Animation How Prions Arise

Check Your Progress 3.4

1. List the roles of proteins in living organisms.
2. Describe how two amino acids are combined to form a polypeptide.
3. Summarize the differences among primary, secondary, teriary, and quaternary structure.
4. Describe the consequences of incorrect protein folding.

1 TSEs: transmissible spongiform encephalopathies.

3.5 Nucleic Acids

Learning Outcomes

Upon completion of this section, you should be able to

1. Distinguish between a nucleotide and nucleic acid.
2. Compare the structure and function of DNA and RNA nucleic acids.
3. Explain how ATP is able to store energy.

Each cell has a storehouse of information that specifies how a cell should behave, respond to the environment, and divide to make new cells. **Nucleic acids,** which are polymers of nucleotides, store information, include instructions for life, and conduct chemical reactions. **DNA** (**deoxyribonucleic acid**) is one type of nucleic acid that not only stores information about how to copy, or replicate, itself but also specifies the order in which amino acids are to be joined to make a protein.

RNA (**ribonucleic acid**) is another diverse type of nucleic acid that has multiple uses. Messenger RNA (mRNA) is a temporary copy of a gene in the DNA that specifies what the amino acid sequence will be during the process of protein synthesis. Transfer RNA (tRNA) is also necessary in synthesizing proteins, and it helps translate the sequence of nucleic acids in a gene into the correct sequence of amino acids during protein synthesis. Ribosomal RNA (rRNA) works as an enzyme to form the peptide bonds between amino acids in a polypeptide. A wide range of other RNA molecules also perform important functions within the cell.

Not all nucleotides are made into DNA or RNA polymers. Some nucleotides are directly involved in metabolic functions in cells. For example, some are components of **coenzymes,** nonprotein organic molecules that help regulate enzymatic reactions. **ATP** (**adenosine triphosphate**) is a nucleotide that stores large amounts of energy needed for synthetic reactions and for various other energy-requiring processes in cells.

Structure of DNA and RNA

Every **nucleotide** is comprised of three types of molecules: a pentose sugar, a phosphate (phosphoric acid), and a nitrogen-containing base (Fig. 3.18*a*). In DNA, the pentose sugar is deoxyribose, and in RNA the pentose sugar is ribose. A difference in the structure of these 5-carbon sugars accounts for their respective names, because, as you might guess, deoxyribose lacks an oxygen atom found in ribose (Fig. 3.18*b*).

Both DNA and RNA contain combinations of four nucleotides (Fig. 3.18*c*), but these differ somewhat between the two nucleic acids (Table 3.4). Nucleotides that have a base with a single ring are called pyrimidines, and nucleotides with a double ring are called purines. In DNA, the pyrimidine bases are cytosine and thymine; in RNA, the pyrimidine bases are cytosine and uracil. Both DNA and RNA contain the purine bases adenine and guanine. These molecules are called bases because their presence raises the pH of a solution. Table 3.4 summarizes the differences between DNA and RNA.

Animation DNA Structure

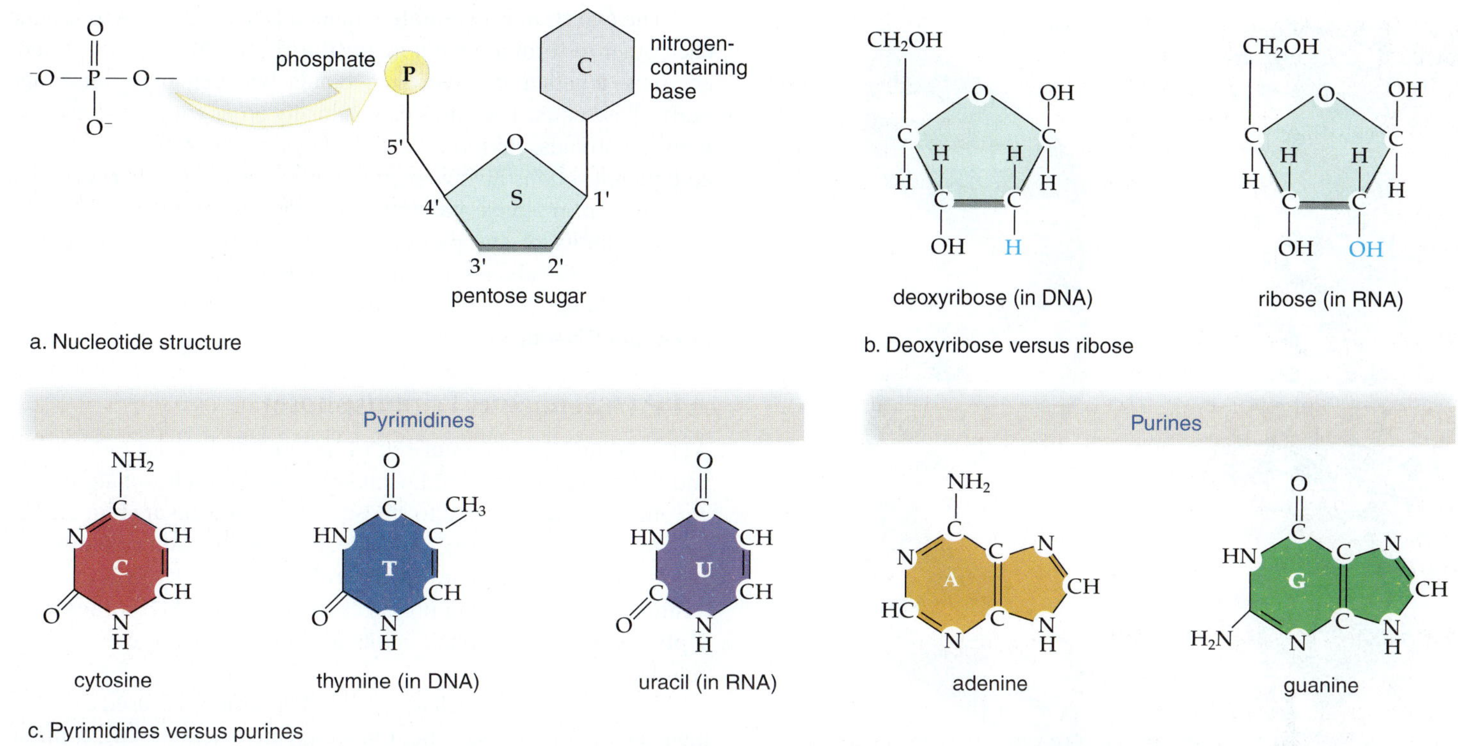

Figure 3.18 Nucleotides. **a.** A nucleotide consists of a pentose sugar, a phosphate molecule, and a nitrogen-containing base. **b.** DNA contains the sugar deoxyribose, and RNA contains the sugar ribose. **c.** DNA contains the pyrimidines C and T and the purines A and G. RNA contains the pyrimidines C and U and the purines A and G.

Table 3.4 DNA Structure Compared to RNA Structure

	DNA	RNA
Sugar	Deoxyribose	Ribose
Bases	Adenine, guanine, thymine, cytosine	Adenine, guanine, uracil, cytosine
Strands	Double-stranded with base pairing	Usually single-stranded
Helix	Yes	No

Nucleotides are joined into a DNA or an RNA polymer by a series of dehydration reactions. The resulting polymer is a linear molecule called a strand, in which the backbone is made up of an alternating series of sugar-phosphate-sugar-phosphate molecules. The bases project to one side of the backbone. Nucleotides are joined in an order specified by the strand they are copied from. DNA is double-stranded, and RNA is single-stranded (Fig. 3.19).

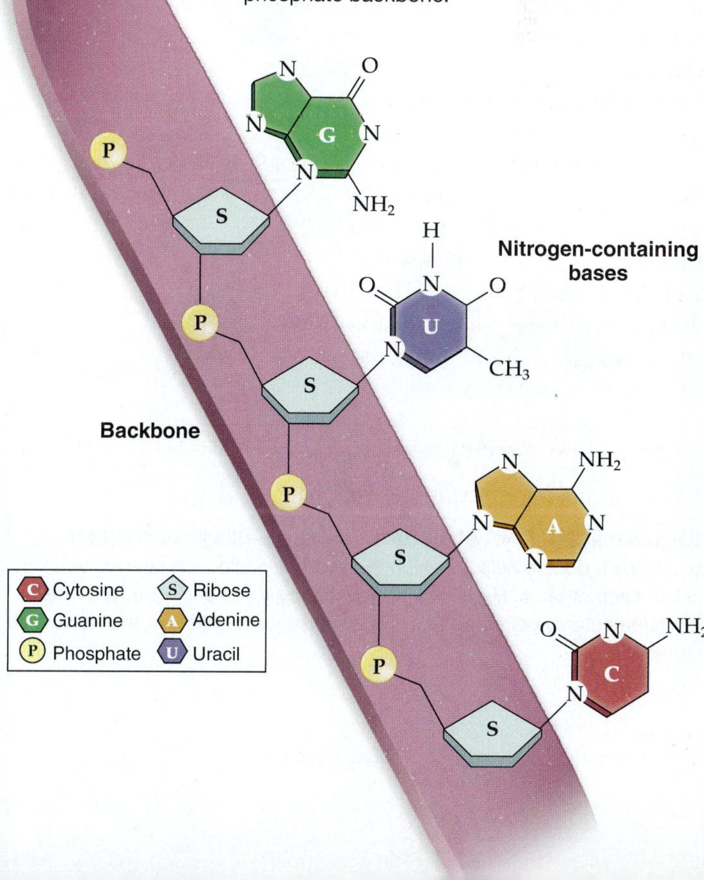

Figure 3.19 RNA structure. RNA is a single-stranded polymer of nucleotides. When the nucleotides join, the phosphate group of one is bonded to the sugar of the next. The bases project out to the side of the resulting sugar–phosphate backbone.

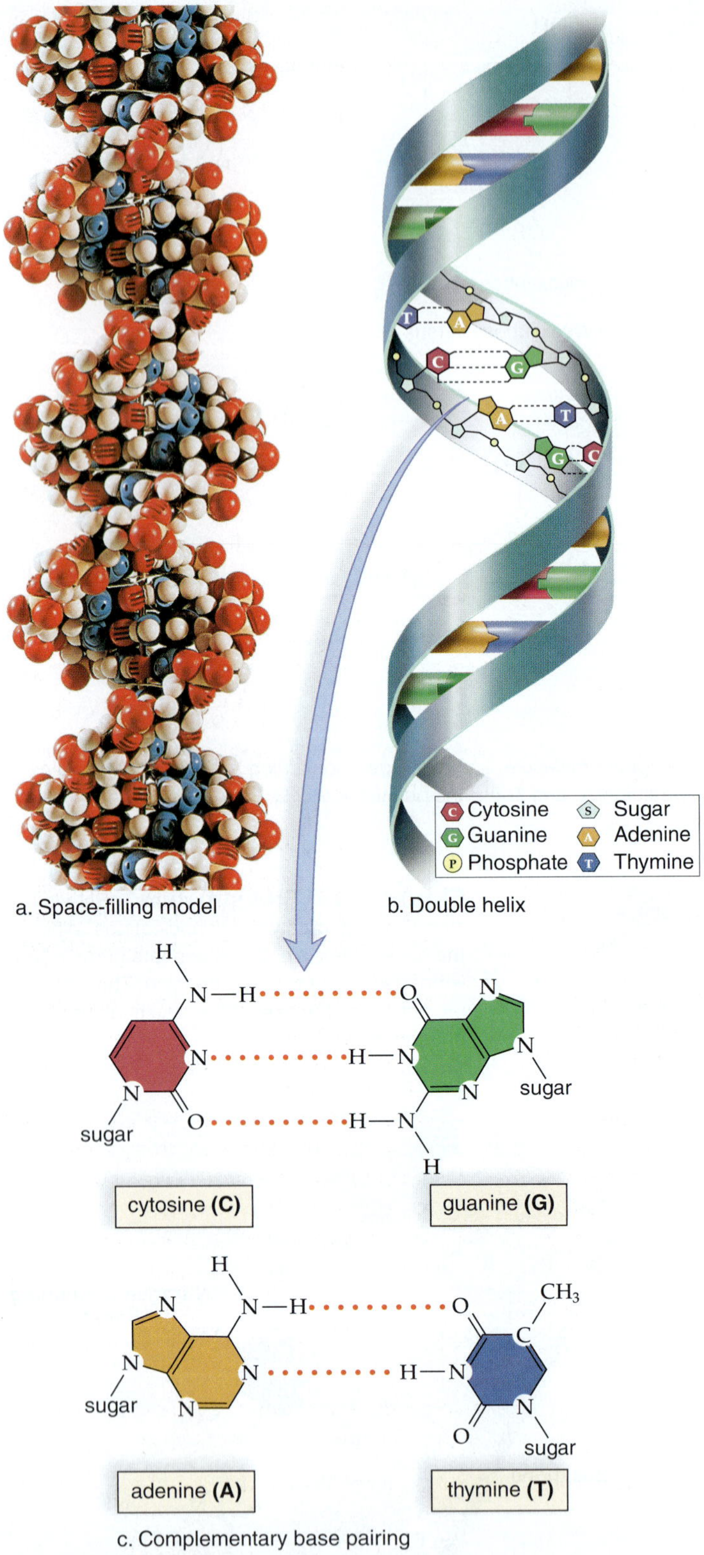

Figure 3.20 DNA structure. **a.** Space-filling model of DNA. **b.** DNA is a double helix in which the two polynucleotide strands twist about each other. **c.** Hydrogen bonds (dotted lines) occur between the complementarily paired bases: C is always paired with G, and A is always paired with T.

The two strands in double-stranded DNA usually twist around each other to form a double helix (Fig. 3.20*a, b*). The two strands are held together by hydrogen bonds between pyrimidine and purine base pairs. The bases can be in any order within a strand, but between strands, thymine (T) is always paired with adenine (A), and guanine (G) is always paired with cytosine (C). This is called **complementary base pairing.** Therefore, regardless of the order or the quantity of any particular base pair, the number of purine bases (A + G) always equals the number of pyrimidine bases (T + C) (Fig. 3.20*c*). We will take a closer look at the structure of DNA and RNA in Chapter 12.

ATP (Adenosine Triphosphate)

ATP is a nucleotide comprised of adenine and ribose (adenosine) and three phosphates (triphosphate). The three phosphate groups are attached together and to ribose, the pentose sugar (Fig. 3.21).

ATP is a high-energy molecule, because the last two phosphate bonds are unstable and are easily broken. In cells, hydrolysis of the terminal phosphate bond produces the molecule **ADP (adenosine diphosphate)**, a phosphate molecule Ⓟ, and lots of energy to do cellular work.

The energy that is released by ATP hydrolysis is used to power many cellular processes, including enzyme reactions, cell communication, and cell division. ATP hydrolysis is chemically favored, because ADP and Ⓟ are more stable than the original ATP molecule. Even though the third phosphate bond is broken, it is the whole molecule that releases energy.

In many cases, the hydrolysis of the ATP nucleotide is coupled to chemically unfavorable reactions in cells to allow these reactions to proceed. For example, key steps in the synthesis of macromolecules, such as carbohydrates and proteins, are able to proceed because the energy from ATP breakdown is used to pay the energy costs of the chemical reaction. ATP also supplies the energy for muscle contraction and nerve impulse conduction. Just as you spend money when you pay for a product or service, cells "spend" ATP when they need something. That's why ATP is called the energy currency of cells.

MP3 Nucleic Acids

Check Your Progress 3.5

1. Examine how a nucleic acid stores information.
2. Describe the three components of a nucleotide.
3. Evaluate the properties of ATP that make it an ideal carrier of energy.

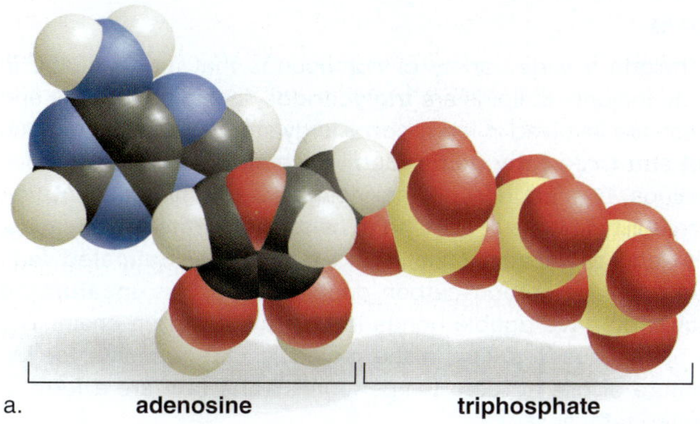

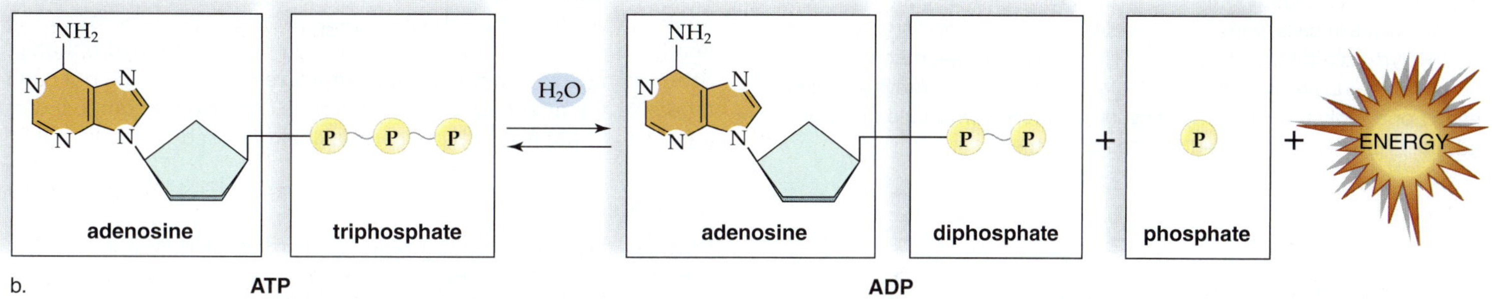

Figure 3.21 ATP. ATP, the universal energy currency of cells, is composed of adenosine and three phosphate groups. **a.** Space-filling model of ATP. **b.** When cells require energy, ATP becomes ADP + Ⓟ, and energy is released.

Connecting *the* Concepts *to the* Themes

Evolution

- Living cells are built from carbohydrates, lipids, proteins, and nucleic acids.
- DNA in genes specifies the amino acid sequences found in all proteins. DNA sequence changes can affect secondary, tertiary, and sometimes quaternary protein structures and therefore function.
- The function of a macromolecule is directly related to its structure.
- Mutations are changes in an organism's normal DNA or protein sequences and may increase, decrease, or have no effect on its biological functions.

Nature of Science

- The term organic can refer to food products that have been grown without the use of chemicals or have been minimally processed, to molecules that contain carbon and hydrogen, to living organisms, or to anything that has been previously alive, such as the plants and animals that make up fossil fuels.
- The breakdown of organic molecules releases carbon dioxide into the atmosphere.
- An understanding of organic molecules plays an important role in establishing guidelines for human health.

Biological Systems

- Carbon is central to the chemistry of life. The intake and synthesis of biomolecules is necessary to maintain cellular and life systems.
- Most biomolecules are simply polymers of small organic molecules. Monomers of sugars, amino acids, and nucleotides undergo dehydration reactions to become carbohydrates, proteins, and nucleic acids, respectively. Fats are composed of fatty acids and glycerol. Organisms must break down polymers to obtain the monomers needed for the specific structures in their bodies.

MEDIA STUDY TOOLS http://connect.mheducation.com

SMARTBOOK® Study smart. SmartBook helps you maximize your study time by identifying what you know and don't know. The recharge feature in SmartBook helps you prepare for the test by identifying the material that you are most likely to forget.

connect BIOLOGY Find even more resources to learn the chapter concepts including animations, tutorial videos, and interactive practice questions.

MP3 Files	Animations	Tutorial
3.1 Organic Molecules **3.2** Carbohydrates **3.3** Lipids **3.4** Proteins **3.5** Nucleic Acids	**3.4** Protein Denaturation • How Prions Arise **3.5** DNA Structure	**3.4** Levels of Protein Organization

SUMMARIZE

3.1 Organic Molecules

The chemistry of carbon, also called **organic chemistry,** accounts for the diversity of **organic** molecules found in living organisms. **Inorganic chemistry** is the study of nonliving matter. Carbon can bond with as many as four other atoms. It can also bond with itself to form both chains and rings. Differences in the carbon skeleton and attached **functional groups** cause biomolecules to have different chemical properties. Isomers are organic molecules with identical formulas but different structures. Some functional groups are hydrophobic, and others are hydrophilic.

The chemical properties of a molecule determine how it interacts with other molecules and its role in the cell. There are four classes of biomolecules in cells: carbohydrates, lipids, proteins, and nucleic acids.

Polymers are formed by linking monomers. For each bond formed during a **dehydration reaction,** a molecule of water is removed, and for each bond broken during a **hydrolysis reaction,** a molecule of water is added. Enzymes are involved in these processes.

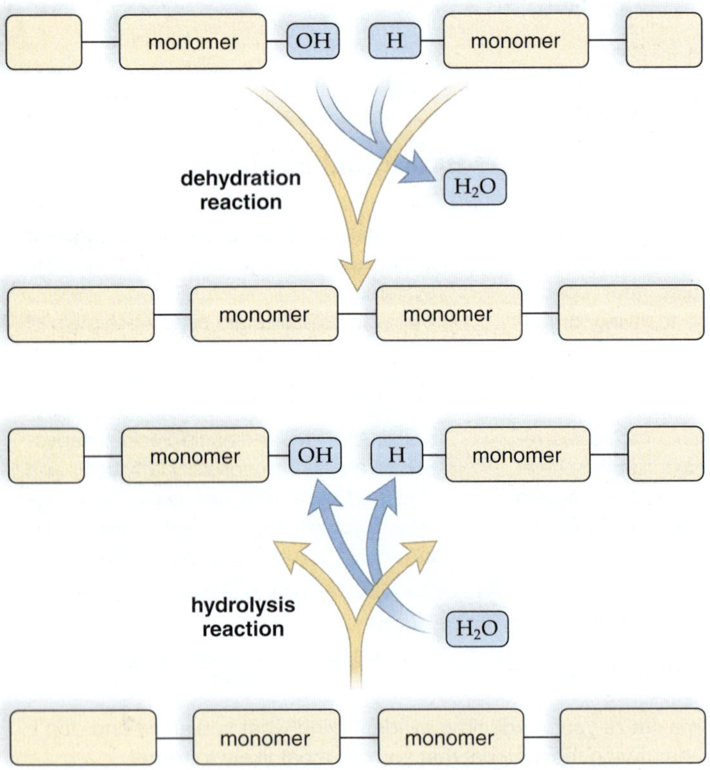

3.2 Carbohydrates

Monosaccharides, disaccharides, and polysaccharides are all **carbohydrates.** Therefore, the term carbohydrate includes both the monomers (e.g., glucose) and the polymers (e.g., starch, glycogen, and cellulose). **Monosaccharides** include **glucose,** a **hexose** sugar, and **ribose** and **deoxyribose, pentose** sugars. **Dissacharides** consist of two monosaccharides. **Polysaccharides** may serve as energy-storage molecules, such as **starch** and **glycogen,** or as structural molecules, such as **cellulose, chitin,** and **peptidoglycan.**

3.3 Lipids

Lipids include a wide variety of compounds that are insoluble in water. The majority of lipids are **triglycerides,** including the **fats** and oils. These are involved in long-term energy storage and contain one **glycerol** and three **fatty acids.** Both glycerol and fatty acids have polar groups, but fats and oils are nonpolar, and this accounts for their insolubility in water. Fats tend to contain **saturated fatty acids,** and oils tend to contain unsaturated fatty acids. Saturated fatty acids do not have carbon–carbon double bonds, but **unsaturated fatty acids** do have double bonds in their hydrocarbon chain. The double bond causes a kink in the molecule that accounts for the liquid nature of oils at room temperature. **Trans fats** are a form of unsaturated fatty acid.

A **phospholipid** replaces one of the fatty acids with a phosphate group. In water, phospholipids form a bilayer, because the head of each molecule is hydrophilic and the tails are hydrophobic. **Steroids** have the same four-ring structure as cholesterol, but each differs by the attached functional groups. **Waxes** are composed of a fatty acid with a long hydrocarbon chain bonded to an alcohol.

3.4 Proteins

Proteins carry out many diverse functions in cells and organisms, including support, metabolism, transport, defense, regulation, and motion. Proteins are polymers of **amino acids.**

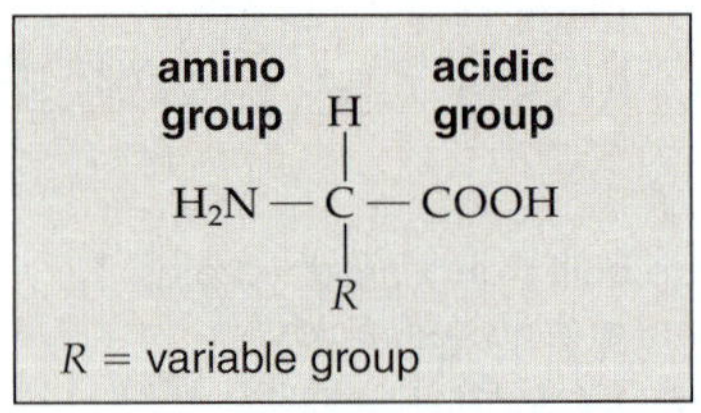

A **peptide** consists of two amino acids joined by a **peptide bond.** A **polypeptide** is a long chain of amino acids. There are 20 different amino acids in cells, and they differ only by their *R* groups. Whether or not the *R* groups are hydrophilic or hydrophobic helps determine the structure, and therefore the function, of the protein. A polypeptide has up to four levels of structure: The primary level is the linear sequence of the amino acids, which is determined by the DNA; the secondary level contains α helices and β (pleated) sheets held in place by hydrogen bonding between amino acids along the polypeptide chain; and the tertiary level is the final folded polypeptide, which is held in place by internal bonding and hydrophobic interactions between *R* groups. Proteins that contain more than one polypeptide have a quaternary level of structure as well.

Some proteins serve as enzymes, which regulate and carry out body functions. As with other proteins, the shape of an enzyme is important to its function. Both high temperatures and drastic pH change can cause proteins to **denature,** lose their shape, and decrease their function. Some proteins, called prions, may cause other proteins to fold incorrectly, thus altering their function.

3.5 Nucleic Acids

The **nucleic acids DNA (deoxyribonucleic acid)** and **RNA (ribonucleic acid)** are linear polymers of nucleotides. Each nucleotide has three components: a 5-carbon sugar, a phosphate (phosphoric acid), and a nitrogen-containing base.

DNA, which contains the sugar deoxyribose, phosphate, and nitrogen-containing bases, is the genetic material that stores information for its own replication and specifies the order in which amino acids are sequenced in proteins. DNA uses mRNA to direct protein synthesis. DNA is a double-stranded helix that has complementary base pairing between the strands. This means that A pairs with T and C pairs with G through hydrogen bonding. RNA is single-stranded, contains the sugar ribose and phosphate, and has the same bases as DNA except for uracil. There are many different types of RNA.

ATP (adenosine triphosphate) is a nucleotide that, with its unstable phosphate bonds, stores energy to do cellular work. Hydrolysis of ATP to **ADP (adenosine diphosphate)** releases energy needed by the cell to make a product or conduct metabolism. Some nucleic acids also act as **coenzymes** in metaboic reactions.

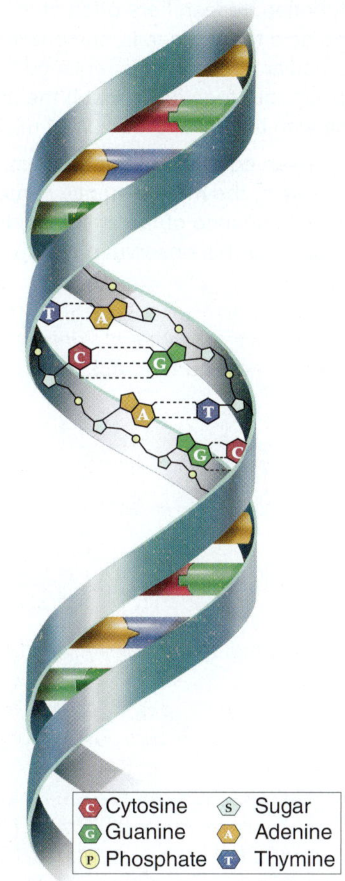

ASSESS

Choose the best answer for each question.

3.1 Organic Molecules

1. A hydrophilic group is
   a. attracted to water.
   b. a polar and/or an ionized group.
   c. found at the end of fatty acids.
   d. the opposite of a hydrophobic group.
   e. All of these are correct.
2. Which of these is not a characteristic of carbon?
   a. forms four covalent bonds
   b. bonds with other carbon atoms
   c. is sometimes ionic
   d. can form long chains
   e. sometimes shares two pairs of electrons with another atom
3. Which of the following reactions combines two monomers to produce a polymer?
   a. dehydration
   b. hydrolysis
   c. phosphorylation
   d. None of the above are correct.

3.2 Carbohydrates

4. The monomers of the carbohydrates are the
   a. polysaccharides.
   b. disaccharides.
   c. monosaccharides.
   d. waxes.
5. Which of the following polysaccharides is used as an energy-storage molecule in plants?
   a. glycogen
   b. chitin
   c. starch
   d. cellulose
6. Fructose and galactose are both isomers of
   a. glycogen.
   b. glucose.
   c. starch.
   d. maltose.

3.3 Lipids

7. A fatty acid is unsaturated if it
   a. contains hydrogen.
   b. contains carbon–carbon double bonds.
   c. contains a carboxyl (acidic) group.
   d. is bound to a glycerol.
8. Which of the following is incorrect regarding phospholipids?
   a. The heads are polar.
   b. The tails are nonpolar.
   c. They contain a phosphate group in place of one fatty acid.
   d. They are energy-storage molecules in the cell.
9. A lipid that contains four fused carbon rings is a
   a. triglyceride.
   b. wax.
   c. phospholipid.
   d. steroid.

3.4 Proteins

10. The chemical differences between one amino acid and another is due to which of the following?
    a. amino group
    b. carboxyl group
    c. *R* group
    d. peptide bond
    e. carbon atoms
11. Which of the following levels of protein structure is determined by interactions of more than one polypeptide chain?
    a. primary
    b. secondary
    c. tertiary
    d. quaternary
12. Which of the following is formed by the linking of two amino acids?
    a. a peptide bond
    b. a functional group
    c. quaternary structure
    d. an ionic bond

3.5 Nucleic Acids

13. Which of the following is incorrect regarding nucleotides?
    a. They contain a sugar, a nitrogen-containing base, and a phosphate group.
    b. They are the monomers of fats and polysaccharides.
    c. They join together by alternating covalent bonds between the sugars and phosphate groups.
    d. They are present in both DNA and RNA.
14. Which of the following is correct regarding ATP?
    a. It is an amino acid.
    b. It has a helical structure.
    c. It is a high-energy molecule that can break down to ADP and phosphate.
    d. It is a nucleotide component of DNA and RNA.
15. Which of the following is correct concerning an RNA molecule?
    a. It contains the sugar ribose.
    b. It may contain uracil as a nitrogen-containing base.
    c. It contains a phosphate molecule.
    d. All of the above are correct.

ENGAGE

LEARNSMART LABS®

The following LearnSmart Labs contain exercises that are related to the content of this chapter:

- Chemical Composition of Cells

THINKING SCIENTIFICALLY

1. The seeds of temperate plants tend to contain unsaturated fatty acids, while the seeds of tropical plants tend to have saturated fatty acids. a. How would you test your hypothesis? b. Assuming your hypothesis is supported, give an explanation.
2. Chemical analysis reveals that an abnormal form of an enzyme contains a polar amino acid at the location where the normal form has a nonpolar amino acid. Formulate a testable hypothesis concerning the abnormal enzyme.
3. In order to understand the relationship between enzyme structure and function, researchers often study mutations that swap one amino acid for another. In one enzyme, function is retained if a particular amino acid is replaced by one that has a nonpolar *R* group, but function is lost if the amino acid is replaced by one with a polar *R* group. Why might that be?
4. Scientists have observed that, although two species might have the same protein, the more distantly related they are, the more likely the sequence of the amino acids has changed. What might account for this observation? Why would this be an important study?

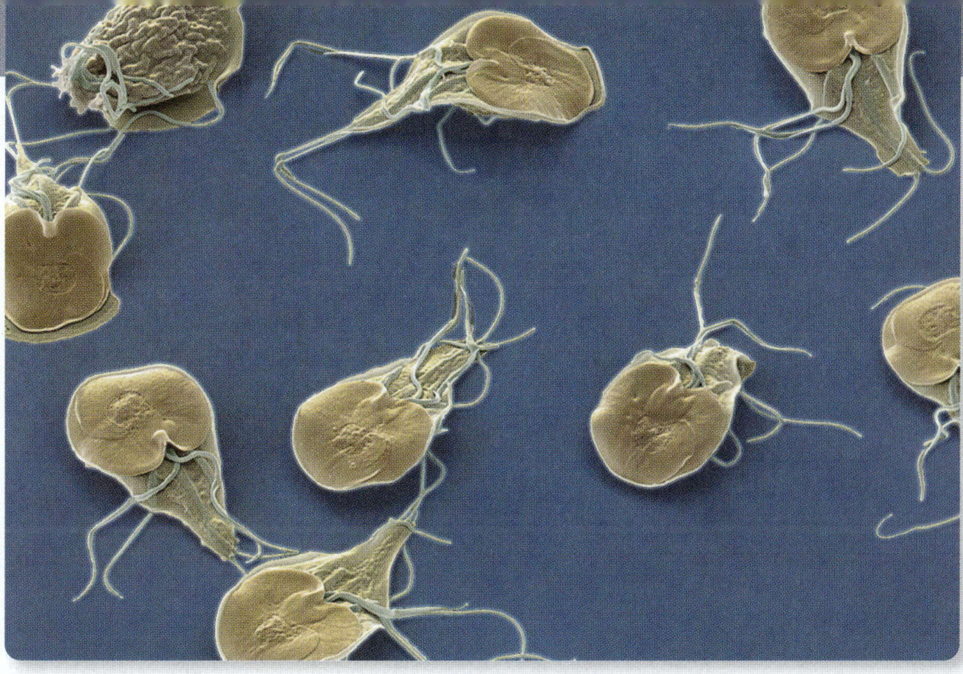

Electron micrograph of *Giardia lamblia,* a cause of diarrhea.

4

Cell Structure and Function

The Dutch shopkeeper Antonie van Leeuwenhoek (1632–1723) may have been the first person to see living cells. Using a microscope he built himself, he looked at everything possible, from the plaque between his teeth to his own feces. During one of these observations, he discovered "animalcules a moving prettily. Their bodies were somewhat longer than broad, and the belly, which was flat-lie, furnished with sundry little paws. . . ." In this way, Antonie van Leeuwenhoek reported seeing the parasite *Giardia lamblia* (also known as *Giardia intestinalis*). We now know that *Giardia* is a cause of some forms of diarhhea, especially in water supplies that have been contaminated by fecal material. And it is very common; up to 20% of the world's population may be infected with *Giardia*. While *Giardia* are single-celled parasites, and humans consist of trillions of cells, the cells of both of these organisms share may similar characteristics.

In this chapter, you will see that cells are the fundamental building blocks of organisms, organized to carry out basic metabolic functions and adapt to changing environmental conditions. The presentation concentrates on the generalized bacterial, animal, or plant cell; however, all cells are specialized in particular ways.

As you read through this chapter, think about the following questions:

1. How is a cell more than simply the sum of its macromolecular parts?
2. What characteristics enable cells to be alive and allow them to self-replicate?
3. How are cells able to metabolize and respond to environmental changes so quickly?

Chapter Outline

4.1 Cellular Level of Organization 58
4.2 Prokaryotic Cells 62
4.3 Introduction to Eukaryotic Cells 64
4.4 The Nucleus and Ribosomes 67
4.5 The Endomembrane System 69
4.6 Microbodies and Vacuoles 72
4.7 The Energy-Related Organelles 73
4.8 The Cytoskeleton 75

Before You Begin

Before beginning this chapter, take a few moments to review the following discussions.

Table 3.1 What role do functional groups serve in biological molecules?

Sections 3.2 to 3.5 What macromolecules are needed to construct a cell?

Figures 3.7, 3.12, 3.17, and 3.21 How does the structure of a macromolecule affect its function?

Following the Themes

CHAPTER 4 CELL STRUCTURE AND FUNCTION

UNIT 1 THE CELL

Evolution	All cells are produced from existing cells, creating an unbroken lineage back to cells from early Earth.
Nature of Science	Understanding how cells function allows us to discover ways to treat cell-based diseases.
Biological Systems	Cells play an important role at all levels of biological organization, from tissues to ecosystems.

4.1 Cellular Level of Organization

Learning Outcomes

Upon completion of this section, you should be able to

1. Understand that cells are the basic unit of life.
2. List the basic principles of the cell theory.
3. Recognize how the surface-area-to-volume ratio limits the size of a cell.

Cells are the basic units of life. All of the chemistry and biomolecules we have discussed to this point are necessary but insufficient on their own to support life. It is only when these components are brought together and organized into a cell that life is possible.

All organisms are made up of cells. When we observe plants, animals, and other organisms, it is important to realize that what we are seeing is a collection of cells that work together in a highly organized, regulated manner and thus conduct the business of life. Figure 4.1 shows the connection between whole organisms and their component cells. Although the cellular basis of life is clear to us now, scientists were unaware of this fact as recently as 200 years ago. The link between cells and life became clear to microscopists during the 1830s.

The **cell** is the smallest unit of living matter. The collective work of the nineteenth-century scientists Robert Brown (1773–1858), Matthias Schleiden (1804–1881), and Theodor Schwann (1810–1882) helped determine that plants and animals are composed of cells. Further work by the German physician Rudolph Virchow (1821–1902) showed that cells self-reproduce and that "every cell comes from a preexisting cell." Today, we know that various illnesses of the body, such as diabetes and prostate cancer, are due to cellular malfunction. Countless scientific investigations since that time verify these initial findings. From these results, we can infer that all life on Earth today came from cells in ancient times, and that all cells are related in some way. In reality, a continuity of cells has been present from generation to generation, even back to the very first cell (or cells) in the history of life.

Today, some life-forms exist as single cells, whereas others are complex, interconnected systems of cells. When single-celled organisms reproduce, a single cell divides and becomes two new organisms. When multicellular organisms grow, many cells divide. The presence of many cells allows some to specialize to do particular jobs within the multicellular organism, including the cells that create genetic variation through sexual reproduction.

The work of Schleiden, Schwann, and Virchow helped created the **cell theory.** It states that

1. All organisms are composed of cells.
2. Cells are the basic units of structure and function in organisms.
3. Cells come only from preexisting cells because cells are self-reproducing.

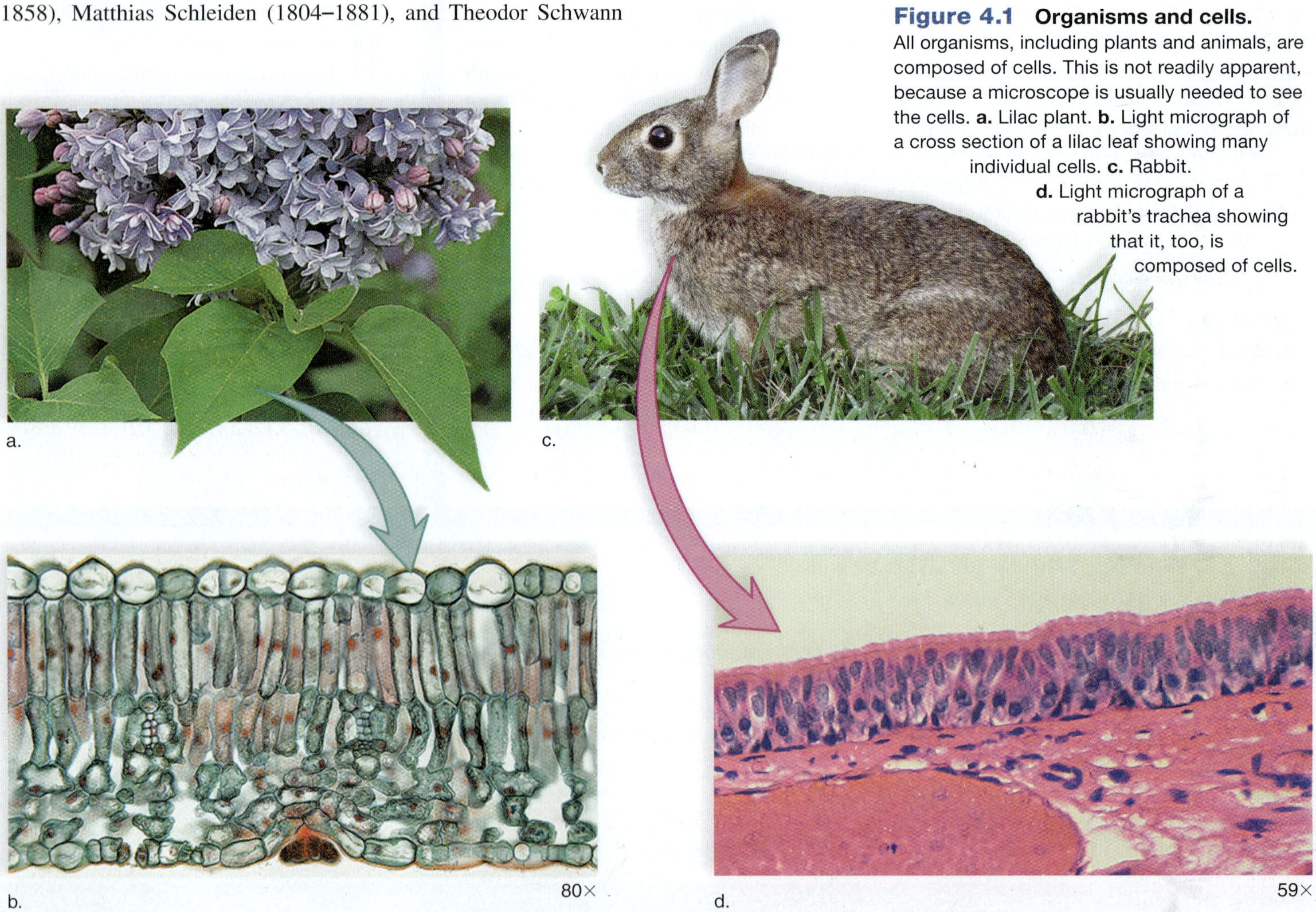

Figure 4.1 Organisms and cells. All organisms, including plants and animals, are composed of cells. This is not readily apparent, because a microscope is usually needed to see the cells. **a.** Lilac plant. **b.** Light micrograph of a cross section of a lilac leaf showing many individual cells. **c.** Rabbit. **d.** Light micrograph of a rabbit's trachea showing that it, too, is composed of cells.

Cell Size

Although they range in size, cells are generally quite small. A frog's egg, at about 1 millimeter (mm) in diameter, is large enough to be seen by the human eye. But most cells are far smaller than 1 mm; some are even as small as 1 micrometer (μm)—one-thousandth of a millimeter. Cell inclusions and macromolecules are smaller than a micrometer and are measured in terms of nanometers (nm).

Because of their size, very small biological structures can only be viewed with microscopes, which magnify a visual image. Figure 4.2 shows the visual range of the eye, light microscope, and electron microscope; the discussion of microscopy in the Nature of Science feature, "Microscopy Today," on page 60 explains why the electron microscope allows us to see so much more detail than the light microscope does.

Why are cells so small? To answer this question, consider that a cell is a system by itself; as such, it needs a surface area large enough to allow adequate nutrients to enter and for wastes to be eliminated. Small cells, not large cells, are likely to have an adequate surface area for exchanging wastes for nutrients. As cells increase in size, the surface area becomes inadequate to exchange the materials that the volume of the cell requires.

Figure 4.3 illustrates that dividing a large cube into smaller cubes provides a lot more surface area per volume. This relationship is called the **surface-area-to-volume ratio.** Calculations show that a 1-cm cube has a surface-area-to-volume ratio of 6:1, whereas a 4-cm cube has a surface-area-to-volume ratio of 1.5:1. In general, a higher surface-area-to-volume ratio increases the efficiency of transporting materials into and out of the cell.

A mental image might help you visualize the importance of surface-area-to-volume ratios and why this relationship favors smaller cells. Imagine a small room and a large room filled with people. The small room, which holds 20 people, has only two doors, and the large room, which holds 80 people, has four doors. If a fire occurred in both rooms, it would be faster to get the people out of the smaller room, because it has the more favorable ratio of doors to people. Similarly, a small cell size is more advantageous for exchanging molecules because of its greater surface-area-to-volume ratio.

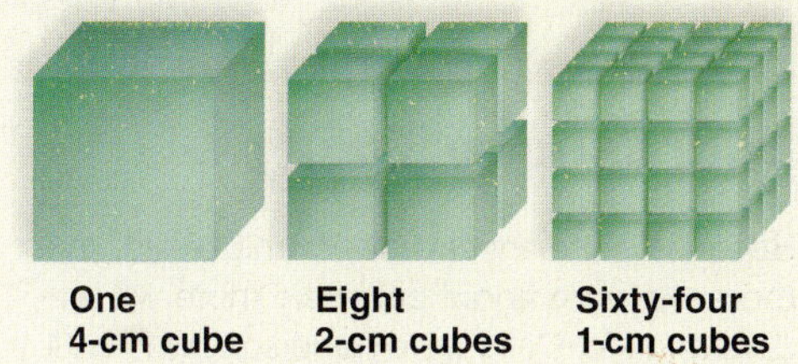

	One 4-cm cube	Eight 2-cm cubes	Sixty-four 1-cm cubes
Total surface area (height × width × number of sides × number of cubes)	96 cm^2	192 cm^2	384 cm^2
Total volume (height × width × length × number of cubes)	64 cm^3	64 cm^3	64 cm^3
Surface area/Volume of each cube (surface area ÷ volume)	1.5:1	3:1	6:1

Figure 4.3 Surface-area-to-volume relationships. As cell size decreases from 4 cm^3 to 1 cm^3, the surface-area-to-volume ratio increases.

Check Your Progress 4.1

1. Explain why cells are alive but macromolecules are not.
2. State the components of the cell theory.
3. Explain why a large surface-area-to-volume ratio is needed for the proper functioning of cells.

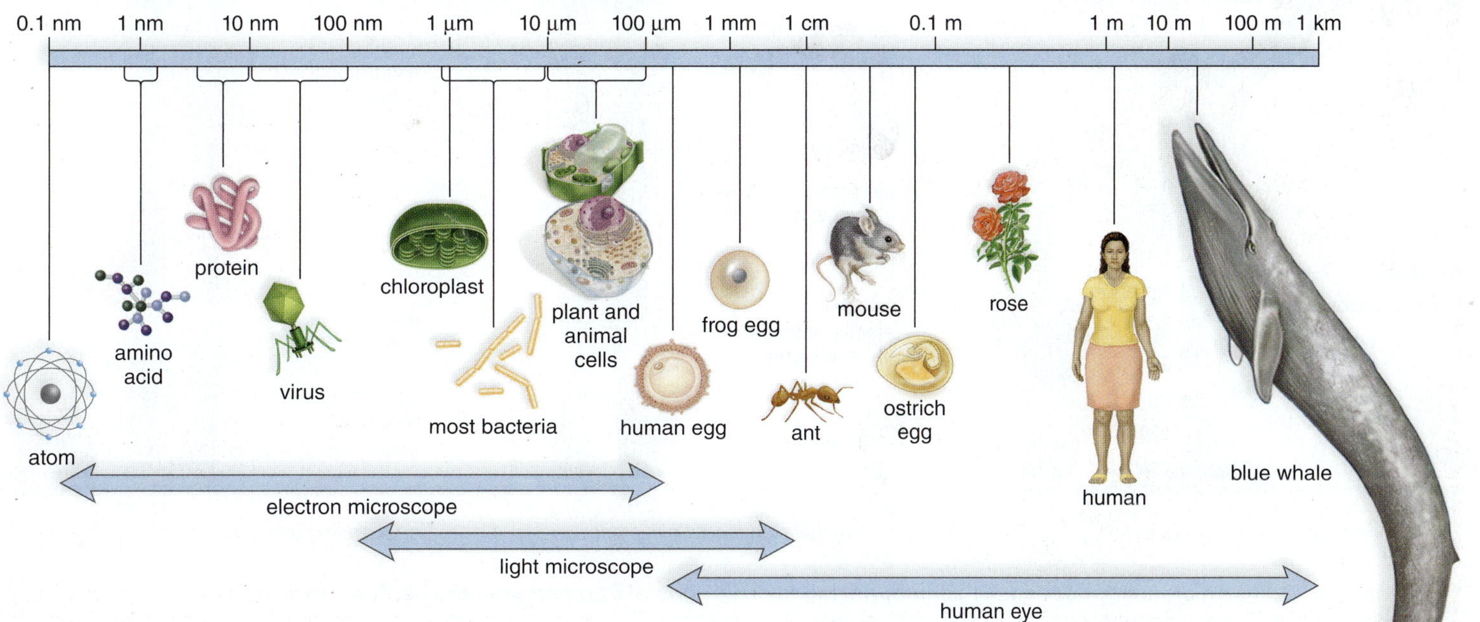

Figure 4.2 The sizes of various objects. The unassisted human eye can usually see macroscopic organisms and a few large cells. Microscopic cells are visible with the light microscope, but not in much detail. An electron microscope is necessary to see organelles in detail and to observe viruses and molecules. In the metric system (see back endsheet), each higher unit is ten times greater than the preceding unit. (1 meter = 10^2 cm = 10^3 mm = 10^6 μm = 10^9 nm)

THEME Nature of Science

Microscopy Today

Because cells are the basic unit of life, the more we learn about cells, the more we understand life. Cells were not discovered until the seventeenth century, when the microscope was invented. Since that time, various types of microscopes have been developed for studying cells and their components.

Many times when scientists don't have suitable tools to investigate natural phenomena, they invent them. Microscopes have given scientists a deeper look into how life works than is possible with the naked eye. Today, there are many types of microscopes. A *compound light microscope* uses a set of glass lenses to focus light rays passing through a specimen to produce an image that is viewed by the human eye. A *transmission electron microscope* (*TEM*) uses a set of electromagnetic lenses to focus electrons passing through a specimen to produce an image, which is projected onto a fluorescent screen or photographic film. A *scanning electron microscope* (*SEM*) uses a narrow beam of electrons to scan over the surface of a specimen that is coated with a thin metal layer. Secondary electrons given off by the metal are detected and used to produce a three-dimensional image on a television screen. Figure 4A shows these three types of microscopic images.

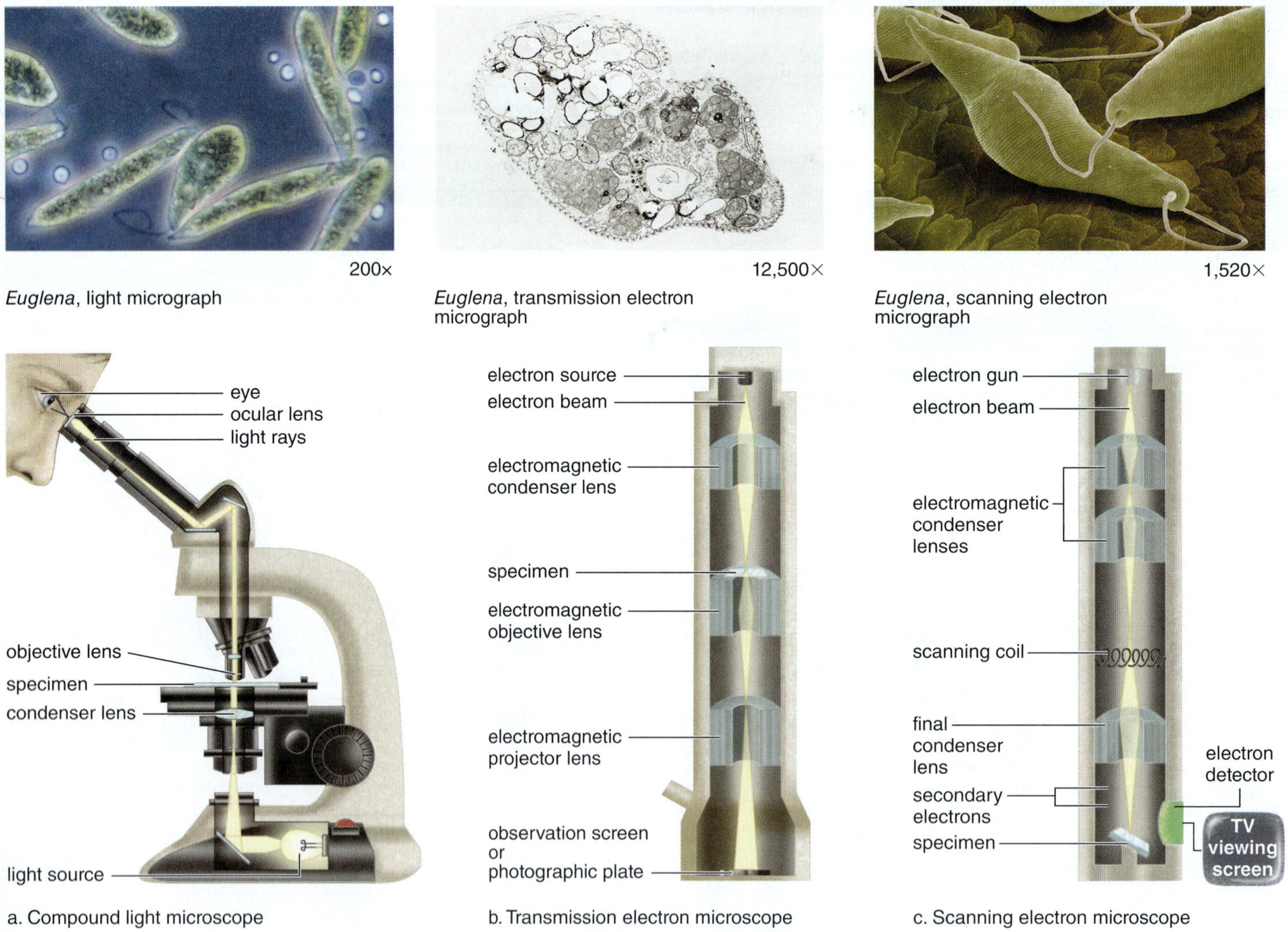

Figure 4A Diagram of microscopes with accompanying micrographs of *Euglena gracilis*. **a.** The compound light microscope and **(b)** the transmission electron microscope provide an internal view of an organism. **c.** The scanning electron microscope provides an external view of an organism.

Magnification, Resolution, and Contrast

Magnification is the ratio between the size of an image and its actual size. Electron microscopes magnify to a greater extent than do compound light microscopes. A light microscope can magnify objects about a thousand times, but an electron microscope can magnify them hundreds of thousands of times. The difference lies in the means of illumination. The path of light rays and electrons moving through space is wavelike, but the wavelength of electrons is much shorter than the wavelength of light. This difference in wavelength accounts for the electron microscope's greater magnifying capability and its greater ability to distinguish between two points (resolving power).

Resolution is the minimum distance between two objects that allows them to be seen as two separate objects. A microscope with poor resolution might enable a student to see only one cellular granule, while the microscope with the better resolution would show two granules next to each other. The greater the resolving power, the greater the detail seen.

If oil is placed between the sample and the objective lens of the compound light microscope, the resolving power is increased, and if ultraviolet light is used instead of visible light, it is also increased. But typically, a light microscope can resolve down to 0.2 μm, while the transmission electron microscope can resolve down to 0.0002 μm. If the resolving power of the average human eye is set at 1.0, then the typical compound light microscope is about 500, and the transmission electron microscope is 100,000 (Fig. 4A*b*).

The ability to make out, or resolve, a particular object can depend on *contrast,* a difference in the shading of an object compared to its background. Higher contrast is often achieved by staining cells with colored dyes (light microscopy) or with electron-dense metals (electron microscopy), which make them easier to see. Optical methods such as phase contrast and differential interference contrast (Fig. 4B) can also be used to improve contrast. Using fluorescently tagged antibodies can also help us visualize subcellular components such as specific proteins (see Fig. 4.19).

Illumination, Viewing, and Recording

Light rays can be bent (refracted) and brought to focus as they pass through glass lenses, but electrons do not pass through glass. Electrons have a charge that allows them to be brought into focus by electromagnetic lenses. The human eye uses light to see an object but cannot use electrons for the same purpose. Therefore, electrons leaving the specimen in the electron microscope are directed toward a screen or a photographic plate that is sensitive to their presence. Humans can view the image on the screen or photograph.

A major advancement in illumination has been the introduction of *confocal microscopy,* which uses a laser beam scanned across the specimen to focus on a single shallow plane within the cell. The microscopist can "optically section" the specimen by focusing up and down, and a series of optical sections can be combined in a computer to create a three-dimensional image, which can be displayed and rotated on the computer screen.

An image from a microscope may be recorded by placing a television camera where the eye would view the image. The television camera converts the light image into an electronic image, which can be entered into a computer. In *video-enhanced contrast microscopy,* the computer makes the darkest areas of the original image much darker and the lightest areas of the original much lighter. The result is a high-contrast image with deep blacks and bright whites. Even more contrast can be introduced by the computer if shades of gray are replaced by colors.

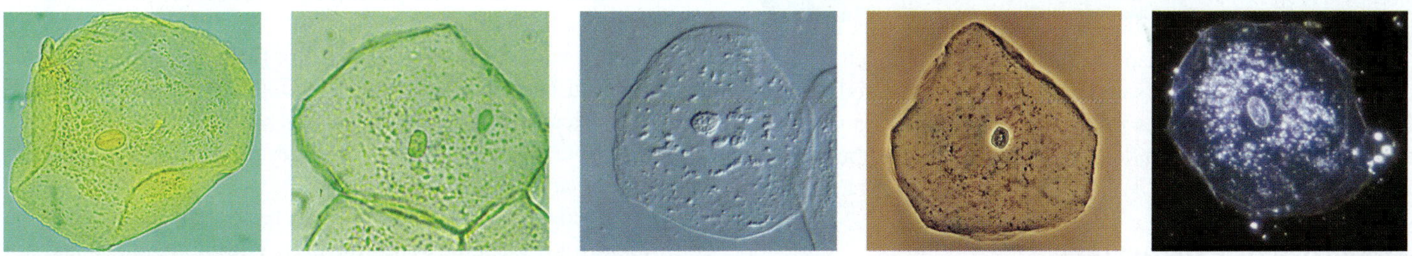

250×

Bright-field. Light passing through the specimen is brought directly into focus. Usually, the low level of contrast within the specimen interferes with viewing all but its largest components.

225×

Bright-field (stained). Dyes are used to stain the specimen. Certain components take up the dye more than other components, and therefore contrast is enhanced.

160×

Differential interference contrast. Optical methods are used to enhance density differences within the specimen so that certain regions appear brighter than others. This technique is used to view living cells, chromosomes, and organelle masses.

600×

Phase contrast. Density differences in the specimen cause light rays to come out of "phase." The microscope enhances these phase differences so that some regions of the specimen appear brighter or darker than others. The technique is widely used to observe living cells and organelles.

600×

Dark-field. Light is passed through the specimen at an oblique angle so that the objective lens receives only light diffracted and scattered by the object. This technique is used to view organelles, which appear quite bright against a dark field.

Figure 4B Photomicrographs of cheek cells. Bright-field microscopy is the most common form used with a compound light microscope. Other types of microscopy include differential interference contrast, phase contrast, and dark-field.

4.2 Prokaryotic Cells

Learning Outcomes

Upon completion of this section, you should be able to

1. Examine the evolutionary relatedness of prokaryotes, eukaryotes, and archaeans.
2. Describe the fundamental components of a bacterial cell.

Fundamentally, two different types of cells exist in nature. **Prokaryotic cells** (Gk. *pro,* "before"; *karyon,* "kernel, nucleus") lack a membrane-bound nucleus. **Eukaryotic cells** (Gk. *eu,* "true") possess a nucleus. The bacteria and archaeans were once thought to be closely related because of their similar size and shape. Comparisons of DNA and RNA sequences now show archaeans to be biochemically distinct from either the bacteria or the eukaryotes. These comparisons also suggest that the archaeans are more closely related to the eukaryotes than the bacteria. These comparisons also help define the three domains of life that were presented in Chapter 1. Two of the three domains, the Eubacteria and Archaea, are prokaryotic cells, while all eukaryotic cells are assigned to domain Eukarya.

Prokaryotes as a group are one of the most abundant and diverse life-forms on Earth, and they are present in great numbers in the air, water, and soil, as well as living in and on other organisms. Although they are structurally less complicated than eukaryotes, their metabolic capabilities as a group far exceed those of eukaryotes. Prokaryotes are an extremely successful group of organisms whose evolutionary history dates back to the first cells on Earth.

Bacteria are well known because they cause some serious diseases, such as tuberculosis, anthrax, tetanus, throat infections, and gonorrhea. But many species of bacteria are important to the environment, because they decompose the remains of dead organisms and contribute to ecological cycles. Bacteria also assist humans in still another way—we use them to manufacture all sorts of products, from industrial chemicals to foodstuffs and drugs. For example, today we know how to place human genes in certain cultured bacteria so that they can produce human insulin, a necessary hormone for the treatment of diabetes.

The Structure of Prokaryotes

Prokaryotes are quite small; an average size is 1.1–1.5 μm wide and 2.0–6.0 μm long. The majority of the prokaryotes have one of these basic shapes:

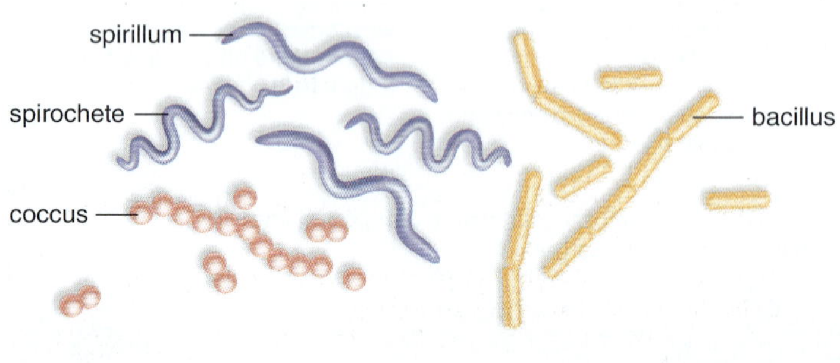

A rod-shaped bacterium is called a **bacillus,** while a spherical-shaped bacterium is a **coccus.** Both of these can occur as pairs or chains, and in addition, cocci can occur as clusters. Some long rods are twisted into spirals, in which case they are **spirilla** if they are rigid or **spirochetes** if they are flexible.

Animation
Prokaryotic Cell Shape

Figure 4.4 shows the generalized structure of a bacterium. "Generalized" means that not all bacteria have all the structures depicted, and some have more than one of each. Also, for the sake of discussion, we divide the organization of bacteria into the cell envelope, the cytoplasm, and the external structures.

Cell Envelope

In bacteria, the **cell envelope** includes the plasma membrane, the cell wall, and the glycocalyx. The **plasma membrane** is a phospholipid bilayer with embedded proteins:

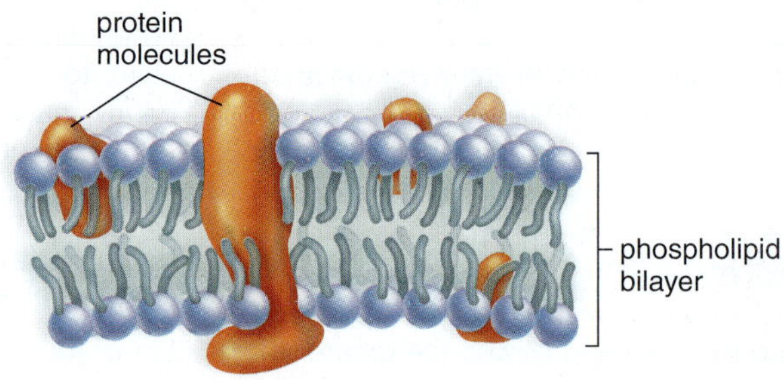

The plasma membrane has the important function of regulating the entrance and exit of substances into and out of the cytoplasm. Regulating the flow of materials into and out of the cytoplasm is necessary in order to maintain its normal composition.

In prokaryotes, the plasma membrane can form internal pouches called *mesosomes.* Mesosomes most likely increase the internal surface area for the attachment of enzymes that are carrying on metabolic activities.

The **cell wall** maintains the shape of the cell, even if the cytoplasm should happen to take up an abundance of water. The cell wall of a bacterium contains peptidoglycan, a complex molecule containing a unique amino disaccharide and peptide fragments.

Animation
Cell Wall Antibiotics

The **glycocalyx** is a layer of polysaccharides that lies outside the cell wall in some bacteria. When the layer is well organized and not easily washed off, it is called a **capsule.** A slime layer, on the other hand, is not well organized and is easily removed. The glycocalyx aids against drying out and helps bacteria resist a host's immune system. It also helps bacteria attach to almost any surface.

Cytoplasm

The **cytoplasm** is a semifluid solution composed of water and inorganic and organic molecules encased by a plasma membrane. Among the organic molecules are a variety of enzymes, which speed the many types of chemical reactions involved in metabolism.

While prokaryotes lack a membrane-bound nucleus, their DNA is located in a region of the cytoplasm called the **nucleoid.** Furthermore, eukaryotic cells typically have multiple chromosomes, but prokaryotes have a single, coiled chromosome. Many prokaryotes also have extrachromosomal pieces of circular DNA called

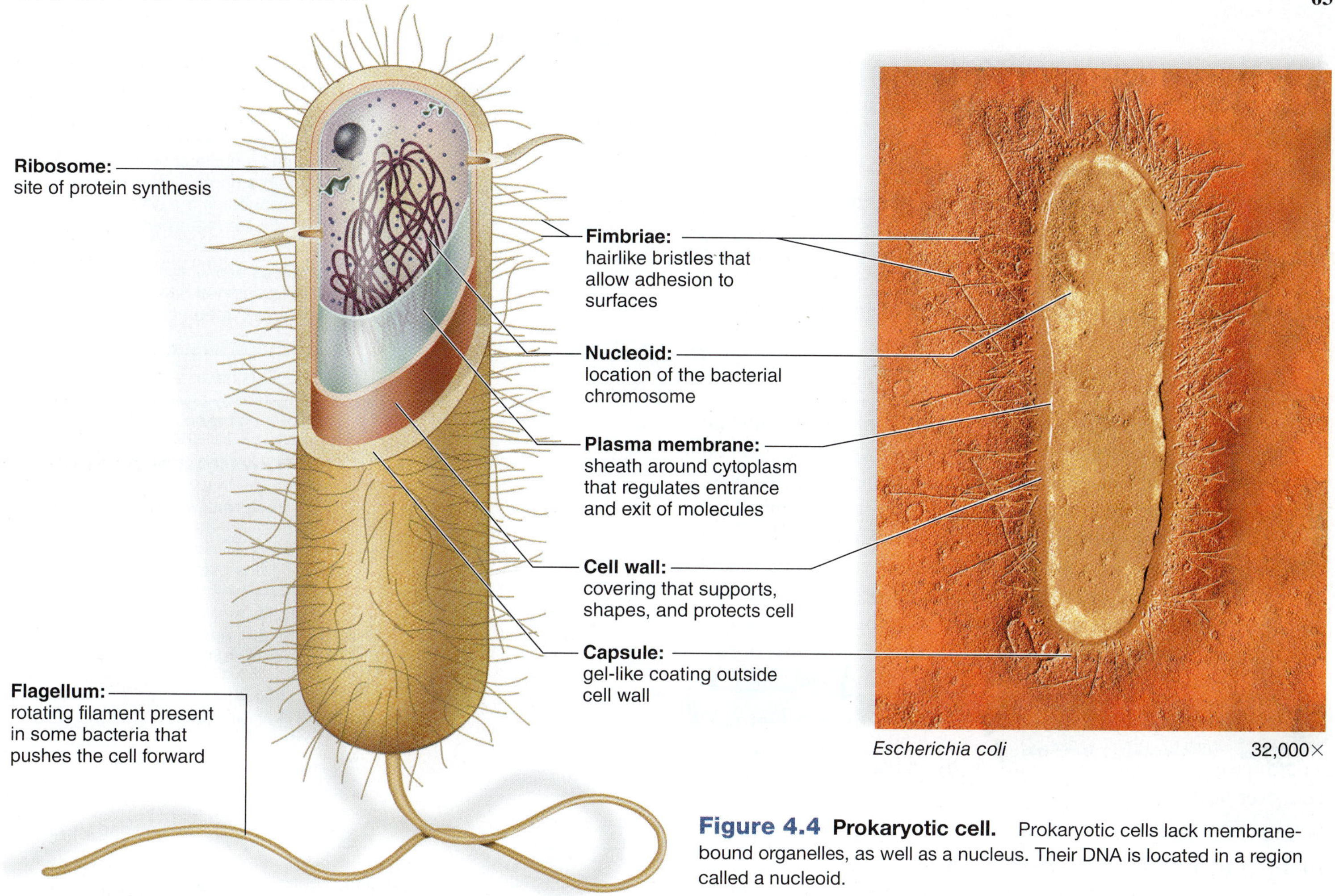

Figure 4.4 Prokaryotic cell. Prokaryotic cells lack membrane-bound organelles, as well as a nucleus. Their DNA is located in a region called a nucleoid.

plasmids. Plasmids are routinely used in biotechnology laboratories as a vector to transport DNA into a bacterium (see Chapter 14). Procedures such as this are possible because all life on Earth is constructed from the same four DNA nucleotides: A, G, C, and T. Biotechnology plays an important role in the production of new medicines and many of the commerical products we use every day.

The many proteins encoded by the prokaryotic DNA are synthesized on tiny structures in the cytoplasm called **ribosomes.** A prokaryotic cell contains thousands of ribosomes that are similar in shape and function but are smaller than eukaryotic ribosomes. Like their eukaryotic counterparts, prokaryotic ribosomes still contain RNA and protein in two subunits.

There is a tremendous amount of metabolic diversity in the prokaryotes. Some prokaryotes carry out metabolism in the same manner as animals (by ingesting other organisms), but the **cyanobacteria** are a form of bacteria that are capable of photosynthesis in the same manner as plants. These organisms live in water, in ditches, on buildings, and on the bark of trees. Their cytoplasm contains extensive internal membranes called **thylakoids** (Gk. *thylakon,* "small sac"), where chlorophyll and other pigments absorb solar energy for the production of carbohydrates. Cyanobacteria are called the blue-green bacteria, because some have a pigment that adds a shade of blue to the cell, in addition to the green color of chlorophyll. The cyanobacteria release oxygen as a by-product of photosynthesis, and ancestral cyanobacteria were some of the earliest photosynthesizers on Earth. Many sources of evidence show that the composition of the early Earth's atmosphere was changed by the addition of oxygen.

External Structures

The external structures of a prokaryote, namely the flagella, fimbriae, and conjugation pili, are made of protein. Motile prokaryotes can propel themselves in water by the means of appendages called **flagella** (usually 20 nm in diameter and 1–70 nm long). The prokaryotic flagellum is one of the great wonders of nature, and it consists of a filament, a hook, and a basal body. The basal body is a series of rings anchored in the cell wall and membrane. Unlike the flagellum of the eukaryotes, which has a whiplike motion, the flagellum of a prokaryote rotates 360 degrees. Sometimes flagella occur only at the two ends of a cell, and sometimes they are dispersed randomly over the surface. The number and location of flagella can be used to help distinguish different types of prokaryotes.

Fimbriae are small, bristlelike fibers that sprout from the cell surface. They are not involved in locomotion; instead, fimbriae are involved in attaching prokaryotes to a surface. **Conjugation pili** are rigid, tubular structures used by prokaryotes to pass DNA from cell to cell. Prokaryotes reproduce asexually by binary fission, but they can exchange DNA by way of the conjugation pili. They can also take up DNA from the external medium or by way of viruses.

Check Your Progress 4.2

1. Explain the major differences between a prokaryotic and eukaryotic cell.
2. Describe the functions of the bacterial cell envelope, cytoplasm, and external structures.

4.3 Introduction to Eukaryotic Cells

Learning Outcomes

Upon completion of this section, you should be able to

1. Describe how the endosymbiotic theory explains eukaryotic cell structure.
2. Summarize the functions of the organelles in a eukaryotic cell.
3. Compare and contrast the structure of animal and plant cells.

Eukaryotic cells, like prokaryotic cells, have a plasma membrane that separates the contents of the cell from the environment and that regulates the passage of molecules into and out of the cytoplasm. The plasma membrane is a phospholipid bilayer with embedded proteins. What distinguishes eukaryotic cells from prokaryotic cells is the presence of a nucleus and internal membrane-bound compartments, called **organelles.** Nearly all organelles are surrounded by a membrane with embedded proteins, many of which are enzymes. These enzymes make products specific to that organelle, but their action benefits the whole cell system. Each organelle carries out specialized functions, which together allow the cell to be more efficient and successful. These features would have given the new cell a selective advantage over other cells.

MP3 Cellular Organelles

Origin of the Eukaryotic Cell

The fossil record, which is based on the remains of ancient life, suggests that the first cells were prokaryotes. Therefore, scientists believe that eukaryotic cells evolved from prokaryotic cells. Biochemical data suggest that eukaryotes are more closely related to the archaea than the bacteria. The eukaryotic cell probably evolved from a prokaryotic cell in stages. The distinguishing characteristic of the eukaryotic cell, the nucleus, is believed to have evolved due to the invagination of the plasma membrane (Fig. 4.5). The same process also explains the origin of organelles such as the endoplasmic reticulum and the Golgi apparatus.

There is strong evidence that the origin of the energy organelles occurred when a larger eukaryotic cell engulfed smaller prokaryotic cells. Observations in the laboratory indicate that an amoeba infected with bacteria can become dependent upon them. Some investigators believe mitochondria and chloroplasts are derived from prokaryotes that were taken up by larger cells (Fig. 4.5). Perhaps mitochondria were originally aerobic heterotrophic bacteria and chloroplasts were originally cyanobacteria. The eukaryotic host cell would have benefited from an ability to utilize oxygen or synthesize organic food when, by chance, the prokaryote was taken up and not destroyed. After the prokaryote entered the host cell, the two would have begun living together cooperatively. This proposal is known as the **endosymbiotic theory** (*endo-*, "in"; *symbiosis*, "living together"). Some of the evidence supporting this hypothesis is as follows:

- Mitochondria and chloroplasts are similar to bacteria in size and in structure.
- Both organelles are surrounded by a double membrane—the outer membrane may be derived from the engulfing vesicle, and the inner one may be derived from the plasma membrane of the original prokaryote.
- Mitochondria and chloroplasts contain a limited amount of genetic material and divide by splitting. Their DNA (deoxyribonucleic acid) is a circular loop like that of prokaryotes.
- Although most of the proteins within mitochondria and chloroplasts are now produced by the eukaryotic host, they do have their own ribosomes and they do produce some proteins. Their ribosomes resemble those of prokaryotes.
- The RNA (ribonucleic acid) base sequence of the ribosomes in chloroplasts and mitochondria also suggests a prokaryotic origin of these organelles.

Animation Endosymbiosis

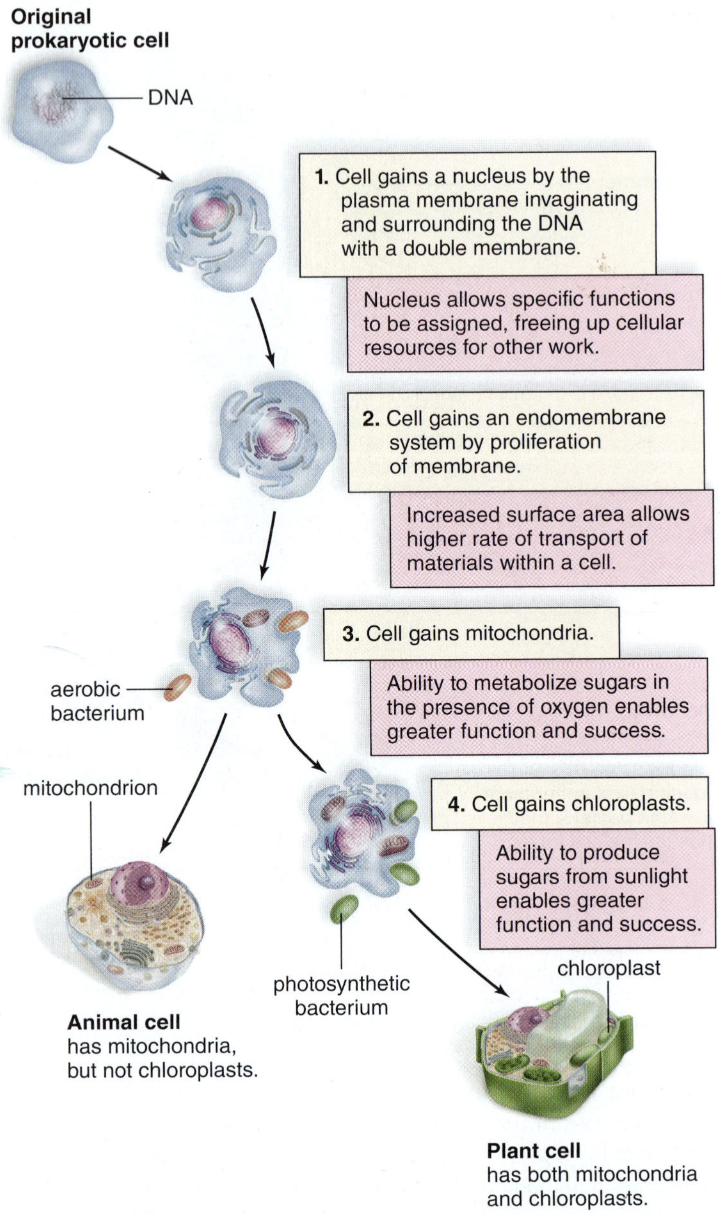

Figure 4.5 Origin of organelles. Invagination of the plasma membrane could have created the nuclear envelope and an endomembrane system that involves several organelles. The endosymbiotic theory states that mitochondria and chloroplasts were independent prokaryotes that took up residence in a eukaryotic cell. Endosymbiosis was a first step toward the origin of the eukaryotic cell during the evolutionary history of life.

Tutorial Endosymbiosis

Figure 4.6 Typical animal cell. Micrograph of a human cell (*right*) and drawing of a generalized animal cell (*below*).

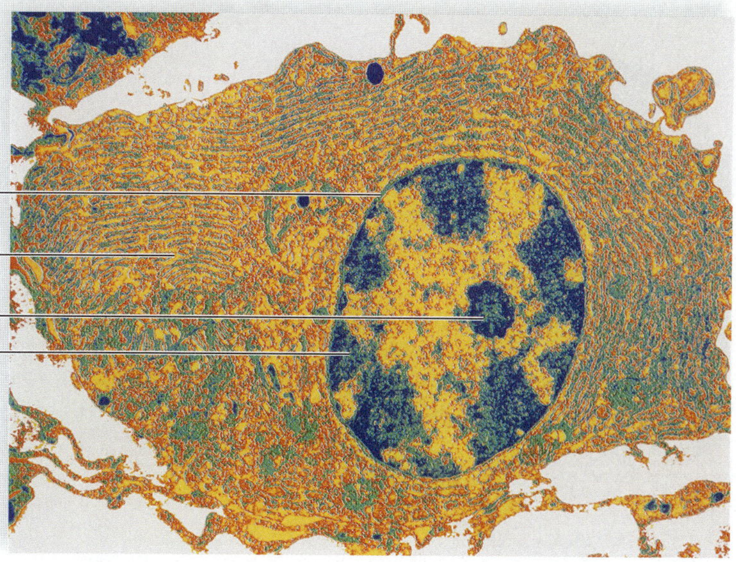

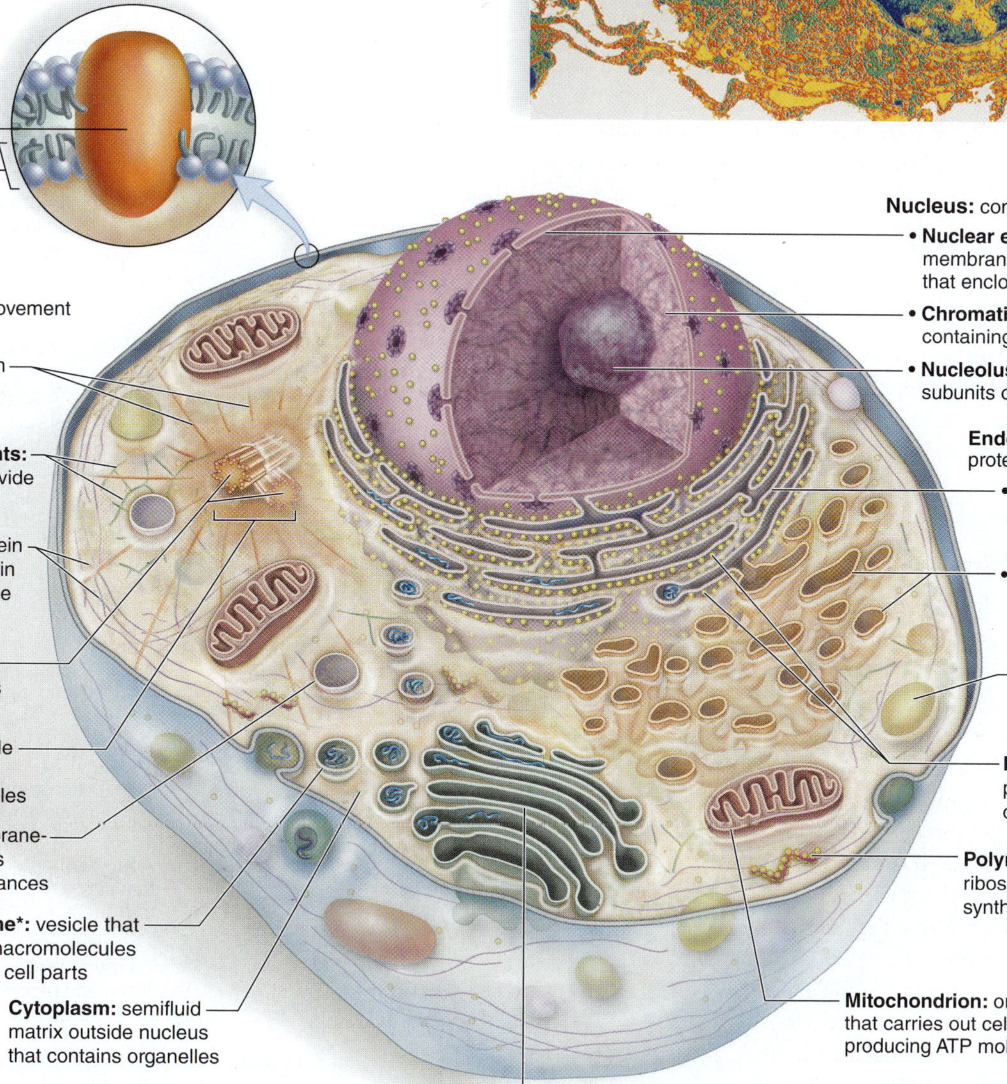

*not commonly found in plant cells
**not found in plant cells

Structure of a Eukaryotic Cell

Figures 4.6 and 4.7 show general features of fully evolved, present-day animal and plant cells. Specialized cells, as opposed to generalized cells, do not necessarily contain all the structures depicted and may have more or fewer copies of any particular organelle, depending on their particular function. These generalized depictions of plant and animal cells are useful for study purposes. A baseline understanding of cell structure and function will be helpful when you study the function of specialized cells later in this text. Overall, the cell can be seen as a system of interconnected organelles that work together to metabolize, regulate, and conduct life processes. For example, the nucleus is a compartment that houses the genetic material within eukaryotic chromosomes and contains hereditary information. The nucleus communicates with ribosomes in the cytoplasm, and the organelles of the endomembrane system—notably the endoplasmic reticulum and the Golgi apparatus—communicate with one another.

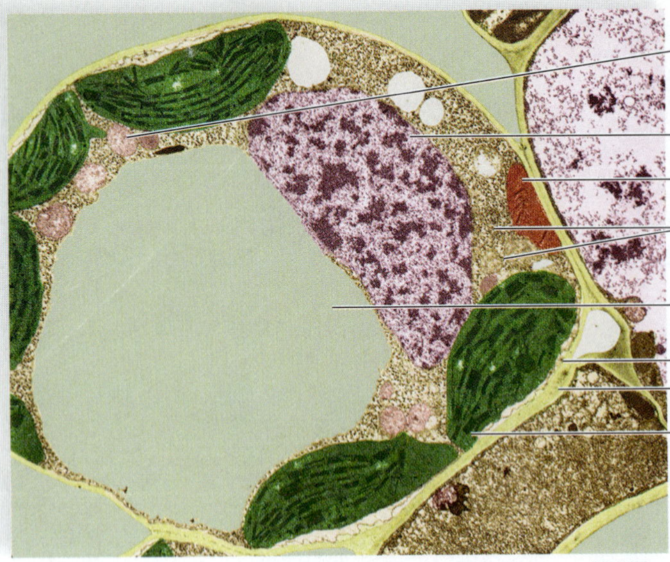

Figure 4.7 Typical plant cell. False-colored micrograph of a young plant cell (*left*) and drawing of a generalized plant cell (*below*).

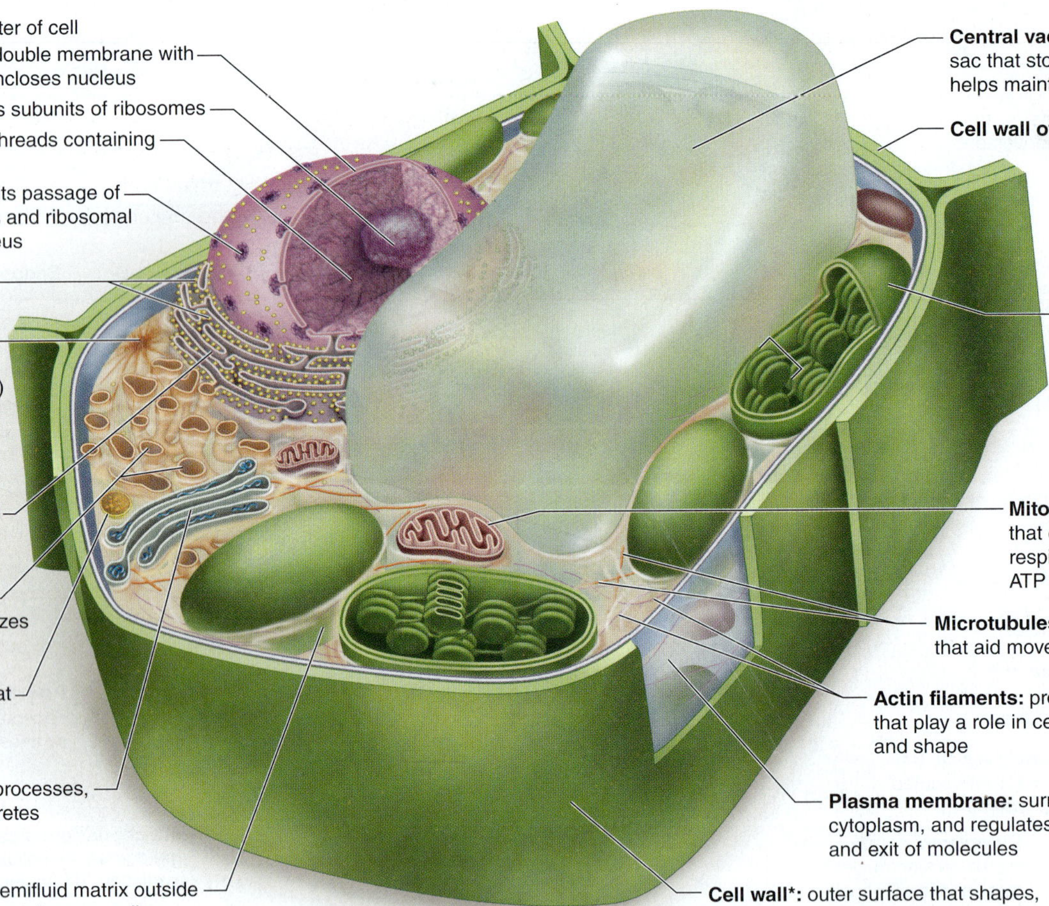

*not found in animal cells

Production of specific molecules takes place inside or on the surface of organelles. As mentioned, enzymes embedded in the organelles' membranes make these molecules. These products are then transported around the cell by transport **vesicles,** membranous sacs that enclose the molecules and keep them separate from the cytoplasm. For example, the endoplasmic reticulum communicates with the Golgi apparatus by means of transport vesicles. Communication with the energy-related organelles—mitochondria and chloroplasts—is less obvious, but it does occur, because they import particular molecules from the cytoplasm.

Vesicles move around by means of an extensive network or lattice of protein fibers called the **cytoskeleton,** which also maintains cell shape and assists with cell movement. The protein fibers serve as tracks for the transport vesicles that are taking molecules from one organelle to another. Organelles are also moved from place to place using this transport system. Think of the cytoskeleton as a three-dimensional road system inside cells used to transport important cargo from place to place. The cytoskeleton is discussed in detail later in this chapter.

In addition to the plasma membrane, some eukaryotic cells, notably plant cells and those of fungi and many protists, have a cell wall. A plant cell wall contains cellulose and, therefore, has a different composition from the bacterial cell wall.

Cells can vary the proportion of organelles they have, depending on the specialized function of the cell. For example, a liver cell whose function is partly to detoxify drugs and other ingested compounds contains a greater proportion of smooth endoplasmic reticulum, the organelle that accomplishes that task. A nerve cell, whose job is to conduct electrical signals across long distances, contains more plasma membrane relative to other cells. Other cells may specialize so extensively that they completely lose an organelle, like a red blood cell that ejects its nucleus to increase the surface area needed to carry oxygen in the blood.

Check Your Progress 4.3

1. Summarize the benefits of compartmentalization found in cells.
2. Examine why organelles increase cell efficiency and function.
3. Explain the origins of the nucleus, chloroplast, and mitochondria of eukaryotic cells.

4.4 The Nucleus and Ribosomes

Learning Outcomes

Upon completion of this section, you should be able to

1. Describe the structure and function of the nucleus.
2. Describe the flow of information from DNA to a protein.
3. Explain the role of ribosomes in protein synthesis.

The **nucleus** is essential to the life of a eukaryotic cell. It contains the genetic information that is passed on from cell to cell and from generation to generation. It specifies the information that ribosomes use to carry out protein synthesis. It also contains instructions for copying itself.

The Nucleus

The nucleus, which has a diameter of about 5 μm, is a prominent structure in the eukaryotic cell (Fig. 4.8). It generally appears as an oval structure located near the center of most cells. Some cells, such as skeletal muscle cells, can have more than one nucleus. The interior of the nucleus contains a semifluid matrix called the **nucleoplasm.** The nucleus is the command center of the cell. It contains **chromatin**

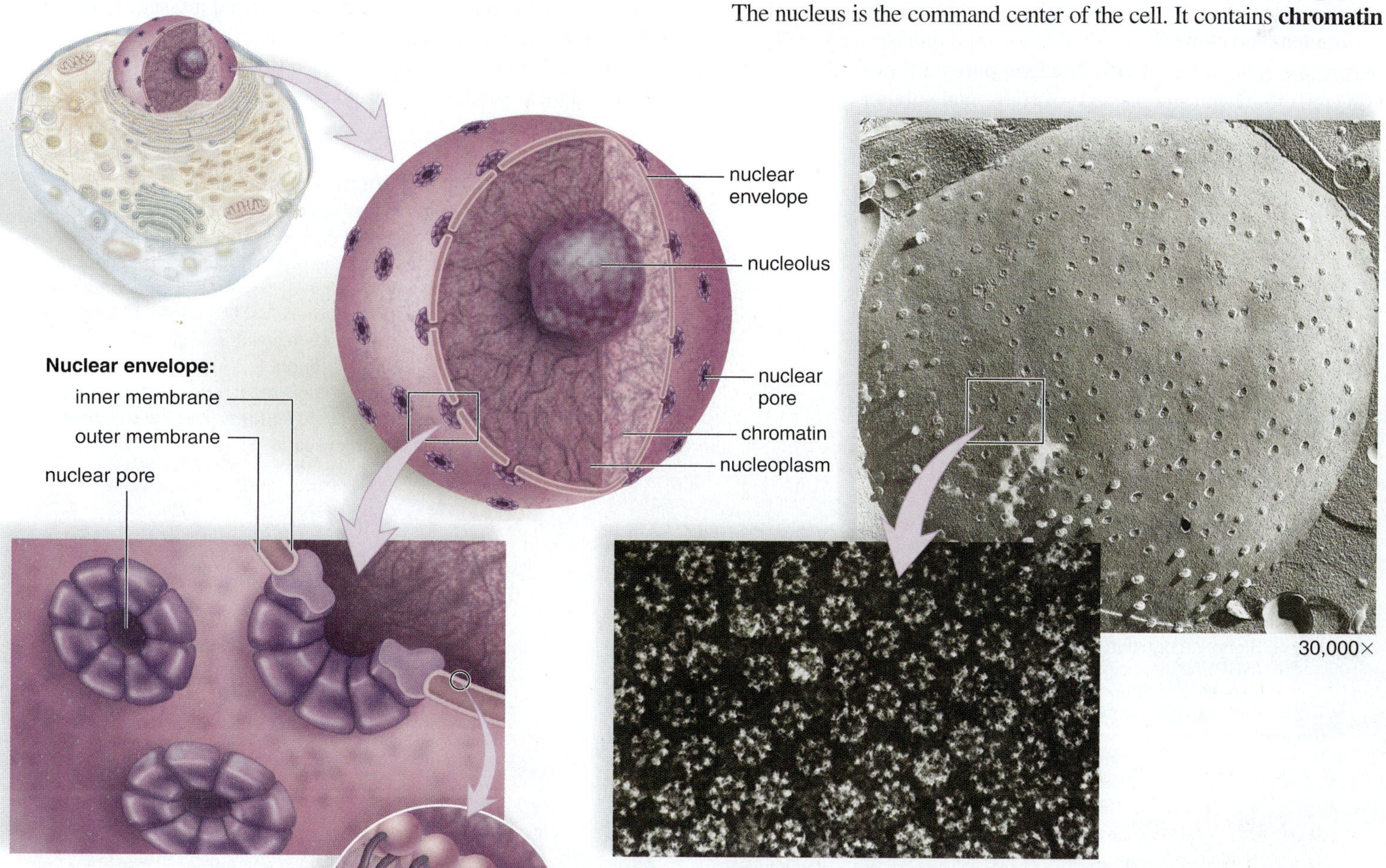

Figure 4.8 Anatomy of the nucleus. The nucleus contains chromatin. The nucleolus is a region of chromatin where ribosomal RNA is produced and ribosomal subunits are assembled. The nuclear envelope contains pores, as shown in the larger micrograph of a freeze-fractured nuclear envelope. Each pore is lined by a complex of eight proteins, as shown in the smaller micrograph and drawing. Nuclear pore complexes serve as passageways for substances to pass into and out of the nucleus.

(Gk. *chroma,* "color"), which is a combination of proteins and nucleic acids. Chromatin looks grainy, but actually it is a network of strands that condenses and undergoes coiling into rodlike structures called **chromosomes,** just before the cell divides. The chromosomes are the carriers of genetic information. This information is organized on the chromosome as **genes,** the basic units of heredity. All the cells of an individual contain the same number of chromosomes, and the mechanics of nuclear division ensure that each daughter cell receives the normal number of chromosomes, except for the egg and sperm, which usually have half this number.

Three types of ribonucleic acid (RNA) are produced in the nucleus: *ribosomal RNA (rRNA), messenger RNA (mRNA),* and *transfer RNA (tRNA).* Ribosomal RNA is produced in the **nucleolus,** a dark region of chromatin where rRNA joins with proteins to form the subunits of ribosomes. Ribosomes are small bodies in the cytoplasm that facilitate protein synthesis. Messenger RNA, a mobile molecule, acts as an intermediary for DNA, a sedentary molecule, which specifies the sequence of amino acids in a protein. Transfer RNA participates in the assembly of amino acids into a polypeptide by recognizing both mRNA and amino acids during protein synthesis.

The nucleus is important to cell structure and function, because it specifies the code to make proteins. Although the nucleus is physically separated from the cytoplasm by a double membrane known as the **nuclear envelope,** it is still able to communicate with the cytoplasm through nuclear pores. **Nuclear pores** are of sufficient size (100 nm) to permit the passage of ribosomal subunits and mRNA out of the nucleus into the cytoplasm, as well as the passage of proteins from the cytoplasm into the nucleus. High-resolution electron micrographs show that nonmembrane components associated with the pores form a nuclear pore complex. Nuclear pore complexes act as gatekeepers to regulate what goes into and out of a nucleus.

Ribosomes

Ribosomes are particles where protein synthesis occurs. A large and small ribosomal subunit, each comprised of a mix of proteins and rRNA, are necessary components of a functional ribosome. In eukaryotes, ribosomes are 20 nm by 30 nm, and in prokaryotes they are slightly smaller. The number of ribosomes in a cell varies depending on its functions; for example, pancreatic cells and those of other glands have many ribosomes because they produce secretions that contain proteins.

In eukaryotic cells, some ribosomes occur freely within the cytoplasm, either singly or in groups called **polyribosomes,** whereas others are attached to the endoplasmic reticulum (ER), a membranous system of flattened saccules (small sacs) and tubules (see section 4.5). In the nucleus, the information within a gene is copied into mRNA, which is exported through a nuclear pore complex into the cytoplasm. Ribosomes receive the mRNA, which carries a coded message from DNA indicating the correct sequence of amino acids in a particular protein. Proteins synthesized by cytoplasmic ribosomes are used in the cytoplasm, and those synthesized by attached ribosomes end up in the ER.

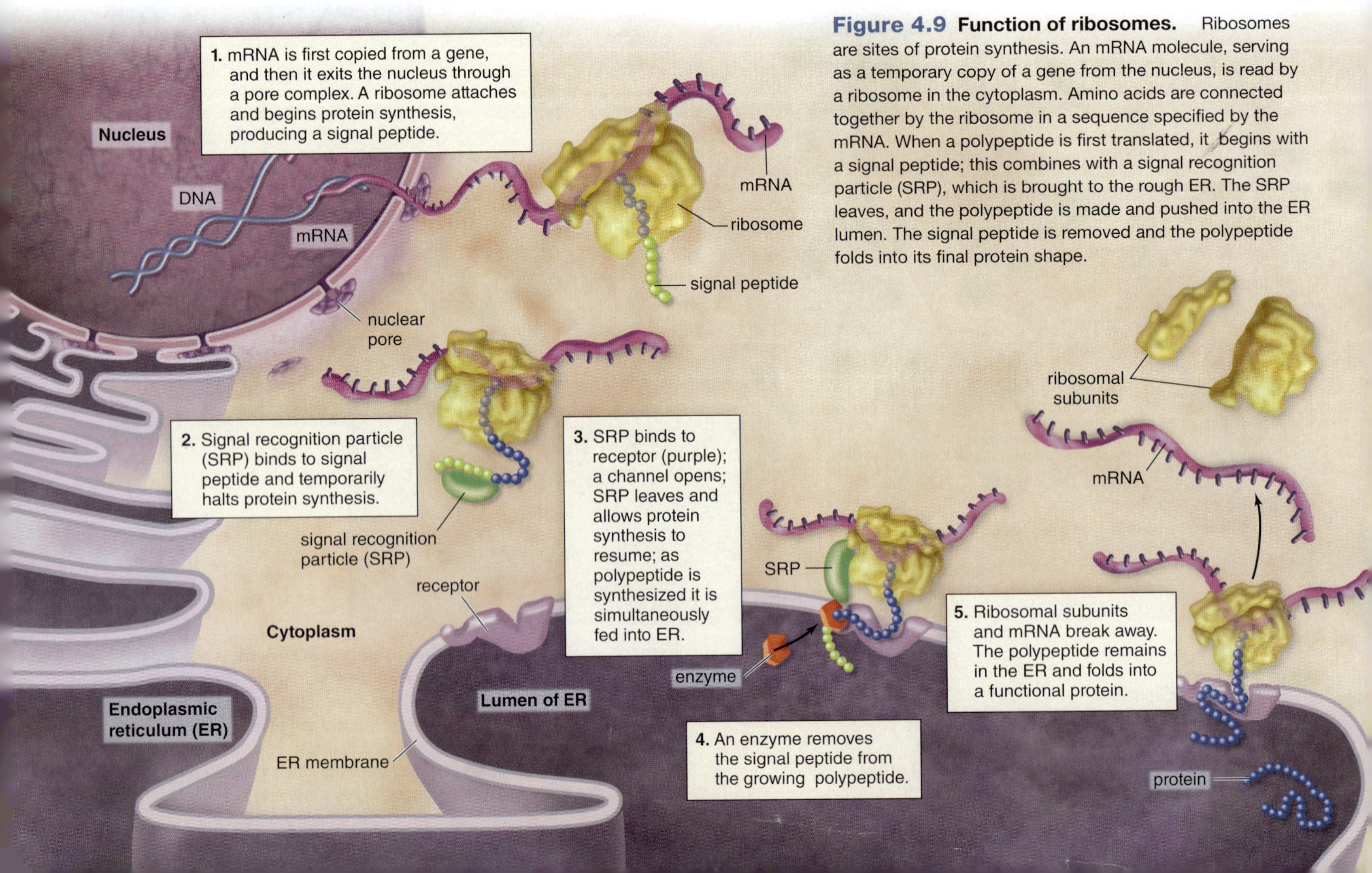

Figure 4.9 Function of ribosomes. Ribosomes are sites of protein synthesis. An mRNA molecule, serving as a temporary copy of a gene from the nucleus, is read by a ribosome in the cytoplasm. Amino acids are connected together by the ribosome in a sequence specified by the mRNA. When a polypeptide is first translated, it begins with a signal peptide; this combines with a signal recognition particle (SRP), which is brought to the rough ER. The SRP leaves, and the polypeptide is made and pushed into the ER lumen. The signal peptide is removed and the polypeptide folds into its final protein shape.

What causes a ribosome to bind to the endoplasmic reticulum? Binding occurs only if the protein being synthesized by a ribosome begins with a sequence of amino acids called a *signal peptide.* The signal peptide binds a particle (signal recognition particle, SRP), which then binds to a receptor on the ER. Once the protein enters the ER, an enzyme cleaves off the signal peptide, and the protein ends up within the lumen (interior) of the ER, where it folds into its final shape (Fig. 4.9).

The sequence of DNA being transcribed into mRNA, and this in turn being translated into a protein, occurs in all living cells, at least during some point in their lifespan. Because of its universality, the DNA–mRNA–protein sequence of events is termed the *central dogma of molecular biology* (see Chapter 12).

Check Your Progress **4.4**

1. Distinguish between the chromatin and chromosomes within the nucleus.
2. Explain the importance of the nuclear pores.
3. Describe the sequence of events that transfers information from a gene to a functional protein.

4.5 The Endomembrane System

Learning Outcomes

Upon completion of this section, you should be able to

1. Explain the importance of the endomembrane system in cellular function.
2. Examine how the ER, Golgi apparatus, and lysosomes differ from one another.
3. Describe how endomembrane vesicles are able to fuse with organelles.

The **endomembrane system** consists of the nuclear envelope, the membranes of the endoplasmic reticulum, the Golgi apparatus, and several types of vesicles. This system compartmentalizes the cell so that particular enzymatic reactions are restricted to specific regions and overall cell efficiency is increased. The vesicles transport molecules from one part of the system to another.

Endoplasmic Reticulum

The **endoplasmic reticulum (ER)** (Gk. *endon,* "within"; *plasma,* "something molded"; L. *reticulum,* "net"), consisting of a complicated system of membranous channels and saccules (flattened vesicles), is physically continuous with the nuclear envelope (Fig. 4.10). The ER consists of rough ER and smooth ER, which have different structures and functions.

Rough ER is studded with ribosomes on the side of the membrane that faces the cytoplasm, giving it the capacity to produce proteins. Inside its lumen, the rough ER allows proteins to fold and take on their final three-dimensional shape. The rough ER also contains enzymes that can add carbohydrate (sugar) chains to proteins, forming glycoproteins that are important in many cell functions.

Smooth ER, which is continuous with the nuclear envelope and the rough ER, does not have attached ribosomes. Certain organs contain cells with an abundance of smooth ER, depending on the organ's function. In some organs, increased smooth ER helps produce more lipids. For example, in the testes, smooth ER produces testosterone, a steroid hormone. In the liver, smooth ER helps detoxify drugs. The smooth ER of the liver increases in quantity when a person consumes alcohol or takes barbiturates on a regular basis. Regardless of functional differences, both rough and smooth ER form vesicles that transport molecules to other parts of the cell, notably the Golgi apparatus.

The Golgi Apparatus

The **Golgi apparatus** is named for Camillo Golgi (1843–1926), who discovered its presence in cells in 1898. The Golgi apparatus

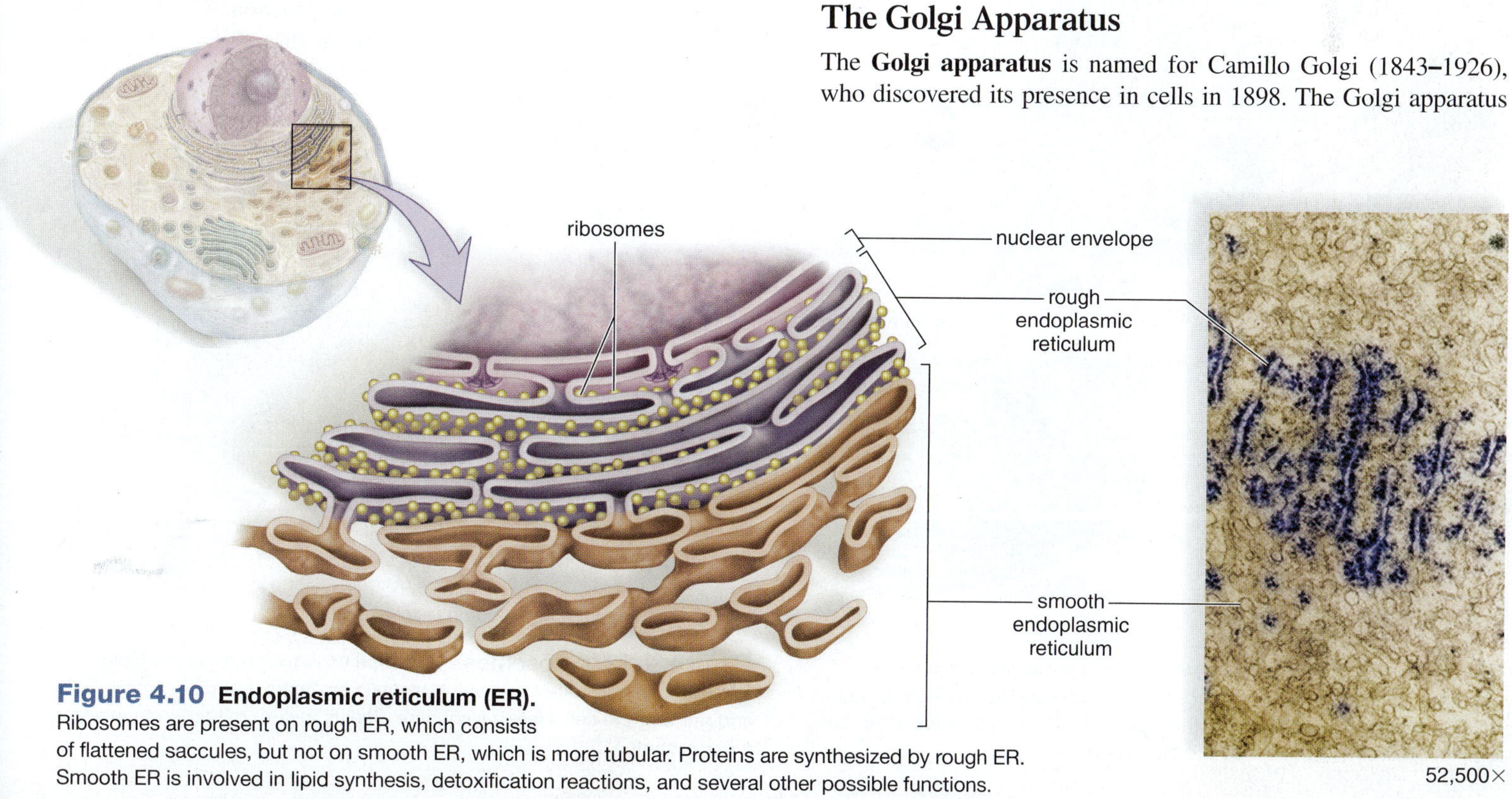

Figure 4.10 Endoplasmic reticulum (ER). Ribosomes are present on rough ER, which consists of flattened saccules, but not on smooth ER, which is more tubular. Proteins are synthesized by rough ER. Smooth ER is involved in lipid synthesis, detoxification reactions, and several other possible functions.

typically consists of a stack of 3 to 20 slightly curved, flattened saccules whose appearance can be compared to a stack of pancakes (Fig. 4.11). In animal cells, one side of the stack (the cis, or inner, face) is directed toward the ER, and the other side of the stack (the trans, or outer, face) is directed toward the plasma membrane. Vesicles can frequently be seen at the edges of the saccules.

Protein-filled vesicles that bud from the rough ER and lipid-filled vesicles that bud from the smooth ER are received by the Golgi apparatus at its inner face. These substances are altered as they move through the saccules. For example, the Golgi apparatus contains enzymes that modify the carbohydrate chains first attached to proteins in the rough ER. It can modify one sugar into another sugar on glycoproteins. In some cases, the modified carbohydrate chain serves as a signal molecule or molecular address label that determines the protein's final destination in the cell.

The Golgi apparatus sorts the modified molecules and packages them into vesicles that depart from the outer face. These vesicles may be transported to various locations within the cell, depending on their molecular address labels. In animal cells, some of these vesicles are lysosomes, which are discussed next. Other vesicles may return to the ER or proceed to the plasma membrane, where they merge and discharge their contents to the outside of the cell by exocytosis.

Lysosomes

Lysosomes (Gk. *lyo,* "loose"; *soma,* "body") are membrane-bound vesicles produced by the Golgi apparatus. They have a very low pH and store powerful hydrolytic-digestive enzymes in an inactive state. Lysosomes act much like your stomach in that they assist in digesting material taken into the cell. They also destroy nonfunctional organelles and portions of cytoplasm (Fig. 4.12).

Materials can be taken into a cell by vesicle or vacuole formation at the plasma membrane. When a lysosome fuses with either, the lysosomal enzymes are activated and digest the material into simpler subunits that are exported into the cytoplasm and recycled by other cell processes. White blood cells, specialized to protect the body from foreign entities, are well known for engulfing pathogens (e.g., disease-causing viruses and bacteria), which are then broken down in lysosomes. White blood cells have a greater proportion of lysosomes than other cells, because their specialized function is the digestion of foreign bodies.

A number of human lysosomal storage diseases are due to a missing lysosomal enzyme. In Tay-Sachs disease, the missing enzyme digests a fatty substance that helps insulate nerve cells and increases their efficiency. The fatty substance accumulates in so many storage bodies that nerve cells die off. Affected individuals appear normal at birth but begin to develop neurological problems at 4 to 6 months of age. Eventually, the child suffers cerebral degeneration, slow paralysis, blindness, and loss of motor function. Children with Tay-Sachs disease live only about

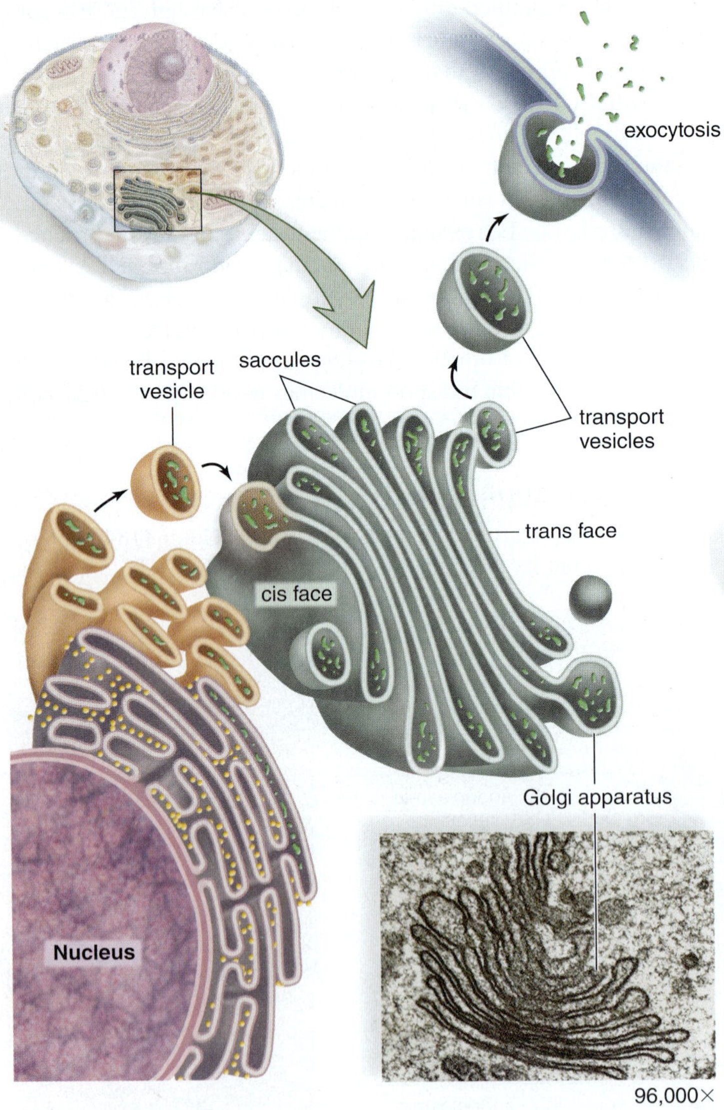

Figure 4.11 Golgi apparatus. The Golgi apparatus is a stack of flattened, curved saccules. It processes proteins and lipids and packages them in transport vesicles that either distribute these molecules to various locations within the cell or secrete them externally.

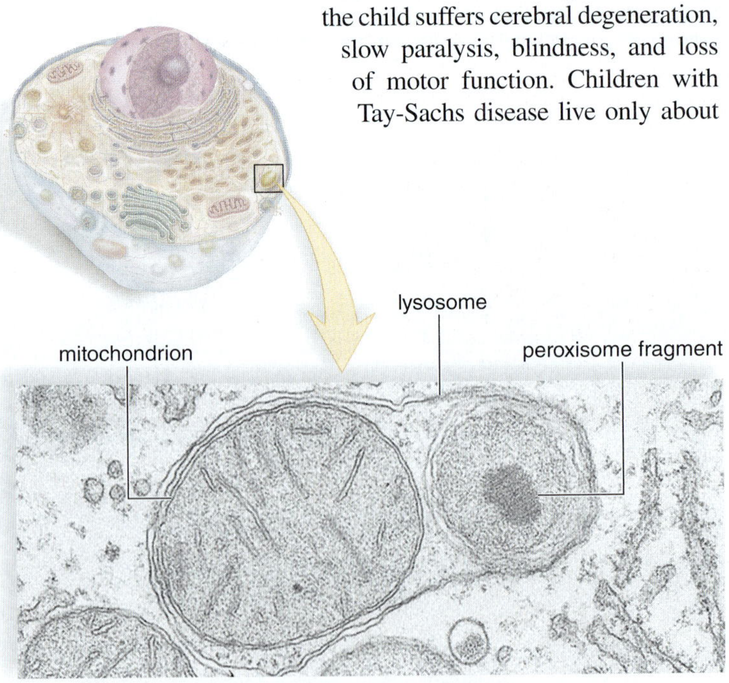

Figure 4.12 Lysosomes. Lysosomes, which bud off the Golgi apparatus in cells, are filled with hydrolytic enzymes that digest molecules and parts of the cell. Here a lysosome digests a worn mitochondrion and a peroxisome.

3 to 4 years. The use of gene therapy (see Chapter 14) to provide the enzyme to the cells may be able to treat Tay-Sachs disease.

Animation Lysosomes

Endomembrane System Summary

You have seen that the endomembrane system is a series of membranous organelles that work together and communicate by means of transport vesicles. The endoplasmic reticulum (ER) and the Golgi apparatus are essentially flattened saccules, and lysosomes are specialized vesicles.

Organelles within the endomembrane system can interact because their membranes readily fuse together, and because membrane-associated proteins enable communication and specialized functions. Figure 4.13 shows how the components of the endomembrane system work together. Products of both rough ER and smooth ER are carried in transport vesicles to the Golgi apparatus, where they are further modified. Using signaling sequences and molecular address labels, the Golgi apparatus sorts these products and packages them into vesicles that transport them to various cellular destinations. Secretory vesicles take the proteins to the plasma membrane, where they exit the cell by exocytosis. For example, secretion into ducts occurs when the mammary glands produce milk or the pancreas produces digestive enzymes.

In animal cells, the Golgi apparatus also produces lysosomes that contain stored hydrolytic enzymes. Lysosomes fuse with incoming vesicles from the plasma membrane and digest macromolecules taken into a cell.

Check Your Progress 4.5

1. Contrast the structure and functions of rough and smooth endoplasmic reticulum.
2. Describe the relationship between the components of the endomembrane system.
3. Examine how cellular function would be affected if the Golgi apparatus ceased to function.

Figure 4.13 Endomembrane system. The organelles in the endomembrane system work together to carry out the functions noted. Plant cells do not have lysosomes, nor do they have incoming and outgoing (secretory) vesicles.

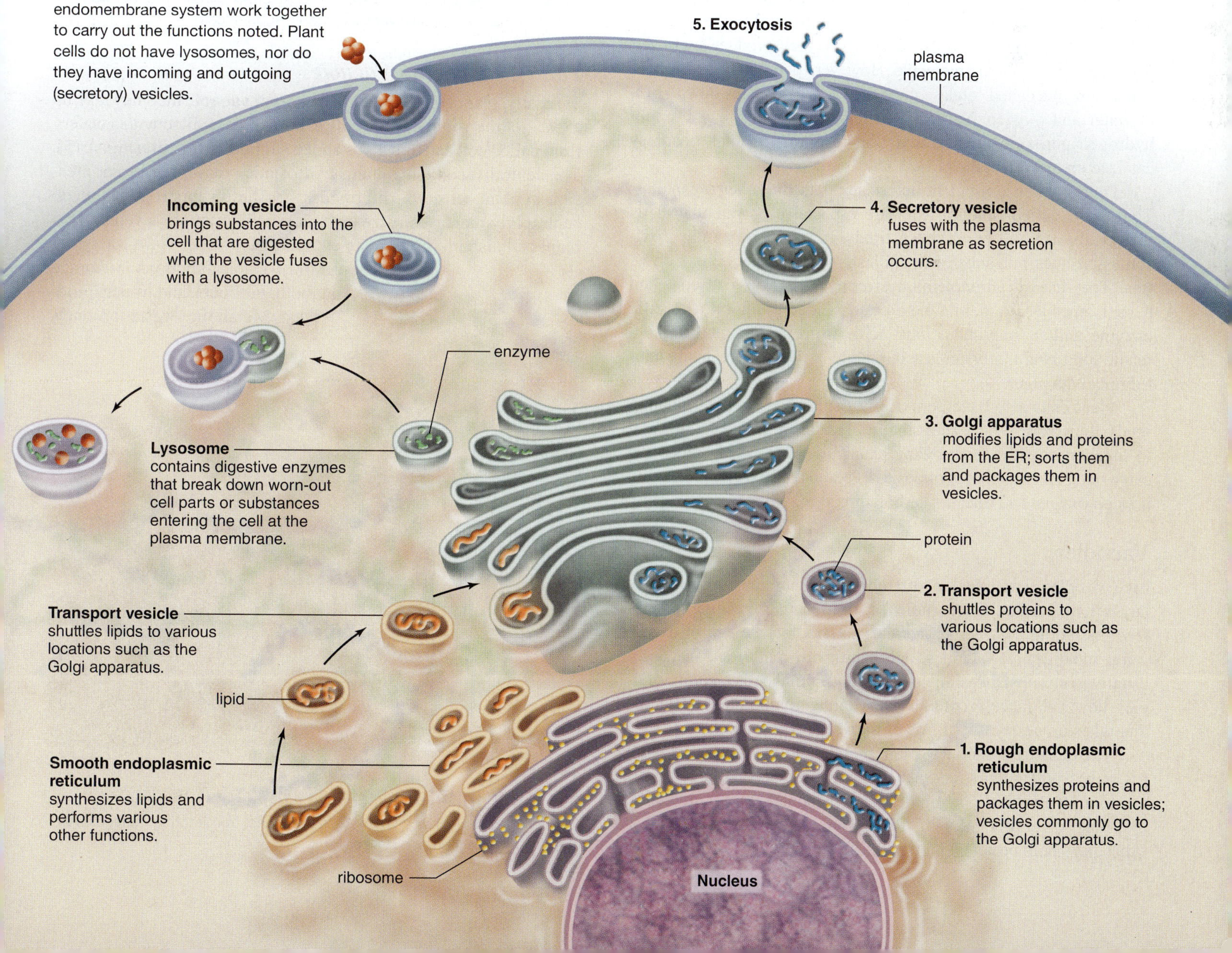

4.6 Microbodies and Vacuoles

Learning Outcomes

Upon completion of this section, you should be able to

1. Describe the role of peroxisomes and vacuoles in cell function.
2. Contrast peroxisomes and vacuoles with endomembrane organelles.

Eukaryotic cells contain a variety of membrane-bound vesicles, called **microbodies,** that contain specialized enzymes to perform specific metabolic functions. One example is the peroxisome. In addition, cells may contain large storage areas called vacoules.

Peroxisomes

Peroxisomes are membrane-bound vesicles that enclose enzymes that are involved in the breakdown of fatty acids. Unlike the enzymes of lysosomes, which are loaded into the vesicle by the Golgi apparatus, the enzymes in peroxisomes are synthesized by free ribosomes and transported into a peroxisome from the cytoplasm. As the enzymes within the peroxisome oxidize fatty acids, they produce hydrogen peroxide (H_2O_2), a toxic molecule. However, peroxisomes also contain an enzyme called catalase that immedidately breaks down H_2O_2 to water and oxygen. You can see this reaction when you apply hydrogen peroxide to a wound; the resulting bubbles occur as catalase breaks down the H_2O_2.

Peroxisomes are metabolic assistants to the other organelles. They have varied functions but are especially prevalent in cells that synthesize and break down lipids. In the liver, some peroxisomes produce bile salts from cholesterol, and others break down fats. The disease adrenoleukodystrophy (ALD) is caused when peroxisomes lack a membrane protein needed to import a specific enzyme and/or long-chain fatty acids from the cytoplasm. As a result, long-chain fatty acids accumulate in the brain, causing neurological damage.

Plant cells also have peroxisomes (Fig. 4.14). In germinating seeds, they oxidize fatty acids into molecules that can be converted to sugars needed by the growing plant. In leaves, peroxisomes can carry out a reaction that is opposite to photosynthesis—the reaction uses up oxygen and releases carbon dioxide.

Vacuoles

Like vesicles, **vacuoles** are membranous sacs, but vacuoles are larger than vesicles. The vacuoles of some protists are quite specialized, including contractile vacuoles for ridding the cell of excess water and digestive vacuoles for breaking down nutrients. Vacuoles usually store substances. In general, few animal cells contain vacuoles; however, fat cells contain a very large, lipid-engorged vacuole that takes up nearly two-thirds of the volume of the cell!

Vacuoles are essential to plant function. Plant vacuoles contain not only water, sugars, and salts but also water-soluble pigments and toxic molecules. The pigments are responsible for many of the red, blue, or purple color of flowers and some leaves. The toxic substances help protect a land plant from herbivorous animals.

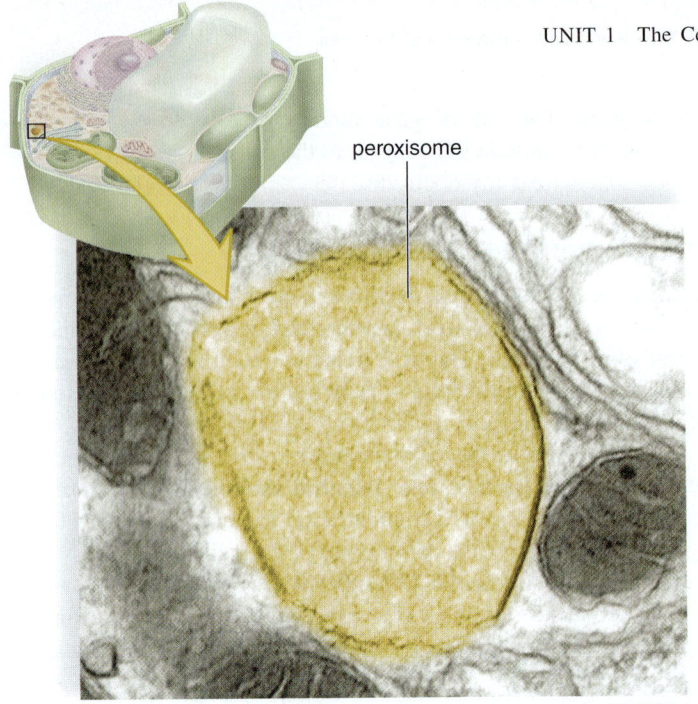

Figure 4.14 Peroxisomes. Peroxisomes contain one or more enzymes that can oxidize various organic substances.

Plant Cell Central Vacuole

Typically, plant cells have a large **central vacuole** that may take up to 90% of the volume of the cell. The vacuole is filled with a watery fluid called cell sap that gives added support to the cell (Fig. 4.15). The central vacuole maintains hydrostatic pressure or turgor pressure in plant cells, which provides structural support. A plant cell can rapidly increase in size by enlarging its vacuole. Eventually, a plant cell also produces more cytoplasm.

The central vacuole functions in storage of both nutrients and waste products. Metabolic waste products

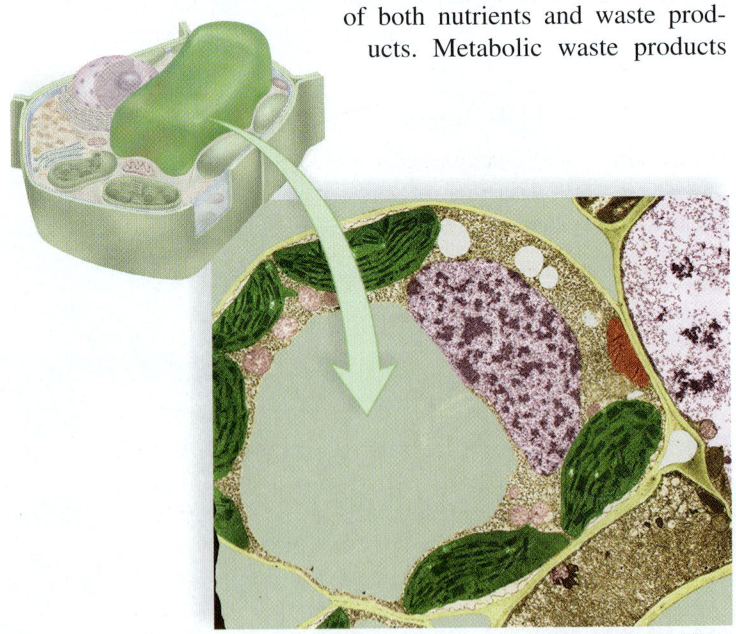

Figure 4.15 Plant cell central vacuole. The large central vacuole of plant cells has numerous functions, from storing molecules to helping the cell increase in size.

are pumped across the vacuole membrane and stored permanently in the central vacuole. As organelles age and become nonfunctional, they fuse with the vacuole, where digestive enzymes break them down. This is a function analogous to that carried out by lysosomes in animal cells.

Check Your Progress 4.6

1. Compare the structure and functions of a peroxisome with those of a lysosome.
2. Distinguish between where peroxisome and lysosome proteins are produced.

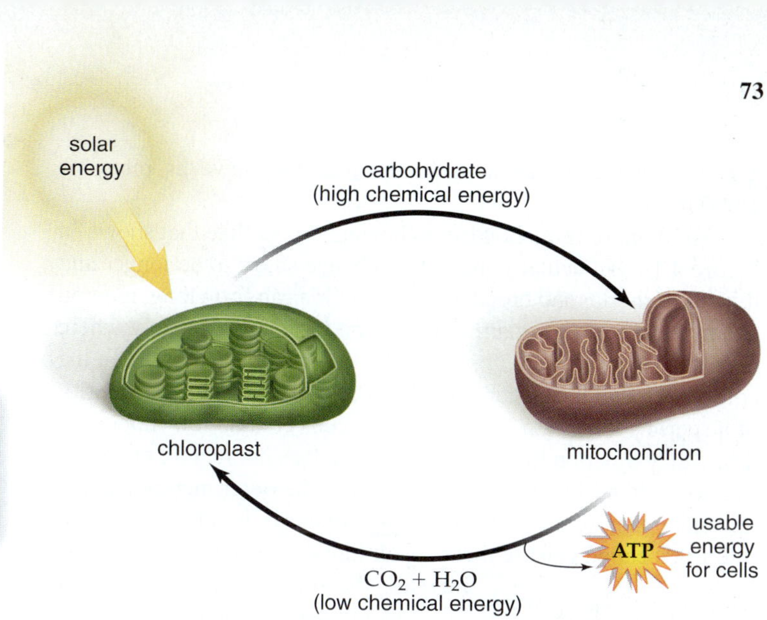

Figure 4.16 Energy-producing organelles. Chloroplasts use sunlight to produce carbohydrates, which in turn are used by the mitochondria. The mitochondria then produce carbon dioxide and water, which is in turn used by the chloroplasts.

4.7 The Energy-Related Organelles

Learning Outcomes

Upon completion of this section, you should be able to

1. Distinguish between the functions of chloroplasts and mitochondria in a cell.
2. Describe the internal structure of mitochondria and chloroplasts.

Life is possible only because a constant input of energy maintains the structure of cells. Chloroplasts and mitochondria are the two eukaryotic membranous organelles that specialize in converting energy to a form that can be used by the cell. Although animal cells contain only mitochondria, plant cells contain both mitochondria and chloroplasts.

During *photosynthesis,* **chloroplasts** (Gk. *chloros,* "green"; *plastos,* "formed, molded") use solar energy to synthesize carbohydrates, which serve as organic nutrient molecules for plants and all life on Earth. Photosynthesis can be represented by this equation:

solar energy + carbon dioxide + water → carbohydrate + oxygen

Plants, algae, and cyanobacteria are capable of conducting photosynthesis in this manner, but only plants and algae have chloroplasts, because they are eukaryotes.

In *cellular respiration,* **mitochondria** (sing., mitochondrion) break down carbohydrate-derived products to produce ATP (adenosine triphosphate). Cellular respiration can be represented by this equation:

carbohydrate + oxygen → carbon dioxide + water + energy

Here the word *energy* stands for ATP molecules. When a cell needs energy, ATP supplies it. The energy of ATP is used to drive synthetic reactions, active transport, and all energy-requiring processes in cells. Figure 4.16 provides a summary of the interactions between these two energy-producing organelles.

Chloroplasts

Some algal cells have only one chloroplast, while some plant cells have as many as a hundred. Chloroplasts can be quite large, being twice as wide and as much as five times the length of a mitochondrion.

Chloroplasts have a three-membrane system (Fig. 4.17). They are surrounded by a double membrane, which includes an outer membrane and an inner membrane. The double membrane encloses the semifluid **stroma,** which contains enzymes and **thylakoids,** disklike sacs formed from a third chloroplast membrane. A stack of thylakoids is a **granum.** The lumens of the thylakoids are believed to form a large, internal compartment called the thylakoid space. Chlorophyll and the other pigments that capture solar energy are located in the thylakoid membrane, and the enzymes that synthesize carbohydrates are located outside the thylakoid in the fluid of the stroma.

The endosymbiotic theory holds that chloroplasts are derived from a photosynthetic bacterium that was engulfed by a eukaryotic cell (see Fig. 4.5). This certainly explains why a chloroplast is surrounded by a double membrane—one membrane is derived from the vesicle that brought the prokaryote into the cell, while the inner membrane is derived from the prokaryote. The endosymbiotic theory is also supported by the finding that chloroplasts have their own prokaryotic-type chromosome and ribosomes, and they produce some of their own enzymes even today.

Animation Endosymbiosis

Other Types of Plastids

A chloroplast is a type of plastid. **Plastids** are plant organelles that are surrounded by a double membrane and have varied functions. *Chromoplasts* contain pigments that result in a yellow, orange, or red color. Chromoplasts are responsible for the color of autumn leaves, fruits, carrots, and some flowers. *Leucoplasts* are generally colorless plastids that synthesize and store starches and oils. A microscopic examination of potato tissue reveals a number of leucoplasts.

Mitochondria

Nearly all eukaryotic cells, and certainly all plant and algal cells in addition to animal cells, contain mitochondria. Even though mitochondria are smaller than chloroplasts, they can usually be seen using a light microscope. The number of mitochondria can vary depending on the metabolic activities and energy needed within a cell. Some cells, such as liver cells, may have as many as 1,000 mitochondria.

We think of mitochondria as having a shape like that shown in Figure 4.18, but actually they often change shape to be longer and thinner or shorter and broader. Mitochondria can form long, moving chains, or they can remain fixed in one location—typically where energy is most needed. For example, they are packed between the contractile elements of cardiac cells and wrapped around the interior of a sperm's flagellum. In contrast, fat cells contain few mitochondria—they function in fat storage, which does not require energy.

Mitochondria have two membranes, the outer membrane and the inner membrane. The inner membrane is highly convoluted into folds called **cristae** that project into the matrix. These cristae increase the surface area of the inner membrane so much that in a liver cell they account for about one-third the total membrane in the cell. The inner membrane encloses a semifluid **matrix,** which contains mitochondrial DNA and ribosomes. Again, the presence of a double membrane and mitochondrial genes is consistent with the endosymbiotic theory regarding the origin of mitochondria, which was illustrated in Figure 4.5.

Animation
Endosymbiosis

Mitochondria are often called the powerhouses of the cell because they produce most of the ATP utilized by the cell. Within the matrix of the mitochondria is a highly concentrated mixture of enzymes that break down carbohydrates and other nutrient molecules. These reactions supply the chemical energy needed for a chain of proteins on the inner membrane to create the conditions that allow ATP synthesis to take place. The entire process, which also involves the cytoplasm, is called *cellular respiration,* because oxygen is used and carbon dioxide is given off, as shown at the beginning of this section.

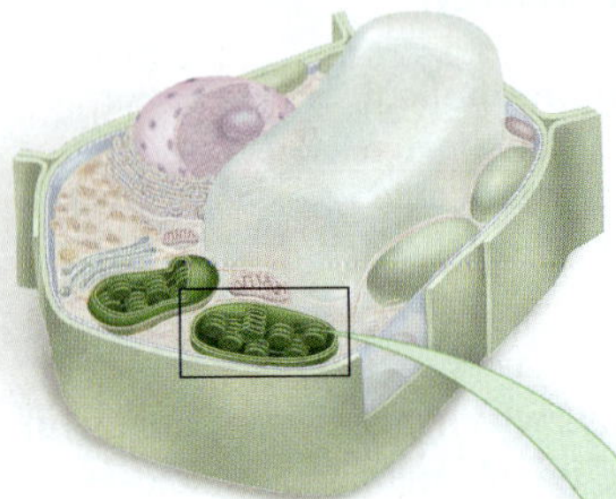

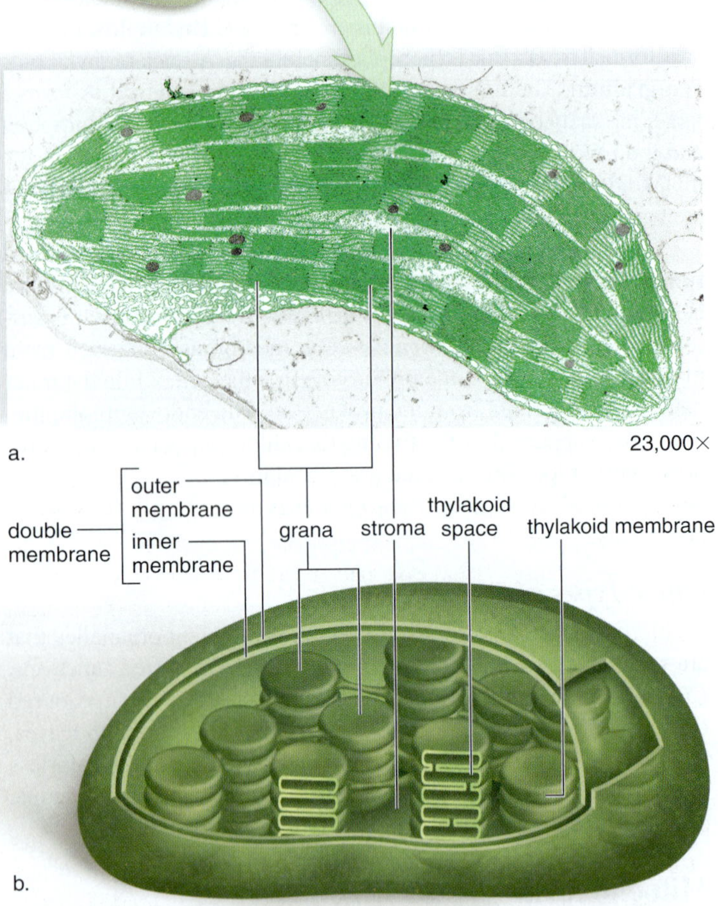

Figure 4.17 Chloroplast structure. Chloroplasts carry out photosynthesis. **a.** Electron micrograph of a longitudinal section of a chloroplast. **b.** Generalized drawing of a chloroplast in which the outer and inner membranes have been cut away to reveal the grana, each of which is a stack of membranous sacs called thylakoids. In some grana, but not all, thylakoid spaces are interconnected.

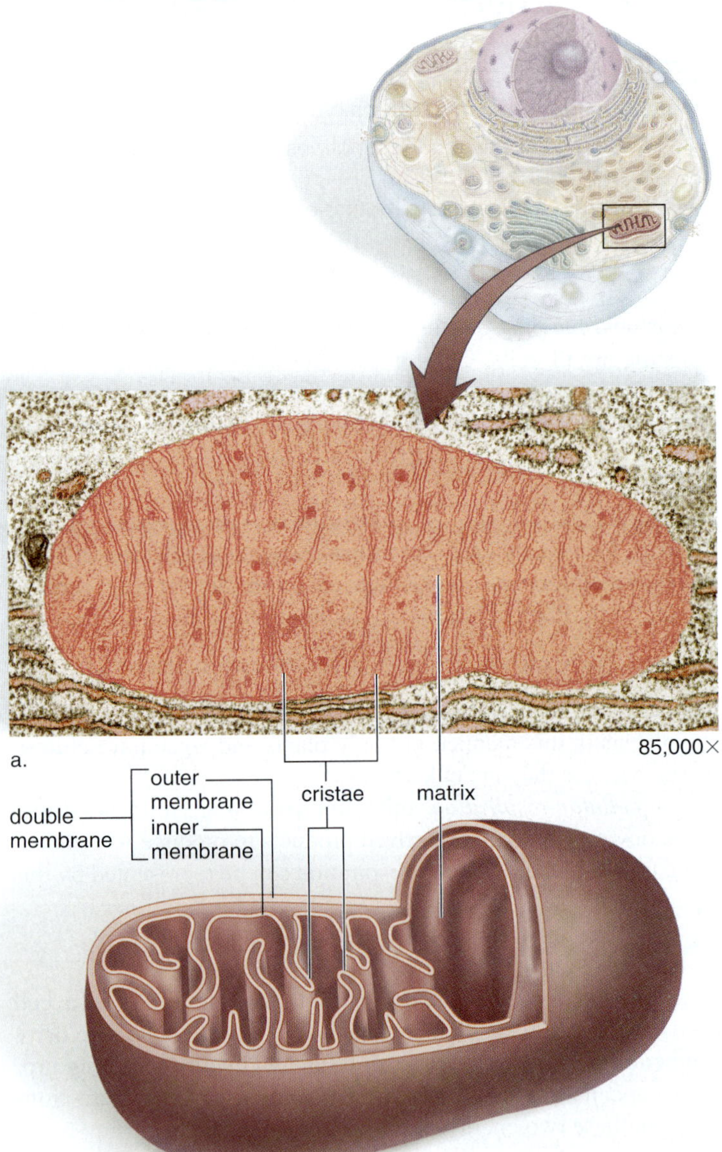

Figure 4.18 Mitochondrion structure. Mitochondria are involved in cellular respiration. **a.** Electron micrograph of a longitudinal section of a mitochondrion. **b.** Generalized drawing in which the outer membrane and portions of the inner membrane have been cut away to reveal the cristae.

Mitochondrial Diseases

So far, dozens of different mitochondrial diseases that affect the brain, muscles, kidneys, heart, liver, eyes, ears, or pancreas have been identified. The common factor among these genetic diseases is that the patient's mitochondria are unable to completely metabolize organic molecules to produce ATP. As a result, toxins accumulate inside the mitochondria and the body. The toxins can be free radicals (substances that readily form harmful compounds when they react with other molecules), and these compounds damage mitochondria over time. In the United States, between 1,000 and 4,000 children per year are born with a mitochondrial disease. In addition, it is possible that many diseases of aging are due to malfunctioning mitochondria.

Check Your Progress 4.7

1. Summarize the roles of mitochondria and chloroplasts in the cell.
2. Discuss the evidence that chloroplasts and mitochondria are derived from ancient bacteria.
3. Explain why chloroplasts and mitochondria contain complex internal membrane structures.

4.8 The Cytoskeleton

Learning Outcomes

Upon completion of this section, you should be able to

1. Compare the structure and function of actin filaments, intermediate filaments, and microtubules.
2. Describe how motor molecules interact with cytoskeletal elements to produce movement.
3. Explain the diverse roles of microtubules within the cell.

Cells are exposed to many physical forces. Cell shape, movement, and internal transport all require structural support, provided by the cytoskeleton. The protein components of the cytoskeleton (Gk. *kytos,* "cell") interconnect and extend from the nucleus to the plasma membrane in eukaryotic cells. Prior to the 1970s, it was believed that the cytoplasm was an unorganized mixture of organic molecules. Then, high-voltage electron microscopes, which can penetrate thicker specimens, showed instead that the cytoplasm is highly organized. The technique of immunofluorescence microscopy identified the makeup of the protein components within the cytoskeletal network (Fig. 4.19).

The cytoskeleton contains actin filaments, intermediate filaments, and microtubules, which maintain cell shape and allow the cell and its organelles to move. Therefore, the cytoskeleton is often compared to the bones and muscles of an animal. However, the cytoskeleton is dynamic; it can rearrange its protein components as necessary in response to changes in internal and external environments. A number of different mechanisms appear to regulate this process, including protein phosphatases, which remove phosphates from proteins and bring about assembly, and protein kinases, which phosphorylate proteins and lead to disassembly.

Actin Filaments

Actin filaments (formerly called microfilaments) are long, extremely thin, flexible fibers (about 7 nm in diameter) that occur in bundles or meshlike networks. Each actin filament contains two chains of globular actin monomers twisted about one another in a helical manner.

Actin filaments provide structural support as a dense, complex web just under the plasma membrane, to which they are anchored by special proteins. Sometimes, actin filaments can dynamically rearrange themselves and facilitate cellular movement, such as when an amoeba moves over a surface with pseudopods (L. *pseudo,* "false"; *pod,* "feet"), or when intestinal cell microvilli lengthen and shorten into the gut lumen (the space where ingested food is processed). In plant cells, actin filaments form the tracks along which chloroplasts circulate in a particular direction in a process called cytoplasmic streaming.

Actin filaments move the cell and its organelles by interacting with *motor molecules,* which are proteins that can attach, detach, and reattach farther along an actin filament. The motor molecule myosin uses ATP to pull actin filaments along in this way. Myosin has both a head and a tail. In muscle cells, the tails of several myosin molecules are joined to form a thick filament. In nonmuscle cells, cytoplasmic myosin tails are bound to membranes, but the heads still interact with actin:

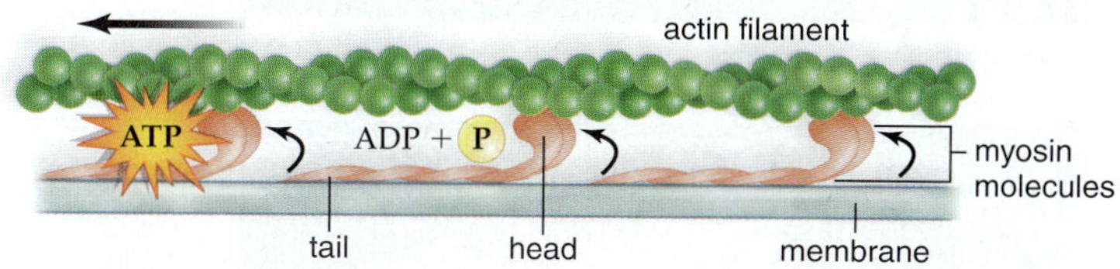

During animal cell division, the two new cells form when actin, in conjunction with myosin, pinches off the cells from one another.

Intermediate Filaments

Intermediate filaments (8–11 nm in diameter) are so named because they are intermediate in size between actin filaments and microtubules. They form a ropelike assembly of fibrous polypeptides, but the specific filament type varies according to the tissue. Some intermediate filaments support the nuclear envelope, whereas others support the plasma membrane and take part in the formation of cell-to-cell junctions. In the skin, intermediate filaments made of the protein keratin give great mechanical strength to skin cells. Like other cytoskeletal components, intermediate filaments are highly dynamic and disassemble when phosphate is added to them by a kinase.

Microtubules

Microtubules (Gk. *mikros,* "small") are small, hollow cylinders about 25 nm in diameter and from 0.2 to 25 μm in length. They are made of a globular protein called tubulin, which is of two types called α and β. Alpha tubulin has a slightly different amino acid sequence than β tubulin. When assembly occurs, α and β tubulin molecules come together as dimers, and the dimers arrange themselves in rows. Microtubules have 13 rows of tubulin dimers, surrounding what appears in electron micrographs to be an empty central core.

Figure 4.19 The cytoskeleton. The cytoskeleton maintains a cell's shape and allows its parts to move. Three types of protein components make up the cytoskeleton. They can be detected in cells by using labeling and fluorescence microscopy. **a.** *Left to right:* Cells showing a twisted double chain of actin filaments (green fibers). The giant cells of the green alga *Chara* use actin filaments to move organelles within the cell. **b.** *Left to right:* Animal cells showing fibrous, ropelike intermediate filaments (blue fibers). A peacock's colorful feathers are strengthened by intermediate filaments. **c.** *Left to right:* Animal cells showing hollow microtubules made of tubulin dimers (orange fibers). A chameleon's skin cells use microtubules to move pigment granules around so that they take on the color of their environment.

Microtubule assembly is under the regulatory control of a microtubule-organizing center (MTOC). In most eukaryotic cells, the main MTOC is in the **centrosome** (Gk. *centrum,* "center"), which lies near the nucleus. Microtubules radiate from the centrosome, helping to maintain the shape of the cell and acting as tracks along which organelles can be moved. Whereas the motor molecule myosin is associated with actin filaments, the motor molecules kinesin and dynein are associated with microtubules:

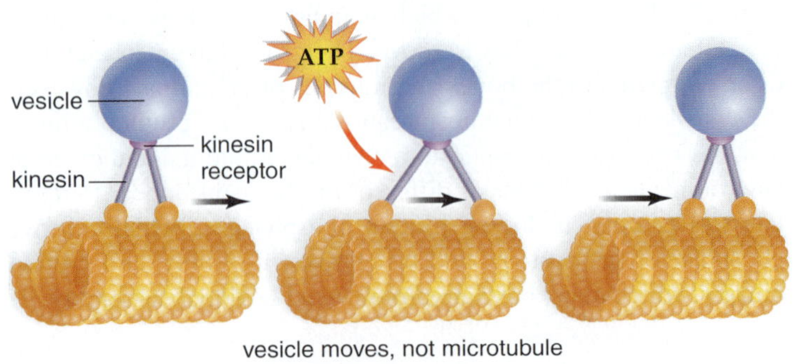

There are different types of kinesin proteins, each specialized to move one kind of vesicle or cellular organelle. Kinesin moves vesicles or organelles in an opposite direction from dynein. Cytoplasmic dynein is closely related to the molecule dynein found in flagella.

Before a cell divides, microtubules disassemble and then reassemble into a structure called a spindle, which distributes chromosomes in an orderly manner. At the end of cell division, the spindle disassembles, and microtubules reassemble once again into their former array. Plants have evolved various types of poisons that prevent them from being eaten by herbivores. One of these, colchicine, is a plant poison that binds tubulin and blocks the assembly of microtubules.

Centrioles

Centrioles are short cylinders with a 9 + 0 pattern of microtubule triplets—nine sets of triplets are arranged in an outer ring, but the center of a centriole does not contain a microtubule. In animal cells and most protists, a centrosome contains two centrioles lying at right angles to each other. A centrosome, as mentioned previously, is the major microtubule-organizing center for the cell. Therefore, it is possible that centrioles are also involved in the process by which microtubules assemble and disassemble.

Before an animal cell divides, the centrioles replicate, and the members of each pair are at right angles to one another (Fig. 4.20). Then each pair becomes part of a separate centrosome. During cell division, the centrosomes move apart and most likely function to organize the mitotic spindle. In any case, each new cell has its own centrosome and pair of centrioles. Plant and fungal cells have the equivalent of a centrosome, but this structure does not contain centrioles, suggesting that centrioles are not necessary to the assembly of cytoplasmic microtubules.

A *basal body* is a structure that lies at the base of cilia and flagella and may direct the organization of microtubules within these structures. In other words, a basal body may do for a cilium or flagellum what the centrosome does for the cell. In cells with cilia and flagella, centrioles are believed to give rise to basal bodies.

Cilia and Flagella

Cilia (L. *cilium,* "eyelash, hair") and **flagella** (L. *flagello,* "whip") are hairlike projections that can move either in an undulating fashion, like a whip, or stiffly, like an oar. In free cells, cilia (or flagella) move the cell through liquid. For example, single-celled paramecia are organisms that move by means of cilia, whereas sperm cells move by means of flagella. If the cell is attached to other cells, cilia (or flagella) are capable of moving liquid over the cell. The cells that line our upper respiratory tract have cilia that sweep debris trapped within mucus back up into the throat, where it can be swallowed or expelled. This action helps keep the lungs clean.

In eukaryotic cells, cilia are much shorter than flagella, but they have a similar construction. Both are membrane-bound cylinders

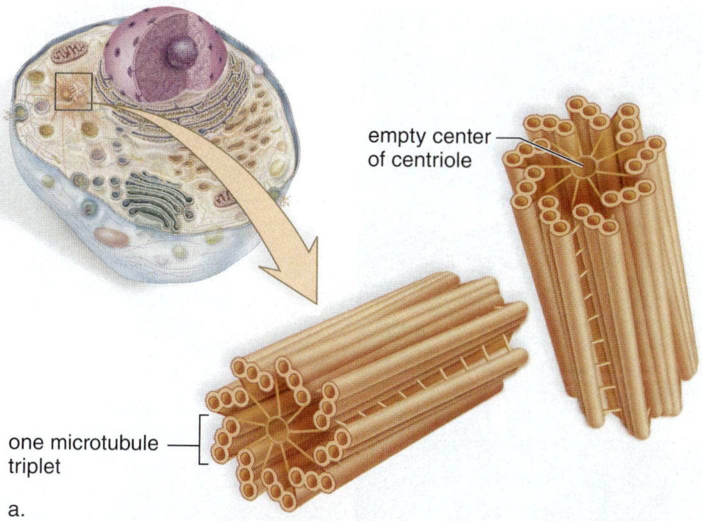

a.

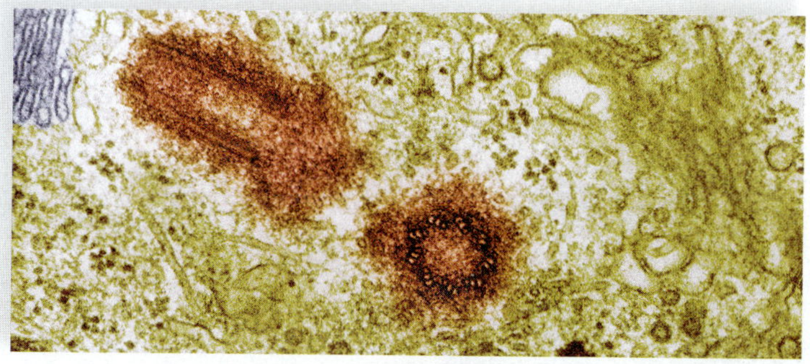

b. one centrosome: one pair of centrioles

Figure 4.20 Centrioles. **a.** The centrosome of an animal cell contains two centrioles positioned at right angles to each other. **b.** A micrograph of one centrosome containing two centrioles.

enclosing a matrix area. In the matrix are nine microtubule doublets arranged in a circle around two central microtubules; this is called the 9 + 2 pattern of microtubules (Fig. 4.21). Cilia and flagella move when the microtubule doublets slide past one another using motor molecules.

As mentioned, each cilium and flagellum has a basal body lying in the cytoplasm at its base. Basal bodies have the same circular arrangement of microtubule triplets as centrioles and are believed to be derived from them. It is possible that basal bodies organize the microtubules within cilia and flagella, but this idea is not supported by the observation that cilia and flagella grow by the addition of tubulin dimers to their tips.

Check Your Progress 4.8

1. Differentiate between the components of the cytoskeleton and how they provide support to the cell.
2. Explain how ATP is used to produce movement in a cell.
3. Describe the role of motor molecules and microtubules in cilia and flagella.

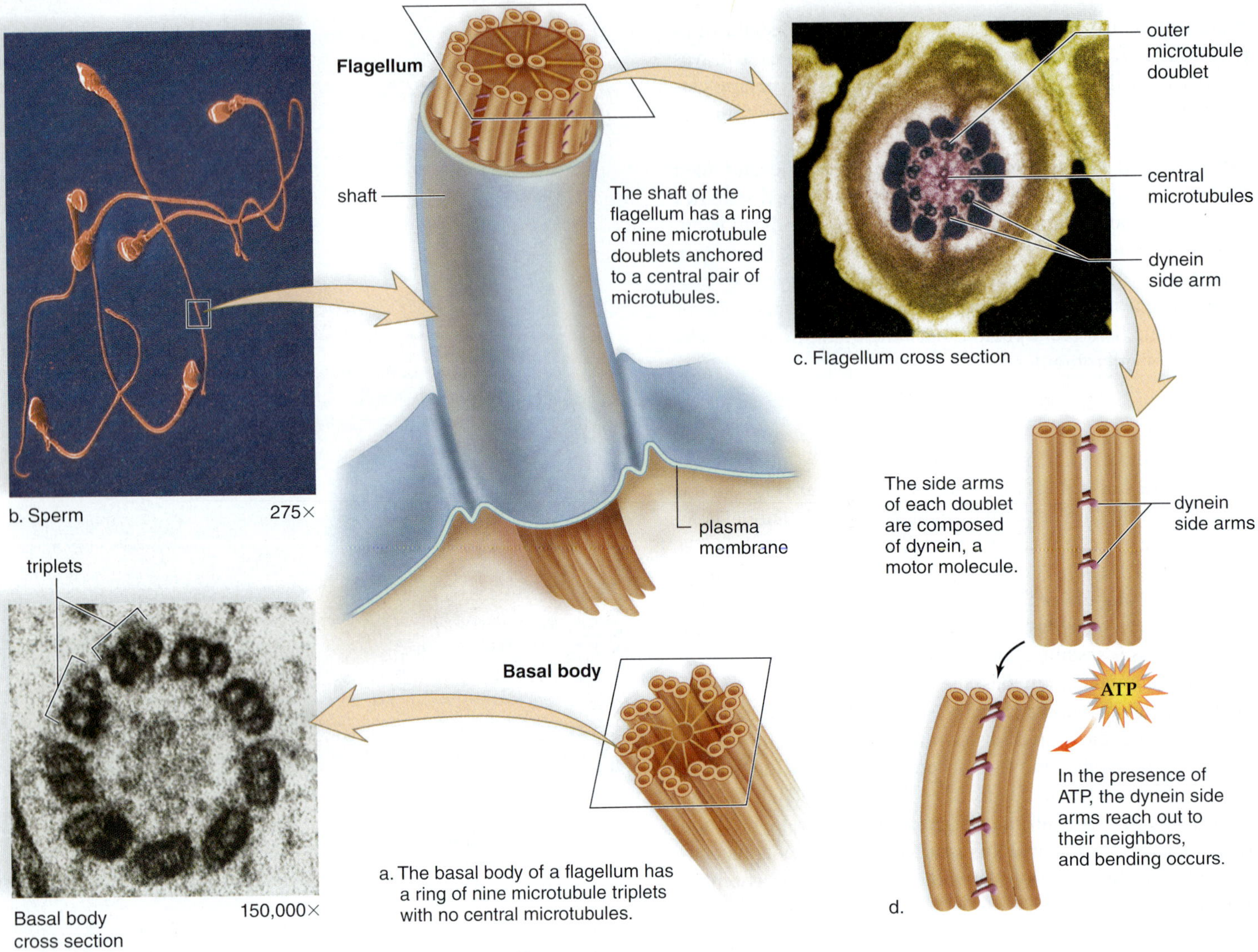

Figure 4.21 Structure of a flagellum. **a.** The basal body of a flagellum has a 9 + 0 pattern of microtubule triplets. Notice the ring of nine triplets, with no central microtubules. **b.** In sperm, the shaft of the flagellum has a 9 + 2 pattern (a ring of nine microtubule doublets surrounds a central pair of microtubules). **c.** In place of the triplets seen in a basal body, a flagellum's outer doublets have side arms of dynein, a motor molecule. **d.** In the presence of ATP, the dynein side arms reach out and attempt to move along their neighboring doublet. Because of the radial spokes connecting the doublets to the central microtubules and motor molecules, bending occurs.

Connecting *the* Concepts *with the* Themes

Evolution

- Cells self-replicate. All modern-day cells are descendants of the first cells on Earth.
- Endosymbiosis gave previously separate prokaryotic cells a metabolic advantage, which allowed them to become more numerous.
- Over time, specialized functions became associated with separate, membrane-bound spaces within the cell. This compartmentalization allowed previously incompatible chemical reactions to occur simultaneously, making cells more efficient.

Nature of Science

- Our knowledge of cell anatomy has been gathered by studying micrographs of cells. Cytologists (biologists who study cells) have arrived at a picture of generalized cells.
- Microscopy includes techniques that provide additional detail and insight into cell structure and function.
- Scientific studies of cellular function provide insight into many human diseases.

Biological Systems

- Cells are integrated systems of assembled macromolecules that collectively support life. Organelles are assemblies that carry out specialized functions within the cell.
- All cellular systems must function normally to maintain the metabolic activity necessary for life. Cellular systems must also be able to adapt to rapidly changing environmental conditions.

MEDIA STUDY TOOLS http://connect.mheducation.com

SMARTBOOK® Study smart. SmartBook helps you maximize your study time by identifying what you know and don't know. The recharge feature in SmartBook helps you prepare for the test by identifying the material that you are most likely to forget.

connect | BIOLOGY Find even more resources to learn the chapter concepts including animations, tutorial videos, and interactive practice questions.

Animations	Tutorial	MP3 Files
4.2 Prokaryotic Cell Shape • Cell Wall Antibiotics **4.3** Endosymbiosis **4.5** Lysosomes **4.7** Endosymbiosis	**4.3** Endosymbiosis	**4.3** Cellular Organelles

SUMMARIZE

4.1 Cellular Level of Organization

The **cell theory** states that all organisms are composed of cells and that all cells come from preexisting cells. **Cells** are very small (measured in micrometers) and must remain small in order to have an adequate **surface-area-to-volume ratio.** The plasma membrane regulates exchange of materials between the cell interior and the external environment.

4.2 Prokaryotic Cells

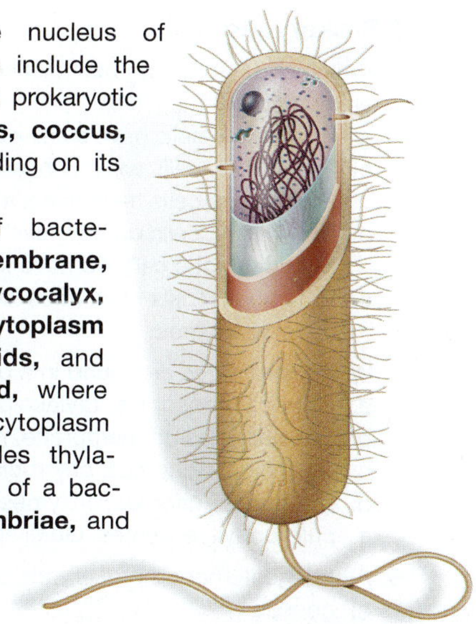

Prokaryotic cells lack the nucleus of **eukaryotic cells.** Prokaryotes include the bacteria and the archaeans. A prokaryotic cell may be called a **bacillus, coccus, spirilla,** or **spirochete** depending on its shape.

The **cell envelope** of bacteria includes a **plasma membrane,** a **cell wall,** and an outer **glycocalyx,** also called a **capsule.** The **cytoplasm** contains **ribosomes, plasmids,** and a region called the **nucleoid,** where the DNA may be found. The cytoplasm of cyanobacteria also includes thylakoids. The external structures of a bacterium include the **flagella, fimbriae,** and **conjugation pili.**

4.3 Introduction to Eukaryotic Cells

Eukaryotic cells are much larger than prokaryotic cells, and they contain compartmentalized structures called **organelles,** each of which has a specific structure and function (Table 4.1) that increases cell efficiency. The **endosymbiotic theory** helps explain the evolutionary origins of many membrane-bound organelles. Most membranous organelles are in constant communication.

Eukaryotic cells also contain a variety of transport **vesicles** that move around the cell using the proteins of the **cytoskeleton.**

4.4 The Nucleus and Ribosomes

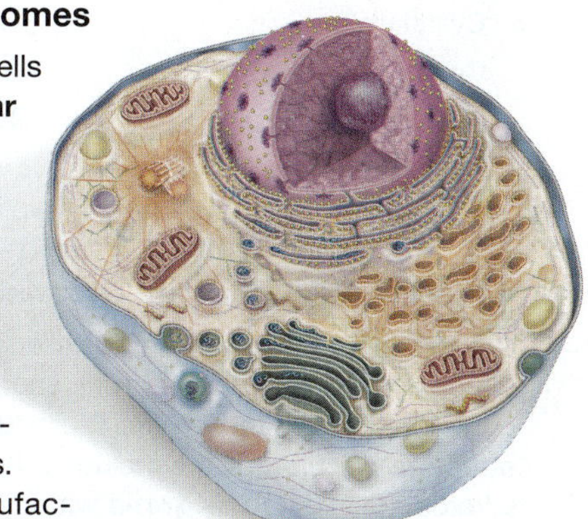

The **nucleus** of eukaryotic cells is surrounded by a **nuclear envelope** containing **nuclear pores** that regulate transport between the cytoplasm and the **nucleoplasm.** The nucleus contains **chromatin** (proteins and nucleic acids) that is organized into **chromosomes. Genes** on the chromosomes contain specific instructions for traits.

Ribosomes are manufactured in the **nucleolus** of the nucleus. **Ribosomes** are organelles that function in protein synthesis. In order to make a protein, mRNA is copied exactly from the DNA, processed, and exits the nucleus through a nuclear pore. After a ribosome attaches to an mRNA, most of the time this assembly goes to the rough ER to make a protein. Ribosomes may be connected as groups, called **polyribosomes.**

4.5 The Endomembrane System

The **endomembrane system** includes the endoplasmic reticulum (ER), the Golgi apparatus, the lysosomes (in animal cells), and transport vesicles. Newly produced proteins made in the **rough ER** are modified before they are packaged in transport vesicles, many of which go to the Golgi apparatus. The **smooth ER** has various metabolic functions, depending on the cell type, but it generally makes lipids that are carried by vesicles to different locations, particularly the **Golgi apparatus.** The Golgi apparatus modifies, sorts, and repackages proteins as well as processes lipids. Some proteins are tagged for transport to different cellular destinations; others are secreted from the cell. **Lysosomes** are produced by the Golgi apparatus and contain digestive enzymes.

4.6 Microbodies and Vacuoles

Cells contain numerous vesicles, **microbodies,** and **vacuoles.** Microbodies are vesicles with specific metabolic functions. **Peroxisomes** are microbodies that are involved in the metabolism of long-chain fatty acids. The large **central vacuole** in plant cells functions in storage and in the breakdown of molecules and cell parts.

Table 4.1 Comparison of Prokaryotic Cells and Eukaryotic Cells

	Prokaryotic Cells (1–20 μm in diameter)	Eukaryotic Cells (10–100 μm in diameter)	
		Animal	Plant
Cell wall	Usually (peptidoglycan)	No	Yes (cellulose)
Plasma membrane	Yes	Yes	Yes
Nucleus	No	Yes	Yes
Nucleolus	No	Yes	Yes
Ribosomes	Yes (smaller)	Yes	Yes
Endoplasmic reticulum	No	Yes	Yes
Golgi apparatus	No	Yes	Yes
Lysosomes	No	Yes	Not usually
Mitochondria	No	Yes	Yes
Chloroplasts	No	No	Yes
Peroxisomes	No	Usually	Usually
Cytoskeleton	No	Yes	Yes
Centrioles	No	Yes	No
9 + 2 cilia or flagella	No	Often	No (in flowering plants) Yes (sperm of bryophytes, ferns, and cycads)

4.7 The Energy-Related Organelles

Cells require a constant input of energy to maintain their structure. **Chloroplasts** are **plastids** that capture the energy of the sun and conduct photosynthesis, which produces carbohydrates. The internal structure of chloroplasts includes the **stroma** and **thylakoids.** Thylakoids are stacked as **grana** within the chloroplast.

Carbohydrate-derived products are broken down in **mitochondria** in the presence of oxygen via cellular respiration, and ATP is produced as a result. The internal structure of a mitochondrion includes the **cristae** and the **matrix.**

4.8 The Cytoskeleton

The cytoskeleton contains actin filaments, intermediate filaments, and microtubules. These maintain cell shape and help transport organelles from place to place within the cell. **Actin filaments** interact with motor molecules to allow a range of functions from muscular contraction to cellular division. **Intermediate filaments** support the nuclear and plasma membranes and participate in the cell-to-cell junctions that produce tissues. **Microtubules** radiate out from the **centrosome** and are present in **centrioles, cilia,** and **flagella.** They serve as an internal transport system along which vesicles and other organelles move.

ASSESS

Choose the best answer for each question.

4.1 Cellular Level of Organization

1. The surface-area-to-volume ratio defines what aspect of a cell?
   a. whether it is eukaryotic or prokaryotic
   b. whether it is plant or animal
   c. its size
   d. its ability to move
2. The cell theory states that
   a. cells are the basic units of life.
   b. all organisms are composed of cells.
   c. all cells come from preexisting cells.
   d. All of these are correct.

4.2 Prokaryotic Cells

3. Which of the following best distinguishes a prokaryotic cell from a eukaryotic cell?
   a. Prokaryotic cells have a cell wall, but eukaryotic cells never do.
   b. Prokaryotic cells are much larger than eukaryotic cells.
   c. Prokaryotic cells have flagella, but eukaryotic cells do not.
   d. Prokaryotic cells do not have a membrane-bound nucleus, but eukaryotic cells do have such a nucleus.
   e. Prokaryotic cells have ribosomes, but eukaryotic cells do not have ribosomes.
4. Which structures are found in a prokaryotic cell?
   a. cell wall, ribosomes, thylakoids, chromosome
   b. cell wall, plasma membrane, nucleus, flagellum
   c. nucleoid, ribosomes, chloroplasts, capsule
   d. plasmid, ribosomes, enzymes, DNA, mitochondria
   e. chlorophyll, enzymes, Golgi apparatus, plasmids
5. A spherical-shaped prokaryotic cell is called a
   a. coccus.
   b. spirochete.
   c. bacillus.
   d. None of these are correct.

4.3 Introduction to Eukaryotic Cells

6. Which organelle most likely originated by invagination of the plasma membrane?
   a. mitochondria
   b. flagella
   c. nucleus
   d. chloroplasts
   e. All of these are correct.

7. Which of the following organelles contains its (their) own DNA, suggesting they were once independent prokaryotes?
   a. Golgi apparatus
   b. mitochondria
   c. chloroplasts
   d. ribosomes
   e. Both b and c are correct.

4.4 The Nucleus and Ribosomes

8. Which of these is not found in the nucleus?
   a. functioning ribosomes
   b. chromatin that condenses to chromosomes
   c. nucleolus that produces rRNA
   d. nucleoplasm instead of cytoplasm
   e. all forms of RNA
9. The _____ is(are) responsible for protein synthesis in a cell.
   a. chromatin
   b. chromosomes
   c. ribosomes
   d. nucleoplasm
10. Which of the following terms indicates the basic unit of hereditary information?
    a. gene
    b. chromosome
    c. chromatin
    d. nucleoplasm

4.5 The Endomembrane System

11. Vesicles from the rough ER most likely are on their way to
    a. the peroxisomes.
    b. the lysosomes.
    c. the Golgi apparatus.
    d. the plant cell vacuole only.
    e. the location suitable to their size.
12. Lysosomes function in
    a. protein synthesis.
    b. processing and packaging.
    c. intracellular digestion.
    d. lipid synthesis.
    e. production of hydrogen peroxide.
13. Which of the following is reponsible for the synthesis of proteins that are being exported from the cell?
    a. smooth ER
    b. rough ER
    c. lysosome
    d. peroxisome

4.6 Microbodies and Vacuoles

14. Vesicles with specific metabolic functions in a cell are called
    a. the cytoskeleton.
    b. centrioles.
    c. ribosomes.
    d. microbodies.
15. These microbodies break down fatty acids and contain catalase to break down hydrogen peroxide.
    a. lysosome
    b. central vacuole
    c. peroxisome
    d. chromatin

4.7 The Energy-Related Organelles

16. Mitochondria
    a. are involved in cellular respiration.
    b. break down ATP to release energy for cells.
    c. are present in animal cells but not plant cells.
    d. All of these are correct.
17. Which organelle releases oxygen?
    a. ribosome
    b. Golgi apparatus
    c. chloroplast
    d. smooth ER
18. Which of the following would not be found in a chloroplast?
    a. grana
    b. thylakoids
    c. cristae
    d. stroma

4.8 The Cytoskeleton

19. Which of these is not true?
    a. Actin filaments are located under the plasma membrane.
    b. Microtubules are organized by centrosomes.
    c. Intermediate filaments are associated with the nuclear envelope.
    d. Motor molecules move materials along intermediate filaments.
20. Cilia and flagella
    a. have a 9 + 0 pattern of microtubules, the same as basal bodies.
    b. contain myosin that pulls on actin filaments.
    c. are organized by basal bodies derived from centrioles.
    d. are constructed similarly in prokaryotes and eukaryotes.
    e. Both a and c are correct.

ENGAGE

LEARNSMART LABS®

The following LearnSmart Labs contain exercises that are related to the content of this chapter:

- Microscopy Biology
- Cell Anatomy

THINKING SCIENTIFICALLY

1. The protists that cause malaria contribute to infections associated with AIDS. Scientists have discovered that an antibiotic that inhibits prokaryotic enzymes will kill the parasite, because it is effective against the plastids in the protist. What can you conclude about the origin of the plastids?
2. In the 1958 movie *The Blob,* a giant, single-celled alien creeps and oozes around, attacking and devouring helpless humans. Why couldn't there be a real single-celled organism as large as the Blob?
3. Calculate the surface-area-to-volume ratio of a 1-mm cube and a 2-mm cube. Which has the smaller ratio?
4. *Giardia lamblia* is a protist that is commonly known for contaminating water supplies and causing diarrhea. While *Giardia* is a eukaryote, its cells lack mitochondria. What would be the overall effect on a eukaryotic cell if it lacked mitochondria? How might have *Giardia* adapted for this potential difficulty?

5

Membrane Structure and Function

The burning sensation of a chili pepper is caused by interactions of chemicals with the membranes of cells.

Chapter Outline

5.1 Plasma Membrane Structure and Function 83

5.2 Passive Transport Across a Membrane 88

5.3 Active Transport Across a Membrane 91

5.4 Modification of Cell Surfaces 95

Before You Begin

Before beginning this chapter, take a few moments to review the following discussions.

Section 3.3 How does the structure of a phospholipid make it an ideal molecule for the plasma membrane?

Section 3.4 How does a protein's shape relate to its function?

Figures 4.6 and 4.7 What are the key features of animal and plant cells?

Have you ever bitten into a hot pepper and had the sensation that your mouth was on fire? This is because the chili pepper plant produces a chemical, called capsaicin, that binds to a protein in the plasma membrane of pain receptors in your mouth. In the membrane are channel proteins that allow the movement of calcium ions across the membrane. When these channels are open, movement of the calcium ions into the cell causes the pain receptor to send a signal to the brain. The brain then interprets this signal as a burning sensation. These channels may also be triggered by temperature, an acidic pH, and heat. As long as the capsaicin is present, the pathway will remain active and signals will be sent to the brain. So the quickest way to alleviate the pain is to remove the capsaicin and close the channel protein. Unfortunately, since capsaicin is lipid-soluble, drinking cool water does very little to alleviate the pain. However, drinking milk, or eating bread or rice, often helps remove the capsaicin. Often the first bite is the worst, since the capsaicin causes an initial opening of all the channels simultaneously. The receptors can become desensitized to capsaicin, which is why later bites of the same pepper don't produce the same results.

In this chapter, we will explore not only how cells move materials in and out but also the basic properties of energy and how cells use metabolic pathways and enzymes to conduct the complex reactions needed to sustain life.

As you read through this chapter, think about the following questions:

1. How do transport and channel proteins function in a plasma membrane?
2. What type of transport are the calcium channels performing?

Following *the* Themes

CHAPTER 5 MEMBRANE STRUCTURE AND FUNCTION

UNIT 1 THE CELL

Evolution	The plasma membrane and internal membrane-bound organelles have evolved to support specialized functions for cells.
Nature of Science	Research in membranes and associated proteins leads to new discoveries in medicine and disease treatment.
Biological Systems	Membranes play an important role in cellular communication and response to environmental stimuli.

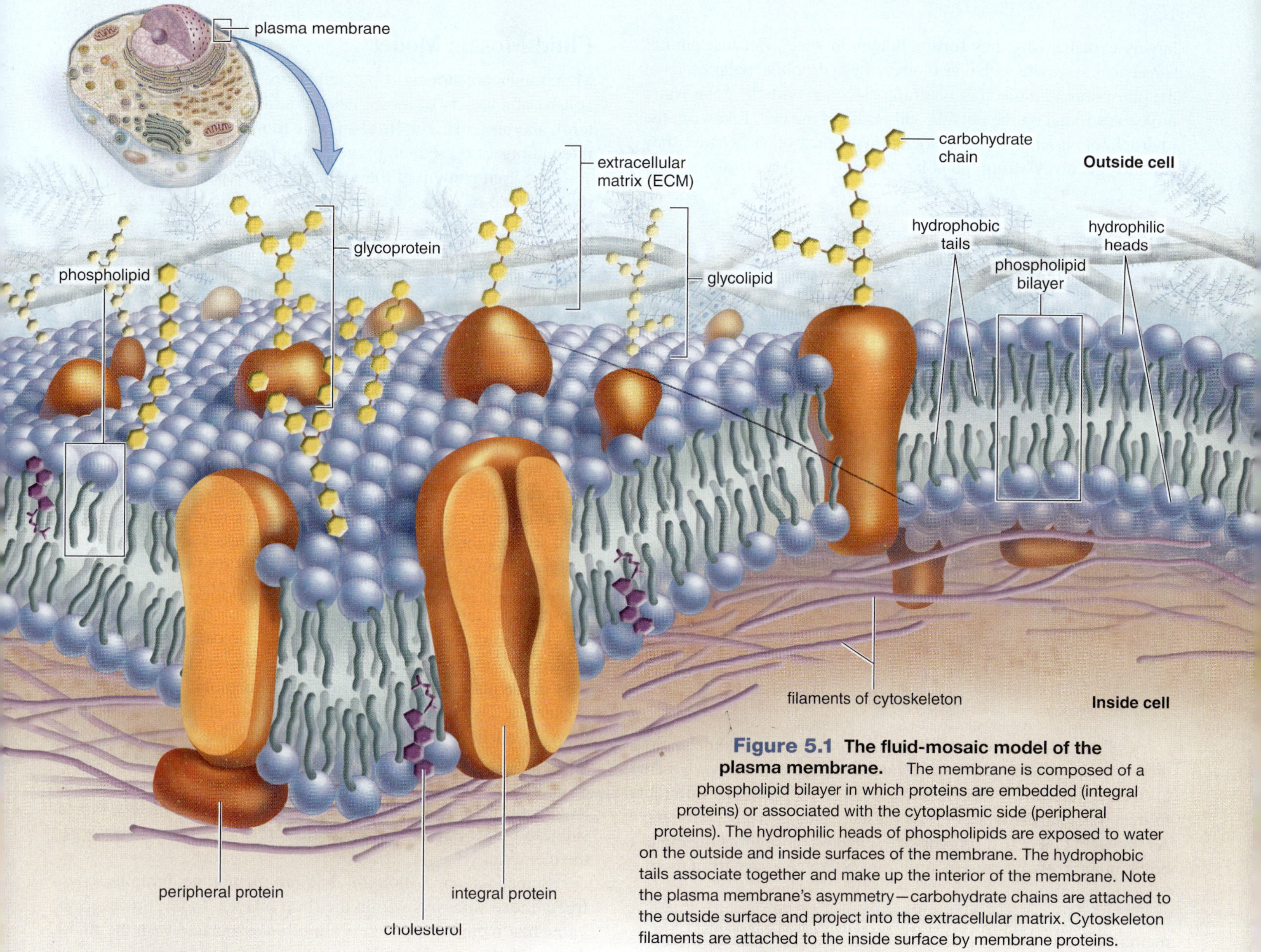

Figure 5.1 The fluid-mosaic model of the plasma membrane. The membrane is composed of a phospholipid bilayer in which proteins are embedded (integral proteins) or associated with the cytoplasmic side (peripheral proteins). The hydrophilic heads of phospholipids are exposed to water on the outside and inside surfaces of the membrane. The hydrophobic tails associate together and make up the interior of the membrane. Note the plasma membrane's asymmetry—carbohydrate chains are attached to the outside surface and project into the extracellular matrix. Cytoskeleton filaments are attached to the inside surface by membrane proteins.

5.1 Plasma Membrane Structure and Function

Learning Outcomes

Upon completion of this section, you should be able to

1. Distinguish between the different structural components of membranes.
2. Describe the nature of the fluid-mosaic model as it relates to membrane structure.
3. Describe the diverse role of proteins in membranes.
4. Explain why the plasma membrane exhibits selective permeability.

The ability to create compartments is a key feature of cells. Membranes, made of a phospholipid bilayer, create separation between the cell and the external environment as well as compartments within the cell itself. Having separate spaces allows multiple, sometimes incompatible, chemical processes to occur simultaneously. This "division of labor" allows cells to operate more efficiently and respond to changing environmental conditions.

Components of the Plasma Membrane

The structure of a typical animal cell's plasma membrane is depicted in Figure 5.1. In addition to the phospholipid bilayer, membrane components include protein molecules that are either partially or wholly embedded in the bilayer. Cholesterol is another lipid found in the animal plasma membrane; related steroids are found in the plasma membrane of plants. As we will see, cholesterol helps modify the fluidity of the membrane over a range of temperatures.

MP3 Membrane Structure

Recall that a phospholipid is an *amphipathic molecule,* meaning that it has both a hydrophilic (water-loving) region and a hydrophobic (water-fearing) region. The amphipathic nature of phospholipids largely explains why they form a bilayer in water. Because similar substances associate with one another, the hydrophilic polar heads of the phospholipid molecules naturally associate with the polar water molecules found on the outside and inside of the cell. Likewise, the hydrophobic nonpolar tails associate with each other because they want to "get away" from the polar water.

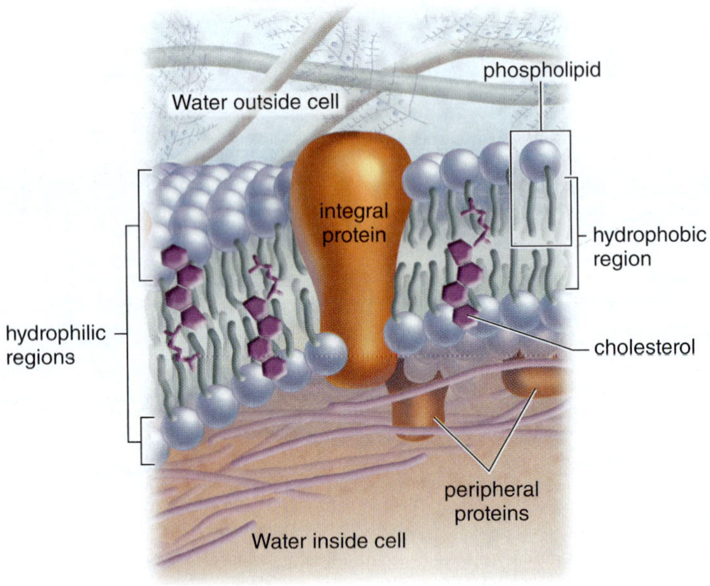

Cell membranes are highly similar in the types of molecules they contain, which makes them interchangeable and allows them to fuse together fairly easily. What makes one membrane different from another are the types of proteins integrated into the membrane. As shown in Figure 5.1, proteins are scattered throughout the membrane in an irregular pattern, and this pattern can vary from membrane to membrane.

Electron micrographs can be used to study the nature of many membrane proteins. A research method called freeze-fracture freezes, and then splits, the membrane so that the upper and lower layers separate. The proteins remain intact and go with one of the layers. The embedded proteins are called *integral proteins,* whereas the proteins that occur only on the cytoplasmic side of the membrane are called *peripheral proteins*.

Some integral proteins protrude from only one surface of the bilayer, but most span the membrane, with a hydrophobic core region that associates with the nonpolar core of the membrane. Hydrophilic ends of integral proteins protrude from both surfaces of the bilayer, interacting with polar water molecules. Integral proteins can be held in place by attachments to protein fibers of the cytoskeleton (inside) and fibers of the extracellular matrix (outside). Only animal cells have an extracellular matrix (ECM), which contains various protein fibers and very large, complex carbohydrate molecules. The ECM, which is discussed in greater detail in section 5.4, has a number of functions, from lending external support to the plasma membrane to assisting in communication between cells.

Fluid-Mosaic Model

Membranes are not rigid but rather are flexible structures. They consist of a variety of molecules, including phospholipids, cholesterol, and proteins. The **fluid-mosaic model** is used to describe the interactions of these membrane components.

The lipid content of the membrane is responsible for its fluidity. Cells are flexible because the phospholipid bilayer is fluid. At body temperature, the phospholipid bilayer of the plasma membrane has the consistency of olive oil. The greater the concentration of unsaturated fatty acid residues, the more fluid the bilayer. In each monolayer, the fatty acid tails jostle around, and an entire phospholipid molecule can move sideways at a rate averaging about 2 μm—the length of a prokaryotic cell—per second. Although it is possible for phospholipid molecules to flip-flop from one monolayer to the other, they rarely do so, because this would require the hydrophilic head to move through the hydrophobic center of the membrane. However, at times special proteins help the phospholipids flip.

The presence of cholesterol molecules prevents the plasma membrane from becoming too fluid at higher temperatures and too solid at lower temperatures. At higher temperatures, cholesterol stiffens the membrane and makes it less fluid than it would otherwise be. At lower temperatures, cholesterol helps prevent the membrane from freezing by not allowing contact between certain phospholipid tails.

A plasma membrane is considered a mosaic because of the presence of many proteins. The number and kinds of proteins can vary in the plasma membrane and in the membranes of the various organelles. The position of these proteins can shift over time, unless they are anchored to another structure, such as the cytoskeleton. Experiments have been conducted in which the proteins were tagged prior to allowing mouse and human cells to fuse. An hour after fusion, the proteins from each cell type were completely mixed, suggesting that at least some proteins are able to move sideways in the membrane.

Scientists once thought that all membrane proteins could freely move sideways within the fluid bilayer. Today, however, we know that membrane proteins are often associated with the ECM, the cytoskeleton, or both. These connections hold a protein in place and partially anchor the otherwise fluid phospholipid bilayer.

It should be noted that the two sides of the membrane are not identical. Carbohydrate chains (see below) are attached only to molecules on the outside surface, and peripheral proteins occur on one surface or the other. Thus, the membrane is said to be asymmetrical.

3D Animation Membrane Transport: Lipid Bilayer

Glycoproteins and Glycolipids

Phospholipids and proteins that have attached carbohydrate (sugar) chains are called **glycolipids** and **glycoproteins,** respectively. The carbohydrate (sugar) chains on a cell's exterior can be highly diverse. The chains can vary in the number and sequence of sugars, and in whether the chain is branched. Each cell within an individual has its own "fingerprint" because of these chains. For this reason, glycolipids and glycoproteins play an important role in cellular identification. As you probably know, transplanted tissues are often rejected by the recipient. Rejection occurs because the immune system is able to detect that the foreign tissue's cells do not have the appropriate carbohydrate chains to be recognized as self. In humans, carbohydrate chains are also the basis for the A, B, and O blood groups.

In animal cells, the carbohydrate chains attached to proteins give the cell a "sugar coat," more properly called a *glycocalyx*. The glycocalyx protects the cell and has various other functions, including cell-to-cell adhesion, reception of signaling molecules, and cell-to-cell recognition.

The Functions of the Proteins

Although the protein components of plasma membranes differ depending on the type of cell and the processes it is undergoing, several types of proteins are likely to be routinely present:

Channel proteins Channel proteins are involved in passing molecules through the membrane. They form a channel that allows a substance to simply move from one side to the other (Fig. 5.2*a*). For example, a channel protein allows hydrogen ions to flow across the inner mitochondrial membrane. Without this movement of hydrogen ions, ATP would never be produced.

Carrier proteins Carrier proteins are also involved in passing molecules through the membrane. They receive a substance and change their shape, and this change moves the substance across the membrane (Fig. 5.2*b*). A carrier protein transports sodium and potassium ions across the plasma membrane of a nerve cell. Without this carrier protein, nerve impulse conduction would be impossible.

Cell recognition proteins Cell recognition proteins are glycoproteins (Fig. 5.2*c*). Among other functions, these proteins help the body recognize when it is being invaded by pathogens, so that an immune response can occur. Without this recognition, pathogens would be able to freely invade the body and hinder its function.

Receptor proteins Receptor proteins have a shape that allows only a specific molecule to bind to it (Fig. 5.2*d*). The binding of this molecule causes the protein to change its shape and thereby bring about a cellular response. The coordination of the body's organs is totally dependent on such signaling molecules. For example, the liver stores glucose after it is signaled to do so by insulin.

Enzymatic proteins Some plasma membrane proteins are enzymes that carry out metabolic reactions directly (Fig. 5.2*e*). Without these enzymes, some of which are attached to the various membranes of the cell, a cell would never be able to perform the chemical reactions needed to maintain its metabolism.

Junction proteins Proteins are involved in forming various types of junctions between animal cells (Fig. 5.2*f*). Signaling molecules that pass through gap junctions allow the cilia of cells that line the respiratory tract to beat in unison.

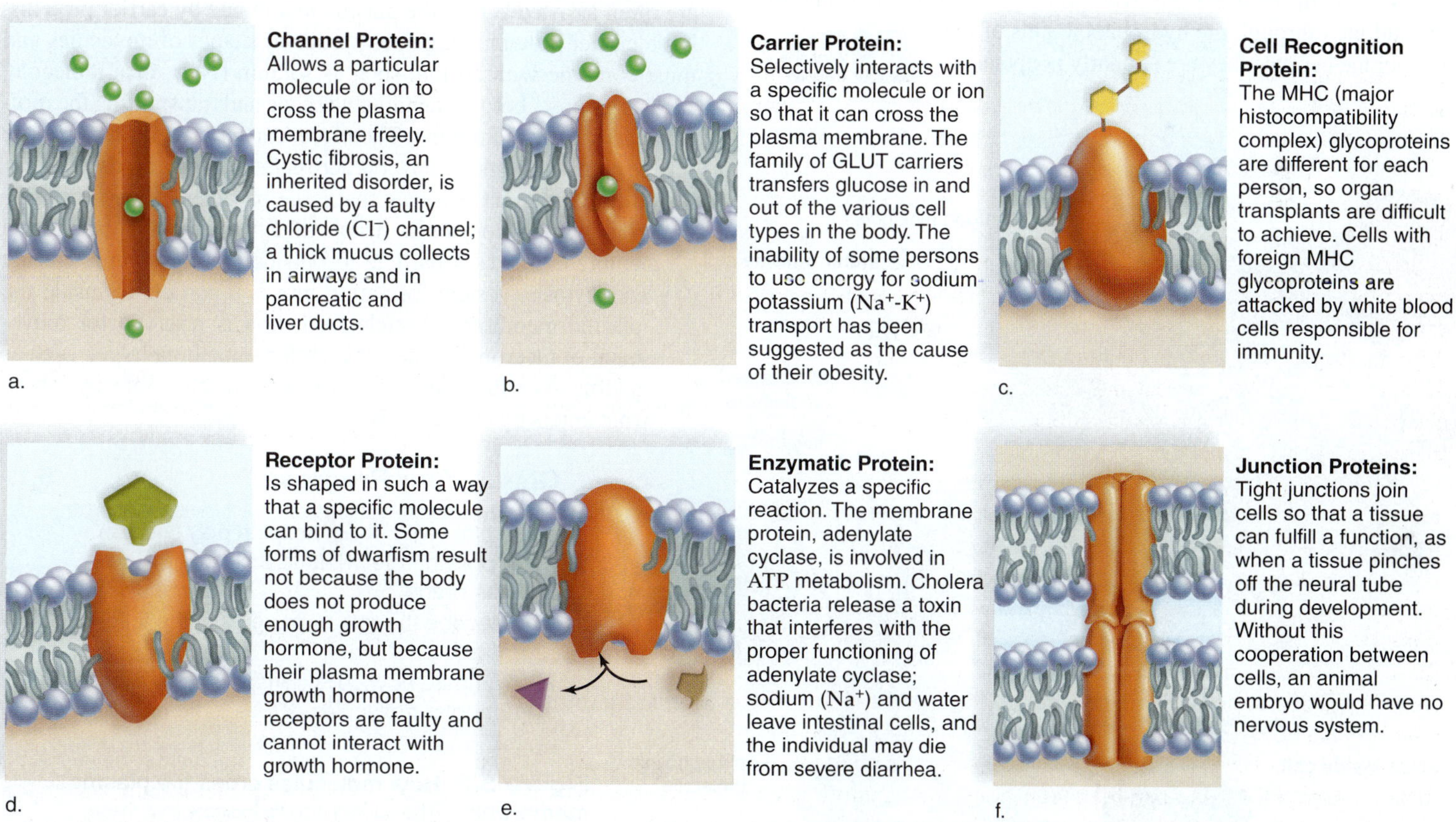

a. **Channel Protein:** Allows a particular molecule or ion to cross the plasma membrane freely. Cystic fibrosis, an inherited disorder, is caused by a faulty chloride (Cl^-) channel; a thick mucus collects in airways and in pancreatic and liver ducts.

b. **Carrier Protein:** Selectively interacts with a specific molecule or ion so that it can cross the plasma membrane. The family of GLUT carriers transfers glucose in and out of the various cell types in the body. The inability of some persons to use energy for sodium-potassium (Na^+-K^+) transport has been suggested as the cause of their obesity.

c. **Cell Recognition Protein:** The MHC (major histocompatibility complex) glycoproteins are different for each person, so organ transplants are difficult to achieve. Cells with foreign MHC glycoproteins are attacked by white blood cells responsible for immunity.

d. **Receptor Protein:** Is shaped in such a way that a specific molecule can bind to it. Some forms of dwarfism result not because the body does not produce enough growth hormone, but because their plasma membrane growth hormone receptors are faulty and cannot interact with growth hormone.

e. **Enzymatic Protein:** Catalyzes a specific reaction. The membrane protein, adenylate cyclase, is involved in ATP metabolism. Cholera bacteria release a toxin that interferes with the proper functioning of adenylate cyclase; sodium (Na^+) and water leave intestinal cells, and the individual may die from severe diarrhea.

f. **Junction Proteins:** Tight junctions join cells so that a tissue can fulfill a function, as when a tissue pinches off the neural tube during development. Without this cooperation between cells, an animal embryo would have no nervous system.

Figure 5.2 Membrane protein diversity. These are some of the functions performed by proteins found in the plasma membrane.

Table 5.1 Passage of Molecules into and out of the Cell

Name	Direction	Requirement	Examples
Diffusion	Toward lower concentration	Concentration gradient	Lipid-soluble molecules, gases
Facilitated transport	Toward lower concentration	Channels or carrier and concentration gradient	Some sugars, amino acids
Active transport	Toward higher concentration	Carrier plus energy	Sugars, amino acids, ions
Bulk transport	Toward outside or inside	Vesicle utilization	Macromolecules

Permeability of the Plasma Membrane

The plasma membrane regulates the passage of molecules into and out of the cell. This function is critical because the cell must maintain its normal composition under changing environmental conditions. The plasma membrane is essential because it is **selectively permeable,** allowing only certain substances into the cell while keeping others out.

Molecules that can freely cross a membrane generally require no energy to do so. Substances that are hydrophobic and therefore similar to the phospholipid center of the membrane are able to diffuse across membranes at no energy cost. Polar molecules, however, are chemically incompatible with the center of the membrane and so require an expenditure of energy to drive their transport.

Table 5.1 and Figure 5.3 examine which types of molecules can passively cross a membrane (no energy required) and which may require transport by a carrier protein and/or an expenditure of energy. In general, small, noncharged molecules, such as carbon dioxide, oxygen, glycerol, and alcohol, can freely cross the membrane. They are able to slip between the hydrophilic heads of the phospholipids and pass through the hydrophobic tails of the membrane because they are similarly nonpolar.

MP3 Membrane Transport

These molecules follow their **concentration gradient** as they move from an area where their concentration is high to an area where their concentration is low. Consider that a cell is always using oxygen when it carries on cellular respiration. The internal consumption of oxygen results in a low cellular concentration. Because oxygen concentration is higher outside than inside the cell, oxygen tends to move across the membrane into the cell. The concentration of carbon dioxide, on the other hand, is highest inside the cell, because it is produced during cellular respiration. Therefore, carbon dioxide tends to move with its concentration gradient from inside to outside the cell.

Water, a polar molecule, would not be expected to readily cross the primarly nonpolar membrane. However, scientists have discovered that the majority of cells have channel proteins, called **aquaporins,** that allow water to cross a membrane more quickly than expected. Aquaporins also allow cells to equalize water pressure differences between their interior and exterior environments, so that their membranes don't burst from environmental pressure changes.

Ions and polar molecules, such as glucose and amino acids, can slowly cross a membrane. To move as quickly as is necessary, they are often assisted across the plasma membrane by carrier proteins. Each carrier protein recognizes particular shapes of molecules and must combine with an ion, such as sodium (Na^+), or a molecule, such as glucose, before changing its shape and transporting the molecule across the membrane. Therefore, carrier proteins are specific for the substances they transport across the plasma membrane.

Bulk transport is a way that large particles can exit or enter a cell. During exocytosis, fusion of a vesicle with the plasma membrane moves a particle to outside the membrane. During endocytosis, vesicle formation moves a particle to inside the plasma membrane. Vesicle formation is reserved for movement of macromolecules or even for something larger, such as a virus. As with many other processes, a cell is selective about what enters by endocytosis.

Check Your Progress 5.1

1. Explain why phospholipids play such an important role in the structure of the cell membrane.
2. Describe the role of proteins in the fluid-mosaic model.
3. Compare how cells transport polar and nonpolar molecules across a membrane.

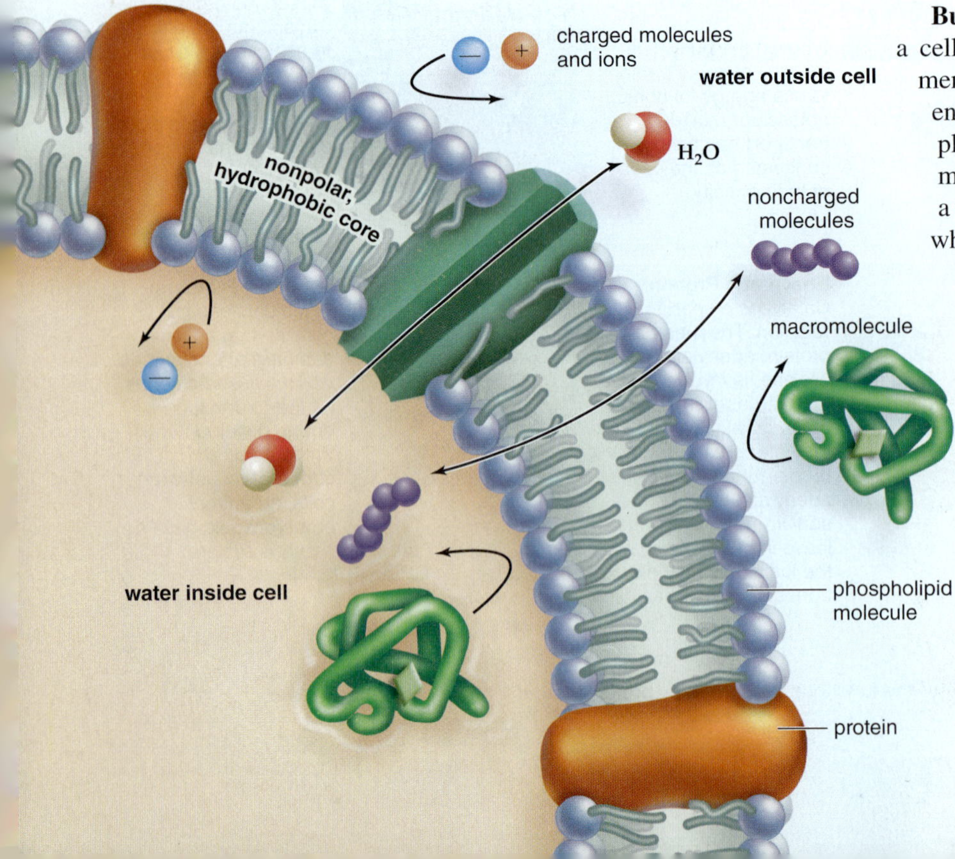

Figure 5.3 How molecules cross the plasma membrane. The curved arrows indicate that these substances cannot passively cross the plasma membrane, and the long, back-and-forth arrows indicate that these substances can diffuse across the plasma membrane.

THEME Biological Systems

How Cells Talk to One Another

All organisms are comprised of cells that are able to sense and respond to specific signals in their environment. A bacterium that lives in your body responds to signaling molecules when it finds food and escapes immune cells in order to stay alive. Signaling helps the bread mold that grows on stale bread detect an opposite mating strain to begin its sexual life cycle. Similarly, the cells of a developing embryo respond to signaling molecules as they move to specific locations and become specific tissues (Fig. 5A*a*).

In newborn animals, internal signals such as hormones are essential to ensure that specific tissues develop when and how they should. In plants, external signals, such as a change in the amount of light, tell them when it is time to resume growth or to flower. Internal signaling molecules enable animals and plants to coordinate their cellular activities, to metabolize, and to better respond in a changing environment. The ability of cells to communicate with one another is an essential part of all biological systems.

Cell Signaling

The cells of a multicellular organism "talk" to one another by using signaling molecules, sometimes called chemical messengers. Some messengers are produced in one location and, in animals, are carried by the circulatory system to various target sites around the body. For example, the pancreas releases a hormone called insulin, which is transported in blood vessels to the liver, and this signal causes the liver to store glucose as glycogen. Failure of the liver to respond appropriately results in a medical condition called diabetes.

In Chapter 9, we are particularly interested in growth factors, which act locally as signaling molecules and cause cells to divide. Overproduction of growth factors can disrupt the balance in cellular systems. If left uncorrected, uncontrolled cell growth and formation of a tumor can result. The importance of cell signaling in regulating cell systems is the focus of much research in cell biology.

Cells respond to only certain signaling molecules. Why? Because they must bind to a receptor protein, and only cells that possess matching receptors can respond to certain signaling molecules. Each cell has a mix of receptors, which gives them the ability to respond differently to a variety of external and internal stimuli. Each cell is also able to balance the relative strength of incoming signals in order to change cellular structure or function. If a minimum level of signaling is not met, the cell dies.

Signaling molecules interacting with their receptor is only the beginning of a complex process of communication that tells the cell how to respond. Once a signaling molecule and receptor interact, a cascade of events occurs that increase, decrease, or otherwise change the signal to elicit a cellular response. This process is called a signal transduction pathway. This pathway is analogous to television transmission: A TV camera (the receptor) views a scene and converts it into electrical signals (transduction pathway) that are understood by the TV receiver in your house, which converts these signals to a picture on your screen (the response). The process in cells is more complicated, because each member of the pathway can activate a number of other proteins. As shown in Figure 5A*b*, the cell response to a transduction pathway can be a change in the shape or movement of a cell, the activation of a particular enzyme, or the activation of a specific gene.

Questions to Consider

1. If your cells needed to respond rapidly to a changing environment, would you want their effect to be short- or long-lived?
2. Given the essential role of signaling in cellular and organismal health, how might diseases arise from signaling errors?

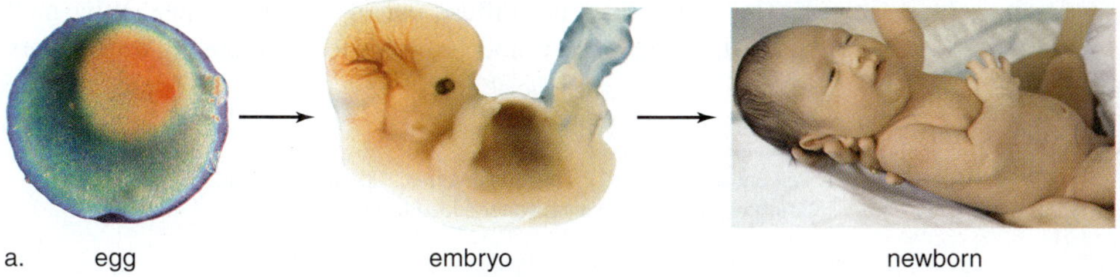

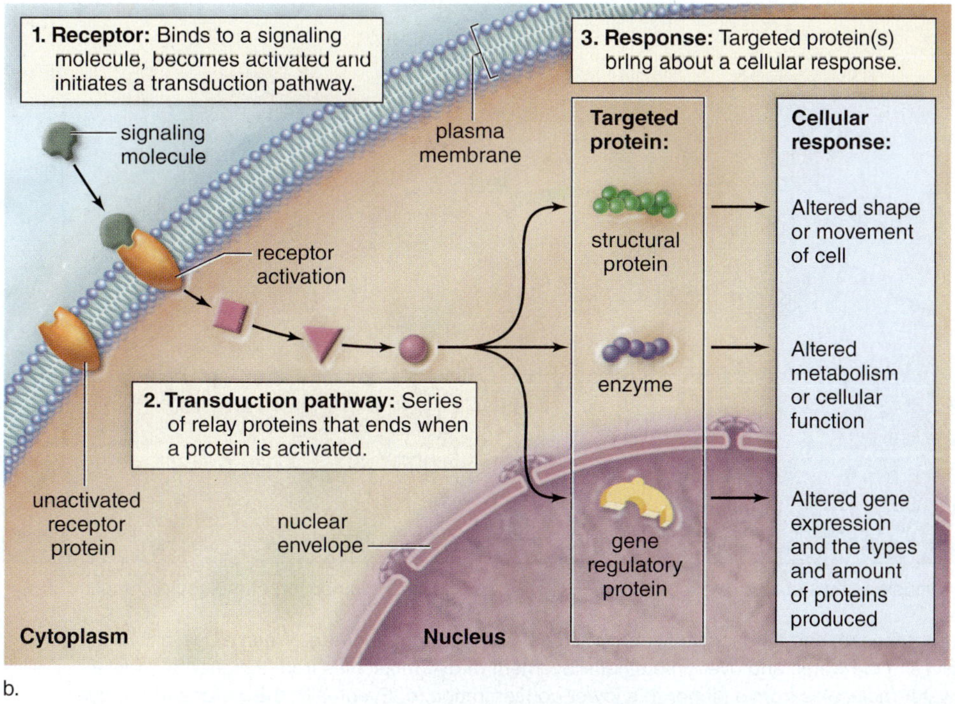

Figure 5A Cell signaling. **a.** The process of signaling helps account for the transformation of an egg into an embryo and then an embryo into a newborn. **b.** The process of signaling involves three steps: binding of the signaling molecule, transduction of the signal, and response of the cell depending on what type of protein is targeted.

5.2 Passive Transport Across a Membrane

Learning Outcomes

Upon completion of this section, you should be able to

1. Compare diffusion and osmosis across a membrane.
2. Describe the role of proteins in the movement of molecules across a membrane.
3. Differentiate among the effects of hypotonic, isotonic, and hypertonic solutions on animal and plant cells.

Diffusion is the movement of molecules from a higher to a lower concentration—that is, down their concentration gradient—until equilibrium is achieved and the molecules are distributed equally. Diffusion is a physical process that results from the random molecular motion that can be observed with any type of molecule. For example, when a crystal of dye is placed in water (Fig. 5.4), the dye and water molecules move in various directions, but their net movement, which is the sum of their motion, is toward the region of lower concentration. Eventually, the dye is dissolved evenly in the water, resulting in equilibrium and a uniformly colored solution.

MP3 Diffusion

A **solution** contains both a **solute,** usually a solid, and a **solvent,** usually a liquid. In this case, the solute is the dye and the solvent is the water molecules. Once the solute and solvent are evenly distributed, they continue to move about, but there is no net movement of either one in any direction.

The chemical and physical properties of the plasma membrane allow only a few types of molecules to enter and exit a cell simply by diffusion. Gases can freely diffuse through the lipid bilayer because they are small and nonpolar; this is the mechanism by which oxygen enters cells and carbon dioxide exits cells. This is also how oxygen diffuses from the alveoli (air sacs) of the lungs into the blood in the lung capillaries (Fig. 5.5). After inhalation (breathing in), the concentration of oxygen in the alveoli is higher than that in the blood; therefore, oxygen diffuses into the blood along its concentration gradient.

Several factors influence the rate of diffusion, including temperature, pressure, electrical currents, and molecular size. For example, as temperature increases, the rate of diffusion increases. The movement of fishes in the tank would also speed the rate of diffusion (see Fig. 5.4).

Animation How Diffusion Works

3D Animation Membrane Transport: Diffusion

Osmosis

The diffusion of water across a selectively permeable membrane from high to low concentration is called **osmosis.** To illustrate osmosis, a thistle tube containing a 10% solute solution[1] is covered at one end by a selectively permeable membrane and then placed in a beaker containing a 5% solute solution (Fig. 5.6*a*). The beaker has a higher concentration of water molecules (lower percentage of solute), and the thistle tube has a lower concentration of water molecules (higher percentage of solute). Diffusion always occurs from higher to lower concentration. Therefore, a net movement of water takes place, moving across the membrane from the beaker to the inside of the thistle tube (Fig. 5.6*b*).

MP3 Osmosis

The solute does not diffuse out of the thistle tube. Why not? Because the membrane is not permeable to the solute. As water enters

1 Percent solutions are grams of solute per 100 ml of solvent. Therefore, a 10% solution is 10 g of sugar with water added to make 100 ml of solution.

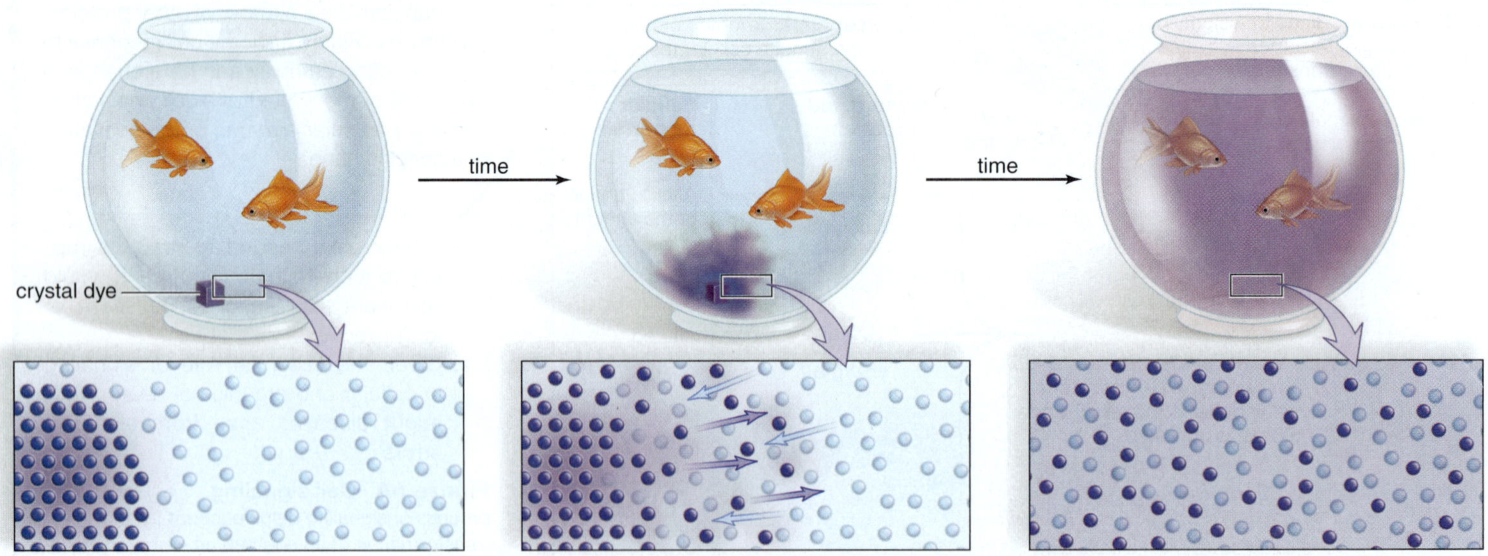

a. Crystal of dye is placed in the water

b. Diffusion of water and dye molecules

c. Equal distribution of molecules results

Figure 5.4 Process of diffusion. Diffusion is spontaneous, and no chemical energy is required to bring it about. **a.** When a dye crystal is placed in water, it is concentrated in one area. **b.** The dye dissolves in the water, and over time a net movement of dye molecules from a higher to a lower concentration occurs. There is also a net movement of water molecules from a higher to a lower concentration. **c.** Eventually, the water and the dye molecules are equally distributed throughout the container.

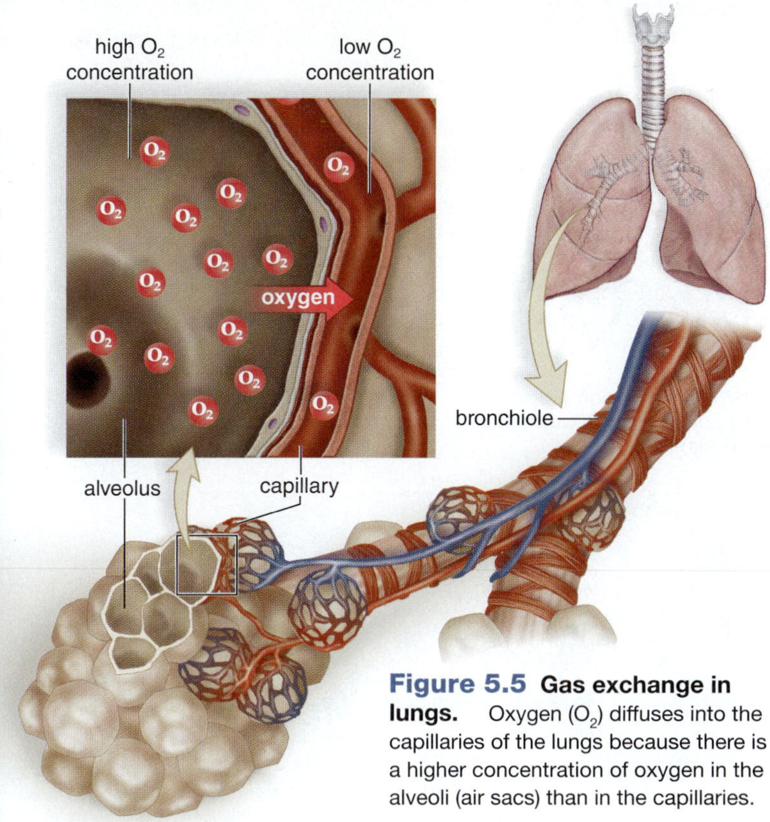

Figure 5.5 Gas exchange in lungs. Oxygen (O_2) diffuses into the capillaries of the lungs because there is a higher concentration of oxygen in the alveoli (air sacs) than in the capillaries.

and the solute does not exit, the level of the solution within the thistle tube rises (Fig. 5.6*c*). In the end, the concentration of solute in the thistle tube is less than 10%. Why? Because there is now less solute per unit volume. And the concentration of solute in the beaker is greater than 5%, because there is now more solute per unit volume.

Water enters the thistle tube due to the osmotic pressure of the solution within the thistle tube until it reaches equilibrium (Fig 5.6*d*). **Osmotic pressure** is the pressure that develops in a system due to osmosis.[2] In other words, the greater the possible osmotic pressure, the more likely it is that water will diffuse in that direction. Due to osmotic pressure, water is absorbed by the kidneys and taken up by capillaries in the tissues. Osmosis also occurs across the plasma membrane, as we'll see next.

Animation How Osmosis Works

Isotonic Solution

In the laboratory, cells are normally placed in **isotonic solutions.** The prefix *iso* means "the same as," and the term **tonicity** refers to the strength of the solution. In an isotonic solution, the solute

2 Osmotic pressure is measured by placing a solution in an osmometer and then immersing the osmometer in pure water. The pressure that develops is the osmotic pressure of a solution.

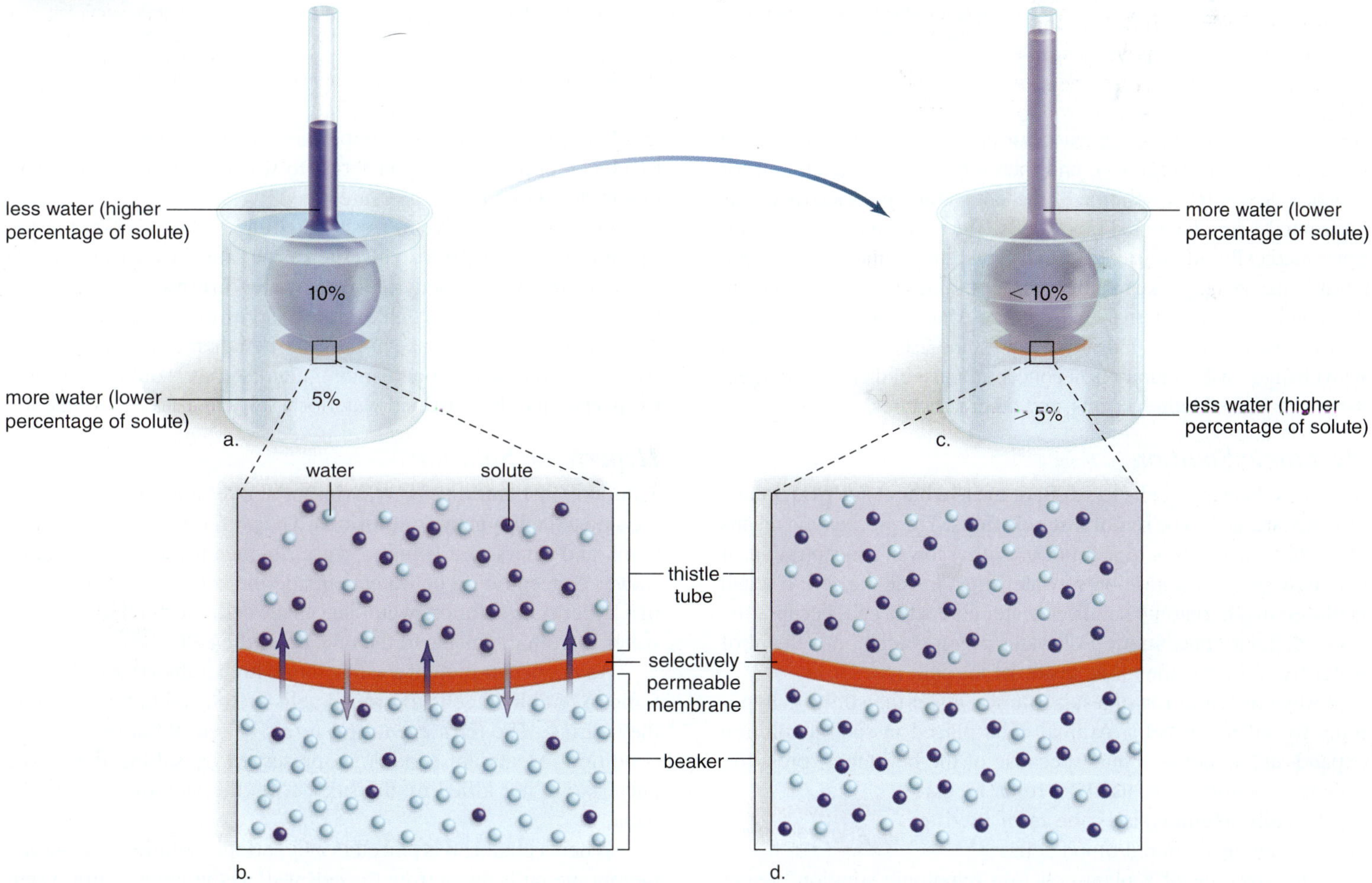

Figure 5.6 Osmosis demonstration. **a.** A thistle tube, covered at the broad end by a selectively permeable membrane, contains a 10% solute solution. The beaker contains a 5% solute solution. **b.** The solute (purple circles) is unable to pass through the membrane, but the water (blue circles) passes through in both directions. There is a net movement of water toward the inside of the thistle tube, where there is a lower percentage of water molecules. **c.** Due to the incoming water molecules, the level of the solution rises in the thistle tube. **d.** Eventually, the concentration of water across the membrane equalizes.

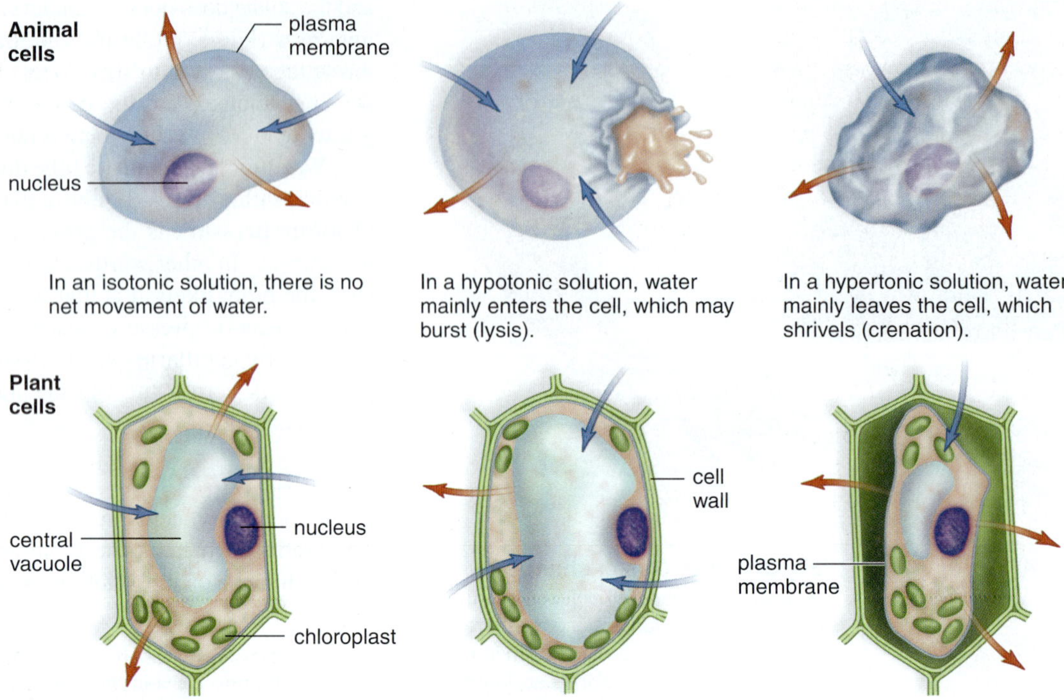

In an isotonic solution, there is no net movement of water.

In a hypotonic solution, water mainly enters the cell, which may burst (lysis).

In a hypertonic solution, water mainly leaves the cell, which shrivels (crenation).

In an isotonic solution, there is no net movement of water.

In a hypotonic solution, vacuoles fill with water, turgor pressure develops, and chloroplasts are seen next to the cell wall.

In a hypertonic solution, vacuoles lose water, the cytoplasm shrinks (plasmolysis), and chloroplasts are seen in the center of the cell.

Figure 5.7 Osmosis in animal and plant cells. The arrows indicate the net movement of water molecules. To determine the net movement of water, compare the number of blue arrows (which are taking water molecules into the cell) versus the number of red arrows (which are taking water out of the cell). In an isotonic solution, a cell neither gains nor loses water; in a hypotonic solution, a cell gains water; and in a hypertonic solution, a cell loses water.

Tutorial Osmosis and Tonicity

concentration and the water concentration both inside and outside the cell are equal (Fig. 5.7), and therefore there is no net gain or loss of water. A 0.9% solution of the salt sodium chloride (NaCl) is known to be isotonic to red blood cells. Therefore, intravenous solutions medically administered usually have this tonicity. Terrestrial animals can usually take in either water or salt as needed to maintain the tonicity of their internal environment. Many animals living in an estuary, such as oysters, blue crabs, and some fishes, are able to cope with changes in the salinity (salt concentrations) of their environment using specialized kidneys, gills, and other structures.

Hypotonic Solution

Solutions that cause cells to swell, or even to burst, due to an intake of water are said to be **hypotonic solutions.** The prefix *hypo* means "less than" and refers to a solution with a lower concentration of solute (higher concentration of water) than inside the cell. If a cell is placed in a hypotonic solution, water enters the cell, because the lower cellular concentration of water prompts a net movement of water from the outside to the inside of the cell.

Any concentration of a salt solution lower than 0.9% is hypotonic to red blood cells. Animal cells placed in such a solution expand and sometimes burst because of the buildup of pressure. The term *cytolysis* is used to refer to disrupted cells. **Hemolysis** is the term used to describe cytolysis in red blood cells.

Animation Hemolysis and Crenation

The swelling of a plant cell in a hypotonic solution creates **turgor pressure.** When a plant cell is placed in a hypotonic solution, the cytoplasm expands, because the large central vacuole gains water and the plasma membrane pushes against the rigid cell wall. Unlike animal cells that have no cell wall, the plant cell does not burst, because the cell wall does not give way. Turgor pressure in plant cells is extremely important to the maintenance of the plant's erect position. If you forget to water your plants, they wilt due to decreased turgor pressure.

Organisms that live in fresh water have to avoid taking in too much water. Many protozoans, such as paramecia, have contractile vacuoles that rid the body of excess water. Freshwater fishes have well-developed kidneys that excrete a large volume of dilute urine. These fish still have to take in salts through their gills. Even though freshwater fishes are good osmoregulators, they would not be able to survive in either distilled water or a salty marine environment.

Hypertonic Solution

Solutions that cause cells to shrink or shrivel due to loss of water are said to be **hypertonic solutions.** The prefix *hyper* means "more than" and refers to a solution with a higher percentage of solute (lower concentration of water) outside the cell. If a cell is placed in a hypertonic solution, water leaves the cell; the net movement of water is from the inside to the outside of the cell.

Any concentration of a salt solution higher than 0.9% is hypertonic to red blood cells. If animal cells are placed in this solution, they shrink. The term **crenation** refers to red blood cells in this condition. Meats are sometimes preserved by salting them. The bacteria are not killed by the salt but by the lack of water in the meat.

When a plant cell is placed in a hypertonic solution, the plasma membrane pulls away from the cell wall as the large central vacuole loses water. This is an example of **plasmolysis,** a shrinking of the cytoplasm due to osmosis. The dead plants you may see along a salted roadside died because they were exposed to a hypertonic solution during the winter. Also, when salt water invades coastal marshes due to storms and human activities, coastal plants die.

Without roots to hold the soil, it washes into the sea, doing away with many acres of valuable wetlands.

Marine animals cope with their hypertonic environment in various ways that prevent them from losing excess water to the environment. Sharks increase or decrease urea in their blood until their blood is isotonic with the environment and, in this way, do not lose too much water. Marine fishes and other types of animals drink no water but excrete salts across their gills. Have you ever seen a marine turtle cry? It is ridding its body of salt by means of glands near the eye.

3D Animation Membrane Transport: Osmosis

Animation Effect of Tonicity on Cells

Facilitated Transport

The plasma membrane impedes the passage of all but a few substances. Yet biologically useful molecules are able to rapidly enter and exit the cell either by way of a channel protein or because of carrier proteins in the membrane. These transport proteins are specific; each can transport only a certain type of molecule or ion across the membrane. How carrier proteins function is not completely understood, but after a carrier combines with a molecule, the carrier is believed to undergo a conformational change in shape that moves the molecule across the membrane. Carrier proteins are utilized for both facilitated transport (movement with concentration gradient; requires no energy) and active transport (movement against concentration gradient; requires energy)(see Table 5.1).

Facilitated transport explains how molecules such as glucose and amino acids are rapidly transported across the plasma membrane. Whereas water moves through a channel protein, the passage of glucose and amino acids is facilitated by their reversible combination with carrier proteins, which transport them through the plasma membrane. These carrier proteins are specific. For example, various sugar molecules of identical size might be present inside or outside the cell, but glucose can cross the membrane hundreds of times faster than the other sugars. As stated earlier, this is the reason the membrane can be called selectively permeable.

A model for facilitated transport (Fig. 5.8) shows that after a carrier has assisted the movement of a molecule to the other side of the membrane, it is free to assist the passage of other solute molecules. Neither diffusion nor facilitated transport requires an expenditure of energy, because the molecules are moving down their concentration gradient.

Animation How Facilitated Diffusion Works

Check Your Progress 5.2

1. Explain why both osmosis and diffusion are passive processes.
2. Describe how a cell would react to a hypertonic or hypotonic solution.
3. Contrast diffusion with facilitated transport.

5.3 Active Transport Across a Membrane

Learning Outcomes

Upon completion of this section, you should be able to

1. Explain how active transport moves substances across a membrane.
2. Compare the energy requirements of passive and active transport.
3. Contrast the bulk transport of large and small substances into a cell.

At times, a cell may need to further increase a concentration gradient across a membrane in order to do more work. The process of **active transport** moves molecules against their concentration

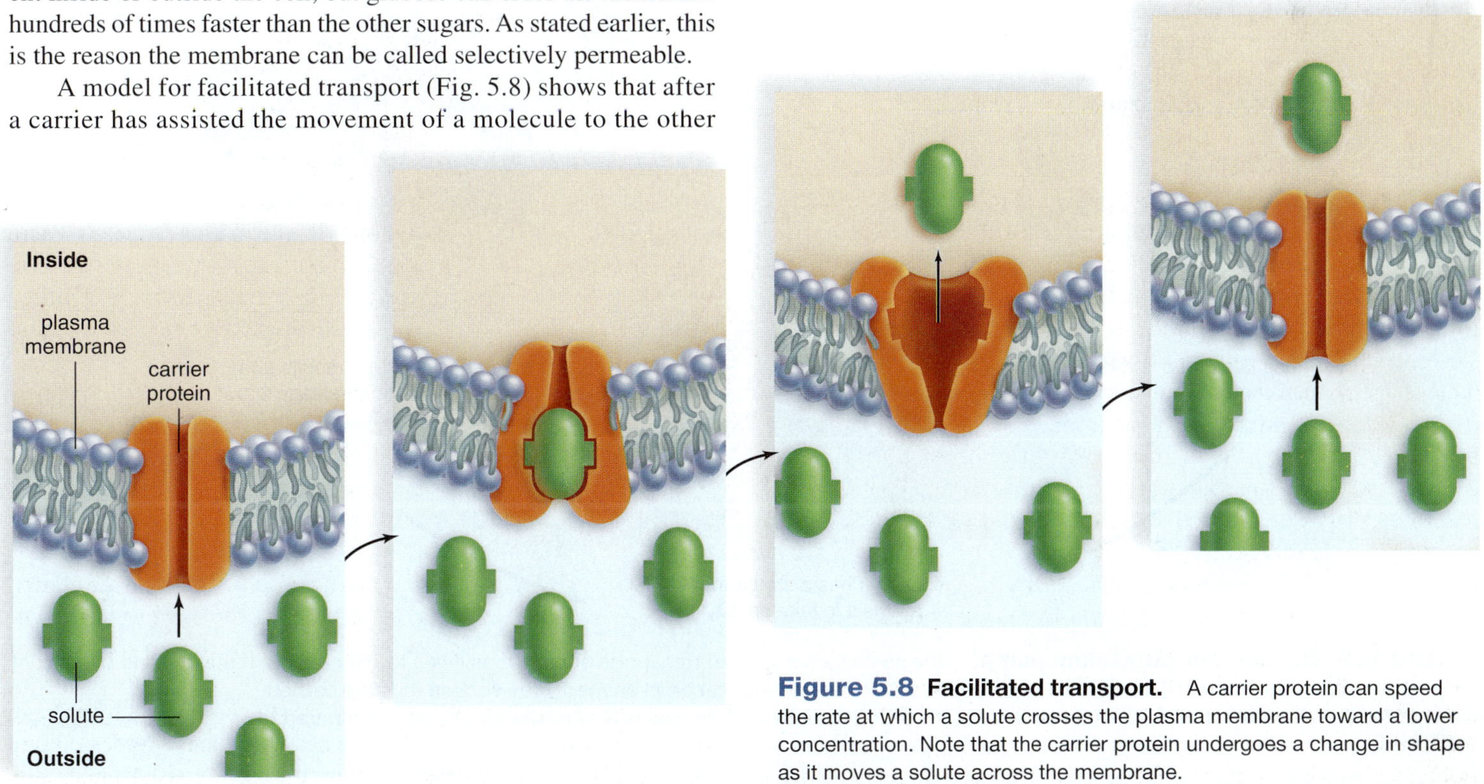

Figure 5.8 Facilitated transport. A carrier protein can speed the rate at which a solute crosses the plasma membrane toward a lower concentration. Note that the carrier protein undergoes a change in shape as it moves a solute across the membrane.

gradient. Active transport requires energy, usually in the form of ATP. For example, iodine collects in the cells of the thyroid gland; glucose is completely absorbed from the gut by the cells lining the digestive tract; and sodium can be almost completely withdrawn from urine by cells lining the kidney tubules. In each of these instances, molecules move from a lower to a higher concentration, exactly opposite the process of diffusion.

Carrier proteins and an expenditure of energy (ATP) are both needed to transport molecules against their concentration gradient. In this case, the ATP is needed for the carrier to combine with the substance to be transported. Therefore, it is not surprising that cells involved primarily in active transport, such as kidney cells, have a large number of mitochondria near membranes where active transport is occurring.

Proteins involved in active transport often are called pumps because, just as a water pump uses energy to move water against the force of gravity, proteins use energy to move a substance against its concentration gradient. One type of pump that is active in all animal cells, but is especially associated with nerve and muscle cells, moves sodium ions (Na^+) to the outside of the cell

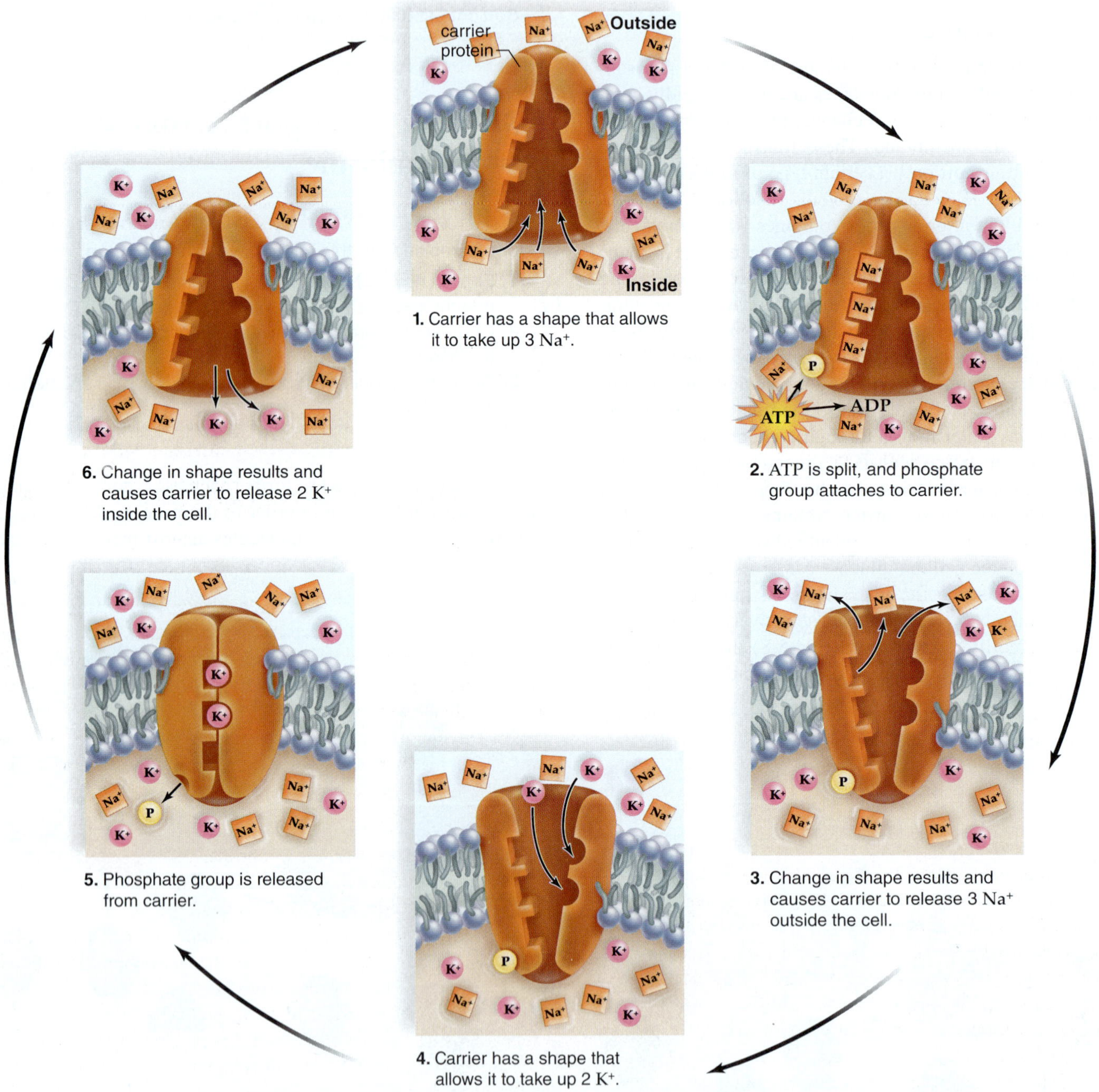

Figure 5.9 The sodium-potassium pump. The same carrier protein transports sodium ions (Na^+) to the outside of the cell and potassium ions (K^+) to the inside of the cell, because it undergoes an ATP-dependent change in shape. Three sodium ions are carried outward for every two potassium ions carried inward; therefore, the inside of the cell is less positively charged compared to the outside.

Tutorial Sodium-Potassium Pump

and potassium ions (K^+) to the inside of the cell. The transport of sodium and potassium are linked together through the same carrier protein, called a **sodium-potassium pump.**

Animation How the Sodium-Potassium Pump Works

The sodium-potassium carrier protein has an initial shape that allows it to bind three sodium ions. Phosphate from an ATP molecule is added to the carrier protein, and it changes shape; this shape change moves sodium across the membrane. The new shape is no longer compatible with binding to the sodium, which falls away.

The new shape, however, is compatible with picking up two potassium ions, which bind to their sites. As the phosphate that was added from ATP in an earlier step leaves, the carrier protein assumes its original shape, and the two potassium ions are released inside the cell (Fig. 5.9). The cotransport of three sodium and two potassium creates not only a solute gradient but also an electrical gradient across the plasma membrane.

Animation Cotransport

The passage of salt (NaCl) across a plasma membrane is of primary importance to most cells. The chloride ion (Cl^-) usually crosses the plasma membrane because it is attracted by positively charged sodium ions (Na^+). First sodium ions are pumped across a membrane, and then chloride ions simply diffuse through channels that allow their passage.

As noted in Figure 5.2*a*, the genetic disorder cystic fibrosis results from a faulty chloride channel protein. When chloride is unable to exit a cell, water stays behind. The lack of water outside the cells causes abnormally thick mucus in the bronchial tubes and pancreatic ducts, thus interfering with the function of the lungs and pancreas.

3D Animation Membrane Transport: Active Transport

Bulk Transport

How do large molecules such as proteins, polysaccharides, or nucleic acids enter and exit a cell? These molecules are too large to be transported by carrier proteins, so they are instead transported into and out of the cell by vesicles. Membrane vesicles formed around macromolecules require an expenditure of cellular energy, but the cost is worth it, because each vesicle keeps its cargo from mixing with molecules within the cytoplasm that could alter the cell's function. Generally, substances can exit a cell through exocytosis, and enter a cell through endocytosis.

Exocytosis

During **exocytosis,** an intracellular vesicle fuses with the plasma membrane as secretion occurs (Fig. 5.10). Hormones, neurotransmitters, and digestive enzymes are secreted from cells in this manner. The Golgi body often produces the vesicles that carry these cell products to the membrane. During exocytosis, the membrane of the vesicle becomes a part of the plasma membrane, because both are nonpolar. Adding additional vesicle membrane to the plasma membrane can enlarge the cell and is a part of growth in some cells. The proteins released from the vesicle may adhere to the cell surface or become incorporated into an extracellular matrix.

Cells of particular organs are specialized to produce and export molecules. For example, pancreatic cells produce digestive enzymes or insulin, and anterior pituitary cells produce growth hormone, among other hormones. In these cells, secretory vesicles accumulate near the plasma membrane, and the vesicles release their contents only when the cell is stimulated by a signal received at the plasma membrane. A rise in blood sugar, for example, signals pancreatic cells to release the hormone insulin. This is called regulated secretion, because vesicles fuse with the plasma membrane only when the needs of the body trigger it to do so.

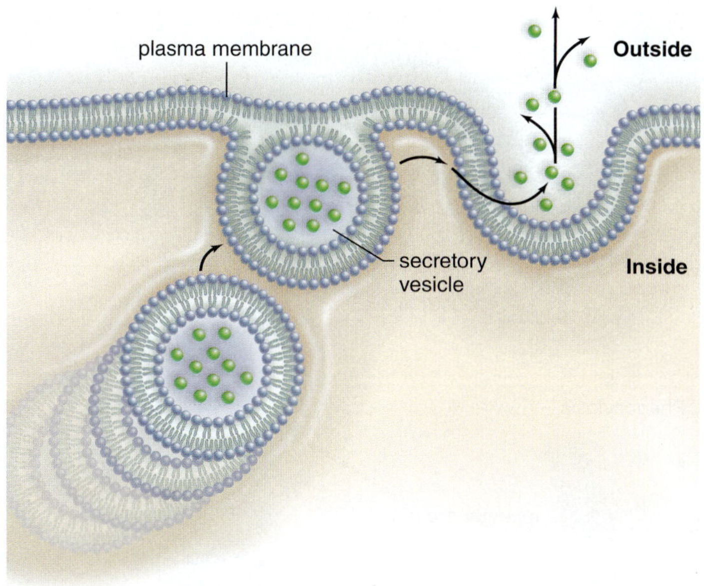

Figure 5.10 **Exocytosis.** Exocytosis secretes or deposits substances on the outside of the cell.

Endocytosis

During **endocytosis,** cells take in substances by forming vesicles around the material. A portion of the plasma membrane invaginates to envelop the substance, and then the membrane pinches off to form an intracellular vesicle. Endocytosis occurs in one of three ways, as illustrated in Figure 5.11. Phagocytosis transports large substances, such as a virus, and pinocytosis transports small substances, such as a macromolecule, into a cell. Receptor-mediated endocytosis is a special form of pinocytosis.

Phagocytosis. When the material taken in by endocytosis is large, such as a food particle or another cell, the process is called **phagocytosis** (Gk. *phagein,* "to eat"). Phagocytosis is common in single-celled organisms, such as amoebas (Fig. 5.11*a*). It also occurs in humans. Certain types of human white blood cells are amoeboid—that is, they are mobile like an amoeba, and they can engulf debris such as worn-out red blood cells or viruses. When an endocytic vesicle fuses with a lysosome, digestion occurs. Later in this text you will see that this process is a necessary and preliminary step toward the development of our immunity to bacterial diseases.

Pinocytosis. **Pinocytosis** (Gk. *pinein,* "to drink") occurs when vesicles form around a liquid or around very small particles (Fig. 5.11*b*). Blood cells, cells that line the kidney tubules or the

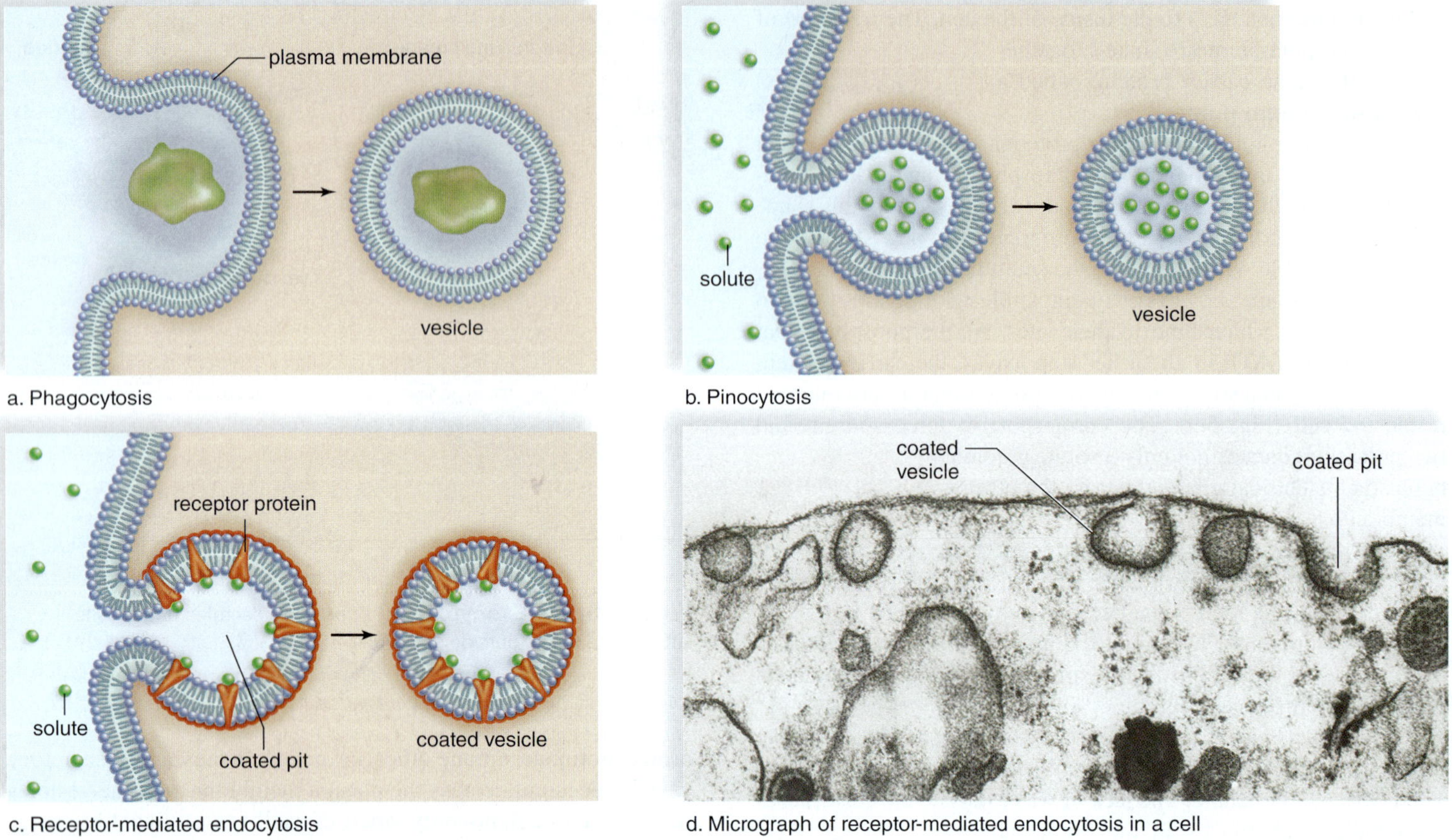

Figure 5.11 Three methods of endocytosis. **a.** Phagocytosis occurs when the substance to be transported into the cell is large. Digestion occurs when the resulting vacuole fuses with a lysosome. **b.** Pinocytosis occurs when a macromolecule, such as a polypeptide, is transported into the cell. **c,d.** Receptor-mediated endocytosis is a form of pinocytosis. Molecules first bind to specific receptor proteins, which migrate to or are already in a coated pit. The coated vesicle that forms contains the molecules and their receptors.

intestinal wall, and plant root cells all use pinocytosis to ingest substances.

Whereas phagocytosis can be seen with the light microscope, the electron microscope must be used to observe pinocytic vesicles, which are no larger than 0.1–0.2 μm. Still, pinocytosis involves a significant amount of the plasma membrane, because it occurs continuously. Cells do not shrink in size, because the loss of plasma membrane due to pinocytosis is balanced by the occurrence of exocytosis.

Receptor-Mediated Endocytosis. **Receptor-mediated endocytosis** is a form of pinocytosis that is quite specific, because it uses a receptor protein to recognize compatible molecules and take them into the cell. Molecules such as vitamins, peptide hormones, or lipoproteins can bind to specific receptors, found in special locations in the plasma membrane (Fig. 5.11*c*). This location is called a coated pit, because there is a layer of protein on the cytoplasmic side of the pit. Once formed, the vesicle is uncoated and may fuse with a lysosome. When empty, a used vesicle fuses with the plasma membrane, and the receptors return to their former location.

Receptor-mediated endocytosis is selective and much more efficient than ordinary pinocytosis. It is involved in uptake and in the transfer and exchange of substances between cells. Such exchanges take place when substances move from maternal blood into fetal blood at the placenta, for example.

Animation Endocytosis and Exocytosis

The importance of receptor-mediated endocytosis is demonstrated by a genetic disorder called familial hypercholesterolemia. Cholesterol is transported in blood by a complex of lipids and proteins called low-density lipoprotein (LDL). Ordinarily, body cells take up LDL when LDL receptors gather in a coated pit. But in some individuals, the LDL receptor is unable to bind properly to the coated pit, and the cells are unable to take up cholesterol. Instead, cholesterol accumulates in the walls of arterial blood vessels, leading to high blood pressure, occluded (blocked) arteries, and heart attacks.

Check Your Progress 5.3

1. Compare facilitated transport with active transport.
2. Explain why active transport requires energy.
3. Summarize why a cell would use bulk transport rather than active transport.

5.4 Modification of Cell Surfaces

Learning Outcomes

Upon completion of this section, you should be able to

1. Explain the role of the extracellular matrix in an animal cell.
2. Compare the structure and function of adhesion, tight, and gap junctions in animals.
3. Explain the role of plasmodesmata in plants.

Most cells do not live isolated from other cells. Rather, they live and interact within an external environment that can dramatically affect cell structure and function. This extracellular environment is made up of large molecules produced by nearby cells and secreted from their membranes. In plants, prokaryotes, fungi, and most algae, the extracellular environment is a fairly rigid cell wall, which is consistent with a somewhat sedentary lifestyle. Animals, which tend to be more active, have a more varied extracellular environment, which can change depending on the tissue type.

Cell Surfaces in Animals

Here we examine two types of animal cell surface features: (1) the extracellular matrix outside cells and (2) the junctions between some types of cells. Both of these can connect to the cytoskeleton and contribute to communication between cells, and therefore tissue formation.

Extracellular Matrix

A protective **extracellular matrix (ECM)** is a meshwork of proteins and polysaccharides in close association with the cell that produced them (Fig. 5.12). Collagen and elastin fibers are two well-known structural proteins in the ECM; collagen resists stretching, and elastin gives the ECM resilience. Fibronectin is an adhesive protein (colored green in Figure 5.12) that binds to a protein, called integrin, in the plasma membrane. Integrins are integral membrane proteins that connect to fibronectin externally and to the actin cytoskeleton internally. Through its connections with both the ECM and the cytoskeleton, integrin plays a role in cell signaling, permitting the ECM to influence the activities of the cytoskeleton and, therefore, the shape and activities of the cell.

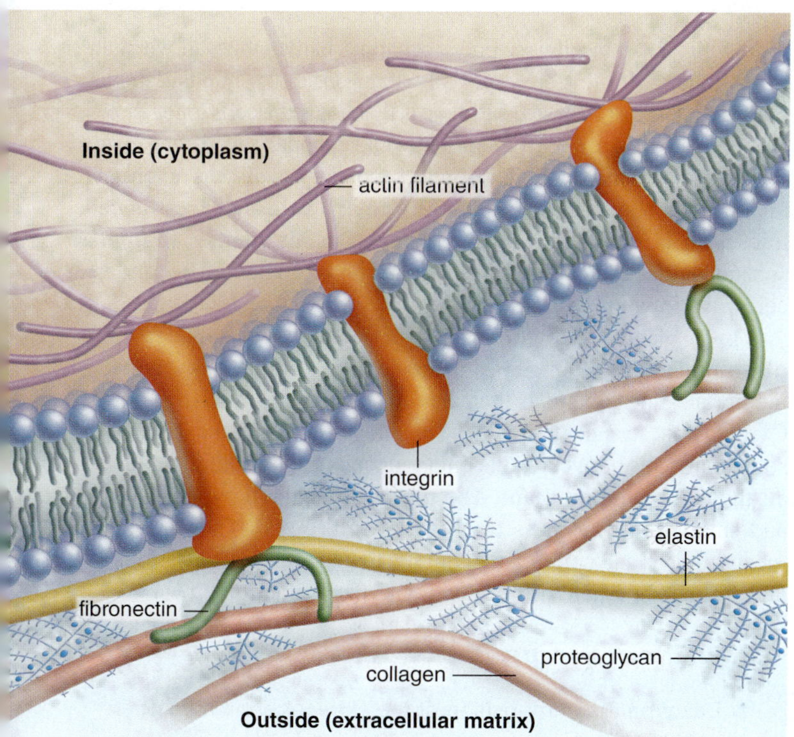

Figure 5.12 Animal cell extracellular matrix. In the extracellular matrix, collagen and elastin have a support function, while fibronectins bind to integrin, and in this way assist communication between the ECM and the cytoskeleton.

Amino sugars in the ECM form multiple polysaccharides that attach to a protein and are, therefore, called proteoglycans. Proteoglycans, in turn, attach to a very long, centrally placed polysaccharide. The entire structure, which looks like an enormous bottle brush, resists compression of the extracellular matrix. Proteoglycans assist cell signaling when they regulate the passage of molecules through the ECM to the plasma membrane, where receptors are located. During development, they help bring about differentiation by guiding cell migration along collagen fibers to specific locations. Thus, the ECM has a dynamic role in all aspects of a cell's behavior.

Later on, in the discussion of tissues (Chapter 31), you'll see that the extracellular matrix varies in quantity and consistency. It can be quite flexible, as in loose connective tissue; semiflexible, as in cartilage; or rock solid, as in bone. The extracellular matrix of bone is hard because, in addition to the components mentioned, mineral salts (notably, calcium salts) are deposited outside the cell.

The proportion of cells to ECM also varies. In the small intestine, for example, epithelial cells constitute the majority of the tissue, and the ECM is a thin sheet beneath the cells. In bone, the ECM makes up most of the tissue, with comparatively fewer cells.

Junctions Between Cells

Certain tissues of vertebrate animals are known to have junctions between their cells that allow them to behave in a coordinated manner. Three types of junctions are shown in Figure 5.13.

Adhesion junctions mechanically attach adjacent cells. One example of an adhesion junction is the *desmosome.* In desmosomes, internal cytoplasmic plaques firmly attach to the intermediate filament cytoskeleton within each cell and are joined between cells by integral membrane proteins called cadherins. The result is a sturdy but flexible sheet of cells. In some organs—such as the heart, stomach, and bladder, where tissues get stretched—desmosomes hold the cells together. At a *hemidesmosome,* the intermediate filaments of the cytoskeleton are attached to the ECM through integrin proteins. Adhesion junctions are the most common type of intercellular junction between skin cells.

Another type of adhesion junction between adjacent cells is the **tight junction,** which brings cells even closer than desmosomes. Tight junction proteins actually connect plasma membranes between adjacent cells together, producing a zipperlike fastening. Tissues that serve as barriers are held together by tight junctions; in the intestine, the digestive juices stay out of the rest of the body, and in the kidneys the urine stays within kidney tubules, because the cells are joined by tight junctions.

A **gap junction** allows cells to communicate. A gap junction is formed when two identical plasma membrane channels join. The channel of each cell is lined by six plasma membrane proteins. A gap junction lends strength to the cells, but it also allows small molecules and ions to pass between them. Gap junctions are

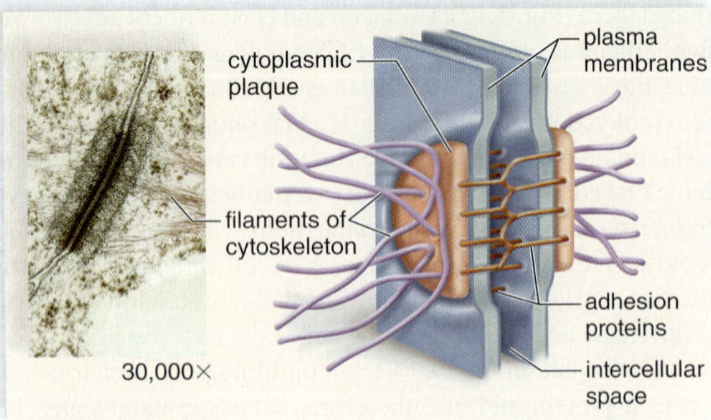

a. Adhesion junction

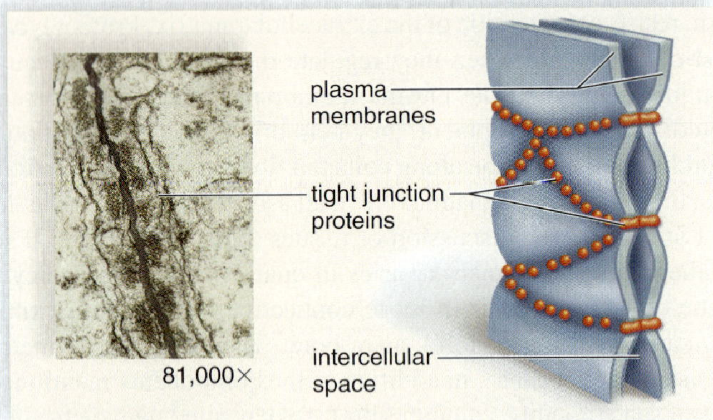

b. Tight junction

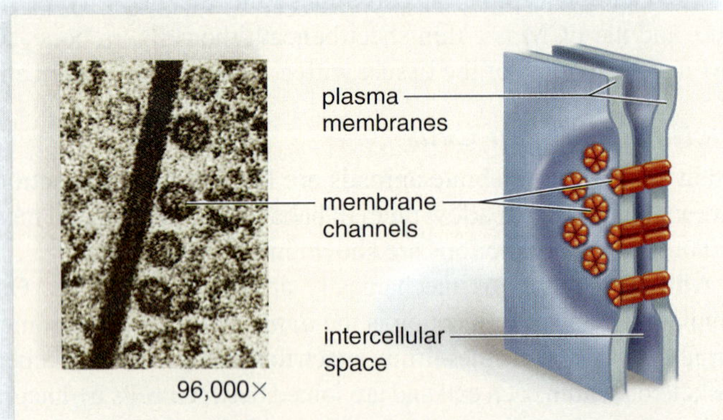

c. Gap junction

Figure 5.13 Junctions between cells of the intestinal wall. **a.** In adhesion junctions such as a desmosome, adhesive proteins connect two cells. **b.** Tight junctions between cells form an impermeable barrier because their adjacent plasma membranes are joined and don't allow molecules to pass. **c.** Gap junctions allow communication between two cells because adjacent plasma membrane channels are joined.

important in heart muscle and smooth muscle, because they permit the flow of ions that is required for the cells to contract as a unit.

Plant Cell Walls

In addition to a plasma membrane, plant cells are surrounded by a porous **cell wall** that varies in thickness, depending on the function of the cell.

All plant cells have a primary cell wall. The primary cell wall contains cellulose fibrils, in which microfibrils are held together by noncellulose substances. Pectins allow the wall to stretch when the cell is growing, and noncellulose polysaccharides harden the wall when the cell is mature. Pectins are especially abundant in the middle lamella, which is a layer of adhesive substances that holds the cells together.

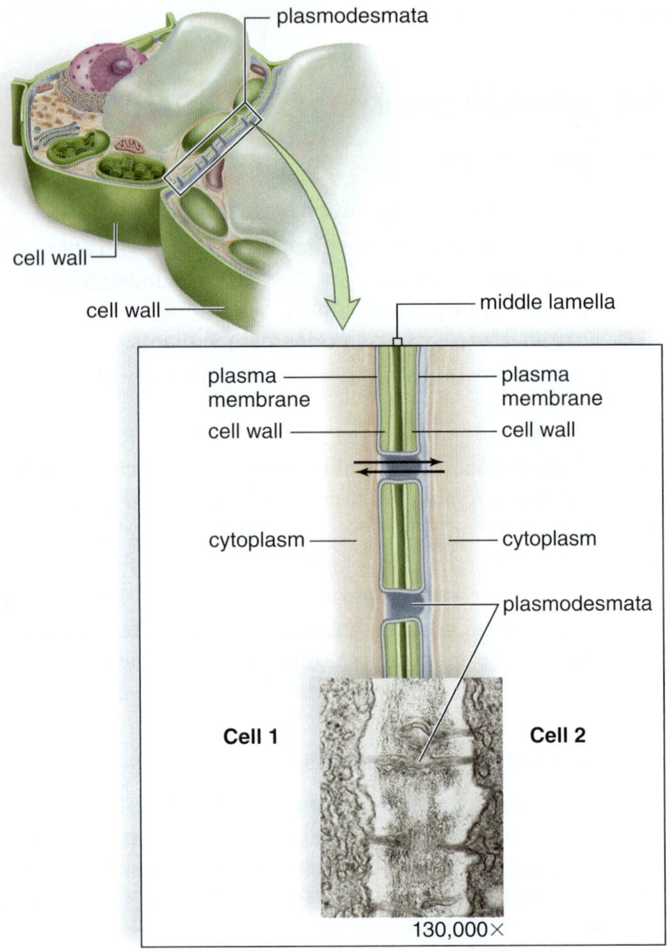

Figure 5.14 Plasmodesmata. Plant cells are joined by membrane-lined channels that contain cytoplasm. Water and small molecules can pass from cell to cell.

Some cells in woody plants have a secondary wall that forms inside the primary cell wall. The secondary wall has a greater quantity of cellulose fibrils than the primary wall, and layers of cellulose fibrils are laid down at right angles to one another. Lignin, a substance that adds strength, is a common ingredient of secondary cell walls in woody plants.

In a plant, the cytoplasm of living cells is connected by **plasmodesmata** (sing., plasmodesma), numerous narrow, membrane-lined channels that pass through the cell wall (Fig. 5.14). Cytoplasmic strands within these channels allow direct exchange of some materials between adjacent plant cells and eventually connect all the cells within a plant. The plasmodesmata allow only water and small solutes to pass freely from cell to cell. This limitation means that plant cells can maintain their own concentrations of larger substances and differentiate into particular cell types.

Check Your Progress 5.4

1. Describe the composition of the extracellular matrix of an animal cell.
2. Explain why a cell would be connected by a tight junction, rather than a gap junction or an adhesion junction.
3. Explain the role of the plasmodesmata in plant cells.

CONNECTING the CONCEPTS with the THEMES

Evolution

- The plasma membrane is appropriately called the gatekeeper of the cell because it maintains the integrity of the cell and stands guard over what enters and leaves.
- Over many generations, membranes have enabled cells to become specialized in their functions because they allow incompatible processes to occur simultaneously. This leads to greater biological fitness within particular environments.
- Glycoproteins present in membranes have also evolved over time, allowing a number of important cellular functions, such as immunity, to arise.

Nature of Science

- The progression in our knowledge about the plasma membrane illustrates how science works. The knowledge we have today will be amended and expanded by new investigations.
- Basic science and membrane research is applied to treating disease and promoting health. To know that the plasma membrane is malfunctioning in a person who has diabetes or cystic fibrosis is a first step toward treating or curing these conditions.
- Cancer is sometimes due to membrane receptor proteins that signal the cell to divide even when it should not. Identification and treatment of cancer can target these receptors.

Biological Systems

- Our ability to understand the functioning of the plasma membrane is dependent on a working knowledge of the molecules that make up the cell and how they interact.
- Membranes that surround various organelles in the cell interact in a coordinated fashion, bringing greater efficiency to the entire cell system.

MEDIA STUDY TOOLS http://connect.mheducation.com

SMARTBOOK® Study smart. SmartBook helps you maximize your study time by identifying what you know and don't know. The recharge feature in SmartBook helps you prepare for the test by identifying the material that you are most likely to forget.

connect BIOLOGY Find even more resources to learn the chapter concepts including animations, tutorial videos, and interactive practice questions.

MP3 Files	Animations	3D Animations	Tutorials
5.1 Membrane Structure • Membrane Transport **5.2** Diffusion • Osmosis	**5.2** How Diffusion Works • How Osmosis Works • Hemolysis and Crenation • Effects of Tonicity on Cells • How Facilitated Diffusion Works **5.3** How the Sodium-Potassium Pump Works • Cotransport • Endocytosis and Exocytosis	**5.1** Membrane Transport: Lipid Bilayer **5.2** Membrane Transport: Diffusion • Membrane Transport: Osmosis **5.3** Membrane Transport: Active Transport	**5.2** Osmosis and Tonicity **5.3** Sodium-Potassium Pump

SUMMARIZE

5.1 Plasma Membrane Structure and Function

The **fluid mosaic model** describes the interactions of the components of the plasma membrane. In the lipid bilayer, phospholipids are arranged with their hydrophilic (polar) heads adjacent to water and their hydrophobic (nonpolar) tails buried in the interior. The lipid bilayer has the consistency of oil but acts as a barrier to the entrance and exit of most biological molecules. Membrane **glycolipids** and **glycoproteins** are involved in marking the cell as belonging to a particular individual and tissue.

A variety of proteins may be embedded within the plasma membrane. The hydrophobic portion of an integral protein lies in the lipid bilayer of the plasma membrane, and the hydrophilic portion lies at the surfaces. Some functions include

- **Channel proteins:** Form a channel through the membrane to move molecules. Examples are the aquaporins that allow the rapid movement of water across a membrane.
- **Carrier proteins:** Interact with a target molecule, causing a change in shape, which allows the movement of the molecule through the membrane
- **Cell recognition proteins:** Glycoproteins involved in cellular identification
- **Receptor proteins:** Detect specific molecules in the environment and invoke a cellular response
- **Enzymatic proteins:** Carry out metabolic reactions within the plasma membrane
- **Junction proteins:** Form connections between cells

The plasma membrane of a cell is **selectively permeable** and isolates the cell from the external environment. In doing so, the membrane establishes a **concentration gradient for** each type of molecule. These molecules are then moved across the membrane by active, passive, or **bulk transport.**

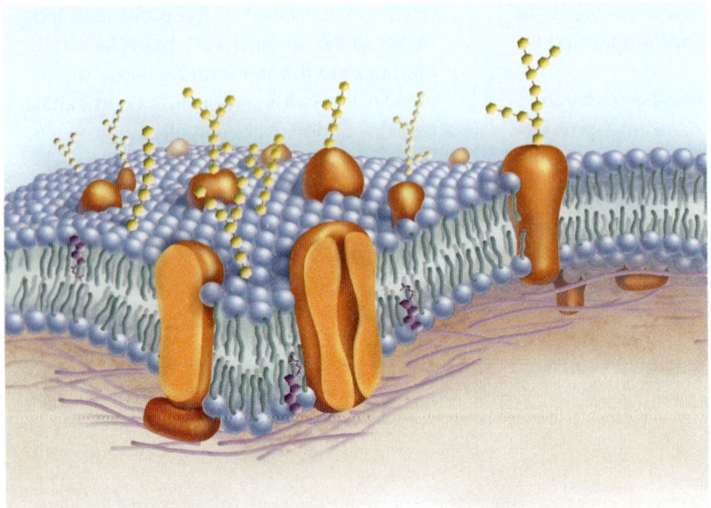

5.2 Passive Transport Across a Membrane

Diffusion is the passive movement of molecules from a higher to lower concentration. In a **solution,** the **solute** diffuses within the **solvent** until it is evenly distributed. Some molecules (lipid-soluble compounds, water, and gases) simply diffuse across the membrane. No metabolic energy is required for diffusion to occur.

The diffusion of water across a selectively permeable membrane is called **osmosis.** During osmosis, **osmotic pressure** directs water to move across the membrane into the area of higher solute (less water) content per volume. **Tonicity** is a measure of the relative solute concentration. When cells are in an **isotonic solution,** they neither gain nor lose water. When cells are in a **hypotonic solution,** they gain water. This accounts for the **turgor pressure** of plants and **hemolysis** in red blood cells. When cells are in a **hypertonic solution,** they lose water, which can cause **crenation** and **plasmolysis** (Table 5.2).

Other molecules are transported across the membrane either by a channel protein or by carrier proteins that span the membrane. During facilitated transport, a substance moves down its concentration gradient. No energy is required.

5.3 Active Transport Across a Membrane

During **active transport,** a carrier protein acts as a pump that causes a substance to move against its concentration gradient. One example is the **sodium-potassium pump** that carries Na^+ to the outside of the cell and K^+ to the inside of the cell. Energy in the form of ATP molecules is required for active transport to occur.

Larger substances can enter and exit a membrane by exocytosis and endocytosis. **Exocytosis** involves secretion. **Endocytosis** includes **phagocytosis, pinocytosis,** and **receptor-mediated endocytosis.** Receptor-mediated endocytosis makes use of receptor proteins in the plasma membrane. Once a specific solute binds to receptors, a coated pit becomes a coated vesicle. After losing the coat, the vesicle can join with the lysosome or, after discharging the substance, the receptor-containing vesicle can fuse with the plasma membrane.

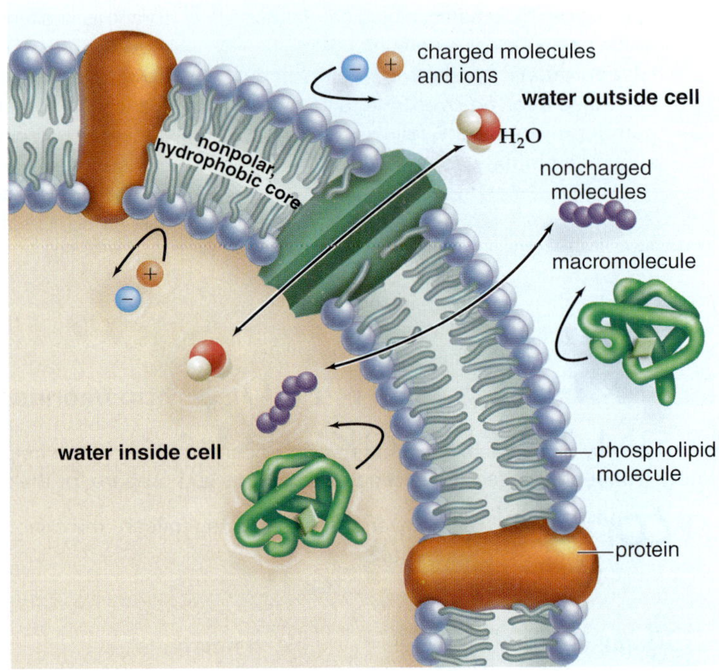

5.4 Modification of Cell Surfaces

Animal cells have an **extracellular matrix** (**ECM**), which influences their shape and behavior. The amount and character of the ECM vary by tissue type. Some animal cells have junction proteins that join them to other cells of the same tissue. **Adhesion junctions** and **tight junctions** help hold cells together; **gap junctions** allow the passage of small molecules between cells.

Plant cells have a freely permeable **cell wall,** with cellulose as its main component. Also, plant cells are joined by narrow, membrane-lined channels, called **plasmodesmata,** that span the cell wall and contain strands of cytoplasm that allow materials to pass from one cell to another.

Table 5.2 Effect of Osmosis on a Cell

	Concentrations			
Tonicity of Solution	Solute	Water	Net Movement of Water	Effect on Cell
Isotonic	Same as cell	Same as cell	None	None
Hypotonic	Less than cell	More than cell	Cell gains water	Swells, turgor pressure
Hypertonic	More than cell	Less than cell	Cell loses water	Shrinks, plasmolysis

ASSESS

Choose the best answer for each question.

5.1 Plasma Membrane Structure and Function

1. In the fluid-mosaic model, the fluid properties are associated with the nature of the ______ and the mosaic pattern is established by the ______.
   a. nucleic acids; phospholipids
   b. phospholipids; embedded proteins
   c. embedded proteins; cholesterol
   d. phospholipids; nucleic acids
2. Which of the following is not a function of proteins present in the plasma membrane?
   a. Proteins assist the passage of materials into the cell.
   b. Proteins interact with and recognize other cells.
   c. Proteins bind with specific hormones.
   d. Proteins carry out specific metabolic reactions.
   e. Proteins produce lipid molecules.
3. The carbohydrate chains projecting from the plasma membrane are involved in
   a. adhesion between cells.
   b. reception of molecules.
   c. cell-to-cell recognition.
   d. All of these are correct.

5.2 Passive Transport Across a Membrane

4. When a cell is placed in a hypotonic solution,
   a. solute exits the cell to equalize the concentration on both sides of the membrane.
   b. water exits the cell toward the area of lower solute concentration.
   c. water enters the cell toward the area of higher solute concentration.
   d. there is no net movement of water or solutes.
5. When a cell is placed in a hypertonic solution,
   a. solute exits the cell to equalize the concentration on both sides of the membrane.
   b. water exits the cell toward the area of lower solute concentration.
   c. water exits the cell toward the area of higher solute concentration.
   d. there is no net movement of water or solute.
6. A plant cell that is placed in a hypertonic solution would experience
   a. crenation.
   b. an increase in turgor pressure.
   c. plasmolysis.
   d. no net changes.
7. Which of the following is incorrect regarding facilitated diffusion?
   a. It is a passive process.
   b. It allows the movement of molecules from areas of low concentration to areas of high concentration.
   c. It may use either channel or carrier proteins.
   d. It allows the rapid transport of glucose across the membrane.

5.3 Active Transport Across a Membrane

8. The sodium-potassium pump
   a. helps establish an electrochemical gradient across the membrane.
   b. concentrates sodium on the outside of the membrane.
   c. uses a carrier protein and chemical energy.
   d. is present in the plasma membrane.
   e. All of these are correct.
9. Which of the following processes is involved in the bulk transport of molecules out of the cell?
   a. phagocytosis
   b. pinocytosis
   c. receptor-mediated endocytosis
   d. exocytosis
   e. None of these are correct.
10. Which process uses special proteins on the surface of the membrane to identify specific molecules for transport into the cell?
    a. phagocytosis
    b. pinocytosis
    c. receptor-mediated endocytosis
    d. exocytosis

5.4 Modification of Cell Surfaces

11. The extracellular matrix
    a. assists in the movement of substances across the plasma membrane.
    b. prevents the loss of water when cells are placed in a hypertonic solution.
    c. has numerous functions that affect the shape and activities of the cell that produced it.
    d. contains the junctions that sometimes occur between cells.
    e. All of these are correct.
12. Which of the following junctions allows for cytoplasm-to-cytoplasm communication between cells?
    a. adhesion junctions
    b. tight junctions
    c. gap junctions
    d. None of these are correct.

ENGAGE

LEARNSMART LABS®

The following LearnSmart Labs contain exercises that are related to the content of this chapter:

- Diffusion
- Osmosis
- How Enzymes Function

THINKING SCIENTIFICALLY

1. As mentioned, cystic fibrosis is a genetic disorder caused by a defective membrane transport protein. The defective protein closes chloride channels in membranes, preventing chloride from being exported out of cells. This results in the development of a thick mucus on the outer surfaces of cells. This mucus clogs the ducts that carry digestive enzymes from the pancreas to the small intestine, clogs the airways in the lungs, and promotes lung infections. Why do you think the defective protein results in a thick, sticky mucus outside the cells, instead of a loose, fluid covering?
2. Winter wheat is planted in the early fall, grows over the winter when the weather is colder, and is harvested in the spring. As the temperature drops, the makeup of the plasma membrane of winter wheat changes. Unsaturated fatty acids replace saturated fatty acids in the phospholipids of the membrane. Why is this a suitable adaptation?

6

Metabolism: Energy and Enzymes

Both the impala and the cheetah depend on metabolic reactions to convert the solar energy captured by photosynthesizers.

Chapter Outline

6.1 Cells and the Flow of Energy 101

6.2 Metabolic Reactions and Energy Transformations 103

6.3 Metabolic Pathways and Enzymes 105

6.4 Oxidation-Reduction Reactions and Metabolism 109

Before You Begin

Before beginning this chapter, take a few moments to review the following discussions.

Section 3.4 What is the relationship between protein structure and function?

Section 3.5 What are the structure and general functions of the nucleic acids?

Section 4.7 What are the roles of the chloroplasts and mitochondria in a cell?

All life on Earth ultimately depends on the flow of energy coming from the sun. Photosynthesizing grasses on an African plain provide impalas with organic building blocks and the energy they need to avoid being caught by a cheetah. Eating impalas provides cheetahs with food and the energy they need to be quick enough to catch impalas.

You, like the cheetah, consume plants and animals that get their energy either directly or indirectly from the sun. Solar energy is concentrated enough to allow plants to photosynthesize and make biological molecules, which in turn provide a continual supply of food for you and other creatures within the biosphere.

In the impala, the cheetah, and your body, metabolic reactions convert food into forms that are needed to power your activities and build muscles, tissues, and organs. Energy is critical to metabolism and enzymatic reactions. Living organisms are bound by certain energy laws. For example, as you digest food and break the bonds in vegetables and meat, some of the energy is used for work, and some energy escapes into the environment as heat.

Many of these metabolic reactions would either not occur or occur too slowly without the help of metabolic assistants called enzymes. Without enzymes, living organisms would not be able to carry out many of the general characteristics of life. As we will see, enzymes play an important role in how our cells, and ultimately our bodies, function.

As you read through this chapter, think about the following questions:

1. What general laws apply to energy?
2. What forms of energy are used by a cell?
3. How is energy used to drive biological processes within cells?

Following *the* Themes

CHAPTER 6 METABOLISM: ENERGY AND ENZYMES

UNIT 1 THE CELL

Evolution	Cells have evolved to metabolize energy in order to support the basic characteristics of life.
Nature of Science	Understanding the principles of metabolism helps us understand how cells and our bodies function.
Biological Systems	Energy flows through all biological systems, creating the ability to do work, although eventually all energy is lost as heat.

6.1 Cells and the Flow of Energy

Learning Outcomes

Upon completion of this section, you should be able to

1. Compare potential and kinetic energy.
2. Describe the first and second laws of thermodynamics.
3. Examine how the organization and structure of living organisms are related to heat and entropy.

To maintain their structural organization and carry out metabolic activities, cells—and organisms comprised of cells—need a constant supply of energy. **Energy** is defined as the ability to do work or bring about a change. The general characteristics of life, including growth, development, metabolism, and reproduction, all require energy.

Organic nutrients, made by photosynthesizing producers (algae, plants, and some bacteria), directly provide organisms with energy by capturing energy from sunlight. Considering that producers use light energy to produce organic nutrients, the majority of life on Earth is ultimately dependent on solar energy.

Forms of Energy

Energy occurs in two forms: kinetic and potential energy. **Kinetic energy** is the energy of motion, as when water flows over a waterfall, a ball rolls down a hill, or a moose walks through grass. **Potential energy** is stored energy whose capacity to accomplish work is not being used at the moment. The food we eat has potential energy, because the energy stored in chemical bonds can be converted into various types of kinetic energy. Food is a form of potential energy called **chemical energy,** because it is composed of organic molecules, such as carbohydrates, proteins, and fat. When a moose walks, it converts chemical energy into a type of kinetic energy called **mechanical energy** (Fig. 6.1).

Two Laws of Thermodynamics

In nature, energy flows in biological systems. Figure 6.1 illustrates the flow of energy in a terrestrial ecosystem. Plants capture only a small portion of solar energy, and much of it dissipates as **heat.** When plants photosynthesize and then make use of the food they produce, more heat results. Even with this considerable heat loss, there is enough remaining to sustain a moose and the other organisms in an ecosystem. As organisms metabolize nutrient molecules, all the captured solar energy eventually dissipates as heat. Therefore, we can see that energy flows through the ecosystem and does not cycle within it.

Two **laws of thermodynamics,** formulated by early energy researchers, explain why energy flows through ecosystems and through cells:

> The first law of thermodynamics—the law of conservation of energy—states that energy cannot be created or destroyed, but it can be changed from one form to another.

When leaf cells photosynthesize, they use solar energy to form carbohydrate molecules from carbon dioxide gas and water. Carbohydrates are energy-rich molecules, because they have many bonds that store energy; carbon dioxide and water are energy-poor molecules, because of the relative lack of bonds. Not all of the captured solar energy becomes carbohydrates; some becomes heat.

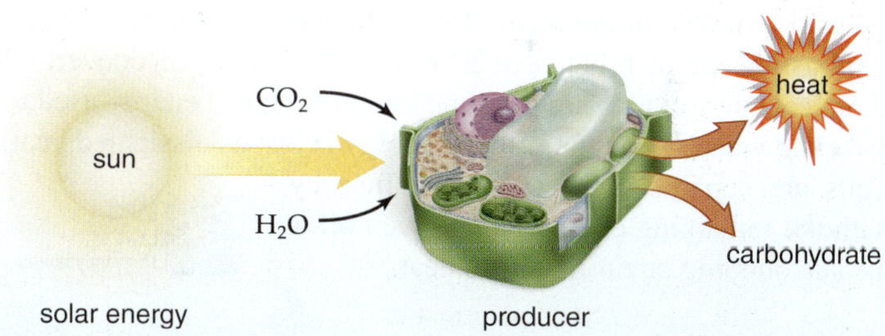

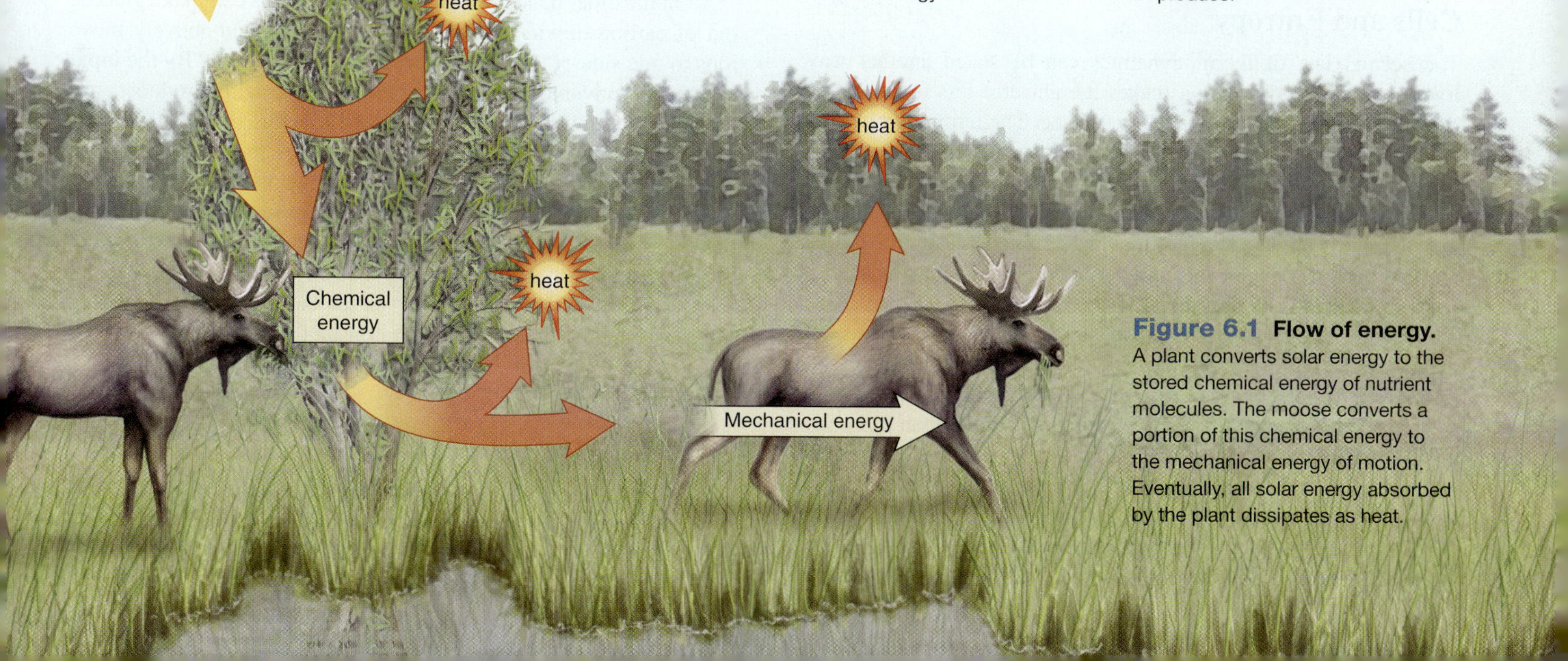

Figure 6.1 Flow of energy. A plant converts solar energy to the stored chemical energy of nutrient molecules. The moose converts a portion of this chemical energy to the mechanical energy of motion. Eventually, all solar energy absorbed by the plant dissipates as heat.

Obviously, plant cells do not create the energy they use to produce carbohydrate molecules; that energy comes from the sun. Is any energy destroyed? No, because the heat the plant cells give off is also a form of energy. Similarly, as a moose walks, it uses the potential energy stored in carbohydrates to kinetically power its muscles. As its cells use this energy, none is destroyed, but each energy exchange produces some heat, which dissipates into the environment.

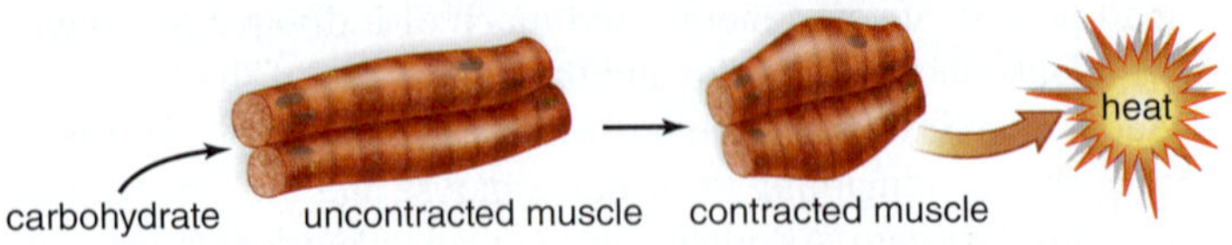

The second law of thermodynamics therefore applies to living systems:

> The second law of thermodynamics states that energy cannot be changed from one form to another without a loss of usable energy.

In our example, this law is upheld because some of the solar energy taken in by the plant and some of the chemical energy within the nutrient molecules taken in by the moose becomes heat. When heat dissipates into the environment, it is no longer usable—that is, it is not available to do work. Each energy transformation moves us closer to a condition where all usable forms of energy become heat that is lost to the environment. Heat that dissipates into the environment cannot be captured and converted to one of the other forms of energy.

As a result of the second law of thermodynamics, no process requiring a conversion of energy is ever 100% efficient. Much of the energy is lost in the form of heat. In automobiles, the internal combustion engine is between 20% and 30% efficient in converting chemical energy stored in gasoline into mechanical energy used to drive the wheels. The majority of energy is lost as dissipated heat. Cells are capable of about 40% efficiency, with the remaining energy being given off to the surrounding environment as heat.

MP3 Laws of Thermodynamics

Cells and Entropy

The second law of thermodynamics can be stated another way: Every energy transformation makes the universe less organized, or structured, and more disordered, or chaotic. The term **entropy** (Gk. *entrope,* "a turning inward") is used to indicate the relative amount of disorganization. Because the processes that occur in cells are energy transformations, the second law means that every process that occurs in cells always does so in a way that increases the total entropy of the universe. The second law means that each cellular process makes less energy available to do useful work in the future.

Figure 6.2 shows two processes that occur in cells. The second law of thermodynamics tells us that glucose tends to break apart into carbon dioxide and water over time. Why? Because glucose is more organized and structured, and therefore less stable, than its breakdown products. Also, hydrogen ions on one side of a membrane tend to move to the other side unless they are prevented from doing so, because when they are distributed randomly, entropy has increased. As an analogy, you know from experience that a neat room is more organized but less stable than a messy room, which is disorganized but more stable. Energy is required to return a messy room to a more organized, or neat, state.

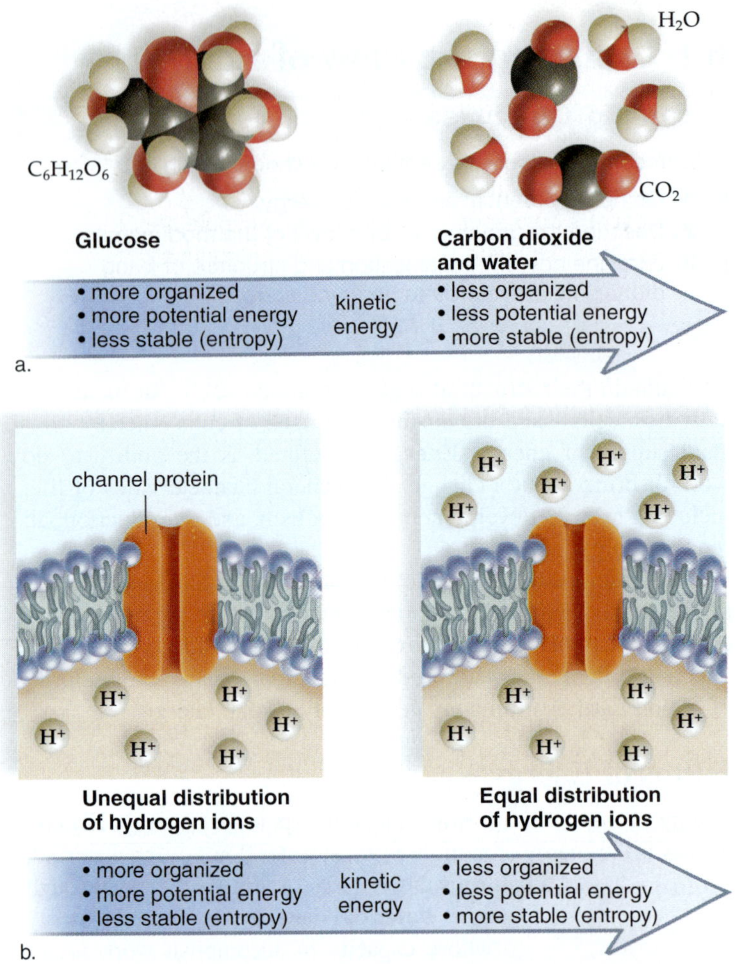

Figure 6.2 Cells and entropy. The second law of thermodynamics tells us that (**a**) glucose, which is more organized, tends to break down to carbon dioxide and water, which are less organized. **b.** Similarly, hydrogen ions (H^+) on one side of a membrane tend to move to the other side, so that the ions are randomly distributed. Both processes result in an increase in entropy.

On the other hand, you know that some cells can make glucose out of carbon dioxide and water, and all cells can actively move ions to one side of the membrane. How do they do it? By the input of energy from an outside source. Photosynthesizing producers use energy from sunlight to create organized structure in biological molecules. Organisms that consume producers are then able to use this potential energy to kinetically drive their own metabolic processes. Thus, living organisms depend on a constant supply of energy ultimately provided by the sun. The ultimate fate of all solar energy in the biosphere is to become randomized in the universe as heat. A living cell can function because it serves as a temporary repository of order, purchased at the cost of a constant flow of energy.

Check Your Progress 6.1

1. Provide an example of a conversion from potential to kinetic energy.
2. Summarize how the first and second laws of thermodynamics relate to cells.
3. Explain the importance of entropy to a living system.

6.2 Metabolic Reactions and Energy Transformations

Learning Outcomes

Upon completion of this section, you should be able to

1. Explain how the ATP cycle involves both endergonic and exergonic reactions.
2. Describe how energy is stored in a molecule of ATP.
3. Examine how cells use ATP to drive energetically unfavorable reactions.

All living organisms maintain their structure and function through chemical reactions. **Metabolism** is the sum of all the chemical reactions that occur in a cell. **Reactants** are substances that participate in a reaction, while **products** are substances that form as a result of a reaction. In the reaction A + B $\rightarrow$ C + D, A and B are the reactants, while C and D are the products. Whether a reaction occurs spontaneously—that is, without an input of energy—depends on how much energy is left after the reaction. Using the concept of entropy, or disorder, a reaction occurs spontaneously if it increases the entropy of the universe.

In cell biology, which occurs on a small scale, we are less concerned about the entire universe, which is vast. In such specific instances, cell biologists use the concept of free energy instead of entropy. **Free energy** (also called "delta G," or ΔG) is the amount of energy left to do work after a chemical reaction has occurred. The change in free energy after a reaction occurs is determined by subtracting the free energy content of the reactants from that of the products. A negative result ($-\Delta G$) means that the products have less free energy than the reactants, and the reaction will occur spontaneously. In our reaction, if C and D have less free energy than A and B, then the reaction occurs without additional input of energy.

Metabolism includes both spontaneous reactions and energy-requiring reactions. **Exergonic reactions** are spontaneous and release energy, while **endergonic reactions** require an input of energy to occur. In the body, many reactions, such as protein and carbohydrate synthesis, are endergonic. For these nonspontaneous reactions to occur during metabolism, they must be coupled with exergonic reactions, such that a net spontaneous reaction results. Many biological processes use ATP as an energy carrier between exergonic and endergonic reactions.

ATP: Energy for Cells

ATP (**adenosine triphosphate**) is the common energy currency of cells; when cells require energy, they use ATP. A sedentary oak tree, a flying bat, and a human require vast amounts of ATP. The more active the organism, the greater the demand for ATP. However, cells do not keep a large store of ATP molecules on hand. Instead, they constantly regenerate ATP using **ADP** (**adenosine diphosphate**) and inorganic phosphate, Ⓟ. This is called the ATP cycle (Fig. 6.3). This cycle is powered by the breakdown of glucose and other biomolecules during cellular respiration. However, according to the second law of thermodynamics, this process is not very efficient. Only 39% of the free energy stored in the chemical bonds of a glucose molecule is transformed to ATP; the rest is lost as heat.

There are many biological advantages to the use of ATP as an energy carrier in living systems. ATP provides a common and universal energy currency because it can be used in many different types of reactions. Also, when ATP is converted to energy, ADP, and Ⓟ, the amount of energy released is sufficient to efficiently power most biological functions. In addition, ATP breakdown can be coupled to endergonic reactions in such a way that it minimizes energy loss.

Structure of ATP

ATP is a nucleotide composed of the nitrogen-containing base adenine and the 5-carbon sugar ribose (together called adenosine) and three phosphate groups. The three phosphates of ATP repel each other, creating instability and potential energy (Fig. 6.3). ATP is called a "high-energy" molecule, because a phosphate group can be easily removed. Under cellular conditions, the amount of energy released when ATP is hydrolyzed to ADP + Ⓟ is about 7.3 kcal per **mole.** A mole is a unit of measurement in chemistry that is equal to the molecular weight of a molecule expressed in grams.

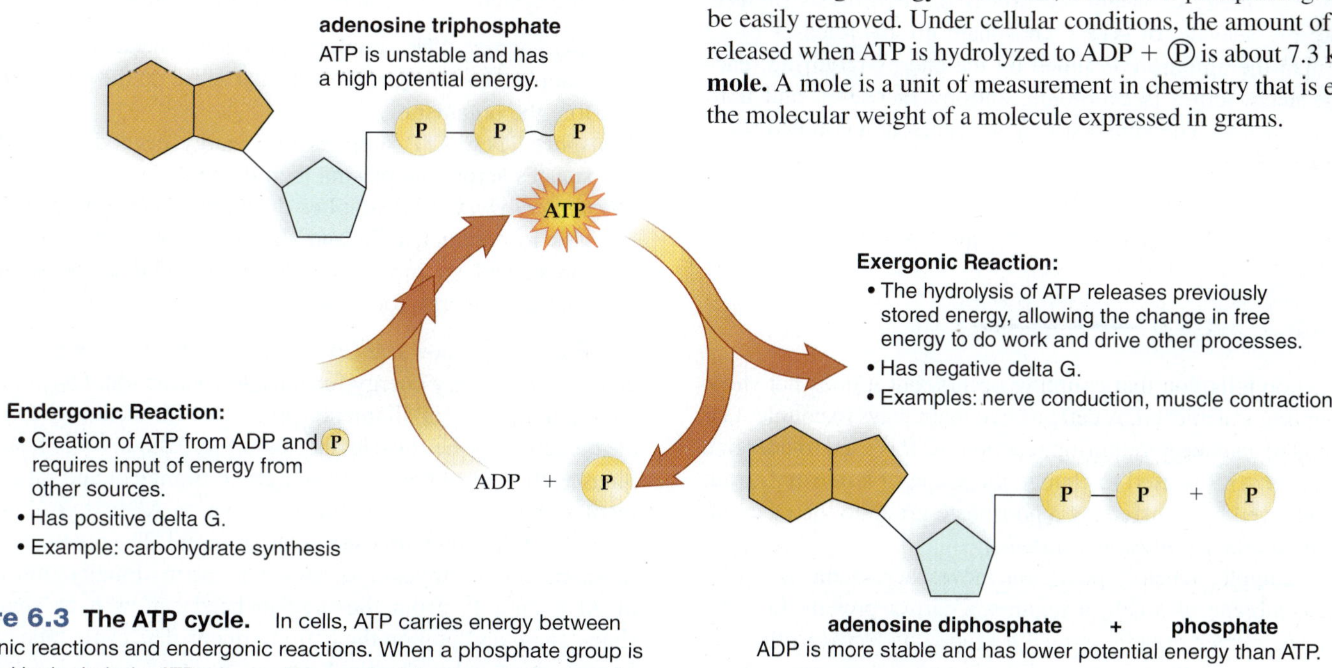

Figure 6.3 The ATP cycle. In cells, ATP carries energy between exergonic reactions and endergonic reactions. When a phosphate group is removed by hydrolysis, ATP releases the appropriate amount of energy for most metabolic reactions.

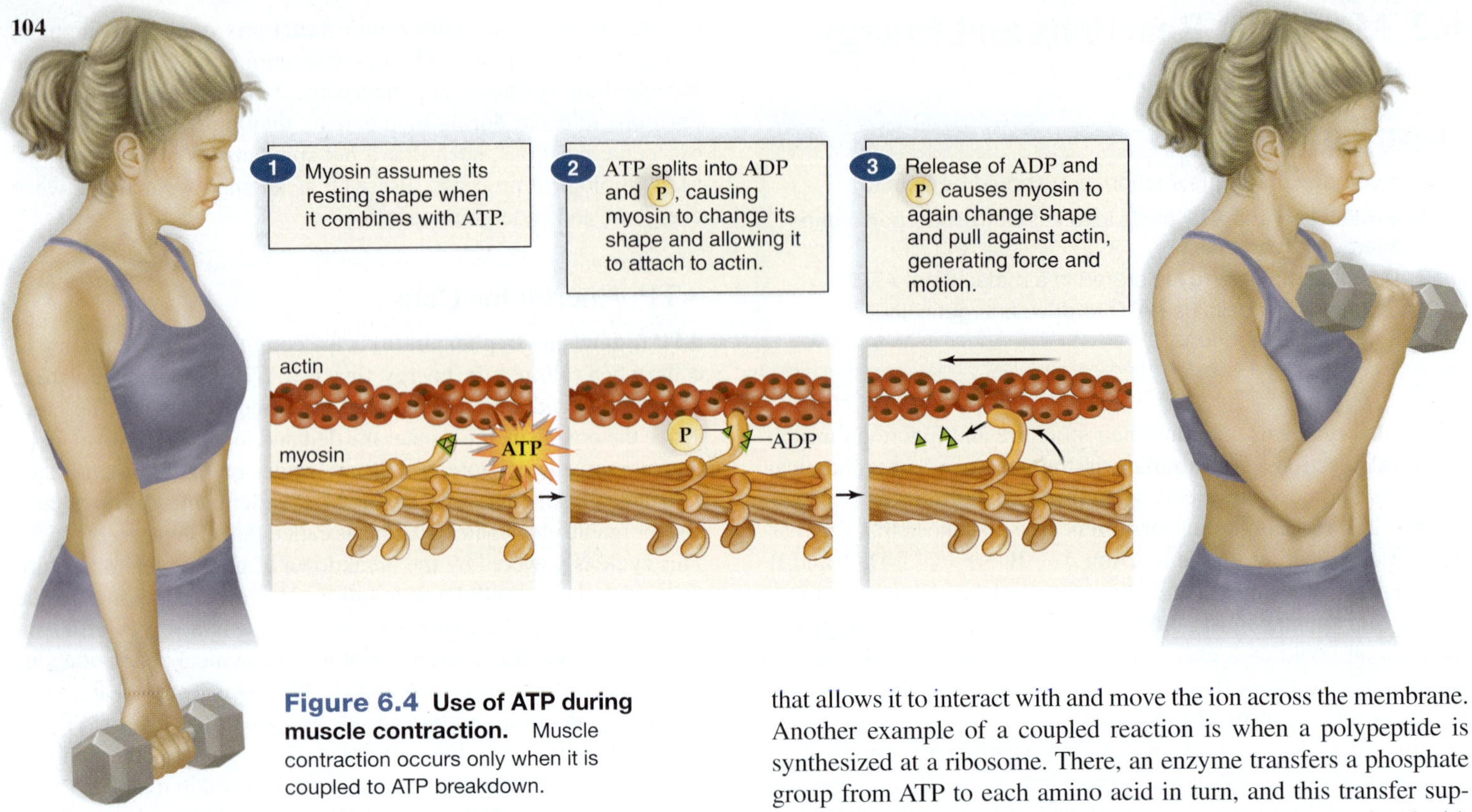

Figure 6.4 Use of ATP during muscle contraction. Muscle contraction occurs only when it is coupled to ATP breakdown.

Coupled Reactions

How can the energy released by ATP hydrolysis be transferred to an endergonic reaction that requires energy and therefore would not ordinarily occur? In other words, how does ATP act as a carrier of chemical energy? How can that energy be transferred efficiently to an energetically unfavorable reaction?

The answer is that ATP breakdown is *coupled* to the energy-requiring reaction, such that both the energetically favorable and unfavorable reactions occur in the same place, at the same time. Usually, the energy-releasing reaction is the hydrolysis of ATP. Because the cleavage of ATP's phosphate groups releases more energy than the amount consumed by the energy-requiring reaction, the net reaction is exergonic, entropy increases, and both reactions proceed. The simplest way to represent a coupled reaction is like this:

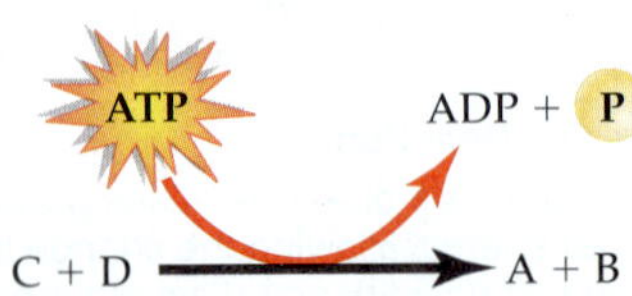

This reaction tells you that coupling occurs, but it does not show how coupling is achieved. A cell has two main ways to couple ATP hydrolysis to an energy-requiring reaction: ATP is used to energize a reactant, or ATP is used to change the shape of a reactant. Both can be achieved by transferring a phosphate group to the reactant, so that the product is *phosphorylated.*

For example, when a polar ion moves across the nonpolar plasma membrane of a cell, it requires a carrier protein. In order to make the carrier protein assume a shape conducive to the ion, ATP is hydrolyzed; then, instead of the last phosphate group floating away, an enzyme attaches it to a carrier protein. The negatively charged phosphate causes the protein to undergo a change in shape that allows it to interact with and move the ion across the membrane. Another example of a coupled reaction is when a polypeptide is synthesized at a ribosome. There, an enzyme transfers a phosphate group from ATP to each amino acid in turn, and this transfer supplies the energy needed to overcome the energy cost associated with bonding one amino acid to another.

Through coupled reactions, ATP drives forward the energetically unfavorable processes that must occur to create the high degree of order and structure essential for life. Macromolecules must be made and organized to form cells and tissues; the internal composition of the cell and the organism must be maintained; and movement of cellular organelles and the organism must occur if life is to continue.

Functions of ATP in Cells

In living systems, ATP can be used for the following:

Chemical work. ATP supplies the energy needed to synthesize macromolecules (anabolism) that make up the cell, and therefore the organism.

Transport work. ATP supplies the energy needed to pump substances across the plasma membrane.

Mechanical work. ATP supplies the energy needed to permit muscles to contract, cilia and flagella to beat, chromosomes to move, and so forth. In most cases, ATP is the immediate source of energy for these processes.

Figure 6.4 shows an example of how ATP hydrolysis provides the necessary energy for muscle contraction. During muscle contraction, myosin filaments pull actin filaments to the center of the cell, and the muscle shortens. First, the myosin head combines with ATP (three connected green triangles) and takes on its resting shape. Next, ATP breaks down to ADP (two connected green triangles) plus Ⓟ (one green triangle). The resulting change in shape allows myosin to attach to actin. Finally, the release of ADP and Ⓟ from the myosin head causes it to change its shape again and pull on the actin filament. The cycle then repeats. During this cycle, chemical energy has been transformed to mechanical energy, and entropy has increased.

Animation Breakdown of ATP and Cross-Bridge Movement

Check Your Progress 6.2

1. Explain why ATP is an effective short-term energy storage molecule.
2. Summarize the ATP cycle.
3. Examine how transferring a phosphate from ATP changes a molecule's structure and function.

6.3 Metabolic Pathways and Enzymes

Learning Outcomes

Upon completion of this section, you should be able to

1. Explain the purpose of a metabolic pathway and how enzymes help regulate it.
2. Recognize how enzymes influence the activation energy rates of a chemical reaction.
3. Distinguish between conditions and factors that affect an enzyme's rate of reaction.

The chemical reactions that constitute metabolism would not easily occur without the use of organic catalysts called enzymes. An **enzyme** is a protein molecule that speeds a chemical reaction without itself being affected by the reaction. Enzymes allow reactions to occur under mild conditions, and they regulate metabolism, partly by eliminating nonspecific side reactions.

Not all enzymes are proteins. **Ribozymes,** which are made of RNA instead of proteins, can also serve as biological catalysts. Ribozymes are involved in the synthesis of RNA and the synthesis of proteins at ribosomes.

Chemical reactions do not occur haphazardly in healthy cells; they are usually part of a **metabolic pathway,** a series of linked reactions. Metabolic pathways begin with a particular reactant and end with a final product. Many specific steps can be involved in a metabolic pathway, and each step is a chemical reaction catalyzed by an enzyme.The reactants in an enzymatic reaction are called the **substrates** for that enzyme. The substrates for the first reaction are converted into products, and those products then serve as the substrates for the next enzyme-catalyzed reaction. One reaction leads to the next reaction in an organized, highly regulated manner.

This arrangement makes it possible for one pathway to interact with several others, because different pathways may have several molecules in common. Also, metabolic pathways are useful for releasing and capturing small increments of molecular energy rather than releasing it all at once. Ultimately, enzymes in metabolic pathways enable cells to regulate and respond to changing environmental conditions.

The diagram below illustrates a simple metabolic pathway:

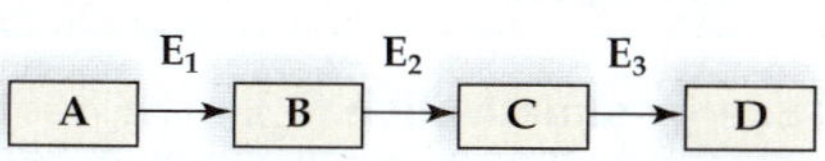

In this diagram, A is the substrate for E_1, and B is the product. Now B becomes the substrate for E_2, and C is the product. This process continues until the final product, D, forms. Any one of the molecules (A–D) in this metabolic pathway could also be a reactant in another pathway. Many of the metabolic pathways in living organisms are highly branched, and interactions between metabolic pathways are very common. It is important to note that each step in the metabolic pathway can be regulated because each step requires an enzyme. The specificity of enzymes allows the regulation of metabolism. The presence of particular enzymes helps determine which metabolic pathways are operative. In addition, some substrates can produce more than one type of product, depending on which pathway is open to them. Therefore, which enzyme is present determines which product is produced, as well as determining the direction of metabolism, without several alternative pathways being activated. As we will see, the ability to regulate these pathways gives our cells fine control over how they respond in a changing environment and helps maximize cell efficiency.

Animation Biochemical Pathways

Enzyme-Substrate Complex

In most instances, only one small part of the enzyme, called the **active site,** associates directly with the substrate (Fig. 6.5).

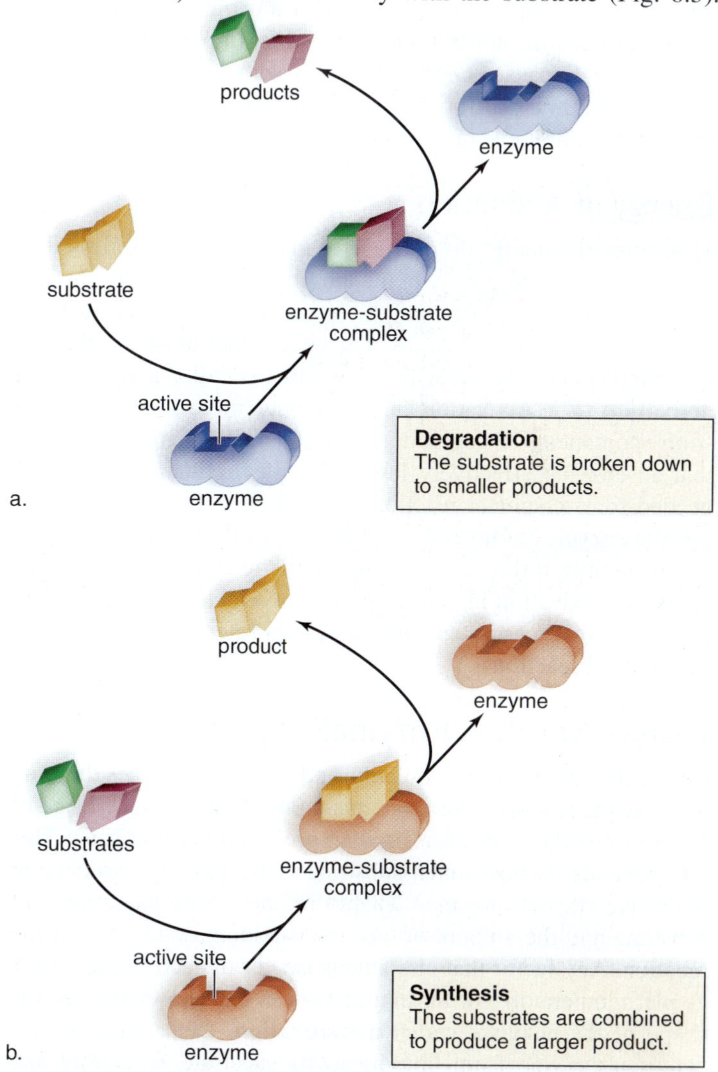

Figure 6.5 Enzymatic actions. Enzymes have an active site where the substrate(s) specifically fit together, so that the reaction will occur. Following the reaction, the product or products are released, and the enzyme is free to act again. Certain enzymes carry out (**a**) degradation, and others carry out (**b**) synthesis.

In the active site, the enzyme and substrate are positioned in such a way that they more easily fit together, seemingly as a key fits a lock. However, an active site differs from a lock and key because it undergoes a slight change in shape to accommodate the substrate(s). This is called the **induced fit model,** because the enzyme is induced to undergo a slight alteration to achieve optimum fit for the substrates.

The change in shape of the active site facilitates the reaction that now occurs. After the reaction has been completed, the product or products are released, and the active site returns to its original state, ready to bind to another substrate molecule. Only a small amount of enzyme is actually needed in a cell, because enzymes are not used up by the reaction; they merely enable it to happen more quickly.

Some enzymes do more than simply form a complex with their substrate(s); they participate in the reaction. Trypsin digests protein by breaking peptide bonds. The active site of trypsin contains three amino acids with *R* groups that actually interact with members of the peptide bond—first to break the bond and then to introduce the components of water. This illustrates that the formation of the enzyme-substrate complex is very important in speeding the reaction. Because enzymes bind only with their substrates, they are sometimes named for their substrates and usually end in *-ase*. For example, lipase is involved in hydrolyzing lipids.

Energy of Activation

Molecules frequently do not react with one another unless they are activated in some way. In the lab, for example, in the absence of an enzyme, molecules may be heated in order to increase the number of effective collisions. The energy that must be added to cause molecules to react with one another is called the **energy of activation** (E_a). Activation energy is essential to keep molecules from spontaneously degrading within the cell. Figure 6.6 shows that an enzyme effectively lowers E_a, thus reducing the energy needed for a chemical reaction to begin. It is important to note that the enzyme has no effect on the energy content of the product; rather, it only influences the rate of the reaction. Reducing the energy of activation increases the rate at which the reaction may occur. For this reason, enzymes are often referred to as catalysts of chemical reactions.

Factors Affecting Enzymatic Speed

Generally, enzymes work quickly, and in some instances they can increase the reaction rate more than 10 million times. The rate of a reaction is the amount of product produced per unit time. This rate depends on how much substrate is available to associate at the active sites of enzymes. Therefore, increasing the amount of substrate and the amount of enzyme can increase the rate of the reaction. Any factor that alters the shape of the active site—such as pH, temperature, or an inhibitor—can cause a change in the shape of the enzyme, called **denaturation.** Denatauration prevents an enzyme from binding to its substrate efficiently and thus can decrease the rate of a reaction. Thus, enzymes require specific conditions to be met in order to be fully operational. In fact, some enzymes require additional molecules called cofactors, which help speed the rate of the reaction, because they help bind the substrate to the active site, or they participate in the reaction at the active site.

Substrate Concentration

Molecules must collide to react. Generally, enzyme activity increases as substrate concentration increases, because there are more collisions between substrate molecules and the enzyme. As more substrate molecules fill active sites, more product results per unit of time. But when the active sites are filled almost continuously with substrate, the rate of the reaction can no longer increase. Maximum rate has been reached.

Just as the amount of substrate can increase or limit the rate of an enzymatic reaction, so the amount of active enzyme can also increase or limit the rate of an enzymatic reaction. Therefore, sufficient concentrations of substrate and enzymes are necessary to achieve maximum reaction rate.

Optimal pH

Each enzyme also has an optimal pH at which the reaction rate is highest. Figure 6.7 shows the optimal pH for the enzymes pepsin and trypsin. At their respective pH values, each enzyme can maintain its normal structural configuration, which enables optimum function. The globular structure of an enzyme is dependent on interactions, such as hydrogen bonding, between *R* groups. A change in pH can alter the ionization of these side chains, causing the enzyme to denature. Under extreme conditions of pH, the enzyme loses its structure and becomes inactive.

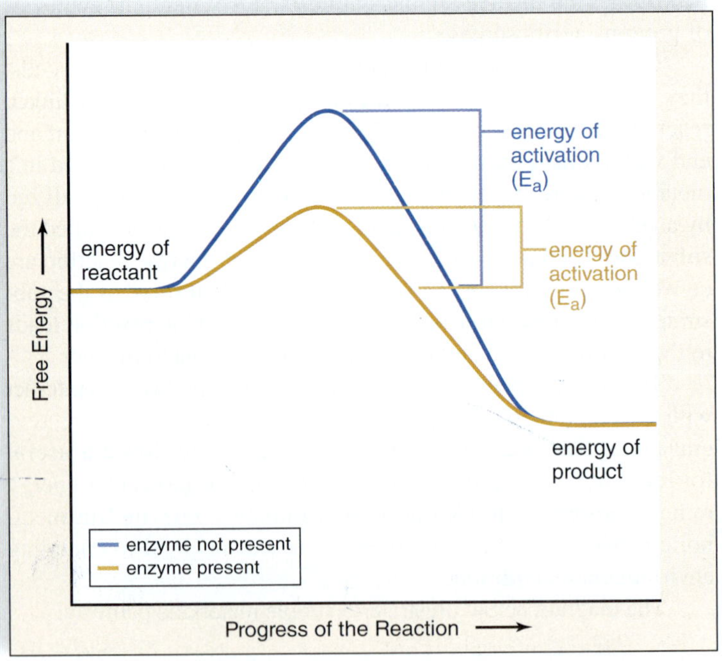

Figure 6.6 Energy of activation (E_a). Enzymes speed the rate of reactions, because they lower the amount of energy required for the reactants to activate. Even spontaneous reactions like this one, in which the energy of the product is less than the energy of the reactant, speed up when an enzyme is present.

Tutorial Energy of Activation

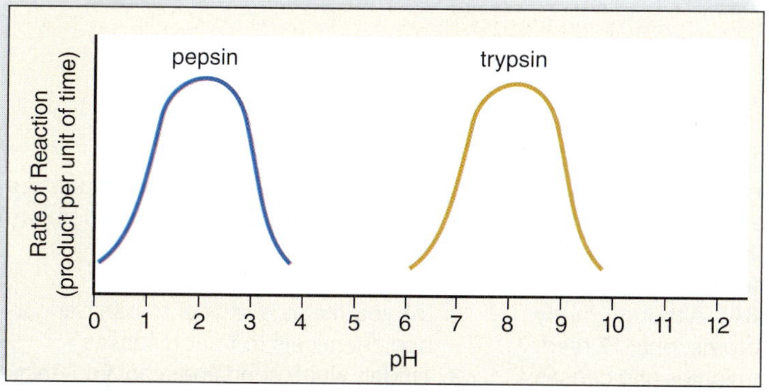

Figure 6.7 The effect of pH on rate of reaction. The optimal pH for pepsin, an enzyme that acts in the stomach, is about 2, while the optimal pH for trypsin, an enzyme that acts in the small intestine, is about 8. Enzyme shape is best maintained at the optimal pH, which allows it to function best and bind with its substrates.

Temperature

Typically, as temperature rises, enzyme activity increases (Fig. 6.8*a*). This occurs because warmer temperatures cause more effective collisions between enzyme and substrate. The body temperature of an animal seems to affect whether it is normally active or inactive (Fig. 6.8*b, c*). It has been suggested that mammals are more prevalent today than reptiles because they maintain a warm internal temperature that allows their enzymes to work at a rapid rate.

In the laboratory and in your body, if the temperature rises beyond a certain point, enzyme activity eventually levels out and then declines rapidly, because the enzyme is denatured. Exceptions to this generalization do occur. For example, some prokaryotes can live in hot springs because their enzymes do not denature. These organisms are responsible for the brilliant colors of the hot springs. Another exception involves the coat color of animals. Siamese cats have inherited a mutation that causes an enzyme to be active only at cooler body temperatures. The enzyme's activity causes the cooler regions of the body—the face, ears, legs, and tail—to be dark in color (Fig. 6.9). The coat color pattern in several other animals can be explained similarly.

Enzyme Cofactors and Coenzymes

Many enzymes require the presence of an inorganic ion or a nonprotein organic molecule at the active site in order to work properly; these necessary ions or molecules are called **cofactors** (Fig. 6.10). The inorganic ions include metals such as copper, zinc, or iron. The nonprotein organic molecules are called **coenzymes.** These cofactors participate in the reaction and may even accept or contribute atoms to the reactions. Examples of these are **NAD^+ (nicotinamide adenine dinucleotide)**, **FAD (flavin adenine dinucleotide)**, and **$NADP^+$ (nicotinamide adenine dinucleotide phosphate)**, each of which plays a significant role in either cellular respiration or photosynthesis.

Vitamins are often components of coenzymes. **Vitamins** are relatively small, organic molecules that are required in trace amounts in our diet and in the diets of other animals for synthesis of coenzymes. The vitamin becomes part of a coenzyme's molecular structure. For example, the vitamin niacin is part of the coenzyme NAD^+ and riboflavin (B_2) is a component of the coenzyme FAD. If a vitamin is not available, enzymatic activity will decrease, and the result will be a vitamin-deficiency disorder. In humans, a niacin deficiency results in a skin disease called pellagra, and riboflavin deficiency results in cracks at the corners of the mouth.

a. Rate of reaction as a function of temperature

b. Body temperature of ectothermic animals often limits rates of reactions.

c. Body temperature of endothermic animals promotes rates of reactions.

Figure 6.8 The effect of temperature on rate of reaction. **a.** Usually, the rate of an enzymatic reaction doubles with every 10°C rise in temperature. This enzymatic reaction is maximum at about 40°C; then it decreases until the reaction stops altogether, because the enzyme has become denatured. **b.** The body temperature of ectothermic animals, such as iguanas, which take on the temperature of their environment, often limits rates of reactions. **c.** The body temperature of endothermic animals, such as polar bears, promotes rates of reaction.

THEME Nature of Science

Enzyme Inhibitors Can Spell Death

Cyanide gas was formerly used to execute people. How did it work? Cyanide can be fatal, because it binds to a mitochondrial enzyme necessary for the production of ATP. MPTP (1-methyl-4-phenyl-1,2,3.6-tetrahydropyridine) is another enzyme inhibitor that stops mitochondria from producing ATP. The toxic nature of MPTP was discovered in the early 1980s, when a group of intravenous drug users in California suddenly developed symptoms of Parkinson disease, including uncontrollable tremors and rigidity. All of the drug users had injected a synthetic form of heroin that was contaminated with MPTP. Parkinson disease is characterized by the death of brain cells, the very ones that are also destroyed by MPTP.

Sarin is a chemical that inhibits an enzyme at neuromuscular junctions, where nerves stimulate muscles. When the enzyme is inhibited, the signal for muscle contraction cannot be turned off, so the muscles are unable to relax and become paralyzed. Sarin can be fatal if the muscles needed for breathing become paralyzed. In 1995, terrorists released sarin gas on a subway in Japan (Fig. 6A). Although many people developed symptoms, only 17 died.

A fungus that contaminates and causes spoilage of sweet clover produces a chemical called warfarin. Cattle that eat the spoiled feed die from internal bleeding, because warfarin inhibits a crucial enzyme for blood clotting. Today, warfarin is widely used as a rat poison. Unfortunately, it is not uncommon for warfarin to be mistakenly eaten by pets and even very small children, with tragic results. Many people are prescribed a medicine called Coumadin to prevent inappropriate blood clotting. For example, those who have received an artificial heart valve need such a medication. Coumadin contains a nonlethal dose of warfarin.

These examples all show how our understanding of science can have positive or negative consequences and emphasize the role of ethics in scientific investigation.

Questions to Consider

1. Do you feel it is ethical to use dangerous chemicals to treat diseases?
2. Under what conditions do you think they should be used?

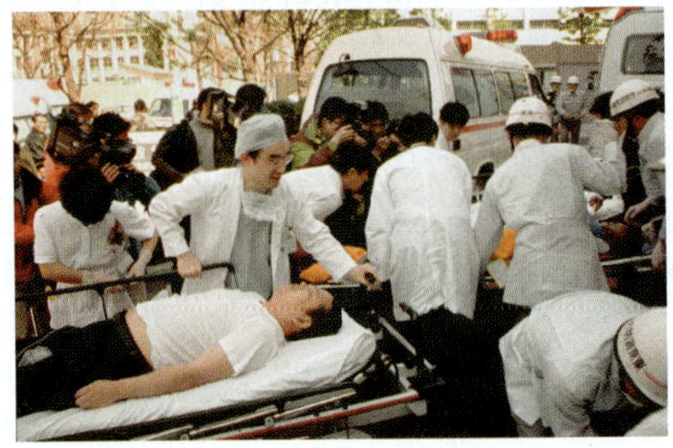

Figure 6A Sarin gas. The aftermath when sarin, a nerve gas that results in the inability to breathe, was released by terrorists in a Japanese subway in 1995.

Figure 6.9 The effect of temperature on enzymes. Siamese cats have inherited a mutation that causes an enzyme to be active only at cooler body temperatures. Therefore, only certain regions of the body are dark in color.

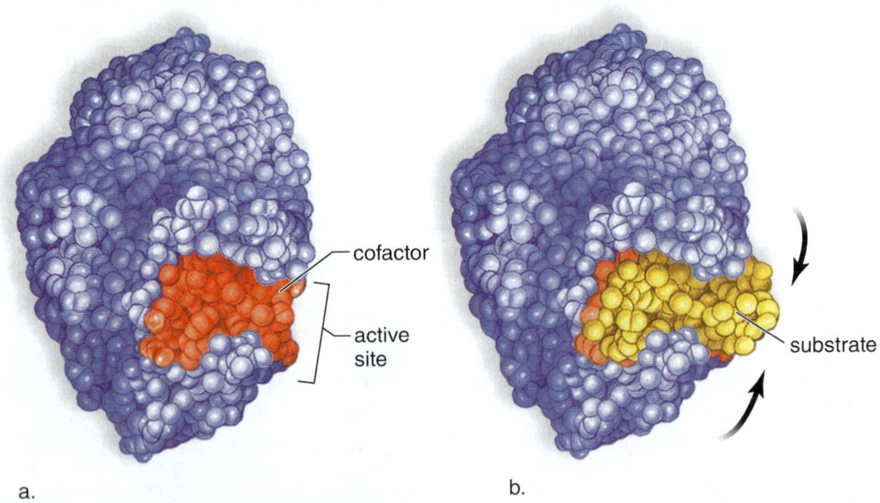

Figure 6.10 Cofactors at active site. **a.** Cofactors, including inorganic ions and organic coenzymes, may participate in the reaction at (**b**) the active site.

Enzyme Inhibition

Sometimes it is necessary to limit the activity of an enzyme. **Enzyme inhibition** occurs when a molecule (the inhibitor) binds to an enzyme and decreases its activity. In Figure 6.11, F is the end product of a metabolic pathway that can act as an inhibitor. This type of inhibition is beneficial, because once sufficient end product is present, inhibiting further production can conserve raw materials and energy.

Figure 6.11 also illustrates **noncompetitive inhibition,** because the inhibitor (F, the end product) binds to the enzyme E_1 at a location other than the active site. The site is called an **allosteric site.** When an inhibitor is at the allosteric site, the active site of the enzyme changes shape, which in turn changes its function.

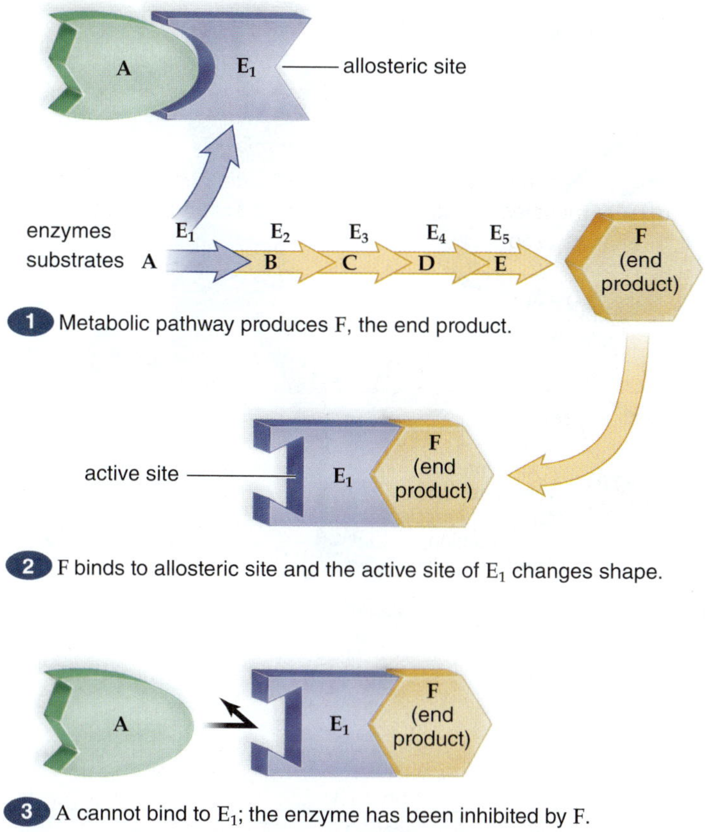

Figure 6.11 Noncompetitive inhibition of an enzyme. In the pathway, A–E are substrates, E_1–E_5 are enzymes, and F is the end product of the pathway that inhibits enzyme E_1. This negative feedback is useful, because it prevents wasteful production of product F when it is not needed.

In Figure 6.11, the enzyme E_1 is inhibited, because it is unable to bind to A, its substrate. The inhibition of E_1 means that the metabolic pathway is inhibited and no more end product is produced, until conditions change and more end product is needed.

In contrast to noncompetitive inhibition, **competitive inhibition** occurs when an inhibitor and the substrate compete for the active site of an enzyme. Product forms only when the substrate, not the inhibitor, is at the active site. In this way, the amount of product is regulated.

Normally, enzyme inhibition is reversible, and the enzyme is not damaged by being inhibited. When enzyme inhibition is irreversible, the inhibitor permanently inactivates or destroys an enzyme.

Check Your Progress 6.3

1. Explain how enzymes are involved in metabolic pathways.
2. Describe how an enzyme interacts with a substrate to reduce the energy of activation.
3. List the environmental conditions that may influence enzyme activity.

6.4 Oxidation-Reduction Reactions and Metabolism

Learning Outcomes

Upon completion of this section, you should be able to

1. Explain how the reactions for photosynthesis and cellular respiration represent oxidation-reduction reactions.
2. Summarize the relationship between the metabolic reactions of photosynthesis and cellular respiration.

In the next two chapters, you will explore two important metabolic pathways: cellular respiration (see Chapter 7) and photosynthesis (see Chapter 8). Both of these pathways are based on the use of special enzymes to facilitate the movement of electrons. The movement of these electrons plays a major role in the energy-related reactions associated with these pathways.

Oxidation-Reduction Reactions

When oxygen (O) combines with a metal such as iron or magnesium (Mg), oxygen receives electrons and becomes an ion that is negatively charged. The metal loses electrons and becomes an ion that is positively charged. When magnesium oxide (MgO) forms, it is appropriate to say that magnesium has been oxidized. On the other hand, oxygen has been reduced, because it has gained negative charges (i.e., electrons). Reactions that involve the gain and loss of electrons are called oxidation-reduction reactions. Sometimes, the terms *oxidation* and *reduction* are applied to other reactions, whether or not oxygen is involved. In a discussion of metabolic reactions, oxidation represents the loss of electrons, and reduction is the gain of electrons. In the reaction Na + Cl → NaCl, sodium has been oxidized (loss of electron), and chlorine has been reduced (gain of electrons). Because oxidation and reduction go hand-in-hand, the entire reaction is called a **redox reaction.** One easy way to remember what is happening in redox reactions is to remember the term OIL RIG:

OIL RIG

Oxidation **I**s **L**oss   **R**eduction **I**s **G**ain

The terms *oxidation* and *reduction* also apply to covalent reactions in cells. In this case, however, oxidation is the loss of hydrogen atoms (e^- + H^+), and reduction is the gain of hydrogen atoms. Notice that when a molecule loses a hydrogen atom it has lost an electron, and when a molecule gains a hydrogen atom it has gained an electron. This form of oxidation-reduction is exemplified in the overall equations for photosynthesis and cellular respiration.

Chloroplasts and Photosynthesis

The chloroplasts in plants capture solar energy and use it to convert water and carbon dioxide into a carbohydrate. Oxygen is a

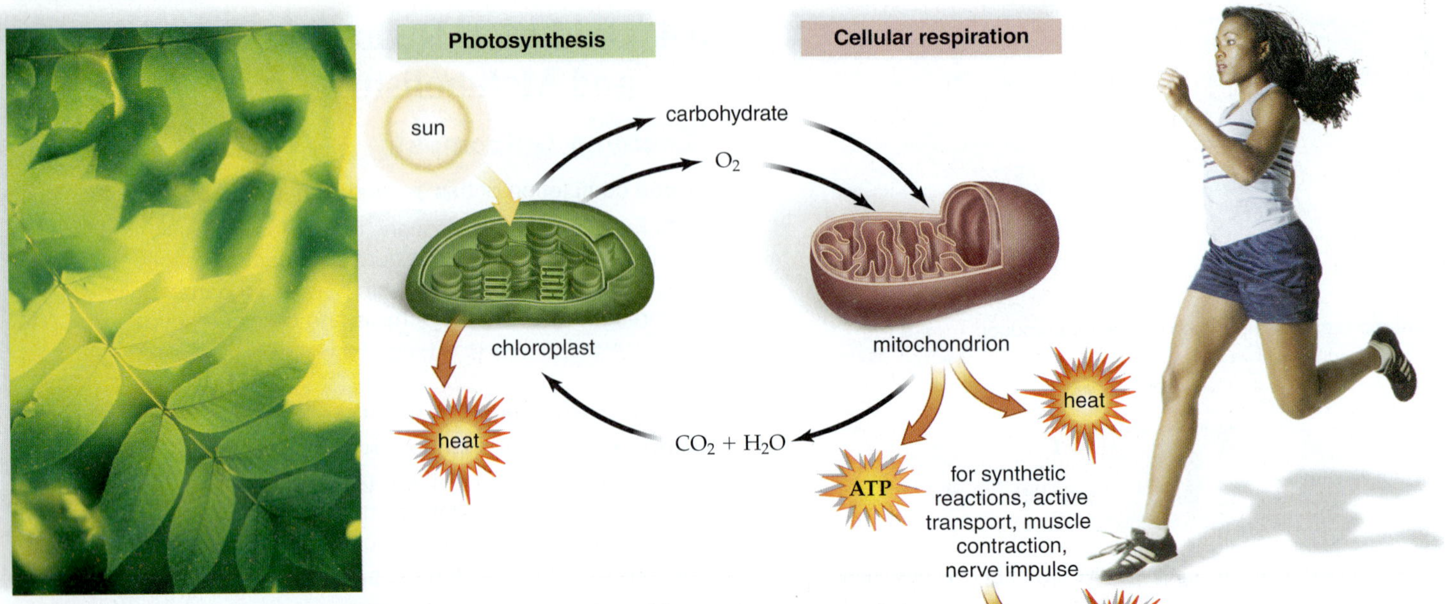

Figure 6.12 Relationship of chloroplasts to mitochondria. Chloroplasts produce energy-rich carbohydrate. Carbohydrate is broken down in mitochondria, and the energy released is used for the buildup of ATP. Mitochondria can also respire molecules derived from fats and amino acids for the buildup of ATP. Usable energy is lost as heat due to the energy conversions of photosynthesis, cellular respiration, and the use of ATP in the body.

by-product that is released (Fig. 6.12, *left*). The overall equation for photosynthesis can be written like this:

$$\underset{\text{}}{\text{energy}} + \underset{\text{carbon dioxide}}{6\,CO_2} + \underset{\text{water}}{6\,H_2O} \longrightarrow \underset{\text{glucose}}{C_6H_{12}O_6} + \underset{\text{oxygen}}{6\,O_2}$$

This equation shows that during photosynthesis hydrogen atoms are transferred from water to carbon dioxide as glucose forms. In this reaction, therefore, carbon dioxide has been reduced and water has been oxidized. It takes energy to reduce carbon dioxide to glucose, and this energy is supplied by solar energy. Chloroplasts are able to capture solar energy and convert it to the chemical energy of ATP, which is used along with hydrogen atoms to reduce carbon dioxide.

The reduction of carbon dioxide to form a mole of glucose stores 686 kcal in the chemical bonds of glucose. This is the energy that living organisms utilize to support themselves only because carbohydrates (and other nutrients) can be oxidized in mitochondria.

Mitochondria and Cellular Respiration

Mitochondria, present in both plants and animals, oxidize carbohydrates and use the released energy to build ATP molecules (Fig. 6.12, *right*). Cellular respiration therefore consumes oxygen and produces carbon dioxide and water, the very molecules taken up by chloroplasts. The overall equation for cellular respiration is the opposite of the one we used to represent photosynthesis:

$$\underset{\text{glucose}}{C_6H_{12}O_6} + \underset{\text{oxygen}}{6\,O_2} \longrightarrow \underset{\text{carbon dioxide}}{6\,CO_2} + \underset{\text{water}}{6\,H_2O} + \text{energy}$$

In this reaction, glucose has lost hydrogen atoms (been oxidized), and oxygen has gained hydrogen atoms (been reduced). When oxygen gains electrons, it becomes water. The complete oxidation of a mole of glucose releases 686 kcal of energy, and some of this energy is used to synthesize ATP molecules. If the energy within glucose were released all at once, most of it would dissipate as heat instead of some of it being used to produce ATP. Instead, cells oxidize glucose step by step. The energy is gradually stored and then converted to that of ATP molecules, which is used in animals in the many ways listed in Figure 6.12.

Figure 6.12 shows us very well that chloroplasts and mitochondria are involved in a cycle. Carbohydrate produced within chloroplasts becomes a fuel for cellular respiration in mitochondria, while carbon dioxide released by mitochondria becomes a substrate during photosynthesis in chloroplasts. These organelles are involved in a redox cycle, because carbon dioxide is reduced during photosynthesis and carbohydrate is oxidized during cellular respiration. Note that energy does not cycle between the two organelles; instead, it flows from the sun through each step of photosynthesis and cellular respiration until it eventually becomes unusable heat as ATP is used by the cell.

Check Your Progress 6.4

1. Compare the role of carbon dioxide in photosynthesis and cellular respiration.
2. Distinguish how energy from electrons is used to establish an electrochemical gradient in chloroplasts and mitochondria.

Connecting *the* Concepts *with the* Themes

Evolution

- Almost all forms of life have evolved to use ATP as an energy source to drive metabolic reactions. ATP is called the universal energy "currency" of life.
- Cellular respiration in mitochondria transforms the chemical energy of carbohydrates into that of ATP molecules. ATP is spent when it is hydrolyzed, and the resulting energy is coupled to an endergonic reaction.
- All cells are continually making and breaking down ATP. If ATP is lacking, the organism dies.

Nature of Science

- An understanding of the laws of energy helps explain why cells must invest energy to stay alive.
- Since many metabolic pathways are interconnected, an understanding of how enzymes function in a pathway is important in the development of treatments for diseases.

Biological Systems

- All cells use energy. Energy is the ability to do work, to bring about change, and to make things happen. The metabolic pathways inside cells use the chemical energy of ATP to synthesize molecules, cause muscle contraction, and even allow you to read these words.
- A metabolic pathway consists of a series of individual chemical reactions, each with its own enzyme. The cell can regulate the activity of the many hundreds of different enzymes taking part in cellular metabolism.
- Enzymes are proteins, and as such they are sensitive to environmental conditions, including pH, temperature, and even certain pollutants, as will be discussed in later chapters.

MEDIA STUDY TOOLS http://connect.mheducation.com

SMARTBOOK® Study smart. SmartBook helps you maximize your study time by identifying what you know and don't know. The recharge feature in SmartBook helps you prepare for the test by identifying the material that you are most likely to forget.

connect BIOLOGY Find even more resources to learn the chapter concepts including animations, tutorial videos, and interactive practice questions.

Animations	Tutorials	MP3 Files
6.2 Breakdown of ATP and Cross-Bridge Movement **6.3** Biochemical Pathways	**6.3** Energy of Activation	**6.1** Laws of Thermodynamics

SUMMARIZE

6.1 Cells and the Flow of Energy

Energy is the ability to do work or bring about a change. Energy may exist in several forms. **Kinetic energy** is the energy of motion, and **potential energy** is stored energy. **Chemical energy** is a form of potential energy that is stored within the chemical bonds of molecules.

Two **laws of thermodynamics** are basic to understanding energy in biological systems. The first law of thermodynamics states that energy cannot be created or destroyed but can only be transferred or transformed. The second law of thermodynamics states that one usable form of energy cannot be completely converted into another usable form, such as **heat.** As a result of these laws, we know that the **entropy** of the universe is increasing and that only a flow of energy from the sun maintains the organization of living organisms.

6.2 Metabolic Reactions and Energy Transformations

The term **metabolism** encompasses all the chemical reactions occurring in a cell. Considering individual reactions, only those that result in a negative **free-energy** difference—that is, the **products** have less usable energy than the **reactants**—occur spontaneously. Such reactions, called exergonic reactions, release energy.

An **endergonic** reaction, which requires an input of energy, occurs only when coupled to an exergonic process. The energy released from the exergonic reaction must exceed the net energy cost of the endergonic reaction in order for the overall reaction to proceed. Hydrolysis of ATP, an **exergonic** reaction, is commonly used to drive energetically unfavorable metabolic reactions.

The energy currency of the cell is **ATP** (**adenosine triphosphate**). As cells use ATP for cellular reactions, **ADP** (**adenosine diphosphate**) is generated. Metabolic reactions in the cell can break down glucose and other macromolecules to release energy and regenerate ATP.

6.3 Metabolic Pathways and Enzymes

A **metabolic pathway** is a series of reactions that proceeds in an orderly, step-by-step manner. **Enzymes** speed reactions by lowering the **energy of activation** when they form a complex with their **substrates.** The **induced fit model** explains how the substrate interacts with the **active site** of the enzyme. The three-dimensional shape of the enzyme is important. When **denaturation** occurs, an enzyme may not be able to interact as efficiently with the substrate. Most enzymes are proteins, but some, called **ribozymes,** may be RNA molecules.

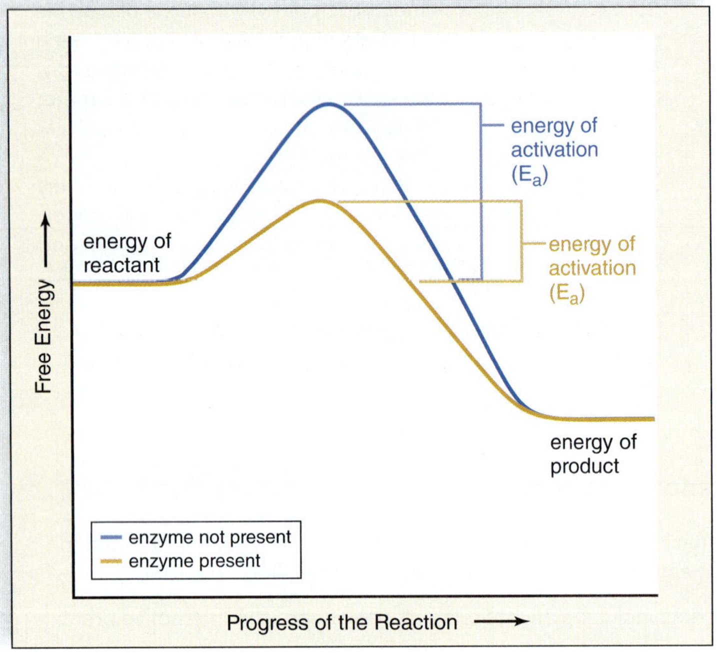

Generally, enzyme activity increases as substrate concentration increases; once all active sites are filled, maximum reaction rate has been achieved. Environmental factors, such as temperature or pH, can affect the shape of an enzyme and therefore its function. Many enzymes need **cofactors** or coenzymes to carry out their reactions. **Vitamins** are often used to build **coenzymes,** such as **NAD^+, FAD^+,** and **$NADP^+$**.

Enzyme inhibition, usually in the form of **feedback inhibition,** regulates the activity of many metabolic pathways. In addition, **noncompetitive inhibition,** where a molecule binds to an allosteric site on the enzyme, may influence enzyme behavior. In **competitive inhibition,** an inhibitor interferes with the binding of the substrate in the active site of an enzyme.

6.4 Oxidation-Reduction Reactions and Metabolism

The overall equation for photosynthesis is the opposite of that for cellular respiration. Both processes involve oxidation-reduction reactions.

> **OIL RIG**
>
> **O**xidation **I**s **L**oss **R**eduction **I**s **G**ain

Redox reactions are a major way in which energy is transformed in cells.

During photosynthesis, carbon dioxide is reduced to glucose, and water is oxidized. Glucose formation requires energy, and this energy comes from the sun. Chloroplasts capture solar energy and convert it to the chemical energy of ATP molecules, which are used along with hydrogen atoms to reduce carbon dioxide to glucose. During cellular respiration, glucose is oxidized to carbon dioxide, and oxygen is reduced to water. This reaction releases energy, which is used to synthesize ATP molecules in all types of cells. Energy flows through all living organisms. Photosynthesis is a metabolic pathway in chloroplasts that converts solar energy to the chemical energy within carbohydrates, and cellular respiration is a metabolic pathway completed in mitochondria that converts this energy into that of ATP molecules. Eventually, the energy within ATP molecules becomes heat.

Cellular respiration is an aerobic process that requires oxygen and gives off carbon dioxide. Cellular respiration involves oxidation. The air we inhale contains the oxygen, and the food we digest after eating contains the carbohydrate glucose needed for cellular respiration.

ASSESS

Choose the best answer for each question.

6.1 Cells and the Flow of Energy

1. The fact that energy transformations increase the amount of entropy is the basis of which of the following?
   a. cell theory
   b. first law of thermodynamics
   c. second law of thermodynamics
   d. oxidation-reduction reactions
2. The energy stored in the carbon-carbon bonds of glucose is an example of _____ energy.
   a. kinetic
   b. potential
   c. chemical
   d. mechanical
   e. Both b and c are correct.
3. During energy transformations, the majority of energy is converted to
   a. chemical bonds.
   b. heat.
   c. ATP.
   d. glucose molecules.

6.2 Metabolic Reactions and Energy Transformations

4. Exergonic reactions
   a. are spontaneous.
   b. have a negative delta G value.
   c. release energy.
   d. All of these are correct.
5. Which of the following is incorrect regarding ATP?
   a. It is the energy currency of the cell.
   b. It is stable.
   c. It is recycled using ADP and inorganic phosphate.
   d. Cells keep only small amounts of ATP on hand.
6. The sum of all the chemical reactions in a cell is called
   a. free energy.
   b. entropy.
   c. metabolism.
   d. oxidation-reduction reactions.

6.3 Metabolic Pathways and Enzymes

7. Which of the following is incorrect regarding the active site of an enzyme?
   a. is unique to that enzyme
   b. is the part of the enzyme where its substrate can fit
   c. can be used over and over again
   d. is not affected by environmental factors, such as pH and temperature
8. Which of the following environmental conditions may have an influence on enzyme activity?
   a. substrate concentration
   b. temperature
   c. pH
   d. All of these are correct.
9. In which of the following does an inhibitor bind to an allosteric site on an enzyme?
   a. competitive inhibition
   b. noncompetitive inhibition
   c. redox reactions
   d. None of the above are correct.
10. Enzymes catalyze chemical reactions by which of the following?
    a. lowering the energy of activation in the reaction
    b. raising the energy of activation in the reaction
    c. increasing entropy
    d. increasing the free energy of the products

6.4 Oxidation-Reduction Reactions and Metabolism

11. The gain of electrons by a molecule is called
    a. inhibition.
    b. entropy.
    c. oxidation.
    d. reduction.
12. In which of the following processes is carbon dioxide reduced to form carbohydrate?
    a. cellular respiration
    b. noncompetitive inhibition
    c. photosynthesis
    d. induced fit model

ENGAGE

LEARNSMART LABS®

The following LearnSmart Labs contain exercises that are related to the content of this chapter:

- How Enzymes Function

THINKING SCIENTIFICALLY

1. Some coupled reactions in cells, including many involved in protein synthesis, use the nucleotide GTP as an energy source instead of ATP. What would be the advantage of using GTP instead of ATP as an energy source for these cellular reactions?
2. Entropy is increased when nutrients break down, so why are enzymatic metabolic pathways required for cellular respiration?
3. Why would you expect glucose storage as glycogen to be an energy-requiring process?
4. How might a knowledge of the structure of an active site of an enzyme allow you to build a drug to regulate a metabolic pathway?

7

Photosynthesis

Biofuels may one day come from modifying the process of photosynthesis.

Chapter Outline

7.1 Photosynthetic Organisms 115

7.2 The Process of Photosynthesis 117

7.3 Plants Convert Solar Energy 119

7.4 Plants Fix Carbon Dioxide 123

7.5 Other Types of Photosynthesis 125

Before You Begin

Before beginning this chapter, take a few moments to review the following discussions.

Figure 6.1 How does energy flow in biological systems?

Section 6.3 What role do enzymes play in regulating metabolic processes?

Section 6.4 How are redox reactions and membranes used to conduct cellular work?

Photosynthesis is a powerful process. Plants, algae, and some bacteria carry out a series of chemical reactions that harness CO_2 from the air, and they combine water from the soil with sunlight to create the molecules that living organisms rely on—oxygen, carbohydrates, oils, and amino acids.

Photosynthesis can also be the key to solving our world's fuel crisis. Plant researchers are tweaking the basic chemistry of photosynthesis to create commercially important oils and fuels. One example is work being done with *Camelina,* a drought-resistant, oilseed crop. Scientists are modifying how *Camelina* captures sunlight by genetically engineering the plant so that the leaves at the top are lighter, allowing sunlight to pass through to the lower leaves, improving the efficiency of photosynthesis. Another goal improves the absorption of CO_2, providing the raw materials for oil production, which are precursors for potential biofuels.

Other researchers are focusing on terpene—another end result of photosynthesis. Terpene is a high-energy organic molecule, produced by pine trees, that makes turpentine. Ongoing research aims to increase terpene production and process this to make a domestic source of diesel and aviation biofuels. In the future, you may be on a commercial flight where the meal providing fuel for your body and the diesel fueling the airplane can both trace their origins to a photosynthesizing plant.

As you read through this chapter, think about the following questions:

1. What raw materials do plants require, and what do they produce?
2. How do some plants modify photosynthesis in extreme conditions?

Following *the* Themes

CHAPTER 7 PHOTOSYNTHESIS

UNIT 1 THE CELL

Evolution	Plants have adapted and spread to nearly every environment and climate on Earth.
Nature of Science	Researchers have created new plant varieties that produce more food and commercially useful products.
Biological Systems	Most ecosystems on Earth depend on photosynthesizing autotrophs as the basis of their food webs.

7.1 Photosynthetic Organisms

Learning Outcomes

Upon completion of this section, you should be able to

1. Explain how autotrophs are able to produce their own food.
2. Describe the components of a chloroplast.
3. Compare the roles of oxygen and carbon dioxide in autotrophs and heterotrophs.

Photosynthesis converts solar energy into the chemical energy of a carbohydrate. Photosynthetic organisms, including plants, algae, and cyanobacteria, are called **autotrophs,** because they produce their own food (Fig. 7.1*a*). It has been estimated that all of the world's green organisms together produce between 100 billion and 200 billion metric tons of sugar each year. Imagine enough sugar cubes to re-create the volume of 2 million Empire State Buildings!

No wonder photosynthetic organisms are able to sustain themselves and all other living organisms on Earth. With few exceptions, it is possible to trace any food chain back to plants and algae. In other words, producers, which have the ability to synthesize carbohydrates, feed not only themselves but also consumers, which must take in preformed organic molecules. Collectively, consumers are called **heterotrophs.** Both autotrophs and heterotrophs use organic molecules produced by photosynthesis as a source of building blocks for growth and repair and as a source of chemical energy for cellular work.

Photosynthesizers also produce copious amounts of oxygen gas (O_2) as a by-product. Oxygen, required by organisms for cellular respiration, rises high into the atmosphere, where it forms an ozone shield that filters out ultraviolet radiation and makes terrestrial life possible.

Oil and coal provide about 90% of the energy needed to power vehicles, factories, computers, and a multitude of electrically energized appliances. The energy within that oil and coal was originally captured from the sun by plants and algae growing millions of years ago—thus the name "fossil fuels." Today's trees are also commonly used as fuel. Fermentation of plant materials produces ethanol, which can be used to fuel automobiles directly or as a gasoline additive.

The products of photosynthesis are critical to humankind in a number of other ways. They serve as a source of building materials, fabrics, paper, and pharmaceuticals. Of course, we also appreciate green plants for the simple beauty of an orchid blossom, the scent of a rose, or the majesty of the forests.

a. *Oscillatoria* 40× Kelp Sequoias

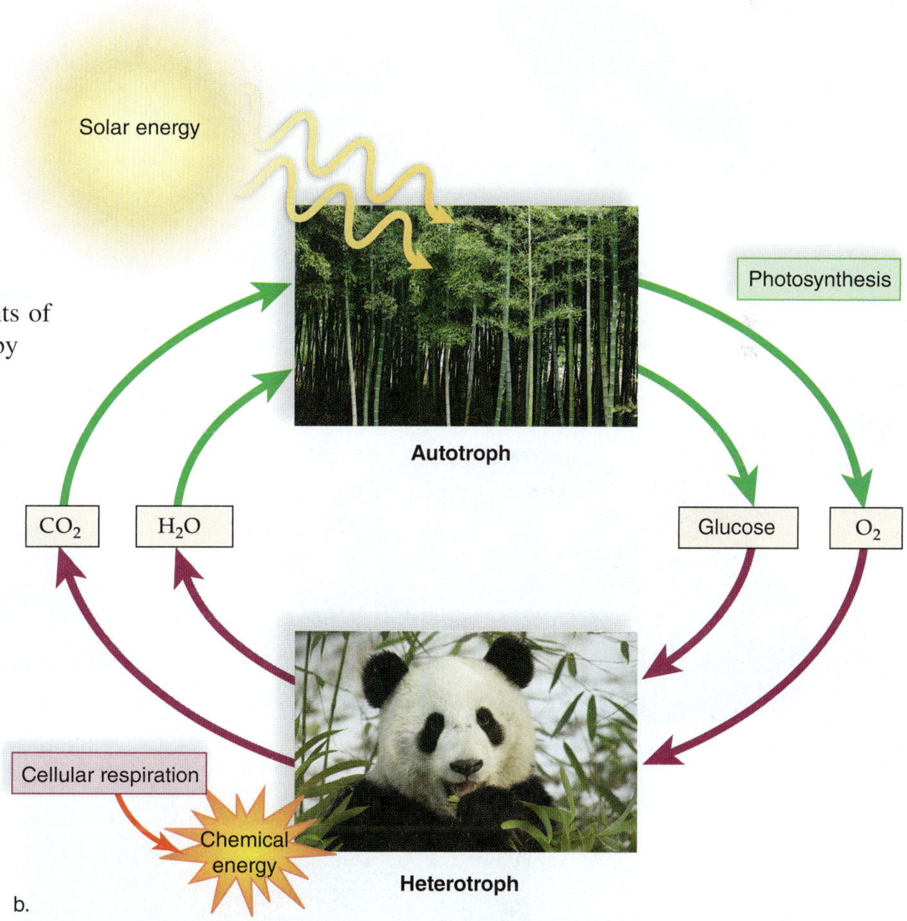

b.

Figure 7.1 Autotrophs and the relationship to heterotrophs. **a.** Photosynthetic organisms (autotrophs) include cyanobacteria (*left*); algae, such as kelp (*middle*); and plants (*right*). **b.** Photosynthetic organisms harness the energy from the sun and provide gases and nutrients for heterotrophs. Heterotrophs generate chemical energy and produce carbon dioxide and water.

Photosynthesis in Flowering Plants

Photosynthesis takes place in the green portions of plants. The leaves of a flowering plant contain mesophyll tissue, in which cells are specialized for photosynthesis (Fig. 7.2). The raw materials for photosynthesis are water and carbon dioxide. The roots of a plant absorb water, which then moves in vascular tissue up the stem to a leaf by way of the leaf veins. Carbon dioxide in the air enters a leaf through small openings called **stomata** (sing., stoma). After entering a leaf, carbon dioxide and water diffuse into **chloroplasts** (Gk. *chloros,* "green"; *plastos,* "formed, molded"), the organelles that carry on photosynthesis.

A double membrane surrounds a chloroplast, and its semifluid interior is called the **stroma** (Gk. *stroma,* "bed, mattress"). A different membrane system within the stroma forms flattened sacs called **thylakoids** (Gk. *thylakos,* "sack"), which in some places are stacked to form **grana** (sing., granum). The space of each thylakoid is thought to be connected to the space of every other thylakoid within a chloroplast, thereby forming an inner compartment within chloroplasts, called the thylakoid space. Overall, chloroplast membranes provide a tremendous surface area for photosynthesis to occur.

The thylakoid membrane contains **chlorophyll** and other pigments that are capable of absorbing the solar energy that drives photosynthesis. The stroma contains an enzyme-rich solution, where carbon dioxide is first attached to an organic compound and then reduced to a carbohydrate.

3D Animation
Photosynthesis: Structure of a Chloroplast

Humans and other respiring organisms release carbon dioxide into the air. Some of the same carbon dioxide molecules enter a leaf through the stoma and are converted to carbohydrate. Carbohydrate, in the form of glucose, is the chief source of chemical energy for most organisms. Thus, an interdependent relationship exists between organisms that make their own food (autotrophs) and those that consume their food (heterotrophs) (see Fig. 7.1*b*).

Check Your Progress 7.1

1. Describe three major groups of photosynthetic organisms.
2. Distinguish the part of a chloroplast that absorbs solar energy from the part that forms a carbohydrate.

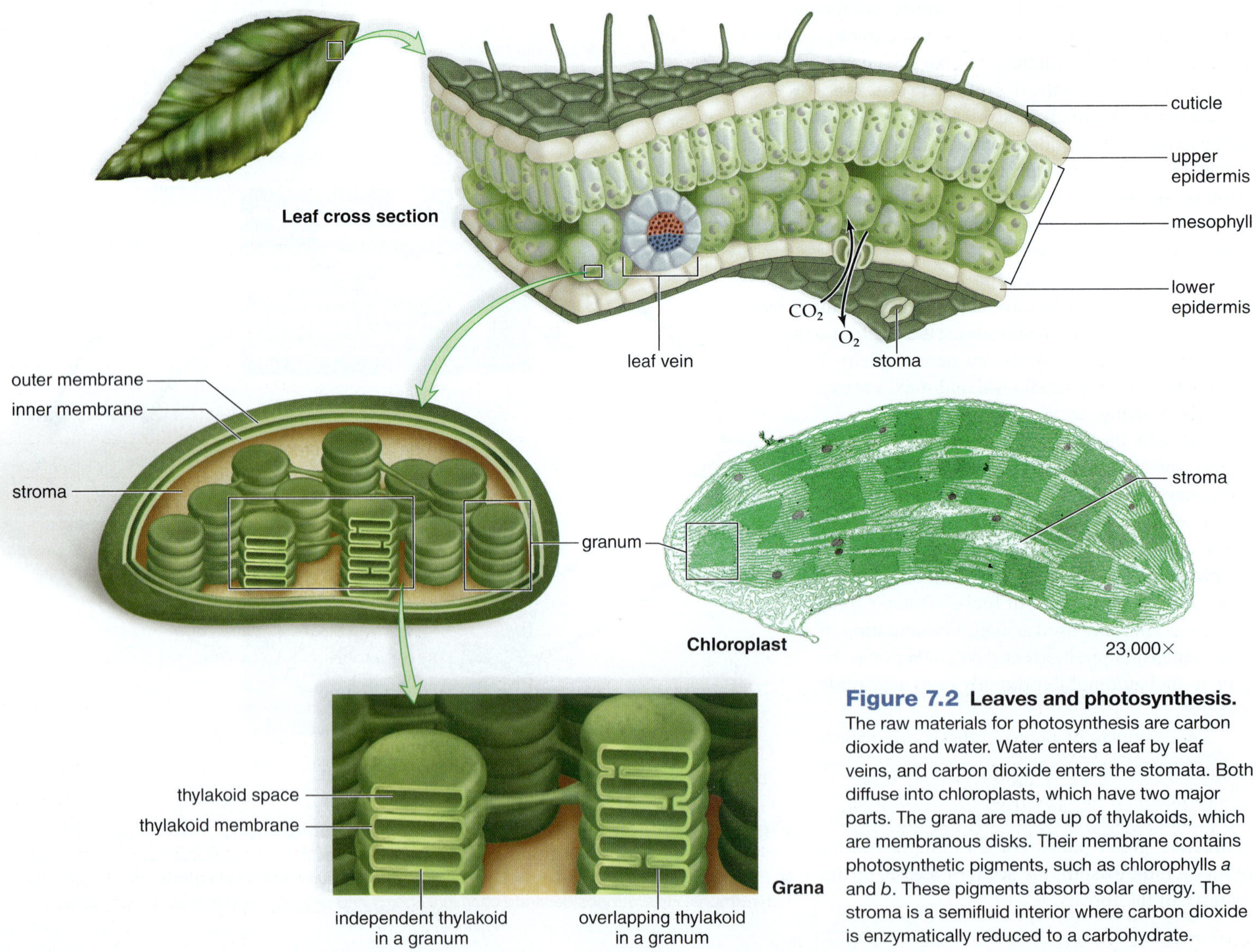

Figure 7.2 Leaves and photosynthesis. The raw materials for photosynthesis are carbon dioxide and water. Water enters a leaf by leaf veins, and carbon dioxide enters the stomata. Both diffuse into chloroplasts, which have two major parts. The grana are made up of thylakoids, which are membranous disks. Their membrane contains photosynthetic pigments, such as chlorophylls *a* and *b*. These pigments absorb solar energy. The stroma is a semifluid interior where carbon dioxide is enzymatically reduced to a carbohydrate.

7.2 The Process of Photosynthesis

Learning Outcomes

Upon completion of this section, you should be able to

1. Describe the overall process of photosynthesis.
2. Compare energy input and output of the light reaction.
3. Compare carbon input and output of the Calvin cycle reaction.

Figure 7.3 Photosynthesis releases oxygen. Bubbling indicates that the aquatic plant *Elodea* releases O_2 gas when it photosynthesizes.

The overall process of photosynthesis can be represented by an equation:

$$6\,CO_2 + 12\,H_2O \xrightarrow{\text{solar energy}} 6\,(CH_2O) + 6\,H_2O + 6\,O_2$$

In this equation, (CH_2O) represents carbohydrate. If the equation were multiplied by 6, the carbohydrate would be $C_6H_{12}O_6$, or glucose.

The overall equation implies that photosynthesis involves oxidation-reduction (redox) and the movement of electrons from one molecule to another. Recall that oxidation is the loss of electrons, and reduction is the gain of electrons. In living organisms, as discussed in Chapter 6, the electrons are very often accompanied by hydrogen ions, so that oxidation is the loss of hydrogen atoms ($H^+ + e^-$) and reduction is the gain of hydrogen atoms. This simplified rewrite of the equation makes it clear that carbon dioxide has been reduced and water has been oxidized:

Reduction ($CO_2 \rightarrow (CH_2O)$)

$$CO_2 + H_2O \xrightarrow{\text{solar energy}} (CH_2O) + O_2$$

Oxidation ($H_2O \rightarrow O_2$)

It takes hydrogen atoms and a lot of energy to reduce carbon dioxide. From your study of energy and enzymes in Chapter 6, you might expect that solar energy is not used directly during photosynthesis; rather, it is converted to ATP molecules. ATP is the energy currency of cells and, when cells need something, they spend ATP. In this case, solar energy is used to generate the ATP needed to reduce carbon dioxide to a carbohydrate. Of course, this carbohydrate represents the food produced by plants, algae, and cyanobacteria that feeds the biosphere.

The Role of NADP+/NADPH

A review of section 6.4 will also lead you to suspect that the electrons needed to reduce carbon dioxide are carried by a coenzyme. $NADP^+$ is the coenzyme of oxidation-reduction (redox coenzyme) active during photosynthesis. When $NADP^+$ is reduced, it has accepted two electrons and one hydrogen atom, and when NADPH is oxidized, it gives up its electrons:

$$NADP^+ + 2\,e^- + H^+ \longrightarrow NADPH$$

What molecule supplies the electrons that reduce $NADP^+$ during photosynthesis? Put a sprig of *Elodea* in a beaker and supply it with light, and you will observe a bubbling (Fig. 7.3). The bubbling occurs because the plant is releasing oxygen as it photosynthesizes.

A famous experiment performed by C. B. van Niel of Stanford University found that the oxygen given off by photosynthesizers comes from water. Van Niel performed two separate experiments. When an isotope of oxygen, ^{18}O, was a part of water, the O_2 given off by the plant contained ^{18}O. When ^{18}O was a part of carbon dioxide supplied to a plant, the O_2 given off by a plant did not contain the ^{18}O. Why not? Because the oxygen in carbon dioxide doesn't come from water; it comes from the air. This was the first step toward discovering that water splits during photosynthesis. When water splits, oxygen is released and the hydrogen atoms ($H^+ + e^-$) are taken up by $NADP^+$. Later, NADPH reduces carbon dioxide to a carbohydrate.

Two Sets of Reactions

How does photosynthesis occur? The process can be divided into two stages, the light reactions and the Calvin cycle reactions. The term *photosynthesis* comes from the associations between these two stages: The prefex *photo* refers to the light reactions that capture the waves of sunlight needed for the *synthesis* of carbohydrates occurring in the Calvin cycle. The light reactions take place on thylakoids, and the Calvin cycle takes place in the stroma.

Light Reactions

The **light reactions** are so named because they occur only when the sun is out. The green pigment chlorophyll, present in thylakoid membranes, is largely responsible for absorbing the solar energy that drives photosynthesis.

During the light reactions, solar energy energizes electrons, which move down an electron transport chain (see Fig. 6.12). As the electrons move down the chain, energy is released and captured to produce ATP molecules. Energized electrons are also taken up by $NADP^+$, which is reduced and becomes NADPH. This equation can be used to summarize the light reactions, because during the light reactions solar energy is converted to chemical energy:

$$\text{solar energy} \longrightarrow \text{chemical energy (ATP, NADPH)}$$

Calvin Cycle Reactions

The **Calvin cycle reactions** are named for Melvin Calvin, who in 1961 received a Nobel Prize in Chemistry for discovering the enzymatic reactions that reduce carbon dioxide to a carbohydrate in the stroma of chloroplasts (Fig. 7.4). The enzymes that speed the reduction of carbon dioxide during both day and night are located in the semifluid substance of the chloroplast stroma.

During the Calvin cycle reactions, CO_2 is taken up and then reduced to a carbohydrate that can later be converted to glucose. This equation can be used to summarize the Calvin cycle reactions, because during these reactions, the ATP and NADPH formed during the light reactions are used to reduce carbon dioxide:

$$\underset{\text{(ATP, NADPH)}}{\text{chemical energy}} \longrightarrow \underset{\text{(carbohydrate)}}{\text{chemical energy}}$$

Figure 7.4 Melvin Calvin. Melvin Calvin, a chemist, is most noted for his work using a carbon 14 isotope to follow the route that carbon travels through a plant during photosynthesis.

Summary

Figure 7.5 summarizes our discussion so far and shows that during the light reactions, (1) solar energy is absorbed, (2) water is split so that oxygen is released, and (3) ATP and NADPH are produced.

During the Calvin cycle reactions, (1) CO_2 is absorbed and (2) reduced to a carbohydrate (CH_2O) by utilizing ATP and NADPH from the light reactions (bottom set of red arrows). The top set of red arrows takes ADP + Ⓟ and $NADP^+$ back to light reactions, where they become ATP and NADPH once more, so that carbohydrate production can continue.

Check Your Progress 7.2

1. Explain how redox reactions are used in photosynthesis.
2. Describe the role of enzymes during photosynthesis.

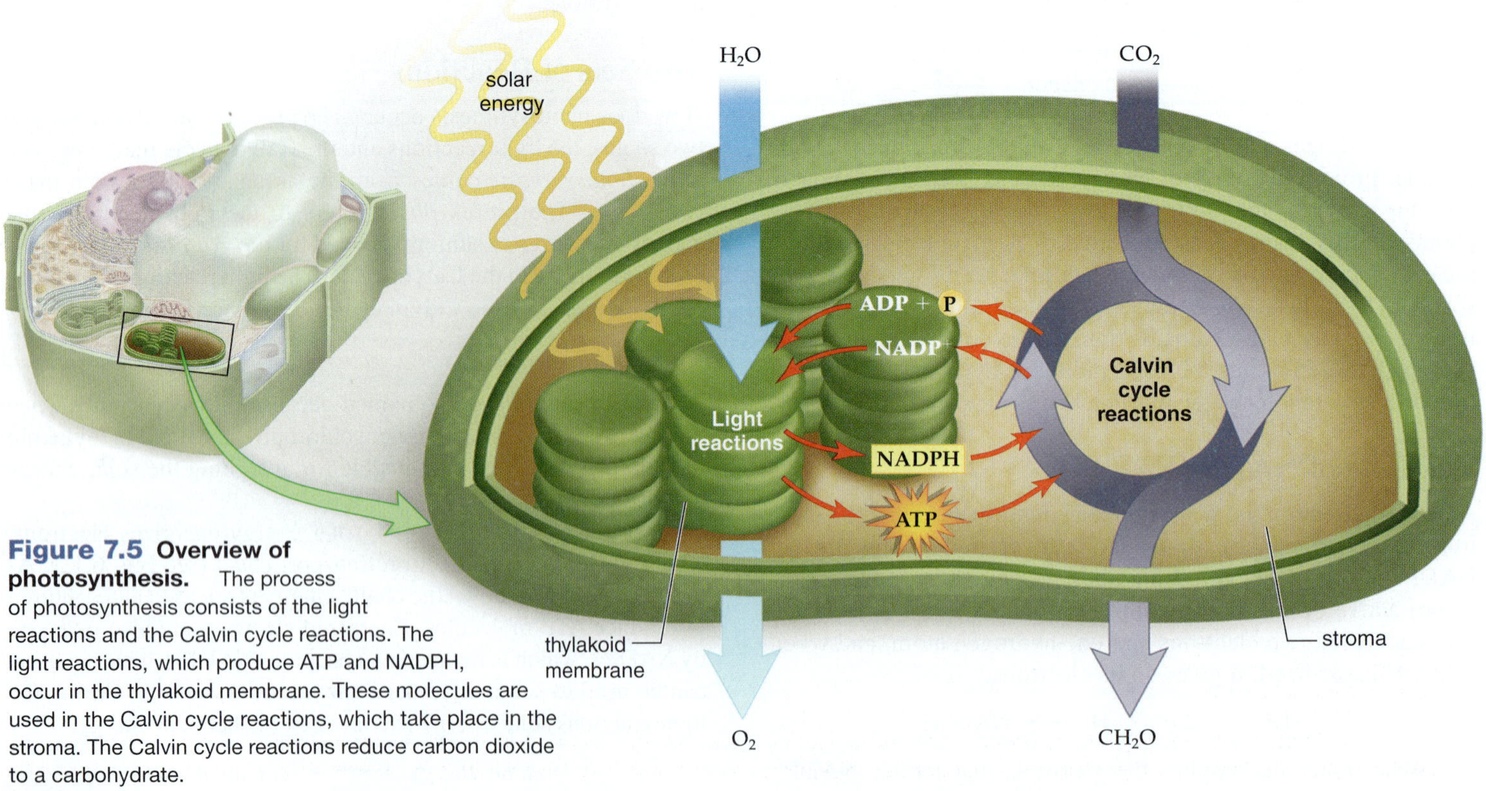

Figure 7.5 Overview of photosynthesis. The process of photosynthesis consists of the light reactions and the Calvin cycle reactions. The light reactions, which produce ATP and NADPH, occur in the thylakoid membrane. These molecules are used in the Calvin cycle reactions, which take place in the stroma. The Calvin cycle reactions reduce carbon dioxide to a carbohydrate.

7.3 Plants Convert Solar Energy

Learning Outcomes

Upon completion of this section, you should be able to

1. Describe the relationship between wavelength and energy in the electromagnetic spectrum.
2. Explain the role of photosynthetic pigments in harnessing solar energy.
3. Examine how ATP and NADPH are produced from redox reactions and membrane gradients.

Solar energy can be described in terms of its wavelength and its energy content. Figure 7.6*a* shows the types of radiant energy from the shortest wavelength, gamma rays, to the longest, radio waves. Most of the radiation reaching the Earth is within the visible-light range. Higher-energy wavelengths are screened out by the ozone layer in the atmosphere before they reach the Earth's surface, and lower-energy wavelengths are screened out by water vapor and carbon dioxide. Because visible light is the most prevalent in the environment, organisms have evolved to use these wavelengths. For example, human eyes have cone cells that respond to color wavelengths, and plants have pigments that are energized by most of the same wavelengths (Fig. 7.6).

3D Animation Photosynthesis: Properties of Light

Pigments and Photosystems

Pigment molecules absorb wavelengths of light. Most pigments absorb only some wavelengths; they reflect or transmit the other wavelengths. The pigments in chloroplasts are capable of absorbing various portions of visible light. This is called their **absorption spectrum.**

Photosynthetic organisms differ in the type of chlorophyll they contain. In plants, chlorophyll *a* and chlorophyll *b* play prominent roles in photosynthesis. **Carotenoids** play an accessory role. Both chlorophylls *a* and *b* absorb violet, blue, and red light better than the light of other colors. Because green light is transmitted and reflected by chlorophyll, plant leaves appear green to us. In short, plants are green because they do *not* use the green wavelength! The carotenoids, which are shades of yellow and orange, are able to absorb light in the violet-blue-green range. These pigments become noticeable in the fall when chlorophyll breaks down.

How do you determine the absorption spectrum of pigments? To identify the absorption spectrum of a particular pigment, a purified sample is exposed to different wavelengths of light inside an instrument called a spectrophotometer. A spectrophotometer measures the amount of light that passes through the sample, and from this it is possible to calculate how much was absorbed. The amount of light absorbed at each wavelength is plotted on a graph, and the result is a record of the pigment's absorption spectrum (Fig. 7.6*b*). Notice the low absorbance reading for the green and yellow wavelengths and recall why plants are green.

A **photosystem** consists of a pigment complex (molecules of chlorophyll *a,* chlorophyll *b,* and the carotenoids) and electron acceptor molecules within the thylakoid membrane. The pigment complex serves as an "antenna" for gathering solar energy.

Electron Flow in the Light Reactions

The light reactions utilize two photosystems, called photosystem I (PS I) and photosystem II (PS II). The photosystems are named for the order in which they were discovered, not for the order in which they occur in the thylakoid membrane or participate in the photosynthetic process.

During the light reactions, electrons usually, but not always, follow a **noncyclic pathway** that begins with photosystem II (Fig. 7.7). The pigment complex absorbs solar energy, which is then passed from one pigment to the other until it is concentrated in a particular pair of chlorophyll *a* molecules, called the *reaction center.* Electrons (e^-) in the reaction center become so energized that they escape from the reaction center and move to nearby electron acceptor molecules.

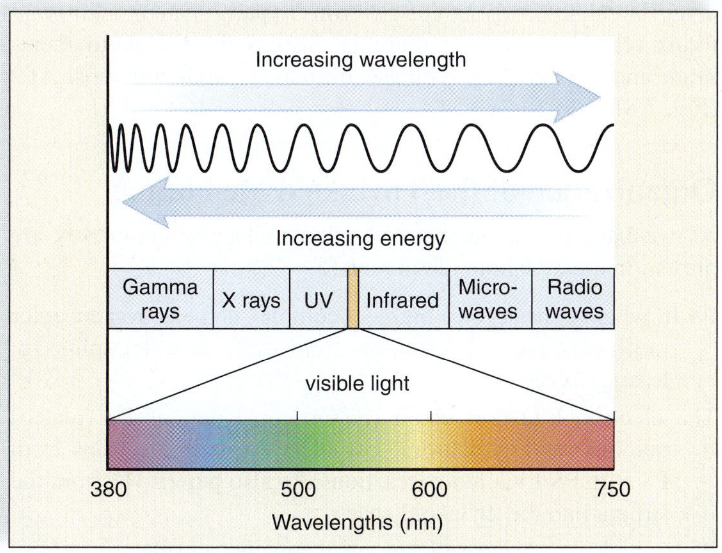

a. The electromagnetic spectrum includes visible light.

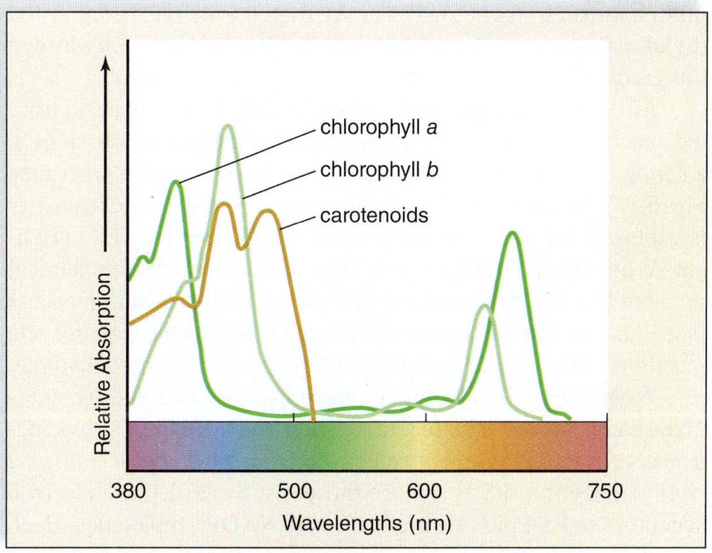

b. Absorption spectrum of photosynthetic pigments.

Figure 7.6 Photosynthetic pigments and photosynthesis. **a.** The wavelengths in visible light differ according to energy content and color. **b.** The photosynthetic pigments in chlorophylls *a* and *b* and the carotenoids absorb certain wavelengths within visible light. This is their absorption spectrum.

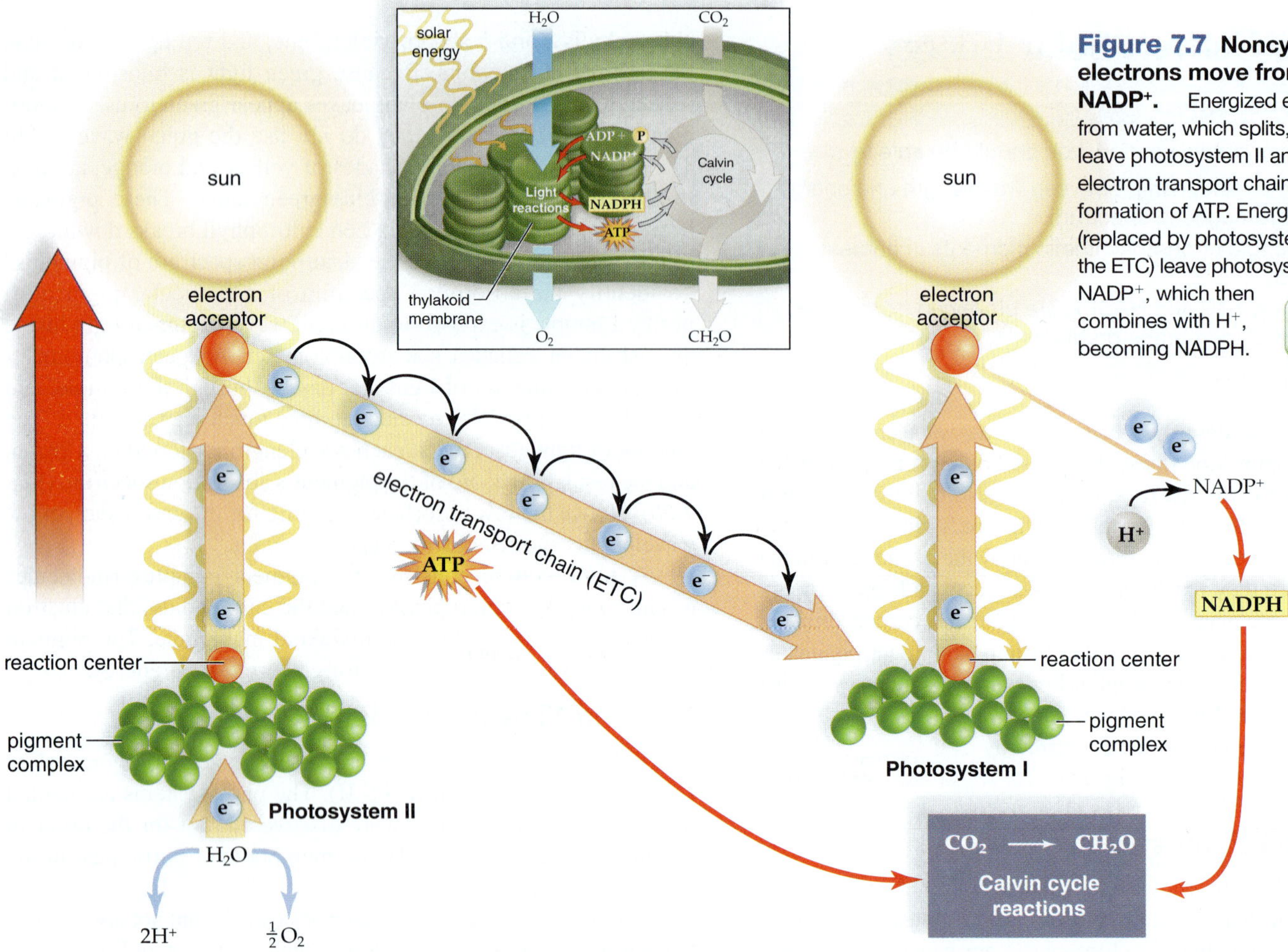

Figure 7.7 Noncyclic pathway: electrons move from water to NADP⁺. Energized electrons (replaced from water, which splits, releasing oxygen) leave photosystem II and pass down an electron transport chain, leading to the formation of ATP. Energized electrons (replaced by photosystem II by way of the ETC) leave photosystem I and pass to $NADP^+$, which then combines with H^+, becoming NADPH.

Tutorial Noncyclic Photosynthesis

PS II would disintegrate without replacement electrons, and these are removed from water, which splits, releasing oxygen to the atmosphere. Notice that with the loss of electrons, water has been oxidized and that the oxygen released during photosynthesis does come from water. Many organisms, including plants themselves and humans, use this oxygen within their mitochondria to make ATP. The hydrogen ions (H^+) stay in the thylakoid space and contribute to the formation of a hydrogen ion gradient.

An electron acceptor sends energized electrons, received from the reaction center, down an **electron transport chain (ETC)**, a series of carriers that pass electrons from one to the other (see Fig. 6.13). As the electrons pass from one carrier to the next, energy is captured and stored in the form of a hydrogen ion (H^+) gradient. When these hydrogen ions flow down their electrochemical gradient through ATP synthase complexes, ATP production occurs (see Fig. 7.9). Notice that this ATP is then used by the Calvin cycle reactions in the stroma to reduce carbon dioxide to a carbohydrate.

When the PS I pigment complex absorbs solar energy, energized electrons leave its reaction center and are captured by electron acceptors. (Low-energy electrons from the *electron transport chain* adjacent to PS II replace those lost by PS I.) The electron acceptors in PS I pass their electrons to $NADP^+$ molecules. Each $NADP^+$ accepts two electrons and an H^+ to become reduced and forms NADPH. This NADPH is then used by the Calvin cycle reactions in the stroma along with ATP in the reduction of carbon dioxide to a carbohydrate.

3D Animation Photosynthesis: Light-Dependent Reactions

ATP and NADPH are not made in equal amounts during the light reactions, and more ATP than NADPH is required during the Calvin cycle. Where does this extra ATP come from? Every so often, an electron moving down the noncyclic pathway is rerouted back to an earlier point in the electron transport chain. The **cyclic pathway,** which occurs in many prokaryotes, and at high oxygen levels in eukaryotes**,** enables electrons to participate in additional redox reactions, moving more H^+ across the thylakoid membrane and through ATP synthase, ultimately producing more ATP (Fig. 7.8).

Organization of the Thylakoid Membrane

As we have discussed, the following molecular complexes are present in the thylakoid membrane (Fig. 7.9):

PS II, which consists of a pigment complex and electron acceptor molecules, receives electrons from water as water splits, releasing oxygen.

The electron transport chain (ETC), consisting of Pq (plastoquinone) and cytochrome complexes, carries electrons from PS II to PS I via redox reactions. Pq also pumps H^+ from the stroma into the thylakoid space.

PS I, which also consists of a pigment complex and electron acceptor molecules, is adjacent to NADP reductase, which reduces $NADP^+$ to NADPH.

The **ATP synthase** complex, which has a channel and a protruding ATP synthase, is an enzyme that joins ADP + Ⓟ.

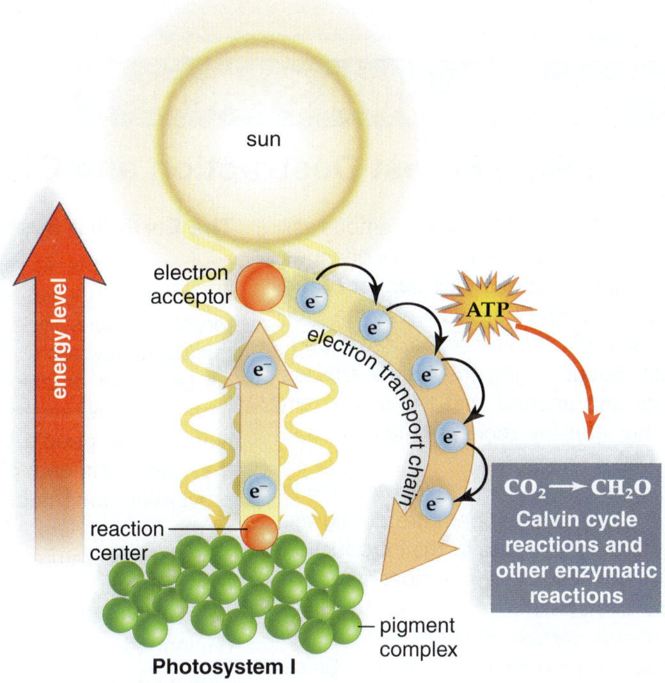

Figure 7.8 Cyclic electron pathway. Electrons leave and return to photosystem I. Energized electrons leave the photosystem I reaction center and are taken up by an electron acceptor, which passes them down an electron transport chain before they return to photosystem I. Only ATP production occurs as a result of this pathway.

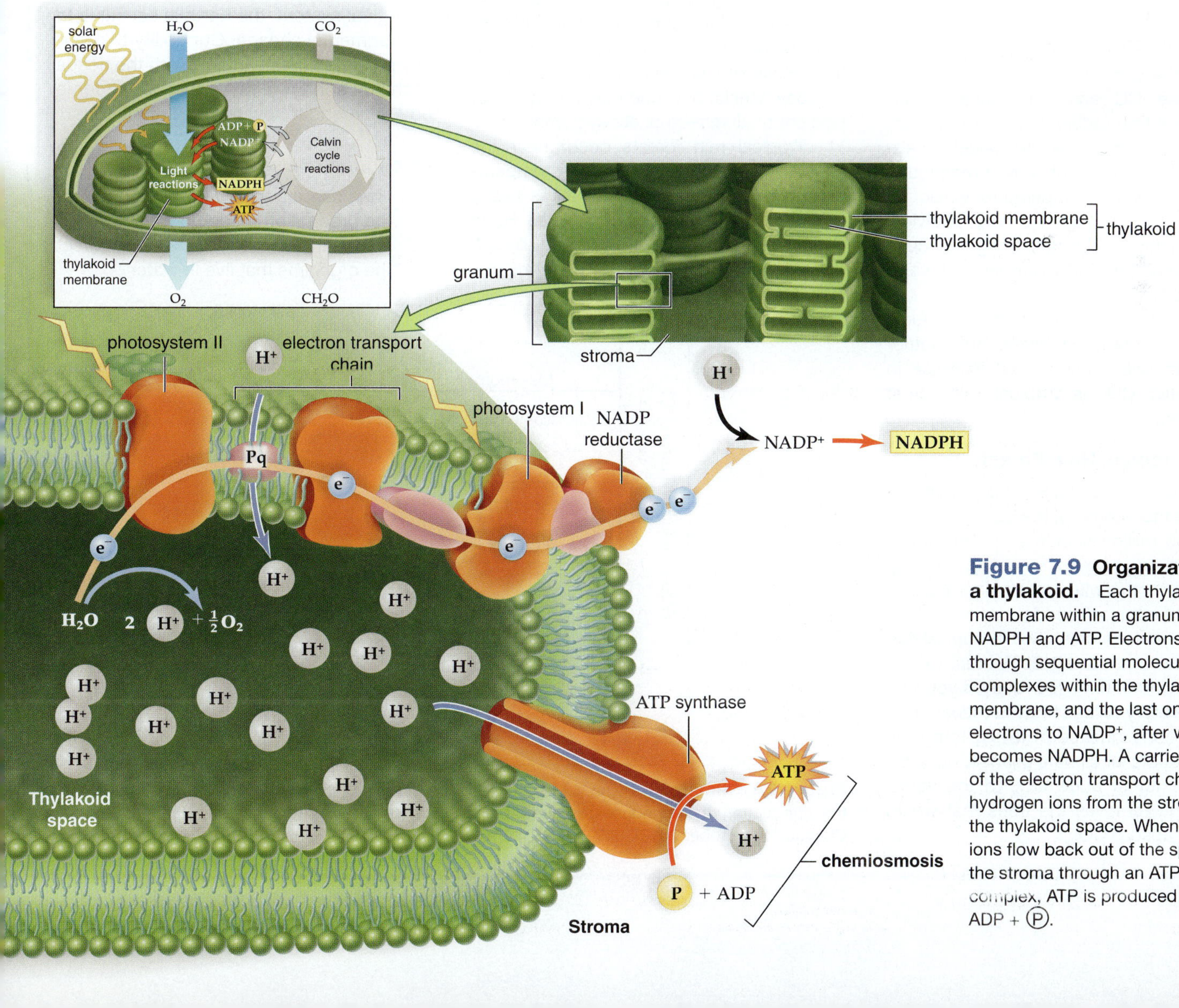

Figure 7.9 Organization of a thylakoid. Each thylakoid membrane within a granum produces NADPH and ATP. Electrons move through sequential molecular complexes within the thylakoid membrane, and the last one passes electrons to $NADP^+$, after which it becomes NADPH. A carrier at the start of the electron transport chain pumps hydrogen ions from the stroma into the thylakoid space. When hydrogen ions flow back out of the space into the stroma through an ATP synthase complex, ATP is produced from ADP + Ⓟ.

THEME Biological Systems

Tropical Rain Forest Destruction and Climate Change

Leonardo DiCaprio not only is a famous actor but also strives to make global changes through his foundation. One aspect of the Leonardo DiCaprio Foundation is the protection of tropical rain forests. Most people think about saving the fragile species of plants and animals that live in the rain forest, but globally there is a larger issue at hand.

Climate change is an expected rise in the average global temperature during the twenty-first century due to the introduction of certain gases into the atmosphere. For at least a thousand years prior to 1850, atmospheric carbon dioxide (CO_2) levels remained fairly constant at 0.028%. Since the 1850s, when industrialization began, the amount of CO_2 in the atmosphere has increased to 0.038% (Fig. 7A).

Role of Carbon Dioxide

In much the same way as the panes of a greenhouse, CO_2 and other gases in our atmosphere trap radiant heat from the sun. Therefore, these gases are called greenhouse gases. Without any greenhouse gases, the Earth's temperature would be about 33°C cooler than it is now. Likewise, increasing the concentration of greenhouse gases makes the Earth warmer and water more acidic.

Certainly, the burning of fossil fuels adds CO_2 to the atmosphere. But another factor that contributes to an increase in atmospheric CO_2 is tropical rain forest destruction.

Role of Tropical Rain Forests

Many scientists consider tropical rain forests to be the "lungs" of the Earth. Between 10 and 30 million hectares of rain forests are lost every year to ranching, logging, mining, and otherwise developing areas of the forest for human needs.

Each year, deforestation in tropical rain forests accounts for 10–20% of all CO_2 in the atmosphere. With your body, if you lose lung capacity, you lose body function. Similarly, the consequence of losing forests is greater trouble for climate change, because burning a forest adds CO_2 to the atmosphere and removes the trees that would ordinarily absorb CO_2.

The Earth Is a System

Carbon dioxide is removed from the air via photosynthesis, which takes place in forests, oceans, and other terrestrial and marine ecosystems. In fact, photosynthesis produces organic matter, which is estimated to be several hundred times the mass of the people living on Earth. Thus, these environments act as a sink for CO_2, preventing too much from accumulating in the atmosphere, where CO_2 can affect global temperatures and bring about climate change.

Despite their reduction in size from an original 15% to less than 5% of land surface today, tropical rain forests make a substantial contribution to global CO_2 removal. They are a critical element of the Earth's systems and, like any biological system, are essential for normal, healthy function. Tropical rain forests contribute greatly to the uptake of CO_2 and the productivity of photosynthesis, because they are the most efficient of all terrestrial ecosystems.

Tropical rain forests occur near the equator. They can exist wherever temperatures are above 26°C and rainfall is heavy (100–200 cm per year) and regular. Huge trees with buttressed trunks and broad, undivided, dark-green leaves predominate. Nearly all land plants in a tropical rain forest are woody, and woody vines are abundant.

It might be hypothesized that an increased amount of CO_2 in the atmosphere would cause photosynthesis to increase in the remaining portion of the forest. Recent studies, however, are showing that the opposite is true. Too much CO_2 can decrease photosynthesis, because increased temperatures can reduce water and mineral availability. Scientists working with wheat showed a decrease in the production of nitrogen-containing compounds; another study showed increased herbivory as plants were unable to produce their defense toxins at higher temperatures.

These and other studies show that, for the Earth, as for any biological system, equilibrium is necessary for health. As a biological system, the Earth is sensitive to environmental change. Our ability to properly balance human activity with the needs of the biosphere requires that we become educated about how the Earth functions.

Questions to Consider

1. How can a rise in temperature affect the production of food crops?
2. How can increased CO_2 levels affect the organims that live in water?

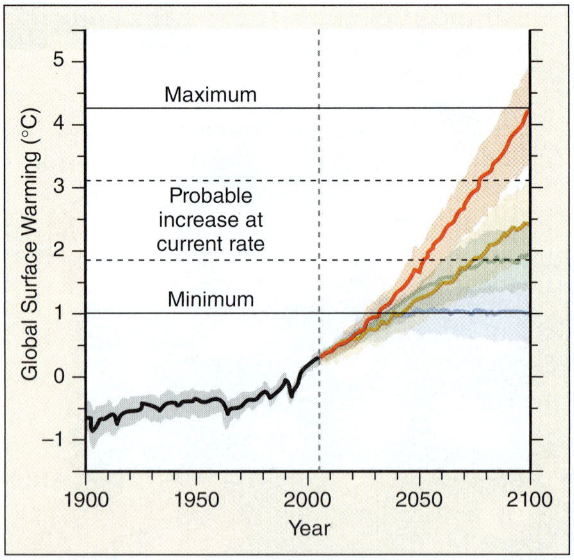

Figure 7A Climate change. Mean global temperature change is expected to rise due to the introduction of greenhouse gases into the atmosphere. (*Source:* nature.com: "Nature Climate Change," 3 [October 2012]: 369–73, doi:10.1038/nclimate1716.)

ATP Production

The thylakoid space acts as a reservoir for many hydrogen ions (H^+). First, each time water is oxidized, two H^+ remain in the thylakoid space. Second, as the electrons move from carrier to carrier via redox reactions along the electron transport chain, the electrons give up energy, which is used to pump H^+ from the stroma into the thylakoid space. Therefore, there are more H^+ in the thylakoid space than in the stroma. This difference and the resulting flow of H^+ (often referred to as protons in this context) from high to low concentration provide kinetic energy that allows an ATP synthase complex enzyme to enzymatically produce ATP from ADP + Ⓟ. This method of producing ATP is called **chemiosmosis,** because ATP production is tied to the establishment of an H^+ gradient (see Fig. 6.13).

Animation ATP Production in the Electron Transport Chain

Animation Proton Pump

Check Your Progress 7.3

1. Distinguish visible light from the electromagnetic spectrum.
2. Describe the movement of electrons from water to $NADP^+$ in the light reactions.

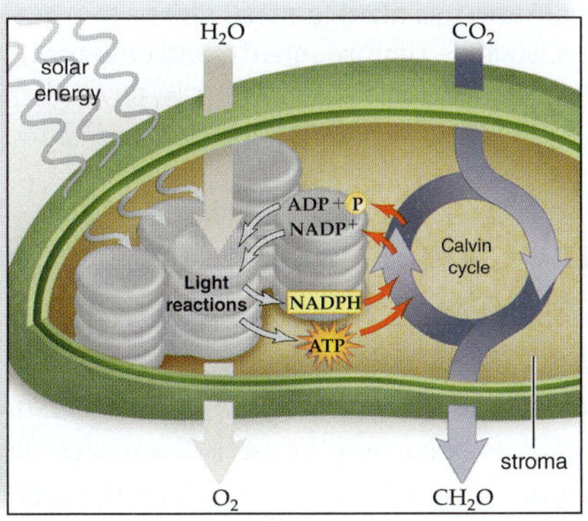

7.4 Plants Fix Carbon Dioxide

Learning Outcomes

Upon completion of this section, you should be able to

1. Describe the three steps of the Calvin cycle and when ATP and/or NADPH is needed.
2. Evaluate the significance of RuBP carboxylase enzyme to photosynthesis.
3. Explain how glyceraldehyde-3-phosphate (G3P) is used to produce other necessary plant molecules.

During the light reactions, the high-energy molecules ATP and NADPH were produced. The Calvin cycle, another series of chemical reactions, will use these high-energy molecules for an amazing process—carbon dioxide fixation. Carbon dioxide in its gas form is all around us in our atmosphere. We and other respiring organisms release it as waste during cellular respiration. Unfortunately, CO_2 is unattainable by heterotrophs—we cannot harness or extract CO_2 from the air and then use those carbon atoms to make sugar. Plants, and other autotrophs, can take the carbon from CO_2 gas and convert, or "fix," it in the bonds of a carbohydrate. The word *fixation* is not limited to photosynthesis. As you will learn in later chapters, some bacteria can undergo fixation by removing nitrogen from the air and fixing it into organic molecules.

The Calvin cycle is a series of reactions that can occur in the dark, but it uses the products of the light reactions to reduce carbon dioxide captured from the atmosphere to a carbohydrate. The Calvin cycle has three steps: (1) carbon dioxide fixation, (2) carbon dioxide reduction, and (3) regeneration of RuBP (Fig. 7.10).

Metabolites of the Calvin Cycle	
RuBP	ribulose-1,5-bisphosphate
3PG	3-phosphoglycerate
BPG	1,3-bisphosphoglycerate
G3P	glyceraldehyde-3-phosphate

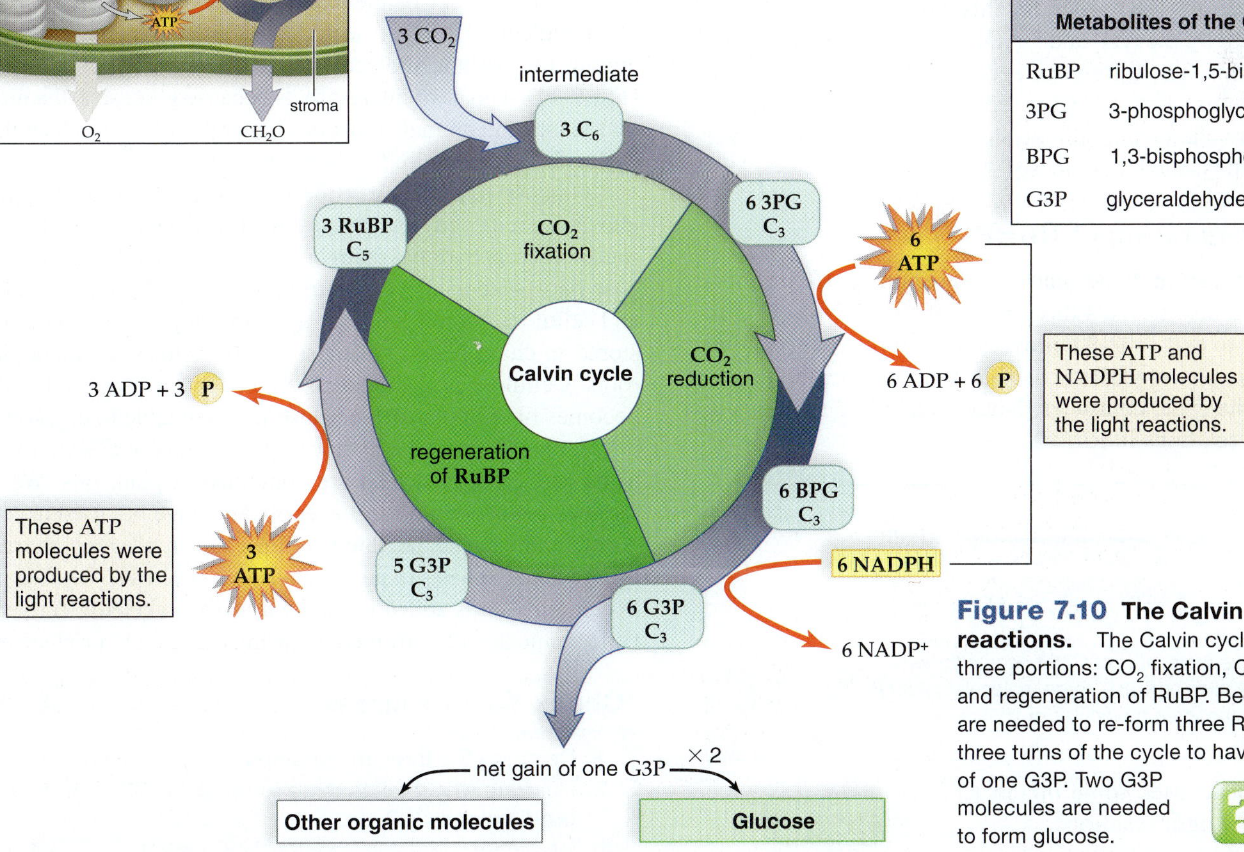

Figure 7.10 The Calvin cycle reactions. The Calvin cycle is divided into three portions: CO_2 fixation, CO_2 reduction, and regeneration of RuBP. Because five G3P are needed to re-form three RuBP, it takes three turns of the cycle to have a net gain of one G3P. Two G3P molecules are needed to form glucose.

Tutorial Calvin Cycle

Step 1: Fixation of Carbon Dioxide

Carbon dioxide fixation is the first step of the Calvin cycle. During this reaction, a molecule of carbon dioxide from the atmosphere is attached to RuBP (ribulose-1,5-bisphosphate), a 5-carbon molecule. The result is one 6-carbon molecule, which splits into two 3-carbon molecules.

The enzyme that speeds this reaction, called **RuBP carboxylase,** is a protein that makes up about 20–50% of the protein content of chloroplasts. The reason for its abundance may be that it is unusually slow—it processes only a few molecules of substrate per second compared to thousands per second for a typical enzyme—so there has to be a lot of it to keep the Calvin cycle going.

Step 2: Reduction of Carbon Dioxide

The first 3-carbon molecule in the Calvin cycle is called 3PG (3-phosphoglycerate). Each of two 3PG molecules undergoes reduction to G3P (glyceraldehyde-3-phosphate) in two steps:

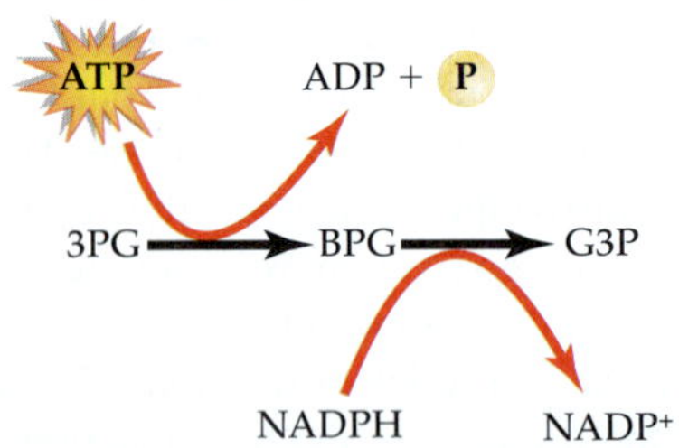

As 3PG becomes G3P, ATP becomes ADP + P, and NADPH becomes $NADP^+$.

Energy and electrons are needed for this reduction reaction, and they are supplied by the ATP and NADPH that were made during the light reactions. The difference between 3PG, BPG, and G3P (all with 3 carbons) is that G3P is reduced, has more electrons, and is now more chemically able to store energy and form larger organic molecules, such as glucose.

Step 3: Regeneration of RuBP

Notice that the Calvin cycle reactions in Figure 7.10 are multiplied by 3 because it takes three turns of the Calvin cycle to allow one G3P to exit. Why? For every three turns of the Calvin cycle, five molecules of G3P are used to re-form three molecules of RuBP, and the cycle continues. Notice that 5 × 3 (carbons in G3P) = 3 × 5 (carbons in RuBP):

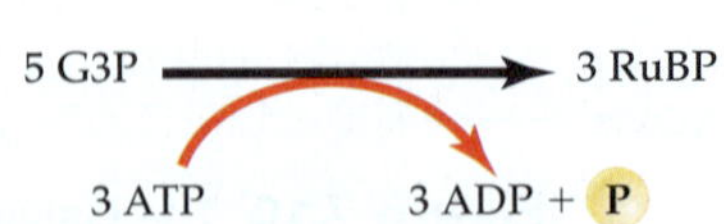

As five molecules of G3P become three molecules of RuBP, three molecules of ATP become three molecules of ADP + P.

This reaction also uses some of the ATP produced by the light reactions.

3D Animation Photosynthesis: Calvin Cycle

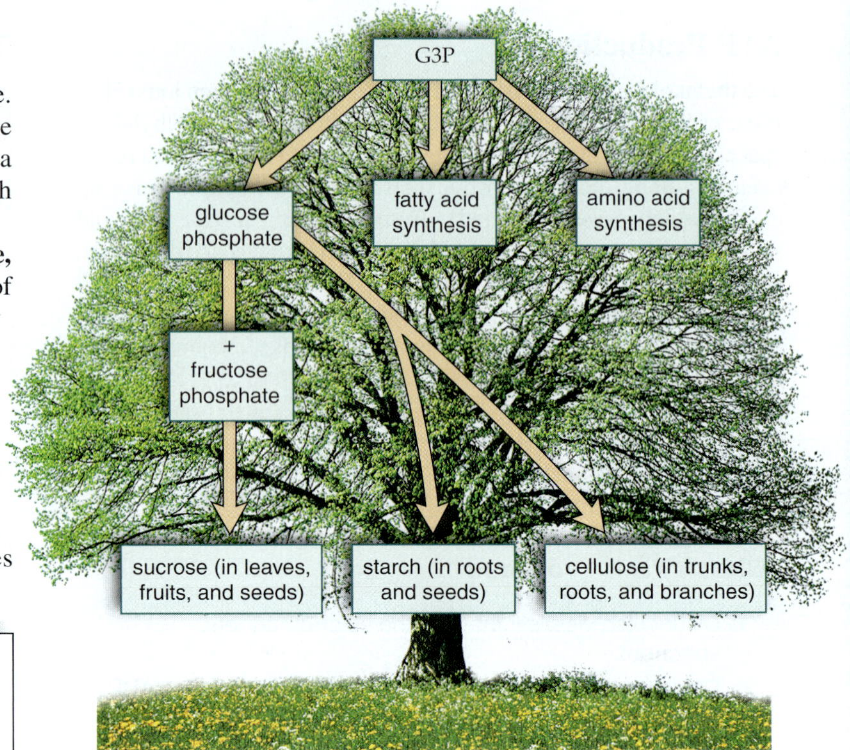

Figure 7.11 Fate of G3P. G3P is the first reactant in a number of plant cell metabolic pathways. From this starting point, different carbohydrates can be produced, such as sucrose, starch, and cellulose. Fatty acid synthesis leads to triglycerides making up plant oils, and production of amino acids allows the plant to make proteins.

The Importance of the Calvin Cycle

G3P is the product of the Calvin cycle that can be converted to other molecules a plant needs. Notice that glucose phosphate is among the organic molecules that result from G3P metabolism (Fig. 7.11). This is of interest to us because glucose is the molecule that plants and animals most often metabolize to produce the ATP molecules they require for their energy needs.

Glucose phosphate can be combined with fructose (and the phosphate removed) to form sucrose, the molecule that plants use to transport carbohydrates from one part of the plant to the other. Glucose phosphate is also the starting point for the synthesis of starch and cellulose. Starch is the storage form of glucose. Some starch is stored in chloroplasts, but most starch is stored in amyloplasts in roots. Cellulose is a structural component of plant cell walls and becomes fiber in our diet, because we are unable to digest it.

A plant can use the hydrocarbon skeleton of G3P to form fatty acids and glycerol, which are combined in plant oils. We are all familiar with corn oil, sunflower oil, and olive oil, used in cooking. As mentioned in the beginning of the chapter, researchers are modifying photosynthesis to produce oils that could also be used as fuel. When nitrogen is added to the hydrocarbon skeleton derived from G3P, amino acids are formed, allowing the plant to produce protein.

Check Your Progress 7.4

1. Describe the three major steps of the Calvin cycle.
2. Illustrate why it takes three turns of the Calvin cycle to produce one G3P.

7.5 Other Types of Photosynthesis

Learning Outcomes

Upon completion of this section, you should be able to

1. Compare the internal location of photosynthesis in C_3 and C_4 plants.
2. Contrast C_3/C_4 modes of photosynthesis with CAM photosynthesis.
3. Explain how different ways of achieving photosynthesis allow plants to adapt to particular environments.

The majority of plants, such as azaleas, maples, and tulips, carry on photosynthesis as previously described and are called **C_3 plants** (Fig. 7.12*a*). C_3 plants use the enzyme RuBP carboxylase to fix CO_2 to RuBP in mesophyll (photosynthetic) cells. The first detected molecule following fixation is the 3-carbon molecule 3PG:

$$\text{RuBP} + CO_2 \xrightarrow{\text{RuBP carboxylase}} 2\ \text{3PG}$$

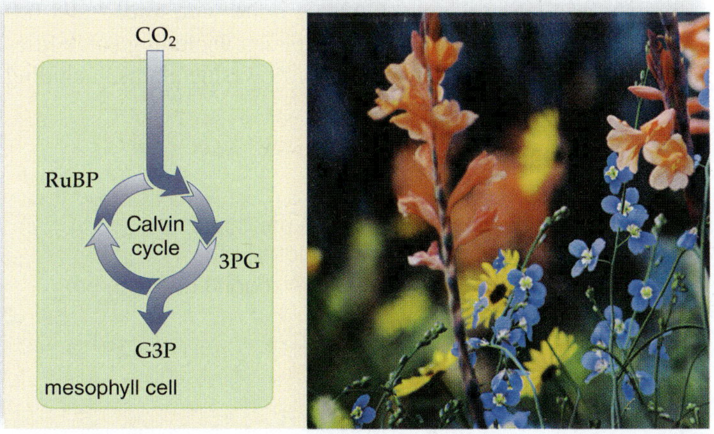

a. CO_2 fixation in a C_3 plant, wildflowers

b. CO_2 fixation in a C_4 plant, corn, *Zea mays*

Figure 7.12 Carbon dioxide fixation in C_3 and C_4 plants.
a. In C_3 plants, CO_2 is taken up by the Calvin cycle directly in mesophyll cells. **b.** C_4 plants form a C_4 molecule in mesophyll cells prior to releasing CO_2 to the Calvin cycle in bundle sheath cells.

As shown in Figure 7.2, leaves have small openings called stomata, through which water can leave and carbon dioxide (CO_2) can enter. If the weather is hot and dry, the stomata close, conserving water. (Water loss might cause the plant to wilt and die.) Now the concentration of CO_2 decreases in leaves, while O_2, a by-product of photosynthesis, increases. When O_2 rises in C_3 plants, RuBP carboxylase combines it with RuBP instead of CO_2. The result is one molecule of 3PG and the eventual release of CO_2. This is called **photorespiration,** because in the presence of light (*photo*), oxygen is taken up and CO_2 is released (*respiration*).

An adaptation called C_4 photosynthesis enables some plants to avoid photorespiration.

C_4 Photosynthesis

In a C_3 plant, the mesophyll cells contain well-formed chloroplasts and are arranged in parallel layers. In a C_4 leaf, the bundle sheath cells, as well as the mesophyll cells, contain chloroplasts. Further, the mesophyll cells are arranged concentrically around the bundle sheath cells:

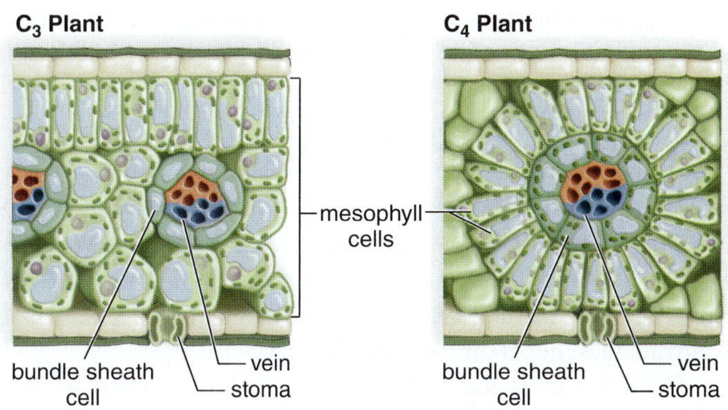

C_4 plants fix CO_2 to PEP (phosphoenolpyruvate, a C_3 molecule) using the enzyme PEP carboxylase (PEPCase). The result is oxaloacetate, a C_4 molecule:

$$\text{PEP} + CO_2 \xrightarrow{\text{PEPCase}} \text{oxaloacetate}$$

In a C_4 plant, CO_2 is taken up in mesophyll cells, and then malate, a reduced form of oxaloacetate, is pumped into the bundle sheath cells (Fig. 7.12*b*). Only here does CO_2 enter the Calvin cycle.

Because it takes energy to pump molecules, you would think that the C_4 pathway would be disadvantageous. Yet in hot, dry climates, the net photosynthetic rate of C_4 plants, such as sugarcane, corn, and Bermuda grass, is about two to three times that of C_3 plants (e.g., wheat, rice, and oats). Why do C_4 plants enjoy such an advantage? The answer is that they can avoid photorespiration, discussed previously. Photorespiration is wasteful, because it is not part of the Calvin cycle. Photorespiration does not occur in C_4 leaves because PEPCase, unlike RuBP carboxylase, does not combine with O_2. Even when stomata are closed, CO_2 is delivered to the Calvin cycle in the bundle sheath cells.

When the weather is moderate, C_3 plants ordinarily have the advantage, but when the weather becomes hot and dry, C_4 plants have the advantage, and we can expect them to predominate. In the early summer, C_3 plants such as Kentucky bluegrass and creeping bent grass predominate in lawns in the cooler parts of the United States, but by midsummer, crabgrass, a C_4 plant, begins to take over.

CAM Photosynthesis

CAM stands for crassulacean-acid metabolism; the Crassulaceae is a family of flowering succulent (water-containing) plants, such as a jade plant, that live in warm, dry regions of the world. CAM was first discovered in these plants, but now it is known to be prevalent among other groups of plants as well, such as pineapples.

Whereas a C_4 plant represents partitioning in space—carbon dioxide fixation occurs in mesophyll cells, while the Calvin cycle occurs in bundle sheath cells—CAM is partitioning by the use of time. During the night, CAM plants use PEPCase to fix some CO_2, forming C_4 molecules, which are stored in large vacuoles in mesophyll cells. During the day, C_4 molecules (malate) release CO_2 to the Calvin cycle when NADPH and ATP are available from the light reactions (Fig. 7.13). The primary advantage for this partitioning again has to do with the conservation of water. CAM plants open their stomata only at night; therefore, only at that time does atmospheric CO_2 enter the plant. During the day, the stomata close; this conserves water, but CO_2 cannot enter the plant. Photosynthesis in a CAM plant is minimal, because a limited amount of CO_2 is fixed at night, but it does allow CAM plants to live under stressful conditions.

CO_2 fixation in a CAM plant, pineapple, *Ananas comosus*

Figure 7.13 Carbon dioxide fixation in a CAM plant. CAM plants, such as pineapple, fix CO_2 at night, forming a C_4 molecule that is released to the Calvin cycle during the day.

Photosynthesis and Adaptation to the Environment

The different types of photosynthesis give us an opportunity to consider that organisms are metabolically adapted to their environment. Each method of photosynthesis has its advantages and disadvantages, depending on the climate.

C_4 plants most likely evolved in, and are adapted to, areas of high light intensities, high temperatures, and limited rainfall. C_4 plants, however, are more sensitive to cold, and C_3 plants do better than C_4 plants below 25°C. CAM plants, on the other hand, compete well with either type of plant when the environment is extremely arid. Surprisingly, CAM is quite widespread and has evolved into 23 families of flowering plants, including some lilies and orchids! And it is found among nonflowering plants, including some ferns and cone-bearing trees.

Check Your Progress 7.5

1. Describe some plants that use a method of photosynthesis other than C_3 photosynthesis.
2. Explain why C_4 photosynthesis is advantageous in hot, dry conditions.

Connecting *the* Concepts *with the* Themes

Evolution

- Plants have evolved to capture solar energy and store it in carbon-based organic nutrients. These are passed on to organisms that have evolved to feed on plants, in turn to organisms that have evolved to feed on the plant-eaters, and so on in a food web.
- Plants are called autotrophs because they make their own organic food. Heterotrophs are organisms that take in organic food made by other organisms.

Nature of Science

- The amount of carbon dioxide in the atmosphere is increasing steadily, in part because of burning fossil fuels. This buildup of carbon dioxide causes global climate change, and we can use the tools of science to understand the scope of this change.
- Scientists are learning how to manipulate photosynthesis to create commercially important products such as oils and turpentines.

Biological Systems

- Autotrophs take in carbon dioxide when they photosynthesize. Carbon dioxide is returned to the atmosphere when autotrophs and heterotrophs carry on cellular respiration. In this way, carbon atoms cycle through living organisms.
- Energy does not cycle; therefore, all life is dependent on the ability of plants to capture solar energy and produce carbohydrate molecules from CO_2.

MEDIA STUDY TOOLS http://connect.mheducation.com

Animations	3D Animations	Tutorials
7.3 ATP Production in the Electron Transport Chain • Proton Pump	**7.1** Photosynthesis: Structure of a Chloroplast **7.3** Photosynthesis: Properties of Light • Photosynthesis: Light-Dependent Reactions **7.4** Photosynthesis: Calvin Cycle	**7.3** Noncyclic Photosynthesis **7.4** Calvin Cycle

SUMMARIZE

7.1 Photosynthetic Organisms

Photosynthesis produces carbohydrates and releases oxygen, both of which are used by the majority of living organisms. Cyanobacteria, algae, and plants are **autotrophs** and carry out photosynthesis. **Heterotrophs** consume the products of photosynthesis. In plants, gases enter and exit through **stomata,** and photosynthesis takes place in **chloroplasts.** A chloroplast is enclosed by a double membrane and contains two main components: the semifluid **stroma** and the membranous **grana** made up of **thylakoids.** Thylakoids contain chlorophyll—a pigment that captures solar energy.

7.2 The Process of Photosynthesis

The overall equation for photosynthesis shows that it is a redox reaction. Carbon dioxide is reduced, and water is oxidized. During photosynthesis, the **light reactions** take place in the thylakoid membranes, and the **Calvin cycle reactions** take place in the stroma.

7.3 Plants Convert Solar Energy

Photosynthesis uses solar energy in the visible-light range. Specifically, **chlorophylls** *a* and *b* absorb violet, blue, and red wavelengths best and reflect green light, whereas the **carotenoids** absorb violet-blue-green light and reflect yellow-to-orange light. Specific wavelengths absorbed by pigments determine a pigment's **absorption spectrum.**

The light reactions contain two **photosystems** (**PS**), which are pigment complexes that capture solar energy. In the light reactions, a **noncyclic pathway** has an electron flow that begins when solar energy enters PS II and energizes chlorophyll *a* electrons. The oxidation (splitting) of water replaces these electrons in the reaction-center chlorophyll *a* molecules. Oxygen is released to the atmosphere, and hydrogen ions (H^+) remain in the thylakoid space. Electrons are ultimately passed to PS I via an **electron transport chain** (**ETC**), which pumps hydrogen ions across the thylakoid membrane and results in **chemiosmosis**—a gradient used to make ATP via **ATP synthase.** Light-energized electrons from PS I are captured by $NADP^+$, which combines with H^+ from the stroma to become NADPH. **The cyclic pathway** has an electron flow in the light reactions that pumps additional hydrogen ions and contributes to ATP production.

7.4 Plants Fix Carbon Dioxide

The energy yield of the light reactions is stored in ATP and NADPH. These molecules are used by the Calvin cycle reactions for **carbon dioxide fixation**—a reduction of CO_2 to carbohydrate, namely G3P, which is then converted to all the organic molecules a plant needs.

During the first stage of the Calvin cycle, the enzyme **RuBP carboxylase** fixes CO_2 to RuBP, producing a 6-carbon molecule that immediately breaks down to two C_3 molecules. During the second stage, CO_2 (incorporated into an organic molecule) is reduced to carbohydrate (CH_2O). This step requires the NADPH and some of the ATP from the light reactions. For every three turns of the Calvin cycle, the net gain is one G3P molecule; the other five G3P molecules are used to re-form three molecules of RuBP, which also requires ATP. It takes two G3P molecules to make one glucose molecule.

7.5 Other Types of Photosynthesis

Plants have adapted ways other than the **C_3 plant** process to photosynthesize in various environments. In C_4 plants, carbon dioxide is first fixed in mesophyll cells via PEPCase, is transported to a different location in bundle sheath cells, and is then released to the Calvin cycle. PEPCase has an advantage over RuBP carboxylase, because it doesn't participate in **photorespiration. C_4 plants** avoid the photorespiration complication by separating where carbon fixation occurs from where the Calvin cycle occurs.

CAM plants, which live in hot, dry environments, cannot leave their stomata open during the day, or they will die from loss of water. CAM plants fix carbon only at night, conserving water. Stores of CO_2 are released to the Calvin cycle during the day, when photosynthesis is possible. CAM plants avoid drying out by separating when they bring in carbon dioxide from when they release it to the Calvin cycle.

ASSESS

Choose the best answer for each question.

7.1 Photosynthetic Organisms

1. All of the following are examples of organisms that can photosynthesize EXCEPT
   a. cyanobacteria.
   b. pine trees.
   c. cacti.
   d. mushrooms.
   e. algae.
2. Carbon dioxide enters leaves through a small opening called the
   a. stoma.
   b. stroma.
   c. thylakoid.
   d. granum.
   e. mesophyll.

7.2 The Process of Photosynthesis

3. The function of light reactions is to
   a. obtain CO_2.
   b. make carbohydrate.
   c. convert light energy into a usable form of chemical energy.
   d. regenerate RuBP.
4. The Calvin cycle reactions
   a. produce carbohydrate.
   b. convert one form of chemical energy into a different form of chemical energy.
   c. regenerate more RuBP.
   d. use the products of the light reactions.
   e. All of these are correct.

7.3 Plants Convert Solar Energy

5. The final acceptor of electrons during the light reactions of the noncyclic electron pathway is
   a. PS I.
   b. PS II.
   c. ATP.
   d. $NADP^+$.
   e. water.
6. The oxygen given off by photosynthesis comes from
   a. H_2O.
   b. CO_2.
   c. glucose.
   d. RuBP.
7. Chemiosmosis
   a. depends on complexes in the thylakoid membrane.
   b. depends on an electrochemical gradient.
   c. depends on a difference in H^+ concentration between the thylakoid space and the stroma.
   d. results in ATP formation.
   e. All of these are correct.

7.4 Plants Fix Carbon Dioxide

For questions 8–10, indicate whether the statement is true (T) or false (F).

8. RuBP carboxylase is the enzyme that fixes carbon dioxide to RuBP in the Calvin cycle. _______
9. When 3PG becomes G3P during the light reactions, carbon dioxide is reduced to carbohydrate. _______
10. NADPH and ATP cycle between the Calvin cycle and the light reactions constantly. _______
11. The ATP and NADPH from the light reactions are used to
    a. split water.
    b. cause RuBP carboxylase to fix CO_2.
    c. re-form the photosystems.
    d. cause electrons to move along their pathways.
    e. convert 3PG to G3P.

7.5 Other Types of Photosynthesis

12. CAM photosynthesis
    a. is the same as C_4 photosynthesis.
    b. is an adaptation to cold environments in the Southern Hemisphere.
    c. is prevalent in desert plants that close their stomata during the day.
    d. occurs in plants that live in marshy areas.
    e. stands for "chloroplasts and mitochondria."
13. C_4 photosynthesis
    a. is the same as C_3 photosynthesis, because it takes place in chloroplasts.
    b. occurs in plants whose bundle sheath cells contain chloroplasts.
    c. takes place in plants such as wheat, rice, and oats.
    d. is an advantage when the weather is hot and dry.
    e. Both b and d are correct.

ENGAGE

LEARNSMART LABS®

The following LearnSmart Labs contain exercises that are related to the content of this chapter:

- Photosynthesis

THINKING SCIENTIFICALLY

1. In the image below, the alga is the autotroph and undergoes photosynthesis, whereas the bacteria (small, black dots) are the heterotrophs. Explain why the bacteria are clustered at the far ends after the alga was exposed to different color wavelengths of light.

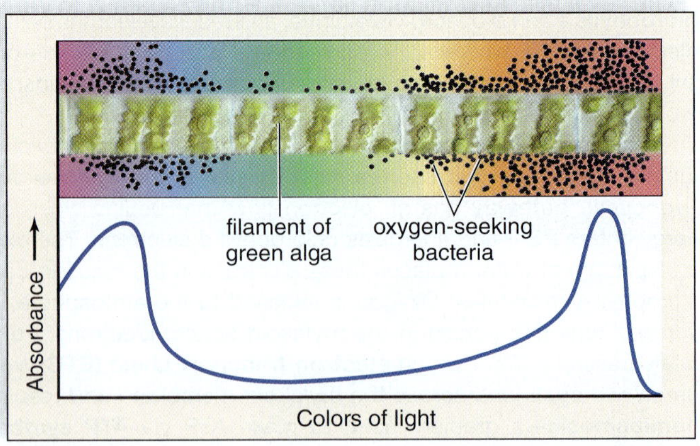

2. A Belgian doctor, Jan Baptista van Helmont (1580–1644) planted a small willow tree in a pot of soil. He weighed the tree and the soil. The tree was watered for 5 years and weighed 74.4 kg more than when he began, and the soil lost 57 g of mass. Explain what accounts for the plant's increase in biomass.

Every cell of this rock climber is manufacturing and using ATP.

8

Cellular Respiration

Chapter Outline

8.1 Overview of Cellular Respiration 130

8.2 Outside the Mitochondria: Glycolysis 132

8.3 Outside the Mitochondria: Fermentation 134

8.4 Inside the Mitochondria 136

8.5 Metabolism 141

Before You Begin

Before beginning this chapter, take a few moments to review the following discussions.

Figure 6.3 How does an ATP molecule store energy?

Section 6.4 How are high-energy electrons used to make energy for cellular work?

Figure 7.5 Where does the glucose that we metabolize come from?

A rock climber, a bacterium moving through a solution, an ocelot climbing a tree, or a snail moving slowly to hide under a rock—each, including the tree, is making and using ATP. ATP is an ancient "molecular fossil." Its molecular structure, plus its presence in the first cell or cells that arose on Earth, accounts for its being the universal energy currency of cells.

ATP is unique among the cell's storehouse of chemicals; amino acids join to make a protein, and nucleotides join to make DNA or RNA, but ATP is singular and works alone. Whether you go skiing, take an aerobics class, or just hang out, ATP molecules provide the energy needed for nerve conduction, muscle contraction, and any other cellular process that requires energy. Cellular respiration, by which cells harvest the energy of organic compounds and convert it to ATP molecules, is the topic of this chapter. It's a process that requires many steps and involves the cytoplasm and the mitochondria, the powerhouses of the cell.

As you read through this chapter, think about the following questions:

1. How does the ATP molecule store chemical energy needed to run biological processes?
2. How are enzymes involved in regulating energy metabolism?
3. If nearly all life on Earth uses ATP, what does that indicate about its origins and biological importance?

Following *the* Themes

CHAPTER 8 CELLULAR RESPIRATION

UNIT 1 THE CELL

Evolution	Cellular respiration is the process by which the majority of the life on Earth generates ATP for cellular processes.
Nature of Science	An understanding of cellular respiration is useful in exploring the nature of human diseases and nutritional needs.
Biological Systems	The ATP produced at the cellular level is used to power all of the activities of an organism.

8.1 Overview of Cellular Respiration

Learning Outcomes

Upon completion of this section, you should be able to

1. Describe the overall reaction for glucose breakdown and show that it is a redox reaction.
2. Examine the role of the NADH and $FADH_2$ redox reactions in cellular respiration.
3. Summarize the phases of cellular respiration.

Cellular respiration is the process by which cells acquire energy by breaking down nutrient molecules produced by photosynthesizers. Cellular respiration requires oxygen (O_2) and gives off carbon dioxide (CO_2), which, in effect, is the opposite of photosynthesis. In fact, it is the reason any animal, such as an ocelot or a human, breathes (Fig. 8.1) and why plants require a supply of oxygen. This chemical interaction between animals and plants is important, because animals, like humans, breathe the oxygen made by photosynthesizers. Most often, cellular respiration involves the complete breakdown of glucose to carbon dioxide and water (H_2O):

$$\underset{\text{glucose}}{C_6H_{12}O_6} + 6\,O_2 \longrightarrow 6\,CO_2 + 6\,H_2O + \textbf{energy}$$

Oxidation ($C_6H_{12}O_6 \rightarrow 6\,CO_2$); Reduction ($6\,O_2 \rightarrow 6\,H_2O$)

This equation shows that cellular respiration is an oxidation-reduction reaction. Recall that oxidation is the loss of electrons and reduction is the gain of electrons (see section 6.4); therefore, glucose has been oxidized and O_2 has been reduced. Also remember that a hydrogen atom consists of a hydrogen ion plus an electron ($H^+ + e^-$). Therefore, when hydrogen atoms are removed from glucose, so are electrons; similarly, when hydrogen atoms are added to oxygen, so are electrons.

Glucose is a high-energy molecule, but its breakdown products, CO_2 and H_2O, are low-energy molecules. Therefore, as the equation shows, energy is released. This is the energy that will be used to produce ATP molecules. The cell carries out cellular respiration in order to build up ATP molecules.

The pathways of cellular respiration allow the energy within a glucose molecule to be released slowly, so that ATP can be produced gradually. Cells would lose a tremendous amount of energy if glucose breakdown occurred all at once—most of the energy would become nonusable heat. The step-by-step breakdown of glucose to CO_2 and H_2O usually produces a maximum yield of 36 to 38 ATP molecules, dependent on the conditions to be discussed later. The energy in these ATP molecules is equivalent to about 39% of the energy that was available in glucose. Even though it might seem less efficient, this conversion is more efficient than many others; for example, only between 20% and 30% of the energy within gasoline is converted to the motion of a car.

MP3 Cellular Respiration

NAD⁺ and FAD

Cellular respiration involves many individual metabolic reactions, each one catalyzed by its own enzyme. Enzymes of particular significance are those that use **NAD⁺**, a coenzyme of oxidation-reduction (sometimes called a redox coenzyme). When a metabolite is oxidized, NAD^+ accepts two electrons plus a hydrogen ion (H^+), and NADH results. The electrons received by NAD^+ are high-energy electrons that are usually carried to the electron transport chain (see Fig. 6.12):

$$NAD^+ + 2\,e^- + H^+ \longrightarrow NADH$$

NAD^+ can oxidize a metabolite by accepting electrons and can reduce a metabolite by giving up electrons. Only a small amount of NAD^+ needs to be present in a cell, because each NAD^+ molecule is used over and over again. **FAD,** another coenzyme of

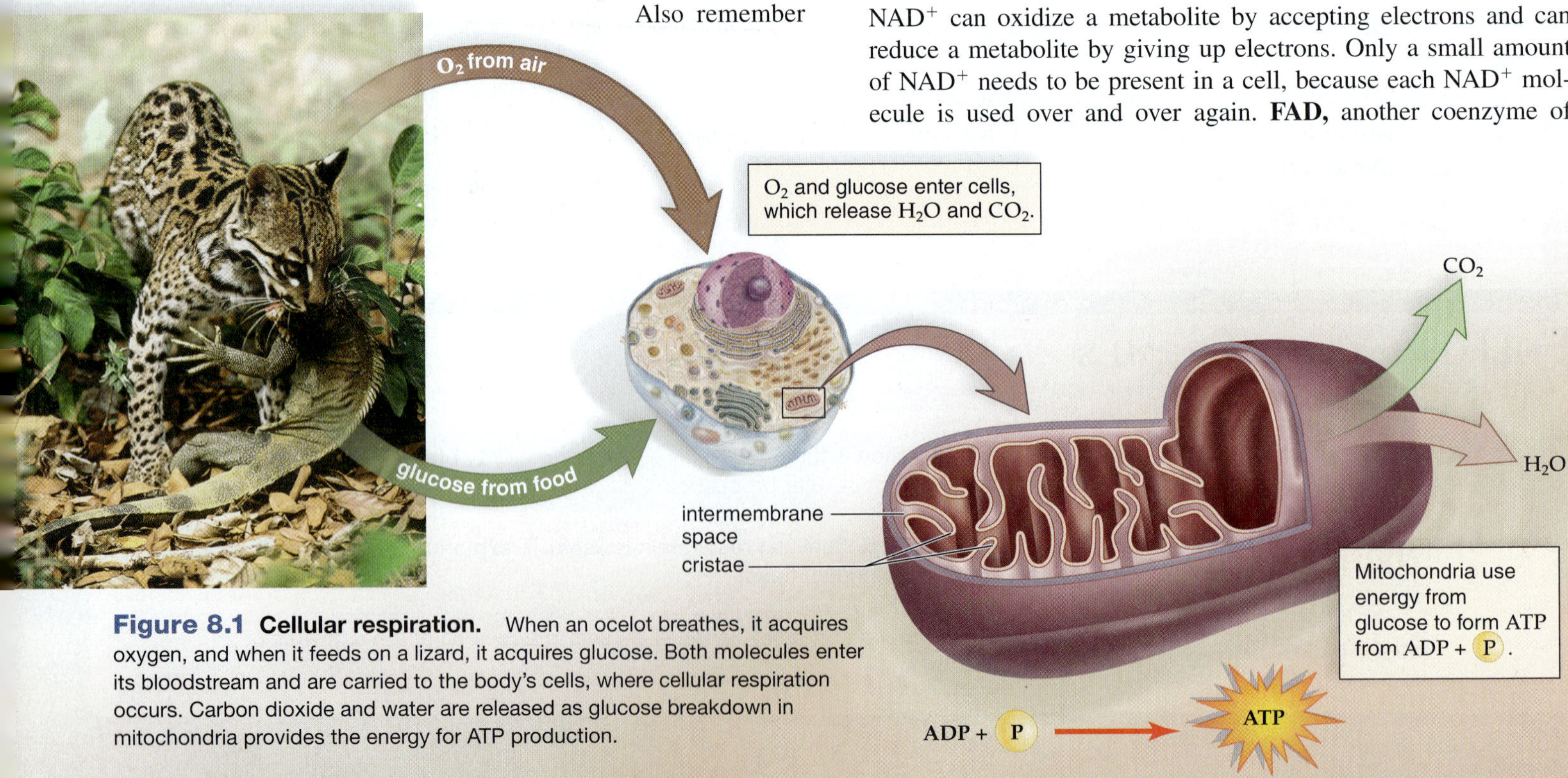

Figure 8.1 Cellular respiration. When an ocelot breathes, it acquires oxygen, and when it feeds on a lizard, it acquires glucose. Both molecules enter its bloodstream and are carried to the body's cells, where cellular respiration occurs. Carbon dioxide and water are released as glucose breakdown in mitochondria provides the energy for ATP production.

oxidation-reduction, is sometimes used instead of NAD^+. FAD accepts two electrons and two hydrogen ions (H^+) to become $FADH_2$.

Animation How the NAD^+ Works

Phases of Cellular Respiration

Cellular respiration involves four phases: glycolysis, the preparatory reaction, the citric acid cycle, and the electron transport chain (Fig. 8.2). Glycolysis takes place outside the mitochondria and does not require the presence of oxygen. Therefore, glycolysis is **anaerobic.** The other phases of cellular respiration take place inside the mitochondria, where oxygen is the final acceptor of electrons. Because they require oxygen, these phases are called **aerobic.**

During these phases, notice where CO_2 and H_2O, the end products of cellular respiration, and ATP, the main outcome of respiration, are produced.

- *Glycolysis* (Gk. *glycos,* "sugar"; *lysis,* "splitting") is the breakdown of glucose (a 6-carbon molecule) to two molecules of pyruvate (two 3-carbon molecules). Oxidation results in NADH and provides enough energy for the net gain of two ATP molecules.
- The *preparatory* (*prep*) *reaction* takes place in the matrix of the mitochondrion. Pyruvate is broken down from a 3-carbon (C_3) to a 2-carbon (C_2) acetyl group, and a 1-carbon CO_2 molecule is released. Since glycolysis ends with two molecules of pyruvate, the prep reaction occurs twice per glucose molecule.
- The *citric acid cycle* also takes place in the matrix of the mitochondrion. Each 2-carbon acetyl group matches up with a 4-carbon molecule, forming two 6-carbon citrate molecules. As citrate bonds are broken and oxidation occurs, NADH and $FADH_2$ are formed, and two CO_2 per citrate are released. The citric acid cycle is able to produce one ATP per turn. Because two acetyl groups enter the cycle per glucose molecule, the cycle turns twice.
- The *electron transport chain* (*ETC*) is a series of carriers on the cristae of the mitochondria. NADH and $FADH_2$ give up their high-energy electrons to the chain. Energy is released and captured as the electrons move from a higher-energy to a lower-energy state during each redox reaction. Later, this energy is used for the production of between 32 and 34 ATP by chemiosmosis. After oxygen receives electrons at the end of the chain, it combines with hydrogen ions (H^+) and becomes water (H_2O).

Pyruvate, the end product of glycolysis, is a pivotal metabolite; its further treatment depends on whether oxygen is available. If oxygen is available, pyruvate enters a mitochondrion and is broken down completely to CO_2 and H_2O, as shown in the cellular respiration equation (page 130). If oxygen is not available, pyruvate is further metabolized in the cytoplasm by an anaerobic process called **fermentation.** Fermentation results in a net gain of only two ATP per glucose molecule.

Check Your Progress 8.1

1. Describe how the formula for cellular respiration includes both oxidation and reduction reactions.
2. Explain why NAD^+ and FAD are needed during cellular respiration.
3. Describe the four phases of complete glucose breakdown, including which release CO_2 and which produce H_2O.

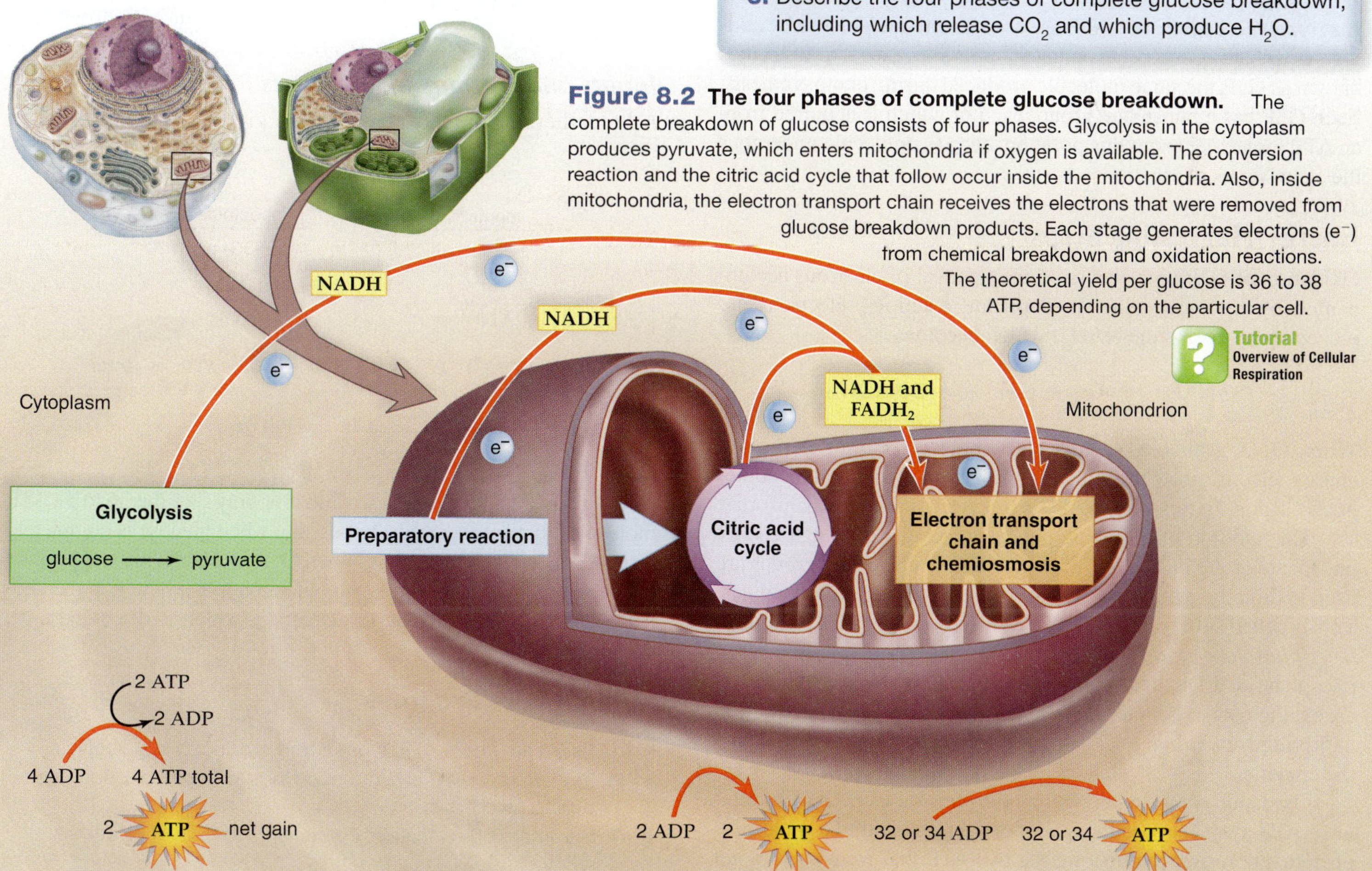

Figure 8.2 The four phases of complete glucose breakdown. The complete breakdown of glucose consists of four phases. Glycolysis in the cytoplasm produces pyruvate, which enters mitochondria if oxygen is available. The conversion reaction and the citric acid cycle that follow occur inside the mitochondria. Also, inside mitochondria, the electron transport chain receives the electrons that were removed from glucose breakdown products. Each stage generates electrons (e^-) from chemical breakdown and oxidation reactions. The theoretical yield per glucose is 36 to 38 ATP, depending on the particular cell.

Tutorial Overview of Cellular Respiration

8.2 Outside the Mitochondria: Glycolysis

Learning Outcomes

Upon completion of this section, you should be able to

1. Describe the role of glycolysis in cellular respiration.
2. List the inputs and outputs of glycolysis.
3. Explain how energy-investment and energy-harvesting steps of glycolysis result in two net ATP.

Glycolysis, which takes place within the cytoplasm outside the mitochondria, is the breakdown of C_6 (6-carbon) glucose to two C_3 (3-carbon) pyruvate molecules. Since glycolysis occurs universally in organisms, it most likely evolved before the citric acid cycle and the electron transport chain. This may be why glycolysis occurs in the cytoplasm and does not require oxygen. There was no free oxygen in Earth's early atmosphere.

Glycolysis is a series of ten reactions, and just as you would expect for a metabolic pathway, each step has its own enzyme. The pathway can be conveniently divided into the energy-investment step and the energy-harvesting steps. During the energy-investment step, ATP is used to "jump-start" glycolysis. During the energy-harvesting steps, four total ATP are made, producing two net ATP overall.

Animation How Glycolysis Works

Energy-Investment Step

As glycolysis begins, two ATP are used to activate glucose by adding phosphate. Glucose eventually splits into two C_3 molecules known as G3P, the same molecule produced during photosynthesis. Each G3P has a phosphate group, each of which is acquired from an ATP molecule. From this point on, each C_3 molecule undergoes the same series of reactions.

Energy-Harvesting Steps

Oxidation of G3P now occurs by the removal of electrons accompanied by hydrogen ions. In duplicate reactions, electrons are picked up by coenzyme NAD^+, which becomes

$$2\,NAD^+ + 4\,e^- + 2\,H^+ \longrightarrow 2\,NADH$$

When O_2 is available, each NADH molecule carries two high-energy electrons to the electron transport chain and becomes NAD^+ again. In this way, NAD^+ is recycled and used again.

The addition of inorganic phosphate results in a high-energy phosphate group on each C_3 molecule. These phosphate groups are used to directly synthesize two ATP in the later steps of glycolysis. This is called **substrate-level ATP synthesis,** also called *substrate-level phosphorylation,* because an enzyme passes a high-energy phosphate to ADP, and ATP results (Fig. 8.3). Notice that this is an example of a coupled reaction: An energy-releasing reaction is driving forward an energy-requiring reaction on the surface of the enzyme.

Oxidation occurs again, but by the removal of H_2O. Substrate-level ATP synthesis occurs again per each C_3, and two molecules of pyruvate result. Subtracting the two ATP that were used to get started, and the four ATP produced overall, there is a net gain of two ATP from glycolysis (Fig. 8.4).

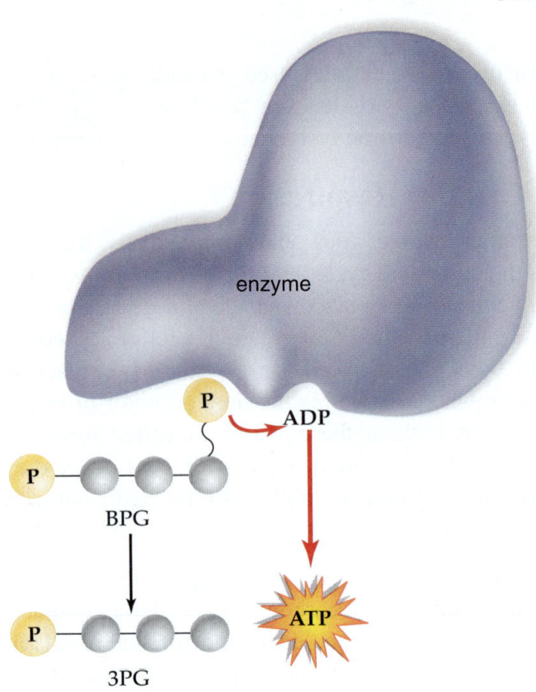

Figure 8.3 Substrate-level ATP synthesis. Substrates participating in the reaction are oriented on the enzyme. A phosphate group is transferred to ADP, producing one ATP molecule. During glycolysis (see Fig. 8.4), BPG is a C_3 substrate (each gray ball is a carbon atom) that gives up a phosphate group to ADP. This reaction occurs twice per glucose molecule.

3D Animation Cellular Respiration: Glycolysis

Inputs and Outputs of Glycolysis

All together, the inputs and outputs of glycolysis are as follows:

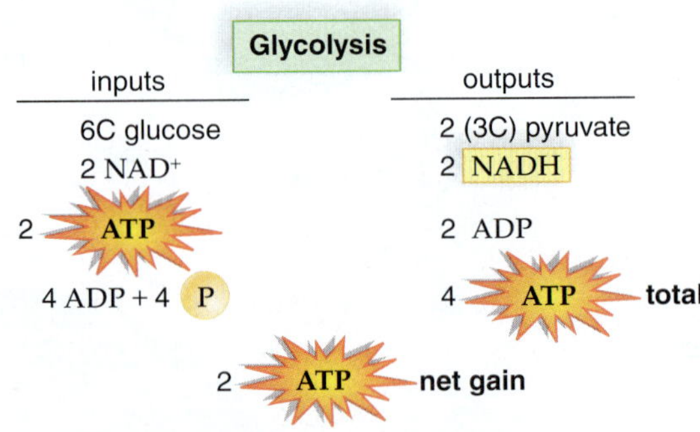

Notice that, so far, we have accounted for only 2 of the 36 to 38 ATP molecules that are theoretically possible when glucose is completely broken down to CO_2 and H_2O. When O_2 is available, the end product of glycolysis, pyruvate, enters the mitochondria, where it is metabolized. If O_2 is not available, fermentation, which is discussed next, occurs.

Check Your Progress 8.2

1. Examine where ATP is used and produced in glycolysis.
2. Explain how ATP is produced from ADP and phosphate during glycolysis.
3. Summarize the location, inputs, and outputs of glycolysis.

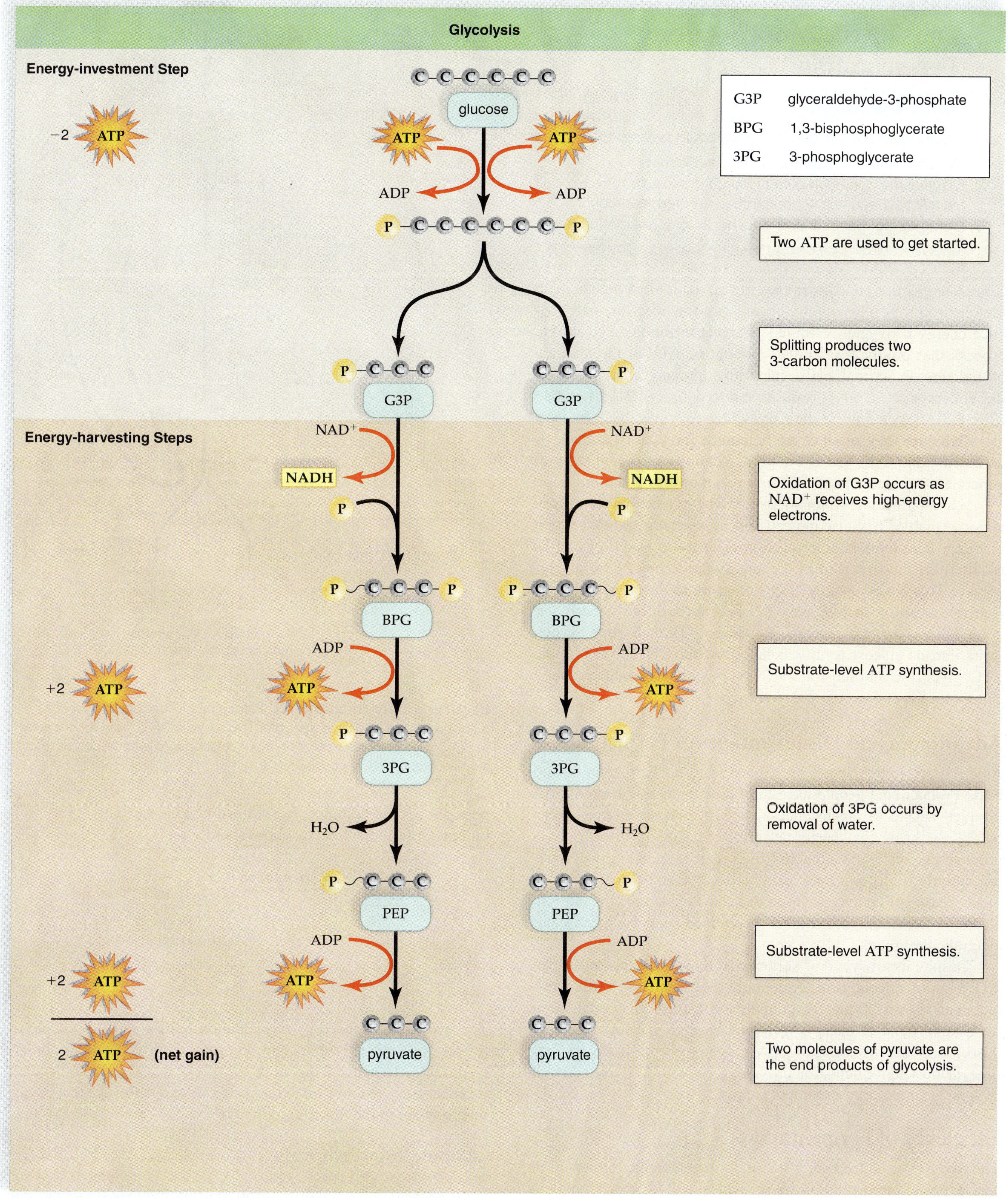

Figure 8.4 Glycolysis. This metabolic pathway begins with C_6 glucose (each gray ball is a carbon atom) and ends with two C_3 pyruvate molecules. Net gain of two ATP molecules can be calculated by subtracting those expended during the energy-investment step from those produced during the energy-harvesting steps. Each of the ten steps is catalyzed by a specialized enzyme.

8.3 Outside the Mitochondria: Fermentation

Learning Outcomes

Upon completion of this section, you should be able to

1. Summarize the two fermentation pathways.
2. Discuss the conditions under which organisms may switch between cellular respiration and fermentation.
3. Compare the benefits and drawbacks of fermentation.

Complete glucose breakdown requires an input of oxygen to keep the electron transport chain working. So how does the cell produce energy if oxygen is limited? **Fermentation** is an anaerobic process that produces a limited amount of ATP in the absence of oxygen. In animal cells, including human cells, pyruvate, the end product of glycolysis, is reduced by NADH to lactate (Fig. 8.5). Depending on their particular enzymes, bacteria vary as to whether they produce an organic acid, such as lactate, or an alcohol and CO_2. Yeasts are good examples of organisms that generate ethyl alcohol and CO_2 as a result of fermentation.

Why is it beneficial for pyruvate to be reduced when oxygen is not available? Because the cell still needs energy when oxygen is absent. The fermentation reaction regenerates NAD^+, which is required for the first step in the energy-harvesting phase of glycolysis. This NAD^+ is now "free" to return to the earlier reaction (see return arrow in Fig. 8.5) and become reduced once more. Although this process generates much less ATP than when oxygen is present and glucose is fully metabolized into CO_2 and H_2O in the ETC, glycolysis and substrate-level ATP synthesis produce enough energy for the cell to continue working.

Advantages and Disadvantages of Fermentation

As discussed in the Nature of Science feature, "Fermentation and Food Production," people have long used anaerobic bacteria that produce lactate to create cheese, yogurt, and sauerkraut—even before we knew that bacteria were responsible! Other bacteria produce chemicals of industrial importance, including isopropanol, butyric acid, proprionic acid, and acetic acid when they ferment. Yeasts, of course, are used to make breads rise. In addition, alcoholic fermentation is utilized to produce wine, beer, and other alcoholic beverages.

Despite its low yield of only two ATP made by substrate-level ATP synthesis, lactic acid fermentation is essential to certain animals and tissues. Typically, animals use lactic acid fermentation for a rapid burst of energy, such as a cheetah chasing a gazelle. Also, when muscles are working vigorously over a short period of time, lactic acid fermentation provides them with ATP, even though oxygen is temporarily in limited supply.

Efficiency of Fermentation

The two ATP produced per glucose during alcoholic fermentation and lactic acid fermentation are equivalent to 14.6 kcal. Complete glucose breakdown to CO_2 and H_2O represents a possible energy yield of 686 kcal per molecule. Therefore, the efficiency of fermentation is only 14.6 kcal/686 kcal × 100, or 2.1% of the total

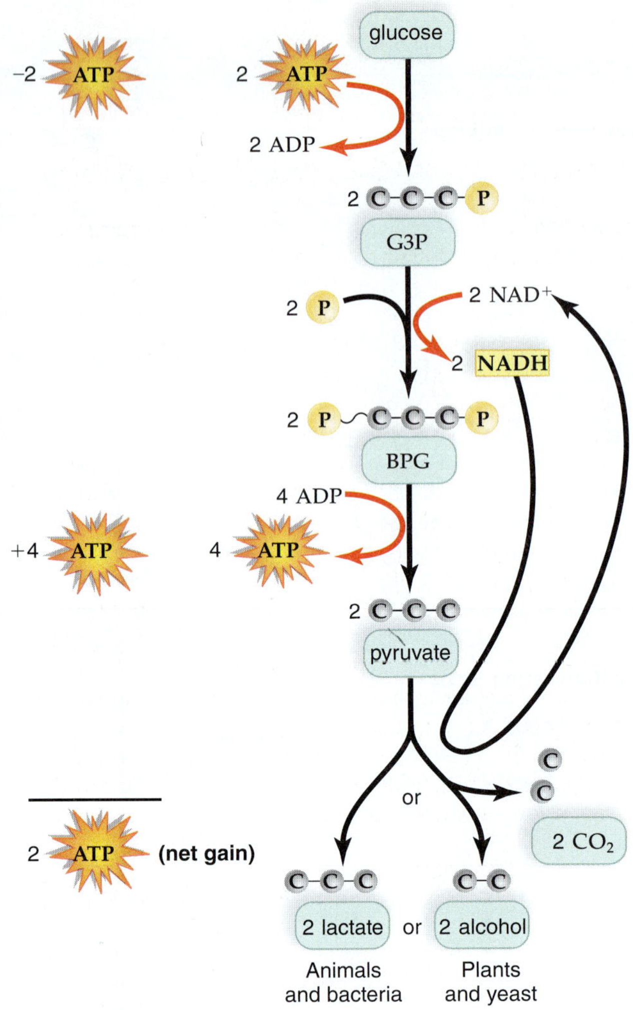

Figure 8.5 Fermentation. Fermentation consists of glycolysis followed by a reduction of pyruvate. This recycles NAD^+ and it returns to the glycolytic pathway to pick up more electrons. As with glycolysis, each step is catalyzed by a specialized enzyme.

possible for the complete breakdown of glucose. The inputs and outputs of fermentation are shown here:

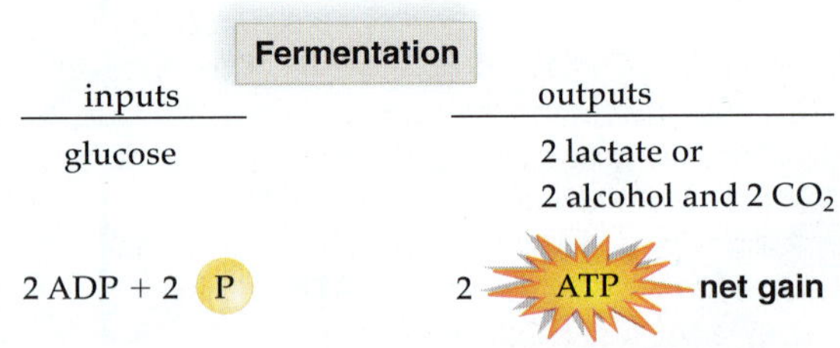

The two ATP produced by fermentation fall far short of the theoretical 36 to 38 ATP molecules that may be produced by cellular respiration. To achieve this number of ATP per glucose molecule, it is necessary to move on to the reactions and pathways that occur with oxygen in the mitochondria.

Check Your Progress 8.3

1. Explain fermentation's role in NAD^+ regeneration.
2. Summarize the two forms of fermentation.
3. List the advantages and disadvantages of fermentation.

THEME Nature of Science

Fermentation and Food Production

At the grocery store, you will find such items as bread, yogurt, soy sauce, pickles, and maybe even beer or wine (Fig. 8A). These are just a few of the many foods that are produced when microorganisms ferment (break down sugar in the absence of oxygen). Foods produced by fermentation last longer, because the fermenting organisms have removed many of the nutrients that would attract other organisms. The products of fermentation can even be dangerous to the very organisms that produced them, as when yeasts are killed by the alcohol they produce.

Yeast Fermentation

Baker's yeast, *Saccharomyces cerevisiae,* is added to bread for the purpose of leavening—the dough rises when the yeasts give off CO_2. The ethyl alcohol produced by the fermenting yeast evaporates during baking. The many varieties of sourdough breads obtain their leavening from a starter composed of fermenting yeasts along with bacteria from the environment. Depending on the community of microorganisms in the starter, the flavor of the bread may range from sour and tangy, as in San Francisco–style sourdough, to a milder taste, such as that produced by most Amish friendship bread recipes.

Ethyl alcohol in beer and wine is produced when yeasts ferment carbohydrates. When yeasts ferment fruit carbohydrates, the end result is wine. If they ferment grain, beer results. A few specialized varieties of beer, such as traditional wheat beers, have a distinctive, sour taste, because they are produced with the assistance of lactic acid–producing bacteria, such as those of the genus *Lactobacillus.* Stronger alcoholic drinks (e.g., whiskey and vodka) require distillation to concentrate the alcohol content.

Bacteria that produce acetic acid, including *Acetobacter aceti,* spoil wine. These bacteria convert the alcohol in wine or cider to acetic acid (vinegar). Until the renowned nineteenth-century scientist Louis Pasteur invented the process of pasteurization, acetic acid bacteria commonly caused wine to spoil. Although today we generally associate the process of pasteurization with making milk safe to drink, it was originally developed to reduce bacterial contamination in wine, so that limited acetic acid would be produced. The discovery of pasteurization is another example of how the pursuit of scientific knowledge can positively affect our lives.

Bacterial Fermentation

Yogurt, sour cream, and cheese are produced through the action of various lactic acid bacteria that cause milk to sour. Milk contains lactose, which these bacteria use as a carbohydrate source for fermentation. Yogurt, for example, is made by adding lactic acid bacteria, such as *Streptococcus thermophilus* and *Lactobacillus bulgaricus,* to milk and then incubating it to encourage the bacteria to convert the lactose. During the production of cheese, an enzyme called rennin must also be added to the milk to cause it to coagulate and become solid.

Old-fashioned brine cucumber pickles, sauerkraut, and kimchi are pickled vegetables produced by the action of acid-producing, fermenting bacteria that can survive in high-salt environments. Salt is used to draw liquid out of the vegetables and to aid in their preservation.

The bacteria need not be added to the vegetables, because they are already present on the surfaces of the plants.

Soy Sauce Production

Soy sauce is traditionally made by adding a mold, *Aspergillus,* and a combination of yeasts and fermenting bacteria to soybeans and wheat. The mold breaks down starch, supplying the fermenting microorganisms with sugar they can use to produce alcohol and organic acids.

As you can see from each of these examples, fermentation is a biologically and economically important process that scientists use for the betterment of our lives.

Questions to Consider

1. How would the production of fermentation products differ from that of other food products?
2. What products of fermentation do you use on a daily basis?

Figure 8A Products from fermentation. Fermentation of different carbohydrates by microorganisms like bacteria and yeast helps produce the products shown.

8.4 Inside the Mitochondria

Learning Outcomes

Upon completion of this section, you should be able to

1. Explain the fate of each carbon during the complete aerobic metabolism of glucose.
2. Contrast substrate-level phosphorylation and chemiosmosis as methods of ATP synthesis.
3. Describe how electron energy from redox reactions is used to create a proton gradient.

The preparatory (prep) reaction, the citric acid cycle, and the electron transport chain, which are needed for the complete breakdown of glucose, take place within the mitochondria. A **mitochondrion** has a double membrane with an intermembrane space (between the outer and inner membrane). Cristae are folds of inner membrane that jut out into the matrix, the innermost compartment, which is filled with a gel-like fluid (Fig. 8.6). Like a chloroplast, a mitochondrion is highly structured, so we would expect reactions to be located in particular parts of this organelle.

The enzymes that speed the prep reaction and the citric acid cycle are arranged in the matrix, and the electron transport chain is located in the cristae in a very organized manner. Most of the ATP from cellular respiration is produced in mitochondria; therefore, mitochondria are often called the powerhouses of the cell.

The Preparatory Reaction

The **preparatory (prep) reaction** is so called because it converts products from glycolysis into products that enter the citric acid cycle. In this reaction, the C_3 pyruvate is converted to a C_2 acetyl group and CO_2 is given off. This is an oxidation reaction in which electrons are removed from pyruvate by NAD^+ and NADH is formed. One prep reaction occurs per pyruvate, so the prep reaction occurs twice per glucose molecule:

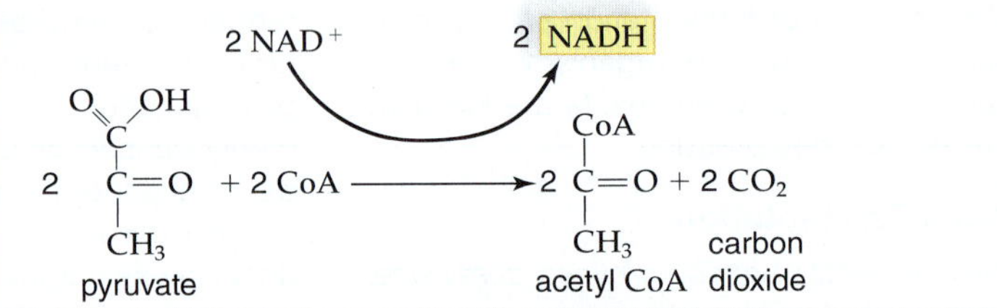

2 pyruvate + 2 CoA ⟶ 2 acetyl—CoA + 2 carbon dioxide

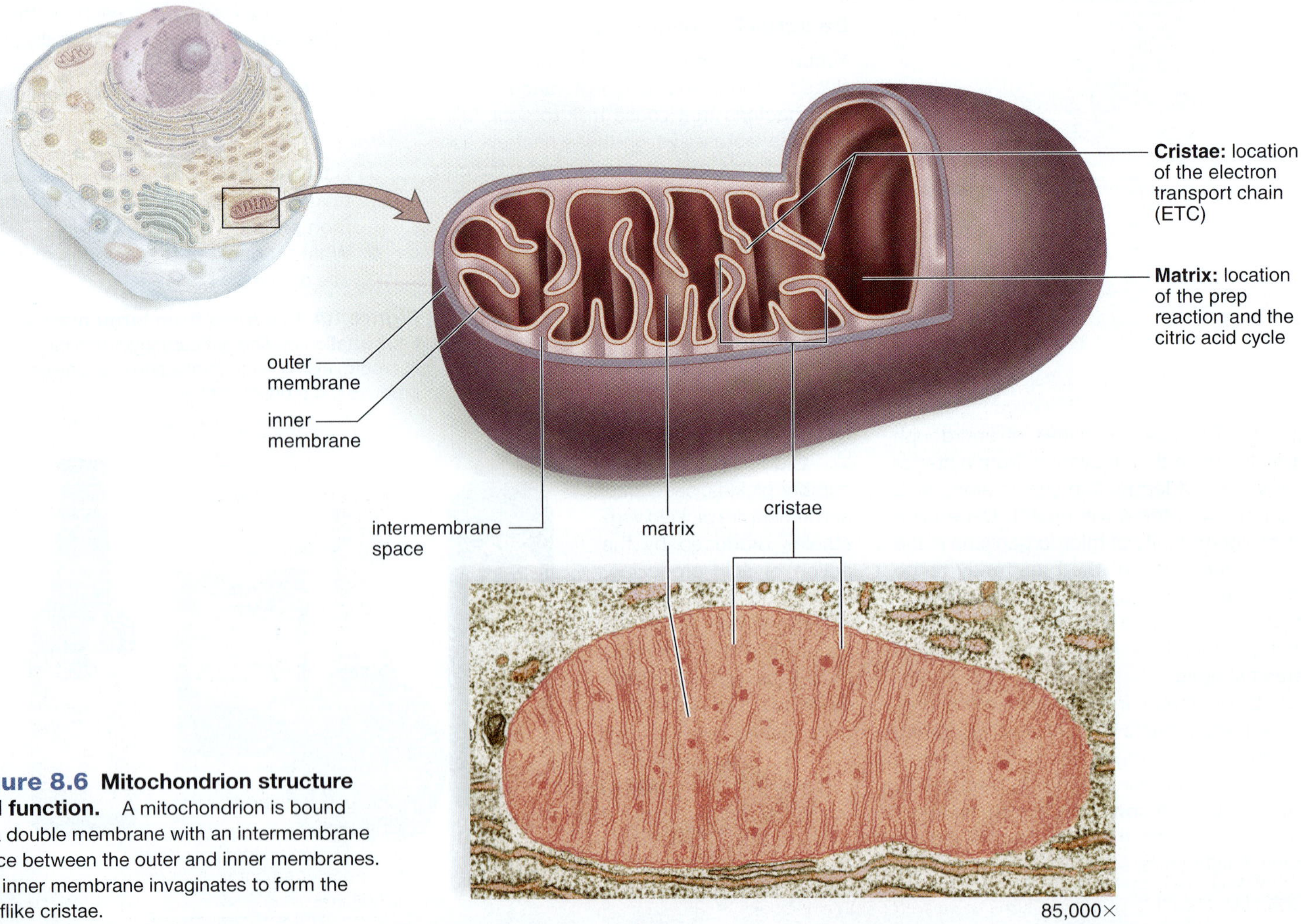

Figure 8.6 Mitochondrion structure and function. A mitochondrion is bound by a double membrane with an intermembrane space between the outer and inner membranes. The inner membrane invaginates to form the shelflike cristae.

The C_2 acetyl group is combined with a molecule known as CoA. CoA will carry the acetyl group to the citric acid cycle in the mitochondrial matrix. The two NADH carry electrons to the electron transport chain. What about the CO_2? In vertebrates, such as ourselves, CO_2 freely diffuses out of cells into the blood, which transports it to the lungs, where it is exhaled.

The Citric Acid Cycle

The **citric acid cycle,** also called the Krebs cycle, is a cyclical metabolic pathway located in the matrix of mitochondria (Fig. 8.7). At the start of the citric acid cycle, the (C_2) acetyl group carried by CoA joins with a C_4 molecule, and a C_6 citrate molecule results. During the cycle, oxidation occurs when electrons are accepted by NAD^+ in three instances and by FAD in one instance. Therefore, three NADH and one $FADH_2$ are formed as a result of one turn of the citric acid cycle. Also, the acetyl group received from the prep reaction is oxidized to two CO_2 molecules. Substrate-level ATP synthesis is also an important event of the citric acid cycle. In substrate-level ATP synthesis, you will recall, an enzyme passes a high-energy phosphate to ADP, and ATP results.

Animation How the Krebs Cycle Works

Because the citric acid cycle turns twice for each original glucose molecule, the inputs and outputs of the citric acid cycle per glucose molecule are as follows:

Citric acid cycle

inputs	outputs
2 (2C) acetyl groups	4 CO_2
6 NAD^+	6 NADH
2 FAD	2 $FADH_2$
2 ADP + 2 P	2 ATP

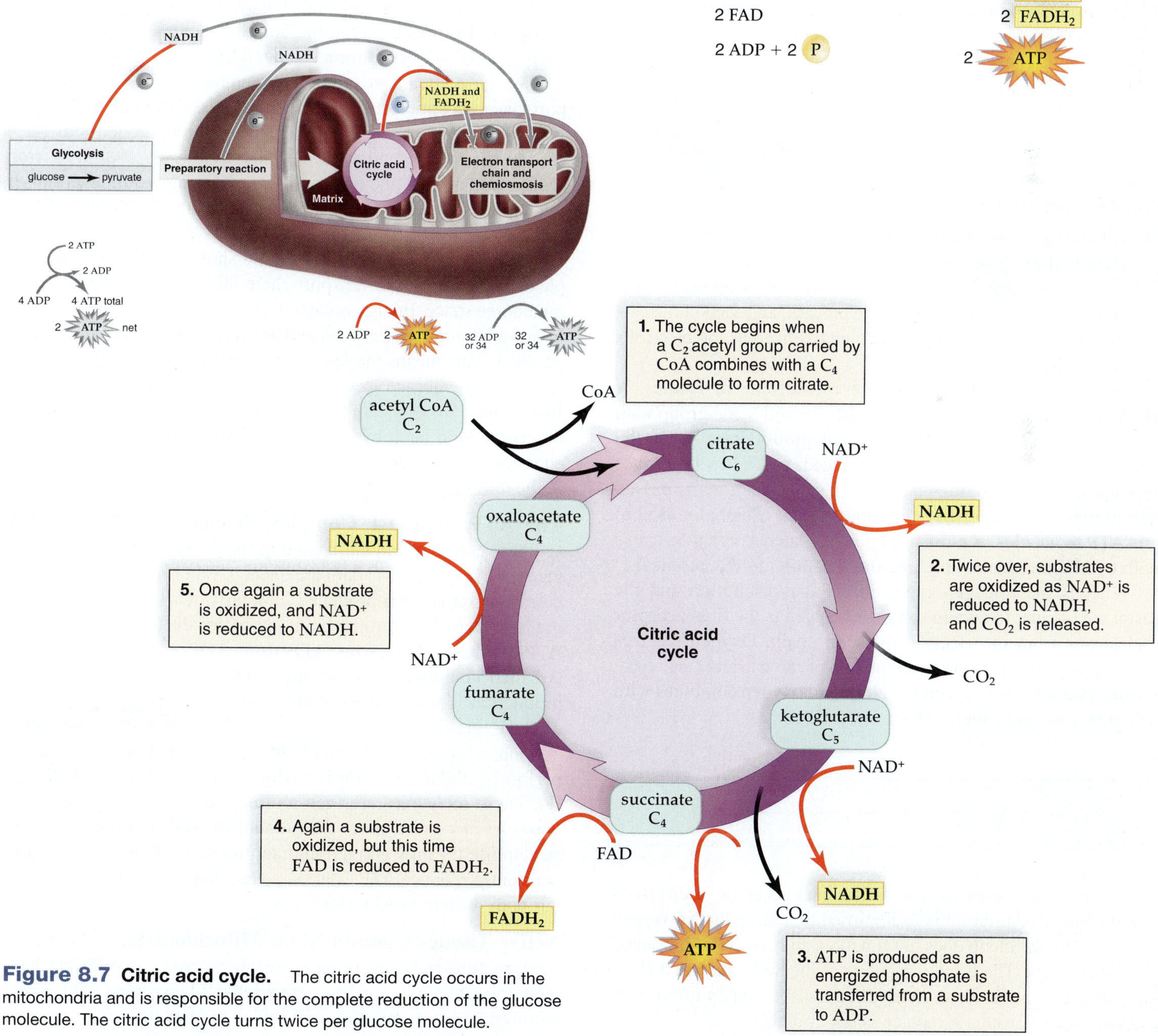

Figure 8.7 Citric acid cycle. The citric acid cycle occurs in the mitochondria and is responsible for the complete reduction of the glucose molecule. The citric acid cycle turns twice per glucose molecule.

Production of CO_2

The six carbon atoms originally located in a glucose molecule have now become CO_2. The prep reaction produces the first two CO_2, and the citric acid cycle produces the last four CO_2 per glucose molecule. We have already mentioned that this is the CO_2 humans and other animals breathe out.

Thus far, we have broken down glucose to CO_2 and hydrogen atoms. Recall that, as bonds are broken and glucose gets converted to CO_2, energy in the form of high-energy electrons is released. NADH and $FADH_2$ capture those high-energy electrons and carry them to the electron transport chain, as discussed next.

3D Animation
Cellular Respiration: Citric Acid Cycle

Electron Transport Chain

The **electron transport chain** (**ETC**), located in the cristae of the mitochondria and the plasma membrane of aerobic prokaryotes, is a series of carriers that pass electrons from one to the other. The high-energy electrons that enter the electron transport chain are carried by NADH and $FADH_2$. Figure 8.8 is arranged to show that high-energy electrons enter the chain and low-energy electrons leave the chain.

Animation
Electron Transport System and ATP Synthesis

Members of the Chain

When NADH gives up its electrons, it becomes oxidized to NAD^+, and when $FADH_2$ gives up its electrons, it becomes oxidized to FAD. The next carrier gains the electrons and is reduced. This oxidation-reduction reaction starts the process, and each of the carriers, in turn, becomes reduced and then oxidized as the electrons move down the chain.

Many of the redox carriers are cytochrome molecules. A **cytochrome** is a protein that has a tightly bound heme group with a central atom of iron, the same as hemoglobin does. When the iron accepts electrons, it becomes reduced, and when iron gives them up, it becomes oxidized. As the pair of electrons is passed from carrier to carrier, energy is captured and eventually used to form ATP molecules. A number of poisons, such as cyanide, cause death by binding to and blocking the function of cytochromes.

What is the role of oxygen in cellular respiration and the reason we breathe to take in oxygen? Oxygen is the final acceptor of electrons from the electron transport chain. Oxygen receives the energy-spent electrons from the last of the carriers (i.e., cytochrome oxidase). After receiving electrons, oxygen combines with hydrogen ions, and water forms:

$$\tfrac{1}{2}O_2 + 2e^- + 2H^+ \longrightarrow H_2O$$

The critical role of oxygen as the final acceptor of electrons during cellular respiration is exemplified by noting that if oxygen is not present, the chain does not function, and no ATP is produced by mitochondria. The limited capacity of the body to form ATP in a way that does not involve the electron transport chain means that death eventually results if oxygen is not available.

Cycling of Carriers

When NADH delivers high-energy electrons to the first carrier of the electron transport chain, enough energy has been captured by the time the electrons are received by O_2 to permit the production of three ATP molecules. When $FADH_2$ delivers high-energy electrons to the electron transport chain, two ATP are produced.

Once NADH has delivered electrons to the electron transport chain and has become NAD^+, it is able to return and pick up more hydrogen atoms. The reuse of coenzymes increases cellular efficiency, because the cell does not have to constantly make new NAD^+; it simply recycles what is already there.

The ETC Pumps Hydrogen Ions. Essentially, the electron transport chain consists of three protein complexes and two carriers. The three protein complexes are the NADH-Q reductase complex, the cytochrome reductase complex, and the cytochrome oxidase complex. The two other carriers that transport electrons between the complexes are coenzyme Q and cytochrome *c* (Fig. 8.8).

Animation
Proton Pump

We have already seen that the members of the electron transport chain accept electrons, which they pass from one to the other via redox reactions. So what happens to the hydrogen ions (H^+) carried by NADH and $FADH_2$? The complexes of the electron transport chain use the energy released during redox reactions to pump these hydrogen ions from the matrix into the intermembrane space of a mitochondrion.

The vertical arrows in Figure 8.8 show that the protein complexes of the electron transport chain all pump H^+ into the intermembrane space. Energy obtained from electron passage is needed, because H^+ ions are pumped and actively transported against their gradient. This means the few H^+ ions in the matrix will be moved to the intermembrane space, where there are already many H^+ ions. Just as the walls of a dam hold back water, allowing it to collect, so do cristae hold back hydrogen ions. Eventually, a strong electrochemical gradient develops; about ten times as many H^+ are found in the intermembrane space as are present in the matrix.

The ATP Synthase Complex Produces ATP. The ATP synthase complex can be likened to the gates of a dam. When the gates of a hydroelectric dam are opened, water rushes through, and electricity (energy) is produced. Similarly, when H^+ flows down a gradient from the intermembrane space into the matrix, the enzyme ATP synthase synthesizes ATP from ADP + Ⓟ. This process is called **chemiosmosis,** because ATP production is tied to the establishment of an H^+ gradient.

3D Animation
Cellular Respiration: Electron Transport Chain

Once formed, ATP moves out of mitochondria and is used to perform cellular work, during which it breaks down to ADP and Ⓟ. These molecules are then returned to mitochondria for recycling. At any given time, the amount of ATP in a human would sustain life for only about a minute; therefore, ATP synthase must constantly produce ATP. It is estimated that mitochondria produce our body weight in ATP every day.

Active Tissues Contain More Mitochondria. Active tissues, such as muscles, require greater amounts of ATP and have more mitochondria than less active cells. When a burst of energy is required, however, muscles still utilize fermentation.

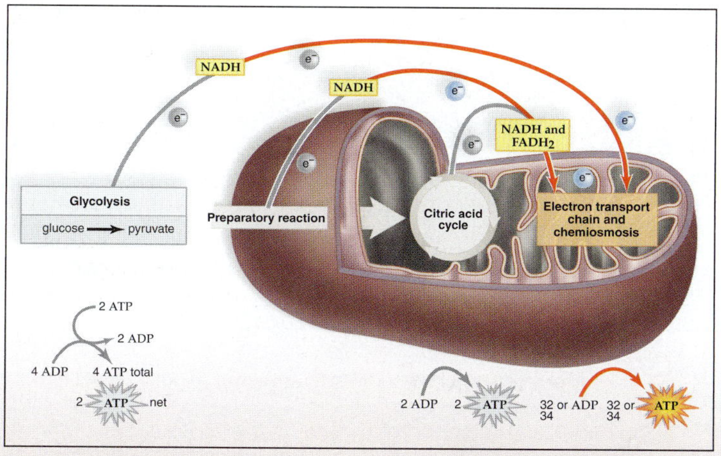

Figure 8.8 Organization and function of the electron transport chain. The electron transport chain is located in the mitochondrial cristae. NADH and $FADH_2$ take electrons to the electron transport chain. As electrons move from one protein complex to the other via redox reactions, energy is used to pump hydrogen ions (H^+) from the matrix into the intermembrane space. As hydrogen ions flow down a concentration gradient from the intermembrane space into the mitochondrial matrix, ATP is synthesized by the enzyme ATP synthase. For every pair of electrons that enters by way of NADH, three ATP result. For every pair of electrons that enters by way of $FADH_2$, two ATP result. Oxygen, the final acceptor of the electrons, becomes a part of water. ATP leaves the matrix by way of a channel protein.

Tutorial Electron Transport Chain

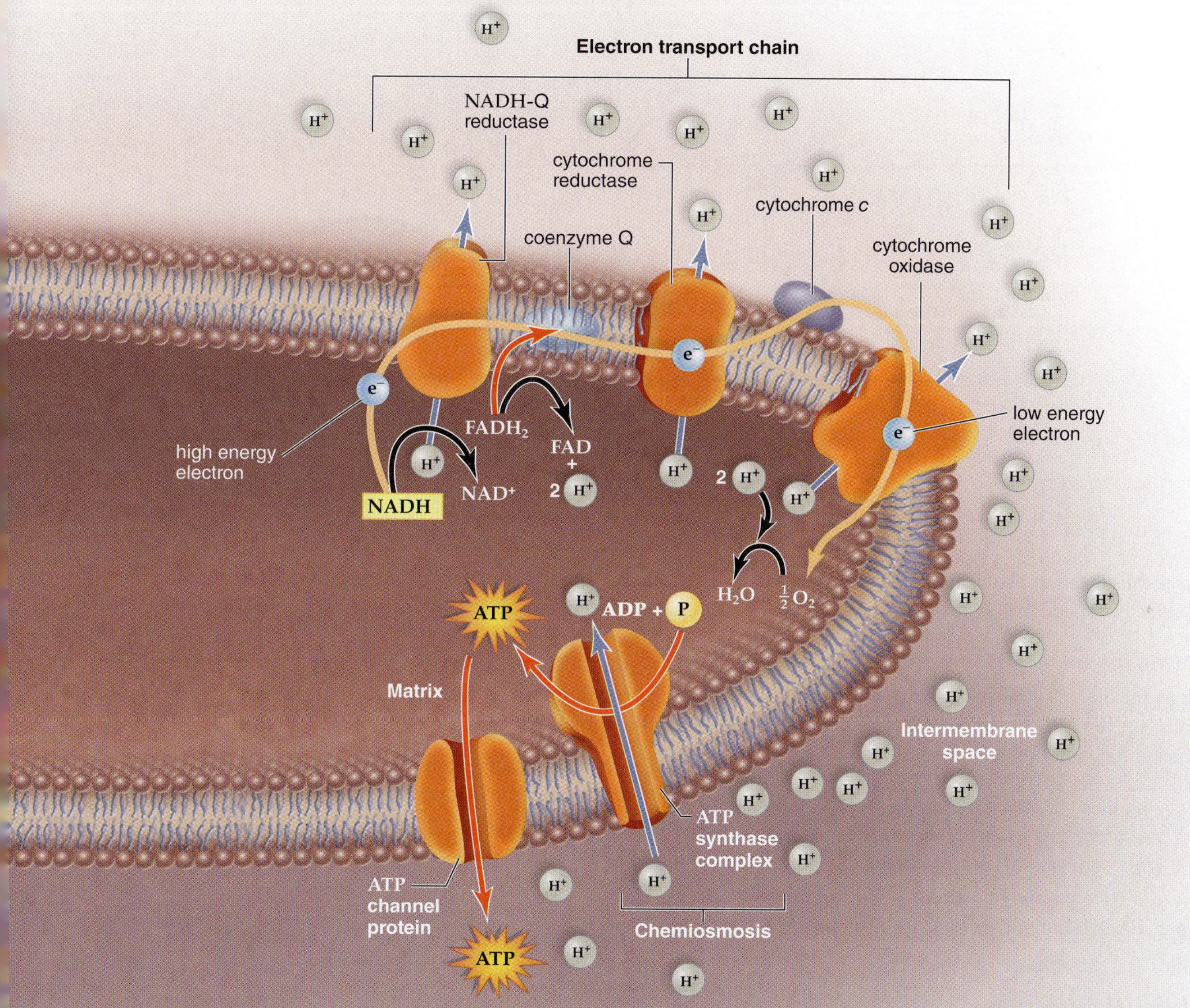

As an example of the relative amounts of ATP, consider that the dark meat of chickens, namely the thigh meat, contains more mitochondria than the white meat of the breast. This suggests that chickens mainly walk or run, rather than fly, about the barnyard.

Energy Yield from Glucose Metabolism

Figure 8.9 calculates the theoretical ATP yield for the complete breakdown of glucose to CO_2 and H_2O during cellular respiration. Notice that the diagram includes the number of ATP produced directly by glycolysis and the citric acid cycle (to the *left*), as well as the number produced as a result of electrons passing down the electron transport chain (to the *right*). A maximum of 32 to 34 ATP molecules may be produced by the electron transport chain.

Substrate-Level ATP Synthesis

Per glucose molecule, there is a net gain of two ATP from glycolysis, which takes place in the cytoplasm. The citric acid cycle, which occurs in the matrix of mitochondria, accounts for two ATP per

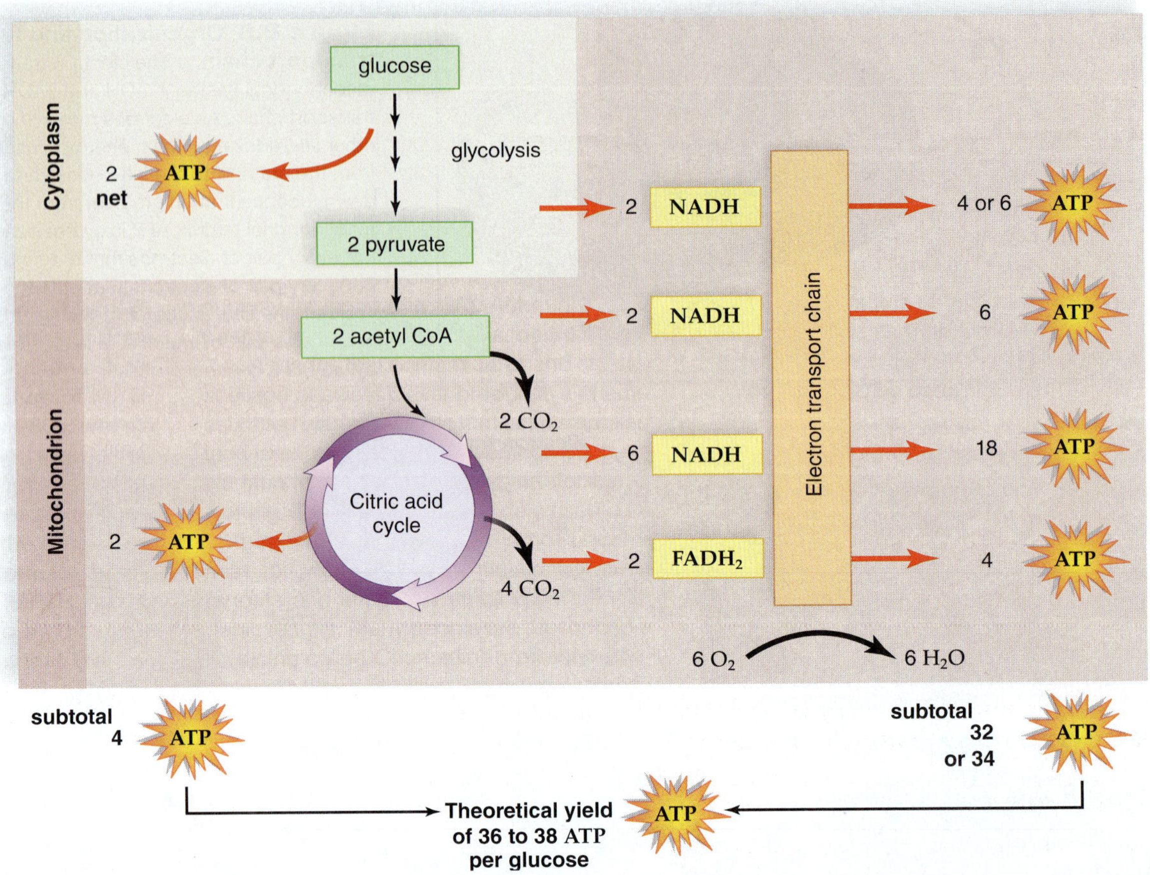

Figure 8.9 Accounting of energy yield per glucose molecule breakdown. Substrate-level ATP synthesis during glycolysis and the citric acid cycle accounts for 4 ATP. The electron transport chain accounts for 32 or 34 ATP, making the theoretical grand total of ATP between 36 and 38 ATP. Other factors may reduce the efficiency of cellular respiration. For example, cells differ as to the delivery of the electrons from NADH generated outside the mitochondria. If they are delivered by a shuttle mechanism to the start of the electron transport chain, 6 ATP result; otherwise, 4 ATP result.

glucose molecule. This means that a total of four ATP are formed by substrate-level ATP synthesis outside the electron transport chain.

ETC and Chemiosmosis

Most ATP is produced by the electron transport chain and chemiosmosis. Per glucose molecule, ten NADH and two $FADH_2$ take electrons to the electron transport chain. For each NADH formed *inside* the mitochondria by the citric acid cycle, three ATP result, but for each $FADH_2$, only two ATP are produced. Figure 8.8 explains the reason for this difference: $FADH_2$ delivers its electrons to the transport chain after NADH, and therefore these electrons do not participate in as many redox reactions and don't pump as many H^+ as NADH. Therefore, $FADH_2$ cannot account for as much ATP production.

Efficiency of Cellular Respiration

Figure 8.9 provides the theoretical ATP for each stage of cellular respiration. However, we know now that cells rarely ever achieve these theoretical values. Several factors can lower the ATP yield for each molecule of glucose entering the pathway:

- In some cells, NADH cannot cross mitochondrial membranes, but a "shuttle" mechanism allows its electrons to be delivered to the electron transport chain inside the mitochondria. The cost to the cell is one ATP for each NADH that is shuttled to the ETC. This reduces the overall count of ATP produced as a result of glycolysis, in some cells, to four instead of six ATP.
- At times, cells need to expend energy to move ADP molecules and pyruvate into the cell and to establish protein gradients in the mitochondria.

There is still considerable research into the precise ATP yield per glucose molecule. However, most estimates place the actual yield at around 30 ATP per glucose. Using this number we can calculate that only between 32 and 39 percent of the available energy is usually transferred from glucose to ATP. The rest of the energy is lost in the form of heat.

3D Animation Cellular Respiration: Summary of Cellular Respiration

In the next section, we consider how cellular respiration fits into metabolism as a whole.

Check Your Progress 8.4

1. Explain when carbon is converted from glucose into carbon dioxide during cellular respiration.
2. Examine which processes during glucose breakdown produce the most ATP.
3. Compare the function of the mitochondrial inner membrane to a hydroelectric dam.

8.5 Metabolism

Learning Outcomes

Upon completion of this section, you should be able to

1. Compare the pathways of carbohydrate, fat, and protein catabolism.
2. Explain how the structure of mitochondria and chloroplasts enables a flow of energy through living organisms.

Key metabolic pathways routinely draw from pools of particular substrates needed to synthesize or degrade larger molecules. Substrates like the end product of glycolysis, pyruvate, exist as a pool that is continuously affected by changes in cellular and environmental conditions (Fig. 8.10). Degradative reactions, termed **catabolism,** that break down molecules must be dynamically balanced with constructive reactions, or **anabolism.** For example, catabolic breakdown of fats will occur when insufficient carbohydrate is present; this breakdown adds to the **metabolic pool** of pyruvate. When energy needs to be stored as fat, pyruvate is taken from the pool. This dynamic balance of catabolism and anabolism is essential to optimal cellular function.

Catabolism

We already know that glucose is broken down during cellular respiration. However, other molecules like fats and proteins can also be broken down as necessary. When a fat is used as an energy source, it breaks down to glycerol and three fatty acids. As Figure 8.10 indicates, glycerol can be converted to pyruvate and enter glycolysis. The fatty acids are converted to 2-carbon acetyl CoA that enters the citric acid cycle. An 18-carbon fatty acid results in nine acetyl CoA molecules. Calculation shows that respiration of these can produce a total of 108 ATP molecules. This is why fats are an efficient form of stored energy—the three long fatty acid chains per fat molecule can produce considerable ATP when needed.

Proteins are less frequently used as an energy source, but they are available as necessary. The carbon skeleton of amino acids can enter glycolysis, be converted to acetyl groups, or enter the citric acid cycle at some other juncture. The carbon skeleton is produced in the liver when an amino acid undergoes **deamination,** or the removal of the amino group. The amino group becomes ammonia (NH_3), which enters the urea cycle and becomes part of urea, the primary excretory product of humans. Just where the carbon skeleton begins degradation depends on the length of the *R* group, since this determines the number of carbons left after deamination.

Anabolism

We have already seen that the building of new molecules requires ATP produced during breakdown of molecules. These catabolic reactions also provide the basic components used to build new molecules. For example, excessive carbohydrate intake can result in the formation of fat. Extra G3P from glycolysis can be converted to glycerol, and acetyl groups from glycolysis can be joined to form fatty acids, which in turn are used to synthesize fat.

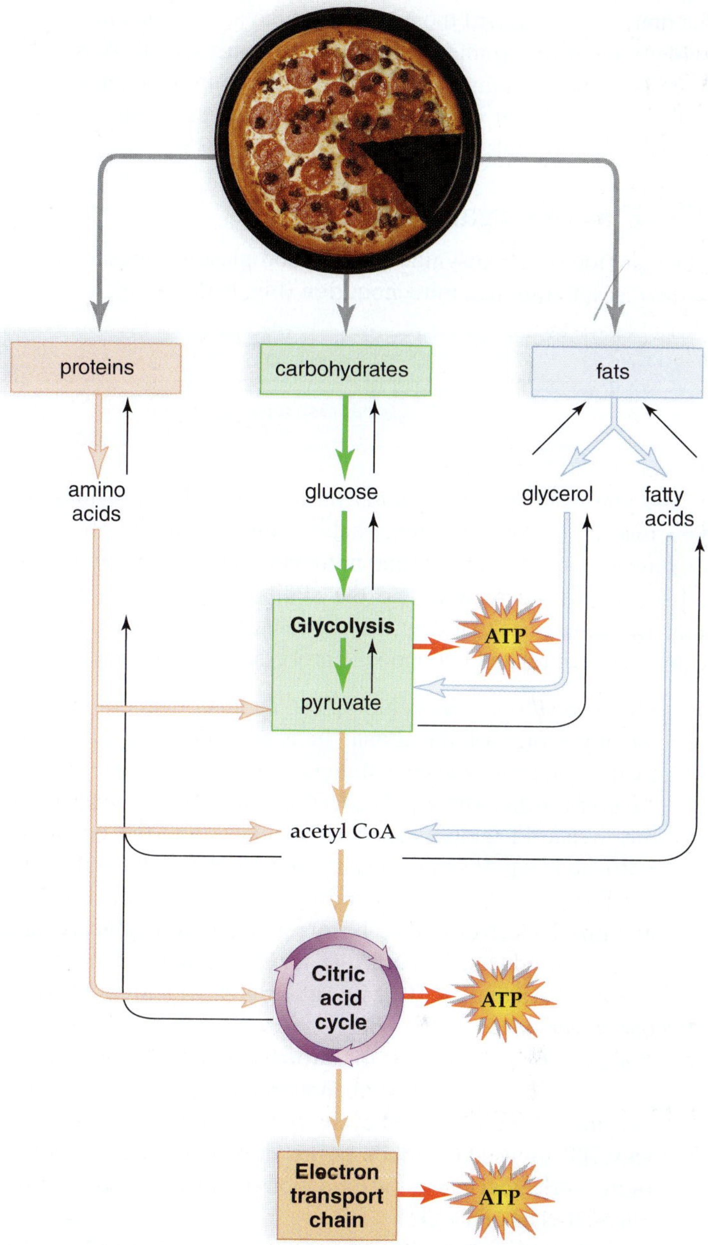

Figure 8.10 The metabolic pool concept. Carbohydrates, fats, and proteins can be used as energy sources, and their monomers (carbohydrates and proteins) or subunits (fats) enter degradative pathways at specific points. Catabolism produces molecules that can also be used for anabolism of other compounds.

This explains why you gain weight from eating too much candy, ice cream, or cake.

Some substrates of the citric acid cycle can be converted to amino acids through transamination—the transfer of an amino group to an organic acid, forming a different amino acid. Plants are able to synthesize all of the amino acids they need. Animals, however, lack some of the enzymes necessary for synthesis of all amino acids. Adult humans, for example, can synthesize 11 of the common amino acids, but they cannot synthesize the other 9. The amino acids that cannot be synthesized must be supplied by the diet; they are called the essential amino acids. The amino acids that can be synthesized are called nonessential. It is quite possible for animals to suffer from protein deficiency if their diets do not contain adequate quantities of all the essential amino acids.

The Energy Organelles Revisited

The equation for photosynthesis in a chloroplast is opposite that of cellular respiration in a mitochondrion (Fig. 8.11):

$$\text{energy} + 6\,CO_2 + 6\,H_2O \underset{\text{cellular respiration}}{\overset{\text{photosynthesis}}{\rightleftarrows}} C_6H_{12}O_6 + 6\,O_2$$

While you were studying photosynthesis and cellular respiration, you may have noticed a remarkable similarity in the structural organization of chloroplasts and mitochondria. Through evolution, all organisms are related, and the similar organization of these organelles suggests that they may be related also. The two organelles carry out related but opposite processes:

1. *Use of membrane.* In a chloroplast, an inner membrane forms the thylakoids of the grana. In a mitochondrion, an inner membrane forms the convoluted cristae.
2. *Electron transport chain (ETC).* An ETC is located on the thylakoid membrane of chloroplasts and the cristae of mitochondria. In chloroplasts, the electrons passed down the ETC have been energized by the sun; in mitochondria, energized electrons have been removed from glucose and glucose products. In both, the ETC establishes an electrochemical gradient of H^+ with subsequent ATP production by chemiosmosis.
3. *Enzymes.* In a chloroplast the stroma contains the enzymes of the Calvin cycle, and in mitochondria the matrix contains the enzymes of the citric acid cycle. In the Calvin cycle, NADPH and ATP are used to reduce carbon dioxide to a carbohydrate. In the citric acid cycle, the oxidation of glucose products produces NADH and ATP.

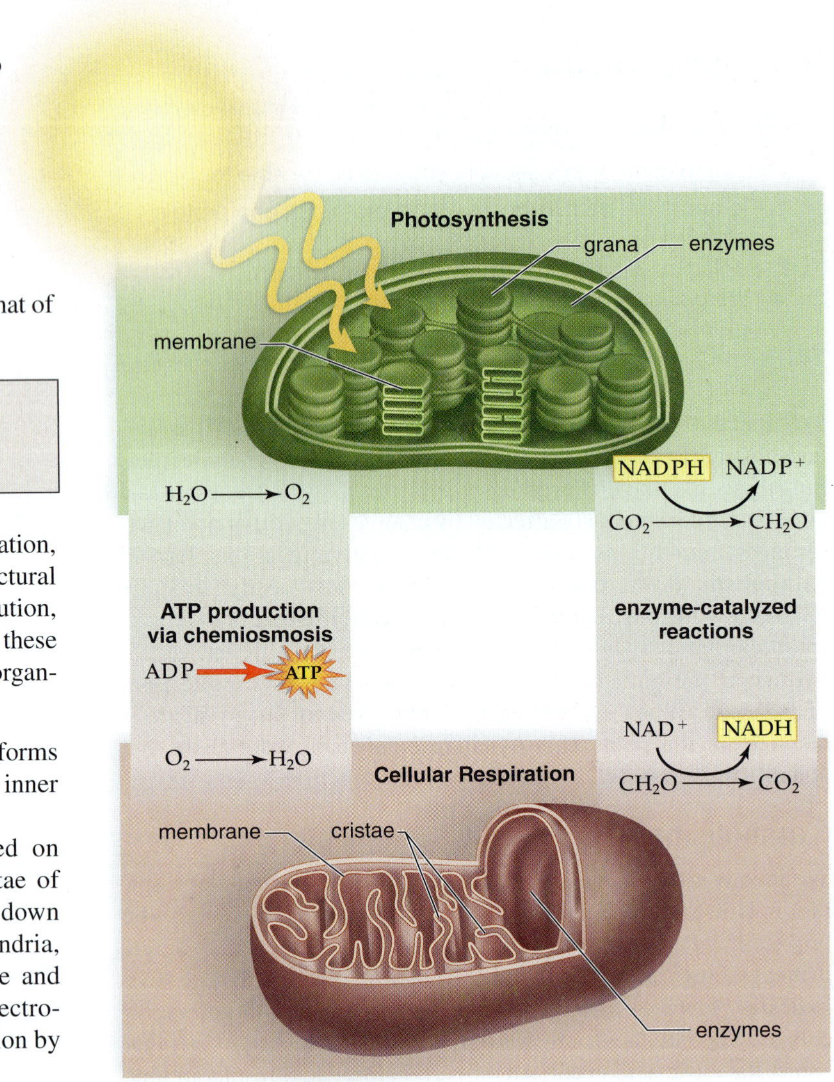

Figure 8.11 Photosynthesis versus cellular respiration. In photosynthesis (*top*), water is oxidized and oxygen is released; in cellular respiration (*bottom*), oxygen is reduced to water. Both processes have an electron transport chain located within membranes (the grana of chloroplasts and the cristae of mitochondria), where ATP is produced by chemiosmosis. Both have enzyme-catalyzed reactions within the semifluid interior. In photosynthesis, CO_2 is reduced to a carbohydrate; in cellular respiration, a carbohydrate is oxidized to CO_2.

Flow of Energy

The ultimate source of energy for producing a carbohydrate in chloroplasts is the sun; the ultimate goal of cellular respiration in a mitochondrion is the conversion of carbohydrate energy into that of ATP molecules. Therefore, energy flows from the sun, through chloroplasts to carbohydrates, and then through mitochondria to ATP molecules.

This flow of energy maintains biological organization at all levels from molecules to organisms to ultimately the biosphere. In keeping with the energy laws, some energy is lost with each chemical transformation, and eventually the solar energy captured by plants is lost in the form of heat. Therefore, all life depends on a continual input of solar energy.

Although energy flows through organisms, chemicals cycle within natural systems. Aerobic organisms utilize the carbohydrate and oxygen produced by chloroplasts to generate energy within the mitochondria to sustain life. Likewise, the carbon dioxide produced by mitochondria returns to chloroplasts to be used in the manufacture of carbohydrates, producing oxygen as a by-product. Therefore, chloroplasts and mitochondria are instrumental in not only allowing a flow of energy through living organisms but also permitting a cycling of chemicals.

Check Your Progress 8.5

1. Evaluate how catabolism and anabolism are balanced within a cell.
2. Compare the structure and function of chloroplasts and mitochondria.

Connecting *the* Concepts *with the* Themes

Evolution

- "Structure equals function" is a concept exemplified by chloroplasts and mitochondria. Their membranous structure has evolved over millions of years and is well suited to the isolation of enzymatic reactions in the interior from complexes located on the membrane.
- The majority of organisms on the planet use cellular respiration to generate ATP molecules.

Nature of Science

- Understanding the processes of energy metabolism, such as cellular respiration, may have positive effects, such as helping people with chronic fatigue, or negative effects, such as overconsumption of substances like energy drinks.
- An understanding of fermentation pathways allows the production of a number of food products.

Biological Systems

- Chloroplasts and mitochondria play a significant role in metabolism, and their enzyme-requiring pathways permit a flow of energy through all living organisms.
- The energy transformations that take place in these organelles result in a loss of energy in the form of heat. Therefore, all organisms are in need of a constant supply of energy, which they get from their food.
- Food is ultimately produced by plants, which have the ability to capture solar energy. Photosynthesizing organisms form the basis of most food chains on Earth.

MEDIA STUDY TOOLS http://connect.mheducation.com

SMARTBOOK® Study smart. SmartBook helps you maximize your study time by identifying what you know and don't know. The recharge feature in SmartBook helps you prepare for the test by identifying the material that you are most likely to forget.

connect BIOLOGY Find even more resources to learn the chapter concepts including animations, tutorial videos, and interactive practice questions.

MP3 Files	Animations	3D Animations	Tutorials
8.1 Cellular Respiration	**8.1** How the NAD^+ Works **8.2** How Glycolysis Works **8.4** How the Krebs Cycle Works • Electron Transport System and ATP Synthesis • Proton Pump	**8.2** Cellular Respiration: Glycolysis **8.4** Cellular Respiration: Citric Acid Cycle • Cellular Respiration: Electron Transport Chain • Cellular Respiration: Summary of Cellular Respiration	**8.1** Overview of Cellular Respiration **8.4** Electron Transport Chain

SUMMARIZE

8.1 Overview of Cellular Respiration

Cellular respiration is an **aerobic** pathway, during which glucose is completely broken down to CO_2 and H_2O; it consists of four phases: glycolysis, the prep reaction, the citric acid cycle, and the passage of electrons along the electron transport chain. Oxidation of substrates involves the removal of hydrogen atoms ($H^+ + e^-$), usually by redox coenzymes. **NAD^+** becomes NADH, and **FAD** becomes $FADH_2$. In the absence of oxygen, the **anaerobic** pathway of glycolysis and **fermentation** produces small amounts of ATP.

8.2 Outside the Mitochondria: Glycolysis

Glycolysis, the breakdown of glucose to two molecules of pyruvate, is a series of enzymatic reactions that occurs in the cytoplasm and is anaerobic. Breakdown releases enough energy to immediately give a net gain of two ATP by **substrate-level ATP synthesis** and the production of two NADH.

8.3 Outside the Mitochondria: Fermentation

Fermentation involves glycolysis followed by the reduction of pyruvate by NADH either to lactate (animals) or to alcohol (yeast) and carbon dioxide (CO_2). The reduction process "frees" NAD^+ so that it can accept more hydrogen atoms from glycolysis.

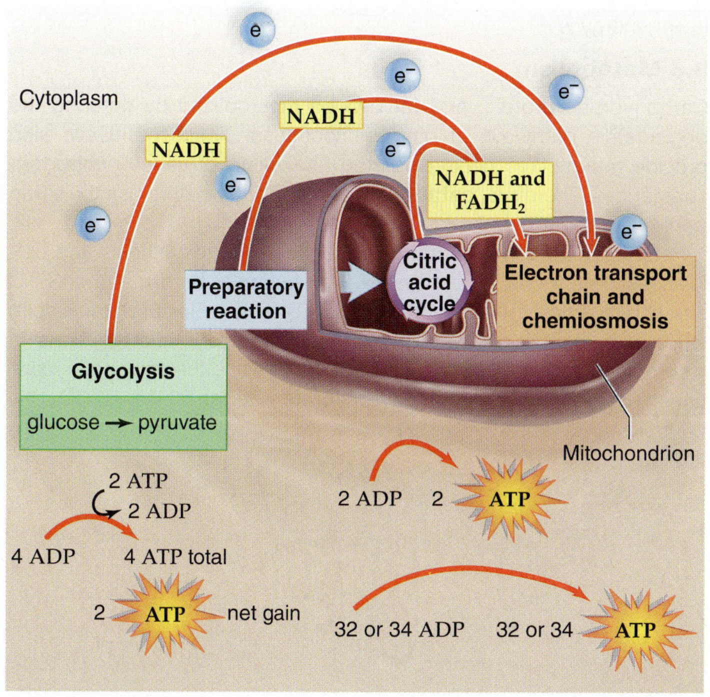

Although fermentation results in only two ATP molecules, it still serves a purpose. Many of the products of fermentation are used in the baking and brewing industries. In vertebrates, it provides a quick burst of ATP energy for short-term, strenuous muscular activity. The accumulation of lactate puts the individual in oxygen debt, because oxygen is needed when lactate is completely metabolized to CO_2 and H_2O.

8.4 Inside the Mitochondria

When oxygen is available, pyruvate from glycolysis enters the **mitochondrion,** where the **prep reaction** takes place. During this reaction, oxidation occurs as CO_2 is removed from pyruvate. NAD^+ is reduced, and CoA receives the C_2 acetyl group that remains. Because the reaction must take place twice per glucose molecule, two NADH result.

The acetyl group enters the **citric acid cycle,** a cyclical series of reactions located in the mitochondrial matrix. Complete oxidation follows, as two CO_2 molecules, three NADH molecules, and one $FADH_2$ molecule are formed. The cycle also produces one ATP molecule. The entire cycle must turn twice per glucose molecule.

The final stage of glucose breakdown involves the **electron transport chain,** located in the cristae of the mitochondria. The electrons received from NADH and $FADH_2$ are passed down a chain of carriers until they are finally received by oxygen, which combines with H^+ to produce water. As the electrons pass down the chain, proteins called **cytochromes** capture energy for use in ATP production.

The cristae of mitochondria contain complexes of the electron transport chain that not only pass electrons from one to the other but also pump H^+ into the intermembrane space, setting up an electrochemical gradient. When H^+ flows down this gradient through an ATP synthase complex, energy is captured and used to form ATP molecules from ADP and Ⓟ. This is ATP synthesis by **chemiosmosis.**

Theoretically, 36 to 38 ATP are produced by complete glucose breakdown. Four are the result of substrate-level ATP synthesis and the rest are produced as a result of the electron transport chain. For most NADH molecules that donate electrons to the electron transport chain, three ATP molecules are produced. However, in some cells, each NADH formed in the cytoplasm results in only two ATP molecules, because a shuttle, rather than NADH, takes electrons through the mitochondrial membrane. $FADH_2$ results in the formation of only two ATP, because its electrons enter the electron transport chain at a lower energy level.

8.5 Metabolism

Carbohydrate, protein, and fat can be metabolized by entering the degradative pathways at different locations. These pathways also provide metabolites needed for the anabolism of various important substances. Therefore, **catabolism** and **anabolism** both use the same pools of metabolites.

Similar to the **metabolic pool** concept, photosynthesis and cellular respiration can be compared. For example, both utilize an ETC and chemiosmosis. As a result of the ETC in chloroplasts, water is split, while in mitochondria, water is formed. The enzymatic reactions in chloroplasts reduce CO_2 to a carbohydrate, while the enzymatic reactions in mitochondria oxidize carbohydrate with the release of CO_2.

ASSESS

Choose the best answer for each question.

8.1 Overview of Cellular Respiration

1. The metabolic process that produces the most ATP molecules is
   a. glycolysis.
   b. the citric acid cycle.
   c. the electron transport chain.
   d. fermentation.
2. Which one of these pathways would not be active in an aerobic condition?
   a. glycolysis
   b. electron transport chain
   c. citric acid cycle
   d. fermentation
   e. All of these would be active.
3. The reduction of NAD^+ produces
   a. acetyl CoA.
   b. pyruvate.
   c. NADH.
   d. oxygen gas.

8.2 Outside the Mitochondria: Glycolysis

4. During glycolysis, what is the net production of ATP per glucose molecule?
   a. 0
   b. 1
   c. 2
   d. 8
   e. 32
5. The process of glycolysis occurs where in the cell?
   a. chloroplasts
   b. mitocondrion
   c. nucleus
   d. cytoplasm
6. Which of the following is not produced by glycolysis?
   a. NADH
   b. pyruvate
   c. ATP
   d. $FADH_2$

8.3 Outside the Mitochondria: Fermentation

7. Which of these is not true of fermentation?
   a. There is a net gain of only two ATP per glucose.
   b. It occurs in cytoplasm.
   c. NADH donates electrons to the electron transport chain.
   d. It begins with glucose.
   e. It is carried on by yeast.
8. Fermentation is primarily involved in the recycling of
   a. ADP.
   b. NAD^+.
   c. oxygen.
   d. glucose.

8.4 Inside the Mitochondria

9. The greatest contributor of electrons to the electron transport chain is
   a. oxygen.
   b. glycolysis.
   c. the citric acid cycle.
   d. the prep reaction.
   e. fermentation.
10. Which of these is not true of the citric acid cycle?
    a. The citric acid cycle includes the prep reaction.
    b. The citric acid cycle produces ATP by substrate-level ATP synthesis.
    c. The citric acid cycle occurs in the mitochondria.
    d. The citric acid cycle produces two ATP per glucose molecule.
11. Which of these is not true of the electron transport chain?
    a. The electron transport chain is located on the cristae of the mitochindria.
    b. The electron transport chain produces more NADH than any metabolic pathway.
    c. The electron transport chain contains cytochrome molecules.
    d. The electron transport chain ends when oxygen accepts electrons.
12. The oxygen required by cellular respiration is reduced and becomes part of which molecule?
    a. ATP
    b. H_2O
    c. pyruvate
    d. CO_2

8.5 Metabolism

13. Fatty acids are broken down to
    a. pyruvate molecules, which take electrons to the electron transport chain.
    b. acetyl groups, which enter the citric acid cycle.
    c. amino acids, which excrete ammonia.
    d. glycerol, which is found in fats.
    e. All of these are correct.
14. Which of the following is not common to the chloroplast and mitochondria?
    a. Both use membranes to establish gradients.
    b. Both have an ETC.
    c. Both use sunlight as a source of energy.
    d. Both use a variety of enzymes.

ENGAGE

LEARNSMART LABS®

The following LearnSmart Labs contain exercises that are related to the content of this chapter:

- Cellular Respiration

THINKING SCIENTIFICALLY

1. You are able to extract mitochondria from a cell and remove the outer membrane. You want to show that the mitochondria can still produce ATP if placed in the right solution. The solution should be isotonic, but at what pH? Why?
2. You are working with acetyl CoA molecules that contain only radioactive carbon. They are incubated with all the components of the citric acid cycle long enough for one turn of the cycle. Examine Figure 8.7 and explain why the carbon dioxide given off is radioactive.
3. Bacteria do not have mitochondria, yet they contain an electron transport chain. Where may this be located in the bacteria?
4. Cyanide is known to be an inhibitor of the electron transport chain. It functions by inhibiting one of the cytochrome enzymes. How, then, would this cause death to the individual?

UNIT 2

Genetic Basis of Life

This unit provides an opportunity to become acquainted with the basics of cellular reproduction and Mendelian and molecular genetics. Cellular reproduction is one of the basic characteristics of life: The processes of mitosis and meiosis allow organisms to reproduce, grow, and repair damaged tissues. An understanding of cellular reproduction has led to the study of stem cells, which have the potential to offer treatments for many human diseases.

Mendelian genetics explains the patterns of inheritance that are founded in the process of meiosis. Among the many applications of knowledge of these patterns is the ability to predict the chances of having a child with a specific genetic disorder. An understanding of molecular genetics has led to the development of DNA technologies that have the potential to cure genetic diseases and produce crops to feed an ever-increasing human population.

The field of genetics is making progress in other areas, too. We are beginning to understand how cell division is regulated by numerous genes, and how a failure of these regulatory mechanisms may lead to cancer. Many other human diseases are the result of mutations in genes as well. Thus, at every turn, it is clear that you can't fully appreciate the happenings of the twenty-first century without a knowledge of genetics, and this is your chance to become a part of the action.

Unit Outline

Chapter 9 The Cell Cycle and Cellular Reproduction 147

Chapter 10 Meiosis and Sexual Reproduction 166

Chapter 11 Mendelian Patterns of Inheritance 186

Chapter 12 Molecular Biology of the Gene 207

Chapter 13 Regulation of Gene Expression 228

Chapter 14 Biotechnology and Genomics 244

Unit Learning Outcomes

The learning outcomes for this unit focus on three major themes in the life sciences.

Evolution	Explain how the process of meiosis introduces the variation necessary for evolutionary change.
Nature of Science	Discuss how an understanding of cellular reproduction and molecular biology can be used to treat human disease.
Biological Systems	Understand how the information contained within the DNA is responsible for the physical characteristics of an organism.

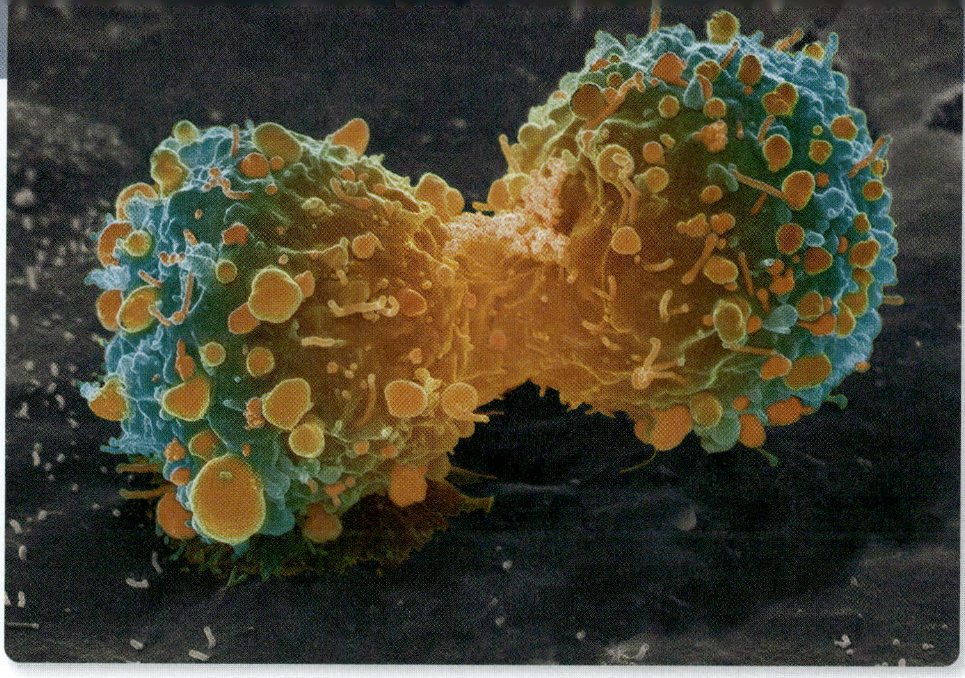

A cell may become cancerous when the regulation of cell division fails.

9

The Cell Cycle and Cellular Reproduction

The process of cell division is highly regulated. In humans, life begins as a single cell, yet in a very short period of time the process of cell division produces trillions of cells, each specialized for a particular function. Over 200 different types of cells are found in the human body; although each is specialized, they all work together in harmony.

But what happens when the regulation of cell division fails? In the United States this year, over 76,000 individuals will be diagnosed with melanoma, a form of skin cancer, and around 9,500 people will die from this disease. In many instances of melanoma, exposure to ultraviolet radiation (UV) from the sun has caused a mutation in the regulatory mechanisms of the cell cycle. Without proper regulation, cell division occurs continuously, a characteristic of cancer. For melanoma, this loss of cell cycle control results from a mutation in a gene known as *CDKN2A.* This gene is an example of a tumor suppressor gene, one of the key regulatory mechanisms of the cell cycle. In this chapter we describe the process of cell division, how it is regulated, and how cancer may develop when regulatory mechanisms malfunction.

As you read through this chapter, think about the following questions:

1. What is the normal sequence of events in the process of cellular reproduction?
2. What are the roles of the checkpoints in a cell cycle?
3. How do tumor suppressor genes regulate the cell cycle?

Chapter Outline

9.1 The Cell Cycle 148

9.2 The Eukaryotic Chromosome 151

9.3 Mitosis and Cytokinesis 152

9.4 The Cell Cycle and Cancer 158

9.5 Prokaryotic Cell Division 161

Before You Begin

Before beginning this chapter, take a few moments to review the following discussions.

Section 3.5 What is the role of the DNA in a cell?

Sections 4.2 and 4.3 What are the major differences between prokaryotic and eukaryotic cells?

Section 4.8 What is the role of the cytoskeleton in a eukaryotic cell?

Following *the* Themes

CHAPTER 9 THE CELL CYCLE AND CELLULAR REPRODUCTION

UNIT 2 GENETIC BASIS OF LIFE

Evolution	As eukaryotic cells became increasingly complex, a similarly complex series of events evolved to separate their multiple chromosomes, located within a membrane-bound nucleus, into new daughter cells.
Nature of Science	By studying the regulatory mechanisms of the cell cycle, scientists are able to gather a deeper understanding of why cancer occurs.
Biological Systems	Cell division must be regulated for the healthy growth of tissues and organisms.

9.1 The Cell Cycle

Learning Outcomes

Upon completion of this section, you should be able to

1. List the stages of interphase, and describe the major events that occur during each stage in preparation for cell division.
2. List the checkpoints that regulate the progression of cells through the cell cycle.
3. Explain the mechanisms within the G_1 cell cycle checkpoint that evaluate growth signals, determine nutrient availability, and assess DNA integrity.

The **cell cycle** is an orderly set of stages that takes place between the time a eukaryotic cell divides and the time the resulting daughter cells also divide. When a cell is going to divide, it grows larger, the number of organelles doubles, and the amount of DNA doubles as DNA replication occurs. The two portions of the cell cycle are interphase, which includes a number of stages, and the mitotic stage, when mitosis and cytokinesis occur.

Interphase

As Figure 9.1 shows, most of the cell cycle is spent in **interphase.** This is the time when a cell performs its usual functions, depending on its location in the body. The amount of time the cell takes for interphase varies widely. Embryonic cells complete the entire cell cycle in just a few hours. For adult mammalian cells, interphase lasts for about 20 hours, which is 90% of the cell cycle. In the past, interphase was known as the resting stage. However, today it is known that interphase is very busy, and that preparations are being made for mitosis. Interphase consists of three stages, referred to as G_1, S, and G_2.

3D Animation
Cell Cycle and Mitosis: Interphase

G_1 Stage

Cell biologists named the stage before DNA replication G_1, and they named the stage after DNA replication G_2. G stood for "gap," but now that we know how metabolically active the cell is, it is better to think of G as standing for "growth." During G_1, the cell recovers from the previous division. The cell grows in size, increases the number of organelles (such as mitochondria and ribosomes), and accumulates materials that will be used for DNA synthesis. Otherwise, cells are constantly performing their normal daily functions during G_1, including communicating with other cells, secreting substances, and carrying out cellular respiration.

Some cells, such as nerve and muscle cells, typically do not complete the cell cycle and are permanently arrested. These cells exit interphase and enter a stage called G_0. While in the G_0 stage, the cells continue to perform normal, everyday processes, but no preparations are being made for cell division. Cells may not leave the G_0 stage without proper signals from other cells and other parts of the body. Thus, completion of the cell cycle is very tightly controlled.

S Stage

Following G_1, the cell enters the S stage, when DNA synthesis, or replication, occurs. At the beginning of the S stage, each chromosome is composed of one DNA double helix. Following DNA replication, each chromosome is composed of two identical DNA double helix molecules. Each double helix is called a **chromatid,** and the two identical chromatids are referred to as **sister chromatids.** The sister chromatids remain attached until they are separated during mitosis.

G_2 Stage

Following the S stage, G_2 is the stage from the completion of DNA replication to the onset of mitosis. During this stage, the cell synthesizes the proteins that will assist cell division. For example, it makes the proteins that form microtubules. Microtubules are used during the mitotic stage to form the mitotic spindle that is critical during M stage.

M (Mitotic) Stage

Following interphase, the cell enters the M (for *mitotic*) stage. This cell division stage includes **mitosis** (nuclear division) and **cytokinesis** (division of the cytoplasm). During mitosis, daughter

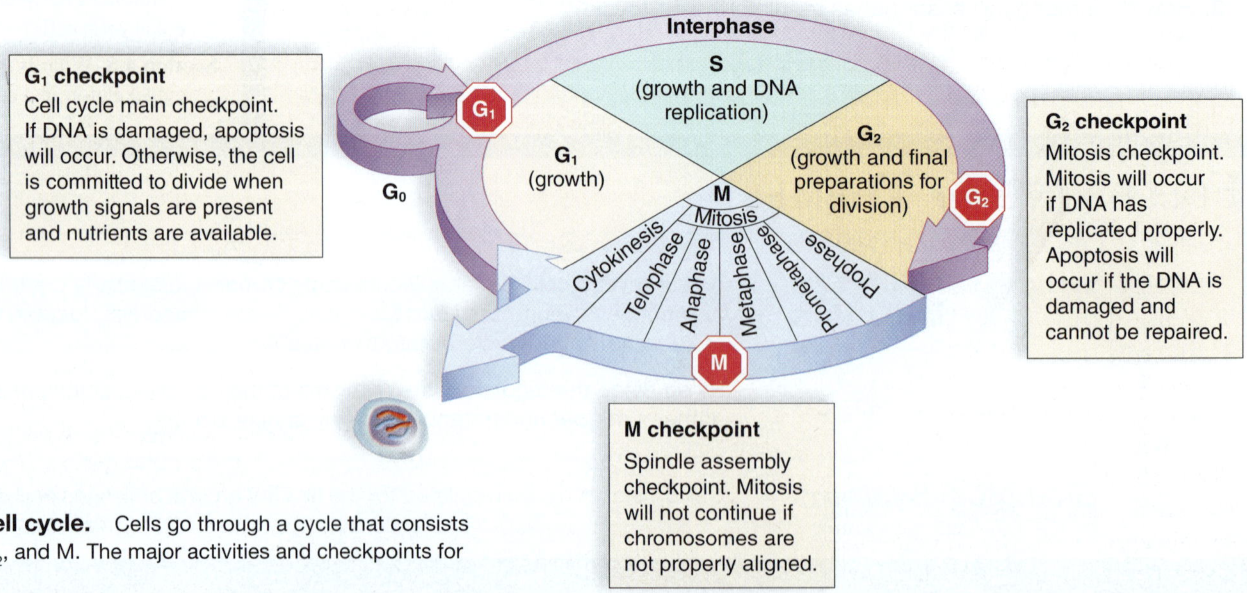

Figure 9.1 The cell cycle. Cells go through a cycle that consists of four stages: G_1, S, G_2, and M. The major activities and checkpoints for each stage are given.

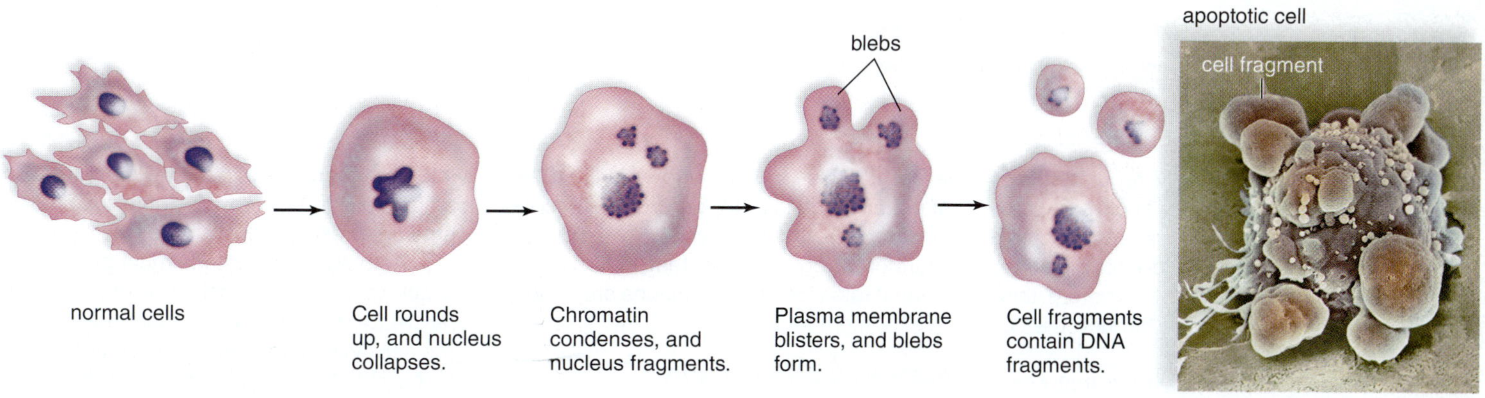

Figure 9.2 Apoptosis. Apoptosis is a sequence of events that results in a fragmented cell. The fragments are phagocytized (engulfed) by white blood cells and neighboring tissue cells.

chromosomes are distributed by the **mitotic spindle** to two daughter nuclei. When division of the cytoplasm is complete, two daughter cells are present.

Control of the Cell Cycle

A **signal** is an agent that influences the activities of a cell. **Growth factors** are signaling proteins received at the plasma membrane. Even cells arrested in G_0 will finish the cell cycle if stimulated to do so by growth factors. In general, signals ensure that the cell cycle stages follow one another in the normal sequence.

Animation Cell Proliferation Signaling Pathway

Cell Cycle Checkpoints

The red stop signs in Figure 9.1 represent three checkpoints at which the cell cycle either stops or continues on, depending on the internal signals received. Researchers have identified a family of internal signaling proteins, called **cyclins,** that increase and decrease as the cell cycle continues. Specific cyclins must be present for the cell to proceed from the G_1 stage to the S stage and from the G_2 stage to the M stage.

As discussed in the Nature of Science feature, "The G_1 Checkpoint," on page 150, the primary checkpoint of the cell cycle is the G_1 checkpoint. In mammalian cells, the signaling protein p53 stops the cycle at the G_1 checkpoint when DNA is damaged. In the name p53, p stands for *p*rotein and 53 represents its molecular weight in kilodaltons. First, p53 attempts to initiate DNA repair, but rising levels of p53 can bring about **apoptosis,** which is programmed cell death (Fig. 9.2). Another protein, called RB, is responsible for interpreting growth signals and nutrient availability signals. RB stands for *r*etino*b*lastoma, a cancer of the retina that occurs when the *RB* gene undergoes a mutation.

The cell cycle may also stop at the G_2 checkpoint if DNA has not finished replicating. This checkpoint prevents the initiation of the M stage before completion of the S stage. If DNA is physically damaged, such as from exposure to solar radiation or X-rays, the G_2 checkpoint also offers the opportunity for DNA to be repaired.

Another cell cycle checkpoint occurs during the mitotic stage. The cycle stops if the chromosomes are not properly attached to the mitotic spindle. Normally, the mitotic spindle ensures that the chromosomes are distributed accurately to the daughter cells.

3D Animation Cell Cycle and Mitosis: Checkpoints

Apoptosis

Apoptosis is often defined as programmed cell death, because the cell progresses through a typical series of events that bring about its destruction (Fig. 9.2). The cell rounds up, causing it to lose contact with its neighbors. The nucleus fragments, and the plasma membrane develops blisters. Finally, the cell fragments are engulfed by white blood cells and/or neighboring cells.

A remarkable finding of the past few years is that the enzymes that bring about apoptosis, called *caspases,* are always present in the cell. The enzymes are ordinarily held in check by inhibitors, but they can be unleashed by either internal or external signals.

Apoptosis and Cell Division. In living systems, opposing events keep the body in balance and maintain homeostasis. Cell division and apoptosis are two opposing processes that keep the number of cells in the body at an appropriate level. Cell division increases and apoptosis decreases the number of **somatic** (body) **cells.** Both are normal parts of growth and development. An organism begins as a single cell that repeatedly divides to produce many cells, but eventually some cells must die for the organism to take shape. For example, when a tadpole becomes a frog, the tail disappears as apoptosis occurs. In a human embryo, the fingers and toes are at first webbed, but then they are usually freed from one another as a result of apoptosis.

Cell division occurs during your entire life. Even now, your body is producing thousands of new red blood cells, skin cells, and cells that line your respiratory and digestive tracts. Also, if you suffer a cut, cell division repairs the injury. Apoptosis occurs all the time, too, particularly if an abnormal cell that could become cancerous appears or a cell becomes infected with a virus. Death through apoptosis prevents a tumor from developing and helps limit the spread of viruses.

Check Your Progress 9.1

1. List, in order, the stages of the cell cycle and briefly summarize what is happening at each stage.
2. Explain what conditions might cause a cell to halt the cell cycle and state briefly where in the cycle this would occur.
3. Discuss how apoptosis represents a regulatory event of the cell cycle.

THEME **Nature of Science**

The G_1 Checkpoint

Cell division is very tightly regulated, so that only certain cells in an adult body are actively dividing. After cell division occurs, cells enter the G_1 stage. Upon completing G_1, they will divide again, but before this happens they have to pass through the G_1 checkpoint.

The G_1 checkpoint ensures that conditions are right for making the commitment to divide by evaluating the meaning of growth signals, determining the availability of nutrients, and assessing the integrity of DNA. Failure to meet any one of these criteria results in a cell's halting the cell cycle and entering G_0 stage, or undergoing apoptosis if the problems are severe.

Evaluating Growth Signals

Multicellular organisms tightly control cell division, so that it occurs only when needed. Signaling molecules, such as hormones, may be sent from nearby cells or distant tissues to encourage or discourage cells from entering the cell cycle. Such signals may cause a cell to enter a G_0 stage, or complete G_1 and enter the S stage. Growth signals that promote cell division cause a cyclin-dependent-kinase (CDK) to add a phosphate group to the RB protein, a major regulator of the G_1 checkpoint.

Ordinarily, a protein called E2F is bound to RB, but when RB is phosphorylated, its shape changes and it releases E2F. Now, E2F binds to DNA, activating certain genes whose products are needed to complete the cell cycle (Fig. 9A*a*). Likewise, growth signals prompt cells that are in G_0 stage to reenter the G_1 stage, complete it, and enter the S stage. If growth signals are sufficient, a cell passes through the G_1 checkpoint and cell division occurs.

Determining Nutrient Availability

Just as experienced hikers ensure that they have sufficient food for their journey, a cell ensures that nutrient levels are adequate before committing to cell division. For example, scientists know that starving cells in culture enter G_0. At that time, phosphate groups are removed from RB (see reverse arrows in Fig. 9A*a*); RB does not release E2F; and the proteins needed to complete the cell cycle are not produced. When nutrients become available, CDKs bring about the phosphorylation of RB, which then releases E2F (see forward arrows in Fig. 9A*a*). After E2F binds to DNA, the proteins needed to complete the cell cycle are produced. Therefore, you can see that cells do not commit to divide until conditions are conducive for them to do so.

Assessing DNA Integrity

For cell division to occur, DNA must be free of errors and damage. The p53 protein is involved in this quality control function. Ordinarily, p53 is broken down because it has no job to do. In response to DNA damage, CDK phosphorylates p53 (Fig. 9A*b*). Now, the molecule is not broken down as usual, and instead its level in the nucleus begins to rise. Phosphorylated p53 binds to DNA; certain genes are activated; and DNA repair proteins are produced. If the DNA damage cannot be repaired, p53 levels continue to rise, and apoptosis is triggered. If the damage is successfully repaired, p53 levels fall, and the cell is allowed to complete G_1 stage—as long as growth signals and nutrients are present, for example.

Actually, many criteria must be met for a cell to commit to cell division, and the failure to meet any one of them may cause the cell cycle to be halted and/or apoptosis to be initiated. The G_1 checkpoint is currently an area of intense research, because understanding it holds the key to possibly curing cancer and to unleashing the power of normal, healthy cells to regenerate tissues, which could be used to cure many other human conditions.

Questions to Consider

1. What is the potential effect of an abnormally high level of a growth hormone on the regulation of the cell cycle?
2. Why might some cancers be associated with a mutation in the gene encoding the p53 protein?

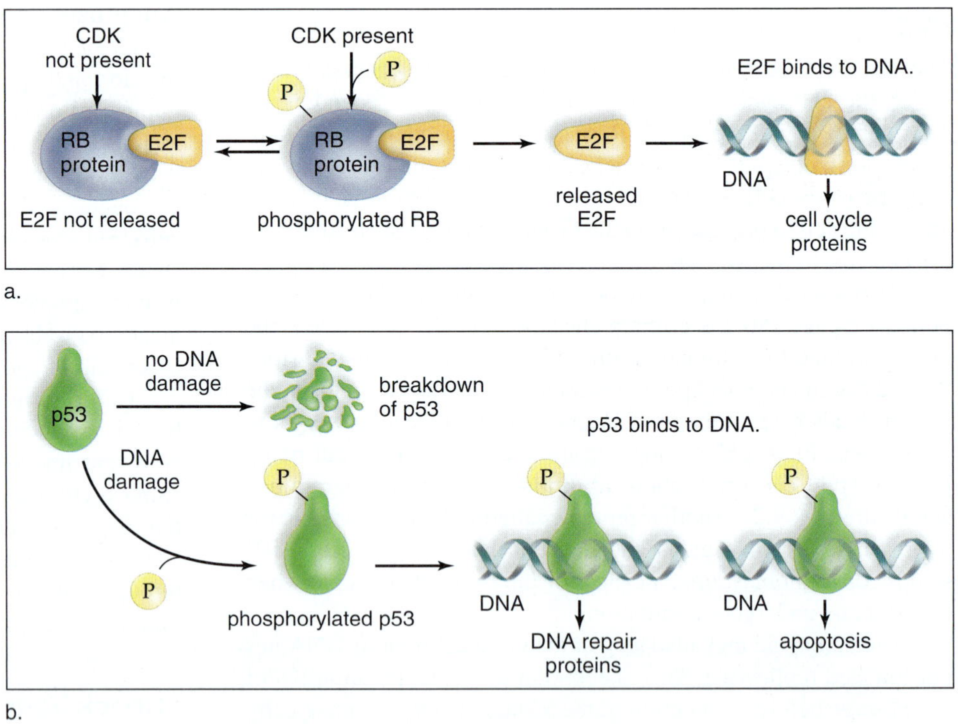

Figure 9A Regulation of the G_1 checkpoint. **a.** When CDK (cyclin-dependent-kinase) is not present, RB retains E2F. When CDK is present, a phosphorylated RB releases E2F, and after it binds to DNA, the proteins necessary for completing cell division are produced. **b.** If DNA is damaged, p53 is not broken down; instead, it is involved in producing DNA repair enzymes and in triggering apoptosis when repair is impossible.

9.2 The Eukaryotic Chromosome

Learning Outcomes

Upon completion of this section, you should be able to

1. Explain how DNA becomes sufficiently compacted to fit inside a nucleus.
2. Distinguish between euchromatin and heterochromatin.

Cell biologists and geneticists have been able to construct detailed models of how chromosomes are organized. A eukaryotic chromosome contains a single double helix DNA molecule, but it is composed of more than 50% protein. Some of these proteins are concerned with DNA and RNA synthesis, but a large majority, termed histones, play primarily a structural role.

The five primary types of histone molecules are designated H1, H2A, H2B, H3, and H4 (see Fig. 13.5*b*). Remarkably, the amino acid sequences of H3 and H4 vary little between organisms. For example, the H4 of peas is only two amino acids different from the H4 of cattle. This similarity suggests that few mutations in the histone proteins have occurred during the course of evolution and that the histones, therefore, have essential functions for survival.

A human cell contains at least 2 m of DNA, yet all of this DNA is packed into a nucleus that is about 6 μm in diameter. The histones are responsible for packaging the DNA so that it can fit into such a small space. First, the DNA double helix is wound at intervals around a core of eight histone molecules (two copies each of H2A, H2B, H3, and H4), giving the appearance of a string of beads (Fig. 9.3*a*). Each bead is called a nucleosome, and the nucleosomes are said to be joined by "linker" DNA.

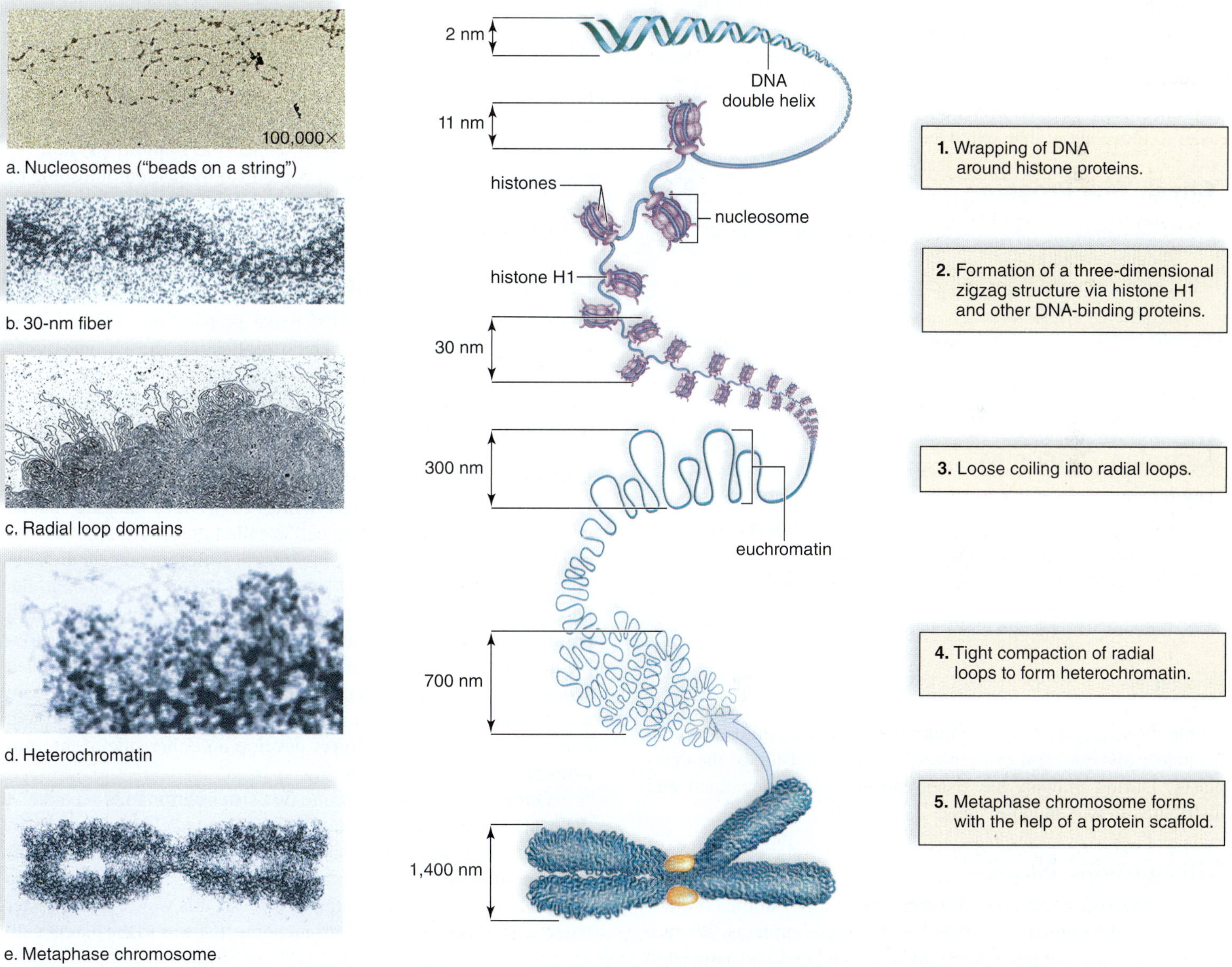

Figure 9.3 Structure of the eukaryotic chromosome. Eukaryotic cells contain nearly 2 m of DNA, yet they must pack it all into a nucleus that is around 6 μm in diameter. Thus, the DNA is compacted by winding it around DNA-binding proteins, called histones, to make nucleosomes. The nucleosomes are further compacted into a zigzag structure, which is then folded upon itself many times to form radial loops, which is the usual compaction state of euchromatin. Heterochromatin is further compacted by scaffold proteins, and further compaction can be achieved prior to mitosis and meiosis.

This string is compacted by folding into a zigzag structure, further shortening the DNA strand (Fig. 9.3*b*). Histone H1 appears to mediate this coiling process. The fiber then loops back and forth into radial loops (Fig. 9.3*c*). This loosely coiled **euchromatin** represents the active chromatin containing genes that are being transcribed. The DNA of euchromatin may be accessed by RNA polymerase and other factors that are needed to promote transcription. In fact, recent research seems to indicate that regulating the level of compaction of the DNA is an important method of controlling gene expression in the cell.

Under a microscope, one often observes dark-stained fibers within the nucleus of the cell. These areas within the nucleus represent a more highly compacted form of the chromosome called **heterochromatin** (Fig. 9.3*d*). Most chromosomes exhibit both levels of compaction in a living cell, depending on which portions of the chromosome are being used more frequently. Heterochromatin is considered inactive chromatin, because the genes contained on it are infrequently transcribed, if at all.

Prior to cell division, a protein scaffold helps further condense the chromosome into a form that is characteristic of metaphase chromosomes (Fig. 9.3*e*). No doubt, compact chromosomes are easier to move about than extended chromatin.

Check Your Progress 9.2

1. Summarize the differences between euchromatin and heterochromatin.
2. List the stages of chromosome compacting, starting with a single DNA strand.

9.3 Mitosis and Cytokinesis

Learning Outcomes

Upon completion of this section, you should be able to

1. Explain how the cell prepares the chromosomes and centrosomes prior to nuclear division.
2. Summarize the major events that occur during mitosis and cytokinesis.
3. Discuss why human stem cells continuously conduct mitosis.

As mentioned, cell division in eukaryotes involves mitosis, which is nuclear division, and cytokinesis, which is division of the cytoplasm. During mitosis, the sister chromatids are separated and distributed to two daughter cells.

Chromosome Number

As we observed in the previous section, the DNA in the chromosomes of eukaryotes is associated with various proteins. When a eukaryotic cell is not undergoing division, the DNA and associated proteins are located within **chromatin,** which has the appearance of a tangled mass of thin threads. Before mitosis begins, chromatin becomes highly coiled and condensed, and it is easy to see the individual chromosomes.

Table 9.1 Diploid Chromosome Numbers of Some Eukaryotes

Type of Organism	Name of Chromosome	Chromosome Number
Fungi	*Saccharomyces cerevisiae* (yeast)	32
Plants	*Pisum sativum* (garden pea)	14
	Solanum tuberosum (potato)	48
	Ophioglossum vulgatum (southern adder's tongue fern)	1,320
Animals	*Drosophila melanogaster* (fruit fly)	8
	Homo sapiens (human)	46
	Carassius auratus (goldfish)	94

When the chromosomes are visible, it is possible to photograph and count them. Each species has a characteristic chromosome number (Table 9.1). This is the full, or **diploid (2n)**, number (Gk. *diplos,* "twofold"; *-eides,* "like") of chromosomes that is found in all cells of the individual. The diploid number includes two chromosomes of each kind. Most somatic cells of animals are diploid. Half the diploid number, called the **haploid (n)** number (Gk. *haplos,* "simple, single"), contains only one chromosome of each kind. The gametes of animals (egg and sperm) are examples of haploid cells.

Preparations for Mitosis

During interphase, a cell must make preparations for cell division. These arrangements include replicating the chromosomes and duplicating most cellular organelles, including the centrosome, which will organize the spindle apparatus necessary for the movement of chromosomes.

Chromosome Duplication

During mitosis, a 2n nucleus divides to produce daughter nuclei that are also 2n. The dividing cell is called the *parent cell,* and the resulting cells are called the *daughter cells*. Before nuclear division takes place, DNA replicates, duplicating the chromosomes in the parent cell. This occurs during the S stage of interphase. Now each chromosome has two identical double helical molecules. Each double helix is a *chromatid,* and the two identical chromatids are called *sister chromatids* (Fig. 9.4). Sister chromatids are constricted and attached to each other at a region called the **centromere.** Protein complexes called **kinetochores** develop on either side of the centromere during cell division.

During nuclear division, the two sister chromatids separate at the centromere, and in this way each duplicated chromosome gives rise to two daughter chromosomes. Each daughter chromosome has only one double helix molecule. The daughter chromosomes are distributed equally to the daughter cells. In this way, each daughter nucleus gets a copy of each chromosome that was in the parent cell.

Division of the Centrosome

The **centrosome** (Gk. *centrum,* "center"; *soma,* "body"), the main microtubule-organizing center of the cell, also divides before mitosis begins. Each centrosome in an animal cell contains a pair of

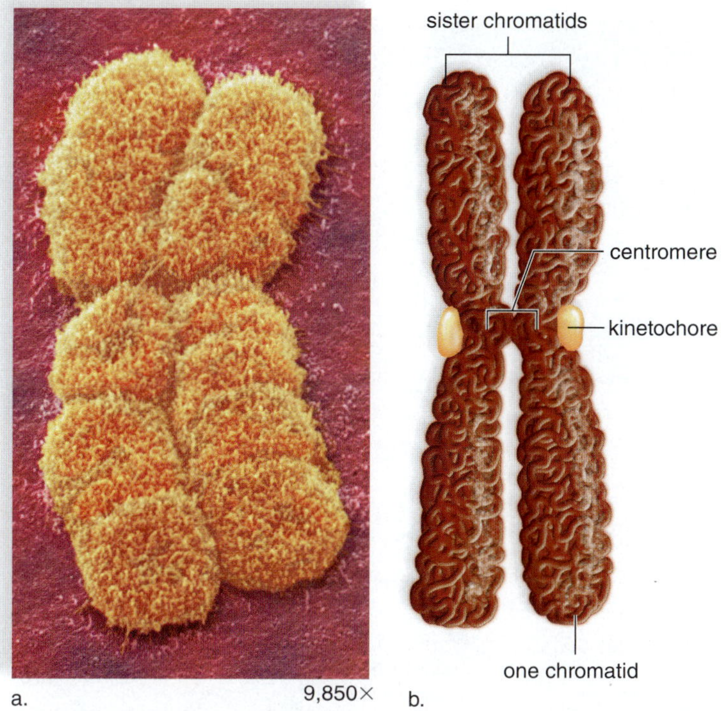

Figure 9.4 Duplicated chromosomes. A duplicated chromosome contains two sister chromatids, each with a copy of the same genes. **a.** Electron micrograph of a highly coiled and condensed chromosome, typical of a nucleus about to divide. **b.** Diagrammatic drawing of a condensed chromosome. The chromatids are held together at a region called the centromere.

barrel-shaped organelles called **centrioles.** Centrioles are not found in plant cells.

The centrosomes organize the mitotic spindle, which contains many fibers, each of which is composed of a bundle of microtubules. Microtubules are hollow cylinders made up of the protein tubulin. They assemble when tubulin subunits join, and when they disassemble, tubulin subunits become free once more. The microtubules of the cytoskeleton disassemble when spindle fibers begin forming. Most likely, this provides tubulin for the formation of the spindle fibers, or it may allow the cell to change shape as needed for cell division.

Phases of Mitosis

Mitosis is a continuous process that is arbitrarily divided into five phases for convenience of description: prophase, prometaphase, metaphase, anaphase, and telophase (Fig. 9.5).

Animation Mitosis

MP3 Mitosis

Prophase

It is apparent during **prophase** that nuclear division is about to occur, because chromatin has condensed and the chromosomes are visible. Recall that DNA replication occurred during interphase, and therefore the *parental chromosomes are already duplicated and composed of two sister chromatids held together at a centromere.* Counting the number of centromeres in diagrammatic drawings gives the number of chromosomes for the cell depicted.

During prophase, the nucleolus disappears and the nuclear envelope fragments. The spindle begins to assemble as the two centrosomes migrate away from one another. In animal cells, an array of microtubules radiates toward the plasma membrane from the centrosomes. These structures are called *asters.* It is thought that asters brace the centrioles during later stages of cell division. Notice that the chromosomes have no particular orientation, because the spindle has not yet formed.

Prometaphase (Late Prophase)

During **prometaphase,** preparations for sister chromatid separation are evident. Kinetochores appear on each side of the centromere, and these attach sister chromatids to the *kinetochore spindle fibers.* These fibers extend from the poles to the chromosomes, which will soon be located at the center of the spindle.

The kinetochore fibers attach the sister chromatids to opposite poles of the spindle, and the chromosomes are pulled first toward one pole and then toward the other before the chromosomes come into alignment. Notice that even though the chromosomes are attached to the spindle fibers in prometaphase, they are still not in alignment.

Metaphase

During **metaphase,** the centromeres of chromosomes are now in alignment on a single plane at the center of the cell. The chromosomes usually appear as a straight line across the middle of the cell when viewed under a light microscope. An imaginary plane that is perpendicular and passes through this circle is called the *metaphase plate.* It indicates the future axis of cell division.

Several nonattached spindle fibers, called *polar spindle fibers,* reach beyond the metaphase plate and overlap. A cell cycle checkpoint, the M checkpoint, delays the start of anaphase until the kinetochores of each chromosome are attached properly to spindle fibers and the chromosomes are properly aligned along the metaphase plate.

Anaphase

At the start of **anaphase,** the two sister chromatids of each duplicated chromosome separate at the centromere, giving rise to two daughter chromosomes. Daughter chromosomes, each with a centromere and single chromatid composed of a single double helix, appear to move toward opposite poles. Actually, the daughter chromosomes are being pulled to the opposite poles as the kinetochore spindle fibers disassemble at the region of the kinetochores.

Even as the daughter chromosomes move toward the spindle poles, the poles themselves are moving farther apart, because the polar spindle fibers are sliding past one another. Microtubule-associated proteins, such as the motor molecules kinesin and dynein, are involved in the sliding process. Anaphase is the shortest phase of mitosis.

Telophase

During **telophase,** the spindle disappears as new nuclear envelopes form around the daughter chromosomes. Each daughter nucleus contains the same number and kinds of chromosomes as the original parent cell. Remnants of the polar spindle fibers are still visible between the two nuclei.

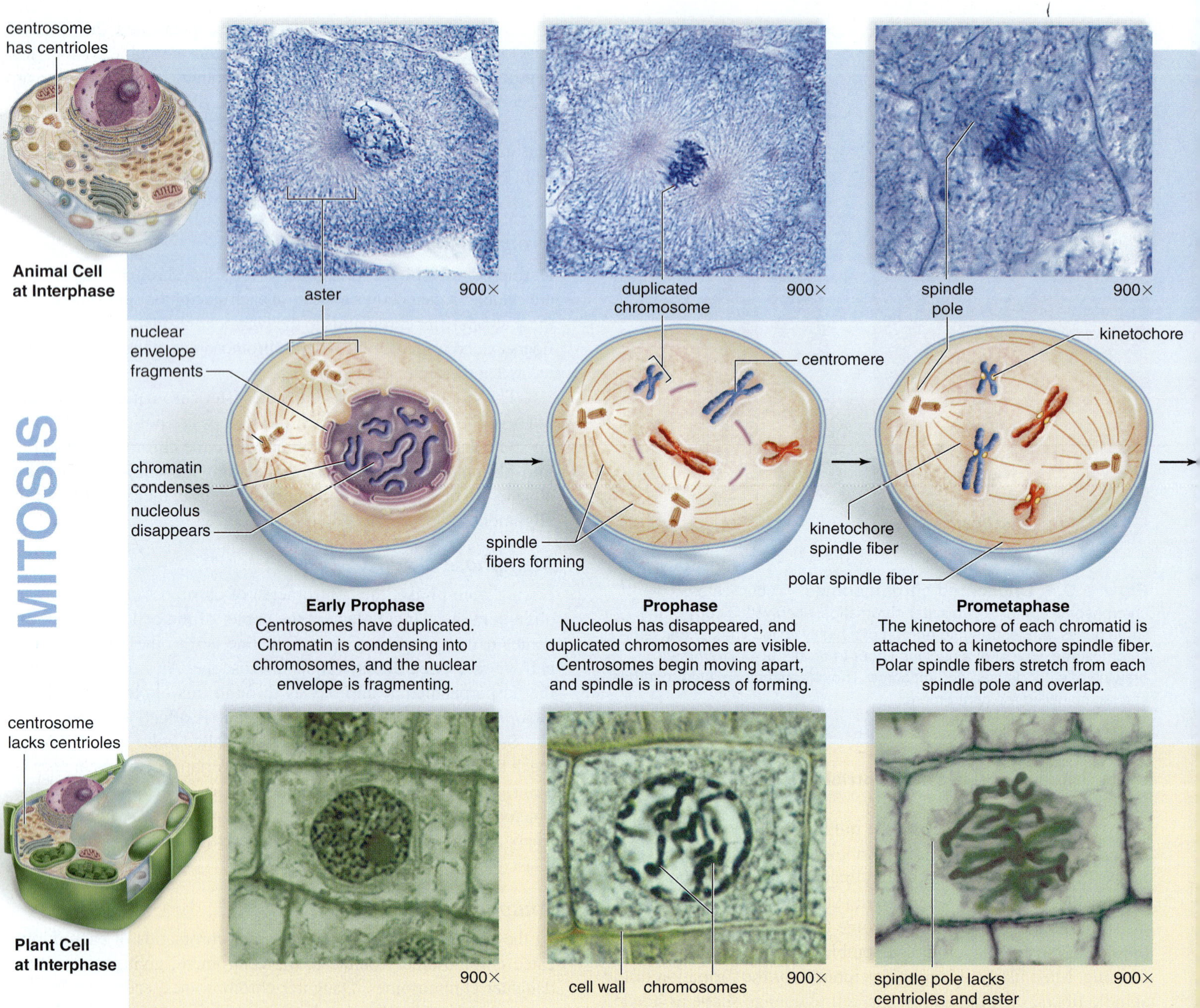

Figure 9.5 Phases of mitosis in animal and plant cells. The blue chromosomes were inherited from one parent, the red from the other parent.

Tutorial Mitosis

The chromosomes become more diffuse chromatin once again, and a nucleolus appears in each daughter nucleus. Division of the cytoplasm requires cytokinesis, which is discussed in the next section.

3D Animation Cell Cycle and Mitosis: Mitosis

Cytokinesis in Animal and Plant Cells

As mentioned previously, cytokinesis is division of the cytoplasm. Cytokinesis accompanies mitosis in most cells, but not all. When mitosis occurs but cytokinesis doesn't occur, the result is a multinucleated cell. For example, you will see in Chapter 27 that the embryo sac in flowering plants is multinucleated.

Division of the cytoplasm begins in anaphase, continues in telophase, but does not reach completion until the following interphase begins. By the end of mitosis, each newly forming cell has received a share of the cytoplasmic organelles that duplicated during interphase. Cytokinesis proceeds differently in plant and animal cells because of differences in cell structure.

Animation Cytokinesis

Cytokinesis in Animal Cells

In animal cells a **cleavage furrow,** which is an indentation of the membrane between the two daughter nuclei, forms just as anaphase draws to a close. By that time, the newly forming cells have received a share of the cytoplasmic organelles that duplicated during the previous interphase.

The cleavage furrow deepens when a band of actin filaments, called the contractile ring, slowly forms a circular constriction

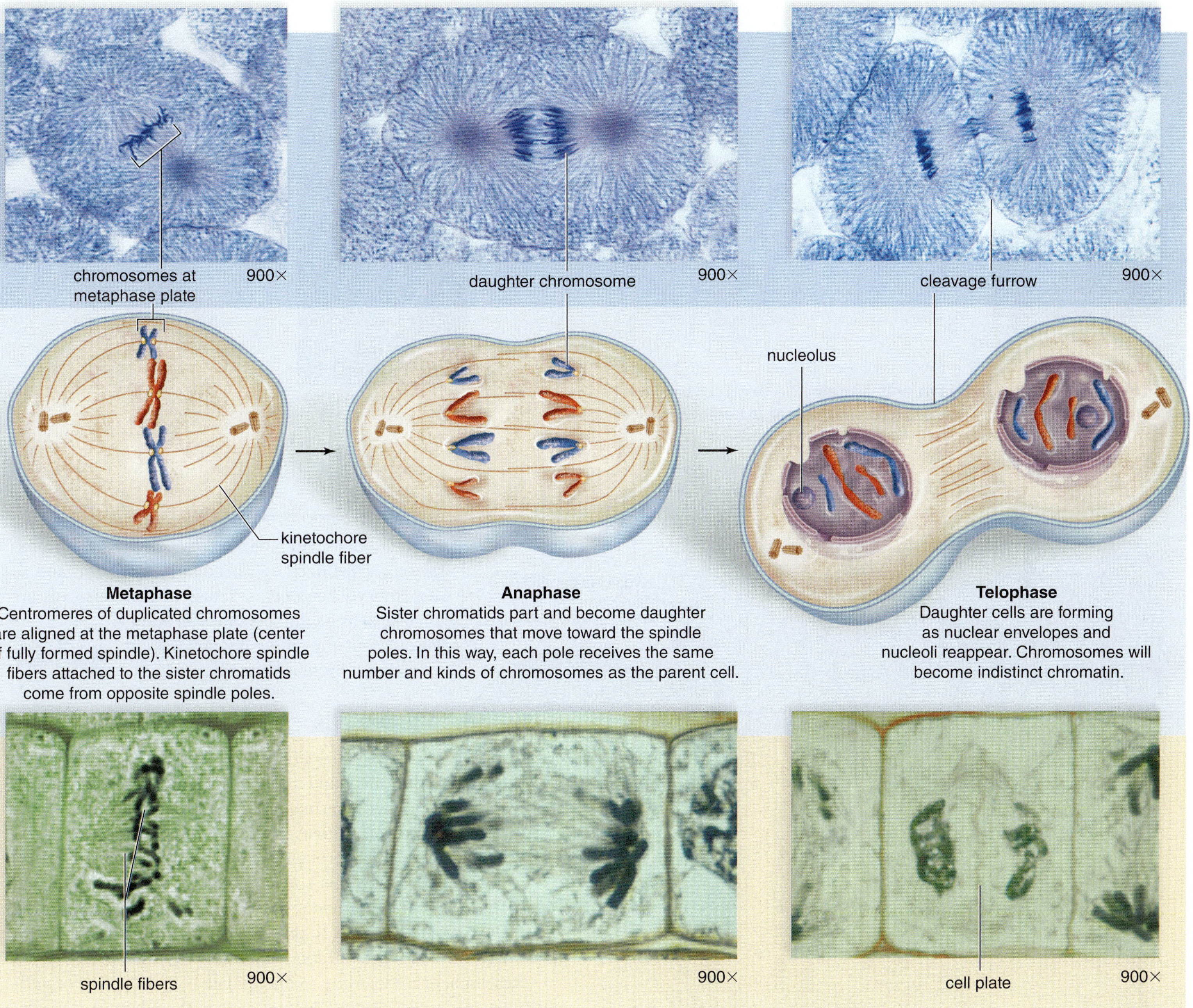

Metaphase
Centromeres of duplicated chromosomes are aligned at the metaphase plate (center of fully formed spindle). Kinetochore spindle fibers attached to the sister chromatids come from opposite spindle poles.

Anaphase
Sister chromatids part and become daughter chromosomes that move toward the spindle poles. In this way, each pole receives the same number and kinds of chromosomes as the parent cell.

Telophase
Daughter cells are forming as nuclear envelopes and nucleoli reappear. Chromosomes will become indistinct chromatin.

between the two daughter cells. The action of the contractile ring can be likened to pulling a drawstring ever tighter about the middle of a balloon. As the drawstring is pulled tight, the balloon constricts in the middle as the material on either side of the constriction gathers in folds. These folds are represented by the longitudinal lines in Figure 9.6.

A narrow bridge between the two cells can be seen during telophase, and then the contractile ring continues to separate the cytoplasm until there are two independent daughter cells (Fig. 9.6).

Cytokinesis in Plant Cells

Cytokinesis in plant cells occurs by a process different from that seen in animal cells (Fig. 9.7). The rigid cell wall that surrounds plant cells does not permit cytokinesis by furrowing. Instead, cytokinesis in plant cells involves the building of new cell walls between the daughter cells.

Cytokinesis is apparent when a small, flattened disk appears between the two daughter plant cells near the site where the metaphase plate once was. In electron micrographs, it is possible to see that the disk is at right angles to a set of microtubules that radiate outward from the forming nuclei. The Golgi apparatus produces vesicles, which move along the microtubules to the region of the disk. As more vesicles arrive and fuse, a cell plate can be seen. The **cell plate** is simply a newly formed plasma membrane that expands outward until it reaches the old plasma membrane and fuses with this membrane.

The new membrane releases molecules that form the new plant cell walls. These cell walls, known as primary cell walls, are later strengthened by the addition of cellulose fibrils. The space between the daughter cells becomes filled with middle lamella, which cements the primary cell walls together.

3D Animation
Cell Cycle and Mitosis: Cytokinesis

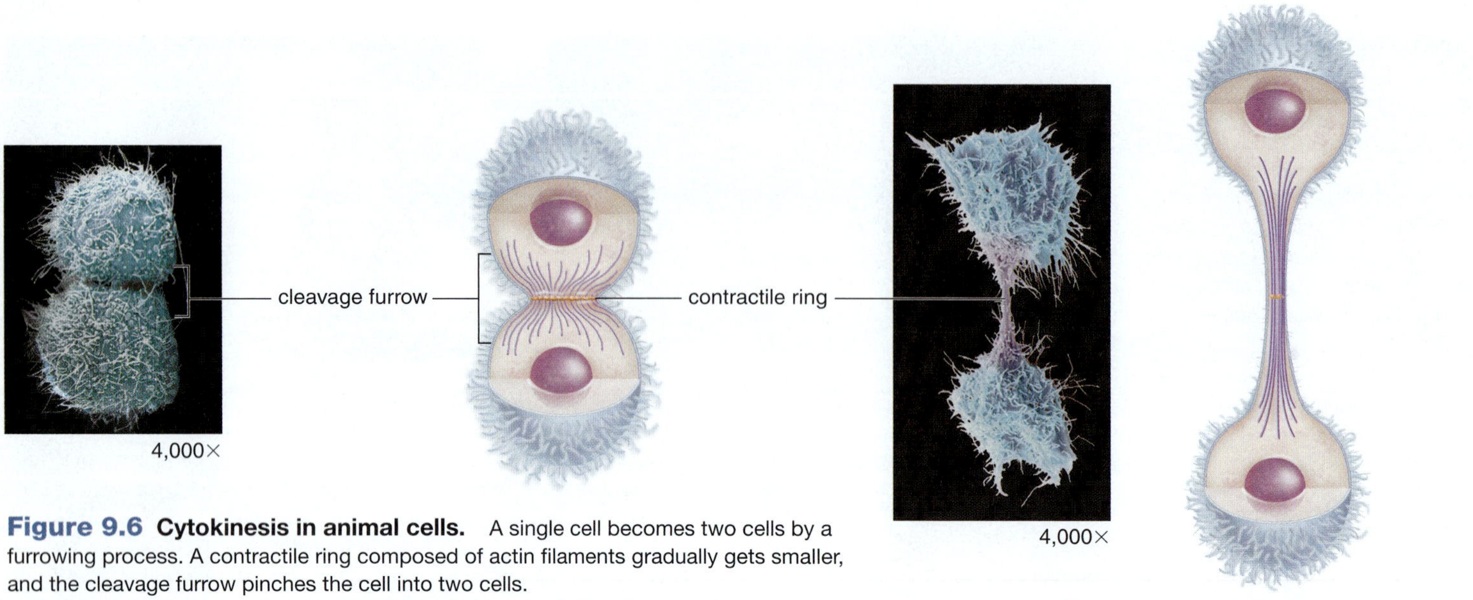

Figure 9.6 Cytokinesis in animal cells. A single cell becomes two cells by a furrowing process. A contractile ring composed of actin filaments gradually gets smaller, and the cleavage furrow pinches the cell into two cells.

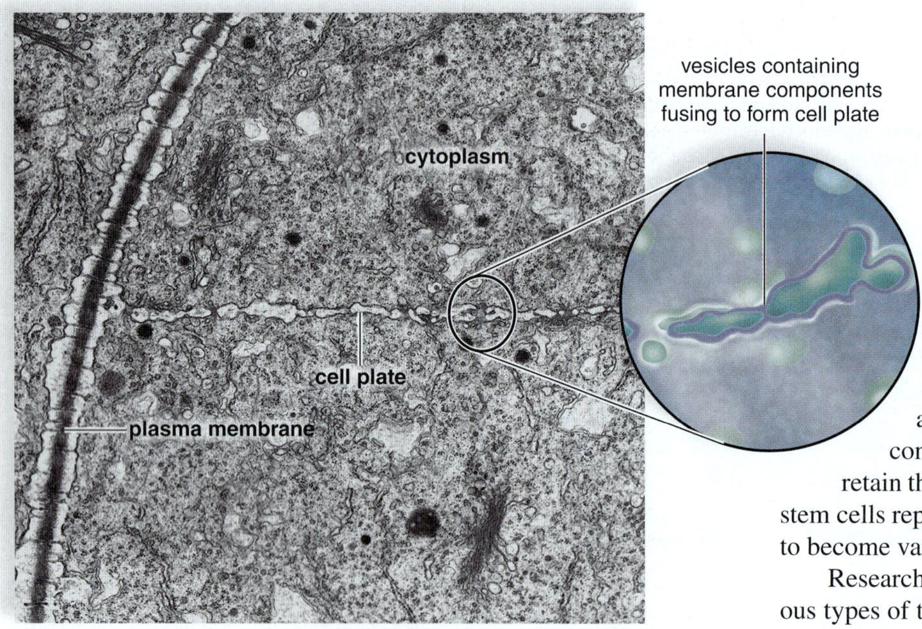

Figure 9.7 Cytokinesis in plant cells. During cytokinesis in a plant cell, a cell plate forms midway between two daughter nuclei and extends to the plasma membrane.

The Functions of Mitosis

Mitosis permits growth and repair. In both plants and animals, mitosis is required during development as a single cell develops into an individual. In plants, the individual could be a fern or daisy, while in animals, the individual could be a grasshopper or a human.

In flowering plants, meristematic tissue retains the ability to divide throughout the life of a plant. Meristematic tissue at the shoot tip accounts for an increase in the height of a plant for as long as it lives. Then, too, lateral meristem accounts for the ability of trees to increase their girth each growing season.

In humans and other mammals, mitosis is necessary as a fertilized egg becomes an embryo and as the embryo becomes a fetus. Mitosis also occurs after birth as a child becomes an adult. Throughout life, mitosis allows a cut to heal or a broken bone to mend.

Video Mitosis

Stem Cells

Earlier, you learned that the cell cycle is tightly controlled, and that most cells of the body at adulthood are permanently arrested in the G_0 stage. However, mitosis is needed to repair injuries, such as a cut or a broken bone. Many mammalian organs contain stem cells (often called adult stem cells) that retain the ability to divide. As one example, red bone marrow stem cells repeatedly divide to produce millions of cells that go on to become various types of blood cells.

Researchers are learning to manipulate the production of various types of tissues from adult stem cells in the laboratory. If successful, these tissues could be used to cure illnesses. As discussed in the Nature of Science feature, "Reproductive and Therapeutic Cloning," **therapeutic cloning,** which is used to produce human tissues, can begin with either adult stem cells or embryonic stem cells. Embryonic stem cells can also be used for **reproductive cloning,** the production of a new individual.

Check Your Progress 9.3

1. Describe the major events that occur during each phase of mitosis.
2. Summarize the differences between cytokinesis in animal and plant cells and explain why the differences are necessary.
3. Discuss the importance of stem cells in the human body.

THEME Nature of Science

Reproductive and Therapeutic Cloning

Our knowledge of how the cell cycle is controlled has yielded major technological breakthroughs, including reproductive cloning—the ability to clone an adult animal from a normal body cell—and therapeutic cloning, which allows the rapid production of mature cells of a specific type. Both types of cloning are a direct result of recent discoveries about how the cell cycle is controlled.

Reproductive cloning, or the cloning of adult animals, was once thought to be impossible, because investigators found it difficult to have the nucleus of an adult cell "start over" with the cell cycle, even when it was placed in an egg cell that had had its own nucleus removed.

In 1997, Dolly the sheep demonstrated that reproductive cloning is indeed possible. The donor cells were starved before the cell's nucleus was placed in an enucleated egg. This caused them to stop dividing and go into a G_0 (resting) stage, and this made the nuclei amenable to cytoplasmic signals for initiation of development (Fig. 9B*a*). This advance has made it possible to clone all sorts of farm animals that have desirable traits and even to clone rare animals that might otherwise become extinct. Despite the encouraging results, however, there are still obstacles to be overcome, and a ban on the use of federal funds in experiments to clone humans remains firmly in place.

In therapeutic cloning, however, the objective is to produce mature cells of various cell types rather than an individual organism. The purpose of therapeutic cloning is (1) to learn more about how specialization of cells occurs and (2) to provide cells and tissues that could be used to treat human illnesses, such as diabetes, or major injuries like strokes or spinal cord injuries.

There are two possible ways to carry out *therapeutic cloning.* The first way is to use exactly the same procedure as reproductive cloning, except that *embryonic stem cells* (*ESCs*) are separated and each is subjected to a treatment that causes it to develop into a particular type of cell, such as red blood cells, muscle cells, or nerve cells (Fig. 9B*b*). Some have ethical concerns about this type of therapeutic cloning, which is still experimental, because if the embryo were allowed to continue development, it would become an individual.

The second way to carry out therapeutic cloning is to use *adult stem cells.* Stem cells are found in many organs of the adult's body; for example, the bone marrow has stem cells that produce new blood cells. However, adult stem cells are limited in the number of cell types they may become. Nevertheless, scientists are beginning to overcome this obstacle. In 2006, by adding just four genes to adult skin stem cells, Japanese scientists were able to coax the cells, called fibroblasts, into becoming induced pluripotent stem cells (iPS), a type of stem cell that is similar to an ESC. The researchers were then able to create heart and brain cells from the adult stem cells. Other researchers have used this technique to reverse Parkinson-like symptoms in rats.

Although questions exist on the benefits of iPS cells, these advances demonstrate that scientists are actively investigating methods of overcoming the current limitations and ethical concerns of using embryonic stem cells.

Questions to Consider

1. How might the study of therapeutic cloning benefit scientific studies of reproductive cloning?
2. What types of diseases might not be treatable using therapeutic cloning?

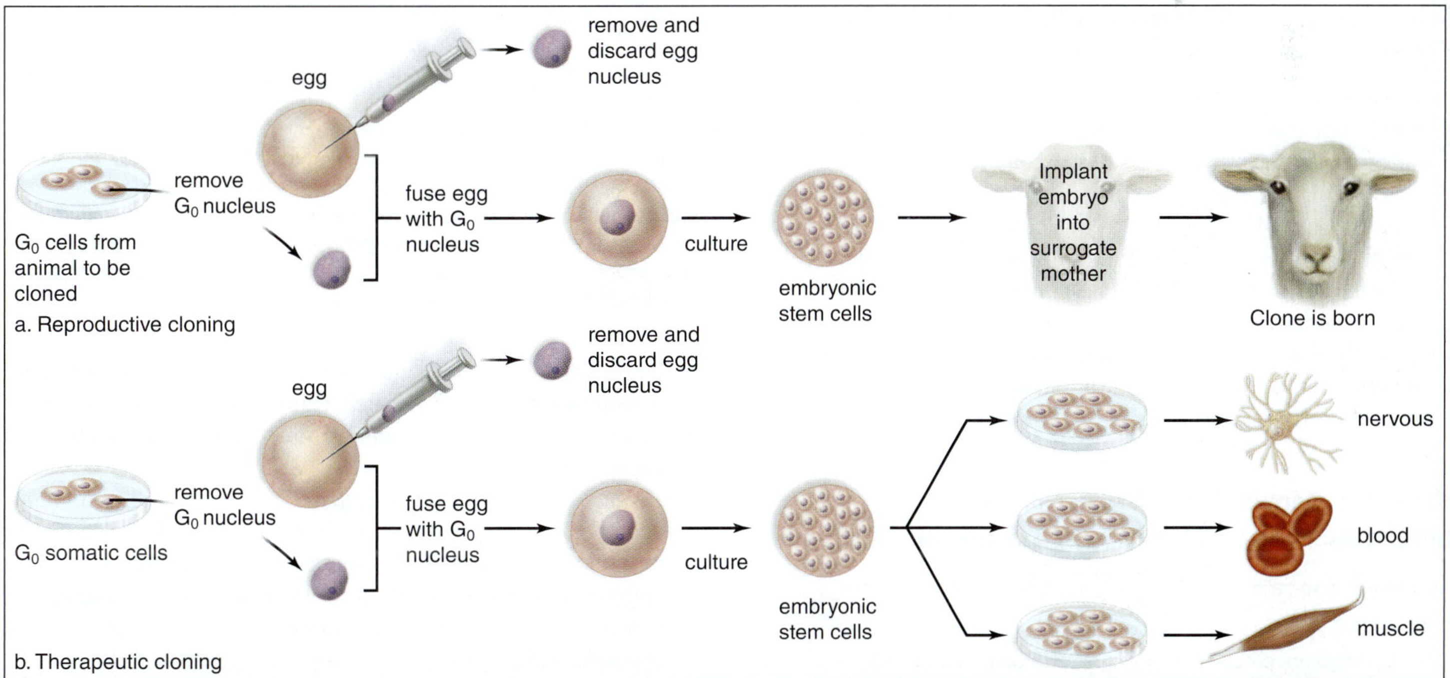

Figure 9B Two types of cloning. **a.** The purpose of reproductive cloning is to produce an individual that is genetically identical to the one that donated a nucleus. The nucleus is placed in an enucleated egg, and, after several mitotic divisions, the embryo is implanted into a surrogate mother for further development. **b.** The purpose of therapeutic cloning is to produce specialized tissue cells. A nucleus is placed in an enucleated egg, and after several mitotic divisions, the embryonic cells (called embryonic stem cells) are separated and treated to become specialized cells.

9.4 The Cell Cycle and Cancer

Learning Outcomes

Upon completion of this section, you should be able to

1. Describe the basic characteristics of cancer cells.
2. Explain the difference between a benign and malignant tumor.
3. Distinguish between the roles of the tumor suppressor genes and proto-oncogenes in the regulation of the cell cycle.

Cancer is a cellular growth disorder that occurs when cells divide uncontrollably. Although causes widely differ, most cancers are the result of accumulating mutations that ultimately cause a loss of control of the cell cycle.

Although cancers vary greatly, they usually follow a common multistep progression (Fig. 9.8). Most cancers begin as an abnormal cell growth that is **benign,** or not cancerous, and usually does not grow larger. However, additional mutations may occur, causing the abnormal cells to fail to respond to inhibiting signals that control the cell cycle. When this occurs, the growth becomes **malignant,** meaning that it is cancerous and possesses the ability to spread.

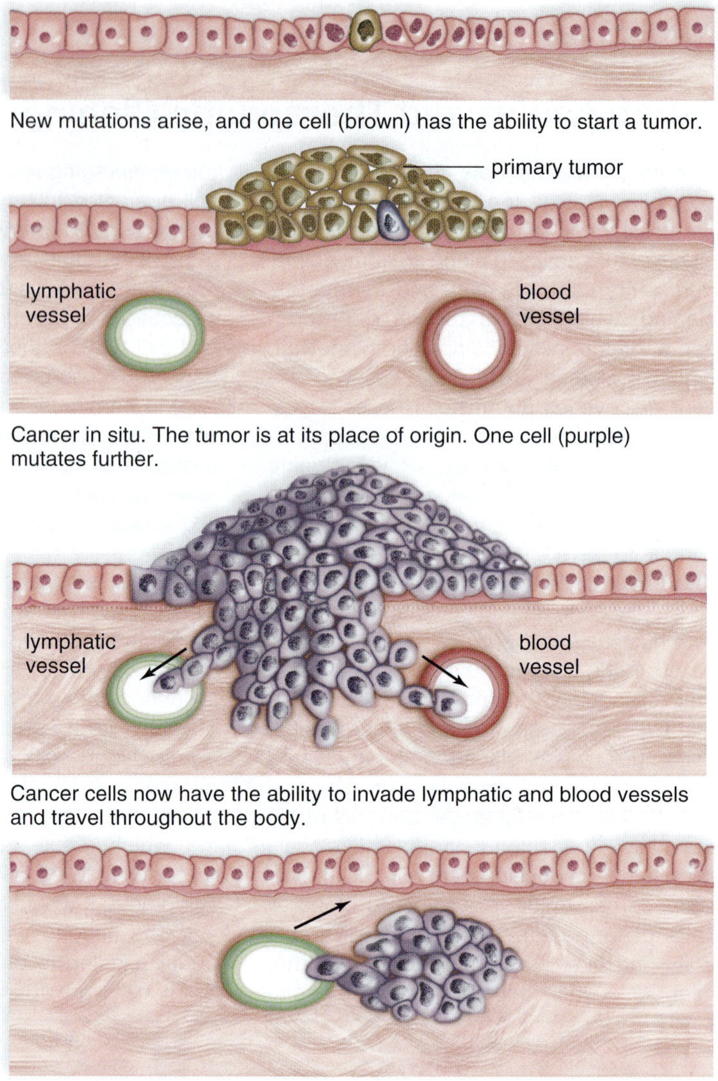

New mutations arise, and one cell (brown) has the ability to start a tumor.

Cancer in situ. The tumor is at its place of origin. One cell (purple) mutates further.

Cancer cells now have the ability to invade lymphatic and blood vessels and travel throughout the body.

New metastatic tumors are found some distance from the primary tumor.

Figure 9.8 Progression of cancer. The development of cancer requires a series of mutations, leading first to a localized tumor and then to metastatic tumors. With each successive step toward cancer, the most genetically altered and aggressive cell becomes the dominant type of tumor. The cells take on characteristics of embryonic cells; they are not differentiated; they can divide uncontrollably; and they are able to metastasize and spread to other tissues.

Characteristics of Cancer Cells

The development of cancer is gradual. A mutation in a cell may cause it to become precancerous, but many other regulatory processes within the body prevent it from becoming cancerous. In fact, it may be decades before a cell possesses most or all of the characteristics of a cancer cell (Table 9.2 and Fig. 9.8). Although cancers vary greatly, cells that possess the following characteristics are generally recognized as cancerous:

Cancer cells lack differentiation. Cancer cells are not specialized and do not contribute to the functioning of a tissue. Although cancer cells may still possess many of the characteristics of surrounding normal cells, they usually look distinctly abnormal. Normal cells can enter the cell cycle about 50 times before they are incapable of dividing again. Cancer cells can enter the cell cycle an indefinite number of times and, in this way, seem immortal.

Cancer cells have abnormal nuclei. The nuclei of cancer cells are enlarged and may contain an abnormal number of chromosomes. Extra copies of one or more chromosomes may be present. Often, there are also duplicated portions of some chromosomes present, which causes gene amplification, or extra copies of specific genes. Some chromosomes may also possess deleted portions.

Table 9.2 Cancer Cells Versus Normal Cells

Cancer Cells	Normal Cells
Nondifferentiated cells	Differentiated cells
Abnormal nuclei	Normal nuclei
Do not undergo apoptosis	Undergo apoptosis
No contact inhibition	Contact inhibition
Disorganized, multilayered	One organized layer
Undergo metastasis	Remain in original tissue

Cancer cells do not undergo apoptosis. Ordinarily, cells with damaged DNA undergo apoptosis, or programmed cell death. The immune system can also recognize abnormal cells and trigger apoptosis, which normally prevents tumors from developing. Cancer cells fail to undergo apoptosis even though they are abnormal cells.

Cancer cells form tumors. Normal cells anchor themselves to a substratum and/or adhere to their neighbors. They exhibit contact inhibition—in other words, when they come in contact with a neighbor, they stop dividing. Cancer cells have lost all restraint and do not exhibit contact inhibition. The abnormal cancer cells pile on top of one another and grow in multiple layers, forming a **tumor.** During carcinogenesis, the most aggressive cell becomes the dominant cell of the tumor.

Cancer cells undergo metastasis and angiogenesis. Additional mutations may cause a benign tumor, which is usually contained within a capsule and cannot invade adjacent tissue, to become malignant, and spread throughout the body, forming new tumors distant from the primary tumor. These cells now produce enzymes that they normally do not express, allowing tumor cells to invade underlying tissues. Then, they travel through the blood and lymph, to start tumors elsewhere in the body. This process is known as **metastasis.**

Tumors that are actively growing soon encounter another obstacle—the blood vessels supplying nutrients to the tumor cells become insufficient to support the rapid growth of the tumor. In order to grow further, the cells of the tumor must receive additional nutrition. Thus, the formation of new blood vessels is required to bring nutrients and oxygen to support further growth. Additional mutations occurring in tumor cells allow them to direct the growth of new blood vessels into the tumor in a process called **angiogenesis.** Some modes of cancer treatment are aimed at preventing angiogenesis from occurring.

Origin of Cancer

Normal growth and maintenance of body tissues depend on a balance between signals that promote and inhibit cell division. When this balance is upset, conditions such as cancer may occur. Thus, cancer is usually caused by mutations affecting genes that directly or indirectly affect this balance, such as those shown in Figure 9.9. The following two types of genes are usually affected:

1. **Proto-oncogenes** code for proteins that promote the cell cycle and prevent apoptosis. They are often likened to the gas pedal of a car, because they cause the cell cycle to speed up.
2. **Tumor suppressor genes** code for proteins that inhibit the cell cycle and promote apoptosis. They are often likened to the brakes of a car, because they cause the cell cycle to go more slowly or even stop.

Proto-oncogenes Become Oncogenes

Proto-oncogenes are normal genes that promote progression through the cell cycle. They are often at the end of a *stimulatory pathway* extending from the plasma membrane to the nucleus. A stimulus, such as an injury, results in the release of a growth factor that binds to a receptor protein in the plasma membrane. This sets in motion a whole series of enzymatic reactions leading to the activation of genes that promote the cell cycle, both directly and indirectly. Proto-oncogenes include the receptors and signal molecules that make up these pathways.

When mutations occur in proto-oncogenes, they become **oncogenes,** or cancer-causing genes. Oncogenes are under constant stimulation and keep on promoting the cell cycle regardless of circumstances. For example, an oncogene may code for a faulty receptor in the stimulatory pathway such that the cell cycle is stimulated, even when no growth factor is present! Or an oncogene may either specify an abnormal protein product or produce abnormally high levels of a normal product that stimulate the cell cycle to begin or to go to completion. As a result, uncontrolled cell division may occur.

Researchers have identified perhaps 100 oncogenes that can cause increased growth and lead to tumors. The oncogenes most frequently involved in human cancers belong to the *ras* gene family. Mutant forms of the *BRCA1* (*br*east *ca*ncer predisposition gene *1*) oncogene are associated with certain hereditary forms of breast and ovarian cancer.

Tumor Suppressor Genes Become Inactive

Tumor suppressor genes, on the other hand, directly or indirectly inhibit the cell cycle and prevent cells from dividing uncontrollably. Some tumor suppressor genes prevent progression of the cell cycle when DNA is damaged. Other tumor suppressor genes may promote apoptosis as a last resort.

A mutation in a tumor suppressor gene is much like brake failure in a car; when the mechanism that slows down and stops cell division does not function, the cell cycle accelerates and does not halt. Researchers have identified about a half-dozen tumor suppressor genes. Among these are the *RB* and *p53* genes that code for the RB and p53 proteins. The Nature of Science feature, "The G_1 Checkpoint," on page 150 discusses the function of these proteins in controlling the cell cycle. The *RB* tumor suppressor gene was discovered when the inherited condition retinoblastoma was being studied, but malfunctions of this gene have now been identified in many other cancers as well, including breast, prostate, and bladder cancers. The *p53* gene turns on the expression of other genes that inhibit the cell cycle. The p53 protein can also stimulate apoptosis. It is estimated that over half of human cancers involve an abnormal or deleted *p53* gene.

Animation How Tumor Suppressor Genes Block Cell Division

Other Causes of Cancer

As mentioned previously, cancer develops when the delicate balance between promotion and inhibition of cell division is tilted toward uncontrolled cell division. Other mutations may occur within a cell that affect this balance. For example, while a mutation affecting the cell's DNA repair system will not immediately cause cancer, it leads to a much greater chance of a mutation occurring within a proto-oncogene or tumor suppressor gene. And in some cancer cells, mutation of the telomerase enzyme that regulates the length of **telomeres,** or the ends of chromosomes, causes the telomeres to remain at a constant length. Because cells with shortened telomeres normally stop dividing, keeping the telomeres at a constant length allows the cancer cells to continue dividing over and over again.

Animation Telomerase Function

Check Your Progress 9.4

1. List the major characteristics of cancer cells that distinguish them from normal cells.
2. Distinguish between a malignant and benign tumor.
3. Compare and contrast the effect on the cell cycle of (a) a mutation in a proto-oncogene and (b) a mutation in a tumor suppressor gene.

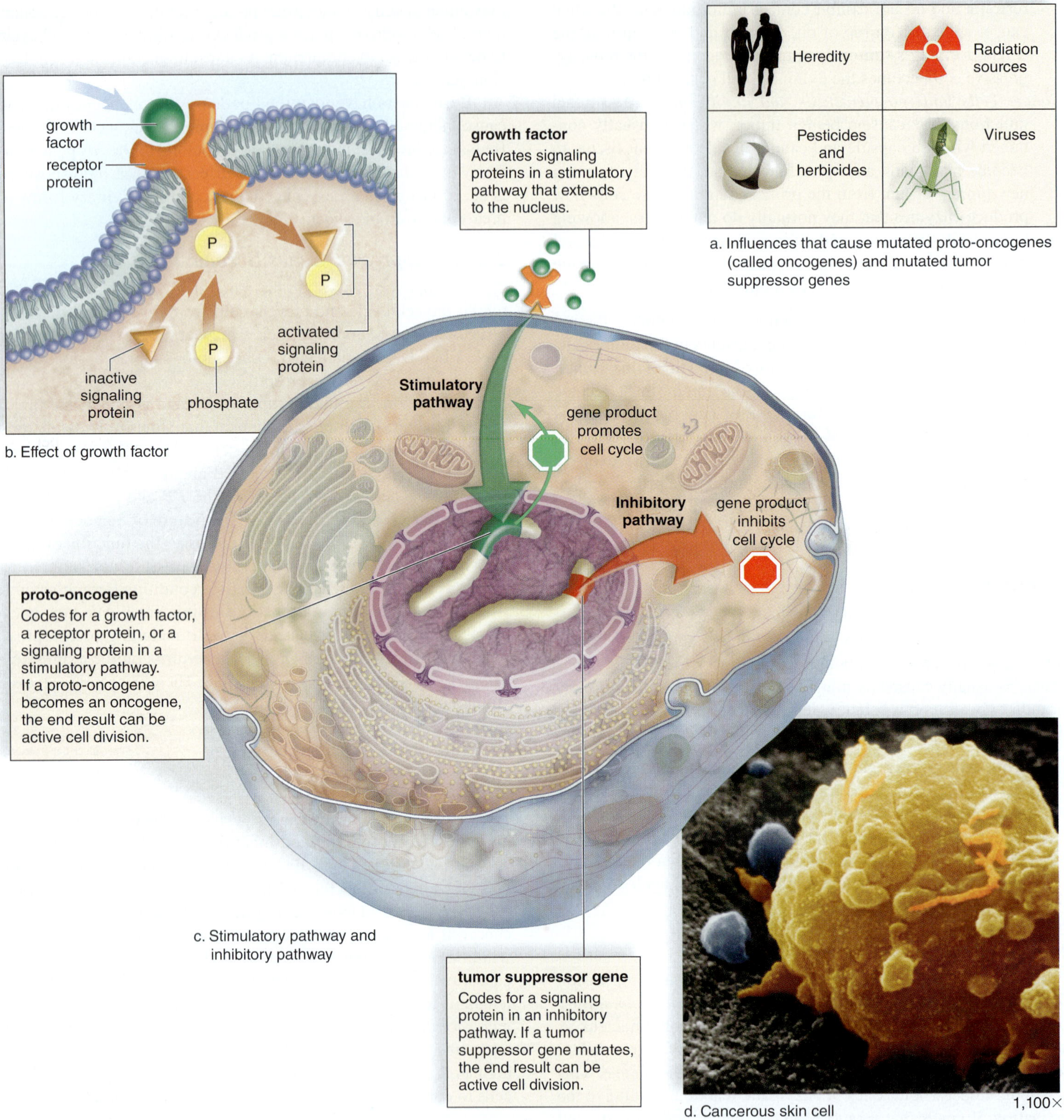

Figure 9.9 Causes of cancer. **a.** Mutated genes that cause cancer can be due to the influences noted. **b.** A growth factor that binds to a receptor protein initiates a reaction that triggers a stimulatory pathway. **c.** A stimulatory pathway that begins at the plasma membrane turns on proto-oncogenes. The products of these genes promote the cell cycle and double back to become part of the stimulatory pathway. When proto-oncogenes become oncogenes, they are turned on all the time. An inhibitory pathway begins with tumor suppressor genes, whose products inhibit the cell cycle. When tumor suppressor genes mutate, the cell cycle is no longer inhibited. **d.** Cancerous skin cell.

9.5 Prokaryotic Cell Division

Learning Outcomes

Upon completion of this section, you should be able to

1. Distinguish between the structures of a prokaryotic and eukaryotic chromosome.
2. Describe the events that occur during binary fission.

Cell division in single-celled organisms, such as prokaryotes, produces two new individuals. This is **asexual reproduction** in which the offspring are genetically identical to the parent. In prokaryotes, reproduction consists of duplicating the single chromosome and distributing a copy to each of the daughter cells. Unless a mutation has occurred, the daughter cells are genetically identical to the parent cell.

The Prokaryotic Chromosome

Prokaryotes (bacteria and archaea) lack a nucleus and other membranous organelles found in eukaryotic cells. Still, they do have a chromosome, which is composed of DNA and a limited number of associated proteins. The single chromosome of prokaryotes contains just a few proteins and is organized differently than eukaryotic chromosomes. A eukaryotic chromosome has many more associated proteins than does a prokaryotic chromosome.

In electron micrographs, the bacterial chromosome appears as an electron-dense, irregularly shaped region called the **nucleoid** (L. *nucleus,* "nucleus, kernel"; Gk. *-eides,* "like"), which is not enclosed by a membrane. When stretched out, the chromosome is seen to be a circular loop with a length that is up to about a thousand times the length of the cell. Special enzymes and proteins help coil the chromosome so that it will fit within the prokaryotic cell.

Binary Fission

Prokaryotes reproduce asexually by binary fission. The process is termed **binary fission** because division (fission) produces two (binary) daughter cells that are identical to the original parent cell. Before division takes place, the cell enlarges, and after DNA replication occurs, there are two chromosomes. These chromosomes attach to a special plasma membrane site and separate by an elongation of the cell that pulls them apart. During this period, a new plasma membrane and cell wall develop and grow inward to divide the cell. When the cell is approximately twice its original length, the new cell wall and plasma membrane for each cell are complete (Fig. 9.10).

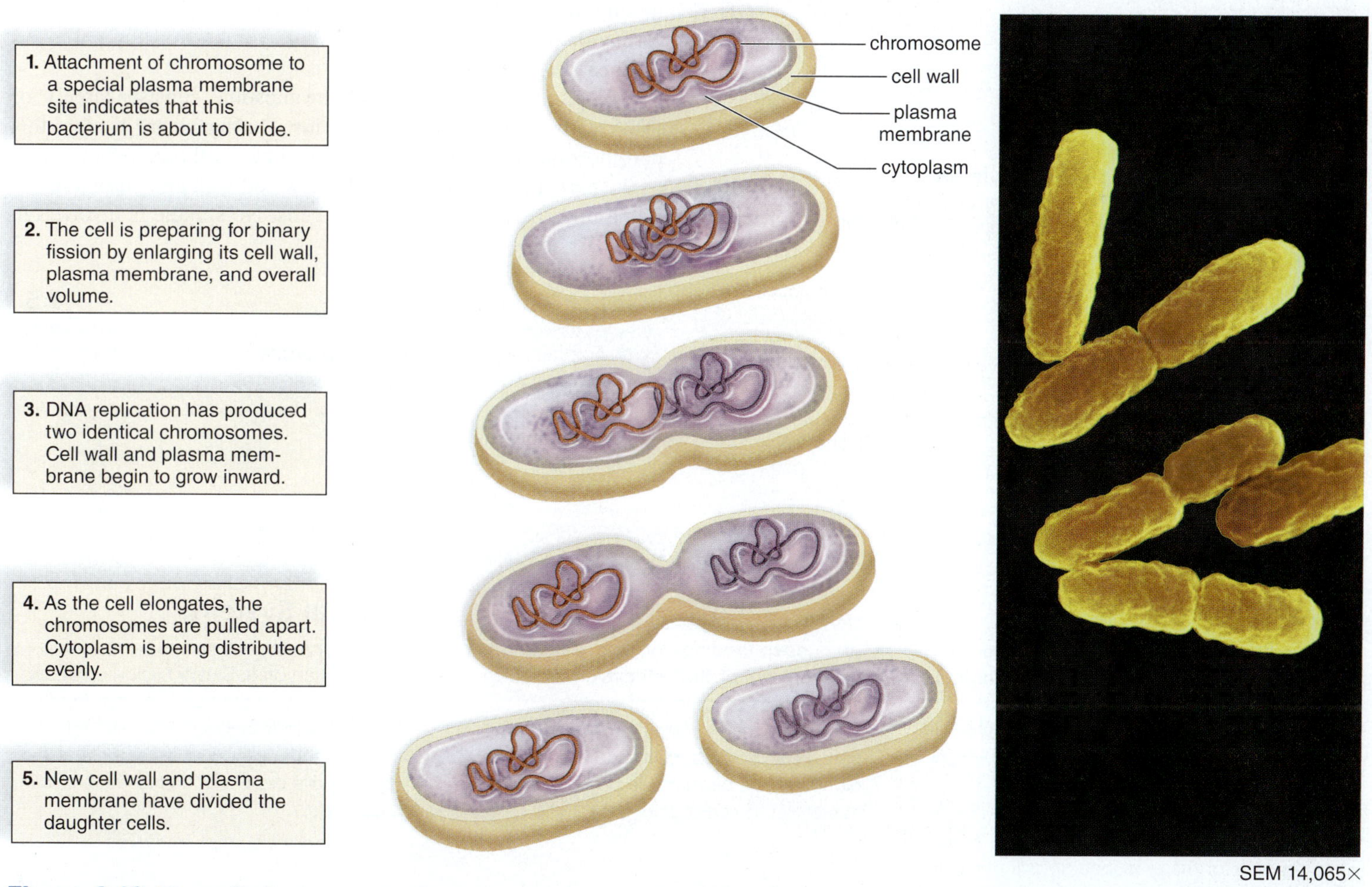

Figure 9.10 Binary fission. First, DNA replicates, and as the cell lengthens, the two chromosomes separate and the cells become divided. The two resulting bacteria are identical.

Escherichia coli, which lives in our intestines, has a generation time (the time it takes the cell to divide) of about 20 minutes under favorable conditions. In about 7 hours, a single cell can increase to over 1 million cells! The division rate of other bacteria varies depending on the species and conditions.

Animation Binary Fission

Comparing Prokaryotes and Eukaryotes

Both binary fission and mitosis ensure that each daughter cell is genetically identical to the parent cell. The genes are portions of DNA found in the chromosomes.

Prokaryotes (bacteria and archaea), protists (many algae and protozoans), and some fungi (yeasts) are single-celled. Cell division in single-celled organisms produces two new individuals:

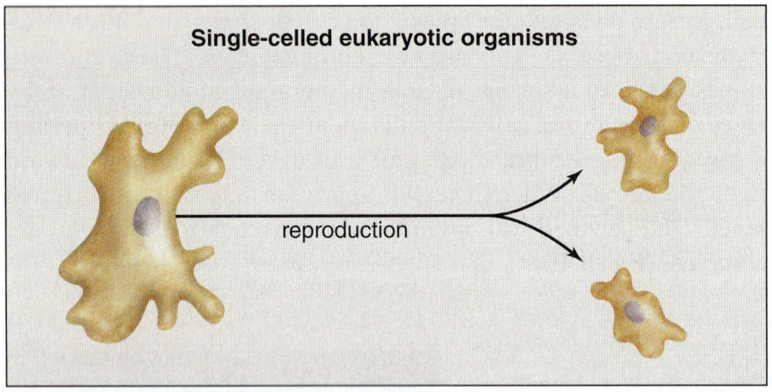

This is a form of asexual reproduction because one parent has produced identical offspring (Table 9.3).

Table 9.3 Functions of Cell Division

Type of Organism	Cell Division	Function
Prokaryotes		
Bacteria and archaea	Binary fission	Asexual reproduction
Eukaryotes		
Protists and some fungi (yeast)	Mitosis and cytokinesis	Asexual reproduction
Other fungi, plants, and animals	Mitosis and cytokinesis	Development, growth, and repair

In multicellular fungi (molds and mushrooms), plants, and animals, cell division is part of the growth process. It produces the multicellular form we recognize as the mature organism. Cell division is also important in multicellular forms for renewal and repair:

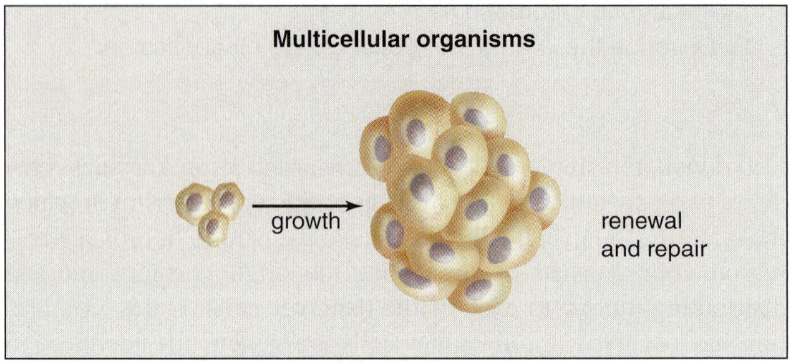

The chromosomes of eukaryotic cells are composed of DNA and many associated proteins. The histone proteins organize a chromosome, allowing it to extend as chromatin during interphase and to coil and condense just prior to mitosis. Each species of multicellular eukaryotes has a characteristic number of chromosomes in the nuclei. As a result of mitosis, each daughter cell receives the same number and kinds of chromosomes as the parent cell. The spindle, which appears during mitosis, is involved in distributing the daughter chromosomes to the daughter nuclei. Cytokinesis, either by the formation of a cell plate (plant cells) or by furrowing (animal cells), is division of the cytoplasm.

In prokaryotes, the single chromosome consists largely of DNA with a few associated proteins. During binary fission, this chromosome duplicates, and each daughter cell receives one copy as the parent cell elongates, and a new cell wall and plasma membrane form between the daughter cells. No spindle is involved in binary fission.

Check Your Progress 9.5

1. Explain how binary fission in prokaryotes differs from mitosis and cytokinesis in eukaryotes.
2. Describe the structure of a prokaryotic and a eukaryotic chromosome.

CONNECTING *the* CONCEPTS *with the* THEMES

Evolution

- Eukaryotic cells have evolved the process of mitosis to sort chromosomes into daughter cells.
- The stages of mitosis are the same for all eukaryotic cells, suggesting a common evolutionary lineage.
- Plants and animals have evolved different strategies for cytokinesis that compensate for the presence or lack of a cell wall.

Nature of Science

- An understanding of the regulatory mechanisms of the cell cycle has led to insight into the causes of cancer.
- The advances in stem cell technologies, specifically reproductive and therapeutic cloning, have been made possible by research on the cell cycle.

Biological Systems

- In prokaryotic organisms and single-celled eukaryotes, binary fission provides for cellular reproduction and the formation of new individuals.
- In eukaryotic organisms, mitosis provides for the growth of tissues and repair of damage.

MEDIA STUDY TOOLS http://connect.mheducation.com

SMARTBOOK® Study smart. SmartBook helps you maximize your study time by identifying what you know and don't know. The recharge feature in SmartBook helps you prepare for the test by identifying the material that you are most likely to forget.

connect BIOLOGY Find even more resources to learn the chapter concepts including animations, tutorial videos, and interactive practice questions.

MP3 Files	Animations	3D Animations	Tutorials	Videos
9.3 Mitosis	**9.1** Cell Proliferation Signaling Pathway **9.3** Mitosis • Cytokinesis **9.4** How Tumor Suppressor Genes Block Cell Division • Telomerase Function **9.5** Binary Fission	**9.1** Cell Cycle and Mitosis: Interphase • Cell Cycle and Mitosis: Checkpoints **9.3** Cell Cycle and Mitosis: Mitosis • Cell Cycle and Mitosis: Cytokinesis	**9.3** Mitosis	**9.3** Mitosis

SUMMARIZE

9.1 The Cell Cycle

The **cell cycle** of a eukaryotic cell includes (1) interphase and (2) a mitotic stage that consists of **mitosis** and **cytokinesis. Interphase,** in turn, is composed of three stages: G_1 (growth as certain organelles double), S (the synthesis stage, where the **chromatids** are duplicated, forming the **sister chromatids)**, and G_2 (growth as the cell prepares to divide). Cells that are no longer dividing are arrested in a G_0 state. During the mitotic stage (M), the chromosomes are sorted by the **mitotic spindle** into two daughter cells, resulting in a full complement of chromosomes. During cytokinesis, the cytoplasm is divided between the new daughter cells

Signals, such as **growth factors** and **cyclin,** are involved in regulating the cell cycle. The cell cycle is regulated by several checkpoints—the G_1 checkpoint, the G_2 checkpoint prior to the M stage, and the M stage checkpoint, or spindle assembly checkpoint, immediately before anaphase. The G_1 checkpoint ensures that conditions are favorable and that the proper signals are present, and it checks the DNA for damage. If the DNA is damaged beyond repair, **apoptosis** may occur. Cell division and apoptosis are two opposing processes that keep the number of **somatic cells** in balance.

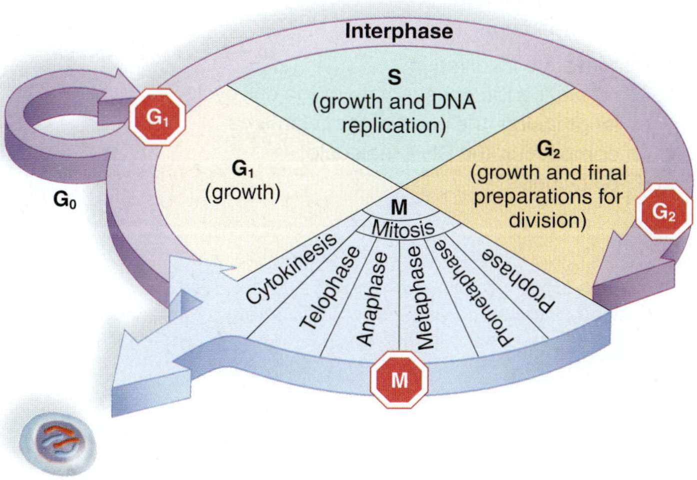

9.2 The Eukaryotic Chromosome

Eukaryotic cells contain nearly 2 m of DNA, yet they must pack it all into a nucleus approximately 6 μm in diameter. Thus, the DNA is compacted by winding it around DNA-binding proteins, called histones, to make nucleosomes. The nucleosomes are further compacted into a zigzag structure, which is then folded upon itself many times to form radial loops, which is the usual compaction state of **euchromatin. Heterochromatin** is further compacted by scaffold proteins, and further compaction can be achieved prior to mitosis and meiosis.

9.3 Mitosis and Cytokinesis

Interphase represents the portion of the cell cycle between nuclear divisions, and during this time, preparations are made for cell division. These preparations include duplication of most cellular contents, including the centrosome, which organizes the mitotic spindle. The DNA is duplicated during S stage, at which time the chromosomes, which consist of a single chromatid each, are duplicated. The G_2 checkpoint ensures that DNA has replicated properly. This results in a nucleus containing the same number of chromosomes, with each now consisting of two chromatids attached at the **centromere.** During interphase, the chromosomes are not distinct and are collectively called chromatin. Each eukaryotic species has a characteristic number of chromosomes. The total number is called the **diploid** number, and half this number is the **haploid** number.

Among eukaryotes, cell division involves both mitosis (nuclear division) and cytokinesis (division of the cytoplasm). As a result of mitosis, the chromosome number stays constant, because each chromosome is duplicated and gives rise to two daughter chromosomes, which consist of a single chromatid each.

Mitosis (nuclear division)

- **Prophase**—The nucleolus disappears, the nuclear envelope fragments, and the spindle forms between centrosomes. The chromosomes condense and become visible under a light microscope. In animal cells, asters radiate from the **centrioles** within the **centrosomes.** Plant cells generally lack **centrioles** and, therefore, asters. Even so, the mitotic spindle forms.
- **Prometaphase** (late prophase)—The **kinetochores** of sister chromatids attach to kinetochore spindle fibers extending from

opposite poles. The chromosomes move back and forth until they are aligned at the metaphase plate.

- **Metaphase**—The spindle is fully formed, and the duplicated chromosomes are aligned at the metaphase plate. The spindle consists of polar spindle fibers that overlap at the metaphase plate and kinetochore spindle fibers that are attached to chromosomes. The M stage checkpoint, or spindle assembly checkpoint, must be satisfied before progressing to the next phase.
- **Anaphase**—Sister chromatids separate, becoming daughter chromosomes that move toward the poles. The polar spindle fibers slide past one another, and the kinetochore spindle fibers disassemble. Cytokinesis by furrowing begins.
- **Telophase**—Nuclear envelopes re-form, chromosomes begin changing back to chromatin, the nucleoli reappear, and the spindle disappears. Cytokinesis continues and is complete by the end of telophase.

Cytoplasmic division

- Cytokinesis in animal cells involves the formation of a **cleavage furrow** that divides the cytoplasm. Cytokinesis in plant cells involves the formation of a **cell plate,** from which the plasma membrane and cell wall are completed.
- In the body, mitosis also produces undifferentiated cells called **stem cells,** that are involved in the production of tissues. **Therapeutic cloning** and **reproductive cloning** both utilize stem cells.

9.4 The Cell Cycle and Cancer

The development of **cancer** is primarily due to the mutation of genes involved in control of the cell cycle. Cancer cells lack differentiation, have abnormal nuclei, do not undergo apoptosis, form tumors, and undergo metastasis and angiogenesis. Cancer often follows a progression in which mutations accumulate, gradually causing uncontrolled growth and the development of a **tumor**. Tumors may be either **benign** (not cancerous) or **malignant** (cancerous). Malignant tumors may undergo both **metastasis** and **angiogenesis.**

Proto-oncogenes stimulate the cell cycle after they are turned on by environmental signals, such as growth factors. Oncogenes are mutated proto-oncogenes that stimulate the cell cycle without need of environmental signals. **Tumor suppressor genes** inhibit the cell cycle. Mutated tumor suppressor genes no longer inhibit the cell cycle, allowing unchecked cell division.

In some cancer cells, the presence of the telomerase enzyme may stop the **telomeres** from shortening with each cell division, causing the cell to proceed through the cell cycle.

9.5 Prokaryotic Cell Division

Binary fission (in prokaryotes) and mitosis (in single-celled eukaryotic protists and fungi) allow for **asexual reproduction.** Mitosis in multicellular eukaryotes is primarily for the purpose of development, growth, and repair of tissues.

The prokaryotic chromosome has a few proteins and a single, long loop of DNA located in the **nucleoid** region of the cytoplasm. When **binary fission** occurs, the chromosome attaches to the inside of the plasma membrane and replicates. As the cell elongates, the chromosomes are pulled apart. Inward growth of the plasma membrane and formation of new cell wall material divide the cell in two.

ASSESS

Choose the best answer for each question.

9.1 The Cell Cycle

For questions 1–4, match each stage of the cell cycle to its correct description.

Key:

a. G_1 stage b. S stage
c. G_2 stage d. M (mitotic) stage

1. At the end of this stage, each chromosome consists of two attached chromatids.
2. During this stage, daughter chromosomes are distributed to two daughter nuclei.
3. The cell doubles its organelles and accumulates the materials needed for DNA synthesis.
4. The cell synthesizes the proteins needed for cell division.
5. Which is not true of the cell cycle?
   a. The cell cycle is controlled by internal/external signals.
   b. Cyclin is a signaling molcule that increases and decreases as the cycle continues.
   c. DNA damage can stop the cell cycle at the G_1 checkpoint.
   d. Apoptosis occurs frequently during the cell cycle.

9.2 The Eukaryotic Chromosome

6. The diploid number of chromosomes
   a. is the 2n number.
   b. is in a parent cell and therefore in the two daughter cells following mitosis.
   c. varies according to the particular organism.
   d. is present in most somatic cells.
   e. All of these are correct.
7. The form of DNA that contains genes that are actively being transcribed is called
   a. histones.
   b. telomeres.
   c. heterochromatin.
   d. euchromatin.
8. Histones are involved in
   a. regulating the checkpoints of the cell cycle.
   b. lengthening the ends of the telomeres.
   c. compacting the DNA molecule.
   d. cytokinesis.

9.3 Mitosis and Cytokinesis

9. At the metaphase plate during metaphase of mitosis, there are
   a. single chromosomes.
   b. duplicated chromosomes.
   c. G_1 stage chromosomes.
   d. always 23 chromosomes.
10. During which mitotic phases are duplicated chromosomes present?
   a. all but telophase
   b. prophase and anaphase
   c. all but anaphase and telophase
   d. only during metaphase at the metaphase plate
   e. Both a and b are correct.
11. Which of these is paired incorrectly?
   a. prometaphase—the kinetochores become attached to spindle fibers
   b. anaphase—daughter chromosomes migrate toward spindle poles
   c. prophase—the nucleolus disappears and the nuclear envelope disintegrates
   d. metaphase—the chromosomes are aligned in the metaphase plate
   e. telophase—a resting phase between cell division cycles

9.4 The Cell Cycle and Cancer

12. Which of the following is not characteristic of cancer cells?
   a. Cancer cells often undergo angiogenesis.
   b. Cancer cells tend to be nonspecialized.
   c. Cancer cells undergo apoptosis.
   d. Cancer cells often have abnormal nuclei.
   e. Cancer cells can metastasize.
13. Which of the following statements is true?
   a. Proto-oncogenes cause a loss of control of the cell cycle.
   b. The products of oncogenes may inhibit the cell cycle.
   c. Tumor suppressor gene products inhibit the cell cycle.
   d. A mutation in a tumor suppressor gene may inhibit the cell cycle.

9.5 Prokaryotic Cell Division

14. In contrast to a eukaryotic chromosome, a prokaryotic chromosome
   a. is shorter and fatter.
   b. has a single loop of DNA.
   c. never replicates.
   d. contains many histones.
   e. All of these are correct.
15. Which of the following is the term used to describe asexual reproduction in a single-celled organism?
   a. cytokinesis
   b. mitosis
   c. binary fission
   d. All of these are correct.

ENGAGE

LEARNSMART LABS®

The following LearnSmart Labs contain exercises that are related to the content of this chapter:

- Mitosis & Meiosis

THINKING SCIENTIFICALLY

1. After DNA is duplicated in eukaryotes, it must be bound to histones. This requires the synthesis of hundreds of millions of new protein molecules. With reference to Figure 9.1, when in the cell cycle would histones be made?
2. The survivors of the atomic bombs that were dropped on Hiroshima and Nagasaki have been the subjects of long-term studies of the effects of ionizing radiation on cancer incidence. The frequencies of different types of cancer in these individuals varied across the decades. In the 1950s, high levels of leukemia and cancers of the lung and thyroid gland were observed. The 1960s and 1970s brought high levels of breast and salivary gland cancers. In the 1980s, rates of colon cancer were especially high. Why do you suppose the rates of different types of cancer varied across time?
3. BPA is a chemical compound that has historically been used in the manufacture of plastic products. However, cells often mistake BPA compounds for hormones that accelerate the cell cycle. Because of this, BPA is associated with an increased risk of certain cancers.
   a. How might BPA interact with the cell cycle and its checkpoints?
   b. Why do you think that very small concentrations of BPA might have a large effect on the cell?
4. Why is it advantageous that checkpoints evolved during the cell cycle?

10

Meiosis and Sexual Reproduction

Meiosis is the process that produces the majority of genetic variation between individuals.

Chapter Outline

10.1 Overview of Meiosis 167

10.2 Genetic Variation 169

10.3 The Phases of Meiosis 172

10.4 Meiosis Compared to Mitosis 174

10.5 The Cycle of Life 176

10.6 Changes in Chromosome Number and Structure 177

Before You Begin

Before beginning this chapter, take a few moments to review the following discussions.

Section 3.5 How is genetic information stored in nucleic acids?

Section 4.8 What are microtubules, and how do they interact with chromosomes?

Section 9.3 How are eukaryotic chromosomes organized and replicated prior to cell division?

What are the chances that there will ever be another human like you on this planet? Because of meiosis, reproduction produces offspring that are genetically different from the parents. Meiosis introduces an enormous amount of diversity; in humans, more than 70 trillion different genetic combinations are possible from the mating of two individuals! In other words, meiosis ensures that, statistically, it is unlikely that anyone will ever be genetically the same as you.

In animals, meiosis begins the process that produces cells called gametes, which play an important role in sexual reproduction. In humans, sperm are the male gametes, and eggs are the female gametes. Males and females differ in the ways they form gametes. While meiosis in the two sexes is very similar, there are some important differences in how they occur. One major difference pertains to the age at which the process begins and ends. In males, sperm production does not begin until puberty but then continues throughout a male's lifetime. In females, the process of producing eggs has started before the female is born and ends around menopause. Another difference concerns the number of gametes that can be produced. In males, sperm production is unlimited, whereas females produce only one egg a month.

In this chapter, you will see how meiosis is involved in providing the variation so important in the production of gametes. We will focus on the role of meiosis in animals but will return to how plants use meiosis in Chapter 23.

As you read through this chapter, think about the following questions:

1. Why is meiosis sometimes referred to as reduction division?
2. What is the role of meiosis in introducing new variation?

Following *the* Themes

CHAPTER 10 MEIOSIS AND SEXUAL REPRODUCTION

UNIT 2 GENETIC BASIS OF LIFE

Theme	
Evolution	The variation introduced during meiosis plays an important role in evolutionary change.
Nature of Science	Changes in chromosome structure and number can have consequences for an individual's physiology.
Biological Systems	The variation introduced by meiosis at a cellular level affects all levels of an organism's physiology.

10.1 Overview of Meiosis

Learning Outcomes

Upon completion of this section, you should be able to

1. Contrast haploid and diploid chromosome numbers.
2. Explain what is meant by *homologous chromosomes.*
3. Summarize the process by which meiosis reduces the chromosome number.

In sexually reproducing organisms, **meiosis** (Gk. *mio,* "less"; *-sis,* "act or process of") is the type of nuclear division that reduces the chromosome number from the diploid (2n) number (Gk. *diplos,* "twofold") to the haploid (n) number (Gk. *haplos,* "single"). The **diploid (2n)** number refers to the total number of chromosomes, which exists in two sets. The **haploid (n)** number of chromosomes is half the diploid number, or a single set of chromosomes. In humans, meiosis reduces the diploid number of 46 chromosomes to the haploid number of 23 chromosomes.

Gametes, or reproductive cells (in animals, these are the sperm and egg), usually have the haploid number of chromosomes. In **sexual reproduction,** haploid gametes, which are produced during meiosis, subsequently merge into a diploid cell called a **zygote.** In plants and animals, the zygote undergoes development to become an adult organism.

Meiosis is necessary in sexually reproducing organisms, because the diploid number of chromosomes has to be reduced by half in each of the parents in order to produce diploid offspring. Otherwise, the number of chromosomes would double with each new generation. Within a few generations, the cells of an animal would be nothing but chromosomes! For example, in humans with a diploid number of 46 chromosomes, in five generations the chromosome number would increase to 1,472 chromosomes (46×2^5). In ten generations, this number would increase to a staggering 47,104 chromosomes (46×2^{10}). The Belgian cytologist (a biologist that studies cells) Pierre-Joseph van Beneden (1809–1894), was one of the first to observe that gametes have a reduced chromosome number. When studying the roundworm *Ascaris,* he noticed that the sperm and egg each contain only two chromosomes, while the zygote and subsequent embryonic cells always have four chromosomes.

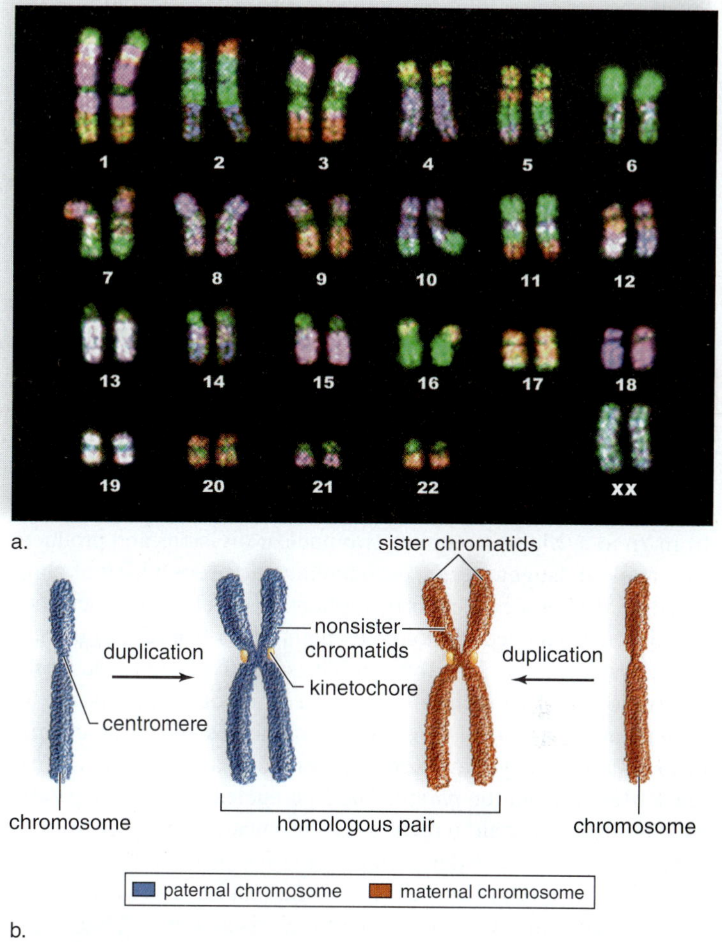

Figure 10.1 Homologous chromosomes. In diploid body cells, the chromosomes occur in pairs called homologous chromosomes. **a.** In this micrograph of stained chromosomes from a human cell, the pairs have been numbered 1–22 and XX. Note that chromosome pairs 1–22 are autosomes, coding for nonsex traits, whereas the XX pair includes the sex chromosomes and helps determine human gender. **b.** These chromosomes are duplicated, and each chromosome in the homologous pair is composed of two chromatids. The sister chromatids contain exactly the same genes; the nonsister chromatids contain genes for the same traits (e.g., type of hair, color of eyes), but one may have DNA that codes for trait variations, such as dark hair versus light hair.

Homologous Pairs of Chromosomes

In diploid body cells, the chromosomes occur in pairs. Figure 10.1*a,* a pictorial display of human chromosomes, called a karyotype, shows the chromosomes arranged according to pairs. The members of each pair are called homologous chromosomes. **Homologous chromosomes,** or **homologues** (Gk. *homologos,* "agreeing, corresponding"), look alike; they have the same length and centromere position. When stained, homologues have a similar banding pattern, because they contain genes for the same traits in the same order in the same locations on both chromosomes in the homologous pair. But while homologous chromosomes have genes for the same traits, such as finger length, the DNA (deoxyribonucleic acid) sequence for the gene on one homologue may code for short fingers and the gene at the same location on the other homologue may code for long fingers. Alternate forms of a gene (as for long fingers and short fingers) are called **alleles.** The DNA sequences of alleles are highly similar, but they are different enough to produce alternative physical traits, such as long or short fingers.

To properly produce a haploid number of chromosomes in gametes, you first have to double the amount of DNA. The chromosomes in Figure 10.1*a* are duplicated as they would be just before nuclear division. Recall that during the S stage of the cell cycle, DNA replicates and the chromosomes become duplicated. The results of the duplication process are depicted in Figure 10.1*b.* When duplicated, a chromosome is composed of two identical parts called sister chromatids, each containing one DNA double helix molecule. The *sister chromatids* are held together at a common region called the centromere.

Why does the zygote have paired chromosomes? One member of a homologous pair was inherited from the male parent, and the other was inherited from the female parent when the haploid sperm and egg fused together. In Figure 10.1*b* and throughout this chapter, the paternal chromosome is colored blue, and the maternal chromosome is colored red. *However, this is simply a method of tracking chromosomes in diagrams. Since chromosomes do not have color, geneticists generally use chromosome length and centromere location to identify homologues.* You will see shortly how meiosis reduces the chromosome number. Whereas the zygote and body cells have homologous pairs of chromosomes, the gametes have only one chromosome of each kind—derived from either the paternal or the maternal homologue.

Meiosis Is Reduction Division

The central purpose of meiosis is to reduce the chromosome number from 2n to n. Meiosis requires two nuclear divisions and produces four haploid daughter cells, each having one of each kind of chromosome. The process begins by replicating the chromosomes, then splitting the matched homologous pairs to go from 2n to n chromosomes during the first division. The second division reduces the amount of DNA in n chromosomes to an amount appropriate for each gamete. Once the DNA has been replicated and chromosomes become a pair, they may exchange genes, creating a genetic mixture different from the parent. The first nuclear division separates each homologous pair, reducing the chromosome number from 2n to n. Even though each daughter cell now has n chromosomes, each chromosome still has a sister chromatid, making a second nuclear division necessary. The end result of meiosis is four gametes with n chromosomes.

MP3 Meiosis

Figure 10.2 presents an overview of meiosis, indicating the two nuclear divisions, meiosis I and meiosis II. Prior to meiosis I, DNA replication has occurred; therefore, each chromosome has two sister chromatids. During meiosis I, something new happens that does not occur in mitosis. The homologous chromosomes come together and line up side by side, forming a **synaptonemal complex.** This process is called **synapsis** (Gk. *synaptos,* "united, joined together") and results in a **bivalent** (L. *bis,* "two"; *valens,* "strength")—that is, two homologous chromosomes that stay in close association during the first two phases of meiosis I. Sometimes the term *tetrad* (Gk. *tetra,* "four") is used instead of *bivalent,* because, as you can see, a bivalent contains four chromatids. Chromosomes may recombine or exchange genetic information during this association (see section 10.2).

Following synapsis, homologous pairs align at the metaphase plate, and then the members of each pair separate. This separation means that only one duplicated chromosome from each homologous pair reaches a daughter nucleus, reducing the chromosome number from 2n to n. It is important for each daughter nucleus to have a member from each pair of homologous chromosomes, because only in that way can there be a copy of each kind of chromosome in the daughter nuclei. Notice in Figure 10.2 that two possible combinations of chromosomes in the daughter cells are shown: short red with long blue and short blue with long red. Knowing that all daughter cells have to have one short chromosome and one long chromosome, what are the other two possible combinations of chromosomes for these cells?

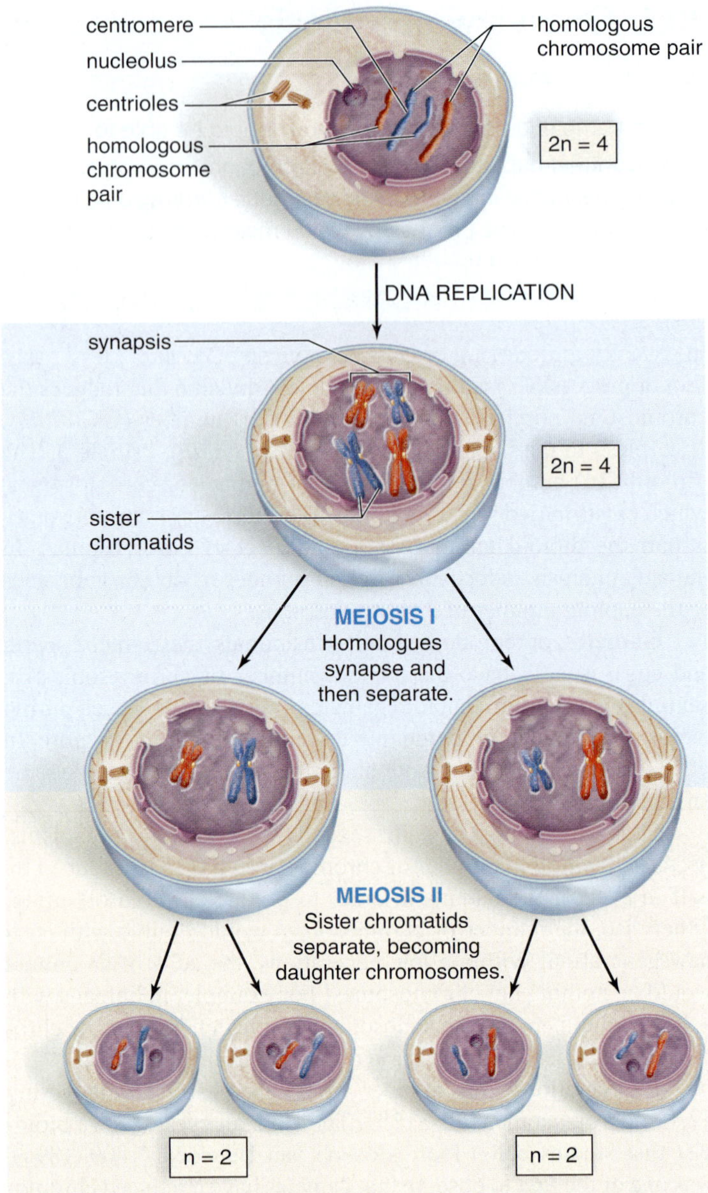

Figure 10.2 Overview of meiosis. Following DNA replication, each chromosome is duplicated and consists of two chromatids. During meiosis I, homologous chromosomes pair and separate. During meiosis II, the sister chromatids of each duplicated chromosome separate. At the completion of meiosis, there are four haploid daughter cells. Each daughter cell has one of each kind of chromosome.

Notice that DNA replication occurs only once during meiosis; no replication is needed between meiosis I and meiosis II, because the chromosomes are already duplicated; they already have two sister chromatids. During meiosis II, the sister chromatids separate, becoming daughter chromosomes that move to opposite poles. The chromosomes in each of the four daughter cells now contain only one DNA double helix molecule in the form of a haploid chromosome.

The number of centromeres can be counted to verify that the parent cell has the diploid number of chromosomes. At the end of meiosis I, the chromosome number has been reduced, because there are half as many centromeres present, even though each chromosome still consists of two chromatids each. Each daughter cell that forms has the haploid number of chromosomes. At the end of meiosis II, sister chromatids separate, and each daughter cell that forms still contains the haploid number of chromosomes, each consisting of a single chromatid.

Animation How Meiosis Works

Fate of Daughter Cells

In the plant life cycle, the daughter cells become haploid spores that germinate to become a haploid generation. This generation then produces the gametes by mitosis. The plant life cycle is studied in Chapter 23. In the animal life cycle, the daughter cells become the gametes, either sperm or eggs. The body cells of an animal normally contain the diploid number of chromosomes due to the fusion of sperm and egg during fertilization. If meiotic events go wrong, the gametes can contain the wrong number of chromosomes or altered chromosomes. This possibility and its consequences are discussed in section 10.6.

Check Your Progress 10.1

1. Describe what is meant by *a homologous pair of chromosomes*.
2. Examine how chromosome number changes during meiosis I and meiosis II.
3. Explain the purpose of a bivalent in chromosome pairing.

10.2 Genetic Variation

Learning Outcomes

Upon completion of this section, you should be able to

1. Understand the importance of genetic variation to evolutionary change.
2. Explain how crossing-over contributes to genetic variation.
3. Examine how independent assortment contributes to genetic variation.

We have seen that meiosis provides a way to keep the chromosome number constant generation after generation. Without meiosis, the chromosome number of the next generation would continually increase. The events of meiosis also help ensure that genetic variation occurs with each generation.

Genetic variation is essential for a species to be able to evolve and adapt in a changing environment. Asexually reproducing organisms, such as the prokaryotes, depend primarily on mutations to generate variation among offspring. This is sufficient for their survival, because they produce great numbers of offspring very quickly. Although mutations also occur among sexually reproducing organisms, the reshuffling of genetic material during sexual reproduction ensures that offspring will have a different combination of genes than their parents. Meiosis brings about genetic variation in two key ways: crossing-over and independent assortment of homologous chromosomes.

Genetic Recombination

Crossing-over is an exchange of genetic material between nonsister chromatids of a bivalent during meiosis I. In humans, it is estimated that an average of two to three crossovers occur between the nonsister chromatids during meiosis. At synapsis, homologues line up side by side, and a nucleoprotein lattice appears between them (Fig. 10.3). This lattice holds the bivalent

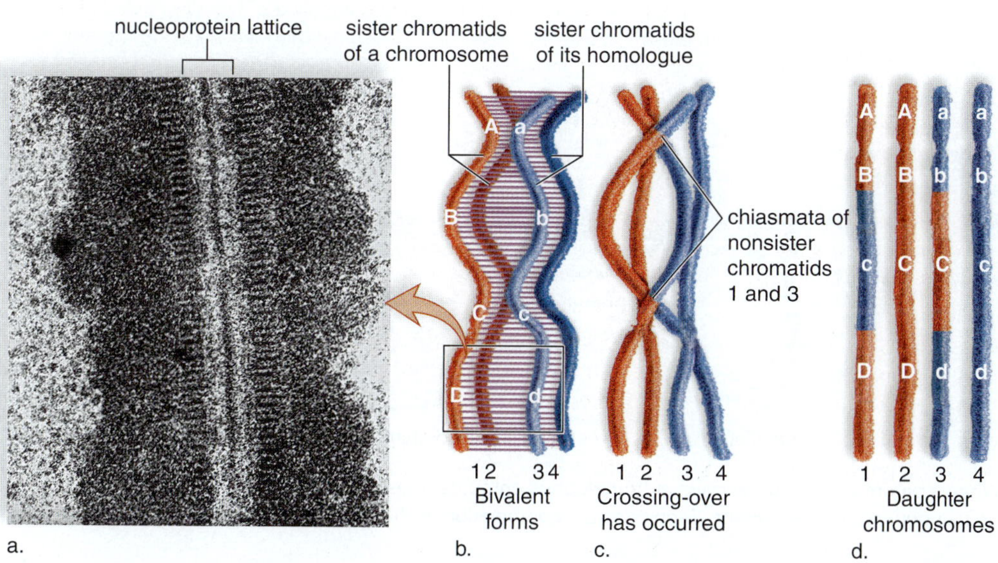

Figure 10.3 Crossing-over during meiosis I. **a.** The homologous chromosomes pair up, and a nucleoprotein lattice develops between them. This is an electron micrograph of the lattice. It "zippers" the members of the bivalent together, so that corresponding genes on paired chromosomes are in alignment. **b.** This visual representation shows only two places where nonsister chromatids 1 and 3 have come in contact. **c.** Chiasmata indicate where crossing-over has occurred. The exchange of color represents the exchange of genetic material. **d.** Following meiosis II, daughter chromosomes have a new combination of genetic material due to crossing-over, which occurred between nonsister chromatids during meiosis I.

together in such a way that the DNA of the duplicated chromosomes of each homologue pair is aligned. This ensures that the genes contained on the nonsister chromatids are directly aligned. Now crossing-over may occur. As the lattice breaks down, homologues are temporarily held together by *chiasmata* (sing., chiasma), regions where the nonsister chromatids are attached due to DNA strand exchange and crossing-over. After exchange of genetic information between the nonsister chromatids, the homologues separate and are distributed to different daughter cells.

To appreciate the significance of crossing-over, keep in mind that the members of a homologous pair can carry slightly different instructions, or alleles, for the same genetic traits. In the end, due to a swapping of genetic material during crossing-over, the chromatids held together by a centromere are no longer identical. Therefore, when the chromatids separate during meiosis II, some of the daughter cells receive daughter chromosomes with recombined alleles. Due to **genetic recombination,** the offspring have a different set of alleles, and therefore genes, than their parents. This increases the genetic variation of the offspring.

Animation Meiosis and Crossing-Over

Independent Assortment of Homologous Chromosomes

During **independent assortment,** the homologous chromosome pairs separate independently, or randomly. When homologues align at the metaphase plate, the maternal or paternal homologue may be oriented toward either pole. Figure 10.4 shows the possible chromosome orientations for a cell that contains only three pairs of homologous chromosomes. Once all possible alignments of independent assortment are considered for these three pairs, the result will be 2^3, or 8, combinations of maternal and paternal chromosomes in the resulting gametes from this cell, simply due to independent assortment of homologues.

Animation Random Orientation of Chromosomes During Meiosis

Significance of Genetic Variation

In humans, who have 23 pairs of chromosomes, the possible chromosomal combinations in the gametes is a staggering 2^{23}, or 8,388,608. The variation that results from meiosis is enhanced by **fertilization,** the union of the male and female gametes. The chromosomes donated by the parents are combined, and in humans, this means that there are potentially $(2^{23})^2$, or 70,368,744,000,000, different chromosome combinations in the zygote. This number assumes that there was no crossing-over between the nonsister chromatids prior to independent assortment. If a single crossing-over event occurs, then $(4^{23})^2$, or 4,951,760,200,000,000,000,000,000,000, genetically different zygotes are possible for every couple. Keep in mind that crossing-over can occur several times in each chromosome!

The staggering amount of genetic variation achieved through meiosis is particularly important to the long-term survival of a species, because it increases genetic variation within a population. The process of sexual reproduction brings about genetic rccombinations among members of a population.

Asexual reproduction passes on exactly the same combination of chromosomes and genes. Asexual reproduction may be advantageous if the environment remains unchanged. However, if the environment changes, genetic variability among offspring introduced by sexual reproduction may be advantageous. Under the new conditions, some offspring may have a better chance of survival and reproductive success than others in a population. For example, suppose the ambient temperature were to rise due to climate change. This change in the environment could place demands on the physiology of an organism. For example, an animal with less fur, or reduced body fat, could have an advantage over other individuals of its generation.

In a changing environment, sexual reproduction, with its reshuffling of genes due to meiosis and fertilization, might give a few offspring a better chance to survive and reproduce, thereby increasing the possibility of passing on their genes to the next generation.

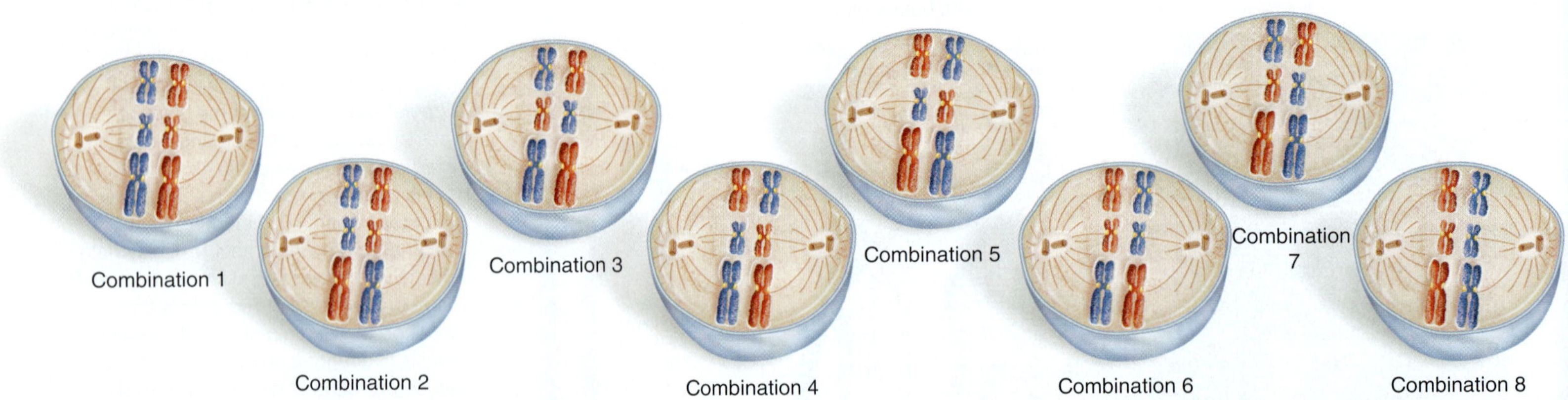

Figure 10.4 Independent assortment. When a parent cell has three pairs of homologous chromosomes, there are 2^3, or 8, possible chromosome alignments at the metaphase plate due to independent assortment. Each possible combination is shown, one in each cell.

THEME Nature of Science

Meiosis and the Parthenogenic Lizards

The process of crossing-over plays an important role in the generation of genetic variation for sexually reproducing species. For most species that undergo asexual reproduction, fast generation times and mutation allow for the species to introduce enough variation to respond to environmental changes. But what about species, such as the whiptail lizard shown in Figure 10A, that undergo parthenogenesis? Parthenogenesis is the production of new individuals from unfertilized eggs. It is not uncommon in the animal kingdom; many arthropods, lizards, fish, and salamanders are known to be parthenogenic.

Parthenogenesis is a form of asexual reproduction; only one parent, the female, contributes genetic information to the next generation. But typically, these species do not have the short generation times of other asexual organisms, such as bacteria. At least on the surface, parthenogenesis would seem to limit the amount of genetic variation in the species, and thus reduce the ability of the species to respond to changes in its environment. While some species, such as honeybees, avoid this problem by switching between parthenogenesis and sexual reproduction, truly parthenogenic species appear to be at an evolutionary disadvantage.

Researchers from a team at the Howard Hughes Medical Institute discovered that in a parthenogenic species of lizards (whiptail lizards, genus *Aspidoscelis*) there is a variation in the normal process of meiosis. In most cases, crossing-over during meiosis occurs between the nonsister chromatids of homologous chromosomes. However, in the whiptail lizard, crossing-over occurs between the sister chromatids.

Figure 10A In parthenogenic species, such as the whiptail lizard, variations in the process of meiosis allow the species to increase genetic variation with each generation.

How does this happen? To make this possible, the species doubles the number of chromosomes prior to meiosis—effectively making an additional copy of the genome and forming a pair of homologous chromosomes from a single parent. This doubling allows the reduction division in meiosis to produce diploid (2n) gametes, a requirement for many species that undergo parthenogenesis. Then, the species allows for crossing-over to occur between the sister chromatids themselves. Since there are always slight differences in the sister chromatids (they are never truly identical), small amounts of variation are maintained in the genome, and this is passed on to the next generation. The amount of genetic variation may be small, but this variation in meiosis allows for some level of genetic recombination, thus providing genetic variation to the species.

Questions to Consider

1. Does this process produce the same amount of genetic variation as would occur in normal sexual reproduction?
2. How would you test to determine the amount of genetic variation produced by parthenogenic species?

Check Your Progress 10.2

1. Describe the two main ways in which meiosis contributes to genetic variation.
2. Examine how many combinations of chromosomes are possible in the gametes in a cell with four pairs of homologous chromosomes.
3. Evaluate why meiosis and sexual reproduction are important in responding to the changing environment.

10.3 The Phases of Meiosis

Learning Outcomes

Upon completion of this section, you should be able to

1. Describe the phases of meiosis and the major events that occur during each phase.
2. Understand how meiosis reduces the chromosome number from diploid to haploid.

Meiosis consists of two unique, consecutive cell divisions, meiosis I and meiosis II. DNA is replicated in S phase of the cell cycle prior to meiosis I but not meiosis II. Both meiosis I and meiosis II contain a prophase, metaphase, anaphase, and telophase.

Animation Stages of Meiosis

MP3 Meiosis

Prophase I

It is apparent during prophase I that nuclear division is about to occur, because a spindle forms as the centrosomes migrate away from one another. The nuclear envelope fragments, and the nucleolus disappears.

The homologous chromosomes, each having replicated during S phase of the cell cycle, consist of two sister chromatids. The homologous chromosomes undergo synapsis to form bivalents. At this time, crossing-over may occur between the nonsister chromatids (see Fig. 10.3). As described earlier, crossing-over increases the genetic diversity of the daughter cells, because after crossing-over, the sister chromatids are no longer identical.

Throughout prophase I, the homologous chromosomes have been condensing, so that by now they have the appearance of compacted metaphase chromosomes.

Metaphase I

During metaphase I, the bivalents held together by chiasmata (see Fig. 10.3) have moved toward the metaphase plate (equator of the spindle). Metaphase I is characterized by a fully formed spindle and alignment of the bivalents at the metaphase plate. As in mitosis, kinetochores are seen, but the two kinetochores of a duplicated chromosome are attached to the same kinetochore spindle fiber.

Bivalents independently align themselves at the metaphase plate of the spindle. Either the maternal or paternal homologue of each bivalent may be oriented toward either pole of the cell. The orientation of one bivalent is not dependent on the orientation of the other bivalents. This independent assortment of chromosomes contributes to the genetic variability of the daughter cells, because all possible combinations of chromosomes can occur in the daughter cells.

Anaphase I

During anaphase I, the homologues of each bivalent separate and move to opposite poles, but sister chromatids do not separate. This splitting of the homologous pair reduces the chromosome number from 2n to n. However, each chromosome still has two chromatids (Fig. 10.5).

Telophase I

Completion of telophase I is not necessary during meiosis. That is, the spindle disappears, but new nuclear envelopes need not form before the daughter cells proceed to meiosis II. Also, this phase may or may not be accompanied by cytokinesis, which is separation of the cytoplasm. Notice in Figure 10.5 that the cells have different chromosome combinations than the original parent cell (not all of the combinations are shown in Fig. 10.5). The cells exiting telophase I are also haploid compared to the diploid parent cell.

3D Animation Meiosis: Interphase and Meiosis I

Animation Meiosis I

Interkinesis

Following telophase, the cells enter interkinesis, a short rest period prior to beginning the second nuclear division, meiosis II. The process of **interkinesis** is similar to interphase between mitotic divisions, except that DNA replication does not occur, because the chromosomes are already duplicated.

Meiosis II and Gamete Formation

At the beginning of meiosis II, the two daughter cells contain the haploid number of chromosomes, or one chromosome from each homologous pair. Note that these chromosomes still consist of duplicated sister chromatids at this point. During metaphase II, the chromosomes align at the metaphase plate, but they do not align in homologous pairs, as in meiosis I, because only one chromosome of each homologous pair is present (Fig. 10.5). Thus, the alignment of the chromosomes at the metaphase plate is similar to what is observed during mitosis.

During anaphase II, the sister chromatids separate, becoming daughter chromosomes that are not duplicated. These daughter chromosomes move toward the poles. At the end of telophase II and cytokinesis, there are four haploid cells. Because of crossing-over of chromatids during meiosis I, each gamete most likely contains chromosomes with a mixture of maternal and paternal genes.

As mentioned, following meiosis II, the haploid cells become gametes in animals (see section 10.5). In plants, they become *spores*, reproductive cells that develop into new multicellular structures without the need to fuse with another reproductive cell. The multicellular structure is the haploid generation, which produces gametes. The resulting zygote develops into a diploid generation. Therefore, plants have both haploid and diploid phases in their life cycle, and plants are said to exhibit an *alternation of generations* (see Chapter 23). In most fungi and algae, the zygote undergoes meiosis, and the daughter cells develop into new individuals. Therefore, the organism is always haploid (see Chapters 21 and 22).

Animation Meiosis II

3D Animation Meiosis: Meiosis II

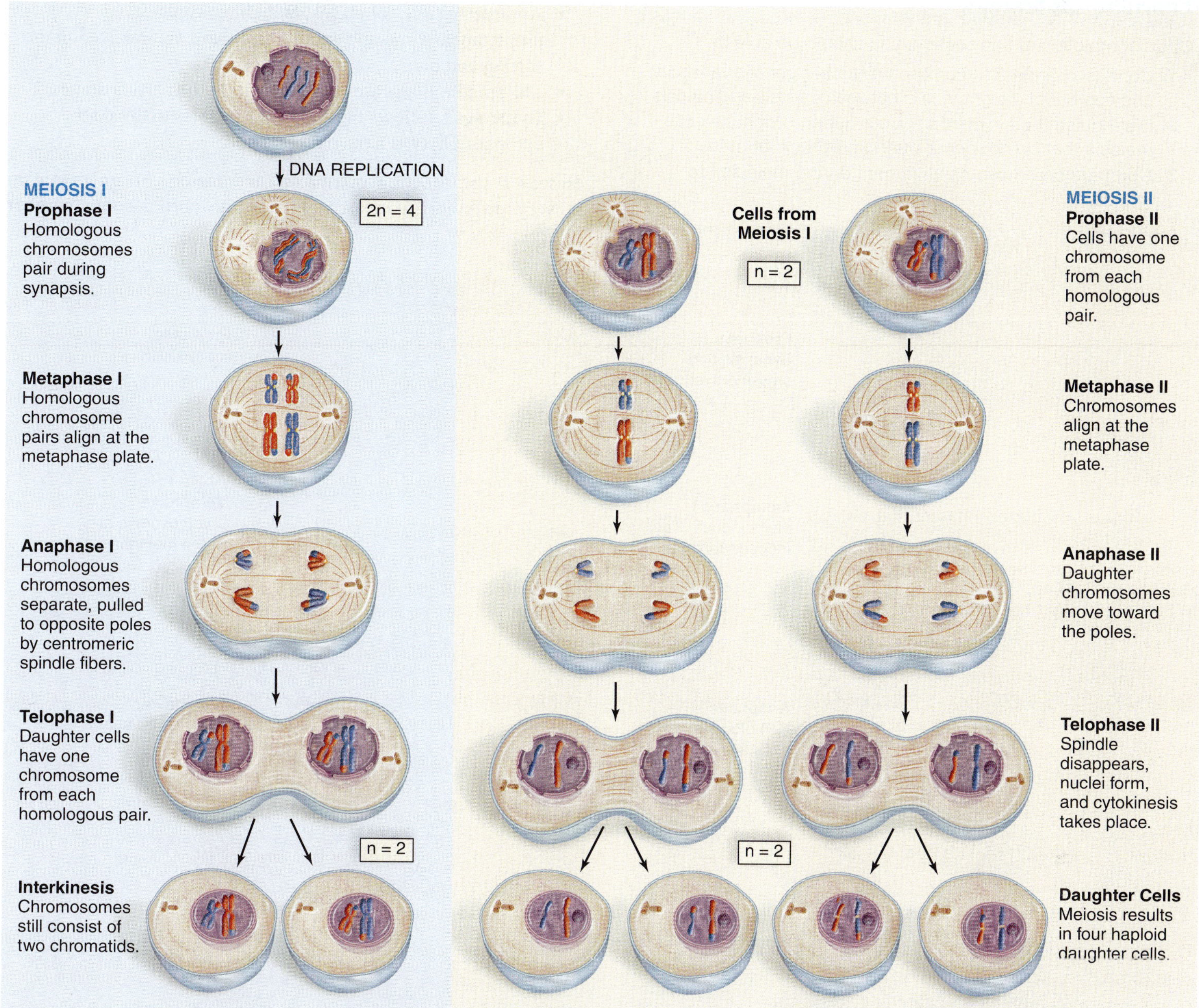

Figure 10.5 Stages of Meiosis. When diploid homologous chromosomes pair during meiosis I, crossing-over occurs, as represented by the exchange of color. Pairs of homologous chromosomes separate during meiosis I, and chromatids separate, becoming haploid daughter chromosomes, with two copies of each during meiosis II. Following meiosis II and the separation of sister chromatids, four haploid daughter cells are produced.

Tutorial
Meiosis

Check Your Progress 10.3

1. Describe the differences between the chromosomal combinations of a cell at metaphase I and metaphase II of meiosis.
2. Explain what would cause daughter cells following meiosis II to contain identical chromosomes or nonidentical chromosomes.
3. Examine what could happen if homologous chromosomes lined up top to bottom instead of side by side during meiosis I.

10.4 Meiosis Compared to Mitosis

Learning Outcomes

Upon completion of this section, you should be able to

1. Contrast changes in chromosome number, genetic variability, and number of daughter cells between meiosis and mitosis.
2. Distinguish the events that occur during prophase I of meiosis that do not occur during prophase of mitosis.
3. Compare chromosome alignment during meiosis I to mitosis.

There are many similarities between the processes of mitosis. In both processes:

- An orderly series of stages, including prophase, prometaphase, metaphase, and telophase are involved in the sorting and divison of the chromosomes.
- The spindle fibers are active in sorting the chromosomes.
- Cytokinesis follows the end of the process to divide the cytoplasm between the daughter cells.

However, the function of mitosis and meiosis in an organism is very different. Mitosis maintains the chromosome number

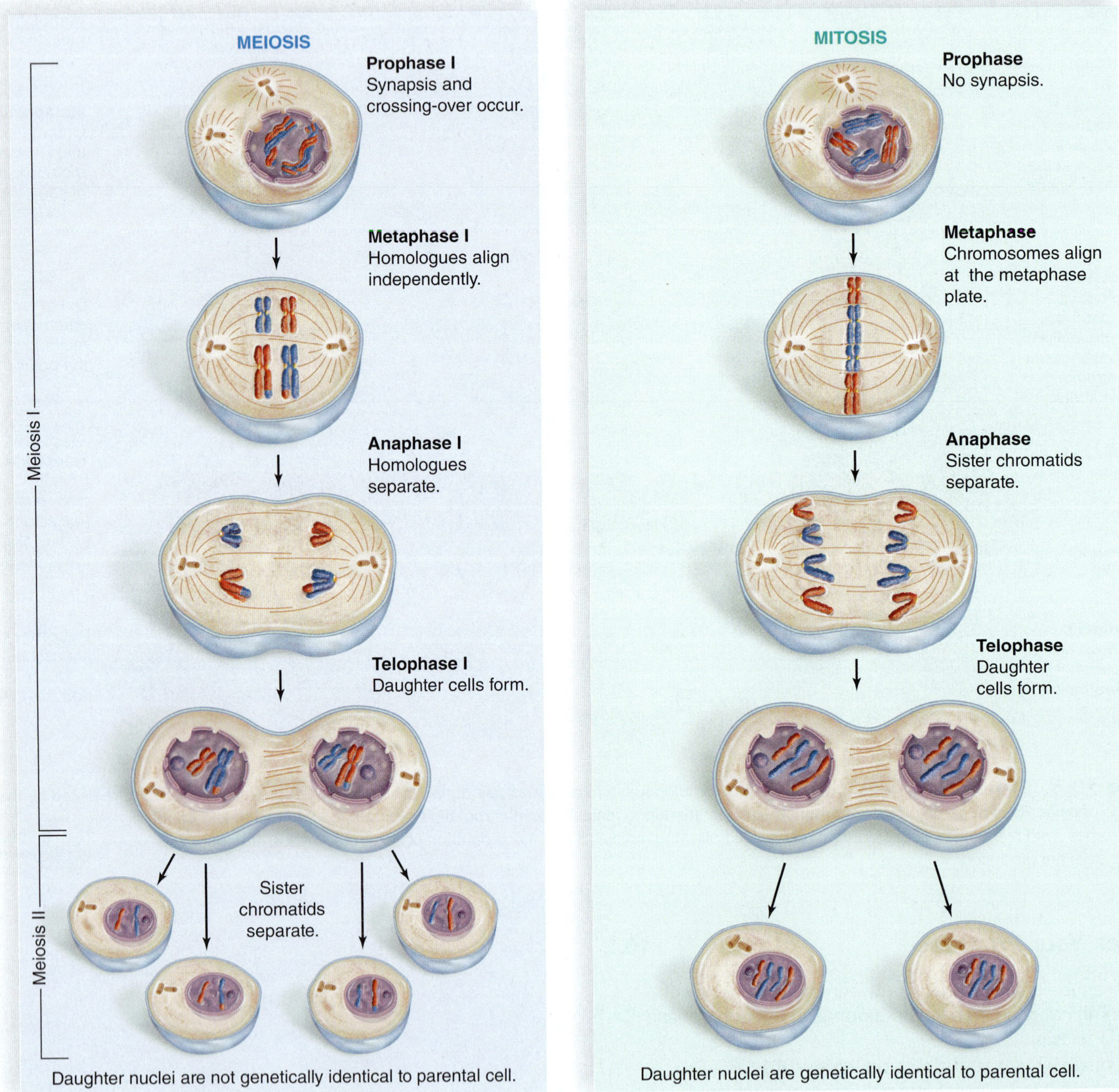

Figure 10.6 Meiosis I compared to mitosis. Why does meiosis produce daughter cells with half the number of chromosomes, whereas mitosis produces daughter cells with the same number of chromosomes as the parent cell? Compare metaphase I of meiosis to metaphase of mitosis. Only in metaphase I of meiosis are the homologous chromosomes paired at the metaphase plate. Members of homologous chromosome pairs separate during anaphase I, and therefore the daughter cells are haploid. The exchange of color between nonsister chromatids represents the crossing-over that occurred during meiosis I. The blue chromosomes were inherited from the paternal parent, and the red chromosomes were inherited from the maternal parent.

between the cells, whereas meiosis is often referred to as reduction division.

Figure 10.6 compares meiosis and mitosis. Several of the fundamental differences between the two processes include:

- Meiosis requires two nuclear divisions, but mitosis requires only one nuclear division.
- Meiosis produces four daughter nuclei. Following cytokinesis, there are four daughter cells. Mitosis followed by cytokinesis results in two daughter cells.
- Following meiosis, the four daughter cells are haploid and have half the chromosome number as the diploid parent cell. Following mitosis, the daughter cells have the same chromosome number as the parent cell.
- Following meiosis, the daughter cells are genetically identical neither to each other nor to the parent cell. Following mitosis, the daughter cells are genetically identical to each other and to the parent cell.

In addition to the fundamental differences between meiosis and mitosis, two specific differences between the two types of nuclear divisions can be categorized. These differences involve occurrence and process.

Animation Comparison of Meiosis and Mitosis

Occurrence

Meiosis occurs only at certain times in the life cycle of sexually reproducing organisms. In humans, meiosis occurs only in the reproductive organs and produces the gametes. Mitosis is more common, because it occurs in all tissues during growth and repair.

Process

We now compare the processes of both meiosis I and meiosis II to mitosis.

Meiosis I Compared to Mitosis

Notice that these events distinguish meiosis I from mitosis:

- During prophase I, bivalents form and crossing-over occurs. These events do not occur during mitosis.
- During metaphase I of meiosis, bivalents independently align at the metaphase plate. The paired chromosomes have a total of four chromatids each. During metaphase in mitosis, individual chromosomes align at the metaphase plate. They each have two chromatids.
- During anaphase I of meiosis, homologues of each bivalent separate, and duplicated chromosomes (with centromeres intact) move to opposite poles. During anaphase of mitosis, sister chromatids separate, becoming daughter chromosomes that move to opposite poles.

Meiosis II Compared to Mitosis

The events of meiosis II are similar to those of mitosis, except that in meiosis II the nuclei contain the haploid number of chromosomes.

Table 10.1 Meiosis I Compared to Mitosis

Meiosis I	Mitosis
Prophase I Pairing of homologous chromosomes	**Prophase** No pairing of chromosomes
Metaphase I Bivalents at metaphase plate	**Metaphase** Duplicated chromosomes at metaphase plate
Anaphase I Homologues of each bivalent separate, and duplicated chromosomes move to poles	**Anaphase** Sister chromatids separate, becoming daughter chromosomes that move to the poles
Telophase I Two haploid daughter cells, not identical to the parent cell	**Telophase** Two diploid daughter cells, identical to the parent cell

Table 10.2 Meiosis II Compared to Mitosis

Meiosis II	Mitosis
Prophase II No pairing of chromosomes	**Prophase** No pairing of chromosomes
Metaphase II Haploid number of duplicated chromosomes at metaphase plate	**Metaphase** Diploid number of duplicated chromosomes at metaphase plate
Anaphase II Sister chromatids separate, becoming daughter chromosomes that move to the poles	**Anaphase** Sister chromatids separate, becoming daughter chromosomes that move to the poles
Telophase II Four haploid daughter cells, not genetically identical	**Telophase** Two diploid daughter cells, identical to the parent cell

In mitosis, the original number of chromosomes is maintained. Meiosis II produces two daughter cells from each parent cell that completes meiosis I, for a total of four daughter cells. These daughter cells contain the same number of chromosomes as they did at the end of meiosis I. Tables 10.1 and 10.2 compare meiosis I and II to mitosis.

Check Your Progress 10.4

1. Compare chromosome alignment between metaphase I of meiosis and metaphase of mitosis.
2. Explain how meiosis II is more similar to mitosis than to meiosis I.

10.5 The Cycle of Life

Learning Outcomes

Upon completion of this section, you should be able to

1. Contrast the life cycle of plants with the life cycle of animals.
2. Describe spermatogenesis and oogenesis in humans.

The term **life cycle** refers to all the reproductive events that occur from one generation to the next similar generation. In animals, including humans, the individual is always diploid, and meiosis produces the gametes, the only haploid phase of the life cycle (Fig. 10.7). In contrast, plants have a haploid phase that alternates with a diploid phase. The haploid generation, known as the **gametophyte,** may be larger or smaller than the diploid generation, called the **sporophyte.**

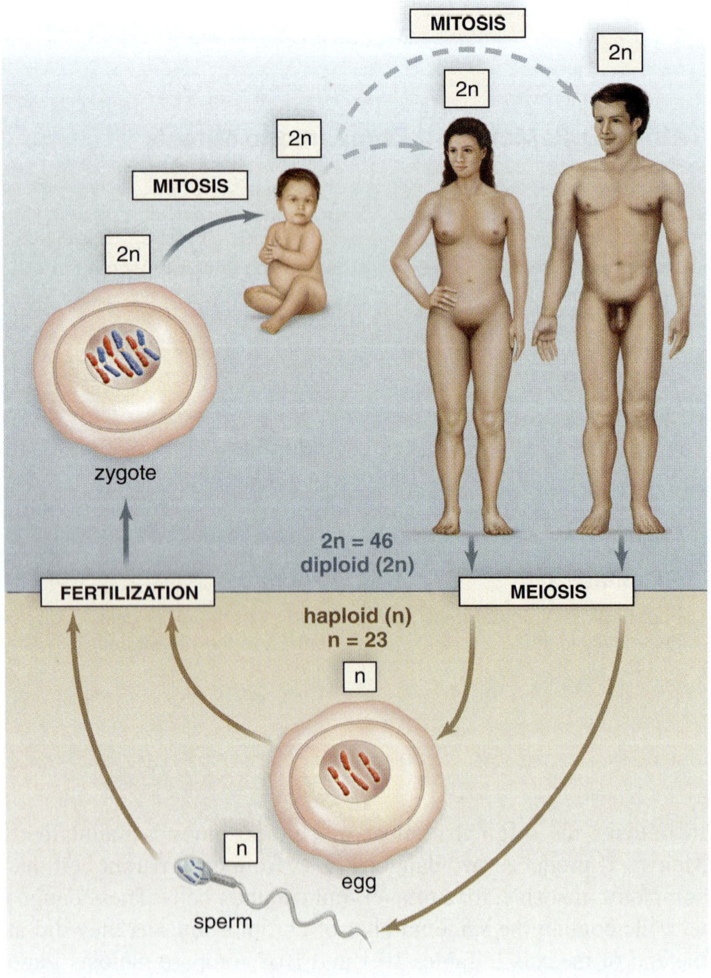

Figure 10.7 Life cycle of humans. Meiosis in males is a part of sperm production, and meiosis in females is a part of egg production. When a haploid sperm fertilizes a haploid egg, the zygote is diploid. The zygote undergoes mitosis as it develops into a newborn child. Mitosis continues throughout life during growth and repair.

Mosses growing on bare rocks and forest floors are the haploid generation, and the diploid generation is short-lived. In most fungi and algae, the zygote is the only diploid portion of the life cycle, and it undergoes meiosis. Therefore, the black mold that grows on bread and the green scum that floats on a pond are haploid.

The majority of plant species, including pine, corn, and sycamore, are usually diploid, and the haploid generation is short-lived. In plants, algae, and fungi, the haploid phase of the life cycle produces gamete nuclei without the need for meiosis, because it has occurred earlier.

Animals are diploid, and meiosis occurs during the production of gametes, called **gametogenesis.** In males, meiosis is a part of **spermatogenesis** (Gk. *sperma,* "seed"), which occurs in the testes and produces sperm. In females, meiosis is a part of **oogenesis** (Gk. *oon,* "egg"), which occurs in the ovaries and produces eggs. A sperm and an egg join at fertilization, restoring the diploid chromosome number. The resulting zygote undergoes mitosis during development of the fetus. After birth, mitosis is involved in the continued growth of the child and the repair of tissues at any time.

Spermatogenesis and Oogenesis in Humans

In human males, spermatogenesis occurs within the testes; in females, oogenesis occurs within the ovaries.

Spermatogenesis

The testes contain stem cells called spermatogonia. These cells keep the testes supplied with primary spermatocytes that undergo spermatogenesis, as described in Figure 10.8, *top.* Primary spermatocytes with 46 chromosomes undergo meiosis I to form two secondary spermatocytes, each with 23 duplicated chromosomes. Secondary spermatocytes undergo meiosis II to produce four spermatids with 23 daughter chromosomes. Spermatids then differentiate into viable sperm (spermatozoa). Upon sexual arousal, the sperm enter ducts and exit the penis upon ejaculation.

Animation Spermatogenesis

Oogenesis

The ovaries contain stem cells, called oogonia, that produce many primary oocytes with 46 chromosomes during fetal development. They even begin oogenesis, but only a few continue when a female has become sexually mature. The result of meiosis I is two haploid cells with 23 chromosomes each (Fig. 10.8, *bottom*). One of these cells, termed the secondary **oocyte,** receives almost all the cytoplasm. The other is a **polar body** that may either disintegrate or divide again.

The secondary oocyte begins meiosis II but stops at metaphase II. Then the secondary oocyte leaves the ovary and enters the uterine tube, where sperm may be present. If no sperm are in the uterine tube, or if a sperm does not enter the secondary oocyte, it eventually disintegrates without completing meiosis. If a sperm does enter the oocyte, some of its contents trigger the completion of meiosis II in the secondary oocyte, and another polar body forms.

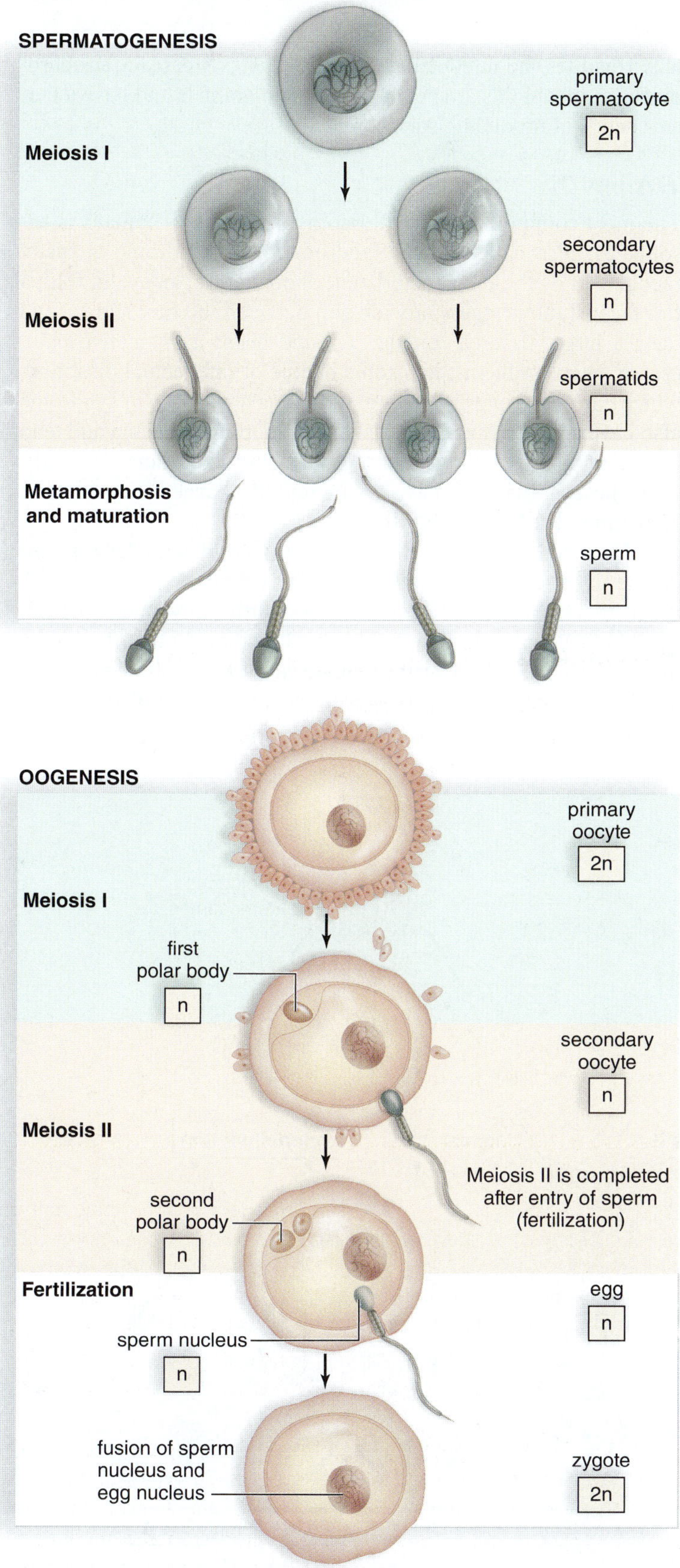

Figure 10.8 Spermatogenesis and oogenesis in mammals. Spermatogenesis produces four viable sperm, whereas oogenesis produces one egg and at least two polar bodies. In humans, both sperm and egg have 23 chromosomes each; therefore, following fertilization, the zygote has 46 chromosomes.

At the completion of oogenesis, following entrance of a sperm, there is one egg and two or three polar bodies. The polar bodies are a way to "dispose of" chromosomes while retaining much of the cytoplasm in the egg. Cytoplasmic molecules and organelles are needed by a developing embryo following fertilization. Some zygote components, such as the centrosome, are contributed by the sperm.

The mature egg has 23 chromosomes, but the zygote formed when the sperm and egg nuclei fuse has 46 chromosomes. Therefore, fertilization restores the diploid number of chromosomes. The production of haploid gametes and subsequent fusion of those gametes into a diploid zygote complete a human life cycle.

Check Your Progress 10.5

1. Describe where cells that undergo meiosis are located in humans.
2. Compare the number of gametes produced during oogenesis and spermatogenesis in humans.

10.6 Changes in Chromosome Number and Structure

Learning Outcomes

Upon completion of this section, you should be able to

1. Distinguish between euploidy and aneuploidy.
2. Explain how nondisjunction can cause monosomy and trisomy aneuploidy.
3. Describe human diseases caused by changes in the number of sex chromosomes.
4. Characterize how changes in chromosome structure can lead to human diseases.

We have seen that crossing-over creates variation within a population and is essential for the normal separation of chromosomes during meiosis. Furthermore, the proper separation of homologous chromosomes during meiosis I and the separation of sister chromatids during meiosis II are essential for the maintenance of normal chromosome numbers in living organisms. Although meiosis almost always proceeds normally, a failure of chromosomes to separate, or **nondisjunction,** may occur, resulting in a gain or loss of chromosomes. Errors in crossing-over may result in extra or missing parts of chromosomes.

Aneuploidy

The correct number of chromosomes in a species is known as **euploidy.** A change in the chromosome number resulting from nondisjunction during meiosis is called **aneuploidy.** Aneuploidy is seen in both plants and animals. Monosomy and trisomy are two aneuploid states.

Monosomy (2n − 1) occurs when an individual has only one of a particular type of chromosome when he or she should have two. **Trisomy** (2n + 1) occurs when an individual has three of a particular type of chromosome when he or she should have two. Both monosomy and trisomy are the result of nondisjunction during mitosis or meiosis. *Primary nondisjunction* occurs during meiosis I when both members of a homologous pair go into the same daughter cell (Fig. 10.9*a*). *Secondary nondisjunction* occurs during meiosis II when the sister chromatids fail to separate and both daughter chromosomes go into the same gamete (Fig. 10.9*b*).

Notice that when secondary nondisjunction occurs, there are two normal gametes and two aneuploid gametes. In contrast, when primary nondisjunction occurs, no normal gametes are produced. Therefore, primary nondisjunction tends to have more deleterious effects than secondary nondisjunction.

In animals, monosomies and trisomies of nonsex, or autosomal, chromosomes are generally lethal, but a trisomic individual is more likely to survive than a monosomic one. In humans, only three autosomal trisomic conditions are known to be viable beyond birth: trisomy 13, 18, and 21. Only trisomy 21 is viable beyond early childhood and is characterized by a distinctive set of physical abnormalities and intellectual disabilities. In comparison, sex chromosome aneuploids are better tolerated in animals and have a better chance of producing survivors.

Trisomy 21

The most common autosomal trisomy seen among humans is trisomy 21, also called Down syndrome. This syndrome is easily recognized by these characteristics: short stature; an eyelid fold; a flat face; stubby fingers; a wide gap between the first and second toes; a large, fissured tongue; a round head; a distinctive palm crease; heart problems; and some degree of intellectual disability, which can sometimes be severe. Individuals with Down syndrome also have a greatly increased risk of developing leukemia and tend to age rapidly, resulting in a shortened life expectancy. In addition, these individuals have an increased chance of developing Alzheimer disease later in life.

The chances of a woman having a child with Down syndrome increase rapidly with age. In women ages 20 to 30, the incidence of trisomy 21 is 1 in 1,400 births; in women 30 to 35, the incidence is about 1 in 750 births. It is thought that the longer the oocytes are stored in the female, the greater the chances of nondisjunction occurring. However, even though an older woman

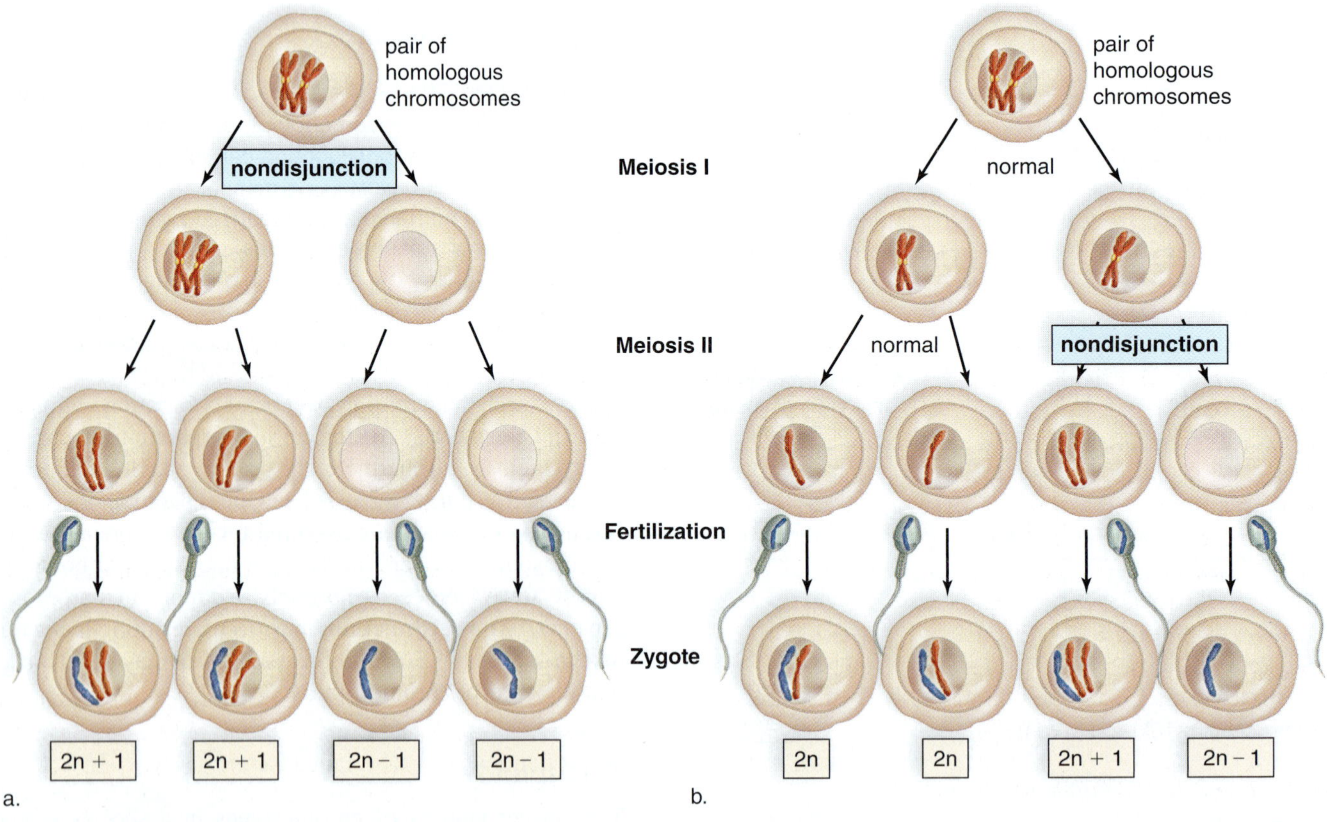

Figure 10.9 Nondisjunction of chromosomes during oogenesis, followed by fertilization with normal sperm. **a.** Nondisjunction can occur during meiosis I (primary nondisjunction) and results in abnormal eggs that also have one more or one less than the normal number of chromosomes. Fertilization of these abnormal eggs with normal sperm results in a zygote with abnormal chromosome numbers. 2n = diploid number of chromosomes. **b.** Nondisjunction can also occur during meiosis II (secondary nondisjunction) if the sister chromatids separate but the resulting daughter chromosomes go into the same daughter cell. Then the egg will have one more or one less than the usual number of chromosomes. Fertilization of these abnormal eggs with normal sperm produces a zygote with abnormal chromosome numbers.

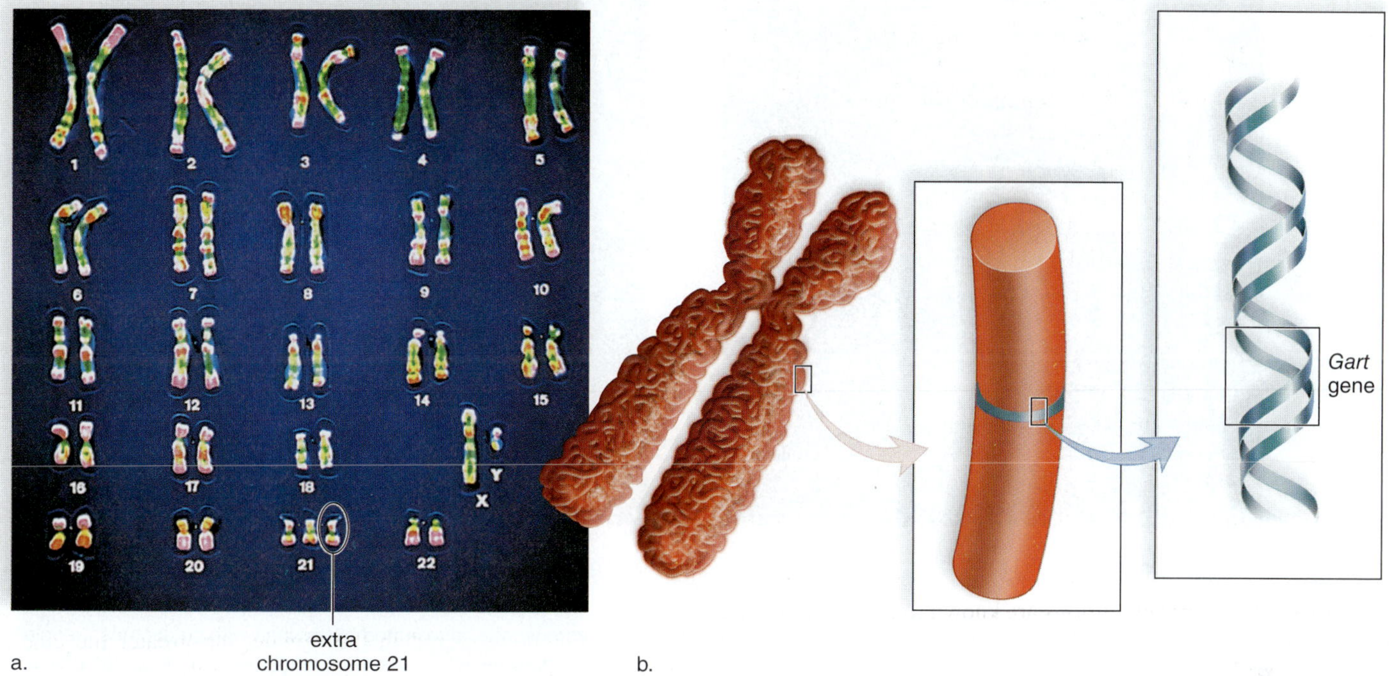

Figure 10.10 Trisomy 21. Persons with Down syndrome, or trisomy 21, have an extra chromosome 21. **b.** The karyotype of an individual with Down syndrome shows three copies of chromosome 21. Therefore, the individual has three copies instead of two copies of each gene on chromosome 21. An extra copy of the *Gart* gene, which leads to high levels of purine in the blood, accounts for many of the characteristics of Down syndrome.

is more likely to have a Down syndrome child, most babies with Down syndrome are born to women younger than age 40, because this is the age group having the most babies. Furthermore, research indicates that in 23% of the cases studied, the sperm contributed the extra chromosome. A **karyotype,** a visual display of the chromosomes arranged by size, shape, and banding pattern, may be performed to identify babies with Down syndrome and other aneuploid conditions (Fig. 10.10).

The genes that cause Down syndrome are located on the long arm of chromosome 21 (Fig. 10.10), and extensive investigative work has been directed toward discovering the specific genes responsible for the characteristics of the syndrome. Thus far, investigators have discovered several genes that may account for various conditions seen in persons with Down syndrome. For example, they have located the genes most likely responsible for the increased tendency toward leukemia, cataracts, and an accelerated rate of aging. Researchers have also discovered that an extra copy of the *Gart* gene causes an increased level of purines in the blood, a finding associated with intellectual disability. One day, it may be possible to control the expression of the *Gart* gene even before birth, so that at least this symptom of Down syndrome does not appear.

Changes in Sex Chromosome Number

An abnormal sex chromosome number is the result of inheriting too many or too few X or Y chromosomes. Nondisjunction during oogenesis or spermatogenesis can result in gametes with an abnormal number of sex chromosomes. However, extra copies of the sex chromosomes are much more easily tolerated in humans than are extra copies of autosomes.

A person with Turner syndrome (XO) is a female, and a person with Klinefelter syndrome (XXY) is a male. However, deletion of the *SRY* gene on the short arm of the Y chromosome results in Swyer syndrome, or an "XY female." Individuals with Swyer syndrome lack a hormone called testis-determining factor, which plays a critical role in the development of male genitals. Furthermore, movement of this gene onto the X chromosome may result in de la Chapelle syndrome, or an "XX male." Men with de la Chapelle syndrome exhibit undersized testes, sterility, and rudimentary breast development. Together, these observations suggest that in humans the presence of the *SRY* gene, not the number of X chromosomes, determines maleness. In its absence, a person develops as a female.

Why are newborns with an abnormal sex chromosome number more likely to survive than those with an abnormal autosome number? Because females have two X chromosomes and males have only one, we might expect females to produce twice the amount of each gene from this chromosome, but both males and females produce roughly the same amount. In reality, both males and females only have one functioning X chromosome. In females, and in males with extra X chromosomes, any additional X chromosomes become an inactive mass called a **Barr body,** named after Murray Barr, the person who discovered it. This inactivation provides a natural method for gene dosage compensation of the sex chromosomes and explains why extra sex chromosomes are more easily tolerated than extra autosomes.

Turner Syndrome. From birth, an XO individual with Turner syndrome has only one sex chromosome, an X; the O signifies the

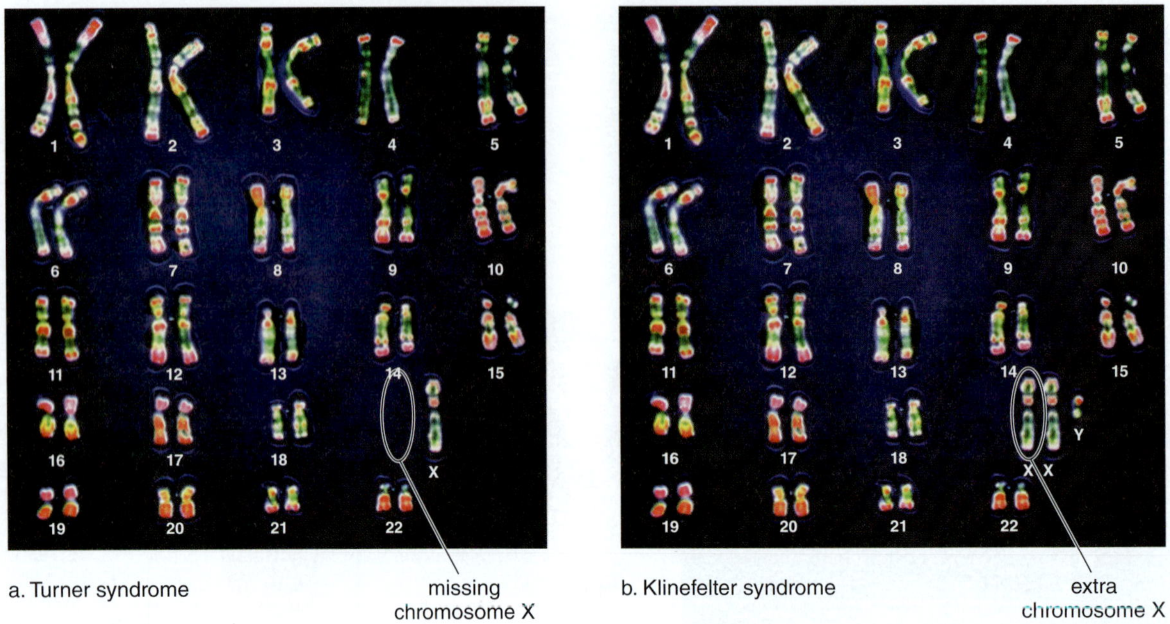

Figure 10.11 Abnormal sex chromosome number. Nondisjunction of sex chromosomes is tolerated better than with autosomes. People with (**a**) Turner syndrome, who have only one X chromosome, as shown, and (**b**) Klinefelter syndrome, who have more than one X chromosome plus a Y chromosome.

absence of a second sex chromosome (Fig. 10.11*a*). Therefore, the nucleus does not contain a Barr body. The approximate incidence is 1 in 10,000 females.

Turner females are short, with a broad chest and widely spaced nipples. These individuals also have a low posterior hairline and neck webbing. The ovaries, oviducts, and uterus are very small and underdeveloped. Turner females do not undergo puberty or menstruate, and their breasts do not develop. However, some have given birth following in vitro fertilization using donor eggs. They usually are of normal intelligence and can lead fairly normal lives if they receive hormone supplements.

Klinefelter Syndrome. A male with Klinefelter syndrome has two or more X chromosomes in addition to a Y chromosome (Fig. 10.11*b*). The extra X chromosomes become Barr bodies. The approximate incidence for Klinefelter syndrome is 1 in 500 to 1,000 males.

In Klinefelter males, the testes and prostate gland are underdeveloped and facial hair is lacking. They may exhibit some breast development. Affected individuals have large hands and feet and very long arms and legs. They are usually slow to learn but do not have an intellectual disability unless they inherit more than two X chromosomes. No matter how many X chromosomes there are, an individual with a Y chromosome is a male.

While males with Klinefelter syndrome exhibit no other major health abnormalities, they have an increased risk of some disorders, including breast cancer, osteoporosis, and lupus, which disproportionately affect females. Although men with Klinefelter syndrome typically do not need medical treatment, some have found that testosterone therapy helps increase muscle strength, sex drive, and concentration ability. Testosterone treatment, however, does not reverse the sterility associated with Klinefelter syndrome due to incomplete testicle development.

Poly-X Females. A poly-X female, sometimes called a superfemale, has more than two X chromosomes and, therefore, extra Barr bodies in the nucleus. Females with three X chromosomes have no distinctive phenotype aside from a tendency to be tall and thin. Although some have delayed motor and language development, as well as learning problems, most poly-X females do not have an intellectual disability. Some may have menstrual difficulties, but many menstruate regularly and are fertile. Children usually have a normal karyotype. The incidence for poly-X females is about 1 in 1,500 females.

Females with more than three X chromosomes occur rarely. Unlike XXX females, XXXX females are usually tall and have a severe intellectual disability. Various physical abnormalities are seen, but they may menstruate normally.

Jacobs Syndrome. XYY males, termed Jacobs syndrome, can result only from nondisjunction during spermatogenesis. Among all live male births, the frequency of the XYY karyotype is about 1 in 1,000. Affected males are usually taller than average, suffer from persistent acne, and tend to have speech and reading problems, but they are fertile and may have children. Despite the extra Y chromosome, there is no difference in behavior between XYY and XY males.

Changes in Chromosome Structure

Changes in chromosome structure are another type of chromosomal mutation. Some, but not all, changes in chromosome structure can be detected microscopically. Various agents in the environment, such as radiation, certain organic chemicals, or even viruses, can cause chromosomes to break. Ordinarily, when breaks occur in chromosomes, the two broken ends reunite to give the same sequence of genes. Sometimes, however, the broken ends of one or more chromosomes do not rejoin in the same pattern as before, and the results are various types of chromosomal mutations.

Changes in chromosome structure include deletions, duplications, translocations, and inversions of chromosome segments. A **deletion** occurs when an end of a chromosome breaks off or when two simultaneous breaks lead to the loss of an internal segment (Fig. 10.12*a*). Even when only one member of a pair of chromosomes is affected, a deletion often causes abnormalities.

Animation Changes in Chromosome Structure

A **duplication** is the presence of a chromosomal segment more than once in the same chromosome (Fig. 10.12*b*). Duplications may or may not cause visible abnormalities, depending on the size of the duplicated region. An **inversion** has occurred when a segment of a chromosome has been turned around 180° (Fig. 10.12*c*). Most individuals with inversions exhibit no abnormalities, but this reversed sequence of genes can result in duplications or deletions being passed on to their children, as shown in Figure 10.13.

A **translocation** is the movement of a chromosome segment from one chromosome to another, nonhomologous chromosome. The translocation shown in Figure 10.12*d* is *balanced,* meaning that there is a reciprocal swap of one piece of the chromosome for the other. Often, there are no visible effects of the swap, but if the individual has children, they receive one normal copy of the chromosome from the normal parent and one of the abnormal

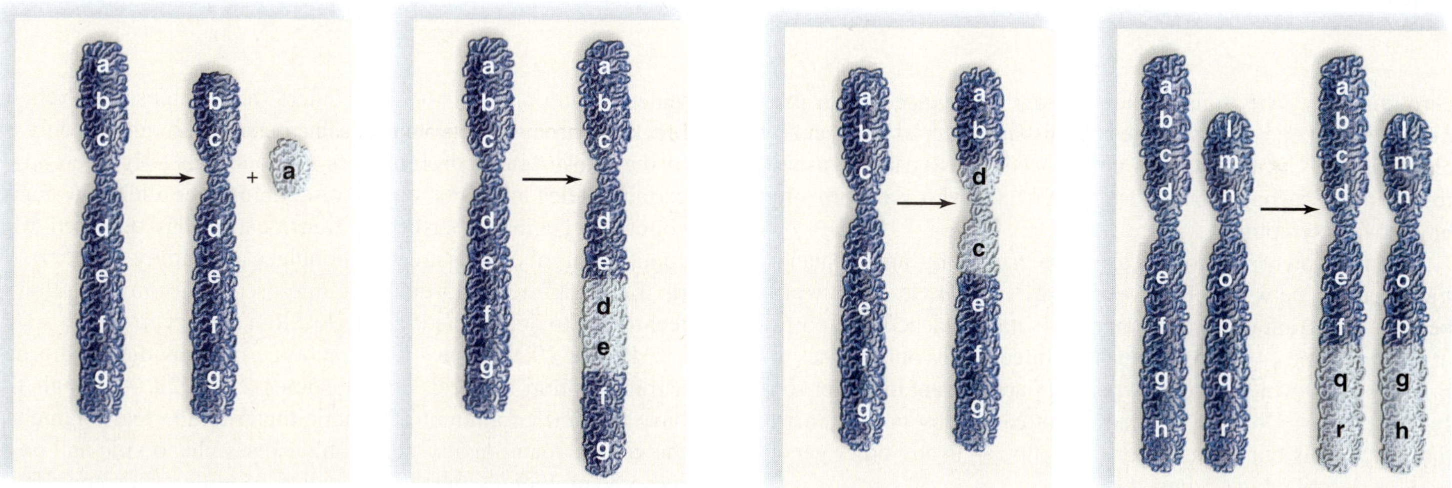

a. Deletion b. Duplication c. Inversion d. Translocation

Figure 10.12 Types of chromosomal mutations. **a.** Deletion is the loss of a chromosome piece. **b.** Duplication occurs when the same piece is repeated within the chromosome. **c.** Inversion occurs when a piece of chromosome breaks loose and then rejoins in the reversed direction. **d.** Translocation is the exchange of chromosome pieces between nonhomologous pairs.

a. b.

Figure 10.13 Deletion. **a.** When chromosome 7 loses an end piece, the result is Williams syndrome. **b.** These children, although unrelated, have the same appearance, health, and behavioral problems.

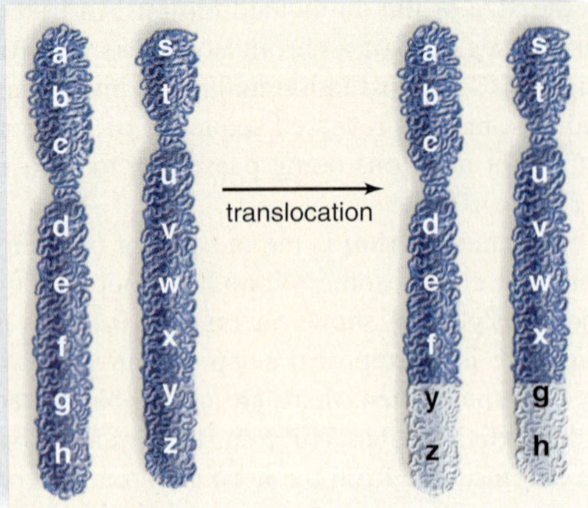

a.

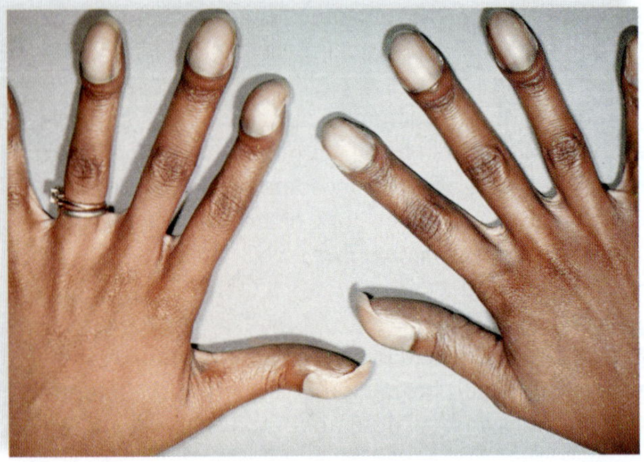

b.

Figure 10.14 Translocation. a. When chromosomes 2 and 20 exchange segments, (b) Alagille syndrome, with distinctive body features, sometimes results because of organ malfunction caused by the chromosome 20 translocation.

chromosomes. The translocation is now *unbalanced*, with extra material from one chromosome and missing material from another chromosome. Embryos with unbalanced translocations usually result in miscarriage, but those individuals who are born often have severe symptoms.

Some Down syndrome cases are caused by an unbalanced translocation between chromosomes 21 and 14. In other words, because a portion of chromosome 21 is now attached to a portion of chromosome 14, the individual has three copies of the genes that bring about Down syndrome when they are present in triplet copy. In these cases, Down syndrome is not caused by nondisjunction during meiosis but is passed on normally, as is any other genetic trait, as described in Chapter 11.

Human Syndromes

Changes in chromosome structure occur in humans and lead to various syndromes, many of which are just now being discovered. Sometimes changes in chromosome structure can be detected in humans by doing a karyotype. They may also be discovered by studying the inheritance pattern of a disorder in a particular family.

Deletion Syndromes. Williams syndrome occurs when chromosome 7 loses a tiny end piece (see Fig. 10.13). Children who have this syndrome look like pixies, with turned-up noses, wide mouths, a small chin, and large ears. Although their academic skills are poor, they exhibit excellent verbal and musical abilities. The gene that governs the production of the protein elastin is missing, and this affects the health of the cardiovascular system and causes their skin to age prematurely. Such individuals are very friendly but need an ordered life, perhaps because of the loss of a gene for a protein that is normally active in the brain.

Cri du chat ("cat's cry") syndrome is seen when chromosome 5 is missing an end piece. The affected individual has a small head, an intellectual disability, and facial abnormalities. Abnormal development of the glottis and larynx results in the most characteristic symptom—the infant's cry resembles that of a cat.

Translocation Syndromes. A person who has both of the chromosomes involved in a translocation has the normal amount of genetic material and is healthy, unless the chromosome exchange breaks an allele into two pieces. The person who inherits only one of the translocated chromosomes no doubt has only one copy of certain alleles and three copies of certain other alleles. A genetic counselor begins to suspect a translocation has occurred when spontaneous abortions are commonplace and family members suffer from various syndromes. A microscopic technique allows a technician to determine if a translocation has occurred.

Figure 10.14 shows the phenotype of individuals who have a translocation between chromosomes 2 and 20. Although they have the normal amount of genetic material, the rearrangment of the genetic material also commonly causes phenotypic and physiological problems, collectively called Alagille syndrome. People with this syndrome ordinarily have a deletion on chromosome 20 (Fig. 10.14*a*), which can lead to a congenital heart condition called tetralogy of fallot, which produces digital clubbing of the fingers (Fig. 10.14*b*). Liver problems are also common in Alagille syndrome. The symptoms of Alagille syndrome range from mild to severe, so some people may not be aware they have the syndrome until after they've had children.

Translocations can also be responsible for a variety of other disorders, including certain types of cancer. In the 1970s, new staining techniques identified that a translocation from a portion of chromosome 22 to chromosome 9 was responsible for many cases of chronic myelogenous leukemia. This translocated chromosome was called Philadelphia chromosome. In Burkitt lymphoma, a cancer common in children in equatorial Africa, a large tumor develops from lymph glands in the region of the jaw. This disorder involves a translocation from a portion of chromosome 8 to chromosome 14.

Check Your Progress 10.6

1. Explain the kinds of changes in chromosome number that can be caused by nondisjunction in meiosis.
2. Examine why sex chromosome aneuploidy is more common than autosome aneuploidy.
3. Compare structural changes between an inversion and a translocation.

CONNECTING *the* CONCEPTS *with the* THEMES

Evolution

- Meiosis is a key process in sexual reproduction, with both evolutionary costs and benefits.
- Although mutations during meiosis can produce faulty gametes, greater genetic diversity is also possible, and this diversity can benefit the species.
- Understanding chromosomal behavior during meiosis is critical to understanding how genes segregate during gamete formation and how this contributes to patterns of inheritance.

Nature of Science

- Researching meiosis provides a deeper understanding of chromosome transmission, leading to the possibility of treating genetic diseases, such as Down syndrome, prior to birth.
- Greater understanding of how chromosomes match up and exchange genetic information during meiosis can help us understand one of the foundations of genetic diversity.

Biological Systems

- Meiosis and mitosis use similar processes to separate chromosomes.
- Like the cell cycle and mitosis, meiosis is tightly controlled. Regulatory mechanisms ensure that homologous chromosomes first pair and then separate during the first division and that sister chromatids do not separate until the second division.
- Meiosis occurs only in certain types of cells during a restricted period of an organism's life span.

MEDIA STUDY TOOLS http://connect.mheducation.com

SMARTBOOK® Study smart. SmartBook helps you maximize your study time by identifying what you know and don't know. The recharge feature in SmartBook helps you prepare for the test by identifying the material that you are most likely to forget.

connect BIOLOGY Find even more resources to learn the chapter concepts including animations, tutorial videos, and interactive practice questions.

MP3 Files	Animations	3D Animations	Tutorials
10.1 Meiosis **10.3** Meiosis	**10.1** How Meiosis Works **10.2** Meiosis and Crossing-Over • Random Orientation of Chromosomes During Meiosis **10.3** Stages of Meiosis • Meiosis I • Meiosis II **10.4** Comparison of Meiosis and Mitosis **10.5** Spermatogenesis **10.6** Changes in Chromosome Structure	**10.3** Meiosis: Interphase and Meiosis I • Meiosis: Meiosis II	**10.3** Meiosis

SUMMARIZE

10.1 Overview of Meiosis

Meiosis reduces the chromosome number of a cell from its **diploid (2n)** number to its **haploid (n)** number. The process ensures that the chromosome number in offspring stays constant generation after generation. In many species, including animals, **meiosis** is associated with the production of **gametes** for sexual reproduction. Gametes are haploid; on fertilization, they restore the diploid chromosome number in the **zygote.**

The nucleus contains pairs of chromosomes, called **homologous chromosomes** (homologues). **Homologues** contain similar genes, but these genes may have different variations, called **alleles.**

Meiosis requires two cell divisions and results in four daughter cells. Replication of DNA takes place before meiosis begins. During meiosis I, a **synaptonemal complex** forms, allowing the homologues to undergo **synapsis,** resulting in a **bivalent.** The bivalent chromosomes align independently at the metaphase plate. The daughter cells receive one member of each pair of homologous chromosomes. There is no replication of DNA during interkinesis, the pause between meiosis I and II. During meiosis II, the sister chromatids separate, becoming daughter chromosomes that move to opposite poles, as they do in mitosis. The four daughter cells contain the haploid number of chromosomes and only one of each kind.

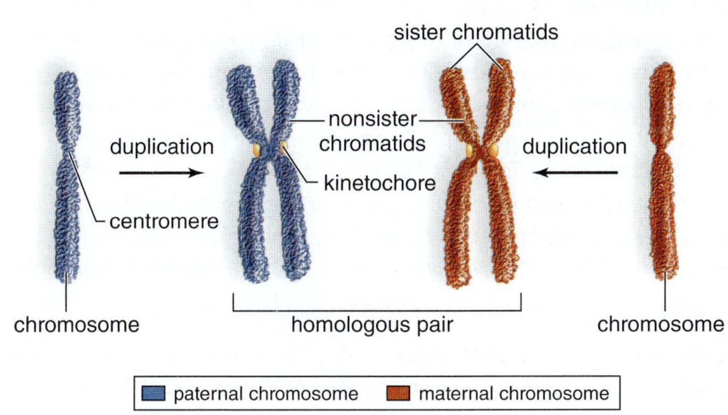

10.2 Genetic Variation

Sexual reproduction ensures that the offspring have a different genetic makeup than the parents and increases the ability of a species to survive. Meiosis contributes to genetic variability in two ways: crossing-over and **independent assortment** of the homologous chromosomes. When homologous chromosomes lie side by side during synapsis, **genetic recombination** occurs, and nonsister chromatids may exchange genetic material. Due to crossing-over, the chromatids that separate during meiosis II have a different combination of genes. When the homologous chromosomes align at the metaphase plate during metaphase I, either the maternal or the paternal chromosome can be facing either pole. Therefore, there will be all possible combinations of chromosomes in the gametes. Random **fertilization** of an egg by a sperm further increases the level of genetic variation.

10.3 The Phases of Meiosis

Meiosis I, which splits pairs of homologous chromosomes and reduces the chromosome number from 2n to n, is divided into four phases:

- Prophase I—bivalents form, and crossing-over occurs as chromosomes condense; the nuclear envelope fragments.
- Metaphase I—bivalents independently align at the metaphase plate.
- Anaphase I—homologous chromosomes separate, and duplicated chromosomes move to poles.
- Telophase I—nuclei become haploid, having received one duplicated chromosome from each homologous pair. **Interkinesis** follows telophase I.

Meiosis II, which reduces the amount of DNA in half from previously replicated haploid chromosomes, is divided into four phases:

- Prophase II—chromosomes condense, and the nuclear envelope fragments.
- Metaphase II—the haploid number of still duplicated chromosomes align at the metaphase plate.
- Anaphase II—sister chromatids separate, becoming daughter chromosomes that move to the poles.
- Telophase II—four haploid daughter cells are genetically different from the parent cell.

10.4 Meiosis Compared to Mitosis

Meiosis and mitosis can be compared in this manner:

Meiosis I	Mitosis
Prophase	
Pairing of homologous chromosomes	No pairing of chromosomes
Metaphase	
Bivalents at metaphase plate	Duplicated chromosomes at metaphase plate
Anaphase	
Homologous chromosomes separate and move to poles	Sister chromatids separate, becoming daughter chromosomes that move to the poles
Telophase	
Daughter nuclei have the haploid number of chromosomes	Daughter nuclei have the parent cell chromosome number

Meiosis II is like mitosis, except that the nuclei are haploid.

10.5 The Cycle of Life

Meiosis occurs in any **life cycle** that involves sexual reproduction. In the animal life cycle, only the gametes are haploid; in plants, the life cycle alternates between **gametophyte** and **sporophyte** generations. Meiosis produces spores, which develop into a multicellular haploid adult that produces the gametes. In single-celled protists and fungi, the zygote undergoes meiosis, and spores become a haploid adult that gives rise to gametes.

During the life cycle of humans and other animals, meiosis is involved in **gametogenesis,** which includes spermatogenesis and oogenesis. Whereas **spermatogenesis** produces four sperm per meiosis, **oogenesis** produces one egg and two or three nonfunctional **polar bodies.** The polar bodies are inactive cells. When a sperm fertilizes an egg, the zygote has the diploid number of chromosomes. Mitosis, which is involved in growth and repair, also occurs during the life cycle of all animals.

10.6 Changes in Chromosome Number and Structure

Euploidy represents the correct number of chromosomes in each cell of the species. **Nondisjunction** during meiosis I or meiosis II may result in **aneuploidy** (extra or missing copies of chromosomes). A **karyotype** may be used to identify if an aneuploid condition exists. **Monosomy** occurs when an individual has only one of a particular type of chromosome ($2n - 1$) and is usually lethal; **trisomy** occurs when an individual has three of a particular type of chromosome ($2n + 1$). Down syndrome is a well-known trisomy in humans resulting from an extra copy of chromosome 21.

Aneuploidy of the sex chromosomes is tolerated more easily than aneuploidy of the autosomes. Turner syndrome, Klinefelter syndrome, poly-X females, and Jacobs syndrome are examples of sex chromosome aneuploidy. **Barr bodies** help identify that extra X chromosomes exist in a cell.

Abnormalities in crossing-over may result in **deletions, duplications, inversions,** and **translocations** within chromosomes. Many human syndromes, including Williams syndrome, cri du chat syndrome, and Alagille syndrome, result from changes in chromosome structure.

ASSESS

Choose the best answer for each question.

10.1 Overview of Meiosis

1. If a parent cell has 16 chromosomes, then each of the daughter cells following meiosis will have
   a. 48 chromosomes.
   b. 32 chromosomes.
   c. 16 chromosomes.
   d. 8 chromosomes.
2. A bivalent is
   a. a homologous chromosome.
   b. the paired homologous chromosomes.
   c. a duplicated chromosome composed of sister chromatids.
   d. the two daughter cells after meiosis I.
   e. the two centrioles in a centrosome.
3. The synaptonemal complex
   a. forms during prophase I of meiosis.
   b. allows synapsis to occur.
   c. forms between homologous chromosomes.
   d. All of these are correct.

10.2 Genetic Variation

4. Crossing-over occurs between
   a. sister chromatids of the same chromosome.
   b. two different kinds of bivalents.
   c. two different kinds of chromosomes.
   d. nonsister chromatids of a bivalent.
   e. two daughter nuclei.
5. Which of the following occurs at metaphase I of meiosis?
   a. independent assortment
   b. crossing-over
   c. interkinesis
   d. formation of new alleles

10.3 The Phases of Meiosis

6. At the metaphase plate during metaphase I of meiosis, there are
   a. chromosomes consisting of one chromatid.
   b. unpaired duplicated chromosomes.
   c. bivalents.
   d. homologous pairs of chromosomes.
   e. Both c and d are correct.
7. At the metaphase plate during metaphase II of meiosis, there are
   a. chromosomes consisting of one chromatid.
   b. unpaired duplicated chromosomes.
   c. bivalents.
   d. homologous pairs of chromosomes.
   e. Both c and d are correct.
8. During which phase of meiosis do homologous chromosomes separate?
   a. prophase I
   b. telophase I
   c. anaphase I
   d. anaphase II

10.4 Meiosis Compared to Mitosis

9. Mitosis ________ chromosome number, whereas meiosis ________ the chromosome number of the daughter cells.
   a. maintains; increases
   b. increases; maintains
   c. increases; decreases
   d. maintains; decreases

For questions 10–13, match the statements that follow to the items in the key. Answers may be used more than once, and more than one answer may be used.

Key:

a. mitosis
b. meiosis I
c. meiosis II
d. Both meiosis I and meiosis II are correct.
e. All of these are correct.

10. Involves pairing of duplicated homologous chromosomes
11. A parent cell with five duplicated chromosomes will produce daughter cells with five chromosomes consisting of one chromatid each.
12. Nondisjunction may occur, causing abnormal gametes to form.
13. Involved in growth and repair of tissues

10.5 The Cycle of Life

14. Polar bodies are formed during the process of
    a. spermatogenesis.
    b. gametophyte formation.
    c. sporophyte formation.
    d. oogenesis.
    e. None of these are correct.
15. In humans, gametogenesis results in the formation of
    a. diploid egg and sperm cells.
    b. gametophytes.
    c. sporophytes.
    d. haploid egg and sperm cells.
    e. a zygote.

10.6 Changes in Chromosome Number and Structure

16. Nondisjunction during meiosis I of oogenesis will result in eggs that have
    a. the normal number of chromosomes.
    b. one too many chromosomes.
    c. one less than the normal number of chromosomes.
    d. Both b and c are correct.
17. In which of the following is genetic material moved between nonhomologous chromosomes?
    a. insertion
    b. nondisjunction
    c. deletion
    d. translocation
    e. inversion
18. Which of the following is not an aneuploid condition?
    a. Turner syndrome
    b. Down syndrome
    c. Alagille syndrome
    d. Klinefelter syndrome

ENGAGE

LEARNSMART LABS®

The following LearnSmart Labs contain exercises that are related to the content of this chapter:

- Mitosis & Meiosis

THINKING SCIENTIFICALLY

1. Why is the first meiotic division considered to be the reduction division for chromosome number?
2. Although most men with Klinefelter syndrome are infertile, some are able to father children. It was found that most fertile individuals with Klinefelter syndrome exhibit mosaicism, in which some cells are normal (46, XY) but others contain the extra chromosome (47, XXY). How might this mosaicism come about? What effects might result?
3. A man has a balanced translocation between chromosomes 2 and 6. If he reproduces with a normal woman, might the child have the same translocation? Why or why not?
4. In a karyotype, larger chromosomes are given lower numbers, so chromosome 1 is larger than chromosome 21. Trisomy conditions, such as trisomy 21, are usually only viable in the smaller chromosomes. Why do you think that this is the case?

11

Mendelian Patterns of Inheritance

Chapter Outline

11.1 Gregor Mendel 187

11.2 Mendel's Laws 188

11.3 Mendelian Patterns of Inheritance and Human Disease 194

11.4 Beyond Mendelian Inheritance 198

Before You Begin

Before beginning this chapter, take a few moments to review the following discussions.

Chapter 2 What are the roles of proteins and nucleic acids in a cell?

Section 10.3 How are chromosomes segregated during the process of meiosis?

Figure 10.9 What genetic changes are possible in gametes when chromosomes fail to segregate properly?

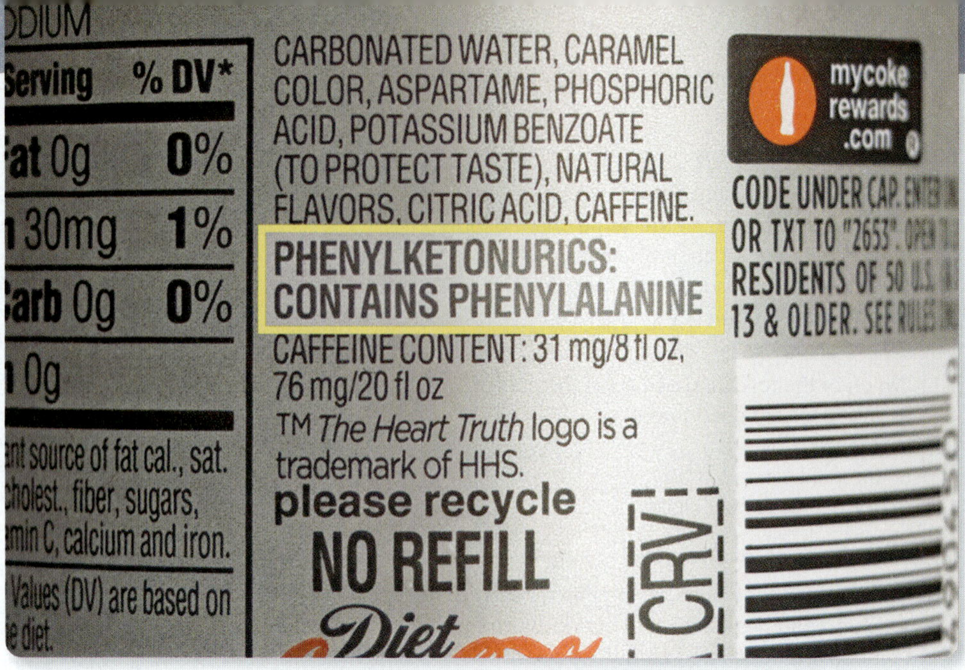

The inability to process phenylalanine is an example of a genetic disorder.

Have you ever looked at the back of a can of diet soda and wondered what the warning "Phenylketonurics: Contains Phenylalanine" means? Phenylalanine is an amino acid that is found in many foods and the artificial sweetener aspartame. For most people, phenylalanine does not present any problems, since any excess is broken down by enzymes in the body. However, some people, called phenylketonurics, lack a functional copy of this enzyme and thus are unable to break down the phenylalanine. The excess may accumulate in the body, causing a variety of nervous system disorders. An estimated 1 in 10,000 people in the United States are phenylketonurics.

Like the rest of us, you are the product of your family tree. The DNA you inherit from your parents directly affects the proteins that enable your body to function properly. Rare genetic disorders, such as phenylketonuria, pique our curiosity about how traits are inherited from one generation to the next. The process of meiosis can be used to predict the inheritance of a trait, and the genetic diversity produced through meiosis can sometimes lead to phenylketonuria.

Through the patterns of inheritance first described by Mendel, you will learn that certain traits, such as phenylketonuria, are recessive and that it takes two nonfunctional copies of that gene before you are affected. This chapter will introduce you to observable patterns of inheritance, including some human genetic disorders, such as phenylketonuria.

As you read through this chapter, think about the following questions:

1. What patterns of inheritance apply to phenylketonuria?
2. How does meiosis help predict the probability of producing gametes and inheriting a trait?

Following *the* Themes

CHAPTER 11 MENDELIAN PATTERNS OF INHERITANCE

UNIT 2 GENETIC BASIS OF LIFE

Evolution	Inheritance of genes within a population is a cornerstone of a species' ability to change over time.
Nature of Science	Gregor Mendel's scientific approach allowed him to establish the basic principles of heredity.
Biological Systems	Inheriting abnormal genes can affect many aspects of body function.

11.1 Gregor Mendel

Learning Outcomes

Upon completion of this section you should be able to

1. Describe how Mendel's scientific approach enabled his genetic experiments to be successful.
2. Contrast blending and the particulate concept of inheritance.

The science of genetics explains the stability of inheritance (why you are human, as are your parents) as well as variations between offspring from one generation to the next (why you have a different combination of traits than your parents). Virtually every culture in history has attempted to explain observed inheritance patterns. An understanding of these patterns has always been important to agriculture, animal husbandry (the science of breeding animals), and medicine.

The Blending Concept of Inheritance

When Gregor Mendel began his work, most plant and animal breeders acknowledged that both sexes contribute equally to a new individual. They thought that parents of contrasting appearance always produced offspring of intermediate appearance. This concept, called the *blending concept of inheritance,* meant that a cross between plants with red flowers and plants with white flowers would yield only plants with pink flowers. When red and white flowers reappeared in future generations, the breeders mistakenly attributed this to instability in the genetic material.

The blending concept of inheritance offered little help to Charles Darwin, the father of evolution (see Chapter 15). Darwin's theory of natural selection was based on the fact that populations possessed variation that allowed for certain individuals to have a selective advantage. According to the blending concept, over time variation would decrease as individuals became more alike in their traits.

Mendel's Particulate Theory of Inheritance

Gregor Mendel was an Austrian monk who developed a *particulate theory of inheritance* after performing a series of ingenious experiments in the 1860s (Fig. 11.1*a*). Mendel studied science and mathematics at the University of Vienna, and at the time of his research in genetics, he was a substitute natural science teacher at a local high school.

Mendel was a successful scientist for several reasons. First, he was one of the first scientists to apply mathematics to biology. Most likely his background in mathematics prompted him to apply statistical methods and the laws of probability to his breeding experiments. He was also a careful, deliberate scientist who followed the scientific method very closely and kept very detailed, accurate records. He prepared for his experiments carefully and conducted many preliminary studies with various animals and plants.

Mendel's theory of inheritance is called a particulate theory because it is based on the existence of minute particles, or hereditary units, we now call genes. Inheritance involves the reshuffling of the same genes from generation to generation. The two laws he proposed, the law of segregation and the law of independent assortment, which we will discuss shortly, describe the behavior of these particulate units of heredity as they are passed from one generation to the next. While Mendel did not know of DNA or genetic material, his theories have been well supported by countless experiments of geneticists and molecular biologists.

a. b.

Figure 11.1 Gregor Mendel, 1822–1884. **a.** Mendel grew and tended the pea plants (**b**) he used for his experiments. His experimental approach allowed him to develop several laws of inheritance.

Mendel Worked with the Garden Pea

Mendel's preliminary experiments prompted him to choose the garden pea, *Pisum sativum* (Fig. 11.1*b*), as his experimental organism. The garden pea was a good choice for many reasons. The plants were easy to cultivate and had a short generation time. Although peas normally self-pollinate (pollen only goes to the same flower), they could be cross-pollinated by hand by transferring pollen from the anther (male part of a flower) to the stigma (female part of a flower).

Many varieties of peas were available, and Mendel chose 22 for his experiments. When these varieties self-pollinated, over generations they became *true-breeding*—meaning that all the offspring were the same and exactly like the parent plants. Unlike his predecessors, Mendel studied the inheritance of relatively simple and discrete traits that were not subjective and were easy to observe, such as seed shape, seed color, and flower color. In his crosses, Mendel observed that the offspring did not possess intermediate charactersitics but, rather, were similar in appearance to one of the parents. As we will see, this disproved the blending concept and supported the particulate theory of inheritance.

Check Your Progress 11.1

1. Explain the difference between the particulate theory of inheritance and the blending concept.
2. Explain why the garden pea was a good choice for Mendel's experiments.

11.2 Mendel's Laws

Learning Outcomes

Upon completion of this section, you should be able to

1. Explain Mendel's law of segregation and law of independent assortment.
2. Compare and contrast dominant alleles with recessive alleles and their relation to genotype and phenotype.
3. Use a Punnett square and the law of probability to predict the chances of producing gametes and offspring.

After ensuring that his pea plants were true-breeding—for example, that his tall plants always had tall offspring and his short plants always had short offspring—Mendel was ready to perform cross-pollination experiments (Fig. 11.2). These crosses allowed Mendel to formulate his law of segregation.

Law of Segregation

For these initial experiments, Mendel chose varieties that differed in only one trait (e.g., plant height). If the blending theory of inheritance were correct, the cross should yield plants with an intermediate appearance of medium height compared to the parents, which were all tall or all short.

Mendel's Experimental Design and Results

Mendel called the original, true-breeding all tall or all short parents the *P generation.* The first generation of offspring were called the F_1, or *filial generation* (L. *filius,* "sons and daughters") (Fig. 11.3). He performed reciprocal crosses: First he dusted the pollen of tall plants onto the stigmas of short plants, and then he dusted the pollen of short plants onto the stigmas of tall plants. In both cases, all F_1 offspring resembled the tall parent.

Certainly, these results were contrary to those predicted by the blending theory of inheritance. Rather than being intermediate, all the F_1 plants were tall and resembled only one parent. Did these results mean that the other characteristic (shortness) had disappeared permanently? Apparently not, because when Mendel

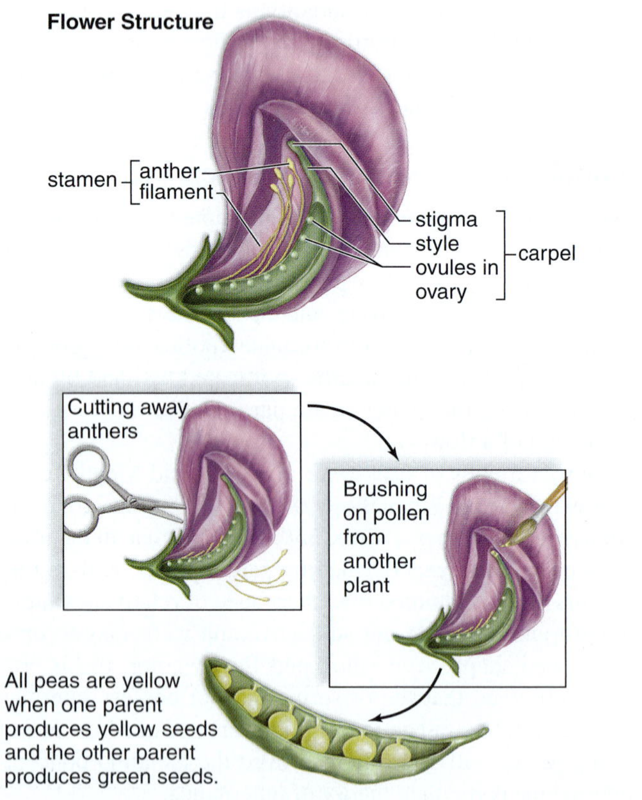

Figure 11.2 Garden pea anatomy. In the garden pea, *Pisum sativum,* pollen grains produced in the anther contain sperm, and ovules in the ovary contain eggs. When Mendel performed crosses, he brushed pollen from one plant onto the stigma of another plant. This cross-pollination allowed sperm to fertilize eggs and ovules to develop into seeds (peas). The open pod shows the seed color trait that resulted from a cross between plants with yellow seeds and plants with green seeds.

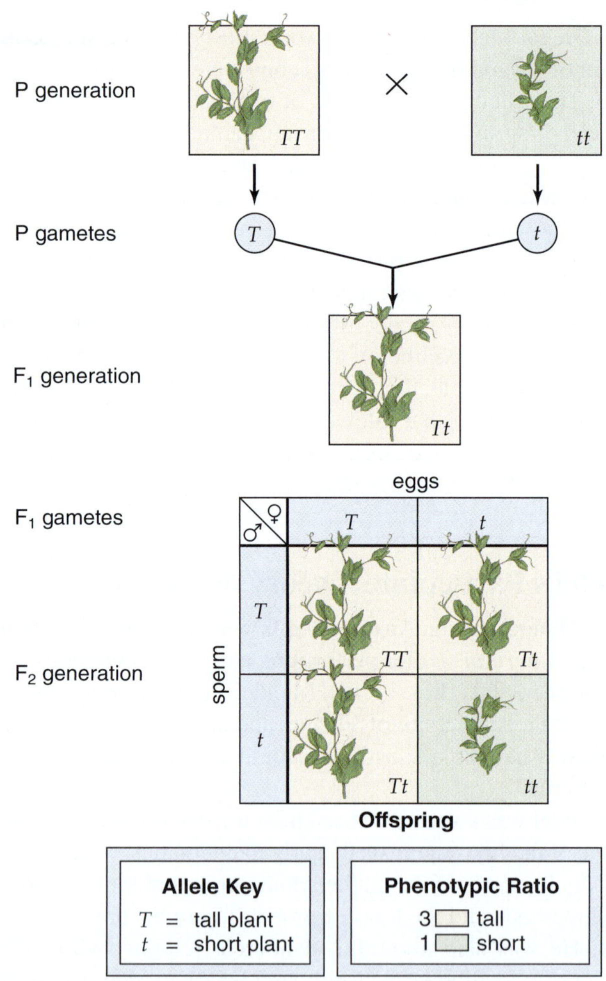

Figure 11.3 Monohybrid cross done by Mendel. The P generation pea plants differ in only one trait—length of the stem. The F_1 generation plants are all tall, but the factor for short has not disappeared, because 1⁄4 of the F_2 generation plants are short. The 3:1 ratio allowed Mendel to deduce that individuals have two discrete and separate genetic factors for each trait.

allowed the F_1 plants to self-pollinate, three-fourths of the next generation of offspring, or *F_2 generation,* were tall and one-fourth were short, a 3:1 ratio (Fig. 11.3).

Mendel inferred that the F_1 plants were able to pass on a factor for shortness—it didn't disappear; it just skipped a generation. Because the F_1 plants were tall but clearly still contained the shortness characteristic, Mendel deduced that tallness was dominant to shortness (Fig. 11.3).

Mendel counted many plants in his plant height and other experiments. When he allowed the F_1 pea plants (which were all tall but carried the characteristic for shortness) to self-fertilize and produce offspring, he counted a total of 1,064 plants, of which 787 were tall and 277 were short. This type of experiment is called a **monohybrid cross** (L. *mono,* "single"; *hybrida,* "mixture"), because it is a cross of a single trait (plant height) with organisms that are a hybrid (tall and short characteristics). In fact, in all monohybrid crosses that he performed for the traits shown in Figure 11.4, he found a 3:1 ratio in the F_2 generation. The characteristic for shortness that had disappeared in the F_1 generation reappeared in one-fourth of the F_2 offspring. In a monohybrid cross of two heterozygotes, assuming a simple dominant/recessive relationship, the expected phenotypic ratio is 3:1.

Mendel's Conclusion

Mendel's mathematical approach led him to interpret his results differently than previous breeders. He knew that the same ratio was obtained among the F_2 generation time and time again when he did a monohybrid cross involving one of the seven traits he was studying (Fig. 11.4). Eventually, Mendel arrived at this explanation: A 3:1 ratio among the F_2 offspring was possible if (1) the F_1 parents contained two separate copies of each hereditary factor, one of these being dominant and the other recessive; (2) the factors separated when the gametes were formed, and each gamete carried only one copy of each factor; and (3) random fusion of all possible gametes occurred upon fertilization. Only in this way could shortness reoccur in the F_2 generation. Thinking this,

Trait	Characteristics: Dominant		Characteristics: Recessive		F_2 Results: Dominant	F_2 Results: Recessive	F_2 Results: Ratio
Stem length	Tall		Short		787	277	2.84:1
Pod shape	Inflated		Constricted	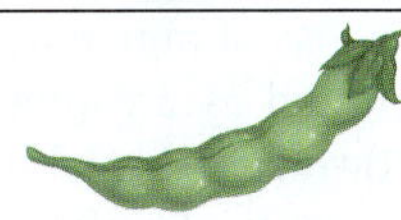	882	299	2.95:1
Seed shape	Round		Wrinkled	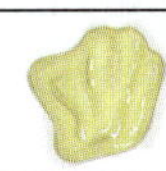	5,474	1,850	2.96:1
Seed color	Yellow		Green		6,022	2,001	3.01:1
Flower position	Axial		Terminal		651	207	3.14:1
Flower color	Purple		White		705	224	3.15:1
Pod color	Green		Yellow	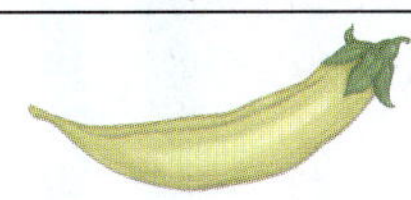	428	152	2.82:1
				Totals:	14,949	5,010	2.98:1

Figure 11.4 Relationship between observed phenotype and F_2 offspring. Mendel was fortunate in choosing the pea plant, because the traits he observed were quite distinct and easily classified. After crossing F_1 hybrids and counting hundreds of F_2 pea plants for each trait, Mendel discovered that each showed a 3:1 ratio.

Mendel arrived at the first of his laws of inheritance—the law of segregation—which is a cornerstone of his particulate theory of inheritance:

The **law of segregation** states the following:

- Each individual has two factors for each trait.
- The factors segregate (separate) during the formation of the gametes.
- Each gamete contains only one factor from each pair of factors.
- Fertilization gives each new individual two factors for each trait.

Mendel's Cross as Viewed by Modern Genetics

We now know that the traits Mendel studied are controlled by single genes. These genes occur on a homologous pair of chromosomes at a particular location, called the gene **locus** (Fig. 11.5). Alternative versions of a gene are called **alleles** (Gk. *allelon,* "reciprocal, parallel"). A **dominant allele** will mask the expression of a **recessive allele** when they are together in the same organism. The word *dominant* is not meant to imply that the dominant allele is better or stronger than the recessive allele. In both cases, these alleles represent DNA sequences that code for proteins. Often, the dominant allele codes for the protein associated with the normal function of the trait within the cell (such as the production of pigment), while the recessive allele represents a "loss of function," meaning that it codes for a protein that has an altered function or no function within the cell (such as a loss of pigment).

In many cases, the dominant allele is identified by a capital letter, the recessive allele by the same letter but lowercase. Usually, the first letter designating a trait is chosen to identify the allele. Using the plant height example, there is an allele for tallness (*T*) and an allele for shortness (*t*).

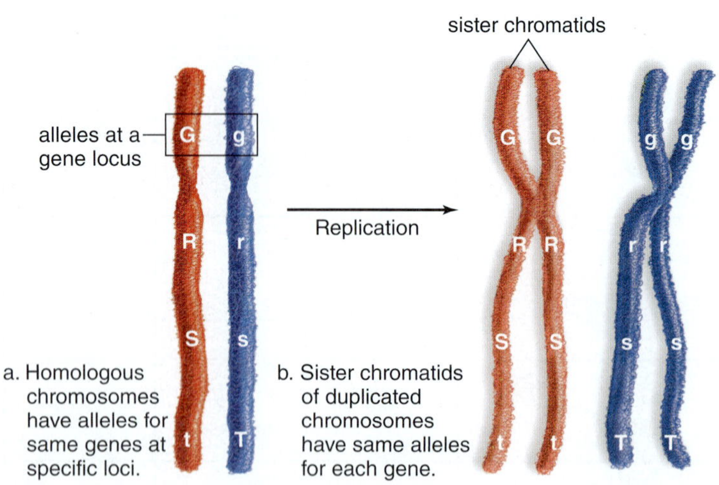

Figure 11.5 Homologous chromosomes. **a.** The letters represent alleles—that is, alternative forms of a gene. Each allelic pair, such as *Gg* or *Tt*, is located on homologous chromosomes at a particular physical location called a gene locus. **b.** Sister chromatids carry the same alleles in the same order. Proteins made from each allele determine the observable traits.

One way to view the outcome of a genetic cross is to use a Punnett square (see Fig. 11.3), in which all possible types of sperm are lined up vertically and all possible types of eggs are lined up horizontally (or vice versa), and every possible combination of gametes occurs within the squares. In Mendel's cross, the original parents (P generation) were true-breeding; therefore, the tall plants had two alleles for tallness (*TT*), and the short plants had two alleles for shortness (*tt*). When an organism has two identical alleles, as these had, we say it is **homozygous** (Gk. *homo,* "same"). Because the parents were homozygous, all gametes produced by the tall plant contained the allele for tallness (*T*), and all gametes produced by the short plant contained an allele for shortness (*t*).

After cross-pollination between different pea plants, all the individuals of the resulting F_1 generation had one allele for tallness and one for shortness (*Tt*). When an organism has two different alleles at a gene locus, we say that it is **heterozygous** (Gk. *hetero,* "different"). Although the plants of the F_1 generation had one of each type of allele, they were all tall. The allele that is expressed in a heterozygous individual is the dominant allele. The allele that is not expressed in a heterozygote is the recessive allele. This explains why shortness, the recessive trait, skipped a generation in Mendel's experiment.

In Chapter 10, we observed that meiosis is the type of cell division that reduces the chromosome number from diploid (2n) to haploid (n). During meiosis I, the members of bivalents (homologous chromosomes, each having sister chromatids) separate. This means that the two alleles for each gene separate from each other during meiosis (see Fig. 11.7). Therefore, the process of meiosis gives an explanation for Mendel's law of segregation, as well as why only one allele for each trait is in a gamete.

Continuing with the discussion of Mendel's cross (see Fig. 11.3), the F_1 plants produce gametes in which 50% have the dominant allele *T* and 50% have the recessive allele *t*. During the process of fertilization, we assume that all types of sperm (*T* or *t*) have an equal chance to fertilize all types of eggs (*T* or *t*). When this occurs, such a monohybrid cross always produces a 3:1 (dominant-to-recessive) ratio among the offspring. Figure 11.4 gives Mendel's results for several monohybrid crosses, and you can see that the results were always close to 3:1.

Genotype Versus Phenotype

It is obvious from our discussion that two organisms with different allelic combinations for a trait can have the same outward appearance. For example, pea plants with both the *TT* and *Tt* combinations of alleles are tall. For this reason, it is necessary to distinguish between the alleles present in an organism and the appearance of that organism.

The word **genotype** (Gk. *genos,* "birth, origin") refers to the alleles an individual receives at fertilization. Genotype may be indicated by letters or by short, descriptive phrases and represents the DNA sequence for a particular gene. Genotype *TT* is called homozygous dominant, and genotype *tt* is called homozygous recessive. Genotype *Tt* is called heterozygous. These refer to the different ways that alleles can be combined in a cell.

The word **phenotype** (Gk. *phaino,* "appear") refers to the physical appearance of an individual, which is determined by the proteins produced by the corresponding alleles. A homozygous dominant (*TT*) individual and a heterozygous (*Tt*) individual both show the dominant phenotype and are tall, because they make fully functional proteins that build the tall trait, while a homozygous recessive individual that shows the recessive phenotype and makes less or nonfunctional protein for that trait is short. Thus, the DNA that makes up the genotype produces the proteins that make up the phenotype.

Mendel's Law of Independent Assortment

Mendel performed a second series of crosses in which true-breeding pea plants differed in two traits. For example, he crossed tall plants having green pods with short plants having yellow pods (Fig. 11.6). The F_1 plants showed both dominant characteristics. As before, Mendel then allowed the F_1 plants to self-pollinate. This F_1 cross is known as a **dihybrid cross** (L. *di,* "two"), because the plants are hybrid in two ways. Two possible results could occur in Mendel's F_2 generation:

1. If the dominant factors (*TG*) always segregated into the F_1 gametes together, and the recessive factors (*tg*) always stayed together, then there would be two phenotypes among the F_2 plants—tall plants with green pods and short plants with yellow pods.
2. If the four factors segregated into the F_1 gametes independently, then there would be four phenotypes among the F_2 plants—tall plants with green pods, tall plants with yellow pods, short plants with green pods, and short plants with yellow pods.

Figure 11.6 shows that Mendel observed four phenotypes among the F_2 plants, supporting the second hypothesis. This is how Mendel formulated his second law of heredity—the law of independent assortment:

> The **law of independent assortment** states the following:
> - Each pair of factors segregates (assorts) independently of the other pairs.
> - All possible combinations of factors can occur in the gametes.

Each chromosome carries a large number of alleles; however, the law of independent assortment applies only to alleles on different chromosomes.

We know that the process of meiosis explains why the F_1 plants produced every possible type of gamete and, therefore, four phenotypes appeared among the F_2 generation of Mendel's plants. Figure 11.7 shows a parent cell with two homologous pairs of chromosomes, with alleles *Aa* on one pair and *Bb* on the other pair. Following duplication of the chromosomes during interphase, the parent cell undergoes meiosis I. At metaphase I, the homologous pairs line up independently of one another, such that the chromosomes with *A* alleles have an equal chance of lining up with the *B* alleles or the *b* alleles. The subsequent segregation of the homologous pairs during anaphase I reduces the chromosome number from 2n to n. Because *A* alleles can be sorted with *B* or *b*, and so can the *a* allele, it is possible to create gametes with *AB, Ab, aB,* and *ab* allele combinations with equal probability.

Animation: Random Orientation of Chromosomes During Meiosis

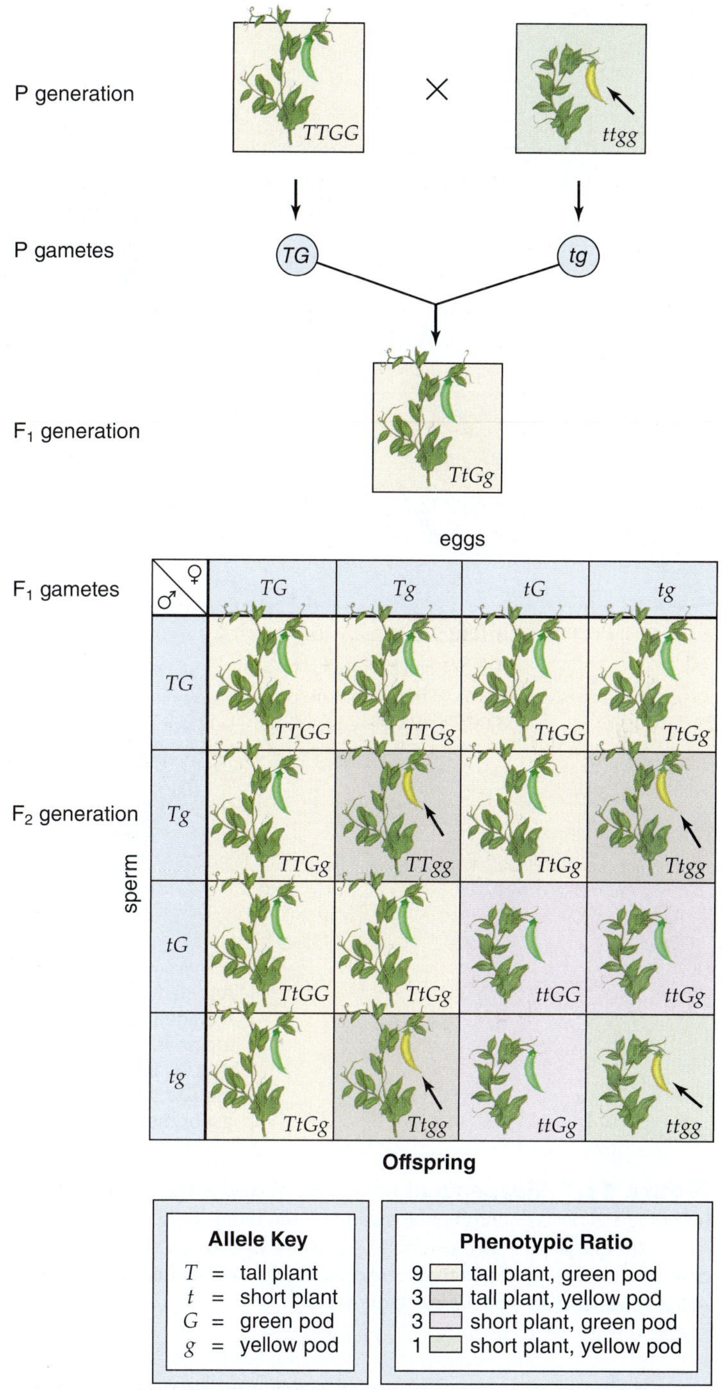

Figure 11.6 Dihybrid cross done by Mendel. P generation plants differ in two traits: length of the stem and color of the pod. The F_1 generation shows only the dominant traits, but all possible phenotypes appear among the F_2 generation, because the F_1 parents are hybrids. The 9:3:3:1 ratio allowed Mendel to deduce that factors segregate into gametes independently of other factors.

Tutorial: Dihybrid Cross

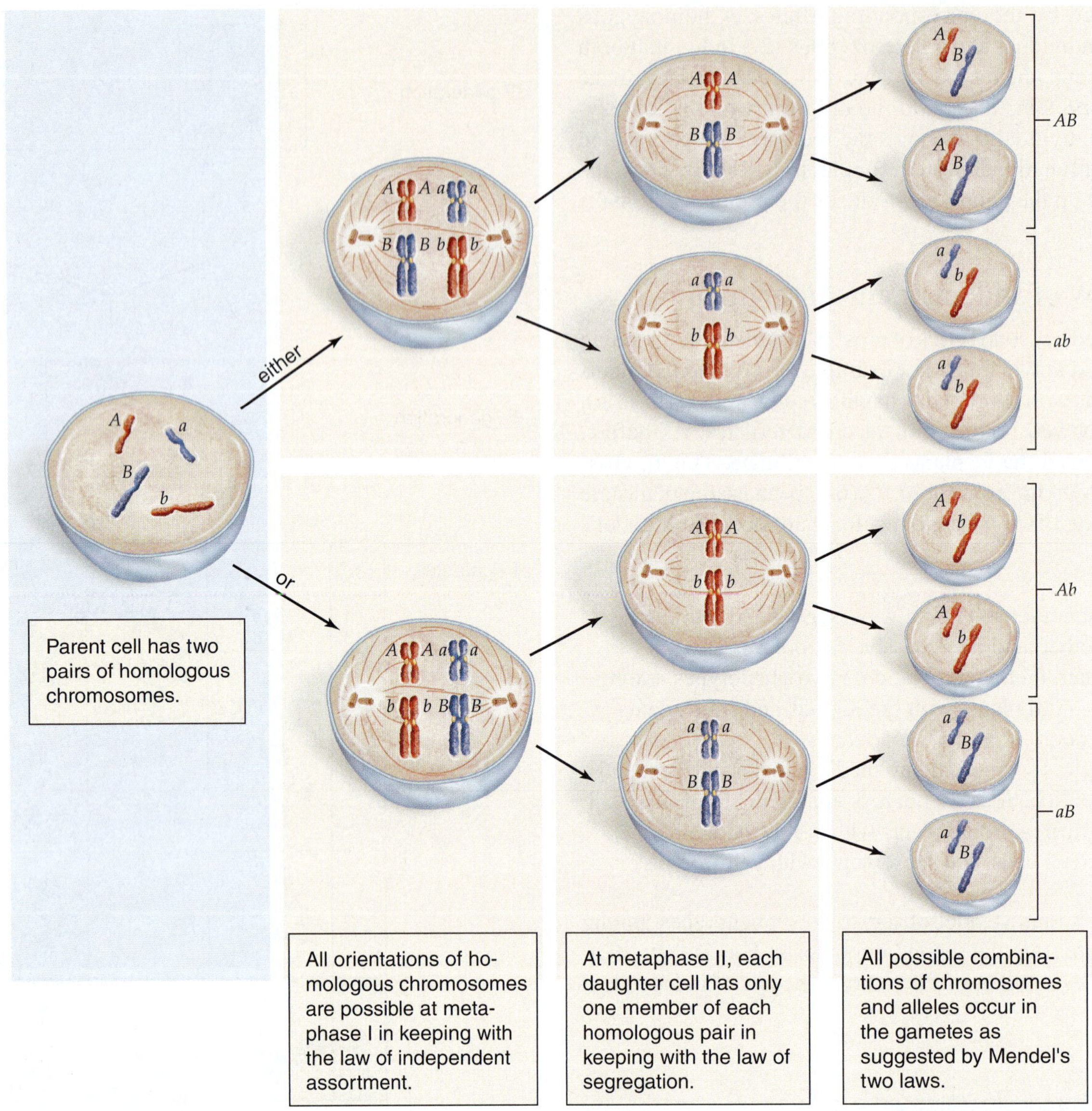

Figure 11.7 Independent assortment and segregation during meiosis. Mendel's laws hold because of the events of meiosis. The homologous pairs of chromosomes line up randomly at the metaphase plate during meiosis I. It doesn't matter which member of a homologous pair faces which spindle pole. In this example, *A* alleles can segregate *B* or *b* alleles. Likewise, *a* alleles can segregate with *B* or *b* alleles. Therefore, the homologous chromosomes, and alleles they carry, segregate independently during gamete formation. All possible combinations of chromosomes and alleles—that is, *AB, Ab, aB, and ab*—occur in the gametes.

The same rule of independent assortment applies for the pea plant example in Figure 11.6. In that case, the possible gametes are the two dominants (such as *TG*), the two recessives (such as *tg*), and the ones that have a dominant and a recessive (such as *Tg* and *tG*). Regardless of whether we are using the A and B chromosome or the T and G chromosome examples, when all possible sperm have an opportunity to fertilize all possible eggs, *the expected phenotypic ratio of a dihybrid cross is always 9:3:3:1.*

Mendel and the Laws of Probability

The diagram we have been using to calculate the results of a cross is called a **Punnett square** (see Figs. 11.3 and 11.6). In a Punnett square, all possible types of sperm are lined up vertically and all possible types of eggs are lined up horizontally (or vice versa), and every possible combination of gametes occurs within the squares. This gives us the ability to easily calculate the chances, or the probability, of genotypes and phenotypes among the offspring. Like flipping a coin, an offspring of the cross illustrated in the Punnett square in Figure 11.8 has a 50% (or ½) chance of receiving an *E* for unattached earlobe or an *e* for attached earlobe from each parent:

The chance of $E = 1/2$
The chance of $e = 1/2$

How likely is it that an offspring will inherit a specific set of two alleles, one from each parent? The *product rule* of probability tells

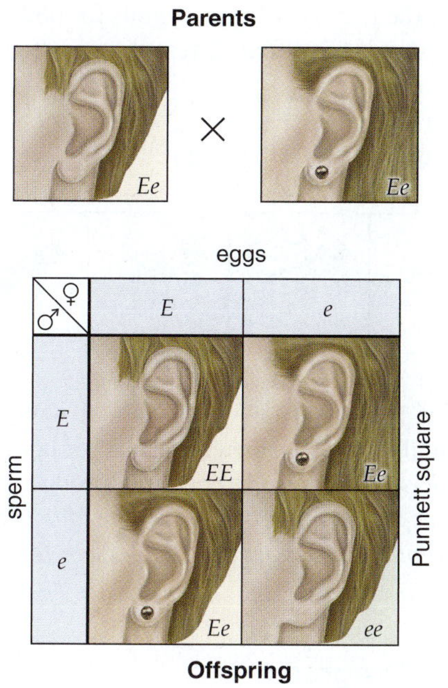

Allele Key

E = unattached earlobes
e = attached earlobes

Phenotypic Ratio

3 unattached earlobes
1 attached earlobes

Figure 11.8 Punnett square. A Punnett square can be used to calculate probable results—in this case, a 3:1 phenotypic ratio.

us that we have to multiply the chances of independent events to get the answer:

1. The chance of *EE* = ½ × ½ = ¼
2. The chance of *Ee* = ½ × ½ = ¼
3. The chance of *eE* = ½ × ½ = ¼
4. Thc chancc of *ee* = ½ × ½ = ¼

The Punnett square does this for us, because we can easily see that each of these is ¼ of the total number of squares.

How do we get the phenotypic results? The *sum rule* of probability tells us that when the same event can occur in more than one way, we can add the results. Because 1, 2, and 3 all result in unattached earlobes, we add them up to know that the chance of unattached earlobes is ¾, or 75%. The chance of attached earlobes is ¼, or 25%. The Punnett square doesn't do this for us—we have to add the results ourselves.

The statement "Chance has no memory" is important when considering inheritance across offspring. Every time a couple produces an offspring, the child has the same chances of inheriting the different allele combinations. Thus, for a heterozygous (*Ee*) couple, each child has a 25% chance of having attached (*ee*) earlobes. Inheriting a recessive trait may not seem significant if we are considering earlobes. However, it becomes important when we consider a recessive genetic disorder such as cystic fibrosis, a debilitating respiratory illness. For a heterozygous couple, there is a 25% chance that their child will inherit two recessive alleles and exhibit the disease. And because each child is an independent event, it is possible that all their children—or none of them—will exhibit cystic fibrosis.

We can use the product rule and the sum rule of probability to predict the results of a dihybrid cross, such as the one shown in Figure 11.6. The Punnett square carries out the multiplication, and we add the results to find that the phenotypic ratio is 9:3:3:1. We expect the same results for every dihybrid cross. Therefore, it is not necessary to do a Punnett square over and over again for either a monohybrid or a dihybrid cross. Instead, we can simply remember the probable results of 3:1 and 9:3:3:1. But we have to remember that the 9 represents the two dominant phenotypes together, the 3s are a dominant phenotype with a hidden recessive, and the 1 stands for the double recessive phenotype.

This tells us the probable phenotypic ratio among the offspring, but not the chances for each possible phenotype. Because the dihybrid Punnett square has 16 squares, the chances are 9/16 for the two dominants together, 3/16 for the dominants with each recessive, and 1/16 for the two recessives together.

Mendel counted the results of many similar crosses to get the probable results, and in the laboratory we, too, have to count the results of many individual crosses to get the probable results for a monohybrid or a dihybrid cross. Why? Consider that each time you toss a coin, you have a 50% chance of getting heads or tails. If you toss the coin only a couple of times, you might very well have heads or tails both times. However, if you toss the coin many times, your results are more likely to approach 50% heads and 50% tails.

Testcrosses

To confirm that the F_1 plants of Mendel's one-trait crosses were, in fact, heterozygous, he crossed his F_1 generation tall pea plants with true-breeding short (homozygous recessive) plants; such a mating is termed a **testcross.** These crosses provided Mendel with further support for his law of segregation.

For the cross in Figure 11.9, Mendel reasoned that half the offspring should be tall and half should be short, producing a 1:1 phenotypic ratio. His results supported the hypothesis that alleles segregate when gametes are formed. In Figure 11.9*a*, the homozygous recessive parent can produce only one type of gamete—*t*—and so the Punnett square has only one column. The use of one column signifies that all the gametes carry a *t. The expected phenotypic ratio for this type of one-trait cross (heterozygous × recessive) is always 1:1.*

One-Trait Testcross

Today, a one-trait testcross is used to determine if an individual with the dominant phenotype is homozygous dominant (e.g., *TT*) or heterozygous (e.g., *Tt*). Because both of these genotypes produce the dominant phenotype, it is not possible to determine the genotype by observation. Figure 11.9*b* shows that if the individual is homozygous dominant, all the offspring will be tall. Each parent has only one type of gamete and, therefore, a Punnett square is not required to determine the results.

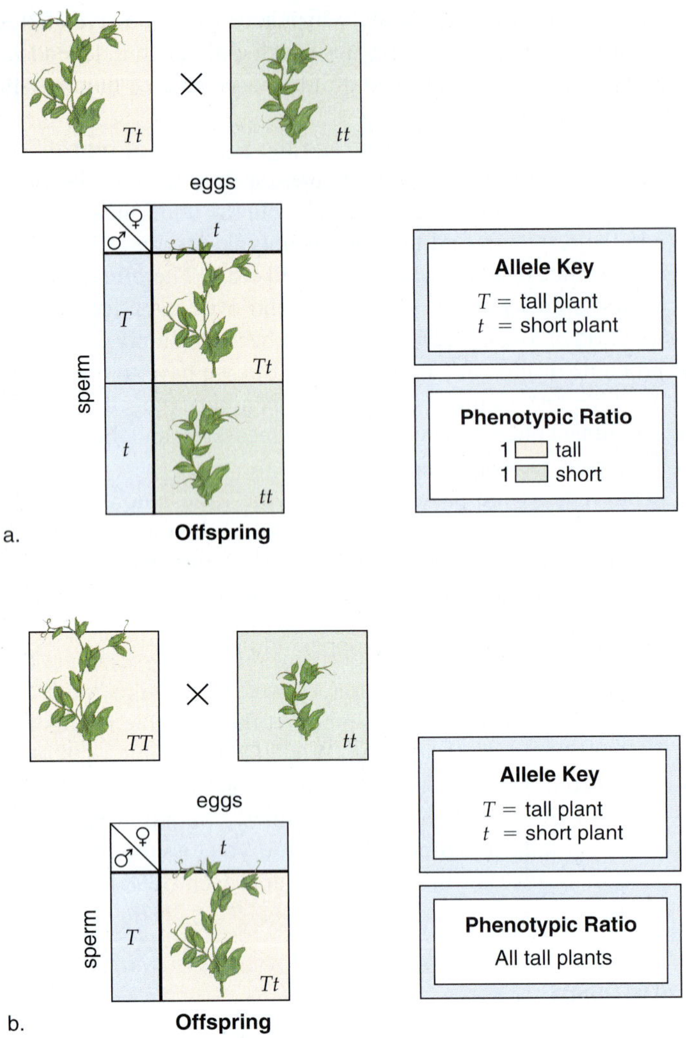

Figure 11.9 One-trait testcrosses. **a.** One-trait testcross when the individual with the dominant phenotype is heterozygous. **b.** One-trait testcross when the individual with the dominant phenotype is homozygous.

Two-Trait Testcross

When doing a two-trait testcross, an individual with the dominant phenotype is crossed with one having the recessive phenotype. Suppose you are working with fruit flies in which

L = long wings	G = gray bodies
l = vestigial (short) wings	g = black bodies

You wouldn't know by examination whether the fly on the left was homozygous or heterozygous for wing and body color. To find out the genotype of the test fly, you cross it with the one on the right. You know by examination that this vestigial-winged and black-bodied fly is homozygous recessive for both traits.

If the test fly is homozygous dominant for both traits with the genotype *LLGG*, it will form only one gamete: *LG*. Therefore, all the offspring from the proposed cross will have long wings and a gray body.

However, if the test fly is heterozygous for both traits with the genotype *LlGg*, it will form four different types of gametes:

Gametes: *LG* *Lg* *lG* *lg*

and can have four different offspring:

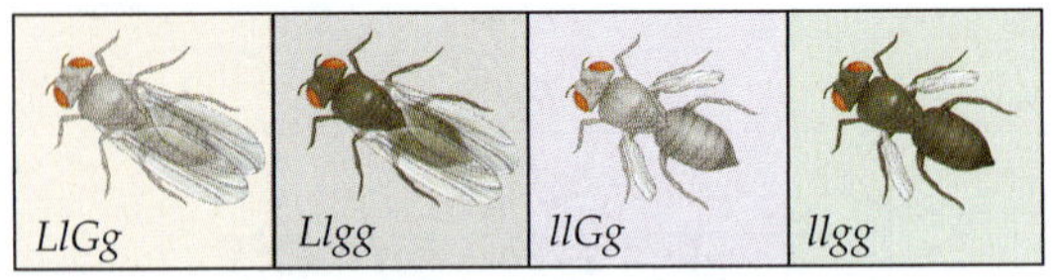

The presence of the offspring with vestigial wings and a black body shows that the test fly is heterozygous for both traits and has the genotype *LlGg*. Otherwise, it could not produce this offspring. In general, *the expected phenotypic ratio for this type of two-trait cross (heterozygous for two traits × recessive for both traits) is always 1:1:1:1.*

Check Your Progress 11.2

1. Summarize how Mendel's laws of independent assortment relate to the process of meiosis.
2. Explain why the *Tt* and *TT* genotypes both have the same phenotype.
3. Calculate the probability of producing an *Aabb* individual from an *AaBb* × *AaBb* cross.

11.3 Mendelian Patterns of Inheritance and Human Disease

Learning Outcomes

Upon completion of this section, you should be able to

1. Distinguish between an autosomal dominant and an autosomal recessive pattern of inheritance.
2. Identify the pattern of inheritance for selected single-gene human disorders.

Many traits and disorders in humans, as well as other organisms, are genetic in origin and follow Mendel's laws. These traits are often controlled by a single pair of alleles on the autosomal chromosomes. An **autosome** is any chromosome other than a sex (X or Y) chromosome. In section 11.4, we will explore patterns of inheritance associated with the sex chromosomes.

Autosomal Patterns of Inheritance

When a genetic disorder is autosomal dominant, the normal allele (*a*) is recessive, and an individual with the alleles *AA* or *Aa* has the disorder. When a genetic disorder is autosomal recessive, the normal allele (*A*) is dominant, and only individuals with the alleles *aa* have the disorder. A pedigree shows the pattern of inheritance for a particular condition; genetic counselors can use a pedigree to determine whether a condition is dominant or recessive. Consider these two possible patterns of inheritance:

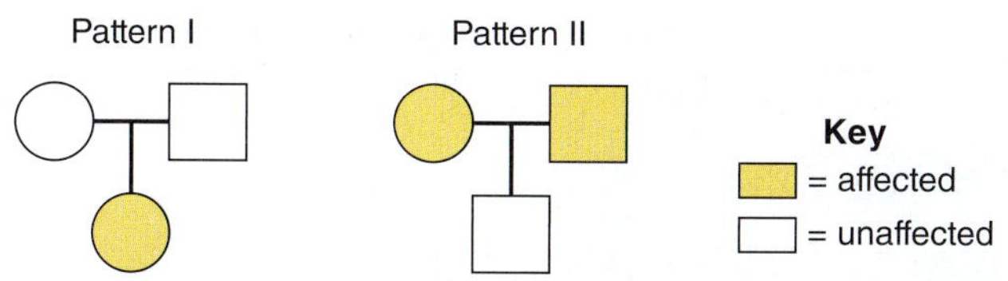

In a pedigree, males are designated by squares and females by circles. Shaded circles and squares are the affected individuals. The shaded boxes do not indicate whether the condition is dominant or recessive, only that the individual exhibits the trait. A line between a square and a circle represents a union. In the patterns above, a vertical line leads to a single child. If there are more children, they are lined up horizontally. In pattern I, the child is affected, but neither parent is; this can happen if the condition is recessive and both parents are *Aa*. Notice that the parents are **carriers,** because they appear normal (do not express the trait) but are capable of having a child with the genetic disorder. In pattern II, the child is unaffected, but the parents are affected. This can happen if the condition is dominant and the parents are *Aa*.

Figure 11.10 shows other ways to recognize an autosomal recessive pattern of inheritance, and Figure 11.11 identifies the characteristics of an autosomal dominant pattern of inheritance. In these pedigrees, generations are indicated by Roman numerals on the left side. Notice in the third generation of Figure 11.10 that two closely related individuals have produced three children, two of which have the affected phenotype. In this case, a double line denotes consanguineous reproduction, or inbreeding, which is reproduction between two closely related individuals. This illustrates that inbreeding significantly increases the chances of children inheriting two copies of a potentially harmful recessive allele.

Autosomal Recessive Disorders

In humans, a number of autosomal recessive disorders have been identified. In this section, we discuss methemoglobinemia, cystic fibrosis, and phenylketonuria.

Methemoglobinemia

Methemoglobinemia is a relatively harmless disorder that results from an accumulation of methemoglobin in the blood. Hemoglobin, the main oxygen-carrying protein in the blood, is usually converted at a slow rate to an alternate form called methemoglobin. Unlike hemoglobin, which is bright red when carrying oxygen, methemoglobin has a bluish color, similar to that of oxygen-poor blood. Although this process is harmless, individuals with methemoglobinemia are unable to clear the abnormal blue protein from their blood, causing their skin to appear bluish-purple (Fig. 11.12).

Methemoglobinemia was documented for centuries, but its exact cause and genetic link remained mysterious until a persistent and determined physician solved the age-old mystery by doing blood tests and pedigree analysis involving a family known as the "blue Fugates" of Troublesome Creek, Kentucky. Enzyme tests indicated that the blue Fugates lacked the enzyme diaphorase, coded for by a gene on chromosome 22. The enzyme normally converts methemoglobin back to hemoglobin.

The physician treated the disorder in a simple but rather unconventional manner. He injected the Fugates with a dye called

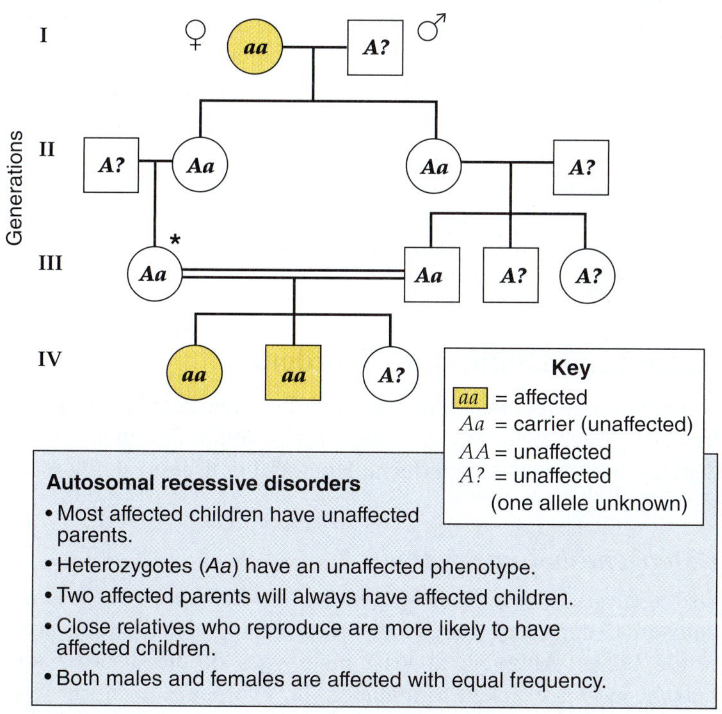

Autosomal recessive disorders

- Most affected children have unaffected parents.
- Heterozygotes (*Aa*) have an unaffected phenotype.
- Two affected parents will always have affected children.
- Close relatives who reproduce are more likely to have affected children.
- Both males and females are affected with equal frequency.

Figure 11.10 Autosomal recessive pedigree. The list gives ways to recognize an autosomal recessive disorder. How would you know the individual at the asterisk is heterozygous? (See Appendix A for the answer.)

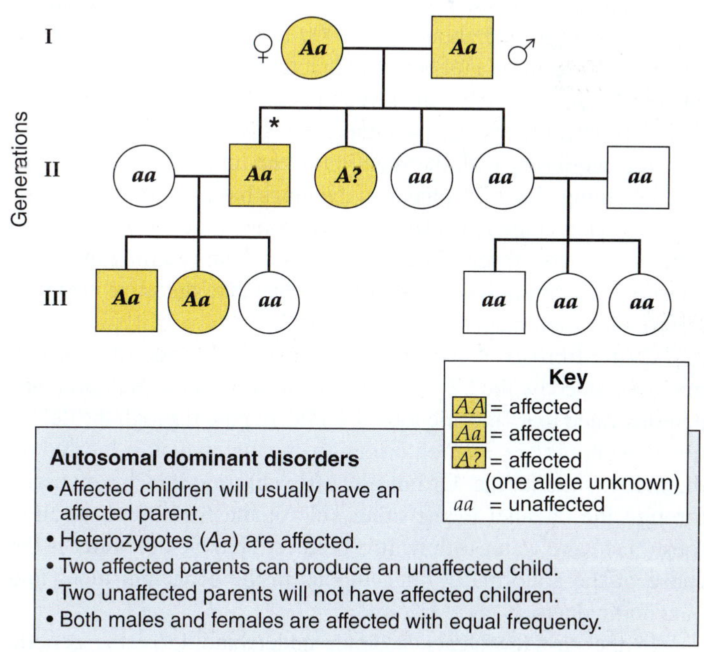

Autosomal dominant disorders

- Affected children will usually have an affected parent.
- Heterozygotes (*Aa*) are affected.
- Two affected parents can produce an unaffected child.
- Two unaffected parents will not have affected children.
- Both males and females are affected with equal frequency.

Figure 11.11 Autosomal dominant pedigree. The list gives ways to recognize an autosomal dominant disorder. How would you know the individual at the asterisk is heterozygous? (See Appendix A for the answer.)

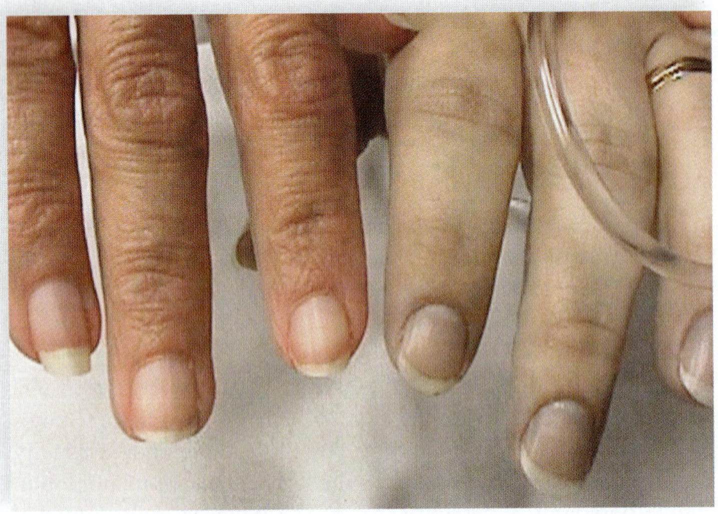

Figure 11.12 Methemoglobinemia. The hands of the woman on the right appear blue due to chemically induced methemoglobinemia.

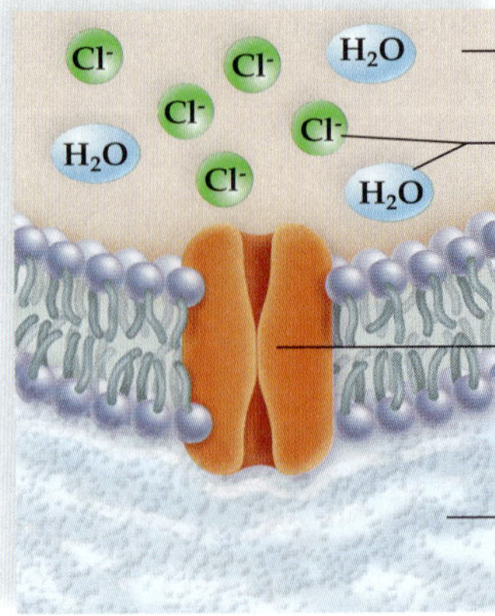

Figure 11.13 Cystic fibrosis. Cystic fibrosis is due to a faulty protein that is supposed to regulate the flow of chloride ions into and out of cells through a channel protein.

methylene blue. This unusual dye can donate electrons to other compounds, successfully converting the excess methemoglobin back into normal hemoglobin. The results were striking but immediate—the patients' skin quickly turned pink after treatment. A pedigree analysis of the Fugates indicated that the trait was common in the family because so many members carried the recessive allele.

Cystic Fibrosis

Cystic fibrosis (CF) is the most common lethal genetic disease among Caucasians in the United States (Fig. 11.13). About 1 in 20 Caucasians is a carrier, and about 1 in 2,000 newborns has the disorder. CF patients exhibit a number of characteristic symptoms, the most obvious being extremely salty sweat. In children with CF, the mucus in the bronchial tubes and pancreatic ducts is particularly thick and viscous, interfering with the function of the lungs and pancreas. To ease breathing, the thick mucus in the lungs has to be loosened periodically, but still the lungs frequently become infected. The clogged pancreatic ducts prevent digestive enzymes from reaching the small intestine, and to improve digestion, patients take digestive enzymes mixed with applesauce before every meal.

Cystic fibrosis is caused by a defective chloride ion channel that is encoded by the *CFTR* allele on chromosome 7. Research has demonstrated that chloride ions (Cl^-) fail to pass through the defective version of the *CFTR* chloride ion channel, which is located on the plasma membrane. Ordinarily, after chloride ions have passed through the channel to the other side of the membrane, sodium ions (Na^+) and water follow. It is believed that lack of water is the cause of the abnormally thick mucus in the bronchial tubes and pancreatic ducts.

In the past few years, a better understanding of the genetic basis of CF, coupled with new treatments, has raised the average life expectancy for CF patients to around 35 years, depending on the severity of the disease. Advances in gene therapy, or the replacement of the faulty allele with a good copy (see section 14.3), is showing considerable promise as a method of treating CF.

Interestingly, the mutated *CFTR* allele is believed to have persisted in the human population as a means of surviving potentially fatal diseases. Individuals who are heterozygous have an increased level of protection against diseases such as cholera. This is called a heterozygote advantage, which we will explore further in Chapter 16.

Phenylketonuria

Phenylketonuria (PKU) is an autosomal recessive metabolic disorder that affects nervous system development. Affected individuals lack the enzyme needed for normal metabolism of the amino acid phenylalanine; therefore, it appears in the urine and the blood. Newborns are routinely tested in the hospital for elevated levels of phenylalanine in the blood. If an elevated level is detected, the newborn will develop normally if placed on a diet low in phenylalanine, which must be continued until the brain is fully developed, around the age of 7, or severe intellectual disabilities will develop. Some doctors recommend that the diet continue for life, but in any case, a pregnant woman with phenylketonuria must be on the diet to protect her unborn child.

Autosomal Dominant Disorders

A number of autosomal dominant disorders have been identified in humans. Three relatively well-known autosomal dominant disorders are osteogenesis imperfecta, Huntington disease, and hereditary spherocytosis.

Osteogenesis Imperfecta

Osteogenesis (L. *os,* "bone"; *genesis,* "origin") imperfecta is an autosomal dominant genetic disorder that results in weakened, brittle bones. Although at least nine types of the disorder are known, most are linked to mutations in two genes necessary for the synthesis of type I collagen, one of the most abundant proteins in the human body. Collagen has many roles, including providing strength and rigidity to bone and forming the framework for most of the body's tissues. Osteogenesis imperfecta leads to a defective collagen I that causes the bones to be brittle and weak. Because the mutant collagen can cause structural defects even when combined with normal collagen I, osteogenesis imperfecta is generally considered to be dominant.

Osteogenesis imperfecta, which has an incidence of approximately 1 in 5,000 live births, affects all racial groups similarly and has been documented since as long as 300 years ago. Some historians think that the Viking chieftain Ivar Ragnarsson, who was known as Ivar the Boneless and was often carried into battle on a shield, had this condition. In most cases, the diagnosis is made in young children who visit the emergency room frequently due to broken bones. Some children with the disorder have an unusual blue tint in the sclera, the white portion of the eye; reduced skin elasticity; weakened teeth; and occasionally heart valve abnormalities. Currently, the disorder is treatable with a number of drugs that help increase bone mass, but these drugs must be taken long-term.

Huntington Disease

Huntington disease is a neurological disorder that leads to progressive degeneration of brain cells. The disease is caused by a mutated copy of the gene for a protein called huntingtin. Most patients appear normal until they are of middle age and have already had children, who may later also be stricken. Occasionally, the first sign of the disease appears during the teen years or even earlier. There is no effective treatment, and death comes 10 to 15 years after the onset of symptoms.

Several years ago, researchers found that the gene for Huntington disease is located on chromosome 4. They developed a test to detect the presence of the gene. However, few people want to know they have inherited the gene, because there is no cure. At least now we know that the disease stems from a mutation that causes the huntingtin protein to have too many copies of the amino acid glutamine. The normal version of huntingtin has stretches of between 10 and 25 glutamines. If huntingtin has more than 36 glutamines, it changes shape and forms large clumps inside neurons. Even worse, it attracts and causes other proteins to clump with it. One of these proteins, called CBP, which helps nerve cells survive, is inactivated when it clumps with huntingtin. Researchers hope to combat the disease by boosting CBP levels.

Hereditary Spherocytosis

Hereditary spherocytosis is an autosomal dominant genetic blood disorder that results from a defective copy of the *ankyrin-1* gene, found on chromosome 8. The protein encoded by this gene serves as a structural component of red blood cells and is responsible for maintaining their disklike shape. The abnormal spherocytosis protein is unable to perform its usual function, causing the affected person's red blood cells to adopt a spherical rather than disklike shape. As a result, the abnormal cells are fragile and burst easily, especially under osmotic stress. Enlargement of the spleen is also commonly seen in people with the disorder.

With an incidence of approximately 1 in 5,000, hereditary spherocytosis is one of the most common hereditary blood disorders. Roughly one-fourth of these cases result from new mutations and are not inherited from either parent. Hereditary spherocytosis exhibits incomplete penetrance, so not all individuals who inherit the mutant allele will actually show the trait. The cause of incomplete penetrance in these cases and others remains poorly understood.

Check Your Progress 11.3

1. Summarize how to distinguish an autosomal recessive disorder from an autosomal dominant disorder using a pedigree.
2. Construct a pedigree of Ivar Ragnarsson's family tree, assuming that his mother, and both her parents, were normal and that Ivar's father's father had osteogenesis imperfecta (mother was normal).

11.4 Beyond Mendelian Inheritance

Learning Outcomes

Upon completion of this section, you should be able to

1. Explain the inheritance pattern of traits when more than two alleles for the trait exist.
2. Contrast incomplete dominance and incomplete penetrance.
3. Describe the effects of pleiotropy on phenotypic traits.
4. Explain the concept of polygenic and multifactorial traits.
5. Understand how X-linked inheritance differs from autosomal inheritance.

Mendelian genetics can be applied to complex patterns of inheritance, such as multiple alleles, incomplete dominance, pleiotropy, and polygenic inheritance.

Multiple Allelic Traits

When a trait is controlled by **multiple alleles,** the gene exists in several allelic forms within a population. For example, although a person's ABO blood type is controlled by a single gene pair, three possible alleles within the human population determine blood type. Each person receives two of these alleles (one from each parent) to determine the presence or absence of antigens on his or her red blood cells.

I^A = A antigen on red blood cells
I^B = B antigen on red blood cells
i = Neither A nor B antigen on red blood cells

The possible phenotypes and genotypes for blood type are as follows:

Phenotype	*Genotype*
A	I^AI^A, I^Ai
B	I^BI^B, I^Bi
AB	I^AI^B
O	ii

The inheritance of the ABO blood group in humans is also an example of **codominance,** because both I^A and I^B are fully expressed in the presence of the other. A person who inherits chromosomes with I^A and I^B alleles will make fully functional A and B protein, and because these alleles are codominant, the resulting mixture of AB protein will give the red blood cell an AB phenotype. On the other hand, both I^A and I^B are dominant over *i*. Therefore, two genotypes are possible for type A blood, and two genotypes are possible for type B blood.

We can use a Punnett square to confirm that reproduction between a heterozygote with type A blood and a heterozygote with type B blood can result in any one of the four blood types. Such a cross makes it clear that an offspring can have a different blood type than either parent. For this reason, rather than blood type, DNA fingerprinting, also called DNA profiling (see Chapter 14) is used to identify the parents of an individual.

Incomplete Dominance and Incomplete Penetrance

Incomplete dominance is exhibited when a heterozygote has an intermediate phenotype between that of either homozygote. In a cross between a true-breeding, red-flowered four-o'clock plant strain and a true-breeding, white-flowered strain, the offspring have pink flowers. Although this outcome might appear to be an example of the blending theory of inheritance, it is not. Although the phenotypes have blended, the individual alleles are not altered. How do we know? When the pink plants self-pollinate, the offspring plants have a phenotypic ratio of 1 red-flowered : 2 pink-flowered : 1 white-flowered. The reappearance of the three phenotypes in this generation makes it clear that we are still dealing with a single pair of alleles (Fig. 11.14).

Incomplete dominance in four-o'clocks actually has more to do with the amount of pigment protein produced in the plant cells: A double dose of pigment results in red flowers; a single dose of pigment results in pink flowers; and a lack of any pigment produces white flowers.

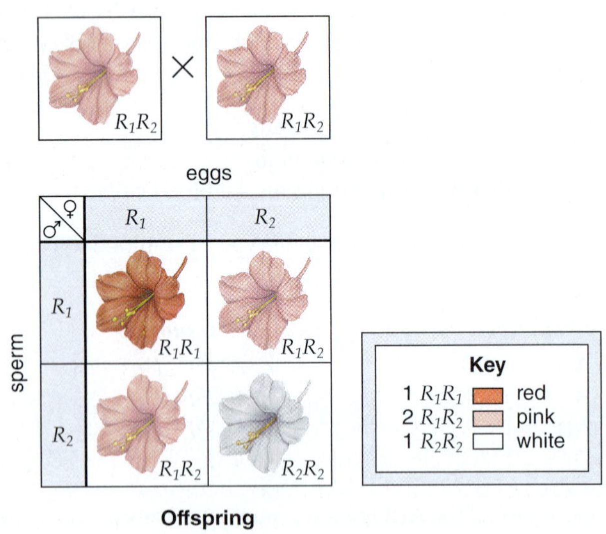

Figure 11.14 Incomplete dominance. When pink four-o'clocks self-pollinate, the results show three phenotypes. This is possible only if the pink parents had an allele for red pigment (R_1) and an allele for no pigment (R_2). Note that alleles involved in incomplete dominance are both given a capital letter.

Human Examples of Incomplete Dominance

In humans, familial hypercholesterolemia (FH) is an example of incomplete dominance. An individual with two alleles for this disorder develops fatty deposits in the skin and tendons and may have a heart attack as a child. An individual with one normal allele and one *FH* allele may suffer a heart attack as a young adult, and an individual with two normal alleles does not have the disorder.

Perhaps the inheritance pattern of other human disorders should be considered one of incomplete dominance. To detect the carriers of cystic fibrosis, for example, it is customary to determine the amount of cellular activity of the gene. When the activity is one-half that of the dominant homozygote, the individual is a carrier, even though the individual does not exhibit the genetic disease. In other words, at the level of gene expression, the homozygotes and heterozygotes differ in the same manner as four-o'clock plants.

A dominant allele may not always lead to the dominant phenotype in a heterozygote, even when the alleles show a true dominant/recessive relationship. The dominant allele in this case does not always determine the phenotype of the individual, so we describe these traits as showing **incomplete penetrance.** In other words, just because a person inherits a dominant allele doesn't mean he or she will fully express the gene or show the dominant phenotype. Many dominant alleles exhibit varying degrees of penetrance.

The best-known example of incomplete penetrance is polydactyly, the presence of one or more extra digits on the hands, the feet, or both. Polydactyly is inherited in an autosomal dominant manner; however, not all individuals who inherit the dominant allele exhibit the trait. The reasons for this are not clear, but expression of polydactyly may require additional environmental factors or be influenced by other genes, as discussed later.

Pleiotropic Effects

Pleiotropy occurs when a single mutant gene affects two or more distinct and seemingly unrelated traits. For example, persons with Marfan syndrome have disproportionately long arms, legs, hands, and feet; a weakened aorta; poor eyesight; and other characteristics (Fig. 11.15). All of these characteristics are due to the production of abnormal connective tissue.

Marfan syndrome has been linked to a mutated gene (*FBN1*) on chromosome 15 that ordinarily specifies a functional protein called fibrillin. Fibrillin is essential for the formation of elastic fibers in connective tissue. Without the structural support of normal connective tissue, the aorta can burst, particularly if the person is engaged in a strenuous sport, such as volleyball or basketball. Flo Hyman may have been the best American woman volleyball player ever, but she fell to the floor and died at the age of 31, because her aorta gave way during a game. Now that coaches are aware of Marfan syndrome, they are on the lookout for it among very tall basketball players.

Many other disorders, including porphyria and sickle-cell disease, are examples of pleiotropic traits. Porphyria is caused by a chemical insufficiency in the production of hemoglobin, the pigment that makes red blood cells red. The symptoms of porphyria

Figure 11.15 Marfan syndrome. Marfan syndrome illustrates the multiple effects a single gene can have. Marfan syndrome is due to any number of defective connective tissue defects.

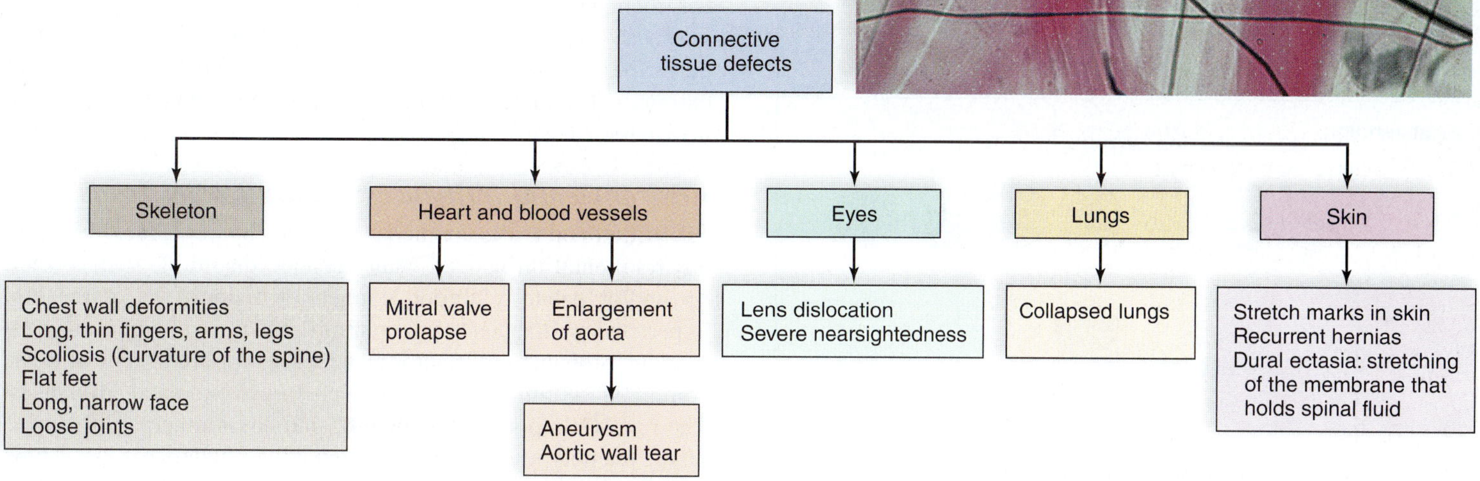

are photosensitivity, strong abdominal pain, port-wine-colored urine, and paralysis in the arms and legs. Many members of the British royal family in the late 1700s and early 1800s suffered from this disorder, which can lead to epileptic convulsions, bizarre behavior, and coma.

In a person suffering from sickle-cell disease (Hb^SHb^S), the cells are sickle-shaped. The underlying mutation is in a gene that codes for a type of polypeptide chain in hemoglobin. Of 146 amino acids, the gene mutation changes only one amino acid, but the result is a less-soluble polypeptide chain that stacks up and causes red blood cells to be sickle-shaped. The abnormally shaped sickle cells slow down blood flow and clog small blood vessels. In addition, sickled red blood cells have a shorter life span than normal red blood cells. Affected individuals may exhibit a number of symptoms, including severe anemia, physical weakness, poor circulation, impaired mental function, pain and high fever, rheumatism, paralysis, spleen damage, low resistance to disease, and kidney and heart failure. All of these effects are due to both the tendency of sickled red blood cells to break down and the resulting decreased oxygen-carrying capacity of the blood, which damage the body.

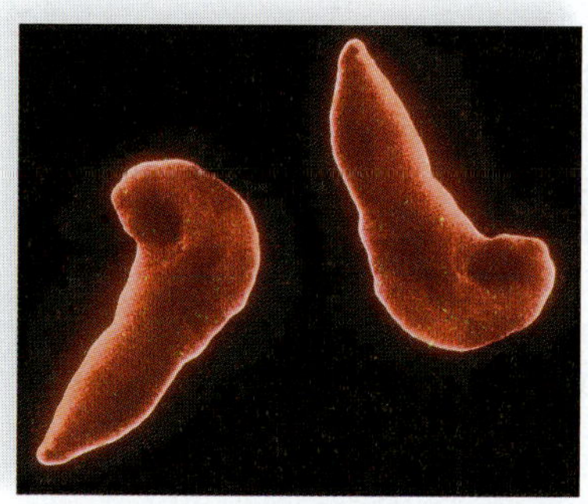

1,600×, colorized SEM

Sickled red blood cell

Although sickle-cell disease is a devastating disorder, from an evolutionary perspective it provides heterozygous individuals with a survival advantage. People who have sickle-cell trait are resistant to the protozoan parasite that causes malaria. The parasite spends part of its life cycle in red blood cells, feeding on hemoglobin, but it cannot complete its life cycle when sickle-shaped cells form and break down earlier than usual. Because of this survival benefit, the sickle-cell allele has been maintained in the human population over evolutionary time (see Chapter 16).

Polygenic Inheritance

Polygenic inheritance (Gk. *poly,* "many"; L. *genitus,* "producing") occurs when a trait is governed by two or more sets of alleles. Examples include human height, skin color, and the prevalence of diabetes. The individual has a copy of all allelic pairs, possibly located on many different pairs of chromosomes. Each dominant allele has a quantitative effect on the phenotype, and these effects are additive. Therefore, a population is expected to exhibit continuous phenotypic variations, such as a wide variation in human height and weight. In Figure 11.16, a cross between genotypes *AABBCC* and *aabbcc* yields F_1 hybrids with the genotype *AaBbCc*. A range of genotypes and phenotypes results in the F_2 generation that can be depicted as a bell-shaped curve (Fig. 11.16).

Skin Color

Skin color is the result of pigmentation produced by skin cells called melanocytes, and over 100 different genes influence skin color. It is an example of a polygenic trait that is likely controlled by many pairs of alleles, which results in a range of phenotypes. The vast majority of people have skin colors in the middle range, whereas fewer people have skin colors in the extreme range.

Even so, we will use the simplest model and we will assume that skin has only three pairs of alleles (*Aa, Bb,* and *Cc*) and that each capital letter contributes pigment to the skin. When a very dark person reproduces with a very light person, the children have medium-brown skin. When two people with the genotype *AaBbCc* reproduce with one another, individuals may range in skin color

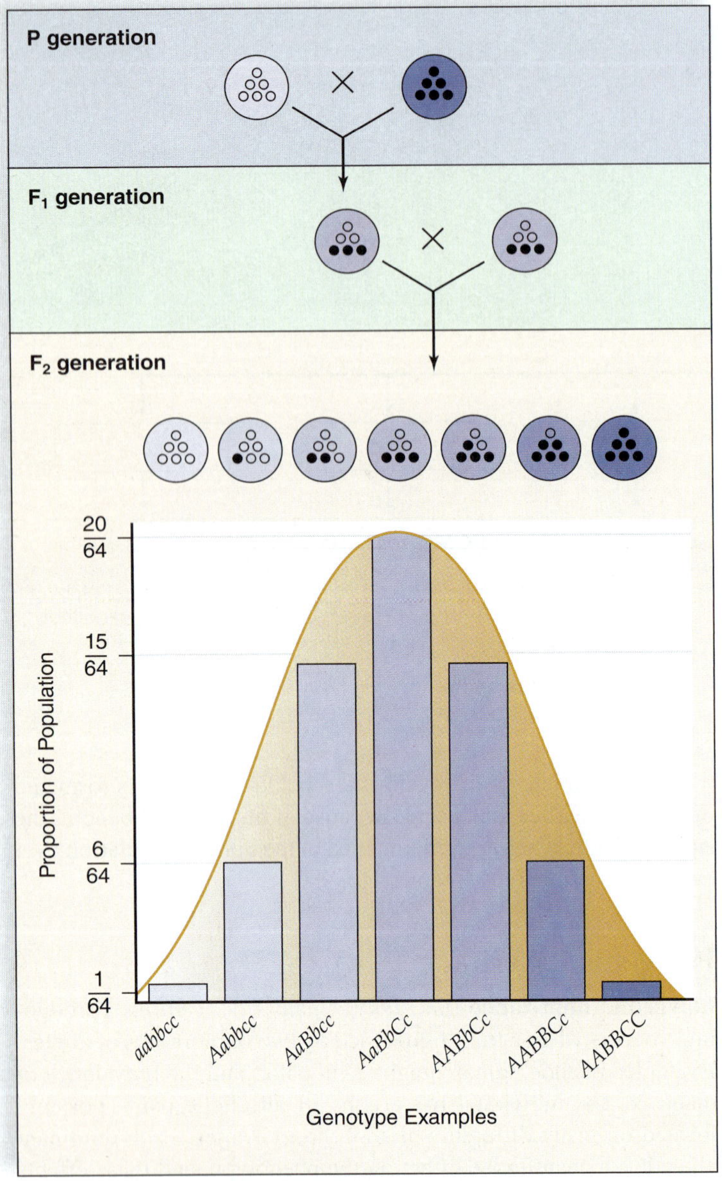

Figure 11.16 Polygenic inheritance. In polygenic inheritance, a number of pairs of genes control the trait. *Above:* Black dots and intensity of blue shading stand for the number of dominant alleles. *Below:* Orange shading shows the degree of environmental influences.

from very dark to very light. The distribution of these phenotypes typically follows a bell-shaped curve, meaning that few people have the extreme phenotypes and most people have the phenotype that lies in the middle. A bell-shaped curve is a common identifying characteristic of a polygenic trait (Fig. 11.17).

However, skin color is also influenced by the sunlight in the environment. Notice again that a range of phenotypes exists for each genotype. For example, individuals who are *AaBbCc* may vary in their skin color, even though they possess the same genotype, and several possible phenotypes fall between the two extremes. The interaction of the environment with polygenic traits is discussed next.

Environmental Influences: Multifactorial Traits

Multifactorial traits are those controlled by polygenes subject to environmental influences. Many genetic disorders, such as cleft lip and/or palate, clubfoot, congenital dislocations of the hip, hypertension, diabetes, schizophrenia, and even allergies and cancers, are probably multifactorial, because they are likely due to the combined action of many genes plus environmental influences. The relative importance of genetic and environmental influences on the phenotype can vary, and often it is a challenge to determine how much of the variation in the phenotype may be attributed to each factor. This is especially true in complex polygenic traits for which there may be an additive effect of multiple genes on the phenotype. If each gene has several alleles, and each allele responds slightly differently to environmental factors, then the phenotype can vary considerably.

Multifactorial traits are a challenge for drug manufacturers, since they must determine the response to a new drug based on genetic factors (for example, the ethnic background of the patient) and environmental factors (such as diet). Temperature is an environmental factor that can influence the phenotypes of plants and animals. Primroses have white flowers when grown above 32°C but red flowers when grown at 24°C.

The coats of Himalayan rabbits are darker in color at the ears, nose, paws, and tail. Himalayan rabbits are known to be homozygous for the allele *ch,* which is involved in the production of melanin. Experimental evidence suggests that the enzyme encoded by this gene is active only at a low temperature and that, therefore, black fur occurs only at the extremities, where body heat is lost to

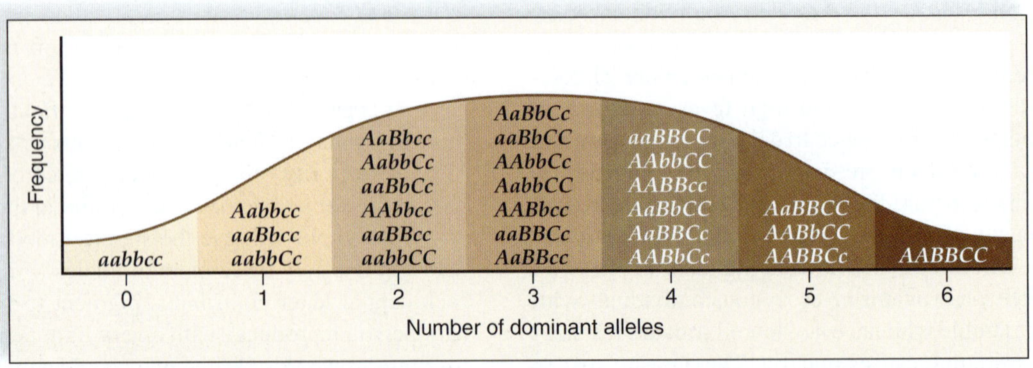

Figure 11.17 Skin color, a polygenic trait. Skin color is controlled by many pairs of alleles, which result in a range of phenotypes. The vast majority of people have skin colors in the middle range, whereas fewer people have skin colors in the extreme range.

the environment. When the animal is placed in a warmer environment, new fur on these body parts is light in color.

Many investigators are trying to determine what percentage of various traits is due to nature (inheritance) and what percentage is due to nurture (the environment). Some studies use twins separated since birth, because if identical twins in different environments share the same trait, the trait is most likely inherited. Identical twins are more similar in their intellectual talents, personality traits, and levels of lifelong happiness than are fraternal twins separated at birth. Biologists conclude that all behavioral traits are partly heritable, and that genes exert their effects by acting together in complex combinations susceptible to environmental influences.

X-linked Inheritance

The X and Y chromosomes in mammals determine the gender of the individual. Females are XX, and males are XY. These chromosomes carry genes that control development; in particular, if the Y chromosome contains an *SRY* gene, the embryo becomes a male. The term **X-linked** is used for genes that have nothing to do with gender yet are carried on the X chromosome. The Y chromosome does not carry these genes and indeed carries very few genes.

This type of inheritance was discovered in the early 1900s by a group at Columbia University headed by Thomas Hunt Morgan. Morgan performed experiments with fruit flies, *Drosophila melanogaster.* Fruit flies are even better subjects for genetic studies than garden peas. They can be easily and inexpensively raised in simple laboratory glassware; after mating, females lay hundreds of eggs during their lifetimes; and the generation time is short, taking only about 10 days from egg to adult. Fruit flies have a sex chromosome pattern similar to that of humans, and therefore Morgan's experiments with X-linked genes apply directly to humans.

Morgan's Experiment

Morgan took a newly discovered mutant male with white eyes and crossed it with a red-eyed female:

	♀		♂
P	red-eyed	×	white-eyed
F_1	red-eyed		red-eyed

From these results, he knew that red eyes are the dominant characteristic and white eyes are the recessive characteristic. He then crossed the F_1 flies. In the F_2 generation, there was the expected 3 red-eyed : 1 white-eyed ratio, but it struck him as odd that all the white-eyed flies were males:

	♀		♂
$F_1 \times F_1$	red-eyed	×	red-eyed
F_2	red-eyed		1 red-eyed : 1 white-eyed

Obviously, a major difference between the male flies and the female flies was their sex chromosomes. Could it be possible that an allele for eye color was on the Y chromosome but not on the X? This idea could be quickly discarded, because usually females have red eyes, and they have no Y chromosome. Perhaps an allele for eye color was on the X, but not on the Y, chromosome. Figure 11.18 indicates that this explanation would match the results obtained in the experiment. These results support the chromosome theory of inheritance by showing that the behavior of a specific allele corresponds exactly with that of a specific chromosome—the X chromosome in *Drosophila.*

Notice that X-linked alleles have a different pattern of inheritance than alleles that are on the autosomes, because the Y chromosome is lacking for these alleles, and the inheritance of a Y chromosome cannot offset the inheritance of an X-linked recessive allele. For the same reason, males always receive an X-linked recessive mutant allele from the female parent—they receive only the Y chromosome from the male parent, and therefore sex-linked recessive traits appear much more frequently in males than in females.

Solving X-linked Genetics Problems

Recall that when solving autosomal genetics problems, the allele key and genotypes can be represented as follows:

Allele key	***Genotypes***
L = long wings	LL, Ll
l = short wings	ll

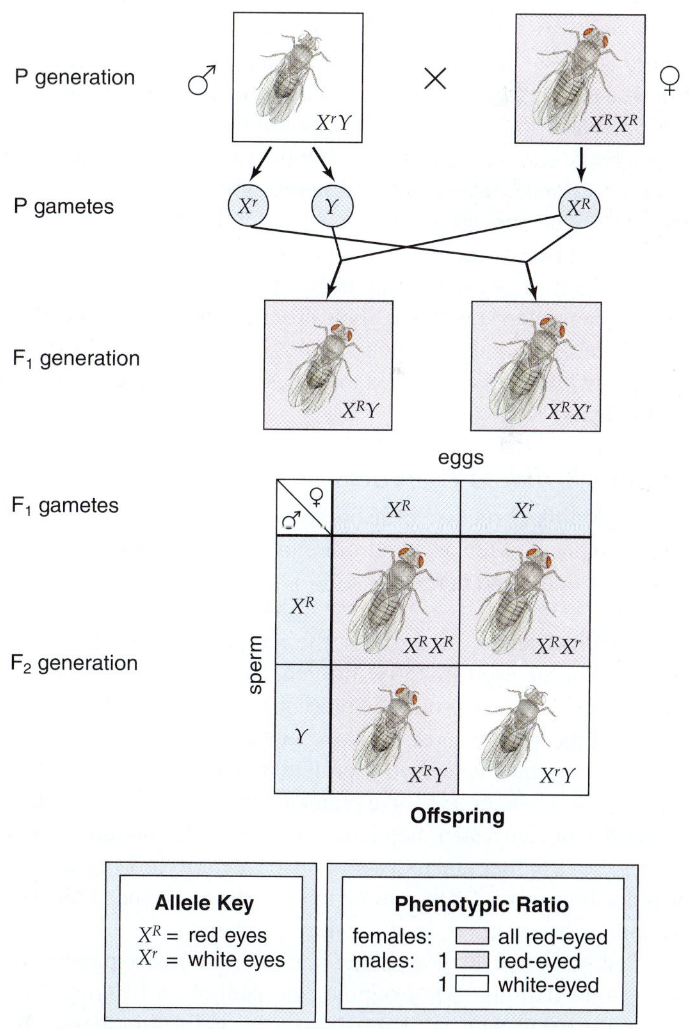

Figure 11.18 X-linked inheritance. Once researchers deduced that the alleles for red/white eye color are on the X chromosome in *Drosophila,* they were able to explain their experimental results. Males with white eyes in the F_2 generation inherit the recessive allele only from the female parent; they receive a Y chromosome lacking the allele for eye color from the male parent.

When predicting inheritance of sex-linked traits, however, it is necessary to indicate the sex chromosomes of each individual. As noted in Figure 11.18, however, the allele key for an X-linked gene shows an allele attached to the X:

Allele key

X^R = red eyes

X^r = white eyes

The possible genotypes and phenotypes in both males and females are as follows:

Genotype	*Phenotype*
X^RX^R	red-eyed female
X^RX^r	red-eyed female
X^rX^r	white-eyed female
X^RY	red-eyed male
X^rY	white-eyed male

Notice that there are three possible genotypes for females but only two for males. Females can be heterozygous X^RX^r, in which case they are carriers. Carriers usually do not show a recessive abnormality, but they are capable of passing on a recessive allele for an abnormality. But unlike autosomal traits, males cannot be carriers for X-linked traits; if the dominant allele is on the single X chromosome, they show the dominant phenotype, and if the recessive allele is on the single X chromosome, they show the recessive phenotype. For this reason, males are considered **hemizygous** for X-linked traits, because a male possesses only one allele for the trait and, therefore, expresses whatever allele is present on the X chromosome.

We know that male fruit flies have white eyes when they receive the mutant recessive allele from the female parent. What is the inheritance pattern when females have white eyes? Females can have white eyes only when they receive a recessive allele from both parents.

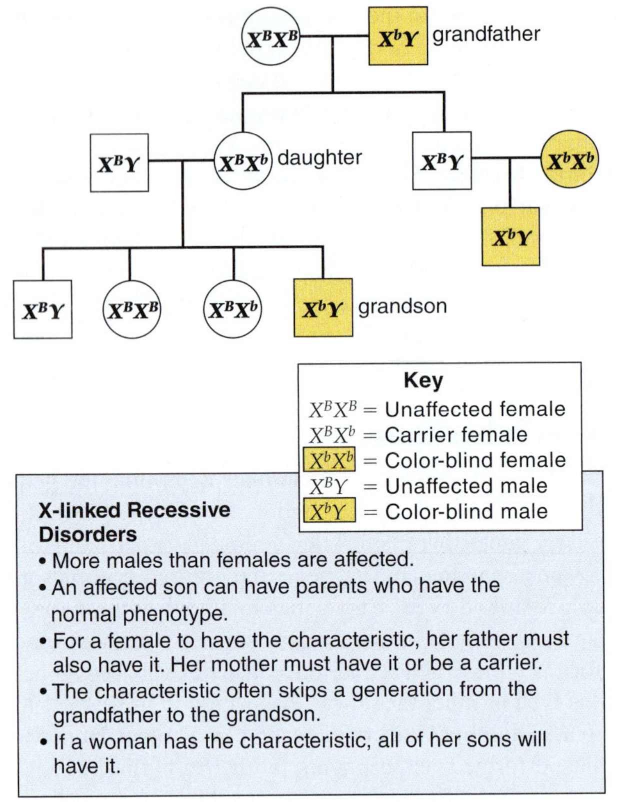

Figure 11.19 X-linked recessive pedigree. This pedigree for color blindness exemplifies the inheritance pattern of an X-linked recessive disorder. The list gives various ways of recognizing the X-linked recessive pattern of inheritance.

Human X-linked Disorders

Several X-linked recessive disorders occur in humans, including color blindness, Menkes syndrome, muscular dystrophy, adrenoleukodystrophy, and hemophilia.

Color Blindness. In humans, the receptors for color vision in the retina of the eyes are three different classes of cone cells. Only one type of pigment protein is present in each class of cone cell; there are blue-sensitive, red-sensitive, and green-sensitive cone cells. The allele for the blue-sensitive protein is autosomal, but the alleles for the red- and green-sensitive pigments are on the X chromosome. About 8% of Caucasian men have red-green color blindness. Most of these see brighter greens as tans, olive greens as browns, and reds as reddish browns. A few cannot tell reds from greens at all. They see only yellows, blues, blacks, whites, and grays.

Pedigrees can also reveal the unusual inheritance pattern seen in sex-linked traits. For example, the pedigree in Figure 11.19 shows the usual pattern of inheritance for color blindness. More males than females have the trait, because recessive alleles on the X chromosome are expressed in males. The disorder often passes from grandfather to grandson through a carrier daughter.

Menkes Syndrome. Menkes syndrome, or kinky hair syndrome, is caused by a defective allele on the X chromosome. Normally, the gene product controls the movement of the metal copper into and out of cells. The symptoms of Menkes syndrome are due to an accumulation of copper in some parts of the body, and the lack of the metal in other parts.

Symptoms of Menkes syndrome include poor muscle tone, seizures, abnormally low body temperature, skeletal anomalies, and the characteristic brittle, steely hair associated with the disorder. Although the condition is relatively rare, affecting approximately 1 in 100,000, mostly males, the prognosis for people with Menkes syndrome is poor, and most individuals die within the first few years of life. In recent years, some people with Menkes syndrome have been treated with injections of copper directly underneath the skin, but with mixed results, and treatment must begin very early in life to be effective.

Muscular Dystrophy. Muscular dystrophy, as the name implies, is characterized by a wasting away of the muscles. The most common form, Duchenne muscular dystrophy, is X-linked and occurs in about 1 out of every 3,600 male births. Symptoms such as waddling gait, toe walking, frequent falls, and difficulty in rising may appear as soon as the child starts to walk. Muscle weakness intensifies until the individual is confined to a wheelchair. Death usually occurs by age 20; therefore, affected males are rarely fathers. The recessive allele remains in the population through passage from carrier mother to carrier daughter.

The allele for Duchenne muscular dystrophy has been isolated, and it has been discovered that the absence of a protein called dystrophin causes the disorder. Much investigative work has determined that dystrophin is involved in the release of calcium from

THEME **Nature of Science**

Hemophilia and the Royal Families of Europe

About 1 in 10,000 males is a hemophiliac. There are two common types of hemophilia: Hemophilia A is due to the absence or minimal presence of a clotting factor known as factor VIII, and hemophilia B is due to the absence of clotting factor IX. Hemophilia is called the bleeder's disease because the affected person's blood either does not clot or clots very slowly. Although hemophiliacs bleed externally after an injury, they also bleed internally, particularly around joints. Hemorrhages can be stopped with transfusions of fresh blood (or plasma) or concentrates of the clotting protein. Also, clotting factors are available as biotechnology products.

The pedigree in Figure 11A shows why hemophilia is often referred to as "the royal disease." Queen Victoria of England, who reigned from 1837 to 1901, was the first of the royals to carry the gene. From her, the disease eventually spread to the Prussian, Spanish, and Russian royal families. In that era, monarchs arranged marriages between their children to consolidate political alliances. This practice allowed the gene for hemophilia to spread throughout the royal families. It is assumed that a spontaneous mutation arose either in Queen Victoria after her conception or in one of the gametes of her parents. However, in the book *Queen Victoria's Gene* by D. M. Potts, the author postulates that Edward Augustus, Duke of Kent, may not have been Queen Victoria's father. Potts suggests that Victoria may have instead been the illegitimate child of a hemophiliac male.

Of Queen Victoria's 26 grandchildren, 4 grandsons had hemophilia and 4 granddaughters were carriers. Because none of Queen Victoria's brothers and sisters were affected, it seems that the faulty allele she carried arose by mutation either in Victoria or in one of her parents. Her carrier daughters, Alice and Beatrice, introduced the allele into the ruling houses of Russia, Prussia, and Spain, respectively. Alexi, the last heir to the Russian throne before the Russian Revolution, was a hemophiliac. There are no hemophiliacs in the present British royal family, because Victoria's eldest son, King Edward VII, did not receive the allele.

Questions to Consider

1. How may a pedigree pattern be used to determine if a disease is autosomal dominant or autosomal recessive?
2. Assume that the mutation for hemophilia did not originate with Victoria. What does this tell you about the genotypes of her parents?

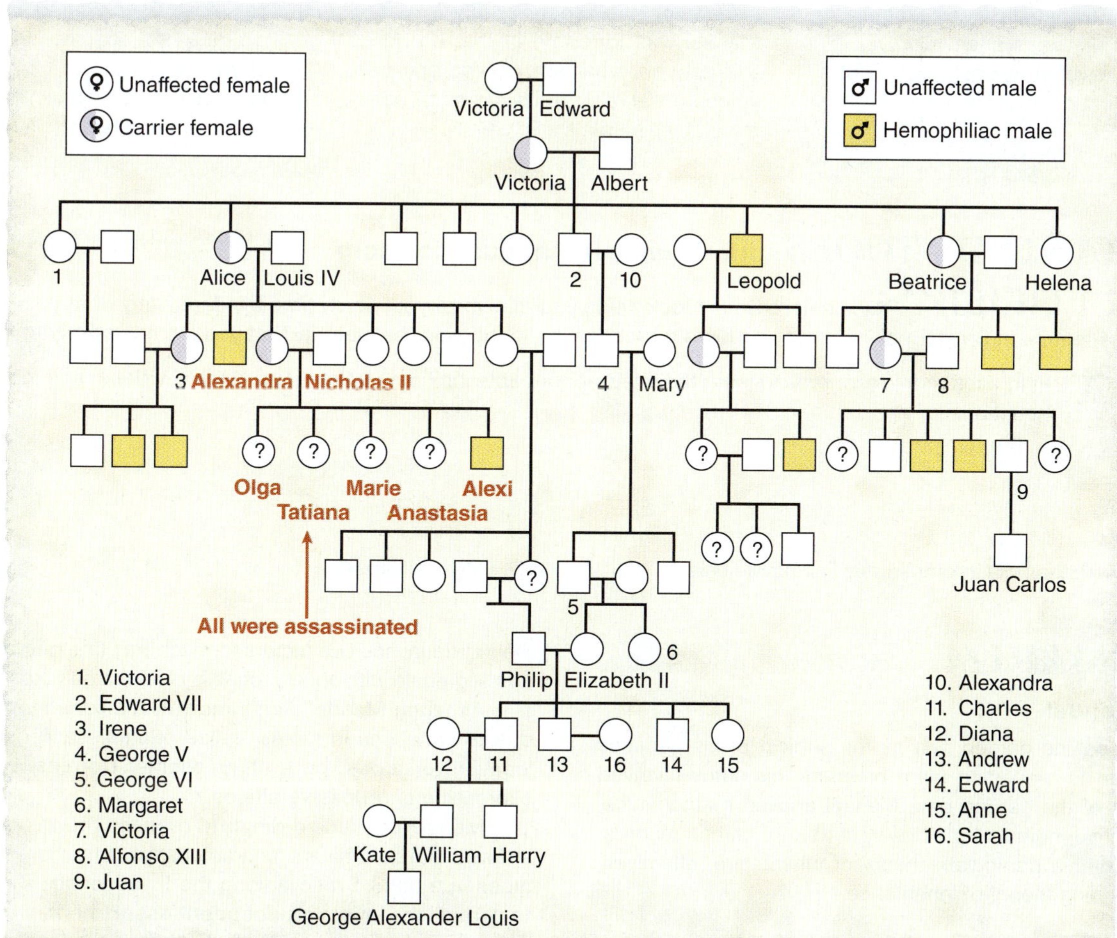

Figure 11A Hemophilia: an X-linked trait. Queen Victoria was a carrier, so each of her sons had a 50% chance of having the disorder, and each of her daughters had a 50% chance of being a carrier. This pedigree shows only the affected descendants.

the sarcoplasmic reticulum in muscle fibers. The lack of dystrophin causes calcium to leak into the cell, which promotes the action of an enzyme that dissolves muscle fibers. When the body attempts to repair the tissue, fibrous tissue forms, and this cuts off the blood supply, so more and more cells die.

A test is now available to detect carriers of Duchenne muscular dystrophy. Also, various treatments have been tried. Immature muscle cells can be injected into muscles, and for every 100,000 cells injected, dystrophin production occurs in 30–40% of muscle fibers. The allele for dystrophin has been inserted into thigh muscle cells, and about 1% of these cells then have produced dystrophin.

Adrenoleukodystrophy. Adrenoleukodystrophy, or ALD, is an X-linked recessive disorder due to the failure of a carrier protein to move either an enzyme or a very long-chain fatty acid (24–30 carbon atoms) into peroxisomes. As a result, these fatty acids are not broken down, and they accumulate inside the cell; the result is severe nervous system damage.

Children with ALD fail to develop properly after age 5, lose adrenal gland function, exhibit very poor coordination, and show a progressive loss of hearing, speech, and vision. The condition is usually fatal, with no known cure, but the onset and severity of symptoms in patients not yet showing symptoms may be mitigated by treatment with a mixture of lipids derived from olive oil. The disease was made well known by the 1992 movie *Lorenzo's Oil,* detailing a mother's and father's determination to devise a treatment for their son who was suffering from ALD.

Check Your Progress 11.4

1. Summarize why incomplete dominance does not support blending.
2. Summarize how to identify an X-linked trait from an autosomal trait.
3. Explain how a trait may be both polygenic and multifactorial.

CONNECTING *the* CONCEPTS *with the* THEMES

Evolution

- The persistence of genetic diseases over evolutionary time may confer some biological benefit to humans. For example, heterozygous sickle-cell anemia individuals are more resistant to malaria, one of the world's most deadly diseases.

Nature of Science

- A good experimental design and careful data analysis allowed Mendel to discover his laws of inheritance.
- Mendel was fortunate in that he chose to study an organism, the garden pea, whose observable traits are often determined by a single allelic pair.

Biological Systems

- With the help of Mendelian genetics, pedigree analysis, and statistics, scientists have been able to link many human diseases to specific genes on certain chromosomes.
- Genetic diseases may have multiple influences on the phenotype of an organism.

MEDIA STUDY TOOLS http://connect.mheducation.com

SMARTBOOK® Study smart. SmartBook helps you maximize your study time by identifying what you know and don't know. The recharge feature in SmartBook helps you prepare for the test by identifying the material that you are most likely to forget.

connect BIOLOGY Find even more resources to learn the chapter concepts including animations, tutorial videos, and interactive practice questions.

Animations	Tutorials
11.2 Random Orientation of Chromosomes During Meiosis	**11.2** Dihybrid Cross

SUMMARIZE

11.1 Gregor Mendel

Gregor Mendel used the garden pea as the subject in his genetics studies. In contrast to preceding plant breeders, his study involved nonblending traits of the garden pea. Mendel applied mathematics, followed the scientific method very closely, and kept careful records. His results supported a particulate theory of inheritance, effectively disproving the blending theory of inheritance.

11.2 Mendel's Laws

When Mendel crossed heterozygous plants with other heterozygous plants, called a **monohybrid cross,** he found that the recessive **phenotype** reappeared in about 1/4 of the F_2 plants; there was a 3:1 phenotypic ratio. This allowed Mendel to propose his law of segregation, which states that the individual has two factors for each trait (the **genotype**), and the factors segregate with equal probability into the gametes.

Although Mendel didn't know it, the variations in the traits were due to variations in genes, called **alleles.** Each gene has a specific location, or **locus,** on a chromosome. **Dominant alleles** mask the expression of **recessive alleles.**

Mendel conducted **dihybrid crosses,** in which the F_1 individuals showed both dominant characteristics, but there were four phenotypes in a 9:3:3:1 ratio among the F_2 offspring. This allowed Mendel to deduce the **law of independent assortment,** which states that the members of one pair of factors separate independently of those of another pair. Therefore, all possible combinations of parental factors can occur in the gametes.

The laws of probability can be used to calculate the expected phenotypic ratio of a cross. A large number of offspring must be counted in order to observe the expected results, and to ensure that all possible types of sperm have fertilized all possible types of eggs, as is done in a **Punnett square.** The Punnett square uses the product law of probability to arrive at possible genotypes among the offspring, and then the sum law can be used to arrive at the phenotypic ratio.

Mendel also crossed the F_1 plants having the dominant phenotype with homozygous recessive plants. The 1:1 results indicated that the recessive factor was present in these F_1 plants (they were heterozygous). Today, we call this a testcross, because it is used to test whether an individual showing the dominant characteristic is **homozygous** dominant or **heterozygous.** The two-trait testcross allows an investigator to test whether an individual showing two dominant characteristics is homozygous dominant for both traits or for one trait only or is heterozygous for both traits.

11.3 Mendelian Patterns of Inheritance and Human Disease

Studies have shown that many human traits and genetic disorders can be explained on the basis of simple Mendelian inheritance. When studying human genetic disorders, biologists often construct pedigrees to show the pattern of inheritance of a characteristic within a family. The particular pattern indicates the manner in which a characteristic is inherited, such as **autosomal** dominant or autosomal recessive.

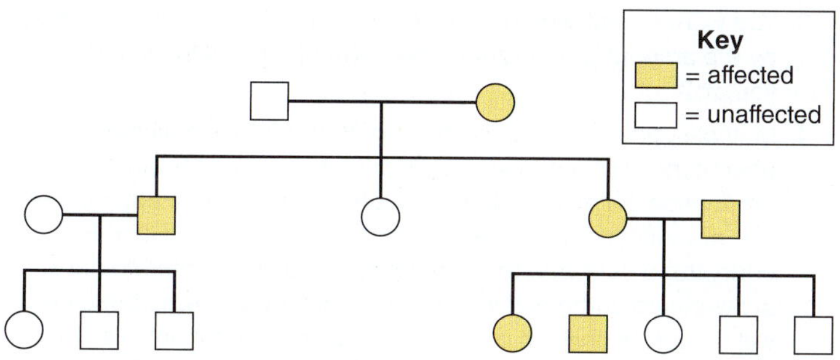

- Autosomal recessive: An individual must possess two copies of the recessive allele to express the trait. Heterozygous individuals may be **carriers** for the trait.
- Autosomal dominant: An individual needs to possess only a single copy of the dominant allele to express the trait. Homozygous dominant and heterozygous individuals both express the trait.

11.4 Beyond Mendelian Inheritance

Other patterns of inheritance have been discovered since Mendel's original contribution. For example, some genes have **multiple alleles,** although each organism has only two alleles, as in the inheritance of blood type in humans. Inheritance of blood type also illustrates **codominance.** With **incomplete dominance,** the phenotype of F_1 individuals is intermediate between the parent phenotypes; this does not support the blending theory, because the parent phenotypes reappear in F_2. With incomplete penetrance, some traits that are dominant may not be expressed due to unknown reasons.

In **pleiotropy,** one gene has multiple effects, as in Marfan syndrome and sickle-cell disease. **Polygenic traits** are controlled by several genes that have an additive effect on the phenotype, resulting in quantitative variations within a population. A bell-shaped curve is seen, because environmental influences bring about many intervening phenotypes, as in the inheritance of height in humans. Skin color and eye color are also examples of **multifactorial traits** (multiple genes plus the environment).

In *Drosophila*, as in humans, the sex chromosomes determine the sex of the individual, with XX being female and XY being male. Experimental support for the chromosome theory of inheritance came when Morgan and his group were able to determine that the gene for a trait unrelated to sex determination, the white-eyed allele in *Drosophila*, is on the X chromosome.

Alleles on the X chromosome are called **X-linked** alleles. Therefore, when doing X-linked genetics problems, it is the custom to indicate the sexes by using sex chromosomes and to indicate the alleles by superscripts attached to the X. The Y is blank because it does not carry these genes. Color blindness, Menkes syndrome, adrenoleukodystrophy, and hemophilia are X-linked recessive disorders in humans.

ASSESS

Choose the best answer for each question.

11.1 Gregor Mendel

1. Mendel's work supported which of the following?
   a. blending theory of inheritance
   b. particulate theory of inheritance
   c. theory of acquired characteristics
   d. All of these are correct.
2. Mendel's success was based on
   a. use of the pea plant as a model organism.
   b. his ability to apply statistics to his studies.
   c. careful planning of his experiments.
   d. All of these are correct.

11.2 Mendel's Laws

3. The law of segregation states all of the following except
   a. factors separate during formation of the gametes.
   b. each individual has two factors for each trait.
   c. gametes contain a single factor for each trait.
   d. factors assort independently of each other by meiosis.
4. In peas, yellow seed (*Y*) is dominant over green seed (*y*). In the F_2 generation of a monohybrid cross that begins when a dominant homozygote is crossed with a recessive homozygote, you would expect
   a. three plants with yellow seeds to every plant with green seeds.
   b. plants with one yellow seed for every green seed.
   c. only plants with the genotype *Yy*.
   d. only plants that produce yellow seeds.
   e. Both c and d are correct.
5. In guinea pigs, smooth coat (*S*) is dominant over rough coat (*s*), and black coat (*B*) is dominant over white coat (*b*). In the cross *SsBb* × *SsBb*, how many of the offspring will have a smooth black coat, on average?
   a. $1/4$
   b. about $9/16$
   c. $1/16$
   d. $6/16$
   e. $2/6$
6. In horses, *B* = black coat, *b* = brown coat, *T* = trotter, and *t* = pacer. A black trotter that has a brown pacer offspring has which of the following genotypes?
   a. *BT*
   b. *BbTt*
   c. *bbtt*
   d. *BBtt*
   e. *BBTT*

11.3 Mendelian Patterns of Inheritance and Human Disease

7. Which of the following is not correct for an autosomal recessive pedigree?
   a. Males inherit the trait 50% of the time.
   b. Heterozygous individuals are carriers.
   c. Only homozygous recessive individuals express the trait.
   d. Homozygous dominant individuals and heterozygotes have the same phenotype.
8. Cystic fibrosis is an example of a/an ________ trait.
   a. autosomal dominant
   b. autosomal recessive
   c. X-linked
   d. incomplete dominant
9. When analyzing a pedigree, you notice that two unaffected parents have produced a child that is affected by the trait. This suggests which of the following patterns of inheritance?
   a. autosomal dominant
   b. X-linked
   c. autosomal recessive
   d. All of these are correct.

11.4 Beyond Mendelian Inheritance

For questions 10–14, match the statements to the items in the key.

Key:

a. multiple alleles
b. polygenic trait
c. pleiotropic gene
d. incomplete dominance

10. A single gene produces a variety of phenotypes.
11. Multiple genes are involved, and the distribution resembles a bell-shaped curve.
12. In humans, there are three possible alleles at the chromosomal locus that determine blood type.
13. A cross of two heterozygous individuals produces an intermediate phenotype.
14. The environment may influence the phenotypic distribution of the trait.

ENGAGE

LEARNSMART LABS®

The following LearnSmart Labs contain exercises that are related to the content of this chapter:

- Mendelian Genetics
- Human Genetics

THINKING SCIENTIFICALLY

1. In tomatoes, red fruit (*R*) is dominant over yellow fruit (*r*), and tallness (*T*) is dominant over shortness (*t*). A plant that is *RrTT* is crossed with a plant that is *rrTt.* What are the chances of an offspring possessing both recessive traits?
2. Assume that you are working in a genetics lab with *Drosophila melanogaster,* a model organism for genetic research. You want to determine whether a newly found *Drosophila* characteristic is dominant or recessive. Explain how you would construct such a cross, and the expected outcomes that would indicate dominance. Why would working with *Drosophila* be easier than working with humans?
3. You want to test whether the leaf pattern of a plant is influenced by the amount of fertilizer in the environment. What would you do?
4. Multiple gene pairs may also interact to produce a single phenotype. In peas, genes *C* and *P* are required for pigment production in flowers. Gene *C* codes for an enzyme that converts a compound into a colorless intermediate product. Gene *P* codes for an enzyme that converts the colorless intermediate product into anthocyanin, a purple pigment. A flower, therefore, will be purple only if it contains at least one dominant allele for each of the two genes (*C*__ *P*__). What phenotypic ratio would you expect in the F_2 generation following a cross between two double heterozygotes (*CcPp*)?

The diversity of life is dependent on variations in genes.

12

Molecular Biology of the Gene

All life on Earth contains the four bases of DNA—A, G, C, and T. Different combinations of these bases make up genes, and every species has its own unique collection and combination of genes. These bases are molecules and code for biological "parts," such as the proteins that make skin, bones, eyes, leaves, and flowers.

MC1R is one of several genes that contribute to skin, hair, and eye color in humans, and studies of this gene have led to major insights into the evolution of human skin pigmentation. *MC1R* is a specific sequence of genetic instructions found in the nucleus of every cell in the human body. The gene is conserved every time a cell divides.

Some of the cells in skin change and become specialized pigment-making cells called melanocytes. All humans have similar numbers of melanocytes, but obviously not all humans have the same skin color. Why would someone have lighter skin or darker skin than another person?

Humans have variation in their *MC1R* and other pigment genes that stem from their ancestry. A few differences in the placement of an adenine (A) or a cytosine (C) in the gene, for example, can alter gene expression—when and how much of the melanin is produced. So although everyone has melanin genes that become replicated with each cell division, each individual's gene expression is determined from the information that DNA transmits to the protein-making machinery.

As you read through this chapter, think about the following questions:

1. How does the flow of genetic information from DNA to protein to trait work?
2. What mechanisms are in place to ensure that genetic information is accurately expressed?

Chapter Outline

12.1 The Genetic Material 208

12.2 Replication of DNA 211

12.3 The Genetic Code of Life 216

12.4 First Step: Transcription 218

12.5 Second Step: Translation 220

Before You Begin

Before beginning this chapter, take a few moments to review the following discussions.

Figures 3.18 and 3.20 What are the components of a nucleotide and the structure of the DNA molecule?

Section 11.2 What is the relationship between an allele and a gene?

Section 11.2 What is the relationship between genotype and phenotype?

Following *the* Themes

CHAPTER 12 MOLECULAR BIOLOGY OF THE GENE

UNIT 2 GENETIC BASIS OF LIFE

Evolution	Changes in DNA sequence produce different proteins that change organismal structure and function.
Nature of Science	Research in gene expression enables scientists to discover new, effective ways to treat disease.
Biological Systems	The flow of genetic information in biological systems is highly regulated and responsive to environmental changes.

12.1 The Genetic Material

Learning Outcomes

Upon completion of this section, you should be able to

1. Describe the properties a substance must possess in order to serve as the genetic material.
2. Examine how historical researchers demonstrated that DNA is the genetic material.
3. Explain the chemical structure of DNA as defined by the Watson and Crick model.

The middle of the twentieth century was an exciting period of scientific discovery. On one hand, geneticists were busy determining that *DNA (deoxyribonucleic acid)* is the genetic material of all living organisms. On the other hand, biochemists were in a frantic race to describe the structure of DNA. The classic experiments performed during this era set the stage for an explosion in our knowledge of modern molecular biology.

When researchers began their work, they knew that the genetic material must be

1. Able to *store information* that pertains to the development, structure, and metabolic activities of the cell or organism
2. Stable, so that it *can be replicated* with high accuracy during cell division and be transmitted from generation to generation
3. Able to *undergo rare changes,* called mutations, that provide the genetic variability required for evolution to occur

This chapter will show, as the researchers of the twentieth century did, that DNA can fulfill these functions.

Transformation of Bacteria

During the late 1920s, the bacteriologist Frederick Griffith (1879–1941) was attempting to develop a vaccine against *Streptococcus pneumoniae* (pneumococcus), which causes pneumonia in mammals. In 1931, he performed a classic experiment with the bacterium. He noticed that when these bacteria are grown on culture plates, some, called S strain bacteria, produce shiny, smooth colonies and others, called R strain bacteria, produce colonies that have a rough appearance. Under the microscope, S strain bacteria have a capsule (mucous coat) that makes them smooth, but R strain bacteria do not.

When Griffith injected mice with the S strain of bacteria, the mice died, but when he injected mice with the R strain, the mice did not die (Fig. 12.1). In an effort to determine whether the capsule alone was responsible for the virulence (ability to kill) of the S strain bacteria, he injected mice with heat-killed S strain bacteria. The mice did not die.

Finally, Griffith injected the mice with a mixture of heat-killed S strain and live R strain bacteria. Most unexpectedly, the mice died—and living S strain bacteria were recovered from the bodies! Griffith concluded that some substance necessary for the bacteria to produce a capsule and be virulent must have passed from the dead S strain bacteria to the living R strain bacteria, so that the R strain bacteria were *transformed* (Fig. 12.1*d*). This change in the phenotype of the R strain bacteria must have been due to a change in their genotype. Indeed, couldn't the transforming substance that passed from S strain to R strain be genetic material? Reasoning such as this prompted investigators at the time to begin looking for the transforming substance to determine the chemical nature of the genetic material.

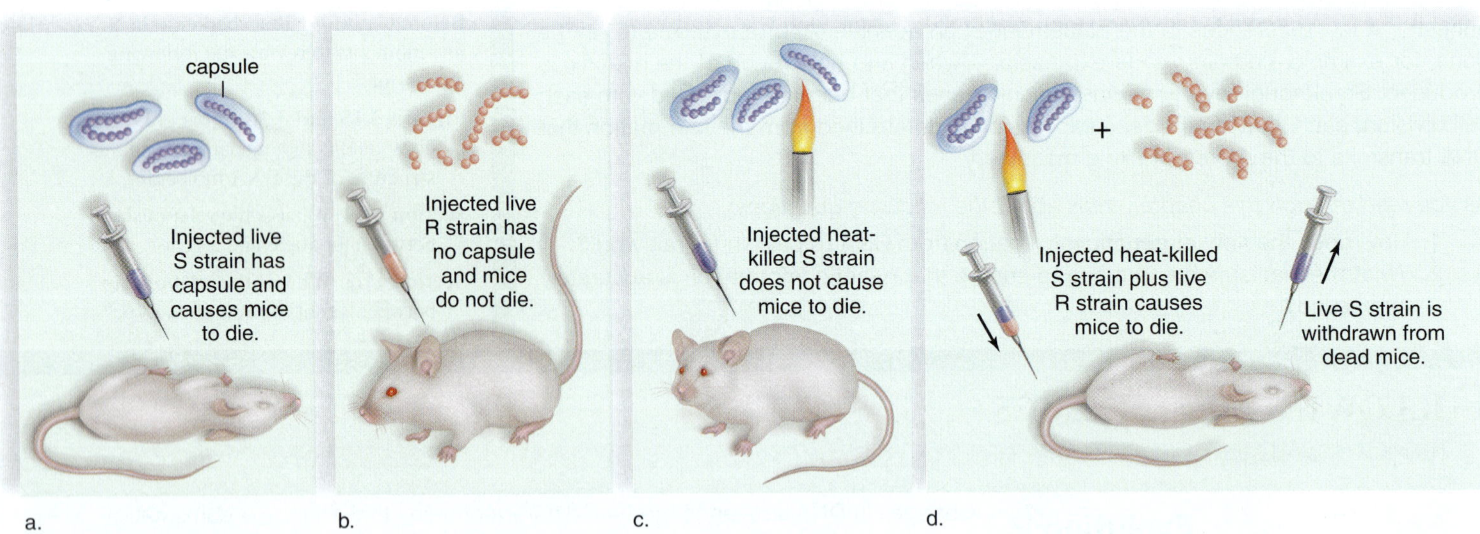

Figure 12.1 Griffith's transformation experiment. **a.** Encapsulated S strain is virulent and kills mice. **b.** Nonencapsulated R strain is not virulent and does not kill mice. **c.** Heat-killed S strain bacteria do not kill mice. **d.** If heat-killed S strain and R strain are both injected into mice, they die, because the R strain bacteria have been transformed into the virulent S strain.

DNA: The Transforming Substance

By the time the next group of investigators, led by Oswald Avery (1877–1955) in the 1940s, began their work, it was known that the genes are on the chromosomes and that the chromosomes contain both proteins and nucleic acids. Investigators were having a very heated debate about whether protein or DNA was the genetic material. Many thought that the protein component of chromosomes must be the genetic material because proteins contain up to 20 different amino acids that can be sequenced in any particular way. On the other hand, nucleic acids—DNA and RNA—contain only four types of nucleotides as basic building blocks. Some argued that DNA did not have enough variability to be able to store information and be the genetic material.

In 1944, after 16 years of research, Oswald Avery and his coinvestigators, Colin MacLeod and Maclyn McCarty, published a paper demonstrating that the transforming substance that allows *Streptococcus* to produce a capsule and be virulent is DNA. This meant that DNA is the genetic material. Here is what they found out:

1. DNA from S strain bacteria causes R strain bacteria to be transformed, so that they can produce a capsule and be virulent.
2. The addition of DNase, an enzyme that digests DNA, prevents transformation from occurring. This supports the hypothesis that DNA is the genetic material.
3. The molecular weight of the transforming substance is large. This suggests the possibility of genetic variability.
4. The addition of enzymes that degrade proteins has no effect on the transforming substance, nor does RNase, an enzyme that digests RNA. This shows that neither protein nor RNA is the genetic material.

These experiments showed that DNA is the transforming substance and, therefore, the genetic material. Although some scientists remained skeptical, many felt that the evidence for DNA being the genetic material was overwhelming.

An experiment by Alfred Hershey and Martha Chase in the early 1950s helped to firmly establish DNA as the genetic material. Hershey and Chase used a virus called a T phage, composed of radioactively labeled DNA and capsid coat proteins, to infect *E. coli* bacteria. They discovered that the radioactive tracers for DNA, but not protein, ended up inside the bacterial cells, causing them to become transformed. Since only the genetic material could have caused this transformation, Hershey and Chase determined that DNA must be the genetic material (Fig. 12.2).

Animation Hershey and Chase Experiment

The Structure of DNA

By the early 1950s, DNA was widely accepted as the genetic material of all living organisms. However. the structure of DNA was not known. How can a molecule with only four different nucleotides produce the great diversity of life on Earth?

To understand the structure of DNA, we need to understand how the bases in DNA are composed. Investigators knew that DNA contains four different types of nucleotides: two with *purine* bases, **adenine** (**A**) and **guanine** (**G**), which have a double ring; and two with *pyrimidine* bases, **thymine** (**T**) and **cytosine** (**C**), which have a single ring (Fig. 12.3*a, b*). Erwin Chargaff used new chemical techniques developed in the 1940s to analyze in detail the base content of DNA.

A sample of Chargaff's data is seen in Figure 12.3*c*. You can see that while some species—*E. coli* and *Zea mays* (corn), for example—do

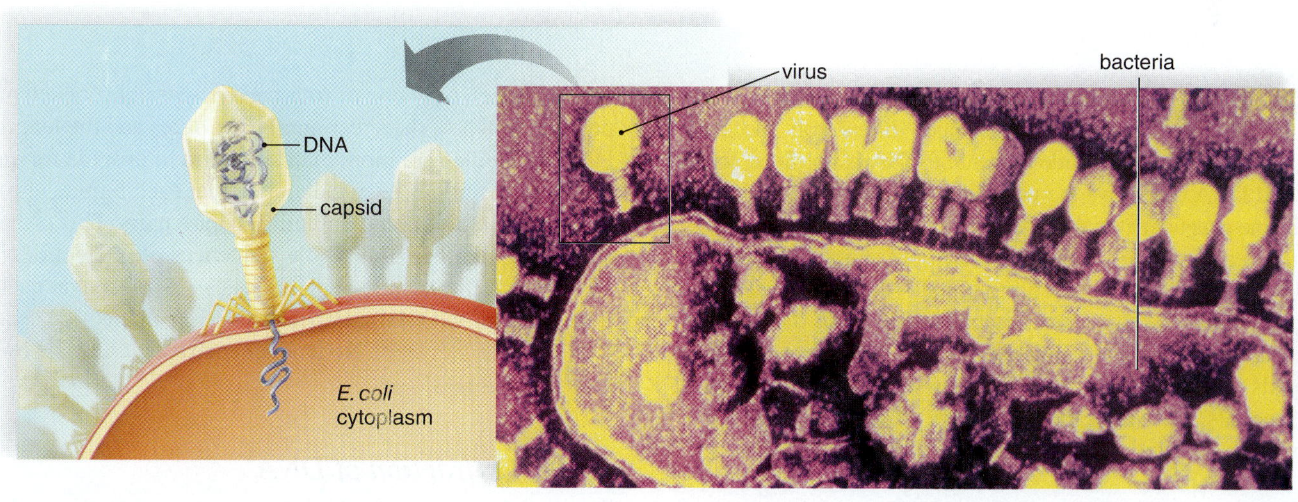

Figure 12.2 The tools of Hershey and Chase. **a.** Drawing of a T phage showing the protein coat (capsid) and DNA. **b.** An electron micrograph showing T phage infecting bacteria. The DNA was labeled with radioactivity, allowing Hershey and Chase to follow its progress into a bacterial cell and eventually provide a blueprint to make new phages.

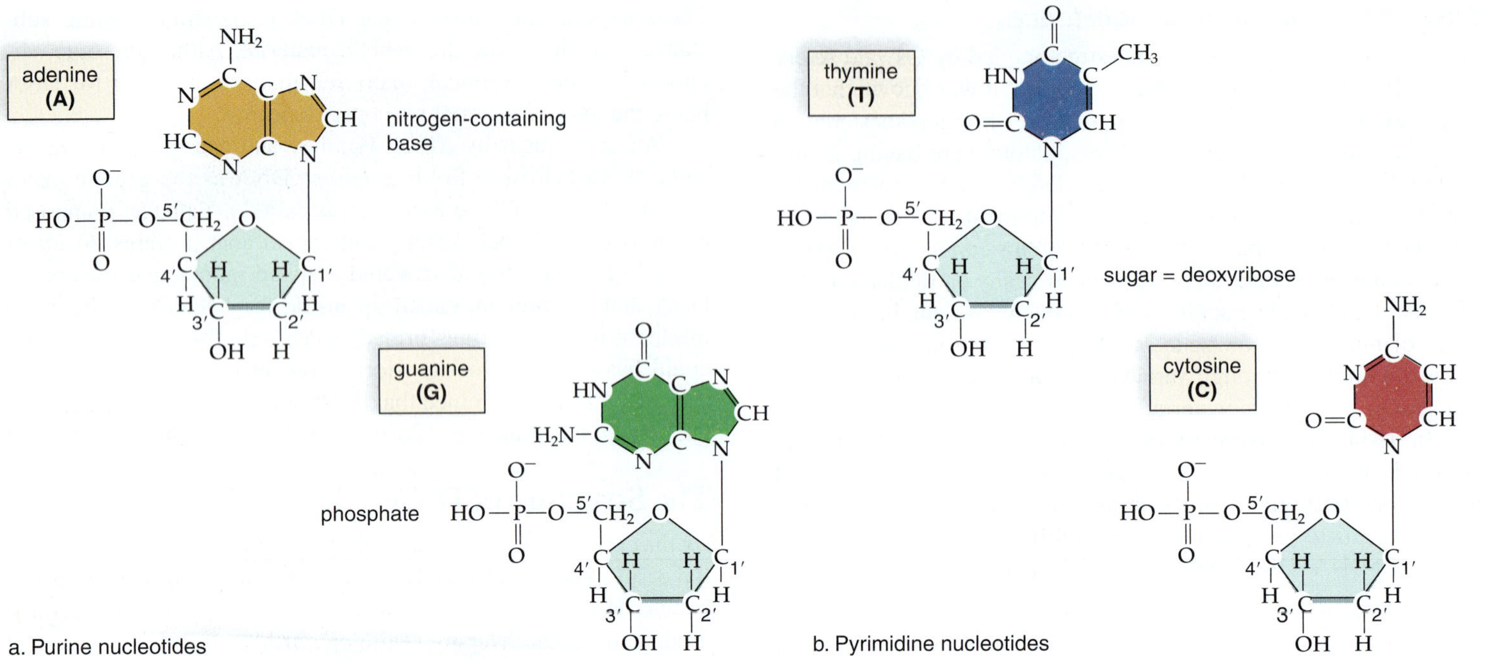

DNA Composition in Various Species (%)				
Species	**A**	**T**	**G**	**C**
Homo sapiens (human)	31.0	31.5	19.1	18.4
Drosophila melanogaster (fruit fly)	27.3	27.6	22.5	22.5
Zea mays (corn)	25.6	25.3	24.5	24.6
Neurospora crassa (fungus)	23.0	23.3	27.1	26.6
Escherichia coli (bacterium)	24.6	24.3	25.5	25.6
Bacillus subtilis (bacterium)	28.4	29.0	21.0	21.6

c. Chargaff's data

Figure 12.3 Nucleotide composition of DNA. All nucleotides contain phosphate, a 5-carbon sugar, and a nitrogen-containing base. In DNA, the sugar is called deoxyribose, because it lacks an oxygen atom in the 2′ position, compared to ribose. The nitrogen-containing bases are **(a)** the purines adenine and guanine, which have a double ring, and **(b)** the pyrimidines thymine and cytosine, which have a single ring. **c.** Chargaff's data show that the DNA of various species differs. For example, in humans the A and T percentages are about 31%, but in fruit flies these percentages are about 27%.

have approximately 25% of each type of nucleotide, most do not. Further, the percentage of each type of nucleotide differs from species to species. Therefore, the nucleotide content of DNA is not fixed across species, and DNA does have the *variability* between species required for it to be the genetic material.

Within each species, however, DNA was found to have the *constancy* required of the genetic material—that is, all members of a species have the same base composition. Also, the percentage of A always equals the percentage of T, and the percentage of G equals the percentage of C. It follows that if the percentage of A + T equals 40%, then the percentage of G + C equals 60%. These relationships are called Chargaff's rules.

Chargaff's rules:

1. The amount of A, T, G, and C in DNA varies from species to species.
2. In each species, the amount of A = T and the amount of G = C.

Although only one of four bases is possible at each nucleotide position in DNA, the sheer number of bases and the length of most DNA molecules are more than sufficient to provide for variability. For example, it has been calculated that each human chromosome typically contains about 140 million base pairs. This provides for a staggering number of possible sequences of nucleotides. Because any of the four possible nucleotides can be present at each nucleotide position, the total number of possible nucleotide sequences is $4^{(140 \times 10^6)}$, or $4^{140,000,000}$. No wonder each species has its own unique base percentages!

X-Ray Diffraction of DNA

Rosalind Franklin (Fig. 12.4*a*), a researcher at King's College in London, studied the structure of DNA using X-rays. She found that if a concentrated, viscous solution of DNA is made, it can be separated into fibers. Under the right conditions, the fibers are enough like a crystal (a solid substance whose atoms are arranged in a definite manner) that when X-rayed, an X-ray diffraction pattern results (Fig. 12.4*b*).

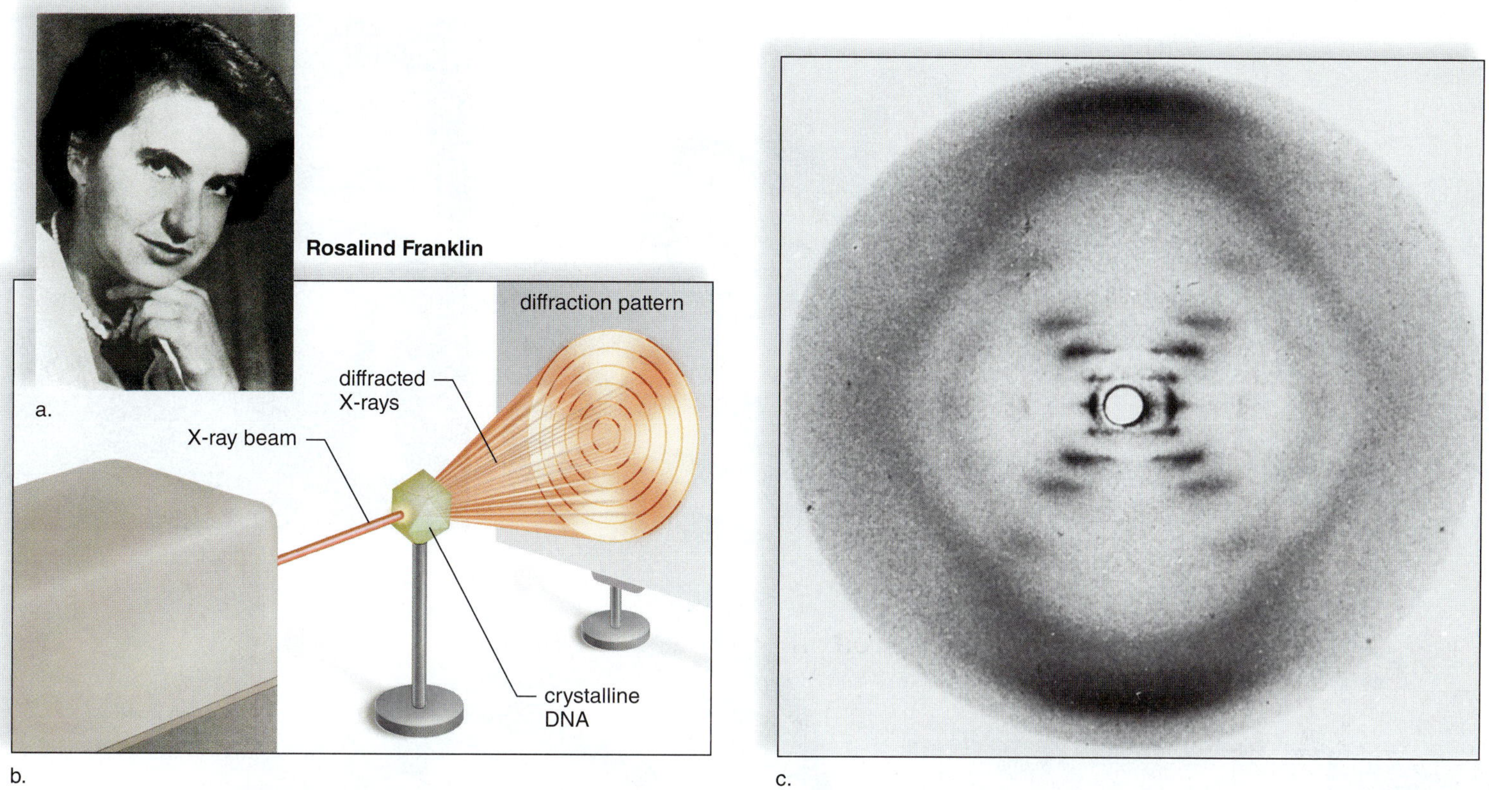

Figure 12.4 X-ray diffraction of DNA. **a.** Rosalind Franklin (1920–1958). **b.** When a crystal is X-rayed, the way in which the beam is diffracted reflects the pattern of the molecules in the crystal. The closer together two repeating structures are in the crystal, the farther from the center the beam is diffracted. **c.** The diffraction pattern of DNA produced by Rosalind Franklin. The crossed (X) pattern in the center told investigators that DNA is a helix, and the dark portions at the top and the bottom told them that some feature is repeated over and over. Watson and Crick determined that this feature was the hydrogen-bonded bases.

The X-ray diffraction pattern of DNA shows that DNA is a double helix. The helical shape is indicated by the crossed (X) pattern in the center of the photograph in Figure 12.4*c*. The dark portions at the top and bottom of the photograph indicate that some portion of the helix is repeated. Maurice H. F. Wilkins, a colleague of Franklin's, showed one of her crystallographic patterns to James Watson, who immediately grasped its significance.

Animation DNA Structure

The Watson and Crick Model

James Watson, an American, was on a postdoctoral fellowship at Cavendish Laboratories in Cambridge, England, when he began to work with the biophysicist Francis H. C. Crick. Using the data provided from X-ray diffraction and other sources, they constructed a model of DNA, for which they received a Nobel Prize in 1962.

Based on previous work of other scientists, Watson and Crick knew that DNA is a polymer of nucleotides, but they did not know how the nucleotides were arranged within the molecule. However, they deduced that DNA is a **double helix** with sugar-phosphate backbones on the outside and paired bases on the inside. This arrangement fits the mathematical measurements provided by Franklin's X-ray diffraction data for the spacing between the base pairs (0.34 nm) and for a complete turn of the double helix (3.4 nm).

According to Watson and Crick's model, the two DNA strands of the double helix are *antiparallel,* meaning that the sugar-phosphate groups that are chained together to make each strand are oriented in opposite directions. As seen in Figure 12.5, each nucleotide possesses a phosphate group located at the 5′ position of the sugar. Nucleotides are joined together by linking the 5′ phosphate of one nucleotide to a free hydroxyl (–OH) located at the 3′ position on the sugar of the preceding nucleotide, giving the molecule directionality. *Antiparallel* simply means that while one DNA strand runs 5′ to 3′, the other strand runs in a parallel but opposite direction.

This model also agreed with Chargaff's rules, which state that A = T and G = C. Figure 12.5 shows that A is hydrogen-bonded to T, and G is hydrogen-bonded to C. This **complementary base pairing** means that a purine (large, two-ring base) is always bonded to a pyrimidine (smaller, one-ring base). This antiparallel pairing arrangement of the two strands ensures that the bases are oriented properly, so that they can interact. The consistent spacing between the two strands of the DNA was detected by Franklin's X-ray

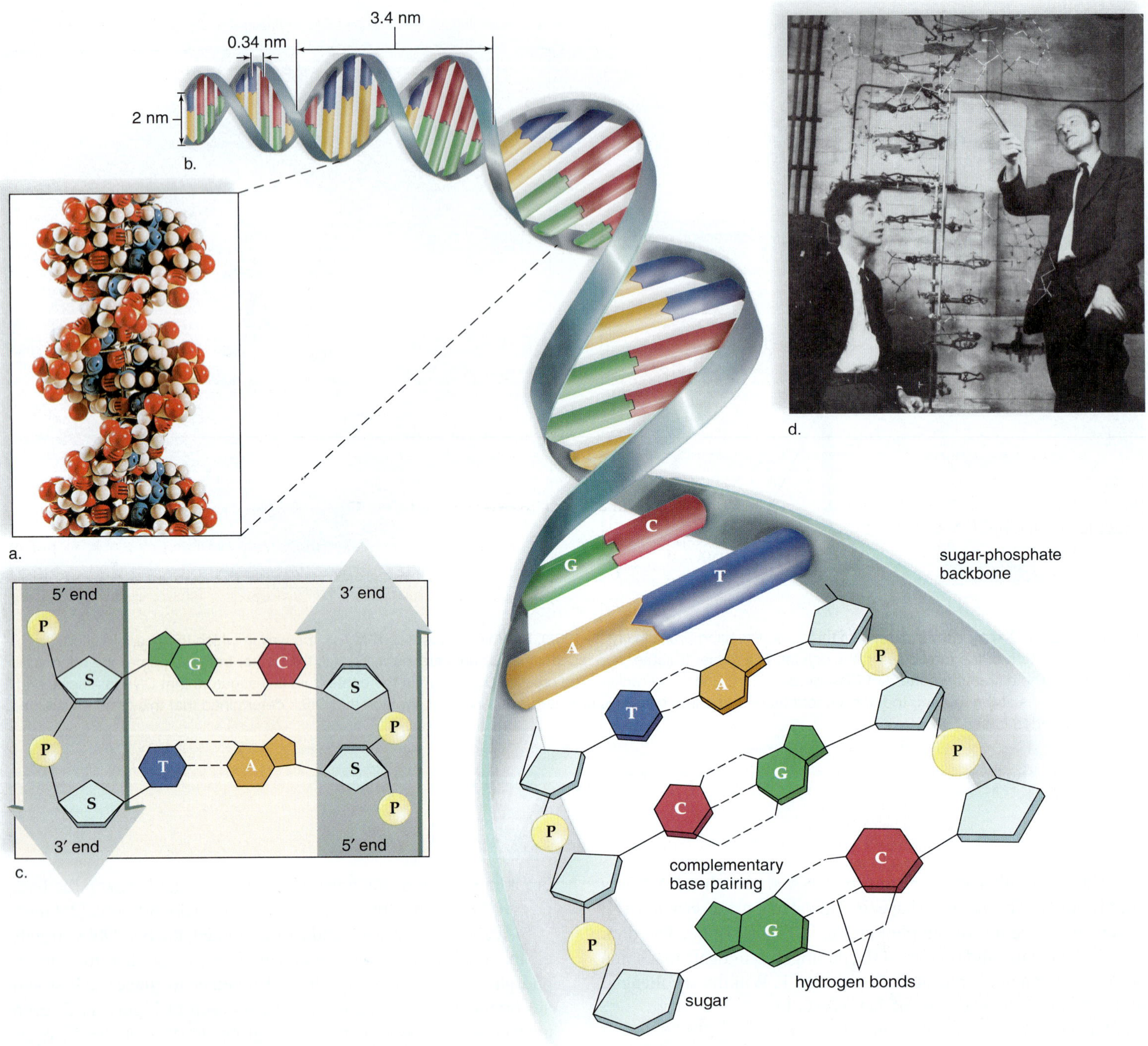

Figure 12.5 Watson and Crick model of DNA.
a. Space-filling model of DNA. **b.** The double helix molecules. **c.** The two strands of the molecule are antiparallel. The direction of the strand is said to be 5′ to 3′ when going from top to bottom. The 5′ end of a DNA strand has a phosphate group, while the 3′ end has an –OH group (not shown). **d.** James Watson (*left*) and Francis Crick (*right*) deduced the molecular configuration of DNA.

diffraction pattern, because two pyrimidines together are too narrow, and two purines together are too wide.

The information stored within DNA must always be read in the 5′ to 3′ direction. Thus, a DNA strand is usually replicated in a 5′ to 3′ direction.

3D Animation DNA Replication: DNA Structure

Check Your Progress 12.1

1. Explain the major features of DNA structure.
2. Explain the roles of Erwin Chargaff and Rosalind Franklin in elucidating the final structure of DNA.

12.2 Replication of DNA

Learning Outcomes

Upon completion of this section, you should be able to

1. Explain why the replication of DNA is semiconservative.
2. Describe the enzymes and proteins involved in DNA replication.
3. Contrast DNA replication in eukaryotes and prokaryotes.

The term **DNA replication** refers to the process of copying a DNA molecule. Following replication, there is usually an exact copy of the parental DNA double helix. As soon as Watson and Crick developed their double helix model, they commented, "It has not escaped our notice that the specific pairing we have postulated immediately suggests a possible copying mechanism for the genetic material."

A **template** is most often a mold used to produce a shape complementary to itself. During DNA replication, each DNA strand of the parental double helix serves as a template for a new strand in a daughter molecule. DNA replication is termed **semiconservative replication,** because each daughter DNA double helix contains an old strand from the parental DNA double helix and a new strand. In Figure 12.6, the backbones of the parental DNA molecule are blue. Following replication, the daughter molecules each have a green backbone (new strand) and a blue backbone (old strand). Because A pairs with T, and G pairs with C, a daughter DNA double helix has the same sequence of bases as the parental DNA double helix had originally.

DNA replication requires three main steps: unwinding, complementary base pairing, and joining. At the molecular level, several enzymes and proteins participate in the synthesis of the new DNA strands (Fig. 12.7 and Table 12.1).

Animation DNA Replication

Unwinding

A **DNA helicase** enzyme unwinds DNA and separates the parental strands. This creates two replication forks that move away from each other. These separated strands now become the template to create two new DNA molecules. DNA is chemically stable as a helix, but not as single strands. **Single-stranded binding proteins (SSB)** attach to newly separated DNA and prevent it from re-forming the helix so replication can occur.

Table 12.1 Proteins Involved in DNA Replication

Protein Name	Function
DNA helicase	Separates double-stranded DNA into single strands
Single-stranded binding protein (SSB)	Binds to single-stranded DNA and prevents it from re-forming a double helix
DNA primase	Synthesizes short RNA primers
DNA polymerase	Synthesizes DNA in the leading and lagging strands, removes RNA primers filling the gaps with more DNA, and proofreads newly made DNA
DNA ligase	Covalently attaches adjacent Okazaki fragments in the lagging strand

Complementary Base Pairing

DNA replication needs a *primer,* a short strand of RNA, to put in place before replication can begin. **DNA primase** places short primers on the strands to be replicated. DNA polymerase recognizes this RNA target and begins DNA synthesis, allowing new nucleotides to form complementary base pairs with the old strand and connecting the new nucleotides together in a chain. DNA polymerase also proofreads the strands and can correct any mistakes.

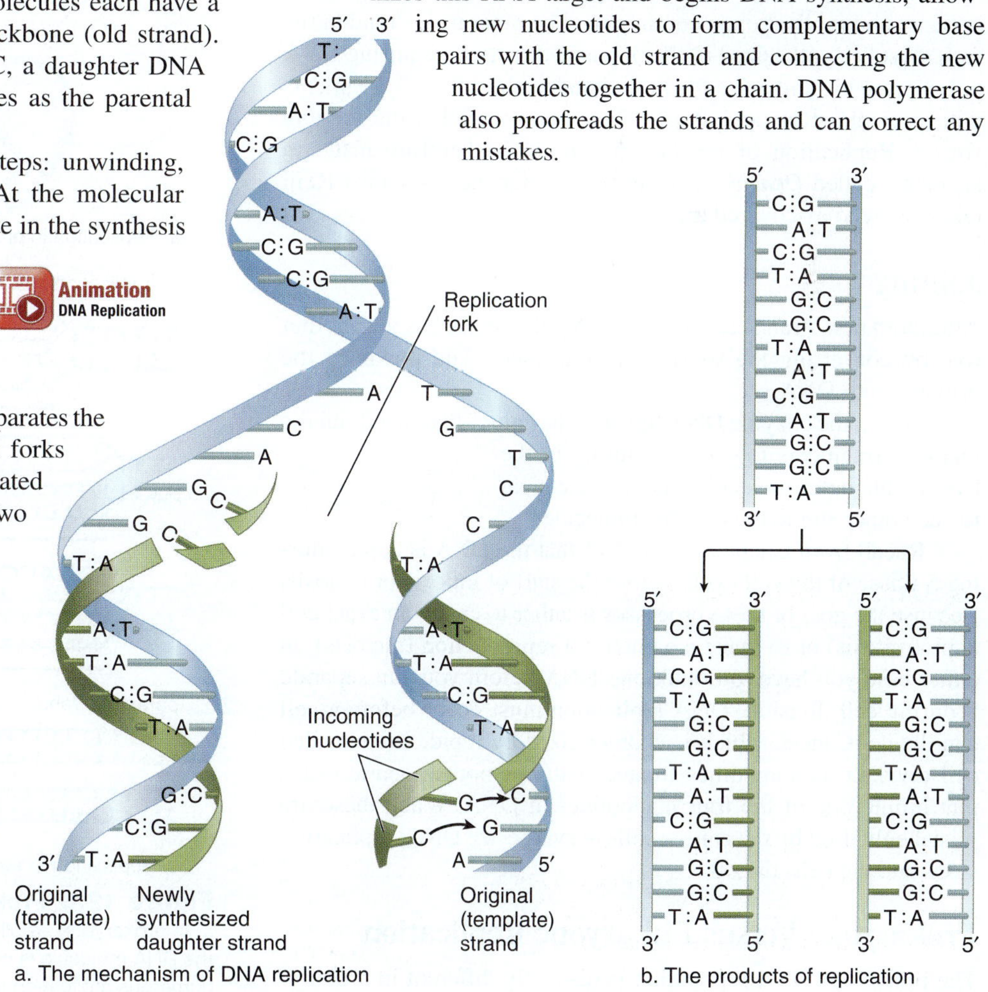

Figure 12.6 Semiconservative replication (simplified). After the DNA double helix unwinds, each parental strand serves as a template for the formation of the new daughter strands. Complementary free nucleotides hydrogen bond to a matching base (e.g., A with T; G with C) in each parental strand and are joined to form a complete daughter strand. Two helices, each with a daughter and parental strand, are produced following replication.

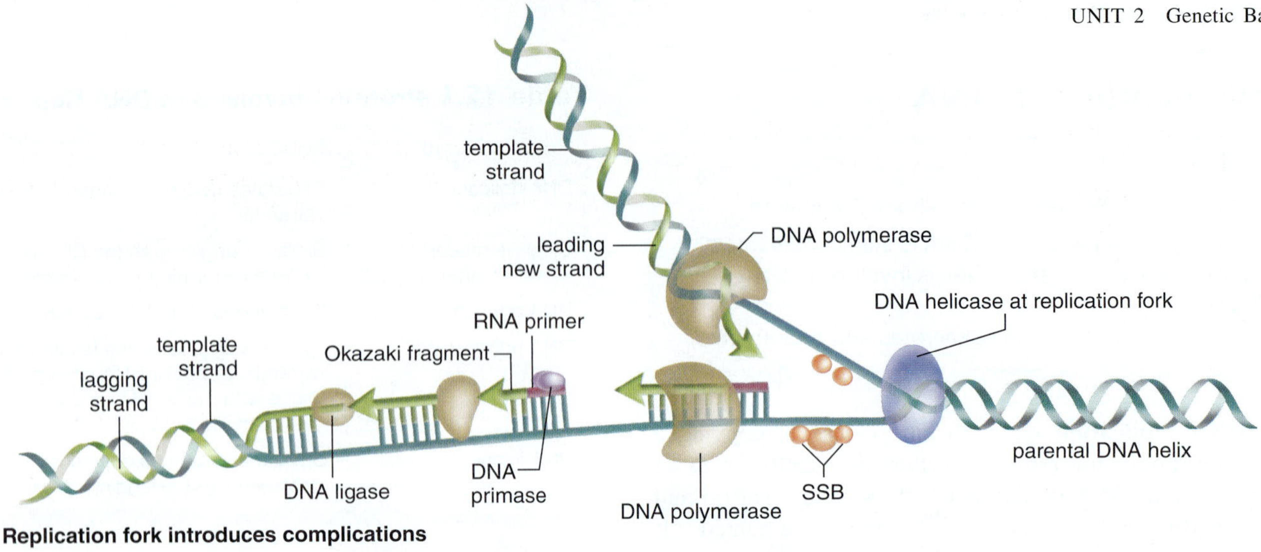

Figure 12.7 Enzymes in DNA replication. The major enzymes involved in DNA replication. Note that the synthesis of the new DNA molecules occurs in opposite directions due to the orientation of the original DNA strands.

Tutorial DNA Replication

The parental strands are antiparallel to each other, and each of the new daughter strands must also be antiparallel to its matching parental strand—which creates a problem. DNA can only be synthesized in a 5′ to 3′ direction (see Fig. 12.5*c*). One strand, the *leading strand,* is exposed so that synthesis in a 5′ to 3′ direction is easier and replication is continuous. The other new strand in the fork must be synthesized in the opposite direction, requiring DNA polymerase to synthesize the new strand in short 5′ to 3′ segments with periodic starts and stops. This strand is called the *lagging strand.* Replication of the lagging strand is therefore made in segments called *Okazaki fragments,* after Japanese scientist Reiji Okazaki, who discovered them.

Joining

After both new strands are made, DNA polymerase has yet another role by converting the short RNA sequences, laid down by the primase, into DNA.

Finally, the enzyme **DNA ligase** is the "glue" that mends all the Okazaki fragments together, resulting in the two double helix molecules that are identical to each other and to the original molecule.

3D Animation DNA Replication

Recall from Chapters 10 and 11 that the DNA is copied during S phase of the cell cycle before the start of mitosis or meiosis. Because the goal of these processes is either to create an exact cell copy (mitosis) or to make a gamete for reproduction (meiosis), in either case you have to double the DNA before you can separate it during cell division. DNA replication must occur before a cell can divide. Cancer, which is characterized by rapid, uncontrolled cell division, is sometimes treated with chemotherapeutic drugs that mimic one of the four nucleotides in DNA. When these are mistakenly used by the cancer cells to synthesize DNA, replication stops and the cells die off.

Prokaryotic Versus Eukaryotic Replication

The process of DNA replication is distinctly different in prokaryotic and eukaryotic cells, although many of these organisms' basic functions are similar (Fig. 12.8).

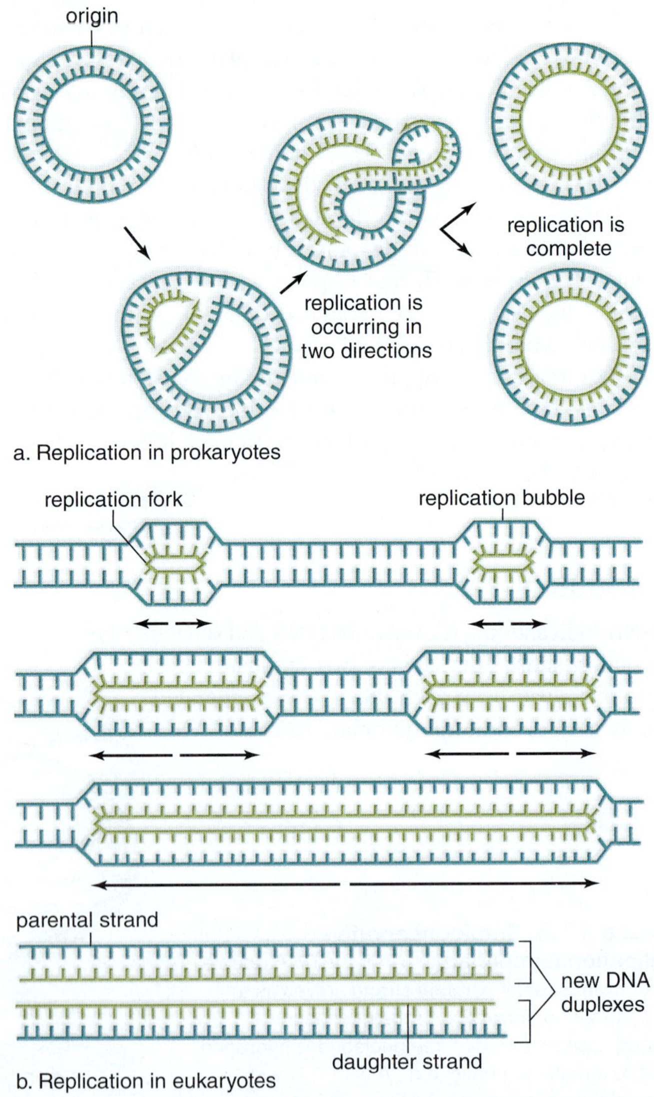

Figure 12.8 Prokaryotic versus eukaryotic replication. **a.** In prokaryotes, replication can occur in two directions at once, because the DNA molecule is circular. **b.** In eukaryotes, replication occurs at numerous replication bubbles, each with two forks. The forks move away from each other until they meet again and the two new daughter helices have been completed.

Prokaryotic DNA Replication

Bacteria have a single circular loop chromosome, whose DNA must be replicated before the cell divides. In some circular DNA molecules, replication moves around the DNA molecule in one direction only. In others, as shown in Figure 12.8*a,* replication occurs in two directions. The process always occurs in the 5′ to 3′ direction.

The process begins at the *origin of replication,* a specific site on the bacterial chromosome. The strands are separated and unwound, and a DNA polymerase enzyme binds to each side of the opening and begins the copying process. When the two DNA polymerases meet at a termination region, replication is halted, and the two copies of the chromosome are separated.

Bacterial cells require about 40 minutes to replicate the complete chromosome. Because bacterial cells are able to divide as often as once every 20 minutes, it is possible for a new round of DNA replication to begin even before the previous round is completed!

Eukaryotic DNA Replication

In eukaryotes, DNA replication begins at numerous origins of replication along the length of the linear chromosome, and replication bubbles spread bidirectionally until they meet. Notice in Figure 12.8*b* that there is a V shape wherever DNA is being replicated. This is called a **replication fork.**

The chromosomes of eukaryotes are long, making replication a more time-consuming process. Eukaryotes replicate their DNA at a slower rate—500 to 5,000 base pairs per minute—but there are many individual origins of replication to accelerate the process. Therefore, eukaryotic cells complete the replication of the diploid amount of DNA (in humans, over 6 billion base pairs) in a matter of hours!

The linear chromosomes of eukaryotes also pose another problem: DNA polymerase is unable to replicate the ends of the chromosomes. The ends of eukaryotic chromosomes are composed of telomeres, which are short DNA sequences that are repeated over and over. Telomeres are not copied by DNA polymerase; rather, they are added by an enzyme called telomerase, which adds the correct number of repeats after the chromosome is replicated. In stem cells, this process preserves the ends of the chromosomes and prevents the loss of DNA after successive rounds of replication. Unregulated telomerase activity can negatively affect cell function, as seen with uncontrolled cell division in cancer cells.

Accuracy of Replication

A DNA polymerase is very accurate and makes a mistake approximately once per 100,000 base pairs at most. This error rate, however, would result in many errors accumulating over the course of several cell divisions. DNA polymerase is also capable of checking for accuracy, or proofreading the daughter strand it is making. It can recognize a mismatched nucleotide and remove it from a daughter strand by reversing direction and removing several nucleotides. Once it has removed the mismatched nucleotide, it changes direction again and resumes making DNA. Overall, the error rate for the bacterial DNA polymerase is only 1 in 100 million base pairs!

Check Your Progress 12.2

1. Explain the three major steps in DNA replication.
2. Explain why replication must occur differently on the leading and lagging strands.
3. Compare DNA replication in prokaryotes and eukaryotes.

12.3 The Genetic Code of Life

Learning Outcomes

Upon completion of this section, you should be able to

1. Explain the function of transcription and translation.
2. Explain how the mRNA nucleotides determine the sequence of amino acids in a polypeptide.

Evidence began to mount in the 1900s that metabolic disorders can be inherited. An English physician, Sir Archibald Garrod, called them "inborn errors of metabolism." Investigators George Beadle and Edward Tatum, working with red bread mold, proposed what they called the "one gene, one enzyme hypothesis," based on the observation that a defective gene caused a defective enzyme.

This and many other examples illustrate the flow of genetic information from DNA to RNA to protein to an observed trait. We now turn our attention to the transfer of information from DNA to RNA, the next component in the system.

RNA Carries the Information

Like DNA, *RNA* (*ribonucleic acid*) is a polymer composed of nucleotides. The nucleotides in RNA, however, contain the sugar ribose and the bases adenine (A), cytosine (C), guanine (G), and **uracil** (U). In RNA, the base uracil replaces the thymine found in DNA. Finally, RNA is single-stranded and does not form a double helix in the same manner as DNA (Table 12.2 and Fig. 12.9).

There are three major classes of RNA. Each class has a unique size, shape, and function in protein synthesis.

Messenger RNA (mRNA) takes a message from DNA in the nucleus to the ribosomes in the cytoplasm.
Transfer RNA (tRNA) transfers amino acids to the ribosomes.
Ribosomal RNA (rRNA), along with ribosomal proteins, makes up the ribosomes, where polypeptides are synthesized.

Table 12.2 RNA Structure Compared to DNA Structure

	RNA	DNA
Sugar	Ribose	Deoxyribose
Bases	Adenine, guanine, uracil, cytosine	Adenine, guanine, thymine, cytosine
Strands	Single-stranded	Double-stranded with base pairing
Helix	No	Yes

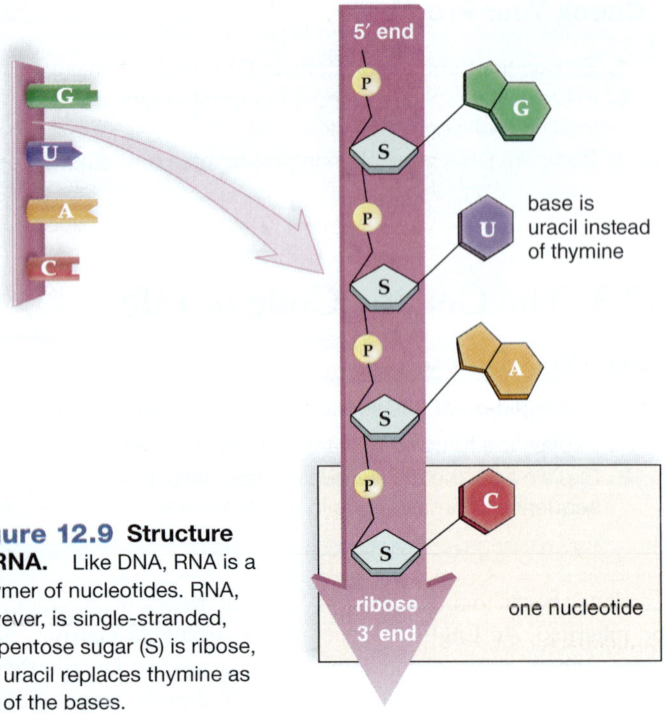

Figure 12.9 Structure of RNA. Like DNA, RNA is a polymer of nucleotides. RNA, however, is single-stranded, the pentose sugar (S) is ribose, and uracil replaces thymine as one of the bases.

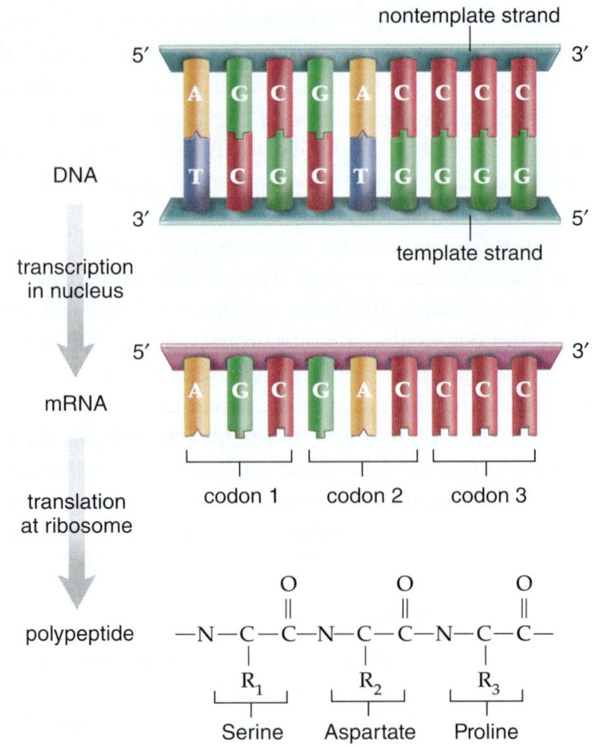

Figure 12.10 The flow of genetic information in a cell. One strand of DNA acts as a template for mRNA synthesis, and the sequence of bases in mRNA determines the sequence of amino acids in a polypeptide.

Tutorial
Overview of Gene Expression

The Genetic Code

In the genetic flow of information, two major steps are needed to convert the information stored in DNA into a protein that supports body function (Fig. 12.10). First, the DNA undergoes **transcription** (L. *trans,* "across"; *scriptio,* "a writing"), a process by which an RNA molecule is produced based on a DNA template. DNA is transcribed, or copied base by base, into mRNA, tRNA, and rRNA.

Second, during **translation** (L. *trans,* "across"; *latus,* "carry or bear"), the mRNA transcript is read by a ribosome and converted into the sequence of amino acids in a polypeptide. Like a translator who understands two languages, the cell changes a nucleotide sequence into an amino acid sequence. Together, the flow of information from DNA to RNA to protein to trait is known as the **central dogma** of molecular biology.

Now that we know that the DNA sequence within a gene is transcribed into an RNA molecule and, for genes that code for proteins, the mRNA sequence determines the sequence of amino acids in a protein, it becomes necessary to identify the specific **genetic code** for each of the 20 amino acids found in proteins. Although scientists knew that DNA somehow directed protein production, they did not initially know specifically how the code was translated. This discovery was made in the 1960s.

Finding the Genetic Code

Logically, the genetic code would have to be at least a **triplet code;** that is, each coding unit, or **codon,** would need to be made up of three nucleotides. The reason is that fewer nucleotides would not provide sufficient variety to encode 20 different amino acids.

In 1961, Marshall Nirenberg and J. Heinrich Matthei performed an experiment that laid the groundwork for cracking the genetic code. First, they found that a cellular enzyme could be used to construct a synthetic RNA (one that does not occur in cells), and then they found that the synthetic RNA polymer could be translated in a test tube that contained the cytoplasmic contents of a cell. Their first synthetic RNA was composed only of uracil, and the protein that resulted was composed only of the amino acid phenylalanine. Therefore, the mRNA codon for phenylalanine was known to be UUU. Later, they were able to translate just three nucleotides at a time; in that way, it was possible to assign an amino acid to each of the mRNA codons (Fig. 12.11).

Like the periodic table and other major works, the genetic code seen in Figure 12.11 is a masterpiece of scientific discovery, because it is a key that unlocks the very basis of biological life. Here are some of its features:

1. The genetic code is *degenerate.* This term means that most amino acids have more than one codon; leucine, serine, and arginine have six different codons, for example. The degeneracy (redundancy) of the code helps protect against potentially harmful mutations.
2. The genetic code is *unambiguous.* Each triplet codon has only one meaning.
3. The code has *start and stop signals.* There is only one start signal, but there are three stop signals.

The Code Is Universal

With a few exceptions, the genetic code (Fig. 12.11) is universal to all living organisms. In 1979, however, researchers discovered

First Base	Second Base				Third Base
	U	C	A	G	
U	UUU phenylalanine	UCU serine	UAU tyrosine	UGU cysteine	U
	UUC phenylalanine	UCC serine	UAC tyrosine	UGC cysteine	C
	UUA leucine	UCA serine	UAA **stop**	UGA **stop**	A
	UUG leucine	UCG serine	UAG **stop**	UGG tryptophan	G
C	CUU leucine	CCU proline	CAU histidine	CGU arginine	U
	CUC leucine	CCC proline	CAC histidine	CGC arginine	C
	CUA leucine	CCA proline	CAA glutamine	CGA arginine	A
	CUG leucine	CCG proline	CAG glutamine	CGG arginine	G
A	AUU isoleucine	ACU threonine	AAU asparagine	AGU serine	U
	AUC isoleucine	ACC threonine	AAC asparagine	AGC serine	C
	AUA isoleucine	ACA threonine	AAA lysine	AGA arginine	A
	AUG (start) methionine	ACG threonine	AAG lysine	AGG arginine	G
G	GUU valine	GCU alanine	GAU aspartate	GGU glycine	U
	GUC valine	GCC alanine	GAC aspartate	GGC glycine	C
	GUA valine	GCA alanine	GAA glutamate	GGA glycine	A
	GUG valine	GCG alanine	GAG glutamate	GGG glycine	G

Figure 12.11 Messenger RNA codons. Notice that in this chart each of the codons (in boxes) is composed of three letters representing the first base, second base, and third base. For example, find the box where C for the first base and A for the second base intersect. You will see that U, C, A, or G can be the third base. The bases CAU and CAC are codons for histidine; the bases CAA and CAG are codons for glutamine.

that the genetic code used within the mitochondria, chloroplasts, and some archaebacteria, differs slightly from the more familiar genetic code.

The universal nature of the genetic code provides strong evidence that all living organisms share a common evolutionary heritage. Because the same genetic code is used by all living organisms, it is possible to transfer genes from one organism to another. Many commercial and medicinal products, such as human insulin, can be produced in this manner. The Nature of Science feature, "Moving Genes Between Species: Green Fluorescent Protein and Cells," on page 219 demonstrates that the gene for GFP could be transferred from jellyfish to a number of other organisms to cause a fluorescent green color. This is made possible only because the genetic code is universal.

Check Your Progress 12.3

1. Examine the flow of genetic information in a cell.
2. Describe the three major classes of RNA; what is the function of each class?
3. Explain why the genetic code is said to be degenerate.

12.4 First Step: Transcription

Learning Outcomes

Upon completion of this section, you should be able to

1. Distinguish among the events of transcription that occur during formation of an mRNA molecule.
2. Describe how eukaryotic mRNA molecules are processed and exported to the cytoplasm.

During *transcription,* a segment of the DNA serves as a template for the production of an RNA molecule. Although mRNA, tRNA, and rRNA are all produced by transcription, we focus here on transcription to make mRNA, the type of RNA that eventually leads to building a protein.

MP3 Protein Synthesis

Animation Stages of Transcription

Messenger RNA Is Produced

The sequences of bases in a gene are transcribed into an mRNA molecule based on complementary base pairing: The T base in the DNA pairs with A in the mRNA, G with C, and A with U (note that uracil replaces T in the newly formed mRNA) (Fig. 12.12). When a gene is transcribed, a segment of the DNA helix unwinds and unzips, and complementary RNA nucleotides pair with DNA nucleotides of the strand opposite the gene. This strand is known as the *template strand;* the other strand is the gene strand. An **RNA polymerase** joins the nucleotides together in the 5′ to 3′ direction. Like DNA polymerase, an RNA polymerase adds a nucleotide only to the 3′ end of the polymer under construction.

Transcription begins when RNA polymerase attaches to a region of DNA called a promoter (Fig. 12.12). A **promoter** defines the start of transcription, the direction of transcription, and the strand to be transcribed. The binding of RNA polymerase to the promoter is the *initiation* of transcription. The RNA-DNA association is not as stable as the two strands in the DNA helix. Therefore, only the newest portion of an RNA molecule that is associated with RNA polymerase is bound to the DNA, and the rest dangles off to the side.

Elongation of the mRNA molecule occurs as the RNA polymerase reads down the DNA template strand in a 5′ to 3′ direction and continues until RNA polymerase comes to a DNA stop sequence, where *termination* occurs. The stop sequence causes RNA polymerase to stop transcribing the DNA and to release the mRNA molecule, now called an **mRNA transcript.**

It is not necessary for RNA polymerase to finish making one mRNA transcript before it starts another. As long as they have access to the gene's promoter, many RNA polymerase molecules can be working one after the other to produce mRNA transcripts at the same time (Fig. 12.13*a*). This allows the cell to produce many thousands of copies of the same mRNA molecule, and eventually many copies of the same protein, within a shorter period of time than if a single mRNA copy were used to direct protein synthesis. This ability to rapidly express the gene enables the cell (and the organism) to better respond to changing environmental conditions and have a greater chance at survival.

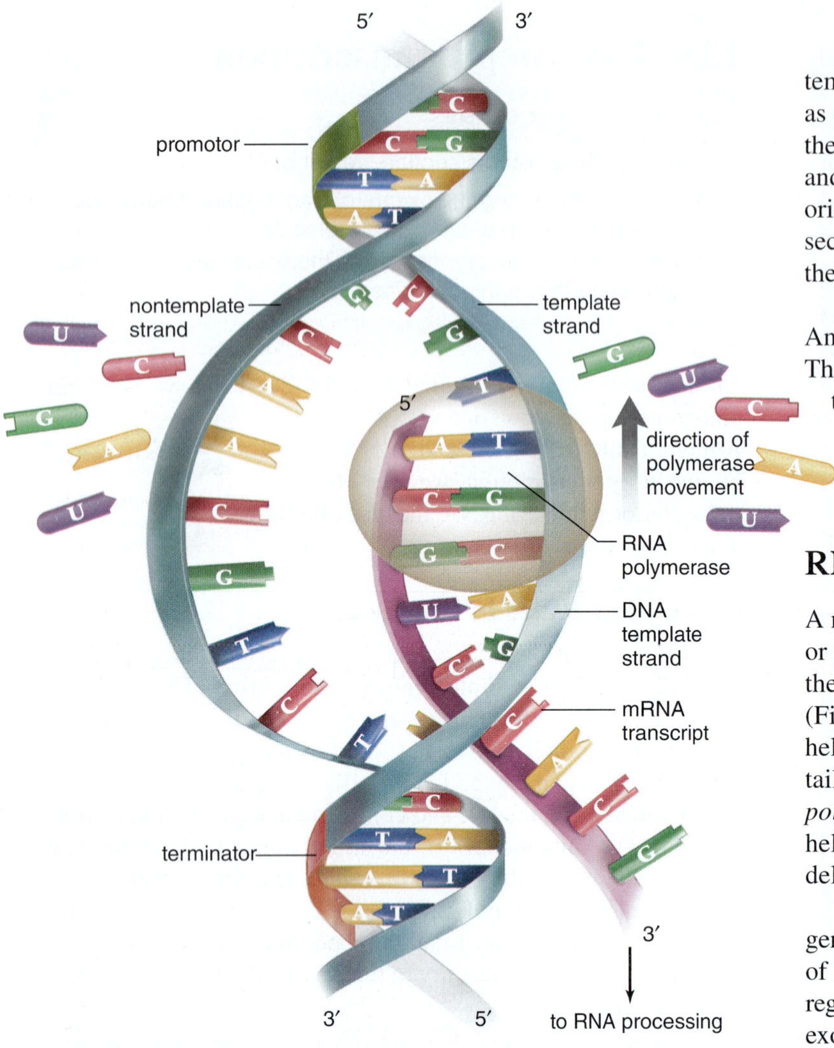

Figure 12.12 Transcription. During transcription, the complementary mRNA is made from a DNA template. At the point of attachment of RNA polymerase, the DNA helix unwinds and unzips, and complementary RNA nucleotides are joined together. After RNA polymerase has passed by, the DNA strands rejoin and the mRNA transcript dangles to the side.

Note that, for a given gene, either strand of the DNA can be a template strand. In Figure 12.12, RNA polymerase uses one strand as the template, but for another gene, the opposite strand may be the template. Assuming both genes are on the same chromosome, and therefore the same piece of DNA, can you think what the orientation of the template strand in the second gene might be? (*Hint:* Consider the directionality of RNA polymerase.)

3D Animation
Molecular Biology of the Gene: Transcription

Some species of *Amanita* mushrooms, such as the Destroying Angel and Death Cap, are so named because of their toxic nature. The *Amanita* produces a toxin called α-amanitin, which inhibits the function of RNA polymerase (Fig. 12.13*b*). The *Amanita* mushrooms are responsible for 95% of all mushroom poisonings and can often lead to a quick death due to the destruction of a person's liver.

RNA Molecules Undergo Processing

A newly formed RNA transcript, called a pre-mRNA, is modified or processed before leaving the eukaryotic nucleus. For example, the molecule receives a cap at the 5′ end and a tail at the 3′ end (Fig. 12.14). The *cap* is a modified guanine (G) nucleotide that helps tell a ribosome where to attach when translation begins. The tail consists of a chain of 150–200 adenine (A) nucleotides. This *poly-A tail* facilitates the transport of mRNA out of the nucleus, helps initiate loading of ribosomes and the start of translation, and delays degradation of mRNA by hydrolytic enzymes.

When the mRNA is first made by RNA polymerase from the gene, it is in a rough form. Called pre-mRNA, it contains a mix of **exons** (protein-coding regions) and **introns** (non-protein-coding regions), particularly in multicellular eukaryotes. Because only the exons of the pre-mRNA will be contained in the mature mRNA, the introns, which occur in between the exons, must be spliced out. An easy way to remember this is with the phrase "exons are *expressed* and introns are *in the way*."

In lower eukaryotes, introns are removed by "self-splicing"—that is, the intron itself has the capability of enzymatically splicing

Figure 12.13 RNA polymerase. **a.** Numerous RNA transcripts extend from a horizontally oriented gene in an amphibian egg cell. The strands get progressively longer because transcription begins to the left. **b.** The α-amanitin toxin from *Amanita* mushrooms inhibits the activity of (**c**) RNA polymerase.

THEME **Nature of Science**

Moving Genes Between Species: Green Fluorescent Protein and Cells

Most cells lack any significant pigmentation. Thus, cell biologists frequently rely on dyes to produce enough contrast to resolve (view) organelles and other cellular structures. The first of these dyes was developed in the nineteenth century from chemicals used to stain clothes in the textile industry. Since then, significant advances have occurred in the development of cellular stains.

In 2008, three scientists—Martin Chalfie, Roger Y. Tsien, and Osamu Shimomura—earned the Nobel Prize in Chemistry or Medicine for their work with a protein called *green fluorescent protein,* or *GFP.* GFP is a bioluminescent protein found in the jellyfish *Aequorea victoria*, commonly called the crystal jelly (Fig. 12A*a*). The crystal jelly is a native of the West Coast of the United States. This jellyfish is normally transparent, but when disturbed it releases a fluorescent protein called aequorin, which fluoresces with a green color. The scientists were able to isolate the fluorescent protein from the jellyfish and develop it as a molecular tag. The molecular tag works by inserting the *GFP* gene just after the promoter region of another gene in a different organism. RNA polymerase binds to the promoter and initiates transcription of that gene. If the *GFP* gene is inserted correctly, it can be expressed (glow) in organisms other than the crystal jellyfish.

These tags can be generated for almost any protein within the cell, revealing not only its cellular location but also how its distribution within the cell may change as a result of a response to its environment. Figure 12A*b* shows fluorescent zebra fish (GloFish®) first developed to glow in the presence of water pollution. They are now widely available for purchase as pets. Figure 12A*c* shows how a GFP-labeled antibody can be used to identify the cellular location of the actin proteins in a human cell. Actin is one of the prime components of the cell's microfilaments, which in turn are part of the cytoskeleton of the cell. This image shows the distribution of actin in a human cell.

Questions to Consider

1. Discuss how a researcher might use a GFP-labeled protein to study cancer.
2. Should this technology be applied to any type of pet?

a. b. c.

Figure 12A GFP as molecular tags. **a.** The jellyfish *Aequorea victoria*. **b.** GFP and other fluorescent proteins used to modify fish. **c.** Human cells tagged with a GFP-labeled antibody to the actin protein.

itself out of a pre-mRNA. In higher eukaryotes, the RNA splicing is done by spliceosomes, which contain *small nuclear RNAs* (*snRNAs*). By means of complementary base pairing, snRNAs are capable of identifying the introns to be removed. A spliceosome utilizes a **ribozyme** (enzyme made of RNA rather than just protein) to cut and remove the introns. Following splicing of the exons together and the addition of the 5′ cap and 3′ poly-A tail, an mRNA is ready to leave the nucleus and be translated into a protein.

Animation How Spliceosomes Process RNA

Function of Introns

For many years, scientists thought that introns were simply wasted space within genes. Now, we realize they serve several key functions in the cell. The presence of introns allows a cell to choose which exons will go into a particular mRNA (see Chapter 13). Just because an mRNA has all the exons in its pre-mRNA doesn't mean they will all make it to the final product. For example, if a gene has three exons, then depending on cell need and environmental conditions, it may produce an mRNA with exons 1 and 2 only, or 1 and 3 only, or 1, 2, and 3. This ability is called *alternative mRNA splicing,* and it increases the flexibility and efficiency of the cell. The snRNAs of the spliceosomes that excise the introns play an important role in alternative splicing in eukaryotes.

Some introns give rise to *microRNAs* (*miRNAs*), which are small molecules involved in regulating the translation of mRNAs. These molecules bind with the mRNA through complementary base pairing and, in that way, prevent translation from occurring.

3D Animation Molecular Biology of the Gene: mRNA Modifications

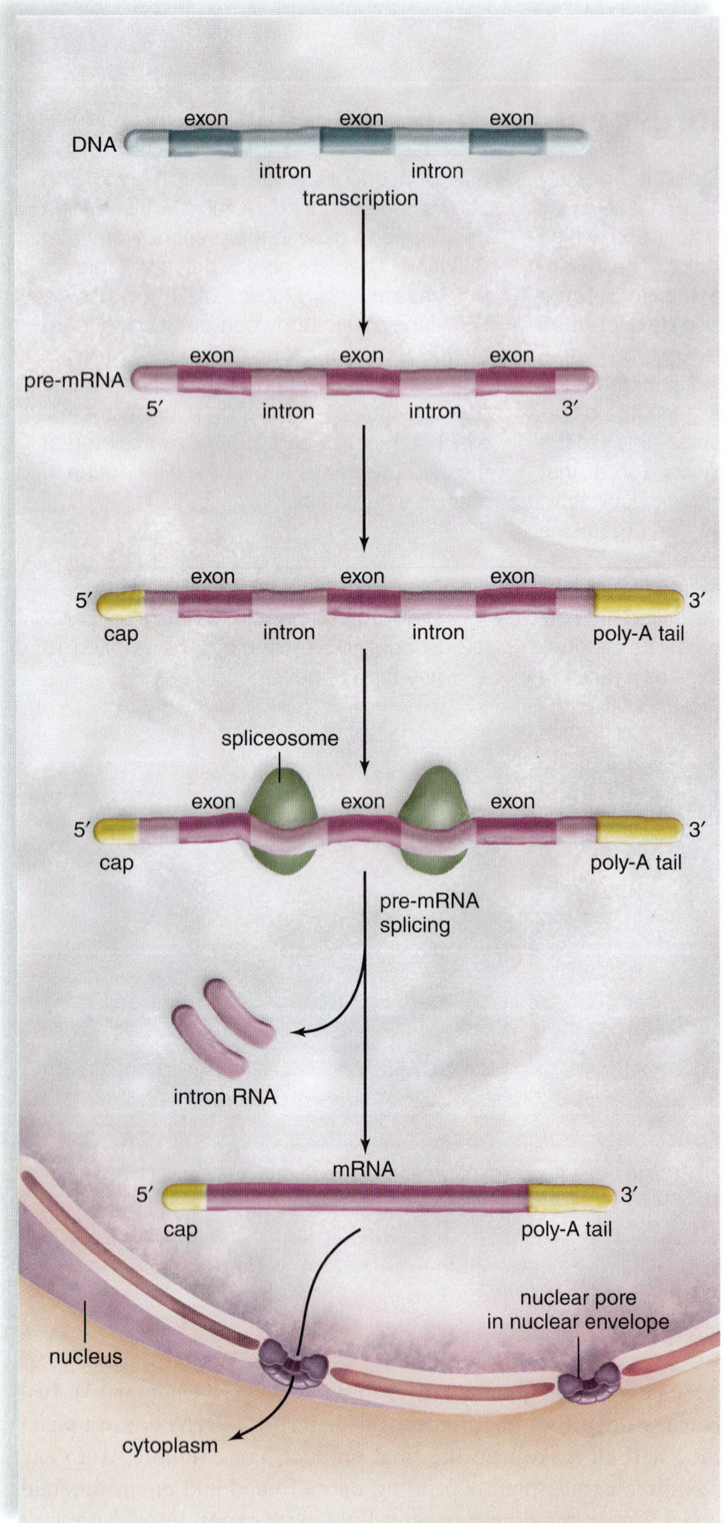

Figure 12.14 Messenger RNA (mRNA) processing in eukaryotes. DNA contains both exons (protein-coding sequences) and introns (non-protein-coding sequences). Both of these are transcribed and are present in pre-mRNA. During processing, a cap and a poly-A tail (a series of adenine nucleotides) are added to the molecule. Also, introns get cut out and the exons get spliced together by complexes called spliceosomes. Once processing is complete, the mRNA molecule is ready to leave the nucleus.

It is also possible that the presence of introns encourages crossing-over during meiosis, and this permits a phenomenon termed *exon shuffling,* which can play a role in the evolution of new genes.

Animation Exon Shuffling

Check Your Progress 12.4

1. Explain the role of RNA polymerase.
2. Describe the three major modifications that occur during the processing of an mRNA.
3. Distinguish between the introns and exons of a gene.
4. Explain the potential evolutionary benefits of alternative mRNA splicing.

12.5 Second Step: Translation

Learning Outcomes

Upon completion of this section, you should be able to

1. Describe the roles of mRNA, tRNA, and rRNA in translating the genetic code.
2. Examine the stages of translation and the events that occur during each stage.

Translation, which takes place in the cytoplasm of eukaryotic cells, is the second step needed to express a gene into a protein. During translation, the sequence of codons (nucleotide triplets) in the mRNA is read by a ribosome, which connects the sequence of amino acids dictated by the mRNA into a polypeptide. The process is called translation because it requires the conversion of information from a nucleic acid language (DNA and RNA) into an amino acid language (protein).

MP3 Translation

The Role of Transfer RNA

Transfer RNA (tRNA) molecules transfer amino acids to the ribosomes. A tRNA molecule is a single-stranded nucleic acid that doubles back on itself to create regions where complementary bases are hydrogen-bonded to one another. The structure of a tRNA molecule is generally drawn as a flat cloverleaf (Fig. 12.15*a*), but a space-filling model shows the molecule's actual three-dimensional shape (Fig. 12.15*b*).

There is at least one tRNA molecule for each of the 20 amino acids found in proteins. The amino acid binds to the 3′ end. The opposite end of the molecule contains an **anticodon,** a group of three bases that is complementary and antiparallel to a specific mRNA codon. For example, a tRNA that has the anticodon 5′ AAG 3′ binds to the mRNA codon 5′ CUU 3′ and carries the amino acid leucine. In the genetic code, 61 codons specify amino acids; the other 3 serve as stop sequences (see Fig. 12.11).

Approximately 40 different tRNA molecules are found in most cells. There are fewer tRNAs than codons, because some tRNAs can pair with more than one codon. In 1966, Francis Crick observed this phenomenon and called it the *wobble hypothesis.* He stated that the first two positions in a tRNA anticodon pair obey the A–U/G–C configuration rule. However, the third position can be variable.

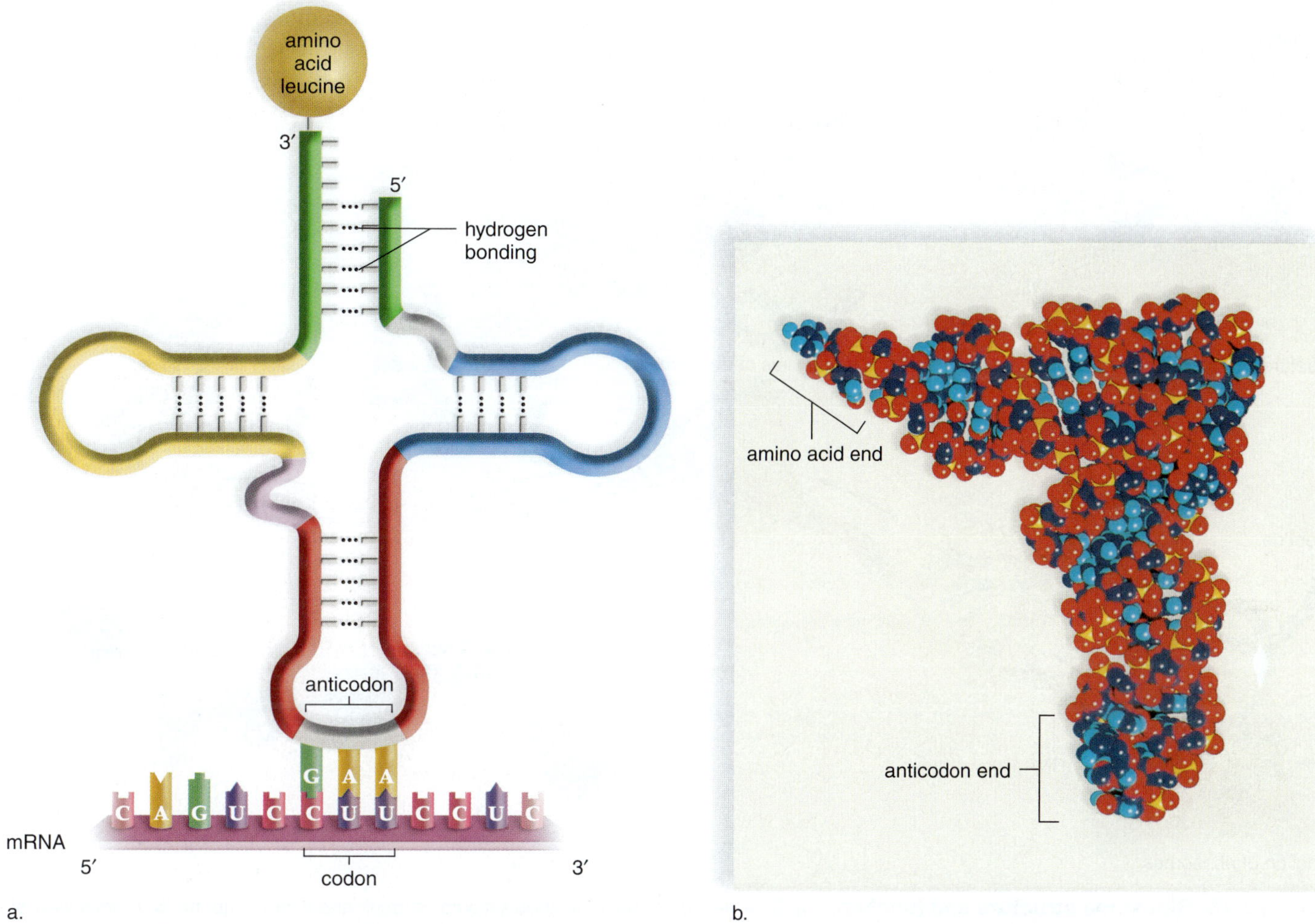

Figure 12.15 Structure of a transfer RNA (tRNA) molecule. **a.** Complementary base pairing indicated by hydrogen bonding occurs between nucleotides within the molecule, and this causes it to form its characteristic loops. The anticodon that base-pairs with a particular messenger RNA (mRNA) codon occurs at one end of the folded molecule; the other two loops help hold the molecule at the ribosome. An appropriate amino acid is attached at the 3′ end of the molecule in the cytoplasm by a tRNA charging enzyme. For this mRNA codon and tRNA anticodon, the specific amino acid is leucine. **b.** Space-filling model of tRNA molecule.

Some tRNA molecules can recognize as many as four separate codons differing only in the third nucleotide. The wobble effect helps ensure that despite changes in DNA base sequences, the resulting sequence of amino acids will produce a correct protein. This is one of the reasons the genetic code is said to be degenerate.

How does the correct amino acid become attached to the correct tRNA molecule? This task is carried out by amino acid–charging enzymes, generically called aminoacyl-tRNA synthetases. Just as a key fits a lock, each enzyme has a recognition site for a particular amino acid to be joined to a specific tRNA. For example, leucine-tRNA synthetase attaches the leucine amino acid to a tRNA with the correct anticodon. This is an energy-requiring process that uses ATP. A tRNA with its amino acid attached is termed a *charged tRNA.* Once the amino acid–tRNA complex is formed, it is added to the large pool of charged tRNAs that exist in the cytoplasm, where it can now be accessed by a ribosome during protein synthesis.

The Role of Ribosomal RNA

As with so many cellular structures, the structure of a ribosome is essential to its function.

Structure of a Ribosome

In eukaryotes, ribosomal RNA (rRNA) is produced from a DNA template in the nucleolus of a nucleus. The rRNA is packaged with a variety of proteins into two ribosomal subunits, one of which is larger than the other. The subunits then move separately through nuclear envelope pores into the cytoplasm, where they join together at the start of translation (Fig. 12.16*a*). Once translation begins, ribosomes can remain in the cytoplasm, or they can become attached to endoplasmic reticulum.

Function of a Ribosome

Both prokaryotic and eukaryotic cells contain thousands of ribosomes per cell, because they play such a significant role in protein synthesis. Ribosomes have a binding site for mRNA and three binding sites for transfer RNA (tRNA) molecules (Fig. 12.16*b*). The tRNA binding sites facilitate complementary base pairing between tRNA anticodons and mRNA codons. The large ribosomal subunit has enzyme activity from rRNA (a ribozyme) that creates the peptide bond between adjacent amino acids. This peptide bond is created many times to produce a polypeptide, which

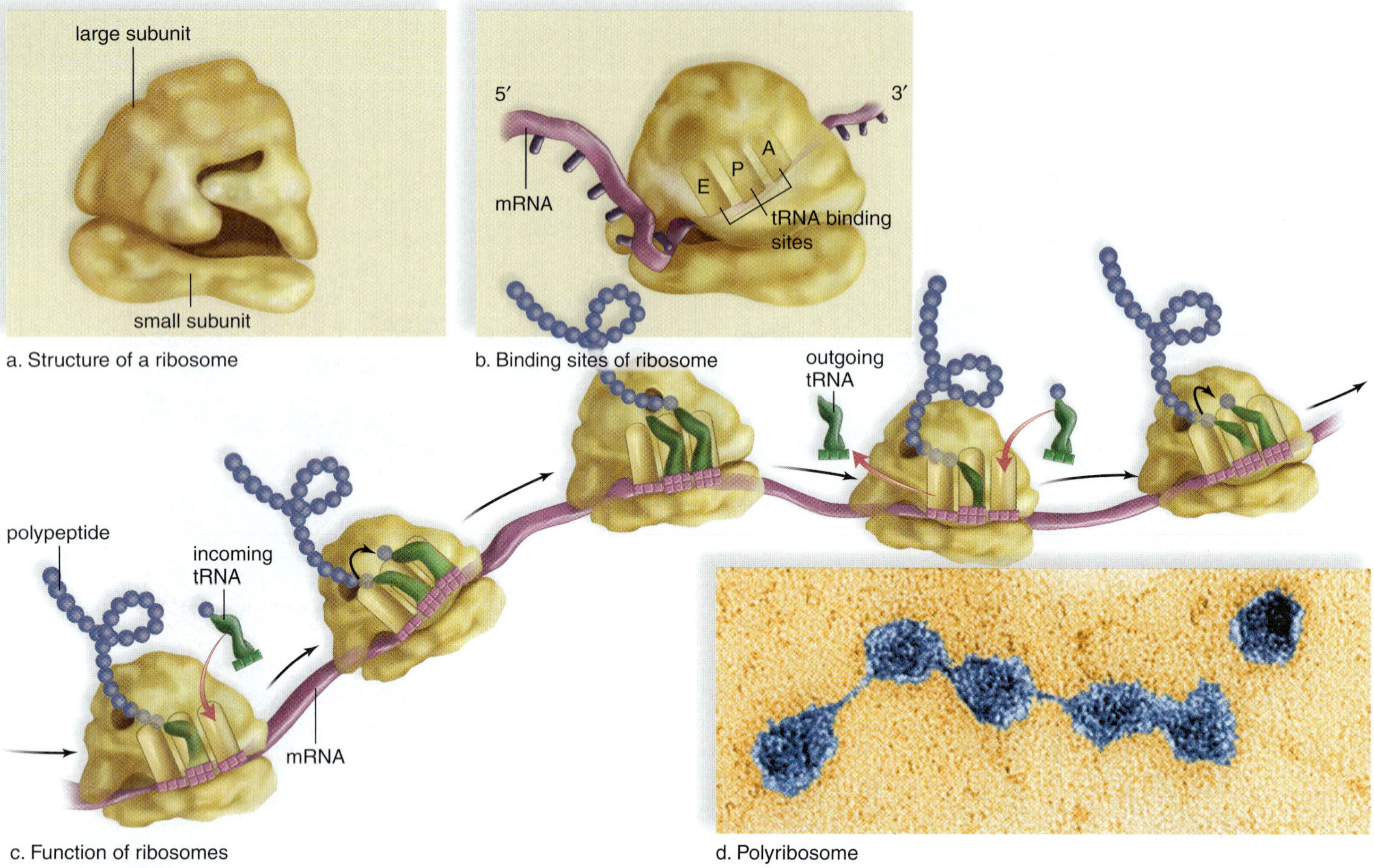

Figure 12.16 Ribosome structure and function. **a.** Side view of a ribosome shows a small subunit and a large subunit. **b.** Frontal view of a ribosome shows its binding sites. mRNA is bound to the small subunit, and the large subunit has three binding sites for tRNAs. **c.** Overview of protein synthesis. The tRNA bearing the growing polypeptide passes the entire chain to the new amino acid carried by the tRNA occupying the A site. The ribosome shifts, and freed of its burden, the "empty" tRNA exits. The new peptide-bearing tRNA moves over one binding site, making the A site accessible once again to a new tRNA. This cycle is repeated until the ribosome reaches the termination codon. **d.** Electron micrograph of a polyribosome, a number of ribosomes all translating the same mRNA molecule.

in turn folds into its three-dimensional shape and becomes a protein.

When a ribosome moves down an mRNA molecule, the polypeptide increases by one amino acid at a time (Fig. 12.16*c*). Translation terminates at a stop codon. Once translation is complete, the polypeptide dissociates from the translation complex and folds into its normal shape. Recall from Chapter 3 that a polypeptide twists and bends into a definite shape based on the makeup of its amino acids. This folding process begins as soon as the polypeptide emerges from a ribosome. Chaperone molecules that are often present in the cytoplasm and the ER ensure that protein folding proceeds as it should. For proteins that contain more than one polypeptide, each subunit is folded first, and then subunits join together into a final, functional protein complex.

Like RNA polymerase during transcription, multiple ribosomes often attach and translate the same mRNA at one time. As soon as the initial portion of mRNA has been translated by one ribosome, and the ribosome has begun to move down the mRNA, another ribosome can attach to the mRNA. The entire complex of mRNA and multiple ribosomes is called a **polyribosome** (Fig. 12.16*d*), and it greatly increases the efficiency of translation.

Translation Requires Three Steps

During translation, the codons of an mRNA base-pair with the anticodons of tRNA molecules carrying specific amino acids. The order of the codons determines the order of the tRNA molecules at a ribosome and the corresponding sequence of amino acids in a polypeptide. The process of translation must be extremely orderly, so that the amino acids of a polypeptide are sequenced correctly. Even a single amino acid change has the potential to dramatically affect a protein's function, as is the case with individuals who carry the alleles for sickle-cell disease.

Protein synthesis involves three steps: initiation, elongation, and termination. Enzymes are required for each of the three steps to function properly. The first two steps, initiation and elongation, require energy.

Animation How Translation Works

Initiation

Initiation is the step that brings all the translation components together. Proteins called initiation factors are required to assemble the small ribosomal subunit, mRNA, initiator tRNA, and the large ribosomal subunit for the start of protein synthesis.

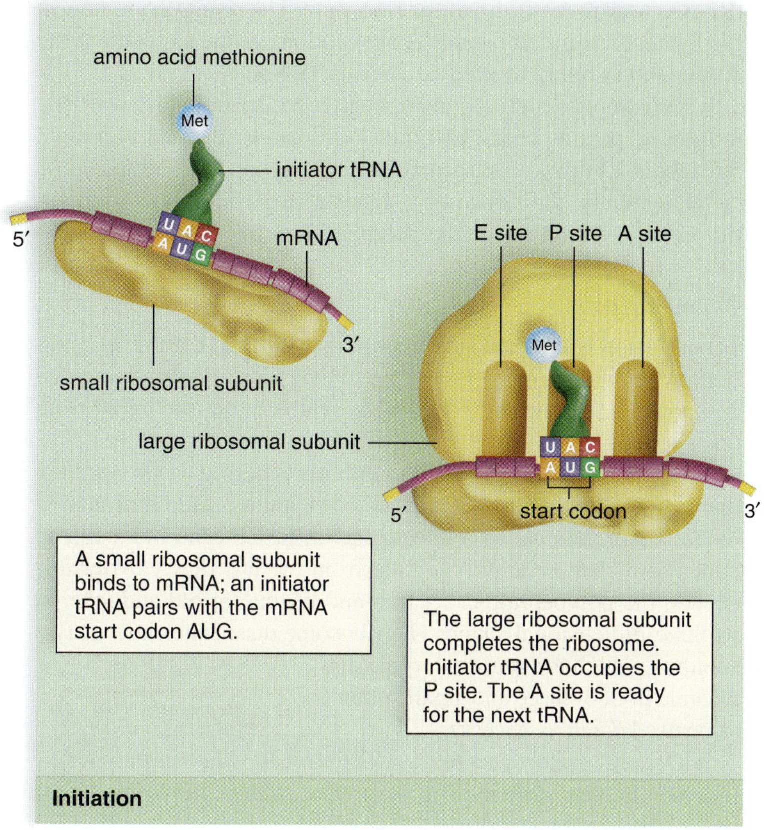

Figure 12.17 Initiation. In prokaryotes, participants in the translation process assemble as shown. The first amino acid is typically a special form of methionine.

Initiation is shown in Figure 12.17. In prokaryotes, a small ribosomal subunit attaches to the mRNA in the vicinity of the *start codon* (AUG). The first, or initiator, tRNA pairs with this codon. Then, a large ribosomal subunit joins to the small subunit (Fig. 12.17). Although similar in many ways, initiation in eukaryotes is much more complex.

As already discussed, a ribosome has three binding sites for tRNAs. One of these is called the E (for "exit") site, second is the P (for "peptide") site, and the third is the A (for "amino acid") site. The initiator tRNA binds to the P site, even though it carries only the amino acid methionine (see Fig. 12.11). The A site is where tRNA carrying the next amino acid enters the ribosome, and the E site is for any tRNAs that are leaving a ribosome. Following initiation, translation continues with elongation and then termination.

Elongation

Elongation is the stage during protein synthesis when a polypeptide increases in length one amino acid at a time. In addition to the necessary tRNAs, elongation requires elongation factors, which facilitate the binding of tRNA anticodons to mRNA codons within a ribosome.

Elongation is shown in Figure 12.18, where a tRNA with an attached peptide is already at the P site, and a tRNA carrying its appropriate amino acid is just arriving at the A site. Once a ribosome has verified that the incoming tRNA matches the codon and is firmly in place at the A site, the entire growing peptide will be transferred to the amino acid on the tRNA in the A site. A ribozyme, an rRNA-based enzyme that is a part of the large ribosomal subunit, uses energy to transfer the growing peptide and create a new peptide bond. Following peptide bond formation, the peptide is one amino acid longer than it was before. Next, *translocation* occurs: The ribosome moves forward, and the peptide-bearing

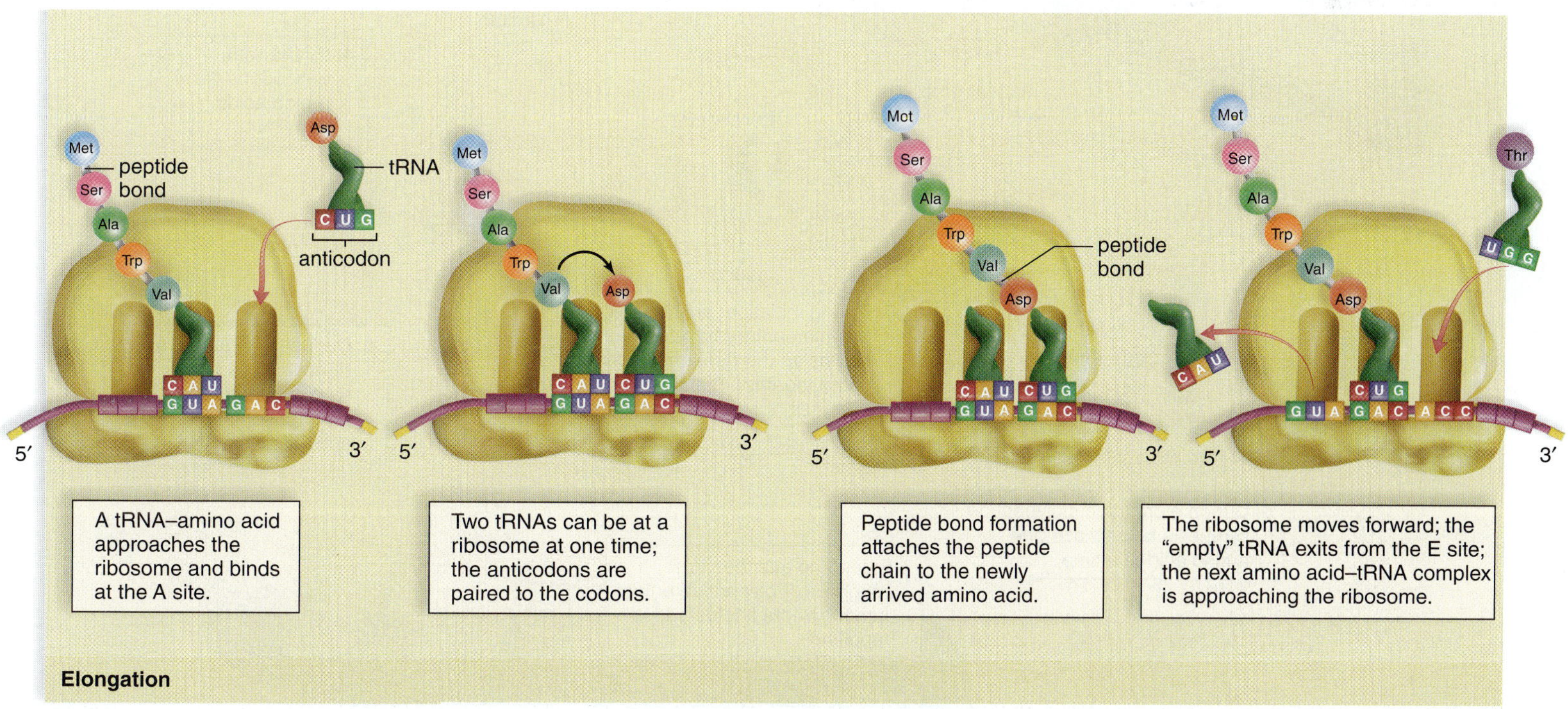

Figure 12.18 Elongation. Note that a polypeptide is already at the P site. During elongation, polypeptide synthesis occurs as amino acids are added one at a time to the growing chain.

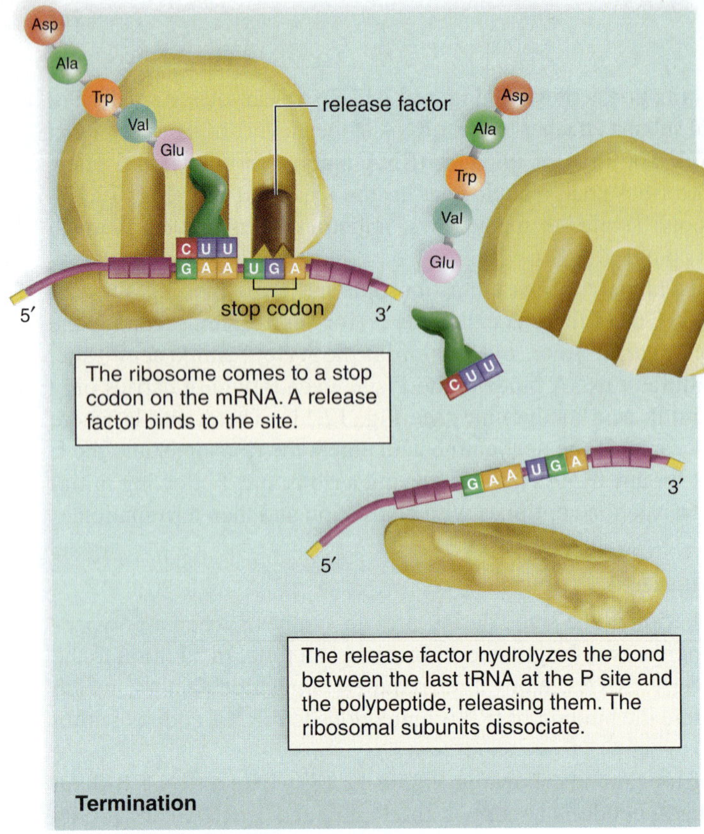

Figure 12.19 Termination. During termination, the finished polypeptide is released, as are the mRNA and the last tRNA.

tRNA is now in the P site of the ribosome. The spent tRNA, now at the E site, exits the ribosome. A new codon is now exposed at the A site and is ready to receive another tRNA.

The complete cycle—complementary base pairing of new tRNA, transfer of peptide chain, and translocation—is repeated at a rapid rate (about 15 times each second in the bacterium *Escherichia coli*).

Eventually, the ribosome reaches a stop codon, and termination occurs, during which the polypeptide is released.

Termination

Termination is the final step in protein synthesis. During termination, as shown in Figure 12.19, the polypeptide and the assembled components that carried out protein synthesis are separated from one another.

Termination of polypeptide synthesis occurs at a *stop codon*—that is, a codon that does not code for an amino acid. Termination requires a protein called a release factor, which can bind to a stop codon and cleave the polypeptide from the last tRNA. After this occurs, the polypeptide is set free and begins to fold and take on its three-dimensional shape. The ribosome dissociates into its two subunits, which are returned to the cytoplasmic pool of large and small subunits, to be used again as necessary.

3D Animation Molecular Biology of the Gene: Translation

Overall, proteins do the work of the cell, whether they reside in a membrane within the cell or are free in the cytoplasm. A new

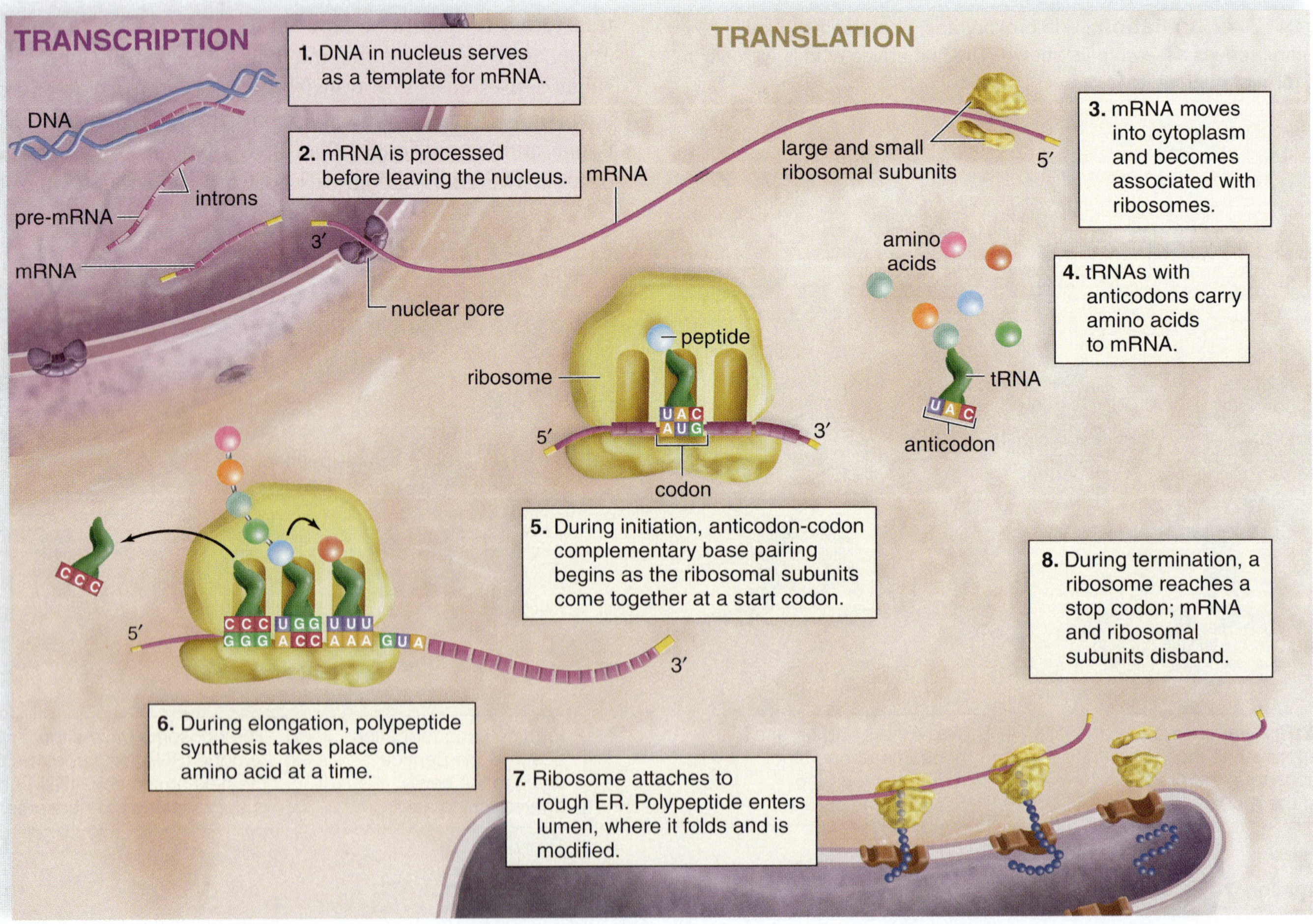

Figure 12.20 Summary of protein synthesis in eukaryotes.

field of biology called **proteomics** is dedicated to understanding the structure of proteins and how they function in metabolic pathways. One of the important goals of proteomics is to understand how proteins are modified in the endoplasmic reticulum and the Golgi apparatus.

Gene Expression

A gene has been expressed once its product, a protein (or an RNA), is made and is operating in the cell. For a protein, gene expression requires transcription and translation (Fig. 12.20), and it requires that the protein be active, as discussed in Chapter 13.

Translation occurs at ribosomes. Some ribosomes (polyribosomes) remain free in the cytoplasm, and some become attached to rough ER. The first few amino acids of a polypeptide act as a signal peptide that indicates where the polypeptide belongs in the cell or if it is to be secreted from the cell. Polypeptides that are to be secreted enter the lumen of the ER by way of a channel and are then folded and further processed by the addition of sugars, phosphates, or lipids. Transport vesicles carry the proteins between organelles and to the plasma membrane as appropriate for that protein.

Check Your Progress 12.5

1. Explain the role of transfer RNA in translation.
2. Describe how the structure of a ribosome contributes to polypeptide synthesis.
3. Examine the events that occur during the three major steps of translation.

CONNECTING *the* CONCEPTS *with the* THEMES

Evolution

- Genes specify proteins through the steps of transcription and translation. During transcription, a strand of DNA is used as a template for the production of an mRNA molecule.
- The actions of RNA polymerase, the enzyme that carries out transcription, and DNA polymerase, which is required for DNA replication, are similar enough to suggest that both enzymes evolved from a common ancestral enzyme.
- RNA molecules play a prominent role in the regulation of the genome. Some believe this is evidence that RNA may have preceded DNA in the evolutionary history of cells.

Nature of Science

- Early investigators who did their work between 1950 and 1990 came to the conclusion that DNA is organized into discrete units called genes.
- Scientists are now discovering that the rest of the DNA that does not specify proteins may also have valuable functions.
- Many biologists believe that we need a new definition of a gene that recognizes that much of our DNA results in RNA molecules rather than protein products.

Biological Systems

- Both protein-coding and non-protein-coding DNA provide the blueprint for building and developing an entire organism.
- Just as a blueprint is useless without a team of engineers, architects, and construction workers to execute it, the expression of genes requires a large system of proteins and other factors to control it.
- The regulation of gene expression contributes to the great complexity and diversity of living organisms.

MEDIA STUDY TOOLS http://connect.mheducation.com

SMARTBOOK® Study smart. SmartBook helps you maximize your study time by identifying what you know and don't know. The recharge feature in SmartBook helps you prepare for the test by identifying the material that you are most likely to forget.

connect BIOLOGY Find even more resources to learn the chapter concepts including animations, tutorial videos, and interactive practice questions.

MP3 Files	Animations	3D Animations	Tutorials
12.4 Protein Synthesis **12.5** Translation	**12.1** Hershey and Chase Experiment • DNA Structure **12.2** DNA Replication **12.4** Stages of Transcription • How Spliceosomes Process RNA • Exon Shuffling **12.5** How Translation Works	**12.1** DNA Replication: DNA Structure **12.2** DNA Replication **12.4** Molecular Biology of the Gene: Transcription • Molecular Biology of the Gene: mRNA Modifications **12.5** Molecular Biology of the Gene: Translation	**12.2** DNA Replication **12.3** Overview of Gene Expression

SUMMARIZE

12.1 The Genetic Material

Early work illustrated that DNA is the hereditary material. Griffith injected strains of pneumococcus into mice and observed that when heat-killed S strain bacteria were injected along with live R strain bacteria, virulent S strain bacteria were recovered from the dead mice. Griffith said that the R strain had been transformed by some substance passing from the dead S strain to the live R strain. Twenty years later, Avery and his colleagues reported that the transforming substance was DNA.

To study the structure of DNA, Chargaff performed a chemical analysis of DNA and found that **adenine (A)** = **thymine (T)** and **guanine (G)** = **cytosine (C)**, and that the amount of purine equals the amount of pyrimidine. Franklin prepared an X-ray photograph of DNA that showed it is a **double helix,** has repeating structural features, and has certain dimensions. Watson and Crick built a model of DNA in which the sugar-phosphate molecules made up the sides of a twisted ladder and the **complementary base pairs** were the rungs of the ladder.

12.2 Replication of DNA

The Watson and Crick model immediately suggested a method by which DNA could be replicated. Basically, the two strands unwind and unzip, and each parental strand acts as a **template** for a new (daughter) strand, resulting in **semiconservative replication.** In the end, each new helix is like the other and like the parental helix.

DNA helicase opens the double-stranded DNA, and **single-stranded binding proteins** ensure the strands stay apart. **DNA primase** helps initiate replication by adding short primers, which are recognized by **DNA polymerase,** and begins replication. **DNA replication** on the leading strand is continuous, unlike synthesis of the lagging strand, which requires making many fragments called Okazaki fragments. Fragments are mended by **DNA ligase,** and the final result is two helices identical to the parental DNA. Replication in prokaryotes typically proceeds in both directions from one point of origin to a termination region until there are two copies of the circular chromosome. Replication in eukaryotes has many points of origin and many bubbles (places where the DNA strands are separating and replication is occurring). Replication occurs at the ends of the bubbles—at **replication forks.**

12.3 The Genetic Code of Life

The **central dogma** of molecular biology says that the flow of genetic information is from DNA to RNA to protein to traits. RNA is a nucleic acid that uses the nucleotide **uracil (U)** instead of thymine (T). **Messenger RNA, transfer RNA,** and **ribosomal RNA** are all specialized RNAs needed to make proteins. More specifically, (1) DNA is a template for its own replication and for RNA formation during **transcription,** and (2) the sequence of nucleotides in mRNA directs the correct sequence of amino acids of a polypeptide during **translation.**

The **genetic code** is a **triplet code,** and each **codon** consists of three bases. The code is degenerate—that is, more than one codon exists for most amino acids. There are also one start and three stop codons. The genetic code is considered universal, but there are a few exceptions.

12.4 First Step: Transcription

Transcription to produce messenger RNA (mRNA) begins when **RNA polymerase** attaches to the **promoter** of a gene. Elongation occurs until RNA polymerase reaches a stop sequence. An **mRNA transcript** is made, then processed, following transcription. A cap is put onto the 5′ end, a poly-A tail is put onto the 3′ end. **Exons** are kept and **introns** are removed in eukaryotes by spliceosomes containing **ribozymes.**

Small nuclear RNAs (snRNAs) present in spliceosomes help identify the introns to be removed. These snRNAs play a role in alternative mRNA splicing, which allows a single eukaryotic gene to code for different proteins, depending on which segments of the gene serve as introns and which serve as exons.

Some introns serve as microRNAs (miRNAs), which help regulate the translation of mRNAs. Research is now directed at discovering the many ways small RNAs influence the production of proteins in a cell.

12.5 Second Step: Translation

Translation requires mRNA, transfer RNA (tRNA), and ribosomal RNA (rRNA). Each tRNA has an **anticodon** at one end and an amino acid at the other; amino acid–charging enzymes ensure that the correct amino acid is attached to the correct tRNA. When tRNAs bind with their codon at a ribosome, the amino acids are correctly sequenced in a polypeptide according to the order predetermined by DNA.

In the cytoplasm, many ribosomes move along the same mRNA at a time. Collectively, these are called a **polyribosome.**

Translation requires these steps: During **initiation,** mRNA, the first (initiator) tRNA, and the two subunits of a ribosome all come together in the proper orientation at a start codon. During **elongation,** as the tRNA anticodons bind to their codons, the growing peptide chain is transferred by peptide bonding to the next amino acid in a polypeptide. During **termination** at a stop codon, the polypeptide is cleaved from the last tRNA. The ribosome now dissociates. The field of **proteomics** studies how proteins are made, function, and are modified by other organelles.

ASSESS

Choose the best answer for each question.

12.1 The Genetic Material

1. Transformation occurs when
   a. DNA is transformed into RNA.
   b. DNA is transformed into protein.
   c. bacteria cannot grow on penicillin.
   d. organisms receive foreign DNA and thereby acquire a new characteristic.
2. The double helix model of DNA resembles a twisted ladder in which the rungs of the ladder are
   a. a purine paired with a pyrimidine.
   b. A paired with G and C paired with T.
   c. sugar-phosphate paired with sugar-phosphate.
   d. a 5′ end paired with a 3′ end.
   e. Both a and b are correct.
3. If 30% of an organism's DNA is thymine, then
   a. 70% is purine.
   b. 20% is guanine.
   c. 30% is adenine.
   d. 70% is pyrimidine.
   e. Both b and c are correct.
4. If the sequence of bases in one strand of DNA is 5′ TAGCCT 3′, then the sequence of bases in the other strand is
   a. 3′ TCCGAT 5′.
   b. 3′ ATCGGA 5′.
   c. 3′ TAGCCT 5′.
   d. 3′ AACGGUA 5′.

12. 2 Replication of DNA

5. DNA replication is said to be semiconservative because
   a. one of the new molecules conserves both of the original DNA strands.
   b. the new DNA molecule contains two new DNA strands.
   c. both of the new molecules contain one new strand and one old strand.
   d. DNA polymerase conserves both of the old strands.
6. The enzyme responsible for separating double-stranded DNA into single-stranded DNA is
   a. DNA helicase.
   b. DNA primase.
   c. DNA polymerase.
   d. DNA ligase.

12.3 The Genetic Code of Life

7. The central dogma of molecular biology
   a. states that DNA is a template for all RNA production.
   b. states that DNA is a template only for DNA replication.
   c. states that translation precedes transcription.
   d. states that RNA is a template for DNA replication.
   e. All of these are correct.
8. If the sequence of DNA on the template strand of a gene is AAA, the mRNA codon produced by transcription will be ______ and will specify the amino acid ______ .
   a. AAA, lysine
   b. AAA, phenylalanine
   c. TTT, arginine
   d. UUU, phenylalanine
   e. TTT, lysine

12.4 First Step: Transcription

9. If the sequence of bases in the coding strand of a DNA molecule is TAGC, then the sequence of bases in the mRNA will be
   a. AUCG.
   b. TAGC.
   c. UAGC.
   d. ATCG.
10. The portion of the mRNA transcript that gets removed during RNA processing is the
   a. exons.
   b. introns.
   c. poly-A tails.
   d. 5′ caps.
   e. spliceosomes.

12.5 Second Step: Translation

11. During protein synthesis, an anticodon on transfer RNA (tRNA) pairs with
   a. DNA nucleotide bases.
   b. ribosomal RNA (rRNA) nucleotide bases.
   c. messenger RNA (mRNA) nucleotide bases.
   d. other tRNA nucleotide bases.
   e. All of these are correct.
12. This is a segment of a DNA molecule. What are (a) the RNA codons, (b) the matching tRNA anticodons, and (c) the sequence of amino acids in the eventual protein?

ENGAGE

LEARNSMART LABS®

The following LearnSmart Labs contain exercises that are related to the content of this chapter:

- DNA & Biotechnology

THINKING SCIENTIFICALLY

1. How would you test a hypothesis that a genetic condition, such as cancer, is due to mistakes in transcription and translation?
2. What kind of genes do you think would be included in the category of "housekeeping" genes?
3. DNA replication is fast, virtually error-free, and coordinated with cell division. Discuss which of these three features you think is the most important.
4. A new virus has recently been discovered that infects human blood cells. The virus can be grown in the laboratory using cultured blood cells as host cells. Design an experiment using a radioactive label that would tell you if the virus contains DNA or RNA.

13

Regulation of Gene Expression

The health of an individual may be affected by social status. Researchers are using female rhesus macaques, seen here, to study the effects of gene expression in social groups.

Chapter Outline

13.1 Prokaryotic Regulation 229

13.2 Eukaryotic Regulation 232

13.3 Gene Mutations 238

Before You Begin

Before beginning this chapter, take a few moments to review the following discussions.

Figure 12.10 What is the central dogma of biology?

Section 12.4 How is an mRNA transcript made?

Section 12.5 What is the role of translation in gene expression?

Social status can affect what genes are turned on and off in fish, honeybees, and insects. A recent study found a similar effect in macaque monkeys. Scientists interested in gene regulation worked with 49 female rhesus macaques housed at the Yerkes National Primate Research Center at Emory University. Females were used because they exhibit a sustained dominant hierarchy in the wild compared to their male counterparts. At the center, new enclosures were created to house the females in a group to create a "social unit." The females that had been in the social unit earlier had a higher-ranking social status, and the females introduced later were lower-ranking.

Blood samples taken from the macaques showed that high-ranking females had stronger immune systems with a higher production of white blood cells, while lower-ranking females had fewer white blood cells. The DNA of high-ranking versus low-ranking females was also different in how certain genes were expressed—some genes were turned on or off.

Gene expression is the information in DNA transcribed and translated into a protein. How does a cell know when to undergo gene expression? Gene expression needs to be controlled or regulated, and as you will see in this chapter, there are many mechanisms that do this.

The study with the macaques shows that although DNA can be transcribed and translated into a protein, many factors regulate when and how this happens. For female macaques, social status can control gene expression and in turn benefit or harm an individual.

As you read through this chapter, think about the following questions:

1. How does gene regulation differ between prokaryotes and eukaryotes?
2. How might mutations influence the ability of a cell to regulate gene expression?

Following *the* Themes

UNIT 2 GENETIC BASIS OF LIFE — CHAPTER 13 REGULATION OF GENE EXPRESSION	
Evolution	Whereas prokaryotic gene regulation operates primarily at the level of the gene, eukaryotes have evolved mechanisms to regulate gene expression at multiple levels.
Nature of Science	By understanding how cells regulate gene expression, it is possible to better understand the basis of many human diseases.
Biological Systems	Mutations in the genetic material and epigenetics may have a profound impact on the function of a cell or the health of an organism.

13.1 Prokaryotic Regulation

Learning Outcomes

Upon completion of this section, you should be able to

1. Describe the structure of an operon and state the role of each component of the operon.
2. Explain how the *trp* and *lac* operons of prokaryotes are regulated.
3. Distinguish between a repressible operon and an inducible operon.

Because their environment is ever changing, bacteria do not always need to express their entire complement of enzymes and proteins. In 1961, French microbiologists François Jacob and Jacques Monod showed that *Escherichia coli* is capable of regulating the expression of its genes. They observed that the genes in a metabolic pathway, called **structural genes,** are grouped on a chromosome and transcribed at the same time. Jacob and Monod, therefore, proposed the **operon** (L. *opera,* "works") model to explain gene regulation in prokaryotes. They later received a Nobel Prize for their investigations.

An operon (Fig. 13.1) typically includes the following parts:

Regulator gene—Normally located outside the operon, this codes for a DNA-binding protein that acts as a **repressor.** The repressor controls whether the operon is active or not.

Promoter—A short sequence of DNA where RNA polymerase first attaches to begin transcription of the grouped genes. Basically, a promoter signals the start of the operon and the location where transcription begins.

Operator—A short portion of DNA located before the structural genes. If a repressor is attached to the operator, then transcription cannot occur; conversely, if a repressor is not attached,

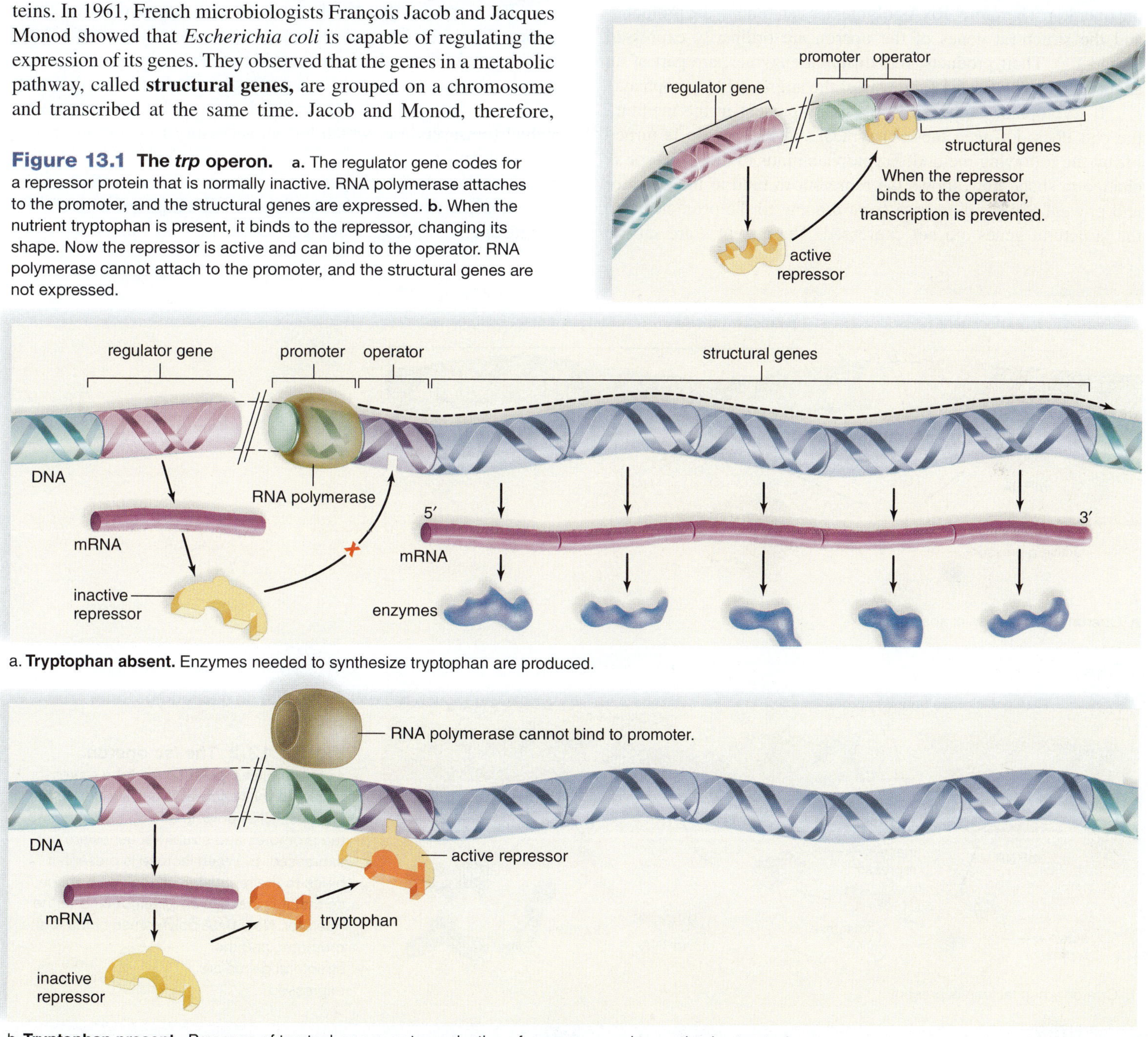

Figure 13.1 The *trp* operon. **a.** The regulator gene codes for a repressor protein that is normally inactive. RNA polymerase attaches to the promoter, and the structural genes are expressed. **b.** When the nutrient tryptophan is present, it binds to the repressor, changing its shape. Now the repressor is active and can bind to the operator. RNA polymerase cannot attach to the promoter, and the structural genes are not expressed.

a. **Tryptophan absent.** Enzymes needed to synthesize tryptophan are produced.

b. **Tryptophan present.** Presence of tryptophan prevents production of enzymes used to synthesize tryptophan.

then transcription can occur. In this way, the operator controls transcription of structural genes.

Structural genes—These genes code for the enzymes and proteins that are involved in the metabolic pathway of the operon. The structural genes are transcribed as a unit.

Next, we will briefly review the findings of Jacob and Monod in their studies of two *E. coli* operons: the *trp* operon and the *lac* operon.

The *trp* Operon

Tryptophan is an essential amino acid synthesized by the enzymes coded for in the *trp* operon. Many investigators, including Jacob and Monod, found that some operons in *E. coli* usually exist in the "on" rather than "off" condition. For example, in the *trp* operon, the regulator codes for a repressor that ordinarily is unable to attach to the operator. Therefore, RNA polymerase can bind to the promoter, and the structural genes of the operon are ordinarily expressed (Fig. 13.1). Their products, five different enzymes, are part of an anabolic pathway for the synthesis of the amino acid tryptophan.

If tryptophan happens to be already present in the medium, the cell does not need these enzymes, and the operon is turned off by the following method. Tryptophan binds to the repressor. A change in shape now allows the repressor to bind to the operator and prevent RNA polymerase from binding to the promoter, and the structural genes are not expressed. The enzymes are said to be repressible (can be turned "off"), and the entire unit is called a *repressible operon*. Tryptophan is called the **corepressor.** Repressible operons are usually involved in anabolic pathways that synthesize a substance needed by the cell.

Animation The Tryptophan Repressor

The *lac* Operon

Bacteria metabolism is remarkably efficient; if proteins or enzymes are needed for metabolism, then the structural genes are expressed. If no metabolism is necessary, then genes are not expressed. For example, if the milk sugar lactose is not present, there is no need to express genes for enzymes involved in lactose catabolism. But when only lactose is present, the cell immediately begins to make the three enzymes needed for lactose metabolism.

The enzymes that break down lactose are encoded by three genes (Fig. 13.2): One gene is for an enzyme called β-galactosidase, which breaks down the disaccharide lactose to glucose and galactose; a second gene codes for a permease that facilitates the entry of lactose into the cell; and a third gene codes for an enzyme called transacetylase, which has an accessory function in lactose metabolism.

The three structural genes are adjacent to one another on the chromosome and are under the control of a single promoter and a single operator. The regulator gene codes for a *lac* operon repressor

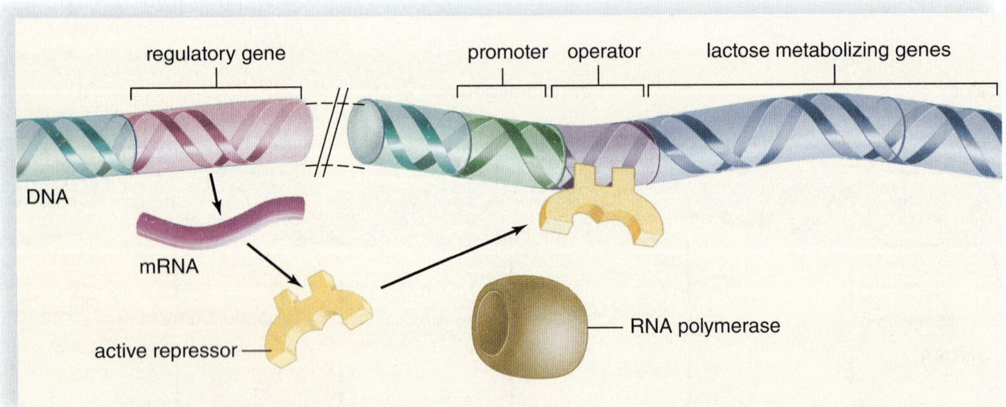

a. Operon when lactose is absent.

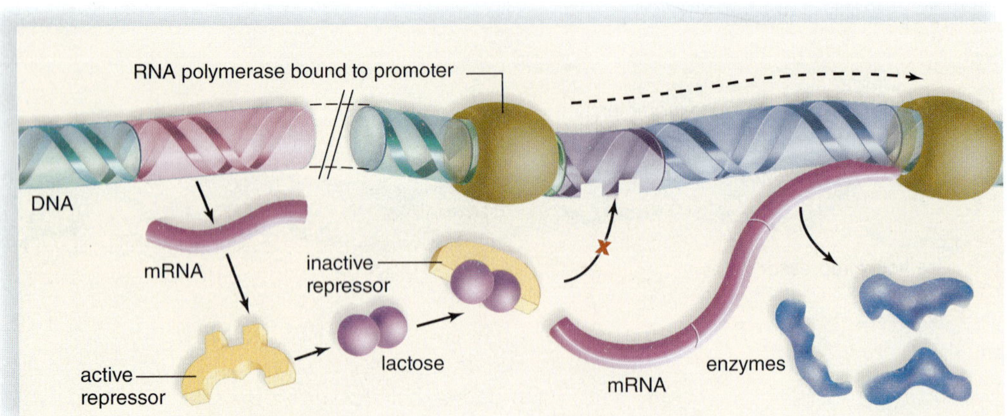

b. Operon when lactose is present.

Figure 13.2 The *lac* operon.
a. The regulator gene codes for a repressor that is normally active. When it binds to the operator, RNA polymerase cannot attach to the promoter, and structural genes are not expressed. **b.** When lactose is present, it binds to the repressor, changing its shape, so that it is inactive and cannot bind to the operator. Now RNA polymerase binds to the promoter, and the structural genes are expressed.

Tutorial *lac* Operon

that ordinarily binds to the operator and prevents transcription of the three genes. When only lactose (more correctly, allolactose, an isomer formed from lactose) is present, lactose binds to the repressor, and the repressor undergoes a change in shape that prevents it from binding to the operator. Because the repressor is unable to bind to the operator, RNA polymerase is better able to bind to the promoter. After RNA polymerase carries out transcription, the three enzymes of lactose metabolism are synthesized.

Because the presence of lactose brings about expression of genes, it is called an **inducer** of the *lac* operon: The enzymes are said to be inducible enzymes (can be turned "on"), and the entire unit is called an *inducible operon.* Inducible operons are usually found in catabolic pathways that break down a nutrient. Why is that beneficial? Because these enzymes need to be active only when the nutrient is present.

Further Control of the lac *Operon*

If both glucose and lactose are present, then *E. coli* preferentially breaks down glucose. The bacterium has a way to ensure that the lactose operon is fully turned on only when glucose is absent. A molecule called *cyclic AMP (cAMP)* accumulates when glucose is absent. Cyclic AMP, which is derived from ATP, has only one phosphate group, which is attached to ribose at two locations:

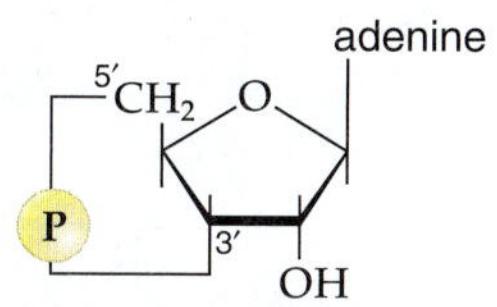

cyclic AMP
(cAMP)

Cyclic AMP binds to a molecule called a *catabolite activator protein (CAP),* and the complex attaches to a CAP binding site next to the *lac* promoter. When CAP binds to DNA, DNA bends, exposing the promoter to RNA polymerase. RNA polymerase is now better able to bind to the promoter, so that the *lac* operon structural genes are transcribed, leading to their expression (Fig. 13.3).

When glucose is present, there is little cAMP in the cell; CAP is inactive, and the lactose operon does not function maximally. CAP affects other operons as well and takes its name for activating the catabolism of various other metabolites when glucose is absent. A cell's ability to encourage the metabolism of lactose and other metabolites when glucose is absent provides a backup system for survival when the preferred energy source, glucose, is absent.

The CAP protein's regulation of the *lac* operon is an example of positive control. Why? When this molecule is active, it promotes the activity of an operon. The use of repressors, on the other hand, is an example of negative control, because when active they shut down an operon. A positive control mechanism allows the cell to fine-tune its response. In the case of the *lac* operon, the operon is only maximally active when glucose is absent and lactose is present. If both glucose and lactose are present, the cell preferentially metabolizes glucose.

Animation Combination of Switches: The *lac* Operon

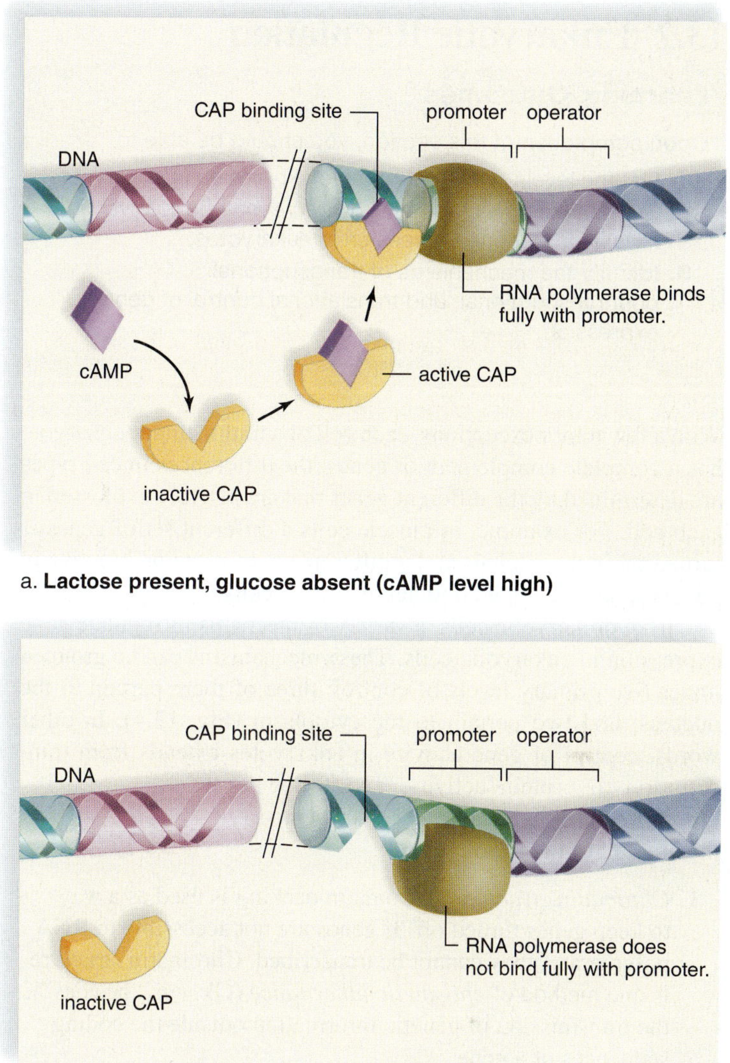

a. **Lactose present, glucose absent (cAMP level high)**

b. **Lactose present, glucose present (cAMP level low)**

Figure 13.3 Action of CAP. When active CAP binds to its site on DNA, the RNA polymerase is better able to bind to the promoter, so that the structural genes of the *lac* operon are expressed. **a.** CAP becomes active in the presence of cAMP, a molecule that is prevalent when glucose is absent. Therefore, transcription of lactose enzymes increases, and lactose is metabolized. **b.** If glucose is present, CAP is inactive, and RNA polymerase does not completely bind to the promoter. Therefore, transcription of lactose enzymes decreases, and less metabolism of lactose occurs.

Check Your Progress 13.1

1. Explain the difference between the roles of the promoter and operator of an operon.
2. Summarize how gene expression differs in an inducible operon versus a repressible operon, and give an example of each.
3. Describe the difference between positive control and negative control of gene expression.
4. Explain which operon discussed in this section is catabolic and which operon is anabolic.

13.2 Eukaryotic Regulation

Learning Outcomes

Upon completion of this section, you should be able to

1. List the levels of control of gene expression in eukaryotes.
2. Summarize how chromatin structure may be involved in regulation of gene expression in eukaryotes.
3. Identify the mechanisms of transcriptional, posttranscriptional, and translational control of gene expression.

With a few minor exceptions, each cell of a multicellular eukaryote has a complete complement of genes; the differences in cell types are determined by the different genes that are actively expressed in each cell. For example, in muscle cells a different set of genes is turned on in the nucleus and a different set of proteins is active in the cytoplasm, compared to nerve or liver cells.

Like prokaryotic cells, a variety of mechanisms regulate gene expression in eukaryotic cells. These mechanisms can be grouped under five primary levels of control; three of them pertain to the nucleus, and two pertain to the cytoplasm (Fig. 13.4). In other words, control of gene activity in eukaryotes extends from transcription to protein activity. The following types of control in eukaryotic cells can modify the amount of the gene product:

Animation Control of Gene Expression in Eukaryotes

1. *Chromatin structure:* Chromatin packing is used as a way to keep genes turned off. If genes are not accessible to RNA polymerase, they cannot be transcribed. Chromatin structure is one method of *epigenetic inheritance* (Gk. *epi,* "besides"), the transmission of genetic information outside the coding sequences of a gene.
2. *Transcriptional control:* The degree to which a gene is transcribed into mRNA determines the amount of gene product. In the nucleus, transcription factors may promote or repress transcription, the first step in gene expression.
3. *Posttranscriptional control:* Posttranscriptional control involves mRNA processing and how fast mRNA leaves the nucleus.
4. *Translational control:* Translational control occurs in the cytoplasm and affects when translation begins and how long it continues. Small RNA molecules (siRNA) are known to regulate translation. In addition, any condition that can cause the persistence of the 5′ cap and 3′ poly-A tail can affect the length of translation. Excised introns may also have effects on the life span of mRNA.
5. *Posttranslational control:* Posttranslational control, which also takes place in the cytoplasm, occurs after protein synthesis. Only a functional protein is an active gene product.

We now explore each of these types of control in greater depth.

Chromatin Structure

The DNA in eukaryotes is always associated with a variety of proteins, and together they make up a stringy material called **chromatin.** Chromatin is most evident in the nucleus during interphase of the cell cycle.

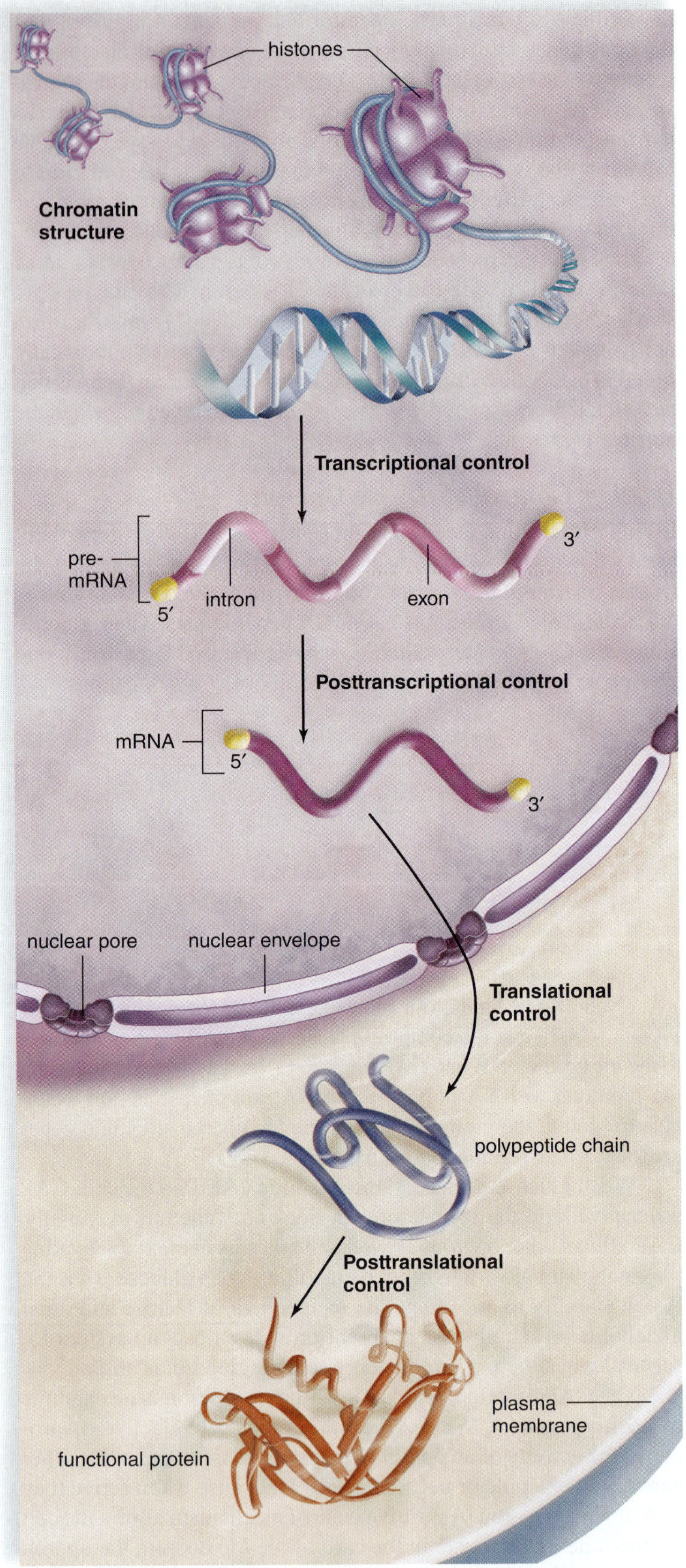

Figure 13.4 Levels at which control of gene expression occurs in eukaryotic cells. The five levels of control are (1) chromatin structure, (2) transcriptional control, and (3) posttranscriptional control, which occur in the nucleus; and (4) translational and (5) posttranslational control, which occur in the cytoplasm.

In Chapter 9 you learned that one class of these DNA-associated proteins consists of the histones. Histones play an important role in the compaction of the DNA (see Fig. 9.3), as well as in eukaryotic gene regulation. Without histones, the DNA would not fit inside the nucleus. Each human cell contains around 2 meters of DNA, yet the nucleus is only 5 to 8 micrometers (μm) in diameter.

The degree to which chromatin is compacted greatly affects the accessibility of the chromatin to the transcriptional machinery of the cell, and thus the expression levels of the genes. Active genes in eukaryotic cells are associated with more loosely packed chromatin called *euchromatin,* while the more tightly packed DNA, called *heterochromatin,* contains mostly inactive genes. Under a microscope, the more densely compacted heterochromatin stains darker than euchromatin (Fig. 13.5*a*).

What regulates whether chromatin exists as heterochromatin or euchromatin? In Chapter 9 you learned that a *nucleosome* consists of a portion of DNA wrapped around a group of histone molecules. Histone molecules have *tails,* strings of amino acids that extend beyond the main portion of a nucleosome (Fig. 13.5*b*). In heterochromatin, the histone tails tend to bear methyl groups (—CH_3); in euchromatin, the histone tails tend to be acetylated and have attached acetyl groups (—$COCH_3$).

Histones regulate accessibility to DNA; euchromatin becomes genetically active when histones no longer block access to DNA. When DNA in euchromatin is transcribed, a group of proteins called the *chromatin remodeling complex* pushes aside, or *unpacks,* the histone portion of a nucleosome, so that access to DNA is not blocked and transcription can begin (Fig. 13.5*c*). After unpacking occurs, many decondensed loops radiate from the central axis of the chromosome. These chromosomes have been named lampbrush chromosomes, because their feathery appearance resembles the brushes that were once used to clean kerosene lamps.

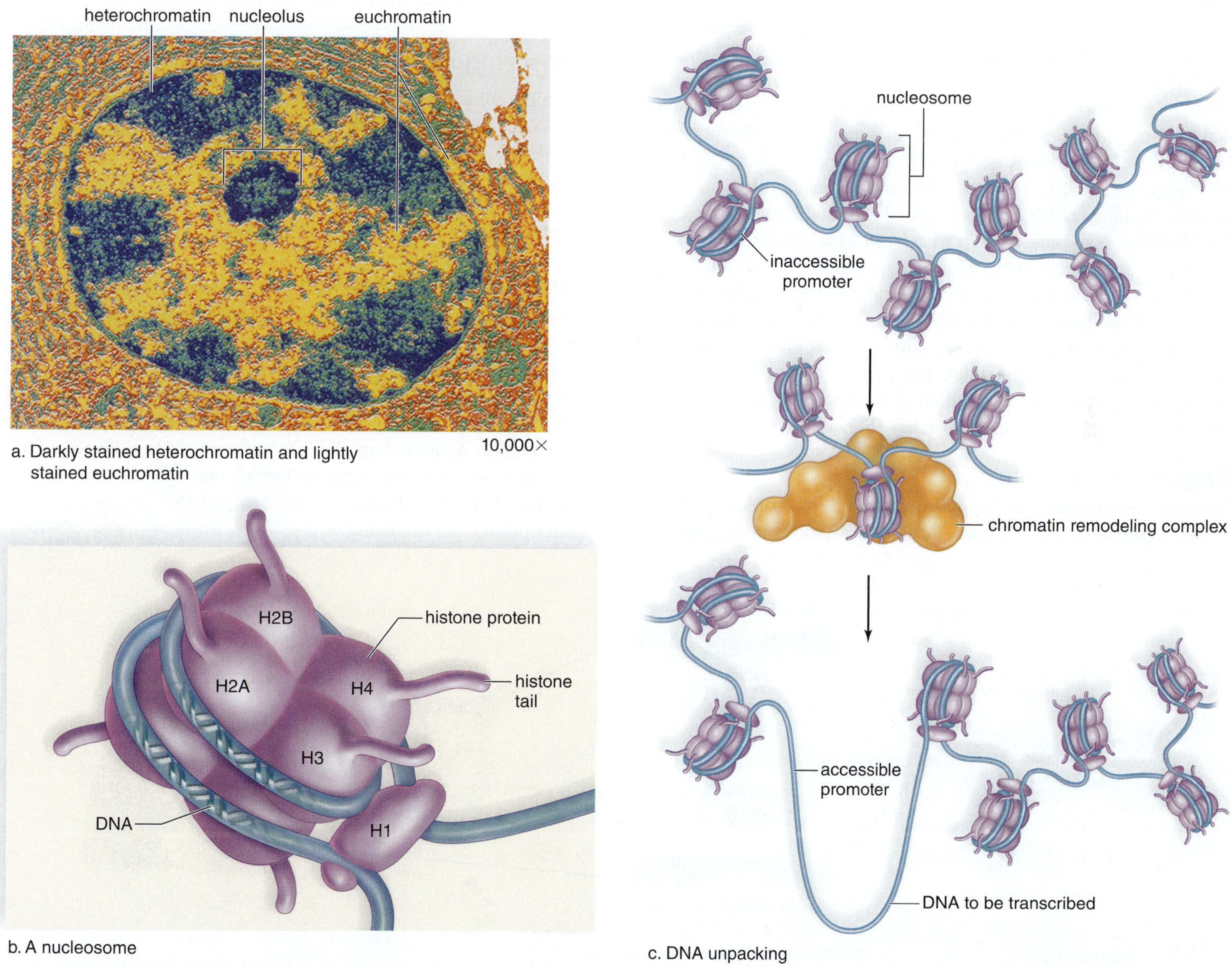

a. Darkly stained heterochromatin and lightly stained euchromatin

b. A nucleosome

c. DNA unpacking

Figure 13.5 Chromatin structure regulates gene expression. **a.** A eukaryotic nucleus contains highly condensed heterochromatin (darkly stained) and euchromatin (lightly stained), which is not as condensed. **b.** Nucleosomes ordinarily prevent access to DNA, so that transcription cannot take place. If histone tails are acetylated, access can be achieved; if the tails are methylated, access is more difficult. **c.** A chromatin remodeling complex works on euchromatin to make the DNA available and thus the promoter accessible for transcription.

In addition to physically moving nucleosomes aside to expose promoters, chromatin remodeling complexes may also affect gene expression by adding acetyl or methyl groups to histone tails.

Heterochromatin Is Not Transcribed

In general, highly condensed heterochromatin is inaccessible to RNA polymerase, and the genes contained within are seldom or never transcribed. A dramatic example of heterochromatin is the **Barr body** in mammalian females, first mentioned in Chapter 10. This small, darkly staining mass of condensed chromatin adhering to the inner edge of the nuclear membrane is an inactive X chromosome. To compensate for the fact that female mammals have two X chromosomes (XX), whereas males have only one (XY), one of the X chromosomes in the cells of female embryos undergoes inactivation. The inactive X chromosome does not produce gene products, allowing both males and females to produce the same amount of gene product from a single X chromosome.

Animation X-Inactivation

How do we know that Barr bodies are inactive X chromosomes that are not producing gene products? In a heterozygous female, 50% of the cells have one X chromosome active, and 50% have the other X chromosome active. The body of a heterozygous female is therefore a mosaic, with "patches" of genetically different cells. Investigators have discovered that human females who are heterozygous for an X-linked recessive form of ocular albinism have patches of pigmented and nonpigmented cells at the back of the eye.

As other examples, women who are heterozygous for X-linked hereditary absence of sweat glands have patches of skin lacking sweat glands. And the female calico cat exhibits a difference in X-inactivation in its cells. In these cats, an allele for black coat color is on one X chromosome, and a corresponding allele for orange coat color is on the other. The patches of black and orange in the coat can be related to which X chromosome is in the Barr bodies of the cells found in the patches (Fig. 13.6).

Epigenetic Inheritance

Histone modification is sometimes linked to a phenomenon termed **epigenetic inheritance,** in which variations in the pattern of inheritance are not due to changes in the sequence of the DNA nucleotides. For example, when histones are methylated, sometimes the DNA itself becomes methylated as well.

During *genomic imprinting,* either the mother's or the father's gene (but not both) is methylated during gamete formation. If an inherited allele is highly methylated, the gene is not expressed, even if it is a normal gene in every other respect. For traits that exhibit genomic imprinting, the expression of the gene depends on whether the unmethylated allele was inherited from the mother or the father.

The term *epigenetic inheritance* is now used broadly for other inheritance patterns that do not depend on the genes themselves. Epigenetic inheritance explains unusual inheritance patterns and may play an important role in growth, aging, and cancer. As discussed in the Biological Systems feature, "Same but Not the Same—the Role of Epigenetics," on page 237, researchers are hopeful that it will be easier to develop drugs to modify this level of inheritance, rather than trying to change the DNA itself.

Transcriptional Control

Although eukaryotes have various levels of genetic control (see Fig. 13.4), **transcriptional control** remains the most critical of these levels. The first step toward transcription is availability of DNA, which involves chromatin structure. Transcriptional control also involves the participation of transcription factors, activators, and repressors.

Transcription Factors, Activators, and Repressors

Although some operons like those of prokaryotic cells have been found in eukaryotic cells, transcription in eukaryotes is still controlled by DNA-binding proteins. Every cell contains many different types of **transcription factors,** proteins that help regulate transcription by assisting the binding of the RNA polymerase to the promotor. A cell has many different types of transcription factors, and a variety of transcription factors may be active at a single promoter. Thus, the absence of one can prevent transcription from occurring.

Animation Transcription Factors

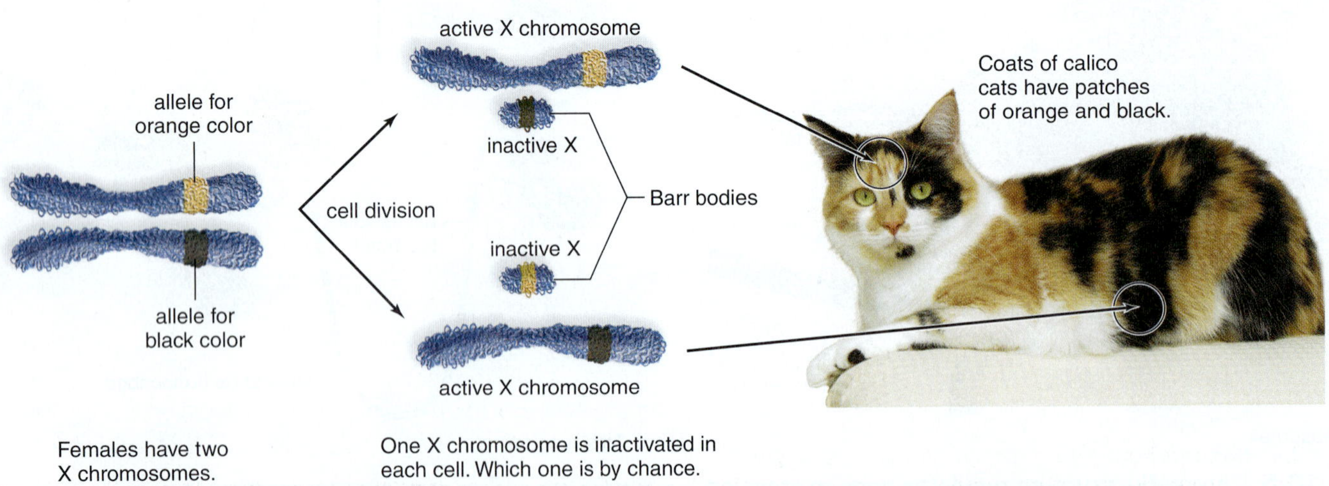

Figure 13.6 X-inactivation in mammalian females. In cats, the alleles for black or orange coat color are carried on the X chromosomes. Random X-inactivation occurs in females. Therefore, in heterozygous females, 50% of the cells have an allele for black coat color and 50% of the cells have an allele for orange coat color. The result is calico cats that have coats with patches of both black and orange.

Even if all the transcription factors are present, transcription may not begin without the assistance of a DNA-binding protein called a **transcription activator.** These bind to regions of DNA called **enhancers,** which may be located some distance from the promoter. A hairpin loop in the DNA brings the transcription activators attached to the enhancer into contact with the transcription factor complex (Fig. 13.7). Likewise, the binding of repressors within the promoter may prohibit the transcription of certain genes. Most genes are subject to regulation by both activators and repressors.

The promoter structure of eukaryotic genes is often very complex, and a large variety of regulatory proteins may interact with each other and with transcription factors to affect a gene's transcription level. Mediator proteins act as a bridge between transcription factors and transcription activators at the promoter. Now RNA polymerase can begin the transcription process (Fig. 13.7). Such protein-to-protein interactions are a hallmark of eukaryotic gene regulation. Together, these mechanisms can fine-tune a gene's transcription level in response to a large variety of conditions. For example, all the cells in a corn plant contain the gene for the pigment anthocyanin, but where and when anthocyanin is made is transcriptionally controlled. UV light induces anthocyanin production in the leaves where it is controlled by one set of transcription factors. Later, organ development cues anthocyanin production in the kernels controlled by a different set of transcription factors.

Posttranscriptional Control

Posttranscriptional control of gene expression occurs in the nucleus and includes alternative mRNA splicing and controlling the speed with which mRNA leaves the nucleus.

Recall that during pre-mRNA splicing, introns (noncoding regions) are excised and exons (expressed regions) are joined together to form an mRNA (see Fig. 12.14). When introns are removed from pre-mRNA, differential splicing of exons can occur, and this affects gene expression. For example, an exon that is normally included in an mRNA transcript may be skipped, and it is excised along with the flanking introns (Fig. 13.8). The resulting mature mRNA has an altered sequence, and the protein it encodes is altered. Sometimes introns remain in an mRNA transcript; when this occurs, the protein-coding sequence is also changed.

Examples of alternative pre-mRNA splicing abound. Both the hypothalamus and the thyroid gland produce a protein hormone called calcitonin, but the mRNA that leaves the nucleus is not the same in both types of cells. This results in the thyroid's releasing a slightly different version of calcitonin than does the hypothalamus.

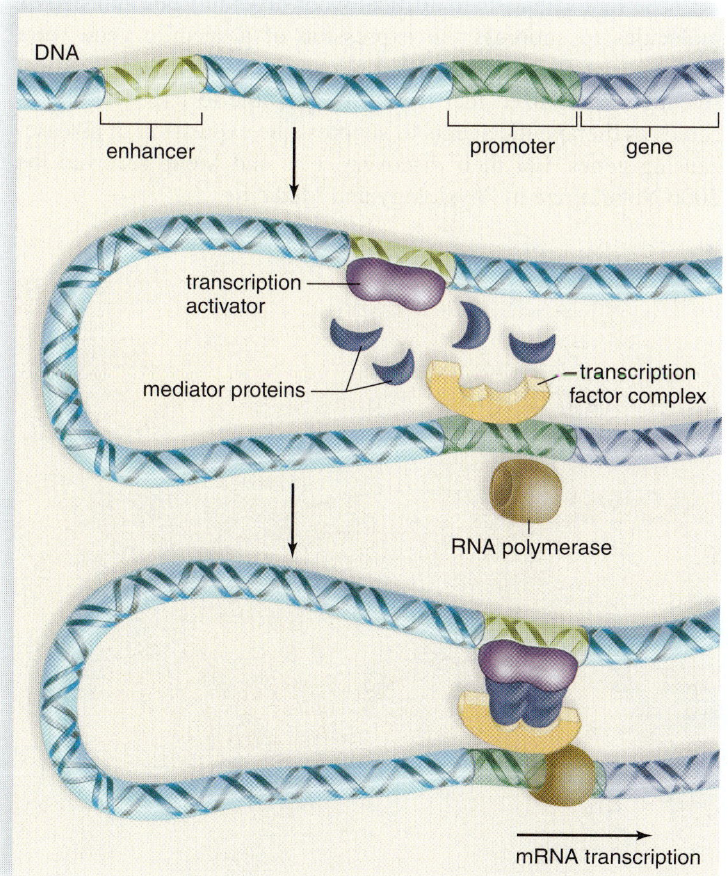

Figure 13.7 Eukaryotic transcription factors. Transcription in eukaryotic cells requires that transcription factors bind to the promoter and transcription activators bind to an enhancer. The enhancer may be far from the promoter, but the DNA loops and mediator proteins act as a bridge joining activators to factors. Only then does transcription begin.

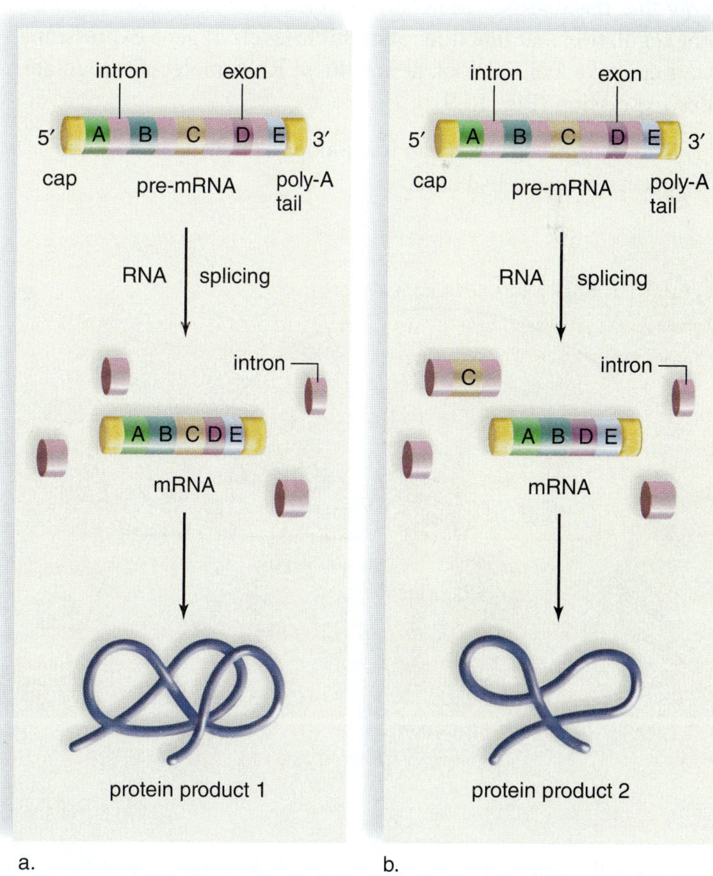

Figure 13.8 Alternative processing of pre-mRNA. Because the pre-mRNAs are processed differently in these two cells (**a** and **b**), distinct proteins result. This is a form of posttranscriptional control of gene expression.

Evidence of alternative mRNA splicing is found in other cells, such as those that produce neurotransmitters, muscle regulatory proteins, and antibodies.

Alternative pre-mRNA splicing allows humans and other complex organisms to recombine their genes in novel ways to create the great variety of proteins found in these organisms. Researchers are busy determining how small nuclear RNAs (snRNAs) affect the splicing of pre-mRNA. Alternative mRNA splicing can also result in the inclusion of an intron that brings about destruction of the mRNA before it leaves the nucleus.

Further posttranscriptional control of gene expression is achieved by modifying the speed of transport of mRNA from the nucleus into the cytoplasm. Evidence indicates there is a difference in the length of time it takes various mRNA molecules to pass through a nuclear pore, affecting the amount of gene product realized per unit of time following transcription.

Small RNA (sRNA) Molecules Regulate Gene Expression

For a long time, scientists were faced with a mystery: A cell appeared to contain vastly more DNA than was needed to account for the number of expressed proteins. The DNA that was not expressed as proteins was initially termed "junk" DNA, but recently scientists have begun to understand the role of this DNA in the cell. Although only about 1.5% of the transcribed DNA codes for protein, the remainder is used to form small RNA (sRNA) molecules. We now know that these sRNA molecules represent an important form of gene regulation that functions at multiple levels of gene expression.

Let's take a closer look at how these RNA molecules regulate gene expression (Fig. 13.9).

1. The transcribed RNA can form loops as hydrogen bonding occurs between its bases.
2. The double-stranded RNA (dsRNA) is diced up by enzymes in the cell to form sRNA molecules.
3. Some of these sRNA molecules regulate transcription, while others are involved in the regulation of translation. Various ways have been found by which sRNA may regulate gene expression. sRNA molecules have been known to alter the compaction of DNA, so that some genes are inaccessible to the transcription machinery of the cell.
4. Small RNAs are the source of **microRNAs** (**miRNAs**), small snippets of RNA that can bind to and disable the translation of mRNA in the cytoplasm.
5. Small RNAs are also the source of **small-interfering RNAs** (**siRNAs**) that join with an enzyme (an RNA-induced silencing complex, or RISC) to form an active silencing complex. This activated complex targets specific mRNAs in the cell for breakdown, preventing them from being expressed.

By using a combination of miRNA and siRNA molecules, a cell can fine-tune the amount of product being expressed from a gene, much as a dimmer switch on a light regulates the brightness of the room. Because both miRNA and siRNA molecules interfere with the normal gene expression pathways, the process is often referred to as **RNA interference.**

Animation RNA Interference

The first scientists to artifically construct miRNA and siRNA molecules to suppress the expression of a specific gene were Andrew Fire and Craig Mello. Following this discovery, medical scientists recognized that it may be possible to use sRNA molecules as therapeutic agents to suppress the expression of disease-causing genes. For their discovery, Fire and Mello received the 2006 Nobel Prize in Physiology and Medicine.

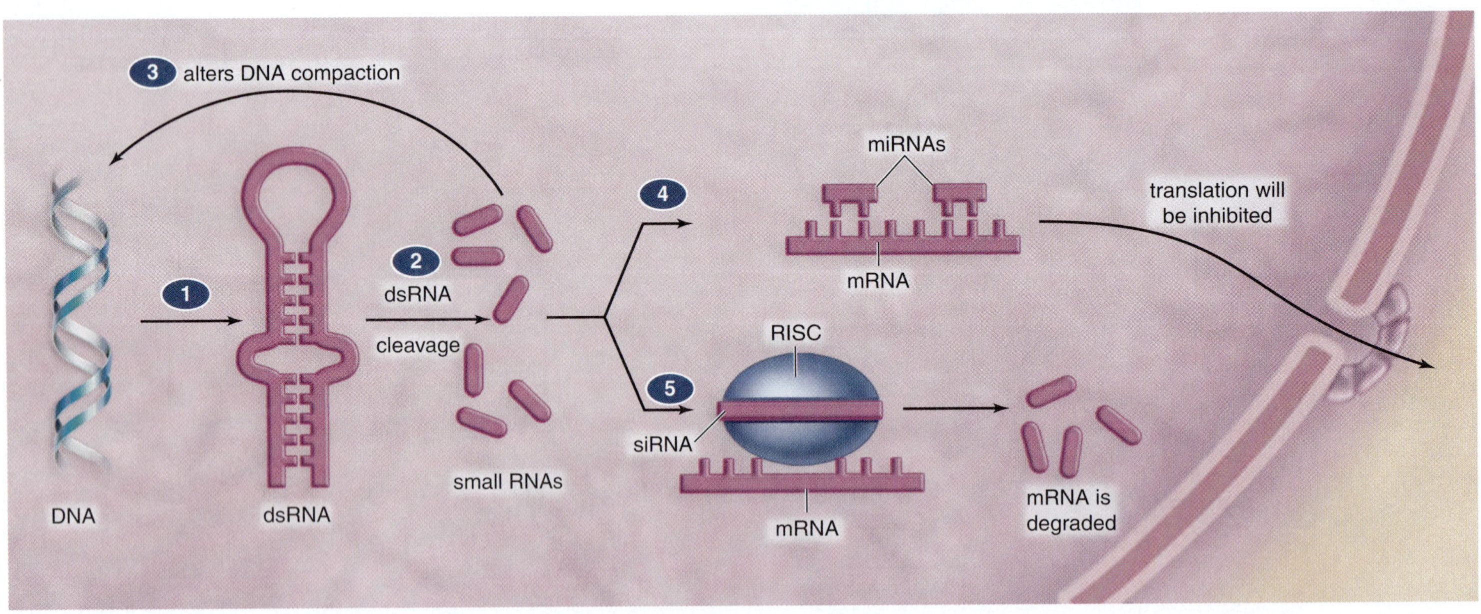

Figure 13.9 Function of small RNA molecules. Transcription of the DNA (1) may lead to looped and double-stranded RNA (dsRNA). (2) The cleavage of the dsRNA produces many small RNA (sRNA) molecules. (3) An sRNA can double-back to increase DNA compaction, or it may become an miRNA or siRNA. (4) miRNA reduces translation by binding to complementary mRNA molecules. (5) siRNA forms a complex with RISC, which then degrades any mRNA with a sequence of bases that are complementary to the siRNA.

THEME Biological Systems

Same but Not the Same—the Role of Epigenetics

Mia and Emma are identical twins in their early 20s. They both have a dimpled chin and blonde hair, and they wear the same-size clothes and shoes. As little girls, their parents emphasized their similarities by dressing them the same and giving them both the same opportunities to play piano and do gymnastics. As teenagers, things began to change. Their clothing styles were different—Mia preferred the current trends, whereas Emma loved black clothing. Mia was also more outgoing and popular; Emma was more reserved and thoughtful.

How is it possible that two people with the same genes and raised alike can be so different? Many scientists attribute a person's outcome to two factors: nature and nurture. Nature, your genes, gives you traits for eye color, hair color, and blood type. Nurture is based on your lifestyle and environment, including diet, rearing, and education. But is there a third force at work that can affect a person's overall health and well-being? Researchers working with identical twins believe there is a bridge between nature and nurture in the form of epigenetics (Gk. *epi-*, "upon, over").

The specific chemical reactions, or epigenetic "tags," can come in different forms but are often associated with DNA methylation, in which a methyl group attaches to the cytosine base of DNA (Fig. 13B). With a methyl group attached, transcription cannot occur. The methyl group interferes with transcription factors and other proteins in the transcription machinery, thereby silencing or weakening a gene. Over time, the differences in these tags accumulate, making twins increasingly different from each other (Fig. 13C).

Epigenetics are heritable changes in gene expression without changing the DNA sequence. Chemical reactions due to environmental exposure influence how genes are turned off or on, how they are weakened or strengthened, how they change our immune systems, and how they build muscle, brains, and all other body parts. Identical twins present a unique opportunity to study epigenetics, because they are clones resulting from a split in a single fertilized egg (Fig. 13A). Assuming a similar upbringing, their gradual differences over time can therefore be attributed to their disparate control of genes.

Epigenetics has important implications for medicine. The appearance of tags on genes helps scientists discover the cause of some illnesses that cannot be explained by DNA or genetic mutations alone. Identical twins discordant (different) for autism, psychiatric disorders, and cancer have been shown to have different DNA methylation on certain genes.

In addition, the epigenetic changes are reversible. A study using rats showed that rat pups that are licked and nurtured by their mothers become calm adults. Rat pups that are not nurtured are anxious. Injecting a calm rat with a drug that adds methyl groups creates an anxious rat. Conversely, injecting an anxious rat with a different drug that removes methyl groups creates a calm rat. In drug development, epigenetic medicines could be used to correct or reverse the particular effect of a tag.

Questions to Consider

1. How does epigenetics affect transcription and translation?
2. What lifestyle choices most likely negatively impact a person's epigenetics?

Figure 13A Identical twins. Identical twins come from a single fertilized egg that splits in two. Their genes are the same.

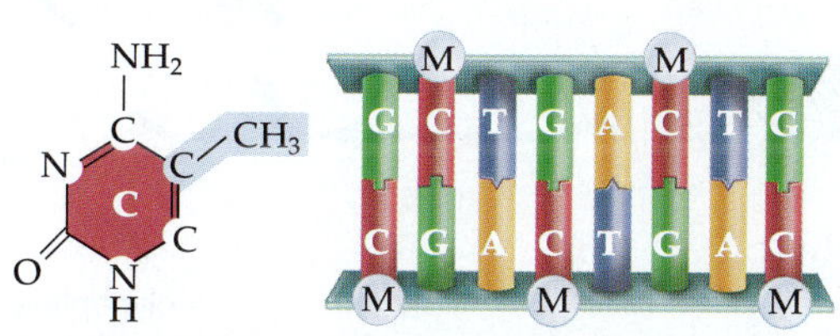

DNA methylation is the addition of a methyl group (M) to the DNA base cytosine (C).

Figure 13B DNA methylation. DNA is methylated when a methyl group attaches to the cytosine nucleotide.

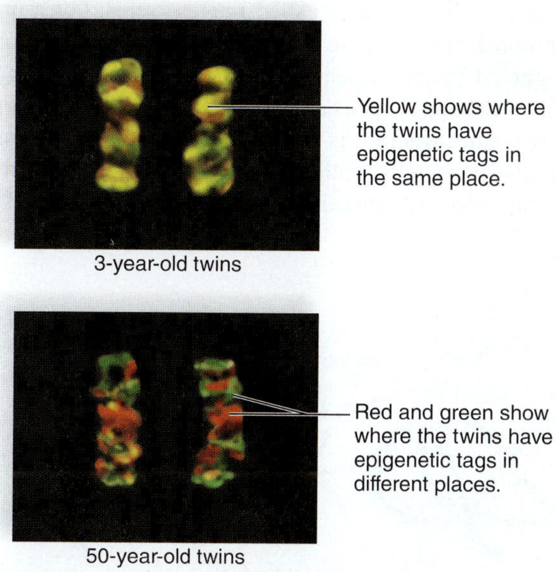

Figure 13C Comparison of twin's chromosomes. One twin's epigenetic tags are dyed green, and the other twin's tags are dyed red. An overlap in green and red shows up as yellow. The 50-year-old twins have more epigenetic tags in different places than do the 3-year-old twins.

Translational Control

Translational control begins when the processed mRNA molecule reaches the cytoplasm and before there is a protein product. Translational control involves the activity of mRNA for translation at the ribosome.

The presence or absence of the 5′ cap and the length of the poly-A (adenine nucleotide) tail at the 3′ end of a mature mRNA transcript can determine whether translation takes place and how long the mRNA is active. The long life of mRNAs that code for hemoglobin in mammalian red blood cells is attributed to the persistence of their 5′ end caps and their long 3′ poly-A tails. Therefore, any condition that affects the length of the poly-A tail or leads to removal of the cap may trigger the destruction of an mRNA.

Posttranslational Control

Posttranslational control begins once a protein has been synthesized and has become active. Posttranslational control represents the last chance a cell has for influencing gene expression.

If all the proteins produced by a cell during its lifetime remained in the cell, serious problems would arise. Thus, proteins are continually being synthesized and then degraded.

Proteins only needed for a short time can be altered chemically, leaving them nonfunctional. Proteins may not be folded correctly or they may change shape over time, leading them to behave erratically or stick to one another and form aggregates. In fact, a number of neurodegenerative diseases, such as Alzheimer dementia, Parkinson disease, and mad cow disease, are related to proteins that aggregate, forming plaques in the brain. Thus, in addition to normal turnover of proteins, cells need a way to get rid of old, unused, and incorrectly folded proteins.

Just how long a protein remains active in a cell is usually regulated by the use of **proteases,** enzymes that break down proteins. To protect the cell, proteases are typically confined to the lysosomes or special structures called **proteasomes.** For a protein to enter a proteasome, it has to be tagged with a signaling protein that is recognized by the proteasome cap (Fig. 13.10). When the cap recognizes the tag, it opens and allows the protein to enter the core of the structure, where it is digested to peptide fragments. Notice that proteasomes help regulate gene expression because they help control the amount of protein product in the cytoplasm.

Check Your Progress 13.2

1. Describe the five levels of genetic control in eukaryotes.
2. Explain how chromatin structure influences gene expression.
3. Discuss how small RNA molecules and proteasomes regulate gene expression.

13.3 Gene Mutations

Learning Outcomes

Upon completion of this section, you should be able to

1. Distinguish between spontaneous and induced mutations.
2. Identify how mutations influence protein structure.
3. Summarize how mutations may cause cancer.

A **gene mutation** is a permanent change in the sequence of bases in DNA. The effect of a DNA base sequence change on protein activity can range from no effect to complete inactivity. Germ-line mutations are those that occur in sex cells and can be passed to subsequent generations. Somatic mutations occur in body cells and, therefore, may affect only a small number of cells in a tissue. Somatic mutations are not passed on to future generations, but they can lead to the development of cancer.

Causes of Mutations

Some mutations are spontaneous—they happen for no apparent reason—whereas others are induced by environmental influences. In most cases, **spontaneous mutations** arise as a result of abnormalities in normal biological processes. **Induced mutations** may result from exposure to toxic chemicals or radiation, which induce (cause) changes in the base sequence of DNA.

Spontaneous Mutations

Spontaneous mutations can be associated with any number of normal processes. For example, a movable piece of DNA, termed a *transposon,* may jump from one location to another, disrupting

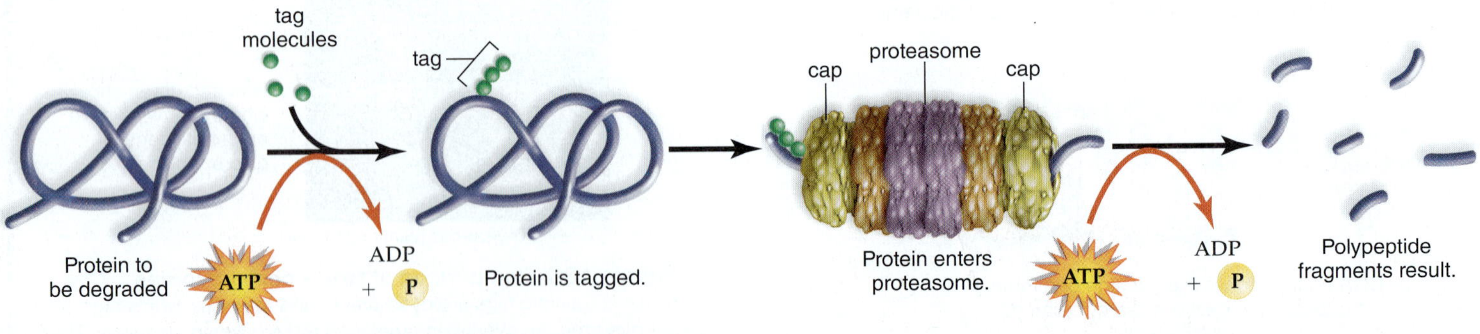

Figure 13.10 Proteasomes degrade proteins in a cell. Proteins to be degraded are first tagged with a signaling molecule. They then enter the proteasome, where they are broken down to polypeptide fragments.

one or more genes and leading to an abnormal product (see Chapter 14). On rare occasions, a base in DNA can undergo a chemical change that leads to a mispairing during replication. A subsequent base-pair change may be carried forth in future generations. Spontaneous mutations due to DNA replication errors, however, are rare. DNA polymerase, the enzyme that carries out replication, proofreads the new strand against the old strand and detects any mismatched nucleotides, and each is usually replaced with a correct nucleotide. In the end, only about one mistake occurs for every 1 billion nucleotide pairs replicated.

Induced Mutations

Induced mutations are caused by **mutagens,** environmental factors that can alter the base composition of DNA. Among the best-known mutagens are radiation and organic chemicals. Many mutagens are also **carcinogens** (cancer-causing mutagens).

Chemical mutagens are present in many sources, including some of the food we eat and many industrial chemicals. The mutagenic potential of AF-2, a food additive once widely used in Japan, and of safrole, a flavoring agent once used to flavor root beer, caused them to be banned. Surprisingly, many naturally occurring substances—like aflatoxin, produced in moldy grain and peanuts (and present in peanut butter at an average level of 2 parts per billion), and acrylamide, a natural product found in french fries—are also suspected mutagens.

Tobacco smoke contains a number of organic chemicals that are known carcinogens, and it is estimated that one-third of all cancer deaths can be attributed to smoking. Lung cancer is the most frequent lethal cancer in the United States, and smoking is implicated in the development of cancers of the mouth, larynx, bladder, kidney, and pancreas. The greater the number of cigarettes smoked per day, the earlier the habit starts, and the higher the tar content, the greater is the possibility of these cancers. When smoking is combined with drinking alcohol, the risk of these cancers increases even more.

Scientists use the Ames test for mutagenicity to hypothesize that a chemical can be carcinogenic (Fig. 13.11). In the Ames test, a histidine-requiring strain of bacteria is exposed to a chemical. If the chemical is mutagenic, the bacteria can grow without histidine. A large number of chemicals used in agriculture and industry give a positive Ames test. Examples are ethylene dibromide (EDB), which is added to leaded gasoline (to vaporize lead deposits in the engine and send them out the exhaust), and ziram, which is used to prevent fungal disease on crops. Some drugs, such as isoniazed (used to prevent tuberculosis), are mutagenic according to the Ames test.

Aside from chemicals, certain forms of radiation, such as X-rays and gamma rays, are called ionizing radiation, because they create free radicals, ionized atoms with unpaired electrons. Free radicals react with and alter the structure of other molecules, including DNA. Ultraviolet (UV) radiation is easily absorbed by the pyrimidines in DNA. Wherever there are two thymine molecules next to one another, ultraviolet radiation may cause them to bond together, forming *thymine dimers.* A kink results in the DNA. Usually, these dimers are removed by **DNA repair enzymes,** which constantly monitor DNA and fix any irregularities. One enzyme excises a portion of DNA that contains the dimer; another makes a new section by using the other strand as a template; and still another seals the new section in place.

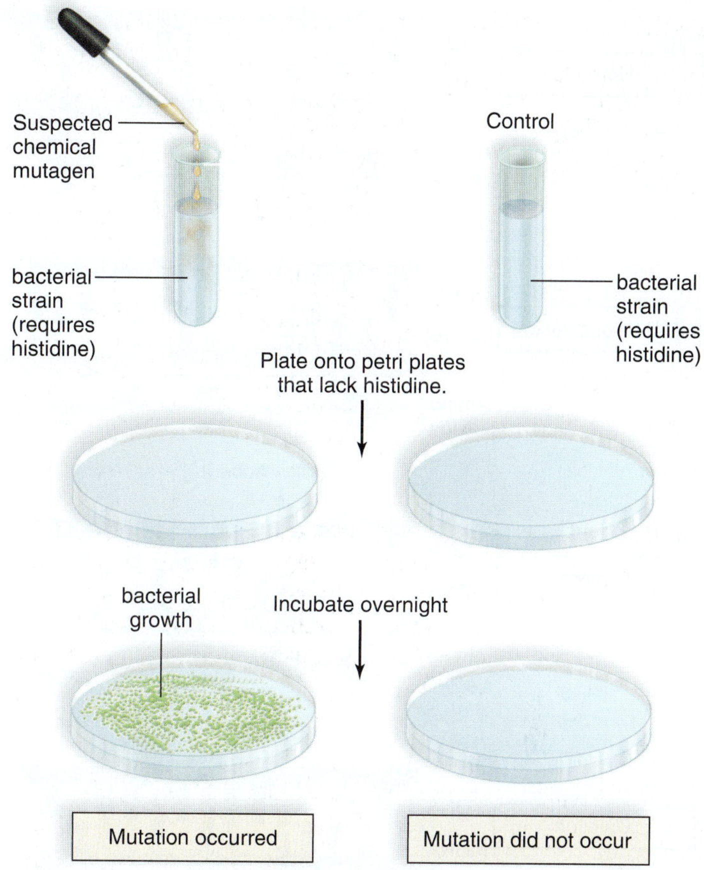

Figure 13.11 The Ames test for mutagenicity. A bacterial strain that requires histidine as a nutrient is exposed to a suspected chemical mutagen, but a control is not exposed. The bacteria are plated on a medium that lacks histidine; only the bacteria exposed to the chemical show growth. A mutation allowed the bacteria to grow; therefore, the chemical can be carcinogenic.

The importance of these repair enzymes is exemplified by individuals with the condition known as xeroderma pigmentosum. They lack some of the repair enzymes, and as a consequence, these individuals have a high incidence of skin cancer because of the large number of mutations that accumulate over time. Also, repair enzymes can fail, as when skin cancer develops because of excessive sunbathing or prolonged exposure to X-rays.

Effect of Mutations on Protein Activity

Point mutations involve a change in a single DNA nucleotide. That change alters transcription and possibly changes the specific amino acid. One type of point mutation is a *base substitution* resulting in one DNA nucleotide being replaced with another incorrect nucleotide. Notice the base difference in the second row of Figure 13.12*a* and how it changes the resultant amino acid sequence. Sometimes a base substitution has little or no effect on the final protein produced, but in some cases early stop codons can be introduced or coding for the wrong amino acid can severely alter the protein shape. Such is the case with the genetic disorder sickle-cell disease (Fig. 13.12*b*). In this gene, there is a base substitution that alters

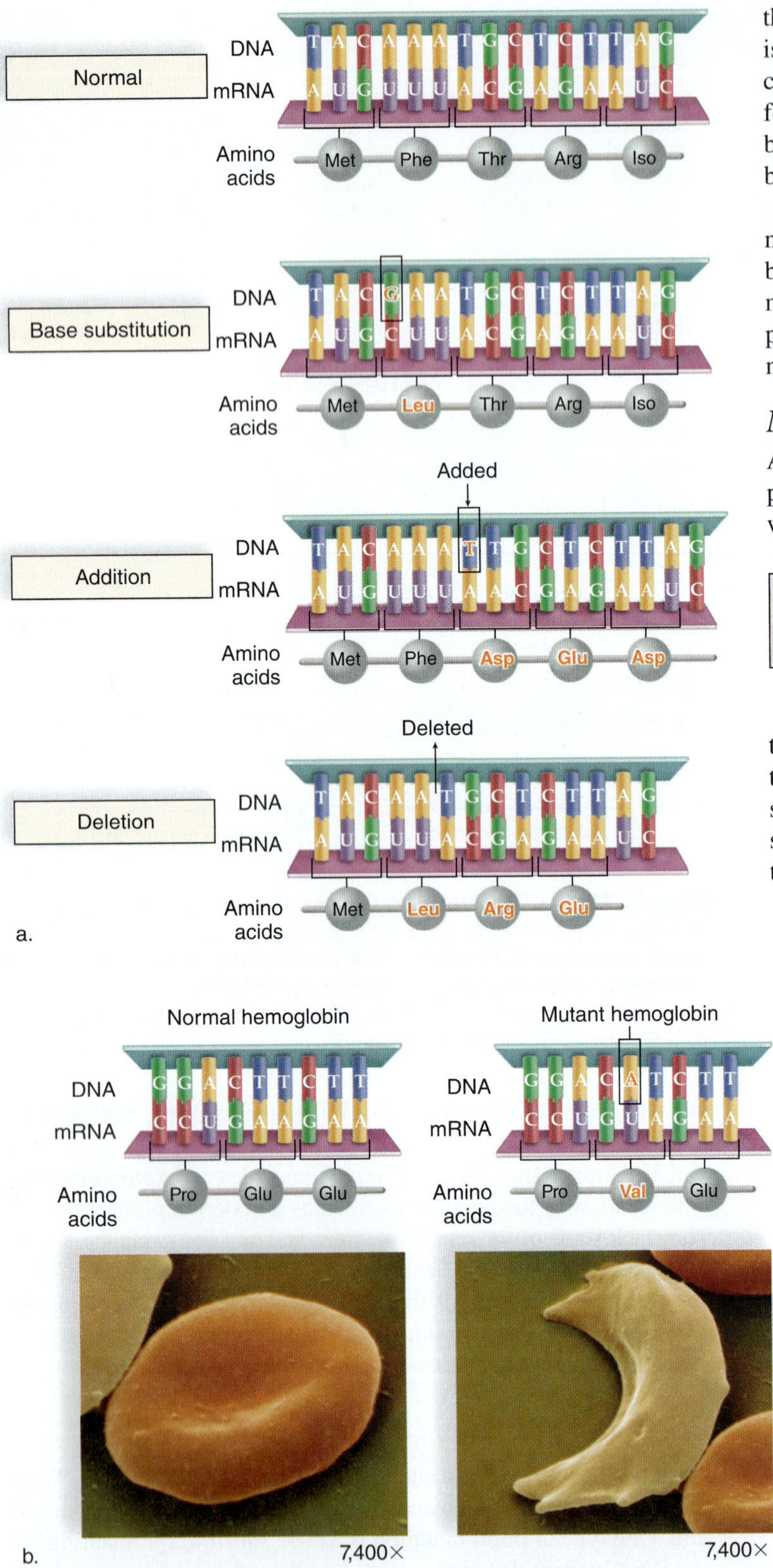

Figure 13.12 Point mutations. The effect of a point mutation can vary. **a.** Starting at the *top:* Normal sequence of bases results in a normal sequence of amino acids; next, a base substitution can result in the wrong amino acid; in the final two rows, an addition or deletion can result in a frameshift mutation altering all the codons downstream of the point mutation. **b.** Due to a base substitution in the hemoglobin gene, the DNA now codes for valine instead of glutamic acid, and the result is that normal red blood cells become sickle-shaped.

the mRNA codon for glutamic acid. Instead, the codon for valine is present, altering the final shape of hemoglobin—the protein that carries oxygen in the blood. The abnormal hemoglobin molecules form semirigid rods, and the red blood cells become sickle-shaped, resulting in decreased blood flow through tiny blood vessels.

Animation Mutation by Base Substitution

Frameshift mutations occur most often when one or more nucleotides are either added or deleted from DNA (Fig. 13.12*a*, bottom two lines). Because all the codons downstream of the mutation are now shifted, the result is a completely new sequence of codons, yielding a nonfunctional protein.

Animation Addition and Deletion Mutations

Nonfunctional Proteins

A single nonfunctioning protein can have a dramatic effect on the phenotype, because enzymes are often a part of metabolic pathways. One metabolic pathway in cells is as follows:

$$\underset{\text{(phenylalanine)}}{\text{A}} \xrightarrow{E_A} \underset{\text{(tyrosine)}}{\text{B}} \xrightarrow{E_B} \underset{\text{(melanin)}}{\text{C}}$$

If a faulty code for enzyme E_A is inherited, a person is unable to convert the molecule A to B. Phenylalanine builds up in the system, and the excess causes an intellectual disability and the other symptoms of the genetic disorder phenylketonuria (PKU). In the same pathway, if a person inherits a faulty code for enzyme E_B, then B cannot be converted to C, and the individual is an albino.

A rare condition called androgen insensitivity is due to a faulty receptor for androgens, which are male sex hormones, such as testosterone. In a male with this condition, plenty of testosterone is present in the blood, but the cells are unable to respond to it. Female instead of male external genitals form, and female instead of male secondary sex characteristics occur at puberty. The individual, who appears to be a normal female, may be prompted to seek medical advice when menstruation never occurs. The karyotype is that of a male rather than a female, and the individual does not have the internal sexual organs of a female.

Mutations Can Cause Cancer

It is estimated that one in three people will develop cancer at some time in their lives. Of these affected individuals, one-third of the females and one-fourth of the males will die due to cancer. In the United States, the three deadliest forms of cancer are lung cancer, colon and rectal cancer, and breast cancer.

The development of cancer involves a series of accumulating mutations that can be different for each type of cancer. As discussed in Chapter 9, tumor suppressor genes ordinarily act as brakes on cell division, especially when it begins to occur abnormally. Proto-oncogenes stimulate cell division but are usually turned off in fully differentiated, nondividing cells. When proto-oncogenes mutate, they become oncogenes that are active all the time. Carcinogenesis begins with the loss of tumor suppressor gene activity and/or the gain of oncogene activity. When tumor suppressor genes are inactive and oncogenes are active, cell division occurs uncontrollably, because a cell signaling pathway that reaches from the plasma membrane to the nucleus no longer functions as it should (Figs. 13.13 and 13.14).

It often happens that tumor suppressor genes and proto-oncogenes code for transcription factors or proteins that control transcription factors. As we have seen, transcription factors are a part of the rich and diverse types of mechanisms that control gene expression in cells. They are of fundamental importance to DNA replication and repair, cell growth and division, control of apoptosis, and cellular differentiation. Therefore, it is not surprising that inherited or acquired defects in transcription factor structure and function contribute to the development of cancer.

For example, the tumor suppressor gene called *p53* is more frequently mutated in human cancers than is any other known gene. It has been found that the p53 protein acts as a transcription factor, and as such it is involved in turning on the expression of genes whose products are cell cycle inhibitors (see Chapter 9). *p53* also promotes apoptosis (programmed cell death) when it is needed. The retinoblastoma protein (RB) controls the activity of a transcription factor for cyclin D and other genes whose products promote entry into the S stage of the cell cycle. When the tumor suppressor gene *p16* mutates, the RB protein is always available, and the result is too much active cyclin D in the cell.

Mutations in many other genes also contribute to the development of cancer. Several proto-oncogenes code for ras proteins, which are needed for cells to grow, to make new DNA, and to not grow out of control. A point mutation is sufficient to turn a normally functioning *ras* proto-oncogene into an oncogene, and abnormal growth results.

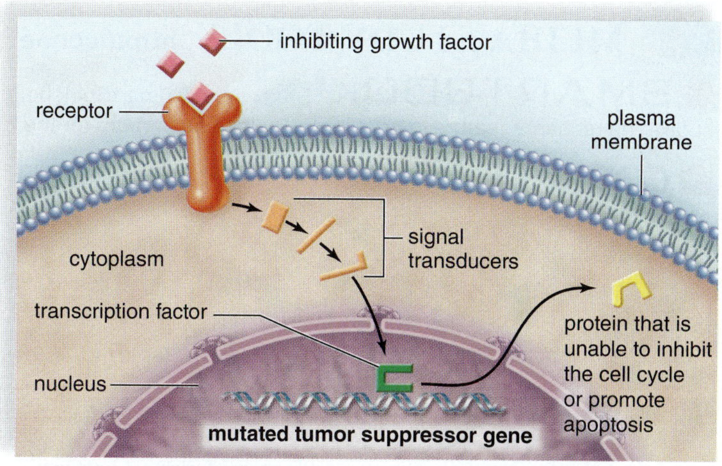

Figure 13.13 Cell signaling pathway that stimulates a mutated tumor suppressor gene. A mutated tumor suppressor gene codes for a product that directly or indirectly stimulates the cell cycle.

Tutorial Tumor Suppressor Genes

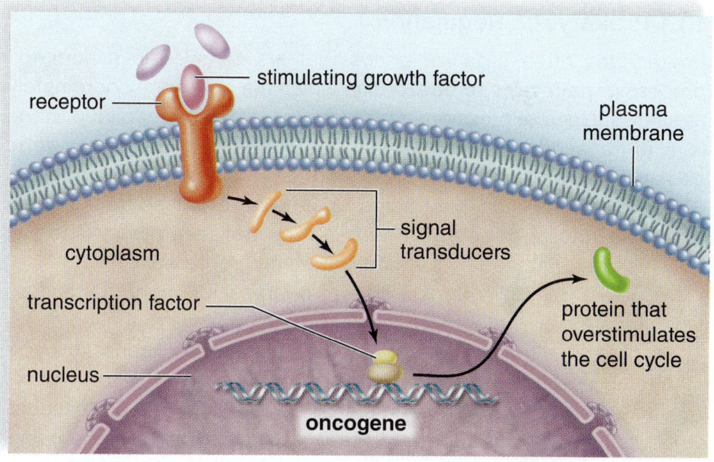

Figure 13.14 Cell signaling pathway that stimulates an oncogene. An oncogene codes for a product that either directly or indirectly overstimulates the cell cycle.

Tutorial Proto-oncogenes

Check Your Progress 13.3

1. List some common causes of spontaneous and induced mutations.
2. Explain how a frameshift mutation may disrupt a gene's function.
3. Discuss how a mutation in a tumor suppressor gene and in proto-oncogenes disrupts the cell cycle.

Connecting *the* Concepts *with the* Themes

Evolution

- In prokaryotes, regulation of gene expression occurs primarily through the regulation of the operon.
- Eukaryotes have evolved a variety of regulatory mechanisms that allow them to fine-tune gene expression and produce a large number of proteins from a relatively small number of genes.
- All living organisms are subject to genetic mutations, or changes, in the base sequence of DNA. DNA repair mechanisms have evolved to reduce the impact of mutations. Mutation is a source of genetic variability that is acted on by natural selection, allowing for evolutionary change over time.

Nature of Science

- By studying the structure of the prokaryotic operon, scientists have developed a basic understanding of gene regulation that allows them to better understand gene expression in eukaryotes.
- The discovery of small RNA molecules has caused scientists to rethink the concept of "junk" DNA. New findings may lead to the development of new treatments for human diseases.

Biological Systems

- In eukaryotes, gene regulation may occur at a number of different levels, all of which influence the amount of end-product protein found in a cell.
- Mutations cause changes in the DNA and may have a significant influence on the structure and function of the cell, as is the case with cancer.
- Epigenetics are heritable changes to the way genes are controlled without changing the DNA sequence itself.
- Changes can be due to environmental exposure and are reversible.

MEDIA STUDY TOOLS http://connect.mheducation.com

SMARTBOOK® Study smart. SmartBook helps you maximize your study time by identifying what you know and don't know. The recharge feature in SmartBook helps you prepare for the test by identifying the material that you are most likely to forget.

connect BIOLOGY Find even more resources to learn the chapter concepts including animations, tutorial videos, and interactive practice questions.

Animations

13.1 The Tryptophan Repressor • Combination of Switches: The *lac* Operon
13.2 Control of Gene Expression in Eukaryotes • X-Inactivation • Transcription Factors • RNA Interference
13.3 Mutation by Base Substitution • Addition and Deletion Mutations

Tutorials

13.1 *lac* Operon
13.3 Tumor Suppressor Genes • Proto-oncogenes

SUMMARIZE

13.1 Prokaryotic Regulation

Prokaryotes often organize genes that are involved in a common process or pathway into **operons** in which the genes are coordinately regulated. Gene expression in prokaryotes is usually regulated at the level of transcription. The operon model states that a **regulator gene** codes for a **repressor.** When the repressor binds to the **operator,** RNA polymerase is unable to bind to the **promoter,** and transcription of the **structural genes** of the operon cannot take place. Operons may also be regulated by both activators and repressors.

The *trp* operon is an example of a repressible operon, because when tryptophan, the **corepressor,** is present, it binds to the repressor. The repressor is then able to bind to the operator, and transcription of structural genes does not take place.

The *lac* operon is an example of an inducible operon, because when lactose, the **inducer,** is present, it binds to the repressor. The repressor is unable to bind to the operator, and transcription of structural genes takes place if glucose is absent.

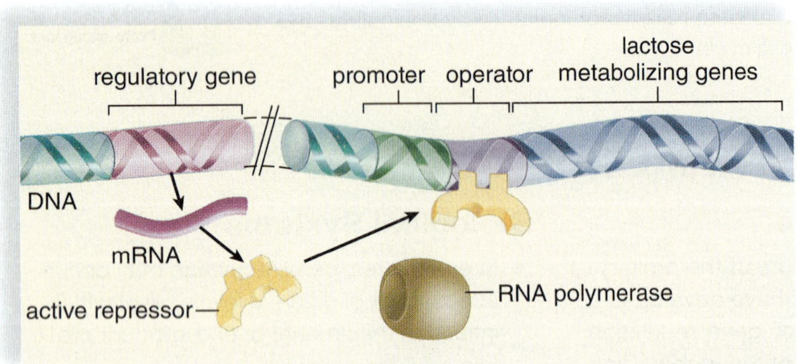

Both the *lac* and *trp* operons exhibit negative control, because a repressor is involved; however, some pathways also provide examples of positive control. The structural genes in the *lac* operon are not maximally expressed unless glucose is absent and lactose is present. At that time, cAMP attaches to a molecule called CAP, and then CAP binds to a site next to the promoter. Now RNA polymerase is better able to bind to the promoter, and transcription occurs.

13.2 Eukaryotic Regulation

The following levels of control of gene expression are possible in eukaryotes: chromatin structure, **transcriptional control,** posttranscriptional control, translational control, and posttranslational control.

Chromatin structure helps regulate transcription. Highly condensed heterochromatin is genetically inactive, as exemplified by **Barr bodies.** Less-condensed euchromatin is genetically active, as exemplified by lampbrush chromosomes in vertebrates.

Regulatory proteins called **transcription factors,** as well as DNA sequences called **enhancers,** play a role in **transcriptional control** in eukaryotes. Transcription factors bind to the promoter, and **transcription activators** bind to an enhancer. Small RNA molecules, such as **microRNAs** and **small-interfering RNAs,** are involved in **RNA interference** and play a role in gene expression.

Posttranscriptional control is achieved by creating variations in messenger RNA (mRNA) splicing, which may yield multiple mRNA messages from the same gene, and by altering the speed with which a particular mRNA molecule leaves the nucleus.

Translational control affects mRNA translation and the length of time it is translated, primarily by altering the stability of an mRNA.

Posttranslational control affects whether or not an enzyme is active and how long it is active. **Proteasomes** are specialized structures containing **proteases** that break down protein molecules and participate in the regulation of gene expression.

13.3 Gene Mutations

In molecular terms, a gene is a sequence of DNA nucleotide bases, and a **gene mutation** is a change in this sequence. Mutations can be **spontaneous,** occurring for no apparent reason, or **induced** and due to environmental **mutagens,** such as radiation and organic chemicals. **Carcinogens** are mutagens that cause cancer. **DNA repair enzymes** monitor and fix irregularities in DNA, notably thymine dimers that come from UV exposure.

Point mutations can have a range of effects, depending on the particular codon change. Sickle-cell disease is an example of a point mutation that greatly changes the activity of the affected gene. **Frameshift mutations** result when one or more bases are added or deleted, and the result is usually a nonfunctional protein. Most cases of cystic fibrosis are due to a frameshift mutation. Nonfunctional proteins can affect the phenotype drastically, as in a form of albinism that is due to a single faulty enzyme, and androgen insensitivity, which is due to a faulty receptor for testosterone.

Cancer is often due to an accumulation of genetic mutations among genes that code for regulatory proteins. The cell cycle occurs inappropriately when proto-oncogenes become oncogenes and tumor suppressor genes are no longer effective. Mutations that affect transcription factors and other regulators of gene expression are frequent causes of cancer.

ASSESS

Choose the best answer for each question.

13.1 Prokaryotic Regulation

1. In regulation of the *lac* operon, when lactose is present and glucose is absent,
   a. there is a low level of cAMP present.
   b. there is a high level of cAMP present.
   c. transcription of structural genes occurs.
   d. transcription of lactose occurs.
   e. Both b and c are correct.
2. In regulation of the *trp* operon, when tryptophan is present,
   a. the repressor is able to bind to the operator.
   b. the repressor is unable to bind to the operator.
   c. transcription of the repressor in inhibited.
   d. transcription of the structural genes, operator, and promoter occurs.
3. In operon models, the function of the promoter is to
   a. code for the repressor protein.
   b. bind with RNA polymerase.
   c. bind to the repressor.
   d. code for the regulator gene.
4. Label this diagram of an operon.

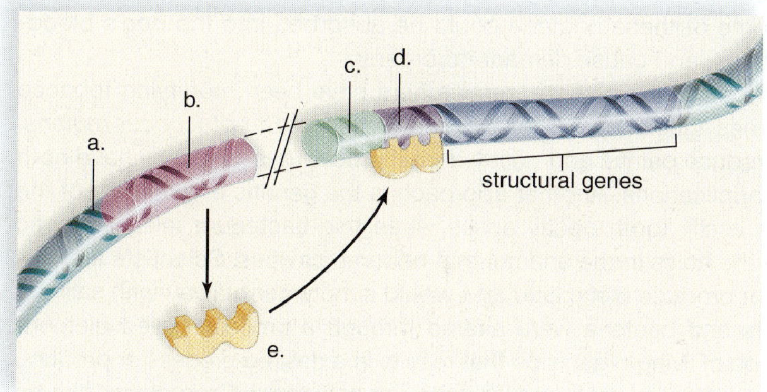

13.2 Eukaryotic Regulation

5. Which of the following regulate(s) gene expression in the eukaryotic nucleus?
   a. posttranslational control
   b. transcriptional control
   c. translational control
   d. posttranscriptional control
   e. Both b and d are correct.
6. Which of the following mechanisms may create multiple mRNAs from the same gene?
   a. posttranslational control
   b. alternative mRNA splicing
   c. binding of a transcription factor
   d. chromatin remodeling
   e. miRNAs
7. Translational control of gene expression occurs within the
   a. nucleus.
   b. cytoplasm.
   c. nucleolus.
   d. mitochondria.
8. Barr bodies are
   a. genetically active X chromosomes in males.
   b. genetically inactive X chromosomes in females.
   c. genetically active Y chromosomes in males.
   d. genetically inactive Y chromosomes in females.

13.3 Gene Mutations

9. A mutation in a DNA molecule involving the replacement of one nucleotide base pair with another is called a(n)
   a. frameshift mutation.
   b. transposon.
   c. deletion mutation.
   d. base substitution.
   e. insertion mutation.
10. THE COW ATE THE HAY. If the letter C is deleted from this sentence, shifting the reading frame, we read THE OWA TET HEH AY. Which of the following mutations *best* explains this example?
    a. spontaneous mutation
    b. frameshift mutation
    c. induced mutation
    d. base substitution
    e. transposon
11. A cell is cancerous. You might find an abnormality in
    a. a proto-oncogene.
    b. a tumor suppressor gene.
    c. regulation of the cell cycle.
    d. tumor cells.
    e. All of these are correct.
12. A tumor suppressor gene
    a. inhibits cell division.
    b. opposes oncogenes.
    c. prevents cancer.
    d. is subject to mutations.
    e. All of these are correct.

ENGAGE

THINKING SCIENTIFICALLY

1. New findings indicate that mutations outside of genes may cause disease, such as in some cases of Hirschsprung disease and multiple endocrine neoplasia. Explain how such a mutation might alter the expression of a gene.
2. From an experimental point of view, is it better to use haploid or diploid organisms for mutagen testing? Think about the Ames test when thinking about your answer.
3. Discuss the times in a person's life when it would be most important to avoid mutagens. Which parts of a person's body should be most protected from mutagens?

14

Biotechnology and Genomics

Chapter Outline

14.1 DNA Cloning 245
14.2 Biotechnology Products 247
14.3 Gene Therapy 250
14.4 Genomics 251

Before You Begin

Before beginning this chapter, take a few moments to review the following discussions.

Section 12.1 What is the basic structure of a DNA molecule?

Section 12.4 What is the difference between an intron and an exon in a gene?

Section 13.2 What is the role of microRNA molecules in a cell?

Combating tooth decay with traditional methods may one day be a thing of the past.

Dental disease in dogs is one of the most common medical conditions seen by veterinarians. Over 80% of all dogs over the age of 3 are estimated to have some form of gum or tooth decay. The oral cavity of dogs is a perfect incubator for many kinds of bacteria, and as these bacteria multiply, they form an invisible layer called plaque. Toxins produced by some of these bacteria could be absorbed into the dog's bloodstream, produce infections, and cause damage to organs.

In an effort to combat dental disease, researchers have been modifying tobacco plants to produce vaccines against the decay-causing bacterium *Streptococcus mutans*. These vaccines could reduce painful and costly dental procedures, and they have both veterinary and human applications. Another approach is the genetic engineering of the *Streptococcus mutans* itself. Tooth decay arises when this bacterium releases lactic acid on teeth, creating the holes in the enamel that become cavities. Scientists are creating strains that cannot produce lactic acid and would simply wash away with saliva.

The tobacco plants and bacteria were altered through a process called biotechnology—the manipulation of living organisms that results in a desired process or product. Other bacteria can be modified to make insulin or to eat pollutants. Crop plants can be altered to have higher vitamin content, or to be more resistant to pests. Even animals are being genetically modified and cloned to produce specific proteins in their milk. Gene therapies also exist in humans, where a DNA change can cure or slow down a disease.

As you read through this chapter, think about the following questions:

1. What procedures are used to introduce a gene into a bacterium, a plant, or an animal?
2. What is the difference between transgenic organisms and gene therapy?

Following the Themes

CHAPTER 14 BIOTECHNOLOGY AND GENOMICS

UNIT 2 GENETIC BASIS OF LIFE

Evolution	The comparison of the genomes of humans and model organisms, such as the mouse, is providing insights into the evolution of our species.
Nature of Science	The development of recombinant DNA technology has enabled scientists to produce genetically modified and transgenic organisms that benefit human society.
Biological Systems	Recombinant DNA technology may be used to modify individual cells, as is the case in gene therapy, or produce organisms that assist in the cleanup of polluted ecosystems.

14.1 DNA Cloning

Learning Outcomes

Upon completion of this section, you should be able to

1. Describe the steps involved in making a recombinant DNA molecule.
2. Explain the purpose of the polymerase chain reaction (PCR).
3. Identify how PCR may be used to analyze DNA.

In biology, **cloning** is the production of genetically identical copies of DNA, cells, or organisms through some asexual means. When an underground stem or root sends up new shoots, the resulting plants are clones of one another. The members of a bacterial colony on a petri dish are clones, because they all came from the division of a single original cell. Human identical twins are also considered clones. Early in embryonic development the cells separate, and each becomes a complete individual.

DNA cloning can be done to produce many identical copies of the same gene—that is, for the purpose of **gene cloning.** Scientists clone genes for a number of reasons. They might want to determine the difference in base sequence between a normal gene and a mutated gene. Or they might use the genes to genetically modify organisms in a beneficial way. When cloned genes are used to modify a human, the process is called **gene therapy.** Otherwise, the organisms are called **transgenic organisms** (L. *trans,* "across, through"; Gk. *genic,* "producing"). Transgenic organisms are frequently used today to produce a product desired by humans.

Recombinant DNA (rDNA) technology and the polymerase chain reaction (PCR) are two procedures that scientists can use to clone DNA.

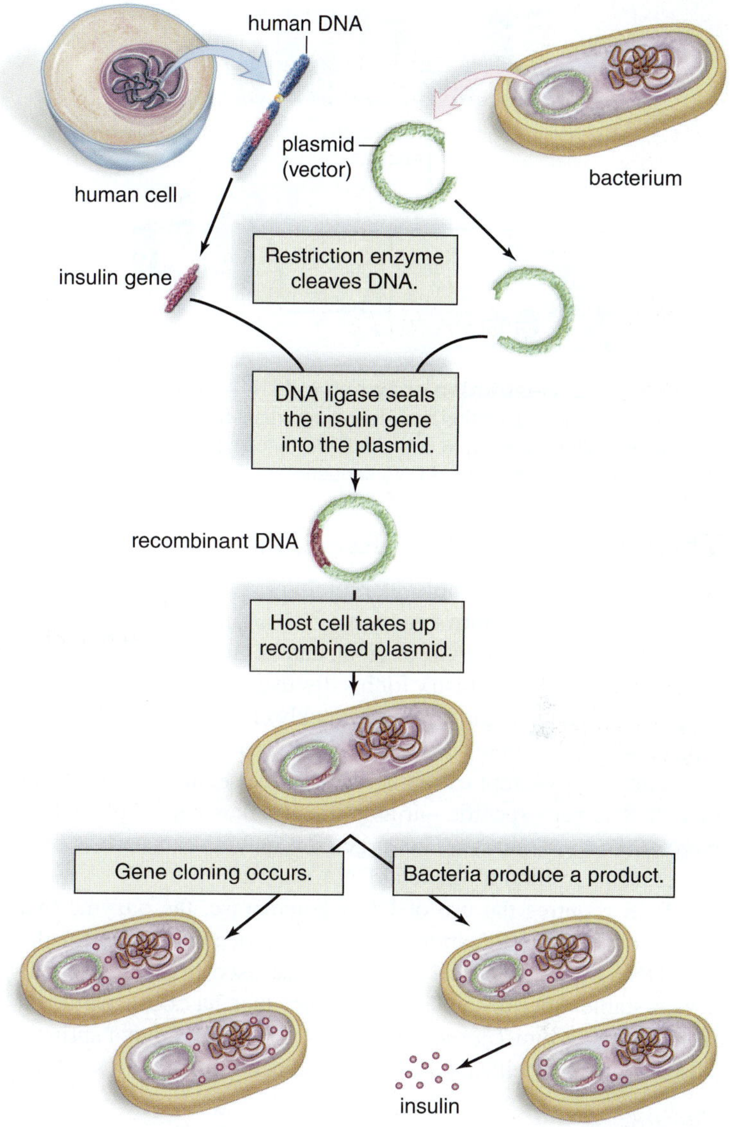

Figure 14.1 Cloning a human gene. This figure shows the basic steps in the cloning of a human gene. Human DNA and plasmid DNA are cleaved by a specific type of restriction enzyme. Then the human DNA, perhaps containing the insulin gene, is spliced into a plasmid by the enzyme DNA ligase. Gene cloning is achieved after a bacterium takes up the plasmid. If the gene functions normally, as expected, the product (e.g., insulin) may also be retrieved.

Recombinant DNA Technology

Recombinant DNA (rDNA) contains DNA from two or more different sources, such as a human cell and a bacterial cell, as shown in Figure 14.1. To make rDNA, a technician needs a **vector** (L. *vehere,* "to carry") by which rDNA will be introduced into a host cell. One common vector is a plasmid. **Plasmids** are small accessory rings of DNA found in bacteria; they were first discovered in the bacterium *Escherichia coli* (*E. coli*). The ring is not part of the main bacterial chromosome; it replicates on its own and can be easily removed from or introduced into a bacterial cell.

Two enzymes are needed to introduce foreign DNA into vector DNA: (1) a **restriction enzyme,** which cleaves (cuts) DNA, and (2) an enzyme called **DNA ligase** (L. *ligo,* "bind"), which seals DNA into an opening created by the restriction enzyme. Hundreds of restriction enzymes occur naturally in bacteria, and they cut up any viral DNA that enters the cell. They are called restriction enzymes because they *restrict* the growth of viruses. Scientists take advantage of these enzymes and use them as molecular scissors to cleave any piece of DNA at a specific site.

Animation Restriction Endonucleases

Notice that the restriction enzyme creates a puzzlelike gap in the DNA (Fig. 14.2), into which a piece of foreign DNA can be placed if its ends are complementary to those exposed by the restriction enzyme. The single-stranded, but complementary, ends of the two DNA molecules are called "sticky ends" because they can bind a piece of foreign DNA by complementary base-pairing. Sticky ends facilitate the insertion of foreign DNA into vector DNA as long as both are cleaved by the same restriction enzyme.

Next, genetic engineers use the enzyme DNA ligase to seal the foreign piece of DNA into the vector. DNA splicing is now complete; an rDNA molecule has been prepared (see Fig. 14.1). Bacterial cells take up recombinant plasmids, when they are treated to make their plasma membranes more permeable. Thereafter, as the plasmid replicates, DNA is cloned.

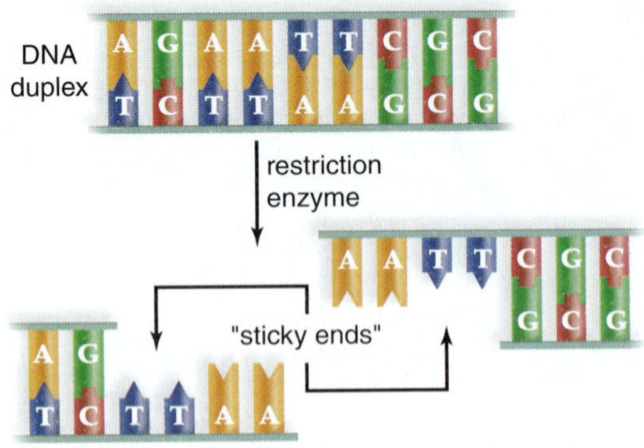

Figure 14.2 Restriction enzymes cut DNA at specific locations. Each restriction enzyme recognizes a specific sequence of nucleotides. After the enzyme cuts the DNA, "sticky ends" may be formed that are useful in the cloning of DNA sequences.

The Polymerase Chain Reaction

Another revolution in molecular biology was the development of the **polymerase chain reaction (PCR)**. American biochemist Kary Mullis developed PCR in 1983, and in 1993, he was awarded the Nobel Prize in Chemistry for his discovery. PCR can accelerate the pace of genetic engineering by quickly creating many clones of a piece of DNA without first inserting it into a plasmid. The process mimics DNA replication in the cell (see section 12.2), except that PCR is very specific—it amplifies (makes copies of) only a targeted DNA sequence. The targeted sequence can be less than one part in a million of the total DNA sample!

PCR requires the use of DNA polymerase, the enzyme that carries out DNA replication, and a supply of nucleotides for the new DNA strands. The DNA polymerase used in the reaction is a heat-stable (thermostable) polymerase that has been extracted from the bacterium *Thermus aquaticus,* which lives in hot springs. The enzyme can withstand the high temperature used to separate double-stranded DNA; therefore, replication does not have to be interrupted by the need to add more enzyme. PCR is a chain reaction because the targeted DNA is repeatedly replicated as long as the process continues. The colors in Figure 14.3 distinguish the old strand from the new DNA strand, but keep in mind that all the newly synthesized strands are identical (clones). Notice that the amount of DNA doubles with each replication cycle.

Animation Polymerase Chain Reaction

Analyzing DNA

DNA amplified by PCR can be analyzed for various purposes. For example, mitochondrial DNA taken from modern living populations was used to decipher the evolutionary history of human populations. For identification purposes, DNA taken from a corpse burned beyond recognition can be matched to that on the bristles of the person's toothbrush!

Analysis of DNA following PCR has undergone improvements over the years. At first, the entire genome was treated with restriction enzymes, resulting in a unique collection of different-sized fragments, because each person has his or her own restriction enzyme sites. A process called **gel electrophoresis,** which separates DNA fragments according to their size, was then employed; the result of fragment sorting was a pattern of distinctive bands that identified the person.

DNA fingerprinting (also called **DNA profiling**) is a technology that can identify and distinguish among individuals based on variations in their DNA. Like the human fingerprint, the DNA of each individual is different and can be used for identification. When subjected to DNA fingerprinting, selected fragments of chromosomal DNA produce a series of bands on a gel (Fig. 14.4*a*). The unique pattern of these bands is usually a distinguishing feature of each individual.

In the past two decades, the technique of DNA fingerprinting has become automated and is now done using PCR, which amplifies

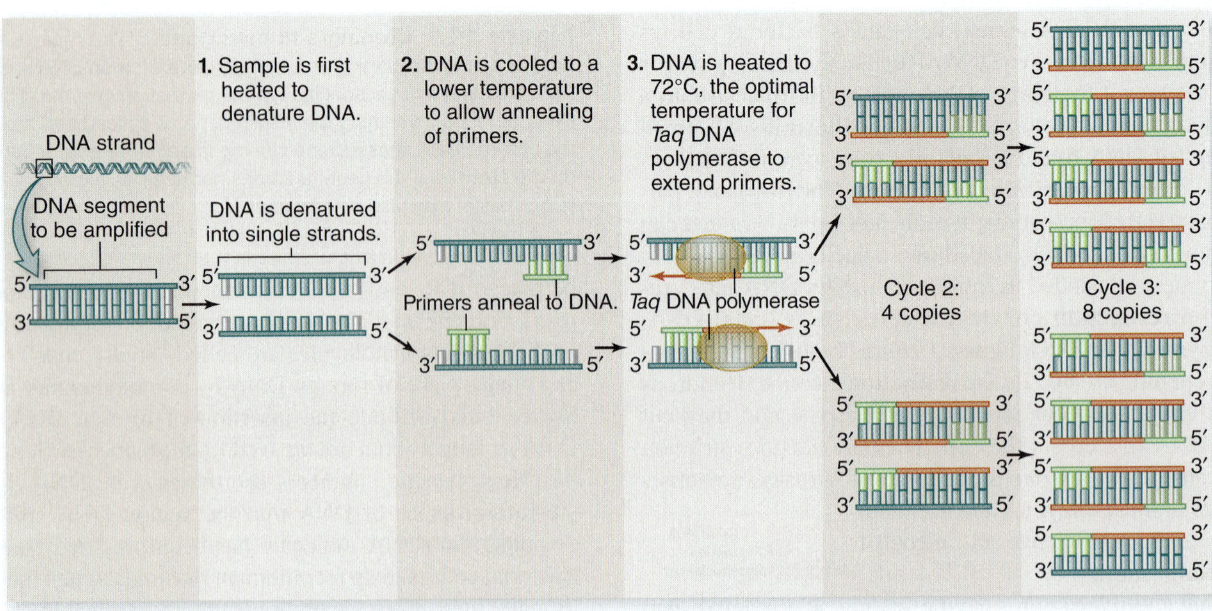

Figure 14.3 Polymerase chain reaction (PCR). PCR allows the production of many identical copies of a specific sequence of DNA in a laboratory setting.

Tutorial Polymerase Chain Reaction

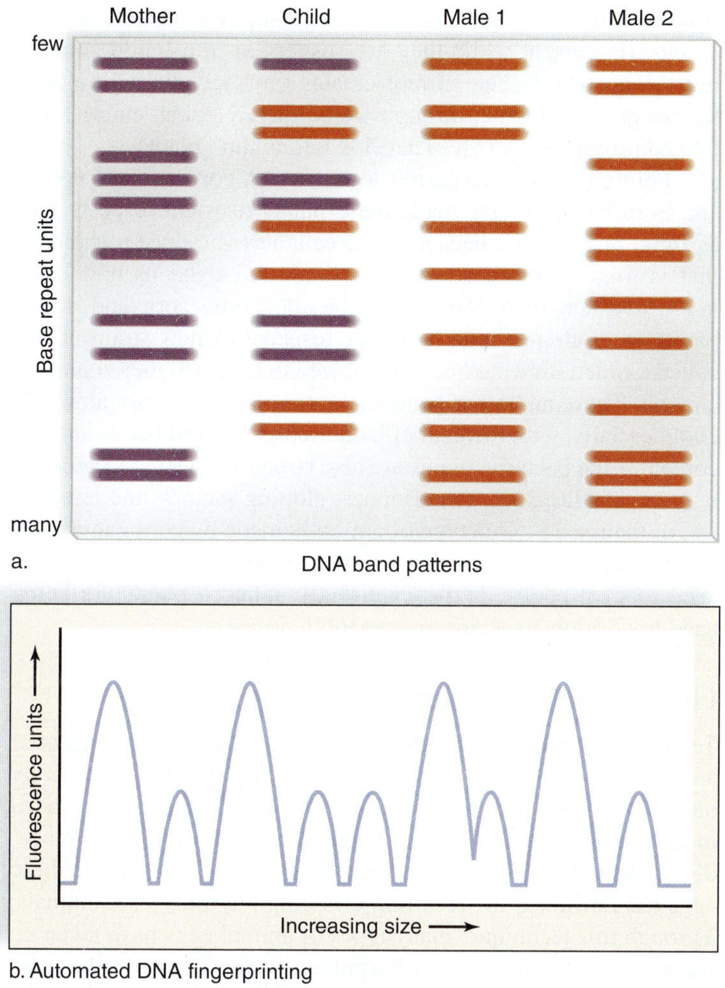

Figure 14.4 DNA fingerprinting using STRs to establish paternity. **a.** In this method, DNA fragments containing STRs are separated by gel electrophoresis. Male 1 is the father. **b.** Each person's DNA fingerprint (only one is shown) can also be printed out by a machine that detects fluorescence.

short tandem repeat sequences (**STRs**)—short DNA sequences that are repeated many times in a row. Such tandem repeat sequences, which are noncoding regions of chromosomal DNA, are found at specific locations in the genomes of all species. The number of repeats at each location tends to vary from one individual to the next. For example, humans have the sequence GATA on chromosome 7. One person may have this sequence 10 times, while another may have 15 repeats. The person with the higher number of repeats will have a larger DNA fragment.

The newest method of producing DNA fingerprints does away with the need to use gel electrophoresis: The DNA fragments are fluorescently labeled. A laser then excites the fluorescent STRs, and a detector records the amount of emission for each DNA fragment in terms of peaks and valleys. Therefore, the greater the fluorescence, the greater the number of repeats at a location. The printout, such as the one shown in Figure 14.4*b*, is the DNA fingerprint, and each person has his or her own unique printout. Currently, the Federal Bureau of Investigation (FBI) uses 13 STR locations on various chromosomes that are now routinely used in the identification of individuals in the United States.

Other Applications

Applications of PCR are limited only by the imagination.

- A viral infection, a genetic disorder, or cancer can be confirmed when the DNA tested matches that of a known virus or mutated gene.
- DNA fingerprinted from blood or tissues at a crime scene has been successfully used in screening suspects, convicting criminals, and exonerating those wrongly convicted.
- DNA fingerprinting through STR profiling has been extensively used to identify the victims of natural disasters, such as tsunamis in Indonesia and Japan.
- Relatives can be found, paternity suits can be settled (Fig. 14.4*a*), genetic disorders can be detected, and illegally poached ivory and fish can be recognized using this technology (see Chapter 19).
- PCR has also shed new light on evolutionary studies by comparing extracted DNA from ancient specimens with that of living organisms or by comparing DNA of two living species to study their genetic relatedness.

Check Your Progress 14.1

1. Explain the purpose of restriction enzymes in creating an rDNA molecule.
2. Contrast how gene cloning is different between recombinant technology and PCR.
3. Explain how DNA fingerprinting distinguishes individuals.

14.2 Biotechnology Products

Learning Outcomes

Upon completion of this section, you should be able to

1. Identify the benefits of genetically modified bacteria, plants, and animals to human society.
2. Describe the steps involved in the production of a transgenic animal.

Today, transgenic bacteria, plants, and animals are often called **genetically modified organisms** (**GMOs**), and the products they produce are called **biotechnology products.**

Genetically Modified Bacteria

Many uses have been found for genetically modified bacteria, besides the production of proteins. Biotechnology products from bacteria include insulin, clotting factor VIII, human growth hormone, t-PA (tissue plasminogen activator), and hepatitis B vaccine.

Animation Early Genetic Engineering Experiment

Transgenic bacteria have many other uses as well. Some have been produced to promote the health of plants. For example, bacteria that normally live on plants and encourage the formation of ice crystals have been changed from frost-plus to frost-minus bacteria. As a result, new crops, such as frost-resistant strawberries, are being developed. Also, a bacterium that normally colonizes the roots of corn plants has now been endowed with genes (from another bacterium) that code for an insect toxin. The toxin protects the roots from insects.

Bacteria can be selected for their ability to degrade a particular substance, and this ability can then be enhanced by bioengineering. *Bioremediation* is the process that uses microorganisms or other organisms, such as plants, to detoxify pollutants in the environment. For instance, naturally occurring bacteria that eat oil have been genetically engineered to clean up beaches (Fig. 14.5) after oil spills, such as the 2010 Deep Water Horizon spill in the Gulf of Mexico. Bacteria can also remove sulfur from coal before it is burned and help clean up toxic waste dumps. One such strain was given genes that allowed it to clean up levels of toxins that would have killed other strains. Further, these bacteria were given "suicide" genes that caused them to self-destruct when the job had been accomplished.

Organic chemicals are often synthesized by having catalysts act on precursor molecules or by using bacteria to carry out the synthesis. Today, it is possible to go one step further and manipulate the genes that code for these enzymes. For instance, biochemists discovered a strain of bacteria that is especially good at producing phenylalanine, an organic chemical needed to make aspartame, the dipeptide sweetener better known as NutraSweet®. They isolated, altered, and formed a vector for the appropriate genes, so that various other bacteria could be genetically engineered to produce phenylalanine.

Genetically Modified Plants

Techniques have been developed to introduce foreign genes into immature plant embryos or into plant cells called *protoplasts* that have had the cell wall removed. The protoplasts are treated with an electric current while they are suspended in a liquid containing foreign DNA. The current creates tiny, self-sealing holes in the plasma membrane, through which the DNA can enter. These treated protoplasts go on to develop into mature plants.

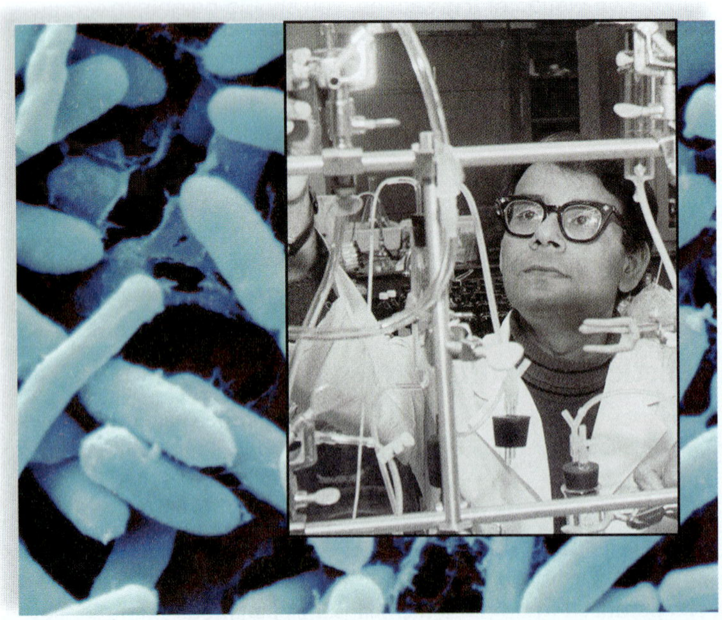

Figure 14.5 Bioremediation. Bacteria capable of decomposing oil have been engineered and patented by researchers such as Dr. Chakrabarty.

Foreign genes transferred to strains of cotton, corn, potato, and even bananas have made these plants resistant to pests such as fungi and insects, because their cells now produce a chemical that is toxic to the pest species. Similarly, soybeans have been made resistant to a common herbicide. Some corn and cotton plants are both pest- and herbicide-resistant. A new strain of rice called Golden Rice has been engineered to have a higher vitamin A content. These and other genetically modified crops are now sold commercially, with the goal of increased yields and better nutrient content. Like bacteria, plants are also being engineered to produce human proteins, such as hormones, clotting factors, and antibodies, in their seeds. One type of antibody made by corn can deliver radioisotopes to tumor cells; another, made by soybeans, can be used to treat genital herpes. Currently, tobacco plants are being used to develop a vaccine against tooth decay.

Genetically Modified Animals

Techniques have been developed to insert genes into the eggs of animals. It is possible to microinject foreign genes into eggs by hand, but another method uses vortex mixing. The eggs are placed in an agitator with DNA and silicon-carbide needles, and the needles make tiny holes through which the DNA can enter. When these eggs are fertilized, the resulting offspring are transgenic animals. Through this technique, many types of animal eggs have taken up the gene for bovine growth hormone (bGH). The procedure has been used to produce larger fishes, cows, pigs, rabbits, and sheep.

Gene pharming, the use of transgenic farm animals to produce pharmaceuticals, is being pursued by a number of firms. Genes that code for therapeutic and diagnostic proteins are incorporated into an animal's DNA, and the proteins appear in the animal's milk. Trials are under way for drugs that treat cystic fibrosis, cancer, blood diseases, and other disorders. Figure 14.6*a* outlines the procedure for producing transgenic mammals: DNA containing the gene of interest is injected into donor eggs. Following in vitro fertilization, the zygotes are placed in host females, where they develop. After female offspring mature, the product is secreted in their milk.

Cloning Transgenic Animals

For many years, researchers believed that adult vertebrate animals could not be cloned, because cloning requires that all the genes of an adult cell be turned on if development is to proceed normally. This had long been thought impossible.

In 1997, however, Scottish scientists announced that they had produced a cloned sheep, which they called Dolly. Since then, calves, goats, pigs, rabbits, and even cats have also been cloned. The techniques can be applied to produce populations of transgenic animals.

As shown in Figure 14.6*b*, after enucleated eggs from a donor are microinjected with 2n nuclei from a single transgenic animal, they are coaxed to begin development in vitro. Development

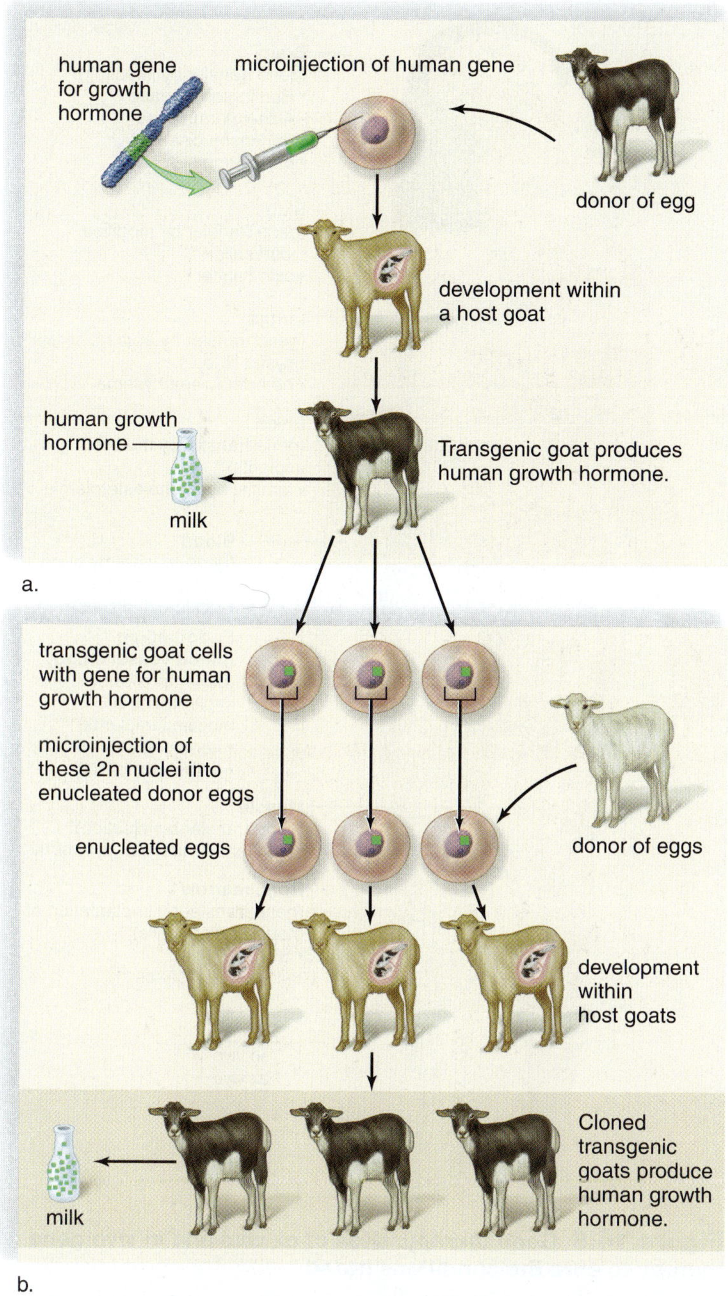

Figure 14.6 Transgenic mammals produce a product. This figure illustrates the basic procedure for generating a transgenic animal. **a.** A bioengineered egg develops in a host to create a transgenic goat, which produces a biotechnology product in its milk. **b.** Nuclei from the transgenic goat are transferred into donor eggs, which develop into cloned transgenic goats.

continues in host females until the clones are born. The female clones have the same product in their milk as does the original transgenic animal. Now that scientists have a way to clone animals, this procedure will undoubtedly be used routinely to procure biotechnology products. However, animal cloning is a difficult process with a low success rate (usually 1 or 2 viable embryos per 100 attempts). The vast majority of cloning attempts are unsuccessful, resulting in the early death of the clone.

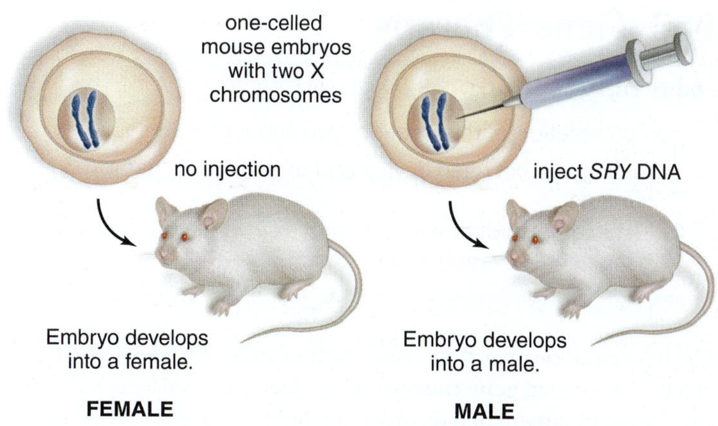

Figure 14.7 Experimental use of mice. Bioengineered mice showed that maleness is due to *SRY* DNA.

Compared with the production of proteins in bacteria, one advantage of molecular pharming is that certain proteins are more likely to function properly when expressed in mammals. This may be due to specific protein folding or other modifications that occur in mammals but not in bacteria. In addition, certain proteins may be degraded rapidly or folded improperly when expressed in bacteria. Furthermore, the yield of recombinant proteins in milk can be quite large. Each dairy cow, for example, produces about 10,000 liters of milk per year. In some cases, a transgenic cow can produce appoximately 1 gram per liter (g/L) of the transgenic protein in its milk.

Applications of Transgenic Animals

Researchers are using transgenic mice for many different research projects. Figure 14.7 shows how this technology has demonstrated that a section of DNA called *SRY* (*s*ex determining *r*egion of the *Y* chromosome) produces a male animal. The *SRY* gene was cloned, and then one copy was injected into single-celled mouse embryos. Injected embryos developed into males, but any that were not injected developed into females.

Eliminating a gene is another way to study a gene's function. A *knockout mouse* has had both alleles of a gene removed or made nonfunctional. For example, scientists have constructed a knockout mouse lacking the *CFTR* gene, the same gene mutated in cystic fibrosis patients. The mutant mouse has a phenotype similar to that of a human with cystic fibrosis and can be used to test new drugs for the treatment of the disease.

Check Your Progress 14.2

1. List some of the beneficial applications of transgenic bacteria, animals, and plants.
2. Distinguish between a transgenic animal and a cloned animal.

14.3 Gene Therapy

Learning Outcomes

Upon completion of this section, you should be able to

1. Distinguish between in vivo and ex vivo gene therapy in humans.
2. List examples of how in vivo and ex vivo gene therapy has been used to treat human disease.

The manipulation of an organism's genes can be extended to humans in a process called **gene therapy.** Gene therapy is an accepted therapy for the treatment of a disorder and has been used to cure inborn errors of metabolism, as well as to treat more generalized disorders, such as cardiovascular disease and cancer (Fig. 14.8).

Viruses genetically modified to be safe can be used to transport a normal gene into the body. Sometimes the gene is injected directly into a particular region of the body. In the following sections, we discuss examples of **ex vivo gene therapy,** in which the gene is inserted into cells that have been removed and then returned to the body, and **in vivo gene therapy,** in which the gene is delivered directly into the body.

> *ex vivo* – a process that takes place outside of a living organism
> *in vivo* – a process that takes place inside of a living organism

Ex Vivo Gene Therapy

Children who have SCID (severe combined immunodeficiency) lack the enzyme ADA (adenosine deaminase), which is involved in the maturation of immune cells. Therefore, these children are prone to constant infections and may die unless they receive treatment. To carry out gene therapy, bone marrow stem cells are removed from the bone marrow of the patient and are infected with a virus that carries a normal gene for the enzyme into their DNA. Then the cells are returned to the patient, where it is hoped they will divide to produce more blood cells with the same genes.

One of the earliest uses of ex vivo gene therapy was for familial hypercholesterolemia, a condition that develops when liver cells lack a receptor protein for removing cholesterol from the blood. The high levels of blood cholesterol make the patient subject to fatal heart attacks at a young age. In this procedure, a small portion of the liver was surgically excised and then infected with a virus containing a normal gene for the receptor before being returned to the patient. Patients experienced lowered serum cholesterol levels following this procedure. Scientists are investigating using ex vivo gene therapy to treat other human diseases, including some forms of hemophilia.

In Vivo Gene Therapy

Cystic fibrosis patients lack a gene that codes for a transmembrane carrier of the chloride ion. They often suffer from numerous and potentially deadly infections of the respiratory tract. In gene therapy trials, the gene needed to cure cystic fibrosis is sprayed into the nose or delivered to the lower respiratory tract by adenoviruses. Another method of delivery is to enclose the gene in a lipid

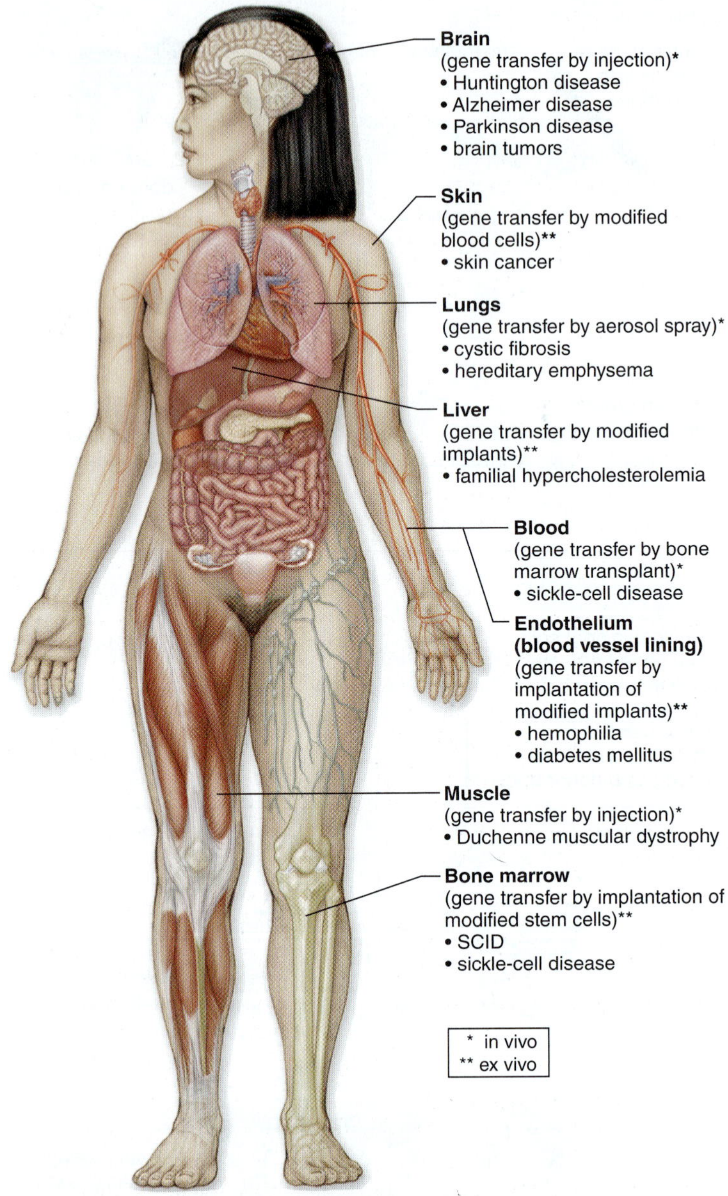

Figure 14.8 Gene therapy: sites of ex vivo and in vivo gene therapy to cure the conditions noted.

globule called a liposome. So far, this treatment has resulted in limited success, but recent advances in the use of lentiviral vectors is promising. Lentiviruses have a long incubation period and have been shown to be effective in infecting lung tissue.

In cancer patients, genes are being used to make healthy cells more tolerant of, and tumors more vulnerable to, chemotherapy. The gene *p53* brings about apoptosis, and there is much interest in introducing it into cancer cells that no longer have the gene and in that way killing them off.

Check Your Progress 14.3

1. Describe the methods that are being used to introduce genes into humans for gene therapy.
2. Discuss an example of ex vivo and of in vivo gene therapy.

14.4 Genomics

Learning Outcomes

Upon completion of this section, you should be able to

1. Distinguish among the sciences of genomics, proteomics, and bioinformatics.
2. Identify the function of repetitive elements, transposons, and unique noncoding DNA sequences in the human genome.
3. Explain how DNA microarrays are used in the study of genomics.

In the preceding century, researchers discovered the structure of DNA, how DNA replicates, and how DNA and RNA are involved in the process of protein synthesis. Genetics in the twenty-first century concerns **genomics,** the study of genomes—our complete genetic makeup and that of other organisms. Knowing the sequence of bases in genomes is the first step, and thereafter we want to understand the function of our genes and their introns, as well as the intergenic sequences. The enormity of the task can be appreciated by knowing that there are approximately 6 billion nucleotides in the 2n human genome. Many other organisms have a larger number of protein-coding genes but fewer noncoding regions compared to the human genome.

Sequencing the Genome

We now know the order of the base pairs in the human genome. This feat, which has been likened to arriving at the periodic table of the elements in chemistry, was accomplished by the **Human Genome Project (HGP)**, a 13-year effort that involved both university and private laboratories around the world.

In the beginning, investigators developed a laboratory procedure that would allow them to decipher a short sequence of base pairs, and then instruments became available that could carry out sequencing automatically. Over the 13-year span, DNA sequencers were constantly improved, and now modern instruments can automatically analyze up to 2 million base pairs of DNA in a 24-hour period.

Sperm DNA was the material of choice for analysis, because it has a much higher ratio of DNA to protein than other types of cells. (Recall that sperm do provide both X and Y chromosomes.) However, white blood cells from female donors were also used in order to include female-originated samples. The male and female donors were of European, African, American (both North and South), and Asian ancestry.

Many small regions of DNA that vary among individuals, termed polymorphisms, were identified during the HGP. Most of these are *single nucleotide polymorphisms (SNPs)*; they vary by only one nucleotide. Many SNPs have no effect; others may contribute to enzymatic differences affecting the phenotype. It's possible that certain SNP patterns change an individual's susceptibility to disease and alter his or her response to medical treatments (see Chapter 16).

Determining the number of genes in the human genome required a number of techniques, many of which relied on identifying RNAs in cells and then working backward to find the DNA that can pair with each RNA. **Structural genomics**—knowing the sequence of the bases and how many genes we have—is now being followed by functional genomics.

Estimates place the number of human genes between 21,000 and 23,000. The majority of these genes are expected to code for proteins. However, much of the human genome was formerly described as "junk," because it does not specify the order of amino acids in a polypeptide. However, recall from Chapter 12 that it is possible for RNA molecules to have a regulatory effect in cells. We examine this in more detail in the next section.

Structure of the Eukaryotic Genome

Historically, genes were defined as discrete units of heredity that corresponded to a locus on a chromosome (see Fig. 11.4). Prokaryotes typically possess a single circular chromosome with genes that are packed together very closely; eukaryotic chromosomes, in contrast, are much more complex: The genes are seemingly randomly distributed along the length of a chromosome and are fragmented into exons, with intervening sequences called introns scattered throughout the length of the gene (Fig. 14.9).

In general, more complex organisms have more complex genes with more and larger introns. In humans, 95% or more of the average protein-coding gene is introns. Once a gene is transcribed, the introns must be removed and the exons joined together to form a functional mRNA transcript (see Fig. 12.14).

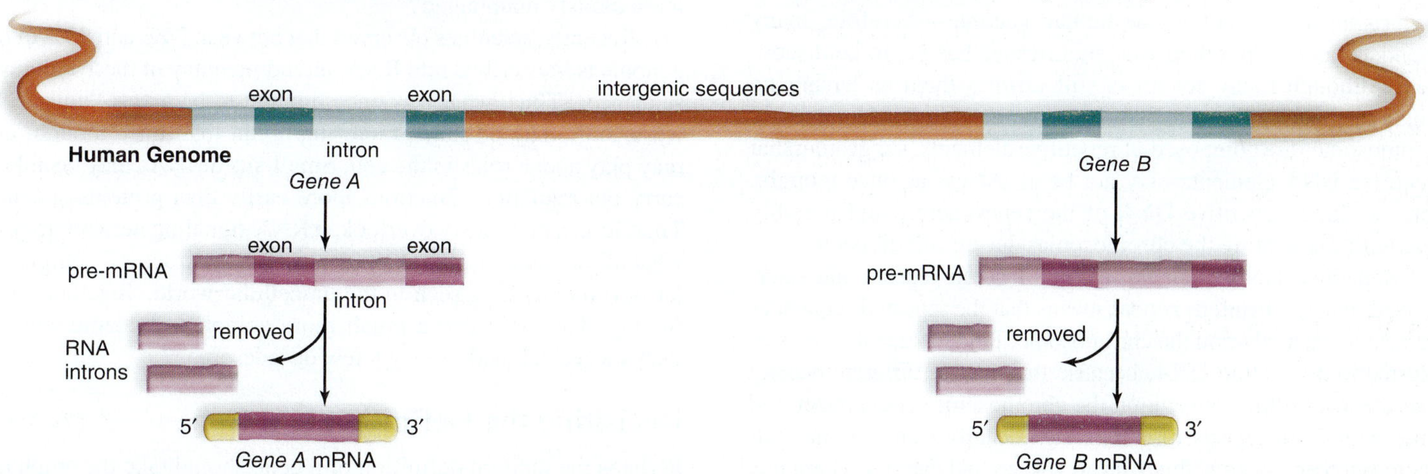

Figure 14.9 Chromosomal DNA. A genome contains protein-coding DNA (exons) and noncoding DNA, including introns (light blue) and other intergenic sequences (red). Only the exons are present in mRNA and specify protein synthesis.

Once regarded as merely intervening sequences, introns are now attracting attention as regulators of gene expression. The presence of introns allows exons to be put together in various sequences, so that different mRNAs and proteins can result from a single gene. Introns might also regulate gene expression and help determine which genes are to be expressed and how they are to be spliced. In fact, entire genes have been found embedded within the introns of other genes.

Intergenic Sequences

DNA sequences occur between genes and are referred to as **intergenic sequences** (Fig. 14.9). In general, as the complexity of an organism increases, so does the proportion of its noncoding DNA sequences. Intergenic sequences are now known to comprise the vast majority of human chromosomes, and protein-coding genes represent only about 1.5–2.0% of our total DNA. The remainder of this DNA, once dismissed as "junk DNA," is now thought to serve many important functions. Several basic types of intergenic sequences are found in the human genome, including (1) repetitive elements, (2) transposons, and (3) unique noncoding DNA. The majority of intergenic sequences belong to this last class.

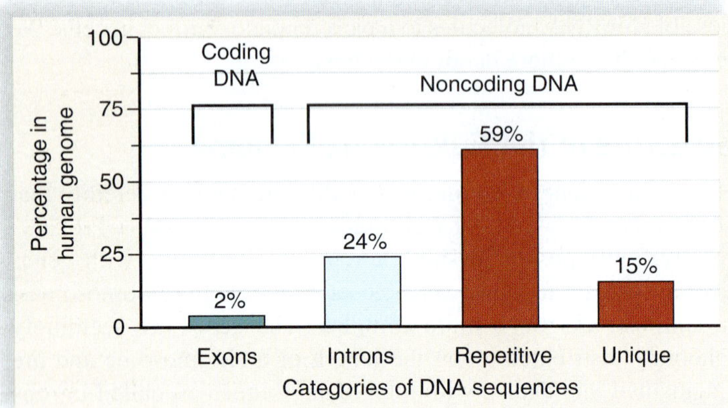

Repetitive DNA Elements

Repetitive DNA elements occur when a sequence of two or more nucleotides is repeated many times along the length of one or more chromosomes. Repetitive elements are very common—comprising nearly half of the human genome—therefore, many scientists believe that their true significance has yet to be discovered. Although many scientists still dismiss them as having no function, others point out that the centromeres and telomeres of chromosomes are composed of repetitive elements, suggesting that repetitive DNA elements may not be as useless as once thought. For one thing, repetitive DNA of the centromere could possibly help with segregating the chromosomes during cell division.

Repetitive DNA elements include tandem repeats and interspersed repeats. **Tandem repeat** means that the repeated sequences are next to each other on the chromosome. Tandem repeats are often referred to as satellite DNA, because they have a different density than the rest of the DNA within the chromosome. The number and types of tandem repeats may vary significantly from one individual to another, making them invaluable as indicators of heritage. One type of tandem repeat sequence, referred to as *short tandem repeats,* or *STRs,* has become a standard method in forensic science for distinguishing one individual from another and for determining familial relationships (see page 247).

The second type of repetitive DNA element is called an **interspersed repeat,** meaning that the repetitions may be placed intermittently along a single chromosome or across multiple chromosomes. For example, a repetitive DNA element, known as the *Alu* sequence, is interspersed every 5,000 base pairs in human DNA and comprises nearly 5–6% of total human DNA. Because of their common occurrence, interspersed repeats are thought to play a role in the evolution of new genes.

Transposons

Transposons are specific DNA sequences that have the remarkable ability to move within and between chromosomes. Their movement to a new location sometimes alters neighboring genes, particularly decreasing their expression. In other words, a transposon sometimes acts as a regulator gene. The movement of transposons throughout the genome is thought to be a driving force in the evolution of living organisms. The *Alu* repetitive element is an example of a transposon. In fact, many scientists now think that many repetitive DNA elements were originally derived from transposons.

Although Barbara McClintock first described these "movable elements" in corn over 60 years ago, it took time for the scientific community to fully appreciate this revolutionary idea. In fact, their significance was only realized within the past few decades. Transposons, sometimes termed "jumping genes," have now been discovered in bacteria, fruit flies, humans, and many other organisms. McClintock received a Nobel Prize in 1983 for her discovery of transposons and for her pioneering work in genetics (Fig. 14.10).

Animation Transposons

Unique Noncoding DNA

Genes constitute an estimated 1.5% of the human genome, and repetitive DNA elements make up about 44%; the function of the remaining half, or *unique noncoding DNA,* remains a mystery. Even though this DNA does not appear to contain any protein-coding genes, it has been highly conserved through evolution. In the many millions of years that separate humans from mice, large tracts of this mysterious DNA have remained almost unchanged. But if this DNA has no relevant function, then why has it been so meticulously maintained?

Recently, scientists observed that between 74% and 93% of the genome is transcribed into RNA, including many of these unknown sequences. Thus, what was once thought to be a vast "junk DNA wasteland" may be much more important than once thought and may play active roles in the cell. Small-sized RNAs may be able to carry out regulatory functions more easily than proteins at times. Therefore, a previously overlooked RNA signaling network may be what allows humans, for example, to achieve structural complexity far beyond anything seen in the unicellular world. Together, these findings have revealed a much more complex, dynamic genome than was envisioned merely a few decades ago.

Revisiting the Definition of a Gene

Perhaps the modern definition of a gene should take the emphasis away from the chromosome and place it on the results of transcription. Previously, molecular genetics considered a gene to

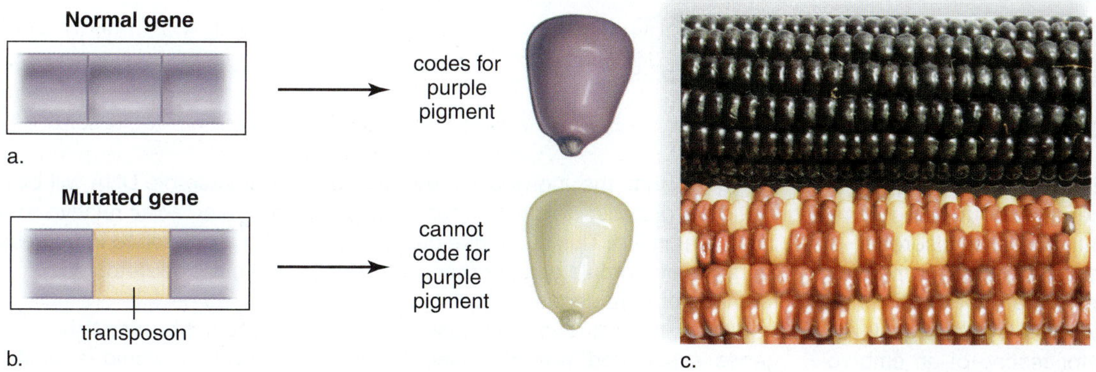

Figure 14.10 Transposons may cause gene mutations. **a.** A purple-coding gene ordinarily codes for a purple pigment. **b.** A transposon "jumps" into the purple-coding gene. This mutated gene is unable to code for purple pigment, and a white kernel results. **c.** Indian corn displays a variety of colors and patterns due to transposon activity.

be a nucleic acid sequence that codes for the sequence of amino acids in a protein. In contrast to this definition, geneticists have known for some time that all three types of RNA are transcribed from DNA, and that these RNAs are useful products. We also know that protein-coding regions can be interrupted by regions that do not code for a protein but do produce RNAs with various functions. This knowledge expands on the central dogma of genetics and recognizes that a gene product need not be a protein, and a gene need not be on one locus on a chromosome. The DNA sequence that results in a gene product can be split and be present on one or several chromosomes. Also, any DNA sequence can result in one or more products. Furthermore, some prokaryotes have RNA genes. In other words, the genetic material need not be DNA. Again, we can view this as a simple expansion of the central dogma of genetics.

Functional and Comparative Genomics

Since we now know the structure of our genome, the emphasis today is on functional genomics and on comparative genomics. The aim of **functional genomics** is to understand the exact role of the genome in cells or organisms.

The Nature of Science feature, "Testing for Genetic Disorders," on page 254, discusses the importance of a new technology that can be used to monitor the expression of thousands of genes simultaneously. **DNA microarrays,** also known as DNA chips or genome chips, contain microscopic amounts of known DNA sequences fixed onto a small glass slide or silicon chip in known locations (see Fig. 14A). The use of a microarray can tell you what genes are turned on in a specific cell or organism at a particular time and under what environmental circumstances. When mRNA molecules of a cell or an organism bind through complementary base pairing with the various DNA sequences on an array, then that gene is active in the cell.

Animation Using a DNA Microarray

DNA microarrays are increasingly available that rapidly identify all the mutations in the genome of an individual. This information is called the person's **genetic profile.** The genetic profile can indicate if any genetic illnesses are likely and what type of drug therapy for an illness might be most appropriate for that individual.

The aim of **comparative genomics** is to compare the human genome to the genome of other organisms, such as the model organisms listed in Table 14.1. Model organisms are used in genetic analysis because they have many genetic mechanisms and cellular pathways in common with each other and with humans. Functional genomics has also been advanced through the study of these genomes.

Much has been learned by genetically modifying mice; however, other model organisms can also be used. Scientists inserted

Table 14.1 Comparison of Sequenced Genomes

Organism	*Homo sapiens* (human) 	*Mus musculus* (mouse) 	*Drosophila melanogaster* (fruit fly) 	*Arabidopsis thaliana* (flowering plant) 	*Caenorhabditis elegans* (roundworm) 	*Saccharomyces cerevisiae* (yeast) 
Estimated Size	3,300 million bases	2,800 million bases	180 million bases	125 million bases	97 million bases	12 million bases
Estimated Number of Genes	~23,000	~23,000	13,600	25,500	21,700	5,700
Chromosome Number	46	40	8	10	12	32

THEME Nature of Science

Testing for Genetic Disorders

Genetic testing is required if prospective parents are concerned about being carriers for autosomal recessive disorders. If a woman is already pregnant, the parents may want to know if the unborn child has a disorder. If the woman is not pregant, the parents may opt for testing of an embryo or egg before she does become pregnant. One way to detect genetic disorders is to test the DNA for mutated genes.

Testing the DNA

DNA testing typically uses procedures that test for a specific genetic marker, or probe the genome for sequences of interest using DNA microarrays. Testing for a genetic marker is similar to the traditional procedure for DNA fingerprinting, as discussed earlier. As an example, consider that individuals with Huntington disease have an abnormality in the sequence of bases at a particular location on a chromosome. An unusually long STR can be found, which in turn causes a frameshift mutation in a gene. The length of the STR can be detected with PCR.

DNA Microarrays

With advances in robotic technology, it is now possible to place the entire human genome onto a single microarray (Fig. 14A). The mRNA from the organism or the cell to be tested is labeled with a fluorescent dye and added to a chip. When the mRNAs bind to the microarray, a fluorescent pattern results and is recorded by a computer. Now the scientist knows what DNA is active in that cell or organism. A researcher can use this method to determine the difference in gene expression between two different cell types, such as between liver cells and muscle cells.

A mutation microarray, the most common type, can be used to generate a person's genetic profile. The microarray contains hundreds to thousands of known diseases-associated mutant gene alleles. Genomic DNA from the individual to be tested is labeled with a fluorescent dye, then added to a microarray. The spots on the microarray fluoresce if the individual's DNA binds to the mutant genes on the chip, indicating that the individual may have a particular disorder or is at risk for developing it later in life. This technique can generate a genetic profile quicker than the older methods of DNA sequencing.

DNA microarrays can also identify genes associated with diseased tissues. The investigator applies the mRNA from normal and abnormal tissue to the microarray. The intensity of fluorescense from a spot on the microarray indicates the amount of mRNA originating from that gene in the diseased tissue relative to the normal tissue. If a gene is activated in the disease, more copies of mRNA will bind to the microarray than the control tissue, and the spot will appear more red than green.

Genomic microarrays are also used to identify links between disease and chromosomal variations. In this case, the chip contains genomic DNA that is cut into fragments. Each spot on the microarray corresponds to a known chromosomal location. Labeled genomic DNA from diseased tissues and control tissues bind to the DNA on the chip, and the fluorescense from both dyes is determined. If the number of copies of any particular target DNA has increased, more sample DNA will bind to that spot on the microarray relative to the control DNA, and a difference in fluorescence of the two dyes will be detected.

Home kits are now available to consumers who want their genetic profile. An individual would submit a DNA sample and receive a report with over 240 possible conditions and traits, including carrier status for various diseases. Currently, the Federal Drug Administration (FDA) is concerned about "false" positives that may cause consumers to undergo unnecessary medical procedures or undue stress and is working with these companies to improve the product and the information relayed to consumers.

Questions to Consider

1. What benefits are there when using a DNA microarray over a genetic marker such as an STR?
2. Why might a researcher want to know what genes are being expressed in different cell types?
3. How might the information from a DNA microarray be used to develop new drugs to treat disease?

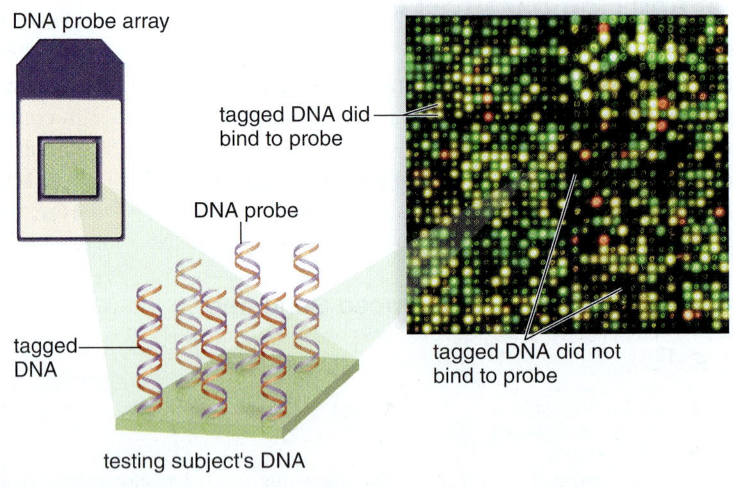

Figure 14A Use of a DNA microarray to test for a genetic disorder. This DNA chip contains rows of DNA sequences for mutations that indicate the presence of particular genetic disorders. If DNA fragments derived from an individual's DNA bind to a sequence representing a mutation on the DNA chip, that sequence fluoresces, and the individual has the mutation.

a human gene associated with early-onset Parkinson disease into *Drosophila melanogaster,* and the flies showed symptoms similar to those seen in humans with the disorder. This outcome suggested that we might be able to use these organisms instead of mice to test therapies for Parkinson disease.

Comparative genomics also offers a way to study changes in a genome over time, because the model organisms have a shorter generation time than humans. Comparing genomes can also help us understand the evolutionary relationships between organisms. One surprising discovery is that the genomes of all vertebrates are highly similar. Researchers were not surprised to find that the genes of chimpanzees and humans are 98% alike, but they did not expect to find that our sequence is also 85% like that of a mouse. Genomic comparisons will likely reveal evolutionary relationships between organisms never previously considered.

Proteomics

The entire collection of a species' proteins is the **proteome.** At first, it may be surprising to learn that the proteome is larger than the genome until we consider all the many regulatory mechanisms, such as alternative pre-mRNA splicing, that increase the number of possible proteins in an organism.

Proteomics is the study of the structure, function, and interaction of cellular proteins. Specific regulatory mechanisms differ between cells, and these differences account for the specialization of cells. One goal of proteomics is to identify and determine the function of the proteins within a particular cell type. Each cell produces thousands of different proteins, which can vary not only between cells but also within each cell, depending on circumstances. Therefore, the goal of proteomics is an overwhelming endeavor. Microarray technology can assist with this project, as can today's supercomputers.

Computer modeling of the three-dimensional shape of cellular proteins is also an important part of proteomics. If the primary structure of a protein is known, it should be possible to predict its final three-dimensional shape, and even the effects of DNA mutations on the protein's shape and function.

The study of protein function is viewed as essential to the discovery of new and better drugs. Also, it may be possible in the future to correlate drug treatment to the particular proteome of the individual to increase its efficiency and decrease side effects.

Bioinformatics

Bioinformatics is the application of computer technologies, specially developed software, and statistical techniques to the study of biological information, particularly databases that contain much genomic and proteomic information (Fig. 14.11). The new, raw data produced by structural genomics and proteomics are stored in databases that are readily available to research scientists. They are called raw data because, as yet, they have little meaning. Functional genomics and proteomics are dependent on computer analysis to find significant patterns in the raw data. For example, BLAST, which stands for *b*asic *l*ocal *a*lignment *s*earch *t*ool, is a computer program that can identify homologous genes among the genomic sequences of model organisms. **Homologous genes** are genes that code for the same proteins, although the base sequences may be slightly different. Finding these differences can help trace the history of evolution among a group of organisms.

Figure 14.11 Bioinformatics. New computer programs are being used to make sense of the raw data generated by genomics and proteomics. Bioinformatics allows researchers to study both functional and comparative genomics in a meaningful way.

Bioinformatics also has various applications in human genetics. For example, researchers found the function of the protein that causes cystic fibrosis by using the computer to search for genes in model organisms that have the same sequence. Because they knew the function of this gene in model organisms, they could deduce the function in humans. This was a necessary step toward possibly developing specific treatments for cystic fibrosis.

The human genome has 3 billion known base pairs, and without the computer it would be almost impossible to make sense of these data. For example, it is now known that an individual's genome often contains multiple copies of a gene. But individuals may differ as to the number of copies—called *copy number variations.* It seems that the number of copies in a genome can be associated with specific diseases. The computer can help make correlations between genomic differences among large numbers of people and disease.

It is safe to say that, without bioinformatics, progress in determining the function of DNA sequences, comparing our genome to model organisms, knowing how genes and proteins interact in cells, and so forth would be extremely slow. The Evolution feature, "Metagenomics," on page 256, discusses the use of bioinformatics in the new field of metagenomics.

Check Your Progress 14.4

1. Distinguish between the genome and the proteome of a cell.
2. Summarize the difference between a short tandem repeat and a transposon.
3. Explain how the use of microarrays and bioinformatics aids in the study of genomics and proteomics.

THEME Evolution

Metagenomics

In Figure 14B, a microbiologist dips a hand in a pond to extract a muddy substance teeming with life. What microorganisms are in this sample? How do the various microbes interact, and how are they all adapted to living in this environment? These are just a few of many questions that can be answered in the field of metagenomics. Metagenomics studies metagenomes—genetic material obtained directly from environmental samples. A broad sample of collected organisms allows investigators to determine evolutionary interactions in a particular environment by revealing the hidden biodiversity of microscopic life.

Traditionally, if a scientist wanted to know what microbial species were present in a sample, he or she would have had to isolate one species from another and be able to culture it in a laboratory to obtain enough DNA for sequencing. Only the most abundant species would be isolated and cultured, resulting in a loss of the true biodiversity that actually existed in the sample. Methods have been developed to address this shortcoming.

Shotgun Sequencing

One of the methods employed in metagenomics is the use of shotgun sequencing, a technique also used in the Human Genome Project. Shotgun sequencing got its name from the broad trajectory of buckshot a shotgun blast produces. Shotgun sequencing can be likened to putting ten copies of this textbook in a shredder, then taking those pieces out and reassembling a complete book. The approach randomly shears DNA, sequences the short fragments, then reassembles these sequences into the correct order in what is called a consensus sequence (Fig. 14C).

When studying microbial biodiversity, the scientist feeds the sequence information into a computer, and bioinformatics software begins the filtering process. Eukaryotic DNA can be identified and removed, leaving the microbial DNA behind for analysis. The biggest challenge scientists face working in metagenomics is the enormous size of the fragmented data. Genes isolated from the microorganims in the human gut revealed 3.3 million genes, requiring close to 570 gigabases of sequence data. Collecting and analyzing data sets of this size is a difficult computational challenge. Many of these challenges are being met by software developers working in bioinformatics.

Studies Using Metagenomics

Investigators working with the San Diego Zoo were interested in the gut microbes that inhabit various species of mammals. They asked, does each type of mammal have its own community of microbes? The study analyzed the fecal DNA from 34 mammalian species, including humans, and determined what microbes lived in the gut of each species. Most of the fecal samples were obtained from zoo animals with closely monitored diets, as well as humans who kept a strict food diary. The results showed that the *diet* of mammals, not the specific species of that mammal, determines what microorganims live in the gut. The collection of gut microbes is conserved across mammalian species, depending on what they eat. If a mammal is an herbivore, one set of microbes exists in the gut compared with the mammals that are omnivores or carnivores.

Metagenomics can also be used as a diagnostic tool to discover causative agents of diseases. Researchers working on a disease affecting boa constrictors suspected that a virus was to blame. DNA from affected and healthy snakes was obtained, and all the snake DNA was filtered out using a computer. The resulting DNA sequences revealed arenaviruses and a virus that was a hybrid of two previously identified viruses. The results not only confirmed that the snake disease was viral but also had larger implications for viral evolution—arenaviruses were only previously known to infect mammals and had never been identified in reptiles. This discovery opened up many questions about host range, evolution, and mechanisms of pathogenesis of viruses.

Questions to Consider

1. Why is metagenomics used versus traditional genomic techniques?
2. Why is shotgun sequencing used in this field of study?
3. Why is metagenomics closely tied to evolutionary biology?

Figure 14B Obtaining an environmental sample. Scientists interested in microbial diversity can collect a sample of mixed species and use metagenomics to analyze what species and genes are present.

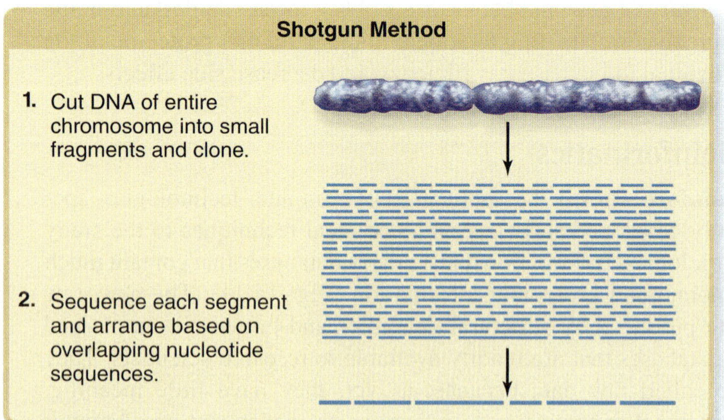

Figure 14C Shotgun sequencing method. The metagenome of an environmental sample is fragmented into small clones and sequenced. Computers assemble and analyze DNA sequences.

Connecting the Concepts with the Themes

Evolution

- Repetitive elements and transposons, once considered to be "junk DNA," are now believed to play a role in genome evolution.
- The field of comparative genomics and metagenomics is yielding valuable new insights into the relationships among species, impacting taxonomy and evolutionary biology.

Nature of Science

- Basic research into the nature and organization of genes in various organisms allowed geneticists to produce recombinant DNA molecules and transgenic organisms.
- Using model organisms as a guide, scientists are able to better understand the role of the elements of the genome.

Biological Systems

- Recombinant DNA technology is being used to generate genetically modified bacteria that can assist in protecting plants from pest species or restoring a polluted ecosystem.
- Techniques such as recombinant DNA technology, PCR, and DNA microarrays are allowing scientists to better understand the structure of the genome and to develop new treatments, such as gene therapy, for a number of human diseases.

Media Study Tools http://connect.mheducation.com

SMARTBOOK® Study smart. SmartBook helps you maximize your study time by identifying what you know and don't know. The recharge feature in SmartBook helps you prepare for the test by identifying the material that you are most likely to forget.

connect | BIOLOGY Find even more resources to learn the chapter concepts including animations, tutorial videos, and interactive practice questions.

Animations	Tutorials
14.1 Restriction Endonucleases • Polymerase Chain Reaction **14.2** Early Genetic Engineering Experiment **14.4** Transposons • Using a DNA Microarray	**14.1** Polymerase Chain Reaction

Summarize

14.1 DNA Cloning

Cloning produces genetically identical copies of DNA, cells, or organisms. **Gene cloning** results when a gene is isolated and many copies are produced. The gene can be studied in the laboratory or inserted into a bacterium, a plant, or an animal, creating **transgenic organisms.** Then, this gene may be transcribed and translated to produce a protein, which can become a commercial product or a medicine. When a gene is inserted into a human, the process is called **gene therapy.**

Two methods are currently available for making copies of DNA: **recombinant DNA** technology and the **polymerase chain reaction** (**PCR**). Recombinant DNA contains DNA from two different sources. A **restriction enzyme** is used to cleave **plasmid** DNA and to cleave foreign DNA. The resulting "sticky ends" facilitate the insertion of foreign DNA into **vector** DNA. The foreign gene is sealed into the vector DNA by **DNA ligase.** Both bacterial plasmids and viruses can be used as vectors to carry foreign genes into bacterial host cells.

PCR uses the enzyme DNA polymerase to quickly make multiple copies of a specific piece (target) of DNA. PCR is a chain reaction, because the targeted DNA is replicated over and over again. Analysis of DNA segments, called **DNA gel electrophoresis,** following PCR has all sorts of uses from assisting genomic research to doing **DNA fingerprinting** for the purpose of identifying individuals and confirming paternity. DNA fingerprinting can be accomplished by taking advantage of **short tandem repeats** (**STR**) present in the genome of all organisms.

14.2 Biotechnology Products

Transgenic organisms, also called **genetically modified organisms** (**GMOs**), have had a foreign gene inserted into them. Genetically modified bacteria, agricultural plants, and farm animals now produce **biotechnology products** of interest to humans, such as hormones and vaccines. Bacteria usually secrete the product. The seeds of plants and the milk of animals contain the product.

Transgenic bacteria have also been engineered to promote the health of plants, perform bioremediation, extract minerals, and produce chemicals. Transgenic crops, engineered to resist herbicides and pests, are commercially available. Transgenic animals have been given various genes for **gene pharming,** in particular the one for bovine growth hormone (bGH). Cloning of animals is now possible.

14.3 Gene Therapy

Gene therapy, by either ex vivo or in vivo methods, is used to correct the genotype of humans and to cure various human ills. **Ex vivo gene**

therapy has been used to treat diseases such as SCID. **In vivo gene therapy** is being employed in the fight against cancer, cystic fibrosis, and cardiovascular disease.

14.4 Genomics

Researchers in the field of **genomics** now know the sequence of all the base pairs along the length of the human chromosomes. This achievement, known as the **Human Genome Project** (**HGP**), helped researchers working in **structural genomics** identify around 21,000 human genes that code for proteins; the rest of our DNA consists of regions that do not code for a protein and make up the **intergenic sequences.**

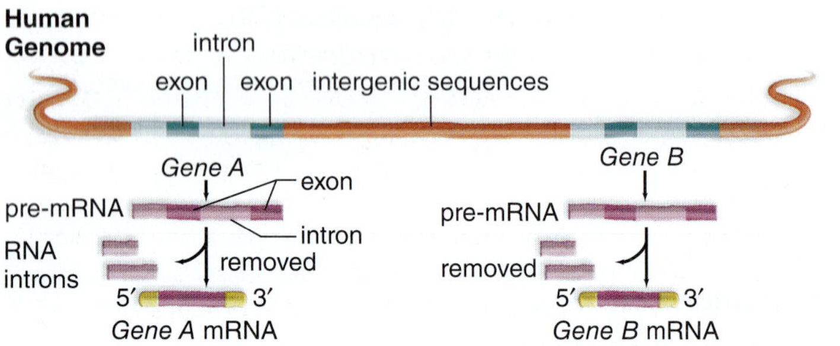

Genes constitute only 1.5% of the human genome. The rest of this DNA is surprisingly more active than once thought. About half of this DNA consists of **repetitive DNA elements,** which may be **tandem repeats** or **interspersed repeats** found throughout several chromosomes. Some of this DNA is made up of mobile DNA sequences called **transposons,** which are a driving evolutionary force within the genome. The role of the remaining portion of the genome is actively being investigated, but it is believed that these DNA sequences may play an important role in regulation of gene expression, thus challenging the classical definition of a gene. **Functional genomics** aims to understand the function of protein-coding regions and noncoding regions of our genome. To that end, researchers are utilizing tools such as **DNA microarrays.** Microarrays can also be used to create an individual's **genetic profile,** which is becoming helpful in predicting illnesses and reactions to particular medications.

Comparative genomics has revealed that little difference exists between the DNA sequence of our bases and those of many other organisms. Genome comparisons have revolutionized the understanding of evolutionary relationships by revealing previously unknown similarities between organisms.

Proteomics studies genes that code for proteins. A species' entire collection of proteins is the **proteome.** Genes that code for the same proteins even though their DNA is different are **homologous genes,** and computers are instrumental in finding these similar genes. **Bioinformatics** is the use of computers and specialized software programs to process and analyze large amounts of genetic data.

ASSESS

Choose the best answer for each question.

14.1 DNA Cloning

1. Restriction enzymes in bacterial cells are ordinarily used
   a. during DNA replication.
   b. to degrade the bacterial cell's DNA.
   c. to degrade viral DNA that enters the cell.
   d. to attach pieces of DNA together.
2. Using the key, put the phrases in the correct order to form a plasmid-carrying recombinant DNA.

Key:
1. use restriction enzymes
2. use DNA ligase
3. remove plasmid from parent bacterium
4. introduce plasmid into new host bacterium

   a. 1, 2, 3, 4 c. 3, 1, 2, 4
   b. 4, 3, 2, 1 d. 2, 3, 1, 4
3. The polymerase chain reaction
   a. uses RNA polymerase.
   b. takes place in huge bioreactors.
   c. uses a temperature-insensitive enzyme.
   d. makes lots of nonidentical copies of DNA.
   e. All of these are correct.

14.2 Biotechnology Products

4. Bacteria are able to successfully transcribe and translate human genes because
   a. both bacteria and humans contain plasmid vectors.
   b. bacteria can replicate their DNA, but humans cannot.
   c. human and bacterial ribosomes are vastly different.
   d. the genetic code is nearly universal.
5. Which of these is an incorrect statement?
   a. Bacteria usually secrete the biotechnology product into the medium.
   b. Plants are being engineered to have human proteins in their seeds.
   c. Animals are engineered to have a human protein in their milk.
   d. Animals can be cloned, but plants and bacteria cannot.
6. Which is not a correct association with regard to bioengineering?
   a. plasmid as a vector—bacteria
   b. protoplast as a vector—plants
   c. RNA virus as a vector—human stem cells
   d. All of these are correct.

14.3 Gene Therapy

7. When a cloned gene is used to modify a human disease, the process is called
   a. bioremediation.
   b. gene therapy.
   c. genetic profiling.
   d. gene pharming.
   e. DNA fingerprinting.
8. Which of the following delivery methods is not used in gene therapy?
   a. virus
   b. nasal sprays
   c. liposomes
   d. electric currents

14.4 Genomics

9. A genetic profile can
   a. assist an individual in maintaining good health.
   b. be accomplished utilizing bioinformatics.
   c. show how many genes are normal.
   d. be accomplished utilizing a microarray.
   e. Both a and d are correct.

10. Which is a true statement?
    a. Genomics would be slow going without bioinformatics.
    b. Genomics is related to the field of proteomics.
    c. Genomics has now moved on to functional and comparative genomics.
    d. Genomics shows that we are related to all other organisms tested so far.
    e. All of these are correct.
11. Which of the following was used to find the function of the cystic fibrosis gene?
    a. microarray
    b. proteomics
    c. comparative genomics and bioinformatics
    d. sequencing of the gene
12. Repetitive DNA elements
    a. may be tandem or spread across several chromosomes.
    b. are found in centromeres and telomeres.
    c. make up nearly half of human chromosomes.
    d. may be present as just a few or many thousands of copies.
    e. All of these are correct.
13. Bioinformatics can
    a. assist genomics and proteomics.
    b. compare our genome to that of a primate.
    c. depend on computer technology.
    d. match genes with proteins.
    e. All of these are correct.
14. Proteomics is used to discover
    a. what genes are active in what cells.
    b. what proteins are active in what cells.
    c. the structure and function of proteins.
    d. how proteins interact.
    e. All except a are correct.

ENGAGE

LEARNSMART LABS®

The following LearnSmart Labs contain exercises that are related to the content of this chapter:

- DNA & Biotechnology

THINKING SCIENTIFICALLY

1. Transposons are considered by many researchers to have played a major role in the evolution of life on Earth. Explain how the movement of genetic material within the genome may produce some organisms with a selective advantage.
2. The Nature of Science feature "Moving Genes Between Species: Green Fluorescent Protein and Cells" feature in Chapter 12 on page 219 describes the use of green fluorescent proteins that can be used to light up regions of interest in many organisms and cell types. Considering this is done with the commercially available GloFish®, how comfortable are you with genetically modified pets created to have interesting phenotypes?
3. Human cloning has been banned around the globe. Considering the discussion on animal cloning, why do you think this ban exists?
4. We can use transgenic viruses to infect humans and help treat genetic disorders. Viruses can be genetically modified to contain "normal" human genes. In this type of gene therapy, a person is infected with a particular virus, which then delivers the "normal" human gene to cells by infecting them. What are some pros and cons of viral gene therapy?
5. Many human proteins, such as hemoglobin, are only functional as an assembly of multiple subunits (Fig. 3.16). Assembly of these functional units occurs within the endoplasmic reticulum and Golgi apparatus of a eukaryotic cell. Discuss what limitations, if any, exist to the large-scale production of genetically engineered hemoglobin.

UNIT 3

Evolution

Evolution may seem like a foreign topic to you, but it need not be. Evolution is the change in the gene pool of populations over time. Changes in the gene pool can happen when mutations produce new gene variants, when individuals move into or out of a population, or because of external forces such as natural disasters. If populations change enough over long periods of time, then new populations can branch from the old. Over time, an evolutionary tree, with many branches, forms.

As with any tree, any branch can be traced back to the root, or the ancestral population that gave rise to all others. We can create a family tree of our ancestors as well, to our great-grandparents and beyond. So, too, can all life trace its ancestry to the first living cells. A remarkable finding in recent times has been that some of our genes are the same as those of single-celled organisms. This is strong evidence for the shared ancestry of all life on Earth, from single-celled bacteria to multicelled plants and animals.

Darwin proposed that populations change in response to pressures from the environment. Some individuals are better able than others to survive and reproduce in a particular environment. Thus, the environment is the force that determines which individuals contribute offspring to the next generation. Environmental conditions are widely variable, and this variation provides the template by which natural selection shapes the evolution of populations. From carnivorous plants to bats and whales, all life has undergone natural selection in response to the environment; the result is a vast Tree of Life with many branches, all of which can be traced to a common ancestor of life on Earth.

Unit Outline

Chapter 15 Darwin and Evolution 261

Chapter 16 How Populations Evolve 279

Chapter 17 Speciation and Macroevolution 296

Chapter 18 Origin and History of Life 317

Chapter 19 Taxonomy, Systematics, and Phylogeny 337

Unit Learning Outcomes

The learning outcomes for this unit focus on three major themes in the life sciences.

Evolution	Evaluate why Darwin's theory of evolution by natural selection is considered a unifying theory of biology.
Nature of Science	Examine the methods by which scientists obtain evidence in support of evolutionary change.
Biological Systems	Describe how evolution occurs at the molecular, organismal, and population levels of biology.

Tiktaalik had a wristlike forelimb and other features that tell us it was an early intermediate between fish and four-limbed animals.

15

Darwin and Evolution

Chapter Outline

15.1 History of Evolutionary Thought 262

15.2 Darwin's Theory of Evolution 265

15.3 Evidence for Evolution 270

In 2004, a fossil "fishapod" was discovered in the Canadian Arctic. *Tiktaalik roseae* is a 375-million-year-old fossil that looks like a cross between an ancient fish and the first four-legged animals, called *tetrapods*. This unique fossil exhibits a number of transitional features that are both fishlike and tetrapod-like. It had fins, scales, and gills like a fish; but it also had a flexible neck, a flat head, and a wristlike forelimb similar to that of a modern tetrapod. *Tiktaalik* lived in wet, swampy areas, and a wrist may have been advantageous for moving along the bottom of shallow pools and rivers. Transitional fossils, such as *Tiktaalik,* support Darwin's theory that all animals descended from a common ancestor.

The evidence for evolution is not limited to fossils such as *Tiktaalik*—Darwin's theory of evolution is supported by over 150 years of biogeographical, biochemical, developmental, and genetic evidence. Evolution not only is evident in the study of fossils over long periods of time but can be witnessed, in action, over very short periods, such as days, weeks, and years. In this chapter we first take a look at the history of evolutionary thought, beginning with the history of ideas that influenced Darwin as he made his observations. Then we trace Darwin's trip around the world, examining the evidence that allowed him to develop his theory of evolution by natural selection. Finally, we explore the modern evidence that supports Darwin's theory.

As you read through this chapter, think about the following questions:

1. How do the features of fossil organisms tell us something about an organism's behavior and the environment in which it lived?
2. How does Darwin's theory of natural selection explain the intermediate features of modern and fossil organisms, such as *Tiktaalik?*

Before You Begin

Before beginning this chapter, take a few moments to review the following discussions.

Section 1.2 What does Darwin mean by evolution as "descent with modification"?

Section 1.3 What is a scientific theory?

Section 12.3 What is the genetic basis of inheritance?

Following *the* Themes

CHAPTER 15 DARWIN AND EVOLUTION

UNIT 3 EVOLUTION

Evolution	Darwin's theory of natural selection proposes that all life on Earth descends from a common ancestor.
Nature of Science	A scientific theory, such as Darwin's theory of evolution by natural selection, is supported by abundant evidence.
Biological Systems	Evolution by natural selection comes about from interaction between an organism and its environment.

15.1 History of Evolutionary Thought

Learning Outcomes

Upon completion of this section, you should be able to

1. Describe Darwin's trip aboard the HMS *Beagle* and some of the observations he made.
2. Identify historical figures and their viewpoints before and during the development of Darwin's theory.

In December 1831, a new chapter in the history of biology had its humble origins. A 22-year-old naturalist, Charles Darwin (1809–82), set sail on a journey of a lifetime aboard the British naval vessel HMS *Beagle* (Fig. 15.1). Darwin's primary mission on his journey around the world was to serve as the ship's naturalist—to collect and record the geological and biological diversity he saw during the voyage.

As Darwin set sail on the HMS *Beagle,* he was a supporter of the long-held idea that species had remained unchanged since the time of creation. The view of the fixity of species was forged from deep-seated religious beliefs, not by experimentation and observation of the natural world. During the 5-year voyage of the *Beagle,* Darwin's observations challenged his belief that species do not change over time—in fact, his observations of geological formations and species variation led him to propose a process by which species arise and change. This process—**evolution** (L. *evolutio,* "an unrolling")—proposed that genetic change occurs in a species over time, which leads to their genetic and phenotypic differences. This process is due to natural, not supernatural, forces.

This new view was not readily accepted by Darwin's peers, but it gained gradual credibility as a result of a scientific and intellectual revolution that began in Europe in the late 1800s. Today, 150 years since Darwin first published his idea of natural selection, the principle he proposed has been subjected to rigorous scientific tests—so much so that it is now considered one of the unifying theories of biology. Darwin's theory of evolution by natural selection explains both the unity and the diversity of life on Earth, how all living organisms share a common ancestor, and how species adapt to various habitats and ways of life.

Although many have believed that Darwin forged this change in worldview by himself, numerous biologists during the preceding century and some of Darwin's contemporaries had a large influence on Darwin as he developed his theory. The European scientists of the eighteenth and nineteenth centuries were keenly interested in understanding the nature of biological diversity. This was a time of exploration and discovery as the natural history of new lands was being mapped and documented. Shipments of strange plants and animals from newly explored regions were arriving in England to be identified and described by biologists—it was a time of rapid expansion of an understanding of the Earth's biological diversity. In this atmosphere of discovery, Darwin's theory first took root and grew.

Mid-Eighteenth-Century Influences

Many of the beliefs of the eighteenth century, although consistent with Judeo-Christian teachings about special creation, can be traced to the works of the Greek philosophers Plato (427–347 B.C.) and Aristotle (384–322 B.C.). Plato said that every species on Earth has a perfect, or "essential," form and species variation is imperfection of this essential form. Aristotle saw that organisms vary in complexity and can be arranged based on their order of increasing complexity.

Georges-Louis Leclerc (1707–88), better known as Count Buffon, was a naturalist who worked most of his life writing a 44-volume natural history series that described all known plants and animals. He provided evidence of evolution and proposed various causes, such as environmental influence and the struggle for existence. Buffon's support of evolution seemed to waiver, and often he professed to believe in special creation and the fixity of species.

Taxonomy, the science of classifying organisms, was an important endeavor during the mid-eighteenth century. Chief among taxonomists was Carolus Linnaeus (1707–78), who developed the binomial system of nomenclature (a two-part name for a species, such as *Homo sapiens*) and a system of classification for living organisms. Linnaeus, like other taxonomists of his time, believed in the fixity of species, that each species had an "ideal" form. He also believed in the *scala naturae,* a sequential ladder of life where the simplest beings occupy the lowest rungs and the most complex and spiritual beings—the angels, humans, and then God—occupy the two highest rungs.

Biologists of this time used comparative anatomy, the evaluation of similar structures across a variety of species, to classify organisms into groups. By the late eighteenth century, scientists had discovered fossils and knew that they were the remains of plants and animals from the past. Explorers traveled the world and brought back newly discovered **extant** (still in existence) and fossil organisms to be compared to known living species. At first, scientists believed that each type of fossil had a living descendant, but eventually some fossils did not seem to match well with known species. Baron Georges Cuvier (1769–1832) was the first to suggest that some species known only from the fossil record had become extinct.

Erasmus Darwin (1731–1802), Charles Darwin's grandfather, was a physician and a naturalist. His writings on both botany and zoology contained comments and footnotes that suggested the possibility of evolution. He based his conclusions on changes in animals during development, animal breeding by humans, and the presence of **vestigial structures** (L. *vestigium,* "trace, footprint")—anatomical structures that apparently functioned in an ancestor but have since lost most or all of their function in a descendant. Like Buffon, Erasmus Darwin thought that species might evolve, but he offered no mechanism by which this change might occur.

Late-Eighteenth/Early-Nineteenth-Century Influences

Baron Georges Cuvier, a distinguished zoologist, used comparative anatomy to develop a system of classifying animals. He also founded the science of **paleontology** (Gk. *palaios,* "old"; *ontos,* "having existed"; *-logy,* "study of"), the study of fossils, and was quite skilled at using fossil bones to deduce the structure of an animal.

Figure 15.1 Voyage of the HMS *Beagle*. **a.** Young Charles Darwin in 1831 at 22 years old. He did not publish his authoritative book, *On the Origin of Species,* until 1859. **b.** A map of Darwin's journey aboard the HMS *Beagle*. **c.** Along the east coast of South America, he noted a bird called the rhea, which looks like an African ostrich. **d.** In the Patagonian Desert, he observed a rodent that resembles the European rabbit. **e.** In the Andes Mountains, he observed fossils in rock layers. **f.** In the tropical rain forest, he found an abundant diversity of life. **g.** On the Galápagos Islands, he saw marine iguanas with blunt snouts suited for eating algae growing on rocks.

Cuvier was a staunch advocate of the fixity of species and special creation, but his studies revealed that the assembly of fossil varieties changed suddenly between different layers of sediment, or **strata,** within a geographic region. He reconciled his beliefs with his observations by proposing that sudden changes in fossil variation could be explained by a series of local catastrophes, or mass extinctions, followed by repopulation by species from surrounding areas. The result of these catastrophes was a turnover in the assembly of life-forms that occupied a particular region over time. Some of Cuvier's followers suggested that there had been worldwide catastrophes and God had created new sets of species to repopulate the world. This explanation of the history of life came to be known as **catastrophism.**

Jean-Baptiste de Lamarck (1744–1829) was the first biologist to offer a testable hypothesis that explained how evolution occurs via adaptation to the environment. Lamarck's ideas about descent were entirely different from those of Cuvier. After studying the succession of fossilized life-forms in the Earth's strata, Lamarck proposed that more complex organisms are descended from less complex organisms. He mistakenly concluded, however, that increasing complexity is the result of a natural motivating force—a striving for perfection—that is inherent in all living organisms.

To explain the process of adaptation to the environment, Lamarck proposed the idea of **inheritance of acquired characteristics**—that the environment can produce physical changes in an organism during its lifetime that are inheritable. One example he gave—and for which he is most famous—is that the long neck of giraffes developed over time because their necks grew longer as they stretched to reach food in tall trees, and this longer neck was then passed on to their offspring (Fig. 15.2). His hypothesis of inheritance of acquired characteristics has never been supported by experimentation. The molecular mechanism of inheritance explains why—phenotypic changes acquired during an organism's lifetime do not result in genetic changes that can be passed to subsequent generations.

In the eighteenth century, geologist James Hutton (1726–97) proposed a theory of slow, uniform geological change. Charles Lyell (1797–1875), the foremost geologist of Darwin's time, made Hutton's ideas popular in his book *Principles of Geology,* published in 1830. Hutton explained that the Earth is subject to slow but continuous cycles of rock formation and erosion, not shaped by sudden catastrophes. He proposed that erosion produces dirt and rock debris that is washed into the rivers, transported to the oceans, and deposited in thick layers, which are converted over time into sedimentary rock. These layers of sedimentary rocks, which often contain fossils, are then uplifted from below sea level to form land during geological upheavals.

Hutton concluded that extreme geological changes can be explained by slow, natural processes, given enough time. Lyell went on to propose the theory of **uniformitarianism,** which states that the natural processes witnessed today are the same processes that occurred in the past. Hutton's general ideas about slow and continual geological change are still accepted today, although modern geologists realize that rates of change have not always been uniform through history. Darwin was not taken by the idea of uniform change, but he was convinced, as was Lyell, that the Earth's massive geological changes are the result of extremely slow processes, and that the Earth, therefore, must be very old.

Thomas Malthus (1766–1834) was an economist who studied the factors that influence the growth and decline of human populations. In 1798 Malthus published *An Essay on the Principle of Population,* in which he proposed that the size of human populations is limited only by the quantity of resources, such as food, water, and shelter, available to support it. He related famine, war, and epidemics to the problem of populations overstretching their limited resources. Darwin, after reading Malthus's essay in 1838, applied similar principles to animal populations—that is, animals tend to produce more offspring than can survive, and competition for limited resources in the environment is the element that determines survival. Darwin thus used Malthus's principle to formulate his idea of natural selection.

Figure 15.2 A comparison of Lamarck's and Darwin's theories of evolution. **a.** Jean-Baptiste de Lamarck's proposal of the inheritance of acquired characteristics. **b.** Charles Darwin's theory of natural selection.

Check Your Progress 15.1

1. Define *catastrophism* and identify who proposed this idea.
2. Evaluate Lamarck's idea of "inheritance of acquired characteristics" as an explanation of biological diversity.
3. Construct a timeline of the history of evolutionary thought. Include major contributors and a brief description of each contribution along the timeline.

15.2 Darwin's Theory of Evolution

Learning Outcomes

Upon completion of this section, you should be able to

1. Summarize the process of evolution by natural selection.
2. List examples of the evidence Darwin gathered from fossils and biogeography that supported his growing idea of shared ancestry.
3. Give examples of how the mechanisms of evolutionary change can be identified and studied.

When Darwin signed on as the naturalist aboard the HMS *Beagle,* he possessed a suitable background for the position. Since childhood, he had been a devoted student of nature and a collector of insects. At age 16, Darwin was sent to medical school to follow in the footsteps of his grandfather and father. However, he did not take to the study of medicine, so his father encouraged him to enroll in the School of Divinity at Christ's College at Cambridge, with the intent of becoming a clergyman.

While at Christ's College, Darwin attended many lectures on biology and geology to satisfy his interest in natural science. During this time, he became the protégé and friend of the botanist John Henslow (1796–1861), from whom he gained skills in the identification and collection of plants. Darwin gained valuable experience in geology in the summer of 1831 by conducting fieldwork with Adam Sedgewick (1785–1873), one of the founders of modern geology. Shortly after Darwin was awarded his BA, Henslow recommended him to serve, without pay, as naturalist aboard the HMS *Beagle,* which was to explore the Southern Hemisphere.

The voyage was to take 2 years—but ended up taking 5 years—and the ship was to traverse the Southern Hemisphere (see Fig. 15.1). Along the way, Darwin encountered species that were very different from those of his native England. As part of his duties as the ship's naturalist, Darwin began to gather evidence that organisms are related through descent with modification from a common ancestor, and that adaptation to various environments results in diversity. Darwin also began contemplating the "mystery of mysteries," the origin of new species.

Observations of Change over Time

On his trip, Darwin observed massive geological changes firsthand. When he explored what is now Argentina, he saw raised beaches for great distances along the coast. Many of the raised beaches had exposed layers of sediment that contained a variety of fossilized shells and bones of extinct mammals. Darwin collected fossil remains of an armadillo-like animal (*Glyptodon*), the size of a small, modern-day car, and a giant ground sloth, *Mylodon darwinii,* the largest of which stood nearly 3 m tall (Fig. 15.3). Darwin also observed marine shells high in the cliffs of the impressive Andes Mountains, which suggested to him that the Earth is very old. Once Darwin accepted the possibility that the Earth must be very old, he began to think that there would have been enough time for descent with modification to occur. Therefore, living forms could be descended from extinct forms known only from the fossil record. It would seem that species were not fixed; instead, they changed over time.

Figure 15.3 Fossils of extinct mammals Darwin found during his exploration of South America. **a.** A giant, armadillo-like glyptodont, *Glyptodon,* is known only by the study of its fossil remains. Darwin came to the conclusion that this extinct animal must be related to living armadillos. The glyptodont weighed 2,000 kg. **b.** Darwin also observed the fossil remains of an extinct giant ground sloth, *Mylodon.*

Biogeographical Observations

Biogeography (Gk. *bios,* "life"; *geo,* "earth"; *grapho,* "writing") is the study of the geographical distribution of organisms throughout the world. The distribution of species and the makeup of species groups in different regions provide hints about past geological events, such as the movement of continents and the formation of volcanic islands, and about ecological change, such as glaciation and river formation.

As Darwin explored the Southern Hemisphere, he compared the animals of South America to those with which he was familiar. He noticed that although the animals in South America were different from those in Europe, similar environments on each continent had similar-looking animals. For example, instead of rabbits, he found the Patagonian cavy in the grasslands of South America. The Patagonian cavy has long legs and ears but the face of a guinea pig, a rodent also native to South America (Fig. 15.4). Did the Patagonian cavy resemble a rabbit because the two types of animals were adapted to the same type of environment? Both animals ate grass, hid in bushes, and moved rapidly using long hind legs. Did the Patagonian cavy have the face of a guinea pig because of having an ancestor in common with guinea pigs?

Figure 15.4 The European hare (head only), and the Patagonian cavy.

As he sailed southward along the eastern coast of South America, Darwin saw how similar species replaced one another. For example, the greater rhea (an ostrichlike bird) found in the north was replaced by the lesser rhea in the south. Therefore, Darwin reasoned that related species could be modified according to environmental differences (i.e., northern vs. southern latitudes). When he explored the Galápagos Islands, he found further evidence of this phenomenon.

The Galápagos Islands are a small group of volcanic islands formed 965 km off the western coast of South America. These islands are too far from the mainland for most terrestrial animals and plants to colonize, yet life is present there. The types of plants and animals Darwin found there were slightly different from species he had observed on the mainland, and even more important, they varied from island to island. Where did the animals and plants inhabiting these islands come from? Why were these species different from those on the mainland, and why were different species found on each island?

For example, each of the Galápagos Islands seemed to have its own type of tortoise, and Darwin began to wonder whether this difference was correlated with variation in vegetation among the islands (Fig. 15.5). Long-necked tortoises seemed to inhabit only dry areas where low-growing vegetation was scarce but tall cacti were abundant. In moist regions with relatively abundant ground foliage, short-necked tortoises were found. Had an ancestral tortoise from the mainland of South America given rise to these different types, each adapted to take advantage of food sources in its unique environment?

One of Darwin's most famous observations from the Galápagos Islands was his study of the finches. Darwin almost overlooked the finches because of their nondescript nature compared with many of the other animals in the Galápagos. At the time, Darwin did not recognize that these birds were all finches, because they were very different from the familiar finches from England. However, these birds would eventually play a major role in the formation of his thoughts about geographic barriers and their contribution to the origin of new species. Upon returning to England, Darwin had John Gould identify the birds. Gould identified the birds as "a series of ground finches . . . an entirely new group," and they exhibited significant variation in beak size and shape (see Fig. 15.9).

Today, many more Galápagos finches have been identified—there are ground-dwelling finches with beaks adapted to eating seeds, tree-dwelling finches with beaks sized according to their insect prey, and a cactus-eating finch with a more pointed beak it uses to punch holes in cactus fruit to extract pulp (see Fig. 15.9). The most unusual of the finches is a woodpecker-type finch. This bird has a sharp beak to chisel through tree bark but lacks the long tongue characteristic of a true woodpecker, which probes for insects. To compensate for this, the bird carries a twig or cactus spine in its beak and uses it to poke into crevices. Once an insect emerges, the finch drops this tool and seizes the insect with its beak (for more information on the shape of finch beaks, see the Chapter 17 Nature of Science feature, "Genetic Basis of Beak Shape in Darwin's Finches").

Later, Darwin speculated whether these different species of finches could have descended from a mainland finch species. In other words, he wondered if a finch from South America was the common ancestor of all the types on the Galápagos Islands. Perhaps new species had arisen because the geographic distance between the islands

a. b.

Figure 15.5 Galápagos tortoises. Darwin wondered whether the Galápagos tortoises were descended from a common ancestor. **a.** The tortoises with dome shells and short necks feed at ground level on islands with enough rainfall to support grasses. **b.** Those with shells that flare up in the front have long necks and live on arid islands, where they feed on tall, treelike cacti.

isolated populations of birds long enough for them to evolve independently. And perhaps the present-day species had resulted from accumulated changes occurring within each of these isolated populations.

Video
Finches Natural Selection

Natural Selection and Adaptation

Upon returning to England, Darwin began to reflect on the voyage of the HMS *Beagle* and to collect additional evidence in support of his ideas about how organisms adapt to the environment. Darwin concluded early on that species change over time and are not fixed entities crafted by a creator. However, he did not yet have a mechanism to explain how change could happen in existing species and how new species could arise.

By 1842, Darwin had fully developed his idea of natural selection as a mechanism for evolutionary change. In 1858, Alfred Russel Wallace (1823–1913) sent an essay to Darwin in which he proposed a similar concept. Eight years of collecting and identifying thousands of species new to science in the Malay Archipelago had helped Wallace formulate his views on evolutionary change. The idea of natural selection was first presented to the Linnean Society of London in 1858 as a pair of essays by Darwin and Wallace.

Natural selection is a process based on the following observations:

- Organisms exhibit variation that can be passed from one generation to the next—that is, they have heritable variation.
- Organisms compete for available resources.
- Individuals within a population differ in terms of their reproductive success.
- Organisms become adapted to conditions as the environment changes.

We consider each of these characteristics in detail in the sections that follow.

Organisms Have Heritable Variation

Darwin emphasized that the members of a population vary in their functional, physical, and behavioral characteristics (Fig. 15.6). Before Darwin, variations were viewed as imperfections that should be ignored because they were not important to the description of "fixed" species (see section 15.1). In contrast, Darwin emphasized that variation is required for the process of natural selection to operate. He suspected that a mechanism of inheritance existed, but he did not have the evidence we have today.

Now we know that genes are the unit of heredity and, along with the environment, determine the phenotype of an organism. Random mutations have been shown to be a source of new genetic variation in a population. Genetic variation can be harmful, helpful, or neutral (have no effect at all) to survival and reproduction. Genetic variation arises by chance and for no particular purpose, and new variation is as likely to be harmful as helpful or neutral to the organism. However, harmful variation is eliminated from the population by natural selection, because individuals with these mutations often do not survive or reproduce. Beneficial or neutral variation can be maintained in a population. Natural selection ignores neutral variation. But a beneficial mutation increases the probability that an individual with this mutation will tend to have

Figure 15.6 Variation in a population. Variation in populations, such as that seen in human populations, is required for natural selection to result in adaptation to the environment.

greater reproductive success. Biologists have abundant evidence that natural selection operates on heritable variation already present in a population's gene pool, and that this selection process is random—it has no goal of "improvement" in anticipation of future environmental changes.

Organisms Compete for Resources

Darwin applied Malthus's treatise on human population growth to animal populations. He realized that if all offspring born to a population were to survive, insufficient resources would be available to support the growing population. He calculated the reproductive potential of elephants, assuming an average life span of 100 years and a breeding span of 30–90 years. Given these assumptions, a single female probably bears no fewer than six young, and if all these young survive and continue to reproduce at the same rate, then after only 750 years the descendants of a single pair of elephants would number about 19 million! Obviously, no environment has the resources to support an elephant population of this magnitude, and no such elephant population has ever existed. This overproduction potential of a species is often referred to as the geometric ratio of increase.

Organisms Differ in Reproductive Success

Some individuals have favorable traits that enable them to better compete for limited resources. The individuals with favorable traits acquire more resources than the individuals with less favorable traits and can devote more energy to reproduction. Darwin called this ability to have more offspring *differential reproductive success.*

Fitness is the reproductive success of an individual relative to other members of a population. The most fit individuals are the ones that capture a larger amount of resources and convert these resources into a larger number of viable offspring. Because organisms vary, as do the conditions of local environments, fitness is influenced by different factors for different populations. For example, among western diamondback rattlesnakes (*Crotalus atrox*) living on dark-colored lava flows, the most fit are those that are black in color. But among those living on desert soil, the most fit are those with the typical light coloring with brown blotching. Background matching helps an animal both capture prey and avoid being captured; therefore, it is expected to lead to survival and increased fitness.

Natural selection occurs because certain members of a population happen to have a variation that allows them to survive and reproduce to a greater extent than do other members. For example, a variation in a desert plant that reduces water loss is beneficial; and a mutation in a wild dog that increases its sense of smell helps it find prey.

Organisms Become Adapted

An **adaptation** is any evolved trait that helps an organism be more suited to its environment. Adaptations are especially recognizable when unrelated organisms living in a particular environment display similar characteristics. For example, manatees, penguins, and sea turtles all have flippers, which help them move through the water. In Chapter 1, we saw other ways in which penguins are adapted to their environment. Similarly, a Venus flytrap, a plant that lives in the nitrogen-poor soil of a bog, is able to obtain nitrogen-containing nutrients because it has specialized leaves adapted to catching and digesting flies.

Such adaptations to specific environments result from natural selection. Differential reproduction generation after generation can cause adaptive traits to increase in frequency in each succeeding generation. Evolution includes other processes in addition to natural selection (see Chapter 16), but natural selection is the only process that results in adaptation to the environment.

We Can Observe Selection at Work

Darwin noted that humans can artificially modify desired traits in plants and animals by selectively breeding certain individuals. For example, the diversity of domestic dogs has resulted from prehistoric humans selectively breeding wolves with particular traits, such as hair length, height, and guarding behavior. This type of human-controlled breeding to increase the frequency of desired traits is called **artificial selection.** Artificial selection, like natural selection, is possible only because the original population exhibits a variety of characteristics, allowing humans to select the traits they prefer. For dogs, artificial selection has produced the many breeds of dogs we see today, all of which are descended from the wolf (Fig. 15.7).

As another example, several varieties of vegetables can be traced to a single ancestor. Chinese cabbage, brussels sprouts, and kohlrabi are all derived from a single species, *Brassica oleracea* (Fig. 15.8). Modern corn, or maize, has a wild ancestor called teosinte. Teosinte looks very different from the corn we grow for food; teosinte has 5–12 kernels in a single row, and a hard, thick outer shell encases each kernel, making it difficult to use as a food source (see Chapter 24). Strong evidence from archaeology and genetics supports the hypothesis that prehistoric humans, selecting for softer shells, more kernels, and other desirable traits, produced modern corn. Darwin surmised that if humans could create such a wide variety of organisms by artificial selection, then natural selection could also produce diversity, but with the environment, not humans, as the force selecting for particular traits.

Figure 15.7 Artificial selection. All dogs, *Canis lupus familiaris,* are descended from the gray wolf, *Canis lupus,* which began to be domesticated about 14,000 years ago. The process of selective breeding by humans has led to extreme phenotypic differences among breeds.

The Galápagos finches have beaks adapted to the food they eat, with different species of finches found on each island (Fig. 15.9). Today, many investigators, including Peter and Rosemary Grant of Princeton University, are documenting natural selection as it occurs on the Galápagos Islands. In 1973, the Grants began a study of the various finches on Daphne Major, an island near the center of the Galápagos Islands. The weather on this island swings widely back and forth from wet years to dry years.The Grants found that the

Figure 15.8 Artificial selection of plants. The vegetables Chinese cabbage, brussels sprouts, and kohlrabi are derived from wild mustard, *Brassica oleracea.* Darwin described artificial selection as a model by which to understand natural selection. With natural selection, however, the environment—not human selection—provides the selective force.

Figure 15.9 Galápagos finches. Each of the 13 species of finches has a beak adapted to a particular way of life. For example, (**a**) the heavy beak of the large ground-dwelling finch (*Geospiza magnirostris*) is suited to a diet of large seeds; (**b**) the beak of the warbler-finch (*Certhidea olivacea*) is suited to feeding on insects found among ground vegetation or caught in the air; and (**c**) the longer beak, somewhat decurved, and the split tongue of the cactus-finch (*Cactornis scandens*) are suited to extracting the flesh of cactus fruit.

beak size of the medium ground finch, *Geospiza fortis,* adapted to each weather swing, generation after generation (Fig. 15.10). These finches like to eat small, tender seeds that require a smaller beak. When the weather turns dry, they must eat larger, drier seeds, which are harder to crush. The birds that have a larger beak depth have an advantage during the dry periods, and they have more offspring. Therefore, among the next generation of *G. fortis* birds, the mean, or average, beak size has more depth than the previous generation (for more information on the shape of finch beaks, see the Chapter 17 Nature of Science feature, "Genetic Basis of Beak Shape in Darwin's Finches"). The Grants' research demonstrates that evolutionary change can sometimes be observed within the timeframe of a human lifespan, rather than over thousands of years.

Recent advances in biotechnology have produced a set of new and revolutionary tools to document phenotype evolution at the level of the gene. As one example, Sean Carroll (University of Wisconsin, Madison) studies the genes that determine variation in the color patterns on the wings of fruit flies. In one species of fly, *Drosophila biarmipes,* a close relative of the common fruit fly, *D. melanogaster,* males have a black spot on the top, forward edge of the wing (Fig. 15.11). This spot is part of the male fly's courtship dance. Carroll's research shows that the spot on the wing of the male fly has evolved from a few simple mutations that have changed how a wing gene is switched on and off during development. A few simple mutations in the DNA code were enough to produce a change in the wing color pattern of *D. biarmipes.* This

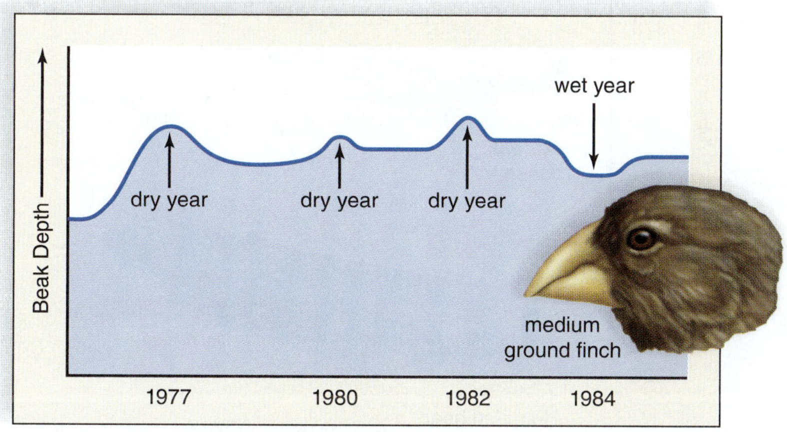

Figure 15.10 Evolution in action. The average beak depth of medium ground finches varies from generation to generation according to the weather. The weather affects the hardness and size of seeds on the islands, and different beak depths are better suited to eating different types of seeds. Average beak features were observed to change many times over a period of a decade. This is one way in which evolution by natural selection has been observable over a short period of time.

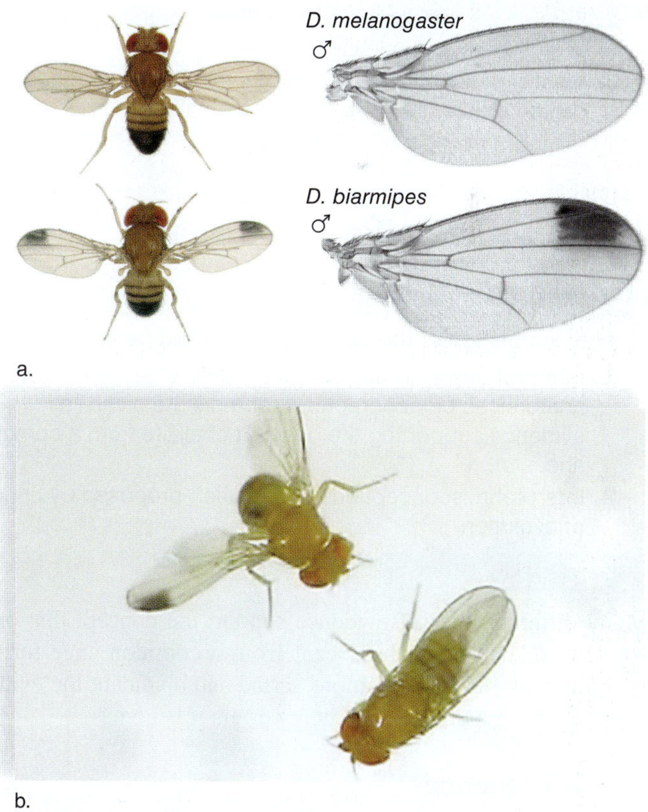

Figure 15.11 Wing spot in *Drosophila biarmipes.*
a. Male *D. biarmipes* have a spot on the leading edge of their wings that is not present in their close relative, *D. melanogaster.*
b. *D. biarmipes* males flash the spot on their wings to attract the attention of females during a courtship dance.

study and others like it demonstrate how new traits can evolve as a result of only a few changes in the DNA code that regulate a gene. In the case of *D. biarmipes,* natural selection, in the form of female mate choice, favors the evolution of males with spotted wings.

Industrial melanism is a common example of how natural selection can shape a trait in a population. Prior to the Industrial Revolution in Great Britain, light-colored peppered moths, *Biston betularia,* were more common than dark-colored peppered moths. It was estimated that only 10% of the moth population was dark at this time. With the advent of industry and an increase in pollution, the number of dark-colored moths exceeded 80% of the moth population. After legislation to reduce pollution, a dramatic reversal in the ratio of light-colored moths to dark-colored moths occurred. In 1994, one collecting site recorded a drop in the frequency of dark-colored moths to 19%, from a high of 94% in 1960. (We revisit this example in Chapter 16, where we discuss evolution of populations.)

The rise in bacterial resistance to antibiotics has occurred within the past 30 years or so. Resistance is an expected way of life now, not only in medicine but also in agriculture. New chemotherapeutic and HIV drugs are required because of the resistance of cancer cells and HIV, respectively. Also, pesticides and herbicides have created resistant insects and weeds.

Check Your Progress 15.2

1. List the three categories of observations made by Darwin that support evolution by natural selection.
2. Summarize the components of Darwin's theory of evolution by natural selection.
3. Identify several mechanisms of evolutionary change that can be studied.

15.3 Evidence for Evolution

Learning Outcomes

Upon completion of this section, you should be able to

1. Interpret one example from each area of study—fossil, anatomical, biogeographical, and biochemical—as evidence supporting the descent of all life from a common ancestor.
2. Interpret misconceptions of evolution proposed by critics of evolution.

Many different lines of evidence support the concept that organisms are related through descent from a common ancestor. This is significant, because the more varied and abundant the evidence supporting a hypothesis, the more certain it becomes.

Fossil Evidence

Fossils are the remains and traces of past life or any other direct evidence of past life. Traces include trails, footprints, burrows, worm casts, or even preserved droppings. Usually when an organism dies, the soft parts are either consumed by scavengers or decomposed by bacteria. Occasionally, the organism is buried quickly and in such a way that decomposition is never completed or is completed so slowly that the soft parts leave an imprint of their structure—for example, animals or plants trapped in a landslide or mudflow. Most fossils, however, consist only of hard parts, such as shells, bones, or teeth, because these are usually not consumed or destroyed.

Transitional fossils bear a resemblence to two groups that in the present are classified separately. They often represent the intermediate evolutionary forms of life in transition from one type to another, or a common ancestor of these types. Transitional fossils allow us to retrace the evolution of organisms over relatively long periods of time.

In 2004, a team of paleontologists discovered fossilized remains of *Tiktaalik roseae,* nicknamed the "fishapod" because it is the transitional form between fish and four-legged animals, the tetrapods (see chapter opener). *Tiktaalik* fossils are estimated to be 375 million years old and are from a time when the transition from fish to tetrapods is likely to have occurred. As expected of an intermediate fossil, *Tiktaalik* has a mix of fishlike and tetrapod-like features that illustrate the steps in the evolution of tetrapods from a fishlike ancestor (Fig. 15.12). For example, *Tiktaalik* has a very fishlike set of gills and fins, with the exception of the pectoral,

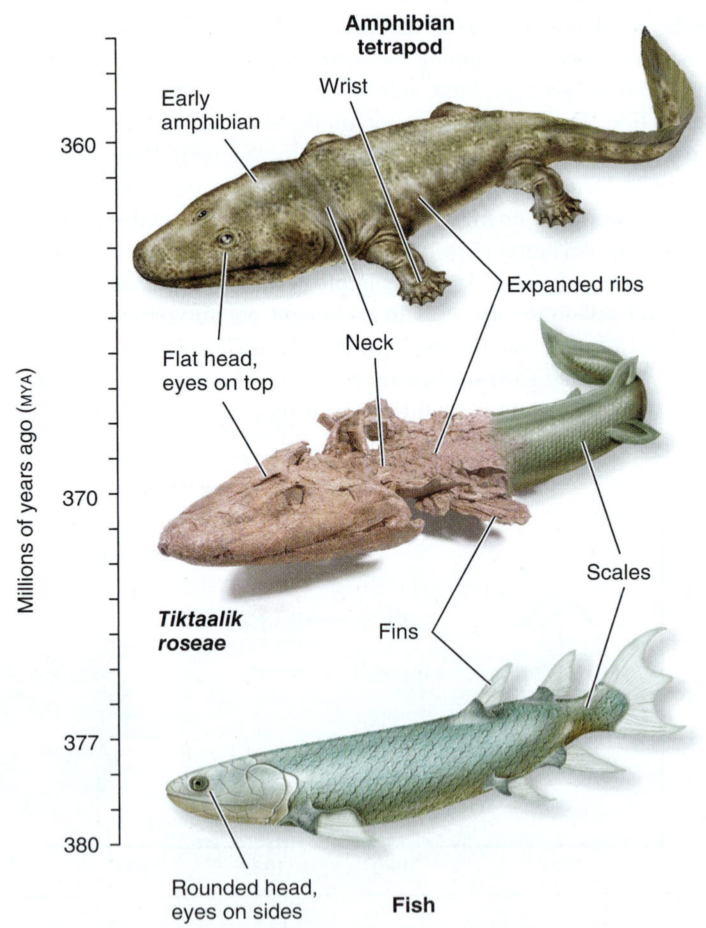

Figure 15.12 Transitional fossils. *Tiktaalik roseae* has a mix of fishlike and tetrapod-like features. Fossils such as *Tiktaalik* provide evidence that the evolution of new groups involves the modification of preexisting features in older groups. The evolutionary transition from one form to another, such as from a fish to a tetrapod, can be gradual, with intermediate forms having a suite of adapted, fully functional features.

or front fins, which have the beginnings of wrist bones similar to those of a tetrapod (see the photo at the beginning of this chapter). Unlike a fish, *Tiktaalik* had a flat head, flexible neck, eyes on the top of its head (like a crocodile), and interlocking ribs that suggest it had lungs. These transitional features suggest that it had the ability to push itself along the bottom of shallow rivers and see above the surface of the water—features that would come in handy in the river habitat where it lived.

Even in Darwin's day, scientists knew of *Archaeopteryx,* which is an intermediate between dinosaurs and birds. Progressively younger fossils than *Archaeopteryx* have been found: The skeletal remains of *Sinornis* suggest it had wings that could fold against its body like those of modern birds, and its grasping feet had an opposable toe—but it still had a tail. Another fossil, *Confuciusornis,* had the first toothless beak. A third fossil, called *Iberomesornis,* had a breastbone to which powerful flight muscles could attach. Such fossils show how the species of today evolved.

Fossils have been discovered that support the hypothesis that whales had terrestrial ancestors. *Ambulocetus natans* (meaning "the walking whale that swims") was the size of a large sea lion, with broad, webbed feet on its forelimbs and hindlimbs that enabled it to both walk and swim. It also had tiny hoofs on its toes and the primitive skull and teeth of early whales (Fig. 15.13). Modern whales still have a vestigial hindlimb consisting of only a few bones that are very reduced in size. As the ancestors of whales adopted an increasingly aquatic lifestyle, the location of the nasal opening underwent a transition—from the tip of the snout, as in *Ambulocetus;* to midway between the tip of the snout and the skull in *Basilosaurus;* to the very top of the head in modern whales (Fig. 15.13). An older fossil, *Pakicetus,* was primarily terrestrial yet had the dentition of an early whale. A younger fossil, *Rodhocetus,* had reduced hindlimbs that would have been no help for either walking or swimming but may have been used for stabilization during mating.

The origin of mammals is also well documented. The synapsids, an early amniote group, gave rise to the premammals. Slowly, mammal-like organisms acquired features that enabled them to breathe and eat at the same time, a muscular diaphragm and rib cage that helped them breathe efficiently. The earliest true mammals were shrew-sized creatures, which have been unearthed in fossil beds about 200 million years old. (We return to the topic of mammalian evolution in Chapter 29.)

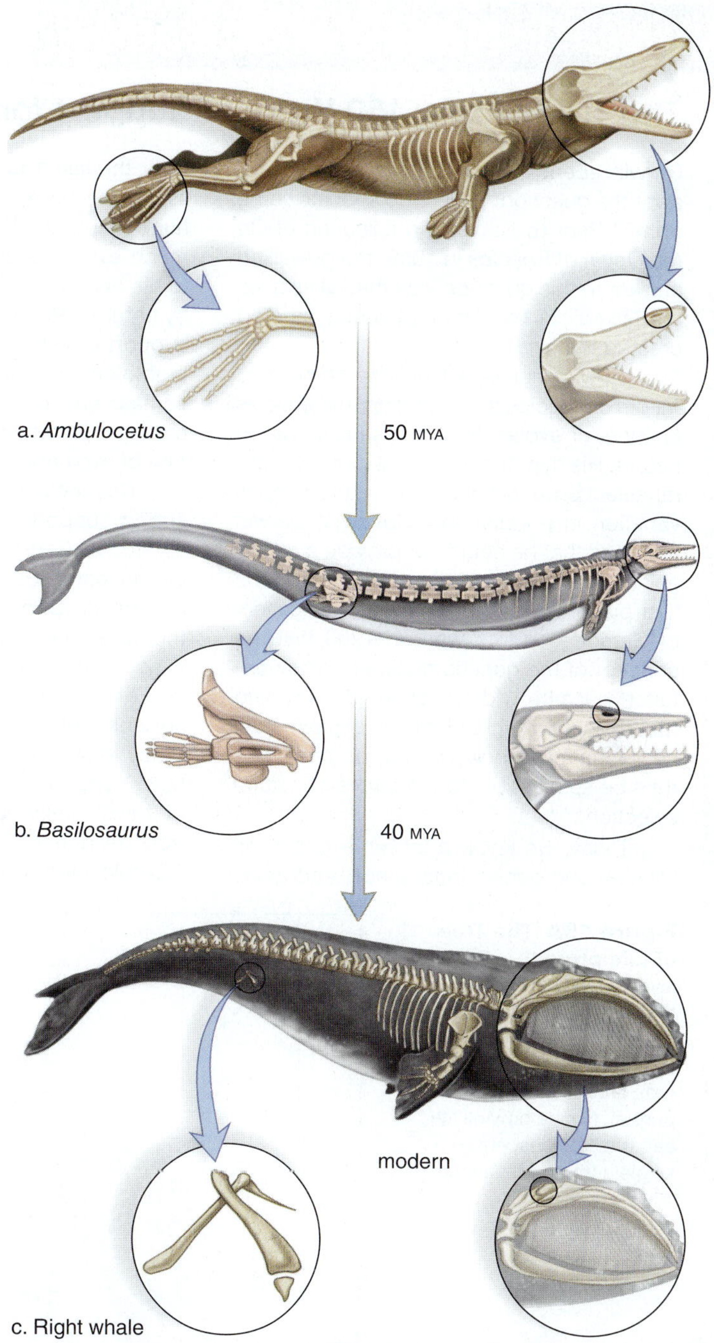

Figure 15.13 Anatomical transitions during the evolution of whales. Transitional fossils, such as *Ambulocetus* and *Basilosaurus,* support the hypothesis that modern whales evolved from terrestrial ancestors that walked on four limbs. These fossils show a gradual reduction in the hindlimb and a movement of the nasal opening from the tip of the nose to the top of the head—both adaptations to living in water.

Biogeographical Evidence

We described in section 15.2 the biogeographical observations that Darwin made during his voyage on the HMS *Beagle.* We noted that in cases where geography separates continents, islands, and seas, we might expect a different mix of plants and animals. For example, during his travels, Darwin observed that South America lacked rabbits, even though the environment was quite suitable for them. He concluded there were no rabbits in South America because rabbits evolved somewhere else and had no means of reaching South America. Instead, a different animal, the Patagonian cavy, occupied the environmental niche that rabbits held elsewhere.

In addition, Darwin noted that the different species of finches on the Galápagos Islands were not found on mainland South America. One reasonable explanation is that an ancestral population of finches from the mainland reached the Galápagos Islands and over time evolved into the different species found on each isolated island.

In the history of the Earth, South America, Antarctica, and Australia were originally connected (see Fig. 18.16). Marsupials, mammals in which females have an external body pouch where their young complete development, have evolved from egg-laying mammalian ancestors. Today marsupials are endemic to South America

THEME **Nature of Science**

The Tree of Life: 150 Years of Support for the Theory of Evolution by Natural Selection

Darwin spent his adult life striving to answer the question "Where do species come from?" Prior to Darwin's publication of *On the Origin of Species* in 1859, the prevailing answer to this question was that all species are "fixed" in their current state, as God created them.

Darwin challenged the idea of the fixed nature of species by proposing that species change, or evolve, in response to forces in nature. His hypothesis of evolution by natural selection explained how nature shapes variation in populations. However, Darwin admitted that he could not provide a mechanism to explain how diversity arises in the first place. It was not until the rediscovery of Gregor Mendel's work in 1900 that the concept of the genetic basis of trait inheritance became widely accepted, providing the missing mechanism to explain how new variation in populations can arise, and then be susceptible to the forces of natural selection.

Today, we know a lot more about the links among genes, inheritance, and traits. Over the last 150 years since Darwin published his book, scientists have amassed huge amounts of support for his ideas—so much evidence, in fact, that we now refer to his hypothesis as the theory of evolution by natural selection. A lot of the evidence in support of Darwin's theory has come from biomolecules—such as DNA, chromosomes, and proteins—that are compared among different species to look for a signature of evolution.

This molecular evidence has provided strong support for Darwin's proposal that all life on Earth can be traced to a single ancestor. Early in the development of his theory, Darwin kept notebooks of his thoughts. One notebook (Notebook B) contains the first known representation of life on Earth as a tree. This was a revolutionary concept at the time, but today evolutionary biologists have constructed thousands of trees similar to Darwin's from evidence provided by biomolecules and fossils.

Recently, a group of scientists has initiated a project to construct the largest evolutionary tree of all—the Tree of Life (Fig. 15A). The Tree of Life project is a collaborative effort to determine how all life on Earth is related and descended from a common ancestor. To date, the Tree of Life contains hundreds of species from all domains of life, and it is growing as more species are added. This project provides an incredible amount of support for Darwin's theory of evolution by natural selection.

Questions to Consider

1. Why was the idea of life as a tree so controversial during Darwin's time?
2. How does the tree of life support Darwin's theory that all life on Earth is descended from a common ancestor?

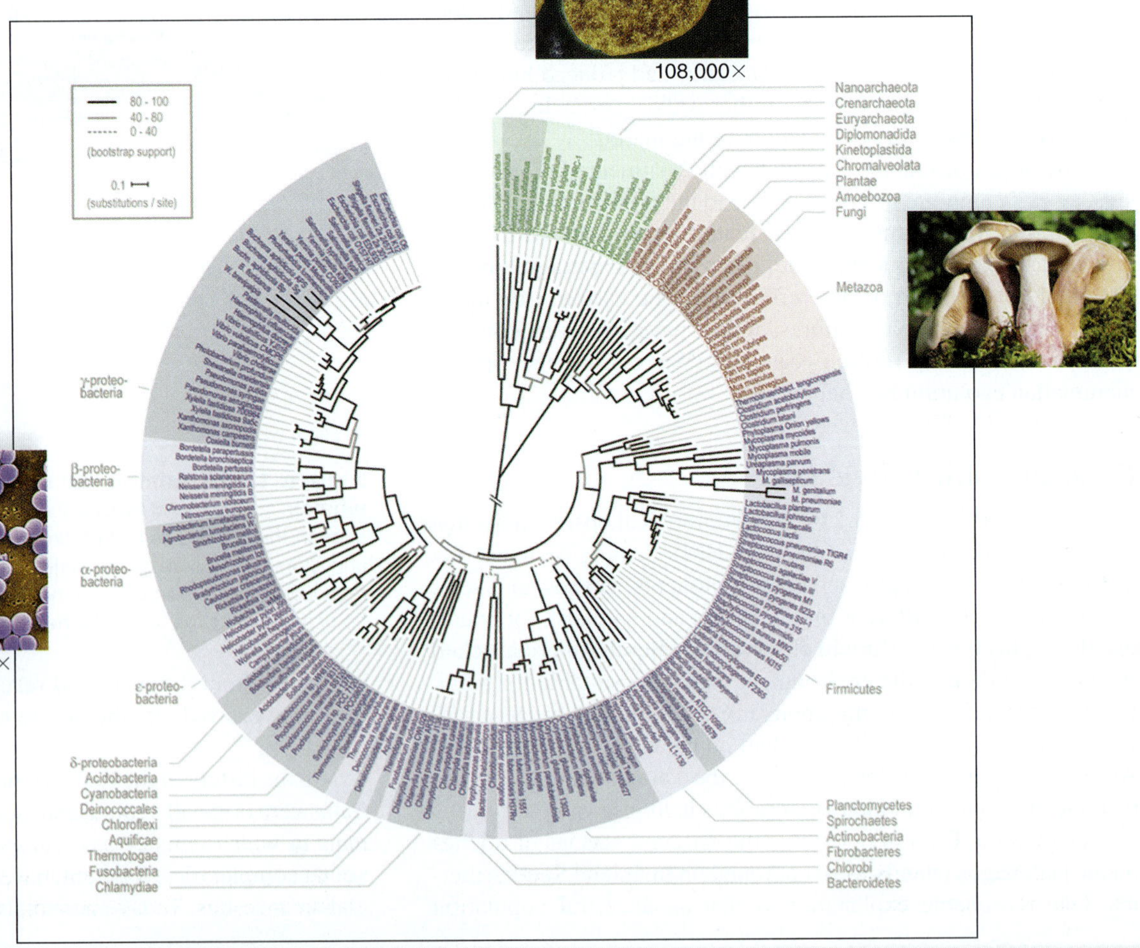

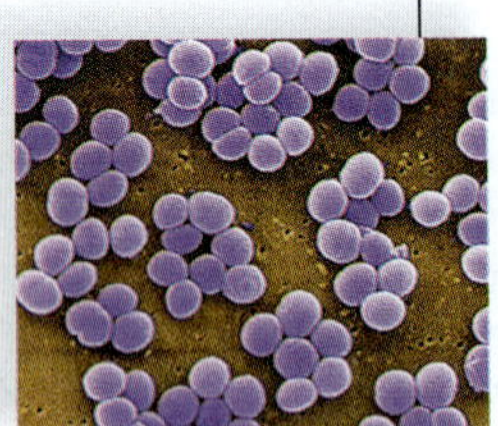

Figure 15A The Tree of Life project. This level of resolution shows the division of life into the three domains: Archaea (in green), Bacteria (in red), and Eukarya (in purple). Notice how all life can trace its descent to a single common ancestor.

and Australia. What is now Australia separated and drifted away from the other landmasses. The marsupials then diversified into many different forms suited to various environments (Fig. 15.14).

Marsupials were free to diversify because few, if any, placental mammals were present in Australia. In placental mammals, young complete their development inside the mother's uterus, nourished by the placenta (see Chapter 29). Where placental mammals are abundant, marsupials are not as diverse due to competition. After the formation of the Isthmus of Panama, placental mammals were able to migrate into South America. As a result, marsupial mammals were outcompeted by placental mammals, and the diversity of marsupials in South America declined greatly.

Biogeographical differences, therefore, provided evidence that variability in a single, ancestral population can lead to adaptation to different environments through the forces of natural selection. Competition for resources appears to provide some of the pressure that leads to diversification.

Anatomical Evidence

Darwin was able to show how descent from a common ancestor can explain anatomical similarities among organisms. Vertebrate forelimbs are used for flight (birds and bats), orientation during swimming (whales and seals), running (horses), climbing (arboreal lizards), and swinging from tree branches (monkeys). However, all vertebrate forelimbs contain the same sets of bones organized in similar ways, despite their dissimilar functions (Fig. 15.15). The most plausible explanation for this unity is that this basic forelimb plan was present in a common vertebrate ancestor. This plan was then modified independently in all descendants as each continued along its own evolutionary pathway.

Structures that are anatomically similar because they are inherited from a common ancestor are called **homologous.** In contrast, **analogous** structures serve the same function but originated independently in different groups of organisms that do not share a common ancestor. The wings of birds and insects are analogous structures. Thus, homologous, not analogous, structures are evidence for a common ancestry of particular groups of organisms.

As mentioned earlier, *vestigial structures* are anatomical features that are fully developed in one group of organisms but are reduced and may have no function in related groups. Most birds, for example, have well-developed wings used for flight, while some species have greatly reduced wings and do not fly. Similarly, snakes and whales have no use for hindlimbs, yet some species have remnants of a pelvic girdle and hindlimbs. Both the tail bone and wisdom teeth are examples of human structures that have no apparent function in our species today. Vestigial structures occur because organisms inherit their anatomy from their ancestors, and therefore their anatomy carries traces of their evolutionary history.

The homology shared by vertebrates is observable during their embryological development (Fig. 15.16). At some time during development, all vertebrates have a postanal tail and exhibit paired pharyngeal (throat) pouches supported by cartilaginous arches. In fishes and amphibian larvae, these pouches develop into functioning gills. In humans, the first pair of pouches and arches becomes

Sugar glider, *Petaurus breviceps*, is a tree-dweller and resembles the placental flying squirrel.

The Australian wombat, *Vombatus*, is nocturnal and lives in burrows. It resembles the placental woodchuck.

Kangaroo, *Macropus*, is an herbivore that inhabits plains and forests. It resembles the placental Patagonian cavy of South America.

Figure 15.14 Biogeography. Each type of marsupial in Australia is adapted to a different way of life. All the marsupials in Australia presumably evolved from a common ancestor that entered Australia some 60 million years ago.

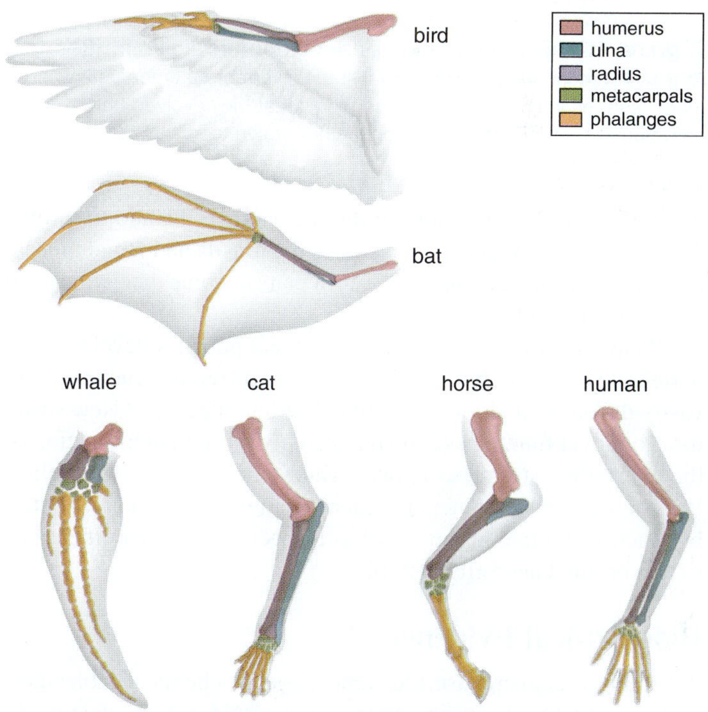

Figure 15.15 Significance of homologous structures. Although the specific design details of vertebrate forelimbs are different, the same bones are present (they are color-coded). Homologous structures provide evidence of a common ancestor.

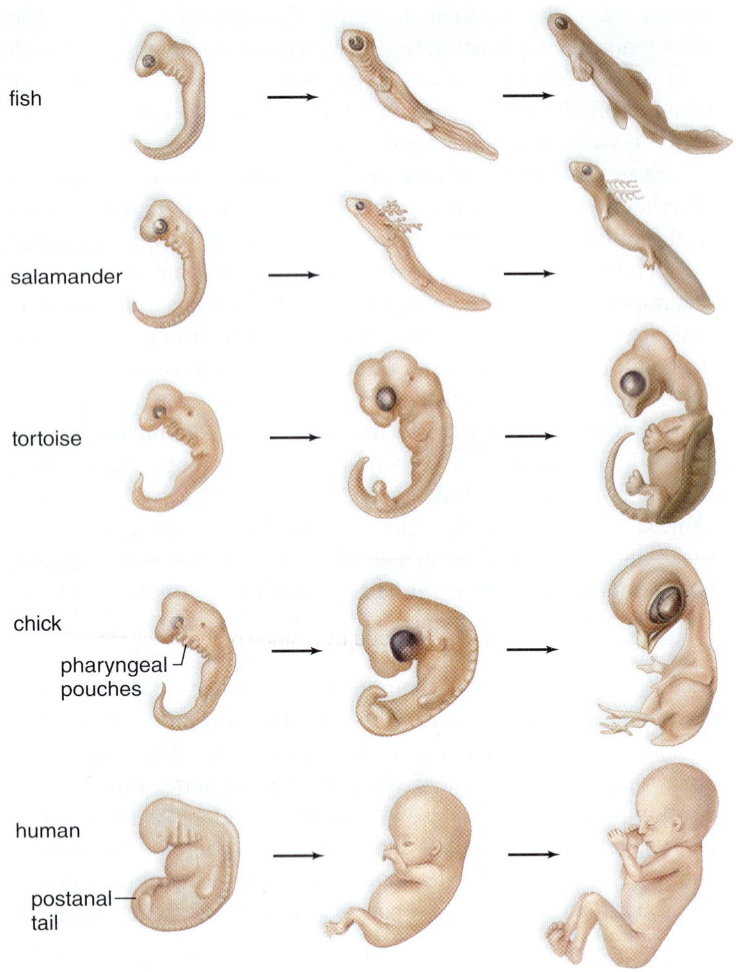

Figure 15.16 Significance of developmental similarities. At these comparable developmental stages, vertebrate embryos have many features in common, which suggests that they evolved from a common ancestor. (These embryos are not drawn to scale.)

the jawbones, the cavity of the middle ear, and the auditory tube. The second pair of pouches becomes the tonsils and facial muscle and nerves, while the third and fourth pairs become the thymus and parathyroid glands.

Why do structures such as pharyngeal pouches develop in all vertebrate embryos, but then become very different structures with vastly different functions in adults of different groups? New structures or novel functions can originate only through modification of the preexisting structures in one's ancestors. All vertebrates inherited the same developmental pattern from their common ancestor, but each vertebrate group now has a specific set of modifications to this original ancestral pattern.

Biochemical Evidence

All living organisms use the same basic biochemical molecules, including DNA (deoxyribonucleic acid), RNA (ribonucleic acid), and ATP (adenosine triphosphate). We can deduce from this that these molecules were present in the first living cell or cells from which life as we know it today has arisen.

Organisms use a triplet nucleic-acid code in their DNA to encode for 1 of 20 amino acids that will form their proteins. Because the sequences of DNA bases in the genomes of many organisms are now known, clear evidence is available that humans have some genes in common with much simpler organisms, such as prokaryotes. Because the genetic code is universal in living organisms, it is possible to insert a human gene into the genome of a bacterium. The bacterium will then produce the human protein that the gene encodes for.

Also, the sequence of amino acids of some proteins is similar across the tree of life. The sequence of amino acids in the human version of cytochrome *c,* a protein essential to cellular respiration, is remarkably similar to that of yeast (Fig. 15.17). The number of differences between the cytochrome *c* amino acid sequence in humans and other organisms increases with the distance in time since they shared a common ancestor—monkey cytochrome *c* differs from that of humans by only 1 amino acid, from that of a duck by 11 amino acids, and from that of yeast by 51 amino acids.

Animation Evolution of Homologous Genes

Evidence from Developmental Biology

The study of the evolution of development has discovered that many developmental genes are shared among all animals ranging from worms to humans. It appears that life's vast diversity has come about by a set of regulatory genes that control the activity of other genes involved in development.

For example, *Hox,* or ***homeobox,*** genes orchestrate the development of the body plan in all animals, from invertebrates (such as sea anemones and fruit flies) to humans (see the Evolution feature, "Evolution of the Animal Body Plan," in Chapter 28). All animals share a *Hox* gene common ancestor, but the number and type of *Hox* genes vary among animal groups. This variation in *Hox* genes is responsible, at least in part, for the wide range of body plans seen in animals. For example, a change in the timing and duration of the expression of *Hox* genes that control the number and type of vertebrae can produce the spinal column of a chicken or the longer spinal column of a snake. Thus, simple changes in how genes are controlled can have profound effects on the phenotype of organisms.

Criticisms of Evolution

Evolution is no longer considered a hypothesis. It is one of the great unifying theories of biology. Evolution is not "just a theory"—in science, the word *theory* is reserved for those concepts that are supported by a large number of observations (see section 1.4). The theory of evolution has the same status in biology that the theory of heredity has in genetics. However, some people propose mechanisms other than evolution to explain the origin of new species. These alternatives, founded in religious philosophy, cannot be tested, so they are not accepted as scientific evidence.

Many misconceptions about evolution, and the scientific process in general, are commonly used to challenge the legitimacy of the theory of evolution. Following are a few examples of these misconceptions, each followed by a brief scientific explanation.

1. Evolution is a theory about how life originated.

   Evolutionary biologists are concerned with how the diversity of life emerged *following* the origin of life. Certainly, the study of the origin of life is interesting to evolutionary biologists, but this is not the focus of their research.

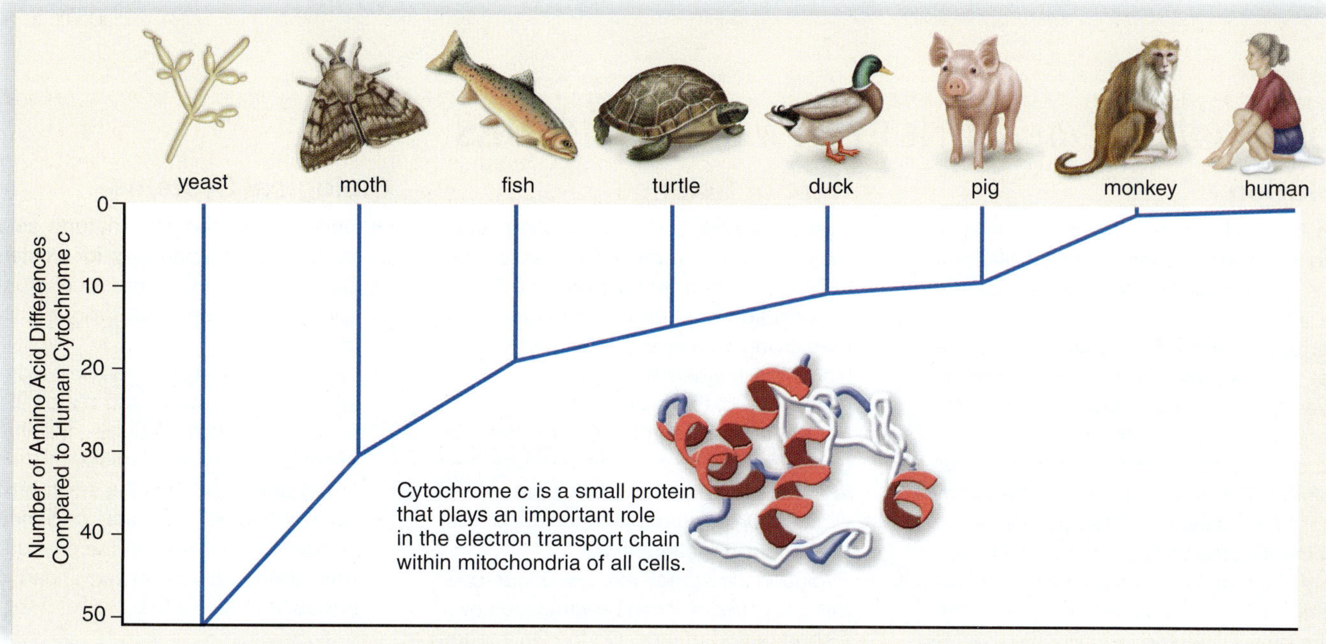

Figure 15.17 Significance of biochemical differences. The branch points in this diagram indicate the number of amino acids that differ between human cytochrome *c* and the organisms depicted. These biochemical data are consistent with those provided by a study of the fossil record and comparative anatomy.

2. There are no transitional fossils.

   Biologists do not expect that all transitional forms have been preserved in the fossil record. In fact, scientists *predict* that not all transitional forms will be discovered. The reason is that a series of events must occur before a fossil can be found. First, the organism must have perished in an area that favors the preservation of its skeletal remains. Soft tissue remains are rarely fossilized; thus, many species will not leave fossil remains at all. Second, scientists have to locate and uncover those remains. This is like finding a needle in a haystack!

   Despite this, scientists have unearthed an array of transitional fossils. For example, the fossil record clearly demonstrates transitional forms in the evolution of the whale from its terrestrial ancestor (see Fig. 15.13). Also, a series of intermediate fossils illustrates the transition from fish to tetrapods, such as *Tiktaalik* (see Fig. 15.12).

3. Evolution proposes that life changed as a result of random events; clearly, traits are too complex to have originated "by chance."

   "Chance" does play a role in evolution, but this is only part of the story. Mutation, the process that *creates* new variation in populations, occurs randomly. However, natural selection, the process that *shapes* variation, is not random. Natural selection can act only on the variation that is present in a population, and it is thus constrained by changes that have occurred in the past. Complex structures, such as the vertebrate eye, did not evolve as a single, functioning unit with all parts intact. Complexity is the result of millions of years of modifications to preexisting traits, each of which provided a useful function at the time. For example, the bacterial flagellum—the long, tail-like propeller of some bacteria—contains a complex, microscopic, rotary motor made from an assembly of many proteins. In a different group of bacteria, a simpler version of this protein assembly does not function as a rotary motor but as a "syringe," called an injectisome, which the bacteria use to "inject" eukaryotic cells with toxins. Scientists hypothesize that the flagellum evolved over time via the addition of proteins to a preexisting, simpler structure, like an injectisome. Both the injectisome and the flagellum, composed of subsets of the same proteins, are totally functional, even though one is less complex than the other.

4. Evolution is not observable or testable; thus, it is not science.

   Evolution is both observable and testable. Recently, scientists discovered that there are genes that encode more than one type of trait. New variation can arise from small changes to single genes. Several studies show how traits in populations change in response to environmental changes (see section 15.2). Other branches of science figure out how things work by accumulating evidence from the real world. Particle physicists cannot see the electrons in an atom. Geologists cannot directly observe the past. But like evolutionary biologists, these scientists can learn a lot about the world by gathering evidence from multiple sources. Evolution has 150 years of such supporting evidence from a wide variety of scientific disciplines.

Check Your Progress 15.3

1. Define *transitional fossils* and provide one example.
2. Summarize the differences between homologous and analogous vestigial structures, as well as what each tells us about common ancestry.
3. Explain how biomolecules support the theory of evolution by natural selection.
4. State one misconception about the theory of evolution and explain why it is incorrect.

Connecting the Concepts with the Themes

Evolution

- All life on Earth has the same building blocks of inheritance, namely DNA, and has in common many proteins essential to life.
- Vertebrate embryos develop the same set of features early in development, even though these features develop into very different structures in the adult.
- All animals have genes in common that control the development of the body plan. *Hox* genes orchestrate the development of the body plan in all animals.
- The Tree of Life project has collected information on hundreds of organisms. Anatomy, DNA, and behavior are used to trace all of life back to a single common ancestor.

Nature of Science

- Fossils provide us with a glimpse of life in the past. Transitional fossils have been discovered that support the theory of evolution via gradual changes from preexisting forms, such as alteration of hindlimbs in whale ancestors.
- At the level of the gene, small changes in the DNA sequence of switches that turn genes "on" and "off" can produce new features, such as the black spot on the wings of *Drosophila biarmipes* males that play a role in mating rituals.
- Evolution does not always occur over millions of years; it can be witnessed over a short period of time. On the Galápagos Islands, a shift in the average beak depth of ground finches can be observed in relatively short periods of time.

Biological Systems

- Darwin proposed that natural selection is, in essence, a struggle for existence.
- Organisms tend to produce more offspring than can be supported by the environment.
- All living organisms require resources such as food, water, and mates in order to survive and reproduce—the intensity of competition is determined by the availability of resources in the environment.
- Natural selection operates on variation in populations. Change in the environment, both long- and short-term, can cause populations to evolve.

MEDIA STUDY TOOLS http://connect.mheducation.com

SMARTBOOK® Study smart. SmartBook helps you maximize your study time by identifying what you know and don't know. The recharge feature in SmartBook helps you prepare for the test by identifying the material that you are most likely to forget.

connect® BIOLOGY Find even more resources to learn the chapter concepts including animations, tutorial videos, and interactive practice questions.

Animations	Videos
15.3 Evolution of Homologous Genes	**15.2** Finches Natural Selection

SUMMARIZE

15.1 History of Evolutionary Thought

In general, the pre-Darwinian worldview was different from the post-Darwinian worldview.

A century before Darwin's trip, most biologists believed in the fixity, or unchanging nature, of species. **Evolution,** the change in species over time, was not widely accepted. As explorers traveled the world and brought back **extant** species, questions began to arise. Erasmus Darwin, Charles's grandfather, suggested the possibility of evolution based on the presence of **vestigial structures.** Linnaeus, the originator of taxonomy, thought that each species had a place in the *scala naturae* and that classification should describe the fixed features of species. Georges Cuvier and Jean-Baptiste de Lamarck, contemporaries of Darwin in the late eighteenth century, differed sharply on evolution. To explain the fossil record of a region, Cuvier, the founder of **paleontology,** proposed that changes in the makeup of fossils in the **strata** of the Earth could be explained by regional catastrophes, or extinctions, followed by repopulation from other regions. Cuvier's idea became known as **catastrophism.** Lamarck was in support of the idea that species can change as they become adapted to their environments. However, he suggested the **inheritance of acquired characteristics** as a mechanism for evolutionary change. We now know that only traits encoded in our genes are heritable.

Charles Lyell, the foremost geologist of Darwin's time, made popular James Hutton's theory of slow, uniform geological change. Lyell's theory of **uniformitarianism** stated that geological processes today are the same as those in the past. Darwin's observations of geology led him to support Lyell's proposal and that the Earth therefore, must be very old.

Thomas Malthus, in his influential work *Essay on the Principle of Population,* proposed that population growth is limited by the availability of resources. Darwin applied this idea to his theory of natural selection.

15.2 Darwin's Theory of Evolution

Charles Darwin formulated hypotheses concerning evolution after taking a 5-year voyage as a naturalist aboard the ship HMS *Beagle*. His hypotheses were that descent with modification from a common ancestor does occur and that natural selection results in adaptation to the environment.

Darwin's study of **biogeography,** including the animals of the Galápagos Islands, led him to conclude that biological diversity arises from adaptation to the environment, which can eventually lead to the formation of new species.

Natural selection is a mechanism of evolutionary change. Members of a population have heritable variations, compete for resources, differ in reproductive success, and adapt to new environmental conditions. **Fitness** is the reproductive success of an organism. **Adaptations**

are evolved traits that increase an organism's chances of survival. Darwin observed selection at work as humans used **artificial selection** to increase the frequency of desired traits in various organisms.

15.3 Evidence for Evolution

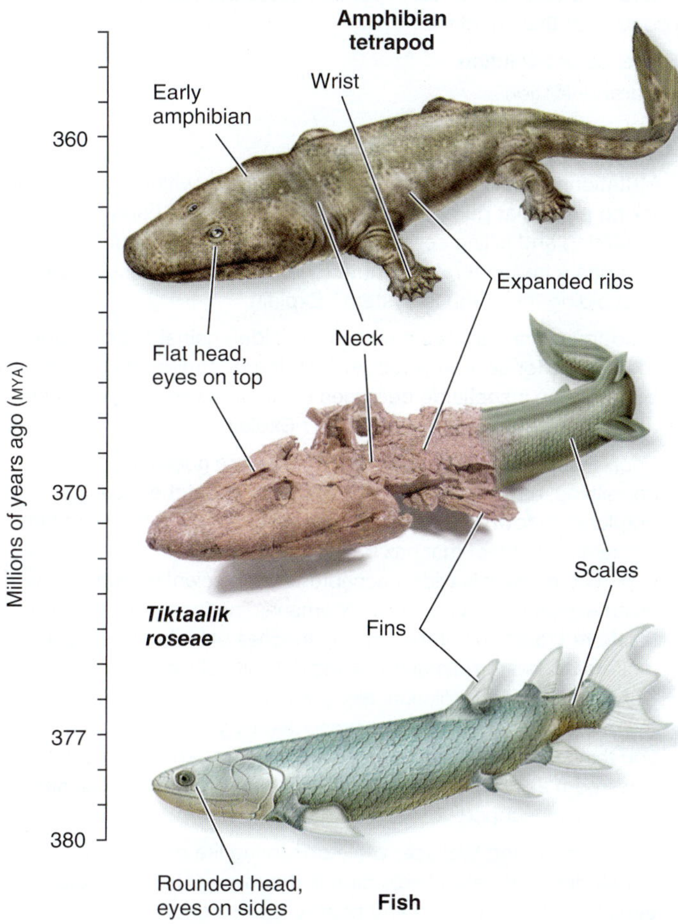

The theory that all organisms share a common ancestor is supported by many lines of evidence, including **fossils,** anatomy, biochemistry, and development. The fossil record gives us a snapshot of the history of life that allows us to trace the descent of a particular group. **Transitional fossils** bear a resemblence to two groups that are classified separately. These allow us to retrace the evolution of organisms over time.

Biogeography is the study of the range and distribution of plants and animals in different places throughout the world and how, and when, they came to be distributed as they are today. Therefore, a different mix of plants and animals might be expected in cases where geography separates continents, islands, and seas.

A comparison of the anatomy and the development of organisms suggests that all life on Earth is closely related. All organisms have certain biochemical molecules and a body plan encoded in genes shared in common, suggesting relatedness. **Homologous structures** are anatomically similar due to inheritance from a common ancestor. **Analogous structures** serve the same function but originated independently in different groups. ***Homeobox*** genes orchestrate the development of the body plan in all animals.

Alternatives to evolution have been proposed to explain the origin of new species. These alternatives are untestable, and thus are unscientific, explanations founded in religious or spiritual philosophy.

Today, the theory of evolution is one of the great unifying theories of biology, because it has been supported by 150 years of scientific evidence.

ASSESS

Choose the best answer for each question.

15.1 History of Evolutionary Thought

1. Which of these pairs is mismatched?
   a. Charles Darwin—natural selection
   b. Linnaeus—classified organisms according to the *scala naturae*
   c. Cuvier—series of catastrophes explains the fossil record
   d. Lamarck—uniformitarianism
   e. Thomas Malthus—principles of population
2. Which scientist's information on geology helped support the observations Darwin made during his voyage aboard the HMS *Beagle*?
   a. Erasmus Darwin
   b. Georges Cuvier
   c. Aristotle
   d. Carolus Linnaeus
   e. Charles Lyell
3. According to the theory of inheritance of acquired characteristics,
   a. if a man loses his hand, then his children will also be missing a hand.
   b. changes in phenotype are passed on by way of the genotype to the next generation.
   c. organisms are able to bring about a change in their phenotype.
   d. evolution is striving toward improving particular traits.
   e. All of these are correct.
4. Why was it helpful to Darwin to learn that Lyell thought the Earth was very old?
   a. An old Earth has more fossils than a new Earth.
   b. It meant there was enough time for evolution to have occurred slowly.
   c. There was enough time for the same species to spread out into all continents.
   d. Darwin said that artificial selection occurs slowly.
   e. All of these are correct.

15.2 Darwin's Theory of Evolution

5. The distribution of organisms across Earth is known as
   a. biogeography.
   b. uniformitarianism.
   c. paleontology.
   d. evolution.
   e. strata.
6. Organisms
   a. compete with other members of their species.
   b. differ in fitness.
   c. are adapted to their environment.
   d. are related by descent from common ancestors.
   e. All of these are correct.
7. Which of the following is not a component of evolution by natural selection?
   a. Variation exists within a population.
   b. Organisms compete for resources.
   c. Individuals have different reproductive success.
   d. Organisms adapt due to environmental changes.
   e. Species are fixed.

8. A key biogeographical observation made by Darwin was
   a. the variation that existed among the Galápagos Island finches.
   b. the similarity between the Galápagos Island tortoises.
   c. the differences between the Galápagos Island hare and cavy.
   d. the difference in the neck lengths of giraffes.
   e. All of these are correct.
9. If evolution occurred, we would expect different biogeographical regions with similar environments to
   a. all contain the same mix of plants and animals.
   b. each have its own specific mix of plants and animals.
   c. have plants and animals with similar adaptations.
   d. have plants and animals with different adaptations.
   e. Both b and c are correct.
10. Humans are able to cause organisms to evolve through which mechanism?
    a. artificial selection
    b. natural selection
    c. fitness
    d. paleontology
    e. uniformitarianism

15.3 Evidence for Evolution

For questions 11–15, match the evolutionary evidence in the key to the description. Choose more than one answer if correct.

Key:

a. biogeographical evidence
b. fossil evidence
c. biochemical evidence
d. anatomical evidence
e. developmental evidence

11. It's possible to trace the evolutionary ancestry of a species.
12. A group of related species has homologous structures.
13. The same types of molecules are found in all living organisms.
14. All vertebrate embryos have pharyngeal pouches.
15. Transitional fossils have been found between some major groups of organisms.

ENGAGE

LEARNSMART LABS®

The following LearnSmart Labs contain exercises that are related to the content of this chapter:

- Evidence of Evolution
- Natural Selection

THINKING SCIENTIFICALLY

1. Mutations occur at random and can increase within a population for no particular purpose. Our immune system is capable of detecting and killing certain viruses. Would a virus, such as HIV, that has a frequent rate of mutation be more or less successful in avoiding the immune system? Explain.
2. A cotton farmer applies a new insecticide against the boll weevil to his crop for several years. At first, the treatment is successful, but then the insecticide becomes ineffective and the boll weevil rebounds. Has evolution occurred? Explain.
3. People often use the word *theory* to refer to a guess about something, as in "the police have a theory about the recent burglaries in town." In the scientific sense, the word *theory* means a scientific concept that has been supported by observation and experiments, so it is widely accepted by the scientific community as an explanation of natural phenomena. The theory of evolution is referred to as one of the unifying theories of biology, because it offers the best explanation for the diversity of life. Criticisms of Darwin's theory of evolution, along with alternative "theories" about the diversity of life, have been presented by various groups. Do you think that these alternatives should be presented in a science classroom as an equal to the theory of evolution? Using scientific information, support your answer.
4. Both Darwin and Wallace, while observing life on islands, concluded that natural selection is the mechanism for biological evolution. The Hawaiian and nearby islands once had at least 50 species of honeycreepers, and they lived nowhere else on Earth. Natural selection occurs everywhere and in all species. What characteristics of islands allow the outcome of natural selection to be so obvious?

MRSA (methicillin-resistant *Staphylococcus aureus*) has evolved a resistance to various types of antibiotics. MRSA can cause life-threatening infections of the bloodstream, pneumonia, and infections of surgical sites.

16

How Populations Evolve

Chapter Outline

16.1 Genes, Populations, and Evolution 280

16.2 Natural Selection 286

16.3 Maintenance of Diversity 290

MRSA (methicillin-resistant *Staphylococcus aureus*) is a strain of a common bacterium that can be deadly because of its resistance to several types of antibiotics. The bacterium is often found on the skin and in the noses of healthy people, but *S. aureus* can cause "staph" infections in scrapes, cuts, and open wounds. Normally, regular hand washing with soap and warm water prevents most staph infections. Until recently, severe staph infections were found primarily in hospitals. However, MRSA may now be found in prisons and schools, especially in gymnasiums and locker rooms.

Use and overuse of antibiotics have resulted in the evolution of resistant bacterial strains, such as MRSA. Although we tend to think of evolution as happening over long timescales, human activities can accelerate the process of evolution quite rapidly. In fact, evolution of resistance to the antibiotic methicillin occurred in just 1 year! Antibiotic-resistant strains of bacteria are expensive and difficult to treat.

Some scientists believe that these "superbugs" will be a far bigger threat to human health than emerging diseases such as H1N1 and AIDS. However, our understanding of evolutionary biology is helping to combat the superbugs. Antibiotic resistance is an example of why evolution is important in people's everyday lives. In this chapter, you will learn about evidence that indicates evolution has occurred and about how the evolutionary process works.

As you read through the chapter, think about the following questions:

1. In the case of antibiotic resistance, what type of selection is operating?
2. Discuss strategies that may be used to help reduce the rate at which antibiotic-resistant bacteria evolve.

Before You Begin

Before beginning this chapter, take a few moments to review the following discussions.

Section 1.1 What is a population?

Section 11.2 How is a Punnett square used to estimate genotype frequencies?

Section 11.2 What is an allele?

Following *the* Themes

CHAPTER 16 HOW POPULATIONS EVOLVE

UNIT 3 EVOLUTION

Evolution	Microevolution, or evolution within populations, is measured as a change in allele frequencies over generations.
Nature of Science	Population geneticists calculate allele and genotype frequencies to characterize how populations evolve.
Biological Systems	Microevolution and the environment are closely linked, such that a phenotype may be adaptive in one environment, but not in another.

16.1 Genes, Populations, and Evolution

Learning Outcomes

Upon completion of this section, you should be able to

1. Explain how evolution in populations is related to a change in allele frequencies.
2. List the five conditions necessary to maintain Hardy-Weinberg equilibrium.
3. Apply the Hardy-Weinberg principle to estimate equilibrium genotype frequencies.
4. Describe the agents of evolutionary change.

A triathlete who spends months in Denver gradually gets used to being at high altitude. Part of the reason is that the number of oxygen-carrying red blood cells has increased in response to the oxygen-poor environment. Many traits can change temporarily in response to a varying environment. The color change in the fur of an Arctic fox from brown to white in winter, the increased thickness of your dog's fur in cold weather, or the darkening of your skin when exposed to the sun lasts only for a season.

These are not evolutionary changes. Changes to traits over an individual's lifetime are not evidence that an individual has evolved, because these traits are not heritable. In order for traits to evolve, they must have the ability to be passed on to subsequent generations. Evolution is about change in a trait within a population over many generations. A **population** is defined as a group of organisms of a single species living together in the same geographic area.

Darwin observed that populations, not individuals, evolve, but he could not explain *how* traits change over time. Now we know that genes interact with the environment to determine traits—the diversity of a population is linked to the genetic diversity of individuals within that population. Because genes and traits are linked, evolution is really about genetic change—or more specifically, *evolution is the change in allele frequencies in a population over time.*

Several mechanisms can cause a population to evolve—to change allele frequencies—over generations. **Microevolution** pertains to evolutionary change within populations. In this chapter, we use the peppered moth example from Chapter 15 to examine how populations evolve over time.

Microevolution in the Peppered Moth

Population genetics, as its name implies, is the field of biology that studies the diversity of populations at the level of the gene. Population geneticists are interested in how genetic diversity in populations changes over generations, as well as in the forces that cause populations to evolve. Population geneticists study microevolution by measuring the diversity of a population in terms of allele and genotype frequencies over generations.

You may recall from Chapter 10 that diploid organisms, such as moths, carry two copies of each chromosome, with one copy of each gene on each chromosome. A single gene can come in many forms, or alleles, that encode variations of a single trait. In the peppered moth, a single gene for body color has two alleles, *D* (dark color) and *d* (light color), with *D* dominant to *d* (Fig. 16.1).

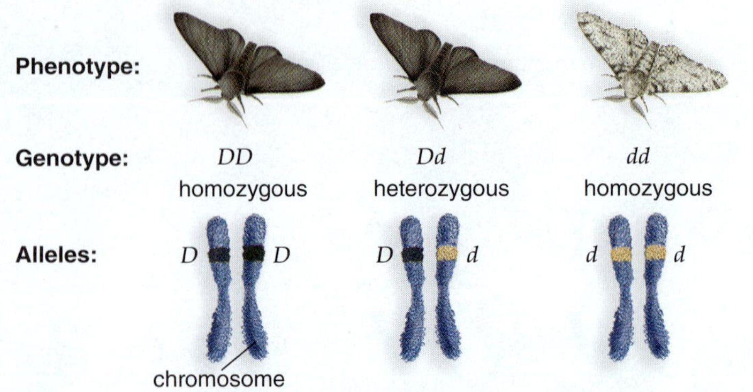

Figure 16.1 The genetic basis of body color in the peppered moth. Light or dark body color in the peppered moth is determined by a gene with two alleles, *D* and *d*. Genotypes *DD* and *Dd* produce dark body color, and *dd* produces light body color. The *D* and *d* alleles are variants of a gene at a particular locus on a chromosome.

We know that with only two alleles, there are three possible genotypes (the combinations of alleles in an individual) for the color gene in the peppered moth: *DD* (homozygous dominant), *Dd* (heterozygous), or *dd* (homozygous recessive). *DD* or *Dd* genotypes produce dark moths, and the *dd* genotype produces light moths (Fig. 16.1).

Allele Frequencies

Suppose that a population geneticist collected a population of 25 moths, some dark and some light, from a forest outside London (Fig. 16.2). In this population you would expect to find a mixture of *D* and *d* alleles in the **gene pool**—the alleles of all genes in all individuals in a population (Fig. 16.2). The population geneticist ran tests to determine the alleles present in each moth. Of the 50 alleles in the peppered moth gene pool (2 alleles × 25 moths), 10 were *D* and 40 were *d*. Thus, the frequency of the *D* and the *d* alleles would be 10/50, or 0.20, and 40/50, or 0.80 (Fig. 16.2). The **allele frequency,** as illustrated in this example, is the percentage of each allele in a population's gene pool.

Notice that the frequencies of *D* and *d* add up to 1. This relationship is true of the sum of allele frequencies in a population for any gene of any diploid organism. This relationship is described by the expression $p + q = 1$, where p is the frequency of one allele, in this case *D*, and q is the frequency of the other allele, *d* (Fig. 16.2).

For the next three seasons, samples of 25 moths were collected from the same forest, and the allele frequencies were always the same: 0.20 *D* and 0.80 *d*. Because allele frequencies in this population did not change over generations, we could conclude, with regard to color, that this population has not evolved.

Hardy-Weinberg Equilibrium

A population in which allele frequencies do not change over time, such as in the moth population just described, is said to be in *genetic equilibrium,* or **Hardy-Weinberg equilibrium (HWE)**—a stable, nonevolving state. Hardy-Weinberg equilibrium is derived from the work of British mathematician Godfrey H. Hardy and German physician Wilhelm Weinberg, who in 1908 developed a mathematical model to estimate genotype frequencies of a population that

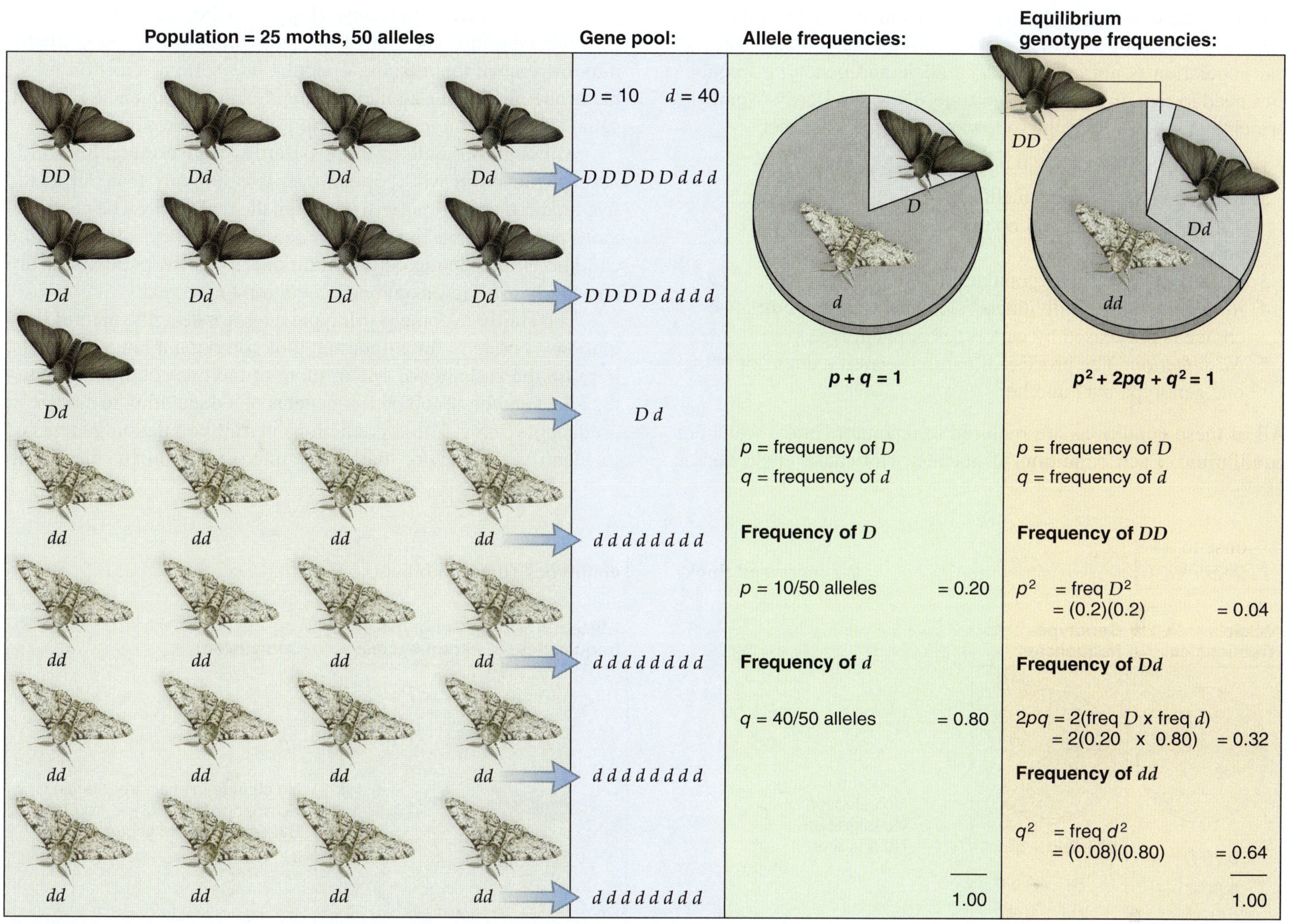

Figure 16.2 How Hardy-Weinberg equilibrium is estimated. A population of 25 moths contains a gene pool of *D* and *d* alleles. The frequencies of the *D* and *d* alleles can be estimated from the gene pool. Under Hardy-Weinberg equilibrium, the frequencies of *D* and *d* alleles should produce in the next generation a predictable frequency of genotypes that can be calculated with the Hardy-Weinberg principle.

Tutorial Hardy-Weinberg Equilibrium

is in genetic equilibrium. Their mathematical model, called the **Hardy-Weinberg principle,** proposes that the genotype frequencies of a nonevolving population can be described by the expression $p^2 + 2pq + q^2$, again with p and q representing the frequency of alleles *D* and *d*. Recall that in our moth population, $D = p$ and $d = q$, so that D^2 is the frequency of the *DD* genotype, $2Dd$ is the frequency of the *Dd* genotype, and d^2 is the frequency of the *dd* genotype (Fig. 16.2).

A simple Punnett square, first described in Chapter 11, is another way to illustrate how the Hardy-Weinberg principle explains the genotype frequencies of a population. The frequency of the *D* and *d* alleles in the gametes (sperm and egg) in this population would be the same as the allele frequencies, so that 20% of alleles in eggs and sperm will be *D,* and 80% of alleles in eggs and sperm will be *d* (Fig. 16.3). The genotype frequencies from the Punnett square match those predicted by the Hardy-Weinberg principle—that is, the frequency of *DD, Dd,* and *dd* is explained by the expression $D^2 + 2Dd + d^2$ (see Figs. 16.2 and 16.3).

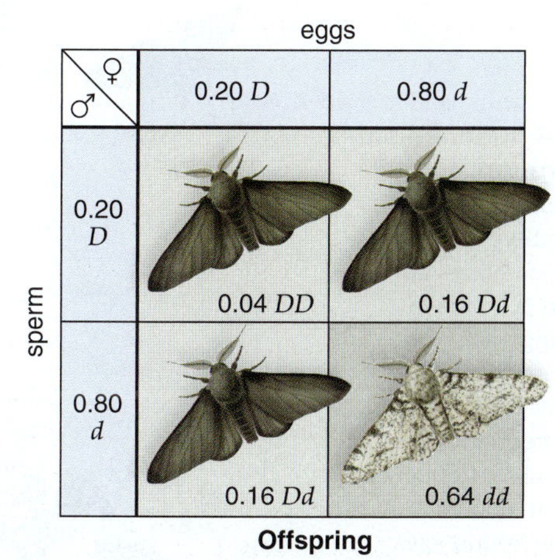

Figure 16.3 Calculation of Hardy-Weinberg equilibrium from gamete frequencies. Notice that the results of the Punnett square match those shown in Figure 16.2.

Tutorial Hardy-Weinberg Equilibrium

The allele and genotype frequencies of the moth population we are considering follow the Hardy-Weinberg principle only if the population is not evolving. The allele and genotype frequencies need to remain constant over time. Thus, the Hardy-Weinberg principle applies only if the following conditions are met.

1. *No mutation:* No new alleles can arise by mutation.
2. *No migration:* No new members (and their alleles) can join the population, and no existing members can leave the population.
3. *Large gene pool:* The population is very large.
4. *Random mating:* Individuals select mates at random; mate choice is not biased by genotypes or phenotypes.
5. *No selection:* The process of natural selection does not favor one genotype over another.

All of these conditions are required to maintain Hardy-Weinberg equilibrium. Each condition, if not met, can cause allele and/or genotype frequencies to change (Fig. 16.4). For example, individuals migrating into and out of the population would bring alleles into, or remove them from, a population. Natural selection might favor one allele over another, changing allele frequencies, so that some alleles are more or less common than others (Fig. 16.4).

Although possible in theory, Hardy-Weinberg equilibrium is never achieved in wild populations. It is unlikely that all of the five required conditions will be met in the real world. The peppered moth population we are using as an example obeys all five of the conditions for genetic equilibrium, but in reality, populations are constantly evolving from one generation to the next.

The Hardy-Weinberg principle does not describe natural populations, but it is an important tool for population geneticists, because the violation of one or more of the five conditions causes the allele and/or genotype frequencies of a population to change in predictable ways. These predictions permit population geneticists to identify the factors that cause microevolution by measuring

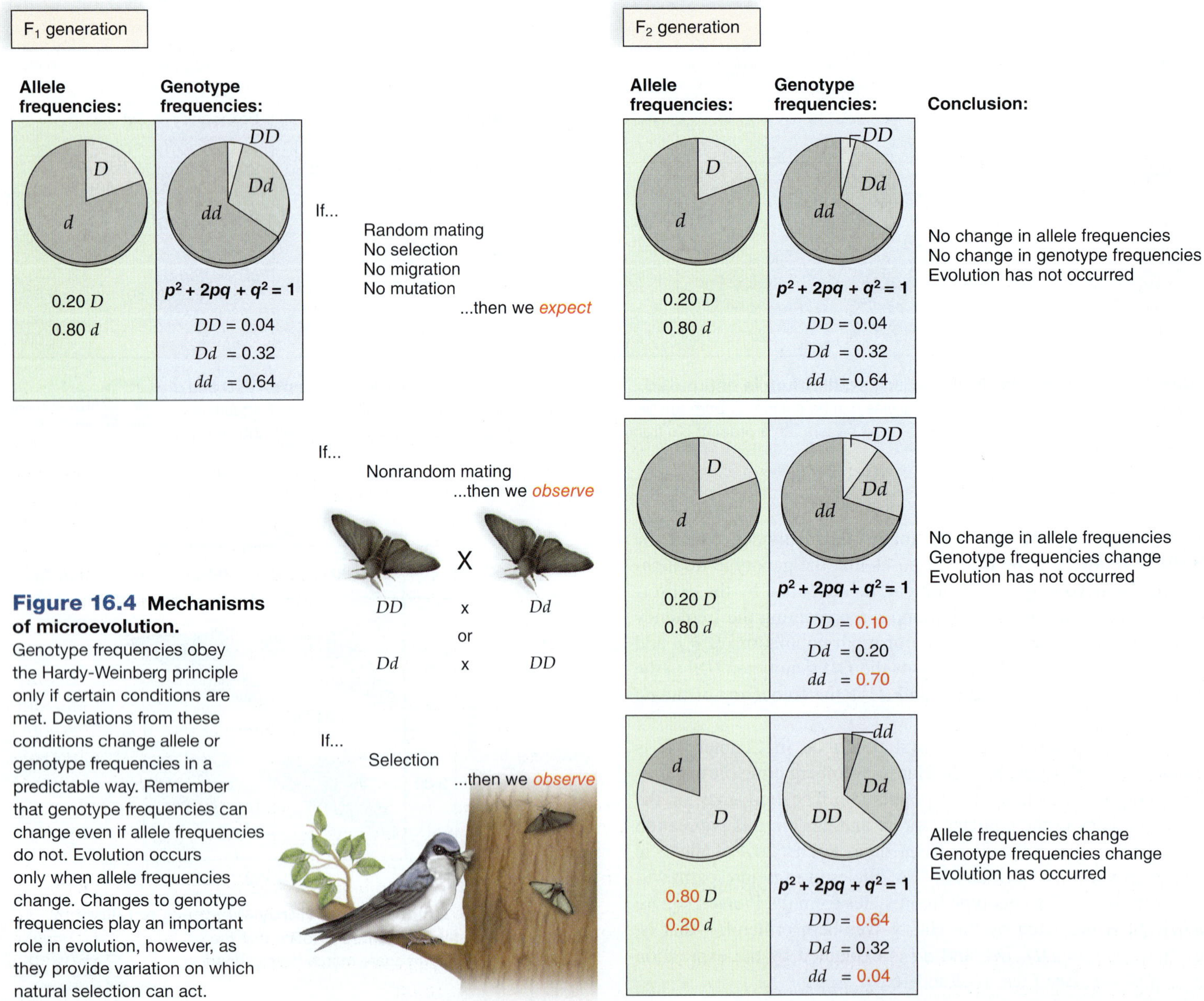

Figure 16.4 Mechanisms of microevolution. Genotype frequencies obey the Hardy-Weinberg principle only if certain conditions are met. Deviations from these conditions change allele or genotype frequencies in a predictable way. Remember that genotype frequencies can change even if allele frequencies do not. Evolution occurs only when allele frequencies change. Changes to genotype frequencies play an important role in evolution, however, as they provide variation on which natural selection can act.

Table 16.1 Hardy-Weinberg Proportions Are Used to Determine If Evolution Has Occurred

HWE Condition	Deviation from Condition	Effect of Deviation on Population	Expected Deviation from HWE	Evolution Occurred?
Random mating	Nonrandom mating  	Allelles do not assort randomly	Change in genotype frequencies	No
No selection	Selection 	Certain alleles are selected for or against	Change in allele frequencies	Yes
No mutation	Mutation	Addition of new alleles	Change in allele frequencies	Yes
No migration	Immigration or emigration 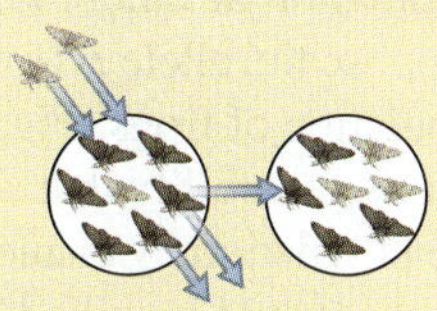	Individuals carry alleles into, or out of, the population	Change in allele frequencies	Yes
Large population (no genetic drift)	Small population (genetic drift) 1. Bottleneck effect 2. Founder effect 	Loss of allele diversity; some alleles may disappear	Change in allele frequencies	Yes

how allele and genotype frequencies of a population are different from those of a population in Hardy-Weinberg equilibrium (Table 16.1).

Microevolution and Hardy-Weinberg Equilibrium

We can modify the peppered moth scenario to illustrate how the Hardy-Weinberg principle can be used to determine whether, and how, microevolution has occurred. Population genetic theory predicts that Hardy-Weinberg equilibrium can be interrupted by deviation from any of its five conditions. Therefore, we predict that evolutionary change can be caused by mutation, migration, small population size, nonrandom mating, or natural selection.

It is important to remember that evolution is measured as a change in allele frequencies from one generation to another. Thus, we start with allele and genotype frequencies of an F_1 generation, then remeasure the same frequencies in the F_2 generation.

Mutation

A *mutation* is a change to the DNA sequence, which can serve as a source of new genetic variation. Most mutations occur because of errors made to the DNA sequence during DNA replication. The rate of mutations is generally very low, on the order of one mutation per 100,000 cell divisions. Mutations may also occur because of exposure to *mutagens,* chemical or physical agents that cause changes to the DNA code, as described in Chapter 13.

Not all mutations affect the genetic equilibrium of a population. Most mutations that occur during DNA replication and from mutagens are repaired by cellular repair mechanisms. Those mutations that are not repaired can affect the gene pool of a population only if they *are* transmitted from the F_1 to subsequent generations. In other words, the mutations must be carried by the gametes of individuals that successfully reproduce. In addition, mutation is a random process, and the mutation must occur in a gene such that the result is a change in the frequency of the gene's alleles. Thus, it is safe to say that *for any particular gene,* mutation, although possible, is not a major force for evolutionary change, because it generally results in only small changes in the allele frequency of a single allele.

In the peppered moth, a hypothetical example of evolution by mutation could be a mutation that changed a single *D* allele to a *d* allele in gametes of a member of the F_1 generation. This change would alter the frequency of *D* and *d* alleles in the F_2 generation gene pool (Fig. 16.4).

Although inherited mutations are rare, the sum of the effect of mutations is essential to evolution. The reason is that a single organism has many thousands of genes, and each gene can have many different alleles, so that even if the frequency of heritable mutation in a particular allele of a single gene is very low, *over thousands of genes and alleles,* mutation can have a significant impact on the evolution of a population.

Animation Mutation by Base Substitution

For example, suppose that the peppered moth has 30,000 genes in its genome. If each gene has only two alleles, then a single moth could carry 60,000 different alleles. In a large population, it is likely that a moth would carry at least one new allele that arose because of mutation. These new alleles are sources of new genetic variation, which is essential for evolutionary processes to work. Without new mutations, evolution could not occur, because natural selection must have variation on which to act.

Migration

Gene flow is the movement of alleles between populations. Gene flow occurs when plants or animals migrate, or more specifically their gametes move, between populations. When gene flow brings a new or rare allele into a population, the allele frequency in the next generation changes. Gene flow in plants may result when the pollen from one plant fertilizes a plant in another population (Fig. 16.5).

In the peppered moth example, the movement of a dark- or light-colored individual into a population would introduce one of the three genotypes: *DD, Dd,* or *dd.* Thus, the numbers of the *D* and *d* alleles would change. In the sample group of moths, the addition of a *Dd* moth would change the number of moths to 26, the total number of alleles to 52, the number of *D* alleles to 11, and the number of *d* alleles to 41. The frequency of *D* would now be 11/52, or 0.21, instead of 0.20; the frequency of *d* would change accordingly, to 0.79. A similar change in allele frequencies would occur if a moth left the population, taking with it its *D* or *d* gametes (see Fig. 16.4).

The amount of gene flow between populations depends on several factors, including the distance between populations, the ability of individuals or their gametes to move between populations, and behavior that determines whether an individual will migrate and mate. When gene flow continuously occurs between populations, the gene pools of each population become more and more similar over time until they appear as if they were a single population. In contrast, if migration between populations does not occur, the gene pools of the populations become more and more different over time. The differences in the genetic makeup of these populations can eventually become so large that they become **reproductively isolated**—or incapable of interbreeding. Reproductive isolation is the first stage in the formation of new species, which is covered in Chapter 17.

Small Population Size

Genetic drift refers to changes in the allele frequencies of a gene pool due to chance events. Such events remove individuals, and their genes, from a population at random—without regard for genotype or phenotype. Suppose the allele *B* (for brown) occurs in 10% of the members in a population of frogs (Fig. 16.6). If a storm kills half the frogs, there is a good possibility that some of the *B* alleles would be removed from the population by chance. However, it is also possible that only green frogs would be killed, and then the allele frequency of *B* would be increased after the storm.

In the equilibrium peppered moth population of 25 individuals, 20% of the alleles are *D,* and 80% are *d.* Suppose that a heavy storm kills 5 of the 25 moths, or 20% of the alleles in the gene pool. To simulate the random loss of individuals from the storm, imagine that all of the 25 moths are in a canvas bag. To simulate the storm, 5 moths are removed at random from the bag. The remaining 20 moths in the bag represent the population after the storm. The phenotype, genotype, and allele frequencies are then determined for the 20 moths remaining after the storm. Overall, suppose we find that the loss of the 5 moths has increased the frequency of the *D* allele by 2.5% and decreased the frequency of

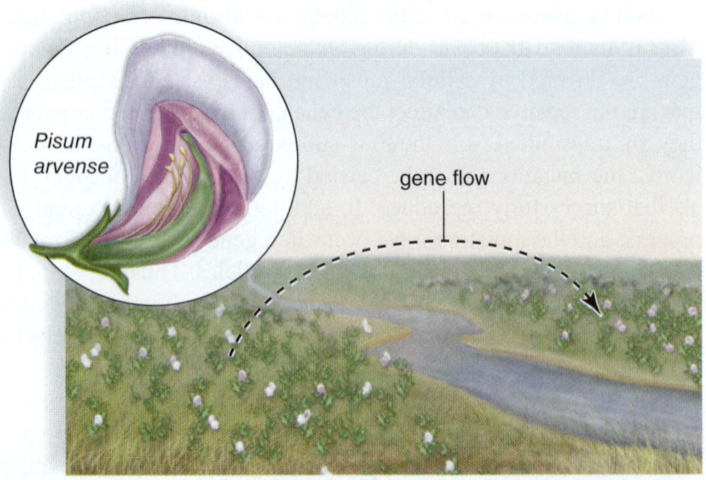

Figure 16.5 Gene flow. Occasional cross-pollination between two different populations of *Pisum arvense* is an example of gene flow.

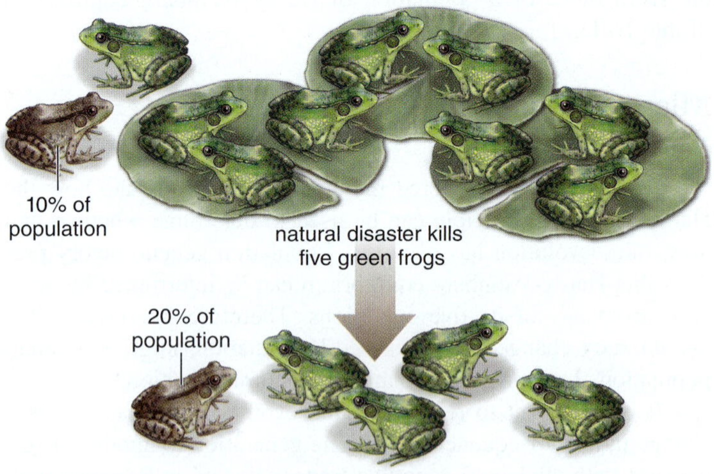

Figure 16.6 Genetic drift. Genetic drift occurs when, by chance, only certain members of a population (in this case, green frogs) reproduce and pass on their alleles to the next generation. A natural disaster can cause the allele frequencies of the next generation's gene pool to be different from those of the previous generation. Genetic drift can be a powerful force for evolutionary change, especially in small populations.

the *d* allele by 2.5%. The *D* and *d* alleles have changed frequency, and we would conclude that the population has evolved due to genetic drift.

Although genetic drift occurs in populations of all sizes, the gene pool of a smaller population is likely to be more affected by genetic drift. For example, in the population of 25 moths, removing 5 moths reduced the size of the gene pool by 20% (10 alleles from a gene pool of 50). But if the storm had removed 5 moths from a population of 500, or 10 alleles out of a pool of 1,000, only 1% of the gene pool would have been eliminated—with less effect on allele frequency. Thus, the smaller the population, the more genetic drift impacts allele frequencies.

In some cases, a large population can suddenly become very small, such as a *bottleneck,* when natural disasters strike and significantly reduce a population. When this occurs, the effects of genetic drift can be significant. A **bottleneck effect** is a type of genetic drift in which the loss of genetic diversity is due to natural disasters (e.g., hurricane, earthquake, or fire), disease, overhunting, overharvesting, or habitat loss. A **founder effect,** another type of genetic drift, is similar to a bottleneck effect except that genetic variation is lost when a few individuals break away from a large population to found a new population.

The outcome of a founder or a bottleneck effect is a small population with a gene pool made up of a random assortment of alleles from the original population. In some cases, alleles disappear altogether. The gene pool of the small population is often much different from the gene pool of the original population (Fig. 16.7). The greater the reduction in population size, the greater the effects on allele frequencies. Thus, genetic drift is one of the more powerful forces for evolution of small populations.

Animation Simulation of Genetic Drift

Another by-product of a very small population is a higher than normal occurrence of **inbreeding,** or mating between relatives, because after a few generations only a few, if any, unrelated mates are available. Unlike genetic drift, *inbreeding alone does not affect the frequency of alleles* and thus does not cause a population to evolve. Nevertheless, inbreeding can have a significant impact on the genotypes, and thus the phenotypes, of individuals, sometimes with unfortunate consequences. Of particular concern are rare recessive disorders, which emerge more frequently in inbred populations (see the Nature of Science feature, "Inbreeding in Populations," on page 290).

Nonrandom Mating

Hardy-Weinberg equilibrium requires individuals in a population to mate randomly. Nonrandom mating alone does not cause allele frequencies to change. Nonrandom mating, however, does affect *how the alleles in the gene pool assort into genotypes,* thus affecting the phenotypes in a population. Inbreeding, such as that in the Pingelapese population described in the Nature of Science feature, is an example of nonrandom mating.

In a randomly mating population, the alleles in the gene pool assort at random. When mating is nonrandom, gametes, and thus alleles, assort according to mating behavior. For example, another type of **nonrandom mating,** called **assortative mating,** occurs when individuals choose a mate with a preferred trait, such as a particular coat color, feather length, or body size. Assortative mating brings together alleles for these traits more often than would happen by chance. The result is that certain genotypes are more frequent than others.

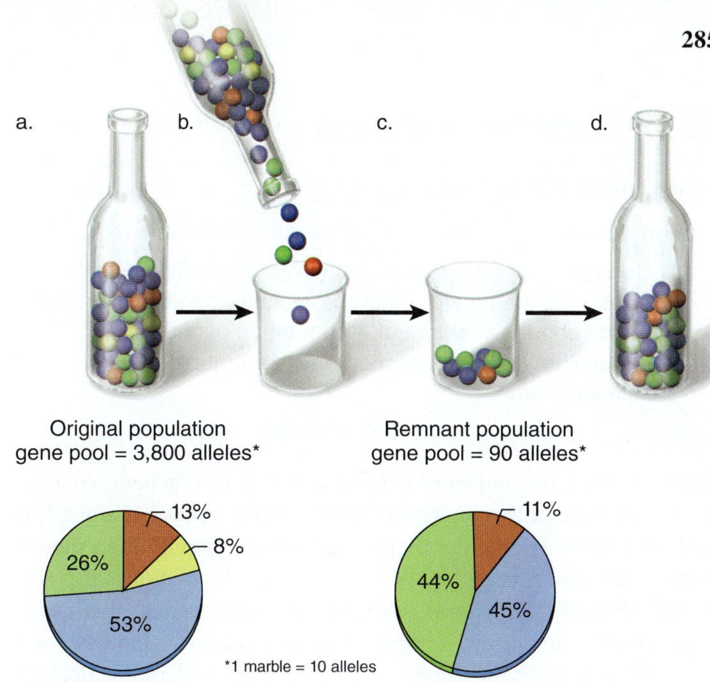

Figure 16.7 Consequences of bottleneck and founder effects. **a.** The gene pool of a large population contains four different alleles, represented by colored marbles in a bottle, each with a different frequency. **b.** A population bottleneck occurs. The marbles, or alleles, that exit the bottle must pass through the narrow neck into the cup. The new gene pool will have a fraction of the alleles from the original population. **c.** The gene pool of the new population has changed from the original. Some alleles are in high frequency, while some are not present. **d.** A founder event is the same as a bottleneck, except that in the founder event the original population still exists.

Although nonrandom mating does not, in itself, cause a population to evolve, it can play an important role in the evolution of a population. For example, in the Pingelapese population of the island of Pingelap, nonrandom mating resulted in an increased frequency of colorblindness.

Natural Selection

A population in Hardy-Weinberg equilibrium has phenotypes that are equally likely to survive and reproduce. One genotype does not have an advantage over another. But in nature some phenotypes do have a reproductive advantage. Individuals who have an advantageous phenotype often pass on the allele for this trait to their offspring. Over time, selection for this advantageous trait increases the frequency of the alleles associated with it, while other alleles decrease.

Natural selection is the foundation of Darwin's theory of evolution. In the next section, we discuss in detail how natural selection works within populations.

Check Your Progress 16.1

1. List the five conditions necessary for Hardy-Weinberg equilibrium, and describe what happens to allele frequencies in a population if these conditions are not met.
2. Estimate the equilibrium genotype frequencies from a population with allele frequencies $p = 0.10$, $q = 0.90$.

16.2 Natural Selection

Learning Outcomes

Upon completion of this section, you should be able to

1. Compare stabilizing, directional, and disruptive selection.
2. Determine the type of natural selection operating on a trait by the change in shape of a phenotype distribution.
3. Explain how sexual selection drives adaptation for increased fitness.

In this chapter, we consider natural selection in a genetic context. Many traits are **polygenic** (controlled by many genes). If a graph is constructed for a particular trait in a population, it will often show a wide distibution of variation. When this range of variation is exposed to the environment, natural selection favors the variant that is the most adaptive under the present environmental conditions. Natural selection acts much the same way as a governing board that decides which students will be admitted to a college. Some students will be favored and allowed to enter, while others will be rejected and not allowed to enter. Of course, in the case of natural selection, the chance to reproduce is the prize awarded.

Types of Natural Selection

Investigators have defined three general types of natural selection: stabilizing selection, directional selection, and disruptive selection (Fig. 16.8). **Stabilizing selection** occurs when an intermediate phenotype is the most adaptive for the given environmental conditions. With stabilizing selection, extreme phenotypes are selected against, and the intermediate phenotype is favored. As an example, consider that when Swiss starlings lay four or five eggs, more young survive than when the female lays more or fewer than this number. Genes determining physiological characteristics, such as the production of yolk, and behavioral characteristics, such as how long the female mates, are involved in determining clutch size.

Human birth weight is another example of stabilizing selection. Over many years, hospital data have shown that human infants born with an intermediate birth weight (3–4 kg) have a better chance of survival than those at either extreme (either much less or much greater than average). When a baby is small, its systems may not be fully functional; when a baby is large, it may have experienced a difficult delivery. Stabilizing selection reduces the variability in birth weight in human populations (Fig. 16.9).

Directional selection occurs when an extreme phenotype is favored, and the distribution curve shifts toward one of the extremes. Over time, directional selection changes the frequency of a phenotype within a population. Such a shift can occur when a population is adapting to a changing environment. For example, the modern horse, *Equus,* showed a gradual increase in body size as the environment changed from forest to grassland (Fig. 16.10). The ancestor of the modern horse, *Hyracotherium,* was around the size of a dog and lived in forested environments of the Eocene. Its smaller body and low-crowned teeth were well suited to finding refuge among trees and eating leaves. In the Miocene, grasslands began to replace forests. Modern horses are much larger than their ancestors, and they are adapted for speed and long-distance

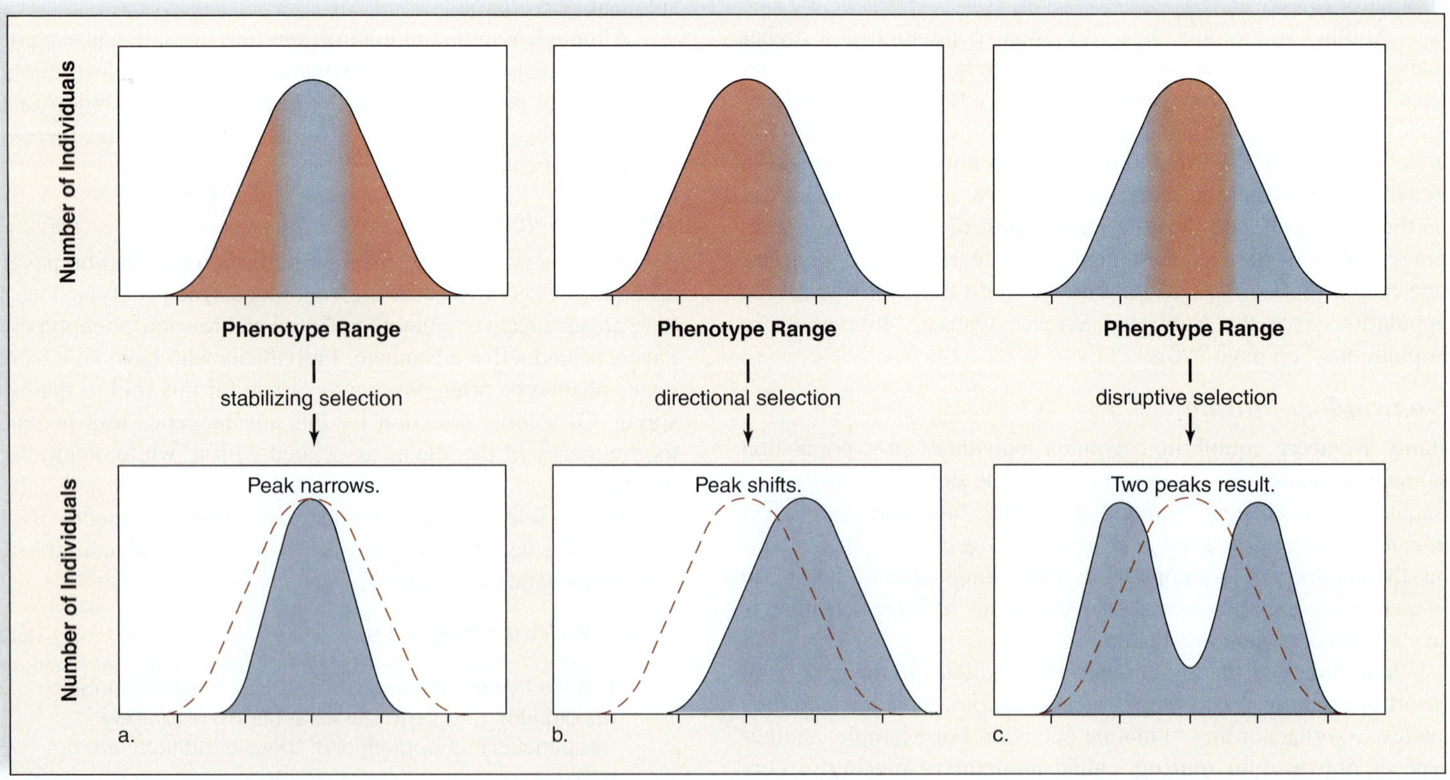

Figure 16.8 Three types of natural selection. Natural selection shifts the average value of a phenotype over time. **a.** During stabilizing selection, the intermediate phenotype increases in frequency; (**b**) during directional selection, an extreme phenotype is favored, which changes the average phenotype value; and (**c**) during disruptive selection, two extreme phenotypes are favored, creating two new average phenotype values, one for each phenotype.

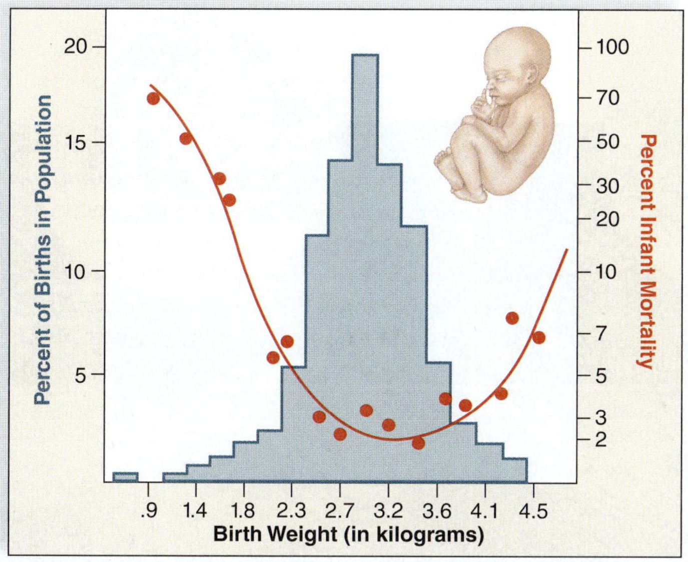

Figure 16.9 Human birth weight. The birth weight (blue) is influenced by the mortality rate (red).

movements in a grassland environment. Long legs provide strength and speed for running, and durable teeth are suited for grinding grasses.

Disruptive selection is found when two or more extreme phenotypes are favored over the intermediate phenotype. For example, British land snails have a wide habitat range that includes low-vegetation areas (grass fields and hedgerows) and forests. In forested areas, thrushes feed mainly on light-banded snails, and the snails with dark shells become more prevalent. In low-vegetation areas, thrushes feed mainly on snails with dark shells, and light-banded snails become more prevalent. Therefore, these two distinctly different phenotypes are found in the population (Fig. 16.11).

Sexual Selection

Sexual selection refers to adaptive changes in males and females that lead to an increased ability to secure a mate. Sexual selection in males may result in an increased ability to compete with other males for a mate, while females may select a male with the best **fitness** (the ability to produce surviving offspring). In that way, the female increases her own fitness. Sexual selection is often considered a form of natural selection, because it affects fitness.

Female Choice

Females produce few eggs compared to a male's production of sperm, so choosing the best mate becomes important. In a study of satin bowerbirds, two opposing hypotheses regarding female choice were tested.

1. *Good genes hypothesis:* Females choose mates on the basis of traits that improve the chance of survival.
2. *Runaway hypothesis:* Females choose mates on the basis of traits that improve male appearance. (The term *runaway* pertains to the possibility that over generations the trait will become exaggerated in the male until its mating benefit is checked by the trait's unfavorable survival cost.)

As investigators observed the behavior of satin bowerbirds, they discovered that aggressive males were usually chosen as mates by females. It could be that inherited aggressiveness improves the chance of survival, or aggressive males might be good at stealing blue feathers from other males—females prefer blue feathers as bower decorations. Therefore, the data did not clearly support either hypothesis.

The Raggiana Bird of Paradise exhibits remarkable **sexual dimorphism,** meaning that males and females differ in size and other traits. The males are larger than the females and have beautiful orange flank plumes. In contrast, the females are drab

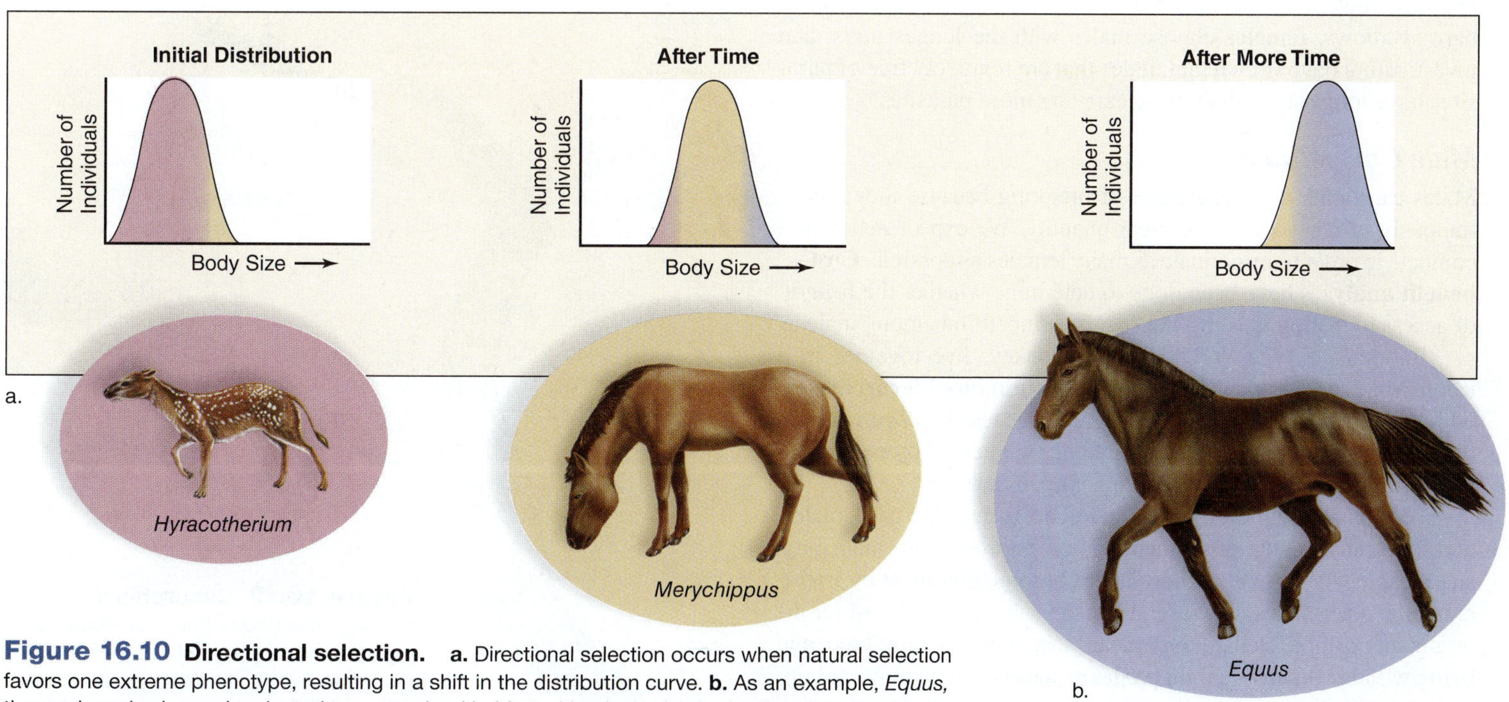

Figure 16.10 Directional selection. **a.** Directional selection occurs when natural selection favors one extreme phenotype, resulting in a shift in the distribution curve. **b.** As an example, *Equus,* the modern-day horse, is adapted to a grassland habitat with a larger body size than its ancestor, *Hyracotherium,* which was adapted to a forest habitat.

Figure 16.11 Disruptive selection. **a.** Disruptive selection favors two or more extreme phenotypes. **b.** Today, British land snails comprise mainly two different phenotypes, each adapted to a different habitat. Snails with dark shells are more prevalent in forested areas, light-banded snails in areas with low-lying vegetation.

(Fig. 16.12). Female choice can explain why male birds are more ornate than females. Consistent with the two hypotheses, it is possible that the remarkable plumes of the male signify health and vigor to the female. Or perhaps females' choice of flamboyant males gives their sons an increased chance of being selected by females.

Some investigators have hypothesized that extravagant male features might indicate males that are relatively parasite-free. In barn swallows, females choose males with the longest tails, and investigators have shown that males that are relatively free of parasites have longer tails than those carrying more parasites.

Male Competition

Males can father a large number of offspring because they continuously produce sperm in great quantity. We expect males to compete in order to inseminate as many females as possible. **Cost–benefit analyses** have been done to determine whether the *benefit* of access to mating is worth the *cost* of competition among males.

Baboons, a type of Old World monkey, live together in a troop. Males and females have separate **dominance hierarchies,** in which a higher-ranking animal has greater access to resources than does a lower-ranking animal. Dominance is decided by confrontations, resulting in one animal giving way to the other.

Baboons are dimorphic; the males are larger than the females, and they can threaten other members of the troop with their long, sharp canine teeth. One or more males become dominant by frightening the other males. However, the male baboon pays a cost for his dominant position. Being larger means that he needs more food, and being willing and able to fight predators means that he may get hurt, and so forth. There is often a reproductive benefit to his behavior. Dominant males tend to be the first to monopolize females when

Figure 16.12 Dimorphism. In the Raggiana Bird of Paradise, *Paradisaea raggiana,* males have brilliantly colored plumage brought about by sexual selection. The drab females tend to choose flamboyant males as mates.

they are most fertile. Nevertheless, there may be other ways to father offspring. A male may act as a helper to a female and her offspring; then, the next time she is in estrus, she may mate preferentially with him instead of with a dominant male. Or subordinate males may form a friendship group that opposes a dominant male, making him give up a receptive female.

A **territory** is an area that is defended against competitors. Scientists are able to track an animal in the wild to determine its home range, or territory. **Territoriality** includes the type of defensive behavior needed to defend a territory. Olive baboons travel within a home range, foraging for food each day and sleeping in trees at night. Dominant males decide where and when the troop will move. If the troop is threatened, dominant males protect the troop as it retreats and attack intruders when necessary.

Vocalization and displays, rather than outright fighting, may be sufficient to defend a territory (Fig. 16.13). In songbirds, for example, males use singing to announce their willingness to defend a territory. Other males of the species become reluctant to make use of the same area.

Red deer stags (males) on the Scottish island of Rhum compete to be the harem master of a group of hinds (females) that mate only with them. The reproductive group occupies a territory that the harem master defends against other stags. Harem masters first attempt to repel challengers by roaring. If the challenger remains, the two lock antlers and push against one another (Fig. 16.14). If the challenger then withdraws, the master pursues him for a short distance, roaring the whole time. If the challenger wins, he becomes the harem master.

A harem master can father two dozen offspring at most, because he is at the peak of his fighting ability for only a short time. And there is a cost to being a harem master. Stags must be large and powerful in order to fight; therefore, they grow faster and have less body fat. During bad times, they are more likely to die of starvation, and in general, they have shorter lives. Harem master behavior will persist in the population only if its cost (reduction in the potential number of offspring because of a shorter life) is lower than its benefit (increased number of offspring due to harem access).

Figure 16.13 A male olive baboon displaying full threat. In olive baboons, *Papio anubis*, males are larger than females and have enlarged canines. Competition between males establishes a dominance hierarchy for the distribution of resources.

Check Your Progress 16.2

1. Recognize the difference between a population undergoing stabilizing, directional, or disruptive selection.
2. Construct an example of a phenotype distribution curve for a population under directional selection that is increasing in body size.
3. Explain why sexual selection is a form of natural selection.

a.

b.

Figure 16.14 Competition between male red deer. Male red deer, *Cervus elaphus*, compete for a harem within a particular territory. **a.** Roaring alone may frighten off a challenger, but (**b**) outright fighting may be necessary, and the victor is most likely the stronger of the two animals.

THEME Nature of Science

Inbreeding in Populations

One of the requirements of a population in Hardy-Weinberg equilibrium is that mates are chosen at random—that is, without preference for a particular trait. Most populations, however, do not meet this requirement, because some traits are more attractive in a mate than others. Humans, for example, select mates based on a set of traits that we find appealing. This type of assortative mating based on trait preference is common in many species. Inbreeding is a unique form of nonrandom, or assortative, mating in which individuals mate with close relatives, such as cousins.

One consequence of inbreeding is an increase in the frequency of homozygous genotypes in a population. For the most part, an increase in homozygotes does not have a large detrimental effect, especially if the population is very large. But when populations are small, inbreeding can have a major impact on the health of a population. Many human diseases are caused by the inheritance of two recessive alleles, such that the disease appears only in persons who are homozygous recessive for the disease-causing allele. In very small populations, the probability that individuals carrying the recessive allele will mate increases, because the mating pool is very small. In this case, inbreeding significantly increases the frequency of homozygous recessive genotypes, and thus those afflicted with a disease.

Human cultures tend to have social rules that discourage inbreeding, but in very small populations, such as those following a bottleneck or founder event, inbreeding is sometimes unavoidable. One example of the effects of inbreeding on human populations is the occurrence of a rare form of non-sex-linked colorblindness, achromatopsia (Fig. 16A). People who are achromatic are completely color-blind. Normal human color vision is possible because of cells in the back of the eye called cones. Those with achromatopsia do not have any cone cells, because they are homozygous for a rare recessive allele that prevents cone cells from developing. Complete colorblindness is so common on Pingelap, a small island in the Pacific Ocean, that it is considered part of everyday life for most families.

Experts propose that the high frequency of achromatopsia appeared on Pingelap following a severe population bottleneck in 1775, when a typhoon killed 90% of the inhabitants. Approximately 20 people survived the typhoon. Four generations after the typhoon struck, achromatopsia began to appear frequently in the population. Figure 16B shows how inbreeding in a population can produce homozygous recessive genotypes. Geneticists explain that the high frequency of this rare genetic disorder on Pingelap is consistent with a large degree of intermarriage among relatives, or inbreeding, following the typhoon.

Mwanenised, a male survivor of the typhoon, was a carrier, or a heterozygote, for the achromatopsia allele. He had ten children, which was a large proportion of the first generation of Pingelapese after the typhoon. Geneticists would predict that on average 50% of his children would have been homozygous for

Figure 16A Achromatopsia. Complete achromatopsia is a recessive genetic disorder that causes complete colorblindness. People with complete achromatopsia see no color, only shades of gray. This image shows how a person with this form of colorblindness would see the world.

16.3 Maintenance of Diversity

Learning Outcomes

Upon completion of this section, you should be able to

1. List two examples of how diversity is maintained in populations.
2. Describe why heterozygote advantage is a form of stabilizing selection.

Diversity can be maintained in a population thru any number of ways. Mutations create new alleles, and sexual reproduction recombines alleles due to meiosis and fertilization. Genetic drift also occurs, particularly in small populations, and the result may be contrary to adaptation to the environment.

Natural Selection

The process of natural selection itself causes imperfect adaptation to the environment. First, it is important to realize that evolution doesn't start from scratch. Just as you can only bake a cake with the ingredients available to you, evolution is constrained by the available diversity. Lightweight titanium bones might benefit birds, but their bones contain calcium and other minerals, the same as other reptiles'. When you mix the

the normal allele, and 50% would have been heterozygous carriers of the color-blind allele. Thus, none of Mwanenised's children would have been color-blind (Fig. 16B). However, intermarriage of successive generations would unavoidably have brought together men and women who were carriers for the achromatopsia allele. Thus, in subsequent generations the homozygous recessive genotype, and complete colorblindness, did appear due to inbreeding within a small population (Fig. 16B). On Pingelap, an increase in colorblindness was observed by the fourth generation.[1]

The effect of inbreeding can be long-term, especially if a population remains relatively small and isolated. Today, 1 in 12 Pingelapese suffers from achromatopsia. On Pingelap it is 3,000 times more frequent than in the United States, where it occurs in approximately 1/40,000 births.

[1]See Sacks, Oliver. 1998. *The Island of the Colorblind.* (Vintage/Anchor Books, New York.) An interesting, nontechnical account of the Pingelapese.

Questions to Consider

1. What might have happened to the color-blind allele in the Pingelapese population following the typhoon if Mwanenised had had no children? Only five children?
2. Would you predict that the Pingelapese population is in Hardy-Weinberg equilibrium? How would you measure this?
3. Has the Pingelapese population evolved? Explain. (Hint: Does inbreeding cause a change in allele frequencies?)

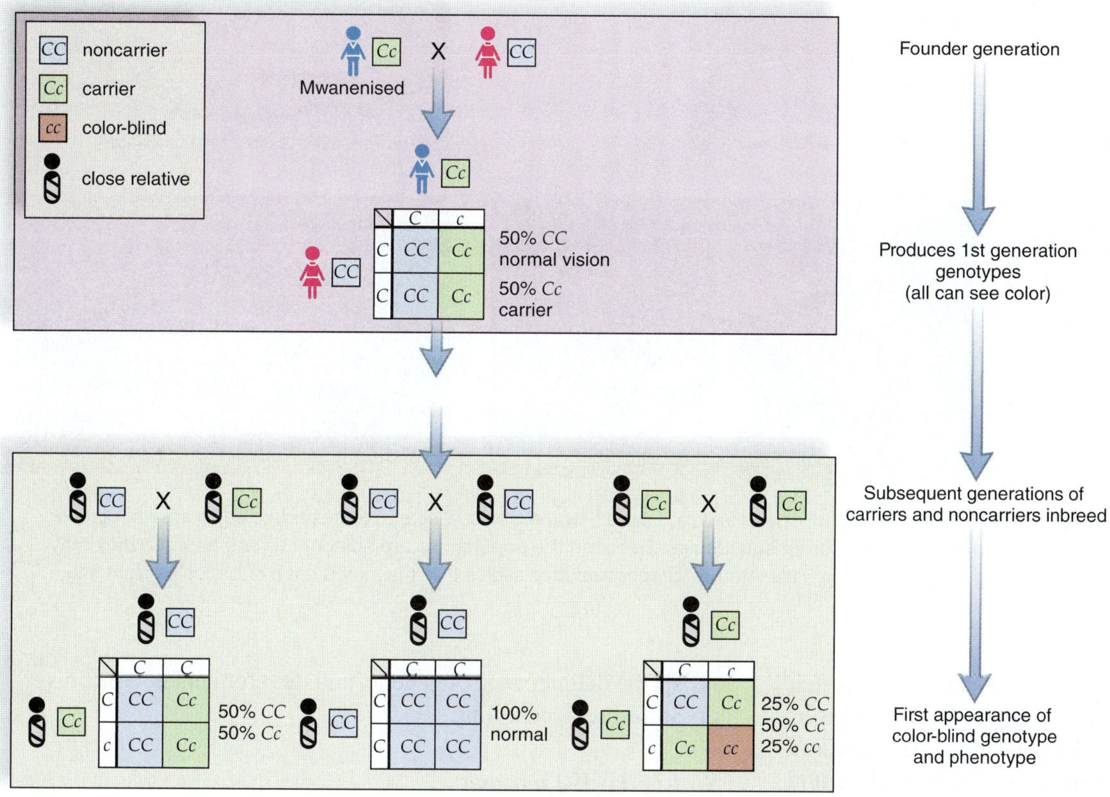

Figure 16B Results of inbreeding in a small population. On the Island of Pingelap, a high level of incidental inbreeding in a small population has resulted in a higher than expected occurrence of complete achromatopsia, a recessive genetic disorder.

ingredients for a cake, you probably follow the same steps taught to you by your elders. Similarly, the processes of development restrict the emergence of novel features. This is why the wing of a bird has the same bones as those of other vertebrate forelimbs.

Imperfections are common because of necessary compromises. The success of humans is attributable to their dexterous hands, but the spine is subject to injury, because the vertebrate spine did not originally support the body in an erect position in our ancestors. A feature that evolves has a benefit that outweighs the cost. For example, the benefit of freeing the hands must have outweighed the cost of spinal injuries from assuming an erect posture. We should also consider that sexual selection has a reproductive benefit, but not necessarily an adaptive benefit.

Second, the environment plays a role in maintaining diversity. It's easy to see that disruptive selection in an environment that differs widely can promote polymorphisms within the population (see Fig. 16.11). Then, too, if a population occupies a wide range, as shown in Figure 16.15, it may have several subpopulations designated as subspecies because of recognizable differences. (Subspecies are given a third name in addition to the usual binomial name.) Each subspecies is partially adapted to its own environment and can serve as a reservoir for a different combination of alleles that flow from one group to the next when adjacent subspecies interbreed.

Figure 16.15 Subspecies help maintain diversity. Each subspecies of rat snake (*Pantherophis obsoleta*) represents a separate population of snakes. Each subspecies has a reservoir of alleles different from another subspecies. Because the populations are adjacent to one another, they may interbreed, and therefore, gene flow may occur among the populations. This interbreeding introduces alleles that may keep each subspecies from fully adapting to its environment.

The environment also includes specific selecting agents that help maintain diversity. We have already seen how insectivorous birds can help maintain the frequencies of both light-colored and dark-colored moths, depending on the color of background vegetation. Some predators have a search image that causes them to select the most common phenotype among their prey. This promotes the survival of the rare forms and helps maintain variation. Or an herbivore can oscillate in its preference for food. In Figure 15.10, we observed that the average beak size of the medium ground finch on the Galápagos Islands depended on the available food supply. In times of drought, when only large seeds were available, birds with larger beaks were favored. In this case, we can clearly see that maintenance of variation among a population's members has survival value for the species.

Video Finches Natural Selection

Heterozygote Advantage

Heterozygote advantage occurs when the heterozygote is favored over the two homozygotes. In this way, heterozygote advantage assists the maintenance of genetic, and therefore phenotypic, diversity in future generations.

Sickle-Cell Disease

Sickle-cell disease can be a devastating condition. Patients can have severe anemia, physical weakness, poor circulation, impaired mental function, pain and high fever, rheumatism, paralysis, spleen damage, low resistance to disease, and kidney and heart failure. In these individuals, the red blood cells are sickle-shaped and tend to pile up and block flow through tiny capillaries. The condition is due to an abnormal form of hemoglobin (*Hb*), the molecule that carries oxygen in red blood cells. People with sickle-cell disease (Hb^SHb^S) tend to die early and leave few offspring, due to hemorrhaging and organ destruction.

Interestingly, however, geneticists studying the distribution of sickle-cell disease in Africa have found that the recessive allele (Hb^S) has a higher frequency in regions where the disease malaria is also prevalent (blue region in Fig. 16.16). Malaria is caused by a protozoan parasite that lives in and destroys the red blood cells of

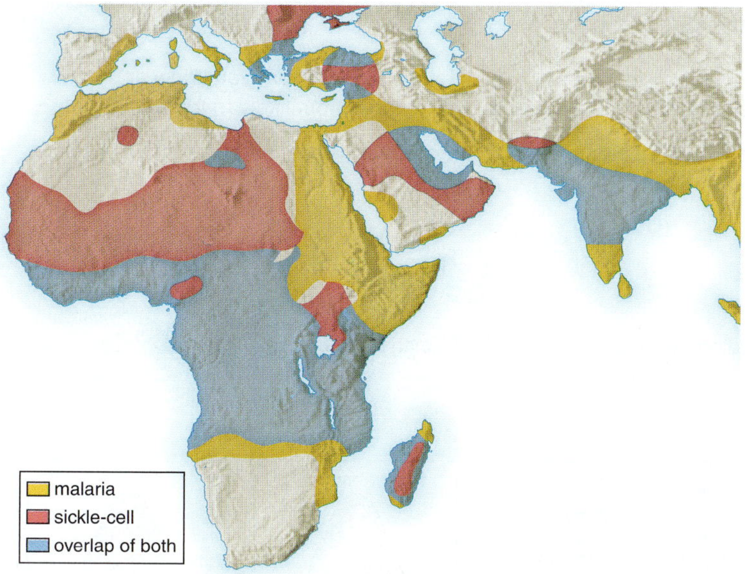

Genotype	Phenotype	Result
Hb^AHb^A	Normal	Dies due to malarial infection
Hb^AHb^S	Sickle-cell trait	Lives due to protection from both
Hb^SHb^S	Sickle-cell disease	Dies due to sickle-cell disease

Figure 16.16 Sickle-cell disease. Sickle-cell disease is more prevalent in areas of Africa where malaria is more common.

the normal homozygote (Hb^AHb^A). Individuals with this genotype also have fewer offspring, due to an early death or to debilitation caused by malaria.

Heterozygous individuals (Hb^AHb^S) have an advantage, because they don't die from sickle-cell disease, and they don't die from malaria. The parasite causes any red blood cell it infects to become sickle-shaped. Sickle-shaped red blood cells lose potassium, and this causes the parasite to die. Heterozygote advantage causes all three alleles to be maintained in the population. As long as the protozoan that causes malaria is present in the environment, it is advantageous to maintain the recessive allele.

Heterozygote advantage is also an example of stabilizing selection, because the genotype Hb^AHb^S is favored over the two extreme genotypes, Hb^AHb^A and Hb^SHb^S. In the parts of Africa where malaria is common, 1 in 5 individuals is heterozygous (has sickle-cell trait) and survives malaria, while only 1 in 100 is homozygous and dies of sickle-cell disease. In the United States, where malaria is not prevalent, the frequency of the Hb^S allele is declining among African Americans, because the heterozygote has no particular advantage in this country.

Cystic Fibrosis

Stabilizing selection is also thought to have influenced the frequency of other alleles. Cystic fibrosis is a debilitating condition that leads to lung infections and digestive difficulties. In this instance, the recessive allele, common among individuals of northwestern European descent, causes the person to have a defective plasma membrane protein. The agent that causes typhoid fever can use the normal version of this protein, but not the defective one, to enter cells. Here again, heterozygote superiority caused the recessive allele to be maintained in the population.

Check Your Progress 16.3

1. Identify the ways in which diversity is maintained in a population.
2. Demonstrate how sickle-cell disease is an example of stabilizing selection. Do the same for cystic fibrosis.

CONNECTING *the* CONCEPTS *with the* THEMES

Evolution

- Evolution occurs when allele frequencies in a population change over time. The Hardy-Weinberg principle measures whether a population is evolving.
- Small populations are especially vulnerable to genetic drift—chance events that, in a small population, can change allele frequencies.
- Some human diseases, such as sickle-cell disease, are maintained in populations in certain environments due to stabilizing selection, because they provide an advantage against other diseases.
- Sexual selection is a type of natural selection that plays a large role in the evolution of "sexy" traits in populations.

Nature of Science

- Population geneticists can measure the frequencies of genotypes in a population and compare them to Hardy-Weinberg genotype frequencies to determine if, and how, a population is evolving.
- Human diseases, such as HIV/AIDS, sickle-cell disease, and cystic fibrosis, are now better understood at the genetic level due to the study of microevolution.

Biological Systems

- Natural selection acts on trait variation, and trait variation is determined by genes. Whether or not a trait gives an advantage depends on the environment. Thus, genes, traits, the environment, and natural selection are involved in microevolution.
- Population dynamics, including social interactions among individuals, must be considered to understand the evolutionary forces at work in a population.

MEDIA STUDY TOOLS http://connect.mheducation.com

SMARTBOOK® Study smart. SmartBook helps you maximize your study time by identifying what you know and don't know. The recharge feature in SmartBook helps you prepare for the test by identifying the material that you are most likely to forget.

connect BIOLOGY Find even more resources to learn the chapter concepts including animations, tutorial videos, and interactive practice questions.

Animations	Tutorial	Video
16.1 Mutation by Base Substitution • Simulation of Genetic Drift	**16.1** Hardy-Weinberg Equilibrium	**16.3** Finches Natural Selection

SUMMARIZE

16.1 Genes, Populations, and Evolution

The phenotype of an individual can change during his or her lifetime, but this is not evolution. Evolution is about the change in the **allele frequency** of a **population** over many generations, not within the individual. **Microevolution** pertains to evolutionary change within populations.

Population genetics studies microevolution by measuring the allele frequency over generations. A population in which allele frequencies do not change over time is said to be in **Hardy-Weinberg equilibrium**—a stable, nonevolving state. The Hardy-Weinberg equilibrium is a constancy of **gene-pool** allele frequencies that remains stable from generation to generation if certain conditions are met. The **Hardy-Weinberg principle,** $p^2 + 2pq + q^2$, can measure the genotype frequencies of a nonevolving population. The conditions are no mutations, no gene flow, random mating, no genetic drift, and no selection. Because these conditions are rarely met, a change in allele frequencies is likely.

When gene-pool frequencies change, microevolution occurs. Deviations from Hardy-Weinberg equilibrium allow us to detect microevolutionary shifts. Mutation occurs when the DNA sequence has changed. **Gene flow** is the movement of alleles between populations. If populations are **reproductively isolated,** there is a greater chance that gene flow will be restricted. **Genetic drift** is when chance events cause allele frequencies to change. Both the **bottleneck effect** and **founder effect** result from the loss of genetic variation within the population. **Inbreeding** can occur as a consequence of genetic drift. **Nonrandom mating,** also called **assortative mating,** occurs when individuals are selective about their mates. The manner in which allele frequencies deviate from Hardy-Weinberg equilibrium indicates which of these processes is causing the population to evolve (see Table 16.1).

16.2 Natural Selection

Most of the traits of evolutionary significance are **polygenic,** controlled by many genes. Natural selection favors the most adaptive variant for a given environment. Three types of natural selection occur: (1) **stabilizing selection**—the intermediate variation is the most adaptive, as is found in human birth weight; (2) **directional selection**—either of the extreme phenotypes is favored, as when body size increases over time; and (3) **disruptive selection**—two or more extreme phenotypes are adaptive; the curve forms two peaks, as when British land snails have one of two different banding patterns of shell color.

Sexual selection is about reproductive success, or **fitness.** Males produce many sperm and compete to inseminate females. Females produce few eggs and are selective about their mates. Traits that promote reproductive success, such as **sexual dimorphism,** are shaped by sexual selection. A **cost-benefit analysis** helps a male determine if it is worth competing for mates. **Dominance hierarchies** provide dominant

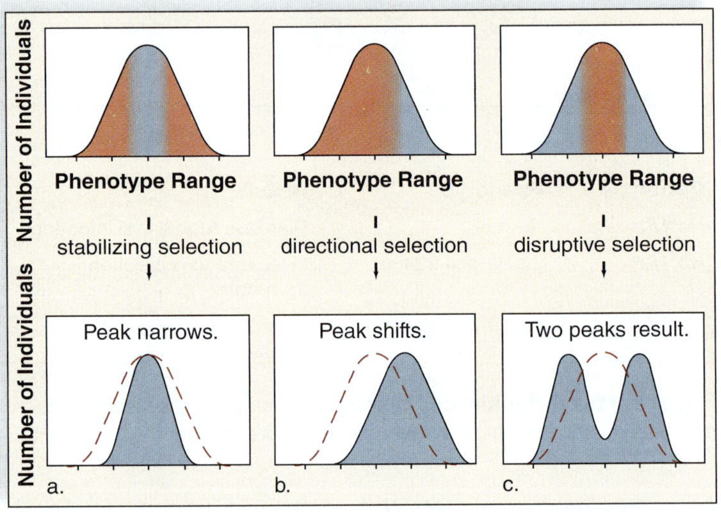

males greater reproductive opportunities than lower-ranking males. A **territory** is defended with specific behaviors known as **territoriality.**

Biological differences between the sexes may promote certain mating behaviors because they increase fitness.

16.3 Maintenance of Diversity

Despite constant natural selection, genetic diversity is maintained. Mutations and recombination still occur; gene flow among small populations can introduce new alleles; and natural selection itself sometimes results in variation. In sexually reproducing diploid organisms, the heterozygote acts as a repository for recessive alleles whose frequency is low. In regard to sickle-cell disease, the heterozygote is more fit in areas where malaria occurs; this is known as the **heterozygote advantage.**

ASSESS

Choose the best answer for each question.

16.1 Genes, Populations, and Evolution

1. Assuming a Hardy-Weinberg equilibrium, 21% of a population is homozygous dominant, 50% is heterozygous, and 29% is homozygous recessive. What percentage of the next generation is predicted to be homozygous recessive?
   a. 21%
   b. 50%
   c. 29%
   d. 42%
   e. 58%

2. A human population has a higher than usual percentage of individuals with a genetic disorder. The most likely explanation is
   a. mutations and gene flow.
   b. mutations and natural selection.
   c. nonrandom mating and founder effect.
   d. nonrandom mating and gene flow.
   e. All of these are correct.
3. The offspring of better-adapted individuals are expected to make up a larger proportion of the next generation. The most likely explanation is
   a. mutations and nonrandom mating.
   b. gene flow and genetic drift.
   c. mutations and natural selection.
   d. mutations and genetic drift.
4. When a population is small, there is a greater chance of
   a. gene flow.
   b. genetic drift.
   c. natural selection.
   d. mutations.
   e. sexual selection.
5. Which of the following cannot occur if a population is to maintain an equilibrium of allele frequencies?
   a. People leave one country and relocate in another.
   b. A disease wipes out the majority of a herd of deer.
   c. Members of an Indian tribe allow only the two tallest people in the tribe to marry each spring.
   d. Large black rats are the preferred males in a population of rats.
   e. All of these are correct.
6. Which of the following applies to the Hardy-Weinberg expression: $p^2 + 2pq + q^2$?
   a. Knowing either p^2 or q^2, you can calculate all the other frequencies.
   b. It applies to Mendelian traits that are controlled by one pair of alleles.
   c. $2pq$ = heterozygous individuals
   d. It can be used to determine the genotype and allele frequencies of the previous and the next generations.
   e. All of these are correct.
7. Following genetic drift,
   a. genotype and allele frequencies would not change.
   b. genotype and allele frequencies would change.
   c. adaptation would occur.
   d. the population would have more phenotypic variation but less genotypic variation.

16.2 Natural Selection

8. Which of the following is an example of stabilizing selection?
   a. Over time, *Equus* developed strength, intelligence, speed, and durable grinding teeth.
   b. British land snails mainly have two different phenotypes.
   c. Swiss starlings usually lay four or five eggs, thereby increasing their chances of more offspring.
   d. Drug resistance increases with each generation; the resistant bacteria survive, and the nonresistant bacteria get killed off.
   e. All of these are correct.
9. One way for disruptive selection to occur is if
   a. the population contains diversity.
   b. the environment contains diversity.
   c. pollution is present.
   d. natural selection occurs.
   e. All of these are correct.
10. In some bird species, the female chooses a mate that is most similar to her in size. This supports
    a. the good genes hypothesis.
    b. the runaway hypothesis.
    c. the sexual dimorphism hypothesis.
    d. All of these hypotheses could be true.
    e. None of these hypotheses are true.
11. A red deer harem master typically dies earlier than other males because
    a. he will likely be expelled from the herd and cannot survive alone.
    b. he will be more prone to disease because he interacts with so many animals.
    c. he needs more food than other males.
    d. he is apt to place himself between a predator and the herd to protect the herd.
    e. he inherited the genetics for a shorter lifespan.

16.3 Maintenance of Diversity

12. The continued occurrence of sickle-cell disease with malaria in parts of Africa is due to
    a. continual mutation.
    b. gene flow between populations.
    c. relative fitness of the heterozygote.
    d. disruptive selection.
    e. protozoan resistance to DDT.
13. Which genotype is characteristic of the heterozygote advantage?
    a. *HH* b. *Hh* c. *hh* d. *ta* e. *HT*

ENGAGE

LEARNSMART LABS®

The following LearnSmart Labs contain exercises that are related to the content of this chapter:

- Natural Selection

THINKING SCIENTIFICALLY

1. A farmer uses a new pesticide. He applies the pesticide as directed by the manufacturer and loses about 15% of his crop to insects. A farmer in the next state learns of these results, uses three times as much pesticide, and loses only 3% of her crop to insects. Each farmer follows this pattern for 5 years. At the end of 5 years, the first farmer is still losing about 15% of his crop to insects, but the second farmer is losing 40% of her crop to insects. How can these observations be interpreted on the basis of natural selection? Should pesticide application be regulated by the government to prevent insects from evolving resistance?
2. You are observing a grouse population in which two feather phenotypes are present in males. One is relatively dark and blends into shadows well, and the other is relatively bright and more obvious to predators. The females are uniformly dark-feathered. Observing the frequency of mating between females and the two types of males, you have recorded the following:

   matings with dark-feathered males: 13

   matings with bright-feathered males: 32

   Propose a hypothesis to explain why females apparently prefer bright-feathered males. What selective advantage might there be in choosing a male with alleles that make it more susceptible to predation? What data would help test your hypothesis?
3. If $p^2 = 0.36$, what percentage of the population has the recessive phenotype, assuming a Hardy-Weinberg equilibrium?
4. If 1% of a human population has the recessive phenotype, what percentage has the dominant phenotype, assuming a Hardy-Weinberg equilibrium?

17

Speciation and Macroevolution

Chapter Outline

17.1 How New Species Evolve 297
17.2 Modes of Speciation 303
17.3 Principles of Macroevolution 308

Before You Begin

Before beginning this chapter, take a few moments to review the following discussions.

Sections 13.1 and 13.2 What is gene regulation, and how are genes regulated?

Section 15.3 What is the genetic, fossil, and morphological evidence for evolution?

Section 16.1 What processes drive evolution within populations (microevolution)?

The micro-frog, *Microhyla nepenthicola,* uses the pitcher plant, *Nepenthes ampullaria,* as a nest and nursery for its young.

In 2007, a new species of micro-frog, *Microhyla nepenthicola,* was discovered in the forests of Borneo. Only a fraction of the size of a penny, this 12-mm frog mates and lays eggs in a pitcher plant, *Nepenthes ampullaria,* that lives on the forest floor. The pitcher plant collects water, in which the frog tadpoles develop into adult frogs. For a long time, *M. nepenthicola* was considered to be a juvenile of a different species, but a closer look revealed a suite of unique features that allow it to scale the waxy, smooth walls of the pitcher plant. The origin of *M. nepenthicola* has been shaped by a tightly knit relationship with *N. ampullaria*, its nest and nursery.

In this chapter we take microevolution one step further and look at how a population, over time, accumulates differences large enough to become a new species. The origin of species is the key to understanding the origin of the diversity of all life on Earth.

As you read through this chapter, think about the following questions:

1. How do scientists determine whether an organism is a new species?
2. What processes drive the evolution of new species? Are they different from those that drive the evolution of populations?
3. What can the fossil record tell us about the origin and extinction of species over time?

Following *the* Themes

CHAPTER 17 SPECIATION AND MACROEVOLUTION

UNIT 3 EVOLUTION

Evolution	Macroevolution, or the origin of new species, results from the accumulation of microevolutionary change over time.
Nature of Science	Evolutionary geneticists research and document the genetic basis of phenotypic changes that can lead to speciation.
Biological Systems	New species can arise from specialization within a particular habitat or microhabitat.

17.1 How New Species Evolve

Learning Outcomes

Upon completion of this section, you should be able to

1. Compare and contrast microevolution and macroevolution.
2. Identify and compare features of prezygotic and postzygotic reproductive isolation.
3. Explain three ways that species are defined.

In Chapter 16, we examined microevolution, or small-scale evolution within populations. Microevolution is measured as change in allele frequencies in a population over generations. In this chapter, we turn our attention to **macroevolution,** evolution on a large scale. The history of life on Earth is a part of macroevolution. Macroevolution involves **speciation,** or the splitting of one species into two or more species. The same microevolutionary mechanisms that are at play within populations—genetic drift, natural selection, mutation, and migration—are also at play during macroevolution. Thus, microevolution and macroevolution are the result of the same processes, differing only in the scale at which they occur. Macroevolution is the result of the accumulation of microevolutionary change that results in the formation of new species (see the Evolution feature, "The Anatomy of Speciation," page 300).

Species originate, adapt to their environment, and then may become extinct. In fact, much of the biodiversity that has existed on Earth is now extinct. For example, mammals experienced many periods of speciation in the past that resulted in high levels of diversity; but the majority of those species are now extinct. Without the continuous origin and extinction of species, life on Earth would not have the ever-changing history that is found in the fossil record.

Darwin devoted his life to understanding the "mystery of mysteries"—how new species originate. Scientists have learned a lot about this mystery since Darwin's time, including some of the processes that cause species to form. In this chapter we take a closer look at what constitutes a species and how new species evolve.

What Is a Species?

When you take a walk in a forest, you see a lot of different "types" of plants and animals. If you've had a biology class, you would probably call these different types "species." Although you may not be able to identify each species, you would intuitively recognize many of these organisms as different because of their appearance. Many differences are obvious; for example, clearly an oak tree is a different species from a squirrel. If you look a little closer, you might recognize two different kinds of oak trees, one with large acorns and one with small acorns and with leaves of a slightly different shape. Whether these two oaks are considered different species or variations of the same species depends on which species definition is used.

Scientists define species based on many types of evidence. When presented with two organisms, a **taxonomist,** a scientist who classifies organisms into groups, makes a working hypothesis about whether they are different species based on the evidence, such as their external features. In this manner, each species that is defined is a hypothesis about how the Earth's tree of life is organized.

As with any hypothesis, the addition of new information can result in the redefinition of species. For example, Linnaeus (the father of taxonomy) once hypothesized that birds and bats should be together in the same group, because both have wings and fly. We now know that birds and bats are very different organisms on separate branches of the tree of life.

How species are defined is an exciting area of study, because all of the diversity of life on Earth has originated from the evolution of new species. Up until now, we have defined a species as a type of living organism, but in this chapter we characterize species in more depth. First, we examine the major species concepts, or the different ways in which a species can be defined, and then we look at some of the mechanisms by which new species originate.

Morphological Species Concept

Linnaeus identified new species by differences in their appearance, or **morphology.** In the **morphological species concept,** species are distinguished from each other by one or more distinct physical characteristics called **diagnostic traits.** It turns out that Linnaeus was very adept at recognizing species, and many of the morphological species he defined have held up to 200 years of scrutiny.

But the morphological species concept has some disadvantages that Linnaeus could not have predicted. Bacteria and other microorganisms do not have many measurable traits. Also, similarities and differences between organisms can be very subtle and sometimes misleading. Some organisms look so similar that they appear to be the same species. **Cryptic species** are species that look almost identical but are very different in other traits, such as habitat use or courtship behaviors. For example, species of leopard frogs in North America are very difficult to distinguish in appearance, but the males of each species have a unique courtship call (Fig. 17.1).

The morphological species concept is useful for paleontologists as a way to define fossil species based only on traits that are preserved in the fossil record. Unfortunately, since fossils do not provide information about color, the anatomy of soft tissues, or behavioral traits, they are of limited value. However, subtle differences in skeletal features can be used to diagnose the differences between species.

Evolutionary Species Concept

The **evolutionary species concept** relies on identification of certain morphological traits to distinguish one species from another. It was proposed to explain speciation in the fossil record. In addition, the evolutionary species concept, as its name implies, requires that the members of a species share a distinct evolutionary pathway. That is, small, transitional changes in a trait are not used to define new species, because these transitional forms are part of the same evolutionary pathway. However, abrupt changes in traits indicate the evolution of a new species in the fossil record. Consider that the species depicted in Figure 17.2 are part of the evolutionary history of *Orcinus orca,* a toothed whale. These species can be recognized individually by differences in diagnostic traits (hindlimbs), but collectively they share an evolutionary pathway distinct from those of other whale species.

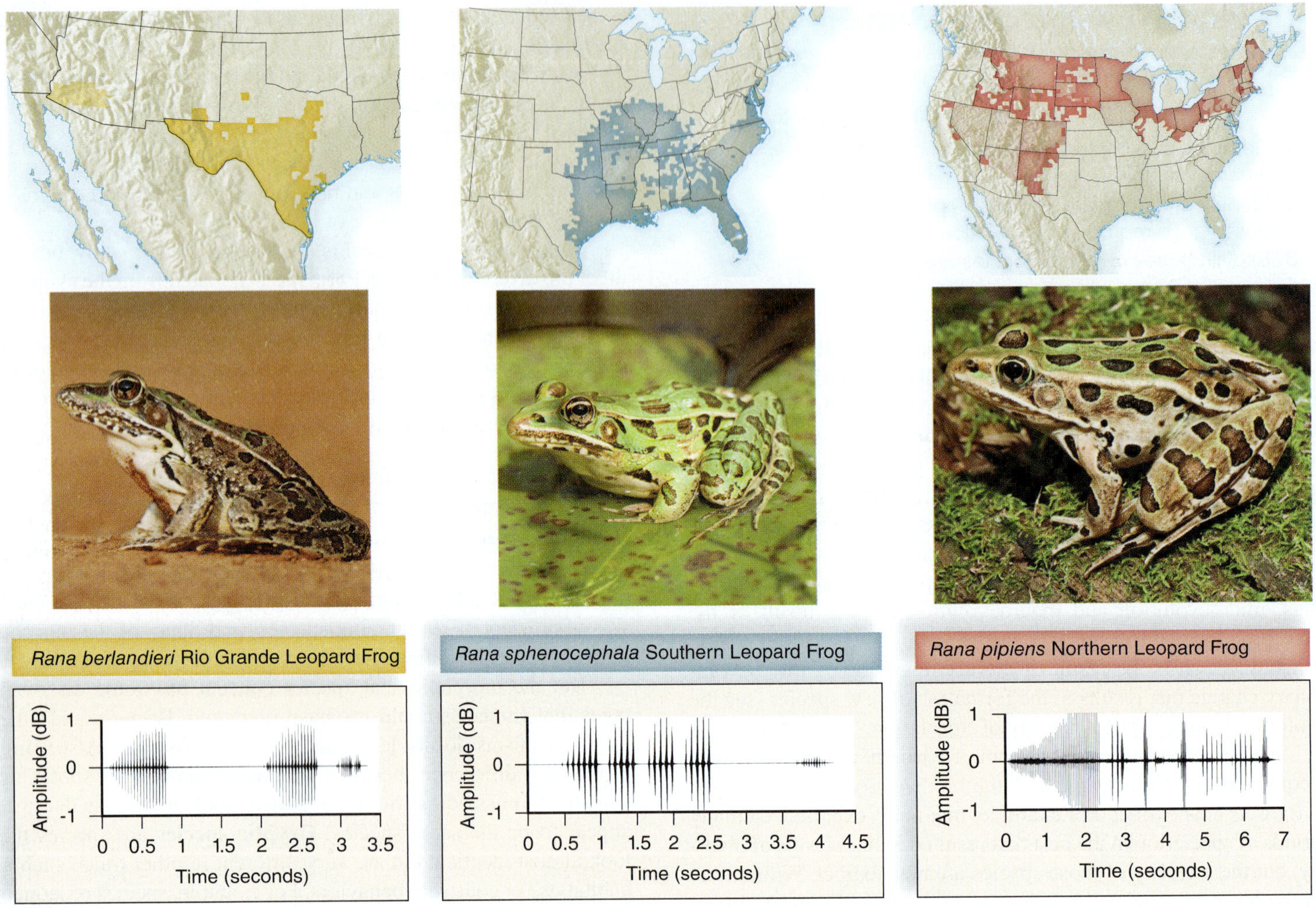

Figure 17.1 Cryptic species of leopard frogs. Leopard frogs are common throughout North America and were once considered to be a single species. Further investigation has revealed that there are at least three species, the Rio Grande, southern, and northern leopard frogs. Although they look similar, they are reproductively isolated, because each has a unique mating call.

Phylogenetic Species Concept

In the **phylogenetic species concept,** an evolutionary "family tree"—or phylogeny—is used to identify species based on a common ancestor—that is, a single ancestor for two or more different groups (see the Evolution feature, "The Anatomy of Speciation," on page 300). For you and your cousins, your shared grandmother is a common ancestor. Similarly, groups of organisms have a common ancestor.

According to the phylogenetic species concept, a species is the smallest set of interbreeding organisms—usually a population—that shares a common ancestor. In a phylogeny, a branch that contains all the descendants of a common ancestor is said to be **monophyletic.** Monophyly is the main criterion for defining species in the phylogenetic species concept.

One advantage of the phylogenetic species concept is that it does not rely only on morphological traits to define a species. The nucleotide sequence of a region of an organism's DNA can be compared to identify the individual A, C, G, or T nucleotide differences that are characteristic of a species. Thus, species of microorganisms and cryptic species can be identified with the phylogenetic species concept, because traits other than morphology can be diagnostic. One example is the giraffe, which has several regional populations distributed around Africa that are distinguishable only by a unique spot shape (Fig. 17.3). Historically, each population was considered as members of a single species, *Giraffa camelopardalis*. A recent phylogeny based on DNA data hypothesizes that each regional population represents a monophyletic

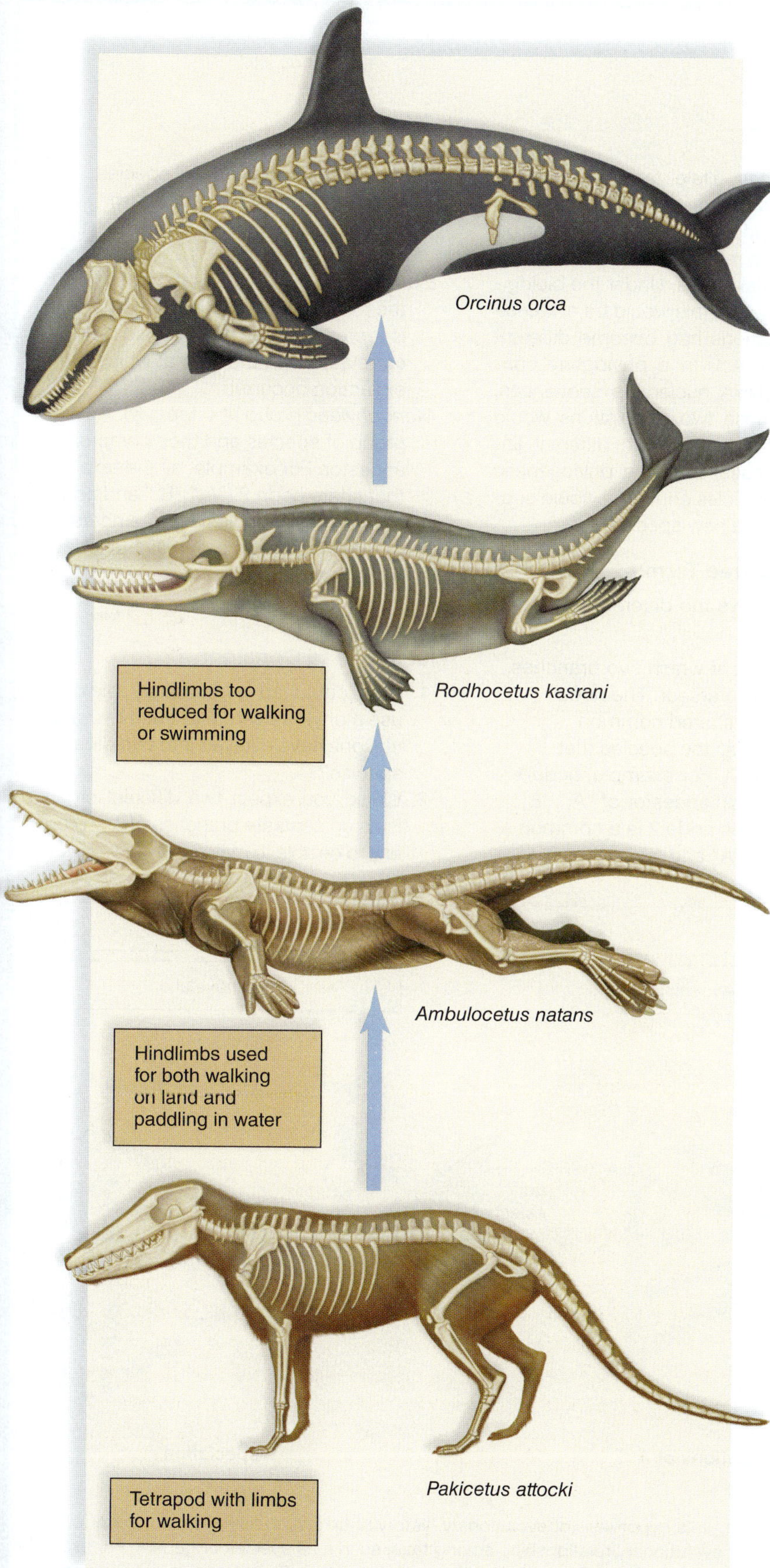

Figure 17.2 Evolutionary species concept. Diagnostic traits can be used to distinguish these species known only from the fossil record. Such traits no doubt would include the anatomy of the limbs.

group, and thus each population should be recognized as an individual species (Fig. 17.3).

Biological Species Concept

The **biological species concept** relies primarily on reproductive isolation to identify different species. The most important criterion, according to the biological species concept, is **reproductive isolation**—physiological, behavioral, and genetic processes that inhibit interbreeding. Specifically, if organisms cannot mate and produce offspring in nature, or if their offspring are sterile, they are defined as different species.

Although useful, the biological species concept often cannot be tested in nature, because many potential species do not overlap in their distribution and thus do not have an opportunity to determine whether they are reproductively isolated. Furthermore, the biological species concept cannot be applied to asexually reproducing living organisms or fossils. The benefit of the concept is that, when applicable, it confirms the lack of gene flow—the best indicator that two populations are following independent evolutionary pathways. For example, a group of birds collectively called the flycatchers all look very similar, but they do not reproduce with one another; therefore, they are separate species. Like the leopard frogs (see Fig. 17.1), not only do they live in different habitats but each group has a unique courtship song.

Reproductive Isolating Mechanisms

For two species to remain separate, populations must be reproductively isolated—that is, gene flow must not occur between them. Reproductive barriers that prevent successful reproduction are called isolating mechanisms (Fig. 17.4). Reproductive isolation can occur either before or after fertilization. Reproductive isolation before fertilization is called prezygotic isolation; after fertilization, postzygotic isolation. A **zygote** is the first cell that results when a sperm fertilizes an egg.

Prezygotic Isolating Mechanisms

Prezygotic isolating mechanisms prevent reproductive attempts or make it unlikely that fertilization will be successful if mating occurs. These isolating mechanisms make it highly unlikely that **hybridization,** or the mating between two species, will occur. Various types of isolating mechanisms can occur between species.

Habitat isolation When two species occupy different habitats, even within the same geographic range, they are less likely to meet and attempt to reproduce. This is one of the reasons that flycatchers do not mate. In tropical rain forests, many animal species are restricted to a particular level of the forest canopy, and in this way they are isolated from similar species.

Temporal isolation Several related species can live in the same locale, but if each reproduces at a different time of year, they do not attempt to mate. Five species of frogs of the genus *Rana* are all found at Ithaca, New York

THEME Evolution

The Anatomy of Speciation

The primary goal of evolutionary biology is to infer the processes of evolution that produce the patterns of diversity seen in nature. More simply put, the evolutionary biologist is interested in explaining biodiversity. At the heart of this endeavor is the phylogeny, the most important tool of the evolutionary biologist, because it represents the history of evolution among organisms. Evolution is considered one of the unifying theories of biology; thus, understanding how to interpret a phylogeny is fundamental to understanding how evolution works. This guide to the anatomy of a phylogeny should help you with the interpretation of a phylogeny.

Macroevolution is about the origin of new species. Darwin's theory proposes that all life on Earth shares a common ancestor. This means that a species originates from evolutionary changes to preexisting species. In this manner, all life on Earth can be envisioned as a large tree of life, with many branches, or species, radiating from it (see the Nature of Science feature in Chapter 15).

Microevolution is about populations, whereas macroevolution is about the origin of new species (Fig. 17A*b*). However, both are governed by the same processes. Different populations of a single species can accumulate genetic differences as microevolution shapes allele frequencies over time. As more and more genetic differences accumulate, a population may no longer be able to recognize members of another population as potential mates. Under the biological species concept, this would be evidence that the populations had become different species (Fig. 17A*c*). In a phylogeny constructed from DNA nucleotide sequences, for example, these two populations would likely be represented by two different lineages. In this case, both the phylogenetic and biological species concepts would support the origin of a new species.

Phylogenetic Tree Terms

Figure 17A*a* shows the different parts of a phylogeny.

Node: The point at which two branches, or lineages, intersect. The node represents a shared common ancestor for all the species that branch from it. For example, node 1 is the common ancestor of "A," "B," and "C," while node 2 is a common ancestor of "A" and "B."

Root: The point to which all species in the phylogeny can trace their ancestry; the origin of their shared common ancestry

Extinction: A taxon that is represented in the fossil record, but is now extinct, is represented by a shortened branch correlated with the time at which the extinction occurred.

Monophyletic group (monophyly): A group of species and their common ancestor. For example, all the taxa that share node 2 ("A," "B," and one extinct fossil species) are members of a monophyletic group (lightshaded rectangle), also called a *clade*.

Animation Phylogenetic Trees

Questions to Consider

1. How could an evolutionary biologist use a phylogeny to find out if two populations have evolved into two different species?
2. Would you expect two different organisms on separate branches of a phylogeny to be able to mate? Why or why not?

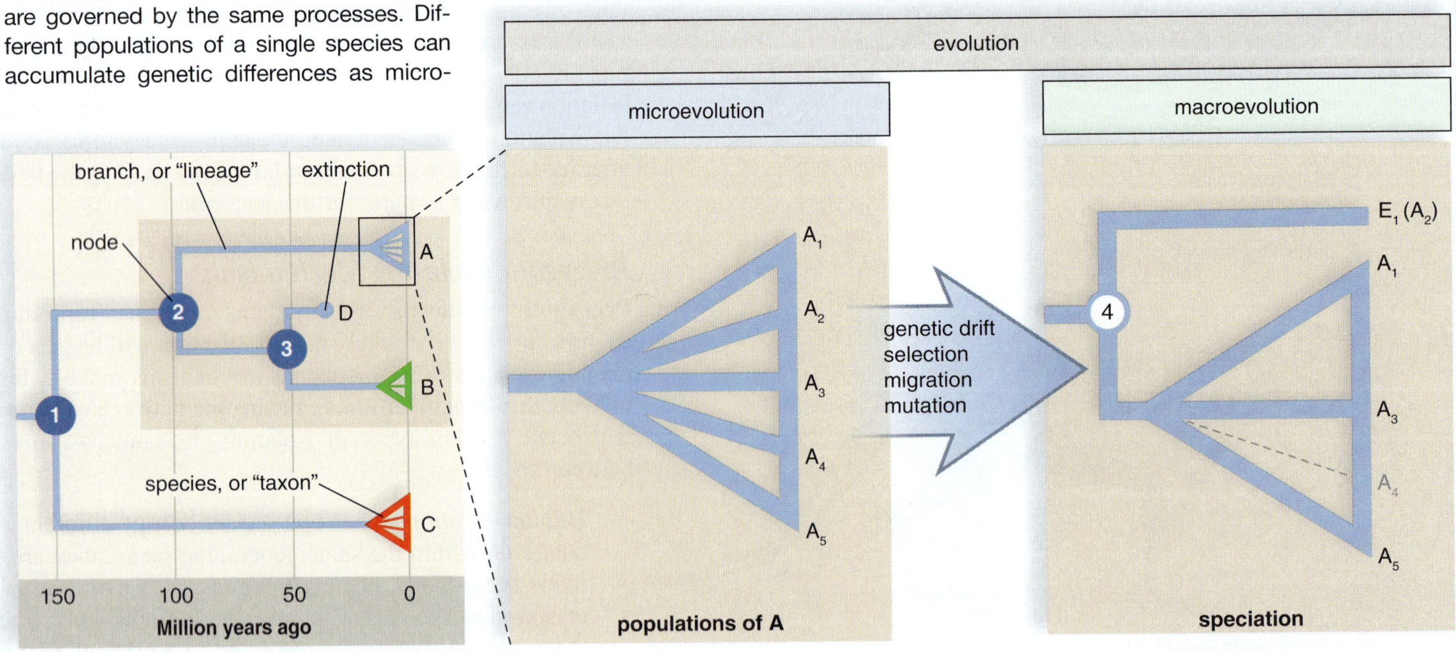

a. b. c.

Figure 17A Anatomy of speciation. A phylogeny, or "family tree," is a hypothesis of evolutionary history of taxa such as species or genera. **a.** A phylogeny has many parts, each of which tells us something about evolutionary relationships among taxa, such as a species or genera. For example, species "A" is more closely related to "B" than to "C," because "A" and "B" share a more recent common ancestor (node 2). **b.** Species "A" is comprised of many populations (A_1–A_5) that are all part of the same branch of the phylogeny. Microevolution occurs at the level of the population. **c.** Microevolution and macroevolution are governed by the same processes, that is genetic drift, selection, migration, and mutation, but at different scales. Speciation is the result of the accumulation of microevolutionary change in a population over time. Eventually, microevolution can result in the divergence of a population (such as A_2) to form a new species (E_1). Another outcome of evolution is the extinction of populations (A_4) and species (D).

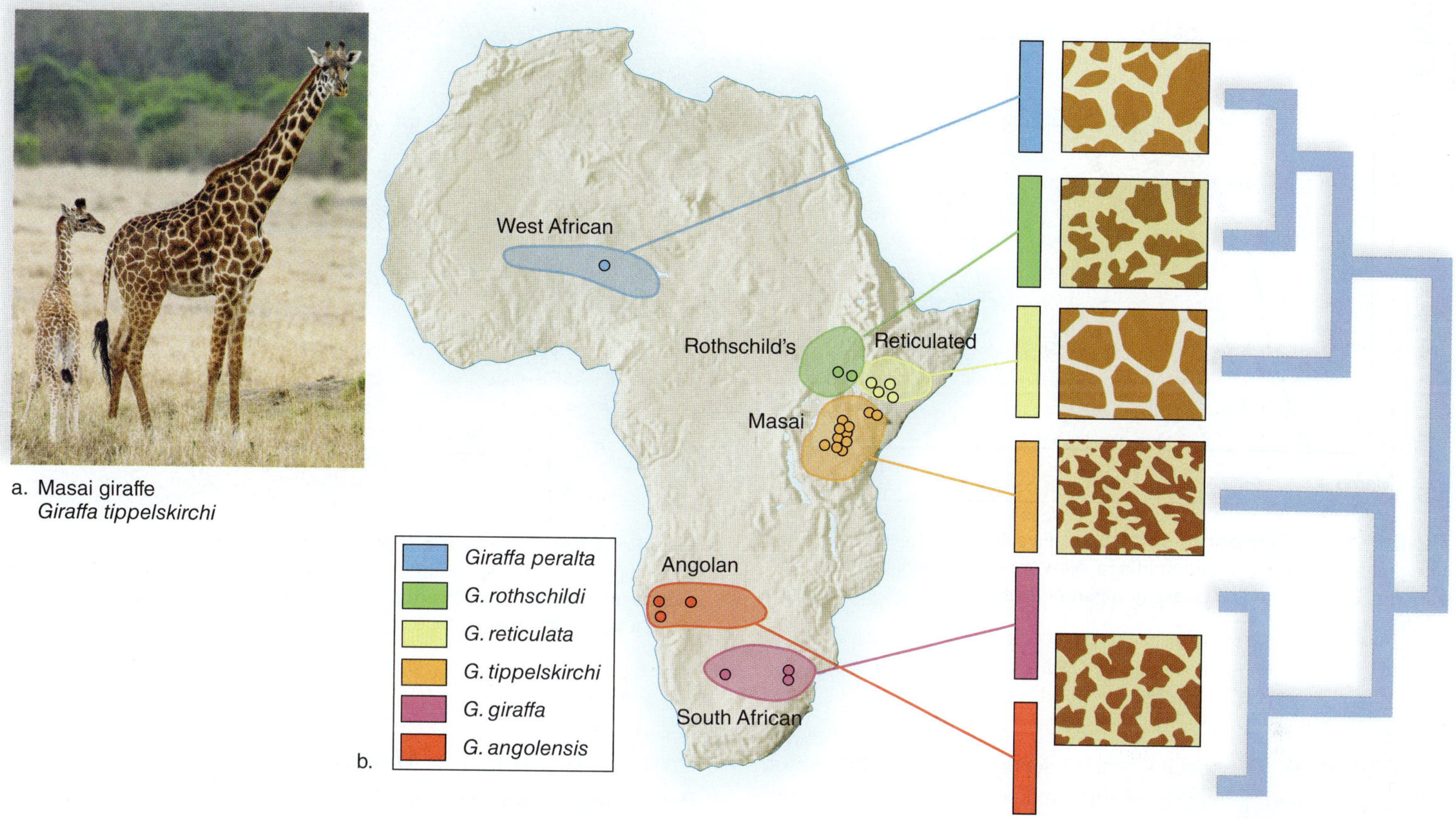

Figure 17.3 The phylogenetic species concept defines species from an evolutionary tree. **a.** *Giraffa tippelskirchi* is known for its unique ragged-edge spot pattern. Historically, regional variations in spot pattern were used to define what were once considered multiple populations of the same species, *Giraffa camelopardalis*. **b.** Recent studies show that each regional population represents a unique evolutionary branch of the giraffe family tree. According to the phylogenetic species concept, each of these branches should be recognized as one of six unique giraffe species.

Figure 17.4 Reproductive barriers. Prezygotic isolating mechanisms prevent mating attempts or a successful outcome, should mating take place. No zygote ever forms. Postzygotic isolating mechanisms prevent the zygote from developing—or should an offspring result, it is not fertile.

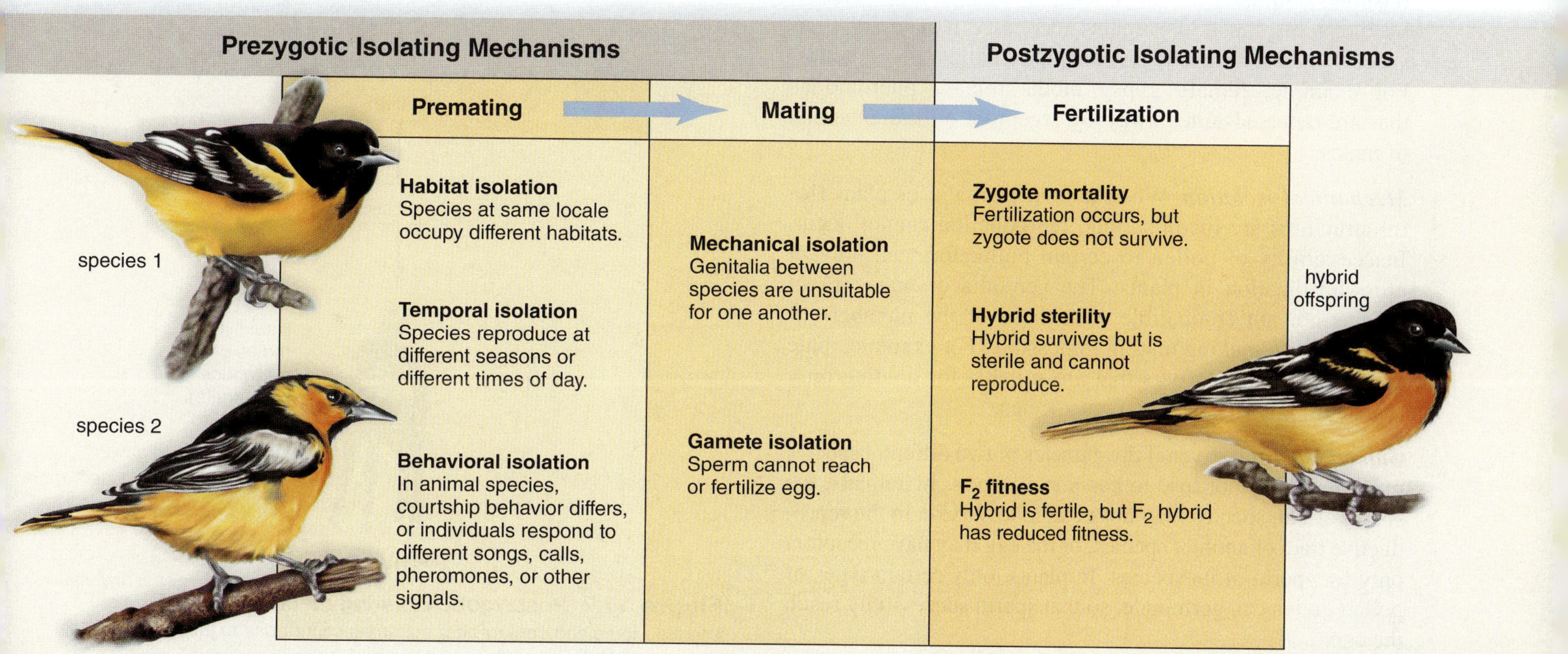

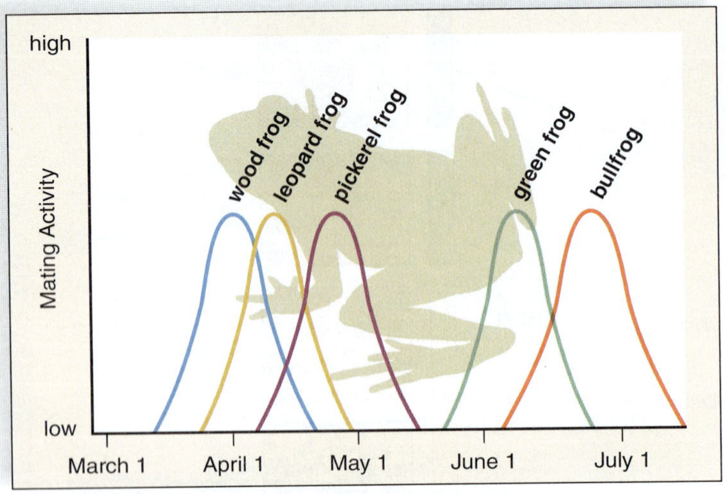

Figure 17.5 Temporal isolation. Five species of frogs of the genus *Rana* are all found at Ithaca, New York. The species remain separate due to breeding peaks at different times of the year, as indicated by this graph.

Figure 17.6 Prezygotic isolating mechanism. An elaborate courtship display allows the blue-footed boobies of the Galápagos Islands to select a mate. The male lifts up his feet in a ritualized manner that shows off their bright blue color.

(Fig. 17.5). The species remain separate because the period of peak mating activity differs, and so do the breeding sites. For example, wood frogs breed in woodland ponds or shallow water, leopard frogs in lowland swamps, and pickerel frogs in streams and ponds on high ground. Having different dispersal times often helps prevent fertilization of the gametes from different species.

Behavioral isolation Many animal species have courtship patterns that allow males and females to recognize one another. The male blue-footed boobie in Figure 17.6 does a special courtship dance. Male fireflies are recognized by females of their species by the pattern of their flashings; similarly, female crickets recognize male crickets by their chirping. Many males recognize females of their species by sensing chemical signals called pheromones. For example, female gypsy moths release pheromones that are detected miles away by receptors on the antennae of males.

Mechanical isolation When animal genitalia or plant floral structures are incompatible, reproduction cannot occur. Inaccessibility of pollen to certain pollinators can prevent cross-fertilization in plants. The genitalia of many insect species are not compatible with those of the members of other species, making mating impossible. For example, male dragonflies have claspers that are suitable for holding only female dragonflies of their own species.

Gamete isolation Even if the gametes of two different species meet, they may not fuse to become a zygote. In animals, the sperm of one species may not be able to survive in the reproductive tract of another species, or the egg may have receptors only for sperm of its species. In plants, only certain types of pollen grains can germinate, so that sperm successfully reach the egg.

Postzygotic Isolating Mechanisms

Postzygotic isolating mechanisms operate after the formation of a zygote. These mechanisms prevent hybrid offspring from developing and reproducing. If a hybrid is born, it is often infertile (Fig. 17.7)

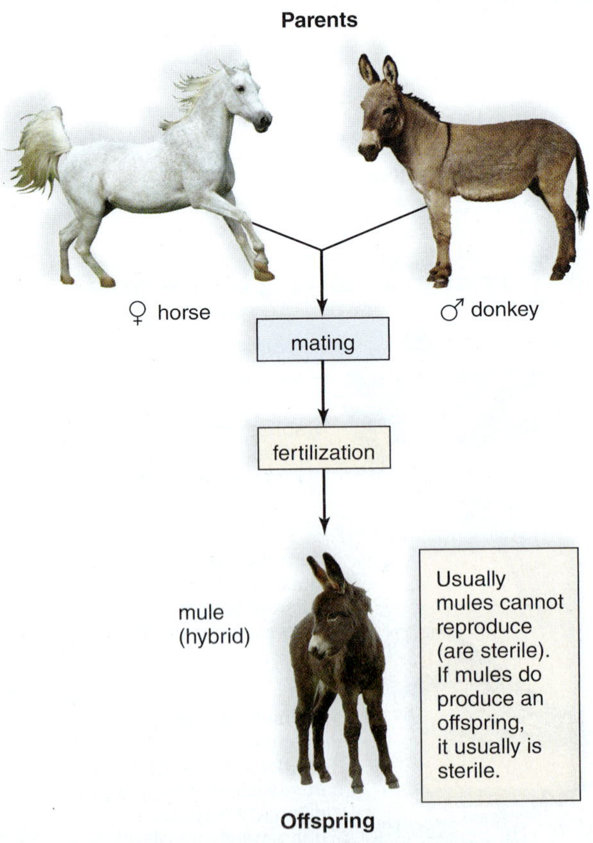

Figure 17.7 Postzygotic isolating mechanism. Mules are infertile. Horse and donkey chromosomes cannot pair to produce gametes.

and cannot reproduce. Either way, the genes of the parents are unable to be passed on to succesive generations.

Hybrid inviability A hybrid zygote may die, because it is not viable. A zygote with two different chromosome sets may fail to go through mitosis properly and not develop. The developing embryo may receive incompatible instructions from the maternal and paternal genes, so that it cannot develop properly.

Hybrid sterility The hybrid zygote may develop into a sterile adult. As is well known, a cross between a female horse and a male donkey produces a mule. Mules are usually sterile and cannot reproduce (Fig. 17.7). Sterility of hybrids generally results from complications in meiosis that lead to an inability to produce viable gametes. Similarly, a cross between a cabbage and a radish produces offspring that cannot form gametes, most likely because the cabbage chromosomes and the radish chromosomes cannot align during meiosis. On the rare occasion that two hybrids mate and produce offspring, the F_2 hybrids have a reduced fitness.

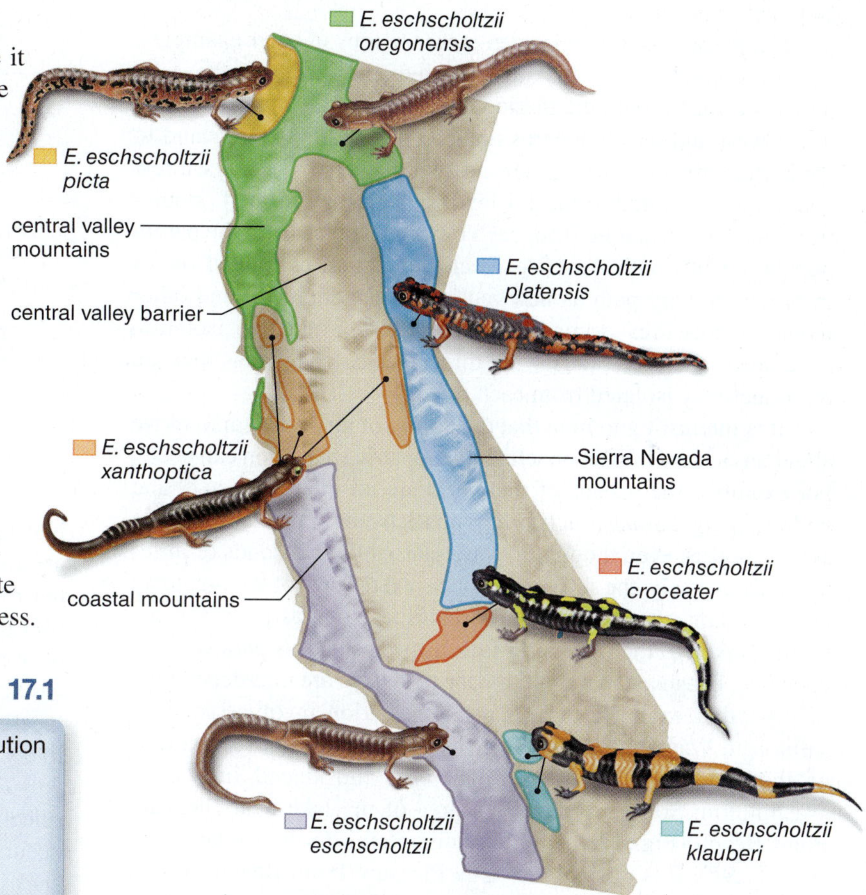

Figure 17.8 Allopatric speciation in progress among *Ensatina* salamanders. The Central Valley of California is reproductively separating a range of populations of *Ensatina eschscholtzii* that are all descended from the same northern ancestral species.

Check Your Progress 17.1

1. Identify the various factors that can lead to microevolution and macroevolution.
2. List three species concepts, and explain the main requirements of each.
3. Explain how frogs that look similar but have different courtship calls can be classified as different species.

17.2 Modes of Speciation

Learning Outcomes

Upon completion of this section, you should be able to

1. Define two modes of speciation and give examples of each.
2. Identify an example of adaptive radiation.
3. Distinguish between coevolution and convergent evolution.

Speciation is the splitting of one species into two or more species, or the transformation of one species into new species over time. One type of speciation requires populations to be physically isolated from one another, while the other main type does not.

Geographic isolation is helpful, because it allows populations to continue on their own evolutionary path. This can eventually cause the population to be reproductively isolated from other species and from one another. Once reproductive isolation has begun, it can be reinforced by the evolution of more traits that prevent breeding with related species. Geographic isolation can occur repeatedly, so one ancestral species can give rise to several other species.

Allopatric Speciation

In 1942, Ernst Mayr, an evolutionary biologist, published the book *Systematics and the Origin of Species*, in which he proposed the biological species concept and a process by which speciation could occur. This process, termed **allopatric speciation** (Gk. *allo,* "other"; *patri,* "fatherland"), is the eventual result of populations that have become separated by a geographic or other type of physical barrier. Mayr said that when populations of a species become geographically isolated, microevolutionary processes, such as genetic drift and natural selection, alter the gene pool of each population independently. If the differences between the groups become large enough, reproductive isolation may occur, resulting in the formation of new species.

Examples of Allopatric Speciation

Figure 17.8 features an example of allopatric speciation that has been extensively studied in southern California. An ancestral population of *Ensatina* salamanders lives in the Pacific Northwest. Members of this ancestral population migrated southward, establishing a series of populations. Each population was exposed to unique selective pressures along the coastal mountains and the Sierra Nevada mountains. Due to the presence of the Central Valley of California, gene flow rarely occurs between the eastern populations and the western populations. Genetic differences increased from north to south, resulting in two distinct forms of *Ensatina* salamanders in southern California that differ dramatically in color and rarely interbreed.

Geographic isolation is even more obvious in other examples. The green iguana of South America is hypothesized to be the common ancestor for both the marine iguana on the Galápagos Islands to the west and the rhinoceros iguana on Hispaniola, an island to the north. Green iguanas are strong swimmers, so by chance a few could have migrated to these islands, where they formed populations that were separate from each other as well as from the parent population in South America. Each population continued on its own evolutionary path as new mutations, genetic drift, and other selection pressures occurred. Eventually, reproductive isolation developed, and the results were three species of iguanas that are reproductively isolated from each other.

It is interesting to note that the ability of an organism to move about has a large impact on whether allopatric speciation can occur. For example, the oceans of the world are all interconnected, and wide-ranging animals, such as humpback whales, are members of a single species, even though they are seasonally thousands of miles apart. Conversely, the scale of distance and the size of the organism are important, too. Many small organisms, such as parasites, are tightly linked to their hosts. In fact, a species and its parasites can coevolve, because their evolutionary pathways are interdependent.

Another example of allopatric speciation involves sockeye salmon in Washington State. In the 1930s and 1940s, hundreds of thousands of sockeye salmon were introduced into Lake Washington. Some colonized an area of the lake near Pleasure Point Beach (Fig. 17.9*a*). Others migrated into the Cedar River (Fig. 17.9*b*). It is possible to tell a Pleasure Point Beach salmon from a Cedar River salmon because of differences in size and shape due to the demands of reproduction. Males in rivers where the waters are fast-moving tend to be more slender than those at the beach. Sockeye salmon turn sideways into the strong current as part of their mating ritual, and a male with a slender body is better able to perform this maneuver. In contrast, the females in rivers tend to be larger than those at the beach. This larger body helps them dig slightly deeper nests in the gravel beds on the river bottom. Deeper nests are not disturbed by river currents and remain warm enough for eggs to survive and hatch. These differences have resulted in reproductive isolation between these two populations. Not all river salmon remain near the beach their whole lives; in fact, a third of the sockeye males in Pleasure Point grew up in the river population. But the two populations do not interbreed because of the difference in size and shape between the males and females in both populations.

Reinforcement of Reproductive Isolation

As seen in sockeye salmon and other animals, independent evolution of populations can result in reproductive isolation. Another example is seen among *Anolis* lizards, in which males court females by extending a colorful flap of skin, called a "dewlap." The dewlap must be seen in order to attract mates. Populations of *Anolis* in a dim forest tend to evolve a light-colored dewlap, while populations in open habitats tend to evolve dark-colored ones. This change in dewlap color causes the populations to be reproductively isolated, because females distinguish males of their species by their dewlaps.

As populations become reproductively isolated, postzygotic isolating mechanisms may arise before prezygotic isolating mechanisms. As we have seen, when a horse and a donkey reproduce, the hybrid is not fertile. Therefore, the process of natural selection tends to favor variations that would prevent the production of hybrids that are unable to reproduce. Indeed, natural selection would favor the continual development of prezygotic isolating mechanisms until the two populations were completely reproductively isolated.

The term **reinforcement** is given to the process of natural selection that "reinforces" reproductive isolation. Reinforcement occurs when two populations, formerly of the same species, come back in contact after being isolated. These two species are not able to reproduce when they come in contact, because they no longer recognize each other as mates. An example of reinforcement has been seen in the pied and collared flycatchers of the Czech Republic and Slovakia, where both species occur in close proximity. Only here have the pied flycatchers evolved a different coat color than the collared flycatchers. The difference in color reinforces the choice to mate with their own species.

Sympatric Speciation

Speciation that occurs in the absence of a geographic barrier is termed **sympatric speciation** (Gk. *sym,* "together"; *patri,* "fatherland"). Sympatric speciation is more difficult to observe in nature, because no physical barrier prevents mating between populations, as in allopatric speciation. Some of the best examples of sympatric speciation in nature have involved divergence in diet, microhabitat, or both. In these cases, a new species evolves when a population becomes specialized to live in a different microhabitat.

One example is the midas and arrow cichlid fishes that live in a small lake in Nicaragua. The midas cichlid colonized the lake and occupied its usual rocky, coastal habitat. Over time, a new species of cichlid, the arrow cichlid, evolved from a population of the midas cichlid that adapted to living and feeding in an open water habitat. The partitioning of lake habitats and dietary preferences resulted in a shift in body size, jaw morphology, and tooth size and shape. Now the midas and arrow cichlids are two distinct species.

a. Sockeye salmon at Pleasure Point Beach, Lake Washington.

b. Sockeye salmon in Cedar River. The river connects with Lake Washington.

Figure 17.9 Allopatric speciation among sockeye salmon. In Lake Washington, salmon that matured (**a**) at Pleasure Point Beach do not reproduce with those that matured (**b**) in the Cedar River. The females from Cedar River are noticeably larger and the males are more slender than those from Pleasure Point Beach, and these shapes help them reproduce in the river.

Sympatric speciation involving **polyploidy** (a chromosome number beyond the diploid [2n] number) is well documented in plants. A polyploid plant can reproduce with itself, but it produces only sterile offspring when mated with 2n individuals, because not all the chromosomes are able to pair during meiosis. Two types of polyploidy are known: autoploidy and alloploidy.

Autoploidy occurs when a diploid plant produces diploid gametes due to nondisjunction during meiosis (see Fig. 10.10). If this diploid gamete fuses with a haploid gamete, a triploid plant results. A triploid (3n) plant is sterile and cannot produce offspring, because the chromosomes cannot pair during meiosis. Humans have found a use for sterile plants, because they produce fruits without seeds. If two diploid gametes fuse, the plant is a tetraploid (4n) and the plant is fertile, as long as it reproduces with another of its own kind. The fruits of polyploid plants are much larger than those of diploid plants. The huge strawberries of today are produced by octaploid (8n) plants.

Alloploidy (Gk. *allo,* "other") occurs when two different but related species of plants hybridize. Hybridization is then followed by a doubling of the chromosomes. For example, the California wildflower *Clarkia concinna* is a diploid plant with 14 chromosomes (seven pairs). A related species, *C. virgata,* is a diploid plant with 10 chromosomes (five pairs). A hybrid of these two species is not fertile, because 7 chromosomes from one plant cannot pair evenly with 5 chromosomes from the other plant (Fig. 17.10). However, if the chromosome number doubles in the hybrid, the chromosomes can pair during meiosis, resulting in a fertile plant. The species *C. pulchella* could have arisen this way. Recent molecular data tell us that polyploidy is common in plants and makes a significant contribution to the evolution of new plant species.

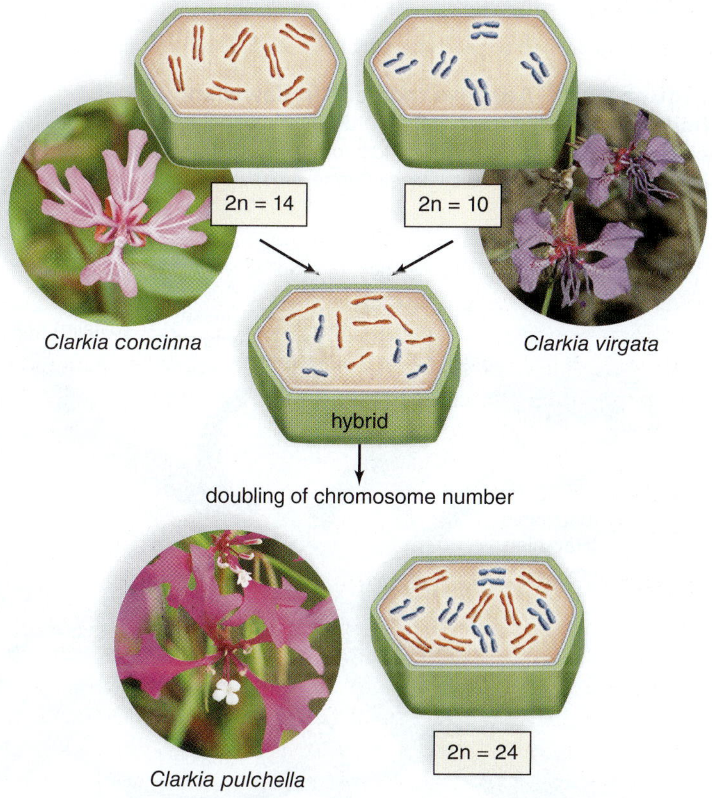

Figure 17.10 Alloploidy produces a new species. Reproduction between two species of *Clarkia* results in a sterile hybrid. Doubling of the chromosome number results in a fertile third *Clarkia* species that can reproduce with itself only.

Adaptive Radiation

Adaptive radiation is a type of speciation that occurs when a single ancestral species rapidly gives rise to a variety of new species as each adapts to a specific environment. Many instances of adaptive radiation involve sympatric speciation following the removal of a competitor, a predator, or a change in the environment. When competition is reduced, it results in **ecological release.** This is an opportunity for a species to expand its use of resources within habitats that now have less competition. Ecological release provides an opportunity for new species to originate as populations become specialized to newly available microhabitats. Allopatric speciation can also cause a population to undergo adaptive radiation.

Examples of Adaptive Radiation

Darwin proposed that a small population of ancestral finches colonized the Galápagos Islands and their descendants spread out to occupy various niches. Geographic isolation of the various finch populations caused their gene pools to become isolated. Because of natural selection, each population adapted to a particular habitat on its island. In time, the many populations became so genotypically different that now, when by chance they reside on the same island, they do not interbreed and are therefore separate species. During mating, female finches use beak shape to recognize members of the same species, and suitors with the wrong type of beak are rejected. The genetic mechanisms by which beak shape changed are described in the Nature of Science feature, "Genetic Basis of Beak Shape in Darwin's Finches," on page 308.

Video Finches Adaptive Radiation

Similarly, on the Hawaiian Islands, a wide variety of honeycreepers are descended from a common goldfinchlike ancestor, which arrived there from Asia or North America about 5 million years ago. Today, honeycreepers have a range of beak sizes and shapes for feeding on various food sources, including seeds, fruits, flowers, and insects (Fig. 17.11). An example of adaptive radiation among plants is the silversword alliance. This group includes plants adapted to both moist and dry environments and even lava fields.

Figure 17.11 Adaptive radiation in Hawaiian honeycreepers. A single ancestral species of goldfinchlike birds colonized the Hawaiian Islands and gave rise through adaptive radiation to more than 20 species of honeycreepers.

Adaptive radiation has occurred throughout the history of life on Earth when a group of organisms exploited new environments. With the demise of the dinosaurs about 66 million years ago, mammals underwent adaptive radiation as they exploited environments previously occupied by the dinosaurs. Mammals diversified in just 10 million years to include the early representatives of all the mammalian orders, including hoofed mammals (e.g., horses and pigs), aquatic mammals (e.g., whales and seals), primates (e.g., lemurs and monkeys), flying mammals (e.g., bats), and rodents (e.g., mice and squirrels). A changing world presented new environmental habitats and new food sources. Insects fed on flowering plants and, in turn, became food for mammals. Primates lived in trees, where fruits were available.

Convergent Evolution

Convergent evolution is said to occur when a biological trait evolves in two unrelated species as a result of exposure to similar environments. Dolphins and tuna do not share a recent common ancestor, yet they both have a dorsal fin for swimming. The ability to swim has evolved in dolphins and tuna independently and, so, has resulted in two different, although similar-looking, solutions to the requirements of moving through water with the greatest amount of efficiency. On both dolphins and fish, the dorsal fin helps keep the animals upright while swimming and prevents them from rolling over. The dolphin's dorsal fin is composed of fibrous connective tissue and blood vessels, whereas the tuna's is composed of bony spines, with skin covering the spines and joining them together.

Traits that evolve convergently in two unrelated lineages because of a response to a similar lifestyle or habitat are said to be **analogous**—such as the dorsal fin of dolphins and tuna. The opposite of analogous is **homologous**—traits that are similar because they evolved from a common ancestor (see Fig. 15.15). For example, the wings of butterflies are homologous to the wings of moths, because both are members of the same lineage of insects called the Lepidoptera. All Lepidoptera evolved from a common winged ancestor.

Recent examples of convergent evolution involve adaptive radiations of species in similar, but unconnected, habitats. Lake Malawi and Lake Tanganyika are two African Rift Valley lakes (Fig. 17.12). Within each lake is a set of 200–500 species of cichlid fishes that have adapted to feed on prey in a particular microhabitat of the lake. Each lake has fish that are adapted to feeding on the sandy bottom as well as species that feed along the rocky shore. Diet specialization has produced a suite of different jaw and tooth shapes and sizes, each adapted to a particular food type. The outcome is an amazing example of convergent evolution. Each lake's assemblage of cichlids has evolved independently of the other, yet if you compare the assemblage in each lake, you find amazing similarities in coloration, body shape, and size and shape of jaws and teeth (Fig. 17.12). Convergent evolution is apparent in the pairing of Lake Malawi and Lake Tanganyika species that have the same features. This is evidence for the evolution of independently derived features that make cichlids adapted to forage in similar habitats.

Video Cichlid Specialization

Figure 17.12 Convergent evolution of African lake fish. Cichlids exhibit remarkable evolutionary convergence. In Lake Malawi and Lake Tanganyika, very similar sets of body shapes and sizes have evolved independently of each other, with each type adapted to feed on a different type of food source. Although they appear morphologically similar, all the cichlids from Lake Malawi are more closely related to one another than to any species within Lake Tanganyika.

Check Your Progress 17.2

1. During the last Ice Age, deer mice in Michigan became separated by a large glacial lake and are now two different species. Identify the mode of speciation.
2. List the evidence you would need to show that the five species of big cats, *Panthera leo* (lion), *P. tigris* (tiger), *P. pardus* (leopard), *P. onca* (jaguar), and *P. uncia* (snow leopard) are an example of an adaptive radiation.
3. Predict the outcome of convergent evolution on the variety of cichlid fish in a newly discovered African Rift Valley lake compared to other lakes with similar microhabitats.

THEME **Nature of Science**

Genetic Basis of Beak Shape in Darwin's Finches

Darwin's finches are a famous example of how many species originate from a common ancestor. Over time, each species of finch on the Galápagos Islands adapted to a unique way of life, with beak size and shape related to their diets. Ground finches have thick, short beaks adept at crushing hard seeds. Cactus finches have long, thin beaks well suited to probing flowers and the fruit of cacti. The warbler finch feeds on both seeds and insects and has a thin, short beak useful for a mixed diet. Multiple sources of

Figure 17B Genetic basis of finch beak size and shape. Bone morphogenetic protein 4 and calmodulin genes regulate the depth and length of the beaks of Darwin's finches. Increases or decreases in gene activity work together to fine-tune beak morphology.

17.3 Principles of Macroevolution

Learning Outcomes

Upon completion of this section, you should be able to

1. Distinguish between the gradualistic and the punctuated equilibrium models of evolution.
2. Explain how gene expression can influence speciation.
3. Identify how macroevolution is not goal-oriented.

Many evolutionary biologists hypothesize, as Darwin did, that macroevolution occurs gradually. After all, natural selection can only do so much to bring about change in each generation. The gradual evolution of new species is the basis of the *gradualistic model* of evolution. This model proposes that speciation occurs after populations become isolated, with each group continuing slowly on its own evolutionary pathway. The proponents of the gradualistic model often show the history of groups of organisms by drawing the type of diagram shown in Figure 17.13*a*. Note that in this diagram an ancestral species has given rise to two separate evidence in DNA sequences and morphology support the hypothesis that Darwin's finches are closely related to one another (Fig. 17B).

Video Finches Adaptive Radiation

The differences in beak shape have been recorded by decades of research. Without any additional information, scientists proposed, as did Darwin, that there must be a genetic explanation for the difference in beak shape among species. In 2006, the genes that are responsible for the variation in finch beak shape were discovered. These findings are direct evidence for the mechanism of macroevolution.

Two genes control beak depth and shape. The gene for bone morphogenetic protein 4 (*BMP4*) determines how deep, or tall, the beak will be. The gene for calmodulin (*CaM*) regulates how long a beak will grow. For example, a high level of *BMP4* creates a deep, wide beak. A high level of *CaM* produces a long beak. In Darwin's finches, a combination of *BMP4* and *CaM* determines overall beak shape (Fig. 17B-ii). The degree of expression of each gene will affect how the beak develops in the embryo.

The cactus finch, for example, has a low level of *BMP4* and a high level of *CaM* expression, which produces a shallow, long beak (Fig. 17B-iii, -iv). In contrast, the ground finch has the opposite pattern, with a high level of *BMP4* and low level of *CaM* expression, producing a short, deep beak (Fig. 17B-iii, -iv). One of the most interesting findings of this research is that evolution of beak shape did not require changes to the *BMP4* or the *CaM* gene. An increase or decrease in the expression of these genes during embryo development is enough to change beak shape!

The ability of *BMP4* and *CaM* to affect beak morphology is not limited to Darwin's finches. Variation in beak shape was reproduced in chicken embryos (Figure 17C). Using molecular tools, finch *BMP4* and *CaM* genes were expressed in the beaks of developing chicken embryos. The expression of *BMP4* caused the chick beaks to deepen, and *CaM* expression caused their beaks to get longer (Figure 17C*b, c*). Overall, this is strong evidence for the genetic basis of macroevolution.

Questions to Consider

1. How might small, or microevolutionary, changes in *BMP4* and *CaM* in finch populations have resulted in new species of finch?
2. Use Darwin's theory of evolution by natural selection to explain the relationship between finch beak shape and diet on the Galápagos Islands.

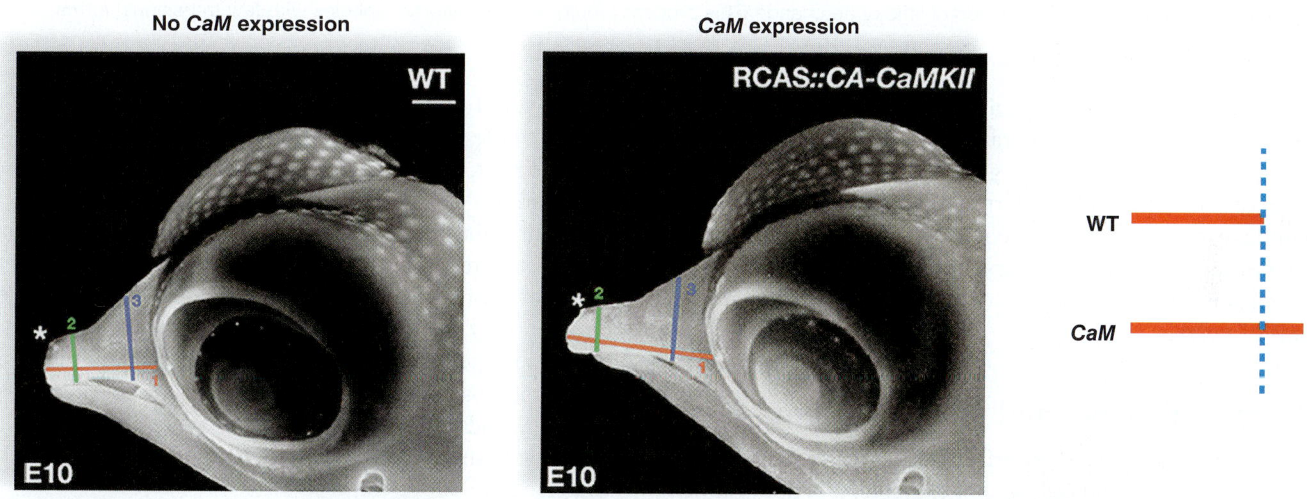

a. b. c.

Figure 17C Expression of *BMP4* and *CaM* in chicks. Chicken embryos were genetically modified to express *CaM* in their beaks during development. **a.** The normal, or wild type (WT), chick did not have *CaM* expression in the beak during development and the length of the beak was normal. **b.** The *CaM* chick produced an elongated beak. **c.** A side-by-side comparison of the length of WT and *CaM* beaks. Similarly, a separate study of *BMP4* expression in chick embryos produced deeper beaks compared to the WT embryos.

species, represented by a slow change in plumage color. The gradualistic model suggests that it is difficult to indicate when speciation occurred, because there would be so many transitional links.

After studying the fossil record, some paleontologists tell us that species can appear quite suddenly, and then they remain essentially unchanged during a period of stasis (sameness) until they either undergo extinction or evolve in response to changes in the environment. Based on these findings, they have developed a *punctuated equilibrium model* to explain the fluctuating pace of evolution. This model says that the assembly of species in the fossil record can be explained by periods of equilibrium, or stasis, punctuated (interrupted) by periods of rapid, abrupt speciation, or change. Figure 17.13*b* shows this way of representing the history of evolution over time.

A strong argument can be made that it is not necessary to choose between these two models of evolution, and that both could very well assist us in interpreting the fossil record. In other words, some fossil species may fit one model, and some may fit the other model. In a stable environment, a species may be kept in equilibrium by stabilizing selection for a long period. If the

Figure 17.13 Gradualistic and punctuated equilibrium models. **a.** Under the gradualistic model, new species evolve from a series of small changes that occur constantly over time. This process brings about a lot of transitional forms. **b.** Under the punctuated model, new species evolve from a series of abrupt, rapid changes after a period of little or no change. This process would result in different species with few transitional forms.

environment changes slowly, a species may be able to adapt gradually. If environmental change is rapid, a new species may arise suddenly before the parent species goes on to extinction. The differences between all possible patterns of evolutionary change are rather subtle, especially when we consider that because geologic time is measured in millions of years, the "sudden" appearance of a new species in the fossil record could actually represent many thousands of years.

Developmental Genes and Macroevolution

Investigators have discovered genes that can bring about radical changes in body shapes and organs. For example, it is now known that the *Pax6* gene is involved in eye formation in all animals, and that homeotic (*Hox*) genes determine the location of repeated structures in all vertebrates (see Chapter 42).

Scientists are working on understanding how evolution could have produced the myriad of animals in the history of life. They are trying to determine how genetic changes brought about such major differences in form. It has been suggested since the time of Darwin that the answer must involve the processes that shape development. In 1917, D'Arcy Thompson asked us to imagine an ancestor in which all parts are developing at a particular rate. A change in gene expression could stop a developmental process or could continue it beyond its normal time. For instance, if the growth of limb bones were stopped early, the result would be shorter limbs, and if it were extended, the result would be longer limbs compared to those of an ancestor. If the whole period of growth were extended, a larger animal would result, accounting for why some species of horses are so large today.

Using the modern techniques of cloning and manipulating genes, investigators have indeed discovered genes whose differences in expression (the timing and location in the body where proteins they encode are synthesized) can bring about changes in body shapes and organs. This result suggests that these genes must date back to a common ancestor that lived more than 600 MYA, and that despite millions of years of divergent evolution, all animals share the same control switches for development (see Chapter 28).

Development of the Eye

The animal kingdom contains many different types of eyes, which were once thought to require their own sets of genes. Flies, crabs, and other arthropods have compound eyes that have hundreds of individual visual units. Humans and other vertebrates have the same camera-type eye with a single lens, as do squids and octopuses. Humans are not closely related to either flies or squids, so it would seem as if all three types of animals evolved "eye" genes separately. Research has shown that this is not the case.

In 1994, Walter Gehring and his colleagues at the University of Basel, Switzerland, discovered that a gene called *Pax6* is required for eye formation in all animals (Fig. 17.14). Mutations in

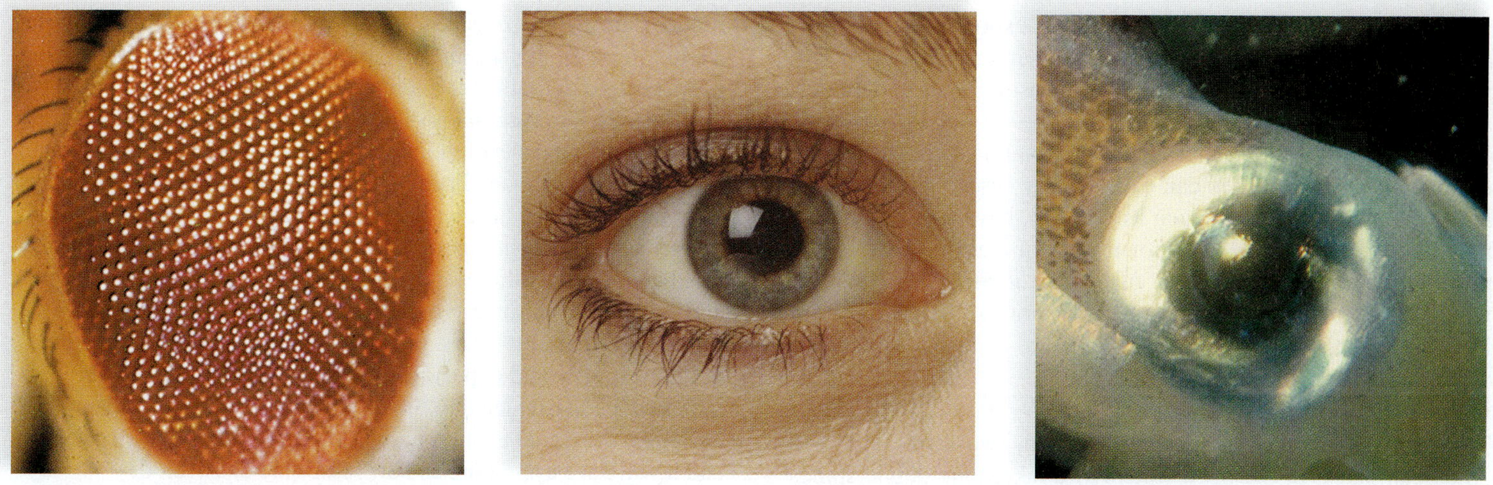

Figure 17.14 ***Pax6*** **gene and eye development.** *Pax6* is involved in eye development in a fly, a human, and a squid.

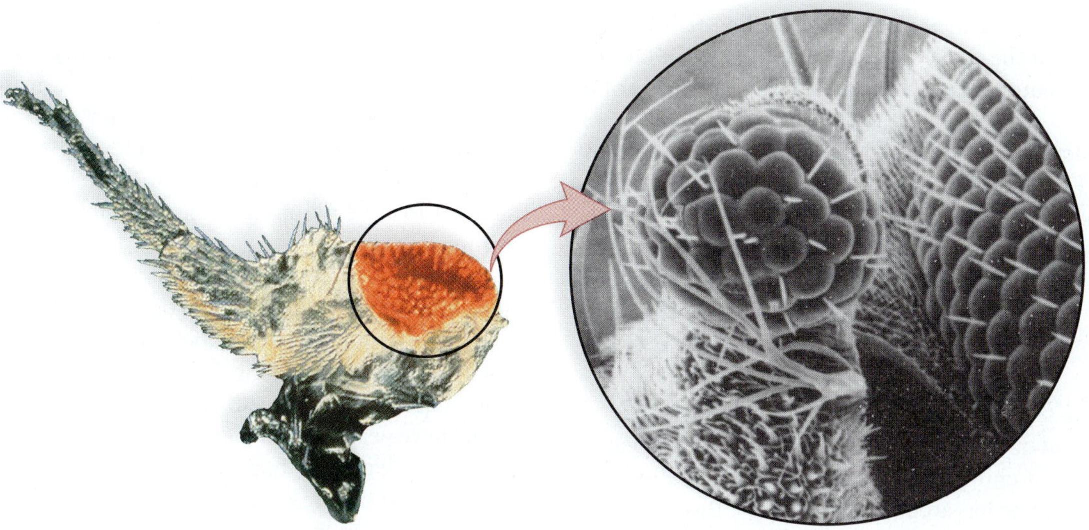

Figure 17.15 Study of ***Pax6*** **gene.** The mouse *Pax6* gene makes a compound eye on the leg of a fruit fly.

the *Pax6* gene lead to the failure of eye development in both people and mice, and remarkably the mouse *Pax6* gene can cause an eye to develop on the leg of a fruit fly (Fig. 17.15).

Development of Limbs

Wings and arms are very different, but both humans and birds express the *Tbx5* gene in developing limb buds. *Tbx5* encodes a protein that is a transcription factor that turns on the genes needed to make a limb during development. Birds and humans both express *Tbx5* but differ in which genes Tbx5 turns on. Perhaps in an ancestral tetrapod the Tbx5 protein triggered the transcription of only one gene. This could have led to the evolution of limb formation in vertebrates such that changes in the genes regulated by Tbx5 and other transcription factors contributed to the variation in tetrapod limb structure. Therefore, subtle changes in gene control can have profound effects on body shape. This might explain the abundance of variation seen in plant and animal shape and form.

There is also a question of timing. Changing the timing of gene expression, as well as which genes are expressed, can result in a dramatic change in shape.

Development of Overall Shape

Vertebrates have repeating segments, as exemplified by the vertebral column. Changes in the number of segments can lead to changes in overall shape. In general, *Hox* genes control the number and appearance of repeated structures along the main body axes of vertebrates. Shifts in when *Hox* genes are expressed in embryos are responsible for the reason a snake has hundreds of rib-bearing vertebrae and essentially no neck, in contrast to other vertebrates, such as a chick.

Hox genes have been found in all animals. Shifts in the expression of these genes can explain why insects have just six legs but other arthropods, such as crayfish, have ten. In general, the study of *Hox* genes has shown how animal diversity is due to variations in the expression of ancient genes, rather than to wholly new and different genes (see the Nature of Science feature in Chapter 28).

Pelvic-Fin Genes

The three-spined stickleback fish occurs in two forms in North American lakes. In the open waters of a lake, long pelvic spines help protect the stickleback from being eaten by large predators. But on the lake bottom, long pelvic spines are a disadvantage, because dragonfly larvae grab young sticklebacks by their spines and feed on them.

The presence of short spines in bottom-dwelling fish can be traced to a reduction in the development of the pelvic-fin bud in the embryo, and this reduction is due to the altered expression of a gene called *Pitx1*.

Hindlimb reduction has occurred during the evolution of other vertebrates. The hindlimbs became greatly reduced in size as whales and manatees evolved from land-dwelling ancestors into fully aquatic forms (see Fig. 15.13). Similarly, legless lizards have evolved many times. The stickleback study has shown how natural selection can lead to major skeletal changes in a relatively short time.

Human Evolution

The sequencing of genomes has shown that our DNA base sequence is very similar to that of chimpanzees, mice, and indeed all vertebrates. The human genome has around 23,000 genes. Based on this knowledge and the work just described, investigators no longer expect to find new genes to account for the evolution of humans. Instead, they predict that differential gene expression, new functions for "old" genes, or both will explain how humans evolved. We discuss the details of human evolution in Chapter 30.

As with all genes, mutations of developmental genes occur by chance, and it is this random process that creates variation. Without variation, evolution cannot occur. Even though mutation is random, natural selection is not a random process. Rather, natural selection acts on the variation that is present, in a way that favors the survival of advantageous traits, in a particular environment at a particular time. This should not be misinterpreted as evidence that evolution is directed or "works" toward an end goal. Evolution is a perpetual process that shapes variation from generation to generation. Where this process leads is unpredictable and depends on a complicated array of external forces. In the next section, we observe that evolution is not directed toward any particular end.

Macroevolution Is Not Goal-Oriented

The evolution of the horse, *Equus,* has been studied since the 1870s, and at first the ancestry of this genus seemed to represent a model for gradual, directed evolution toward the "goal" of the modern horse. Three trends were particularly evident during the evolution of the horse: an increase in overall size, toe reduction, and a change in tooth size and shape.

By now, however, many more fossils have been found, making it easier to tell that the evolutionary history of the horse is complicated by the presence of many lineages that evolved, went extinct, and thus were not on the lineage that led to the modern horse. The evolutionary tree of the horse in Figure 17.16 is an oversimplification, because it is based only on the evidence from the few fossils that we have, and there are likely many more fossils yet to be discovered. If the evolution of the horse were directed toward the "goal" of the modern horse, we would expect to see a single branch on this tree with intermediate fossils leading directly from the ancestor to the horse. However, the actual evolutionary tree of *Equus* has many branches, and it will have even more as new fossils are discovered.

Each of the ancestral species of the horse was adapted to its environment. Adaptation occurs only because the members of a population with an advantage are able to have more offspring than other members. Natural selection is opportunistic, not goal-directed.

Fossils named *Hyracotherium* have been designated as the first probable members of the horse family, living about 57 MYA. These animals had a wooded habitat, ate leaves and fruit, and were about the size of a dog. Their short legs and broad feet with several toes would have allowed them to scamper from thicket to thicket to avoid predators. *Hyracotherium* was obviously well adapted to its environment, because this genus survived for 20 million years.

The family tree of *Equus* indicates that speciation, diversification, and extinction are common occurrences during a species' existence. The first adaptive radiation of horses occurred about 35 MYA. The weather was becoming drier, and grasses were evolving. Eating grass requires tougher teeth, and an increase in size and longer legs would have permitted greater speed to escape predators. The second adaptive radiation of horses occurred about 15 MYA and included *Merychippus* as a representative of those groups that were speedy grazers living on the open plain. By 10 MYA, the horse family was quite diversified. Some species were large forest browsers, some were small forest browsers, and others were large plains grazers. Many species had three toes, but some had one strong toe. (The hoof of the modern horse includes only one toe.)

Modern horses evolved about 4 MYA from ancestors with features that were adaptive for living on an open plain, such as large size, long legs, hoofed feet, and strong teeth. The other groups of horses prevalent at the time became extinct, no doubt for complex reasons. Humans have corralled modern horses for various purposes, and this makes it difficult to realize that the traits of a modern horse are adaptive for living in a grassland environment.

Check Your Progress 17.3

1. Explain how the punctuated equilibrium model provides an alternative explanation of the theory of catastrophism proposed by Cuvier (see Chapter 15).
2. Discuss how the study of developmental genes supports the possibility of rapid speciation in the fossil record.
3. Identify the developmental genes that influence macroevolution.

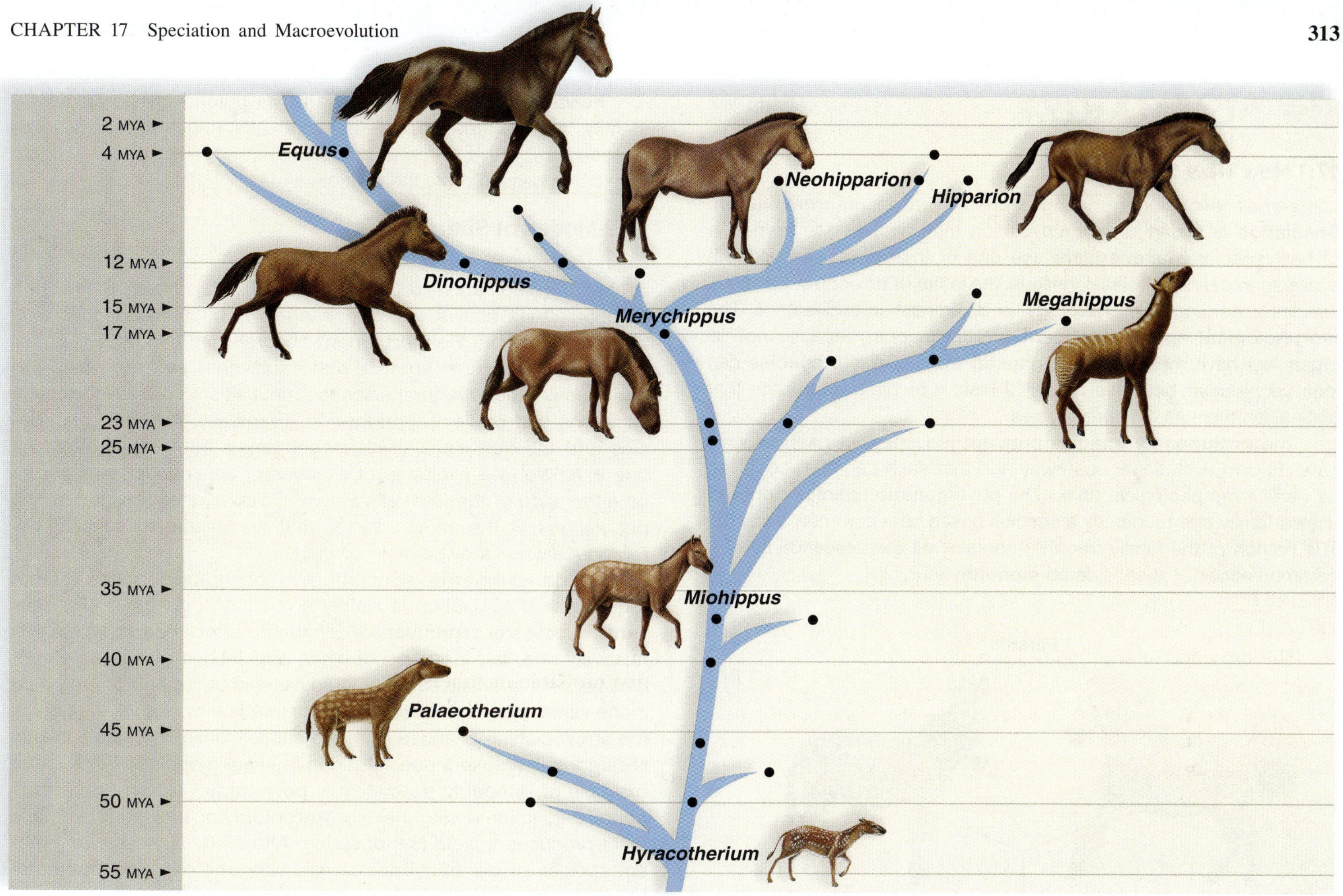

Figure 17.16 Simplified family tree of *Equus*. Every dot represents a genus.

Connecting the Concepts with the Themes

Evolution

- Microevolution is change in the allele frequencies of populations over time. Macroevolution is what happens when microevolutionary change accumulates until speciation occurs.
- Evolution does not have direction. Variation arises by random mutation, not because a particular solution is "needed." Natural selection "weeds out" variation in a particular environment at a particular time.

Nature of Science

- Molecular biologists are now able to document shared ancestry of species at the level of DNA sequences.
- All life on Earth shares some developmental genes in common. These universal genes support the theory that macroevolution is the source of biodiversity.

Biological Systems

- An ecological or physical barrier promotes allopatric speciation. Microhabitat specialization drives sympatric speciation.
- Microhabitat specialization is a common characteristic of adaptive radiation.
- In similar environments, similar features may evolve in unrelated organisms through convergent evolution.

MEDIA STUDY TOOLS http://connect.mheducation.com

SMARTBOOK® Study smart. SmartBook helps you maximize your study time by identifying what you know and don't know. The recharge feature in SmartBook helps you prepare for the test by identifying the material that you are most likely to forget.

connect BIOLOGY Find even more resources to learn the chapter concepts including animations, tutorial videos, and interactive practice questions.

Animation	Videos
17.1 Phylogenetic Trees	**17.2** Finches Adaptive Radiation • Cichlid Specialization

SUMMARIZE

17.1 How New Species Evolve

Large-scale evolutionary changes are known as **macroevolution.** **Speciation** is a part of macroevolution that involves the formation of new species. **Taxonomists** use one of three species concepts available to define species. Linnaeus, the father of taxonomy, identified species according to **morphology,** or differences in appearance. The **morphological species concept** has limitations, because not all organisms have measurable **diagnostic traits. Cryptic species** can look very similar but have other life history or behavioral traits that distinguish them as different species.

The **evolutionary species concept** requires that each species have its own evolutionary pathway and that each can be recognized by certain morphological traits. The **phylogenetic species concept** uses a family tree to identify a species based on a common ancestor. The branch of the family tree that contains all the descendants of a common ancestor is considered **monophyletic.**

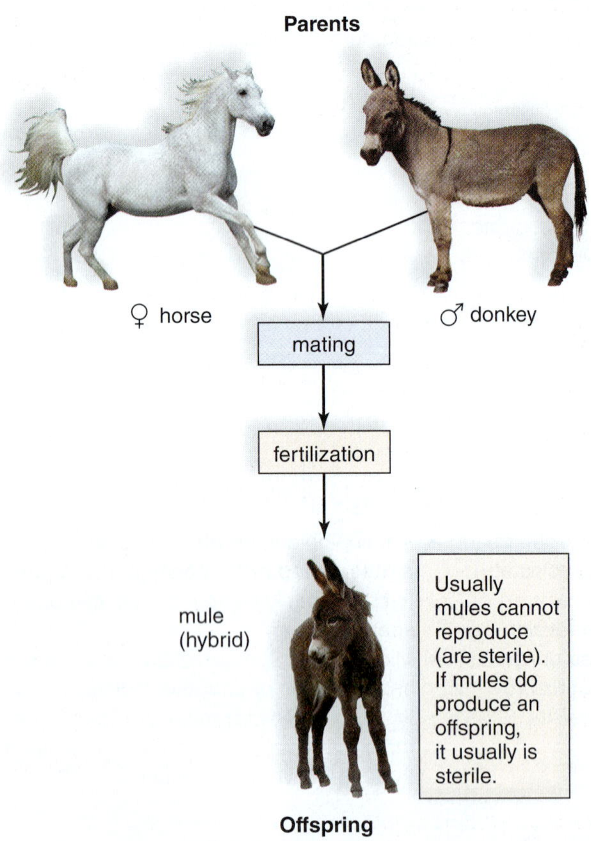

The **biological species concept** identifies species based on whether two or more populations experience **reproductive isolation** that inhibits inbreeding.

Prezygotic isolation occurs when reproductive isolation is prior to fertilization, whereas postzygotic isolation occurs after fertilization. A **zygote** is the first cell that forms after fertilization.

Prezygotic isolating mechanisms (habitat, temporal, behavior, mechanical, and gamete isolation) prevent mating from being attempted or prevent fertilization from being successful if mating is attempted. These types of isolating mechanisms prevent **hybridization,** or the mating of two different species.

Postzygotic isolating mechanisms (hybrid inviability, hybrid sterility, and F_2 fitness) prevent hybrid offspring from surviving or reproducing. These mechanisms prevent the genes of the parents from being passed on to the next generation.

17.2 Modes of Speciation

During **allopatric speciation,** a geographic or physical barrier causes the population to become reproductively isolated. Isolation of populations allows genetic changes to accumulate over time via microevolution. **Reinforcement** occurs when two related populations come back in contact with each other and are unable to reproduce. Various natural selection mechanisms will favor genetic variations that prevent hybridization. Eventually, the ancestral species and the new species no longer breed with one another. As one example, the evolution of a series of salamander subspecies on either side of the Central Valley of California has resulted in two populations of the same species that are unable to successfully reproduce when they come in contact.

During **sympatric speciation,** a geographic barrier is not required, and speciation is simply a change in genotype that prevents successful reproduction. Sympatric speciation in animals is relatively rare, but it can occur when populations of the same species become specialized on a particular subhabitat and/or food item in the same geographic area. For example, a new species of cichlid, the arrow cichlid, diverged from the midas cichlid in a single lake in Nicaragua, because it specialized in feeding in open water. Another example of sympatric speciation is **polyploidy** in plants as a result of nondisjunction during meiosis. **Autoploidy** occurs when a diploid plant produces triploid (3n) offspring. **Alloploidy** is when two different species of plants hybridize, increasing the chromosomal numbers beyond normal for either species.

Adaptive radiation is a type of speciation that occurs when a single ancestral species rapidly gives rise to a radiation of new species as each adapts to a specific environment. Many instances of adaptive radiation involve sympatric speciation following **ecological release.** Ecological release is the reduction of competition that provides opportunity for new species to originate as populations become specialized in newly available microhabitats. Finches on the Galápagos Islands, the honeycreepers of the Hawaiian Islands, and the rise of mammals after the extinction of dinosaurs are examples of adaptive radiation events. Allopatric speciation can also lead to adaptive radiation.

Convergent evolution has occurred when the same biological trait has evolved in two unrelated species as a result of exposure to similar environmental conditions. The wings of birds and bats are examples of convergent evolution. **Analogous** traits are traits found in unrelated lineages that perform a similar function. **Homologous** traits are similar traits found in similar species due to a common ancestor.

An adaptive radiation accompanied by convergent evolution can produce similar assemblages of morphological types in geographically isolated, but similar, environments. Lake Tanganyika and Lake Malawi contain assemblages of cichlids that are similar as a result of convergent evolution during two independent adaptive radiations in each lake.

17.3 Principles of Macroevolution

Macroevolution is evolution of new species and higher levels of classification. The fossil record gives us a view of life across millions of years. The hypothesis that species evolve gradually is now being challenged by the hypothesis that speciation can also occur rapidly. In that case, the fossil record could show periods of stasis interrupted by spurts of change—that is, a punctuated equilibrium. Transitional fossils would be expected with gradual change, but not with punctuated equilibrium. It is most likely that species have evolved as the result of a combination of punctuated equilibirum and gradualism. Environmental factors would also have contributed to the process.

It could be that both models are seen in the fossil record but rapid change can occur by differential expression of regulatory genes. The same regulatory gene (*Pax6*) controls the development of both the camera-type and the compound-type eyes. The *Tbx5* gene controls the development of limbs, whether the wing of a bird or the leg of a tetrapod. *Hox* genes control the number and appearance of a repeated structure along the main body axes of vertebrates. The same pelvic-fin genes control the development of a pelvic girdle. Variation in the expression of *BMP4* and *CaM* produces different beak shapes in each of Darwin's finches. Changing the timing of gene expression, as well as which genes are expressed, can result in dramatic changes in shape.

Speciation, diversification, and extinction are seen during the evolution of *Equus*. These three processes are commonplace throughout the evolutionary history of a species and illustrate that macroevolution is not goal-directed. The life we see about us represents adaptations to particular environments. Such adaptations have changed in the past and will change in the future.

ASSESS

Choose the best answer for each question.

17.1 How New Species Evolve

1. Which of the following events is part of macroevolution?
   a. speciation
   b. mutation
   c. gene flow
   d. natural selection
   e. All of these are correct.
2. A biological species
   a. always looks different from other species.
   b. always has a different chromosome number than that of other species.
   c. is reproductively isolated from other species.
   d. never occupies the same niche as other species.
   e. does not survive long in the wild.
3. Which of these is a prezygotic isolating mechanism?
   a. habitat isolation
   b. temporal isolation
   c. hybrid sterility
   d. zygote mortality
   e. Both a and b are correct.
4. Male moths recognize females of their species by sensing chemical signals called pheromones. This is an example of
   a. gamete isolation.
   b. habitat isolation.
   c. behavorial isolation.
   d. mechanical isolation.
   e. temporal isolation.
5. Which of these is an example of mechanical isolation?
   a. Sperm cannot reach or fertilize an egg.
   b. Courtship patterns differ.
   c. Organisms live in different locales.
   d. Organisms reproduce at different times of the year.
   e. Genitalia are unsuitable to each other.

17.2 Modes of Speciation

6. Complete the following diagram illustrating allopatric speciation by using these phrases: genetic changes (used twice), geographic barrier, species 1, species 2, species 3.

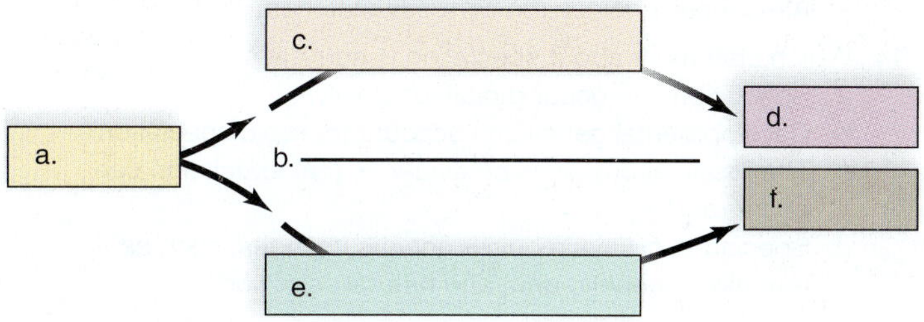

7. The creation of new species due to geographic barriers is called
   a. isolation speciation.
   b. allopatric speciation.
   c. allelomorphic speciation.
   d. sympatric speciation.
   e. symbiotic speciation.
8. The many species of Galápagos finches are each adapted to eating different foods. This is the result of
   a. gene flow.
   b. adaptive radiation.
   c. sympatric speciation.
   d. genetic drift.
   e. All of these are correct.
9. Allopatric, but not sympatric, speciation requires
   a. reproductive isolation.
   b. geographic isolation.
   c. spontaneous differences in males and females.
   d. prior hybridization.
   e. rapid rate of mutation.
10. Which of the following are an example of analogous structures?
    a. wings of birds and bats
    b. hooves of horses and deer
    c. wings of moths and butterflies
    d. leaves on oak trees and maples
    e. arms of primates and cats

17.3 Principles of Macroevolution

11. Which of the following is not part of the gradualistic model of evolution?
    a. isolation of the population
    b. a slow change of the population
    c. the presence of fossils
    d. an ancestral population that gives rise to two separate populations
    e. a period of rapid change
12. A change in expression of the *Tbx5* gene can lead to which of the following problems?
    a. malformation of the legs
    b. incorrect development of the eye
    c. lack of development of neck vertebrae in a snake
    d. development of six legs in a grasshopper
    e. a shorter dorsal spine in stickleback fish

13. Which gene is incorrectly matched to its function?
    a. *Hox*—body shape
    b. *Pax6*—body segmentation
    c. *Tbx5*—limb development
    d. *Pitx1*—pelvic-fin development
    e. All of these choices are correctly matched.
14. Which statement about speciation is not true?
    a. Speciation can occur rapidly or slowly.
    b. Developmental genes can account for rapid speciation.
    c. The fossil record gives no evidence that speciation can occur rapidly.
    d. Speciation always requires genetic changes, such as mutations, genetic drift, and natural selection.
15. In the evolution of the modern horse, which of the following was the goal of the evolutionary process?
    a. large size
    b. single toe
    c. change in tooth size
    d. change in tooth shape
    e. None of these are correct.

ENGAGE

LEARNSMART LABS®

The following LearnSmart Labs contain exercises that are related to the content of this chapter:

- Natural Selection

THINKING SCIENTIFICALLY

1. You want to decide what definition of a species to use in a study. What are the advantages and disadvantages of one based on DNA sequences as opposed to the evolutionary and biological species concept?
2. At a recent school board meeting, a member of the audience stated that there is not enough evidence to support how various species have arisen. Due to this lack of evidence, he or she thinks that students should be presented with alternative ideas to the theory of evolution. Indicate the scientific lines of evidence that are used to support the process of speciation, using the modern horse as an example.
3. Explain the evolutionary mechanisms behind the difference in size between the sockeye salmon of Pleasure Point Beach, Lake Washington, and those of Cedar River, Washington.
4. Over time, enough genetic variations can develop within a population to cause it to undergo speciation. Identify the various mechanisms that will prevent different species from being able to reproduce succesfully. Which of these mechanisms is the most influential in keeping species separate?

Many transitional forms—such as *Microraptor,* shown here in reconstruction—indicate a link between dinosaurs and birds.

18

Origin and History of Life

Today, paleontologists are setting the record straight about dinosaurs. It now appears that some dinosaurs nested in the same manner as some species of birds! Bowl-shaped nests containing dinosaur eggs have been found in Mongolia, Argentina, and the United States. These nests contain fossilized eggs and bones along with eggshell fragments. From this evidence, it seems that baby dinosaurs stayed in the nest after hatching until they were big enough to walk around and fragment the eggshells. The spacing between the nests of *Maiasaura* (meaning "good mother lizard" in Greek) suggests that this dinosaur fed its young. The remains of an enormous group of about 10,000 *Maiasaura* found in Montana is further evidence that this dinosaur was indeed social in its behavior.

Maiasaura and *Microraptor,* the winged gliding dinosaur shown here, provide structural and behavioral evidence of the link between dinosaurs and birds. In this chapter, we trace the origin of life before considering the history of life.

As you read through this chapter, think about the following questions:

1. Does the theory of evolution explain the origins of life? Why or why not?
2. How is the fossil record used to measure the age of life on Earth?
3. What types of scientific evidence, including the fossil record, tell us something about the history of life on Earth?

Chapter Outline

18.1 Origin of Life 318
18.2 History of Life 323
18.3 Geological Factors That Influence Evolution 332

Before You Begin

Before beginning this chapter, take a few moments to review the following discussions.

Section 1.1 What are the basic characteristics that define life?

Section 2.1 What are chemical isotopes?

Section 15.3 How do the fossil and biogeographical records provide evidence for the evolution of new organisms?

Following *the* Themes

CHAPTER 18 ORIGIN AND HISTORY OF LIFE

UNIT 3 EVOLUTION

Evolution	The theory of evolution explains how life on Earth became diverse after it originated. The history of life can be summarized by the macroevolutionary changes shown in the fossil record.
Nature of Science	The absolute age of fossils can be calculated from the rate of decay of radioisotopes. Radiometric methods can determine the age of rocks as old as 4.5 billion years.
Biological Systems	Biogeography, the interface between geology and biology, is used to reconstruct the evolutionary history of life by comparing the geological history of Earth with the distribution of living organisms.

18.1 Origin of Life

Learning Outcomes

Upon completion of this section, you should be able to

1. List and describe the four stages of the origin of life.
2. Differentiate between the stages of chemical and biological evolution.
3. Describe a protocell membrane structure and its importance to the evolution of the first living cell.
4. Summarize at least one hypothesis that explains each of the four stages of the origin of life.

At the heart of Darwin's theory of evolution by natural selection is the principle of common ancestry, that all life on Earth can be traced back to a single ancestor. This ancestor, also called the **last universal common ancestor** (**LUCA**), is common to all organisms that live, and have lived, on Earth since life began (Fig. 18.1). Darwin thought that perhaps the first living organism arose in a "warm little pond with all sorts of ammonia and phosphoric salts, light, heat, electricity, etc." What Darwin unknowingly described were some of the conditions of ancient Earth that gave rise to the first forms of life.

In Chapter 1 we considered the characteristics shared by all living organisms. Living organisms acquire energy through metabolism, or the chemical reactions that occur within cells. Living organisms also respond to and interact with their environment, self-replicate, and are subject to the forces of natural selection that drive adaptation to the environment. The molecules of living organisms, called **biomolecules,** are organic molecules. The first living organisms on Earth would have had all these characteristics, yet early Earth was very different from the Earth we know today, and it consisted mainly of inorganic substances. The challenge is to determine how life started in this inorganic "warm little pond."

Advances and discoveries in chemistry, evolutionary biology, paleontology, microbiology, and other branches of science have helped scientists develop new, and test old, hypotheses about the origin of life. These studies contribute to an ever-growing body of scientific evidence that life originated 3.5–4 billion years ago (BYA) from nonliving matter in a series of four stages:

Stage 1. *Organic monomers.* Simple organic molecules, called monomers, evolved from inorganic compounds prior to the existence of cells. Amino acids, the basis of proteins, and nucleotides, the building blocks of DNA and RNA, are examples of organic monomers.

Stage 2. *Organic polymers.* Organic monomers were joined, or polymerized, to form organic polymers, such as DNA, RNA, and proteins.

Stage 3. *Protocells.* Organic polymers became enclosed in a membrane to form the first cell precursors, called *protocells* or *probionts.*

Stage 4. *Living cells.* Probionts acquired the ability to self-replicate, as well as other cellular properties.

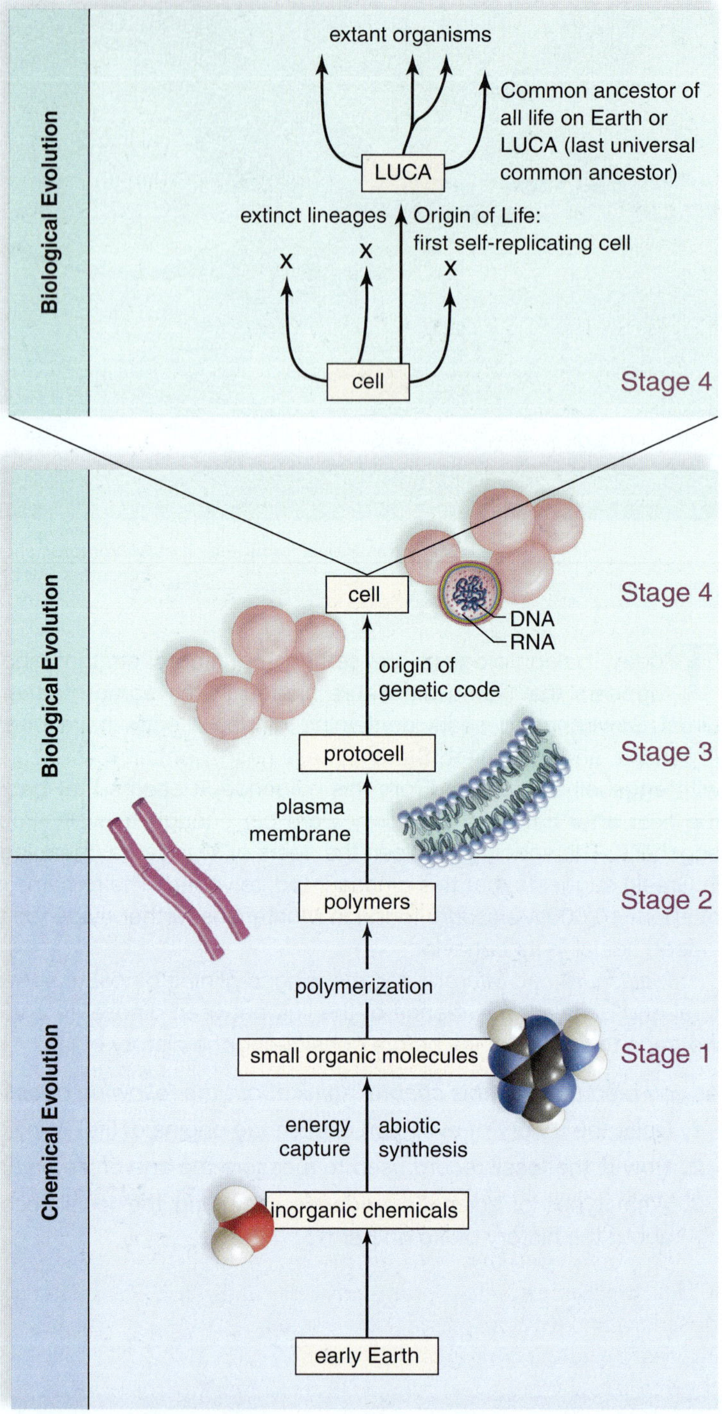

Figure 18.1 Stages of the origin of life. The first organic molecules (*bottom*) originated from chemically altered inorganic molecules present on early Earth (Stage 1). More complex organic macromolecules were synthesized to create polymers (Stage 2), which were then enclosed in a plasma membrane to form the protocells, or probionts (Stage 3). The protocell underwent biological evolution to produce the first true, self-replicating, living cell (Stage 4). This first living cell underwent continued biological evolution (*top*), with a single surviving lineage, the LUCA, which became the common ancestor of all life on Earth.

Scientists have performed experiments to test hypotheses at each stage of the origin of life. Stages 1–3 involve the processes of a "chemical evolution," before the origin of life (Fig. 18.1). Stage 4 is when life first evolved through the processes of "biological evolution" (Fig. 18.1). In this chapter we examine the hypotheses and supporting scientific evidence for each stage of chemical and biological evolution, or **abiogenesis,** the origin of life from nonliving matter.

Stage 1: Evolution of Monomers

In the 1920s, Alexander Oparin, a notable biochemist, and J. B. S. Haldane, a geneticist and evolutionary biologist, independently proposed a hypothesis about the chemical origin of life. They proposed that the first stage in the origin of life was the evolution of simple organic molecules, or monomers (Fig. 18.1), from the inorganic compounds that were present in the Earth's early atmosphere. The Oparin-Haldane hypothesis, sometimes called the **primordial soup hypothesis,** proposes that early Earth had very little oxygen (O_2) but instead was made up of water vapor (H_2O), hydrogen gas (H_2), methane (CH_4), and ammonia (NH_3).

Methane and ammonia are reducing agents, because they readily donate their electrons. In the absence of oxygen, their reducing capability is powerful. **Abiotic synthesis** is the process of chemical evolution, forming organic molecules from inorganic materials. The early Earth's reducing atmosphere could have driven this process. Note that *oxidation* used in this context refers to a chemical redox reaction (see section 6.4), not to oxygen gas.

Primordial Soup Hypothesis and the Miller-Urey Experiment

In 1953, Stanley Miller, under the mentorship of Harold Urey, performed an experiment to test Oparin and Haldane's hypothesis of early chemical evolution (Fig. 18.2). The energy sources on early Earth included heat from volcanoes and meteorites, radioactivity from isotopes, powerful electric discharges in lightning, and solar radiation, especially ultraviolet radiation.

For his experiment, Miller placed a mixture of methane (CH_4), ammonia (NH_3), hydrogen (H_2), and water (H_2O) in a closed system, heated the mixture, and circulated it past an electric spark (simulating lightning). After a week's run, Miller discovered that a variety of amino acids and other organic acids had been produced.

The Miller-Urey experiment has been tested and re-examined over the decades since it was first performed. Other investigators have achieved similar results as Miller by using other, less-reducing combinations of gases dissolved in water. In 2008, a group of scientists examined 11 vials of compounds produced from variations of the Miller-Urey experiment and found a greater variety of organic molecules than Miller reported, including all 22 amino acids.

Animation Miller-Urey Experiment

Recent evidence suggests that nitrogen gas (N_2), not ammonia (NH_3), would have been abundant in the primitive atmosphere. The scarcity of ammonia challenged the Miller-Urey experiment as a valid test of early Earth conditions. However, later experiments showed that ammonia could have been produced when various mixes of iron-nickel sulfides catalyzed the change of N_2 to NH_3. A laboratory test of this hypothesis worked perfectly. Under conditions simulating that of deep-ocean thermal vents (Fig. 18.3), 70% of various nitrogen sources were converted to ammonia within 15 minutes. Thus, the early formation of organic monomers could have been supported by the production of ammonia from these vents.

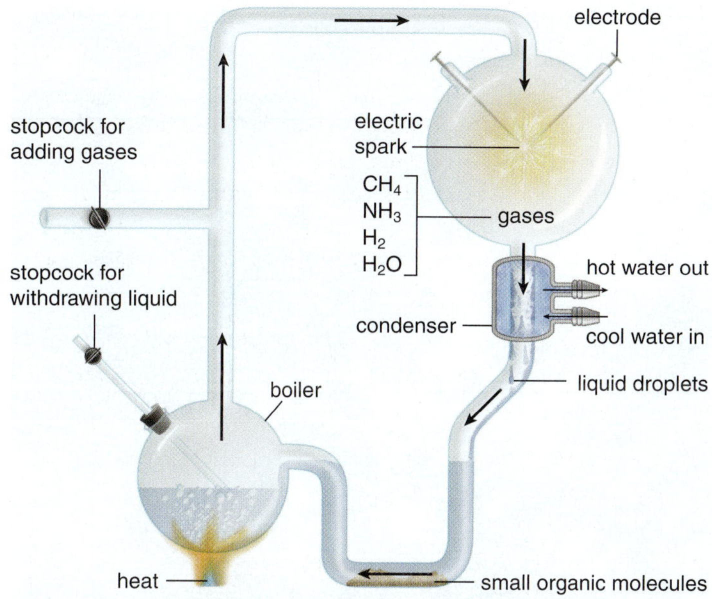

Figure 18.2 Stanley Miller's experiment. Gases that were thought to be present in the early Earth's atmosphere were admitted to the apparatus, circulated past an energy source (electric spark), and cooled to produce a liquid that could be withdrawn. Upon chemical analysis, the liquid was found to contain various small, organic molecules, which could serve as monomers for large, cellular polymers.

If early atmospheric gases did react with one another to produce small organic compounds, neither oxidation (no free oxygen was present) nor decay (no bacteria existed) would have destroyed these molecules, and rainfall would have washed them into the ocean, where they would have accumulated for hundreds of millions of years. Therefore, the oceans would have been a thick, warm organic soup—much like Darwin's "warm little pond."

Iron-Sulfur World Hypothesis

The Oparin-Haldane hypothesis was an important contribution to our understanding of the early stages of life's origins, but other hypotheses have also been proposed and tested. In the late 1980s, biochemist Günter Wächtershäuser proposed that thermal vents at the bottom of the Earth's oceans provided all the elements and conditions necessary to synthesize organic monomers. According to his **iron-sulfur world** hypothesis, dissolved gases emitted from thermal vents, such as carbon monoxide (CO), ammonia, and hydrogen sulfide, would pass over iron and nickel sulfide minerals that are present at thermal vents. The iron and nickel sulfide molecules would act as catalysts to drive the chemical evolution from inorganic to organic molecules.

Figure 18.3 Chemical evolution at hydrothermal vents. Minerals that form at deep-sea hydrothermal vents can catalyze the formation of ammonia and even monomers of larger organic molecules that occur in cells.

Extraterrestrial Origins Hypothesis

Comets and meteorites have constantly pelted the Earth throughout history. In recent years, scientists have confirmed the presence of organic molecules in some meteorites. Many scientists, perhaps the foremost being Chandra Wickramasinghe, feel that these organic molecules could have seeded the chemical origin of life on early Earth. Others even hypothesize that bacterium-like cells evolved first on another planet and then were carried to Earth. A meteorite from Mars labeled ALH84001 landed on Earth some 13,000 years ago. When experts examined it, they found tiny rods similar in shape to fossilized bacteria. This hypothesis continues to be investigated.

Stage 2: Evolution of Polymers

Within a cell's cytoplasm, organic monomers join to form polymers in the presence of enzymes—such as the synthesis of protein polymers from amino acids by ribosomes. Enzymes themselves are proteins, but proteins were not present on early Earth. The challenge is to determine how the first organic polymers formed if enzymes were not present.

Iron-Sulfur World Hypothesis

The iron-sulfur world hypothesis of Günter Wächtershaüser and the research of Claudia Huber have shown that organic molecules will react and amino acids will form peptides in the presence of iron-nickel sulfides under conditions found at thermal vents. The inorganic iron-nickel sulfides have a charged surface that attracts amino acids and binds them together to make proteins.

Protein-First Hypothesis

Sidney Fox has shown that amino acids polymerize abiotically when exposed to dry heat. He suggests that once amino acids were present in the oceans, they could have collected in shallow puddles along the rocky shore line. Then, the heat of the sun could have caused them to form **proteinoids,** small polypeptides that have some catalytic properties.

The formation of proteinoids has been simulated in the laboratory. When placed in water, proteinoids form **microspheres** (Gk. *mikros,* "small, little"; *sphaera,* "ball"), structures composed only of protein that have many properties of a cell. It's possible that even newly formed polypeptides had enzymatic properties. Some may have been more enzymatically active than others, perhaps giving them a selective advantage. If a certain level of enzyme activity provided an advantage over others, this would have set the stage for natural selection to shape the evolution of these first organic polymers. Those that evolved to be part of the first cell or cells would have had a selective advantage over those that did not become part of a cell.

Fox's **protein-first hypothesis** assumes that protein enzymes arose prior to the first DNA molecule. Thus, the genes that encode proteins followed the evolution of the first polypeptides.

RNA-First Hypothesis

The **RNA-first hypothesis** suggests that only the macromolecule RNA was needed to progress toward formation of the first cell or cells. Thomas Cech and Sidney Altman shared a Nobel Prize in 1989 for their discovery that RNA can be both a substrate and an enzyme. Some viruses today have RNA genes; therefore, the first genes could have been RNA. It would seem, then, that RNA could have carried out the processes of life commonly associated today with DNA and proteins. Those who support this hypothesis say that it was an "RNA world" some 4 BYA.

Stage 3: Evolution of Protocells

Before the first true cell arose, a **protocell (protobiont)** would have emerged—a structure that is characterized by having an outer membrane. After all, life requires chemical reactions to take place within a boundary, protecting them from disruption of conditions.

The Plasma Membrane

The plasma membrane of a cell provides a boundary between the inside of the cell and its outside world. This membrane is critical to the proper regulation and maintenance of cellular activities, and therefore the evolution of a membrane was a critical step in the origin of life.

The modern plasma membrane is made up of phospholipids assembled in a bilayer (Fig. 18.4). The first plasma membranes were likely made up of fatty acids, which are smaller than phospholipids but like phospholipids have a hydrophobic "tail" and a hydrophilic "head" (Fig. 18.4). Fatty acids are one

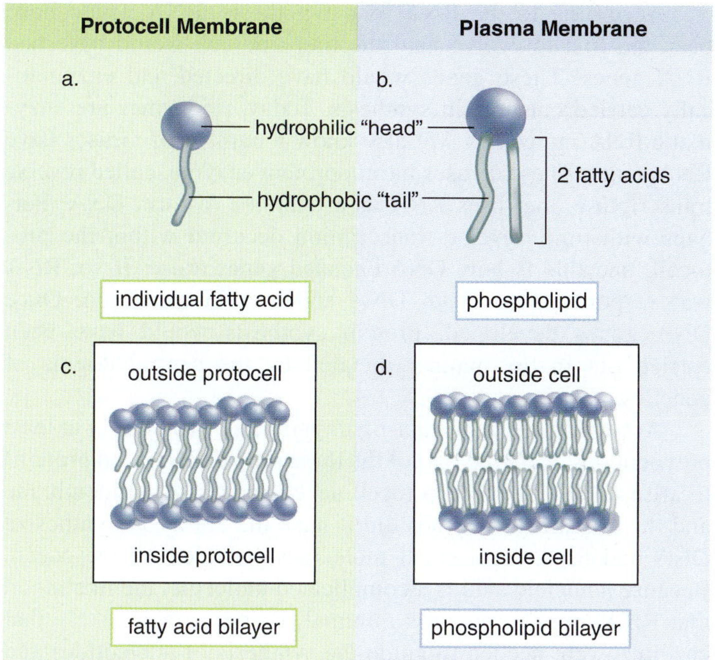

Figure 18.4 A comparison of protocell and modern plasma membranes. The first lipid membrane was likely made of a single layer of fatty acids. These first protocell membranes and the modern plasma membrane have features in common. **a.** Individual fatty acids have a single fatty acid chain with a hydrophilic head and a hydrophobic tail. **b.** Phospholipids of modern plasma membranes are made of two hydrophobic fatty acid chains (the "tails") attached to a hydrophilic head. **c.** Protocell membranes were likely made up of a bilayer of fatty acids, with hydrophilic heads pointing outward and hydrophobic tails pointing inward. **d.** Modern cells are organized in a similar fashion. Both bilayers create a semipermeable barrier between the inside and outside of a cell.

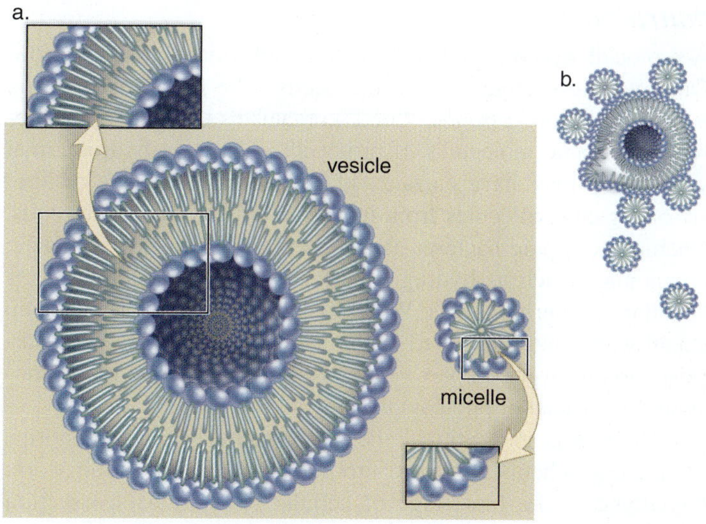

Figure 18.5 Structure and growth of the first plasma membrane. The first plasma membrane was likely made of a fatty acid bilayer, similar to that seen in vesicles. Protocells, the ancestor of modern cells, are thought to have had this type of membrane. **a.** A cross section of a vesicle reveals the fatty acid bilayer of a vesicle membrane. Micelles are spherical droplets formed by a single layer of fatty acids, and they are much smaller than vesicles. **b.** Under proper conditions, micelles can merge to form vesicles. As micelles are added to the growing vesicle, individual fatty acids flip their hydrophobic heads toward the inside and outside of the vesicle and their hydrophilic tails toward each other. This process forms a bilayer of fatty acids.

of the organic polymers that could have formed from chemical reactions at deep-water thermal vents early in the history of life. In water, fatty acids assemble into small spheres called **micelles** (Fig. 18.5). A micelle is a single layer of fatty acids organized with their heads pointing out and tails pointing toward the center of the sphere.

Under appropriate conditions, micelles can merge to form **vesicles.** Vesicles are larger than micelles and are surrounded by a bilayer (two layers) of fatty acids (Fig. 18.5), similar to the *phospholipid bilayer* of modern plasma membranes. An important feature of a vesicle lipid bilayer is that the individual fatty acids can flip between the two layers, which helps move select molecules, such as amino acids, from the outside to the inside of the vesicle. The first protocell would likely have been a type of vesicle with this type of fatty acid bilayer membrane (Fig. 18.5). Interestingly, if lipids are made available to protein microspheres, lipids tend to become associated with microspheres, producing a lipid-protein membrane. Lipid-protein microspheres share some interesting properties with modern cells: They resemble bacteria, they have an electrical potential difference, and they divide and perhaps are subject to selection.

In the 1920s, Alexander Oparin demonstrated that under appropriate conditions of temperature, ionic composition, and pH, concentrated mixtures of macromolecules tend to give rise to complex units called **coacervate droplets.** Coacervate droplets have a tendency to absorb and incorporate various substances from the surrounding solution. Eventually, a semipermeable-type boundary, also a trait of the modern plasma membrane, may form about the droplet.

In the early 1960s, biophysicist Alec Bangham of the Animal Physiology Institute in Cambridge, England, discovered that when he extracted lipids from egg yolks and placed them in water, the lipids naturally organized themselves into double-layered bubbles roughly the size of a cell known as **liposomes** (Gk. *lipos,* "fat"; *soma,* "body").

Later, biophysicist David Deamer, of the University of California, and Bangham realized that liposomes might have provided life's first membranous boundary. Perhaps liposomes with a phospholipid membrane engulfed early molecules that had enzymatic, even replicative, abilities. The liposomes would have protected the molecules from their surroundings and concentrated them, so that they could react (and evolve) quickly and efficiently. These investigators called this the **membrane-first hypothesis,** meaning that the first cell had to have a plasma membrane before any of its other parts. Perhaps the first membrane formed in this manner.

Nutrition

A protocell would have had to acquire nutrition in order to grow. One hypothesis suggests that protocells were heterotrophic (Gk. *hetero,* "different"; *trophe,* "food") organisms that consumed preformed organic molecules. If protocells evolved at hydrothermal vents, they may have carried out chemosynthesis—the synthesis of organic molecules from inorganic molecules and nutrients. Chemoautotrophic bacteria obtain energy by oxidizing inorganic compounds, such as hydrogen sulfide (H_2S), a molecule that is abundant at thermal vents. When hydrothermal vents in the deep ocean were first discovered in the 1970s, investigators were surprised to discover complex vent ecosystems supported by chemosynthesis instead of photosynthesis.

Glycolysis is a critical metabolic pathway that transforms high-energy chemical bonds into energy for a cell to do work. Glycolysis is the first stage of cellular respiration (see section 8.2), and it occurs outside of the mitochondria. ATP (adenosine triphosphate) is the most important energy-carrying molecule in living organisms. In the early stages of the origin of life, ATP would have been available to protocells in a preformed state. Over time, the ATP would have been transformed to ADP as it was used up. Natural selection would have favored the evolution of ATP/ADP recycling as a means to provide a renewable energy supply to the first cells. Modern cells have a way to produce ATP from ADP and inorganic phosphate via substrate-level phosphorylation (see Fig. 8.3) during glycolysis.

The greatest amount of ATP is synthesized by oxidative phosphorylation—in particular, via the electron transport chain (see section 8.2). Because all life on Earth uses ATP to fuel cellular metabolism, the evolution of a means of synthesizing ATP must have occurred very early in the history of life. The first bacteria evolved in the oxygen-poor environment of early Earth (Table 18.1). Thus, ATP was likely synthesized first by fermentation (see section 8.3). The evolution of oxidative phosphorylation in eukaryotes provided an advantage, because it greatly increased the amount of ATP synthesized per unit of energy. Mitochondria share a common ancestor with a group of bacteria that synthesize ATP via an electron transport chain. Oxidative phosphorylation is possible in eukaryotes because mitochondria provide an electron-transport-chain ATP factory.

At first the protocell must have had limited ability to break down organic molecules. Scientists speculate that it took millions of years for glycolysis to evolve completely. Some evidence suggests that microspheres from which protocells may have evolved have some catalytic ability. Oparin's coacervates incorporate enzymes if they are available in the medium.

Stage 4: Evolution of a Self-replication System

Today's cell is able to carry on protein synthesis in order to produce the enzymes that allow DNA to replicate. The central dogma of genetics states that DNA directs protein synthesis and that information flows from DNA to RNA to protein. It is possible that this sequence developed in stages.

According to the RNA-first hypothesis, RNA would have been the first to evolve, and the first true cell would have had RNA genes. These genes would have directed and enzymatically carried out protein synthesis. Today, ribozymes are enzymatic RNA molecules. We also know a number of viruses have RNA genes. These viruses have a protein enzyme, called reverse transcriptase, that uses RNA as a template to form DNA. Perhaps with time, reverse transcription occurred within the protocell, and this is how DNA-encoded genes arose. If so, RNA was responsible for both DNA and protein formation. Once DNA genes developed, protein synthesis would have been carried out in the manner dictated by the central dogma of genetics.

According to the protein-first hypothesis, proteins, or at least polypeptides, were the first of the three (DNA, RNA, and protein) to arise. Only after the protocell developed a plasma membrane and the necessary enzymes did it have the ability to synthesize DNA and RNA from small molecules provided by the ocean. Because a nucleic acid is a complicated molecule, the likelihood that RNA arose *de novo* is minimal. It seems more likely that enzymes were needed to guide the synthesis of nucleotides and then nucleic acids. Similar to the RNA-first hypothesis, once the information was contained within genes then protein synthesis would occur.

Cairns-Smith proposed that polypeptides and RNA evolved simultaneously. Therefore, the first true cell would have contained RNA genes that could have replicated because of the presence of proteins. This eliminates the baffling chicken-and-egg paradox of whether proteins or RNA came first. It means, however, that these two events would have had to happen at the same time.

After DNA formed, the genetic code had to evolve before DNA could store genetic information. The present genetic code is subject to fewer errors than a million other possible codes. Also, the present code is among the best at minimizing the effect of mutations. A single-base change in a present codon is likely to result in the substitution of a chemically similar amino acid, resulting in minimal changes in the final protein. This evidence suggests that the genetic code did undergo a natural selection process before finalizing into today's code.

Check Your Progress 18.1

1. Identify the stages of the origin of life hypothesis.
2. List two different hypotheses that explain the origin of organic molecules from inorganic matter, and identify the key features of these hypotheses.
3. Compare and contrast the features of the protocell membrane with a modern plasma membrane.

a.

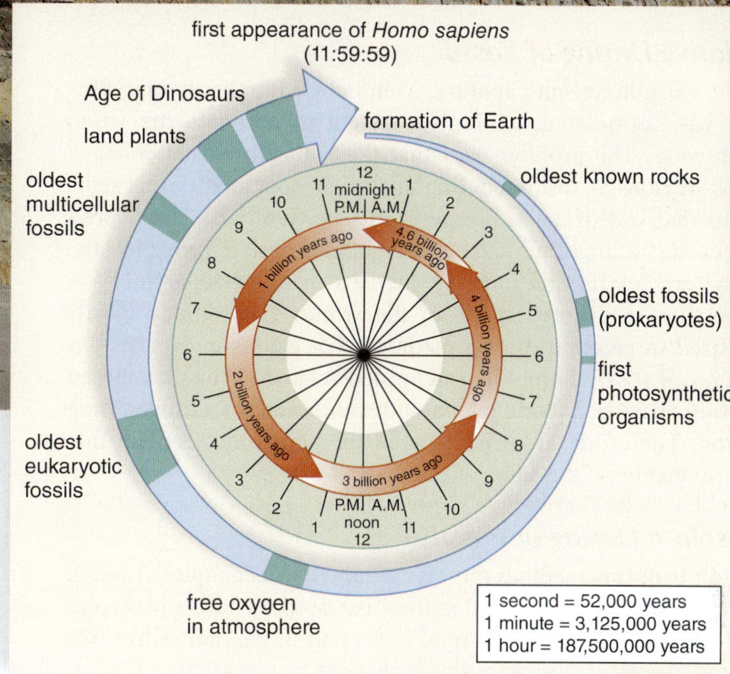

b.

Figure 18.6 The history of life. **a.** Strata, layers of sedimentary rock, are one source of fossils that provide information about the history of life. **b.** The blue ring of this diagram shows the history of life as it would be measured on a 24-hour timescale starting at midnight. (The red ring shows the actual years going back in time to 4.6 BYA.) The fossil record suggests that a very large portion of life's history was devoted to the evolution of unicellular organisms. The first multicellular organisms do not appear in the fossil record until just before 8 P.M., and humans are not on the scene until less than a second before midnight. BYA = billion years ago

18.2 History of Life

Learning Outcomes

Upon completion of this section, you should be able to

1. Explain the processes of relative and absolute dating of fossils.
2. List three sources of evidence that support the endosymbiotic theory of organelle evolution.
3. Discuss when and where the first multicellular organisms evolved.
4. Describe in chronological order the periods of Earth's history, and identify one major biological event that took place in each.

Macroevolution is the origin of new species and other taxonomic groups. The fossil record documents the history of macroevolution over long periods of time.

Fossils Tell a Story

Fossils (L. *fossilis,* "dug up") are the remains and traces of evidence of past life. Recall from Chapter 15 that fossils consist mainly of hard parts, such as shells, bones, or teeth, because these are usually not consumed or destroyed. However, some fossilized traces can be trails, footprints, or the impressions of soft body parts. **Paleontology** (Gk. *palaios,* "old, ancient"; *-logy,* "study of") is the study of the fossil record that results in knowledge about the history of life, ancient climates, and environments.

The great majority of fossils are found embedded in or recently eroded from sedimentary rock. **Sedimentation** (L. *sedimentum,* "a settling") is the gradual settling of particles of eroded and weathered rock and soil, called silt, that are carried by moving water. Silt is deposited gradually, forming layers of particles that vary in size and composition called **sediment.** Sediment becomes a **stratum** (pl., strata), a recognizable layer of sediment in a stratigraphic sequence (Fig. 18.6*a*).

The law of superposition states that a given stratum is considered to be older than the one above it and younger than the one immediately below it. However, the layers of sedimentation can be disturbed by geological forces, and such disturbances can complicate the interpretation of the stratigraphic sequence. Figure 18.6*b* shows the history of the Earth as if it had occurred during a 24-hour time span that starts at midnight. The actual years are shown on an inner ring of the diagram. This figure illustrates dramatically that only single-celled organisms were present during most (about 80%) of the history of the Earth.

If the Earth formed at midnight, prokaryotes did not appear until about 5 A.M., eukaryotes are present at approximately 4 P.M., and multicellular forms do not appear until around 8 P.M. Invasion of the land doesn't occur until about 10 P.M., and humans don't appear until 30 seconds before the end of the day.

This timetable has been worked out by studying the fossil record. In addition to sedimentary fossils, more recent fossils can be found in tar, ice, bogs, and amber. Shells, bones, leaves, and even footprints can also be found in the fossil record (Fig. 18.7).

Relative Dating of Fossils

In the early nineteenth century, even before the theory of evolution was formulated, geologists sought to correlate the strata worldwide. The problem was that the nature of strata changes across regions. A stratum in England might contain different sediments than one of the same age in Russia. Geologists discovered that each stratum of the same age contained certain **index fossils.** These are fossils that identify deposits made at apparently the same time in different parts of the world. These index fossils are used in **relative dating** methods. For example, a particular species of fossil ammonite (an animal related to the chambered nautilus) has been found over a wide range and for a limited time period. Therefore, all strata around the world that contain this fossil must be of the same age.

Absolute Dating of Fossils

Absolute dating methods rely on **radiometric** techniques to assign an actual date to a fossil. All radioactive isotopes have a particular half-life, the length of time it takes for half of the radioactive isotope to change into another stable element.

Radiocarbon dating uses the radioactive decay of ^{14}C, a rare carbon isotope. Over time, ^{14}C will stabililize into ^{14}N (see section 2.1). This process of radioactive decay—that is, when ^{14}C becomes ^{14}N—occurs at a constant rate. Willard Libby won the Nobel Prize in Chemistry for his discovery in 1949 that the half-life of ^{14}C is 5,730 years. This means that half the original ^{14}C in organic matter decays to ^{14}N in this period of time. All living organisms have in them about the same ratio of ^{12}C to ^{14}C. The absolute date of a fossil can be determined by comparing the ^{12}C to ^{14}C ratio of a fossil to that of a living organism. This comparison can be used to calculate the age of the fossil.

Radiocarbon dating is accurate only for fossils up to approximately 100,000 years old. The amount of ^{14}C radioactivity in older fossils is so low that it cannot be used to measure the age of a fossil accurately. ^{14}C is the only radioactive isotope contained within organic matter, but it is possible to use others to date rocks, and from that to infer the age of a fossil contained in the rock. Potassium-argon (K-Ar) dating can be used to date rocks that are 100,000 to 4.5 billion years old, and thus it can be used to indirectly estimate the age of fossils well beyond the limits of radiocarbon dating.

The Precambrian Time

Geologists have devised the **geologic timescale,** which divides the history of the Earth into eras and then periods and epochs (Table 18.1). The geologic timescale was derived from the accumulation of data from the age of fossils in strata all over the world. Biologists traditionally begin the geologic timescale when life

Figure 18.7 Fossils. Fossils are the remains of living organisms from the past. They can be impressions left in rocks, footprints, mineralized bones, shells, or any other evidence of life-forms that lived in the past.

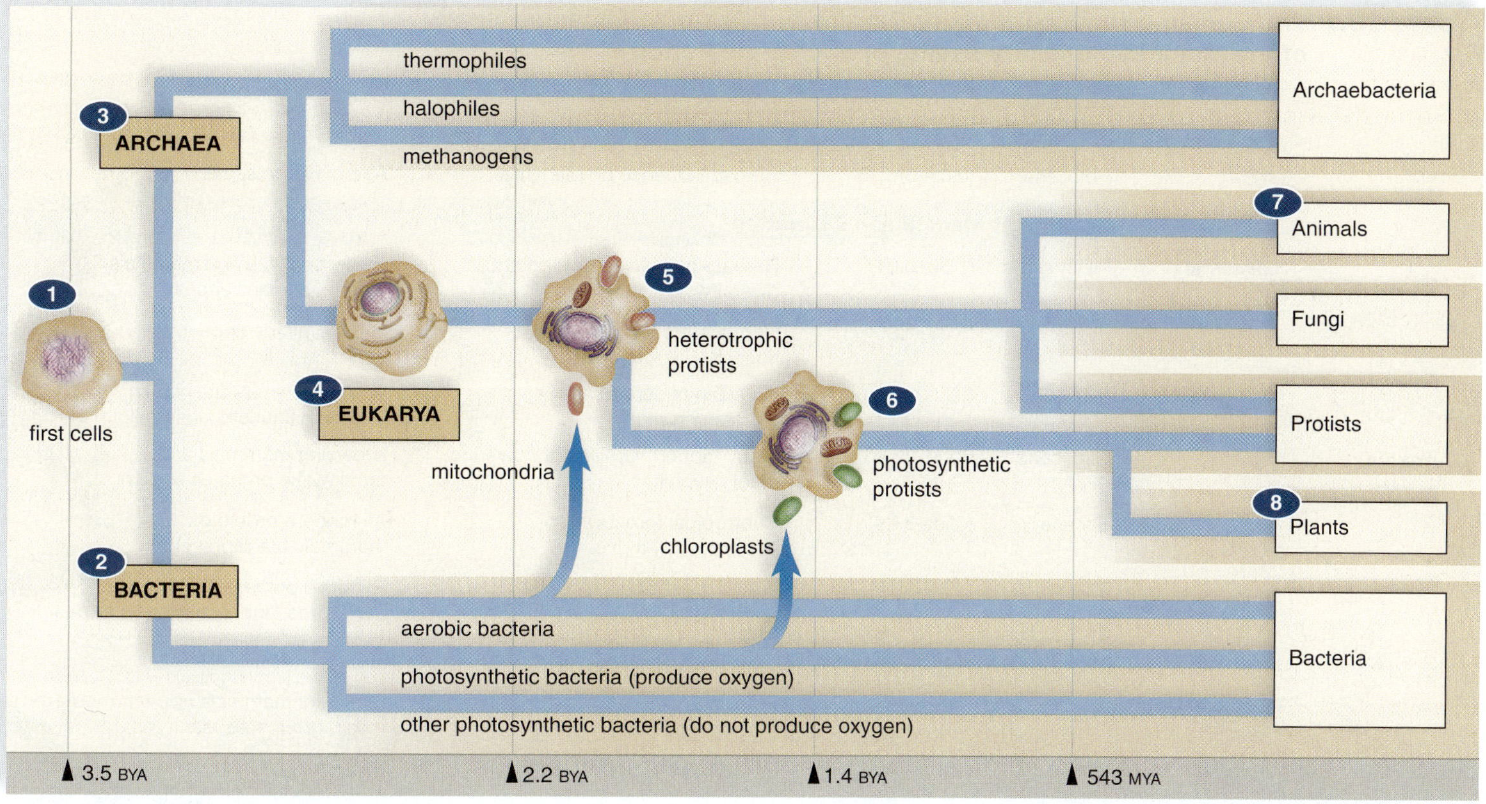

Figure 18.8 The tree of life. During the Precambrian time, (1) the first cell or cells gave rise to (2) bacteria and (3) archaea; (4) the first eukaryotic cell evolved from archaea. (5) Heterotrophic protists arose when eukaryotic cells gained mitochondria by engulfing aerobic bacteria, and (6) photosynthetic protists arose when these cells gained chloroplasts by engulfing photosynthetic bacteria. (7) Animals (and fungi) evolved from heterotrophic protists, and (8) plants evolved from photosynthetic protists. BYA = billion years ago; MYA = million years ago

began during Precambrian time (Table 18.1). The Precambrian is a very long period of time, comprising about 87% of the geologic timescale. During this time, life arose and the first cells came into existence (Fig. 18.8).

The first modern-type cells were probably prokaryotes, which first appeared approximately 3.5 BYA (Table 18.1). Prokaryotes are relatively simple cells—they do not have a nucleus or membrane-bound organelles. Prokaryotes can live in the most inhospitable of environments, such as hot springs, salty lakes, and oxygen-free swamps—all of which may typify habitats on early Earth. The cell wall, plasma membrane, RNA polymerase, and ribosomes of one lineage of prokaryotes, the archaea, are more like those of eukaryotes than those of bacteria (see Chapter 20).

The first identifiable fossils are those of complex prokaryotes. Chemical fingerprints of complex cells are found in sedimentary rocks from southwestern Greenland, dated at 3.8 BYA. The oldest prokaryotic fossils have been discovered in western Australia. These 3.46-billion-year-old microfossils resemble today's cyanobacteria, prokaryotes that carry on photosynthesis in the same manner as plants. At that time, only volcanic rocks jutted above the waves, and there were as yet no continents. Strange-looking boulders, called **stromatolites,** littered beaches and shallow waters (Fig. 18.9*a*). Living stromatolites can still be found along Australia's western coast.

The outer surface of a stromatolite is alive with cyanobacteria. The cyanobacteria in ancient stromatolites added oxygen to the atmosphere (Fig. 18.9*b*). By 2.0 BYA, the presence of oxygen made most environments unsuitable for anaerobic prokaryotes. Photosynthetic cyanobacteria and aerobic bacteria proliferated as new metabolic pathways evolved. Due to the presence of oxygen, the atmosphere became an oxidizing one instead of a reducing one. Oxygen in the upper atmosphere forms ozone (O_3), which filters out the ultraviolet (UV) rays of the sun. Before the formation of the **ozone shield,** the amount of ultraviolet radiation reaching the Earth could have helped create organic molecules, but it would have destroyed any land-dwelling organisms. Once the ozone shield was in place, living organisms were sufficiently protected and able to live on land.

Eukaryotic Cells Arise

The eukaryotic cell, which originated around 2.1 BYA, obtains energy from cellular metabolism in the presence of oxygen. These cells contain a nucleus as well as other membranous organelles. The eukaryotic cell acquired its organelles gradually. The nucleus may have developed by an invagination of the plasma membrane. The ancestors of the mitochondria in eukaryotic cells were free-living bacteria that synthesized ATP via an electron transport chain, and chloroplasts were free-living photosynthetic prokaryotes. The **endosymbiotic theory** states

Table 18.1 **The Geologic Timescale: Major Divisions of Geologic Time and Some of the Major Evolutionary Events of Each Time Period**

Era	*Period*	*Epoch*	*Million Years Ago (MYA)*	*Plant Life*	*Animal Life*
		Holocene	(0.01–0)	Human influence on plant life	Age of *Homo sapiens*
		Significant Mammalian Extinction			
	Quaternary	Pleistocene	(1.80–0.01)	Herbaceous plants spread and diversify.	Presence of Ice Age mammals. Modern humans appear.
		Pliocene	(5.33–1.80)	Herbaceous angiosperms flourish.	First hominids appear.
		Miocene	(23.03–5.33)	Grasslands spread as forests contract.	Apelike mammals and grazing mammals flourish; insects flourish.
Cenozoic	Tertiary	Oligocene	(33.9–23.03)	Many modern families of flowering plants evolve.	Browsing mammals and monkeylike primates appear.
		Eocene	(55.8–33.9)	Subtropical forests with heavy rainfall thrive.	All modern orders of mammals are represented.
		Paleocene	(65.5–55.8)	Flowering plants continue to diversify.	Primitive primates, herbivores, carnivores, and insectivores appear.
		Mass Extinction: 50% of All Species, Dinosaurs and Most Reptiles			
	Cretaceous		(145.5–65.5)	Flowering plants spread; conifers persist.	Placental mammals appear; modern insect groups appear.
Mesozoic	Jurassic		(199.6–145.5)	Flowering plants appear.	Dinosaurs flourish; birds appear.
		Mass Extinction: 48% of All Species, Including Corals and Ferns			
	Triassic		(251–199.6)	Forests of conifers and cycads dominate.	First mammals appear; first dinosaurs appear; corals and molluscs dominate seas.
		Mass Extinction ("The Great Dying"): 83% of All Species on Land and Sea			
Paleozoic	Permian		(299–251)	Gymnosperms diversify.	Reptiles diversify; amphibians decline.
	Carboniferous		(359.2–299)	Age of great coal-forming forests; ferns, club mosses, and horsetails flourish.	Amphibians diversify; first reptiles appear; first great radiation of insects.
		Mass Extinction: Over 50% of Coastal Marine Species, Corals			
	Devonian		(416–359.2)	First seed plants appear. Seedless vascular plants diversify.	First insects and first amphibians appear on land.
	Silurian		(443.7–416)	Seedless vascular plants appear.	Jawed fishes diversify and dominate the seas.
		Mass Extinction: Over 57% of Marine Species			
	Ordovician		(488.3–443.7)	Nonvascular land plants appear on land.	First jawless and then jawed fishes appear.
	Cambrian		(542–488.3)	Marine algae flourish.	All invertebrate phyla present; first chordates appear.
			630	Soft-bodied invertebrates.	
			1,000	Protists diversify.	
Precambrian Time			2,100	First eukaryotic cells.	
			2,700	O_2 accumulates in atmosphere.	
			3,500	First prokaryotic cells.	
			4,570	Earth forms.	

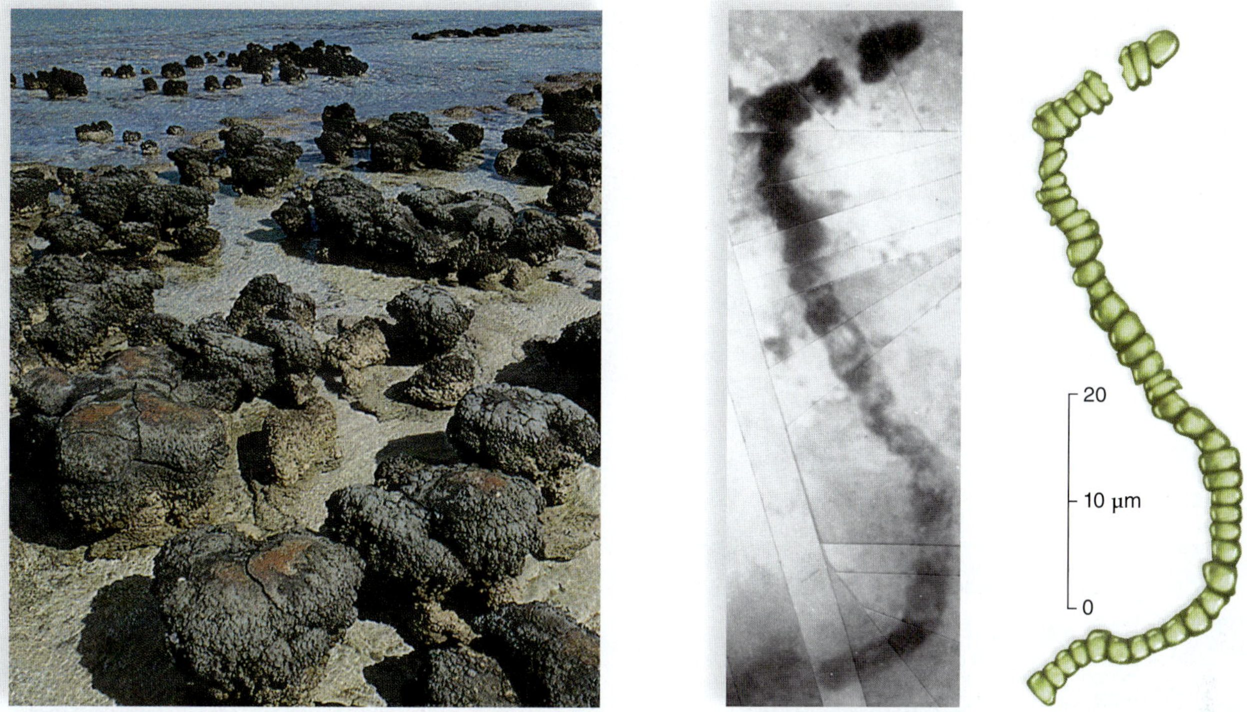

a. Stromatolites

b. *Primaevifilum*

Figure 18.9 Prokaryote fossils of the Precambrian. **a.** Stromatolites date back to 3.46 BYA. Living stromatolites are located in shallow waters off the shores of western Australia and in other tropical seas. **b.** The prokaryotic microorganism *Primaevifilum* was found in fossilized stromatolites.

that a nucleated cell engulfed various prokaryotes, which became the cell's organelles. The nucleated cell and the engulfed bacteria coevolved the ability to synthesize ATP via oxidative phosphorylation (see Fig. 18.8). The evidence for the endosymbiotic theory is as follows.

1. The present-day mitochondria and chloroplasts have a size that lies within the range of that for bacteria.
2. Mitochondria and chloroplasts have their own DNA and make some of their own proteins. (The DNA of the nucleus also codes for some of the mitochondrial proteins.)
3. The mitochondria and chloroplasts divide by binary fission, as bacteria do.
4. The outer membrane of mitochondria and chloroplasts differ—the outer membrane resembles that of a eukaryotic cell and the inner membrane resembles that of a bacterial cell.

It has been suggested that flagella (and cilia) also arose by endosymbiosis. First, slender, undulating prokaryotes could have attached themselves to a host cell to take advantage of food leaking from the host's plasma membrane. Eventually, these prokaryotes adhered to the host cell and became the flagella and cilia we know today. The first eukaryotes were single-celled, the same as prokaryotes.

Animation Endosymbiosis

Multicellularity Arises

Fossils of multicellular protists at least as old as 1.4 BYA have been found in Arctic Canada. It's possible that the first multicellular organisms practiced sexual reproduction. Among today's protists we find colonial forms in which some cells are specialized to produce the gametes needed for sexual reproduction.

Separation of germ cells, which produce gametes, from somatic cells may have been an important first step toward the evolution of the Ediacaran invertebrates, which appeared about 630 MYA and died out about 545 MYA. The first fossils of these organisms were found in the Ediacara Hills of Australia. Since then, similar fossils have been discovered on a number of other continents.

Many of the fossils, dated 630–545 MYA, are thought to be of soft-bodied invertebrates (animals without a vertebral column) that most likely lived on mudflats in shallow marine waters. Some may have been mobile, but others were large, immobile, bizarre creatures resembling spoked wheels, corrugated ribbons, and lettucelike fronds. All were flat and probably had two tissue layers; few had any type of skeleton (Fig. 18.10). They apparently had no mouths; perhaps they absorbed nutrients from the sea or had photosynthetic organisms living on their tissues. With few exceptions, this group disappeared from the fossil record at 545 MYA, but it may have given rise to modern cnidarians and related animals.

The Paleozoic Era

The Paleozoic era lasted about 300 million years. Even though the era was quite short compared to the Precambrian, three major mass extinctions occurred during this era (Table 18.1). An **extinction** is the total disappearance of all the members of a species or higher taxonomic group. **Mass extinctions** are the disappearance

Figure 18.10 Ediacaran fossils. The Ediacaran invertebrates lived from about 600 to 545 MYA. They were all flat, soft-bodied invertebrates. **a.** Classified as *Charniodiscus,* this filter-feeder historically was identified as a sea pen, although that classification is being questioned. **b.** Classified as *Dickinsonia,* these fossils are often interpreted to be segmented worms. However, in the opinion of some, they may be cnidarian polyps.

Figure 18.11 Sea life of the Cambrian period. The animals depicted here are found as fossils in the Burgess Shale, a formation of the Rocky Mountains of British Columbia, Canada. Some lineages represented by these animals are still evolving. *Marrella* has been classified as an arthropod, *Thaumaptilon* is considered to be an early sea pen. *Vauxia* is believed to be a sponge and *Wiwaxia* a segmented worm.

of a large number of species or higher taxonomic groups within an interval of just a few million years (see page 333).

Cambrian Animals

The seas of the Cambrian period, which began at about 542 MYA, teemed with invertebrate life (Fig. 18.11). Life became so abundant that scientists refer to this period in Earth's history as the Cambrian explosion. All of today's groups of animals can trace their ancestry to this time, and perhaps earlier, according to new molecular clock data. A *molecular clock* is based on the principle that mutations in certain parts of the genome occur at a fixed rate and are not tied to natural selection. Therefore, the number of DNA base-pair differences tells how long two species have been evolving separately.

Animals that lived during the Cambrian possessed protective outer skeletons known as exoskeletons. These hard-body parts were fossilized more readily than Precambrian soft-bodied organisms. Thus, fossils are more abundant during the Cambrian. Cambrian seafloors were dominated by now-extinct trilobites, which had thick, jointed armor covering them from head to tail. Trilobites, a common Cambrian fossil, are classified as arthropods, a major phylum of animals today.

Invasion of Land

Life first began to move out of the ocean and onto land around 500 MYA. The process of adapting to a land-based environment occurred over a long period of time. Different life-forms, such as

Figure 18.12 Swamp forests of the Carboniferous period. **a.** Vast swamp forests of treelike club mosses and horsetails dominated the land during the Carboniferous period (Table 18.1). The air contained insects with wide wingspans, such as the predecessors of dragonflies. Amphibians diversified greatly in this environment. **b.** Dragonfly fossil from the Carboniferous period. **c.** Modern-day dragonfly.

plants, invertebrates, and vertebrates, had independent histories, and each group colonized the land at different times.

Plants. During the Ordovician period, algae, which were abundant in the seas, most likely began to take up residence in bodies of fresh water. Eventually, algae invaded damp areas on land. The first land plants were nonvascular (did not possess water-conducting tissues), similar to the mosses and liverworts that survive today. The lack of water-conducting tissues limited the height of these plants to a few centimeters. Although the Ordovician evidence is scarce, spore fossils from this time support this hypothesis.

Fossils of seedless vascular plants (those having tissue for water and organic nutrient transport) date back to the Silurian period. They later flourished in the warm swamps of the Carboniferous period. Club mosses, horsetails, and seed ferns were the trees of that time, and they grew to enormous size. A wide variety of smaller ferns and fernlike plants formed an underbrush (Fig. 18.12*a*).

Invertebrates. The jointed appendages and exoskeleton of arthropods are adapted to life on land. Various arthropods—spiders, centipedes, mites, and millipedes—all preceded the appearance of insects on land. Insects, also arthropods, enter the fossil record in the Carboniferous period. One fossil dragonfly from the Carboniferous had a wingspan of nearly a meter. The evolution of wings provided advantages that allowed insects to radiate into the most diverse and abundant group of animals today. Flying provides a way to escape predators, find food, and disperse to new territories.

Vertebrates. Vertebrates are animals with a vertebral column. The vertebrate line of descent began in the early Ordovician period with the evolution of jawless fishes. Jawed fishes appeared later in the Silurian period. Fishes are ectothermic (see section 29.3) aquatic vertebrates that have gills, scales, and fins. The cartilaginous and ray-finned fishes made their appearance in the Devonian period, which is called the Age of Fishes.

At that time, the seas contained giant predatory fishes covered with protective armor made of external bone. Sharks cruised up deep, wide rivers, and smaller, lobe-finned fishes lived at the river's edge in waters too shallow for large predators. Fleshy fins helped the small fishes push aside debris or hold their place in strong currents, and the fins may have allowed these fishes to venture onto land and lay their eggs safely in inland pools. A large amount of data tells us that lobe-finned fishes were ancestral to the amphibians and to modern-day lobe-finned fishes.

Amphibians are thin-skinned vertebrates that are not fully adapted to life on land, particularly because they must return to water to reproduce. The Carboniferous swamp forests provided the water they needed, and amphibians adaptively radiated into many different sizes and shapes. Some superficially resembled alligators and were covered with protective scales; others were small and snakelike; and a few were larger plant-eaters. The largest measured 6 m from snout to tail. The Carboniferous period is called the Age of Amphibians.

Video Amphibian Origin

The process that turned the great Carboniferous forests into the coal we use today to fuel our modern society started during the Carboniferous period. The weather turned cold and dry, and this brought an end to the Age of Amphibians. A major mass extinction event occurred at the end of the Permian period, bringing an end to the Paleozoic era and setting the stage for the Mesozoic era.

The Mesozoic Era

Although a severe mass extinction occurred at the end of the Paleozoic era, the evolution of certain types of plants and animals continued into the beginning of the Mesozoic era. Nonflowering seed plants (collectively called gymnosperms) evolved and spread during the Paleozoic, becoming the dominant plant life. Cycads are short and stout, with palmlike leaves that produce large cones. Cycads and related plants

Figure 18.13 Dinosaurs of the late Cretaceous period. *Parasaurolophus walkeri,* although not as large as other dinosaurs, was one of the largest plant-eaters of the late Cretaceous period. The crest atop its head was about 2 m long, and it may have been used as a way to regulate body temperature or as a way to communicate by making booming calls. Also living at that time were the rhinolike dinosaurs represented here by *Triceratops* (*left*), another herbivore.

were so prevalent during the Triassic and Jurassic periods that these periods are sometimes called the Age of Cycads. Reptiles can be traced back to the Permian period of the Paleozoic era. Unlike amphibians, reptiles can thrive in a dry climate, because they have scaly skin and lay a shelled egg, which hatches on land. Reptiles underwent an adaptive radiation during the Mesozoic era to produce forms that lived in the sea and on the land, as well as forms that could fly. One group of reptiles, the therapsids, had several mammalian skeletal traits.

During the Jurassic period, large, flying reptiles called pterosaurs ruled the air, and giant marine reptiles with paddlelike limbs ate fishes in the sea. But on land, dinosaurs dominated while the evolving mammals remained small and less conspicuous.

Although the average size of the dinosaurs was about that of a crow, many giant species developed. The gargantuan *Apatosaurus* and the armored, tractor-sized *Stegosaurus* fed on cycad seeds and conifer trees. The size of a dinosaur such as *Apatosaurus* is hard for us to imagine. It was 4.5 m tall at the hips and 27 m in length, and it weighed about 40 tons. A large body size is beneficial in that, being ectothermic (cold-blooded), the surface-area-to-volume ratio was favorable for retaining heat. Some data also suggest that dinosaurs were endothermic (warm-blooded).

During the Cretaceous period, great herds of rhinolike dinosaurs, *Triceratops,* roamed the plains, as did *Tyrannosaurus rex. T. rex* was a carnivore, perhaps a part-time scavenger, filling the same ecological role that lions do today. *Parasaurolophus* was a long-crested, duck-billed dinosaur (Fig. 18.13). The long, hollow crest was bigger than the rest of its skull and may have functioned as a thermoregulatory organ, or as a resonating chamber for making booming calls during mating or helping members of a herd locate each other. In comparison to *Apatosaurus, Parasaurolophus* was small. It was less than 3 m tall at the hips and weighed only about 3 tons. Still, it was one of the largest plant-eaters of the late Cretaceous period and fed on pine needles, leaves, and twigs. *Parasaurolophus* was easy prey for large predators; its main defense would have been running away in large herds.

At the end of the Cretaceous period, the dinosaurs became victims of a mass extinction, which is discussed on page 333.

One group of bipedal dinosaurs, called theropods, includes the Tyrannosaurs and various raptors, such as *Velociraptor*. This group most likely gave rise to the birds, whose fossil record includes the famous *Archaeopteryx*.

Up until 1999, Mesozoic mammalian fossils largely consisted of teeth. This changed when a fossil found in China was dated at 120 MYA and named *Jeholodens.* The animal, identified as a mammal, apparently looked like a long-snouted rat. Surprisingly, *Jeholodens* had sprawling hindlimbs as do reptiles, but its forelimbs were under the belly, as in today's mammals.

Small mammals appeared early in the Mesozoic and were very diverse in the Jurassic and early Cretaceous periods. By the end of the Cretaceous, the common ancestors of each of the modern orders of mammals had evolved. However, by the late Cretaceous, the majority of the Jurassic and early Cretaceous mammalian diversity had gone exinct.

The Cenozoic Era

Classically, the Cenozoic era is divided into the Tertiary and Quaternary periods. Another scheme, dividing the Cenozoic into the Paleogene and the Neogene periods, is gaining popularity. This

Figure 18.14 Mammals of the Oligocene epoch. The artist's representation of these mammals and their habitat vegetation is based on fossil remains.

Figure 18.15 Woolly mammoth of the Pleistocene epoch. Woolly mammoths were animals that lived along the borders of continental glaciers.

new system divides the epochs differently. In any case, we are currently living in the Holocene epoch.

Mammalian Diversification

At the end of the Mesozoic era, mammals began an adaptive radiation into the many habitats now left vacant by the demise of the dinosaurs. Mammals are endotherms that have hair to help keep body heat from escaping. Their name refers to the presence of mammary glands, which produce milk to feed their young. Mammals were very diverse in the Mesozoic, but by the start of the Paleocene epoch, the majority of Mesozoic mammal diversity had gone extinct.

By the end of the Eocene epoch, all the modern orders of mammals had evolved—representatives of modern mammal lineages were the only mammals left in existence. Mammals underwent adaptive radiation into a number of environments. Several species of mammals, including the bats, took to the air. Ancestors of modern whales, dolphins, and manatees began their return to an aquatic lifestyle. On land, herbivorous hoofed mammals populated the forests and grasslands and were preyed upon by carnivorous mammals. Many of the types of herbivores and carnivores of the Oligocene epoch are extinct today (Fig. 18.14).

Evolution of Primates

The ancestors of modern primates appeared during the Eocene epoch about 55 MYA. The first primates were small, squirrel-like animals. Flowering plants (collectively called angiosperms) were already diverse and plentiful by the Cenozoic era. Many species of tree-sized angiosperms had evolved by the Eocene. Some primates adapted to living in trees, where protection from predators and food in the form of fruit was plentiful.

Ancestral apes appeared during the Oligocene epoch. These primates were adapted to living in the open grasslands and savannas. Apes diversified during the Miocene and Pliocene epochs and gave rise to the first hominids, a group that includes humans. Many of the skeletal differences between apes and humans relate to the fact that humans walk upright. Exactly what caused humans to adopt bipedalism is still being debated.

The world's climate became progressively colder during the Tertiary period. The Quaternary period begins with the Pleistocene epoch, which is known for multiple ice ages in the Northern Hemisphere. During periods of glaciation, snow and ice covered about one-third of the land surface of the Earth. The Pleistocene epoch was an age of not only humans but also mammalian **megafauna**—giant ground sloths, beavers, wolves, bison, woolly rhinoceroses, mastodons, and mammoths (Fig. 18.15). Humans are still in existence, but the oversized mammals have gone extinct. Mammalian megafauna had died out by the end of the Pleistocene. One hypothesis is that human hunting was at least partially responsible for the extinction of these animals. Climate change is another hypothesis.

Check Your Progress 18.2

1. Determine the age of a fossil that contains 25% of its original ^{14}C.
2. Identify two features of organelles that support the endosymbiotic theory of organelle evolution.
3. List the sequence of events in the Precambrian that led to the evolution of heterotrophic and photosynthetic protists.
4. List the major events in the history of life that occurred during each of the Earth's eras.

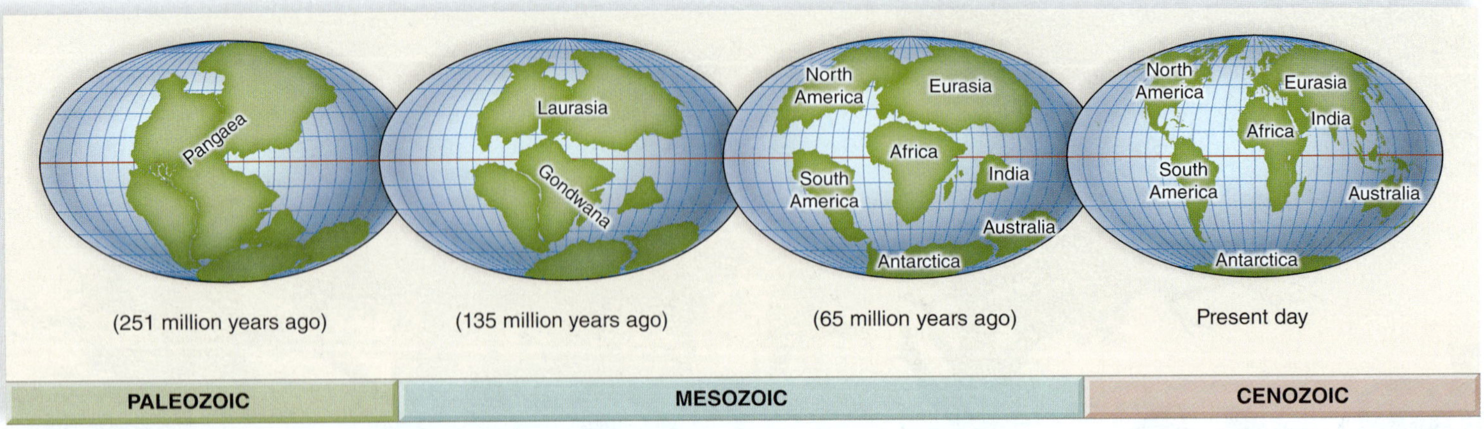

Figure 18.16 Continental drift. About 251 MYA, all the continents were joined into a supercontinent called Pangaea. During the Mesozoic era, the joined continents of Pangaea began moving apart, forming two large continents called Laurasia and Gondwana. Then, all the continents began to separate. Presently, North America and Europe are drifting apart at a rate of about 2 cm per year.

18.3 Geological Factors That Influence Evolution

Learning Outcomes

Upon completion of this section, you should be able to

1. Identify who proposed the theory of continental drift, as well as the evidence provided to support this theory.
2. Describe plate tectonics and how it explains the drifting of continents.
3. Interpret biogeographical and geological evidence that supports continental drift.
4. Describe the geological points at which mass extinctions occurred and give one proposed cause for each extinction event.

In the past, it was thought that the Earth's crust was immobile, that the continents had always been in their present positions, and that the ocean floors were only a catch basin for the debris that washed off the land. But in 1920, Alfred Wegener, a German meteorologist, presented data from a number of disciplines to support his hypothesis of continental drift.

Continental Drift

Continental drift was finally confirmed in the 1960s, establishing that the continents are not fixed; instead, their positions and the positions of the oceans have changed over time (Fig. 18.16). During the Paleozoic era, the continents joined to form one supercontinent, which Wegener called Pangaea (Gk. *pangea,* "all lands"). First, Pangaea divided into two large subcontinents, called Gondwana and Laurasia, and then these also split to form the continents of today. Presently, the continents are still drifting in relation to one another.

Continental drift explains why the coastlines of several continents are roughly mirror images of each other—for example, the outline of the west coast of Africa matches that of the east coast of South America. The same geological structures are also found in many of the areas where the continents touched. For example, based on geological data, a single mountain range runs through South America, Antarctica, and Australia.

Continental drift also explains the unique distribution patterns of several fossils. Fossils of the seed ferns (*Glossopteris*) have been found on all the continents, which supports the theory that all continents were once joined (Fig. 18.17). Similarly, the fossil reptile *Cynognathus* is found in Africa and in South America, and an assemblage of early mammal-like reptiles, the *Lystrosaurus* group, was discovered in Antarctica, far from Africa and Southeast Asia, where it also occurs.

With mammalian fossils, the situation is different: Australia, South America, and Africa share lineages of mammals that were widely distributed across Gondwana. For example, each continent has representatives of the rodent lineage. However, each continent also has its own unique mammal lineages that evolved in isolation after the continents separated.

The mammalian biological diversity of today's world is the result of isolated evolution on separate continents. For example, as mentioned in Chapter 15, marsupials are endemic only to South America and Australia, which were connected when the landmasses

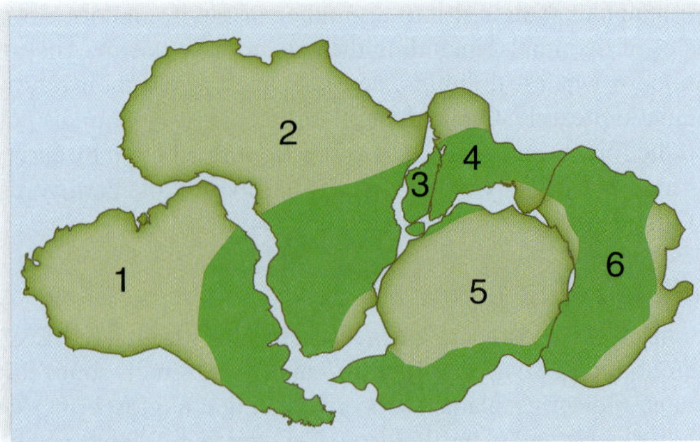

Figure 18.17 Biogeography of *Glossopteris* supports continental drift. Fossils of the fern *Glossopteris* (represented by the dark green shading) are distributed throughout what was Gondwanaland.

were a supercontinent (see Fig. 18.16). The fossil record suggests that marsupials were abundant in South America up until about 7 MYA. Marsupials are more prevalent in the Australian biogeographical region than in South America. This is most likely due to marsupials evolving in the Americas and reaching Australia when the southern continents were still joined. Once Australia separated off, marsupials were able to diversify without competition from the placental mammals that evolved elsewhere. Today, the only placental mammals native to Australia are bats and rodents. The rest of the placental mammals were introduced to the continent relatively recently. In contrast, placental mammals are prevalent in the Americas, and much of the marsupial diversity has disappeared on this continent.

Plate Tectonics

The answer to why continents drift has been suggested through a branch of geology known as **plate tectonics** (Gk. *tektos,* "fluid, molten, able to flow"). This idea shows that the Earth's crust is fragmented into slablike plates that float on a lower, hot mantle layer. The continents and the ocean basins are a part of these rigid plates, which move like conveyor belts. At ocean ridges, seafloor spreading occurs as molten mantle rock rises and material is added to the ocean floor. Seafloor spreading causes the continents to move a few centimeters a year, on the average. At *subduction zones,* the forward edge of a moving plate sinks into the mantle and is destroyed, forming deep-ocean trenches bordered by volcanoes or volcanic island chains.

The Earth isn't getting bigger or smaller, so the amount of oceanic crust being formed is as much as that being destroyed. When two continents collide, the result is often a mountain range; for example, the Himalayas resulted when India collided with Eurasia. The place where two plates meet and scrape past one another is called a *transform boundary*. The San Andreas fault in southern California is at a transform boundary, and the movement of the two plates is responsible for the many earthquakes in that region. Earthquakes leave visible evidence of the movement of plates at transform boundaries. In rare cases, the movement of the plates has been witnessed by people as it occurs—roads diverge, the ground uplifts, and cracks emerge during seismic activity. Examples of this were seen in the massive earthquake that took place in Japan in 2011.

Mass Extinctions

At least five mass extinctions have occurred throughout history: at the ends of the Ordovician, Devonian, Permian, Triassic, and Cretaceous periods (Fig. 18.18 and Table 18.1). Mass extinction can be caused by cataclysmic events or by a more gradual process brought on by environmental change.

These ideas were brought to the fore when Walter and Luis Alvarez proposed in 1977 that the Cretaceous extinction, during which the dinosaurs died out, was due to a bolide. A **bolide** is a large, crater-forming projectile that impacts Earth. The Alvarezes found that Cretaceous clay contains an abnormally high level of iridium, an element that is rare in the Earth's crust but more common in asteroids and meteorites. Walter Alvarez stated that the "10^8-megaton impact of the comet (or asteroid) which ended the Cretaceous was . . . the equivalent of the explosion of 10,000 times the entire nuclear arsenal of the world."[1] An impact of this magni-

[1]Alvarez, Walter. 2008. *T. rex and the Crater of Doom.* (Princeton University Press, Princeton, N.J.).

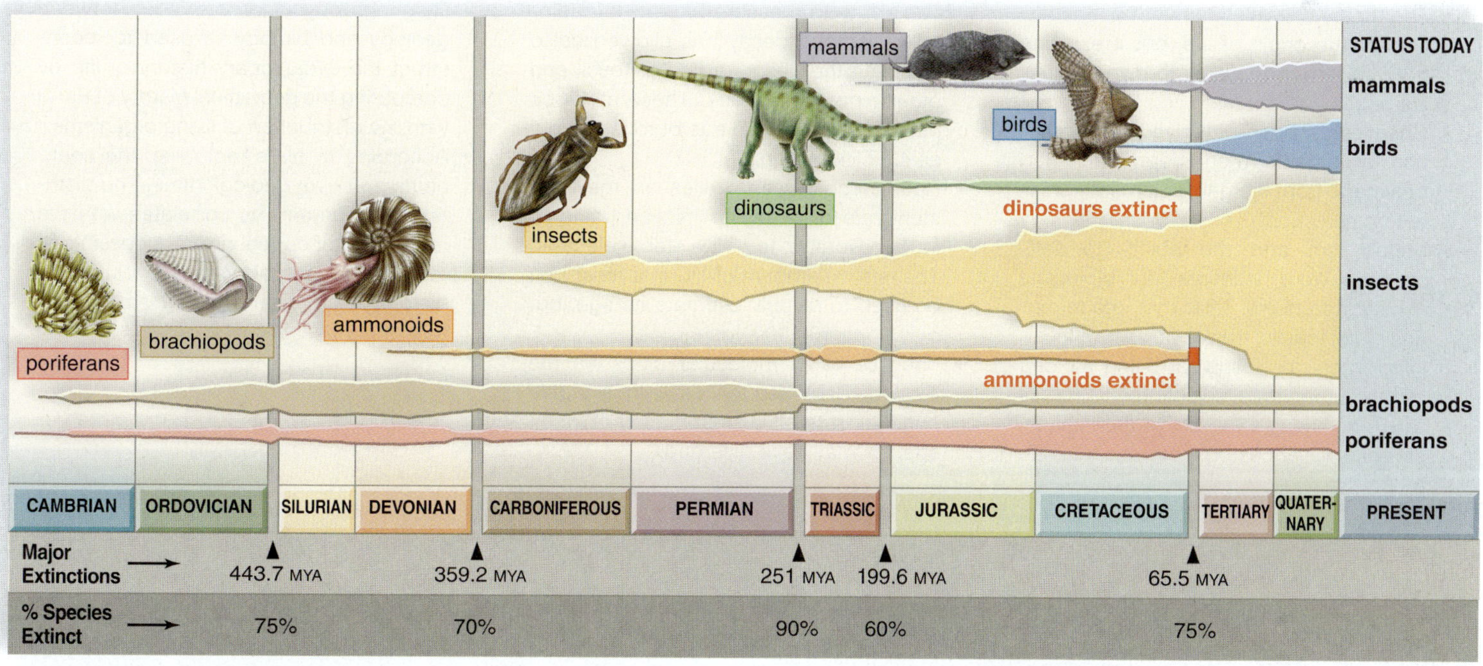

Figure 18.18 Mass extinctions. Five significant mass extinctions and their effects on the abundance of certain forms of marine and terrestrial life. The thickness of the horizontal, colored bars indicates the varying abundance of each life-form considered. MYA=million years ago

tude would have blasted into the atmosphere a layer of ash and soot so expansive it would have blocked the sun.

A crater that could have been caused by a meteorite impact of this magnitude was found in the Caribbean–Gulf of Mexico region on the Yucatán peninsula. A layer of soot alongside the iridium has dated the impact to the end of the Cretaceous and beginning of the Tertiary periods. This crater is known as the Chicxulub crater. Certainly, continental drift contributed to the Ordovician extinction. This extinction occurred after Gondwana arrived at the South Pole. Immense glaciers, which drew water from the oceans, chilled even once-tropical land. Marine invertebrates and coral reefs, which were especially hard hit, didn't recover until Gondwana drifted away from the pole and warmth returned.

The mass extinction at the end of the Devonian period saw an end to 70% of marine invertebrates. Helmont Geldsetzer of Canada's Geological Survey notes that iridium has also been found in Devonian rocks in Australia, suggesting that a bolide event was involved. Some scientists believe that this mass extinction could have been due to the movement of Gondwana back to the South Pole.

The extinction at the end of the Permian period was quite severe; 90% of species disappeared. The latest hypothesis attributes the Permian extinction to excess carbon dioxide. When Pangaea formed, there were no polar ice caps to initiate ocean currents. The lack of ocean currents caused organic matter to stagnate at the bottom of the ocean. Then, as the continents drifted into a new configuration, ocean circulation switched back on. Then, the extra carbon on the seafloor was swept up to the surface, where it became carbon dioxide, a deadly gas for sea life. The trilobites became extinct, and the crinoids (sea lilies) barely survived. Excess carbon dioxide on land led to climate change, which altered the pattern of vegetation. Areas that were wet and rainy became dry and warm, and vice versa. Burrowing animals that could escape land surface changes seemed to have the best chance of survival.

The extinction at the end of the Triassic period is another that has been attributed to the environmental effects of a meteorite collision with Earth. Central Quebec has a crater half the size of Connecticut that some believe is the impact site. The dinosaurs may have benefited from this event, because this is when the first of the gigantic dinosaurs took charge of the land. A second wave occurred in the Cretaceous period, but it ended in dinosaur extinction, as discussed previously.

Check Your Progress 18.3

1. Explain why continents drift, and summarize the geological evidence to support plate tectonics.
2. Defend the theory of continental drift with evidence from the biogeographical distribution of organisms.
3. List the mass extinction events that have occurred on Earth, and describe the hypotheses that explain the cause of each.

CONNECTING *the* CONCEPTS *with the* THEMES

Evolution

- The theory of evolution is not needed to explain the origin of life, but it explains how life on Earth became diverse after life began.
- Chemical evolution involved the formation of organic monomers, and then polymers, from inorganic elements present on early Earth.
- Metabolism and the ability to self-replicate were necessary for biological evolution to begin. After that, the formation of protocells, true cells, and multicellular organisms followed, being molded by forces of natural selection.
- Only one lineage, the LUCA (last universal common ancestor), gave rise to all life on Earth.

Nature of Science

- The ability to measure the rate of radioactive isotope decay has allowed us to calculate the absolute age of fossil and other types of remains. These methods have determined the age of rocks as old as 4.5 billion years.
- Radiocarbon dating relies on the constant rate of decay of ^{14}C, an unstable carbon isotope, to ^{14}N, a stable element. The relative amount of ^{14}C in a fossil can be used to reliably estimate its age, but only up to 100,000 years.
- The use of a "molecular clock" in the form of DNA sequences with a stable mutation rate allows estimation of when species may have diverged.

Biological Systems

- Biogeography, the interface between geology and biology, is used to reconstruct the evolutionary history of life by comparing the geological history of Earth with the distribution of living organisms.
- According to plate tectonics, the continents drift over geologic time. The distribution of organisms correlates with the movement of continents. It is possible to test hypotheses about the evolutionary history of organisms based on the history of continent movement.

MEDIA STUDY TOOLS http://connect.mheducation.com

SMARTBOOK® Study smart. SmartBook helps you maximize your study time by identifying what you know and don't know. The recharge feature in SmartBook helps you prepare for the test by identifying the material that you are most likely to forget.

connect | BIOLOGY Find even more resources to learn the chapter concepts including animations, tutorial videos, and interactive practice questions.

Animations	Video
18.1 Miller-Urey Experiment **18.2** Endosymbiosis	**18.2** Amphibian Origin

SUMMARIZE

18.1 Origin of Life

The **last universal common ancestor** (**LUCA**) is believed to be the ancestor of all life on Earth. **Biomolecules** of living organisms are believed to have formed from nonliving matter or the process of **abiogenesis.** The Oparian-Haldane hypothesis, or **primordial soup hypothesis,** proposes the conditions of early Earth that would have allowed an **abiotic synthesis** or chemical evolution of organic molecules from inorganic ones either in the atmosphere or at hydrothermal vents. These monomers may have joined together to form polymers at thermal vents according to the **iron-sulfur world hypothesis.** If abiogenesis occurred on land, the sun could have caused **protenoids** to form, which could have turned into **microspheres** when placed in water. The **protein-first hypothesis** assumes that proteins arose prior to DNA. The **RNA-first hypothesis** assumes that RNA evolved first.

The aggregation of polymers inside a fatty acid plasma membrane produced a **protocell** (**protobiont**) having some enzymatic properties such that it could grow. The basic assemblage of fatty acids into a small sphere is called a **micelle,** which can merge with others to form a **vesicle. Coacervate droplets** are complex units formed from concentrated mixtures of macromolecules under appropriate conditions. A collection of lipids in solution will form bubbles called **liposomes,** leading researchers to formulate the **membrane-first hypothesis.** Biological evolution began with the origin of self-replicating molecules, such as DNA and RNA.

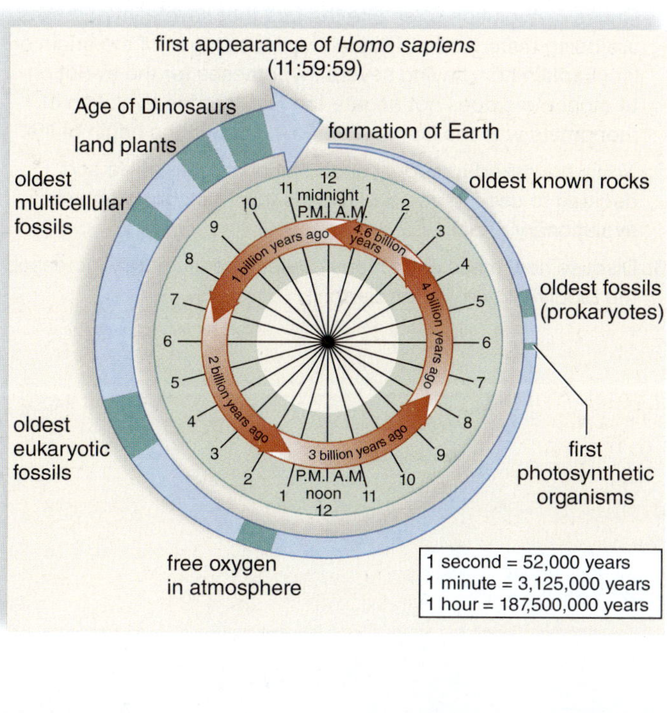

18.2 History of Life

Fossils provide evidence about the life that has existed on Earth. **Paleontology** is the study of the fossil record. **Sedimentation** is the accumulation of particles carried by moving water that leads to the formation of **sediment.** Layers of sediment accumulate to form **strata. Index fossils** are often found in the stratum and used as a form of **relative dating. Absolute dating** is accomplished by using **radiometric** techniques, such as **radiocarbon dating** to assign an actual age to a fossil. The accumulation of these data has enabled geologists to establish the **geologic timescale,** which divides the Earth into eras, periods, and epochs.

The oldest prokaryotic fossils are cyanobacteria, dated about 3.5 BYA, found in strange boulder forms called **stromatolites.** They were the first organisms to add oxygen to the atmosphere, thus forming the early **ozone shield.** The **endosymbiotic theory** states that a nucleated cell engulfed various prokaryotes, forming the first eukaryotic cell about 2.2 BYA. Multicellular animals (the Ediacaran animals) did not evolve until about 600 MYA.

The Paleozoic era is characterized by the disappearance of all members of a species, or **extinction.** More specifically, there were three **mass extinctions** during this time. A rich animal fossil record starts at the Cambrian period of the Paleozoic era. The evolution of the exoskeleton seems to explain the increased number of fossils at this time. Predation pressures may also have selected for the evolution of a hard external enclosure like an exoskeleton. The fishes were the first vertebrates to diversify and become dominant. Amphibians are descended from lobe-finned fishes.

Plants invaded the land from water during the Ordovician period. The swamp forests of the Carboniferous period contained seedless vascular plants, insects, and amphibians. This period is sometimes called the Age of Amphibians.

The Mesozoic era was the Age of Cycads and Reptiles. Mammals and birds evolved from reptilian ancestors. During this era, dinosaurs of enormous size were present. By the end of the Cretaceous period, the dinosaurs were extinct.

The Cenozoic era is divided into the Tertiary period and the Quaternary period. The Tertiary is associated with the adaptive radiation of mammals and flowering plants that formed vast tropical forests. The Quaternary is associated with the evolution of mammalian **megafauna** as well as the primates. Grasslands were replacing forests, and this put pressure on primates, which were adapted to living in trees, to respond to an ever-expanding terrestrial habitat. The result may have been the evolution of humans—primates who left the trees.

18.3 Geological Factors That Influence Evolution

The continents are on massive plates that move, carrying the land with them. **Continental drift** helps explain the distribution pattern of today's land organisms. **Plate tectonics** is a branch of geology that helps explain the movement of the Earth's plates, which contributes to continental drift. Five mass extinctions have played a dramatic role in the history of life on Earth. It has been suggested that the extinction at the end of the Cretaceous period was caused by the impact of a large **bolide,** or crater-forming meteorite. Evidence indicates that other extinctions have a similar cause as well. It has also been suggested that tectonic, oceanic, and climatic fluctuations, particularly due to continental drift, can bring about mass extinctions.

ASSESS

Choose the best answer for each question.

18.1 Origin of Life

For questions 1–5, match the statements with events in the key. Answers may be used more than once.

Key:

a. early Earth
b. monomers evolve
c. polymers evolve
d. protocell evolves
e. self-replication system evolves

1. The heat of the sun could have caused amino acids to form proteinoids.
2. In a liquid environment, phospholipid molecules automatically form a membrane.
3. As the Earth cooled, water vapor condensed, and subsequent rain produced the oceans.
4. Miller's experiment shows that under the right conditions, inorganic chemicals can react to form small organic molecules.
5. An abiotic synthesis may have occurred at hydrothermal vents.
6. Evolution of the DNA ⟶ RNA ⟶ protein system was a milestone, because the protocell could now
   a. be a heterotrophic fermenter.
   b. pass on genetic information.
   c. use energy to grow.
   d. take in preformed molecules.
   e. All of these are correct.

18.2 History of Life

7. Fossils
   a. are the remains and traces of past life.
   b. can be dated absolutely according to their location in strata.
   c. are usually remains of soft tissues, such as muscle.
   d. have been found for all types of animals except humans.

For questions 8–12, match the phrases with the divisions of geologic time in the key. Answers may be used more than once.

Key:

a. Cenozoic era
b. Mesozoic era
c. Paleozoic era
d. Precambrian time

8. dinosaur diversity, evolution of birds and mammals
9. contains the Carboniferous period
10. prokaryotes abound; eukaryotes evolve and become multicellular
11. mammalian diversification
12. invasion of land

18.3 Geological Factors That Influence Evolution

13. Continental drift helps explain
    a. mass extinctions.
    b. the distribution of fossils on the Earth.
    c. geological upheavals, such as earthquakes.
    d. climatic changes.
    e. All of these are correct.
14. Which scientist proposed the theory of continental drift?
    a. Alfred Wegener
    b. Stanley Miller
    c. Thomas Cech
    d. Claudia Huber
    e. David Deamer
15. Mass extinctions occurred during all of the following periods except which one?
    a. Devonian
    b. Permian
    c. Triassic
    d. Cretaceous
    e. Carboniferous

ENGAGE

THINKING SCIENTIFICALLY

1. Critics of evolution often cite the fact that several hypotheses are being tested to help explain various stages of the origin of life. Explain how having several hypotheses for the evolution of monomers does not negate the fact that the evolution of monomers was the first stage of evolution in the origin of life.
2. You were asked to supply an evolutionary tree of life and decided to use Figure 18.8. How is this tree consistent with evolutionary principles?
3. Discuss how mass extinctions could have ultimately increased the biodiveristy of life on Earth.

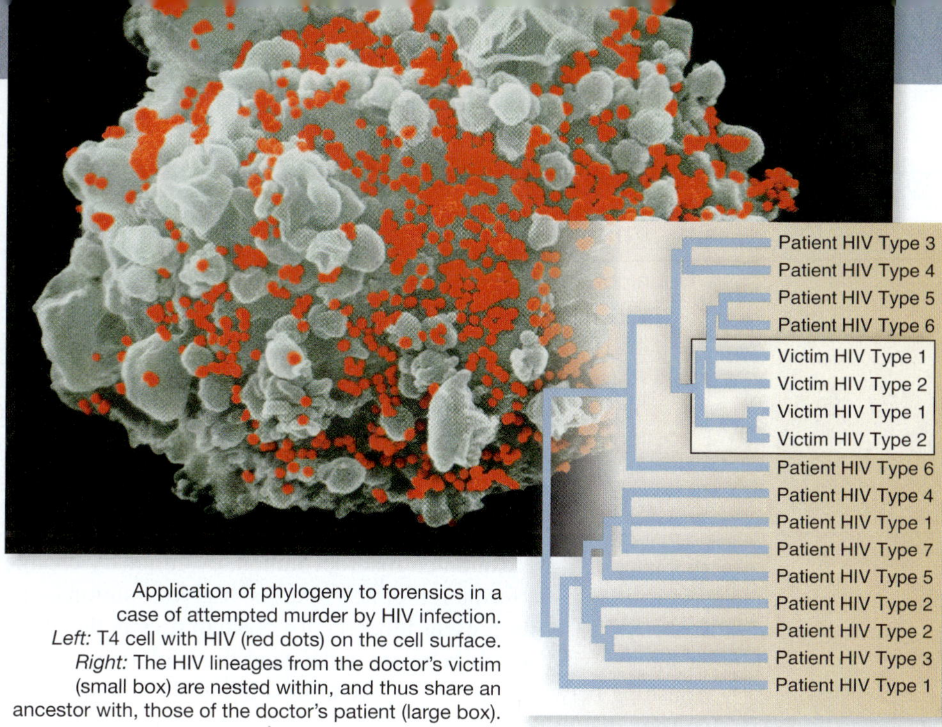

Application of phylogeny to forensics in a case of attempted murder by HIV infection. *Left:* T4 cell with HIV (red dots) on the cell surface. *Right:* The HIV lineages from the doctor's victim (small box) are nested within, and thus share an ancestor with, those of the doctor's patient (large box).

19

Taxonomy, Systematics, and Phylogeny

Chapter Outline

19.1 Systematic Biology 338
19.2 The Three-Domain System 341
19.3 Phylogeny 344

Molecular technology offers a powerful new set of tools to address issues important to human society. The ability to identify the genetic profile of an organism has helped address problems in conservation, agriculture, medicine, and forensic science. In the courtroom, a segment of a gene from HIV was used to convict a gastroenterologist for attempted murder of his girlfriend when he purposely injected her with HIV disguised in a vitamin B_{12} shot. A phylogeny, or "evolutionary tree," of the HIV gene traced the girlfriend's HIV to one of the doctor's patients with HIV/AIDS. The doctor was convicted and sentenced to a 50-year prison sentence for attempted murder. Systematic biology uses phylogenies, in a similar way, to trace the origin of biodiversity back to a common ancestor.

In this chapter we examine how systematic biologists use traits to reconstruct evolutionary history and classify organisms according to shared ancestry.

As you read through this chapter, think about the following questions:

1. How are the study of macroevolution and the study of systematic biology interrelated?
2. Why is it important to classify biodiversity, and why is evolutionary history important to this classification?
3. What is a phylogeny? What does it represent, and what can it tell about the history of life on Earth?

Before You Begin

Before beginning this chapter, take a few moments to review the following discussions.

Section 1.4 What is biodiversity?

Section 12.3 Why is DNA called the "genetic code of life"?

Section 17.2 How does adaptive radiation result in the formation of new species?

Following *the* Themes

Chapter 19 Taxonomy, Systematics, and Phylogeny

Unit 3 Evolution

Evolution	Systematic biology reconstructs evolutionary history as a tree showing common ancestors and their descendants.
Nature of Science	The tree of life is a phylogeny, or an evolutionary tree, that represents the best hypothesis of the evolutionary history of life on Earth.
Biological Systems	The evolutionary relationship among living organisms, in the form of a phylogeny, serves as the framework for addressing issues in conservation, medicine, and agriculture.

19.1 Systematic Biology

Learning Outcomes

Upon completion of this section, you should be able to

1. Differentiate among taxonomy, classification, and systematic biology.
2. Reconstruct the levels of the Linnaean classification hierarchy.
3. Identify the genus and species of an organism from its scientific name.

In Chapter 17 we examined macroevolution, evolutionary change that results in the formation of new species. Macroevolution is the source of past and present biodiversity. **Systematics** is the study of biodiversity, which helps us understand the evolutionary relationships between species. **Systematic biology** is a quantitative science that uses **traits** of living and fossil organisms to infer the relationships among organisms over time.

The Field of Taxonomy

Taxonomy (Gk. *tasso,* "arrange, classify"; *nomos,* "usage, law") is the branch of systematic biology that identifies, names, and organizes biodiversity into related categories (Fig. 19.1). A **taxon** (pl. taxa) is the general name for a group containing an organism or a group of organisms that exhibits a set of shared traits. **Classification** is the process of naming and assigning organisms or groups of organisms to a taxon. For example, the taxon Vertebrata contains organisms with a bony spinal column. As another example, the taxon Canidae contains the wolf (*Canis lupus*) and its close relative, the domestic dog (*Canis familiaris*).

Taxonomists, scientists who study taxonomy, strive to classify all of the life on Earth. The methods used to classify living organisms have changed throughout history. The Greek philosopher Aristotle (384–322 BCE) was interested in taxonomy, and he sorted organisms into groups, such as horses, birds, and oaks, based on a set of shared traits. Similarly, early taxonomists after Aristotle relied on physical traits to classify organisms. This method proved to be problematic, however, because many features of organisms were similar not because they shared a common ancestor but because of convergent evolution (see Chapter 17). For example, animals that have wings could be grouped into a single taxonomic group, but birds, bats, and beetles, all of which have wings, are profoundly different in many other ways.

Today, taxonomists attempt to classify organisms into **natural groups,** groupings of organisms that represent a shared evolutionary history. Natural groups are classified by using a set of traits to construct a **phylogeny,** or evolutionary "family tree," that represents the evolutionary history of taxa. This evolutionary history is then used to classify taxa based on shared ancestry.

Figure 19.1 Classifying organisms. How would you name and classify these organisms? After naming them, how would you assign each to a particular group? Based on what criteria? An artificial system would not take into account how they might be related through evolution, as would a natural system.

a.

b. *Lilium canadense*

c. *Lilium bulbiferum*

Figure 19.2 Carolus Linnaeus. a. Linnaeus was the father of taxonomy and devised the binomial system of naming and classifying organisms. His original name was Karl von Linne, but he later latinized it because of his fascination with scientific names. Linnaeus was particularly interested in classifying plants. **b, c.** Both of these two lilies are species in the same genus, *Lilium.*

Advances in DNA technology allow modern systematic biologists to compare traits other than external features to classify organisms. For example, a phylogeny of animals constructed from DNA sequences clearly shows that beetles, birds, and bats have wings that evolved at different times in the history of life (see Fig. 19.9 for an example of DNA sequence differences). This means that birds, bats, and beetles do not share a common ancestor with wings. Rather, wings originated three times independently, on three different branches of the tree of life, as a result of convergent evolution.

Linnaean Taxonomy

The classification hierarchy that taxonomists use today was created by Carolus Linnaeus (1707–78), the father of modern taxonomy (see section 15.1). Linnaeus's system was developed as a way to organize biodiversity. In the mid-eighteenth century, Europeans traveled to distant parts of the world and described, collected, and sent back to Europe examples of plants and animals they had not encountered before. During this time of discovery, Linnaeus developed **binomial nomenclature,** part of his classification system in which each species receives a unique two-part Latin name (Fig. 19.2).

As an example, *Lilium bulbiferum* and *Lilium canadense* are two different species of lily. The first word, *Lilium,* is the genus (pl., genera), a classification category that can contain many species. The second word, known as the **specific epithet,** refers to one species within that genus. The specific epithet sometimes tells us something descriptive about the organism. Notice that the scientific name is in italics. The species name is designated by the full binomial name—in this case, either *Lilium bulbiferum* or *Lilium canadense.* The specific epithet without the genus gives no clue as to species—just as a house number alone without the street name is useless for finding an address. The genus name can be used alone, however, to refer to a group of related species.

Scientific names are derived in a number of ways. Some scientific names are descriptive—for example, *Acer rubrum* for the red maple (*Acer,* "maple"; *rubrum,* "red"). Other scientific names include geographic descriptions, such as *Alligator mississippiensis* for the *American alligator.* Scientific names can also include eponyms (derived from the name of a person), such as the owl mite *Strigophilus garylarsonii* (named after the cartoonist Gary Larson). Many scientific names are derived from mythical characters, such as *Iris versicolor,* named for Iris, the goddess of the rainbow.

Scientists use Latin, rather than common names, to describe organisms for a variety of reasons. First, a common name varies from country to country because of language differences. Second, even people who speak the same language sometimes use different common names to describe the same organism. For example, bowfin, grindle, choupique, and cypress trout all refer to the same common fish, *Amia calva.* Furthermore, between countries, the same common name is sometimes given to different organisms. A robin (*Erithacus rubecula*) in England is very different from a robin (*Turdus migratorius*) in the United States. Latin, however, is a universal language that not too long ago was well known by most scholars, many of whom were physicians or clerics. When scientists throughout the world use scientific binomial names, they avoid the confusion caused by common names.

Linnaean Classification Hierarchy

The binomial nomenclature system of Linnaeus is used to classify species. Today, taxonomists use a nested, hierarchical set of categories to classify organisms (Fig. 19.3). Each taxon is given a name and a rank according to which of the following set of major taxonomic groups it belongs: **species, genus, family, order, class, phylum, kingdom,** and more recently the higher taxonomic category of **domain.** The organisms that fill a particular classification category are distinguishable from other organisms by their shared set of traits.

Organisms in the same domain have general traits in common, whereas those in the same species have quite specific traits in common. For example, the kingdom Animalia includes all animals. Within the kingdom Animalia is the phylum Chordata, a taxonomic group that contains only those animals with a spinal cord. Within the phylum Chordata is the class Mammalia, which contains animals that have spinal cords (phylum Chordata) and, among other characteristics, mammary glands. The species is the most exclusive

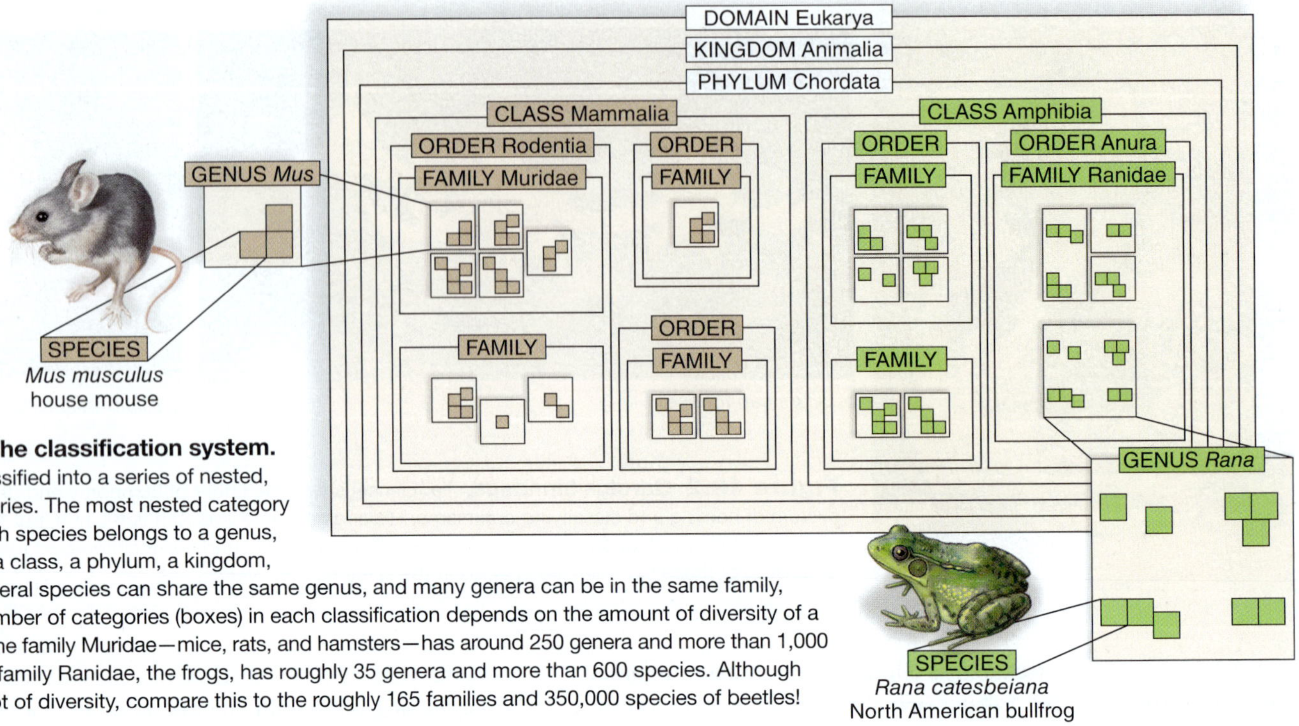

Figure 19.3 The classification system. Organisms are classified into a series of nested, hierarchical categories. The most nested category is the species. Each species belongs to a genus, a family, an order, a class, a phylum, a kingdom, and a domain. Several species can share the same genus, and many genera can be in the same family, and so on. The number of categories (boxes) in each classification depends on the amount of diversity of a particular group. The family Muridae—mice, rats, and hamsters—has around 250 genera and more than 1,000 species, while the family Ranidae, the frogs, has roughly 35 genera and more than 600 species. Although this seems like a lot of diversity, compare this to the roughly 165 families and 350,000 species of beetles!

of all the categories, as it contains only a single type of organism. The house mouse, *Mus musculus,* is a single species of mouse in the family Muridae, a family in the order Rodentia (one of several orders in the class Mammalia) (Fig. 19.3).

The classification hierarchy is very useful, because it allows scientists to organize the diversity of life, but it is important to remember that the hierarchy was created by scientists and thus does not represent any special relationship among organisms in nature. For example, grouping monkeys and apes into the order Primates tells scientists something about their evolutionary history, but being in the same order does not mean much to the genetics or behavior of monkeys or apes. The classification hierarchy, if based on natural groups, should be viewed as the best working hypothesis of evolutionary relationships.

As with any scientific hypothesis, the hierarchy, and the placement of organisms within it, is revised with the addition of new information. That is why you can find the same organism classified differently in older textbooks or even among taxonomists. This uncertainty is part of the nature of science and is one of its greatest principles.

Linnaeus set the standard of binomial nomenclature in the mid-1700s, but there were no official rules for classifying organisms until the mid-1800s. During this 100-year period, problems with name confusion arose when scientists in different parts of the world, who did not communicate with each other (remember, there was no Internet!), began to give the same Latin names to different species or to develop their own version of Linnaeus's classification system.

The first attempt to make standardized rules of **nomenclature,** the procedure of assigning scientific names to taxonomic groups, occurred in 1842 at a meeting of scientists, among them Charles Darwin. In 1961, after 120 more years of revisions, the International Code of Zoological Nomenclature (ICZN) was officially accepted as the universal guide for naming animals. Today, the International Commission on Zoological Nomenclature is the keeper of the ICZN, and all scientists of the world use its rules to classify animal groups. The International Code of Botanical Nomenclature is responsible for overseeing the naming of plants and certain groups of algae and fungi.

Despite a universal set of rules and over 250 years of taxonomy, only a fraction of the estimated 30 million or more species now living on Earth have been classified. Some taxonomic groups are classified more completely than others; naming of the birds and mammals may be complete, but there are millions of insects and microorganisms that remain to be discovered and classified.

The task of identifying and naming the species of the world is a daunting one. A new fast and efficient way of identifying species based on their DNA is described in the Nature of Science feature, "DNA Barcoding of Life," on page 342. This method, called DNA "barcoding," compares a short fragment of DNA sequence from an unknown organism to a large database of sequences from known organisms. The similarities and differences between the nucleotide sequence of the unknown organism and the sequences in the database can help determine which taxonomic group the organism likely belongs to or if it is something new to science.

DNA barcoding does not always get the taxonomy correct, however, and for this reason it has been criticized by some taxonomists as being too simplistic. Nevertheless, it is a potentially powerful way to rapidly and inexpensively catalog at least a portion of the world's biodiversity.

Check Your Progress 19.1

1. Describe the relationship among classification, systematic biology, and taxonomy.
2. Use the Linnaean classification system to fully classify the human species *Homo sapiens.*
3. Explain why the grouping together of birds and bats based on having wings does not represent a natural group.

19.2 The Three-Domain System

Learning Outcomes

Upon completion of this section, you should be able to

1. List the three domains of life.
2. Summarize two characteristics that define each of the three domains.

From Aristotle's time to the middle of the twentieth century, biologists recognized only two kingdoms: kingdom Plantae (plants) and kingdom Animalia (animals). Plants were literally organisms that were planted (immobile), whereas animals were animated (moved about). In the 1880s, a German scientist, Ernst Haeckel, proposed adding a third kingdom: The kingdom Protista (protists) included single-celled microscopic organisms but not multicellular, largely macroscopic ones.

In 1969, R. H. Whittaker expanded the classification system to the **five-kingdom system:** Monera, Protista, Fungi, Plantae, and Animalia. Organisms were placed in these kingdoms based on the type of cell (prokaryotic or eukaryotic), complexity (single-celled or multicellular), and type of nutrition. Kingdom Monera contained all the prokaryotes, which are organisms that lack a membrane-bound nucleus. These single-celled organisms were collectively called the bacteria. The other four kingdoms contain types of eukaryotes, which we describe later.

Defining the Domains

In the late 1970s, Carl Woese and his colleagues at the University of Illinois were studying the relationships among the prokaryotes by comparing the nucleotide sequences of ribosomal RNA (rRNA). Woese found that the rRNA sequences of prokaryotes that lived at high temperatures or produced methane are quite different from that of all the other types of prokaryotes and from all eukaryotes. Therefore, he proposed that there are two groups of prokaryotes (rather than one group, the Monera, in the five-kingdom system). Further, Woese found that these two groups of prokaryotes have rRNA sequences so fundamentally different from each other that they should be assigned to separate domains, the category of classification that is higher than the kingdom category. The two designated domains are **domain Bacteria** and **domain Archaea.** He placed the eukaryotes in a third domain, the **domain Eukarya.**

The phylogenetic tree shown in Figure 19.4 is based on Woese's rRNA sequencing data. The data suggested that both bacteria and archaea evolved early in the history of life from the last universal common ancestor (LUCA) (see Fig. 18.1). Later, the eukarya diverged from the archaea lineage. This means that the bacteria are members of the oldest lineage of living organisms on Earth, and the eukaryotes, the youngest lineage, are more closely related to the archaea than to the bacteria (Fig. 19.4).

Animation Three Domains

Domain Bacteria

Bacteria are such a diversified group that they can be found in large numbers in nearly every environment on Earth. The archaea are structurally similar to bacteria but are placed in a separate domain, domain Archaea, because of biochemical differences (Table 19.1). The details of these differences are covered in Chapter 20.

The cyanobacteria are large, photosynthetic prokaryotes. They carry on photosynthesis in the same manner as plants in that they use solar energy to convert carbon dioxide and water to a carbohydrate, in the process giving off oxygen. The cyanobacteria belong to a very ancient lineage of bacteria, and they may have been the

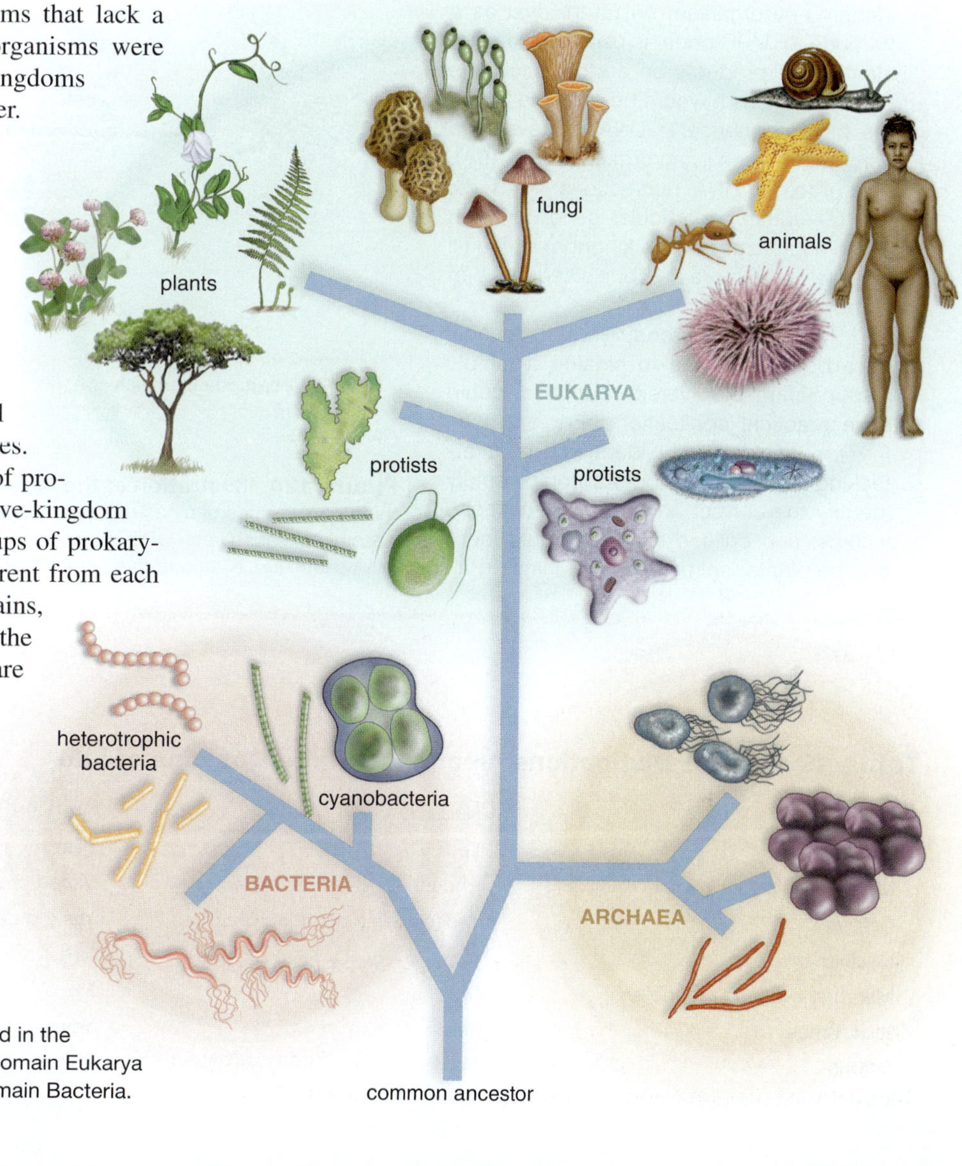

Figure 19.4 A tree of life showing the three domains. Representatives of each domain are depicted in the shaded areas. The tree shows that domain Archaea and domain Eukarya are more closely related to each other than either is to domain Bacteria.

THEME Nature of Science

DNA Barcoding of Life

Traditionally, taxonomists relied on anatomical traits to tell species apart. Although useful, physical features have several limitations when used as the only means to define a species. The identification of anatomical traits often requires the assistance of taxonomic experts who specialize on a particular group of organisms. In recent time, the number of species has far exceeded the number of available experts. This puts the cataloging of the world's biodiversity on a slow path—much slower than the rate at which our natural areas are disappearing. So far, scientists have identified only about 1.8 million species out of a potential 15 million or more.

The Consortium for the Barcode of Life (CBOL) proposes that any scientist, not just taxonomists, could use a sample of DNA to identify any organism on Earth. Just as a barcode, or UPC code, is used as a unique identifier of products on a store shelf, the CBOL suggests it would be possible to use the base sequence in DNA to develop a barcode for each living organism (Fig. 19A). The order of DNA nucleotides—A, T, C, and G—within a particular gene common to the organisms in each kingdom would fill the role of the store barcode's sequence of numbers.

Speedy DNA barcoding would not only be a boon to efforts to catalog a rapidly disappearing biodiversity but also would have practical applications. For example, farmers could readily identify a pest attacking their crops, doctors could rapidly identify the correct antivenin for snakebite victims, and college students could identify the plants, animals, and protists on an

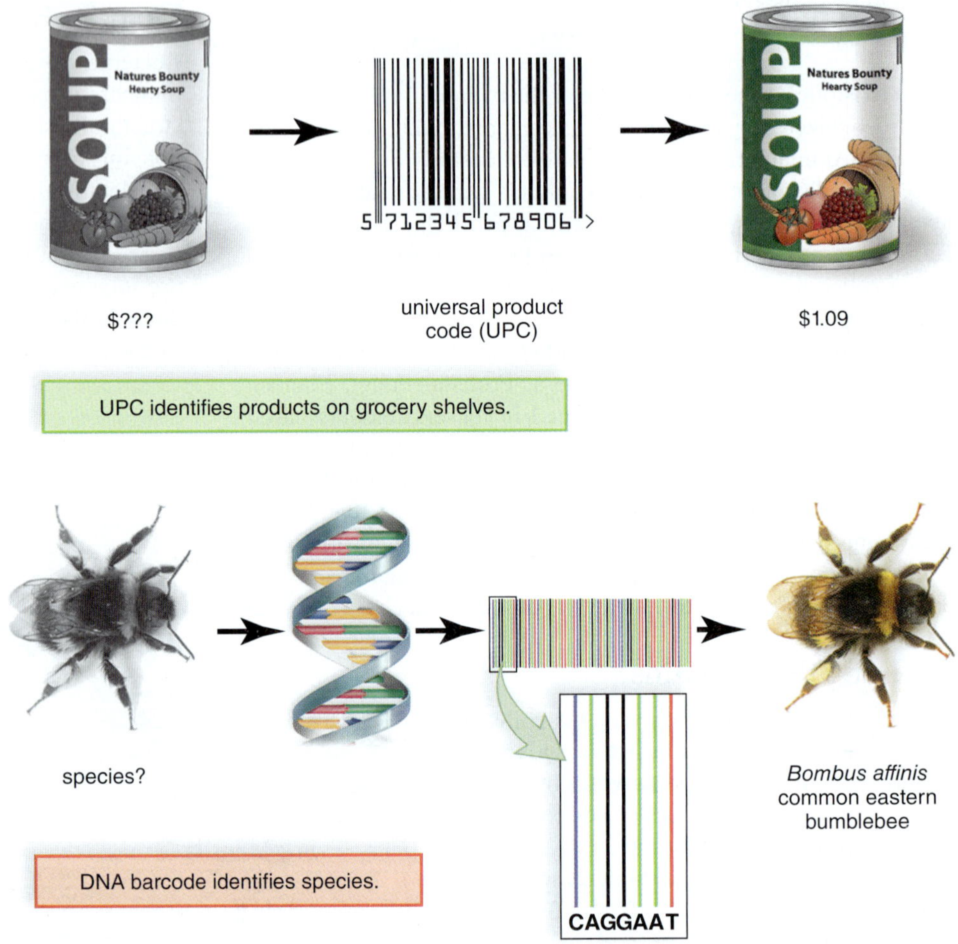

Figure 19A Illustration of the barcoding concept. Much as a barcode (UPC) can identify the type and price of a can of soup, a DNA barcode can be used to identify a species. This is possible because each UPC is unique to a particular product, and each DNA barcode, based on a DNA sequence, is unique to an individual species.

Table 19.1 Major Distinctions Among the Three Domains of Life

	Bacteria	*Archaea*	*Eukarya*
Single-celled	Yes	Yes	Some, many multicellular
Membrane lipids	Phospholipids, unbranched	Varied branched lipids	Phospholipids, unbranched
Cell wall	Yes (contains peptidoglycan)	Yes (no peptidoglycan)	Some yes, some no
Nuclear envelope	No	No	Yes
Membrane-bound organelles	No	No	Yes
Ribosomes	Yes	Yes	Yes
Introns	Some	Some	Yes

ecological field trip. Already, the CBOL has accumulated hundreds of thousands of DNA barcodes representing species across the diversity of life.

The CBOL initiative has the potential to be a powerful tool for conservation biologists and wildlife officials worldwide. A DNA barcode could be used to identify illegal trade in endangered species and for the early detection of invasive species that arrive into other countries as a consequence of global transportation.

In 2008, a pair of New York City high school students found a commercial application for the CBOL database (Fig. 19B). Kate Stoeckle and Louisa Strauss, two Trinity School seniors, did a project on the identification of fishes sold in markets and sushi restaurants in Manhattan, New York. They collected 60 fish samples from four restaurants and ten grocery stores in Manhattan, which they sent off to have the DNA segment, the barcode, sequenced and compared to a global library of fish barcodes representing nearly 5,500 fish species. Their results sent a wave of controversy throughout Manhattan and beyond: two of the four restaurants, and six of the ten grocery stores, sold fish that had been mislabeled. Most of the mislabeled fish were being sold as more expensive species. For example, Mozambique tilapia, a commonly farmed fish selling for \$1.70 per pound wholesale, was being sold as albacore tuna at \$8.50 per pound (Fig. 19C). In one case, they found an endangered fish, the Acadian redfish, being sold as red snapper!

Questions to Consider

1. How might systematic biologists use DNA barcoding to speed up the classification of biodiversity?
2. Propose additional ways that DNA barcoding could aid in managing modern societal problems, such as the conservation of biodiversity, global warming, crime, and disease.

Figure 19B Student Scientists. Two high-school students, Katie Stoeckle (*left*) and Louisa Strauss (*right*) uncovered mislabeled fish in Manhattan.

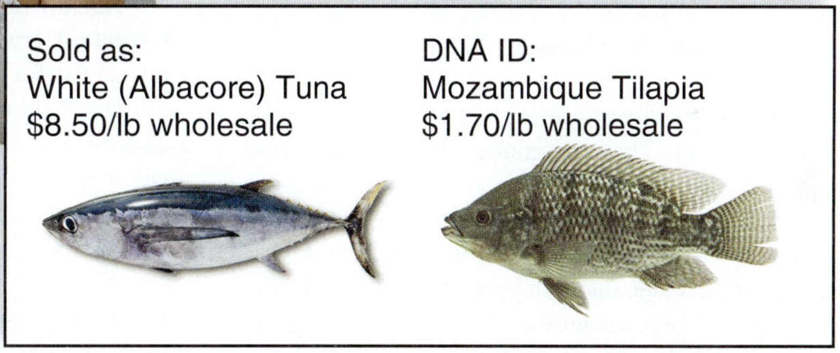

Figure 19C Mislabeled fish. DNA barcoding identified mislabeled fish in some New York City restaurants.

first organisms to contribute oxygen to early Earth's atmosphere. It is possible that they played a role in making the environment hospitable for the evolution of oxygen-using organisms, including animals.

Bacteria have a wide variety of means for obtaining nutrients, but most are heterotrophic. *Escherichia coli,* which lives in the human intestine, is heterotrophic. *Clostridium tetani* (the cause of tetanus), *Bacillus anthracis* (the cause of anthrax), and *Vibrio cholerae* (the cause of cholera) are disease-causing species of bacteria. Heterotrophic bacteria are beneficial in ecosystems, because they break down organic remains. Along with fungi, they keep chemical cycling going, so that plants always have a source of inorganic nutrients.

Domain Archaea

Like bacteria, archaea are prokaryotic single-celled organisms that reproduce asexually. Archaea don't look that different from bacteria under the microscope. The extreme conditions under which many species live has made it difficult to grow them in the laboratory. This may have been the reason that their unique place among the living organisms went unrecognized for a long time.

The archaea are distinguishable from bacteria by a difference in their rRNA nucleotide sequences and by their unique plasma membrane and cell wall chemistry. The chemical nature of the archaeal cell wall is diverse and is different from that of a bacterial cell. The unique cell wall structure of archaea might help them live in the extreme conditions in which they are found. In addition, the branched nature of diverse lipids in the archaeal plasma membrane is very different from that of the bacterial plasma membrane.

The archaea live in all sorts of environments, but they are known for thriving in extreme environments that are thought to be similar to those of the early Earth. For example, the methanogens live in environments without oxygen, such as swamps, marshes, and the guts of animals where methane is abundant. The halophiles thrive in salty environments, such as the Great Salt Lake in Utah, whereas the thermoacidophiles are both high temperature and acid loving. These archaea live in extremely hot, acidic environments, such as hot springs and geysers.

Domain Eukarya

Eukaryotes are single-celled to multicellular organisms whose cells have a membrane-bound nucleus. They also have various organelles, some of which arose through endosymbiosis of other single-celled organisms (see section 18.2). Sexual reproduction is common in eukaryotes, and various types of life cycles are seen.

Later in this text, we study the individual kingdoms that occur within the domain Eukarya. In the meantime, we can note that the protists are a diverse group of single-celled eukaryotes that are hard to classify and define. Some protists have filaments and form colonies or multicellular sheets. Even so, protists do not have true tissues. Nutrition is diverse; some are heterotrophic by ingestion or absorption and some are photosynthetic. Green algae, paramecia, and slime molds are representative protists. There has been considerable debate over the classification of protists, and presently they are placed in six supergroups (see Table 21.1) within the domain Eukarya.

Fungi are eukaryotes that form spores, lack flagella, and have cell walls containing chitin. They are multicellular, with a few exceptions. Fungi are saprotrophic by absorption—they secrete digestive enzymes and then absorb nutrients from decaying organic matter. Mushrooms, molds, and yeasts are representative fungi. Despite appearances, molecular data suggest that fungi and animals are more closely related to each other than either is to plants.

Plants are photosynthetic, multicellular organisms that are primarily adapted to a land environment. They share a common ancestor, which is an aquatic photosynthetic protist. Land plants possess true tissues and have the organ-system level of organization. Examples include cacti, ferns, and cypress trees.

Animals are motile, eukaryotic, multicellular organisms that evolved from a heterotrophic protist. Like land plants, animals have true tissues and the organ-system level of organization. Animals are heterotrophs. Examples include the worms, whales, and insects.

Check Your Progress 19.2

1. List the traits that separate the Archaea from other single-celled organisms.
2. Explain the evidence indicating that fungi are more closely related to animals than to plants.
3. Identify the domain that includes all multicellular organisms.

19.3 Phylogeny

Learning Outcomes

Upon completion of this section, you should be able to

1. Discriminate between ancestral and derived traits.
2. Interpret the evolutionary relationships depicted in a phylogeny.
3. List the types of traits used to construct a phylogeny.

Systematic biologists use characters from the fossil record, comparative anatomy and development, and the sequence, structure, and function of RNA and DNA molecules to construct a phylogeny. Systematic biologists study the evolutionary history of biodiversity, represented by a phylogeny. In essence, systematic biology is the study of the evolutionary history of biodiversity, and a phylogeny is the visual representation of that history.

Interpreting a Phylogeny

Systematic biologists construct a phylogeny from traits that are unique to, and shared by, a taxon and their **common ancestor** (an ancestor to two or more lines of descent). Each branch, or **lineage,** in a phylogeny represents a descendant of a common ancestor. When a new character evolves, a new evolutionary path can begin, or **diverge,** from the old, a new lineage is formed, and a new branch of the phylogeny arises (Fig. 19.5).

Animation Phylogenetic Trees

Not all traits have equal value when making a phylogeny. **Ancestral traits,** or those found in the common ancestor, are not useful for determining the evolutionary relationships of an ancestor's descendants. For example, deer, cattle, monkeys, and apes, all examples of mammals, share a common ancestor that had mammary glands (Fig. 19.5). Because all mammals have mammary glands, this trait is an ancestral trait and thus is not helpful for understanding how deer, cattle, monkeys, and apes are related to each other. In contrast, **derived traits,** or those not found in the common ancestor of a taxonomic group, are the most important traits for clarifying evolutionary relationships. For example, both monkeys and apes have an opposable thumb capable of grasping, a trait not present in the common ancestor of mammals. This shared derived trait places monkeys and apes on a separate lineage of mammals called "primates" (Fig. 19.5).

Whether a trait is derived or ancestral is dependent on whether it is present in the common ancestor, and thus relative to its location within a phylogeny. For example, an opposable thumb is a derived trait for all monkeys and apes when compared to the common ancestor of all mammals. But the opposable thumb, while a derived trait when compared to the mammal common ancestor, is nevertheless an ancestral trait of primates (Fig. 19.5). Similarly, deer and cattle, both artiodactyls, have even-toed hooves, a trait not found in the common ancestor of mammals or in primates. Thus, even-toed hooves is a derived trait when compared to the common ancestor of mammals, but an ancestral trait of artiodactyls (Fig. 19.5). Both even-toed hooves and the opposable thumb suggest that the evolutionary history of artiodactyls and primates became independent as each group diverged from the mammal common ancestor.

Derived traits provide a more detailed phylogeny as they define closer and closer evolutionary relationships. Within the primates, a fully rotating shoulder joint, which allows apes to swing from branch to branch in trees, and the prehensile tail of monkeys are derived traits that define separate ape and monkey branches within the primates. Similarly, horns and antlers are derived traits that divide the cattle and deer into two individual artiodactyl branches (Fig. 19.5).

The hierarchical classification system of Linnaeus defines species as closely related to other species within the same genus. A genus is related to other genera in the same family, and so forth, from order to class to phylum to kingdom to domain (see Fig. 19.3). When we say that two species (or genera, families, etc.) are closely related, we mean that they share a recent common ancestor. For example, all the animals in Figure 19.5 are related, because we can trace their ancestry back to a common ancestor. Taxonomists use the pattern of branching in a phylogeny constructed from an analysis of derived traits to classify taxa into natural groups.

For example, all mammals with even-toed hooves form a single lineage that is assigned to the order Artiodactyla (Fig. 19.5). Furthermore, artiodactyls that have antlers form a lineage within the order Artiodactyla that is classified as the family Cervidae. Antlers are grown only in males during the breeding season, and they can grow quite large and be highly branched. In contrast, artiodactyls that have horns form a lineage within the order Artiodactyla that is classified as the family Bovidae. Unlike antlers, horns are not shed seasonally and are found on both males and females, although they are smaller in females.

Similarly, the order Primates is a lineage of mammals with opposable thumbs. The rotating shoulder of apes and the prehensile tail of monkeys form two independent lineages in the order Primates that are classified as the family Hominidae and family Cebidae, respectively.

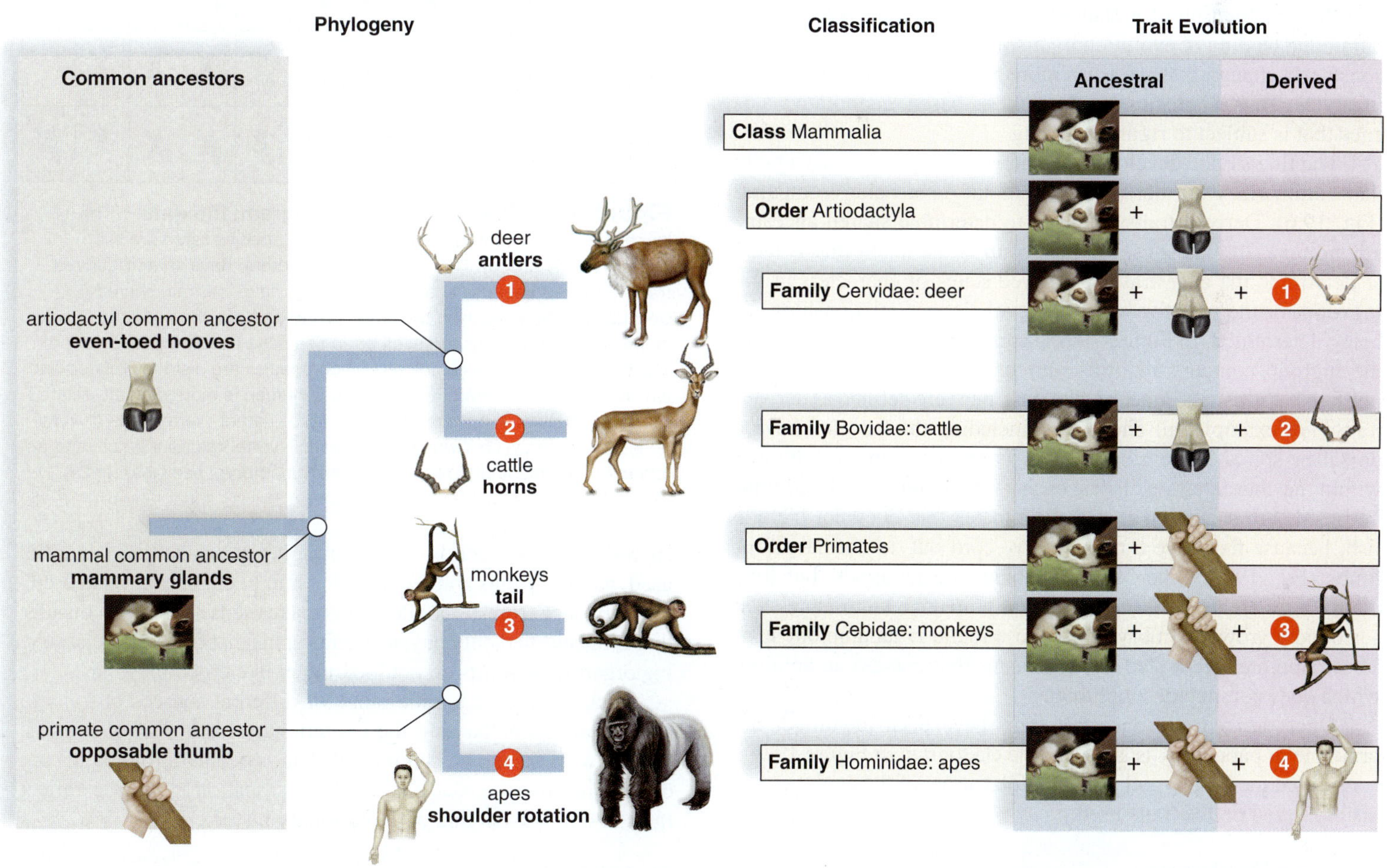

Figure 19.5 The relationship among phylogeny, classification, and traits. The phylogeny is a representation of the evolutionary history of a few members of the class Mammalia. Each common ancestor has a trait that is present in all of its descendants. For example, the red-numbered traits are derived traits present in all descendants of a clade: (1) deer have antlers; (2) cattle have horns; (3) monkeys have tails; (4) apes have full shoulder rotation. As you move toward the tips of the tree, new derived traits provide greater resolution to the classifications. For example, mammals have mammary glands. Artiodactyls have mammary glands and even-toed hooves. Deer have mammary glands, even-toed hooves, and antlers.

Cladistics

When constructing any phylogeny, multiple characteristics from a variety of organisms are compared at the same time (Fig. 19.6). This approach can produce various evolutionary trees, because not all traits are equally useful to the study of evolutionary history. The challenge is determining which phylogeny of the many possible phylogenies is the best hypothesis of evolutionary history. One way to determine the answer is through the use of cladistics.

Cladistics is a method that uses shared, derived traits to develop a hypothesis of evolutionary history. The evolutionary history of derived traits is interpreted into a type of phylogeny constructed with cladistic methods, called a **cladogram.** In a cladogram, a common ancestor and all of its descendant lineages is called a **clade.**

Cladistics applies the principle of **parsimony** (L. *parsimonia,* "frugality, thrift") to a set of traits to construct a cladogram. Parsimony considers the simplest solution to be the "optimal" solution. Thus, the cladogram that represents the simplest evolutionary history—that is, the one that requires the fewest number of evolutionary changes—is considered the best hypothesis based on the traits used to construct the cladogram. As with any hypothesis, a cladogram may change when new traits are discovered and incorporated into the construction of a cladogram. The important point is that our understanding of evolutionary history is a working hypothesis, constantly changing as we learn more about organisms' traits and lifestyles. Thus, cladistics is a hypothesis-based, quantitative science that is subject to rigorous testing.

The first step in developing a cladogram is to construct a table that summarizes the derived traits of the taxa being compared (Fig. 19.6). Derived traits are used to determine shared ancestry among taxa. In cladistics, the **outgroup** is the taxon that is used to determine the ancestral and derived states of characters in the **ingroup,** or the taxa for which the evolutionary relationships are being determined. In Figure 19.6, the outgroup is the lancelet, and the ingroup contains all other vertebrates. Traits present in the ingroup but not in the lancelet (the outgroup) are defined as derived traits. For example, all **chordates,** including the lancelet, have a dorsal or spinal nerve cord, so this is an ancestral trait. Nested within the chordates are clades, each with a uniquely derived trait. Tetrapods are a clade within the chordates that does not include fish, because fish have a dorsal nerve cord but do not have four limbs (Fig. 19.7). Likewise, amphibians are tetrapods, but they do not have an amnion, one of several protective membranes that surround a growing embryo, like those found in an amniotic egg. Thus, amniotes are a clade of organisms that possess an amnion, which does not include amphibians.

Once derived and ancestral traits have been identified, the principle of parsimony is applied. The cladogram in Figure 19.7 is considered the best hypothesis, or explanation, of the evolutionary history based on the traits used.

	Species							
	ingroup							lancelet (outgroup)
Traits	chimpanzee	dog	finch	crocodile	lizard	frog	tuna	
mammary glands	X	X						
hair	X	X						
gizzard			X	X				
epidermal scales			X	X	X			
amniotic egg	X	X	X	X	X			
four limbs	X	X	X	X	X	X		
vertebrae	X	X	X	X	X	X	X	
notochord in embryo	X	X	X	X	X	X	X	X

Figure 19.6 Constructing a cladogram: the data. The lancelet is in the outgroup, and all the other species listed are in an ingroup (study group). The species in the ingroup have shared derived traits—derived because a lancelet does not have the trait, shared because certain species in the study group do have them. All the species in the ingroup have vertebrae, all but a fish have four limbs, and so forth. The shared derived traits indicate which species are distantly related and which are closely related. For example, a human is more distantly related to a fish, with which it shares only one trait—namely vertebrae—than to an iguana, with which it shares three traits—vertebrae, four limbs, and an amniotic egg (the amnion layer protects the embryo; see Chapter 29).

Tracing Phylogeny

Traditionally, systematic biologists relied on morphological data to study evolutionary relationships between taxa. However, morphology can be misleading. For example, recent studies suggest that birds and crocodiles are more closely related to each other than crocodiles are to lizards. The morphological traits that Linnaeus used, namely wings and feathers, led him to classify birds as a different lineage from crocodiles. This makes sense when we consider that Linnaeus had only physical traits at his disposal when classifying organisms—birds do not look much like crocodiles!

Today, we have a wide range of different sources of traits to assist with understanding the evolutionary history of organisms. Armed with these new sources of data, systematic biologists are continually revising the historical classification system to reflect their best understanding of evolutionary history.

Fossil Traits

One of the advantages of fossils is that they can be dated (see Chapter 18), but it is not always possible to tell to which lineage, living or extinct, a fossil is related. At present, paleontologists are discussing whether fossil turtles indicate that turtles are distantly

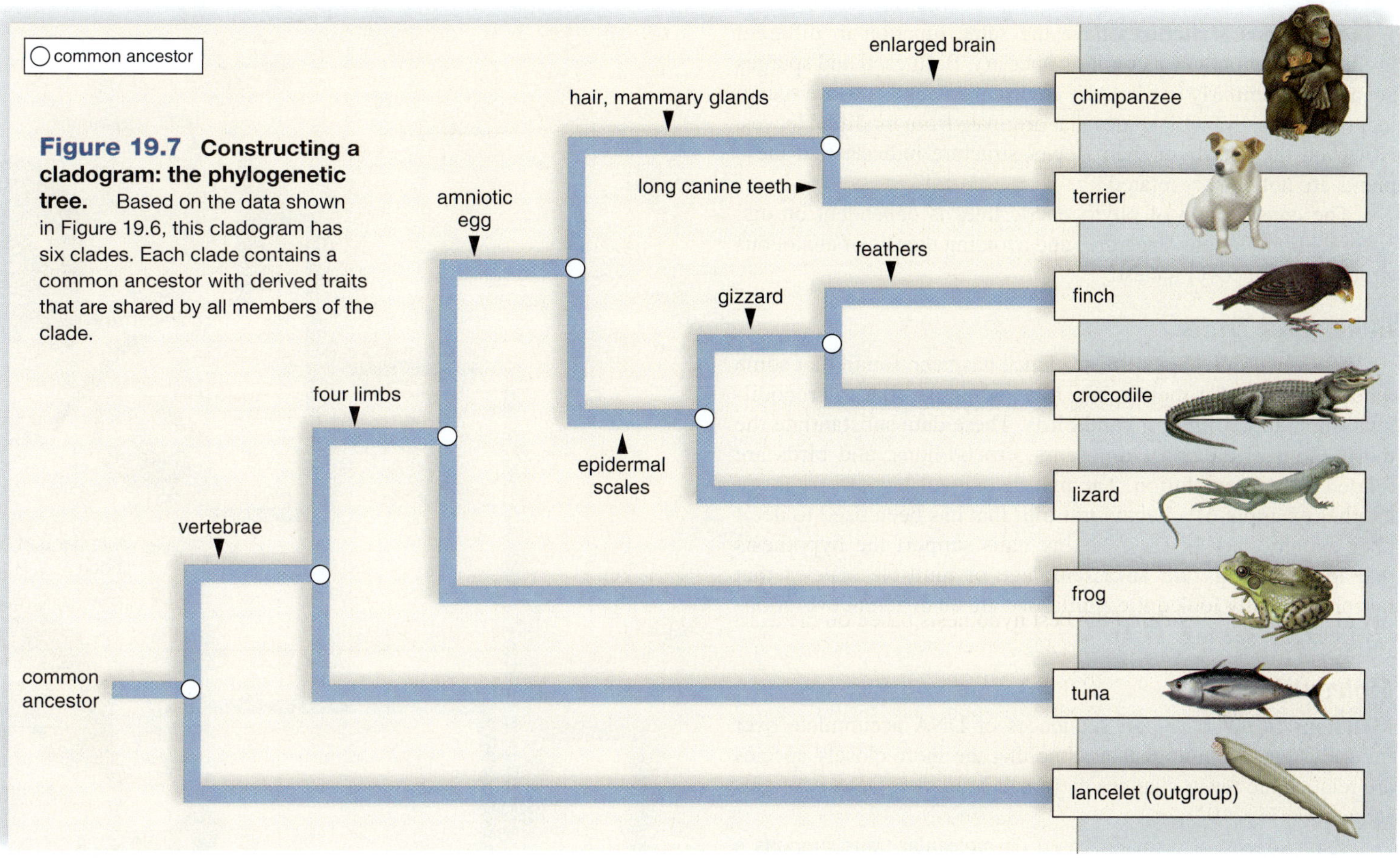

Figure 19.7 Constructing a cladogram: the phylogenetic tree. Based on the data shown in Figure 19.6, this cladogram has six clades. Each clade contains a common ancestor with derived traits that are shared by all members of the clade.

or closely related to crocodiles. On the basis of his interpretation of fossil turtles, Olivier C. Rieppel of the Field Museum of Natural History in Chicago is challenging the conventional interpretation that turtles have traits seen in a common ancestor to all reptiles but are not closely related to crocodiles, which evolved later. His interpretation is being supported by molecular data that show turtles and crocodiles to be closely related (see section 29.5 for an overview of reptile evolution).

If the fossil record were more complete, fewer controversies might arise about the interpretation of fossils. One reason the fossil record is incomplete is that most fossils are formed from harder body parts, such as bones and teeth. Soft parts are usually eaten or decayed before they have a chance to be buried and preserved. This may be one reason it has been difficult to discover when angiosperms (flowering plants) first evolved. A Jurassic fossil recently found may help pinpoint the date of origin (Fig. 19.8). As paleontologists continue to discover new fossils, the fossil record will reveal more traits useful to systematic biologists.

Morphological Traits

Homology (Gk. *homologos,* "agreeing, corresponding") is structural similarity that stems from having a common ancestor. Comparative anatomy, including developmental evidence, provides information regarding homology (see Figs. 15.15 and 15.16).

Homologous structures are similar to each other because of common descent. The forelimbs of vertebrates contain the same bones organized as they were in a common ancestor, despite adaptations to different environments. As Figure 15.15 showed, even though a horse has but a single digit and toe (the hoof), while a bat has four lengthened digits that support its wing, a horse's forelimb and a bat's forelimb contain the same bones.

Deciphering homology is sometimes difficult because of convergent evolution. **Convergent evolution** has occurred when distantly related species have a structure that looks the same only because of adaptation to the same type of environment (see Fig. 17.12 for an example of convergent evolution in fish). Similarity due to convergence is termed **analogy.** The wings of an insect and the wings of a bat are analogous.

Figure 19.8 Ancestral angiosperm. The fossil *Archaefructus liaoningensis,* dated from the Jurassic period, may be the earliest angiosperm to be discovered. Without knowing the anatomy of the first flowering plant, it has been difficult to determine the ancestry of angiosperms.

Analogous structures have the same function in different groups but do not have a common ancestry. Both cacti and spurges are adapted similarly to a hot, dry environment, and both are succulent (thick, fleshy) with spines that originate from modified leaves. However, the details of their flower structure indicate that these plants are not closely related.

The construction of phylogenetic trees is dependent on discovering homologous structures and avoiding the use of analogous structures to uncover ancestry.

Behavioral Traits

As mentioned in Chapter 18, evidence has been found that some dinosaurs cared for their young in a manner similar to crocodilians (including alligators) and birds. These data substantiate the morphological data that dinosaurs, crocodilians, and birds are related through evolution. The mating calls of leopard frogs are another example of a behavioral trait that has been used to decipher evolutionary history. Mating calls support the hypothesis that leopard frogs are an assemblage of multiple species that morphologically look quite similar but are on different evolutionary lineages (see Fig. 17.2).

Molecular Traits

Mutations in the base-pair sequences of DNA accumulate over time. Systematic biologists assume that the more closely species are related, the fewer changes there will be in their DNA base-pair sequences (Fig. 19.9).

A phylogeny of primates based on molecular traits supports a recent common ancestor for humans and chimpanzees (Fig. 19.10). Because DNA codes for amino acid sequences that form proteins, it also follows that, the more closely species are related, the fewer differences there will be in the amino acid sequences within their proteins.

	1	2	3	4	5	6	7	8	9	10	11	12	13	14	15	16	17	18	19	20	21	22	23
Cow	c	c	c	c	g	t	g	g	a	g	g	t	a	c	g	c	t	t	c	a	c	t	c
Pig	c	c	c	c	g	t	g	g	a	g	g	t	g	c	g	c	t	t	c	a	c	t	c
Horse	t	c	c	g	g	t	g	g	a	g	g	t	g	c	g	c	t	t	c	g	c	c	c
Mouse	c	c	c	c	g	t	g	g	a	g	g	t	g	c	g	c	t	t	c	a	c	c	c
Rat	c	c	c	c	g	t	a	g	a	g	g	t	g	c	g	c	t	t	c	a	c	c	c
Dog	c	c	c	t	g	t	g	g	a	g	g	t	c	c	g	c	t	t	c	a	c	c	c
Guinea Pig	c	c	c	t	g	t	g	g	g	g	g	t	g	c	g	c	t	t	c	a	c	c	c
Chimp	c	c	t	g	g	t	g	g	g	g	c	t	a	c	g	c	t	t	c	a	c	c	t
Human	c	c	t	g	g	t	g	g	g	g	g	t	a	c	g	c	t	t	c	a	c	c	t
Orangutan	c	c	c	g	g	t	g	g	g	g	g	t	g	c	g	c	t	t	c	a	c	c	c
Macaque	a	c	c	g	g	t	g	g	g	g	g	t	g	c	g	c	t	t	c	a	c	c	c

Figure 19.9 DNA sequence alignment. The DNA sequences of a small section of a gene are aligned to determine the evolutionary relationships among some mammals. Each nucleotide is aligned in columns and assigned a number (in this case, 1–23). Just as with physical traits, individual nucleotides provide information about how organisms are related. For example, the sixth nucleotide is a T shared among all mammals. Thus, this T is an ancestral trait of the class Mammalia (see Fig. 19.5). Chimpanzees and humans have in common a T at nucleotides 3 and 23, a trait not found in other mammals. Thus, these two nucleotides are shared, derived traits that support a close relationship between chimps and humans. Some nucleotide sequence alignments are thousands of nucleotides long, so computers are used to perform comparisons.

Figure 19.10 A phylogeny determined from molecular data. The relationship of certain primate species based on a study of their genomes. The length of the branches indicates the relative number of nucleotide pair differences that were found between groups. These data, along with knowledge of the fossil record for one divergence, make it possible to suggest a date for the other divergences in the tree.

Advances in molecular biology have made it very quick, easy, and inexpensive to collect nucleotide sequences for many different taxa. Software breakthroughs have made it possible to analyze nucleotide sequences or amino acid sequences quickly and accurately. Archives of DNA sequences from different genes for thousands of organisms are freely available to anyone doing systematic biology research.

Mitochondrial DNA (mtDNA) mutates ten times faster than nuclear DNA. Therefore, when determining the phylogeny of closely related species, investigators often choose to sequence mtDNA instead of nuclear DNA. One such study concerned North American songbirds. Ornithologists believed for a long time that these birds diverged into eastern and western subspecies due to retreating glaciers some 250,000–100,000 years ago. Sequencing of mtDNA allowed investigators to conclude that the two groups of North American songbirds diverged from one another an average of 2.5 million years ago (MYA). Because the old hypothesis based on glaciation is apparently flawed, a new hypothesis is required to explain why eastern and western subspecies arose among these songbirds.

Phylogenetic trees derived from molecular data are routinely used to apply evolutionary theory to all areas of biology. This information has helped human society in areas such as agriculture, medicine, and forensic science. For example, a phylogeny of the HIV types in an individual guides doctors' decisions about antiviral drug treatment. Likewise, phylogenies of insects have assisted agriculturalists in designing effective controls of crop pests.

Protein Comparisons. Before amino acid sequencing became routine, immunological techniques were used to roughly judge the similarity of plasma membrane proteins. In one procedure, antibodies are produced by transfusing a rabbit with the cells of one species. Cells of the second species are exposed to these antibodies, and the degree of the reaction is observed. The stronger the reaction, the more similar are the cells from the two species.

Later, it became customary to use amino acid sequencing to determine the number of amino acid differences in a particular protein. Cytochrome *c* is a protein found in all aerobic organisms, so its sequence has been used to examine evolutionary relationships among many organisms. There are three amino acid differences in the cytochrome *c* of chickens and ducks, but between chickens and humans there are 13 amino acid differences. From these data we can conclude that, as expected, chickens and ducks are more closely related than are chickens and humans. In addition to amino acids, differences in individual nucleic acids in DNA and RNA have become a powerful tool for examining evolutionary relationships.

Molecular Clocks. Some nucleic acid changes are neutral (not under the influence of natural selection) and thus accumulate at a fairly constant rate. These neutral mutations can be used as a kind of **molecular clock** to construct a timeline of evolutionary history. For example, songbird subspecies have mtDNA with 5.1% nucleic acid differences. Researchers know the average rate at which mtDNA nucleotide changes occur, called the mutation rate, measured in the number of mutations per unit of time. The researchers doing comparative mtDNA sequencing used their data as a molecular clock when they equated a 5.1% nucleic acid difference among songbird subspecies to 2.5 MYA.

In Figure 19.10, the researchers used their DNA sequence data to suggest how long the different types of primates have been separate. When the fossil record for one divergence is known, it indicates how long it probably takes for each nucleotide pair difference to occur. When the fossil record and molecular clock data agree, researchers have more confidence that the proposed phylogenetic tree is correct.

Check Your Progress 19.3

1. Interpret the ancestral or derived state of traits relative to their position on the phylogeny in Figure 19.5.
2. Compare two phylogenies of the same set of organisms; one requires 10 evolutionary changes, the other 15. Explain which phylogeny would be the best hypothesis for evolutionary history, and why.
3. Recognize the various traits used to construct a phylogeny.

CONNECTING *the* CONCEPTS *with the* THEMES

Evolution

- Biodiversity can be organized based on a tree, or phylogeny, that represents how organisms are related to each other in an evolutionary sense.
- Systematic biologists use a classification system first derived by Linnaeus to organize biodiversity into groups, called taxa, that reflect shared evolutionary relationships.
- Macroevolution is the sum of microevolutionary changes that lead to the evolution of new lineages of species.
- The tree of life is the best working hypothesis of the evolutionary history of life on Earth.

Nature of Science

- The science of systematic biology investigates the evolutionary relationships of living organisms.
- Linnaeus's system of taxonomy, developed over 200 years ago, is still being used by taxonomists to classify biodiversity, although modern taxonomists use evolutionary biology as a guide.
- Systematic biology is a quantitative science that uses models to reconstruct the best hypothesis of evolutionary history, given the traits available.
- Cladistics employs the principle of parsimony, which considers the simplest explanation to be the best hypothesis.

Biological Systems

- Forensic scientists can use gene phylogenies to identify criminals in cases where physical evidence is lacking, such as in the case of attempted murder with HIV.
- Phylogenies of viral DNA proteins provide doctors with drug treatment regimes that are effective in combating HIV/AIDS.

MEDIA STUDY TOOLS http://connect.mheducation.com

SMARTBOOK® Study smart. SmartBook helps you maximize your study time by identifying what you know and don't know. The recharge feature in SmartBook helps you prepare for the test by identifying the material that you are most likely to forget.

connect BIOLOGY Find even more resources to learn the chapter concepts including animations, tutorial videos, and interactive practice questions.

Animations

19.2 Three Domains
19.3 Phylogenetic Trees

SUMMARIZE

19.1 Systematic Biology

Systematics is dedicated to understanding the evolutionary history of life on Earth. **Systematic biology** is a field of science that uses **traits** of living and fossil organisms to determine relationships. **Taxonomy,** a part of systematics, identifies, names, and organizes biodiversity into various **taxa. Classification** is the process of naming and assigning an organism to a particular taxon. **Taxonomists** strive to classify organisms into **natural groups** in order to construct a **phylogeny,** or evolutionary "family tree." Traditional Linnaean taxonomy uses a **binomial nomenclature** in order to give every species a unique Latin name. The first part is the genus; the second part is the **specific epithet.** Taxonomists use a nested, hierarchical set of categories to classify organisms. There are eight main categories of classification: **species, genus, family, order, class, phylum, kingdom,** and **domain.** The categories are hierarchical such that each higher category is more inclusive than the one below it; species in the same kingdom share general characters, and species in the same genus share quite specific characters. The standardized rules of **nomenclature** are governed by the International Commission on Zoological Nomenclature and the International Code of Botanical Nomenclature.

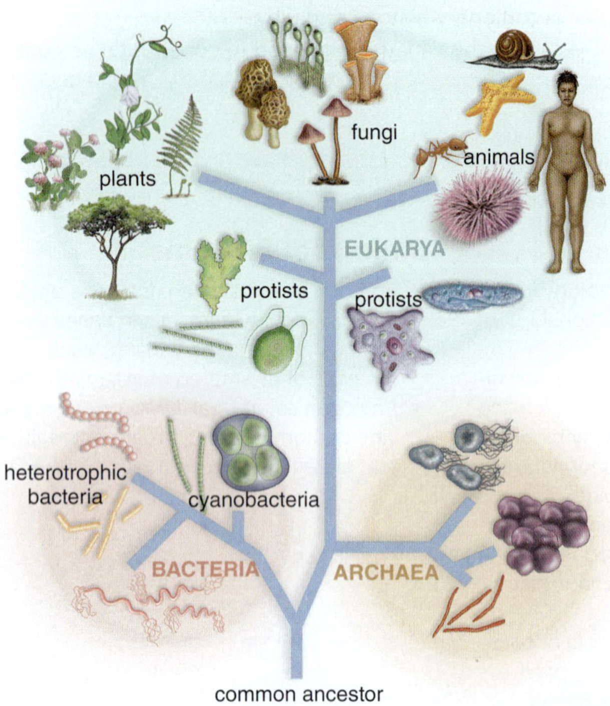

19.2 The Three-Domain System

Over time, the classification system has been expanded in order to include new information. In 1969, the **five-kingdom system** was introduced. Then in the late 1970s, the system was expanded to include three domains. Prokaryotes are placed in **domain Bacteria** or **domain Eukarya** based on molecular information. All of the protists, fungi, plants, and animals are placed in **domain Eukarya.**

19.3 Phylogeny

Phylogenies are constructed from traits that are unique to a taxon and shared by their common ancestor. The branches, or **lineages,** on a phylogeny represent different lines of descent that occur when new traits cause lineages to **diverge. Ancestral traits** are those found in the **common ancestor,** while **derived traits** are unique to a particular taxon and are useful in determining evolutionary relationships.

Cladistics uses shared derived traits to distinguish different groups of species from one another. The phylogeny that results from cladistic analysis is called a **cladogram,** with lineages called **clades.** The principle of **parsimony** is used to select the tree with the simplest explanation of evolutionary history, given the set of traits used. An **outgroup** is the taxon used to determine the ancestral and derived states of characters in the **ingroup,** or taxa for which the evolutionary relationships are being determined. All **chordates** share the ancestral trait of a spinal nerve cord. The most parsimonious cladogram is the one that requires the fewest number of evolutionary steps. Fossil, morphological, behavioral, and molecular traits help systematic biologists study evolutionary relationships. If the fossil record is complete enough, we can sometimes trace a lineage through time.

Homology helps indicate when species share a common ancestor by using **homologous structures. Convergent evolution** occurs when distantly related species have evolved similar structures in response to the environment. This convergence is termed **analogy. Analogous structures** have the same function in different groups but do not share a common ancestor. Changes to the genetic code can also be used to construct a timeline of evolutionary history by establishing a **molecular clock.** DNA nucleotide sequence data are commonly used to help determine evolutionary relationships. When multiple lines of evidence agree, it provides more confidence in the phylogenetic tree.

ASSESS

Choose the best answer for each question.

19.1 Systematic Biology

1. Which of the following is the scientific name of an organism?
   a. *Rosa rugosa*
   b. *Rosa*
   c. *rugosa*
   d. *Rugosa rugosa*
   e. Both a and d are correct.
2. Which of the following describes systematics?
   a. studies evolutionary relationships
   b. includes taxonomy and classification
   c. includes phylogenetic trees
   d. utilizes fossil, morphological, and molecular data
   e. All of these are correct.
3. The classification category below the level of family is
   a. class.
   b. species.
   c. phylum.
   d. genus.
   e. order.

19.2 The Three-Domain System

4. Which of the following are domains? Choose more than one answer if correct.
   a. Bacteria
   b. Archaea
   c. Eukarya
   d. Animals
   e. Plants
5. Which of these characteristics is shared by bacteria and archaea? Choose more than one answer if correct.
   a. presence of a nucleus
   b. absence of a nucleus
   c. presence of ribosomes
   d. absence of membrane-bound organelles
   e. presence of a cell wall
6. Which of the following pairs is mismatched?
   a. fungi—prokaryotic single cells
   b. plants—nucleated
   c. plants—flowers and mosses
   d. animals—arthropods and humans
   e. protists—single-celled eukaryotes
7. Which of the following pairs is mismatched?
   a. fungi—heterotrophic by absorption
   b. plants—usually photosynthetic
   c. animals—rarely ingestive
   d. protists—various modes of nutrition
   e. Both c and d are mismatched.

19.3 Phylogeny

8. Which of the following traits would be considered a derived trait in primates?
   a. shoulder rotation
   b. production of breast milk
   c. hair
   d. opposable thumb
   e. a tail
9. Concerning a phylogenetic tree, which is incorrect?
   a. Dates of divergence are always given.
   b. Common ancestors give rise to descendants.
   c. The more recently evolved organisms are always at the top of the tree.
   d. Ancestors have primitive characteristics.
10. Which pair is mismatched?
    a. homology—character similarity due to a common ancestor
    b. molecular data—matching DNA strands
    c. fossil record—bones and teeth
    d. homology—functions always differ
    e. molecular data—molecular clock
11. The discovery of common ancestors in the fossil record, the presence of homologies, and nucleic acid similarities help scientists decide
    a. how to classify organisms.
    b. how to determine the proper cladogram.
    c. how to construct phylogenetic trees.
    d. how evolution occurred.
    e. All of these are correct.

ENGAGE

THINKING SCIENTIFICALLY

1. Recent DNA evidence suggests to some plant taxonomists that the traditional way of classifying flowering plants is not correct, and that flowering plants need to be completely reclassified. Other botanists disagree, saying it would be chaotic and unwise to disregard the historical classification groups. Argue for and against keeping traditional classification schemes.
2. What data might make you conclude that the eukaryotes should be in more than one domain? What domains would you hypothesize might be required?
3. Because the genomes of chimpanzees and humans are almost identical, and the differences between them are no greater than between any two humans, their classification has been changed. Chimpanzees and humans are placed in the same family and subfamily. They are in different "tribes," which is a rarely used classification category between subfamily and genus.The former classification of chimpanzees and humans placed the two animals in different families. Do you believe the chimpanzees should be classified in the same family and subfamily as humans, or do you prefer the classification used formerly? Which way seems prejudicial? Give your reasons for preferring one method over the other.

UNIT 4

Microbiology and Evolution

Microbes occupy a world unseen by the naked eye. It's richly populated because it includes the viruses, the prokaryotic bacteria and archaea, and the eukaryotic protists and fungi. These organisms occur everywhere from the highest mountain peaks to the deepest ocean trenches and in every type of environment, even those that are extremely hot and acidic.

Although some microbes cause serious diseases in plants and animals, including ourselves, we make use of microbes in innumerable ways. For example, bacteria help us accomplish gene cloning and genetic engineering; make foods and antibiotics; and help dispose of sewage and environmental pollutants. The biosphere is totally dependent on the services of microorganisms.

We are aware of how much we rely on land plants, but we may fail to acknowledge that without microorganisms, land plants could not exist. Decomposing fungi and bacteria make inorganic nutrients available to plants, which they can absorb all the better because their roots are coated with friendly fungi. Photosynthetic bacteria first put oxygen in the atmosphere, and they, along with certain protists, are the producers of food in the oceans.

Microbes are our ancestors. They alone were on Earth for about 2.5 billion years, and single-celled protists gave rise to the animals and plants that populate the macroscopic world. This unit discusses microbes, organisms that contribute so much to our world, even though we cannot see them without the use of a microscope.

Unit Outline

Chapter 20 Viruses, Bacteria, and Archaea 353

Chapter 21 Protist Evolution and Diversity 373

Chapter 22 Fungi Evolution and Diversity 395

Unit Learning Outcomes

The learning outcomes for this unit focus on three major themes in the life sciences.

Evolution	Recognize the evolutionary relationships of the microorganisms and their position in the tree of life.
Nature of Science	Identify how microorganisms are used as models in medical, environmental, and genetics research.
Biological Systems	Explain how microorganisms are essential for a healthy ecosystem.

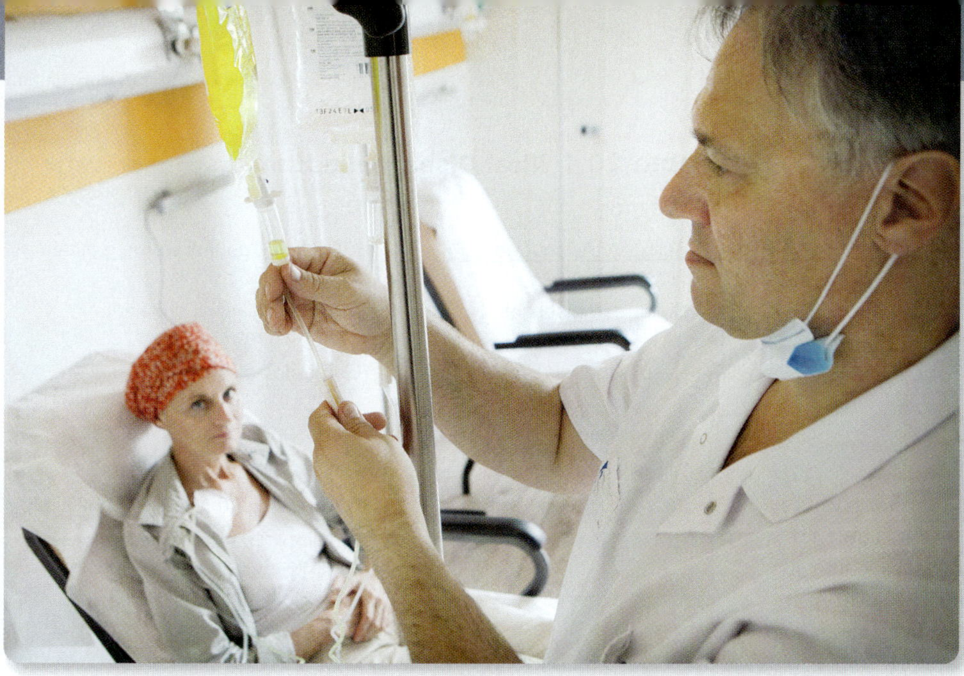

A cancer patient receiving chemotherapy treatment.

20

Viruses, Bacteria, and Archaea

Out of every nine people reading this, three will develop cancer and one will die, unless, that is, a cure or an effective treatment is developed. How does cancer relate to microorganisms? Multiple novel treatments using microorganisms could lead to a cure for cancer. In one experiment, reovirus injected into cancer-stricken mice was found to locate cancer cells, infect them, and destroy the cancer cells as they replicated inside. Most importantly, the virus didn't affect normal cells, as current drug and radiation treatments do. Bacteria are also showing promise in treating tumors, which account for 90% of all cancers. *Clostridium* bacteria are known for causing botulism and tetanus, but scientists are genetically engineering new strains of *Clostridium* that secrete enzymes, toxins, and antibodies to fight cancer cells. The *Clostridium* only colonizes tumors, minimizing the effects on healthy tissues, much as the reovirus does.

For most, the words *virus* and *bacteria* are associated only with causing illness, when in reality less than 1% of microorganisms are harmful. With the perseverance of scientists, some can benefit or even save your life.

As you read through this chapter, think about the following questions:

1. How do viruses and prokaryotic bacteria differ from each other and eukaryotes?
2. How do scientists study viruses and prokaryotes?
3. Although some cause disease, why are microorganisms essential to life?

Chapter Outline

20.1 Viruses, Viroids, and Prions 354

20.2 The Prokaryotes 360

20.3 The Bacteria 363

20.4 The Archaea 368

Before You Begin

Before beginning this chapter, take a few moments to review the following discussions.

Section 1.1 What characteristics are necessary for an organism to be considered "living"?

Section 4.2 What is the structure of a prokaryotic cell?

Section 19.2 What characteristics divide microorganisms into the domains Bacteria and Archaea?

Following *the* Themes

CHAPTER 20 VIRUSES, BACTERIA, AND ARCHAEA

UNIT 4 MICROBIOLOGY AND EVOLUTION

Evolution	All living organisms evolved from the first prokaryotic cells 3.5 billion years ago.
Nature of Science	The study of how microorganisms evolve over time allows researchers to predict, and develop, vaccines and treatments for disease.
Biological Systems	Microorganisms are found in, and are essential to, all environments on Earth.

20.1 Viruses, Viroids, and Prions

Learning Outcomes

Upon completion of this section, you should be able to

1. Identify the basic structures of a virus.
2. Explain the unique characteristics of viruses compared to living cells.
3. Describe the process of viral reproduction.

The term **virus** (L. *virus,* "poison") is associated with a number of plant, animal, and human diseases (Table 20.1). The mere mention of the term brings to mind serious illnesses, such as polio, rabies, and AIDS (acquired immunodeficiency syndrome), as well as formerly common childhood maladies such as measles, chickenpox, and mumps. Viral diseases are of concern to everyone; it is estimated that the average person catches a cold two or three times a year.

Viruses are a biological enigma. They have some characteristics of living organisms, such as a DNA or RNA genome, and the ability to evolve and replicate. However, they can replicate only by using the metabolic machinery of a host cell, they do not have a metabolism, and they do not respond to stimuli. For these reasons, viruses are known as *obligate intracellular parasites* and are either active or inactive, rather than living or nonliving.

Table 20.1 Viral Diseases in Humans

Category	Disease
Sexually transmitted diseases	AIDS (HIV), genital warts, genital herpes
Childhood diseases	Mumps, measles, chickenpox, German measles
Respiratory diseases	Common cold, influenza, severe acute respiratory syndrome (SARS)
Skin diseases	Warts, fever blisters, shingles
Digestive tract diseases	Gastroenteritis, diarrhea
Nervous system diseases	Poliomyelitis, rabies, encephalitis
Other diseases	Smallpox, hemorrhagic fevers, cancer, hepatitis, mononucleosis, yellow fever, dengue fever, conjunctivitis, hepatitis C

Viruses do not fossilize, as other living organisms do, so their history is difficult to study. However, several hypotheses exist regarding their origin and evolution. Hypotheses about the origin of life suggest that proteins or nucleic acids, the two organic molecules present in viruses, were first to evolve. It is possible that viruses arose from these basic polymers at the same time as living cells. Some scientists offer an alternative hypothesis suggesting viruses actually originated after living cells. They propose that viruses were

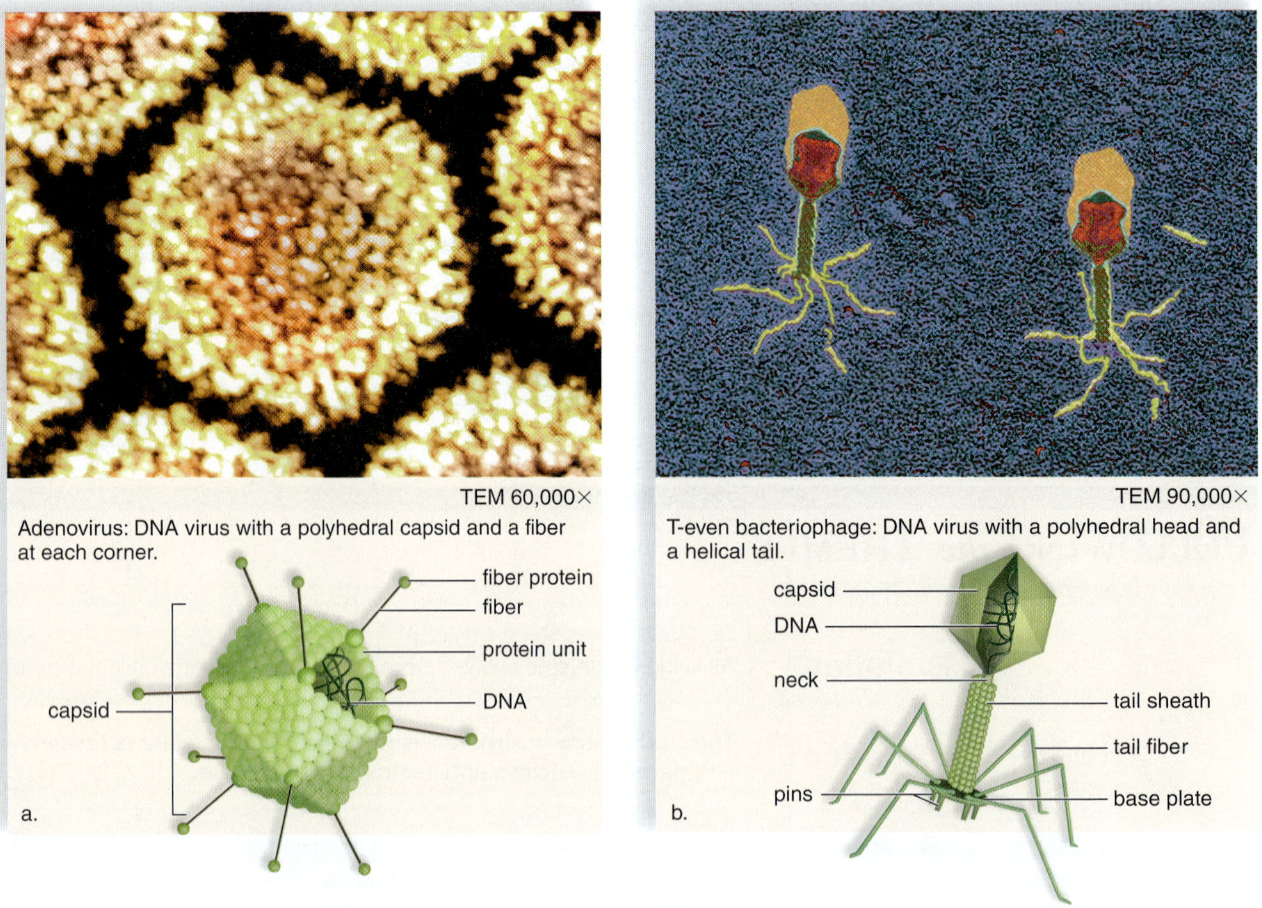

Figure 20.1 Viruses. Despite their diversity, all viruses have an outer capsid composed of protein subunits and a nucleic acid core—composed of either DNA or RNA, but not both. Some types of viruses also have a membranous envelope.

derived from pieces of cell genomes or evolved backwards from living cells, or degenerated, into the simplest possible form required for reproduction.

Discovery of Viruses

Our knowledge of viruses began in 1884, when the French chemist Louis Pasteur (1822–95) suggested that something smaller than a bacterium was the cause of rabies, and he chose the word *virus* from the Latin word meaning poison.

In 1892, Dimitri Ivanowsky (1864–1920), a Russian microbiologist, was studying tobacco mosaic disease, which causes damage to the leaves and fruit of tobacco plants. He noticed that even when an infective extract was filtered through a fine-pore porcelain filter that retained bacteria, the extract still caused disease. This substantiated Pasteur's belief, because it meant that the disease-causing agent was smaller than any known bacterium.

In the twentieth century, electron microscopy was born, and viruses were seen for the first time. By the 1950s, virology had become an active field of research; the study of viruses, and now viroids and prions, has contributed much to our understanding of disease, genetics, and the characteristics of living organisms.

Classification System of Viruses

Because viruses mutate rapidly and are not living organisms, they are difficult to classify and name. However, the International Committee on Taxonomy of Viruses (ICTV) has developed a classification system similar to the system used for living organisms. Viral classification differs because it includes only the taxonomic levels of order, family, genus, and species, not the higher levels of kingdom, phylum, and class. Over 2,500 species of viruses have been identified, including the species that causes the seasonal flu, influenza A, in the *Influenza A* genus and Orthomyxoviridae family.

When a new type of virus emerges within a single species of virus, additional classification levels, such as subtype, are required for clear identification. For example, influenza A is classified into subtypes based on the type of two glycoprotein spikes in its envelope, hemagglutinin (H) and neuraminidase (N). Because there are 16 H-type and 9 N-type spikes, many virus subtypes may evolve. Notable examples are the H5N1 "bird flu" and H1N1 "swine flu."

Structure of Viruses

The size of a virus is comparable to that of a large protein macromolecule, approximately 10–400 nm. Viruses are best studied through electron microscopy (Fig. 20.1). Many viruses can be purified and crystallized, and the crystals can be stored just as chemicals are stored. Still, viral crystals become infectious when the viral particles they contain are given the opportunity to invade a host cell.

Viruses are categorized by (1) their size and shape; (2) their type of nucleic acid, including whether it is single-stranded or

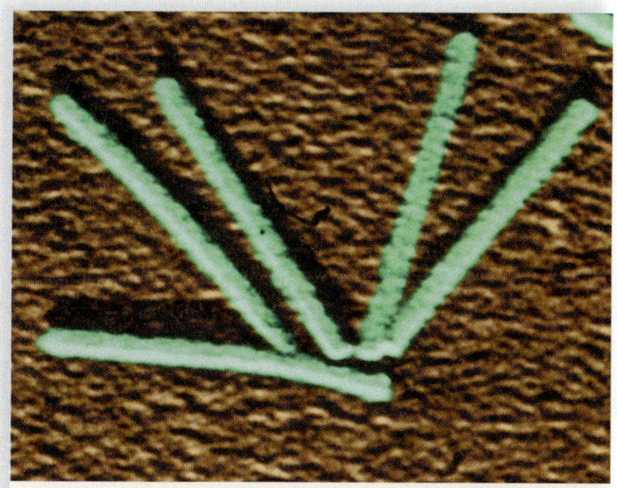

TEM 44,000×

Tobacco mosaic virus: RNA virus with a helical capsid.

c.

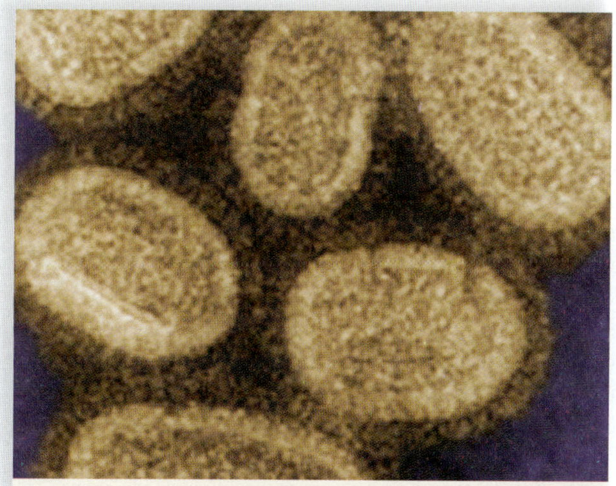

TEM

Influenza virus: RNA virus with a spherical capsid surrounded by an envelope with spikes.

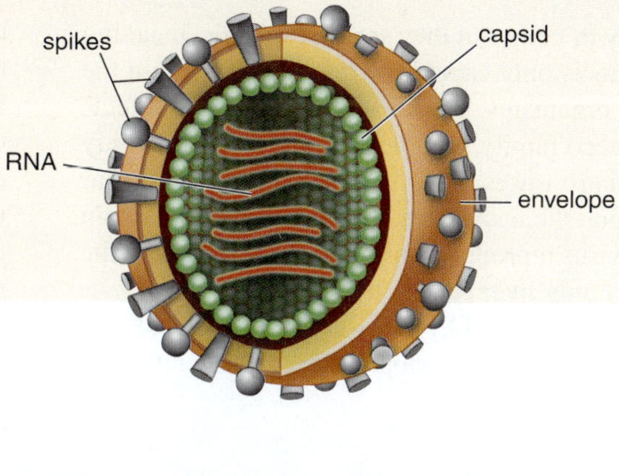

d.

double-stranded; and (3) the presence or absence of an outer envelope. The structure of a virus can be summarized by the following diagram:

- Virus particle
  - Covering
    - Capsid (protein)
    - Envelope (not found in all viruses)
  - Inner core
    - Nucleic acid molecule (DNA or RNA)
    - Various proteins (enzymes)

Viruses vary in shape from threadlike to polyhedral. However, all viruses possess the same basic anatomy: an outer **capsid** composed of protein subunits and an inner core of nucleic acid—either DNA or RNA. A viral genome may have as few as 3 or as many as 100 genes; a human cell, in contrast, contains tens of thousands of genes.

If the viral capsid is the outermost structure on a virus, the virus is said to be naked. Figure 20.1*a, b, c* gives examples of naked viruses. In contrast, enveloped viruses are surrounded by an outer membranous envelope actually derived from the host cell's plasma membrane. Figure 20.1*d* is an example of an enveloped virus. Inserted into the envelope are glycoprotein spikes coded for by the viral genome. These molecules vary among viruses and allow each virus to bind to a new host cell. Aside from its genome, a viral particle may also contain various proteins, especially enzymes such as the polymerases, needed to produce viral DNA and/or RNA.

Parasitic Nature of Viruses

As *obligate intracellular parasites,* viruses cannot replicate outside a living cell. Like prokaryotic and eukaryotic cells, viruses have genetic material. Whereas a cell is capable of copying its own genetic material in order to reproduce, a virus cannot duplicate its genetic material or any of its other components on its own. For a virus to reproduce, it must infect a living cell. Once inside a living cell, the virus "hijacks" the cell's protein synthesis machinery to replicate the nucleic acid and other parts of the virus, including the capsid, viral enzymes, and for some viruses the envelope.

Cells infected by some viruses are killed or damaged by the replicating virus, causing the symptoms associated with viral infections. For example, cells infected by adenovirus in the respiratory tract are lysed when viral replication is complete, leading to conditions such as bronchitis and pneumonia.

Host Specificity

Viruses infect a variety of cells, but they are *host specific,* meaning that any particular virus is only capable of reproducing within the cells of specific living organisms. The tobacco mosaic virus infects only plants in the tobacco family, and the rabies virus infects only mammals. Some human viruses are even specific to a particular tissue. Human immunodeficiency virus (HIV) enters only certain blood cells, the polio virus reproduces in spinal nerve cells, and the hepatitis viruses infect only liver cells.

Host specificity is determined by the structure of molecules in the naked capsid or spikes on an enveloped virus. These attach in a lock-and-key manner with a receptor on the host cell's outer surface. A cell that does not have a receptor to match a virus exactly cannot be infected, because the virus will be unable to enter. Many antiviral medications are effective because they interfere with the lock-and-key attachment of viruses to host cells.

Animation Antiviral Agents

Reproduction of Viruses

Viruses are microscopic pirates, commandeering the metabolic machinery of a host cell during their reproduction cycle, consisting of five steps:

1. *Attachment:* A virus binds to a specific host cell based on the host-specific match between virus surface molecules and host cell receptors.
2. *Penetration:* The host cell engulfs the virus or the virus injects its genome into the cytoplasm.
3. *Biosynthesis:* New viral components, including the capsid subunits, spikes, and copies of the genome, are synthesized using the host's ribosomes, enzymes, transfer RNA (tRNA), and energy.
4. *Maturation:* Viral components are assembled into new viruses.
5. *Release:* New viruses exit the host cell through lysis or budding in order to infect new host cells.

This reproductive cycle is most common, but notable exceptions do occur among different types of viruses, including bacteriophages, animal viruses, and retroviruses.

Reproduction of Bacteriophages

Bacteriophages (Gk. *bacterion,* "rod"; *phagein,* "to eat") are viruses that parasitize bacteria. They have provided a useful model for scientists to study the reproductive cycle of viruses. Figure 20.2 shows two alternative life cycles of bacteriophages, called the lytic cycle and the lysogenic cycle.

In the **lytic cycle,** as shown in Figure 20.2, the five steps of the viral reproduction cycle occur in immediate sequence. The lytic cycle is so named because the bacterial host is lysed at the end of the cycle during release by a virally coded enzyme called lysozyme. In the process, the bacterial cell dies, and several hundred new viral particles are released.

Video Virus Lytic Cycle

When a virus enters the **lysogenic cycle** instead of the lytic cycle, viral reproduction and release of new viruses does not occur immediately, but reproduction may take place in the future. In the meantime, the infecting phage is *latent*—not actively replicating.

Following attachment and penetration, *integration,* instead of biosynthesis, occurs: Viral DNA becomes incorporated into bacterial DNA with no destruction of host DNA. While latent, the viral DNA is called a *prophage.* The prophage is replicated along with the host DNA, and all subsequent cells, called **lysogenic cells,** carry a copy of the prophage genome. Sometime in the future, certain environmental factors, such as ultraviolet radiation, can induce the prophage to re-enter the lytic stage of biosynthesis, followed by maturation and release.

Lysogenic bacterial cells may have distinctive properties due to the prophage genes they carry. The presence of a prophage may cause a bacterial cell to produce a toxin. For example, if the same bacterium that causes strep throat happens to carry a certain prophage, then it will cause scarlet fever, so named because the toxin causes a widespread red skin rash as it spreads through the body. Likewise, diphtheria is caused by a bacterium carrying a prophage. The diphtheria toxin damages the lining of the upper respiratory tract, resulting in the formation of a thick membrane that restricts breathing.

Animation Lambda Phage Replication Cycle

Reproduction of Animal Viruses

Animal viruses reproduce in a manner similar to that of bacteriophages, but various animal viruses have different ways of introducing their genetic material into their host cells. For some enveloped viruses, the process is as simple as attachment and fusion of the spike-studded envelope with the host cell's plasma membrane. Many naked and some enveloped viruses are taken into host cells by endocytosis. Once the virus enters, it is uncoated—that is, the capsid and, if necessary, the envelope are removed. The viral genome, either DNA or RNA, is now free of its covering, and the virus continues the lytic cycle or enters the lysogenic cycle.

Viruses that are highly virulent enter directly into the lytic cycle, causing rapid and severe destruction of host cells. The Ebola virus is highly virulent, and up to 90% of those infected die from the disease within 2–21 days of infection. In contrast, HIV enters first into the lysogenic cycle and thus can lie inactive, or latent, for many years before AIDS symptoms emerge.

Viral release is just as variable as penetration for animal viruses. Some mature viruses are released by budding. During budding, the

Figure 20.2 Lytic and lysogenic cycles of bacteriophages. **a.** In the lytic cycle, viral particles escape when the cell is lysed (broken open). In the lysogenic cycle, viral DNA is integrated into host DNA. At some time in the future, the lysogenic cycle can be followed by the lytic cycle. **b.** Micrograph of bacteriophage viruses attaching to a bacterium, *Escherichia coli*.

Tutorial Viral Life Cycle

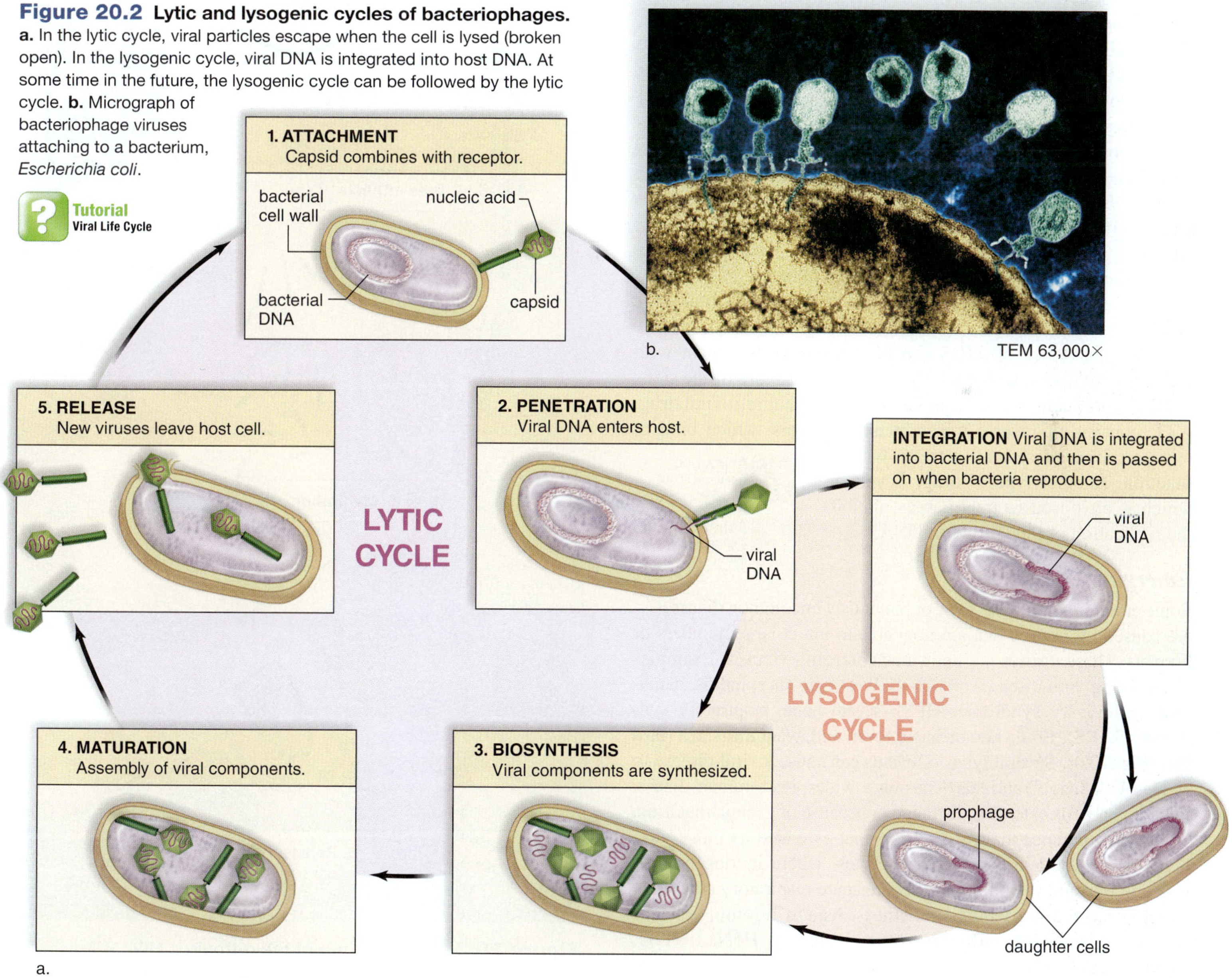

virus picks up its envelope, consisting of lipids, proteins, and carbohydrates, from the host cell. Most enveloped animal viruses acquire their envelope from the plasma membrane of the host cell, but some take envelopes from other membranes, such as the nuclear envelope or Golgi apparatus. Envelope markers, such as the glycoprotein spikes that allow the virus to enter a host cell, are coded for by viral genes. Naked animal viruses are usually released by host cell lysis.

Animation Entry of Virus into Host Cell

Retroviruses. **Retroviruses** (L. *retro,* "backward") are animal viruses with an RNA genome that is converted into DNA within the host cell by an enzyme called **reverse transcriptase.** Figure 20.3 illustrates the reproduction of HIV, a type of retrovirus.

Animation How the HIV Infection Cycle Works

Before a retrovirus can integrate into the host's genome, or use the host cell's machinery to transcribe and translate its proteins, it must first convert its RNA to DNA. First, the enzyme reverse transcriptase synthesizes from its RNA genome a single DNA strand, called cDNA because it is a complementary DNA strand to the viral RNA. The single strand of cDNA is used as a template to make a double-stranded DNA.

Using host enzymes, the double-stranded virus DNA is integrated into the host genome. The viral DNA remains in the host genome and is replicated when host DNA is replicated. When and if this DNA is transcribed, new viruses are produced by the steps already cited: biosynthesis, maturation, and release; in the case of HIV, it is released by budding from the host plasma membrane. As mentioned, HIV can remain latent for many years. Without treatment, the median survival time after HIV infection is 9–11 years.

The emergence of AIDS can be delayed by treatment with antiretroviral drugs, which interfere with one or more of the steps of HIV reproduction. For example, one type of antiretroviral drug, AZT, consists of reverse transcriptase inhibitors, which bind to reverse transcriptase and interfere with its function. Another type of drug, Acyclovir, which is also used to treat herpes, inhibits the replication of the HIV viral DNA.

Animation Treatment of HIV Infection

Animation Replication Cycle of a Retrovirus

Emerging Viruses

Some emerging diseases—new or previously uncommon illnesses—are caused by viruses that are now able to infect large numbers of humans. These viruses are known as **emerging viruses.** Examples of emerging viral diseases are AIDS, West Nile encephalitis, hantavirus pulmonary syndrome (HPS), severe acute respiratory syndrome (SARS), Ebola hemorrhagic fever, and avian influenza (bird flu) (Fig. 20.4). Several types of events can cause a viral disease to suddenly "emerge" and start causing a widespread human illness, including a virus extending its range or because of genetic mutation.

West Nile encephalitis is a virus that extended its range after being transported into the United States, where it took hold in bird and mosquito populations. Severe acute respiratory syndrome (SARS) was transported from Southeast Asia to Toronto, Canada.

New strains of influenza virus, including H5N1, H1N1, and H7N9, are emerging viral diseases because they are created through rapid mutations of flu viruses that once only infected animals. Mutations that occur allow the virus to jump to other species, including humans. Even the seasonal influenza is well known for

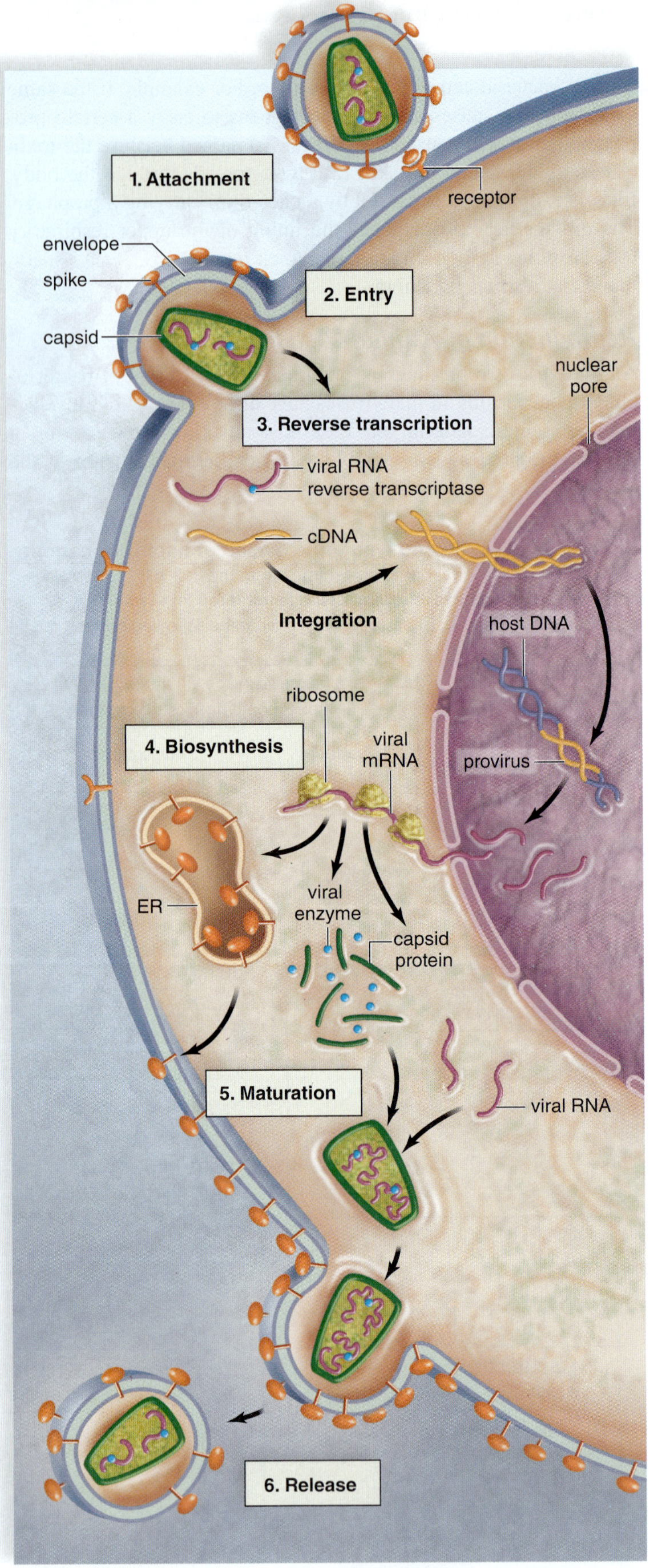

Figure 20.3 Reproduction of the retrovirus HIV. HIV uses reverse transcription to produce a DNA copy (cDNA) of RNA genes; double-stranded DNA integrates into the cell's chromosomes before the virus reproduces and buds from the cell. Reverse transcription is unique to retroviruses.

THEME Nature of Science

Flu Viruses

A flu virus has an H (hemagglutinin) spike and an N (neuraminidase) spike embedded in its plasma membrane (Fig. 20A*a, left*). Its H spike allows the virus to bind to its receptor, and its N spike attacks host plasma membranes in a way that allows mature viruses to exit the cell. At least 16 types of H and 9 types of N spikes exist, allowing different combinations that can infect different hosts. Many of the flu viruses are assigned specific codes based on the type of spike. For example, H5N1 virus gets its name from its variety of H5 spikes and its variety of N1 spikes.

Our immune systems can recognize only the particular variety of H spikes and N spikes they have been exposed to in the past by infection or immunization. When a new flu virus arises, one for which there is little or no immunity in the human population, a flu pandemic (global outbreak) may occur.

Currently, the H7N9 and H5N1 subtypes of flu virus are of great concern because of their potential to reach pandemic proportions. Both viruses are common in wild birds such as waterfowl, and they can readily infect domestic poultry, such as chickens, which is why they are referred to as "bird flu." More pathogenic strains of these viruses have appeared with the ability to spread from birds to humans, causing severe illness and death.

Humans become infected because the viral H spikes can attach to both a bird flu receptor and a human flu receptor. Close contact between domestic poultry and humans is necessary for this to happen, and the virus has rarely been transmitted from one human to another.

At this time, bird flu infects mostly the lungs, or the lower respiratory tract, and transmission by coughing or sneezing is rare. However, a spontaneous mutation in the H spike could enable it to attack the upper respiratory tract, making the virus easily spread from person to person by coughing and sneezing (Fig. 20A*a*). Another possibility is that a combining of spikes could occur in a person who is infected with both the bird flu and the human flu viruses (Fig. 20A*b*).

In both cases, the new virus would likely spread rapidly and be lethal. Currently, there are no available vaccines for an H5N1 or H7N9 virus, and approximately half of infected individuals die.

Questions to Consider

1. How is the host-specific nature of viruses preventing a global pandemic?
2. Compare the names of the H5N1 and the H7N9 viruses. What can their names tell you about differences in their structure?
3. Why do humans need a flu vaccine every year?

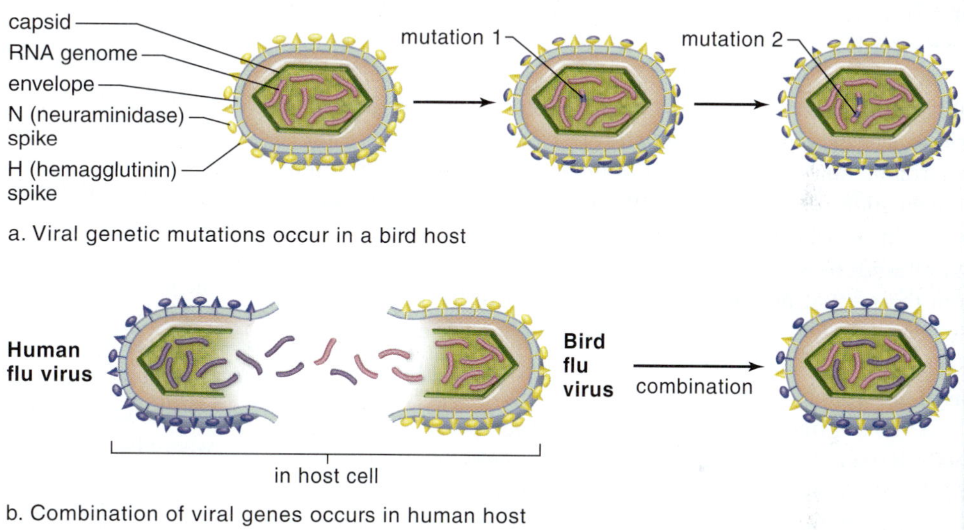

Figure 20A Spikes of bird flu virus. a. Genetic mutations in bird flu viral spikes could allow the virus to infect the human upper respiratory tract. **b.** Alternatively, a combination of bird flu and human spikes could allow the virus to infect the human upper respiratory tract.

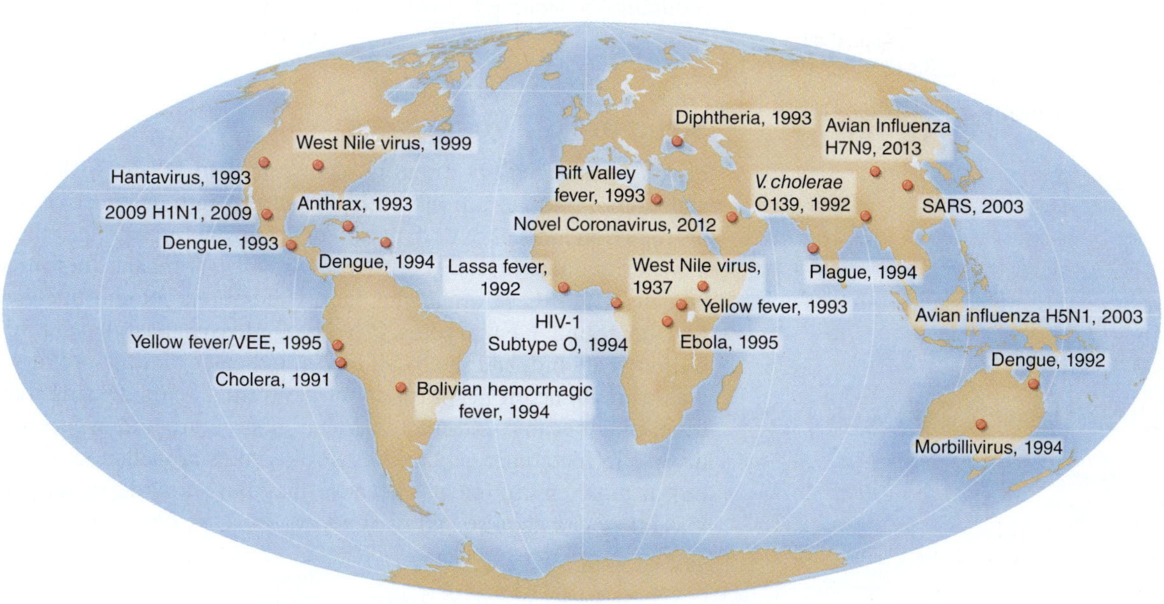

Figure 20.4 Emerging diseases. Emerging diseases, such as those noted here, are new or demonstrate increased prevalence. These disease-causing agents may have acquired new virulence factors, or environmental factors may have encouraged their spread to an increased number of hosts.

mutating, and this is why it is necessary to have a flu shot every year—antibodies generated from last year's shot are not expected to be effective this year.

HIV is also a very rapidly evolving virus, which makes it difficult to find a cure. Viruses such as HIV are often engaged in an "arms race" with an animal's immune system. Its rapid evolution results in many different types of HIV within a patient. Although a new drug or vaccine may work against some, it is very likely that one type within the billions of virus copies will evolve resistance and continue the infection. The rapid evolution of resistance in HIV is why there is currently no cure.

Video Virus Crisis

Viroids and Prions At least a thousand different viruses cause diseases in plants. About a dozen diseases of crops, including potatoes, coconuts, and citrus, have been attributed not to viruses but to **viroids,** which are naked strands of RNA (not covered by a capsid). Like viruses, however, viroids direct the cell to produce more viroids.

A number of fatal brain diseases, known as *transmissible spongiform encephalopathies,* or TSEs, have been attributed to **prions,** a term coined for *proteinaceous infectious* particles. Prions are proteins that normally exist in an animal but have a different conformation, or structure. Like viruses, prions cannot replicate on their own but cause infection by interacting with a normal protein and altering its structure. The process that changes the structure from the normal protein conformation to the prion conformation can be as simple as changing a chemical bond. Once a prion infects tissue, a chain reaction begins that converts normal proteins to prions at an exponential rate.

TSEs are **neurodegenerative diseases,** or those that destroy nerve tissue in the brain. In the brain, prion proteins form clusters that break down normal brain tissue, creating small holes that give the brain a spongy appearance. All TSEs are untreatable and fatal. Mad cow disease, or bovine spongiform encephalopathy (BSE), is a neurodegenerative disease of cattle that is transmissible to humans by eating cattle brains or meat contaminated with prion-infected brain tissue. The discovery of prions began when it was observed that members of a primitive tribe in the highlands of Papua New Guinea died from a disease commonly called kuru (meaning "trembling with fear"). The disease occurred after the individual had participated in the cannibalistic practice of eating a deceased person's brain; the brain was evidently infected with prions.

Animation How Prions Arise

Animation Prion Diseases

Check Your Progress 20.1

1. Describe the structure of a virus.
2. Explain why a virus can only infect specific cells and organisms.
3. Compare the lytic and lysogenic cycles of viral reproduction.
4. Explain, from an evolutionary standpoint, why it is beneficial to a virus if its host lives.

20.2 The Prokaryotes

Learning Outcomes

Upon completion of this section, you should be able to

1. Describe the evolution of prokaryotes.
2. Identify structural features of prokaryotes.
3. Describe at least four ways in which the cells of prokaryotes differ from eukaryotic cells.

Prokaryotes include bacteria and archaea, which are fully functioning, living, single-celled organisms. Because they are microscopic, the prokaryotes were not discovered until the Dutch microscopist Antonie van Leeuwenhoek (1632–1723) first described them, along with many other microorganisms (see the Nature of Science feature, "Microscopy Today," in Chapter 4).

Leeuwenhoek and others after him believed that the "little animals" he observed could arise spontaneously from inanimate matter. *Spontaneous generation,* the idea that living organisms can emerge from nonliving things, was common at the time. When meat spoiled, for example, it was thought that maggots arose from the meat spontaneously.

For about 200 years, scientists carried out various experiments to determine the origin of microorganisms in laboratory cultures. Finally, in about 1850, Louis Pasteur devised an experiment for the French Academy of Sciences (described in Fig. 20.5). It showed that a previously sterilized broth cannot become cloudy with microorganism growth unless it is exposed directly to the air where bacteria are abundant.

Today, we know that bacteria are plentiful in air, water, and soil and that new bacteria arise from the division of preexisting bacteria—not by spontaneous generation. We also know a lot about the structure of bacteria; their membranes, DNA, and proteins; how and where they live; where they get their nutrition; and how they coordinate replication. In the following pages, the general characteristics of prokaryotes are discussed before those specific to the bacteria (domain Bacteria), and then the archaea (domain Archaea) are considered in more detail.

Structure of Prokaryotes

Prokaryotes generally range in size from 1 to 10 μm in length and from 0.7 to 1.5 μm in width. To put this into perspective, an average human is about 1.5 m tall, or 1 million times longer than a bacterium (see Fig. 4.2). The term *prokaryote* means "before a nucleus," and these organisms lack a membrane-bound nucleus like that found in eukaryotes. Prokaryotic fossils exist that are 3.5 billion years old, and the fossil record indicates that the prokaryotes were alone on Earth for about 2.5 billion years. During that time, they became extremely diverse in structure and especially diverse in metabolic capabilities. Prokaryotes are adapted to living in most environments, because they have a wide variety of ways they can acquire and use energy.

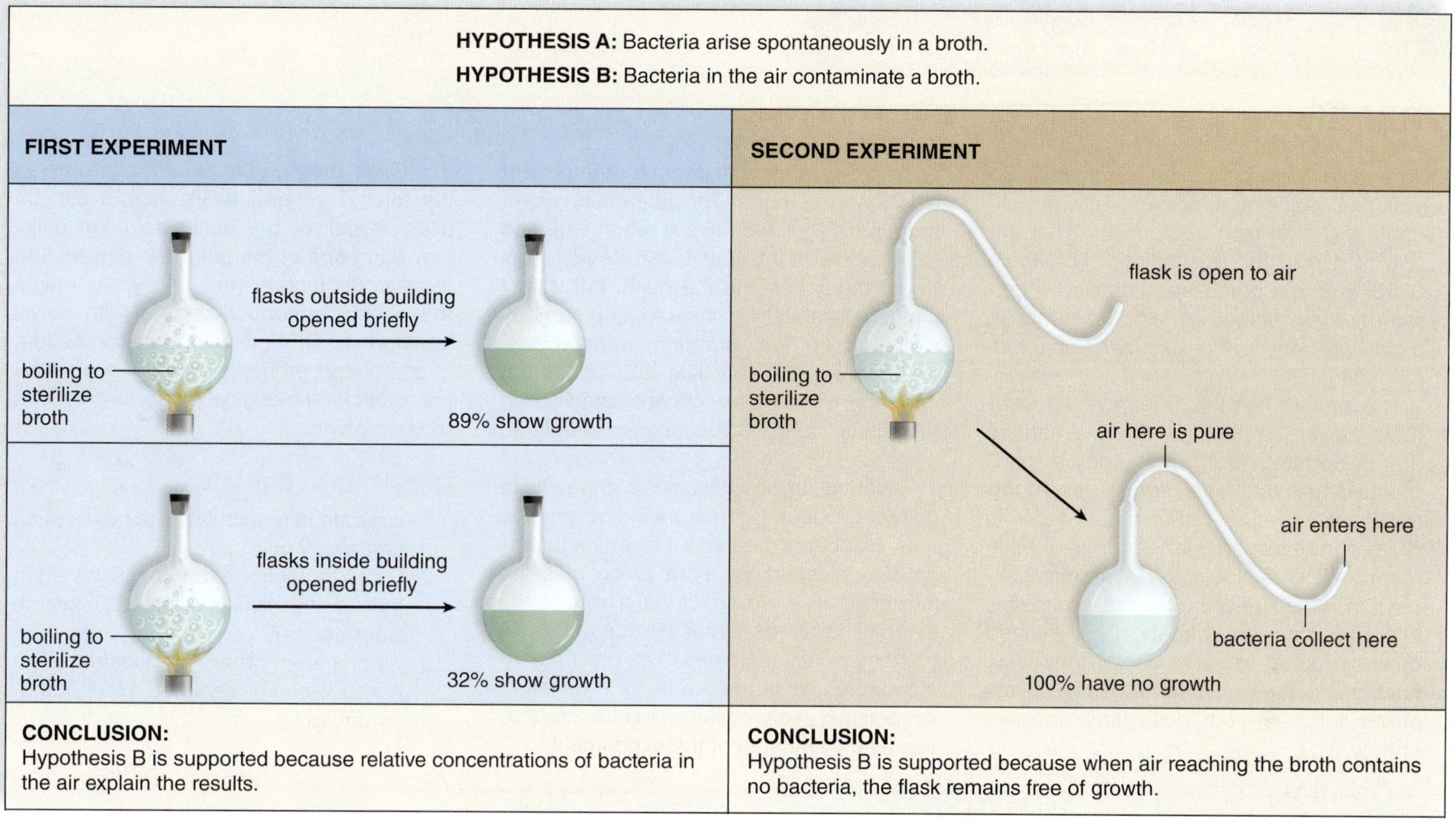

Figure 20.5 Pasteur's experiments. Pasteur disproved the theory of spontaneous generation of microbes by performing these types of experiments.

The organization of a typical prokaryotic cell is illustrated in the following diagram:

- Prokaryotic cell
  - Cell envelope
    - Glycocalyx
    - Cell wall
    - Plasma membrane
  - Cytoplasm
    - Nucleoid
    - Ribosomes
    - Thylakoids (cyanobacteria)
  - Appendages
    - Flagella
    - Conjugation pilus
    - Fimbriae

A prokaryotic cell is surrounded by a plasma membrane and a cell wall situated outside the membrane (Fig. 20.6). The cell wall prevents a prokaryote from bursting or collapsing due to fluctuations in the amount of fluid inside the cell. Yet another layer may exist outside the cell wall, depending on the type of prokaryote.

In many bacteria, the additional cover is a layer of polysaccharides called a *glycocalyx* (Fig. 20.6*a*). A well-organized glycocalyx is called a capsule, whereas a loosely organized one is called a slime layer. Instead of a glycocalyx covering, many bacteria and archaea have a layer comprising protein, or glycoprotein, called an *S-layer.* In parasitic forms of bacteria, these outer coverings help protect the cell from host defenses.

Some prokaryotes move by means of *flagella* (Fig. 20.6*b*). A bacterial flagellum has a filament composed of strands of the protein flagellin wound in a helix. The filament is inserted into a hook anchored by a basal body. The 360-degree rotation of the flagellum causes the cell to spin and move forward. The archaeal flagellum is similar but more slender and apparently lacking a basal body.

Many prokaryotes adhere to surfaces by means of *fimbriae*—short, bristlelike fibers extending from the surface (Fig. 20.6*a*). The fimbriae of the bacterium *Neisseria gonorrhoeae* allow it to attach to host cells and cause gonorrhea, a sexually transmitted disease.

Animation Bacterial Locomotion

A prokaryotic cell lacks the membranous organelles of a eukaryotic cell; instead, the plasma membrane contains many infolds, where metabolic reactions such as respiration and photosynthesis occur. Although prokaryotes do not have a nucleus, they do have a dense area called a *nucleoid,* where a single chromosome consisting of a circular strand of DNA is found. Many prokaryotes also have accessory rings of DNA called *plasmids,* which contain genes for antibiotic resistance, production of toxins, or degradation of chemicals. Plasmids can also be extracted and used to carry foreign DNA into host bacteria during genetic engineering processes, as discussed in the Nature of Science feature, "DIY Bio," on page 362.

Protein synthesis in a prokaryotic cell is carried out by thousands of ribosomes, which are smaller than eukaryotic ribosomes.

THEME Nature of Science

DIY Bio

Picture yourself becoming a scientist. Do you see years of education, a white lab coat, and high-tech equipment? That image of a scientist is changing. For about $250, you can purchase materials to perform a sophisticated experiment to create a glowing plant right in your own home with no experience.

A project like this is part of a movement called DIY biology, where amateurs are performing modern biology experiments in their kitchens, garages, and community lab spaces, rather than academic or corporate research laboratories. A main focus of DIY biologists is synthetic biology, in which new biological processes or organisms are genetically designed or constructed to serve useful purposes, such as creating plants that could replace street lights or your desk lamp to save energy.

One necessary material for the project is a gene that codes for green fluorescent protein (GFP), which glows when exposed to ultraviolet light. You'll also need a tool to get the genes into the plant, but it isn't a mechanical device. Instead, you can buy a culture of *Agrobacterium tumefaciens,* a Gram-negative bacillus soil bacterium commonly used for a genetic engineering technique called *Agrobacterium*-mediated plant transformation.

Natural *Agrobacterium* is pathogenic to many agricultural crops, including grapes, nuts, and beets, because it inserts a tumor-causing plasmid, or extra piece of DNA, into plant cells. *Agrobacterium* used in the synthetic biology process doesn't have the tumor-causing genes in its plasmid; instead, the goal is to add the GFP genes to the plasmid in a process called transformation as the first step of the experiment.

Once the *Agrobacterium* incorporates the foreign genes into its own, it can be used to deliver the genes to plant cells. Flowering tips of the plant are dipped into the *Agrobacterium,* which naturally infects the plant's reproductive cells with newly transformed GFP plasmid. If the procedure is completed properly, some of the seeds produced by the original plant will grow into glowing plants.

Questions to Consider

1. Is it safe to manipulate the genes of an organism?
2. Should scientific experiments be left to trained scientists in regulated research laboratories?
3. What are other applications of *Agrobacterium*-mediated plant transformation?

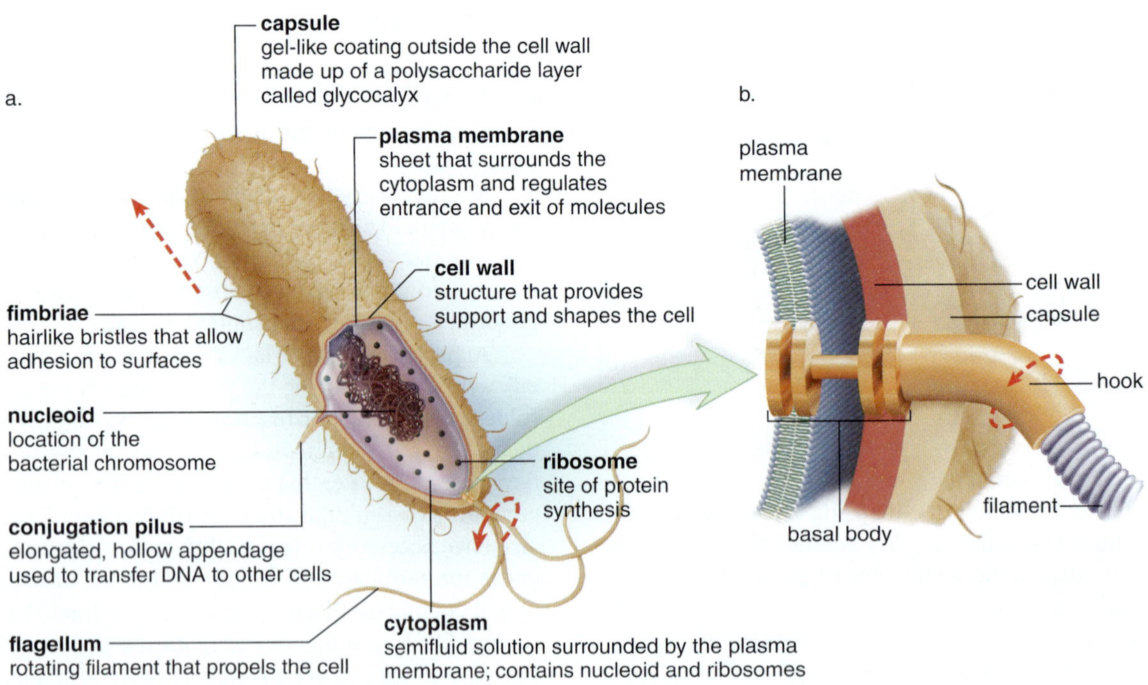

Figure 20.6 Features of prokaryotic cells. **a.** Structural components of a generalized prokaryotic cell. **b.** Each flagellum of a bacterium contains a basal body, a hook, and a filament. The red-dashed arrows indicate that the hook and filament rotate 360 degrees.

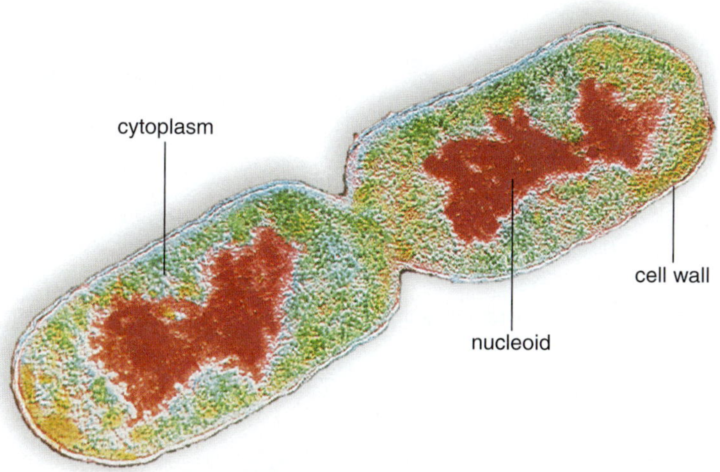

Figure 20.7 Binary fission. When conditions are favorable for growth, prokaryotes divide to reproduce. This is a form of asexual reproduction, because the daughter cells have exactly the same genetic material as the parent cell.

Reproduction in Prokaryotes

Mitosis, which requires the formation of a spindle apparatus, does not occur in prokaryotes. Instead, prokaryotes reproduce asexually by means of **binary fission** (Fig. 20.7).

The single circular chromosome replicates, and then the two copies separate as the cell enlarges. Newly formed plasma membrane and cell wall separate the cell into two cells. Prokaryotes have a generation time as short as 12 minutes under favorable conditions. Mutations are generated and passed on to offspring more quickly than in eukaryotes. Also, prokaryotes are haploid, so mutations are immediately subjected to natural selection, which determines any possible adaptive benefit in the particular environment.

In eukaryotes, genetic recombination occurs as a result of sexual reproduction. Sexual reproduction does not occur among prokaryotes, but three means of genetic recombination have been observed in prokaryotes.

- During **conjugation,** two bacteria are temporarily linked together, often by means of a **conjugation pilus** (see Fig. 20.6*a*). While they are linked, the donor cell passes DNA to a recipient cell in the form of a plasmid.
- **Transformation** occurs when a cell picks up free pieces of DNA secreted by live prokaryotes or released by dead prokaryotes.
- During **transduction,** bacteriophages carry portions of DNA from one bacterial cell to another. Viruses have also been found to infect archaeal cells, so transduction may play an important role in gene transfer for both domains of prokaryotes.

Animation Bacterial Conjugation

Animation Bacterial Transformation

Check Your Progress 20.2

1. Explain the connection between Pasteur's experiment and the sterilization of surgical instruments.
2. Describe the difference between a prokaryote nucleoid and a eukaryote nucleus.
3. Define three ways in which prokaryotes can recombine their genetic material without sexual reproduction.

20.3 The Bacteria

Learning Outcomes

Upon completion of this section, you should be able to

1. Identify the similarities and differences in the cell wall structure of Gram-positive and Gram-negative bacteria.
2. Identify three metabolic types of bacteria and describe how they obtain nutrients from their environments.
3. Describe the unique properties of cyanobacteria.

Bacteria (domain Bacteria) are the more common type of prokaryote. The amount of bacteria on our planet is amazing: Even though we can't see them with our naked eye, the biomass of bacteria on Earth exceeds those of plants and animals combined. They are found in practically every kind of environment on Earth. In this section we consider the bacteria—their characteristics, metabolism, and lifestyle.

Characteristics of Bacterial Cells

Most bacterial cells are protected by a cell wall composed of the unique molecule **peptidoglycan,** a complex of polysaccharides linked by amino acids. Two types of bacteria have been distinguished based on the structure of their cell wall—Gram-positive bacteria and Gram-negative bacteria.

Gram-positive bacteria have a very thick peptidoglycan cell wall relative to thin-walled *Gram-negative bacteria,* as shown in Figure 20.8*b*. In addition, Gram-negative bacteria are surrounded by a second plasma membrane outside the cell wall, which often blocks antibiotic drugs, making infections difficult to treat.

A procedure called the Gram stain, shown in Figure 20.8*a,* lends its name to the types of bacteria, because it distinguishes their cell wall type based on the color bacteria appear after staining. Gram-positive bacteria appear dark purple after the Gram stain process and Gram-negative bacteria appear red, providing a useful first step for identifying unknown bacteria causing an infection.

Animation Gram Stain

Bacteria can also be described in terms of their three basic cell shapes (Fig. 20.9):

- Spirilli (sing., spirillum), spiral-shaped or helical-shaped
- Bacilli (sing., bacillus), rod-shaped
- Cocci (sing., coccus), round or spherical

In addition to cell wall type and shape, bacteria can be characterized by their growth arrangement. For example, *staph* arrangement describes clusters of cells (e.g., staphylococcus), *strept* arrangement describes chains (e.g., streptobacillus), and *diplo* arrangement describes pairs (e.g., diplococcus).

Describing the characterstics of bacterial cells aids in the identification of species such as *Streptococcus pyogenes,* a Gram-positive streptococcus that causes strep throat, and *Neisseria gonorrhoeae,* a Gram-negative diplococcus responsible for gonorrhea.

Bacterial Metabolism

Bacteria are astoundingly diverse in terms of their metabolic lifestyles. With respect to basic nutrient requirements, bacteria are not

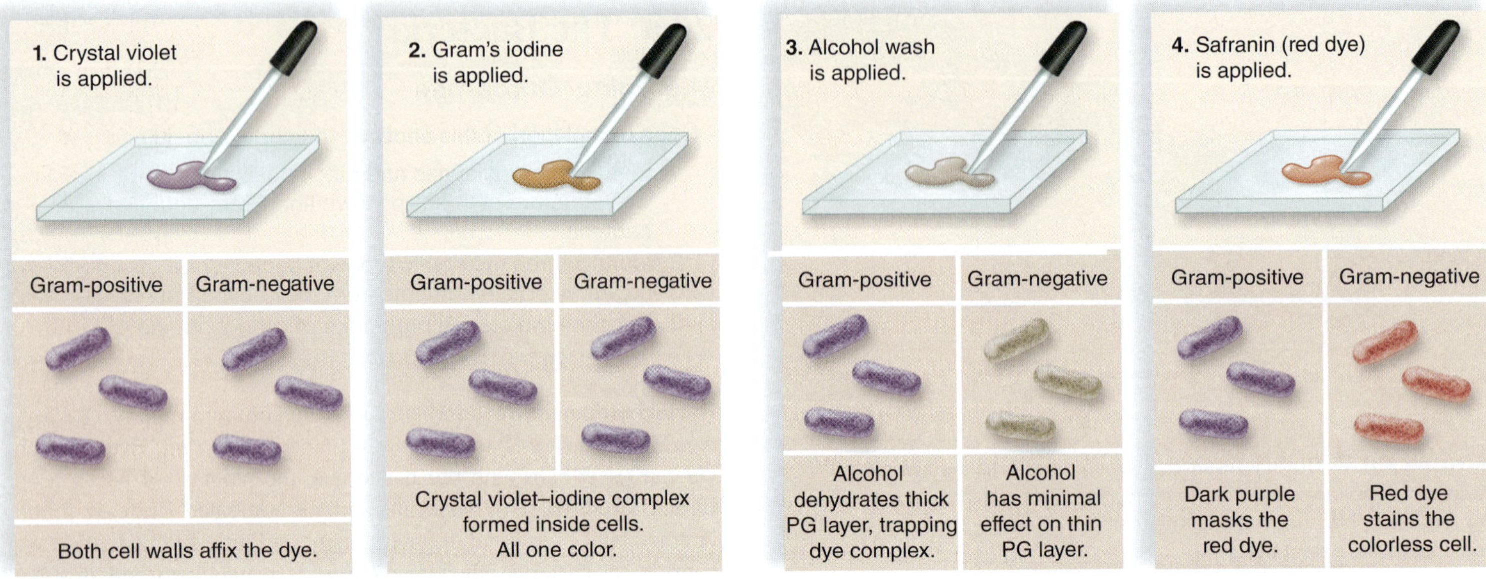

a.

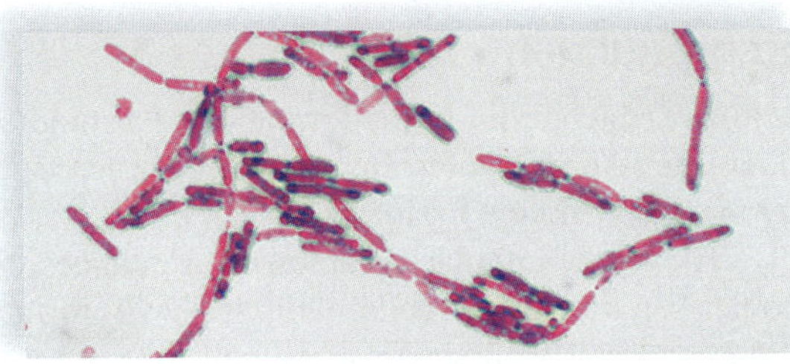

b. 1,000×

Figure 20.8 Gram staining. **a.** The thick peptidoglycan (PG) layer encasing Gram-positive bacteria traps crystal violet dye, so the bacteria appear purple after the Gram stain. Because Gram-negative bacteria have much less peptidoglycan (located between the plasma membrane and an outer membrane), they do not retain the crystal violet dye and so exhibit the red counterstain (usually a safranin dye). **b.** A micrograph showing the results of a Gram stain with both Gram-positive and Gram-negative cells.

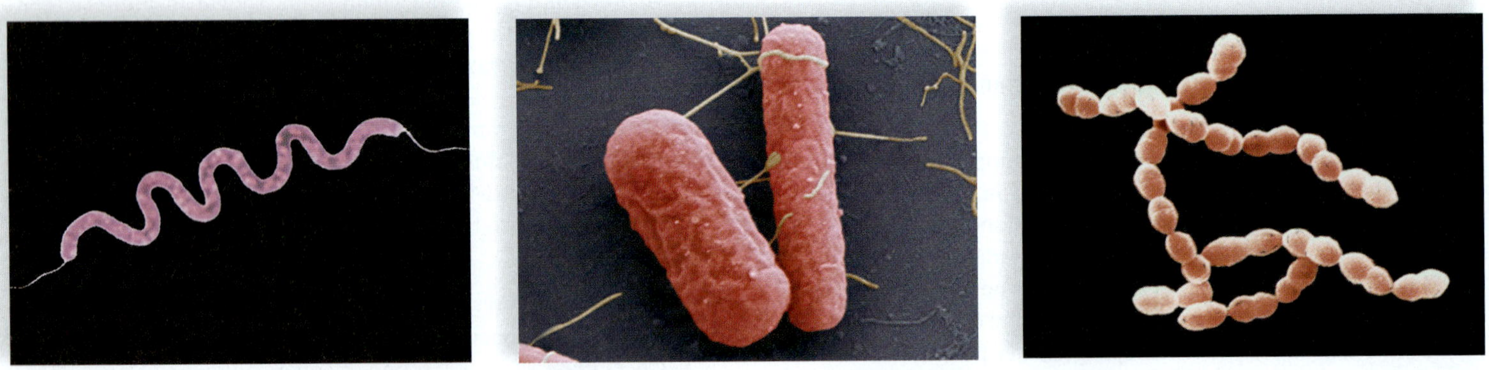

a. Spirillum: *Spirillum volutans* LM 400×

b. Bacilli: *Escherichia coli* SEM 13,300×

c. Cocci: *Streptococcus thermophilus* SEM 6,250×

Figure 20.9 Diversity of bacteria. **a.** Spirillum, a spiral-shaped bacterium. **b.** Bacilli, rod-shaped bacteria. **c.** Cocci, round bacteria.

much different from other organisms. One difference, however, concerns the need for oxygen. Most bacteria are aerobic and, like animals, require a constant supply of oxygen to carry out cellular respiration. Other bacteria, called **facultative anaerobes,** are able to grow in either the presence or the absence of gaseous oxygen. Some bacteria are **obligate anaerobes** and are unable to grow in the presence of free oxygen. A few serious illnesses—such as botulism, gas gangrene, and tetanus—are caused by anaerobic bacteria that infect oxygen-free environments in the human body, such as in the intestine or in deep puncture wounds.

Autotrophic Bacteria

Bacteria called **photoautotrophs** (Gk. *photos,* "light"; *auto,* "self"; *trophe,* "food") are photosynthetic (for a review of photosynthesis, see section 7.2). They use solar energy to reduce carbon dioxide to organic compounds. There are two types of photoautotrophic bacteria:

those that perform anoxygenic photosynthesis and those that perform oxygenic photosynthesis. Their characteristics are shown here:

Photoautotrophic Bacteria	
Anoxygenic Photosynthesis	**Oxygenic Photosynthesis**
- Does not produce O_2 - Photosystem I only - Unique type of chlorophyll called bacteriochlorophyll	- Produces O_2 - Photosystems I and II

Green sulfur bacteria and some purple bacteria carry out anoxygenic photosynthesis. These bacteria usually live in anaerobic (oxygen-poor) conditions, such as the muddy bottom of a marsh. They cannot photosynthesize in the presence of oxygen, and they do not emit oxygen. In contrast, the cyanobacteria (see Fig. 20.12) contain chlorophyll *a* and carry on oxygenic photosynthesis, just as algae and plants do; that is, they reduce carbon dioxide to organic compounds and give off oxygen as a by-product.

Bacteria called **chemoautotrophs** (Gk. *chemo*, "pertaining to chemicals"; *auto*, "self"; *trophe*, "food") carry out chemosynthesis. They oxidize inorganic compounds such as hydrogen gas, hydrogen sulfide, and ammonia to obtain the necessary energy to reduce CO_2 to an organic compound. The nitrifying bacteria oxidize ammonia (NH_3) to nitrites (NO_2^-) and nitrites to nitrates (NO_3^-). Their metabolic abilities keep nitrogen cycling through ecosystems. Other bacteria oxidize sulfur compounds. They live in environments such as deep-sea vents 2.5 km below sea level.

The organic compounds produced by such bacteria and archaea support the growth of a community of organisms found at vents (see page 369). This discovery lends support to the suggestion that the first cells originated at deep-sea vents.

Heterotrophic Bacteria

Bacteria called **chemoheterotrophs** (*hetero*, "different") obtain carbon and energy in the form of organic nutrients produced by other living organisms. For example, parasitic bacteria feed on the tissues and fluids of their living host.

In many ecosystems, chemoheterotrophic bacteria called **saprotrophs** serve as *decomposers* that break down organic matter from dead organisms. Probably no natural organic molecule exists that cannot be digested by at least one prokaryotic species, and this plays a critical role in recycling matter and making inorganic molecules available to photosynthesizers.

The metabolic capabilities of chemoheterotrophic bacteria have long been exploited by humans. Bacteria are used commercially to produce chemicals such as ethyl alcohol, acetic acid, butyl alcohol, and acetones. Bacterial action is also involved in the production of butter, cheese, sauerkraut, rubber, silk, coffee, and cocoa. Even antibiotics are produced by some bacteria.

Symbiotic Relationships

Bacteria (and archaea) form **symbiotic relationships** (Gk. *sym*, "together"; *bios*, "life") in which two different species live together in an intimate way.

- In **mutualism,** both species benefit from the association.
- In **commensalism,** only one species benefits, whereas the other is unaffected.
- In **parasitism,** one species benefits while harming the other.

Mutualistic bacteria live in human intestines, where they release vitamins K and B_{12}, which we can use to help produce blood components. In the stomachs of cows and goats, mutualistic prokaryotes digest cellulose, enabling these animals to feed on grass. Mutualistic bacteria live in the root nodules of soybean, clover, and alfalfa plants, where they reduce atmospheric nitrogen (N_2) to ammonia, a process called nitrogen fixation (Fig. 20.10). Plants are unable to fix atmospheric nitrogen, leaving bacteria their only source for usable nitrogen.

Commensalism often occurs when one population modifies the environment in such a way that a second population benefits. Obligate anaerobes can live in our intestines only because the bacterium *Escherichia coli* uses up the available oxygen.

Parasitic bacteria cause diseases and therefore are called **pathogens;** a few are listed in Table 20.2. In some cases, the growth of microbes themselves does not cause disease; what they release is the pathological portion. When Gram-negative bacteria are killed by an antibiotic, their outer plasma membrane releases a substance called lipopolysaccharide, which acts as a *superantigen* to overstimulate the immune response. The result may be a high fever and a severe drop in blood pressure, leading to shock and possibly death.

When someone steps on a rusty nail, *Clostridium tetani* bacteria can be injected deep into damaged tissue and produce a **toxin** that causes the disease tetanus. The bacteria never leave the site of the wound, but the tetanus toxin they produce does move throughout the body. This toxin prevents the relaxation of muscles. In time, the body contorts, because all the muscles have contracted. Eventually, suffocation occurs.

Figure 20.10 Nodules of a legume. Some free-living bacteria carry on nitrogen fixation; however, bacteria of the genus *Rhizobium* invade the roots of legumes, with the resultant formation of nodules. Here the bacteria convert atmospheric nitrogen to an organic nitrogen the plant can use. These are nodules on the roots of a broad bean plant (*Vicia* sp.)

Table 20.2 Bacterial Diseases in Humans

Category	Disease
Sexually transmitted diseases	Syphilis, gonorrhea, chlamydia
Respiratory diseases	Strep throat, scarlet fever, tuberculosis, pneumonia, Legionnaires disease, whooping cough, inhalation anthrax
Skin diseases	Erysipelas, boils, carbuncles, impetigo, acne, infections of surgical or accidental wounds and burns, leprosy (Hansen disease)
Digestive tract diseases	Gastroenteritis, food poisoning, dysentery, cholera, peptic ulcers, dental caries
Nervous system diseases	Botulism, tetanus, leprosy, spinal meningitis
Systemic diseases	Plague, typhoid fever, diphtheria
Other diseases	Tularemia, Lyme disease

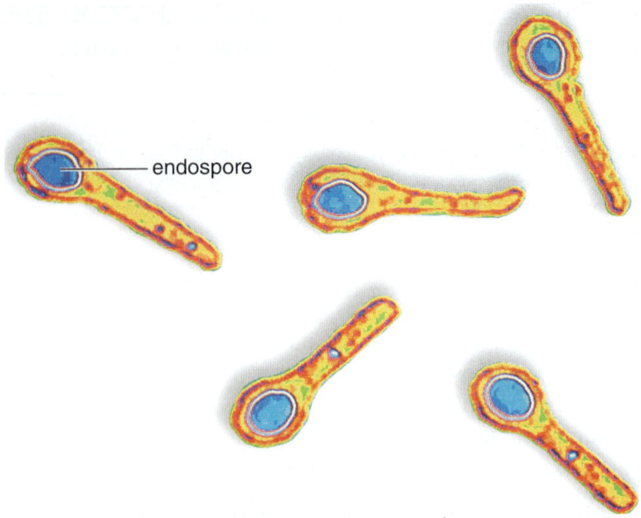

Figure 20.11 The endospore of *Clostridium tetani*. *C. tetani* produces a terminal endospore that causes it to have a drumstick appearance. If endospores gain access to a wound, they germinate and release bacteria that produce a neurotoxin. The patient develops tetanus, a progressive rigidity that can result in death; immunization can prevent tetanus.

Fimbriae allow a pathogen to bind to certain cells, and their specificity determines which organs or cells of the body are its host. Like many bacteria that cause dysentery (severe diarrhea), *Shigella dysenteriae* is able to stick to the intestinal wall. In addition, *S. dysenteriae* produces a toxin, called Shiga toxin, that increases the potential for fatality. Also, invasive mechanisms that give a pathogen the ability to move through tissues and into the bloodstream result in a more medically significant disease than if it were localized. Usually, a person can recover from food poisoning caused by *Salmonella.* But some strains of *Salmonella* have virulence factors—including a needle-shaped toxin secretion apparatus—that allow the bacteria to penetrate the lining of the colon and move beyond this organ. Typhoid fever, a life-threatening disease, can then result.

Some of the deadliest pathogens form **endospores** (Gk. *endon,* "within"; *spora,* "seed") when faced with unfavorable environmental conditions. A portion of the cytoplasm and a copy of the chromosome become dehydrated and are then encased by a heavy, protective endospore coat (Fig. 20.11). In some bacteria, the rest of the cell deteriorates, and the endospore is released.

Endospores survive in the harshest of environments—desert heat and dehydration, boiling temperatures, polar ice, and extreme ultraviolet radiation. They also survive for very long periods. When anthrax endospores 1,300 years old germinate, they can still cause a severe infection (usually seen in cattle and sheep). Humans also fear a deadly but uncommon type of food poisoning called botulism, which is caused by the germination of endospores inside cans of food. To germinate, the endospore absorbs water and grows out of the endospore coat. In a few hours' time, it becomes a typical bacterial cell, capable of reproducing once again by binary fission. Endospore formation is not a means of reproduction, but it does allow the survival and dispersal of bacteria to new places.

Antibiotics

A number of antibiotic compounds are active against bacteria and are widely prescribed. Most antibacterial compounds fall within two classes, those that inhibit protein biosynthesis and those that inhibit cell wall biosynthesis. Two types of antibiotics, erythromycin and tetracyclines, inhibit bacterial protein synthesis by affecting bacterial ribosomes because they function somewhat differently than eukaryotic ribosomes. Antibiotics that inhibit cell wall biosynthesis generally block the formation of peptidoglycan, a compound necessary to maintain the integrity of bacterial cell walls. Penicillin, ampicillin, and fluoroquinolone (e.g., Cipro) inhibit bacterial cell wall biosynthesis without harming animal cells.

Animation Antibiotic Inhibition of Protein Synthesis

Antibiotics are heavily prescribed to treat infection, often when not needed. One outcome of excessive and improper use of antibiotics has been increasing bacterial resistance to antibiotics. Genes conferring resistance to antibiotics can be transferred between infectious bacteria by transformation, conjugation, or transduction. When penicillin was first introduced, less than 3% of *Staphylococcus aureus* strains were resistant to it. Now, because of selective advantage, 90% or more are resistant to penicillin and, increasingly, to methicillin, an antibiotic developed in 1957. A strain of methicillin-resistant *S. aureus* (MRSA) is responsible for many difficult-to-treat infections that have become a threat to human health. MRSA is common in hospitals and nursing facilities, where it causes problems for those with a greater risk of infection, especially those with open wounds and weakened immune systems.

Cyanobacteria

Cyanobacteria are Gram-negative bacteria with a number of unusual traits. They photosynthesize in the same manner as plants and are believed to be responsible for first introducing oxygen into the primitive atmosphere. Formerly, the cyanobacteria were called blue-green algae and were classified with eukaryotic algae, but now they are classified as prokaryotes. Cyanobacteria can have other pigments that mask the color of chlorophyll, so that they appear red, yellow, brown, or black, rather than only blue-green (Fig. 20.12).

Cyanobacterial cells are rather large, ranging from 1 to 50 μm in width. They can be single-celled, colonial, or filamentous. Cyanobacteria lack any visible means of locomotion, although some glide when in contact with a solid surface and others oscillate (sway back and forth). Some cyanobacteria have a special advantage, because they possess heterocysts, which are thick-walled cells without nuclei, where nitrogen fixation occurs. The ability to photosynthesize and to fix atmospheric nitrogen (N_2) means that their nutritional requirements are minimal. They can serve as food for heterotrophs in ecosystems.

Cyanobacteria are common in fresh and marine waters, in soil, and on moist surfaces, but they are also found in harsh habitats, such as hot springs. They are symbiotic with a number of organisms, including liverworts, ferns, and even at times invertebrates such as corals. In association with fungi, they form **lichens** that can grow on rocks. In a lichen, the cyanobacterium mutualistically provides organic nutrients to the fungus, while the fungus possibly protects and furnishes inorganic nutrients to the cyanobacterium. It is also possible that the fungus is parasitic on the cyanobacterium. Lichens help transform rocks into soil; other forms of life then may follow. It is hypothesized that cyanobacteria were the first colonizers of land during the course of evolution.

Cyanobacteria are ecologically important in still another way. If care is not taken in disposing of industrial, agricultural, and human wastes, phosphates drain into lakes and ponds, resulting in a "bloom" of these organisms. The surface of the water becomes turbid, and light cannot penetrate to lower levels. When a portion of the cyanobacteria die off, the decomposers feeding on them use up the available oxygen, causing fish to die from lack of oxygen.

Check Your Progress 20.3

1. Describe how the peptidoglycan layer is different in Gram-positive and Gram-negative cells.
2. Explain how autotrophic and heterotrophic bacteria differ.
3. Construct a hypothesis about how cyanobacteria may have affected the atmosphere of early Earth.

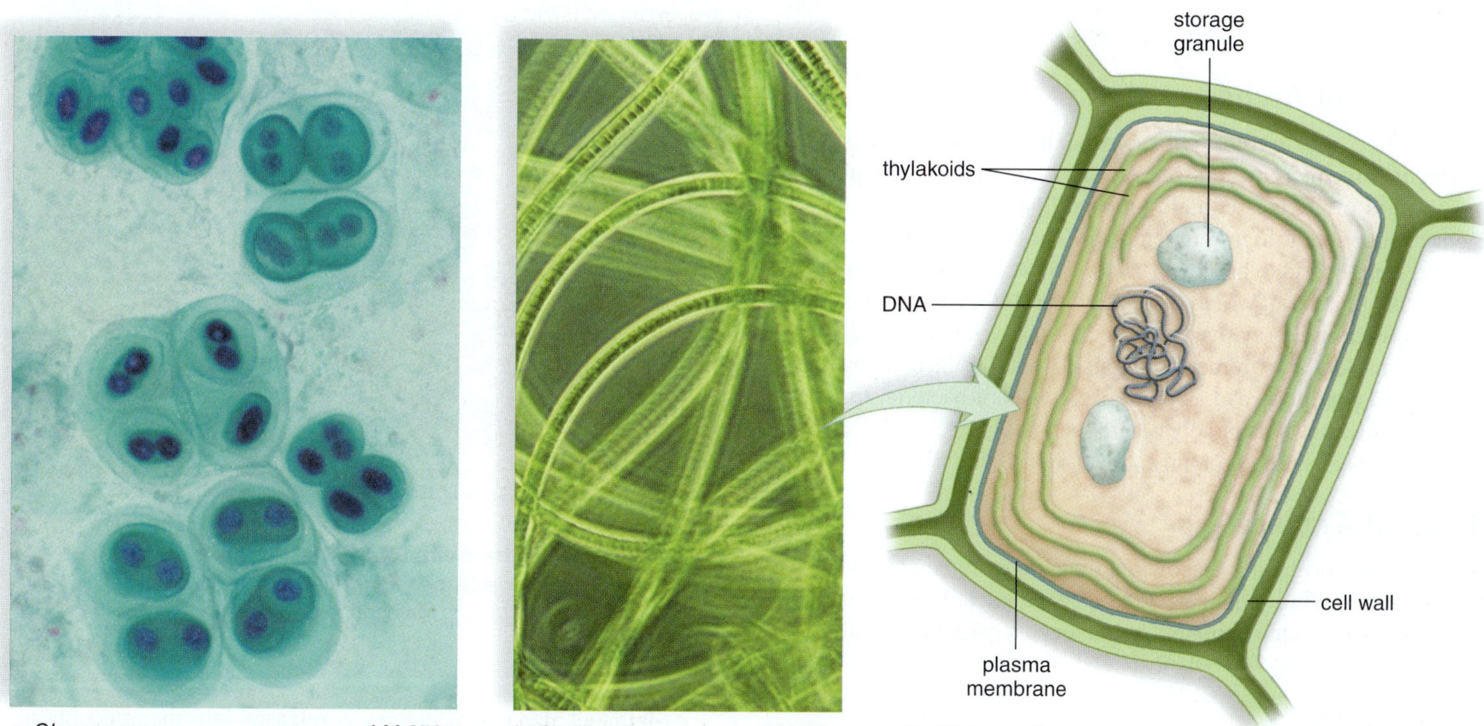

Figure 20.12 Diversity among the cyanobacteria. **a.** In *Gloeocapsa,* single cells are grouped in a common gelatinous sheath. **b.** Filaments of cells occur in *Oscillatoria.* **c.** One cell of *Oscillatoria*.

20.4 The Archaea

Learning Outcomes

Upon completion of this section, you should be able to

1. List the biochemical characteristics that distinguish archaea from bacteria and eukaryotes.
2. Describe three different types of archaea and the habitats in which they are found.
3. Explain two ways in which archaea metabolize inorganic compounds in extreme environments.

At one time, **archaea** (domain Archaea) were considered to be a unique group of bacteria. Archaea came to be viewed as a distinct domain of organisms in 1977, when Carl Woese and George Fox discovered that the rRNA of archaea has a different sequence of bases than the rRNA of bacteria. They chose rRNA because of its involvement in protein synthesis—any changes in rRNA sequence probably occur at a slow, steady pace as evolution occurs.

As discussed in Chapter 19, it is proposed that the tree of life contains three domains: Archaea, Bacteria, and Eukarya. Because archaea and some bacteria are found in extreme environments (hot springs, thermal vents, salt basins), they may have diverged from a common ancestor relatively soon after life began. Then later, the eukarya diverged from the archaeal line of descent. In other words, the eukarya are more closely related to the archaea than to the bacteria. Archaea and eukarya share some of the same ribosomal proteins (not found in bacteria), initiate transcription in the same manner, and have similar types of tRNA.

Structure of Archaea

Archaea are prokaryotes with biochemical characteristics that distinguish them from both bacteria and eukaryotes. The plasma membranes of archaea contain unusual lipids that allow many of them to function at high temperatures. The lipids of archaea contain glycerol linked to branched-chain hydrocarbons, in contrast to the lipids of bacteria, which contain glycerol linked to fatty acids.

The archaea also evolved diverse cell wall types, which facilitate their survival under extreme conditions. The cell walls of archaea do not contain peptidoglycan as do the cell walls of bacteria. In some archaea, the cell wall is largely composed of polysaccharides, and in others, the wall is pure protein. A few have no cell wall.

Types of Archaea

Archaea were originally discovered living in extreme environmental conditions. Three main types of archaea are still distinguished based on their unique habitats: methanogens, halophiles, and thermoacidophiles (Fig. 20.13).

Methanogens

The **methanogens** (methane makers) are obligate anaerobes found in environments such as swamps, marshes, and the intestinal tracts of animals. Methanogenesis, the ability to form methane (CH_4), is a type of metabolism performed only by some archaea. Methanogens are chemoautotrophs, using hydrogen gas (H_2) to reduce carbon dioxide (CO_2) to methane and couple the energy released to ATP production.

Figure 20.13 Extreme habitats. **a.** Halophilic archaea can live in salt lakes. **b.** Thermoacidophilic archaea can live in the hot springs of Yellowstone National Park. **c.** Methanogens live in swamps and in the guts of animals.

Methane, also called biogas, is released into the atmosphere, where it contributes to the greenhouse effect and climate change. About 65% of the methane in our atmosphere is produced by these methanogenic archaea.

Methanogenic archaea may help us anticipate what life may be like on other celestial bodies. Consider, for instance, the unusual microbial community residing in the Lidy Hot Springs of eastern Idaho. The springs, which originate 200 m (660 feet) beneath the Earth's surface, are lacking in organic nutrients but rich in H_2. Scientists have found the springs to be inhabited by vast numbers of microorganisms; over 90% are archaea, and the overwhelming majority are methanogens. The researchers who first investigated the Lidy Hot Springs microbes point out that similar methanogenic communities may someday be found beneath the surfaces of Mars and Europa (one of Jupiter's moons). Because hydrogen is the most abundant element in the universe, it would be readily available for use by methanogens everywhere.

Halophiles

The **halophiles** require high salt concentrations (usually 12–15%; the ocean, in contrast, is about 3.5%). They have been isolated from highly saline environments in which few organisms are able to survive, such as the Great Salt Lake in Utah, the Dead Sea, solar salt ponds, and hypersaline soils.

These archaea have evolved a number of mechanisms to survive in environments that are high in salt. To prevent osmotic water loss to the hypertonic environment, halophiles increase solutes such as chloride ions, potassium ions, and organic molecules within the cell, creating an internal environment more isotonic to the outside salt water. This survival ability benefits the halophiles, as they do not have to compete with as many microorganisms as they would encounter in a more moderate environment.

These organisms are aerobic chemoheterotrophs; however, some species can carry out a unique form of photosynthesis if their oxygen supply becomes scarce, as commonly occurs in highly saline conditions. Instead of chlorophyll, these halophiles use a purple pigment called bacteriorhodopsin to capture solar energy for use in ATP synthesis. Interestingly, most halophiles are so adapted to a high-saline environment that they perish if placed in a solution with a low salt concentration (such as pure water).

Thermoacidophiles

The third major type of archaea are the **thermoacidophiles.** These archaea thrive in extremely hot, acidic environments, such as hot springs, geysers, submarine thermal vents, and volcanoes. They are chemoautotrophic anaerobes that use hydrogen (H_2) as the electron donor, and sulfur (S) or sulfur compounds as terminal electron acceptors, for their electron transport chains. Hydrogen sulfide (H_2S) and protons (H^+) are common products.

Recall that the greater the concentration of protons, the lower (and more acidic) the pH. Thus, it is not surprising that thermoacidophiles grow best at extremely low pH levels, between pH 1 and 2. Due to the unusual lipid composition of their plasma membranes, thermoacidophiles survive best at temperatures above 80°C; some can even grow at 105°C (remember that water boils at 100°C)!

Archaea in Moderate Habitats

Although archaea are capable of living in extremely stressful conditions, they are found in all moderate environments as well. For example, some archaea have been found living in symbiotic relationships with animals, including sponges and sea cucumbers. Such relationships are sometimes mutualistic or even commensalistic, but there are no parasitic archaea—that is, they are not known to cause infectious diseases.

The roles of archaea in activities such as nutrient cycling are still being explored. For example, a group of nitrifying marine archaea has recently been discovered. Some scientists think that these archaea may be major contributors to the supply of nitrite in the oceans. Nitrite can be converted by certain bacteria to nitrate, a form of nitrogen that can be used by plants and other producers to construct amino acids and nucleic acids. Archaea have also been found inhabiting lake sediments, rice paddies, and soil, where they are likely to be involved in nutrient cycling.

Check Your Progress 20.4

1. Identify the differences between archaea and bacteria.
2. List the three types of archaea distinguished by their unique habitats.
3. Archaea are thought to be closely related to eukaryotes. Explain the evidence that supports this possibility.

CONNECTING *the* CONCEPTS *with the* THEMES

Evolution

- Bacteria and archaea are single-celled organisms that originated at least 3.5 billion years ago. Many are able to live in extreme habitats that are similar to those present on early Earth.
- Bacteria reproduce rapidly and as a result can undergo selection and evolution quickly; an example is the development of antibiotic resistance.
- Virus genomes mutate rapidly, and this helps them evade the attempts of the human immune system to recognize and destroy them.

Nature of Science

- Virology is the study of the structure, reproduction, and transmission of viruses; this science has increasing importance in fighting and preventing viral diseases.
- Properties of bacteria are manipulated to synthesize a variety of products, from medicines and foods to silk and rubber.

Biological Systems

- Saprotrophic bacteria are essential to the decomposition of organic material.
- Mutualistic bacteria that live in the root nodules of soybean, clover, and alfalfa plants allow these plants to fix nitrogen.
- In the human gut environment, bacteria are critical to the digestion of food and absorption of nutrients.

MEDIA STUDY TOOLS http://connect.mheducation.com

SMARTBOOK® Study smart. SmartBook helps you maximize your study time by identifying what you know and don't know. The recharge feature in SmartBook helps you prepare for the test by identifying the material that you are most likely to forget.

connect BIOLOGY Find even more resources to learn the chapter concepts including animations, tutorial videos, and interactive practice questions.

Animations	Tutorial	Videos
20.1 Antiviral Agents • Lambda Phage Replication Cycle • Entry of Virus into Host Cell • How the HIV Infection Cycle Works • Treatment of HIV Infection • Replication Cycle of a Retrovirus • How Prions Arise • Prion Diseases **20.2** Bacterial Locomotion • Bacterial Conjugation • Bacterial Transformation **20.3** Gram Stain • Antibiotic Inhibition of Protein Synthesis	**20.1** Viral Life Cycle	**20.1** Virus Lytic Cycle • Virus Crisis

SUMMARIZE

20.1 Viruses, Viroids, and Prions

Viruses are nonliving, obligate intracellular parasites that can only replicate within a living cell.

Classification of viruses is difficult, because they are nonliving and mutate rapidly.

All viruses have at least two parts: an outer **capsid** composed of protein subunits and an inner core of nucleic acid, either DNA or RNA. Some also have an outer membranous envelope that contains glycoprotein spikes used for attachment to host cells.

Viruses are host-specific, meaning they can only attach to and infect the cells of certain organisms. To reproduce, a virus hijacks the host cell, forcing it to make new virus copies during the viral replication cycle.

The **lytic cycle** of viral replication follows five steps:

1. *Attachment:* The viral capsid or spikes lock to a host cell receptor.
2. *Penetration:* The viral genome and certain enzymes enter the cell.
3. *Biosynthesis:* Host cell enzymes, ribosomes, and ATP are used to create new viral components.
4. *Maturation:* Viral components are assembled into new viruses.
5. *Release:* New viruses exit the host cell through lysis or budding.

Some virus genomes enter the **lysogenic cycle** after penetration and are integrated into the host cell's DNA, creating a **lysogenic cell** that passes copies of the viral genome to its daughter cells. Later (possibly years later), the viral genome will continue the lytic cycle to create new virus particles.

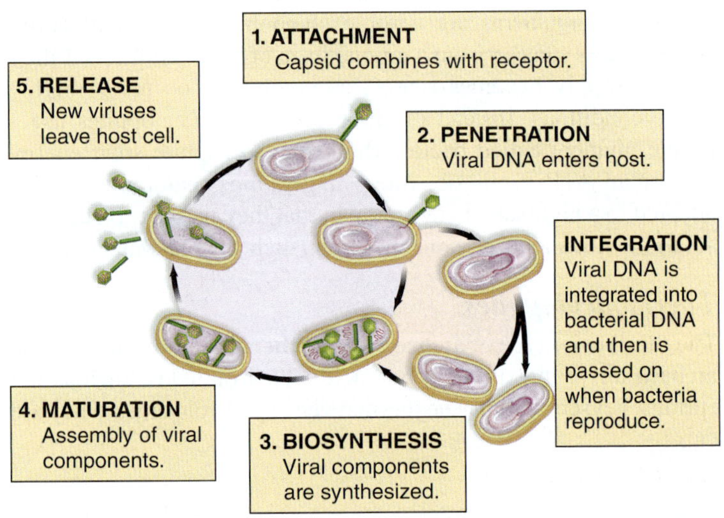

Replication varies among viruses. **Bacteriophages** are viruses that infect prokaryotes and lyse the host cell during release. Viruses that infect animal cells need to be uncoated to free the genome from the capsid, and either budding or lysis releases the viral particles from the cell. **Retroviruses,** like HIV, use **reverse transcriptase** to produce single-stranded DNA from viral RNA, which is then copied and integrated into the host DNA. HIV, once unknown, is an example of an **emerging virus** able to infect a large number of people.

Viroids are naked strands of RNA (not covered by a capsid) that can cause disease in plants. **Prions** are protein molecules that have a misshapen tertiary structure. Prions cause **neurodegenerative diseases,** such as TSEs in humans and mad cow disease in cattle.

20.2 The Prokaryotes

The **bacteria** (domain Bacteria) and **archaea** (domain Archaea) are **prokaryotes.** Prokaryotes were the first living cells, lack a nucleus, and do not have cytoplasmic organelles.

Prokaryotes reproduce asexually by **binary fission.** Their chief method for achieving genetic variation is mutation and genetic recombination by means of **conjugation** (using a **conjugation pilus**), **transformation,** and **transduction.**

20.3 The Bacteria

Bacteria (domain Bacteria) are the more prevalent type of prokaryote, found in nearly every type of environment on Earth.

Bacteria possess a cell wall made of **peptidoglycan,** a complex of polysaccharides linked by amino acids. Gram staining highlights the difference between cell wall types. Gram-positive bacteria possess a thick peptidoglycan cell wall, and Gram-negative bacteria have a thinner cell wall surrounded by a second plasma membrane. Bacterial cells occur in three basic shapes: spiral-shaped (spirillum), rod-shaped (bacillus), and round (coccus).

Prokaryotes differ in their need (and tolerance) for oxygen. There are aerobic prokaryotes requiring oxygen, **facultative anaerobes** capable of living with or without oxygen, and **obligate anaerobes** unable to survive in the presence of oxygen.

Autotrophic prokaryotes gain energy through photosynthesis (**photoautotrophs**) or oxidation of inorganic compounds (**chemoautotrophs**). **Chemoheterotrophic** prokaryotes gain organic molecules from living (parasites) or dead organisms (**saprotrophs**). Saprotrophs are absolutely essential to the cycling of nutrients in ecosystems.

Many bacteria exist in **symbiotic relationships** with other organisms, including **lichens,** a symbiotic growth of cyanobacteria and fungi. In **mutualism,** both species benefit, such as nitrogen-fixing bacteria living on the roots of legumes. **Commensal** relationships benefit one organism but have no afffect on the other. Some bacteria are **pathogens,** existing in **parasitic,** disease-causing relationships with other living organisms. Production of **endospores, toxins,** and superantigens increases their ability to cause disease.

Cyanobacteria are unique, because they photosynthesize as plants do, may grow in colonies or long filaments, and are able to fix their own nitrogen.

20.4 The Archaea

The archaea (domain Archaea) are a second type of prokaryote. On the basis of rRNA sequencing, the archaea appear to be more closely related to the eukarya than to the bacteria.

Archaea do not have peptidoglycan in their cell walls, as do the bacteria, and they share more biochemical characteristics with the eukarya than do bacteria.

Three types of archaea live under harsh conditions, such as anaerobic marshes (**methanogens**), salty lakes (**halophiles**), and hot sulfur springs (**thermoacidophiles**). Archaea are also found in moderate environments.

ASSESS

Choose the best answer for each question.

20.1 Viruses, Viroids, and Prions

1. Viruses are considered nonliving because
   a. they do not locomote.
   b. they cannot reproduce independently.
   c. their nucleic acid does not code for protein.
   d. they are single-celled.
2. Which of these are found in all viruses?
   a. envelope, nucleic acid, capsid
   b. DNA, RNA, and proteins
   c. capsid and a nucleic acid
   d. proteins, nucleic acids, carbohydrates, and lipids
   e. tail fibers, spikes, and rod shape
3. The envelope of an animal virus is derived from the ________________ of its host cell.
   a. cell wall
   b. membrane
   c. glycocalyx
   d. receptors
4. What is the correct order of the stages in the lytic cycle of viral reproduction?
   a. penetration, attachment, uncoating, biosynthesis, release
   b. attachment, release, penetration, biosynthesis, maturation, lysogeny
   c. attachment, penetration, biosynthesis, maturation, release
   d. maturation, release, attachment, penetration, biosynthesis

20.2 The Prokaryotes

5. Which is not true of prokaryotes?
   a. They are living cells.
   b. They lack a nucleus.
   c. They all are parasitic.
   d. They are both archaea and bacteria.
   e. They evolved early in the history of life.
6. Which of these is most apt to be a prokaryotic cell wall function?
   a. transport
   b. motility
   c. support
   d. adhesion
7. Prokaryotes reproduce by means of
   a. the lytic cycle.
   b. binary fission.
   c. mitosis.
   d. meiosis.
   e. the lysogenic cycle.

20.3 The Bacteria

8. How do Gram-negative bacteria differ from Gram-positive bacteria?
   a. They are not pathogenic.
   b. Their peptidoglycan cell wall is thinner.
   c. Gram-negative bacteria cannot be Gram stained.
   d. They possess a single plasma membrane.
   e. Gram-negative bacteria do not photosynthesize.

9. The cell wall of bacteria is composed of
   a. lipopolysaccharide.
   b. phospholipids.
   c. cellulose.
   d. peptidoglycan.
   e. protein capsomers.
10. Facultative anaerobes
    a. require a constant supply of oxygen.
    b. are killed in an oxygenated environment.
    c. do not always need oxygen.
    d. are photosynthetic but do not give off oxygen.
11. Chemoautotrophic prokaryotes
    a. are chemosynthetic.
    b. use the rays of the sun to acquire energy.
    c. oxidize inorganic compounds to acquire energy.
    d. are always bacteria, not archaea.
12. A prokaryote that gains organic molecules from other living organisms is a
    a. photoautotroph.
    b. photoheterotroph.
    c. chemoautotroph.
    d. chemoheterotroph.
13. Gram-negative bacteria that first introduced oxygen into the atmosphere through photosynthesis are
    a. methanogens.
    b. pathogens.
    c. lichens
    d. cyanobacteria.

20.4 The Archaea

14. Archaea are distinct from bacteria because
    a. some archaea form methane.
    b. each have different rRNA sequences.
    c. archaea do not have peptidoglycan cell walls.
    d. archaea are more closely related to eukaryotes.
    e. All of these are correct.
15. Which of these archaea would live at a deep-sea vent?
    a. thermoacidophile
    b. halophile
    c. methanogen
    d. parasitic forms

ENGAGE

THINKING SCIENTIFICALLY

1. Most viral infections, including the common cold, chickenpox, mononucleosis, and Ebola, are untreatable. Few antiviral drugs exist, and those in use may have severe side effects on the patient. Why are viral infections so difficult to treat, and why do the drugs cause side effects? Are there steps in viral reproduction that new antiviral medications could target?
2. Antibiotic medications work by targeting specific structures and functions in bacterial cells. Side effects on the patient are usually minimal, because their eukaryotic cells do not possess the same structures and characteristics as the prokaryotic pathogens. What structures or functions of the prokaryotic cell would serve as good targets for new antibiotics?
3. We know that bacteria and archaea are very diverse in their metabolic capabilities, such as the consumption of organic waste and carbon dioxide and the creation of others such as oxygen, alcohol, methane, nitrates, and antibiotics. Amazingly, these known capabilities come from only the 1–10% of prokaryotes that have been identified. Scientists hypothesize that at least 90% are undiscovered and have potential metabolic capabilities we could harness to benefit human life. If you were a scientist, what metabolic capability of bacteria would you try to discover to solve a human or environmental problem?

Beaches are closed to protect swimmers from possible amoeba infections.

21

Protist Evolution and Diversity

Protected from the heat wave in an air-conditioned lab, a pathologist pulls a petri plate covered in *E. coli* bacteria from an incubator. A boy has suddenly passed away, and a sample of his brain tissue is in the center of the plate. Radiating from the tissue are tiny pathways cut through the film of bacteria, pathways characteristic of the "brain-eating amoeba" *Naegleria fowleri.* The tiny protists must have been the cause of death, and they were feeding their way through the *E. coli,* just as they had fed through the boy's brain. Five days ago, the boy suddenly experienced a high fever, stiff neck, and vomiting. Doctors suspected bacterial meningitis, but antibiotics didn't help, and he died before more tests could be performed. But how was the boy infected? At the start of the oppressive heat wave, he and his family had gone to the lake to cool off. Stirred up from the bottom by the boy's boisterous play, the amoeba likely entered his nose during a handstand attempt and burrowed toward his brain, virtually guaranteeing his death. Fortunately, *N. fowleri* is just one rare species among thousands of other beneficial protists that live around us, on us, and in us. In the same lake where the boy was infected, there are critical species producing oxygen, recycling nutrients, and forming the base of the aquatic food chain.

As you read through this chapter, think about the following questions:

1. What explains the amazing diversity of protists?
2. How do protists affect other organisms and the environment?
3. How do microorganisms, such as protists, impact our health and welfare?

Chapter Outline

21.1 General Biology of Protists 374
21.2 Supergroup Archaeplastida 377
21.3 Supergroup Chromalveolata 380
21.4 Supergroup Excavata 385
21.5 Supergroups Amoebozoa, Opisthokonta, and Rhizaria 389

Before You Begin

Before beginning this chapter, take a few moments to review the following discussions.

Section 4.3 How does the endosymbiotic theory explain the origin of energy-producing organelles in the eukaryotic cell?

Section 4.4 What is the basic structure of a eukaryotic cell?

Section 6.4 How are eukaryotic organelles involved in the production and flow of energy in a cell?

Following *the* Themes

CHAPTER 21 PROTIST EVOLUTION AND DIVERSITY

UNIT 4 MICROBIOLOGY AND EVOLUTION

Evolution	Protists represent the oldest branch of eukaryotes in the tree of life. Endosymbiosis was a key step in the evolution of the first eukaryotes billions of years ago.
Nature of Science	Many protists serve as model organisms critical to many areas of scientific research, including ecosystem ecology, medicine, and epidemiology.
Biological Systems	Protists are at the base of many food chains, serving as the foundation for all life in an ecosystem.

21.1 General Biology of Protists

Learning Outcomes

Upon completion of this section, you should be able to

1. Explain the origin of eukaryotic organelles.
2. Assign protists into one of two groups based on mode of nutrition.
3. Understand that protists represent multiple evolutionary lineages.

Protists are the simplest, but most diverse, of the eukaryotes. Most are single-celled, but some exist as colonies of cells or are multicellular. As eukaryotes, protists have membranous organelles, such as mitochondria and plastids, that serve as the energy centers of the cell. (Section 4.7 describes the structure and function of mitochondria and plastids.)

The endosymbiotic theory proposes that eukaryotic cells acquired mitochondria and plastids, including chloroplasts, by engulfing a free-living bacterium that developed a symbiotic relationship within the host cell, a process termed **endosymbiosis** (see section 4.3). Mitochondria were derived first from the endosymbiosis of an aerobic bacterium, and chloroplasts were derived later from the endosymbiosis of a cyanobacterium (see Fig. 4.5). Much of the endosymbiotic bacteria's genomes have been incorporated into the genome of the host cell and now complement the life processes of the host.

Animation Endosymbiosis

Characteristics of Protists

Protists vary in size from microscopic algae and protozoans to kelp that can exceed 200 m in length. Kelp, a brown alga, is multicellular; *Volvox,* a green alga, is colonial; *Spirogyra,* also a green alga, is filamentous. Most protists are single-celled, but despite their small size they have attained a high level of complexity. The amoeboids and ciliates possess unique organelles—their contractile vacuole is an organelle that assists in water regulation.

Protists are sometimes grouped according to how they acquire organic nutrients. The algae are a diverse group of photoautotrophic protists that synthesize organic compounds via photosynthesis. Protozoans are a group of heterotrophic protists that obtain organic compounds from the environment. Some protozoans, such as *Euglena,* are **mixotrophic,** meaning they are able to combine autotrophic and heterotrophic nutritional modes.

Protists reproduce sexually and asexually. Asexual reproduction by mitosis is the norm in protists. Sexual reproduction generally occurs only when environmental conditions are unfavorable. Protists can form spores or **cysts,** which are dormant phases of the protist life cyle, that can survive until favorable conditions return. Parasitic protists form cysts for the transfer to a new host.

Many protists cause diseases in humans, but many others have significant ecological importance. Aquatic photoautotrophic protists produce oxygen and are the foundation of the food chain in both freshwater and saltwater ecosystems. They are a part of **plankton,** organisms suspended in the water that serve as food for heterotrophic protists and animals. Interestingly, whales, the largest animals in the sea, feed on plankton, one of the smallest.

Evolution and Diversity of Protists

Protists were once classified together as a single kingdom. Recently, DNA evidence has suggested that protists are not **monophyletic**—that they do not all belong to the same evolutionary lineage. In fact, protists and other eukaryotes, including the plants, fungi, and animals, are currently classified into six supergroups (Table 21.1). A **supergroup** is a high-level taxonomic group below domain and above kingdom. Each supergroup represents a separate evolutionary lineage and can be summarized as follows:

- Archaeplastida—red and green algae (also includes land plants); they have plastids
- Chromalveolata—brown and golden brown algae, water molds, and the alveolates; most have plastids
- Excavata—zooflagellates, often with distinctive oral grooves
- Amoebozoa—protozoans that move via pseudopods
- Opisthokonta—single-celled and multicellular protists, including choanoflagellates (also includes animals and fungi)
- Rhizaria—foraminiferans and radiolarians

The DNA evidence supports these multiple protist lineages, but the relationships among the lineages are difficult to decipher. Protist lineages are very long and old, dating back to the origin of the first eukaryotes. As lineages stretch back in time, we can be less and less certain about how they are related to each other, just as the history of humans is less complete the further back in time we look. New research in the evolution of protists has helped clarify some of the evolutionary relationships among eukaryote lineages (Fig. 21.1), but much research still needs to be done.

In the following sections, we examine the various supergroups into which protists and other eukaryotes have been placed based on our current understanding.

Check Your Progress 21.1

1. Explain how mitochondria and chloroplasts originated in eukaryotic cells.
2. Describe how algae and protozoans are nutritionally different from one another.
3. Identify the eukaryote supergroups that (1) include both plants and protists and (2) include fungi, animals, and protists.

Table 21.1 Protist Diversity

Supergroup	Members	Distinguishing Features	Ecological Roles	Human Interactions
Archaeplastids 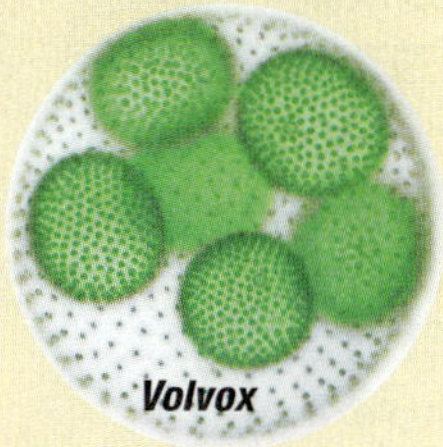	Green algae, red algae, charophytes **Other Members** Plants	Plastids; single-celled, colonial, and multicellular	Algae are primary producers for aquatic food webs	Biofuels, food, gelatin-like agar
Chromalveolates 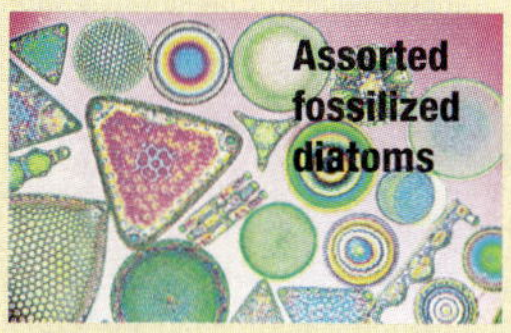	Stramenopiles: brown algae, diatoms, golden brown algae, water molds Alveolates: ciliates, apicomplexans, dinoflagellates	Most with plastids; single-celled and multicellular Alveoli support plasma membrane; single-celled	Kelp forests support diverse marine environments, dinoflagellates support coral growth	Fertilizer, toxic red tides, malaria, toxoplasmosis
Excavates 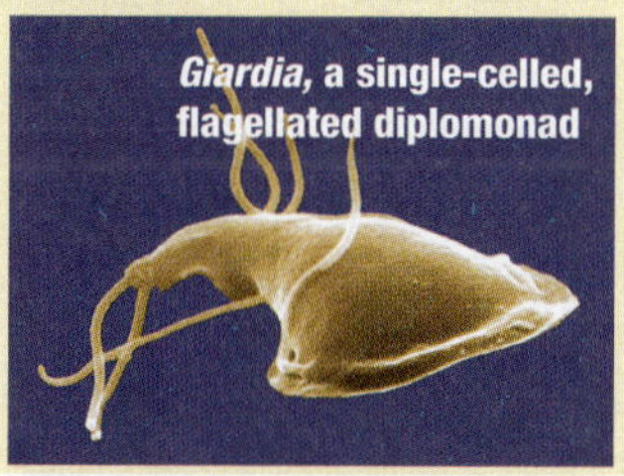	Euglenids, kinetoplastids, parabasalids, diplomonads	Feeding groove; unique flagella; single-celled	Live in the guts of termites where they assist in the breakdown of cellulose	STD-trichomoniasis, giardiasis, African sleeping sickness
Amoebozoans 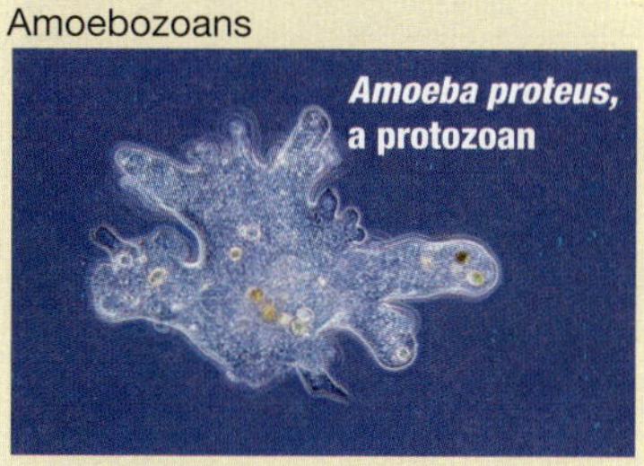	Amoeboids, plasmodial and cellular slime molds	Pseudopods; single-celled	Important terrestrial decomposers	*Entamoeba histolyca*—dysentery, *Naegleria fowleri*—brain-eating amoeba
Rhizaria 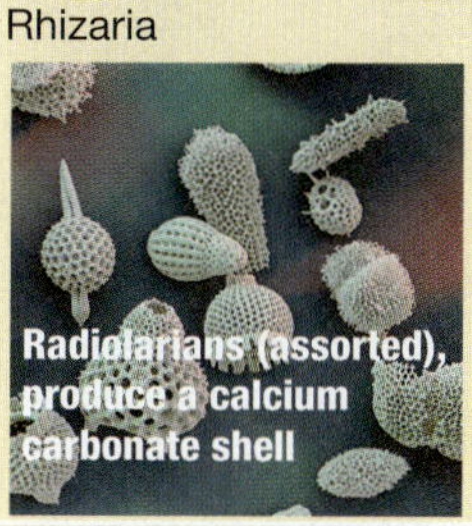	Foraminiferans, radiolarians	Thin pseudopods; some with tests; single-celled	Bioindicators, produce carbonates for coral reef development	Fossils indicate oil deposits and provide an index for geological dating
Opisthokonts 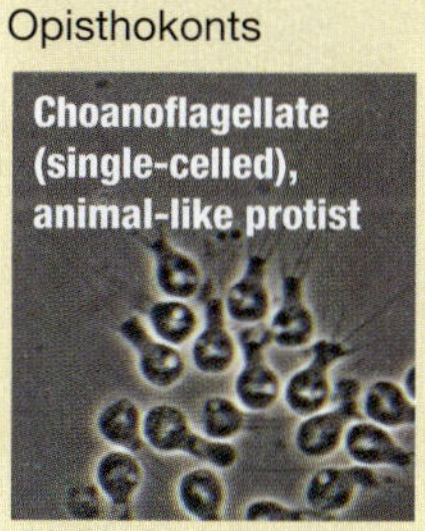	Choanoflagellates, nucleariids **Other Members** Animals, fungi	Some with flagella; single-celled and colonial	Participate in aquatic carbon cycling by feeding on aquatic bacteria and detritus	Close relationship to animals allows insight into early animal evolution

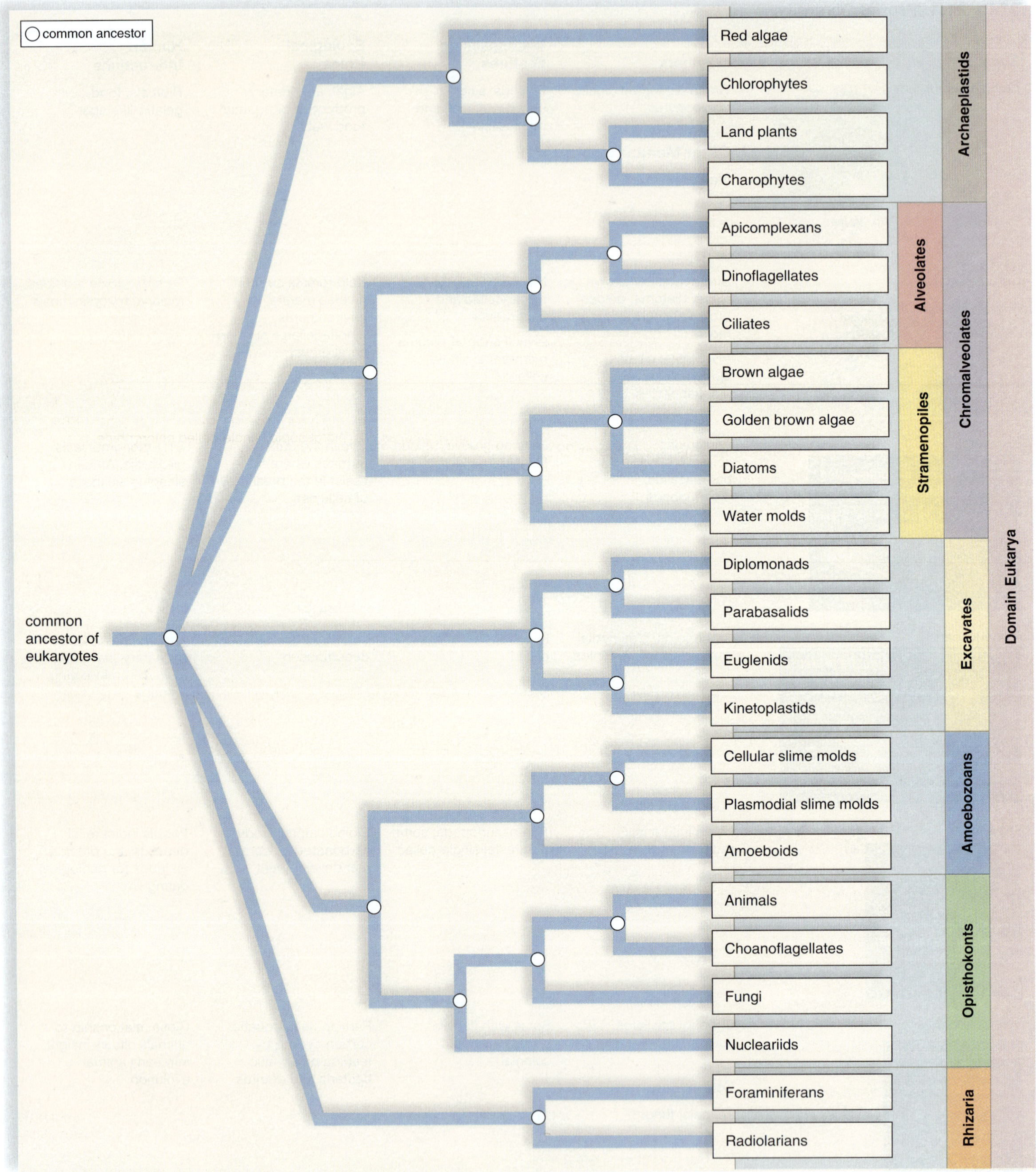

Figure 21.1 Evolutionary relationships between the eukaryotic supergroups. Molecular data are used to determine the relatedness of the supergroups and their constituents. This is a simplified tree that does not include all members of each supergroup.

21.2 Supergroup Archaeplastida

Learning Outcomes

Upon completion of this section, you should be able to

1. Identify the distinguishing characteristics of archaeplastida.
2. Describe the life cycles of archaeplastida.

The **archaeplastids** (Gk. *archeos,* "ancient"; *plastikos,* "moldable") include land plants and other photosynthetic organisms, such as green and red algae, that have plastids derived from endosymbiotic cyanobacteria (see Fig. 4.5).

Green Algae

The **green algae** are protists that contain both chlorophylls *a* and *b.* They inhabit a variety of environments, including oceans, fresh water, snowbanks, the bark of trees, and the backs of turtles. Some of the 8,000 species of green algae also form symbiotic relationships with plants, animals, and fungi in lichens (see Chapter 22).

Green algae occur in many different forms. The majority are single-celled; however, filamentous and colonial forms exist. Seaweeds are multicellular green algae that resemble lettuce leaves. Despite the name, green algae are not always green; some have additional pigments that give them an orange, red, or rust color.

Biologists propose that land plants are closely related to the green algae, because both land plants and green algae have chlorophylls *a* and *b,* a cell wall that contains cellulose, and food reserves made of starch. Molecular data suggest that the green algae are subdivided into two groups, the **chlorophytes** and the **charophytes.** Charophytes are thought to be the green algae group most closely related to land plants.

Chlorophytes

Chlamydomonas is a tiny, photoautotrophic chlorophyte that inhabits still, freshwater pools. Its fossil ancestors date back over a billion years. The anatomy of *Chlamydomonas* is best seen in an electron micrograph, because it is less than 25 μm long (Fig. 21.2). It has a defined cell wall and a single, large, cup-shaped chloroplast that contains a *pyrenoid,* a dense body where starch is synthesized. In many species, a bright red, light-sensitive eyespot helps guide individuals toward light for photosynthesis.

When conditions are favorable—that is, proper nutrients and sunlight are available—*Chlamydomonas* exists as haploid cells. These haploid cells, called *vegetative cells,* have two long, whiplike flagella projecting from the end that operate with a breaststroke-like motion. *Chlamydomonas* vegetative cells often reproduce asexually by mitosis. Each mitosis event produces two haploid daughter cells, and as many as 16 daughter cells can form inside the parent cell wall (Fig. 21.3). Each daughter cell then secretes a cell wall and acquires flagella. The fully formed, functional haploid daughter cells emerge from within the parent by secreting an enzyme that digests the parent cell wall.

When growth conditions are unfavorable, *Chlamydomonas* reproduces sexually. Two haploid vegetative cells of two different mating types come into contact and fuse to form a diploid zygote. A heavy wall forms around the zygote, and it becomes a zygospore, which

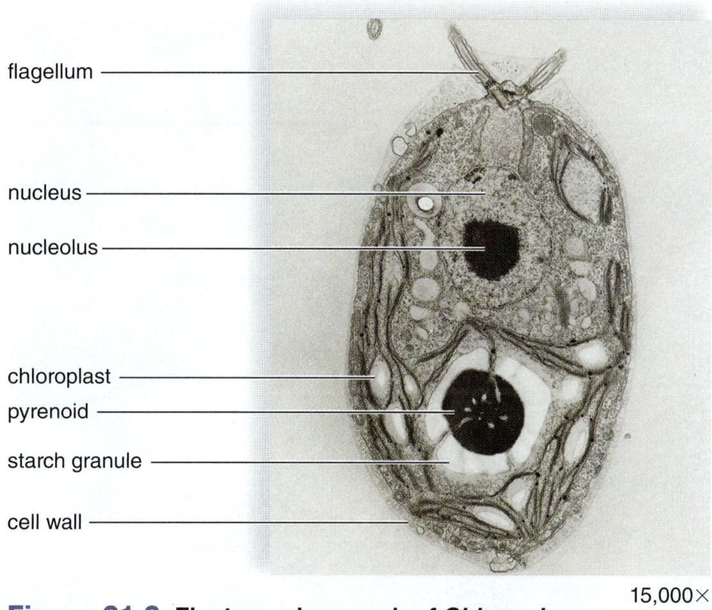

Figure 21.2 Electron micrograph of *Chlamydomonas.* *Chlamydomonas* is a microscopic, single-celled chlorophyte.

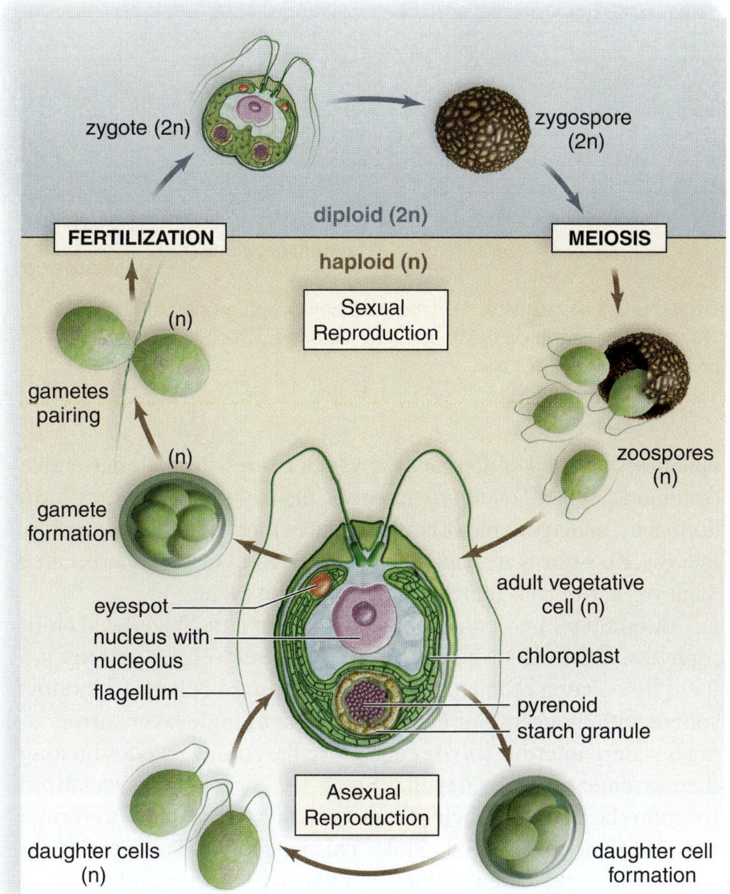

Figure 21.3 Haploid life cycle of *Chlamydomonas.* *Chlamydomonas* reproduction is an example of the haploid life cycle common to algae. During asexual reproduction, all structures are haploid; during sexual reproduction, meiosis follows the zygospore stage, which is the only diploid part of the cycle.

Figure 21.4 ***Volvox.*** *Volvox* is a colonial chlorophyte. The adult *Volvox* colony often contains daughter colonies, which are asexually produced by special cells.

undergoes a period of dormancy in which it is resistant to unfavorable conditions. When conditions improve, the zygospore emerges from dormancy, undergoes meiosis and produces four haploid zoospores by meiosis. **Zoospores** are haploid flagellated spores that grow to become adult vegetative cells, thus completing the life cycle.

A number of colonial forms occur among the flagellated chlorophytes. *Volvox* is a well-known colonial green alga. A **colony** is a loose association of independent cells. A *Volvox* colony is a hollow sphere with thousands of cells arranged in a single layer surrounding a watery interior. *Volvox* cells move the colony by coordinating the movement of their flagella. Some *Volvox* cells are specialized for reproduction, and each of these can divide asexually to form a new daughter colony (Fig. 21.4). This daughter colony resides for a time within the parent colony, but then it escapes by releasing an enzyme that dissolves away a portion of the parent colony.

Ulva is a multicellular chlorophyte; it's called sea lettuce, because it lives in the sea and has a leafy appearance (Fig. 21.5*a*). The body of *Ulva* is two cells thick and can be as much as a meter long. *Ulva* has an alternation-of-generations life cycle (Fig. 21.5*b*) like that of land plants.

Charophytes

The charophytes are filamentous algae. **Filaments** (L. *filum,* "thread") are end-to-end chains of cells. Charophytes have both branched and unbranched filaments. Charophytes often grow on aquatic flowering plants. Others attach to rocks or other objects under water or are suspended in the water column.

Spirogyra is an example of an unbranched charophyte (Fig. 21.6) found in green masses on the surfaces of ponds and streams. It has ribbonlike, spiraled chloroplasts. *Spirogyra* undergoes sexual reproduction via **conjugation** (L. *conjugalis,* "pertaining to marriage"), a temporary union during which the cells exchange genetic material. Two haploid filaments line up parallel to each other, and the cell contents of one filament move into the cells of the other filament, forming diploid zygospores. Diploid zygospores survive the winter, and in the spring they undergo meiosis to produce new haploid filaments.

a. *Ulva,* one individual

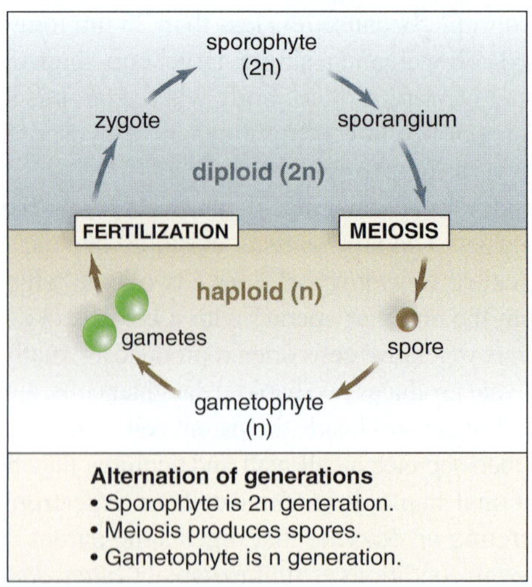

b. Alternation-of-generations life cycle

Figure 21.5 ***Ulva.*** **a.** *Ulva* is a multicellular chlorophyte known as sea lettuce. A single *Ulva* individual has a flat, leaflike appearance. **b.** Alternation-of-generations life cycle.

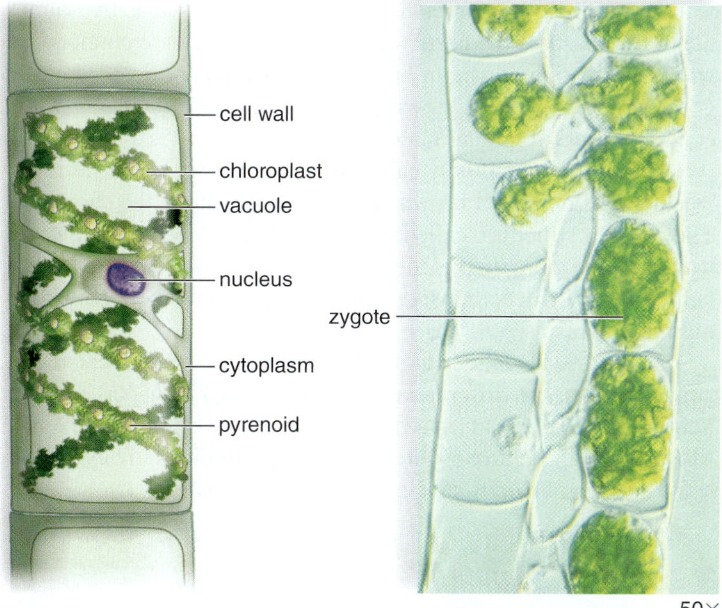

Figure 21.6 ***Spirogyra.*** **a.** *Spirogyra* is an unbranched charophyte in which each cell has a ribbonlike chloroplast. **b.** During conjugation, the cell contents of one filament enter the cells of another filament. Zygote formation follows.

Chara (Fig. 21.7) is a charophyte that lives in freshwater lakes and ponds. It is commonly called a stonewort, because it is encrusted with calcium carbonate deposits. The main strand of the alga, which can be over a meter long, is a single-file strand of very long cells anchored by rhizoids, which are colorless, hairlike filaments. Only the cell at the upper end of the main strand produces new cells. Whorls of branches occur at multicellular nodes, regions between the giant cells of the main strand. Each of the branches is also a single-file thread of cells (Fig. 21.7*b*).

Figure 21.7 ***Chara.*** *Chara* is an example of a stonewort, a charophyte that shares a common ancestor with land plants.

Figure 21.8 Red algae. Red algae are multicellular seaweeds, represented by *Rhodoglossum affine.*

Male and female multicellular reproductive structures grow at the nodes, and in some species they occur on separate individuals. The male structure produces flagellated sperm, and the female structure produces a single egg. The gametes fuse to produce a diploid zygote, which is retained until it is enclosed by tough walls.

DNA sequencing data suggest that among green algae the stoneworts are most closely related to land plants.

Red Algae The **red algae** are multicellular seaweeds that possess red and blue **accessory pigments,** which transfer energy from absorbed light to the photopigment chlorophyll during photosynthesis (Fig. 21.8). These algae live in warm seawater, some at depths exceeding 70 m. Their accessory pigments allow them to absorb the wavelengths of light that penetrate into deep water.

Most of the more than 7,000 species of red algae are much smaller and more delicate than brown algae, but some species can exceed a meter in length. Red algae can be filamentous but most have feathery, flat, or, ribbonlike branches. Coralline red algae have cell walls that contain calcium carbonate, a mineral that contributes to the growth of coral reefs.

Red algae are economically important. Agar is a gelatin-like product made primarily from the algae *Gelidium* and *Gracilaria.* Agar is used commercially to make capsules for vitamins and drugs, as a material for making dental impressions, and as a base for cosmetics. In the laboratory, agar is a solidifying agent for a bacterial culture medium. When purified, it becomes the gel for electrophoresis, a procedure that separates proteins or nucleotides. Agar is also used in food preparation as an antidrying agent for baked goods and to make jellies and desserts set rapidly.

Carrageenan, extracted from various red algae, is an emulsifying agent for the production of chocolate and cosmetics. *Porphyra,* another red alga, is the basis of a billion-dollar aquaculture industry in Japan. The reddish-black wrappings around sushi rolls consist of processed *Porphyra* blades.

Check Your Progress 21.2

1. What are the unique features of green algae?
2. How does reproduction occur within the archaeplastids?

21.3 Supergroup Chromalveolata

Learning Outcomes

Upon completion of this section, you should be able to

1. Identify the characteristics of chromalveolates.
2. Identify unique species of chromalveolates.
3. Describe the life cycle of *Plasmodium*.

The **chromalveolates** (Gk. *chroma,* "color"; L. *alveolus,* "hollow") include two large subgroups: the stramenopiles and the alveolates.

Stramenopiles

The **stramenopiles** include the brown algae, diatoms, golden brown algae, and water molds.

Brown Algae

The **brown algae** have chlorophylls *a* and *c* in their chloroplasts and an accessory carotenoid pigment that gives them their characteristic brown color. Food reserves are stored as a carbohydrate called *laminarin.* The brown algae range from small forms with simple filaments to large, multicellular forms that may reach 100 m in length (Fig. 21.9). The vast majority of the 1,800 species live in cold ocean waters.

The multicellular brown algae are seaweeds that live along the rocky coasts in the north temperate zone. They are pounded by waves as the tide comes in and are exposed to dry air as the tide goes out. They dry out slowly, however, because their cell walls contain a water-retaining material.

Laminaria, commonly called kelp, and *Fucus,* known as rockweed, are examples of brown algae that grow along the shoreline. They have a structure, called a holdfast, that allows them to cling to rocks. The giant kelps *Macrocystis* and *Nereocystis* form dense kelp forests in deeper water that provide food and habitat for marine organisms.

Laminaria is unique among the protists in that members of this genus show tissue differentiation—that is, they transport organic nutrients by way of a tissue that resembles phloem in land plants. Most brown algae have an alternation-of-generations life cycle, but some species of *Fucus* have an exclusively sexual life cycle.

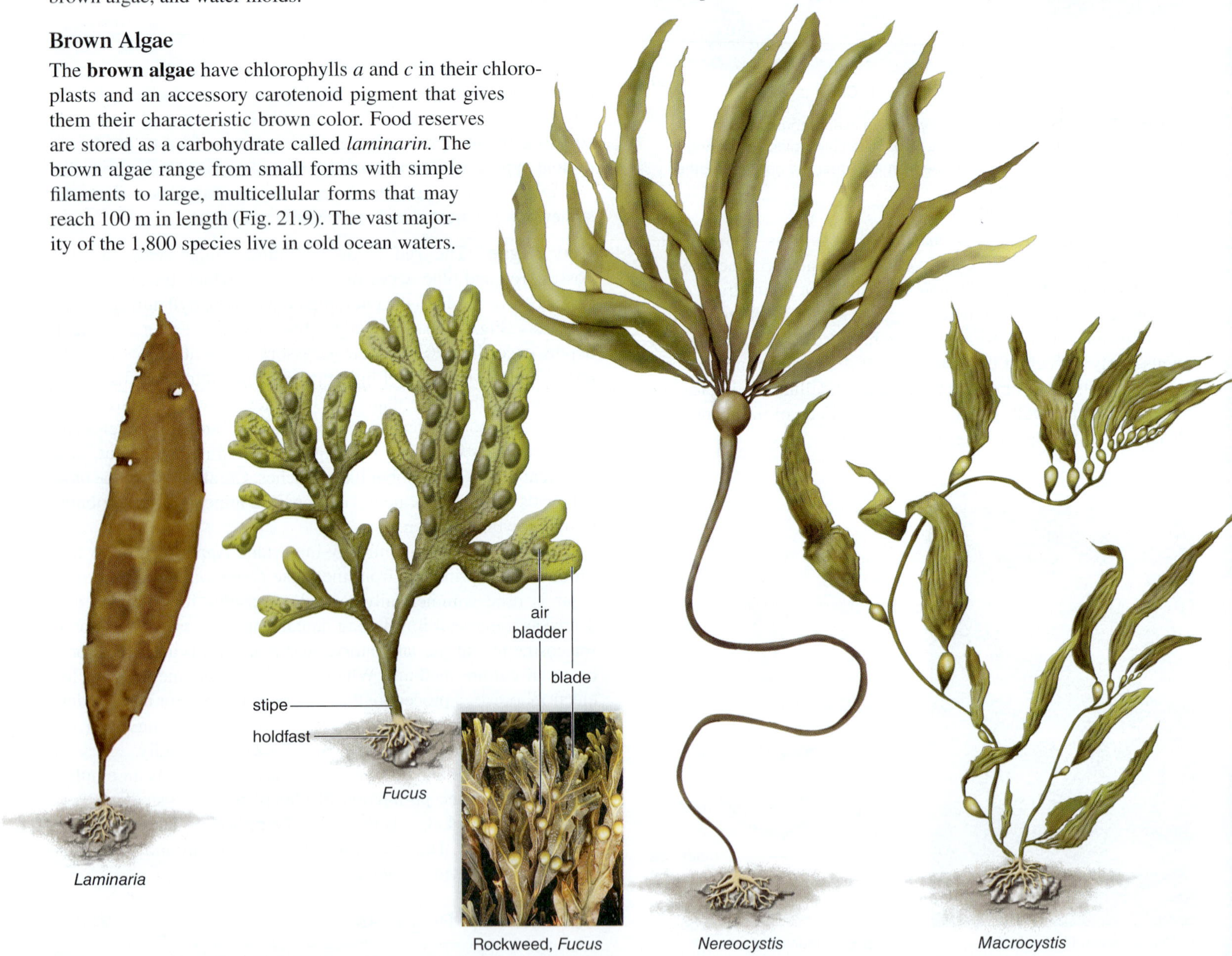

Figure 21.9 Brown algae. *Laminaria* and *Fucus* are seaweeds known as kelps. They live along rocky coasts of the north temperate zone. The other brown algae featured, *Nereocystis* and *Macrocystis,* form spectacular underwater "forests" at sea.

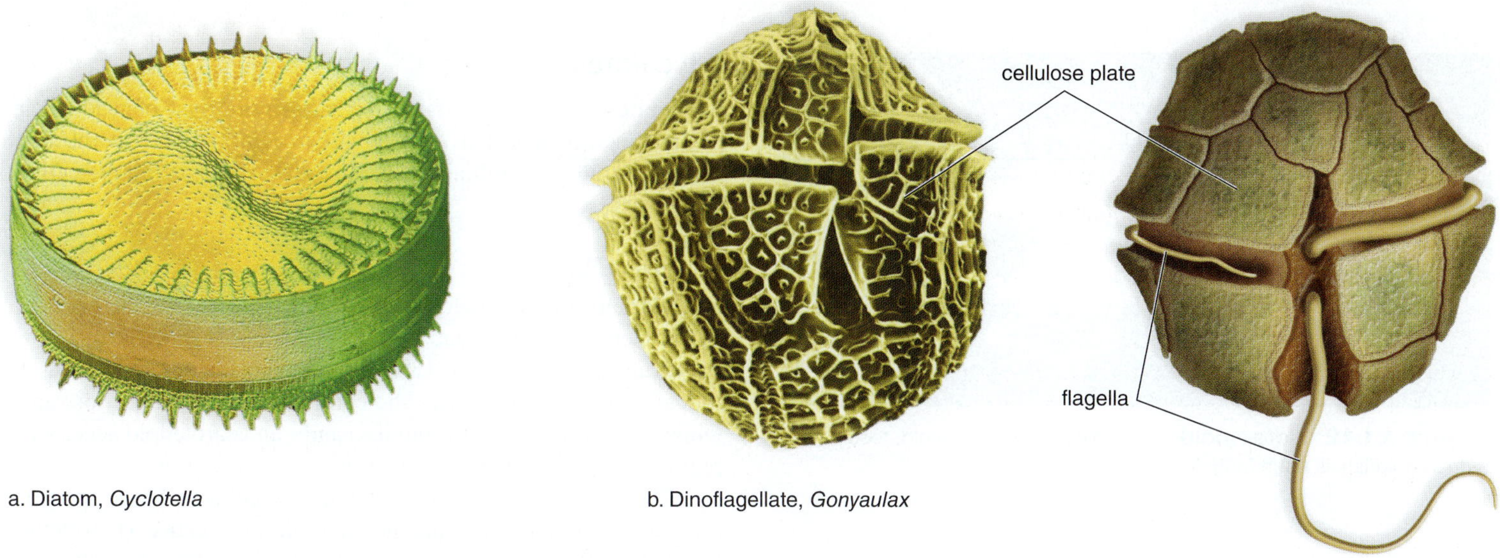

Figure 21.10 Diatoms and dinoflagellates. **a.** Diatoms are stramenopiles of various colors, but even so their chloroplasts contain a unique golden brown pigment (*fucoxanthin*), in addition to chlorophylls *a* and *c*. The beautiful pattern results from markings on the silica-embedded wall. **b.** Dinoflagellates are alveolates, such as *Gonyaulax,* with cellulose plates.

Brown algae are harvested for human food and for fertilizer. *Macrocystis* is the source of algin, a pectinlike material that is added to ice cream, sherbet, cream cheese, and other products to give them a stable, smooth consistency.

Diatoms

A **diatom** is a tiny, single-celled stramenopile with an ornate silica shell. The shell is made up of upper and lower shelves, called valves, that fit together (Fig. 21.10*a*). Diatoms have a carotenoid accessory pigment, which gives them an orange-yellow color. Diatoms make up a significant part of plankton, which serves as a source of oxygen and food for heterotrophs in both freshwater and marine ecosystems.

Diatoms reproduce asexually and sexually. Asexual reproduction occurs by diploid parents undergoing mitosis to produce two diploid daughter cells. Each time a diatom reproduces asexually, the size of the daughter cells decreases until diatoms are about 30% of their original size. At this point, they begin to reproduce sexually. The diploid cell produces gametes by meiosis. Gametes fuse to produce a diploid zygote, which grows and then divides via mitosis to produce new diploid diatoms of normal size.

The valves of diatoms are covered with a great variety of striations and markings that form beautiful patterns when observed under the microscope. These are actually depressions or pores through which the organism makes contact with the outside environment. The remains of diatoms, called diatomaceous earth, accumulate on the ocean floor and are mined for use as filtering agents, soundproofing materials, and gentle polishing abrasives, such as those found in silver polish and toothpaste.

Golden Brown Algae

The **golden brown algae** derive their distinctive color from yellow-brown carotenoid accessory pigments. The cells of these single-celled or colonial protists typically have two flagella with tubular hairs, a characteristic of stramenopiles. Golden brown algae cells may be naked, covered with organic or silica scales, or enclosed in a secreted, cagelike structure called a lorica. Many golden brown algae, such as *Ochromonas* (Fig. 21.11), are mixotrophs, capable of photosynthesis as well as phagocytosis. Like diatoms, the golden brown algae contribute to freshwater and marine phytoplankton.

Water Molds

Water molds are funguslike protists, often seen as furry growths on their food source. In spite of their common name, some water molds live on land and parasitize insects and plants. Over 700 species of water molds have been described. A water mold, *Phytophthora infestans,* was responsible for the 1840s potato famine in Ireland,

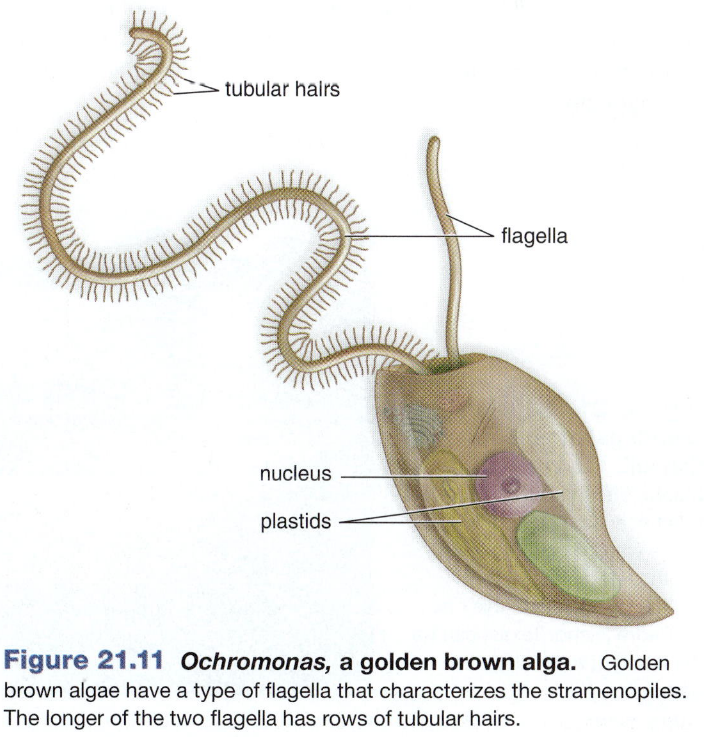

Figure 21.11 ***Ochromonas,* a golden brown alga.** Golden brown algae have a type of flagella that characterizes the stramenopiles. The longer of the two flagella has rows of tubular hairs.

Figure 21.12 Water mold. *Saprolegnia,* a water mold, feeding on a dead goldfish, is not a fungus.

and another, *Plasmopara viticola,* for the downy mildew of grapes that ravaged the vineyards of France in the 1870s. However, most water molds are saprotrophic and live off dead organic matter. Another well-known water mold is *Saprolegnia,* which is often seen as a cottonlike white mass on dead organisms (Fig. 21.12).

Water molds, also called oomycetes ("egg fungi"), used to be grouped with fungi, because they are similar to fungi in many ways. The funguslike water molds have a filamentous body similar to the hyphae of fungi, but their cell walls are composed of cellulose instead of chitin. The life cycle of water molds also differs from that of the fungi.

Water molds are diploid, but they can reproduce both asexually and sexually. During asexual reproduction, water molds produce flagellated, motile, diploid zoospores inside structures called *sporangia* (*zoosporangia*). During sexual reproduction, structures called oogonia, which produce haploid eggs, and antheridia, which produce haploid sperm, form at the tips of filaments. Sperm and eggs come together when the antheridium and oogonium come in contact and sperm is inserted into the oogonium. Eggs and sperm fuse and produce diploid zoospores, which emerge from the oogonium.

Alveolates

Alveolates have alveoli (small air sacs) lying just beneath their plasma membranes that are thought to lend support to the cell surface or aid in membrane transport. Alveolates are single-celled organisms.

Dinoflagellates

Dinoflagellates are single-celled, photoautotrophic algae encased by protective cellulose and silicate plates (see Fig. 21.10). Dinoflagellates typically have two flagella: one flagellum acts as a rudder, and the other causes the cell to spin as it moves forward. Dinoflagellates have chlorophylls *a* and *c* and carotenoid accessory pigments that give them a yellow-green to brown color.

There are over 2,000 species of dinoflagellates. Some species, such as *Noctiluca,* are capable of bioluminescence (producing light). Dinoflagellates are a component of plankton and thus an important source of food for small animals in the ocean.

Zooxanthellae are dinoflagellates that form symbiotic relationships with invertebrates such as corals and other protozoans. Zooxanthellae are endosymbionts, living within the bodies of their hosts. Endosymbiotic dinoflagellates lack cellulose plates and flagella. Corals (see Chapter 28), members of the animal kingdom, usually contain large numbers of zooxanthellae, which provide their animal hosts with organic nutrients. In return, the corals provide the zooxanthellae with shelter, nutrients, and protection.

Dinoflagellates are one of the most important groups of primary producers in the marine ecosystem. Under unusually high nutrient conditions, populations of dinoflagellates and other algae can undergo a population explosion, called an algal bloom. During an algal bloom, a single milliliter of water can contain more than 30,000 algae. Some algal blooms caused by dinoflagellates are so large that they turn the water brown or red because of the high density of the algae (Fig. 21.13*a*). These colorful algal blooms, called **red tides,** are sometimes so extensive that they can be seen from space. The dinoflagellate *Alexandrium catanella* can cause a harmful algal bloom (HAB), because it secretes saxitoxin, a neurotoxin that is responsible for neurotoxic shellfish poisoning. Massive fish kills can

Figure 21.13 Dinoflagellate bloom and fish kill. **a.** A dinoflagellate bloom, often called a red tide because of the color of the water, occurring in southeastern Alaska. **b.** Fish kills, such as this one on Padre Island, Texas, can be the result of a dinoflagellate bloom.

a. b.

occur as a result of saxitoxin produced during a red tide (Fig. 21.13*b*). Humans who consume shellfish that have fed during an *A. catanella* outbreak can die from paralytic shellfish poisoning, which paralyzes the respiratory organs.

Dinoflagellates usually reproduce asexually. Each daughter cell inherits half of the parent's cellulose plates. During sexual reproduction, the daughter cells act as gametes and fuse to form a diploid zygote. The zygote enters a resting stage until signaled to undergo meiosis. The product of meiosis is a single haploid cell, because the other cells disintegrate.

Ciliates

The **ciliates** are single-celled protists that move by means of cilia. They are the most structurally complex and specialized of all protozoa. Members of the genus *Paramecium* are classic examples of ciliates (Fig. 21.14*a*). Hundreds of cilia project through tiny holes in a semirigid outer covering, or pellicle. *Paramecium* beat their cilia in a coordinated and rhythmic manner to "swim" through their environment. Oval capsules that lie just beneath the pellicle contain **trichocysts.** Upon mechanical or chemical stimulation, trichocysts discharge long, barbed threads, which are useful for defense and for capturing prey. Some trichocysts release poisons.

Most ciliates ingest their food. *Paramecium* feed by sweeping food particles down a gullet, below which food vacuoles form. Following digestion, the soluble nutrients are absorbed by the cytoplasm, and the nondigestible residue is eliminated through the anal pore.

The ciliates are a diverse group of protozoans that range in size from 10 to 3,000 μm. The majority of the 8,000 species of ciliates are free-living, mobile, and single-celled; however, several parasitic, sessile, and colonial forms exist. *Suctoria* are sessile ciliates that have specialized microtubules for capturing, paralyzing, and ingesting other ciliates. *Stentor* may be the most elaborate ciliate, resembling a giant vase (Fig. 21.14*c*). Cilia at the opening of the "vase" sweep in food particles. They are one of the largest single-celled protozoans, reaching lengths of 2 mm. *Ichthyophthirius* is an ectoparasitic protozoan that causes a common disease in fishes called "ick." If left untreated, it can be fatal.

During asexual reproduction, ciliates divide by transverse binary fission. Ciliates have two types of nuclei: a large macronucleus and one or more small micronuclei. The macronucleus controls the normal metabolism of the cell, whereas the micronuclei participate in reproduction. During sexual reproduction, the micronuclei undergo meiosis. Sexual reproduction involves conjugation (Fig. 21.14*b*), during which two ciliates unite and exchange haploid micronuclei. After conjugation, the macronuclei dissolve and new macronuclei are formed from the fusion of the micronuclei.

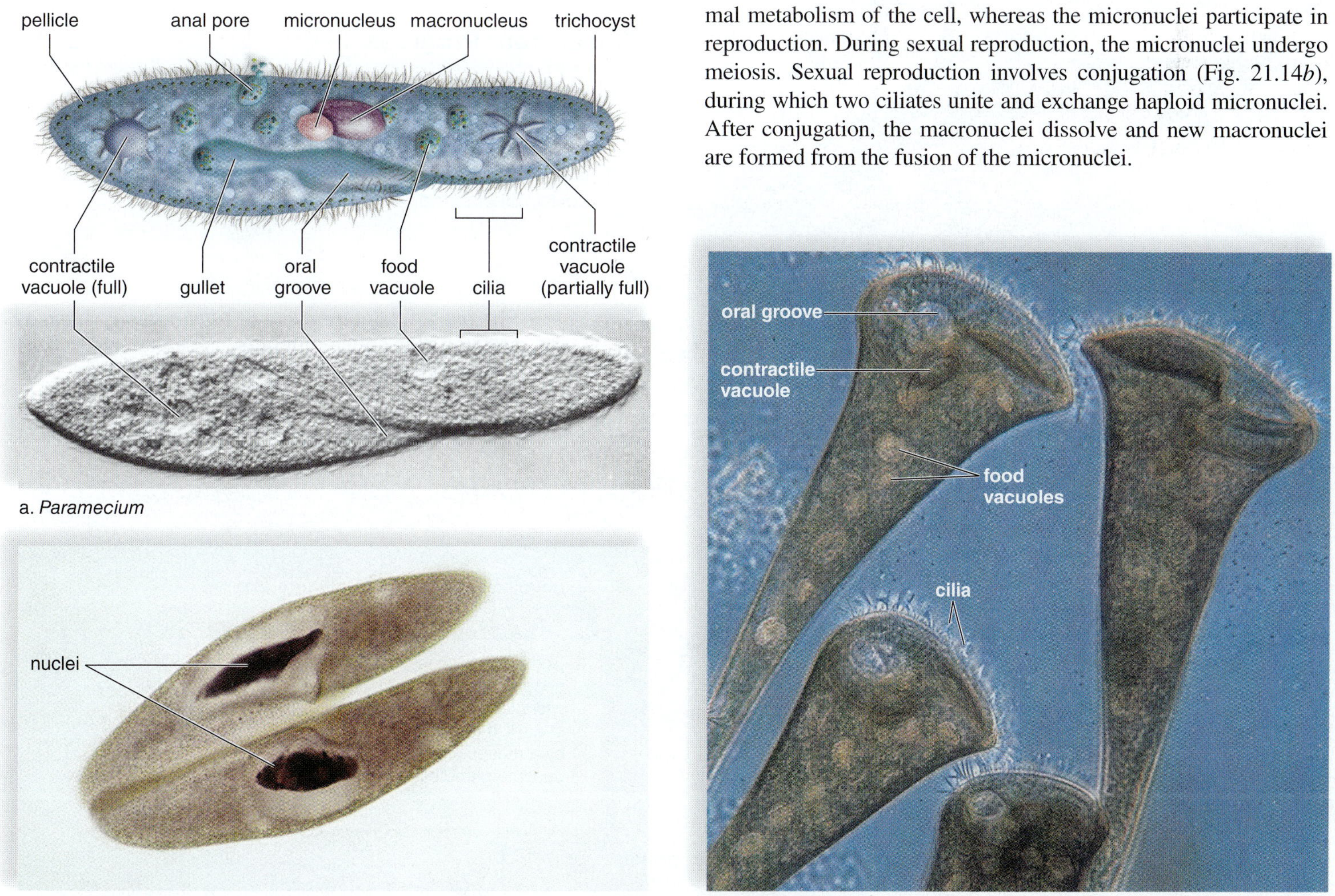

a. *Paramecium*

b. During conjugation two *paramecia* first unite at oral areas

c. *Stentor*

Figure 21.14 Ciliates. Ciliates are the most complex of the protists. **a.** Structure of *Paramecium,* adjacent to an electron micrograph. Note the oral groove, the gullet, and the anal pore. **b.** A form of sexual reproduction called conjugation occurs periodically. **c.** *Stentor,* a large, vase-shaped, freshwater ciliate.

Apicomplexans

The **apicomplexans,** also known as *sporozoans,* include about 4,000 species of nonmotile, parasitic, spore-forming protozoans. Apicomplexans have a unique organelle called an apicoplast, which is used to penetrate a host cell. All apicomplexans are parasites of animals, and some can infect multiple hosts.

Plasmodium, an apicomplexan, is responsible for malaria, a disease that affects over 200 million people each year. *Plasmodium* is a parasitic protozoan that infects human red blood cells. The chills and fever of malaria appear when the infected cells burst and release toxic substances into the blood (Fig. 21.15). In humans, malaria is caused by four distinct members of the genus *Plasmodium. Plasmodium vivax,* the cause of one type of malaria, is the most common.

The transmission cycle of *Plasmodium* involves the *Anopheles* mosquito as an intermediate organism, or **vector,** that transmits the disease between the host and other organisms. The life cycle of *Plasmodium* alternates between a sexual and an asexual phase, dependent upon whether reproduction takes place inside the mosquito (sexual) or the human host (asexual) (Fig. 21.15). Female *Anopheles* mosquitoes acquire protein for the production of eggs by biting humans and other animals.

Despite efforts to control malaria, it kills more than 500,000 people each year. Malaria is most common in densely populated, tropical areas of the globe. Because of its prevalence in these areas, almost 4 billion people, nearly half the world's population, are at risk for the disease.

Apicomplexans include other human parasites. Cyclosporiasis is an infection of the intestine caused by the parasitic apicomplexan *Cyclospora cayetanensis*. The cyclosporin parasite is transmitted by feces-contaminated fresh produce and water. Outbreaks of cyclosporiasis in the United States have been attributed to contaminated fresh raspberries, basil, snow peas, and mesclun lettuce. *Toxoplasma gondii,* another apicomplexan, causes toxoplasmosis that is transmitted to humans from the feces of infected cats. In pregnant women, the parasite can infect the fetus and cause birth defects and intellectual disability; in AIDS patients, it can infect the brain and cause neurological problems.

Check Your Progress 21.3

1. What characteristics separate alveolates from stramenopiles?
2. Which chromalveolates are useful to humans?
3. Summarize the life cycle of *Plasmodium vivax,* one apicomplexan that causes malaria.

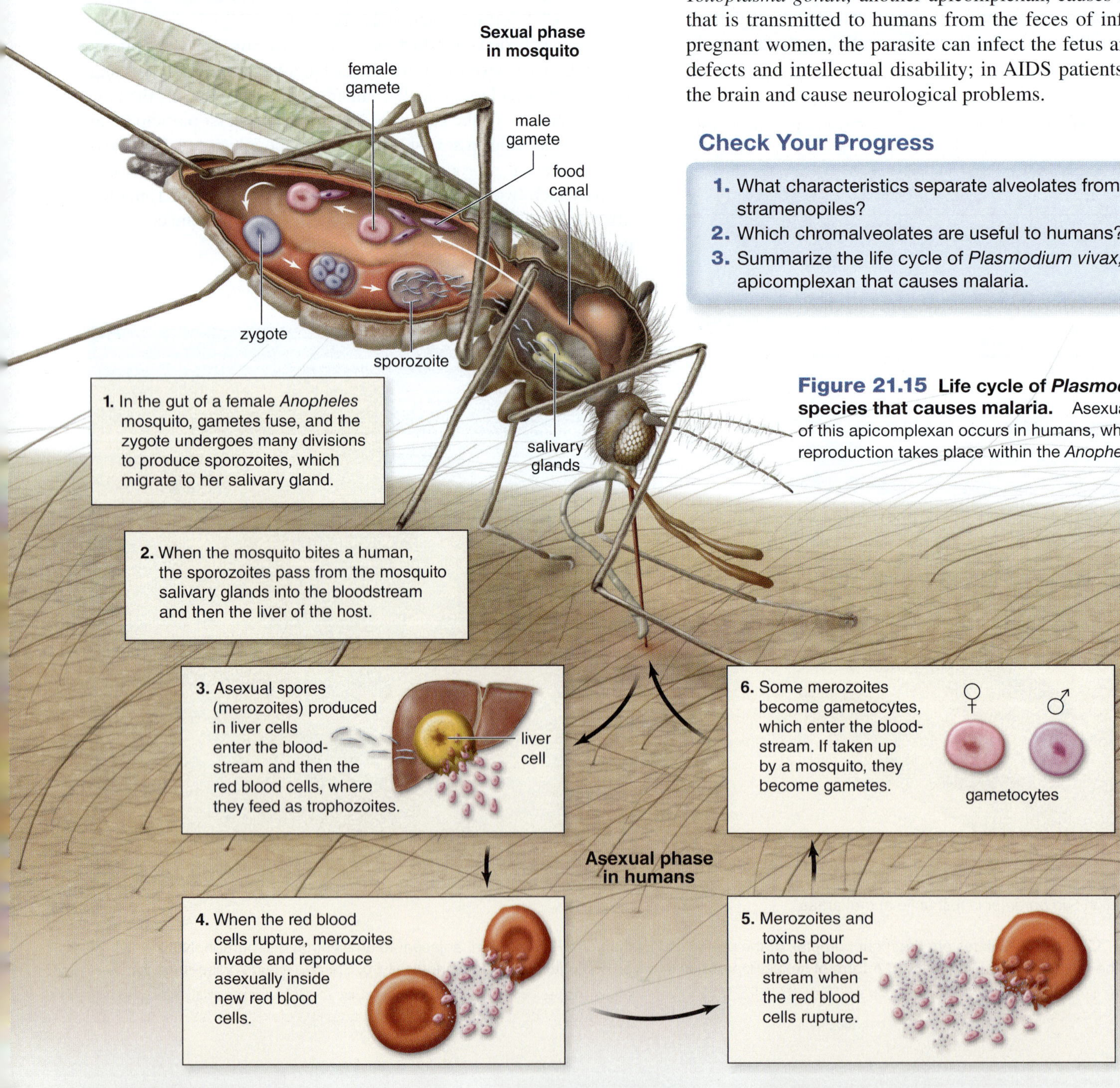

Figure 21.15 Life cycle of *Plasmodium vivax,* a species that causes malaria. Asexual reproduction of this apicomplexan occurs in humans, while sexual reproduction takes place within the *Anopheles* mosquito.

21.4 Supergroup Excavata

Learning Outcomes

Upon completion of this section, you should be able to

1. Describe the distinguishing characteristics of excavates.
2. Identify pathogenic excavate species.

The **excavates** (L. *cavus,* "hollow") have atypical or absent mitochondria and distinctive flagella and/or deep (excavated) oral grooves.

Euglenids

The **euglenids** are small (10–500 μm), single-celled, freshwater organisms. *Euglena deses* is a common inhabitant of freshwater ditches and ponds. Classifying the approximately 1,500 species of euglenids is problematic, because they are very diverse. Some euglenids are mixotrophic, some are photoautotrophic, and others are heterotrophic. One-third of all genera have chloroplasts; the rest do not. Those that lack chloroplasts ingest or absorb their food.

When present, the chloroplasts are surrounded by three rather than two membranes. Carbohydrates are synthesized in a region of the chloroplast called a pyrenoid. Euglenids produce an unusual type of carbohydrate called paramylon.

Euglenids have two flagella, one much longer than the other (Fig. 21.16). The longer flagellum is called a tinsel flagellum, because it has hairs on it. Near the base of the tinsel flagellum is an eyespot that has a photoreceptor capable of detecting light.

Euglenids are bounded by a flexible *pellicle* composed of protein bands lying side by side; this arrangement allows euglenids to assume different shapes. A contractile vacuole rids the body of excess water. Euglenids reproduce by longitudinal cell division, and sexual reproduction is not known to occur.

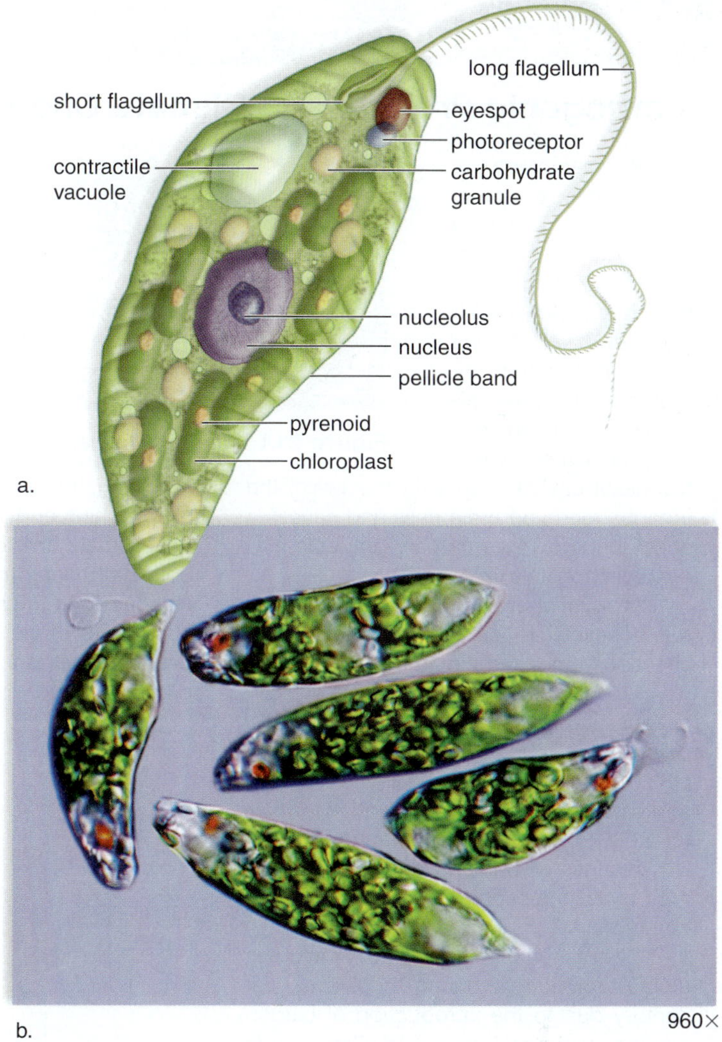

Figure 21.16 ***Euglena.*** **a.** In *Euglena,* a very long flagellum propels the body, which is enveloped by a flexible pellicle. **b.** Micrograph of *Euglena.*

Parabasalids and Diplomonads

Parabasalids and diplomonads are single-celled, flagellated excavates that are endosymbionts of animals. They are able to survive in anaerobic, or low-oxygen, environments. These protozoans lack mitochondria; instead, they rely on fermentation for the production of ATP. A variety of forms are found in the guts of termites, where they assist with the breakdown of cellulose.

Parabasalids have a unique, fibrous connection between the Golgi apparatus and flagella. The most common sexually transmitted disease, trichomoniasis, is caused by the parabasalid *Trichomonas vaginalis.* Infection causes vaginitis in women. The parasite may also infect the male genital tract; however, the male may have no symptoms.

A **diplomonad** (Gk. *diplo,* "double"; *monas,* "unit") cell has two nuclei and two sets of flagella. The diplomonad *Giardia lamblia* forms cysts, which are transmitted by contaminated water. *Giardia* attaches to the human intestinal wall, causing severe diarrhea (Fig. 21.17). This protozoan lives in the digestive tracts of a variety of other mammals as well. Beavers are known to be a reservoir of *Giardia* infection in the mountains of the western United States, and many cases of infection have been acquired by hikers who filled their canteens at a beaver pond.

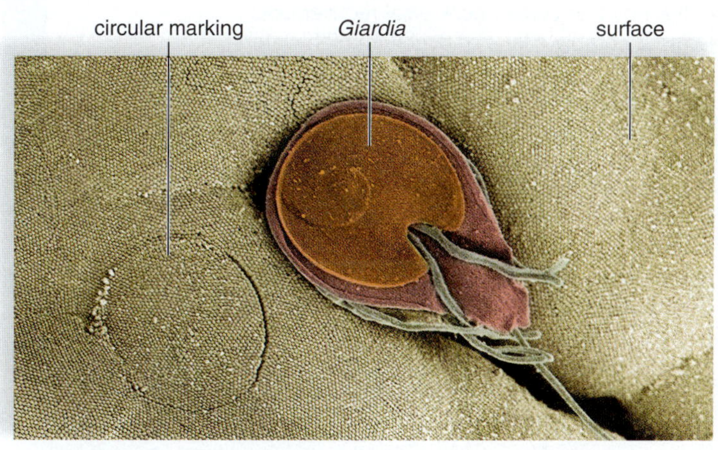

Figure 21.17 ***Giardia lamblia.*** This diplomonad adheres to any surface, including epithelial cells, by means of a sucking disk. Characteristic markings can be seen after the disk detaches.

THEME Biological Systems

Pathogenic Protists and Climate Change

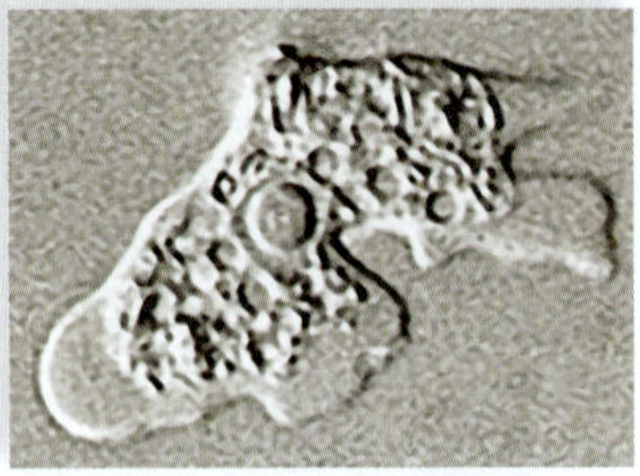

Figure 21A ***Naegleria fowleri.***

At the beginning of the chapter, you were introduced to a young boy who became infected by a lethal protist, *Naegleria fowleri* (Fig. 21A). *N. fowleri* is an amoeba that migrates to the brain after contaminated lake or river water enters the nasal cavities. Once in the brain, the amoeba feeds on brain cells and causes a condition known as primary amoebic meningoencephalitis (PAM).

An overwhelming majority of protists are beneficial to human welfare and not infectious. For example, half of the oxygen we breathe is produced by protists living in oceans and lakes. However, if you're the person who is infected, all of those benefits won't mean much. The key to reducing any type of infection is prevention. Unfortunately, global climate change is driving changes in humans and protists that may make infections like the boy's more common.

Climate change is a rise in average global temperature during the twenty-first century due to the introduction of carbon dioxide, methane, ozone, and nitrous oxide gases to the atmosphere from anthropogenic (human-influenced) sources. Along with increased temperatures, there has been a rise in extreme weather events, such as floods, droughts, and storms. This has led to famine, displacement, and changing behaviors in human populations.

The *N. fowleri* amoeba that infected the boy has certain ecological requirements, most notably a lake or river with high water temperature and low water level (Fig. 21B). Climate change has made those conditions more common. For example, in 2010 and 2012 the first-ever cases of *N. fowleri* infection were documented in Minnesota, over 500 miles north of any other recorded infections (Fig. 21C). Higher than normal summer temperatures allowed the pathogen to survive in a swimming lake and take the lives of two children.

Changing climate influences not only the pathogen but also the human behavior that makes infection more likely. During the summers of 2010 and 2012, in the same heat that allowed *N. fowleri* to survive, large numbers of people visited area lakes and rivers to cool down. More people in the water can create a higher chance of infection, especially for a protist, such as *N. fowleri,* that is stirred up from the bottom by splashing and jumping.

Climate change may continue to drive both pathogens and humans closer to infectious encounters (Fig. 21D). *N. fowleri* is just one example, but climate change has been implicated in increased malaria, dengue fever, food poisoning, toxic tides, and drinking water contamination, all caused by certain protists.

This chapter discusses the characteristics, ecology, and diversity of protists. Understanding those factors and how they will be affected by climate change is the key to preventing increased infections by protists and the loss of beneficial protist species.

Questions to Consider

1. How might climate change affect beneficial protists that produce our oxygen and form the base of the food chain?
2. How might climate change be beneficial for protist species?
3. What life characteristics of protists will be most affected by climate change?

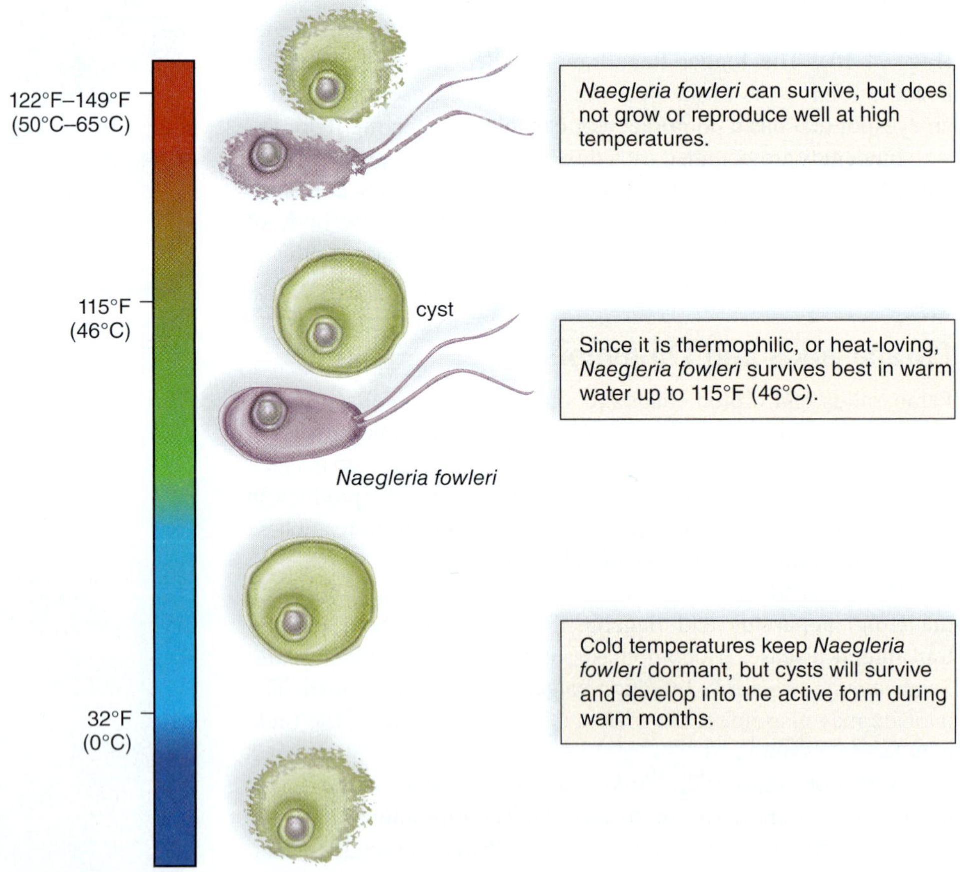

Figure 21B Effects of temperature on *Naegleria fowleri.* Within only a small range of temperature *Naegleria fowleri,* the brain-eating amoeba, will become active.

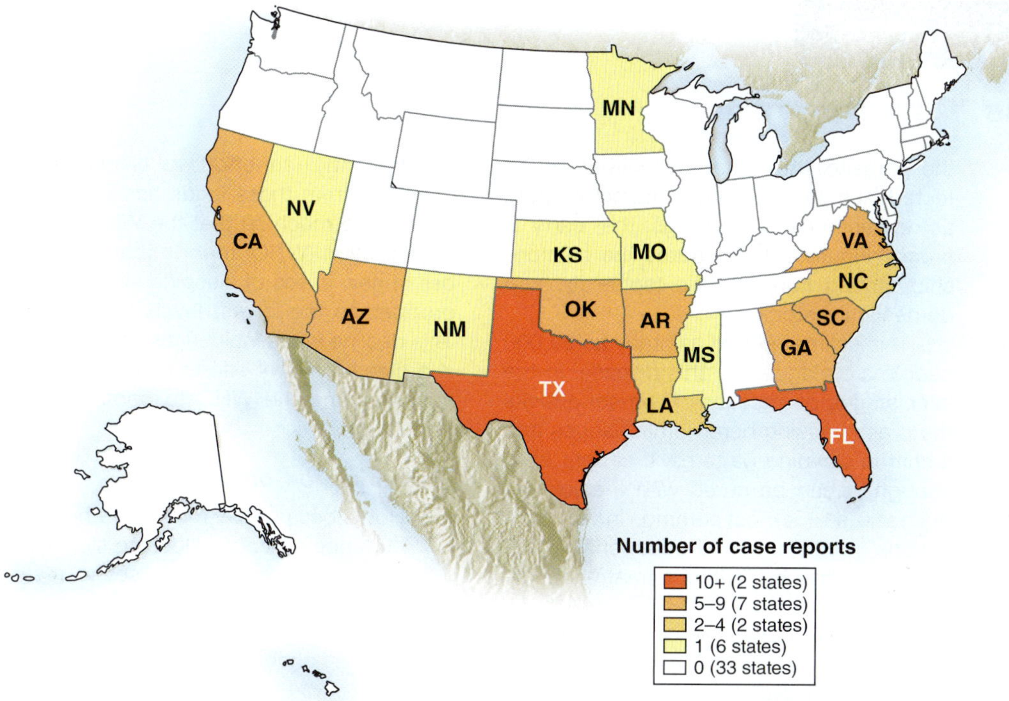

Figure 21C Case reports. Number of case reports of *Naegleria fowleri* infection by state of exposure, United States, 1962–2012.

Source: http://www.cdc.gov/parasites/naegleria/state-map.html

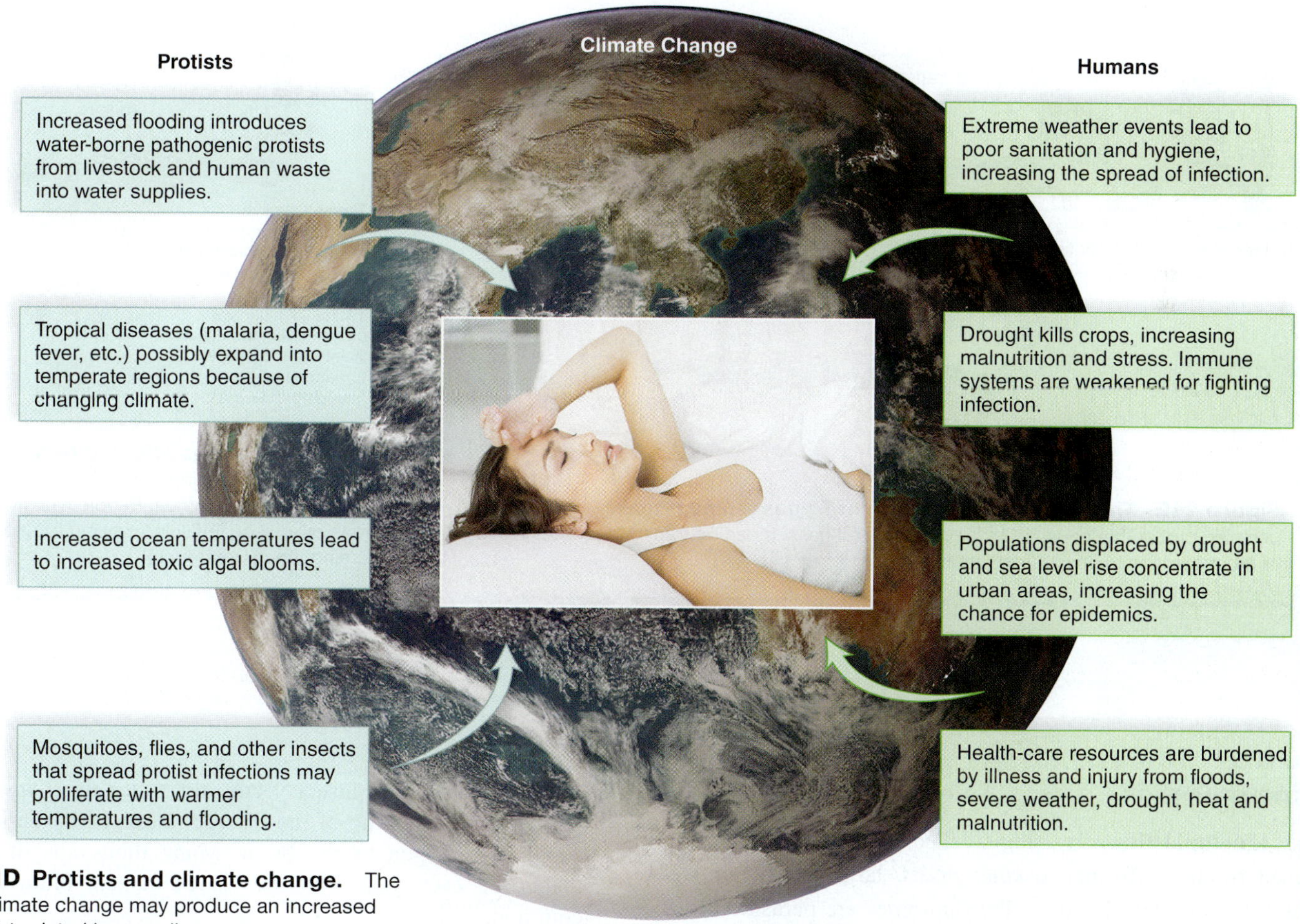

Figure 21D Protists and climate change. The results of climate change may produce an increased rate of protist-related human diseases.

THEME **Biological Systems**

African Sleeping Sickness

The World Health Organization (WHO) recognizes 17 neglected tropical diseases (NTDs) that affect more than 1 billion people exclusively in the most impoverished communities. African sleeping sickness, also called African human trypanosomiasis, is the deadliest of all of the NTDs. It is caused by a type of parasitic, single-celled protist, *Trypanosoma brucei,* which is transmitted to humans in the bite of the blood-sucking tsetse fly (*Glossina*) (Fig. 21E).

The tsetse fly is the transmission vector of the disease, because it carries the trypanosomes and transmits the protists into the human bloodstream while feeding. The flies first become infected with the parasite when they bite either an infected human or another mammal, such as domestic cattle and pigs, that carry *T. brucei.* Figure 21F illustrates the relationship among humans, the tsetse fly, and domestic livestock.

It is estimated that as many as 30,000 people are plagued by this disease. In the later stages of infection, *T. brucei* attacks the brain, causing behavioral changes and a shift in sleeping patterns. If caught early enough, it can be cured with medication, but because it is most common in very poor nations, medication and treatment are difficult to come by. Without treatment, the disease is fatal.

Recently, an effort to eliminate the tsetse fly from these lands has been successful, so much so that the World Health Organization (WHO) reports that the number of new cases of sleeping sickness has dropped to the lowest levels in 50 years. This decline in the incidences of new cases is a direct result of efforts by public health agencies and the WHO to control tsetse populations.

Questions to Consider

1. How does poverty reinforce a high occurrence of African sleeping sickness?
2. Livestock are called "disease reservoirs." Explain why.

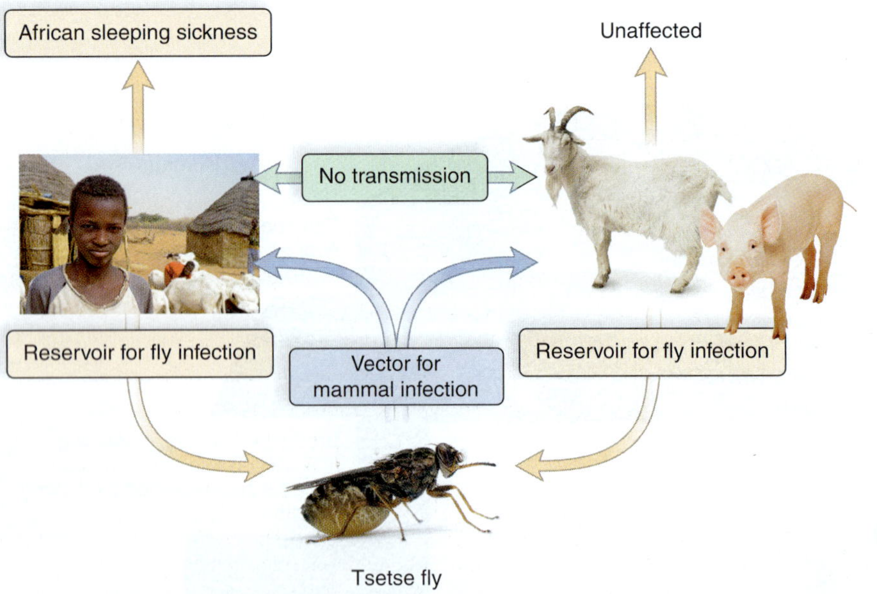

Figure 21E The transmission pattern of *Trypanosoma brucei.* *T. brucei* is a parasitic protist that causes African sleeping sickness. It can be transmitted from livestock or other humans to new human hosts.

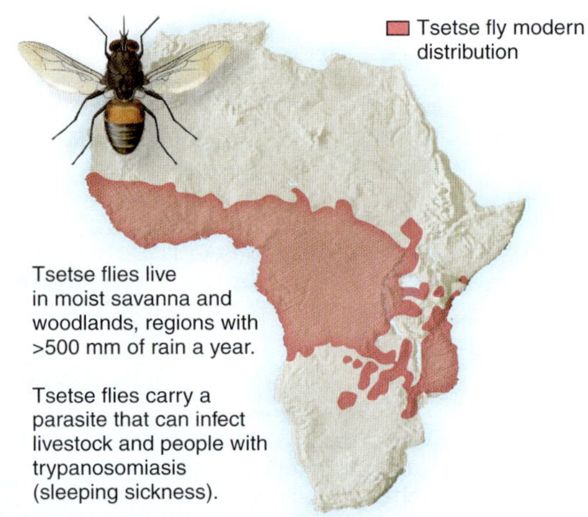

Figure 21F Areas affected by sleeping sickness. Sleeping sickness is found in moist tropical regions of Africa. People who live near wet areas or engage in fishing, raising livestock, and farming are at greater risk.

Kinetoplastids

The **kinetoplastids** are single-celled, flagellated protozoans named for their distinctive *kinetoplasts,* large masses of DNA found in their mitochondria. Trypanosomes are parasitic kinetoplastids that are passed to humans by insect bites. *Trypanosoma brucei* (Fig. 21.18) is the cause of African sleeping sickness (see the Biological Systems feature, "African Sleeping Sickness"). It is transmitted by an insect vector, the tsetse fly (*Glossina*). The lethargy characteristic of the disease is caused by an inadequate supply of oxygen to the brain. Many thousands of cases are diagnosed each year. Fatalities or permanent brain damage is common.

Trypanosoma cruzi causes Chagas disease in humans in Central and South America. Approximately 7 million people are infected with this parasite.

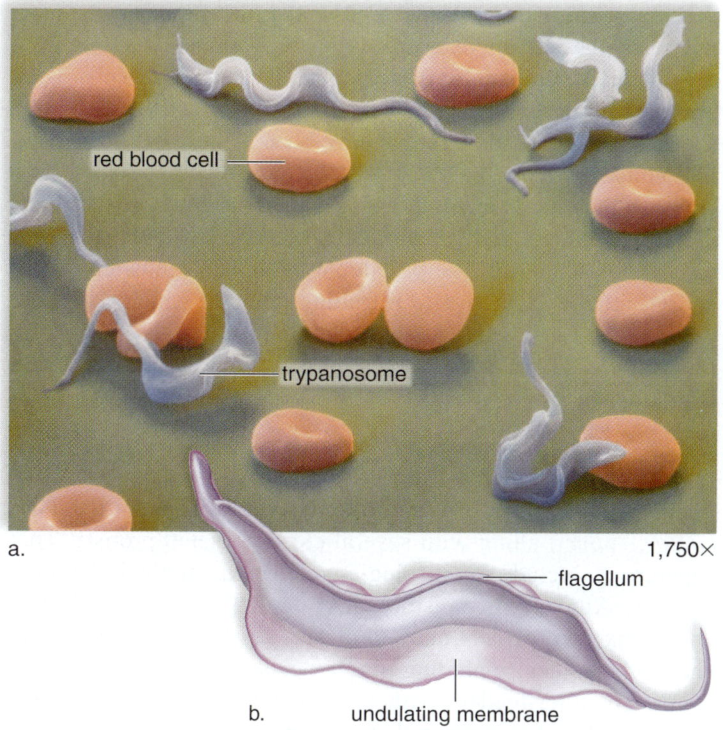

Figure 21.18 ***Trypanosoma brucei.*** **a.** Micrograph of *Trypanosoma brucei,* a causal agent of African sleeping sickness, among red blood cells. **b.** The drawing shows its general structure.

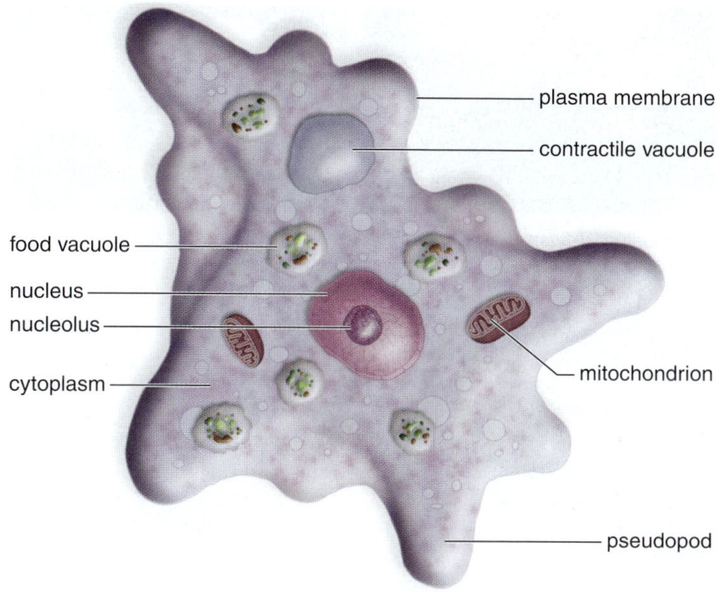

Figure 21.19 ***Amoeba proteus.*** This amoeboid is common in freshwater ponds. Bacteria and other microorganisms are digested in food vacuoles, and contractile vacuoles rid the body of excess water.

Check Your Progress 21.4

1. What are the two distinctive features of the excavates?
2. Match two excavates with the human diseases they cause.

21.5 Supergroups Amoebozoa, Opisthokonta, and Rhizaria

Learning Outcomes

Upon completion of this section, you should be able to

1. Defiine the unique features of amoebozoans.
2. Identify the members of supergroup opisthokonta.
3. Describe the unique properties of rhizaria tests.

Supergroup Amoebozoa

The **amoebozoans** (Gk. *ameibein,* "to change"; *zoa,* "animal") are protozoans that move by **pseudopods** (Gk. *pseudes,* "false"; *podos,* "foot"). Pseudopods form when an amoebozoan's microfilaments contract and extend as the cytoplasm streams toward a particular direction. Amoebozoans usually live in aquatic environments, where they are often a part of the plankton.

Amoeboids

The **amoeboids** are protists that move and ingest their food with pseudopods. Hundreds of species of amoeboids have been identified. *Amoeba proteus* is a commonly studied freshwater amoeba (Fig. 21.19). Amoeboids feed by phagocytosis, by which they engulf their prey with a pseudopod; prey organisms may be algae, bacteria, or other protists. Digestion occurs within a *food vacuole.* Freshwater amoeboids, including *Amoeba proteus,* have contractile vacuoles, which excrete excess water from the cytoplasm.

Video Amoeboid Locomotion

Entamoeba histolytica is a parasitic amoeboid that can live in the human large intestine and cause amoebic dysentery. Infectious *Entamoeba* cysts pass out of the body with infected stool. Others become infected from ingesting water and food contaminated with cysts. Infection of the liver and brain can also be fatal.

Slime Molds

Slime molds are important decomposers that feed on dead plant material. Slime molds were once classified as fungi, but unlike fungi, they lack cell walls, and they have flagellated cells at some time during their life cycle. The vegetative state of the slime molds is mobile and amoeboid. Slime molds produce spores by meiosis; the spores germinate to form gametes. Nearly 1,000 species of slime molds have been identified, including plasmodial and cellular forms.

Video Decomposers

Usually, **plasmodial slime molds** exist as a *plasmodium,* a diploid, multinucleated, cytoplasmic mass enveloped by a slime sheath. This term should not be confused with the genus *Plasmodium,* which is in the alveolate group. This sluglike slime mass consumes decaying plant material as it creeps along a forest floor or an agricultural field (Fig. 21.20).

When conditions are unfavorable, a plasmodium develops many sporangia. A **sporangium** (Gk. *spora,* "seed") is a reproductive structure that produces spores. An aggregate of sporangia is called a fruiting body. The spores produced by a plasmodial slime mold sporangium can survive until moisture is sufficient for them to germinate. In plasmodial slime molds, spores release either a haploid flagellated cell or an amoeboid cell. Eventually, two of the haploid

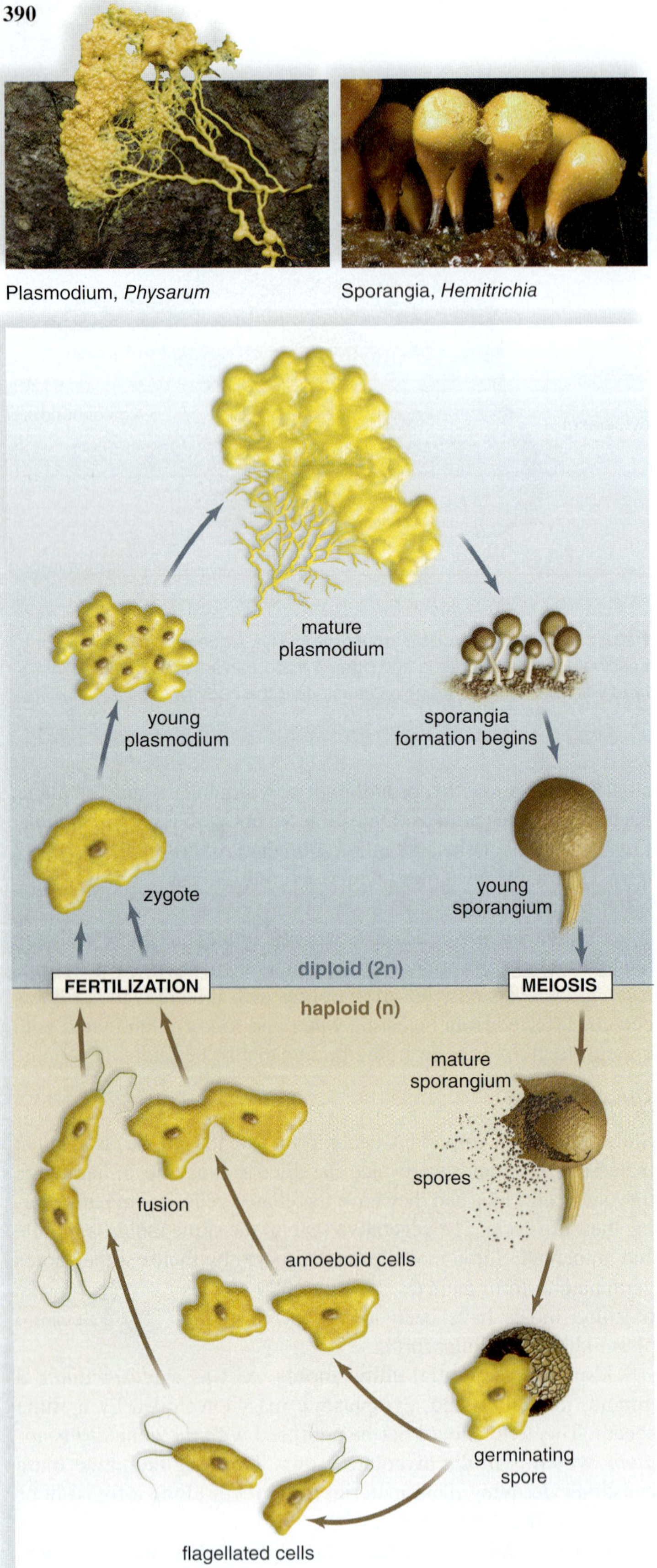

Figure 21.20 Plasmodial slime molds. The diploid adult forms sporangia during sexual reproduction, when conditions are unfavorable to growth. Haploid spores germinate, releasing haploid amoeboid or flagellated cells that fuse.

cells fuse to form a zygote that feeds and grows, producing a multinucleated plasmodium once again.

Cellular slime molds exist as individual amoeboid cells. They are common in soil, where they feed on bacteria and yeasts. Fewer species of cellular slime molds have been identified compared to plasmodial species.

As unfavorable environmental conditions develop, the slime mold cells release a chemical that causes them to aggregate into a pseudoplasmodium. The pseudoplasmodium stage is temporary and eventually gives rise to a fruiting body, in which sporangia produce spores. When favorable conditions return, the spores germinate, releasing haploid amoeboid cells, and the asexual cycle begins again.

Supergroup Opisthokonta

Animals and fungi are **opisthokonts** (Gk. *opisthos,* "behind"; *kontos,* "pole") along with several closely related protists. This supergroup includes both single-celled and multicellular protozoans. Among the opisthokonts are the **choanoflagellates,** animal-like protozoans that are closely related to sponges. The choanoflagellates, including singled-celled as well as colonial forms, are filter-feeders with cells that bear a striking resemblance to the choanocytes that line the inside of sponges (see section 28.2). Each choanoflagellate has a single posterior flagellum surrounded by a collar of slender microvilli. Beating of the flagellum creates a water current that flows through the collar, where food particles are taken in by phagocytosis. Colonial choanoflagellates such as *Codonosiga* (Fig. 21.21*a*) commonly attach to surfaces with a stalk, but sometimes float freely like *Proterospongia* (Fig. 21.21*b*).

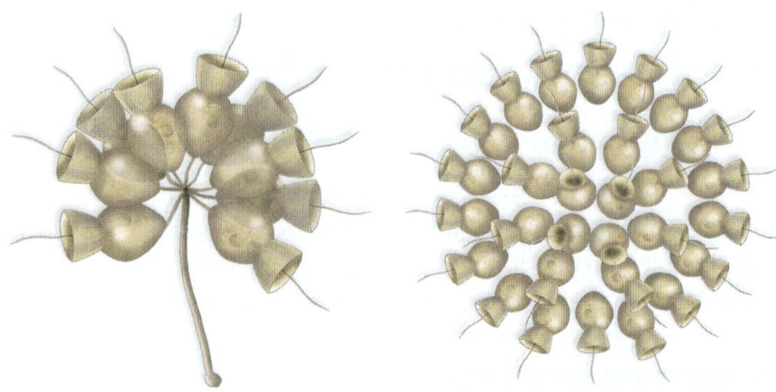

a. *Codonosiga* b. *Proterospongia*

Figure 21.21 Colonial choanoflagellates. **a.** A *Codonosiga* colony can anchor itself with a slender stalk. **b.** A *Proterospongia* colony is unattached.

a. Foraminiferan, *Globigerina,* and the White Cliffs of Dover, England

b. Radiolarian tests SEM 150×

Figure 21.22 Foraminiferans and radiolarians.
a. Pseudopods of a live foraminiferan project through holes in the calcium carbonate shell. Fossilized shells were so numerous they became a large part of the White Cliffs of Dover when a geological upheaval occurred.
b. Skeletal test of a radiolarian. In life, pseudopods extend outward through the openings of the glassy silicon shell.

Nucleariids are opisthokonts with a rounded or slightly flattened cell body and threadlike pseudopods called filopodia.

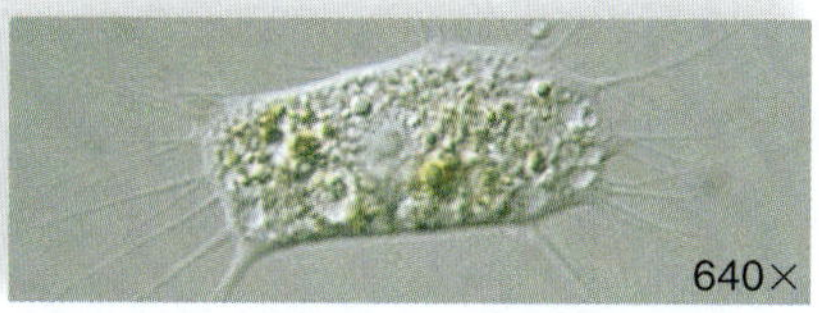

Most feed on algae or cyanobacteria. Although they lack the characteristic cell walls found in fungi, molecular similarities suggest that nucleariids are close relatives of fungi.

Supergroup Rhizaria

The **rhizarians** (Gk. *rhiza,* "root") consist of the **foraminiferans** and the **radiolarians,** organisms with fine, threadlike pseudopods. Although rhizarians were once classified with amoebozoans, they are now assigned to a different supergroup. Foraminiferans and radiolarians both have a skeleton, called a **test,** made of calcium carbonate.

The tests of foraminiferans and radiolarians are intriguing and beautiful. In the foraminiferans, the calcium carbonate test is often multichambered. The pseudopods extend through openings in the test, which covers the plasma membrane (Fig. 21.22*a*). In the radiolarians, the glassy silicon test is internal and usually has a radial arrangement of spines (Fig. 21.22*b*). The pseudopods are external to the test.

The tests of dead foraminiferans and radiolarians form a layer of sediment 700–4,000 m deep on the ocean floor. The presence of tests is used as an indicator of oil deposits on land and sea. Their fossils date as far back as the Precambrian, and they are evidence of the antiquity of the protists. Each geological period has a distinctive form of foraminiferan; thus, foraminiferans can be used as index fossils to date sedimentary rock. Millions of years of foraminiferan deposits formed the White Cliffs of Dover along the southern coast of England. Also, the great Egyptian pyramids are built of foraminiferan limestone. One foraminiferan test found in the pyramids is about the size of an old silver dollar—about an inch and a half across! This species, known as *Nummulites,* produced a flattened, coiled test, and its fossils have been found in deposits worldwide, including central-eastern Mississippi.

Check Your Progress 21.5

1. How do amoebozoans move and feed?
2. Which eukaryotic organisms are classified as opisthokonts?
3. What practical uses have humans found for radiolarian tests?

CONNECTING the CONCEPTS to the THEMES

Evolution

- Endosymbiosis of energy-producing organelles was a key step in the evolution of the first eukaryotes.
- The diversity of protists makes it difficult to determine the evolutionary relationships among main lineages of the eukaryote family tree.
- Morphology and molecular data have assisted with the placement of protists into six main evolutionary lineages, called supergroups.

Nature of Science

- Many protists serve as model organisms critical to many areas of scientific research, including ecosystem ecology, medicine, and epidemiology.
- Research of protist biology leads to the discovery of substances that are useful and economically important.
- Fossil protists provide paleontologists with relative dating of other fossil species.

Biological Systems

- Protists are the cause of many human diseases that affect the ecology and economies of communities.
- Plankton are at the base of the food chain in aquatic systems. A large proportion of plankton is made up of protists.
- An imbalance in the population density of protists can lead to contamination of aquatic ecosystems.
- Protists provide an important service in ecosystems as decomposers.

MEDIA STUDY TOOLS http://connect.mheducation.com

SMARTBOOK® Study smart. SmartBook helps you maximize your study time by identifying what you know and don't know. The recharge feature in SmartBook helps you prepare for the test by identifying the material that you are most likely to forget.

connect BIOLOGY Find even more resources to learn the chapter concepts including animations, tutorial videos, and interactive practice questions.

Animation	Videos
21.1 Endosymbiosis	**21.5** Amoeboid Locomotion • Decomposers

SUMMARIZE

21.1 General Biology of Protists

Protists are generally single-celled (sometimes multicellular) organisms in the domain Eukarya that acquired mitochondria and chloroplasts through **endosymbiosis** of bacteria. While most protists are single-celled, some may be multicellular or exist in colonies or filaments. To obtain nutrients, protists may be autotrophic, heterotrophic, or **mixotrophic.** Reproduction can be sexual, asexual, or even both, and employ unique strategies including alternation of generations or conjugation.

In favorable environmental conditions, protists exhibit asexual reproduction. When faced with an unfavorable environment, they perform sexual reproduction and can produce spores or **cysts** that withstand harsh conditions.

Protists are of great ecological importance, because they are the main producers in aquatic ecosystems, contributing to the community of **plankton.**

Protists, and other eukaryotes, are classified into six **supergroups,** each representing a separate evolutionary lineage. However, protists are not a **monophyletic** group.

21.2 Supergroup Arachaeplastida

Supergroup **Archaeplastida** includes land plants and all green and red algae.

Green algae include the **chlorophytes** and **charophytes** that possess chlorophylls, store energy as starch, and have cell walls made of cellulose.

Chlorophytes include single-celled (*Chlamydomonas*), colonial (*Volvox*), and multicellular *(Ulva)* forms. Chlorophytes form **filaments** and have diverse life cycles, including a haploid life cycle common to algae and alternation of generations similar to land plants (*Ulva*).

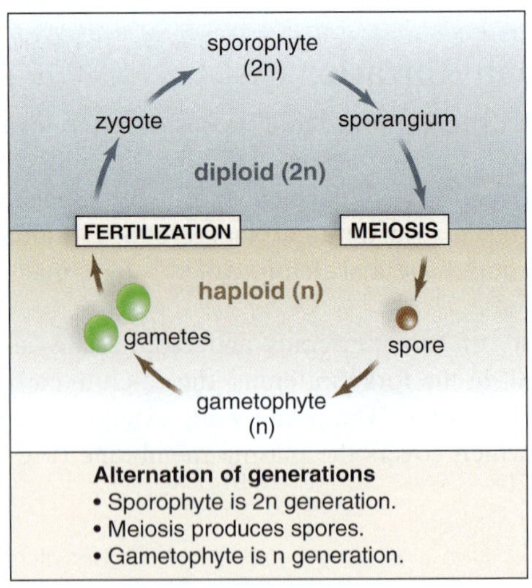

Alternation of generations
- Sporophyte is 2n generation.
- Meiosis produces spores.
- Gametophyte is n generation.

Charophytes include *Spirogyra* as well as stoneworts (*Chara*) and are thought to be the closest living relatives of land plants. Some charophytes, such as *Spirogyra,* reproduce by **conjugation.**

Red algae are economically valuable and contain plastids with pigments that give them a reddish or reddish-brown appearance.

21.3 Supergroup Chromalveolata

Supergroup **Chromalveolata** consists of **stramenopiles** and **alveolates.** Stramenopiles include **brown algae, diatoms, golden brown algae,** and **water molds.**

Brown algae have chlorophylls *a* and *c* plus a brownish carotenoid **accessory pigment.** Seaweeds are economically important, large, complex brown algae. Diatoms and golden brown algae have an outer layer of silica. Water molds are similar to fungi, except that they produce flagellated, diploid **zoospores.**

Alveolates include **dinoflagellates, ciliates,** and **apicomplexans.**

Dinoflagellates can form harmful algal blooms called **red tides.** The ciliates, such as *Paramecium,* move by coordinated movement of their many cilia, capturing prey with specialized barbed threads called **trichocysts.** The apicomplexans are nonmotile parasites that form spores. *Plasmodium,* an apicomplexan that causes malaria, is transmitted between disease reservoirs and hosts by *Anopheles* mosquitoes.

21.4 Supergroup Excavata

Supergroup **Excavata** includes single-celled, motile protists, including **euglenids, parabasalids, diplomonads,** and **kinetoplastids.**

Euglenids are photoautotrophic, flagellated cells with a pellicle instead of a cell wall. Many of the kinetoplastids are parasites, including trypanosomes, which cause Chagas disease and African sleeping sickness when transmitted by the bite of an insect **vector.** Parabasalids and diplomonads, such as *Trichomonas vaginalis* and *Giardia lamblia,* are common parasites of animal hosts.

21.5 Supergroups Amoebozoa, Opisthokonta, and Rhizaria

Supergroup **Amoebozoa** contains **amoeboids** and **slime molds**—protists that use **pseudopods** for motility and feeding.

Amoeboids are heterotrophs that move and feed by **phagocytosis. Plasmodial** and **cellular slime molds** are important decomposers, which produce nonmotile spores during unfavorable conditions using a **sporangium.** They are similar to fungi, except they have an amoeboid stage.

Supergroup **Opisthokonta** includes animals, the animal-like **choanoflagellates,** the fungi, and the funguslike protists called **nucleariids.**

Supergroup **Rhizaria** includes the **foraminiferans** and **radiolarians** that have threadlike pseudopods and skeletons called tests. The tests of foraminiferans and radiolarians form a deep layer of sediment on the ocean floor that can be used as index fossils.

ASSESS

Choose the best answer for each question.

21.1 General Biology of Protists

1. Mitochondria in protists originated from
   a. cyanobacteria.
   b. chloroplasts.
   c. anaerobic bacteria.
   d. aerobic bacteria.
   e. eukaryotic cells.
2. Through what process did eukaryotic cells gain chloroplasts and mitochondria?
   a. engulfment
   b. endosymbiosis
   c. endocytosis
   d. phagocytosis
   e. transformation
3. What feature of protists allow them to survive harsh environmental conditions?
   a. vegetative cells
   b. plankton
   c. cysts
   d. plastids
   e. vacuoles

21.2 Supergroup Archaeplastida

4. Which of these is a green alga?
   a. *Volvox*
   b. *Gelidium*
   c. *Euglena*
   d. *Paramecium*
   e. *Plasmodium*
5. In *Chlamydomonas,* the adult vegetative cell
   a. is haploid.
   b. survives harsh environments.
   c. produces zygospores.
   d. is dormant.
   e. is ciliated.
6. In the life cycle of *Ulva,* which stage is diploid?
   a. spore
   b. sporophyte
   c. gamete
   d. gametophyte

21.3 Supergroup Chromalveolata

7. Kelp represent which type of protist?
   a. archaeplastida
   b. chromalveolata
   c. opisthokonta
   d. excavata
   e. rhizaria
8. What is the main component of a diatom shell?
   a. agar
   b. peptidoglycan
   c. silica
   d. calcium carbonate
   e. paramylon
9. Dinoflagellates
   a. usually reproduce sexually.
   b. have protective cellulose plates.
   c. are insignificant producers of food and oxygen.
   d. have cilia instead of flagella.
   e. tend to be larger than brown algae.
10. Ciliates
    a. move by pseudopods.
    b. are not as varied as other protists.
    c. feed and move using cilia.
    d. do not divide by binary fission.
    e. are closely related to the radiolarians.
11. Apicomplexans are responsible for which disease?
    a. vaginitis
    b. African sleeping sickness
    c. red tide
    d. Chagas disease
    e. malaria

21.4 Supergroup Excavata

12. Protists that may lack mitochondria and possess deep oral grooves are known as
    a. excavates.
    b. apicomplexans.
    c. alveolates.
    d. amoeboids.
    e. ciliates.
13. Which genus of protists causes African sleeping sickness and Chagas disease?
    a. *Plasmodium*
    b. *Trypanosoma*
    c. *Entamoeba*
    d. *Trichomonas*
    e. *Giardia*

21.5 Supergroups Amoebozoa, Opisthokonta, and Rhizaria

14. Plasmodial and cellular _____ are important decomposers that have an amoeboid vegetative state.
    a. brown algae
    b. water molds
    c. filamentous algae
    d. slime molds
    e. red algae
15. Skeletons called _____ are formed by rhizarians and can be used to date fossils.
    a. tests
    b. plankton
    c. cysts
    d. plastids
    e. valves

ENGAGE

THINKING SCIENTIFICALLY

1. While studying a single-celled alga, you discover a mutant in which the daughter cells do not separate after mitosis. This gives you an idea about how filamentous algae may have evolved. You hypothesize that the mutant alga is missing a protein or making a new form of a protein. How might each lead to a filamentous appearance?
2. You are trying to develop a new anti-termite chemical that will not harm environmentally beneficial insects. Because termites are adapted to eat only wood, they will starve if they cannot digest this food source. Termites have two symbiotic partners: the protozoan *Trichonympha collaris* and the bacteria it harbors that actually produce the enzyme that digests the wood. Knowing this, how might you prevent termite infestations without targeting the termites directly?
3. Certain protist species cause human diseases that are difficult to treat and are often deadly. How does the evolutionary relatedness of humans and protists explain the difficulty of treating a person infected with a pathogenic protist?
4. Convergent evolution is a process by which unrelated species evolve similar adaptations to their environment. For example, both birds and bats have independently evolved the ability to fly instead of inheriting the trait from a common ancestor. In protists, an example is parasitism. Certain apicomplexans in the chromalveolata supergroup and parabasilids in the excavata supergroup have evolved the ability to parasitize human hosts. What are other examples of traits that have developed through convergent evolution within the protists?

Mushroom® Packaging is an ecologically sustainable insulation and packaging product derived entirely from fungal mycelium and other organic materials.

22

Fungi Evolution and Diversity

Fungi were the first eukaryotes to invade land. They are ancient organisms, with deep, underground networks of mycelium, or "fungal roots," that service the ecosystem by recycling organic debris back into topsoil. The part of the fungi that we see, and sometimes eat, is the fruiting body, or mushroom, which is the center of reproduction for many species.

The medicinal properties of fungi have been known for centuries—distillates from fungi have antimicrobial and antiviral properties, which are just now being tapped by pharmaceutical research. Fungi also hold the potential to replace the plastics and styrofoam packing materials on which we depend so heavily. For example, the eco-minded company Ecovative Design has developed a compound, called Mushroom® Packaging, entirely from the mycelium of fungi and agricultural waste products. This packaging can be grown into any shape, creating custommade home insulation and packaging materials for any type of product. Mushroom® Packaging requires much less energy to create than styrofoam or plastic and is completely organic, nontoxic, and compostable.

Products such as antibiotics, antivirals, and Mushroom® Packaging, derived from the natural processes of fungi, have the potential to revolutionize human society and our impact on the environment.

As you read through this chapter, think about the following questions:

1. Why is it important to human health to preserve fungi biodiversity?
2. How might the commercial production of materials from fungi reduce our impact on the environment?
3. How would ecosystems be impacted if fungi were to go extinct?

Chapter Outline

22.1 Evolution and Characteristics of Fungi 396

22.2 Diversity of Fungi 398

22.3 Symbiotic Relationships of Fungi 405

Before You Begin

Before beginning this chapter, take a few moments to review the following discussions.

Section 9.3 What is the difference between a haploid and a diploid cell?

Section 10.1 Why is meiosis essential to sexual reproduction?

Chapter 21 What supergroup of protists is most closely related to the fungi?

Following *the* Themes

CHAPTER 22 FUNGI EVOLUTION AND DIVERSITY

UNIT 4 MICROBIOLOGY AND EVOLUTION

Evolution	Fungi are more closely related to animals than to any other group of eukaryote; both animals and fungi belong in the supergroup Opisthokonta.
Nature of Science	Fungi have properties that make them an important resource for the study of medicine, genetics, and molecular biology, as well as for industrial applications.
Biological Systems	Fungi have huge networks of mycelium beneath the soil that serve as nature's recycling center; decomposition by fungi is a critical ecosystem service.

22.1 Evolution and Characteristics of Fungi

Learning Outcomes

Upon completion of this section, you should be able to

1. Identify two traits that are similar between animals and fungi.
2. Distinguish among fungi that are haploid, dikaryotic, or diploid.
3. Define and identify the structural features of fungi.

The **fungi** include over 100,000 species of mostly multicellular eukaryotes that share a common mode of nutrition. Mycologists, scientists who study fungi, expect this number of species to increase to in the millions in the future as new species are discovered and molecular biology techniques improve.

Plants are autotrophs and create their own food. Conversely, fungi, like animals, are heterotrophs and consume preformed organic matter. Animals, however, are heterotrophs that ingest food, whereas fungi are **saprotrophs** that absorb food. Their cells send out digestive enzymes into the immediate environment that break down dead and decaying organic matter. The resulting nutrient molecules are then absorbed by the fungus cells.

Video Decomposers

Evolution of Fungi

Figure 22.1 illustrates the evolutionary relationships among the six groups of fungi we will be discussing. The evolutionary tree is a hypothesis about how these groups are related. The Microsporidia and chytrids are different from all other fungi, because they are single-celled. The chytrids are aquatic and have flagellated spores and gametes. Our description of fungal structure applies best to the zygospore fungi, sac fungi, and club fungi. The AM fungi are notable, because they exist only as mycorrhizae in symbiotic association with plant roots.

Protists evolved some 1.5 BYA (billion years ago). Plants, animals, and fungi can all trace their ancestry to protists, but molecular data tell us that animals and fungi shared a more recent common ancestor than animals and plants. Therefore, animals and fungi, both in the supergroup Opisthokonta, are more closely related to each other than either is to plants (Fig. 22.1). The common ancestor of animals and fungi was most likely an aquatic, flagellated, single-celled protist. Multicellular forms evolved sometime after animals and fungi split into two different lineages.

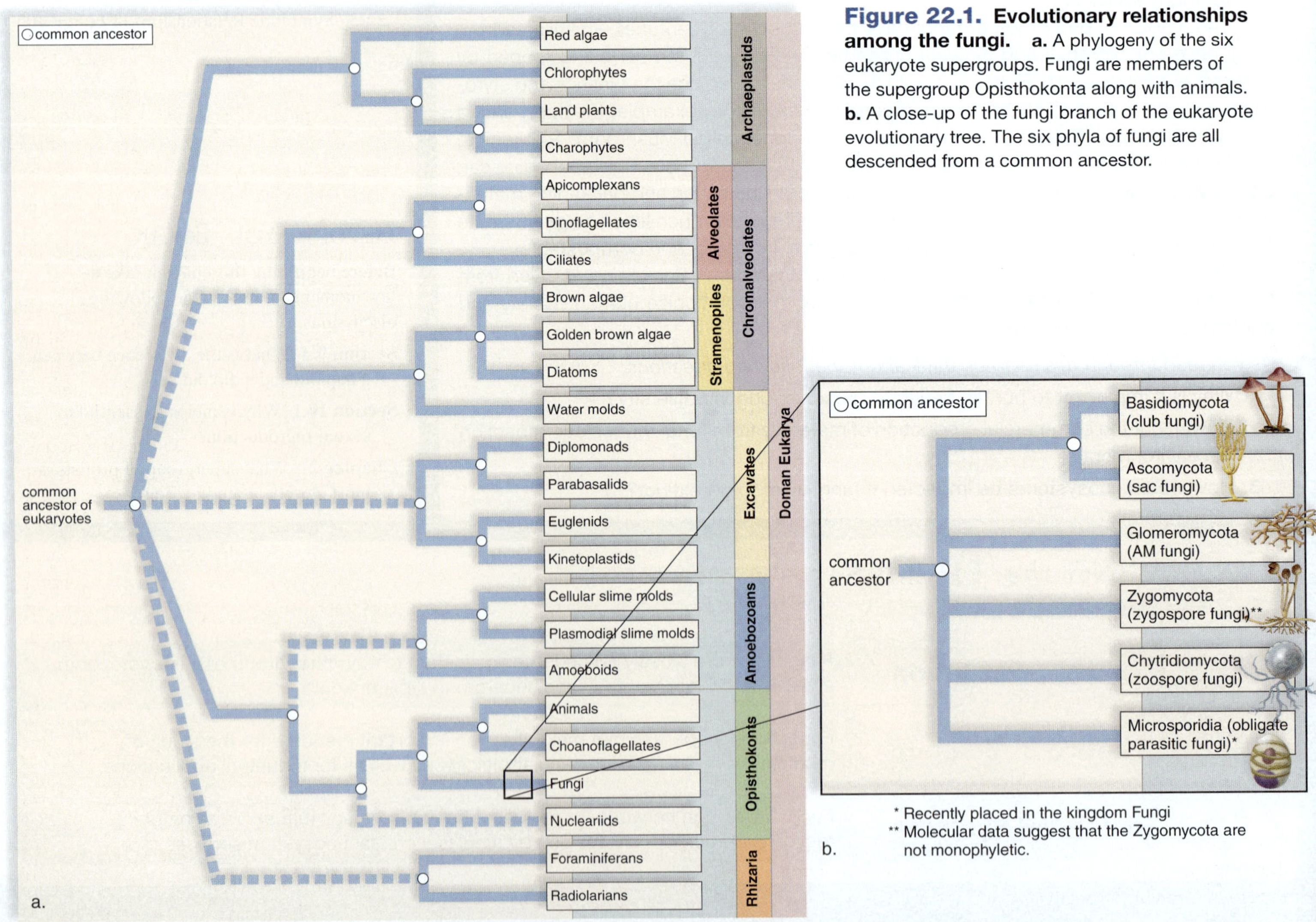

Figure 22.1. Evolutionary relationships among the fungi. **a.** A phylogeny of the six eukaryote supergroups. Fungi are members of the supergroup Opisthokonta along with animals. **b.** A close-up of the fungi branch of the eukaryote evolutionary tree. The six phyla of fungi are all descended from a common ancestor.

a. Fungal mycelia. b. Individual strands of hyphae. 170× c. Septate and aseptate hyphae.

Figure 22.2 Mycelia and hyphae of fungi. **a.** Fungal mycelium made from hyphae, growing as a white mass on strawberries. **b.** Hyphal filaments growing on the surface of a plant. **c.** Hyphae are either septate (do have cross-walls) or aseptate (do not have cross-walls.)

Fungi do not fossilize well, so it is difficult to estimate from the fossil record when they first evolved. The earliest known fossil fungi are dated at 460 MYA (million years ago), but fungi probably evolved a lot earlier. While animals were still swimming in the seas during the Silurian, plants were beginning to live on the land, and they brought fungi with them. Mycorrhizae are evident in plant fossils, also some 460 MYA. Perhaps fungi were instrumental in the colonization of land by plants. Much of the fungal diversity most likely had its origin in an adaptive radiation when organisms began to colonize land.

Structure of Fungi

Some fungi, including the yeasts, are single-celled organisms; however, the vast majority of species are multicellular. The body of most fungi is a multicellular structure known as a mycelium (Fig. 22.2*a*). A **mycelium** (Gk. *mycelium,* "fungus filaments") is a network of fungal filaments; these filaments are called **hyphae** (Gk. *hyphe,* "web"). Hyphae give the mycelium quite a large surface-to-volume ratio, and this maximizes the absorption of nutrients into the body of a fungus (Fig. 22.2*b*).

Hyphae grow from their tips, and in some fungi, **septa** (sing., septum), or walls of tissue, are formed behind the growing tip, partitioning the hyphae into individual cells. Fungi that have septa in their hyphae are called **septate** (L. *septum,* "fence, wall"). Pores in the septa allow cytoplasm and sometimes even organelles to pass freely from one cell to another. The septa that separate reproductive cells, however, are completely closed in all fungal groups. **Aseptate** fungi are not divided into cells, and many nuclei are present in the cytoplasm of a single hypha (Fig. 22.2*c*). Some hyphae are capable of penetrating rigid substances, such as plant tissues. When a fungus reproduces, a specific portion of the mycelium becomes a reproductive structure, which is nourished by the rest of the mycelium.

Fungal cells are quite different from plant cells, because they lack chloroplasts and have a cell wall made of chitin, not cellulose. **Chitin,** like cellulose, is a polymer of glucose organized into microfibrils, but each glucose molecule of chitin has a nitrogen-containing amino group attached to it. Chitin is also found in the exoskeleton of arthropods, a major phylum of animals that includes the insects and crustaceans. Unlike plants that store energy as starch, fungi store energy as glycogen, the same molecule that animals use to store energy. Except for the aquatic chytrids, fungi are not motile. Terrestrial fungi lack basal bodies (see section 4.8) and do not have flagella at any stage in their life cycle. They move toward a food source by hyphae growing toward it. Growing hyphae can cover as much as a kilometer a day!

Reproduction of Fungi

Both sexual and asexual reproduction occur in fungi. Sexual reproduction of terrestrial fungi occurs in these stages:

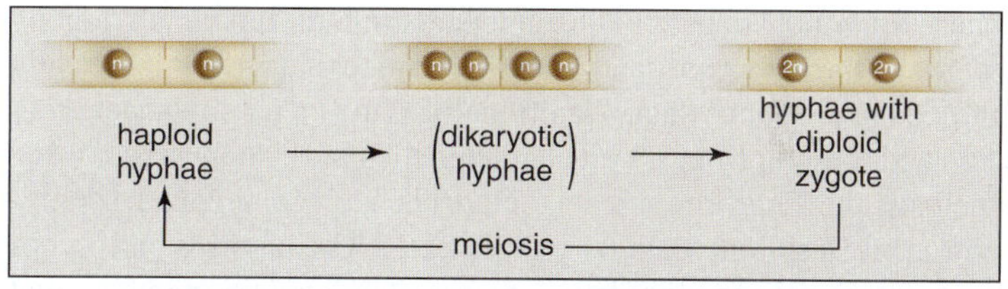

The relative length of time of each phase varies with the species.

During sexual reproduction, hyphae (or a portion thereof) from two different mating types make contact and fuse. In some species, the nuclei from the two mating types fuse immediately. In other species, the nuclei pair up but do not fuse for days, months, or even years. The nuclei continue to divide in such a way that every cell (in septate hyphae) has at least one of each nucleus. A hypha that contains paired haploid nuclei is said to be n + n, or **dikaryotic** (Gk. *dis,* "two"; *karyon,* "nucleus"). When the nuclei do eventually fuse, the zygote undergoes meiosis prior to spore formation.

How can fungi ensure that the offspring will be dispersed to new locations? As an adaptation to life on land, fungi usually produce nonmotile, windblown spores during both sexual and asexual reproduction. A **spore** is a reproductive cell that develops into a new organism without the need to fuse with another reproductive cell (see Fig. 22.5*b*). A large mushroom may produce billions of spores within a few days. When a spore lands on an appropriate food source, it germinates and begins to grow.

Asexual reproduction usually involves the production of spores by a specialized part of a single mycelium. Alternatively, asexual reproduction can occur by fragmentation—a portion of a mycelium begins a life of its own. Also, single-celled yeasts reproduce asexually by **budding;** a small cell forms and gets pinched off as it grows to full size (see Fig. 22.5*a*).

Check Your Progress 22.1

1. Describe how animals and fungi differ with respect to nutritional mode.
2. Explain how fungal cells differ from plant cells.
3. Describe the function of a fungal spore.

22.2 Diversity of Fungi

Learning Outcomes

Upon completion of this section, you should be able to

1. List the major phyla of fungi.
2. Summarize the life cycle typical of fungi in each of the six phyla.
3. Identify one benefit and one disadvantage of human and fungi interactions.

In 1969 R. H. Whittaker classified fungi as a separate group from protists, plants, animals, and prokaryotic organisms. He based his reasoning on the observation that fungi are the only type of multicellular organism to be saprotrophic. However, fungi are now considered to be the closest multicellular relative of animals. Both fungi and animals are placed in the eukaryote supergroup Opisthokonta, which also includes certain heterotrophic protists (see Fig. 22.1).

The phylogenetic relationships among fungi have been the cause of much debate. The understanding of fungal phylogeny is going through rapid and exciting changes, aided by increasing molecular sequence data. Traditionally, four fungal phyla were recognized, based primarily on characteristics of the cells undergoing meiosis: the Chytridiomycota ("chytrids"), Zygomycota, Ascomycota, and Basidiomycota (see Table 22.1). The chytrids and Zygomycota are not monophyletic—that is, they contain members that did not all originate from a common ancestor. In addition to these four, we describe two other groups of fungi, the Glomeromycota, or AM fungi, and the Microsporidia, a medically important group of organisms recently moved into the kingdom Fungi (see Fig. 22.1).

Microsporidians Are Parasitic Fungi

The single-celled **Microsporidia** are obligate, intracellular, animal parasites, most often seen in insects but also found in vertebrates, such as fish, rabbits, and humans (Fig. 22.3*a*). Biologists once believed that Microsporida were an ancient line of protist due in part to their lack of mitochondria. However, genome sequencing of the microsporidian *Encephalitozoon cuniculi* revealed genes that are mitochondrial, leading to the hypothesis that this organism once had a mitochondrion, which later became greatly reduced. In addition, microsporidians have the smallest known eukaryotic genome. New sequence information places *E. cuniculi* and other microsporida in fungi, rather than with the protists.

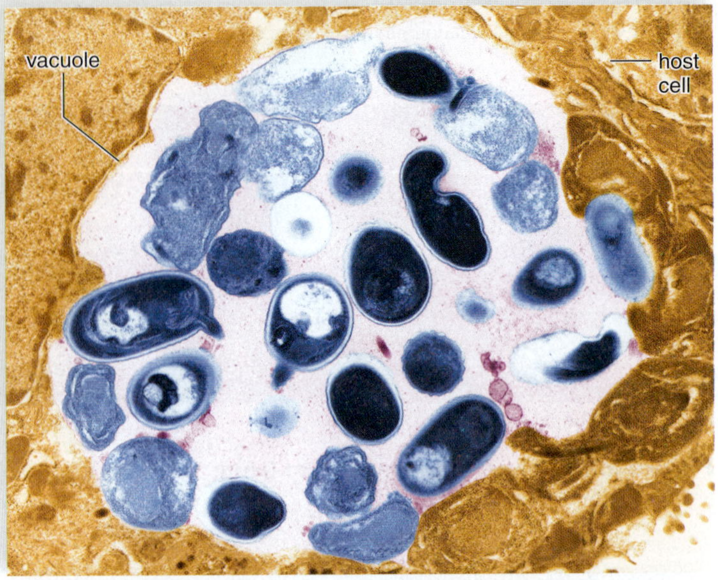

a. A microsporidian *Encephalitozoon cuniculi* infection. 16,000×

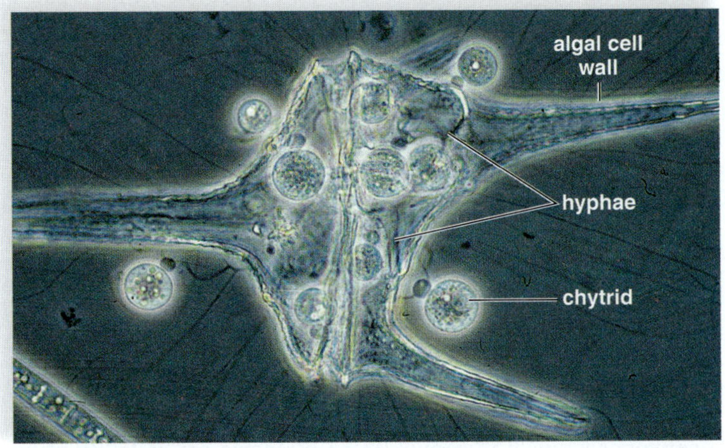

b. Chytrids parasitizing a protist.

Figure 22.3 The Microsporidia and Chytridiomycota. **a.** Microsporidian *Encephalitozoon cuniculi* spores (stained blue) grow and develop within a vacuole inside a host cell. **b.** The chytrids on this dinoflagellate are absorbing nutrients meant for their host. They will produce flagellated zoospores, which will go on to parasitize other protists.

E. cuniculi and other Microsporidia commonly cause diseases in immunocompromised patients, such as those receiving organ transplants or individuals with AIDS. Microsporidians infect their hosts with spores that contain a polar tube. This tube extrudes the contents of the spore into intestinal or neuronal cells and

Table 22.1 Features of the Fungi Phyla

Phylum (common name)	Reproduction	Key Features	Examples
Basidiomycota (club fungi)	Basidiospores, sexual	Basidiocarp fruiting body (mushroom)	Most edible mushrooms, common button mushroom (*Agaricus bisporus*)
Ascomycota (sac fungi)	Conidiospores, asexual; ascus with spores, sexual	Ascocarp fruiting body (mushroom), yeasts, molds	Morel mushroom (*Morchella*), cup fungus (*Sarcoscypha*)
Glomeromycota (AM fungi)	Spores, asexual	Arbuscules, symbiotic with plants	*Glomerales*
Zygomycota (zygospore fungi)	Zygospores, sexual; sporangiospores, asexual	Sporangia, gametangia; saprotrophic; feed on animal remains or bakery goods	Black bread mold (*Rhizopus stolonifer*)
Chytridiomycota (chytrid fungi)	Zoospores with flagella; most asexual; some alternation of generations	Single-celled, simplest fungi, aquatic	*Chytriomyces hyalinus*

remains hidden in a vacuole, causing the human host to suffer with diarrhea and neurodegenerative diseases. The parasitic nature of these organisms makes understanding the phylogenetic placement important in identifying effective disease treatments.

Chytrids Are Aquatic Fungi

The Chytridiomycota, or **chytrids,** include about 750 species of the simplest fungi, which may resemble the first fungi to have evolved. Some chytrids are single cells; others form branched aseptate hyphae. Chytrids are unique among fungi, because they have flagellated gametes and spores, a feature consistent with their aquatic lifestyle, although some also live in moist soil. The placement of the flagella in their spores, called **zoospores,** suggests a shared ancestry of fungi and choanoflagellates (see Table 21.1).

Most chytrids reproduce asexually through the production of zoospores within a single cell. The zoospores grow into new chytrids. However, some have an alternation-of-generations life cycle, much like that of green plants and certain algae (see Fig. 21.5), which is quite uncommon among fungi.

Chytrids play a role in the decay and digestion of dead aquatic organisms, but some are parasites of living plants, animals, and protists (Fig. 22.3*b*). The parasitic chytrid *Batrachochytrium dendrobatidis* has in the last 30 years decimated populations of over 200 species of frogs resulting in a staggering loss of amphibian biodiversity. The chytrid infects the frog's skin, causing a normally permeable layer to thicken, inhibiting oxygen and water intake. The electrolyte imbalances result in cardiac arrest and death.

Zygomycota Produce Zygospores

The Zygomycota, or **zygospore fungi,** include approximately 1,060 species of fungi. Zygospore fungi live off plant and animal remains in the soil or in bakery goods in the pantry. Some are parasites of soil protists, worms, and insects such as the housefly.

The black bread mold, *Rhizopus stolonifer,* is a common example of this phylum. *Rhizopus* has both a sexual and an asexual phase in its life cycle (Fig. 22.4). The body of this fungus is composed of mostly aseptate hyphae that can specialize to perform various tasks. *Rhizopus* exhibits three kinds of specialized hyphae:

- *stolons*—horizontal hyphae that exist on the surface of the bread.
- *rhizoids*—hyphae that grow into bread, anchor the mycelium, and carry out digestion.
- *sporangiophores*—aerial hyphae that bear sporangia.

A **sporangium** (pl., sporangia) is a capsule that produces haploid spores called *sporangiospores* during the asexual phase of reproduction (Fig. 22.4).

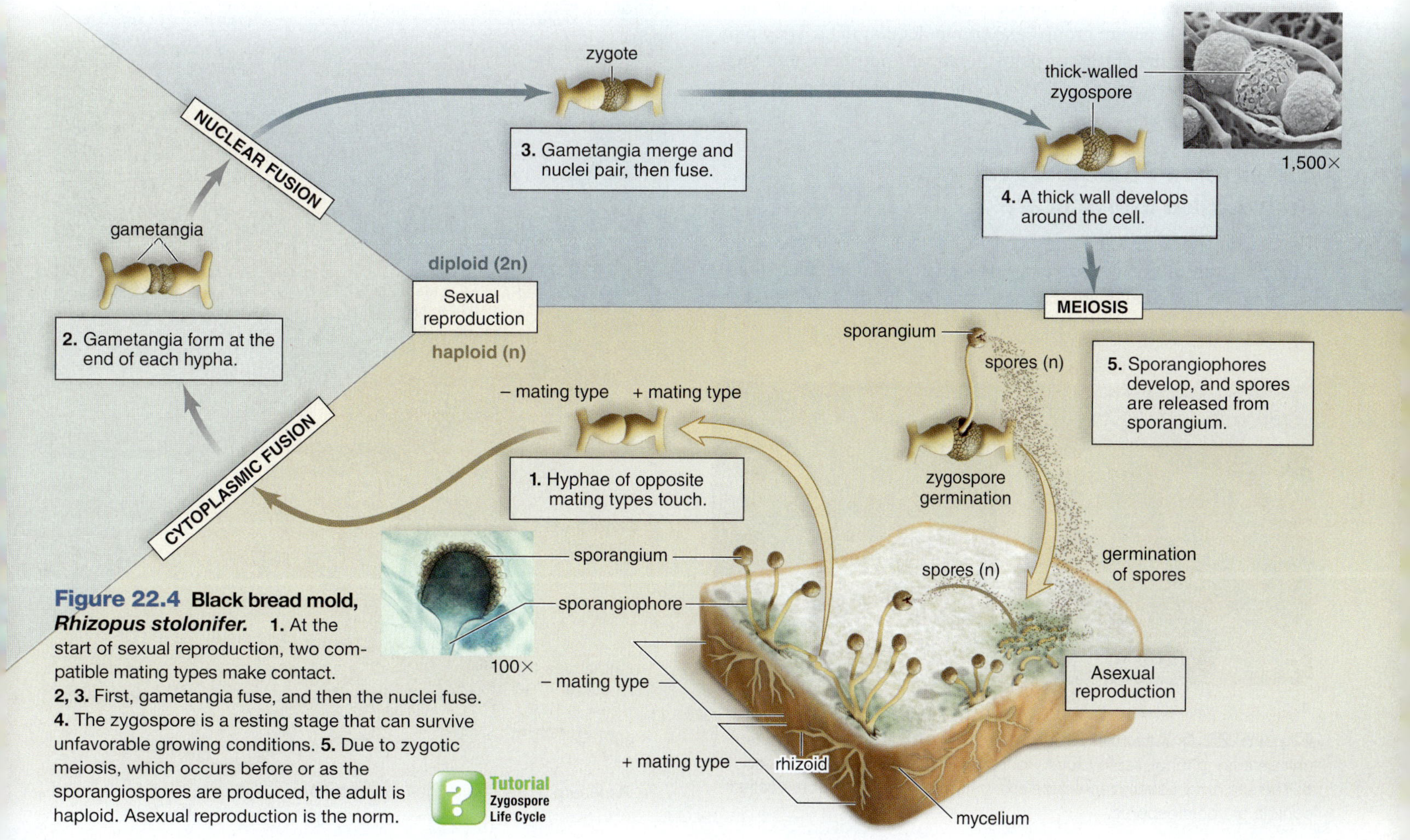

Figure 22.4 Black bread mold, *Rhizopus stolonifer.* **1.** At the start of sexual reproduction, two compatible mating types make contact. **2, 3.** First, gametangia fuse, and then the nuclei fuse. **4.** The zygospore is a resting stage that can survive unfavorable growing conditions. **5.** Due to zygotic meiosis, which occurs before or as the sporangiospores are produced, the adult is haploid. Asexual reproduction is the norm.

Tutorial Zygospore Life Cycle

Zygospores are diploid spores produced during sexual reproduction. Two mating types of hyphae, termed plus (+) and minus (−), are chemically attracted to one another, and they grow toward each other until they touch (Fig. 22.4). The ends of the touching hyphae swell as nuclei are directed toward, and enter, the tips of the hyphae. Cross-walls then develop a short distance behind the swollen end of each hypha, forming an isolated capsule called a *gametangium* (pl., gametangia). The gametangia of each hypha merge, and the nuclei fuse to form a diploid zygote.

A thick wall develops around the zygote, which is now called a **zygospore.** The zygospore undergoes a period of dormancy before new haploid sporangiospores are formed by meiosis. Sporangiophores with sporangia at their tips germinate from the zygospore, and many sporangiospores are released. The spores, dispersed by air currents, give rise to new haploid mycelia, which will continue the sexual phase of the life cycle. Spores from black bread mold have been found in the air above the North Pole, in the tropics, and far out at sea.

Glomeromycota Are Important Symbiotic Fungi

The Glomeromycota, or **AM fungi,** are a relatively small group (230 species) of fungi. The name AM stands for *arbuscular mycorrhizal* fungi. Arbuscules are branching invaginations that the fungus makes when it invades plant roots.

Mycorrhizae (see section 22.3) are a mutualistic association of plants and fungi, and AM fungi are the most common fungi to form symbiotic relationships with plants. The AM fungi were classified with the zygospore fungi for a long time, but they are now recognized as a separate group based on molecular data. They play a critical role in the ability of plants to absorb nutrients with their roots. The majority of plants have a mutually beneficial relationship, or symbiosis, with AM fungi.

Ascomycota Produce a Fruiting Body Called an Ascocarp

The Ascomycota, or **sac fungi,** consist of about 64,000 species of fungi that can reproduce asexually or sexually. Multicellular **molds** and single-celled **yeasts** are the most common morphological types.

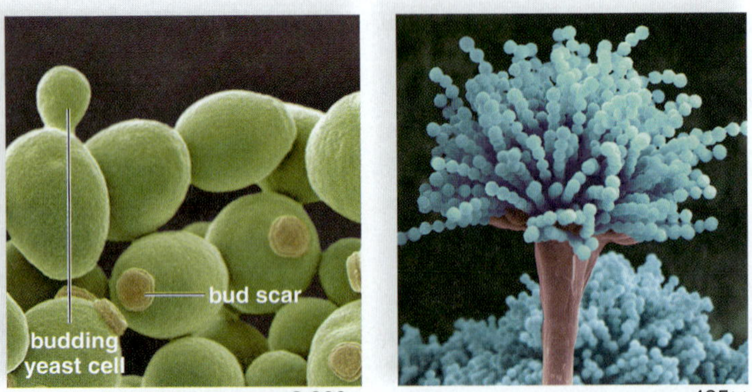

a. 3,000×   b. 425×

Figure 22.5 Asexual reproduction in sac fungi. **a.** Yeasts, unique among fungi, reproduce by budding, leaving a bud scar behind. **b.** The sac fungi usually reproduce asexually by producing spores called conidia or conidiospores.

The sac fungi play an essential role in recycling by digesting materials that do not easily decompose, such as cellulose, lignin, and collagen; some species have even been known to consume jet fuel and wall paint.

Members of this phylum are ecologically important, as some have beneficial relationships with plants, algae, and some animals; others have parasitic relationships with these organisms, too. The Ascomycota have an important economic role. Some species, mainly yeasts, are used to produce foods, while others cause food spoilage and can produce toxins. Fungal infections are of medical importance, as is the production of antibiotics from sac fungi.

Reproduction

Asexual reproduction is most common among these fungi. Yeasts usually reproduce by asexual budding, in which a small cell forms and pinches off as it grows to full size (Fig. 22.5*a*). Molds tend

Figure 22.6 Sexual reproduction in sac fungi. The sac fungi reproduce sexually by producing asci, in fruiting bodies called ascocarps. **a.** In the ascocarp of cup fungi, dikaryotic hyphae terminate, forming the asci, where meiosis follows nuclear fusion and spore formation takes place. **b.** In morels, the asci are borne on the ridges of pits.

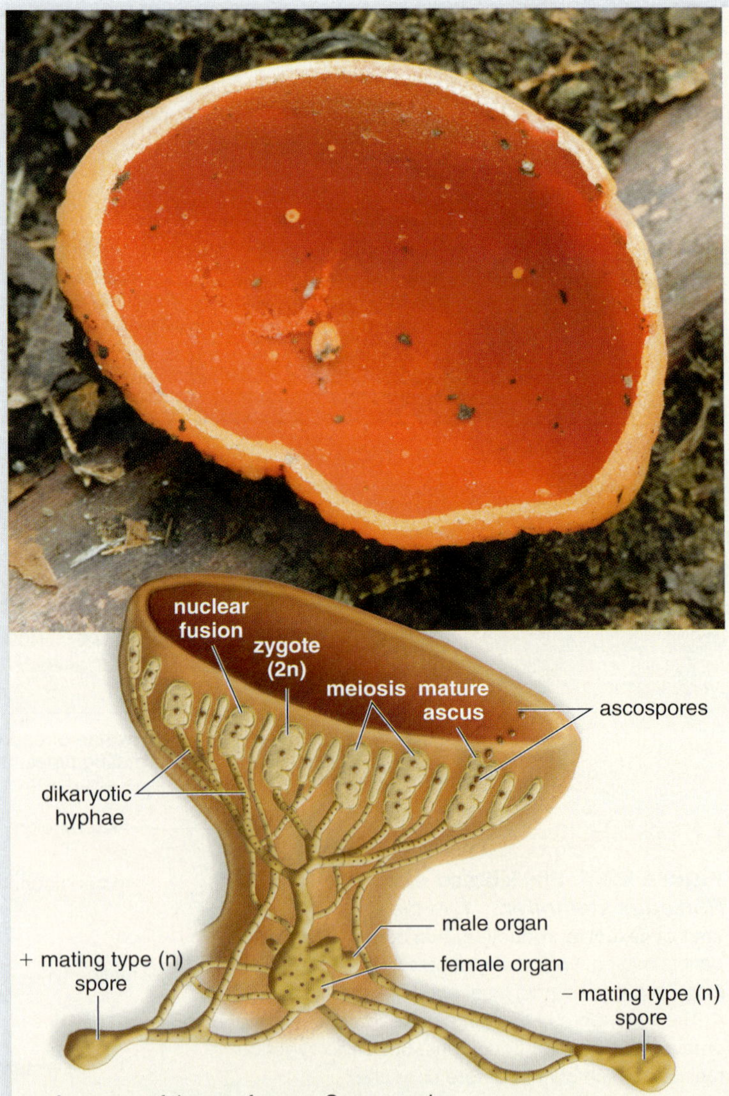

a. Ascocarp of the cup fungus, *Sarcoscypha*

to produce spores called conidia, or **conidiospores,** which vary in size and shape. Conidiospores can develop at the tips of aerial hyphae called *conidiophores* (Fig. 22.5*b*). When released, the conidiospores are dispersed by wind.

Sexual reproduction takes place in a **fruiting body.** A fruiting body affords an evolutionary advantage, as it is formed to produce and enhance the release of spores. The name ascomycota refers to the **ascus** (pl., asci) (Gk. "bag, sac"), a fingerlike sac that develops during sexual reproduction. In some instances, the asci are surrounded and protected by sterile hyphae within a fruiting body, called an ascocarp. *Ascocarps* can have different shapes; they can be cup-shaped or stalked with saclike pits (Fig. 22.6*a, b*). Within the ascus, the hyphae are dikaryotic, and two nuclei fuse to create diploid cells that in turn undergo meiosis to become new haploid cells called *ascospores*. Ascospores are released and dispersed by the wind after bursting from a swollen ascus.

Symbiotic Relationships

The Ascomycota have many mutualistic relationships with plants and animals. For example, leaf cutter ants and some species of termites keep fungal "gardens" within their colonies (Fig. 22.7*a*). The insects feed and protect the fungus, and in turn the fungus provides food for developing larvae. Section 22.3 explores another beneficial partnership of sac fungi in lichens and mycorrhizae.

Fungal parasites of plants exist throughout the Fungi. Sac fungi in particular are responsible for wiping out elm trees in the United States by inducing Dutch elm disease. A species called *Geomyces destructans* has decimated bat populations all along the east coast of the United States from Georgia to Canada. The disease is called white nose syndrome because of the characteristic fungi growing on infected bats' muzzles. This infection disturbs hibernating bats by waking them up during the winter months, causing them to look for food and starve to death (Fig. 22.7*b*).

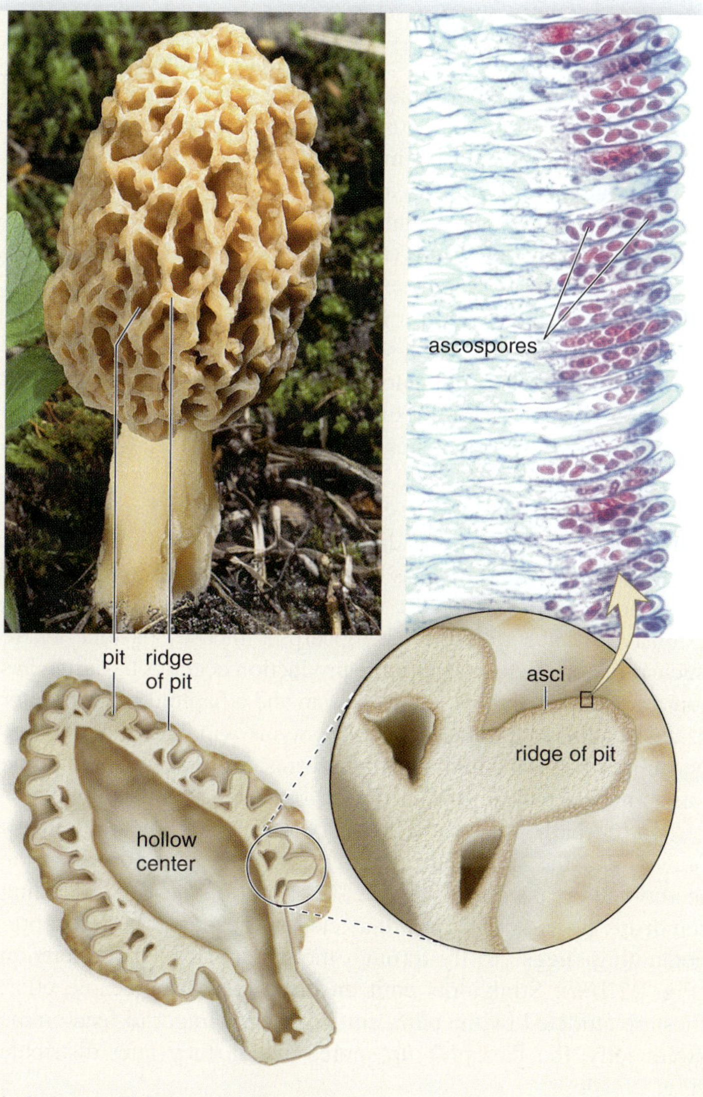

b. Ascocarp of the morel, *Morchella*

Figure 22.7 Symbiotic relationships with sac fungi.
a. Leaf-cutter ants farm sac fungi and feed it to developing larvae in a mutualistic relationship. **b.** White nose syndrome is cause by a parasitic fungus that disrupts hibernating bats and forms the characteristic mold on their muzzles.

Food Industry

The yeast *Saccharomyces cerevisiae* is widely used in the baking and brewing industries. Through the process of fermentation (see section 8.3), yeast can produce ethanol and carbon dioxide gas. Bread rises because of the gas given off, and drinks such as beer, wine, and liquor obtain their alcohol by fermenting yeast. The genus *Aspergillus,* a mold, is used to produce soy sauce and the Japanese food miso by fermentation of soybeans. In addition, the food industry uses *Aspergillus* to produce citric acid and gallic acid. These serve as additives during the manufacture of a wide variety of products, including foods, inks, medicines, dyes, plastics, toothpaste, soap, and chewing gum. The cheese industry uses the mold *Penicillium* to make blue cheese, Roquefort, Camembert, Brie, and Stilton cheeses.

Molds can also be detrimental to the food industry. Molds accelerate spoilage and can release toxins on food. Peanuts and other grains are susceptible to colonization by certain species of *Aspergillus* that produce aflatoxin, which can damage an animal's liver. The U.S. FDA has established guidelines on the amount of aflatoxin allowed in food consumed by humans and livestock.

Importance for Humans

Yeasts can be harmful to humans. *Candida albicans* is a yeast that is normally found on the body. When the balance of *Candida* and other microorganisms is disturbed, the yeast proliferates, causing vaginal infections or oral thrush in newborns (Fig. 22.8*a*). Moldlike fungi cause infections of the skin called tineas. Athlete's foot, jock itch, and ringworm are all tineas characterized by itching, peeling skin, or a raised inflammation (Fig. 22.8*b*).

Patients with AIDS or other immunocompromised conditions are particularly susceptible to fungal infections. The strong similarities between fungal and human cells make it difficult to design fungal medications that do not harm patients. Researchers target and exploit any biochemical difference they can discover, such as focusing on the cell wall or fungal-specific steroid production. Fungicides are developed and applied to crops, used on skin to battle tineas, or taken internally to fight systemic fungal infections.

One positive medical benefit is that many sac fungi have antibiotic and antimicrobial properties. Certain members of the genus *Penicillium* can be used to produce the antibiotic penicillin, which is a treatment for bacterial infections; another produces cyclosporine, which controls the immune system to prevent organ rejection after a transplant.

Many geneticists use Ascomycetes as model organisms. With genomic information available, *Aspergillus nidulans* serves as a safe model for more dangerous related species; *Neurospora crassa* was used to develop the "one gene, one enzyme hypothesis" described in Chapter 12, and *Saccharomyces* is a single-celled eukaryote with genes similar to those of other eukaryotic organisms, and they can be easily manipulated in a lab.

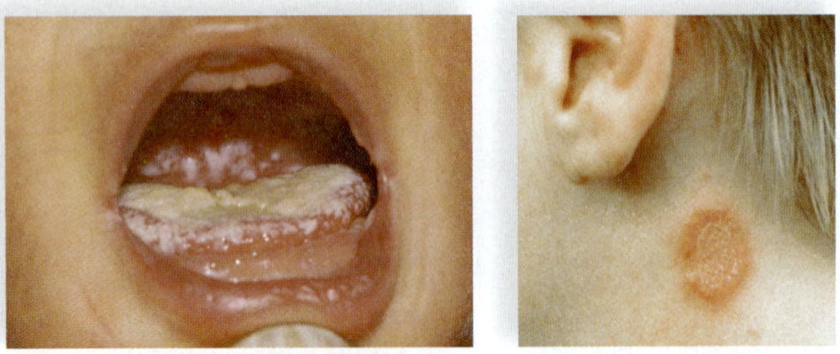

a. b.

Figure 22.8 Fungal infections. Oral thrush **(a)** and the ringworm tinea **(b)** are fungal infections.

Basidiomycota Produce a Fruiting Body Called a Basidium

The Basidiomycotan, or **club fungi,** consist of over 31,000 species. Mushrooms, toadstools, puffballs, shelf fungi, jelly fungi, bird's-nest fungi, and stinkhorns are basidiomycetes. In addition, fungi that cause plant diseases, such as the smuts and rusts, are placed in this phylum. Several mushrooms, such as the portabella and shiitake mushrooms, are savored as foods by humans. Approximately 75 species of basidiomycetes are considered poisonous. The poisonous "death cap" mushroom is discussed in the Biological Systems feature, "Deadly Fungi," on page 404.

Biology of Club Fungi

The body of a basidiomycete is a mycelium composed of septate hyphae. Most members of this phylum are saprotrophs, feeding on dead and decaying organic matter, although several parasitic species obtain nutrition from living hosts.

Although club fungi occasionally produce conidia asexually, they usually reproduce sexually. Their formal name, Basidiomycota, refers to the **basidium** (L. *basidi,* "small pedestal"), a club-shaped structure in which spores called basidiospores develop. Basidia are located within a fruiting body called a *basidiocarp,* which we recognize as a mushroom (Fig. 22.9). Prior to the formation of a basidiocarp, haploid hyphae of opposite mating types meet and fuse, producing a dikaryotic mycelium. The dikaryotic mycelium continues its existence year after year, even for hundreds of years on occasion.

Mushrooms are composed of tightly packed hyphae whose walled-off ends become basidia. In gilled mushrooms, the basidia are located on radiating folds called gills. In shelf fungi and pore mushrooms (Fig. 22.10*a, b*), the basidia terminate in tubes. The extensive surface area of a basidiocarp is lined by basidia, where nuclear fusion, meiosis, and spore production occur. A basidium has four projections, into which cytoplasm and a haploid nucleus enter as the basidiospore forms. Basidiospores are windblown; when they germinate, a new haploid mycelium forms. It is estimated that some large mushrooms can produce up to 40 million spores per hour.

In puffballs, spores are produced inside parchmentlike membranes, and the spores are released through a pore or when the membrane breaks down (Fig. 22.10*c*). In bird's-nest fungi, falling raindrops provide the force that causes the nest's basidiospore-containing "eggs" to fly through the air and land on vegetation (Fig. 22.10*d*). Stinkhorns emit an incredibly disagreeable odor; flies are attracted by the odor, and when they linger to feed on the sweet jelly, the flies pick up spores, which they later distribute (Fig. 22.10*e*).

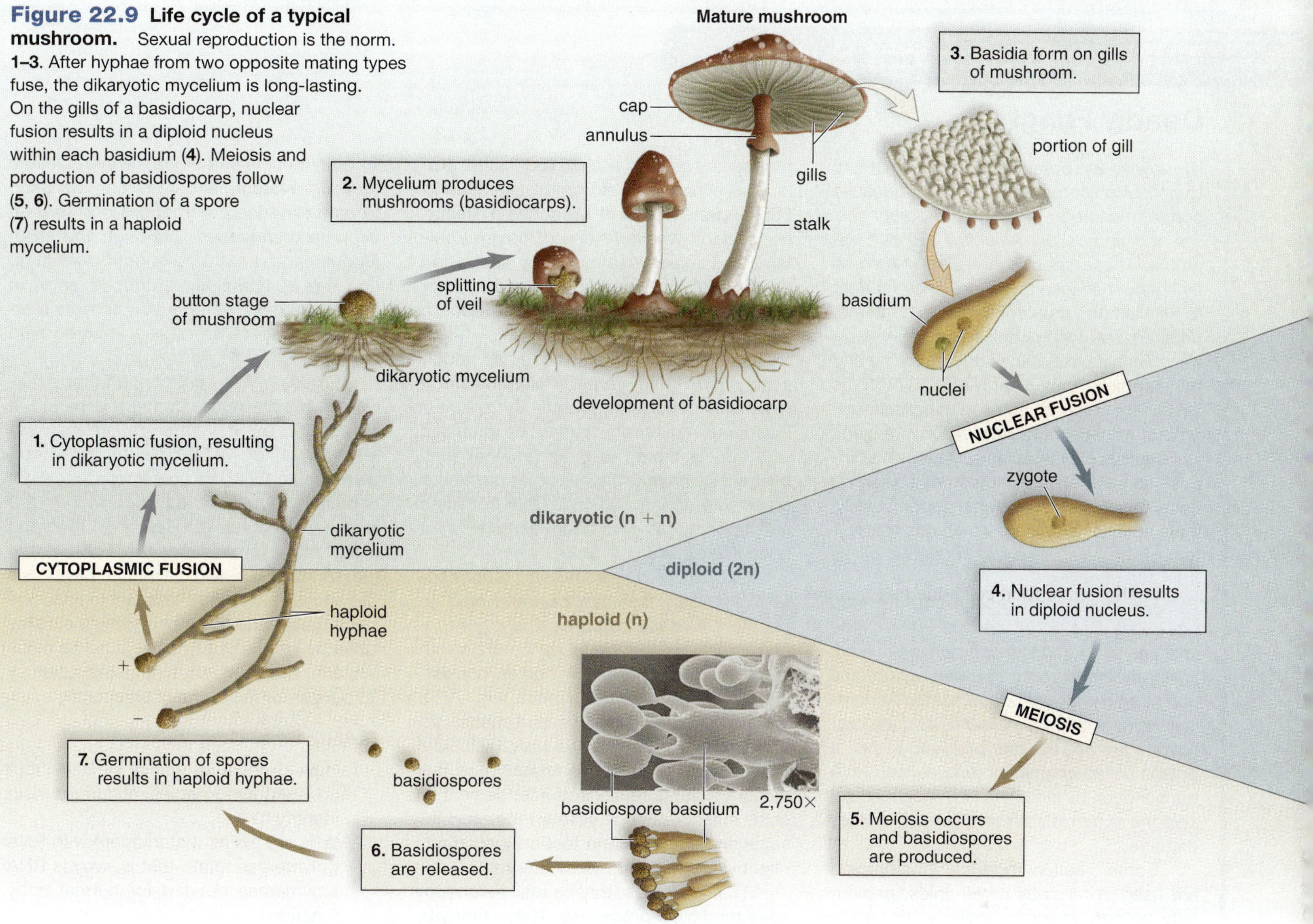

Figure 22.9 Life cycle of a typical mushroom. Sexual reproduction is the norm. **1–3**. After hyphae from two opposite mating types fuse, the dikaryotic mycelium is long-lasting. On the gills of a basidiocarp, nuclear fusion results in a diploid nucleus within each basidium (**4**). Meiosis and production of basidiospores follow (**5, 6**). Germination of a spore (**7**) results in a haploid mycelium.

a. Shelf fungus

b. Pore mushroom, *Boletus*

c. Puffball

d. Bird's nest fungi

e. Stinkhorn

Figure 22.10 Club fungi. **a.** A shelf fungus. **b.** Fruiting bodies of *Boletus*. This mushroom is not gilled; instead, it has basidia-lined tubes that open on the undersurface of the cap. **c.** In puffballs, the spores develop inside an enclosed fruiting body. **d.** In bird's-nest fungi, spores are contained within white packets that are ejected when hit with a raindrop. **e.** Flies are attracted to the sticky spore mass produced by stinkhorns.

THEME **Biological Systems**

Deadly Fungi

It is unwise and potentially fatal for amateurs to collect mushrooms in the wild, because certain mushroom species are poisonous. The red and yellow *Amanitas* are one example. These species are also known as fly agaric, because they were once thought to kill flies (the mushrooms were gathered, crushed, and then sprinkled into milk to attract flies). Its toxins include muscarine and muscaridine, which produce symptoms similar to those of acute alcoholic intoxication. In 1–6 hours, the victim staggers, loses consciousness, and becomes delirious, sometimes suffering from hallucinations, manic conditions, and stupor. Luckily, it also causes vomiting, which rids the system of the poison, so death occurs in less than 1% of cases.

The death cap mushroom, *Amanita phalloides* (Fig. 22A), causes 90% of the fatalities attributed to mushroom poisoning. When this mushroom is eaten, symptoms don't begin until 10–12 hours later. Abdominal pain, vomiting, delirium, and hallucinations are not the real problem; rather, a poison called amanitin interferes with RNA transcription by inhibiting RNA polymerase, and the victim dies from liver and kidney damage.

Some hallucinogenic mushrooms are used in religious ceremonies, particularly among Mexican Indians. *Psilocybe mexicana* contains a chemical called psilocybin, which is a structural analogue of LSD and mescaline. It produces a dreamlike state in which visions of colorful patterns and objects seem to fill up space and dance past in endless succession. Other senses are also sharpened to produce a feeling of "intense reality."

The only reliable way to tell a nonpoisonous mushroom from a poisonous one is to be able to correctly identify the species. Poisonous mushrooms cannot be identified with simple tests, such as whether they peel easily, have a bad odor, or blacken a silver coin during cooking. Only consume mushrooms that have been identified by a bona fide expert!

Like club fungi, some sac fungi contain chemicals that can be dangerous to people. *Claviceps purpurea,* the ergot fungus, infects rye and replaces the grain with ergot—hard, purple-black bodies consisting of tightly cemented hyphae (Fig. 22B). When ground with the rye and made into bread, the fungus releases toxic alkaloids, which cause the disease ergotism. In humans, vomiting, feelings of intense heat or cold, muscle pain, a yellow face, and lesions on the hands and feet are accompanied by hysteria and hallucinations.

The alkaloids that cause ergotism have medicinal properties. They stimulate smooth muscle and block the sympathetic nervous system, and they can be used to cause uterine contractions and to treat migraine headaches. Although the ergot fungus can be cultured in petri dishes, no one has successfully produced ergot in the laboratory. The only way to obtain ergot is to collect it from an infected field of rye.

Ergotism was common in Europe during the Middle Ages. During this period, it was known as St. Anthony's Fire and was responsible for 40,000 deaths in an epidemic in AD 994. We now know that ergot contains lysergic acid, from which LSD is easily synthesized. Based on recorded symptoms, some historians believe that those individuals who claimed to have been "bewitched" in Salem, Massachusetts, during the seventeenth century were actually suffering from ergotism. The ensuing mass hysteria, however, led to the execution of 20 people for the crime of witchcraft.

Questions to Consider

1. How might a sequence of DNA from an unknown species of fungus help identify it?
2. Why are toxins that interfere with RNA polymerase fatal—that is, why is RNA polymerase needed for normal body function?

Figure 22A A poisonous mushroom species, *Amanita phalloides.*

Figure 22B Ergot infection of rye, caused by *Claviceps purpurea.*

Smuts and Rusts

Smuts and rusts are club fungi that parasitize cereal crops, such as corn, wheat, oats, and rye. They are of great economic importance because of the crop losses they cause every year. Smuts and rusts don't form basidiocarps, and their spores are small and numerous, resembling soot. Some smuts enter seeds and exist inside the plant, becoming visible only near maturity. Other smuts externally infect plants. In corn smut, the mycelium grows between the corn kernels and secretes substances that cause the development of tumors on the ears of corn (Fig. 22.11*a*).

The life cycle of rusts requires alternate hosts, and one way to keep them in check is to eradicate the alternate host. Wheat rust (Fig. 22.11*b*) is also controlled by producing new and resistant strains of wheat. The process is continuous, because rust can mutate to cause infection once again.

a. Corn smut, *Ustilago*

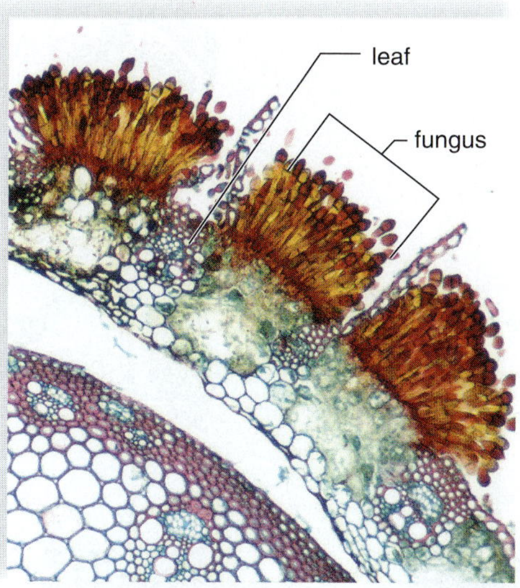

b. Wheat rust, *Puccinia*.

Figure 22.11 Smuts and rusts. a. Corn smut. **b.** Micrograph of wheat rust.

Check Your Progress 22.2

1. Compare and contrast Microsporidia and chytrids.
2. Discuss the evolutionary advantage of a fruiting body and how spore dispersal is accomplished in different phyla of fungi.
3. Identify three fungal infections of plants or animals and the fungi responsible for the infections.

22.3 Symbiotic Relationships of Fungi

Learning Outcomes

Upon completion of this section, you should be able to

1. Summarize the association that occurs between a fungus and its photosynthesizing partner in lichens.
2. Define *mycorrhizae.*
3. Explain the mutualistic relationship between mycorrhizae and plants.

Several instances in which fungi are parasites of plants and animals were mentioned earlier in this chapter. Two other symbiotic associations are of interest: lichen associations and mycorrhizae.

Lichens

Lichens are an association between a fungus, usually a sac fungus, and a cyanobacterium or a green alga. As one example, a crustose lichen has a body consisting of three layers. The fungus forms a thin, tough upper layer and a loosely packed lower layer, which shield the photosynthetic cells in the middle layer (Fig. 22.12*a*). Specialized fungal hyphae, which penetrate or envelop the photosynthetic cells, transfer nutrients directly to the rest of the fungus. Lichens can reproduce asexually by releasing fragments that contain hyphae and an algal cell. In fruticose lichens, the sac fungus reproduces sexually (Fig. 22.12*b*).

In the past, lichens were assumed to be mutualistic relationships in which the fungus received nutrients from the algal cells and the algal cells were protected from desiccation by the fungus. Actually, lichens may involve a controlled form of parasitism of the algal cells by the fungus, with the algae not benefiting at all from the association. This idea is supported by experiments in which the fungal and algal components are removed and grown separately. The algae grow faster when they are alone than when they are part of a lichen. In contrast, it is difficult to cultivate the fungus, which does not naturally grow alone. The different lichen species are identified according to their fungal partners.

Three types of lichens are recognized by shape. Compact crustose lichens are often seen on bare rocks or tree bark; fruticose lichens are shrublike; and foliose lichens are leaflike (Fig. 22.12*c*). Lichens are efficient at acquiring nutrients and moisture, and therefore they can survive in areas of low moisture and low temperature, as well as in areas with poor or no soil. They produce and improve

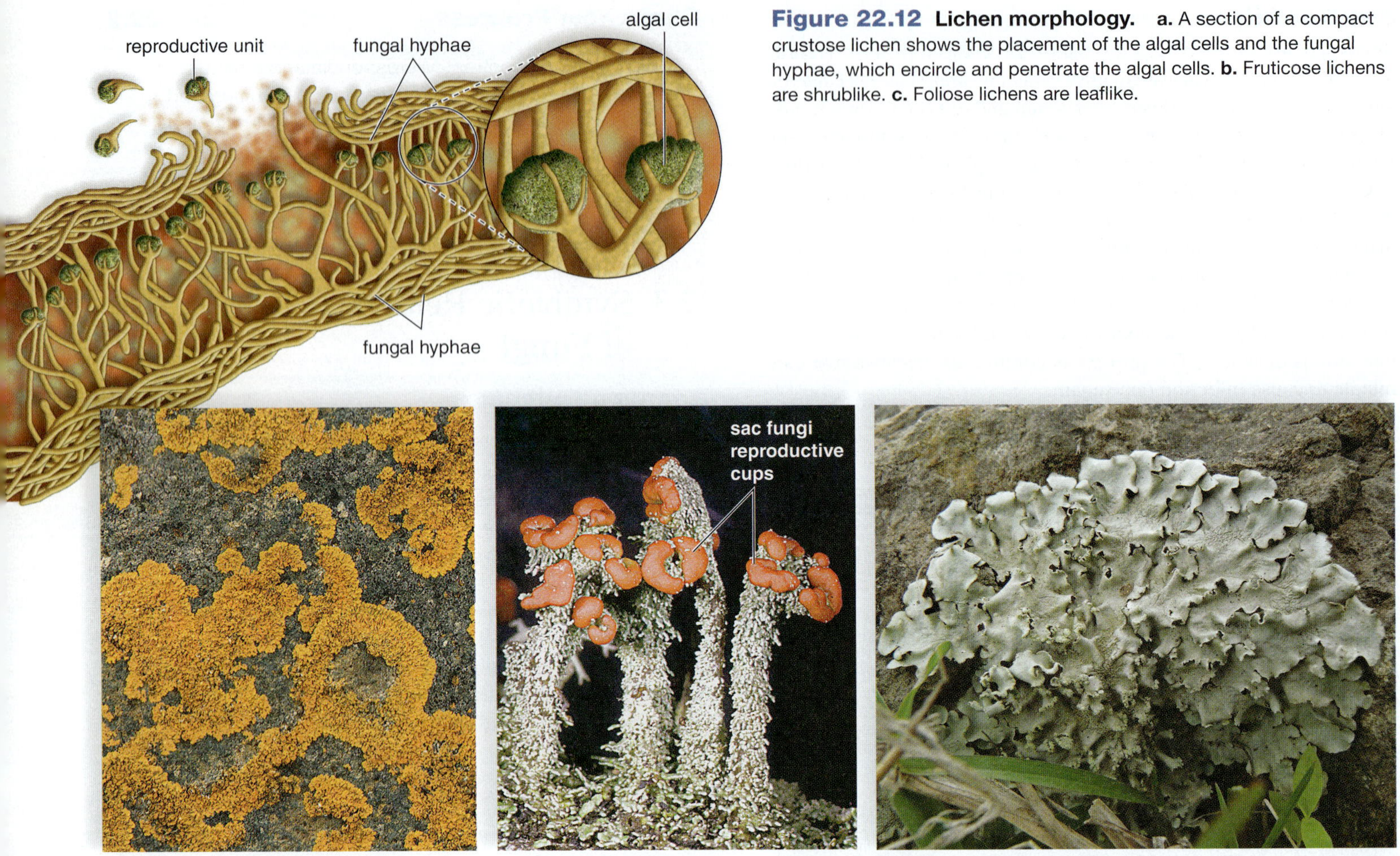

a. Crustose lichen, *Xanthoria*

b. Fruticose lichen, *Lobaria*

c. Foliose lichen, *Parmelia*

Figure 22.12 Lichen morphology. **a.** A section of a compact crustose lichen shows the placement of the algal cells and the fungal hyphae, which encircle and penetrate the algal cells. **b.** Fruticose lichens are shrublike. **c.** Foliose lichens are leaflike.

the soil, thus making it suitable for plants to invade the area. Unfortunately, lichens also take up pollutants, and they cannot survive where the air is polluted. Therefore, lichens can serve as air pollution sensors.

Mycorrhizae

Mycorrhizae (Gk. *mykes,* "fungus"; *rhizion,* "little root") are mutualistic relationships between soil fungi and the roots of most plants. Plants whose roots are invaded by mycorrhizae grow more successfully in poor soils—particularly soils deficient in phosphates—than do plants without mycorrhizae (Fig. 22.13).

The fungal partner, either a glomerulomycete or a sac fungus, may enter the cortex of roots but does not enter the cytoplasm of plant cells. Ectomycorrhizae form a mantle that is exterior to the root, and they grow between cell walls. Endomycorrhizae, such as the AM fungi mentioned earlier, penetrate only the cell walls. The presence of the fungus gives the plant a greater absorptive surface for the intake of minerals. The fungus also benefits from the association by receiving carbohydrates from the plant. As mentioned, even the earliest fossil plants have mycorrhizae associated with them. It would appear, then, that mycorrhizae helped plants adapt to and flourish on land.

The truffle, an underground fungus that is a gourmet delight and that has an ascocarp somewhat prunelike in appearance, is a

Figure 22.13 Plant growth experiment. A bean plant without mycorrhizae (*left*) grows poorly compared to another plant infected with mycorrhizae (right).

mycorrhizal sac fungus living in association with oak and beech tree roots. In the past, the French used pigs ("truffle-hounds") to sniff out and dig up truffles, but now they have succeeded in cultivating truffles by inoculating the roots of seedlings with the proper mycelium. In addition to providing water and minerals, some black truffle species, such as *Tuber melanosporum,* engage in chemical warfare against other plants, fungi, and even bacteria. This truffle creates a burnt "dead zone" around its partner tree, so that the tree has no competition (Fig. 22.14).

Figure 22.14 A black truffle. The truffle *Tuber melanosporum* (inset) eliminates competition for its mycorrhizal partner creating a "dead zone" around the tree.

Check Your Progress 22.3

1. Identify the components of a lichen.
2. Explain how a lichen reproduces.
3. Summarize the symbiotic relationship between mycorrhizae and plants.

Connecting the Concepts with the Themes

Evolution

- Fungi were the first multicellular eukaryotes to evolve, and molecular data suggest a close relationship between animals and fungi.
- The earliest known fossil fungi date back to 460 MYA, but fungi have probably existed longer than this.
- As fungi became more adapted to live on land, spore dispersal mechanisms improved, such as the development of fruiting bodies.

Nature of Science

- Fungi produce substances that kill bacteria cells; many antibiotics are derived from fungi, including penicillin.
- Fungi serve as model organisms to study genetics and molecular biology; model organisms are used to develop new antifungal drugs.
- Fungi consume hydrocarbons and convert them into fungal sugars; fungi break down oil and may serve as an important bioremediation tool in the future.

Biological Systems

- Fungi exhibit symbiotic relationships with plants and animals that can be mutualistic or parasitic.
- Fungi help decompose organic matter into topsoil that supports plant life. Lichens can live on bare rock, breaking it down and making minerals available to other organisms.
- Mycorrhizae facilitate a plant's uptake of nutrients through association with their roots.

MEDIA STUDY TOOLS http://connect.mheducation.com

SMARTBOOK® Study smart. SmartBook helps you maximize your study time by identifying what you know and don't know. The recharge feature in SmartBook helps you prepare for the test by identifying the material that you are most likely to forget.

connect | BIOLOGY Find even more resources to learn the chapter concepts including animations, tutorial videos, and interactive practice questions.

Tutorial	Video
22.2 Zygospore Life Cycle	**22.1** Decomposers

SUMMARIZE

22.1 Evolution and Characteristics of Fungi

Fungi are multicellular, **saprotrophic** eukaryotes. After external digestion, they absorb the resulting nutrient molecules. Saprotrophic fungi aid the cycling of organic molecules in ecosystems by decomposing dead remains. Some fungi are parasitic, especially on plants, and others are mutualistic with plant roots and algae.

The body of a fungus is composed of thin filaments called **hyphae,** which collectively are termed a **mycelium.** The cell wall contains **chitin,** and the energy reserve is glycogen. With the notable exception of the chytrids, which have flagellated spores and gametes, fungi do not have flagella at any stage in their life cycle. **Aseptate** hyphae have no cross-walls; **septate** hyphae have cross-walls called **septa,** but pores allow the cytoplasm and even organelles to pass through.

Fungi produce **spores** during both asexual and sexual reproduction. Yeasts reproduce asexually by **budding.** During sexual reproduction in some multicellular fungi, hyphae tips fuse, so that **dikaryotic** (n + n) hyphae usually result, depending on the type of fungus. Following nuclear fusion, zygotic meiosis occurs during the production of the sexual spores.

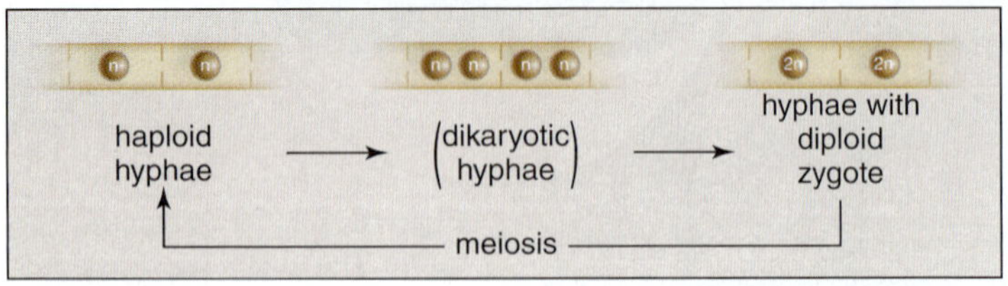

22.2 Diversity of Fungi

Four fungal phyla are traditionally described: the Chytridiomycota (chytrids), Zygomycota (zygospore fungi), Ascomycota (sac fungi), and Basidiomycota (club fungi). Newer data include the Glomeromycota (AM fungi) and Microsporidia.

Microsporidia are single-celled intracelluar parasites found in insects and vertebrates, including humans.

The **chytrids** are predominately aquatic, single-celled fungi with motile **zoospores** and gametes. Some chytrids have an alternation-of-generations life cycle similar to that of plants and certain algae. There are also some multicelluar, hyphae-forming chytrids.

The **zygospore fungi** are multicellular fungi with aseptate hyphae. During sexual reproduction they have a dormant stage consisting of a thick-walled **zygospore.** Meiosis occurs in the diploid zygospore. The zygospore germinates a **sporangium** with haploid sporangiospores. The **AM fungi** were once classified with the zygospore fungi but are now viewed as a distinct group. AM fungi exist in mutualistic associations with the roots of most land plants.

The **sac fungi** are septate, and during sexual reproduction a saclike **ascus** produces spores. Asci are sometimes located in **fruiting bodies** called ascocarps. Asexual reproduction, which is dependent on the production of **conidiospores,** is more common. Sac fungi include various **yeasts** and **molds,** some of which cause disease in plants and animals, including humans.

The **club fungi** are septate, and during sexual reproduction, club-shaped structures called **basidia** produce spores. Basidia are located in fruiting bodies called basidiocarps. Club fungi have a prolonged dikaryotic stage. A dikaryotic mycelium periodically produces fruiting bodies. Mushrooms and puffballs are examples of club fungi.

22.3 Symbiotic Relationships of Fungi

Lichens are an association between a fungus, usually a sac fungus, and a cyanobacterium or a green alga. Traditionally, this association was considered mutualistic, but experimentation suggests a controlled parasitism by the fungus on the alga. Lichens can live in extreme environments and on bare rocks; they allow other organisms to colonize these harsh environments, and together they eventually form soil.

The term **mycorrhizae** refers to an association between a fungus and the roots of a plant. The fungus helps the plant absorb minerals, and the plant supplies the fungus with carbohydrates.

ASSESS

Choose the best answer for each question.

22.1 Evolution and Characteristics of Fungi

1. Which of the following represents an organism that uses decomposition as its mode of nutrition?
   a. parasite
   b. saprotroph
   c. autotroph
   d. All of these are correct.
2. Hyphae are generally characterized by
   a. strong, impermeable walls.
   b. rapid growth.
   c. large surface area.
   d. pigmented cells.
   e. Both b and c are correct.
3. A fungal spore
   a. contains an embryonic organism.
   b. germinates directly into an organism.
   c. is always windblown, because none are motile.
   d. is most often diploid.
   e. Both b and c are correct.

22.2 Diversity of Fungi

4. Which of the following is not a characteristic of Microsporida?
   a. parasitic
   b. lives in host cells
   c. multicellular
   d. can infect invertebrates like insects
   e. can infect vertebrates like humans
5. Which of the following is not a characteristic of chytrids?
   a. They have flagellated spores.
   b. They have flagellated gametes.
   c. They can live on very dry land.
   d. They can be single cells.
   e. They can be animal parasites.
6. Label this diagram of black bread mold structure and asexual reproduction.

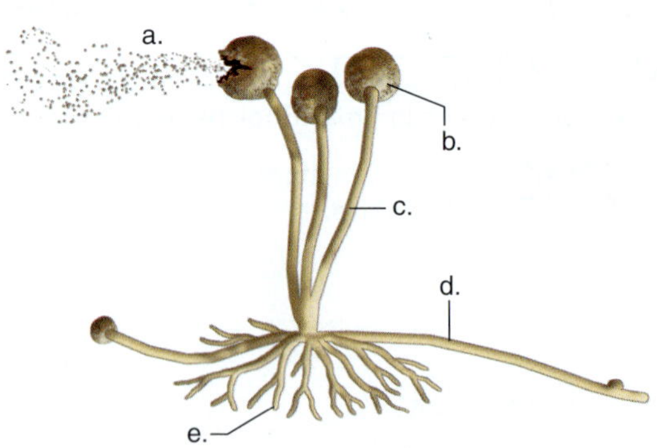

7. Which part of the life cycle of black bread mold is the resting stage that can survive unfavorable conditions?
   a. mycelium
   b. zygospore
   c. gametangia
   d. zygote
   e. rhizoid
8. Tineas are
   a. plant parasites.
   b. skin infections.
   c. deep systemic infections.
   d. frog and bat parasites.
   e. None of these are correct.
9. In an ascocarp,
   a. there are only fertile hyphae.
   b. ascospores are formed.
   c. a sperm fertilizes an egg.
   d. hyphae do not have chitinous walls.
   e. conidiospores form.
10. In which fungus is the dikaryotic stage longer-lasting?
   a. zygospore fungus
   b. sac fungus
   c. club fungus
   d. chytrids

22.3 Symbiotic Relationships of Fungi

11. Compact lichens are called _____; leaflike lichens are called _____; and shrublike lichens are _____.
   a. crustose; foliose; fruticose.
   b. foliose; crustose; fruticose.
   c. fruitcose; foliose; crustose.
   d. foliose; fruitcose; crustose.
12. Mycorrhizae
   a. are a type of lichen.
   b. are mutualistic relationships.
   c. help plants gather solar energy.
   d. help plants gather inorganic nutrients.
   e. Both b and d are correct.

ENGAGE

THINKING SCIENTIFICALLY

1. A fine line seems to exist between symbiosis and parasitism when you examine the relationships between fungi and plants. What hypotheses could explain how different selective pressures may have caused particular fungal species to adopt one or the other relationship? Under what circumstances might a mutualistic relationship evolve between fungi and plants? Under what circumstances might a parasitic relationship evolve?
2. Many fungal infections of humans are considered to be opportunistic, meaning that the fungi that are normally in our environment suddenly become infectious on or in the body. From the fungal "point of view," what unique challenges would be encountered when trying to survive on human skin? What about inside human lungs?
3. The importance of fungi in the evolution of terrestrial life is typically understated. Evaluate the importance of fungi in the colonization of land.
4. You are looking through a seed catalog and discover you have a choice between pea seeds that are treated with a fungicide and those that are not. Discuss one reason you might choose the treated seed and one reason you might prefer the untreated seed.

UNIT 5

Plant Evolution and Biology

Life would be impossible without plants. Plants provide nearly all living organisms with the food and oxygen needed for survival. They can be a source of fuel to heat our homes, as well as providing fibers to make our clothes. The frames of our houses and the furniture inside them are often made of wood and other plant products. We even use plants to make the paper for our magazines and textbooks.

Because plants and our everyday lives are intertwined, it's a good idea to know the principles of plant biology—plant structure and function. Plants have the same characteristics of life as animals do, and most reproduce sexually, as we do. Life began in the sea, and plants invaded the terrestrial environment before animals, as you would expect, because animals are so dependent on plants. Throughout the course of primate and human evolution, we have been dependent on plants for the majority of our nutrients. Even today, people rely on plants to provide the bulk of their caloric intake. These chapters introduce you to the biology of plants, those incredible organisms that keep the biosphere functioning as it should.

Unit Outline

Chapter 23 Plant Evolution and Diversity 411

Chapter 24 Flowering Plants: Structure and Organization 435

Chapter 25 Flowering Plants: Nutrition and Transport 456

Chapter 26 Flowering Plants: Control of Growth Responses 476

Chapter 27 Flowering Plants: Reproduction 495

Unit Learning Outcomes

The learning outcomes for this unit focus on three major themes in the life sciences.

Evolution	Identify the key structural innovations that occurred during the evolution of various plant groups.
Nature of Science	Explain how humans have manipulated plants to better serve our needs.
Biological Systems	Assess how the structure and response of plants to the environment enable them to survive.

Produce aisle in a grocery store.

23

Plant Evolution and Diversity

As you wander down the produce aisle of the grocery store, you can't help but notice the large diversity of fruits and vegetables. What is even more amazing is the fact that despite the diversity of shapes, colors, textures, and tastes, all fruits and vegetables have evolved from a common ancestor. In many respects, plants owe their characteristics to their evolutionary past. Plants, like other organisms, have evolved to adapt to a wide variety of environments over millions of years. But plants have also been influenced by human activity. For thousands of years, humans have been directing the evolution of plants as a result of artificial selection and genetic engineering.

This chapter traces the evolutionary history of plants from their green algal ancestor to the various groups we depend on today for our survival. During the course of this evolutionary journey, some groups have proved to be more successful than others due to various structural adaptations. In this chapter, we discuss the similarities and differences among the major plant groups and the evolutionary adaptations that have contributed to their success. We pay special attention to the reasons behind the tremendous success of the flowering plants.

As you read through this chapter, think about the following questions:

1. What environmental challenges did plants have to overcome in order to survive on land?
2. Why are angiosperms more widespread than all other groups of plants?
3. What characteristics are unique to each of the major groups of plants?

Chapter Outline

23.1 Ancestry and Features of Land Plants 412

23.2 Evolution of Bryophytes: Colonization of Land 415

23.3 Evolution of Lycophytes: Vascular Tissue 417

23.4 Evolution of Pteridophytes: Megaphylls 419

23.5 Evolution of Seed Plants: Full Adaptation to Land 423

Before You Begin

Before beginning this chapter, take a few moments to review the following discussions.

Figure 4.7 What cellular structures are unique to plants?

Figure 10.5 What is the end result of meiosis?

Section 16.2 What is the role of natural selection in the evolutionary process?

Following *the* Themes

CHAPTER 23 PLANT EVOLUTION AND DIVERSITY

UNIT 5 PLANT EVOLUTION AND BIOLOGY

Evolution	Key characteristics evolved in each group of plants that enabled them to be more successful than the earlier groups.
Nature of Science	Humans use plants in a variety of ways that have made us dependent upon them for our survival.
Biological Systems	Each group of plants possesses key structural adaptations that enable them to perform their functions.

23.1 Ancestry and Features of Land Plants

Learning Outcomes

Upon completion of this section, you should be able to

1. Compare and contrast algae with land plants.
2. List the traits that enabled plants to adapt to life on land.
3. Evaluate the differences in the alternation of generations of land plants.

Plants are multicellular, photosynthetic eukaryotes whose evolution is marked by adaptations to a land existence. Plant systematists use molecular and morphological information to classify plants. In this textbook, the plants (listed in Table 23.1) will be described as nonvascular (bryophytes), seedless vascular (lycophytes, ferns, and fern allies), and vascular seed plants (gymnosperms and angiosperms). Phylogenetic information suggests that the modern plant groups arose in a particular sequence (Fig. 23.1). Liverworts diverged first, mosses diverged next, and hornworts seem to be closely related to the vascular plants. The vascular plants are distinguished by tissues that conduct food and water and provide structural support. Lycophytes are then related to all the other modern vascular plants; ferns arose next, then the seed plants.

Plants are thought to have evolved from a freshwater green alga. In this section, we consider similarities between green algae and plants, as well as the different adaptations plants have to facilitate life on land.

Video Plants

The Ancestry of Plants

The evolutionary history of plants begins in the water. Most likely, land plants evolved from a form of freshwater green algae some 590 MYA (million years ago). Green algae are not plants, but like plants they (1) contain chlorophylls *a* and *b* and various accessory pigments, (2) store excess carbohydrates as starch, and (3) have cellulose in their cell walls. One major difference, however, between algae and terrestrial plants is that plants not only protect the zygote but also protect and nourish the resulting embryo—an important feature that separates land plants from green algae (Fig. 23.2).

The commonality with green algae, as well as a comparison of DNA and RNA base sequences, suggests that land plants are most closely related to a particular group of freshwater green algae known as **charophytes.** Molecular data show that charophytes and land plants are in the same clade and form a monophyletic group. Although *Spirogyra* (see Fig. 21.6) is a charophyte, the ancestors of land plants were more closely related to the charophytes *Chara* and *Coleochaete,* shown in Figure 23.3.

Chara are commonly known as stoneworts, because they are encrusted with calcium carbonate deposits. The body consists of a single file of very long cells anchored in mud by thin filaments. Whorls of branches occur at regions called *nodes,* located between the cells of the main axis. Male and female reproductive structures grow at the nodes. *Coleochaete* appears disklike, but the body is actually composed of long, branched filaments of cells. Most important to the evolution of plants, charophytes protect the zygote.

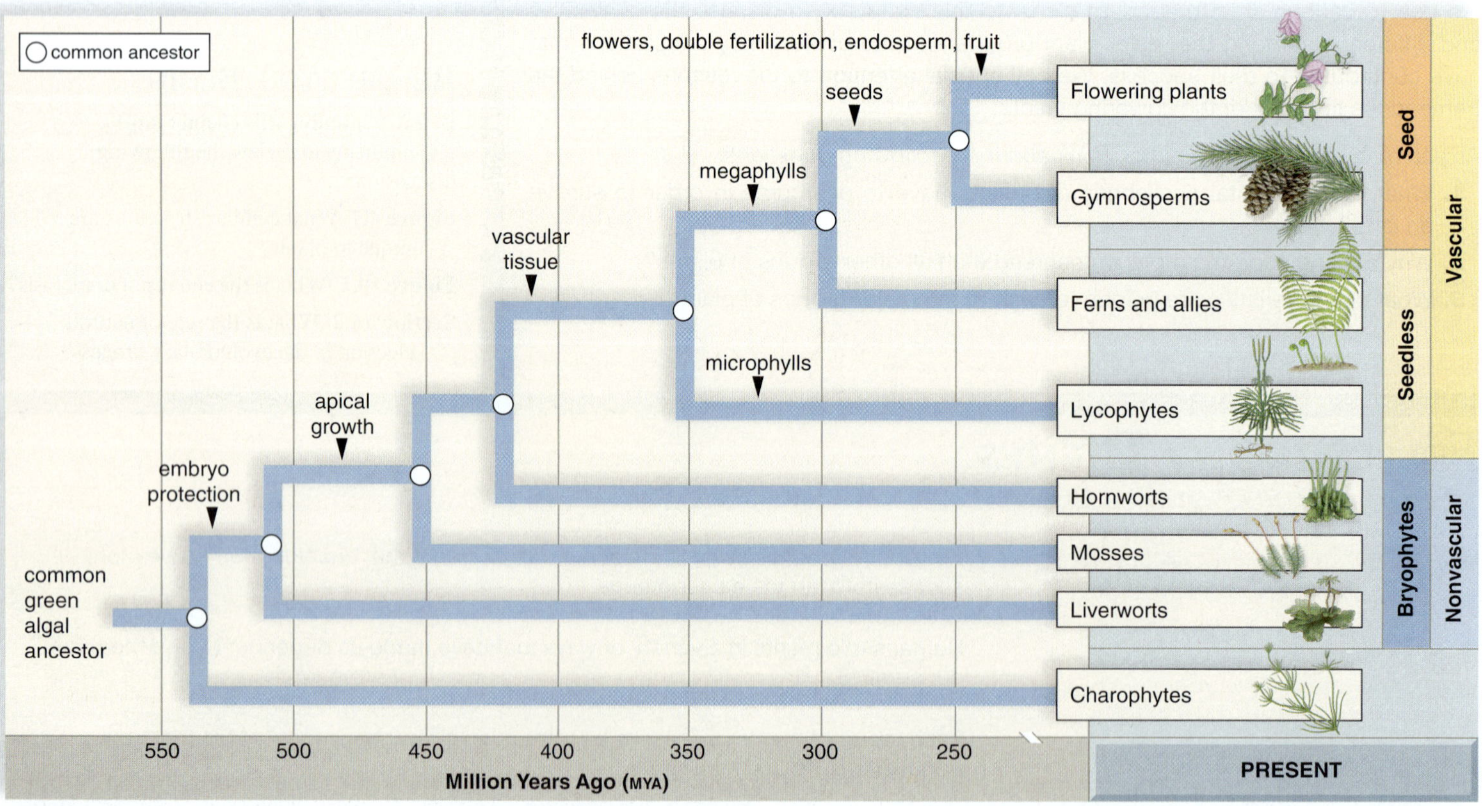

Figure 23.1 Evolutionary history of plants. The evolution of plants involves several significant innovations.

Table 23.1 Key Groups of Plants and Their Features

DOMAIN: Eukarya
KINGDOM: Plantae

CHARACTERISTICS
Multicellular, usually with specialized tissues; photosynthesizers that became adapted to living on land; most have alternation-of-generations life cycle.

Charophytes
Live in water; haploid life cycle; share certain traits with the land plants.

LAND PLANTS
Alternation-of-generations life cycle; protect a multicellular sporophyte embryo; gametangia produce gametes; apical tissue produces complex tissues; waxy cuticle prevents water loss.

Bryophytes (liverworts, hornworts, mosses)
Low-lying, nonvascular plants that prefer moist locations: Dominant gametophyte produces flagellated sperm; unbranched, dependent sporophyte produces windblown spores.

VASCULAR PLANTS (lycophytes, ferns and their allies, seed plants)
Dominant, branched sporophyte has vascular tissue: Lignified xylem transports water, and phloem transports organic nutrients; typically has roots, stems, and leaves; and gametophyte is eventually dependent on sporophyte.

Lycophytes (club mosses)
Leaves are microphylls with a single, unbranched vein; sporangia borne on sides of leaves produce windblown spores; independent and separate gametophyte produces flagellated sperm.

Ferns and Allies (pteridophytes)
Leaves are megaphylls with branched veins; dominant sporophyte produces windblown spores in sporangia borne on leaves; and independent and separate gametophyte produces flagellated sperm.

SEED PLANTS (gymnosperms and angiosperms)
Leaves are megaphylls; dominant sporophyte produces heterospores that become dependent male and female gametophytes. Male gametophyte is pollen grain and female gametophyte occurs within ovule, which becomes a seed.

Gymnosperms (cycads, ginkgoes, conifers, gnetophytes)
Usually large; cone-bearing; existing as trees in forests. Sporophyte bears pollen cones, which produce windblown pollen (male gametophyte), and seed cones, which produce seeds.

Angiosperms (flowering plants)
Diverse; live in all habitats. Sporophyte bears flowers, which produce pollen grains, and bear ovules within ovary. Following double fertilization, ovules become seeds that enclose a sporophyte embryo and endosperm (nutrient tissue). Fruit develops from ovary.

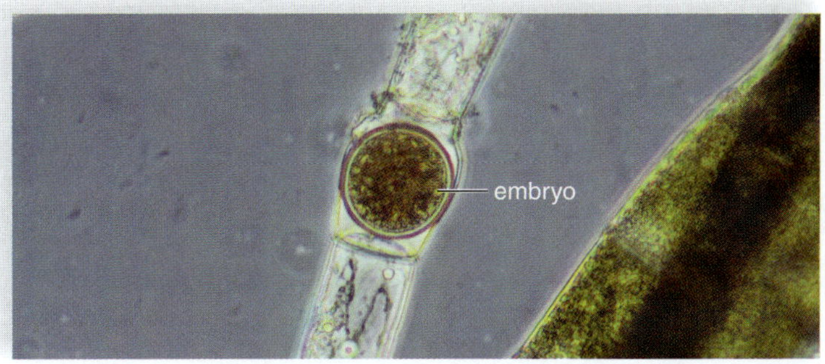

a. Embryo of the green alga *Oedogonium*. 100×

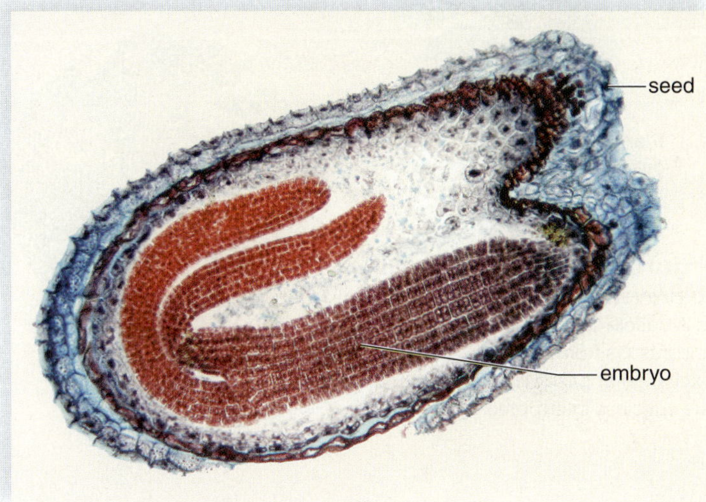

b. Embryo from the land plant *Capsella*. 60×

Figure 23.2 Embryo difference in green algae and land plants. **a.** The filamentous green alga Oedogonium with an unprotected embryo. **b.** A seed from the shepherd's purse plant, *Capsella,* showing the embryo protected and nourished.

Figure 23.3 Charophytes. The charophytes (represented here by *Chara* and *Coleochaete*) are the green algae most closely related to the land plants.

Adaptation to Land

Recall what it is like swimming in a pool. Like algae, you are weightless, surrounded by water and protected from the sun. However, all that changes when you leave the water. When they moved from water to land, plants needed mechanisms to deal with water loss, gravity, and sun exposure. The following features are evolutionary adaptions of most terrestrial plants.

1. Unlike algae, plants living on land have a limited amount of water available to them. As an adaptation to living on land, most plants are protected from desiccation (drying out) by

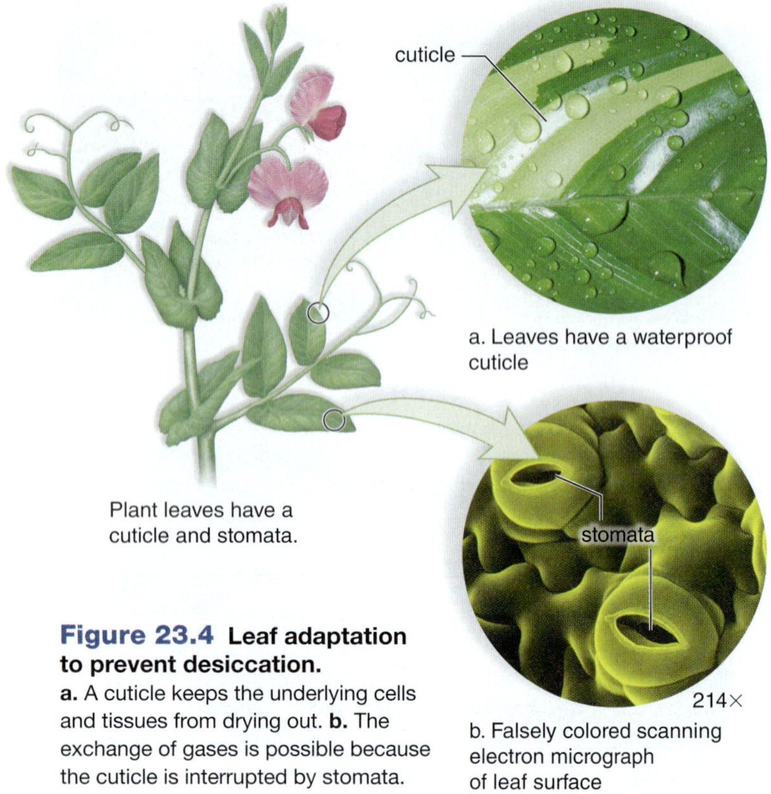

Figure 23.4 Leaf adaptation to prevent desiccation. **a.** A cuticle keeps the underlying cells and tissues from drying out. **b.** The exchange of gases is possible because the cuticle is interrupted by stomata.

a waxy covering called a **cuticle** (Fig. 23.4*a*). The cuticle covers all exposed surfaces and is relatively impermeable. The problem is that this limits gas exchange for photosynthesis and cellular respiration. The solutions are tiny openings found mainly on the underside of leaves called **stomata** (sing., stoma) (Fig. 23.4*b*). These openings, or pores, allow for gas exchange but can also allow water vapor to escape. Chapter 26 describes how stomata can be closed at times to limit water loss.

2. Moving water within plants is a challenge that increases with plant size. Picture the height of moss compared to the height of a redwood tree. The different members of the land plants can be distinguished based on the presence or absence of **tracheids,** specialized cells with proteins that resist gravity and facilitate the upward transport of water and minerals.
3. Compared to green algae, the bodies of land plants are mostly composed of three-dimensional tissues. A *tissue* is an association of many cells of the same type that can lead to specialized structures, such as organs. Tissues provide land plants with an increased ability to avoid water loss at their surfaces, because bodies composed of tissues have a lower surface-area-to-volume ratio than do branched filaments, as seen in *Chara* (see Fig. 23.3).
4. Plants living on land are exposed to higher intensities of UV rays than are aquatic algae. Exposure to ultraviolet light can increase the chance of mutations, but having a diploid genome can hide the effect of a single, deleterious allele. All terrestrial plants have both a haploid and a diploid generation, and the evolutionary shift toward a more dominant diploid generation allows for greater genetic variability in land plants.

Alternation of Generations

In the life cycle (see Fig. 10.7) of humans, and other animals, the diploid stage is multicellular and the haploid stage is composed of single-celled gametes. Plants also have a diploid and a haploid stage, but the significant difference is that both of these stages contain multicellular structures. Therefore, all land plants exhibit an **alternation of generations,** meaning that the plant has two alternating forms in the course of its life cycle. The two types of multicellular bodies are (1) the diploid, spore-producing **sporophyte** generation and (2) the haploid, gamete-producing **gametophyte** generation (Fig. 23.5).

The sporophyte produces a specialized structure called a **sporangium,** where meiosis takes place to produce haploid spores. A **spore** is a reproductive cell that develops into a new organism without the need to fuse with another reproductive cell. The spore then undergoes mitosis and becomes a new multicellular gametophyte.

The gametophyte is a multicellular haploid structure that will develop both male and female gamete-producing regions called *gametangia*. Best seen in the moss life cycle in Figure 23.9, male gametangia are called **antheridia;** they produce sperm. The female gametangia are called **archegonia;** they produce eggs. Sperm and egg will fuse, become a *zygote,* and begin the diploid part of the life cycle.

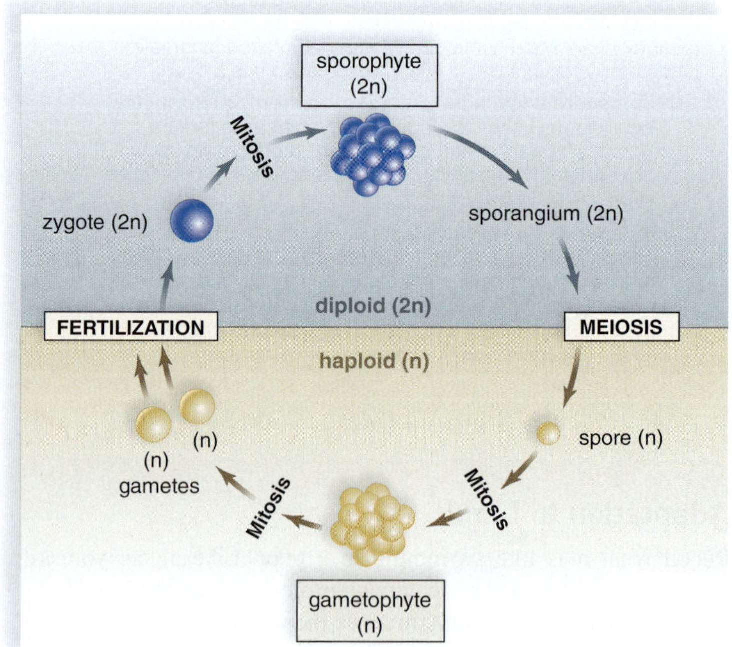

Figure 23.5 Alternation of generations in land plants. The zygote develops into a multicellular 2n generation, and meiosis produces spores in multicellular sporangia. The gametophyte generation produces gametes within multicellular gametangia.

Tutorial Alternation of Generations

Figure 23.6 Reduction in the size of the gametophyte. Notice the reduction in the size of the gametophyte and the increase in the size of the sporophyte among these representatives of today's land plants. This trend occurred as these plants became adapted for life on land. In the moss and fern, spores disperse the gametophyte. In gymnosperms and angiosperms, seeds disperse the sporophyte.

Dominant Generation

Land plants differ as to which generation is dominant—that is, more conspicuous. In plants such as mosses, the gametophyte is dominant, but in plants such as ferns, pine trees, and peach trees, the sporophyte is dominant (Fig. 23.6). In the history of land plants, a vascular system evolves only in the sporophyte; therefore, the shift to sporophyte dominance is an adaptation to life on land. Notice that as the sporophyte gains in dominance, the gametophyte becomes microscopic. The gametophyte also becomes dependent on the sporophyte.

Check Your Progress 23.1

1. Compare and contrast the traits of charophytes and land plants.
2. List the adaptations that led to a land existence for plants.
3. Identify the role of each generation in the alternation-of-generations life cycle.

23.2 Evolution of Bryophytes: Colonization of Land

Learning Outcomes

Upon completion of this section, you should be able to

1. List the traits that classify a plant as a bryophyte.
2. Compare the three groups of bryophytes.
3. Identify the key structures and stages in the life cycle of a moss.

The **bryophytes**—the liverworts, hornworts (Anglo-Saxon *wort,* "herb"), and mosses—were the first plants to colonize land. They only superficially appear to have roots, stems, and leaves, because, by definition, true roots, stems, and leaves must contain vascular tissue. **Vascular tissue** is specialized for the transport of water and organic nutrients throughout the body of a plant. Bryophytes, which lack vascular tissue, are often called the **nonvascular plants.** Vascular tissue also provides support to the plant body, so bryophytes typically are low-lying; some mosses reach a maximum height of only about 20 cm.

The fossil record contains some evidence that the various bryophytes evolved during the Ordovician period (488.3–443.7 MYA). An incomplete fossil record makes it difficult to tell how closely related the various bryophytes are. Molecular data, in particular, suggest that these plants have individual lines of descent, as shown in Figure 23.1, and that they do not form a monophyletic group. The observation that today's mosses have a rudimentary form of vascular tissue suggests that they are more closely related to vascular plants than to the hornworts and liverworts.

Bryophytes do share other traits with the vascular plants. For example, they have an alternation-of-generations life cycle, and they have the traits listed in Table 23.1. Their bodies are covered by a cuticle, which is interrupted in hornworts and mosses by stomata, and they have apical tissue that produces complex tissues. However, bryophytes are the only land plants in which the gametophyte is dominant (Fig. 23.6). Antheridia produce flagellated sperm, which means they need a film of moisture in order to swim to eggs located inside archegonia. The bryophytes' lack of vascular tissue and the presence of flagellated sperm mean that you are apt to find bryophytes in moist locations. Some bryophytes compete well in harsh environments because they can reproduce asexually.

Bryophytes contribute to the lush beauty of forests. They can convert mountain rocks to soil, hold moisture and metals, and resist desiccation. Scientists working in genetic engineering are interested in bryophytes' ability to resist chemical reagents, decay, and herbivory. The idea is that these traits could be transferred to other plants.

Liverworts

Liverworts are divided into two groups—the thallose liverworts with flattened bodies, known as a thallus; and the leafy liverworts, which superficially resemble mosses. The name liverwort

Figure 23.7 Liverwort, *Marchantia.* **a.** Gemmae can detach and start a new plant. **b.** Antheridia are present in disk-shaped structures. **c.** Archegonia are present in umbrella-shaped structures.

refers to the lobes of the thallus, which to some resemble the lobes of the liver. The majority of liverwort species are the leafy types.

The liverworts in the genus *Marchantia* have a thin thallus, about 30 cells thick in the center. Each branched lobe of the thallus is approximately a centimeter in length; the upper surface is divided into diamond-shaped segments with a small pore, and the lower surface bears numerous hairlike extensions called **rhizoids** (Gk. *rhizion,* dim. of "root") that project into the soil (Fig. 23.7). Rhizoids serve in anchorage and limited absorption.

Marchantia species reproduce both asexually and sexually. Gemmae cups on the upper surface of the thallus contain gemmae, groups of cells that detach from the thallus and can start a new plant. Sexual reproduction depends on disk-headed stalks that bear antheridia and on umbrella-headed stalks that bear archegonia. Following fertilization, tiny sporophytes, composed of a foot, a short stalk, and a capsule, begin growing within archegonia. Windblown spores are produced within the capsule.

Hornworts

The **hornwort** gametophyte usually grows as a thin rosette or ribbonlike thallus between 1 and 5 cm in diameter. Although some species of hornworts live on trees, most live in moist, well-shaded areas. They photosynthesize but also have a symbiotic relationship with cyanobacteria, which, unlike plants, can fix nitrogen from the air.

The small sporophytes of a hornwort resemble tiny green broom handles rising from a thin gametophyte, usually less than 2 cm in diameter (Fig. 23.8). Like the gametophyte, a sporophyte can photosynthesize, although it has only one chloroplast per cell. A hornwort can bypass alternation of generations by reproducing asexually through fragmentation.

Mosses

Mosses are the largest group of nonvascular plants, with over 15,000 species. There are three distinct groups of mosses: peat mosses, granite mosses, and true mosses. Although most

Figure 23.8 Hornwort, *Anthoceros* sp. The "horns" of a hornwort are sporophytes that grow continuously from a base anchored in gametophyte tissue.

prefer damp, shaded locations in the temperate zone, some survive in deserts, and others inhabit bogs and streams. In forests, they frequently form a mat that covers the ground and rotting logs. In dry environments, they may become shriveled, turn brown, and look completely dead. As soon as it rains, however, the plant becomes green and resumes metabolic activity. The Nature of Science feature, "Bryophytes—Frozen in Time," on page 418 describes the powerful regenerative abilities of these plants.

Members of the moss genus *Sphagnum* (peat moss) have great commercial and ecological importance to humans. The cell walls of peat moss have a tremendous ability to absorb water, which is why it is used in gardening to improve the water-holding capacity of the soil. One percent of the Earth's surface is peatlands, where dead *Sphagnum* accumulates but does not decay.

Figure 23.9 describes the life cycle of a typical temperate-zone moss. The gametophyte of mosses begins as an algalike, branching filament of cells, the protonema, which precedes and produces upright, leafy shoots that sprout rhizoids. The shoots bear either antheridia or archegonia. The dependent sporophyte consists of a foot, which is enclosed in female gametophyte tissue; a stalk; and an upper capsule, which contains the sporangium, where spores are produced. A moss sporophyte is always attached to the gametophyte. At first, the sporophyte is green and photosynthetic; at maturity, it is brown and nonphotosynthetic. In some species, the sporangium can produce as many as 50 million spores. The spores disperse the new gametophyte generation.

Check Your Progress 23.2

1. Explain the various methods of bryophyte reproduction.
2. List the characteristics that enabled the bryophytes to successfully colonize land.
3. Explain which portion of the bryophyte life cycle is the most dominant.

23.3 Evolution of Lycophytes: Vascular Tissue

Learning Outcomes

Upon completion of this section, you should be able to

1. List the unique structural adaptations found in the lycophytes.
2. Recognize the features present in *Cooksonia* that make it a vascular plant.

Today, **vascular plants** dominate the natural landscape in nearly all terrestrial habitats. Vascular plants can achieve great heights, because they have roots that absorb water from the soil and a

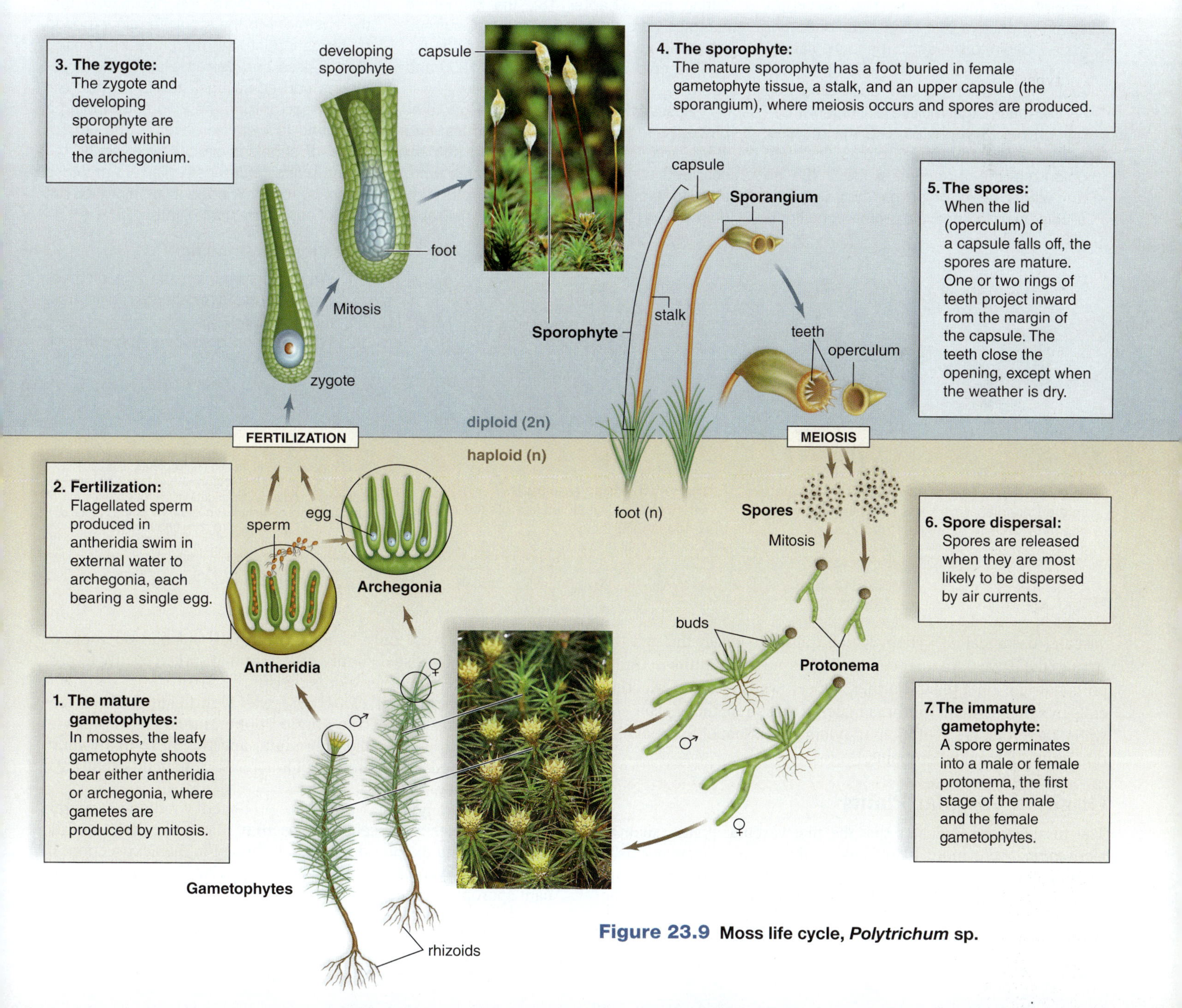

Figure 23.9 **Moss life cycle, *Polytrichum* sp.**

THEME **Nature of Science**

Bryophytes—Frozen in Time

Glaciers, huge masses of ancient ice, are modern-day time machines. Glaciologists, scientists who study glaciers, are able to unlock the mysteries of our prehistoric world. Glaciers can tell how the atmosphere was and what kinds of plants and animals lived thousands of years ago. Glaciology can reveal geologic and climactic processes, such as global climate change. Melting, or *retreating,* glaciers, in particular, offer a unique opportunity for discovery as organisms are liberated from their frozen entombment. Much interest has been in uncovering ancient microbes that, once unfrozen, begin to live again (become viable).

Recently, scientists working in the Arctic regions of Canada uncovered and revived much larger multicellular organisms—bryophytes. Bryophytes, like the mosses, are ancient nonvascular plants that first colonized the terrestrial world. With no true stems, roots, or leaves, their simplicity allows them to survive extreme conditions. Modern-day bryophytes are known to lie dormant when dry and come back to life when water is available, but it had never been known if bryophytes could live after many years of being frozen.

The researchers[1] working on the Teardrop Glacier in the Canadian Arctic archipelago (Fig. 23A) collected dead-looking bryophytes at the edges of the retreating glacier with the goal of answering two questions: (1) How old are these plants? (2) Are they still viable? Using radiocarbon dating (see Chapter 18), the bryophyte samples were determined to be about 400 years old and present during the Little Ice Age—a cold period between AD 1550 and 1850.

Researchers ground up the gametophyte tissue samples (rhizoid and leaf and stem structures), placed them in nutrient soil, and watered them. The plants were able to grow and differentiate into new plants (Fig. 23B), exhibiting remarkable totipotency of bryophytes in extreme conditions. *Totipotency* (Chapter 27) means that plant cells have the amazing genetic capability of becoming an entire plant; although it has been observed many times with a wide range of plants, it has never been shown to happen after a long period (more than 400 years) of being frozen.

This discovery reveals a whole new view of glaciers. Glaciers were once thought to have simply just scraped the landscape as they moved, crushing everything beneath them. We now know that there is more to the story. Terrain exposed by retreating ice can no longer be considered lifeless and barren. Rather, old bryophytes can come back to life and be the key to recolonizing the land surrounding a retreating glacier. Moreover, in the current world of shrinking biological diversity, the frozen world underneath glaciers provides a genetic reservoir not only of microbes but also of early terrestrial plants.

Questions to Consider

1. Why is determining the age of bryophytes found at retreating glaciers useful information?
2. What role do bryophytes play in their environment?

Figure 23A Teardrop Glacier. Located in the Canadian Arctic, this retreating glacier is the recovery site of ancient bryophytes.

Figure 23B Moss regrowth. Moss grown from a 400-year-old sample found at the edge of the glacier.

[1] La Farge, C., Williams, K. H., England, J. H. 2013. Regeneration of Little Ice Age bryophytes emerging from a polar glacier; implications of totipotency in extreme environments. *Proceedings of the National Academy of Science, 110* (24): 9839–9844.

vascular tissue called **xylem,** which transports water through the stem to the leaves. Another conducting tissue called **phloem** transports nutrients in a plant. Further, the cell walls of the conducting cells in xylem contain **lignin,** a material that strengthens plant cell walls; therefore, the evolution of xylem was essential to the evolution of upright and taller plants.

Origin of Vascular Plants

The fossil record indicates that the first vascular plants, such as *Cooksonia,* were more likely bushes than trees. *Cooksonia* is a rhyniophyte, a group of vascular plants that flourished during the Silurian period (443.7–416 MYA) but are now extinct. The rhyniophytes were only about 6.5 cm tall and had no roots or leaves. They consisted simply of a stem that forked evenly to produce branches ending in sporangia (Fig. 23.10).

The branching of *Cooksonia* was significant for two reasons. First, instead of the single sporangium in a bryophyte, this plant produced many sporangia, and therefore many more spores. Second, branching is characteristic of plants that have vascular tissue.

The sporangia of *Cooksonia* produced windblown spores, which classifies it as a **seedless vascular plant,** like the rest of the lycophytes and the pteridophytes, discussed in the next section. In addition, the lycophytes and all other vascular plants have a dominant sporophyte generation.

Figure 23.10 A *Cooksonia* fossil. The upright branches of a *Cooksonia* fossil, no more than a few centimeters tall, terminated in sporangia.

Lycophytes

Like the stem of early vascular plants, the first **lycophytes** also had leaves and roots. The leaves are called **microphylls,** because they had only one strand of vascular tissue. Microphylls most likely evolved as simple side extensions of the stem (see Fig 23.12*a*). Roots evolved simply as lower extensions of the stem; the organization of vascular tissue in the roots of lycophytes today is much as it was in the stems of fossil vascular plants—the vascular tissue is centrally placed.

Today's lycophytes, commonly called the club mosses, include around 1,200 species in three main groups: the ground pines (*Lycopodium*), spike mosses (*Selaginella*), and quillworts (*Isoetes*). Figure 23.11 shows the structure of *Lycopodium;* beginning at the top, notice the cone-shaped structure called the **strobili** (sing., strobilus [Gr. *strobilos,* "pinecone"]). The strobili resemble a club or heavy stick, accounting for the common name, club mosses. The sporangia are borne on the strobili, where meiosis takes place and spores are produced. Next, the aerial stem and microphylls contain vascular tissue, as does the underground stem, called the **rhizome.** The roots develop and branch from the rhizome.

Thus far, sporangia have been described as spore-producing structures. Some lycophytes and other vascular plants produce spores that grow into one type of gametophyte, and they are called **homosporous** (see Fig. 23.16); other vascular plants produce two types of spores, and they are **heterosporous.** The two types of spores are called microspores and megaspores. **Microspores** grow into a male gametophyte, and **megaspores** grow into a female gametophyte (see Fig. 23.18.)

Check Your Progress 23.3

1. Name two features of lycophytes that increase survival on land.
2. Explain how xylem contributes to an upright body plan.
3. Define the terms *homosporous* and *heterosporous.*

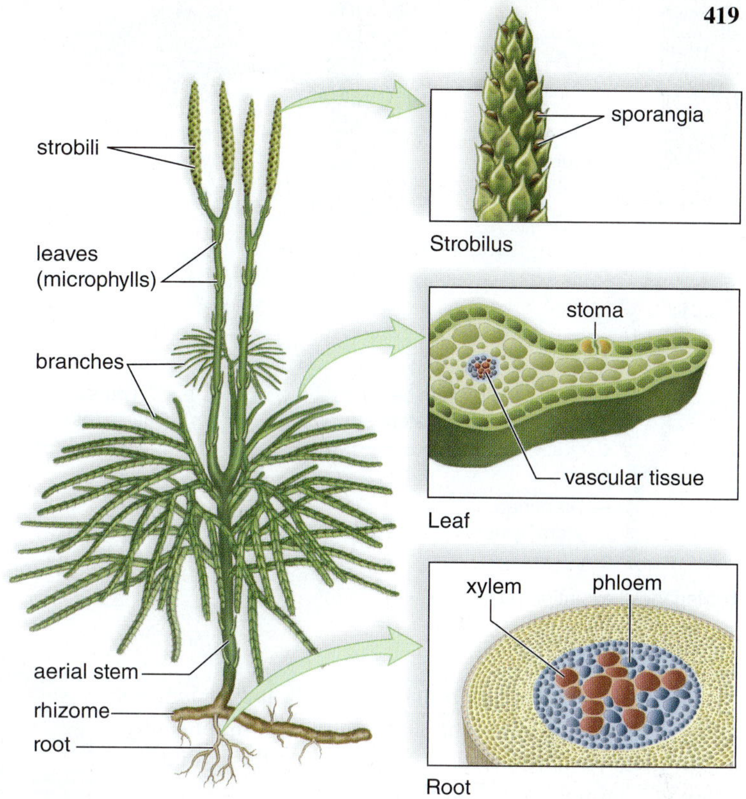

Figure 23.11 Ground pine, *Lycopodium.* The *Lycopodium* sporophyte develops an underground rhizome system. A rhizome is an underground stem. This rhizome produces true roots along its length.

23.4 Evolution of Pteridophytes: Megaphylls

Learning Outcomes

Upon completion of this section, you should be able to

1. Identify three types of pteridophytes.
2. Compare and contrast microphylls and megaphylls.
3. Identify the components of the ferns' life cycle.

Pteridophytes is a broad term used to describe a group of seedless vascular plants including the ferns and their allies, the horsetails and whisk ferns. Both pteridophytes and the seed plants have megaphylls. **Megaphylls** are broad leaves with several strands of vascular tissue. Figure 23.12*a* shows the difference between microphylls and megaphylls, and Figure 23.12*b* shows how megaphylls could have evolved.

Megaphylls, which evolved about 370 MYA, allow plants to efficiently collect solar energy, leading to the production of more food and the possibility of producing more offspring than plants without megaphylls. Therefore, the evolution of megaphylls made plants more fit. Recall that fitness, in an evolutionary sense, is judged by the number of living offspring an organism produces relative to others of its own kind.

The pteridophytes, like the lycophytes, were dominant from the late Devonian period through the Carboniferous period. Today, the lycophytes are quite small, but some of the extinct relatives of today's club mosses were 35 m tall and dominated the Carboniferous

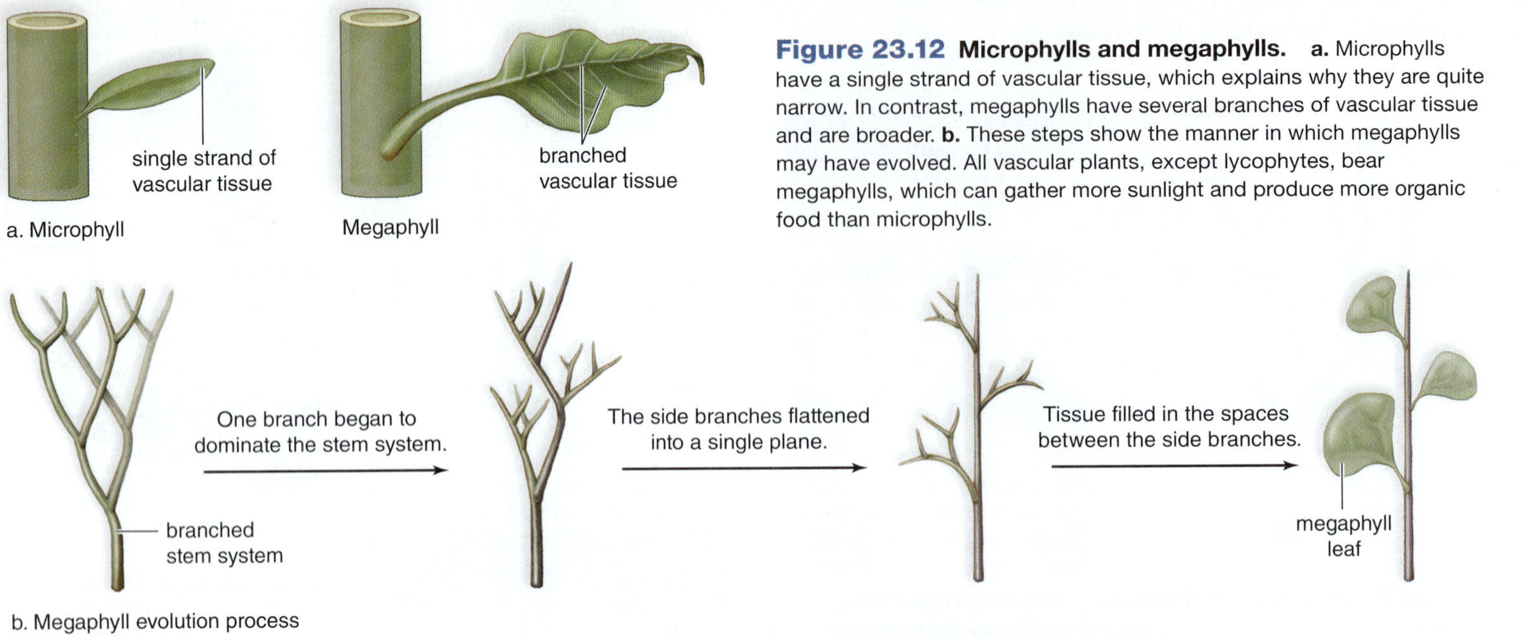

Figure 23.12 Microphylls and megaphylls. **a.** Microphylls have a single strand of vascular tissue, which explains why they are quite narrow. In contrast, megaphylls have several branches of vascular tissue and are broader. **b.** These steps show the manner in which megaphylls may have evolved. All vascular plants, except lycophytes, bear megaphylls, which can gather more sunlight and produce more organic food than microphylls.

swamps. The horsetails (around 18 m tall) and ancient tree ferns (8 m or more in height) also contributed significantly to the great swamp forests of the time (see the Evolution feature "Carboniferous Forests").

Figure 23.13 Horsetail, *Equisetum*. Whorls of branches and tiny leaves are at the nodes of the stem. Spore-producing sporangia are borne in strobili.

Horsetails

Today, **horsetails** consist of one genus, *Equisetum,* and approximately 25 species of distinct seedless vascular plants. Most horsetails inhabit wet, marshy environments around the globe. About 300 MYA, horsetails were dominant plants and grew as large as modern trees. Today, horsetails have a rhizome that produces hollow, ribbed aerial stems and reach a height of 1.3 m (Fig. 23.13).

Equisetum can be branched or unbranched with leaves that may have been megaphylls at one time but now are reduced and form whorls at the nodes. The skirtlike, slender, green side branches make the plant bear a resemblance to a horse's tail. Many horsetails have strobili at the tips of all stems; others send up buff-colored stems that bear the strobili. The spores germinate into inconspicuous and independent gametophytes. *Equisetum* spores are sensitive to humidity and are "spring-loaded." When conditions are right, the spores have been known to be ejected and travel upwards of 2 miles in an hour.

The stems are tough and rigid because of silica deposited in cell walls. Early Americans, in particular, used horsetails for scouring pots and called them "scouring rushes." Today, they are still used as ingredients in a few abrasive powders.

Figure 23.14 Whisk fern, *Psilotum*. *Psilotum* has no leaves—the branches carry on photosynthesis. The sporangia are yellow.

Whisk Ferns

Whisk ferns are represented by the genera *Psilotum* and *Tmesipteris,* which are native to tropical and subtropical regions. The two *Psilotum* species resemble a whisk broom (Fig. 23.14), because they have no leaves. A horizontal rhizome gives rise to aboveground stems that repeatedly fork. The pumpkin-shaped sporangia are borne on short side branches. The two or three species of *Tmesipteris* have appendages that some maintain are reduced megaphylls.

THEME Evolution

Carboniferous Forests

Our industrial society runs on fossil fuels, such as coal. The term *fossil fuel* might seem odd at first until one realizes that it refers to the remains of organic material from ancient times. During the Carboniferous period more than 300 million years ago, a great swamp forest (Fig. 23C) encompassed what is now northern Europe, the Ukraine, and the Appalachian Mountains in the United States. The weather was warm and humid, and the plants grew very tall. These were not plants that would be familiar to us; instead, they were related to various groups of plants that are less well known—the lycophytes, horsetails, and ferns.

Lycophytes today may stand as high as 30 cm, but their ancient relatives were 35 m tall and 1 m wide. The stroboli were up to 30 cm long, and some had leaves more than 1 m long. Horsetails, too—at 18 m tall—were giants compared to today's specimens. Tree ferns were also taller than the tree ferns found in the tropics today. The progymnosperms, including "seed ferns," were significant plants of a Carboniferous swamp.

The amount of biomass in a Carboniferous swamp forest was enormous, and occasionally the swampy water rose, covering the plants that had died. Dead plant material under water does not decompose well. The partially decayed remains became covered by sediment, which changed them into sedimentary rock. Exposed to significant amounts of pressure and over long periods of time, the organic material became coal. This process continued for millions of years, resulting in immense deposits of coal. Geologic upheavals raised the deposits to the levels where they can be mined today.

With a change of climate, many of the plants of the Carboniferous period became extinct. Some of their smaller herbaceous relatives have evolved and survived to our time. We owe the industrialization of today's society to these ancient forests.

Questions to Consider

1. How might a geologist use this information to look for new sources of coal?
2. Why is coal not considered to be a renewable resource?

Figure 23C Swamp forest of the Carboniferous period. Nonvascular plants, early vascular plants, and early gymnosperms dominated the swamp forests of the Carboniferous period (below). Among the early gymnosperms were the seed ferns, so named because their leaves looked like fronds, as shown in a micrograph of fossil remains (right).

Ferns

Ferns are a widespread group of plants (~11,000 species) that are well known for their attractiveness. Ferns are most abundant in warm, moist, tropical regions, but they can also be found in temperate regions and as far north as the Arctic Circle. Several species live in dry, rocky places, whereas others have adapted to an aquatic life. Ferns range in size from tiny aquatic species less than 1 cm in diameter to modern giant tropical tree ferns that exceed 20 m in height. Recall that unlike lycophytes, ferns have megaphylls, which are more commonly referred to as **fronds.** The leatherleaf fern (used in flower arrangements) has fronds that are broad, with subdivided leaflets; those of a tree fern can be about 1.4 meters long; and those of the hart's tongue fern are straplike and leathery (Fig. 23.15). Sporangia are often located in clusters, called **sori** (sing., *sorus*), on the undersides of the fronds, where they may be shielded by thin, protective structures called *indusia* (sing., *indusium*) (Fig. 23.16).

Figure 23.15 Diversity of ferns. Structural adaptations found in a variety of ferns.

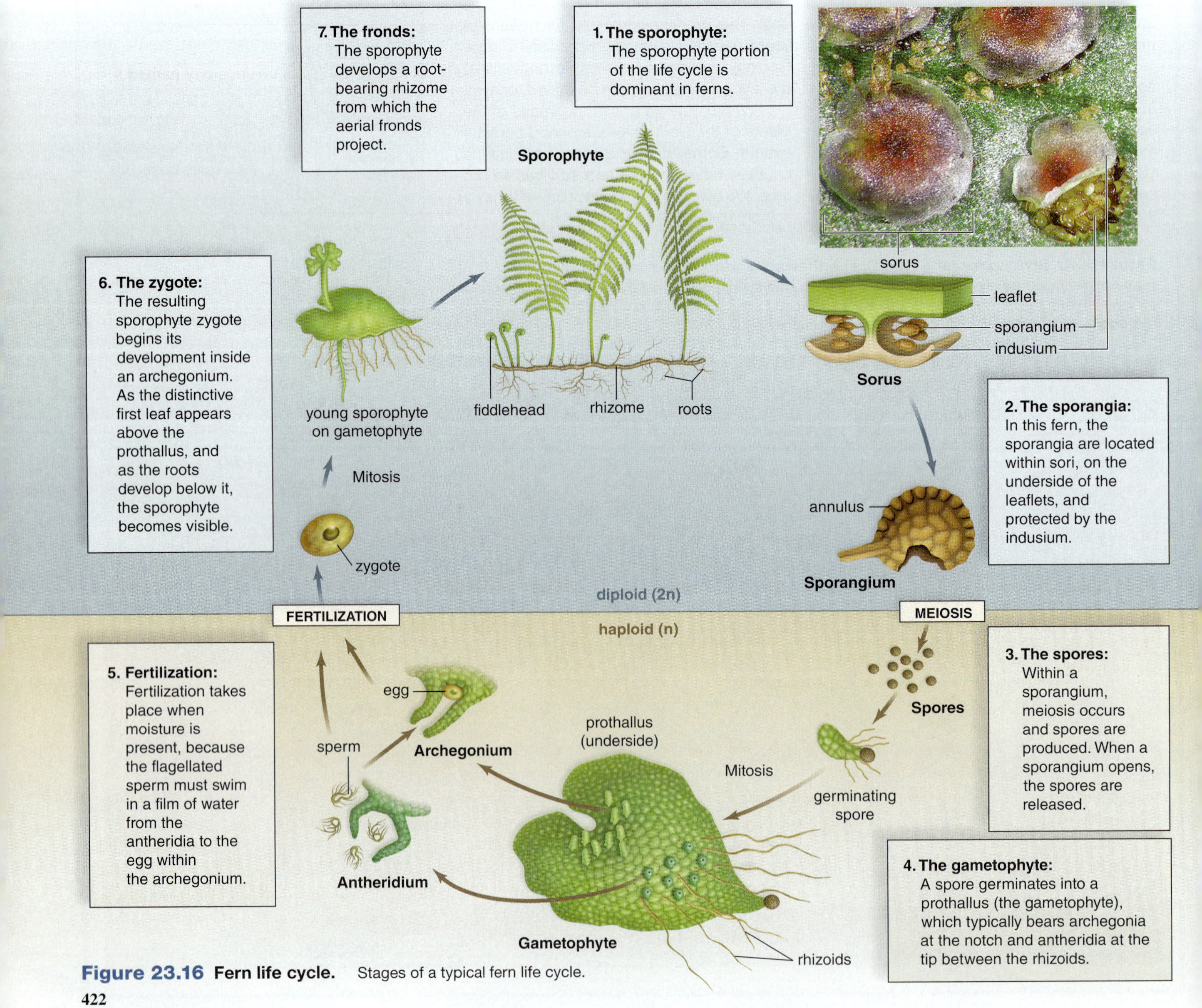

Figure 23.16 Fern life cycle. Stages of a typical fern life cycle.

The life cycle of a typical temperate-zone fern, shown in Figure 23.16, applies in general to the other types of vascular seedless plants. The dominant sporophyte produces windblown spores by meiosis within sporangia. The windblown spores disperse, land, and germinate into the gametophyte, the generation that lacks vascular tissue. The separate, heart-shaped gametophyte produces flagellated sperm, which swim in a film of water from the antheridium to the egg within the archegonium, where fertilization occurs. Eventually, the gametophyte disappears and the sporophyte is independent. Both generations of a fern are considered to be independent of one another. Once established, some ferns, such as the bracken fern, can spread into drier areas, because their rhizomes, which grow horizontally in the soil, produce new plants.

Ferns have much economic value and are frequently used by florists in decorative bouquets and as ornamental plants in the home and garden. Wood from tropical tree ferns is often used as a building material, because it resists decay, particularly by termites. Ferns, especially the ostrich fern, are used as food—in the northeastern United States, many restaurants feature fiddleheads (that season's first growth) as a special treat. Ferns also have medicinal value; many Native Americans use them as an astringent during childbirth to stop bleeding, and the maidenhair fern is the source of a cold medicine.

Check Your Progress 23.4

1. Compare the life cycle of a fern to that of a moss.
2. Describe the structures found on a fern megaphyll.
3. Explain the sequence of events in the haploid portion of the fern life cycle.

23.5 Evolution of Seed Plants: Full Adaptation to Land

Learning Outcomes

Upon completion of this section, you should be able to

1. Compare and contrast the differences between seed and seedless plants.
2. Describe different groups of gymnosperms.
3. Identify the key components of the gymnosperm and angiosperm life cycles.

Seed plants are vascular plants that use seeds during the dispersal stage of their life cycle. **Seeds** contain a sporophyte embryo and stored food within a protective seed coat. The seed coat and stored food allow an embryo to survive harsh conditions during long periods of dormancy (arrested state) until environmental conditions become favorable for growth. When a seed germinates, the stored food is a source of nutrients for the growing seedling. The survival of seeds largely accounts for the dominance of seed plants today.

Like a few of the seedless vascular plants, seed plants are heterosporous (have microspores and megaspores), but their innovation was to retain the spores, and not release them into the environment (see Fig. 23.18). As mentioned earlier in this chapter, the microspores become male gametophytes; these are called **pollen grains. Pollination** occurs when a pollen grain is brought into contact with the female gametophyte by wind or a pollinator. Then, sperm move toward the female gametophyte through a growing **pollen tube.** A megaspore develops into a female gametophyte within an **ovule,** which becomes a seed following fertilization.

Note that because the whole male gametophyte (a pollen grain) moves to the female gametophyte, rather than just the sperm (as in seedless plants), no external water is needed to accomplish fertilization. In essence, the less a plant relies on water for reproduction, the farther that plant can radiate onto drier land, take advantage of new resources, and become more abundant.

The two groups of seed plants alive today are gymnosperms and angiosperms. In **gymnosperms** (mostly cone-bearing seed plants), the ovules are not completely enclosed by sporophyte tissue at the time of pollination. In **angiosperms** (flowering plants), the ovules are completely enclosed within diploid sporophyte tissue (ovary), which becomes a fruit.

The first type of seed plant was a woody plant that appeared during the Devonian period; it has been erroneously named a seed fern. The seed ferns of the Devonian were not ferns at all; they were progymnosperms. It's possible that these were the type of progymnosperm that gave rise to today's gymnosperms and angiosperms. All gymnosperms are still woody plants, but whereas the first angiosperms were woody, many today are nonwoody. Progymnosperms, including seed ferns, were part of the Carboniferous swamp forests (see the Evolution feature, "Carboniferous Forests," on page 421).

Gymnosperms

The four groups of living gymnosperms (Gk. *gymnos,* "naked"; *sperma,* "seed") are conifers, cycads, ginkgoes, and gnetophytes. Since their seeds are not enclosed by fruit, gymnosperms have "naked seeds." Today, living gymnosperms are classified into more than 1,000 species; the conifers are more plentiful than the other types of gymnosperms.

Conifers

Conifers consist of about 630 species of trees, many evergreen, including pines, spruces, firs, cedars, hemlocks, redwoods, cypresses, yews, and junipers. The name *conifers* signifies plants that bear **cones,** but other gymnosperm phyla are also cone-bearing. The coastal redwood (*Sequoia sempervirens*), a conifer native to northwestern California and southwestern Oregon, is the tallest living vascular plant and may attain nearly 100 m in height. Another conifer, the bristlecone pine (*Pinus longaeva*) of the White Mountains of California, is the oldest living tree; one individual is estimated to be 4,900 years of age.

Vast areas of northern temperate regions are covered in evergreen coniferous forests (Fig. 23.17). The tough, needlelike leaves of pines conserve water, because they have a thick cuticle and recessed stomata. Note that in the life cycle of the pine (Fig. 23.18) the sporophyte is dominant, pollen grains are windblown, and the seed is what is dispersed. Conifers are **monoecious,** which means that a single plant carries both male and female reproductive structures.

Figure 23.17 Conifers. **a.** Pine trees are the most common of the conifers. The pollen cones (male) are smaller than the seed cones (female) and produce pollen. Other conifers include **(b)** the spruces, which make beautiful Christmas trees, and **(c)** the cypress, which can be found in northern temperate regions and is a popular landscape tree.

a. Pine

b. Spruce

c. Juniper

Pines, in particular, are well known for their beauty and pleasant smell; they make attractive additions to parks and gardens. Pines have medicinal value in that the needles and bark are rich in vitamins A and C and can be used to create teas that ease symptoms of a cold or cough. Large pine seeds, called pine nuts, are sometimes harvested for use in cooking, and pine oil is used to scent a number of household and personal care products, such as room sprays and deodorant.

Cycads

Cycads include 10 genera and 320 species of distinctive gymnosperms. The cycads are native to tropical and subtropical forests. *Zamia pumila,* found in Florida, is the only species of cycad native to North America. Cycads are commonly used in landscaping. One species, *Cycas revoluta,* referred to as the sago palm, is a common landscaping plant. Their large, finely divided leaves grow in clusters at the top of the stem, and therefore they resemble palms or ferns, depending on their height. The trunk of a cycad is unbranched, even if it reaches a height of 15–18 m, as is possible in some species.

Unlike conifers, cycads have pollen and seed cones on separate plants. The cones, which grow at the top of the stem surrounded by the leaves, can be huge—more than a meter long with a weight of 40 kg (Fig. 23.19*a*). Cycads exhibit the life cycle of a gymnosperm, except they are pollinated by insects rather than by wind. Also, the pollen tube bursts in the vicinity of the archegonium, and multiflagellated sperm swim to reach an egg.

Cycads were plentiful in the Mesozoic era at the time of the dinosaurs, and it's likely that dinosaurs fed on cycad seeds. Now, cycads are in danger of extinction, because they grow very slowly.

Ginkgoes

Although **ginkgoes** are plentiful in the fossil record, they are represented today by only one surviving species of tree, the *Ginkgo biloba*. Ginkgoes are **dioecious,** which means that a single plant produces either male or female reproductive structures, but not both (Fig. 23.19*b*). The fleshy seeds, which ripen in the fall, give off such a foul odor that male trees are usually preferred for planting. Ginkgo trees are resistant to pollution and do well along city

a. An African cycad

b. *Ginkgo biloba*, a native of China

c. *Ephedra*, a type of gnetophyt

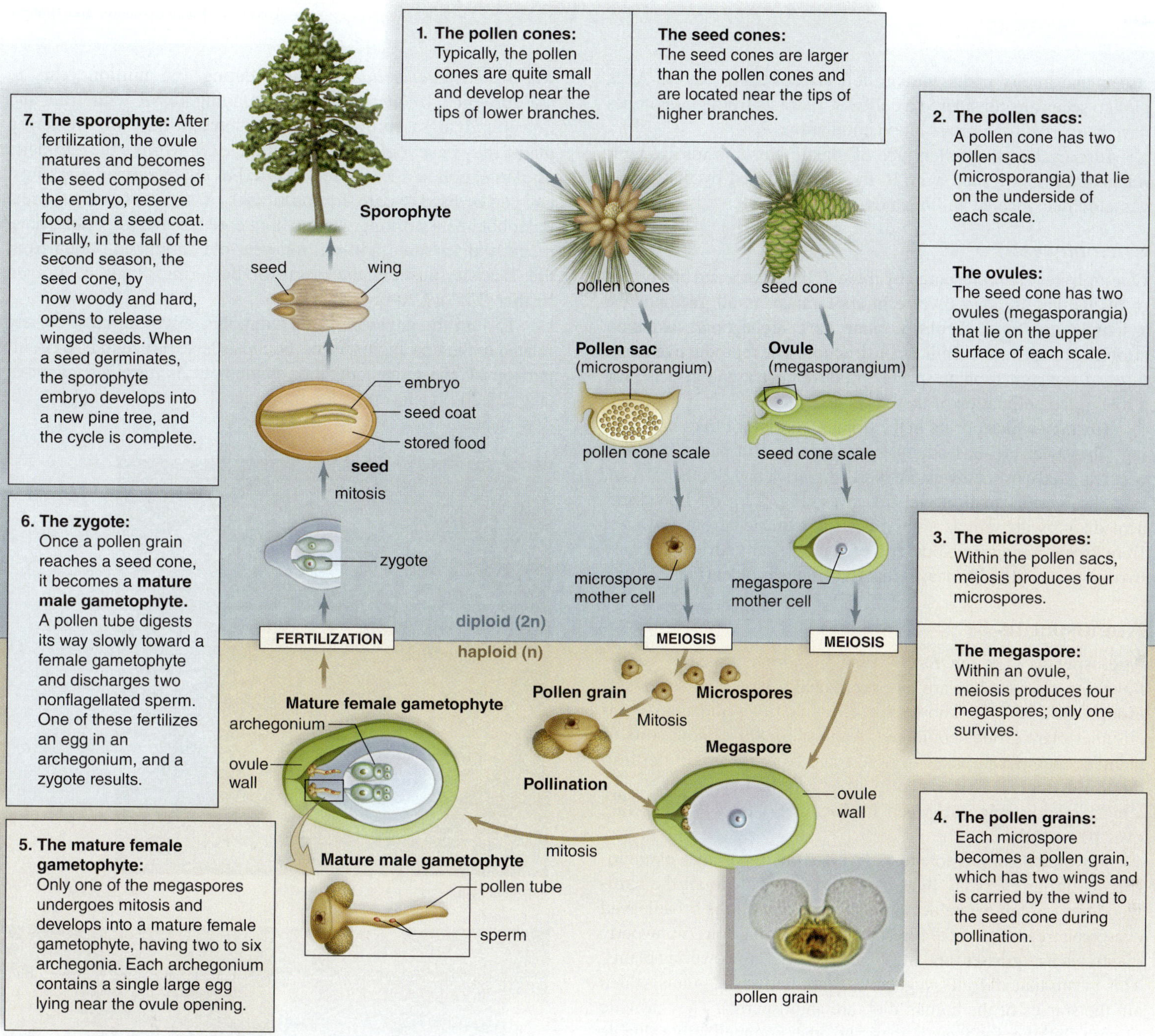

Figure 23.18 **Pine life cycle.** Stages of a typical conifer life cycle.

d. *Welwitschia*, a type of gnetophyte

Figure 23.19 **Three groups of gymnosperms.** **a.** Cycads may resemble ferns or palms, but they are cone-producing gymnosperms. **b.** A ginkgo tree has broad leaves and fleshy seeds borne at the end of stalklike megasporophylls. **c.** *Ephedra,* a type of gnetophyte, is a branched shrub. This specimen produces pollen in microsporangia. **d.** *Welwitschia,* another type of gnetophyte, produces two straplike leaves that fray and split.

streets and in city parks. Ginkgo is native to China, and in Asia, ginkgo seeds are considered a delicacy. Extracts from ginkgo trees have been used to improve blood circulation.

Like cycads, the pollen tube of ginkgo bursts to release multiflagellated sperm that swim to the egg produced by the female gametophyte, located within an ovule.

Gnetophytes

Gnetophytes are represented by three living genera and 70 species of plants that are very diverse in appearance. In all gnetophytes, xylem is structured similarly, none have archegonia, and their strobili (cones) have a similar construction. The reproductive structures of some gnetophyte species produce nectar, and insects play a role in the pollination of these species.

Gnetum, which occurs in the tropics, consists of trees or climbing vines with broad, leathery leaves arranged in pairs. *Ephedra,* occurring only in southwestern North America and Southeast Asia, is a shrub with small, scalelike leaves (Fig. 23.19*c*). Ephedrine, a medicine with serious side effects, is extracted from *Ephedra. Welwitschia,* living in the deserts of southwestern Africa, has only two enormous, straplike leaves that fray as it gets older (Fig. 23.19*d*).

Angiosperms

Angiosperms (Gk. *angion,* "vessel"; *sperma,* "seed") are the flowering plants. They are an exceptionally large and successful group, with 250,000 known species—six times the number of all other plant groups combined. Angiosperms live in all sorts of habitats, from fresh water to desert and from the frigid north to the torrid tropics. They range in size from the tiny, almost microscopic duckweed to *Eucalyptus* trees over 100 m tall.

Video Plants

It would be impossible to exaggerate the importance of angiosperms in our everyday lives. Angiosperms include all the hardwood trees of temperate deciduous forests and all the broadleaved evergreen trees of tropical forests. Also, all herbaceous (nonwoody plants, such as grasses) and most garden plants are flowering plants. This means that all fruits, vegetables, nuts, herbs, and grains, which are the staples of the human diet, are angiosperms. They provide us with clothing, food, medicines, and other commercially valuable products. Over the past 12,000 years, humans have also artificially selected plants to serve us better, resulting in the cultivation of plants as food crops and for other uses. The Evolution feature, "Evolutionary History of Maize," describes corn (*Zea mays*) as an example.

The flowering plants are called angiosperms because their ovules, unlike those of gymnosperms, are always enclosed within diploid tissues. In the Greek derivation of their name, *angio* ("vessel") refers to the ovary, which develops into a fruit, a unique angiosperm feature.

Origin and Radiation of Angiosperms

Plants in general provide an incomplete fossil record, because unlike bones and teeth, the soft parts of plants are often eaten or decayed before they can be preserved. This has created a vigorous debate among paleobotanists on the origins of flowering plants. Until recently, the earliest known angiosperm fossil was *Archaefructus liaoningensis,* dating 135 million years to the Jurassic. *Archaefructus* is a stunning fossil with fruit and flowers, giving a glimpse into what some of the early flowering plants may have looked like (see Fig. 19.8). Unlike the soft parts of plants, pollen does preserve well, has characteristic markings, and can be used to date the appearance of a seed plant in an area. Paleobotanists working with drilling cores from Switzerland have recently discovered 240-million-year-old fossilized pollen from the Triassic, pushing the origin of flowering plants back even further (Fig. 23.20*a*).

To find the angiosperm of today that might be most closely related to the first angiosperms, botanists have turned to DNA comparisons. Gene-sequencing data singled out *Amborella trichopoda* (Fig. 23.20*b*) as having ancestral traits. This small, woody shrub,

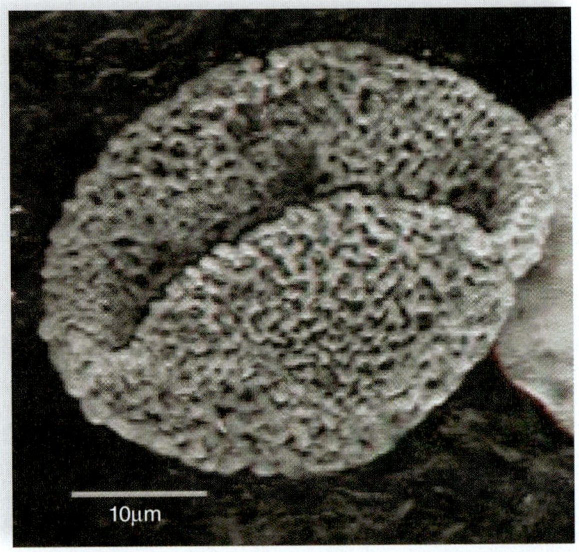

a. Fossil pollen.

b. *Amborella trichopoda*

Figure 23.20 Evolution of flowering plants. **a.** Flowering plant pollen from the Triassic. **b.** Analysis of molecular data suggests that the plant seen here, *Amborella trichopoda,* is the most closely related to the first flowering plants.

THEME Evolution

Evolutionary History of Maize

Approximately 12,000 years ago, humans began shifting away from a hunter-gatherer way of life and moving toward an agriculturally based one. As early groups of humans began to spend more time in the same location, and sedentary populations grew, they required larger and more readily available food sources. Over time, these early farmers increasingly selected plants that produced greater yields of food with less investment of time and energy.

Maize, which we commonly call corn, was first cultivated in Central America nearly 10,000 years ago in the highlands of Mexico. By the time Europeans were exploring Central America in the 1500s, over 300 varieties were being cultivated (Fig. 23D). All of these varieties can trace their ancestral lineage back to a wild Mexican grass known as teosinte.

Among the earliest structural changes that occurred during corn domestication was the loss of *shattering,* the process by which ears of wild corn break apart and disperse their grains (Fig. 23E). A mutation likely caused the ears of some corn plants to remain intact, a trait that is disadvantageous in nature but beneficial to humans. Nonshattering ears are much easier for humans to harvest.

How did our ancestors artificially select for desirable traits? Hunter-gatherers often followed seasonal migration routes between base camps, returning to the same locations year after year. The disturbance of the natural vegetation at these sites would have provided an ideal site for the colonization of species that were being cultivated by these groups. If next season's seeds were collected from the plants that had the most desirable traits, then over time the frequency of plants with these traits would increase. A modern example, showing how quickly artificial selection can change corn, was conducted in 1896 by agricultural scientists selecting for high oil content of corn kernels. Only the top 20% of oil-producing individuals of each generation were allowed to reproduce, and after 90 generations, the average oil content of the corn kernels had increased 450%.

Harvesting plants with specific traits, however, can decrease the genetic diversity within the population. This decline, in turn, produces a genetic bottleneck, in which various traits are lost while others increase in frequency, giving rise to a change in the phenotypic frequency within the population (see Chapter 16).

It was the practice of creating corn hybrids in the 1920s that significantly increased yields, as well as the genetic diversity of cultivars. Modern-day corn produces larger grain, generally a more robust plant, an increased amount of apical growth, and fruit with a higher protein and sugar content. In fact, corn has changed so much through artificial selection that domesticated corn can no longer successfully reproduce without human intervention.

Currently, the world's growing population, the loss of farmland, and erratic climate patterns are straining global food production. The production of corn must now rely on molecular breeding and the use of transgenes. Information gained from genomic studies can be used to customize corn varieties that increase yields under the growing environmental pressures of drought, pests, and microbial pathogens. Genetic engineering has opened up new windows of evolutionary options for the future of maize and, with that, the future of the humans who rely on this plant.

Figure 23D Varieties of corn. Corn can be cultivated to have a specific color, texture, and sugar content.

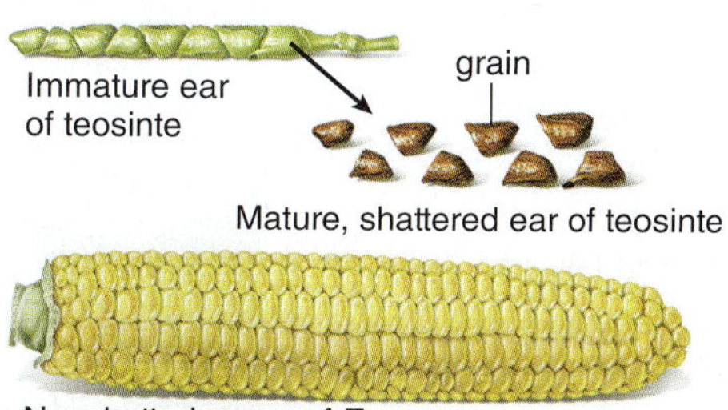

Figure 23E Ears and grain of modern corn and its ancestor, teosinte. Domesticated corn ears are larger than those of the ancestral grass teosinte. In addition, corn fruits are nonshattering, softer, and more edible than are grains of teosinte.

Questions to Consider

1. What potentially beneficial traits may have been lost during artificial selection in maize?
2. What other food plants might have undergone a similar form of natural selection?

with small, cream-colored flowers, lives only on the island of New Caledonia in the South Pacific. Its flowers are about 4–8 mm wide, and the petals and sepals look the same; therefore, they are called tepals. Plants bear either male or female flowers, with a variable number of stamens or carpels.

Although *A. trichopoda* may not be the original angiosperm species, it is sufficiently close that much may be learned from studying its reproductive biology. Botanists hope that this knowledge will help them understand the early adaptive radiation of angiosperms during the Tertiary period. The gymnosperms were abundant during the Mesozoic era but declined during the mass extinction that occurred at the end of the Cretaceous period (145.5–65.5 MYA). Angiosperms survived and went on to become the dominant plants during modern times.

Monocots and Eudicots

Most flowering plants belong to one of two classes. These classes are the **Monocotyledones,** often shortened to simply the **monocots** (about 65,000 species), and the **Eudicotyledones,** shortened to **eudicots** (about 175,000 species). The term *eudicot* (meaning "true dicot") is more specific than the term *dicot.* It was discovered that some of the plants formerly classified as dicots diverged before the evolutionary split that gave rise to the two major classes of angiosperms. These earlier-evolving plants are not included in the designation eudicots.

Cotyledons (Gk. *kotyledon,* "cuplike cavity") are the "seed leaves" containing the nutrients that nourish the plant embryo. If a seed has one cotyledon, it is a monocot, and if a seed has two cotyledons, it is a eudicot (see Fig. 27.8). Several common monocots are corn, tulips, pineapples, and sugarcane; common eudicots include cacti, strawberries, dandelions, poplars, and beans. Table 23.2 lists several fundamental features of monocots and eudicots.

The Flower

Although **flowers** vary widely in appearance (Fig. 23.21), most have certain structures in common (Fig. 23.22).

1. The **sepals,** collectively called the calyx, protect the flower bud before it opens. The sepals may drop off or may be colored like the petals. Usually, however, sepals are green and remain attached to the flower stalk.
2. The **petals,** collectively called the corolla, are quite diverse in size, shape, and color. The petals are often used to attract a particular pollinator.
3. Next are the **stamens.** Each stamen consists of two parts: first, a slender stalk, called a filament, that holds up a second structure, a saclike container called the anther. Pollen grains develop from microspores produced within the anther.
4. At the very center of a flower is the **carpel,** a vaselike structure with three major regions: the stigma, an enlarged, sticky knob; the style, a slender stalk; and the **ovary,** an enlarged base that encloses one or more ovules. The ovule becomes the seed, and the ovary becomes the fruit. Fruit is often instrumental in the distribution of seeds.

Note that not all flowers have all these parts (Table 23.3). A flower is said to be *complete* if it has all four parts; otherwise, it is *incomplete.*

Table 23.2 Key Features of Monocots and Eudicots

Monocots	Eudicots
One cotyledon	Two cotyledons
Flower parts in threes or multiples of three	Flower parts in fours or fives or multiples of four or five
Pollen grain with one pore	Pollen grain with three pores
Usually herbaceous	Woody or herbaceous
Usually parallel venation	Usually net venation
Scattered bundles in stem	Vascular bundles in a ring
Fibrous root system	Taproot system

a. Monocot

b. Eudicot

Figure 23.21 Flower diversity. Regardless of size and shape, flowers, such as this **(a)** monocot, and **(b)** eudicot, share certain features.

Flowering Plant Life Cycle

Figure 23.23 depicts the life cycle of a typical flowering plant. Like the gymnosperms, flowering plants are heterosporous, producing two types of spores. A megaspore located in an ovule within an ovary of a carpel develops into an egg-bearing female gametophyte, called the embryo sac. In most angiosperms, the embryo sac has seven cells; one of these is an egg, and another contains two polar nuclei. They are called the polar nuclei because they came from opposite ends of the embryo sac.

Microspores, produced within anthers, become pollen grains that, when mature, are male gametophytes with sperm. The mature

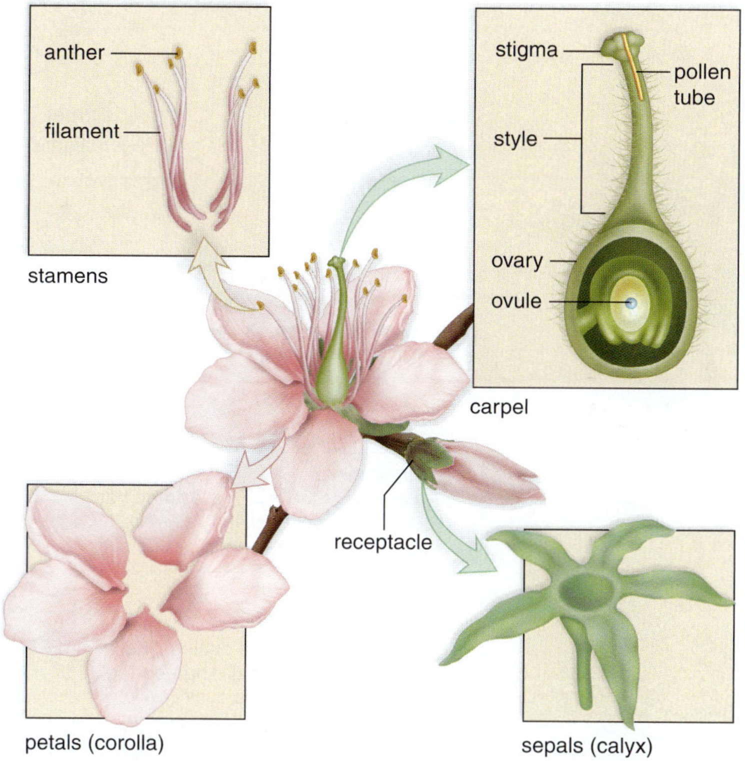

Figure 23.22 Generalized flower. A flower has four main parts: sepals, petals, stamens, and carpels. A stamen has an anther and filament. A carpel has a stigma, style, and ovary. An ovary contains ovules.

Table 23.3 Other Flower Terminology

Term	Type of Flower
Complete	All four parts (sepals, petals, stamens, and carpels) present
Incomplete	Lacks one or more of the four parts
Perfect	Has both stamens and (a) carpel(s)
Imperfect	Has stamens or (a) carpel(s), but not both
Inflorescence	A cluster of flowers
Composite	Appears to be a single flower but consists of a group of tiny flowers

male gametophyte consists of only three cells: the tube cell and two sperm cells.

During pollination, pollen is transported from the anther to the stigma of a carpel. Here, the tube cell produces a pollen tube that carries the two sperm to the micropyle (small opening) of an ovule. Flowering plants undergo **double fertilization:** One sperm unites with an egg, forming a diploid zygote, and the other unites with polar nuclei, forming a triploid endosperm nucleus (see Chapter 27).

Ultimately, the ovule becomes a seed that contains the embryo (the sporophyte of the next generation) and stored food enclosed within a seed coat (see Fig. 23.2*b*). Endosperm in some seeds is absorbed by the cotyledons, whereas in other seeds endosperm is digested as the seed matures.

A **fruit** is derived from an ovary, and in some instances it is an accessory part of the flower. Some fruits, such as apples and tomatoes, provide a moist, fleshy covering; other fruits, such as pea pods and acorns, provide a dry covering for seeds.

Flowers and Diversification

Flowers are involved in the production and development of spores, gametophytes, gametes, and embryos enclosed within seeds. Successful completion of sexual reproduction in angiosperms requires the effective dispersal of pollen and then seeds. The various ways pollen and seeds can be dispersed have resulted in many different types of flowers (see Chapter 27).

Wind-pollinated flowers are usually not showy, whereas many insect- and bird-pollinated flowers are colorful. Night-blooming flowers attract nocturnal mammals or insects; these flowers are usually aromatic and white or cream-colored.

Although some flowers disperse their pollen by wind, many are adapted to attract specific pollinators, such as bees, wasps, flies, butterflies, moths, and even bats, that carry pollen from one flower to another flower of the same type. For example, glands in the region of the ovary produce nectar, a nutrient that is gathered by pollinators as they go from flower to flower. Bee-pollinated flowers are usually blue or yellow and have ultraviolet shadings that lead the pollinator to the location of the nectar. The mouthparts of bees are fused into a long tube, which is able to obtain nectar from the base of the flower. In another example, instead of beautiful colors and temping nectar, some species of *Ophyrys* orchids use "sexual mimicry" to attract their wasp pollinators. The orchid flowers look like a female wasp and emit pheromones to attract male wasps. The males will engage in "pseudocopulation" with the "female" and end up with pollen attached to its head. Frustrated, the male wasp leaves the flower and attempts to mate with another flower. The male then deposits pollen to the second flower thereby completing the cross pollination that the orchid intended.

Video Pollinators

The fruits of flowers protect and aid in the dispersal of seeds. Dispersal occurs when seeds are transported by wind, gravity, water, and animals to another location. Fleshy fruits may be eaten by animals, which transport the seeds to a new location and then deposit them when they defecate. Because animals live in particular habitats and/or have particular migration patterns, they are apt to deliver the fruit-enclosed seeds to a suitable location for seed germination (when the embryo begins to grow again) and development of the plant.

Video Dung Seed Dispersal

Video Fruit Bat Seed Dispersal

Check Your Progress 23.5

1. List the life cycle changes that have enabled pines to better adapt to life on land.
2. Compare and contrast the four types of gymnosperms.
3. List the functions of the key structures required for angiosperm reproduction.

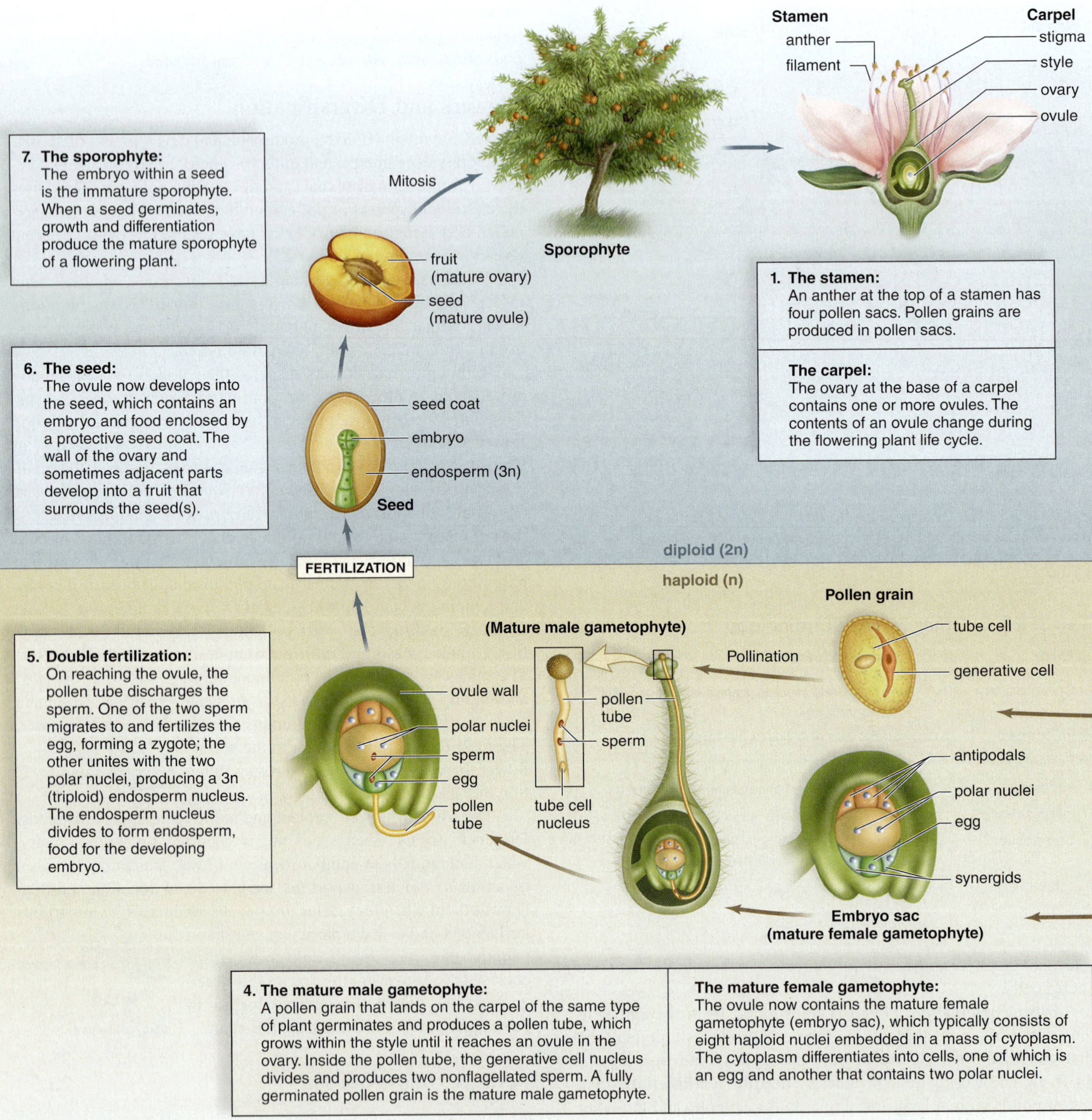

Figure 23.23 Flowering plant life cycle. The parts of the flower involved in reproduction are the stamens and the carpel. Reproduction has been divided into significant stages of female gametophyte development, male gametophyte development, and important stages of sporophyte development.

Tutorial Angiosperm Life Cycle

2. The pollen sacs: In pollen sacs (microsporangia) of the anther, meiosis produces microspores.	**The ovules:** In an ovule (megasporangium) within an ovary, meiosis produces four megaspores.

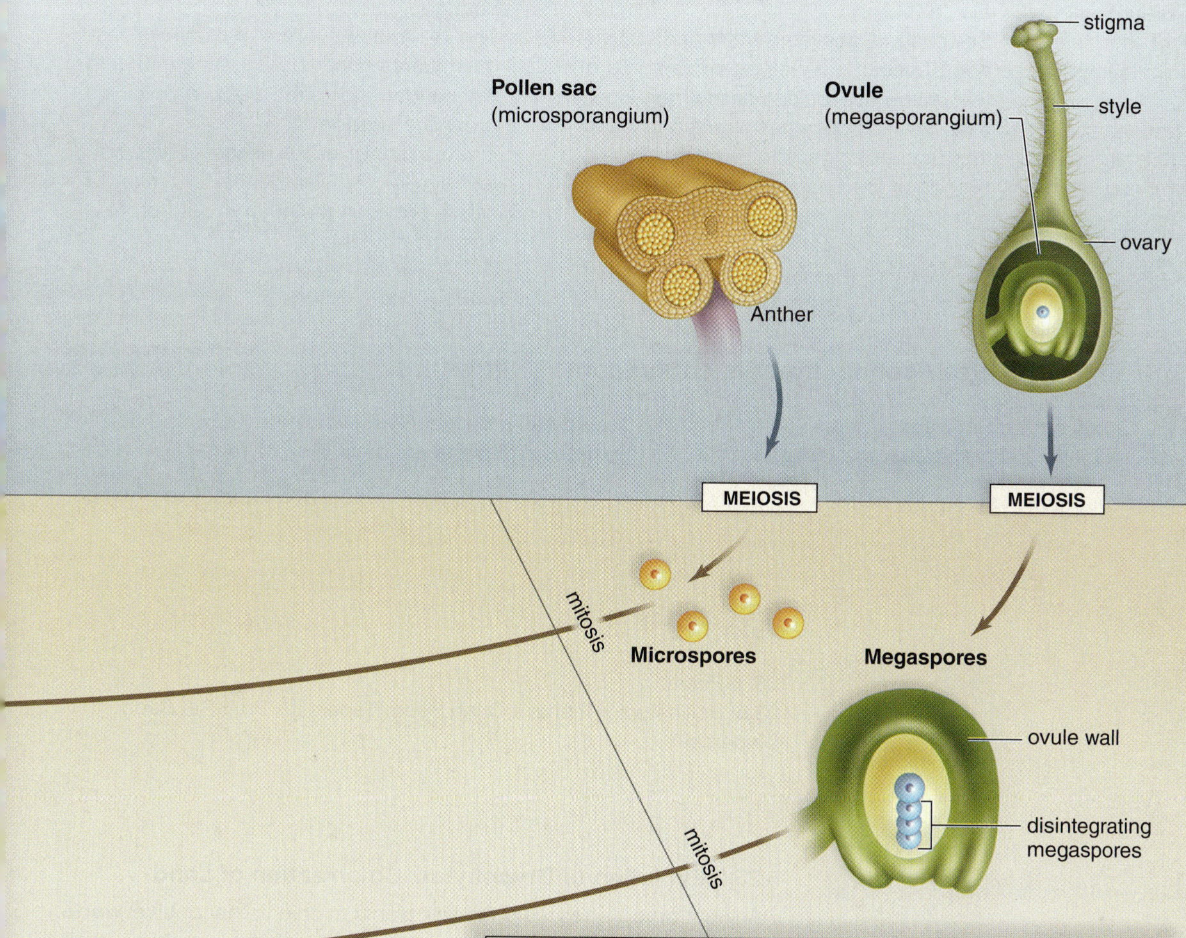

3. The microspores:
Each microspore in a pollen sac undergoes mitosis to become an immature pollen grain with two cells: the tube cell and the generative cell. The pollen sacs open, and the pollen grains are windblown or transported by an animal carrier, usually to other flowers. This is known as pollination.

The megaspores:
Inside the ovule of an ovary, three megaspores disintegrate, and only the remaining one undergoes mitosis to become a female gametophyte.

CONNECTING the CONCEPTS with the THEMES

Evolution

- Several evolutionary modifications were necessary for plants' successful invasion of land. The forces of natural selection determined which traits were most suitable to terrestrial life.
- Successful life on land requires protection from dessication, a structural framework to support the plant body, the ability to move water and nutrients throughout the plant, and a diploid part of the life cycle to deal with increased mutations with sun exposure.
- A land-based lifestyle allows greater exposure to the energy of sunlight, giving an advantage to any photosynthetic organism that can make use of it.

Nature of Science

- Better technology has allowed paleobotanists to study plant fossils and determine the appearance and/or arrival of a plant group in geologic time.
- Plants are a major part of the human diet, and over the past 10,000 years of human history, people have been a force for artificial selection of today's food-producing plants.
- Selection of desirable traits and cultivation of plants has helped solve some of the challenges of human hunger. However, this selection has resulted in a limited amount of genetic diversity in many of today's food plants.

Biological Systems

- Each modern group of land plants exhibits increasingly well-adapted features for existence on land.
- The flagellated sperm of the nonvascular plants enables male gametes to swim through surface water to mate with the female gametophyte; however, these plants are therefore restricted to moist habitats.
- The xylem and phloem of seedless vascular plants conduct water and nutrients throughout the plant body, as well as providing support.
- The megaphyll leaves of seed plants collect sunlight efficiently; the formation of seeds provides the embryo with protection and nutrients.

MEDIA STUDY TOOLS http://connect.mheducation.com

SMARTBOOK® Study smart. SmartBook helps you maximize your study time by identifying what you know and don't know. The recharge feature in SmartBook helps you prepare for the test by identifying the material that you are most likely to forget.

connect BIOLOGY Find even more resources to learn the chapter concepts including animations, tutorial videos, and interactive practice questions.

Tutorials	Videos
23.1 Alternation of Generations **23.5** Angiosperm Life Cycle	**23.1** Plants **23.5** Pollinators • Plants • Dung Seed Dispersal • Fruit Bat Seed Dispersal

SUMMARIZE

23.1 Ancestry and Features of Land Plants

Land **plants** evolved from a common ancestor with multicellular, freshwater algae about 590 MYA. The **charophytes** *Chara* and *Coleochaete* are green algae related to the ancestors of plants and like other algae have characteristics in common with plants, such as photosynthetic pigments, the presence of starch, and cell wall structure.

Plants evolved modifications for life on land, such as a waxy **cuticle** to combat desiccation and **stomata** for gas exchange. Specialized cells called **tracheids** facilitate the movement of water within a plant, and the formation of tissues and organs reduces surface water loss. The **alternation-of-generations** life cycle contains a multicellular diploid stage that can be beneficial if mutations are caused by UV light. The **sporophyte** is the diploid portion of the life cycle; here, specialized structures called **sporangia** produce haploid spores. **Spores** divide by mitosis and become the gametophyte. The haploid stage is the multicellular **gametophyte,** with male and female gametangia called the **antheridia** and **archegonia.** Sperm and egg are produced, fuse, and form a diploid zygote.

23.2 Evolution of Bryophytes: Colonization of Land

Ancient **bryophytes** were the first plants to colonize land. **Liverworts, hornworts,** and **mosses** are examples of bryophytes, which lack well-developed **vascular tissue** and are therefore called the **nonvascular plants.** Without true roots, **rhizoids** instead anchor these plants to a substrate. The sporophyte is nutritionally dependent on the gametophyte, because the gametophyte is larger and photosynthetic. The life cycle of mosses demonstrates the reproductive strategies of flagellated sperm and dispersal by means of windblown spores.

23.3 Evolution of Lycophytes: Vascular Tissue

Vascular plants, such as rhyniophytes, evolved during the Silurian period. The sporophytes contain two types of conducting tissues. **Phloem** is specialized to move organic nutrients, whereas **xylem** is specialized to conduct water and dissolved minerals with the help of **lignin** in xylem cells. **Lycophytes** are descended from these first plants. They contain vascular tissue, produce spores, and are called the **seedless vascular plants.** Ancient lycophytes were the first plants to have leaves known as **microphylls,** and modern lycophytes have conelike structures called **strobili,** where sporangia form. An underground stem called a **rhizome** has extensions of roots. Seedless vascular plants can produce one type of spore, and be **homosporous,** or two types of spores, and be **heterosporous.** The two types of spores made by hetersporous plants are **microspores** and **megaspores.**

23.4 Evolution of Pteridophytes: Megaphylls

In the **pteridophytes** (**ferns** and their allies, **horsetails** and **whisk ferns**) and the lycophytes, the sporophyte is the dominant stage of the life cycle and is separate from the tiny gametophyte. Windblown spores are the dispersal agents for these plants. The ferns found on Earth today have an obvious **megaphyll;** horsetails and whisk ferns have reduced megaphylls.

Fern megaphylls are called **fronds,** and there is much diversity of shape and size of these fronds. Most fronds have sporangia on the underside, found in clusters called **sori.** Established ferns can often reproduce asexually by growing new plants from the underground rhizome.

23.5 Evolution of Seed Plants: Full Adaptation to Land

Seed plants also have an alternation of generations, but they are heterosporous, producing both microspores and megaspores. Microspores become the windblown or animal-transported male gametophytes—the **pollen grains. Pollination** involves pollen grains carrying and delivering sperm via a **pollen tube** to the female gametophyte (megaspore), which is the **ovule.** Following fertilization, the ovule becomes the **seed,** which contains a sporophyte embryo. Fertilization no longer requires external water, and sexual reproduction is fully adapted to the terrestrial environment.

The **gymnosperms** (**cone**-bearing plants) evolved from woody seed ferns during the Devonian period. The **conifers,** represented by the pine tree, exemplify the traits of these plants. Other gymnosperms are the **cycads, ginkgoes,** and **gnetophytes.** Gymnosperms have "naked seeds," because they are not enclosed by fruit, and can be **monoecious** (having both male and female structures on the same plant) or **dioecious** (being either male or female but not both).

A woody shrub, *Amborella trichopoda,* has been identified as most closely related to the common ancestor for the angiosperms. **Angiosperms** belong to two classes: **Monocotyledones** (**monocots**) with one **cotyledon** and **Eudicotyledones** (**eudicots**) with two cotyledons. The reproductive organs are found in **flowers.** Most flowers have **sepals, petals, stamens,** and **carpels.** After fertilization, the ovules become seeds, which are located in the **ovary.** Seeds are the result of **double fertilization,** with two sperm yielding the embryo and the endosperm to feed the embryo. This ultimately becomes the **fruit.** Therefore, angiosperms have "covered seeds."

In many angiosperms, pollen is transported from flower to flower by various pollinators. Both flowers and fruits are found only in angiosperms and may account for the extensive colonization of terrestrial environments by the flowering plants.

ASSESS

Choose the best answer for each question.

23.1 Ancestry and Features of Land Plants

1. Which of these are characteristics of land plants?
   a. multicellular with specialized tissues and organs
   b. photosynthetic and contain chlorophylls *a* and *b*
   c. protect the developing embryo from desiccation
   d. have an alternation-of-generations life cycle
   e. All of these are correct.

2. Label this diagram of alternation-of-generations life cycle.

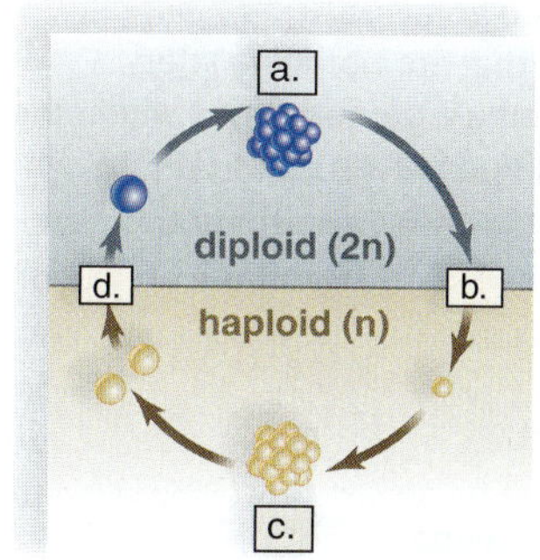

23.2 Evolution of Bryophytes: Colonization of Land

3. In the moss life cycle, the sporophyte
   a. consists of leafy, green shoots.
   b. is the heart-shaped prothallus.
   c. consists of a foot, a stalk, and a capsule.
   d. is the dominant generation.
   e. All of these are correct.
4. In bryophytes, sperm usually move from the antheridium to the archegonium by
   a. swimming.
   b. flying.
   c. insect pollination.
   d. wind pollination.
   e. bird pollination.

23.3 Evolution of Lycophytes: Vascular Tissue

5. A small, upright plant that resembles a tiny, upright pine tree with club-shaped strobili and microphylls is a
   a. whisk fern.
   b. lycophyte.
   c. conifer.
   d. horsetail.
   e. fern.
6. Microphylls
   a. have a single strand of vascular tissue.
   b. evolved before megaphylls.
   c. evolved as extensions of the stem.
   d. are found in lycophytes.
   e. All of these are correct.

23.4 Evolution of Pteridophytes: Megaphylls

7. How are ferns different from mosses?
   a. Only ferns produce spores as dispersal agents.
   b. Ferns have vascular tissue.
   c. In the fern life cycle, the gametophyte and sporophyte are both independent.
   d. Ferns do not have flagellated sperm.
   e. Both b and c are correct.
8. Ferns have
   a. a dominant gametophyte generation.
   b. vascular tissue.
   c. seeds.
   d. Both a and b are correct.
   e. Choices a, b, and c are correct.

23.5 Evolution of Seed Plants: Full Adaptation to Land

9. Which of these is found in seed plants?
   a. complex vascular tissue
   b. pollen grains that are not flagellated
   c. retention of female gametophyte within the ovule
   d. roots, stems, and leaves
   e. All of these are correct.
10. Gymnosperms
    a. have flowers.
    b. are eudicots.
    c. are monocots.
    d. produce fruit.
    e. reproduce by seeds.
11. In the life cycle of the pine tree, the ovules are found on
    a. needlelike leaves.
    b. seed cones.
    c. pollen cones.
    d. root hairs.
    e. All of these are correct.
12. Which of these pairs is mismatched?
    a. anther—produces microspores
    b. carpel—produces pollen
    c. ovule—becomes seed
    d. ovary—becomes fruit
    e. flower—is a reproductive structure

ENGAGE

THINKING SCIENTIFICALLY

1. Using as many terms as necessary (from both X and Y axes), fill in the proposed phylogenetic tree for vascular plants.

	ferns	conifers	ginkgoes	monocots	eudicots
vascular tissue	X	X	X	X	X
seed plants		X	X	X	X
naked seeds		X	X		
needlelike leaves		X			
fan-shaped leaves			X		
enclosed seeds				X	X
one embryonic leaf				X	
two embryonic leaves					X

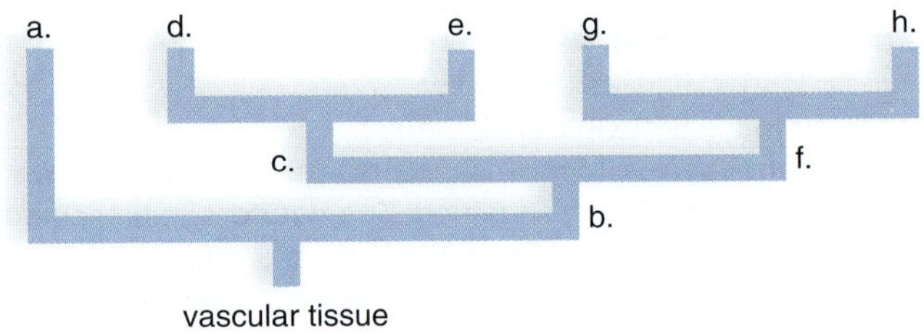

2. Using Figure 23.1, distinguish between the (a) microphyll and the (b) megaphyll clade.
3. Compare and contrast the plant alternation-of-generations life cycle with the human life cycle (Fig. 10.7)
4. Would you expect the sporophyte generation of a moss or a fern to have more mitotic divisions? Why?
5. Imagine hypothetical moss and fern trees, each 10 m tall. Which would face the greater barriers to sexual reproduction? Why?
6. What types of environmental stresses are placed on plants living on dry land, and how do plants overcome these stresses?

The various organs of the neem tree are used for medicinal and agricultural applications.

24

Flowering Plants: Structure and Organization

Chapter Outline

24.1 Cells and Tissues of Flowering Plants 436

24.2 Organs of Flowering Plants 440

24.3 Organization and Diversity of Roots 442

24.4 Organization and Diversity of Stems 445

24.5 Organization and Diversity of Leaves 450

Before You Begin

Before beginning this chapter, take a few moments to review the following discussions.

Figure 4.7 What cellular structures are necessary for plant cells to function?

Section 7.2 What are the reactants, intermediates, and end products of photosynthesis?

Section 23.5 What structural features helped promote angiosperm success?

If you were to walk into a health-food store looking for neem products, you would find oils, toothpaste, soap, and facial washes. At a plant nursery, you would find neem insecticide and fungicide. But what exactly is neem? The neem tree (*Azadirachta indica*) is found in India and Pakistan and is one of the oldest and widely used plants in the world. It is locally known as the "village pharmacy," and the United Nations declared neem the "tree of the 21st century" due to its many uses in health and agriculture.

Neem's nontraditional medical uses are many. Bark and roots act as an analgesic and diuretic, and provide flea and tick protection to dogs. The sap, or phloem, effectively treats skin diseases, such as psoriasis. Gum, a sticky substance exuded by stems, is used to combat scabies (an itch mite) and surface wounds. And in some countries, the tree's woody twigs aid dental hygiene, in the form of homemade toothbrushes.

The neem tree also plays an important role in agriculture and pest control. In India and Pakistan, for example, neem leaves have traditionally been mixed with stored grains or placed in drawers with clothing, keeping insects at bay. Not surprisingly, the U.S. Department of Agriculture funds a significant amount of research on neem, due to its potential as an all-natural insecticide and fungicide. In this chapter, we consider the structure of roots, stems, and leaves and discuss the specialized cells and tissues that make up these organs.

As you read through this chapter, think about the following questions:

1. How do each of the vegetative plant organs contribute to the success of flowering plants?
2. How do modifications of vegetative organs increase fitness?

Following *the* Themes

Chapter 24 Flowering Plants: Structure and Organization

Unit 5 Plant Evolution and Biology

Theme	
Evolution	A number of structural adaptations were necessary for angiosperm evolution.
Nature of Science	Human life is inextricably linked to plants, and we continue to find a wide variety of uses for plants.
Biological Systems	The structural and physiological systems of angiosperms have enabled them to become the dominant form of plant life on Earth today.

24.1 Cells and Tissues of Flowering Plants

Learning Outcomes

Upon completion of this section, you should be able to

1. Explain how plant cells are different from animal cells.
2. Define *apical meristem* and describe where on a plant it is found.
3. Identify the three types of tissue found in angiosperms.
4. Recognize the differences in the location, structure, and function among various angiosperm tissues.

The body of a plant is organized in a similar fashion to the body of an animal. As in animals, a *cell* is a basic unit of life. A *tissue* is composed of specialized cells that perform a particular function, and an *organ* is a structure made up of multiple tissues. In Chapter 4, you learned about eukaryotic cell features common to both plant and animal cells. Table 24.1 highlights some differences between plant and animal cells.

When a plant embryo first begins to develop, the first cells are called **meristem** cells. Like animal stem cells, plant meristem cells are undifferentiated cells that can divide indefinitely and give rise to many types of differentiated cells (Fig. 24.1). As new cells are produced, they are small and boxlike, with a large nucleus and tiny vacuoles. As these cells mature, they assume many different shapes and sizes, each related to the cell's ultimate function.

When a seed germinates and an embryo grows, meristem tissue is present at the tips, or apices, of the young plant and are called **apical meristems.** Apical meristems in turn give rise to three specialized meristems that create the differing plant tissues (see Fig. 23.14):

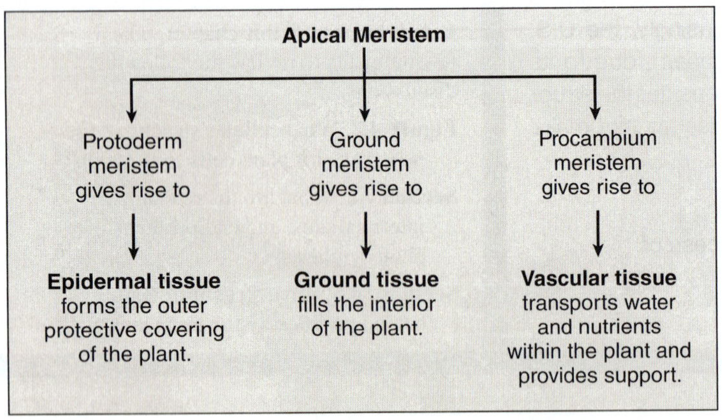

Another type of meristem, mainly found in monocots, such as grasses, is the *intercalary meristem,* found at the node of the plant instead of the tip. This is why you can cut your lawn, yet the grass continues to grow. In addition, plants not only grow from their apices but also can grow wide. The **vascular cambium** is another type of meristem, which gives rise to new vascular tissue called *secondary growth*. Secondary growth causes a plant to increase in girth.

Video Seedling Growth

Table 24.1 Comparison of Plant and Animal Cells

Plant Cell	Animal Cell
Cell wall is present.	No cell wall is present.
Cells connect with plasmodesmata.	Cells connect with various junctions.
Cell division involves a cell plate.	Cell division involves a cleavage furrow.
No centrioles are present during mitosis.	Centrioles are present during mitosis.
Plastids are present.	No plastids are present.
Vacuoles are large.	Vacuoles are small or absent.

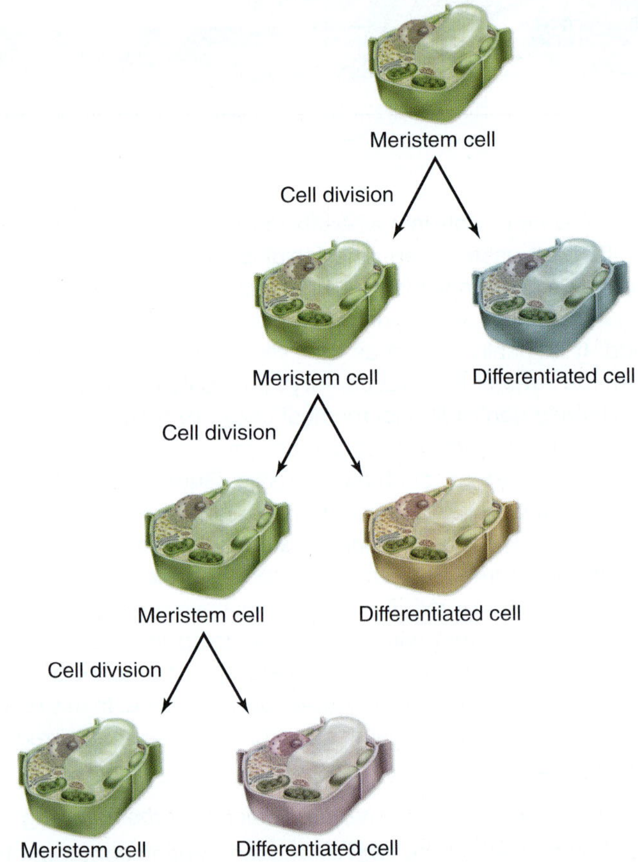

Figure 24.1 Meristem cell division. Meristem cells are located in new and developing parts of a plant. Meristem cells divide to give rise to a differentiating daughter cell and a cell that persists as a meristem cell.

Epidermal Tissue

The entire body of both nonwoody (herbaceous) and young woody plants contains closely packed epidermal cells called the **epidermis** (Gk. *epi,* "over"; *derma,* "skin"). The walls of epidermal cells that are exposed to air are covered with a waxy **cuticle** (L. *cutis,* "skin") to minimize water loss. The cuticle also protects against bacteria and other organisms that might cause disease.

In roots, certain epidermal cells have long, slender projections called **root hairs** (Fig. 24.2*a*). The hairs increase the surface area

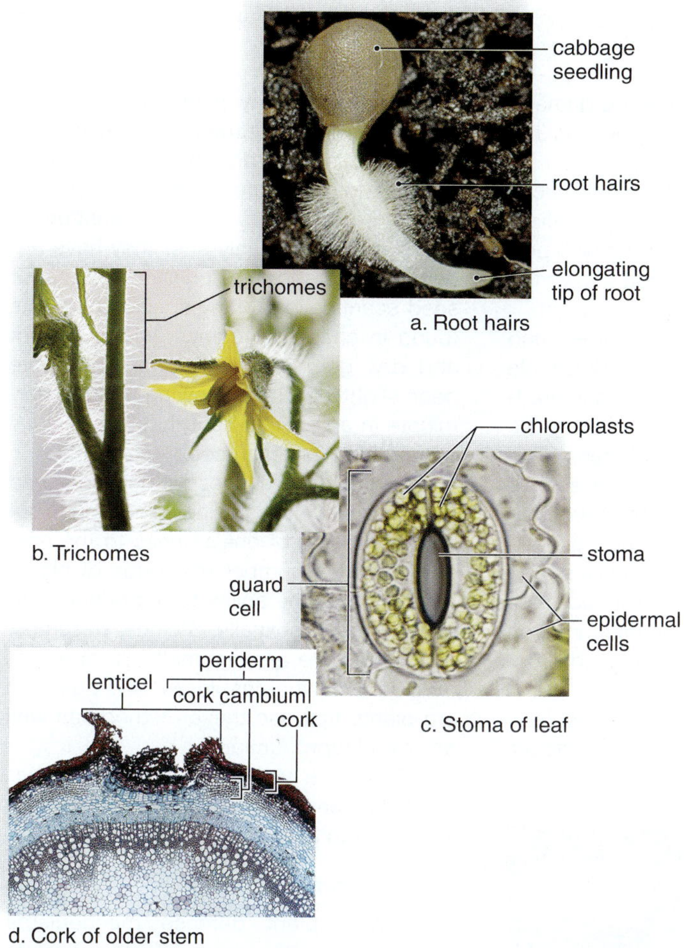

Figure 24.2 Modifications of epidermal tissue. **a.** Root epidermis has root hairs to absorb water. **b.** Trichomes are hairlike epidermal extensions on fruit, leaves, or stems that can protect from herbivory. **c.** Leaf epidermis contains stomata (sing., stoma) for gas exchange. **d.** Periderm includes cork and cork cambium. Lenticels in cork are important in gas exchange.

of the root for absorption of water and minerals, as well as anchoring the plant to various substrates.

On stems, leaves, and reproductive organs, epidermal cells produce hairs, called **trichomes,** that have two important functions: to protect the plant from too much sun and moisture loss and to discourage herbivory (plant eating) (Fig. 24.2*b*). Sometimes trichomes, particularly glandular ones, help protect a plant from herbivores by producing a toxic substance. For example, under the slightest pressure, the stiff trichomes of the stinging nettle lose their tips, forming "hypodermic needles" that inject an intruder with a stinging secretion. (See the Evolution feature, "Survival Mechanisms of Plants," on page 438 for a discussion of some of the defenses that have evolved in angiosperms.)

In leaves, the lower epidermis of eudicots and both surfaces of monocots contain specialized cells called *guard cells* (Fig. 24.2*c*). Guard cells, which are epidermal cells with chloroplasts, surround microscopic pores called **stomata** (sing., stoma). When the stomata are open, gas exchange and water loss occur.

In plants with wood, the epidermis of the stem is replaced by **cork** cells. At maturity, cork cells can be sloughed off (Fig. 24.2*d*). New cork cells are made by a meristem called **cork cambium.** This entire cork area of the plant is called the **periderm** (Gk. *peri,* "around"; *derma,* "skin"). As the new cork cells mature, they increase slightly in volume, and their walls become encrusted with suberin, a lipid material, so that they are waterproof and chemically inert. These nonliving cells protect the plant and help it resist fungal, bacterial, and animal attacks. Some cork tissues, notably from the cork oak (*Quercus suber*), are used commercially for bottle corks and other products.

Notice in Figure 24.2*d* how the cork cambium overproduces cork in certain areas of the stem surface, causing ridges and cracks to appear. These features on the surface are called *lenticels.* Lenticels are the site of gas exchange between the interior of a stem and the air.

Ground Tissue

Ground tissue forms the bulk of a flowering plant; it contains parenchyma, collenchyma, and sclerenchyma cells (Fig. 24.3). **Parenchyma** cells are the most abundant and correspond best to the typical plant cell. These are the least specialized of the cell types and are found in all the organs of a plant. They may contain chloroplasts and carry on photosynthesis, or they may contain colorless plastids that store the products of photosynthesis. A juicy bite from an apple yields mostly storage parenchyma cells. Parenchyma cells line the connected air spaces of a water lily and other aquatic plants. Parenchyma cells can divide and give rise to

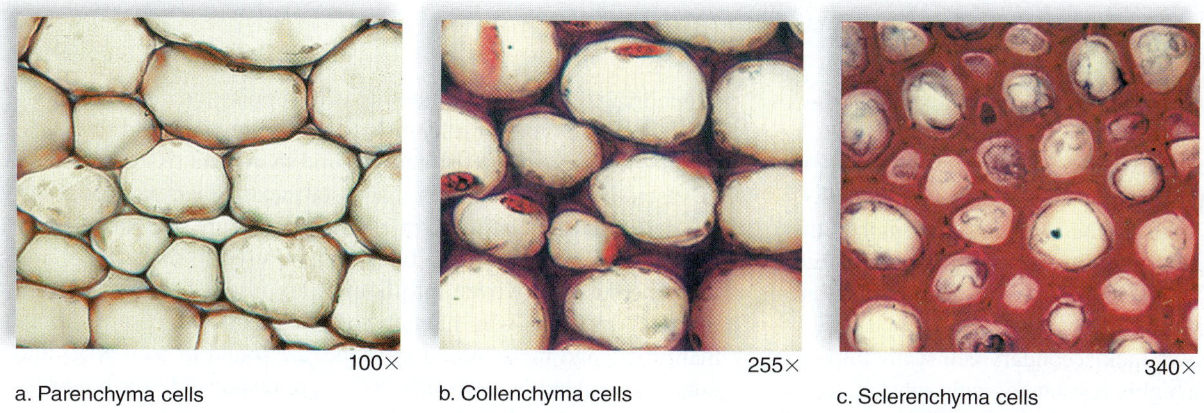

Figure 24.3 Ground tissue cells. **a.** Parenchyma cells are the least specialized of the plant cells. **b.** Collenchyma cells. Notice how much thicker and irregular the walls are compared to those of parenchyma cells. **c.** Sclerenchyma cells are dead and have very thick walls (stained in red)—their only function is to give strong support.

THEME Evolution

Survival Mechanisms of Plants

Plants first made their appearance on land approximately 450 million years ago. Since then, they have evolved a wide variety of mechanisms in order to survive and have established the base of terrestrial ecosystems.

Some groups of plants employ defensive strategies in an attempt to deter predation (Fig. 24A). A defensive strategy is a mechanism that has arisen through a process of natural selection in which the members of a group that possess the strategy compete better than those without it. The more successful competitors usually have a greater opportunity to pass on their genes.

Thorns and spines like those found on black locust trees and cacti are often used to repel large herbivores, but they are generally ineffective against the smaller herbivores. If a tree does become injured, the tracheids and vessel elements of xylem immediately plug up with chemicals that block them off above and below the site of the injury. This response prevents the damage from spreading to other locations on the tree.

Other plants produce toxins or sticky secretions in an attempt to deter predation. Anyone who has come in contact with poison ivy or the sap of a pine tree knows the effectiveness of this defense mechanism.

Cellulose is the polysaccharide found in the cell walls of plants. This indigestible substance makes it difficult for many predators to obtain nutrients from eating the leaves, discouraging continued predation.

Dormancy enables plants to survive in environments that have seasonal conditions that do not allow year-round growth. Deciduous trees shed their leaves and transfer their nutrients into their root systems in response to the decrease in light, temperature, and moisture levels that occurs during the fall.

Seed dormancy allows the next generation to wait until growing conditions are optimal before germinating and competing for resources. Many plant life cycles are timed so that the seeds are produced during the summer, sit dormant throughout the winter, and germinate the following spring.

In some plants, the germination of seeds is triggered only after they have undergone some form of physical trauma, or seed scarification. Many species of plants found in chaparral regions, which are hot and dry, germinate only after they have been slightly burned, allowing them to germinate in an environment that has minimal competition for resources.

Evolutionary success is not measured by the survival of a single individual but by the passing of one's genes to the next generation. A number of groups of plants have evolved the ability to reproduce both sexually and asexually. Stolons, rhizomes, and tubers are asexual methods of reproduction. Strawberries, irises, and potatoes are plants that use these methods as well as sexual reproduction.

The wide variety of survival mechanisms has ensured that plants will be present on Earth for a very long time.

Questions to Consider

1. Which plant defense mechanisms would be the most effective against large predators? Small predators?
2. How would the suppression of fires in a chaparral region impact the plant diversity?

a.

b.

Figure 24A Survival mechanisms used by plants. Plants employ a wide variety of mechanisms to ensure their survival. **a.** Thorns of a locust tree. **b.** Poison ivy contains a toxin within the leaves.

more specialized cells, as when roots develop from stem cuttings placed in water.

Collenchyma cells are like parenchyma cells except they have thicker primary walls. The thickness is uneven and usually involves the corners of the cell. Collenchyma cells often form bundles just beneath the epidermis and give flexible support to immature regions of a plant body. The familiar strands in celery stalks are composed mostly of collenchyma cells.

Sclerenchyma cells have thick secondary cell walls impregnated with **lignin,** which is a highly resistant organic substance that makes the walls tough and hard. Most sclerenchyma cells are dead at maturity; their primary function is to support the mature regions of a plant. Two types of sclerenchyma cells are *fibers* and *sclereids.* Although fibers are occasionally found in ground tissue, most are in vascular tissue, which is discussed next. Fibers are long and slender and may be grouped in bundles, which are sometimes commercially important. Hemp fibers can be used to make rope, and flax fibers can be woven into linen. Flax fibers, however, are not lignified, which is why linen is soft. Sclereids, which are shorter than fibers and more varied in shape, are found in seed coats and nutshells. Sclereids, or "stone cells," are responsible for the gritty texture of pears and the hardness of nuts and peach pits.

Vascular Tissue

There are two types of **vascular tissue. Xylem** transports water and minerals from the roots to the leaves, and **phloem** transports sucrose and other organic compounds, usually from the leaves to the roots. Both xylem and phloem are considered complex tissues, because they are composed of two or more kinds of cells. In the roots, the vascular tissue is located in the **vascular cylinder;** in the stem, it forms **vascular bundles;** and in the leaves, it is found in **leaf veins.**

Xylem contains two types of conducting cells: tracheids and vessel elements, which are modified sclerenchyma cells (Fig. 24.4). Both types of conducting cells are hollow and dead, but the **vessel elements** are larger, may have perforation plates in their end walls, and are arranged to form a continuous vessel for water and mineral transport.

The elongated **tracheids,** with tapered ends, form a less obvious means of transport, but water can move across the end walls and side walls, because there are *pits,* or depressions, where the secondary wall does not form. In addition to vessel elements and tracheids, xylem can contain sclerenchyma fibers that lend additional support as well as parenchyma cells that store various substances. Vascular rays, which are flat ribbons or sheets of parenchyma cells located between rows of tracheids, conduct water and minerals across the width of a plant.

The conducting cells of phloem are specialized parenchyma cells called **sieve-tube members,** arranged to form a continuous sieve tube (Fig. 24.5). Sieve-tube members contain cytoplasm but no nuclei. The term *sieve* refers to a cluster of pores in the end walls, which is known as a sieve plate. Each sieve-tube member has a **companion cell,** which contains a nucleus. The two are connected by numerous plasmodesmata, and the nucleus of the companion cell may control and maintain the life of both cells. The companion cells are also believed to be involved in the transport function of phloem. Sclerenchyma fibers also lend support to phloem.

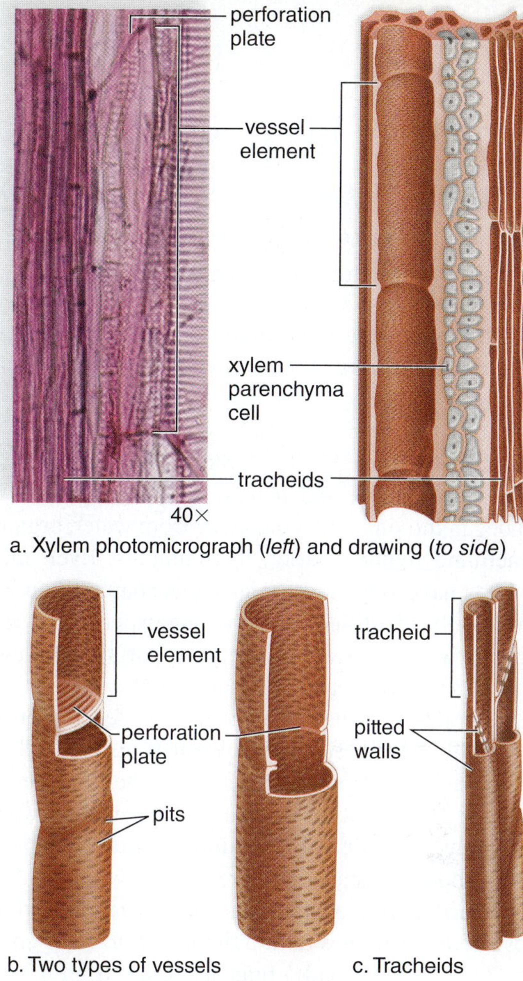

Figure 24.4 Xylem structure. **a.** Photomicrograph of xylem vascular tissue and drawing showing general organization of xylem tissue. **b.** Drawing of two types of vessels (composed of vessel elements)—the perforation plates differ. **c.** Drawing of tracheids.

Check Your Progress 24.1

1. List the three specialized meristems that arise from the apical meristem.
2. List the three specialized tissues in angiosperms and the cells that make up these tissues.
3. Compare the transport functions of xylem and phloem.

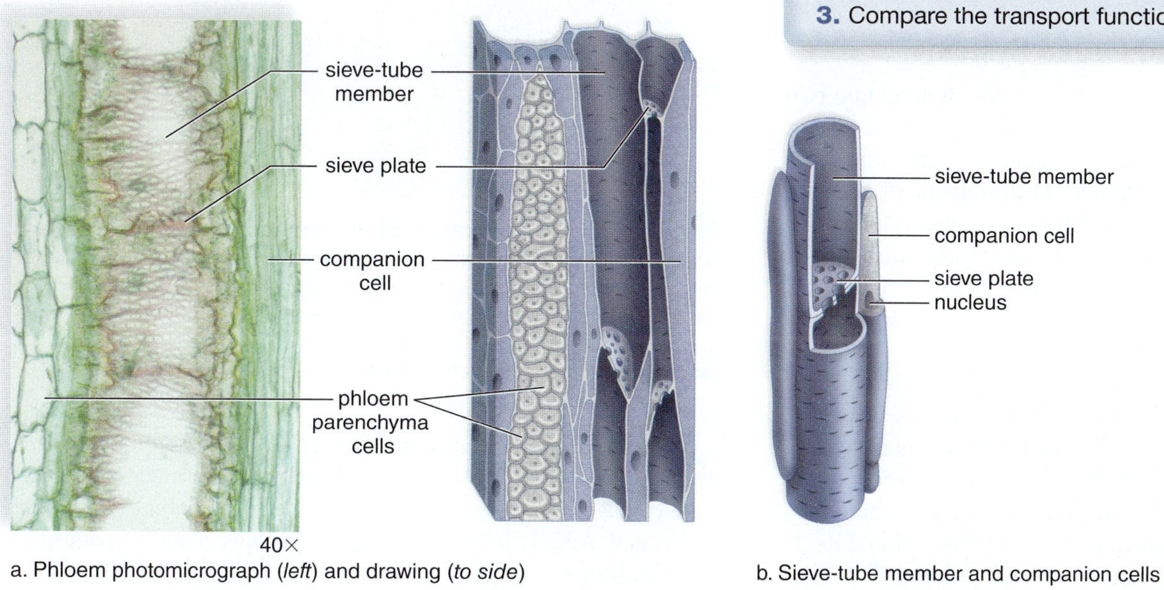

Figure 24.5 Phloem structure. **a.** Photomicrograph of phloem vascular tissue and drawing showing general organization of phloem tissue. **b.** Drawing of sieve tube (composed of sieve-tube members) and companion cells.

24.2 Organs of Flowering Plants

Learning Outcomes

Upon completion of this section, you should be able to

1. Compare the structure and function of roots, stems, and leaves.
2. List and describe the key features of monocots and eudicots.
3. Explain the difference between annual and perennial plants.

As discussed in Chapter 23, the earliest plants were simple and lacked true stems, roots, and leaves. As plants gained vascular tissue and began moving onto land away from water, organs developed to facilitate living in drier environments. Even though all vascular plants have vegetative organs, this chapter focuses on the organs commonly identified with the angiosperms, or flowering plants. **Vegetative organs** are all the plant parts except the reproductive structures of flowers, fruits, and seeds.

A flowering plant, whether a cactus, a water lily, or an apple tree, has a shoot system and root system (Fig. 24.6). The **shoot system** of a plant is composed of the **stem,** the branches, and the leaves. A stem supports the leaves in a way that exposes each one to as much sunlight as possible. In addition, the stem transports materials between roots and leaves and produces new tissue. At the end of a stem, a **terminal bud** contains an apical meristem and produces new leaves and other tissues during the initial *primary growth* of a plant (Fig. 24.7). Lateral (side) branches grow from a lateral bud located at the angle where a leaf joins a stem. A **node** occurs where a leaf or leaves are attached to the stem, and an **internode** is the region between nodes. Vascular tissue transports water and minerals from the roots through the stem to the leaves and transports the products of photosynthesis, usually in the opposite direction. The **root system** simply consists of the roots. The root tip also contains an apical meristem and results in primary growth downward.

Ultimately, the three vegetative organs—the root, the stem, and the leaf—perform functions that allow a plant to live and grow. Flowers and fruit are reproductive organs and will be discussed in Chapter 27.

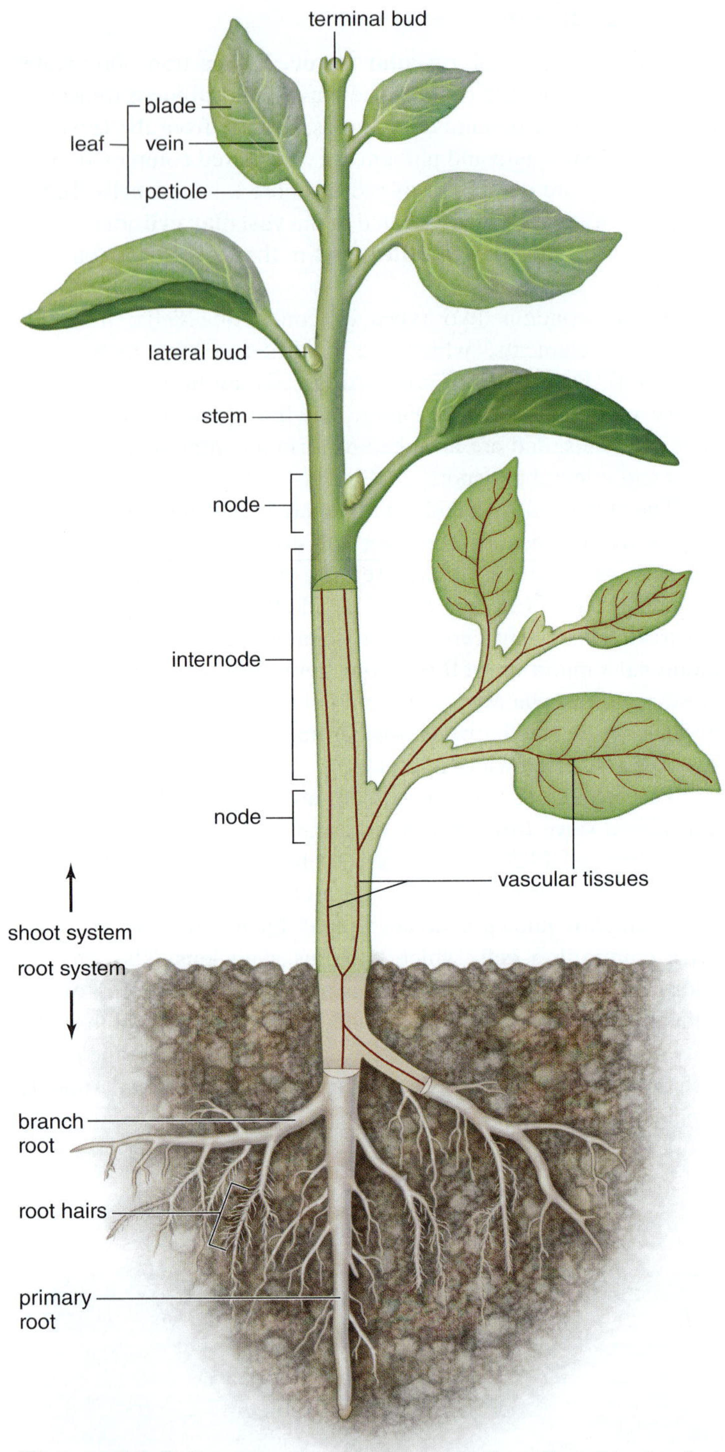

Figure 24.6 Organization of a plant body. The body of a plant consists of a root system and a shoot system. The shoot system contains the stem and leaves, two types of plant vegetative organs. Axillary buds can develop into branches of stems or flowers, the reproductive structures of a plant. The root system is connected to the shoot system by vascular tissue (brown) that extends from the roots to the leaves.

Variation of Organs and Organ Systems

As described in Chapter 23, flowering plants are divided into two groups, depending on the number of *cotyledons,* or seed leaves, in the embryonic plant. Plants with a seed containing only one cotyledon are referred to as monocotyledons, or **monocots.** Plants with seeds that contain two cotyledons are known as eudicotyledons, or **eudicots** (Fig. 24.8). Cotyledons of eudicots supply nutrients for seedlings, but the cotyledons of monocots act as transfer tissue, and the nutrients are derived from the endosperm before the true leaves begin photosynthesizing.

The vascular (transport) tissue is organized differently in monocots and eudicots. In the monocot root, vascular tissue occurs in a ring; in the monocot stem, the vascular bundles, which contain vascular tissue surrounded by a bundle sheath, are scattered. In the eudicot root, the xylem, which transports water and minerals, is star-shaped, and the phloem, which transports organic nutrients, is located between the points of the star. In a eudicot stem, the vascular bundles occur in a ring.

Leaf veins are vascular bundles within a leaf. Monocot leaves usually have a smooth margin (edge) and parallel venation, whereas eudicot leaf margins can be smooth but also commonly seen as lobed or serrated. Additionally, eudicot leaf veins have a netlike pattern.

Adult monocots and eudicots also have structural differences in the number of flower parts. Monocots have their flower parts

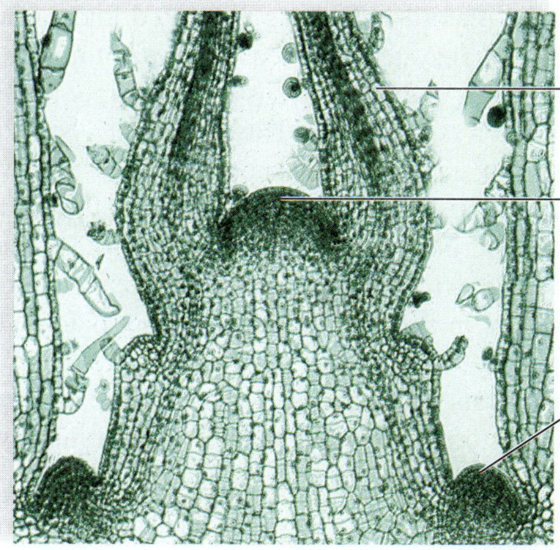

Figure 24.7 A terminal bud. A terminal bud contains an apical meristem and produces new leaves and other tissues during primary growth. Lateral branches grow from the lateral bud.

arranged in multiples of three, whereas eudicots have their flower parts arranged in multiples of four or five. Monocot pollen grains have one pore, and eudicot pollen grains usually have three pores. Recall from Chapter 23 that it was fossilized pollen characteristic of flowering plants that helped date the appearance of these plants in geologic time.

Although the distinctions between monocots and eudicots may seem of limited importance, they do, in fact, affect many aspects of their structure. The eudicots are the larger group and include some of our most familiar flowering plants—from dandelions to oak trees. The monocots include grasses, lilies, orchids, and palm trees, among others. Some of our most significant food sources are monocots, including rice, wheat, and corn.

Vegetative Organs and Plant Life Cycles

Annual plants live for only one growing season, *biennial* plants for two growing seasons, and **perennial** plants for three or more seasons. Perennials, such as certain roses, irises, and potatoes expend energy making vegetative structures, such as wood, bulbs, underground tubers, and leaf buds. These modified organs help the plant survive year after year. Annual plants, such as sunflowers and peas, produce enough vegetative structures to support flower and seed production. After seeds are produced and dispersed, the entire plant dies.

The difference comes down to two important flower-inducing genes, *LEAFY* and *Apetala 1*. In the shoot system, the shoot apical meristem divides, and vegetative organs, such as leaves, are made. When the *Apetala 1* gene turns on (with the help of the *LEAFY* gene), the apical meristem stops vegetative growth and switches to flower production. The timing of these genetic switches determines whether a plant is an annual or a perennial. Scientists have been able to block flower-inducing genes in annual plants and induce them to grow like perennials by diverting energy into the production of vegetative organs.

Check Your Progress 24.2

1. Describe how the plant body is organized.
2. List the three vegetative organs in a plant, and state their major functions.
3. Compare and contrast the structures of monocots and eudicots.

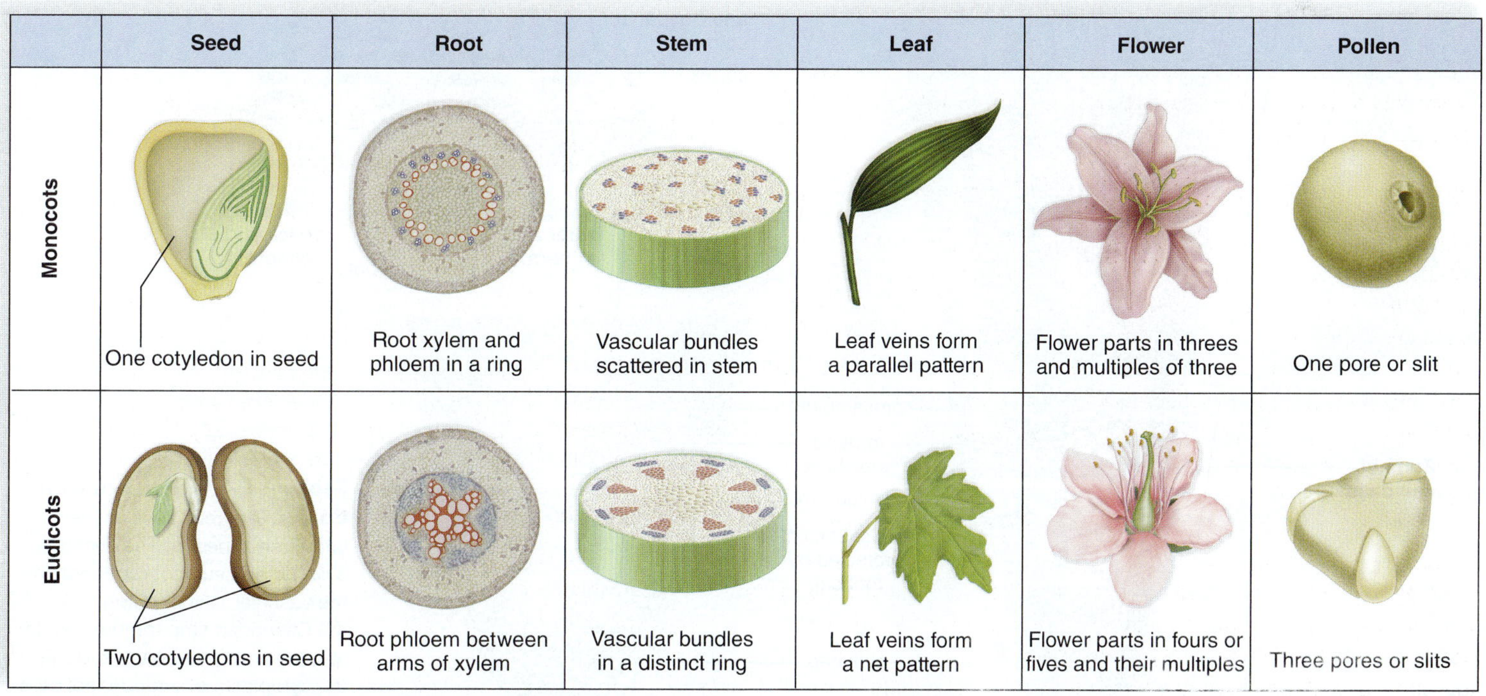

Figure 24.8 Flowering plants are either monocots or eudicots. Six features, illustrated here, are used to distinguish monocots from eudicots: the number of cotyledons; the arrangement of vascular tissue in roots, stems, and leaves; the number of flower parts, and the differences in pollen structure.

24.3 Organization and Diversity of Roots

Learning Outcomes

Upon completion of this section, you should be able to

1. Describe the tissue types that are found in each zone of a root.
2. Identify the structural differences between the roots of monocots and eudicots.
3. Describe the various specializations and symbiotic relationships that lead to root diversity.

The root system in the majority of plants is located underground. As a rule of thumb, the root system is at least equivalent in size and extent to the shoot system. An apple tree has a much larger root system than a corn plant, for example. The extensive root system of a plant anchors it in the soil, gives the plant support, and absorbs water and minerals. The cylindrical shape of a root allows it to penetrate the soil as it grows and permits water to be absorbed from all sides. The absorptive capacity of a root is dependent on its many branches, which all bear root hairs in a zone near the tip. Root hairs, which are projections from epidermal root-hair cells, are the structures that absorb water and minerals. In addition, roots produce hormones and can function in the storage of carbohydrates. Carrots and sweet potatoes are examples of storage roots.

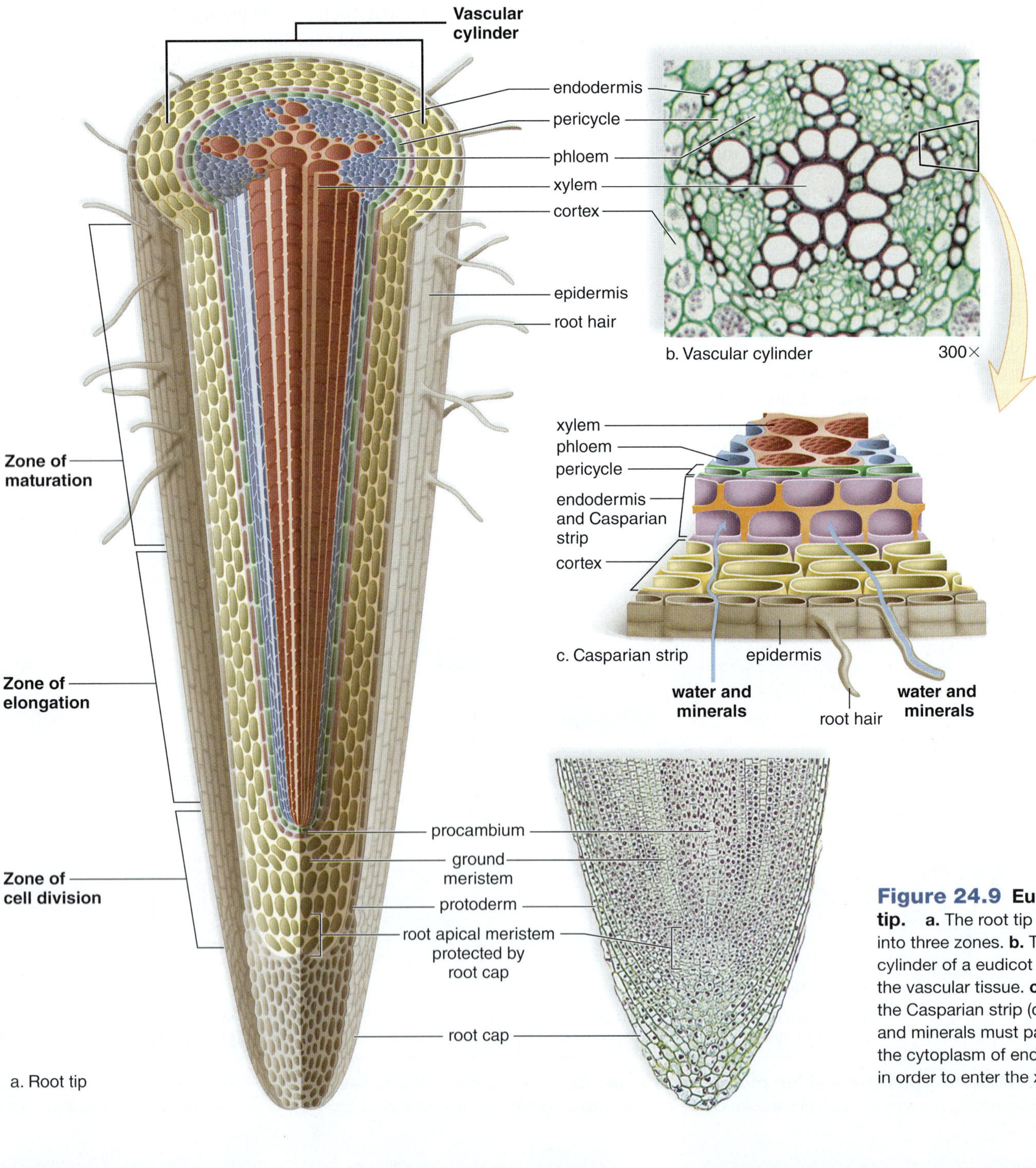

Figure 24.9 Eudicot root tip. **a.** The root tip is divided into three zones. **b.** The vascular cylinder of a eudicot root contains the vascular tissue. **c.** Because of the Casparian strip (orange), water and minerals must pass through the cytoplasm of endodermal cells in order to enter the xylem.

Cells and Tissues of a Eudicot Root

The longitudinal section of a eudicot root (Fig. 24.9*a*) reveals zones where cells are in various stages of differentiation as primary growth occurs. The **root apical meristem** is protected by the **root cap.** Root cap cells have to be replaced constantly, because they get ground off by rough soil particles as the root grows. Notice that the root apical meristem is in the *zone of cell division,* which continuously provides new cells to the region above—the zone of elongation. In the *zone of elongation,* the cells grow longer as they become specialized. The *zone of maturation,* which contains fully differentiated cells, is recognizable because here root hairs are found on many of the epidermal cells.

Figure 24.9*a* also shows a cross section of a root at the zone of maturation. These specialized tissues are identifiable:

Epidermis The epidermis forms the outer layer of the root and consists of only a single layer of cells. The majority of epidermal cells are thin-walled and rectangular, but in the zone of maturation, many epidermal cells have root hairs. These can project as far as 5–8 mm into the soil particles.

Cortex Moving inward, next to the epidermis are the large, thin-walled parenchyma cells that make up the cortex of the root. The cortex mainly functions in food storage. These irregularly shaped cells are loosely packed, making it possible for absorbed water and minerals to weave their way in between the cells or travel through the cells toward the middle of the root.

Endodermis (Gk. *endon,* "within"; *derma,* "skin") The endodermis is the next single layer of cells between the cortex and the inner vascular cylinder. The endodermal cells have a very important role in that they regulate the water and ions that can move toward the vascular tissue. Notice the movement of water in Figure 24.9*c*. In one option, water can move in between the cells of the cortex but is suddenly blocked at the endodermis. Like a tight rubber band, the endodermal cells are fitted with a **Casparian strip** made up of the protein lignin and a waterproof lipid substance called *suberin.* This strip prevents the passage of water and mineral ions *between* cell walls; instead, water and ions are forced *through* the endodermal cells, where biological regulation occurs.

Pericycle The pericycle, the first layer of cells within the vascular cylinder, can continue to divide and is the starting point where branch, or lateral, roots develop (Fig. 24.10).

Vascular tissue The main portion of the vascular cylinder contains xylem and phloem. The xylem appears star-shaped in eudicots, because several arms of tissue radiate from a common center (see Fig. 24.9*b*). The phloem is found in separate regions between the arms of the xylem.

Organization of Monocot Roots

Monocot roots have the same growth zones and undergo the same secondary growth as eudicot roots. However, the organization of their tissues is slightly different. The ground tissue of a monocot root's **pith** is centrally located and is surrounded by a vascular ring composed of alternating xylem and phloem bundles (Fig. 24.11). Monocot roots also have pericycle, endodermis, cortex, and epidermis.

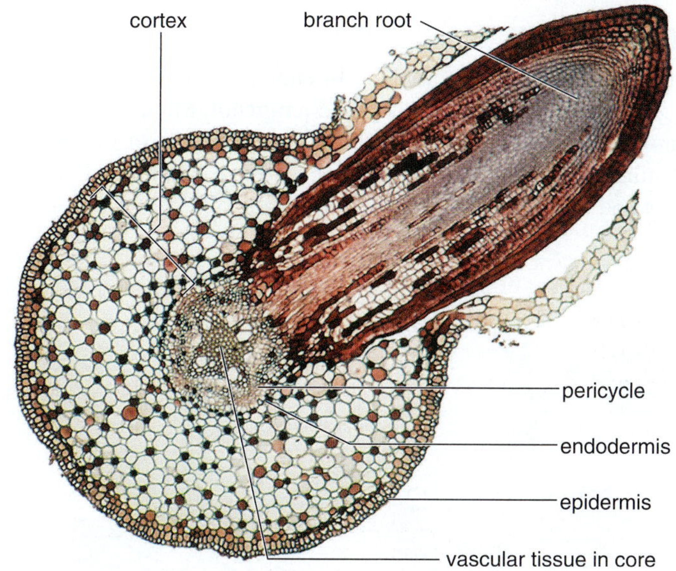

Figure 24.10 Branching of a eudicot root. This cross section of a willow, *Salix,* shows the origination and growth of a branch root from the pericycle.

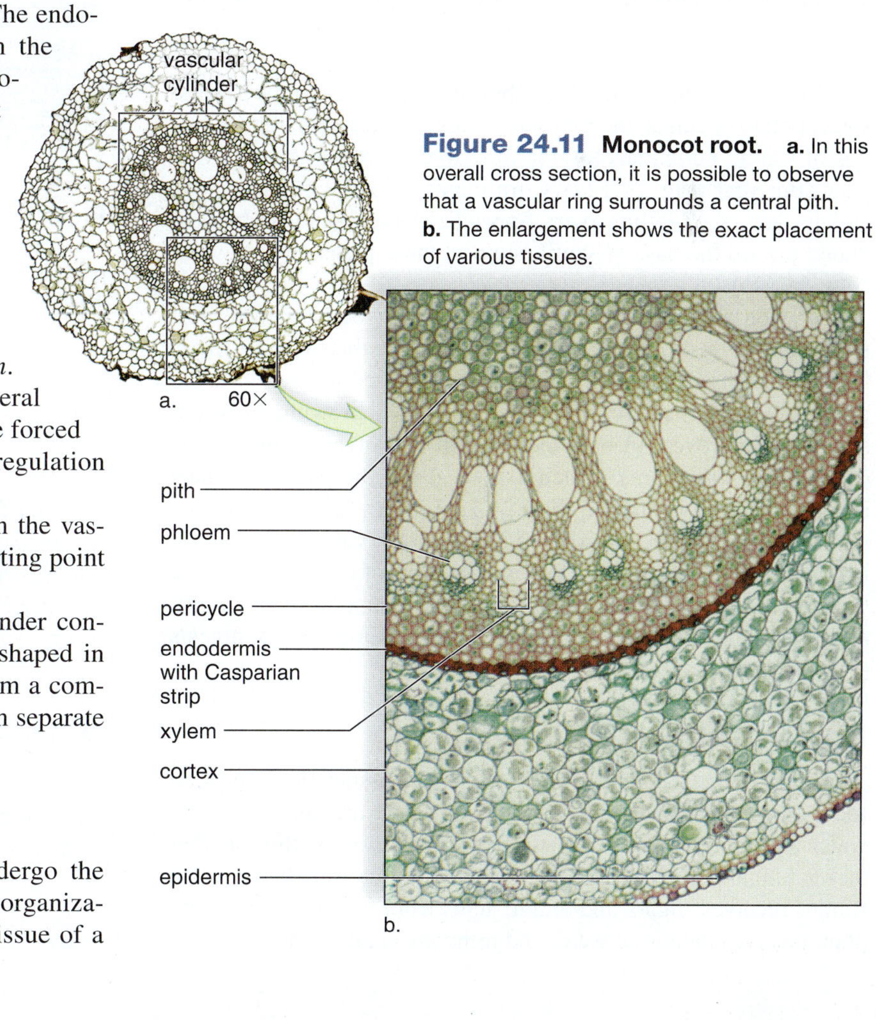

Figure 24.11 Monocot root. a. In this overall cross section, it is possible to observe that a vascular ring surrounds a central pith. **b.** The enlargement shows the exact placement of various tissues.

Root Diversity

Roots tend to be of two varieties. In eudicots, the first (primary) root grows straight down and is called a **taproot.** Monocots have a **fibrous root** system, which may have large numbers of fine roots of similar diameter. Many mature plants have a combination of taproot and fibrous root systems.

Some plants have roots that possess a variety of adaptations to better perform their functions. Root adaptations can improve anchorage to the ground or the storage of carbohydrates. Other root systems need to increase absorption of water, minerals, oxygen, or sunlight.

Most roots store some food, but in certain plants, such as the sweet potato or carrot, the roots are enlarged and store large quantities of starch. Perennials rely on this stored carbohydrate to grow year after year (Fig. 24.12*a*). Corn plants are straight, top-heavy, and in danger of falling over. Specialized corn prop roots, produced toward the base of stems, support these plants in high wind (Fig. 24.12*b*).

As you will learn in Chapter 25, roots need oxygen to perform cellular respiration. Roots normally get their oxygen from the air pockets in the soil, but oxygen becomes scarce when the plant grows exclusively in water. Large trees growing in water, such as mangrove and bald cypress, have evolved spongy roots called *pneumatophores,* which extend above the water's surface and enhance gas exchange (Fig. 24.12*c*). Many *epiphytes* (plants that live in or on trees) have aerial roots for a variety of reasons. English ivy uses aerial roots to climb the bark of trees. Orchids use their roots to capture moisture in the air and support their weight, and in some instances, green aerial roots perform photosynthesis (Fig. 24.12*d*).

Figure 24.12 Root specializations. **a.** Sweet potato plants have food storage roots. **b.** Prop roots are specialized for support. **c.** The pneumatophores of this tree allow it to acquire oxygen even though it lives in water. **d.** The aerial roots of orchids offer physical support, water and nutrient uptake, and in some cases even photosynthesis.

Root Symbiotic Relationships

As described in Chapter 22, *mycorrhizae* are associations between roots and fungi. Plants that have mycorrhizae are able to extract water and minerals from the soil better than those with roots that lack a fungal partner. This relationship is mutualistic, because the fungus receives sugars and amino acids from the plant, while the plant receives increased water and minerals via the fungus.

Peas, beans, and other legumes have **root nodules,** where nitrogen-fixing bacteria live. Plants cannot extract nitrogen from the air, but the bacteria within the nodules can take up and reduce atmospheric nitrogen. The plant gets a source of nitrogen from the bacteria, and the bacteria receive carbohydrates from the plant.

Animation Root Nodule Formation

Check Your Progress 24.3

1. Explain the relationship between the root apical meristem and the root cap.
2. List the function of the endodermis and the Casparian strip in a root.
3. Describe the advantages of some root specializations.

24.4 Organization and Diversity of Stems

Learning Outcomes

Upon completion of this section, you should be able to

1. Identify the anatomical structures of a woody twig.
2. Recognize the differences in the arrangement of vascular tissue between a herbaceous dicot and a monocot stem.
3. Describe how secondary growth of a woody stem results in the various tissues within it.
4. Characterize the variations of stem diversity.

A stem is the main axis to a plant's shoot system. It carries leaves and flowers and supports the plant's weight. Stiff stems rising upward against gravity is an ancient adaptation that allowed plants to move into terrestrial habitats.

The anatomy of a woody twig helps us review the organization of a stem (Fig. 24.13). The **terminal bud** contains the shoot tip protected by modified leaves called bud scales. Each spring when growth resumes, bud scales fall off and leave a scar. Each bud-scale scar indicates 1 year of growth. Leaf scars and bundle scars mark the locations of leaves that have dropped. Dormant axillary buds that will give rise to branches or flowers are also found here.

As seasonal growth resumes, the apical meristem at the shoot tip produces new cells, which increase the height of the stem. The **shoot apical meristem** is protected within the terminal bud, where leaf primordia (immature leaves) envelop it (Fig. 24.14). The leaf primordia mark the locations of nodes; the portion of stem in between nodes is an internode. As a stem grows, the internodes increase in length.

In addition to leaf primordia, the three specialized types of primary meristem (see section 24.1) develop from a shoot apical meristem (Fig. 24.14*b*). These primary meristems contribute to the length of a shoot. The *protoderm,* the outermost primary meristem, gives rise to the epidermis. The *ground meristem* produces two tissues composed of parenchyma cells: the pith and the cortex. The *procambium* (see Fig. 24.13*a*) produces the first xylem cells, called primary xylem, and the first phloem cells, called primary phloem.

Differentiation continues as certain cells become the first tracheids or vessel elements of the xylem within a vascular bundle. The first sieve-tube members of a vascular bundle do not have companion cells and are short-lived (some live only a day before being replaced). Mature vascular bundles contain fully differentiated xylem, phloem, and a lateral meristem called **vascular cambium.** Vascular cambium is discussed more fully later in this section.

Herbaceous Stems

Basil, dandelions, and tulips are all examples of plants with **herbaceous stems** (L. *herba,* "vegetation, plant") that exhibit mostly primary growth—no wood or bark. The outermost tissue

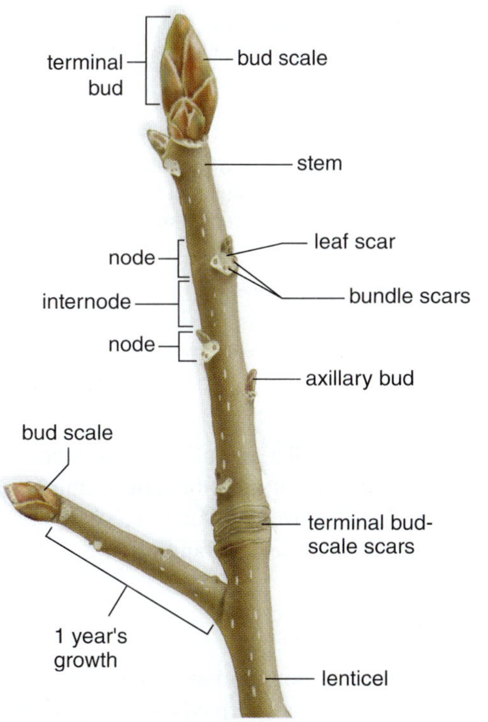

Figure 24.13 Woody twig. The major parts of a stem are illustrated by a woody twig collected in winter.

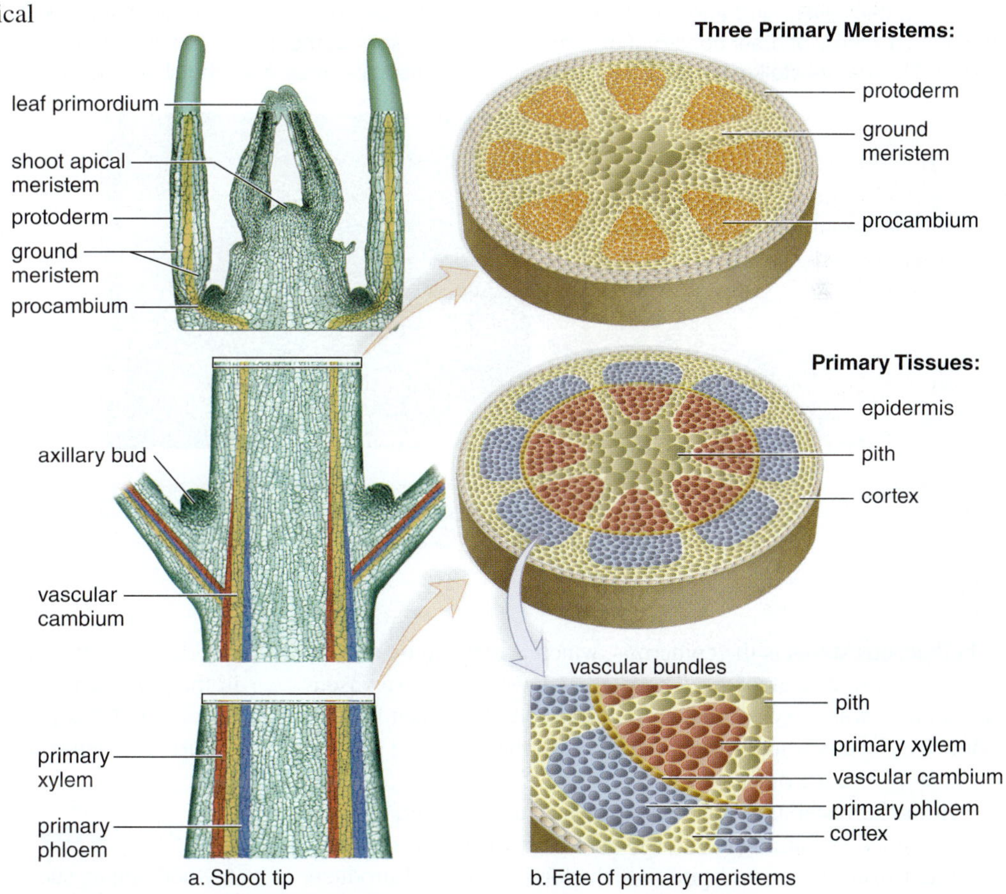

Figure 24.14 Shoot tip and primary meristems. a. The shoot apical meristem within a terminal bud is surrounded by leaf primordia. **b.** The shoot apical meristem produces the primary meristems: Protoderm gives rise to epidermis; ground meristem gives rise to pith and cortex; and procambium gives rise to vascular tissue, including primary xylem, primary phloem, and vascular cambium.

THEME **Nature of Science**

The Many Uses of Bamboo

Because of its resilience, amazing rate of growth, and ability to grow in a variety of climates, bamboo is quickly becoming a valuable crop. Certain varieties of bamboo are capable of growing up to a foot per day and can reach their full height within 1 year in the right conditions.

Bamboo is classified as a grass that contains varieties ranging in height from 1 foot to over 100 feet. Globally, there are over 1,400 species of bamboo. Approximately 900 can be found in tropical climates, and the remaining 500 are in temperate environments. Several varieties of bamboo are even native to the eastern and southeastern United States. Currently, most bamboo is grown in and exported from China.

Bamboo is recognized for its versatility as a building material, toilet paper, commercial food product, and clothing material. It can be processed into roofing material, flooring (Fig. 24B), and support beams, as well as a variety of other construction materials. The mature stalks can be used as support columns in "green" construction. It can also be used to replace the steel reinforcing rods that are typically used in concrete-style construction. Bamboo products are three times harder than oak.

Although not used extensively in the commercial food market, bamboo does have a variety of culinary uses. The shoots are often used in Asian dishes as a type of vegetable (Fig. 24B). Shoots can be boiled and added to a variety of dishes or eaten raw. In China, bamboo is used to make certain alcoholic drinks; other Asian countries make bamboo soups, pancakes, and broths. The hollow bamboo stalk can serve as a container to cook rice and soups, giving the foods a subtle but distinctive taste. Another use is modifying bamboo into cooking and eating utensils, most notably chopsticks.

Clothing products are now being made out of bamboo. Clothing made of bamboo fibers is reported to be very light and extremely soft. Bamboo also has the ability to wick moisture away from the skin, making it ideal to wear during exercise. Several lines of baby clothing are being made from bamboo.

Bamboo is being recognized as one of the most eco-friendly crops. It requires lower amounts of chemicals or pesticides to grow. It removes nearly five times more greenhouse gases and produces nearly 35% more oxygen than an equivalent stand of trees. Harvesting can be done from the second or third year of growth through the fifth to seventh year of growth. Because bamboo is a perennial, a stand can regrow after yearly harvesting, instead of requiring replanting. Greenhouses with cloned bamboo produced from tissue culture will speed up the production of this plant even more.

The bamboo-goods industry started increasing in popularity in the United States during the mid-1990s and is now a multibillion-dollar industry. Its versatility, hardiness, and ease of growing and a greater awareness of environmental issues are quickly making bamboo an ideal natural product, which may someday replace metal, wood, and plastics.

Questions to Consider

1. What might be some drawbacks to planting bamboo for human uses?
2. Why hasn't bamboo become a more popular crop in the United States?

Figure 24B The many uses of bamboo. Bamboo is quickly becoming a multifunctional product in today's society. Uses range from building materials to food to clothing.

of herbaceous stems is the epidermis, which is covered by a waxy cuticle to prevent water loss. These stems have distinctive vascular bundles, where xylem and phloem are found. In each bundle, xylem is typically found toward the inside of the stem, and phloem is found toward the outside.

In the herbaceous eudicot stem, such as a sunflower, the vascular bundles are arranged in a distinct ring where the cortex is separated from the central pith, which stores water and products of photosynthesis (Fig. 24.15). The cortex is sometimes green and carries on photosynthesis.

In a monocot stem, such as a corn stalk, the vascular bundles are scattered throughout the stem, and often the cortex and pith are not clearly distinguishable (Fig. 24.16). The stems of one monocot in the grass family, bamboo, have been of great benefit in human history, and this plant continues to be useful to us today (see the Nature of Science feature, "The Many Uses of Bamboo").

Woody Stems

A woody plant, such as an oak tree, has both primary and secondary tissues. Primary tissues are thee new tissues formed each year from the primary meristems. Secondary tissues develop during the first and subsequent years of growth from lateral meristems, forming the vascular cambium and cork cambium. *Primary growth,* which

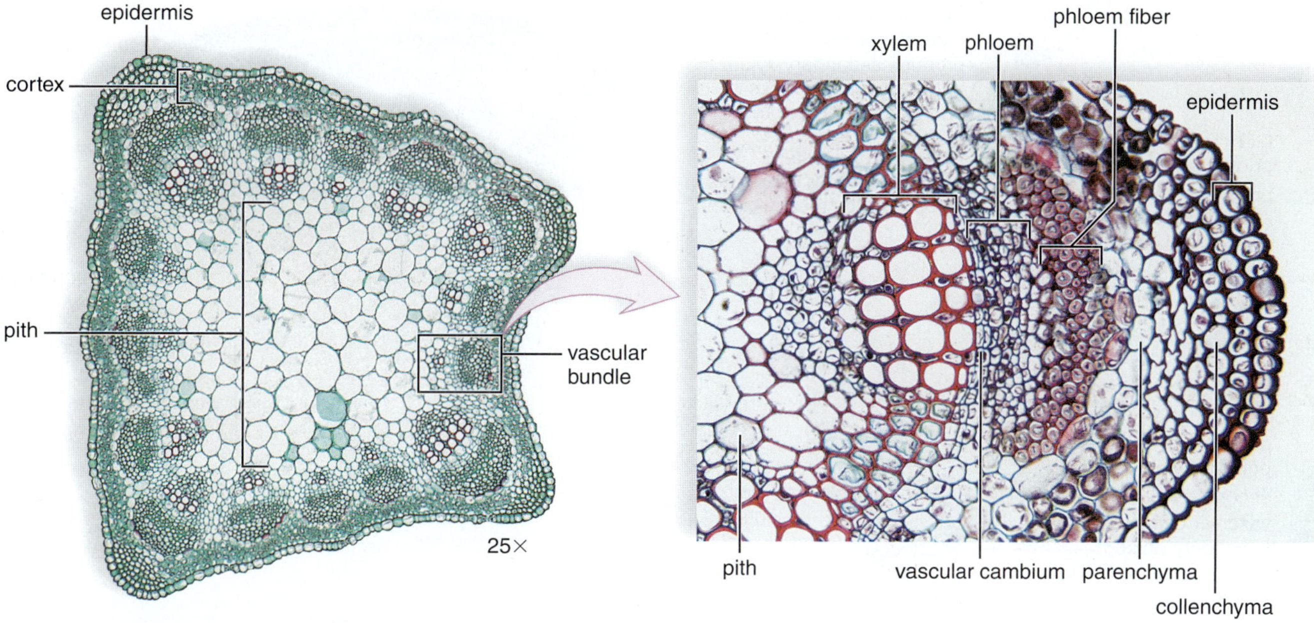

Figure 24.15 **Herbaceous eudicot stem.**

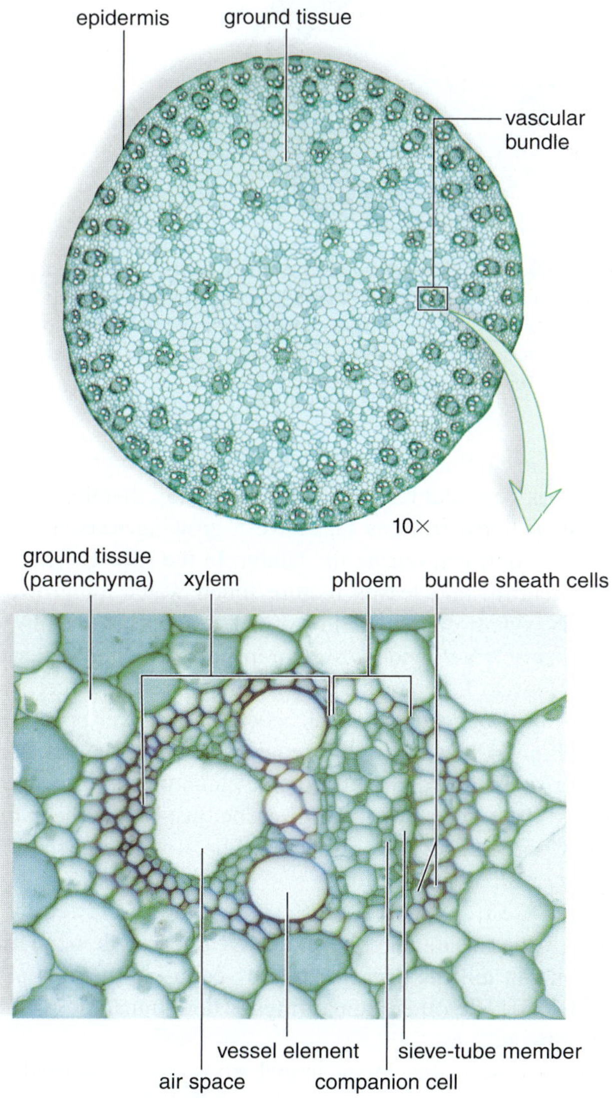

Figure 24.16 **Monocot stem.**

occurs in all plants, increases the length of plant stems and roots; *secondary growth,* which occurs only in conifers and woody eudicots, increases the girth of trunks, stems, branches, and roots.

Trees and shrubs undergo secondary growth because of a change in the location and activity of vascular cambium (Fig. 24.17). In herbaceous plants, vascular cambium is present between the xylem and phloem of each vascular bundle. In woody plants, the vascular cambium develops to form a ring of meristem that divides parallel to the surface of the plant and produces new xylem toward the inside and phloem toward the outside on a yearly basis.

Eventually, a woody eudicot stem has an entirely different organization than that of a herbaceous eudicot stem. A woody stem forms three distinct areas: the bark, the wood, and the pith. Vascular cambium occurs between the bark and the wood, which causes woody plants to increase in girth. Cork cambium, occurring first beneath the epidermis, is instrumental in the production of cork in woody plants.

Also notice in Figure 24.17 the *xylem rays* and *phloem rays* that are visible in the cross section of a woody stem. Rays consist of parenchyma cells that permit lateral conduction of nutrients from the pith to the cortex, as well as some storage of food. A phloem ray can vary in width and is a continuation of a xylem ray.

Bark

The **bark** of a tree contains cork, cork cambium, cortex, and phloem. It is very harmful to remove the bark of a tree, because without phloem, organic nutrients cannot be transported. Although new phloem tissue is produced each year by vascular cambium, it does not build up in the same manner as xylem. In North America, herbivores, such as beavers, elk, and porcupines, eat bark and inadvertently girdle trees. Girdling involves removing bark from around the tree, inevitably leading to the death of the tree. Many forest management agencies wrap the trunks of trees to protect them from animal damage.

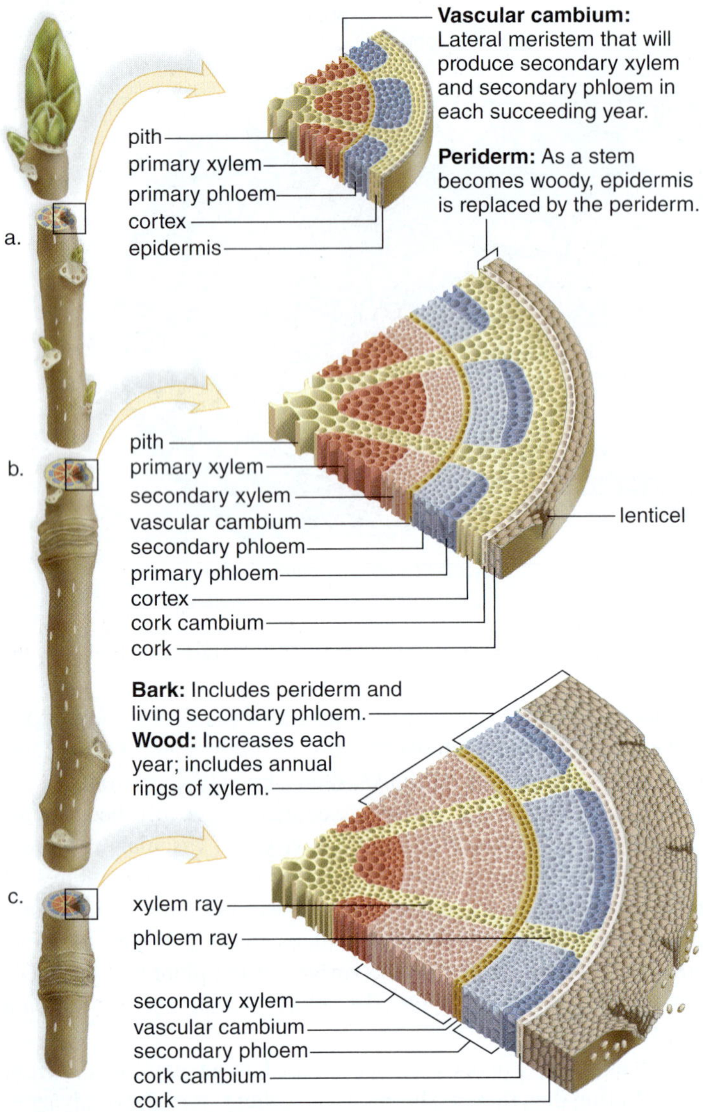

Figure 24.17 Diagrams of secondary growth of stems. **a.** Diagram showing eudicot herbaceous stem just before secondary growth begins. **b.** Diagram showing that secondary growth has begun. Periderm has replaced the epidermis. Vascular cambium produces secondary xylem and secondary phloem each year. **c.** Diagram showing a 2-year-old stem. The primary phloem and cortex will eventually disappear, and only the secondary phloem (within the bark) produced by vascular cambium will be active that year. Secondary xylem builds up to become the annual rings of a woody stem.

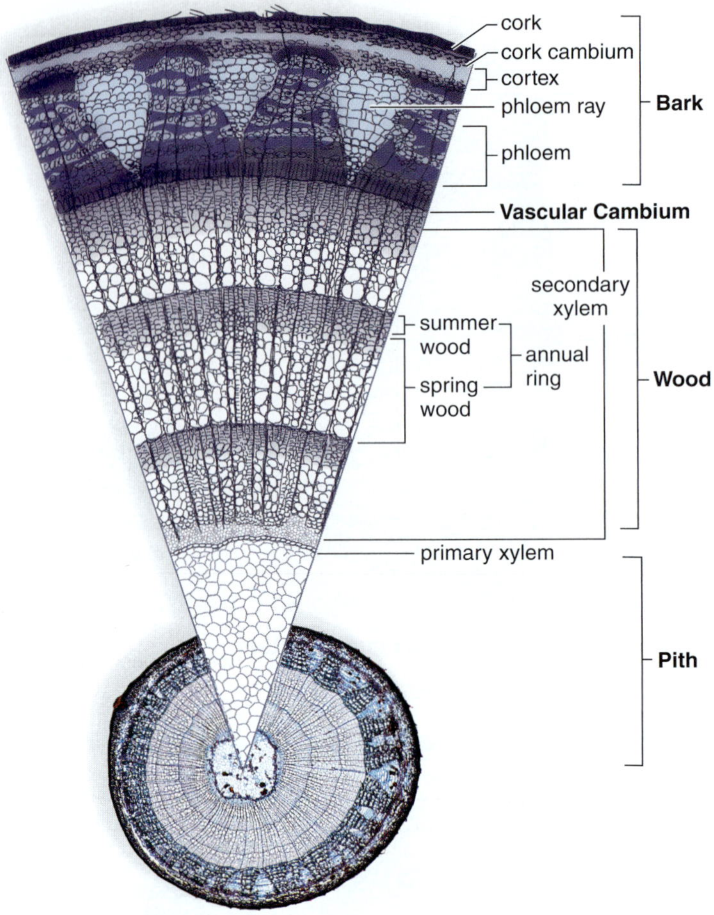

Figure 24.18 Three-year-old woody twig. The buildup of secondary xylem in a woody stem results in annual rings, which tell the age of the stem. The rings can be distinguished because each one begins with spring wood (large vessel elements) and ends with summer wood (smaller and fewer vessel elements).

In bark, the region of active cell division occurs at the **cork cambium.** When cork cambium first begins to divide, it produces tissue that disrupts the epidermis and replaces it with **cork** cells (see Fig. 24.2*d*). Recall that cork cells are impregnated with suberin, a waxy layer that makes them waterproof but also causes them to die, and that suberin also makes up the Casparian strip in roots. In a woody stem, gas exchange is impeded, except at lenticels, which are pockets of loosely arranged cork cells not impregnated with suberin.

Wood

When a plant first begins growing, the xylem is made by the apical meristem. Later, as the plant matures, xylem is made by the vascular cambium and is called secondary xylem. **Wood** is secondary xylem that builds up year after year, thereby increasing the girth of trees. In trees that have a growing season, vascular cambium is dormant during the winter. In the spring, when moisture is plentiful and leaves require much water for growth, the secondary xylem contains wide vessel elements with thin walls. In this *spring wood,* wide vessels transport sufficient water to the growing leaves. Later in the season, moisture is scarce, and the wood at this time, called *summer wood,* has a lower proportion of vessels (Fig. 24.18). Strength is required, because the tree is growing larger, and summer wood contains numerous, thick-walled tracheids. At the end of the growing season, just before the cambium becomes dormant again, only heavy fibers with especially thick secondary walls may develop.

When the trunk of a tree has spring wood followed by summer wood, the two together make up 1 year's growth, or an **annual ring.** You can tell the age of a tree by counting the annual rings (Fig. 24.19*a*). The outer annual rings, where transport occurs, are called sapwood.

In older trees, the inner annual rings, called heartwood, no longer function in water transport. The cells become plugged with deposits, such as resins, gums, and other substances that inhibit the growth of bacteria and fungi. Heartwood may help support a

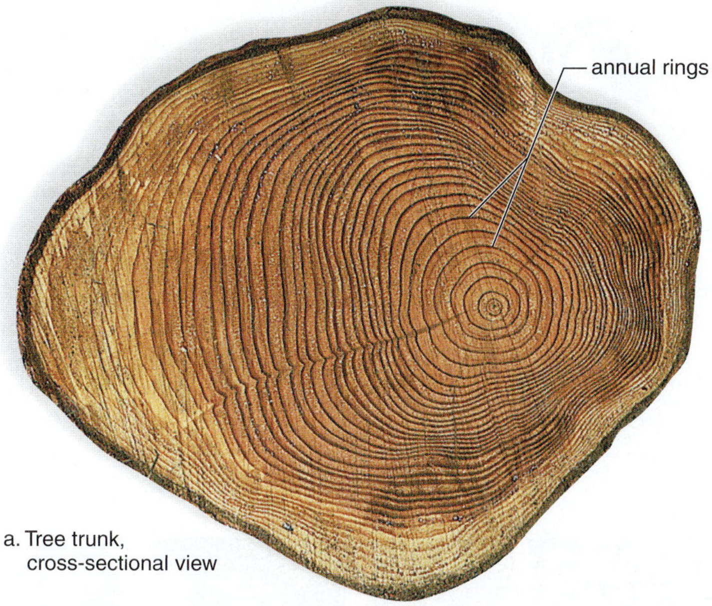

a. Tree trunk, cross-sectional view

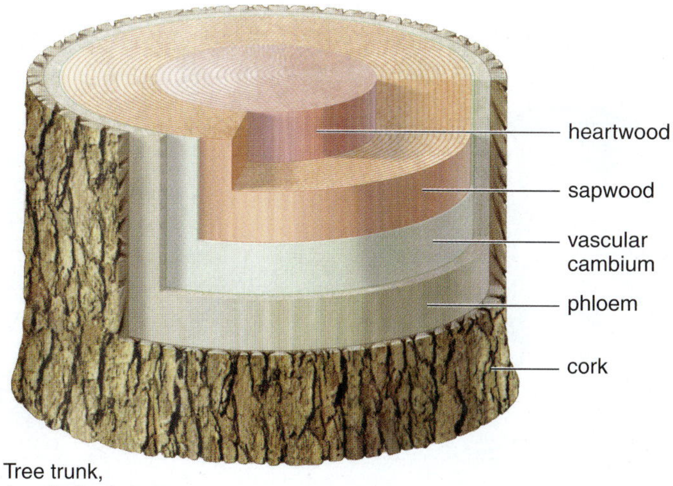

b. Tree trunk, longitudinal view

Figure 24.19 Tree trunk. **a.** A cross section of a 39-year-old larch, *Larix decidua.* The xylem within the darker heartwood is inactive; the xylem within the lighter sapwood is active. **b.** The relationship of bark, vascular cambium, and wood is retained in a mature stem. The pith has been buried by the growth of layer after layer of new secondary xylem.

tree, although some trees stand erect and live for many years after the heartwood has rotted away. Figure 24.19*b* shows the layers of a woody stem in relation to one another.

The annual rings are used to tell the age of a tree as well as the historical record of tree growth. For example, if rainfall and other conditions have been extremely favorable during a season, the annual ring may be wider than usual. If the tree has been shaded on one side by another tree or a building, the rings may be wider on the sunnier side.

Advantages and Disadvantages of Woody Plants

What are the evolutionary benefits of woody plants? With adequate rainfall, woody plants can grow taller and have more growth, because they have adequate vascular tissue to support and service their leaves. Furthermore, a long life may mean more opportunity to reproduce.

However, it takes energy to produce secondary growth and to prepare the body for winter if the plant lives in the temperate zone. Also, woody plants need more defense mechanisms, because a long-lived plant is likely to be attacked by herbivores and parasites. Trees usually do not reproduce until after they have grown for several seasons, by which time they may have been attacked by predators or been infected with a disease. In certain habitats, it is more advantageous for a plant to put most of its energy into producing a large number of seeds rather than being woody.

Stem Diversity

There are many plants whose ground level or belowground stems are often mistaken for roots. The diversity of some of these stems is illustrated in Figure 24.20. Horizontal stems, called *stolons* or runners, produce new plants where nodes touch the ground. The strawberry plant is a common example of this type of stem, which functions in vegetative reproduction.

Rhizomes are underground horizontal stems; they may be long and thin, as in sod-forming grasses, or thick and fleshy, as in irises. Rhizomes survive the winter and contribute to asexual reproduction, because each node bears a bud. Some rhizomes have enlarged portions called tubers, which function in food storage. Potatoes are tubers, and the potato "eyes" are buds that mark the nodes.

Corms are bulbous, underground stems that lie dormant during the winter, just as rhizomes do. They also produce new plants the next growing season. Gladiolus corms are referred to as bulbs by laypersons, but botanists reserve the term *bulb* for a structure composed of modified leaves attached to a short, vertical stem. An onion is a bulb.

Aboveground vertical stems can also be modified. For example, cacti have succulent stems specialized for water storage, and the tendrils of grape plants (which are stem branches) allow them to climb. The morning glory and its relatives have stems that twine around support structures. Such tendrils and twining shoots help plants expose their leaves to the sun.

Humans make use of stems in many ways. The stem of the sugarcane plant is a primary source of table sugar; cinnamon and the drug quinine are derived from the bark of *Cinnamomum verum* and various *Cinchona* species, respectively; and wood is necessary for the production of paper and building materials and is used as fuel in many parts of the world.

Check Your Progress 24.4

1. Describe the transport tissues that are found in a vascular bundle.
2. Compare the arrangements of the vascular bundles in monocot stems and eudicot stems.
3. Contrast primary growth with secondary growth.
4. List the components of bark.
5. Compare the features in the annual rings of spring wood and summer wood.

Figure 24.20 Stem diversity. **a.** A strawberry plant has aboveground, horizontal stems called stolons. Every other node produces a new shoot system. **b.** The underground, horizontal stem of an iris is a fleshy rhizome. **c.** The underground stem of a potato plant has enlargements called tubers. We call the tubers potatoes. **d.** The corm of a gladiolus is a stem covered by papery leaves.

24.5 Organization and Diversity of Leaves

Learning Outcomes

Upon completion of this section, you should be able to

1. Identify the structures and functions of various leaf tissues.
2. Describe the many forms of leaf diversity.

Leaves are the part of a plant that generally carries on the majority of photosynthesis, a process that requires water, carbon dioxide, and sunlight. Leaves receive water from the root system by way of the stem.

The size, shape, color, and texture of leaves are highly variable. These characteristics are fundamental in plant identification. The leaves of some aquatic duckweeds are less than 1 mm in diameter, while some palms have leaves that exceed 6 m in length. The shape of leaves can vary from cactus spines to deeply lobed white oak leaves. Leaves can exhibit a variety of colors from various shades of green to deep purple. The texture of leaves varies from smooth and waxy, like a magnolia, to coarse, like a sycamore. Plants that bear leaves the entire year are called **evergreens,** and those that lose all their leaves at the end of their growing season are called **deciduous.**

Broad and thin plant leaves have the maximum surface area for the absorption of carbon dioxide and the collection of solar energy needed for photosynthesis. Unlike stems, leaves are almost never woody. With few exceptions, their cells are living, and the bulk of a leaf contains photosynthetic tissue.

The wide portion of a foliage leaf is called the **blade.** The **petiole** is a stalk that attaches the blade to the stem (Fig. 24.21). The upper acute angle between the petiole and stem is the leaf axil, where the axillary bud is found.

Leaf Morphology

Figure 24.22 shows a cross section of a typical eudicot leaf of a temperate-zone plant. At the top and bottom are layers of epidermal tissue that often bear trichomes, protective hairs often modified as glands that secrete irritating substances. These features help deter insects from eating the leaf. The epidermis characteristically has an outer, waxy cuticle that helps keep the leaf from drying out. The cuticle also prevents gas exchange, because it is not gas permeable. However, the lower epidermis of eudicot and both surfaces of monocot leaves contain stomata that allow gases to move into and out of the leaf. Water loss also occurs at stomata, but each stoma has two

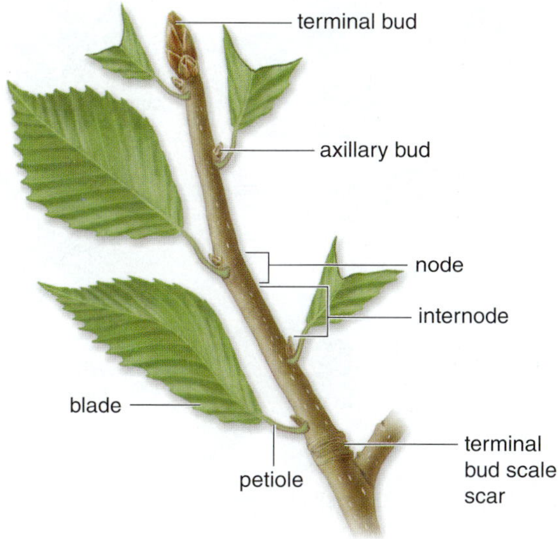

Figure 24.21 Twig with leaves. The major parts of a leaf and attachments are illustrated by a woody twig collected in summer.

guard cells that regulate its opening and closing, and stomata close when the weather is hot and dry.

The body of a leaf is composed of **mesophyll** (Gk. *mesos,* "middle"; *phyllon,* "leaf") tissue. Most eudicot leaves have two distinct regions: **palisade mesophyll,** containing tightly packed, elongated cells, and **spongy mesophyll,** containing irregular cells bounded by air spaces. There are important reasons why palisade mesophyll and spongy mesophyll need to look so different from each other. The long cylindrical cells of the palisade mesophyll are close together, packed with chloroplasts, and are most responsible for photosynthesis. Although close together, the palisade cells are separated from each other by a tiny film of water, which increases the surface area for slow-moving carbon dioxide molecules to diffuse in.

The spongy mesophyll is made up of cells that are loosely packed, irregularly shaped, and have a higher volume of air spaces. This structure increases the chance that carbon dioxide, which does not dissolve easily, will enter the leaf and stay there long enough to diffuse into the palisade layer.

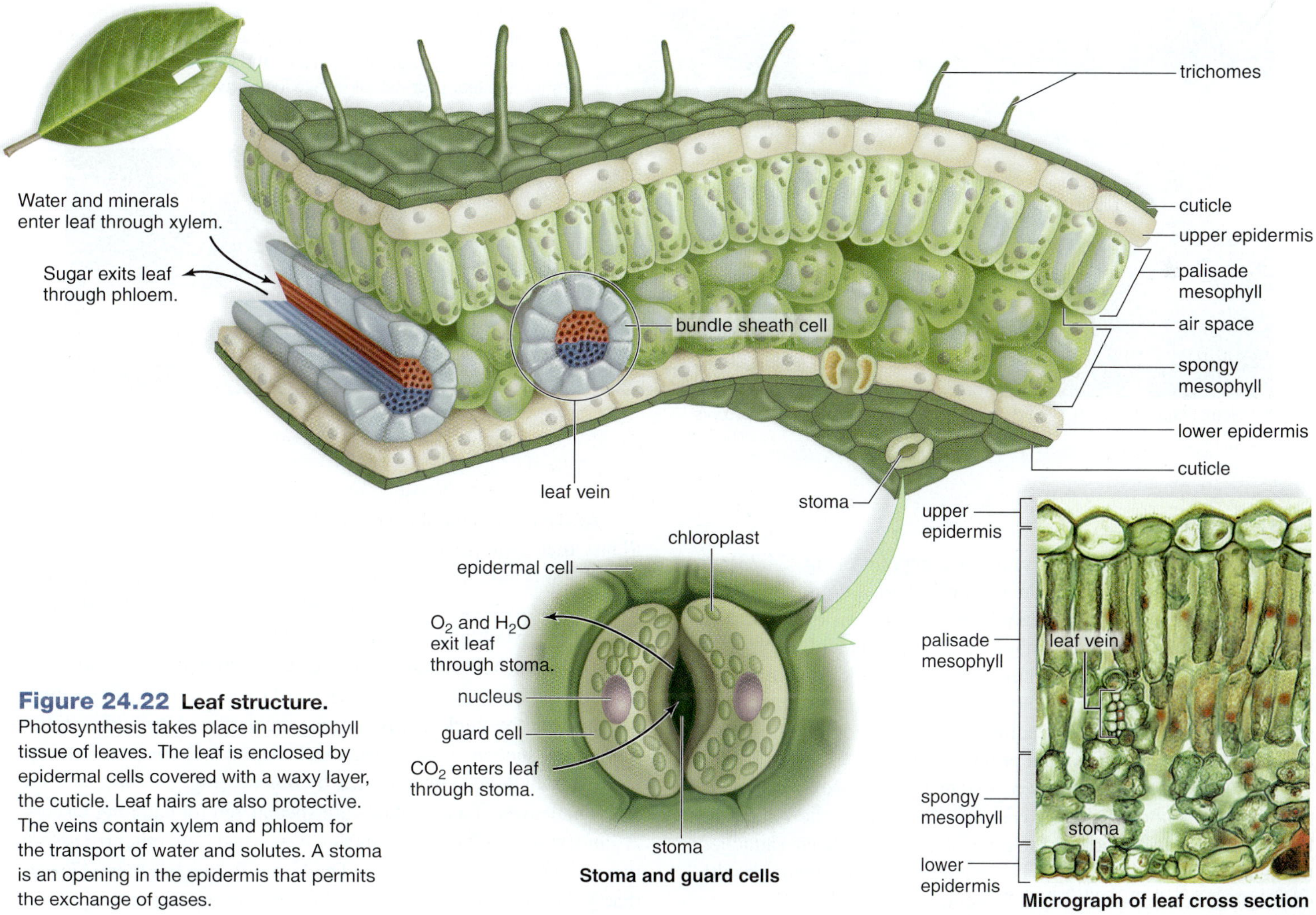

Figure 24.22 Leaf structure. Photosynthesis takes place in mesophyll tissue of leaves. The leaf is enclosed by epidermal cells covered with a waxy layer, the cuticle. Leaf hairs are also protective. The veins contain xylem and phloem for the transport of water and solutes. A stoma is an opening in the epidermis that permits the exchange of gases.

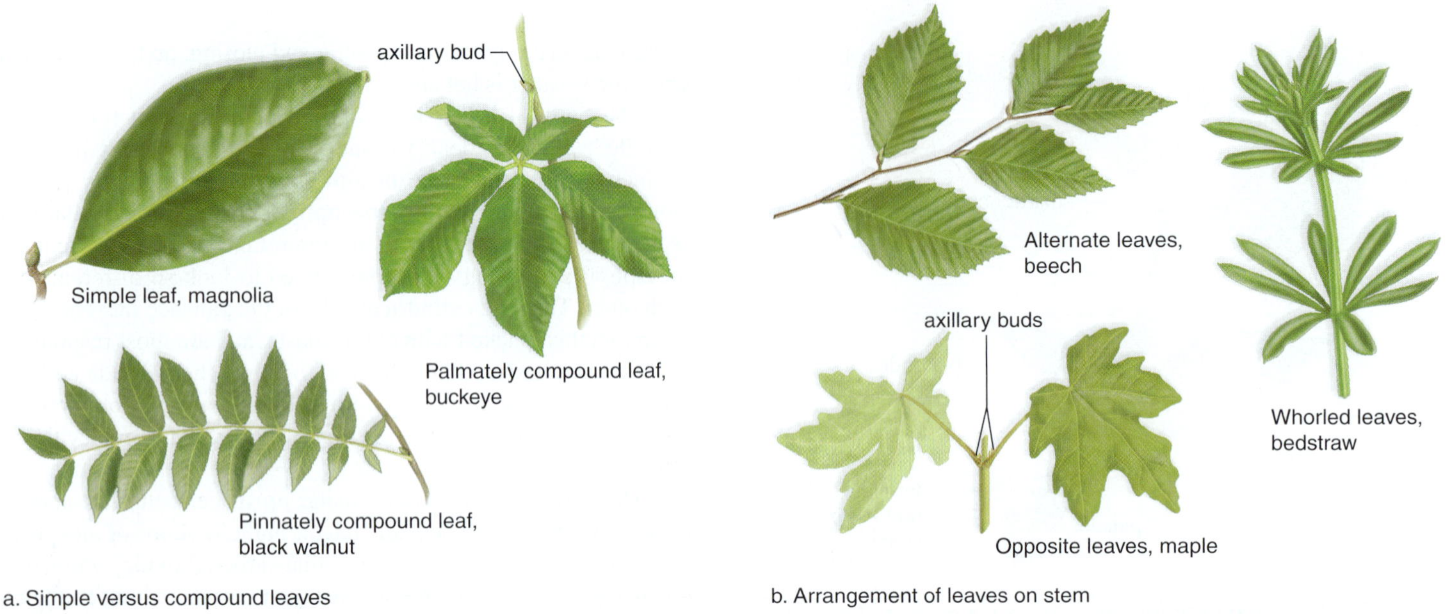

Figure 24.23 Classification of leaves. **a.** Leaves are either simple or compound, being either pinnately compound or palmately compound. Note the one axillary bud per compound leaf. **b.** Leaf arrangement on a stem can be alternate, opposite, or whorled.

a. Cactus, *Opuntia* b. Cucumber, *Cucumis* c. Venus flytrap, *Dionaea*

Figure 24.24 Leaf diversity. **a.** The spines of a cactus plant are modified leaves that protect the fleshy stem from animal predation. **b.** The tendrils of a cucumber are modified leaves that attach the plant to a physical support. **c.** The modified leaves of the Venus flytrap serve as a trap for insect prey. When triggered by an insect, the leaf snaps shut. Once shut, the leaf secretes digestive juices, which break down the soft parts of the insect's body.

Leaf Diversity

The blade of a leaf can be simple or compound (Fig. 24.23). A simple leaf has a single blade in contrast to a compound leaf, which is divided in various ways into leaflets. An example of a plant with simple leaves is a magnolia, and plants with compound leaves include the buckeye and black walnut. In addition, the black walnut has pinnately compound leaves with leaflets occurring in pairs, while the buckeye has palmately compound leaves with all the leaflets attached to a single point.

Leaves can be arranged on a stem in three ways: alternate, opposite, or whorled. The leaves are alternate in the American beech; in a maple, the leaves are opposite, with two leaves attached to the same node. Bedstraw has a whorled leaf arrangement with several leaves originating from the same node.

Leaves are adapted to various environmental conditions. Plants that grow in shade tend to have broad, wide leaves, and desert plants tend to have reduced leaves with sunken stomata. The spines of a cactus are actually modified leaves attached to the succulent (water-containing) stem (Fig. 24.24*a*).

An onion bulb is made up of leaves surrounding a short stem. In a head of cabbage, large leaves overlap one another. The petiole of a leaf can be thick and fleshy, as in celery and rhubarb. Climbing leaves, such as those of peas and cucumbers, are modified into tendrils, which can attach to nearby objects (Fig. 24.24*b*).

The leaves of a few plants are specialized for catching insects. A sundew has sticky trichomes that trap insects and other trichomes that secrete digestive enzymes. The Venus flytrap has hinged leaves that snap shut and interlock when an insect triggers sensitive trichomes that project from inside the leaves (Fig. 24.24*c*). Certain leaves of a pitcher plant resemble a pitcher and have downward-pointing hairs that lead insects into a pool of digestive enzymes secreted by trichomes. Carnivorous plants commonly grow in marshy regions, where the supply of soil nitrogen is severely limited. The digested insects provide the plants with a source of organic nitrogen.

Video Carnivorous Plant

Check Your Progress 24.5

1. Explain how the cuticle, stomata, and trichomes protect the leaf.
2. Compare the structural and functional differences between the palisade and spongy mesophyll.
3. Give examples of different types of leaves and their functions.

Connecting the Concepts with the Themes

Evolution

- A number of novel features evolved in the angiosperms that improved their fitness. The evolution of the seed was one of the greatest keys to angiosperm success, along with the production of epidermis, vascular tissue, leaves with trichomes, cuticle, stomata, and reproduction utilizing flowers.
- Growth from apical meristems produces an increase in shoot and root length, allowing increased access to sunlight and nutrients. Growth in the primary meristems produces an increase in girth, which allows for stronger stems and trunks, which provide support for the plant.

Nature of Science

- Humans rely on angiosperms for a vast number of uses, including food, shelter, and fuel.
- The medicinal properties of plants have been known for millennia, and modern science is investigating plant compounds for use as drugs and nutrients.
- Scientists understand the genetic switches that stop organ development and can use this knowledge to manipulate plants to continue vegetative growth.

Biological Systems

- Angiosperms' development of root and shoot systems, with separate but intertwined functions, has helped them become the dominant form of plant life.
- The evolution of the leaf as a primary photosynthetic organ has allowed efficient gathering of sunlight, enabling flowering plants, in particular, to put more resources into growth, production of offspring, and provision of nutrients for offspring in the form of fruits.

MEDIA STUDY TOOLS http://connect.mheducation.com

SMARTBOOK® Study smart. SmartBook helps you maximize your study time by identifying what you know and don't know. The recharge feature in SmartBook helps you prepare for the test by identifying the material that you are most likely to forget.

connect | BIOLOGY Find even more resources to learn the chapter concepts including animations, tutorial videos, and interactive practice questions.

Animation	Videos
24.3 Root Nodule Formation	**24.1** Seedling Growth **24.5** Carnivorous Plant

SUMMARIZE

24.1 Cells and Tissues of Flowering Plants

Flowering plants contain **apical meristems** along with three types of primary **meristems.** The protoderm produces the **epidermal tissue.** Within the roots, the epidermal cells can produce a waxy **cuticle, root hairs,** and **trichomes** and, in the leaves, the **epidermis** contains the **stoma.** In a woody stem, the epidermis is replaced by the **periderm** layer, made up of the **cork** and **cork cambium.**

The ground meristem produces **ground tissue,** making up the bulk of a plant. Ground tissue can have living **parenchyma** cells, supportive **collenchyma** cells, and strong, dead **sclerenchyma** cells lined with **lignin.**

The procambium produces **vascular tissue.** Vascular tissue is located in the **vascular cylinder** of roots; the **vascular bundles** in stems; and **leaf veins** in leaves.

The vascular tissue consists of **xylem** and **phloem.** The xylem contains two types of conducting cells: **vessel elements** and **tracheids.** Vessel elements form a continuous pipeline from the roots to the leaves.

The elongated tracheids contain tapered ends, which allow water to move through pits in the end walls and sidewalls. Xylem transports water and minerals.

The phloem contains sieve tubes made up of **sieve-tube members,** each of which is associated with a **companion cell.** The phloem transports sucrose and other organic compounds, including plant hormones.

24.2 Organs of Flowering Plants

Flowering plants have three main **vegetative organs.** The **root** anchors the plant and absorbs water and minerals, as well as storing the products of photosynthesis. **Stems** produce new tissues, support the leaves, transport minerals from the **root system** to the leaves and back, and store plant products. The tip of the stem contains the **terminal bud,** and **nodes** are where leaves attach to stems. Between these regions are the **internodes.** Leaves are specialized for gas exchange and carry on the majority of photosynthesis within the plant. Of the vegetative organs, leaves and stems make up the **shoot system.**

Flowering plants are classified into the **monocots** and the **eudicots.** This distinction is based on the number of cotyledons in the seed; the arrangement of the vascular tissue in the roots, stems, and leaves; the number of flowering parts; and the structure of pollen.

Plant life cycles depend on gene switches that promote vegetative growth or flower growth. A plant in which flower production is dominant and that lives only one growing season is an **annual;** a plant that produces vegetative structures for multiple growing seasons is a **perennial.**

24.3 Organization and Diversity of Roots

The root tip has three main zones: the zone of cell division (contains the **root apical meristem**), the zone of elongation, and the zone of maturation. At the very tip of the root is a **root cap,** which protects the growing root from damage.

A cross section of an herbaceous eudicot root reveals the **epidermis,** which functions in protection; the **cortex,** which stores food; the **endodermis,** which along with the **Casparian strip** regulates the movement of water and minerals; and the vascular cylinder, which is composed of a **pericycle,** then the vascular tissue. Within the vascular cylinder of a eudicot, the xylem appears star-shaped; the phloem is found in separate regions between the points of the star. In contrast, a monocot root has a ring of vascular tissue with alternating bundles of xylem and phloem surrounding the **pith.**

Roots tend to be either **taproots** (one dominant root) or **fibrous roots** with many similarly sized roots. Roots can be highly diverse and include specializations such as storage functions, prop roots, aerial roots, and pneumatophores. Most flowering plant roots have mycorrhizal associations with fungi, which increase water and mineral uptake, and the legume plants house nitrogen-fixing bacteria in **root nodules.**

24.4 Organization and Diversity of Stems

The activity of the **shoot apical meristem** accounts for the primary growth of a stem. The **terminal bud** contains internodes and leaf primordia at the nodes. The lengthening of the internodes allows for stem growth.

A cross section of an **herbaceous** eudicot stem reveals an epidermis, a cortex, vascular bundles in a ring, and an inner pith. Monocot stems have scattered vascular bundles and a cortex and pith that are not well defined.

Secondary growth of a woody stem is due to the **vascular cambium,** which produces new xylem and phloem on an annual basis. The **cork cambium** produces new **cork** cells when needed. Cork is part of the **bark** and replaces the epidermis in woody plants. In a cross section of a woody stem, all the tissue outside the vascular cambium is bark. It consists of secondary phloem, cork cambium, and cork. **Wood** consists of secondary xylem that builds up year after year and forms **annual rings.**

24.5 Organization and Diversity of Leaves

Leaves are the main part of the plant responsible for photosynthesis. The wide portion of a leaf is the **blade,** and the blade attaches to the stem with a **petiole.** Plants that keep their leaves all year-round are **evergreens,** while plants that drop their leaves in a growing season are **deciduous.**

The ground tissue of a leaf is made up of **mesophyll** cells. These cells contain chloroplasts and undergo photosynthesis. The **palisade mesophyll** makes up the upper layer; the **spongy mesophyll** is the lower layer. The spongy mesophyll has many air spaces for gas exchange, and its lower epidermal layer contains the stoma. Leaves can be specialized to protect against herbivores, as in a cactus, or catch insects, as found in the carnivorous plants.

ASSESS

Choose the best answer for each question.

24.1 Cells and Tissues of Flowering Plants

1. New plant cells originate from the
   a. parenchyma.
   b. collenchyma.
   c. sclerenchyma.
   d. base of the shoot.
   e. apical meristem.
2. Meristem tissue that gives rise to epidermal tissue is called
   a. procambium.
   b. ground meristem.
   c. epiderm.
   d. protoderm.
   e. periderm.
3. Which of these cell types is dead at maturity?
   a. parenchyma
   b. meristem
   c. epidermis
   d. companion cell
   e. sclerenchyma

24.2 Organs of Flowering Plants

4. All of the following are vegetative organs except
   a. leaves.
   b. herbaceous stems.
   c. woody stems.
   d. seeds.
   e. roots.
5. Which of these is an incorrect contrast between monocots (stated first) and eudicots (stated second)?
   a. one cotyledon—two cotyledons
   b. leaf veins parallel—net veined
   c. pollen with three pores—pollen with one pore
   d. flower parts in threes—flower parts in fours or fives
   e. All of these are correct contrasts.

24.3 Organization and Diversity of Roots

6. Roots
   a. are the primary site of photosynthesis.
   b. give rise to new leaves and flowers.
   c. have a thick cuticle to protect the epidermis.
   d. absorb water and nutrients.
   e. contain spores.

7. Root hairs are found in the zone of
   a. cell division.
   b. elongation.
   c. maturation.
   d. apical meristem.
   e. All of these are correct.
8. The Casparian strip is found
   a. between all epidermal cells.
   b. between xylem and phloem cells.
   c. surrounding endodermal cells.
   d. within the secondary wall of parenchyma cells.
   e. in both endodermis and pericycle.

24.4 Organization and Diversity of Stems

9. Monocot stems have
   a. vascular bundles arranged in a ring.
   b. vascular cambium.
   c. scattered vascular bundles.
   d. a cork cambium.
   e. a distinct pith and cortex.
10. Between the bark and the wood in a woody stem, there is a layer of meristem called
    a. cork cambium.
    b. vascular cambium.
    c. apical meristem.
    d. the zone of cell division.
    e. procambium preceding bark.
11. Which of these is a stem?
    a. carrot
    b. stolon of strawberry plants
    c. spine of cacti
    d. sweet potato
    e. pneumatophore

24.5 Organization and Diversity of Leaves

12. Which part of a leaf carries on most of the photosynthesis of a plant?
    a. vascular bundle
    b. mesophyll
    c. epidermal layer
    d. guard cells
    e. trichomes
13. How are compound leaves distinguished from simple leaves?
    a. Compound leaves do not have axillary buds at the base of leaflets.
    b. Compound leaves are smaller than simple leaves.
    c. Simple leaves are usually deciduous.
    d. Compound leaves are found only on pine trees.
    e. Simple leaves are found only in gymnosperms.
14. Label this leaf using these terms: leaf vein, lower epidermis, palisade mesophyll, spongy mesophyll, and upper epidermis.

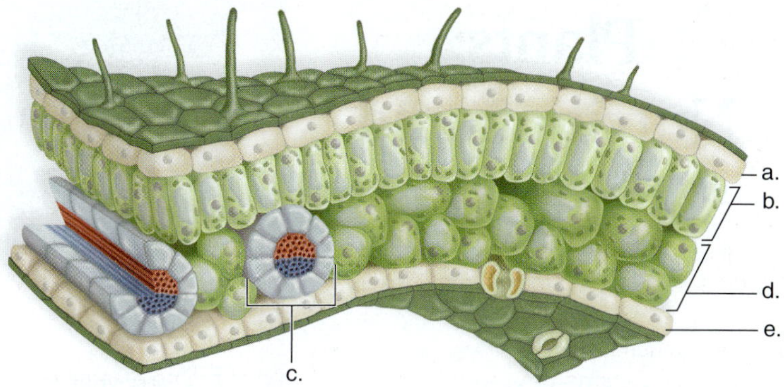

ENGAGE

THINKING SCIENTIFICALLY

1. Using an electron microscope, how might you confirm that a companion cell communicates with its sieve-tube member?
2. Design an experiment that tests the hypothesis that new plants arise at the nodes of a stolon according to environmental conditions (temperature, water, and sunlight).
3. If you were given an unfamiliar vegetable, how could you tell if it was a root or a stem, based on its external features and microscopic examination of its cross section?
4. Many potato tubers harvested from wet soil have large lenticels. What is the adaptive significance of this?

25

Flowering Plants: Nutrition and Transport

Roses can be artificially multicolored by taking advantage of their transport systems.

Chapter Outline

25.1 Plant Nutrition and Soil 457

25.2 Water and Mineral Uptake 461

25.3 Transport Mechanisms in Plants 465

Before You Begin

Before beginning this chapter, take a few moments to review the following discussions.

Section 2.3 Which properties of water are essential for the conduction of water from the root system to the leaves?

Section 5.2 How do diffusion and osmosis affect how water, minerals, and nutrients move in a plant?

Section 22.3 Which organisms have evolved symbiotic relationships with flowering plants?

Every year, we see a dazzling array of floral creations at weddings and for occasions such as Valentine's Day. Blue carnations, green daisies, and multicolored roses are artificially colored to increase the variations. Florists have learned how to alter flower color by using the plant's natural conducting system.

To accomplish the artificial color change, the florist needs the flower and its stem. The stem is cut under water to prevent air bubbles from getting trapped within the conduction tubes of the stem. An air bubble will block the transport of fluid up the stem. For one color, the flowers are placed in a vase of water containing dye. In the case of multicolored roses, the stems are cut into four quadrants, and each quadrant is placed in dye. The larger commercial suppliers of these special roses inject dye into different sections of the stem. The dye is then transported up the stem and into the flower due to water potential and the cohesion of water. A wide variety of floral colors can be created by using dye and the natural conducting system within the plant.

In this chapter, we explore how plants use water to conduct essential nutrients throughout their systems—from the highest leaves to the tips of the deepest roots.

As you read through this chapter, think about the following questions:

1. Which nutrients are essential for plant growth?
2. What structures enable plants to absorb water and minerals from the soil?
3. Why does fluid "leak" from a branch when it is cut?

Following *the* Themes

CHAPTER 25 FLOWERING PLANTS: NUTRITION AND TRANSPORT

UNIT 5 PLANT EVOLUTION AND BIOLOGY

Evolution	Vascular plants have evolved symbiotic relationships with a variety of species. These relationships have enabled them to become the dominant form of plant life on Earth today.
Nature of Science	Research has shown that plants that have nutrient deficiencies will not fully develop. Biotechnology helps with understanding how nutrients are conducted through a plant.
Biological Systems	Flowering plants use xylem and phloem to conduct nutrients and water throughout the plant body.

25.1 Plant Nutrition and Soil

Learning Outcomes

Upon completion of this section, you should be able to

1. Identify the macronutrients and micronutrients that plants require.
2. Describe how and why mineral ions enter a plant through the roots.
3. Explain a simplified soil profile.

Plant nutrition is the study of how a plant gains and uses mineral nutrients from the soil. Nutrients are elements such as nitrogen and calcium and are obtained primarily as inorganic ions. In traditional farming, crop plants absorb inorganic nutrients from the soil; then humans and other animals consume them. Leftover crop residue and the human and animal manure return the nutrients to the soil. In essence, mineral nutrients continually cycle through all organisms and enter the biosphere predominantly through the root systems of plants. For this reason, many plant scientists refer to plants as the "miners" of the Earth's crust.

Water is an essential nutrient for a plant, but much of the water entering a plant evaporates at the leaves. The photosynthetic tissues carry out photosynthesis, a process that uses carbon dioxide and gives off oxygen. Roots, like all plant organs, carry on cellular respiration, a process that uses oxygen and gives off carbon dioxide (Fig. 25.1). Just as roots require water and minerals, it is important that roots always have a continuous supply of oxygen.

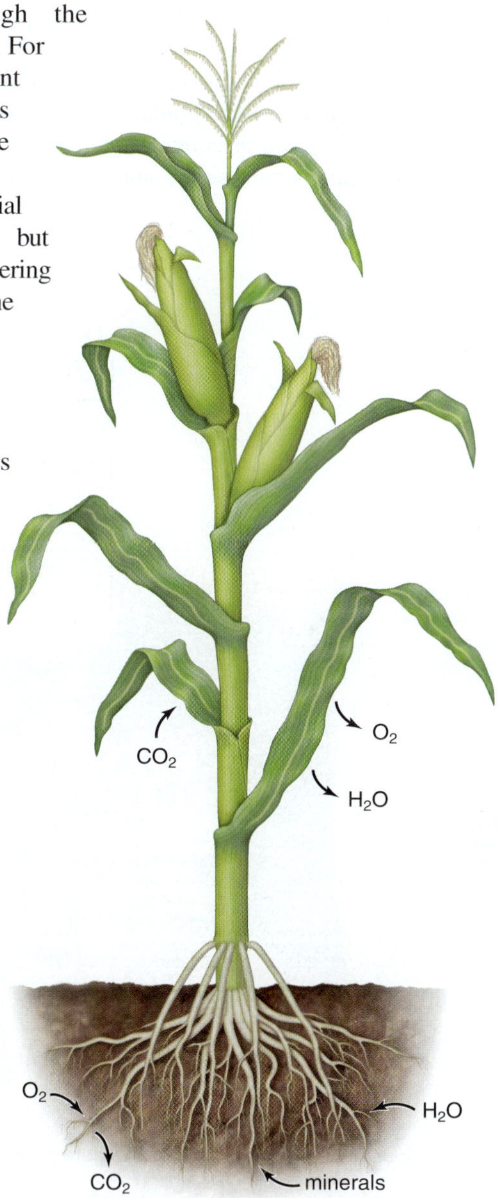

Figure 25.1 Overview of plant nutrition. Carbon dioxide, which enters leaves, and water, which enters roots, are combined during photosynthesis to form carbohydrates, with the release of oxygen from the leaves. Root cells, and all other plant cells, carry on cellular respiration, which uses oxygen and gives off carbon dioxide. Aside from the elements carbon, hydrogen, and oxygen, plants require nutrients, which are absorbed as minerals by the roots.

Essential Inorganic Nutrients

Approximately 95% of a typical plant's dry weight (weight excluding free water) is carbon, hydrogen, and oxygen. Why? Because these are the elements found in most organic compounds, such as carbohydrates. Carbon dioxide (CO_2) supplies the carbon, and water (H_2O) supplies the hydrogen and oxygen found in the organic compounds of a plant.

In addition to carbon, hydrogen, and oxygen, plants require certain other nutrients, which the roots absorb as minerals. A **mineral** is an inorganic substance, usually containing two or more elements. Why do plants need minerals from the soil? In plants, nitrogen is a major component of nucleic acids and proteins, magnesium is a component of chlorophyll, and iron is a building block of cytochrome molecules.

The major **essential nutrients** for plants are listed in Table 25.1. A nutrient is essential if (1) it has an identifiable role, (2) no other nutrient can substitute for it and fulfill the same role, and (3) a deficiency of this nutrient disrupts plant function and metabolism, causing a plant to die without completing its life cycle. Essential nutrients are divided into **macronutrients** and **micronutrients** according to their relative concentrations in plant tissue.

Beneficial nutrients are another category of elements taken up by plants. Beneficial nutrients either are required for growth or enhance the growth of a particular plant. Horsetails require silicon as a mineral nutrient, and sugar beets show enhanced growth in the

Table 25.1 Some Essential Inorganic Nutrients in Plants

Elements	Form in Which Element Is Absorbed
Macronutrients	
Carbon (C)	CO_2
Hydrogen (H)	H_2O
Oxygen (O)	O_2
Phosphorus (P)	$H_2PO_4^-$, HPO_4^{2-}
Potassium (K)	K^+
Nitrogen (N)	NO_3^-, NH_4^+
Sulphur (S)	SO_4^{2-}
Calcium (Ca)	Ca^{2+}
Magnesium (Mg)	Mg^{2+}
Micronutrients	
Iron (Fe)	Fe^{2+}, Fe^{3+}
Boron (B)	BO_3^{3-}, $B_4O_7^{2-}$
Manganese (Mn)	Mn^{2+}
Copper (Cu)	Cu^{2+}
Zinc (Zn)	Zn^{2+}
Chlorine (Cl)	Cl^-
Molybdenum (Mo)	MoO_4^{2-}

presence of sodium. Nickel is a beneficial mineral nutrient in soybeans when root nodules are present. Aluminum is used by some ferns, and selenium, which is often fatally poisonous to livestock, is utilized by canola plants.

Determination of Essential Nutrients

Soil is a complex medium, which makes it difficult to figure out what missing nutrient may be affecting a plant's growth. The preferred method for determining the mineral requirements of a plant was developed at the end of the nineteenth century by the German plant physiologists Julius von Sachs (1832–97) and Wilhelm Knop (1817–91). This method is called hydroponics (Gk. *hydrias,* "water"; *ponos,* "hard work"). **Hydroponics** allows plants to grow well if they are supplied with all the nutrients they need; this method provided proof that plants can fulfill all their needs with simply sunlight, water, and minerals (Fig. 25.2). In order to test for specific nutrient deficiencies, an investigator omits a particular mineral from the liquid medium and observes the effect on plant growth. If growth suffers, it can be concluded that the omitted mineral is an essential nutrient. Aside from generally stunted growth, there are characteristic symptoms for some of the most common nutrient deficiencies (Fig. 25.3*a*).

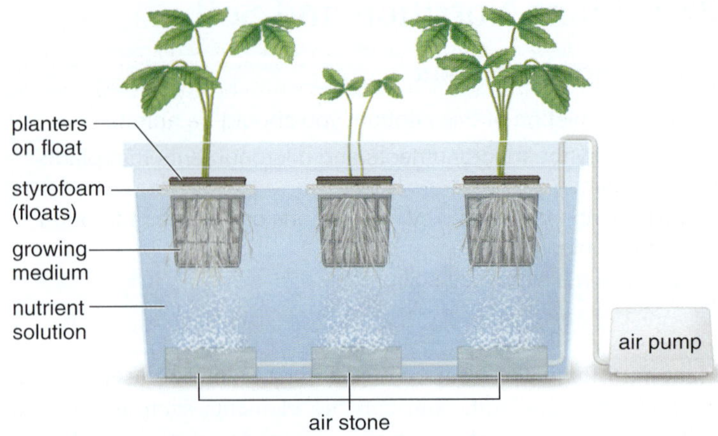

Figure 25.2 Hydroponics. Normally, soil provides nutrients and support, but both of these functions can be replaced in a hydroponics system to maximize growth. In this example, plants are suspended and roots are bathed in a nutrient bath.

Farmers and home gardeners often supplement the soil with fertilizers to avoid nutrient deficiencies. Most mixed fertilizer packaging includes three numbers, referred to as the *NPK ratio* (Fig. 25.3*b*). This ratio describes the percentage by weight of nitrogen (N), phosphorus (P), and potassium (K) contained in the fertilizer mix. For example, if a 100-pound bag of fertilizer has an NPK ratio of 18-24-6 it contains 18 pounds of nitrate, 24 pounds of phosphate (which contains phosphorus), 6 pounds of potash (which contains potassium), and 52 pounds of filler.

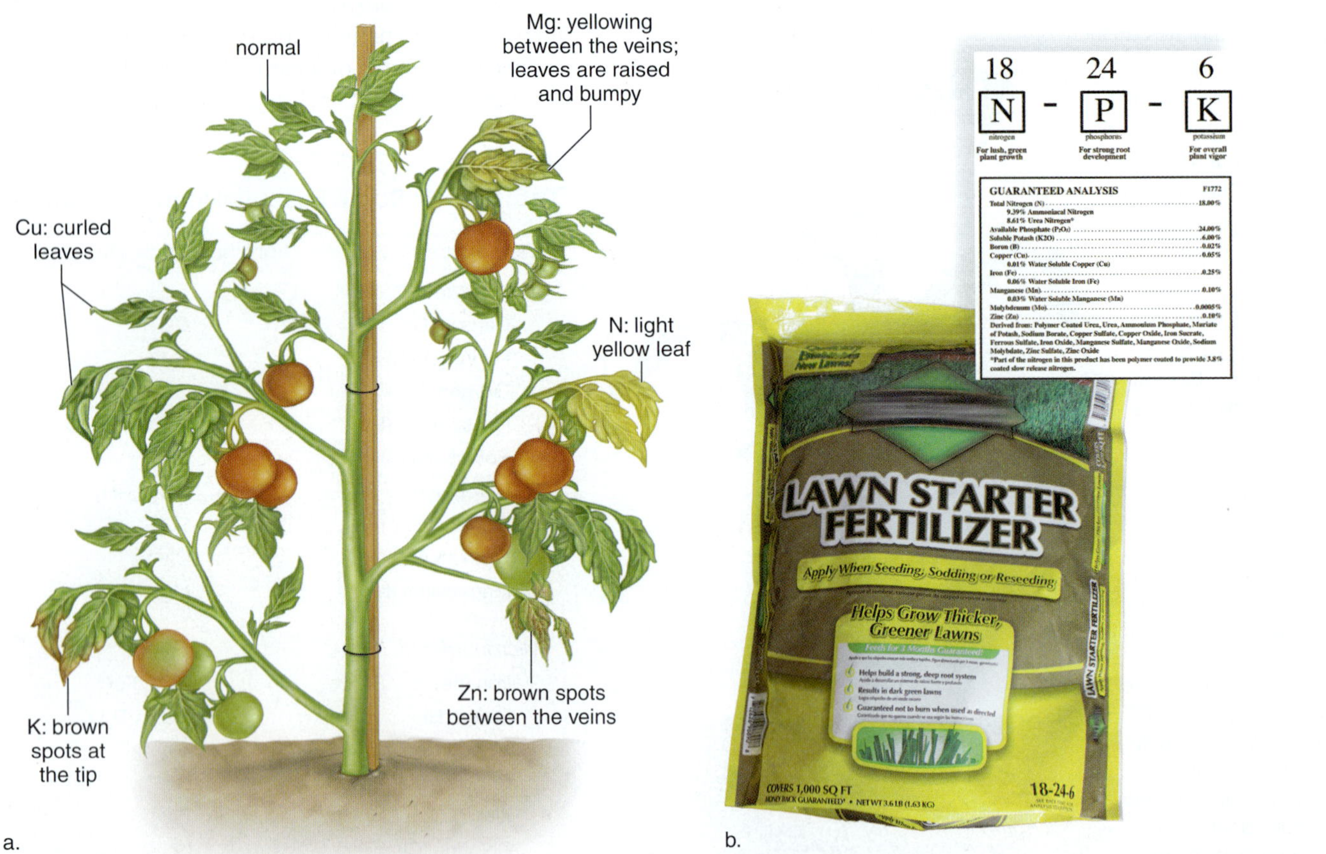

Figure 25.3 Nutrient deficiencies and fertilizer. **a.** Characteristics of various mineral deficiencies in a tomato plant. **b.** Typical packaging for store-bought fertilizer showing the NPK ratio of 18-24-6.

Soil

Soil is a mixture of mineral particles, decaying organic material, living organisms, air, and water, which together support the growth of plants. All of the essential nutrients, the water, and most of the oxygen the plant requires are absorbed from the soil by the roots. It would not be an exaggeration to say that terrestrial life is dependent on the quality of the soil and its ability to provide plants with the nutrients they need.

Soil Formation

Soil is created when rock is weathered (broken down). Weathering first gradually breaks down rock to rubble and then to smaller particles of sand, silt, and clay. Mechanical weathering includes the forces of wind, rain, the freeze-thaw cycle of ice, and the grinding of rock on rock by the action of glaciers or river flow. Chemical weathering can come in the form of acid rain, the formation of iron oxide, or degradation by lichens and mosses, which can live on bare rock. Lichens are so effective in breaking down rock that they pose a great threat to historic castles and monuments made of stone.

Over the weathered rock layer is decaying organic matter called humus. **Humus** supplies nutrients to plants, and the acidity of decomposition releases minerals from rock.

Building soil takes a long time. Under ideal conditions, depending on the type of parent material (the original rock) and the various processes at work, a centimeter of soil may take 15 years to develop.

The Nutritional Function of Soil

In a good agricultural soil, mineral particles, organic matter, and living organisms come together in such a way that there are spaces for air and water (Fig. 25.4). It is best if the soil contains particles of different sizes, because only then can spaces for air be present. It is within these air spaces that roots take up oxygen. Ideally, water clings to particles by capillary action and does not fill the spaces. Flooding or overwatering of plants fills the air spaces with excess water. The plant is therefore deprived of oxygen, cannot undergo cellular respiration, and dies.

Soil Particles. Soil particles vary in size: Sand particles are the largest (0.05–2.0 mm in diameter); silt particles have an intermediate size (0.002–0.05 mm); and clay particles are the smallest (less than 0.002 mm). Soils are a mixture of these three types of particles. Because sandy soils have many large particles, they have large spaces, and the water drains readily between the particles. In contrast to sandy soils, a soil composed mostly of clay particles has small spaces, which fill completely with water.

Most likely, you have experienced the feel of sand and clay: Sand, having no moisture, flows right through your fingers, while clay clumps together in one large mass because of its water content. The ideal soil for agriculture is called *loam* soil; it combines the aeration provided by sand with the mineral- and water-retention capacity of silt and clay.

It is also important that soils have a healthy balance of clay particles and humus. Clay and humus are negatively charged and will bind to positively charged minerals, such as calcium (Ca^{2+}) and potassium (K^+), preventing these minerals from being washed away by leaching. Through a process called **cation exchange,** hydrogen ions (and other positive ions) switch places with a positively charged mineral ion, and the root takes up the needed mineral nutrient (Fig. 25.4). The better this exchange, the healthier the soil. In soil science, soils are rated by an index called the cation exchange capacity (CEC), which indicates the availability of negative charge sites able to bind positive cations. As expected, the CEC of sandy soils is less than that of soils with a higher quantity of clay and humus mixed in.

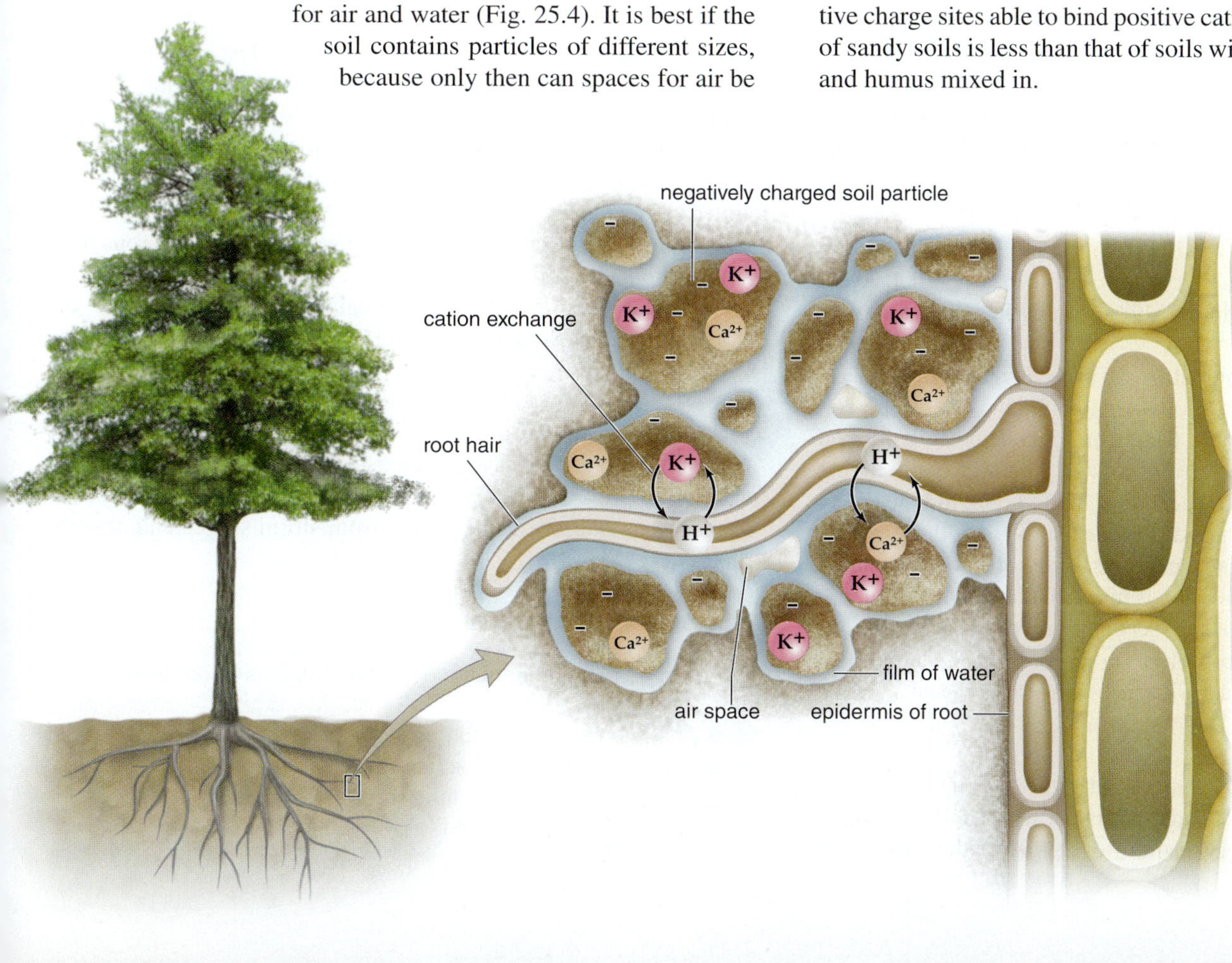

Figure 25.4 Components of healthy soil. Variously sized soil particles promote air spaces and water retention. Negatively charged clay particles bind positively charged minerals, such as Ca^{2+} and K^+. Plants extract these minerals by exchanging H^+ for them.

Humus. Humus, which mixes with the top layer of soil particles, increases the benefits of soil. Plants do well in soils that contain 10–20% humus.

Humus causes soil to have a loose, crumbly texture that allows water to soak in without doing away with air spaces. After a rain, the presence of humus decreases the chances of runoff. Humus swells when it absorbs water and shrinks as it dries. This action helps aerate soil.

Soil that contains humus is nutritious for plants. Humus is acidic; therefore, it retains positively charged minerals until plants take them up. When the organic matter in humus is broken down by bacteria and fungi, inorganic nutrients are returned to plants. Although soil particles are the original source of minerals in soil, the recycling of nutrients, as you know, is a major characteristic of ecosystems.

Living Organisms. Small plants play a major role in the formation of soil from bare rock. Due to the process of succession (see Fig. 45.13), larger plants eventually become dominant in certain ecosystems. The roots of larger plants penetrate soil even to the bedrock layer. This action slowly opens up soil layers, allowing water, air, and animals to follow.

A wide variety of animals dwell in the soil, at least part of the time. The largest of them, such as snakes, moles, badgers, and rabbits, disturb and mix soil by burrowing. Smaller animals, such as earthworms, ingest fine soil particles and deposit them on the surface as worm casts. Earthworms also loosen and aerate the soil. A range of small soil animals, including mites, springtails, and millipedes, help break down leaves and other plant remains by eating them. Soil-dwelling ants construct tremendous colonies with massive chambers and tunnels. These ants also loosen and aerate the soil.

The microorganisms in soil, such as protozoans, fungi, and bacteria, are responsible for the final decomposition of organic remains in humus to inorganic nutrients. As mentioned, plants are unable to make use of atmospheric nitrogen (N_2), and soil bacteria play an important nutrient role because they make nitrate available to plants.

Figure 25.5 Simplified soil profile. The top layer (A horizon) contains most of the humus; the next layer (B horizon) accumulates materials leached from the A horizon; and the lowest layer (C horizon) is composed of weathered parent material. Erosion removes the A horizon, a primary source of humus and minerals in soil.

Soil Profiles

A **soil profile** is a vertical section from the ground surface to the unaltered rock below. Usually, a soil profile has parallel layers known as **soil horizons.** Mature soil generally has three horizons (Fig. 25.5).

Because the parent material (rock) and climate (e.g., temperature and rainfall) differ in various parts of the biosphere, the soil profile varies according to the particular ecosystem. Soils formed in grasslands tend to have a deep A horizon built up from decaying grasses over many years, but because of limited rain, little leaching into the B horizon has occurred. In forest soils, both the A and B horizons have enough inorganic nutrients to allow for root growth. In tropical rain forests, the A horizon is more shallow than the generalized profile, and the B horizon is deeper, signifying that leaching is more extensive. Since the topsoil of a rain forest lacks nutrients, it can support crops for only a few years before it is depleted.

Soil Erosion

Soil erosion occurs when water or wind carries soil away to a new location. Soil erosion is the leading cause of water pollution in the United States and is a direct result of overgrazing, the clearing of land for urbanization and roads, and plowing. The uppermost soil layer, topsoil, is difficult to replace, yet erosion is removing it faster than ever before. At least 40% of the world's farmland is seriously degraded, and fertile soil in an area equal to the size of Texas is lost annually. Farmland erosion may be reduced by practices that are aimed at increasing soil organic matter. Effective practices include crop rotation, reduced tillage, and the use of companion crops that hold and nourish soil.

Coastal erosion, often a dramatic result of hurricane or storm surges, can reshape a coastal landscape. One example is in Pacifica, California, where urban development occurred on cliffs that were undercut and collapsed after a major storm. In recent years, there has been a net loss of beach area as a result of an intensification of storms and increases in sea level. Trees are critical for the prevention of coastal erosion; they not only hold soil in place but also break the waves from storm surges and absorb their energy.

Check Your Progress 25.1

1. Name the elements that make up the bulk of commercial fertilizers.
2. List the benefits of humus in the soil.
3. Explain how cation exchange works.
4. Identify the benefits of leaving the remains of the previous year's crops in the field to overwinter.

25.2 Water and Mineral Uptake

Learning Outcomes

Upon completion of this section, you should be able to

1. Choose the correct order of mineral uptake across the plasma membrane within a plant cell wall.
2. Describe the mutualistic relationships that assist plants in acquiring nutrients from the soil.

There are two main ways that water and dissolved minerals can enter a root. As seen in Figure 25.6, in pathway A, water weaves its way in between cells, diffusing from the porous cell wall of one cell to the cell wall of adjacent cells. Entry of water past the cell wall is blocked at the *Casparian strip,* a waxy layer that surrounds endodermal cells (see Fig. 24.8). Here the water is forced to enter endodermal cells through the plasma membrane. In pathway B, water travels from the plasma membrane of one cell to the plasma membrane of another cell, connected by openings called plasmodesmata (see Fig. 5.15). Regardless of the pathway, water enters root cells when *osmotic pressure* in the root tissues is lower than that of the soil solution. That is to say, if there is more water outside the root, and less water inside the root, then water moves in by osmosis, causing osmotic pressure.

Mineral Uptake

In contrast to the passive movement of water, minerals can be taken up by passive or active transport. Plants possess an astonishing ability to concentrate minerals until they are many times more concentrated in the plant than in the surrounding medium. The concentration of certain minerals in roots is as much as 10,000 times greater than in the surrounding soil. The presence of the Casparian strip prevents the backflow of minerals and allows the plant to maintain a higher mineral concentration in root xylem than can be found in the surrounding soil. The Nature of Science feature, "Plants Can Be Used for Cleaning and Discovery of Minerals," on page 462 explains how this ability of plants can be exploited for environmental cleanup and mineral exploration.

By what mechanism do minerals cross plasma membranes? As it turns out, the energy of ATP is involved, but only indirectly. Recall that plant cells absorb minerals in the ionic form: Nitrogen is absorbed as nitrate (NO_3^-), phosphorus as phosphate (HPO_4^{2-}), potassium as potassium ions (K^+), and so forth. Ions cannot cross the plasma membrane, because they are unable to enter the nonpolar portion of the lipid bilayer. Plant physiologists know that plant cells expend energy to actively take up and concentrate mineral ions. If roots are deprived of oxygen or are poisoned so that cellular respiration cannot occur, mineral ion uptake is diminished.

As shown in Figure 25.6*b,* a plasma-membrane pump, called a proton pump, hydrolyzes ATP and uses the energy released to

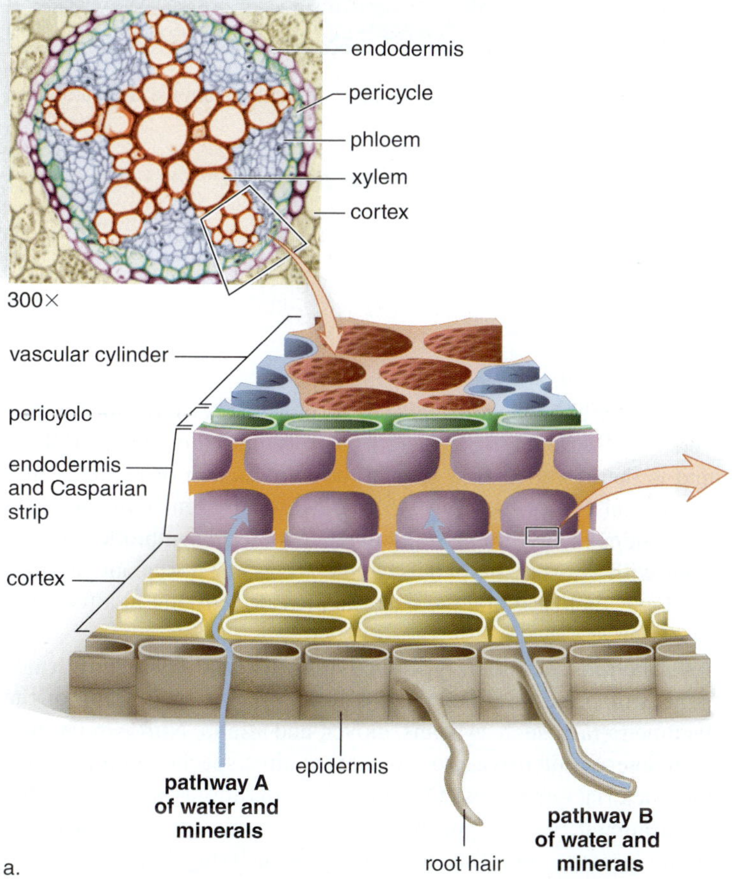

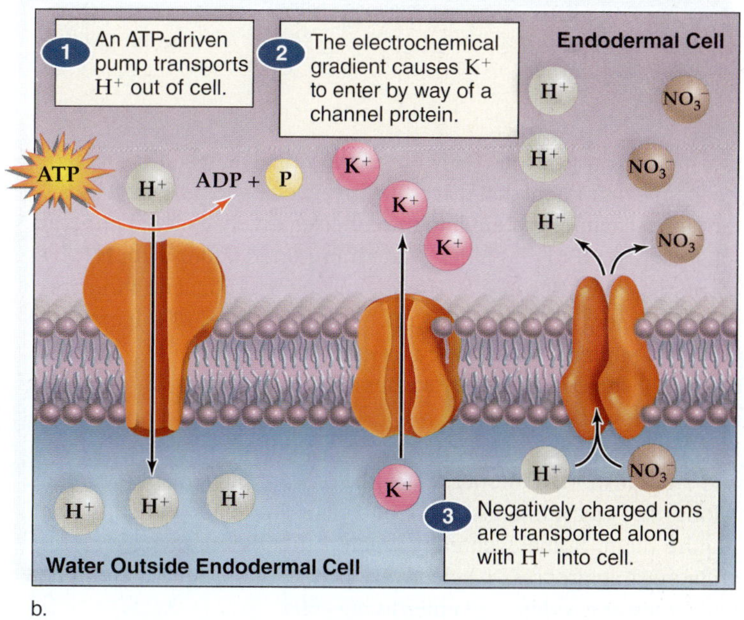

Figure 25.6 Water and mineral uptake. **a.** Pathways of water and minerals. Water and minerals can travel via porous cell walls but then must enter endodermal cells because of the Casparian strip (pathway A). Alternatively, water and minerals can enter root hairs and move from cell to cell (pathway B). **b.** Transport of minerals across an endodermal plasma membrane. **1** An ATP-driven pump removes hydrogen ions (H^+) from the cell. **2** This establishes an electrochemical gradient that allows potassium (K^+) and other positively charged ions to cross the membrane via a channel protein. **3** Negatively charged mineral ions (e.g., NO_3^-) can cross the membrane by way of a carrier when they co-transport with hydrogen ions (H^+), which are diffusing down their concentration gradient.

THEME Nature of Science

Plants Can Be Used for Cleaning and Discovery of Minerals

Phytoremediation uses plants such as mulberry, poplar, and canola to clean up environmental pollutants. The genetic makeup of these plants allows them to absorb, store, degrade, or transform substances that normally kill or harm other plants and animals. "It's an elegantly simple solution to pollution problems," says Louis Licht, who runs Ecolotree, an Iowa City phytoremediation company.

The idea behind phytoremediation is not new; scientists have long recognized certain plants' abilities to absorb and tolerate toxic substances. But the idea of using these plants on contaminated sites has gained support just in the last 25 years. Different plants work on different contaminants. The mulberry bush, for instance, is effective on industrial sludge; some grasses attack petroleum wastes; and sunflowers remove lead.

The plants clean up sites in different ways depending on the substance involved. If it is an organic contaminant, such as spilled oil, the plants, or the

Figure 25A Poplar trees cleaning up nitrogen. Poplars are able to remove large amounts of nitrogen from runoff.

transport hydrogen ions (H^+) out of the cell. The result is an electrochemical gradient that drives positively charged ions such as K^+ through a channel protein into the cell. Negatively charged mineral ions, such as NO_3^- and HPO_4^{2-}, are transported, along with H^+, by carrier proteins. Because H^+ is moving down its concentration gradient, no energy is required. Notice that this model of mineral ion transport in plant cells is based on *chemiosmosis,* the establishment of an electrochemical gradient to perform work.

Animation Proton Pump

Following their uptake by root cells, minerals move into xylem and are transported into leaves by the upward movement of water. Along the way, minerals can exit xylem and enter the cells that require them. Some eventually reach leaf cells. In any case, minerals must again cross a selectively permeable plasma membrane when they exit xylem and enter living cells.

Adaptations of Roots for Mineral Uptake

Two mutualistic relationships assist roots in obtaining mineral nutrients. Root nodules involve a mutualistic relationship with bacteria, and mycorrhizae are a mutualistic relationship with fungi.

Bacterial nitrogen fixation is responsible for most of the conversion of nitrogen from the air ($N \equiv N$) into ammonium (NH_4^+) and is therefore the first step in the introduction of nitrogen into ecological cycles (see Fig. 45.24). Most nitrogen-fixing bacteria live independently in the soil, but some do form symbiotic associations with a host plant. In these associations, the host plant provides food and shelter, while the bacteria provide nitrogen in a form the plant can use.

The most common types of symbiosis occur between various genera of bacteria collectively called **rhizobia** and plants of the legume family, such as beans, clover, and alfalfa. Nitrogen fixation is an energy-intensive process that requires special conditions for the bacteria's enzymes. One of those conditions is an anaerobic environment, as the presence of oxygen disrupts the nitrogen-fixing process. For this reason, nitrogen-fixing bacteria live in organs called **root nodules** (Fig. 25.7), where oxygen levels are maintained high enough for cellular respiration but low enough so as to not inactivate important nitrogen-fixing enzymes. In addition, large-scale farming of legume crops often depletes the native populations of rhizobia, and farmers must often supplement with pellets containing these bacteria.

microbes around their roots, break down the substance. The remnants can be either absorbed by the plant or left in the soil or water. For an inorganic contaminant, such as cadmium or zinc, the plants absorb the substance and trap it. The plants must then be harvested and disposed of, or processed to reclaim the trapped contaminant.

Poplars Take Up Excess Nitrates

Most trees planted along the edges of farms are intended to break the wind, but another use of poplars is to remove excess minerals from runoff. The poplars act as vacuum cleaners, sucking up nitrate-laden runoff from a fertilized cornfield before this runoff reaches a nearby brook—and perhaps other waters (Fig. 25A). Nitrate runoff into the Mississippi River from Midwest farms is a major cause of the large "dead zone" of oxygen-depleted water that develops each summer in the Gulf of Mexico.

Canola Plants Take Up Selenium

Canola plants (*Brassica rapus* and *B. napa*) are grown in California's San Joaquin Valley to soak up excess selenium in the soil to help prevent an environmental catastrophe like the one that occurred there in the 1980s. Back then, irrigated farming caused naturally occurring selenium to rise to the soil surface. When excess water was pumped onto the fields, some selenium flowed off into drainage ditches, eventually ending up in Kesterson National Wildlife Refuge. The selenium in ponds at the refuge accumulated in plants and fish and subsequently deformed and killed waterfowl.

Eucalyptus Trees Reveal Hidden Gold

The ability of plants to take up minerals, called *biogeochemical absorption,* may not only clean up toxic messes but also serve as a valuable beacon for desirable minerals. Gold, for example, is an element for which worldwide discoveries are down by 45%. Normally, prospectors drill in suspected areas, test soil samples, and disturb the ecosystem in promising areas with no guarantees of success.

In Australia, scientists have found that *Eucalyptus* trees growing over deep deposits of gold have leaves with high concentrations of this sought-after element. Gold is toxic to plants, and when drawn up in the soil through the xylem, gold accumulates in the outermost regions of plants—their leaves and bark. Leaves and bark sampled from various eucalyptus trees were taken to the lab, where x-ray and chemical analyses revealed the levels of gold. The Australian scientists were able to show that the trees growing *directly* over a 35-meter-deep gold deposit were the samples with the unusually high gold readings.

Questions to Consider

1. What happens to the pollutants when the plant dies?
2. Why would one plant be more adapted at absorbing a particular pollutant than another plant?
3. What are the ecological and economic benefits of using plants for gold prospecting?

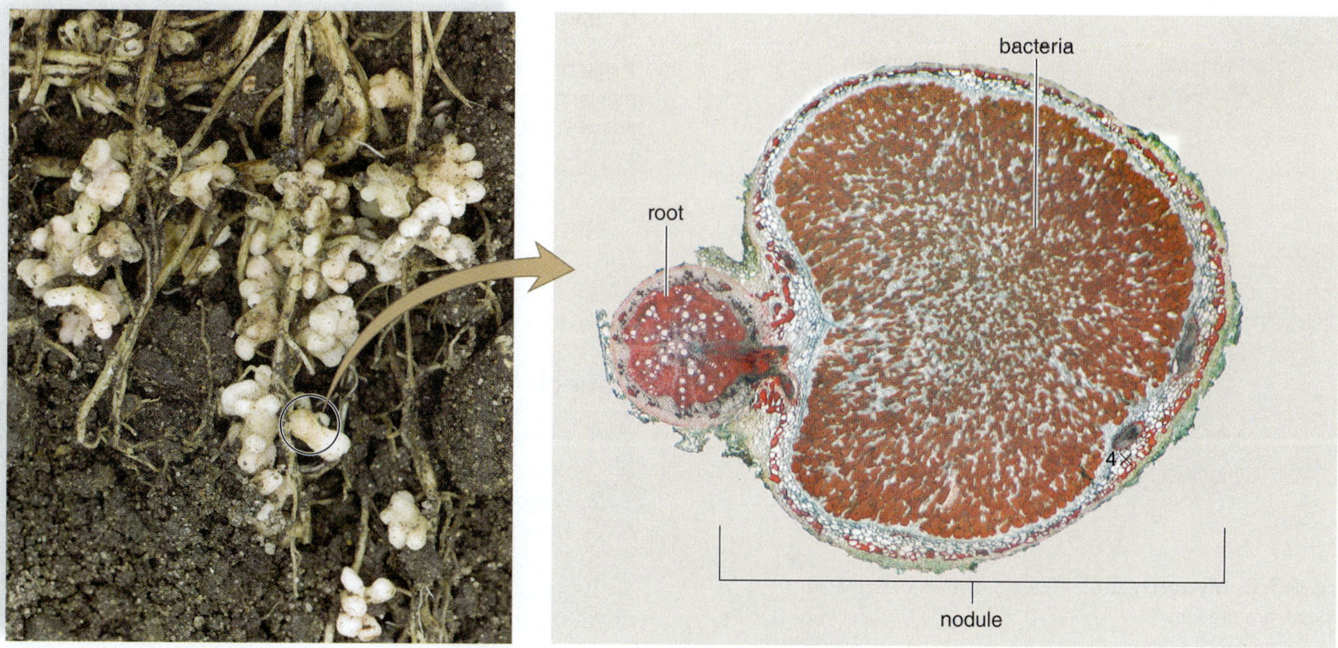

Figure 25.7 Root nodules. Nitrogen-fixing bacteria live in nodules on the roots of plants, particularly legumes.

The second type of mutualistic relationship, called **mycorrhizae,** involves fungi and almost all plant roots (Fig. 25.8) (see section 22.3). Only a small minority of plants do not have mycorrhizae, and these plants are most often limited as to the environment in which they can grow. The fungus increases the surface area available for mineral and water uptake and breaks down organic matter in the soil, releasing nutrients the plant can use. In return, the root furnishes the fungus with sugars and amino acids. Plants are extremely dependent on mycorrhizae. Orchid seeds, which are quite small and contain limited nutrients, do not germinate until a mycorrhizal fungus has invaded their cells.

Animation Root Nodule Formation

Other means of acquiring nutrients also occur. Parasitic plants, such as dodders, broomrapes, and pinedrops, send out rootlike projections, called haustoria, that tap into the xylem and phloem of the host stem (Fig. 25.9*a*). Carnivorous plants, such as the Venus flytrap and sundew, obtain some nitrogen and minerals when their leaves capture and digest insects (Fig. 25.9*b*).

Figure 25.8 Mycorrhizae. Plant growth is better when mycorrhizae are present.

a. Dodder, *Cuscuta* sp.

b. Cape sundew, *Drosera capensis*

Figure 25.9 Other ways to acquire nutrients. **a.** Some plants, such as the dodder, are parasitic. **b.** Some plants, such as the sundew, are carnivorous.

Check Your Progress 25.2

1. Explain the role of the endodermis and Casparian strip in concentrating minerals in a plant.
2. Describe the relationship of nitrogen-fixing bacteria with a host plant.
3. Explain how both partners benefit from a mycorrhizal association.

25.3 Transport Mechanisms in Plants

Learning Outcomes

Upon completion of this section, you should be able to

1. Describe the relationship between water potential and root pressure.
2. Identify the properties of water that influence the upward movement of water in flowering plants.
3. Explain how environmental factors influence the opening and closing of stomata.
4. List the correct sequence of events for the movement of water in xylem, and sucrose in phloem.

Flowering plants are well adapted to living in a terrestrial environment. Their leaves, which carry on photosynthesis, are positioned to catch the rays of the sun, because they are held aloft by the stem (see Fig. 25.11). Carbon dioxide enters leaves at the stomata, but water, the other main requirement for photosynthesis, is absorbed by the roots. Water must be transported from the roots through the stem to the leaves.

Reviewing Xylem and Phloem Structure

Vascular plants have a transport tissue, called *xylem,* that moves water and minerals from the roots to the leaves. Xylem, with its strong-walled, nonliving cells, gives trees much-needed internal support. Xylem contains two types of conducting cells: tracheids and vessel elements (Fig. 25.10*a*).

- *Tracheids* are tapered at both ends. The ends overlap with those of adjacent tracheids, and pits allow water to pass from one tracheid to the next.
- *Vessel elements* are long and tubular with perforation plates at each end. Vessel elements placed end to end form a completely hollow pipeline from the roots to the leaves.

The process of photosynthesis results in sugars, which are used as a source of energy and building blocks for other organic molecules throughout a plant. *Phloem* is the type of vascular tissue that transports organic nutrients to all parts of the plant. Roots buried in the soil cannot carry on photosynthesis, but they require a source of energy, so that they can carry on cellular metabolism. In flowering plants, phloem consists of two types of cells: sieve-tube members and companion cells (Fig. 25.10*b*).

- *Sieve-tube members* are the conducting cells of phloem. The end walls are called sieve plates and have numerous pores; strands of cytoplasm extend from one sieve-tube member to another through the pores. Sieve-tube members lack nuclei.
- *Companion cells,* which do have nuclei, provide proteins to sieve-tube members.

In this way, sieve-tube members form a continuous *sieve tube* for organic nutrient transport throughout the plant.

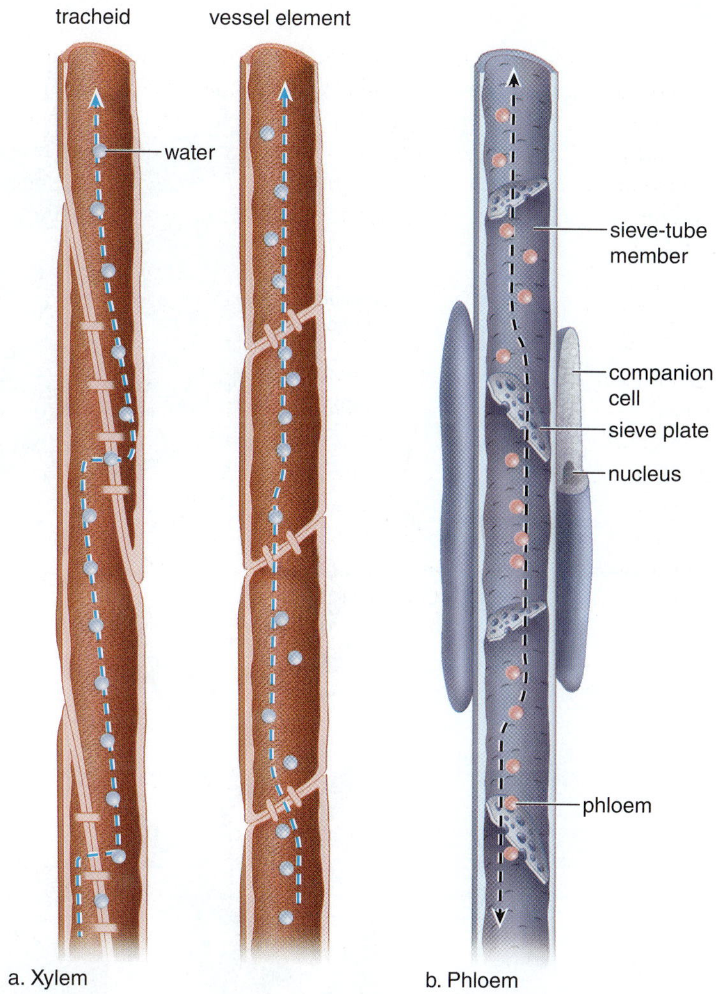

Figure 25.10 Organization of conducting cells in xylem and phloem. The specialized cells making up xylem and phloem form a series of interconnected and parallel cells that form a pathway for the movement of water and carbohydrates in a plant. Xylem cells **(a)** move water from the bottom to the top. Phloem cells **(b)** can move sucrose and other materials in any direction.

The Role of Water Potential

Knowing that vascular plants are structured in a way that allows materials to move from one part to another does not tell us the mechanisms by which these materials move. Plant physiologists have performed numerous experiments to determine how water and minerals rise to the tops of very tall trees in xylem, and how organic nutrients move in the opposite direction in phloem. We might expect that these processes are mechanical and based on the properties of water, because water is a large part of both *xylem sap* and *phloem sap*.

In living systems, water molecules diffuse freely across plasma membranes from the area of higher concentration to the area of lower concentration. Plant scientists prefer describing the movement of water in terms of **water potential:** Water always flows passively from the area of higher water potential to the area of lower water potential (Fig. 25.11). As can be seen in the Biological Systems feature, "The Concept of Water Potential," on page 467, the concept of water potential involves water pressure in addition to osmotic pressure.

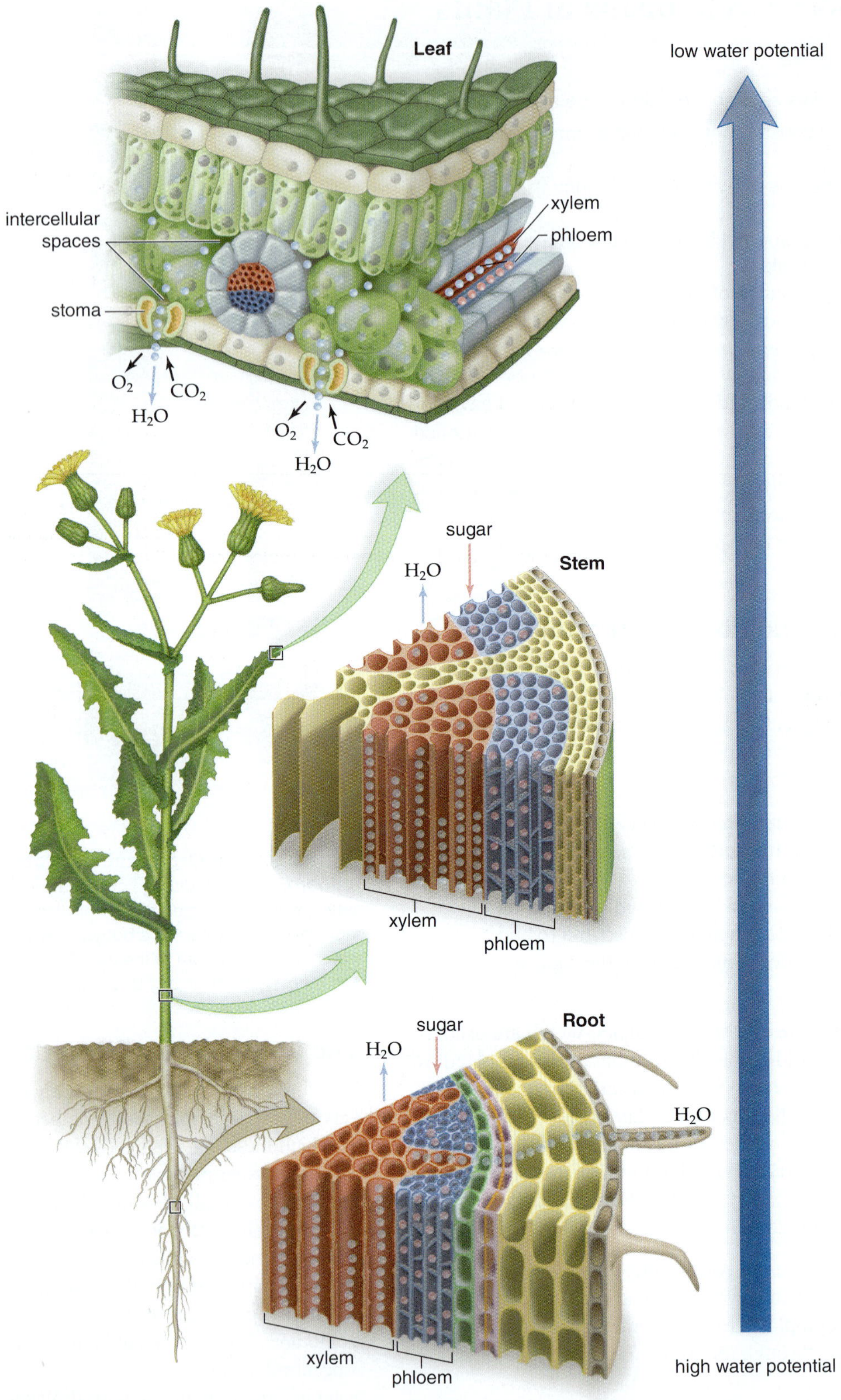

Figure 25.11 Plant transport and water potential. Vascular tissue in plants includes xylem, which transports water and minerals from the roots to the leaves, and phloem, which transports organic nutrients, often in the opposite direction. Notice that xylem and phloem are continuous from the roots through the stem to the leaves, which are the vegetative organs of a plant. Water potential is higher at the roots as water moves in by osmosis. Water potential is lower at the leaves as water escapes through stomata.

THEME Biological Systems

The Concept of Water Potential

As you learned in Chapter 6, potential energy is stored energy. Potential energy can exist in an object's position, such as that of a boulder at the top of a hill (mechanical energy), or in chemical bonds, such as the bonds between phosphate groups in ATP (chemical energy). Kinetic energy is the energy of motion; it is energy actively engaged in doing work. A boulder rolling down a hillside is exhibiting kinetic energy, as is an enzyme reaction that breaks a bond, converting ATP to ADP and releasing energy in the process.

Water potential is defined as the mechanical energy of water. Just like the boulder, water at the top of a waterfall has a higher water potential than water at the bottom of the waterfall. As illustrated by this example, water moves from a region of higher water potential to a region of lower water potential.

In terms of cells, two factors usually determine water potential, which in turn determines the direction in which water moves across a plasma membrane. These factors concern differences in

1. Water pressure across a membrane
2. Solute concentration across a membrane

Pressure potential is the effect that pressure has on water potential. Water moves across a membrane from an area of higher pressure to an area of lower pressure. The higher the water pressure, the higher the water potential, and vice versa. Pressure potential is the concept that best explains the movement of sap in xylem and phloem.

Osmotic potential, in contrast, takes into account the effects of solutes on the movement of water. The presence of solutes restricts the movement of water, because water tends to engage in molecular interactions with solutes, such as hydrogen bonding. Water therefore tends to move across a membrane from an area of lower solute concentration to an area of higher solute concentration. The lower the concentration of solutes, the higher the water potential, and vice versa.

It is not surprising that increasing the water pressure can counter the effects of solute concentration. This situation is common in plant cells. As water enters a plant cell by osmosis because of the higher solute concentration, water pressure increases inside the cell—a plant cell has a strong cell wall that allows water pressure to build up. When the pressure potential inside the cell balances the osmotic potential outside the cell, the flow of water in and out becomes the same.

Pressure potential that increases due to osmosis is often called *turgor pressure*. Turgor pressure is critical, because plants depend on it to maintain the turgidity of their bodies (Fig. 25B). The cells of a wilted plant have insufficient turgor pressure, and the plant droops as a result.

Questions to Consider

1. What variables will restrict the movement of water across the plasma membrane?
2. What structures are necessary for a plant to maintain turgidity?
3. What environmental conditions might cause a plant to lose its turgidity?

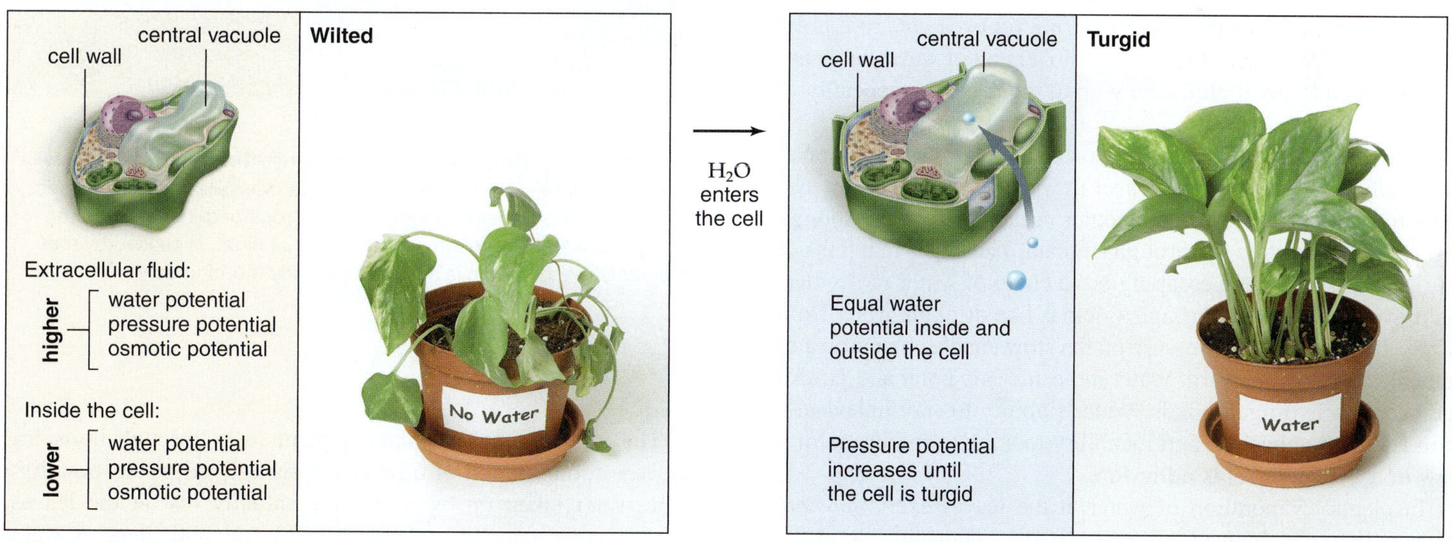

a. Plant cells need water.

b. Plant cells are turgid.

Figure 25B Water potential and turgor pressure. Water flows from an area of higher water potential to an area of lower water potential. **a.** The cells of a wilted plant have a lower water potential; therefore, water enters the cells. **b.** Equilibrium is achieved when the water potential is equal inside and outside the cell. Cells are now turgid, and the plant is no longer wilted.

Chemical properties of water are also important in the movement of xylem sap. The polarity of water molecules and the hydrogen bonding between water molecules allow water to fill xylem cells.

Water Transport

Recall that minerals accumulate at high concentrations beyond the endodermis in the root xylem tissue. This solute concentration difference results in the continuous movement of water into the root, creating **root pressure.** For example, if the stem of a young plant is cut, the cut end will often "leak" xylem sap. If a glass tube is sealed over the cut end, the sap will rise, because root pressure raises the water level in the glass tube (Fig. 25.12*a*). During the day, root pressure is not as obvious, because water is being drawn out from the leaves. At night, water continues to enter the roots, but evaporation slows down at the surface of the leaves. This results in a phenomenon called guttation. **Guttation** occurs when drops of water are forced out of vein endings along the edges of leaves (Fig. 25.12*b*). This morning "dew" effect is the direct result of root pressure.

Cohesion-Tension Model of Xylem Transport

Water that enters xylem must be transported against gravity to all parts of the plant. Transporting water can appear to be a daunting task, especially for plants such as redwood trees, which can exceed 90 m (almost 300 ft) in height.

The **cohesion-tension model** of xylem transport, outlined in Figure 25.13, describes how water and minerals travel upward in xylem yet requires no expenditure of energy by the plant. To understand how it works, one must start at the bottom of Figure 25.13 in the soil. Recall that there is a higher water potential in the soil and a lower water potential in the plant. Water will move into the root by osmosis. All of the water entering roots creates root pressure, which is helpful for the upward movement of water but is not nearly enough to get it all the way up to the leaves—especially in a tall tree.

Transpiration is the phenomenon that explains how water can completely resist gravity and travel upward. Focusing on the top of the tree, notice the water molecules escaping from the spongy mesophyll and into the air through stomata. The key is that it is not just one water molecule escaping but a chain of water molecules. The movement of water through xylem is like drinking water from a straw. Drinking exerts pressure on the straw, and a chain of water molecules is drawn upward. Water molecules are polar and "stick" together with hydrogen bonds. Water's ability to stay linked in a chain is called **cohesion,** and its ability to stick to the inside of a straw or a xylem vessel is **adhesion.**

In plants, evaporation of water at the leaves exerts *tension,* which pulls on a chain of water molecules. Transpiration is the constant tugging or pulling of the **water column** from the top due to evaporation. Cohesion of water molecules and adhesion to the inside of a xylem vessel facilitate this process. As transpiration occurs, the water column is pulled upward—first within the leaf, then from the stem, and finally from the roots. In addition, unlike other plant cells, xylem vessels offer a simple pipeline, with reinforced lignified walls and low resistance for the movement of water.

3D Animation Plant Transport: Water Transport in Xylem

a.

b.

Figure 25.12 Root pressure and guttation. **a.** Root pressure, as measured in this experiment, is a positive pressure potential caused by the entrance of water into root cells. **b.** Drops of guttation water on the edges of a leaf. Guttation, which occurs at night, is caused by root pressure. Often, guttation is mistaken for early morning dew.

The total amount of water a plant loses through transpiration over a long period of time is surprisingly large. At least 90% of the water taken up by roots is eventually lost at the leaves. A single corn plant loses between 135 and 200 liters of water through transpiration during the growing season. A typical tree loses 400 liters of water per day! On a global scale, plant transpiration has enormous effects on climate. For example, an estimated one-half to three-quarters of the rainfall received by the Amazon rain forest originates from water vapor of transpiring plants, often visible as a mist (Fig. 25.14). The evaporation of large amounts of water from plant surfaces dissipates heat and explains how plants cool themselves and their environments.

Opening and Closing of Stomata

The way water is transported in plants has an important consequence. When a plant is under water stress, the stomata close. Now the plant loses little water, because the leaves are protected against water loss by the waxy *cuticle* of the upper and lower epidermis. When stomata are closed, however, carbon dioxide cannot enter the leaves, and many plants are unable to photosynthesize efficiently. Photosynthesis, therefore, requires an abundant supply of water, so that stomata remain open, allowing carbon dioxide to enter.

Each *stoma,* a small pore in the leaf epidermis, is bordered by **guard cells.** When water enters the guard cells and turgor pressure increases, the stoma opens; when water exits the guard cells and turgor pressure decreases, the stoma closes. Notice in Figure 25.15 that the guard cells are attached to each other at their ends and that the inner walls are thicker than the outer walls. When water enters, a guard cell's radial expansion is restricted because of cellulose microfibrils in the walls, but lengthwise expansion of the outer walls is possible. When the outer walls expand lengthwise, they buckle out from the region of their attachment, and the stoma opens.

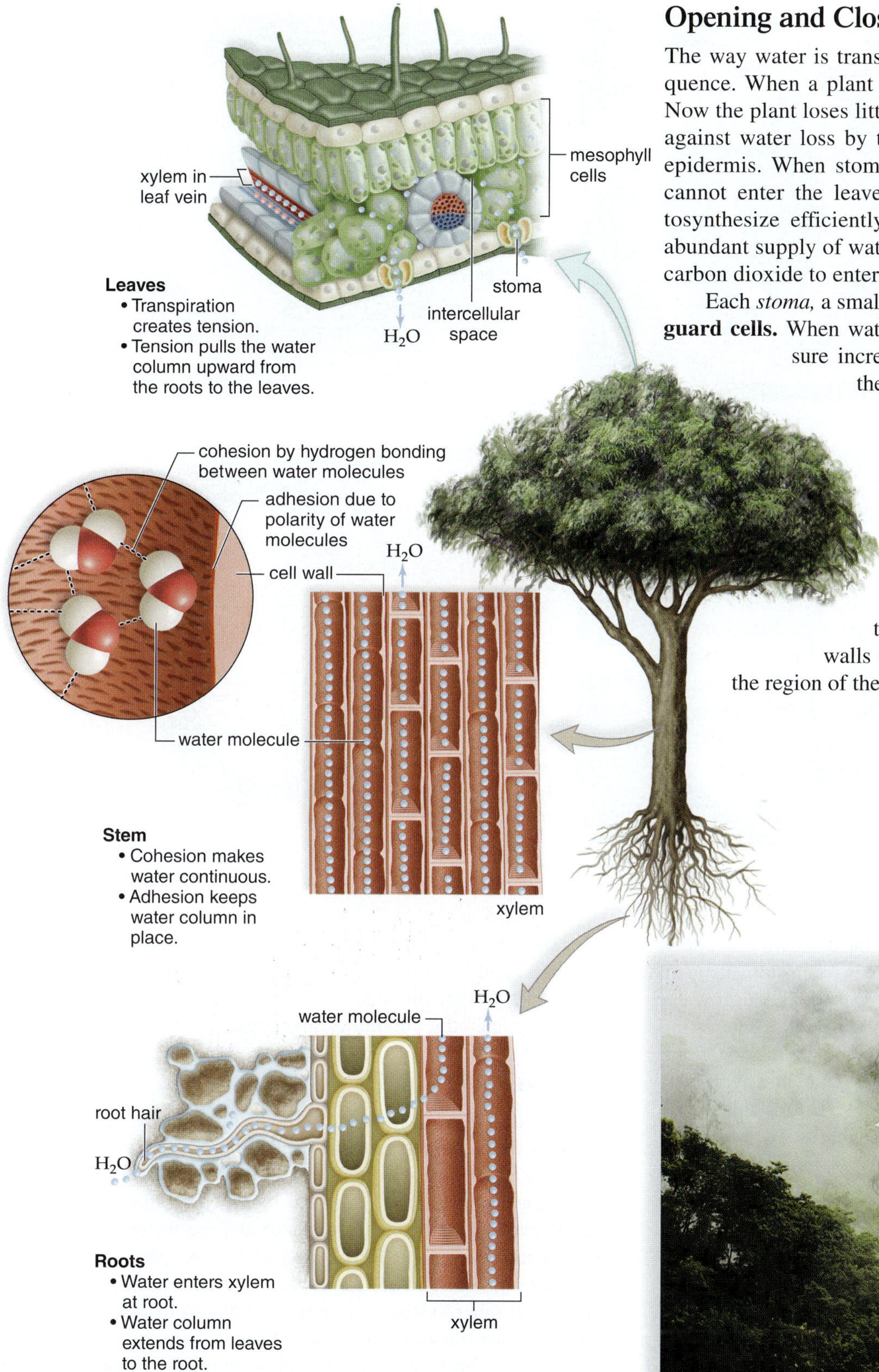

Figure 25.13 Cohesion-tension model of xylem transport. Tension created by evaporation (transpiration) at the leaves pulls water along the length of the xylem—from the roots to the leaves.

Figure 25.14 Plant-transpired mist rising from a tropical rain forest. Plants transpire enormous amounts of water, creating water vapor. Water vapor is an important source of rainfall, and the process of evaporative cooling is responsible for cooling the plants and affecting climate.

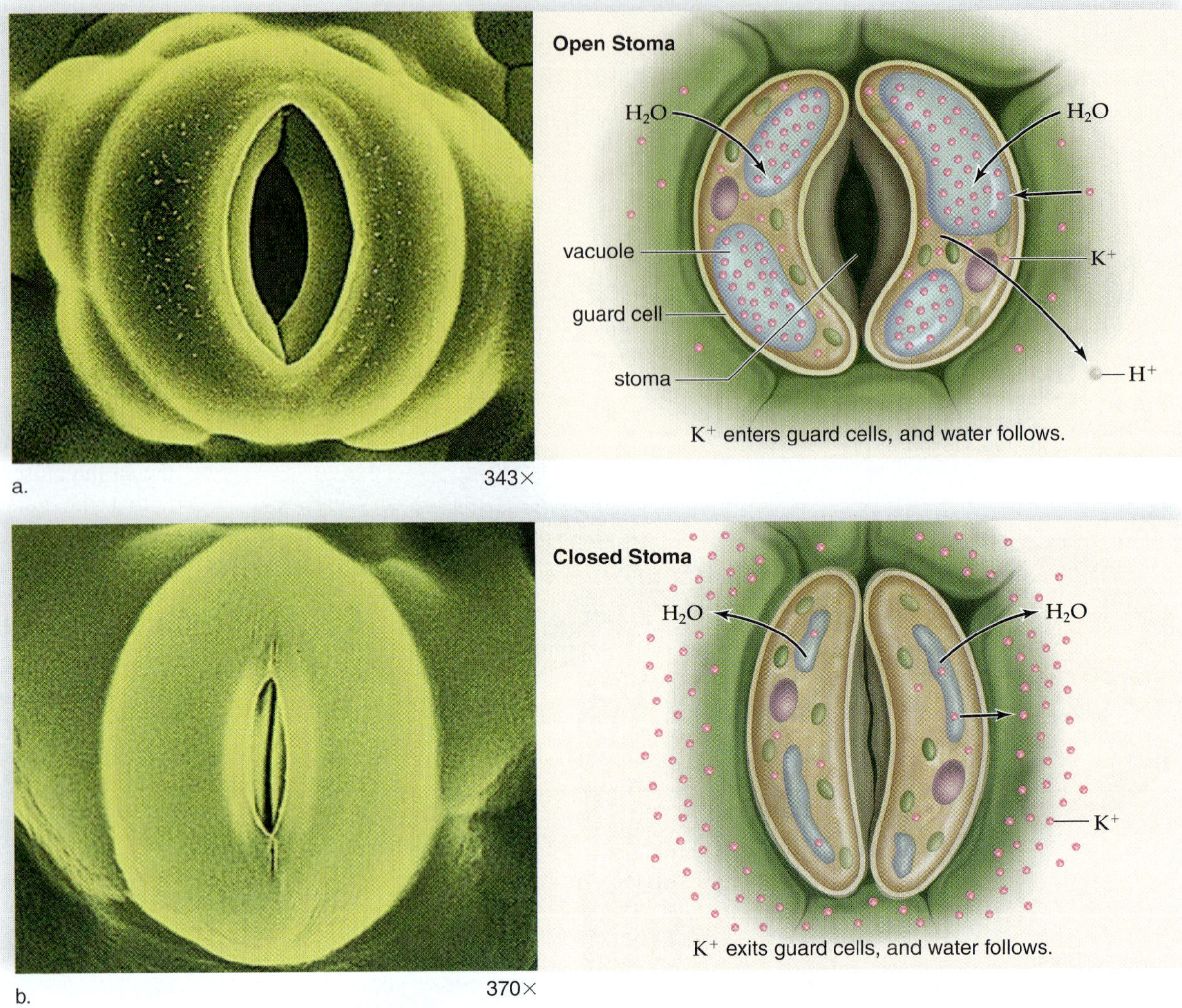

Figure 25.15 Opening and closing of stomata. **a.** A stoma opens when turgor pressure increases in guard cells due to the entrance of K^+ followed by the entrance of water. **b.** A stoma closes when turgor pressure decreases due to the exit of K^+ followed by the exit of water.

Since about 1968, plant physiologists have known that potassium ions (K^+) accumulate within guard cells when stomata open. In other words, active transport of K^+ into guard cells causes water to follow by osmosis and stomata to open. Another interesting observation is that hydrogen ions (H^+) accumulate outside guard cells as K^+ moves into them. A proton pump run by the hydrolysis of ATP transports H^+ to the outside of the cell. This establishes an electrochemical gradient that allows K^+ to enter by way of a channel protein (see Fig. 25.6*b*).

The blue-light component of sunlight has been found to regulate the opening and closing of stomata. Evidence suggests that a flavin pigment absorbs blue light, and then this pigment sets in motion the cytoplasmic response that leads to activation of the proton pump. In a similar way, a receptor in the plasma membrane of guard cells could bring about the inactivation of the pump when carbon dioxide (CO_2) concentration rises, as might happen when photosynthesis ceases. Abscisic acid (ABA), which is produced by cells in wilting leaves, can also cause stomata to close (see Chapter 26). Although photosynthesis cannot occur, water is conserved.

If plants are kept in the dark, stomata open and close about every 24 hours, as though they were responding to the presence of sunlight in the daytime and the absence of sunlight at night. The implication is that some sort of internal biological clock must be keeping time. Circadian rhythms (behaviors that occur nearly every 24 hours) and biological clocks are areas of intense investigation. Other factors that influence the opening and closing of stomata include temperature, humidity, and stress.

Organic Nutrient Transport

Mosses, described in Chapter 23, are short, ancient plants with no vascular tissue. Water, minerals, and the products of photosynthesis all move passively from one cell to the next. As plants evolved and moved onto land, there were many challenges for survival. Plants evolved tissues and organs to acquire water and minerals and to collect sunlight for photosynthesis. As plants grew taller, the shoot system and the root system became increasingly separated, and other systems (xylem and phloem tissue) evolved for long-distance travel. Phloem, specifically, is the tissue that translocates (transfers) the products of photosynthesis. Sugar, produced in mature leaves, moves to areas of development and storage, such as young leaves, fruit, and roots.

Role of Phloem

Phloem tissue is typically found external to the xylem in vascular tissues (see Fig. 24.14). In plants with woody stems, phloem makes

Figure 25.16 The movement of phloem from source to sink. **a.** An illustration of an *Arabidopsis* plant where radioactive ^{14}C , in the form of CO_2, was supplied to a mature leaf that can produce sugar. **b.** An X-ray image showing the movement of radioactive phloem sap from the source leaf to sink leaves and roots.

up the inner bark. It is the location of phloem in woody stems that helps explain why **girdling** a tree will cause the tree to die. If a strip of bark is removed from around the tree, then the lower half of the tree is cut off from its supply of sugar.

Interestingly, phloem sap does not move exclusively upward or downward, and it is not defined by gravity, as water is in xylem. In essence, phloem can travel in any direction, but these directions have a beginning, called the **source** (where sugars originate) and an end, or **sink** (where the sugars are unloaded).

Radioactive tracer studies with carbon 14 (^{14}C) have confirmed that phloem transports organic nutrients from source to sink. When ^{14}C-labeled carbon dioxide (CO_2) is supplied to mature leaves, radioactively labeled sugar is soon found moving down the stem into areas that cannot undergo photosynthesis, such as immature leaves and the roots (Fig. 25.16).

As expected, the liquid traveling in phloem is mostly water and sucrose, but other substances travel in the phloem as well, such as amino acids, hormones, RNA, and proteins involved in defense and protection.

Pressure-Flow Model of Phloem Transport

Figure 25.16 shows one of many experiments that plant scientists have conducted proving that what starts at a source can end up in a sink. The question now is, how does phloem sap move from the source to the sink? The translocation of sugar can be explained using the **pressure-flow model.** As mentioned earlier, phloem can travel in any direction. (For simplicity, Fig. 25.18 will later describe the movement of phloem sap from leaves, the source, to roots, the sink.)

Photosynthesizing leaves make sugar, and that sugar is actively transported from cells in the leaf mesophyll into the sieve tubes of phloem. Recall that, like xylem, phloem is a continuous pipeline throughout the plant. Active transport, or *loading,* of sugar into phloem is dependent on an electrochemical gradient established by a proton pump. Sugar is co-transported with hydrogen ions (H^+) that are moving down their concentration gradient (Fig. 25.17).

Next, high concentrations of sugar in the sieve tubes cause water to flow in by osmosis (Fig. 25.18). Like turning the nozzle on a hose, there is an increase in positive pressure as water flows in. The sugar (sucrose) solution, under massive *pressure* at the source, is forced to move by bulk *flow* to areas of lower pressure at the sink, like a root. This step highlights where the pressure-flow model got its name, and indeed, phloem has been measured moving at a velocity of 1 m an hour. The same distance with passive diffusion would take 30 years!

3D Animation **Plant Transport: Translocation in Phloem**

When the sugar arrives at the root, it is transported out of sieve tubes into the root cells. There, the sugar is used for cellular respiration or other metabolic processes. The high concentration of sugar in the root cells causes water to follow by osmosis, where it is later reclaimed by the xylem tissue.

Although leaves are generally the source, storage roots and stems such as carrots, beets, and potatoes are also examples of sources providing much needed sugar during winter or periods of dormancy.

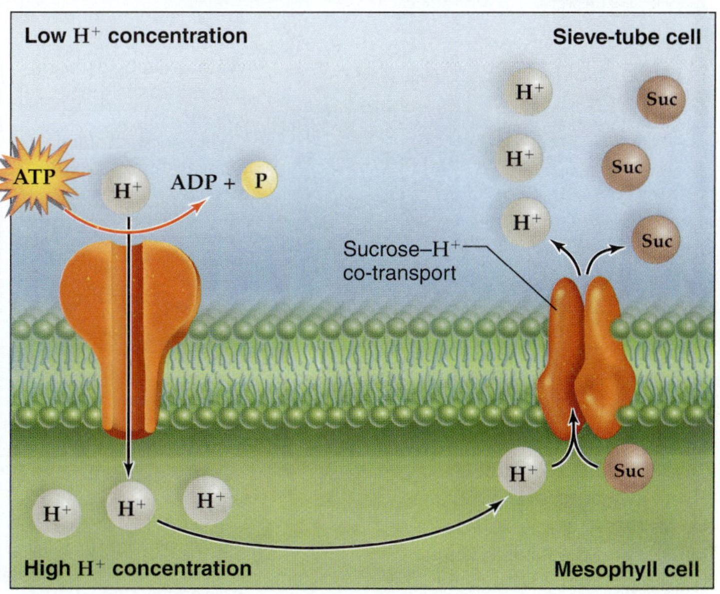

Figure 25.17 Sucrose loading is dependent on a H^+ ion gradient. H^+ ions are actively pumped into the mesophyll cell so they can be co-transported with sucrose into sieve-tube cells.

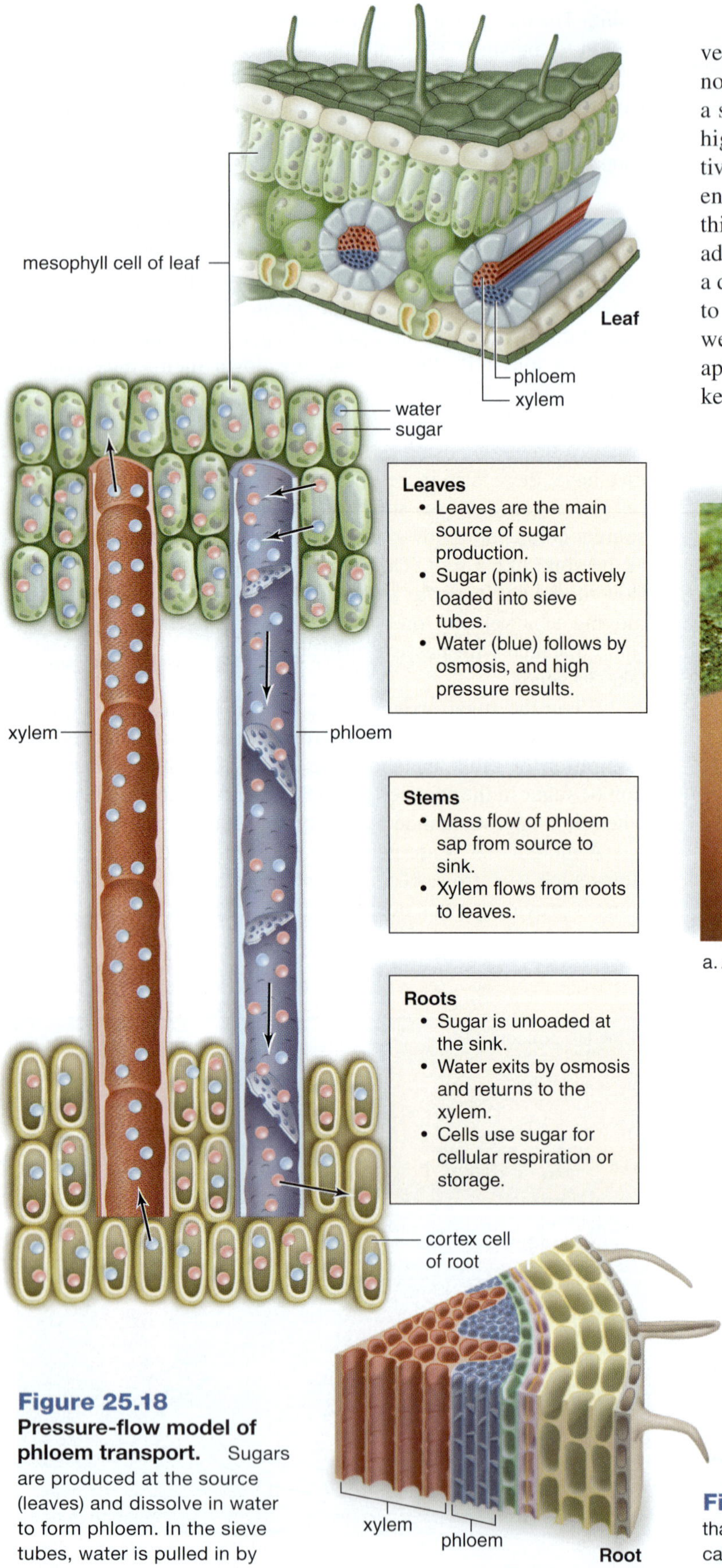

Figure 25.18 Pressure-flow model of phloem transport. Sugars are produced at the source (leaves) and dissolve in water to form phloem. In the sieve tubes, water is pulled in by osmosis. The phloem follows positive pressure and moves toward the sink (root system).

The high pressure of sucrose in phloem has resulted in a very interesting observation of tiny insects called aphids. Aphids, normally a pest in a garden, have a needlelike mouthpart, called a stylet, that can penetrate a stem and tap into a sieve tube. The high-pressure sucrose solution is forced through an aphid's digestive tract very quickly, resulting in a droplet of sucrose on the rear end, called "honeydew" (Fig. 25.19*a*). Many ant species consume this honeydew and, in turn, protect the aphids. Scientists also take advantage of this natural phloem-tapping system by anesthetizing a drinking aphid, removing its body, and using the inserted stylet to collect phloem for analysis (Fig. 25.19*b*). If the researcher were to insert a needle into the stem, the phloem would clot, but aphids have a unique anticlotting property in their saliva that keeps the initial sieve-tube clot from forming.

a. An aphid feeding on a plant stem

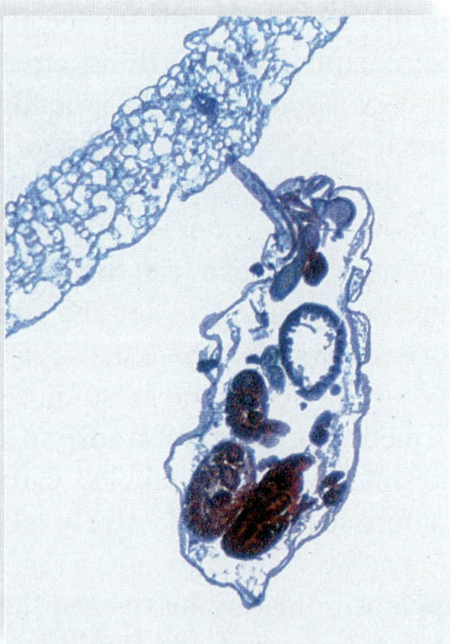

b. Aphid stylet in place

Figure 25.19 Acquiring phloem sap. Aphids are small insects that remove nutrients from phloem by means of a needlelike mouthpart called a stylet. **a.** Excess phloem sap appears as a droplet after passing through the aphid's body. **b.** Micrograph of a stylet in plant tissue. When an aphid is cut away from its stylet, phloem sap becomes available for collection and analysis.

Check Your Progress 25.3

1. Describe how water flows upward against gravity.
2. Identify the cohesion and adhesion properties of water that pertain to water transport.
3. Describe the process in which sugars move from source to sink in a plant.

CONNECTING *the* CONCEPTS *with the* THEMES

Evolution

- Through the process of natural selection, vascular plants have evolved a vascular system that provides efficient absorption and delivery of water via xylem and distribution of sugars via phloem.
- The coevolution of symbiotic associations with fungi and nitrogen-fixing bacteria have enhanced plants' success.

Nature of Science

- In addition to water, plants require minerals from the soil, such as phosphorus, nitrogen, and calcium. Active mechanisms have evolved to concentrate essential nutrients inside the plant cells.
- The ability to concentrate minerals and other molecules allows some plants to remove pollutants from soil and water, aiding in environmental remediation.

Biological Systems

- The separation of the two conduction systems allows plants to move different nutrients simultaneously. Water can continue to flow upward in xylem sap, while sugars flow downward in phloem sap toward sinks, such as the roots.
- Plant conduction systems take advantage of the cohesive and adhesive properties of water to move fluids as an unbroken column through conducting cells.
- The mechanisms that control the opening and closing of stomata enable plants to regulate loss of water, even at the expense of photosynthesis.

MEDIA STUDY TOOLS http://connect.mheducation.com

SMARTBOOK® Study smart. SmartBook helps you maximize your study time by identifying what you know and don't know. The recharge feature in SmartBook helps you prepare for the test by identifying the material that you are most likely to forget.

connect BIOLOGY Find even more resources to learn the chapter concepts including animations, tutorial videos, and interactive practice questions.

Animations	3D Animations
25.2 Proton Pump • Root Nodule Formation	**25.3** Plant Transport: Water Transport in Xylem • Plant Transport: Translocation in Phloem

SUMMARIZE

25.1 Plant Nutrition and Soil

Plants need various **essential nutrients** called **minerals.** Essential nutrients needed in large quantities are called **macronutrients;** those needed in smaller quantities are **micronutrients.** Carbon, hydrogen, and oxygen make up 95% of a plant's dry weight. The other necessary nutrients are taken up by the roots as mineral ions. **Hydroponics** is a way of growing plants in water, but it is also useful in figuring out mineral deficiencies. Mineral deficiencies can be avoided with the use of fertilizer.

Plant life is dependent on **soil,** which is formed by the mechanical and chemical weathering of rock. Soil is a mixture of mineral particles, **humus,** living organisms, air, and water. Soil particles are of three types, from the largest to the smallest: sand, silt, and clay. Loam, which contains about equal proportions of all three types, retains water but still has air spaces. Humus contributes to the texture of soil and its ability to provide inorganic nutrients to plants. **Cation exchange** is the chemical process by which minerals are absorbed by plant roots. Topsoil (a **soil horizon** of a **soil profile**) contains humus, and this is the layer that is lost by **soil erosion.**

25.2 Water and Mineral Uptake

Water, along with minerals, can enter a root by passing between the porous cell walls until it reaches the Casparian strip, after which it passes through an endodermal cell before entering xylem. Water can also enter root hairs and then pass through the cells of the cortex and endodermis to reach xylem.

Mineral ions cross plasma membranes by a chemiosmotic mechanism. A proton pump transports H^+ out of the cell. This establishes an electrochemical gradient that causes positive ions to flow into the cells. Negative ions are carried across the membrane when H^+ moves along its concentration gradient.

Plants have a number of adaptations that assist them in acquiring nutrients. Legumes have **root nodules** infected with **rhizobia,** which makes nitrogen compounds available to these plants. Many other plants have **mycorrhizae,** or fungus associated with the root system. The fungus gathers nutrients from the soil, while the root provides the fungus with sugars and amino acids.

25.3 Transport Mechanisms in Plants

As an adaptation to life on land, plants have a vascular system that transports water and minerals from the roots to the leaves. Sugars are then transported in the opposite direction. Vascular tissue includes xylem and phloem.

In xylem, vessels composed of vessel elements aligned end to end form an open pipeline from the roots to the leaves. The difference in **water potential** accounts for the upward movement of water in a plant. At night, **root pressure** can build in the root, causing **guttation.** However, this does not contribute significantly to xylem transport.

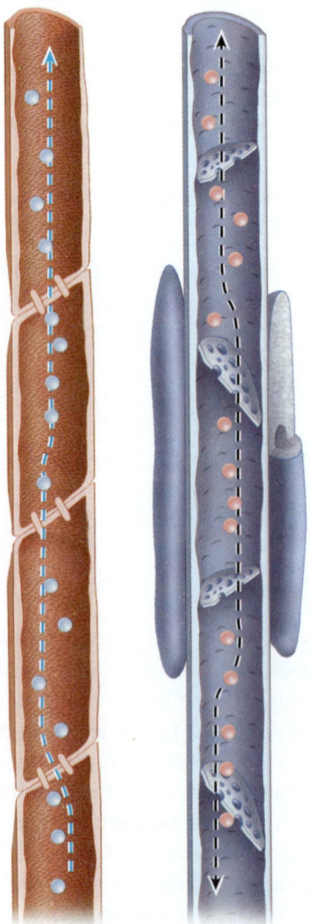

The **cohesion-tension model** of xylem transport states that **transpiration** creates a tension, which pulls water upward in xylem from the roots to the leaves. This means of transport works only because water molecules are **cohesive** with one another, forming a **water column,** and **adhesive** with xylem walls.

Most of the water taken in by a plant is lost through stomata by transpiration. Only when there is plenty of water do stomata remain open, allowing carbon dioxide to enter the leaf and photosynthesis to occur.

Stomata open when **guard cells** take up water. The guard cells stretch lengthwise, because the microfibrils in their walls prevent lateral expansion. Water enters the guard cells after potassium ions (K^+) have entered, causing these cells to buckle outward. Light signals stomata to open, and a high carbon dioxide (CO_2) level may signal stomata to close. Abscisic acid produced by wilting leaves also signals for closure.

An early observation for the function of phloem came from the effects of girdling a tree. **Girdling** involves the removal of bark, which contains phloem, and once removed would kill a tree.

In phloem, sieve tubes composed of sieve-tube members aligned end to end form a continuous pipeline from the leaves to the roots. Sieve-tube members have sieve plates through which strands of cytoplasm extend through plasmodesmata from one member to the other.

The **pressure-flow model** of phloem transport proposes that a positive pressure drives phloem contents in sieve tubes. Sucrose is actively transported into sieve tubes—by a chemiosmotic mechanism—at a **source,** and water follows by osmosis. The resulting increase in pressure creates a flow that moves water and sucrose to a **sink.** A sink can be at the roots or at any other part of the plant that requires organic nutrients.

ASSESS

Choose the best answer for each question.

25.1 Plant Nutrition and Soil

1. A nutrient element is considered essential if
   a. plant growth increases with a reduction in the concentration of the element.
   b. plants die in the absence of the element.
   c. plants can substitute a similar element for the missing element with no ill effects.
   d. the element is a positive ion.
2. Humus
   a. supplies nutrients to plants.
   b. is basic in its pH.
   c. is found in the deepest soil horizons.
   d. is inorganic in origin.
3. Soils rich in which type of mineral particle will have a high water-holding capacity?
   a. sand
   b. silt
   c. clay
   d. All soil particles hold water equally well.
4. Negatively charged clay particles attract
   a. K^+.
   b. NO_3^-.
   c. Ca^{2+}.
   d. Both a and b are correct.
   e. Both a and c are correct.

25.2 Water and Mineral Uptake

5. The Casparian strip affects
   a. how water and minerals move into the vascular cylinder.
   b. vascular tissue composition.
   c. how soil particles function.
   d. cation exchange.
   e. Both a and c are correct.
6. Nitrogen-fixing bacteria
   a. reduce atmospheric nitrogen.
   b. can live independently in the soil.
   c. are usually associated with legume plants.
   d. break N ≡ N bonds.
   e. All of these are correct.
7. Which of the following is not an adaptation by plants to obtain minerals?
   a. specialized leaves that capture insects
   b. parasitic plants, such as a dodder
   c. mycorrhizal associations
   d. nitrogen-fixing bacteria in root nodules
   e. broad leaves in low sunlight

25.3 Transport Mechanisms in Plants

8. The opening in the leaf that allows gas and water exchange is called the
   a. lenticel.
   b. hole.
   c. stoma.
   d. guard cell.
   e. accessory cell.
9. Stomata are usually open
   a. at night, when the plant requires a supply of oxygen.
   b. during the day, when the plant requires a supply of carbon dioxide.
   c. day or night if there is excess water in the soil.
   d. during the day, when transpiration occurs.
   e. Both b and d are correct.
10. What role do cohesion and adhesion play in xylem transport?
    a. Like transpiration, they create a tension.
    b. Like root pressure, they create a positive pressure.
    c. Like sugars, they cause water to enter xylem.
    d. They create a continuous water column in xylem.
    e. All of these are correct.
11. After sucrose enters sieve tubes,
    a. it is removed by the source.
    b. water follows passively by osmosis.
    c. it is driven by active transport to the source, which is usually the roots.
    d. stomata open so that water flows to the leaves.
    e. All of these are correct.
12. The pressure-flow model of phloem transport states that
    a. phloem content always flows from the leaves to the root.
    b. phloem content always flows from the root to the leaves.
    c. water flow takes sucrose from a source to a sink.
    d. water pressure creates a flow of water toward the source.
    e. Both c and d are correct.

ENGAGE

THINKING SCIENTIFICALLY

1. Design an experiment to determine whether calcium, nitrogen, or phosphate is an essential plant nutrient. State the possible results.
2. Explain why this experiment supports the hypothesis that transpiration can cause water to rise to the tops of tall trees.

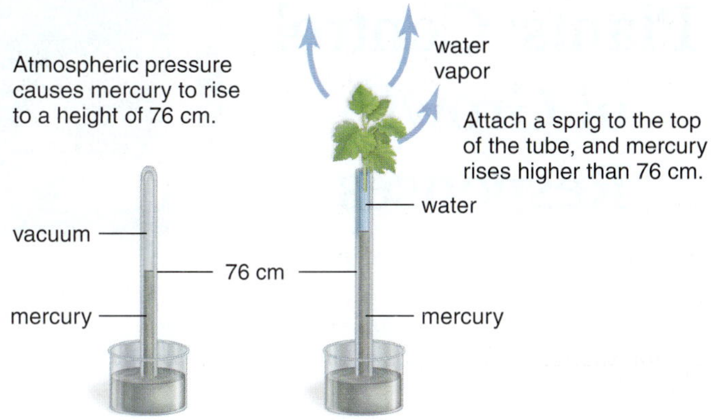

3. *Welwitschia* is a genus of plant that lives in the Namib and Mossamedes Deserts in Africa. Annual rainfall averages only 2.5 cm (1 inch) per year. *Welwitschia* plants contain a large number of stomata (22,000 per cm^2), which remain closed most of the time. Can you suggest how a large number of stomata would be beneficial to these desert plants?
4. Explain why solution flows from the left bulb to the right bulb and how this experiment relates to the pressure-flow model of phloem transport.

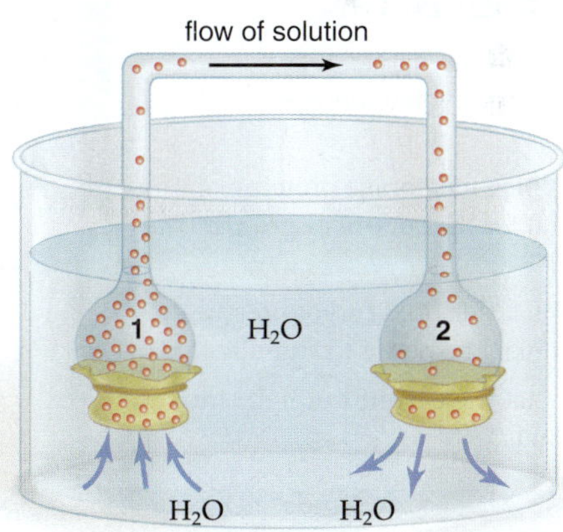

26

Flowering Plants: Control of Growth Responses

A photograph of sunflowers uniformly tracking the movement of the sun.

Chapter Outline

26.1 Plant Hormones 477

26.2 Plant Growth and Movement Responses 485

26.3 Plant Responses to Phytochrome 489

Before You Begin

Before beginning this chapter, take a few moments to review the following discussions.

Figure 5.2 Which membrane proteins are essential for the control of growth responses?

Figure 5.7 What role does turgor pressure play in the plant response to stimuli?

Figure 7.6 What wavelengths of light are used by plants?

The observation that sunflowers track the sun as it moves through the sky is a striking example of a flowering plant's ability to respond to environmental stimuli. Other responses to light can take longer than sun-tracking, because they involve hormones and an alteration in growth. For example, flowering plants will exhibit a bend toward the light within a few hours, because a hormone produced by the growing tip has moved from the sunny side to the shady side of the stem. Hormones also help flowering plants respond to stimuli in a coordinated manner. In the spring, seeds germinate and growth begins if the soil is warm enough to contain liquid water. In the fall, when temperatures drop, shoot and root apical growth ceases. Some plants also flower according to the season. The pigment phytochrome is instrumental in detecting the photoperiod and bringing about changes in gene expression, which determine whether a plant flowers or does not flower.

Plant defenses include physical barriers, chemical toxins, and even mutualistic animals. This chapter discusses the variety of ways flowering plants can respond to their environment, including other organisms.

As you read through this chapter, think about the following questions:

1. Which hormones are essential for plant growth?
2. What environmental stimuli trigger the various plant responses?

Following *the* Themes

CHAPTER 26 FLOWERING PLANTS: CONTROL OF GROWTH RESPONSES

UNIT 5 PLANT EVOLUTION AND BIOLOGY

Evolution	Plants have evolved responses to specific stimuli that increase their chances of survival.
Nature of Science	Scientific research has found a variety of human applications for a number of plant hormones.
Biological Systems	Plants respond to stimuli by using specific signal transduction pathways.

26.1 Plant Hormones

Learning Outcomes

Upon completion of this section, you should be able to

1. Explain the role of hormones when plant cells utilize signal transduction to respond to stimuli.
2. Compare and contrast the effects of auxins, gibberellins, cytokinins, abscisic acid, and ethylene on plant growth and development.

All organisms are capable of responding to environmental stimuli. Being able to respond to stimuli is a beneficial adaptation, because it leads to organisms' longevity and ultimately to the survival of the species. Flowering plants perceive and react to a variety of environmental stimuli. Some examples include light, gravity, carbon dioxide levels, pathogen infection, drought, and touch. Their responses can be short-term, as when stomata open and close in response to light levels, or long-term, as when plants respond to gravity with the downward growth of the root and the upward growth of the stem.

Although we think of responses in terms of a plant structure, the mechanism that brings about a response occurs at the cellular level. Research has shown that plant cells respond to stimuli by utilizing **signal transduction,** the binding of a molecular "signal" that initiates and amplifies a cellular response. You first encountered the concept of signal transducers in Chapter 13 in the description of tumor suppressor genes and oncogenes (see Figs. 13.13 and 13.14).

Notice in Figure 26.1 that signal transduction involves the following.

Receptors—proteins activated by a specific signal. Receptors can be located in the plasma membrane, the cytoplasm, the nucleus, or even the endoplasmic reticulum. A receptor that responds to light has a pigment component. For example, the phytochrome receptor protein has a region that is sensitive to red light, and the phototropin receptor protein has a region that is sensitive to blue light.

Transduction pathway—a series of relay proteins or enzymes that pass a signal until it reaches the machinery of the cell. In some instances, a bound receptor immediately communicates with the transduction pathway, and in other instances, a second messenger, such as Ca^{2+}, initiates the response.

Animation Second Messengers

Cellular response—the result of the transduction pathway. Very often, the response is either the transcription of particular genes or the end product of an activated metabolic pathway. The cellular response brings about the overall visible change, such as stomata closing or a stem that turns toward the light.

What role do plant **hormones** play in the ability of flowering plants to respond to a stimulus? The answer is that they serve as chemical signals that coordinate cell responses. These molecules

Figure 26.1 Signal transduction in plants. (1) The hormone auxin enters the cell and is received by a receptor in the nucleus. This complex alters gene expression. (2) A light receptor in the plasma membrane is sensitive to and activated by blue light. Activation leads to stimulation of a transduction pathway that ends with gene expression changes. (3) When attacked by an herbivore, the flowering plant produces defense hormones that bind to a plasma membrane receptor. Again, the transduction pathway results in a change in gene expression.

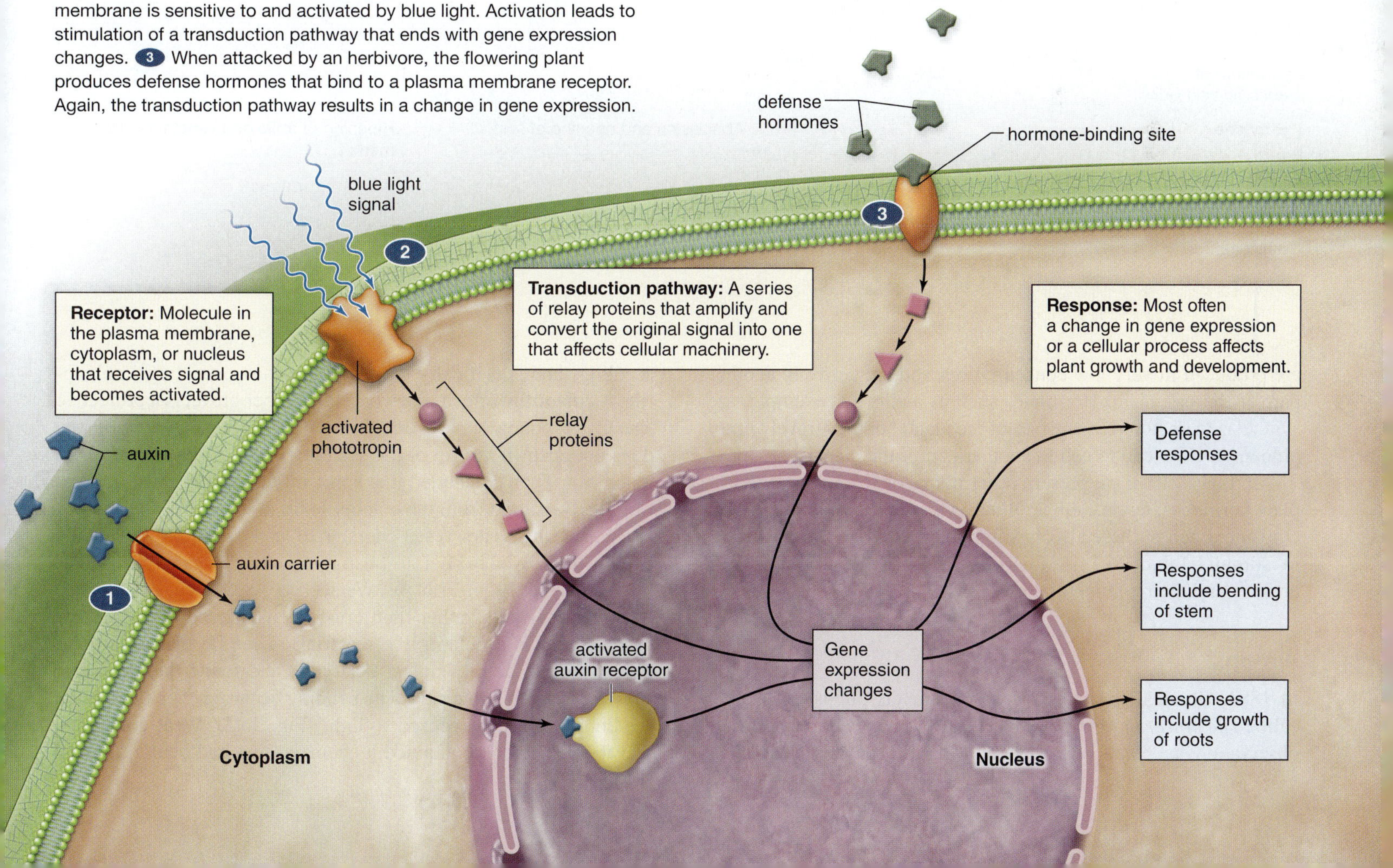

Table 26.1 Plant Hormones

Hormone	Plant use	Commercial use
Auxin (indoleacetic acid; IAA) 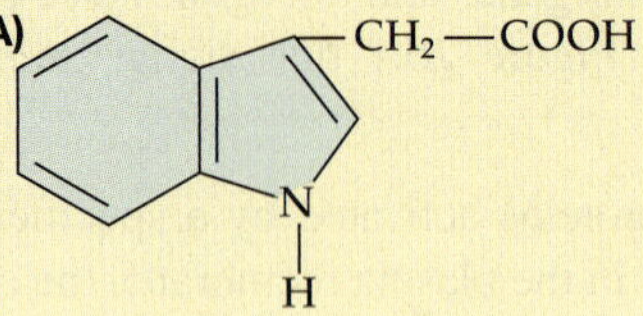 Structure of indoleacetic acid (IAA)	Growth of roots and fruits, prevents the loss of leaves	Induce fruit production without pollination, used in commercial herbicides (2,4-D)
Gibberellins (gibberellic acid; GA_3) 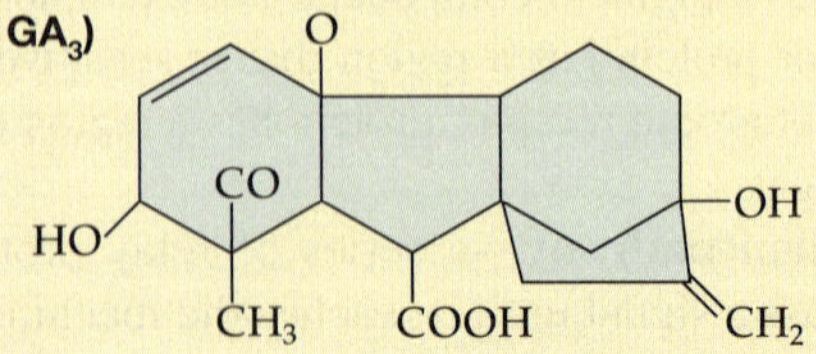 Structure of gibberellic acid (GA_3)	Stem elongation	Increase growth and size of the plants, breaks the dormancy cycle
Cytokinins (zeatin) 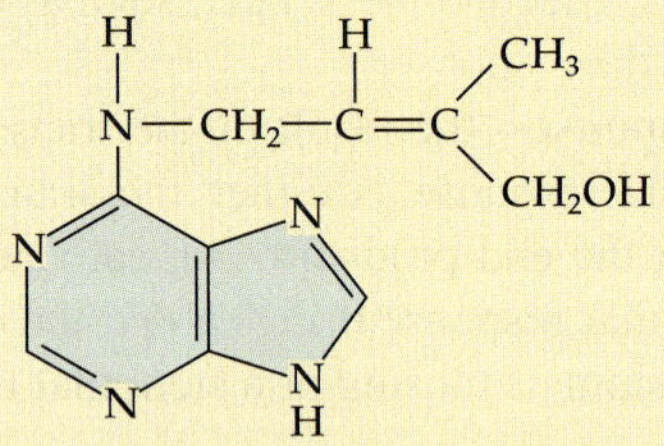 Structure of zeatin	Cell division, prevents senescence	Prolongs the shelf life of flowers and vegetables
Abscisic acid (abscisic acid; ABA) 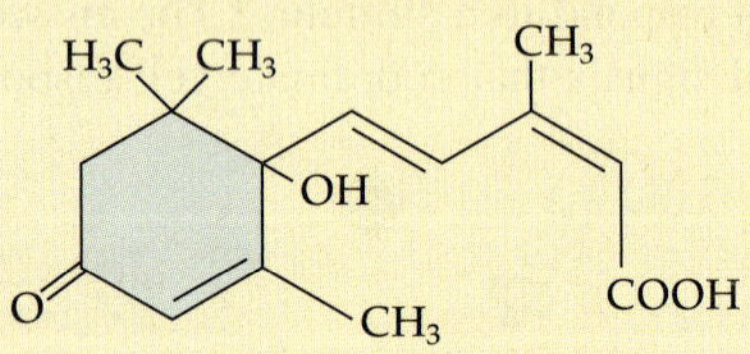 Structure of abscisic acid (ABA)	Initiates and maintains seed and bud dormancy	Commercial thinning of fruits to promote growth in the remaining fruit
Ethylene 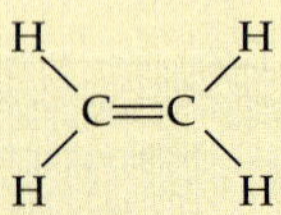 Structure of ethylene	Abscission and ripening of fruit	Ripening of fruits and vegetables for market

are produced in very low concentrations and are active in another part of the organism. Hormones such as auxin, for example, are synthesized or stored in one part of the plant, but they travel within phloem or from cell to cell to another part of the plant.

In this section, we present descriptions of five major types of plant hormones: auxins, gibberellins, cytokinins, abscisic acid, and ethylene. Each of these affects different aspects of plant responses to stimuli. Table 26.1 summarizes these five hormones, their actions, and their commercial uses.

Auxins

In 1881, Charles Darwin and his son Francis published a book called *The Power of Movement in Plants*. Here they described their observation that plants bend toward the light. The Darwins used oat seedlings and specifically looked at the young seedlings' coleoptile. A **coleoptile,** much like wearing a rubber glove, is a protective sheath for young leaves. Bending toward light, or *phototropism,* does not occur if the tip of the seedling is cut off or covered by a black cap. They concluded that some influence that causes curvature is transmitted from the coleoptile tip to the rest of the shoot.

For many years researchers tried and failed to isolate the chemical involved in phototropism by crushing and analyzing coleoptile tips. In 1926, Frits W. Went (1903–1990) cut off the tips of coleoptiles and, rather than crush them, placed them on agar blocks (agar is a gelatin-like material). Then he placed an agar block to one side of a tipless coleoptile and found that the shoot curved away from that side. The bending occurred even though the seedlings were not exposed to light (Fig. 26.2). Went concluded that the agar block contained a chemical that had been produced

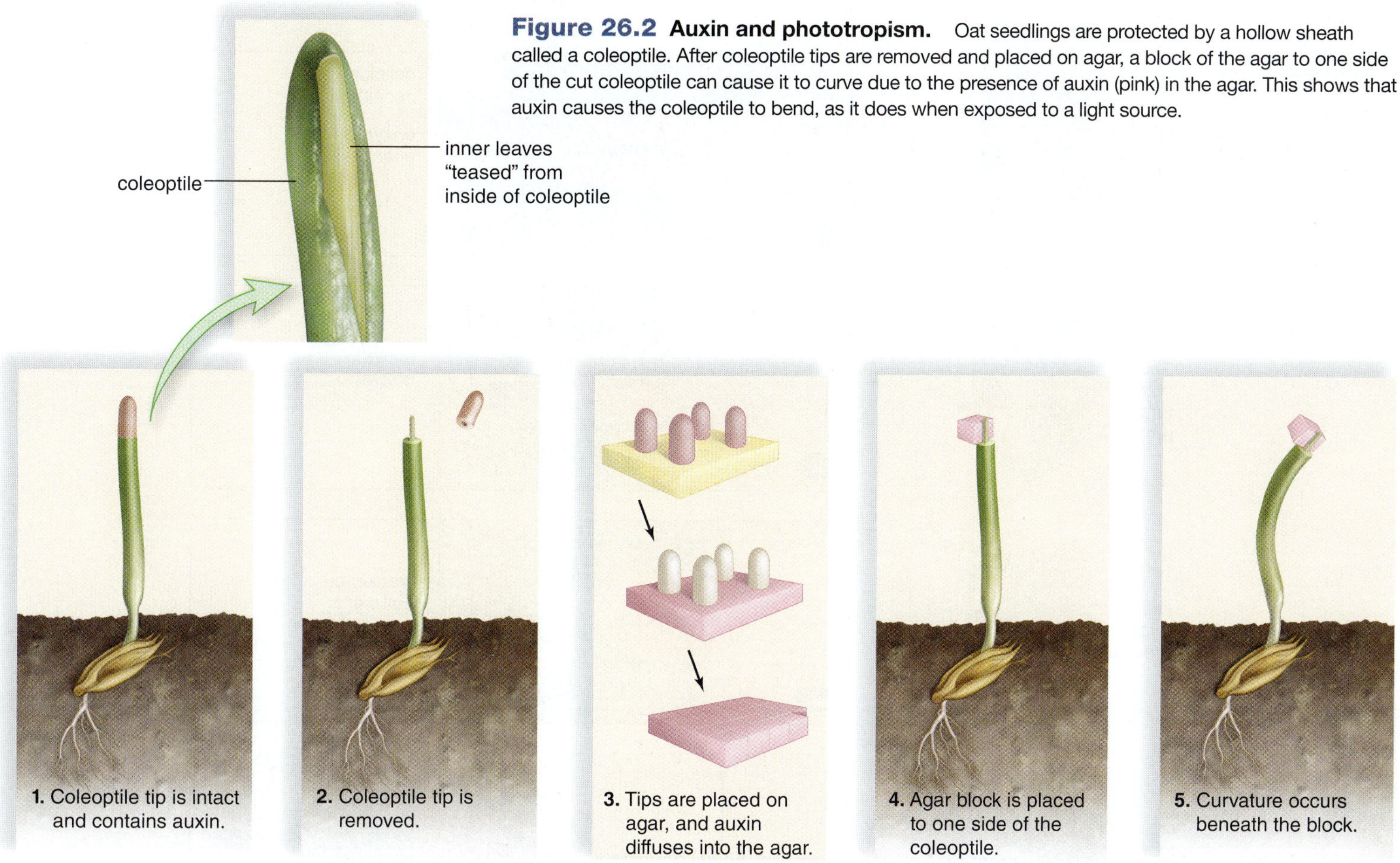

Figure 26.2 Auxin and phototropism. Oat seedlings are protected by a hollow sheath called a coleoptile. After coleoptile tips are removed and placed on agar, a block of the agar to one side of the cut coleoptile can cause it to curve due to the presence of auxin (pink) in the agar. This shows that auxin causes the coleoptile to bend, as it does when exposed to a light source.

by the coleoptile tips. This chemical, he decided, had caused the shoots to bend. He named the chemical substance **auxin** after the Greek word *auximos,* which means "promoting growth." Since then, it has been determined that auxins are also produced in shoot apical meristems (see Fig. 24.7), young leaves, flowers, fruits, and at lower levels in the root apical meristem. There are many variations of auxin, but the most common naturally occurring form is indoleacetic acid (IAA).

How Auxins Cause Stems to Bend

When a stem is exposed to unidirectional light, auxin moves to the shady side, where it enters the nuclei of shady cells and attaches to a receptor (Fig. 26.3*a*). Next, hydrogen ions are pumped into the cell wall, creating an acidic environment. The acid triggers enzymes to dismantle the cellulose fibers in the cell wall, resulting in a weakened wall. To test the hypothesis that acid weakens walls, scientists have neutralized the acid in shady cells and the result was cells that would not grow.

A plant with a cell wall that is weak and loose will result in a decrease in turgor pressure. The water potential inside the cell is now less than outside the cell, and as you learned in Chapter 25, water moves from areas of high potential to low potential. Water flows into the cell from other parts of the plant, attempting to restore turgor pressure once again. The pressure of the water begins to stretch the wall, and the plant cell responds by rebuilding a now longer wall, resulting in cell growth. Without the rebuilding, the cell would burst. Overall, the role of auxin is to cause the wall to weaken, so that it can be stretched and rebuilt even larger. The end result of these activities is elongation of the stem, because only the cells on the shady side are getting larger, bending the stem toward the light (Fig. 26.3*b*).

Auxins Affect Growth and Development

As mentioned, auxin causes phototropism, maximizing exposure of a plant to the sun. In addition, auxin is responsible for a process called *gravitropism* where, after the direction of gravity has been detected by a flowering plant, auxin moves to the lower surface of roots and stems. Thereafter, roots curve downward and stems curve upward. Phototropism and gravitropism are discussed at more length on page 486.

This versatile hormone is also responsible for a phenomenon called **apical dominance.** Experienced gardeners know that, to produce a bushier plant, they must remove the terminal bud (see Fig. 24.6). In a plant that has not been altered, auxin produced in the apical meristem of the terminal bud is transported downward, inhibiting the growth of lateral or axillary buds. Release from apical dominance occurs when pruning (cutting) removes the shoot tip. Then the axillary buds grow and the plant takes on a fuller appearance. Interestingly, if auxin were to be applied to the broken terminal stem, apical dominance would be restored.

Auxin causes the growth of roots and fruits and prevents the loss of leaves and fruit. The application of an auxin paste to a stem

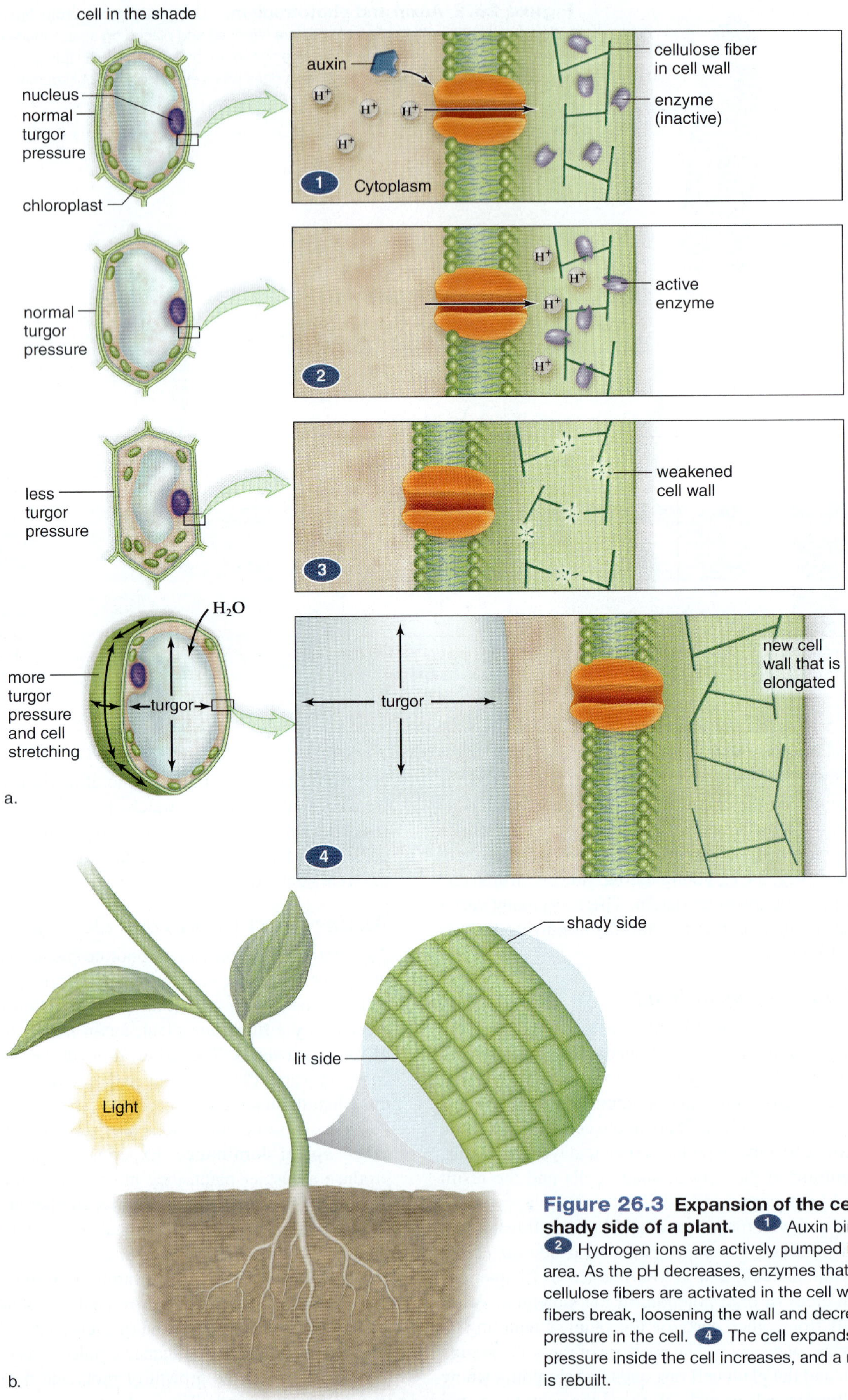

Figure 26.3 Expansion of the cell wall on the shady side of a plant. 1 Auxin binds to receptors. 2 Hydrogen ions are actively pumped into the cell wall area. As the pH decreases, enzymes that can break down cellulose fibers are activated in the cell wall. 3 Cellulose fibers break, loosening the wall and decreasing turgor pressure in the cell. 4 The cell expands as turgor pressure inside the cell increases, and a new, longer wall is rebuilt.

cutting causes adventitious roots to develop more quickly than they would otherwise. Auxin production by seeds promotes the growth of fruit. As long as auxin is concentrated in leaves or fruits rather than in the stem, leaves and fruits do not drop off. Therefore, trees can be sprayed with auxin to keep mature fruit from falling to the ground.

Auxins Have Many Commerical Uses

The most common naturally occurring auxin, IAA, has a relatively simple chemical structure (Table 26.1), allowing it to be easily copied and altered into various synthetic forms. Synthetic auxins are used today in a number of applications. These auxins are sprayed on plants, such as tomatoes, to induce the development of fruit without pollination, creating seedless varieties. Synthetic auxins have been used as herbicides to control broadleaf weeds, such as dandelions and other plants. These substances have little effect on grasses. Agent Orange is a powerful synthetic auxin that was used in extremely high concentrations to defoliate the forests of Vietnam during the Vietnam War. This powerful auxin proved to be carcinogenic and harmed many of the local people.

Gibberellins

Gibberellins were discovered in 1926 when a Japanese scientist was investigating a fungal disease of rice plants called "foolish seedling disease." Rapid stem elongation weakened the plants and caused them to collapse. The fungus infecting the plants produced an excess of a chemical called gibberellin, named after the fungus *Gibberella fujikuroi*. It wasn't until 1956 that a form of gibberellin, now known as gibberellic acid, was isolated from a flowering plant rather than from a fungus. We now know of about 136 gibberellins, and the most common of these is gibberellic acid, GA_3 (the subscript designation distinguishes it from other gibberellins). Young leaves, roots, embryos, seeds, and fruits are places where natural gibberellins can be found.

Gibberellins Have Commercial Uses

When gibberellins are applied externally to plants, the most obvious effect is stem elongation (Fig. 26.4*a*). Gibberellins can cause dwarf plants to grow, cabbage plants to become 2 m tall, and bush beans to become pole beans.

Gibberellins induce growth in a variety of crops, such as apples, cherries, and sugarcane. A notable example is their use on many of the table grapes grown in the United States. Commercial grapes are a genetically seedless variety that would naturally produce small fruit on very small bunches. Treating with GA_3 substitutes for the presence of seeds, which would normally be the source of endogenous gibberellins for fruit growth. Applications of GA_3 increase both fruit stem length and fruit size (Fig. 26.4*b*).

Dormancy is a period of time when plant growth is suspended. Gibberellins can break the dormancy of buds and seeds. Therefore, application of gibberellins is one way to hasten the development of a flower bud. When gibberellins break the dormancy of barley seeds, a large, starchy endosperm is broken down into sugars to provide energy for the growing seedling. This occurs because amylase, an enzyme that breaks down starch, makes its appearance. In the brewing industry, the production of beer relies on the breakdown of starch. Gibberellins are added to barley seeds, so that they artificially break dormancy and provide sugar for the fermentation process.

a. b.

Figure 26.4 Gibberellins cause stem elongation. **a.** The plant on the right was treated with gibberellins; the plant on the left was not treated. **b.** The grapes are larger on the right, because gibberellins caused an increase in the space between the grapes, allowing them to grow larger.

Cytokinins

Cytokinins were discovered as a result of attempts to grow plant tissues and organs in culture vessels in the 1940s. It was found that cell division occurred when coconut milk (a liquid endosperm) and yeast extract were added to the culture medium. Although the specific chemicals responsible could not be isolated at the time, they were collectively called cytokinins, because, as you may recall, *cytokinesis* means "division of the cytoplasm." A naturally occurring cytokinin was not isolated until 1967 and was called zeatin, because it came from corn (*Zea mays*) (Table 26.1).

Cytokinins Promote Cell Division and Organ Formation

Cytokinins influence plant growth by promoting cell division in all tissues of growing plants. In addition, plant organ formation is influenced by cytokinins and its interaction with auxin. Furthermore, cytokinins and auxins are different from all other plant hormones in that they are required for embryo survival. Researchers have tested and are aware that the ratio of auxin and cytokinin and the acidity of the culture medium determine whether plant cells will form an undifferentiated mass of cells, called a *callus,* or a mass of cells with roots, leaves, or flowers (Fig. 26.5). Cytokinins are also responsible for root nodule formation (housing nitrogen-fixing bacteria), as well as gall formation on wounded trees. A gall is a tumorlike growth caused by infections from bacteria, fungi, insects, or nematodes. These organisms can disrupt normal cytokinin function and result in a plant growing uncontrollably in small areas.

Cytokinins Prevent Senescence

The aging of plants is called **senescence.** During senescence, large molecules within the leaf are broken down and transported to other parts of the plant. Senescence does not always affect the entire plant at once; for example, as some plants grow taller, they

a. b. c. d.

Figure 26.5 The interaction of cytokinins and auxins in organ development. Tissue culture experiments have revealed that auxin and cytokinin interact to affect differentiation during development. **a.** In tissue culture that has the usual amounts of these two hormones, tobacco cells develop into a callus of undifferentiated tissue. **b.** If the ratio of auxin to cytokinin is appropriate, the callus produces roots. **c.** Change the ratio, and vegetative shoots and leaves are produced. **d.** Yet another ratio causes floral shoots. It is now clear that each plant hormone rarely acts alone; it is the relative concentrations of both hormones that produce an effect.

Figure 26.6 Leaves change colors when cytokinin levels are low. Seasonal changes signal a drop in cytokinin production in deciduous plants, causing the senescence of leaves.

naturally lose their lower leaves. In the autumn, low levels of cytokinin cause leaves to change color and eventually die (Fig. 26.6). Interestingly, it has been found that senescence of leaves can be prevented by the application of cytokinins. Some varieties of lettuce have been genetically modified to produce cytokinins at the onset of aging. The modified lettuce heads stay fresher longer and avoid brown and wilting leaves.

Abscisic Acid

Abscisic acid (ABA) is a hormone produced in the chloroplast and is derived from carotenoid pigments. Abscisic acid is sometimes called the stress hormone, because it initiates and maintains seed and bud dormancy and brings about the closure of stomata. It was once believed that ABA functioned in **abscission,** the dropping of leaves, fruits, and flowers from a plant. But although the external application of ABA promotes abscission this hormone is no longer believed to function naturally in this process. Instead, the hormone ethylene seems to bring about abscission.

ABA Promotes Dormancy

Recall that dormancy is a period of low metabolic activity and arrested growth. Dormancy occurs when a plant organ readies itself for adverse conditions by ceasing to grow. For example, it is believed that the hormone ABA travels from leaves to vegetative buds in the fall, and thereafter these buds are converted to winter buds. A winter bud is covered by thick, hardened bud scales (see Fig. 24.13). A reduction in the level of ABA and an increase in the level of gibberellins are believed to break seed and bud dormancy. Then seeds germinate, and buds send forth leaves.

Figure 26.7 shows what can happen if a plant becomes insensitive to ABA. In normal corn, a home gardener can leave a cob on the stalk to dry, then collect the dry kernels to plant the following year. ABA will keep the dry kernels from germinating until conditions are right. In the ABA-insensitive mutant corn, the kernels exhibit *vivipary*—an early break in dormancy and germination while still on the cob. The corn seedlings growing from the cob in Figure 26.7 are germinating in the wrong season and will most likely die.

Figure 26.7 Dormancy and germination. Image of viviparous mutant of maize (Indian corn) showing germination on the cob due to reduced sensitivity to abscisic acid. Red arrows indicate emerging seedlings.

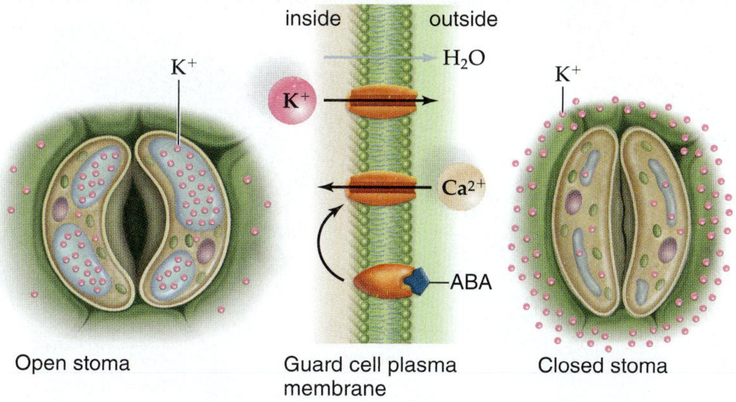

Figure 26.8 Abscisic acid promotes closure of stomata. The stoma is open (*left*). When ABA (the first messenger) binds to its receptor in the guard cell plasma membrane, the second messenger (Ca^{2+}) enters (*middle*). Now, K^+ channels open, and K^+ exits the guard cells. After K^+ exits, so does water. The stoma closes (*right*).

Figure 26.9 Ethylene and abscission. Normally, there is no abscission when a holly twig is placed under a glass jar for a week. When an ethylene-producing ripe apple is also under the jar, abscission of the holly leaves occurs.

ABA Closes Stomata

The reception of abscisic acid brings about the closing of stomata when a plant is under water stress, as described in Figure 26.8. Investigators have also found that ABA induces rapid depolymerization of actin filaments and formation of a new type of actin that is randomly oriented throughout the cell. This change in actin organization may also be part of the transduction pathways involved in stomata closure.

Ethylene

Ethylene ($H_2C = CH_2$) is a gas formed from the amino acid methionine. This hormone is involved in abscission (dropping leaves and fruit) and the ripening of fruits.

Ethylene Causes Abscission

The absence of auxin, and perhaps gibberellin, probably initiates abscission. But once abscission has begun, ethylene stimulates certain enzymes, such as cellulase, which helps cause leaf, fruit, or flower drop. In Figure 26.9, a ripe apple, which gives off ethylene, is under the bell jar on the right, but not under the bell jar on the left. As a result, only the holly plant on the right loses its leaves.

Ethylene Ripens Fruit

In the early 1900s, it was common practice to prepare citrus fruits for market by placing them in a room with a kerosene stove. Only later did researchers realize that an incomplete combustion product of kerosene, ethylene, ripens fruit. It does so by increasing the activity of enzymes that soften fruits. For example, in addition to stimulating the production of cellulase, it promotes the activity of enzymes that produce the flavor and smell of ripened fruits and breaks down chlorophyll, inducing the color changes associated with fruit ripening.

Ethylene moves freely through a plant by diffusion, and because it is a gas, ethylene also moves freely through the air. That is why a basket of ripening apples can induce ripening of a bunch of bananas some distance away. Ethylene is released at the site of a plant wound due to physical damage or infection (which is why one rotten apple spoils the whole bushel).

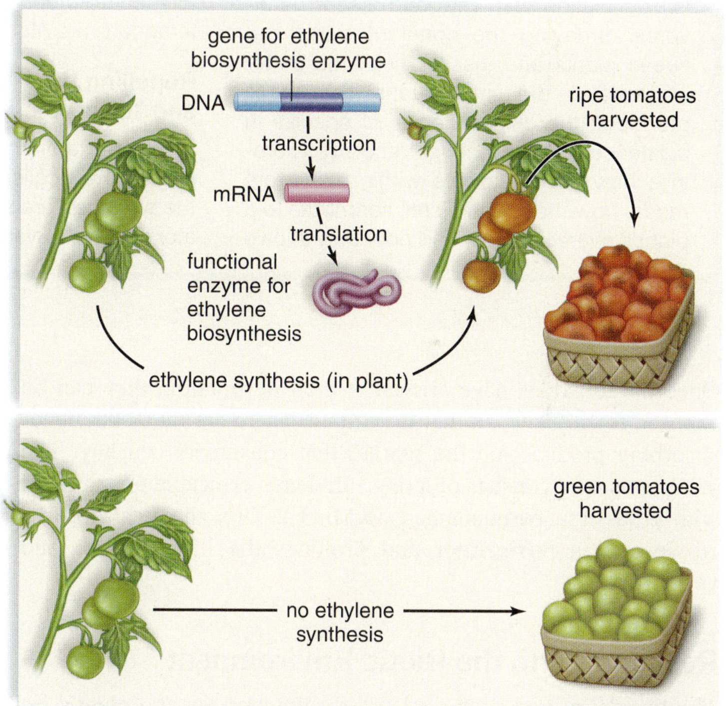

Figure 26.10 Ethylene and fruit ripening. Wild-type tomatoes (*top*) ripen on the vine after producing ethylene. Tomatoes (*bottom*) are genetically modified to produce no ethylene and stay green for shipping.

The use of ethylene in agriculture is extensive. It is used to hasten the ripening of green fruits, such as melons and honeydews, and it is applied to citrus fruits to attain pleasing colors before being put out for sale. Normally, tomatoes ripen on the vine, because the plants produce ethylene. Today, tomato plants can be genetically modified to not produce ethylene. This facilitates shipping, because green tomatoes are not subject to as much damage (Fig. 26.10).

THEME Evolution

The Chemical Ecology of Plants

Because plants are rooted to the ground, they are unable to escape from herbivores, pathogens, or even competing plants in the area. By producing a variety of chemical defenses, plants have overcome these constraints. With various organic chemicals, plants can attract mycorrhizal partners, pollinators, and the enemies of herbivores. They also repel herbivores, pathogens, and competing plants.

Chemical ecology is the study of the interaction between chemical signals, plants, animals, and the environment in which they live. Chemical ecology brings together scientists from many different fields, such as entomology, chemistry, and plant biology, who work together to study the complex chemical communication systems that occur in nature.

The most common research focus is coevolution of plants and insect herbivores. Studying the constant battle between plants and insects helps us better understand the interactions that have produced the diverse range of species in existence today. Chemical ecology examines how the chemicals within plants are made, how these chemicals contribute to a plant's overall fitness, and how they evolve in response to environmental pressures. In this feature, four examples of chemical interactions are highlighted.

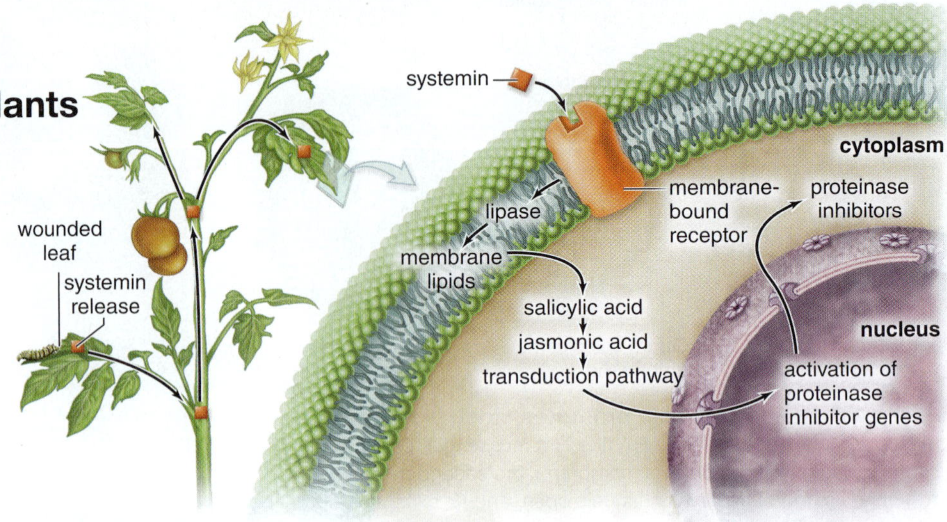

Figure 26A Defense response in tomato. Wounded leaves produce systemin, which travels in phloem to all parts of a plant, where it binds to cells that have a systemin receptor. These cells then produce jasmonic acid, a molecule that initiates a transduction pathway, which leads to the production of proteinase inhibitors, which limit insect feeding.

Repelling the Herbivore and Telling the Neighbors

After a predator chews a tomato plant leaf, a small protein called *systemin* is produced in the wounded area in response to the predator's saliva. Systemin is part of a signal transduction pathway that produces other defense compounds, such as *jasmonic acid* and *salicylic acid*. These defense compounds travel in phloem, become widely distributed throughout the plant, and activate the gene expression of proteinase inhibitors. When the next predator begins to eat the same plant, it will be poisoned or repelled by the bad taste of these inhibitors (Fig. 26A). In addition, many of the defense

Once the tomatoes have arrived at their destination, they can be exposed to ethylene, so that they ripen. There are many ethylene-absorbing products on the market that consumers can buy. The product usually consists of a tiny, ethylene-permeable pouch filled with potassium permanganate ($KMnO_4$). This chemical absorbs ethylene in a refrigerator and prolongs the life of fruits and vegetables.

Responding to the Biotic Environment

The hormones just discussed mostly function for the plant's response to abiotic stimuli, such as light, oxygen, water, pH, and temperature. Plants must also have an arsenal of chemicals to handle biotic stimuli, such as herbivory, parasitism, and competition from other plants.

A plant's epidermis and bark do a good job of discouraging attackers. But, unfortunately, herbivores have ways around a plant's first line of defense. A fungus can invade a leaf via the stomata and set up shop inside the leaf, where it feeds on nutrients meant for the plant. Underground nematodes have sharp mouthparts to break through the epidermis of a root and establish a parasitic relationship. Tiny insects called aphids have piercing mouthparts that allow them to tap into the phloem of a nonwoody stem. These examples illustrate why plants need a variety of defenses that are not dependent on its outer surface. The primary metabolites of plants, such as sugars and amino acids, are necessary to the normal workings of a cell. Plants also produce molecules termed **secondary metabolites** as a defense, or survival, mechanism. Secondary metabolites were once thought to be waste products, but now we know that they are part of a plant's arsenal to prevent predation or discourage competition. The Evolution feature, "The Chemical Ecology of Plants," explains the field of chemical ecology and how plants have evolved various chemical messages to make them more successful in a particular situation.

Check Your Progress 26.1

1. Explain how hormones assist in bringing about responses to stimuli.
2. Describe how auxin causes a plant to bend toward light.
3. Explain why abscisic acid is sometimes referred to as a stress hormone.

compounds are volatile (evaporate easily) and can stimulate defenses in nearby plants.

Attracting the Enemy's Enemy

The wild tobacco (*Nicotiana attenuate*) plant of the Southwest United States and Mexico produces nicotine to poison herbivores. Unfortunately for the plant, some herbivores, such as the hawkmoth caterpillar, have become resistant to nicotine and decimate the weedy shrubs. Recently, researchers from the Max Planck Institute for Chemical Ecology in Germany found a new way the plant is ridding itself of the hawkmoth caterpillar larva. Dubbed "poison lollipops," the trichomes of some plants produce a sugary substance irresistible to hungry caterpillars (Fig 26B). The sugar contains a volatile substance, which in turn makes the caterpillar smell good to predatory ants, which grab the caterpillars and carry them off to their nests.

Figure 26B Trichomes. Trichomes, or "poison lollipops" contain sugar to attract caterpillars. Other chemicals increase the predation of the caterpillars.

Smelling Your Prey

The dodder vine (*Cuscuta pentagona*) is a parasitic plant that winds itself around a host plant, inserts sharp pegs called haustoria, and feeds on the host's xylem and phloem (see page 464). Noticing that dodders seemed to have a preference for certain host plants, scientists hypothesized that dodders could "smell" their preferred meal. To test this, a dodder seedling was placed in a pot between a tomato and a wheat plant (Fig. 26C). After vacillating for several hours, the dodder chose to latch onto the tomato plant. A subsequent experiment confirmed that the attractant was the volatile chemical produced by the tomato. The tomato "perfume" was isolated and placed near the dodder. The dodder "smelled" its preferred prey and moved in that direction.

Keeping Others Away

Some chemical toxins protect plants from other plants. Black walnut trees (*Juglans nigra*) have roots that secrete a chemical toxin that blocks the germination of nearby seeds and inhibits the growth of neighboring plants. This strategy minimizes shading and competition for nutrients while maximizing exposure to the sun.

Questions to Consider

1. How have humans taken advantage of the chemicals that plants produce for their own defenses?
2. How can you test if a defense response is pathogen-specific?
3. During the domestication of crops, humans have intentionally or inadvertently selected for lower levels of toxic compounds. Explain why this type of selection would have occurred.

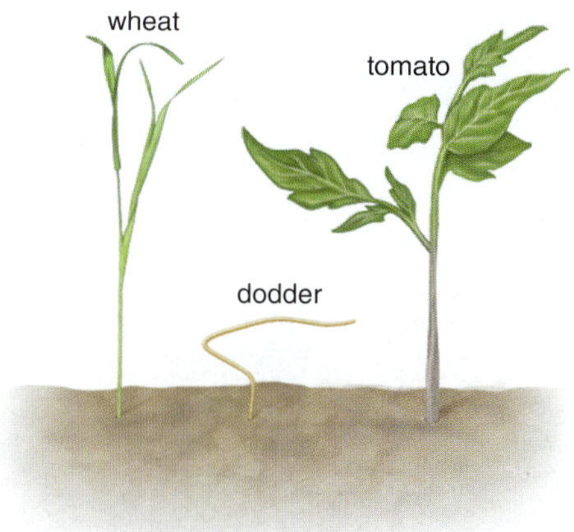

Figure 26C Dodder plant's response to multiple hosts. When placed between two possible hosts, the dodder will choose the host that will sustain the parasite the most.

26.2 Plant Growth and Movement Responses

Learning Outcomes

Upon completion of this section, you should be able to

1. Explain the difference between a tropism and a turgor movement.
2. Describe how a photoreceptor works in phototropism.
3. Explain why a shoot grows upward and a root grows downward.
4. List the internal stimuli involved in closing a Venus flytrap.

All living organisms respond to stimuli and exhibit movement. Plant movements are slow and difficult to notice unless seen in time-lapse video or demonstrated experimentally. Plant movement responses to stimuli can be internal—such as changes in turgor pressure, electrical impulses, or the action of hormones—or external—such as responses to sunlight, water, oxygen, gravity, and barriers such as rocks.

Recall from Figure 26.1 that when there is a stimulus, whether internal or external, the first step is *reception* of the stimulus. The next step is *transduction,* meaning that the stimulus has been changed into a form that is meaningful to the plant. Finally, a *response* is made, usually by the plant's genes. Animals and plants go through the same sequence of events when they respond to a stimulus; however, in the case of plants, no nerves are present—instead, chemical signals are released, and binding of these signals brings about transduction and response.

In this section, we consider plant tropisms—responses caused by external stimuli—and turgor movements—responses caused by internal stimuli. Note that tropisms are growth movements, and turgor movements are nongrowth movements.

Movement Caused By External Stimuli

Growth toward or away from a unidirectional stimulus is called a **tropism** (Gk. *tropos,* "turning"). *Unidirectional* means that the stimulus is coming from only one direction instead of multiple directions. Growth toward a stimulus is called a *positive tropism,* and growth away from a stimulus is called a *negative tropism.* Tropisms are due to differential growth—one side of an organ elongates faster than the other, and the result is a curving toward or away from the stimulus.

A number of tropisms have been observed in plants. The three best-known tropisms are phototropism (light), thigmotropism (touch), and gravitropism (gravity).

Phototropism: a movement in response to a light stimulus
Thigmotropism: a movement in response to touch
Gravitropism: a movement in response to gravity

Other tropisms include chemotropism (chemicals), traumotropism (trauma), skototropism (darkness), and aerotropism (oxygen).

Phototropism

A potted plant left in the open with sunlight on all sides will grow and develop vertically. However, if a potted plant is placed on a sunny windowsill with unidirectional light, the stems will begin to bend toward the light (see Fig. 26.12*a*). Positive **phototropism** of stems occurs because the cells on the shady side of the stem elongate due to the presence of auxin. Plant growth that curves away from light is called negative phototropism. Roots, depending on the species examined, are either insensitive to light or exhibit negative phototropism.

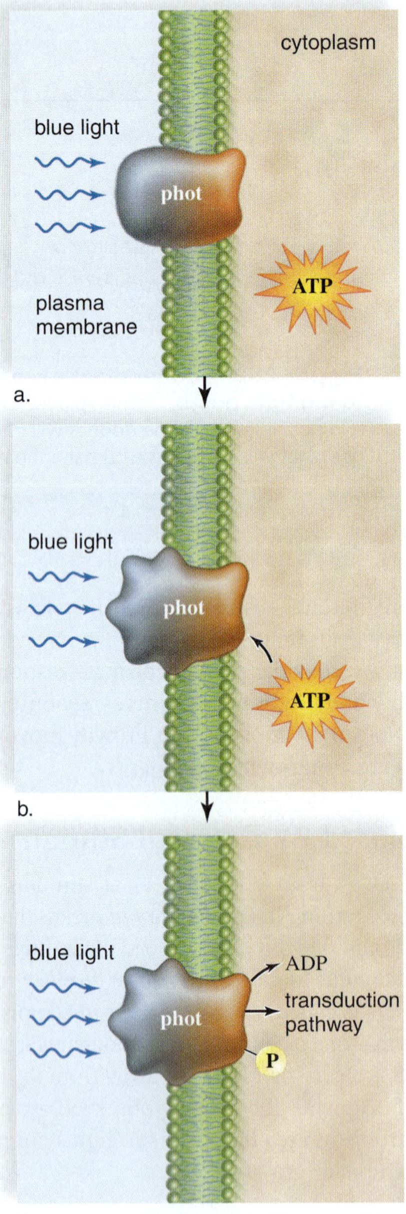

Figure 26.11 Phototropin. In the presence of blue light, (**a**) a photoreceptor called phototropin (phot) is activated (**b**) and becomes phosphorylated (**c**). A transduction pathway begins, leading to the accumulation of auxin.

Through the study of mutant *Arabidopsis* plants (see the Nature of Science feature, "Why So Many Scientists Work with *Arabidopsis*," on page 488), plant scientists know that phototropism occurs because plants have membrane receptors that respond to wavelengths of light; these receptors are called *photoreceptors*. Photoreceptors are proteins embedded with pigment molecules that, in the case of phototropism, respond to blue wavelengths (400–500 nm) of light. Figure 26.11 describes the steps initiating the signal transduction pathway that eventually leads to elongation of cells and the bending of a plant.

When blue light wavelengths are absorbed (Fig. 12.11*a*), the pigment portion of the photoreceptor, called *phototropin* (*phot*), changes its shape (Fig. 12.11*b*). This shape change results in the transfer of a phosphate group from ATP to the protein portion of the photoreceptor (Fig. 26.11*c*). The phosphorylated photoreceptor triggers a transduction pathway that leads to the entry of auxin into the cell (see Fig. 26.3).

Thigmotropism

Thigmotropism (Gk. *thigma,* "touch"; *tropos,* "turning") is a response to touch from another plant, an animal, rocks, or the wind. An example of this response is the coiling of tendrils or the stems of plants, such as the stems of runner bean and morning glory plants (Fig. 26.12*b*).

a. Phototropism

b. Thigmotropism

Figure 26.12 Phototropism and thigmotropism. **a.** The stem of the plant curves toward the light, exhibiting positive phototropism. **b.** The stem of a runner bean plant, *Phaseolus,* coiling around a pole illustrates thigmotropism.

A flowering plant grows straight up and down until it touches something. Then the cells in contact with an object, such as a pole or an underground rock, grow less while those on the opposite side elongate. Thigmotropism can be quite rapid; tendrils have been observed to encircle an object within 10 minutes. Several minutes of touching can bring about a response that lasts for several days. The response isn't always immediate—tendrils touched in the dark will respond once they are illuminated. ATP rather than light initiates the response. It is possible that auxin and ethylene play a role in the process, since they are capable of inducing the curvature of tendrils even in the absence of touch.

Thigmomorphogenesis is when a plant changes its overall shape due to an environmental touch stimulus, such as a barrier, wind, or rain. For example, when a storm blows across a field, plants in the field respond to these and other mechanical stresses by increasing production of fibers and collenchyma tissue (see Chapter 24). Cell elongation is inhibited, building shorter, sturdier plants. A tree growing in a windy location often has a shorter, thicker trunk than the same type of tree growing in a more protected location. Even simple mechanical stimulation, such as rubbing a plant with a stick, can inhibit cellular elongation and produce a sturdier plant with increased amounts of support tissue.

Gravitropism

Gravitropism is the effect of gravity on plant growth. When a seed germinates, the embryonic shoot exhibits negative gravitropism by growing upward *against* gravity. Increased auxin concentration of the lower side of the young stem causes the cells in that area to grow more than the cells on the upper side, resulting in growth upward. The embryonic root exhibits positive gravitropism by growing *with* gravity downward into the soil (Fig. 26.13*a*).

Charles and Francis Darwin, in addition to studying coleoptiles, studied roots and discovered that if the root cap is removed, roots no longer respond to gravity. Since then, investigators have developed an explanation as to how root cells know which way is down. The root cap contains specialized cells filled with starch granules called **statoliths.** The statoliths are found within organelles called *amyloplasts.* Like marbles in a bag, statoliths settle to the bottom of a cell and put pressure on the other organelles, thus signaling the downward direction (Fig. 26.13*b*). This signal influences the placement of auxin and instructions for growth that follow.

If you place a potted plant on its side, the signals from the statoliths (and light) will change, and auxin will redistribute, causing the roots to bend downward and the shoot to grow upward. If a sideways plant is put in a clinostat, it will grow straight, because gravity will be negated if the plant is in constant motion (Fig. 26.13*c*).

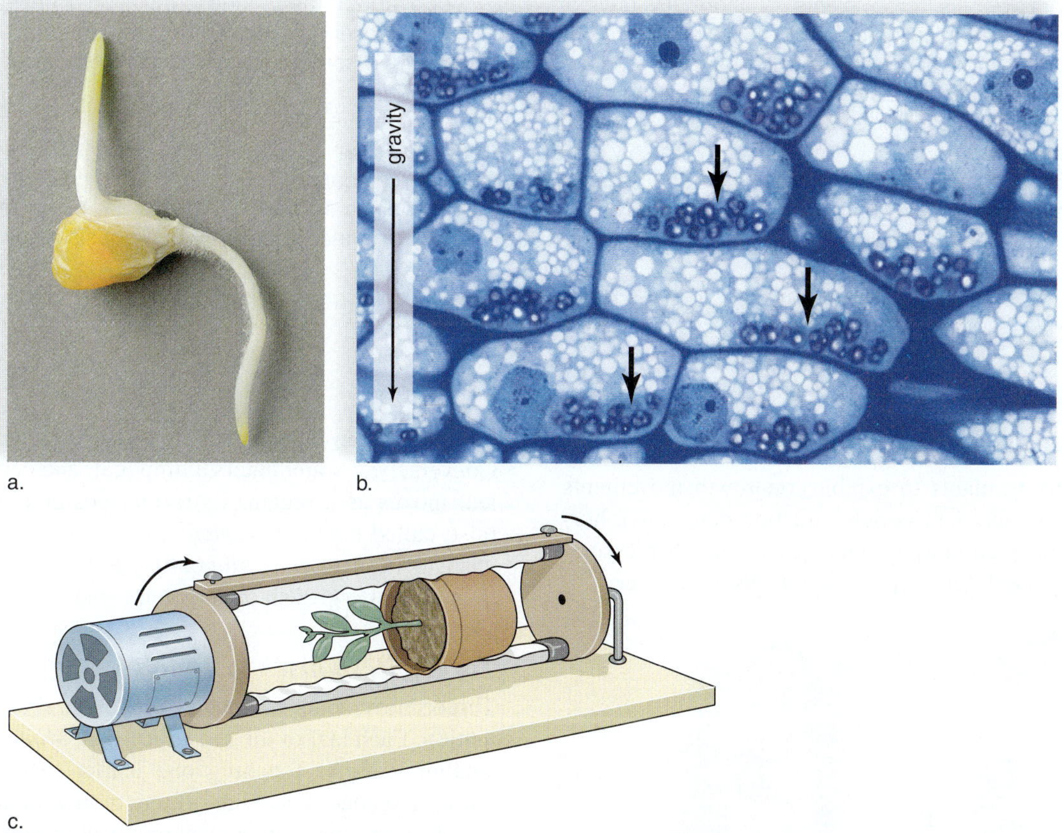

Figure 26.13 Gravitropism. **a.** This corn seed was germinated in a sideways orientation and in the dark. The shoot is growing upward (negative gravitropism) and the root downward (positive gravitropism.) **b.** Sedimentation of statoliths (see arrows), which are starch granules, is thought to explain how roots perceive gravity. **c.** A clinostat, a tool used by plant biologists to negate the effects of gravity. Plants are slowly rotated so that the statoliths do not settle to the bottom of cells. Typical bending of shoots and roots in response to gravity does not occur.

THEME Nature of Science

Why So Many Scientists Work with *Arabidopsis*

Arabidopsis thaliana is a small flowering plant related to cabbage and mustard plants (Fig. 26D). *Arabidopsis* has no commercial value—in fact, it is a weed. However, it has become a model organism for the study of plant molecular genetics, including signal transduction. Unlike the crop plants used formerly, *Arabidopsis* has characteristics that make it an ideal model organism.

- It is small, so many hundreds of plants can be grown in a small amount of space.
- Generation time is short. It takes only 5–6 weeks for plants to mature, and each one produces about 10,000 seeds.
- It can be self-pollinated or cross-pollinated. This feature facilitates the production of strains with multiple mutations.
- It has a relatively small genome: 5 chromosomes, 125 million base pairs of DNA, and 25,500 genes.

In contrast to *Arabidopsis,* crop plants, such as corn, have generation times of at least several months, and they require a great deal of field space for a large number to grow. Crop plants also have much larger genomes than *Arabidopsis*. For comparison, the genome sizes for rice (*Oryza sativa*), wheat (*Triticum aestivum*), and corn (*Zea mays*) are 420 million, 16 billion, and 2.5 billion base pairs, respectively. However, crop plants have about the same number of functional genes as *Arabidopsis,* and they occur in the same sequence. Therefore, knowledge of the *Arabidopsis* genome can be used to help locate specific genes in the genomes of other plants.

The creation of *Arabidopsis* mutants plays a significant role in discovering what each of its genes does. For example, if a mutant plant lacks stomata, then we know that the affected gene influences the formation of stomata.

The application of *Arabidopsis* genetics to other plants has been demonstrated. For example, one of the mutant genes that alters the development of flowers has been cloned and introduced into tobacco plants, where, as expected, it causes sepals and stamens to appear where petals would normally develop. Knowledge about the development of flowers in *Arabidopsis* can have far-ranging applications. It will someday lead to more productive crops.

A study of the *Arabidopsis* genome will undoubtedly promote plant molecular genetics in general. And because *Arabidopsis* is a model organism, genetic findings from this plant may have applications to humans, just as Mendel's work with pea plants led to the formulation of genetic laws. It's far easier to study signal transduction in *Arabidopsis* cells than in human cells.

Arabidopsis thaliana

Figure 26D Photograph of *Arabidopsis thaliana.* *Arabidopsis* consists of a flat rosette of leaves, from which grows a short flower stalk. Many investigators have turned to this weed as an experimental organism to study the actions of genes, including those that control growth and development.

Questions to Consider

1. Why is *Arabidopsis* a better study organism than cabbage or mustard?
2. What are the potential applications of *Arabidopsis* research?

Movement Caused By Internal Stimuli

Internal signals causing plants to exhibit nongrowth movements could be the result of electrical impulses (action potentials), hormonal action, or most commonly, changes in turgor pressure. Recall that a plant cell exhibits turgor when it fills with water:

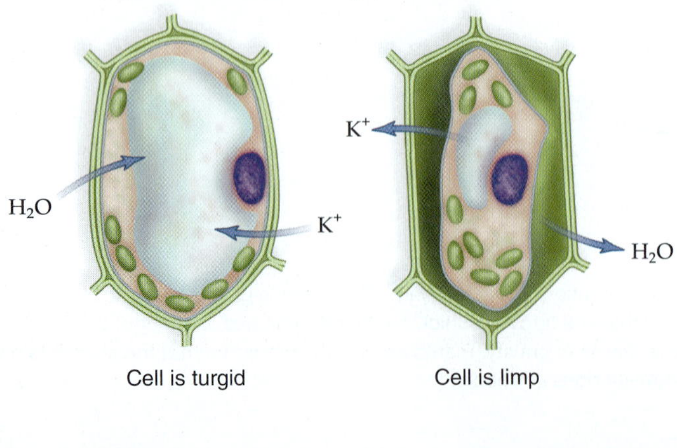

In general, if water exits many cells of a leaf, the leaf goes limp. Conversely, if water enters a limp leaf, and cells exhibit turgor, the leaf moves as it regains its former position. **Turgor movements** (also called *nastic movements*) are dependent on turgor pressure changes in plant cells. In contrast to tropisms, turgor movements do not involve growth of an organ toward or away from a stimulus.

Turgor Responses to Touch

Turgor movements can result from touch, shaking, or thermal stimulation. Each leaf of the sensitive plant, *Mimosa pudica,* has many smaller leaflets. Touching one leaflet collapses the whole leaf within 1 second, knocking off any munching insect (Fig. 26.14).

The structure of this plant involved in controlling turgor movement is a thickening called a *pulvinus* at the base of each leaflet. A leaf folds and the petiole drops when the cells in the lower half of the pulvinus, called the motor cells, lose potassium ions (K^+), then water. When the pulvinus cells lose turgor, the leaflets of the

Figure 26.14 Turgor movement. A leaf of the sensitive plant, *Mimosa pudica,* before and after it is touched.

leaf collapse. An electrical mechanism may cause the response to move from one leaflet to another. The speed of an electrical charge transmission has been measured at about 1 cm/sec.

A Venus flytrap closes its trap in less than 1 second when three hairs inside the trap, called trigger hairs, are touched by an insect. When the trigger hairs are stimulated by the insect, an electrical charge is transmitted throughout the lobes of a leaf. Exactly what causes this electrical charge is being studied. Two possible explanations have been suggested: (1) Cells located near the outer region of the lobes rapidly secrete hydrogen ions into their cell walls, loosening them and allowing the walls to swell rapidly by osmosis or (2) cells in the inner portion of the lobes and the midrib rapidly lose ions, leading to a loss of water by osmosis and collapse of these cells. In any case, it appears that turgor movements are involved.

Video Plant Tactile Response

Sleep movements, discussed later, are another example of turgor movement caused by circadian rhythms and the effects of photosensitive pigments.

Check Your Progress 26.2

1. Distinguish between a positive and a negative tropism.
2. Explain how root cells determine the downward direction.
3. List examples of internal and external stimuli.
4. Explain how changes in turgor pressure can cause a leaf to collapse.

26.3 Plant Responses to Phytochrome

Learning Outcomes

Upon completion of this section, you should be able to

1. Explain the conversion cycle of phytochrome.
2. Compare flowering in short-day/long-night plants and in long-day/short-night plants as it relates to photoperiodism.
3. Describe phytochrome's role in plant spacing.
4. Describe circadian rhythms and the connection to phytochrome.

Plants are aware of changes in light in two different ways: (1) Plants can sense if it is daytime, nighttime, or dawn/dusk and respond by adjusting metabolic processes such as photosynthesis, and (2) plants can sense the time of year, affecting seasonal responses for processes such as germination or flowering.

The ability of plants to sense these changes in light lies mostly with the absorption of blue and red wavelengths of the visible light spectrum (see Fig. 7.6.) by the light-sensing pigments phototropin (discussed earlier) and phytochrome. The following discussion will focus on phytochrome and its involvement in both day/night cycles and seasonal cycles.

Phytochrome

Phytochrome (Gk. *phyton,* "plant"; *chroma,* "color") is a blue-green leaf pigment that is present in the cytoplasm of plant cells. A phytochrome molecule is composed of two identical proteins (Fig. 26.15). Each protein has a larger portion in which a light-sensitive region is located. The smaller portion is a kinase that can link light absorption with a transduction pathway within the cytoplasm. Phytochrome can be said to act as a light switch, because, like a light switch, it can be in the down (inactive) position or in the up (active) position.

Animation Phytochrome Signaling

Phytochrome can distinguish between red wavelengths (650–680 nm) present during the daytime and far-red wavelengths (710–740 nm) present at dawn or dusk. These two ranges of red wavelengths cause the phytochrome protein to interconvert between two forms:

P_r (phytochrome red) absorbs red light and is converted into P_{fr} in the daytime.

P_{fr} (phytochrome far-red) absorbs far-red light and is converted into P_r in the evening.

Functions of Phytochrome

The $P_r \rightarrow P_{fr}$ conversion cycle is known to control various growth functions in plants. P_{fr} promotes seed germination, inhibits shoot elongation, promotes flowering, and affects plant spacing and accumulation of chlorophyll. The effects of phytochrome on flowering, germination, and plant spacing will be discussed in this section.

Flowering and Photoperiodism As just noted, many physiological changes in flowering plants are related to a seasonal change in day length. A physiological response prompted by changes in the length of day or night in a 24-hour daily cycle is called **photoperiodism** (Gk. *photos,* "light"; *periodus,* "completed course"). In some plants, photoperiodism influences flowering; for example, violets and tulips flower in the spring, and asters and goldenrod flower in the fall. Photoperiodism requires the participation of a biological clock (discussed later) and the activity of phytochrome.

Flowering plants can be divided into three groups on the basis of their flowering status.

1. **Short-day plants** flower when the day length is shorter than a *critical length.* (Examples are cocklebur, goldenrod, poinsettia, and chrysanthemum.)
2. **Long-day plants** flower when the day length is longer than a critical length. (Examples are wheat, barley, rose, iris, clover, and spinach.)
3. **Day-neutral plants** are not dependent on day length for flowering. (Examples are tomato and cucumber.)

The criterion for designating plants as short-day or long-day is not an absolute number of hours of light but a critical number that either must be or cannot be exceeded. Spinach is a long-day plant that has a critical day length of 14 hours; ragweed is a short-day plant with the same critical length. Spinach, however, flowers in the summer when the day length increases to 14 hours or more, and ragweed flowers in the fall, when the day length shortens to 14 hours or fewer. In addition, some plants require a specific sequence of day lengths in order to flower.

Soon after the three groups of flowering plants were distinguished, researchers began to experiment with artificial lengths of light and dark that did not necessarily correspond to a normal 24-hour day. These investigators discovered that the cocklebur, a short-day plant, does not flower if a required long dark period is interrupted by a brief flash of white light. (Interrupting the light period with darkness has no effect.) In contrast, a long-day plant does flower if an overly long dark period is interrupted by a brief flash of white light. They concluded that the length of the dark period, not the length of the light period, controls flowering. Of course, in nature, short days always go with long nights, and vice versa.

To recap, let's consider Figure 26.16.

- Cocklebur is a short-day plant (Fig. 26.16a, *left*). First, when the night is longer than a critical length, cocklebur flowers. Conversely, the plant does not flower when the night is shorter than the critical length. Cocklebur also does not flower if the longer-than-critical-length night is interrupted by a flash of light.
- Clover is a long-day plant (Fig. 26.16*b, right*). When the night is shorter than a critical length, clover flowers. Conversely, the plant does not flower when the night is longer than a critical length. Finally, unlike the cocklebur, clover does flower when a slightly longer-than-critical-length night is interrupted by a flash of light.

Commercial florists and nursery owners have made extensive use of photoperiods, manipulating with artificial light the flowering times of poinsettias, some lilies, and other plants in order to have them flower at times of biggest demand, such as Christmas and Mother's Day.

Phytochrome and Germination The presence of P_{fr} indicates to some seeds that sunlight is present and conditions are favorable

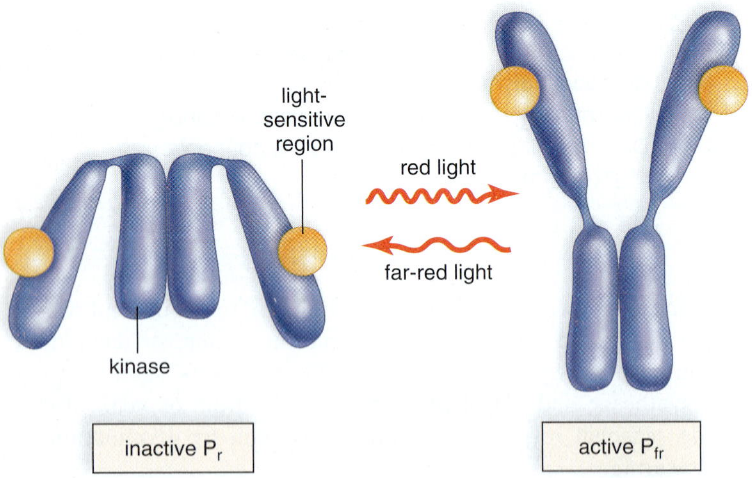

Figure 26.15 Phytochrome conversion cycle. The inactive form of phytochrome (P_r) is converted to the active form P_{fr} in the presence of red light, which is prevalent in daylight. P_{fr} is involved in various plant responses, such as seed germination, shoot elongation, and flowering. P_{fr} is converted to P_r whenever light is limited, such as in the shade or during the night.

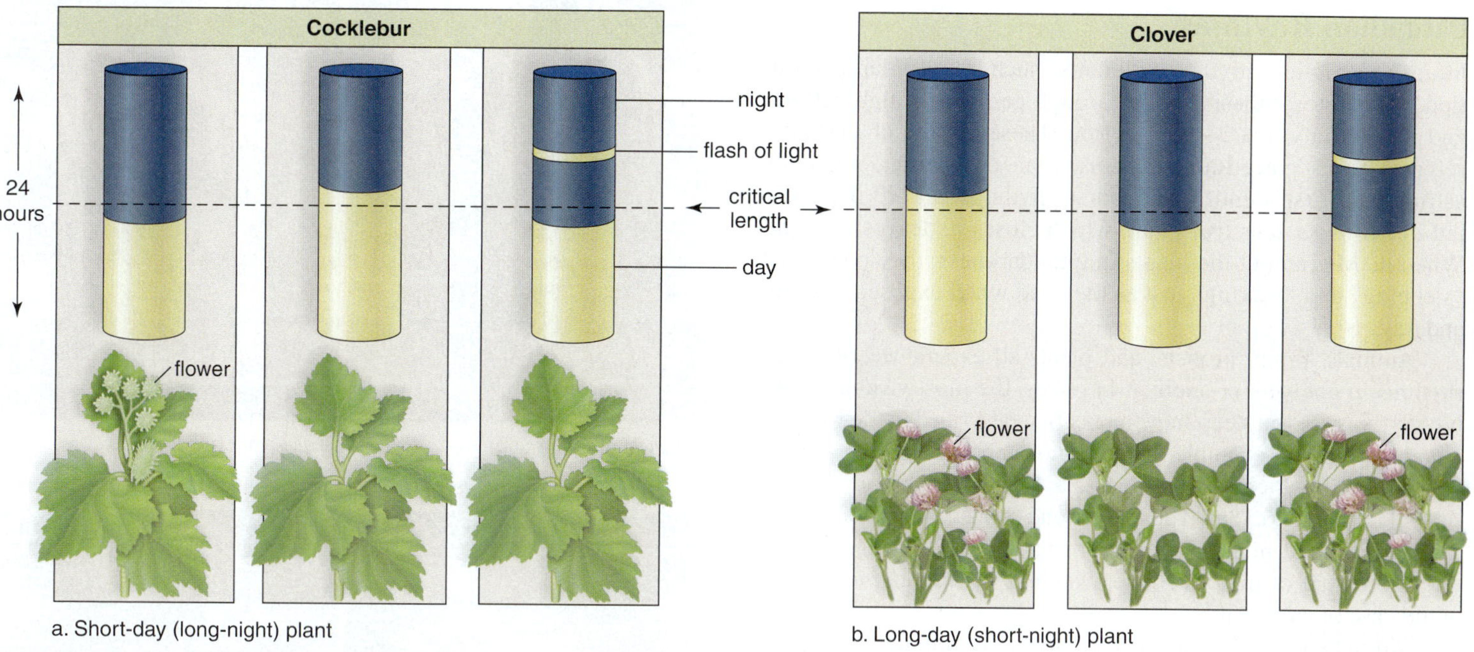

Figure 26.16 Photoperiodism and flowering. **a.** Short-day plant. When the day is shorter than a critical length, this type of plant flowers. The plant does not flower when the day is longer than the critical length. It also does not flower if the longer-than-critical-length night is interrupted by a flash of light. **b.** Long-day plant. The plant flowers when the day is longer than a critical length. When the day is shorter than a critical length, this type of plant does not flower. However, it does flower if the slightly longer-than-critical-length night is interrupted by a flash of light.

for germination. This explains why some seeds, such as those from snapdragons, poppies, and petunias, must be only partly covered with soil when planted. The seeds of the birch tree need 8 consecutive days of 10 hours of sunlight before they will germinate. Germination of other seeds, such as those of the mustard plant *Arabidopsis* or corn, is inhibited by light, so they must be planted deeper.

When a deeply planted seed germinates, it uses the food reserves stored within that seed to grow roots and elongate the shoot until it can push its way out from the soil. Notice the shoot in Figure 26.13*a*. The stem is elongated and yellow, as this seed has been grown in the dark, simulating growth in soil. The first flash of dim light instantly changes the developmental instructions for seedlings, and a new set of physical changes occur called *photomorphogenesis*. The presence of P_{fr} indicates that sunlight is available, and the seedlings begin to grow normally—the leaves expand and become green and the stem begins branching (see Fig. 27.11).

Next time you are in a produce aisle, notice two vegetables that are grown in the absence of light—white asparagus and alfalfa sprouts. These vegetables are simply elongated stems that are yellow, or **etiolated**—that is, the shoot increases in length, and the leaves remain small and yellow (Fig. 26.17). Plants that are grown in the dark do not receive a different set of instructions from their phytochrome pigments and thus never exhibit photomorphogenesis.

Figure 26.17 Phytochrome controls shoot elongation and chlorophyll production. **a.** If red light is prevalent, as it is in bright sunlight, normal growth occurs. **b.** If far-red light is prevalent, as it is in the shade, etiolation occurs. These effects are due to phytochrome.

Phytochrome and Competition Plant spacing is another interesting function of phytochrome. A store-bought seed packet will always have instructions on spacing seeds to be planted. In nature, red and far-red light also signal spacing. Leaf shading increases the amount of far-red light relative to red light. Plants measure the amount of far-red light bounced back to them from neighboring plants. The closer together plants are, the more far-red relative to red light they perceive and the more likely they are to grow taller—a strategy for outcompeting others for sunlight.

Circadian Rhythms

Many metabolic activities in plants, such as cellular respiration and photosynthesis, cycle through periods of high activity and low activity in a 24-hour period. These cycling changes are referred to as **circadian rhythms.** Jean de Mairan, a French astronomer, first identified circadian rhythms in 1729. He studied the *Mimosa* sensitive plant, which closes its leaves at night. When de Mairan put the plants in total darkness, they continued "sleeping" and "waking" just as they had when exposed to night and day.

Animals, fungi, protists, and plants all experience circadian rhythms in one way or another. In plants, the most visually striking rhythms are the sleep movements first described by de Mairan. Another common example of sleep movements occurs in a plant called *Oxalis,* which is often sold as a "shamrock" before St. Patrick's Day (Fig. 26.18*a*). The leaves (and flowers) open during the day and close at night. This movement is due to changes in the turgor pressure of motor cells in swellings called *pulvini,* located at the base of each leaf.

Morning glory (*Ipomoea leptophylla*) is a plant that opens its flowers in the early part of the day and closes them at night (Fig. 26.18*b*). In most plants, stomata open in the morning and close at night, and some plants secrete nectar at the same time of the day or night.

In the natural environment, circadian rhythms are entrained to a daily cycle through the action of phytochrome and blue-light receptors. *Entrainment* means to be synchronized to light at daybreak. Overall, to qualify as a circadian rhythm, the activity must (1) occur every 24 hours; (2) take place whether or not the day/night lighting is present; and (3) be able to be reset if external cues are provided. For example, if you take a transcontinental flight, you will likely suffer jet lag at the destination, because your body will still be attuned to the day/night pattern of its previous environment. But after several days, you probably will adjust and will be able to go to sleep and wake up according to your new time.

Biological Clock

The internal mechanism by which a circadian rhythm is maintained in the absence of appropriate environmental stimuli is termed a **biological clock.** If organisms are sheltered from environmental stimuli, their biological clock keeps the circadian rhythms going, but the cycle extends. In plants with sleep movements, the sleep cycle changes to 26 hours when the plant is kept in constantly dim light, as opposed to 24 hours when in traditional day/night conditions. Therefore, it is suggested that biological clocks are synchronized by external stimuli to 24-hour rhythms.

As previously mentioned, the length of daylight compared to the length of darkness, called the photoperiod and influenced by phytochrome, sets the clock. Temperature has little or no effect. This synchronization with light is adaptive, because the photoperiod indicates seasonal changes better than temperature changes. Spring and fall, in particular, can have both warm and cold days.

Figure 26.18 Circadian rhythms. a. The leaves of the *Oxalis* plant fold every 24 hours at night. **b.** The flowers of the morning glory, *Ipomoea leptophylla,* close at night. **c.** Graph of circadian rhythm exhibited by morning glory plant. The period is the time between peaks in a repeating cycle.

Work with *Arabidopsis* (see the Nature of Science feature, "Why So Many Scientists Work with *Arabidopsis*," on page 488) and other organisms suggests that the biological clock involves the transcription of a small number of "clock genes." Although circadian rhythms are outwardly very similar in all species, the clock genes that have been identified are not the same, because biological clocks have evolved several times in different organisms to perform similar tasks.

Check Your Progress 26.3

1. Describe the phytochrome protein and its conversion between two forms.
2. Explain why a long-day plant still flowers if the long day is interrupted by a period of darkness.
3. Describe the various roles of phytochrome.
4. Explain the criteria for circadian rhythms.

CONNECTING *the* CONCEPTS *with the* THEMES

Evolution

- The majority of flowering plants are anchored to one spot; they must therefore have evolved chemical signals to respond to environmental conditions.
- Successful defenses that have evolved have been those that act as deterrents to predation, invasion, and competition.

Nature of Science

- Scientific study of plant hormones and their actions has allowed agricultural researchers to enhance food production and preservation.
- Studies of mutations in the model organism *Arabidopsis* have application to other commercial plants.
- Chemical ecologists study plant chemical interactions with other organisms.

Biological Systems

- Signal transduction in plants via plant hormones has many similarities with the actions of animal hormones.
- Plant responses to stimuli follow a precise series of steps that if interrupted can significantly alter the original response.
- Absorption of light with photoreceptors allows a plant to perceive its environment.

MEDIA STUDY TOOLS http://connect.mheducation.com

SMARTBOOK® Study smart. SmartBook helps you maximize your study time by identifying what you know and don't know. The recharge feature in SmartBook helps you prepare for the test by identifying the material that you are most likely to forget.

connect BIOLOGY Find even more resources to learn the chapter concepts including animations, tutorial videos, and interactive practice questions.

Animations	Video
26.1 Second Messengers **26.3** Phytochrome Signaling	**26.2** Plant Tactile Response

SUMMARIZE

26.1 Plant Hormones

Like animals, flowering plants use a **signal transduction pathway** when they respond to a stimulus. The process involves **receptor** activation, **transduction** of the signal by relay proteins, and a **cellular response,** which can consist of the turning on of a gene or an enzymatic pathway. There are five main plant **hormones** that are often the activation or the response of these pathways.

Early studies using **coleoptiles** helped discover and define the role of **auxin.** Auxin-controlled cell elongation is involved in phototropism and gravitropism. When a plant is exposed to light, auxin moves laterally from the bright to the shady side of a stem. Auxin is also responsible for the **apical dominance** of a plant.

Gibberellin causes stem elongation between nodes and breaks bud and seed **dormancy.** After this hormone binds to a plasma membrane receptor, a DNA-binding protein activates a gene leading to the production of amylase.

Cytokinins cause cell division, the effects of which are especially obvious when plant tissues are grown in culture. The absence of cytokinins results in aging, or **senescence.**

Abscisic acid (**ABA**) and **ethylene** are two plant growth inhibitors. ABA is well known for causing stomata to close, and ethylene is known for causing **abscission** and causing fruits to ripen.

Secondary metabolites are other chemicals produced by plants to combat biotic stimuli, such as herbivory and parasitism.

26.2 Plant Growth and Movement Responses

When flowering plants respond to stimuli, growth, movement, or both occur. **Tropisms** are growth responses toward or away from unidirectional stimuli. The positive **phototropism** of stems results in a bending toward light, and the negative **gravitropism** of stems results in a bending away from the direction of gravity. Roots that bend toward the direction of gravity show positive gravitropism. Positive gravitropism of roots is due to **statoliths** settling to the bottom of root cap cells. **Thigmotropism** occurs when a plant part makes contact with an object, as when tendrils coil about a pole.

Turgor movements, or nastic movements, are not directional. Due to turgor pressure changes, some plants respond to touch and some perform sleep movements.

26.3 Plant Responses to Phytochrome

Phytochrome is a pigment that is involved in **photoperiodism,** the ability of plants to sense the length of the day and night during a 24-hour period. This sense leads to seed germination, shoot elongation, and flowering during favorable times of the year. The conformation, and activity, of phytochrome is influenced by daylight. Phytochrome in the P_{fr} form leads to a biological response, such as flowering.

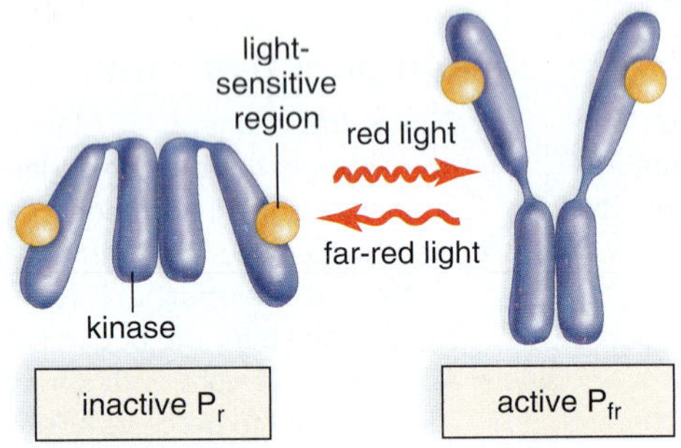

Short-day plants flower only when the days are shorter than a critical length, and **long-day plants** flower only when the days are longer than a critical length. Research has shown that actually it is the length of darkness that is critical. Interrupting the dark period with a flash of white light prevents flowering in a short-day plant and induces flowering in a long-day plant. **Day-neutral plants** are not affected by periods of dark and light. Phytochrome is also involved in plant spacing, germination, and stem elongation. A plant grown in darkness will exhibit **etiolation,** where stems continue to elongate and remain yellow.

Plants exhibit **circadian rhythms,** which are believed to be controlled by a **biological clock.** The sleep movements of *Oxalis* plants, the closing of stomata, and the daily opening of certain flowers have a 24-hour cycle.

ASSESS

Choose the best answer for each question.

26.1 Plant Hormones

1. During which step of signal transduction is a second messenger released into the cytoplasm?
   a. reception
   b. response
   c. transduction
   d. final step

For questions 2–8, match each statement with a hormone in the key. Answers can be used more than once.

KEY:
a. auxin
b. gibberellin
c. cytokinin
d. ethylene
e. abscisic acid

2. It is present in a gaseous form.
3. Grapes can grow larger and exhibit stem elongation.
4. Stomata close when a plant is water-stressed.
5. Stems bend toward the sun.
6. It inhibits plant senescence.
7. It causes leaf abscission.
8. It is responsible for apical dominance.
9. Primary metabolites are needed for ______, while secondary metabolites are produced for ______.
   a. growth, signal transduction
   b. normal cell functioning, defense
   c. defense, growth
   d. signal transduction, normal cell functioning

26.2 Plant Growth and Movement Responses

10. The sensors in the cells of the root cap are called
    a. mitochondria.
    b. central vacuoles.
    c. statoliths.
    d. chloroplasts.
    e. intermediate filaments.
11. A student places 25 corn seeds in a large pot and allows the seeds to germinate in total darkness. Which of the following growth or movement activities would the seedlings exhibit?
    a. gravitropism, as the roots grow down and the shoots grow up
    b. phototropism, as the shoots search for light
    c. thigmotropism, as seedlings move toward the edge of the pot
    d. All of these are possible.
12. The snapping of a Venus flytrap is an example of a
    a. phototropism.
    b. gravitropism.
    c. circadian rhythm.
    d. turgor movement.
    e. thigmotropism.

26.3 Plant Responses to Phytochrome

13. A plant requiring a dark period of at least 14 hours will
    a. flower if a 14-hour night is interrupted by a flash of light.
    b. not flower if a 14-hour night is interrupted by a flash of light.
    c. not flower if the days are 14 hours long.
    d. not flower if the nights are longer than 14 hours.
    e. Both b and c are correct.
14. Which of the following is false about the phytochrome protein?
    a. There are two convertible forms.
    b. It primarily absorbs blue wavelengths of light.
    c. It is involved in seed germination.
    d. It is involved in plant sensing of spacing.
15. A morning glory flower opens during the day and closes at night. This response is due to its
    a. phototropism.
    b. circadian rhythm.
    c. hours in the darkness.
    d. temperature.

ENGAGE

THINKING SCIENTIFICALLY

1. You hypothesize that abscisic acid (ABA) is responsible for the turgor pressure changes that permit a plant to track the sun (see the chapter-opening photograph). What observations could you make to support your hypothesis?
2. You formulate the hypothesis that the negative gravitropic response of stems is greater than the positive phototropism of stems. How would you test your hypothesis?
3. Farmers who grow crops that are planted as seedlings may prepare them for their transition from the greenhouse to the field by brushing them gently every day for a few weeks. Why is this beneficial?
4. Because diverse plants exude volatile compounds in response to herbivore or pathogen attack, some experts have written about "talking trees." Explain why someone would use this adage.

A commercial tulip field in the Netherlands.

27

Flowering Plants: Reproduction

Chapter Outline

27.1 Sexual Reproductive Strategies 496

27.2 Seed Development 502

27.3 Fruit Types and Seed Dispersal 504

27.4 Asexual Reproductive Strategies 507

Bright petals, enticing nectar, and volatile perfume make flowers the perfect ornament to attract pollinators. But insects, birds, and bats are not the only animals attracted to these plant organs—humans have found flowers irresistible for thousands of years. In the 1600s, with the development of greenhouses, the cut flower industry became established in the Netherlands with tulips and other flowers. Even today, tulip festivals and tourism thrive every spring in the Netherlands.

In the United States, the cut flower industry followed European settlers, and with the invention of refrigerated trucks, states such as California, Colorado, and Florida quickly dominated the fresh-cut flower industry. The flower industry spread to South America, where growing conditions were favorable and labor was relatively cheap. Currently, the "big three"—roses, carnations, and chrysanthemums—are mostly grown and imported from Colombia and Ecuador. The United States has had to reinvent itself by growing specialty flowers and unusual varieties, such as dahlias, lilies, and zinnias.

Even though flowers are viewed for their beauty and the positive feelings they evoke, remember that a flower is a reproductive organ. It is where genetic recombination takes place and sperm meets egg. Flowers produce the seeds that become our trees, our crops, and our garden landscape. Flowers produce fruit, which also sustains and nourishes us.

As you read through this chapter, think about the following questions:

1. How are flowering plants adapted to a terrestrial lifestyle?
2. What would happen to the diversity of flowering plants if a large percentage of insect species were to disappear in a given community?
3. What are the advantages of sexual reproduction versus asexual reproduction in flowering plants?

Before You Begin

Before beginning this chapter, take a few moments to review the following discussions.

Section 10.3 What role does meiosis play in sexual reproduction?

Section 14.2 How is biotechnology changing how plants are reproduced?

Table 18.1 How long have angiosperms and modern insects been coevolving?

Following *the* Themes

CHAPTER 27 FLOWERING PLANTS: REPRODUCTION

UNIT 5 PLANT EVOLUTION AND BIOLOGY

Evolution	The majority of flowering plants owe their success to the animal pollinators with which they have coevolved.
Nature of Science	Plants rely on wind, water, and animals for pollination. Ecologists are studying the potential consequences for the flowering plants if the population of specific animal pollinators were to decrease.
Biological Systems	The life cycle of flowering plants is dependent on successful pollination, fertilization, and seed germination.

27.1 Sexual Reproductive Strategies

Learning Outcomes

Upon completion of this section, you should be able to

1. Identify the key components of the flowering plant life cycle.
2. Recognize the functions of the key flower parts.
3. Compare the male and female gametophytes of flowering plants.
4. Describe the process of double fertilization in flowering plants.

Sexual reproduction in plants is advantageous, because it generates variation among the offspring through the process of meiosis and fertilization. In a changing environment, a new variation may be better adapted for survival and reproduction than either parent.

Life Cycle Overview

When plants reproduce sexually, they undergo alternation of generations, in which they alternate between two multicellular stages, one diploid and one haploid.

To review, let's consider Figure 27.1 by beginning with the flowering sporophyte and moving clockwise. In flowering plants, the diploid sporophyte is dominant, and it is the portion of the life cycle that bears flowers. Next, a **flower,** which is the reproductive structure of angiosperms, produces two types of spores by meiosis, microspores and megaspores. The haploid **microspore** (Gk. *mikros,* "small, little") undergoes mitosis and becomes a pollen grain, which is either windblown or carried by an animal to the vicinity of the female gametophyte. A pollen grain is the male gametophyte. In the meantime, the haploid **megaspore** (Gk. *megas,* "great, large") undergoes mitosis and becomes the female gametophyte. The female gametophyte is an embryo sac within the ovule that is within an ovary. At maturity, a pollen grain contains nonflagellated sperm, which travel by way of a pollen tube to the embryo sac. Once a sperm fertilizes an egg, the zygote becomes an embryo, still within an ovule. The ovule develops into a **seed,** which contains the embryo and stored food surrounded by a seed coat. The ovary becomes a fruit, which aids in dispersing the seeds. When a seed germinates, a new sporophyte emerges and through mitosis and growth becomes a mature organism.

Notice that the sexual life cycle of flowering plants is adapted to a land-based existence. The microscopic female gametophytes develop completely within the sporophyte and are thereby protected from desiccation. Pollen grains (male gametophytes) are not released until they develop a thick wall. Unlike the mosses, no external water is needed to bring about fertilization in flowering plants. Instead, the pollen tube provides passage for a sperm to reach an egg. Following fertilization, the embryo and its stored food are enclosed within a protective seed coat until external conditions are favorable for germination.

Flowers

The flower is unique to angiosperms (Fig. 27.2). Aside from producing the spores and protecting the gametophytes, flowers often

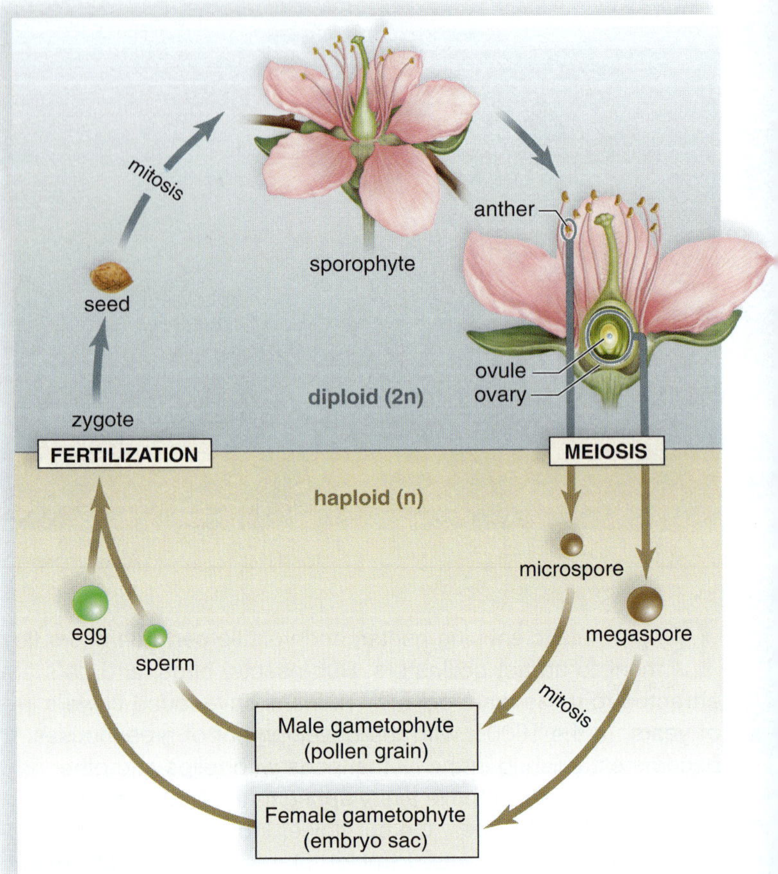

Figure 27.1 Sexual reproduction in flowering plants. The sporophyte bears flowers. The flower produces microspores within anthers and megaspores within ovules by meiosis. A megaspore becomes a female gametophyte, which produces an egg within an embryo sac, and a microspore becomes a male gametophyte (pollen grain), which produces sperm. Fertilization results in a zygote. A seed contains an embryo and stored food within a seed coat. After dispersal, a seed becomes a new sporophyte plant.

Tutorial Alternation of Generations

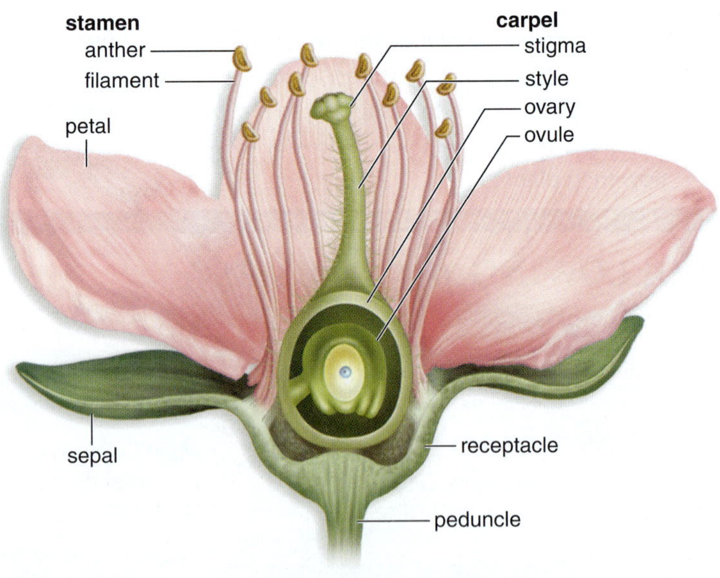

Figure 27.2 Anatomy of a flower. A complete flower has all flower parts: sepals, petals, stamens, and at least one carpel.

a. Daylily, *Hemerocallis* sp.

b. Festive azalea, *Rhododendron* sp.

Figure 27.3 Monocot versus eudicot flowers. **a.** Monocots, such as daylilies, have flower parts in threes. In particular, note the three petals and three sepals, both of which are colored in this flower. **b.** Azaleas are eudicots. They have flower parts in fours or fives; note the five petals of this flower. p = petal, s = sepal

attract pollinators, which aid in transporting pollen from plant to plant. Flowers also produce the fruits that enclose the seeds.

The evolution of the flower was a major factor leading to the success of angiosperms, with over 250,000 species. Flowering is often a response to environmental signals, such as the length of the day or time of year (see Chapter 26). In many plants, a flower develops when shoot apical meristem that previously formed leaves suddenly stops producing leaves and starts producing a flower enclosed within a bud. In other plants, axillary buds develop directly into flowers. In monocots, flower parts occur in multiples of three; in eudicots, flower parts occur in multiples of four or five (Fig. 27.3).

Flower Structure

The typical eudicot flower has four whorls of modified leaves attached to a receptacle at the end of a flower stalk called a peduncle.

1. The **sepals** are the most leaflike of all the flower parts and are usually green. Sepals protect the bud as the flower develops within. Collectively, the sepals are called the **calyx.**
2. An open flower next has a whorl of **petals,** whose color accounts for the attractiveness of many flowers. The size, shape, and color of petals are attractive to specific pollinators. Some wind-pollinated flowers have no petals at all. Collectively, the petals are called the **corolla.**
3. **Stamens** are the "male" portions of flowers. Each stamen contains two parts: the slender stalk called the *filament* (L. *filum,* "thread") and the saclike *anther* supported by the filament. Pollen grains develop from the microspores produced in the anther.
4. At the very center of a flower is the **carpel,** a vaselike structure that represents the "female" portion of the flower. A carpel usually has three parts: The **style** is a slender stalk that supports the **stigma;** an enlarged, sticky knob; and the **ovary,** an enlarged base that encloses one or more *ovules* (see Fig. 27.2).

Ovules (L. *ovulum,* "little egg") play a significant role in the production of megaspores and therefore female gametophytes, as described shortly.

A flower can have a single carpel or multiple carpels. Sometimes several carpels are fused into a single structure, in which case the ovary is termed compound; it has several chambers, each of which contains ovules. For example, an orange develops from a compound ovary, and every section of the orange is a chamber.

Variations in Flower Structure

Not all flowers have sepals, petals, stamens, or carpels. Those that do are said to be *complete,* and those that do not are said to be *incomplete.* Flowers that have both stamens and carpels are called *perfect* (bisexual) flowers; flowers with only stamens and those with only carpels are *imperfect* (unisexual) flowers.

If both staminate (male) flowers and carpellate (female) flowers occur on a single plant, the plant is *monoecious* (Gk. *monos,* "one"; *oikos,* "home, house") (Fig. 27.4*a*). If staminate and carpellate flowers occur on separate plants, the plant is *dioecious.* Holly trees are dioecious; if you hope to have holly berries, you need one plant with staminate flowers and another plant with carpellate flowers (Fig. 27.4*b*).

Life Cycle in Detail

In all land plants, the sporophyte produces haploid spores by meiosis. The haploid spores grow and develop into haploid **gametophytes,** which produce gametes by mitotic division. Flowering plants, however, are *heterosporous*—they produce microspores and megaspores. Microspores become mature male gametophytes (sperm-bearing pollen grains), and megaspores become mature female gametophytes (egg-bearing embryo sacs).

Development of Male Gametophyte

Microspores are produced in the anthers of flowers (Fig. 27.5). An anther has four pollen sacs, each containing many microspore mother cells. A microspore mother cell undergoes meiosis to produce four haploid microspores. In each, the haploid nucleus divides mitotically, followed by unequal cytokinesis, and the result is two cells enclosed by a finely sculptured wall. This structure, called

a.

b.

Figure 27.4 Monoecious and dioecious plants.
a. The flowers of the mature *Jatropha* plant are monoecious, having both male and female flowers on the same plant. **b.** Holly trees are dioecious; berries are produced only by female plants, and pollen only by male plants.

Figure 27.5 Life cycle of flowering plants. *Development of gametophytes (far page):* A pollen sac in the anther contains microspore mother cells, which produce microspores by meiosis. A microspore develops into a pollen grain, which germinates and has two sperm. An ovule in an ovary contains a megaspore mother cell, which produces a megaspore by meiosis. A megaspore develops into an embryo sac containing seven cells, one of which is an egg. *Development of sporophyte (this page):* A pollen grain contains two sperm by the time it germinates and forms a pollen tube. During double fertilization, one sperm fertilizes the egg to form a diploid zygote, and the other fuses with the polar nuclei to form a triploid (3n) endosperm cell. A seed contains the developing sporophyte embryo plus stored food.

Tutorial
Angiosperm Life Cycle

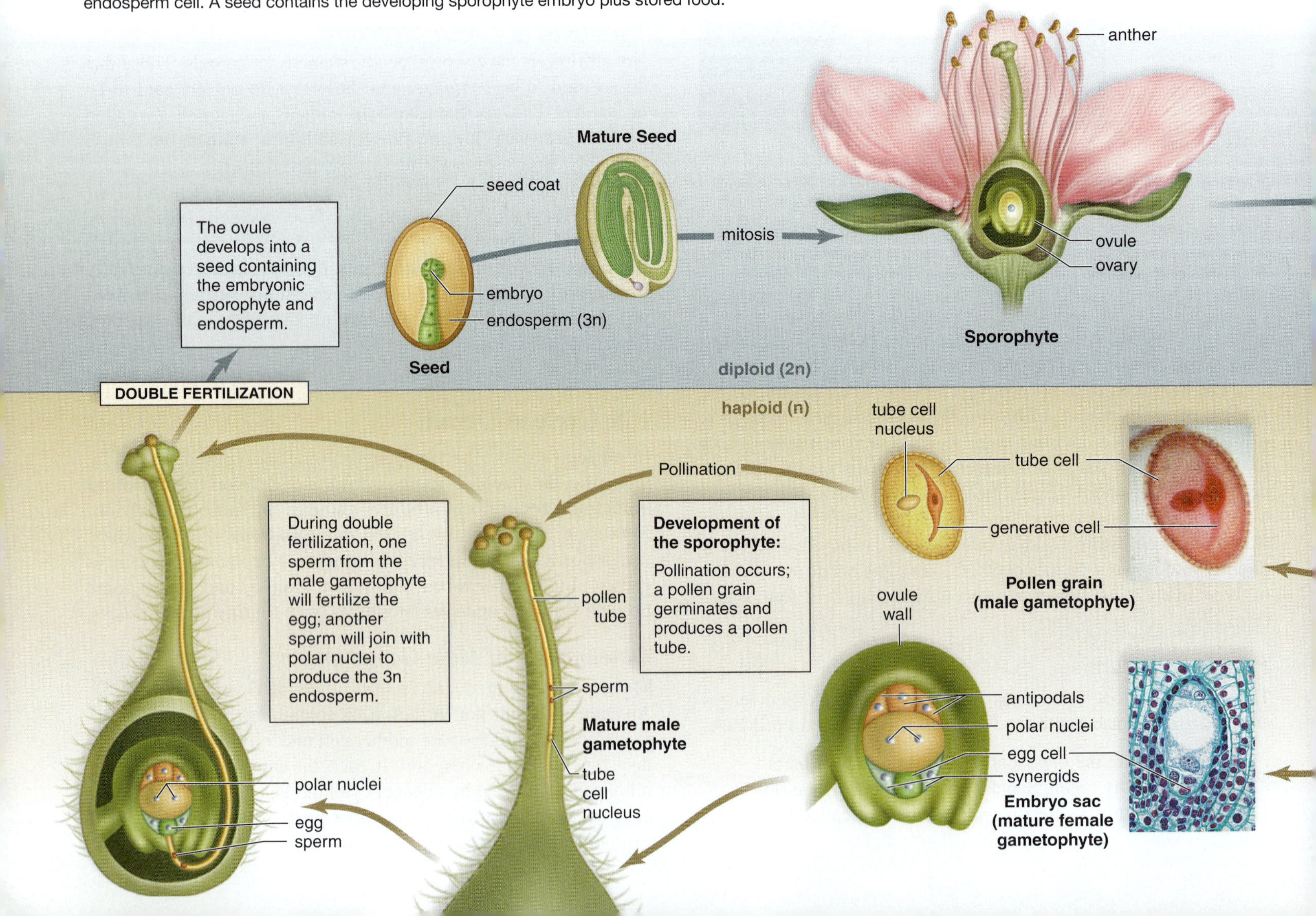

the **pollen grain,** is at first an immature *male gametophyte* that consists of a tube cell and a generative cell. The larger tube cell will eventually produce a *pollen tube.* The smaller generative cell divides mitotically either now or later to produce two sperm. Once these events have taken place, the pollen grain has become the mature male gametophyte.

Development of Female Gametophyte

The ovary contains one or more ovules. An ovule has a central mass of parenchyma cells almost completely covered by layers of tissue called integuments, except where there is an opening called a micropyle. One parenchyma cell enlarges to become a megaspore mother cell, which undergoes meiosis, producing four haploid megaspores (Fig. 27.5). Three of these megaspores are nonfunctional; one is functional. In a typical pattern, the nucleus of the functional megaspore divides mitotically until there are eight nuclei in the *female gametophyte.* When cell walls form later, there are seven cells, one of which is binucleate.

The female gametophyte, also called the **embryo sac,** consists of these seven cells:

- One egg cell, associated with two synergid cells
- One central cell, with two polar nuclei
- Three antipodal cells

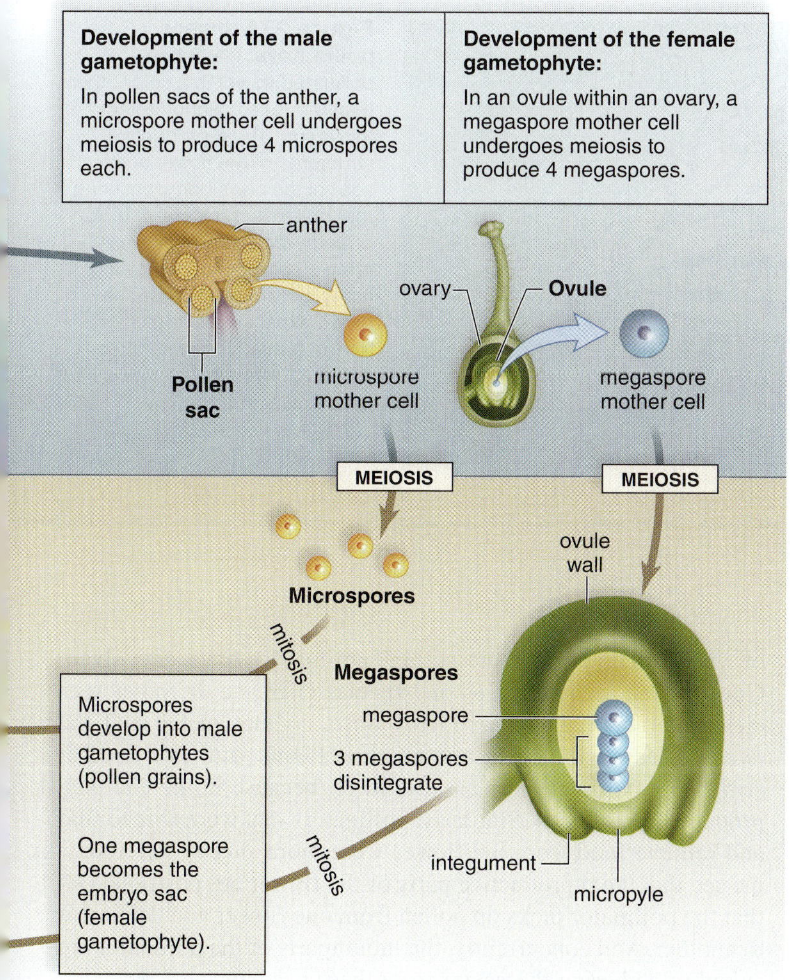

Development of New Sporophyte

The walls separating the pollen sacs in the anther break down when the pollen grains are ready to be released. **Pollination** is simply the transfer of pollen from an anther to the stigma of a carpel. Self-pollination occurs if the pollen is from the same plant, and cross-pollination occurs if the pollen is from a different plant of the same species.

When a pollen grain lands on the stigma of the same species, it germinates, forming a pollen tube (Fig. 27.5). The germinated pollen grain, containing a tube cell and two sperm, is the mature male gametophyte. As it grows, the pollen tube passes between the cells of the stigma and the style to reach the micropyle, a pore of the ovule. When the pollen tube reaches the micropyle, **double fertilization** occurs: One of the sperm unites with the egg to form a 2n zygote; however, the second sperm unites with the two polar nuclei centrally placed in the embryo sac to form a 3n endosperm nucleus. This latter fertilization is unique to angiosperms. The endosperm nucleus eventually develops into the **endosperm** (Gk. *endon,* "within"; *sperma,* "seed"), a nutritive tissue that the developing embryonic sporophyte will use as an energy source.

Now the ovule begins to develop into a seed. One important aspect of seed development is formation of the seed coat from the ovule wall. A mature seed contains the embryo, stored food, and the seed coat (see Fig. 27.8).

Cross-Pollination

Some species of flowering plants, such as grasses and grains, rely on wind pollination (Fig. 27.6), as do the gymnosperms, the other type of seed plant. Much of the plant's energy goes into making

Figure 27.6 Pollination. **a.** Grass releasing pollen. **b.** Pollen grains of Canadian goldenrod, *Solidago canadensis.* **c.** Pollen grains of asparagus, *Asparagus officinalis.* The shape and pattern of pollen grain walls are quite distinctive, and experts can use them to identify the genus, and even sometimes the species, that produced a particular pollen grain. Pollen grains have strong walls resistant to chemical and mechanical damage; therefore, they frequently become fossils.

THEME Evolution

Plants and Their Pollinators

Plants and their pollinators have adapted to one another. They have a mutualistic relationship in which each benefits—the plant uses its pollinator to ensure that cross-pollination takes place, and the pollinator uses the plant as a source of food. This mutualistic relationship came about through the process of coevolution—that is, the interdependency of the plant and the pollinator is the result of suitable changes in the structure and function of each.

The evidence for coevolution is observational. For example, floral coloring and odor are suited to the sense perceptions of the pollinator; the mouthparts of the pollinator are suited to the structure of the flower; the type of food provided is suited to the nutritional needs of the pollinator; and the pollinator forages at the time of day that specific flowers are open.

Bee- and Wasp-Pollinated Flowers

There are 20,000 known species of bees that pollinate flowers. The best-known pollinators are the honeybees (Fig. 27A*a*). As noted in the text, bee eyes see ultraviolet (UV) wavelengths. Therefore, bee-pollinated flowers are usually brightly colored and are predominantly blue or yellow; they are not entirely red. They may also have ultraviolet shadings called nectar guides, which highlight the portion of the flower that contains the reproductive structures.

The mouthparts of bees are fused into a long tube that contains a tongue. This tube is an adaptation for sucking up nectar provided by the plant, usually at the base of the flower. Bees also collect pollen as a food.

Bee flowers are delicately sweet and fragrant, advertising that nectar is present. The nectar guides often point to a narrow floral tube large enough for the bee's feeding apparatus but too small for other insects to reach the nectar. Bee-pollinated flowers are sturdy and may be irregular in shape, because they often have a landing platform where the bee can alight. The flower structure requires the bee to brush up against the anther and stigma as it moves toward the floral tube to feed.

One type of orchid, *Ophrys,* has evolved a unique adaptation. The flower resembles a female wasp, and when the male of that species attempts to copulate with the flower, the flower spring loads pollen on the wasp's head. When the frustrated wasp attempts to "copulate" with another flower, the pollen is perfectly positioned to come in contact with the stigma of the second flower.

Moth- and Butterfly-Pollinated Flowers

Contrasting moth- and butterfly-pollinated flowers emphasizes the close adaptation

a. b.

Figure 27A Insect pollinators. **a.** A bee-pollinated flower is a color other than red (bees cannot detect this color). The reproductive structures of the flower brush up against the bee's body, ensuring that pollen is transferred. **b.** A butterfly-pollinated flower is often a composite, containing many individual flowers. The broad expanse provides room for the butterfly to land, after which it lowers its proboscis into each flower in turn.

pollen to ensure that some pollen grains actually reach a stigma. The amount of pollen successfully transferred is staggering: A single corn plant may produce 20–50 million pollen grains a season. In corn, the flowers tend to be monoecious, and clusters of tiny male flowers move in the wind, freely releasing pollen into the air.

Most angiosperms rely on animals—insects (e.g., bumblebees, flies, butterflies, and moths), birds (e.g., hummingbirds), or mammals (e.g., bats)—to carry out pollination. The use of animal pollinators is unique to flowering plants, and it helps account for why these plants are so successful on land.

By the time flowering plants appear in the fossil record, around 240 MYA, insects had long been present. For millions of years, plants and their animal pollinators have coevolved. **Coevolution** means that as one species changes, the other species undergoes adaptation in response, so that in the end the two species are suited to one another. Plants with flowers that attracted a pollinator had an advantage, because in the end they produced more seeds. Similarly, pollinators that were able to find and remove food from the flower were more successful. Today, we see that the reproductive parts of the flower are positioned so that the pollinator picks up pollen from one flower and delivers it to another. And concurrently, the mouthparts of the pollinator are suited to gathering the nectar from these particular plants.

Video Pollinators

between pollinator and flower. Both moths and butterflies have a long, thin, hollow proboscis, but they differ in other characteristics. Moths usually feed at night and have a well-developed sense of smell. The flowers they visit are visible at night, because they are lightly shaded (white, pale yellow, or pink), and they have strong, sweet perfume, which helps attract moths. Moths hover when they feed, and their flowers have deep tubes with open margins that allow the hovering moths to reach the nectar with their long proboscis.

Butterflies, in comparison, are active in the daytime and have good vision but a weak sense of smell. Their flowers have bright colors—even red, because butterflies can see the color red—but the flowers tend to be odorless. Unable to hover, butterflies need a place to land. Flowers that are visited by butterflies often have flat landing platforms (Fig. 27A*b*). Composite flowers (composed of a compact head of numerous individual flowers) are especially favored by butterflies. Each flower has a long, slender floral tube, accessible to the long, thin butterfly proboscis.

Bird- and Bat-Pollinated Flowers

In North America, the most well-known bird pollinators are the hummingbirds. These small animals have good eyesight but do not have a well-developed sense of smell. Like moths, they hover when they feed. Typical flowers pollinated by hummingbirds are red, with a slender floral tube and margins that are curved back and out of the way. And although they produce copious amounts of nectar, the flowers have little odor. As a hummingbird feeds on nectar with its long, thin beak, its head comes in contact with the stamens and pistil (Fig. 27B*a*).

Bats are adapted to gathering food in various ways, including feeding on the nectar and pollen of plants. Bats are nocturnal and have an acute sense of smell. Those that are pollinators also have keen vision and a long, extensible, bristly tongue. Typically, bat-pollinated flowers open only at night and are light-colored or white. They have a strong, musky smell similar to the odor that bats produce to attract one another. The flowers are generally large and sturdy and are able to hold up when a bat inserts part of its head to reach the nectar. While the bat is at the flower, its head becomes dusted with pollen (Fig. 27B*b*).

Coevolution

How did this coevolution of plants and pollinators come about? Some 250 million years ago, when seed plants were just beginning to evolve and insects were not as diverse as they are today, wind alone carried pollen. Wind pollination, however, is a hit-or-miss affair. Perhaps beetles feeding on vegetative leaves were the first insects to carry pollen directly from plant to plant by chance. Because flowers undergoing direct cross-fertilization would likely produce more fruit, natural selection favored flowers with features that would attract pollinators.

As cross-fertilization continued, more and more flower variations likely developed, and pollinators became increasingly adapted to specific angiosperm species. Today, there are some 250,000 species of flowering plants and over 900,000 species of insects. This diversity suggests that the success of angiosperms has contributed to the success of insects, and vice versa.

Questions to Consider

1. What are the potential consequences if honeybees were to go extinct?
2. How does coevolution cause two species to change over time?

Figure 27B Bird and bat pollinators.
a. Hummingbird-pollinated flowers are curved back, allowing the bird to insert its beak to reach the rich supply of nectar. While doing this, the bird's forehead and other body parts touch the reproductive structures. **b.** Bat-pollinated flowers are large, sturdy flowers that can take rough treatment. Here the head of the bat is positioned so that its bristly tongue can lap up nectar.

a. b.

One well-studied example of coevolution has occurred between bees and the plants they pollinate. Bee-pollinated flowers tend to be yellow, blue, or white, because these are the colors bees can see. Bees respond to ultraviolet (UV) markings, called nectar guides, that help them locate nectar. Humans cannot detect light in the ultraviolet range, but bees are sensitive to UV. A bee has a feeding proboscis of the right length to collect nectar from certain flowers and a pollen basket on its hind legs that allows it to carry pollen back to the hive.

Because many fruits and vegetables are dependent on bee pollination, many people have great concern today that the number of bees is declining due to disease and the use of pesticides. More examples of the coevolution between plants and their pollinators are given in the Evolution feature, "Plants and Their Pollinators."

Check Your Progress 27.1

1. Compare the development and structure of the male gametophyte with those of the female gametophyte.
2. Describe the products of double fertilization in angiosperms.
3. Explain how a flowering plant may coevolve in response to an increase in the body size of its pollinator.

27.2 Seed Development

Learning Outcomes

Upon completion of this section, you should be able to

1. Describe the stages of eudicot development.
2. Compare the main parts of a monocot seed to those of a eudicot.

Development of the embryo within the seed is the next event in the life cycle of the angiosperm. Plant growth and development involves cell division, cell elongation, and differentiation of cells into tissues, which then develop into organs. **Development** is a programmed series of stages from a simple to a more complex form. Cellular **differentiation,** or specialization of structure and function, occurs as development proceeds.

Stages of Eudicot Development

Figure 27.7 shows the stages of development for a eudicot embryo.

Zygote and Proembryo Stages

Immediately after double fertilization, the zygote and the endosperm become visible. The zygote is small with dense cytoplasm (Fig. 27.7*a*). Next, the zygote divides repeatedly in different planes, forming several cells called a proembryo. Also present is an elongated structure called a suspensor. The suspensor transfers and produces nutrients from the endosperm, which allows the embryo to grow (Fig. 27.7*b*).

Globular Stage

During the globular stage, the proembryo is largely a ball of cells. The root-shoot axis of the embryo is already established at this stage. The embryonic cells near the suspensor will go on to become a root, while those at the other end will ultimately become a shoot (Fig. 27.7*c*).

The outermost cells of the plant embryo will become the dermal tissue. These cells divide with their cell plate perpendicular to the surface; therefore, they produce a single outer layer of cells. Recall that dermal tissue protects the plant from desiccation and includes the stomata, which open and close to facilitate gas exchange and minimize water loss.

The Heart Stage and Torpedo Stage Embryos

The embryo has a heart shape when the **cotyledons,** or seed leaves, appear because of local, rapid cell division (Fig. 27.7*d*). As the embryo continues to enlarge and elongate, it takes on a torpedo shape. Now the root and shoot apical meristems are distinguishable. The shoot apical meristem is responsible for aboveground growth, while the root apical meristem is responsible for underground growth. Ground meristem gives rise to the bulk of the embryonic interior, which is now present (Fig. 27.7*e*).

Figure 27.7 Development of a eudicot embryo. After fertilization, the eudicot embryo undergoes multiple stages of differentiation. **a.** In the zygote stage, the nourishing endosperm and zygote are visible. **b.** After multiple cell divisions, a stalklike structure called a suspensor develops in the proembryo. **c.** The axis for shoot and root development is established, and the embryo takes on a globelike appearance. **d.** The heart stage is indicative of the formation of cotyledons. **e.** In the torpedo stage, the shoot, root, and ground meristems are present and actively dividing. **f.** In the mature embryo, a rudimentary shoot and root develop with continued division from the apical meristems. The integuments of the original ovule harden, becoming the seed coat.

The Mature Embryo

In the mature embryo, the *epicotyl* is the portion between the cotyledons that contributes to shoot development. The *plumule* is found at the tip of the epicotyl and consists of the shoot tip and a pair of small leaves. The *hypocotyl* is the portion below the cotyledon(s). It contributes to stem development and terminates in the radicle or embryonic root.

The cotyledons are quite noticeable in a eudicot embryo and may fold over. Procambium is located at the core of the embryo and is destined to form the future vascular tissue responsible for water and nutrient transport.

As the embryo develops, the integuments of the ovule become the seed coat. The seed coat encloses and protects the embryo and its food supply (Fig. 27.7*f*).

Monocot Versus Eudicot Seeds

Monocots, unlike eudicots, have only one cotyledon. Another important difference between monocots and eudicots is the manner in which nutrient molecules are stored in the seed. In monocots, the cotyledon, in addition to storing certain nutrients, absorbs other nutrient molecules from the endosperm and passes them to the embryo. In eudicots, the cotyledons usually store all the nutrient molecules the embryo uses. Therefore, in Figure 27.7, you can see that the endosperm seemingly disappears. Actually, it has been taken up by the two cotyledons.

Figure 27.8 contrasts the structure of a bean seed (eudicot) and a corn kernel (monocot). The size of seeds may vary from the dust-sized seeds of orchids to the 27-kg seed of double coconuts.

Check Your Progress 27.2

1. Identify the origin of each of the three parts of a seed.
2. Explain why the seed coat and the embryo are both 2n.
3. Describe the structure and function of the cotyledon.

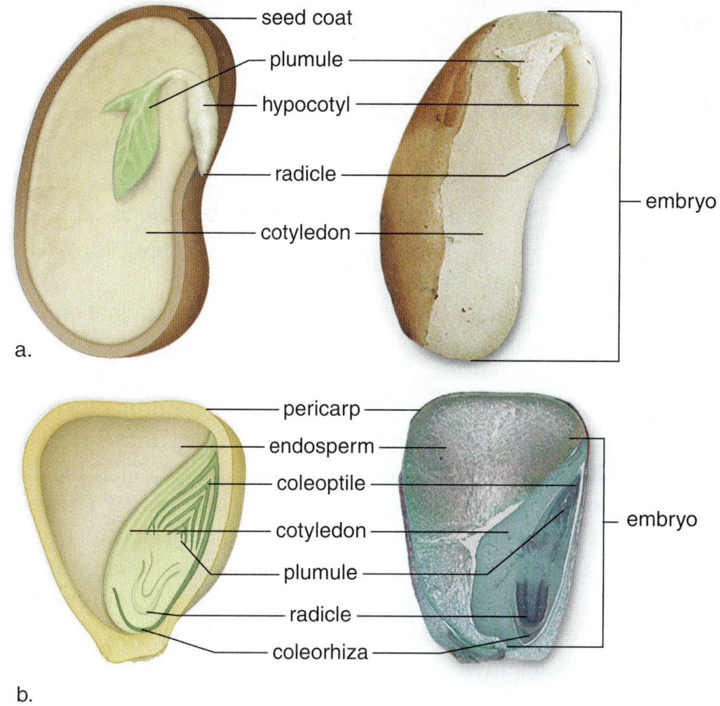

Figure 27.8 Monocot versus eudicot. **a.** In a bean seed (eudicot), the endosperm has disappeared; the bean embryo's cotyledons take over food storage functions. The hypocotyl becomes the shoot system, which will include the plumule (first leaves). The radicle becomes the root system. **b.** The corn kernel (monocot) has endosperm that is still present at maturity. The coleoptile is a protective sheath for the shoot system; the coleorhiza similarly protects the future root system. The pericarp of the fruit develops from the ovary wall.

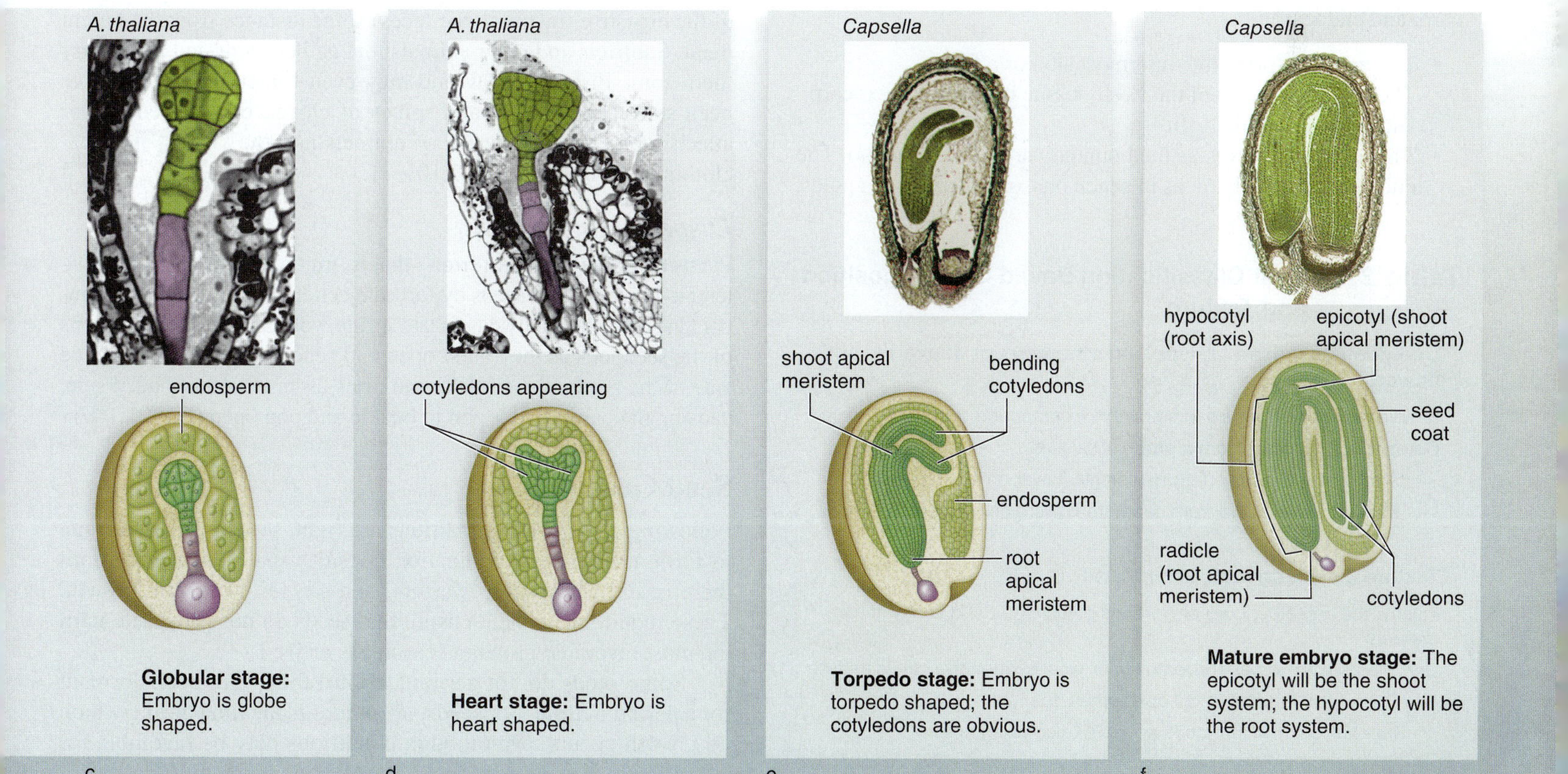

27.3 Fruit Types and Seed Dispersal

Learning Outcomes

Upon completion of this section, you should be able to

1. Identify examples of fleshy and dry fruits that are simple, compound, aggregate, or accessory fruits.
2. Explain the sequence of events during seed germination.

Most people are unfamiliar with the botanical definition of a fruit (Table 27.1). A **fruit** in botany is a mature ovary that can also contain other flower parts, such as the receptacle. This means that pea pods, tomatoes, and what are usually called winged maple seeds are actually fruits (Fig. 27.9). Fruits protect and help disperse seeds. Some fruits are better at one function than the others. The fruit of a peach protects the seed well, but the pit may make it difficult for germination to occur. Peas easily escape from pea pods, but once they are free, they are protected only by the seed coat.

Kinds of Fruits

Fruits can be simple or compound (Fig. 27.9). A *simple fruit* is derived from a single ovary that can have one or several chambers (Fig. 27.9*a–d*). A *compound fruit* is derived from several groups of ovaries (Fig. 27.9*e–f*). If a single flower has multiple ovaries, as in a blackberry, then it produces an *aggregate fruit* (Fig. 27.9*e*). In contrast, a pineapple comes from many individual ovaries. Because the flowers have only one receptacle, the ovaries fuse to form a large, *multiple fruit* (Fig. 27.9*f*).

As a fruit develops, the ovary wall thickens to become the *pericarp,* which can have as many as three layers: exocarp, mesocarp, and endocarp.

- The *exocarp* forms the outermost skin of a fruit.
- The *mesocarp* is often the fleshy tissue between the exocarp and endocarp of the fruit.
- The *endocarp* serves as the boundary around the seed(s). The endocarp may be hard, as in peach pits, or papery, as in apples.

Table 27.1 Fruit Classification Based on Composition and Texture

Composition (based on type and arrangement of ovaries and flowers)
Simple: develops from a simple ovary or compound ovary
Compound: develops from a group of ovaries
Aggregate: ovaries are from a single flower on one receptacle
Multiple: ovaries are from separate flowers on a common receptacle
Texture (based on mature pericarp)
Fleshy: the entire pericarp or portions of it are soft and fleshy at maturity
Dry: the pericarp is papery, leathery, or woody when the fruit is mature
Dehiscent: the fruit splits open when ripe
Indehiscent: the fruit does not split open when ripe

Some fruits, such as legumes and cereal grains of wheat, rice, and corn, are *dry fruits.* The fruits of grains can be mistaken for seeds, because a dry pericarp adheres to the seed within. Legume fruits such as the pea pod (Fig. 27.9*c*) are dehiscent, which means they split open when ripe. Grains are indehiscent—they do not split open. Humans gather grains before they are released from the plant and then process them to acquire their nutrients.

You are probably more familiar with fleshy fruits, such as the peach and tomato. In these fruits, the mesocarp is well developed.

Dispersal of Fruits

Generally, it is beneficial for plants to diperse their fruits away from the parent plant, so that seedlings do not have to compete with the parent for nutrients. Fruits may drift or be blown to new locations by wind or forced ejection or may be carried away by animals.

Dispersal by Air

Many dry fruits are dispersed by wind. Woolly hairs, plumes, and wings are all adaptations for this type of dispersal. The somewhat heavier dandelion fruit uses a tiny "parachute" for dispersal (Fig. 27.10*a*). Milkweed pods split open to release seeds that float away on puffy, white threads. The winged fruit of a maple tree has been known to travel up to 10 km from its parent.

Dispersal by Animals

Ripe, fleshy, colorful fruits, such as peaches and cherries, often attract animals and provide them with food (Fig. 27.10*b*). Their hard endocarp protects the seed, so it can pass through the digestive system of an animal and remain unharmed. As the flesh of a tomato is eaten, the small size of the seeds and the slippery seed coat mean that tomato seeds rarely get crushed by the teeth of animals. The seeds swallowed by birds and mammals are defecated (passed out of the digestive tract with the feces) some distance from the parent plant. Squirrels and other animals that gather seeds and fruits bury them some distance away and may even forget where they have been stored. The hooks and spines of clover, bur, and cocklebur attach a dry fruit to the fur of animals and the clothing of humans (Fig. 27.10*b*, *c*).

Video Fruit Bat Seed Dispersal

Dispersal by Ejection

Plants that have dehiscent fruit—that is, fruit with slits or openings—tend to disperse their seeds by forced ejection. If you walk by a trumpet vine on a hot, sunny day in late spring, you can hear the explosions of the seedpods as they burst open and send their seeds flying. The side of the seedpod facing the sun dries faster than the shaded side, causing the partially dry pod to buckle and pop open (Fig. 27.10*d*).

Seed Germination

Following dispersal, if conditions are right, seeds may **germinate** to form a seedling. Germination doesn't usually take place until there is sufficient water, warmth, and oxygen to sustain growth. These requirements help ensure that seeds do not germinate until the most favorable growing season has arrived.

Some seeds do not germinate until they have been dormant for a period of time. For seeds, *dormancy* is the time during which no growth occurs, even though conditions may be favorable for

a. A drupe is a fleshy fruit with a pit containing a single seed produced from a simple ovary.

b. A berry is a fleshy fruit having seeds and pulp produced from a compound ovary.

c. A legume is a dry dehiscent fruit produced from a simple ovary.

d. A samara is a dry indehiscent fruit produced from a simple ovary.

e. An aggregate fruit contains many fleshy fruits produced from simple ovaries of the same flower.

f. A multiple fruit contains many fused fruits produced from simple ovaries of individual flowers.

Figure 27.9 Fruits. **a.** A peach is a drupe. **b.** A tomato is a true berry. **c.** A pea is a legume. **d.** The fruit of a maple tree is a samara. **e.** A raspberry is an aggregate fruit. **f.** A pineapple is a multiple fruit.

a.

b.

c.

d.

Figure 27.10 Methods of fruit and seed dispersal. **a.** Dandilion fruit float away in air currents. **b.** When birds eat fleshy fruits, seeds pass through their digestive system. **c.** Burdock, a dry fruit, clings to the fur of animals. **d.** Dry pods split open, forcibly ejecting their seeds.

growth. In the temperate zone, seeds often have to be exposed to a period of cold weather before dormancy is broken. Fleshy fruits (e.g., apples, pears, oranges, and tomatoes) contain inhibitors, so that germination does not occur while the fruit is still on the plant. For seeds to take up water, bacterial action and even fire may be needed. Once water enters, the seed coat bursts and the seed germinates.

If the two cotyledons of a bean seed are parted, the rudimentary plant with immature leaves is exposed (Fig. 27.11*a*). As the eudicot seedling starts to form, the root emerges first. The shoot is hook-shaped to protect the immature leaves as they emerge from the soil. The cotyledons provide the new seedlings with enough energy for the stem to straighten and the leaves to grow. As the mature leaves of the plant begin photosynthesizing, the cotyledons shrivel up.

A kernel of corn is actually a fruit, and therefore its outer covering is the pericarp and seed coat combined (Fig. 27.11*b*). Inside is the single cotyledon. Also, the immature leaves and the radicle are covered, respectively, by a coleoptile and a coleorhiza. These sheaths are discarded as the root grows directly downward into the soil and the shoot of the seedling begins to grow directly upward.

Check Your Progress 27.3

1. Compare the structure and dispersal methods of dry and fleshy fruits.
2. Compare the protective methods of monocot and eudicot seedlings used to protect their first true leaves.

Figure 27.11 Eudicot and monocot seed germination. **a.** Bean (eudicot) seed germinates, revealing two cotyledons. **b.** Corn (monocot) seed germinates, with an emerging coleoptile.

27.4 Asexual Reproductive Strategies

Learning Outcomes

Upon completion of this section, you should be able to

1. Identify the asexual methods of reproduction in plants.
2. Describe how tissue culture can be used to clone plants with desirable traits.

Asexual reproduction is the production of an offspring from a single parent. Because the offspring is genetically identical to the parent, it is called a **clone.** Asexual reproduction is less complicated in plants, because pollination and seed production are not required. Therefore, it can be advantageous when the parent is already well adapted to a particular environment and the production of genetic variations is not an apparent necessity. In addition, clones are desirable for plant sellers, because they know exactly how the plant will grow and look.

Asexual Reproduction from Stems, Roots, and Cuttings

Figure 27.12 features a strawberry plant that has produced a stolon. *Stolons* are horizontal stems that can be seen because they run aboveground. As you know, the nodes of stems are regions where new growth can occur. In the case of stolons, a new shoot system appears above the node, and a new root system appears below the node. The larger plant on the *left* is the parent plant, and the smaller plant on the *right* is the asexual clone that has arisen at a node. The characteristics of new offspring produced by stolons are identical to those of the parent plant.

Rhizomes are underground stems that produce new plants asexually. Irises are examples of plants that have no aboveground stem because their main stem is a rhizome that grows horizontally underground. As with stolons, new plants arise at the nodes of a rhizome. White potatoes are expanded portions of a rhizome branch called tubers, and each "eye" is a bud that can produce a new potato plant if it is planted with a portion of the swollen

Figure 27.12 Asexual reproduction in plants. Meristem tissues at nodes can generate new plants, as when the stolons of strawberry plants, *Fragaria,* give rise to new plants.

Figure 27.13 Asexual reproduction through tissue culture. **a.** Meristem tissue is placed on sterile media and an undifferentiated mass, called a callus, develops. **b.** From the callus, organogenesis takes place and leaves or roots develop. **c.** The callus develops into a plantlet. **d.** Plantlets can be stored, then shipped in sterile containers and transferred to soil for growth into adult plants.

tuber. Sweet potatoes, by contrast, are modified roots; they can be propagated by planting sections of the root, which are called slips.

Tissue Culture of Plants

One of the major disadvantages of most asexual propagation techniques is that they also propagate pathogens. Plant pathogens can be viruses, bacteria, or fungi; clones created from an infected parent will also be infected. However, it is possible to maintain plants in a disease-free status if clones from an uninfected parent are made in sterile conditions through tissue culture. Hence, **tissue culture** is simply plant propagation done in a laboratory under sterile conditions.

The key to plant tissue culture, as opposed to animal cell culture, is the totipotency of plant cells. **Totipotency** is the ability of individual plant cells to develop into an entire plant.

Techniques for tissue culture vary, but most begin with cells from the *meristem* of the parent plant. Meristematic cells, which are already actively dividing, are grown on an agar medium in flacks, tubes, or petri dishes. The medium contains auxin, cytokinins, various nutrients, and water. Initially, a mass of undifferentiated (unspecialized) cells, called a callus, forms (Fig. 27.13*a*). As described in Chapter 26, the addition of different combinations of auxin and cytokinins will signal *organogenesis* (organ formation) and initiate the formation of roots and shoots, until a fully developed plantlet is formed (Fig. 27.13*b,c*). Plantlets can be shipped in their sterile containers to growers for transplantation into a greenhouse or field.

Tissue culture is an important technique for propagating many fruits and vegetables. A notable example is the banana, which is a sterile hybrid that cannot produce seeds. The only way to produce this commercially important fruit is through tissue culture. As another example, asparagus is a dioecious plant, and all commercial stalks are male. The female stalks favor the production of flowers and are undesirable for eating. Tissue culture is a more efficient means of producing disease-free male asparagus for growers.

Many botanical gardens and universities use tissue culture for plant conservation. The Atlanta Botanical Garden has a tissue culture lab that propagates native species of orchids and lilies. The propagated plants are shared with local nurseries, resulting in fewer plants being removed from the wild by collectors. Tissue culture of rare species is also used to help increase the population in the wild, as they are replanted in native habitats (Fig. 27.14).

Figure 27.14 Tissue culture for plant conservation. This rare Kentucky ladyslipper orchid is grown in a tissue culture lab and will be replanted into native habitats.

Cell Suspension Culture

A technique called **cell suspension culture** allows scientists to extract chemicals (secondary metabolites) from plant cells, which may have been genetically modified. This technique allows scientists to avoid overcollection of wild plants from their natural environments. These cells produce the same chemicals that the plant produces. For example, cell suspension cultures of *Cinchona ledgeriana* produce quinine, which is used to treat leg cramping, a major symptom of malaria. Several *Digitalis* species produce digitalis, digitoxin, and digoxin, which are useful in the treatment of heart disease.

Check Your Progress 27.4

1. Identify the possible benefits of asexual reproduction.
2. Describe methods of asexual reproduction in wild plants.
3. Explain the benefits of plant tissue culture.

Connecting the Concepts with the Themes

Evolution

- The evolution of flowers enabled plants to increase cross-pollination and genetic diversity.
- The evolution of fruits enabled plants to better protect and disperse their seeds.
- Flowering plants and the animals that pollinate them have coevolved, each adapting to the changes that have occurred in the other.

Nature of Science

- People use plants for a variety of purposes, such as food, building materials, and even medicines.
- Research into plant reproduction has facilitated breeding. Genetic engineering has also allowed the production of plants more suited to human needs.

Biological Systems

- Double fertilization, unique to flowering plants, produces a 2n zygote, which will become a new plant embryo, and a 3n endosperm that supports the embryo.
- The flowering plant life cycle, which does not require water for gamete transfer, is adapted for a land-based existence.

MEDIA STUDY TOOLS http://connect.mheducation.com

SMARTBOOK® Study smart. SmartBook helps you maximize your study time by identifying what you know and don't know. The recharge feature in SmartBook helps you prepare for the test by identifying the material that you are most likely to forget.

connect BIOLOGY Find even more resources to learn the chapter concepts including animations, tutorial videos, and interactive practice questions.

Tutorials	Videos
27.1 Alternation of Generations • Angiosperm Life Cycle	**27.1** Pollinators **27.3** Fruit Bat Seed Dispersal

SUMMARIZE

27.1 Sexual Reproductive Strategies

Flowering plants exhibit an alternation-of-generations life cycle. **Flowers** borne by the sporophyte produce **microspores** and **megaspores** by meiosis. Microspores develop into a male **gametophyte,** and megaspores develop into a female gametophyte. Following fertilization, the sporophyte is enclosed within a **seed** covered by the fruit.

The flowering plant life cycle is adapted to a land existence. The microscopic gametophytes are protected from desiccation by the sporophyte; the pollen grain has a protective wall; and fertilization does not require external water. The seed has a protective seed coat, and seed germination does not occur until conditions are favorable.

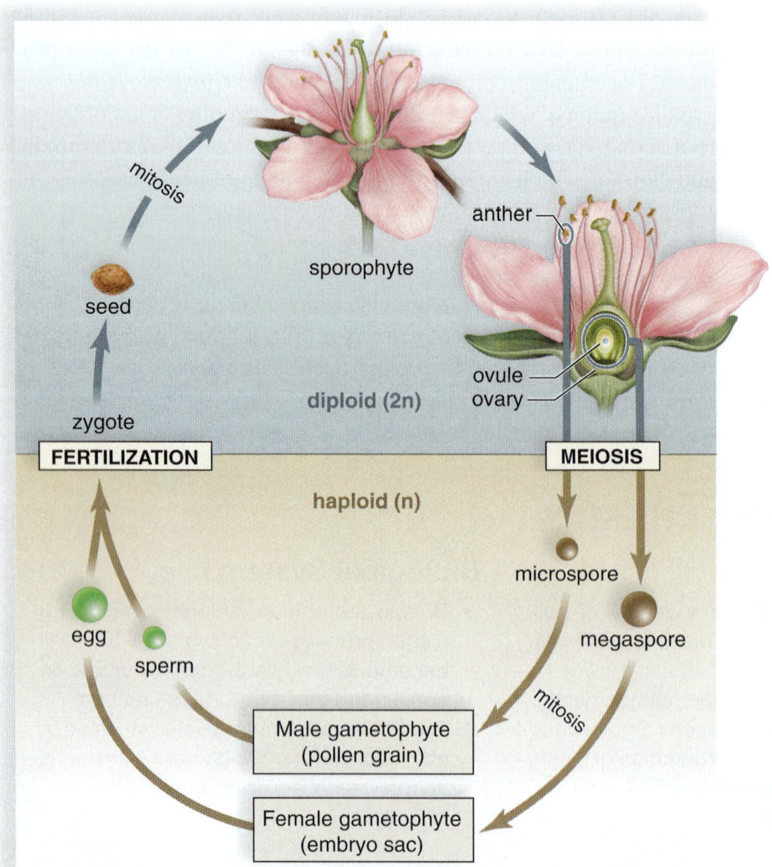

A typical flower has several parts: **Sepals,** which are usually green, form an outer whorl called a **calyx; petals,** often colored, are the next whorl and are called a **corona;** and **stamens,** each having a filament and an anther, form a whorl around the base of at least one **carpel.** The carpel, in the center of a flower, consists of a **stigma,** a **style,** and an **ovary.** The ovary contains **ovules.**

The anthers contain microspore mother cells, which divide meiotically to produce four haploid microspores. These divide into a two-celled **pollen grain,** which is the male gametophyte. One of these cells will divide later to become two sperm cells. After pollination, the pollen grain germinates, and as the pollen tube grows, the sperm cells travel to the embryo sac. **Pollination** is simply the transfer of pollen from anther to stigma.

Each ovule contains a megaspore mother cell, which divides meiotically to produce an egg cell. The egg cell is in the **embryo sac,** which is the female gametophyte.

Flowering plants undergo **double fertilization.** One sperm nucleus unites with the egg nucleus, forming a 2n zygote, and the other unites with the polar nuclei of the central cell, forming a 3n endosperm cell.

After fertilization, the **endosperm** cell divides to form multicellular endosperm. The zygote becomes the sporophyte embryo. The ovule matures into the seed (its integuments become the seed coat). The ovary becomes the fruit.

Coevolution occurs when specific traits or behaviors evolve in one species in response to changes that have occurred in another species. The coevolution of flowering plants with animal pollinators may help explain the success of flowering plants on land. Flowering plants would have gained an increased chance of seed production. The pollinators would have gained more feeding opportunities. In the end, both species became dependent on each other for success.

27.2 Seed Development

After fertilization, the cells in the ovule undergo **differentiation** and **development** into a seed. As the ovule is becoming a seed, the zygote is becoming an embryo. After the first several divisions, it is possible to discern the embryo and the suspensor. The suspensor attaches the embryo to the ovule and supplies it with nutrients.

The eudicot embryo becomes first heart-shaped and then torpedo-shaped. Once you can see the two **cotyledons,** it is possible to distinguish the shoot tip and the root tip, which contain the apical meristems. In eudicot seeds, the cotyledons frequently absorb the endosperm.

27.3 Fruit Types and Seed Dispersal

The seeds of flowering plants are enclosed by **fruits.** Simple fruits are derived from a single ovary (which can be simple or compound). Some simple fruits are fleshy, such as a peach or an apple. Others are dry, such as peas, nuts, and grains. Compound fruits consist of aggregate fruits, which develop from a number of ovaries of a single flower, and multiple fruits, which develop from a number of ovaries of separate flowers.

Flowering plants have several ways to disperse seeds. Seeds may be blown by the wind, be attached to animals that carry them away, or be eaten by animals that defecate them some distance away.

Prior to **germination,** you can distinguish the two cotyledons and plumule of a bean seed (eudicot). The plumule is the shoot that bears leaves. Also present are the epicotyl, the hypocotyl, and the radicle. In a corn kernel (monocot), the endosperm, the cotyledon, the plumule, and the radicle are visible.

27.4 Asexual Reproductive Strategies

Many flowering plants undergo **asexual reproduction,** producing **clones,** as when the nodes of stems (either aboveground or underground) give rise to entire plants or when roots produce new shoots.

The production of clonal plants utilizing **tissue culture** for many common fruits and vegetables is common. Plant cells are **totipotent,** so adult plants can be made from a few parent cells. Leaf, stem, and root culture can result in **cell suspensions** that allow plant chemicals to be produced in large tanks.

ASSESS

Choose the best answer for each question.

27.1 Sexual Reproductive Strategies

1. In plants,
   a. a gamete becomes a gametophyte.
   b. a spore becomes a sporophyte.
   c. both sporophyte and gametophyte produce spores.
   d. only a sporophyte produces spores.
   e. Both a and b are correct.
2. The flower part that contains ovules is the
   a. carpel. b. stamen. c. sepal. d. petal.
3. The megaspore and the microspore mother cells
   a. both produce pollen grains.
   b. both divide meiotically.
   c. both divide mitotically.
   d. produce pollen grains and embryo sacs, respectively.
   e. All of these are correct.
4. A pollen grain is
   a. a 3n structure.
   b. a diploid structure.
   c. first a diploid and then a haploid structure.
   d. first a haploid and then a diploid structure.
   e. the male gametophyte.
5. Bat-pollinated flowers
   a. are colorful.
   b. are open throughout the day.
   c. are strongly scented.
   d. have little scent.
   e. Both b and c are correct.

27.2 Seed Development

6. Which of these pairs is incorrectly matched?
   a. polar nuclei—plumule
   b. egg and sperm—zygote
   c. ovule—seed
   d. ovary—fruit
   e. stigma—carpel
7. *Heart, torpedo,* and *globular* refer to
   a. embryo development.
   b. sperm development.
   c. female gametophyte development.
   d. seed development.
   e. Both b and d are correct.
8. A seed contains
   a. a seed coat.
   b. an embryo.
   c. stored food.
   d. cotyledon(s).
   e. All of these are correct.

27.3 Fruit Types and Seed Dispersal

9. Fruits
   a. nourish embryo development.
   b. help with seed dispersal.
   c. signal gametophyte maturity.
   d. attract pollinators.
   e. stay where they are produced.
10. Which of the following is a simple, dry, and dehiscent fruit?
   a. apple
   b. tomato
   c. walnut
   d. pineapple
   e. legume pod
11. Fruit dispersal can be accomplished by
   a. wind.
   b. forced ejection.
   c. mammals.
   d. birds.
   e. All of these are correct.

27.4 Asexual Reproductive Strategies

12. Asexual reproduction produces
   a. seeds.
   b. embryos.
   c. pollen.
   d. clones.
   e. ovules.
13. Plant tissue culture takes advantage of
   a. a difference in flower structure.
   b. sexual reproduction.
   c. gravitropism.
   d. phototropism.
   e. totipotency.
14. Cell suspension culture is used to
   a. extract chemicals.
   b. clone vegetative organs.
   c. produce pollen.
   d. increase genetic diversity.

ENGAGE

THINKING SCIENTIFICALLY

1. You notice that a type of wasp has been visiting a flower type in your garden. What web/library research would allow you to hypothesize that this wasp is a pollinator for this flower type?
2. The pollinator for a very rare plant has become extinct. **a.** What laboratory technique would you use to prevent the plant from also becoming extinct? **b.** How might you improve the hardiness of the plant?
3. In most parts of the world, commerical potato crops are produced asexually by planting tubers. However, in some regions, such as Southeast Asia and the Andes, some potatoes are grown from true seeds. Discuss the advantages and disadvantages of growing potatoes from true seeds.

UNIT 6

Animal Evolution and Diversity

In his book *On the Origin of Species,* Charles Darwin notes that while the planet Earth has cycled year after year around the sun, "endless forms most beautiful and most wonderful" have appeared and will keep on appearing. This statement can certainly be applied to the evolution of animals, whose variety seems without bounds. The fossil record even reveals a myriad of animals that are extinct today, of which the dinosaurs are only the most well known. The search for food, shelter, and mates under a variety of environmental conditions can explain why the diversity of animals is so great.

Despite their diversity, animals have an unbroken thread of unity, provided by evolution from a common ancestor. At the biosphere level, animals are heterotrophic consumers that require a constant supply of food, ultimately supplied by the autotrophs at the base of the food chain. At the organismal level, their eukaryotic cells usually form tissues and organs with specialized functions. At the molecular level, animals share a common chemistry, including a genetic code that we now know reveals how the many groups of animals are related. In this unit of the text, we concentrate on the characteristics of animals, their origin, and their evolution as revealed by molecular genetics.

Unit Outline

Chapter 28 Invertebrate Evolution 513

Chapter 29 Vertebrate Evolution 544

Chapter 30 Human Evolution 564

Unit Learning Outcomes

The learning outcomes for this unit focus on three major themes in the life sciences.

Evolution	Describe the position of humans in relation to the other animals in the tree of life.
Nature of Science	Explain why, from an evolutionary perspective, animal model organisms can be used in human biomedical research.
Biological Systems	Identify how the environment has shaped the evolution of the animal body plan.

Bees such as these are involved in the reproduction of plants around the world.

28

Invertebrate Evolution

What is happening to the bees? Honeybees first evolved around 130 million years ago, during the Cretaceous period. Flowering plants present at that time developed nectar that contained sugars to attract bees, encouraging them to feed on the flowers, spreading pollen. Humans have been keeping bees for thousands of years, to produce honey and wax. However, over the past several years, bee populations around the world have been decreasing, with as many as one-third of bees dying each winter. More than 10 million beehives have been wiped out since 2006. Although the cause of this "colony collapse disorder (CCD)" remains unclear, scientists have uncovered evidence that exposure to the pesticides and fungicides used in agriculture reduces bees' ability to resist a common single-celled parasite, *Nosema ceranae.* Other researchers speculate that the causes of CCD include mites, a newly identified virus, altered bee genetics, and diesel fumes masking the normal odors of flowers that attract bees.

Although beekeepers are generally able to replace their bee colonies, it has been noted that bee pollination is involved in about one in three bites of food we eat. If bees were to disappear from the planet, it is not likely that humans would run out of food, but many of our favorite fruits and vegetables would become rare. This story of dying bees illustrates how important invertebrates can be to the health of humans as well as the planet.

As you read through this chapter, think about the following questions:

1. Where do invertebrates fit into the evolutionary history of eukaryotes? Of animals?
2. What evolutionary advantage might there be to invertebrates having multiple developmental stages, each with a different body form, habitat, and lifestyle?

Chapter Outline

28.1 Evolution of Animals 514
28.2 The Simplest Invertebrates 521
28.3 Diversity Among the Lophotrochozoans 524
28.4 Diversity of the Ecdysozoans 532
28.5 Invertebrate Deuterostomes 539

Before You Begin

Before beginning this chapter, take a few moments to review the following discussions.

Figure 18.8 When did the lineage of animals first appear in the tree of life?

Figure 21.1 Where do animals fit into the evolutionary history of eukaryote supergroups?

Section 21.1 Which single-celled eukaryotes are thought to be the common ancestor of all animals?

Following *the* Themes

CHAPTER 28 INVERTEBRATE EVOLUTION

UNIT 6 ANIMAL EVOLUTION AND DIVERSITY

Evolution	According to the colonial flagellate hypothesis, animals evolved from an invertebrate, single-celled, colonial, choanoflagellate-like protist.
Nature of Science	The phylogeny of invertebrates is a hypothesis of evolutionary history, which like all scientific hypotheses is subject to revision with the addition of new data.
Biological Systems	Invertebrates are widely successful animals that have adapted to a wide range of habitats and lifestyles, including parasitism, which in some cases has a direct impact on human health.

28.1 Evolution of Animals

Learning Outcomes

Upon completion of this section, you should be able to

1. List three common characteristics of animals that are not found in other multicellular eukaryotes.
2. Summarize the colonial flagellate hypothesis as it relates to the origin of animals.
3. Distinguish among the different body plans of animals.
4. Compare protostome and deuterostome development.

The traditional five-kingdom classification placed animals in the kingdom Animalia. The modern three-domain system places animals in the domain Eukarya. Within the Eukarya, they are placed in the supergroup Opisthokonta (see Chapter 21), along with fungi and certain protozoans (notably, the choanoflagellates). In this section, we consider what characteristics distinguish animals from other eukaryotes and look at how these traits evolved.

Animation Three Domains

Characteristics of Animals

Animals, fungi, and plants are all multicellular eukaryotes, but unlike plants, which make their food through photosynthesis, fungi and animals are heterotrophs and must acquire nutrients from an external source. Animals differ from fungi, which are saprotrophs, because animals ingest (eat) food and digest it internally, while fungi dissolve food externally and absorb nutrients. Animal cells also lack a cell wall, which is common to both the plants and fungi, although the cell walls in fungi and plants are made of different organic molecules. In general, animals are mobile, have nerves and muscles, and reproduce sexually. However, the tremendous diversity of animals makes assigning specific characteristics to all animals difficult.

Animals have a variety of life cycles. Many reproduce sexually, some reproduce asexually, and some combine both life cycles. While many animals have a diploid life cycle, other life cycles, such as haploid or haplodiploid, are also represented. Animals undergo a series of developmental stages to produce an organism that has specialized tissues that carry on specific functions.

Video Tadpole Development

Muscle and nerve tissues, both characteristic of animals, allow motility and a variety of flexible movements. In turn, motility enables animals to search actively for food and to prey on other organisms. Coordinated movements also allow animals to seek mates, shelter, and a suitable climate—behaviors that have allowed animals to live in all habitats and to become vastly diverse.

Animals are descended from a single common ancestor, thus forming a single lineage on the tree of life. Within the animal lineage are two main branches, the vertebrates and the invertebrates. **Vertebrates** are animals that at some stage of their lives have a spinal cord (backbone) running down the center of the back, whereas **invertebrates,** the topic of this chapter, do not have a backbone. Figure 28.1 illustrates some common features of animals, including invertebrates.

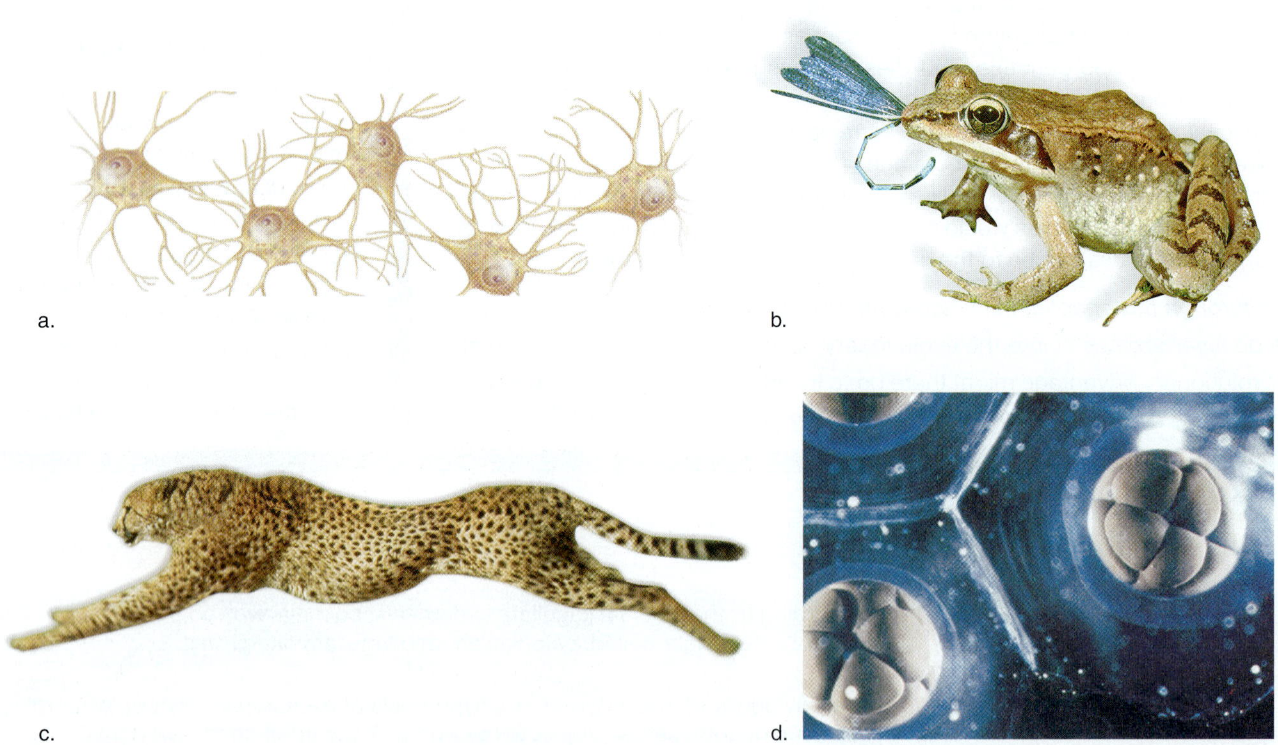

Figure 28.1 Common features of animals. **a.** Animals are multicellular, and all animal cells lack a cell wall. **b.** Animals are heterotrophs, meaning they obtain nutrition from external sources. **c.** Animals are typically motile, due to their well-developed nervous and muscular systems. **d.** Most animals reproduce sexually, beginning life as a 2n zygote, which undergoes development to produce a multicellular organism that has specialized tissues.

Figure 28.2 The colonial flagellate hypothesis. The numbered statements explain the hypothesis by which motile flagellates may undergo evolution into a multicellular organism with specialized cells and more than a single cell layer.

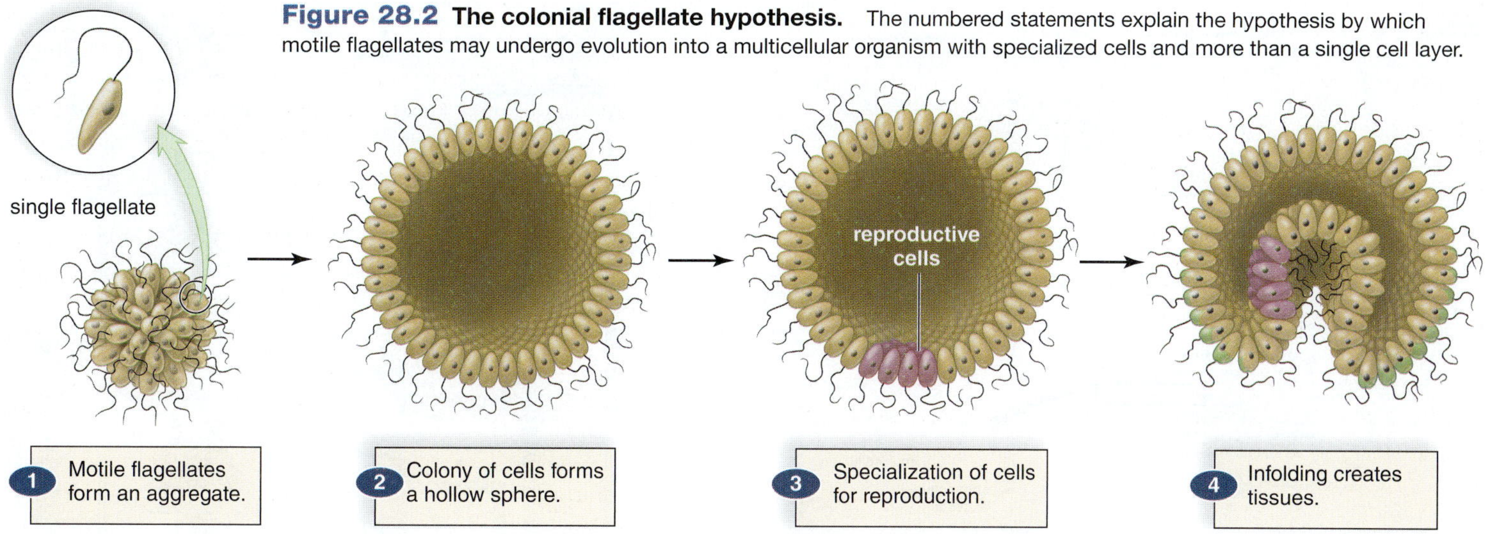

Ancestry of Animals

In Chapter 23, we discussed evidence that plants most likely share a protist (green algae) ancestor with the charophytes. What about animals? The available evidence suggests they also evolved from a protist, most likely a motile protozoan. The **colonial flagellate hypothesis** states that animals are descended from an ancestor that resembled a hollow, spherical colony of flagellated cells.

Among the protists, the **choanoflagellates** most likely resemble the last single-celled ancestor of animals, and molecular data tell us that they are the closest living relatives of animals. A choanoflagellate is a single cell, 3–10 μm in diameter, with a flagellum surrounded by a collar of 30–40 microvilli. Movement of the flagellum creates water currents, which pull the protist along. As water moves through the microvilli, they engulf bacteria and debris from the water. Choanoflagellates can also exist as a colony of cells (see Fig. 21.21). Several can be found together at the end of a stalk or simply clumped together, like a bunch of grapes.

Figure 28.2 shows how the process of transition from colonial flagellates to multicellular animals might have begun with an aggregate of a few flagellated cells. From there, a larger number of cells could have formed a hollow sphere. Individual cells within the colony would have become specialized for particular functions, such as reproduction. Two tissue layers could have arisen by an infolding of certain cells into a hollow sphere. Tissue layers arise in this manner during the development of animals today. The colonial flagellate hypothesis is also attractive because of its implications regarding animal symmetry, which is discussed shortly.

Evolution of Body Plans

All of the various animal body plans we see today were present by the Cambrian period, roughly 500 million years ago (see Table 18.1). How could such diversity have arisen within a relatively short period of geologic time? As an animal develops, there are many possible outcomes regarding the number, position, size, and patterns of its body parts. Different combinations could have led to the great variety of animal forms in the past and present. Scientists now think that slight shifts in the DNA code and the expression of genes called *Hox* (homeotic) genes are responsible for the major differences between animals that arise during development, as described in the Evolution feature, "Evolution of the Animal Body Plan," on pages 518–519.

The Phylogenetic Tree of Animals

The fossil record of the early evolution of animals is very sparse. Therefore, the best hypothesis of the evolution of animals, represented by the phylogenetic tree of animals shown in Figure 28.3, is based on a combination of morphological characters of living and fossil organisms, developmental homologies, and molecular (DNA) characters. The more closely related two organisms are, the more similar their DNA sequences. The addition of molecular data has resulted in an updated phylogenetic tree that is quite different from hypotheses that were historically based solely on morphological characteristics.

Refer to the phylogeny in Figure 28.3 as we discuss the anatomical characteristics that are shared among animals of each branch of the animal family tree.

Type of Symmetry

Symmetry (Gk. *syn,* "together"; *metron,* "measure") is a pattern of similarity that is observed in objects; three types of symmetry exist in the animal kingdom. **Asymmetry,** or lack of symmetry, is seen in sponges that have no particular pattern to body shape (see Fig. 28.6).

The cnidarians and comb jellies exhibit **radial symmetry**—their bodies are organized circularly, similar to a wheel, such that any longitudinal cut through the central point produces two identical halves (Fig. 28.4*a*). Many adult and immature, or larval, forms of animals are radially symmetrical. Some radially symmetrical animals are mobile and others are immobile, or sessile, when attached to a substrate. Radial symmetry allows an organism to extend out in all directions from one center. Floating animals with radial symmetry, such as jellyfish, also have this feature.

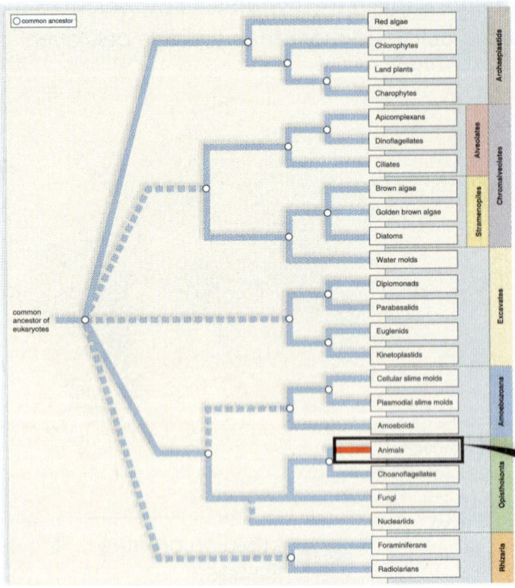

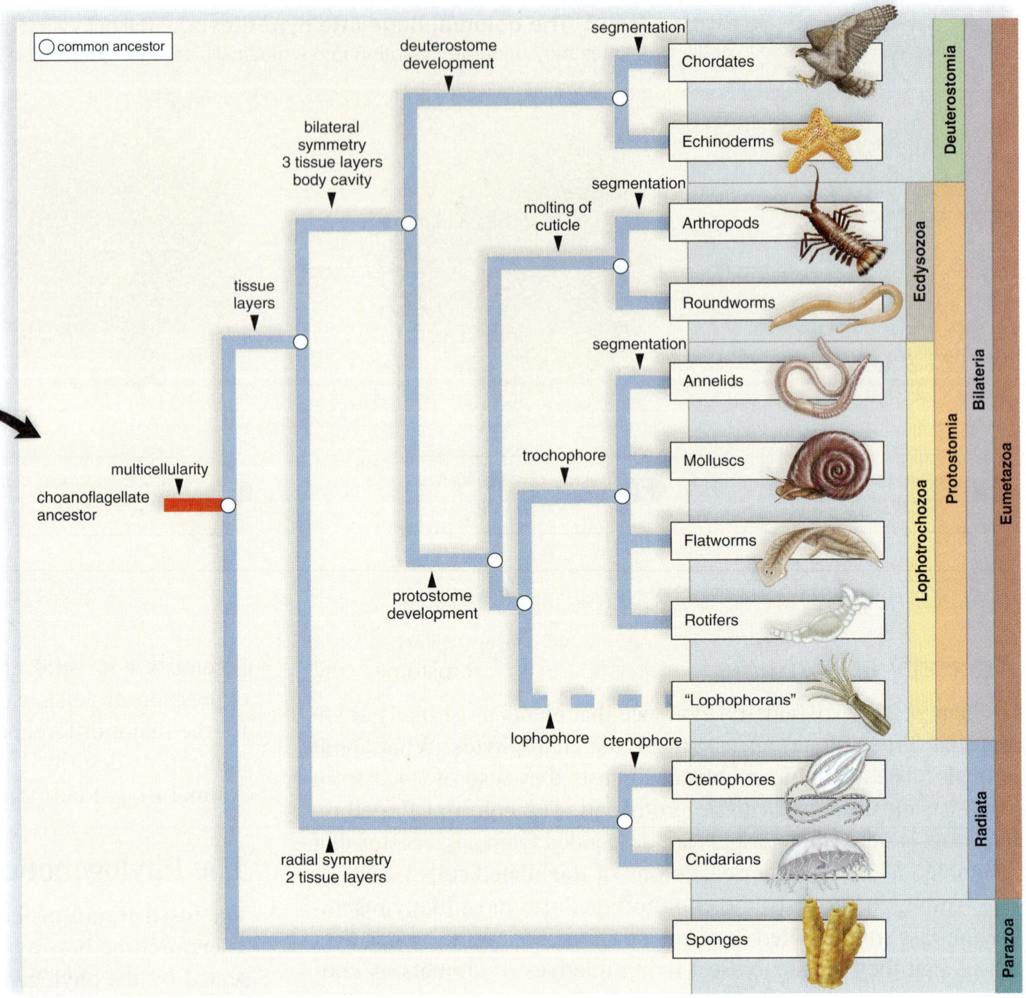

Figure 28.3 Phylogenetic tree of animals. All animal phyla living today are most likely descended from a colonial flagellated protist living about 600 MYA. This phylogenetic (evolutionary) tree uses morphological and molecular data (rRNA sequencing) to determine which phyla are most closely related to one another. The "lophophorans" are members of the Lophotrochozoa that filter-feed with a fanlike lophophore. A dotted line indicates that the evolutionary relationships of the "lophophoran" lineage are uncertain.

The rest of the animals exhibit **bilateral symmetry** as adults—they have definite left and right halves, and only a single longitudinal cut down the centerline of the animal produces two equal halves (Fig. 28.4*b*). Bilaterally symmetrical animals have defined anterior and posterior ends, and forward movement is guided with the anterior end. The colonial flagellate hypothesis, mentioned earlier, is attractive because it implies that radial symmetry preceded bilateral symmetry in animal history, and a phylogeny supports this hypothesis.

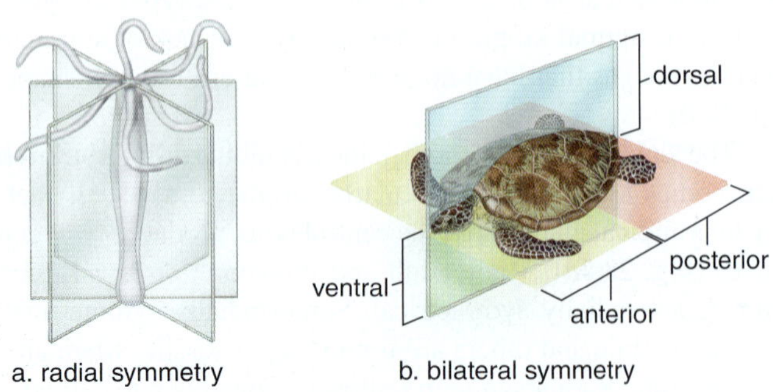

Figure 28.4 Symmetry in animals The body plans of animals can be **(a)** radially symmetrical or **(b)** bilaterally summetrical.

During the evolution of animals, bilateral symmetry was accompanied by **cephalization,** localization of a brain and specialized sensory organs at the anterior end of an animal. This development was critical to an animal's ability to engage in directed movement—toward food or mates and away from danger.

Embryonic Development

The simplest animals are the sponges. Like all animals, sponges are multicellular. However, they do not have truly specialized tissues. True tissues develop in the more complex animals as they undergo embryological development. The first tissue layers that appear are called **germ layers,** and they give rise to the organs and organ systems of complex animals.

Animals such as the cnidarians, which as embyros have only two tissue layers—the ectoderm and endoderm—are termed *diploblastic.* Diploblastic animals develop tissues, but no specialized organs. The animals that develop specialized organs are termed *triploblastic,* because as embryos they have three tissue layers—the ectoderm, mesoderm, and endoderm. All animals except the sponges (Parazoa) and cnidarians and comb jellies (Radiata) are triploblastic.

Notice in the phylogenetic tree (see Fig. 28.3) the triploblastic animals are either **protostomes** (Gk. *proto,* "first"; *stoma,* "mouth") or **deuterostomes** (Gk. *deuter,* "second"). These terms have to do with whether the mouth or the anus develops first in the embryo. In protostomes, the mouth develops prior to the anus, whereas in deuterostomes, the anus develops prior to the mouth. Most invetebrates are protostomes, but all vertebrates, including humans, are deuterostomes. Figure 28.5 shows that protostome and deuterostome development are differentiated by three major events: cleavage, blastula formation, and coelom development.

Cleavage. The first developmental event after fertilization is **cleavage,** cell division without cell growth. In protostomes, spiral cleavage occurs, and daughter cells sit in grooves formed by the previous cleavages. The fate of these cells is fixed and determinate in protostomes; each can contribute to development in only one particular way.

In deuterostomes, radial cleavage occurs, and the daughter cells sit right on top of the previous cells. The fate of these cells is indeterminate—that is, if they are separated from one another, each cell can go on to become a complete organism.

Blastula Formation. As development proceeds, a hollow sphere of cells, or **blastula,** forms, and the indentation that follows produces an opening called the blastopore. In protostomes, the mouth appears at or near the blastopore.

In deuterostomes, the anus appears at or near the blastopore, and only later does a second opening form the mouth.

Video Blastocyst Formation

Coelom Development. Certain protostomes and all deuterostomes have a body cavity lined by mesoderm called a **coelom** (Gk. *koiloma,* "cavity"). More specifically, the coelom in these groups is a **true coelom,** because the mesoderm cells line the cavity completely. However, the coelom develops differently in the two groups. In protostomes, the mesoderm arises from cells located near the embryonic blastopore, and a splitting occurs that produces the coelom.

In deuterostomes, the coelom arises as a pair of mesodermal pouches from the wall of the primitive gut. The pouches enlarge until they meet and fuse. To learn more about animal development, see Chapter 42.

In the sections that follow, we will compare the major groups of invertebrate animals. All of these, except the echinoderms, are protostomes. We begin with the simplest multicellular animals, the sponges.

Check Your Progress 28.1

1. State three characteristics that all animals have in common.
2. Explain the colonial flagellate hypothesis about the origin of animals.
3. Compare radial and bilateral symmetry, and provide examples of animals that exemplify each.
4. List two differences between deuterostomes and protostomes.

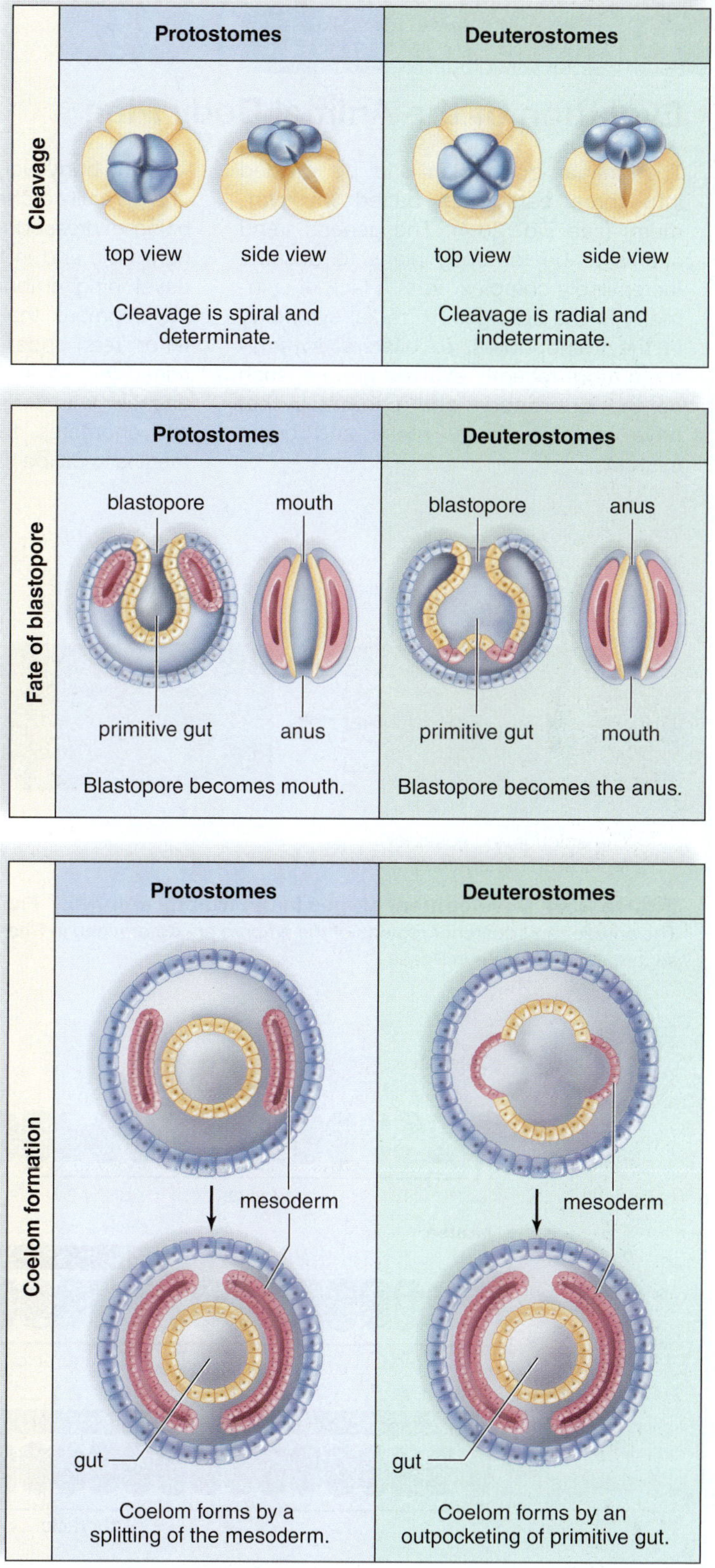

Figure 28.5 Protostomes compared to deuterostomes. *Left:* In the embryo of protostomes, cleavage is spiral—new cells are at an angle to old cells—and each cell has limited potential and cannot develop into a complete embryo; the blastopore is associated with the mouth; and the coelom, if present, develops by a splitting of the mesoderm. *Right:* In deuterostomes, cleavage is radial—new cells sit on top of old cells—and each one can develop into a complete embryo; the blastopore is associated with the anus; and the coelom, if present, develops by an outpocketing of the primitive gut.

Tutorial Protostomes and Deuterostomes

THEME Evolution

Evolution of the Animal Body Plan

The animal body plan can be divided into three categories based on symmetry (see Fig. 28.4). The general trend seems to be for body plans to become increasingly complex, from a lack of symmetry in the sponges, to radial symmetry in the ctenophores, to bilateral symmetry in more recently evolved groups, such as the arthropods and chordates that have multiple tissue types and organ systems.

The body plan of an animal is the result of a carefully orchestrated pattern of genes being expressed (or not expressed) at the right time and in the correct region of the developing embryo. In the first stage of development, the anterior (front) and posterior (rear) ends of the embryo are determined (Fig. 28A). In bilaterally symmetrical animals with segments, such as insects and chordates, the next step in development is to divide the embryo into segments, each of which will become a different part of the body. In fruit flies, genes such as the *gap* and the *pair-rule* genes determine the number of segments. In vertebrates, *FGF8* is one of many genes that determine segmentation pattern.

Once the segmentation pattern is established, *homeotic,* or *HOX,* genes determine the ultimate developmental fate of each segment. *HOX* genes encode homeotic proteins that bind to the regulatory region

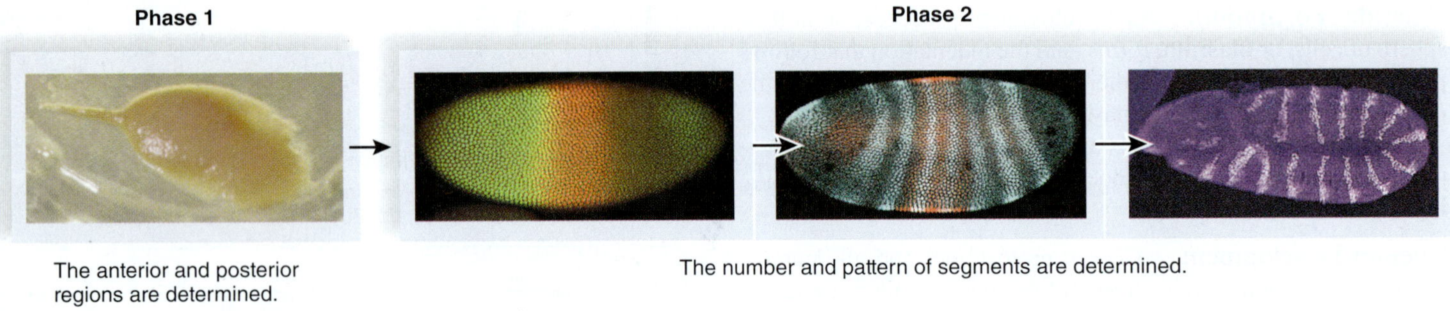

Figure 28A Development stages in the fruit fly embryo. Fruit fly embryos are sectioned and stained during different stages of development. The anterior and posterior regions of the embryo are determined in Phase 1 by genes such as *bicoid.* Genes such as *gap* and *pair-rule* determine the segmentation pattern in Phase 2.

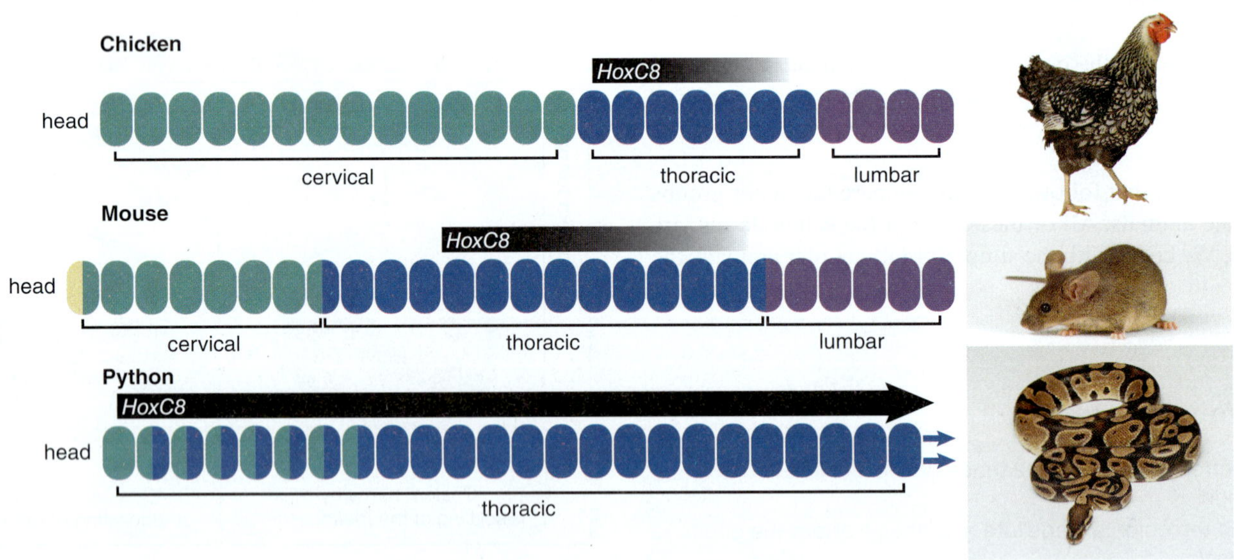

Figure 28B *HOX* genes orchestrate the development of the body plan of animals. In animal development, the number of segments, or vertebrae in vertebrates, is controlled by *HOX* genes. One *HOX* gene, *HOXC8,* determines the number of thoracic vertebrae. Variation in the number of thoracic vertebrae in the spine of a chicken, mouse, and snake is caused by a shift in when, and for how long, *HOXC8* is active during embryonic development. In the chicken, 7 thoracic vertebrae are formed; in the mouse, 12. A large increase in the number of thoracic vertebrae results in the long, thin body plan of a snake.

of genes that determine the body plan during development. Homeotic proteins act as "switches" that control when, where, and for how long a particular developmental gene is active.

Each *HOX* gene orchestrates the developmental fate of a particular region of the body. In mice, *HOXC8* sets the fate of 12 segments to become thoracic vertebrae, while in snakes, *HOXC8* orchestrates the development of hundreds of thoracic vertebrae (Fig. 28B).

HOX genes have played a role in the development of the body plan since the early stages of animal evolution. We know this because *HOX* genes are found in all animals with multiple types of tissue, and there is a shared similarity of *HOX* genes across animal groups (Fig. 28C). For example, the *HOX* genes that determine the fate of the head region have the same evolutionary origin in flies, worms, and mice. Even cnidarians, with a simple body plan, have some *HOX* genes in common with more complex animals. This implies that all *HOX* genes evolved in animals from a common *HOX* gene ancestor. *HOX* genes are found in linear clusters on the same chromosome. However, not all *HOX* genes are found in all animals. Some groups of animals have more than one cluster of genes with duplicate copies of genes within each cluster.

Questions to Consider

1. Why are *HOX* genes evidence for a common ancestor of all life?
2. What would you expect to happen if a particular *HOX* gene, such as *HOXC8*, were not functioning properly?
3. What experiments could you plan to test what a particular *HOX* gene controlled at a particular time of development?

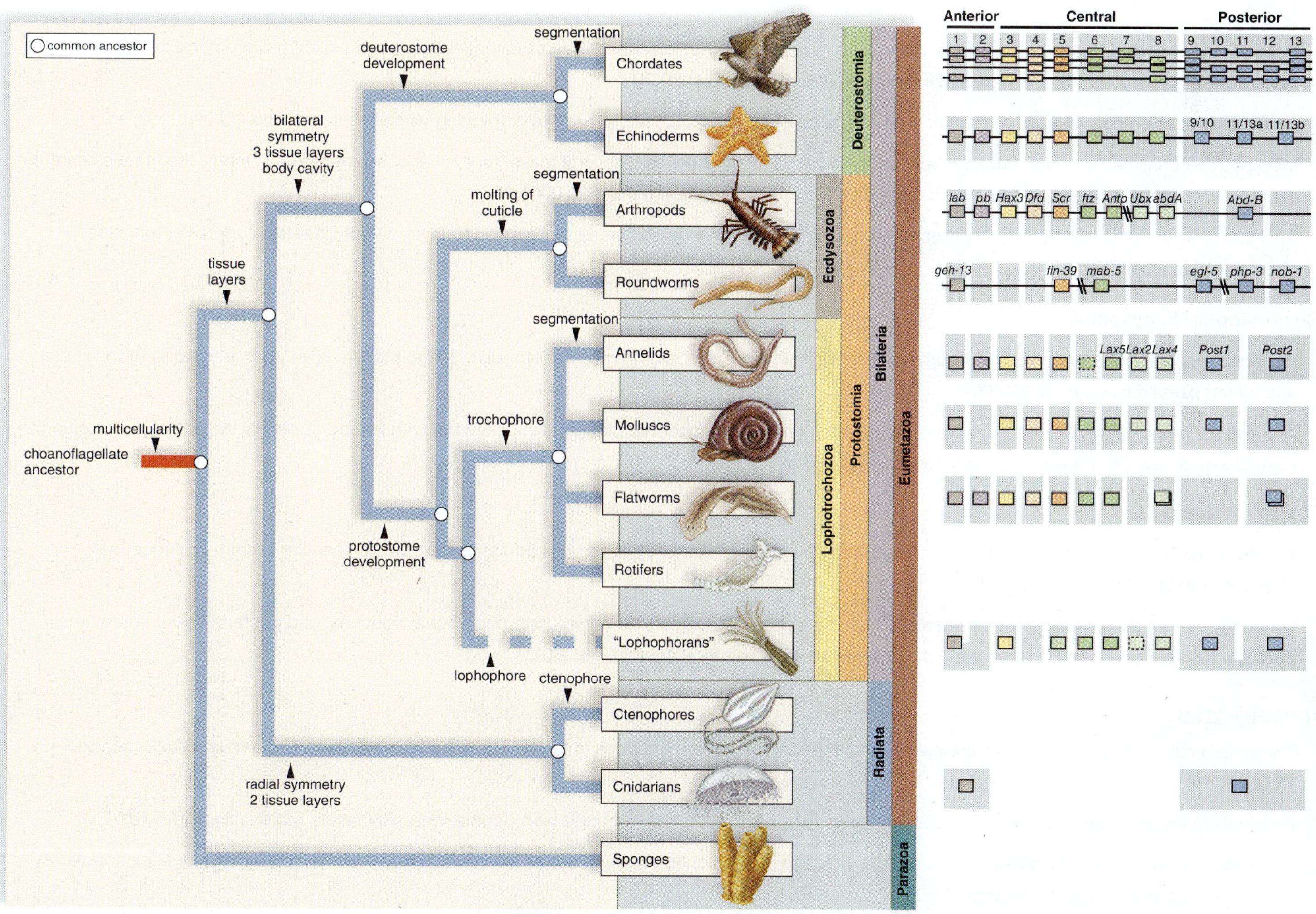

Figure 28C Shared evolutionary history of *HOX* genes in animals. *HOX* genes are found in all groups of animals with multiple tissue layers. A general trend is observed for an increase in the number of *HOX* genes and *HOX* gene clusters and an increase in body plan complexity. Although the number of *HOX* genes varies among animals, there is strong evidence for common ancestry of each *HOX* gene. The color coding indicates that a particular *HOX* gene is shared between lineages of animals.

Table 28.1 The Animal Kingdom

DOMAIN: Eukarya
KINGDOM: Animalia

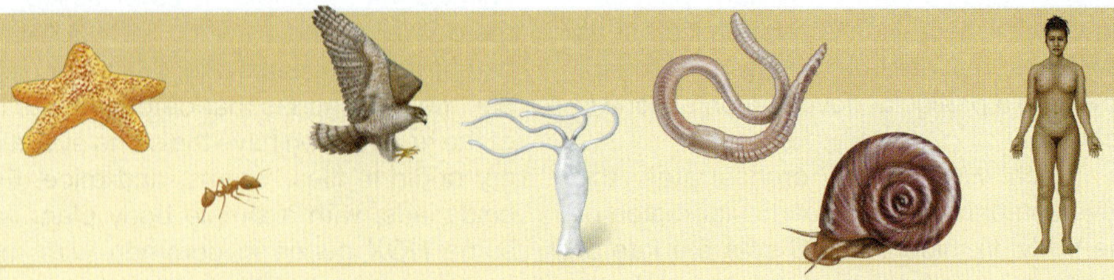

CHARACTERISTICS

Multicellular, usually with specialized tissues; ingest or absorb food; diploid life cycle

INVERTEBRATES

Sponges (bony, glass, spongin): Asymmetrical, saclike body perforated by pores; internal cavity lined by choanocytes; spicules serve as internal skeleton. 5,150*

Radiata

Cnidarians (hydra, jellyfish, corals, sea anemones): Radially symmetrical with two tissue layers; sac body plan; tentacles with nematocysts. 10,000*

Comb jellies: Have the appearance of jellyfish; the "combs" are eight visible longitudinal rows of cilia that can assist locomotion; lack the nematocysts of cnidarians but some have two tentacles. 150*

Protostomia (Lophotrochozoa)

Lophophorates (lampshells, bryozoa): Filter feeders with a circular or horseshoe-shaped ridge around the mouth that bears feeding tentacles. 5,935*

Flatworms (planarians, tapeworms, flukes): Bilateral symmetry with cephalization; three tissue layers and organ systems; acoelomate with incomplete digestive tract that can be lost in parasites; hermaphroditic. 20,000*

Rotifers (wheel animals): Microscopic animals with a corona (crown of cilia) that looks like a spinning wheel when in motion. 2,000*

Molluscs (chitons, clams, snails, squids): Coelom; all have a foot, mantle, and visceral mass; foot is variously modified; in many, the mantle secretes a calcium carbonate shell as an exoskeleton; true coelom and all organ systems. 110,000*

Annelids (polychaetes, earthworms, leeches): Segmented with body rings and setae; cephalization in some polychaetes; hydroskeleton; closed circulatory system. 16,000*

Protostomia (Ecdysozoa)

Roundworms (*Ascaris,* pinworms, hookworms, filarial worms): Pseudocoelom and hydroskeleton; complete digestive tract; free-living forms in soil and water; parasites common. 25,000*

Arthropods (crustaceans, spiders, scorpions, centipedes, millipedes, insects): Chitinous exoskeleton with jointed appendages; undergoes molting; insects—most have wings—are most numerous of all animals. 1,000,000*

Deuterostomia

Echinoderms (sea stars, sea urchins, sand dollars, sea cucumbers): Radial symmetry as adults; unique water-vascular system and tube feet; endoskeleton of calcium plates. 7,000*

Chordates (tunicates, lancelets, vertebrates): All have notochord, dorsal tubular nerve cord, pharyngeal pouches, and postanal tail at some time; contains mostly vertebrates in which notochord is replaced by vertebral column. 56,000*

VERTEBRATES

Fishes (jawless, cartilaginous, bony): Endoskeleton, jaws, and paired appendages in most; internal gills; single-loop circulation; usually scales. 31,000*

Amphibians (frogs, toads, salamanders): Jointed limbs; lungs; three-chambered heart with double-loop circulation; moist, thin skin. 5,670*

Reptiles (snakes, turtles, crocodiles): Amniotic egg; rib cage in addition to lungs; three- or four-chambered heart typical; scaly, dry skin; copulatory organ in males and internal fertilization. 7,000*

Birds (songbirds, waterfowl, parrots, ostriches): Endothermy, feathers, and skeletal modifications for flying; lungs with air sacs; four-chambered heart. 8,600*

Mammals (monotremes, marsupials, eutherians): Hair and mammary glands. 4,500*

*Number of species.

28.2 The Simplest Invertebrates

Learning Outcomes

Upon completion of this section, you should be able to

1. Explain why sponges are considered to be the simplest animals.
2. Discuss how a sponge respires, feeds, and reproduces.
3. Compare the anatomical features of comb jellies to those of cnidarians, such as hydras.

Table 28.1 summarizes the classification of organisms found in kingdom Animalia. Sponges, cnidarians, and comb jellies represent the most ancient and the simplest animals.

Sponges

All animals are multicellular; **sponges** (phylum Porifera [L. *porus,* "pore"; *ferre,* "to bear"]) are the only animals to lack true tissues and to have only a cellular level of organization. They have only a few cell types, and they lack the nerve and muscle cells seen in more complex animals. Molecular data place them at the base of the evolutionary tree of animals (see Fig. 28.3).

The saclike body of a sponge is perforated by many pores (Fig. 28.6). Sponges are aquatic, largely marine animals that vary greatly in size, shape, and color. However, they all have a canal system of pores of varying complexity that allows water to move through their bodies.

The interior of the canals is lined with flagellated cells that resemble choanoflagellates. In a sponge, these cells are called collar cells, or choanocytes. The beating of the flagella produces water currents, which flow through the pores into the central cavity and out through the osculum, the upper opening of the body. Even a simple sponge only 10 cm tall is estimated to filter as much as 100 liters of water each day. All of a sponge's cells are able to acquire the oxygen they need for cellular respiration by diffusion from this water.

A sponge is a sessile filter feeder, also called a suspension feeder, because it filters suspended particles from the water by means of a straining device—in this case, the pores of the walls and the microvilli making up the collar of collar cells. Microscopic food particles that pass between the microvilli are engulfed by the collar cells and digested by them in food vacuoles.

The skeleton of a sponge prevents the body from collapsing. All sponges contain tough fibers made of spongin, a modified form of collagen; a natural bath sponge is the dried spongin skeleton from which all living tissue has been removed. Today, however, commercial "sponges" are usually synthetic.

Typically, the endoskeleton of a sponge also contains small, needle-shaped structures called **spicules.** Traditionally, the type of spicule has been used to classify sponges; there are bony, glass, and spongin sponges. The success of sponges—they have existed longer than any other animal group—is due in part to their spicules. They have few predators, because a mouth full of spicules is an unpleasant experience. Some sponges also produce toxic substances that discourage predators.

Sponges can reproduce both asexually and sexually. They reproduce asexually by fragmentation or by budding. During budding, a small protuberance appears and gradually increases in size until a complete organism forms. Budding produces colonies of sponges that can become quite large. During sexual reproduction, eggs and sperm are released into the central cavity, and the zygote develops into a flagellated larva, which may swim to a new location.

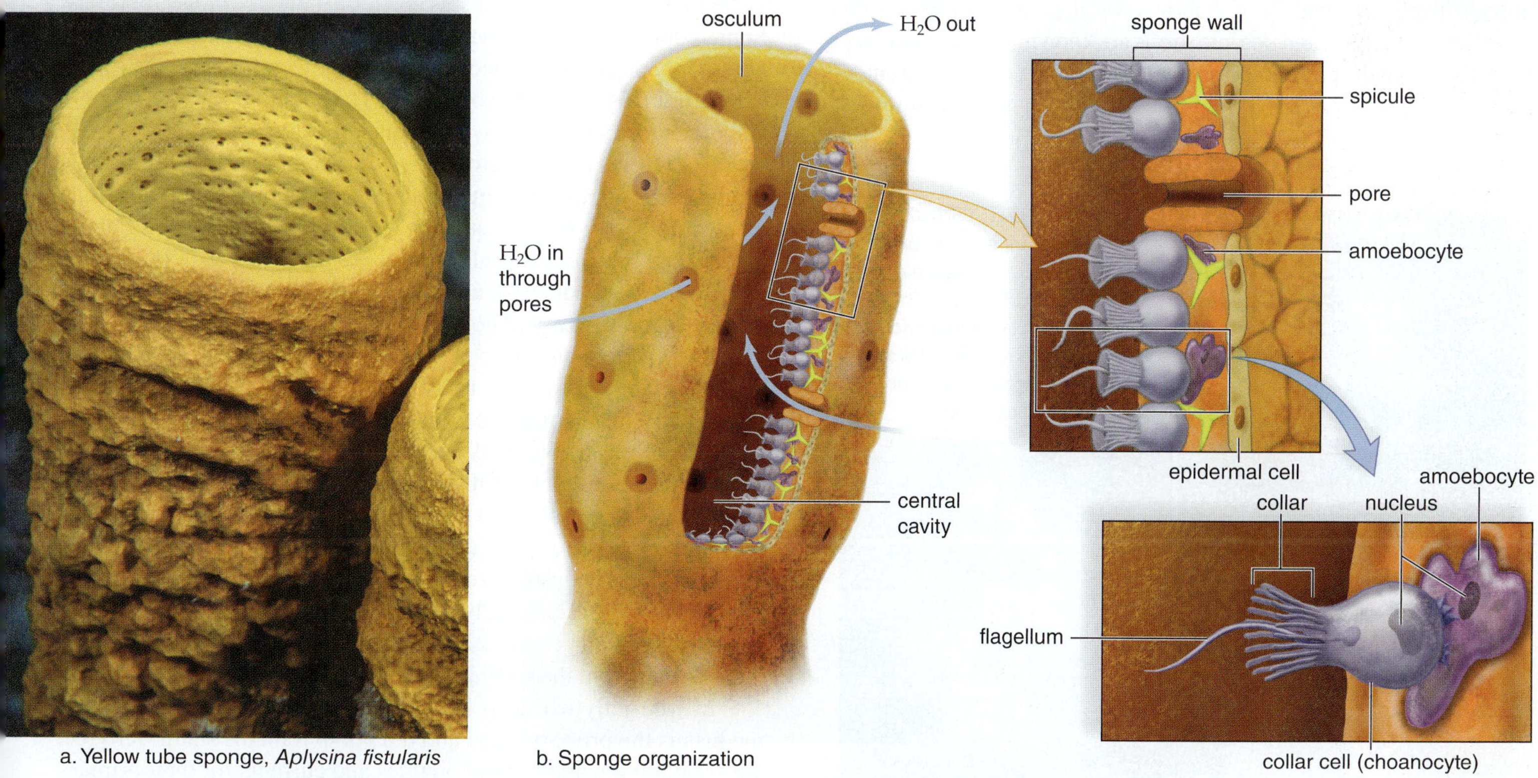

a. Yellow tube sponge, *Aplysina fistularis*

b. Sponge organization

Figure 28.6 Simple sponge anatomy.

Like many relatively simple organisms, sponges are capable of regeneration, or growth of a whole from a small part. Moreover, if the cells of a sponge are mechanically separated, they will reassemble into a complete and functioning organism.

Comb Jellies and Cnidarians

These two groups of animals have true tissues, and as embryos, they have two germ layers, ectoderm and endoderm. They are radially symmetrical as adults.

Comb Jellies

Comb jellies (phylum Ctenophora) (Fig. 28.7*a*) are solitary, mostly free-swimming marine invertebrates that are usually found in warm waters. Ctenophores represent the largest of these animals; they are propelled by beating cilia and range in size from a few centimeters to 1.5 m in length. Their body is made up of a transparent, jellylike substance called **mesoglea.** Most ctenophores do not have stinging cells and capture their prey by using sticky, adhesive cells called colloblasts. Some ctenophores are *bioluminescent,* meaning that they are capable of producing their own light.

Because sponges are the simplest animals, traditional phylogenetic trees (such as Fig. 28.3) indicate that they are also the oldest group of animals. Recent studies, however, suggest that comb jellies contain DNA sequences that are more ancient than those of sponges, indicating that comb jellies may have been the first animals to evolve. Because animals typically evolve to become more complex over time, further research is needed to solve this mystery.

Video Ctenophores

Cnidarians

Cnidarians (phylum Cnidaria) (Fig. 28.7*b*) are tubular or bell-shaped animals that reside mainly in shallow coastal waters; however, some freshwater, brackish, and oceanic forms are known. The term *cnidaria* is derived from the presence of specialized stinging cells called cnidocytes. Each cnidocyte has a fluid-filled capsule called a **nematocyst** (Gk. *nema,* "thread"; *kystis,* "bladder") that contains a long, spirally coiled, hollow thread. When the trigger of the cnidocyte is touched, the nematocyst is discharged. Some threads merely trap a prey; others have spines that penetrate and inject paralyzing toxins.

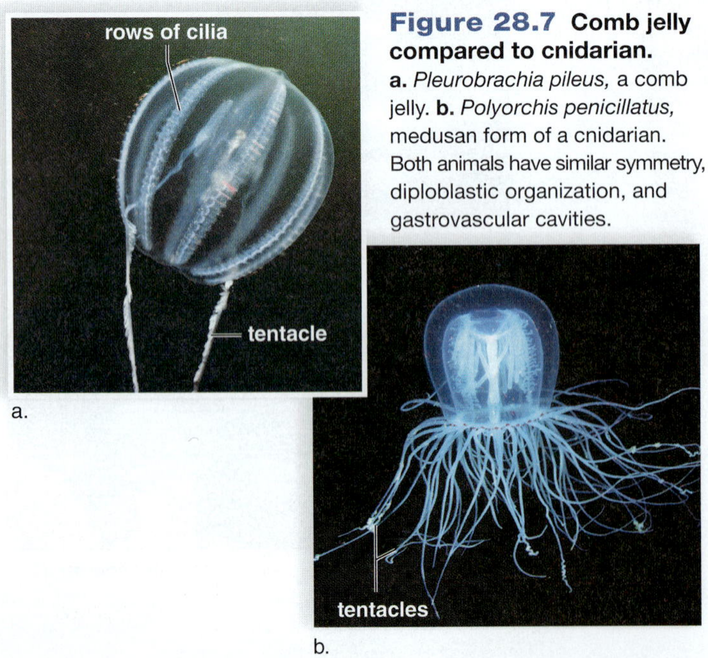

Figure 28.7 Comb jelly compared to cnidarian.
a. *Pleurobrachia pileus,* a comb jelly. **b.** *Polyorchis penicillatus,* medusan form of a cnidarian. Both animals have similar symmetry, diploblastic organization, and gastrovascular cavities.

The body of a cnidarian is a two-layered sac. The outer tissue layer is a protective epidermis derived from ectoderm. The inner tissue layer, which is derived from endoderm, secretes digestive juices into the internal cavity, called the **gastrovascular cavity** (Gk. *gastros,* "stomach") because it functions in the digestion of food and circulation of nutrients. The fluid-filled gastrovascular cavity also serves as a supportive **hydrostatic skeleton,** so called because it offers some resistance to the contraction of muscle but permits flexibility. The two tissue layers are separated by mesoglea.

Two basic body forms are seen among cnidarians. The mouth of a **polyp** is directed upward, while the mouth of a jellyfish, or **medusa,** is directed downward. The bell-shaped medusa has more mesoglea than a polyp, and the tentacles are concentrated on the margin of the bell.

At one time, both body forms may have been a part of the life cycle of all cnidarians. When both are present, the animal is dimorphic: The sessile polyp stage produces medusae by asexual budding, and the motile medusan stage produces egg and sperm. In some cnidarians, one stage is dominant and the other is reduced; in other species, one form is absent altogether.

Cnidarian Diversity. Sea anemones (Fig. 28.8*a*) are sessile polyps that live attached to submerged rocks, timbers, or other substrate. Most sea anemones range in size from 0.5 to 20 cm in length and 0.5 to 10 cm in diameter and are often colorful. Their upward-turned oral disk contains the mouth and is surrounded by a large number of hollow tentacles containing nematocysts.

Corals (Fig. 28.8*b*) resemble sea anemones encased in a calcium carbonate (limestone) house. The coral polyp can extend into the water to feed on microorganisms and retreat into the house for safety. Some corals are solitary, but the vast majority live in colonies that vary in shape from rounded to branching. Corals are the animals that build coral reefs. They exhibit elaborate geometric designs and stunning colors.

Video Coral Reef Spawning

Coral reefs are built from the slow accumulation of limestone produced by corals. Over hundreds of years, this accumulation can result in massive structures, such as the Great Barrier Reef along the eastern coast of Australia. Coral reef ecosystems are very productive, and a diverse group of marine life calls the reef home.

Video Coral Reef Ecosystems

The hydrozoans have a dominant polyp stage. *Hydra* (see Fig. 28.9) is a hydrozoan, and so is a Portuguese man-of-war. You might think the Portuguese man-of-war is an odd-shaped medusa, but actually it is a colony of polyps (Fig. 28.8*c*). The original polyp becomes a gas-filled float that provides buoyancy, keeping the colony afloat. Other polyps, which bud from this one, are specialized for feeding or for reproduction. A long, single tentacle armed with numerous nematocysts arises from the base of each feeding polyp. Swimmers who accidentally come upon a Portuguese man-of-war can receive painful, even serious, injuries from these stinging tentacles.

Video Portuguese Man-of-War

In true jellyfish, also known as sea jellies (Fig. 28.8*d*), the medusa is the primary stage and the polyp remains small. Jellyfish are zooplankton and depend on tides and currents for their primary means of movement. They feed on a variety of invertebrates and fishes and are themselves food for marine animals.

a. Sea anemone, *Condylactis*

b. Cup coral, *Tubastrea*

c. Portuguese man-of-war, *Physalia*

d. Jellyfish, *Aurelia*

Figure 28.8 Cnidarian diversity. **a.** The sea anemone, which is sometimes called the flower of the sea, is a solitary polyp. **b.** Corals are colonial polyps residing in a calcium carbonate or proteinaceous skeleton. **c.** The Portuguese man-of-war is a colony of modified polyps and medusae. **d.** Jellyfish, *Aurelia*.

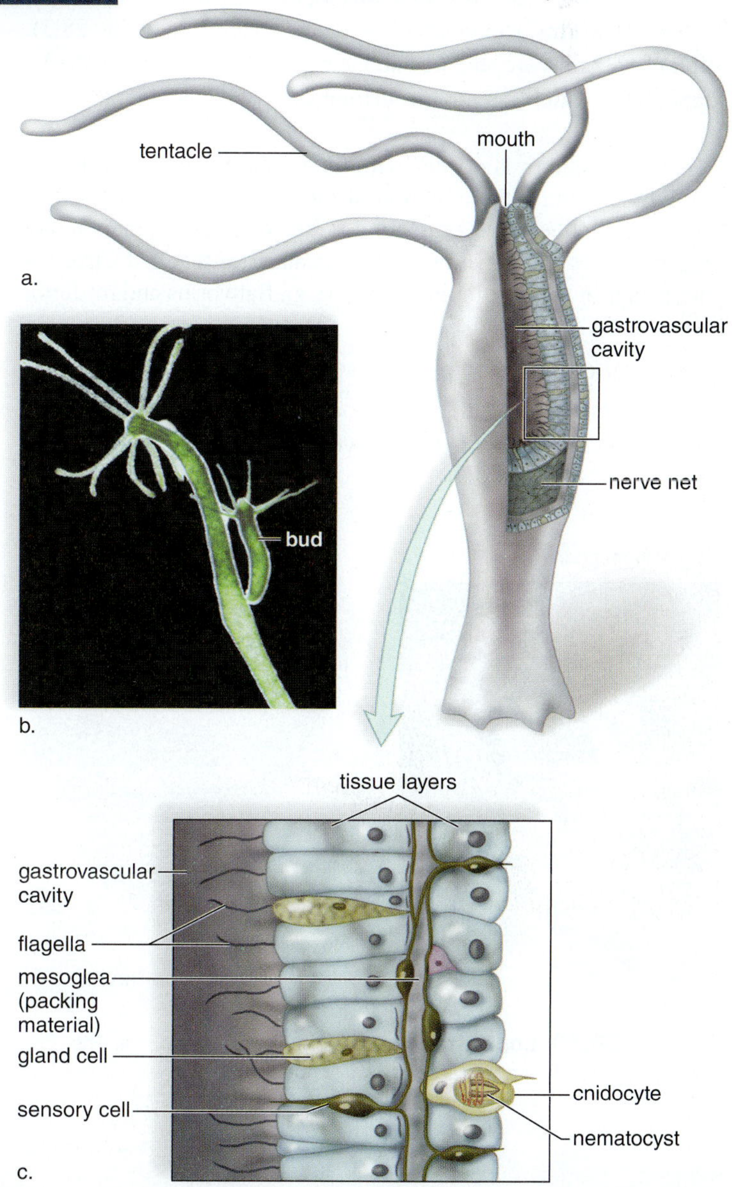

Figure 28.9 Anatomy of *Hydra.* **a.** The body of *Hydra* is a small, tubular polyp that reproduces asexually by forming outgrowths called buds. **b.** The buds develop into a complete animal. **c.** The body wall contains two tissue layers separated by mesoglea. Cnidocytes are cells that contain nematocysts.

A Typical Cnidarian: Hydra. Hydras are often studied as an example of a cnidarian. Hydras are likely to be found attached to underwater plants or rocks in most lakes and ponds. The body of a hydra is a small, tubular polyp about one-quarter of an inch in length. The only opening (which serves as both mouth and anus) is in a raised area surrounded by four to six tentacles that contain a large number of nematocysts.

Figure 28.9 shows the microscopic anatomy of *Hydra.* The cells of the epidermis are termed epitheliomuscular cells because they contain muscle fibers. Also present in the epidermis are nematocyst-containing cnidocytes and sensory cells that make contact with the nerve cells within a **nerve net.** These interconnected nerve cells allow the transmission of impulses in several directions at once. The body of a hydra can contract or extend, and the tentacles that ring the mouth can reach out and grasp prey and discharge nematocysts.

Hydras reproduce asexually by forming buds, small outgrowths that develop into a complete animal and then detach. Interstitial cells of the epidermis are capable of becoming other types of cells, such as an ovary and/or a testis. When hydras reproduce sexually, sperm from a testis swim to an egg within an ovary. The embryo is encased within a hard, protective shell that allows it to survive until conditions are optimum for it to emerge and develop into a new polyp.

Like the sponges, cnidarians have great regenerative powers, and hydras can grow an entire organism from a small piece.

Check Your Progress 28.2

1. List three ways in which cnidarians are more complex than sponges.
2. Summarize how a sponge obtains nutrients.
3. Describe the medusa and polyp body forms of a cnidarian.
4. Explain how a cnidarian, such as a jellyfish, stings its prey.

28.3 Diversity Among the Lophotrochozoans

Learning Outcomes

Upon completion of this section, you should be able to

1. List the basic features of lophotrochozoans.
2. Describe the basic anatomy and physiology of a planarian.
3. Summarize the steps in the life cycles of *Schistosoma* and *Taenia*.
4. Identify morphological features of molluscs, bivalves, rotifers, and annelids.

The **lophotrochozoa** are a diverse group of protostomes. These animals are bilaterally symmetrical during at least one stage of their development. As embryos, they have three germ layers, and as adults they have the organ level of organization. Some have a true coelom (see section 28.1), as exemplified by the annelid worms.

The lophotrochozoans can be divided into two groups: the **lophophorans** (Gk. *lophos,* "crest"), such as bryozoans, phoronids, and brachiopods, and the **trochozoans** (Gk. *trochos,* "wheel"), such as flatworms, rotifers, molluscs, and annelids (see Fig. 28.3). All lophophorans are aquatic and have a feeding apparatus called a lophophore, which is a mouth surrounded by ciliated, tentacle-like structures (Fig. 28.10*a*).

A trochophore is a free-swimming, marine larva with bands of cilia that control the direction of movement (Fig. 28.10*b*). The trochozoans either have a trochophore stage of development today (e.g., molluscs and annelids) or had an ancestor that had a trochophore stage at some point in the past (e.g., flatworms and rotifers).

Lophophorans

Traditionally, the "lophophorans" were considered a lineage of protostomes that included the bryozoans and brachiopods, but recent evidence from analysis of the small subunit of ribosomal RNA (rRNA) (see Fig. 12.16) and other genes suggest that the evolutionary relationships among the lophophorans within the Protostomia are more complex. However, for our purposes, we can represent the "lophophorans" in the traditional sense as a single lineage (see Fig. 28.3), containing three closely related groups, the bryozoans, the brachiopods, and the phoronids (Fig. 28.11).

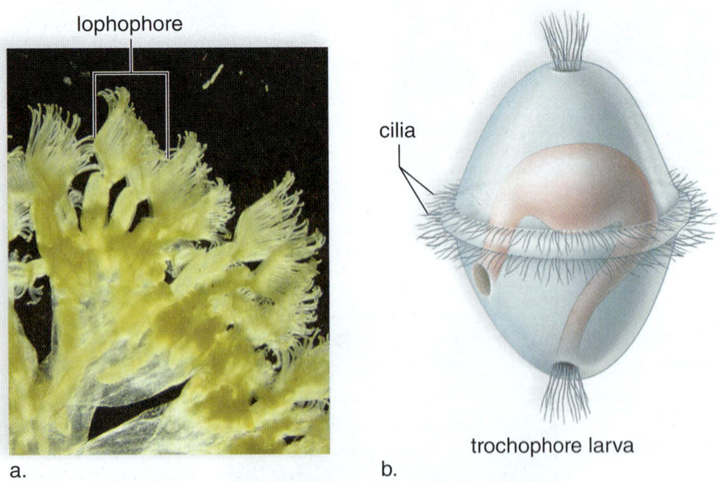

Figure 28.10 Lophotrochozoan characteristics. a. The lophophore feeding apparatus. **b.** A trochophore larva.

Figure 28.11 Lophophorans. A bryozoan (*Canda* sp.) with a horseshoe-shaped tentacle crown.

Bryozoans (phylum Bryozoa) are aquatic, colonial lophophorans. Colonies are made up of individuals called zooids. Zooids are not independent animals but, rather, single members of a colony that cooperate as a single organism. Some zooids specialize in feeding and they filter particles from the water with the lophophore; some specialize in reproduction; and some can perform both functions. Zooids coordinate functions within a colony by communicating through chemical signals. Zooids have protective exoskeletons, which they use to attach to substrates, including the bottoms of ships, where they cause a nuisance by increasing drag and impeding maneuverability.

Brachiopods (phylum Brachiopoda) are a small group of lophophorans that have two hinged shells, much as molluscs do—but instead of having a left and right shell, they have a top and a bottom shell. Brachiopods affix themselves to hard surfaces with a muscular pedicle. Like other lophophorans, brachiopods use their lophophore to feed by filtering particles from the water.

Phoronids (phylum Phoronida) live inside a long tube formed from their own chitinous secretions. The tube is buried in the ground and their lophophore extends from it, but it can retract very quickly when needed. Only about 15 species of phoronids exist worldwide.

Trochozoans

Flatworms (phylum Platyhelminthes) are trochozoans with an extremely flat body. Like the cnidarians, flatworms have a sac body plan with only one opening, the mouth. Organisms with a single opening are said to have an *incomplete digestive tract,* whereas a *complete digestive tract* has two openings. Flatworms have no body cavity; instead, the third germ layer, mesoderm, fills the space between their organs. Among flatworms, planarians are free-living; flukes and tapeworms are parasitic.

Free-living Flatworms

Planarians are a group of nonparasitic, free-living flatworms. *Dugesia* is a planarian that lives in freshwater lakes, streams, and ponds, where it feeds on small living or dead organisms. A planarian captures food by wrapping itself around the prey, entangling it in slime, and pinning it down. Then, a muscular pharynx is extended through the mouth and a sucking motion takes pieces of the prey into the pharynx. The pharynx leads into a three-branched gastrovascular cavity, in which digestion is both extracellular and intracellular (Fig. 28.12*a*). The digestive system delivers nutrients and oxygen to the cells; the animal has no circulatory system or respiratory system. Waste molecules exit through the mouth.

Planarians have a well-developed excretory system (Fig. 28.12*b*). The excretory organ functions in osmotic regulation, as well as in water excretion. The organ consists of a series of interconnecting canals that run the length of the body on each side. Bulblike structures containing cilia are at the ends of the side branches of the canals. The cilia move back and forth, taking water into the canals that empty at pores. The excretory system often functions as an osmotic-regulating system. The beating of the cilia reminded an early investigator of the flickering of a flame, so the excretory organ of the flatworm is called a flame cell.

Planarians usually reproduce sexually. They are monoecious (**hermaphroditic**), which means that they possess both male and female sex organs and gametes in a single individual (Fig. 28.12*c*). The worms cross-fertilize when the penis of one is inserted into the genital pore of the other. During this process, each planarian gives and receives sperm. The fertilized eggs are enclosed in a cocoon and hatch in 2 or 3 weeks as tiny worms.

Planarians also can reproduce asexually via regeneration. The tail portion of the planaria breaks off, and each part grows into a new worm. Because planarians have the ability to regenerate, they have been the subject of a field of research called regenerative medicine. Many animals, including humans, do not have the ability to regenerate parts of the body after amputation. The study of how planarians are able to regenerate may lead to advances in regenerative medicine for humans and other animals.

The nervous system of planarians is called a *ladder-type,* because the two lateral nerve cords plus transverse nerves look like a ladder (Fig. 28.12*d*). Paired **ganglia,** or collections of nerve cells, function as a primitive brain.

Planarians are bilaterally symmetrical and exhibit cephalization. The head of a planarian is bluntly arrow-shaped, with lateral extensions, called auricles, that contain chemosensory cells and tactile cells they use to detect potential food sources and enemies. They do not have complex eyes but, rather, two pigmented, light-sensitive eyespots on the top of the head that make the worm look "cross-eyed" (Fig. 28.12*e*).

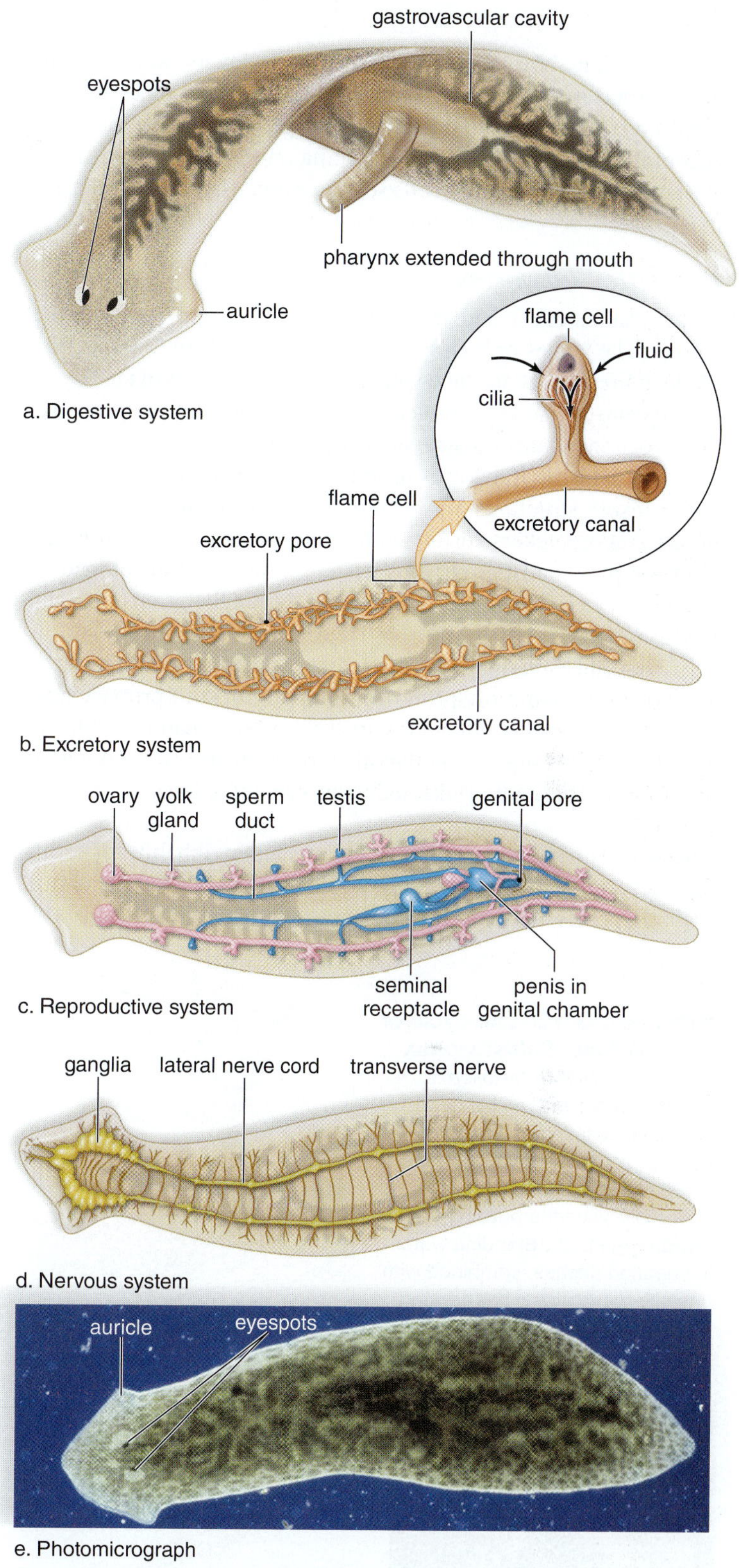

Figure 28.12 Planarian anatomy. **a.** When a planarian extends the pharynx, food is sucked up into a gastrovascular cavity, which branches throughout the body. **b.** The excretory system with flame cells is shown in detail. **c.** The reproductive system (shown in pink and blue) has both male and female organs. **d.** The nervous system has a ladderlike appearance. **e.** The photomicrograph shows that a planarian, *Dugesia,* is bilaterally symmetrical and has a head region with eyespots.

Planarians have three kinds of muscle layers that allow for quite varied movement: an outer circular layer, an inner longitudinal layer, and a diagonal layer. In larger forms, locomotion is accomplished by the movement of cilia on the ventral and lateral surfaces. Numerous gland cells secrete a mucus, on which the animal glides.

Parasitic Flatworms

Flukes (trematodes) and tapeworms (cestodes) are parasitic flatworms. The bodies of both groups are highly evolved for a parasitic mode of life. Flukes and tapeworms feed on nutrients provided by the host and are covered by a protective tegument, which is a specialized body covering that is resistant to host digestive juices.

The parasitic flatworms have lost cephalization, and the head with sensory structures has been replaced by an anterior end with hooks and/or suckers for attachment to the host. They no longer hunt for prey, and the nervous system is not well developed. In contrast, a well-developed reproductive system helps ensure transmission to a new host.

Both flukes and tapeworms utilize a secondary, or intermediate, host to transmit offspring from primary host to primary host. The primary host is infected with the sexually mature adult; the secondary host contains the larval stage or stages. Several human diseases are caused by fluke and tapeworm infections.

Flukes. Flukes are named for the organ they inhabit; for example, there are liver, lung, and blood flukes (Fig. 28.13). The almost 11,000 species have an oval to more elongated, flattened body about 2.5 cm long. At the anterior end is an oral sucker surrounded by sensory papillae and at least one other sucker, used for attachment to a host.

Schistosomiasis is a serious disease caused by a genus of blood fluke, *Schistosoma,* which occurs predominantly in the Middle East, Asia, and Africa. The World Health Organization also lists schistosomiasis as one of several neglected tropical diseases (NTDs) that afflict the poor people in impoverished nations (see the Biological Systems feature, "African Sleeping Sickness," in Chapter 21). In schistosomiasis, female flukes deposit their eggs in small blood vessels close to the lumen of the human intestine, and the eggs make their way into the digestive tract by a slow migratory process (Fig. 28.13). After the eggs pass out with the feces, they hatch into tiny larvae that swim about in rice paddies and elsewhere until they enter a particular species of snail. Within the snail, asexual reproduction occurs; sporocysts, which are spore-containing sacs, eventually produce new larval forms that leave the snail. If the larvae penetrate the skin of a human, they begin to mature in the liver and implant themselves in the blood vessels of the small intestine.

The flukes and their eggs can cause dysentery, anemia, bladder inflammation, brain damage, and severe liver complications. Infected persons usually die of secondary diseases brought on by their weakened condition. It is estimated that over 200 million people worldwide are afflicted with this disease.

The Chinese liver fluke, *Clonorchis sinensis,* is a parasite of cats, dogs, pigs, and humans; it requires two secondary hosts: a snail and a fish. The adults reside in the liver and deposit their eggs in bile ducts, which carry them to the intestines for elimination in

Figure 28.13 Life cycle of a blood fluke, *Schistosoma.* **a.** Micrograph of *Schistosoma*. **b.** *Schistosomiasis,* an infection of humans caused by the blood fluke *Schistosoma,* is an extremely prevalent disease in Egypt—especially since the building of the Aswan High Dam. Standing water in irrigation ditches, combined with unsanitary practices, has created the conditions for widespread infection.

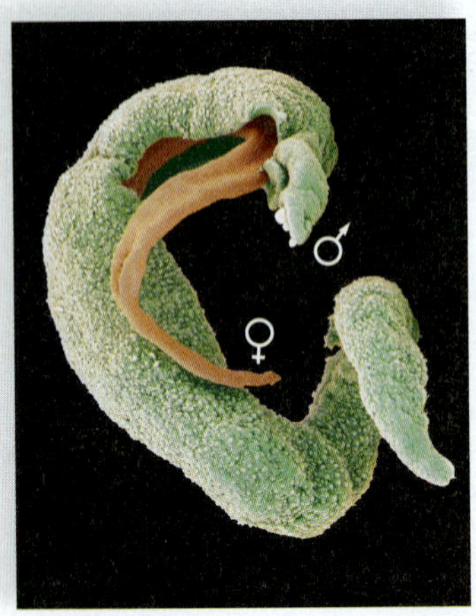

a.

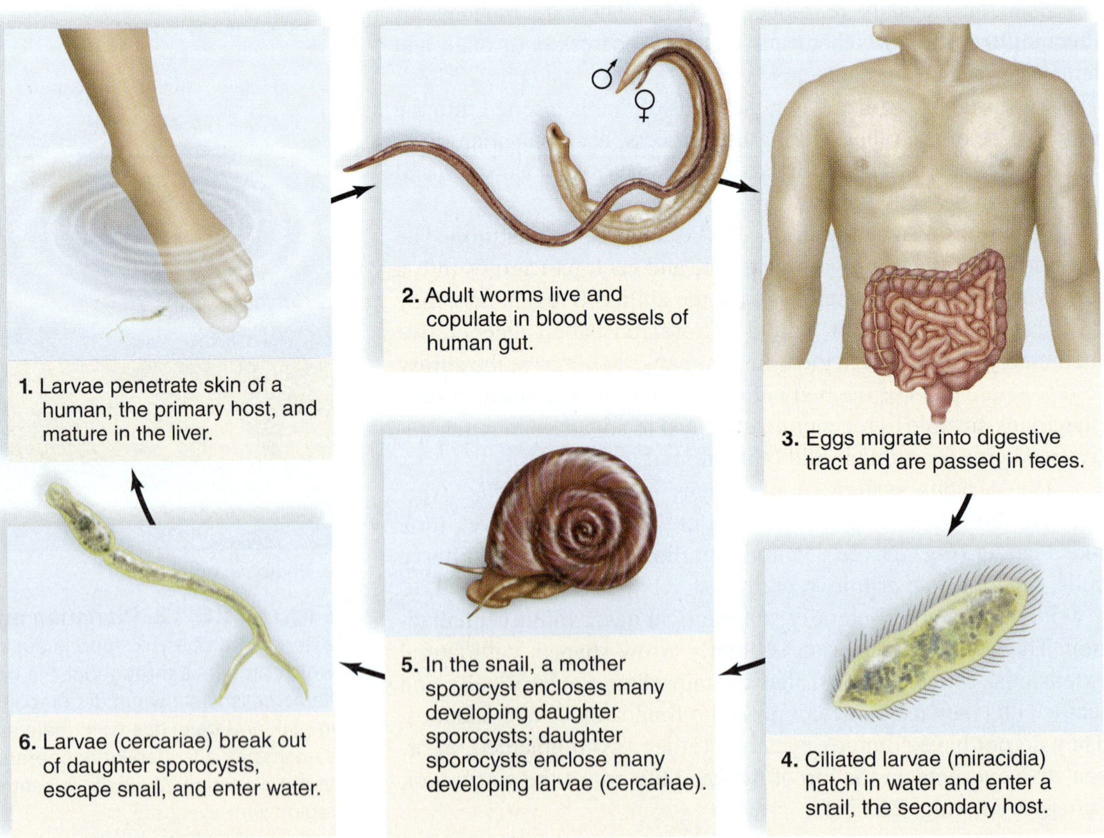

b.

feces. Nonhuman species generally become infected through the fecal route, but humans usually become infected by eating raw fish. A heavy *Clonorchis* infection can cause severe cirrhosis of the liver and death.

Tapeworms. Tapeworms vary in length from a few millimeters to nearly 20 m. They have a highly modified head region called the **scolex,** which contains hooks for attachment to the intestinal wall of the host and suckers for feeding. Tapeworms are hermaphrodites; behind the scolex is a series of reproductive units, called **proglottids,** that contain a full set of female and male sex organs. The number of proglottids may vary depending on the species.

After fertilization, the organs within a proglottid disintegrate and the proglottids become gravid, or full of mature eggs. Gravid proglottids may contain 100,000 eggs. Once mature, the eggs of some species may be released through a pore in the proglottid into the host's intestine, where they exit with feces. In other species, the gravid proglottids break off and are eliminated with the feces. These "segments" can be mistaken for fly maggots—for example, in dog or cat feces—but tapeworm segments are flatter and appear in fresh samples, whereas fly eggs take up to a day to hatch.

Most tapeworms have complicated life cycles, which usually involve several hosts. Figure 28.14 illustrates the life cycle of the pork tapeworm, *Taenia solium,* which has the human as the primary host and the pig as the secondary host. After a pig feeds on feces-contaminated food, the larvae are released. They burrow through the intestinal wall and travel in the bloodstream to finally lodge and encyst in muscle. This **cyst** is a small, hard-walled structure that contains a larva called a bladder worm. Humans become infected with tapeworms when they eat infected meat that has not been thoroughly cooked.

In impoverished nations of the world, more than 50 million people suffer from cysticercosis, an advanced tapeworm infection that interferes with the uptake of nutrients and is therefore particularly serious for growing children. After tainted meat is eaten, the bladder worms break out of the cysts, attach themselves to the intestinal wall, and grow to adulthood. Then the cycle begins again. Generally, tapeworm infections cause diarrhea, weight loss, and fatigue in the primary host.

Rotifers

Rotifers (phylum Rotifera) are trochozoans related to the flatworms. Antonie van Leeuwenhoek (1632–1723), the inventor of the microscope, viewed microscopic rotifers and called them the "wheel animalcules." Rotifers have a crown of cilia, known as the corona, on their heads (Fig. 28.15). When in motion, the corona, which looks like a spinning wheel, serves as an organ of locomotion and directs food into the mouth.

The approximately 2,000 species primarily live in fresh water; however, some marine and terrestrial forms exist. The majority of rotifers are transparent, but some are very colorful. Many species of rotifers can desiccate during harsh conditions and remain dormant for lengthy periods of time.

Molluscs

The **molluscs** (phylum Mollusca) are the second most numerous group of animals, with over 100,000 named species. They inhabit marine, freshwater, and terrestrial habitats. This diverse phylum includes chitons, limpets, slugs, snails, abalones, conchs, nudibranchs, clams, oysters, scallops, squid, and octopuses. Molluscs

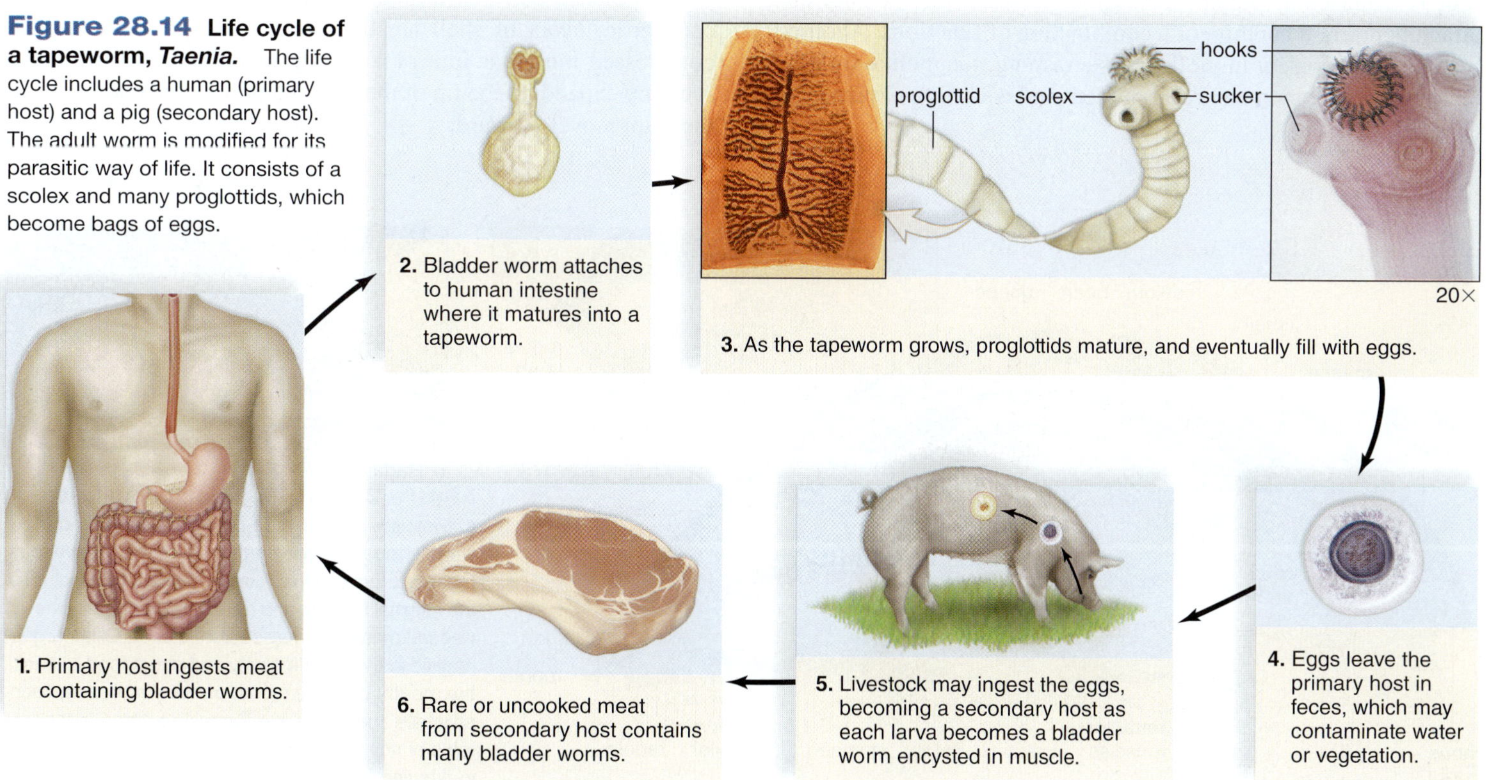

Figure 28.14 Life cycle of a tapeworm, *Taenia.* The life cycle includes a human (primary host) and a pig (secondary host). The adult worm is modified for its parasitic way of life. It consists of a scolex and many proglottids, which become bags of eggs.

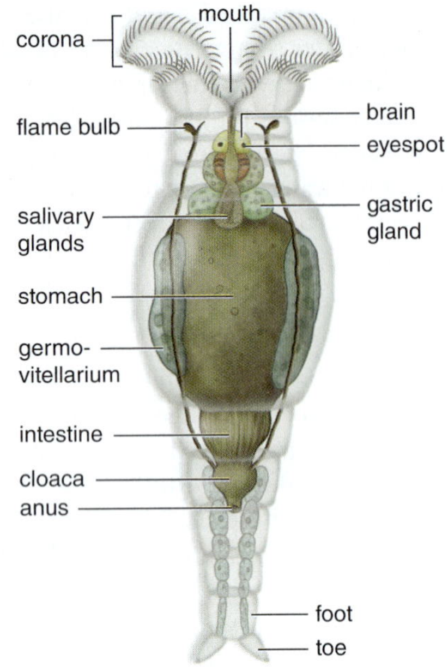

Figure 28.15

Rotifer. Rotifers are microscopic animals only 0.1–3 mm in length. The beating of cilia on two lobes at the anterior end of the animal gives the impression of a pair of spinning wheels.

vary in size from microscopic to the giant squid, which can attain lengths of over 20 m and weigh over 450 kg. The group includes herbivores, carnivores, filter feeders, and parasites.

Although diverse, molluscs share a three-part body plan consisting of the visceral mass, mantle, and foot (Fig. 28.16*a*). The visceral mass contains the internal organs, including a highly specialized digestive tract, paired kidneys, and reproductive organs. The **mantle** is a covering that lies to either side of, but does not completely enclose, the visceral mass. It may secrete a shell and/or contribute to the development of gills or lungs. The space between the folds of the mantle is called the mantle cavity. The foot is a muscular organ that may be adapted for locomotion, attachment, food capture, or a combination of functions. Another feature often present in molluscs is a rasping, tonguelike *radula,* an organ that bears many rows of teeth and is used to obtain food (Fig. 28.16*b*).

The true coelom is reduced in molluscs and largely limited to the region around the heart. Most molluscs have an open circulatory system. The heart pumps blood, more properly called hemolymph, through vessels into sinuses (cavities) collectively called a **hemocoel** (Gk. *haima,* "blood"; *koiloma,* "cavity"). Blue hemocyanin, rather than red hemoglobin, is the oxygen-carrying pigment. Nutrients and oxygen diffuse into the tissues from these sinuses instead of being carried into the tissues by capillaries, the microscopic blood vessels present in animals with closed circulatory systems.

The nervous system of molluscs consists of several ganglia connected by nerve cords. The amount of cephalization and sensory organs varies from nonexistent in clams to complex in squid and octopuses. The molluscs also exhibit variation in mobility. Oysters are sessile, snails are extremely slow moving, and squid are fast-moving, active predators.

Bivalves

Clams, oysters, shipworms, mussels, and scallops are all **bivalves** (class Bivalvia), with a two-part shell that is hinged and closed by one or two powerful adductor muscles (Fig. 28.17). Bivalves have no head, no radula, and very little cephalization. Clams use their hatchet-shaped foot for burrowing in sandy or muddy soil, and mussels use their foot to produce threads that attach them to nearby objects. Scallops both burrow and swim; rapid clapping of the valves releases water in spurts and causes the animal to move forward in a jerky fashion for a few feet.

Video
Clam Locomotion

In freshwater clams such as *Anodonta* (Fig. 28.17*c*), the shell, secreted by the mantle, is composed of protein and calcium carbonate with an inner layer, called mother of pearl. If a foreign body is placed between the mantle and the shell, pearls form as concentric layers of shell are deposited about the particle. The compressed muscular foot of a clam projects ventrally from the shell; by expanding the tip of the foot and pulling the body after it, the clam moves forward.

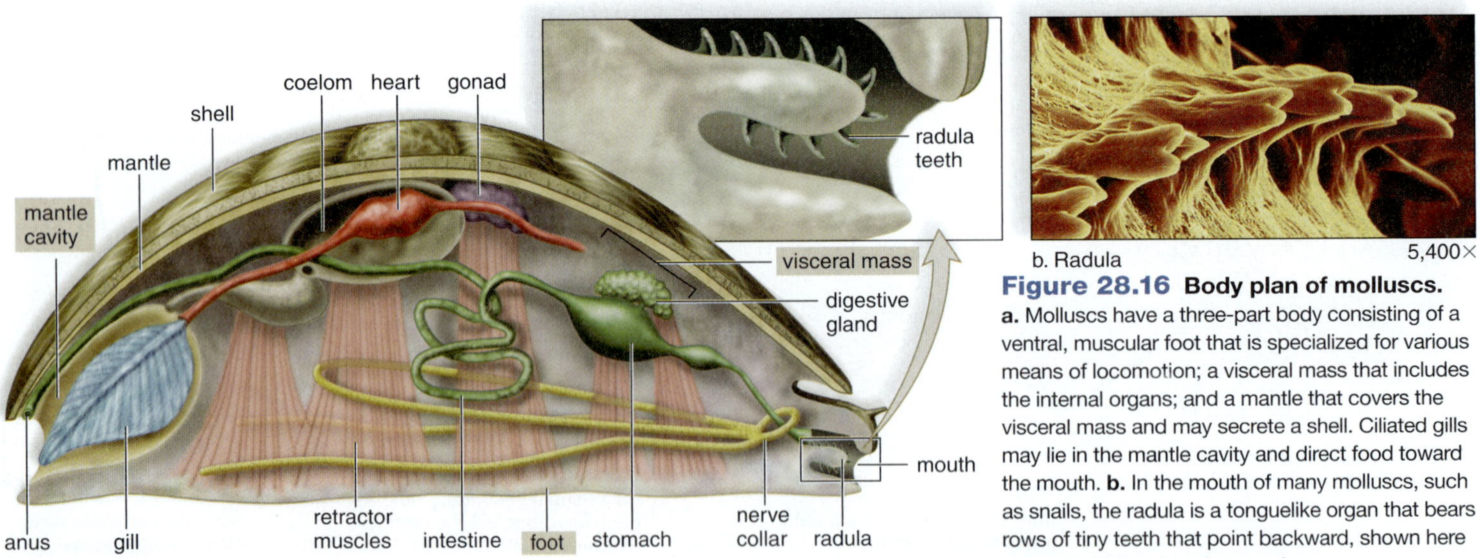

a. Generalized molluscan anatomy

b. Radula

Figure 28.16 Body plan of molluscs. **a.** Molluscs have a three-part body consisting of a ventral, muscular foot that is specialized for various means of locomotion; a visceral mass that includes the internal organs; and a mantle that covers the visceral mass and may secrete a shell. Ciliated gills may lie in the mantle cavity and direct food toward the mouth. **b.** In the mouth of many molluscs, such as snails, the radula is a tonguelike organ that bears rows of tiny teeth that point backward, shown here in drawing (a) and a micrograph.

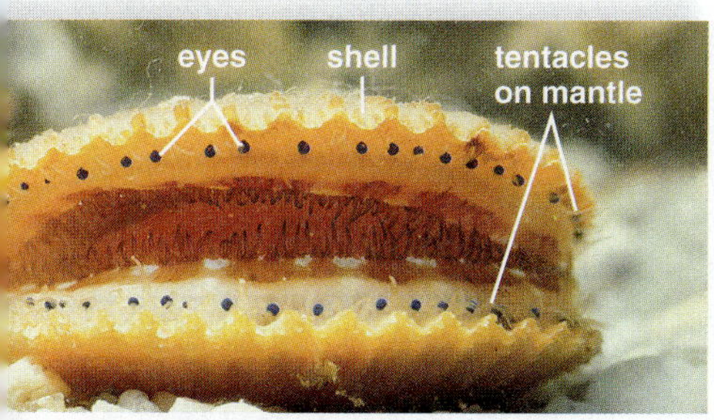

a. Scallop, *Pecten* sp.

b. Mussels, *Mytilus edulis*

Figure 28.17 Bivalve diversity. Bivalves have a two-part shell. **a.** Scallops clap their valves and swim by jet propulsion. This scallop has sensory organs consisting of blue eyes and tentacles along the mantle edges. **b.** Mussels form dense beds in the intertidal zone of northern shores. **c.** In this drawing of a clam, the mantle has been removed from one side. Follow the path of food from the incurrent siphon to the gills, the mouth, the stomach, the intestine, the anus, and the excurrent siphon. Locate the three ganglia: anterior, foot, and posterior. The heart lies in the reduced coelom.

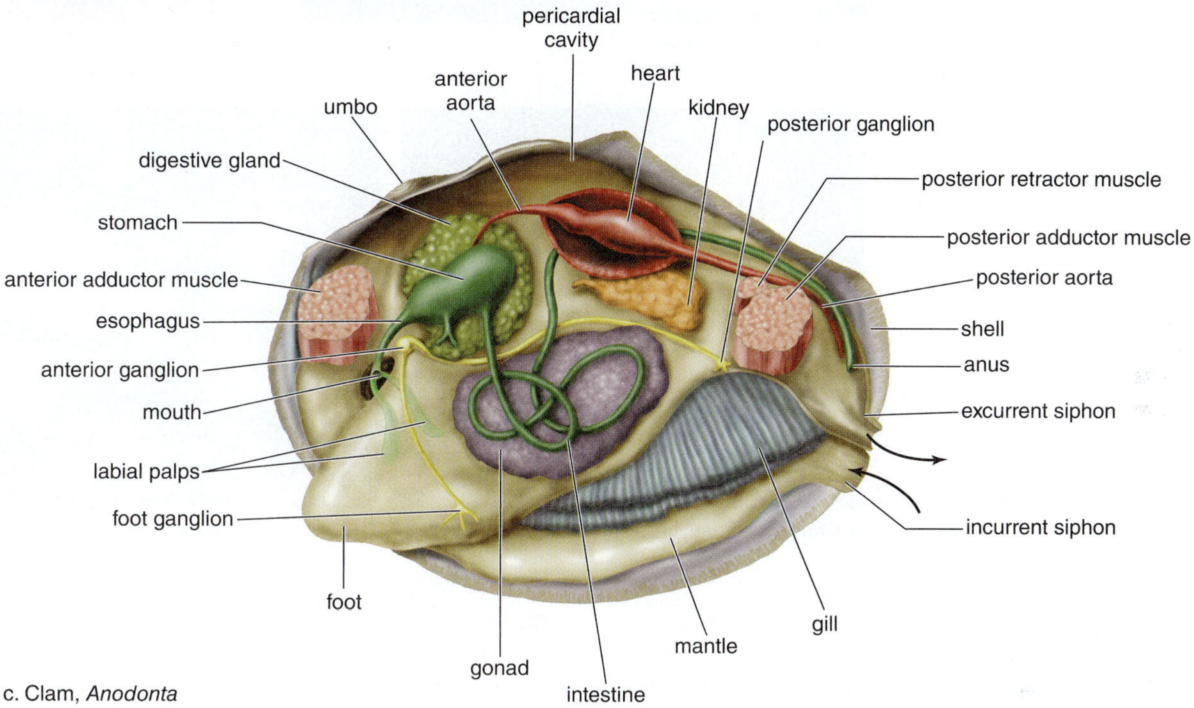

c. Clam, *Anodonta*

Within the mantle cavity, the ciliated gills hang down on either side of the visceral mass. The beating of the cilia causes water to enter the mantle cavity by way of the incurrent siphon and to exit by way of the excurrent siphon. The clam is a filter feeder; small particles in this constant stream of water adhere to the gills, and ciliary action sweeps them toward the mouth.

The mouth leads to a stomach and then to an intestine, which coils about in the visceral mass before going right through the heart and ending in an anus. The anus empties at the excurrent siphon. An accessory organ of digestion, called a digestive gland, is also present. The heart lies just below the hump of the shell within the pericardial cavity, the only remains of the coelom. The circulatory system is open; the heart pumps hemolymph into vessels that open into the *hemocoel*. The nervous system is composed of three pairs of ganglia (located anteriorly, posteriorly, and in the foot), which are connected by nerves.

Two excretory kidneys lie just below the heart and remove waste from the pericardial cavity for excretion into the mantle cavity. The clam excretes ammonia (NH_3), a toxic substance that requires the excretion of water at the same time.

In freshwater clams, the sexes are separate, and fertilization is internal. Fertilized eggs develop into specialized larvae and are released from the clam. Some larvae attach to the gills of a fish and become a parasite before they sink to the bottom and develop into a clam. Certain clams and annelids have the same type of larva, namely the trocophore larva, and this reinforces the evolutionary relationship between molluscs and annelids.

Other Molluscs

The **gastropods** (Gk. *gastros,* "stomach"; *podos,* "foot"), the largest class of molluscs, include slugs, snails, whelks, conchs, limpets, and nudibranchs (Fig. 28.18*a–c*). Most are marine; however, slugs and garden snails are adapted to terrestrial environments (Fig. 28.18*c*).

Gastropods have an elongated, flattened foot, and most, except for slugs and nudibranchs, have a one-piece, coiled shell that protects the visceral mass. The anterior end bears a well-developed head region with a cerebral ganglion and eyes on the ends of tentacles. Land snails are hermaphroditic; when two snails meet, they shoot calcareous darts into each other's body wall as a part of premating behavior. Then, each inserts a penis into the vagina of the other to provide sperm for the future fertilization of eggs, which are deposited in the soil. Development proceeds directly without the formation of larvae.

Cephalopods (Gk. *kaphale,* "head"; *podos,* "foot") range in length from 2 cm to 20 m, as in the giant squid, *Architeuthis.*

Figure 28.18 Gastropod and cephalopod diversity.
a, b. Gastropods have the three parts of a mollusc, and the foot is muscular, elongated, and flattened. **c, d.** Cephalopods have tentacles and/or arms in place of a head. Speed suits their predatory lifestyle, and only the chambered nautilus has a shell.

a. Flamingo tongue shell, *Cyphoma gibbosum*

b. Land snail, *Helix aspersa*

c. Chambered nautilus, *Nautilus belauensis*

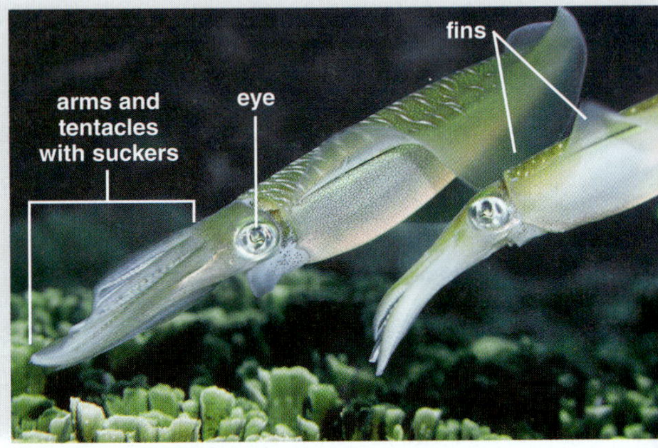

d. Bigfin reef squid, *Sepioteuthis lessoniana*

Cephalopod means head-footed; both squids and octopuses can squeeze their mantle cavity so that water is forced out through a funnel, propelling them by jet propulsion (Fig. 28.18*d*). Also, the tentacles and arms that circle the head capture prey by adhesive secretions or by suckers. A powerful, parrotlike beak is used to tear prey apart. They have well-developed sense organs, including eyes that are similar to those of vertebrates and focus like a camera. Cephalopods, particularly octopuses, have well-developed brains and show a remarkable capacity for learning.

Annelids

Annelids (phylum Annelida [L. *anellus,* "little ring"]), which are sometimes called the segmented worms, vary in size from microscopic to tropical earthworms that can be over 4 m long. The most familiar members of this group are earthworms, marine worms, and leeches.

Annelids are the only trochozoan with segmentation and a well-developed coelom. **Segmentation** is the repetition of body parts along the length of the body. The well-developed coelom is fluid-filled and serves as a supportive *hydrostatic skeleton.* A hydrostatic skeleton, along with partitioning of the coelom, permits independent movement of each body segment. Instead of just burrowing in the mud, an annelid can crawl on a surface.

Setae (L. *seta,* "bristle") are bristles that protrude from the body wall, can anchor the worm, and help it move. The *oligochaetes* are annelids with few setae, and the *polychaetes* are annelids with many setae.

Earthworms

The common earthworm, *Lumbricus terrestris,* is an oligochaete (Fig. 28.19). Earthworm setae protrude in pairs directly from the surface of the body. Locomotion, which is accomplished section by section, uses muscle contraction and the setae. When longitudinal muscles contract, segments bulge and their setae protrude into the soil; then, when circular muscles contract, the setae are withdrawn, and these segments move forward.

Earthworms reside in soil where there is adequate moisture to keep the body wall moist for gas exchange. They are scavengers that feed on leaves or any other organic matter conveniently taken into the mouth along with dirt. Segmentation and a complete digestive tract have led to increased specialization of digestive system components. Food drawn into the mouth by the action of the muscular pharynx is stored in a crop and ground up in a thick, muscular gizzard (Fig. 28.19*a*). Digestion and absorption occur in a long intestine, whose dorsal surface has an expanded region called a typhlosole, which increases the surface for absorption. Waste is eliminated through the anus.

Earthworm segmentation, which is obvious externally, is also internally evidenced by septa that occur between segments. The long, ventral nerve cord leading from the brain has ganglionic swellings and lateral nerves in each segment. The excretory system consists of paired **nephridia** (Gk. *nephros,* "kidney"), which are coiled tubules in each segment (Fig. 28.19*b*). A nephridium has two openings: One is a ciliated funnel that collects coelomic fluid, and the other is an exit in the body wall. Between the two openings is a convoluted region where waste material is removed from the blood vessels about the tubule.

Annelids have a *closed circulatory system,* which means that the blood is always contained in blood vessels that run the length of the body. Oxygenated blood moves anteriorly in the dorsal blood vessel, which connects to the ventral blood vessel by five pairs of muscular vessels called "hearts." Pulsations of the dorsal blood vessel and the five pairs of hearts are responsible for blood flow. As the ventral vessel takes the blood toward the posterior regions of the worm's body, it gives off branches in every segment.

Earthworms are *hermaphroditic;* the male organs are the testes, the seminal vesicles, and the sperm ducts, and the female organs are the ovaries, the oviducts, and the seminal receptacles. During

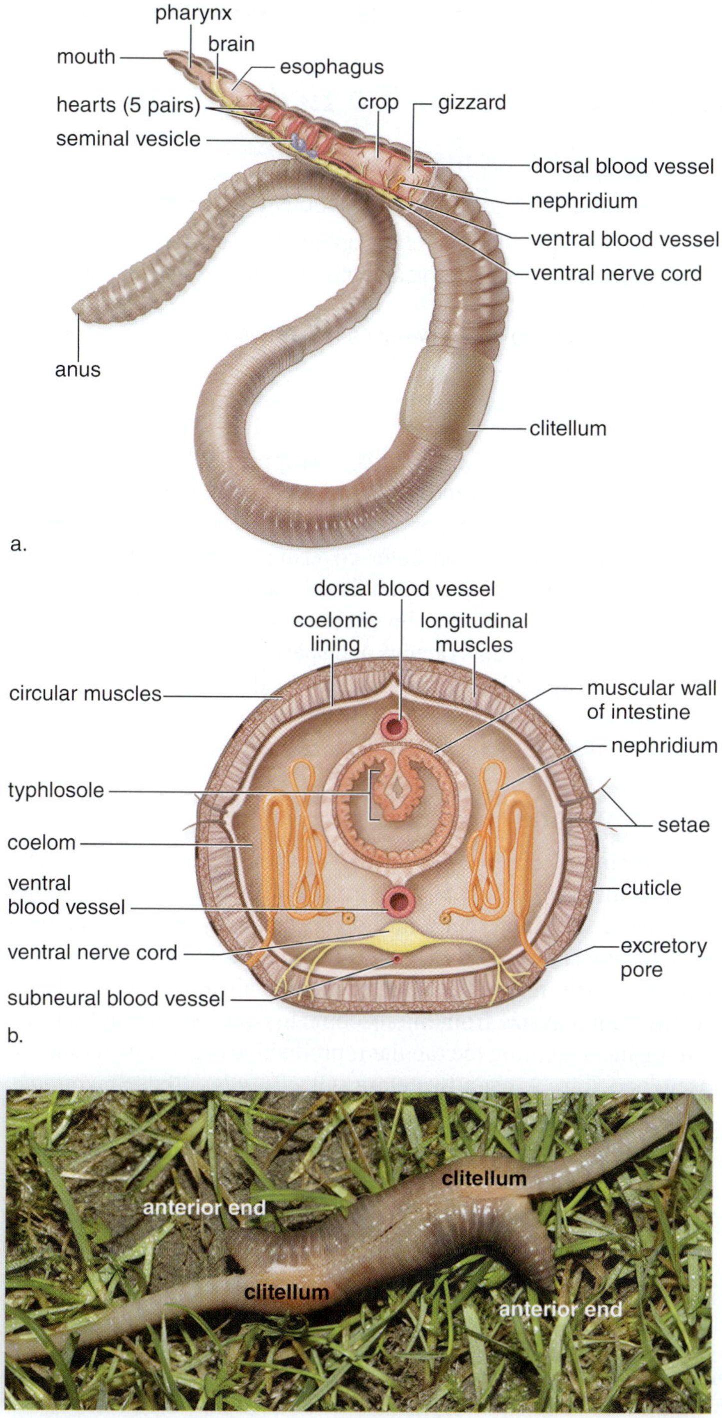

Figure 28.19 Earthworm, *Lumbricus terrestris*. **a.** In the longitudinal section, note the specialized parts of the digestive tract. **b.** In cross section, note the spacious coelom, the paired setae and nephridia, and a ventral nerve cord that has branches in each segment. **c.** When earthworms mate, they are held in place by a mucus secreted by the clitellum. The worms are hermaphroditic, and when mating, sperm pass from the seminal vesicles of each to the seminal receptacles of the other.

mating, two worms lie parallel to each other, facing in opposite directions (Fig. 28.19*c*). The fused midbody segment, called a *clitellum*, secretes mucus, protecting the sperm from drying out as they pass between the worms. After the worms separate, the clitellum of each produces a slime tube, which is moved along over the anterior end by muscular contractions. As it passes, eggs and the sperm received earlier are deposited, and fertilization occurs. The slime tube then forms a cocoon to protect the hatched worms as they develop. There is no larval stage in earthworms.

Other Annelids

Approximately two-thirds of annelids are marine polychaetes. Polychaetes have setae in bundles on parapodia (Gk. *para,* "beside"; *podos,* "foot"), which are paddlelike appendages found on most segments. These are used not only in swimming but also as respiratory organs, where the expanded surface area allows for exchange of dissolved gases. Some polychaetes are free-swimming, but the majority live in crevices or burrow into the ocean bottom. Clam worms, such as *Nereis* (Fig. 28.20*a*), are predators. They prey on crustaceans and other small animals, which are captured by a pair of strong, chitinous jaws that extend with a part of the pharynx when the animal is feeding. Associated with its way of life, *Nereis* is cephalized, having a head region with eyes and other sense organs.

Another group of marine polychaetes are sessile tube worms, with *radioles* (ciliated mouth appendages) used to gather food (Fig. 28.20*b*). Christmas tree worms, fan worms, and featherduster

Figure 28.20 Annelid diversity. **a.** Clam worms are predaceous polychaetes that undergo cephalization. Note also the parapodia, which are used for swimming and as respiratory organs. **b.** Christmas tree worms (a type of tube worm) are sessile feeders whose radioles (ciliated mouth appendages) spiral, as shown here. **c.** Medical leeches are blood suckers.

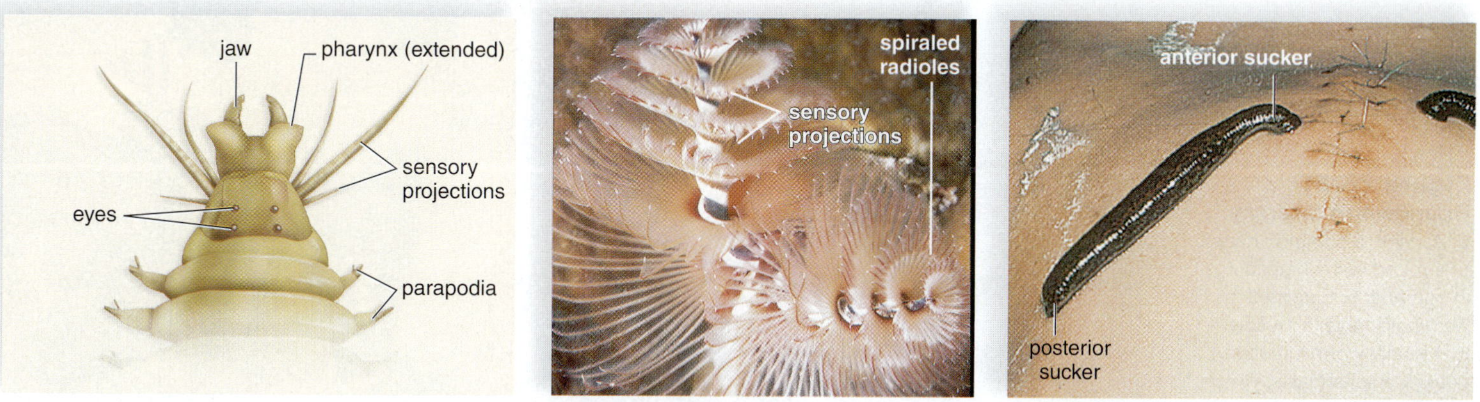

a. Clam worm, *Nereis succinea*

b. Christmas tree worm, *Spirobranchus giganteus*

c. Medicinal leech, *Hirudo medicinalis*

worms all have radioles. In featherduster worms, the beautiful radioles cause the animal to look like an old-fashioned feather duster.

Polychaetes have breeding seasons, and only during these times do the worms have sex organs. In *Nereis,* many worms simultaneously shed a portion of their bodies containing either eggs or sperm, and these float to the surface, where fertilization takes place. The zygote rapidly develops into a trochophore larva.

Leeches are annelids that normally live in freshwater habitats. They range in size from less than 2 cm to the medicinal leech, which can be 20 cm in length. They exhibit a variety of patterns and colors but most are brown or olive green. The body of a leech is flattened dorsoventrally. They have the same body plan as other annelids, but they have no setae and each body ring has several transverse grooves.

While some are free-living, most leeches are fluid feeders that attach themselves to open wounds. Among their modifications are two suckers—a small, oral one around the mouth and a posterior one. Some bloodsuckers, such as the medicinal leech, can cut through tissue. Leeches are able to keep blood flowing by means of hirudin, a powerful anticoagulant in their saliva. Research suggests that leech saliva contains a chemical with local anesthetic properties, which helps explan why the bite of a leech is usually not painful. Medicinal leeches have been used for centuries in blood-letting and other procedures. Today, they are used in reconstructive surgery for severed digits and in plastic surgery (Fig. 28.20*c*).

Check Your Progress 28.3

1. List the characteristics that unite the flatworms, molluscs, and annelids.
2. Compare the features of the flatworm, mollusc, and annelid body cavity, digestive tract, and circulatory system.
3. Describe the life cycle of two lophotrochozoan parasites.
4. List three phyla that feed with lophophores.

28.4 Diversity of the Ecdysozoans

Learning Outcomes

Upon completion of this section, you should be able to

1. Identify the characteristics unique to ecdysozoans.
2. Compare the anatomical features of roundworms and arthropods.
3. Describe five characteristics responsible for the success of the arthropods.

The **ecdysozoans** (Gk. *ecdysis,* "stripping off") include the roundworms and arthropods. They are one of the major groups of animals, which also contains the largest number of species. Ecdysozoans construct an outer covering called a **cuticle,** which protects and supports the animal and is periodically shed to allow growth. Unlike many other invertebrate species that reproduce by releasing large amounts of eggs and sperm into their aqueous environment, many ecdysozoans have evolved separate sexes, which come together and copulate to deliver sperm into the female body, or to directly fertilize eggs as they are released.

Roundworms

Roundworms (phylum Nematoda) are nonsegmented worms that are prevalent in almost any environment. Around 20,000 species have been identified, but a million species may exist. Nematodes have been characterized as the most numerous multicellular organisms on Earth, mainy because they are so abundant in soil. These worms range in size from microscopic to over 1 m in length. Internal organs, including the tubular reproductive organs, lie within the pseudocoelom. A **pseudocoelom** (Gk. *Pseudes,* "false"; *koiloma,* "cavity") is a body cavity that is incompletely lined by mesoderm. In other words, mesoderm occurs inside the body wall but not around the digestive cavity (gut).

Although many nematodes are free-living, others are important parasites of plants and animals. One free-living species, *Caenorhabditis elegans,* has been used extensively as a model animal to study genetics and developmental biology.

a. b. c.

Figure 28.21
Roundworm diversity.
a. *Ascaris,* a common cause of a roundworm infection in humans. **b.** Encysted *Trichinella* larva in muscle. **c.** A filarial worm infection causes elephantiasis, which is characterized by a swollen body part when the worms block lymphatic vessels.

Parasitic Roundworms

The CDC estimates that at any one time, up to a billion people are infected with nematodes of the genus *Ascaris*. Other *Ascaris* species can infect domestic animals and a number of other vertebrates. Though relatively rare in the United States, ascariasis is common in tropical countries, especially in areas with poor sanitation.

As in other nematodes, *Ascaris* males tend to be smaller (15–31 cm long) than females (20–49 cm long) (Fig. 28.21*a*). A typical female *Ascaris* produces over 200,000 eggs daily. The eggs are eliminated with the host's feces and can remain viable in the soil for many months. Eggs enter the human body via uncooked vegetables, soiled fingers, or ingested fecal material and hatch in the intestines. The juveniles make their way into the veins and lymphatic vessels and are carried to the heart and lungs. From the lungs, the larvae travel up the trachea, where they are swallowed and eventually reach the intestines. There, the larvae mature and begin feeding on intestinal contents.

The symptoms of an *Ascaris* infection depend on the stage of the infection. Most infected people have surprisingly few symptoms, but larval *Ascaris* in the lungs can cause coughing or bloody sputum. In the intestines, *Ascaris* can cause malnutrition, blockage of the bile and pancreatic ducts, and poor health.

Trichinosis, caused by the nematode *Trichinella spiralis,* is a serious infection that humans can contract when they eat rare pork containing encysted *Trichinella* larvae. After maturation, the female adult burrows into the wall of the small intestine and produces living offspring, which are carried by the bloodstream to the skeletal muscles, where they encyst (Fig. 28.21*b*). Heavy infections can be painful and lethal.

Filarial worms, another type of roundworm, cause various diseases. For example, mosquitoes can transmit the larvae of a parasitic filarial worm to dogs. Because the worms live in the heart and the arteries that serve the lungs, the infection is called *heartworm disease.* The condition can be fatal; therefore, heartworm medicine is recommended as a preventive measure for all dogs.

Elephantiasis is a disease of humans caused by another filarial worm, *Wuchereria bancrofti.* Restricted to tropical areas of Africa, this parasite uses mosquitoes as a vector to transmit the disease to humans. Because the adult filarial worms reside in lymphatic vessels, the removal of lymph fluid is impeded, and the limbs of an infected person may swell to a monstrous size, resulting in elephantiasis (Fig. 28.21*c*). Elephantiasis is treatable in its early stages but usually not after scar tissue has blocked lymphatic vessels. The World Health Organization lists elephantiasis as one of the 16 neglected tropical diseases (NTDs), that affect impoverished nations. It is estimated that 120 million people in subtropical and tropical areas of the world have filariasis.

Pinworms are the most common nematode parasite in the United States, occurring most commonly in children. The adult parasites live in the cecum and large intestine. Females migrate to the anal region at night and lay their eggs. Scratching the resultant itch can contaminate hands, clothes, and bedding. The eggs are swallowed, and the life cycle begins again. Fortunately, once diagnosed (usually by microscopic examination of sticky tape used to collect eggs from the skin around the anus), the infection is easily cured.

Arthropods

The **arthropods** (phylum Arthropoda [Gk. *arthron,* "joint"; *podos,* "foot"]) are a very large group of protostomes that have exoskeletons and jointed appendages. The phylum Arthropoda includes insects, crustaceans, millipedes, centipedes, spiders, and scorpions. The parasitic mite measures less than 0.1 mm in length, while the Japanese crab measures up to 4 m in length. Arthropods, which also occupy every type of habitat, are considered the most successful group of all the animals.

Characteristics of Arthropods

The remarkable success of arthropods is dependent on five characteristics: an exoskeleton, segmentation, a well-developed nervous system, a variety of respiratory systems, and a life cycle that includes metamorphosis.

Exoskeleton. Arthropods feature a rigid, jointed **exoskeleton** (Fig. 28.22*a, b*). The exoskeleton is composed primarily of chitin,

Figure 28.22 Arthropod skeleton and eye. **a.** The joint in an arthropod skeleton is a region where the cuticle is thinner and not as hard as the rest of the cuticle. The direction of movement is toward the flexor muscle or the extensor muscle, whichever one has contracted. **b.** The exoskeleton is secreted by the epidermis and consists of the endocuticle; the exocuticle, hardened by the deposition of calcium carbonate; and the epicuticle, a waxy layer. Chitin makes up the bulk of the exo- and endocuticles. **c.** Because the exoskeleton is nonliving, it must be shed through a process called molting for the arthropod to grow. **d.** Arthropods have a compound eye that contains many individual units, each with its own lens and photoreceptors.

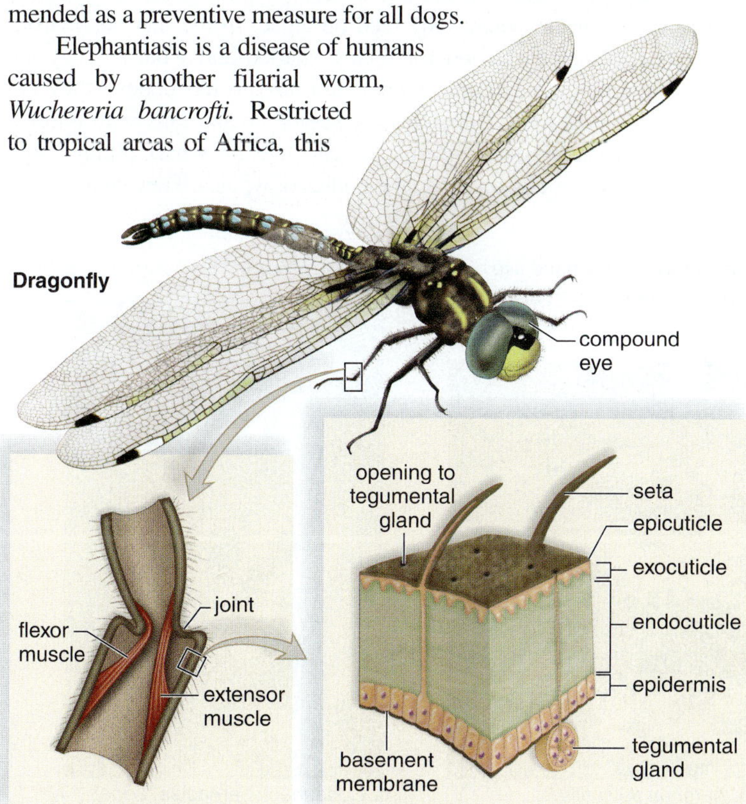

a. Joint movement

b. Exoskeleton composition

c. Molting

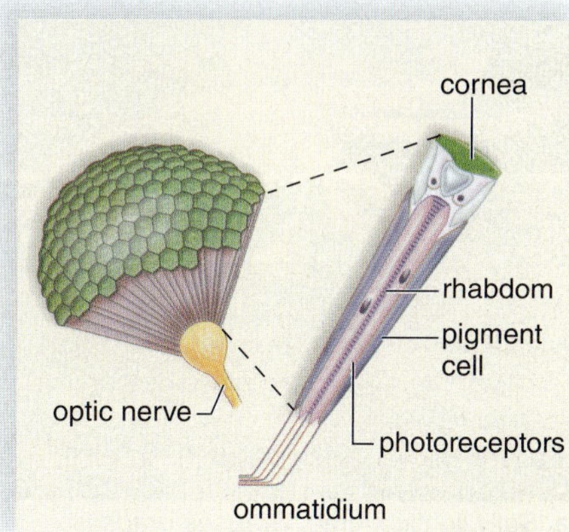

d. Compound eye

a strong, flexible, nitrogenous polysaccharide. The exoskeleton serves many functions, including protection, attachment for muscles, locomotion, and prevention of desiccation. However, because it is hard and nonexpandable, arthropods must molt, or shed, the exoskeleton as they grow larger. Arthropods have this in common with other ecdysozoans.

Before molting, the body secretes a new, larger exoskeleton, which is soft and wrinkled, underneath the old one. After enzymes partially dissolve and weaken the old exoskeleton, the animal breaks it open and wriggles out. The new exoskeleton then quickly expands and hardens (Fig. 28.22*c*).

Segmentation. Segmentation is readily apparent, because each segment has a pair of jointed appendages, even though certain segments are fused into a head, a thorax, and an abdomen. The jointed appendages of arthropods are basically hollow tubes moved by muscles. Typically, the appendages are highly adapted for a particular function, such as food gathering, reproduction, or locomotion. In addition, many appendages are associated with sensory structures and used for tactile purposes.

Nervous System. Arthropods have a brain and a ventral nerve cord. The head bears various types of sense organs, including eyes of two types—simple and compound. The compound eye is composed of many complete visual units, each of which operates independently (Fig. 28.22*d*). The lens of each visual unit focuses an image on the light-sensitive membranes of a small number of photoreceptors within that unit. The simple eye, like that of vertebrates, has a single lens that brings the image to focus onto many receptors, each of which receives only a portion of the image.

In addition to sight, many arthropods have well-developed touch, smell, taste, balance, and hearing. Arthropods display many complex behaviors and methods of communication.

Variety of Respiratory Organs. Marine forms use gills, which are vascularized, highly convoluted, thin-walled tissue specialized for gas exchange. Terrestrial forms have book lungs (e.g., spiders) or air tubes called tracheae (L. *trachea,* "windpipe"). Tracheae serve as a rapid way to transport oxygen directly to the cells.

Metamorphosis. Many arthropods undergo a drastic change in form and physiology that occurs as an immature stage, or larva, becomes an adult. This change is termed **metamorphosis** (Gk. *meta,* "implying change"; *morphe,* "shape, form"). Among arthropods, the larva eats different food and lives in a different environment than does the adult. For example, larval crabs live among and feed on plankton, while adult crabs are bottom dwellers that catch live prey or scavenge dead organic matter. Among insects, such as butterflies, the caterpillar feeds on leafy vegetation, while the adult feeds on nectar. The result is reduced competition for resources, increasing the animal's overall fitness.

While keeping these five arthropod features in mind, let's look at representatives of this large and varied phylum.

Crustaceans

The name **crustacean** is derived from their hard, crusty exoskeleton, which contains calcium carbonate in addition to the typical chitin. Although crustaceans are extremely diverse, the head usually bears a pair of compound eyes and five pairs of appendages. The first two pairs, called antennae and antennules, lie in front of the mouth and have sensory functions. The other three pairs (mandibles and first and second maxillae) lie behind the mouth and are mouthparts used in feeding. Biramous (Gk. *bi-,* "two") appendages on the thorax and abdomen are segmentally arranged; one branch is the gill branch, and the other is the leg branch.

The majority of crustaceans live in marine and aquatic environments (Fig. 28.23). Decapods, which are the most familiar and numerous crustaceans, include lobsters, crabs, crayfish, hermit crabs, and shrimp. These animals have a thorax that bears five pairs of walking appendages. Typically, the gills are positioned above the walking legs. The first pair of walking legs may be modified as claws.

Copepods and krill are small, free-swimming crustaceans that live in the water, where they feed on algae. In the marine environment, they serve as food for fishes, sharks, and whales. They are so numerous that, despite their small size, some believe they are harvestable as food. Barnacles are sessile crustaceans with a thick, heavy, protective shell. Barnacles can live on wharf pilings, ship hulls, seaside rocks, and even the bodies of whales. They begin life as

Figure 28.23 Crustacean diversity. Crabs **(a)** and crayfish **(b)** are decapods—they have five pairs of walking legs. Crabs have a reduced abdomen compared to crayfish. The gooseneck barnacle **(c)** is attached to an object by a long stalk.

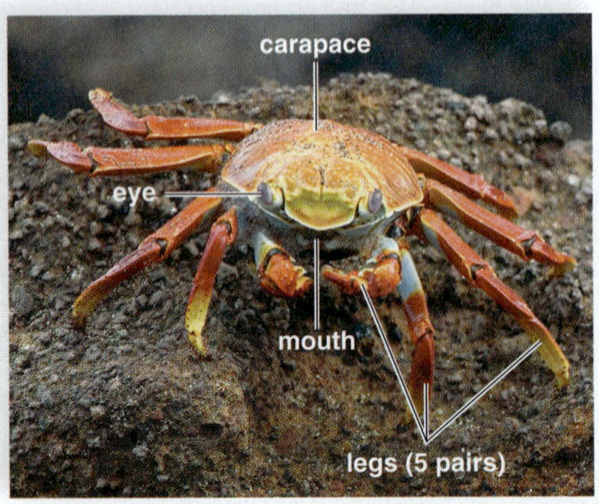

a. Sally lightfoot crab, *Grapsus grapsus*

b. European crayfish, *Actacus fluviatilis*

c. Gooseneck barnacles, *Lepas anatifera*

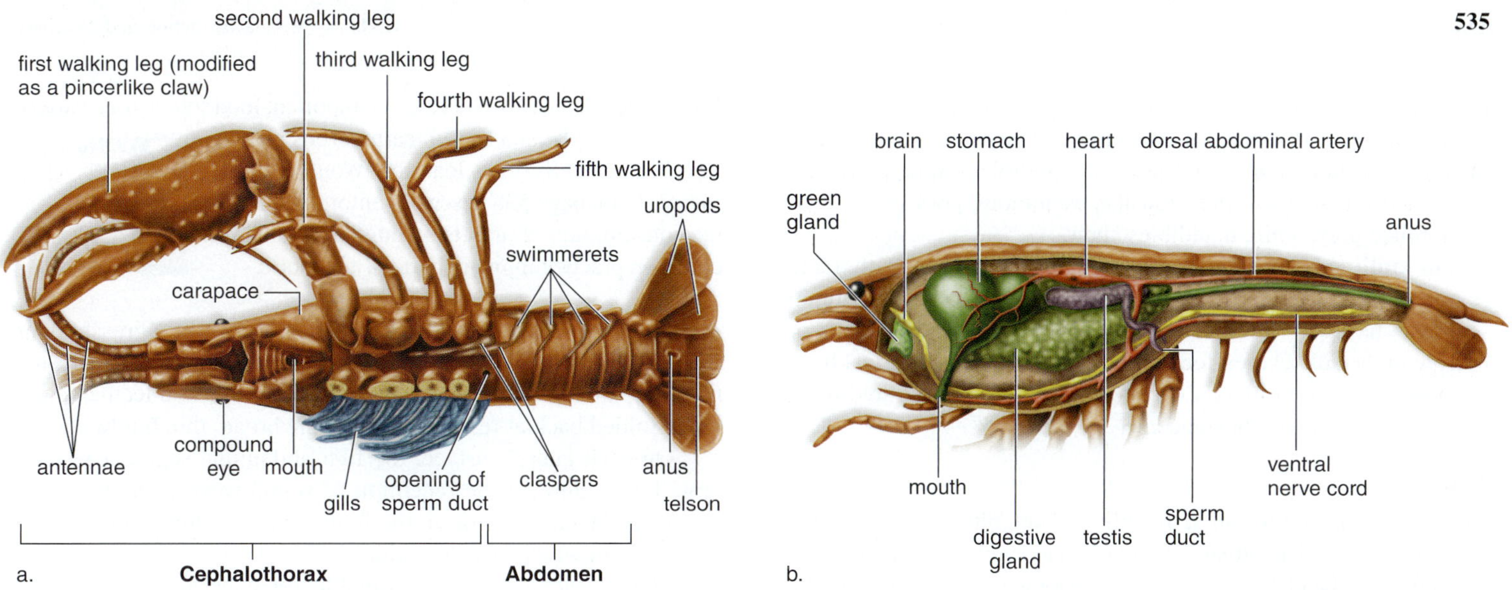

Figure 28.24 Male crayfish, *Cambarus.* **a.** Externally, it is possible to observe the jointed appendages, including the swimmerets, and the walking legs, which include the claws. These appendages, plus a portion of the carapace, have been removed from the right side, so that the gills are visible. **b.** Internally, the parts of the digestive system are particularly visible. The circulatory system can also be clearly seen. Note the ventral nerve cord.

free-swimming larvae, but they undergo a metamorphosis that transforms their swimming appendages to cirri, feathery structures that are extended and allow them to filter feed when they are submerged.

A Typical Crustacean: Crayfish. Figure 28.24*a* gives a view of the external anatomy of the crayfish. The head and thorax are fused into a *cephalothorax* (Gk. *kephale,* "head"; *thorax,* "breastplate"), which is covered on the top and sides by a nonsegmented carapace. The abdominal segments are equipped with swimmerets, small, paddlelike structures. The first two pairs of swimmerets in the male, known as claspers, are quite strong and are used to pass sperm to the female. The last two segments bear the uropods and the telson, which make up a fan-shaped tail.

Ordinarily, a crayfish lies in wait for prey. It faces out from an enclosed spot with the claws extended, the antennae moving about. The claws seize any small animal, either dead or living, that happens by and carry it to the mouth. When a crayfish moves about, it generally crawls slowly, but it may swim rapidly by using its heavy abdominal muscles and tail.

The respiratory system consists of gills that lie above the walking legs protected by the carapace. As shown in Figure 28.24*b,* the digestive system includes a stomach, which is divided into two main regions: an anterior portion called the gastric mill, equipped with chitinous teeth to grind coarse food; and a posterior region, which acts as a filter to prevent coarse particles from entering the digestive glands, where absorption takes place.

Green glands lying in the head region, anterior to the esophagus, excrete metabolic wastes through a duct that opens externally at the base of the antennae. The coelom, which is well developed in the annelids, is reduced in the arthropods and is composed chiefly of the space about the reproductive system. A heart pumps hemolymph containing the blue respiratory pigment hemocyanin into a *hemocoel* consisting of sinuses (open spaces), where the hemolymph flows about the organs. As in the molluscs, this is an open circulatory system because blood is not contained within blood vessels.

The crayfish nervous system is well developed. Crayfish have a brain and a ventral nerve cord that passes posteriorly. Along the length of the nerve cord, periodic ganglia give off lateral nerves. Sensory organs are well developed. The compound eyes are on the ends of movable eyestalks. These eyes are accurate and can detect motion and respond to polarized light. Other sensory organs include tactile antennae and chemosensitive setae. Crayfish also have statocysts, which serve as organs of equilibrium.

The sexes are separate in the crayfish, and the gonads are located just ventral to the pericardial cavity. In the male, a coiled sperm duct opens to the outside at the base of the fifth walking leg. Sperm transfer is accomplished by the modified first two swimmerets of the abdomen. In the female, the ovaries open at the bases of the third walking legs. A stiff fold between the bases of the fourth and fifth pairs serves as a seminal receptacle. Following fertilization, the eggs are attached to the swimmerets of the female. Young hatchlings are miniature adults, and no metamorphosis occurs.

Centipedes and Millipedes

The centipedes and millipedes are known for their many legs (Fig. 28.25*a*). In **centipedes** ("hundred-leggers"), each of their many body segments has a pair of walking legs. The approximately 3,000 species prefer to live in moist environments, such as under

Figure 28.25 Centipede and millipede. **a.** A centipede has a pair of appendages on almost every segment. **b.** A millipede has two pairs of legs on most segments.

logs, in crevices, and in leaf litter, where they are active predators on worms, small crustaceans, and insects. The head of a centipede includes paired antennae and jawlike mandibles. Appendages on the first trunk segment are clawlike, venomous jaws that kill or immobilize prey, while mandibles chew.

In **millipedes** ("thousand-leggers"), each of four thoracic segments bears one pair of legs (Fig. 28.25*b*), while abdominal segments have two pairs of legs. Millipedes live under stones or burrow in the soil as they feed on leaf litter. Their cylindrical bodies have a tough, chitinous exoskeleton. Some secrete hydrogen cyanide, a poisonous substance.

Insects

Insects are adapted for an active life on land, although some have secondarily invaded aquatic habitats. The body of an insect is divided into a head, a thorax, and an abdomen. The head bears the sense organs and mouthparts (Fig. 28.26). The thorax bears three pairs of legs and possibly one or two pairs of wings; and the abdomen contains most of the internal organs. Wings enhance an insect's ability to survive by providing a way of escaping enemies, finding food, facilitating mating, and dispersing offspring.

Many insects, such as butterflies, undergo *complete metamorphosis,* involving drastic changes in form. At first, the animal is a wormlike larva (caterpillar) with chewing mouthparts. It then forms a case, or cocoon, about itself and becomes a pupa. During this stage, the body parts are completely reorganized; the adult then emerges from the cocoon. As mentioned earlier, this life cycle allows the larvae and adults to use different food sources.

Insects show remarkable behavioral adaptations. Bees, wasps, ants, termites, and other colonial insects have complex societies. Insects themselves also serve as an important food source for a variety of other animals (and certain carnivorous plants). The Biological Systems feature, "Would You Eat Insects?," on page 538 describes entomophagy, or the consumption of insects by humans, which is a common practice in many parts of the world.

Video
Thorn Tree Ants

Video
Ant-Caterpillar Mutualism

A Typical Insect: Grasshopper. In the grasshopper (Fig. 28.27), the third pair of legs is suited to jumping. This insect has two pairs of wings. The forewings are tough and leathery, and when folded back at rest they protect the broad, thin hindwings.

On each lateral surface, the first abdominal segment bears a large tympanum for the reception of sound waves. The posterior region of the exoskeleton in the female has an ovipositor, used to dig a hole in which eggs are laid.

The digestive system is suitable for a herbivorous diet. In the mouth, food is broken down mechanically by mouthparts and enzymatically by salivary secretions. Food is temporarily stored in the crop before passing into the gizzard, where it is finely ground. Digestion is completed in the stomach, and nutrients are absorbed into the hemocoel from outpockets called gastric ceca (sing., cecum). A cecum is a cavity that is open at one end only.

The excretory system consists of *Malpighian tubules,* which extend into a hemocoel and collect nitrogenous wastes, which are concentrated and excreted into the digestive tract. The formation of a solid nitrogenous waste—uric acid—conserves water.

The respiratory system begins with openings in the exoskeleton called spiracles. From here, air enters small tubules called **tracheae** (Fig. 28.27*a*). The tracheae branch and rebranch, finally ending in moist areas where the actual exchange of gases takes place. No

Mealybug, order Homoptera
Beetle, order Coleoptera
Leafhopper, order Homoptera
Head louse, order Anoplura
Wasp, order Hymenoptera
Dragonfly, order Odonata

Figure 28.26 Insect diversity.

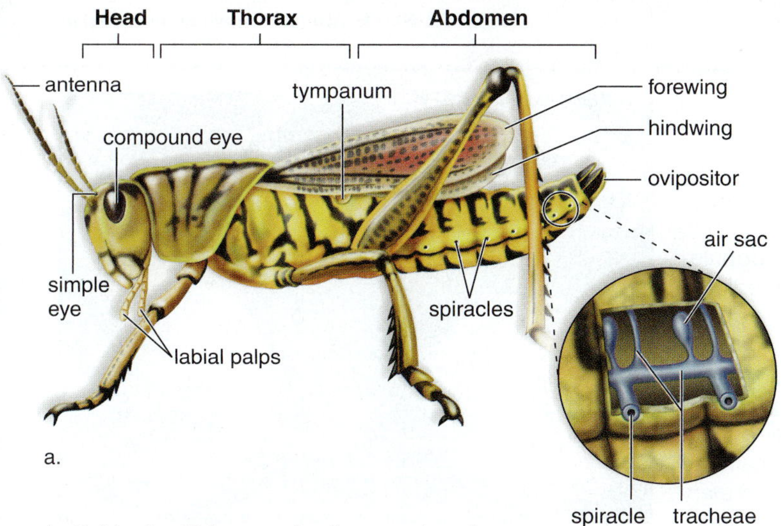

a.

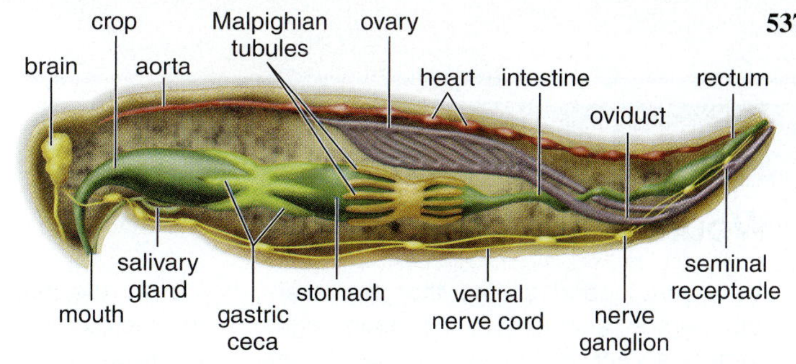

b.

Figure 28.27 Female grasshopper, *Romalea*. **a.** Externally, the body of a grasshopper is divided into three sections and has three pairs of legs. The tympanum receives sound waves, and the jumping legs and the wings are for locomotion. **b.** Internally, the digestive system is specialized. The Malpighian tubules excrete a solid, nitrogenous waste (uric acid). A seminal receptacle receives sperm from the male, which has a penis.

individual cell is very far from a site of gas exchange. The movement of air through this complex of tubules is not a passive process; air is pumped through by a series of bladderlike structures (air sacs) attached to the tracheae near the spiracles. Air enters the anterior four spiracles and exits by the posterior six spiracles. Breathing by tracheae may account for the small size of insects (most are less than 6 cm in length), because the tracheae are so tiny and fragile that they would be crushed by any amount of weight.

The circulatory system contains a slender, tubular heart, which lies against the dorsal wall of the abdominal exoskeleton and pumps hemolymph into the hemocoel, where it circulates before returning to the heart again. The hemolymph lacks a respiratory pigment; therefore it is colorless and transports mainly nutrients and wastes. The highly efficient tracheal system transports respiratory gases.

Grasshoppers undergo *incomplete metamorphosis,* a gradual change in form as the animal matures. The immature grasshopper, called a nymph, is recognizable as a grasshopper, even though it differs in body proportions from the adult.

Chelicerates

The **chelicerates** live in terrestrial, aquatic, and marine environments. The first pair of appendages is the pincerlike chelicerae, used in feeding and defense. The second pair is the pedipalps, which can have various functions. A cephalothorax (fused head and thorax) is followed by an abdomen that contains internal organs.

Horseshoe crabs of the genus *Limulus* are familiar along the East Coast of North America (Fig. 28.28*a*). The body is covered by exoskeletal shields. The anterior shield is a horseshoe-shaped carapace, which bears two prominent, compound eyes. Ticks, mites, scorpions, spiders, and harvestmen are all arachnids. Over 25,000 species of mites and ticks have been classified, some of which are parasitic on a variety of other animals. Ticks are ectoparasites of various vertebrates, and they are carriers for such diseases as Rocky Mountain spotted fever and Lyme disease. When not attached to a host, ticks hide on plants and in the soil.

Video
Lyme Disease

Scorpions can be found on all continents except Antarctica (Fig. 28.28*b*). North America is home to approximately 1,500 species. Scorpions are nocturnal and spend most of the day hidden under a log or a rock. Their pedipalps are large pincers, and their long abdomen ends with a stinger, which contains venom.

Currently, over 35,000 species of spiders have been classified (Fig. 28.28*c*). Spiders, the most familiar chelicerates, have a narrow waist that separates the cephalothorax from the abdomen. Spiders do not have compound eyes; instead, they have numerous simple eyes that perform a similar function. The chelicerae are modified as fangs, with ducts from poison glands, and the pedipalps are used to hold, taste, and chew food. The abdomen often contains silk glands, and spiders spin a web in which to trap their prey.

Check Your Progress 28.4

1. Name two ways in which the roundworms are anatomically similar to the arthropods.
2. List two ways that crustaceans are adapted to an aquatic life and insects are adapted to living on land.
3. Describe the features chelicerates have in common.

a. Horseshoe crab, *Limulus* b. Kenyan giant scorpion, *Pandinus* c. Black widow spider, *Latrodectus*

Figure 28.28 Chelicerate diversity. **a.** Horseshoe crabs are common along the East Coast. **b.** Scorpions are more common in tropical areas. **c.** The black widow spider is a poisonous spider that spins a web.

THEME **Biological Systems**

Would You Eat Insects?

An urban legend claims that the average person accidentally swallows eight spiders a year while sleeping. No matter how often this myth is repeated, however, there have been no studies performed to confirm it, and, in fact, it is unlikely that most of us ever swallow a spider, whether we're sleeping or awake. However, we do accidentally consume a much larger quantity of insects each year than we might imagine.

Despite our use of pesticides and even genetically resistant crops, insects are commonly present in our food. In its "Defect Levels Handbook," the United States Food and Drug Administration has set some limits on how many insect parts are allowable:

- Broccoli, frozen—60 insects or mites per 100 g
- Chocolate—60 insect fragments per 100 g
- Hops (to make beer)—2,500 aphids per 10 g
- Macaroni—225 insect fragments per 225 grams
- Mushrooms—20 maggots per 100 g
- Peanut butter—30 insect fragments per 100 g

Most of these limits are justified for "aesthetic" reasons, meaning anything more than that might be noticeable enough to gross out the average consumer, but that there are no health consequences associated with the consumption of insects. In fact, it's been estimated that each of us probably ingests 1 or 2 pounds of insect parts each year, without ever noticing.

Sometimes the use of insect products is intentional, however. In 2012, Starbucks announced it would stop using a dye made from the crushed bodies of Peruvian cochineal bugs, to color certain beverages and desserts, after vegan customers raised concerns. In contrast, humans have widely accepted the use of honey as a sweetener for thousands of years, even though you might hesitate if someone at the health food store offered to sweeten your tea with a sweet, sticky, yellowish-brown fluid make from fluid that bees regurgitate after feeding on nectar (also known as honey)!

As the human population increases to a predicted 9 billion by 2050, global food demand is expected to double, and it is unlikely that conventional agricultural methods will be able to meet this need. In particular, the ways we produce our dietary protein have a huge impact, as more people worldwide increase their consumption of beef, pork, and chicken. The livestock industry currently uses 70% of available agricultural land and accounts for 20% of greenhouse gas production. In a 2013 report, the United Nations Food and Agriculture Organization recommends considering the potential of insects as a logical food source for the rapidly growing human population.

Figure 28D Using insects for food. Insects being offered for human consumption in Bangkok, Thailand.

Although insects remain decidedly unpopular in the United States and most European countries, over 2 billion people include insects as a part of their regular diet (Fig. 28D). Out of about 1.3 million identified species of insects, approximately 1,900 species are considered edible. Beetles are most commonly consumed, followed by caterpillars, bees, wasps, and ants. Compared to cattle, insects produce nine times as much protein for the same amount of feed. Insects are also lower in fat and a great source of vitamins, minerals, and essential fatty acids. They require less energy to farm and can be cultivated at a much higher density than conventional livestock, and most people don't have the ethical concerns about the treatment of insects that many feel about raising conventional livestock in concentrated animal feeding operations ("factory farms"). Finally, insects are generally resistant to the types of microbial pathogens (e.g., *E. coli, Salmonella*) that seem to be an increasing problem with intensively raised conventional livestock.

A variety of companies are planning strategies to take advantage of the likely increase in entomophagy, or consumption of insects, in our near future. By 2020, a London-based company called Ento plans on introducing its stylish insect-based products, first into local markets and festivals, then restaurants and supermarkets. Bug Muscle targets bodybuilders and other athletes as the main consumers of its grasshopper- and cricket-based nutritional supplements; Hotlix has produced insect-containing candies for 25 years; and Tiny Farms will sell you kits to begin growing your own edible bugs at home.

You might also be surprised to learn that gourmet dishes featuring insects are already on the menus of some acclaimed restaurants: a grasshopper taco at Oyamel in Washington, DC; fried bamboo caterpillars at Sticky Rice Thai in Chicago; and Singapore-style scorpions at Typhoon in Santa Monica. You can also order insects for home cooking and snacking from a variety of online vendors. The next time you are offered an insect-based food (and it is likely that you will be), why not give it a try?

Questions to Consider

1. Many people find the idea of eating insects to be disgusting. However, what one person experiences as disgusting may not have the slightest effect on someone else. How can you explain this?
2. Our reaction to certain "disgusting" things seems to be innate—we are born with it. Can you think of why our tendency to avoid certain "disgusting" stimuli might be biologically based?
3. So, would you eat insects?

28.5 Invertebrate Deuterostomes

Learning Outcomes

Upon completion of this section, you should be able to

1. List the two major groups of animals in the Deuterostomia.
2. Identify the major morphological structures of the sea star, an echinoderm.
3. Describe how sea stars move, feed, and reproduce.

Deuterostomes include invertebrate echinoderms, such as sea stars (also known as starfish) and chordates. Molecular data tell us that echinoderms and chordates are closely related. Morphological data indicate that these two groups share the deuterostome pattern of development (see Fig. 28.5). The echinoderms and a few chordates are invertebrates; most of the chordates are vertebrates.

Echinoderms

Echinoderms (phylum Echinodermata [Gk. *echinos,* "spiny"; *derma,* "skin"]) are primarily bottom-dwelling marine animals. They range in size from brittle stars less than 1 cm in length to giant sea cucumbers over 2 m long (Fig. 28.29*c*).

The most striking feature of echinoderms is their five-pointed radial symmetry, as illustrated by a sea star (Fig. 28.29*b*). Although echinoderms are radially symmetrical as adults, their larvae are free-swimming filter feeders with bilateral symmetry. Echinoderms have an endoskeleton of spiny, calcium-rich plates called ossicles (Fig. 28.29*a*). Another innovation of echinoderms is their unique **water vascular system** consisting of canals and appendages that function in locomotion, feeding, gas exchange, and sensory reception.

The more familiar of the echinoderms are the Asteroidea, containing the sea stars (Fig. 28.29*a, b*); the Holothuroidea, including the sea cucumbers, which have long, leathery bodies in the shape

a. Sea star anatomy

b. Red sea star, *Mediastar*

c. Sea apple, *Pseudocolochirus*

d. Purple sea urchin, *Strongylocentrotus*

Figure 28.29 Echinoderms. **a.** Sea star (starfish) anatomy. Like other echinoderms, sea stars have a water vascular system that begins with the sieve plate and ends with expandable tube feet. **b.** The red sea star *Mediastar* uses the suction of its tube feet to open a clam, a primary source of food. **c.** Sea cucumber (*Pseudocolochirus*). **d.** Sea urchin (*Strongylocentrotus*).

of a cucumber (Fig. 28.29*c*); and the Echinoidea, including the sea urchin and sand dollar, both of which use their spines for locomotion, defense, and burrowing (Fig. 28.29*d*). Less familiar are the Ophiuroidea, which includes the brittle stars, with a central disk surrounded by radially flexible arms; and the Crinoidea, the oldest group, which includes the stalked feather stars and the motile feather stars.

A Typical Echinoderm: Sea Star

Sea stars number about 1,600 species commonly found along rocky coasts, where they feed on clams, oysters, and other bivalve molluscs. Various structures project through the body wall: (1) Spines from the endoskeletal plates offer some protection; (2) pincerlike structures around the bases of spines keep the surface free of small particles; and (3) skin gills, tiny, fingerlike extensions of the skin, are used for respiration. On the oral surface, each arm has a groove lined by little *tube feet* (Fig. 28.29*a*).

To feed, a sea star positions itself over a bivalve and attaches some of its tube feet to each side of the shell. By working its tube feet in alternation, it gradually pulls the shell open. A very small crack is enough for the sea star to evert its cardiac stomach and push it through the crack, so that it contacts the soft parts of the bivalve. The stomach secretes enzymes, and digestion begins, even while the bivalve is attempting to close its shell. Later, partly digested food is taken into the sea star's body, where digestion continues in the pyloric stomach using enzymes from the digestive glands found in each arm. A short intestine opens at the anus on the aboral side (side opposite the mouth).

In each arm, the well-developed coelom contains a pair of digestive glands and gonads (either male or female) that open on the aboral surface by very small pores. The nervous system consists of a central nerve ring, which gives off radial nerves into each arm. A light-sensitive eyespot is at the tip of each arm.

Locomotion depends on the water vascular system. Water enters this system through a structure on the aboral side called the *sieve plate,* or *madreporite* (Fig. 28.29*a*). From there it passes through a stone canal to a ring canal, which circles around the central disc, and then to a radial canal in each arm. From the radial canals, many lateral canals extend into the tube feet, each of which has an ampulla. Contraction of an ampulla forces water into the tube foot, expanding it. When the foot touches a surface, the center is withdrawn, giving it suction so that it can adhere to the surface. By alternating the expansion and contraction of the tube feet, a sea star moves slowly along.

Echinoderms do not have a respiratory, excretory, or circulatory system. Fluids within the coelomic cavity and the water vascular system carry out many of these functions. For example, gas exchange occurs across the skin gills and the tube feet. Nitrogenous wastes diffuse through the coelomic fluid and the body wall. Cilia on the coelom and other structures keep the coelomic fluid moving.

Sea stars reproduce asexually and sexually. If the body is fragmented, each fragment can regenerate a whole animal as long as a portion of the central disc is present. Sea stars spawn and release either eggs or sperm at the same time. The bilaterally symmetrical larvae undergo metamorphosis to become radially symmetrical adults.

Video Sea Urchin Reproduction

Check Your Progress 28.5

1. Explain why echinoderms and chordates are now considered to be closely related.
2. Delineate the evidence that supports the evolution of echinoderms from bilaterally symmetrical animals.
3. Describe the location and function of skin gills, tube feet, and the stomach.
4. Explain the functions of the water vascular system in sea stars.

CONNECTING *the* CONCEPTS *with the* THEMES

Evolution

- Some eukaryote colonial protists have individuals that specialize in particular functions; the first multicellular animals may have evolved from colonial flagellate specialization.
- Tissues could have evolved from infolding of specialized colonial cells.
- The evolutionary history of invertebrates is complex because of the diversity of body shapes and lifestyles, which are sometimes present in multiple stages of development.

Nature of Science

- Traditional phylogenies of invertebrates, based on morphological traits, are being challenged by new data from multiple genes.
- Each new phylogeny is the best current hypothesis supported by both morphological and molecular data.

Biological Systems

- Parasitic invertebrates are adapted to life inside a host—in some cases, humans—but they also require a life stage inside one or more different intermediate hosts.
- Human invertebrate parasites are responsible for many diseases with high incidence in impoverished nations, because living and environmental conditions favor the spread of these diseases. These diseases have been listed by the World Health Organization as neglected tropical diseases (NTDs).

MEDIA STUDY TOOLS http://connect.mheducation.com

SMARTBOOK® Study smart. SmartBook helps you maximize your study time by identifying what you know and don't know. The recharge feature in SmartBook helps you prepare for the test by identifying the material that you are most likely to forget.

connect BIOLOGY Find even more resources to learn the chapter concepts including animations, tutorial videos, and interactive practice questions.

Animation	Tutorial	Videos
28.1 Three Domains	**28.1** Protostomes and Deuterostomes	**28.1** Tadpole Development • Blastocyst Formation **28.2** Ctenophores • Coral Reef Spawning • Coral Reef Ecosystems • Portuguese Man-of-War **28.3** Clam Locomotion **28.4** Thorn Tree • Ant-Caterpillar Mutualism Ants • Lyme Disease **28.5** Sea Urchin Reproduction

SUMMARIZE

28.1 Evolution of Animals

Animals are multicellular organisms that are heterotrophic and ingest their food. Typically, they are diploid, they move by means of contracting fibers, and they are able to reproduce sexually. Animals can be **vertebrates,** which have a backbone, or **invertebrates,** which do not.

The **colonial flagellate hypothesis** proposes that animals evolved from a protist that resembled the **choanoflagellates** of today. The traditional evolutionary tree of invertebrate animals was based only on morphological characters. A modern phylogeny constructed with molecular and morphological data presents an alternative evolutionary history.

The first developmental event after fertilization is **cleavage,** or cell division without growth. Next, a hollow ball of cells called a **blastula** appears, forming a body cavity that becomes the **coelom.** In animals with a **true coelom,** mesoderm completely lines the coelom cavity. Most animals have a particular **symmetry,** or pattern, to their body shape, although a few exhibit **asymmetry.** Animal bodies that are organized in a circular fashion have a **radial symmetry;** those with definite right and left halves have **bilateral symmetry.** Many body plans include **cephalization,** or location of a brain and specialized sensory organs at the anterior end.

All animals except sponges, cnidarians, and comb jellies are triploblastic, meaning their tissues develop from three embryonic **germ layers.** Triploblastic animals are either **deuterostomes** or **protostomes,** which differ in the development of the coelom and whether the blastopore gives rise to the mouth (protostomes) or the anus (deuterostomes).

28.2 The Simplest Invertebrates

Sponges, which may be the oldest lineage of animals, resemble colonial protozoans. They have a cellular level of organization, lack tissues, and are asymmetrical. Sponges are sessile filter feeders that depend on a flow of water through the body to acquire food. Sponges can be classified by their type of **spicules,** which are small, needle-shaped structural components of the endoskeleton.

Comb jellies and cnidarians are diploblastic. They are aquatic and radially symmetrical, with bands of cilia that propel them through the water. The body of a comb jelly is composed largely of jellylike **mesoglea.**

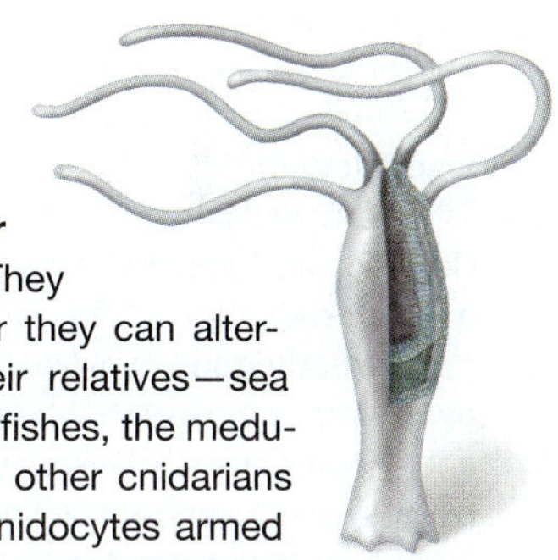

Cnidarians have a sac body plan. They have an internal **gastrovascular cavity** and a **hydrostatic skeleton.** They exist as either a **polyp** or a **medusa,** or they can alternate between the two. Hydras and their relatives—sea anemones and corals—are polyps; in jellyfishes, the medusan stage is most common. Hydras and other cnidarians possess tentacles to capture prey and cnidocytes armed with **nematocysts** to stun it. A **nerve net** coordinates movements.

28.3 Diversity Among the Lophotrochozoans

There are two major types of **Lophotrochozoa. Lophophorans,** which include the **bryozoans, brachiopods,** and **phoronids,** feed with a feathery lophophore that filters microorganisms from water. **Trocozoans** have a trocophore larva, either in living forms or in their ancestors.

Free-living **flatworms** (planarians) have three tissue layers and no coelom. They are **hermaphroditic,** and usually reproduce sexually. Planarians have muscles and a ladder-type nervous system with paired **ganglia** serving as the brain. They take in food through an extended pharynx, leading to a gastrovascular cavity that extends through the body. An osmotic-regulating organ contains flame cells.

Flukes and tapeworms are common parasites of vertebrate animals. Flukes have two suckers by which they attach to and feed from their hosts. Tapeworms have a **scolex** with hooks and suckers for attaching to the host's intestinal wall. The body of a tapeworm is made up of **proglottids,** which, when mature, contain thousands of eggs. If these eggs are taken up by pigs or cattle, for example, larvae become encysted in their muscles. If humans eat this meat, they, too, may become infested with a tapeworm. Tapeworm infection is a serious concern in impoverished nations.

Rotifers are microscopic and aquatic and have a corona that resembles a spinning wheel when in motion.

The body of a **mollusc** typically contains a visceral mass, a **mantle,** and a foot. Many also have a head and a radula for feeding. The nervous system consists of several ganglia connected by nerve cords. They have a reduced coelom and an open circulatory system with hemolymph that flows into a **hemocoel.** Clams (**bivalves**) are adapted to a sedentary coastal life, snails (**gastropods**) to life on land, and squids (**cephalopods**) to life in the sea.

Annelids are worms that exhibit **segmentation.** They have a well-developed coelom that acts as a hydrostatic skeleton, a closed circulatory system, a ventral solid nerve cord, **setae** for movement, and paired **nephridia.** Earthworms are oligochaetes ("few bristles") that use the body wall for gas exchange. Polychaetes ("many bristles") are marine worms that may be predators, with a definite head region, or filter feeders, with ciliated tentacles to filter food from the water. Leeches also belong to this phylum.

28.4 Diversity of the Ecdysozoans

Ecdysozoans are protostomes that produce a **cuticle,** which is periodically shed. **Roundworms,** which have a **pseudocoelom,** are usually small and very diverse; they are present almost everywhere in great numbers. Many have adapted to a parasitic lifestyle; examples in humans include *Ascaris, Trichinella,* and filarial worms. These diseases are more common in developing tropical and subtropical regions where sanitation is poor.

Many biologists consider the **arthropods** to be the most successful and varied group of animals. Their success is largely attributable to a flexible **exoskeleton,** specialized body regions, and jointed appendages. Also important are a high degree of cephalization and a variety of respiratory organs, and many undergo some form of **metamorphosis.** The hard exoskeleton requires a molt at different stages of growth. Crustaceans, insects, and chelicerates are representative groups of arthropods.

Crustaceans (crayfish, lobsters, shrimps, copepods, krill, and barnacles) have a head that bears compound eyes, antennae, antennules, and mouthparts. Crayfish have an open circulatory system, respiration by gills, and a ventral solid nerve cord. **Centipedes** and **millipedes** are known for their many legs.

Insects include butterflies, grasshoppers, bees, and beetles. The anatomy of the grasshopper is adapted to life on land. Like other insects, grasshoppers have wings and three pairs of legs attached to the thorax. They have a tympanum for sound reception, a digestive system specialized for a grass diet, Malpighian tubules for excretion of solid nitrogenous waste, **tracheae** for respiration, internal fertilization, and incomplete metamorphosis.

Chelicerates (horseshoe crabs, spiders, scorpions, ticks, and mites) have chelicerae, pedipalps, and four pairs of walking legs attached to a cephalothorax.

28.5 Invertebrate Deuterostomes

Echinoderms (sea stars, sea urchins, sea cucumbers, and sea lilies) share the deuterostome pattern of development with the chordates. They are radially symmetrical as adults but not as larvae, and they have internal, calcium-rich plates with spines. Typical of echinoderms, **sea stars** have tiny skin gills, a central nerve ring with branches, and a **water vascular system** for locomotion. Each arm of a sea star contains branches from the nervous, digestive, and reproductive systems.

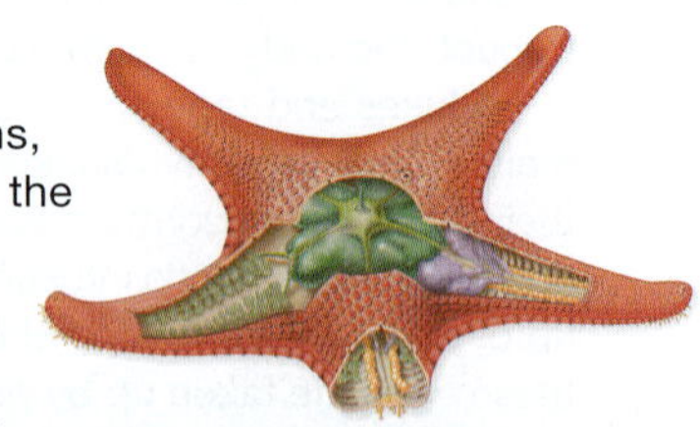

ASSESS

Choose the best answer for each question.

28.1 Evolution of Animals

1. Which of these is not a characteristic of animals?
   a. heterotrophic
   b. ingest their food
   c. cells lack cell walls
   d. have chlorophyll
   e. multicellular
2. The phylogenetic tree of animals shows that
   a. rotifers are closely related to flatworms.
   b. both molluscs and annelids are protostomes.
   c. some animals have radial symmetry.
   d. sponges were the first to evolve from an ancestral protist.
   e. All of these are correct.
3. Which of these does not pertain to a protostome?
   a. spiral cleavage
   b. blastopore is associated with the anus
   c. coelom, splitting of mesoderm
   d. annelids, arthropods, and molluscs
   e. mouth is associated with first opening

28.2 The Simplest Invertebrates

4. Which features of a sponge are mismatched?
   a. buds—asexual reproduction
   b. collar cells—flagellated
   c. osculum—upper opening
   d. spicules—sexual reproduction
   e. spongin—support
5. Comb jellies are most closely related to
   a. cnidarians.
   b. sponges.
   c. flatworms.
   d. roundworms.
   e. rotifers.
6. Which of these pairs is mismatched?
   a. sponges—spicules
   b. cnidarians—nematocysts
   c. cnidarians—polyp and medusa
   d. hydra—nerve net
   e. comb jellies—proglottids

28.3 Diversity Among the Lophotrochozoans

7. Flukes and tapeworms
   a. show cephalization.
   b. have well-developed reproductive systems.
   c. have well-developed nervous systems.
   d. are free-living.
   e. are generally microscopic.
8. Which of these best shows that snails are not closely related to crayfish?
   a. Snails are terrestrial, and crayfish are aquatic.
   b. Snails have a broad foot, and crayfish have jointed appendages.
   c. Snails are hermaphroditic, and crayfish have separate sexes.
   d. Snails are insects, but crayfish are fishes.
   e. Snails are bivalves, and crayfish are chelicerates.

9. Segmentation in the earthworm is not exemplified by
   a. body rings.
   b. coelom divided by septae.
   c. setae on most segments.
   d. nephridia interior in most segments.
   e. tympanum exterior to segments.

28.4 Diversity of the Ecdysozoans

10. Roundworms lack which feature?
    a. periodically shed their cuticle
    b. pseudocoelom
    c. segmentation
    d. some are parasitic
    e. All of these are correct.
11. Label this diagram.

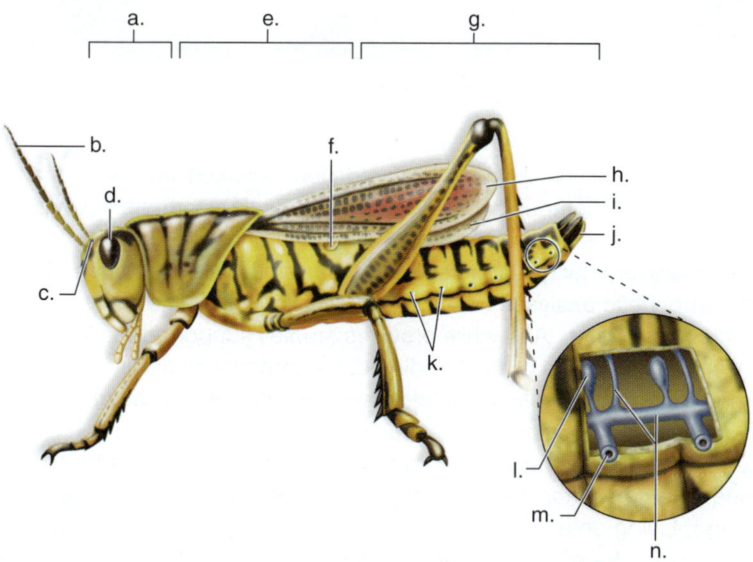

12. Which characteristic(s) account(s) for the success of arthropods?
    a. jointed exoskeleton
    b. well-developed nervous system
    c. segmentation
    d. respiration adapted to environment
    e. All of these are correct.

28.5 Invertebrate Deuterostomes

13. The two phyla in the Deuterostomia are
    a. Arthropoda and Nematoda.
    b. Brachiopoda and Platyhelminthes.
    c. Cestoda and Rotifera.
    d. Cnidaria and Ctenophora.
    e. Echinodermata and Chordata.
14. Which is not a characteristic feature of sea stars?
    a. coelom
    b. eyespot
    c. skin gills
    d. trachea
    e. tube feet
15. The water vascular system of sea stars functions in
    a. circulation of gasses and nutrients.
    b. digestion.
    c. locomotion.
    d. reproduction.
    e. Both a and c are correct.

ENGAGE

THINKING SCIENTIFICALLY

1. Multicellular animals are thought to have evolved from single-celled protists resembling the choanoflagellates that exist today. Describe some specific advantages to multicellularity that gave animals an advantage over their single-celled relatives.
2. Cnidarians such as *Hydra* have only one opening, which must serve as both mouth and anus. In contrast, most animal species have a one-way, tubelike digestive system, with separate openings for intake and output. Describe some specific advantages of the latter arrangement.
3. Many species of invertebrates (e.g., flatworms, roundworms, arthropods) are parasites of other animals. List several advantages, as well as challenges, inherent in a parasitic lifestyle.
4. It was recently discovered that the filarial worm *Wuchereria bancrofti,* which causes elephantiasis, itself carries bacterial symbionts of the genus *Wolbachia.* Studies have suggested that the bacteria are essential for reproduction and survival of the worm. What does this suggest regarding potential treatments for elephantiasis?

29

Vertebrate Evolution

New Guinea singing dogs, *Canis familiaris hallstromi.*

Chapter Outline

29.1 The Chordates 545

29.2 The Vertebrates 547

29.3 The Fishes 548

29.4 The Amphibians 551

29.5 The Reptiles 553

29.6 The Mammals 559

Before You Begin

Before beginning this chapter, take a few moments to review the following discussions.

Section 15.3 How does *Archaeopteryx,* and similar transitional fossils, support the close relationship between birds and dinosaurs?

Table 18.1 When in the history of life on Earth did the vertebrates first appear?

Figure 21.1 Which eukaryotes are the closest relatives of vertebrate animals?

Dogs are probably our most familiar and beloved domesticated animals, "man's best friend." They are our companions as well as our working partners in war, law enforcement, herding, service, rescue, and therapy. But when and where did domestic dogs first evolve? Scientists have long suspected that *Canis familiaris* evolved from *Canis lupus,* the gray wolf, but genetic analyses now indicate that modern dogs evolved from a species of wolf that is now extinct. And while previous studies supported the idea that dogs were first domesticated in China or the Middle East, comparisons of mitochondrial DNA isolated from modern dogs, wolves, and ancient dog fossils now indicate that modern dogs most likely originated in Europe, at least 16,000 years ago.

The New Guinea singing dog is named for its habit of howling at different pitches. It is also called Stone Age dog, after the stone tool–using people who took it to New Guinea 6,000 years ago. Living on an island, it has remained geographically and reproductively isolated ever since. Therefore, it might be a "living fossil," an organism with the same genes and characteristics as its original ancestor. If so, the New Guinea singing dog offers an opportunity to study what the first domesticated dog was like. However, like many other vertebrates, which are the topic of this chapter, the New Guinea singing dog is threatened with extinction. Expeditions into the highlands of New Guinea have yielded only a few droppings, tracks, and haunting howls in the distance.

As you read through this chapter, think about the following questions:

1. When in the history of life did vertebrates evolve?
2. What characteristics set vertebrates apart from other animals?
3. How are vertebrates important to the day-to-day life of humans?

Following *the* Themes

Chapter 29 Vertebrate Evolution

Unit 6 Animal Evolution and Diversity

Evolution	Vertebrates share a common chordate ancestor that had a set of basic characteristics, including a nerve cord.
Nature of Science	The study of vertebrate evolution is key to solving important issues related to the Earth's changing environment and other aspects of human health and welfare.
Biological Systems	Vertebrates live in a wide diversity of habitats, where they contribute to the structure of the ecosystem.

29.1 The Chordates

Learning Outcomes

Upon completion of this section, you should be able to

1. Identify the four basic characteristics of a chordate.
2. Name the two groups of nonvertebrate chordates.
3. Describe two features of each of the two groups of nonvertebrate chordates.

Chordates (phylum Chordata), like the echinoderms (discussed in Chapter 28), are deuterostomes. In contrast to the exoskeleton of many invertebrates, however, most chordates have an internal skeleton made of bone and/or cartilage, to which the muscles are attached. This arrangement allows the chordates to enjoy a greater freedom of movement and enables many to attain a larger body size than invertebrates.

Most chordates are vertebrates, meaning they have a vertebral column, and most of this chapter is devoted to this very familiar group of animals. First, though, it is important to recognize the nonvertebrate chordates, some of which likely resemble the ancestors of all vertebrate animals, including humans.

Characteristics of Chordates

All chordates have four basic characteristics that appear at some point during development (Fig. 29.1). Although all of these characteristics were likely present in the adult stage of the common ancestor of chordates, not all of the traits are present in the adults of modern chordates. However, the presence of these characteristics during the development of all chordates indicates their evolutionary ties to a single common ancestor. These four characteristics are:

1. **Notochord** (Gk. *notos,* "back"; *chorde,* "string"). A dorsal supporting rod, the **notochord** is located just below the nerve cord. The majority of vertebrates have an embryonic notochord that is replaced by the vertebral column during embryonic development.
2. **Dorsal tubular nerve cord.** In contrast to the arthropods, which have a ventral nerve cord, chordates have a tubular cord situated dorsally. The anterior portion becomes the brain in most chordates. In vertebrates, the nerve cord, often called the spinal cord, is protected by vertebrae.

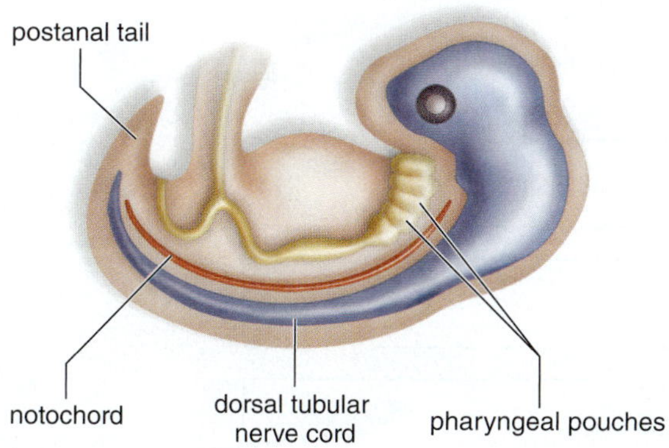

Figure 29.1 Characteristics of the chordates. Chordates have four distinctive traits. The notochord is present in all chordates.

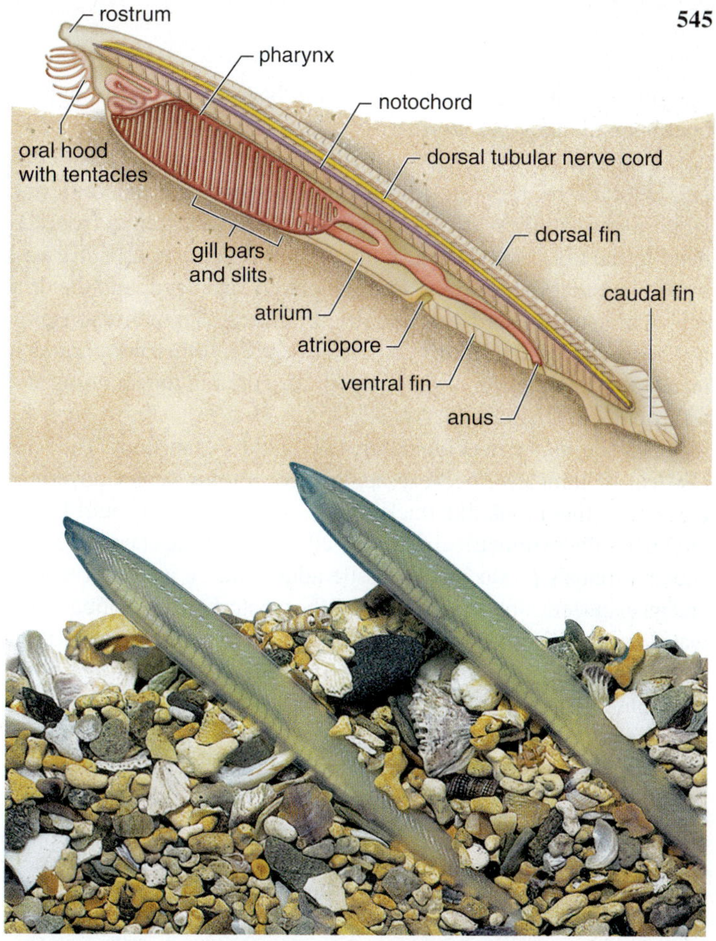

Figure 29.2 Lancelet, *Branchiostoma.* Lancelets are filter feeders. Water enters the mouth and exits at the atriopore after passing through the gill slits.

3. **Pharyngeal pouches.** These are seen only during embryonic development in most vertebrates. In the nonvertebrate chordates, the fishes, and amphibian larvae, the pharyngeal pouches become functioning **gills** (respiratory organs). Water passing into the mouth and the pharynx goes through the gill slits, which are supported by gill arches. In terrestrial vertebrates, the pouches are modified for various purposes. For example, in humans the first pair of pouches become the auditory tubes.
4. **Postanal tail.** A tail—in the embryo, if not in the adult—extends beyond the anus. In some chordates, including humans, the tail disappears during embryonic development.

Nonvertebrate Chordates

The nonvertebrate chordates do not have a spine made of bony vertebrae. They are divided into two groups: the cephalochordates and the urochordates.

Lancelets (genus *Branchiostoma,* previously called *Amphioxus*) are **cephalochordates** (Gk. *kephalo,* "head"). These marine chordates, which are only a few centimeters long, are named for their resemblance to a lancet—a small, two-edged surgical knife (Fig. 29.2). Lancelets are found in the shallow water along most coasts, where they usually lie partly buried in the sandy or muddy bottom with only their mouth and gill apparatus exposed. They feed on microscopic particles, which they filter out of a constant stream of water that enters the mouth and passes through the gill slits into a chamber, or *atrium,* before exiting through an opening called an *atriopore*.

Lancelet adults possess all four general chordate characteristics; therefore, they are important in comparative anatomy and evolutionary studies. In lancelets, the notochord extends from the head to the tail. In addition, segmentation is present, as illustrated by the fact that the muscles are segmentally arranged, and the dorsal tubular nerve cord has periodic branches. Segmentation may not be an important feature in lancelets, but in other vertebrates it leads to specialization of parts, as we observed in the annelids and arthropods in Chapter 28.

Sea squirts, or **urochordates,** live on the ocean floor. They are also called tunicates, because adults have a tunic (outer covering) that makes them look like thick-walled, squat sacs. Sea squirt larvae are bilaterally symmetrical and have all four chordate characteristics. Metamorphosis produces the sessile adult with an incurrent siphon and an excurrent siphon (Fig. 29.3). When disturbed, they often squirt water from their excurrent siphon, a habit that is reflected in their name. The pharynx is lined by numerous cilia, whose beating creates a current of water that moves into the pharynx and out the numerous gill slits, the only chordate characteristic that remains in the adult.

Many evolutionary biologists hypothesize that the sea squirts are directly related to the vertebrates. It has been suggested that a larva with the four chordate characteristics may have become sexually mature without developing the other adult sea squirt characteristics. Over time, it may have evolved into a fishlike vertebrate. Figure 29.4 shows how the main groups of chordates may have evolved.

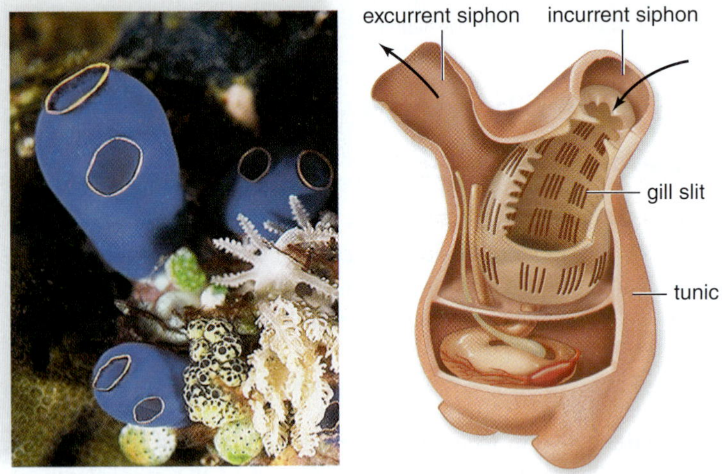

Figure 29.3 Sea squirt, *Halocynthia.* Note that the only chordate characteristic remaining in the adult is gill slits.

Check Your Progress 29.1

1. Discuss how humans, as chordates, possess all four characteristics either as embryos or adults.
2. Explain why adult sea squirts are classified as chordates, although they look like thick-walled, squat sacs.

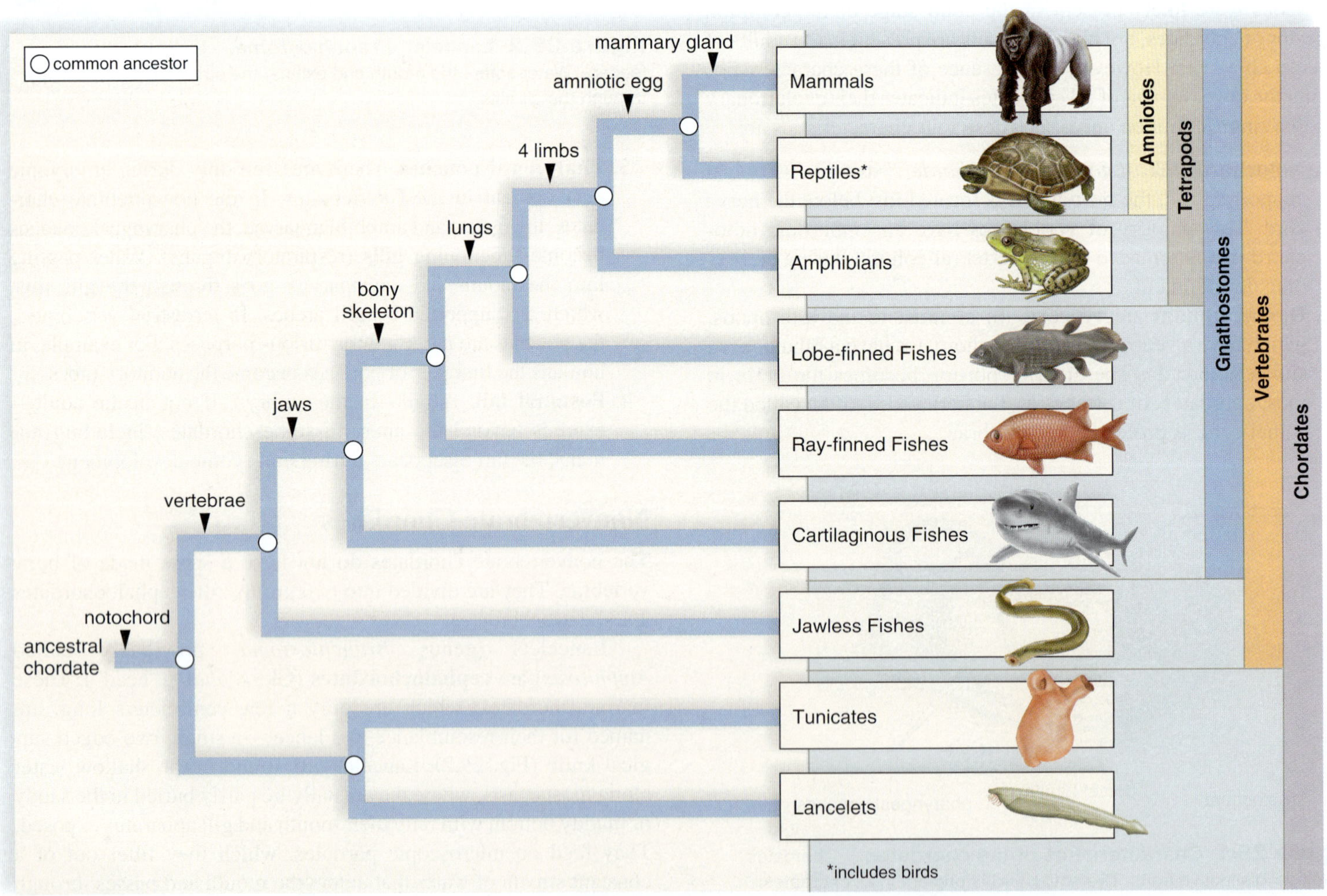

Figure 29.4 Phylogenetic tree of the chordates. For each branch of the tree, new characteristics seen in the common ancestor of each group are noted.

29.2 The Vertebrates

Learning Outcomes

Upon completion of this section, you should be able to

1. Describe the four characteristics that are unique to vertebrates.
2. Explain how the terms *tetrapod, gnathostome,* and *amniote* relate to vertebrate evolution.
3. Identify the geologic era and periods in which chordates and the first vertebrates appear.

Vertebrates are chordates with vertebral columns and certain other features that distinguish them from the nonvertebrate chordates.

Characteristics of Vertebrates

As embryos, vertebrates have the four chordate characteristics. In addition, vertebrates have these features:

Vertebral column. The embryonic notochord is generally replaced by a vertebral column composed of individual vertebrae (Fig. 29.5). Remnants of the notochord give rise to the intervertebral discs, which are compressible, cartilaginous pads between the vertebrae. The vertebral column, which is a part of the flexible but strong endoskeleton, gives evidence that vertebrates are segmented.

Skull. The main axis of the internal skeleton consists of not only the vertebral column but also a skull, which encloses and protects the brain. During vertebrate evolution, the brain has increased in complexity, and specialized regions have developed to carry out specific functions.

The high degree of cephalization is accompanied by complex sense organs. The eyes develop as outgrowths of the brain. The ears are primarily equilibrium devices in aquatic vertebrates, but they also function as sound-wave receivers in land vertebrates. In addition, many vertebrates possess well-developed senses of smell and taste.

Endoskeleton. The vertebrate skeleton (either cartilage or bone) is a living tissue that grows with the animal. It protects internal organs and serves as a place of attachment for muscles. Together, the skeleton and muscles form a system that permits rapid and efficient movement. Two pairs of appendages are characteristic. Fishes typically have pectoral and pelvic fins, while terrestrial tetrapods have four limbs.

Internal organs. Vertebrates have a large coelom and a complete digestive tract. The blood in vertebrates is contained entirely within blood vessels and is therefore a closed circulatory system. The respiratory system consists of gills or lungs, which obtain oxygen from the environment. The kidneys are important excretory and water-regulating organs that conserve or rid the body of water as necessary. The sexes are generally separate, and reproduction is usually sexual.

Vertebrate Evolution

The Paleozoic era is distinguished by the arising of the chordates and first vertebrates. Chordates appear on the scene suddenly at the start of the Cambrian period, 542 MYA. By the end of the Ordovician period, which followed, both jawless and jawed fishes had appeared; during this period, nonvascular plants moved onto the land.

Even though we do not know the precise origin of vertebrates, we can trace their evolutionary history, as shown in Figure 29.4. (See also Table 18.1 to review evolutionary events on the geologic time scale.) The earliest vertebrates were fishes, organisms that are abundant today in both marine and freshwater habitats. A few of today's fishes lack jaws and have to suck and otherwise engulf their prey. Most fishes have jaws, which are a more efficient means of grasping and eating prey. The jawed fishes and all the other vertebrates are **gnathostomes**—animals with jaws.

Jawed fishes had dominated the seas by the Silurian period, which followed the Ordovician. Some of these had not only jaws but also a bony skeleton, lungs, and fleshy fins. These characteristics were preadaptive for a land existence, and the amphibians, the first vertebrates to live on land, had evolved from these fishes by the Devonian period, 416 MYA. The amphibians were the first vertebrates to have limbs. The terrestrial vertebrates are **tetrapods** (Gk. *tetra,* "four"; *podos,* "foot"), because they have four limbs. Some, such as the snakes, no longer have four limbs, but their evolutionary ancestors did have four limbs.

Many amphibians, such as the frog, reproduce in an aquatic environment. This means that, in general, amphibians are not fully adapted to living on land. Reptiles are fully adapted to life on land, because, among other features, they produce an amniotic egg. The amniotic egg is so named because the embryo is

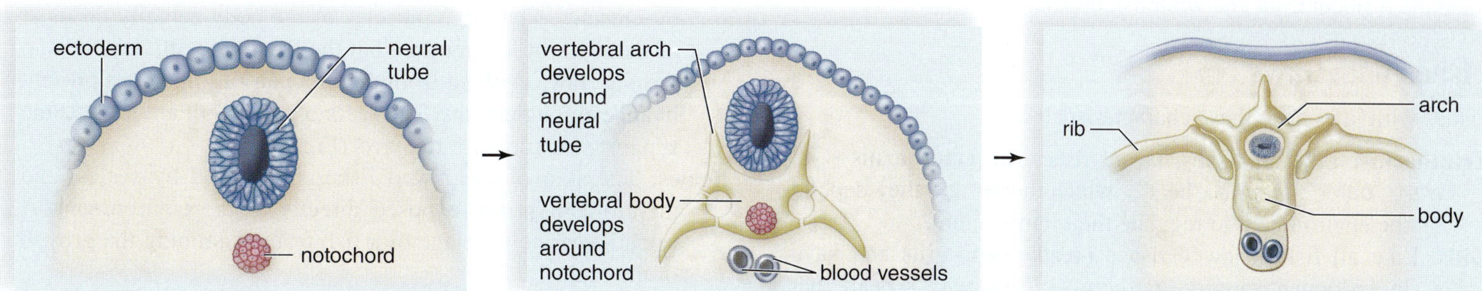

Figure 29.5 Replacement of notochord by the vertebrae. During vertebrate development, the vertebrae replace the notochord and surround the neural tube. The result is the flexible vertebral column, which protects the nerve cord.

surrounded by an amniotic membrane that encloses the amniotic fluid. Therefore, **amniotes,** animals that exhibit an amniotic membrane, develop within an aquatic environment, but one of their own making. In placental mammals, such as ourselves, the fertilized egg develops inside the female, where it is surrounded by an amniotic membrane.

A watertight skin, which is seen among the reptiles and mammals, is also a good feature to have when living on land.

Check Your Progress 29.2

1. Describe the advantages of an endoskeleton.
2. Explain how four legs would be useful in terrestrial environments.

29.3 The Fishes

Learning Outcomes

Upon completion of this section, you should be able to

1. List several features of jawless fishes.
2. Describe four characteristics shared by all jawed fishes.
3. Compare and contrast cartilaginous fishes and bony fishes.
4. Discuss the evolutionary significance of lobe-finned fishes.

Fishes are the largest group of vertebrates, with nearly 28,000 recognized species. They range in size from a few millimeters in length to the whale shark, which may reach lengths of 12 m. The fossil record of the fishes is extensive.

Jawless Fishes

The earliest fossils of Cambrian origin were the small, filter-feeding, jawless, and finless **ostracoderms.** Several groups of ostracoderms developed heavy dermal armor for protection.

Today's **jawless fishes,** or **agnathans,** have a cartilaginous skeleton and persistent notochord. They are cylindrical, are up to a meter long, and have smooth, nonscaly skin (Fig. 29.6). The hagfishes are exclusively marine scavengers that feed on soft-bodied invertebrates and dead fishes. Many species of lampreys are filter feeders, like their ancestors. Parasitic lampreys have a round, muscular mouth used to attach themselves to another fish and suck nutrients from the host's cardiovascular system. The parasitic sea lamprey gained access to the Great Lakes in 1829, and by 1950 it had almost demolished the resident trout and whitefish populations.

Fishes with Jaws

Fishes with jaws have these characteristics:

Ectothermy. Like all fishes, jawed fishes are **ectotherms** (Gk. *ekto,* "outer"; *therme,* "heat"), which means that they depend on the environment to regulate their temperature.

Gills. Like all fishes, jawed fishes breathe with gills and have a single-looped, closed circulatory system with a heart that pumps the blood first to the gills. Then, oxygenated blood passes to the rest of the body.

Figure 29.6 Lamprey, genus *Petromyzon.* Lampreys, which are agnathans, have an elongated, rounded body and nonscaly skin. Note the lamprey's toothed oral disk, which is attached to the aquarium glass.

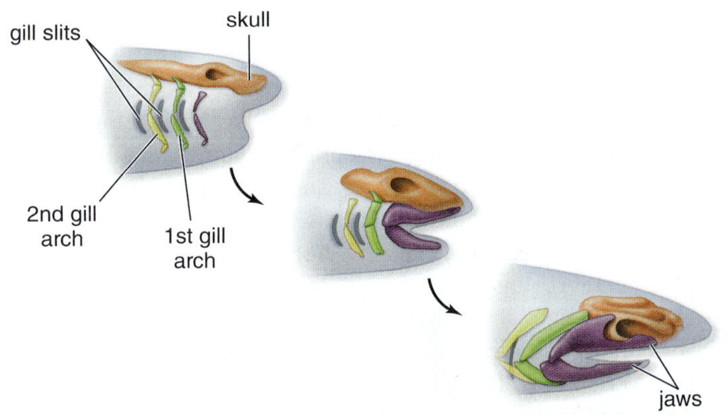

Figure 29.7 Evolution of the jaw. The first jaw evolved from the first and second gill arches of fishes.

Cartilaginous or bony endoskeleton. The endoskeleton of jawed fishes includes the vertebral column, a skull with jaws, and paired pectoral and pelvic **fins,** projections that are controlled by muscles. The large muscles of the body actually do most of the work of locomotion, but the fins help with balance and turning. Jaws evolved from the first pair of gill arches present in ancestral agnathans. The second pair of gill arches became support structures for the jaws (Fig. 29.7).

Scales. The skin of the jawed fishes is covered by scales, and therefore, it is not exposed directly to the environment. A scientist can tell the age of a fish from examining the growth of the scales.

The **placoderms,** extinct jawed fishes of the Devonian period, are probably the ancestors of early sharks and bony fishes. Placoderms

a. Tiger shark, *Galeocerdo cuvier*

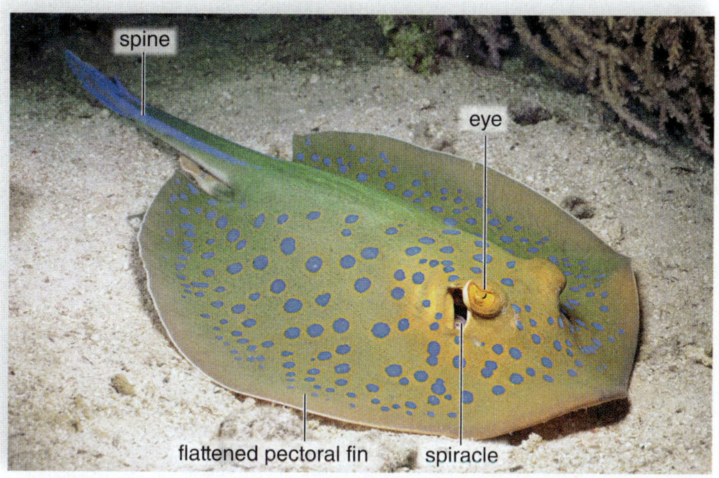

b. Blue-spotted stingray, *Taeniura lymma*

Figure 29.8 Cartilaginous fishes. **a.** Sharks are predators or scavengers that move gracefully through open ocean waters. **b.** Most stingrays grovel in the sand, feeding on bottom-dwelling invertebrates. This blue-spotted stingray is protected by the spine on its whiplike tail.

were armored with heavy, bony plates and had strong jaws. Like modern-day fishes, they also had paired pectoral and pelvic fins (Fig. 29.8).

Cartilaginous Fishes

Sharks, rays, skates, and chimaeras are marine **cartilaginous fishes** (Chondrichthyes). The cartilaginous fishes have a skeleton composed of cartilage instead of bone; have five to seven gill slits on both sides of the pharynx; and lack the gill cover of bony fishes. In addition, many have openings to the gill chambers located behind the eyes called spiracles. Their body is covered with dermal denticles—tiny, teethlike scales that project posteriorly—which is why a shark's skin feels like sandpaper. The menacing teeth of sharks and their relatives are simply larger, specialized versions of these scales. At any one time, a shark such as the great white shark may have up to 3,000 teeth in its mouth, arranged in 6 to 20 rows. Only the first row or two are actively used for feeding; the other rows are replacement teeth.

Three well-developed senses enable sharks and rays to detect their prey. They have the ability to sense electric currents in water—even those generated by the muscle movements of animals. They have a lateral line system, a series of pressure-sensitive cells that lie within canals along both sides of the body, which can sense pressure caused by a fish or other animal swimming nearby. They also have a very keen sense of smell; the part of the brain associated with this sense is very well developed. Sharks can detect about one drop of blood in 115 liters of water.

The largest sharks are filter feeders, not predators. The basking sharks and whale sharks ingest tons of small crustaceans, collectively called krill. Many sharks are fast-swimming predators in the open sea (Fig. 29.8*a*). The great white shark, about 7 m in length, feeds regularly on dolphins, sea lions, and seals. Humans are normally not attacked, except when mistaken for sharks' usual prey. Tiger sharks, so named because the young have dark bands, reach 6 m in length and are unquestionably one of the most predaceous sharks. As it swims through the water, a tiger shark will swallow anything, including rolls of tar paper, shoes, gasoline cans, paint cans, and even human parts. The number of bull shark attacks has increased in the Gulf of Mexico in recent years. This increase is due to more swimmers in the shallow waters and a loss of the sharks' natural food supply.

In rays and skates (Fig. 29.8*b*), the pectoral fins are greatly enlarged into a pair of large, winglike fins, and the bodies are dorsoventrally flattened. The spiracles are enlarged and allow them to move water over the gills while resting on the bottom, where they feed on organisms such as crustaceans, small fishes, and molluscs.

Stingrays have a whiplike tail that has serrated spines with venom glands at the base. This tail spike is a defensive weapon, but it can deliver a harmful or even fatal wound. Manta rays are harmless, oceanic, filter-feeding giants with a fin span of up to 6 m and a weight of 2,000 kg. Sawfish rays are named for their large, protruding anterior "saw." Some species of stingrays can deliver an electric shock. Their large electric organs, located at the base of their pectoral fins, can discharge over 300 volts. Skates resemble stingrays but possess two dorsal fins and a caudal fin.

The chimaeras, or ratfishes, are a group of cartilaginous fishes that live in cold marine waters. They are known for their unusual shape and iridescent colors.

Bony Fishes

The majority of living vertebrates, approximately 25,000 species, are **bony fishes** (Osteichthyes). The bony fishes range in size from gobies, which are less than 7.5 mm long, to the giant sturgeons, which can obtain a length of 4 m.

The majority of fish species are **ray-finned bony fishes** with fan-shaped fins supported by a thin, bony ray. These fishes are the most successful and diverse of all the vertebrates (Fig. 29.9). Some, such as herrings, are filter feeders; others, such as trout, are opportunists; and still others are predaceous carnivores, such as piranhas and barracudas.

Video Cichlid Specialization

Despite their diversity, bony fishes have many features in common. They lack external gill slits; instead, their gills are covered by an *operculum.* Many bony fishes have a **swim bladder,** a gas-filled sac into which they can secrete gases or from which they can absorb gases, altering its pressure. This results in a

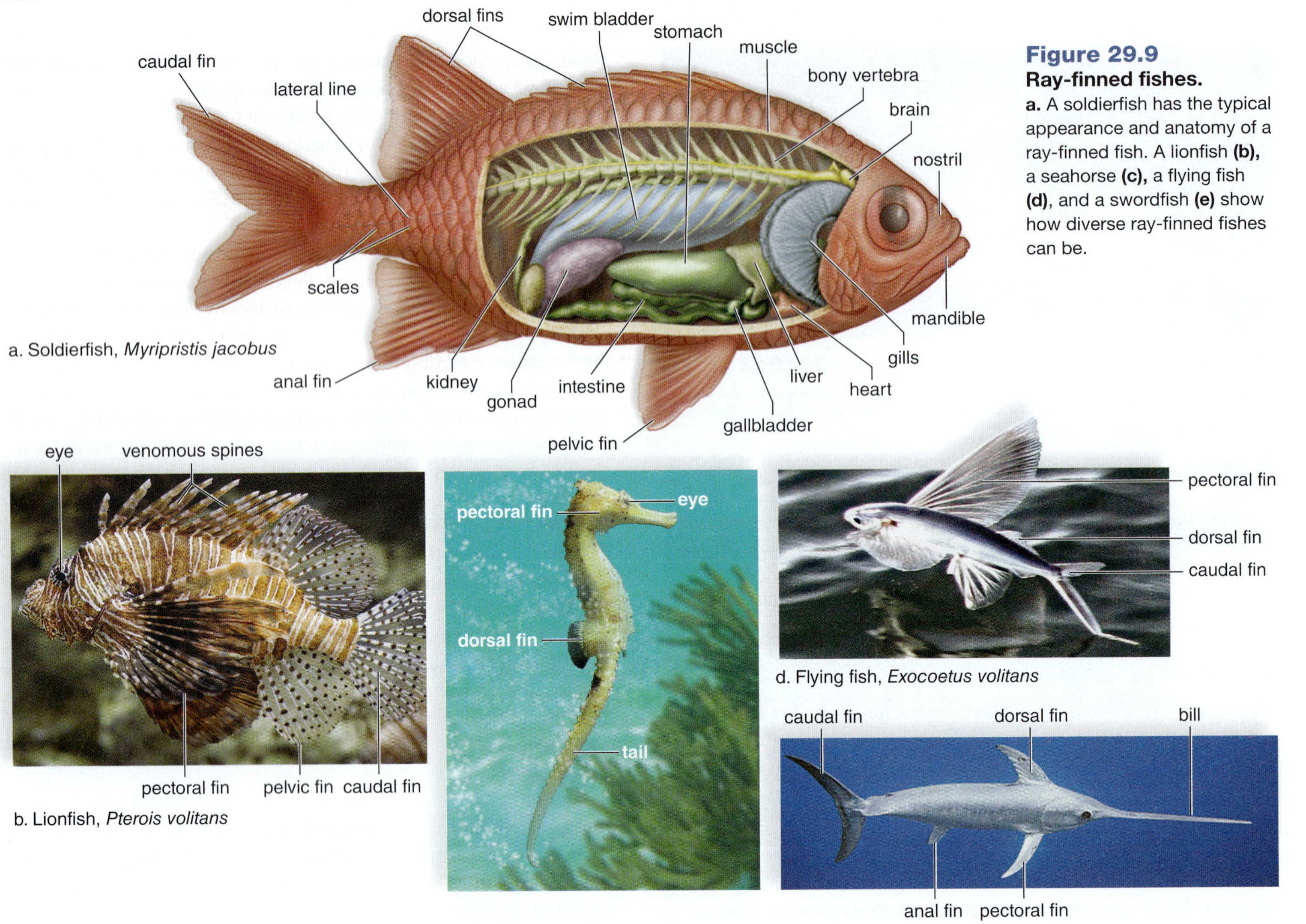

a. Soldierfish, *Myripristis jacobus*

b. Lionfish, *Pterois volitans*

c. Seahorse, *Hippocampus kuda*

d. Flying fish, *Exocoetus volitans*

e. Swordfish, *Xiphias gladius*

Figure 29.9 Ray-finned fishes.
a. A soldierfish has the typical appearance and anatomy of a ray-finned fish. A lionfish **(b),** a seahorse **(c),** a flying fish **(d),** and a swordfish **(e)** show how diverse ray-finned fishes can be.

change in the fishes' buoyancy and, therefore, their depth in the water. Bony fishes have a single-loop, closed cardiovascular system (see Fig. 29.11*a*).

The nervous system and brain in bony fishes are well developed, and complex behaviors are common. Bony fishes have separate sexes, and the majority of species undergo external fertilization after females deposit eggs and males deposit sperm into the water.

Lobe-finned Fishes. **Lobe-finned fishes** possess fleshy fins supported by bones. Ancestral lobe-finned fishes gave rise to modern-day lobe-finned fishes, the lungfishes, and to the amphibians. **Lungfishes** have lungs as well as gills for gas exchange. The lobe-finned fishes and lungfishes are grouped together as the Sarcopterygii. Today, only two species of lobe-finned fishes (the coelacanths) and six species of lungfishes are known. Lungfishes live in Africa, South America, and Australia, either in stagnant fresh water or in ponds that dry up annually.

In 1938, a coelacanth was caught from the deep waters of the Indian Ocean off the eastern coast of South Africa. It took the scientific world by surprise, because these animals were thought to be extinct for 70 million years. Approximately 200 coelacanths have been captured since that time (Fig. 29.10).

Check Your Progress 29.3

1. List and describe the characteristics that fishes have in common.
2. Distinguish between lobe-finned and ray-finned bony fishes.

Figure 29.10 Coelacanth, *Latimeria chalumnae.*
A coelacanth is a lobe-finned fish once thought to be extinct.

29.4 The Amphibians

Learning Outcomes

Upon completion of this section, you should be able to

1. List the seven characteristics that define the amphibians.
2. Describe the features of the three groups of living amphibians.
3. Summarize the two hypotheses that explain the evolution of amphibians from lobe-finned fishes.

Amphibians (class Amphibia [Gk. *amphibios,* "living both on land and in water"]) were abundant during the Carboniferous period and exhibit these characteristics:

Limbs. Typically, amphibians are tetrapods, as mentioned earlier. The skeleton, particularly the pelvic and pectoral girdles, is well developed to promote locomotion.

Smooth and nonscaly skin. The skin, which is kept moist by mucous glands, plays an active role in water balance and respiration, and it can help in temperature regulation when on land through evaporative cooling. A thin, moist skin does mean, however, that most amphibians stay close to water, or else they risk drying out.

Lungs. If lungs are present, they are relatively small, and respiration is supplemented by exchange of gases across the porous skin (called cutaneous respiration).

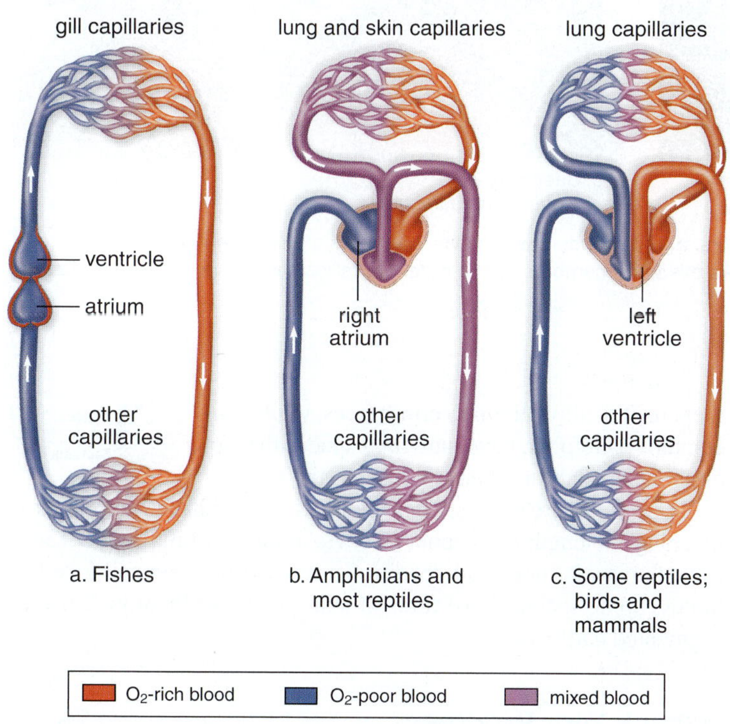

Figure 29.11 Vertebrate circulatory pathways. **a.** The single-loop pathway of fishes has a two-chambered heart. **b.** The double-loop pathway of other vertebrates sends blood to the lungs and to the body. In amphibians and most reptiles, limited mixing of oxygen-rich and oxygen-poor blood takes place in the single ventricle of their three-chambered heart. **c.** The four-chambered heart of some reptiles (crocodilians and birds) and mammals sends only oxygen-poor blood to the lungs and oxygen-rich blood to the body.

Double-loop circulatory pathway. A three-chambered heart, with a single ventricle and two atria, pumps blood to both the lungs and the body (Fig. 29.11*b, c*).

Sense organs. Special senses, such as sight, hearing, and smell, are fine-tuned for life on land. Amphibian brains are larger than those of fish, and the cerebral cortex is more developed. These animals have a specialized tongue for catching prey, eyelids for keeping their eyes moist, and a sound-producing larynx.

Ectothermy. Like fishes, amphibians are ectotherms, but they are able to live in environments where the temperature fluctuates greatly. During winters in the temperate zone, they become inactive and enter torpor. The European common frog can survive in temperatures dropping to as low as −6°C.

Aquatic reproduction. Their name, amphibians, is appropriate because many return to water for the purpose of reproduction. They deposit their eggs and sperm into the water, where external fertilization takes place. Generally, the eggs are protected only by a jelly coat, not by a shell. When the young hatch, they are tadpoles (aquatic larvae with gills) that feed and grow in the water. After amphibians undergo a *metamorphosis* (change in form), they emerge from the water as adults that breathe air. Some amphibians, however, have evolved mechanisms that allow them to bypass this aquatic larval stage and reproduce on land.

Evolution of Amphibians

Amphibians evolved from the lobe-finned fishes with lungs by way of transitional forms. Two hypotheses have been suggested to account for the evolution of amphibians from lobe-finned fishes. Perhaps lobe-finned fishes had an advantage over others because they could use their lobed fins to move from pond to pond. Or perhaps the supply of food on land in the form of plants and insects—and the absence of predators—promoted further adaptations to the land environment. Paleontologists have recently found a well-preserved transitional fossil from the late Devonian period in Arctic Canada that represents an intermediate between lobe-finned fishes and tetrapods with limbs. This fossil, named *Tiktaalik roseae* (see Chapter 15), provides unique insights into how the legs of tetrapods arose (Fig. 29.12).

Video Amphibian Origin

Diversity of Living Amphibians

The amphibians of today occur in three groups: salamanders and newts; frogs and toads; and caecilians. Salamanders and newts have elongated bodies, long tails, and usually two pairs of limbs (Fig. 29.13*a*). Salamanders and newts range in size from less than 15 cm to the giant Japanese salamander, which exceeds 1.5 m in length. Most have limbs that are set at right angles to the body and resemble the earliest fossil amphibians. They move like a fish, with side-to-side, sinusoidal (S-shaped) movements.

Both salamanders and newts are carnivorous, feeding on small invertebrates, such as insects, slugs, snails, and worms. Salamanders practice internal fertilization; in most, males produce a sperm-containing spermatophore, which females pick up with their *cloaca* (the terminal chamber common to the urinary, digestive, and genital tracts). Then, the fertilized eggs are laid in water or on land, depending on the species. Some amphibians, such as

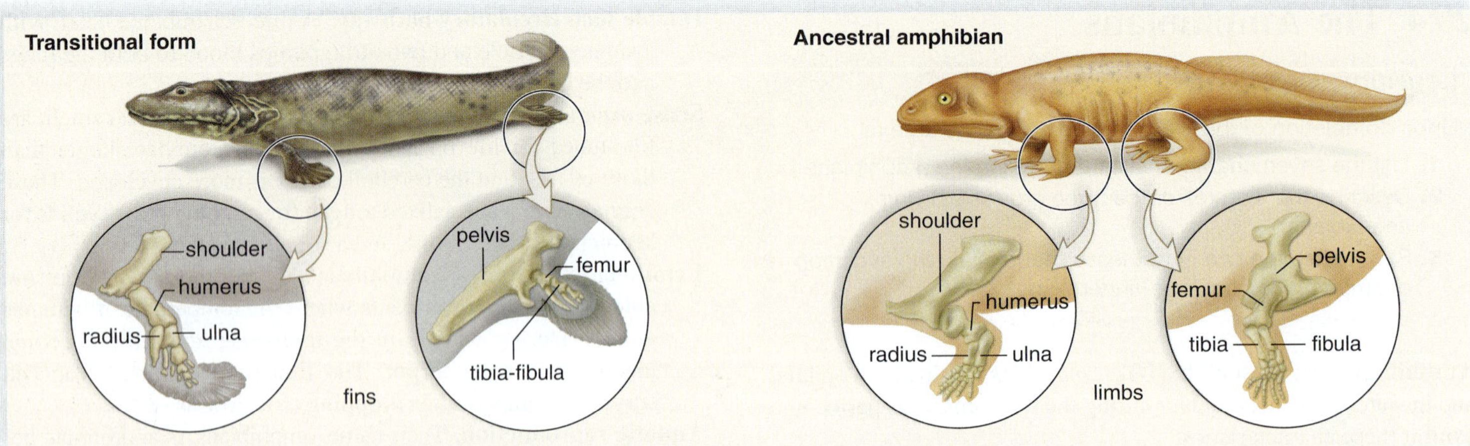

Figure 29.12 Lobe-finned fishes to amphibians. This transitional form links the lobes of lobe-finned fishes to the limbs of ancestral amphibians. Compare the fins of the transitional form (*left*) to the limbs of the ancestral amphibian (*right*).

a. Barred tiger salamander, *Ambystoma tigrinum*

b. Bullfrog, *Rana catesbeiana*

c. Caecilian, *Boulengerula fischeri*

Figure 29.13 Amphibians. Living amphibians are divided into three orders: **a.** Salamanders and newts. Members of this order have a tail throughout their lives and, if present, unspecialized limbs. **b.** Frogs and toads. Like this frog, members of this order are tailless and have limbs specialized for jumping. **c.** The caecilians are wormlike burrowers.

the mudpuppy of eastern North America, remain in the water and retain the gills of the larva.

Frogs and toads, which range in length from less than 1 cm to 30 cm, are common in subtropical to temperate to desert climates around the world. In these animals, which lack tails as adults, the head and trunk are fused, and the long hindlimbs are specialized for jumping (Fig. 29.13*b*). All species are carnivorous and have a tremendous array of specializations, depending on their habitats. Glands in the skin secrete poisons that make the animals distasteful to eat and protect them from microbial infections. Some tropical species with brilliant fluorescent coloration are particularly poisonous, with the bright colors serving as a warning to potential predators. Colombian Indians dip their darts in the deadly secretions of poison-dart frogs (see Fig. 29A*a*). The tree frogs have adhesive toepads that allow them to climb trees, while others, the spadefoots, have hardened spades that act as shovels, enabling them to dig into the soil.

Video Frog Reproduction

Caecilians are legless, often sightless, worm-shaped amphibians that range in length from about 10 cm to more than 1 m (Fig. 29.13*c*). Most burrow in moist soil, feeding on worms and other soil invertebrates. Some species have folds of skin that make them look like a segmented earthworm.

Check Your Progress 29.4

1. List the characteristics that amphibians have in common.
2. Describe the usual life cycle of amphibians.

29.5 The Reptiles

Learning Outcomes

Upon completion of this section, you should be able to

1. List the seven features that define the reptiles.
2. Explain why the reptiles are represented by more than one evolutionary lineage.
3. Define the traits of birds that are related to flight.

The **reptiles** (class Reptilia) are a very successful group of terrestrial animals consisting of more than 17,000 species, including the birds. Reptiles have the following characteristics, showing that they are fully adapted to life on land.

Paired limbs. They have two pairs of limbs, usually with five toes each. Reptiles are adapted for climbing, running, paddling, or flying.

Skin. A thick and dry skin is impermeable to water. Therefore, the skin prevents water loss. In reptiles, the skin is wholly or in part scaly (Fig. 29.14). Many reptiles (e.g., snakes and lizards) molt several times a year.

Efficient breathing. The lungs are more developed than in amphibians. Also in many reptiles, an expandable rib cage assists breathing.

Efficient circulation. The heart prevents mixing of blood. A septum divides the ventricle either partially or completely. If it partially divides the ventricle, the mixing of oxygen-poor blood and oxygen-rich blood is reduced. If the septum is complete, oxygen-poor blood is completely separated from oxygen-rich blood (see Fig. 29.11*c*).

Efficient excretion. The kidneys are well developed. The kidneys excrete uric acid, and therefore less water is required to rid the body of nitrogenous wastes.

Ectothermy. Most reptiles are ectotherms, and this allows them to survive on a fraction of the food per body weight required by birds and mammals. Ectothermic reptiles are adapted behaviorally to maintain a warm body temperature by warming themselves in the sun.

Well-adapted reproduction. Sexes are separate and fertilization is internal. Internal fertilization prevents sperm from drying out when copulation occurs. The **amniotic egg** contains extraembryonic membranes, which protect the embryo, remove nitrogenous wastes, and provide the embryo with oxygen, food, and water (see Fig. 29.16). These membranes are not part of the embryo itself and are disposed of after development is complete. One of the membranes, the amnion, is a sac that fills with fluid and provides a "private pond" within which the embryo develops.

Evolution of Amniotes

An ancestral amphibian gave rise to the amniotes at some point in the Carboniferous period, beginning some 359 MYA. The amniotes include animals now classified as the reptiles (including birds) and the mammals. The embryo of an amniote has extracellular membranes, including an amnion (see Fig. 29.16).

Figure 29.15 shows that the amniotes consist of three lineages: (1) the turtles, in which the skull is **anapsid**—that is, it has no openings behind the orbit, or eye socket; (2) all the other reptiles, including the birds, in which the skull is **diapsid,** or has two openings behind the orbit; and (3) the mammals, in which the skull is **synapsid** and has one opening behind the orbit.

Systemicists are continuing to explore the evolutionary relationships of the reptiles, which now include the birds. For example, the classical view of reptile evolution considers the turtles, or anapsids, as an independent lineage, separate from that of the rest of the reptiles, which are diapsids. The modern view of turtle evolution places turtles within the archosaurs with birds and crocodiles. This would mean that the anapsids are really just highly specialized diapsids. The evidence in support of this modern view is still

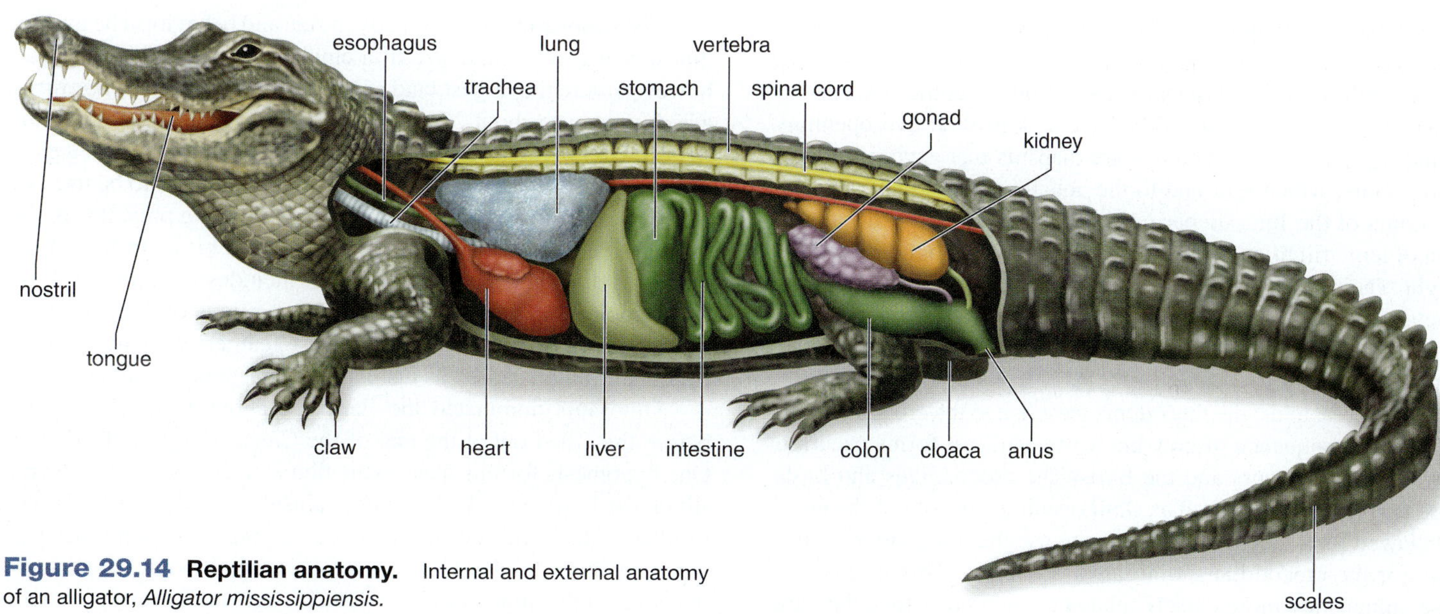

Figure 29.14 Reptilian anatomy. Internal and external anatomy of an alligator, *Alligator mississippiensis.*

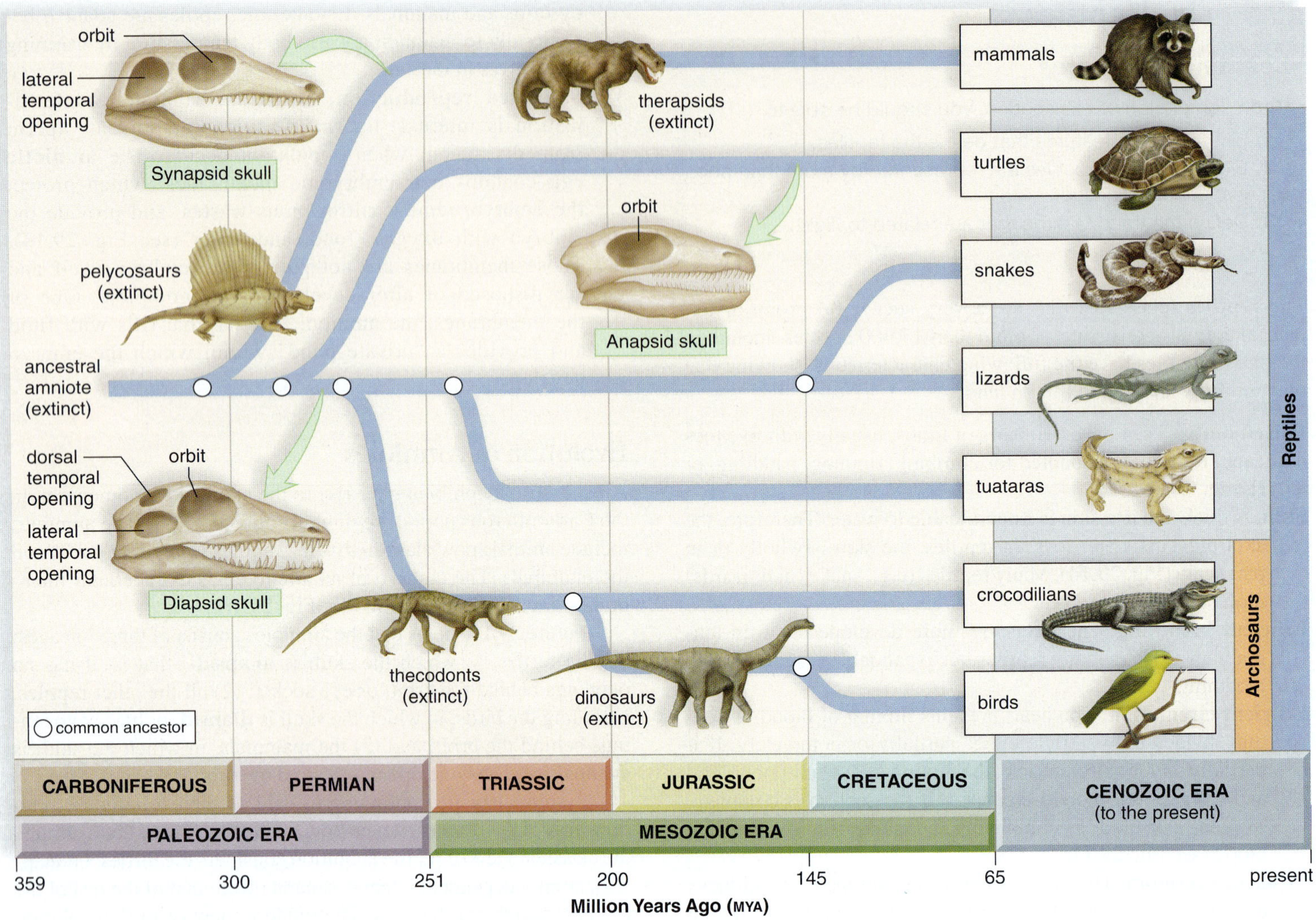

Figure 29.15 Timeline of the evolution of the amniotes. This diagram shows an overview of the presumed evolutionary relationships among amniotes, including the major groups of reptiles. The amniote ancestor evolved in the Paleozoic era. Historically, the openings in the skull were considered evidence that there were two major groups of reptiles, with turtles separate from the other reptile groups. This historical understanding of turtle evolution is presented in this figure. A modern view, not shown, proposes that the turtles are archosaurs along with crocodiles and birds.

being debated, and thus the classical view of reptile evolution is presented here (Fig. 29.15).

According to the classical view, all other reptiles except the turtles are diapsids because they have a skull with two openings behind the eyes. The thecodonts are diapsids that gave rise to the ichthyosaurs, which returned to the aquatic environment, and the pterosaurs of the Jurassic period, which had a keel for the attachment of large flight muscles and air spaces in their bones to reduce weight. Their wings were membranous and supported by elongated bones of the fourth finger. *Quetzalcoatlus,* the largest flying animal ever to live, had an estimated wingspan of nearly 13.5 m (over 40 feet)!

Of interest to us, the thecodonts gave rise to the crocodiles and dinosaurs. A sequence of now known transitional forms occurred between the dinosaurs and the birds. The crocodilians and birds share derived features, such as skull openings in front of the eyes, and clawed feet. It is customary now to use the designation archosaurs for the crocodilians, dinosaurs, and birds. This means that these animals are more closely related to each other than they are to snakes and lizards.

The **dinosaurs** varied greatly in size and behavior. The average dinosaur was about the size of a chicken. Some of the dinosaurs, however, were the largest land animals ever to live. *Brachiosaurus,* a herbivore, was about 23 m long and about 17 m tall. *Tyrannosaurus rex,* a carnivore, was 5 m tall when standing on its hind legs. A bipedal stance freed the forelimbs and allowed them to be used for purposes other than walking, such as manipulating prey. It was also preadaptive for the evolution of wings in the birds. In fact, many fossils of dinosaurs in the family that includes the familiar *T. rex* and velociraptor bear unmistakable impressions of feathers. Prior to the evolution of birds, however, these feathers were not used for flight but, rather, served as insulation against cold weather.

Dinosaurs dominated the Earth for about 170 million years before they died out at the end of the Cretaceous period, 65 MYA. One hypothesis for this mass extinction is that a massive meteorite struck the Earth near the Yucatán Peninsula (see section 18.3). The resultant cataclysmic events disrupted existing ecosystems, destroying many living things. This hypothesis is supported by the presence of a layer of the mineral iridium, which is rare on Earth but common in meteorites, in the late Cretaceous strata.

THEME **Nature of Science**

Vertebrates and Human Medicine

Many pharmaceutical products come from vertebrates, and even some deadly venoms can be sources of beneficial medicines.

Natural Products with Medical Applications

The black-and-white spitting cobra of Southeast Asia paralyzes its victims with a potent venom, which eventually leads to respiratory arrest. However, that venom is also the source of the drug Immunokine, which can inhibit some harmful effects of an overactive immune system. It is approved in Thailand for use in combating the side effects of cancer therapy, and it is being studied for use in treating AIDS, autoimmune diseases, and other disorders.

Although snakebites can be very painful, certain components found in venom actually relieve pain. The black mamba, found mainly in sub-Saharan Africa, is one of the most lethal snakes on Earth. Compounds in its venom called mambalgins, however, block pain signals by inhibiting the flow of certain ions through nerves that carry pain messages. When tested in mice, these compounds were as effective as morphine, with fewer side effects.

Another compound, known as epibatidine, derived from the skin of an endangered Ecuadorian poison-dart frog, is 50–200 times more powerful than morphine in relieving chronic and acute pain, without the addictive properties. Unfortunately, it can also have serious side effects, so companies have synthesized compounds with a similar structure, hoping to improve its safety profile.

Other venoms mainly affect blood clotting. Eptifibatide is derived from the venom of the pigmy rattlesnake, which lives in the southeastern United States. Because it binds to blood platelets and reduces their tendency to clump together, this drug is used to reduce the risk of clot formation in patients at risk for heart attacks. Alternatively, the venom of several pit vipers, such as the copperhead, contains "clot-busting" (thrombolytic) substances, which can be used to dissolve abnormal clots that have already formed.

Sharks produce a variety of chemicals with potentially medicinal properties. Squalamine is a steroidlike molecule that was first isolated from the liver of dogfish sharks. It has broad antimicrobial properties, and it can inhibit the abnormal growth of new blood vessels, which is a factor in cancer and a variety of other diseases. Squalamine is also safe enough to be used in the eye and is currently being tested as a potential treatment for macular degeneration, an eye disease that will affect 3 million Americans by 2020.

Animal Pharming

Some of the most powerful applications of genetic engineering can be found in the development of drugs and therapies for human diseases. In fact, this technology has led to a new industry: animal pharming, which uses genetically altered vertebrates, such as mice, sheep, goats, cows, pigs, and chickens, to produce medically useful pharmaceutical products.

To accomplish this, the human gene for some useful product is inserted into the embryo of a vertebrate. That embryo is implanted into a foster mother, which gives birth to the transgenic animal, which contains genes from the two sources. An adult transgenic vertebrate produces large quantities of the pharmed product in its blood, eggs, or milk, from which the product can be easily harvested and purified.

An example of a pharmed product used in human medicine is human antithrombin. This medication is important in the treatment of individuals who have a hereditary deficiency of this protein and so are at high risk for life-threatening blood clots. Approved by the FDA in 2009, the bioengineered drug, known by the brand name ATryn, is purified from the milk of transgenic goats.

Xenotransplantation

There is an alarming shortage of human donor organs to fill the need for hearts, kidneys, and livers. One solution is xenotransplantation, the transplantation of nonhuman vertebrate tissues and organs into humans. The first such transplant occurred in 1984 when a team of surgeons implanted a baboon heart into an infant, who, unfortunately, lived only 20 days.

Although apes are more closely related to humans, pigs are considered to be the best source for xenotransplants. Pig organs are similar to human organs in size, anatomy, and physiology, and large numbers of pigs can be produced quickly. Most infectious microbes of pigs are unlikely to infect a human recipient. Currently, pig heart valves and skin are routinely used for treatment of humans. Miniature pigs, whose heart size is similar to that of humans, are being genetically engineered to make their tissues less foreign to the human immune system, to minimize rejection.

Questions to Consider

1. Is it ethical to change the genetic makeup of vertebrates in order to use them as drug or organ factories?
2. What are some of the health concerns that may arise due to xenotransplantation?

a. Poison-dart frogs, source of a medicine

b. Pigs, source of organs

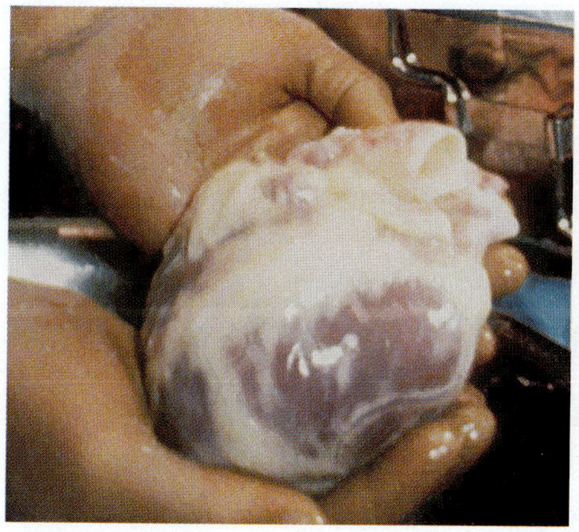

c. Heart for transplantation

Figure 29A Use of other vertebrates for medical purposes. **a.** The poison-dart frog is the source of a pain medication. **b.** Pigs are now being genetically altered to provide a supply of **(c)** hearts for heart transplant operations.

Diversity of Living Reptiles

Living reptiles are represented by turtles, lizards, snakes, tuataras, crocodilians, and birds. Figure 29.16 shows representatives of all but the birds.

Along with tortoises, turtles can be found in marine, freshwater, and terrestrial environments. Most turtles have ribs and thoracic vertebrae that are fused into a heavy shell. They lack teeth but have a sharp beak. The legs of sea turtles are flattened and paddlelike (Fig. 29.16*a*), while terrestrial tortoises have strong limbs for walking.

Lizards have four clawed feet and resemble their prehistoric ancestors in appearance (Fig. 29.16*b*), although some species have lost their limbs and superficially resemble snakes. Typically, they are carnivorous and feed on insects and small animals, including other lizards. Marine iguanas of the Galápagos Islands are adapted to spending time each day at sea, where they feed on sea lettuce and other algae. Chameleons are adapted to live in trees and have long, sticky tongues for catching insects some distance away. They can change color to blend in with their background. Geckos are primarily nocturnal lizards with adhesive pads on their toes. Skinks are common elongated lizards with reduced limbs and shiny scales. Monitor lizards and Gila monsters, despite their names, are generally not a dangerous threat to humans.

Video Basilisk Lizards

Video Leaf-Tailed Gecko

Although most snakes (Fig. 29.16*c*) are harmless, several venomous species, including rattlesnakes, cobras, mambas, and copperheads, have given the whole group a reputation of being dangerous. Snakes evolved from lizards and have lost their limbs as an adaptation to burrowing. A few species such as pythons and boas still possess the vestiges of pelvic girdles. Snakes are carnivorous and have a jaw that is loosely attached to the skull; therefore, they can eat prey that is much larger than their head size. When snakes and lizards flick out their tongues, they are collecting airborne molecules and transferring them to a *Jacobson's organ* at the roof of the mouth and sensory cells on the floor of the mouth. The Jacobson's organ is an olfactory organ for the analysis of airborne chemicals. Snakes possess internal ears that are capable of detecting low-frequency sounds and vibrations. Their ears lack external ear openings.

Video Two-Headed Snake

Video Snake Eating

Two species of tuataras are found in New Zealand (Fig. 29.16*d*). They are lizardlike animals that can attain a length of 66 cm and can live for nearly 80 years. These animals possess a well-developed "third" eye, known as a pineal eye, which is light-sensitive and buried beneath the skin in the upper part of the head. The tuataras are the only member of an ancient group of reptiles that included the common ancestor of modern lizards and snakes.

The majority of crocodilians (including alligators and crocodiles) live in fresh water feeding on fishes, turtles, and terrestrial animals that venture too close to the water. They have long, powerful jaws (Fig. 29.16*e*) with numerous teeth and a muscular tail that serves as both a weapon and a paddle. Male crocodiles and

a. Green sea turtle, *Chelonia mydas*

b. Gila monster, *Heloderma suspectum*

c. Diamondback rattlesnake, *Crotalus atrox*

d. Tuatara, *Sphenodon punctatus*

e. American crocodile, *Crocodylus acutus*

f. Reptile egg

Figure 29.16 Reptilian diversity other than birds. Representative living reptiles include **(a)** green sea turtles, **(b)** the venomous Gila monster, **(c)** the diamondback rattlesnake, and **(d)** the tuatara. **e.** A young crocodile hatches from an egg. The eggshell is leathery and flexible, not brittle, like birds' eggs. **f.** Inside the egg, the embryo is surrounded by three membranes. The chorion aids in gas exchange, the allantois stores waste, and the amnion encloses a fluid that prevents drying out and provides protection. The yolk sac provides nutrients for the embryo.

alligators bellow to attract mates. In some species, the male protects the eggs and cares for the young.

Birds

As already noted, a plentiful fossil record shows that birds evolved from dinosaurs—probably small meat eaters resembling the velociraptors featured in the movie *Jurassic Park.* All **birds** have traits such as the presence of scales (feathers are modified scales), a tail with vertebrae, and clawed feet that show they are indeed reptiles. Interestingly, a close examination of velociraptor fossils has revealed that they also had feathers.

To many people, birds are the most conspicuous, melodic, beautiful, and fascinating group of vertebrates. Birds range in size from the tiny "bee" hummingbird at 1.8 g (less than a penny) and 5 cm long to the ostrich at a maximum weight of 160 kg and a height of 2.7 m.

Nearly every anatomical feature of a bird can be related to its ability to fly (Fig. 29.17). These features are involved in the action of flight, providing energy for flight or the reduction of the bird's body weight, making flight less energetically costly:

Feathers. Soft down keeps birds warm, wing feathers allow flight, and tail feathers are used for steering. A feather is a modified reptilian scale with the complex structure shown in Figure 29.17*a*. Nearly all birds molt (lose their feathers) and replace their feathers about once a year.

Modified skeleton. Unique to birds, the collarbone is fused (the wishbone), and the sternum has a keel (Fig. 29.17*b*). Many other bones are fused, making the skeleton more rigid than the reptilian skeleton. The breast muscles are attached to the keel, and their action accounts for a bird's ability to fly. A horny beak has replaced jaws equipped with teeth, and a slender neck connects the head to a rounded, compact torso.

Modified respiration. In birds, unlike other reptiles, the lobular lungs connect to anterior and posterior air sacs. The presence of these sacs means the air circulates one way through the lungs and gases are continuously exchanged across respiratory

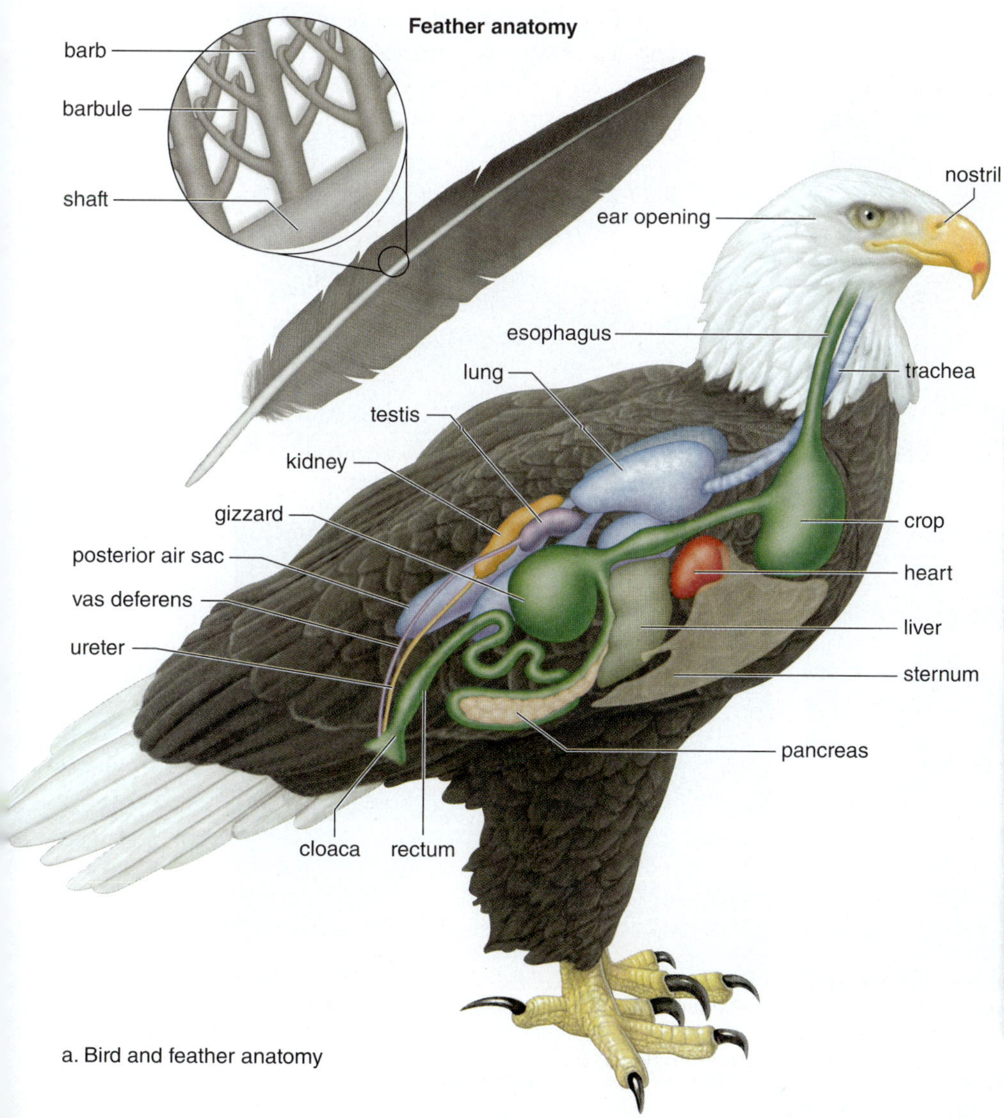

a. Bird and feather anatomy

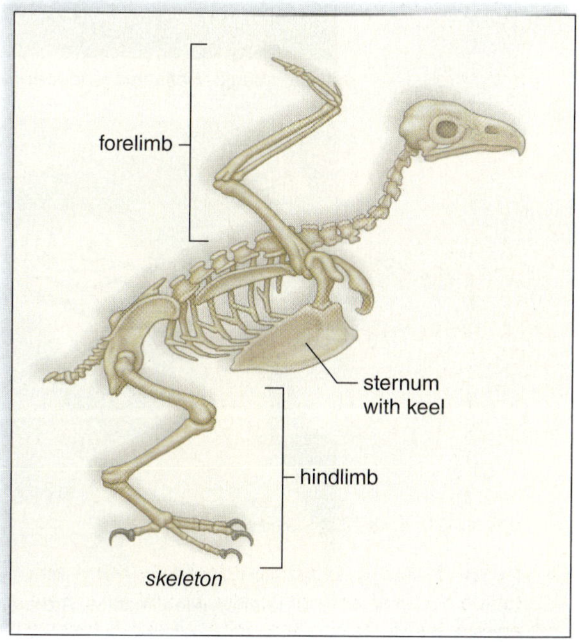

b. Bald eagle, *Haliaetus*

Figure 29.17 Bird anatomy and flight.
a. Bird anatomy. *Top:* In feathers, a hollow central shaft gives off barbs and barbules, which interlock in a latticelike array. *Bottom:* The anatomy of an eagle is representative of bird anatomy. **b.** Bird flight. The skeleton of an eagle shows that birds have a large, keeled sternum to which flight muscles attach. The bones of the forelimb help support the wings.

tissues. Another benefit of air sacs is that they lighten the body and bones for flying. Some of the air sacs are present in cavities within the bones.

Endothermy. Birds, unlike other reptiles, generate internal heat. Many **endotherms** can use metabolic heat to maintain a constant internal temperature. Endothermy may be associated with their efficient nervous, respiratory, and circulatory systems.

Well-developed sense organs and nervous system. Birds have particularly acute vision and well-developed brains. Their muscle reflexes are excellent. An enlarged portion of the brain seems to be the area responsible for instinctive behavior.

A ritualized courtship often precedes mating. Many newly hatched birds require parental care before they are able to fly away and seek food for themselves.

A remarkable aspect of bird behavior is the seasonal migration of many species over long distances. Birds navigate by day and night, whether it is sunny or cloudy, by using the sun and stars and even the Earth's magnetic field to guide them.

Birds are very vocal animals. Their vocalizations are distinctive and so convey an abundance of information.

Diversity of Living Birds

The majority of birds, including eagles, geese, and mockingbirds, have the ability to fly. However, some birds, such as emus, penguins, kiwis, and ostriches, are flightless. Traditionally, birds have been classified according to beak and foot type (Fig. 29.18) and, to some extent, on their habitat and behavior. The birds of prey have notched beaks and sharp talons; shorebirds have long, slender, probing beaks and long, stiltlike legs; woodpeckers have sharp, chisel-like beaks and grasping feet; waterfowl have broad beaks and webbed toes; penguins have wings modified as paddles; songbirds have perching feet; and parrots have short, strong, plierslike beaks and grasping feet.

Video Finches Adaptive Radiation

Video Harris Hawk

Check Your Progress 29.5

1. Contrast the characteristics of crocodilians with those of snakes.
2. Explain what features indicate that birds are reptiles.

a. Bald eagle, *Haliaetus leucocephalus*

b. Pileated woodpecker, *Dryocopus pileatus*

c. Blue-and-yellow macaw, *Ara ararauna*

d. Cardinal, *Cardinalis cardinalis*

Figure 29.18 Bird beaks. **a.** A bald eagle's beak allows it to tear apart prey. **b.** A woodpecker's beak is used to chisel in wood. **c.** A parrot's beak is modified to pry open nuts. **d.** A cardinal's beak allows it to crack tough seeds.

29.6 The Mammals

Learning Outcomes

Upon completion of this section, you should be able to

1. Describe five features of mammals.
2. Discuss the timeline of the evolution of mammals.
3. Identify several features that define each of the three living lineages of mammals.

The **mammals** (class Mammalia) include the largest animal ever to live, the blue whale (130 metric tons); the smallest mammal, the Kitti's bat (1.5 g); and the fastest land animal, the cheetah (110 km/hr). These characteristics distinguish mammals:

Hair. The most distinguishing characteristics of mammals are the presence of hair and milk-producing mammary glands. Hair provides insulation against heat loss, and being endothermic allows mammals to be active even in cold weather. The color of hair can camouflage a mammal and help the animal blend into its surroundings. In addition, hair can be ornamental and can serve sensory functions.

Mammary glands. These glands enable females to feed (nurse) their young without having to leave them to collect food, as birds do. Nursing also creates a bond between mother and offspring that helps ensure parental care while the young are helpless, and it provides antibodies to the young from the mother through the milk.

Skeleton. The mammalian skull accommodates a larger brain relative to body size than does the reptilian skull. Also, mammalian cheek teeth are differentiated as premolars and molars. The vertebrae of mammals are highly differentiated; typically, the middle region of the vertebral column is arched, and the limbs are under the body rather than out to the sides.

Internal organs. Efficient respiratory and circulatory systems ensure a ready oxygen supply to muscles whose contraction produces body heat. Like birds, mammals have a double-loop circulatory pathway and a four-chambered heart. The kidneys are adapted to conserving water in terrestrial mammals. The nervous system of mammals is highly developed. Special senses in mammals are well developed, and mammals exhibit complex behavior.

Internal development. In most mammals, the young are born alive after a period of development in the uterus, a part of the female reproductive tract. Internal development shelters the young and allows the female to move actively about while the young are maturing.

Evolution of Mammals

Mammals share an amniote ancestor with reptiles (see Fig. 29.15). Their more immediate ancestors in the Mesozoic era had a synapsid skull. The first true mammals appeared during the Triassic period, about the same time as the first dinosaurs, and were similar in size to mice. During the reign of the dinosaurs (170 million years), mammals were a minor group that changed little. The common ancestor of all three mammal groups appeared in the late Triassic–early Jurassic period, about 200 MYA (Fig. 29.19). The

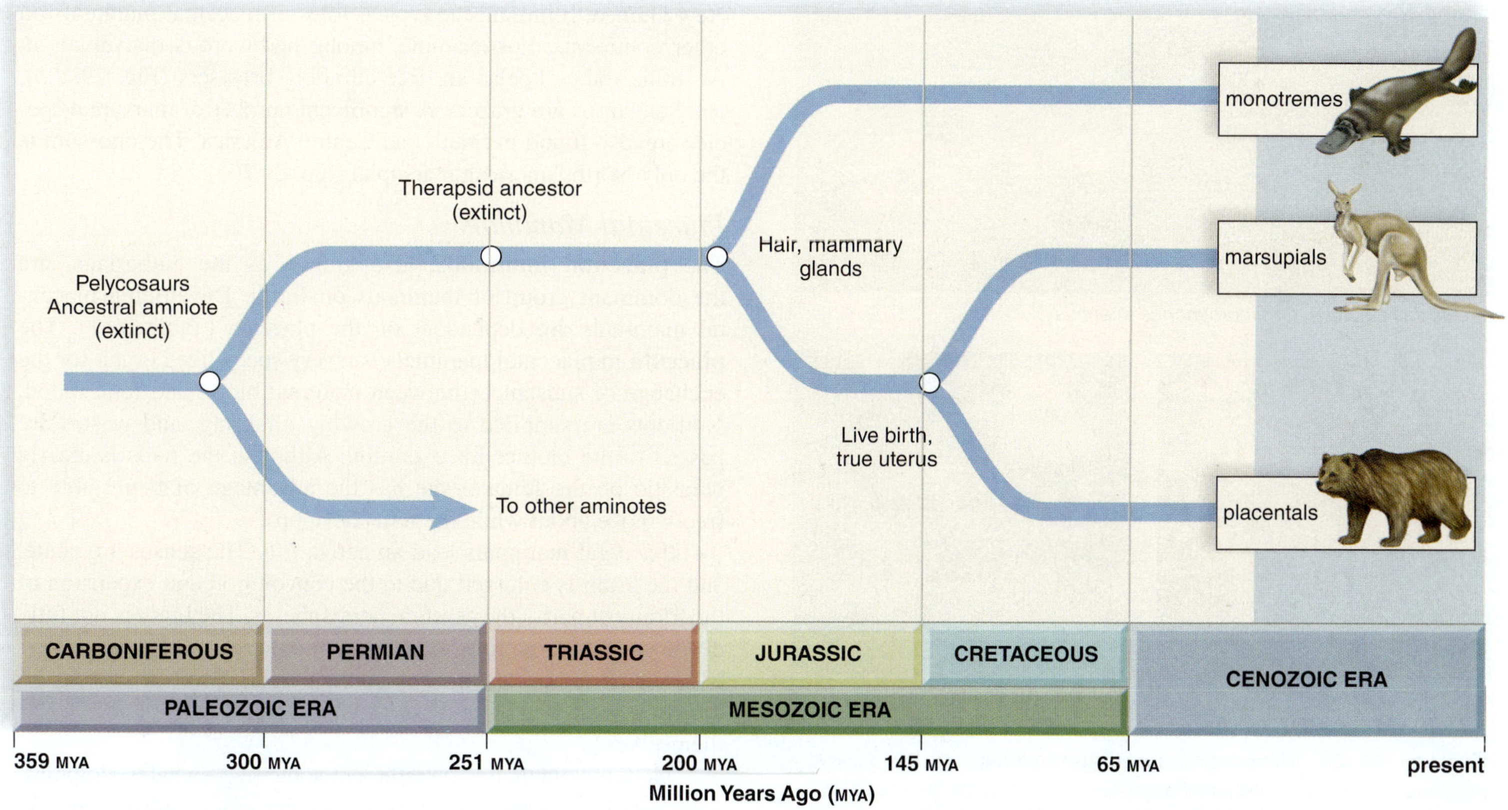

Figure 29.19 Timeline of the evolution of mammal groups. Only three lineages of mammals exist on Earth today, the monotremes, marsupials, and placentals (eutherians). Monotremes are the oldest group of mammals, have hair and mammary glands, but do not give live birth—they lay eggs. The marsupials and placental lineages evolved in the early Cretaceous. Both marsupials and placentals give live birth. In the placental mammals, however, the placenta is more complex and specialized than in the marsupials. Both marsupials and placentals have a true uterus but it is structurally different.

earliest mammalian group is the monotremes. The marsupials probably originated in the Americas and then spread through South America and Antarctica to Australia before these continents separated. Placental mammals, the third branch of the mammalian lineage (Fig. 29.19), originated in Eurasia and spread to the Americas also by land connections that existed between the continents during the Mesozoic. The placental mammals underwent an adaptive radiation into the habitats previously occupied by the dinosaurs.

Table 29.1 Female reproductive traits of mammal lineages.

	Milk	Nipples	Birth	Uterus	Placenta	Birth Canal
Monotremes	Yes	No	Eggs	None	None	Cloaca
Marsupials	Yes	Yes	Live	True (>1)	Simple (yolk)	Vagina (>1)
Placentals	Yes	Yes	Live	True (1)	Complex (tissue)	Vagina (1)

Monotremes

Monotremes (Gk. *monos,* "one"; *trema,* "hole") are egg-laying mammals that include only the duckbill platypus (Fig. 29.20*a*) and two species of echidna, or spiny anteaters. The term *monotreme* refers to the presence of a single urogenital opening, the cloaca, which is a shared excretory and reproductive canal. Monotremes, unlike other mammals, lay hard-shelled amniotic eggs (Table 29.1). No embryonic development occurs inside the female's body. The female duckbill platypus lays her eggs in a burrow in the ground. She incubates the eggs, and after hatching, the young lick up milk that seeps from modified sweat glands on the mother's abdomen.

Echidnas, which feed mainly on termites, have pores that seep milk in a shallow belly pouch formed by skin folds on each side. The egg moves from the cloaca to this pouch, where hatching takes place and the young remain for about 53 days. Then, they stay in a burrow, where the mother periodically visits and nurses them.

a. Duckbill platypus, *Ornithorhynchus anatinus*

b. Koala, *Phascolarctos cinereus*

c. Virginia opossum, *Didelphis virginianus*

Figure 29.20 Monotremes and marsupials. **a.** The duckbill platypus is a monotreme that inhabits Australian streams. **b.** The koala is an Australian marsupial that lives in trees. **c.** The opossum is the only marsupial in North America. The Virginia opossum is found in a variety of habitats.

Marsupials

The **marsupials** (Gk. *marsupium,* "pouch") are also known as the pouched mammals. Marsupials include kangaroos, koalas, Tasmanian devils, wombats, sugar gliders, and opossums. Marsupials have a true uterus (Table 29.1). The embryos of marsupials are nourished by a yolk-based type of placenta inside the female's body, but they are born in a very immature condition. Newborns are typically hairless and have yet to open their eyes, but they crawl up into a pouch on their mother's abdomen. Inside the pouch, they attach to nipples of mammary glands and continue to develop. Frequently, more are born than can be accommodated by the number of nipples, and it's "first come, first served."

Today, marsupial mammals are most abundant in Australia and New Guinea, filling all the typical roles of placental mammals on other continents. For example, among herbivorous marsupials in Australia today, koalas are tree-climbing browsers (Fig. 29.20*b*), and kangaroos are grazers. A significant number of marsupial species are also found in South and Central America. The opossum is the only North American marsupial (Fig. 29.20*c*).

Placental Mammals

The **placental mammals,** also known as the eutherians, are the dominant group of mammals on Earth. Developing placental mammals are dependent on the placenta (Table 29.1). The **placenta** in placental mammals is a very specialized organ for the exchange of substances between maternal blood and fetal blood. Nutrients are supplied to the growing offspring, and wastes are passed to the mother for excretion. Although the fetus is clearly parasitic on the female, she has the advantage of being able to freely move about while the fetus develops.

Placental mammals lead an active life. The senses are acute, and the brain is enlarged due to the convolution and expansion of the foremost part—the cerebral hemispheres. The brain is not fully developed for some time after birth, and there is a long period of dependency on the parents, during which the young learn to take care of themselves.

Video Mom Grizzly Teaches Her Cubs

Most mammals live on land, but some (e.g., whales, dolphins, seals, sea lions, and manatees) are secondarily adapted to live in water, and bats are able to fly. Although bats are the only mammal that can actually fly, three types of placentals can glide: the flying squirrels, scaly-tailed squirrels, and flying lemurs. There are 20 different orders of placental mammals (Table 29.2).

Video
Bat Echolocation

Humans are mammals, and in Chapter 30 we consider the evolutionary history of humans, an enormously successful mammalian group.

Check Your Progress 29.6

1. Identify two traits that are unique to mammals.
2. Describe features that distinguish the three groups of mammals.

Table 29.2 Orders of placental mammals

Order	Examples	Traits
Cetacea	Whales, dolphins	Marine, no fur, streamlined bodies
Artiodactyla	Cattle, deer, antelope	Even-toed ungulates, horns or antlers, ruminants
Perissodactyla	Horses, rhinoceroses	Odd number toes, nonruminants
Carnivora	Dogs, cats, weasels, minks, stoats	Meat eaters, long canines
Pholidota	Pangolins	Scaly armor, no teeth
Chiroptera	Bats	Fly with membranous wings, some echolocate
Erinaceomorpha	Hedgehogs	Spines, insectivores
Soricomorpha	Shrews, moles	Small, high metabolism, insectivorous
Rodentia	Mice, rats, voles, beavers, squirrels	One pair of specialized incisors that grow continuously
Lagomorpha	Rabbits, hares	Two pair of incisors that grow continuously, long hindlimbs
Dermoptera	Colugos ("flying lemurs")	Membrane between hands and feet
Scandentia	Tree shrews	Large forward-facing eyes, grasping hand
Primates	Monkeys, apes	Opposable thumb, developed brains, social
Xenarthra	Sloths, armadillos	Special projections on the spine
Afrocoricida	Golden moles, tenrecs	Insectivores
Macroscelidia	Elephant shrews	Small, long snout, long hindlimbs
Tubulidentata	Aardvarks	Insectivore, long snout, coarse fur
Sirenia	Manatees, dugongs	Marine, long whiskers
Hyracoidea	Hyraxes	Small mammals
Proboscidea	Elephants	Largest land mammals, trunk, tusks

CONNECTING *the* CONCEPTS *with the* THEMES

Evolution

- The presence of a notochord and a dorsal tubular nerve cord, at least during some stage of the life cycle, is a major distinguishing feature of chordates. Vertebrates evolved from a chordate ancestor.
- The appearance of four limbs and the amniotic egg were key events that shaped the evolution of vertebrate diversity.
- Mammals underwent an explosive adaptive radiation following the extinction of the dinosaurs.

Nature of Science

- Many natural products of vertebrates, such as hormones, toxins, and other pharmacologically active molecules, can be used as important human medicines.
- The ability to produce transgenic animals has allowed human genes to be inserted into vertebrate animals. The animal then produces the human gene's product, which can be collected for human use.

Biological Systems

- Vertebrates occupy and contribute to the majority of habitats on Earth.
- In many habitats, particularly terrestrial ones, vertebrates are among the top herbivores and carnivores.

MEDIA STUDY TOOLS http://connect.mheducation.com

SMARTBOOK® Study smart. SmartBook helps you maximize your study time by identifying what you know and don't know. The recharge feature in SmartBook helps you prepare for the test by identifying the material that you are most likely to forget.

connect | BIOLOGY Find even more resources to learn the chapter concepts including animations, tutorial videos, and interactive practice questions.

Videos

29.3 Cichlid Specialization
29.4 Amphibian Origin • Frog Reproduction
29.5 Basilisk Lizards • Leaf-Tailed Gecko • Snake Eating • Two-Headed Snake • Finches Adaptive Radiation • Harris Hawk
29.6 Mom Grizzly Teaches Her Cubs • Bat Echolocation

SUMMARIZE

29.1 The Chordates

At some time in their life history, all **chordates** (sea squirts, lancelets, and vertebrates) have a **notochord,** a dorsal tubular nerve cord, pharyngeal pouches, and a postanal tail. Pharyngeal pouches develop into the **gills** of aquatic vertebrates.

Lancelets (**cephalochordates**) and sea squirts (**urochordates**) are the nonvertebrate chordates. Lancelets are the only chordate to have all four characteristics in the adult stage. Sea squirts lack chordate characteristics (except gill slits) as adults, but they have a larva that could be ancestral to the vertebrates.

29.2 The Vertebrates

As embryos, vertebrates have the four chordate characteristics, plus they develop a vertebral column, a skull, an endoskeleton, and internal organs. The adult vertebral column is derived from the embryonic notochord. Vertebrates have also undergone cephalization, and they have paired appendages, and well-developed internal organs.

Vertebrate evolution is marked by the appearance of vertebrae, jaws, a bony skeleton, lungs, limbs, and the amniotic egg. The **gnathostomes,** or animals with jaws, gave rise to **tetrapods,** or animals with four limbs. **Amniotes** developed an egg surrounded by membranes and amniotic fluid.

29.3 The Fishes

The **fishes** are the largest group of vertebrates. All are ectotherms, meaning their temperature fluctuates with the environment. The earliest known fishes, the **ostracoderms,** lacked jaws and paired muscular appendages (**fins**). They are represented today by the **jawless fishes** (also called **agnathans**), such as hagfishes and lampreys. Ancestral bony fishes like the **placoderms,** which had jaws and fins, gave rise during the Devonian period to two groups: today's **cartilaginous fishes** (skates, rays, and sharks) and the **bony fishes,** including the **ray-finned bony fishes** and the **lobe-finned fishes.** The ray-finned fishes (Actinopterygii) became the most diverse group among the vertebrates. Many of these have a gas-filled **swim bladder** that affects their buoyancy. Ancient lobe-finned fishes (Sarcopterygii) gave rise to the coelacanths, **lungfishes,** and amphibians.

29.4 The Amphibians

Amphibians are ectothermic tetrapods represented primarily today by frogs and salamanders. Most have thin, moist, nonscaly skin, a three-chambered heart, small lungs supplemented by cutaneous respiration, and finely tuned senses. Most frogs and some salamanders return to the water to reproduce and then metamorphose into terrestrial adults.

29.5 The Reptiles

Reptiles (today's alligators and crocodiles, birds, turtles, tuataras, lizards, and snakes) lay a shelled **amniotic egg,** which allows them to reproduce on land. The skull of an amniote may be **anapsid,** with no opening behind the eye socket; **synapsid,** with one opening, or **diapsid,** with two openings. There are two views of reptile evolution. The classical view considers turtles, with an anapsid skull, as having a separate ancestry from the other reptiles. The other reptiles, with a diapsid skull, include the crocodilians, **dinosaurs,** and birds. A more recent view of reptile evolution proposes that turtles are actually very specialized diapsid archosaurs along with birds and crocodiles. **Birds** have reptilian features, including scales (feathers are modified scales), a tail with vertebrae, and clawed feet.

The feathers of birds help them maintain a constant body temperature. Birds are adapted for flight: Their bones are hollow, their shape is compact, their breastbone is keeled, and they have well-developed sense organs. Unlike other reptiles, birds are **endotherms** that maintain a constant internal temperature.

29.6 The Mammals

Mammals share an amniote ancestor with reptiles, but they have a synapsid skull. Mammals remained small and insignificant while the dinosaurs existed, but when dinosaurs became extinct at the end of the Cretaceous period, mammals became the dominant land organisms.

Mammals are vertebrates with hair and mammary glands. Hair helps them maintain a constant body temperature, and the mammary glands allow them to feed and establish an immune system in their young. **Monotremes** lay eggs, while **marsupials** have a pouch in which the newborn crawls and continues to develop. The **placental mammals,** or eutherians, are the most varied and numerous, and they retain offspring inside the female, where they receive nourishment via the **placenta** until birth.

ASSESS

Choose the best answer for each question.

29.1 The Chordates

1. Which of these is not a chordate characteristic?
   a. dorsal supporting rod, the notochord
   b. dorsal tubular nerve cord
   c. pharyngeal pouches
   d. postanal tail
   e. vertebral column

2. Adult sea squirts
   a. do not have all five chordate characteristics.
   b. are also called tunicates.
   c. are fishlike in appearance.
   d. are the first chordates to be terrestrial.
   e. All of these are correct.
3. Label the following diagram of a chordate embryo.

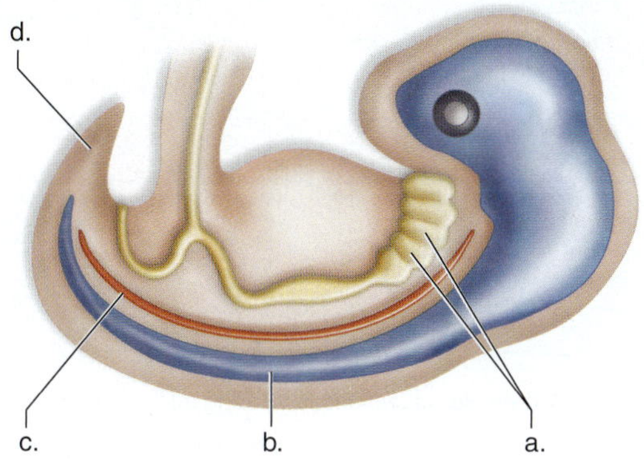

29.2 The Vertebrates

4. Which of these is not characteristic of all vertebrates?
   a. complete digestive system
   b. closed circulatory system
   c. skin with either scales or feathers
   d. endoskeleton made of bone or cartilage
   e. vertebral column
5. The first vertebrates to evolve were
   a. amphibians.
   b. jawed fishes.
   c. jawless fishes.
   d. reptiles.
   e. amniotes.

29.3 The Fishes

6. Which is not a characteristic of jawed fishes?
   a. endothermy
   b. gills
   c. skin covered by scales
   d. two-chambered heart
7. Cartilaginous fishes and bony fishes are different in that only
   a. bony fishes have paired fins.
   b. bony fishes have a keen sense of smell.
   c. bony fishes have an operculum.
   d. cartilaginous fishes have a complete skeleton.
   e. cartilaginous fishes are predaceous.
8. Bony fishes are divided into which two groups?
   a. hagfishes and lampreys
   b. sharks and ray-finned fishes
   c. ray-finned fishes and lobe-finned fishes
   d. jawless fishes and cartilaginous fishes

29.4 The Amphibians

9. Amphibians evolved from what type of ancestral fish?
   a. sea squirts and lancelets
   b. cartilaginous fishes
   c. jawless fishes
   d. ray-finned fishes
   e. lobe-finned fishes
10. Which of these is not a feature of amphibians?
    a. dry skin that resists desiccation
    b. metamorphosis from a swimming form to a land form
    c. small lungs and a supplemental means of gas exchange
    d. reproduction in the water
    e. a single ventricle
11. Which of the following groups has a three-chambered heart?
    a. birds
    b. reptiles other than birds
    c. mammals
    d. amphibians

29.5 The Reptiles

12. Reptiles
    a. were dominant during the Mesozoic era.
    b. include the birds.
    c. lay shelled eggs.
    d. are ectotherms, except for birds.
    e. All of these are correct.
13. The amniotes include all but the
    a. birds.
    b. mammals.
    c. reptiles.
    d. amphibians.
14. A major difference between birds and other reptiles is that birds
    a. produce a shelled egg.
    b. are ectotherms.
    c. are tetrapods.
    d. care for their young.
    e. have air sacs.
15. Which of these does not produce an amniotic egg? Choose more than one answer if correct.
    a. bony fishes
    b. duckbill platypus
    c. snake
    d. robin
    e. frog

29.6 The Mammals

16. The first mammals to evolve were
    a. aquatic.
    b. marsupials.
    c. monotremes.
    d. placental.
    e. primates.
17. Which of these is a true statement?
    a. In all mammals, offspring develop completely within the female.
    b. All mammals have hair and mammary glands.
    c. All mammals have one birth at a time.
    d. All mammals are land-dwelling forms.
    e. All of these are true.

ENGAGE

THINKING SCIENTIFICALLY

1. *Archaeopteryx* was a birdlike reptile that had a toothed beak. Give an evolutionary explanation for the elimination of teeth in a bird's beak.
2. While amphibians have rudimentary lungs, skin is also a respiratory organ. Why would a thin skin be more sensitive to pollution than lungs?
3. Most paleontologists believe that the major function of feathers in dinosaurs was to insulate against the cold. What are some other possible advantages to feathers, compared to scales?
4. As techniques for DNA sequencing become more commonplace, scientists have recently determined the entire genome of exotic vertebrates, such as *Petromyzon marinus* (a lamprey), the West Indian Ocean coelacanth, and the Old World fruit bat. How might this information be helpful in the development of new medical treatments for humans?

30

Human Evolution

Chapter Outline

30.1 Evolution of Primates 565
30.2 Evolution of Humanlike Hominins 569
30.3 Evolution of Early Genus *Homo* 572
30.4 Evolution of Later Genus *Homo* 573

Before You Begin

Before beginning this chapter, take a few moments to review the following discussions.

Section 15.3 What is a transitional fossil?

Section 17.1 What is the difference between the evolutionary species concept and the biological species concept?

Table 29.2 To which order of mammals do the hominins belong?

Comparison of the genomes of ancient and modern *Homo sapiens* is yielding insight into human evolution.

The study of human evolution is being transformed by molecular genetics. Up until a few years ago, scientists uncovered most of what we know about our ancestors by studying fossilized bones, teeth, and artifacts. Based on this type of evidence, researchers have generally hypothesized that humans evolved in a relatively orderly fashion, from more primitive forms to our more advanced modern selves. Analysis of the information contained within our DNA is revealing that our history may be more complicated.

Beginning in 2010, researchers began publishing sequences of DNA obtained from Neandertal bones that were between 38,000 and 45,000 years old. By December 2013, the entire Neandertal genome had been sequenced, allowing interesting comparisons with modern human DNA. For example, we now know that modern humans of European or Asian descent carry about 2% Neandertal DNA, indicating that Neandertals and *Homo sapiens* likely interbred. Some of these Neandertal genes may influence tolerance of cold weather, immune reponses, and the risk of certain diseases. A lack of Neandertal DNA sequences in modern African populations suggests that Neandertals and humans interacted after the latter migrated from Africa to Asia and Europe, around 100,000 years ago.

These discoveries, among others, have renewed interest in discovering the genes that distinguish us from our closest ancestors and, in the process, in developing a greater understanding of the evolutionary history of our species.

As you read through this chapter, think about the following questions:

1. What was the last common ancestor of both Neandertals and *Homo sapiens?*
2. What does the replacement model tell us about the evolution of our species?
3. How might an understanding of evolutionary patterns in other animals help us understand our evolutionary history?

Following the Themes

CHAPTER 30 HUMAN EVOLUTION

UNIT 6 ANIMAL EVOLUTION AND DIVERSITY

Evolution	Like all other organisms, humans have an evolutionary history and are related to all other species through common ancestors.
Nature of Science	By studying the fossil record, and using comparative genomics, scientists are piecing together the story of human evolution.
Biological Systems	The physical and physiological characteristics of the hominins have been shaped by their environment.

30.1 Evolution of Primates

Learning Outcomes

Upon completion of this section, you should be able to

1. Identify the major groups of primates.
2. Discuss the traits common to the primates.
3. Arrange the groups of primates in an evolutionary tree that shows their relationships.

The mammalian order **Primates** (L. *primus,* "first") includes prosimians, monkeys, apes, and humans (Fig. 30.1). Most primates are adapted for an **arboreal** life—that is, for living in trees. The evolution of primates is characterized by trends toward mobile limbs, grasping hands, a flattened face and stereoscopic vision, a large and complex brain, and a reduced reproductive rate. These traits are particularly useful for living in trees.

Mobile Forelimbs and Hindlimbs

Primates tend to have prehensile hands and feet, meaning that they are adapted for grasping and holding. In most primates, flat nails have replaced the claws of ancestral primates, and sensitive pads on the undersides of fingers and toes assist in the grasping of objects. All primates have thumbs, but they are only truly opposable in Old World monkeys, great apes, and humans. Because an **opposable thumb** can touch each of the other fingers, the grip is both powerful and precise (Fig. 30.2). In all but humans, primates with opposable thumbs also have opposable toes.

The evolution of the primate limb was a very important adaptation for their life in trees. Mobile limbs with clawless, opposable digits allow primates to freely grasp and release tree limbs. They also allow primates to easily reach out and bring food, such as fruit, to the mouth.

Stereoscopic Vision

A foreshortened snout and a relatively flat face are also evolutionary trends in primates. These may be associated with a general decline in the importance of smell and an increased reliance on vision. In most primates, the eyes are located in the front, where they can focus on the same object from slightly different angles (Fig. 30.3). The result is **stereoscopic** (three-dimensional) **vision** with good depth perception that permits primates to make accurate judgments about the distance and position of adjacent tree limbs.

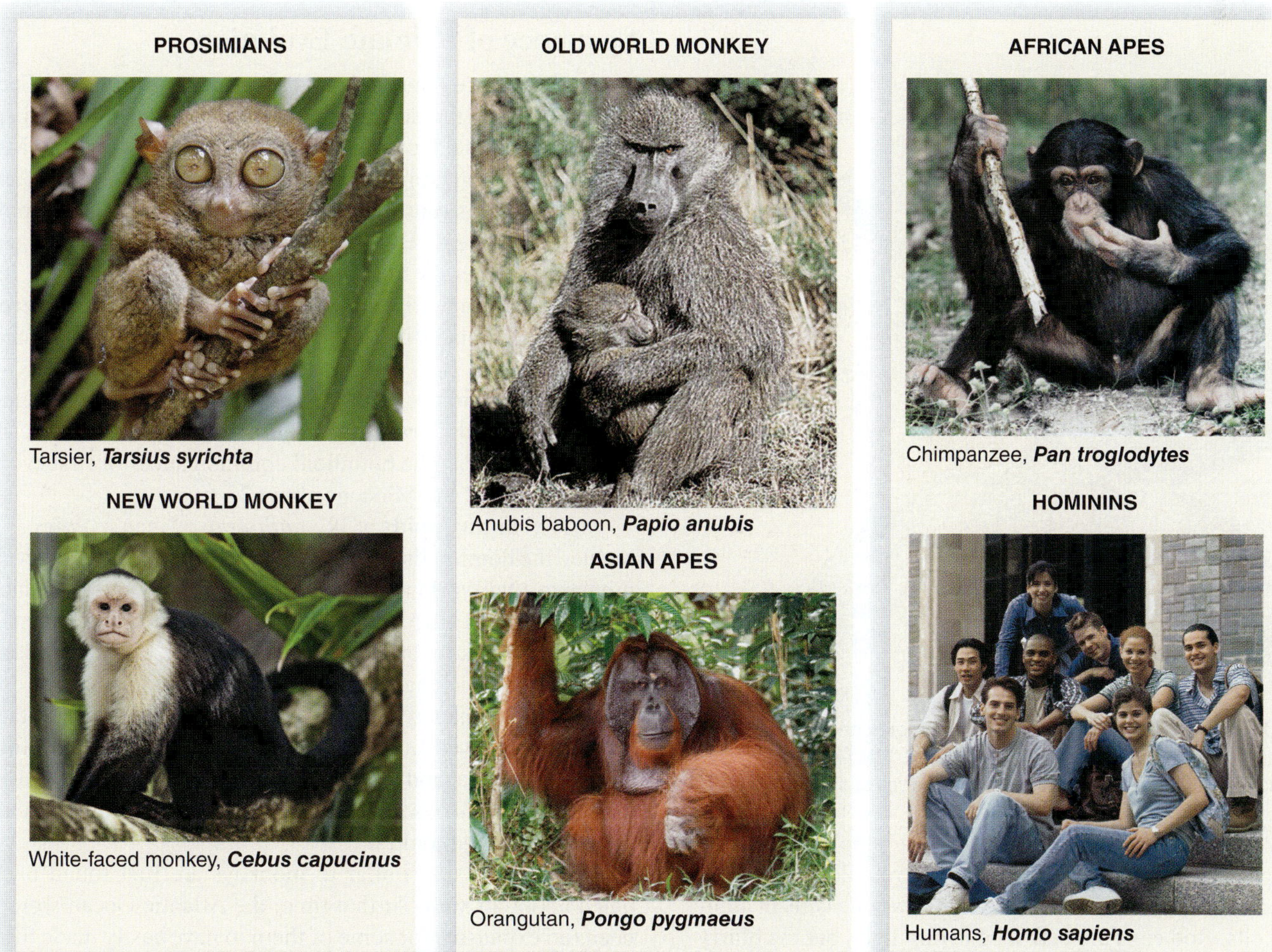

Figure 30.1 Primate diversity. Today's prosimians may resemble the first group of primates to evolve. Modern monkeys are divided into the New World monkeys and the Old World monkeys. The apes can be divided into the Asian apes (orangutans and gibbons) and the African apes (chimpanzees and gorillas). Humans (hominins) are also primates.

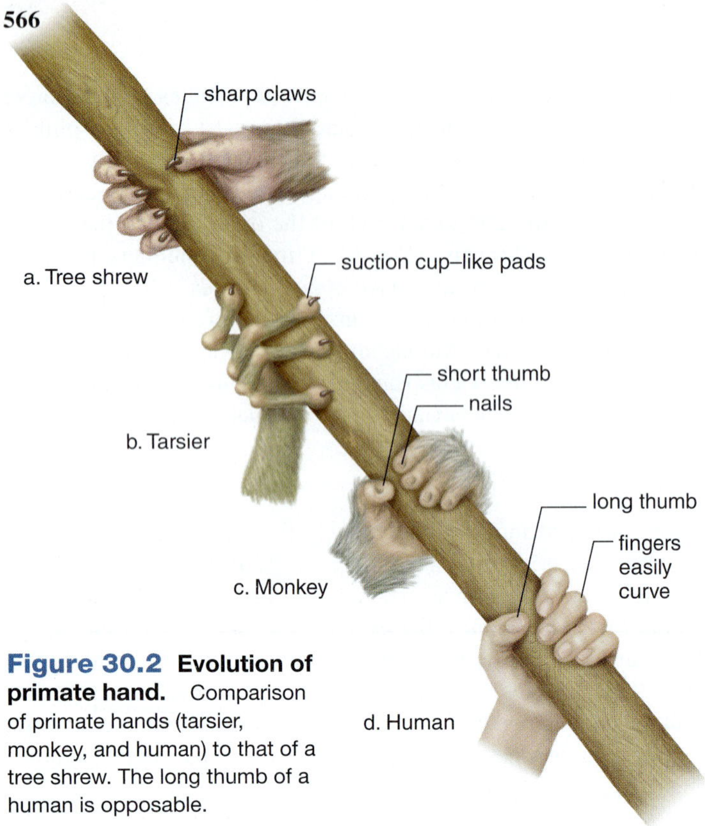

Figure 30.2 Evolution of primate hand. Comparison of primate hands (tarsier, monkey, and human) to that of a tree shrew. The long thumb of a human is opposable.

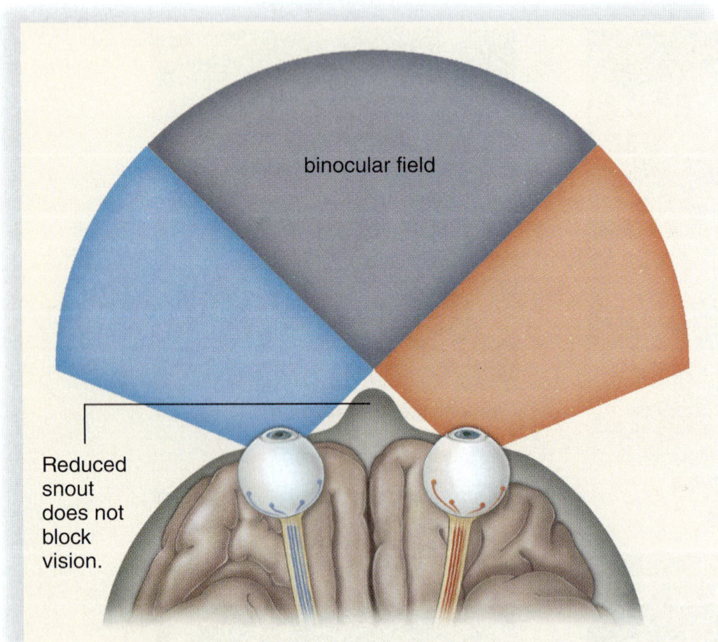

Figure 30.3 Stereoscopic vision. In primates, the snout is reduced, and the eyes are at the front of the head. The result is a binocular field that aids depth perception and provides stereoscopic vision.

Some primates, humans in particular, have color vision and greater visual acuity because the retina contains cone cells in addition to rod cells. Rod cells are activated in dim light, but the blurry image is in shades of gray. Cone cells require bright light, but the image is sharp and in color. The lens of the eye focuses light directly on the fovea, a region of the retina where cone cells are concentrated (see Chapter 38).

Large, Complex Brain

Sense organs are only as beneficial as the brain that processes their input. The evolutionary trend among primates is toward a larger and more complex brain. This is evident when comparing the brains of prosimians, such as lemurs and tarsiers, with those of apes and humans. In apes and humans, the portion of the brain devoted to smell is smaller, and the portions devoted to sight have increased in size and complexity. Also, more of the brain is devoted to controlling and processing information received from the hands and the thumbs. The result is good hand-eye coordination. A larger portion of the brain is devoted to communication skills, which supports primates' tendency to live in social groups.

Reduced Reproductive Rate

One other trend in primate evolution is a general reduction in the rate of reproduction, associated with increased age of sexual maturity and extended lifespans. Gestation is lengthy, allowing time for forebrain development. One birth at a time is the norm in primates; it is difficult to care for several offspring in the trees while moving from limb to limb. The juvenile period of dependency is extended, and there is an emphasis on learned behavior and complex social interactions.

Sequence of Primate Evolution

Figure 30.4 traces the evolution of primates during the Cenozoic era. **Hominins** (the designation that includes humans and species very closely related to humans) first evolved about 5 MYA. Molecular data show that hominins and gorillas are closely related and that these two groups must have shared a common ancestor sometime during the Miocene. Hominins, chimpanzees, and gorillas are now grouped together as **hominines.** The Nature of Science feature, "A Genomic Comparison of *Homo sapiens* and Chimpanzees," on page 568 presents DNA comparisons between modern humans and chimpanzees.

The **hominids** (L. *homo,* "man"; Gk. *eides,* "like") include the hominines and the orangutan. The **hominoids** include the gibbon and the hominids. The hominoid common ancestor first evolved at the beginning of the Miocene about 23 MYA.

The **anthropoids** (Gk. *anthropos,* "man"; *eides,* "like") include the hominoids and the Old World monkeys and New World monkeys. Old World monkeys, native to Africa and Asia, lack prehensile tails and have protruding noses. Some of the better-known Old World monkeys are the baboon, a ground dweller, and the rhesus monkey, which has been used extensively in medical research. New World monkeys, which often have long, prehensile tails and flat noses, evolved in South America and are now also found in Central America and parts of Mexico. Two of the well-known New World monkeys are the spider monkey and the capuchin, the "organ grinder's monkey."

Primate fossils similar to monkeys are first found in Africa, dated about 45 MYA. At that time, the Atlantic Ocean would have been too expansive for some of them to have easily made their way to South America. It is hypothesized that a common ancestor of both the New World and Old World monkeys arose much earlier, when a narrower Atlantic would have made crossing much more reasonable.

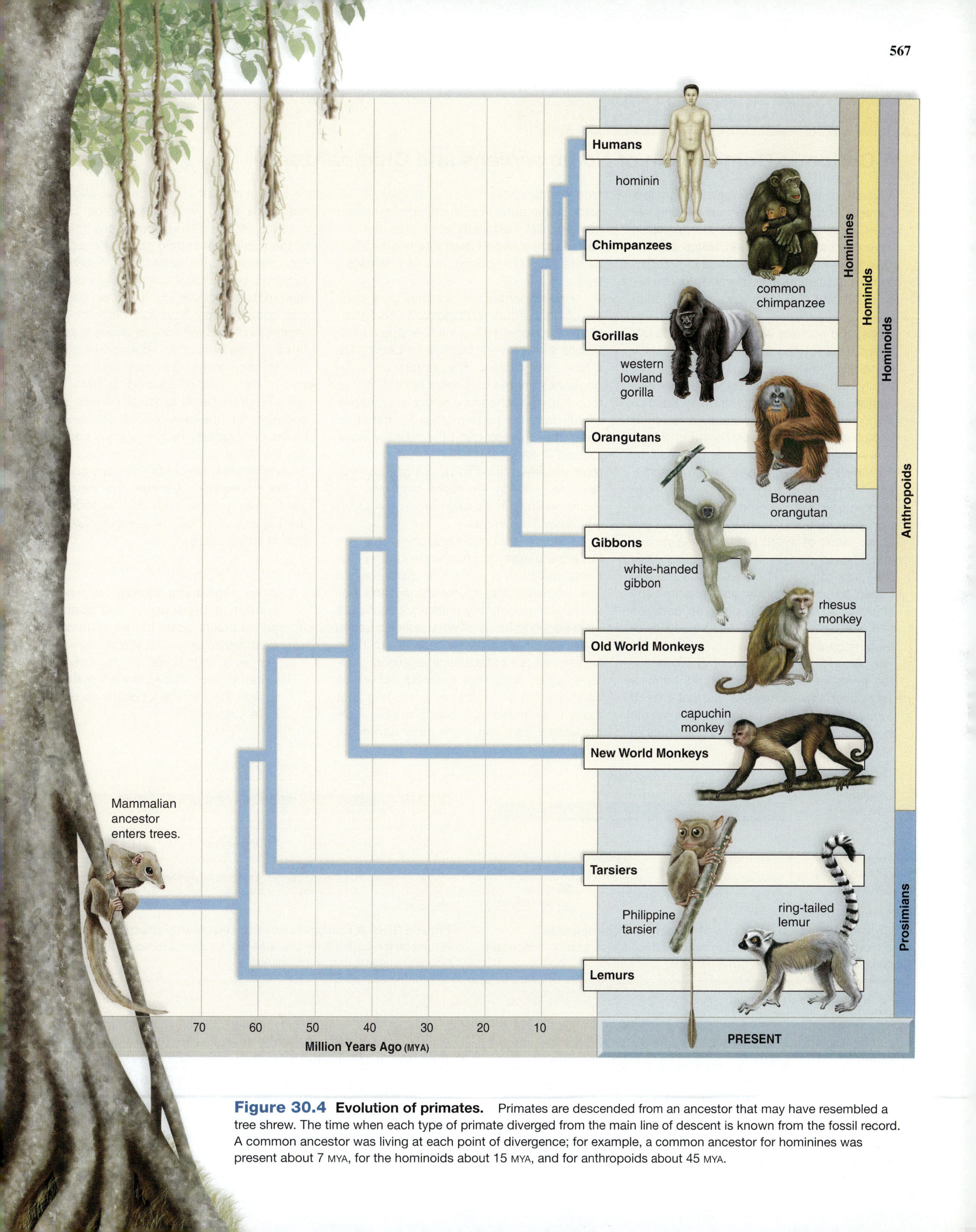

Figure 30.4 Evolution of primates. Primates are descended from an ancestor that may have resembled a tree shrew. The time when each type of primate diverged from the main line of descent is known from the fossil record. A common ancestor was living at each point of divergence; for example, a common ancestor for hominines was present about 7 MYA, for the hominoids about 15 MYA, and for anthropoids about 45 MYA.

THEME **Nature of Science**

A Genomic Comparison of *Homo sapiens* and Chimpanzees

A wealth of genetic evidence suggests that humans and chimpanzees are closely related, despite the fact that chimpanzees have 48 chromosomes and *Homo sapiens* have 46. At first, the difference in chromosome number was considered significant—so significant that for a long time the great apes (chimpanzees, gorillas, and orangutans) and humans were classified into different families. The apes were in the family Pongidae, while the humans were in the family Hominidae. However, in 1991, investigators at Yale University showed that human chromosome 2 is actually a fusion of two chimpanzee chromosomes (Fig. 30A).

Additional evidence was provided by the study of transposons. In Chapter 14, you learned that transposons are mobile elements in the genome. Many copies of transposons are present in the human genome, most of which are no longer mobile and instead remain in one location. They are relics of past retrovirus infections. Because these infections are random, any similarity in transposon patterns between two species can be considered evidence of a common ancestor.

Through the course of many studies, investigators have found that humans and chimpanzees have similar patterns of transposons in their genome. An example is shown in Fig. 30B of an *Alu* element (a type of transposon) in the vicinity of the hemoglobin gene in both chimpanzees and humans. This similarity suggests that the transposon inserted itself into this location before the chimpanzee-human lineages split.

Pseudogenes are nonfunctional copies of genes that were active in the past. Most pseudogenes are inactive due to mutations that prevent them from coding for a functional protein. The pattern of pseudogenes in humans is most similar to that found in the chimpanzees, but it varies slightly from the patterns found in the other great apes. These and other studies have caused a reclassification of the primates most closely related to us. All of the great apes are now in the same family as humans, while chimpanzees are in the same subfamily (Homininae).

Modern genomic data show that the base sequence of chimpanzees and humans differs only by 1.5%. Because we now consider that chimpanzees and humans are genetically similar, geneticists have begun to focus on what specific genes make us different—even though our base sequences are quite similar, significant differences do exist. For example, when we compare the two genomes, we find that many DNA stretches (about 5 million in all) are absent from one or the other genomes.

We know that in the evolution of mammals there was an explosion in the amount of noncoding sequences relative to the number of coding genes. So a common mammalian ancestor may have had a larger quantity of noncoding DNA, and each mammalian ancestor (for example, chimps and humans) may have lost different DNA stretches since their lines of descent separated. These noncoding regions still play a role in gene expression (see Chapter 14), so the stretches of noncoding DNA that were retained by a particular primate may account for the anatomical differences. Evidence suggests that the genes controlling the development of the brain may have been affected the most by the gaps present in the chimpanzee genome compared to the human genome. This may account for the larger size of our brains compared to that of chimpanzees today.

Questions to Consider

1. Which would you consider to have a pattern of transposons and pseudogenes that is closer to that of humans, a prosimian or an Old World monkey?
2. For what additional differences between chimps and humans would you screen the genome for evidence of lost DNA stretches?

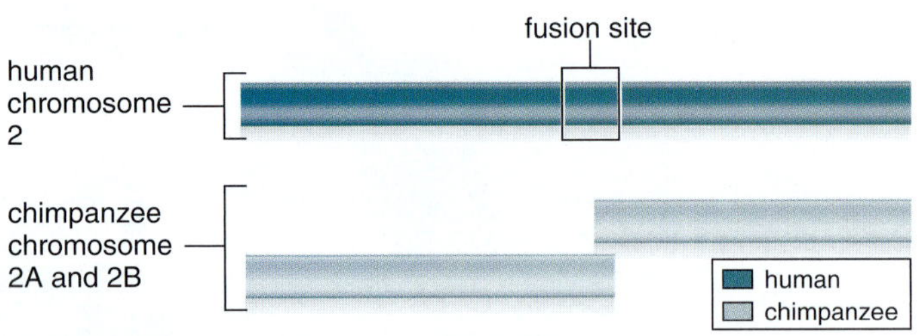

Figure 30A A comparison of human and chimpanzee chromosomes. Human chromosome 2 is a fusion of two chimpanzee chromosomes.

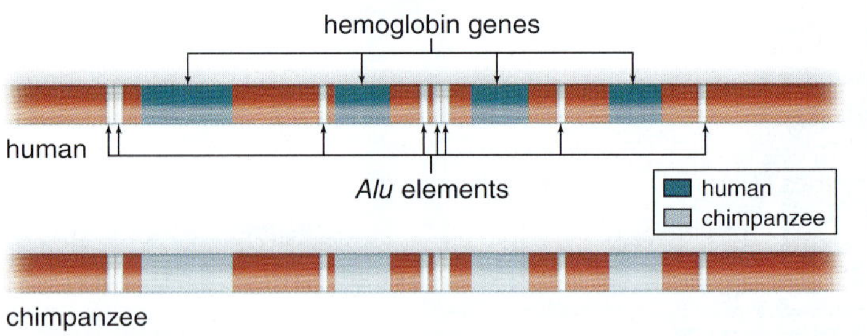

Figure 30B A comparison of transposon patterns. Similarities in transposon patterns, in this case *Alu,* suggest a close common ancestor between humans and chimpanzees.

The transitional link between the monkeys and the hominoids (around 35 MYA) is best represented by a fossil classified as *Proconsul. Proconsul* was about the size of a baboon, and the size of its brain (165 cc) was also comparable. This fossil species didn't have the tail of a monkey (Fig. 30.5), but its limb proportions suggest that it walked as a quadruped on top of tree limbs, as monkeys do. Although primarily a tree dweller, *Proconsul* may have also spent time exploring nearby environs for food.

Proconsul was probably ancestral to the **dryopithecines,** from which the hominoids arose. About 10 MYA, Africarabia (Africa plus the Arabian Peninsula) joined with Asia, and the apes migrated into Europe and Asia. In 1966, Spanish paleontologists announced the discovery of a specimen of *Dryopithecus* dated at 9.5 MYA near Barcelona. The anatomy of these bones clearly indicates that *Dryopithecus* was a tree dweller and locomoted by swinging from branch to branch, as gibbons

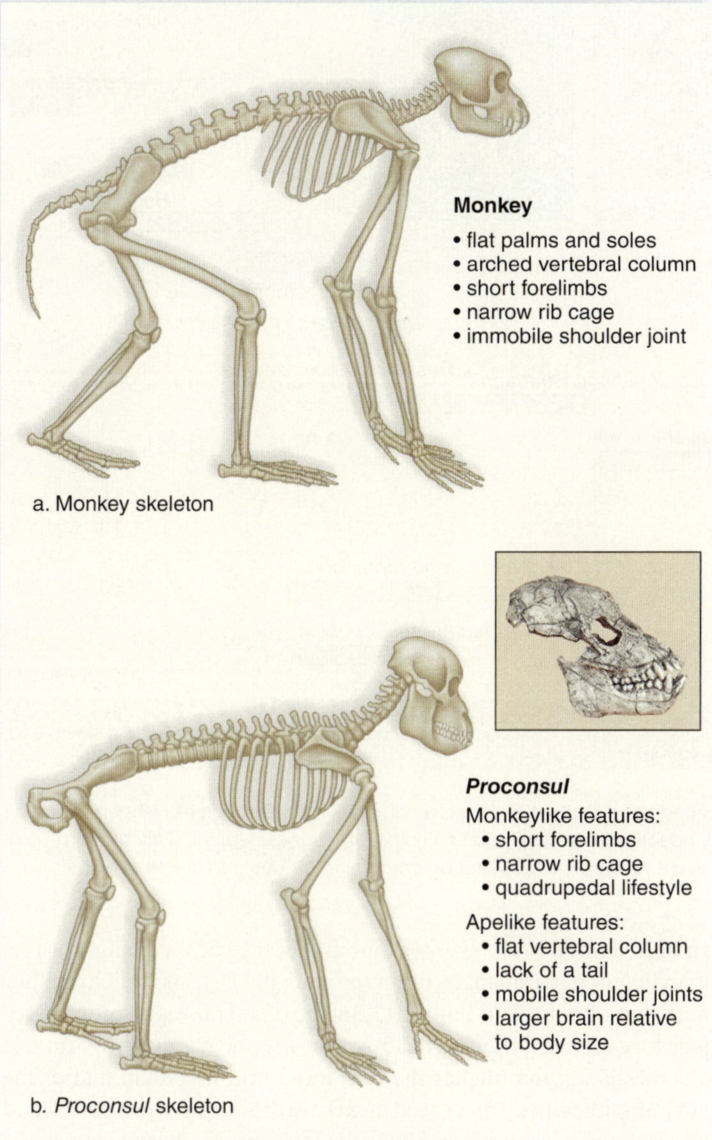

Figure 30.5 Monkey skeleton compared to *Proconsul* skeleton. Comparison of a monkey skeleton **(a)** with that of *Proconsul* **(b)** shows various dissimilarities, indicating that *Proconsul* is more related to today's apes than to today's monkeys.

do today. They did not walk along the top of tree limbs as *Proconsul* did.

Note that **prosimians** (L. *pro,* "before"; *simia,* "ape, monkey"), represented by lemurs and tarsiers, were the first type of primate to diverge from the common ancestor for all the primates. All primates share one common mammalian ancestor, which lived about 55 MYA. This ancestor may have resembled today's tree shrews.

Check Your Progress 30.1

1. Identify primate traits that are adaptive for living in trees.
2. Identify the location of hominins, hominines, hominoids, hominids, anthropoids, and prosimians in the evolutionary tree of the primates.
3. Describe the characteristics that humans share with the chimpanzees.

30.2 Evolution of Humanlike Hominins

Learning Outcomes

Upon completion of this section, you should be able to

1. Explain the significance of bipedalism in hominin evolution.
2. Compare the features of "Ardi" with those of "Lucy."
3. Summarize the importance of ardipithecines and australopithecines in hominin evolution.

The relationship of hominins to the other primates is shown in the classification box below. Molecular data have been used to determine when hominin evolution began. When two lines of descent first diverge from a common ancestor, the genes of the two lineages are nearly identical. But as time goes by, each lineage accumulates genetic changes. Genetic changes compared to the other hominines suggest that hominin evolution began about 5 MYA.

Evolution of Bipedalism

The anatomy of humans is suitable for standing erect and walking on two feet, a characteristic called *bipedalism.* Humans are bipedal, while apes are quadrupedal (walk on all fours). Early humanlike hominins are not in the genus *Homo,* but they are considered closely related to humans, because they exhibit bipedalism. Although bipedalism places stress on the spinal column, the upright posture frees the hands for tool use.

Figure 30.6 provides a timeline of hominin evolution. The orange and green bars signify the early humanlike hominins, a lavendar bar signifies an early species of the genus *Homo*, while the later members of genus *Homo* are in blue. The length of each bar indicates when evidence of the species first appears in the fossil record until the estimated time of its extinction.

CLASSIFICATION

ORDER: Primates

- Adapted to an arboreal life
- Prosimians, Anthropoids

FAMILY: Hominidae (hominids)

SUBFAMILY: Homininae (hominines)

TRIBE*: Hominini (hominins)

Early Humanlike Hominins → *Sahelanthropus*, ardipithecines,

Later Humanlike Hominins → australopithecines

GENUS: *Homo* (humans)

Early *Homo* → *Homo habilis, Homo ergaster, Homo erectus*
Brain size greater than 600 cc; tool use and culture

Later *Homo* → *Homo heidelbergensis, Homo neandertalensis, Homo sapiens*
Brain size greater than 1,000 cc; tool use and culture

*A tribe is a taxonomic level that lies between subfamily and genus.

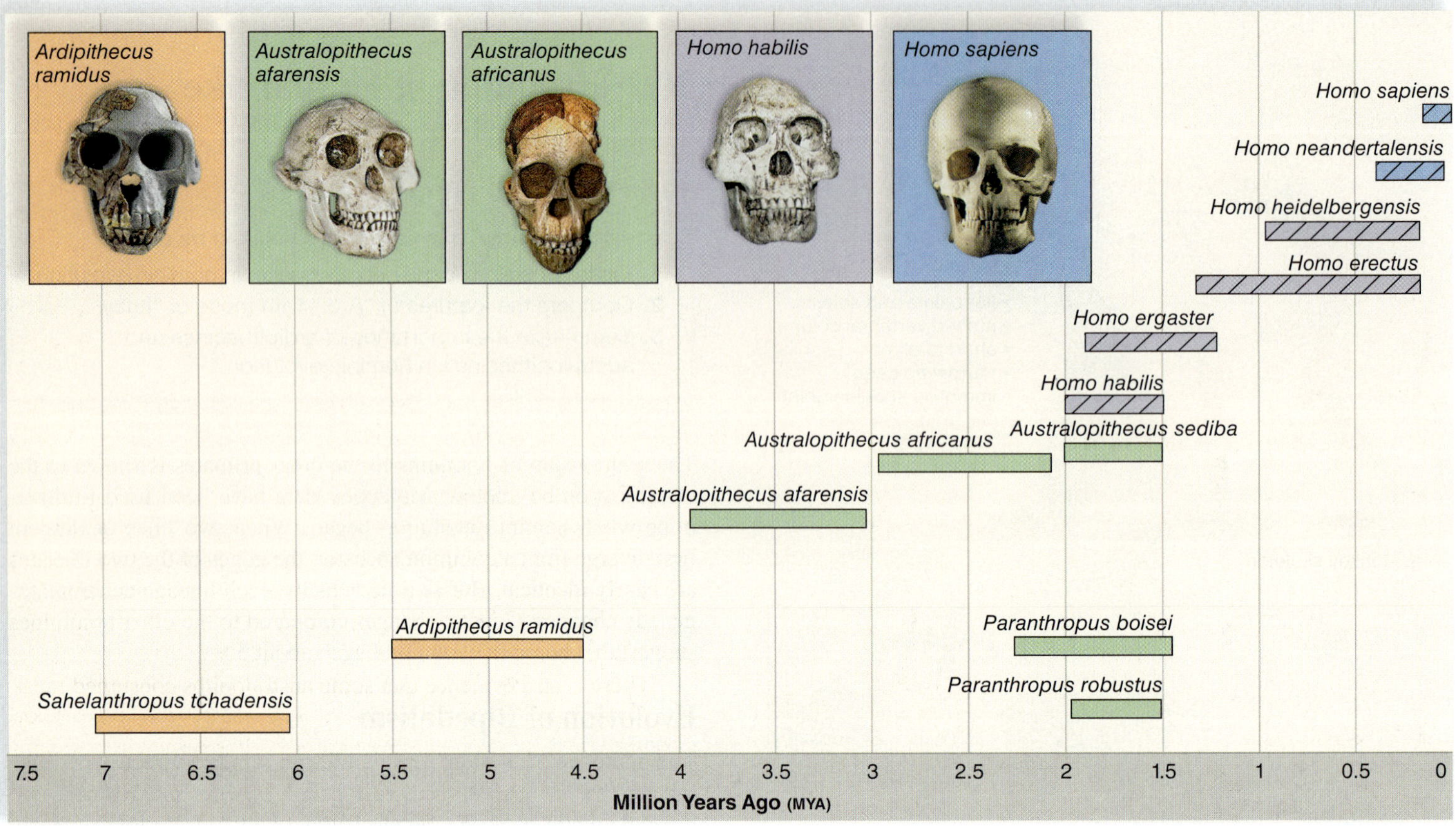

Figure 30.6 Human evolution. Several groups of extinct hominins preceded the evolution of modern humans. These groups have been divided into the early humanlike hominins (orange), later humanlike hominins (green), early species (lavender), and finally the later species (blue). Only modern humans are classified as *Homo sapiens.* The cross marks indicate areas where current research is focusing on combining groups into single species.

Paleontologists have identified several fossils dated around the time of the split between the human and ape lineages. One of these is *Sahelanthropus tchadensis,* a species that lived in West Central Africa between 6 and 7 MYA. Only cranial fragments of this species have been uncovered to date, but the point on the back of the skull where the neck muscles would have attached suggests bipedalism. The brain of *S. tchadensis* was similar in size to that of modern chimpanzees, and some scientists suspect that it may have been a common ancestor to chimpanzees and humans. The skull of this fossil is very similar to that of the ardipithecines, discussed next.

Ardipithecines

Two species of **ardipithecines** have been uncovered, *Ardipithecus kadabba* and *A. ramidus.* Only teeth and a few bone bits have been found for *A. kadabba,* and these have been dated to around 5.6 MYA. A more extensive collection of fossils has been collected for *A. ramidus.* To date, over 100 skeletons, all dated to 4.4 MYA, have been identified from this species; all were collected near a small town in Ethiopia, East Africa. In 2009, scientists announced that they had reconstructed these fossils to form a female fossil called Ardi.

Some of Ardi's features are primitive, like that of an ape such as *Dryopithecus* (see section 30.1), but others are like that of a human. Ardi was about the size of a chimpanzee, standing about 120 cm (4 ft) tall and weighing about 55 kg (110 lb). It appears that males and females were about the same size.

Ardi had a small head compared to the size of her body. The skull has the same features as *Sahelanthropus tchadensis* but is smaller. Ardi's brain size was around 300 to 350 cc, slightly less than that of a chimpanzee brain (around 400 cc), and much smaller than that of a modern human (1,360 cc). The nose and mouth project forward, and the forehead is low with heavy eyebrow ridges, a combination that makes the face more primitive than that of the australopithecines (discussed next). Ardi's teeth were small and like those of an omnivore. She lacked the strong, sharp canines of a chimpanzee, and her diet probably consisted mostly of soft, rather than tough, plant material.

Ardi could walk erect, but she spent a lot of time in trees. Notice in Figure 30.7 that in both the human and Ardi skeletons, the spine exits from the center of the skull rather than toward the rear. Also, the femurs angle inward toward the knees (red arrows). These skeletal features assist in walking erect by placing the trunk's center of gravity squarely over the feet. Also, Ardi's pelvis and hip joints are broad enough to keep her from swaying from side to side (as chimps do) while walking. The knee joint in both humans and Ardi is modified to support the body's weight, because the bones broaden at this joint.

Ardi's feet have a bone, missing in apes, that would keep her feet squarely on the ground, a sure sign that she was bipedal and not a quadruped like the apes. Nevertheless, like the apes, she has an opposable big toe. Opposable toes allow an animal's feet to grab hold of a tree limb.

The wrists of Ardi's hands were flexible, and most likely she moved along tree limbs on all fours, as ancient apes did. Modern apes *brachiate*—use their arms to swing from limb to limb. Ardi did not do this, but her shoulders were flexible enough to allow

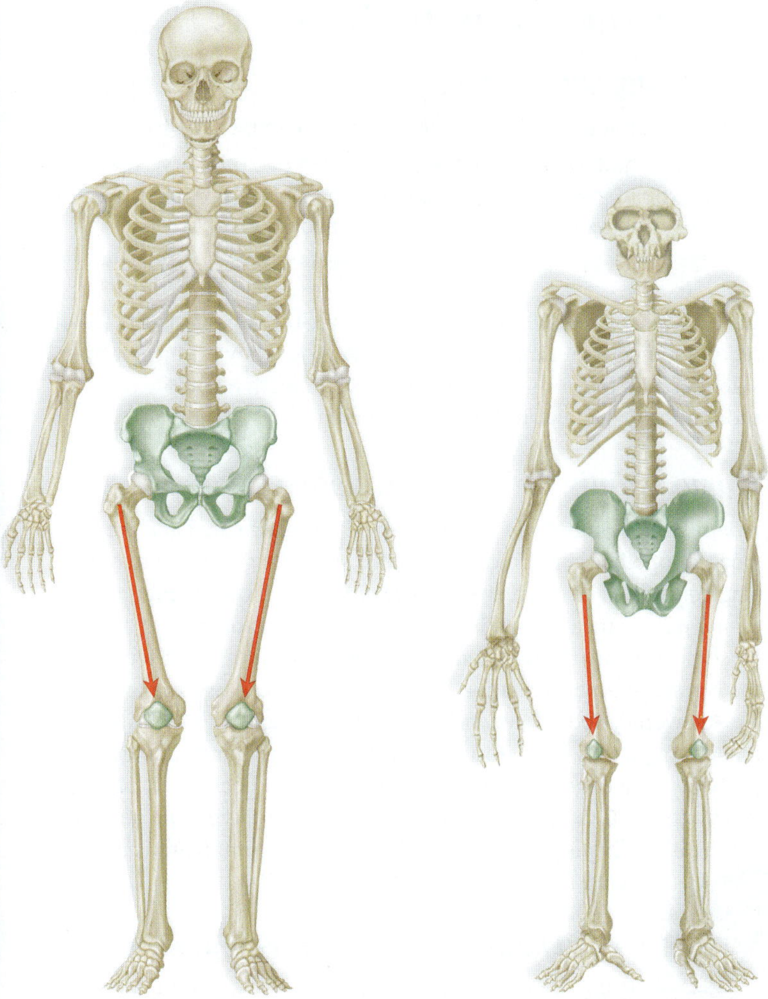

Figure 30.7 Adaptations for walking erect. A human skeleton compared to an ardipithecine (Ardi). In both skeletons, the spine exits from the center of the skull. A broad pelvis (green) causes the femurs to angle (red arrows) toward the broadened knee joints. However, the skeleton of Ardi has an opposable toe, indicating that this species still lived in trees.

her to reach for limbs to the side or over her head. The general conclusion is that Ardi moved carefully in trees. Although the top of her pelvis is like that of a human, and probably served as the attachment of muscles needed for walking, the bottom of the pelvis served as an attachment for the strong muscles needed for climbing trees.

Until recently, it has been suggested that bipedalism evolved when a dramatic change in climate caused the forests of East Africa to be replaced by grassland. However, evidence suggests that Ardi lived in the woods, which questions the advantage that walking erect would have afforded her. Bipedalism does provide an advantage in caring for a helpless infant by allowing it to be carried by hand from one location to another. It is also possible that bipedalism may have benefited the males of the species as they foraged for food on the floor of the forests. More evidence is needed to better understand this mystery, but one thing is clear—the ardipithecines represent a link between our quadruped ancestors and the bipedal hominins.

Australopithecines

The **australopithecines** (called **australopiths** for short) are a group of hominins that evolved and diversified in Africa from 4 MYA until about 1.5 MYA. In Figure 30.6, the australopiths are represented by green-colored bars. The australopiths had a small brain (an apelike characteristic) and walked erect (a humanlike characteristic). Therefore, it seems that not all human characteristics evolved at the same time. This is an example of **mosaic evolution,** meaning that different body parts change at different rates and, therefore, at different times.

Australopiths stood about 100–115 cm in height and had relatively small brains, averaging about 370–515 cc—slightly larger than that of a chimpanzee. Males were distinctly larger than females. Some australopiths were slight of frame and termed *gracile* (slender). Others were *robust* (powerful) and tended to have massive jaws because of their large grinding teeth. Recent changes in the classification of these groups has separated the gracile types into genus *Australopithecus* and the robust types into genus *Paranthropus.*

There is also evidence that some australopiths consumed meat, as indicated by the recent discovery in Ethiopia of two fossilized bones from large mammals dated 3.4 MYA. The bones bear obvious marks indicating the meat was scraped off with stone tools, providing the earliest evidence of australopiths using stone tools, about 1 million years earlier than scientists had previously thought.

East African Australopiths

The most significant fossil from East Africa is from a species of australopiths called *Australopithecus afarensis*. The female specimen of this species, known as Lucy (Fig. 30.8), had a low forehead and a face that projected forward, with large canine teeth. The body of Lucy was broader than that of an ardipithecine. Although the brain size was small (around 400 cc), Lucy's skeleton indicates that she was a biped that stood upright. She stooped a bit like a chimpanzee, and the arms were somewhat proportionally longer than the legs. This suggests brachiation as a possible mode of locomotion in trees. Otherwise, the skeleton was humanlike, even though the pelvis lacked refinements that would have allowed Lucy to walk with a striding gait in a manner similar to modern humans.

Even better evidence of bipedal locomotion comes from a trait of fossilized footprints in Tanzania dated to about 3.7 MYA (Fig. 30.8). The larger footprints are double, as though a smaller being were stepping in the footprints of another, and there are additional footprints off to one side, within hand-holding distance.

Some 30 years after Lucy was discovered, paleontologists discovered a skeleton of a child that has been dated to be 0.1 million years older than Lucy. The face of this skeleton, named Selam but sometimes referred to as "Lucy's baby," looks more like that of an ardipithecine, and the structure of the bones suggests that Selam was not as agile a walker as Lucy.

A. afarensis is a gracile form of australopith and is believed to be ancestral to the robust types found in eastern Africa, *P. boisei* and *P. robustus*. *P. boisei* had a powerful upper body and the largest molars of any hominin. *A. afarensis* is generally considered more directly related to the early members of the genus *Homo* than are the South African species.

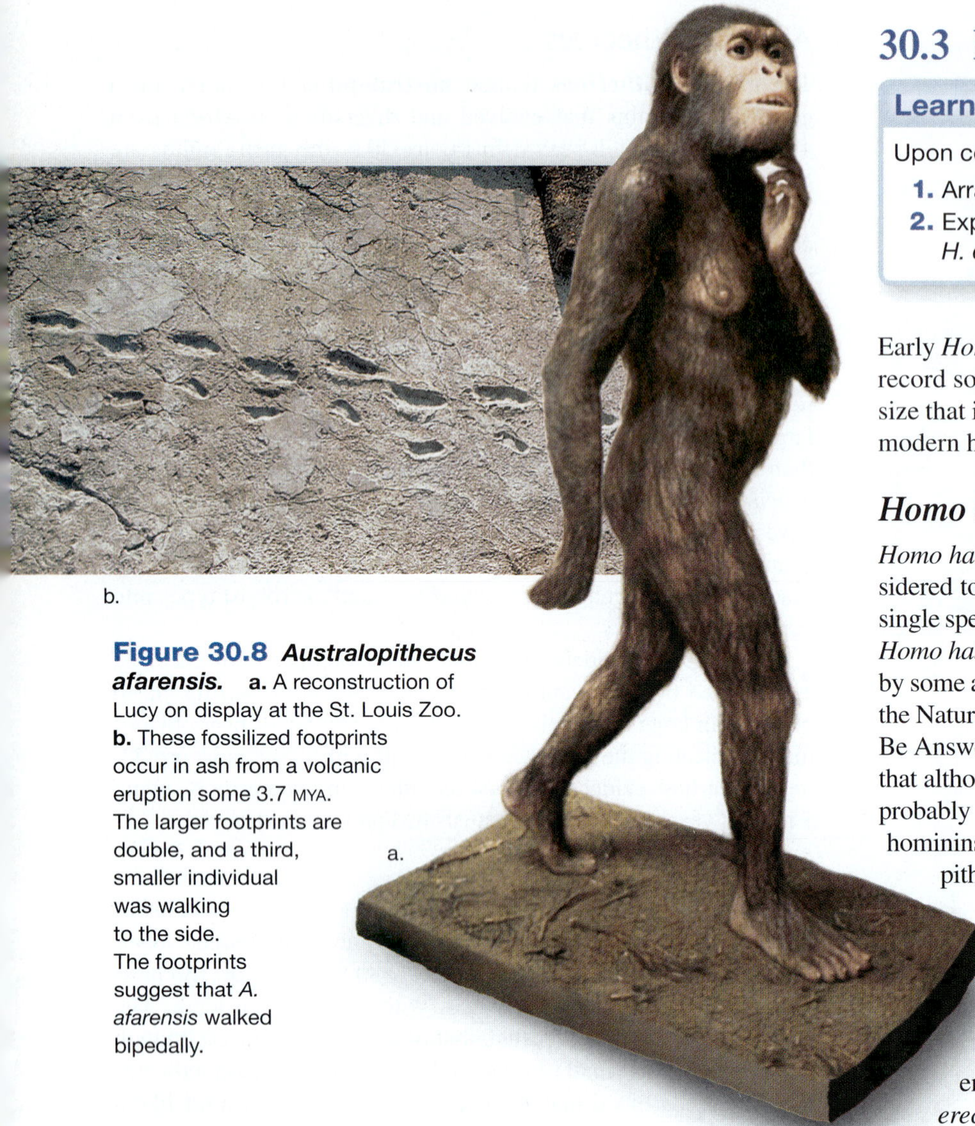

b.

a.

Figure 30.8 ***Australopithecus afarensis.*** **a.** A reconstruction of Lucy on display at the St. Louis Zoo. **b.** These fossilized footprints occur in ash from a volcanic eruption some 3.7 MYA. The larger footprints are double, and a third, smaller individual was walking to the side. The footprints suggest that *A. afarensis* walked bipedally.

South African Australopiths

The first australopith to be discovered was unearthed in southern Africa in the 1920s. This hominin, named *Australopithecus africanus,* is a gracile type. A second southern African specimen called *A. robustus,* discovered in the 1930s, is a robust type that is believed to have had a brain size of around 530 cc.

In 2008, the American anthropologist Lee Berger discovered the bones of a 2-million-year-old australopithecine he named *A. sediba* ("wellspring"). The small brain (500 cc) and long arms suggest that this species is an australopith that climbed trees. However, it also had a humanlike pelvis, nose, and dentition (teeth). A series of papers published in 2013 by the journal *Science* described *A. sediba's* mosaic of primitive and modern traits, leading some to suggest that this species may have been a direct ancestor of the genus *Homo.* The scientific jury is still out on this question.

Check Your Progress 30.2

1. Compare the characteristics of an australopith with those of an ardipithecine.
2. Explain why an understanding of bipedalism and brain size is important in understanding the evolution of the hominins.

30.3 Evolution of Early Genus *Homo*

Learning Outcomes

Upon completion of this section, you should be able to

1. Arrange the early species of *Homo* in evolutionary order.
2. Explain the signficance of *Homo habilis, H. ergaster,* and *H. erectus* in the study of human evolution.

Early *Homo* species (lavender bars in Fig. 30.6) appear in the fossil record somewhat earlier or later than 2 MYA. They all have a brain size that is 600 cc or greater, their jaws and teeth resemble those of modern humans, and tool use is in evidence.

Homo habilis and *Homo rudolfensis*

Homo habilis and *Homo rudolfensis* are closely related and are considered together here. Recent evidence suggests that these may be a single species. For this reason only *H. habilis* is shown in Figure 30.6. *Homo habilis* means "handyman," and these two species are credited by some as being the first hominins to use stone tools, as discussed in the Nature of Science feature, "Some Major Questions Remaining to Be Answered About Human Evolution," on page 576. Most believe that although they appear to have been socially organized, they were probably scavengers rather than hunters. The cheek teeth of these hominins tend to be smaller than even those of the gracile australopiths. This is also evidence that they were omnivorous and ate meat in addition to plant material.

Homo ergaster and *Homo erectus*

Homo ergaster evolved in Africa, most likely from *H. habilis.* Similar fossils found in Asia are different enough to be classified as *Homo erectus* (L. *homo,* "man"; *erectus,* "upright"). These fossils span the dates between 1.9 and 0.3 MYA, and many other fossils belonging to both species have been found in Africa and Asia.

Compared to other early *Homo* species, *H. ergaster* had a larger brain (about 1,000 cc), a rounder jaw, prominent brow ridges, and a projecting nose. This type of nose is adaptive for a hot, dry climate, because it permits water to be removed before air leaves the body. The recovery of an almost complete skeleton of a 10-year-old boy indicates that *H. ergaster* was much taller than the hominins discussed thus far (Fig. 30.9). Males were 1.8 m tall, and females were 1.55 m tall. Indeed, these hominins stood erect and most likely had a striding gait like that of modern humans. The robust and probably heavily muscled skeleton still retained some australopithecine features. Even so, the size of the birth canal in female specimens indicates that infants were born in an immature state that required an extended period of care.

H. ergaster first appeared in Africa but then migrated into Europe and Asia sometime between 2 MYA and 1 MYA. Most likely, *H. erectus* evolved from *H. ergaster* after *H. ergaster* arrived in Asia. In any case, such an extensive population movement is a first in the history of humankind and a tribute to the intellectual and physical skills of these hominins. They also had a knowledge of fire and may have been the first to cook meat.

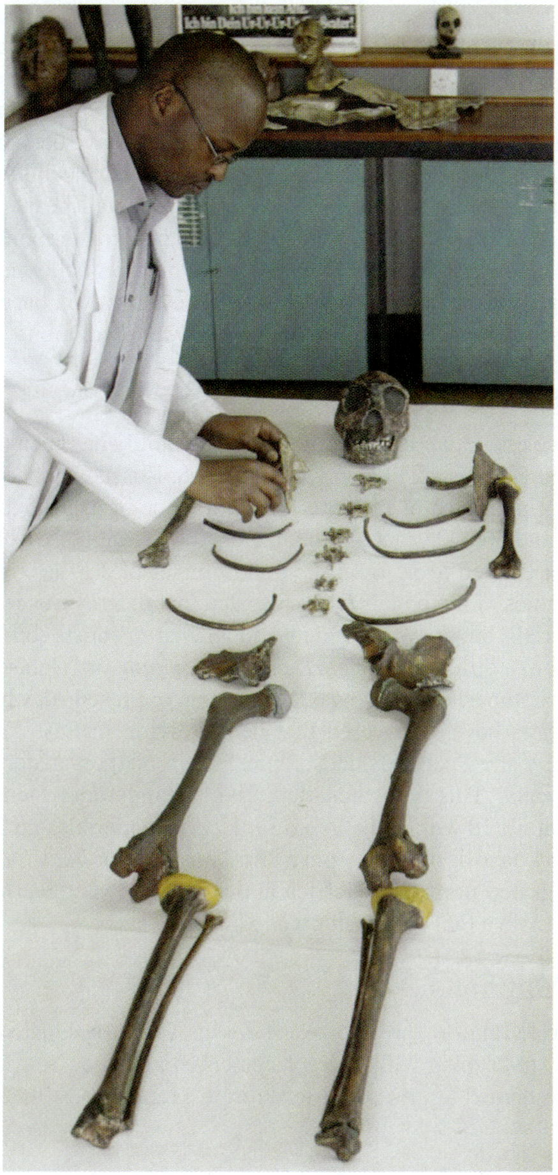

Figure 30.9 ***Homo ergaster.*** This skeleton of a 10-year-old boy who lived 1.6 MYA in eastern Africa shows femurs that are angled because the neck is quite long.

Homo floresiensis

In 2004, scientists announced the discovery of the fossil remains of *Homo floresiensis.* The 18,000-year-old fossil of a 1-m-tall, 25-kg adult female was discovered on the island of Flores in the South Pacific. The specimen was the size of a 3-year-old *Homo sapiens* but possessed a braincase only one-third the size of a modern human. Due to its small size, this species has been nicknamed "Hobbits," after characters in the book by J. R. R. Tolkien. A 2007 study supports the hypothesis that these diminutive hominins evolved from normal-sized, island-hopping *Homo erectus* populations that reached Flores about 840,000 years ago. Apparently, *H. floresiensis* used tools and fire.

Check Your Progress 30.3

1. Discuss the general evolutionary trends in early *Homo* species.
2. Discuss how the evolution of bipedalism and increased brain size probably contributed to *H. ergaster*'s migration from Africa.

30.4 Evolution of Later Genus *Homo*

Learning Outcomes

Upon completion of this section, you should be able to

1. Describe the evidence for the replacement model hypothesis regarding the evolution of later members of the genus *Homo.*
2. Discuss the significance of increased tool use in Cro-Magnons.
3. Summarize how the replacement model explains the major human ethnic groups.

Later *Homo* species are represented by blue-colored bars in Figure 30.6. The evolution of these species from older *Homo* species has been the subject of much debate. Most researchers believe that modern humans (*Homo sapiens*) evolved from *H. ergaster,* but they differ as to the details.

Evolutionary Hypotheses

Many disparate early *Homo* species in Europe are now classified as *Homo heidelbergensis.* Just as *H. erectus* is believed to have evolved from *H. ergaster* in Asia, so *H. heidelbergensis* is believed to have evolved from *H. ergaster* in Europe. Further, for the sake of discussion, *H. ergaster* in Africa, *H. erectus* in Asia, and *H. heidelbergensis* (and *H. neandertalensis*) in Europe can be grouped together as early *Homo* species who lived between 1.5 and 0.25 MYA.

The most widely accepted hypothesis for the evolution of modern humans from archaic humans is referred to as the *replacement model* or *out-of-Africa hypothesis,* which proposes that modern humans evolved from archaic humans only in Africa, and then modern humans migrated to Asia and Europe, where they replaced the early *Homo* species about 100,000 years BP (before the present) (Fig. 30.10).

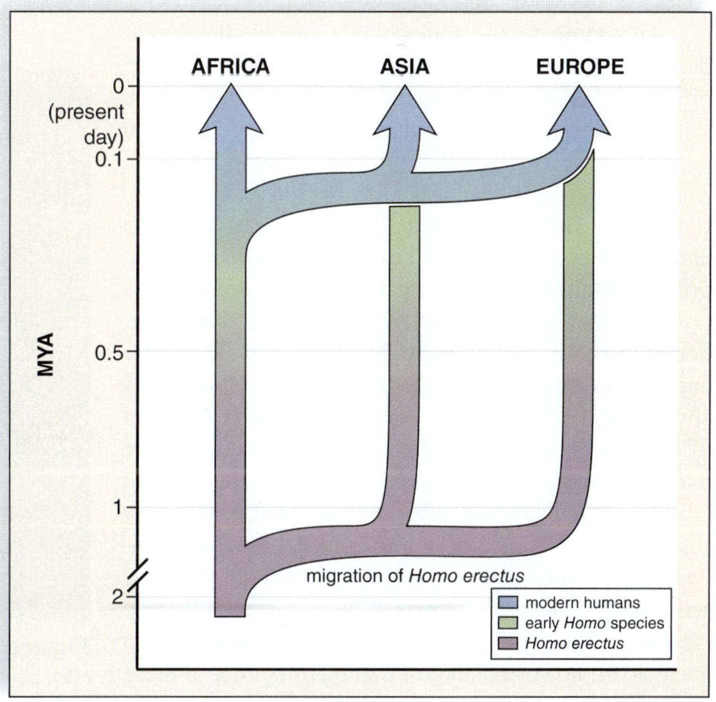

Figure 30.10 Replacement model. Modern humans evolved in Africa and then replaced archaic humans in Asia and Europe.

The replacement model is supported by the fossil record. The earliest remains of modern humans (the Cro-Magnons), dating at least 130,000 years BP, have been found only in Africa. Modern humans are not found in Asia until 100,000 years BP and not in Europe until 60,000 years BP. Until earlier modern human fossils are found in Asia and Europe, the replacement model is supported.

The replacement model is also supported by DNA data. Several years ago, a study showed that the mitochondrial DNA of Africans is more diverse than the DNA of the people in Europe (and the rest of the world). This is significant, because if mitochondrial DNA has a constant rate of mutation, Africans should show the greatest diversity, since modern humans have existed the longest in Africa. Called the "mitochondrial Eve" hypothesis by the press (note that this is a misnomer, because no single ancestor is proposed), the statistics that calculated the date of the African migration were found to be flawed. Still, the raw data—which indicate a close genetic relationship among all Europeans—support the replacement model.

A hypothesis opposing the replacement model does exist. This hypothesis, called the *multiregional continuity hypothesis,* proposes that modern humans arose from archaic humans in essentially the same manner in Africa, Asia, and Europe. The hypothesis is multiregional, because it applies equally to Africa, Asia, and Europe, and it proposes that in these regions genetic continuity will be found between modern populations and early *Homo*. This hypothesis has sparked many innovative studies to test which hypothesis is correct.

Neandertals

The **Neandertals,** *Homo neandertalensis,* are an intriguing species of archaic humans that lived between 200,000 and 30,000 years ago. Neandertal fossils are known from the Middle East and throughout Europe. Neandertals take their name from Germany's Neander Valley, where one of the first Neandertal skeletons, dated some 200,000 years ago, was discovered.

According to the replacement model, the Neandertals were also supplanted by modern humans. Surprisingly, however, the Neandertal brain was, on the average, slightly larger than that of *Homo sapiens* (1,400 cc, compared with 1,360 cc in most modern humans). The Neandertals had massive brow ridges and wide, flat noses. They also had a forward-sloping forehead and a receding lower jaw. Their nose, jaws, and teeth protruded far forward. Physically, the Neandertals were powerful and heavily muscled, especially in the shoulders and neck. The bones of Neandertals were shorter and thicker than those of modern humans. New fossils show that the pubic bone was long compared to that of modern humans. The Neandertals lived in Europe and Asia during the last Ice Age, and their sturdy build could have helped conserve heat. As noted in the story that opened this chapter, interbreeding between Neandertals and *Homo sapiens* may have provided the latter with some genes that increased their tolerance of cold weather, among other traits. Two papers published in early 2014, however, suggest that the offspring of these hybrid matings likely had fertility problems.

Archaeological evidence suggests that Neandertals were culturally advanced. Some Neandertals lived in caves; however, others probably constructed shelters. They manufactured a variety of stone tools, including spear points, which could have been used for hunting, and scrapers and knives, which would have helped in food preparation. They most likely successfully hunted bears, woolly mammoths, rhinoceroses, reindeer, and other contemporary animals. They used and could control fire, which probably helped in cooking frozen meat and in keeping warm. They even buried their dead with flowers and tools and may have had a religion.

Denisovans

Evidence of another member of the human family has been found in a cave in Russia, called Denisova. In 2008, Russian archaeologists working in an area of the cave known to contain deposits between 30,000 and 50,000 years old found a tiny bone fragment, which was later determined to be from the pinkie finger of a 5- to 7-year-old girl. Studies of the bone's DNA revealed that the girl was related to Neandertals, but she was different enough that scientists consider her to be from a different species, *Denisova hominin,* or "Denisovans."

By comparing Denisovan DNA to that of present-day humans, researchers have discovered that people living in East Asia, and especially the Pacific Islands, share 3–5% of their genome with Denisovans. This indicates that, like Neandertals, Denisovans likely interbred with early *Homo sapiens*. Neandertals and Denisovans may have descended from a "first wave" of *H. heidelbergensis* that migrated north from Africa about 200,000 years earlier than the later wave that evolved into *H. sapiens*.

Cro-Magnons

The **Cro-Magnons** are the oldest fossils to be designated *Homo sapiens*. In keeping with the replacement model, the Cro-Magnons, who are named after a fossil location in France, were the modern

Figure 30.11 Cro-Magnons. Cro-Magnon people are the first to be designated *Homo sapiens.* Their tool-making ability and other cultural attributes, such as their artistic talents, are legendary.

humans who entered Asia from Africa about 100,000 years BP and then spread to Europe. They probably reached western Europe about 40,000 years ago.

Cro-Magnons had a thoroughly modern appearance (Fig. 30.11). They had lighter bones, flat high foreheads, domed skulls housing brains of 1,590 cc, small teeth, and a distinct chin. They were **hunter-gatherers** who collected food from the environment rather than domesticating animals and growing food plants. *H. erectus* were also hunter-gatherers, but Cro-Magnons hunted more efficiently.

Tool Use in Cro-Magnons

Cro-Magnons designed and manipulated tools and weapons of increasing sophistication. They made advanced stone tools, including compound tools, as when stone flakes were fitted to a wooden handle. They may have been the first to make knifelike blades and to throw spears, enabling them to kill animals from a distance. They were such accomplished hunters that some researchers believe they may have been responsible for the extinction of many larger mammals, such as the giant sloth, the mammoth, the sabertoothed tiger, and the giant ox, during the late Pleistocene epoch. This event is known as the Pleistocene overkill.

Language and Cro-Magnons

A more highly developed brain may have also allowed Cro-Magnons to perfect a language composed of patterned sounds. Language greatly enhanced the possibilities for cooperation and a sense of cohesion within the small bands that were the predominant form of human social organization, even for the Cro-Magnons.

The Cro-Magnons were highly creative. They sculpted small figurines and jewelry out of reindeer bones and antlers. These sculptures could have had religious significance or may have been seen as a way to increase fertility. The most impressive artistic achievements of the Cro-Magnons were cave paintings, realistic and colorful depictions of a variety of animals, from woolly mammoths to horses, that have been discovered deep in caverns in southern France and Spain. These paintings suggest that Cro-Magnons had the ability to think symbolically, as would be needed in order to speak.

Rise of Agriculture

The Cro-Magnons combined hunting and fishing with gathering fruits, berries, grains, and root crops that grew in the wild. With the rise of agriculture about 10,000 years BP, modern humans are no longer considered Cro-Magnon. However, full dependency on domestic crops and animals did not occur until humans started making tools of bronze (instead of stone), about 4,500 years BP.

Anthropologists previously thought that early humans turned to agriculture because life as a hunter-gatherer had its drawbacks. However, skeletal evidence suggests that early agricultural societies experienced an increase in infectious diseases, malnutrition, and anemia compared to earlier hunter-gatherer groups. Many anthroplogists now think that the change in lifestyle was dictated by extinctions of the large game animals, as well as a general warming of the climate. As the glaciers retreated, fertile soil was deposited into rivers and streams full of fish. In suitable locations, such as the fertile crescent of Mesopotamia, fishing villages may have developed, causing populations to settle in one location. Combined with a beneficial climate, the increase in food supplies would have resulted in an increase in the population, reducing its ability to migrate easily. An increase in agriculture would also have supported the specialization of tasks in the population and enhanced the process of **biocultural evolution,** in which cultural achievements—not individual phenotypes—are influenced by natural selection.

Human Variation

Humans have been widely distributed about the globe ever since they evolved. As with any other species that has a wide geographic distribution, phenotypic and genotypic variations are noticeable between populations. Today, we say that people have different ethnicities (Fig. 30.12*a*).

Figure 30.12 Ethnic groups. **a.** Some of the differences among the prevalent ethnic groups in the United States may be due to adaptations to their original environments. **b.** The Maasai live in East Africa. **c.** Eskimos live near the Arctic Circle.

THEME **Nature of Science**

Some Major Questions Remaining to Be Answered About Human Evolution

Whose Bones Are in the Pit?

In the 1970s, scientists discovered a cave in northern Spain, which came to be known as the Sima de los Huesos, or Pit of Bones, because the remains of at least 28 ancient hominin skeletons have been recovered there. None of the bones show signs of violent trauma, suggesting that the bodies may have been deposited there on purpose. However, at an estimated age of 400,000 years, the bones are far too old to have belonged to any of our more recent ancestors who buried their dead. Researchers compared DNA from a thighbone from this site to that of Neandertals and other early humans and were surprised to find that this fossil is more closely related to Denisovans than to Neandertals. Because Denisovans are thought to have lived far to the east and south of Spain, the researchers speculate that these bones may be those of an as yet unidentified species that may have given rise to the Denisovans, and maybe the Neandertals as well.

Which Human Ancestors Used Complex Tools?

A variety of animals are known to use, and even make, simple tools. Primates are probably the best-known tool users, as Jane Goodall first described in her groundbreaking studies on wild chimpanzees. They modify twigs to use for gathering termites and use stones to open nuts. Birds such as crows and ravens are also good at using sticks and twigs to extract insects from holes, or even dropping nuts in front of moving cars to crack them open.

Even though tool use is not unique to *Homo sapiens,* the evolution of modern humans has been dependent in many ways on our ability to use our hands to design, produce, and use complex tools (Fig. 30C). This process requires not only a highly developed brain but also an anatomical feature of the human hand, in which the third metacarpal bone has a projection known as a styloid process. This helps the thumb and fingers apply more pressure to the palm, providing the hand strength and dexterity needed to make and use complex tools. This feature had been thought to be present only in modern humans and Neandertals, but in 2013 researchers working in Kenya discovered a hand fossil with this feature dated 1.4 MYA. This indicates that complex tool use could have developed at least 500,000 years earlier than previously thought.

What Happened to the Neandertals?

Fossilized remains of *Homo neandertalensis* have been found widely throughout what is now Europe and Asia, suggesting that they were the dominant hominin group residing there for hundreds of thousands of years. About 80,000 to 100,000 years ago, ancestors of modern humans entered Europe from the south. The Neandertals disappeared about 30,000 years ago, but what caused their extinction is still a matter of much scientific debate. Although no one can be sure how much contact occurred between *H. neandertalensis* and *H. sapiens,* recent genetic analyses have shown that interbreeding did occur, so there obviously was some contact. The biologist and author Jared Diamond has suggested that there may have been violent battles between the two groups, which the Neandertals ultimately lost. There is some evidence that conflicts did occur—for example, a 2009 discovery in southern France of a Neandertal jawbone from which the teeth had been manually removed, as if to be worn as a necklace.

The truth may have been less gruesome. New infectious diseases that traveled with *H. sapiens* to Europe could have been difficult for the Neandertal population to resist. The modern humans may have had some other competitive advantage, such as better hunting and gathering of food, depleting available resources. Or Neandertals may have died out for reasons completely unrelated to interactions with their human cousins, such as overhunting of their main food sources or failure to adapt to a changing climate.

Questions to Consider

1. After an interbreeding between two groups, what types of genes are most likely to remain in the gene pool?
2. Besides level of complexity, what else distinguishes tool use by humans vs. other animals?
3. What are some additional factors that could have led to Neandertal extinction?

a.

b.

Figure 30C Some mysteries of human evolution. **a.** Which hominins were the first to make and use complex tools, such as spears for hunting and knives for carving? **b.** What factor(s) caused Neandertals to become extinct? (photograph of a model, constructed using anatomical data)

Evolutionists have hypothesized that human variations evolved as adaptations to local environmental conditions. One obvious difference among people is skin color. A darker skin is protective against the high UV intensity of bright sunlight. On the other hand, a white skin ensures vitamin D production in the skin when the UV intensity is low. Harvard University geneticist Richard Lewontin points out, however, that this hypothesis concerning the survival value of dark and light skin has never been tested.

Two correlations between body shape and environmental conditions have been noted since the nineteenth century. The first, known as Bergmann's rule, states that animals in colder regions of their range have a bulkier body build. The second, known as Allen's rule, states that animals in colder regions of their range have shorter limbs, digits, and ears. Both of these effects help regulate body temperature by increasing the surface-area-to-volume ratio in hot climates and decreasing the ratio in cold climates. For example, Figure 30.12*b, c* shows that the Maasai of East Africa tend to be slightly built with elongated limbs, while the Eskimos, who live in northern regions, are bulky and have short limbs.

Other anatomical differences among ethnic groups, such as hair texture, a fold on the upper eyelid (common in Asian peoples), and the shape of lips, cannot be explained as adaptations to the environment. Perhaps these features became fixed in different populations due simply to genetic drift. As far as intelligence is concerned, no significant disparities have been found among different ethnic groups.

Video High Altitude Peoples

Genetic Evidence for a Common Ancestry

The replacement model for the evolution of humans, discussed earlier in this section, pertains to the origin of ethnic groups. This hypothesis proposes that all modern humans have a relatively recent common ancestor—that is, Cro-Magnon—who evolved in Africa and then spread into other regions. Paleontologists tell us that the variation among modern populations is considerably less than existed among archaic human populations some 250,000 years ago. If so, all ethnic groups evolved from the same single ancestral population.

A comparative study of mitochondrial DNA shows that the differences among human populations are consistent with their having a common ancestor no more than a million years ago. Lewontin has also found that the genotypes of different modern populations are extremely similar. He examined variations in 17 genes, including blood groups and various enzymes, among seven major geographic groups: Europeans (caucasians), black Africans, mongoloids, South Asian Aborigines, Amerinds, Oceanians, and Australian Aborigines. He found that the great majority of genetic variation—85%—occurs within ethnic groups, not between them. In other words, the amount of genetic variation between individuals of the same ethnic group is greater than the variation between any two ethnic groups.

Check Your Progress 30.4

1. Explain how the replacement model explains both the dominance of Cro-Magnon and the formation of human ethnic groups.
2. Discuss what factors led to the development of biocultural evolution as a factor in human evolution.

Connecting *the* Concepts *with the* Themes

Evolution

- An understanding of human evolution focuses on the development of bipedalism and the enlargement of the brain.
- The mobile limbs, stereoscopic vision, large and complex brain, and reduced reproductive rate of primates can be seen as adaptations to a life in the trees.
- Humans and all other primates share a common mammalian ancestor, which first appeared around 55 MYA.

Nature of Science

- Scientists study transitional fossils to better understand the patterns of human evolution.
- Recently developed technologies enable genomic comparison of the primates and provide additional evidence of our evolutionary history.

Biological Systems

- Human evolution has been influenced by the environment in which each species lived and the effects of climate change over time.
- Mosaic evolution has occurred during primate evolution, causing the different systems of the human body to evolve at different rates.
- The evolution of bipedalism resulted in changes in the structure of the primate skeleton, as well as in the muscular system, which supports its motion.

MEDIA STUDY TOOLS http://connect.mheducation.com

SMARTBOOK® Study smart. SmartBook helps you maximize your study time by identifying what you know and don't know. The recharge feature in SmartBook helps you prepare for the test by identifying the material that you are most likely to forget.

connect BIOLOGY Find even more resources to learn the chapter concepts including animations, tutorial videos, and interactive practice questions.

Video

30.4 High Altitude Peoples

SUMMARIZE

30.1 Evolution of Primates

Most **primates** are adapted for an **arboreal** life. The evolution of primates is characterized by trends toward mobile limbs; grasping hands with or without an **opposable thumb;** a flattened face; **stereoscopic vision;** a large, complex brain; and the birth of only one offspring at a time. These traits are particularly useful for living in trees.

The term **hominin** is used for humans and their closely related, but extinct, relatives. A hominin is a member of a larger group called the **hominines,** which also includes the chimpanzees and gorillas. **Hominids, hominoids,** and **anthropoids** are groupings that contain additional, more distant relatives.

The **prosimians** (lemurs, tarsiers) were the first primates to diverge from the common ancestor of all primates. *Proconsul* is a transitional link between monkeys and the **dryopithecines,** from which the hominoids arose.

30.2 Evolution of Humanlike Hominins

Fossil and molecular data tell us humanlike hominins shared a common ancestor with chimpanzees until about 5 MYA, and the split between their lineage and the human lineage occurred around that time.

Adaptations that allow humans to walk erect have resulted in our anatomy differing from that of the apes. In humans, the spinal cord curves and exits from the center of the skull, rather than from the rear of the skull. The human pelvis is broader and more bowl-shaped to place the weight of the body over the legs. Humans use only the longer, heavier lower limbs for walking upright bipedally; in apes, all four limbs are used for walking, and the upper limbs are longer than the lower limbs.

Several early humanlike hominin fossils, such as *Sahelanthropus tchadensis,* have been dated around the time of a shared ancestor for apes and humans (7 MYA). The **ardipithecines** appeared about 4.5 MYA. Ardi (*Ardipithecus ramidus*) is an example of an ardipithecine. All the early humanlike hominins have a chimp-sized braincase but are believed to have walked erect. They may also have been the first hominins to use tools as an aid in the consumption of meat.

Australopithecines (or **australopiths**) lived and diversified in Africa from about 4 MYA–1.5 MYA. They had some apelike and some humanlike characteristics, and thus provide an example of **mosaic evolution.** It is possible that an astralopith is a direct ancestor of humans. These hominins walked upright and had a brain size of 370–515 cc. In southern Africa, hominins classified as australopiths include *Australopithecus africanus,* a gracile form, *Paranthropus robustus,* a robust form, and *Australopithecus sediba*, which had both primitive and modern traits. In eastern Africa, hominins classified as australopiths include *A. afarensis* (Lucy), a gracile form, as well as robust forms. Many of the australopiths coexisted, and one of these species is the probable ancestor of the genus *Homo.*

30.3 Evolution of Early Genus *Homo*

Early *Homo,* such as *Homo habilis* and *Homo rudolfensis,* dated around 2 MYA, is characterized by a brain size of at least 600 cc, a jaw with teeth that resemble those of modern humans, and the use of tools.

Homo ergaster and *Homo erectus* (1.9–0.3 MYA) had a striding gait, made well-fashioned tools, and could control fire. *Homo ergaster* migrated into Asia and Europe from Africa between 2 and 1 MYA. *Homo erectus* evolved in Asia and gave rise to *H. floresiensis.*

30.4 Evolution of Later Genus *Homo*

The replacement model of human evolution says that modern humans originated only in Africa and, after migrating into Europe and Asia, replaced the archaic *Homo* species found there. The multiregional continuity model suggests that modern humans arose in several regions.

The **Neandertals,** a group of archaic humans, lived in Europe and Asia. Their sloping chins, squat frames, and heavy muscles are apparently adaptations to the cold. Another group, the Denisovans, may have evolved from a common ancestor that also gave rise to Neandertals, but the **Denisovans** migrated more to East Asia and the South Pacific Islands. Genetic evidence suggests that both Neandertals and Denisovans interbred with early humans.

Cro-Magnon is a name often given to modern humans. These **hunter-gatherers** used sophisticated tools, and they definitely had a culture, as witnessed by the paintings on the walls of caves. In advanced human populations, specialization of tasks led to **biocultural evolution.**

The human ethnic groups of today differ in ways that can be explained in part by adaptation to the environment. Genetic studies tell us that there are more genetic differences between people of the same ethnic group than between ethnic groups. We are one species.

ASSESS

Choose the best answer for each question.

30.1 Evolution of Primates

1. Which of the following are not classified as primates?
   a. apes
   b. humans
   c. marsupials
   d. prosimians
2. All primates have a(n)
   a. keen sense of smell.
   b. opposable toe.
   c. tail.
   d. thumb.
3. Choose the correct order of primate evolution, from the oldest to the most recent group.
   a. prosimians—anthropoids—hominoids—hominids—hominees
   b. hominees—hominids—hominoids—anthropoids—prosimians
   c. prosimians—anthropoids—hominees—hominids—hominoids
   d. anthropoids—hominees—hominids—hominoids—prosimians

30.2 Evolution of Humanlike Hominins

4. What most likely influenced the evolution of bipedalism?
   a. Humans wanted to stand erect in order to run faster.
   b. Bipedalism facilitates tool use.
   c. Bipedalism facilitates sexual intercourse.
   d. An upright stance exposes more of the body to the sun, and vitamin D production requires sunlight.
   e. All of these are correct.
5. The fossil nicknamed Lucy was a(n)
   a. early *Homo.*
   b. australopith.
   c. ardipithecine.
   d. modern human.
6. A major difference between the ardipithecines and the australopithecines is that the latter
   a. had large brains.
   b. lived only in Australia.
   c. were primarily tree dwellers.
   d. had gracile and robust forms.

30.3 Evolution of Early Genus *Homo*

7. Choose the correct order of evolution of the genus *Homo,* from the oldest to the most recent group.
   a. *H. rudolfensis—H. ergaster—H. erectus—H. sapiens*
   b. *H. ergaster—H. rudolfensis—H. sapiens—H. erectus*
   c. *H. erectus—H. ergaster—H. rudolfensis—H. sapiens*
   d. *H. sapiens—H. erectus—H. ergaster—H. rudolfensis*
8. Which of these pairs is incorrectly matched?
   a. *H. erectus*—made tools
   b. *H. rudolfensis*—ate meat
   c. *H. habilis*—controlled fire
   d. *H. floresiensis*—short stature
9. *H. ergaster* could have been the first *Homo* species to
   a. use and control fire.
   b. migrate out of Africa.
   c. make axes and cleavers.
   d. All of these are correct.

30.4 Evolution of Later Genus *Homo*

10. The most likely direct ancestor of modern *H. sapiens* is
    a. *A. africanus.*
    b. *H. ergaster.*
    c. *H. habilis.*
    d. *H. neandertalensis.*
11. If the replacment (out-of-Africa) model is correct, then
    a. human fossils in China after 100,000 years BP would not be expected to resemble earlier fossils.
    b. human fossils in China after 100,000 years BP would be expected to resemble earlier fossils.
    c. humans did not migrate out of Africa.
    d. Both b and c are correct.
    e. Both a and c are correct.
12. Which of the following descriptions applies to Cro-Magnons?
    a. developed agriculture
    b. made bronze tools
    c. were hunter-gatherers
    d. All of these are correct.
13. Which statement is true regarding the development of variation in human populations (ethnicities)?
    a. Ethnic groups evolved from different, distant ancestors.
    b. Allen's rule suggests that humans living in warmer climates would have shorter limbs and digits.
    c. Differences in hair texture or eyelids are likely to have survival advantage.
    d. More genetic variation occurs within ethnic groups than among different ethnic groups.
14. Complete this diagram of the replacement model by filling in the blanks.

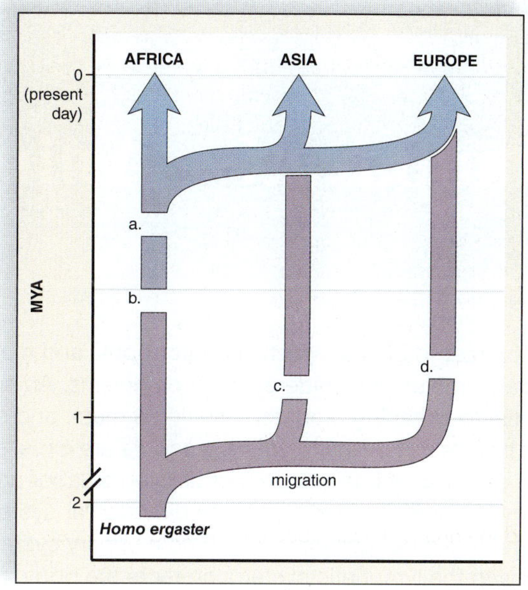

ENGAGE

THINKING SCIENTIFICALLY

1. Bipedalism has many selective advantages. However, there is one disadvantage to walking on two feet: Giving birth to an offspring with a large head through a smaller pelvic opening, which is necessitated by upright posture, is very difficult. This situation results in a high percentage of deaths (of both mother and child) during birth compared to other primates. How do you explain the selection of a trait that is both positive and negative?
2. How might you use biotechnology to show that humans today have Neandertal genes and, therefore, Cro-Magnons and Neandertals interbred with one another?
3. Some modern ethnic groups (white Europeans, Asians) have apparently inherited genes from Neandertals that may influence resistance to cold temperatures. What types of genes might these be?
4. Chimpanzees and humans obviously look and act very differently, but their genomes are remarkably similar. What types of mechanisms could explain how small differences in genotype can result in such large differences in phenotype?

UNIT 7

Comparative Animal Biology

In contrast to plants, which are autotrophic and make their own organic food, animals are heterotrophic and feed on organic molecules made by other organisms. An animal's mobility, which is dependent on nerve and muscle fibers, is essential to escaping predators, finding a mate, and acquiring food. Food is then digested, and the nutrients are distributed to the body's cells. Finally, waste products are expelled.

In complex animals, a distinct division of labor among organs exists, and each of the organ systems is specialized to carry out specific functions. A cardiovascular system transports materials from one body part to another; a respiratory system carries out gas exchange; and a urinary system filters the blood and removes its wastes. The immune system, along with the lymphatic system, protects the body from infectious diseases. The nervous system and endocrine system coordinate the activities of the other systems.

Our comparative study will show how the different organ systems evolved and how they function to maintain homeostasis, the relative constancy of the internal environment.

Unit Outline

Chapter 31 Animal Organization and Homeostasis 581

Chapter 32 Circulation and Cardiovascular Systems 600

Chapter 33 The Lymphatic and Immune Systems 621

Chapter 34 Digestive Systems and Nutrition 641

Chapter 35 Respiratory Systems 659

Chapter 36 Body Fluid Regulation and Excretory Systems 677

Chapter 37 Neurons and Nervous Systems 691

Chapter 38 Sense Organs 714

Chapter 39 Locomotion and Support Systems 733

Chapter 40 Hormones and Endocrine Systems 750

Chapter 41 Reproductive Systems 770

Chapter 42 Animal Development 793

Unit Learning Outcomes

The learning outcomes for this unit focus on three major themes in the life sciences.

Evolution	Explain how various anatomical and physiological adaptations that have evolved in animals confer selective advantages.
Nature of Science	Analyze how specific scientific research studies are advancing our understanding of how animal systems function.
Biological Systems	Review the fundamental structures and functions that are uniquely found in animals.

Astronauts need a special suit to take a walk outside their spacecraft.

31 Animal Organization and Homeostasis

Chapter Outline

31.1 Types of Tissues 582

31.2 Organs, Organ Systems, and Body Cavities 589

31.3 The Integumentary System 591

31.4 Homeostasis 594

Before You Begin

Before beginning this chapter, take a few moments to review the following discussions.

Figure 1.2 What levels of biological organization are found in animals?

Section 24.1 What types of tissues are found in flowering plants?

Section 28.1 Which types of tissues are most characteristic of animals?

In the 2013 movie *Gravity,* medical engineer Ryan Stone (Sandra Bullock) and astronaut Matt Kowalski (George Clooney) are marooned in space, attached only to each other. The fact that an astronaut needs to wear a special suit to survive in space is a reminder that the internal environment of our body functions must stay within normal limits. For example, our enzymes function best at around 37°C; a moderate blood pressure helps blood perfuse the tissues; and a sufficient oxygen concentration facilitates ATP production. This concept is known as homeostasis, a dynamic equilibrium of the internal environment. Swim the English Channel, cross the Sahara Desert by camel, visit the South Pole, or take a space walk—a variety of internal conditions in your body will stay within fairly narrow ranges, as long as you take proper precautions. An astronaut depends on artificial systems in addition to natural systems to maintain homeostasis.

This chapter discusses homeostasis after a look at the body's organization. Just as in other complex animals, each organ system of the human body contains a particular set of organs. For example, the circulatory system contains the heart and blood vessels, and the nervous system contains the brain and nerves. Organs are composed of tissues, and each type of tissue has cells that perform specific functions. We begin the chapter by examining several of the major types of tissues.

As you read through this chapter, think about the following questions:

1. How did the evolution of specialized tissues, organs, and organ systems allow animals to better adapt to their environment?
2. What are some of the most important functions of animal skin?
3. How does the disruption of homeostasis lead to disease?

Following *the* Themes

CHAPTER 31 ANIMAL ORGANIZATION AND HOMEOSTASIS

UNIT 7 COMPARATIVE ANIMAL BIOLOGY

Evolution	The degree of cell specialization seen in tissues, organs, and organ systems provides evolutionary advantages for animals.
Nature of Science	Scientific research is increasing our understanding of how tissues and organs can regenerate.
Biological Systems	Most animals, especially vertebrates, have similar types of tissues, organs, and organ systems.

31.1 Types of Tissues

Learning Outcomes

Upon completion of this section, you should be able to

1. List and describe the four major types of tissues found in animals.
2. Identify the common locations of the various types of animal tissues.
3. Explain how specialization of cells in tissues enhances tissue function.

Like all living organisms, animals are highly organized. Animals begin life as a single cell—a fertilized egg, or *zygote.* The zygote undergoes cell division, producing cells that will eventually form the variety of tissues that make up organs and organ systems. Although all of these cells carry out a number of common functions—such as obtaining nutrients, synthesizing basic cellular constituents, and in most cases reproducing themselves—the cells of multicellular organisms further differentiate, so that they can perform additional, unique functions.

A **tissue** is composed of specialized cells of the same or similar type that perform a common function in the body. The tissues of most complex animals can be categorized into four major types:

1. *Epithelial tissue* covers body surfaces, lines body cavities, and forms glands.
2. *Connective tissue* binds and supports body parts.
3. *Muscular tissue* moves the body and its parts.
4. *Nervous tissue* receives stimuli and transmits nerve impulses.

MP3 Overview of Tissues

Epithelial Tissue

Epithelial tissue, also called *epithelium* (pl., epithelia), consists of tightly packed cells that form a continuous layer. Epithelial tissue covers surfaces and lines body cavities. Usually, it has a protective function, but it can also be modified to carry out secretion, absorption, excretion, and filtration.

Epithelial cells may be connected to one another by three types of junctions composed of proteins (see Fig. 5.13). Regions where proteins join them together are called tight junctions. In the intestine, the gastric juices stay out of the body, and in the kidneys, the urine stays within kidney tubules, because epithelial cells are joined by tight junctions. In the skin, adhesion junctions add strength and allow epithelial cells to stretch and bend, whereas gap junctions are protein channels that permit the passage of molecules between two adjacent cells. (These junctions are described in more detail in section 5.4.)

Epithelial tissues are often exposed to the environment on one side, but on the other side they are attached to a *basement membrane.* The basement membrane is simply a thin layer of various types of proteins that anchors the epithelium to the extracellular matrix, which is often a type of connective tissue. The basement membrane should not be confused with the plasma membrane or with the body membranes we will be discussing.

MP3 Epithelial Tissue

Simple Epithelia

Epithelial tissue is either simple or complex. Simple epithelia have only a single layer of cells (Fig. 31.1) and are classified according to cell type. **Squamous epithelium,** which is composed of flattened cells, lines blood vessels and the air sacs of lungs.

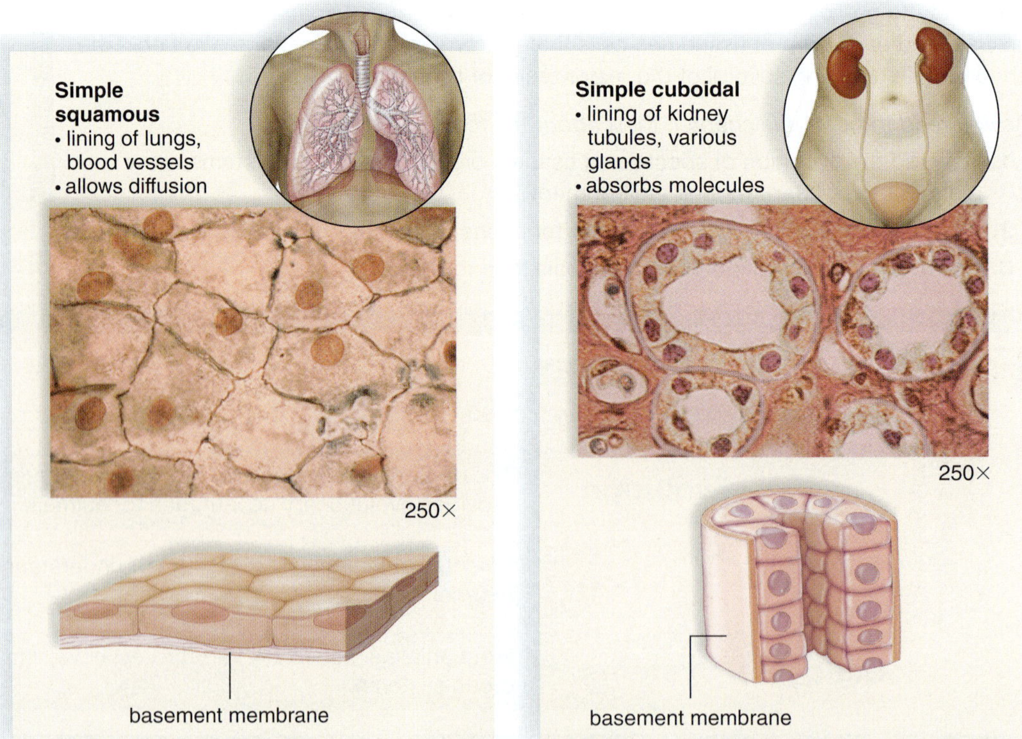

Figure 31.1 Types of epithelial tissues in vertebrates. Basic epithelial tissues found in vertebrates are shown, along with the locations of the tissue and the primary function of the tissue at these locations.

Cuboidal epithelium contains cube-shaped cells and is found lining the kidney tubules and various glands. **Columnar epithelium** has cells resembling rectangular pillars or columns, with nuclei usually located near the bottom of each cell. This epithelium lines the digestive tract, where it efficiently absorbs nutrients from the small intestine because of minute cellular extensions called microvilli. Ciliated columnar epithelium lines the uterine tubes, where it propels the egg toward the uterus.

When an epithelium is pseudostratified, it appears to be layered, but true layers do not exist, because each cell touches the basement membrane. The lining of the windpipe, or trachea, is pseudostratified ciliated columnar epithelium. A secreted covering of mucus traps foreign particles, and the upward motion of the cilia carries the mucus to the back of the throat, where it may be either swallowed or expectorated. Smoking can cause a change in mucus secretion and inhibit ciliary action, resulting in a chronic inflammatory condition called bronchitis.

Stratified Epithelia

Stratified epithelia have layers of cells piled one on top of the other. Only the bottom layer touches the basement membrane. The nose, mouth, esophagus, anal canal, and vagina are all lined with stratified squamous epithelium. As you'll see, the outer layer of skin is also stratified squamous epithelium, but the cells have been reinforced by keratin, a protein that provides strength. Stratified cuboidal and stratified columnar epithelia also occur in the body.

Glandular Epithelia

When an epithelium secretes a product, it is said to be glandular. A **gland** can be a single epithelial cell, as in the case of mucus-secreting goblet cells within the columnar epithelium lining the digestive tract, or a gland can contain many cells. Glands that secrete their product into ducts are called **exocrine glands.**

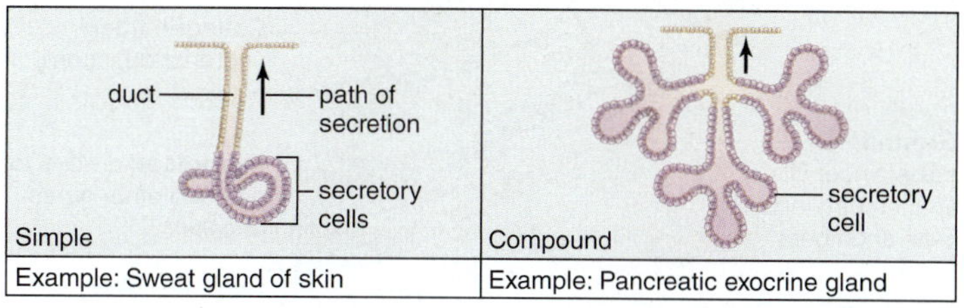

Glands that have no duct are known as **endocrine glands.** Endocrine glands (e.g., pituitary gland and thyroid) secrete hormones internally, so they are transported by the bloodstream (see Chapter 40).

MP3 Epithelial Glands

Connective Tissue

Connective tissue is the most abundant and widely distributed tissue in complex animals. It is quite diverse in structure and function; even so, all types have three components: specialized cells, ground substance, and protein fibers (Fig. 31.2).

The ground substance is a noncellular material that separates the cells and varies in consistency from solid to semifluid to fluid. The fibers[1] are of three possible types. White **collagen fibers** contain collagen, a protein that gives them flexibility and strength. **Reticular fibers** are very thin collagen fibers that are highly branched and form delicate supporting networks. Yellow **elastic fibers** contain elastin, a protein that is not as strong as

[1]In connective tissue, a fiber is a component of the matrix; in muscular tissue, a fiber is a muscle cell; in nervous tissue, a nerve fiber is an axon and its myelin sheath.

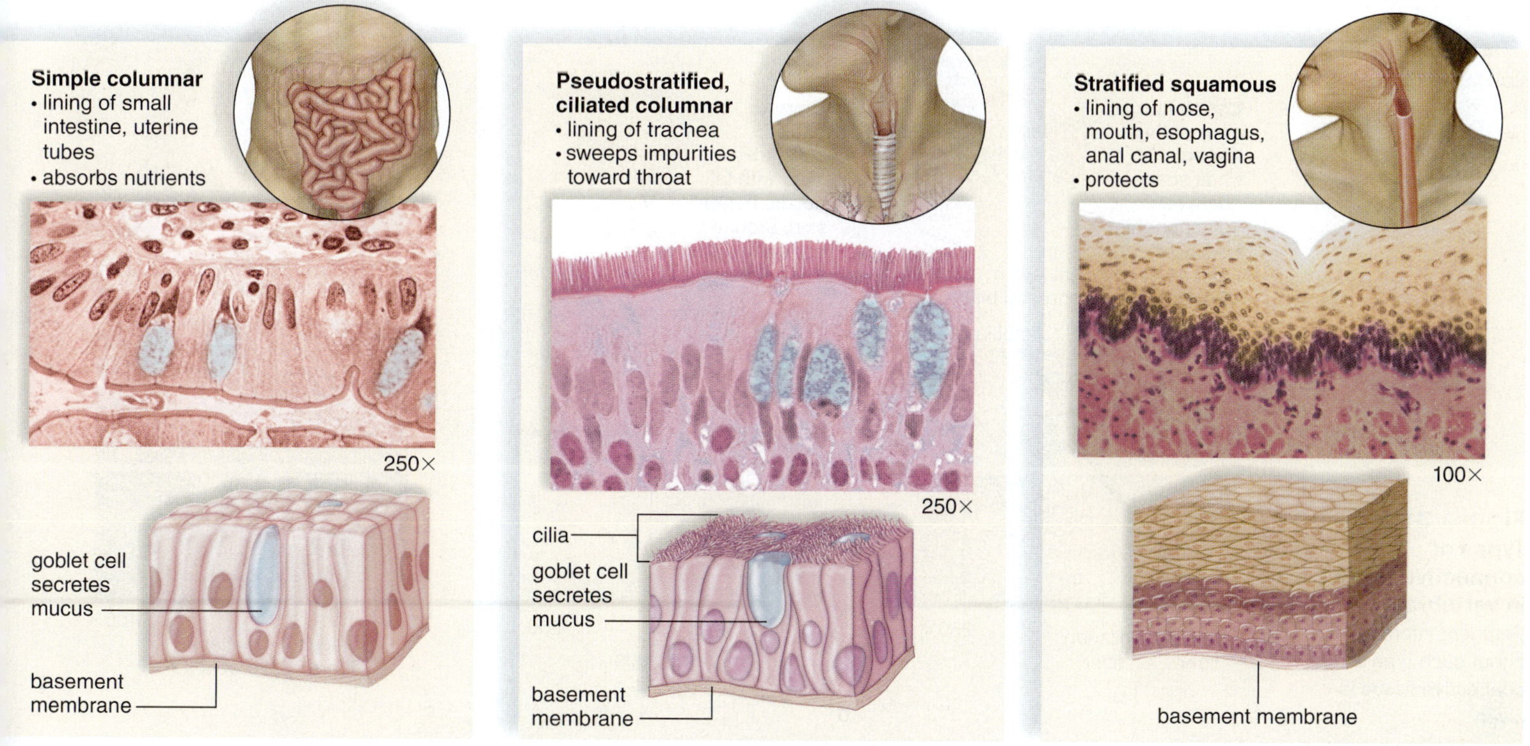

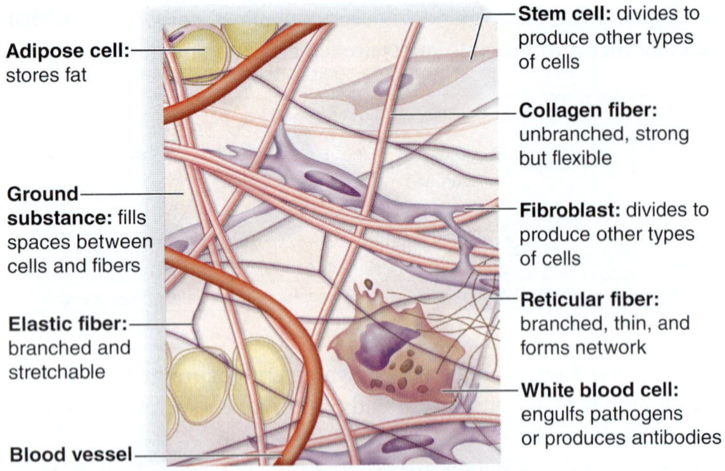

Figure 31.2 **Diagram of fibrous connective tissue.**

collagen but is more elastic. The ground substance plus the fibers together are referred to as the connective tissue *matrix.*

Connective tissue is classified into three major categories: fibrous, supportive, and fluid. Each category includes a number of different types of tissues.

MP3 Connective Tissue

Fibrous Connective Tissue

Both loose fibrous and dense fibrous connective tissues have cells called **fibroblasts** (L. *fibra,* "thread"; Gk. *blastos,* "bud"), located some distance from one another and separated by a jellylike matrix containing white collagen fibers and yellow elastic fibers.

Loose fibrous connective tissue supports epithelium and many internal organs (Fig. 31.3*a*). Its presence in lungs, arteries, and the urinary bladder allows these organs to expand. It forms a protective covering enclosing many internal organs, such as muscles, blood vessels, and nerves.

Adipose tissue serves as the body's primary energy reservoir (Fig. 31.3*b*). It is loose fibrous connective tissue composed mostly of enlarged fibroblasts that store fat. These specialized fibroblasts are called *adipocytes.* Adipose tissue also insulates the body, contributes to body contours, and provides cushioning. In mammals, adipose tissue is found particularly beneath the skin, around the kidneys, and on the surface of the heart.

The number of adipocytes in an individual is fixed. When a person gains weight, the cells become larger, and when weight is lost, the cells shrink. In obese people, the individual cells may be up to five times larger than normal. Most adipose tissue is white, but in newborns and hibernating mammals, some is brown due to an increased number of mitochondria that can produce heat.

Dense fibrous connective tissue contains many collagen fibers that are packed together (Fig. 31.3*c*). This type of tissue has more specific functions than does loose connective tissue. For example, dense fibrous connective tissue is found in **tendons** (L. *tendo,* "stretch"), which connect muscles to bones, and in **ligaments** (L. *ligamentum,* "band"), which connect bones to other bones at joints.

Supportive Connective Tissue

Cartilage and bone are the two main supportive connective tissues that provide structure, shape, protection, and leverage for movement. Generally, cartilage is more flexible than bone, because it lacks mineralization of the matrix.

Cartilage. In **cartilage,** the cells lie in small chambers called lacunae (sing., **lacuna**), separated by a matrix that is solid yet flexible. Unfortunately, because this tissue lacks a direct blood supply, it heals very slowly. There are three types of cartilage, distinguished by the type of fiber in the matrix.

Hyaline cartilage (Fig. 31.3*d*), the most common type of cartilage, contains only very fine collagen fibers. The matrix has

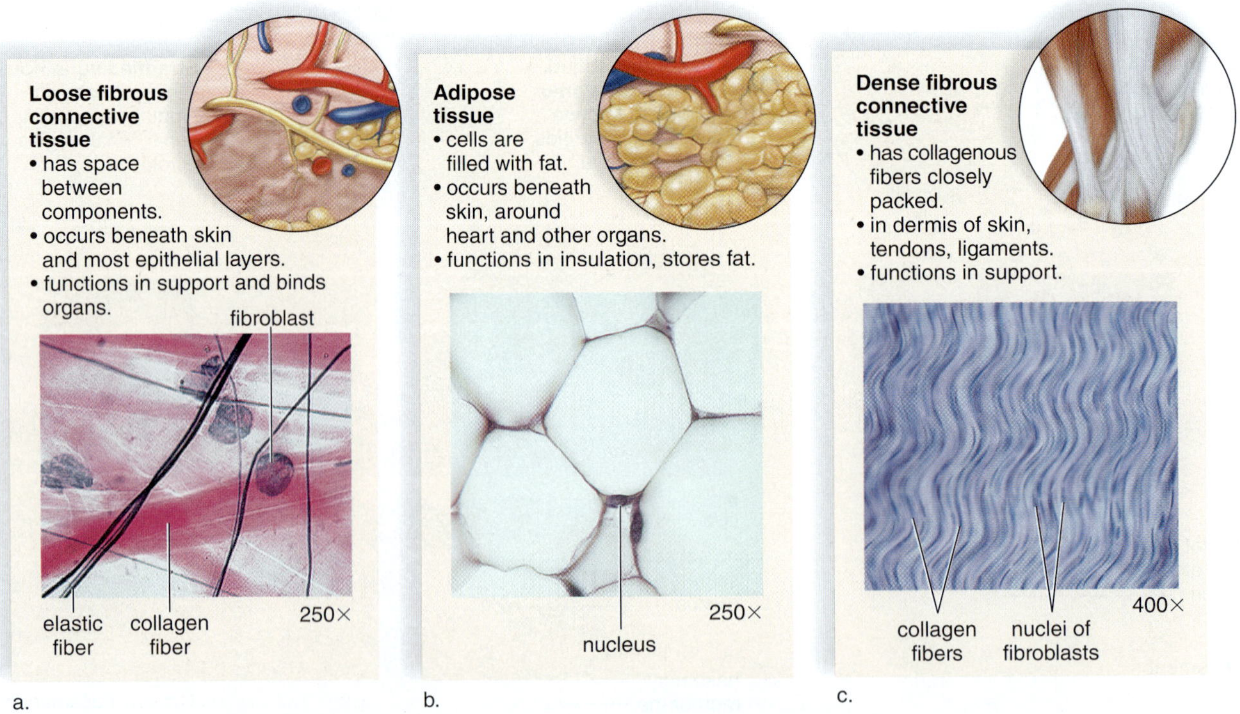

Figure 31.3 **Types of connective tissue in vertebrates.** Pertinent information about each type of connective tissue is given.

a white, translucent appearance. Hyaline cartilage is found in the nose and at the ends of the long bones and the ribs, and it forms rings in the walls of respiratory passages. The fetal skeleton also is made of this type of cartilage. Later, the cartilaginous fetal skeleton is replaced by bone.

Elastic cartilage has more elastic fibers than hyaline cartilage. For this reason, it is more flexible and is found, for example, in the framework of the outer ear.

Fibrocartilage has a matrix containing strong collagen fibers. Fibrocartilage is found in structures that withstand tension and pressure, such as the pads between the vertebrae in the backbone and the wedges in the knee joint.

Bone. Of all the connective tissues, **bone** is the most rigid. It consists of an extremely hard matrix of inorganic salts, notably calcium salts, deposited around protein fibers, especially collagen fibers. The inorganic salts give bones rigidity, and the protein fibers provide elasticity and strength, much as steel rods do in reinforced concrete.

Compact bone makes up the shaft of a long bone (Fig. 31.3*e*). It consists of cylindrical structural units called osteons (Haversian systems). The central canal of each osteon is surrounded by rings of hard matrix. Bone cells are located in spaces, called lacunae, between the rings of matrix. Blood vessels in the central canal carry nutrients that allow bone to renew itself. Thin extensions of bone cells within canaliculi (minute canals) connect the cells to each other and to the central canal. The hollow shaft of long bones, such as the femur (thigh bone), is filled with yellow bone marrow (see Chapter 39).

The ends of a long bone contain spongy bone, which has an entirely different structure. **Spongy bone** contains numerous bony bars and plates, separated by irregular spaces. Although lighter than compact bone, spongy bone is still designed for strength. Just as braces are used for support in buildings, the solid portions of spongy bone follow lines of stress. Spongy bone is also the site of red bone marrow, which is critical to the production of blood cells (see Chapters 32 and 33).

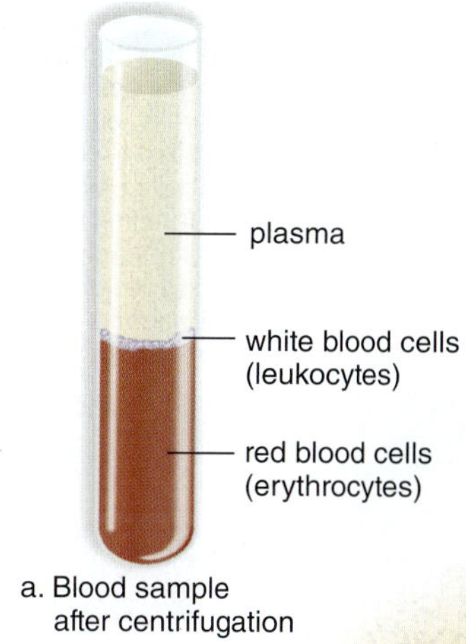

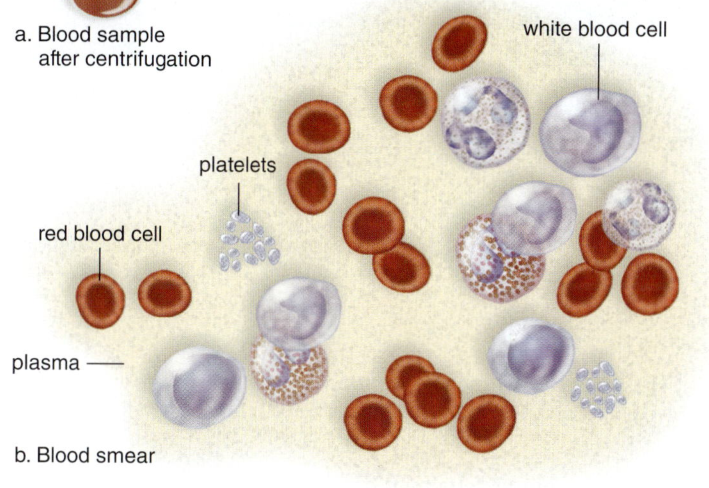

Figure 31.4 Blood, a liquid connective tissue. **a.** Blood is classified as connective tissue because the cells are separated by a matrix—plasma. Plasma, the liquid portion of blood, usually contains several types of cells. **b.** Drawing of the components seen in a stained blood smear: red blood cells, white blood cells, and platelets (which are actually fragments of a larger cell).

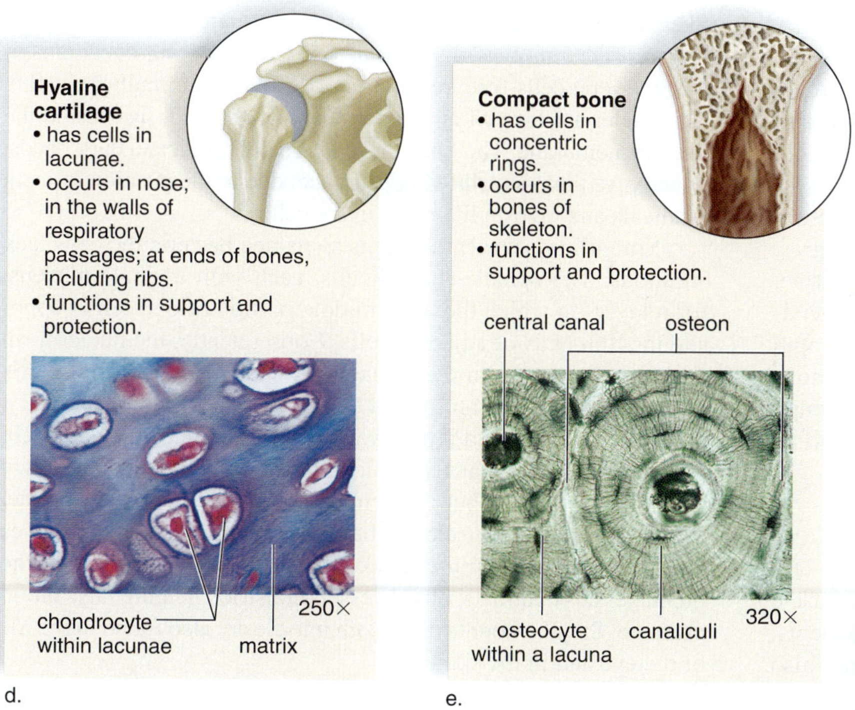

Fluid Connective Tissues

Blood, which consists of formed elements and plasma, is a fluid connective tissue located in blood vessels (Fig. 31.4). Formed elements in the blood consist of the many kinds of blood cells and the platelets.

The systems of the body help keep blood composition and chemistry within normal limits, and blood in turn creates interstitial fluid. Blood transports nutrients and oxygen to interstitial fluid and removes carbon dioxide and other wastes. It helps distribute heat and plays a role in fluid, ion, and pH balance. The formed elements, discussed next, each have specific functions.

MP3 General Functions and Composition of Blood

The **red blood cells** are small, disk-shaped cells without nuclei. The absence of a nucleus makes the cells biconcave. The presence of the red pigment hemoglobin makes them red and in turn makes the blood red. Hemoglobin is composed of four units; each unit is composed of the protein globin and a complex, iron-containing structure called heme. The iron forms a loose association with oxygen, and in this way red blood cells transport oxygen and readily give it up in the tissues.

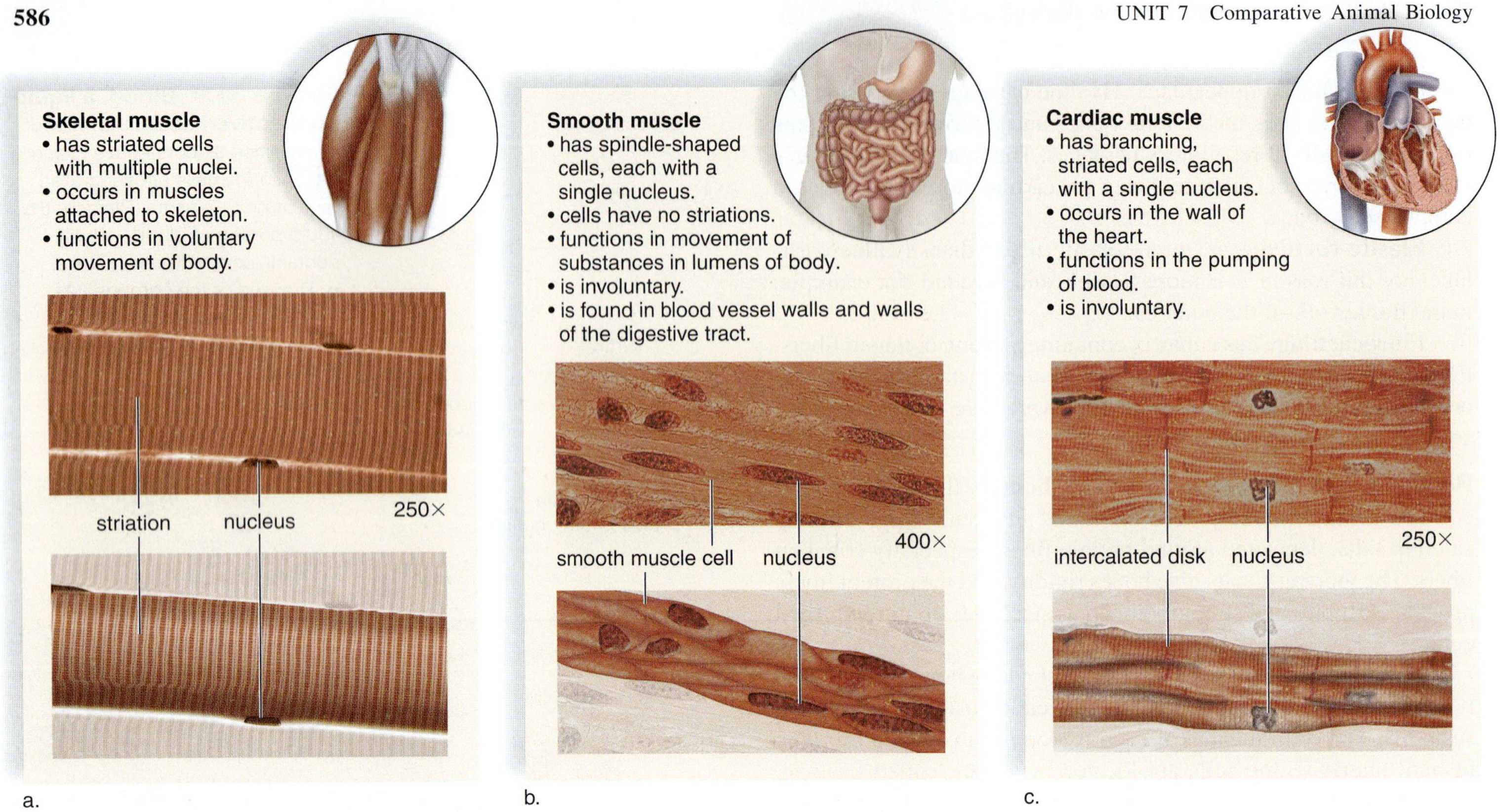

Figure 31.5 Muscular tissue. **a.** Skeletal muscle is voluntary and striated. **b.** Smooth muscle is involuntary and nonstriated. **c.** Cardiac muscle is involuntary and striated. Cardiac muscle cells branch and fit together at intercalated disks.

White blood cells may be distinguished from red blood cells by the fact that they are usually larger, have a nucleus, and without staining would appear translucent. When blood is smeared onto a microscope slide and stained, the nucleus of a white blood cell typically looks blue or purple. White blood cells fight infection, primarily in two ways. Some white blood cells are phagocytic and engulf infectious **pathogens,** while other white blood cells either produce antibodies, molecules that combine with foreign substances to inactivate them, or kill cells outright.

Platelets are not complete cells; rather, they are fragments of large cells present only in bone marrow. When a blood vessel is damaged, platelets help form a plug that seals the vessel, and injured tissues release molecules that help the clotting process.

Lymph is a fluid connective tissue located in lymphatic vessels. Lymphatic vessels absorb excess **interstitial** fluid and return it to the cardiovascular system. Special lymphatic capillaries, called lacteals, also absorb fat molecules from the small intestine. Lymph nodes, composed of fibrous connective tissue plus specialized white blood cells called lymphocytes, occur along the length of lymphatic vessels. These lymphocytes and other cells remove any foreign material from the lymph as it passes through lymph nodes. Lymph nodes may enlarge when these cells respond to an infection.

Muscular Tissue

Muscular (contractile) tissue is composed of cells called muscle fibers. Muscle fibers contain actin filaments and myosin filaments, whose interaction accounts for movement. The muscles are also important in the generation of body heat. There are three distinct types of muscle tissue: skeletal, smooth, and cardiac. Each type differs in appearance, physiology, and function.

MP3 Muscle Tissue

Skeletal muscle, also called voluntary muscle (Fig. 31.5*a*), is attached by tendons to the bones of the skeleton, and when it contracts, body parts move. Contraction of skeletal muscle is under voluntary control and occurs faster than in the other muscle types. Skeletal muscle fibers are cylindrical and quite long—sometimes they run the length of the muscle. They arise during development when several cells fuse, resulting in one fiber with multiple nuclei. The nuclei are located at the periphery of the cell, just inside the plasma membrane. The fibers have alternating light and dark bands, which give them a **striated** appearance, due to the position of actin filaments and myosin filaments in the cell.

Smooth (visceral) muscle is so named because the cells lack striations. The spindle-shaped cells, each with a single nucleus, form layers in which the thick middle portion of one cell is opposite the thin ends of adjacent cells. Consequently, the nuclei form an irregular pattern in the tissue (Fig. 31.5*b*). Smooth muscle is not under voluntary control and therefore is said to be involuntary. Smooth muscle, found in the walls of viscera (intestine, stomach, and other internal organs) and blood vessels, contracts more slowly than skeletal muscle but can remain contracted for a longer time. When the smooth muscle of the intestine contracts, food moves along its lumen (central cavity). When the smooth muscle of the blood vessels contracts, blood vessels constrict, helping raise blood pressure. Small amounts of smooth muscle are also found in the iris of the eye and in the skin.

Cardiac muscle (Fig. 31.5*c*) is found only in the walls of the heart. Its contraction pumps blood and accounts for the heartbeat. Cardiac muscle combines features of both smooth muscle and skeletal muscle. Like skeletal muscle, it has striations, but the contraction of the heart is involuntary for the most part. Cardiac muscle cells also differ from skeletal muscle cells in that they usually have a single, centrally placed nucleus. The cells are branched and seemingly fused one with the other, and the heart appears to be composed of one large, interconnecting mass of muscle cells. Actually, cardiac muscle cells are separate and individual, but they are bound end to end at *intercalated disks,* areas where folded plasma membranes between two cells contain adhesion junctions and gap junctions.

Nervous Tissue

Nervous tissue contains nerve cells called neurons and supporting cells called neuroglia. An average human body has about 1 trillion neurons. The nervous system conveys signals termed nerve impulses throughout the body.

MP3 Nervous Tissue

Neurons

A **neuron** is a specialized cell that has three parts: dendrites, a cell body, and an axon (Fig. 31.6*a*). A dendrite is a process that conducts signals toward the cell body. The cell body contains the major portion of the cytoplasm and the nucleus of the neuron. An axon is a process that typically conducts nerve impulses away from the cell body. Long axons are covered by myelin, a white, fatty substance. The term *fiber* is used here to refer to an axon along with its myelin sheath, if it has one. Outside the brain and spinal cord, fibers bound by connective tissue form **nerves.**

The nervous system has just three functions: sensory input, integration of data, and motor output. Nerves conduct impulses from sensory receptors to the spinal cord and the brain, where integration occurs. The phenomenon called sensation occurs only in the brain, however. Nerves also conduct nerve impulses away from the spinal cord and brain to the muscles and glands, causing them to contract and secrete, respectively. In this way, a coordinated response to the stimulus is achieved.

Neuroglia

In addition to neurons, nervous tissue contains cells called **neuroglia.** In the human brain, these cells outnumber neurons as much as ten to one, and they make up approximately half the volume of the organ. Although the primary function of neuroglia is to support and nourish neurons, recent research has shown that some neuroglia directly contribute to brain function.

Several types of neuroglia are found in the brain. Microglia, astrocytes, and oligodendrocytes are shown in Figure 31.6*a*. Microglia, in addition to supporting neurons, engulf bacterial and cellular debris. Astrocytes provide nutrients to neurons and produce a hormone known as glial cell–derived growth factor, which is being studied as a possible treatment for Parkinson disease and other diseases caused by neuron degeneration. Oligodendrocytes form myelin in the brain.

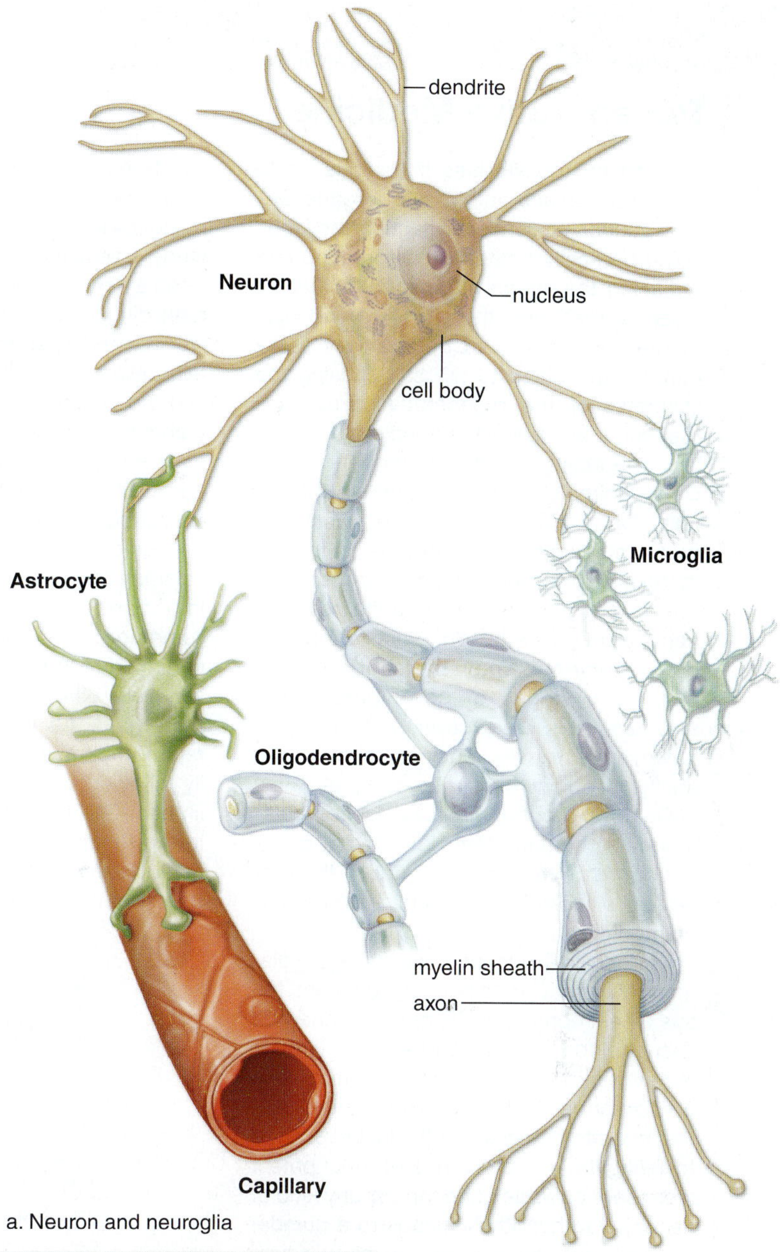

a. Neuron and neuroglia

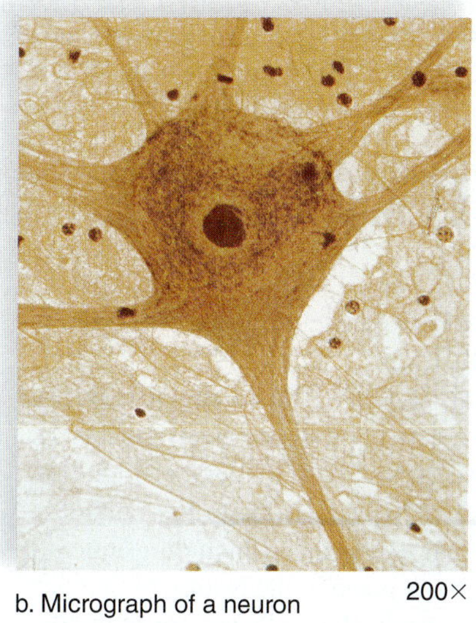

b. Micrograph of a neuron 200×

Figure 31.6 Neurons and neuroglia. Neurons conduct nerve impulses. Neuroglia consist of cells that support and service neurons and have various functions: Microglia become mobile in response to inflammation and phagocytize debris. Astrocytes lie between neurons and a capillary; therefore, nutrients must first pass through astrocytes before entering neurons. Oligodendrocytes form the myelin sheaths around fibers in the brain and spinal cord.

THEME Nature of Science

Regenerative Medicine

Imagine the possibilities if humans could regrow or replace our limbs or organs. Severe injuries or cancers affecting the limbs could be easily treated by amputating and replacing the affected appendage. Deadly diseases that affect the heart, liver, kidneys, or other organs could be relatively easily cured. But scientists are still chasing the "secrets" of the regenerative processes that occur in a few other species.

The ability of salamanders and a few other animals to regenerate amputated tails, limbs, and other organs has fascinated scientists for centuries. Although many invertebrate species can regenerate missing parts (some decapitated snails can regrow a new head!), most vertebrates can regenerate body parts only during embryonic stages of development. The salamander is an exception to this rule, however, with many species retaining the ability to replace lost limbs, tails, and even jaws throughout their lives. Unlike salamanders, humans who lose limbs or organs must either get by without them or rely on artificial limbs, machines, or transplanted organs.

Much like human arms or legs, salamander limbs contain many different tissue types: bone, muscle, nerve, connective tissue, blood vessels, and skin. Until quite recently, most scientists believed that when a salamander limb was severed, the cells that migrate to the site of the injury reverted to being truly undifferentiated, pluripotent stem cells capable of becoming any kind of tissue. That hypothesis leads to a number of questions: How does an injury cause the cells to become "reprogrammed"? And how do the cells know which types of tissues to become?

Some recent research has cast doubt on the idea that the cells that regrow a salamander's limb are true stem cells. In a 2009 study, scientists in a German laboratory devised a new method to track the fate of cells rebuilding the severed limb. The researchers first inserted a gene coding for a jellyfish protein called green fluorescent protein (GFP) into the genome of embryonic axolotl salamanders. The result was transgenic salamanders, in which every cell of the animal gave off a detectable green glow when illuminated with ultraviolet light (Fig. 31A).

Different tissues from these transgenic salamanders could then be transplanted into nontransgenic salamanders of the same species. When the researchers were certain that the transplanted tissues had survived, different limb amputations were performed. As the limb regenerated, the fate of the transplanted cells could be easily followed by examining the regenerating limb for GFP+ (green) cells. This tracking allowed the researchers to answer the question of whether the GFP+ donor cells would be found throughout the regenerating limb, or only in the same type of tissue that was originally transplanted (Fig. 31B).

Somewhat surprisingly, the results showed that most GFP+ tissue types gave rise only to the same tissue type, or to a very limited set of tissues (Fig. 31B). For example, when GFP+ skin was transplanted, GFP+ cells were found only in skin or cartilage, but never in muscle. Similarly, when GFP+ cartilage was transplanted, GFP+ cells were found in cartilage, but never in muscle, and GFP+ muscle tissue never produced cartilage. Additionally, although previous research had suggested that neuroglia-type cells present in the regenerating limb could revert to pluripotent cells capable of forming many types of tissues, in these experiments GFP+ neuroglia-type cells gave rise only to nerve tissue.

Although the ability to regenerate human limbs and organs is still far in the future, the key to teaching human cells how to regenerate may lie in experiments like these. Beyond regenerating limbs and organs, scientists working in the field of regenerative medicine are closing in on

Figure 31A Creating a new type of salamander for limb regeneration research. *Left:* Wild-type axolotl salamander showing normal coloration. *Right:* Researchers used genetic engineering techniques to produce an axolotl whose cells would fluoresce green under UV illumination.

One fundamental difference between neurons and neuroglia is that the neurons of adult animals usually cannot undergo cell division, but neuroglial cells retain this capacity. As a result, the majority of brain tumors in adults involve actively dividing neuroglial cells. Most of these tumors have to be treated with surgery or radiation therapy, because a large number of tight juctions in the epithelial cells of brain capillaries prevent many substances (including anticancer drugs) from entering the brain. As noted in the Nature of Science feature, "Regenerative Medicine," neuroglial cells may promote the regeneration of some types of nerve tissue.

Check Your Progress 31.1

1. List five types of epithelium, and identify an example of where each type can be found.
2. Compare and contrast the three major types of connective tissue.
3. Describe the structure and function of skeletal, smooth, and cardiac muscle.
4. Recall the three parts of a neuron, and explain the function of each.

ways to replace insulin-producing cells in type 1 diabetes, to grow new blood vessels to replace blocked ones in failing hearts, and even to replace damaged nerves as well as bone, muscle, and cartilage. Although a large amount of financial support for this research is needed initially, regenerative medicine offers the potential to cure many conditions that currently can only be managed over the lifetime of a patient.

Questions to Consider

1. Certain primitive invertebrate animals, such as sponges and hydras, can regenerate their entire body from one or a few cells. How does this phenomenon complicate the definition of what constitutes an "animal" versus a single-celled life-form?
2. Even in humans, certain cell types (e.g., neurons) are relatively less likely to regenerate after an injury than other types (e.g., skin). What are some relevant differences between these two types of cells that might explain this observation?

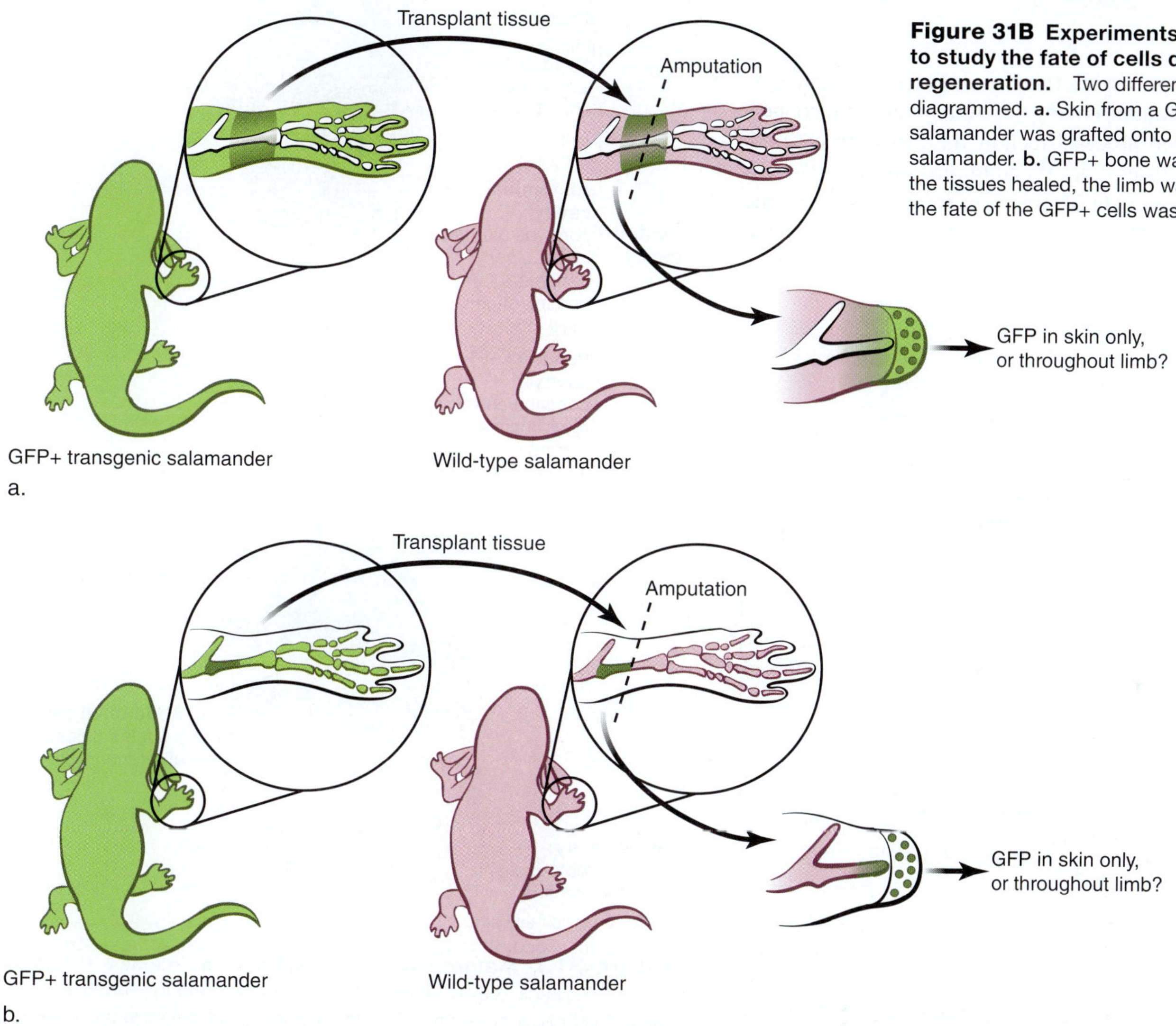

Figure 31B Experiments designed to study the fate of cells during limb regeneration. Two different experiments are diagrammed. **a.** Skin from a GFP+ transgenic salamander was grafted onto a wild-type salamander. **b.** GFP+ bone was transplanted. After the tissues healed, the limb was amputated, and the fate of the GFP+ cells was determined.

31.2 Organs, Organ Systems, and Body Cavities

Learning Outcomes

Upon completion of this section, you should be able to

1. Distinguish among tissues, organs, and organ systems.
2. List the major life processes carried out by each organ system in vertebrate animals.
3. Describe the two main cavities of the human body and the major organs found in each.

Tissues have four main types; however, two or more of these types may be arranged together in a structure, termed an organ, that has a specific function; in turn, several organs may work together in an organ system to accomplish a general process. The evolution of organs and organ systems allowed animals to localize these processes to certain areas of the body and to accomplish them more efficiently.

Organs

We first described the concept of organs in Chapter 24 when we discussed the vegetative and reproductive organs of flowering plants. An **organ** is composed of two or more types of tissues working together to perform a particular function. For example, a kidney is an organ that contains a variety of epithelial and connective tissues, and these tissues are specialized for the function of eliminating waste products from the blood.

Organ Systems

In most animals, individual organs function as part of an organ system. An **organ system** contains many different organs that cooperate to carry out a general process, such as the digestion of food. Similar types of organ systems are found in most invertebrates, as well as in all vertebrate animals. These organ systems carry out the life processes that all of these animals, including humans, must carry out.

Life Processes	Organ Systems
Coordinate body activities	Nervous system Endocrine system
Acquire materials and energy (food)	Skeletal system Muscular system Digestive system
Maintain body shape	Skeletal system Muscular system
Exchange gases	Respiratory system
Transport materials	Cardiovascular system
Eliminate wastes	Urinary system Digestive system
Protect the body from pathogens	Lymphatic system Immune system
Produce offspring	Reproductive system

Body Cavities

Each organ system has a particular distribution within the body. Vertebrates have two main **body cavities:** the smaller dorsal cavity and the larger ventral cavity (Fig. 31.7*a*). The brain and the spinal cord are in the dorsal cavity.

MP3 Body Cavities

During development, the ventral cavity develops from the coelom. In humans and other mammals, the coelom is divided by a muscular diaphragm that assists breathing. The heart and the lungs are located in the upper (thoracic, or chest) cavity (Fig. 31.7*b*). The major portions of the digestive system, including the accessory organs (e.g., the liver and pancreas) are located in the abdominal cavity, as are the kidneys of the urinary system. The urinary bladder, the female reproductive organs, and certain of the male reproductive organs are located in the pelvic cavity.

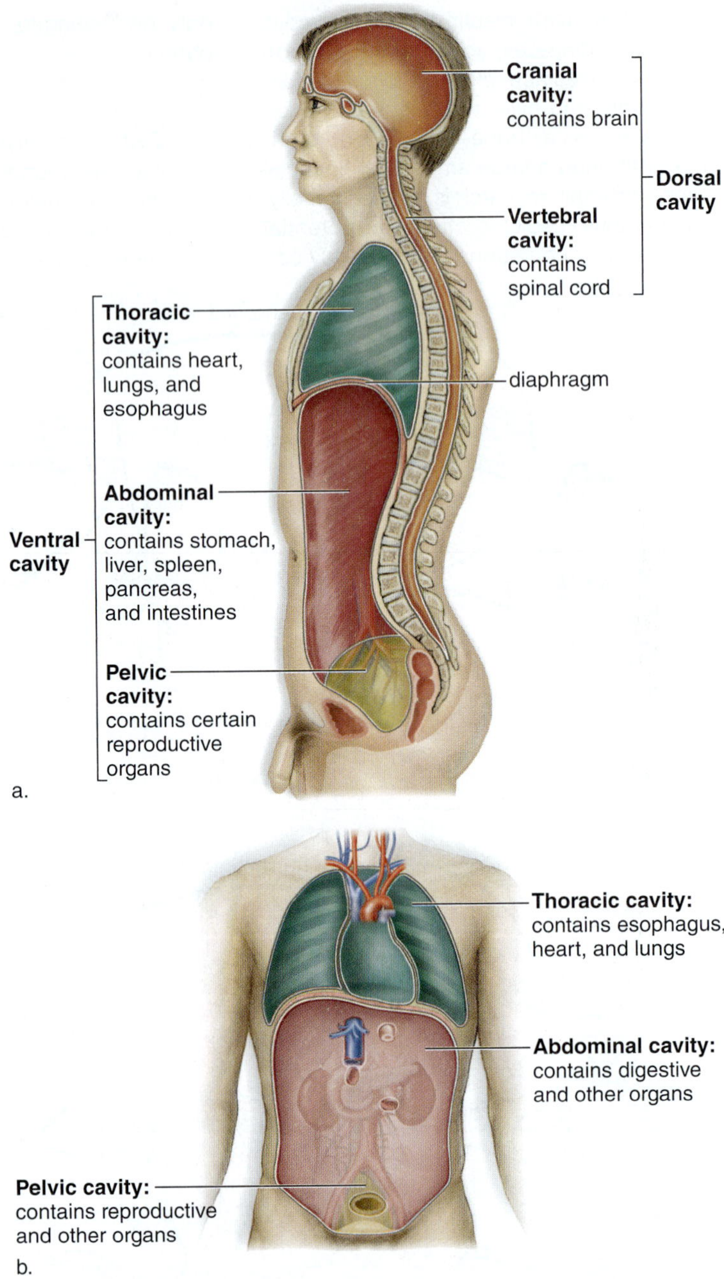

Figure 31.7 Mammalian body cavities. **a.** Side view. The dorsal ("toward the back") cavity contains the cranial cavity and the vertebral canal. The brain is in the cranial cavity, and the spinal cord is in the vertebral canal. The well-developed ventral ("toward the front") cavity is divided by the diaphragm into the thoracic cavity and the abdominopelvic cavity (abdominal cavity and pelvic cavity). The heart and lungs are in the thoracic cavity, and most other internal organs are in the abdominal cavity. **b.** Frontal view of the thoracic cavity.

Check Your Progress 31.2

1. Explain the difference between an organ and an organ system.
2. Identify two organ systems that protect the body from disease.
3. List and locate the two major body cavities in humans, as well as the two cavities found in each of these.

31.3 The Integumentary System

Learning Outcomes

Upon completion of this section, you should be able to

1. Distinguish between the functions of skin that are common to all animals and those that are unique to specific groups.
2. Identify the two main regions of skin and how these differ from the subcutaneous layer.
3. Explain the function of melanocytes in the skin and the effects of UV radiation.
4. Describe the makeup and function of the accessory structures of human skin.

The **integumentary system,** consisting of the skin, its derivatives, and its accessory organs, is the largest and most conspicuous organ system in the body. The skin of an average human covers an area of 21 square feet and accounts for nearly 15% of the body weight.

Derivatives of the skin differ throughout the vertebrate world. Most fishes have protective outgrowths of the skin called scales. Amphibian skin is usually covered with mucous glands. Scales are characteristic of reptiles, and they are found on the legs and feet of birds; however, only birds have feathers, which grow from specialized follicles in the skin. Hair is found only on the skin of mammals, and all mammals (even whales) have hair at some stage of their life.

Functions of Skin

Skin covers the body, protecting underlying parts from dessication, physical trauma, and pathogen invasion. It is also important in regulating body temperature. The skin of small aquatic or semiaquatic animals is often involved in the exchange of gases with the environment (see Chapter 35). In contrast, the dry, scaly skin of reptiles is very poor at gas exchange, but it prevents water loss and thus was probably an important evolutionary adaptation to life on land. Feathers are unique appendages of bird skin that function in insulation, waterproofing, and, of course, flight. Skin is also equipped with a variety of sensory structures that monitor touch, pressure, temperature, and pain. In addition, skin cells manufacture precursor molecules that are converted to vitamin D after exposure to UV light.

MP3 Skin and Its Tissues

Regions of Skin

The **skin** has two main regions: the epidermis and the dermis (Fig. 31.8). A *subcutaneous layer,* also known as the hypodermis, is found between the skin and any underlying structures, such as muscle or bone.

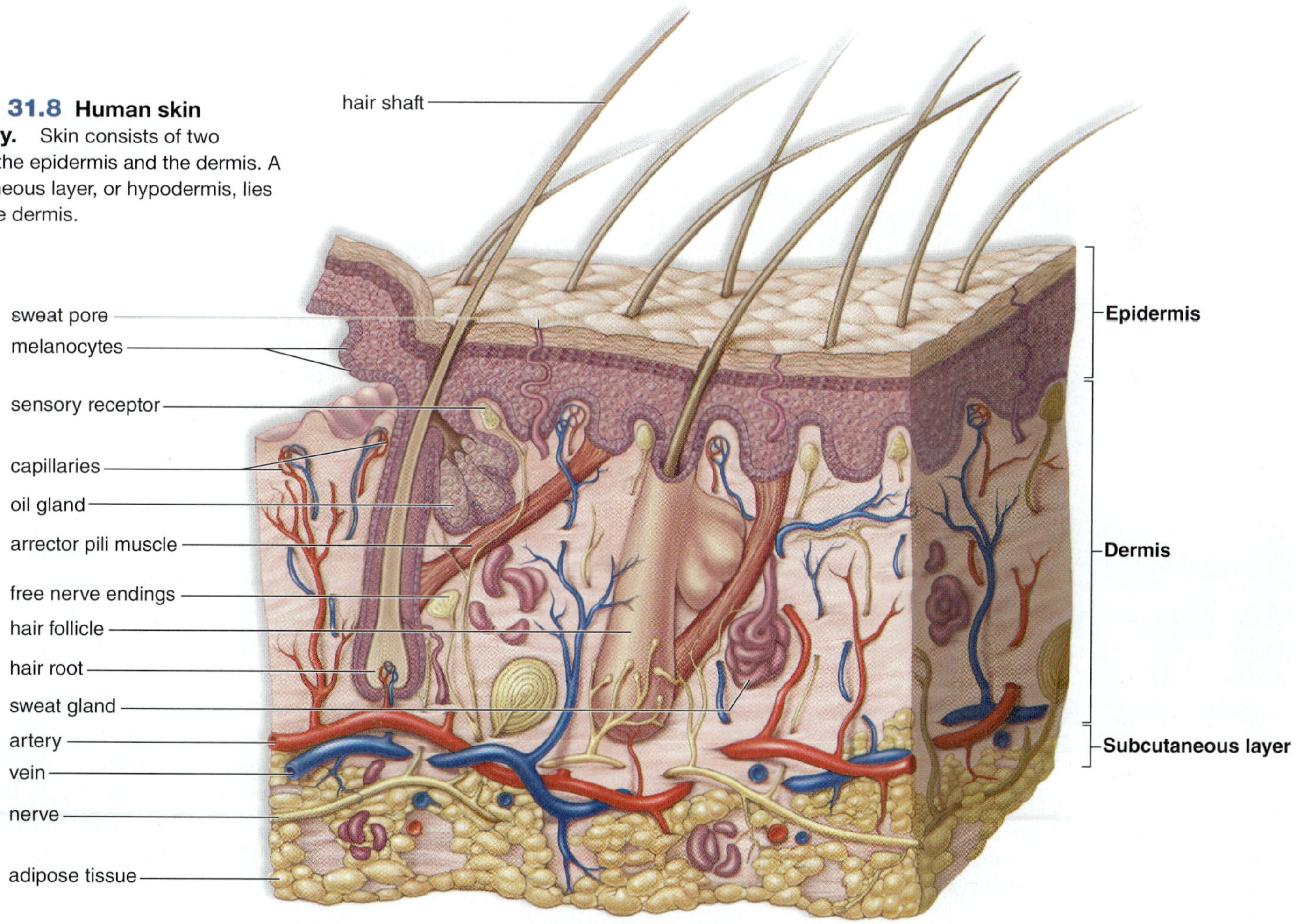

Figure 31.8 Human skin anatomy. Skin consists of two regions, the epidermis and the dermis. A subcutaneous layer, or hypodermis, lies below the dermis.

The Epidermis

The **epidermis** (Gk. *epi,* "over"; *derma,* "skin") is made up of stratified squamous epithelium. Human skin can be described as thin skin or thick skin based on the thickness of the epidermis. Thin skin covers most of the body and is associated with hair follicles, sebaceous (oil) glands, and sweat glands. Thick skin appears in regions of wear and tear, such as the palms of the hands and soles of the feet. Thick skin has sweat glands but no sebaceous glands or hair follicles.

In both types of skin, new cells derived from stem (basal) cells become flattened and hardened as they push to the surface (Fig. 31.9). Hardening takes place because the cells produce keratin, a waterproof protein. It is estimated that 1.5 million of these cells are shed from the human body every day. A thick layer of dead, keratinized cells, arranged in spiral and concentric patterns, forms fingerprints (and toe prints), which are thought to increase friction and aid in gripping objects.

Specialized cells in the epidermis called **melanocytes** produce melanin, the pigment responsible for skin color. The amount of melanin varies throughout the body. It is concentrated in freckles and moles. Tanning occurs after a lighter-skinned person is exposed to sunlight because melanocytes produce more melanin, which is distributed to epidermal cells before they rise to the surface. Although we tend to associate a tan with health, in reality it signifies that the body is trying to protect itself from the dangerous rays of the sun. Some UV radiation seems to be required for good health, however. The Biological Systems feature, "UV Rays: Too Much Exposure or Too Little?," examines some pros and cons of sun exposure.

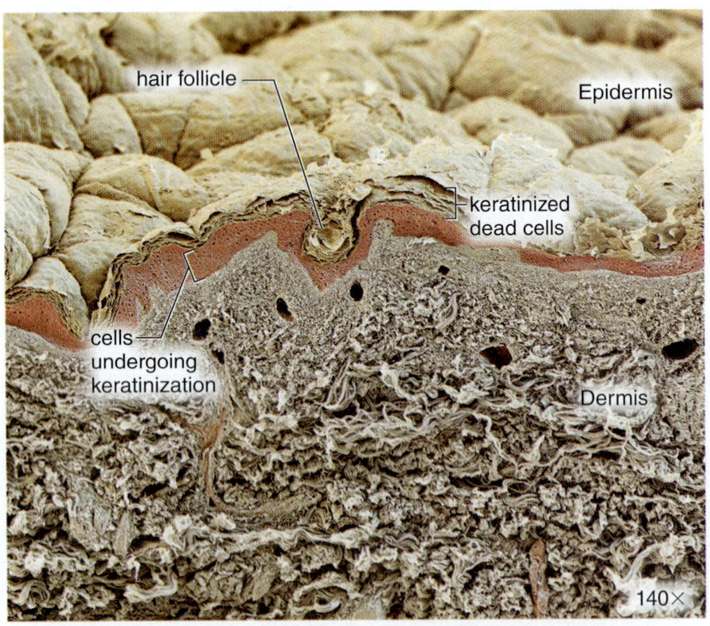

Photomicrograph of skin

Figure 31.9 Layers of skin. The epidermal surface covers and protects the dermis below. A hair follicle and the dermis are seen in cross section. Cells at the dermal-epidermal border are becoming keratinized and pushed to the surface, from which they are eventually shed.

Besides skin cancer, the epidermal layer is susceptible to a number of other diseases. Perhaps more than animals with skin covered by tough scales or a thick fur, human skin can be easily traumatized. It is also prone to a variety of bacterial, fungal, and viral infections. Fungi that grow on keratinized tissue are called dermatophytes, and they include the organisms that cause ringworm and athlete's foot. Warts, caused by infection with the human papillomavirus, are small areas of epidermal proliferation that are generally harmless. Genital warts (see chapter 41), however, can be a more serious problem.

The Dermis

The **dermis** is a region of dense fibrous connective tissue beneath the epidermis. As seen in Figure 31.9, the deeper epidermis forms ridges, which interact with projections of the dermis. The dermis contains collagen and elastic fibers. The collagen fibers are flexible but offer great resistance to overstretching; they prevent the skin from being torn. Stretching of the dermis, as occurs in obesity and pregnancy, can produce stretch marks, or striae.

The elastic fibers maintain normal skin tension but also stretch to allow movement of underlying muscles and joints. (The number of collagen and elastic fibers decreases with age and with exposure to the sun, causing the skin to become less supple and more prone to wrinkling.) The dermis also contains blood vessels that nourish the skin. When blood rushes into these vessels, a person blushes, and when blood is minimal in them, a person turns "blue."

Sensory receptors are specialized nerve endings in the dermis that respond to external stimuli. There are sensory receptors for touch, pressure, pain, and temperature. The fingertips contain the most touch receptors, and these add to our ability to use our fingers for delicate tasks.

The Subcutaneous Layer

Technically speaking, the subcutaneous layer (the hypodermis) beneath the dermis is not a part of skin. It is composed of loose connective tissue and adipose tissue. Subcutaneous adipose tissue helps thermally insulate the body from either gaining heat from the outside or losing heat from the inside. Excessive development of the subcutaneous layer occurs with obesity.

Accessory Structures of Human Skin

Nails, hair, and glands are of epidermal origin, even though some parts of hair and glands are largely found in the dermis.

Nails are a protective covering of the distal part of fingers and toes, collectively called digits. Nails grow from epithelial cells at the base of the nail in the portion called the nail root. The cuticle is a fold of skin that hides the nail root. The whitish color of the half-moon-shaped base, or lunula, results from the thick layer of cells in this area. The cells of a nail become keratinized as they grow out over the nail bed. The appearance of nails can reveal clues about a person's health. For example, in clubbing of the nails, the nails turn down instead of lying flat. This condition is associated with a deficiency of oxygen in the blood.

THEME Biological Systems

UV Rays: Too Much Exposure or Too Little?

The sun is the major source of energy for life on Earth. Without the sun, most organisms would quickly die out. But the sun's energy can also be damaging. In addition to visible light, the sun emits ultraviolet (UV) radiation, which has a shorter wavelength (and thus a higher energy) than visible light. Based on wavelength, UV radiation can be grouped into three types: UVA, UVB, and UVC, but only UVA and UVB reach the Earth's surface. Scientists have developed a UV index to determine how powerful the solar rays are in different U.S. cities. In general, the more southern the city, the higher the UV index and the greater the risk for skin cancer. Maps indicating the daily risk levels for various U.S. regions can be viewed at http://www.weather.gov/os/uv/.

Both UVA and UVB rays can damage skin cells. Tanning occurs when melanin granules increase in keratinized cells at the surface of the skin as a way to prevent further damage by UV rays. Too much exposure to UVB rays can directly damage skin cell DNA, leading to mutations that can cause skin cancer. UVB is also responsible for the pain and redness characteristic of a sunburn. In contrast, UVA rays do not cause sunburn but penetrate more deeply into the skin, damaging collagen fibers and inducing premature aging of the skin, as well as certain types of skin cancer. UVA rays can also induce skin cancer by causing the production of highly reactive chemicals that can indirectly damage DNA.

Skin cancer is the most commonly diagnosed type of cancer in the United States, outnumbering cancers of the lung, breast, prostate, and colon combined. The most common type of skin cancer is basal cell carcinoma (Fig. 31C), which is rarely fatal but can be disfiguring if allowed to grow. Squamous cell carcinoma, the second most common type, is fatal in about 1% of cases. The most deadly form is melanoma, which occurs in adolescents and young adults as well as in older people. If detected early, over 95% of patients survive at least 5 years, but if the cancer cells have already spread throughout the body, only 10–20% can expect to live this long.

Melanoma affects pigmented cells and often has the appearance of an unusual mole (Fig. 31C). Any moles that become malignant are removed surgically. If the cancer has spread, chemotherapy and a number of other treatments are also available. In March 2007, the USDA approved a vaccine to treat melanoma in dogs, which was the first time a vaccine has been approved as a treatment for any cancer in animals or humans. Clinical trials are under way to test a similar vaccine for use in humans. In March 2011, the FDA approved a drug called Yervoy, after it was shown to extend the lives of patients with advanced, late-stage melanoma.

According to the Skin Cancer Foundation, about 90% of nonmelanoma skin cancers, and 65% of melanomas, are associated with exposure to UV radiation from the sun. So how can we protect ourselves? First, try to minimize sun exposure between 10 A.M. and 4 P.M. (when the UV rays are most intense) and wear protective clothing, hats, and sunglasses. Second, use sunscreen. Sunscreens generally do a better job of blocking UVB than UVA rays. In fact, the sun protection factor (SPF) printed on sunscreen labels refers only to the degree of protection against UVB. Many sunscreens don't provide as much protection against UVA, and because UVA doesn't cause sunburn, people may have a false sense of security. Some sunscreens do a better job of blocking UVA—look for those that contain zinc oxide, titanium dioxide, avobenzone, or mexoryl SX. Unfortunately, tanning salons use lamps that emit UVA rays that are two to three times more powerful than the UVA rays emitted by the sun. Because of the potential damage to deeper layers of skin, most medical experts recommend avoiding indoor tanning salons altogether.

Since UV light is potentially damaging, why haven't all humans developed the more protective dark skin that is common to humans living in tropical regions? It turns out that vitamin D is produced in the body only when UVB rays interact with a form of cholesterol found mainly in the skin. This "sunshine vitamin" serves several important functions in the body, including keeping bones strong, boosting the immune system, and reducing blood pressure. Certain foods also contain vitamin D, but it can be difficult to obtain sufficient amounts through diet alone. Therefore, in more temperate areas of the planet, lighter-skinned individuals have the advantage of being able to synthesize sufficient vitamin D. Interestingly, dark-skinned people living in such regions may be at increased risk for vitamin D deficiency.

How much sun exposure is enough in temperate regions of the world? During the summer months, an average fair-skinned person will synthesize plenty of vitamin D after exposure to 10–15 minutes of midday sun. During winter months, however, anyone living north of Atlanta probably receives too few UV rays to stimulate vitamin D synthesis, and therefore he or she must fulfill the requirement through the diet.

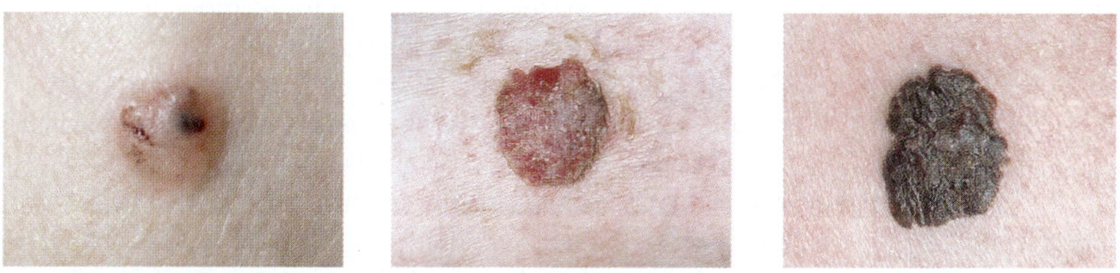

a. Basal cell carcinoma b. Squamous cell carcinoma c. Melanoma

Figure 31C Skin cancer. **a.** Basal cell carcinomas are the most common type of skin cancer. **b.** Squamous cell carcinomas arise from the epidermis, usually in sun-exposed areas. **c.** Malignant melanomas result from proliferation of pigmented cells. Warning signs include changes in size, shape, or color of a normal mole, as well as itching or pain.

Discussion Questions

1. Which is more important to you—having a "healthy-looking" tan now or preserving your skin's health later in life? Why?
2. If caught early, melanoma is rarely fatal. What are some factors that could delay the detection of a melanoma?
3. There is considerable controversy about the level of dietary vitamin D intake that should be recommended. Why is this the case?

Hair follicles begin in the dermis and continue through the epidermis, where the hair shaft extends beyond the skin. Contraction of the arrector pili muscles attached to hair follicles causes the hairs to "stand on end" and goose bumps to develop. Epidermal cells form the root of hair, and their division causes a hair to grow. The cells become keratinized and die as they are pushed farther from the root.

Hair, except for the root, is formed of dead, hardened epidermal cells; the root is alive and resides at the base of a follicle in the dermis. A person's scalp has about 100,000 hair follicles on average. The number of follicles varies from one body region to another. The texture of hair is dependent on the shape of the hair shaft. In wavy hair, the shaft is oval, and in straight hair, the shaft is round. Hair color is determined by pigmentation. Dark hair is due to melanin concentration, and blond hair has scanty amounts of melanin. Red hair is caused mainly by the presence of an iron-containing pigment called pheomelanin. Gray or white hair results from a lack of pigment.

A hair on the scalp grows about 1 mm every 3 days. Interestingly, chemical substances in the body, such as illicit drugs and by-products of alcohol metabolism, are incorporated into growing hair shafts, where they can be detected by laboratory tests. A certain amount of hair loss is inevitable. According to the American Academy of Dermatology, most people lose about 50–100 hairs from their scalp every day, but sometimes hair loss is more dramatic. Sudden hair loss is often due to illness, nutritional deficiency, medications, or pregnancy. A gradual loss of hair that becomes more noticeable each year is usually a hereditary condition called *androgenetic alopecia,* which affects about 80 million American men and women.

Each hair follicle has one or more **oil glands** (**sebaceous glands**), which secrete sebum, an oily substance that lubricates the hair within the follicle and the skin itself. If the oil glands fail to discharge, the secretions collect and form "whiteheads" or "blackheads." The color of blackheads is due to oxidized sebum. Acne is an inflammation of the sebaceous glands that most often occurs during adolescence, when hormonal changes lead to increased oil production.

On average, a person's skin has about 250,000 **sweat glands,** which are present in all regions of skin. A sweat gland is a tubule that begins in the dermis and either opens into a hair follicle or, more often, opens onto the surface of the skin. Sweat glands all over the body play a role in modifying body temperature. When body temperature starts to rise, sweat glands become active. Sweat absorbs body heat as it evaporates. Once the body temperature lowers, sweat glands are no longer active. Other sweat glands occur in the groin and axillary regions and are associated with distinct scents.

Check Your Progress 31.3

1. List one function of skin that is common to all animals and one that is unique to a single group.
2. Compare the structure and function of the epidermal and dermal layers of the skin.
3. Discuss why a dark-skinned individual living in northern Canada might develop bone problems.
4. Describe the structure and functions of nails, hair, sweat glands, and oil glands.

31.4 Homeostasis

Learning Outcomes

Upon completion of this section, you should be able to

1. Define *homeostasis* and explain why it is an essential feature of all living organisms.
2. Evaluate the evolutionary benefits of regulating an internal variable, such as body temperature, versus the cost.
3. Differentiate between positive and negative feedback mechanisms and list one specific example of each in animals.

All organ systems of animals contribute to **homeostasis,** the ability of an organism to maintain a relatively constant internal environment. Although all organisms must carry out some degree of homeostasis, animals vary in the degree to which they regulate these internal variables.

3D Animation Homeostasis

MP3 Homeostasis

Examples of Homeostatic Regulation

With respect to body temperature, all invertebrates, as well as fish, amphibians, and reptiles, are "cold-blooded," or **poikilothermic,** meaning that their body temperature fluctuates depending on the environmental temperature. This approach saves energy, but it also may restrict the ability of these species to live in extremely cold or hot environments.

Birds and mammals tend to be "warm-blooded," or **homeothermic,** and they have mechanisms for regulating their body temperature toward an optimum. This approach is energetically expensive, but it provides the evolutionary advantage of being able to adapt to many different environments.

Homeostasis does not mean a rigid or unvarying stability but, rather, a dynamic fluctuation around a set point. Besides temperature, animal systems regulate pH, salt balance, and the concentrations of many other body constituents, such as glucose, oxygen, CO_2, and various minerals. All organ systems participate in this regulation:

- The digestive system takes in and digests food, providing nutrient molecules to replace those constantly being consumed by the body cells.
- The respiratory system adds oxygen to the blood and removes carbon dioxide to meet body needs.
- The liver removes glucose from the blood and stores it as glycogen; later, glycogen is broken down to supply the needs of body cells. Blood glucose levels remain fairly constant.
- The pancreas secretes insulin in response to elevated blood glucose; insulin helps regulate glycogen storage.
- Under hormonal control, the kidneys excrete wastes and salts, substances that can affect the pH of the blood.

When homeostasis fails, disease or death often results.

Although homeostasis is, to a degree, controlled by hormones, it is ultimately controlled by the nervous system. In humans, the brain contains regulatory centers that control the function of other organs, maintaining homeostasis. These regulatory centers are often a part of negative feedback systems.

Negative Feedback

Negative feedback is the primary homeostatic mechanism that keeps a variable, such as the blood glucose level, close to a particular value, or set point.

A homeostatic mechanism has at least two components: a sensor and a control center. The sensor detects a change in the internal environment; the control center then initiates an action to bring conditions back to normal again. When normal conditions are reached, the sensor is no longer activated. In other words, a negative feedback mechanism is present when the output of the system dampens the original stimulus. As an example, when blood pressure rises, sensory receptors signal a control center in the brain. The center stops sending nerve impulses to the arterial walls, and they relax. Once the blood pressure drops, signals no longer go to the control center.

A home heating system is often used to illustrate how a more complicated negative feedback mechanism works (Fig. 31.10). You set the thermostat at, say, 68°F. This is the *set point*. The thermostat contains a thermometer, a sensor that detects when the room temperature is above or below the set point. The thermostat also contains a control center; it turns the furnace off when the room is warm and turns it on when the room is cool. When the furnace is off, the room cools a bit, and when the furnace is on, the room warms a bit. In other words, typical of negative feedback mechanisms, there is a fluctuation above and below normal.

Human Example: Regulation of Body Temperature

The sensor and the control center for body temperature are located in a part of the brain called the hypothalamus.

Above Normal Temperature. When the body temperature is above normal, the control center directs the blood vessels of the skin to dilate. The result is that more blood flows near the surface of the body, where heat can be lost to the environment. In addition, the nervous system activates the sweat glands, and the evaporation of sweat helps lower body temperature. Gradually, body temperature decreases to 37.0°C (98.6°F). Body temperature does not get colder and colder, because a body temperature below normal brings about a change toward a warmer body temperature.

Below Normal Temperature. When the body temperature falls below normal, the control center directs (via nerve impulses) the blood vessels of the skin to constrict (Fig. 31.11). This action conserves heat. If body temperature falls even lower, the control center sends nerve impulses to the skeletal muscles, and shivering occurs. Shivering generates heat, and gradually body temperature rises to 37.0°C. When the temperature rises to normal, the control center is inactivated.

MP3 Temperature Regulation

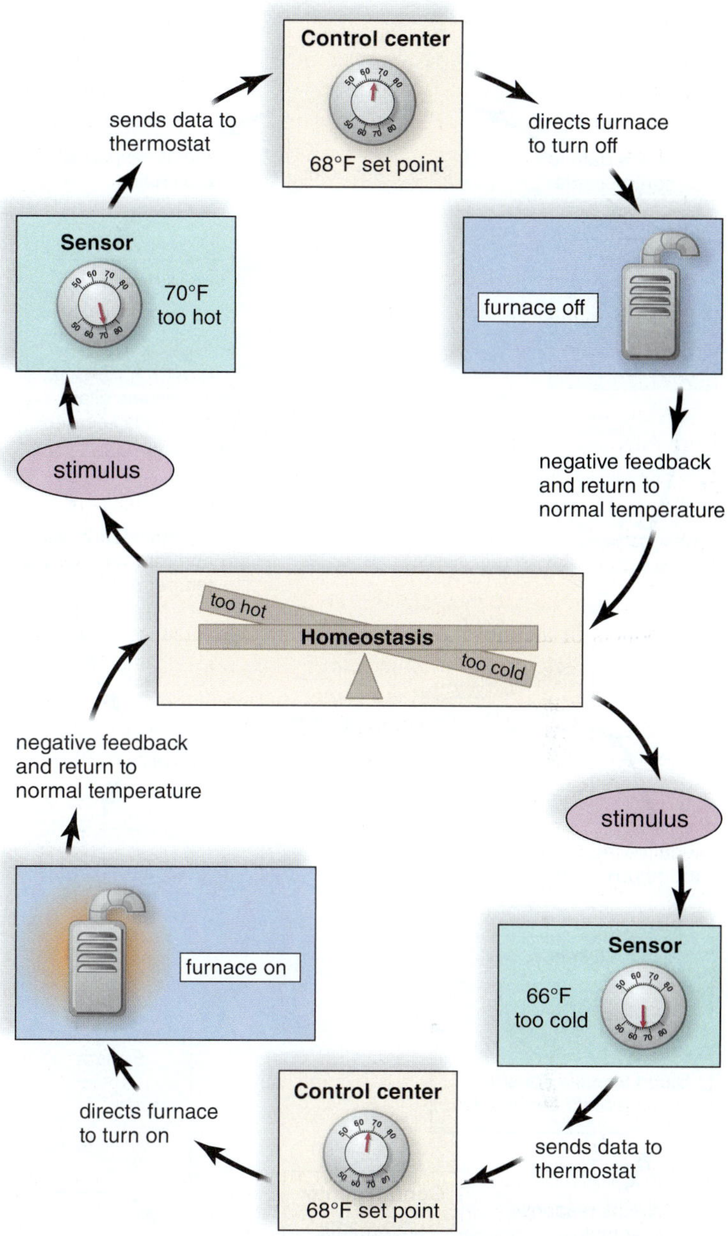

Figure 31.10 Regulation of room temperature using negative feedback. Room temperature is returned to normal when the room becomes too hot (*above*) or too cold (*below*). The thermostat contains both the sensor and the control center. *Above:* The sensor detects that the room is too hot, and the control center turns the furnace off. The room cools, removing the stimulus (negative feedback). *Below:* The sensor detects that the room is too cold, and the control center turns the furnace on. When the temperature returns to normal, the stimulus is no longer present.

Positive Feedback

Positive feedback is a mechanism that brings about a continually greater change in the same direction. Body processes regulated by positive feedback can produce significant change in a relatively short period of time. For example, when a woman is giving birth, the head of the baby begins to press against the cervix (opening to

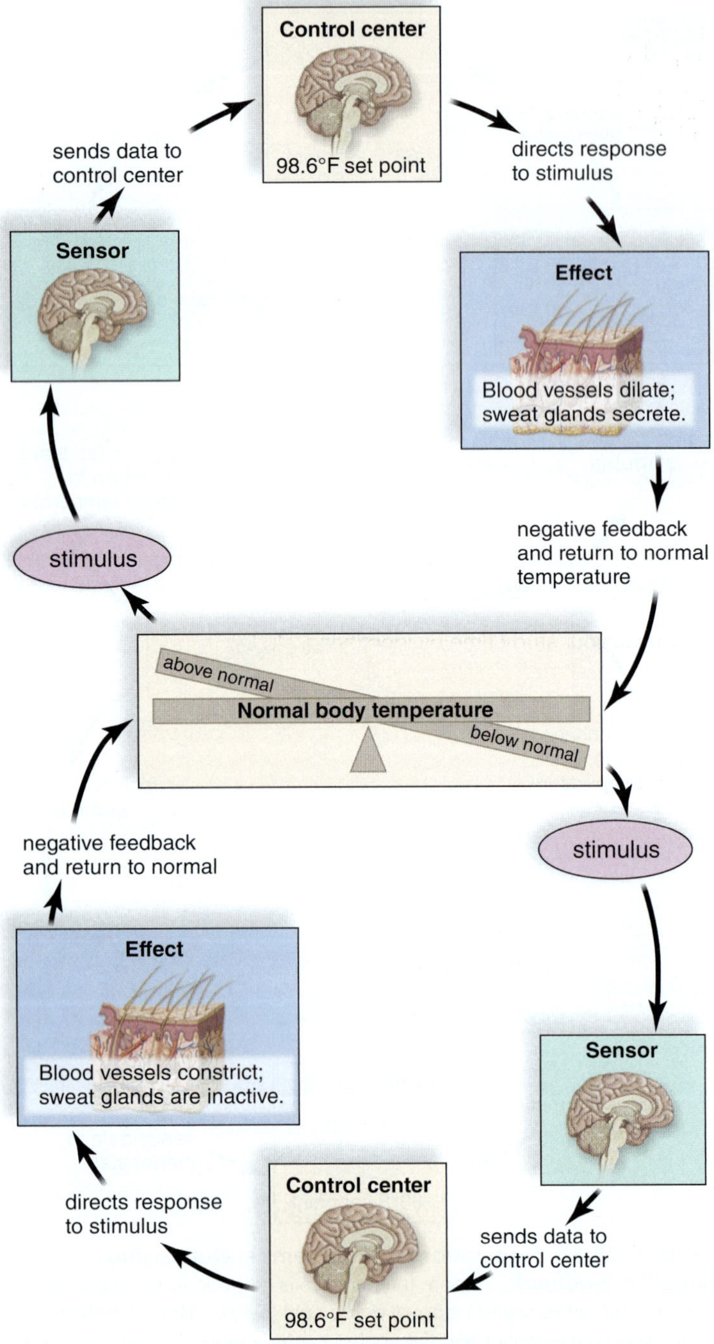

Figure 31.11 Regulation of body temperature by negative feedback. *Above:* When body temperature rises above normal, the hypothalamus senses the change and causes blood vessels to dilate and sweat glands to secrete, so that body temperature returns to normal. *Below:* When body temperature falls below normal, the hypothalamus senses the change and causes blood vessels to constrict. In addition, shivering may occur to bring body temperature back to normal. In this way, the original stimulus is removed (negative feedback).

Tutorial Negative Feedback

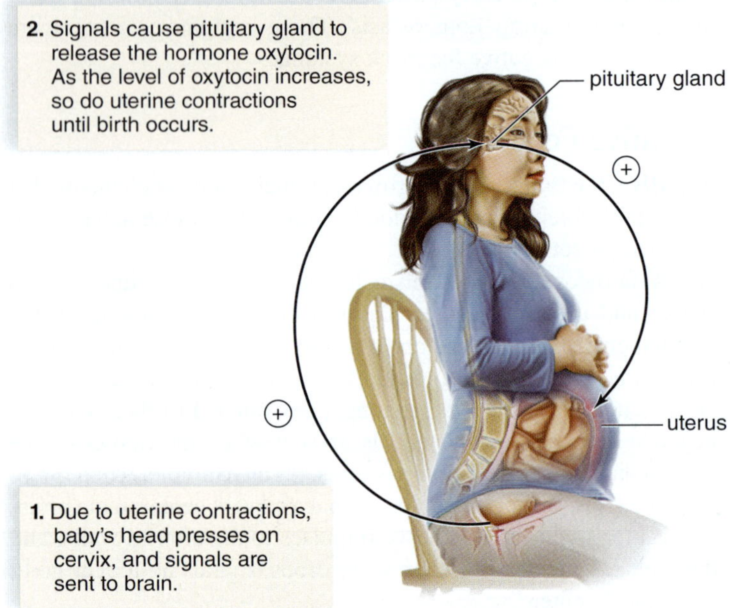

Figure 31.12 Positive feedback. This diagram shows how positive feedback works. The signal causes a change in the same direction until there is a definite cutoff point, such as the birth of a child.

the birth canal) stimulating sensory receptors there (Fig. 31.12). When nerve impulses reach the brain, the brain causes the pituitary gland to secrete the hormone oxytocin. Oxytocin travels in the blood and causes the uterus to contract. As labor continues, the cervix is increasingly more stimulated, and uterine contractions become stronger until birth occurs.

A positive feedback mechanism can be harmful, as when a fever causes metabolic changes that push the fever still higher. At temperatures higher than about 113°F, many cellular proteins begin to denature, and metabolism stops. Still, positive feedback loops such as those involved in childbirth, blood clotting, and the stomach's digestion of protein assist the body in completing a process that has a definite cutoff point.

Animation Feedback Mechanisms

Check Your Progress 31.4

1. Define *homeostasis,* and discuss why is it important to body function.
2. Explain how the circulatory, respiratory, and urinary systems specifically contribute to homeostasis.
3. Discuss how negative feedback is similar to the way a thermostat works.

CONNECTING the CONCEPTS with the THEMES

Evolution

- With the evolution of multicellularity, cell differentiation became possible.
- The development of organs—tissues combined into a structure with a particular function—gave animals the ability to more efficiently localize a function, such as waste elimination (kidneys) and respiration (lungs).
- The evolution of the skin and its derivatives, such as feathers and fur, facilitated movement into new environments and aided homeostasis.

Nature of Science

- Cells of many tissues continue to divide throughout life—for example, skin, hair, and nails. Other tissues are limited in their ability to regenerate.
- Certain animals are able to regenerate various parts of their body. Study of these organisms may someday lead to the ability to stimulate regeneration of human tissues and organs.

Biological Systems

- The organs of complex animals comprise four major types of tissues: epithelial, connective, muscular, and nervous.
- The integumentary system serves many functions for animals, including protection, temperature regulation, and detection of sensations.
- Homeostatic mechanisms, especially negative feedback mechanisms, help ensure optimum internal conditions for the body's metabolic functions.

MEDIA STUDY TOOLS http://connect.mheducation.com

SMARTBOOK® Study smart. SmartBook helps you maximize your study time by identifying what you know and don't know. The recharge feature in SmartBook helps you prepare for the test by identifying the material that you are most likely to forget.

connect BIOLOGY Find even more resources to learn the chapter concepts including animations, tutorial videos, and interactive practice questions.

MP3 Files	Animation	3D Animation	Tutorial
31.1 Overview of Tissues • Epithelial Tissue • Epithelial Glands • Connective Tissue • General Functions and Composition of Blood • Muscle Tissue • Nervous Tissue **31.2** Body Cavities **31.3** Skin and Its Tissues **31.4** Homeostasis • Temperature Regulation	**31.4** Feedback Mechanism	**31.4** Homeostasis	**31.4** Negative Feedback

SUMMARIZE

31.1 Types of Tissues

During development, the zygote divides to produce cells that go on to form **tissues** composed of similar cells specialized for a particular function. Tissues make up organs, and organ systems make up the organism. This sequence describes the levels of organization within an organism. Tissues are categorized into four groups:

- **Epithelial tissue,** which covers the body and lines cavities, is of three types: **squamous epithelium, cuboidal epithelium,** and **columnar epithelium.** In all cases, epithelial tissues are attached to a basement membrane. Each type can be simple or stratified; it can also contain **glands,** which secrete a product. **Endocrine glands,** which secrete hormones, are a specialized example of this.
- **Connective tissue** contains cells and a matrix containing fibers and ground substance. The three types of fibers are **collagen fibers, reticular fibers,** and/or **elastic fibers. Loose fibrous connective tissue** has both collagen and elastic fibers. **Dense fibrous connective tissue,** like that of **tendons** and **ligaments,** contains closely packed collagen fibers. Both loose fibrous and dense fibrous connective tissues contain cells called **fibroblasts. Adipose tissue** contains cells, called adipocytes, that enlarge and store fat. Supportive connective tissue includes **cartilage** and **bone.** Both have cells within **lacunae,** but the matrix for cartilage is more flexible than that for bone, which contains calcium salts. Three types of cartilage are **hyaline cartilage, elastic cartilage,** and **fibrocartilage,** while bone can be either **compact bone** or **spongy bone. Blood** is a fluid connective tissue that contains **red blood cells, white blood cells,** and **platelets. Lymph** is also a fluid connective tissue, which is derived from excess **interstitial fluid** that flows through lymphatic vessels.

- **Muscular** (**contractile**) **tissue** can be smooth or striated, as well as involuntary (smooth and cardiac) or voluntary (skeletal). In humans, **skeletal muscle** is attached to bone, **smooth** (**visceral**) **muscle** is in the wall of internal organs, and **cardiac muscle** makes up the heart.
- **Nervous tissue** has one main type of conducting cell, the **neuron,** and several types of **neuroglia.** The majority of neurons have dendrites, a cell body, and an axon. The brain and spinal cord contain complete neurons, while nerves contain only axons. Axons are specialized to conduct nerve impulses.

31.2 Organs, Organ Systems, and Body Cavities

An **organ** contains two or more tissues. An **organ system** is made up of several organs that function together in a particular life process.

The human body has two main **body cavities.** The dorsal cavity contains the brain and spinal cord. The ventral cavity is divided into the thoracic cavity (containing the heart and lungs) and the abdominal cavity (containing most other internal organs).

31.3 The Integumentary System

The **integumentary system** includes the **skin** and its derivatives, such as scales (in fish, reptiles, and birds), feathers (birds only), and hair (mammals only). This organ system protects underlying tissues against dessication, trauma, and pathogens plus functions in thermoregulation, sensory perception, and vitamin D synthesis.

Skin itself has two regions. **Epidermis** (stratified squamous epithelium) overlies the **dermis** (fibrous connective tissue containing sensory receptors, hair follicles, blood vessels, and nerves). The epidermis contains **melanocytes,** cells that produce a pigment, called melanin, that helps protect against the harmful effects of UV light. A subcutaneous layer is composed of loose connective tissue and adipose tissue. Accessory structures of human skin include **nails, hair follicles, oil glands,** and **sweat glands.**

31.4 Homeostasis

All animals must be capable of maintaining a certain degree of **homeostasis,** or maintainance of a relatively constant internal environment. Animals differ in their approach to regulating certain variables. For example, the body temperature of **poikilothermic** animals varies according to their environment, while **homeothermic** animals maintain an optimal temperature.

All organ systems contribute to homeostasis, but special contributions are made by the liver, which keeps the blood glucose constant, and the kidneys, which regulate the pH. The nervous and hormonal systems regulate the other body systems. Both of these are controlled by **negative feedback** mechanisms, which result in slight fluctuations above and below desired levels. Less commonly, **positive feedback** mechanisms can bring about an increasingly greater change in the same direction.

ASSESS

Choose the best answer for each question

31.1 Types of Tissues

1. Which of these is not a type of epithelial tissue?
   a. simple cuboidal
   b. cartilage
   c. stratified squamous
   d. striated
   e. All of these are epithelial tissue.
2. Tendons and ligaments
   a. are dense fibrous connective tissue.
   b. are associated with the bones.
   c. are found in vertebrates.
   d. contain collagen.
   e. All of these are correct.
3. Which tissue has cells in lacunae?
   a. epithelial
   b. cartilage
   c. bone
   d. smooth muscle
   e. Both b and c are correct.

In questions 4–7, match each description to the type of connective tissue in the key.

Key:

a. loose fibrous
b. adipose tissue
c. hyaline cartilage
d. elastic cartilage

4. occurs in the nose, ends of bones, and walls of respiratory passages
5. occurs in the outer ear
6. allows lungs, arteries, and bladder to expand
7. found mostly beneath skin and around organs
8. Which of these components of blood fights infection?
   a. red blood cells
   b. white blood cells
   c. platelets
   d. hydrogen ions
   e. All of these are correct.

In questions 9–11, match each type of muscle tissue to as many terms in the key as possible.

Key:

a. voluntary
b. involuntary
c. striated
d. nonstriated
e. spindle-shaped cells
f. branched cells
g. long, cylindrical cells

9. skeletal muscle
10. smooth muscle
11. cardiac muscle
12. Give the name, the location, and the function for each of the illustrated tissues in the human body.
    a. type of epithelial tissue ______________
    b. type of muscular tissue ______________
    c. type of connective tissue ______________

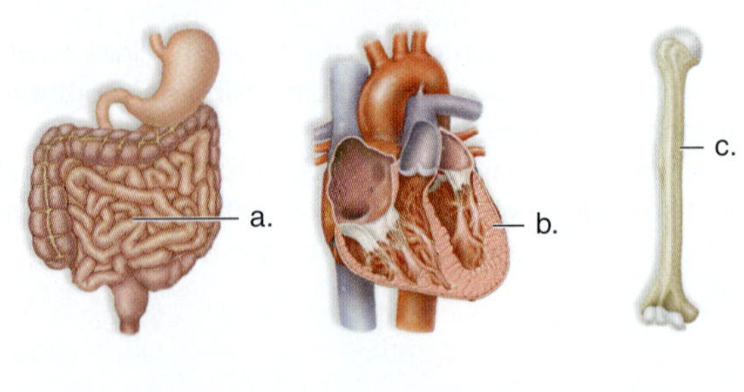

31.2 Organs, Organ Systems, and Body Cavities

13. The ______________ separates the thoracic cavity from the abdominal cavity.
    a. liver
    b. pancreas
    c. diaphragm
    d. pleural membrane
    e. intestine

In questions 14–17, match each organ system(s) to the life process(es) it (they) carries (carry) out.

Key:

a. coordinate body activities
b. exchange gases
c. protect the body from pathogens
d. eliminate wastes

14. urinary and digestive systems
15. lymphatic and immune systems
16. nervous and endocrine systems
17. respiratory system

31.3 The Integumentary System

18. Which of these is a function of skin?
    a. temperature regulation
    b. manufacture of vitamin D
    c. collection of sensory input
    d. protection from invading pathogens
    e. All of these are correct.
19. Which of these correctly describes a layer of the skin?
    a. The epidermis is simple squamous epithelium in which hair follicles develop and blood vessels expand when we are hot.
    b. The subcutaneous layer lies between the epidermis and the dermis. It contains adipose tissue, which keeps us warm.
    c. The dermis is a region of connective tissue that contains sensory receptors, nerve endings, and blood vessels.
    d. The skin has a special layer, still unnamed, in which there are all the accessory structures, such as nails, hair, and various glands.
20. When a person is cold, the superficial blood vessels
    a. dilate, and the sweat glands are inactive.
    b. dilate, and the sweat glands are active.
    c. constrict, and the sweat glands are inactive.
    d. constrict, and the sweat glands are active.
    e. contract, so that shivering occurs.

31.4 Homeostasis

21. Which of these body systems contribute to homeostasis?
    a. digestive and urinary systems
    b. respiratory and nervous systems
    c. nervous and endocrine systems
    d. immune and cardiovascular systems
    e. All of these are correct.
22. Which of these is an example of negative feedback?
    a. Air conditioning switches off when room temperature lowers.
    b. Insulin decreases blood sugar levels after eating a meal.
    c. Heart rate increases when blood pressure drops.
    d. All of these are examples of negative feedback.
23. In a negative feedback mechanism,
    a. the output cancels the input.
    b. there is a fluctuation above and below the average.
    c. there is self-regulation.
    d. a regulatory center communicates with other body parts.
    e. All of these are correct.

ENGAGE

LEARNSMART LABS®

The following LearnSmart Labs contain exercises that are related to the content of this chapter:

- Homeostasis

THINKING SCIENTIFICALLY

1. Many cancers develop from epithelial tissue. These include lung, colon, and skin cancers. What are two attributes of this tissue type that make cancer more likely to develop?
2. Bacterial or viral infections can cause a fever. Fevers occur when the hypothalamus changes its temperature set point. Signaling of the hypothalamus can be direct (from the infectious agent itself) or indirect (from the immune system). Which of these would enable the hypothalamus to respond to the greatest variety of infectious agents? Is there any disadvantage to such a signaling system?
3. The risk for human skin cancer, especially melanoma, rises with sun exposure. However, the amount of overall exposure over the years doesn't seem to be as important as the number of brief, intense exposures. A single, blistering sunburn may be even more dangerous than years of moderate tanning. What are some possible explanations for this?
4. In addition to their major functions, some organ systems play additional homeostatic roles. Describe some additional, "lesser" functions of the nervous system, endocrine system, skeletal system, muscular system, and lymphatic system.

32

Circulation and Cardiovascular Systems

Chapter Outline

32.1 Transport in Invertebrates 601
32.2 Transport in Vertebrates 603
32.3 The Human Cardiovascular System 605
32.4 Blood 613

Before You Begin

Before beginning this chapter, take a few moments to review the following discussions.

Figure 8.1 Animal cells use glucose and oxygen for what specific purpose(s)? Where is carbon dioxide generated?

Figure 29.11 How do the circulatory pathways of fishes, amphibians, reptiles, birds, and mammals resemble each other? How are they different?

Section 31.3 What types of tissues constitute the various parts of the cardiovascular system?

In his late fifties, Denver Broncos football coach John Fox needed surgery to repair a faulty heart valve he had had since birth.

Coaching professional sports is a stressful job, and the 2013 season was a tough one for some NFL head coaches. During the same week in November, Denver Broncos coach John Fox and Houston Texans coach Gary Kubiak were both hospitalized for cardiovascular-related illnesses.

The Broncos coach became dizzy while playing golf and was immediately taken to a local hospital, where he learned that he needed surgery to replace a failing heart valve. As he was recovering after surgery, Fox explained that he had been diagnosed in childhood with an abnormal aortic valve. Blood is ejected through this valve from the left ventricle into the aorta, and then to the rest of the body, with every heartbeat. Over time, the abnormal valve can become scarred and narrowed, eventually interfering with blood flow to the body. Aortic valve abnormalities are actually the most common congenital heart defect, occuring in 1–2% of the population, many of whom never develop symptoms. However, more than 60,000 aortic valve replacements are performed each year in the United States.

In Kubiak's case, the problem was not his heart, but a transient ischemic attack, or "mini stroke." He was able to return to work after about a week, as his body presumably dissolved the clot that temporarily obstructed blood flow to part of his brain. The word *ischemia* refers to a restricted blood supply to any tissue, but the brain is particularly sensitive to being deprived of the oxygen and nutrients that blood distributes throughout the body.

As you read through this chapter, think about the following questions:

1. What are the essential components of any circulatory system, and their functions?
2. Why are the processes that occur in capillaries essential to life?

Following the Themes

CHAPTER 32 CIRCULATION AND CARDIOVASCULAR SYSTEMS

UNIT 7 COMPARATIVE ANIMAL BIOLOGY

Evolution	The circulatory systems of animals show a distinct pattern of evolution—from no discrete system to open systems to closed cardiovascular systems.
Nature of Science	Experimentation and observation have increased our understanding of circulatory systems, in some cases resulting in the ability to treat or prevent cardiovascular disease.
Biological Systems	Circulatory systems carry out critical functions in most animals, helping provide the cells of the body with oxygen and nutrients and removing wastes.

32.1 Transport in Invertebrates

Learning Outcomes

Upon completion of this section, you should be able to

1. Describe the common features that determine why some invertebrates, such as sponges, cnidarians, and flatworms, do not require a circulatory system.
2. Explain two differences between blood and hemolymph.
3. Compare and contrast the open circulatory system of an arthropod with the closed system of an annelid.

All animal cells require a steady supply of oxygen and nutrients, and their waste products must be removed. In most animals, these tasks are facilitated by a **circulatory system,** which moves fluid between various parts of the body. However, some invertebrates, such as sponges, cnidarians (e.g., hydras, sea anemones), and flatworms (e.g., planarians), lack a circulatory system (Fig. 32.1*a, b*). Their thin body wall makes a circulatory system unnecessary.

In hydras, cells either are part of an external layer or line the gastrovascular cavity. Each cell is exposed to water and can independently exchange gases and rid itself of wastes. The cells that line the gastrovascular cavity are specialized to complete the digestive process. They pass nutrient molecules to other cells by diffusion. In planarians, a trilobed gastrovascular cavity branches throughout the small, flattened body. No cell is very far from one of the three digestive branches, so nutrient molecules can diffuse from cell to cell. Similarly, diffusion meets the respiration and elimination needs of the cells.

Pseudocoelomate invertebrates, such as nematodes, use the coelomic fluid of their body cavity for transport purposes. The coelomate echinoderms also rely on movement of coelomic fluid within a body cavity as a circulatory system (Fig. 32.1*c*).

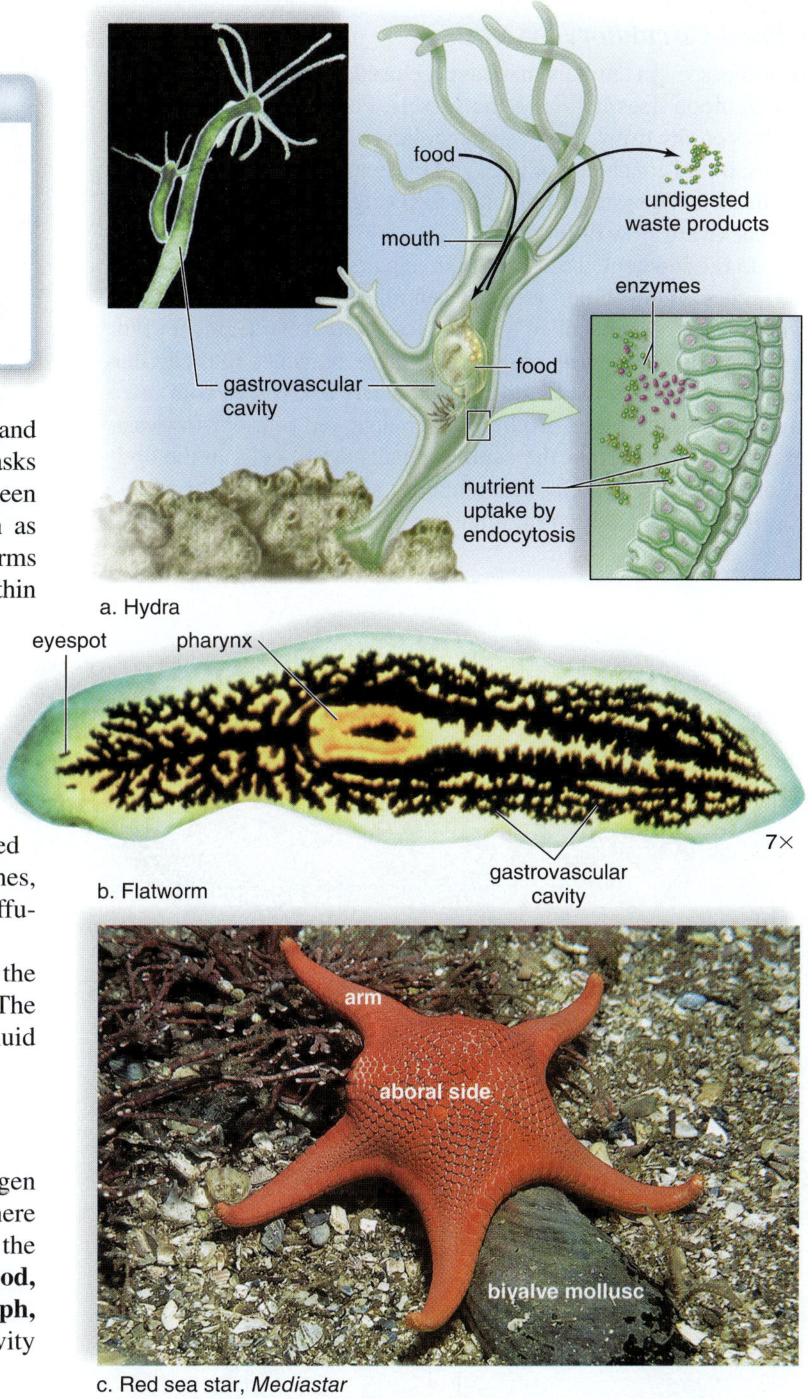

Figure 32.1 Aquatic animals without a circulatory system. **a.** In a hydra, a cnidarian, the gastrovascular cavity makes digested material available to the cells that line the cavity. These cells can also acquire oxygen from the watery contents of the cavity and discharge their wastes there. **b.** In a planarian, a flatworm, the gastrovascular cavity branches throughout the body, bringing nutrients to body cells. **c.** In a sea star, the coelomic fluid distributes oxygen and picks up wastes.

Invertebrates with a Circulatory System

Most invertebrates have a circulatory system that transports oxygen and nutrients, such as glucose and amino acids, to their cells. There it picks up wastes, which are later excreted from the body by the lungs or kidneys. There are two types of circulatory fluids: **blood,** which is always contained within blood vessels, and **hemolymph,** a mixture of blood and interstitial fluid, which fills the body cavity and surrounds the internal organs.

Open Circulatory Systems

Hemolymph is seen in animals that have an **open circulatory system** that consists of blood vessels plus open spaces. Open circulatory systems were likely the first to evolve, as they are present in simpler and evolutionarily older animals. For example, in most molluscs and arthropods, the heart pumps hemolymph via vessels into tissue spaces that are sometimes enlarged into saclike sinuses (Fig. 32.2*a*). Eventually, hemolymph drains back to the heart. In the grasshopper, an arthropod, the dorsal tubular heart pumps hemolymph into a dorsal aorta, which empties into the hemocoel. When the heart contracts, openings called ostia (sing., ostium) are closed; when the heart relaxes, the hemolymph is sucked back into the heart by way of the ostia.

The hemolymph of a grasshopper is colorless, because it does not contain hemoglobin or any other respiratory pigment. It carries nutrients but no oxygen. Oxygen is taken to cells, and carbon dioxide is removed from them, by way of air tubes called tracheae, which are found throughout the body. The tracheae provide efficient transport and delivery of respiratory gases while restricting water loss.

Closed Circulatory Systems

Blood is seen in animals that have a **closed circulatory system,** in which blood does not leave the vessels. For example, in annelids, such as earthworms, and in some molluscs, such as squid and octopuses, blood consisting of cells and plasma (a liquid) is pumped by the heart into a system of blood vessels (Fig. 32.2*b*). Valves prevent the backward flow of blood.

In the segmented earthworm, five pairs of anterior hearts (aortic arches) pump blood into the ventral blood vessel (an artery), which has a branch, called a lateral vessel, in every segment of the worm's body. Blood moves through these branches into capillaries, the thinnest of the blood vessels, where exchanges with interstitial fluid take place. Both gas exchange and nutrient-for-waste exchange occur across the capillary walls. Most cells in the body of an animal with a closed circulatory system are not far from a capillary. In an earthworm, after leaving a capillary, blood moves from small veins into the dorsal blood vessel (a vein). This dorsal blood vessel returns blood to the heart for repumping.

The earthworm has red blood that contains the respiratory pigment hemoglobin. Hemoglobin is dissolved in the blood and is not contained within blood cells. The earthworm has no specialized organ, such as lungs, for gas exchange with the external environment. Gas exchange takes place across the body wall, which must always remain moist for this purpose.

Check Your Progress 32.1

1. List the general functions of all circulatory systems.
2. Explain how blood differs from hemolymph.
3. Regarding oxygen transport, deduce the specific additional step that must occur in animals with a closed circulatory system, compared to those with an open system.

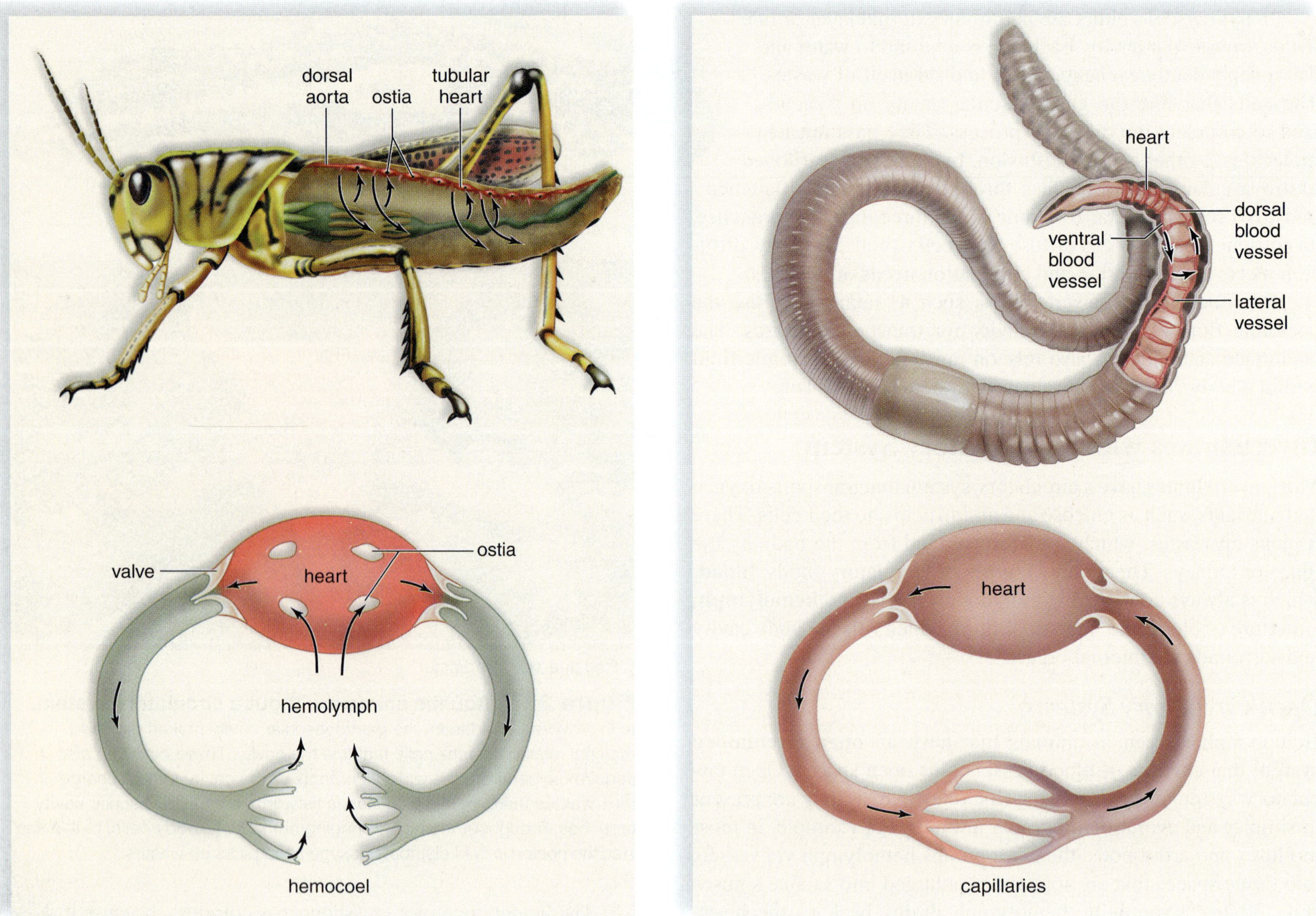

a. Open circulatory system

b. Closed circulatory system

Figure 32.2 Open versus closed circulatory systems. **a.** *Top:* The grasshopper, an arthropod, has an open circulatory system. *Bottom:* A hemocoel is a body cavity filled with hemolymph, which freely bathes the internal organs. The heart, a pump, sends hemolymph out through vessels and collects it through ostia (openings). This open system probably could not supply oxygen to wing muscles rapidly enough. These muscles receive oxygen directly from tracheae (air tubes). **b.** *Top:* The earthworm, an annelid, has a closed circulatory system. The dorsal and ventral blood vessels are joined by five pairs of anterior hearts, which pump blood. *Bottom:* The lateral vessels distribute blood to the rest of the worm.

32.2 Transport in Vertebrates

Learning Outcomes

Upon completion of this section, you should be able to

1. Distinguish among the structure and functions of arteries, veins, and capillaries.
2. Compare the path of blood in animals with a one-circuit circulatory pathway vs. a two-circuit pathway.
3. Identify the number of atria and ventricles in each type of vertebrate animal: fish, amphibians, most reptiles, crocodilians, birds, and mammals.

All vertebrate animals have a closed circulatory system, which is called a **cardiovascular system** (Gk. *kardia,* "heart"; L. *vascular,* "vessel"). It consists of a strong, muscular heart in which the atria (sing., atrium) receive blood and the muscular ventricles pump blood through the blood vessels. There are three kinds of blood vessels: **arteries,** which carry blood away from the heart; **capillaries** (L. *capillus,* "hair"), which exchange materials with **interstitial fluid** (the fluid between the body's cells); and **veins** (L. *vena,* "blood vessel"), which return blood to the heart (Fig. 32.3).

MP3 Classification of Blood Vessels

An artery or a vein has three distinct layers (Fig. 32.3*a, c*). The outer layer consists of fibrous connective tissue, which is rich in elastic and collagen fibers. The middle layer is composed of smooth muscle and elastic tissue. The innermost layer, called the endothelium, is similar to squamous epithelium.

Arteries have thick walls, and those attached to the heart are resilient, meaning that they are able to expand and accommodate the sudden increase in blood volume that results after each heartbeat. **Arterioles** are small arteries whose diameter can be regulated by the nervous and endocrine systems. Arteriole constriction and dilation affect blood pressure in general. The greater the number of vessels dilated, the lower the blood pressure.

Arterioles branch into capillaries, which are extremely narrow, microscopic tubes with a wall composed of only one layer of cells. Capillary beds, which consist of many interconnected capillaries (Fig. 32.4), are so prevalent that in humans almost all cells are within 60–80 µm of a capillary. But only about 5% of the capillary beds are open at the same time. After an animal has eaten, precapillary sphincters relax, and the capillary beds in the digestive tract are usually open. During muscular exercise, the capillary beds of the muscles are open. Capillaries, which are usually so narrow that red blood cells pass through in single file, allow exchange of nutrient and waste molecules across their thin walls.

Venules and veins collect blood from the capillary beds and take it to the heart. First, the venules drain the blood from the capillaries, and then they join to form a vein. The wall of a vein is much thinner than that of an artery, and this may be associated with a lower blood pressure in the veins. Valves within the veins point, or open, toward the heart, preventing a backflow of blood when they close (see Fig. 32.3*c*).

Comparison of Circulatory Pathways

Two different types of circulatory pathways are seen among vertebrate animals. In fishes, blood follows a one-circuit (single-loop) circulatory pathway through the body. The heart has a single atrium and a single ventricle (Fig. 32.5*a*).

The pumping action of the ventricle sends blood under pressure to the gills, where gas exchange occurs. After passing through the gills, blood returns to the dorsal aorta, which distributes blood throughout the body. Veins return oxygen-poor blood to an enlarged chamber called the sinus venosus, which leads to the atrium. The atrium pumps blood back to the ventricle. This single circulatory loop has an advantage in that the gill capillaries receive oxygen-poor blood and the capillaries of the body, called systemic capillaries, receive fully oxygen-rich blood. It is disadvantageous in that after leaving the gills, the blood is under reduced pressure.

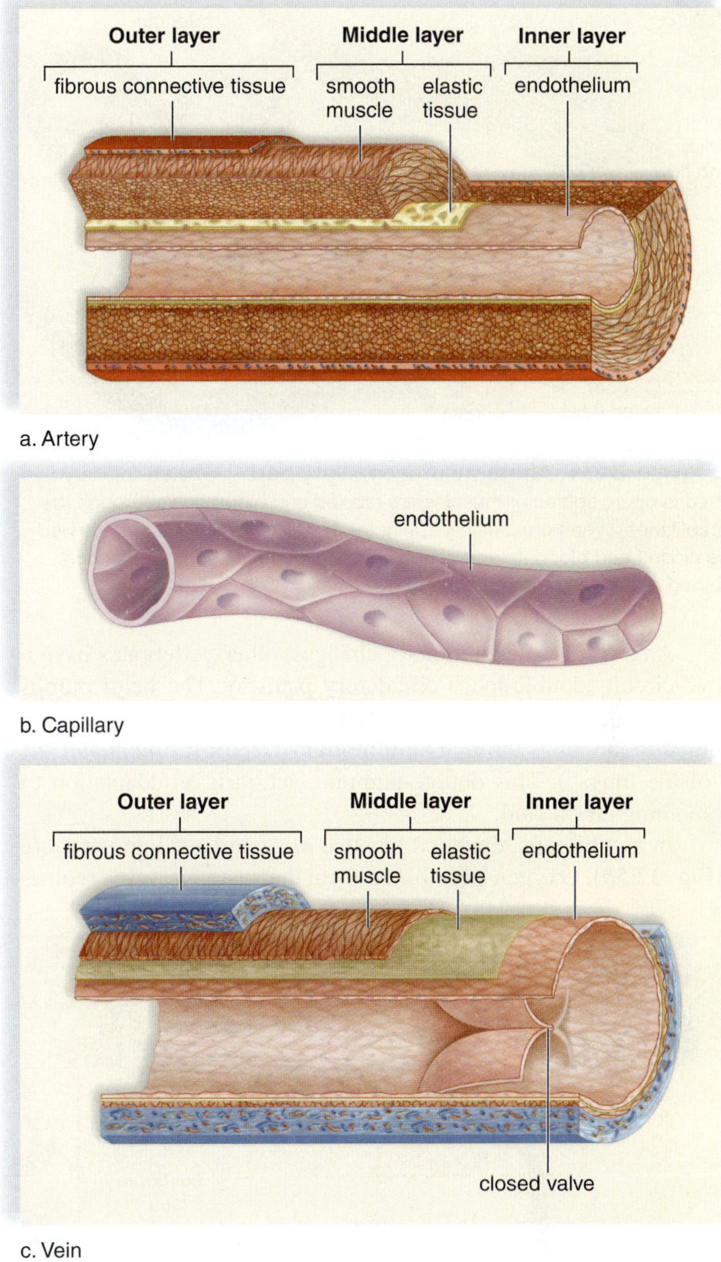

Figure 32.3 Transport in vertebrates. **a.** Arteries have well-developed walls with a thick middle layer of elastic tissue and smooth muscle. **b.** Capillary walls are only one cell thick. **c.** Veins have flabby walls, particularly because the middle layer is not as thick as in arteries. Veins have valves, which ensure one-way flow of blood back to the heart.

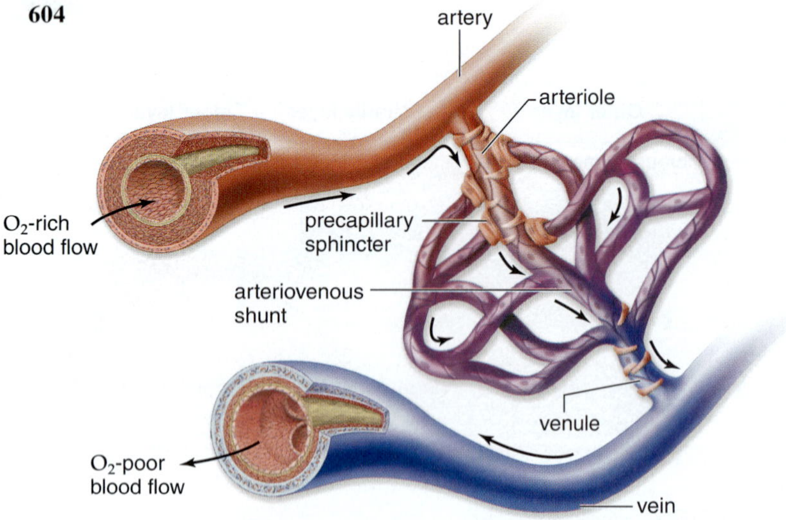

Figure 32.4 Anatomy of a capillary bed. When a capillary bed is open, sphincter muscles are relaxed and blood flows through the capillaries. When precapillary sphincter muscles are contracted, the bed is closed and blood flows through an arteriovenous shunt that carries blood directly from an arteriole to a venule.

As a result of evolutionary changes, other vertebrates have a two-circuit (double-loop) circulatory pathway. The heart pumps blood to the tissues through a **systemic circuit,** as well as pumping blood to the lungs through a **pulmonary circuit** (L. *pulmonarius,* "of the lungs"). This double-pumping action is an adaptation to breathing air on land.

In amphibians, the heart has two *atria* and a single *ventricle* (Fig. 32.5*b*). Oxygen-poor blood from the systemic veins returns to the right atrium. Oxygen-rich blood returning from the lungs passes to the left atrium. Both of the atria empty into the single ventricle. Oxygen-rich and oxygen-poor blood are kept somewhat separate, because oxygen-poor blood is pumped out of the ventricle before the oxygen-rich blood enters. When the ventricle contracts, the division of the main artery also helps keep the blood somewhat separated. More oxygen-rich blood is distributed to the body, and more oxygen-poor blood is delivered to the lungs, and perhaps to the skin, for recharging with oxygen.

In most reptiles, a septum partially divides the ventricle. In these animals, mixing of oxygen-rich and oxygen-poor blood is kept to a minimum. In crocodilians (alligators and crocodiles), the septum completely separates the ventricle. These reptiles have a four-chambered heart. The heart of birds and mammals is also divided into left and right halves (Fig. 32.5*c*). The right ventricle pumps blood to the lungs, and the larger left ventricle pumps blood to the rest of the body. This arrangement provides adequate blood pressure for both the pulmonary and systemic circuits.

Check Your Progress 32.2

1. List and describe the functions of three types of vessels in a cardiovascular system.
2. Explain why veins are the only blood vessels that contain valves.
3. Examine the evolutionary benefits of a two-circuit circulatory pathway compared to a one-circuit pathway, especially for animals that breathe air on land.

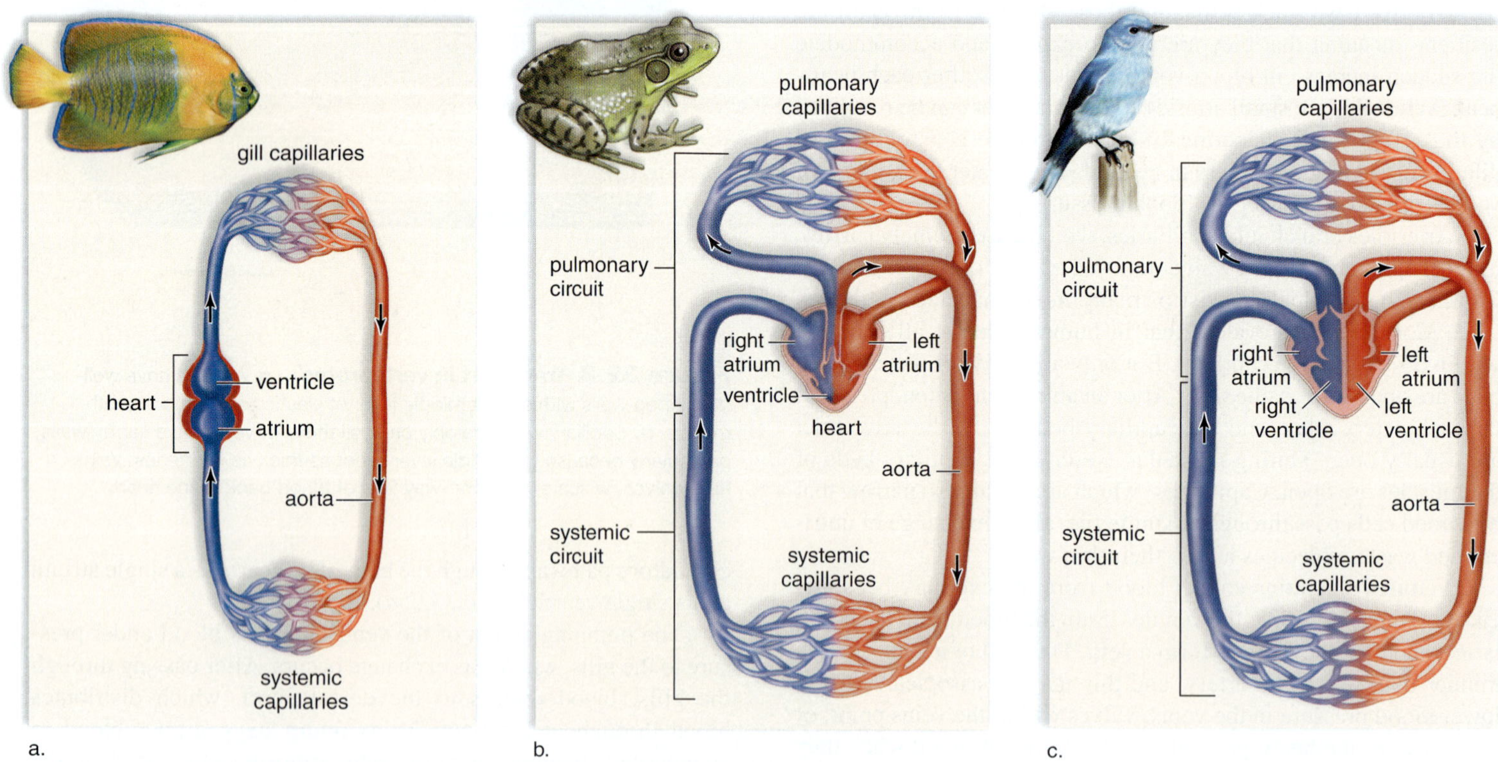

Figure 32.5 Comparison of circulatory pathways in vertebrates. **a.** In fishes, the blood moves in a single circuit. Blood pressure created by the pumping of the heart is dissipated after the blood passes through the gill capillaries. This is a disadvantage of this one-circuit system. **b.** Amphibians and most reptiles have a two-circuit system in which the heart pumps blood to both the pulmonary capillaries in the lungs and the systemic capillaries in the body itself. Although there is a single ventricle, little mixing of oxygen-rich and oxygen-poor blood takes place. **c.** The pulmonary and systemic circuits are completely separate in crocodiles (a reptile) and in birds and mammals, because the heart is divided by a septum into right and left halves. The right side pumps blood to the lungs, and the left side pumps blood to the rest of the body.

32.3 The Human Cardiovascular System

Learning Outcomes

Upon completion of this section, you should be able to

1. List the major components of the human heart, including the four chambers and four valves.
2. Trace the path of blood through the human heart, lungs, and major vessels leading to the lower leg.
3. Discuss how the SA and AV nodes control the contractions of the heart muscle, as well as how these electrical changes result in the characteristic patterns seen in an ECG.
4. Describe the major categories of cardiovascular disease that occur in the United States.

In the cardiovascular system of humans, the pumping of the heart keeps blood moving primarily in the arteries. Skeletal muscle contraction pressing against veins is the main force responsible for the movement of blood in the veins.

The Human Heart

The **heart** is a cone-shaped, muscular organ about the size of a fist (Fig. 32.6). It is located between the lungs directly behind the sternum (breastbone) and is tilted so that the apex (the pointed end) is oriented to the left.

Structure of the Heart

The major portion of the heart, called the myocardium, consists largely of cardiac muscle tissue. The myocardium receives oxygen and nutrients from the coronary arteries, not from the blood it pumps. The muscle fibers of the myocardium are branched and tightly joined to one another at intercalated disks.

The heart lies within the pericardium, a thick, membranous sac that secretes a small quantity of lubricating liquid. The inner surface of the heart is lined with endocardium, a membrane composed of connective tissue and endothelial tissue. The lining is continuous with the endothelium lining of the blood vessels.

Internally, a wall called the **septum** separates the heart into a right side and a left side (Fig. 32.7). The heart has four chambers.

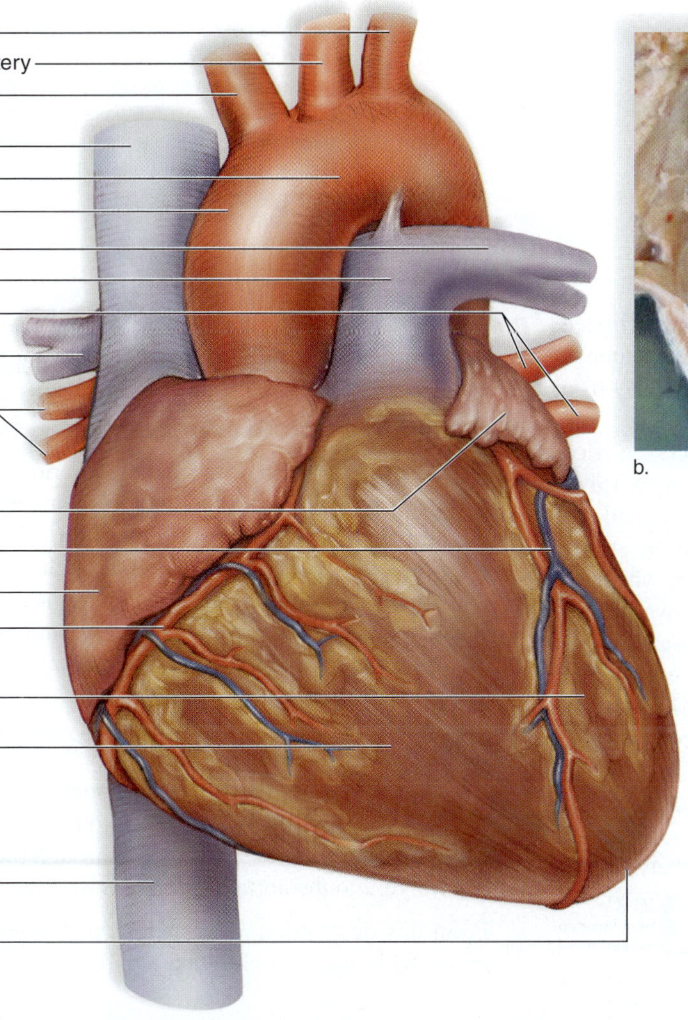

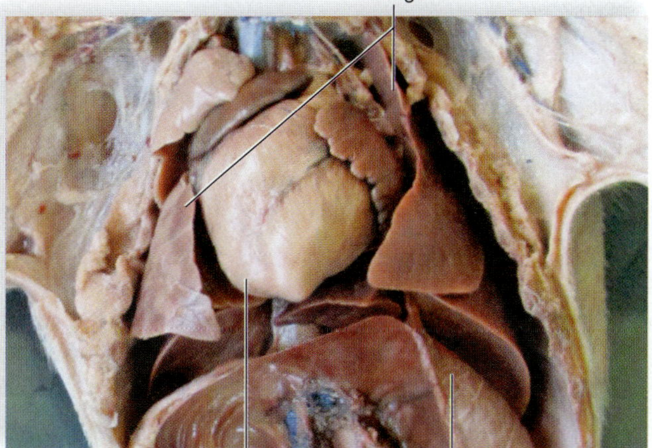

Figure 32.6 External heart anatomy. **a.** The venae cavae and the pulmonary trunk are attached to the right side of the heart. The aorta and the pulmonary veins are attached to the left side of the heart. Blood vessels are colored red if they carry oxygen-rich blood and blue if they carry oxygen-poor blood. **b.** Photograph of a mammalian heart in its natural position in the chest.

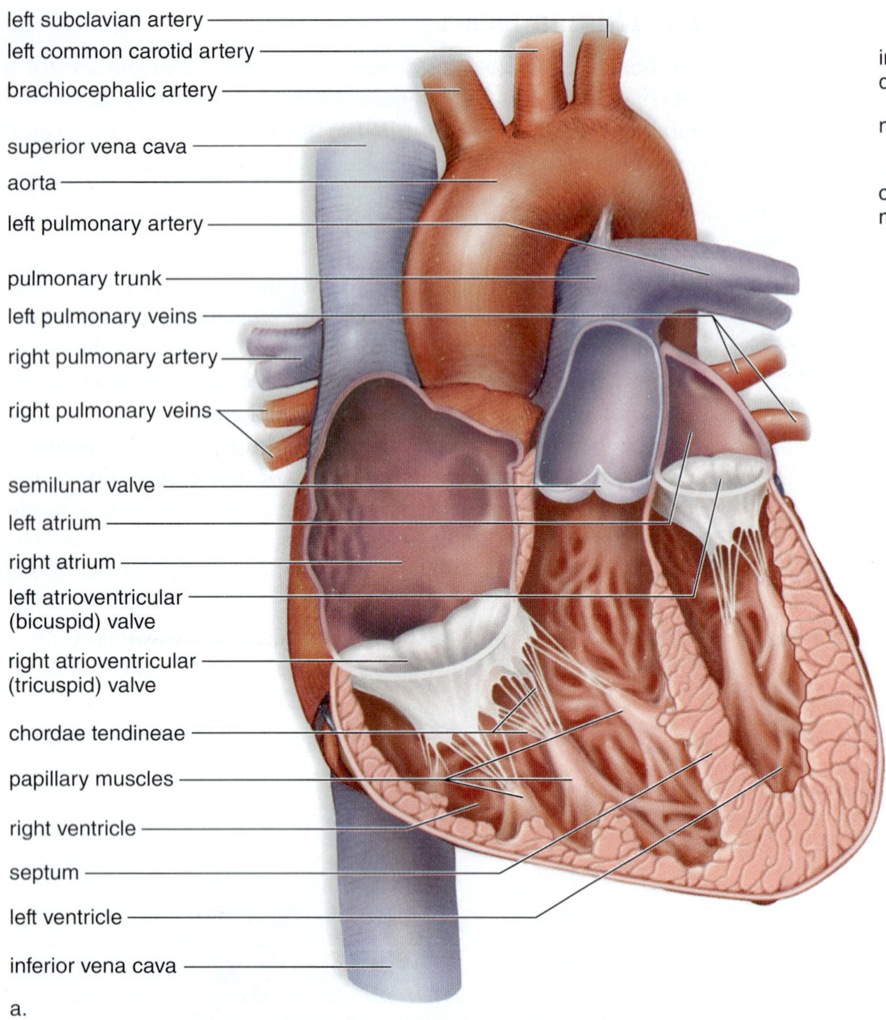

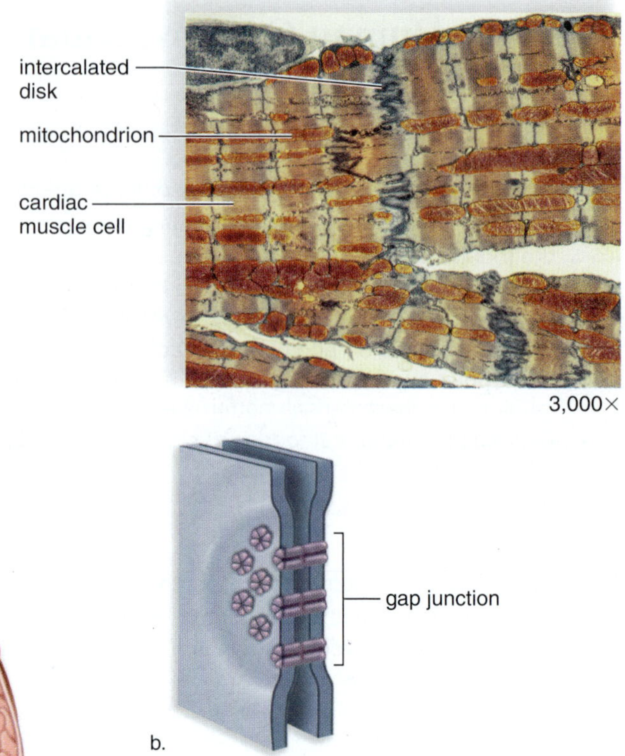

Figure 32.7 Internal view of the heart.
a. The heart has four chambers; the two chambers on the right are separated from the two chambers on the left by a septum. When the atrioventricular valves open, blood passes from the atria to the ventricles, and when the semilunar valves open, blood passes out of the heart. **b.** Intercalated disks contain gap junctions, and these allow muscle cells to contract simultaneously. Desmosomes at the same location allow the cells to bend and stretch.

The two upper, thin-walled atria (sing., **atrium**) have wrinkled, protruding appendages called auricles. The two lower chambers are the thick-walled **ventricles,** which pump the blood away from the heart.

The heart also has four valves, which direct the flow of blood and prevent its backward movement. The two valves that lie between the atria and the ventricles are called the **atrioventricular valves.** These valves are supported by strong, fibrous strings called chordae tendineae. The chordae, which are attached to muscular projections of the ventricular walls, support the valves and prevent them from inverting when the heart contracts. The atrioventricular valve on the right side is called the tricuspid valve, because it has three flaps, or cusps. The valve on the left side is called the bicuspid (or the mitral), because it has two flaps.

The remaining two valves are the **semilunar valves,** whose flaps resemble half-moons, between the ventricles and their attached vessels. The pulmonary semilunar valve lies between the right ventricle and the pulmonary trunk. The aortic semilunar valve lies between the left ventricle and the aorta.

MP3 Heart Structure and Function

Path of Blood Through the Heart

Even though both atria and then both ventricles contract simultaneously due to the presence of intercalated disks (Fig. 32.7*b*), we can trace the path of blood through the heart in the following manner:

- The superior vena cava and the inferior vena cava, which carry oxygen-poor blood that is relatively high in carbon dioxide, empty into the right atrium.
- The right atrium sends blood through an atrioventricular valve (the tricuspid valve) to the right ventricle.
- The right ventricle sends blood through the pulmonary semilunar valve into the pulmonary trunk and the two pulmonary arteries to the lungs.
- Four pulmonary veins, which carry oxygen-rich blood, empty into the left atrium.
- The left atrium sends blood through an atrioventricular valve (the bicuspid, or mitral, valve) to the left ventricle.
- The left ventricle sends blood through the aortic semilunar valve into the aorta and to the rest of the body.

From this description, it is obvious that oxygen-poor blood never mixes with oxygen-rich blood and that blood must go through the lungs in order to pass from the right side to the left side of the heart, as is typical in a double-loop circulatory system. Because the left ventricle has the harder job of pumping blood to the entire body, its walls are thicker than those of the right ventricle, which pumps blood a relatively short distance to the lungs.

People often associate oxygen-rich blood with all arteries and oxygen-poor blood with all veins, but this idea is incorrect: Pulmonary arteries and pulmonary veins are just the reverse. That is why pulmonary arteries are colored blue and pulmonary veins are colored red in Figures 32.6 and 32.7.

The pumping of the heart sends blood out under pressure into the arteries. Because the left side of the heart is the stronger pump, blood pressure is greatest in the aorta. Blood pressure then decreases as the cross-sectional area of arteries and then arterioles increases. Therefore, a different mechanism is needed to move blood in the veins, as we will discuss later.

The Heartbeat

The average human heart contracts, or beats, about 70 times a minute, so each heartbeat lasts about 0.85 second. This adds up to about 100,000 beats per day. Over a 70-year lifespan, the average human heart will have contracted about 2.5 billion times! The term **systole** (Gk. *systole,* "contraction") refers to contraction of the heart chambers, and the word **diastole** (Gk. *diastole,* "dilation, spreading") refers to relaxation of these chambers. Each heartbeat, or **cardiac cycle,** consists of the following phases, which are also depicted in Figure 32.8.

Cardiac Cycle		
Time	**Atria**	**Ventricles**
0.15 sec	Systole	Diastole
0.30 sec	Diastole	Systole
0.40 sec	Diastole	Diastole

First the atria contract (while the ventricles relax), then the ventricles contract (while the atria relax), and then all chambers rest. Note that the heart is in diastole about 50% of the time. The short systole of the atria is appropriate because the atria send blood only into the ventricles. It is the muscular ventricles that actually pump blood out into the cardiovascular system proper.

The volume of blood that the left ventricle pumps per minute into the systemic circuit is called the **cardiac output.** A person with a heartbeat of 70 beats per minute has a cardiac output of 5.25 liters a minute. This is almost equivalent to the amount of blood in the body, and it adds up to about 2,000 gallons a day. During heavy exercise, the cardiac output can increase manyfold.

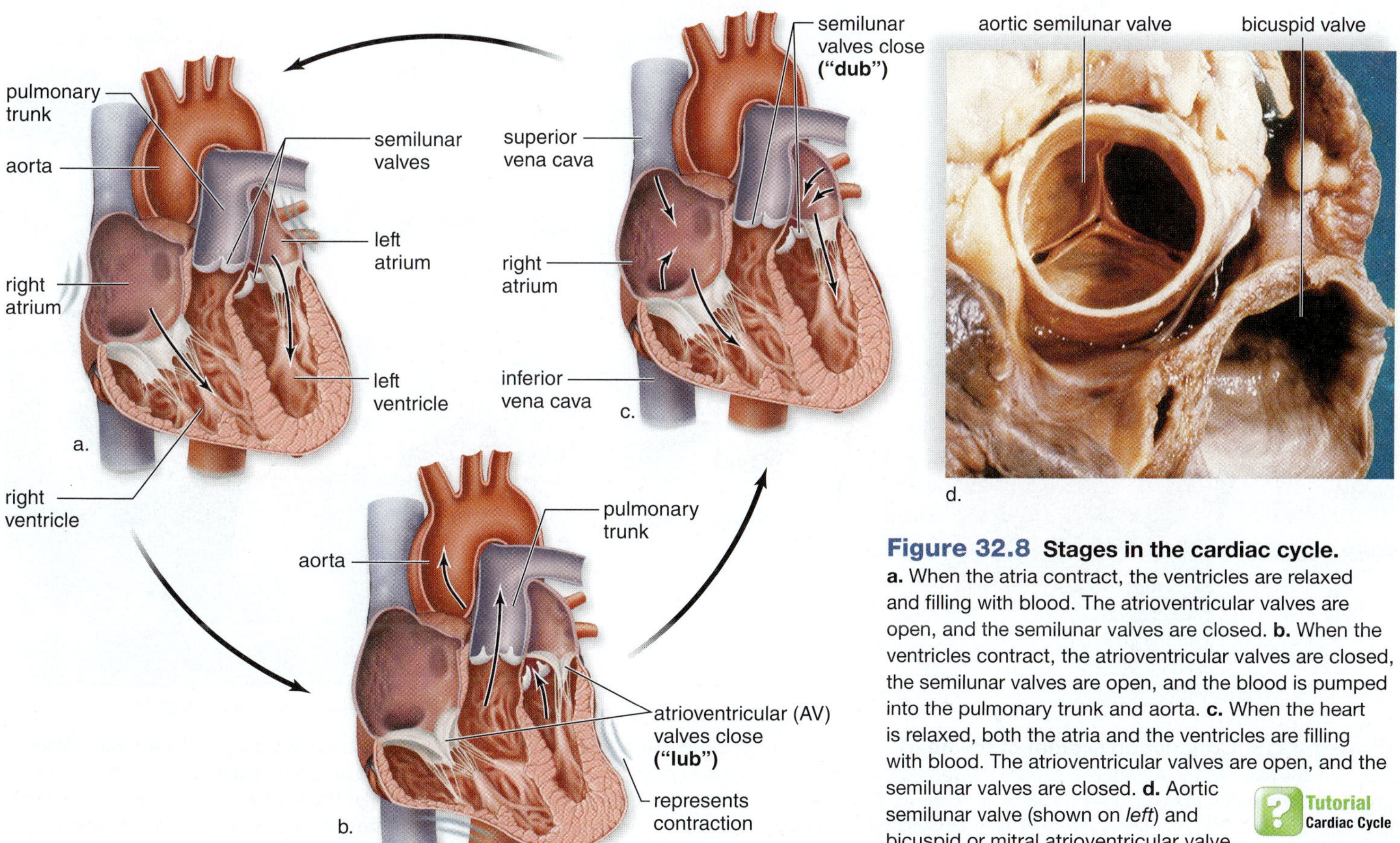

Figure 32.8 Stages in the cardiac cycle.
a. When the atria contract, the ventricles are relaxed and filling with blood. The atrioventricular valves are open, and the semilunar valves are closed. **b.** When the ventricles contract, the atrioventricular valves are closed, the semilunar valves are open, and the blood is pumped into the pulmonary trunk and aorta. **c.** When the heart is relaxed, both the atria and the ventricles are filling with blood. The atrioventricular valves are open, and the semilunar valves are closed. **d.** Aortic semilunar valve (shown on *left*) and bicuspid or mitral atrioventricular valve (shown on *right*).

Tutorial Cardiac Cycle

When the heart beats, the familiar lub-dub sound is heard as the valves of the heart close. The longer and lower-pitched *lub* is caused by vibrations of the heart when the atrioventricular valves close due to ventricular contraction. The shorter and sharper *dub* is heard when the semilunar valves close due to back pressure of blood in the arteries. A heart murmur, a slight slush sound after the lub, is often due to ineffective valves, which allow blood to pass back into the atria after the atrioventricular valves have closed.

The **pulse** is a wave effect that passes down the walls of the arterial blood vessels when the aorta expands and then recoils following ventricular systole. Because there is one arterial pulse per ventricular systole, the arterial pulse rate can be used to determine the heart rate.

The rhythmic contraction of the atria and ventricles is due to the internal (intrinsic) conduction system of the heart. Nodal tissue, which has both muscular and nervous characteristics, is a unique type of cardiac muscle located in two regions of the heart. The *SA* (*sinoatrial*) *node* is found in the upper dorsal wall of the right atrium; the *AV* (*atrioventricular*) *node* is found in the base of the right atrium very near the septum (Fig. 32.9*a*). The SA node initiates the heartbeat about every 0.85 second by automatically sending out an excitation impulse, which causes the atria to contract. Therefore, the SA node is called the **pacemaker,** because it usually keeps the heartbeat regular. When the impulse reaches the AV node, the AV node signals the ventricles to contract by way of large fibers terminating in the more numerous and smaller Purkinje fibers.

Although the heart muscle will contract without any external nervous stimulation, input from the brain can increase or decrease the rate and strength of heart contractions. In addition, the hormones epinephrine and norepinephrine, secreted into the blood by the adrenal glands, also stimulate the heart. When a person is frightened, for example, the heart pumps faster and stronger due to both nervous and hormonal stimulation.

The Electrocardiogram. An **electrocardiogram (ECG)** is a recording of the electrical changes that occur in the myocardium during a cardiac cycle. Body fluids contain ions that conduct electrical currents, and therefore these electrical changes can be detected on the body surface. During an ECG procedure, these changes pass from electrodes placed on the skin through wires to an instrument, generating "waves" that can be traced onto paper. Figure 32.9*b* depicts the pattern that results from a normal cardiac cycle.

MP3 Cardiac Cycle

Animation Cardiac Cycle

When the SA node triggers an impulse, the atrial fibers produce an electrical change called the P wave. The P wave indicates that the atria are about to contract. After that, the QRS complex signals that the ventricles are about to contract and the atria are relaxing. The electrical changes that occur as the ventricular muscle fibers recover produce the T wave.

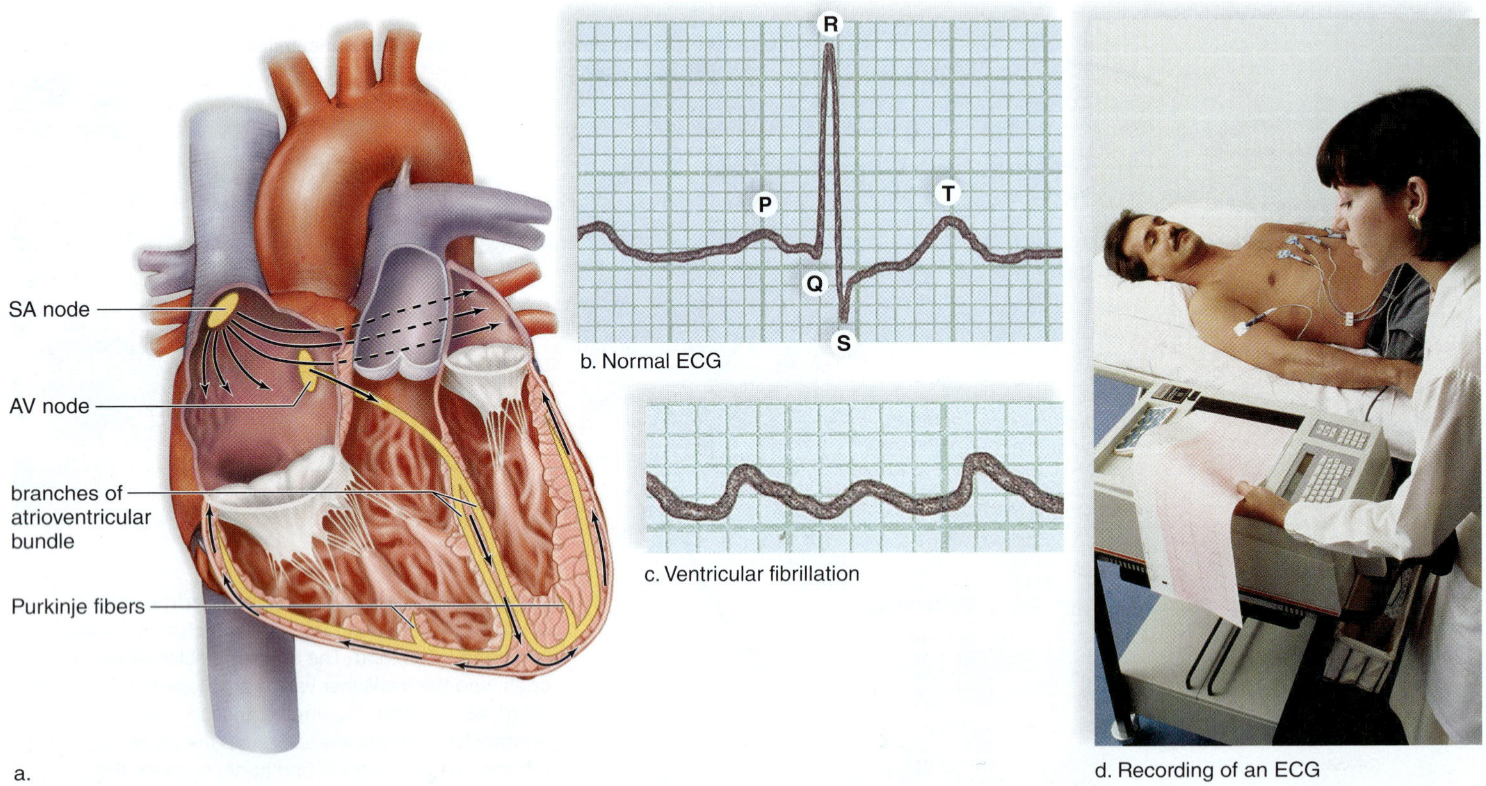

Figure 32.9 Conduction system of the heart. **a.** The SA node sends out a stimulus (black arrows), which causes the atria to contract. When this stimulus reaches the AV node, it signals the ventricles to contract. Impulses pass down the two branches of the atrioventricular bundle to the Purkinje fibers, and thereafter the ventricles contract. **b.** A normal ECG usually indicates that the heart is functioning properly. The P wave occurs just prior to atrial contraction; the QRS complex occurs just prior to ventricular contraction; and the T wave occurs when the ventricles are recovering from contraction. **c.** Ventricular fibrillation produces an irregular electrocardiogram due to irregular stimulation of the ventricles. **d.** The recording of an ECG.

Various types of abnormalities can be detected by an electrocardiogram. One of these, called ventricular fibrillation, is caused by uncoordinated contraction of the ventricles (Fig. 32.9*c*). Ventricular fibrillation is of special interest, because it can be caused by an injury or a drug overdose. It is the most common cause of sudden cardiac death in a seemingly healthy person. When the ventricles are fibrillating, they can be defibrillated by applying a strong electrical current for a short period of time. Then, the SA node may be able to reestablish a coordinated beat. Many public places, and even private homes, have automatic external defibrillators (AEDs). These are small devices that can be used to determine whether a person is suffering from ventricular fibrillation. If so, the AED administers an appropriate electrical shock to the chest.

Animation Conducting System of the Heart

Comparison of Circulatory Circuits

As mentioned, the human cardiovascular system includes two major circular pathways, the pulmonary circuit and the systemic circuit (Fig. 32.10).

The Pulmonary Circuit

In the pulmonary circuit, the path of blood can be traced as follows: Oxygen-poor blood from all regions of the body collects in the right atrium and then passes into the right ventricle, which pumps it into the pulmonary trunk. The pulmonary trunk divides into the right and left pulmonary arteries, which carry blood to the lungs. As blood passes through pulmonary capillaries, carbon dioxide is given off and oxygen is picked up. Oxygen-rich blood returns to the left atrium of the heart, through pulmonary venules that join to form pulmonary veins.

The Systemic Circuit

The **aorta** and the **venae cavae** (sing., vena cava) are the major blood vessels in the systemic circuit. To trace the path of blood to any organ in the body, you need only start with the left ventricle, mention the aorta, the proper branch of the aorta, the organ, and the vein returning blood to the vena cava, which enters the right atrium. In the systemic circuit, arteries contain oxygen-rich blood and have a bright red color, but veins contain oxygen-poor blood and appear dull red or, when viewed through the skin, blue.

The coronary arteries are extremely important because they supply oxygen and nutrients to the heart muscle itself (see Fig. 32.6). The coronary arteries arise from the aorta just above the aortic semilunar valve. They lie on the exterior surface of the heart, where they branch into arterioles and then capillaries. In the capillary beds, nutrients, wastes, and gases are exchanged between the blood and the tissues. The capillary beds enter venules, which join to form the cardiac veins, and these empty into the right atrium.

A **portal system** (L. *porto,* "carry, transport") is a structure in which blood from capillaries travels through veins to reach another set of capillaries, without first traveling through the heart. The hepatic portal system takes blood from the intestines directly to the liver. The liver then performs such functions as metabolizing nutrients and removing toxins (liver functions are explored further in Chapter 34). Blood leaves the liver by way of the hepatic vein, which enters the inferior vena cava.

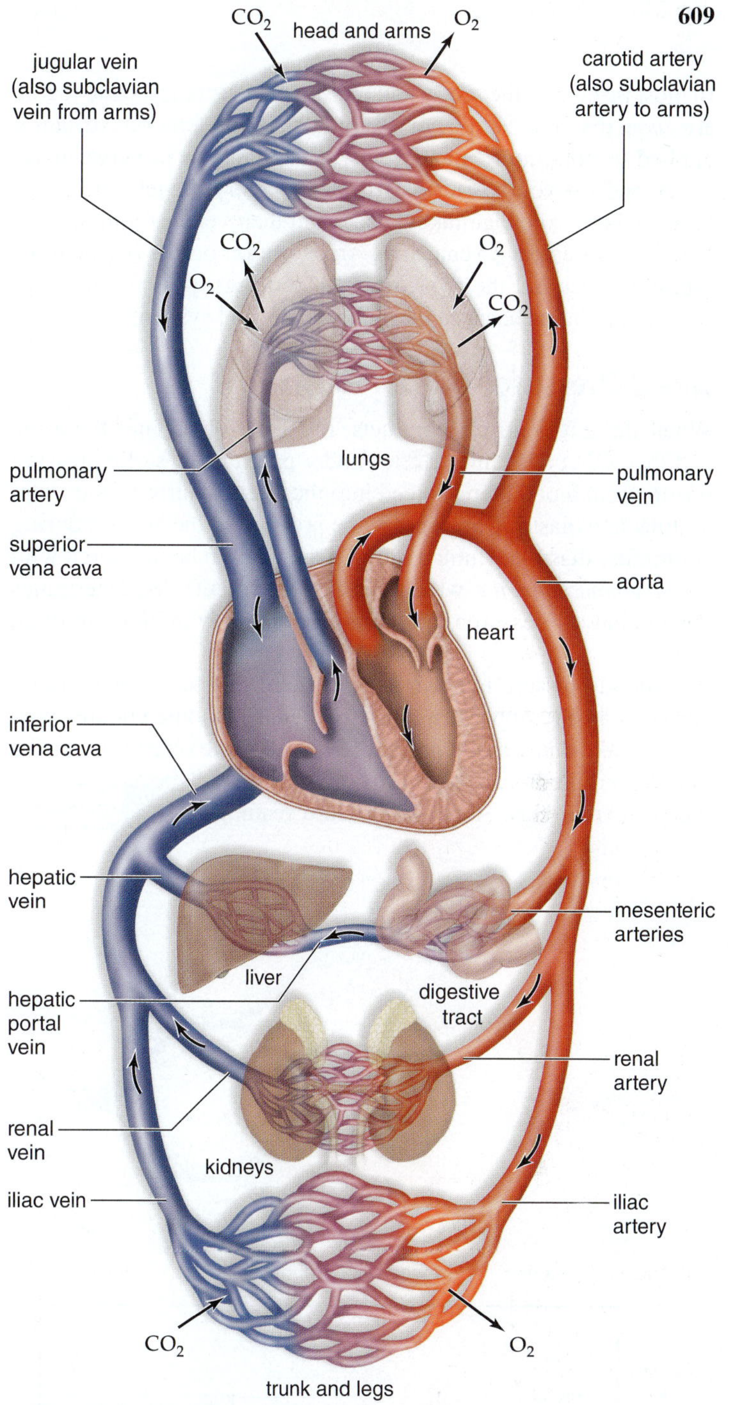

Figure 32.10 Path of blood. When tracing blood from the right to the left side of the heart in the pulmonary circuit, you must mention the pulmonary vessels. When tracing blood from the digestive tract to the right atrium in the systemic circuit, you must mention the hepatic portal vein, the hepatic vein, and the inferior vena cava. The blue-colored vessels carry oxygen-poor blood, and the red-colored vessels carry oxygen-rich blood; the arrows indicate the flow of blood.

Tracing the Path of Blood. Branches from the aorta go to the organs and major body regions. For example, this is the path of blood to and from the lower legs:

> left ventricle—aorta—common iliac artery—femoral artery—lower leg capillaries—femoral vein—common iliac vein—inferior vena cava—right atrium

In most instances, the artery and the vein that serve the same region are given the same name. For example, iliac and femoral are names applied to both arteries and veins. What happens in between the artery and the vein? Arterioles from the artery branch into capillaries, where exchange takes place, and then venules join to form the vein that enters a vena cava. An exception occurs between the digestive tract and the liver, where blood must pass through two sets of capillaries because of the hepatic portal system.

Blood Pressure

When the left ventricle contracts, blood is forced into the aorta and then other systemic arteries under pressure. Systolic pressure results from blood being forced into the arteries during ventricular systole, and diastolic pressure is the pressure in the arteries during ventricular diastole. Human **blood pressure** can be measured with a *sphygmomanometer,* which has a pressure cuff that determines the amount of pressure required to stop the flow of blood through an artery.

Blood pressure is normally measured on the brachial artery, an artery in the upper arm. Today, digital manometers are often used to take one's blood pressure instead of the older type with a dial. Blood pressure is given in millimeters of mercury (mm Hg). A blood pressure reading consists of two numbers—for example, 120/80—that represent systolic and diastolic pressures, respectively.

MP3 Blood Flow and Blood Pressure

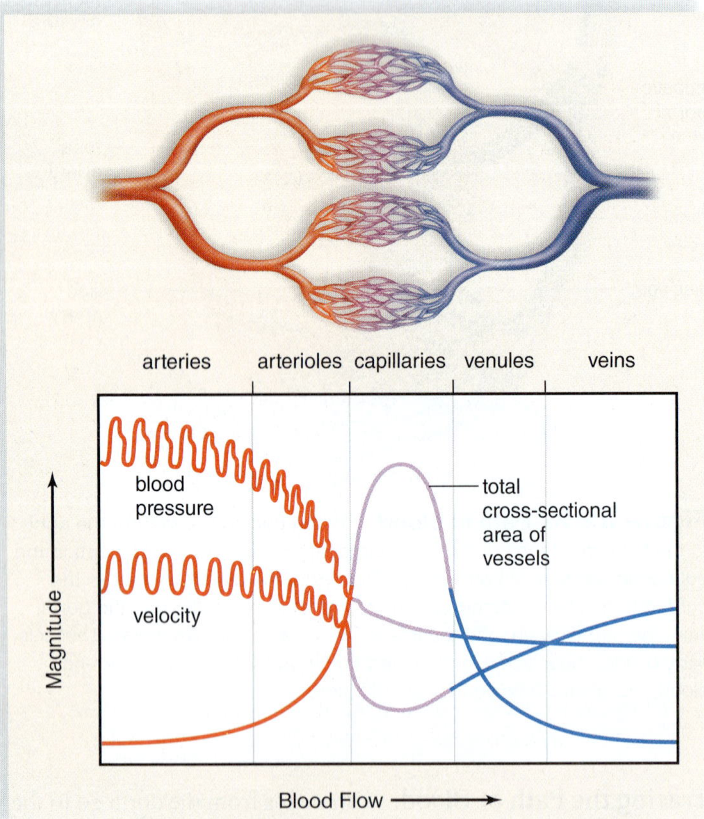

Figure 32.11 Velocity and blood pressure related to vascular cross-sectional area. In capillaries, blood is under minimal pressure and has the least velocity. Blood pressure and velocity drop off, because capillaries have a greater total cross-sectional area than arterioles.

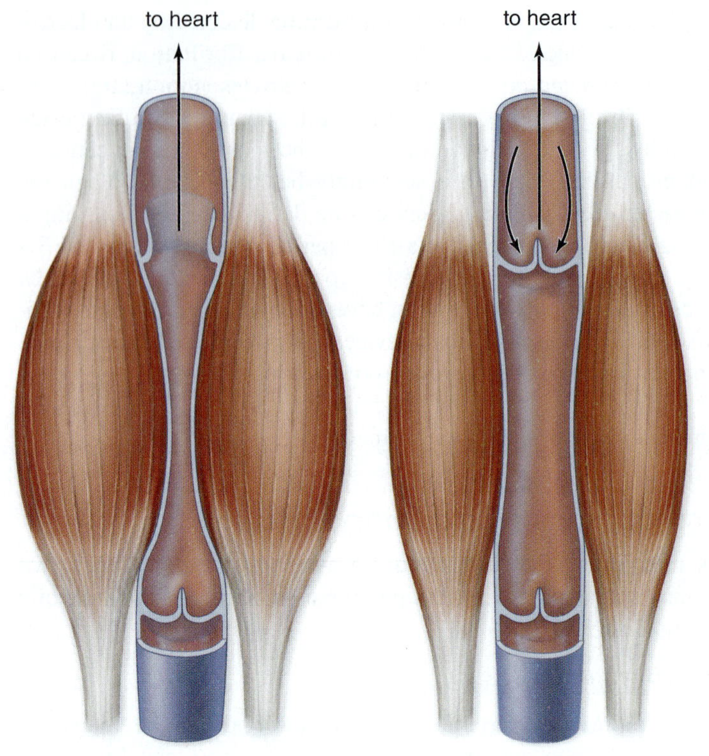

Figure 32.12 Cross section of a valve in a vein. **a.** Pressure on the walls of a vein, exerted by skeletal muscles, increases blood pressure within the vein and forces a valve open. **b.** When external pressure is no longer applied to the vein, blood pressure decreases, and back pressure forces the valve closed. Closure of the valves prevents the blood from flowing in the opposite direction.

As blood flows from the aorta into the various arteries and arterioles, blood pressure falls. Also, the difference between systolic and diastolic pressure gradually diminishes. In the capillaries, there is a slow, fairly even flow of blood. This may be related to the very high total cross-sectional area of the capillaries (Fig. 32.11). It has been calculated that if all the blood vessels in a human body were connected end to end, the total distance would reach around the Earth at the equator two times! Most of this distance would be due to the large number of capillaries.

Blood pressure in the veins is low and is insufficient for moving blood back to the heart, especially from the limbs of the body. Venous return is dependent on three factors:

- Skeletal muscles near veins put pressure on the collapsible walls of the veins, and therefore on the blood contained in these vessels, when they contract.
- Valves in the veins prevent the backward flow of blood, and therefore pressure from muscle contraction moves blood toward the heart (Fig. 32.12). Varicose veins, abnormal dilations in superficial veins, develop when the valves of the veins become weak and ineffective due to a backward pressure of the blood.

- Variations in pressure in the chest cavity during breathing, also known as the *respiratory pump,* cause blood to flow from areas of higher pressure (such as the abdominal cavity) to lower pressure (in the thoracic cavity) during each inhalation.

Cardiovascular Disease

Cardiovascular disease (CVD) is the leading cause of death in most Western countries. According to the American Heart Association, CVD has been the most common cause of death in the United States every year since 1900. The only exception to this statistic was 1918, the worst year of a global influenza pandemic. According to the American Heart Association's Heart Disease and Stroke Statistical Update 2014, about 2,150 Americans die of CVD each day, which is an average of one death every 40 seconds, and about one out of every three deaths overall. The Nature of Science feature, "Recent Findings About Preventing Cardiovascular Disease," on page 612, emphasizes the possible prevention of CVD.

Hypertension

It is estimated that about 30% of Americans suffer from **hypertension,** which is high blood pressure. Another 30% are thought to have a condition called prehypertension, which can lead to hypertension. Under age 45, a reading above 130/90 is hypertensive, and beyond age 45, a reading above 140/95 is hypertensive.

Hypertension is most often caused by a narrowing of arteries due to atherosclerosis (described next). This narrowing causes the heart to work harder to supply the required amount of blood. The resulting increase in blood pressure can damage the heart, arteries, and other organs. Other risk factors that can contribute to hypertension include obesity, smoking, chronic stress, and a high dietary salt intake (which causes retention of fluid). Only about two-thirds of people with hypertension seek medical help for their condition, and it is likely that many people with high blood pressure are unaware of it.

Atherosclerosis

Atherosclerosis is an accumulation of soft masses of fatty materials, particularly cholesterol, beneath the inner linings of arteries (see Fig. 32A). Such deposits are called plaque. As deposits occur, plaque tends to protrude into the lumen of the vessel, interfering with the flow of blood. Plaque can also cause a clot to form on the irregular arterial wall. As long as the clot remains stationary, it is called a *thrombus,* but when and if it dislodges and moves along with the blood, it is called an *embolus*. If thromboembolism is not treated, complications can arise (see the following section).

Atherosclerosis often begins in early adulthood and develops progressively through middle age, but symptoms may not appear until an individual is 50 or older. See the Nature of Science feature, "Recent Findings About Preventing Cardiovascular Disease," on page 612 for some recent recommendations on reducing risk.

Stroke and Heart Attack

Strokes and heart attacks are associated with hypertension and atherosclerosis. A **stroke,** or disruption of blood supply to the brain, often results when a small cranial arteriole bursts or is blocked by an embolus. A lack of oxygen causes a portion of the brain to die, and paralysis or death can result. A person is sometimes forewarned of a stroke by a feeling of numbness in the hands or the face, difficulty speaking, or temporary blindness in one eye.

If a coronary artery becomes partially blocked, the individual may suffer from **angina pectoris,** characterized as a squeezing sensation or a flash of burning. If a coronary artery is completely blocked, perhaps by a thromboembolism, a portion of the heart muscle dies due to a lack of oxygen. This is a myocardial infarction, also called a **heart attack.** It may be necessary to place a stent, or self-expanding wire mesh tube, inside a blocked artery to keep it open. About 700,000 of these stents are placed in U.S. patients every year (Fig. 32.13). If this approach is unsuccessful, a coronary bypass may be required, in which a surgeon replaces one or more blocked coronary arteries with an artery taken from elsewhere in the patient's body. Each year, more than 500,000 of these procedures are performed in the United States.

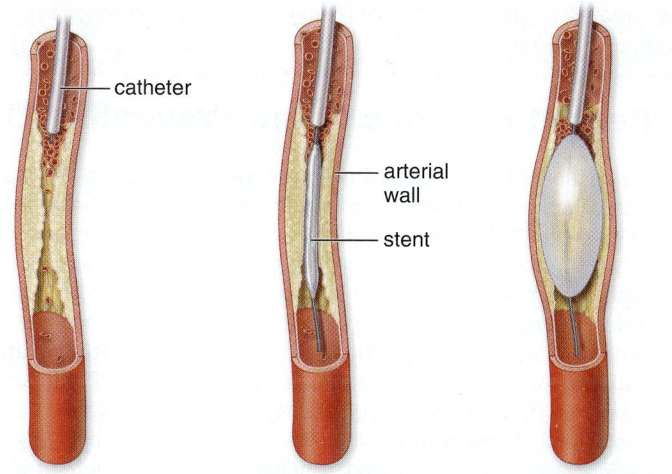

Figure 32.13 Angioplasty with stent placement. **a.** A plastic tube (catheter) is inserted into the coronary artery until it reaches the clogged area. **b.** A metal stent with a balloon inside it is pushed out the end of the plastic tube into the clogged area. **c.** When the balloon is inflated, the vessel opens, and the stent is left in place to keep the vessel open.

Check Your Progress 32.3

1. Name each blood vessel and heart chamber that blood passes through on its way from the venae cavae to the aorta, and identify which artery carries oxygen-poor blood.
2. Explain what specifically causes the sounds of the heartbeat.
3. Discuss why systolic blood pressure is higher than diastolic.
4. Predict what type of conditions might occur as a result of chronic hypertension and plaque.

THEME Nature of Science

Recent Findings About Preventing Cardiovascular Disease

For decades, several factors have been associated with an increased risk of cardiovascular disease (CVD), especially atherosclerosis (Fig. 32A). Some of these cannot be avoided, such as increasing age, male gender, family history of heart disease, and belonging to certain races, including African American, Mexican American, and American Indian. Other risk factors—smoking, obesity, high cholesterol, hypertension, diabetes, physical inactivity—can be avoided or at least affected by changing one's behavior or taking medications. In recent years, however, other factors have been under consideration, such as the following.

- **Alcohol.** Alcohol abuse can destroy just about every organ in the body, the heart included. But recent research suggests that a moderate level of alcohol intake can improve cardiovascular health by improving the blood cholesterol profile, decreasing unwanted clot formation, increasing blood flow in the heart, and reducing blood pressure. According to the American Heart Association, people who consume one or two drinks per day have a 30–50% reduction in cardiovascular disease compared to nondrinkers. However, the maximum protective effect is achieved with only one or two drinks per day—consuming more than that increases the risk of many alcohol-related problems. The American Heart Assocation does *not* recommend that nondrinkers start using alcohol, or that drinkers increase their consumption, based on these findings.
- **Resveratrol.** The "red wine effect," or "French paradox," refers to the observation that levels of CVD in France are relatively low, despite the consumption of a high-fat diet. One possible explanation is that wine is frequently consumed with meals. In addition to its alcohol content, red wine contains especially high levels of antioxidants, including resveratrol. Resveratrol is mainly produced in the skin of grapes, so it is also found in grape juice. Resveratrol supplements are also available at health food stores. The benefits of resveratrol alone are questionable, however, and most controlled studies to date have demonstrated no beneficial effects. The lower incidence of CVD in the French may be due to multiple factors, including lifestyle and genetic differences.
- **Omega-3 fatty acids.** The influence that diet has on blood cholesterol levels has been well studied. It is generally beneficial to minimize our intake of foods high in saturated fat (red meat, cream, and butter) and trans fats (most margarines, commercially baked goods, and deep-fried foods). Replacing these harmful fats with healthier ones, such as monounsaturated fats (olive and canola oils) and polyunsaturated fats (corn, safflower, and soybean oils), is beneficial. In addition, the American Heart Association now recommends eating at least two servings of fish a week, especially salmon, mackerel, herring, lake trout, sardines, and albacore tuna, which are high in omega-3 fatty acids. These essential fatty acids can decrease triglyceride levels, slow the growth rate of atherosclerotic plaque, and lower blood pressure. However, children and pregnant women are advised to limit their fish consumption because of the high levels of mercury contamination in some fishes. For middle-aged and older men and postmenopausal women, the benefits of fish consumption far outweigh the potential risks.

Questions to Consider

1. Would you be helping your health if you decided to eat mackerel every day and drink two glasses of red wine with it? Why or why not?
2. What would be some difficulties in trying to determine the true cause of the "French paradox"?

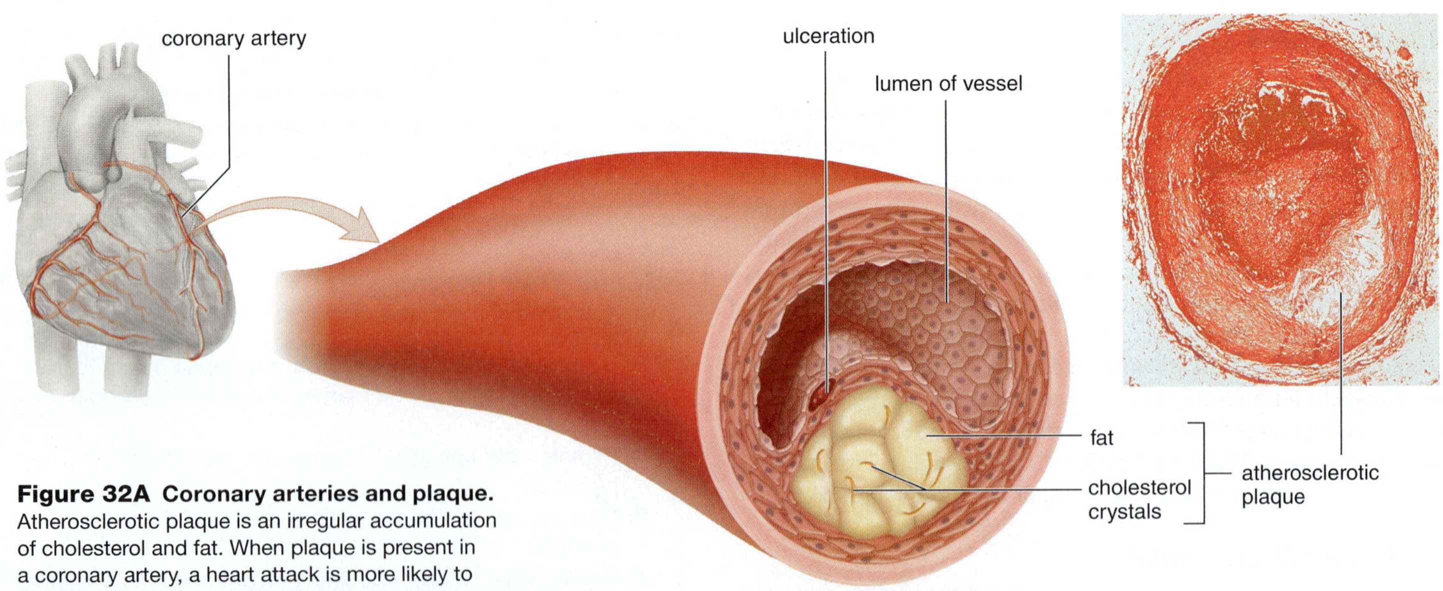

Figure 32A Coronary arteries and plaque. Atherosclerotic plaque is an irregular accumulation of cholesterol and fat. When plaque is present in a coronary artery, a heart attack is more likely to occur because of restricted blood flow.

32.4 Blood

Learning Outcomes

Upon completion of this section, you should be able to

1. List the major types of blood cells and their functions.
2. Identify the major cellular and molecular events that result in a blood clot.
3. Compare and contrast the ABO and Rh blood classification systems.
4. Define *capillary exchange* and describe the two major forces involved.

As discussed in Chapter 31, blood is considered to be a connective tissue with a fluid matrix. In contrast to the hemolymph found in open circulatory systems, blood is normally contained within blood vessels. The blood of mammals has a number of functions that help maintain homeostasis:

- Transporting gases, nutrients, waste products, and hormones throughout the body
- Combating pathogenic microorganisms
- Helping maintain water balance and pH
- Regulating body temperature
- Carrying platelets and factors that ensure clotting to prevent blood loss

Blood has two main portions: a liquid portion, called plasma, and the formed elements, consisting of cells and platelets (Fig. 32.14).

Plasma

Plasma contains many types of molecules, including nutrients, wastes, salts, and hundreds of different types of proteins. Some of these proteins are involved in buffering the blood, effectively keeping the pH near 7.4. They also maintain the blood's osmotic pressure, so that water has an automatic tendency to enter blood capillaries. Several plasma proteins are involved in blood clotting, and others transport large organic molecules in the blood.

Albumin, the most plentiful of the plasma proteins, transports bilirubin, a breakdown product of hemoglobin, and various types of lipoproteins transport cholesterol. Another very significant group of plasma proteins are the **antibodies,** which are proteins produced by the immune system in response to specific pathogens and other foreign materials (see Chapter 33).

Formed Elements

The **formed elements** are of three types: red blood cells, or erythrocytes (Gk. *erythros,* "red"; *kytos,* "cell"); white blood cells, or leukocytes (Gk. *leukos,* "white"); and platelets, or thrombocytes (Gk. *thrombos,* "blood clot").

Red Blood Cells

Red blood cells (RBCs) are small, biconcave disks that at maturity lack a nucleus and contain the respiratory pigment hemoglobin. The average adult human has 5 to 6 million RBCs per cubic millimeter (mm^3) of whole blood, and each one of these cells contains about 250 million hemoglobin molecules. **Hemoglobin** (Gk. *haima,* "blood"; L. *globus,* "ball") contains four globin protein

Plasma	
Type	**Function**
Water (90–92% of plasma)	Maintains blood volume; transports molecules
Plasma proteins (7–8% of plasma)	Maintain blood osmotic pressure and pH
Globulins	Transport; fight infection
Fibrinogen	Blood clotting
Salts (less than 1% of plasma)	Maintain blood osmotic pressure and pH; aid metabolism
Gases (O_2 and CO_2)	Cellular respiration
Nutrients (lipids, glucose, and amino acids)	Food for cells
Wastes (urea and uric acid)	End product of metabolism; excretion by kidneys
Hormones	Aid metabolism

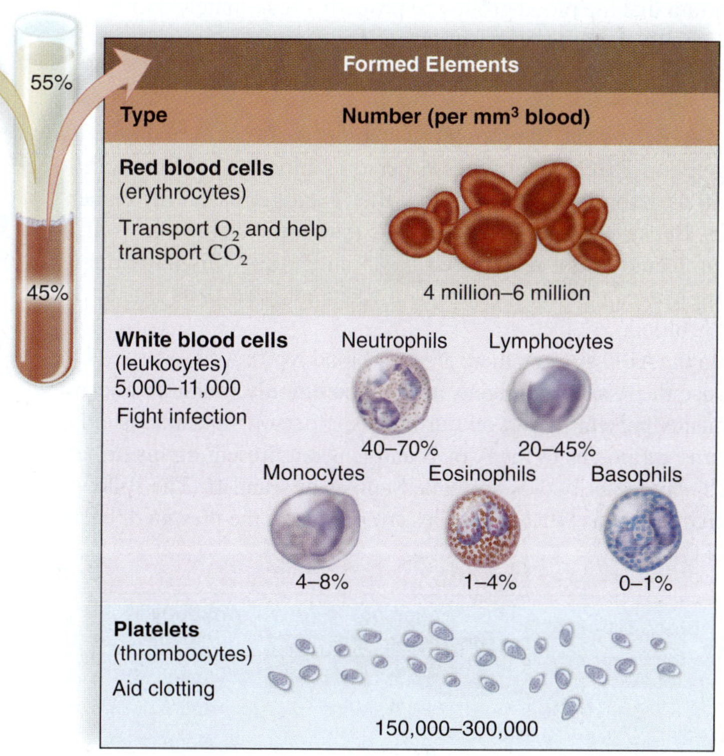

Figure 32.14 **Composition of blood.**

chains, each associated with heme, an iron-containing group. Iron combines loosely with oxygen, and in this way oxygen is carried in the blood. If the number of RBCs is insufficient, or if the cells do not have enough hemoglobin, the individual suffers from anemia and has a tired, run-down feeling.

In adults, RBCs are manufactured in the red bone marrow of the skull, the ribs, the vertebrae, and the ends of the long bones. The hormone erythropoietin, produced by the kidneys, stimulates RBC production. Now available as a drug, erythropoietin is helpful to persons with anemia and is sometimes abused by athletes who want to enhance the oxygen-carrying capacity of their blood.

Before they are released from the bone marrow into blood, RBCs synthesize hemoglobin and lose their nuclei. After living about 120 days, they are destroyed chiefly in the liver and the spleen, where they are engulfed by large phagocytic cells. When RBCs are destroyed, hemoglobin is released. The iron is recovered and returned to the red bone marrow for reuse. The heme portions of the molecules undergo chemical degradation and are excreted by the liver as bile pigments in the bile. The bile pigments are primarily responsible for the color of feces.

Blood Types

The earliest attempts at blood transfusions resulted in illness and even the death of some recipients. Eventually, it was discovered that only certain transfusion donors and recipients are compatible, because red blood cell membranes carry specific proteins or carbohydrates that are antigens to blood recipients. An **antigen** (Gk. *anti,* "against"; L. *genitus,* "forming, causing") is a molecule, usually a protein or carbohydrate, that can trigger a specific immune response. Several groups of RBC antigens exist, the most significant being the ABO and Rh systems. Clinically, it is very important that the blood groups be properly cross-matched to avoid a potentially deadly transfusion reaction.

ABO System

In the ABO system, the presence or absence of type A and type B antigens on RBCs determines a person's blood type. For example, if a person has type A blood, the A antigen is on his or her RBCs. Because it is considered by the immune system to be "self," this molecule is not recognized as an antigen by this individual, although it can be an antigen to a recipient who does not have type A blood.

In the ABO system, there are four blood types: A, B, AB, and O. Because the A and B antigens are also commonly found on microorganisms present in and on our bodies, a person's plasma usually contains antibodies to the A or B antigens not present on his or her RBCs. These antibodies are called anti-A and anti-B. The following chart explains what antibodies are present in the plasma of each blood type:

Blood Type	Antigen on Red Blood Cells	Antibody in Plasma
A	A	Anti-B
B	B	Anti-A
AB	A, B	None
O	None	Anti-A and anti-B

Because type A blood has anti-B but not anti-A antibodies in the plasma, a donor with type A blood can give blood to a recipient with type A blood (Fig. 32.15). However, if type A blood is given to a type B recipient, **agglutination** (Fig. 32.16), the clumping of RBCs, can cause blood to stop circulating in small blood vessels, leading to organ damage.

Theoretically, a person with which blood type can donate to all recipients? The answer is that type O RBCs have no A or B antigens, and this is sometimes called the universal donor type. A person with which blood type can receive blood from any donor? Type AB blood has no anti-A or anti-B antibodies, and thus it is sometimes called the universal recipient. In practice, however, it is not safe to rely solely on the ABO system when matching blood. Instead, samples of the two types of blood are physically mixed, and the result is microscopically examined for agglutination before blood transfusions are done.

An equally important concern when transfusing blood is to make sure that the donor is free from transmissible infectious agents, such as the microbes that cause AIDS, hepatitis, and syphilis.

Rh System

Another important antigen on RBCs is the Rh factor. Eighty-five percent of the U.S. population have this particular antigen on their RBCs and are Rh-positive. Fifteen percent do not have the

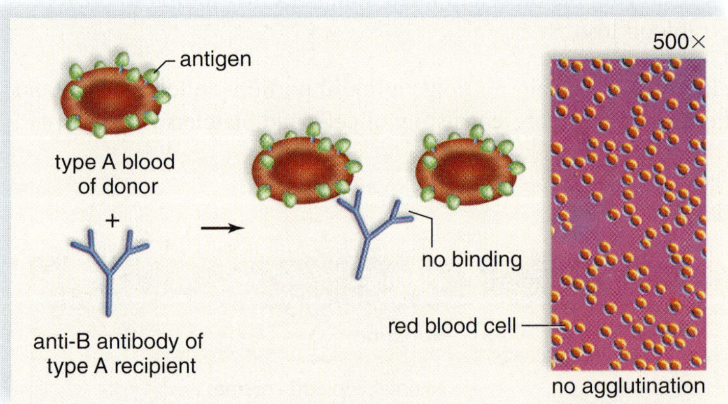

Figure 32.15 Matched blood transfusion. No agglutination occurs when the donor and recipient have the same type blood.

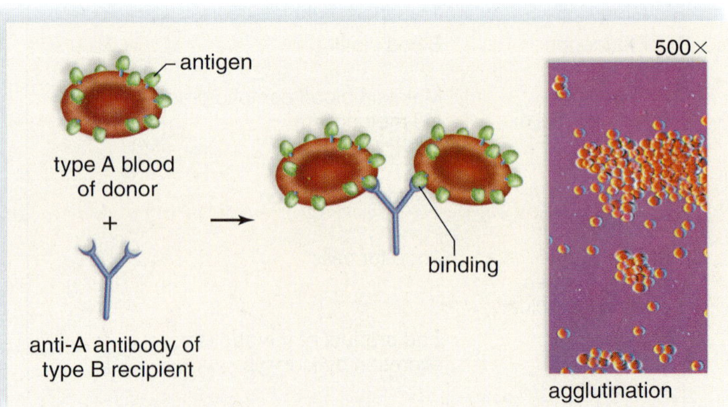

Figure 32.16 Mismatched blood transfusion. Agglutination occurs, because blood type B has anti-A antibodies in the plasma.

antigen and are Rh-negative. The designation of blood type usually includes whether the person has or does not have the Rh factor on the RBCs—for example, type A-positive (A+). Unlike the case with the A and B antigens, Rh-negative individuals normally do not have antibodies to the Rh factor, but they may make them when exposed to the Rh factor.

During pregnancy, if the mother is Rh-negative and the father is Rh-positive, the child may be Rh-positive. If the Rh-positive fetal RBCs leak across the placenta, the mother may produce anti-Rh antibodies. In this or a subsequent pregnancy with another Rh-positive baby, these antibodies may cross the placenta and destroy the child's RBCs. This condition, called hemolytic disease of the newborn (HDN), can be fatal without an immediate blood transfusion after birth.

The problem of Rh incompatibility can be prevented by giving Rh-negative women an Rh immunoglobulin injection toward the end of pregnancy and within 72 hours of giving birth to an Rh-positive child. This treatment contains a relatively low level of anti-Rh antibodies that help destroy any Rh-positive blood cells in the mother's blood before her immune system produces high levels of anti-Rh antibodies.

White Blood Cells

Because they are a critical component of the immune system, the functions of white blood cells are discussed in detail in Chapter 33 and only briefly here. **White blood cells (WBCs)**, or **leukocytes,** differ from RBCs in that they are usually larger, have a nucleus, lack hemoglobin, and without staining appear translucent. With staining, WBCs appear light blue unless they have granules that bind with certain stains (see Fig. 32.14). There are far fewer WBCs than RBCs in the blood, with approximately 5,000–11,000 WBCs per mm^3 in humans.

On the basis of their structure, WBCs can be divided into granular and agranular leukocyte*s*. Within these two categories, five main types of WBCs can be identified.

Granular Leukocytes. The cytoplasm of **granular leukocytes** (neutrophils, eosinophils, and basophils) contains spherical vesicles, or granules, filled with enzymes and proteins, which these cells use to help defend the body against invading microbes and other parasites.

Neutrophils have a multilobed nucleus, resulting in their other name, polymorphonuclear cells. They are the most abundant of the WBCs and are able to squeeze through capillary walls and enter the tissues, where they phagocytize and digest bacteria. The thick, yellowish fluid called *pus* that develops in some bacterial infections contains mainly dead neutrophils that have fought the infection. **Basophil** granules stain a deep blue and contain inflammatory chemicals, such as histamine. The prominent granules of **eosinophils** stain a deep red, and these WBCs are involved in fighting parasitic worms, among other actions.

Agranular Leukocytes. The **agranular leukocytes,** which are also called mononuclear cells, lack obvious granules and include the monocytes and the lymphocytes.

Monocytes are the largest of the WBCs, and they tend to migrate into tissues in response to chronic, ongoing infections, where they differentiate into large phagocytic **macrophages** (Gk. *makros,* "long"; *phagein,* "to eat"). These long-lived cells not only fight infections directly but also release growth factors that increase the production of different types of WBCs by the bone marrow. Some of these factors are available for medicinal use and may be helpful to people with low immunity, such as AIDS patients or people on chemotherapy for cancer. A third function of macrophages is to interact with lymphocytes to help initiate the adaptive immune response (see Chapter 33).

Lymphocytes are the second most common type of WBC in the blood. The two major types of lymphocytes, T cells and B cells, each play a distinct role in adaptive immune responses to specific antigens. One type of T cell, the helper T cell, initiates and influences most of the other cell types involved in adaptive immunity. The other type, the cytotoxic T cell, attacks infected cells that contain viruses. In contrast, the main function of B cells is to produce antibodies. Each B cell produces just one type of antibody, which is specific for one type of antigen. As mentioned earlier in this section, an antigen is a molecule that causes a specific immune response because the immune system recognizes it as "foreign." When antibodies combine with antigens, the complex is often phagocytized by a macrophage. The activities of lymphocytes, along with other aspects of animal immune systems, are discussed in more detail in Chapter 33.

Platelets and Blood Clotting

Platelets (thrombocytes) result from fragmentation of large cells, called *megakaryocytes,* in the red bone marrow. Platelets are produced at a rate of 200 billion a day, and the blood contains 150,000–300,000 per mm^3. These formed elements are involved in blood **clotting,** or coagulation (Fig 32.17).

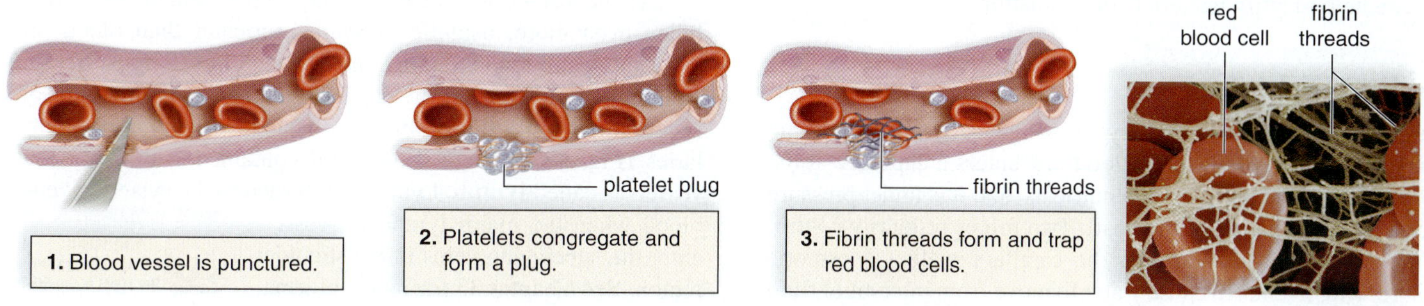

Figure 32.17 Blood clotting. A number of plasma proteins participate in a series of enzymatic reactions that lead to the formation of fibrin threads.

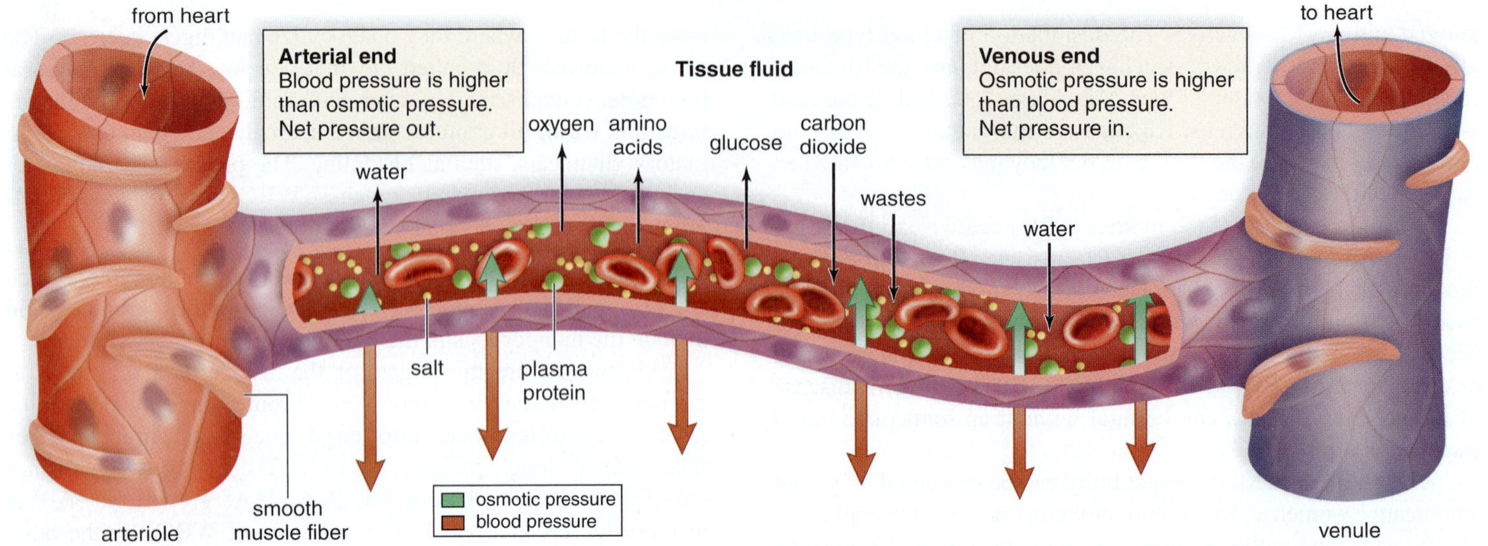

Figure 32.18 Capillary exchange. A capillary, illustrating the exchanges that take place and the forces that aid the process. At the arterial end of a capillary, the blood pressure is higher than the osmotic pressure; therefore, water (H_2O) tends to leave the bloodstream. In the midsection, molecules, including oxygen (O_2) and carbon dioxide (CO_2), follow their concentration gradients. At the venous end of a capillary, the osmotic pressure is higher than the blood pressure; therefore, water tends to enter the bloodstream. Notice that the red blood cells and the plasma proteins are too large to exit a capillary.

Tutorial
Capillary Exchange

When a blood vessel in the body is damaged, platelets clump at the site of the puncture and partially seal the leak. Platelets and the injured tissues release a clotting factor called prothrombin activator, which converts prothrombin in the plasma to thrombin. This reaction requires calcium ions (Ca^{2+}). *Thrombin,* in turn, acts as an enzyme that severs two short amino acid chains from a *fibrinogen* molecule, one of the proteins in plasma. These activated fragments then join end to end, forming long threads of *fibrin*.

Fibrin threads wind around the platelet plug in the damaged area of the blood vessel and provide the framework for the clot. Red blood cells also are trapped within the fibrin threads; these cells make a clot appear red. A fibrin clot is present only temporarily. As soon as blood vessel repair is initiated, an enzyme called *plasmin* destroys the fibrin network and restores the fluidity of plasma.

The Nature of Science feature, "How Horseshoe Crabs Save Human Lives," describes how a clotting reaction in these arthropods can help identify bacterial contamination.

Capillary Exchange

Figure 32.18 illustrates capillary exchange between a systemic capillary and interstitial fluid. Blood that enters a capillary at the arterial end is rich in oxygen and nutrients, and it is under pressure created by the pumping of the heart. Two forces primarily control the movement of fluid through the capillary wall: (1) osmotic pressure, which tends to cause water to move from interstitial fluid into blood, and (2) blood pressure, which tends to cause water to move in the opposite direction. At the arterial end of a capillary, the osmotic pressure of blood (21 mm Hg) is lower than the blood pressure (30 mm Hg). Osmotic pressure is created by the presence of salts and the plasma proteins. Because osmotic pressure is lower than blood pressure at the arterial end of a capillary, water exits a capillary at this end.

MP3
Capillary Exchange and Bulk Flow

Midway along the capillary, where blood pressure is lower, the two forces essentially cancel each other, and there is no net movement of water. Solutes now diffuse according to their concentration gradient: Oxygen and nutrients (glucose and amino acids) diffuse out of the capillary; carbon dioxide and wastes diffuse into the capillary. Red blood cells and almost all plasma proteins remain in the capillaries.

The substances that leave a capillary contribute to interstitial fluid. Because plasma proteins are too large to readily pass out of the capillary, interstitial fluid tends to contain all components of plasma but has much lower amounts of protein.

At the venule end of a capillary, where blood pressure has fallen even more, osmotic pressure is greater than blood pressure, and water tends to move into the capillary. Almost the same amount of fluid that left the capillary returns to it, although some excess interstitial fluid is always collected by the lymphatic capillaries (Fig. 32.19). interstitial fluid contained within lymphatic vessels is called **lymph.** Lymph is returned to the systemic venous blood when the major lymphatic vessels enter the subclavian veins in the shoulder region. See Chapter 33 for more information about the lymphatic system.

Animation
Fluid Exchange Across the Walls of Capillaries

THEME Nature of Science

How Horseshoe Crabs Save Human Lives

Take a walk along a beach on the northeastern U.S. coast and you are likely to see horseshoe crabs (*Limulus polyphemus*) (Fig. 32B). Although not truly "crabs," they are classified in the phylum Arthropoda, along with insects, arachnids, and crustaceans. Because of their prehistoric appearance, horseshoe crabs are sometimes called "living fossils," and, in fact, they were living on Earth before the dinosaurs. They have an open circulatory system, with an elongated heart that pumps hemolymph between the gills and the body, without returning to the heart in between.

Instead of hemoglobin, the hemolymph of horseshoe crabs contains hemocyanin, which binds to oxygen using copper instead of iron, giving the blood a light blue color. The blood also contains amebocytes, cells that are analogous to the neutrophils or macrophages of higher animals, which serve a similar role in protecting against bacterial infections. Oddly enough, these amebocytes turned out to be the key to developing a method for detecting potentially fatal bacterial contamination of medical products such as IV solutions, vaccines, and injectable medications.

In the summer of 1950, a scientist named Frederick Bang was working at the Marine Biological Laboratory in Woods Hole, Massachusetts. Bang was interested in the immune system of primitive organisms, so he chose to inject various types of bacteria into horseshoe crabs to study their immune response. What he found was that injection of any bacteria of the Gram-negative type, or an extract of their cell walls, caused the horseshoe crabs to die quickly, not from the infection but from a massive coagulation of their circulatory fluid.

After many experiments and collaborations with other scientists, Bang developed a test using an extract of the amebocytes, which could be mixed with any sample to determine if that sample contained any contamination by Gram-negative bacteria. If so, the material would clot within 45 minutes. This LAL test, as it is now called, replaced the existing pyrogen test, which required that materials be injected into rabbits, took more time, and was less sensitive to low levels of contamination.

The only drawback of the LAL test is that it requires the removal of hemolymph from the horseshoe crabs. To do this, biomedical companies hire trawlers to catch adult horseshoe crabs. These are taken to a laboratory, then washed; about 30% of the animal's hemolymph is removed from the animal's heart with a large-gauge needle. The hemolymph is then centrifuged to separate the amebocytes, distilled water is added to lyse the cells, and the proteins responsible for the clotting reaction are separated and processed into the product used for the LAL test.

The horseshoe crabs are usually returned to the ocean within 72 hours of bleeding, and studies suggest that most of them survive, perhaps to be caught and bled again. Because 1 quart of hemolymph is worth about $15,000, the companies have good reason to preserve this ancient, fascinating species.

Figure 32B Horseshoe crabs. Horseshoe crabs have lived on the Earth for an estimated 450 million years.

Questions to Consider

1. Compared to hemoglobin, hemocyanins are much larger, free-floating molecules. Why might hemocyanins work better with an open circulatory system, compared to hemoglobin?
2. The amebocytes of horseshoe crabs cause the animal's hemolymph to clot in response to certain bacteria. How could this response be beneficial to the animal?

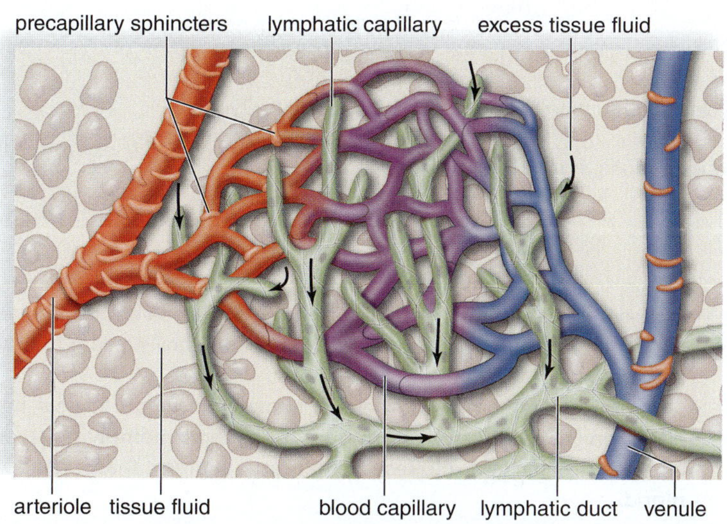

Not all capillary beds are open at the same time. When the precapillary sphincters (circular muscles) shown in Figure 32.4 are relaxed, the capillary bed is open and blood flows through the capillaries. When precapillary sphincters are contracted, blood flows through a shunt that carries blood directly from an arteriole to a venule.

Figure 32.19 Capillary bed. A lymphatic capillary bed lies near a blood capillary bed. When lymphatic capillaries take up excess tissue fluid, it becomes lymph. Precapillary sphincters can shut down a blood capillary, and blood then flows through the shunt.

In addition to nutrients and wastes, the blood distributes heat to body parts. When you are warm, many capillaries that serve the skin are open, and your face is flushed. This helps rid the body of excess heat. When you are cold, skin capillaries close, conserving heat, and your skin takes on a bluish tinge.

Check Your Progress 32.4

1. List the major components of blood and the functions of each.
2. Name the major events, in chronological order, that result in a blood clot.
3. Explain why Rh incompatibility is a problem only when a fetus is Rh-positive and the mother is Rh-negative, but not vice versa.
4. Describe the major factors that affect the rate of capillary exchange.

CONNECTING *the* CONCEPTS *with the* THEMES

Evolution

- Some animals have had no adaptive need to develop a circulatory system, because their structure allows each cell to exchange nutrients and wastes directly with the environment.
- The open circulatory system most likely evolved first. As demands for oxygen and efficient exchange increased with a terrestrial lifestyle, natural selection favored the development of closed systems.
- Within closed systems, the heart evolved from the one atrium–one ventricle heart, seen in most fish, to the four-chambered heart, seen in crocodilians, birds, and mammals.

Nature of Science

- Because the circulatory system's function is so critical, a major research effort has been devoted to understanding and treating cardiovascular diseases.
- Recent advances in cardiovascular disease prevention and treatment range from lifestyle changes to drug therapies and surgical interventions.

Biological Systems

- The cardiovascular system's function is linked with the functions of several other systems: the respiratory system (gas exchange); the urinary system (waste excretion); the digestive system (provision of nutrients); and the immune system (body defenses).
- In more complex animals, circulatory systems also transport many substances, such as hormones, bilirubin, and cholesterol; protect against microbial invaders; and participate in temperature regulation.

MEDIA STUDY TOOLS http://connect.mheducation.com

SMARTBOOK® Study smart. SmartBook helps you maximize your study time by identifying what you know and don't know. The recharge feature in SmartBook helps you prepare for the test by identifying the material that you are most likely to forget.

connect® | BIOLOGY Find even more resources to learn the chapter concepts including animations, tutorial videos, and interactive practice questions.

MP3 Files	Animations	Tutorial
32.2 Classification of Blood Vessels **32.3** Heart Structure and Function • Cardiac Cycle • Blood Flow and Blood Pressure **32.4** Capillary Exchange and Bulk Flow	**32.3** Cardiac Cycle • Conducting System of the Heart **32.4** Fluid Exchange Across the Walls of Capillaries	**32.3** Cardiac Cycle **32.4** Capillary Exchange

SUMMARIZE

32.1 Transport in Invertebrates

Most animals have a **circulatory system** that provides their tissues with oxygen and nutrients, and removes wastes. However, some invertebrates lack such a transport system. The presence of a gastrovascular cavity allows diffusion alone to supply the needs of cells in cnidarians and flatworms. Roundworms make use of their pseudocoelom in the same way that echinoderms use their coelom to circulate materials.

Other invertebrates do have a transport system. Insects have an **open circulatory system** that transports **hemolymph,** and earthworms have a **closed circulatory system** that transports **blood.**

32.2 Transport in Vertebrates

Vertebrates have a closed **cardiovascular system** in which **arteries** carry blood away from the heart, branch into smaller **arterioles,** and then into **capillaries,** where exchange with **interstitial fluid** takes place. **Venules** collect blood from the capillaries, and merge into **veins** that carry blood to the heart.

Fishes have a one-circuit circulatory pathway, because the heart, with the single atrium and ventricle, pumps blood to the gills and then to the body, without a second pass through the heart. The other vertebrates have both pulmonary and systemic circuits. Amphibians have two atria but a single ventricle. Crocodilians, birds, and mammals, including humans, have a heart with two atria and two ventricles, in which oxygen-rich blood is always separate from oxygen-poor blood. Animals with such a two-circuit circulatory pathway have a **systemic circuit** and a **pulmonary circuit.**

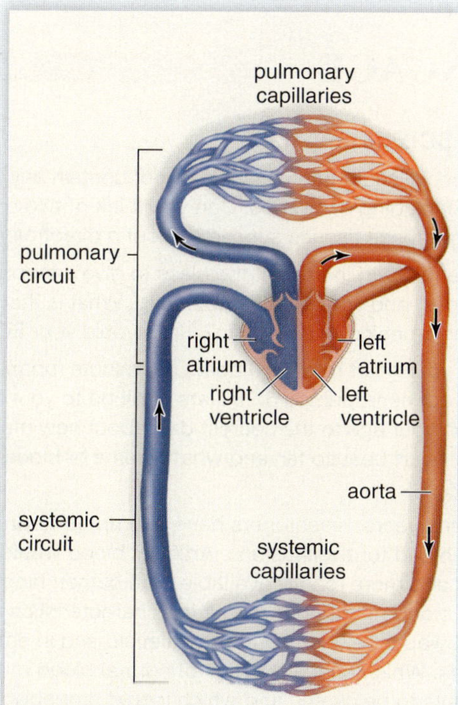

32.3 The Human Cardiovascular System

The human **heart** is largely made of cardiac muscle (myocardium), and it is surrounded by pericardium. The internal chambers of the heart—the left and right **atria** and **ventricles**—are separated by a **septum.**

During **systole,** the chambers contract to pump blood, and during **diastole** they relax, allowing filling. The **cardiac cycle** (heartbeat) in humans begins when the SA (sinoatrial) node (**pacemaker**) causes the two atria to contract, and blood moves through the **atrioventricular valves** to the two ventricles. The SA node also stimulates the AV (atrioventricular) node, which in turn causes the two ventricles to contract. This electrical activity can be measured with an **electrocardiogram (ECG).** Ventricular contraction sends blood through the **semilunar valves** to the pulmonary trunk and the aorta. Then, all chambers rest. The heart sounds, lub-dub, are caused by the closing of the valves, and the wave effect of blood being pumped through arteries can be felt as the **pulse.** The amount of blood pumped by the ventricles each minute is the **cardiac output.**

In the pulmonary circuit, blood travels to and from the lungs. In the systemic circuit, the **aorta** divides into blood vessels that serve the body's cells. The **venae cavae** return oxygen-poor blood to the heart.

In a **portal system,** blood from one set of capillaries (e.g., in the intestine) travels through veins directly to another set of capillaries (e.g., in the liver).

Blood pressure created by the pumping of the heart accounts for the flow of blood in the arteries, but skeletal muscle contraction is largely responsible for the flow of blood in the veins, which have valves preventing a backward flow.

Hypertension and **atherosclerosis** are two circulatory disorders that can lead to **angina pectoris,** a **heart attack,** or **stroke.** Following a heart-healthy diet, getting regular exercise, maintaining a proper weight, and not smoking cigarettes can help protect against the development of these conditions.

32.4 Blood

Blood has two main parts: **plasma** and the formed elements. Plasma is mostly water (90–92%), but it also contains 7–8% proteins (such as albumin and **antibodies**), nutrients, and wastes.

The formed elements include **red blood cells,** white blood cells, and platelets. Red blood cells contain hemoglobin, which functions in oxygen transport.

White blood cells (WBCs) also called **leukocytes,** which include **granular leukocytes** and **agranular leukocytes,** defend the body against infections. Three types of granular leukocytes are the **neutrophils,** which are phagocytes; **basophils,** which are involved in inflammation; and **eosinophils,** which are important in parasitic infections.

The two types of agranular leukocytes are the **monocytes,** which enter tissues to become phagocytic macrophages, and **lymphocytes,** which carry out adaptive (specific) immunity to infection.

The **platelets** and two plasma proteins, prothrombin and fibrinogen, function in blood **clotting,** an enzymatic process that results in fibrin threads. Blood clotting includes three major events: (1) Platelets and injured tissue release prothrombin activator, which (2) enzymatically changes prothrombin to thrombin, which is an enzyme that (3) causes fibrinogen to be converted to fibrin threads.

The ABO blood-typing system is based on the presence or absence of A and B **antigens** on the red blood cells (A and B). If a mismatched transfusion is given, antibodies in the recipient's blood may react to these antigens, causing **agglutination** of red blood cells. A second type of red blood cell antigen is the Rh factor. If an Rh-negative woman becomes pregnant with an Rh-positive fetus, she may produce anti-Rh antibodies that could damage any Rh-positive fetus she carries.

When blood reaches a capillary, water moves out at the arterial end due to blood pressure. At the venous end, water moves in due to osmotic pressure. In between, nutrients diffuse out of, and wastes diffuse into, the capillary according to concentration gradients. Any excess interstitial fluid is absorbed into lymphatic vessels, where it is known as **lymph.**

ASSESS

Choose the best answer for each question.

32.1 Transport in Invertebrates

1. Which animals lack a true circulatory system?
   a. cnidarians
   b. flatworms
   c. nematodes
   d. All of these are correct.
2. Which one of these would you expect to be part of a closed, but not an open, circulatory system?
   a. ostia
   b. capillary beds
   c. hemolymph
   d. heart
   e. All of these are correct.
3. Which animal has a closed circulatory system?
   a. earthworm b. grasshopper c. hydra d. sea star

32.2 Transport in Vertebrates

4. All vertebrates have
   a. a closed circulatory system.
   b. a heart with at least three chambers.
   c. a two-circuit circulatory pathway.
   d. All of these are correct.
5. A major difference between arteries and veins is that
   a. arteries always carry oxygenated blood; veins never do.
   b. arteries carry blood away from the heart; veins return blood.
   c. only arteries have valves.
   d. veins feed blood into capillaries.
6. In which animal does aortic blood have less oxygen than blood in the pulmonary vein?
   a. frog
   b. chicken
   c. monkey
   d. fish
   e. All of these are correct.

32.3 The Human Cardiovascular System

7. In humans, blood returning to the heart from the lungs returns to
   a. the right ventricle.
   b. the right atrium.
   c. the left ventricle.
   d. the left atrium.
   e. both the right and left sides of the heart.
8. All arteries in the body contain oxygen-rich blood, with the exception of the
   a. aorta.
   b. pulmonary arteries.
   c. renal arteries.
   d. coronary arteries.
9. *Systole* refers to the contraction of the
   a. major arteries.
   b. SA node.
   c. atria and ventricles.
   d. major veins.
   e. All of these are correct.
10. Which of these is an incorrect statement concerning the heartbeat?
    a. The atria contract at the same time.
    b. The ventricles relax at the same time.
    c. The atrioventricular valves open at the same time.
    d. The semilunar valves open at the same time.
    e. First the right side contracts, and then the left side contracts.
11. Label this diagram of the heart.

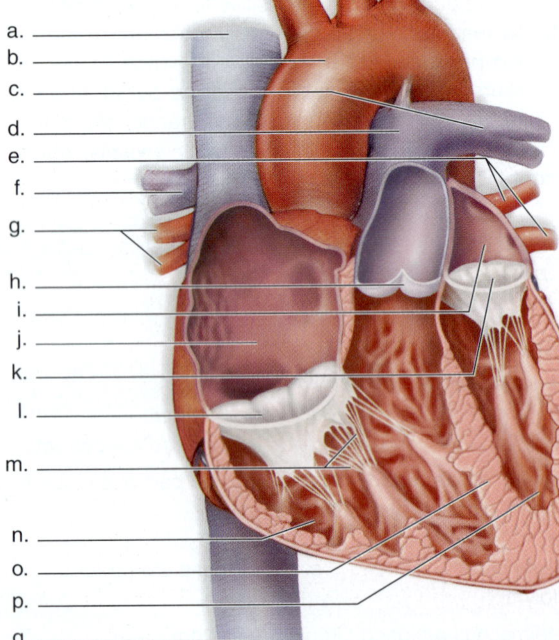

32.4 Blood

12. Which of the following is not a formed element of blood?
    a. leukocyte
    b. eosinophil
    c. fibrinogen
    d. platelet
13. Which of these is an incorrect association?
    a. white blood cells—infection fighting
    b. red blood cells—blood clotting
    c. plasma—water, nutrients, and wastes
    d. red blood cells—hemoglobin
    e. platelets—blood clotting
14. Water enters capillaries on the venous end as a result of
    a. active transport from tissue fluid.
    b. an osmotic pressure gradient.
    c. higher blood pressure on the venous end.
    d. higher blood pressure on the arterial side.
    e. higher red blood cell concentration on the venous end.

ENGAGE

THINKING SCIENTIFICALLY

1. A few specialized human tissues do not contain any blood vessels, including capillaries. Can you think of two or three? How might these tissues survive without a direct blood supply?
2. You have to stand in front of the class to give a report. You are nervous, and your heart is pounding. What is the specific mechanism behind this reaction? How would your ECG appear?
3. Assume your heart rate is 70 beats per minute (bpm), and each minute your heart pumps 5.25 liters of blood to your body. Based on your age to the nearest day, about how many times has your heart beat so far, and what volume of blood has it pumped?
4. For several years, researchers have attempted to produce artificial blood for transfusions. Artificial blood would most likely be safer and more readily available than human blood. While artificial blood might not have all the characteristics of human blood, it would be useful on the battlefield and in emergency situations. Which characteristics of normal blood must artificial blood have to be useful, and which would probably be too difficult to reproduce?

Allergic reactions can produce serious, even life-threatening consequences.

33

The Lymphatic and Immune Systems

Chapter Outline

33.1 Evolution of Immune Systems 622

33.2 The Lymphatic System 623

33.3 Innate Immune Defenses 625

33.4 Adaptive Immune Defenses 628

33.5 Immune System Disorders and Adverse Reactions 636

Before You Begin

Before beginning this chapter, take a few moments to review the following discussions.

Figure 5.3 Which of the major protein functions are most important in the immune system?

Figure 21.19 In what ways do the amoeboid protists resemble macrophages of the mammalian immune system?

Section 31.2 What life processes are carried out by the lymphatic and immune systems?

Kelsey is a junior in college, who will apply to medical school next year. An excellent student, Kelsey also works as a phlebotomist at a local hospital. She must be careful, however, to avoid wearing, or even contacting, latex gloves. She also knows to avoid latex balloons, and she steers clear of certain foods. Kelsey became aware that seemingly harmless items or foods can pose a threat to some people when she developed a life-threatening condition known as anaphylactic shock. A few minutes after eating some guacamole dip at a high school party, Kelsey felt her throat tighten, and her breathing became labored. By the time the ambulance arrived, Kelsey had lost consciousness. Once this severe reaction had been tamed by medications, tests revealed that Kelsey was also allergic to latex, kiwi, and mango. As her medical caregivers explained, all these natural materials contain some common proteins, to which Kelsey had developed an allergic reaction. She now carries an EpiPen and knows how to inject herself with lifesaving medication in case she begins to feel the early symptoms of anaphylaxis.

Allergic reactions illustrate the power of our immune system. Clearly, this power can be harmful in such cases, but our immune system also serves the essential role of protecting us against the vast array of viruses, bacteria, fungi, parasites, and toxins in our environment. In this chapter, we explore that system, along with the lymphatic system, which helps produce and distribute the cells of the immune system.

As you read through this chapter, think about the following questions:

1. What are the most essential components of the immune system?
2. How do the lymphatic and immune systems work together?

Following *the* Themes

CHAPTER 33 THE LYMPHATIC AND IMMUNE SYSTEMS

UNIT 7 COMPARATIVE ANIMAL BIOLOGY

Evolution	Even the simplest of multicellular organisms have developed cells that specialize in immune defenses, which provides clues about how more complex immunity may have evolved.
Nature of Science	Because the immune system is essential to our health, studies of patients with immune system disorders, such as AIDS, have improved our basic understanding of how the immune system functions.
Biological Systems	The lymphatic and immune systems are both essential to homeostasis, with the lymphatic system serving some additional functions unrelated to immunity.

33.1 Evolution of Immune Systems

Learning Outcomes

Upon completion of this section, you should be able to

1. Summarize the evidence suggesting that cellular slime molds can form a rudimentary "immune system."
2. Define *PAMPs* and explain how they enable many animals to identify the presence of harmful microbes.
3. Compare the types of antigens recognized by the innate versus the adaptive immune system.

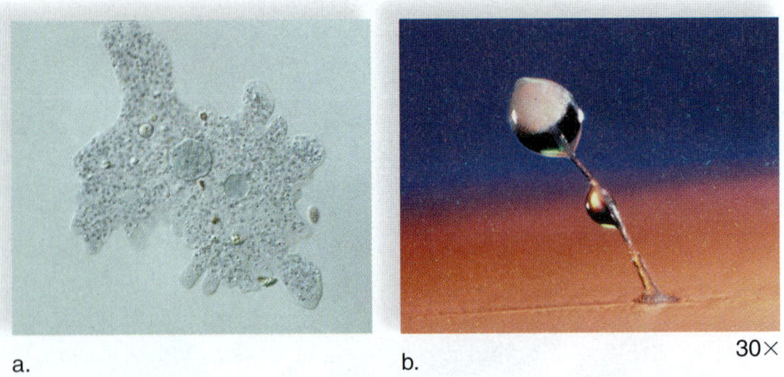

Figure 33.1 Social amoebas (*Dictyostelium discoideum*). **a.** The single-celled form, shown here at a high magnification. **b.** Under certain conditions, thousands of amoebas can form a multicellular slug, in which some cells develop protective functions.

Our **immune system** protects us from all sorts of harmful invaders, including bacterial and viral pathogens, various toxins, and perhaps even the cancerous cells that occasionally arise. It is a very intricate system made up of many components that work together, perhaps rivaling the nervous system in its complexity. But how did such a complicated system first evolve? And which components developed first? Scientists are beginning to answer those questions.

Immunity in Cellular Slime Molds

In August 2007, researchers at Baylor College of Medicine in Houston, Texas, discovered that very simple creatures called cellular slime molds can exhibit signs of a rudimentary immune system. These organisms are commonly known as "social amoebas," because of their unique life cycle. When food is plentiful, these protists live as separate amoeboid cells (Fig. 33.1*a*), ingesting bacteria through phagocytosis and reproducing by binary fission (dividing into identical copies). As the food supply dwindles, many thousands of these amoebas can join together to form a slug 2–4 millimeters long (Fig. 33.1*b*). The slug migrates toward light as a single, multicellular body, then differentiates into a reproductive structure that releases spores, many of which disperse to become new amoebas.

For several years, biologists have known that individual cells within the slug can become specialized to perform various functions, but the Baylor group discovered a new type of cell within the slug that they named *sentinel cells.* These cells circulate throughout the slug, engulfing bacteria and toxins. Eventually, the sentinel cells are sloughed from the body of the slug, thereby "sacrificing themselves" for the good of the organism. Some scientists believe that phagocytic white blood cells in the human body, such as neutrophils and macrophages, may have evolved from these types of cells.

Immunity in *Drosophila*

Although social amoebas may provide an example of how phagocytic cells first became specialized to protect multicellular organisms, the immune system is far more developed in invertebrates, such as the fruit fly, *Drosophila melanogaster.* It was in this well-studied insect that scientists discovered the existence of a group of cellular receptors that could recognize common components found in many pathogenic microbes, but not in the insect's own cells. When these receptors bind to these *pathogen-associated molecular patterns,* or *PAMPs,* they trigger an immune reaction, increasing the odds that the pathogen can be eliminated from the fly.

Examples of PAMPs found on pathogenic microbes include the double-stranded RNA that is produced during the replication cycle of many viruses and certain arrangements of carbohydrates, lipids, or proteins found only on bacterial or fungal cell walls. Receptors for PAMPs have been found in organisms as diverse as fruit flies, plants, and humans, suggesting that they were one of the earliest, and most successful, types of cellular receptors that evolved for the recognition of pathogens. As you will see in section 33.3, these two examples illustrate a type of host defense known as **innate immunity,** which can recognize common microbial invaders very quickly but shows no signs of an increased response on repeated exposure to the same invader.

The Rise of Adaptive Immunity

In addition to innate immunity, vertebrate animals also exhibit **adaptive immunity,** characterized by the production of a very large number of diverse receptors on the surface of specialized white blood cells (such as B and T lymphocytes in humans). These receptors bind very specifically to molecules called **antigens,** much as a key fits a lock. This binding stimulates lymphocytes to divide and become much more numerous, resulting in the characteristic features of adaptive immunity, such as greatly increased responses to specific antigens and immunological memory after an initial exposure to an antigen.

The generation of such a diverse array of antigen receptors depends on a rearrangement of the DNA that codes for these receptors, somewhat like choosing different combinations of cards from a deck. Scientists have now discovered that this process developed quite suddenly in an ancestor that gave rise to the jawed vertebrates—including the cartilaginous fishes (sharks and rays), bony fishes, amphibians, reptiles, birds, and mammals.

The precise mechanism by which this "explosion" of adaptive immunity occurred is still incompletely understood, although it now seems quite likely that it involved the insertion of a small piece of DNA (a transposon, or "jumping gene"; see section 14.4) into a gene coding for a more primitive, less variable antigen receptor—perhaps similar to the receptors for PAMPs, mentioned earlier. In contrast to the relatively "fixed" receptors for PAMPs recognized by the innate immune system, the generation of antigen receptors by gene rearrangement allows the adaptive immune system to be able to respond to new antigens that evolve, for example, in emerging infectious agents. In other words, the vertebrate immune system has evolved an ability to respond to the continuing evolution of pathogenic microbes and other threats to our health.

Check Your Progress 33.1

1. Describe the function of sentinel cells in cellular slime molds.
2. List three specific types of PAMPs found on microbes.
3. Describe three ways that the evolution of receptors for specific antigens increased the effectiveness of the immune system.

33.2 The Lymphatic System

Learning Outcomes

Upon completion of this section, you should be able to

1. Describe three major functions of the lymphatic system.
2. Distinguish between the roles of primary and secondary lymphoid tissues and list examples of each.

The **lymphatic system,** which is closely associated with the cardiovascular and immune systems, includes the lymphatic vessels and the lymphoid organs. It has three main functions that contribute to homeostasis:

- Lymphatic capillaries absorb excess interstitial fluid and return it to the bloodstream.
- In the small intestine, lymphatic capillaries called lacteals absorb fats in the form of lipoproteins and transport them to the bloodstream.
- The lymphoid organs and lymphatic vessels are sites of production and distribution of lymphocytes, which help defend the body against pathogens.

MP3 Lymphatic System

Lymphatic Vessels

Lymphatic vessels form a one-way system, beginning with **lymphatic capillaries**—tiny, closed-ended vessels that are found throughout the body (Fig. 33.2). Lymphatic capillaries take up

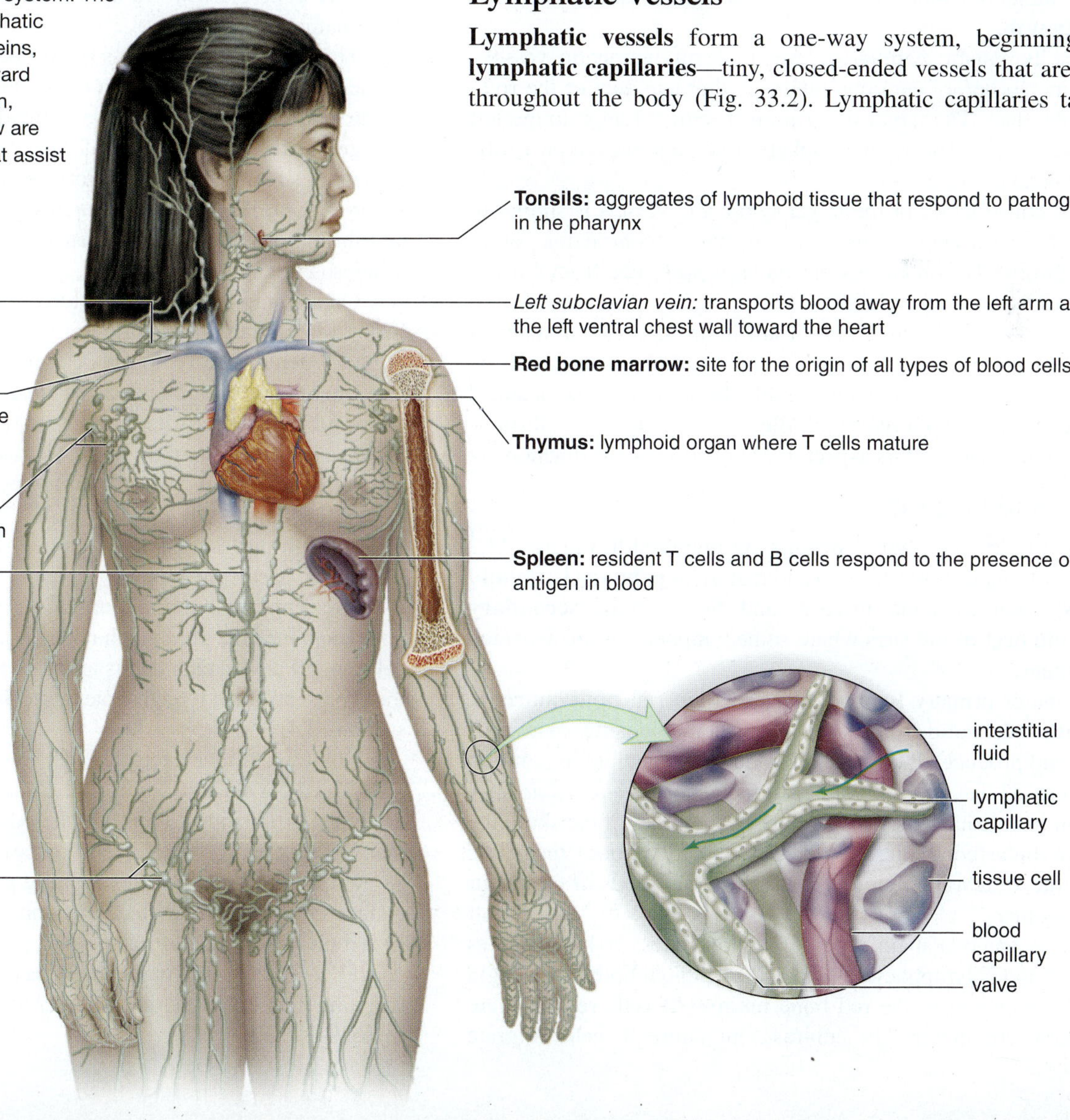

Figure 33.2 Lymphatic system. Lymphatic vessels drain excess fluid from the tissues and return it to the cardiovascular system. The enlargement shows that lymphatic vessels, like cardiovascular veins, have valves to prevent backward flow. The lymph nodes, spleen, thymus, and red bone marrow are the main lymphoid organs that assist immunity.

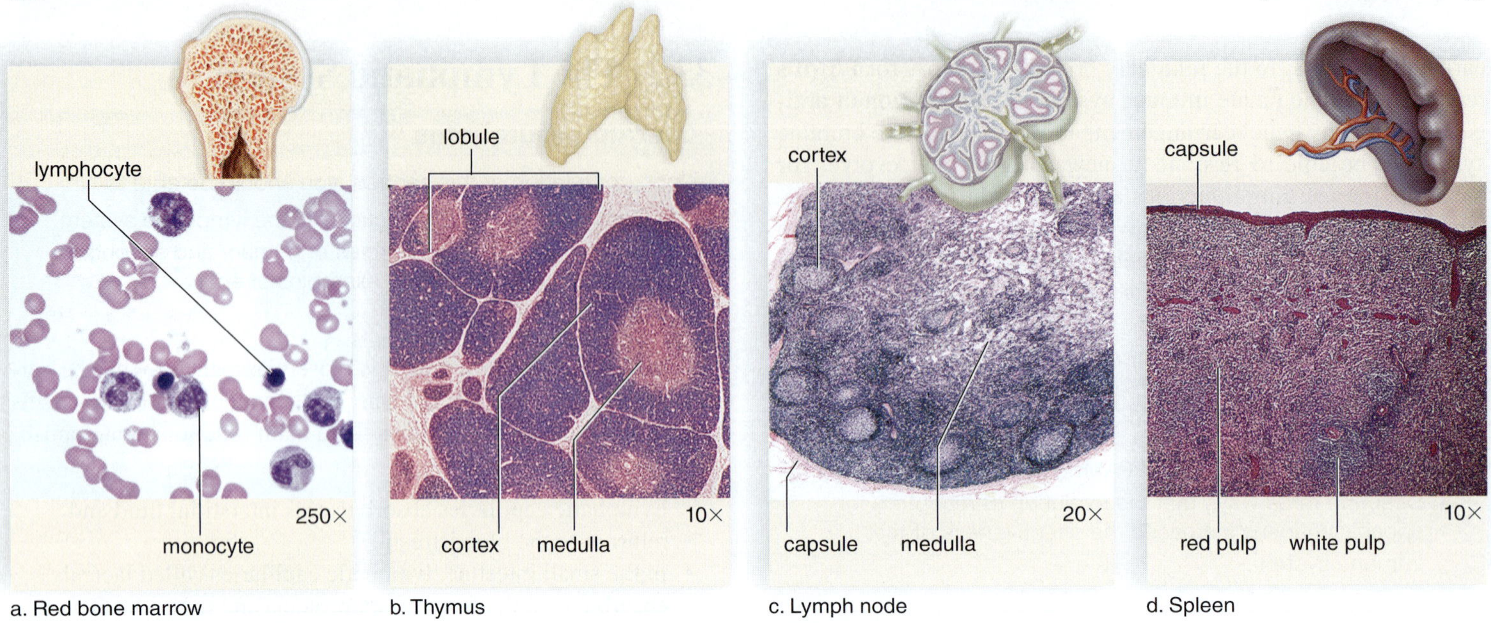

Figure 33.3 The lymphoid organs. **a.** Blood cells, including lymphocytes, are produced in red bone marrow. B cells mature in the bone marrow, but (**b**) T cells mature in the thymus. **c.** Lymph is cleansed in lymph nodes, while (**d**) blood is cleansed in the spleen.

excess interstitial fluid. The fluid inside lymphatic capillaries is called **lymph.**

The lymphatic capillaries join to form lymphatic vessels, which merge before entering either the thoracic duct or the right lymphatic duct. The larger thoracic duct returns lymph to the left subclavian vein. The right lymphatic duct returns lymph to the right subclavian vein.

The construction of the larger lymphatic vessels is similar to that of cardiovascular veins. Skeletal muscle contraction forces lymph through lymphatic vessels, and it is prevented from flowing backward by one-way valves.

A number of diseases may result in an increased amount of fluid leaving the blood capillaries, or an insufficient return of fluid to the blood via the lymphatic vessels. In either case, a localized accumulation of interstitial fluid called *edema* may result, illustrating the importance of this aspect of lymphatic system function.

Lymphoid Organs

The **lymphoid (lymphatic) organs** are reviewed in Figures 33.2 and 33.3. Lymphocytes develop and mature in **primary lymphoid organs,** such as bone marrow and the thymus; **secondary lymphoid organs** are sites where some lymphocytes are activated by antigens.

A major primary lymphoid organ is the **red bone marrow,** a spongy, semisolid, red tissue where hematopoietic stem cells divide and produce all the types of blood cells, including lymphocytes (Fig. 33.3*a*). In a child, most of the bones have red bone marrow, but in an adult, it is present only in the bones of the skull, the sternum (breastbone), the ribs, the clavicle (collarbone), the pelvic bones, the vertebral column, and the proximal heads of the femur and humerus.

There are two main types of lymphocytes: B lymphocytes (B cells) and T lymphocytes (T cells). Although both types begin their development in the red bone marrow, **B cells** remain there until they are mature. In contrast, immature **T cells** migrate from the bone marrow via the bloodstream to the thymus, where they mature.

The soft, bilobed **thymus** is a primary lymphoid organ located in the thoracic cavity between the trachea and the sternum ventral to the heart (see Fig. 33.2). It is in the thymus that T cells learn to recognize the combinations of self-molecules and foreign molecules; this recognition characterizes mature T-cell responses (see section 33.4). The thymus varies in size, but it is largest in children and shrinks as we get older. When well developed, it contains many lobules (Fig. 33.3*b*).

Once lymphocytes are mature, they enter the bloodstream. From there, they frequently migrate into secondary lymphoid organs, such as the lymph nodes and spleen. Here, lymphocytes may encounter foreign molecules or cells; in response, they proliferate and become activated. Activated lymphocytes then reenter the bloodstream, searching for signs of infection or inflammation, like a squadron of highly trained military personnel seeking to destroy a specific enemy.

Lymph nodes are small (about 1–25 mm in diameter), ovoid structures occurring along lymphatic vessels. They are a major type of secondary lymphoid organ. As lymph percolates through the cortex and medulla of a lymph node (Fig. 33.3*c*), resident phagocytic cells engulf any foreign debris and pathogens. These phagocytes can then "present" these foreign materials to T cells in the lymph node (see section 33.4).

Sometimes incorrectly called "lymph glands," lymph nodes are named for their location. For example, inguinal lymph nodes are in the groin and axillary lymph nodes are in the armpits. Physicians often feel for the presence of swollen, tender lymph nodes as evidence that the body is fighting an infection. Unfortunately, cancer cells sometimes enter lymphatic vessels and congregate in lymph nodes. Therefore, when a person undergoes surgery for cancer, it is a common procedure to remove some lymph nodes and examine them to determine whether the cancer has spread to other regions of the body.

The **spleen,** an oval secondary lymphoid organ with a dull purplish color, is located in the upper left side of the abdominal cavity posterior to the stomach. Most of the spleen contains red pulp, which filters and cleanses the blood. Red pulp consists of blood vessels and sinuses, where macrophages remove old and defective blood cells. The spleen also has white pulp, consisting of small areas of secondary lymphoid tissue (Fig. 33.3*d*). Much as the lymph nodes serve as sites for lymphocytes to respond to foreign material from the tissues, the spleen serves a similar role for the blood.

The spleen's outer capsule is relatively thin, and an infection or a trauma can cause the spleen to burst, necessitating surgical removal. Although some of the spleen's functions can be largely replaced by other organs, an individual with no spleen is more susceptible to certain types of infections and may require antibiotic therapy indefinitely.

Patches of lymphoid tissue in the body include the *tonsils,* located in the pharynx; *Peyer patches,* located in the intestinal wall; and the *vermiform appendix,* attached to the cecum. These structures encounter pathogens and antigens that enter the body by way of the mouth.

Check Your Progress 33.2

1. Distinguish between the lymphatic and circulatory systems.
2. Summarize the functions of the lymphatic system.
3. Describe the general location and function of the lymphoid organs.

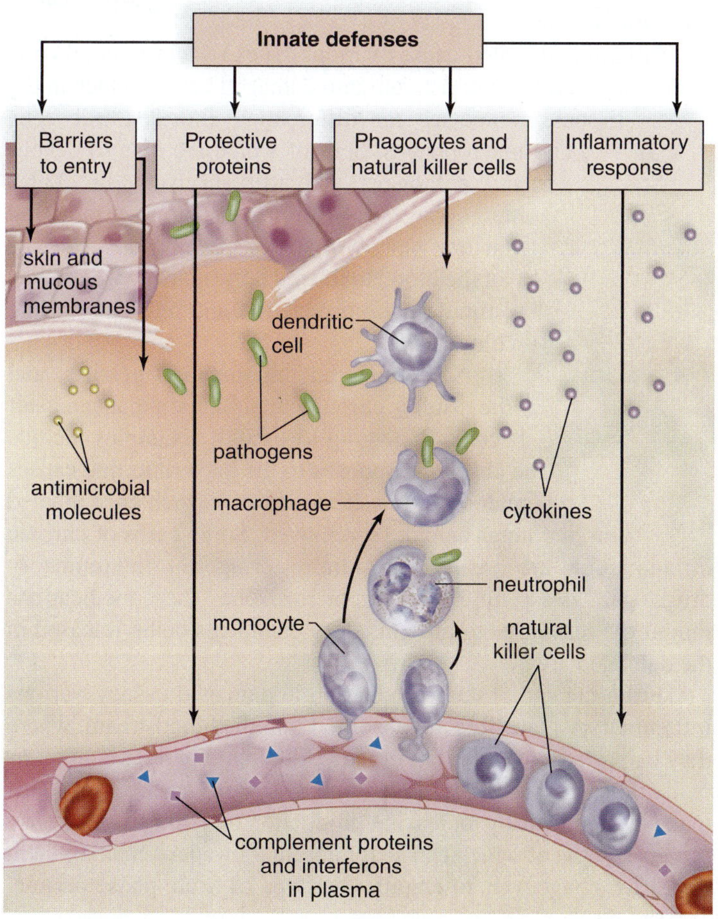

Figure 33.4 Overview of innate immune defenses. Most innate defenses act rapidly to detect and respond to various, highly conserved molecules expressed by pathogens.

33.3 Innate Immune Defenses

Learning Outcomes

Upon completion of this section, you should be able to

1. Define *innate immunity.*
2. Describe four mechanisms of innate immunity and the major tissues, molecules, and/or cells involved.
3. Explain some specific ways that the innate immune system interacts with and influences the adaptive immune system.

We are constantly exposed to microbes, such as viruses, bacteria, and fungi, in our environment. **Immunity** is the capability of removing or killing foreign substances, pathogens, and cancer cells from the body. Mechanisms of *innate immunity* are fully functional without previous exposure to these invaders, while *adaptive immunity* (see section 33.4) is initiated and amplified by exposure.

As summarized in Figure 33.4, innate immune defenses include the following:

- Physical and chemical barriers
- The inflammatory response
- Phagocytes and natural killer cells
- Protective proteins, such as complement and interferons

Innate defenses occur immediately or very shortly after an infection occurs. With innate immunity, there is no recognition that an intruder has attacked before, and therefore no immunological "memory" is present for the attacker.

Physical and Chemical Barriers

Physical barriers to various types of invaders include the skin as well as the mucous membranes lining the respiratory, digestive, and urinary tracts. As you saw in Chapter 31, the outer layers of our skin are composed of dead, keratinized cells that form a relatively impermeable barrier. But when the skin has been injured, one of the first concerns is the possibility of an infection.

The mucus produced by mucous membranes physically ensnares microbes. The upper respiratory tract is lined by ciliated cells that sweep mucus and trapped particles up into the throat, where they can be swallowed or expectorated (coughed out). In addition, various bacteria that normally reside in the intestine and in other areas, such as the vagina, take up nutrients and block binding sites that could be exploited by pathogens.

The secretions of oil glands in human skin also contain chemicals that weaken or kill certain bacteria; saliva, tears, milk, and mucus contain lysozyme, an enzyme that can lyse bacteria; and the stomach has an acidic pH, which inhibits or kills many microbes.

MP3 Barriers and Nonspecific Defenses

Inflammatory Response

When tissues are damaged by a variety of causes, including pathogens, a series of events known as the **inflammatory response**

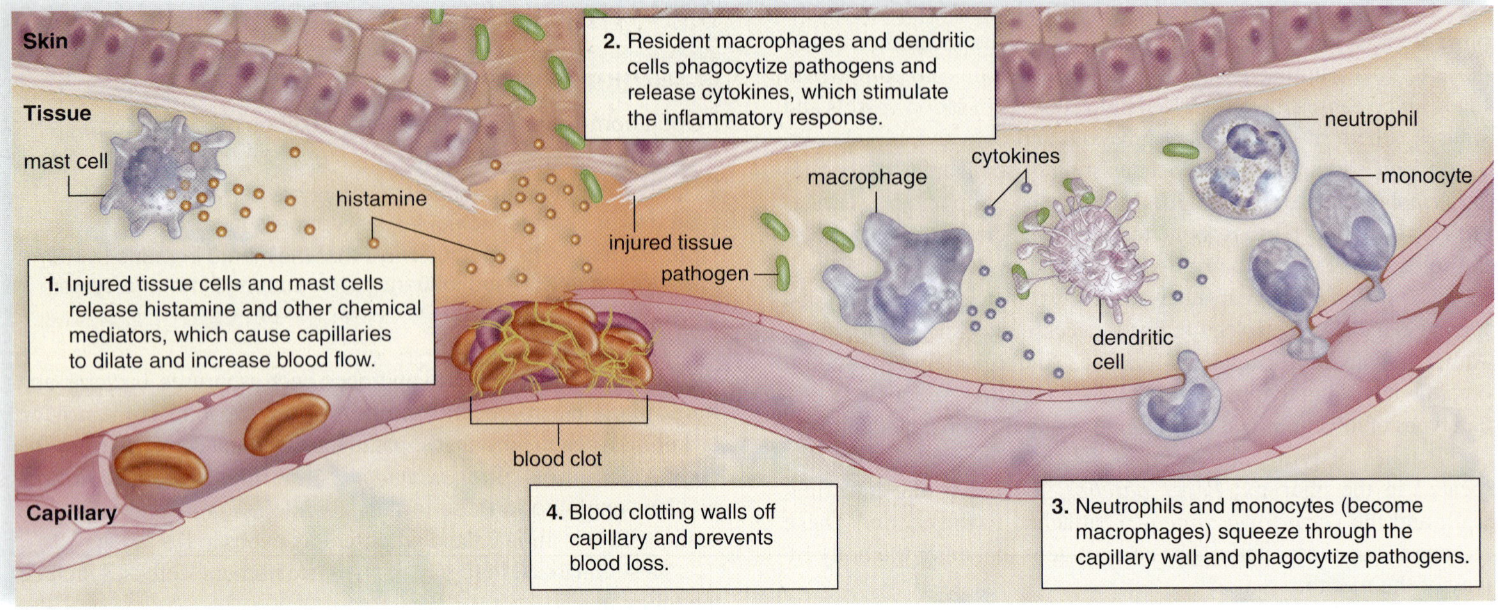

Figure 33.5 Inflammatory response. Due to capillary changes in a damaged area and the release of chemical mediators, such as histamine by mast cells, an inflamed area exhibits redness, heat, swelling, and pain. The inflammatory response can be accompanied by other reactions to the injury. Macrophages and dendritic cells, present in the tissues, phagocytize pathogens, as do neutrophils, which squeeze through capillary walls from the blood. Macrophages and dendritic cells release cytokines, which stimulate the inflammatory and other immune responses. A blood clot can form to seal a break in a blood vessel.

Tutorial Inflammatory Response

occurs. An inflamed area has at least four common signs: redness, heat, swelling, and pain. Most of these signs are due to capillary changes in the damaged area, as illustrated in Figure 33.5. Chemical mediators released by damaged cells, including **histamine** that is mainly secreted by tissue-dwelling cells of the innate immune system called **mast cells,** cause capillaries to dilate and become more permeable. Increased blood flow to the area causes the skin to redden and become warm. Increased permeability of the capillaries allows proteins and fluids to escape into the tissues, resulting in swelling. Various chemicals released by damaged cells stimulate free nerve endings, causing the sensation of pain.

Animation Inflammatory Response

Inflammation also causes various types of white blood cells to migrate from the bloodstream into damaged tissues. Once in the tissues, monocytes can differentiate into dendritic cells and macrophages, both of which are able to devour many pathogens and still survive (Fig. 33.6). Macrophages also release colony-stimulating factors, namely cytokines, which pass by way of the blood to the red bone marrow where they stimulate the production and release of white blood cells.

Sometimes an inflammation persists, and the result is chronic inflammation that can itself become damaging to tissues. Examples include the chronic responses to the bacterium that causes tuberculosis or to asbestos fibers, which once inhaled into the lungs cannot be removed. Some cases of chronic inflammation are treated by administering anti-inflammatory drugs, such as aspirin, ibuprofen, or cortisone. These medications inhibit the responses to inflammatory chemicals being released in the damaged area.

From the site of their production in damaged tissues, various inflammatory mediators are absorbed into the bloodstream, where they can affect several other organs. Although the liver is not normally thought of as a part of the immune system, the liver responds to these chemicals by increasing production of various *acute phase proteins,* some of which can coat microbial invaders, making them easier for phagocytes to engulf. One type of acute phase protein,

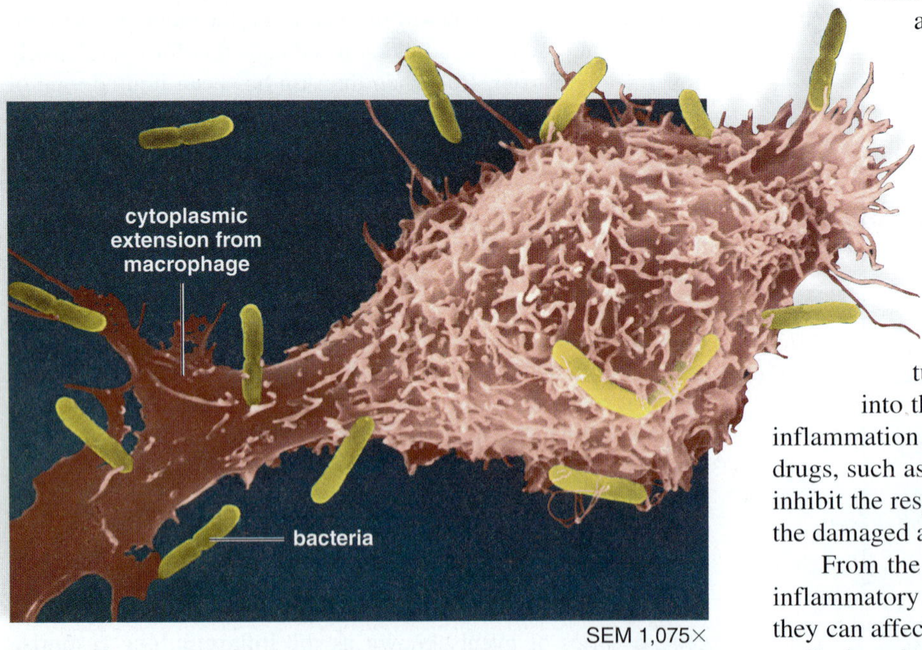

Figure 33.6 Macrophage engulfing bacteria. Monocyte-derived macrophages are the body's scavengers. They engulf microbes and debris in the body's fluids and tissues, as illustrated in this colorized scanning electron micrograph.

called C-reactive protein, is frequently measured to assess levels of inflammation in patients suffering from certain diseases.

Inflammatory chemicals in the blood may also act on the brain to initiate an elevated body temperature, or *fever*. Although the exact function of the body's fever response is unknown, many speculate that certain bacteria or viruses may not survive as well at higher temperatures, or that certain immune mechanisms work better at higher body temperatures. Experimental data support both hypotheses and both, in fact, may be true. In either case, because mild to moderate fever appears to help the body fight off invaders more effectively, the wisdom of using drugs to treat mild fevers can be questioned. However, a body temperature higher than about 107°F can be fatal especially in children, so obviously this situation must be treated as an emergency.

Inflammatory responses are accompanied by other responses to the injury. The clotting system can be activated to seal a break in a blood vessel. Antigens, chemical mediators, dendritic cells, and macrophages move from the damaged tissue via the lymph to the lymph nodes. There, these phagocytes interact with T cells and B cells to activate a specific adaptive response to the infection (see section 33.4). Inflammation also initiates the healing response, in which macrophages play an essential role.

Phagocytes and Natural Killer Cells

Several types of white blood cells are phagocytic, meaning that they can engulf and digest relatively large particles, such as viruses and bacteria. **Neutrophils** are able to leave the bloodstream and phagocytize bacteria in tissues. They have multiple ways of killing bacteria. The cytoplasm of a neutrophil is packed with granules that contain antimicrobial peptides, as well as enzymes that can digest bacteria. Other enzymes inside neutrophil granules generate highly reactive free radicals such as superoxide and hydrogen peroxide, all of which participate in killing engulfed bacteria.

Video Neutrophils

As the infection is being overcome, some neutrophils die. These—along with dead tissue cells, dead bacteria, and living white blood cells—may form pus, a whitish material. The presence of pus usually indicates that the body is trying to overcome a bacterial infection.

Eosinophils can be phagocytic, but they are better known for mounting an attack against animal parasites, such as tapeworms, that are too large to be phagocytized.

As mentioned, the two longer-lived types of phagocytic white blood cells are **macrophages** (Fig. 33.6) and **dendritic cells.** Macrophages are found in all sorts of tissues, whereas dendritic cells are especially prevalent in the skin. Both cell types engulf pathogens, which are then digested and broken down into smaller molecular components. They then travel to lymph nodes, where they stimulate T cells, which are responsible for initiating adaptive immune responses.

Natural killer (NK) cells are large, granular lymphocytes that kill virus-infected cells and cancer cells by cell-to-cell contact. NK cells do their work while adaptive defenses are still mobilizing, and they produce cytokines that promote adaptive immunity.

What makes NK cells attack and kill a cell? NK cells seek out cells that lack a particular type of "self" molecule, called MHC-I (major histocompatibility complex I), on their surface. Because some virus-infected and cancer cells may lack these MHC-I molecules, they may be recognized by NK cells, which kill these cells by inducing them to undergo cellular suicide (apoptosis). Because NK cells do not recognize specific viral or tumor antigens, and do not proliferate when exposed to a particular antigen, their numbers do not increase after stimulation.

Protective Proteins

Complement is composed of a number of blood plasma proteins, produced mainly by the liver, that "complement" certain immune responses. These proteins are continually present in the blood plasma but must be activated by pathogens to exert their effects. The complement system helps destroy pathogens in three ways:

1. *Enhanced inflammation.* Complement proteins are involved in and amplify the inflammatory response, because certain ones can bind to mast cells and trigger histamine release, and others can attract phagocytes to the scene.
2. *Increased phagocytosis.* Some complement proteins bind to the surface of pathogens, increasing the odds that the pathogens will be phagocytized by a neutrophil or macrophage.
3. *Membrane attack complexes.* Certain other complement proteins join to form a membrane attack complex, which produces holes in the surface of some bacteria and viruses. Fluids and salts then enter the bacterial cell or virus to the point that it bursts (Fig. 33.7).

Animation Activation of Complement

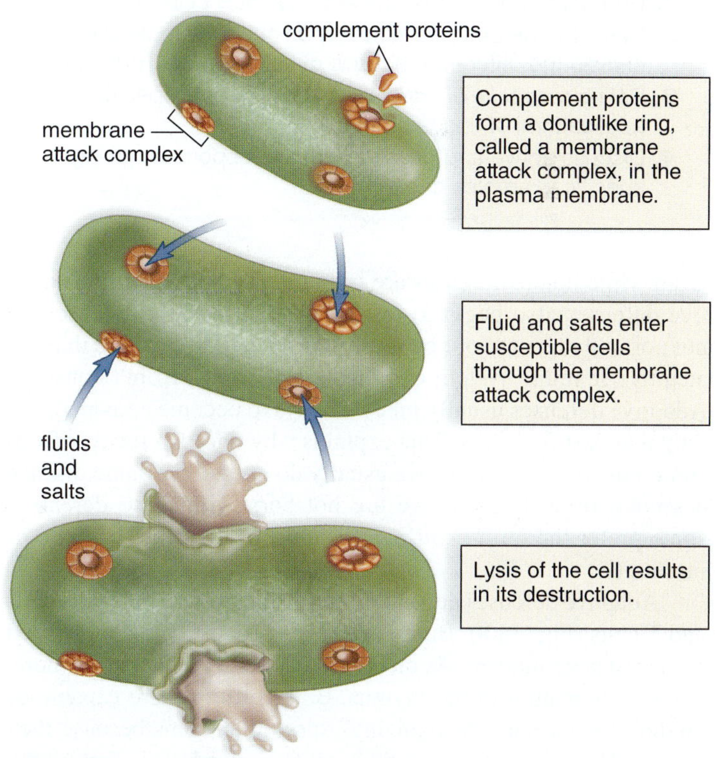

Figure 33.7 Action of the complement system against a bacterium. When complement proteins in the blood plasma are activated by an immune response, they form a membrane attack complex, which makes holes in bacterial cell walls and plasma membranes, allowing fluids and salts to enter until the cell eventually bursts.

Interferons come in several different types, but all are **cytokines,** soluble proteins that affect the behavior of other cells. Most interferons are made by virus-infected cells. They bind to the receptors of noninfected cells, causing them to produce substances that slow cellular metabolism and interfere with viral replication. Interferons are used to treat certain cancers and viral infections, such as hepatitis C.

Animation Antiviral Activity of Interferon

Check Your Progress 33.3

1. List three physical and three chemical barriers.
2. Describe the inflammatory response and explain how this response is beneficial.
3. Name five cell types involved in innate immunity and the major functions of each.
4. Summarize three specific functions of the complement system.

33.4 Adaptive Immune Defenses

Learning Outcomes

Upon completion of this section, you should be able to

1. Compare and contrast the activities of B cells and T cells.
2. Describe the basic structure of an antibody molecule and explain the different functions of IgG, IgA, IgM, and IgE.
3. Define *monoclonal antibodies* and list some specific applications of this technology.
4. Discuss active and passive immune responses, giving specific examples of each.

Even while innate defenses are trying to fight an infection, adaptive defenses also begin to respond. Because these defenses do not normally react to our own cells or molecules, it is said that the adaptive immune system can distinguish "self" from "nonself." Adaptive defenses usually take 5–7 days to become activated, but they may last for years. This explains why once we recover from some infectious diseases, we usually do not get the same disease a second time. Because we are not born with these defenses, some prefer the term *acquired immunity* to describe this type of immunity.

Adaptive defenses depend primarily on the activities of B cells and T cells (Fig. 33.8). Both B cells and T cells are manufactured in the red bone marrow. As mentioned earlier, B cells mature there, but T cells mature in the thymus. Both cell types are capable of binding to and thus "recognizing" specific antigens because they have **antigen receptors** on their plasma membrane. Pathogens, cancer cells, and transplanted tissues and organs bear antigens the immune system usually recognizes as nonself.

During our lifetime, we need a diversity of B cells and T cells to recognize these antigens and protect us against them. Remarkably, diversification occurs during the lymphocyte maturation process to so great an extent that there are specific B cells and/or T cells for almost every possible antigen. Because B and T cells defend us from disease by specifically reacting to antigens, they can be likened to special forces that can attack selected targets without harming nearby residents (uninfected cells).

B Cells and Antibody-Mediated Immunity

The receptor for antigen on the surface of a B cell is called a **B-cell receptor** (**BCR**). B cells are usually activated in a lymph node or the spleen, after their BCRs bind to a specific antigen. Subsequently, the B cell divides by mitosis many times, making many copies (clones) of itself. The **clonal selection theory** states that the antigen receptor of each B cell or T cell binds to only a single type of antigen.

As illustrated in Figure 33.9, many B cells are present, but only those that have BCRs that can combine with the specific

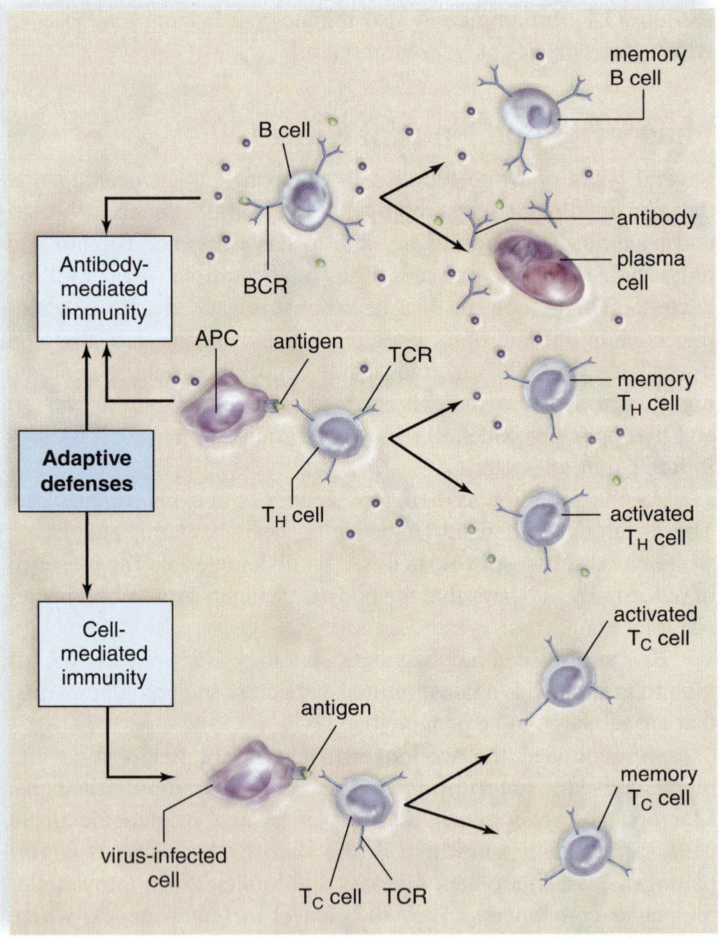

Figure 33.8 Overview of adaptive immune defenses. B cells, helper T (T_H) cells, and cytotoxic T (T_C) cells respond to specific antigens by dividing and differentiating. The BCRs of B cells bind to whole, intact antigens, while the TCRs of T_H and T_C cells only bind to antigens that are processed and presented by MHC proteins on the surface of other cells. When activated by antigens, B cells differentiate into antibody-secreting plasma cells, T_H cells become cytokine-secreting cells, and T_C cells are able to destroy virus-infected or cancer cells. Each cell type also produces memory cells that can respond more quickly to a subsequent exposure to the same antigen.

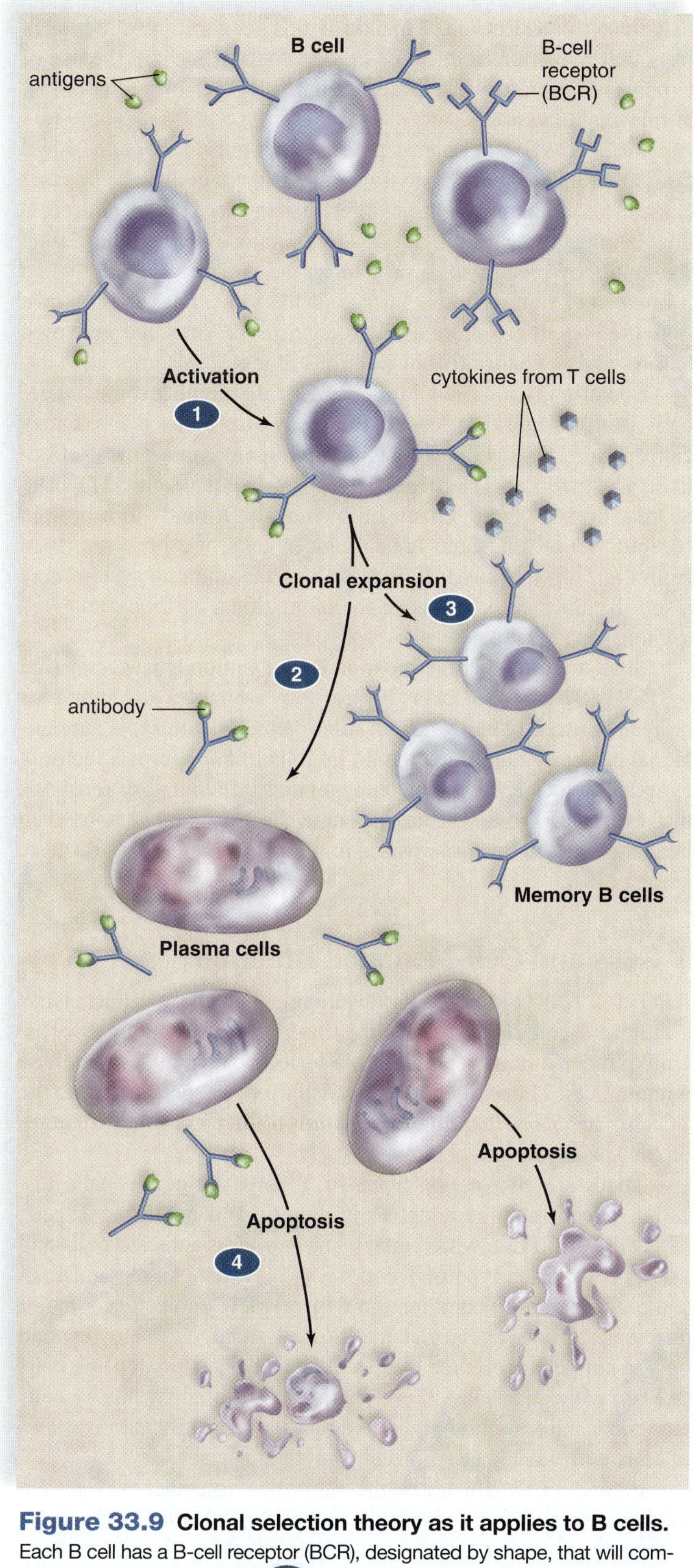

Figure 33.9 Clonal selection theory as it applies to B cells. Each B cell has a B-cell receptor (BCR), designated by shape, that will combine with a specific antigen. (1) Activation of a B cell occurs when its BCR can combine with an antigen (colored green). (2) In the presence of cytokines, the B cell undergoes clonal expansion, producing many plasma cells and memory B cells. Plasma cells secrete antibodies specific to the antigen. (3) Memory B cells immediately recognize the antigen in the future. (4) After the infection passes, plasma cells undergo apoptosis, also called programmed cell death.

Tutorial B Cell Clonal Selection

antigen(s) present go on to divide and produce many new cells. Therefore, the antigen is said to "select" the B cells that will begin dividing. At the same time, cytokines secreted by helper T cells stimulate B cells to differentiate. Many of these B cells become **plasma cells,** which are specialized for secretion of antibodies. Plasma cells are larger than regular B cells because they have extensive rough endoplasmic reticulum for the mass production and secretion of **antibodies** that bind to a specific antigen. Antibodies are the secreted form of the BCR of an activated B cell, and these antibodies react to the same antigen as the original B cell. As the antigen that stimulated the B-cell response is removed from the body, the development of new plasma cells ceases, and those present undergo apoptosis. Other progeny of the dividing B cells become **memory B cells,** so named because these cells always "remember" a particular antigen and make us immune to a particular illness, but not to any other illness.

Defense of the body by B cells is known as **antibody-mediated immunity** (see Fig. 33.8, *top*). It is also called humoral immunity, because these antibodies are present in blood and lymph. (Historically, the term *humor* referred to any fluid normally occurring in the body.)

Structure of Antibodies

The basic unit of antibody structure is a Y-shaped protein molecule with two arms. Each arm has a "heavy" (long) polypeptide chain and a "light" (short) polypeptide chain (Fig. 33.10). These chains have constant (*C*) regions, located at the trunk of the Y, where the sequence of amino acids is set. The variable (*V*) regions at the tips of the Y form two antigen-binding sites, and their shape is specific to a particular antigen. The antigen combines with the antibody at the antigen-binding site in a lock-and-key manner.

Animation Antibody Diversity

The binding of antibodies to an antigen can have several outcomes. Often, the reaction produces a clump of antigens combined with antibodies, termed an *immune complex*. The antibodies in an immune complex are like a beacon, attracting white blood cells that move in for the kill. For example, immune complexes may be engulfed by neutrophils or macrophages, or they may activate NK cells to destroy a cell coated with antibodies. Immune complexes may also activate the complement system. Antibodies may also "neutralize" viruses or toxins by preventing them from binding to specific receptors on cells.

There are several types, or classes, of antibodies, also called **immunoglobulins (Ig)**. The class of antibody is determined by the structure of the antibody's constant region. The major class of antibody found in the blood, called *IgG,* is a single Y-shaped molecule (Fig. 33.10). IgG antibodies are the major type that can cross the placenta from a mother to her fetus, to provide some temporary protection to the newborn. IgG is also found in breast milk, along with a type called IgA. *IgA* is also the main class secreted in milk, tears, and saliva and at mucous membranes. Antibodies of the *IgM* class are pentamers—that is, clusters of five Y-shaped molecules linked together. IgM antibodies are the first antibodies produced during most B-cell responses. As such, their presence is often interpreted as indicating a recent infection. The other major type of antibody is *IgE,* which is mainly bound to receptors on eosinophils and on mast cells in the tissues.

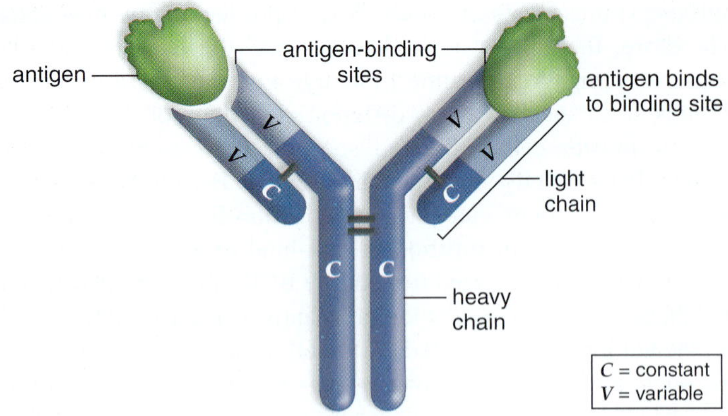

Figure 33.10 Structure of antibodies. An antibody contains two heavy (long) polypeptide chains and two light (short) chains arranged so that there are two variable regions, where a particular antigen is capable of binding with the antibody. The shape of the antigen fits the shape of the binding site.

Monoclonal Antibodies

As has been noted, every plasma cell derived from a single B cell secretes antibodies that bind to a single antigen. This discovery, along with the development of techniques for growing cells in the laboratory, led to the production of **monoclonal antibodies.** Niels Jerne, George Köhler, and César Milstein were awarded the Nobel Prize in 1984 for their development of this technology.

Monoclonal antibodies are produced by cells derived from a single plasma cell; thus, all these antibodies have identical specificity for one antigen. Monoclonal antibodies are typically produced by first immunizing an animal (usually a mouse) with the antigen of interest. The animal is then killed, its spleen is removed, and the spleen cells (including a large number of B cells) are fused with mouse myeloma cells (malignant plasma cells that live and divide indefinitely). The fused cells are called *hybridomas*: *hybrid* because they result from the fusion of two different cells, and *oma* because one of the cells is a cancer cell. The hybridomas are then isolated as individual cells and screened to select only those that are producing the desired monoclonal antibody.

Animation Monoclonal Antibody Production

Research Uses for Monoclonal Antibodies. The ability to quickly produce monoclonal antibodies in the laboratory has made them an important tool for academic research. Monoclonal antibodies are very useful because of their extreme specificity for only a particular molecule. A monoclonal antibody can be used to select out a specific molecule among many others, much like finding needles in a haystack. Then, the target molecule can be purified from all the others that are also present in a sample. In this way, monoclonal antibodies have simplified formerly tedious laboratory tasks.

Medical Uses for Monoclonal Antibodies. Monoclonal antibodies also have many applications in medicine. In one application, they can be used to make quick and certain diagnoses of infections and other conditions. For example, a particular hormone called hCG is present in the urine of a woman only if she is pregnant. An anti-hCG monoclonal antibody can be used to detect this hormone. Thanks to this technology, pregnancy tests that once required a visit to a doctor's office and the use of expensive laboratory equipment can now be performed at home, at minimal expense.

Monoclonal antibodies can be used not only to diagnose infections and illnesses but also to fight them. RSV, a common virus that causes serious respiratory tract infections in very young children, is now being successfully treated with a monoclonal antibody drug. The antibody recognizes a protein on the viral surface, and when it binds very tightly to the surface of the virus, the patient's own immune system can more easily recognize the virus and destroy it before it has a chance to cause serious illness.

Because monoclonal antibodies can distinguish some cancer cells from normal tissue cells, they may also be used to identify cancers at very early stages, when treatment can be most effective. Trastuzumab (Herceptin) is a monoclonal antibody used to treat breast cancer. Given intravenously, it binds to a protein receptor found on some breast cancer cells and prevents them from dividing so quickly. Other cells of the immune system may also kill tumor cells that have the monoclonal antibody attached to their surface.

Since the first therapeutic monoclonal antibody was approved by the FDA in 1986, over 20 are now available, and hundreds more are currently being tested. Adalimumab (Humira) is a monoclonal antibody that binds to and inhibits tumor necrosis factor, a cytokine associated with the exaggerated inflammatory reactions that characterize several autoimmune diseases, and it is used to treat rheumatoid arthritis, psoriatic arthritis, and Crohn's disease, among others.

T Cells and Cell-Mediated Immunity

After a T cell completes its development in the thymus, it has a unique **T-cell receptor (TCR)** similar to the BCR on B cells. Unlike B cells, however, T cells are unable to recognize an antigen without help. The antigen must be displayed, or "presented," to the TCR by an **MHC (major histocompatibility complex) protein** on the surface of another cell.

There are two major types of T cells: **helper T cells** (T_H cells), which regulate adaptive immunity, and **cytotoxic T cells** (T_C cells, or CTLs), which attack and kill virus-infected cells and cancer cells. Each type of T cell has a TCR that can recognize an antigen fragment in combination with an MHC molecule. A major difference, however, is that the T_H cells recognize and respond only to antigens presented by specialized **antigen-presenting cells (APCs)** with MHC class II proteins on their surface, while T_C cells recognize and respond only to antigens presented by various types of cells with MHC class I proteins on their surface.

Figure 33.11 illustrates the important role that APCs such as macrophages and dendritic cells play in stimulating T_H cells. After phagocytizing a pathogen, APCs travel to a lymph node or the spleen, where T_H cells, T_C cells, and B cells congregate. In the meantime, the APC has broken the pathogen apart in a lysosome. A fragment of the pathogen is then displayed in association with an MHC class II protein on the cell's surface, which can bind to and select any T_H cell that has a TCR capable of combining with a particular antigen/MHC combination. This interaction stimulates the

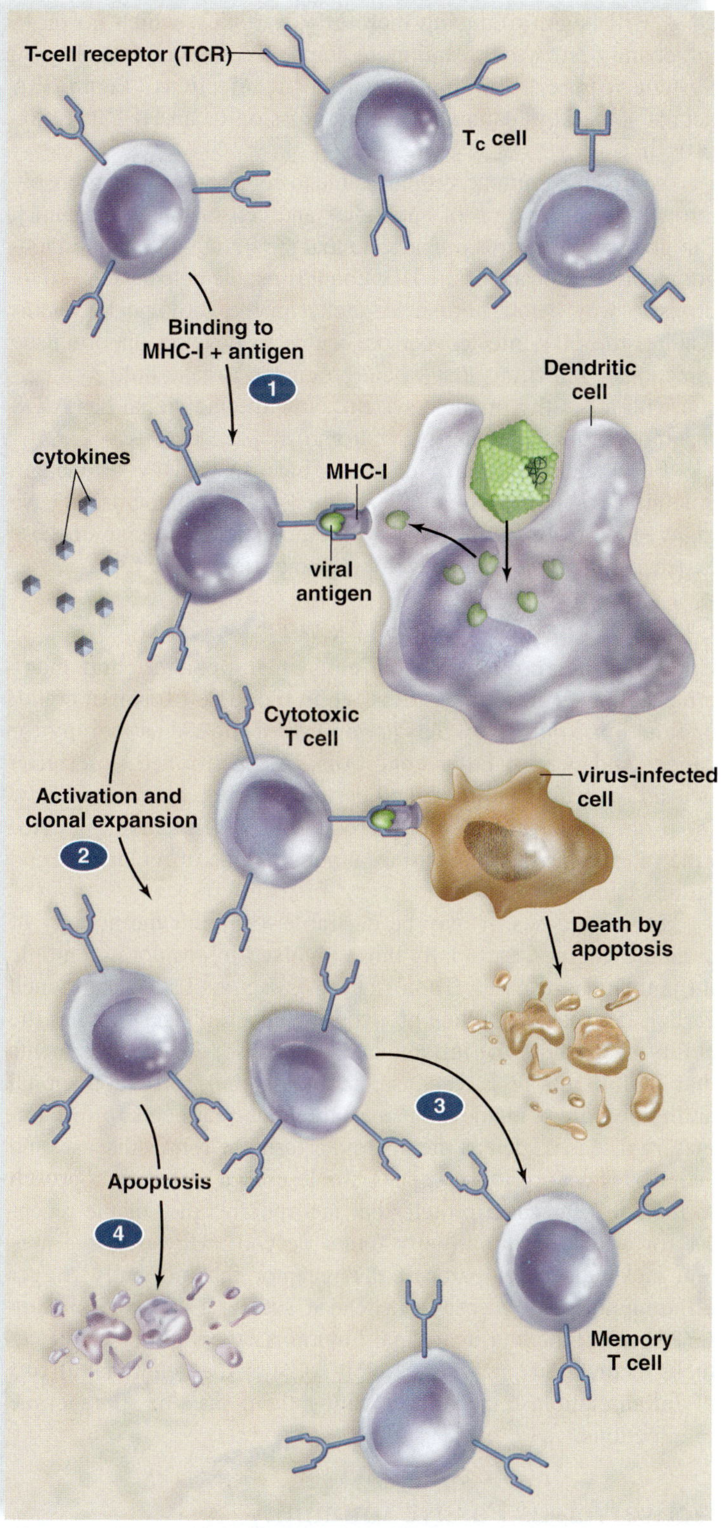

Figure 33.11 Clonal selection theory as it applies to T cells. Each T cell has a T-cell receptor (TCR) designated by a shape that will combine only with a specific antigen. (1) Activation of a T cell occurs when its TCR can combine with an antigen. A dendritic cell presents the antigen (colored green) in the groove of an MHC class I (MHC-I) molecule. (2) The T_C cell subsequently undergoes clonal expansion, and many copies of the same type of T cell are produced. (3) Memory T cells provide protection, should the same antigen enter the body again at a future time. (4) After the immune response has been successful, the majority of T cells undergo apoptosis.

Tutorial
T Cell Clonal Selection

T_H cell to divide, thereby cloning itself. Some of these proliferating T_H cells secrete various cytokines that influence B cells, T_C cells, and other cell types. Many other cloned T_H cells become **memory T cells.** Like memory B cells, memory T cells are long-lived, and their number is far greater than the original number of T cells that could recognize a specific antigen. Therefore, when the same antigen enters the body later on, the immune response may occur so rapidly that no detectable illness occurs. As the antigen is cleared, activated T cells become susceptible to apoptosis, just as plasma cells do during a B-cell response.

By releasing a variety of cytokines, activated helper T cells are largely responsible for **cell-mediated immunity** (see Fig. 33.8, *middle*), which is the destruction or elimination of pathogens and other threats by T_C cells, macrophages, natural killer cells, or other cells. For example, certain cytokines cause T_C cells to proliferate (see Fig. 33.8, *bottom*), while others activate macrophages to more actively seek, engulf, and destroy pathogens. Other cytokines released by T_H cells influence B-cell activities, even though B cells are responsible for antibody-mediated immunity. Cytokines are also used for immunotherapy purposes, as discussed in the next section.

Functions of Cytotoxic T Cells

A major difference in recognition of an antigen by helper T cells and cytotoxic T cells is that helper T cells recognize an antigen only in combination with MHC class II proteins, while T_C cells recognize an antigen only in combination with MHC class I proteins, which are found on almost all types of cells. The outcome of this recognition is also very different—activated T_H cells secrete a variety of cytokines that "help" other cells, but activated T_C cells specialize in killing other cells.

The cytoplasm of a T_C cell contains storage vacuoles that are filled with a chemical called *perforin,* as well as enzymes called *granzymes* (Fig. 33.12). After a T_C cell is activated, it travels via the bloodstream to areas of inflammation in the body. On migrating into the tissues, if the T_C cell recognizes its unique target combination of antigen plus MHC-I protein on a virus-infected cell or a cancer cell, it releases perforin molecules, which perforate the plasma membrane, forming a pore. Granzymes then use the pore to enter the abnormal cell, causing it to undergo apoptosis and die. Once T_C cells have released the perforins and granzymes, they move on to the next target cell. Because of this ability, T_C cells have been referred to as the "serial killers" of the immune system. As with B cells and T_H cells, most activated T_C cells are short-lived, while others become long-lived memory T_C cells, ever ready to defend against the same virus or kill the same type of cancer cell again.

HIV Infection

At the end of 2009, an estimated 33.3 million people worldwide were infected with **human immunodeficiency virus (HIV)**, and more than 25 million people have died of AIDS since 1981. Helper T cells are the major host cell for HIV, but macrophages and dendritic cells can also be infected.

After HIV enters a host cell, it reproduces as was illustrated in Figure 20.4, inserting itself into the host cell DNA, where it will remain for the life of the cell. Perhaps years later, this hidden viral

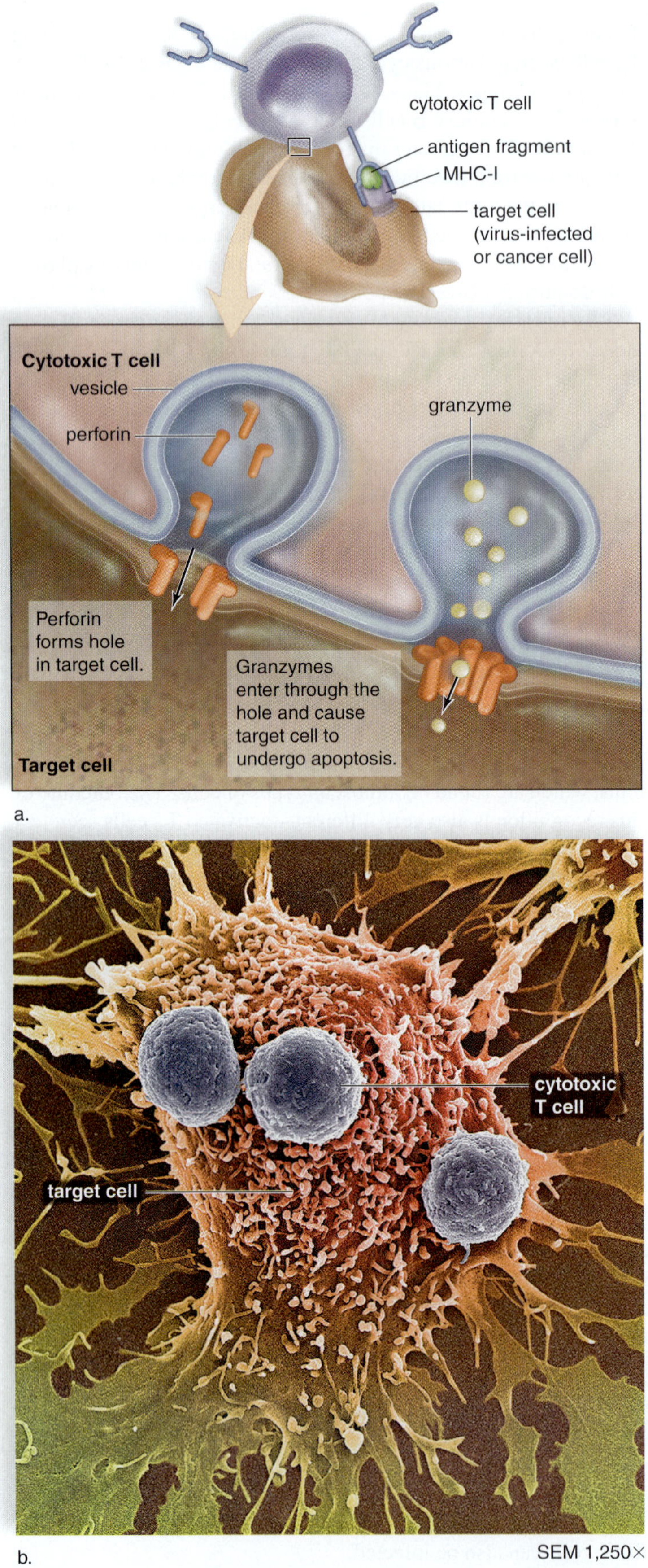

Figure 33.12 Cell-mediated immunity. **a.** How a T cell destroys a virus-infected cell or cancer cell. **b.** The scanning electron micrograph shows cytotoxic T cells attacking and destroying a cancer cell (target cell).

DNA will begin producing more HIV particles, which go on to infect and destroy more and more helper T cells. In the Biological Systems feature, "AIDS and Opportunistic Infections," Figure 33A details some key features of the progression of an HIV infection over time.

At first the immune system is able to control the virus, probably through a combination of antibodies and cell-mediated immunity. But gradually the virus mutates, so that the immune system no longer recognizes it, and as the HIV count rises, the helper T-cell count drops to way below normal. As noted in the Biological Systems feature, the HIV-infected person begins to develop opportunistic infections—infections that would be unable to take hold in a person with a healthy immune system. Now the individual has AIDS (acquired immunodeficiency syndrome).

Fortunately, antiretroviral drugs have made HIV a manageable infection in most treated individuals, although the drugs have many side effects. Unfortunately, there is still a shortage of these expensive drugs in many developing countries.

Cytokines as Therapeutic Agents

Cytokines are produced by T cells, macrophages, and many other cells. Because they affect white blood cell formation and/or function, some cytokines are approved for use as therapy for cancer and certain other conditions. As mentioned, interferon is used to treat hepatitis C and certain cancers, and it may slow the progression of multiple sclerosis. The T-cell activating cytokine interleukin-2 is used to treat some forms of melanoma and kidney cancer.

In some cases, it may be desirable to inhibit the activity of certain cytokines, especially those involved in promoting chronic inflammatory diseases. Tumor necrosis factor (TNF), mentioned earlier, is a major cytokine produced by macrophages that has the ability to promote the inflammatory response. Several treatments that inhibit the TNF response, including anti-TNF monoclonal antibodies, are being developed as potential treatments for inflammatory diseases such as rheumatoid arthritis, Crohn's disease, and asthma. The drug etanercept (Enbrel) uses a different approach to block TNF; it is a protein that incorporates part of the receptor for TNF that is normally found on cells. However, the protein lacks the other parts of the receptor that normally trigger inflammatory responses, so instead it acts as a "decoy," binding to TNF and blocking its normal functions. As would be expected, these medications may also inhibit some of the beneficial effects of inflammation, and patients taking them may be more prone to infections.

Active Versus Passive Immunity

In general, adaptive immune responses can be induced either actively or passively. **Active immunity** occurs when an individual produces his or her own immune response against an antigen. For example, when you catch a cold, you recover because your body produces the T-cell and B-cell responses that eventually clear the offending viruses from your body. Active immunity can also be induced artificially when a person is well, to prevent infection in the future. **Immunization** involves the use of vaccines, substances that contain an antigen to which the immune system

THEME **Biological Systems**

AIDS and Opportunistic Infections

AIDS (acquired immunodeficiency syndrome) is caused by HIV, the human immunodeficiency virus. An HIV infection leads to the eventual destruction of helper T (T_H) cells. HIV kills T_H cells by directly infecting them, and it causes many uninfected T_H cells to die by a variety of mechanisms. Many HIV-infected T_H cells are also killed by the person's own immune system.

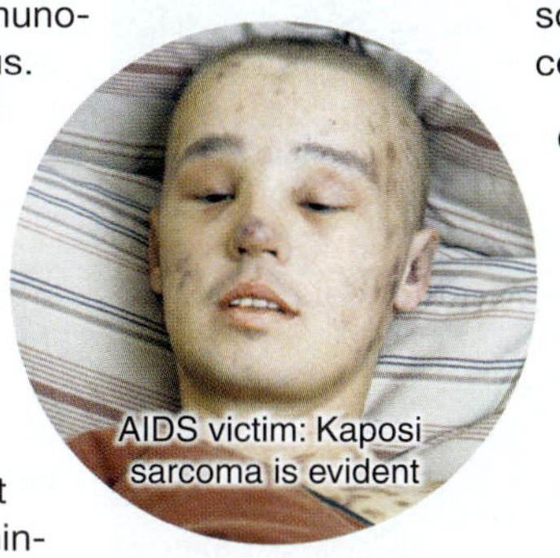

AIDS victim: Kaposi sarcoma is evident

A healthy individual typically has 800–1,000 T_H cells per mm³ of blood (Fig. 33A). After an initial HIV infection, it may take several years for an individual's T_H-cell numbers to drop below 500 cells per mm³ of blood, at which point the HIV-infected individual usually begins to suffer from many unusual types of infections that would not cause disease in a person with a healthy immune system. Such infections are known as opportunistic infections (OIs). The U.S. Centers for Disease Control and Prevention (CDC) has defined three categories of HIV infection, based mainly on the types of OIs seen in the patient. These OIs also tend to be associated with a decreasing number of T_H cells in the blood:

CDC category A (>500 T_H cells/mm³ blood)

- Typically, no OIs seen; may have persistently enlarged lymph nodes

CDC category B (200–499 T_H cells/mm³ blood)

- Shingles, a painful infection with *Varicella zoster* (chickenpox) virus
- Candidiasis, or thrush, a fungal infection of the mouth, throat, or vagina

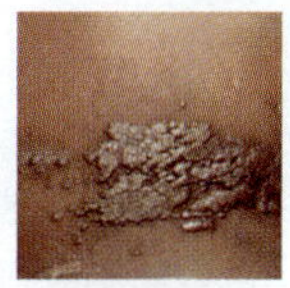

Shingles

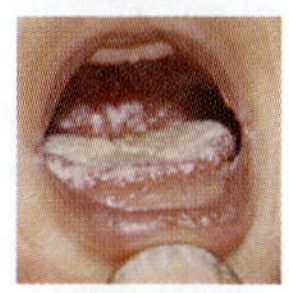

Candidiasis

CDC category C (< 200 T_H cells/mm³ blood)

- Pneumocystis pneumonia, a fungal infection, causing the lungs to become useless as they fill with fluid and debris
- Kaposi sarcoma, a cancer of blood vessels due to human herpesvirus 8, giving rise to reddish-purple, coin-sized spots and lesions on the skin
- Toxoplasmic encephalitis, a protozoan infection characterized by severe headaches, fever, seizures, and coma
- *Mycobacterium avium* complex (MAC), a bacterial infection resulting in persistent fever, night sweats, fatigue, weight loss, and anemia
- Cytomegalovirus, a viral infection that leads to blindness, inflammation of the brain, and throat ulcerations

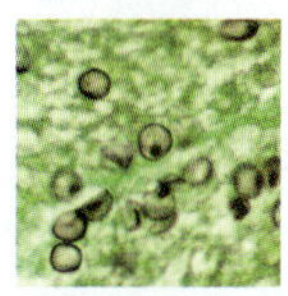

Pneumocystis pneumonia

Thanks to the development of powerful drug therapies that inhibit the life cycle of HIV, people infected with HIV in the United States are suffering a lower incidence of OIs than in the 1980s and 1990s.

Questions to Consider

1. Why has it been so difficult to develop an effective vaccine for HIV?
2. What are some possible differences between the types of OIs typically seen in category B and those seen in category C (i.e., why do shingles and candidiasis occur in B, but others more commonly in C)?
3. What, if any, obligation do relatively wealthy countries, such as the United States, have in providing anti-HIV drugs to poorer countries?

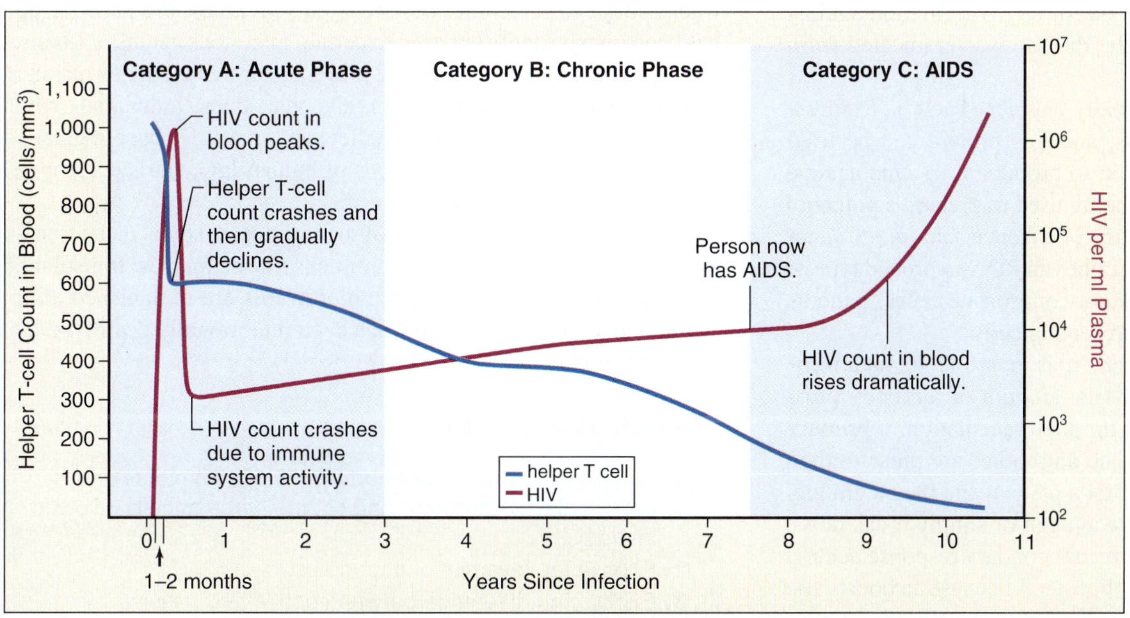

Figure 33A Progression of HIV infection during its three stages, called categories A, B, and C. In category A, the individual may have no symptoms or very mild symptoms associated with the infection. By category B, opportunistic infections—such as candidiasis, shingles, and diarrhea—have begun to occur. Category C is characterized by more severe opportunistic infections and is clinically described as AIDS.

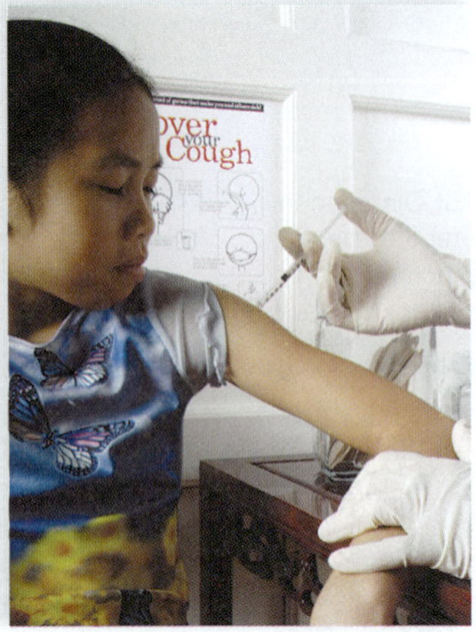

Figure 33.13 Antibody titers. During immunization, the primary response after the first injection of a vaccine is minimal, but the secondary response, which occurs after the second injection, shows a dramatic rise in the amount of antibody present in plasma.

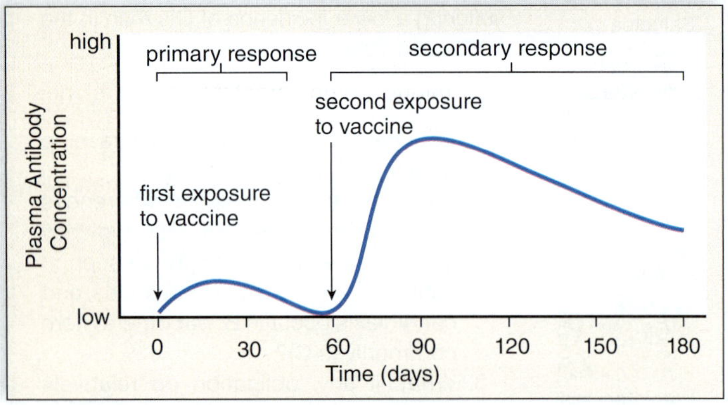

responds (Fig. 33.13). Traditionally, vaccines are the pathogens themselves, or their products, that have been treated, so that they are no longer virulent (able to cause disease). The use of vaccines has been effective in reducing the rates of bacterial diseases such as diphtheria, tetanus, and whooping cough, as well as viral diseases such as measles, mumps, and rubella. In fact, vaccination against smallpox was so successful that the disease was eradicated from the planet in 1977.

Today, it is possible to genetically engineer bacteria to mass-produce a protein from pathogens, and this protein can be used as a vaccine. This method was used to produce a vaccine against hepatitis B, a viral disease, and is being used to prepare a potential vaccine against malaria. The Nature of Science feature, "Cancer Vaccines: Becoming a Reality," describes an FDA-approved type of cancer vaccine that boosts the immunity of prostate cancer patients by stimulating their own antigen-presenting cells.

After most vaccines are given, it is possible to determine the antigen-specific antibody titer (the amount of antibody present in a sample of plasma). After the first vaccination, a primary response occurs. For several days, no antibodies are present; then the titer rises slowly, followed by first a plateau and then a gradual decline as the antibodies bind to the antigen or simply break down (Fig. 33.13). After a second exposure, a secondary response occurs. The second exposure is called a "booster," because it boosts the immune response to a high level. The high levels of antigen-specific T cells and antibodies are expected to prevent disease symptoms if the individual is later exposed to the disease-causing agent. Even years later, if the antigen enters the body, memory B cells can quickly give rise to more plasma cells capable of producing the correct type of antibody.

Figure 33.14 Passive immunity. Breast-feeding is believed to prolong the passive immunity an infant receives from the mother during pregnancy because antibodies are present in the mother's milk.

Passive immunity occurs when an individual receives another person's antibodies or immune cells. The passive transfer of antibodies is a common natural process. For example, newborn infants are passively immune to some diseases, because antibodies have crossed the placenta from the mother's blood. These antibodies soon disappear, however, so that within a few months, infants become more susceptible to infections as their own immune system must now protect them. Breast-feeding may prolong the natural passive immunity an infant receives from its mother, because antibodies are present in the mother's milk (Fig. 33.14).

Even though passively administered antibodies last only a few weeks, they can sometimes be used to prevent illness in a patient who has been unexpectedly exposed to certain infectious agents or toxins. Examples of diseases that may be prevented or treated in this manner include rabies, tetanus, botulism, and snakebites. Individuals with certain types of genetic immunodeficiencies may also benefit greatly from regular intravenous injections of human IgG, extracted from a large, diverse adult population.

Instead of antibodies, cells of the immune system can be transferred into a patient. The best example is a bone marrow transplant, in which stem cells that produce blood cells are replenished after a cancer patient's own marrow has been intentionally destroyed by radiation or chemotherapy.

Check Your Progress 33.4

1. Distinguish between antibody-mediated immunity and cell-mediated immunity, and list the types of cells involved in each.
2. Explain the diversity of antibodies.
3. Explain the difference between active and passive immunity, and list three examples of each.

THEME **Nature of Science**

Cancer Vaccines: Becoming a Reality

Approximately 200 types of human cancer have been identified, all of which involve an uncontrolled proliferation of cells. Some of these cancer cells express unusual molecules on their surface, which may allow the immune system to identify and destroy them. Unfortunately, many cancer cells grow unchecked in the body, apparently avoiding the immune system's many weapons.

Vaccines against infectious disease are an essential component of modern medicine. Most vaccines are administered before a person is exposed to an infection, but in a few cases (e.g., rabies, tetanus) vaccines can be used therapeutically, after exposure to a pathogen. Medical researchers are developing vaccines to treat certain types of cancer, and in April 2010 the U.S. Food and Drug Administration approved the first therapeutic cancer vaccine for use in humans.

Prostate cancer is the second most common type of cancer among men in the United States (behind skin cancer); about 30,000 men died from the disease in 2013. Sipuleucel-T (Provenge) can now be used to treat certain types of advanced prostate cancer.

Provenge is prepared in a very different manner than a typical vaccine (Fig. 33B). First, the patient undergoes a procedure called leukapheresis, in which a catheter is inserted into a large vein and blood flows into a machine that removes some of the white blood cells. Another catheter returns the plasma and red blood cells to the patient. Within the white blood cells are the antigen-presenting cells, such as dendritic cells. These cells are shipped to another lab, where they are exposed to a protein, called PAP, that is commonly expressed on prostate cancer cells. The APCs can take up the PAP molecule and present it on MHC molecules. These APCs are also activated using a cytokine called granulocyte-macrophage-CSF. Three days later, the cells are shipped back and infused into the patient. Typically, three doses of Provenge are administered, each containing a minimum of 50 million stimulated cells, at a cost of about $31,000 per dose.

Once inside the patient, the APCs are able to activate a cytotoxic T-cell response that targets prostate cancer cells for destruction. In a trial of 512 men with advanced prostate cancer, the median survival for patients receiving Provenge was 25.8 months, compared to 21.7 months for those who received a placebo.

Although this approach to cancer therapy is a significant advance, there are also drawbacks, including the fact that a separate vaccine must be prepared for each patient. Other cancer vaccines are currently in clinical testing. Another study showed that prostate cancer patients who received Prostvac, a vaccine containing harmless viruses that express prostate specific antigen (PSA), lived an average of 8.5 months longer than patients who received a placebo injection.[1] Researchers are also making progress on therapeutic vaccines for cancers of the lung, breast, prostate, colon, and skin.

Questions to Consider

1. In what ways is this approach to vaccination similar to vaccination for infectious diseases? In what ways is it different?
2. What are some biological limitations to this approach? In other words, how might some prostate cancers become resistant to this type of cancer treatment?
3. Do you see any ethical issues with the cost of this cancer vaccine? Even if a prostate cancer patient lives only an additional 4 months, is it ever reasonable to place a monetary value on human life?

[1]Kantoff, Philip W., Schuetz, Thomas J., et al. 2010. "Overall Survival Analysis of a Phase II Randomized Controlled Trial of a Poxviral-based PSA-targeted Immunotherapy in Metastatic Castration-resistant Prostate Cancer," *Journal of Clinical Oncology* 28:1099–1105.

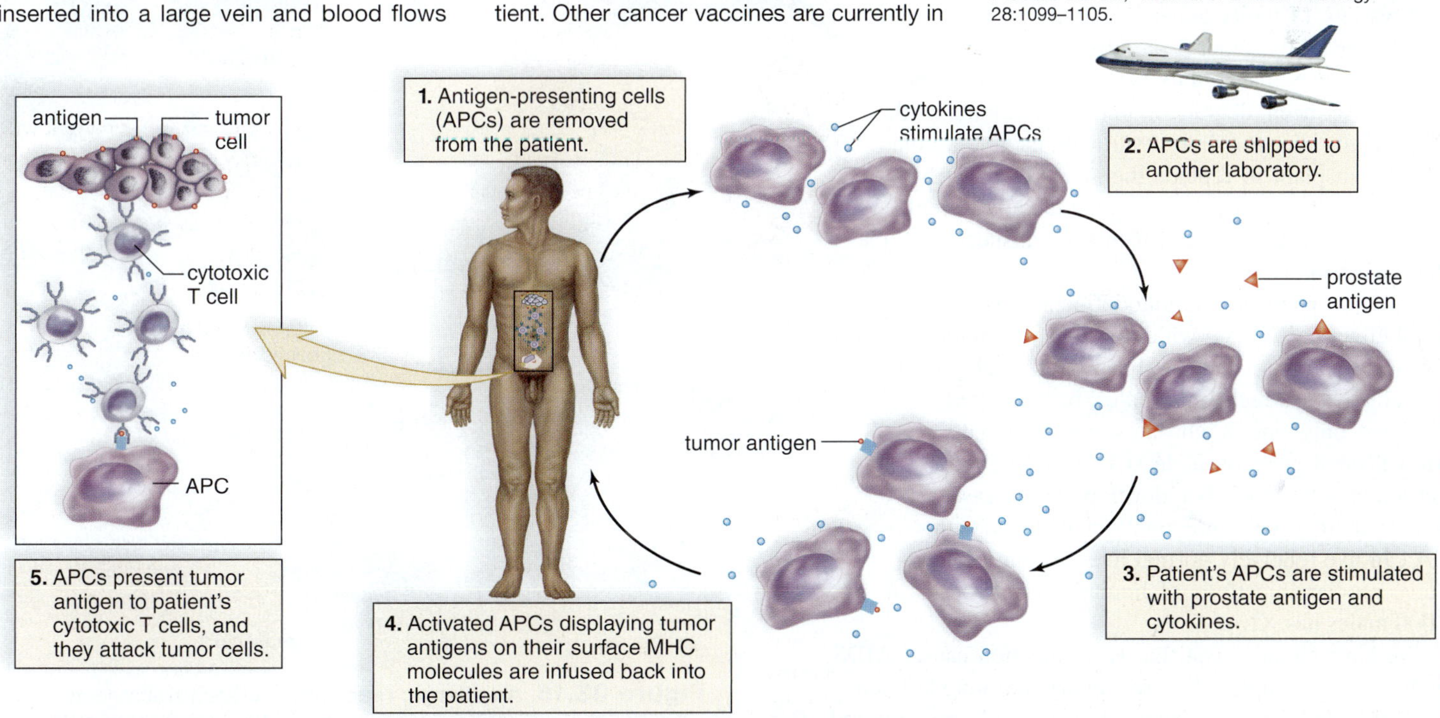

Figure 33B Prostate cancer vaccine. The first therapeutic vaccine for a human cancer involves a complex procedure in which the patient's own antigen-presenting cells (APCs) are removed, shipped to another laboratory, stimulated with cytokines and prostate antigen, then returned to be injected back into the patient.

33.5 Immune System Disorders and Adverse Reactions

Learning Outcomes

Upon completion of this section, you should be able to

1. Describe the two main types of immunodeficiency disorders and provide examples of each.
2. Discuss the most common immunological mechanisms responsible for allergies and how these may be treated.
3. Define *autoimmune disease* and list several specific examples of these diseases.
4. Explain the types of precautions that must be taken when transplanting organs.

The immune system can be thought of as a "double-edged sword." It is essential for our health and survival, as demonstrated by the diseases, some of them fatal, that occur in people who are immunodeficient. In other instances, the immune system may work against the best interests of the body, as occurs in allergies, autoimmune disorders, and rejection of transplanted organs.

Immunodeficiencies

A number of immunodeficiency disorders are known, but all result in some degree of increased susceptibility to infections. Primary immunodeficiencies are genetic, meaning that they are passed from parents to offspring. For example, in severe combined immunodeficiency (SCID), both T cells and B cells are either lacking completely or not functioning well enough to protect the body from a variety of infections that are not a problem for most people. SCID only occurs in about 1 in 500,000 births.

A variety of faulty genes can cause SCID. In most cases, however, by about 3 months of age, when most of the antibodies that infants have obtained from their mother have been degraded, untreated infants with SCID usually die. Possible treatments include a bone marrow transplant to replace the stem cells that form all of our white blood cells and gene therapy to replace the faulty DNA. If these treatments are unsuccessful, the outcome is usually poor.

Another primary immunodeficiency is X-linked agammaglobulinemia (XLA), which is due to a mutated gene on the X chromosome that is needed for proper development of B cells. XLA affects only males, because their cells have only one X chromosome. A female with a normal gene on at least one of her two X chromosomes does not develop the disease. Because their T cells are unaffected, boys with XLA can live relatively normal lives as long as they receive regular injections of human IgG. About 1 in 50,000 males has XLA.

We have already seen that HIV infection causes AIDS, which is an example of a secondary immunodeficiency. These disorders are not genetic but, instead, are acquired after birth. Besides infections, other potential causes of secondary immunodeficiencies include malnutrition, irradiation, certain drugs and toxins, and certain cancers. Some of these can be cured by addressing their cause.

Allergies

Allergies are hypersensitivities to substances, such as pollen, food, or animal hair, that ordinarily would do no harm to the body. The response to these antigens, called allergens, usually includes some degree of tissue damage.

An **immediate allergic response** can occur within seconds of contact with an allergen. The response is caused by antibodies of the IgE class (Table 33.1). IgE antibodies are attached to receptors on the plasma membrane of mast cells in the tissues, as well as to basophils and eosinophils in the blood. When an allergen attaches to these IgE antibodies, the cells release histamine and other substances that bring about the symptoms of an allergy (Fig. 33.15). When an allergen such as pollen is inhaled, histamine stimulates the inflammatory response in the mucous membranes of the nose and eyes typical of *hay fever.* If a person has **asthma,** the airways leading to the lungs constrict, resulting in difficult breathing accompanied by wheezing. An allergen in food typically causes nausea, vomiting, and diarrhea.

As described in the story that opened this chapter, **anaphylactic shock** is an immediate allergic response that occurs after an allergen has entered the bloodstream. Bee stings, foods, various medications, and latex rubber are all known to cause this reaction in some individuals. Anaphylactic shock is characterized by a sudden and life-threatening drop in blood pressure, due to an increased

Figure 33.15 An allergic reaction. An allergen attaches to IgE antibodies, which then cause mast cells to release histamine and other chemicals that are responsible for the allergic reaction.

Table 33.1 Comparison of Immediate and Delayed Allergic Responses

	Immediate Response	Delayed Response
Onset of Symptoms	Takes several minutes	Takes 2 to 3 days
Lymphocytes Involved	B cells	T cells
Immune Reaction	IgE antibodies	Cell-mediated immunity
Type of Symptoms	Hay fever, asthma, and many other allergic responses	Contact dermatitis (e.g., poison ivy)
Therapy	Antihistamine and epinephrine	Cortisone

dilation of the capillaries by histamine throughout the body. The smooth muscle lining the bronchi may also be strongly stimulated to constrict, resulting in an inability to breathe. Injecting epinephrine can counteract this reaction until medical help is available, and some people carry an epinephrine-containing, spring-loaded syringe (sometimes called an EpiPen) for this purpose.

Mild to moderate allergies are usually treated with antihistamines, which compete with histamine for binding to histamine receptors. In more serious cases, injections of the allergen can be given in an effort to stimulate the immune system to produce high quantities of IgG against the allergen. The hope is that these IgG antibodies will combine with the allergen molecules before they have a chance to reach the IgE antibodies. A monoclonal antibody called Xolair is also available; it blocks the binding of IgE to its receptor on inflammatory cells.

A **delayed allergic response** is initiated by memory T cells at the site of allergen contact in the body. The allergic response is regulated by the cytokines secreted by these "sensitized" T cells at the site. A classic example of a delayed allergic response is the skin test for tuberculosis (TB). When the test result is positive, the tissue where the antigen was injected becomes red and hardened. This indicates prior exposure to *Mycobacterium tuberculosis,* the bacterium that causes TB. Contact dermatitis, which occurs when a person's skin reacts to poison ivy, jewelry, cosmetics, or many other substances that touch the skin, is another example of a delayed allergic response.

Autoimmune Diseases

When a person has an **autoimmune disease,** the immune system mistakenly attacks the body's own cells or molecules. Exactly what causes autoimmune diseases is not known. In some cases, there appears to be a genetic tendency to develop autoimmune diseases. Many autoimmune diseases also seem to occur after an individual has recovered from an infection. It is also known that certain antigens of microbial pathogens can resemble antigens found in their host, a phenomenon known as molecular mimicry. A good example of this is rheumatic fever, which sometimes follows an infection with bacteria of the genus *Streptococcus.* Certain proteins in the cell wall of these bacteria are known to resemble proteins in the heart; as a result antibodies formed against the bacterial proteins may cause inflammation of the heart, which can continue even after the bacteria have been cleared from the body.

Some autoimmune diseases affect only specific tissues. Rheumatoid arthritis is a common autoimmune disorder that causes recurring inflammation in synovial joints (Fig. 33.16). Complement proteins, T cells, and B cells all participate in the destruction of the joints, which eventually become immobile. In myasthenia gravis, antibodies interfere with the functioning of neuromuscular junctions, causing muscular weakness. In multiple sclerosis, T cells attack the myelin sheath of nerve fibers, causing a variety of symptoms related to the defective transmission of messages by nerves.

Systemic lupus erythematosus (lupus) is a chronic autoimmune disorder that affects multiple tissues and organs. It is characterized by the production of antibodies that react with the DNA contained in almost every cell of the body. The symptoms vary somewhat, but most patients experience a characteristic skin rash (Fig. 33.17), joint pain, and kidney damage, which may be life-threatening. About 500,000 to 1.5 million people in the United States have lupus, 90% of whom are women of childbearing age. Because little is known about the origin of autoimmune disorders, no cures are available. The symptoms can sometimes be controlled using immunosuppressive drugs, such as cortisone, but these drugs can have serious side effects.

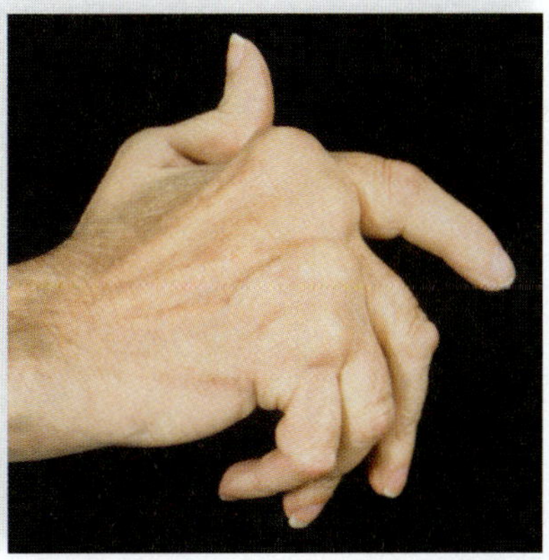

Figure 33.16 Rheumatoid arthritis. Rheumatoid arthritis is caused by recurring inflammation in synovial joints, due to immune system attack.

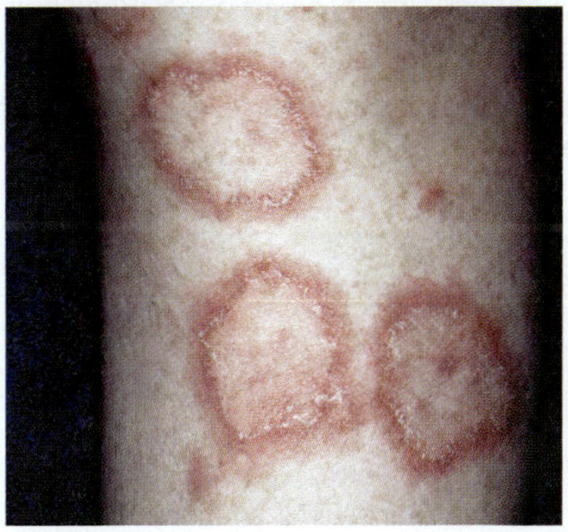

Figure 33.17 Systemic lupus. The characteristic rash of systemic lupus.

Transplant Rejection

Certain organs, such as the skin, heart, and kidneys, could be transplanted relatively easily from one person to another if the body did not attempt to reject them. Unfortunately, like the immune system itself, MHC proteins are a double-edged sword. In addition to their beneficial role in presenting antigens to T cells, MHC proteins are major targets of the immune response during the rejection of a transplanted organ.

The MHC proteins of different individuals differ by the sequence of their amino acids. The immune system will attack any foreign tissue that bears MHC antigens that are different from those of the individual. Ideally, a transplant donor would have exactly the same type of MHC proteins as those of the recipient; however, it is difficult to find a perfect MHC match, but the odds of organ rejection can be reduced by administering immunosuppressive drugs. Two commonly used drugs, cyclosporine and tacrolimus, inhibit the production of certain cytokines by T cells.

Xenotransplantation, the transplantation of animal tissues and organs into humans, is a potential way to solve the shortage of organs from human donors. The pig is the most popular choice as an organ source, because pig organs are generally the right size, and pigs are a common meat source and thus readily available. Genetic engineering can make pig organs less antigenic by removing the MHC antigens. The ultimate goal is to make pig organs as widely accepted as type O blood cells.

Other researchers hope that tissue engineering, including the production of human organs from stem cells, will someday do away with the problem of rejection. Scientists have recently grown new heart valves in the laboratory using stem cells gathered from amniotic fluid following amniocentesis, and surgeons have sucessfully used lab-grown urinary bladder tissue to rebuild defective bladders in human patients.

Check Your Progress 33.5

1. Explain the general type of abnormality that causes most primary immunodeficiencies.
2. Describe the treatments for autoimmune diseases.
3. Define *xenotransplantation.*

CONNECTING *the* CONCEPTS *with the* THEMES

Evolution

- Cells specialized for phagocytizing pathogens can be found in some of the most primitive multicellular organisms.
- Insects such as *Drosophila melanogaster* have a relatively advanced innate immune system, including the ability to recognize pathogens by their expression of pathogen-associated molecular patterns.
- Although the origins of the adaptive immune system are not completely understood, the ability to generate specific antigen receptors occurred early in vertebrate history.

Nature of Science

- Research targeted toward the development of new medical treatments for infectious disease has also resulted in an increased understanding of basic immunological mechanisms.
- Medical investigators hope that cancer vaccines will help trigger the body's adaptive immune response to target cancer cells that otherwise may go undetected.

Biological Systems

- Immune systems in animals can range from a loose association of phagocytic cells to the vastly complicated interactions seen in the mammalian immune system.
- The innate and adaptive immune systems interact with and influence each other in a number of ways to provide both short-term and long-term protection.

MEDIA STUDY TOOLS http://connect.mheducation.com

SMARTBOOK® Study smart. SmartBook helps you maximize your study time by identifying what you know and don't know. The recharge feature in SmartBook helps you prepare for the test by identifying the material that you are most likely to forget.

connect® BIOLOGY Find even more resources to learn the chapter concepts including animations, tutorial videos, and interactive practice questions.

MP3 Files	Animations	Tutorials	Video
33.2 Lymphatic System **33.3** Barriers and Nonspecific Defenses	**33.3** Inflammatory Response • Activation of Complement • Antiviral Activity of Interferon **33.4** Antibody Diversity • Monoclonal Antibody Production	**33.3** Inflammatory Response **33.4** B Cell Clonal Selection • T Cell Clonal Selection	**33.3** Neutrophils

SUMMARIZE

33.1 Evolution of Immune Systems

The **immune system** protects an organism from a variety of threats. From the available evidence, it appears that **innate immunity** is much older than adaptive immunity, because aspects of innate immunity are present in relatively simple multicellular animals. Innate immunity is quite well developed in insects and often involves the recognition of pathogen-associated molecular patterns (PAMPs). Available evidence indicates that **adaptive immunity,** or the ability to generate a diverse array of antigen receptors that can recognize very specific **antigens,** first appeared in an ancestor of the jawed vertebrates.

33.2 The Lymphatic System

The **lymphatic system** includes the **lymphatic vessels** and **lymphoid (lymphatic) organs.** The lymphatic system (1) removes excess interstitial fluid (called **lymph**) collected by **lymphatic capillaries;** (2) absorbs fats from the small intestine; and (3) produces and distributes lymphocytes.

Lymphocytes are produced in the **primary lymphoid organs** (**red bone marrow** and **thymus**), and lymphocytes respond to antigens in **secondary lymphoid organs** (e.g., **lymph nodes** and **spleen**). **T cells** (T lymphocytes) mature in the thymus, while **B cells** (B lymphocytes) mature in the red bone marrow.

33.3 Innate Immune Defenses

Two types of **immunity** protect the body from infections and other threats. Innate defenses are always present, or they occur very soon after exposure to an infection. These include barriers to entry, the **inflammatory response** (which includes **histamine** released from tissue-dwelling **mast cells**), phagocytes (including **neutrophils, eosinophils, macrophages,** and **dendritic cells**) and **natural killer cells,** and protective proteins (including **complement** and **interferons**).

33.4 Adaptive Immune Defenses

Adaptive (also called acquired) immunity involves the B cells and T cells. These cells have unique **antigen receptors** on their surface. According to the **clonal selection theory,** upon binding of its antigen receptor with with a specific antigen, B cells and T cells divide and differentiate. The antigen receptor on a B cell is called a **B-cell receptor** (**BCR**), and clonal selection of B cells forms **plasma cells** and **memory B cells.** Plasma cells secrete **antibodies** and are responsible for **antibody-mediated immunity.**

Antibodies, or **immunoglobulins** (Ig), are Y-shaped molecules that have at least two binding sites for a specific antigen. **Monoclonal antibodies** are produced by cells derived from a single plasma cell fused to a myeloma cancer cell, and have a variety of applications in science and medicine.

The two main types of T cells are **helper T cells** (T_H cells) and **cytotoxic T cells** (T_C cells, or CTLs). For a T_H cell to recognize an antigen, its **T-cell receptor** (**TCR**) must bind to antigen presented by class II **MHC** (**major histocompatibility complex**) **proteins** on the surface of **antigen-presenting cells** (**APCs**). Types of APCs include dendritic cells and macrophages. Thereafter, the activated T cell undergoes clonal expansion, forming activated T cells and **memory T cells** (T_H cells, in this case). Activated T_H cells produce cytokines that affect many other immune cells, resulting in **cell-mediated immunity. HIV** (**human immunodeficiency virus**) infects and destroys T_H cells.

T_C cells recognize antigens presented by MHC class I proteins on the surface of virus-infected or cancer cells. They then kill these cells by releasing perforin and granzymes, inducing apoptosis.

Active immunity occurs as a response to an illness or to **immunization** with a vaccine. **Passive immunity** is needed when an individual is in immediate danger of succumbing to an infectious disease. Passive immunity can occur naturally (as in the transfer of antibodies from mother to infant) or may be used as a medical treatment.

33.5 Immune System Disorders and Adverse Reactions

Immunodeficiencies can be primary (genetic) or secondary (due to some other cause). **Allergies,** as seen in hay fever or in **asthma,** occur when the immune system reacts to substances not normally recognized as dangerous. An **immediate allergic response** is mediated by IgE antibodies, and in its most severe form can result in **anaphylactic shock.** A **delayed allergic response,** such as occurs in the tuberculosis skin test, is mediated by T cells. **Autoimmune disease** occurs when the immune system attacks the body's own cells or tissues. In a transplant rejection, the immune system is usually responding to unmatched MHC proteins on the cells of a donated organ.

ASSESS

Choose the best answer for each question.

33.1 Evolution of Immune Systems

1. Receptors for pathogen-associated molecular patterns, or PAMPS, are found in all but which type of organism?
   a. amoebas
   b. fruit flies
   c. humans
   d. plants
2. Which of these are examples of pathogen-associated molecular patterns?
   a. bacterial cell wall components
   b. double-stranded viral RNA
   c. fungal cell wall components
   d. All of these are correct.
3. True adaptive immunity, involving the production of a large number of diverse antigen receptors, first evolved in
   a. amphibians.
   b. invertebrates.
   c. jawed vertebrates.
   d. mammals.

33.2 The Lymphatic System

4. Like veins, lymphatic vessels
   a. have thick walls of smooth muscle.
   b. contain valves for a one-way flow of fluids.
   c. empty directly into the heart.
   d. receive fluids directly from capillaries.
5. Which of these is a primary lymphoid organ?
   a. lymph nodes b. spleen c. thymus d. tonsils
6. B cells mature within
   a. the lymph nodes.
   b. the spleen.
   c. the thymus.
   d. the bone marrow.
7. The organ that contains red pulp, which filters the blood, and white pulp, where lymphocytes respond to antigens in the blood, is the
   a. bone marrow.
   b. lymph node.
   c. spleen.
   d. thymus.

33.3 Innate Immune Defenses

8. Which of the following is an innate defense against pathogens?
   a. skin
   b. gastric juice
   c. complement
   d. interferons
   e. All of these are correct.

9. Which cell is not a phagocyte?
   a. neutrophil
   b. lymphocyte
   c. dendritic cell
   d. macrophage
10. Natural killer cells
    a. are known for attacking large parasites, such as tapeworms.
    b. engulf pathogens and present antigens to T cells.
    c. recognize the absence of MHC-I molecules on a cell surface.
    d. produce antibodies.
11. Complement
    a. is an innate defense mechanism.
    b. is involved in the inflammatory response.
    c. is a series of proteins present in the plasma.
    d. plays a role in destroying bacteria.
    e. All of these are correct.

33.4 Adaptive Immune Defenses

12. The clonal selection theory says that
    a. an antigen selects certain B cells and suppresses them.
    b. an antigen stimulates the multiplication of B cells that produce antibodies against it.
    c. T cells select those B cells that should produce antibodies, regardless of the antigens present.
    d. T cells suppress all B cells except the ones that should multiply and divide.
    e. Both b and c are correct.
13. Plasma cells are
    a. the same as memory cells.
    b. formed from blood plasma.
    c. B cells that are actively secreting antibody.
    d. inactive T cells carried in the plasma.

For questions 14–17, note the antibody class described (IgA, IgE, IgG, or IgM).

14. The major class of antibody found in the blood
15. The first class of antibodies produced
16. The major class secreted in milk, tears, and saliva
17. The class that is mostly bound to eosinophils and mast cells
18. Which of these statements pertain(s) to T cells?
    a. They have specific receptors.
    b. They recognize antigen presented by MHC proteins.
    c. They are responsible for cell-mediated immunity.
    d. They stimulate antibody production by B cells.
    e. All of these are correct.
19. Active immunity can be produced by
    a. having a disease.
    b. receiving a vaccine.
    c. receiving gamma globulin injections.
    d. Both a and b are correct.
    e. Both b and c are correct.

33.5 Immune System Disorders and Adverse Reactions

20. A child with severe combined immunodeficiency will have low or absent amounts of
    a. antibodies.
    b. B cells.
    c. cell-mediated immunity.
    d. T cells.
    e. All of these are correct.
21. Immediate hypersensitivity occurs after an allergen combines with
    a. IgG antibodies.
    b. IgE antibodies.
    c. IgM antibodies.
    d. IgA antibodies.
22. A positive skin test for tuberculosis is mediated mainly by
    a. histamine.
    b. IgE antibodies.
    c. memory T cells.
    d. neutrophils.
23. Which condition is not an autoimmune disease?
    a. multiple sclerosis
    b. rheumatoid arthritis
    c. systemic lupus erythematosis
    d. transplant rejection

ENGAGE

THINKING SCIENTIFICALLY

1. Some primitive organisms, such as invertebrates, have no lymphocytes and thus lack an adaptive immune system, but they have some components of an innate immune system, including phagocytes and certain protective proteins. What are some general features of innate immunity that make it very valuable to organisms lacking more specific antibody- and cell-mediated responses? What are some disadvantages to having only an innate immune system?
2. Compare the ways in which natural killer cells and cytotoxic T cells recognize the cells they are going to kill. Why does it make good biological sense for NK cells and cytotoxic T cells to recognize, for example, virus-infected cells in these different ways?
3. Why is it that Rh incompatibility can be a serious problem when an Rh-negative mother is carrying an Rh-positive fetus, but ABO incompatibility between mother and fetus is usually no problem? That is, a type A mother can usually safely carry a type B fetus. (Hint: The antibodies produced by an Rh-negative mother against the Rh antigen are usually IgG, whereas the antibodies produced against the A or B antigen are IgM.) Because the Rh antigen obviously serves no vital function (most humans lack it), why do you think it hasn't been completely eliminated during human evolution?
4. The transplantation of organs from one person to another was impossible until the discovery of immunosuppressant drugs. Now, with the use of drugs such as cyclosporine, organs can be transplanted without rejection. Transplant patients must take immunosuppressant drugs for the remainder of their lives. What is a major expected side effect of this lifelong therapy?

Cattle, like sheep, goats, and other ruminants, are able to digest the cellulose found in grasses because of their highly specialized digestive system.

34 Digestive Systems and Nutrition

Humans first domesticated cattle around 8,000 years ago. Cattle are part of a large group of mammals called ruminants, which use a process of digestion that begins when plant material is swallowed and enters a large chamber called the rumen. Here, a rich population of bacteria and other microbes break down the cellulose present in plant material. During this process some solid material is also regurgitated as the cud, which is chewed slowly to break down the plant fibers into a more digestible size.

This ability of ruminants to utilize the cellulose present in grasses and other plants is the main advantage of using these animals as a source of meat for human consumption. Rangeland that is not suitable for growing other kinds of crops can be used to raise cattle (although most beef cattle in the United States are fed grain). However, due largely to the growing human population and high demand for meat in some countries, the total number of domesticated cattle on Earth has more than doubled in the last 40 years to its current estimate of 1.5 billion. Estimates vary, but ruminants account for about 15–20% of the global production of methane, an important contributor to climate change. Most medical experts also believe a diet containing too much red meat is an important factor in major diseases such as atherosclerosis, diabetes, and many cancers. As you will see in this chapter, eating a well-balanced diet is one of the most important things we can do to maintain good health.

As you read through this chapter, think about the following questions:

1. What different types of strategies have animals evolved to efficiently obtain nutrients?
2. In what ways do the types of diets that humans choose to consume play a role in our health, as well as in the quality of our environment?

Chapter Outline

34.1 Digestive Tracts 642
34.2 The Human Digestive System 645
34.3 Digestive Enzymes 651
34.4 Nutrition and Human Health 652

Before You Begin

Before beginning this chapter, take a few moments to review the following discussions.

Chapter 3 What are some structural differences between carbohydrates, lipids, proteins, and nucleic acids?

Figure 6.1 How does energy flow from the sun, into chemical energy, to be ultimately dissipated as heat?

Figure 8.10 How do components of the human diet enter common metabolic pathways?

Following *the* Themes

UNIT 7 COMPARATIVE ANIMAL BIOLOGY

CHAPTER 34 DIGESTIVE SYSTEMS AND NUTRITION	
Evolution	Animals have evolved a wide variety of strategies to acquire nutrients from their environment.
Nature of Science	With the incidence of obesity rising in developed countries, medical researchers seek to learn more about the condition to develop new treatments.
Biological Systems	Like all living organisms, animals require a source of energy and the molecular building blocks needed to construct their components.

34.1 Digestive Tracts

Learning Outcomes

Upon completion of this section, you should be able to

1. Compare the structural features of incomplete versus complete digestive tracts.
2. Describe several examples of animals that are either continuous or discontinuous feeders.
3. Discuss some specific adaptations that are seen in omnivores, herbivores, and carnivores.

A digestive system includes all the organs, tissues, and cells involved in ingesting food and breaking it down into smaller components. Digestion contributes to homeostasis by providing the body with the nutrients needed to sustain the life of cells. A digestive system

1. Ingests food;
2. Breaks food down into small molecules that can cross plasma membranes;
3. Absorbs these nutrient molecules;
4. Eliminates undigestible remains.

A digestive tract, or *gut,* is typically defined as a long tube through which food passes as it is being digested. The majority of animals have some sort of digestive tract, but some (e.g., sponges) have no digestive tract at all. Instead, as water from the aqueous environment flows through the sponge (see Fig. 28.6), food particles are removed by cells that make up the inner lining of the organism. Cells in the sponge called *archaeocytes* may also ingest and distribute food to the rest of the organism.

Incomplete Versus Complete Tracts

An **incomplete digestive tract** has a single opening, usually called a mouth; however, the single opening is used as both an entrance for food and an exit for wastes. Planarians, which are flatworms, have an incomplete tract (Fig. 34.1). It begins with a mouth and muscular pharynx, and then the tract, a gastrovascular cavity, branches throughout the body.

Planarians are primarily carnivorous and feed largely on smaller, aquatic animals, as well as bits of organic debris. When a planarian is feeding, the pharynx actually extends beyond the mouth. The body is wrapped about the prey and the pharynx sucks up small quantities at a time. Digestive enzymes in the tract allow some extracellular digestion to occur. Digestion is finished intracellularly by the cells that line the tract. No cell in the body is far from the digestive tract; therefore, diffusion alone is sufficient to distribute nutrient molecules.

The digestive tract of a planarian is notable for its lack of specialized parts. It is saclike, because the pharynx serves not only as an entrance for food but also as an exit for undigestible material. This use of the same body parts for more than one function tends to minimize the evolution of more specialized parts, such as those seen in complete tracts.

Planarians have some modified parasitic relatives. Tapeworms, which are parasitic flatworms, lack a digestive system. Nutrient

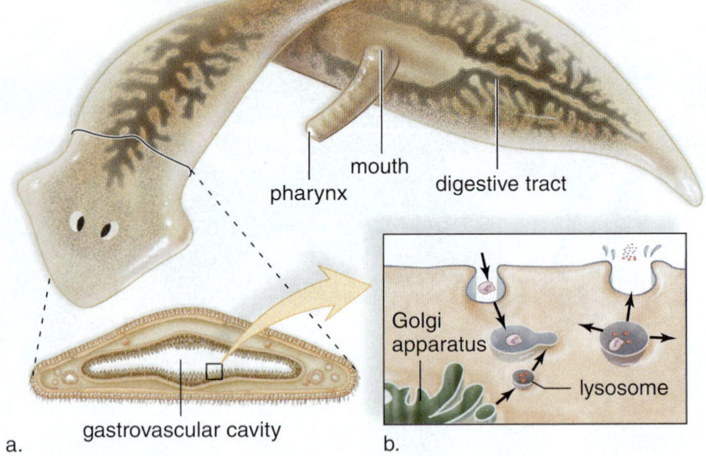

Figure 34.1 Incomplete digestive tract of a planarian. **a.** Planarians, which are flatworms, have a gastrovascular cavity with a single opening that acts as both an entrance and an exit. Like hydras, planarians rely on intracellular digestion to complete the digestive process. **b.** Phagocytosis produces a vacuole, which joins with an enzyme-containing lysosome. The digested products pass from the vacuole into the cytoplasm before any undigestible material is eliminated at the plasma membrane.

molecules are absorbed by the tapeworm from the intestinal juices of the host, which surround the tapeworm's body. The integument and body wall of the tapeworm are highly modified for this purpose. They have millions of microscopic, fingerlike projections that increase the surface area for absorption.

In contrast to planarians, earthworms, which are annelids, have a **complete digestive tract,** meaning that the tract has a mouth and an anus (Fig. 34.2). Earthworms feed mainly on the decayed organic matter found in soil. The muscular pharynx draws in a large amount of soil with a sucking action. Soil then enters the crop, which is a storage area with thin, expansive walls. From there, it goes to the gizzard, where thick, muscular walls crush the food and ingested sand grinds it. Digestion is extracellular within the intestine. The surface area of digestive tracts is often increased for absorption of nutrient molecules, and in earthworms, this is accomplished by an intestinal fold called the *typhlosole.* Undigested remains pass out

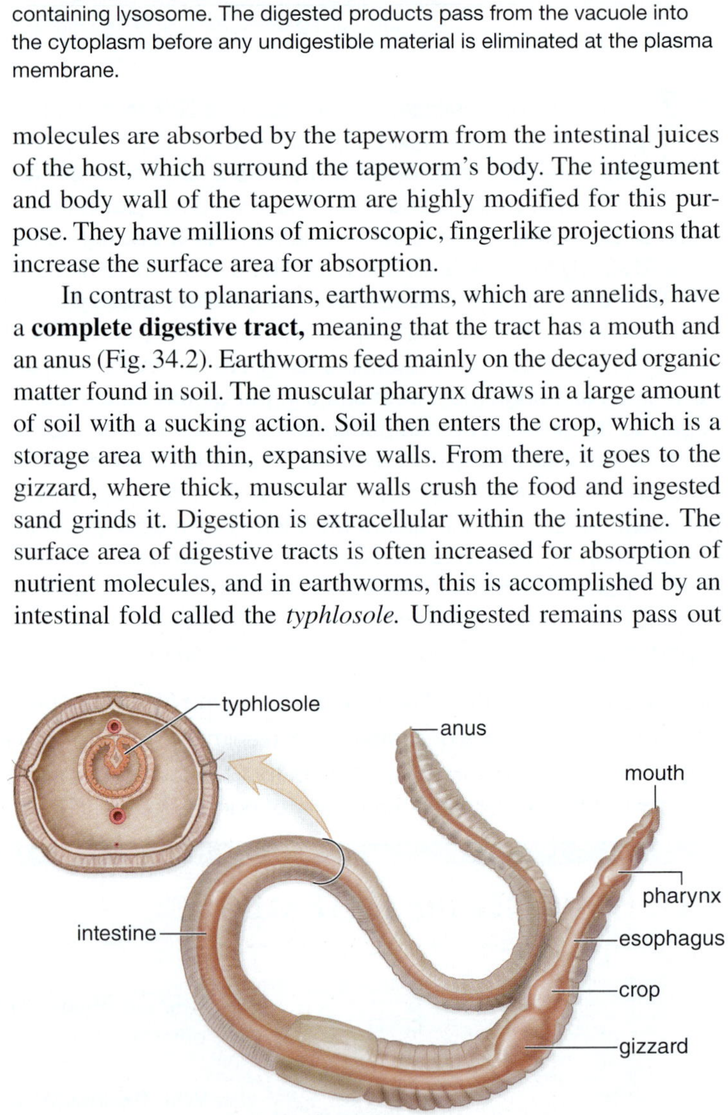

Figure 34.2 Complete digestive tract of an earthworm. Complete digestive tracts have both a mouth and an anus and can have many specialized parts, such as those labeled in this drawing. Also in earthworms, which are annelids, the absorptive surface of the intestine is increased by an internal fold called the typhlosole.

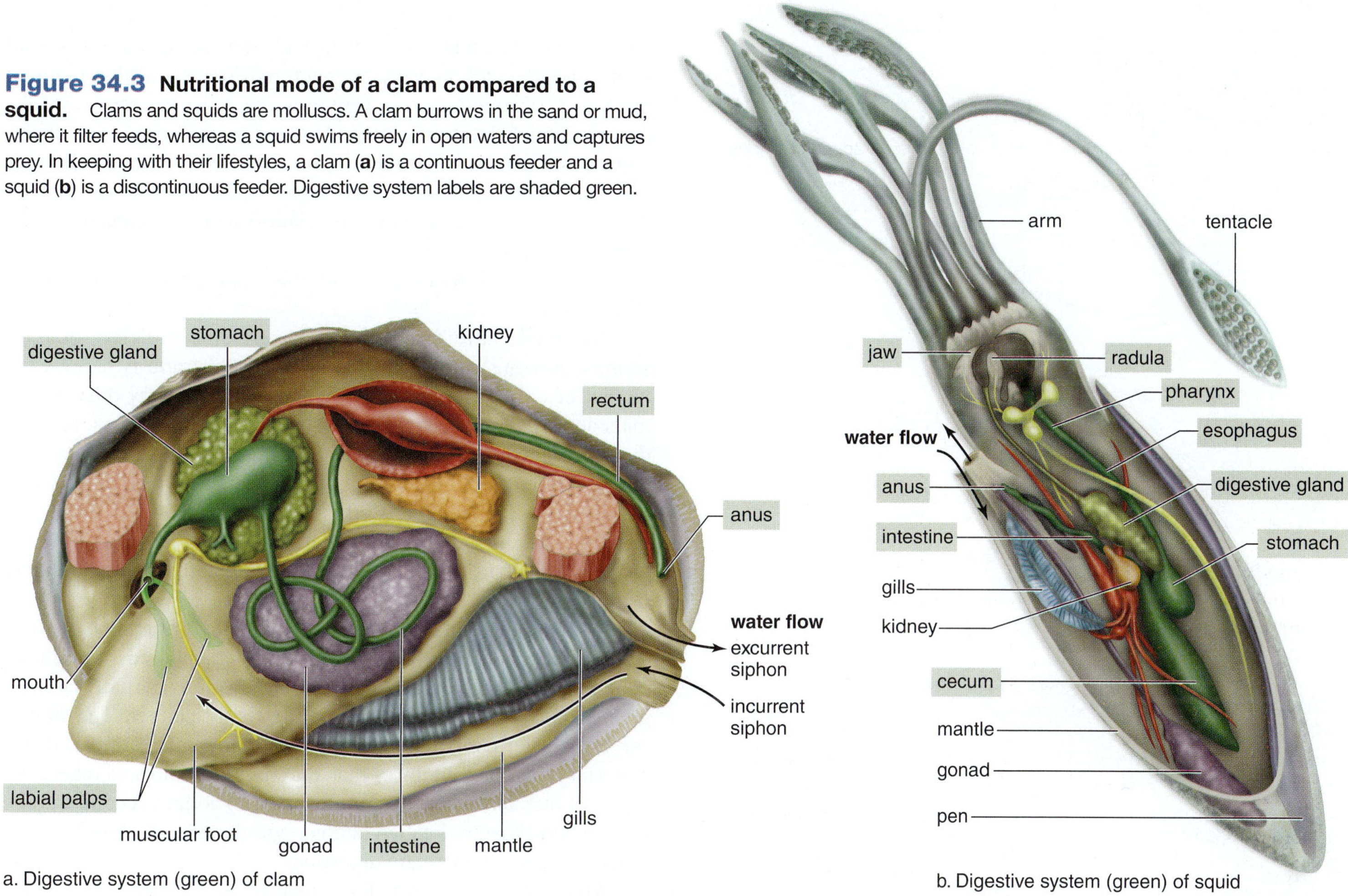

Figure 34.3 Nutritional mode of a clam compared to a squid. Clams and squids are molluscs. A clam burrows in the sand or mud, where it filter feeds, whereas a squid swims freely in open waters and captures prey. In keeping with their lifestyles, a clam (**a**) is a continuous feeder and a squid (**b**) is a discontinuous feeder. Digestive system labels are shaded green.

of the body at the anus. Specialization of parts is obvious in the earthworm, because the pharynx, crop, gizzard, and intestine have particular functions as food passes through the digestive tract.

Continuous Versus Discontinuous Feeders

Some aquatic animals acquire their nutrients by continuously passing water through some type of apparatus that captures food. Clams, which are molluscs, are filter feeders (Fig. 34.3*a*). Water is always moving into the mantle cavity by way of the incurrent siphon (slitlike opening) and depositing particles, including algae, protozoans, and minute invertebrates, on the gills. The size of the incurrent siphon permits the entrance of only small particles, which adhere to the gills. Ciliary action moves suitably sized particles to the labial palps, which force them through the mouth into the stomach. Digestive enzymes are secreted by a large digestive gland, but amoeboid cells throughout the tract are believed to complete the digestive process by intracellular digestion.

Not all filter feeders are relatively small invertebrates. A baleen whale, such as the blue whale, is an active filter feeder. Baleen—a keratinized, curtainlike fringe—hangs from the roof of the mouth and filters small shrimp, called krill, from the water. A baleen whale filters up to a ton of krill every few minutes.

Discontinuous feeders have evolved the ability to store food temporarily while it is being digested, enabling them to spend less time feeding and more time engaging in other activites. Discontinuous feeding requires a storage area for food, which can be a crop, where no digestion occurs, or a stomach, where digestion begins.

Squids, which are molluscs, are discontinuous feeders (Fig. 34.3*b*). The body of a squid is streamlined, and the animal moves rapidly through the water using jet propulsion (forceful expulsion of water from a tubular funnel). The head of a squid is surrounded by ten arms, two of which have developed into long, slender tentacles whose suckers have toothed, horny rings. These tentacles seize prey (fishes, shrimps, and worms) and bring it to the squid's beaklike jaws, which bite off pieces pulled into the mouth by the action of a *radula,* a tonguelike structure. An esophagus leads to a stomach and a cecum (blind sac), where digestion occurs. The stomach, supplemented by the cecum, retains food until digestion is complete.

Adaptations to Diet

Beyond the general categories of continuous versus discontinuous feeders, some animals have further adapted to more specialized diets. Some animals are omnivores; they eat both plants and animals. Others are strict herbivores; they feed only on plants. Still others are strict carnivores; they eat only other animals. Among invertebrates, filter feeders such as clams and tube worms

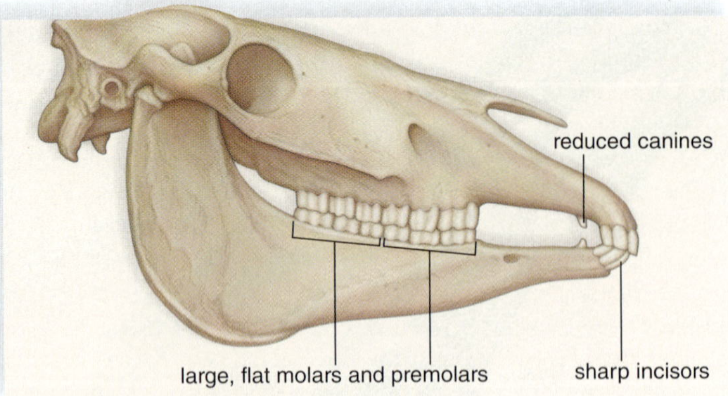

a. Horses are herbivores.

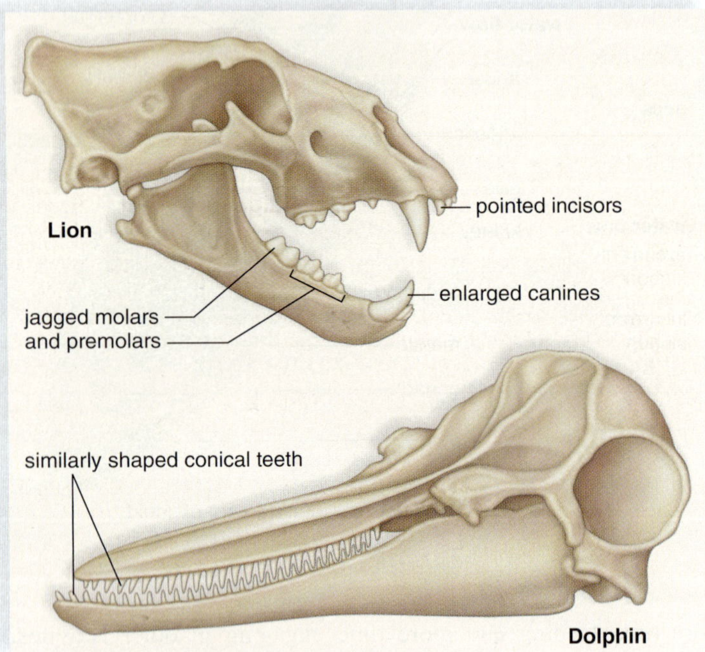

b. Lions and dolphins are carnivores.

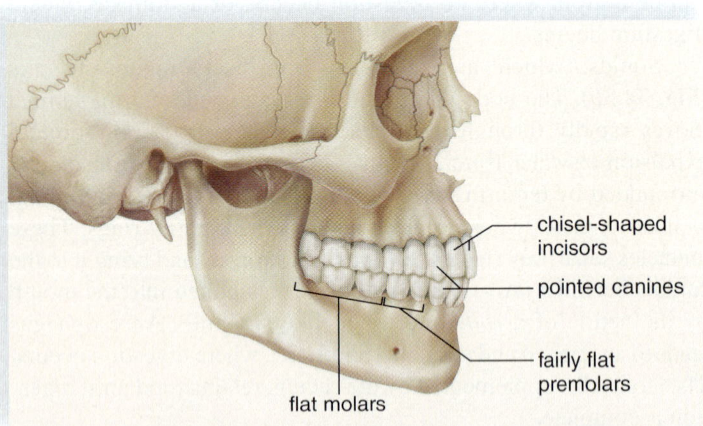

c. Humans are omnivores.

Figure 34.4 Dentition among mammals. **a.** Horses are herbivores and have teeth suitable to clipping and chewing grass. **b.** Lions and dolphins are carnivores. Dentition in a lion is suitable for killing large animals, such as zebras and wildebeests, and tearing apart their flesh. Dentition in a dolphin is suitable to grasping small animals, such as fish, which are swallowed whole. **c.** Humans are omnivores and have teeth suitable to a mixed diet of vegetables and meat.

are omnivores. Land snails, which are terrestrial molluscs, and some insects, such as grasshoppers and locusts, are herbivores. Spiders (arthropods) are carnivores, as are sea stars (echinoderms), which feed on clams. A sea star positions itself above a clam and uses its tube feet to pull the valves of the shell apart (see Fig. 28.29). Then, it everts a part of its stomach to start the digestive process, even while the clam is trying to close its shell. Some invertebrates are cannibalistic. A female praying mantis (an insect), if starved, will feed on her mate as the reproductive act is taking place!

Mammals have also adapted to consume a variety of food sources. Among herbivores, the koala of Australia is famous for its diet of only eucalyptus leaves, and likewise many other mammals are browsers, feeding on bushes and trees. Grazers, such as the horse, feed off grasses. The horse has sharp, even incisors for neatly clipping off blades of grass and large, flat premolars and molars for grinding and crushing the grass (Fig. 34.4*a*). Extensive grinding and crushing disrupts plant cell walls, allowing bacteria located in the part of the digestive tract called the cecum to digest cellulose.

As mentioned in the chapter-opening story, ruminants such as cattle, sheep, and goats have a large, four-chambered stomach. In contrast to horses, they graze quickly and swallow partially chewed grasses into the **rumen,** which is the first chamber. The rumen serves as a fermentation vat, where microorganisms break down material, such as cellulose, that the animal could not otherwise digest. Later on, when the ruminant is no longer feeding, undigested, solid material called cud is regurgitated and chewed again to facilitate more complete digestion.

Many mammals, including dogs, lions, toothed whales, and dolphins, are carnivores. Lions use pointed canine teeth for killing, short incisors for scraping bones, and pointed molars for slicing flesh (Fig. 34.4*b*, *top*). Dolphins and toothed whales swallow food whole without chewing it first; they are equipped with many identical, conical teeth that are used to catch and grasp their slippery prey before swallowing (Fig. 34.4*b*, *bottom*). Meat is rich in protein and fat and is easier to digest than plant material. The intestine of a rabbit, a herbivore, is much longer than that of a similarly sized cat, a carnivore.

Humans, like pigs, raccoons, mice, and most bears, are omnivores. Therefore, the dentition has a variety of specializations to accommodate both a vegetable diet and a meat diet. An adult human has 32 teeth. One-half of each jaw has teeth of four types: two chisel-shaped incisors for shearing; one pointed canine (cuspid) for tearing; two fairly flat premolars (bicuspids) for grinding; and three molars, well flattened for crushing (Fig. 34.4*c*). Omnivores are generally better able to adapt to different food sources, which can vary by location and season.

Check Your Progress 34.1

1. Compare the digestive tract of a planarian with that of an earthworm.
2. Describe some of the limitations of an incomplete digestive tract.
3. Compare the teeth of carnivores to those of herbivores.

34.2 The Human Digestive System

Learning Outcomes

Upon completion of this section, you should be able to

1. List all the major components of the human digestive tract, from the mouth to the anus.
2. Compare and contrast the structural features of the small intestine and the large intestine.
3. Discuss the major functions of the pancreas, liver, and gallbladder.

Humans have a complete digestive tract, which begins with a mouth and ends in an anus. The major structures of the human digestive tract are illustrated in Figure 34.5. The pancreas, liver, and gallbladder are accessory organs that aid digestion.

The digestion of food in humans is an extracellular event and requires a cooperative effort between different parts of the body. Digestion consists of two major stages: mechanical digestion and chemical digestion.

Mechanical digestion involves the physical breakdown of food into smaller particles. This task is accomplished through the chewing of food in the mouth and the physical churning and mixing of food in the stomach and small intestine. Chemical digestion requires enzymes that are secreted by the digestive tract or by accessory glands that lie nearby. Specific enzymes break down particular macromolecules into smaller molecules that can be absorbed.

MP3 An Overview of the Digestive System

Animation Organs of Digestion

Mouth

The **mouth,** or oral cavity, serves as the beginning of the digestive tract. The *palate,* or roof of the mouth, separates the oral cavity

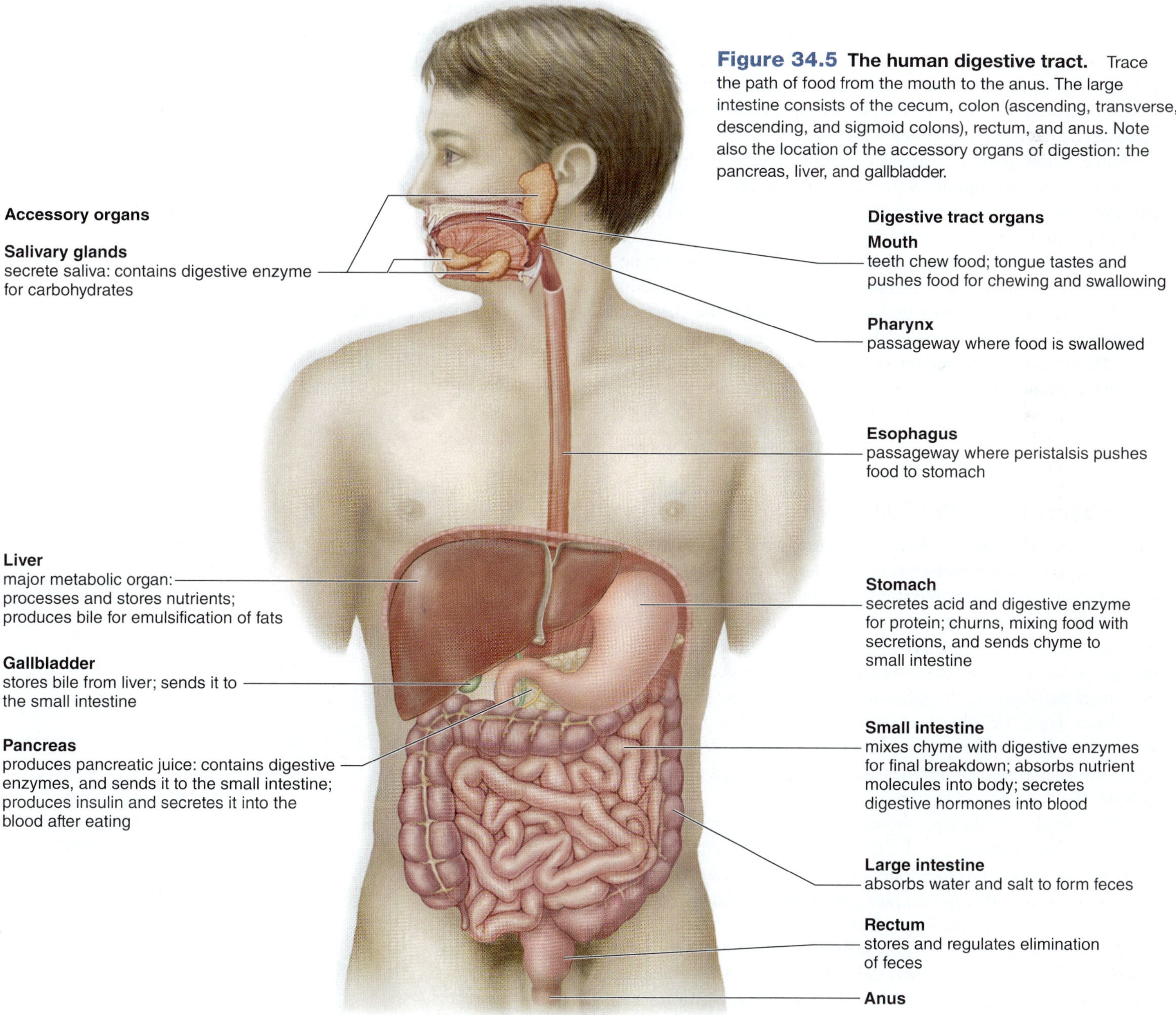

Figure 34.5 The human digestive tract. Trace the path of food from the mouth to the anus. The large intestine consists of the cecum, colon (ascending, transverse, descending, and sigmoid colons), rectum, and anus. Note also the location of the accessory organs of digestion: the pancreas, liver, and gallbladder.

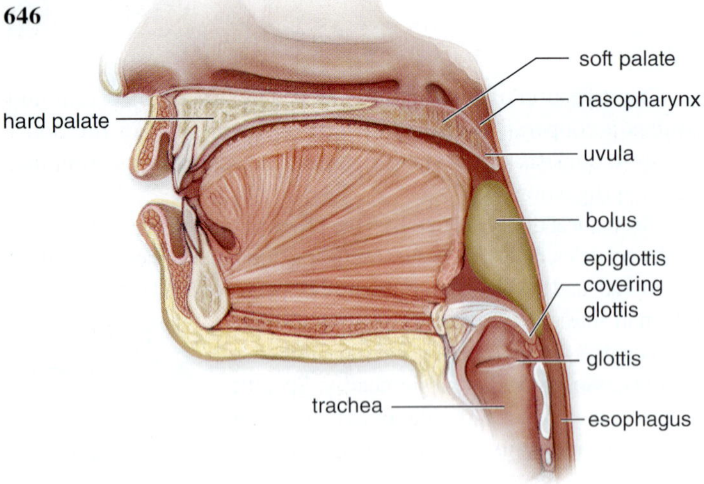

Figure 34.6 Swallowing. Respiratory and digestive passages converge and diverge in the pharynx. When food is swallowed, the soft palate closes off the nasopharynx, and the epiglottis covers the glottis, forcing the bolus to pass down the esophagus. Therefore, a person does not breathe when swallowing.

from the nasal cavity. It consists of the anterior hard palate and the posterior soft palate. The fleshy *uvula* is the posterior extension of the soft palate (Fig. 34.6). The cheeks and lips retain food while it is chewed by the teeth and mixed with saliva.

Three major pairs of **salivary glands** send their juices by way of ducts into the mouth. Saliva contains the enzyme **salivary amylase,** which begins to digest the starch that is present in many foods of plant origin (see section 34.3).

While in the mouth, food is manipulated by a muscular tongue, which has touch and pressure receptors similar to those in the skin. Taste buds, sensory receptors that are stimulated by the chemical composition of food, are also found primarily on the tongue as well as on the surface of the mouth. The tongue, which is composed of striated muscle and an outer layer of mucous membrane, mixes the chewed food with saliva. It then forms this mixture into a mass called a *bolus* in preparation for swallowing.

MP3 Oral Cavity, Esophagus, and the Swallowing Reflex

The Pharynx and the Esophagus

The digestive and respiratory passages come together in the **pharynx** and then separate. The **esophagus** is a tubular structure, about 25 cm in length, that takes food to the stomach. *Sphincters* are muscles that encircle tubes and act as valves; tubes close when sphincters contract, and they open when sphincters relax. The lower gastroesophageal sphincter is located where the esophagus enters the stomach. When food enters the stomach, the sphincter relaxes for a few seconds and then closes again. Heartburn occurs due to acid reflux, when some of the stomach's contents escape into the esophagus. When vomiting occurs, the abdominal muscles and the diaphragm, a muscle that separates the thoracic and abdominal cavities, contract.

When food is swallowed, the soft palate, the rear portion of the mouth's roof, moves back to close off the nasopharynx. A flap of tissue called the epiglottis covers the glottis, or opening into the trachea. Now the bolus must move through the pharynx into the esophagus, because the air passages are blocked (Fig. 34.6).

The central space of the digestive tract, through which food passes as it is digested, is called the **lumen** (Fig. 34.7). From the esophagus to the large intestine, the wall of the digestive tract is composed of four layers. The innermost layer next to the lumen is called the **mucosa.** The mucosa is a type of mucous membrane, and therefore it produces mucus, which protects the wall from the digestive enzymes inside the lumen.

The second layer in the digestive wall is called the **submucosa.** The submucosal layer is a broad band of loose connective tissue that contains blood vessels, lymphatic vessels, and nerves. Lymph nodules, called Peyer patches, are also in the submucosa. Like other secondary lymphoid tissues, they are sites of lymphocyte responses to antigens (see Chapter 33).

The third layer is termed the **muscularis,** and it contains two layers of smooth muscle. The inner, circular layer encircles the tract; the outer, longitudinal layer lies in the same direction as the tract. The contraction of these muscles, which are under involuntary nervous control, accounts for the movement of the gut contents from the esophagus to the rectum by **peristalsis** (Gk. *peri,* "around"; *stalsis,* "compression"), a rhythmic contraction that moves the contents along in various tubular organs (Fig. 34.8).

The fourth layer of the wall is the **serosa,** which secretes a watery fluid that lubricates the outer surfaces of the digestive tract and reduces friction as various parts rub against each other and other organs. The serosa is actually a part of the *peritoneum,* the internal lining of the abdominal cavity.

MP3 Oral Cavity, Esophagus, and the Swallowing Reflex

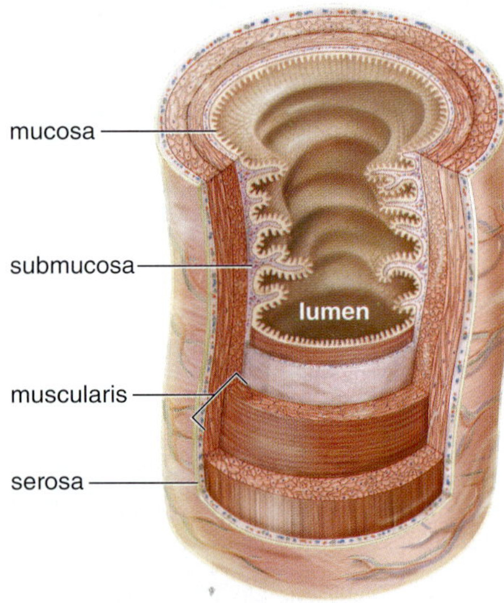

Figure 34.7 Wall of the digestive tract. The esophagus, stomach, small intestine, and large intestine all have a lumen and walls composed of similar layers.

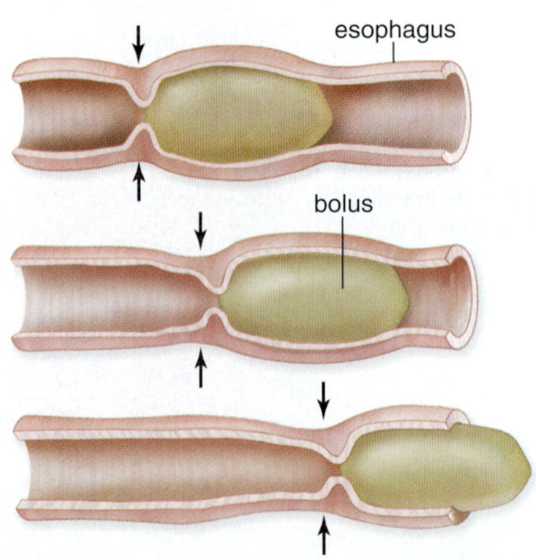

Figure 34.8 Peristalsis in the digestive tract. These three drawings show how a peristaltic wave moves through a single section of the esophagus over time. The arrows point to areas of contraction.

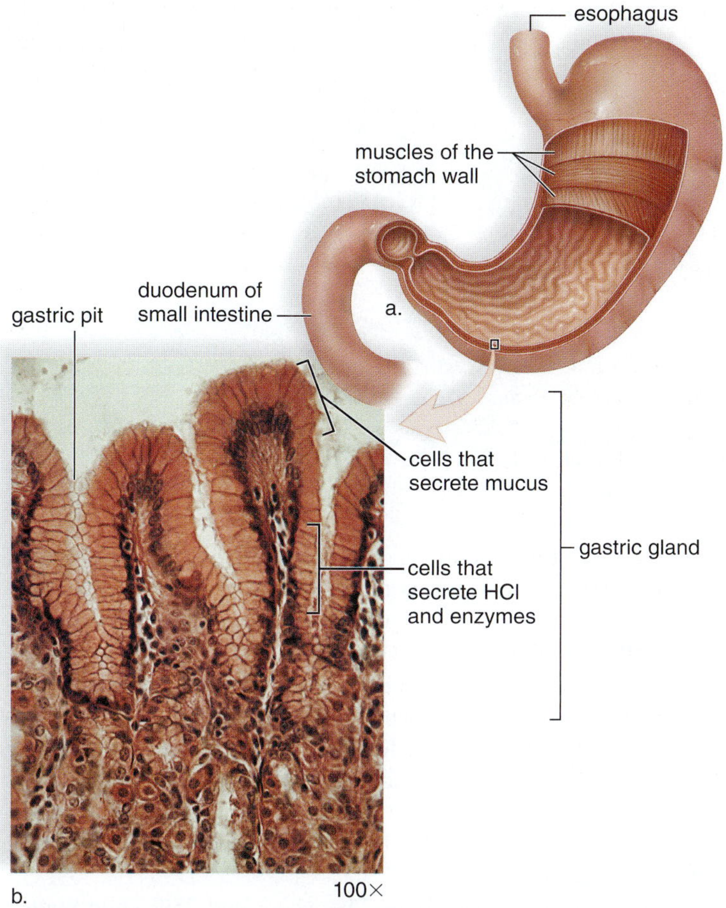

Figure 34.9 Anatomy of the stomach. **a.** The stomach, which has thick walls, expands as it fills with food. **b.** The mucous membrane layer of its walls secretes mucus and contains gastric glands, which secrete a gastric juice active in the digestion of protein.

Stomach

The **stomach** (Fig. 34.9) is a thick-walled, J-shaped organ that lies on the left side of the body beneath the diaphragm. The wall of the stomach has deep folds (rugae), which disappear as the stomach fills to its capacity (approximately 1 liter in humans). Therefore, many animals can periodically eat relatively large meals and spend the rest of their time at other activities.

The stomach is more than a food storage organ, as was discovered by William Beaumont (1785–1853) in the mid-nineteenth century. Beaumont, an American doctor, had a patient who had been shot in the stomach, and when the wound healed, he was left with a fistula, or opening, that allowed Beaumont to look inside the stomach and collect the juices produced by gastric glands. Beaumont was able to determine that the muscular walls of the stomach contract vigorously and mix food with juices that are secreted whenever food enters the stomach. He found that gastric juice contains hydrochloric acid (HCl) and a substance, now called pepsin, that is active in digestion. Beaumont's work pioneered the study of digestive physiology. For similar reasons, modern animal scientists can surgically create fistulas into the rumen of cattle in order to study ruminant nutrition.

The epithelial lining of the stomach has millions of *gastric pits,* which lead into *gastric glands* (Fig. 34.9). The gastric glands produce gastric juice. So much hydrochloric acid is secreted by the gastric glands that the stomach routinely has a pH of about 2.0. Such a high acidity usually is sufficient to kill bacteria and other microorganisms that might be in food. This low pH also stops the activity of salivary amylase, which functions optimally at the near-neutral pH of saliva.

Animation Three Phases of Gastric Secretion

A thick layer of mucus protects the wall of the stomach from enzymatic action. Sometimes, however, gastric acid can leak upwards, through the lower esophageal sphincter, where its acidic pH can irritate the mucosal lining of the esophagus. This gastroesophageal reflux disease (GERD) can cause heartburn and a number of other symptoms. In people with chronic GERD, the epithelial cells of the esophagus may change from stratified squamous (see Fig. 31.1) to a more columnar shape characteristic of the small intestine. This condition, known as "Barrett's esophagus," can lead to esophageal cancer.

Other individuals may develop gastric ulcers, which are areas where the protective epithelial layer of the stomach has been damaged. For many years these stomach ulcers were attributed mainly to stress, but through the work of Australian scientists Barry Marshall and Robin Warren, we now know that they can be caused by an acid-resistant bacterium, *Helicobacter pylori.* Wherever the bacterium attaches to the epithelial lining, the lining stops producing mucus, and the area becomes damaged by acid and digestive enzymes. If the condition is promptly diagnosed, antibiotic treatment is usually curative. Marshall and Warren were awarded the Nobel Prize in Medicine in 2005.

Eventually, food mixing with gastric juice in the stomach contents becomes **chyme,** which has a thick, creamy consistency. At the base of the stomach is a narrow opening controlled by a sphincter. Whenever the sphincter relaxes, a small quantity of chyme passes through the opening into the small intestine. When chyme enters the small intestine, it sets off a neural reflex, which causes the muscles of the sphincter to contract vigorously and close the opening temporarily. Then, the sphincter relaxes again and allows more chyme to enter. The slow manner in which chyme enters the small intestine allows for thorough digestion.

MP3 The Stomach

Small Intestine

The **small intestine** is named for its small diameter (compared to that of the large intestine), but perhaps it should be called the long intestine. The small intestine averages about 6 m in length, compared to the large intestine, which is about 1.5 m in length.

The first 25 cm of the small intestine is called the **duodenum.** A duct brings bile from the liver and gallbladder, and pancreatic juice from the pancreas, into the small intestine (see Fig. 34.12*a*). **Bile** emulsifies fat—emulsification causes fat droplets to disperse in water. The intestine has a slightly basic pH, because pancreatic juice contains sodium bicarbonate ($NaHCO_3$), which neutralizes chyme. The enzymes in pancreatic juice and enzymes produced by the intestinal wall complete the process of food digestion.

It has been estimated that the surface area of the small intestine is approximately that of a tennis court. What factors contribute to increasing its surface area? First, the wall of the small intestine contains fingerlike projections called villi (sing., **villus**), which

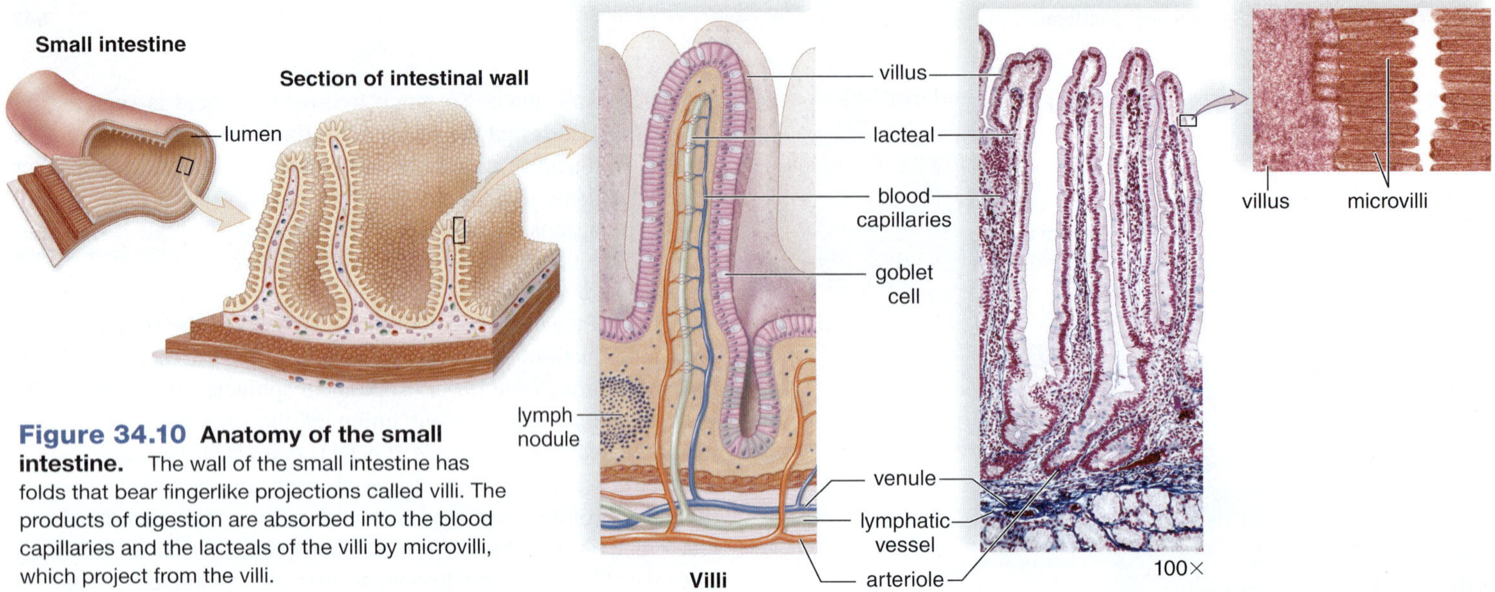

Figure 34.10 Anatomy of the small intestine. The wall of the small intestine has folds that bear fingerlike projections called villi. The products of digestion are absorbed into the blood capillaries and the lacteals of the villi by microvilli, which project from the villi.

give the intestinal wall a soft, velvety appearance (Fig. 34.10). Second, a villus has an outer layer of columnar epithelial cells, and each of these cells has thousands of microscopic extensions called microvilli. Collectively, in electron micrographs, microvilli give the villi a fuzzy border, known as a "brush border." Because the microvilli bear the intestinal enzymes, these enzymes are called brush-border enzymes. The microvilli greatly increase the surface area of the villus for the absorption of nutrients.

Animation Enzyme Action and the Hydrolysis of Sucrose

Nutrients are absorbed into the vessels of a villus, which contains blood capillaries and a lymphatic capillary, called a **lacteal.** Sugars (digested from carbohydrates) and amino acids (digested from proteins) enter the blood capillaries of a villus. Glycerol and fatty acids (digested from fats) enter the epithelial cells of the villi, and within these cells they are joined and packaged as lipoprotein droplets, which enter a lacteal. After nutrients are absorbed, they are eventually carried to all the cells of the body by the bloodstream.

MP3 Absorption of Nutrients and Water

Large Intestine

The **large intestine,** which includes the cecum, colon, rectum, and anus, is larger in diameter (6.5 cm) but shorter in length (1.5 m) than the small intestine. The large intestine absorbs water, salts, and some vitamins. It also stores undigestible material until it is eliminated as feces.

The cecum, which lies below the junction with the small intestine, is the blind end of the large intestine. The cecum has a small projection called the vermiform **appendix** (L. *verm,* "worm"; *form,* "shape"; *append,* "an addition") (Fig. 34.11). The function of the human appendix is unclear, although many experts suggest it may serve as a reservoir for the "good bacteria" that help maintain our intestinal health. In the case of appendicitis, the appendix becomes infected and so filled with fluid that it may burst. If an infected appendix bursts before it can be removed, it can lead to a serious, generalized infection of the abdominal lining called peritonitis.

The colon joins the rectum, the last 20 cm of the large intestine. About 1.5 liters of water enter the digestive tract daily as a result of eating and drinking. An additional 8.5 liters enter the digestive tract each day carrying the various substances secreted by the digestive glands. About 95% of this water is absorbed by the small intestine, and much of the remaining portion is absorbed by the colon. If this water is not reabsorbed, **diarrhea,** the passing of watery feces, can lead to serious dehydration and ion loss, especially in children.

The large intestine has a large population of bacteria, including *Escherichia coli* and perhaps 400 other species. By taking up space and nutrients, these bacteria provide protection against more pathogenic species. They also produce some vitamins—such as vitamin K, which is necessary to blood clotting. Digestive wastes, or feces, eventually leave the body through the **anus,** the opening of the anal canal.

Feces are normally about 75% water and 25% solid matter. Almost one-third of this solid matter is made up of intestinal bacteria. In fact, there are about 100 billion bacteria per gram of feces! The rest of the solids are undigested plant material, fats, waste products (such as bile pigments), inorganic material, mucus, and dead cells from the intestinal lining. The color of feces is the result

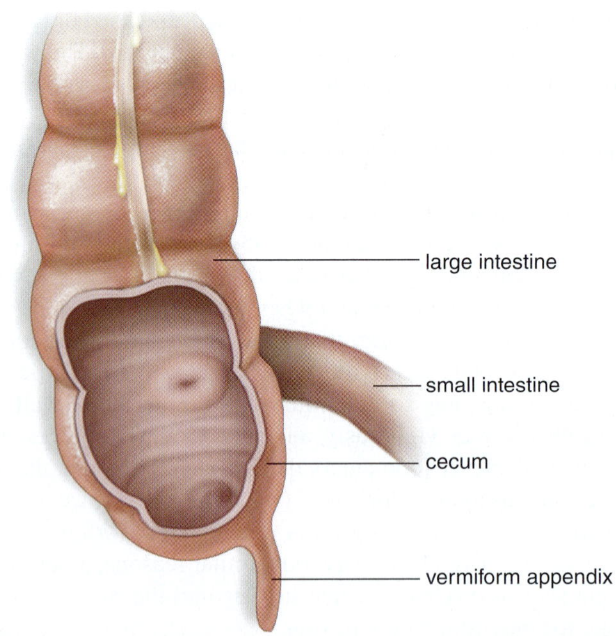

Figure 34.11 Junction of the small intestine and the large intestine.

of bilirubin breakdown and the presence of oxidized iron. The foul odor is the result of bacterial action.

The colon is subject to the development of **polyps,** which are small growths arising from the mucosa. Polyps, whether they are benign or cancerous, can be removed surgically. Some investigators believe that dietary fat increases the likelihood of colon cancer. Dietary fat causes an increase in bile secretion, and it could be that intestinal bacteria convert bile salts to substances that promote the development of colon cancer. Dietary fiber absorbs water and adds bulk, thereby diluting the concentration of bile salts and facilitating the movement of substances through the intestine. Regular elimination reduces the time that the colon wall is exposed to any cancer-promoting agents in feces.

Three Accessory Organs

The pancreas, liver, and gallbladder are accessory digestive organs. Figure 34.12*a* shows how the pancreatic duct from the pancreas and the common bile duct from the liver and gallbladder enter the duodenum.

Pancreas

The **pancreas** lies deep in the abdominal cavity, resting on the posterior abdominal wall. It is an elongated and somewhat flattened organ that has both an endocrine and an exocrine function. As an endocrine gland, it secretes insulin and glucagon, hormones that help keep the blood glucose level within normal limits (see Chapter 40). In this chapter, however, we are interested in its exocrine function. Most pancreatic cells produce pancreatic juice, which contains sodium bicarbonate ($NaHCO_3$) and digestive enzymes for all types of food. Sodium bicarbonate neutralizes acid chyme from the stomach. Pancreatic amylase digests starch, trypsin digests protein, and lipase digests fat.

Liver

The **liver,** which is the largest gland in the body, lies mainly in the upper right section of the abdominal cavity, under the diaphragm (see Fig. 34.5). The liver contains approximately 100,000 lobules, which serve as its structural and functional units (Fig. 34.12*b*). Triads, located between the lobules, consist of a bile duct, which takes bile away from the liver; a branch of the hepatic artery, which brings oxygen-rich blood to the liver; and a branch of the hepatic portal vein, which transports nutrients to the liver from the intestines (see Fig. 34.12). The central veins of lobules enter a hepatic vein. Blood moves from the intestines to the liver via the hepatic portal vein and from the liver to the inferior vena cava via the hepatic veins.

In some ways, the liver acts as the gatekeeper to the blood. As blood in the hepatic portal vein passes through the liver, it removes many toxic substances and metabolizes them. The liver also removes and stores iron and the vitamins A, B_{12}, D, E, and K. The liver makes many of the proteins found in blood plasma and helps regulate the quantity of cholesterol in the blood.

The liver maintains the blood glucose level at about 100 mg/100 mL (0.1%), even though a person eats intermittently. When insulin is present, any excess glucose present in blood is removed and stored by the liver as glycogen. Between meals, glycogen is broken down to glucose, which enters the hepatic veins. In this way, the blood glucose level remains constant.

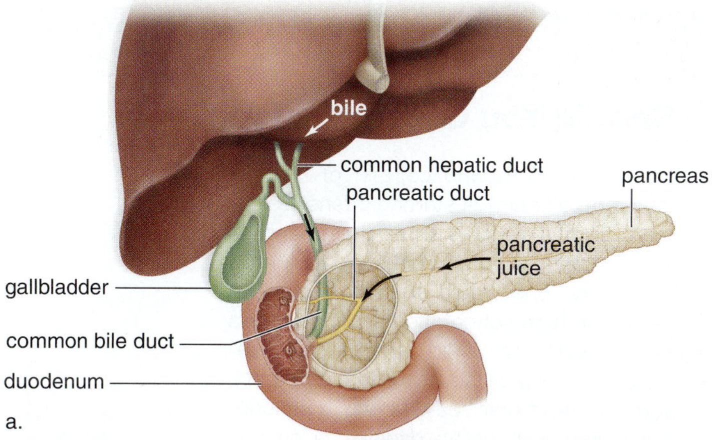

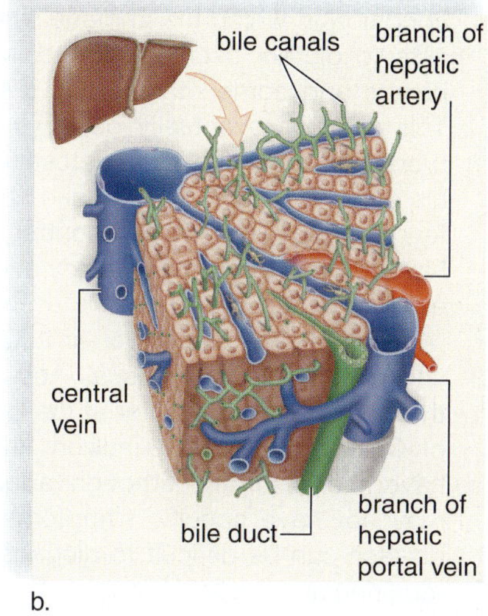

Figure 34.12 Liver, gallbladder, and pancreas. **a.** The liver makes bile, which is stored in the gallbladder and sent (black arrow) to the small intestine by way of the common bile duct. The pancreas produces digestive enzymes that are sent (black arrows) to the small intestine by way of the pancreatic duct. **b.** The liver contains over 100,000 lobules. Each lobule contains many cells that perform the various functions of the liver. They remove and add materials to the blood and deposit bile in a duct.

If the supply of glycogen is depleted, the liver converts glycerol (from fats) and amino acids to glucose molecules. The conversion of amino acids to glucose necessitates deamination, the removal of amino groups. By a complex metabolic pathway, the liver then combines ammonia with carbon dioxide to form urea. Urea is the usual nitrogenous waste product from amino acid breakdown in humans.

The liver produces bile, which is stored in the gallbladder. Bile has a yellowish green color, because it contains the bile pigment *bilirubin,* derived from the breakdown of hemoglobin, the red pigment of red blood cells. Bile also contains bile salts. Bile salts are derived from cholesterol, and they emulsify fat in the small intestine. When fat is emulsified, it breaks up into droplets, providing a much larger surface area, which can be acted upon by a digestive enzyme from the pancreas.

Liver Disorders. Because the liver performs so many vital functions, serious disorders of the liver can be life-threatening. When a person has a liver ailment, a yellowing of the skin and the sclera of the eyes called **jaundice** may occur. Jaundice results when the liver is not helping the body excrete excess bilirubin, which is then deposited in the tissues.

Regardless of the cause, inflammation of the liver is called **hepatitis.** The most common causes of hepatitis are viruses. Hepatitis A virus is usually acquired from food or water that has been contaminated with feces. Hepatitis B, which is usually spread by

THEME Nature of Science

Should You Go Gluten-Free?

Jake is a 280-pound captain of his college football team. A starting offensive lineman, he lifts weights three times a week and keeps in good aerobic shape. Looking at him, no one would suspect that when he was in high school, he had 3 feet of his small intestine removed, after he had experienced weeks of severe abdominal pain, vomiting, and diarrhea. He was eventually diagnosed with celiac disease, a serious condition in which the immune system reacts to gluten, a protein found in wheat, barley, and some other foods (Fig. 34A). This reaction eventually destroys the microvilli that line the small intestine. Since his diagnosis and surgery, Jake has had a few flare-ups, but he can control most of his symptoms by taking medications and eating a gluten-free diet.

Figure 34A People following a strict gluten-free diet must avoid foods like these, which contain wheat, barley, rye, or a number of other grains.

Jake's story is based on an actual case, and one that is not unusual. According to the Celiac Disease Center at the University of Chicago, at least 3 million Americans have the disorder, although about 40% may not have specific symptoms. Celiac disease can be difficult to diagnose, often requiring an intestinal biopsy to confirm the condition, and insurance companies often hesitate to pay for this procedure. As a result, even when a person has symptoms such as chronic diarrhea, it takes an average of 4 years to confirm the diagnosis.

Meanwhile, undiagnosed and untreated celiac disease can contribute to the development of other disorders, such as autoimmune disease, osteoporosis, infertility, and neurological conditions. Complicating matters further, some people who repeatedly test negative for celiac disease may have a different condition called gluten sensitivity. A landmark 2011 study showed that while people with celiac disease had increased production of cytokines associated with the adaptive immune system, those with gluten sensitivity had increased expression of innate immune markers, such as Toll-like receptors (see Chapter 33).

A variety of other studies have shown that the incidence of celiac disease and gluten sensitivity is rising. Some experts think as many 1 in 20 Americans may have health problems due to gluten, ranging from digestive problems to headaches, fatigue, and depression. It's unclear why gluten-related problems seem to be increasing, but changes in agricultural practices may have altered the type or amount of gluten in wheat.

The increased level of concern about gluten has not escaped the attention of the food industry. More than 2,000 gluten-free food products are now available. In 2010 Americans spent $2.6 billion on these foods; in 2013 it was $10.5 billion, with 11% of households purchasing some gluten-free foods. Major League Baseball stadiums are offering gluten-free options; so are the Girl Scouts (chocolate chip shortbread cookies).

So should you go gluten-free? If you are diagnosed with celiac disease or gluten sensitivity, it's very likely your doctor will recommend avoiding gluten. For everyone else, though, it's more complicated. Gluten-free often means more expensive: Gluten-free customers spend an average of $100 per grocery shopping trip versus $33 for others. Also, gluten-free does not necessarily mean healthy—one can eat a gluten-free diet that is rich in sugar and fat, or become so obsessed about avoiding gluten that one becomes nutritionally deficient. It is also possible that people who test negative for celiac disease but claim they feel better after banishing gluten from their diet are simply eating a healthier diet overall, or even benefiting from the placebo effect.

For those with unexplained health problems, many dietary experts recommend giving up gluten for a month, then reintroducing it and seeing how your body responds. They also encourage choosing naturally gluten-free, whole foods—fruits, vegetables, meats, seafood, dairy, nuts, seeds, and grains such as brown rice and quinoa, rather than gluten-free "junk" foods. Gluten-free or not, eating a healthier diet is always a good idea!

Questions to Consider

1. Many foods labeled "gluten-free," when tested at labs, have been found to contain trace amounts of gluten. What are common ways that this sort of contamination could happen?
2. When a person with celiac disease consumes gluten, his or her immune system recognizes and attacks not only the gluten but also an enzyme in the intestinal wall called tissue transglutamine (tTG), eventually resulting in destruction of the microvilli. List some specific immune mechanisms that can destroy tissues (you might want to review Chapter 33).
3. One hypothesis to explain the increasing rates of gluten–related disorders is that the newer, hybrid wheat strains we are eating today contain more gluten. What are some other possible explanations?

sexual contact, can also be spread by blood transfusions or contaminated needles. The hepatitis B virus is more contagious than the AIDS virus, which is spread in the same way. A vaccine is now available for hepatitis B, however. Hepatitis C, which is usually acquired by contact with infected blood and for which no vaccine is available, can lead to chronic hepatitis, liver cancer, and death.

Cirrhosis is another chronic disease of the liver. First, the organ becomes fatty, and then liver tissue is replaced by inactive, fibrous scar tissue. Cirrhosis of the liver is often seen in alcoholics, due to malnutrition and to the excessive amounts of alcohol (a toxin) the liver is forced to break down.

The liver has amazing regenerative powers and can recover if the rate of regeneration exceeds the rate of damage. During liver failure, however, there may not be enough time to let the liver heal itself. Liver transplantation is usually the preferred treatment for liver failure, but currently an estimated 4,000 people in the United States alone are waiting for a liver transplant.

Because the liver serves so many functions, artificial livers have been difficult to develop. One type is a cartridge that contains cultured liver cells (either human or pig). Like kidney dialysis, the patient's blood is passed outside of the body and through an apparatus containing the liver cells, which perform their normal functions, and the blood is returned to the patient.

Progress is also being made in the area of transplanting a smaller number of liver cells, as opposed to the entire organ. These cells can be either grown from stem cells or derived from the livers of donors who have died, but whose livers as entire organs are not suitable for transplantation.

Gallbladder

The **gallbladder** is a pear-shaped, muscular sac attached to the surface of the liver (see Fig. 34.5). About 1,000 ml of bile are produced by the liver each day, and any excess is stored in the gallbladder. Water is reabsorbed by the gallbladder, so that bile becomes a thick, mucuslike material. When bile is needed, the gallbladder contracts, releasing bile into the duodenum via the common bile duct (Fig. 34.12).

The cholesterol content of bile can come out of solution and form crystals called gallstones. These stones can be as small as a grain of sand or as large as a golf ball. The passage of the stones from the gallbladder may block the common bile duct, causing pain as well as possible damage to the liver or pancreas. Then, the gallbladder must be removed.

Check Your Progress 34.2

1. Trace the path of food from the mouth to the large intestine.
2. Describe the likely selective pressures that resulted in the evolution of taste buds.
3. Explain how the stomach, small intestine, and large intestine are each adapted to perform their particular functions.
4. Discuss how each accessory organ contributes to the digestion of food.

34.3 Digestive Enzymes

Learning Outcomes

Upon completion of this section, you should be able to

1. Describe the overall characteristics and functions of digestive enzymes.
2. Compare the specific types of nutrients that are digested in the mouth, stomach, and small intestine.

The various digestive enzymes present in the digestive juices, mentioned earlier, help break down carbohydrates, proteins, nucleic acids, and fats, the major nutritional components of food. Starch is a polysaccharide, and its digestion begins in the mouth. Saliva from the salivary glands has a neutral pH and contains **salivary amylase,** the first enzyme to act on starch.

$$\text{starch} + H_2O \xrightarrow{\text{salivary amylase}} \text{maltose}$$

Maltose molecules cannot be absorbed by the intestine; additional digestive action in the small intestine converts maltose to glucose, which can be absorbed.

Protein digestion begins in the stomach. Gastric juice secreted by gastric glands has a very low pH—about 2.0—because it contains hydrochloric acid (HCl). Pepsinogen, a precursor that is converted to **pepsin** when exposed to HCl, is also present in gastric juice. Pepsin acts on protein to produce peptides.

$$\text{protein} + H_2O \xrightarrow{\text{pepsin}} \text{peptides}$$

Peptides are usually too large to be absorbed by the intestinal lining, but later they are broken down to amino acids in the small intestine.

Starch, proteins, nucleic acids, and fats are all enzymatically broken down in the small intestine. Pancreatic juice, which enters the duodenum, has a basic pH because it contains sodium bicarbonate ($NaHCO_3$). One pancreatic enzyme, **pancreatic amylase,** digests starch (Fig. 34.13*a*).

$$\text{starch} + H_2O \xrightarrow{\text{pancreatic amylase}} \text{maltose}$$

Another pancreatic enzyme, **trypsin,** digests protein (Fig. 34.13*b*).

$$\text{protein} + H_2O \xrightarrow{\text{trypsin}} \text{peptides}$$

Trypsin is secreted as trypsinogen, which is converted to trypsin in the duodenum.

Maltase and peptidases, enzymes produced by the small intestine, complete the digestion of starch to glucose and protein to amino acids, respectively. Glucose and amino acids are small molecules that cross into the cells of the villi and enter the blood (Fig. 34.13*a*, *b*).

Maltose, a disaccharide that results from the first step in starch digestion, is digested to glucose by **maltase.**

$$\text{maltose} + H_2O \xrightarrow{\text{maltase}} \text{glucose} + \text{glucose}$$

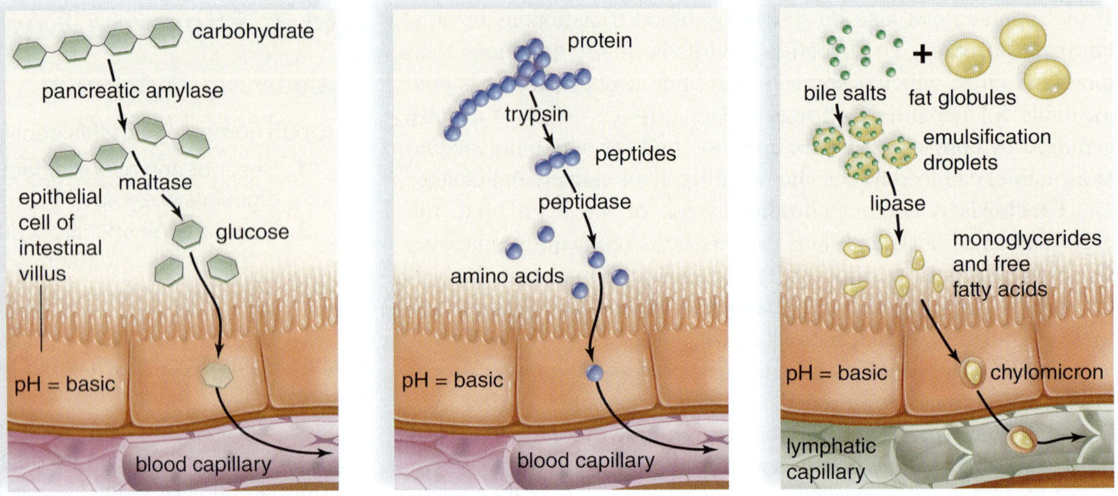

a. Carbohydrate digestion b. Protein digestion c. Fat digestion

Figure 34.13 Digestion and absorption of nutrients. **a.** Starch is digested to glucose, which is actively transported into the epithelial cells of intestinal villi. From there, glucose moves into the bloodstream. **b.** Proteins are digested to amino acids, which are actively transported into the epithelial cells of intestinal villi. From there, amino acids move into the bloodstream. **c.** Fats are emulsified by bile and digested to monoglycerides and fatty acids. These diffuse into epithelial cells, where they recombine and join with proteins to form lipoproteins, called chylomicrons. Chylomicrons enter a lacteal.

The brush border of the small intestine produces other enzymes for digestion of specific disaccharides. The absence of any one of these enzymes can cause illness. For example, approximately 75% of the world's adult human population is estimated to be lactose intolerant, because of a decreased expression of the enzyme lactase beyond the age of childhood. When such a person ingests milk or other products containing lactose, the undigested sugar is fermented by intestinal bacteria, resulting in a variety of unpleasant intestinal symptoms.

Peptides, which result from the first step in protein digestion, are digested to amino acids by **peptidases.**

$$\text{peptides} + H_2O \xrightarrow{\text{peptidases}} \text{amino acids}$$

Lipase, a third pancreatic enzyme, digests fat molecules in fat droplets after they have been emulsified by bile salts.

$$\text{fat} \xrightarrow{\text{bile salts}} \text{fat droplets}$$

$$\text{fat droplets} + H_2O \xrightarrow{\text{lipase}} \text{glycerol} + 3 \text{ fatty acids}$$

Specifically, the end products of lipase digestion are monoglycerides (glycerol + one fatty acid) and fatty acids. These enter the cells of the villi, and within these cells, they are rejoined and packaged as lipoprotein droplets, called chylomicrons. Chylomicrons enter the lacteals (Fig. 34.13*c*).

MP3 Chemical Digestion in the Small Intestine

Check Your Progress 34.3

1. Describe the location(s) in the digestive tract where each of the major types of nutrients is broken down.
2. Explain what final molecule (monomer) results from the digestion of carbohydrates, proteins, and fats.

34.4 Nutrition and Human Health

Learning Outcomes

Upon completion of this section, you should be able to

1. List the major types of nutrients and provide examples of foods that are a good source of each.
2. Describe the connection between a person's diet and the likely development of obesity, type 2 diabetes, and cardiovascular disease.
3. Distinguish among vitamins, coenzymes, and minerals.

This section of the chapter discusses the components of a balanced human diet, as well as some problems that may arise from consuming a poor diet.

Carbohydrates

Carbohydrates are present in food in the form of sugars, starch, and fiber. Fruits, vegetables, milk, and honey are natural sources of sugars. Glucose and fructose are monosaccharide sugars, and lactose (milk sugar) and sucrose (table sugar) are disaccharides. Disaccharides are broken down in the small intestine, and monosaccharides are absorbed into the bloodstream and delivered to cells. Once inside animal cells, monosaccharides are converted to glucose, much of which is used for the production of ATP by cellular respiration (see section 8.1).

Plants store glucose as starch, and animals store glucose as glycogen. Good sources of starch are beans, peas, cereal grains, and potatoes. Starch is digested to glucose in the digestive tract, and excess glucose is stored as glycogen. The human liver and muscles can only store a total of about 600 g of glucose in the form of glycogen; excess glucose is converted into fat and stored in adipose tissues.

Although other animals likewise store glucose as glycogen in liver or muscle tissue (meat), little is left by the time an animal is eaten for food. Except for honey and milk, which contain sugars, animal foods do not contain high levels of carbohydrates.

Fiber includes various undigestible carbohydrates derived from plants. Food sources rich in fiber include beans, peas, nuts, fruits, and vegetables. Whole-grain products are also a good source of fiber and are therefore more nutritious than food products made from refined grains. During *refinement,* fiber as well as vitamins and minerals are removed from grains, so that primarily starch remains. For example, a slice of bread made from whole-wheat flour contains 3 g of fiber; a slice of bread made from refined wheat flour contains less than 1 g of fiber.

Technically, fiber is not a nutrient for humans, because it cannot be digested to small molecules that enter the bloodstream. Insoluble fiber, however, adds bulk to fecal material, which stimulates movement in the large intestine, preventing constipation. Soluble fiber combines with bile acids and cholesterol in the small intestine and prevents them from being absorbed. In this way, high-fiber diets may protect against heart disease. The typical American consumes only about 15 g of fiber each day; the recommended daily intake of fiber is 25 g for women and 38 g for men. To increase your fiber intake, eat whole-grain foods, snack on fresh fruits and raw vegetables, and include nuts and beans in your diet.

If you, or someone you know, has lost weight by following low-carbohydrate diets, you may think "carbs" are unhealthy and should be avoided. According to the American Dietetic Association, however, some low-carbohydrate, high-fat diets are potentially hazardous and have no benefits over well-balanced diets that include the same number of calories. In fact, a recent study of over 4,400 Canadian adults found the lowest risk of obesity in people who consumed about half of their calories from carbohydrates.[1] Evidence also suggests that many Americans are not eating the right kind of carbohydrates. In some countries, the traditional diet is 60–70% high-fiber carbohydrates, and these people have a low incidence of the diseases that plague Americans.

A current controversy in human nutrition is the relative risk of consuming high levels of high-fructose corn syrup (HFCS), compared to other sweeteners. HFCS, or corn sugar, is now the most commonly used sweetening agent, found in soft drinks and a huge variety of foods that end up on our plates. Many websites and a few research studies have suggested that HFCS is a major factor in the rising epidemic of obesity and related diseases, but many nutritionists contend that the *type* of sugar consumed is not as important as the *amount.* As an example, the typical American obtains about one-sixth of his or her daily caloric intake from HFCS and other sugars. It is likely that consuming such a high percentage of "empty calories" in the form of simple sugars is contributing to the increasing incidence of obesity in the United States.

Lipids

Like carbohydrates, *triglycerides* (fats and oils) supply energy for cells, but *fat* is also stored for the long term in the body. Dietary experts generally recommend that people include unsaturated, rather than saturated, fats in their diets (see Fig. 3.10 to review these structures). Two unsaturated fatty acids, alpha-linolenic and linoleic acids (also called omega-3 fatty acids), are *essential* in the diet, meaning that we cannot synthesize them. Delayed growth and skin problems can develop in people who consume an insufficient amount of these essential unsaturated fatty acids, which are found in high amounts in certain fish and in plant oils such as canola and soybean oils.

Another type of lipid, *cholesterol,* is a necessary component of the plasma membrane of all animal cells. It is also a precursor for the synthesis of various compounds, including bile, steroid hormones, and vitamin D. Plant foods do not contain cholesterol, but animal foods such as cheese, egg yolks, liver, and certain shellfish (shrimp and lobster) are rich in cholesterol. Elevated blood cholesterol levels are associated with an increased risk of cardiovascular disease, the number one cause of disease-related death in the United States (described later on).

Animal-derived foods, such as butter, red meat, whole milk, and cheeses, contain saturated fats, which are also associated with cardiovascular disease. Statistical studies suggest that trans fatty acids (trans fats) are even more harmful than saturated fatty acids. Trans fatty acids arise when unsaturated oils are hydrogenated to produce a solid fat, as in shortening and some margarines. Trans fats may reduce the function of the plasma membrane receptors that clear cholesterol from the bloodstream. Trans fats are found in commercially packaged foods, such as cookies and crackers; in commercially fried foods, such as french fries; and in packaged snacks.

Proteins

Dietary *proteins* are digested to amino acids, which cells use to synthesize thousands of different cellular proteins. Of the 20 different amino acids, 8 are *essential amino acids* that normal adult humans cannot synthesize and thus must be present in the diet. Animal products such as beef, pork, poultry, eggs, and dairy products contain all these essential amino acids and are considered "complete" or "high-quality" protein sources.

Most foods derived from plants do not have as much protein per serving as those derived from animals, and some types of plant foods lack one or more of the essential amino acids. For example, the proteins in corn have a low content of the essential amino acid lysine (although high-lysine corn has been produced through genetic engineering technology). Approximately 3% of Americans (and millions of people in other countries) are either vegetarians, who avoid eating animal flesh, or vegans, who avoid consuming any products derived from animals. Neither group needs to rely on animal sources of protein.

To meet their protein needs, vegetarians and vegans can eat grains, beans, and nuts in various combinations. Also, tofu, soymilk, and other foods made from processed soybeans are complete protein sources. A 2009 report from the American Dietetic Association states that "well-planned vegetarian diets are appropriate for individuals during all stages of the life cycle, including pregnancy, lactation, infancy, childhood, and adolescence, and for athletes."[2]

Although a severe deficiency in dietary protein intake can be life-threatening, most Americans probably consume too much protein. Even further, some health food stores are full of protein supplements, aimed mainly at athletes who are trying to build

[1]Merchant, A. T., et al. "Carbohydrate Intake and Overweight-Obesity Among Healthy Adults," *J. Am. Dietetic Assn.* 109: 1165–1172 (2009).

[2]Craig, W. J., and Mangels, A. R. "Position of the American Dietetic Association: Vegetarian Diets," *J. Am. Dietetic Assn.* 109: 1266–1282 (2009).

muscle mass. However, both the American and Canadian Dietetic Associations recommend that even athletes should consume only 1–1.5 grams of protein per kilogram of body weight per day, which is just slightly higher than the 0.8 gram per kilogram recommended for sedentary people. This means an inactive 150-pound person would need to consume only about 60 grams of protein per day, which is about the amount contained in two cheeseburgers.

When amino acids are broken down, the liver removes the nitrogen portion (*deamination*) and uses it to form urea, which is excreted in urine. The water needed for the excretion of urea can cause dehydration when a person is exercising and losing water by sweating. High-protein diets can also increase calcium loss in the urine and encourage the formation of kidney stones. Furthermore, high-protein foods derived from animals often contain a high amount of fat, and some plant proteins may cause problems for those who have immune reactions to gluten (see the Nature of Science feature, "Should You Go Gluten-Free?," on page 650).

Diet and Obesity

As mentioned, the consumption of an excess amount of calories (relative to calories expended) from any source causes storage of these calories in the form of body fat. Obesity can be defined in several ways: (1) a condition in which excess body fat has an adverse effect on normal activity and health; (2) weight over 20% more than the ideal for your height and body build, and (3) a body mass index (BMI) over 30. A person's BMI can be calculated by dividing weight in kilograms by height in meters squared, or by using an online BMI calculator. Most estimates indicate that about 30% of Americans are obese. Obesity raises the risk of many medical conditions, including type 2 diabetes and cardiovascular disease. The seriousness of obesity as a health-care problem is evidenced by the increasing popularity of surgical procedures designed to reduce food consumption (see the Nature of Science feature, "New Approaches to Treating Obesity").

Type 2 Diabetes

Diabetes mellitus occurs when the hormone insulin is not functioning properly, resulting in abnormally high levels of glucose in the blood. This may occur due to a deficiency of insulin secretion by the pancreas, as in type 1 diabetes, or to an inability of cells to respond to insulin (also called insulin resistance), defined as type 2 diabetes. In both types, the excess blood glucose spills over into the urine, leading to increased urination, thirst, and weight loss. Over time, the high levels of blood glucose, and lack of other insulin functions, can lead to damage to blood vessels, nerves, eyes, and kidneys, and even to death. Type 1 diabetes can usually be successfully managed with insulin injections, but type 2 diabetes can be much more resistant to treatment.

Animation
Blood Sugar Regulation in Diabetics

In a 2010 report published in the *Journal of the American Medical Association,* 4,193 adults were studied for an average of 12.4 years, during which 339, or 8.1%, developed diabetes. Among the key findings, people who gained 20 pounds or more after age 50 had three times the risk of developing diabetes, and the risk was four times greater for those with the biggest waist circumferences and highest BMIs. Because we tend to lose muscle and gain fat as we age, the study authors noted the importance of exercise, as opposed to weight loss alone, in maintaining good health.

Cardiovascular Disease

Cardiovascular disease is the leading cause of death in the United States. Heart attacks and strokes often occur when arteries become blocked by plaque, which contains saturated fats and cholesterol. Cholesterol is carried in the blood by two types of lipoproteins: low-density lipoprotein (LDL) and high-density lipoprotein (HDL). LDL molecules are considered "bad," because they are like delivery trucks that carry cholesterol from the liver to the cells and to the arterial walls. HDL molecules are considered "good," because they are like garbage trucks that dispose of cholesterol. HDL transports cholesterol from the cells to the liver, which converts it to bile salts that enter the small intestine.

According to the American Heart Association, diets high in saturated fats, trans fats, and/or cholesterol tend to raise LDL cholesterol levels, while eating unsaturated fats may actually lower LDL cholesterol levels. Furthermore, coldwater fish (e.g., herring, sardines, tuna, and salmon) contain polyunsaturated fatty acids and especially omega-3 fatty acids, which are believed to reduce the risk of cardiovascular disease. However, taking fish oil supplements to obtain omega-3s is not recommended without a physician's approval, because too much of these fatty acids can interfere with normal blood clotting.

The American Heart Association also recommends limiting total cholesterol intake to 300 mg per day. This requires careful selection of the foods we include in our daily diets. For example, an egg yolk contains about 210 mg of cholesterol, which would be two-thirds of the recommended daily intake. Still, this doesn't mean eggs should be eliminated from a healthy diet, because the proteins in them are very nutritious; in fact, most healthy people can eat a couple of whole eggs each week without experiencing an increase in their blood cholesterol levels.

A physician can determine whether blood lipid levels are normal. If a person's cholesterol and triglyceride levels are elevated, modifying the fat content of the diet, losing excess body fat, and exercising regularly can reduce them. If lifestyle changes do not lower blood lipid levels enough to reduce the risk of cardiovascular disease, a physician may prescribe cholesterol-lowering medications.

Vitamins and Minerals

Vitamins are organic compounds other than carbohydrates, fats, and proteins that regulate various metabolic activities and must be present in the diet. Many vitamins are part of coenzymes; for example, niacin is the name for a portion of the coenzyme NAD^+, and riboflavin is a part of FAD. Coenzymes are needed in small amounts, because they are used over and over again in cells. Not all vitamins are coenzymes, however; vitamin A, for example, is a precursor for the pigment that prevents night blindness.

Animation
B Vitamins

It has been known for some time that the absence of a vitamin can be associated with a particular disorder. Vitamins are especially abundant in fruits and vegetables, so it is suggested that we eat about 4 1/2 cups of fruits and vegetables per day. Although many

THEME Nature of Science

New Approaches to Treating Obesity

We all know that the most critical factor in weight gain is consuming more calories than we need for our level of physical activity. But considering the rising rates of obesity in developed countries, and the many associated health risks, researchers are investigating factors and approaches that might help some people reduce their weight.

Adequate Sleep

In 2011, the National Sleep Foundation's Sleep in America poll found that 43% of Americans between the ages of 13 and 64 reported that they rarely or never get a good night's sleep on weeknights. The Centers for Disease Control and Prevention (CDC) website refers to insufficient sleep as "a Public Health Epidemic," partly because accumulating scientific evidence points to a link between declining sleep and increasing obesity rates.

Even a modest amount of sleep deprivation can cause alterations in hormones that control appetite and regulate metabolism. In a 2004 study, 12 healthy college-age males were divided into two groups: one that slept 10 hours a night, and a second that slept only 4 hours. After 2 days, the sleep-deprived subjects averaged an 18% decrease in serum leptin (a hormone that normally suppresses appetite) and a 28% increase in ghrelin (an appetite stimulator). Those with the greatest hormonal differences also reported greater increases in hunger, especially for high-carb foods.[1] Other, larger studies have mostly confirmed these findings, suggesting that adequate sleep may be a significant factor in avoiding obesity.

Anti-Obesity Drugs

In general, two types of drugs are available for the treatment of obesity: fat absorption inhibitors and appetite suppressants. Xenical (Orlistat), taken three times a day with meals, interferes with fat digestion by inhibiting pancreatic lipase. Various studies have shown that xenical is more effective than dietary management alone in promoting weight loss, although the typical patient may lose only a few pounds. Because an increased amount of fat is passing undigested through the digestive tract, some people experience side effects such as abdominal pain, increased frequency of bowel movements, or even fecal incontinence (inability to control fecal release).

Although several appetite suppressor drugs are available, some have been removed from the market due to safety concerns. Research is continuing on identifying strategies to control hunger using drugs that affect various appetite control mechanisms.

Surgical Procedures

Despite advances in understanding the factors behind obesity, many individuals still struggle with this problem. The number of bariatric (weight-loss) surgeries performed per year in the United States increased from about 16,000 in the early 1990s to an estimated 220,000 in 2008. Many types of procedures are performed, but all are intended to reduce the size of the stomach, to decrease the absorption of nutrients, or both. When successful, any procedure that reduces the weight of an obese person to a value closer to normal can lead to a significant reduction in the risk of other health problems. However, as with any major surgery, there is also a substantial risk of harm, including death.

Traditionally, bariatric surgery has been recommended only for people who are morbidly obese (BMI greater than 40) or those who have a BMI greater than 35 but with weight-related medical problems. Many of these procedures require the removal of stomach tissue, or bypassing the stomach altogether. However, less invasive procedures are being developed, and in December 2010 a panel of FDA advisors recommended that a product called Lap-Band be approved for use in patients whose BMI is as low as 30. As shown in Figure 34B, the Lap-Band is an adjustable plastic strip that is inserted into the abdomen and over the top of the stomach via a small incision. Advantages of this procedure over more traditional surgeries are that the tension can be adjusted for greater or less restriction of stomach volume and it can be removed. Many experts seem to believe that gastric banding procedures are not as effective as other types of bariatric surgery, however.

Questions to Consider

1. From an evolutionary perspective, why are humans predisposed to gain weight when food supplies are readily available?
2. Besides diet, exercise, and lack of sleep, what are some some other factors that could cause some people to be more prone to obesity?
3. The average cost for various bariatric surgeries is \$18,000 to \$35,000. With limited health-care dollars to go around, do you think this is a justifiable expense, either practically or ethically?

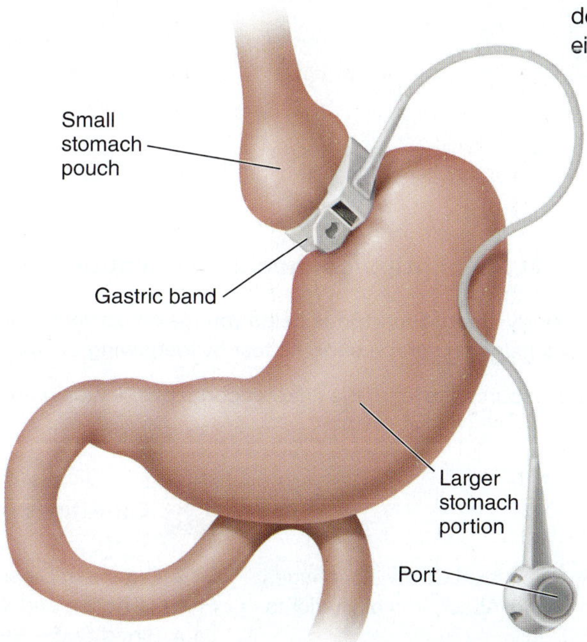

Figure 34B The Lap-Band. The Lap-Band is an inflatable loop that is surgically placed around the upper part of the stomach to reduce the amount of food that can enter. It is approved for use only in obese people for whom other weight-loss therapies have failed.

[1]Spiegel, K., et al. "Sleep Curtailment in Healthy Young Men. . . ." *Ann. Internal Medicine* 141: 846–850 (2004).

foods are now enriched or fortified with vitamins, some individuals are still at risk for vitamin deficiencies, generally as a result of poor food choices.

The body also needs about 20 elements called *minerals* for various physiological functions, including the regulation of biochemical reactions, the maintenance of fluid balance, and as components of certain structures, such as bone. Some individuals (especially women) do not get enough iron, calcium, magnesium, or zinc in their diets. Adult females need more iron in the diet than males (18 mg compared to 10 mg) if they are menstruating each month. Many people take calcium supplements, as directed by a physician, to counteract osteoporosis, a degenerative bone disease that especially affects older women and men. Many people consume too much sodium, even double the amount needed. Excess sodium can cause water retention and contribute to hypertension.

Animation Osteoporosis

Check Your Progress 34.4

1. Review several reasons a diet that includes plenty of vegetables is generally better for you than a diet that includes excess protein.
2. Discuss the relationship among blood cholesterol, saturated fat intake, and cardiovascular disease.
3. Define *vitamin*.

CONNECTING *the* CONCEPTS *with the* THEMES

Evolution

- An incomplete digestive tract fulfills the nutritional needs of some relatively simple animals. The evolution of a complete digestive tract has allowed specialization of function and efficient food processing.
- Discontinuous feeders have developed the ability to store food temporarily while it is being digested, enabling them to spend less time feeding and more time engaging in other activities.
- Many features of animal body structure—including type of dentition, ability to catch prey, and structure of the digestive tract—have evolved as adaptations to a particular diet: herbivore, carnivore, or omnivore.

Nature of Science

- Observational studies of the digestive systems of animals, ranging from the simplest to the most complex, have greatly aided our understanding of digestion and nutrition.
- Basic research on the digestive system has resulted in screenings and treatments for diseases such as ulcers, hepatitis, gallstones, and colon cancer.
- Through research on the causes of obesity, medical scientists are attempting to develop new strategies and medications to help reduce weight.

Biological Systems

- Some animals are simple enough, with no cell located far from their aqueous environment, that their cells can obtain nutrients with no need for a digestive system.
- The human digestive system is a good example of the complexity and specialization of the digestive system of an omnivorous mammal.
- Humans in modern societies have access to types and quantities of food that were unknown to our ancestors, and this may account for the rise of obesity.

MEDIA STUDY TOOLS http://connect.mheducation.com

SMARTBOOK® Study smart. SmartBook helps you maximize your study time by identifying what you know and don't know. The recharge feature in SmartBook helps you prepare for the test by identifying the material that you are most likely to forget.

connect BIOLOGY Find even more resources to learn the chapter concepts including animations, tutorial videos, and interactive practice questions.

MP3 Files	Animations
34.2 An Overview of the Digestive System • Oral Cavity, Esophagus and the Swallowing Reflex • The Stomach • Absorption of Nutrients and Water **34.3** Chemical Digestion in the Small Intestine	**34.2** Organs of Digestion • Three Phases of Gastric Secretion • Enzyme Action and the Hydrolysis of Sucrose **34.4** Blood Sugar Regulation in Diabetics • B Vitamins • Osteoporosis

SUMMARIZE

34.1 Digestive Tracts

A few animals (e.g., sponges) lack a digestive tract; others, such as planarians, have an incomplete digestive tract that has only one opening. An **incomplete digestive tract** has little specialization. Many other animals, such as earthworms, have a complete digestive tract that has both a mouth and an anus. A **complete digestive tract** tends to have specialized regions.

Some animals are continuous feeders (e.g., clams, which are filter feeders); others are discontinuous feeders (e.g., squid). Discontinuous feeders need a storage organ for food.

Most mammals have teeth. Herbivores need teeth that can clip off plant material and grind it up. Also, many herbivores have a **rumen** inhabited by bacteria that can digest cellulose. Carnivores need teeth that can tear and rip animal flesh into pieces. Meat is easier to digest than plant material, so the digestive system of carnivores has fewer specialized regions and the intestine is shorter than that of herbivores.

34.2 The Human Digestive System

In the human digestive system, both mechanical and chemical digestion begin in the **mouth,** where food is chewed and mixed with saliva produced by **salivary glands.** Saliva contains **salivary amylase,** which begins carbohydrate digestion.

Food then passes to the **pharynx** and down the **esophagus** by **peristalsis** to the **stomach.** The stomach stores and mixes food with mucus and gastric juices to produce chyme. Pepsin begins protein digestion in the stomach. Beginning with the esophagus, the wall of the digestive tract has a central space or **lumen,** an inner **mucosa,** then a **submucosa, muscularis,** and outer **serosa.**

Partially digested food (**chyme**) passes into the **small intestine.** The **duodenum** of the small intestine receives bile from the liver and pancreatic juice from the pancreas. **Bile** emulsifies fat and readies it for digestion by pancreatic lipase. The pancreas also produces amylase and proteases. These and other intestinal enzymes finish the process of chemical digestion.

The walls of the small intestine have fingerlike projections called villi (sing. **villus**), where small nutrient molecules are absorbed. Amino acids and glucose enter the blood vessels of a villus. Glycerol and fatty acids are packaged as lipoproteins before entering lymphatic vessels, called **lacteals,** in the villi.

The **large intestine** consists of the cecum, colon, and rectum, which ends at the **anus.** The cecum, a blind pouch at the junction of the small and large intestines, has a small projection called the **appendix,** which sometimes becomes infected and inflamed, necessitating its removal. The large intestine does not produce digestive enzymes; it does absorb water, salts, and some vitamins. Reduced water absorption results in **diarrhea.** The intake of water and fiber helps prevent constipation. **Polyps** in the colon are small growths that can be benign or cancerous.

Three accessory organs of digestion—the **pancreas, liver,** and **gallbladder**—send secretions to the duodenum via ducts. The pancreas produces pancreatic juice, which contains digestive enzymes for carbohydrates, protein, and fat.

The liver produces the bile, which is stored in the **gallbladder.** The liver receives blood from the small intestine by way of the hepatic portal vein. **Hepatitis** and **cirrhosis** are common disorders of the liver, and either can result in **jaundice** due to accumulation of bilirubin.

34.3 Digestive Enzymes

Digestive enzymes are present in digestive juices and break down food into the nutrient molecules glucose, amino acids, fatty acids, and glycerol. **Salivary amylase** and **pancreatic amylase** begin the digestion of starch, and **maltase** completes the digestion of starch to glucose. **Pepsin** and **trypsin** digest protein to peptides, then intestinal **peptidases** digest these to amino acids. Following emulsification by bile, **lipase** digests fat to glycerol and fatty acids.

Each digestive enzyme is present in a particular part of the digestive tract. Salivary amylase functions in the mouth; pepsin functions in the stomach; trypsin, lipase, and pancreatic amylase occur in the intestine along with the various enzymes that digest disaccharides and peptides.

34.4 Nutrition and Human Health

The nutrients released by the digestive process should provide us with an adequate amount of major nutrients, essential amino acids and fatty acids, and all necessary vitamins and minerals.

Carbohydrates are necessary in the diet, but simple sugars and refined starches are not as healthy, because they provide calories but little or no fiber, vitamins, or minerals. Proteins supply us with essential amino acids, but many Americans consume more protein than is healthy. It is also wise to restrict one's intake of meats that are fatty, because animal fats are saturated fats. Unsaturated fatty acids, particularly the omega-3 fatty acids, are protective against cardiovascular disease, whereas saturated fatty acids may lead to plaque formation, which blocks blood vessels.

Obesity is becoming an increasingly serious problem, especially because it is associated with the development of type 2 diabetes and cardiovascular disease.

ASSESS

Choose the best answer for each question.

34.1 Digestive Tracts

1. Archaeocytes are associated with
   a. digestion in bacteria.
   b. the digestive tracts of sponges.
   c. the filter organs of continuous feeders.
   d. the human appendix.
   e. the ingestion and distribution of food in sponges.
2. The typhlosole within the gut of an earthworm compares best to which of these organs in humans?
   a. teeth in the mouth
   b. esophagus in the thoracic cavity
   c. folds in the stomach
   d. villi in the small intestine
   e. the large intestine, because it absorbs water
3. Animals that feed discontinuously
   a. have a digestive tract that permits storage.
   b. are always filter feeders.
   c. exhibit extremely rapid digestion.
   d. have a nonspecialized digestive tract.
   e. usually eat only meat.

34.2 The Human Digestive System

4. Beginning with the mouth and going from a–f, which structure is out of order first when tracing the path of food in humans?
   a. mouth   c. esophagus   e. stomach
   b. pharynx   d. small intestine   f. large intestine
5. Why can a person not swallow food and talk at the same time?
   a. To swallow, the epiglottis must close off the trachea.
   b. The brain cannot control two activities at once.
   c. To speak, air must come through the larynx to form sounds.
   d. A swallowing reflex is only initiated when the mouth is closed.
   e. Both a and c are correct.

6. Which association is incorrect?
   a. mouth—starch digestion
   b. esophagus—protein digestion
   c. small intestine—starch, lipid, protein digestion
   d. liver—bile production
   e. large intestine—water absorption
7. In humans, most of the absorption of the products of digestion takes place across
   a. the squamous epithelium of the esophagus.
   b. the convoluted walls of the stomach.
   c. the fingerlike villi of the small intestine.
   d. the smooth wall of the large intestine.
   e. the lacteals of the lymphatic system.
8. The appendix connects to the
   a. cecum.
   b. small intestine.
   c. esophagus.
   d. large intestine.
   e. liver.
   f. All of these are correct.

34.3 Digestive Enzymes

9. Which of these could be absorbed directly without need of digestion?
   a. glucose
   b. fat
   c. protein
   d. nucleic acid
   e. All of these are correct.
10. Which association is incorrect?
    a. protein—trypsin
    b. fat—lipase
    c. maltose—pepsin
    d. starch—amylase
    e. protein—pepsin
11. Predict and explain the expected digestive results per test tube for this experiment.

Incubator

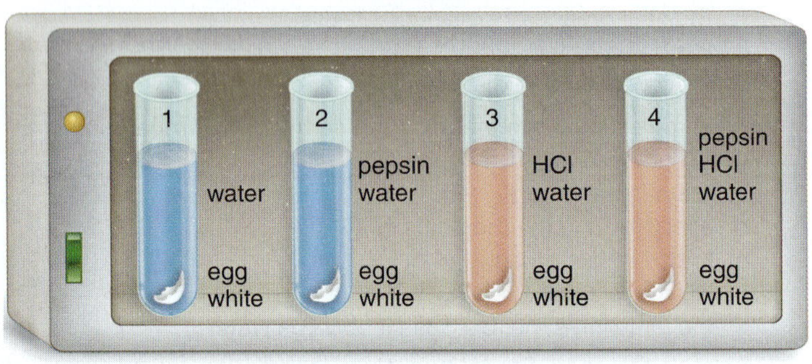

34.4 Nutrition and Human Health

12. Plants generally store their excess glucose as
    a. amino acids.
    b. fiber.
    c. glycogen.
    d. starch.
13. Lipids
    a. are never essential in the diet.
    b. include fats and oils.
    c. contain less energy than carbs.
    d. are found only in animals, not in plants.
14. In order to meet their need for protein, vegetarians
    a. must consume a small quantity of meat.
    b. should consume plenty of fresh fruits.
    c. can eat combinations of grains, beans, and nuts.
    d. should add powdered protein supplements to their diet.

ENGAGE

THINKING SCIENTIFICALLY

1. Imagine you could track a single molecule of carbon that is a part of a glucose molecule that, in turn, is a part of a starch molecule in an apple. Describe what happens to that molecule of carbon, in a chemical sense, after you bite into the apple. What enzymes are involved, and where, as the starch molecule is broken down into simpler carbohydrates, then into glucose, which is absorbed into your bloodstream and eventually into cells, which use it to produce ATP?
2. Snakes often swallow whole animals, a process that takes a long time. Then, snakes spend some time digesting their food. What structural modifications to the digestive tract would allow the slow swallowing and storage of a whole animal to occur? What chemical modifications would be necessary to digest a whole animal?
3. A drug for leukemia is not destroyed in the stomach and is well absorbed by the small intestine. However, the molecular form of the drug collected from the blood is not the same as the form that was swallowed by the patient. What explanation is most likely?
4. Suppose you are taking large doses of creatine, an amino acid supplement advertised for its ability to enhance muscle growth. Because your muscles can grow only at a limited rate, what do you suppose happens to the excess creatine that is not used for the synthesis of new muscle?

David Blaine displays an amazing ability to hold his breath while under water, which he has learned through training. Mammals living in aquatic environments have gained this ability through adaptation.

35

Respiratory Systems

Chapter Outline

35.1 Gas-Exchange Surfaces 660

35.2 Breathing and Transport of Gases 665

35.3 Respiration and Human Health 669

On April 30, 2008, magician David Blaine set a world record by holding his breath for 17 minutes, 4 seconds while submerged in a glass globe filled with cold water. While this was an amazing feat, Blaine may have been aided by an ancient evolutionary adaptation called the "diving response"—simply immerse your face in cold water, and your heart rate decreases, your spleen may contract (to release stored red blood cells), and blood vessels in your extremities constrict. Taken to the extreme, however, this response—along with decreasing oxygen levels—can lead to painful muscle cramping, and even tissue damage.

Despite efforts to push the limits of human physiology, the true breath-holding champions are aquatic mammals, such as the elephant seal, which can dive almost a mile deep and hold its breath for up to 2 hours. These animals benefit from various evolutionary adaptations: They have more red blood cells per body weight than we do; their muscles contain more oxygen-storing proteins; and they have a particularly effective diving response. Research indicates that elephant seals also tolerate exceptionally low levels of oxygen in their blood. Wherever they live, animals have evolved an amazing variety of strategies for delivering oxygen to their cells and removing carbon dioxide.

As you read through this chapter, think about the following questions:

1. What are some possible evolutionary pressures that might explain why a strictly terrestrial species, such as humans, would have a diving response?
2. Considering the adaptations that are required, what kinds of physiological limitations prevent elephant seals from being able to hold their breath for even longer?

Before You Begin

Before beginning this chapter, take a few moments to review the following discussions.

Figure 7.5 During which specific parts of photosynthesis do plants produce oxygen and use carbon dioxide?

Section 8.4 At what point is carbon dioxide produced inside mitochondria, and why is oxygen required?

Figure 32.5 What path does blood travel from the heart to the site of gas exchange in fish, amphibians, and birds?

Following *the* Themes

CHAPTER 35 RESPIRATORY SYSTEMS

UNIT 7 COMPARATIVE ANIMAL BIOLOGY

Evolution	Strategies have evolved in virtually all multicellular animals that allow them to efficiently extract oxygen from their environment and to eliminate carbon dioxide.
Nature of Science	Research studies into respiration are providing the basic biological framework needed for treating respiratory disease and building functioning lung tissues in the laboratory.
Biological Systems	Animal respiratory systems are responsive to changing demands for oxygen. They also play an important role in the interaction of an organism with its environment.

35.1 Gas-Exchange Surfaces

Learning Outcomes

Upon completion of this section, you should be able to

1. Distinguish among ventilation, external respiration, and internal respiration.
2. Compare and contrast the gas-exchange mechanisms of hydras, earthworms, insects, aquatic vertebrates, and terrestrial vertebrates.
3. Trace the path of a molecule of oxygen as it passes from the human nose to an alveolus.

Respiration is the sequence of events that results in gas exchange between the body's cells and the environment. In terrestrial vertebrates, respiration includes these steps:

- **Ventilation** (breathing) includes inspiration (the entrance of air into the lungs) and expiration (the exit of air from the lungs).
- **External respiration** is gas exchange between the air and the blood within the lungs. Blood then transports oxygen from the lungs to the tissues.
- **Internal respiration** is gas exchange between the blood and the interstitial fluid. (The body's cells exchange gases with the interstitial fluid.) The blood then transports carbon dioxide to the lungs.

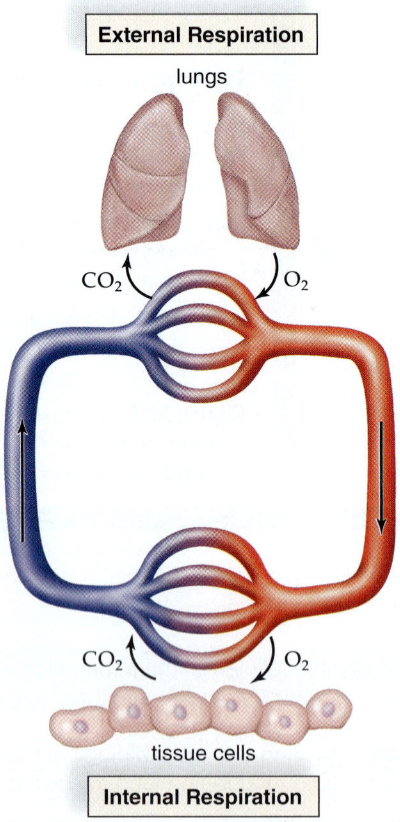

Gas exchange takes place by the physical process of diffusion (see section 5.2). For external respiration to be effective, the gas-exchange region must be (1) moist, (2) thin, and (3) large in relation to the size of the body.

Some animals are small and shaped in a way that allows the surface of the animal to be the gas-exchange surface. Most complex animals have evolved specialized tissues for external respiration, such as gills in aquatic animals and lungs in terrestrial animals. The effectiveness of diffusion is enhanced by vascularization (the presence of many capillaries), and delivery of oxygen to the cells is promoted when the blood contains a respiratory pigment, such as hemoglobin.

Regardless of the particular external respiration surface and the manner in which gases are delivered to the cells, in the end, oxygen enters mitochondria, where cellular respiration takes place (see section 8.1). A rare exception to this was discovered in April 2010, when a team of Italian and Danish deep-sea divers discovered a new species of tiny, jellyfish-like animals called loriciferans living in sediment more than 10,000 feet below the surface of the Mediterranean Sea, a depth that contains almost no oxygen. This discovery represents the first known multicellular animals that do not appear to require oxygen! Subsequent studies have indicated that the cells of these animals may lack mitochondria, but instead contain structures that resemble those used by anaerobic bacteria to undergo cellular respiration in the absence of oxygen. For most animals, however, if internal respiration does not occur, ATP production declines dramatically, and life ceases.

Overview of Gas-Exchange Surfaces

It is more difficult for animals to obtain oxygen from water than from air. Water fully saturated with air contains only a fraction of the amount of oxygen that is present in the same volume of air. Also, water is more dense than air. Therefore, aquatic animals expend more energy carrying out gas exchange than do terrestrial animals. Fish use as much as 25% of their energy output to respire, while terrestrial mammals use only 1–2% of their energy output for that purpose.

Hydras, which are cnidarians, and planarians, which are flatworms, have a large surface area in comparison to their size. This makes it possible for most of their cells to exchange gases directly with the environment. In hydras, the outer layer of cells is in contact with the external environment, and the inner layer can exchange gases with the water in the gastrovascular cavity (Fig. 35.1).

The earthworm is an example of a terrestrial invertebrate that is able to use its body surface for respiration because the capillaries come close to the surface (Fig. 35.2). An earthworm keeps its body surface moist by secreting mucus and by releasing fluids from excretory pores. Further, the worm is behaviorally adapted to remain in damp soil during the day, when the air is driest.

Aquatic invertebrates (e.g., clams and crayfish) and aquatic vertebrates (e.g., fish and tadpoles) have gills that extract oxygen from a watery environment. **Gills** are finely divided, vascularized outgrowths of the body surface or the pharynx (Fig. 35.3*a*). Various mechanisms are used to pump water across the gills, depending on the organism.

Insects have a system of air tubes called **tracheae** through which oxygen is delivered directly to the cells without entering the blood (Fig. 35.3*b*). Air sacs located near the wings, legs, and abdomen act as bellows to help move the air into the tubes through external openings.

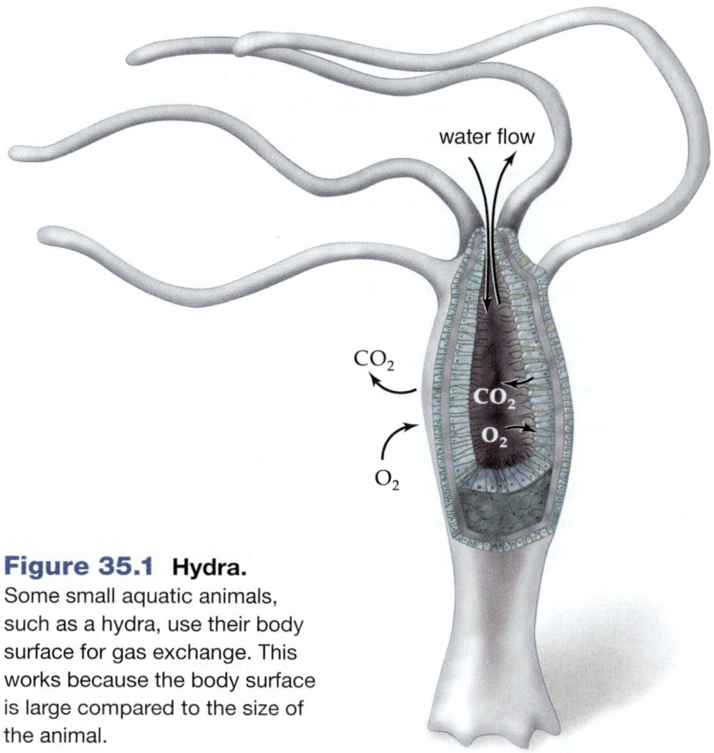

Figure 35.1 Hydra. Some small aquatic animals, such as a hydra, use their body surface for gas exchange. This works because the body surface is large compared to the size of the animal.

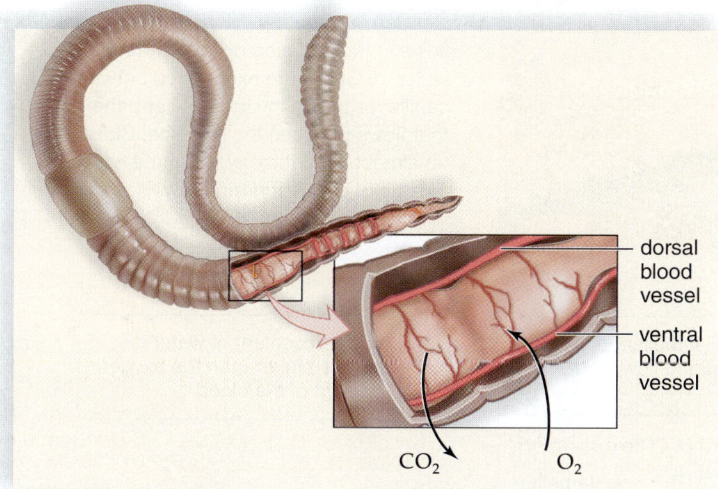

Figure 35.2 Earthworm. An earthworm's entire external surface functions in external respiration.

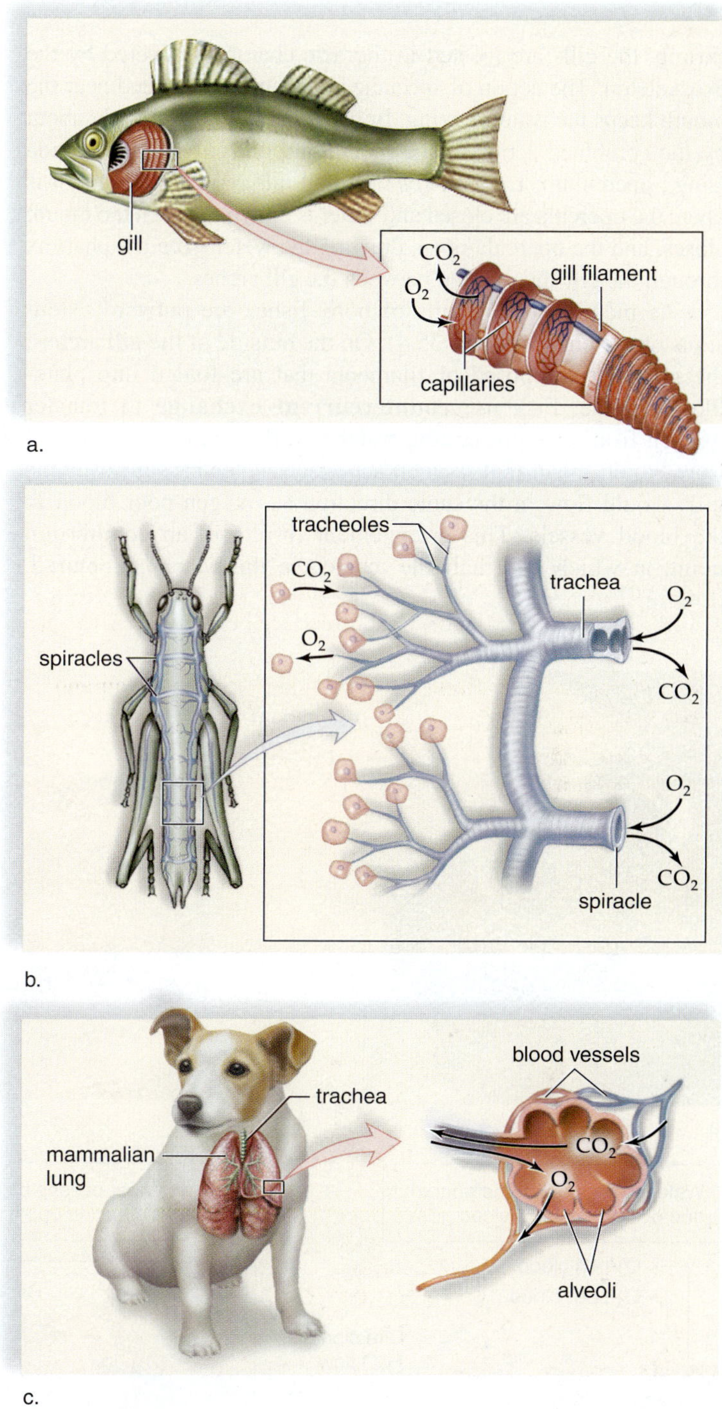

Figure 35.3 Respiratory organs. **a.** Fish have gills to assist with external respiration. **b.** Insects have a tracheal system that delivers oxygen directly to their cells. **c.** Terrestrial vertebrates have lungs with a large total external respiration surface.

Terrestrial vertebrates usually have **lungs,** which are vascularized outgrowths from the lower pharyngeal region. The tadpoles of frogs live in the water and have gills as external respiratory organs, but adult amphibians possess simple, saclike lungs. Most amphibians respire to some extent through the skin, and some salamanders depend entirely on the skin, which is kept moist by mucus produced by numerous glands on the surface of the body.

The lungs of birds and mammals are elaborately subdivided into small passageways and spaces (Fig. 35.3*c*). It has been estimated that human lungs have a total surface area of about 70 square meters, which is about 50 times the skin's surface area. Air is a rich source of oxygen compared to water; however, it does have a drying effect on external respiratory surfaces. A human loses about 350 ml of water per day through respiration when the air has a relative humidity of only 50%. To keep the lungs from drying out, air is moistened as it moves in through the passageways leading to the lungs.

The Gills of a Fish

Animals with gills use various means of ventilation. Among molluscs, such as clams or squids, water is drawn into the mantle cavity, where it passes through the gills. In crustaceans such as crabs and shrimp, the gills are located in thoracic chambers covered by the exoskeleton. The action of specialized appendages located near the mouth keeps the water moving. In fish, ventilation is brought about by the combined action of the mouth and gill covers, or opercula (sing., operculum; L. *operculum,* "small lid"). When the mouth is open, the opercula are closed and water is drawn in. Then the mouth closes, and the opercula open, drawing the water from the pharynx through the gill slits located between the gill arches.

As mentioned, the gills of bony fishes are outward extensions of the pharynx (Fig. 35.4). On the outside of the gill arches, the gills are composed of filaments that are folded into platelike lamellae. Fish use **countercurrent exchange** to transfer oxygen from the surrounding water into their blood. *Con*current flow would mean that oxygen (O_2)–rich water passing over the gills would flow in the same direction as oxygen-poor blood in the blood vessels. This arrangement results in an equilibrium point, at which only half the oxygen in the water is captured. *Counter*current flow, in contrast, means that the two fluids flow in opposite directions. With countercurrent flow, as blood gains oxygen, it always encounters water having an even higher oxygen content. A countercurrent mechanism prevents an equilibrium point from being reached, and about 80–90% of the initial dissolved oxygen in water is extracted.

The Tracheal System of Insects

Arthropods are coelomate animals, but the coelom is reduced and the internal organs lie within a cavity called the hemocoel, because it contains hemolymph, a mixture of blood and lymph (see Chapter 32). Hemolymph flows freely through the hemocoel, making circulation in arthropods inefficient. Many insects are adapted for flight, and their flight muscles require a steady supply of oxygen.

Insects overcome the inefficiency of their blood flow by having a respiratory system that consists of **tracheae,** tiny air tubes

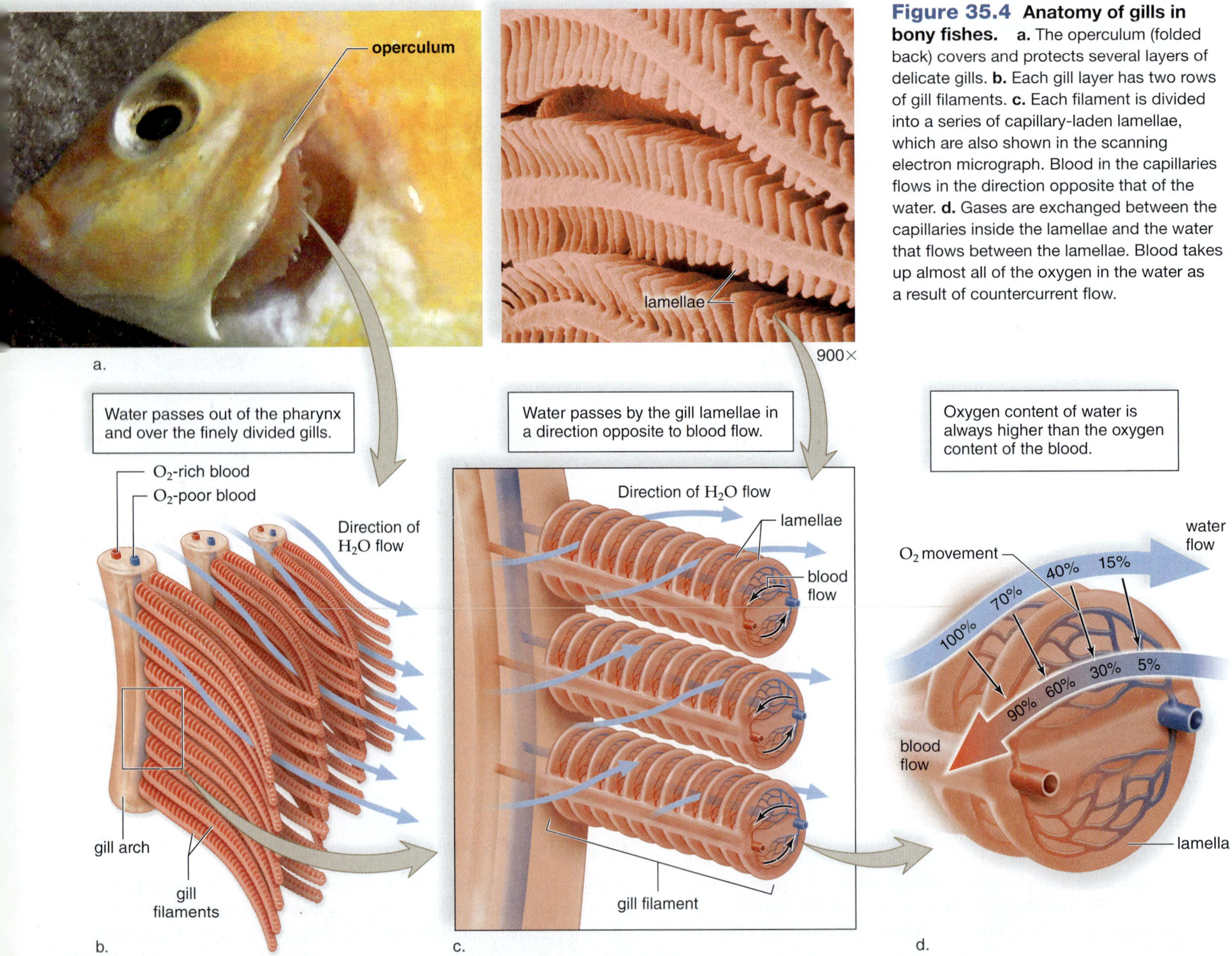

Figure 35.4 Anatomy of gills in bony fishes. **a.** The operculum (folded back) covers and protects several layers of delicate gills. **b.** Each gill layer has two rows of gill filaments. **c.** Each filament is divided into a series of capillary-laden lamellae, which are also shown in the scanning electron micrograph. Blood in the capillaries flows in the direction opposite that of the water. **d.** Gases are exchanged between the capillaries inside the lamellae and the water that flows between the lamellae. Blood takes up almost all of the oxygen in the water as a result of countercurrent flow.

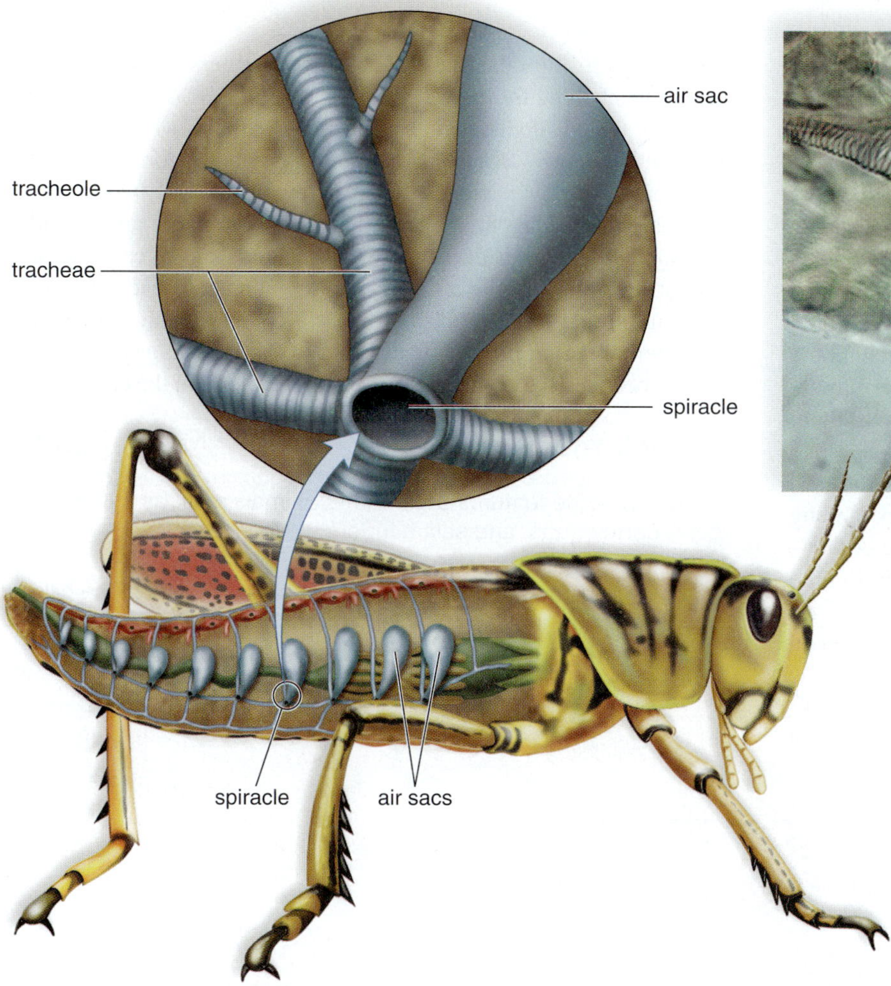

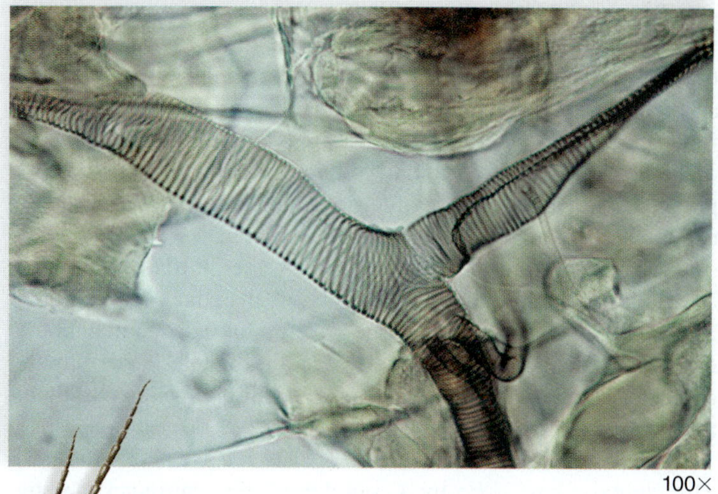

Figure 35.5 Tracheae of insects. A system of air tubes extends throughout the body of an insect, and the tubes carry oxygen to the cells. Air enters the tracheae at openings called spiracles. From here, the air moves to the smaller tracheoles, which take it to the cells, where gas exchange takes place. The photomicrograph shows how the walls of the trachea are stiffened with bands of chitin.

that take oxygen directly to the cells (Fig. 35.5). The tracheae branch into even smaller tubules called tracheoles, which also branch and rebranch until finally the air tubes are only about 0.1 μm in diameter. There are so many fine tracheoles that almost every cell is near one. Also, the tracheoles indent the plasma membrane, so that they terminate close to mitochondria. Therefore, O_2 can flow more directly from a tracheole to mitochondria, where cellular respiration occurs. The tracheae also dispose of CO_2.

The tracheoles are fluid-filled, but the larger tracheae contain air and open to the outside by way of spiracles (Fig. 35.5). Usually, the spiracle has some sort of closing device that reduces water loss, and this may be why insects have no trouble inhabiting drier climates. Scientists have determined that the tracheae can actually expand and contract, thereby drawing air into and out of the system. To improve the efficiency of the tracheal system, many larger insects also have air sacs, which are thin-walled and flexible, located near major muscles. Contraction of these muscles causes the air sacs to empty, and relaxation causes the air sacs to expand and draw in air. This method is comparable to the way that human lungs expand to draw air into them.

Even with all these adaptations, insects still lack the efficient circulatory system of birds and mammals, which is able to pump oxygen-rich blood through arteries to all the cells of the body. This may be why insects have remained relatively small (despite the attempts of science-fiction movies to make us think otherwise).

A tracheal system is an adaptation to breathing air, yet some insect larval stages, and even some adult insects, live in the water. In these instances, the tracheae do not receive air by way of spiracles. Instead, diffusion of oxygen across the body wall supplies the tracheae with oxygen. Mayfly and stonefly nymphs have thin extensions of the body wall called tracheal gills—the tracheae are particularly numerous in this area. This is an interesting adaptation, because it dramatizes that tracheae function to deliver oxygen in the same manner as vertebrate blood vessels.

Some aquatic insects have developed a different strategy. Like most insects, water beetles breathe through spiracles. Because they live in water however, they capture a bubble of air from the surface and carry it with them, exchanging the oxygen inside for CO_2. Water spiders even spin an underwater web, which they fill with air bubbles, forming what some scientists have called an "external lung."

The Lungs of Humans

The human respiratory system includes all of the structures that conduct air in a continuous pathway to and from the lungs (Fig. 35.6*a*). The lungs lie deep in the body, within the thoracic cavity, where they are protected from drying out. As air moves through the nose, the pharynx, trachea, and bronchi to the lungs, it is filtered, so that it is free of debris, warmed, and humidified. By the time the air reaches the lungs, it is at body temperature and saturated with water.

MP3 Respiratory Structure and Function

In the nose, hairs and cilia act as a screening device. In the trachea and the bronchi, cilia beat upward, carrying mucus, dust, and occasionally small bits of food that "went down the wrong way" back into the throat, where the accumulation may be swallowed or expectorated (Fig. 35.6*b*).

The hard and soft palates separate the nasal cavities from the oral cavity, but the air and food passages cross in the **pharynx.** This arrangement may seem inefficient, and there is danger of choking if food accidentally enters the trachea; however, it has the advantage of letting you breathe through your mouth in case your nose is plugged up. In addition, it permits greater intake of air during heavy exercise, when greater gas exchange is required.

Air passes from the pharynx through the **glottis,** which is an opening into the **larynx,** or voice box. At the edges of the glottis are two folds of connective tissue covered by mucous membrane called the **vocal cords.** These flexible and pliable bands vibrate against each other, producing sound when air is expelled past them through the glottis from the larynx.

The larynx and trachea remain open to receive air at all times. The larynx is held open by a complex of nine cartilages, among them the Adam's apple. Easily seen in many men, the Adam's apple resembles a small, rounded apple just under the skin in the front of the neck. The **trachea** (windpipe) is held open by a series of C-shaped, cartilaginous rings that do not completely meet in the rear. When food is being swallowed, the larynx rises, and the glottis is covered by a flap of tissue called the **epiglottis.** A backward movement of the soft palate covers the entrance of the nasal passages into the pharynx. The food then enters the esophagus, which lies behind the larynx.

The trachea divides into two primary **bronchi,** which enter the right and left lungs. Branching continues, eventually forming a great number of smaller passages called **bronchioles.** The two bronchi resemble the trachea in structure, but as the bronchial tubes divide and subdivide, their walls become thinner, and rings of cartilage are absent. Each bronchiole terminates in an elongated space enclosed by a multitude of air pockets, or sacs, called **alveoli,** which fill the internal region of the lungs (Fig. 35.6*c*). Internal gas exchange occurs between the air in the alveoli and the blood in the capillaries.

Check Your Progress 35.1

1. List some features common to animals such as hydras, earthworms, and salamanders, which are able to exchange gases directly with their environment.
2. Explain why the countercurrent flow that occurs in the gills of fish is much more efficient than concurrent flow would be.
3. Describe the role of each of the following in insect respiration: hemolymph, tracheae, tracheoles, spiracles, air sacs, tracheal gills.

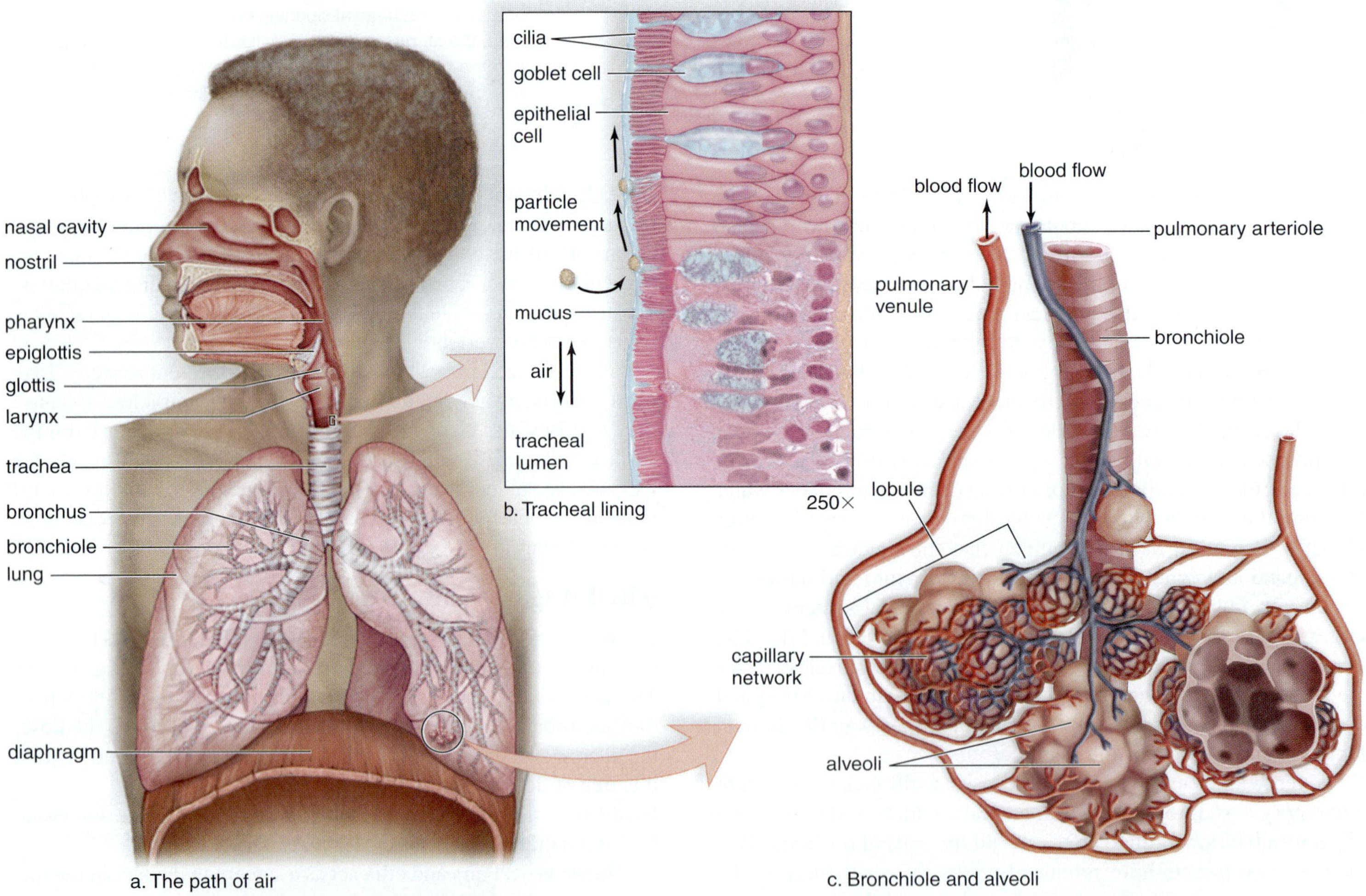

Figure 35.6 The human respiratory tract. **a.** The respiratory tract extends from the nose to the lungs, which are composed of air sacs called alveoli. **b.** The lining of the trachea is a ciliated epithelium with mucus-producing goblet cells. The lining prevents inhaled particles from reaching the lungs: The mucus traps the particles, and the cilia help move the mucus toward the throat, where it can be swallowed or expectorated. **c.** Gas exchange occurs between air in the alveoli and blood within a capillary network that surrounds the alveoli.

35.2 Breathing and Transport of Gases

Learning Outcomes

Upon completion of this section, you should be able to

1. Compare the mechanisms used by amphibians, mammals, and birds to inflate their lungs.
2. Explain how the breathing rate in humans is influenced by both physical and chemical factors.
3. Describe how carbon dioxide (CO_2) is carried in the blood and the effect that blood P_{CO_2} has on blood pH.

During breathing, the lungs are ventilated. Oxygen (O_2) moves into the blood, and carbon dioxide (CO_2) moves out of the blood into the lungs. Blood transports O_2 to the body's cells and CO_2 from the cells to the lungs.

Breathing

Terrestrial vertebrates ventilate their lungs by moving air into and out of the respiratory tract. Amphibians use positive pressure to force air into the respiratory tract. With the mouth and nostrils firmly shut, the floor of the mouth rises and pushes the air into the lungs. Reptiles, birds, and mammals use negative pressure to move air into the lungs and positive pressure to move it out. **Inspiration** (inhalation) is the act of moving air into the lungs, and **expiration** (exhalation) is the act of moving air out of the lungs.

Reptiles have jointed ribs that can be raised to expand the lungs, but mammals have both a rib cage and a diaphragm. The **diaphragm** is a horizontal muscle that divides the thoracic cavity (above) from the abdominal cavity (below). During inspiration in mammals, the rib cage moves up and out, and the diaphragm contracts and moves down (Fig. 35.7*a*). As the thoracic (chest) cavity expands and lung volume increases, air flows into the lungs due to decreased air pressure in the thoracic cavity and lungs. Inspiration is the active phase of breathing in reptiles and mammals.

During expiration in mammals, the rib cage moves down, and the diaphragm relaxes and moves up to its former position (Fig. 35.7*b*). No muscle contraction is required; thus, expiration is the passive phase of breathing in reptiles and mammals. During expiration, air flows out as a result of increased pressure in the thoracic cavity and lungs.

We can compare ventilation in reptiles and mammals to the relationship between air pressure and volume in a flexible

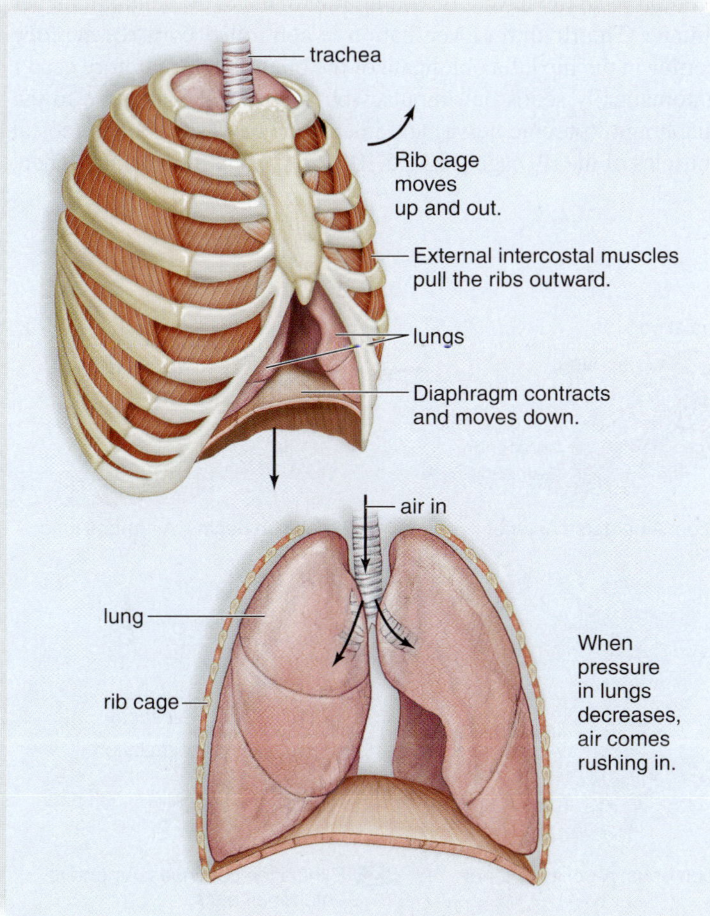

a. Inspiration

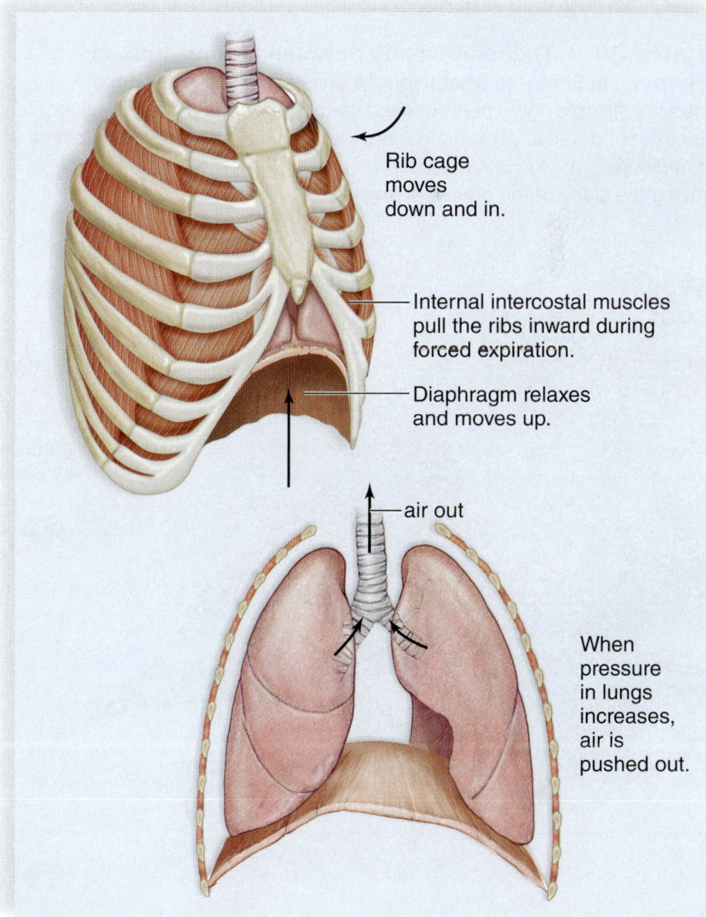

b. Expiration

Figure 35.7 The thoracic cavity during inspiration and expiration. **a.** During inspiration, the thoracic cavity and lungs expand, so that air is drawn in. **b.** During expiration, the thoracic cavity and lungs resume their original positions and pressures. Then, air is forced out.

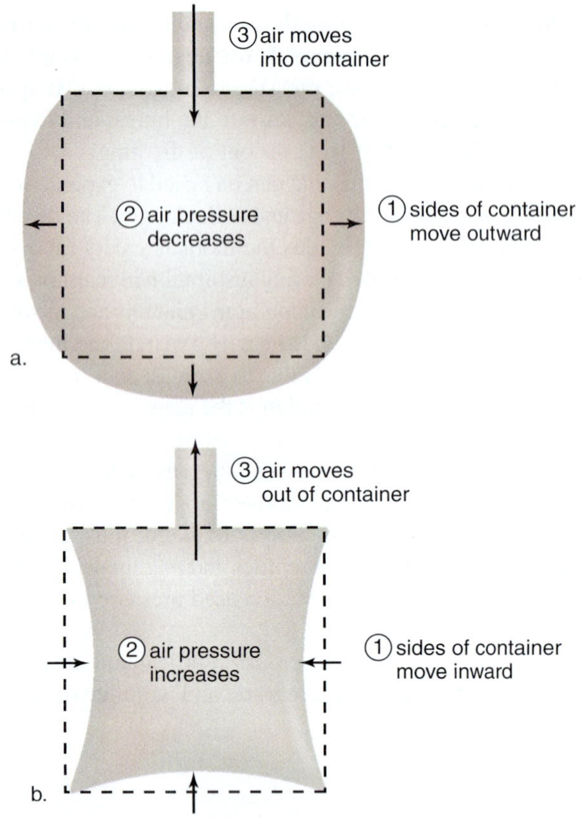

Figure 35.8 The relationship between air pressure and volume. **a.** Similar to what happens during inhalation, when the sides of a flexible container move outward, the volume of the container increases and the air pressure decreases. **b.** During exhalation, increased air pressure in the lungs causes air to flow out, similar to the effects of moving the sides of the container inward.

container (Fig. 35.8). When the sides of the container move outward, air pressure decreases inside the container and air moves in, just as air automatically enters the lungs because the rib cage moves up and out during inspiration. Conversely, if the sides of the container are pressed inward, air flows out because of increased air pressure inside the container. Similarly, air automatically exits the lungs when the rib cage moves down and in during expiration. The analogy is not exact, however, because no force is required for the rib cage to move down, and inspiration is the only active phase of breathing. Forced expiration can occur if we so desire, however.

All terrestrial vertebrates, except birds, use a *tidal ventilation mechanism,* so named because the air moves in and out by the same route. This means that the lungs of amphibians, reptiles, and mammals are not completely emptied and refilled during each breathing cycle. Because of this, the air entering mixes with used air remaining in the lungs. Although this does help conserve water, it also decreases gas-exchange efficiency. In contrast, birds use a *one-way ventilation mechanism* (Fig. 35.9). Incoming air is carried past the lungs by a trachea, which takes it to a set of posterior air sacs. The air then passes forward through the lungs into a set of anterior air sacs. From here, it is finally expelled. Notice that fresh air never mixes with used air in the lungs of birds, thereby greatly improving gas-exchange efficiency.

Modifications of Breathing in Humans

Normally, adults have a breathing rate of 12 to 20 ventilations per minute. The rhythm of ventilation is controlled by a **respiratory center** in the medulla oblongata of the brain. The respiratory center automatically sends out impulses by way of a spinal nerve to the diaphragm (phrenic nerve) and intercostal nerves to the intercostal muscles of the rib cage (Fig. 35.10). Now inspiration occurs. Then,

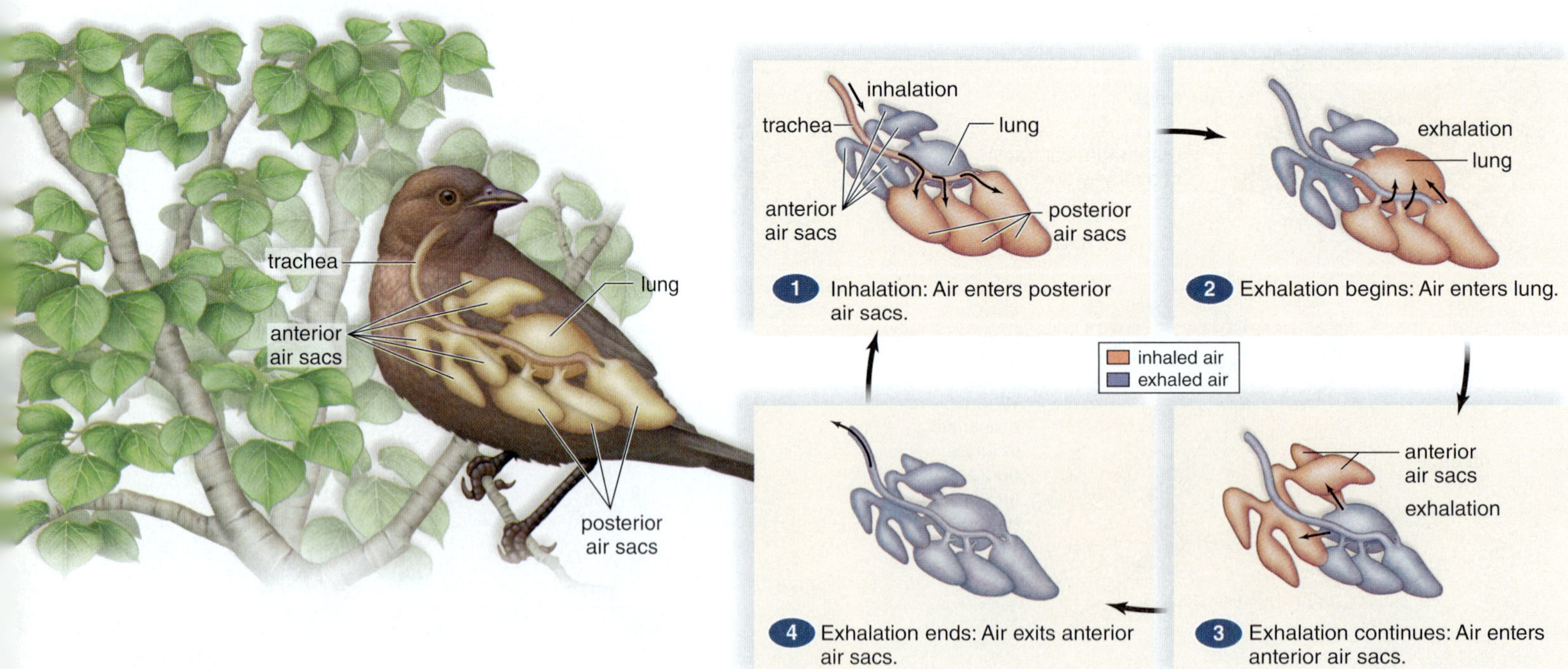

Figure 35.9 Respiratory system in birds. Air sacs are attached to the lungs of birds. These allow birds to have a one-way mechanism of ventilating their lungs.

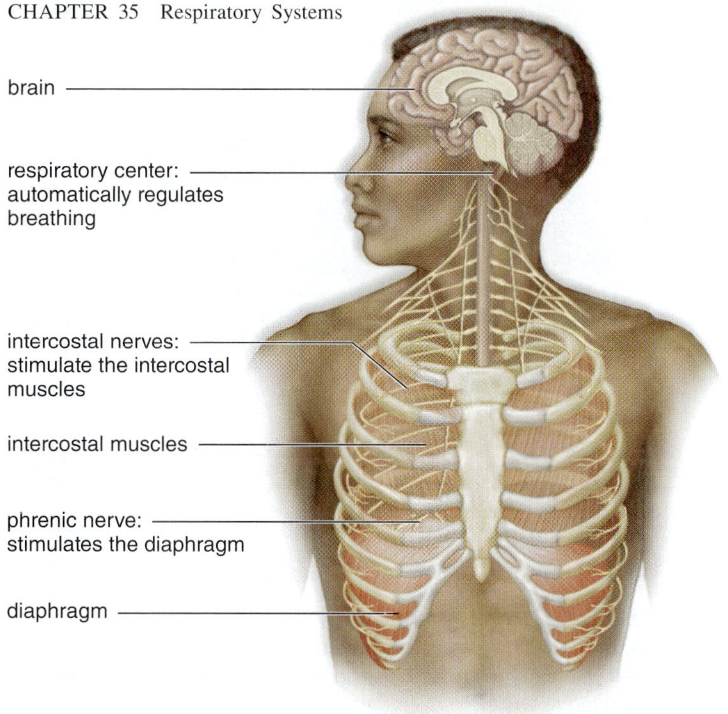

Figure 35.10 Nervous control of breathing. The breathing rate can be modified by nervous stimulation of the intercostal muscles and diaphragm.

when the respiratory center stops sending neuronal signals to the diaphragm and the rib cage, expiration occurs.

Although the respiratory center automatically controls the rate and depth of breathing, its activity can also be influenced by nervous input and chemical input. Following forced inhalation, stretch receptors in the alveolar walls initiate inhibitory nerve impulses that travel from the inflated lungs to the respiratory center. This stops the respiratory center from sending out nerve impulses.

The respiratory center is directly sensitive to the levels of hydrogen ions (H^+). However, when carbon dioxide enters the blood, it reacts with water and releases hydrogen ions. In this way, CO_2 participates in regulating the breathing rate. When hydrogen ions rise in the blood and the pH decreases, the respiratory center increases the rate and depth of breathing. The chemoreceptors in the **carotid bodies,** located in the carotid arteries, and in the **aortic bodies,** located in the aorta, stimulate the respiratory center during intense exercise due to a reduction in pH, or if arterial oxygen decreases to 50% of normal.

MP3 Control of Respiration

Gas Exchange and Transport

Respiration includes the exchange of gases in our lungs, called external respiration, as well as the exchange of gases in the tissues, called internal respiration (Fig. 35.11). The principles of diffusion

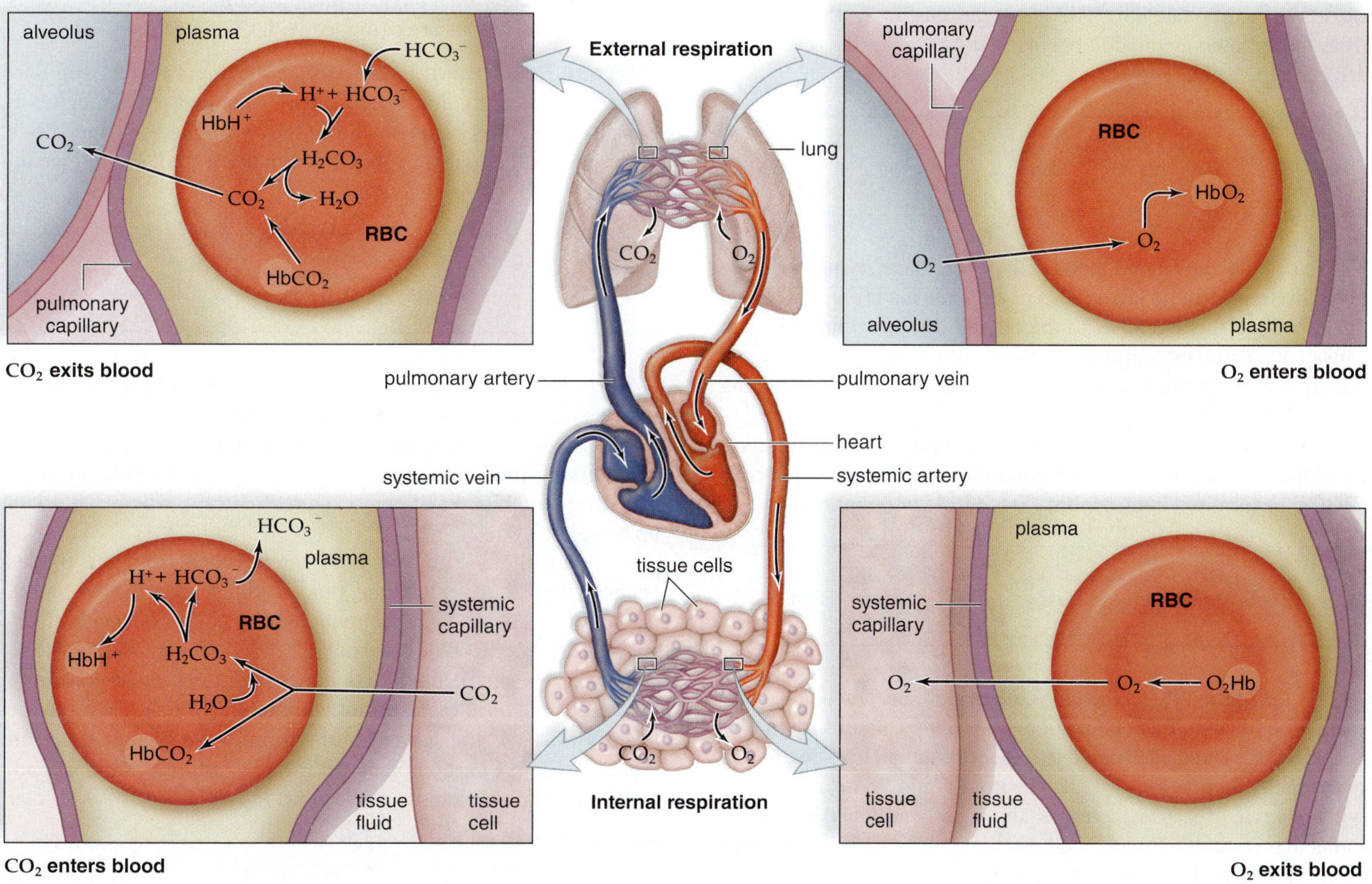

Figure 35.11 External and internal respiration. During external respiration (*top*) in the lungs, carbon dioxide (CO_2) leaves blood, and oxygen (O_2) enters blood. During internal respiration (*bottom*) in the tissues, oxygen leaves blood, and carbon dioxide enters blood.

Tutorial Internal and External Respiration

largely govern the movement of gases into and out of blood vessels in the lungs and in the tissues. Gases exert pressure, and the amount of pressure each gas exerts is called the **partial pressure,** symbolized as P_{O_2} and P_{CO_2}. If the partial pressure of oxygen differs across a membrane, oxygen will diffuse from the higher to the lower pressure. Similarly, carbon dioxide diffuses from the higher to the lower partial pressure.

MP3 Gas Exchange

Ventilation causes the alveoli of the lungs to have a higher P_{O_2} and a lower P_{CO_2} than the blood in pulmonary capillaries, and this accounts for the exchange of gases in the lungs. When blood reaches the tissues, cellular respiration in cells causes the interstitial fluid to have a lower P_{O_2} and a higher P_{CO_2} than the blood in the systemic capillaries, and this accounts for the exchange of gases in the tissues.

Transport of Oxygen and Carbon Dioxide

The transport of O_2 and CO_2 is somewhat different in external respiration than in internal inspiration, although the driving forces of diffusion are the same.

External Respiration. As blood enters the lungs, a small amount of CO_2 is being carried by hemoglobin with the formula $HbCO_2$. Also, some hemoglobin is carrying hydrogen ions with the formula HbH^+. Most of the CO_2 in the pulmonary capillaries is carried as bicarbonate ions (HCO_3^-) in the plasma. As the free CO_2 from the following equation begins to diffuse out, this reaction is driven to the right:

H^+	+	HCO_3^-	$\longrightarrow$	H_2CO_3	$\longrightarrow$	H_2O	+	CO_2
hydrogen ion		bicarbonate ion		carbonic acid		water		carbon dioxide

The reaction occurs in red blood cells, where the enzyme **carbonic anhydrase** speeds the breakdown of carbonic acid (Fig. 35.11, *top left*). Pushing this equation to the far right by breathing fast can cause you to stop breathing for a time; pushing this equation to the left by not breathing is even more temporary, because breathing will soon resume due to the rise in H^+.

Most oxygen entering the pulmonary capillaries from the alveoli of the lungs combines with **hemoglobin** (**Hb**) in red blood cells (RBCs) to form **oxyhemoglobin** (Fig. 35.11, *top right*):

Hb	+	O_2	$\longrightarrow$	HbO_2
deoxyhemoglobin		oxygen		oxyhemoglobin

At the normal P_{O_2} in the lungs, hemoglobin is practically saturated with oxygen. Each hemoglobin molecule contains four polypeptide chains, and each chain is folded around an iron-containing group called **heme** (Fig. 35.12). The iron forms a loose association with oxygen. Because there are about 250 million hemoglobin molecules in each red blood cell, each red blood cell is capable of carrying at least 1 billion molecules of oxygen.

Carbon monoxide (CO) is an air pollutant that is produced by the incomplete combustion of natural gas, gasoline, kerosene, and

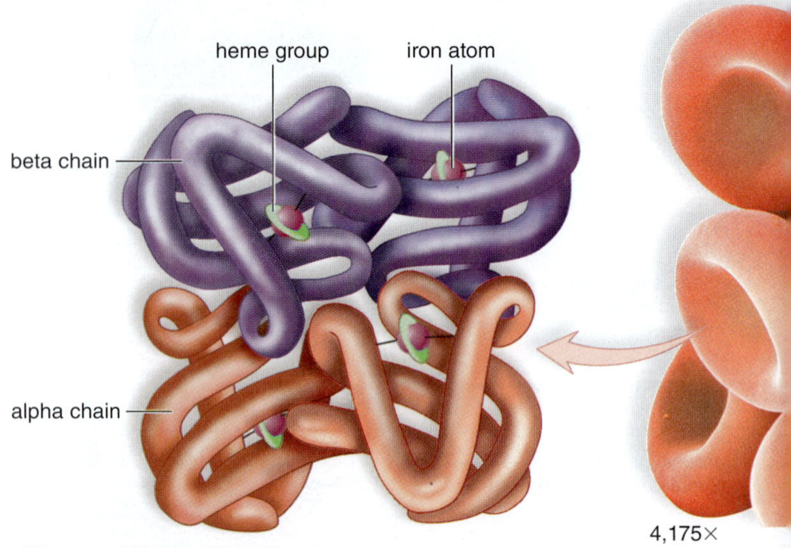

Figure 35.12 Hemoglobin. Hemoglobin consists of four polypeptide chains, two alpha (*red*) and two beta (*purple*), each associated with a heme group. Each heme group contains an iron atom, which can bind to O_2.

even wood and charcoal. Because CO is a colorless, odorless gas, people can be unaware that they are breathing it. But once CO is in the bloodstream, it combines with the iron of hemoglobin 200 times more tightly than oxygen, and the result can be death. This is the reason that homes are equipped with CO detectors.

Internal Respiration. Blood entering the systemic capillaries is a bright red color, because RBCs contain oxyhemoglobin. Because the temperature in the tissues is higher and the pH is lower than in the lungs, oxyhemoglobin has a tendency to give up oxygen:

$$HbO_2 \longrightarrow Hb + O_2$$

Oxygen diffuses out of the blood into the tissues because the P_{O_2} of interstitial fluid is lower than that of blood (Fig. 35.11, *bottom right*). The lower P_{O_2} is due to cells continuously using up oxygen in cellular respiration. After oxyhemoglobin gives up O_2, this oxygen leaves the blood and enters interstitial fluid, where it is taken up by cells.

Carbon dioxide, in contrast, enters blood from the tissues because the P_{CO_2} of interstitial fluid is higher than that of blood. Carbon dioxide, produced continuously by cells, collects in interstitial fluid. After CO_2 diffuses into the blood, it enters the red blood cells, where a small amount combines with the protein portion of hemoglobin to form **carbaminohemoglobin** ($HbCO_2$). Most of the CO_2, however, is transported in the form of the **bicarbonate ion** (HCO_3^-). First, CO_2 combines with water, forming carbonic acid, and then this dissociates to a hydrogen ion (H^+) and HCO_3^-:

CO_2	+	H_2O	$\longrightarrow$	H_2CO_3	$\longrightarrow$	H^+	+	HCO_3^-
carbon dioxide		water		carbonic acid		hydrogen ion		bicarbonate ion

Carbonic anhydrase also speeds this reaction. The HCO_3^- diffuses out of the red blood cells to be carried in the plasma (see Fig. 35.11, *bottom left*).

The release of H^+ from this reaction could drastically change the pH of the blood, which is highly undesirable, because cells require a normal pH in order to remain healthy. However, the H^+ is absorbed by the globin portions of hemoglobin. Hemoglobin that has combined with H^+ is called reduced hemoglobin and has the formula HbH^+. HbH^+ plays a vital role in maintaining the normal pH of the blood. Blood that leaves the systemic capillaries is a dark maroon color, because red blood cells contain reduced hemoglobin.

MP3 Gas Transport

MP3 Gas Exchange During Respiration

Check Your Progress 35.2

1. Describe how the mechanism of ventilation in reptiles and mammals is similar to changing the volume of a flexible container.
2. Explain how the carotid bodies and aortic bodies affect the rate of respiration.
3. Define the role of oxyhemoglobin, reduced hemoglobin, and carbaminohemoglobin in homeostasis.

35.3 Respiration and Human Health

Learning Outcomes

Upon completion of this section, you should be able to

1. Describe several common disorders that mainly affect the upper respiratory tract as well as several that affect the lower respiratory tract.
2. Classify several common respiratory disorders according to whether they are mainly caused by allergies, infections, a genetic defect, or toxin exposure.

The human respiratory tract is constantly exposed to environmental air, which may contain infectious agents, allergens, tobacco smoke, or other toxins. This results in the respiratory tract being susceptible to a number of diseases. Some of the most important of these are summarized here.

Disorders of the Upper Respiratory Tract

The upper respiratory tract consists of the nasal cavities, sinuses, pharynx, and larynx. Because the upper part of the respiratory tract filters out many pathogens and other materials that may be present in the air, it is commonly affected by a variety of infections, which may also spread to the middle ear or the sinuses.

The Common Cold

Most "colds" are relatively mild viral infections of the upper respiratory tract characterized by sneezing, rhinitis (runny nose), and perhaps a mild fever. Most colds last a few days, after which the immune response is able to eliminate the inciting virus. However, since colds are caused by several different viruses, and by several hundred strains of these viruses, we usually have no immunity to the next strain that "goes around," and vaccines are very difficult to develop. As with all viral infections, antibiotics such as penicillin are useless in treating colds.

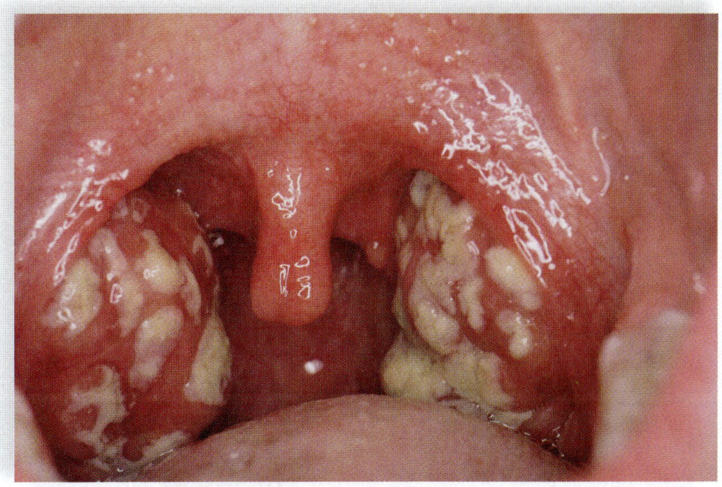

Figure 35.13 Strep throat. Pharyngitis caused by the bacterium *Streptococcus pyogenes* can cause swollen tonsils, as shown here. The whitish patches are areas of pus formation, indicating that white blood cells are fighting the infection.

Strep Throat

Most cases of **pharyngitis,** or inflammation of the pharynx, are caused by viruses, but strep throat is an acute pharyngitis caused by the bacterium *Streptococcus pyogenes*. Typical symptoms include severe sore throat, high fever, and white patches in the tonsillar area (Fig. 35.13). Adults experience about half as many sore throats as do children, who average about five upper respiratory infections per year and about one strep throat infection every 4 years. Many untreated strep infections probably resolve on their own, but some can lead to more serious conditions, such as scarlet fever or rheumatic fever. Fortunately, infection with *S. pyogenes* can be easily and quickly diagnosed with specific laboratory tests, and it is usually curable with antibiotics.

Disorders of the Lower Respiratory Tract

Several common disorders affecting the lower respiratory tract are summarized in Figure 35.14.

Disorders Affecting the Trachea and Bronchi

One of the most obvious and life-threatening disorders that can affect the trachea is choking. The best way for a person without extensive medical training to help someone who is choking is to perform the Heimlich maneuver, which involves grabbing the choking person around the waist from behind and forcefully pulling both hands into his or her upper abdomen to expel whatever is lodged. If this fails, trained medical personnel may be able to quickly insert a breathing tube through an incision made in the trachea. This procedure is called a tracheotomy, and the opening is a *tracheostomy*.

If infections of the upper respiratory tract spread into the lower respiratory tract, *acute bronchitis,* or inflammation of the bronchi,

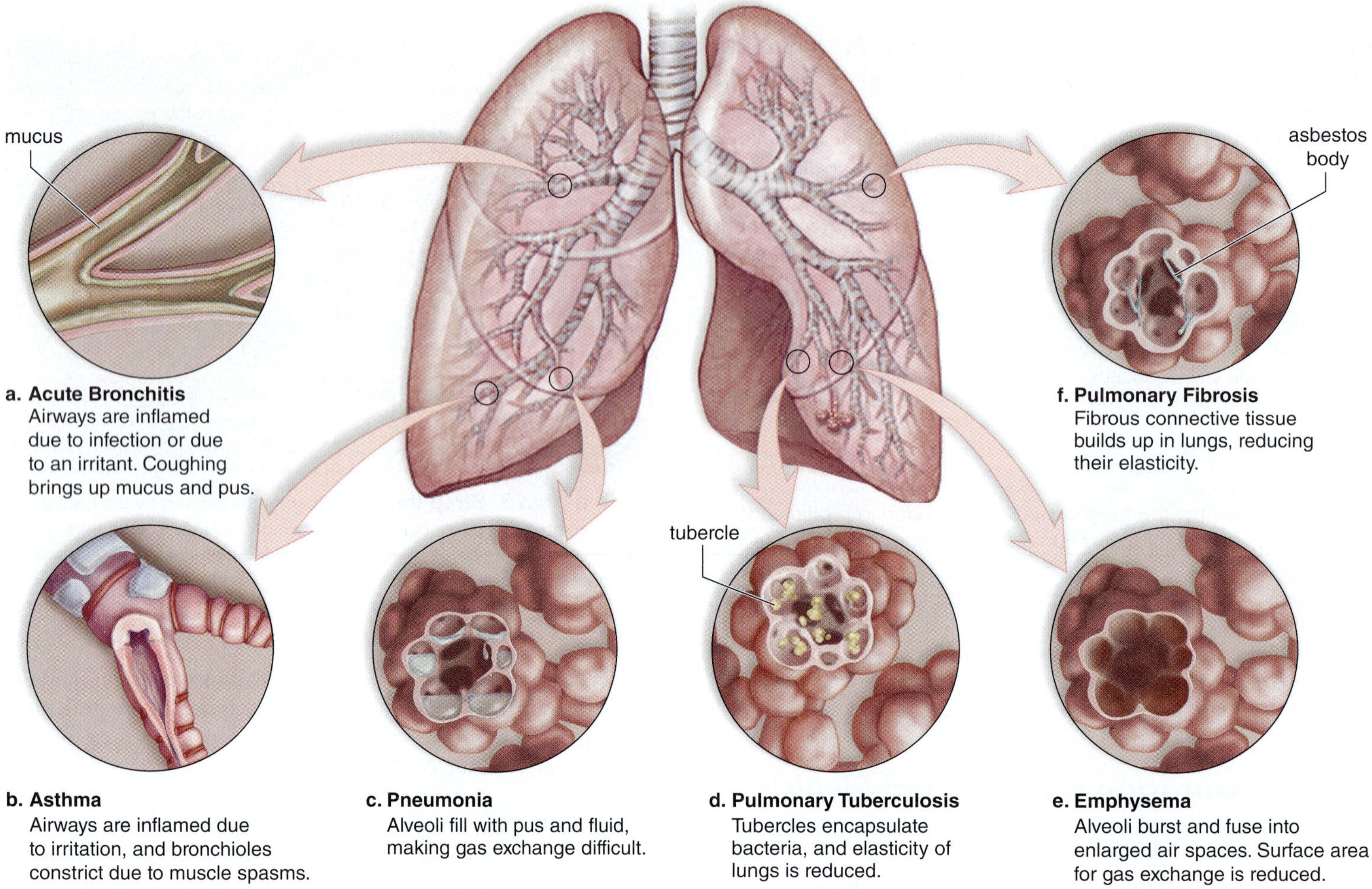

Figure 35.14 Common bronchial and pulmonary diseases. Exposure to infectious pathogens and/or polluted air, including tobacco smoke, causes the diseases and disorders shown here.

often results (Fig. 35.14*a*). Other causes of acute bronchitis include allergic reactions and damage from environmental toxins, such as those present in cigarette smoke. It is estimated that approximately 5% of the U.S. population suffers from a bout of acute bronchitis in any given year. Symptoms include fever, a cough that produces phlegm or pus, and chest pain. Depending on the cause, acute bronchitis may be treatable with antibiotics, or it may resolve with time or progress to more serious conditions.

If the inciting cause (such as smoking) persists, acute bronchitis can develop into *chronic bronchitis,* in which the airways are inflamed and filled with mucus. Over time, the bronchi undergo degenerative changes, including the loss of cilia and their normal cleansing action. Under these conditions, infections are more likely to occur. Smoking and exposure to other airborne toxins are the most frequent causes of chronic bronchitis. Along with emphysema, chronic bronchitis is a major component of *chronic obstructive pulmonary disease* (*COPD*), the fourth leading cause of death in the United States.

Asthma is a disease of the bronchi and bronchioles marked by coughing, wheezing, and breathlessness. The airways are unusually sensitive to various irritants, which include allergens such as pollen, animal dander, dust, and cigarette smoke. Even cold air or exercise can be an irritant.

An asthmatic attack results from inflammation in the airways and the contraction of smooth muscle lining their walls, resulting in a narrowing of the diameter of the airways (Fig. 35.14*b*). All estimates indicate that the incidence of asthma in American children has been increasing steadily since the early 1980s. Possible explanations for this include a more sedentary lifestyle, with more exposure to indoor toxins, and less frequent exposure to beneficial microbes. Asthma is not curable, but it is treatable. Drugs administered by inhalers can help prevent the inflammation and dilate the bronchi.

Disorders Affecting the Lungs

Altogether, the various diseases of the lung cause about 400,000 deaths per year in the United States, and they affect hundreds of millions of people worldwide. Depending on the cause, treatments for lung disease include antibiotics, supplemental oxygen, and anti-inflammatory drugs. For patients with serious lung conditions who have exhausted all other treatment options, a lung transplant may be the best option, but there aren't enough donor lungs to meet the need. The Nature of Science feature, "Artifical Lung Technology," describes some recent advances in artificial lung technology.

Pneumonia is a viral, bacterial, or fungal infection of the lungs in which bronchi and alveoli fill with a discharge, such as pus and fluid (Fig. 35.14*c*). Along with coughing and difficulty breathing, people suffering from pneumonia often have a high fever, sharp chest pain, and a cough that produces thick phlegm or even pus. Several

THEME **Nature of Science**

Artificial Lung Technology

Some organs, such as the kidneys, are relatively easy to transplant from one well-matched individual to another, with a high rate of success. Lungs are more difficult to transplant, however, with a 10-year survival rate of only 10–20%. Some recent research is showing how one day it may be possible to replace diseased lungs with laboratory-grown versions.

In a study at Yale University, scientists anesthetized a group of adult rats, surgically removed their left lungs, and replaced them with tissue-engineered lungs that had been produced in the laboratory. For periods of up to 2 hours, the implanted lung tissue exchanged oxygen and carbon dioxide at similar rates as the natural lungs.[1]

In order to build the artificial rat lungs, the Yale researchers began by removing lungs from adult rats and treating the organs to remove most of the cellular components, while preserving the airways and supporting connective tissue matrix. Next, they placed these decellularized structures, along with various types of cultured cells, into a sterile container designed to mimic some aspects of the fetal environment in which the lungs normally first develop. The scientists found that the cells were able to form much of the lung tissues as well as the blood vessels needed to supply the tissue with blood and transport gases.

Although these results are an important early step toward growing replacement lungs in the lab, there is a long way to go before anyone will contemplate implanting engineered lungs into humans.

In a separate study published on the same day as the Yale study, a Harvard group announced that they had developed a pea-sized device that mimics human lung tissues. Made of human lung cells, a permeable membrane, plus blood capillary cells, all mounted on a microchip (Fig. 35A), the device is able to mimic the function of alveoli. When the researchers placed bacteria on the alveolar side of the device, and white blood cells on the capillary side, the blood cells crossed the membrane, mimicking an immune response.[2] At a minimum, the researchers hope that their "lung-on-a-chip" device can be used for testing certain drugs or the effects of various toxins on the lungs, which might replace much of the animal testing that is currently performed.

Questions to Consider

1. With regard to producing tissue-engineered human lungs, what is the major drawback in the procedure developed by the Yale group?
2. What are some of the aspects of lung structure that make it a more difficult organ to grow in the lab than, for example, a urinary bladder?
3. Besides increased availability, what are two other potential advantages of laboratory-grown lungs (or other tissues) compared to regular donor tissues?

[1]Petersen, T. H., et al. "Tissue-Engineered Lungs for in Vivo Implantation," *Science* 329: 538–541 (2010).

[2]Huh, D., et al. "Reconstituting Organ-Level Lung Functions on a Chip," *Science* 328: 1662–1668 (2010).

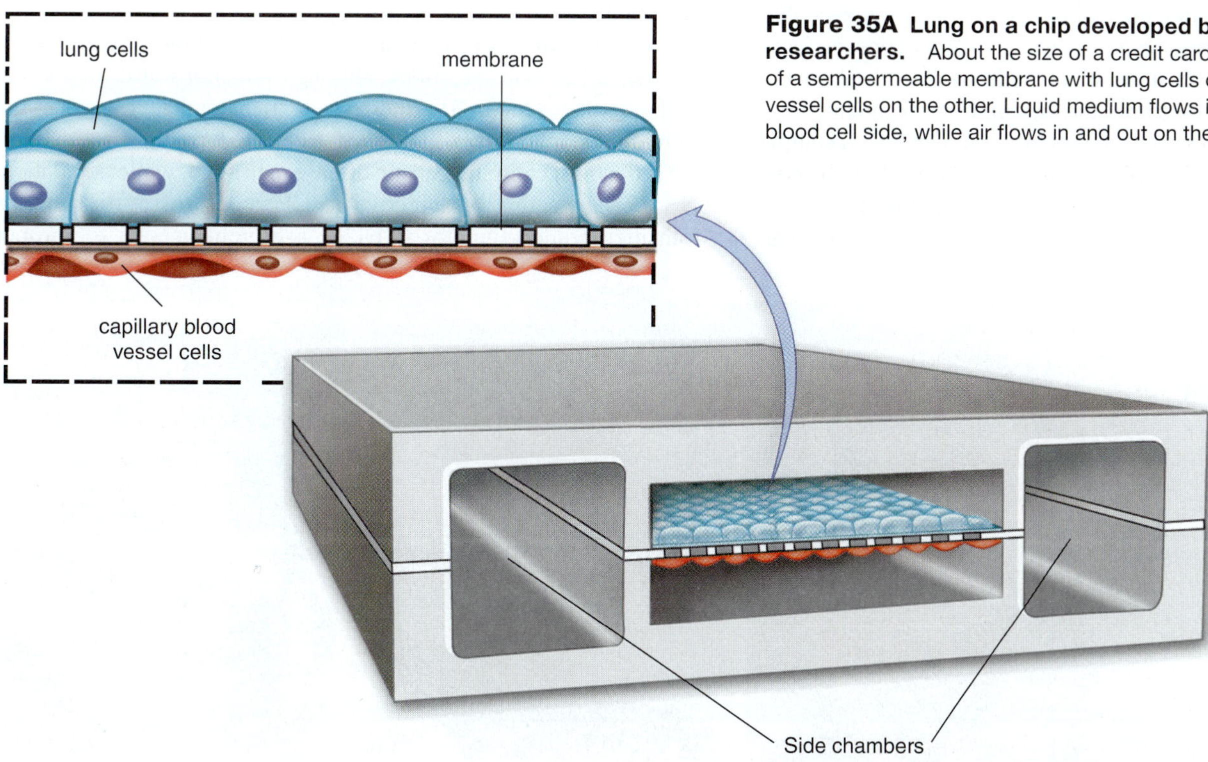

Figure 35A Lung on a chip developed by Harvard researchers. About the size of a credit card, the device consists of a semipermeable membrane with lung cells on one side and blood vessel cells on the other. Liquid medium flows in a channel on the blood cell side, while air flows in and out on the lung cell side.

bacteria can cause pneumonia, as can the influenza virus, especially in the very young, the very old, and people with a suppressed immune system. AIDS patients are subject to a particularly rare form of pneumonia caused by a fungus of the genus *Pneumocystis,* but they suffer from many other types of pneumonias as well.

Pulmonary tuberculosis is caused by the bacterium *Mycobacterium tuberculosis*. Tuberculosis (TB) was a major killer in the United States before the middle of the twentieth century, after which antibiotic therapy brought it largely under control. However, the incidence of TB is rising in certain areas of the world, especially where HIV infection (which reduces immunity to *M. tuberculosis*) is common and treatments are not widely available. According to the Centers for Disease Control and Prevention, approximately one-third of the world's population is infected with *M. tuberculosis,* and TB is the cause of death for as many as half of all persons with AIDS. In 2013, the United States had fewer than 10,000 cases, the lowest number ever recorded, but infection rates remain high in certain ethnic groups and among foreign-born persons.

When tubercle bacilli invade the lung tissue, the cells build a protective capsule around the organisms, isolating them from the rest of the body. This tiny capsule is called a tubercle (Fig. 35.14*d*). If the resistance of the body is high, the imprisoned organisms die, but if the resistance is low, the organisms can escape and spread.

It is possible to tell if a person has ever been exposed to *M. tuberculosis* with a TB skin test, in which a highly diluted extract of the bacterium is injected into the skin of the patient. A person who has never been exposed to the bacterium shows no reaction, but one who has previously been infected develops an area of inflammation that peaks in about 48 hours.

Emphysema is a chronic and incurable lung disorder in which the alveoli are distended and their walls damaged, so that the surface area available for gas exchange is reduced (Fig. 35.14*e*). As mentioned, emphysema often contributes to COPD in smokers. Air trapped in the lungs leads to alveolar damage and a noticeable ballooning of the chest. The elastic recoil of the lungs is reduced, so not only are the airways narrowed but the driving force behind expiration is also reduced. The patient is breathless and may have a cough. Because the surface area for gas exchange is reduced, less oxygen reaches the heart and brain, leaving the person feeling depressed, sluggish, and irritable. Exercise, drug therapy, and supplemental oxygen, along with giving up smoking, may relieve the symptoms and slow the progression of emphysema and COPD (Fig. 35.15*b*).

Inhaling particles such as silica (sand), coal dust, or asbestos can lead to **pulmonary fibrosis,** a condition in which fibrous connective tissue builds up in the lungs. The lungs cannot inflate properly and are always tending toward deflation (Fig. 35.14*f*). Breathing asbestos is also associated with the development of cancer, including a type called mesothelioma. In the United States, the use of asbestos as a fireproofing and insulating agent has been limited since the 1970s; however, many thousands of lawsuits are filed each year by patients suffering from asbestos-related illnesses.

Lung cancer is the leading cause of cancer-related death worldwide. More people die from lung cancer each year than from cancer of the colon, breast, and prostate combined. Lung cancer is more common in men than women, but rates in women have increased in recent years due to an increasing number of women who smoke. Lung cancer rates remain low until about age 40, when they gradually start to rise, peaking at around age 70. Symptoms may include coughing, shortness of breath, blood in the sputum, and chest pain. Many other symptoms can occur if the cancer spreads to other parts of the body, which is common.

Lung cancer may be treated with a combination of surgery, chemotherapy, and radiation. Even with treatment, lung cancer is highly lethal—5-year survival rates range from 15% in the United States to 8% in less developed countries. About 150,000 people in the United States die of lung cancer each year (Fig. 35.15*c*). The American Cancer Society links about 90% of these deaths to smoking. Smoking is also associated with bronchitis, emphysema, heart disease, and other types of cancer. Considering that the nicotine in cigarette smoke is addictive, it is better never to start smoking than to try quitting later on. This advice applies to e-cigarettes, as well (see the Nature of Science feature, "Is 'Vaping' Safer Than Smoking?").

Cystic fibrosis (CF) is an example of a lung disease that is genetic rather than infectious, although infections also play a role in the disease. One in 31 Americans carries the defective gene, but a child must inherit two copies of the gene to have the disease. Still, CF is the most common genetic disease in the U.S. white population.

The gene that is defective in CF codes for cystic fibrosis transmembrane regulator (CFTR), a protein needed for proper

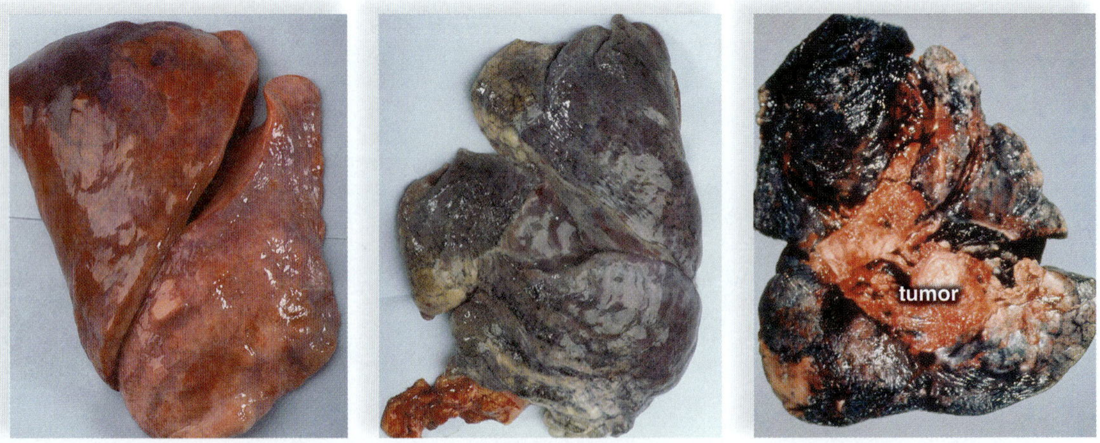

a. Normal lung b. Emphysema c. Lung cancer

Figure 35.15 Smoking and lung disorders. Smoking causes 90% of all lung cancers and is a major cause of emphysema. **a.** Normal lung. **b.** The lung of a person who died from emphysema, shrunken and blackened from trapped smoke. **c.** The lung of a person who died from lung cancer, blackened from smoke except for the presence of the tumor, which is a mass of malformed soft tissue.

THEME Nature of Science

Is "Vaping" Safer Than Smoking?

For at least 50 years, the many health risks of smoking have been clear. Despite this, about 42 million adult Americans are smokers. Once a person starts smoking, the addictive power of nicotine is strong. But why do young people start smoking? Some may want to look mature or "cool," to be accepted by friends, or to rebel against authority. Some smokers believe the habit helps them control their weight; others admit they simply enjoy the "buzz" that nicotine can provide.

Because of the unhealthy side effects of smoking, people often look for alternatives, which explains the growing popularity of electronic cigarettes (e-cigarettes). E-cigarettes are often designed to look like real cigarettes (Fig. 35B), but instead of tobacco, they contain a cartridge filled with an "e-liquid" that consists mainly of nicotine plus propylene glycol or vegetable glycerin. When the device is used, a battery heats the liquid, turning it into a vapor that can be inhaled (explaining the popular term *vaping* for this practice). Often, an LED light at the tip glows, mimicking a lit cigarette. There is no cigarette smell, though, because no tobacco is burning.

Manufacturers claim that the vapor from an e-cigarette is much safer than cigarette smoke. In contrast, many health experts are concerned that the effects of inhaling pure nicotine vapor have not yet been sufficiently studied. The nicotine content of e-liquids can be quite variable, and a variety of contaminants—including metals and trace levels of certain carcinogens—have been detected. Although some smokers say using e-cigarettes helped them quit smoking cigarettes, health authorities worry that people who first become hooked on vaping might "graduate" to smoking.

Another concern is that although companies claim they aren't marketing to children, some nicotine solutions contain flavors—such as butterscotch, chocolate, and even cotton candy—which could appeal to the very young. An April 2014 *New York Times* article titled "Selling a Poison by the Barrel" characterized e-liquids as powerful neurotoxins, which when ingested or even absorbed through the skin can be lethal, and poisonings due to contact with or ingestion of e-liquids are on the rise.

So is vaping safer than smoking? Considering all the health risks associated with smoking, it's unlikely that vaping is worse. Still, until many issues with e-cigarettes can be sorted out, it is clearly best to follow the FDA's advice and avoid vaping as well.

Questions to Consider

1. Compared to most drugs, nicotine is unusual, because at low doses it is mainly a stimulant, but at high doses it has more sedative effects. How might these properties contribute to nicotine's addictive potential?
2. Suppose you've never "vaped" before, and someone offers you an e-cigarette at a party. Would you be tempted to try it? Why or why not?

Figure 35B Electronic cigarettes.

transport of chloride (Cl^-) ions out of the epithelial cells of the lung. Because this also reduces the amount of water transported out of lung cells, the mucus secretions become very sticky and can form plugs that interfere with breathing.

Symptoms of CF include coughing and shortness of breath; part of the treatment involves clearing mucus from the airways by vigorously slapping the patient on the back as well as by administering mucus-thinning drugs. None of these treatments is curative, however, and because the lungs can be severely affected, the median survival age for people with CF is only 30 years. Researchers are attempting to develop gene therapy strategies to replace the faulty *CFTR* gene.

Check Your Progress 35.3

1. Explain why antibiotic drugs such as penicillin are ineffective at treating the common cold.
2. Name two disorders of the lower respiratory tract that mainly cause a narrowing of the airways and two that restrict the lungs' ability to expand.
3. List six illnesses associated with smoking cigarettes.

CONNECTING *the* CONCEPTS *with the* THEMES

Evolution

- Animals that are small and have sufficient body surface area may exchange gases directly with the environment. The evolution of respiratory systems allowed a more efficient exchange of gases.
- Gills evolved in many aquatic animals as an effective way to obtain sufficient oxygen from water.
- A wide variety of respiratory structures have evolved in terrestrial animals, including tracheoles in insects and lungs in some amphibians, reptiles, birds, and mammals.

Nature of Science

- Studies of respiratory systems have provided us with extensive knowledge, which has been applied to respiratory diseases.
- Medical researchers are working to develop new strategies for treating respiratory diseases, including the possibility of replacing diseased lungs with laboratory-engineered lung tissue.

Biological Systems

- The respiratory and cardiovascular systems work together to provide animal tissues with oxygen and to remove carbon dioxide and other wastes.
- Respiration is regulated to adjust to changing needs. In humans, the carotid bodies and aortic bodies detect the lowering of blood pH and oxygen content and increase respiration accordingly.

MEDIA STUDY TOOLS http://connect.mheducation.com

SMARTBOOK® Study smart. SmartBook helps you maximize your study time by identifying what you know and don't know. The recharge feature in SmartBook helps you prepare for the test by identifying the material that you are most likely to forget.

connect BIOLOGY Find even more resources to learn the chapter concepts including animations, tutorial videos, and interactive practice questions.

MP3 Files	Tutorial
35.1 Respiratory Structure and Function **35.2** Control of Respiration • Gas Exchange • Gas Transport • Gas Exchange During Respiration	**35.2** Internal and External Respiration

SUMMARIZE

35.1 Gas-Exchange Surfaces

During **respiration,** gas exchange occurs between an animal's cells and their environment. In terrestrial vertebrates, **ventilation** includes inspiration and expiration. **External respiration** refers to gas exchange between the air and blood, and **internal respiration** is exhange between blood and interstitial fluid.

Some aquatic animals, such as hydras, earthworms, and some amphibians, use their entire body surface for gas exchange. However, most animals have a specialized gas-exchange area. Large aquatic animals usually pass water through **gills.** In bony fishes, blood in the capillaries flows in the direction opposite that of the water. Blood takes up almost all of the oxygen in the water as a result of this **countercurrent exchange.**

For gas exchange on land, insects have **tracheae,** and vertebrates have **lungs.** In insects, air enters the tracheae at openings called spiracles. From there, the air moves to ever smaller tracheoles until gas exchange takes place at the cells themselves. Lungs are found inside the body, where water loss is reduced. To ventilate the lungs, some vertebrates use positive pressure, but most inhale, using muscular contraction to produce a negative pressure that causes air to rush into the lungs. When the breathing muscles relax, air is exhaled.

Birds have a series of air sacs attached to the lungs. When a bird inhales, air enters the posterior air sacs, and when a bird exhales, air moves through the lungs to the anterior air sacs before exiting the respiratory tract. The one-way flow of air through the lungs allows more fresh air to be present in the lungs with each breath, and this leads to greater uptake of oxygen from one breath of air. Mammals have two-way or tidal airflow in and out of the lungs, and as a result some mixing is always occurring between fresh air and previously inhaled air, with less uptake of oxygen than in birds.

In humans, inhaled air passes from the mouth or nasal cavities to the **pharynx,** through the **glottis, larynx, trachea, bronchi,** and **bronchioles,** finally reaching the **alveoli** of the lungs. The glottis contains **vocal cords** that produce sound. Because both air and food pass through the pharynx, the glottis is covered by the **epiglottis** when food is being swallowed.

35.2 Breathing and Transport of Gases

During **inspiration,** air moves into the lungs, where exchange occurs; during **expiration,** air passes in the opposite direction. Humans breathe by negative pressure, as do other mammals. During inspiration, the rib cage goes up and out, and the **diaphragm** lowers. The lungs expand and air rushes in. During expiration, the rib cage goes down and in, and the diaphragm rises. Therefore, air rushes out.

A **respiratory center** in the brain controls the rate and depth of ventilation. The rate of breathing increases when the amount of H^+ and carbon dioxide in the blood rises, as detected by chemoreceptors such as the **aortic bodies** and **carotid bodies.**

Gas exchange in the lungs and tissues is brought about by diffusion, according to the **partial pressure** of each gas. In the lungs, **hemoglobin** (**Hb**) takes up oxygen, forming **oxyhemoglobin,** which transports oxygen in the blood. Within Hb, the oxygen molecules are actually bound to iron, which is a component of **heme.** Carbon dioxide is mainly transported in plasma as the **bicarbonate ion,** but some combines with Hb, forming **carbaminohemoglobin.** Excess hydrogen ions are transported by hemoglobin. The enzyme **carbonic anhydrase** found in red blood cells speeds the formation of the bicarbonate ion.

35.3 Respiration and Human Health

The respiratory tract is susceptible to a wide variety of infections and other disease conditions. Infections of the upper respiratory tract include the common cold, caused by several different viruses, and strep throat, which is a form of **pharyngitis** caused by a bacterium. In the lower respiratory tract, bronchitis is an inflammation of the bronchi, usually due to infections, whereas in **asthma** the inflammation and smooth muscle contraction in the bronchi and bronchioles are due to sensitivity to various irritants. Many important diseases affect the lungs, including **pneumonia, pulmonary tuberculosis, pulmonary fibrosis, emphysema, lung cancer,** and **cystic fibrosis** (**CF**).

ASSESS

Choose the best answer for each question.

35.1 Gas-Exchange Surfaces

1. Label the following diagram depicting respiration in terrestrial vertebrates.

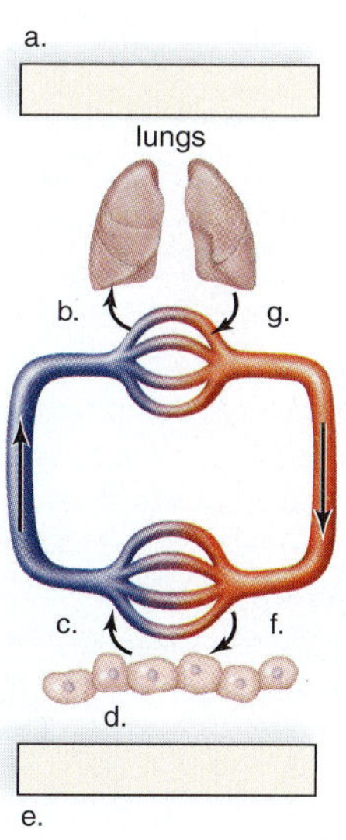

2. Birds have more efficient lungs than humans because the flow of air in birds
   a. is the same during both inspiration and expiration.
   b. travels in only one direction through the lungs.
   c. follows a tidal ventilation pattern.
   d. is not hindered by a larynx.
   e. enters their bones.
3. If the digestive and respiratory tracts were completely separate in humans, there would be no need for
   a. swallowing.
   b. a nose.
   c. an epiglottis.
   d. a diaphragm.
   e. All of these are correct.
4. Which of these is a true statement?
   a. In lung capillaries, carbon dioxide combines with water to produce carbonic acid.
   b. In tissue capillaries, carbonic acid breaks down to carbon dioxide and water.
   c. In lung capillaries, carbonic acid breaks down to carbon dioxide and water.
   d. In tissue capillaries, carbonic acid combines with hydrogen ions to form the carbonate ion.
   e. All of these statements are true.

For questions 5–7, match each description with a structure in the key.

Key:

a. bronchi	d. larynx
b. bronchioles	e. pharynx
c. glottis	f. trachea

5. reinforced tube that connects larynx with bronchi
6. opening into larynx
7. divisions of air tubes that enter lungs

35.2 Breathing and Transport of Gases

8. Which of these is incorrect concerning inspiration?
   a. The rib cage moves up and out.
   b. The diaphragm contracts and moves down.
   c. As pressure in lungs decreases, air comes rushing in.
   d. The lungs expand because air comes rushing in.
9. In humans, the respiratory control center
   a. is stimulated by carbon dioxide.
   b. is located in the medulla oblongata.
   c. controls the rate of breathing.
   d. is stimulated by hydrogen ion concentration.
   e. All of these are correct.
10. Air enters the human lungs because
    a. atmospheric pressure is less than the pressure inside the lungs.
    b. atmospheric pressure is greater than the pressure inside the lungs.
    c. although the pressures are the same inside and outside, the partial pressure of oxygen is lower within the lungs.
    d. the residual air in the lungs causes the partial pressure of oxygen to be less than it is outside.
    e. the process of breathing pushes air into the lungs.
11. Carbon dioxide is carried in the plasma
    a. in combination with hemoglobin.
    b. as the bicarbonate ion.
    c. combined with carbonic anhydrase.
    d. only as a part of interstitial fluid.
    e. All of these are correct.
12. The chemical reaction that converts carbon dioxide to a bicarbonate ion takes place in
    a. the blood plasma.
    b. red blood cells.
    c. the alveolus.
    d. the hemoglobin molecule.

35.3 Respiration and Human Health

13. It is difficult to develop immunity to the common cold because
    a. antibodies don't work against cold viruses.
    b. bacteria that cause colds are resistant to antibiotics.
    c. colds can be caused by hundreds of viral strains.
    d. common cold viruses cause severe immunosuppression.
14. The first action that should be taken when someone is choking on food is to perform
    a. a tracheostomy.
    b. CPR.
    c. the Heimlich maneuver.
    d. X-rays to determine where the food is lodged.
15. Asthma
    a. mainly affects the upper respiratory tract.
    b. is usually caused by an infection.
    c. is considered to be a genetic disorder.
    d. is usually curable.
    e. None of these statements are true.
16. Pulmonary tuberculosis is caused by a
    a. bacterium.
    b. fungus.
    c. protist.
    d. virus.
17. In which chronic lung disease do the alveoli become distended and often fuse into enlarged air spaces?
    a. asthma
    b. cystic fibrosis
    c. emphysema
    d. pulmonary fibrosis

ENGAGE

THINKING SCIENTIFICALLY

1. Why would someone who nearly drowned have a blue tint to his or her skin?
2. Fetal hemoglobin picks up oxygen from the maternal blood. If the oxygen-binding characteristics of hemoglobin in the fetus were identical to the hemoglobin of the mother, oxygen could never be transferred at the placenta to fetal circulation. What hypothesis about the oxygen-binding characteristics of fetal hemoglobin would explain how fetuses get the oxygen they need?
3. Carbon monoxide (CO) binds to hemoglobin about 200 times more strongly than oxygen does, which means less O_2 is being delivered to tissues. What would be the specific cause of death if too much hemoglobin became bound to CO instead of to O_2?
4. As mentioned in the chapter, sometimes it is necessary to install a permanent tracheostomy—for example, in smokers who develop laryngeal cancer. Besides interfering with the ability to speak, what sorts of health problems would you expect to see in an individual with a permanent tracheostomy, and why?

Marine organisms rid the body of excess salt; fishes extrude salt at their gills, and turtles do so near their eyes.

36

Body Fluid Regulation and Excretory Systems

Chapter Outline

36.1 Animal Excretory Systems 678

36.2 The Human Urinary System 681

Before You Begin

Before beginning this chapter, take a few moments to review the following discussions.

Figure 5.7 What occurs when a cell is surrounded by a solution having a higher or lower solute concentration than that inside the cell?

Section 8.5 What happens when proteins are used as an energy source?

Section 34.4 What happens to excess nutrients and minerals that cannot be stored?

If the salt concentration in body fluids is too high, cells shrivel and die. If it is too low, cells swell and rupture. However, animals are found in all sorts of environments, including marine environments that are too salty, freshwater environments that don't have enough salt, and even terrestrial environments that are simply too dry. Animals clearly spend a lot of energy regulating the composition of their body fluids, and chief among the organs that help are the kidneys of the urinary system. Sometimes, animals such as marine birds and reptiles get some assistance from accessory glands. Sea turtles have salt glands above their eyes that, true to their name, rid the body of salt. When the glands excrete a salty solution collected from body fluids, sea turtles appear to cry. Humans lack salt glands and cannot survive after drinking too much salt water, because the kidneys alone can't handle all the salt.

In this chapter, you'll learn how animals maintain their normal water-salt balance while excreting various metabolic wastes and regulating their pH. All these functions are of primary importance to homeostasis and continued good health.

As you read through this chapter, think about the following questions:

1. Besides salt glands, what other types of strategies have animals developed to either conserve or excrete excess salt?
2. If you were stranded on a desert island with plenty of food but only seawater to drink, how long do you think you could survive?
3. By what mechanisms is the human kidney able to regulate the salt concentration of urine it produces?

Following the Themes

CHAPTER 36 BODY FLUID REGULATION AND EXCRETORY SYSTEMS	
UNIT 7 COMPARATIVE ANIMAL BIOLOGY	
Evolution	Many different mechanisms for maintaining water-salt balance and excreting metabolic wastes have evolved, molded by the environments in which animals live.
Nature of Science	By studying many different animals in a variety of environments, scientists have learned how animal excretory systems regulate water-salt balance, excrete metabolic wastes, and maintain a healthy pH.
Biological Systems	The urinary system plays an important role in maintaining pH and ion concentration in vertebrates.

36.1 Animal Excretory Systems

Learning Outcomes

Upon completion of this section, you should be able to

1. Describe the overall, specific functions of animal excretion systems.
2. List the costs and benefits of the excretion of ammonia, urea, or uric acid as nitrogenous waste products.
3. Compare and contrast the excretory organs of earthworms, arthropods, aquatic vertebrates, and terrestrial vertebrates.

An important part of maintaining homeostasis in animals involves **osmoregulation,** or balancing the levels of water and salts in the body. Often, the osmoregulatory system of an animal also removes metabolic wastes from the body, a process called **excretion.**

Nitrogenous Waste Products

The breakdown of nitrogen-containing molecules, such as amino acids and nucleic acids, results in excess nitrogen, which must be excreted. When the body breaks down amino acids to generate energy, or converts them to fats or carbohydrates, the amino ($—NH_2$) groups must be removed, because they are not needed, and they may be toxic at high levels. Depending on the species, this excess nitrogen may be excreted in the form of ammonia, urea, or uric acid. The removal of amino groups from amino acids requires a fairly constant amount of energy; however, the amount of energy required to convert amino groups to ammonia, urea, or uric acid differs, as indicated in Figure 36.1.

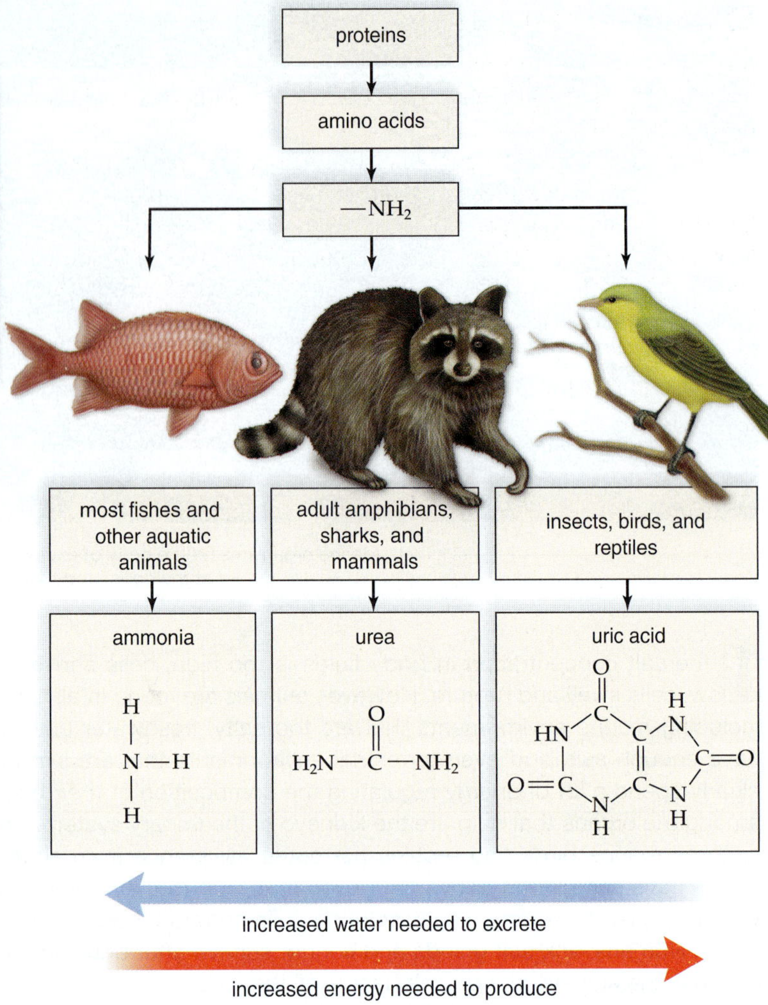

Figure 36.1 Nitrogenous wastes. Proteins are hydrolyzed to amino acids, whose breakdown results in carbon chains and amino groups ($—NH_2$). The carbon chains can be used as an energy source, but the amino groups must be excreted as ammonia, urea, or uric acid.

Ammonia

Amino groups removed from amino acids immediately form **ammonia** (NH_3) by the addition of a third hydrogen ion (H^+). This reaction requires little or no energy. Ammonia is quite toxic, but it can be a nitrogenous excretory product if sufficient water is available to wash it from the body. Ammonia is excreted by most fishes and other aquatic animals whose gills and/or body surfaces are in direct contact with the water of the environment.

Urea

Sharks, adult amphibians, and mammals usually excrete **urea** as their main nitrogenous waste. Urea is much less toxic than ammonia and can be excreted in a moderately concentrated solution. This elimination strategy allows body water to be conserved, an important advantage for terrestrial animals with limited access to water. The production of urea requires the expenditure of energy, however, because it is produced in the liver by a set of energy-requiring enzymatic reactions, known as the urea cycle. In this cycle, carrier molecules take up carbon dioxide and two molecules of ammonia, finally releasing urea.

Uric Acid

Uric acid is synthesized by a long, complex series of enzymatic reactions that requires the expenditure of even more energy than does urea synthesis. Uric acid is not very toxic, and it is poorly soluble in water. Poor solubility is an advantage if water conservation is needed, because uric acid can be concentrated even more readily than can urea.

Uric acid is routinely excreted by insects, reptiles, and birds. In reptiles and birds, a dilute solution of uric acid passes from the kidneys to the *cloaca,* a common reservoir for the products of the digestive, urinary, and reproductive systems. The cloacal contents are refluxed into the large intestine, where water is reabsorbed. The white substance in bird feces is uric acid.

Embryos of reptiles and birds develop inside completely enclosed shelled eggs. The production of insoluble, relatively nontoxic uric acid is advantageous for shelled embryos, because all nitrogenous wastes are stored inside the shell until hatching takes place. For all these reasons, the evolutionary advantages of uric acid production have outweighed the disadvantage of energy expenditure needed for its synthesis.

Humans have retained the ability to produce uric acid, mainly from the breakdown of excess purine and pyrimidine nucleic acids in the diet. Although the exact causes are unknown, in some individuals uric acid builds up in the blood and can precipitate in and around the joints, producing a painful ailment called *gout.*

Excretory Organs Among Invertebrates

Most invertebrates have tubular excretory organs that regulate the water-salt balance of the body and excrete metabolic wastes into the environment.

The planarians, flatworms that live in fresh water, have two strands of branching excretory tubules that open to the outside of the body through excretory pores (Fig. 36.2*a*). Located along the tubules are bulblike *flame cells,* each of which contains a cluster of beating cilia that looks like a flickering flame under the microscope. The beating of flame-cell cilia propels fluid through the excretory tubules and out of the body. The system is believed to function in ridding the body of excess water and in excreting wastes.

The body of an earthworm is divided into segments, and nearly every body segment has a pair of excretory structures called *nephridia.* Each nephridium is a tubule with a ciliated opening and an excretory pore (Fig. 36.2*b*). As fluid from the coelom is propelled through the tubule by beating cilia, its composition is modified. For example, nutrient substances are reabsorbed and carried away by a network of capillaries surrounding the tubule. **Urine** is a liquid that contains metabolic wastes, excreted salts, and water; the urine of an earthworm is passed out of the body via the excretory pore. Although the earthworm is considered a terrestrial animal, it excretes a very dilute urine. Each day, an earthworm may produce a volume of urine equal to 60% of its body weight.

Insects have a unique excretory system consisting of long, thin *Malpighian tubules* attached to the gut. Uric acid is actively transported from the surrounding hemolymph into these tubules, and water follows a salt gradient established by active transport of K^+. Water and other useful substances are reabsorbed at the rectum, but the uric acid leaves the body through the anus. Insects that live in water, or eat large quantities of moist food, reabsorb little water. But insects in dry environments reabsorb most of the water and excrete a dry, semisolid mass of uric acid.

The excretory organs of other arthropods are given different names, although they function similarly. In aquatic crustaceans (e.g., crabs, crayfish), nitrogenous wastes are generally removed by diffusion across the gills. Some crustaceans also possess excretory organs, called *green glands,* in the ventral portion of the head region. Fluid collects within the tubules from the surrounding blood of the hemocoel, but this fluid is modified before it leaves the tubules. The secretion of salts into the tubules regulates the amount of urine excreted.

In shrimp and pillbugs, the excretory organs are located in the maxillary segments and are called *maxillary glands.* Spiders, scorpions, and other arachnids possess *coxal glands,* which are located near one or more appendages and are used for excretion. Coxal glands are spherical sacs resembling annelid nephridia. Wastes are collected from the surrounding blood of the hemocoel and discharged through pores at one to several pairs of appendages.

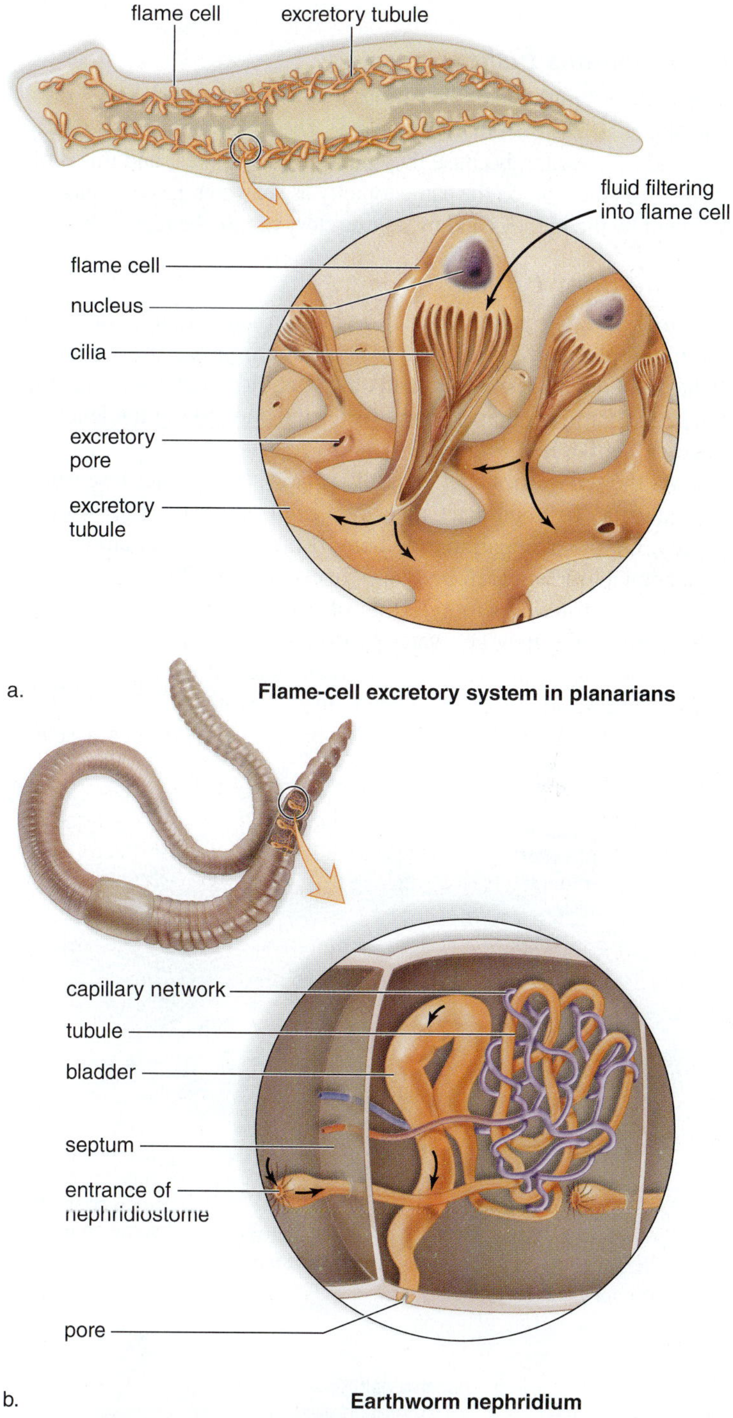

Figure 36.2 Excretory organs in animals. **a.** Two or more tracts of branching tubules run the length of the body and open to the outside by pores. At the ends of side branches are small, bulblike cells called flame cells. **b.** The nephridium has a ciliated opening, the nephridiostome, which leads to a coiled tubule surrounded by a capillary network. Urine can be temporarily stored in the bladder before being released to the outside via a pore called a nephridiopore.

Osmoregulation by Aquatic Vertebrates

In most vertebrates, the kidneys are the most important organs involved in osmoregulation. As described later in this chapter, the kidneys perform several functions critical to homeostasis, including maintaining the balance between water and several types of salts. This is a necessity, because ions such as Na^+, Ca^{2+}, K^+, and PO_4^- greatly affect the workings of the body systems, such as the skeletal, nervous, and muscular systems.

The kidneys produce urine, a liquid that contains a number of different metabolic wastes. The concentration of the urine produced by an animal varies depending on its environment, as well as on factors such as water and salt intake.

Cartilaginous Fishes

The total concentration of the various ions in the blood of sharks, rays, and skates is less than that in seawater. Their blood plasma is nearly isotonic to seawater, because they pump it full of urea, and this molecule gives their blood the same tonicity as seawater. Excess salts are secreted by the kidneys and by an excretory organ, the rectal gland.

Marine Bony Fishes

The marine environment, which is high in salts, is hypertonic to the blood plasma of bony fishes. The common ancestor of marine fishes evolved in fresh water, and only later did some groups invade the sea. Therefore, marine bony fishes must avoid the tendency to become dehydrated (Fig. 36.3*a*).

As the sea washes over their gills, marine bony fishes lose water by osmosis. To counteract this, they drink seawater almost constantly. On the average, marine bony fishes swallow an amount of water equal to 1% of their body weight every hour. This is equivalent to a human drinking about 700 ml of water every hour around the clock. But while they get water by drinking, this habit also causes these fishes to acquire salt. To rid their body of excess salt, they actively transport it into the surrounding seawater at the gills. The kidneys conserve water, and marine bony fishes produce a scant amount of isotonic urine.

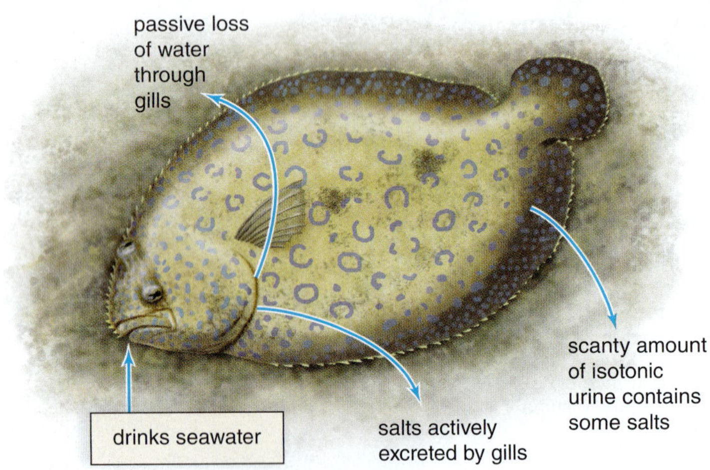

a. Marine bony fish (a flounder)

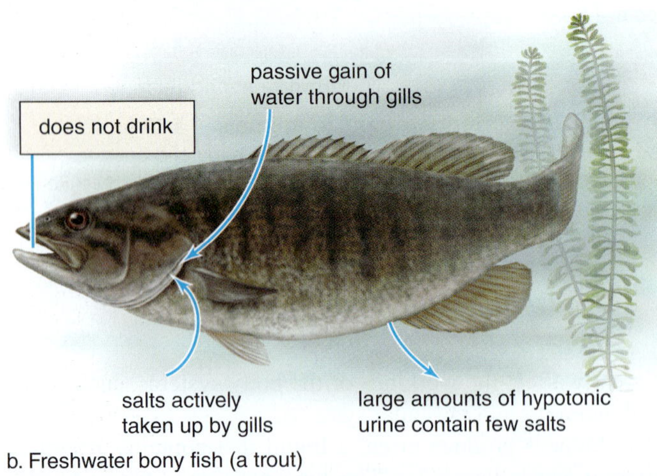

b. Freshwater bony fish (a trout)

Figure 36.3 Body fluid regulation in bony fishes. Marine bony fishes (**a**) use different mechanisms than do freshwater fishes (**b**) to osmoregulate their body fluids.

Freshwater Bony Fishes

The osmotic problems of freshwater bony fishes and the response to their environment are exactly opposite those of marine bony fishes (Fig. 36.3*b*). Freshwater fishes tend to gain water by osmosis across the gills and the body surface. As a consequence, these fishes never drink water. They actively transport salts into the blood across the membranes of their gills. They eliminate excess water by producing large quantities of dilute (hypotonic) urine. They discharge a quantity of urine equal to one-third their body weight each day.

Osmoregulation by Terrestrial Vertebrates

An important evolutionary adaptation that allowed animals to survive on land was the development of a kidney that could produce a concentrated (hypertonic) urine. The need for water conservation is particularly well illustrated in desert mammals, such as the kangaroo rat, as well as in animals that drink seawater.

Kangaroo Rat

Dehydration threatens all terrestrial animals, especially those that live in a desert, as does the kangaroo rat. During daylight hours, kangaroo rats remain in a cool burrow, a behavioral adaptation to conserve water. In addition, the kangaroo rat's nasal passages have a highly convoluted mucous membrane surface that captures condensed water from exhaled air. Exhaled air is usually full of moisture, which is why you can see it on cold winter mornings—the moisture in exhaled air is condensing.

A major adaptation that allows the kangaroo rat to conserve water is the ability to form a very hypertonic urine—20 times more concentrated than its blood plasma. The kidneys of a kangaroo rat are able to accomplish this feat because the structure in their kidneys that is largely responsible for producing concentrated urine, called the loop of the nephron (see Fig. 36.8), is much longer and more efficient than that in most other animals. Also, kangaroo rats produce fecal material that is almost completely dry.

Most terrestrial animals need to drink water at least occasionally to make up for the water lost from the skin and respiratory passages and through urination. However, the kangaroo rat is so adapted to conserving water that it can survive by using metabolic water derived from cellular respiration, and it never drinks water (Fig. 36.4).

Marine Mammals and Sea Birds

Reptiles, birds, and mammals evolved on land, and their kidneys are especially good at conserving water. However, some of these vertebrates have become secondarily adapted to living in or near the sea. They can drink seawater and still manage to survive. If humans drink seawater, we lose more water than we take in just ridding the body of all that salt!

Little is known about how whales manage to get rid of extra salt, but we know that their kidneys are enormous. In some marine animals, however, the kidneys are not efficient enough to secrete all the excess salt. As mentioned in the chapter-opening story, some animals living in high-salt environments have developed specialized glands for excreting these salts. These glands work by actively transporting salt from the blood into the gland, where it can be excreted as a concentrated solution.

In sea birds, salt-excreting glands are located near the eyes. The glands produce a salty solution that is excreted through the

Figure 36.4 Adaptations of a kangaroo rat to a dry environment. The kangaroo rat minimizes water loss through a variety of ways.

Figure 36.5 Adaptations of marine birds to a high salt environment. Many marine birds and reptiles have glands that pump salt out of the body.

nostrils and moves down grooves on their beaks until it drips off (Fig. 36.5). In marine turtles, the salt gland is a modified tear (lacrimal) gland, and in sea snakes, a salivary sublingual gland beneath the tongue gets rid of excess salt. The work of these glands is regulated by the nervous system. Osmoreceptors, perhaps located near the heart, are thought to stimulate the brain, which then directs the gland to excrete salt until the salt concentration in the blood decreases to a tolerable level.

Check Your Progress 36.1

1. Distinguish between osmoregulation and excretion.
2. Describe two advantages of excreting urea instead of ammonia or uric acid.
3. Summarize the strategies used by kangaroo rats to conserve water.

36.2 The Human Urinary System

Learning Outcomes

Upon completion of this section, you should be able to

1. Trace the anatomical path that urine takes from the glomeruli to its exit from the body.
2. Discuss the contributions of glomerular filtration, tubular reabsorption, and tubular secretion to the formation of urine.
3. Summarize the four major functions of human kidneys in maintaining homeostasis.

The major excretory organs of humans, as with most other vertebrates, are the kidneys (Fig. 36.6). The kidneys are the ultimate regulators of blood composition, because they can remove various unwanted products from the body.

Human **kidneys** are bean-shaped, reddish-brown organs, each about the size of a fist. They are located on each side of the vertebral column just below the diaphragm, in the lower back, where they are partially protected by the lower rib cage. The right kidney is slightly lower than the left kidney.

Urine made by the kidneys is conducted from the body by the other organs in the urinary system. Each kidney is connected to a **ureter,** a duct that takes urine from the kidney to the **urinary bladder,** where it is stored until it is voided from the body through the single **urethra.** In males, the urethra passes through the penis; in females, the opening of the urethra is ventral to that of the vagina. No connection exists between the genital (reproductive) and urinary systems in females, but in males the urethra also carries sperm during ejaculation.

MP3 Functional Anatomy of the Urinary System

Kidneys

If a kidney is sectioned longitudinally, three major parts can be distinguished (Fig. 36.7). The *renal cortex,* which is the outer region of a kidney, has a somewhat granular appearance. The *renal medulla* consists of six to ten cone-shaped renal pyramids that lie on the inner side of the renal cortex. The innermost part of the kidney is a hollow chamber called the *renal pelvis.* Urine collects in the renal pelvis and then is carried to the bladder by a ureter.

Nephrons

Microscopically, each kidney is composed of over 1 million tiny tubules called **nephrons** (Gk. *nephros,* "kidney"). The nephrons of a kidney produce urine. Some nephrons are located primarily in the renal cortex, but others dip down into the renal medulla, as shown in Figure 36.7*b*. Each nephron is made of several parts (Fig. 36.8). The blind end of a nephron is pushed in on itself to form a cuplike structure called the **glomerular capsule** (L. *glomeris,* "ball"), also known as Bowman's capsule. The outer layer of the glomerular capsule is composed of squamous epithelial cells; the inner layer is composed of specialized cells that allow easy passage of molecules.

Leading from the glomerular capsule is a portion of the nephron known as the **proximal convoluted tubule** (L. *proximus,* "nearest"), which is lined by cells with many mitochondria and tightly packed microvilli. Then, simple squamous epithelium appears in the

Figure 36.6 The human urinary system. **a.** The kidneys are well supplied with blood, as shown in the angiogram. **b.** Urine is found only in the kidneys, ureters, urinary bladder, and urethra.

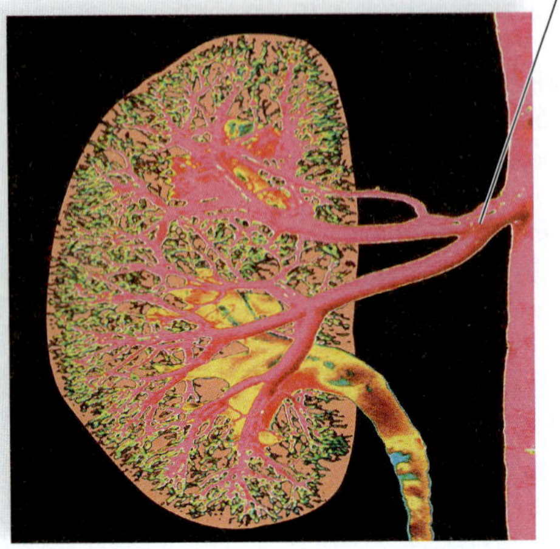

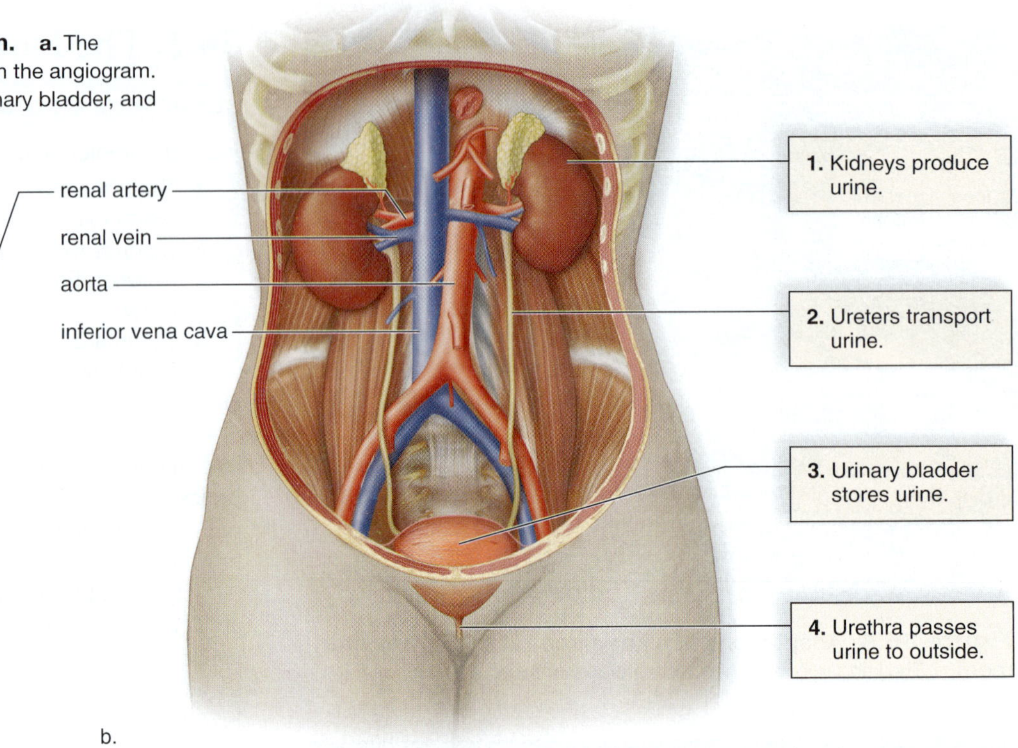

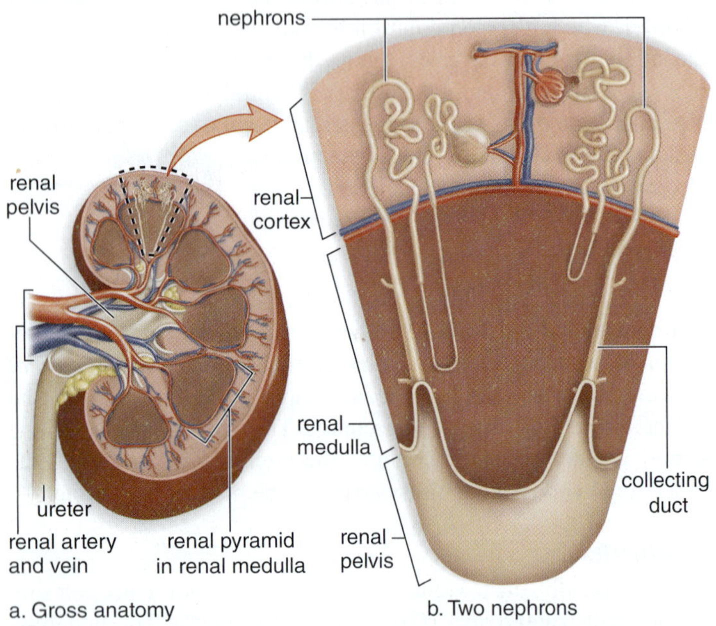

Figure 36.7 Macroscopic and microscopic anatomy of the kidney. **a.** Longitudinal section of a kidney, showing the location of the renal cortex, the renal medulla, and the renal pelvis. **b.** An enlargement of one renal lobe, showing the placement of nephrons.

loop of the nephron (loop of Henle), which has a descending limb and an ascending limb. This is followed by the **distal convoluted tubule** (L. *distantia,* "far"). Several distal convoluted tubules enter one **collecting duct.** The collecting duct transports urine down through the renal medulla and delivers it to the renal pelvis.

Each nephron has its own blood supply (Fig. 36.8). The renal artery branches into numerous small arteries, which branch into arterioles, one for each nephron. Each arteriole, called an afferent arteriole, divides to form a capillary bed, the **glomerulus** (L. *glomeris,* "ball"), which is surrounded by the glomerular capsule. The glomerulus drains into an efferent arteriole, which subsequently branches into a second capillary bed around the tubular parts of the nephron. These capillaries, called peritubular capillaries, lead to venules that join to form veins leading to the renal vein, a vessel that enters the inferior vena cava.

Urine Formation

An average human produces between 1 and 2 liters of urine daily. The fundamental process of urine formation involves initially filtering a large amount of water and a collection of solutes out of the blood, then reabsorbing much of the water, along with other material the body needs to conserve.

MP3 An Overview of Urine Formation

Urine production requires three distinct processes (Fig. 36.9*a*):

1. Glomerular filtration at the glomerular capsule
2. Tubular reabsorption at the convoluted tubules
3. Tubular secretion at the convoluted tubules

Glomerular Filtration

Glomerular filtration (Fig. 36.9*a*) is the movement of small molecules across the glomerular wall into the glomerular capsule as a result of blood pressure. When blood enters the glomerulus, blood pressure is sufficient to cause small molecules, such as water, nutrients, salts, and wastes, to move from the glomerulus to the inside of the glomerular capsule, especially since the glomerular walls are 100 times more permeable than the walls of most capillaries elsewhere in the body. The molecules that leave the blood and enter the glomerular capsule are called the *glomerular filtrate.* Plasma proteins and blood cells are too large to be part of this filtrate, so they remain in the blood as it flows into the efferent arteriole.

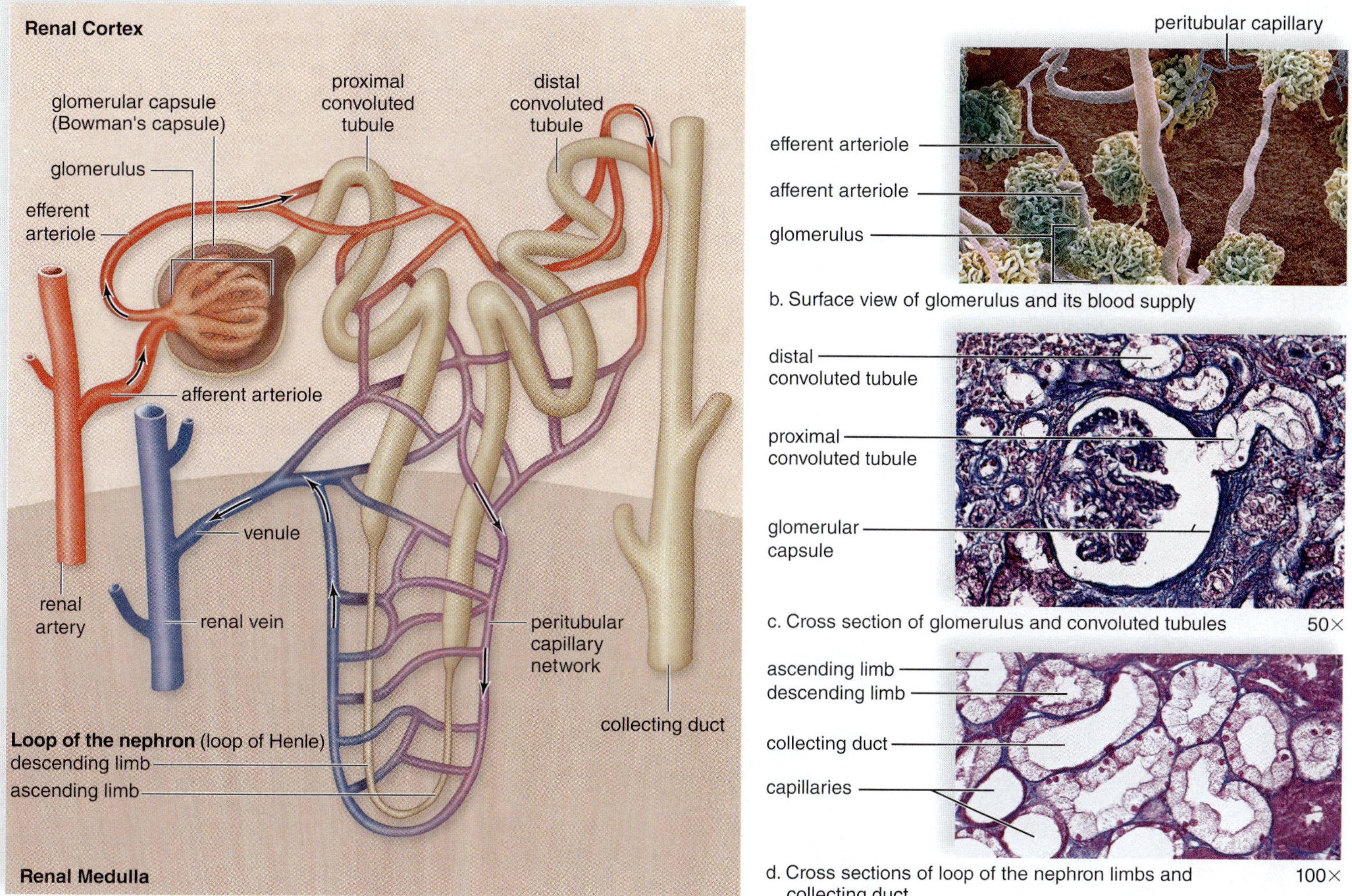

a. A nephron and its blood supply

b. Surface view of glomerulus and its blood supply

c. Cross section of glomerulus and convoluted tubules

d. Cross sections of loop of the nephron limbs and collecting duct.

Figure 36.8 Nephron anatomy. **a.** You can trace the path of blood about a nephron by following the arrows. A nephron is made up of a glomerular capsule, the proximal convoluted tubule, the loop of the nephron, the distal convoluted tubule, and the collecting duct. The micrographs in (**b**), (**c**), and (**d**) show these structures.

Glomerular filtrate is essentially protein-free, but otherwise it has the same composition as blood plasma. If this composition were not altered in other parts of the nephron, death from starvation (loss of nutrients) and dehydration (loss of water) would quickly follow. The total blood volume averages about 5 liters, and this amount of fluid is filtered every 40 minutes. Thus, 180 liters of filtrate are produced daily, some 60 times the amount of blood plasma in the body. Most of the filtered water is obviously quickly returned to the blood, or a person would die from urination. Tubular reabsorption prevents this from happening.

Tubular Reabsorption

Tubular reabsorption (Fig. 36.9*a*) takes place when substances move across the walls of the tubules into the associated peritubular capillary network (Fig. 36.9*a, b*). Here, osmosis comes into play. You may remember that *osmosis* is the diffusion of water down its concentration gradient across a membrane (see section 5.2). *Osmolarity* is a measure of the potential for osmosis; water tends to move from a solution with low osmolarity into a solution with high osmolarity.

The osmolarity of the blood is essentially the same as that of the filtrate within the glomerular capsule, and therefore osmosis of water from the filtrate into the blood cannot yet occur. However, sodium ions (Na^+) are actively pumped into the peritubular capillary, and then chloride ions (Cl^-) follow passively. The osmolarity of the blood then is such that water moves passively from the tubule into the blood. About 60–70% of salt and water are reabsorbed at the proximal convoluted tubule, and 20–25% at the loop of the nephron.

Nutrients such as glucose and amino acids also return to the blood, mostly at the proximal convoluted tubule. This is a selective process, because only molecules recognized by carrier proteins in plasma membranes are actively reabsorbed. The cells of the proximal convoluted tubule have numerous microvilli, which increase the surface area, and numerous mitochondria, which supply the energy needed for active transport (Fig. 36.9*b*).

Glucose is an example of a molecule that ordinarily is reabsorbed completely because the supply of carrier molecules for it is plentiful. However, if the filtrate contains more glucose than there are carriers to handle it, glucose exceeds its renal threshold, or transport maximum. When this happens, the excess glucose in

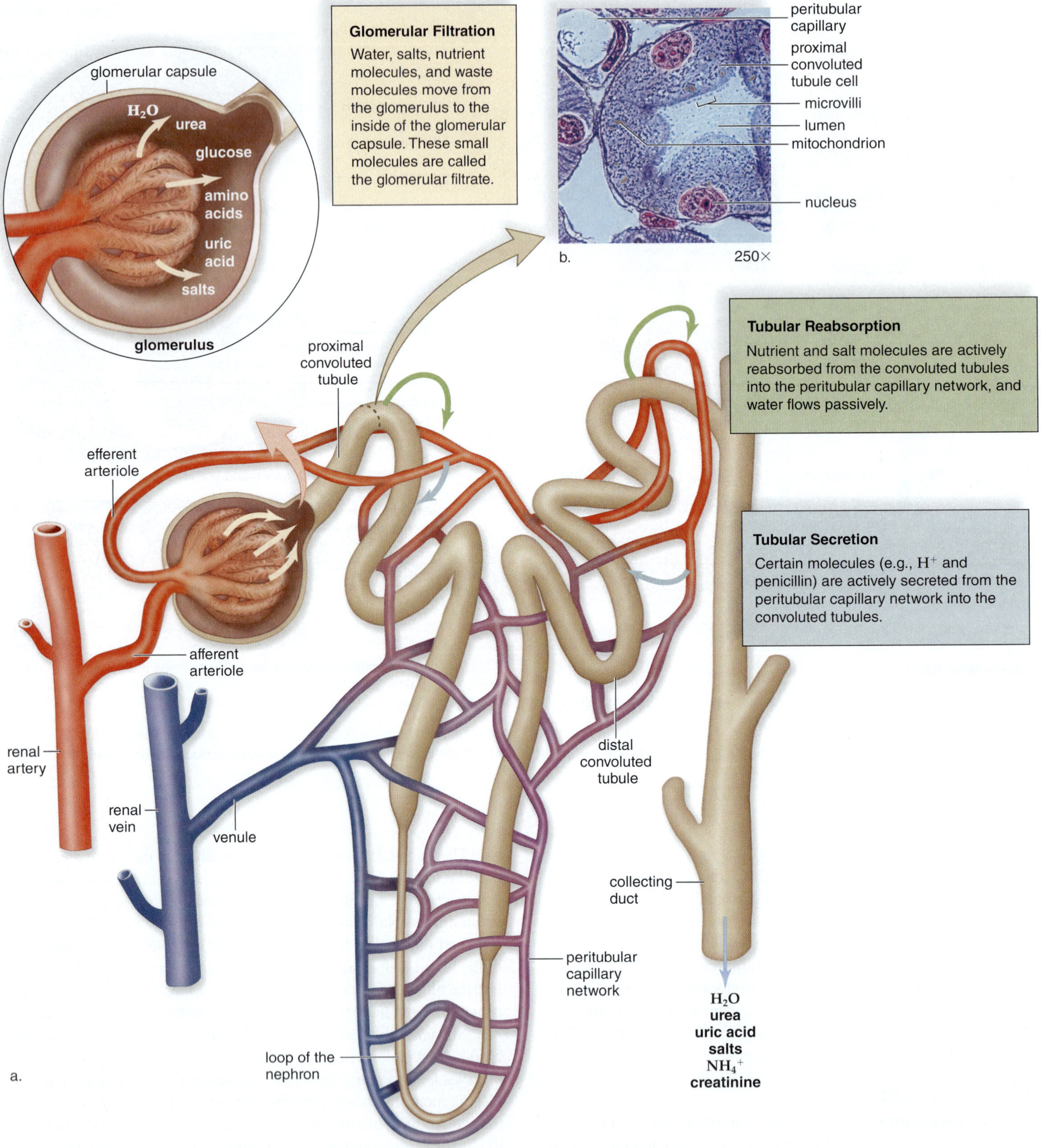

Figure 36.9 Processes in urine formation. **a.** The three main processes in urine formation are described in boxes and color coded to arrows that show the movement of molecules into or out of the nephron at specific locations. In the end, urine is composed of the substances within the collecting duct (blue arrow). **b.** This photomicrograph shows that the cells lining the proximal convoluted tubule have a brush border composed of microvilli, which greatly increases the surface area exposed to the lumen. The peritubular capillary adjoins the cells.

Tutorial
Urine Formation

the filtrate appears in the urine. In diabetes mellitus, an abnormally large amount of glucose is present in the filtrate, because the liver cannot store all the excess glucose as glycogen. The presence of glucose in the filtrate results in less water being absorbed; the increased thirst and frequent urination in untreated diabetics are a result of less water being reabsorbed into the peritubular capillary network.

Urea is an example of a substance that is passively reabsorbed from the filtrate. At first, the concentration of urea within the filtrate is the same as that in blood plasma. But after water is reabsorbed, the urea concentration is greater than that of peritubular plasma. In the end, about 50% of the filtered urea is reabsorbed.

Tubular Secretion

Tubular secretion is the second way substances are removed from blood and added to tubular fluid (Fig. 36.9*a*). Substances such as hydrogen ions, uric acid, salts, ammonia, creatinine, and penicillin are eliminated by tubular secretion. The process of tubular secretion may be viewed as helping rid the body of potentially harmful compounds that were not filtered into the glomerulus.

The Kidneys and Homeostasis

The kidneys are organs of homeostasis for four main reasons:

1. The kidneys *excrete metabolic wastes,* such as urea, which is the primary nitrogenous waste of humans.
2. They *maintain the water-salt balance,* which in turn affects blood volume and blood pressure.
3. Kidneys *maintain the acid-base balance* and therefore the pH balance.
4. They *secrete hormones.*

One hormone secreted by the kidneys, **erythropoietin,** stimulates the stem cells in bone marrow to produce more red blood cells. The Nature of Science feature, "The Misuse of Erythropoietin in Sports" (p. 686), examines the potential abuse of this hormone by endurance athletes. Another substance produced by the kidneys, called renin, is discussed later in this section.

Maintaining the Water-Salt Balance

Most of the water and salt (NaCl) present in filtrate is reabsorbed across the wall of the proximal convoluted tubule. The excretion of a hypertonic urine (one that is more concentrated than blood) is dependent on the reabsorption of water from the loop of the nephron and the collecting duct. During the process of reabsorption, water passes through water channels called **aquaporins,** which were first discovered in 1992.

Loop of the Nephron. A long loop of the nephron, which typically penetrates deep into the renal medulla, is made up of a descending (downward) limb and an ascending (upward) limb. Salt (NaCl) passively diffuses out of the lower portion of the ascending limb, but the upper, thick portion of the limb actively extrudes salt out into the tissue of the outer renal medulla (Fig. 36.10). Less and less salt is available for transport as fluid moves up the thick portion of the ascending limb. Because of these circumstances, an osmotic gradient is created within the tissues of the renal medulla: The concentration of salt is greater in the direction of the inner medulla. Note that water cannot leave the ascending limb, because this limb is impermeable to water.

The innermost portion of the inner medulla has the highest concentration of solutes. This cannot be due to salt, because active transport of salt does not start until fluid reaches the thick portion of the ascending limb. Urea is believed to leak from the lower portion of the collecting duct, and it is this molecule that contributes to the high solute concentration of the inner medulla.

Because of the osmotic gradient within the renal medulla, water leaves the descending limb along its entire length. This is a countercurrent mechanism: As water diffuses out of the descending limb, the remaining fluid within the limb encounters an even greater osmotic concentration of solute; therefore, water continues to leave the descending limb from the top to the bottom. Filtrate within the collecting duct also encounters the same osmotic gradient mentioned earlier (Fig. 36.10). Therefore, water diffuses out of the collecting duct into the renal medulla, and the urine within the collecting duct becomes hypertonic to blood plasma.

Antidiuretic hormone (ADH) is released by the posterior lobe of the pituitary in response to an increased concentration of salts in the blood. To understand the action of this hormone, consider its name. *Diuresis* means increased amount of urine, and

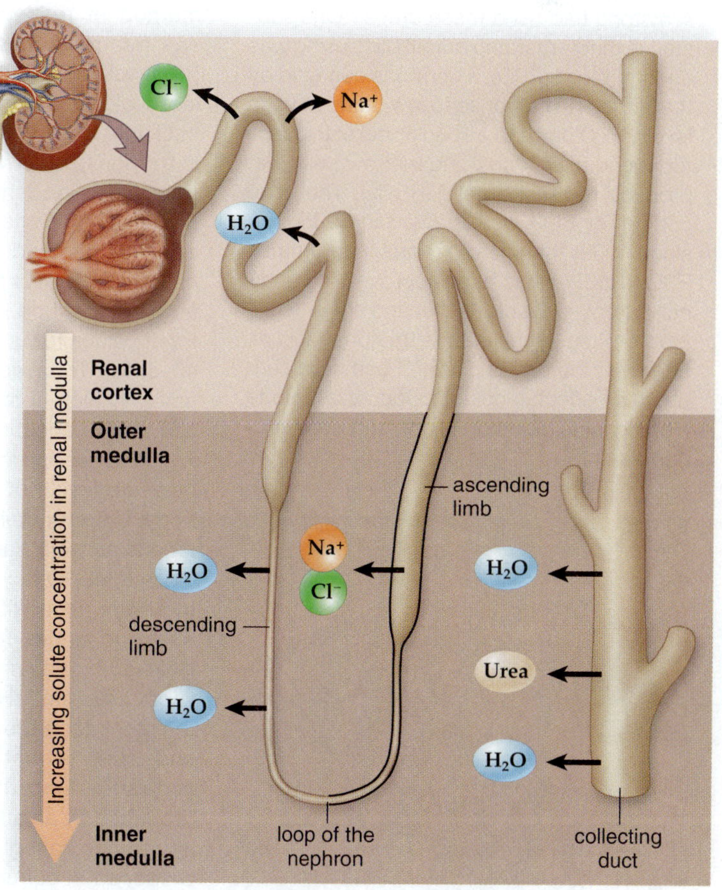

Figure 36.10 Reabsorption of salt and water. Salt (NaCl) diffuses and is actively transported out of the ascending limb of the loop of the nephron into the renal medulla; also, urea leaks from the collecting duct and enters the tissues of the renal medulla. These actions create a hypertonic environment, which draws water out of the descending limb and the collecting duct. This water is returned to the cardiovascular system.

THEME Nature of Science

The Misuse of Erythropoietin in Sports

For almost two decades, Lance Armstrong was a hero to many sports fans (Fig. 36A). There is no disputing that he was a seven-time winner of the Tour de France cycling event, a survivor of testicular cancer, and a sponsor of many charitable causes. However, in 2011 the TV news program *60 Minutes* aired a segment in which Tyler Hamilton, a former teammate of Armstrong's, said he had witnessed Armstrong using performance-enhancing drugs on several occasions. At that time, Armstrong denied those charges, noting that he had tested negative nearly 500 times during his 20-year career. Accusations continued back and forth until a 2013 interview with Oprah Winfrey, in which Armstrong confessed to using performance-enhancing drugs during much of his athletic career.

Why was it so difficult to prove that Armstong had used these drugs? One substance that Armstrong admitted to using was erythropoietin (EPO), a hormone secreted by the kidneys in response to low blood oxygen levels. EPO binds to a specific cellular receptor, found mainly on bone marrow cells that produce red blood cells (RBCs). The gene coding for EPO was isolated in 1985, and since 1989, the use of recombinant human EPO (rHuEPO) has been approved for medical purposes, such as treating the anemia that is often associated with kidney failure. However, the use of rHuEPO can have serious side effects, including increased blood clotting, high blood pressure, and even sudden death.

Figure 36A Lance Armstrong and Tyler Hamilton. Hamilton (*green and white helmet*) says he observed Armstrong (*red, white, and blue helmet*) using performance-enhancing drugs, including rHuEPO.

In addition to its legitimate medical uses, rHuEPO has been misused by athletes in several endurance sports. Research has generally verified that EPO is effective at increasing athletic performance. A small study published in 2007 showed a 13% increase in peak power output and a 54% increase in time to exhaustion in cyclists who used rHuEPO (Fig. 36B). Because this is an unfair advantage, rHuEPO use has been banned by the Tour de France, the Olympics, and other sports organizations.

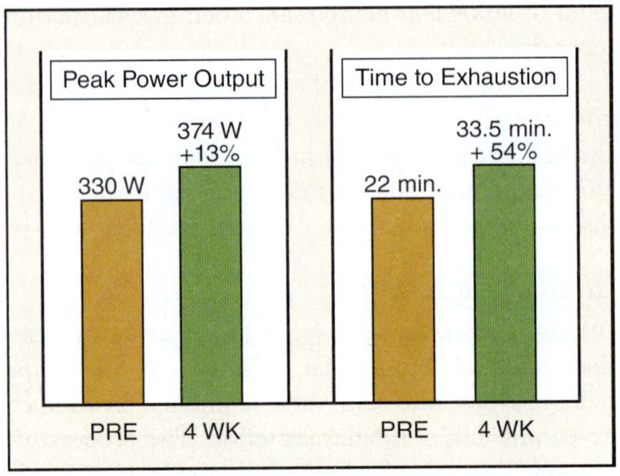

Figure 36B Use of rHuEPO and improved athletic performance. Four weeks of rHuEPO treatments increased the peak power output and prolonged the time that relatively fit cyclists were able to ride (at 80% of their maximum exertion level).

To determine whether an athlete is using rHuEPO, an accurate test is needed. One method of testing athletes for any treatment designed to increase the RBC count (including collecting the athlete's blood, storing it, and transfusing it back into the athlete soon before an event) is to monitor the *hematocrit;* the percentage of the blood that is comprised of cells. Since 1997, sports cycling governing bodies have decreed that any athlete whose hematocrit is over 50% may be suspended from competition. However, normal hematocrit values for men can vary widely, from 40–54%. This means an athlete with a naturally high hematocrit could be unfairly accused of cheating.

A better approach would be to directly measure levels of rHuEPO in an athlete's body. Unfortunately, these tests have limitations as well. Part of the problem is that when rHuEPO is injected, it can persist in the body for as little as 24 hours, but its effect continues for as long as 2 weeks. Unless a test is conducted during the short time that rHuEPO is present, even the most accurate test may miss it. A second issue is that rHuEPO must be distinguished from the EPO that is naturally produced by an athlete's own kidneys, and is also circulating in the blood. rHuEPO is produced in cultured hamster ovary cells, which process the protein differently than human cells do, attaching different sugar molecules to the amino acids. This should result in slight differences in the movement of rHuEPO versus EPO through a gel material that is subjected to an electrical field (i.e., electrophoresis).

Unfortunately, direct testing for rHuEPO may also be unreliable. In one study, eight adult male volunteers received weekly rHuEPO injections, and identical samples of their urine were submitted to two labs that were approved for rHuEPO testing. "Lab A" found that 6 of 16 samples were positive, but "Lab B" concluded that all samples were negative.[1]

Because of these issues, it will be difficult to eliminate the misuse of rHuEPO in endurance sports, especially in participants whose motivation to win overpowers their concern about the risks of abusing rHuEPO.

Questions to Consider

1. Certain kidney tumors secrete large amounts of EPO. What types of symptoms might this cause?
2. Some medical conditions, such as bacterial infections, cause the numbers of white blood cells to increase, perhaps even to twice the normal levels. Why does this have very little effect on the hematocrit?

[1]Lundby, C., et al. "Testing for Recombinant Human Erythropoietin in Urine: Problems Associated with Current Anti-doping Testing," *J. Applied Physiology* 105:417–419 (2008).

antidiuresis means decreased amount of urine. When ADH is present, more water is reabsorbed (blood volume and pressure rise), and a decreased amount of more concentrated urine is produced. One way by which ADH accomplishes this change is by causing the insertion of additional aquaporin water channels into the epithelial cells of the distal convoluted tubule and collecting duct, allowing more water to be reabsorbed.

In practical terms, if an individual does not drink much water on a certain day, the posterior lobe of the pituitary releases ADH, causing more water to be reabsorbed and less urine to form. On the other hand, if an individual drinks a large amount of water and does not perspire much, ADH is not released. More water is excreted, and more urine forms. Diuretics, such as caffeine and alcohol, increase the flow of urine by interfering with the action of ADH. ADH production also decreases at night, an adaptation that allows longer periods of sleep without the need to wake up to urinate.

Hormones Control the Reabsorption of Salt. Usually, more than 99% of the Na^+ filtered at the glomerulus is returned to the blood. Most sodium (67%) is reabsorbed at the proximal convoluted tubule, and a sizable amount (25%) is extruded by the ascending limb of the loop of the nephron. The rest is reabsorbed from the distal convoluted tubule and collecting duct.

Blood volume and pressure are, in part, regulated by salt reabsorption. When blood volume, and therefore blood pressure, is not sufficient to promote glomerular filtration, a cluster of cells near the glomerulus called the *juxtaglomerular apparatus* secretes renin. **Renin** is an enzyme that changes angiotensinogen (a large plasma protein produced by the liver) into angiotensin I. Later, angiotensin I is converted to **angiotensin II,** a powerful vasoconstrictor that also stimulates the adrenal glands, which lie on top of the kidneys, to release aldosterone (Fig. 36.11). **Aldosterone** is a hormone that promotes the excretion of potassium ions (K^+) and the reabsorption of sodium ions (Na^+) at the distal convoluted tubule. The reabsorption of sodium ions is followed by the reabsorption of water. Therefore, blood volume and blood pressure increase.

Atrial natriuretic hormone (**ANH**) is a hormone secreted by the atria of the heart when cardiac cells are stretched due to increased blood volume. ANH inhibits the secretion of renin by the juxtaglomerular apparatus and the secretion of aldosterone by the adrenal cortex. Its effect, therefore, is to promote the excretion of Na^+—that is, natriuresis. When Na^+ is excreted, so is water, and therefore blood volume and blood pressure decrease.

These examples show that the kidneys regulate the water balance in blood by controlling the excretion and reabsorption of ions. Sodium is an important ion in plasma that must be regulated, but the kidneys also excrete or reabsorb other ions, such as potassium ions (K^+), bicarbonate ions (HCO_3^-), and magnesium ions (Mg^{2+}), as needed.

MP3 Water Conservation

Maintaining the Acid-Base Balance

The functions of cells are influenced by pH. Therefore, the regulation of pH is extremely important to good health. The *bicarbonate (HCO_3^-) buffer system* and breathing work together to help maintain the pH of the blood. Central to the mechanism is the following reaction, which you have seen before:

$$H^+ + HCO_3^- \rightleftharpoons H_2CO_3 \rightleftharpoons H_2O + CO_2$$

The excretion of carbon dioxide (CO_2) by the lungs helps keep the pH within normal limits, because when carbon dioxide is exhaled, this reaction is pushed to the right and hydrogen ions are tied up in water. As you learned in Chapter 35, when blood pH decreases, chemoreceptors in the carotid bodies (in the carotid arteries) and in aortic bodies (in the aorta) stimulate the respiratory control center, and the rate and depth of breathing increase. And when blood pH begins to rise, the respiratory control center is depressed, and the amount of bicarbonate ion increases in the blood.

As powerful as this system is, only the kidneys can rid the body of a wide range of acidic and basic substances. The kidneys are slower-acting than the buffer/breathing mechanism, but they have a more powerful effect on pH. For the sake of simplicity, we can think of the kidneys as reabsorbing bicarbonate ions and

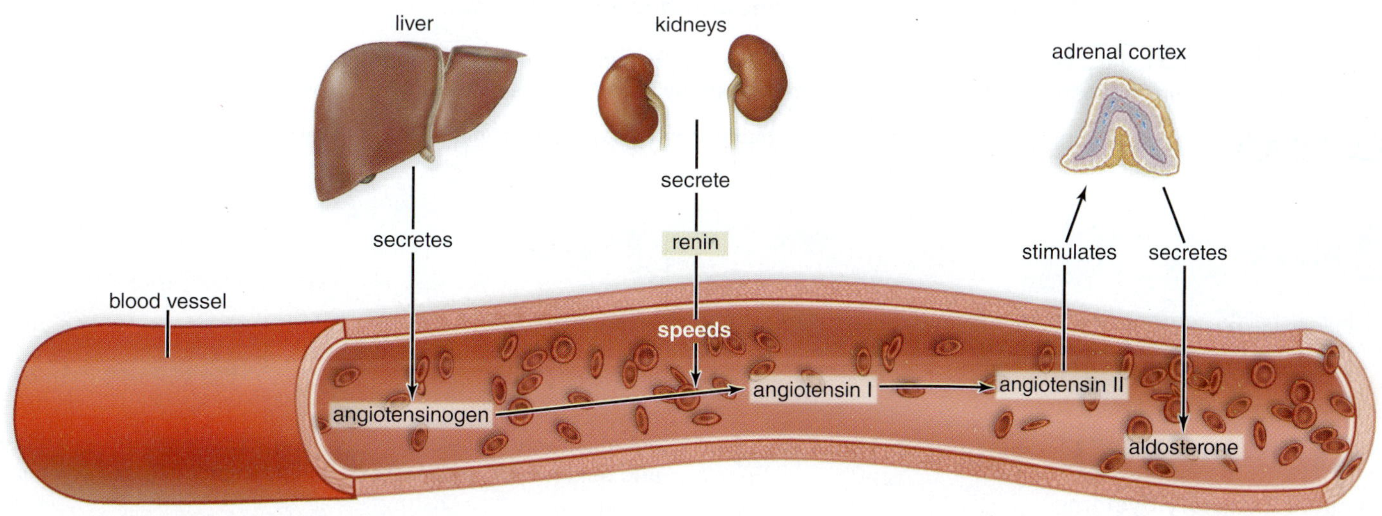

Figure 36.11 The renin-angiotensin-aldosterone system. The liver secretes angiotensinogen into the bloodstream. Renin from the kidneys initiates the chain of events that results in angiotensin II. Angiotensin II acts on the adrenal cortex to secrete aldosterone, which causes reabsorption of sodium ions by the kidneys and a subsequent rise in blood pressure.

excreting hydrogen ions as needed to maintain the normal pH of the blood:

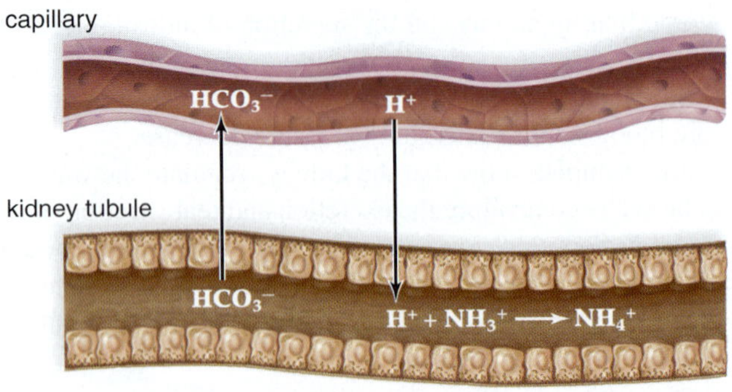

If the blood is acidic, hydrogen ions are excreted and bicarbonate ions are reabsorbed. If the blood is basic, hydrogen ions are not excreted and bicarbonate ions are not reabsorbed. The fact that urine is typically acidic (pH about 6) shows that usually an excess of hydrogen ions are excreted. Ammonia (NH_3) provides a means for buffering these hydrogen ions in urine: ($NH_3 + H^+ \longrightarrow NH_4^+$). Ammonia is produced in tubule cells by the deamination of amino acids. Phosphate provides another means of buffering hydrogen ions in urine.

MP3 Acid-Base Balance

Check Your Progress 36.2

1. Describe which of the four major functions of the human urinary system are accomplished solely by the kidneys, and which are shared with other body systems.
2. Explain the effect of the renin-angiotensin-aldosterone system on water-salt balance.
3. Describe how the kidneys contribute to the maintenance of normal blood pH.

CONNECTING *the* CONCEPTS *with the* THEMES

Evolution

- Proper salt balance has a critical effect on cellular functions, and many adaptations have evolved to maintain it.
- Animals that live in wet or watery environments tend to secrete nitrogenous wastes as ammonia; those in drier environments secrete urea or uric acid, which requires more energy to produce.
- A variety of excretory organs have evolved in animals, from relatively simple, tubular structures in invertebrates to the extremely efficient kidneys found in the kangaroo rat.

Nature of Science

- By studying the excretory systems of many different animals, scientists have come to understand the great variety of evolutionary strategies and mechanisms that regulate the composition of body fluids.
- The production of the recombinant hormone erythropoietin has tremendous benefits but also the potential for misuse.

Biological Systems

- Without removal of wastes and excess salts, an animal would die. In most animals, the kidneys are the main regulators of the internal environment.
- Human kidneys excrete nitrogenous waste (mainly as urea), maintain the salt balance, work with the respiratory system to maintain pH, and secrete hormones that affect water reabsorption, blood volume and blood pressure, and red blood cell production.

MEDIA STUDY TOOLS http://connect.mheducation.com

SMARTBOOK® Study smart. SmartBook helps you maximize your study time by identifying what you know and don't know. The recharge feature in SmartBook helps you prepare for the test by identifying the material that you are most likely to forget.

connect BIOLOGY Find even more resources to learn the chapter concepts including animations, tutorial videos, and interactive practice questions.

MP3 Files	Tutorial
36.2 Functional Anatomy of the Urinary System • An Overview of Urine Formation • Water Conservation • Acid-Base Balance	**36.2** Urine Formation

SUMMARIZE

36.1 Animal Excretory Systems

Osmoregulation, or balancing the body's water and salt levels, includes **excretion** of metabolic wastes, including nitrogenous wastes. Aquatic animals usually excrete **ammonia** (which requires little energy but much water to excrete), and land animals excrete either **urea** or **uric acid** (which requires much energy to produce, but can conserve water).

Most animals have specialized excretory organs. The flame cells of planarians rid the body of excess water. Earthworm nephridia exchange molecules with the blood in a manner similar to that of vertebrate kidneys. Malpighian tubules in insects take up metabolic wastes and water from the hemolymph. Later, the water is absorbed by the gut.

Osmotic regulation is important to animals. Most must balance their water and salt intake and excretion to maintain normal solute and water concentration in body fluids. Marine fishes constantly drink water, excrete salts at the gills, and pass an isotonic **urine.** Freshwater fishes never drink water; they take in salts at the gills and excrete a hypotonic urine.

Some terrestrial animals have adapted to extreme environments. For example, the desert kangaroo rat can survive on metabolic water; marine birds and reptiles have glands that extrude salt.

36.2 The Human Urinary System

The human urinary system includes the **kidneys, ureters, urinary bladder,** and **urethra.** The kidneys serve four basic homeostatic functions: excretion of metabolic waste; maintenance of water-salt balance; maintenance of pH balance; and production of hormones, such as erythropoietin.

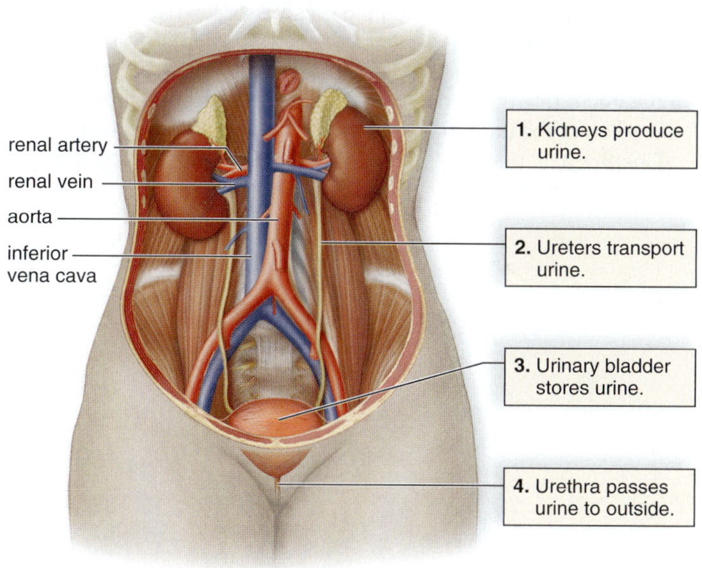

Kidneys are made up of **nephrons,** each of which has several parts. The renal artery branches into many smaller arteries, each of which forms numerous arterioles. Each **afferent arteriole** forms a tuft of capillaries, called the **glomerulus.** From the **glomerular capsule,** a **proximal convoluted tubule** leads to the **loop of the nephron,** which is followed by the **distal convoluted tubule.** Several nephrons enter one **collecting duct,** which transports urine through the renal medulla into the renal pelvis.

Urine formation by a nephron requires three steps: **glomerular filtration,** during which nutrients, water, and wastes enter the nephron's glomerular capsule; **tubular reabsorption,** when nutrients and most water are reabsorbed into the peritubular capillary network; and **tubular secretion,** during which additional wastes are added to the convoluted tubules.

In order to excrete a hypertonic urine, the ascending limb of the loop of the nephron extrudes salt, so that the renal medulla is increasingly hypertonic. Since urea leaks from the lower end of the collecting duct, the inner renal medulla has the highest concentration of solute. Therefore, a countercurrent mechanism ensures that water diffuses out of the descending limb and the collecting duct. During this process, water passes through membrane channels called **aquaporins.**

Three hormones are involved in maintaining the water-salt balance of the blood. **Antidiuretic hormone (ADH),** which makes the collecting duct more permeable to water, is secreted by the posterior pituitary in response to an increase in the osmotic pressure of the blood. **Aldosterone** is secreted by the adrenal cortex after low blood pressure has caused the kidneys to release **renin.** The presence of renin leads to the formation of **angiotensin II,** which causes the adrenal cortex to release aldosterone. Aldosterone causes the kidneys to retain Na^+; therefore, water is reabsorbed and blood pressure rises. **Atrial natriuretic hormone,** in contrast, prevents the secretion of renin and aldosterone.

The kidneys also keep blood pH within normal limits. They reabsorb HCO_3^- and excrete H^+ as needed to maintain the pH at about 7.4. Finally, the kidneys help to regulate red blood cell production by secreting **erythropoietin,** a hormone that is sometimes abused by athletes.

ASSESS

Choose the best answer for each question.

36.1 Animal Excretory Systems

1. Which of these is not correct?
   a. Uric acid is produced from the breakdown of nucleic acids.
   b. Urea is produced from the breakdown of proteins.
   c. Ammonia results from the deamination of amino acids.
   d. All of these are correct.
2. One advantage of the excretion of urea instead of uric acid is that urea
   a. requires less energy than uric acid to produce.
   b. can be concentrated to a greater extent.
   c. is not a toxic substance.
   d. requires no water to excrete.
   e. is a larger molecule.
3. Which of these pairs is mismatched?
   a. insects—excrete uric acid
   b. humans—excrete urea
   c. fishes—excrete ammonia
   d. birds—excrete ammonia
   e. All of these are mismatched.
4. Freshwater bony fishes maintain water balance by
   a. excreting salt across their gills.
   b. periodically drinking small amounts of water.
   c. excreting a hypotonic urine.
   d. excreting wastes in the form of uric acid.
   e. Both a and c are correct.
5. Which of these is not an adaptation that helps kangaroo rats conserve water?
   a. formation of a very hypertonic urine
   b. highly convoluted nasal passages
   c. production of very dry feces
   d. relatively long loops of the nephron in kidneys
   e. secretion of excess salt from glands near eyes

36.2 The Human Urinary System

6. In the path of blood through the human kidney, the blood vessel that follows the renal artery is the
   a. peritubular capillary.
   b. efferent arteriole.
   c. afferent arteriole.
   d. renal vein.

7. Which of these materials is not filtered from the blood at the glomerulus?
   a. water
   b. urea
   c. protein
   d. glucose
8. Excretion of a hypertonic urine in humans is associated best with
   a. the glomerular capsule.
   b. the proximal convoluted tubule.
   c. the loop of the nephron.
   d. the collecting duct.
   e. Both c and d are correct.
9. Which of these causes blood pressure to decrease?
   a. aldosterone
   b. antidiuretic hormone (ADH)
   c. renin
   d. atrial natriuretic hormone (ANH)
10. Label this diagram of a nephron.

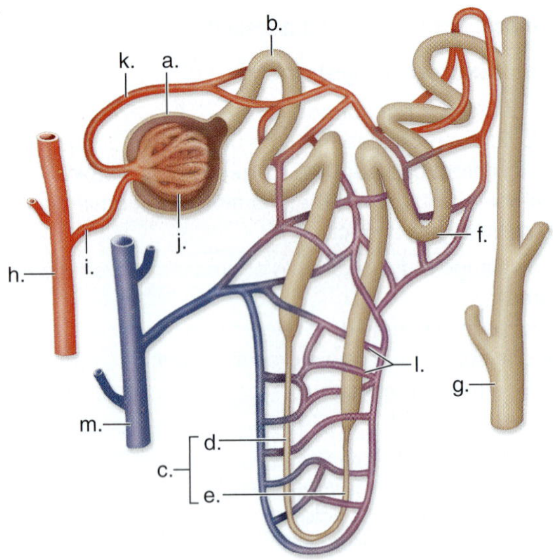

ENGAGE

THINKING SCIENTIFICALLY

1. High blood pressure often is accompanied by kidney damage. In some people, the kidney damage is subsequent to the high blood pressure, but in others the kidney damage is the cause of the high blood pressure. Explain how a low-salt diet would enable you to determine whether the high blood pressure or the kidney damage came first.
2. The renin-angiotensin-aldosterone system can be inhibited in order to reduce high blood pressure. Usually, the angiotensin-converting enzyme, which converts angiotensin I to angiotensin II, is inhibited by drug therapy. Why would this enzyme be an effective point at which to disrupt the system?
3. In diabetes mellitus, low insulin levels can lead to high blood sugar levels. In diabetes insipidus, ADH levels are abnormally low. What would be the effect of each condition on urine production?
4. Approximately one in six adults has some degree of overactive bladder syndrome, in which the muscles of the bladder can contract unexpectedly, causing an urge to urinate. This condition can be treated with a medication called Detrol LA, which blocks the neurotransmitter acetylcholine in the parasympathetic division of the autonomic nervous system. Referring ahead to Chapter 37 if needed, predict some side effects of using this type of medication.

The actor Michael J. Fox is also a leading advocate for Parkinson disease research.

37

Neurons and Nervous Systems

Chapter Outline

37.1 Evolution of the Nervous System 692

37.2 Nervous Tissue 695

37.3 The Central Nervous System 699

37.4 The Peripheral Nervous System 707

In his autobiography, titled *Lucky Man,* actor Michael J. Fox relates a story of how he woke up one morning in 1990 after a night of partying to find that the pinky finger on his left hand was trembling, and wouldn't stop. Initially, the 29-year-old star of the *Back to the Future* movies and the *Family Ties* TV show assumed the strange symptom might be a result of a bad hangover. However, the odd trembling and tingling sensation continued, even though doctors initially couldn't find a cause. About a year later, the actor consulted with a neurologist, who diagnosed Fox with early-onset Parkinson disease (PD). Over 20 years later, although he is still a working actor, Fox now suffers from widespread tremors and difficulty walking and speaking, along with other neurological symptoms.

About 1 million Americans are living with PD, which results from a degeneration of certain neurons in the brain. Other than the distinctive collection of symptoms, there is no definitive test for the disease. There is also no cure, although symptoms can usually be decreased with medication. Seven years after his diagnosis, Fox had an experimental surgical procedure called a thalamotomy, which has successfully reduced the symptoms of some PD patients. However, after a temporary improvement, Fox admitted that his symptoms had returned in full. Meanwhile, he founded the Michael J. Fox Foundation, which has provided over $450 million to fund PD research.

As you read through this chapter, think about the following questions:

1. How did the evolution of the nervous system provide advantages to animals?
2. What specific types of processes occur uniquely in nervous tissues?
3. Why are diseases of the human nervous system generally difficult to treat?

Before You Begin

Before beginning this chapter, take a few moments to review the following discussions.

Figure 5.9 How does the sodium-potassium pump function?

Section 29.1 From what embryonic structures are the brain, spinal cord, and (in vertebrates) vertebral column derived?

Section 30.1 How has the evolution of a large, complex brain allowed humans to become the most dominant species on the planet?

Following *the* Themes

CHAPTER 37 NEURONS AND NERVOUS SYSTEMS

UNIT 7 COMPARATIVE ANIMAL BIOLOGY

Evolution	Nervous systems have evolved in all types of animals except the simplest multicellular animals and range from relatively simple nerve nets to the highly complex human brain.
Nature of Science	Scientific studies have revealed much (but not all) about how the nervous system functions, leading to new treatments for some neurological diseases.
Biological Systems	Although animal nervous systems vary greatly, all are involved in controlling and coordinating body functions and activities.

37.1 Evolution of the Nervous System

Learning Outcomes

Upon completion of this section, you should be able to

1. Compare the nervous systems of cnidarians, planarians, and annelids.
2. Describe the essential features of a typical vertebrate nervous system.
3. Explain the major adaptations that evolved in the brains of mammals.

The nervous system is vital in complex animals, enabling them to seek food and mates and to avoid danger. It ceaselessly monitors internal and external conditions and makes appropriate changes to maintain homeostasis. A comparative study of animal nervous systems shows the evolutionary trends that led to the nervous system of mammals.

Invertebrate Nervous System Organization

The simplest multicellular animals, such as sponges, lack neurons (nerve cells) and therefore have no nervous system. However, their cells can respond to their environment and can communicate with each other, perhaps by releasing calcium or other ions; the most common example is closure of the osculum (central opening) in response to various stimuli.

Hydras, which are cnidarians with the tissue level of organization and radial symmetry, have a **nerve net** composed of neurons in contact with one another and with contractile cells in the body wall (Fig. 37.1*a*). They can contract and extend their bodies, move their tentacles to capture prey, and even turn somersaults. Sea anemones and jellyfish, which are also cnidarians, seem to have two nerve nets: A fast-acting one allows major responses, particularly in times of danger; the slower one coordinates slower and more delicate movements.

Planarians (flatworms) have a nervous organization that reflects their bilateral symmetry. They have a ladderlike nervous system, with two ventrally located lateral or longitudinal nerve cords (bundles of nerves) that extend from the cerebral ganglia to the posterior end of their body. Transverse nerves connect the nerve cords, as well as the cerebral ganglia, to the eyespots. **Cephalization,** or concentration of nervous tissue in the anterior or head region, has occurred. A cluster of neuron cell bodies is called a **ganglion** (pl., ganglia), and the anterior cerebral ganglia of flatworms receive sensory information from photoreceptors in the eyespots and sensory cells in the auricles (Fig. 37.1*b*). The two lateral nerve cords allow a rapid transfer of information from the cerebral ganglia to the posterior end, and the transverse nerves between the nerve cords keep the movement of the two sides coordinated. Bilateral symmetry plus cephalization are two significant trends in the development of a nervous organization that is adaptive for an active way of life.

Annelids (e.g., earthworm) (Fig. 37.1*c*) and arthropods (e.g., crab) (Fig. 37.1*d*) are complex animals with the typical invertebrate nervous system. A brain is present, and a ventral nerve cord has a ganglion in each segment. The brain, which normally receives sensory information, controls the activity of the ganglia and assorted nerves, so that the muscle activity of the entire animal is coordinated.

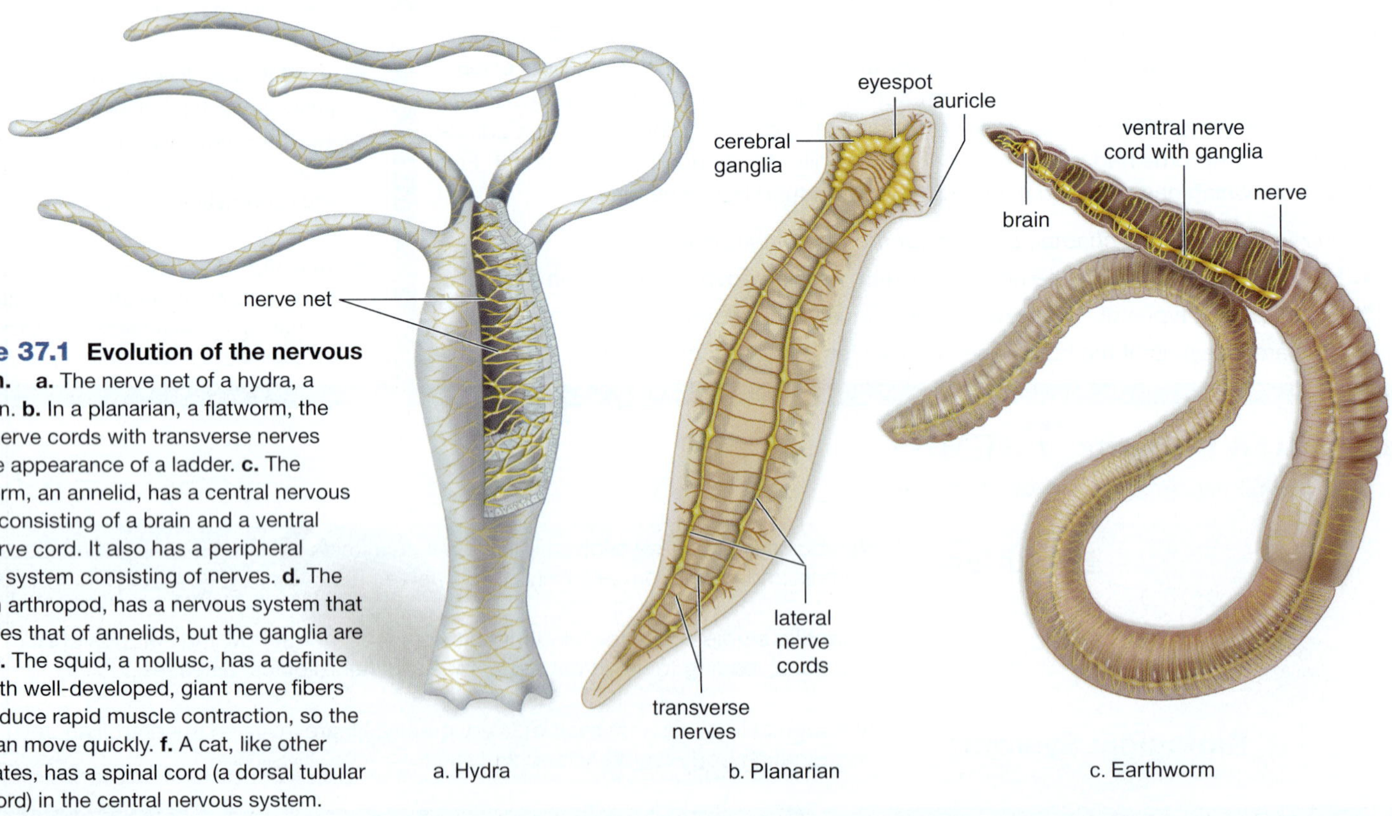

Figure 37.1 Evolution of the nervous system. **a.** The nerve net of a hydra, a cnidarian. **b.** In a planarian, a flatworm, the paired nerve cords with transverse nerves have the appearance of a ladder. **c.** The earthworm, an annelid, has a central nervous system consisting of a brain and a ventral solid nerve cord. It also has a peripheral nervous system consisting of nerves. **d.** The crab, an arthropod, has a nervous system that resembles that of annelids, but the ganglia are larger. **e.** The squid, a mollusc, has a definite brain with well-developed, giant nerve fibers that produce rapid muscle contraction, so the squid can move quickly. **f.** A cat, like other vertebrates, has a spinal cord (a dorsal tubular nerve cord) in the central nervous system.

A group of molluscs called cephalopods (e.g., squid) (Fig. 37.1*e*) show marked cephalization—the anterior end has a well-defined brain and well-developed sense organs, such as eyes. The cephalopods are widely regarded as the most intelligent invertebrates; many are highly social creatures, and some, such as the octopus, have been observed to collect, transport, and assemble coconut shells for later use as a shelter.

Vertebrate Nervous System Organization

Vertebrates have many more neurons than do invertebrates. For example, an insect's entire nervous system contains a total of about 1 million neurons, while a cat's nervous system may contain many thousand times that number (Fig. 37.1*f*). The human cerebral cortex alone contains an estimated 11 billion neurons; some whales have even more.

All vertebrates have a brain that controls the nervous system. It is customary to divide the vertebrate brain into the hindbrain, midbrain, and forebrain (Fig. 37.2), although the relative sizes of the parts vary greatly among species. The hindbrain, the most ancient part of the brain, regulates motor activity below the level of consciousness. For example, the lungs and heart function even when an animal is sleeping. The medulla oblongata contains control centers for breathing and heart rate. Coordination of motor activity associated with limb movement, posture, and balance eventually became centered in the cerebellum.

Several types of paired sensory receptors, including the eyes, ears, and olfactory structures, allow the animal to gather information from the environment. These sense organs are generally located at the anterior end of the animal, because this end is usually the first to enter new environments. The optic lobes are part of the midbrain, which was originally a center for coordinating reflexes involving the eyes and ears. In early vertebrate evolution, the forebrain was concerned mainly with the sense of smell. Beginning with the amphibians and continuing in the other vertebrates, the forebrain processes sensory information. Later, the thalamus evolved to receive sensory input from the midbrain and the hindbrain and to pass it on to the cerebrum, the anterior part of the forebrain in vertebrates. In the forebrain, the hypothalamus is particularly concerned with homeostasis, and in this capacity, the hypothalamus communicates with the medulla oblongata and the pituitary gland.

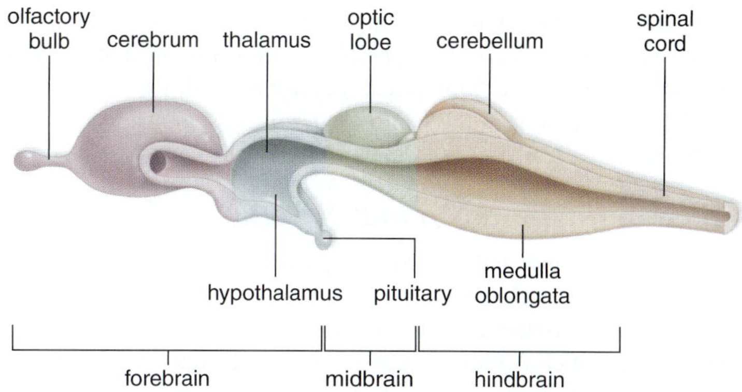

Figure 37.2 Organization of the vertebrate brain. The vertebrate brain is divided into the forebrain, midbrain, and hindbrain.

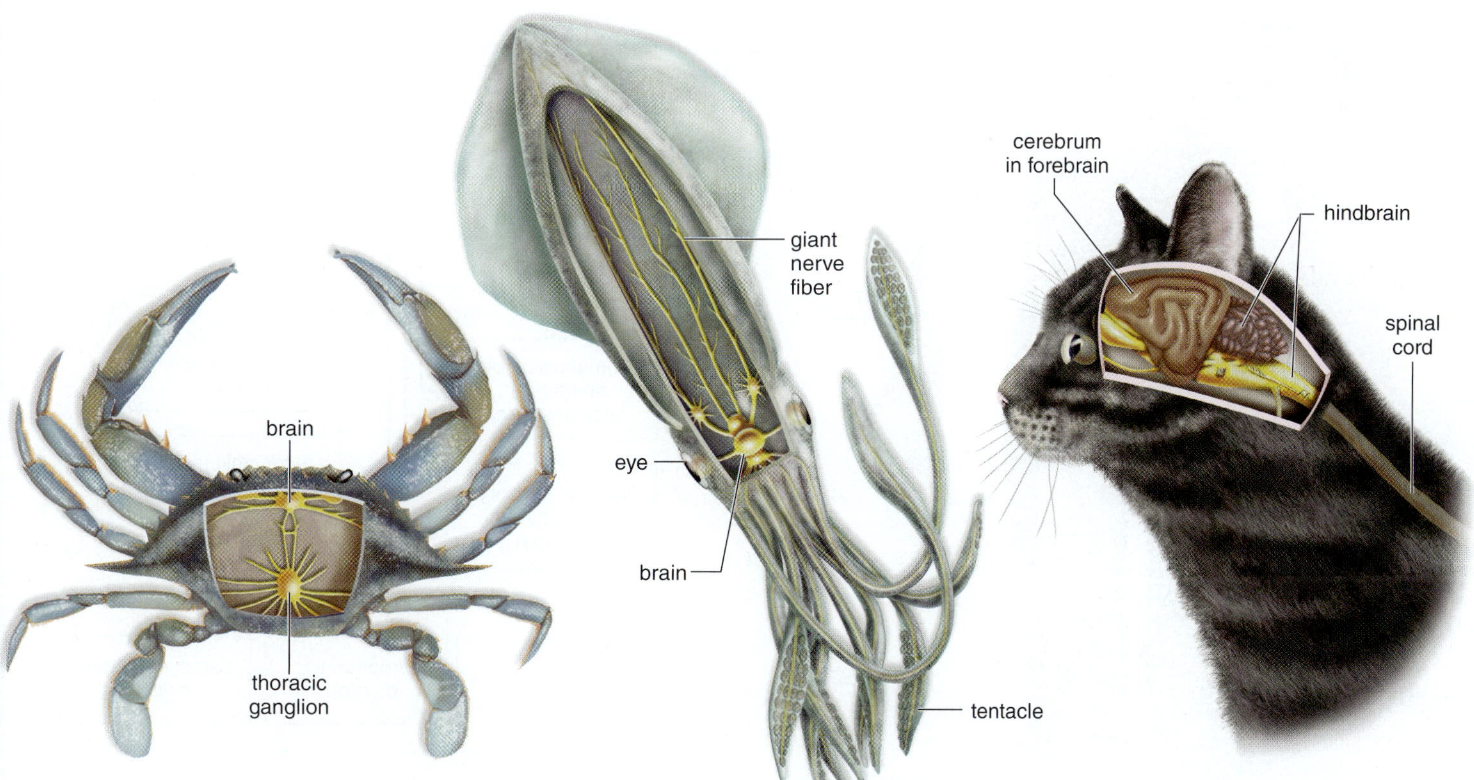

d. Crab

e. Squid

f. Cat

The **central nervous system (CNS)** consists of the brain and spinal cord (Fig. 37.3). The **peripheral nervous system (PNS)** (Gk. *periphereia,* "circumference") consists of all the nerves and ganglia that lie outside the central nervous system. The CNS and PNS are considered in more detail in sections 37.3 and 37.4.

The Mammalian Nervous System

The hindbrain and midbrain of mammals are similar to those of other vertebrates. However, the forebrain of mammals is greatly enlarged, due to the addition of an outermost layer called the *neocortex,* which is seen only in mammals. It functions in higher mental processes, such as spatial reasoning, conscious thought, and language.

Although all mammals have a neocortex, all neocortexes are not the same. Large variation has been observed in the number of crevices and folds, which can greatly increase the surface area and numbers of connections between regions. The frontal lobes (see section 37.3) are especially large and complex in primates, and in humans other parts of the cortex are also enlarged and form very complex connections with other parts of the brain. It is likely that this greatly increased brain capacity allowed mammals, and especially humans, to become increasingly adept at higher mental activities, such as manipulating the environment, complex learning, and anticipating the future, all of which have provided tremendous evolutionary advantages.

Check Your Progress 37.1

1. Define the terms *nerve net, ganglion,* and *brain.*
2. Describe the major functions of the hindbrain, midbrain, and forebrain.
3. Identify the specific location of the more recently evolved parts of the brain, compared to the older parts.

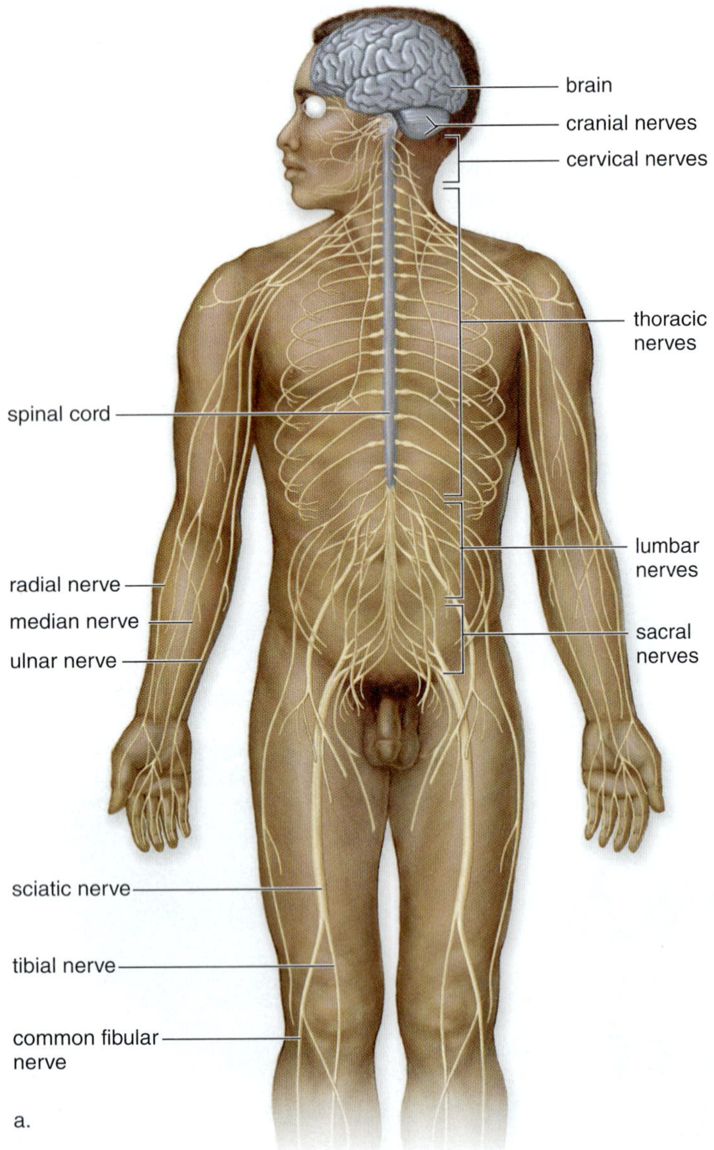

a.

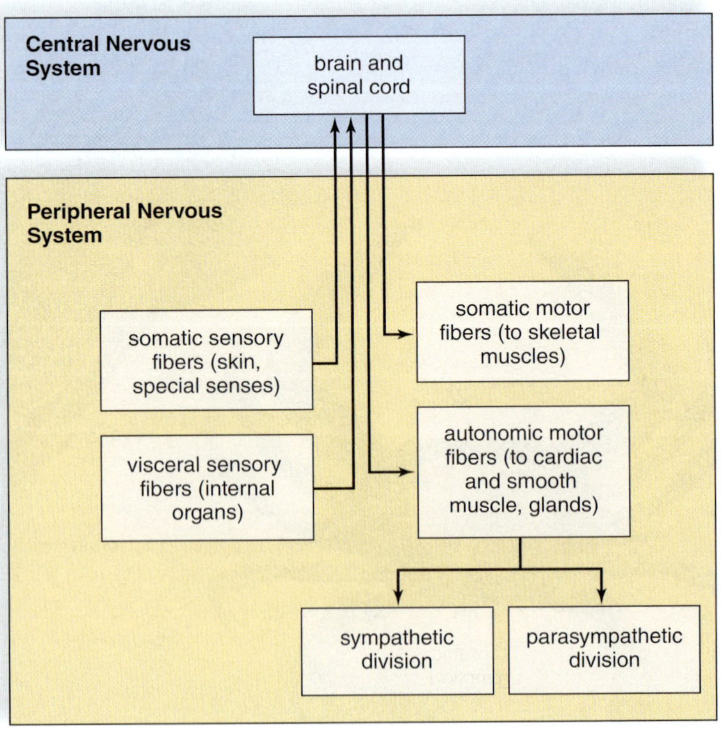

b.

Figure 37.3 Organization of the nervous system in humans. **a.** The central nervous system (CNS) is composed of the brain and spinal cord; the peripheral nervous system (PNS) consists of nerves. **b.** In the somatic system of the PNS, nerves conduct impulses from sensory receptors in the skin and internal organs to the CNS, and motor impulses from the CNS to the skeletal muscles. In the autonomic system, consisting of the sympathetic and parasympathetic divisions, motor impulses travel to smooth muscle, cardiac muscle, and glands.

37.2 Nervous Tissue

Learning Outcomes

Upon completion of this section, you should be able to

1. Describe the basic structure of a neuron and compare the functions of the three types of neurons.
2. Discuss the changes in ion concentrations inside and outside a neuron that result in an action potential.
3. Summarize the role of various neurotransmitters in propagating nerve impulses.

Although complex, nervous tissue is composed of just two principal types of cells. **Neurons,** also known as nerve cells, are the functional units of the nervous system. They receive sensory information, convey the information to an integration center such as the brain, and conduct signals from the integration center to effector structures, such as the glands and muscles. **Neuroglia** serve as supporting cells, providing support and nourishment to the neurons.

MP3 Cells of the Nervous System

Neurons and Neuroglia

Neurons vary in appearance depending on their function and location. They consist of three major parts: a cell body, dendrites, and an axon (Fig. 37.4). The **cell body** contains a nucleus and a variety of organelles. The **dendrites** (Gk. *dendron,* "tree") are short, highly branched processes that receive signals from the sensory receptors or other neurons and transmit them to the cell body. The **axon** (Gk. *axon,* "axis") is the portion of the neuron that conveys information to another neuron or to other cells. Axons can be bundled together to form nerves. For this reason, axons are often called **nerve fibers.** Many axons are covered by a white insulating layer called the **myelin sheath** (Gk. *myelos,* "spinal cord").

Neuroglia, or glial cells, greatly outnumber neurons in the brain. Named for the Greek work for "glue," glial cells were once thought to simply provide structural and nutritional support for neurons. However, some researchers now characterize glial cells as the "supervisors" of the neurons, because some glial cells play an important role in synapse formation and help neurons process information.

There are several types of neuroglia in the CNS, each with some specific known functions. The most numerous type of cell in the brain is the **astrocyte,** which serves many roles in maintaining neuron health and function. **Microglia** are phagocytic cells that help remove bacteria and debris. The myelin sheath is formed from the membranes of tightly spiraled neuroglia. In the CNS, neuroglial cells called **oligodendrocytes** form the myelin sheath. In the PNS, **Schwann cells** perform this function, leaving gaps called **nodes of Ranvier. Ependymal cells** line the ventricles of the brain, where they produce the cerebrospinal fluid. Finally, **satellite cells** surround neuron cell bodies in ganglia (ganglia are discussed in section 37.4), where they participate in responses to injury and inflammation.

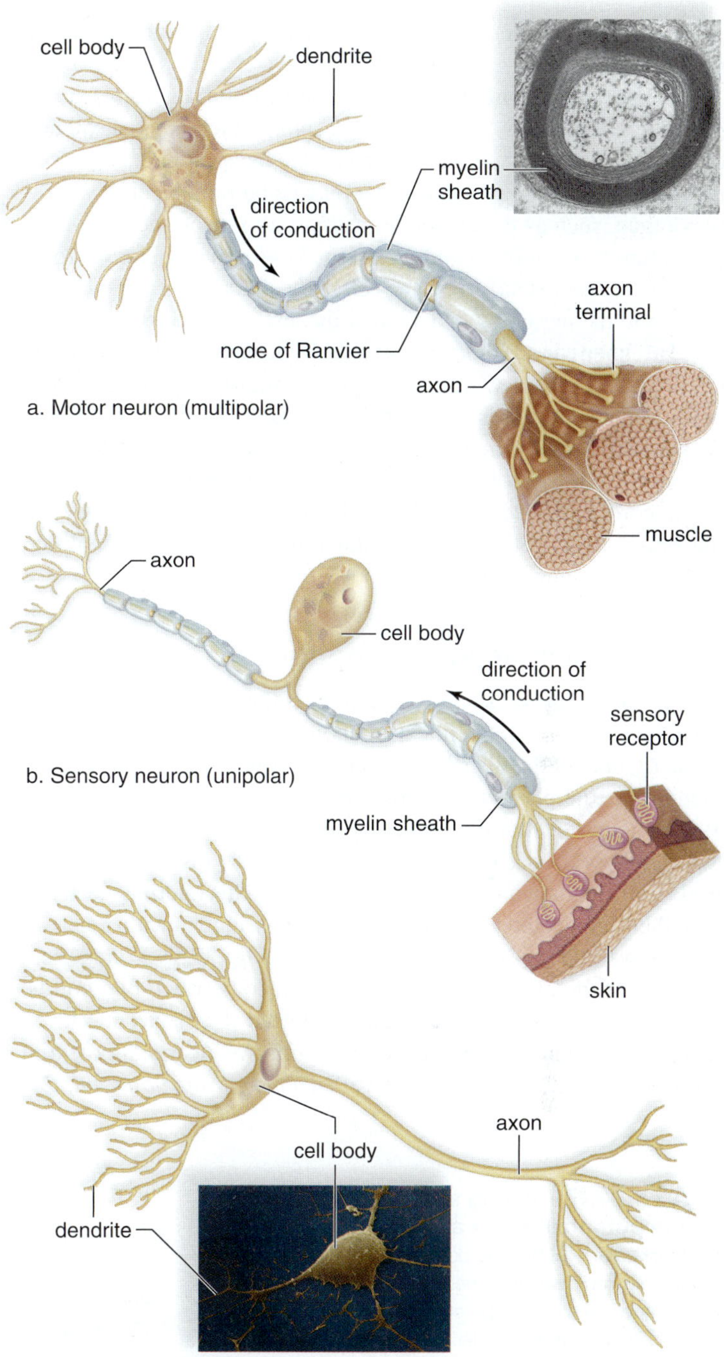

Figure 37.4 Neuron anatomy. a. Motor neuron. Note the branched dendrites and the single, long axon, which branches only near its tip. **b.** Sensory neuron with dendritelike structures projecting from the peripheral end of the axon. **c.** Interneuron (from the cortex of the cerebellum) with very highly branched dendrites.

Types of Neurons

Neurons can be described in terms of their function and shape. **Motor (efferent) neurons** take nerve impulses from the CNS to muscles or glands. Motor neurons are said to have a *multipolar* shape, because they have many dendrites and a single axon (Fig. 37.4*a*). Motor neurons cause muscle fibers to contract or glands to secrete, and therefore they are said to innervate these structures.

Sensory (afferent) neurons take nerve impulses from sensory receptors to the CNS. The sensory receptor, which is the distal end of the long axon of a sensory neuron, may be as simple as a naked nerve ending (a pain receptor), or it may be built into a highly complex organ, such as the eye or ear. Almost all sensory neurons have a structure that is termed *unipolar* (Fig. 37.4*b*). In unipolar neurons, the process that extends from the cell body divides into a branch that extends to the periphery and another that extends to the CNS.

Interneurons (L. *inter,* "between") occur entirely within the CNS. Interneurons, which are typically multipolar (Fig. 37.4*c*), convey nerve impulses between various parts of the CNS. Some lie between sensory neurons and motor neurons; some take messages from one side of the spinal cord to the other or from the brain to the cord, and vice versa. They also form complex pathways in the brain, leading to higher mental functions, such as thinking, memory, and language.

Transmission of Nerve Impulses

In the early 1900s, scientists first hypothesized that the nerve impulse is an electrochemical phenomenon involving the movement of unequally distributed ions on either side of an axonal membrane, the plasma membrane of an axon. It was not until the 1960s, however, that experimental techniques were developed to test this hypothesis. Investigators were able to insert a tiny electrode into the giant axon of the squid *Loligo.* This internal electrode was then connected to a voltmeter, an instrument with a screen that shows voltage differences over time (Fig. 37.5). Voltage is a measure of the electrical potential difference between two points, which in this case is the difference between the electrode placed inside and another placed outside the axon. An electrical potential difference across a membrane is called a *membrane potential.*

Resting Potential

When the axon is not conducting an impulse, the voltmeter records a membrane potential equal to about −70 mV (millivolts), indicating that the inside of the neuron is more negative than the outside (Fig. 37.5*a*). This is called the **resting potential,** because the axon is not conducting an impulse.

The existence of this polarity can be correlated with a difference in ion distribution on either side of the axonal membrane. As Figure 37.5*a* shows, there is a higher concentration of sodium ions (Na^+) outside the axon and a higher concentration of potassium ions (K^+) inside the axon.

The unequal distribution of these ions is due in part to the activity of the *sodium-potassium pump* (described in section 5.3; see Fig. 5.10). This pump is an active transport system in the plasma membrane that pumps three sodium ions out of the axon and two potassium ions into the axon. The pump is always working, because the membrane is somewhat permeable to these ions, and they tend to diffuse toward areas of lesser concentration. Because the membrane is more permeable to potassium than to sodium, there are always more positive ions outside the membrane than inside; this accounts for some of the membrane potential recorded by the voltmeter. The axon cytoplasm also contains large, negatively charged proteins. Altogether, then, the voltmeter records that the resting potential is −70 mV inside the cell.

Animation How the Sodium-Potassium Pump Works

Action Potential

An **action potential** is a rapid change in polarity across a portion of an axonal membrane as the nerve impulse occurs. An action potential involves two types of gated ion channels in the axonal membrane, one that allows the passage of Na^+ and one that allows the passage of K^+. In contrast to ungated ion channels,

Animation Action Potential Propagation

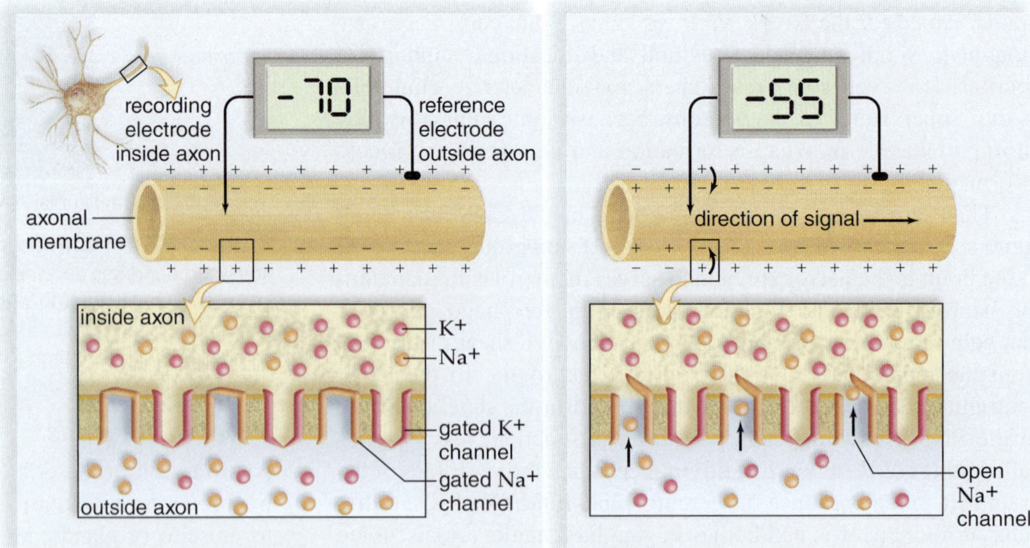

a. Resting potential: Na+ outside the axon, K+ and large anions inside the axon. Separation of charges polarizes the cell and causes the resting potential.

b. Stimulus causes the axon to reach its threshold; the axon potential increases from −70 to −55. The action potential has begun.

Figure 37.5 Resting and action potential of the axonal membrane. **a.** Resting potential. A voltmeter indicates that the axonal membrane has a resting potential of −70 mV. There is a preponderance of Na^+ outside the axon and a preponderance of K^+ inside the axon. The permeability of the membrane to K^+ compared to Na^+ causes the inside to be negative compared to the outside. **b.** During an action potential, depolarization occurs when Na^+ gates open and Na^+ begins to move inside the axon. **c.** Depolarization continues until a potential of +35 mV is reached. **d.** Repolarization occurs when K^+ gates open and K^+ moves outside the axon. **e.** Graph of an action potential.

Tutorial Neuron Action Potentials

which constantly allow ions across the membrane, gated ion channels open and close in response to a stimulus, such as a signal from another neuron.

The *threshold* is the minimum change in polarity across the axonal membrane that is required to generate an action potential. Therefore, the action potential is an all-or-none event. During *depolarization,* the inside of a neuron becomes positive because of the sudden entrance of sodium ions. If threshold is reached, many more sodium channels open, and the action potential begins. As sodium ions rapidly move across the membrane to the inside of the axon, the action potential swings up from −70 mV to +35 mV (Fig. 37.5*c*). This reversal in polarity causes the sodium channels to close and the potassium channels to open. As potassium ions leave the axon, the membrane potential swings down from +35 mV to −70 mV. In other words, a *repolarization* occurs (Fig. 37.5*d*). An action potential takes only 2 msec (milliseconds). To visualize such rapid fluctuations in voltage across the axonal membrane, researchers generally find it useful to plot the voltage changes over time (Fig. 37.5*e*).

Animation Nerve Impulse

Propagation of Action Potentials

In nonmyelinated axons (such as sensory receptors in the skin), the action potential travels down an axon one small section at a time, at a speed of about 1 m/sec (meter per second). In myelinated axons, the gated ion channels that produce an action potential are concentrated at the nodes of Ranvier. *Saltar* in Spanish means "to jump," so this mode of conduction, called **saltatory conduction,** means that the action potential "jumps" from node to node:

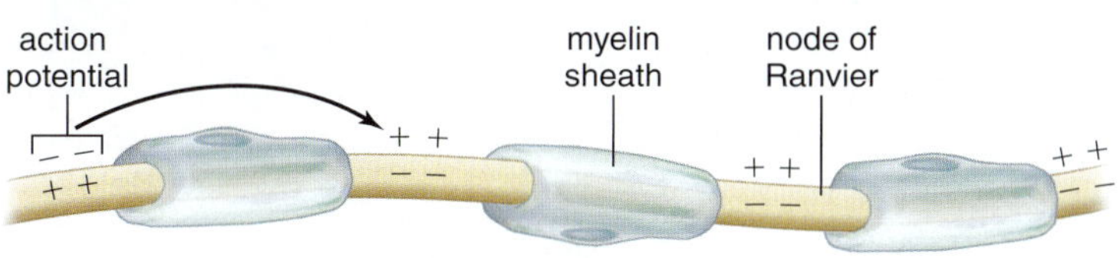

Speeds of 200 m/sec (about 450 miles per hour) have been recorded. As you can see, this speed is considerably greater than the rate of travel in nonmyelinated axons and allows what seems to be an instantaneous response.

As soon as an action potential has moved on, the previous section undergoes a **refractory period,** during which the Na^+ gates are unable to open. Notice, therefore, that the action potential cannot move backward and instead always moves down an axon toward its terminals. The intensity of a signal traveling down a nerve fiber is determined by how many nerve impulses are generated within a given time span.

Transmission Across a Synapse

Every axon branches into many fine endings, each tipped by a small swelling, called an axon terminal (Fig. 37.6). Each terminal lies very close to the dendrite (or the cell body) of another neuron. This region of close proximity is called a **synapse.** At a synapse, the membrane of the first neuron is called the *pre*synaptic membrane, and the membrane of the next neuron is called the *post*synaptic membrane. The small gap between the neurons is called the **synaptic cleft.**

MP3 Synapses

A nerve impulse cannot cross a synaptic cleft. Transmission across a synapse is carried out by molecules called **neurotransmitters,** which are stored in synaptic vesicles. When nerve impulses traveling along an axon reach an axon terminal, gated channels for calcium ions (Ca^{2+}) open, and calcium enters the terminal. This sudden rise in Ca^{2+} stimulates synaptic vesicles to merge with the presynaptic membrane, and neurotransmitter molecules are released into the synaptic cleft. They diffuse across the cleft to the postsynaptic membrane, where they bind with specific receptor proteins.

Depending on the type of neurotransmitter and/or the type of receptor, the response of the postsynaptic neuron can be toward

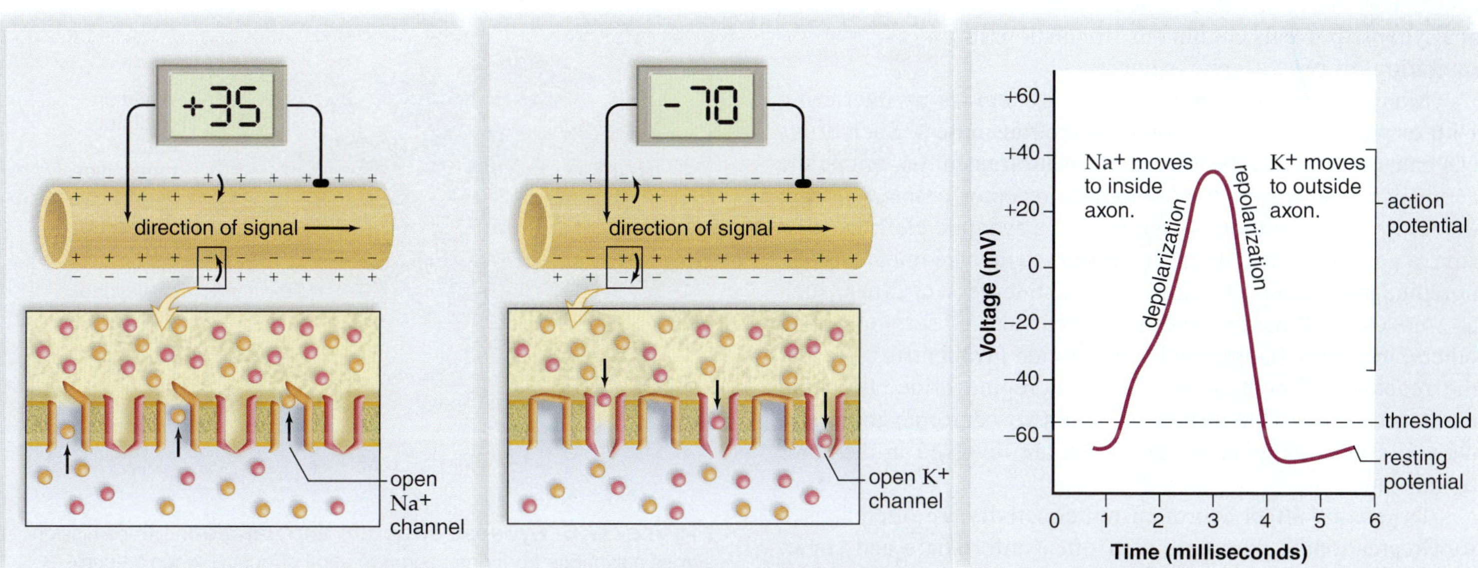

c. Depolarization continues as Na^+ gates open and Na^+ moves inside the axon.

d. Action potential ends: Repolarization occurs when K^+ gates open and K^+ moves to outside the axon. The sodium-potassium pump returns the ions to their resting positions.

e. An action potential can be visualized if voltage changes are graphed over time.

excitation or toward inhibition. Excitatory neurotransmitters that use gated ion channels are fast-acting. Other neurotransmitters affect the metabolism of the postsynaptic cell and therefore are slower-acting.

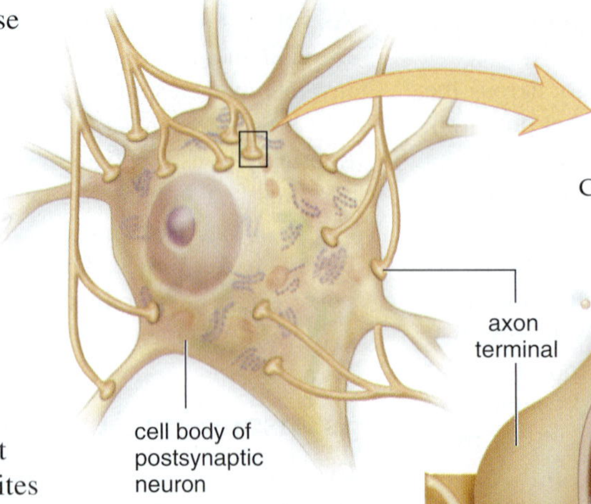

Neurotransmitters

More than 100 substances are known or suspected to be neurotransmitters in both the CNS and the PNS. Many of these can have opposing effects on different tissues. *Acetylcholine* (*ACh*) excites skeletal muscle but inhibits cardiac muscle. It has either an excitatory or an inhibitory effect on smooth muscle or glands, depending on their location. In the CNS, *norepinephrine* is important to dreaming, waking, and mood. *Dopamine* is involved in emotions, learning, and attention, and *serotonin* is involved in thermoregulation, sleeping, emotions, and perception. *Endorphins* are neurotransmitters that bind to natural opioid receptors in the brain. They are associated with the "runner's high" of exercisers, because they also produce a feeling of tranquility. Endorphins are produced by the brain not only when there is physical stress but also when emotional stress is present.

After a neurotransmitter has been released into a synaptic cleft and has initiated a response, it is removed from the cleft. In some synapses, the postsynaptic membrane contains enzymes that rapidly inactivate the neurotransmitter. For example, the enzyme *acetylcholinesterase* (*AChE*) breaks down acetylcholine. In other synapses, the presynaptic cell is responsible for reuptake, a process in which it rapidly reabsorbs the neurotransmitter, possibly for repackaging in synaptic vesicles or for molecular breakdown. The short existence of neurotransmitters at a synapse prevents continuous stimulation (or inhibition) of postsynaptic membranes.

Animation Chemical Synapses

Many drugs affecting the nervous system act by interfering with or potentiating the action of neurotransmitters. Such drugs can enhance or block the release of a neurotransmitter, mimic the action of a neurotransmitter or block the receptor, or interfere with the removal of a neurotransmitter from a synaptic cleft. Depression, a common mood disorder, appears to involve imbalances in norepinephrine and serotonin. Some antidepressant drugs, such as fluoxetine (Prozac), prevent the reuptake of serotonin, and others, including bupropion hydrochloride (Wellbutrin), prevent the reuptake of both serotonin and norepinephrine. Blocking reuptake prolongs the effects of these two neurotransmitters in networks of neurons in the brain that are involved in the emotional state.

Drugs that affect neurotransmitter activity are often abused for "recreational" purposes, with often unfortunate and sometimes deadly results. The Biological Systems feature, "Drugs of Abuse," on page 700 describes a number of these dangerous drugs.

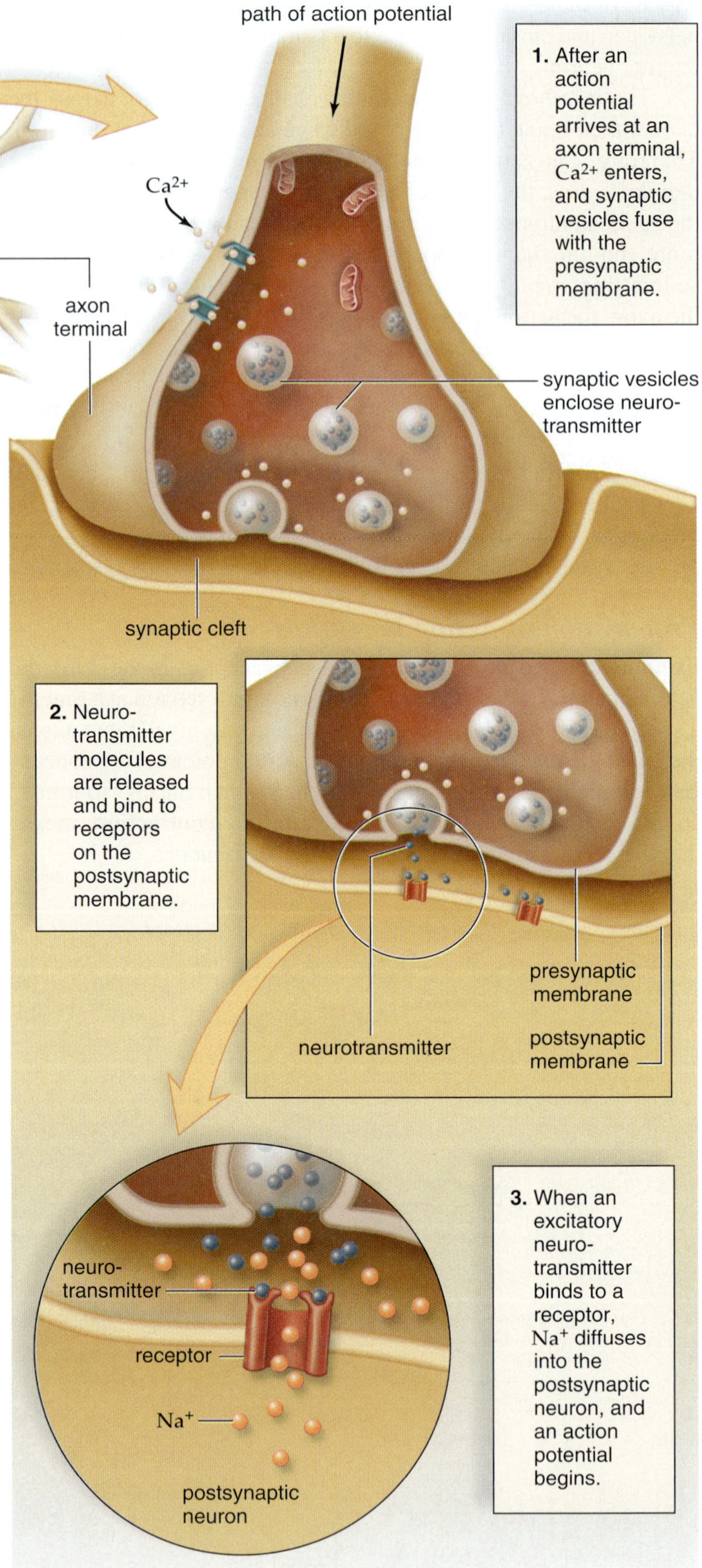

Figure 37.6 Synapse structure and function. Transmission across a synapse from one neuron to another occurs when a neurotransmitter is released at the presynaptic membrane, diffuses across a synaptic cleft, and binds to a receptor in the postsynaptic membrane. An action potential may begin.

Tutorial Synaptic Cleft

Synaptic Integration

A single neuron has many dendrites plus the cell body, and both can have synapses with many other neurons. One thousand to 10,000 synapses per single neuron is not uncommon. Therefore, a neuron is on the receiving end of many excitatory and inhibitory signals. An excitatory signal produces a potential change that causes the neuron to become less polarized, or closer to triggering an action potential. An inhibitory signal causes the neuron to become hyperpolarized, or farther from an action potential.

Neurons integrate these incoming signals, and they do so specifically at the area of the neuron cell body where the axon emerges, called the *axon hillock*. **Integration** is the summing up of excitatory and inhibitory signals (Fig. 37.7). If a neuron receives many excitatory signals (either from different synapses or at a rapid rate from one synapse), chances are the axon will transmit a nerve impulse. In Figure 37.7*b*, the inhbitory signals (shown in blue) are canceling out the excitatory signals, resulting in no nerve impulse.

Check Your Progress 37.2

1. Explain why a nerve impulse travels more quickly down a myelinated axon than down an unmyelinated axon.
2. Describe the movement of specific ions during the generation of a nerve impulse.
3. Analyze how the bite of a black widow spider, which contains a powerful AChE inhibitor, might cause each of the common symptoms of muscle cramps, salivation, fast heart rate, and high blood pressure.

37.3 The Central Nervous System

Learning Outcomes

Upon completion of this section, you should be able to

1. Describe the anatomy of the spinal cord and spinal nerves.
2. List the major regions of the human brain and describe some major functions of each.
3. Compare the causes and types of symptoms seen in some common CNS disorders.

The central nervous system (CNS) consists of the spinal cord and brain. It has three specific functions:

1. *Receives sensory input*—sensory receptors in the skin and other organs respond to external and internal stimuli by generating nerve impulses that travel to the CNS
2. *Performs integration*—the CNS sums up the input it receives from all over the body
3. *Generates motor output*—nerve impulses from the CNS go to the muscles and glands; muscle contractions and gland secretions are responses to stimuli received by sensory receptors

As an example of the operation of the CNS, consider the events that occur as a person raises a glass to the lips. Continuous sensory input to the CNS from the eyes and hand informs the CNS of the position of the glass, and the CNS continually sums up the incoming data before commanding the hand to proceed. At any time, integration with other sensory data might cause the CNS to

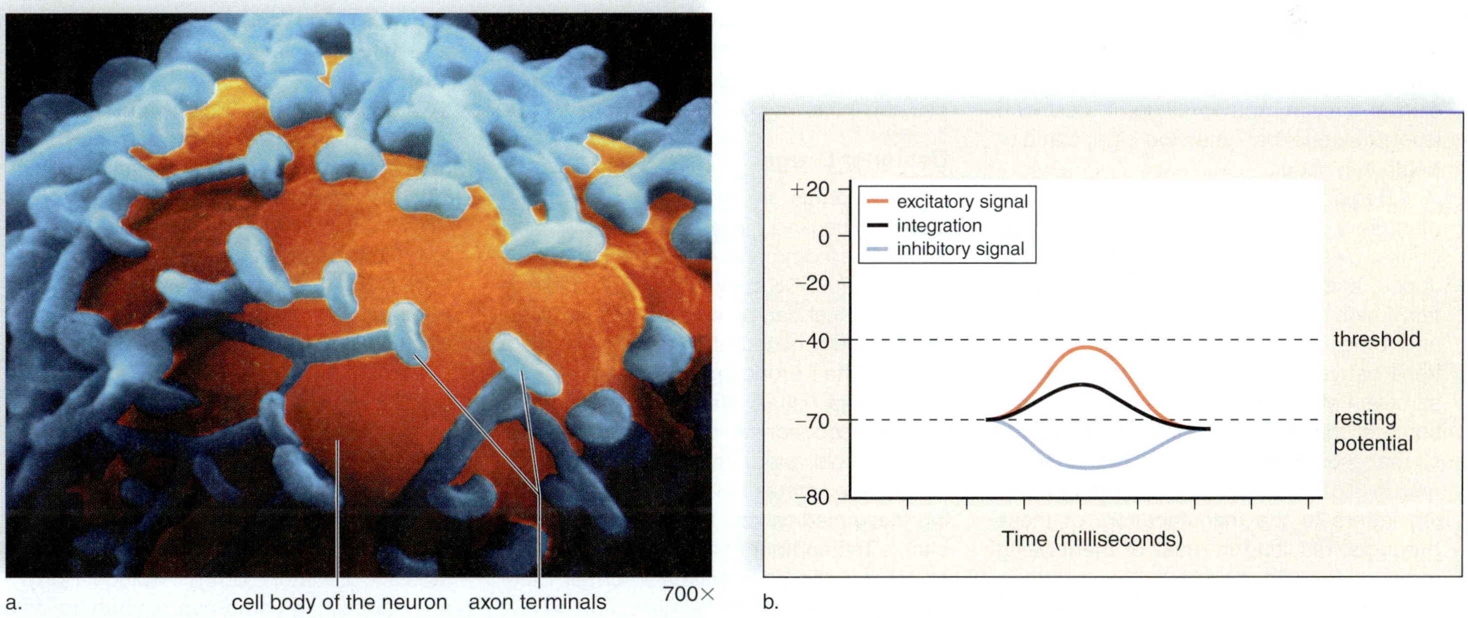

Figure 37.7 Synaptic integration. **a.** Many neurons synapse with a cell body. **b.** Both inhibitory signals (blue) and excitatory signals (red) are summed up in the dendrite and cell body of the postsynaptic neuron. Only if the combined signals cause the membrane potential to rise above threshold does an action potential occur. In this example, threshold was not reached.

THEME Biological Systems

Drugs of Abuse

Drug abuse is apparent when a person takes a drug at a dose level and under circumstances that increase the potential for a harmful outcome. Addiction is present when more and more of the drug is needed to get the same effect and withdrawal symptoms occur when the user stops taking the drug. This is true not only for teenagers and adults but also for newborn babies of mothers who are addicted to drugs.

Alcohol

With the exception of caffeine, alcohol (ethanol) consumption is the most socially accepted form of drug use in the United States, although consuming alcohol is illegal for those under 21. According to a 2011 survey, 39% of all U.S. high school students drank some amount of alcohol, and 22% reported hazardous drinking (five or more drinks in one setting) during the 30 days preceding the survey. Notably, 80% of college-age young adults drink. According to a U.S. government study, drinking in college contributes to an estimated 1,825 student deaths, 690,000 incidents of assault, and over 95,000 cases of sexual assault or date rape each year.

Alcohol acts as a *depressant* on many parts of the brain by increasing the action of GABA, an inhibitory neurotransmitter. Depending on the amount consumed, the effects of alcohol on the brain can lead to a feeling of relaxation, lowered inhibitions, impaired concentration and coordination, slurred speech, and vomiting. If the blood level of alcohol becomes too high, coma or death can occur.

Beginning in about 2005, several manufacturers began selling alcoholic energy drinks. With names such as Four Loco, Joose, and Sparks, these drinks combine fairly high levels of alcohol with caffeine and other ingredients. Although interactions between drugs can be complex, the stimulant effects of caffeine can counteract some of the depressant effects of alcohol, so that users feel able to drink more. Beginning in 2010, the FDA began sending warning letters to the manufacturers of these products, resulting in most of them being removed from the market.

Nicotine

About 23% of U.S. high school students, and 7% of middle school students, reported using tobacco products in 2012. The use of chewing tobacco is about half of what it was in the mid-1990s, but from 2011 to 2012, electronic cigarette use doubled among middle and high school students (see the Nature of Science feature "Is 'Vaping' Safer Than Smoking?" in Chapter 35).

When tobacco is smoked or chewed, nicotine is rapidly delivered throughout the body. It causes a release of epinephrine from the adrenal glands, increasing blood sugar levels and initally causing a feeling of stimulation. As blood sugar falls, depression and fatigue set in, causing the user to seek more nicotine. In the CNS, nicotine stimulates neurons to release dopamine, a neurotransmitter that promotes a temporary sense of pleasure, reinforcing dependence on the drug. About 70% of people who try smoking become addicted.

As mentioned in earlier chapters, smoking is strongly associated with serious diseases of the cardiovascular and respiratory systems. Once addicted, however, only 10–20% of smokers are able to quit. Most medical approaches to quitting involve the administration of nicotine in safer forms, such as skin patches, gum, or a nicotine inhaler, so that withdrawal symptoms can be minimized while dependence is gradually reduced. In a 2011 trial, a vaccine called NicVAX, which stimulates the production of antibodies that prevent nicotine from entering the brain, provided no advantage over a placebo in helping people give up smoking.

Designer Drugs

Designer drugs are those that are synthesized, or "cooked," in laboratories. Methamphetamine (commonly called meth or crank) is a powerful CNS stimulant. It is available as a powder that can be snorted or as crystals (crystal meth or ice) that can be smoked. Meth is often produced in makeshift home laboratories, usually starting with ephedrine or pseudoephedrine, common ingredients in many cold and asthma medicines. As a result, many states have passed laws making these medications more difficult to purchase. The number of toxic chemicals used to prepare the drug makes a former meth lab site hazardous to humans and to the environment. About 11 million people in the United States have used methamphetamine at least once in their lifetime.

The structure of methamphetamine is similar to that of dopamine, and the most immediate effect of taking meth is a rush of euphoria, energy, alertness, and elevated mood. However, this is typically followed by a state of agitation, which in some individuals leads to violent behavior. Chronic use can result in what is called amphetamine psychosis, characterized by paranoia, hallucinations, irritability, and aggressive, erratic behavior.

Ecstasy, "X", and "molly" are common names for MDMA (methylenedioxymethamphetamine), which is chemically similar to methamphetamine. Many users say that X, taken as a pill that looks like an aspirin or candy, increases their feelings of well-being and love for other people. However, it has many of the same side effects as other stimulants; plus, it can interfere with the body's temperature regulation, leading to hyperthermia, high blood pressure, and seizures.

"Bath salts" resemble various inorganic salts that are actually added to bath water, but the effects of smoking, snorting, or injecting these highly addictive substances are anything but relaxing. After an initial euphoria, common side effects of these drugs include high fever, hallucinations, and extreme paranoia. Horrifying stories have also appeared in the news about users inflicting a variety of violent acts on themselves or others.

As recently as 2010, bath salts were legal and available from head shops, convenience stores, and online, with innocuous-sounding brand names such as Bliss, Cloud Nine, and Vanilla Sky. In 2011, the FDA banned two key ingredients—MDPV (methylenedioxypyrovalerone) and mephedrone, but designer drug cookers try to stay ahead of the law by slightly altering the ingredients. As a result, it's difficult to predict what chemicals might be found in bath salts, but it's safe to say that abusing these drugs often does not turn out well.

Date Rape Drugs

Drugs with sedative effects, known as date rape or predatory drugs, include Rohypnol (roofies), gamma-hydroxybutyric acid (GHB), and ketamine ("special K"). Ketamine is a drug that veterinarians sometimes use to perform surgery on animals.

Any of these drugs can be given to an unsuspecting person, who may fall into a dreamlike state and be unable to move and thus vulnerable to sexual assault.

Cocaine and Crack

Cocaine is an alkaloid derived from the shrub *Erythroxylon coca.* Approximately 35 million Americans have used cocaine by sniffing/snorting, injecting, or smoking it. Cocaine is a powerful *stimulant* in the CNS that interferes with the reuptake of dopamine at synapses, increasing overall brain activity. The result is a rush of well-being that lasts 5–30 minutes. This is followed by a crash period, characterized by fatigue, depression, irritability, and lack of interest in sex. In fact, men who use cocaine often become impotent.

Crack is the street name given to cocaine that is processed to a free base form for smoking. The term *crack* refers to the crackling sound heard when the drug is smoked. Smoking allows high doses of the drug to reach the brain rapidly, providing an intense and immediate high, or "rush." Approximately 8 million Americans use crack.

Cocaine is highly addictive; with continued use, the brain makes less dopamine to compensate for a seemingly endless supply. The user experiences withdrawal symptoms and an intense craving for the drug. Over time, the brain of a cocaine user becomes less active (Fig. 37A).

Heroin

Heroin is derived from the resin or sap of the opium poppy plant, which is widely grown—from Turkey to Southeast Asia and parts of Latin America. Drugs derived from opium are called opiates, a class that also includes morphine and oxycodone. The number of heroin users in the United States has nearly doubled in the last few years, partly due to an increased supply.

As with other drugs of abuse, addiction is common. Heroin binds to receptors in the brain that normally bind to endorphins, naturally occurring neurotransmitters that kill pain and produce feelings of tranquility. After heroin is injected, snorted, or smoked, a feeling of euphoria, along with the relief of any pain, occurs within a few minutes. With repeated heroin use, the body's production of endorphins decreases. Tolerance develops, so that the user needs to take more of the drug just to prevent withdrawal symptoms (tremors, restlessness, cramps, vomiting), and the original euphoria is no longer felt.

Long-term users commonly acquire hepatitis, HIV/AIDS, and various bacterial infections due to the use of shared needles, and heavy users may experience convulsions and death by respiratory arrest. Some well-publicized recent cases of heroin overdose include the actors Cory Monteith and Philip Seymour Hoffman.

Heroin addiction can be treated with synthetic opiate compounds, such as methadone or suboxone, that decrease withdrawal symptoms and block heroin's effects. However, methadone itself can be addictive, and methadone-related deaths are on the rise.

Marijuana and K2

The dried flowering tops, leaves, and stems of the marijuana plant, *Cannabis sativa,* contain and are covered by a resin that is rich in THC (tetrahydrocannabinol). Marijuana can be ingested, but usually it is smoked in a cigarette called a "joint," or in pipes or other paraphernalia. An estimated 65 million Americans have used marijuana, making it the most commonly used illegal drug in the United States. As of early 2014, about 20 states have legalized its use for medical purposes, such as treating cancer, AIDS, or glaucoma, and two states (Colorado and Washington) have legalized its recreational use.

Researchers have found that, in the brain, THC binds to a receptor for anandamide, a naturally occurring neurotransmitter that is important for short-term memory processing, and perhaps for feelings of contentment. The occasional marijuana user experiences mild euphoria, along with alterations in vision and judgment. Heavy use can cause hallucinations, anxiety, depression, paranoia, and psychotic symptoms. Some researchers believe that long-term marijuana use leads to brain impairment.

In recent years, awareness has been increasing about a synthetic compound (designer drug) called K2 or spice. Compounds in K2 may be 10–100 times stronger than THC. The chemical is typically sprayed onto a mixture of other herbal products and smoked. However, because there is no regulation of how it is produced, the amount or types of chemicals in K2 can vary greatly. This may account for the several reports of serious medical problems and even deaths in K2 users.

Questions to Consider

1. Suppose a form of heroin had only the desired effects (euphoria and pain relief) with no side effects. Should such a drug be legal for everyone to use?
2. Should medical marijuana be legal for use in all states? If so, how should it be regulated?
3. In November 2010, the U.S. Drug Enforcement Agency banned the sale of five chemicals used to make K2. Is this an overreaction?

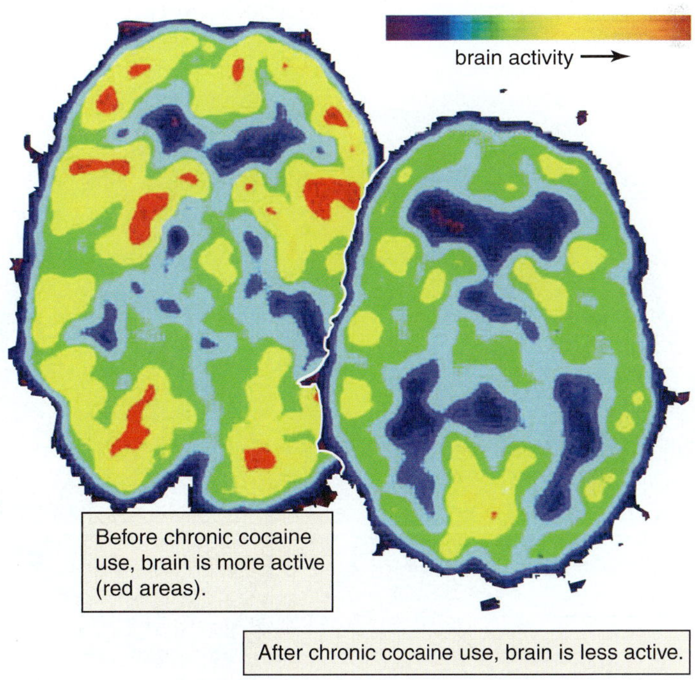

Figure 37A Drug use. Brain activity before and after the use of cocaine.

command a different motion instead. The lips detect the arrival of the glass, passing this information to the CNS, which then directs the actions of drinking.

MP3 Organization of the Nervous System

The spinal cord and the brain are both protected by bone; the spinal cord is surrounded by vertebrae, and the brain is enclosed by the skull. Both the spinal cord and the brain are wrapped in three protective membranes known as **meninges** (Fig. 37.8). The spaces between the meninges are filled with **cerebrospinal fluid,** which cushions and protects the CNS. Cerebrospinal fluid, produced by a type of glial cell, is contained in the central canal of the spinal cord and within the **ventricles** of the brain, which are interconnecting spaces that produce and serve as reservoirs for cerebrospinal fluid. **Meningitis** (inflammation of the meninges) is a serious disorder caused by a number of bacteria or viruses that can invade the meninges.

The Spinal Cord

The **spinal cord** is a bundle of nervous tissue enclosed in the vertebral column (see Fig. 37.12); it extends from the base of the brain to the vertebrae just below the rib cage. The spinal cord has two main functions: (1) it is the center for many **reflex actions,** which are automatic responses to external stimuli, and (2) it provides a means of communication between the brain and the spinal nerves, which leave the spinal cord.

A cross section of the spinal cord reveals that it is composed of a central portion of gray matter and a peripheral region of white matter. The **gray matter** consists of cell bodies and unmyelinated fibers. In cross section, it is shaped like a butterfly, or the letter *H,* with two dorsal (posterior) horns and two ventral (anterior) horns surrounding a central canal. The gray matter contains portions of sensory neurons and motor neurons, as well as short interneurons that connect sensory and motor neurons.

Myelinated long fibers of interneurons that run together in bundles called **tracts** give **white matter** its color. These tracts connect the spinal cord to the brain. They are like a busy superhighway, by which information continuously passes between the brain and the rest of the body. In the dorsal part of the cord, the tracts are primarily ascending, taking information *to* the brain. Ventrally, the tracts are primarily descending, carrying information *from* the brain. Because the tracts cross over at one point, the left side of the brain controls the right side of the body and the right side of the brain controls the left side of the body. Researchers estimate that there are about 100,000 miles of myelinated nerve fibers in the adult human brain.

If the spinal cord is damaged as a result of an injury, paralysis may result. If the injury occurs in the cervical (neck) region, all four limbs are usually paralyzed, a condition known as quadriplegia. If the injury occurs in the thoracic region, the lower body may be paralyzed, a condition called paraplegia.

Other disease processes can cause paralysis. In **amyotrophic lateral sclerosis** (**ALS**), or Lou Gehrig disease, motor neurons in the brain and spinal cord degenerate and die, leaving patients weakened, then paralyzed, then unable to breathe properly. Although there is no cure, some drugs slow the disease, and others are currently in clinical trials.

The Brain

Nerve impulses are the same in all neurons, so how is it that the stimulation of our eyes causes us to see and the stimulation of our ears causes us to hear? Essentially, the central nervous system carries out the function of integrating incoming data. The **brain** allows us to perceive our environment, to reason, and to remember.

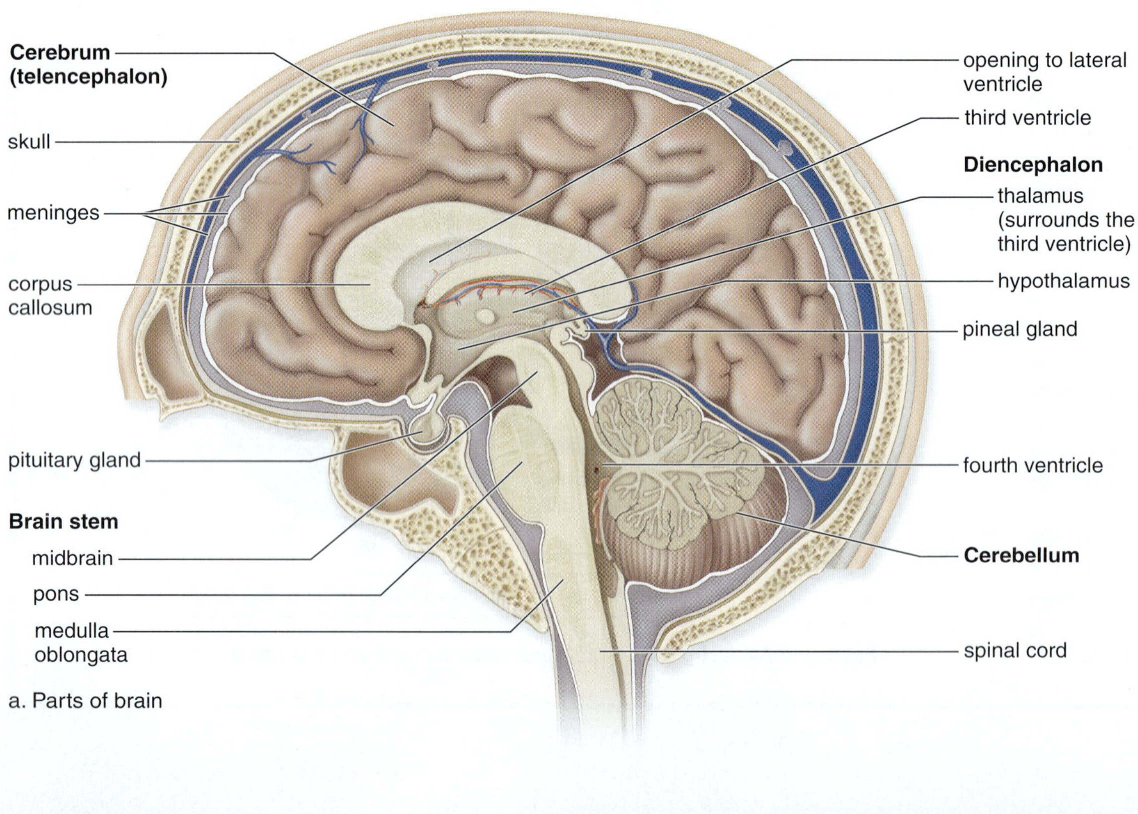

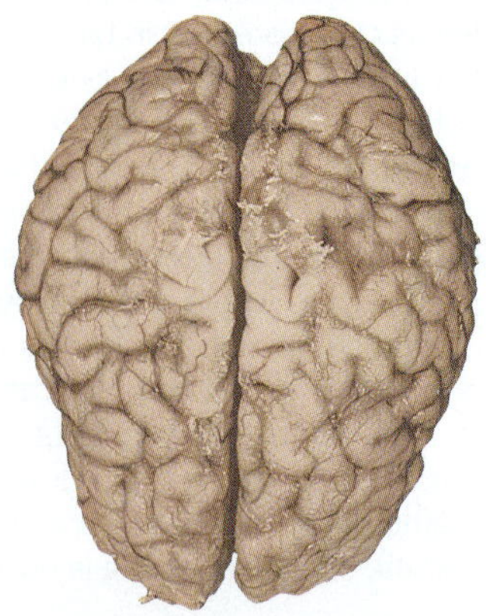

Figure 37.8 The human brain. a. The right cerebral hemisphere is shown, along with other, closely associated structures. The hemispheres are connected by the corpus callosum. **b.** The cerebrum is divided into the right and left cerebral hemispheres.

The exact processes by which the brain generates these higher functions remain largely mysterious.

MP3 The Brain

The Cerebrum

The **cerebrum** is the largest, outermost portion of the brain in humans (Fig. 37.8*a*). The cerebrum is the last center to receive sensory input and carry out integration before commanding voluntary motor responses. It communicates with and coordinates the activities of the other parts of the brain. The cerebrum also contains the two lateral ventricles; the third ventricle is surrounded by the diencephalon, and the fourth ventricle lies between the cerebellum and the pons.

MP3 The Cerebrum

Cerebral Hemispheres. The cerebrum is divided into two halves, called **cerebral hemispheres** (Fig. 37.8*b*). A deep groove called the *longitudinal fissure* divides the cerebrum into the right and left hemispheres. Each hemisphere receives information from and controls the opposite side of the body. Although the hemispheres appear the same, the right hemisphere is associated with artistic and musical ability, emotion, spatial relationships, and pattern recognition. The left hemisphere is more adept at mathematics, language, and analytical reasoning. The two cerebral hemispheres are connected by a bridge of tracts within the *corpus callosum.*

Shallow grooves called *sulci* (sing., *sulcus*) divide each hemisphere into paired lobes (Fig. 37.9). The *frontal lobes* lie toward the front of the hemispheres and are associated with motor control, memory, reasoning, and judgment. For example, if a fire occurs, the frontal lobes enable you to decide whether to exit via the stairs or the window, or if it is winter how to dress if the temperature plummets to subzero. The left frontal lobe contains *Broca's area,* which organizes motor commands to produce speech.

The *parietal lobes* lie posterior to the frontal lobe and are concerned with sensory reception and integration, as well as taste. A *primary taste area* in the parietal lobe accounts for taste sensations.

The *temporal lobes* are located laterally. A primary auditory area in each temporal lobe receives information from the ears. The *occipital lobes* are the most posterior lobes. A *primary visual area* in each occipital lobe receives information from the eyes.

The Cerebral Cortex. The **cerebral cortex** is a thin (less than 5 mm thick) but highly convoluted outer layer of gray matter that covers the cerebral hemispheres. The convolutions increase the surface area of the cerebral cortex. It contains tens of billions of neurons and is the region of the brain that accounts for sensation, voluntary movement, and all the thought processes required for learning, memory, language, and speech.

Two regions of the cerebral cortex are of particular interest. The *primary motor area* is in the frontal lobe just ventral to (before) the central sulcus. Voluntary commands to skeletal muscles begin in the primary motor area, and each part of the body is controlled by a certain section. The size of the section indicates the precision of motor control. For example, controlling the muscles of the face and hands takes up a much larger portion of the primary motor area than controlling the entire trunk. The *primary somatosensory area* is just dorsal to the central sulcus in the parietal lobe. Sensory information from the skin and skeletal muscles arrives here.

When the blood supply to any area of the brain is disrupted, a **stroke** results. Stroke is the third leading cause of death in the United States. The most common type is *ischemic* stroke, in which there is a sudden loss of blood supply to an area of the brain, usually due to arterial blockage or clot formation. The area(s) of the brain affected by a stroke will determine what type of symptoms arise. For example, a stroke that affects only the motor areas of the cerebral

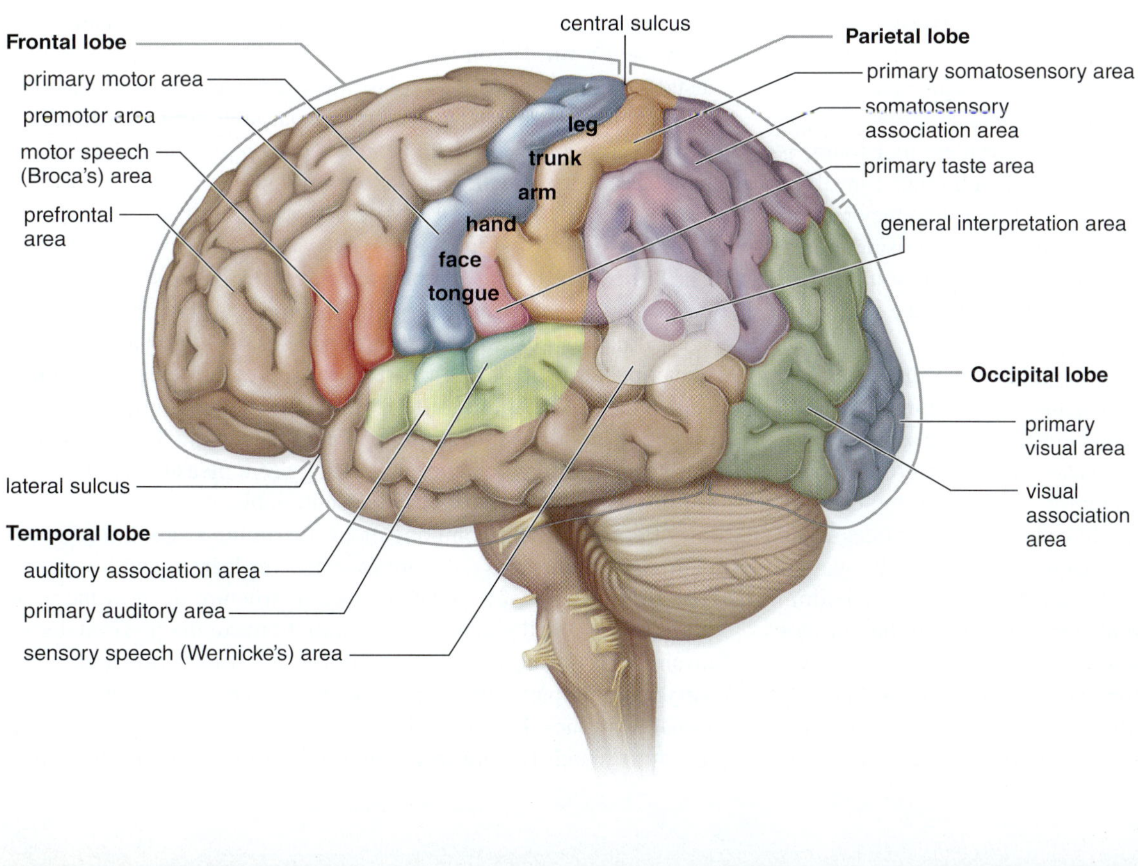

Figure 37.9 The lobes of a cerebral hemisphere. Each cerebral hemisphere is divided into four lobes: frontal, parietal, temporal, and occipital. These lobes contain centers for reasoning and movement, somatic sensing, hearing, and vision, respectively.

cortex might paralyze one side of the body, while a stroke involving Broca's area might render a stroke victim unable to speak.

Although strokes are most common in older people, a 2011 study noted a 51% increase in strokes in men ages 15 through 34, as well as a 17% increase in women the same age. Because many of the risk factors for stroke are similar to those for cardiovascular disease, see the Nature of Science feature "Recent Findings About Preventing Cardiovascular Disease" in Chapter 32 to learn how to reduce your risk.

Basal Nuclei. Although the bulk of the cerebrum is composed of white matter (tracts), masses of gray matter are located deep within the white matter. These so-called **basal nuclei** (basal ganglia) integrate motor commands, ensuring that proper muscle groups are activated or inhibited. As mentioned in the chapter-opening essay about Michael J. Fox, **Parkinson disease (PD)** is a brain disorder characterized by tremors, speech difficulties, and difficulty standing and walking. PD results from a loss of the cells in the basal nuclei that normally produce the neurotransmitter dopamine. See the Nature of Science feature, "An Accidental Experimental Model for Parkinson Disease," on page 706 to learn about the somewhat strange history of research into PD, as well as some treatments for this disease.

Other Parts of the Brain

The hypothalamus and the thalamus are in the *diencephalon,* a region that encircles the third ventricle. The **hypothalamus** forms the floor of the third ventricle. It is an integrating center that helps maintain homeostasis by regulating hunger, sleep, thirst, body temperature, and water balance. The hypothalamus controls the pituitary gland and, thereby, serves as a link between the nervous and endocrine systems (see Chapter 40).

The **thalamus** consists of two masses of gray matter in the sides and roof of the third ventricle. It receives all sensory input except smell. The thalamus integrates this information and sends it on to the appropriate portions of the cerebrum. For this reason, the thalamus is often referred to as the gatekeeper for sensory information en route to the cerebral cortex. The thalamus also participates in higher mental functions, such as memory and emotions.

The **pineal gland,** which secretes the hormone melatonin, is also located in the diencephalon. *Melatonin* is a hormone involved in maintaining a normal sleep-wake cycle. It is sometimes recommended for people suffering from insomnia, but many side effects can occur. Relatively high levels of melatonin are a key ingredient in the "relaxation brownies" that are sold at convenience stores and online; many states are banning their sale, however.

The **cerebellum** lies under the occipital lobe of the cerebrum and is separated from the brain stem by the fourth ventricle. It is the largest part of the hindbrain. The cerebellum receives sensory input from the eyes, ears, joints, and muscles about the present position of body parts, and it receives motor output from the cerebral cortex about where those parts should be located. After integrating this information, the cerebellum sends motor impulses by way of the brain stem to the skeletal muscles. In this way, the cerebellum maintains posture and balance. It also ensures that all the muscles work together to produce smooth, coordinated voluntary movements, such as those in playing the piano or hitting a baseball.

The **brain stem** contains the midbrain, the pons, and the medulla oblongata (see Fig. 37.8). The **midbrain** acts as a relay station for tracts passing between the cerebrum and the spinal cord or cerebellum. The tracts cross in the brain stem, so the right side of the body is controlled by the left portion of the brain and the left side of the body is controlled by the right portion of the brain.

The **pons** (L. *pons,* "bridge") contains bundles of axons that form a bridge between the cerebellum and the rest of the CNS. The pons also works with the medulla oblongata to regulate many basic body functions.

The **medulla oblongata** lies just superior to the spinal cord, and it contains tracts that ascend or descend between the spinal cord and higher brain centers. It regulates the heartbeat, breathing, swallowing, and blood pressure. It also contains reflex centers for vomiting, coughing, sneezing, hiccuping, and swallowing.

The most common neurological disease of young adults is **multiple sclerosis (MS)**. It typically affects myelinated nerves in the cerebellum, brain stem, basal ganglia, and optic nerve. MS is considered an autoimmune disease, in which the patient's own white blood cells attack the myelin, oligodendrocytes, and eventually neurons in the CNS. The word *sclerosis* refers to the multiple scars, or plaques, that can be seen through various types of scans. The myelin damage affects the transmission of nerve impulses, resulting in the most common symptoms: fatigue, vision problems, weakness, numbness, and tingling. Nearly 400,000 people in the United States have MS, and about 10,000 new cases are diagnosed each year, mainly in young adults.

The Reticular Activating System. The **reticular activating system (RAS)** contains the reticular formation, a complex network of nuclei and nerve fibers extending the length of the brain stem (Fig. 37.10). The reticular formation receives sensory signals, which it sends up to higher centers, and motor signals, which it sends to the spinal cord.

The RAS arouses the cerebrum via the thalamus and causes a person to be alert. Apparently, the RAS can filter out unnecessary sensory stimuli, which explains why some individuals can study with the TV on. If you want to awaken the RAS, surprise it with a sudden stimulus, such as splashing your face with cold water; if you want to deactivate it, remove visual and auditory stimuli. A severe injury to the RAS can cause a person to be comatose, from which recovery may be impossible.

The Limbic System

The **limbic system** is a complex group of brain structures that lie just under the cortex, near the thalamus. Although definitions vary somewhat, the limbic system includes the hypothalamus, hippocampus, amygdala, olfactory bulb, and other nearby structures (Fig. 37.11). The limbic system blends higher mental functions and primitive emotions into a united whole. It accounts for why activities such as sexual behavior and eating seem pleasurable and why, for instance, mental stress can cause high blood pressure.

Two significant components of the limbic system, the hippocampus and the amygdala, are essential for learning and memory. The **hippocampus,** a seahorse-shaped structure deep in the temporal lobe, is well situated in the brain to make the prefrontal area aware of past experiences stored in sensory association areas. The **amygdala,** in particular, can cause these experiences to have emotional overtones. For example, the smell of smoke may serve as an alarm to search for fire in the house. The inclusion of the frontal

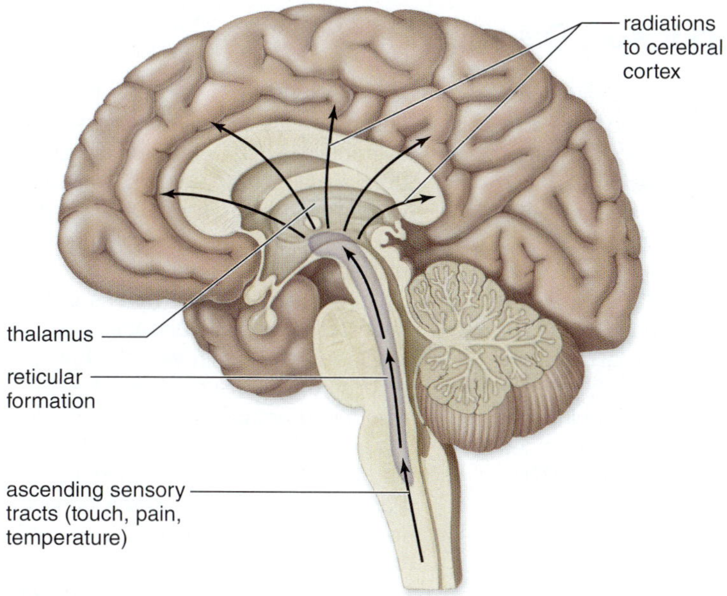

Figure 37.10 The reticular activating system. The reticular formation receives and sends on motor and sensory information to various parts of the CNS. One portion, the reticular activating system (RAS) (arrows), arouses the cerebrum and, in this way, controls alertness versus sleep.

lobe in the limbic system gives us the ability to restrain ourselves from acting out on strong feelings by using reason.

Learning and Memory. **Memory** is the ability to hold a thought in mind or recall events from the past, ranging from a word we learned only yesterday to an early emotional experience that has shaped our lives. Learning takes place when we retain and use memories.

The prefrontal area in the frontal lobe is active during short-term memory, as when we temporarily recall a phone number. However, some phone numbers go into long-term memory. Think of a phone number you know by heart, and see if you can bring it to mind without also thinking about the place or person associated with the number. Most likely you cannot, because typically long-term memory is a mixture of what is called semantic memory (numbers, words, etc.) and episodic memory (persons, events, etc.). Skill memory is a type of memory that can exist independently of episodic memory. Skill memory is being able to perform motor activities, such as riding a bike or playing ice hockey.

What parts of the brain are functioning when you remember something from long ago? The hippocampus gathers long-term memories, which are stored in bits and pieces throughout the sensory association areas, and makes them available to the frontal lobe. Why are some memories so emotionally charged? The amygdala is responsible for fear conditioning and associating danger with sensory information received from the thalamus and the cortical sensory areas.

Several diseases of the brain can affect memory. **Alzheimer disease (AD)** is the most common cause of dementia, or a loss of reasoning, memory, and other higher brain functions, especially in people over age 65. AD patients have abnormal neurons throughout the brain, but especially in the hippocampus and amygdala. These neurons have two abnormalities: (1) plaques, containing a protein called beta amyloid, which accumulate around the axons, and (2) neurofibrillary tangles (bundles of fibrous protein) surrounding the nucleus. The cause of these protein abnormalites is unknown, although several genes that predispose a person to develop AD have been identified. Although no cure is available, most of the drugs that are currently approved to treat symptoms of AD are cholinesterase inhibitors, which effectively increase the levels of acetylcholine in the AD patient's brain. This in turn can improve learning and memory, at least temporarily.

Check Your Progress 37.3

1. Trace the path of a nerve impulse from a stimulus in an internal organ (such as food in the large intestine stimulating peristalsis) to the brain and back.
2. Name the four major lobes of the human brain.
3. List at least two common diseases of the CNS, and describe their symptoms and causes.

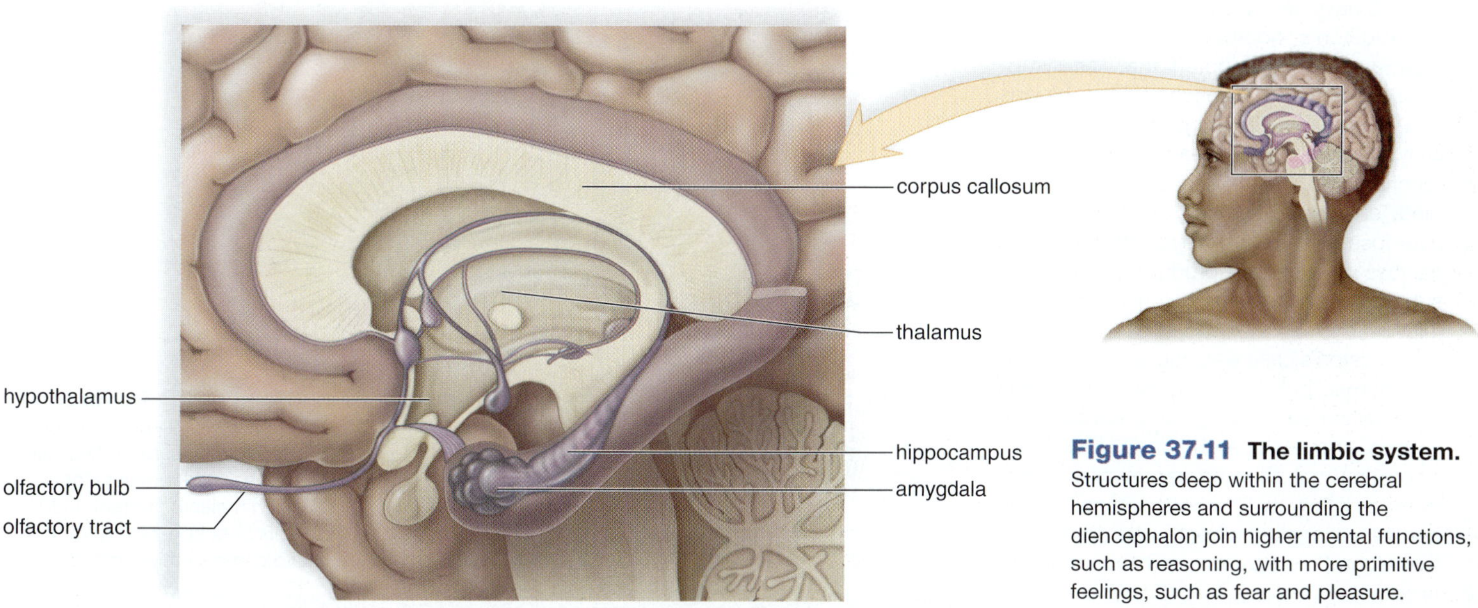

Figure 37.11 The limbic system. Structures deep within the cerebral hemispheres and surrounding the diencephalon join higher mental functions, such as reasoning, with more primitive feelings, such as fear and pleasure.

THEME **Nature of Science**

An Accidental Experimental Model for Parkinson Disease

Parkinson disease (PD) was first described as the "shaking palsy" in 1817 by the English surgeon James Parkinson, for whom the disease was later named. It affects about 1 million people in the United States and over 6 million worldwide. PD is most common in people over 60 and rarely affects those under 40 (the actor Michael J. Fox, described in the chapter-opening story, is one exception). The symptoms of PD include bradykinesis (slow movements), tremors, rigidity of the limbs and trunk, and impaired speech, balance, and coordination. PD is caused by injury to or the death of dopamine-producing neurons in the basal ganglia (specifically, the substantia nigra) deep in the forebrain, which normally help control voluntary movement. The initial events that cause the neuron damage are not well understood, and there is no cure.

Frozen Junkies

In 1976, Barry Kidston was a graduate student in chemistry at the University of Maryland. He wanted to experiment with hard drugs and decided to synthesize a narcotic he had read about in a 1947 scientific paper. The drug, called MPPP, was said to be less addictive than morphine and was technically not illegal (although it is now). Kidston was successful initially and was apparently able to achieve a satisfactory "high" by intravenously injecting himself with the compound, which he had synthesized in a makeshift lab in his parent's basement. His luck ran out, however, when he accidently produced a related compound, called MPTP, instead of MPPP. Soon after injecting the MPTP, Kidston's speech became slurred, he had trouble walking, and within 3 days he could hardly move.

Kidston's doctors were baffled, but after a neurologist noted that his symptoms resembled Parkinson disease, he was treated with anti-PD drugs, and he dramatically improved. For 2 years he was able to function well on medication, but then he died, somewhat ironically, from a cocaine overdose. His autopsy revealed a substantial loss of dopamine-producing cells in the substantia nigra, which is a hallmark of PD (Fig. 37B).

Kidston's case was published in a medical journal, but it was barely noticed until 1981, when six IV drug users turned up at various emergency rooms in the San Francisco area, all showing very similar symptoms, as if they had "turned to stone." Some quick investigative work turned up the fact that all had tried some new heroin that had become available on the street. When some of this material was analyzed, one of the toxicologists remembered seeing the article about Kidston a few years earlier and quickly determined that the drug all the new patients had injected contained MPTP.[1]

Developing Treatments for PD

As news that PD could be induced by MPTP in humans reached the biomedical research community, scientists were quickly able to demonstrate that MPTP could also induce PD in animals such as monkeys and mice. Since the mid-1980s, animals with MPTP-induced PD have been used in hundreds of studies that have advanced our understanding of how PD develops, and they have been instrumental in the development of therapies, such as the following:

- Drugs that increase dopamine. These vary from L-dopa, which is converted into dopamine in the brain, to newer drugs, such as entacapone, which inhibit the normal breakdown of dopamine.
- Surgical treatments. The most common of these is deep brain stimulation, in which an electrode is inserted into an area of the brain called the subthalamic nucleus, which seems to be overactive in PD. Coincidentally, one of the original "frozen addicts," still serving time in prison, had this procedure done and has had very significant improvement in his condition.
- Stem cell transplants. The use of fetal stem cells to replace diseased neurons in PD remains controversial. However, results from animal models are encouraging, and groups such as the California Institute for Regenerative Medicine and the Michael J. Fox Foundation for Parkinson Disease Research continue to support stem cell research as a potential cure for PD.
- Gene therapy. A 2014 study showed significant reduction in the symptoms of 15 PD patients injected with Pro-Savin, which delivers genes for dopamine synthesis directly into the brains of PD patients.

[1]Further details about these cases can be read in Langston, J. W., and Palfreman, J. 1995. *The Case of the Frozen Addicts* (Pantheon, New York). A PBS video of the same name is also available.

Questions to Consider

1. What are some other instances in which accidental findings have resulted in scientific discoveries?
2. Should the families of people like Kidston, who accidentally benefit medicine, be compensated financially?
3. Why is it so difficult to determine the specific causes of PD and other brain diseases?

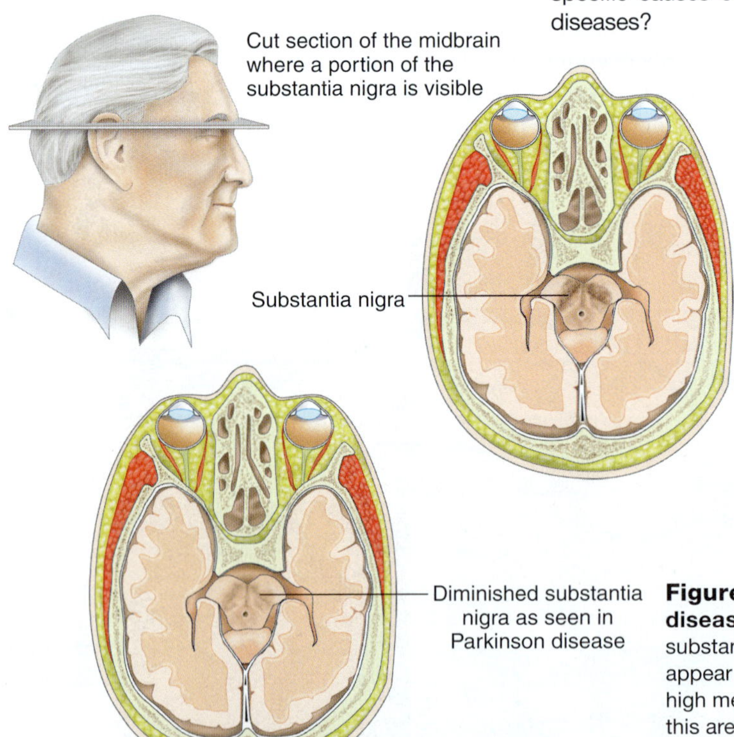

Figure 37B Parkinson disease. Neurons in the substantia nigra of the brain appear dark because of a high melanin content. In PD, this area appears lighter due to a loss of these neurons.

37.4 The Peripheral Nervous System

Learning Outcomes

Upon completion of this section, you should be able to

1. Describe the overall anatomy of the PNS, including the cranial nerves and spinal nerves.
2. Explain how the somatic system differs from the autonomic system.
3. Contrast the functions of the sympathetic and parasympathetic divisions of the autonomic nervous system.

The peripheral nervous system (PNS) lies outside the central nervous system and contains **nerves,** which are bundles of axons. Axons that occur in nerves are also called *nerve fibers.* The cell bodies of neurons are found in the CNS and in ganglia, collections of cell bodies outside the CNS.

MP3 Organization of the Nervous System

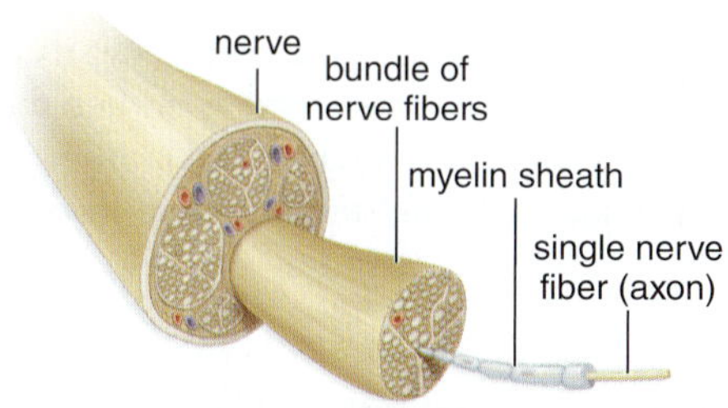

The paired cranial and spinal nerves are part of the PNS. In the PNS, the somatic nervous system has sensory and motor functions that control the skeletal muscles. Ascending tracts carry sensory information to the brain, and descending tracts carry motor commands to the neurons in the spinal cord that control the muscles. The autonomic nervous system controls smooth muscle, cardiac muscle, and the glands. It is further divided into the sympathetic and parasympathetic divisions.

Humans have 12 pairs of **cranial nerves** attached to the brain (Fig. 37.12*a*). Some of these are sensory nerves; they contain only sensory nerve fibers. Some are motor nerves, containing only motor fibers; others are mixed nerves, with both sensory and motor fibers. Cranial nerves are largely concerned with the head, neck, and facial regions of the body. However, the vagus nerve has branches not only to the pharynx and larynx but also to most of the internal organs.

Humans also have 31 pairs of **spinal nerves** (Figs. 37.12*b* and 37.13), which emerge from the spinal cord via two short branches, or roots. The dorsal roots contain axons of sensory neurons, which conduct impulses to the spinal cord from sensory receptors. The cell body of a sensory neuron is in the **dorsal root ganglion.** The ventral roots contain axons of motor neurons, which conduct impulses away from the spinal cord to effectors. These two roots join to form a spinal nerve. All spinal nerves are mixed nerves, containing many sensory and motor fibers.

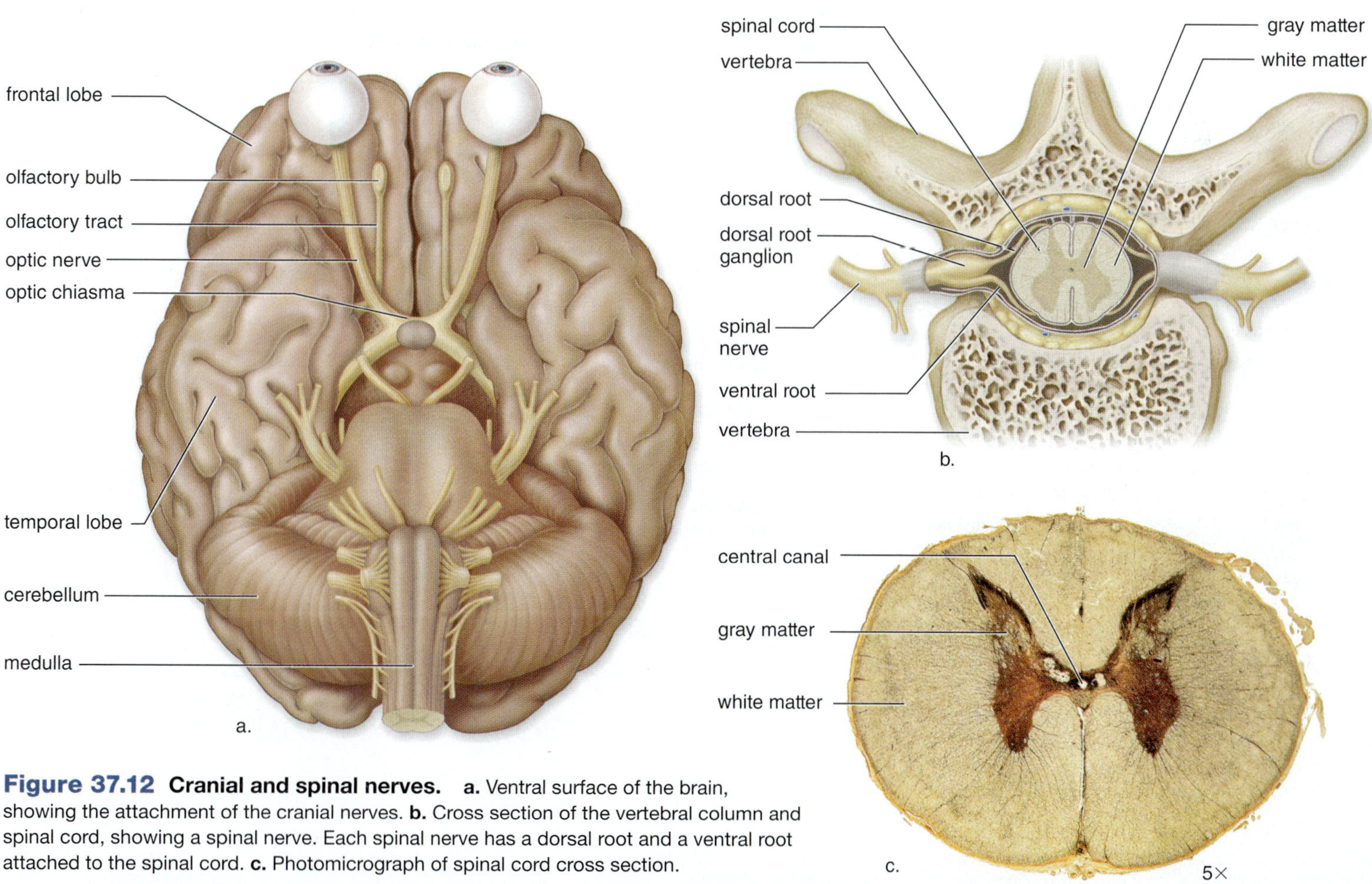

Figure 37.12 Cranial and spinal nerves. **a.** Ventral surface of the brain, showing the attachment of the cranial nerves. **b.** Cross section of the vertebral column and spinal cord, showing a spinal nerve. Each spinal nerve has a dorsal root and a ventral root attached to the spinal cord. **c.** Photomicrograph of spinal cord cross section.

Somatic System

The PNS has two divisions—somatic and autonomic (Table 37.1). The nerves in the **somatic system** serve the skin, joints, and skeletal muscles. Therefore, the somatic system includes nerves with the following functions:

- Take sensory information from external sensory receptors in the skin and joints to the CNS
- Carry motor commands away from the CNS to the skeletal muscles

The neurotransmitter acetylcholine (ACh) is active in the somatic system.

Voluntary control of skeletal muscles always originates in the brain. Involuntary responses to stimuli, called reflex actions, can involve only the spinal cord. Reflexes enable the body to react swiftly to stimuli that could disrupt homeostasis. For example, flying objects cause our eyes to blink, and sharp pins cause our hands to jerk away, even without our having to think about it.

The Reflex Arc. Figure 37.13 illustrates the path of a reflex that involves only the spinal cord. For instance, if your hand touches a sharp pin, **sensory receptors** generate nerve impulses that move along sensory axons through a dorsal root ganglion toward the spinal cord. Sensory neurons that enter the cord dorsally pass signals on to many interneurons in the gray matter of the spinal cord. Some of these interneurons synapse with motor neurons. The short dendrites and the cell bodies of motor neurons are also in the spinal cord, but their axons leave the cord ventrally. Nerve impulses travel along motor axons to an **effector,** which brings about a response to the stimulus. In this case, a muscle contracts, so you withdraw your hand from the pin. (Sometimes an effector is a gland.)

Various other reactions are possible—you will most likely look at the pin, wince, and cry out in pain. This series of responses is explained by the fact that some of the interneurons in the white matter of the cord carry nerve impulses in tracts to the brain. The brain makes you aware of the stimulus and directs subsequent reactions to the situation. You don't feel pain until the brain receives the information and interprets it. Visual information received directly by way of a cranial nerve may make you aware that your finger is bleeding. Then, you might decide to look for a bandage.

Autonomic System

The **autonomic system** of the PNS regulates the activity of cardiac and smooth muscle and glands. It carries out its duties without our awareness or intent. The system is divided into the sympathetic and parasympathetic divisions (Table 37.1). Both of these divisions function automatically and usually in an involuntary manner,

Figure 37.13 A reflex arc showing the path of a spinal reflex. A stimulus (e.g., a sharp pin) causes sensory receptors in the skin to generate nerve impulses that travel in sensory axons to the spinal cord. Interneurons integrate data from sensory neurons and then relay signals to motor axons. Motor axons convey nerve impulses from the spinal cord to a skeletal muscle, which contracts. Movement of the hand away from the pin is the response to the stimulus.

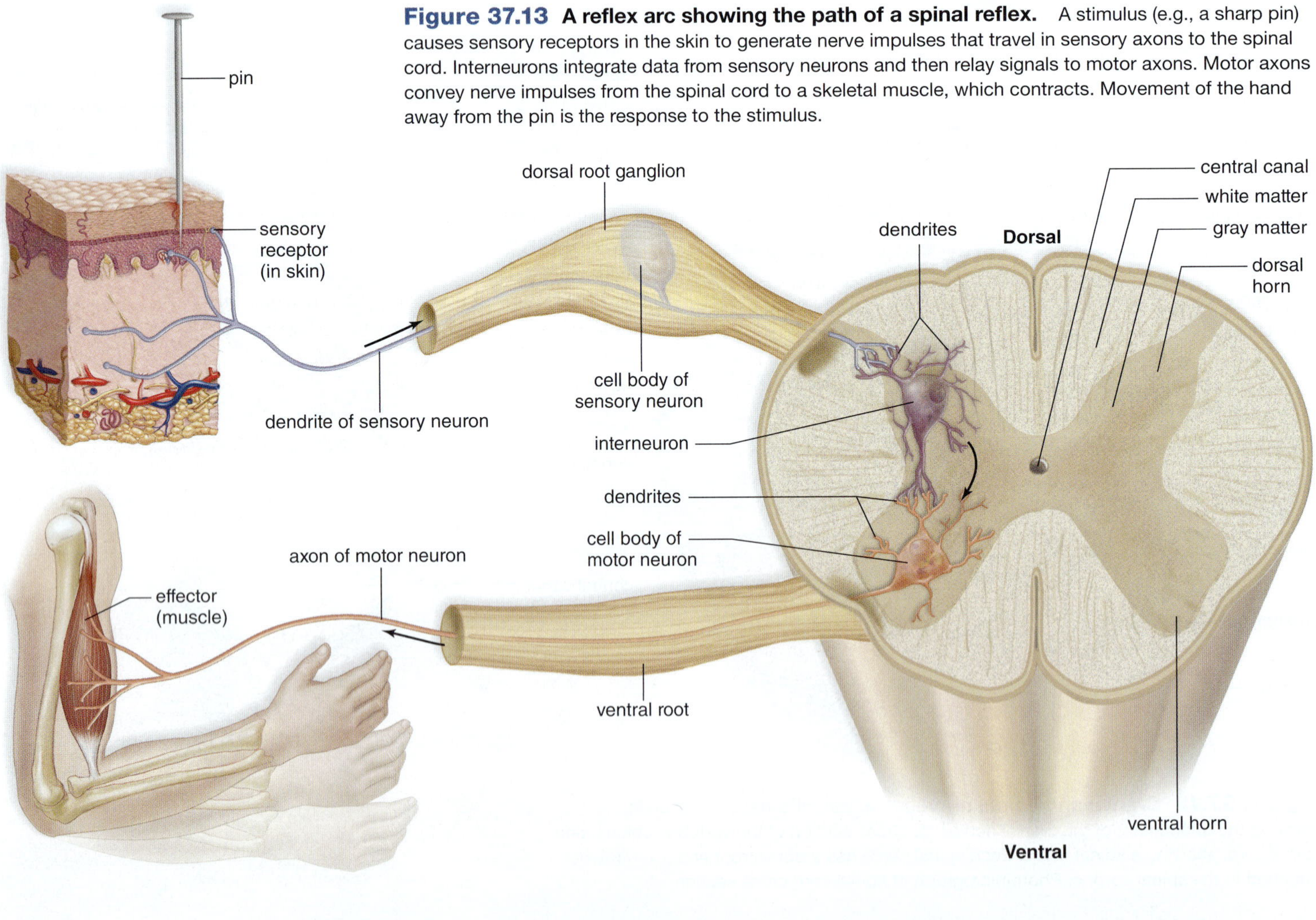

Table 37.1 Comparison of Somatic Motor and Autonomic Motor Pathways

	Somatic Motor Pathway	Autonomic Motor Pathways	
		Sympathetic	**Parasympathetic**
Type of control	Voluntary/involuntary	Involuntary	Involuntary
Number of neurons per message	One	Two (preganglionic shorter than postganglionic)	Two (preganglionic longer than postganglionic)
Location of motor fiber	Most cranial nerves and all spinal nerves	Thoracolumbar spinal nerves	Cranial (e.g., vagus) and sacral spinal nerves
Neurotransmitter	Acetylcholine	Norepinephrine	Acetylcholine
Effectors	Skeletal muscles	Smooth and cardiac muscle, glands	Smooth and cardiac muscle, glands

innervate all internal organs, and use two neurons and one ganglion for each impulse. The first neuron has a cell body within the CNS and a preganglionic fiber. The second neuron has a cell body within the ganglion and a postganglionic fiber.

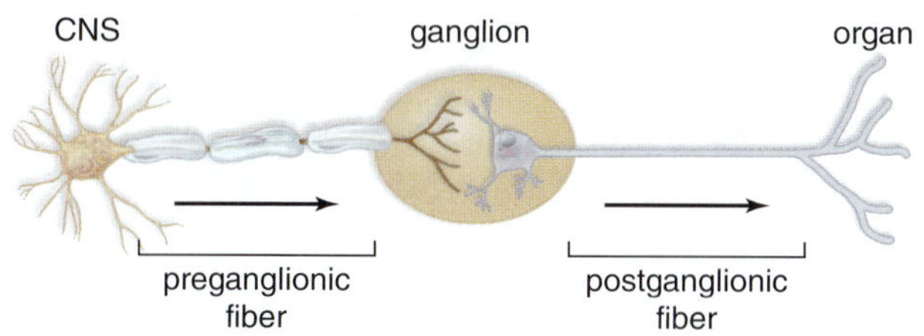

Reflex actions, such as those that regulate blood pressure and breathing rate, are especially important in the maintenance of homeostasis. These reflexes begin when the sensory neurons in contact with internal organs send information to the CNS. They are completed by motor neurons in the autonomic system.

Sympathetic Division

Most preganglionic fibers of the **sympathetic division** arise from the middle, or thoracolumbar, portion of the spinal cord and almost immediately terminate in ganglia lying near the cord. Therefore, in this division, the preganglionic fiber is short, but the postganglionic fiber that makes contact with an organ is long.

The sympathetic division is especially important during emergency situations and is associated with "fight or flight" (Fig. 37.14). In order to fend off a foe or flee from danger, active muscles require a ready supply of glucose and oxygen. The sympathetic division accelerates the heartbeat and dilates the bronchi. At the same time, the sympathetic division inhibits the digestive tract, because digestion is not an immediate necessity if you are under attack. The neurotransmitter released by the postganglionic axon is primarily norepinephrine (NE). Structurally, NE resembles epinephrine (adrenaline), an adrenal medulla hormone that usually increases heart rate and contraction (see Chapter 40).

Parasympathetic Division

The **parasympathetic division** includes a few cranial nerves (e.g., the vagus nerve) and fibers that arise from the sacral (bottom) portion of the spinal cord. Therefore, this division is often referred to as the craniosacral portion of the autonomic system. In the parasympathetic division, the preganglionic fiber is long and the postganglionic fiber is short, because the ganglia lie near or within the organ.

The parasympathetic division, sometimes called the "housekeeper" or "rest and digest division," promotes all the internal responses we associate with a relaxed state; for example, it causes the pupil of the eye to contract, promotes the digestion of food, and slows the heartbeat. The neurotransmitter used by the parasympathetic division is acetylcholine (ACh).

Several disorders can affect the peripheral nerves. Guillain-Barré syndrome (GBS) results from an abnormal immune reaction to one of several types of infectious agents. Antibodies formed against these microbes cross-react with the myelin coating of peripheral nerves, causing demyelination of peripheral nerve axons. The first symptom of GBS is usually weakness in the legs two to four weeks after an infection or immunization. Soon the arms are affected, and in some cases the respiratory muscles may be weakened to the point that mechanical ventilation is required. Fortunately, the inflammation usually subsides in a few weeks, and most patients fully recover within 6 to 12 months. In myasthenia gravis (MG), abnormal antibodies react with the acetylcholine (ACh) receptor at the neuromuscular junction of the skeletal muscles. When an action potential arrives at the synaptic cleft of a peripheral nerve, these antibodies block the normal action of ACh, resulting in muscle weakness. Although there is no cure, MG patients often respond well to drugs that inhibit acetylcholinesterase, an enzyme that normally destroys ACh after it is released into synapses.

Check Your Progress 37.4

1. Review the neurological explanation for the observation that, after you touch a hot stove, you withdraw your hand before you feel any pain.
2. Apply your knowledge of the autonomic nervous system to explain why your stomach may ache if you exercise after a meal.
3. Describe the shift in autonomic system activity that occurs when you are startled from your sleep.

Figure 37.14 Autonomic system structure and function. Sympathetic preganglionic fibers (*left*) arise from the cervical, thoracic, and lumbar portions of the spinal cord; parasympathetic preganglionic fibers (*right*) arise from the cranial and sacral portions of the spinal cord. Each system innervates the same organs but has contrary effects.

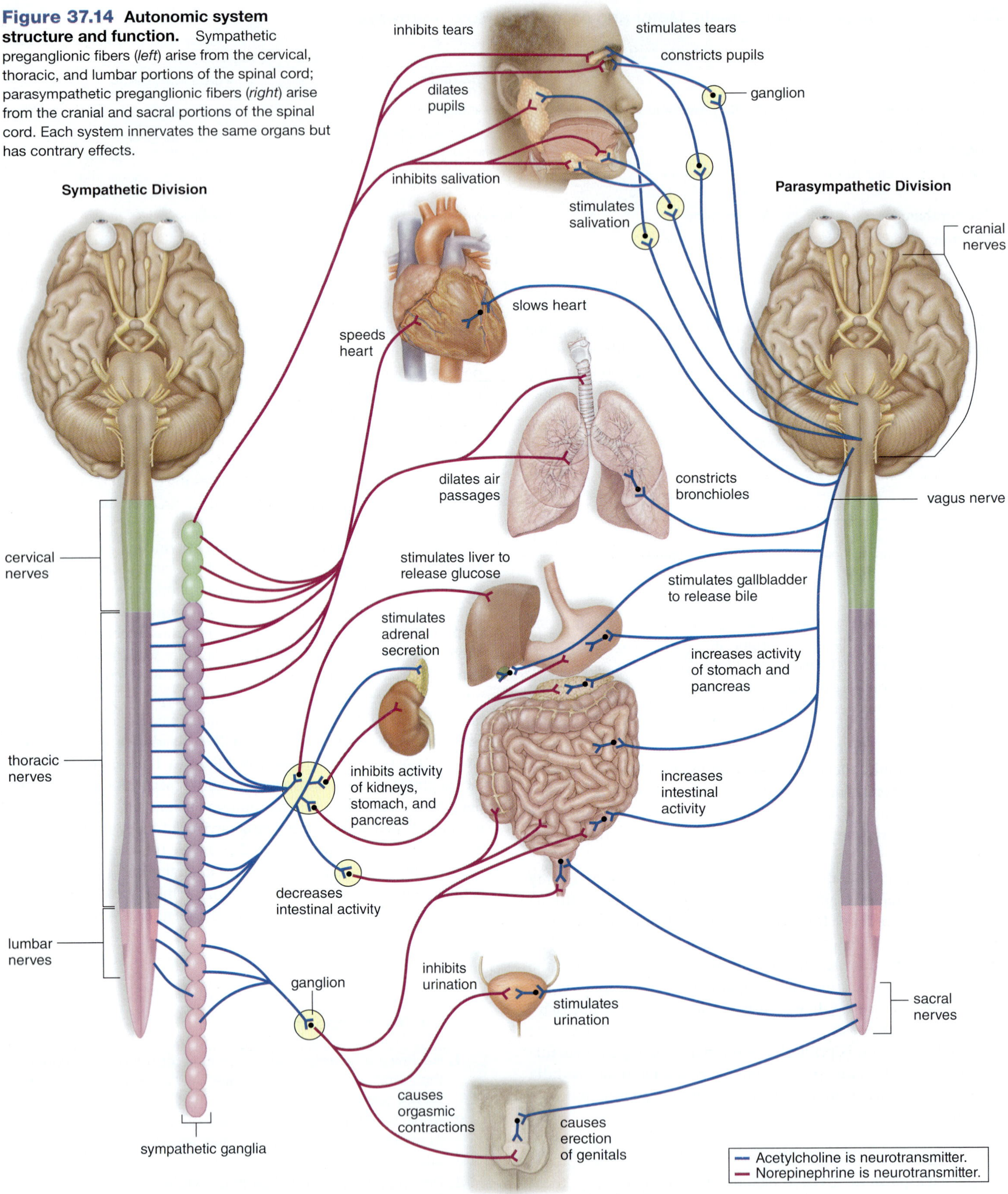

CONNECTING the CONCEPTS with the THEMES

Evolution

- The simplest type of nervous system is the nerve net found in cnidarians, which is a network of neurons in contact with each other and with contractile cells.
- Most invertebrates have a brain and ventral nerve cord with a ganglion in each body segment.
- Vertebrates have a great increase in total number of neurons that make up their nervous system, as well as a well-developed brain and, in the case of mammals, a cerebral cortex.

Nature of Science

- Research studies of the electrical charges that develop across the plasma membrane of the giant axon of squids first helped scientists understand the action potentials generated by neurons.
- The discovery that Parkinson disease could be induced by a potent toxin accidently synthesized in a home lab led to useful animal models for studying PD and the development of new treatments.
- Brain-imaging techniques are useful in examining the effects of illicit drugs, as well as other brain conditions.

Biological Systems

- Nervous tissue is made up of cells called neurons, which transmit nerve impulses, and supporting cells called neuroglia. Transmission of impulses between neurons is usually accomplished by means of chemical neurotransmitters.
- The central nervous system includes the brain and spinal cord; the peripheral nervous system is divided into a somatic system, which controls skeletal muscles, and an autonomic system, which regulates the activity of the heart, smooth muscle, and glands.

MEDIA STUDY TOOLS http://connect.mheducation.com

SMARTBOOK® Study smart. SmartBook helps you maximize your study time by identifying what you know and don't know. The recharge feature in SmartBook helps you prepare for the test by identifying the material that you are most likely to forget.

connect BIOLOGY Find even more resources to learn the chapter concepts including animations, tutorial videos, and interactive practice questions.

MP3 Files	Animations	Tutorials
37.2 Cells of the Nervous System • Synapses **37.3** Organization of the Nervous System • The Brain • The Cerebrum **37.4** Organization of the Nervous System	**37.2** How the Sodium-Potassium Pump Works • Action Potential Propagation • Nerve Impulse • Chemical Synapses	**37.2** Neuron Action Potentials • Synaptic Cleft

SUMMARIZE

37.1 Evolution of the Nervous System

A comparative study of the invertebrates shows a gradual increase in the complexity of the nervous system. Cnidarians have a **nerve net** made up of neurons that communicate with each other and with contractile cells in the body wall. Flatworms have a ladderlike nervous system, with **cephalization** and a cluster of nerve cell bodies, or **ganglion,** at the anterior end. Annelids have a brain, plus a ganglion present in each body segment. Vertebrate nervous systems are much more complex, with a brain and spinal cord making up the **central nervous system** (**CNS**), and an additional **peripheral nervous system** (**PNS**). In mammals the forebrain is greatly enlarged, with an additional outer layer called the neocortex.

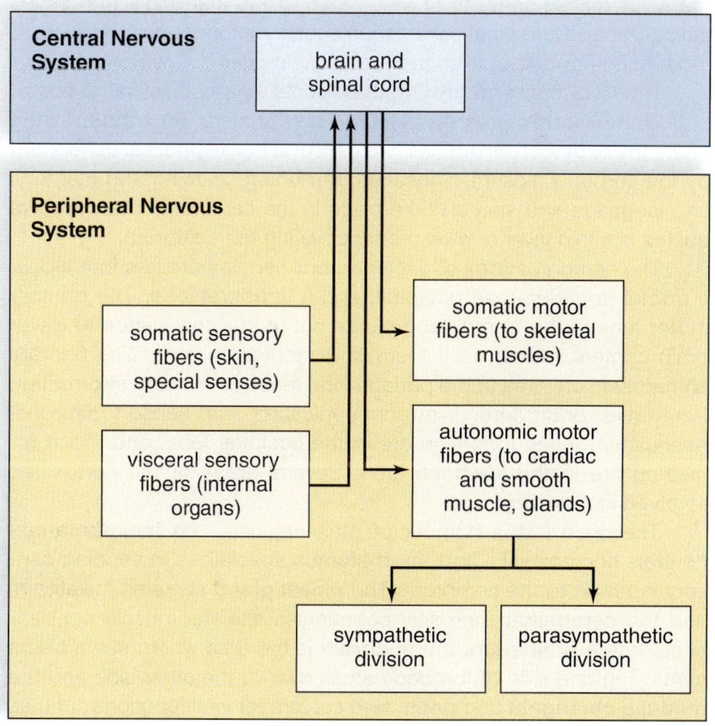

37.2 Nervous Tissue

The anatomical unit of the nervous system is the **neuron,** of which there are three types: sensory neuron, motor neuron, and interneuron. Each of these is made up of a **cell body,** an **axon,** and **dendrites.** Axons are often bundled together as **nerve fibers,** forming nerves, and many axons are covered by a **myelin sheath. Motor (efferent) neurons** take nerve impulses from the CNS to muscles or glands, and **sensory (afferent) neurons** deliver impulses to the CNS from sensory receptors. **Interneurons** convey messages between different areas of the CNS.

Glial cells (**neuroglia**) serve important roles in supporting neuron health and function. The types of glial cells include **astrocytes, microglia, oligodendrocytes, Schwann cells, ependymal cells,** and **satellite cells.**

When an axon is not conducting an **action potential** (nerve impulse), the **resting potential** indicates that the inside of the fiber is negative compared to the outside. The sodium-potassium pump helps maintain this resting potential. When the axon is conducting a nerve impulse, an action potential (a change in membrane potential) travels along the fiber. Depolarization occurs (inside becomes positive) due to the movement of Na^+ to the inside, and then repolarization occurs (inside becomes negative again) due to the movement of K^+ to the outside of the fiber. In myelinated axons, the nerve impulse "jumps" from one nonmyelinated area (node of Ranvier) to the next, a mode known as **saltatory conduction.** Once an action potential occurs, that section of the axon undergoes a brief **refractory period,** during which the sodium gates are unable to open.

Transmission of the nerve impulse from one neuron to another takes place across a **synapse.** Synaptic vesicles usually release a chemical, known as a **neurotransmitter,** into the **synaptic cleft.** The binding of neurotransmitters to receptors in the postsynaptic membrane can either increase the chance of an action potential (stimulation) or decrease the chance of an action potential (inhibition) in the next neuron. A neuron can have as many as 10,000 synapses with other neurons; **integration** is the summing up of these signals.

37.3 The Central Nervous System

The CNS consists of the brain and **spinal cord,** which are both protected by bone and covered by **meninges. Cerebrospinal fluid** fills the spaces between these meninges, as well as the **ventricles** of the brain. When infections invade the meninges, **meningitis** results. The CNS receives and integrates sensory input and formulates motor output. The gray matter of the spinal cord contains neuron cell bodies; the white matter consists of myelinated axons that occur in bundles called tracts. The spinal cord sends sensory information to the brain, receives motor output from the brain, and carries out **reflex actions.**

The CNS contains **grey matter,** containing unmyelinated fibers, and **white matter,** with myelinated fibers that run in **tracts.** In the brain, the outer **cerebrum** has two **cerebral hemispheres,** connected by the corpus callosum. Sensation, reasoning, learning and memory, and language and speech take place in the cerebrum. The **cerebral cortex** is a thin layer of gray matter covering the cerebrum.

The cerebral cortex of each cerebral hemisphere has four lobes: a frontal, a parietal, an occipital, and a temporal lobe. The primary motor area in the frontal lobe sends out motor commands to lower brain centers, which pass them on to motor neurons. The primary somatosensory area in the parietal lobe receives sensory information from lower brain centers in communication with sensory neurons. Association areas for vision are in the occipital lobe, and those for hearing are in the temporal lobe. Several areas of the cortex are involved in **memory.**

The brain has a number of other regions. The **hypothalamus** controls homeostasis, and the **thalamus** specializes in sending sensory input on to the cerebrum. The **pineal gland** secretes melatonin, and the **cerebellum** primarily coordinates skeletal muscle contractions. In the **brainstem,** the **midbrain** is the area where nerve tracts controlling one side of the body cross over to the other side and the **medulla oblongata** and **pons** have centers for vital functions such as breathing and the heartbeat. The **reticular activating system** (**RAS**) is involved in alertness and filtering out irrelevant information. The **limbic system** connects higher reasoning and **memory** with emotions; the **hippocampus** and **amygdala** also play a role.

A number of diseases affect the human nervous system. In **amyotrophic lateral sclerosis** (**ALS**), motor neurons in the brain and spinal cord degenerate and die. When a **stroke** occurs, the blood supply to the brain is disrupted. **Parkinson disease** (**PD**) and **Alzheimer disease** (**AD**) are brain disorders that mainly affect older individuals. **Multiple sclerosis** (**MS**) is an autoimmune disease that affects the myelin sheaths, disrupting nerve transmission.

37.4 The Peripheral Nervous System

The **nerves** of the peripheral nervous system (PNS) lie outside the CNS, and their cell bodies make up ganglia. In humans, these nerves include 12 paired **cranial nerves,** and 31 paired **spinal nerves.** The cell body of a sensory nerve lies in a **dorsal root ganglion.**

The PNS contains the **somatic system** and the **autonomic system.** Reflexes are automatic, and some do not require the involvement of the brain. A simple reflex uses neurons that make up a reflex arc. In the somatic system, a sensory neuron conducts nerve impulses from a **sensory receptor** to an interneuron, which in turn transmits impulses to a motor neuron, which stimulates an **effector** to react.

The motor portion of the somatic system of the PNS controls skeletal muscle; in contrast, the motor portion of the autonomic system controls smooth muscle of the internal organs and glands. The **sympathetic division,** which is often associated with reactions that occur during times of stress, and the **parasympathetic division,** which is often associated with activities that occur during times of relaxation, are both parts of the autonomic system.

ASSESS

Choose the best answer for each question.

37.1 Evolution of the Nervous System

1. Which type of animal has a nerve net?
   a. cnidarians
   b. planarians
   c. sponges
   d. All of these are correct.
2. The most intelligent invertebrates are probably the
   a. annelids.
   b. arthropods.
   c. cephalopods.
   d. None of these can be regarded as intelligent.
3. The part of the brain that is seen only in mammals is the
   a. cerebellum.
   b. cerebrum.
   c. neocortex.
   d. olfactory bulb.

37.2 Nervous Tissue

4. Which of these correctly describes the distribution of ions on either side of an axon when it is not conducting a nerve impulse?
   a. more sodium ions (Na^+) outside and fewer potassium ions (K^+) inside
   b. K^+ outside and Na^+ inside
   c. charged proteins outside; Na^+ and K^+ inside
   d. charged proteins inside
   e. Both a and d are correct.

5. When the action potential begins, sodium gates open, allowing Na^+ to cross the membrane. Now the polarity changes to
   a. negative outside and positive inside.
   b. positive outside and negative inside.
   c. There is no difference in charge between outside and inside.
   d. None of these are correct.
6. Transmission of the nerve impulse across a synapse is accomplished by
   a. the release of Na^+ at the presynaptic membrane.
   b. the release of neurotransmitters at the postsynaptic membrane.
   c. the reception of neurotransmitters at the postsynaptic membrane.
   d. Both a and c are correct.
7. Repolarization of an axon during an action potential is produced by
   a. inward diffusion of Na^+.
   b. active extrusion of K^+.
   c. outward diffusion of K^+.
   d. inward active transport of Na^+.
8. A drug that inactivates acetylcholinesterase
   a. stops the release of ACh from presynaptic endings.
   b. prevents the attachment of ACh to its receptor.
   c. increases the ability of ACh to stimulate postsynaptic cells.
   d. All of these are correct.

37.3 The Central Nervous System

9. Which of these is not a specific function of the CNS?
   a. generate motor output
   b. perform integration
   c. regulate blood sugar
   d. receive sensory input
10. Which of the following is not part of the spinal cord?
    a. grey matter
    b. dorsal horn
    c. association areas
    d. tracts
    e. ventral horn
11. The largest, outermost portion of the brain in humans is the
    a. cerebellum.
    b. cerebrum.
    c. reticular activating system.
    d. thalamus.
12. The limbic system
    a. includes the hippocampus and amygdala.
    b. is responsible for our deepest emotions, including pleasure, rage, and fear.
    c. is not responsible for reason and self-control.
    d. All of these are correct.
13. Which of these pairs is mismatched?
    a. cerebrum—thinking and memory
    b. thalamus—motor and sensory centers
    c. hypothalamus—internal environment regulation
    d. cerebellum—motor coordination
    e. medulla oblongata—fourth ventricle

37.4 The Peripheral Nervous System

14. A spinal nerve takes nerve impulses
    a. to the CNS.
    b. away from the CNS.
    c. both to and away from the CNS.
    d. only inside the CNS.
    e. only from the cerebrum.
15. The autonomic system has two divisions, called the
    a. CNS and PNS.
    b. somatic and skeletal systems.
    c. efferent and afferent systems.
    d. sympathetic and parasympathetic divisions.
16. Which of these statements about autonomic neurons is correct?
    a. They are motor neurons.
    b. Preganglionic neurons have cell bodies in the CNS.
    c. Postganglionic neurons innervate smooth muscles, cardiac muscle, and glands.
    d. All of these are correct.
17. Sympathetic nerve stimulation does not cause
    a. the liver to release glycogen.
    b. the dilation of bronchioles.
    c. the gastrointestinal tract to digest food.
    d. an increase in the heart rate.
18. Label this diagram of a reflex arc.

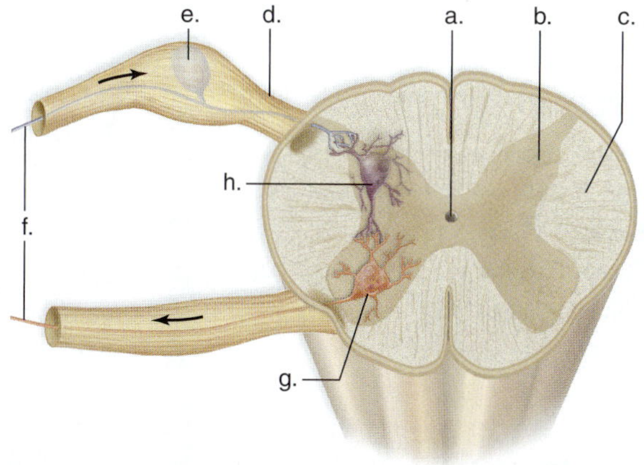

ENGAGE

THINKING SCIENTIFICALLY

1. In individuals with panic disorder, the fight-or-flight response is activated by inappropriate stimuli. How might it be possible to directly control this response in order to treat panic disorder? Why is such control often impractical?
2. A man who lost his leg several years ago continues to experience pain as though it were coming from the missing limb. What hypothesis could explain the neurological basis of this pain?
3. Dopamine is a key neurotransmitter in the brain. Patients with Parkinson disease suffer from a lack of this chemical, whereas the abusers of drugs such as nicotine, cocaine, and methamphetamine enjoy an enhancement of their dopamine activity. Do you think these drugs might be a possible treatment, or even a cure, for Parkinson disease? Why or why not?
4. What are some factors that make brain diseases such as Alzheimer disease, Parkinson disease, and MS so difficult to treat?

38

Sense Organs

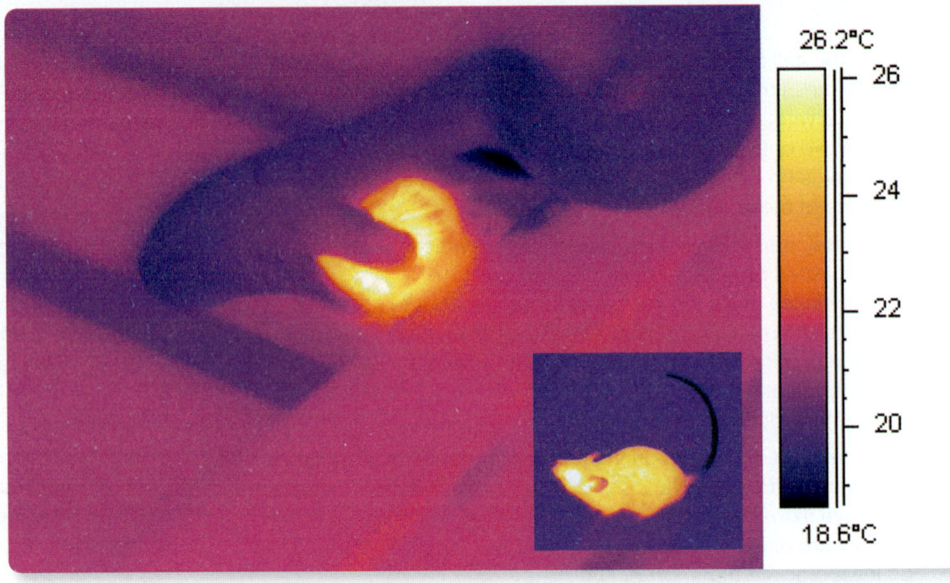

Certain snakes can detect infrared energy emitted by their prey. In this infrared image, the red-orange areas are the warmest, and the blue-black areas are the coldest.

Chapter Outline

38.1 Sensory Receptors 715

38.2 Chemical Senses 716

38.3 Sense of Vision 718

38.4 Senses of Hearing and Balance 724

38.5 Somatic Senses 729

Before You Begin

Before beginning this chapter, take a few moments to review the following discussions.

Sections 37.3 and 37.4 What are the roles of the central and peripheral nervous systems in an animal's responses to its environment?

Figure 37.9 How is the cerebral cortex involved in the processing of sensory information?

Figure 37.13 How do sensory receptors in the skin stimulate a spinal reflex?

Visible light is made up of waves of electromagnetic energy with different wavelengths. As you'll see in this chapter, sensory receptors in the human eye contain pigments that undergo a chemical change when exposed to electromagnetic energy at these various wavelengths, which is then perceived by the brain as vision. Electromagnetic energy with a wavelength at the shortest end of the visible spectrum is perceived as violet; that with a wavelength at the highest end is perceived as red, with other colors in between.

A few types of animals have evolved the ability to detect electromagnetic energy with longer wavelengths, known as the infrared spectrum. Just about any source of heat—the sun, a fire, or a warm body—emits energy in the infrared spectrum. Certain kinds of snakes, such as the pit vipers, have evolved specialized infrared sensory organs. Located in a pit below each eye, these organs are very sensitive to infrared waves emitted by their warm-blooded prey; the photo above, taken using an infrared-sensitive camera, shows how a mouse might "appear" to a snake. Even when placed in total darkness, snakes with this ability can track and find such prey quickly. Although the ability to detect infrared energy with such precision is unusual, all animals rely on many types of sensory systems to maintain homeostasis.

As you read through this chapter, think about the following questions:

1. How does the ability to detect infrared energy provide snakes with a competitive advantage over predators lacking this ability?
2. Of the types of sensory receptors described in this chapter—chemoreceptors, photoreceptors, mechanoreceptors, or thermoreceptors—which is the most necessary for an animal to survive?

Following *the* Themes

CHAPTER 38 SENSE ORGANS

UNIT 7 COMPARATIVE ANIMAL BIOLOGY

Evolution	The sense organs that have evolved in animals are essential to their ability to maintain homeostasis, avoid danger, find food, and locate mates.
Nature of Science	Biomedical scientists are using the knowledge of the transmission of sensory information to restore the abilities of deaf people to hear and of blind people to see.
Biological Systems	Four types of sensory receptors—chemoreceptors, photoreceptors, mechanoreceptors, and thermoreceptors—provide animals with information about their internal and external environments.

38.1 Sensory Receptors

Learning Outcomes

Upon completion of this section, you should be able to

1. Explain the differences among sensory receptors, sensory transduction, and perception.
2. Describe four types of sensory receptors and list examples of each.

In order to survive, animals must be able to maintain homeostasis as well as locate required nutrients, avoid dangers, and learn from previous experiences. These kinds of selective pressures have resulted in the evolution of the many different types of **sensory receptors,** which are specialized cells capable of detecting changes in internal or external conditions, and of communicating that information to the central nervous system.

A sensory receptor is able to convert some type of event, or *stimulus,* into a nerve impulse. This process is known as **sensory transduction.** Some sensory receptors are modified neurons, and others are specialized cells closely associated with neurons.

The plasma membrane of a sensory receptor contains proteins that react to a stimulus. For example, these membrane proteins might be sensitive to temperature or react with a certain chemical. When this happens, ion channels open, and ions flow across the plasma membrane. If the stimulus is sufficient, nerve impulses begin and are carried by a sensory nerve fiber within the PNS to the CNS.

Note that there is no difference between the nerve impulses generated by different types of sensory receptors. All these impulses are simply the action potentials discussed in Chapter 37, regardless of whether they arise in the eyes, ears, nose, mouth, skin, or internal organs. The interpretation of these nerve impulses by appropriate areas of the brain brings about a response that is appropriate for the particular type of stimulus. That is why artificial stimulation of the nerves that normally carry impulses generated in the ear or the eye are interpreted by the brain as sound or light, respectively (see the Nature of Science feature, "Artificial Retinas Come into Focus," on page 723).

What's more, not all of these sensory impulses are received at the conscious levels of the brain—for example, we are not aware of the constant adjustments that are occurring in response to various internal stimuli. Any sensory stimuli of which we become conscious are known as *perceptions*.

Based on the source of the stimulus, sensory receptors can be classified as *interoceptors* or *exteroceptors*. Interoceptors receive stimuli from inside the body, such as changes in blood pressure, blood volume, and the pH of the blood. Interoceptors located within internal organs are sometimes called visceroceptors; those that help maintain muscle tone and posture are proprioceptors (discussed in section 38.5).

A few types of exteroceptors enable an animal to detect information in its environment. **Chemoreceptors** can respond to a diverse range of chemical substances, from oxygen levels in the blood to molecules of food in the mouth or nasal passages. **Electromagnetic receptors** respond to heat or light energy. The infrared sensors of snakes are an example, as are the **photoreceptors** (Gk. *photos,* "light") of the eyes that detect light. **Mechanoreceptors** are stimulated by mechanical forces, usually pressure of some sort. Mechanoreceptors are responsible for detecting changes that are perceived as sound or touch, as well as for maintaining equilibrium (balance). **Thermoreceptors,** located in the hypothalamus and skin, are stimulated by changes in temperature.

Although the extent to which nonhuman animals have perceptions is largely unknown, it is likely that some of them perceive their world in very different ways (Fig. 38.1). As noted in the

Figure 38.1 Perception in the animal kingdom. Animals have a variety of mechanisms by which they perceive their environment. **a.** Some snakes detect thermal energy. **b.** Whales use echolocation. **c.** Dogs detect chemicals in the environment.

chapter-opening essay, some snakes can detect infrared energy that is completely invisible to humans. Bats, dolphins, and whales are capable of *echolocation,* meaning they can produce very high frequency sounds, and then learn about objects in their environment by listening for echoes. Some whales can also hear very low frequency sounds emitted by other whales hundreds of miles away. Dogs have a sense of smell that is more sensitive than that of humans, and they can be trained to detect drugs, human remains, blood, and even bedbugs. Several scientific studies have confirmed that dogs can detect some types of human cancer just by sniffing the appropriate samples. Clearly, the sensory systems are a subject of fascination to biologists, because through them we experience our world.

MP3 Sensations and Receptors

Check Your Progress 38.1

1. Define *sensory transduction*.
2. List three examples of sensory capabilities found in animals that are lacking in humans.

38.2 Chemical Senses

Learning Outcomes

Upon completion of this section, you should be able to

1. Discuss the locations of chemoreceptors in arthropods, crustaceans, and vertebrates.
2. Describe the types and locations of taste receptors in humans.
3. Compare and contrast how the brain receives information about taste versus smell.

Chemoreception is found almost universally in animals and is therefore believed to be the most primitive sense. Chemoreceptors sensitive to certain chemical substances can be important in locating food, finding a mate, and detecting potentially dangerous chemicals in the environment.

The location and sensitivity of chemoreceptors vary throughout the animal kingdom. Although chemoreceptors are present throughout the body of planarians, they are concentrated in the auricles located on the sides of the head. Many insects have taste receptors on their mouthparts, but in the housefly, chemoreceptors are located primarily on the feet. Insects also detect airborne *pheromones,* which are chemical messages passed between individuals.

In crustaceans such as lobsters and crabs, chemoreceptors are widely distributed on their appendages and antennae. Many fish have chemoreceptors scattered over the surface of their skin. Snakes possess *Jacobson's organs,* a pair of pitlike sensory organs in the roof of the mouth. When a snake flicks its forked tongue, scent molecules are carried to the Jacobson's organs, and sensory information is transmitted to the brain for interpretation.

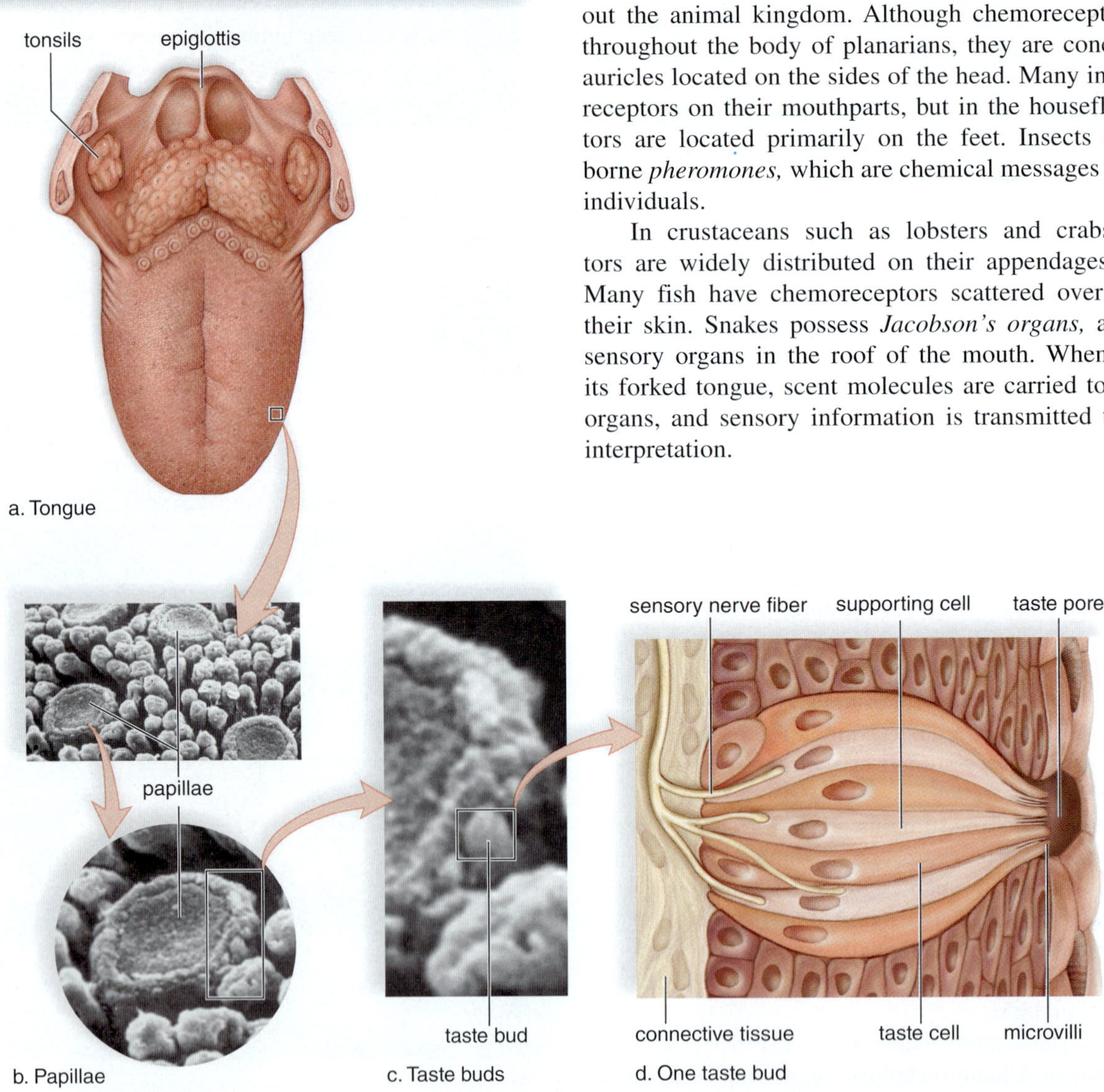

Figure 38.2 Taste buds in humans. **a.** Papillae on the tongue contain taste buds that are sensitive to sweet, sour, salty, bitter, and umami. **b.** Photomicrograph and enlargement of papillae. **c.** Taste buds occur along the walls of the papillae. **d.** Taste cells end in microvilli that bear receptor proteins for certain molecules. When molecules bind to the receptor proteins, nerve impulses are generated and go to the brain, where the sensation of taste occurs.

Sense of Taste in Humans

In adult humans, approximately 3,000 **taste buds** are located primarily on the tongue (Fig. 38.2). Many taste buds lie along the walls of the papillae, the small elevations on the tongue that are visible to the unaided eye. Isolated taste buds are also present on the hard palate, the pharynx, and the epiglottis.

Taste buds on the tongue open at a taste pore. Taste buds have supporting cells and a number of elongated taste cells that end in microvilli. The microvilli, which project into the taste pore, bear receptor proteins for certain molecules. When molecules bind to receptor proteins, nerve impulses are generated in associated sensory nerve fibers. These nerve impulses travel to the brain, where they are interpreted as tastes.

Humans have five main types of taste receptors: sweet, sour, salty, bitter, and umami (Japanese, "savory, delicious"). Foods rich in certain amino acids, such as the common seasoning monosodium glutamate (MSG), as well as certain flavors of cheese, beef broth, and some seafood, produce the taste of umami. Taste buds for each of these tastes are located throughout the tongue, although certain regions may be slightly more sensitive to particular tastes. A food can stimulate more than one of these types of taste buds. The brain appears to survey the overall pattern of incoming sensory impulses and take a "weighted average" of their taste messages as the perceived taste.

Researchers have found chemoreceptors in the human lung that are sensitive only to chemicals that normally taste bitter. These receptors are not clustered in buds, and they do not send taste signals to the brain. Stimulation of these receptors causes the airways to dilate, leading the scientists to speculate about implications for new medications to treat diseases such as asthma.

Sense of Smell in Humans

In humans, the sense of smell, or olfaction, is dependent on between 10 and 20 million **olfactory cells.** These structures are located within olfactory epithelium high in the roof of the nasal cavity (Fig. 38.3). Olfactory cells are modified neurons. Each cell ends in a tuft of about five olfactory cilia, which bear receptor proteins for odor molecules. Each olfactory cell has only 1 out of 1,000 different types of receptor proteins. Nerve fibers from similar olfactory cells lead to the same neuron in the olfactory bulb, an extension of the brain.

An odor contains many odor molecules that activate a characteristic combination of receptor proteins. A rose might stimulate certain olfactory cells, designated by blue and green in Figure 38.3, whereas a gardenia might stimulate a different combination. When the neurons communicate this information via the olfactory tract to the olfactory areas of the cerebral cortex, we perceive that we have smelled a rose or a gardenia.

For decades scientists have estimated that humans can discern only about 10,000 different odors. However, according to a 2014 study in which human volunteers were asked to distinguish between very similar odorant molecules, it is likely that the average human can actually perceive over one trillion different smells!

Have you ever noticed that a certain aroma vividly brings to mind a certain person or place? A whiff of perfume may remind you of someone you knew, or the smell of boxwood may remind you of your grandfather's farm. The olfactory bulbs have direct connections with the limbic system and its centers for emotions and memory. One study found that participants with previous negative experiences of visiting the dentist rated the smell of a chemical

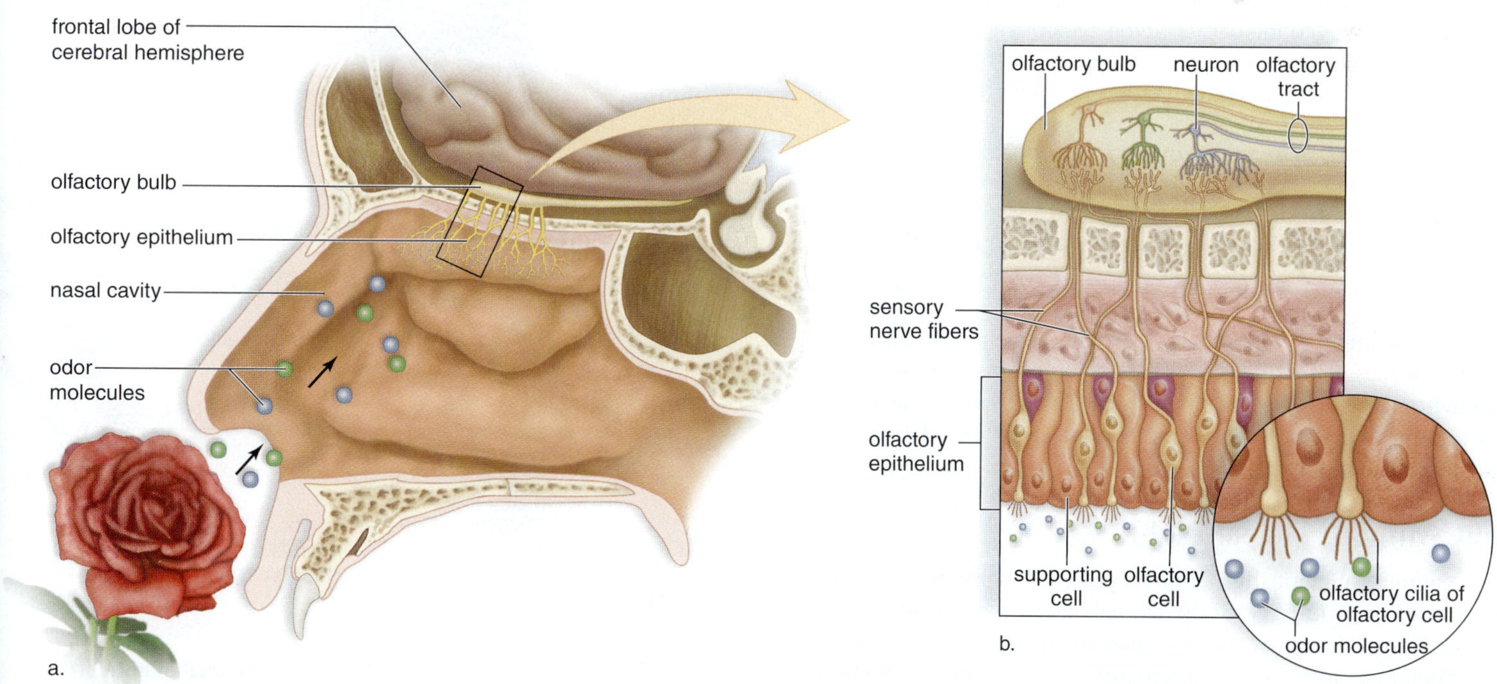

Figure 38.3 Olfactory cell location and anatomy. **a.** The olfactory epithelium in humans is located high in the nasal cavity. **b.** Olfactory cells end in cilia that bear receptor proteins for specific odor molecules. The cilia of each olfactory cell can bind to only one type of odor molecule (signified here by color). For example, if a rose causes olfactory cells sensitive to "blue" and "green" odor molecules to be stimulated, then neurons designated by blue and green in the olfactory bulb are activated. The primary olfactory area of the cerebral cortex interprets the pattern of stimulation as the scent of a rose.

often encountered in dentists' offices as unpleasant, while those lacking such negative experiences rated it as pleasant.

The number of olfactory cells declines with age, and the remaining population of receptors becomes less sensitive. Thus, older people may tend to apply excessive amounts of perfume or aftershave. The ability to smell can also be lost as a result of head trauma, respiratory infection, or brain disease. This condition can become dangerous if these individuals cannot smell spoiled food, a gas leak, or smoke.

Usually, the sense of taste and the sense of smell work together to create a combined effect when interpreted by the cerebral cortex. For example, when you have a cold, you may think food has lost its taste, but most likely you have lost the ability to detect its smell. This method works in reverse also. When you smell something, some of the molecules move from the nose down into the mouth region and stimulate the taste buds there. Therefore, part of what we refer to as smell may, in fact, be taste.

MP3 Taste and Smell

Check Your Progress 38.2

1. Compare and contrast the senses of smell and taste.
2. List the five types of taste receptors in humans.
3. Discuss what could account for how a nerve impulse would be interpreted by the different sense organs.

38.3 Sense of Vision

Learning Outcomes

Upon completion of this section, you should be able to

1. Compare the structure of the compound eyes of arthropods with the camera-type eyes of vertebrates.
2. List all tissues or cell layers through which light passes from when it enters the eye until it is converted to a nerve impulse.
3. Discuss the distinct roles of rod cells, cone cells, and rhodopsin in converting a light stimulus into a nerve impulse.
4. Describe several common disorders affecting vision.

Vision is an important capability for many, but not all, animals. Like the senses of smell and hearing, vision allows us to perceive the environment at a distance, which can have survival value. In this section, we review how animals detect light and how the human eye accomplishes vision.

How Animals Detect Light

As mentioned previously, photoreceptors are sensory receptors that are sensitive to light. Some animals lack photoreceptors and instead depend on senses such as smell and hearing; other animals have photoreceptors but live in environments that do not require them. For example, moles live underground and use their senses of smell and touch rather than eyesight.

Not all photoreceptors form images. The "eyespots" of planarians allow these animals to sense and move away from light. Image-forming eyes are found among four invertebrate groups: cnidarians, annelids, molluscs, and arthropods. Arthropods have **compound eyes** composed of many independent visual units called ommatidia (Gk. *ommation,* dim. of *omma,* "eye"), each possessing all the elements needed for light reception (Fig. 38.4). Both the cornea and the crystalline cone function as lenses to direct light rays toward the photoreceptors. The photoreceptors generate nerve impulses, which pass to the brain by way of optic nerve fibers. The outer pigment cells absorb stray light rays, so that the rays do not pass from one visual unit to the other.

Flies and mosquitoes can see only a few millimeters in front of them, but dragonflies can see small prey insects several meters away. Research has shown that foraging bees use their sense of vision as a sort of "odometer" to estimate how far they have flown from their hive.

Most insects have color vision, but they see a limited number of colors compared to humans. However, many insects can also see some ultraviolet rays, and this enables them to locate the particular parts of flowers, such as nectar guides, that have ultraviolet patterns (Fig. 38.5). Some fishes, all reptiles, and most birds are believed to have color vision, but among mammals, only humans

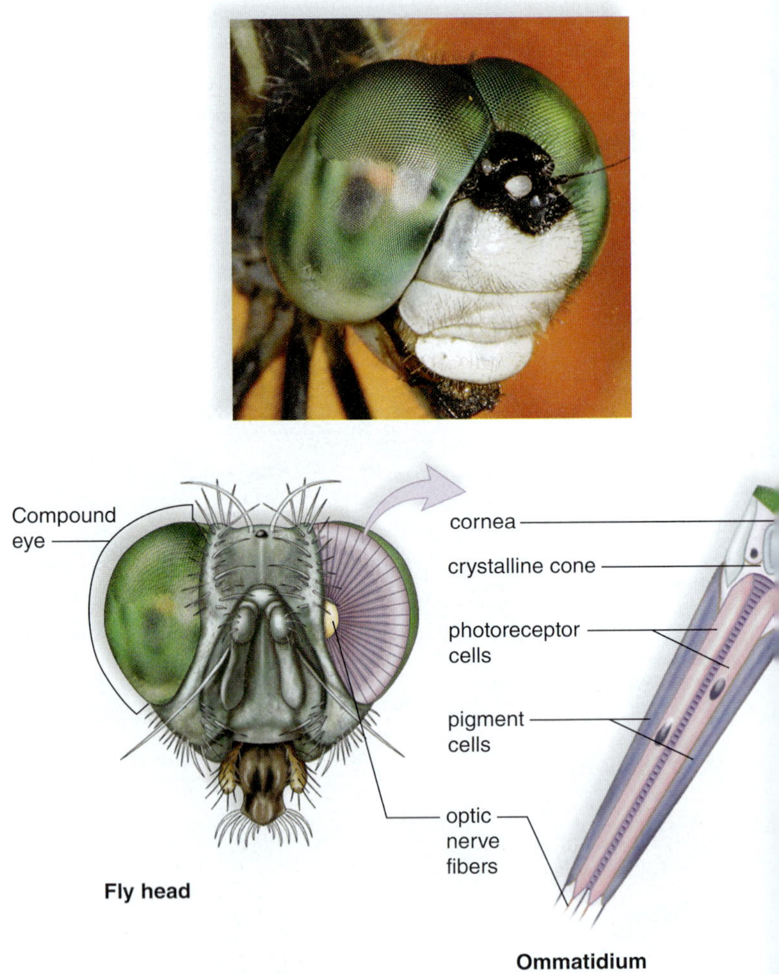

Figure 38.4 Compound eye. Each visual unit of a compound eye has a cornea and a lens, which focus light onto photoreceptors. The photoreceptors generate nerve impulses, which are transmitted to the brain, where interpretation produces a mosaic image.

Figure 38.5 Nectar guides. Evening primrose, *Oenothera,* as seen by humans (*left*) and insects (*right*). Humans see no markings, but insects see distinct lines and central blotches, because their eyes respond to ultraviolet rays. These types of markings, known as nectar guides, often highlight the reproductive parts of flowers, where insects feed on nectar and pick up pollen at the same time.

and other primates have color vision. It would seem, then, that this trait was adaptive for a diurnal habit (active during the day), which accounts for its retention in only a few mammals.

Vertebrates (including humans) and certain molluscs, such as the squid and the octopus, have a **camera-type eye.** Because molluscs and vertebrates are not closely related, this similarity is an example of convergent evolution. A single lens focuses an image of the visual field on photoreceptors, which are closely packed together. In vertebrates, the lens changes shape to aid focusing, but in molluscs the lens moves back and forth. All of the photoreceptors taken together can be compared to a piece of film in a camera. The human eye is more complex than a camera, however, as you will see.

Animals with two eyes facing forward have three-dimensional vision, or **stereoscopic vision.** The visual fields overlap, and each eye is able to view an object from a different angle. Predators tend to have stereoscopic vision, as do humans. Animals with eyes facing sideways, such as rabbits, don't have stereoscopic vision, but they do have **panoramic vision,** meaning that their visual field is very wide. Panoramic vision is useful to prey animals, because it makes it more difficult for a predator to sneak up on them.

Many vertebrates have a membrane in the back of their eye called a *tapetum lucidum,* which reflects light back into the photoreceptor cells of the retina to increase sensitivity to light. This explains the eerie glowing appearance of some animals' eyes at night.

The Human Eye

The human eye, which is an elongated sphere about 2.5 cm in diameter, has three layers: the sclera, the choroid, and the retina (Fig. 38.6). The outer **sclera** is an opaque, white, fibrous layer that covers most of the eye; in front of the eye, the sclera becomes the transparent **cornea,** the window of the eye. A thin layer of epithelial cells forms a mucous membrane called the **conjunctiva,** which covers the surface of the sclera and keeps the eyes moist.

The middle, thin, dark-brown **choroid** layer contains many blood vessels and a brown pigment that absorbs stray light rays. Toward the front of the eye, the choroid thickens and forms the ring-shaped ciliary body and a thin, circular, muscular diaphragm, the iris. The **iris,** the colored portion of the eye, regulates the size of an opening called the pupil. The **pupil,** like the aperture of a camera lens, regulates light entering the eye. The **lens,** which is attached to the ciliary body by ligaments, divides the cavity of the

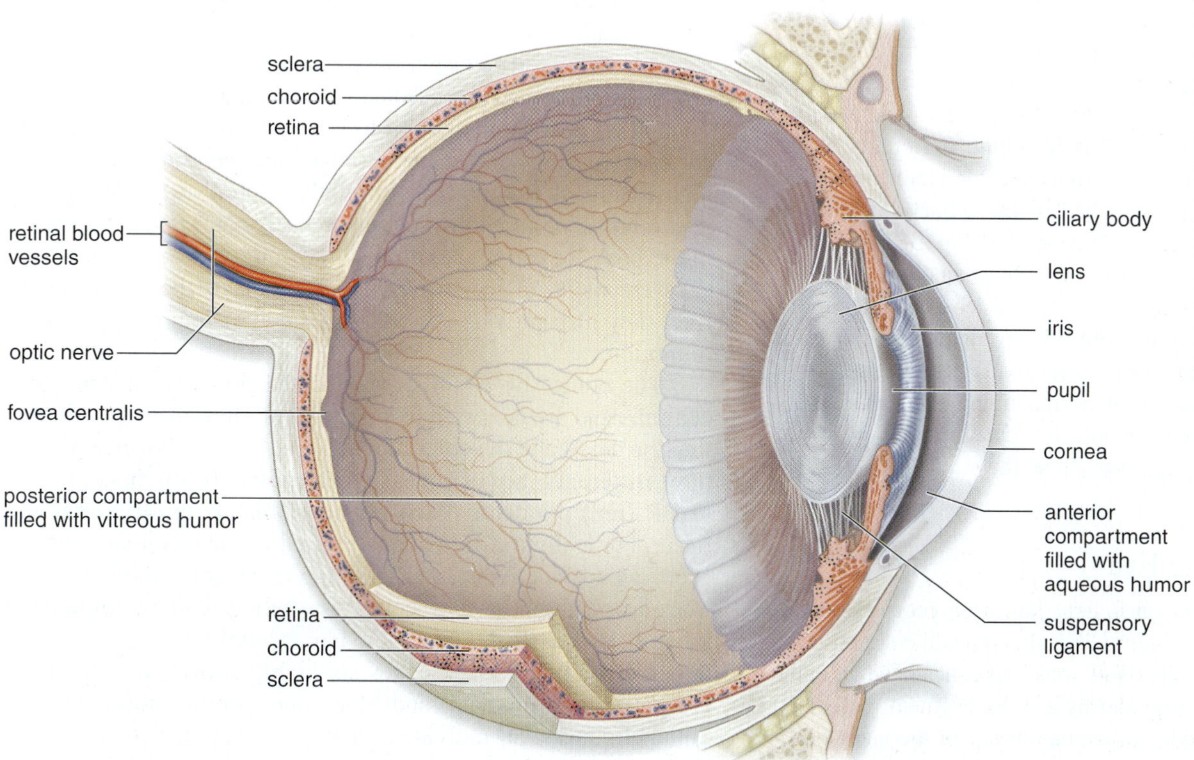

Figure 38.6 Anatomy of the human eye. Notice that the sclera, the outer layer of the eye, becomes the cornea and that the choroid, the middle layer, is continuous with the ciliary body and the iris. The retina, the inner layer, contains the photoreceptors for vision. The fovea centralis is the region where vision is most acute.

eye into two portions and helps form images. A basic, watery solution called aqueous humor fills the anterior compartment between the cornea and the lens. The aqueous humor provides a fluid cushion, as well as nutrient and waste transport, for the eye.

The inner layer of the eye, the **retina,** is located in the posterior compartment. The retina contains photoreceptors called **rod cells** and **cone cells.** The rods are very sensitive to light, but they do not respond to colors; therefore, at night or in a darkened room, we see only shades of gray. Rods are distributed in the peripheral regions of the retina. The cones, which require bright light, are sensitive to different wavelengths of light, and therefore humans have the ability to distinguish colors. The retina has a central region called the **fovea centralis,** where cone cells are densely packed. Light is normally focused on the fovea when we look directly at an object. This is helpful because vision is most acute in the fovea centralis.

Sensory fibers form the optic nerve, which takes nerve impulses to the brain. No rods or cones are present where the optic nerve exits the retina (see Fig. 38.10). Therefore, no vision is possible in this area, and it is termed the **blind spot.** You can detect your own blind spot by putting a dot to the right of center on a piece of paper. Use your right hand to move the paper slowly toward your right eye while you look straight ahead. The dot will disappear at one point—this is your blind spot.

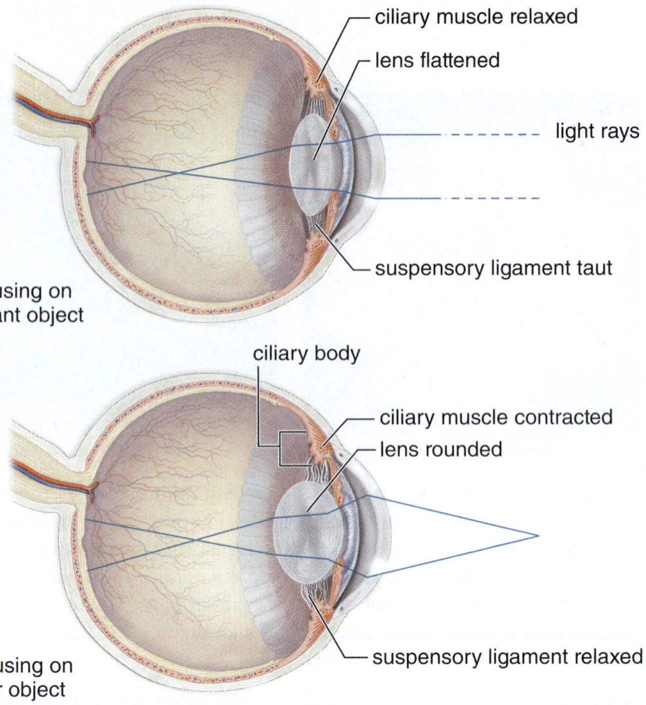

Figure 38.7 Focusing of the human eye. Light rays from each point on an object are bent by the cornea and the lens in such a way that an inverted and reversed image of the object forms on the retina. **a.** When focusing on a distant object, the lens is flat, because the ciliary muscle is relaxed and the suspensory ligament is taut. **b.** When focusing on a near object, the lens accommodates; that is, it becomes rounded, because the ciliary muscle contracts, causing the suspensory ligament to relax.

Focusing of the Eye

When we look directly at something, such as the printed letters on this page, light rays pass through the pupil and are focused on the retina. The image produced is much smaller than the object, because light rays are bent (refracted) when they are brought into focus. The image on the retina is also upside down and reversed from left to right. When information from the retina reaches the brain, it is processed so that we perceive our surroundings in the correct orientation.

Focusing starts at the cornea and continues as the rays pass through the lens. The lens provides additional focusing power as **visual accommodation** occurs for close vision. The shape of the lens is controlled by the **ciliary muscle** within the ciliary body. When we view a distant object, the ciliary muscle is relaxed, causing the suspensory ligaments attached to the ciliary body to be taut; therefore, the lens remains relatively flat (Fig. 38.7*a*). When we view a near object, the ciliary muscle contracts, releasing the tension on the suspensory ligaments, and the lens becomes more round due to its natural elasticity (Fig. 38.7*b*). Because close work requires contraction of the ciliary muscle, it very often causes muscle fatigue known as eyestrain. With normal aging, the lens loses its ability to accommodate for near objects; thus, many people need reading glasses once they reach middle age.

MP3 Sense of Vision

Photoreceptors of the Eye

Sensory transduction occurs once light has been focused on the photoreceptors in the retina. Figure 38.8 illustrates the structure of these rod cells and cone cells. Both rods and cones have an outer segment joined to an inner segment by a stalk. Pigment molecules that react to light are embedded in the membrane of the many disks present in the outer segment. Synaptic vesicles are located at the synaptic endings of the inner segment.

The visual pigment in rods is called rhodopsin. **Rhodopsin** is a complex molecule made up of the protein opsin and a light-absorbing molecule called *retinal,* which is a derivative of vitamin A. When a rod absorbs light, rhodopsin splits into opsin and retinal, leading to a cascade of reactions and the closure of ion channels in the rod cell's plasma membrane. The release of inhibitory transmitter molecules from the rod's synaptic vesicles ceases. Thereafter, nerve impulses go to the visual areas of the cerebral cortex. Rods are very sensitive to light and therefore are suited to night vision. Carrots and other brightly colored vegetables are rich in *carotenoids,* some of which can be easily converted to vitamin A in the body, so eating these foods may improve night vision. Rod cells are plentiful in the peripheral region of the retina; therefore, they also provide us with peripheral vision and perception of motion.

The cones, by contrast, are located primarily in the fovea centralis and are activated by bright light. They allow us to detect the fine detail and the color of an object. Color vision depends on three different kinds of cones, which contain B (blue), G (green), and R (red) pigments. Each pigment is made up of retinal and opsin, but a slight difference is present in the opsin structure of each, which accounts for their individual absorption patterns. Various combinations of cones are believed to be stimulated by in-between shades of color. For example, the color yellow is perceived when green cones are highly stimulated, red cones are partially stimulated, and blue cones are not stimulated. In color blindness, an individual lacks certain visual pigments. As

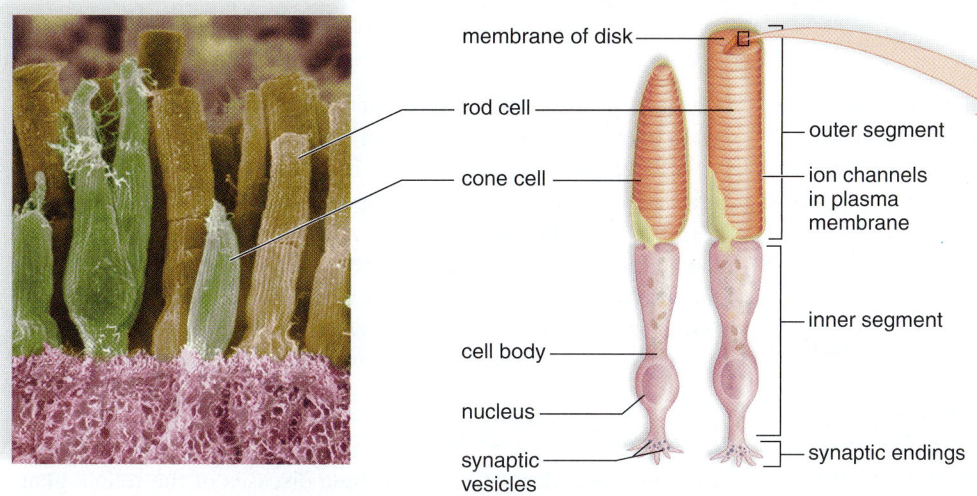

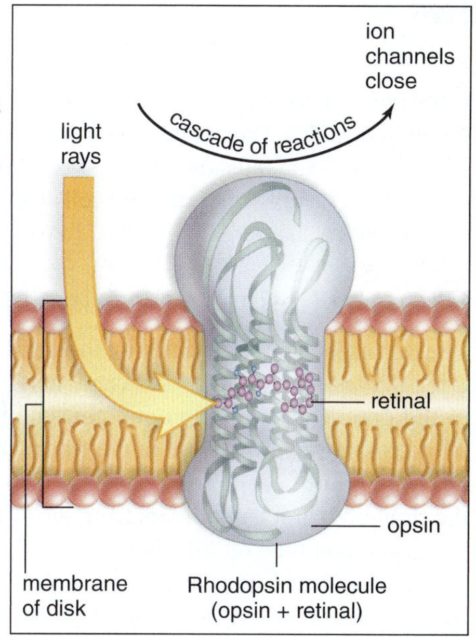

Figure 38.8 Photoreceptors in the eye. The outer segment of rods and cones contains stacks of membranous disks, which contain visual pigments. In rods, the membrane of each disk contains rhodopsin, a complex molecule containing the protein opsin and the pigment retinal. When rhodopsin absorbs light energy, it splits, releasing opsin, which sets in motion a cascade of reactions that cause ion channels in the plasma membrane to close. Thereafter, nerve impulses go to the brain.

indicated in Chapter 11, some forms of color blindness are a sex-linked hereditary disorder.

Integration of Visual Signals in the Retina

The retina has three layers of neurons (Fig. 38.9). The layer closest to the choroid contains the rod cells and cone cells; the middle layer contains bipolar cells; and the innermost layer contains ganglion cells, whose sensory fibers become the optic nerve. Only the rod cells and cone cells are sensitive to light; therefore, before these photoreceptors can be stimulated, light must penetrate through the other cell layers.

The rod cells and cone cells synapse with the bipolar cells, which in turn synapse with ganglion cells that initiate nerve impulses. Notice in Figure 38.9 that there are many more rod cells and cone cells than ganglion cells. In fact, the human retina has about 150 million rod cells, 6 million cone cells, but only 1 million ganglion cells.

The sensitivity of cones versus rods is mirrored by how directly they connect to ganglion cells. As many as 150 rods may excite the same ganglion cell, meaning that each ganglion cell receives signals

Figure 38.9 Structure and function of the retina.
a. The retina is the inner layer of the eye. Rod cells and cone cells, located at the back of the retina nearest the choroid, synapse with bipolar cells, which synapse with ganglion cells. Further, many rod cells share one bipolar cell, but cone cells do not. Certain cone cells synapse with only one ganglion cell. Cone cells, in general, distinguish more detail than do rod cells.
b. Micrograph of the retina.

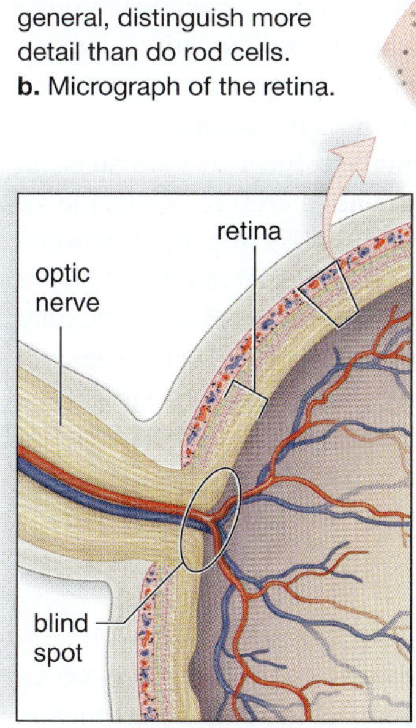

a. Location of retina

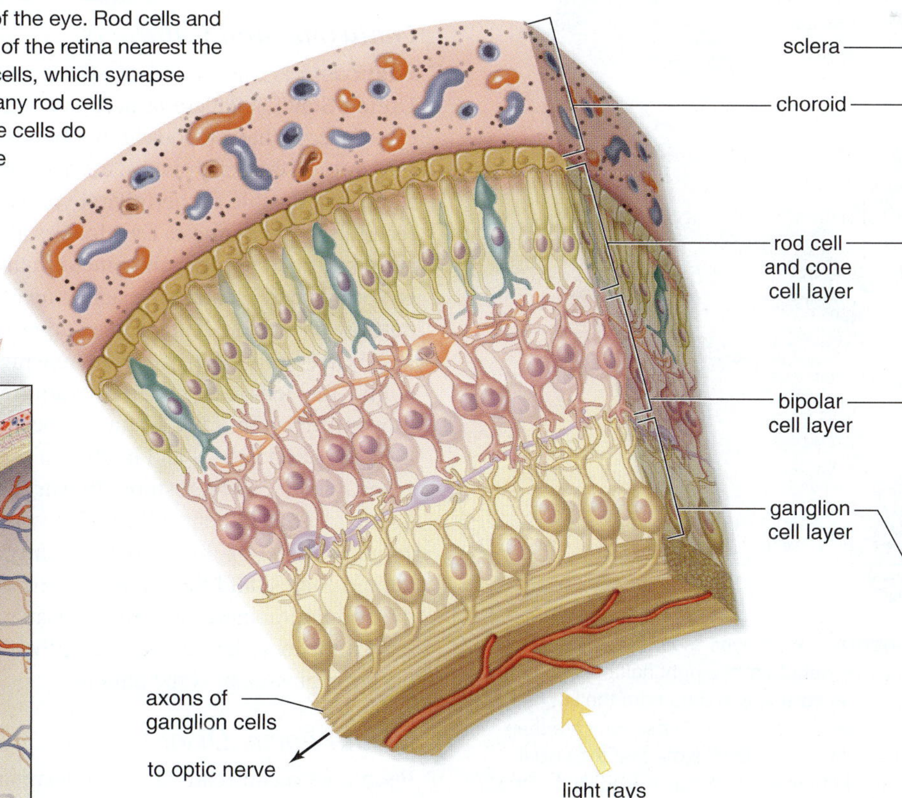

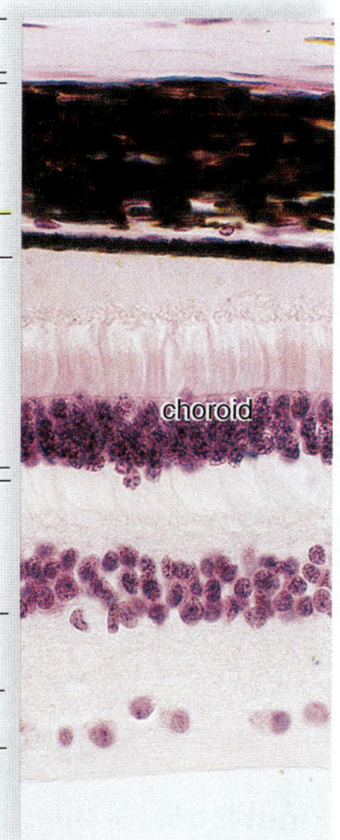

b. Micrograph of retina

from rod cells covering about 1 mm^2 of retina (about the size of a thumb tack hole). Therefore, the stimulation of rods results in vision that is blurred and indistinct. In contrast, some cone cells in the fovea centralis excite only one ganglion cell. This explains why cones provide us with a sharper, more detailed image of an object.

From the Retina to the Visual Cortex

The axons of ganglion cells in the retina assemble to form the optic nerves. The optic nerves carry nerve impulses from the eyes to the optic chiasma. The optic chiasma has an X shape, formed by a crossing-over of optic nerve fibers (Fig. 38.10). Fibers from the right half of each retina converge and continue on together in the right optic tract, and fibers from the left half of each retina converge and continue on together in the left optic tract.

The optic tracts sweep around the hypothalamus, and most fibers synapse with neurons in nuclei (masses of neuron cell bodies) within the thalamus. Axons from the thalamic nuclei form optic radiations that take nerve impulses to the visual area within the occipital lobe. Notice in Figure 38.10 that the image arriving at the thalamus, and therefore the visual area, has been split, because the left optic tract carries information about the right portion of the visual field (shown in green) and the right optic tract carries information about the left portion of the visual field (shown in red). Therefore, the right and left visual areas must communicate with each other for us to see the entire visual field. Also, because the image is inverted and reversed, it must be righted in the brain for us to correctly perceive the visual field.

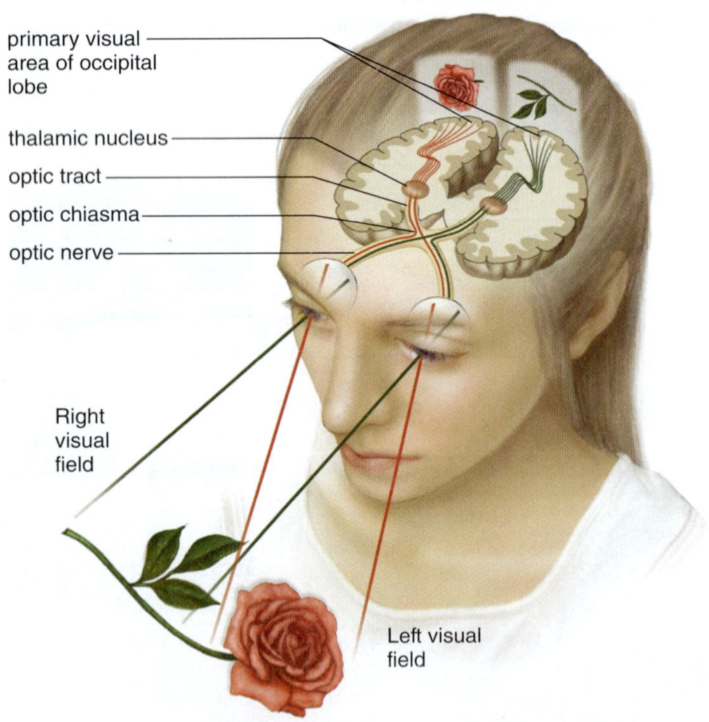

Figure 38.10 Optic chiasma. Both eyes "see" the entire visual field. Because of the optic chiasma, data from the right half of each retina (red lines) go to the right visual cortex, and data from the left half of each retina (green lines) go to the left visual cortex. These data are then combined to allow us to see the entire visual field. Note that the visual pathway to the brain includes the thalamus, which has the ability to filter sensory stimuli.

The most surprising finding has been that the primary visual area acts as a post office, parceling out information regarding color, form, motion, and possibly other attributes to different portions of the adjoining visual association area. Therefore, the brain has taken the visual field apart, even though we see a unified visual field. The visual association areas are believed to rebuild the field and give us an understanding of it at the same time.

Disorders of Vision

Common disorders of vision include diseases of the retina, glaucoma and cataracts, and problems with visual focus.

Diseases of the Retina

Diseases of the retina are the most common cause of blindness in adults. One of these is **diabetic retinopathy,** in which the capillaries to the retina become damaged secondary to diabetes. In **macular degeneration,** the leading cause of blindness for people in the Untied States under the age of 65, capillaries supplying the retinas become damaged, and hemorrhages and blocked vessels can occur. Smokers are 20 times more likely to acquire this disorder than are nonsmokers. If diagnosed early, both of these retinal diseases can be treated, using drugs that are injected directly into the vitreous humor. With **retinal detachment,** the retina peels away from the supportive choroid layer, due to eye trauma or other diseases. A detached retina can often be reattached using a laser. As noted in the Nature of Science feature, "Artificial Retinas Come into Focus," artificial retina technology is becoming an option for some blind people whose visual pathways remain intact.

Glaucoma and Cataracts

Glaucoma is the second most common cause of blindness in the United States. Glaucoma occurs when the drainage system of the eyes fails, so that aqueous humor builds up and increases intraocular pressure. In the early stages, this pressure tends to destroy the nerve fibers responsible for peripheral vision, but untreated glaucoma can result in total blindness. Eye doctors always check intraocular pressure, but the disorder can come on quickly, and any nerve damage is permanent. Treatment usually involves drugs that increase the outflow of aqueous humor, but surgery may be required.

With aging, the eye is increasingly subject to **cataracts,** in which the lens can become opaque and therefore incapable of transmitting light rays. Cataracts occur in 50% of people between the ages of 65 and 74 and in 70% of those ages 75 or older. In most cases, vision can be restored by surgical removal of the unhealthy lens and replacement with a clear plastic artificial lens. Cataract surgery is one of the most frequently performed surgeries in the United States, and the number is increasing as the average age of the population increases. Surgeons may be able to replace cataracts with multifocal lenses that allow the eye to focus, eliminating the need for glasses in some patients.

Visual Focus Disorders

People who can read what are designated as size 20 letters on an optometrist's chart 20 feet away are said to have 20/20 vision.

THEME **Nature of Science**

Artificial Retinas Come into Focus

Over 25 million people worldwide are blind due to diseases of the retinas. In most cases, there are no cures and few effective treatments. However, after years of work, medical researchers have now developed artificial retina technology that is restoring some vision for people who were completely blind.

Retinal diseases such as macular degeneration cause the rods and cones to die but leave the ganglion cells and neurons intact. In 1988, researchers first demonstrated that a blind person could see light if the nerve cells behind the retina were stimulated with an electrical current. Since then, scientists have been working on developing technological approaches that could take the place of the diseased photoreceptors. In March 2011, a California company received approval in Europe to sell the Argus II, a retinal prosthesis designed for implantation in blind patients who still have intact connections between their retinas and brain.

Priced at about $115,000, the device includes a tiny digital camera embedded in a pair of glasses worn by the patient (Fig. 38A). Images from the camera are translated into electrical signals representing patterns of light and dark. These signals are transmitted wirelessly to a receiver implanted above the ear or near the eye. This receiver in turn sends signals via a tiny wire attached to a 1 mm × 1 mm microchip that has been surgically implanted under the retina (Fig. 38A). The chip contains 60 electrodes, which can stimulate the ganglion cells, producing electrical impulses that travel via the optic nerve to the brain, where they are interpreted as vision.

Although it takes time for patients to learn to interpret the patterns of light and dark transmitted by the device, some formerly blind patients have been able to recognize simple objects, see people in front of them and follow their movement, and even read large print slowly. The Argus II device was approved for use in patients in the United States in 2013. Moreover, researchers are already working on upgraded models that incorporate up to 1,500 electrodes, to increase the resolution of the image produced. This is anticipated to be the threshold of information that will be required for formerly blind patients to recognize faces. Other groups are working on artificial retinas that can be implanted inside the eye itself.

Although the amount of information that is transmitted by an artificial retina device is important, other scientists are focusing on the type of information that is transmitted. A 2010 study conducted at Weill Cornell Medical College in New York by investigators Shiela Nirenberg and Chethan Pandarinath demonstrated that a different approach could restore normal vision in blind mice. First, these scientists focused on deciphering the patterns by which the ganglion cells normally transmit information from the photoreceptors to the brain. Next, they tested these codes in blind mice that had been genetically altered so that their ganglion cells expressed a protein called channelrhodopsin, which initiates a nerve impulse when it is exposed to light. That way, the investigators didn't have to implant artificial retinas in the mice but instead exposed the mice to different patterns of flashing light. The results of these experiments showed that the mice receiving light information using the proper code appeared to be able to see at nearly normal levels.

While eliminating the need for surgery is a major advantage of this approach, it is impossible (or at least unethical) to genetically engineer humans to express certain genes. The solution to this problem may lie in gene therapy, in which the ganglion cells of blind patients would be treated so that they express the *channelrhodopsin* gene. If that approach were successful, patients would wear a pair of glasses with an embedded camera, only this camera would emit light that had been translated into the patterns the brain can understand.

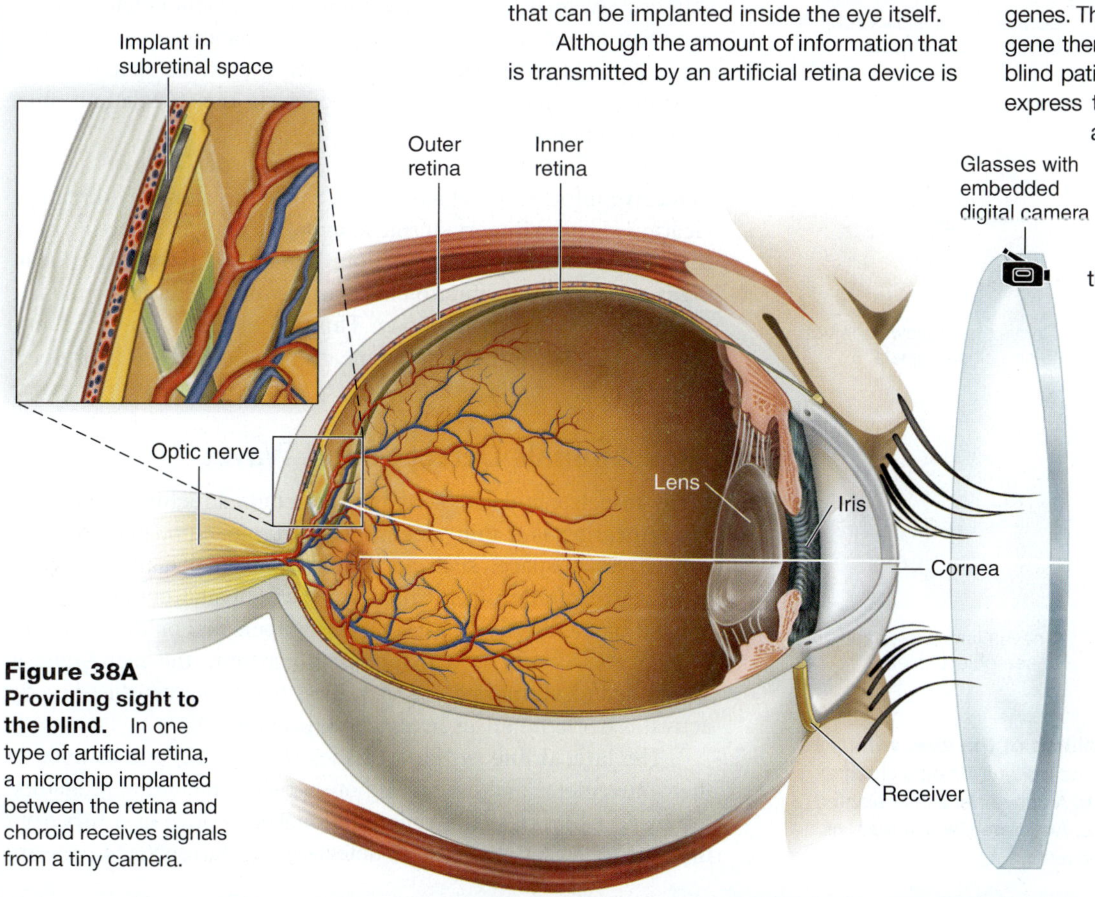

Figure 38A Providing sight to the blind. In one type of artificial retina, a microchip implanted between the retina and choroid receives signals from a tiny camera.

Questions to Consider

1. Although this technology is very exciting, it is also expensive. Do you believe that health insurance companies should pay the $115,000 cost of an artificial retina?
2. Suppose this technology advances to the point where vision can not only be restored but also greatly enhanced. Should it then be legally available to anyone who can afford it—for example, highly paid professional baseball players? What about for children who want to be professional athletes?

Those who cannot read these letters but can focus on close objects are said to be **nearsighted (myopic)**. These individuals often have an elongated eye, and when they attempt to look at a distant object, the image is brought to focus in front of the retina. Concave lenses, which diverge the light rays so that the image can be focused on the retina, usually correct this problem (Fig. 38.11*a*).

Those who can easily read the optometrist's chart but cannot easily focus on near objects are said to be **farsighted (hyperopic)**. They often have a shortened eye, and when they try to view near objects, the image is focused behind the retina. Convex lenses that increase the bending of light rays allow the image to be focused on the retina (Fig. 38.11*b*).

When the cornea or lens is uneven, the image appears fuzzy. This condition, called **astigmatism,** can sometimes be corrected by an unevenly ground lens to compensate for the uneven cornea (Fig. 38.11*c*).

Rather than wear glasses or contact lenses, many people with visual focus problems are now choosing to undergo laser-assisted in situ keratomileusis, or LASIK. First, specialists determine how much the cornea needs to be flattened to achieve optimum vision. Controlled by a computer, the laser then removes this amount of the cornea. As of 2012, about 19 million LASIK procedures have been performed in the United States. Various surveys indicate that about 95% of people who have the LASIK procedure are satisfied with the results.

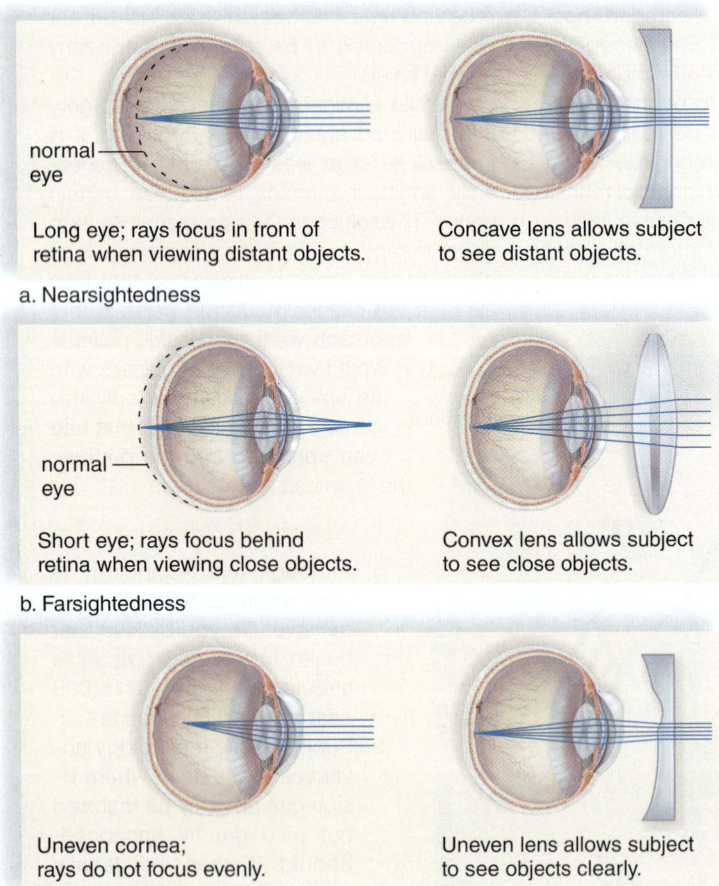

Figure 38.11 Common abnormalities of the eye, with possible corrective lenses. **a.** A concave lens in nearsighted persons focuses light rays on the retina. **b.** A convex lens in farsighted persons focuses light rays on the retina. **c.** An uneven lens in persons with astigmatism focuses light rays on the retina.

Check Your Progress 38.3

1. Compare rods and cones in terms of their main functions, their light sensitivity, and the excitation of ganglion cells.
2. List the three layers of cells in the human retina.
3. Summarize how the shape of the eye can result in nearsightedness or farsightedness, and describe the cause of astigmatism.

38.4 Senses of Hearing and Balance

Learning Outcomes

Upon completion of this section, you should be able to

1. Compare several strategies that different types of animals use to hear sounds.
2. Distinguish among the parts of the human ear that make up the outer, middle, and inner ear.
3. Review the differences between rotational and gravitational equilibrium and how each is accomplished.
4. Describe several common disorders affecting hearing and/or equilibrium.

Mechanoreception is sensing physical contact on the surface of the skin or movement of the surrounding environment (such as sound waves in air or water). The simplest mechanoreceptors are free nerve endings in the skin. At the other end of the spectrum, the most complex mechanoreception occurs in the middle and inner ear of vertebrates.

The evolutionary advantage of hearing is that it allows animals to receive information at a distance, as well as from any direction. Hearing plays an important role in avoiding danger, detecting prey, finding mates, and communication. In the most basic sense, hearing is caused by the vibration in a surrounding medium to resonate some part of an animal's body. This resonance is converted into electrical signals through some means that can then be interpreted by the animal's brain.

How Animals Detect Sound Waves

Many insects can detect sounds. A common structure involved in insect hearing is a thin membrane, or tympanum, that stretches across an air space, such as the tracheae, which also function in insect respiration (see Chapter 35). Tympanal organs are also located on the thorax (chest) of grasshoppers and on the front legs of crickets. Similar to a mammalian eardrum, the membrane is stimulated to vibrate by sound waves, but in insects this directly activates nerve impulses in attached receptor cells.

The **lateral line** system of fishes (Fig. 38.12) guides them in their movements and in locating other fishes, including predators, prey, and mates. Usually running along both sides of a fish from the gills to the tail, the system detects water currents and pressure

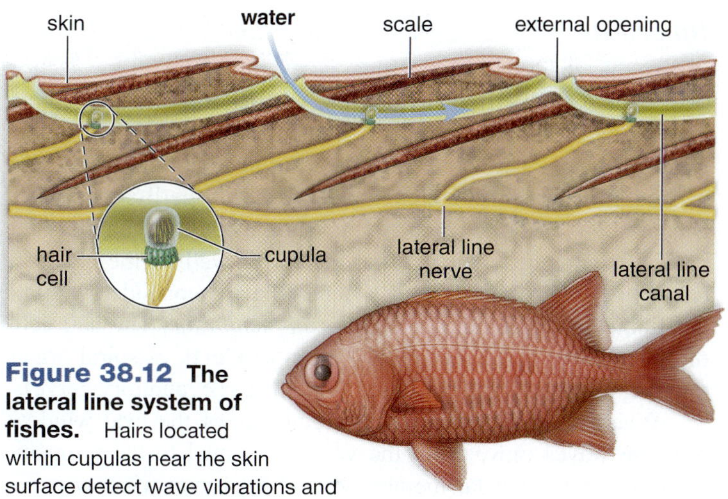

Figure 38.12 The lateral line system of fishes. Hairs located within cupulas near the skin surface detect wave vibrations and currents, helping guide fish movements in order to locate predators and prey, as well as mates.

waves from nearby objects in a manner similar to the sensory receptors in the human ear. Water from the environment enters tiny canals containing hair cells with cilia embedded in a gelatinous cupula. When the cupula bends due to pressure waves, the hair cells initiate nerve impulses.

Most terrestrial vertebrates can hear sound traveling in air, but some, such as amphibians and snakes, are also sensitive to vibrations from the ground, which travel to their inner ear via various parts of their skeleton. A middle ear with an eardrum that transmits sound waves to the inner ear via three small bones is unique to mammals.

The Human Ear

The ear has two sensory functions: hearing and balance (equilibrium). The mechanoreceptors for both of these are located in the inner ear, and each consists of *hair cells* with stereocilia (long microvilli) that are sensitive to mechanical stimulation.

MP3 The Senses of Hearing and Equilibrium

The ear has three distinct divisions: the outer, inner, and middle ear (Fig. 38.13). The **outer ear** consists of the pinna (external "ear") and the auditory canal. The opening of the auditory canal is lined with fine hairs and glands. Glands that secrete earwax, or cerumen, are located in the upper wall of the auditory canal. Earwax helps guard the ear against the entrance of foreign materials, such as air pollutants and microorganisms.

The **middle ear** begins at the **tympanic membrane** (eardrum) and ends at a bony wall containing two small openings covered by membranes. These openings are called the *oval window* and the *round window*. Three small bones are found between the tympanic membrane and the oval window. Collectively called the **ossicles,** individually they are the *malleus* (hammer), the *incus* (anvil), and the *stapes* (stirrup), so named because their shapes resemble these objects. The malleus adheres to the tympanic membrane, and the stapes touches the oval window.

An **auditory tube** (eustachian tube), which extends from each middle ear to the nasopharynx, permits the equalization of air pressure. Chewing gum, yawning, and swallowing in elevators and airplanes help move air through the auditory tubes on ascent and descent. As this occurs, we often hear the ears "pop."

Whereas the outer ear and the middle ear contain air, the inner ear is filled with fluids. Anatomically speaking, the **inner ear** has three areas: the **semicircular canals** and the **vestibule** are both concerned with equilibrium; the **cochlea** is concerned with hearing. The cochlea resembles the shell of a snail, because it spirals.

The Auditory Canal and Middle Ear

The process of hearing begins when sound waves enter the auditory canal. Just as ripples travel across the surface of a pond, sound waves travel by the successive vibrations of molecules. Ordinarily, sound waves do not carry much energy, but when a large number of waves strike the tympanic membrane, it moves back and forth (vibrates) ever so slightly. The malleus then takes the pressure from the inner surface of the tympanic membrane and passes it by means of the incus to the stapes in such a way that the pressure is multiplied about 20 times as it moves. The stapes strikes the membrane of the oval window, causing it to vibrate, and in this way, the pressure is passed to the fluid within the cochlea.

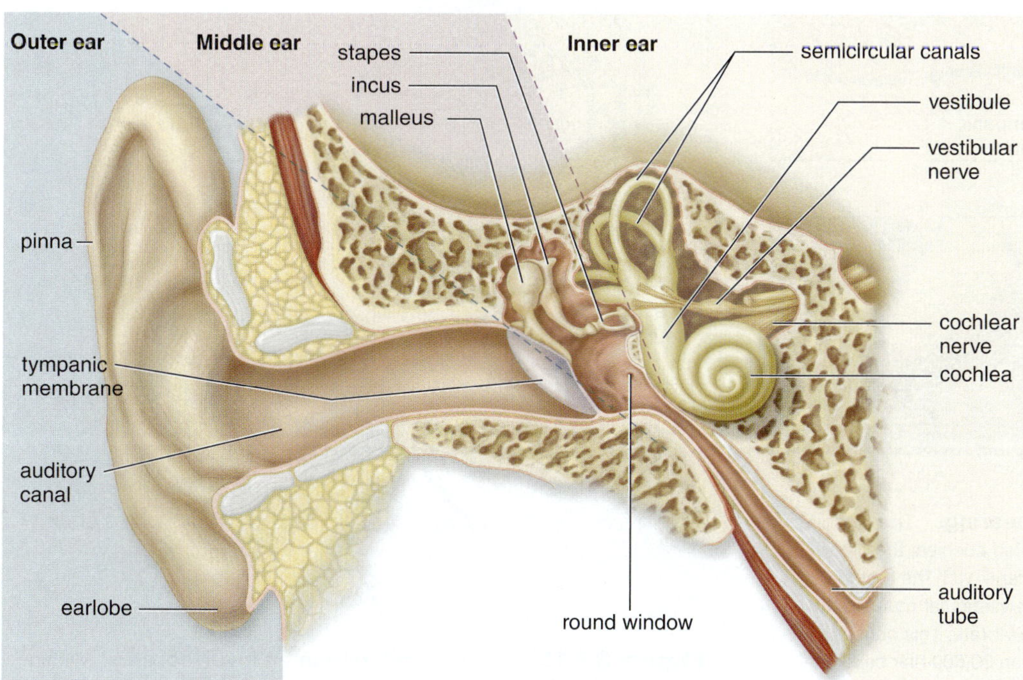

Figure 38.13 Anatomy of the human ear. In the middle ear, the malleus (hammer), the incus (anvil), and the stapes (stirrup) amplify sound waves. In the inner ear, the mechanoreceptors for equilibrium are in the semicircular canals and the vestibule, and the mechanoreceptors for hearing are in the cochlea.

Inner Ear

When the snail-shaped cochlea is examined in cross section (Fig. 38.14), the vestibular canal, the cochlear canal, and the tympanic canal become apparent. The cochlear canal contains endolymph, which is similar in composition to interstitial fluid. The vestibular and tympanic canals are filled with perilymph, which is continuous with the cerebrospinal fluid. Along the length of the basilar membrane, which forms the lower wall of the *cochlear canal,* are little hair cells whose stereocilia are embedded within a gelatinous material called the *tectorial membrane*. The hair cells of the cochlear canal, called the **organ of Corti,** or spiral organ, synapse with nerve fibers of the *cochlear nerve* (auditory nerve).

When the stapes strikes the membrane of the oval window, pressure waves move from the vestibular canal to the tympanic canal across the basilar membrane, and the round window membrane bulges. The basilar membrane moves up and down, and the stereocilia of the hair cells embedded in the tectorial membrane bend. Then, nerve impulses begin in the cochlear nerve and travel to the brain stem. When they reach the auditory areas of the cerebral cortex, they are interpreted as a sound.

Animation Effects of Sound Waves on Cochlear Structures

Each part of the organ of Corti is sensitive to a different wave frequency, or pitch of sound. Near the tip, the organ of Corti responds to low pitches, such as the sound of a tuba, and near the base, it responds to higher pitches, such as that of a bell or whistle. The nerve fibers from each region along the length of the organ of Corti lead to slightly different areas in the brain. The pitch sensation we experience depends on which region of the basilar membrane vibrates and which area of the brain is stimulated.

Volume is a function of the amplitude of sound waves. Loud noises cause the fluid in the vestibular canal to exert more pressure

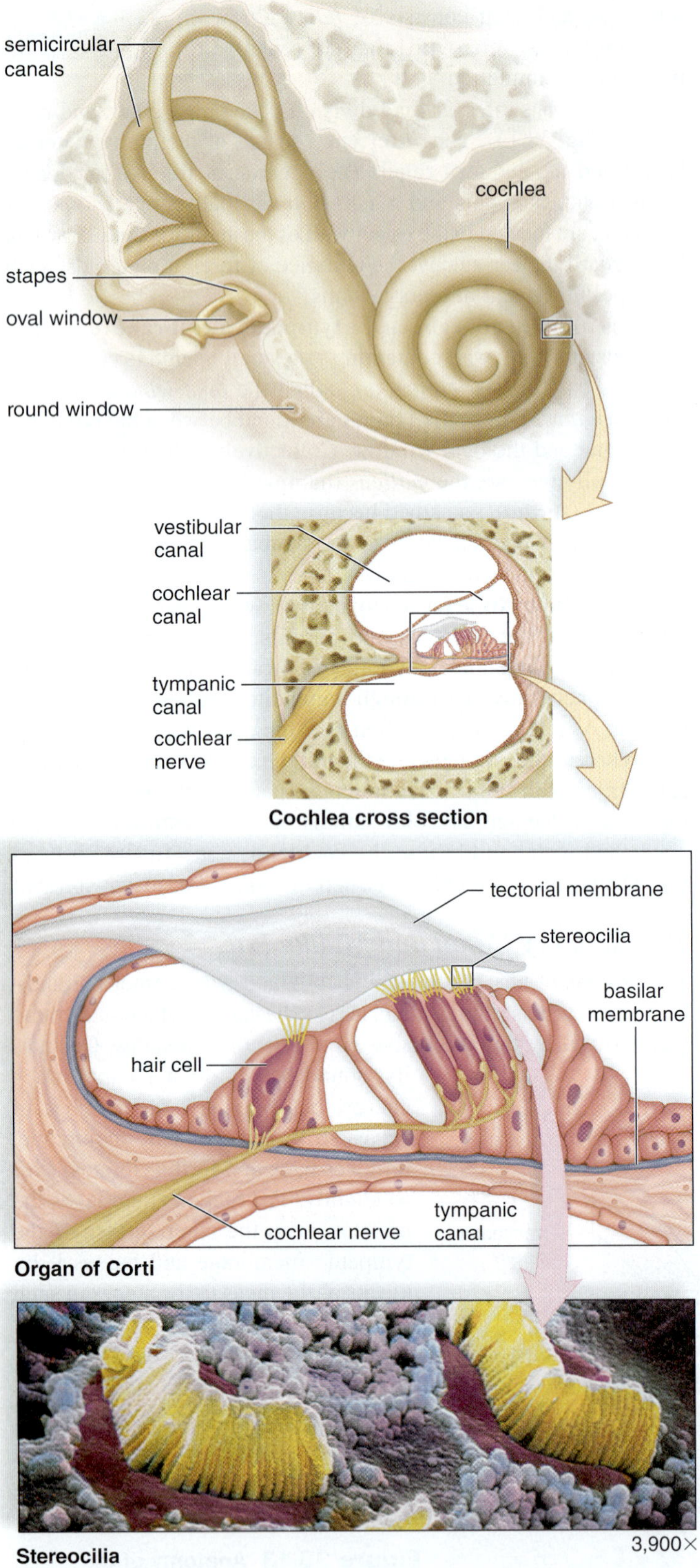

Figure 38.14 Mechanoreceptors for hearing. The organ of Corti is located within the cochlea. In the uncoiled cochlea, the organ consists of hair cells resting on the basilar membrane, with the tectorial membrane above. Pressure waves move from the vestibular canal to the tympanic canal, causing the basilar membrane to vibrate. This causes the stereocilia (or at least a portion of the more than 20,000 hair cells) embedded in the tectorial membrane to bend. Nerve impulses traveling in the cochlear nerve result in hearing.

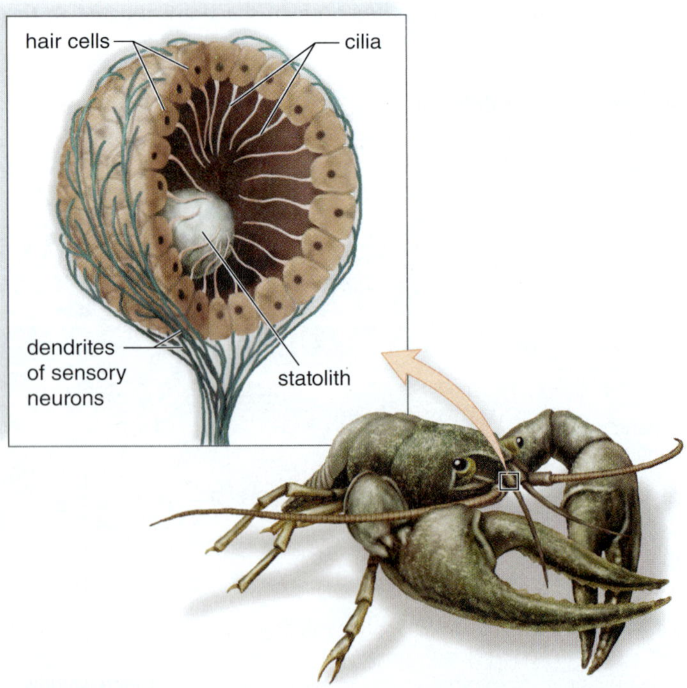

Figure 38.15 Sense of balance in an invertebrate. Within a statocyst, a small particle (the statolith) comes to rest on hair cells and allows a crustacean to sense the position of its head.

and the basilar membrane to vibrate to a greater extent. The resulting increased stimulation is interpreted by the brain as volume. It is believed that the brain interprets the tone of a sound based on the distribution of the hair cells that are stimulated.

Sense of Balance

Gravitational equilibrium organs, called statocysts (Fig. 38.15), are found in cnidarians, molluscs, and crustaceans, which are arthropods. When the head stops moving, a small particle called a statolith stimulates the cilia of the closest hair cells, and these cilia generate impulses that are interpreted as the position of the head.

In the human ear, mechanoreceptors in the semicircular canals detect rotational and/or angular movement of the head (**rotational equilibrium**), while mechanoreceptors in the utricle and saccule detect straight-line movement of the head in any direction (**gravitational equilibrium**).

Rotational Equilibrium

Rotational equilibrium (Fig. 38.16*a*) involves the semicircular canals, which are arranged so that there is one in each dimension of space. The base of each of the three canals, called the ampulla, is slightly enlarged. Little hair cells, whose stereocilia are embedded

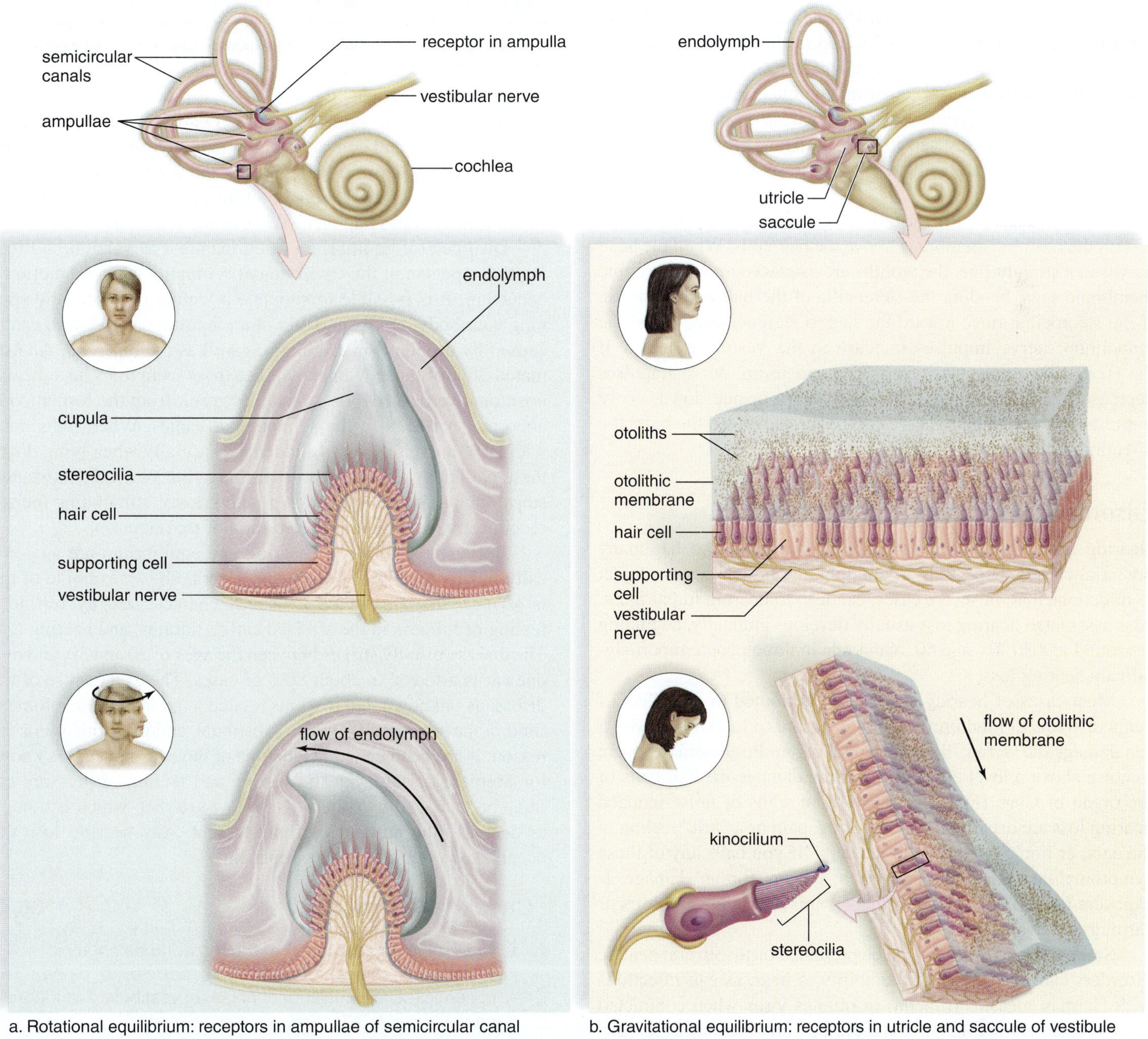

Figure 38.16 Mechanoreceptors for equilibrium. **a.** Rotational equilibrium. The ampullae of the semicircular canals contain hair cells with stereocilia embedded in a cupula. When the head rotates, the cupula is displaced, bending the stereocilia. Thereafter, nerve impulses travel in the vestibular nerve to the brain. **b.** Gravitational equilibrium. The utricle and the saccule contain hair cells with stereocilia embedded in an otolithic membrane. When the head bends, otoliths are displaced, causing the membrane to sag and the stereocilia to bend. If the stereocilia bend toward the kinocilium, the longest of the stereocilia, nerve impulses increase in the vestibular nerve. If the stereocilia bend away from the kinocilium, nerve impulses decrease in the vestibular nerve. This difference tells the brain in which direction the head moved.

within a gelatinous material called a cupula, are found within the ampullae. Because there are three semicircular canals, each ampulla responds to head movement in a different plane of space. As fluid (endolymph) within a semicircular canal flows over and displaces a cupula, the stereocilia of the hair cells bend, and the pattern of impulses carried by the vestibular nerve to the brain changes. The brain uses information from the semicircular canals to maintain equilibrium through appropriate motor output to various skeletal muscles that can right our position in space as need be.

Gravitational Equilibrium

Gravitational equilibrium (Fig. 38.16*b*) depends on the **utricle** and **saccule,** two membranous sacs located in the vestibule. Both of these sacs contain little hair cells, whose stereocilia are embedded within a gelatinous material called an otolithic membrane. Calcium carbonate ($CaCO_3$) granules, or **otoliths,** rest on this membrane. The utricle is especially sensitive to horizontal (back and forth) movements of the head, while the saccule responds best to vertical (up and down) movements.

When the head is still, the otoliths in the utricle and the saccule rest on the otolithic membrane above the hair cells. When the head moves in a straight line, the otoliths are displaced and the otolithic membrane sags, bending the stereocilia of the hair cells beneath. If the stereocilia move toward the largest stereocilium, called the kinocilium, nerve impulses increase in the vestibular nerve. If the stereocilia move away from the kinocilium, nerve impulses decrease in the vestibular nerve. If you are upside down, nerve impulses in the vestibular nerve cease. These data tell the brain the direction of the movement of the head.

Disorders of Hearing and Equilibrium

Hearing loss can develop gradually or suddenly and has many potential causes. Especially in children, the middle ear is subject to infections that, in severe cases, can lead to hearing impairment. Age-associated hearing loss usually develops gradually, beginning at around age 20. By age 60, about one in three people report significant hearing loss.

Most cases of hearing loss can be attributed to the effect of years of frequent (and preventable) exposure to loud noise, which can damage the stereocilia of the spiral organ. Prolonged exposure to noise above a level of 80 decibels can damage the hair cells of the organ of Corti (Fig. 38.17). The first signs of noise-induced hearing loss are usually muffled hearing, pain or a "full" feeling in the ears, or tinnitus (ringing in the ears). If you have any of these symptoms, take steps immediately to prevent further damage. If exposure to loud noise is unavoidable, noise-reducing earmuffs or earplugs are available.

Some types of deafness can be present at birth. Several genetic disorders can interfere with the ability to hear, as can infections with German measles (rubella) or mumps virus when contracted by a woman during her pregnancy. For this reason, every girl should be vaccinated against these viruses before she reaches childbearing age.

People with certain types of deafness may be able to hear again with cochlear implants. A cochlear implant consists of an external device that sits behind the ear and an internal device surgically implanted under the skin. The external part picks up sounds from the environment and converts them into electrical impulses, which are sent directly to different regions of the auditory nerve and then to the brain. As of 2012, about 58,000 adults and 38,000 children in the United States had received cochlear implants.

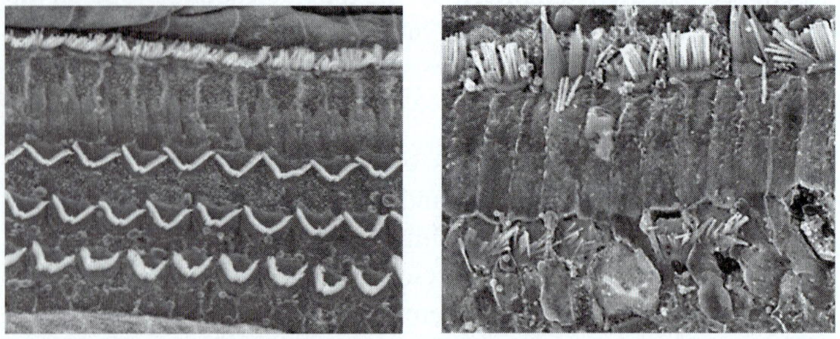

a. b.

Figure 38.17 Studies of hearing loss. a. Microscopic view of normal hair cells in the organ of Corti of a guinea pig. **b.** Damage to these hair cells occurred after 24 hours of exposure to a noise level typical of a rock concert.

Disorders of equilibrium often manifest as **vertigo,** the feeling that a person or the environment is moving when no motion is occurring. It is possible to simulate a feeling of vertigo by spinning your body rapidly and then stopping suddenly. Vertigo can be caused by problems in the brain as well as the inner ear. An estimated 20% of those who experience these symptoms have benign positional vertigo (BPV), which may result from the formation of abnormal particles in the semicircular canals. When individuals with BPV move their head suddenly, especially when lying down, these particles shift like pebbles inside a tire. When the movement stops, the particles tumble down with gravity, stimulating the stereocilia and resulting in the sensation of movement.

Because the senses of hearing and equilibrium are anatomically linked, certain disorders can affect both. An example of this is *Meniere disease,* which is usually characterized by vertigo, a feeling of fullness in the affected ear(s), tinnitus, and hearing loss. The disease usually strikes between the ages of 20 and 50, and only one ear is affected in about 80% of cases. The exact cause of this disease is unknown, but it seems related to an increased volume of fluid in the semicircular canals, vestibule, and/or cochlea. For this reason, it has been called "glaucoma of the ear." No cure is known for Meniere disease, but the vertigo and feeling of pressure can often be managed by adhering to a low-salt diet, which is thought to decrease the amount of fluid present. Any hearing loss that occurs, however, is usually permanent.

Check Your Progress 38.4

1. Determine whether each of the following belongs to the outer, middle, or inner ear: a. ossicles, b. pinna, c. semicircular canals, d. cochlea, e. vestibule, f. auditory canal.
2. List, in order, the structures that must conduct a sound wave from the time it enters the auditory canal until it reaches the cochlea.
3. Identify which structures of the inner ear are responsible for gravitational equilibrium and for rotational equilibrium.

38.5 Somatic Senses

Learning Outcomes

Upon completion of this section, you should be able to

1. Compare and contrast the functions of proprioceptors, cutaneous receptors, and pain receptors.
2. List the specific types of cutaneous receptors that are sensitive to fine touch, pressure, pain, and temperature.

Senses whose receptors are associated with the skin, muscles, joints, and viscera are termed the *somatic senses.* These receptors can be categorized into three types: proprioceptors, cutaneous receptors, and pain receptors. All of these send nerve impulses via the spinal cord to the primary somatosensory areas of the cerebral cortex (see Fig. 37.10).

Proprioceptors

Proprioceptors are mechanoreceptors involved in reflex actions that maintain muscle tone, and thereby the body's equilibrium and posture. For example, proprioceptors called *muscle spindles* are embedded in muscle fibers (Fig. 38.18). If a muscle relaxes too much, the muscle spindle stretches, generating nerve impulses that cause the muscle to contract slightly. Conversely, when muscles are stretched too much, proprioceptors called *Golgi tendon organs,* buried in the tendons that attach muscles to bones, generate nerve impulses that cause the muscles to relax. Both types of receptors act together to maintain a functional degree of muscle tone.

Cutaneous Receptors

As noted in Chapter 31, the skin is composed of an epidermis and a dermis (see Fig. 31.9). The dermis contains many **cutaneous receptors,** which make the skin sensitive to touch, pressure, pain, and temperature.

Four types of cutaneous receptors are sensitive to fine touch. *Meissner corpuscles* and *Krause end bulbs* are concentrated in the fingertips, palms, lips, tongue, nipples, penis, and clitoris. *Merkel disks* are found where the epidermis meets the dermis. A free nerve ending called a *root hair plexus* winds around the base of a hair follicle and fires if a hair is touched.

Two types of cutaneous receptors are sensitive to pressure. *Pacinian corpuscles* are onion-shaped sensory receptors deep inside the dermis. *Ruffini endings* are encapsulated by sheaths of connective tissue and contain lacy networks of nerve fibers.

At least two types of free nerve endings in the epidermis are thermoreceptors. Both cold and warm receptors contain ion channels with activities that are affected by temperature. Cold receptors generate nerve impulses at an increased frequency as the temperature drops; warm receptors increase activity as the temperature rises. Some chemicals (e.g., menthol) can stimulate cold receptors.

Pain Receptors

The skin and many internal organs and tissues have pain receptors, also called free nerve endings or *nociceptors*. Regardless of the cause, damaged cells release chemicals that cause nociceptors to generate nerve impulses, which the brain interprets as pain. Other types of nociceptors are sensitive to extreme temperatures or excessive pressure.

Pain receptors have arisen in evolution because they alert us to potential danger. If you accidentally reach too close to a fire, for example, the reflex action of withdrawing your hand will help protect you from further tissue damage, while the unpleasant sensation your brain perceives will help you remember not to reach too close to a fire again.

Figure 38.18 Muscle spindles and Golgi tendon organs. 1 When a muscle is stretched, muscle spindles send sensory nerve impulses to the spinal cord. 2 Motor nerve impulses from the spinal cord cause slight muscle contraction. 3 When tendons are stretched excessively, Golgi tendon organs cause muscle relaxation.

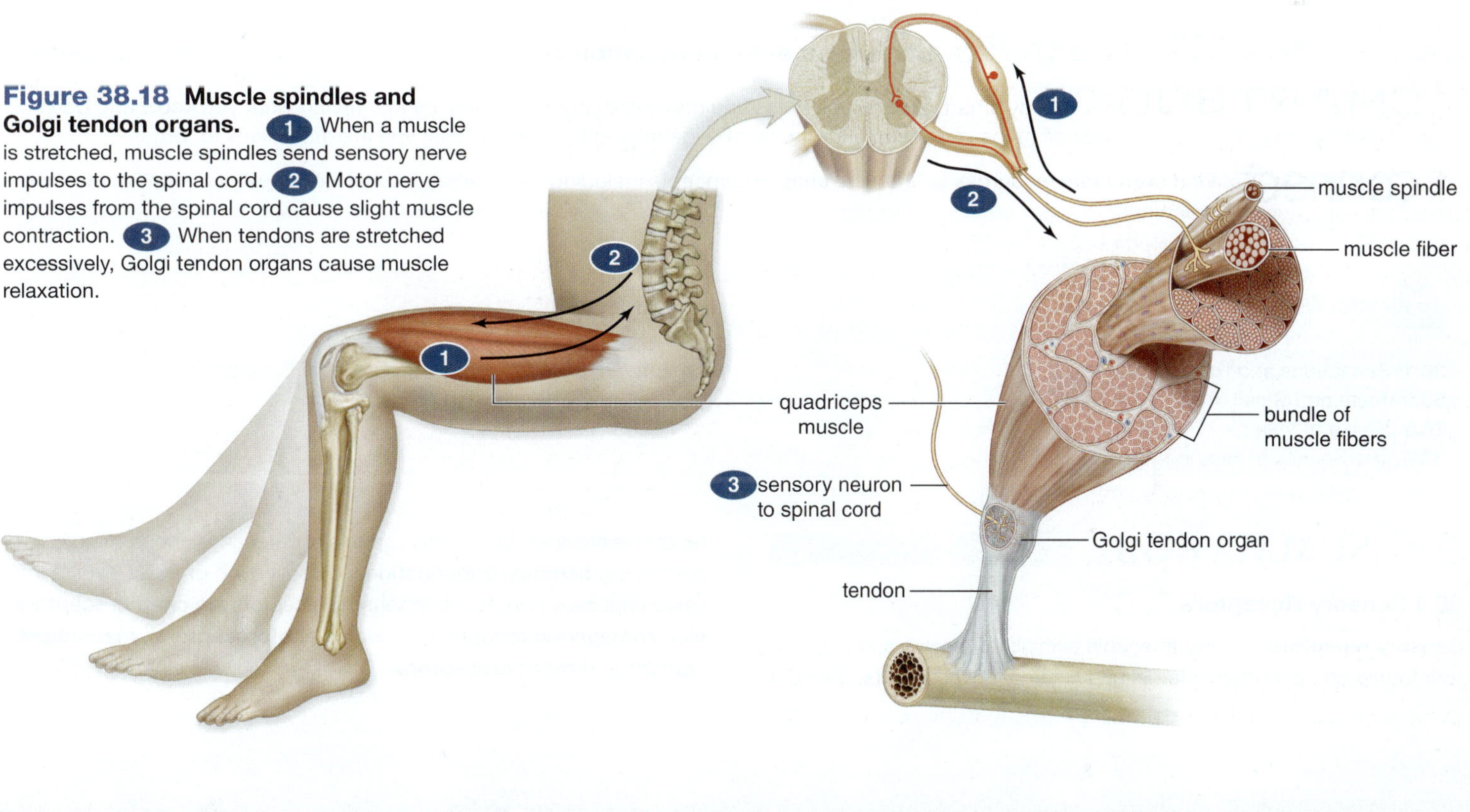

Unfortunately, despite the evolutionary benefit of acute pain, chronic pain often serves no such purpose. To relieve such nonadaptive pain, a variety of painkilling medications, or *analgesics,* have been developed. If the source of the pain is inflammation, anti-inflammatory medications can be used. These include natural anti-inflammatory compounds like corticosteroids, or nonsteroidal anti-inflammatory drugs (NSAIDs) like aspirin or ibuprophen. Each day an estimated 17 million Americans use NSAIDs, which are generally available over-the-counter, and work by inhibiting enzymes that generate inflammatory chemicals called prostaglandins. For more intense pain, opioid medications such as morphine or oxycodone can be prescribed. These stimulate receptors in the brain for another class of naturally occurring analgesics, the *endorphins*.

Check Your Progress 38.5

1. Identify the problems that would likely occur if a person lacked muscle spindles, or nociceptors.
2. In evolutionary terms, assess why cutaneous receptors quickly become adapted to stimuli (e.g., why we don't continue to feel a chair once we settle in), whereas the sense of pain seems to be much less adaptable (e.g., many people suffer from chronic pain).

CONNECTING *the* CONCEPTS *with the* THEMES

Evolution

- Through the evolutionary process, animals tend to rely on those stimuli and senses that are adaptive to their particular environment and way of life.
- In humans and perhaps other animals, impulses received from sensory receptors result in perception, or conscious awareness of events occurring in the external environment.
- Virtually all animals are capable of learning from their sensory experiences to engage in behavior that helps them avoid danger, locate food, and find potential mates.

Nature of Science

- Scientists now have a fairly thorough understanding of how sensory receptors function; however, the processes of sensory transduction and, in particular, perception remain considerably more mysterious.
- Researchers in biomedical science are developing new technologies to enable hearing, in some cases of deafness, and vision, in some cases of blindness.

Biological Systems

- Sensory receptors convey information to the nervous system by converting environmental stimuli into nerve impulses (sensory transduction).
- Depending on the level of stimulus and its source, the nervous system may react by altering heart rate and breathing rate, releasing hormones from glands, and stimulating muscles to perform actions such as fight-or-flight, feeding, or mating behaviors.
- Our knowledge of the world around us is completely dependent on functioning sensory receptors and the interpretation of their signals.

MEDIA STUDY TOOLS http://connect.mheducation.com

SMARTBOOK® Study smart. SmartBook helps you maximize your study time by identifying what you know and don't know. The recharge feature in SmartBook helps you prepare for the test by identifying the material that you are most likely to forget.

connect BIOLOGY Find even more resources to learn the chapter concepts including animations, tutorial videos, and interactive practice questions.

MP3 Files	Animation
38.1 Sensations and Receptors **38.2** Taste and Smell **38.3** Sense of Vision **38.4** The Senses of Hearing and Equilibrium	**38.4** Effects of Sound Waves on Cochlear Structures

SUMMARIZE

38.1 Sensory Receptors

Sensory receptors evolved to enable animals to receive and respond to information about their internal and external environments. Sensory receptors are specialized cells capable of detecting various stimuli, then performing **sensory transduction,** or conversion of those events into nerve impulses. Four types of sensory receptors are **chemoreceptors, electromagnetic receptors** (including **photoreceptors**), **mechanoreceptors** and **thermoreceptors.**

38.2 Chemical Senses

Chemoreception is found universally in animals and is therefore believed to be the most primitive sense. Chemoreceptors are present in a wide variety of locations in animals. Examples of chemoreceptors in humans include **taste buds** and **olfactory cells.**

38.3 Sense of Vision

Most animals have **photoreceptors** that are sensitive to light. Arthropods have **compound eyes;** vertebrates and some molluscs have a **camera-type eye.** Animals with eyes that face sideways have **panoramic vision,** while those with eyes facing forward have **stereoscopic vision.** In humans, vision is dependent on the eye, the optic nerves, and the visual areas of the cerebral cortex. The human eye has three layers. The **sclera** can be seen as the white of the eye; it also becomes the transparent bulge in the front of the eye called the **cornea.** The surface of the sclera is covered by the **conjunctiva.** The middle **choroid** layer contains blood vessels and pigment that absorbs stray light rays. It also forms the **iris,** which regulates the size of the **pupil.** The **lens** helps to form images. The **retina,** or inner layer, contains **rod cells** (sensory receptors for dim light) and **cone cells** (sensory receptors for bright light and color). Many cones are present in the **fovea centralis,** but no rods or cones are in the **blind spot.**

The cornea begins focusing light, and the lens provides further **visual accomodation,** controlled by the **ciliary muscle.** The retina is composed of three layers of cells: the rod and cone layer, the bipolar cell layer, and the ganglion cell layer. When light strikes **rhodopsin** in the membranous disks of rod cells, rhodopsin splits into opsin and retinal, generating nerve impulses that are transmitted from the ganglion cells, via the optic nerve, to the brain. On their way, the optic nerves cross at the optic chiasma and pass through the thalamus before images are interpreted by the visual cortex.

Common causes of blindness include retinal disorders like **diabetic retinopathy, macular degeneration,** and **retinal detachment. Glaucoma** is an abnormal increase in intraocular pressure. **Cataracts,** or opaque lens, affect transmission of light. People with visual focus disorders may be **nearsighted** (**myopic**), **farsighted** (**hyperopic**), or have **astigmatism.**

38.4 Senses of Hearing and Balance

Animals use a variety of strategies to detect sounds or vibrations in their environment. Insects have tympanal organs to detect sounds; fishes have a **lateral line** system to detect vibrations in water. Hearing in humans is dependent on the ear, the cochlear nerve, and the auditory areas of the cerebral cortex. The ear is divided into three parts. The **outer ear** consists of the pinna and the auditory canal, which direct sound waves to the middle ear. The **middle ear** begins with the **tympanic membrane** and contains the **ossicles** (malleus, incus, and stapes). The malleus is attached to the tympanic membrane, and the stapes is attached to the oval window, which is covered by membrane. An **auditory tube** connects each middle ear to the nasopharynx. The **inner ear** contains the **cochlea** for hearing, and the **semicircular canals** and **vestibule** for equilibrium.

Hearing begins when the outer and middle portions of the ear convey and amplify the sound waves that strike the oval window. Its vibrations set up pressure waves within the cochlea, which contains the **organ of Corti,** consisting of hair cells whose stereocilia are embedded in the tectorial membrane. When the stereocilia of the hair cells bend, nerve impulses begin in the cochlear nerve and are carried to the brain.

The ear also contains receptors for our sense of equilibrium. **Rotational equilibrium** is dependent on the stimulation of hair cells in the ampullae of the semicircular canals. **Gravitational equilibrium** relies on the stimulation of hair cells by **otoliths** in the **utricle** and **saccule,** inside the vestibule. Disorders affecting hearing and balance include age-associated hearing loss, **vertigo,** and Meniere disease.

38.5 Somatic Senses

The somatic sensory receptors include proprioceptors, sensory receptors, and pain receptors. **Proprioceptors,** such as muscle spindles and Golgi tendon organs, help maintain equilibrium and posture. **Cutaneous receptors** in the skin detect touch, pressure, pain, and temperature. Pain receptors are free nerve endings that respond to chemicals released by damaged cells.

ASSESS

Choose the best answer for each question.

38.1 Sensory Receptors

1. The conversion of an environmental stimulus into a nerve impulse is called sensory
   a. conduction.
   b. conversion.
   c. transduction.
   d. transformation.
2. The brain is able to interpret nerve impulses coming from sensory receptors as vision, hearing, taste, and so on mainly because
   a. the impulses are of different strength (quantity).
   b. the impulses are of different types (quality).
   c. impulses originating in different sensory organs travel to different areas of the brain.
   d. None of these are correct.
3. Which type of sensory receptor is involved in detecting blood pressure?
   a. chemoreceptors
   b. mechanoreceptors
   c. photoreceptors
   d. thermoreceptors

38.2 Chemical Senses

4. Which association is incorrect?
   a. chemoreceptors on antennae—crustaceans
   b. chemoreceptors on feet—insects
   c. Jacobson's organs in mouth—planarians
   d. taste buds on tongue—humans
5. Which of the following is not a major type of taste receptor in humans?
   a. salty
   b. spicy
   c. sour
   d. sweet
6. In humans, most olfactory cells are located in the
   a. nasal cavity.
   b. nostrils.
   c. pharynx.
   d. roof of the mouth.

38.3 Sense of Vision

7. Which association is incorrect?
   a. camera-type eye—molluscs
   b. compound eye—arthropods
   c. eyespots—planarians
   d. panoramic vision—predators
8. Which of the following is the correct path for light rays entering the human eye?
   a. sclera, retina, choroid, lens, cornea
   b. fovea centralis, pupil, aqueous humor, lens
   c. cornea, pupil, lens, vitreous humor, retina
   d. optic nerve, sclera, choroid, retina, humors
   e. All of these are correct.
9. Label this diagram of the human eye. State a function for each structure labeled.

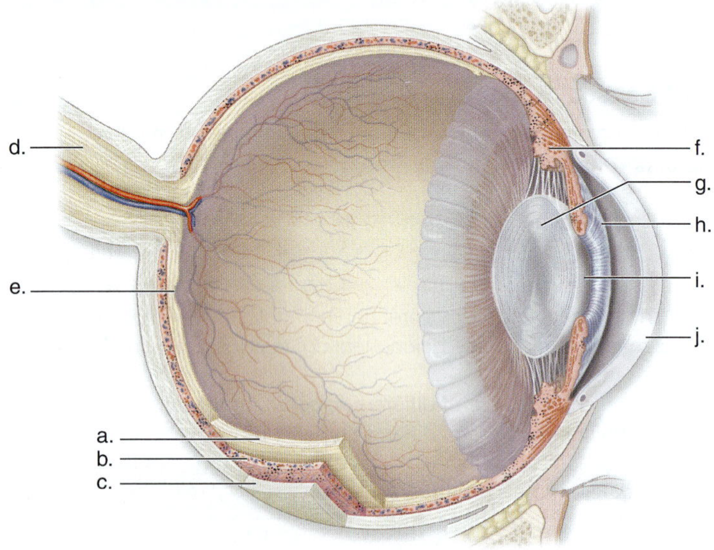

10. To focus on objects that are close to the viewer,
    a. the suspensory ligaments must be pulled tight.
    b. the lens needs to become more rounded.
    c. the ciliary muscle will be relaxed.
    d. the image must focus on the area of the optic nerve.
11. Which abnormality of the eye is correctly matched?
    a. cataracts—cloudy corneas
    b. glaucoma—intraocular pressure is decreased
    c. farsightedness—eyeball is longer than usual
    d. nearsightedness—image focuses behind the retina
    e. astigmatism—cornea or lens is uneven

38.4 Senses of Hearing and Balance

12. Which of the following is not involved with the detection of sound waves?
    a. tympanum of insects
    b. statocysts of crustaceans
    c. lateral line of fishes
    d. skeleton of amphibians and snakes
13. Which one of these wouldn't you mention if you were tracing the path of sound vibrations through the human ear?
    a. auditory canal
    b. tympanic membrane
    c. semicircular canals
    d. cochlea
    e. ossicles
14. Which one of these correctly describes the location of the organ of Corti?
    a. between the tympanic membrane and the oval window in the inner ear
    b. in the utricle and saccule within the vestibule
    c. between the tectorial membrane and the basilar membrane in the cochlear canal
    d. between the outer and inner ear within the semicircular canals
15. Stimulation of hair cells in the semicircular canals results from the movement of
    a. endolymph.
    b. aqueous humor.
    c. basilar membrane.
    d. otoliths.

38.5 Somatic Senses

16. Mechanoreceptors involved in maintaining muscle tone and body posture are called
    a. cutaneous receptors.
    b. pain receptors.
    c. proprioceptors.
    d. statocysts.
17. Which type of cutaneous receptor is sensitive to pressure?
    a. Meissner corpuscle
    b. Merkel disk
    c. Pacinian corpuscle
    d. root hair plexus
18. Nerve impulses generated by nociceptors are interpreted by the brain as
    a. fine touch.
    b. heat.
    c. pain.
    d. pressure.

ENGAGE

THINKING SCIENTIFICALLY

1. Suppose you go to a popular pizza buffet restaurant on a busy Saturday night. As you take your seat in a booth near the kitchen, and later bite into a hot slice of pizza, describe which of your sensory systems are being stimulated, and what types of receptors are involved in each perceived sensation.
2. The density of taste buds on the tongue can vary. Some obese individuals have a lower density of taste buds than usual. Assume that taste perception is related to taste bud density. If so, what hypothesis would you test to see if there is a relationship between taste bud density and obesity?
3. Imagine that you are in a room with very little light, and you are having trouble finding the light switch. You notice that you can see better with your peripheral vision than in the center of your visual field. Why is this the case?
4. Some sensory receptors, such as those for taste, smell, and pressure, readily undergo the process of sensory adaptation, or decreased response to a stimulus. In contrast, receptors for pain are less prone to adaptation. Why does this make good biological sense? What do you think happens to children who are born without the ability to feel pain normally?

Gymnastics requires coordination between the nervous and support systems.

39

Locomotion and Support Systems

Chapter Outline

39.1 Diversity of Skeletons 734

39.2 The Human Skeletal System 736

39.3 The Muscular System 742

Before You Begin

Before beginning this chapter, take a few moments to review the following discussions.

Figure 37.6 What is the function of acetylcholine in the transmission of a nerve impulse to skeletal muscle?

Sections 37.3 and 37.4 How does the primary motor area of the cerebral cortex generate commands to skeletal muscle, and how does the somatic division of the PNS control the muscles?

Chapter 38 How do the various types of sensory receptors provide information and feedback to the brain?

Gabrielle "Gabby" Douglas took her first gymnastics class when she was 2 years old. By age 8, she had won a gymnastics title in Virginia, and she won two gold medals at the 2012 Summer Olympics. When Gabby does a routine, her muscular and skeletal systems are working together under the control of her nervous system. The same is true when eagles fly, fish swim, or animals feed, escape prey, reproduce, or simply play. Although some animals lack muscles and bones, they all use contractile fibers to move about at some stage of their lives. In many invertebrates, muscles push against body fluids located inside either a gastrovascular cavity or a coelom.

Only in vertebrates are muscles attached to a bony endoskeleton. Both the skeletal system and the muscular system contribute to homeostasis. Aside from giving the body shape and protecting internal organs, the skeleton serves as a storage area for inorganic calcium and produces blood cells. The skeleton also protects internal organs while supporting the body against the pull of gravity. While contributing to body movement, the skeletal muscles give off heat, which warms the body. This chapter compares locomotion in animals and reviews the musculoskeletal system of vertebrates.

As you read through this chapter, think about the following questions:

1. What advantages do animals with a skeletal system have over animals that completely lack such a system?
2. How does the nervous system specifically control the skeletal system?
3. How do the sensory systems exert influence over the nervous system and therefore over the muscular and skeletal systems?

Following *the* Themes

CHAPTER 39 LOCOMOTION AND SUPPORT SYSTEMS

UNIT 7 COMPARATIVE ANIMAL BIOLOGY

Evolution	The need for protection, support, and mobility in different environments has led to the evolution of a number of different skeletal systems.
Nature of Science	Investigations into bone and muscle functions have provided an understanding of how animals are adapted to their lifestyles, as well as having many applications to human health.
Biological Systems	In humans, the skeletal system protects the internal organs, stores ions and contributes to homeostasis, produces blood cells for the circulatory system, and works together with the muscular system.

39.1 Diversity of Skeletons

Learning Outcomes

Upon completion of this section, you should be able to

1. Describe a typical hydrostatic skeleton and list some examples of animals that possess one.
2. Discuss some advantages of having an endoskeleton versus an exoskeleton.
3. Provide several examples of how mammalian skeletons are adapted to particular forms of locomotion.

Skeletons serve as support systems for animals, providing rigidity, protection, and surfaces for muscle attachment. Several different kinds of skeletons occur in the animal kingdom. Cnidarians, flatworms, roundworms, and annelids have a hydrostatic skeleton. Typically, molluscs and arthropods have an exoskeleton (external skeleton) composed of calcium carbonate or chitin, respectively. Sponges, echinoderms, and vertebrates possess an internal skeleton, or *endoskeleton.* In echinoderms, the endoskeleton is composed of calcareous plates; in vertebrates, the endoskeleton is composed of cartilage, bone, or both.

Hydrostatic Skeleton

In animals that lack a hard skeleton, a fluid-filled gastrovascular cavity or a fluid-filled coelom can act as a hydrostatic skeleton. A **hydrostatic skeleton** utilizes fluid pressure to offer support and resistance to the contraction of muscles, so that mobility results. As analogies, consider that a garden hose stiffens when filled with water, and that a water-filled balloon changes shape when squeezed at one end. Similarly, an animal with a hydrostatic skeleton can change shape and perform a variety of movements.

Hydras and planarians use their fluid-filled gastrovascular cavity as a hydrostatic skeleton. The tentacles of a hydra also have hydrostatic skeletons, allowing them to be extended to capture food. Roundworms have a fluid-filled pseudocoelom and move in a whiplike manner when their longitudinal muscles contract.

The coelom of annelids, such as earthworms, is segmented and has septa that divide it into compartments (Fig. 39.1). Each segment has its own set of longitudinal and circular muscles and its own nerve supply, so each segment or group of segments may function independently. When circular muscles contract, the segments become thinner and elongate. When longitudinal muscles contract, the segments become thicker and shorten. By alternating circular muscle contraction and longitudinal muscle contraction and by using its setae to hold its position during contractions, the animal moves forward.

Use of Muscular Hydrostats

Even animals that have an exoskeleton or an endoskeleton move selected body parts by means of *muscular hydrostats,* meaning that fluid contained within the individual cells that constitute a muscle assists with movement of that part. Muscular hydrostats are used by clams to extend their muscular foot and by sea stars to extend their tube feet. Spiders depend on them to move their legs, and moths rely on them to extend their proboscis. In vertebrates, movement of an elephant's trunk involves a muscular hydrostat that allows the animal to reach high into trees, pick up a morsel of food off the ground, or manipulate other objects.

Exoskeletons and Endoskeletons

Molluscs and arthropods have a rigid **exoskeleton,** an external covering composed of a stiff material. The strength of an exoskeleton can be improved by increasing its thickness and weight, but this leaves less room for internal organs.

In molluscs, such as snails and clams, a thick and nonmobile calcium carbonate shell is primarily used for protection against the environment and predators. A mollusc's shell can grow as the animal grows.

The exoskeleton of arthropods, such as insects and crustaceans, is composed of chitin, a strong, flexible, nitrogenous polysaccharide. Their exoskeleton protects them against wear and tear, predators, and desiccation (drying out)—an important feature for arthropods that live on land. Working together with muscles, the jointed and movable appendages of arthropods allow them to crawl, fly, and/or swim. Because their exoskeleton is of fixed size, however, arthropods must *molt,* or shed their skeleton, in order to grow (Fig. 39.2).

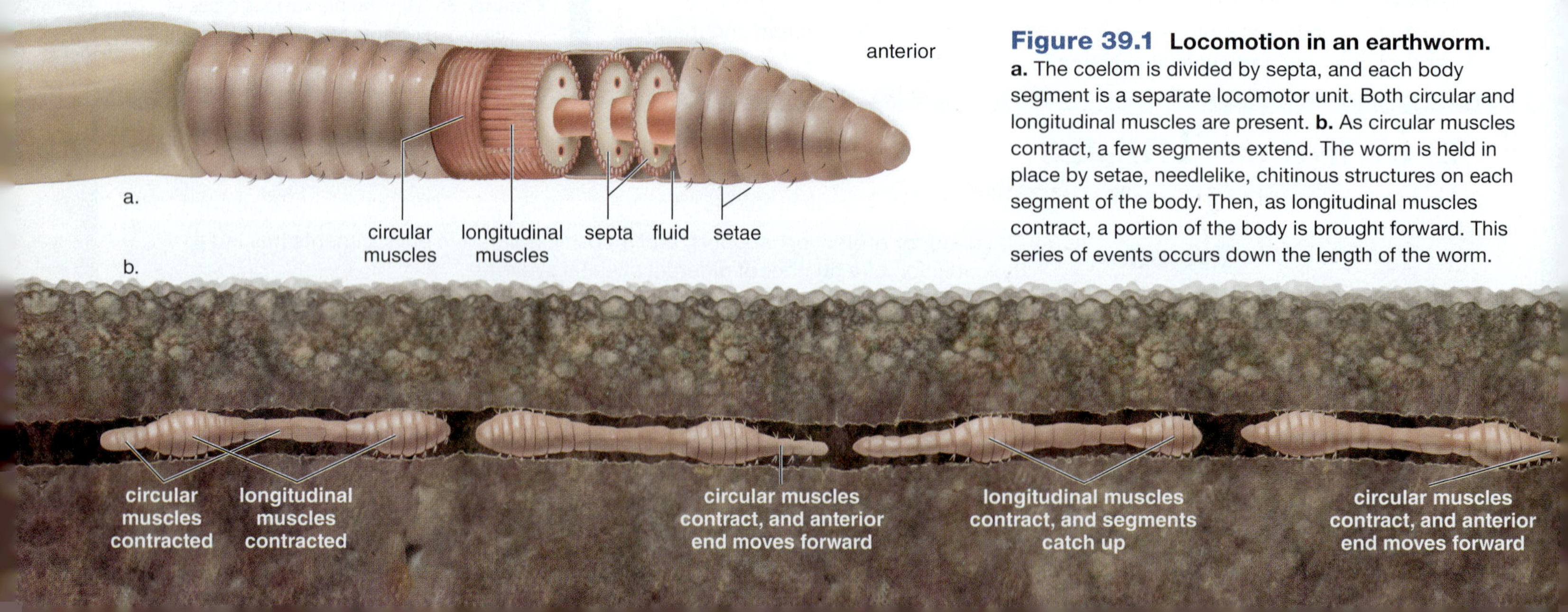

Figure 39.1 Locomotion in an earthworm. **a.** The coelom is divided by septa, and each body segment is a separate locomotor unit. Both circular and longitudinal muscles are present. **b.** As circular muscles contract, a few segments extend. The worm is held in place by setae, needlelike, chitinous structures on each segment of the body. Then, as longitudinal muscles contract, a portion of the body is brought forward. This series of events occurs down the length of the worm.

Figure 39.2 Exoskeleton. Exoskeletons support muscle contraction and prevent drying out. The chitinous exoskeleton of an arthropod is shed as the animal molts; until the new skeleton dries and hardens, the animal is more vulnerable to predators, and muscle contractions may not translate into body movements. In this photo, a cicada has just finished molting.

Both echinoderms and vertebrates have an **endoskeleton,** which is made up of rigid internal structures. The skeleton of echinoderms consists of spicules and plates of calcium carbonate embedded in the living tissue of the body wall. In contrast, the vertebrate endoskeleton is living tissue. Sharks and rays have skeletons composed only of cartilage. Other vertebrates, such as bony fishes, amphibians, reptiles, birds, and mammals, have endoskeletons composed of bone and cartilage.

The advantages of the jointed vertebrate endoskeleton are listed in Figure 39.3. An endoskeleton grows with the animal, so molting is not required. It supports the weight of a large animal without limiting the space for internal organs. An endoskeleton also offers protection to vital internal organs, but it is protected by the soft tissues around it. Injuries to soft tissue are usually easier to repair than injuries to a hard skeleton. Compared to the relatively limited mobility of arthropod appendages, vertebrate limbs are generally more flexible and have different types of joints, allowing for even more complex movements.

The skeletons of mammals come in many sizes and shapes, which are often adapted to a particular mode of locomotion. Aquatic animals such as seals, sea lions, whales, and dolphins have a streamlined, torpedo-shaped skeleton that facilitates movement through water. Many animals that jump, such as kangaroos and rabbits, have a compact skeleton with elongated hindlimbs that propel them forward. Carnivores, such as members of the cat family, walk on their toes, which is an adaptation to running and chasing prey. (Note that when humans run, we push off with our toes to move faster.) Hoofed mammals, such as horses and deer, have evolved long legs and run on the tips of elongated phalanges. The lowest part of each limb of a horse consists entirely of a modified third digit.

Humans are bipedal and walk on the soles of the feet formed by the tarsal and metatarsal bones. This form of locomotion allows the hands to be free and may have evolved from the monkeys' and apes' habit of using only forelimbs as they swing through the branches of trees. Dexterity of hands and feet is actually the ancestral mammalian condition. In humans and apes, the bones of the hands and feet are not fused, and the wrist and ankle can rotate in three dimensions.

Check Your Progress 39.1

1. List the types of skeletons found in animals.
2. Describe the type of support system that makes it possible to stick out your tongue.
3. Explain why an earthworm loses its cylindrical shape when it dies.

Advantages of Jointed Endoskeleton
Can grow with the animal
Supports the weight of large animal
Protects vital internal organs
Is protected by outer tissues
Allows flexible movements

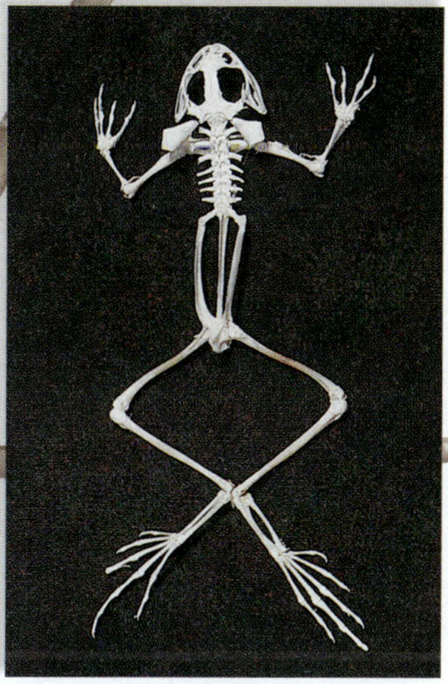

Figure 39.3 The vertebrate endoskeleton. The jointed vertebrate endoskeleton has the advantages listed to the left.

39.2 The Human Skeletal System

Learning Outcomes

Upon completion of this section, you should be able to

1. Review the five major functions of the skeletal system.
2. Describe the macroscopic and microscopic structure of bone.
3. List the major bones that constitute the human axial and appendicular skeletons.

The skeletal system has many functions that contribute to homeostasis:

- *Support of the body.* The rigid skeleton provides an internal framework that largely determines the body's shape.
- *Protection of vital internal organs,* such as the brain, heart, and lungs. The bones of the skull protect the brain; the rib cage protects the heart and lungs. The vertebrae protect the spinal cord.
- *Sites for muscle attachment.* The pull of muscles on the bones makes movement possible. Articulations (joints) occur between all the bones, but we associate body movement particularly with jointed appendages.
- *Storage reservoir for ions.* All bones have a matrix that contains calcium phosphate, a source of calcium ions and phosphate ions in the blood.
- *Production of blood cells.* Blood cells and other blood elements are produced in the red bone marrow of the skull, ribs, sternum, pelvis, and long bones.

In this section, we describe the characteristics of human bones, the components of various regions of the skeleton, and the different types and functions of joints.

Bone Growth and Renewal

During prenatal development, the structures that will form the bones of the human skeleton are composed of cartilage. Because these cartilaginous structures are shaped like the future bones, they provide "models" of these bones. The models are converted to bones as calcium salts are deposited in the matrix (nonliving material), first by the cartilage cells and later by bone-forming cells called **osteoblasts** (Gk. *osteon,* "bone"; *blastos,* "bud"). The conversion of cartilaginous models to bones is called endochondral ossification.

In some cases, ossification occurs without any previous cartilaginous model. This type of ossification occurs in the dermis and forms bones called dermal bones. Examples include the mandible (lower jaw), certain bones of the skull, and the clavicle (collarbone). During intramembranous ossification, fibrous connective tissue membranes give support as ossification begins.

Endochondral ossification of a long bone begins in a region called a primary ossification center, located in the middle of the cartilaginous model. In the primary ossification center, the cartilage is broken down and invaded by blood vessels, and cells in the area mature into bone-forming osteoblasts. Later, secondary ossification centers form at the ends of the model. A cartilaginous *growth plate* remains between the primary ossification center and each secondary center. As long as these plates remain, growth is possible. The rate of growth is controlled by hormones, particularly growth hormone (GH) and the sex hormones. Eventually, the plates become ossified, causing the primary and secondary centers of ossification to fuse, and the bone stops growing.

In the adult, bone is continually being broken down and built up again. Bone-absorbing cells called **osteoclasts** (Gk. *osteon,* "bone"; *klastos,* "broken in pieces") break down bone, remove worn cells, and deposit calcium in the blood. In this way, osteoclasts help maintain the blood calcium level and contribute to homeostasis.

Animation
Bone Growth

Among many other functions, calcium ions play a major role in muscle contraction and nerve conduction. The blood calcium level is closely regulated by the antagonistic hormones parathyroid hormone (PTH) and calcitonin. PTH promotes the activity of osteoclasts, and calcitonin inhibits their activity to keep the blood calcium level within normal limits.

MP3
Calcium Homeostasis

Assuming that the blood calcium level is normal, bone destruction caused by the work of osteoclasts is repaired by osteoblasts. As they form bone, some of these osteoblasts get caught in the matrix (nonliving material) they secrete and are converted to **osteocytes** (Gk. *osteon,* "bone"; *kytos,* "cell"). These cells live within the lacunae of osteons, where they continue to affect the timing and location of bone remodeling.

While a child is growing, the rate of bone formation is greater than the rate of bone breakdown. The skeletal mass continues to increase until ages 20 to 30. After that, the rate of formation and rate of breakdown of bone mass are equal, until ages 40 to 50. Then, reabsorption begins to exceed formation, and the total bone mass slowly decreases.

As people age, an abnormal thinning of the bones called **osteoporosis** can lead to an increased risk of fractures, especially of the wrist, vertebrae, and pelvis. About 10 million people in the United States have osteoporosis, which results in 1.5 million fractures each year.

Animation
Osteoporosis

Women are twice as likely as men to have an osteoporosis-related fracture in their lifetime, and about one in four women will experience such a fracture. This is partly due to the fact that women have about 30% less bone mass than men to begin with. Also, because sex hormones play an important role in maintaining bone strength, women typically lose about 2% of their bone mass each year after menopause. Estrogen replacement therapy has been shown to increase bone mass and reduce fractures, but it can also increase the risk of cardiovascular disease and certain types of cancer. Other strategies for decreasing osteoporosis risk include consuming 1,000–1,500 mg of calcium per day, obtaining sufficient vitamin D, and engaging in regular physical exercise.

Anatomy of a Long Bone

A long bone, such as the humerus, illustrates principles of bone anatomy. When the bone is split open, as in Figure 39.4, the longitudinal section shows that it is not solid but has a cavity called the medullary cavity bounded at the sides by compact bone and at the ends by spongy bone. Beyond the spongy bone, there is a thin shell of compact bone and

MP3
Bone Structure

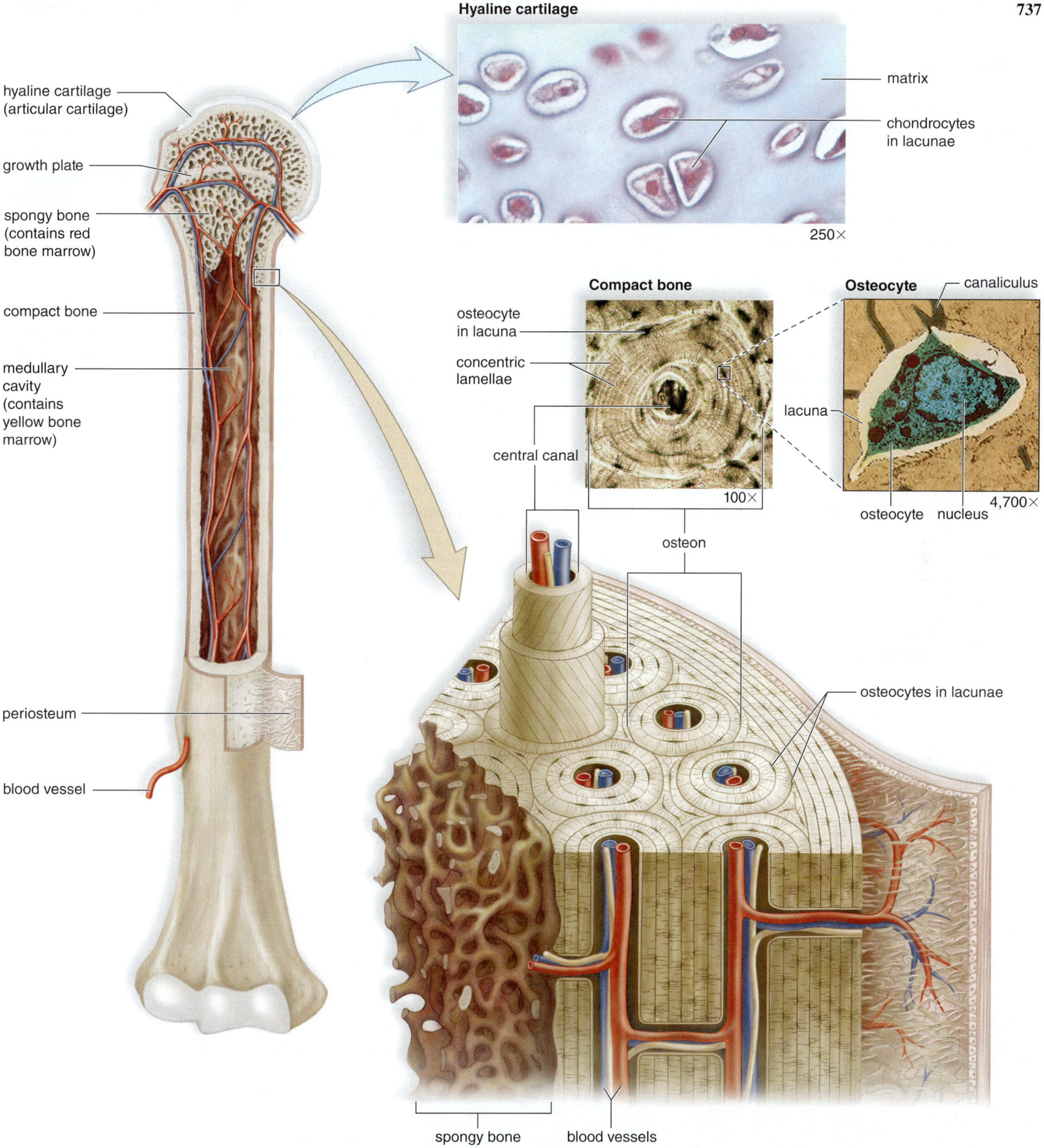

Figure 39.4 Anatomy of a long bone. *Left:* A long bone is encased by fibrous membrane (periosteum), except where it is covered at the ends by hyaline cartilage (see micrograph). Spongy bone located beneath the cartilage may contain red bone marrow. The central shaft contains yellow bone marrow and is bordered by compact bone, which is shown in the enlargement and micrograph (*right*).

finally a layer of hyaline cartilage. The cavity of an adult long bone usually contains yellow bone marrow, which is a fat-storage tissue.

Compact bone contains many osteons (also called Haversian systems), where osteocytes lie in tiny chambers called lacunae. The lacunae are arranged in concentric circles around central canals that contain blood vessels and nerves. The lacunae are separated by a matrix of collagen fibers and mineral deposits, primarily calcium and phosphorus salts.

Spongy bone has numerous bony bars and plates separated by irregular spaces. Although lighter than compact bone, spongy bone is still designed for strength. Just as braces are used for support in buildings, the solid portions of spongy bone follow lines of stress. The spaces in spongy bone are often filled with **red bone marrow,** a specialized tissue that produces blood cells. This is an additional way the skeletal system assists homeostasis. As you know, red blood cells transport oxygen, and white blood cells are a part of the immune system, which fights infection.

The Axial Skeleton

Approximately 206 bones make up a human skeleton. A total of 80 bones make up the **axial skeleton** (L. *axis,* "axis, hinge"; Gk. *skeleton,* "dried body"), which lies in the midline of the body and consists of the skull, the vertebral column, the thoracic cage, the sacrum, and the coccyx (blue labels in Fig. 39.5).

The Skull

The skull, which protects the brain, is formed by the cranium and the facial bones (Fig. 39.6). In newborns, certain bones of the cranium are joined by membranous regions called *fontanels* ("soft spots"), all of which usually close and become **sutures** by the age of 2 years. The bones of the cranium contain the sinuses (L. *sinus,* "hollow"), air spaces lined by mucous membrane that reduce the weight of the skull and give a resonant sound to the voice. Two sinuses, called the mastoid sinuses, drain into the middle ear. Mastoiditis, a condition that can lead to deafness, is an inflammation of these sinuses.

MP3 The Skull

The major bones of the cranium have the same names as the lobes of the brain. On the top of the cranium, the frontal bone forms the forehead, and the parietal bones extend to the sides. Below the much larger parietal bones, each temporal bone has an opening that leads to the middle ear. In the rear of the skull, the occipital bone curves to form the base of the skull. At the base of the skull, the spinal cord passes upward through a large opening, called the *foramen magnum,* and becomes the brain stem.

The temporal and frontal bones are cranial bones that contribute to the face. The sphenoid bones account for the flattened areas on each side of the forehead, which we call the temples. The frontal bone not only forms the forehead but also has supraorbital ridges, where the eyebrows are located. Glasses sit where the frontal bone joins the nasal bones.

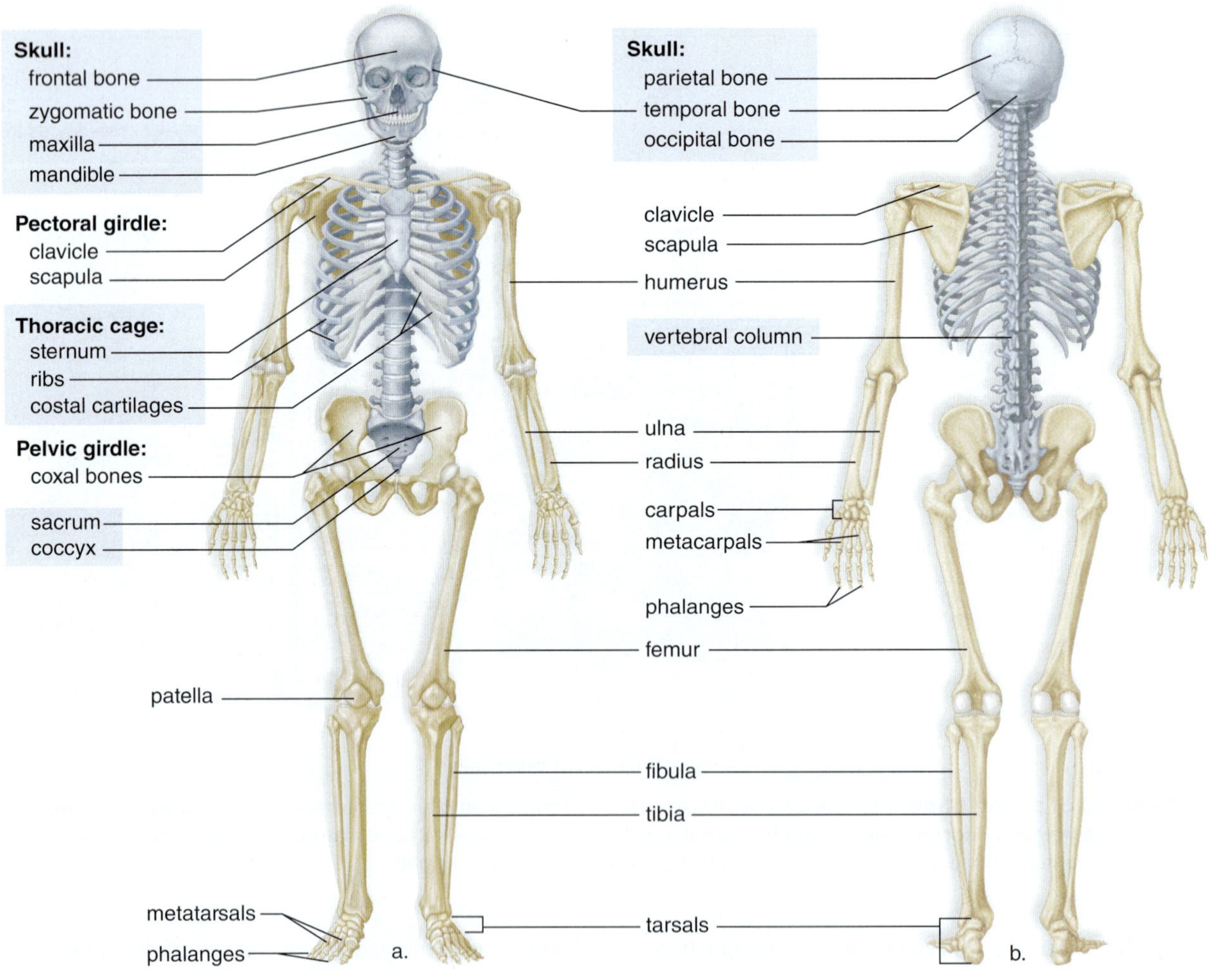

Figure 39.5 The human skeleton. **a.** Anterior view. **b.** Posterior view. The bones of the axial skeleton are in blue, and the rest is the appendicular skeleton.

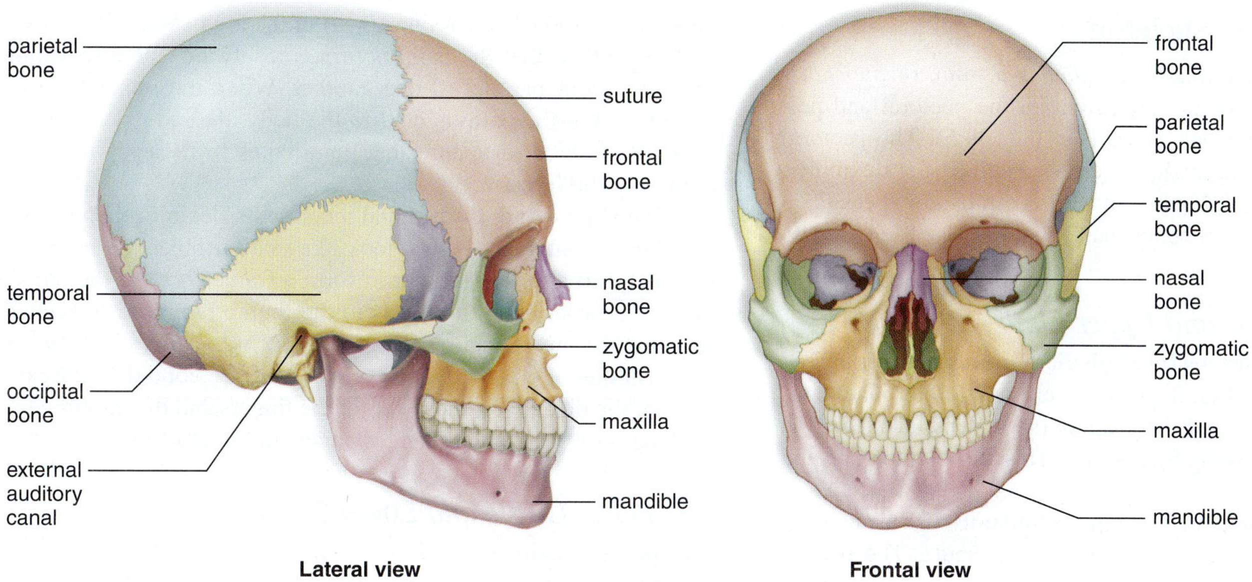

Figure 39.6 **The skull.** The skull consists of the cranium and the facial bones. The frontal bone is the forehead; the zygomatic arches form the cheekbones, and the maxillae form the upper jaw. The mandible has a projection we call the chin.

The most prominent of the facial bones are the mandible, the maxillae, the zygomatic bones, and the nasal bones. The mandible, or lower jaw, is the only freely movable portion of the skull (Fig. 39.6), and its action permits us to chew our food. It also forms the "chin." Tooth sockets are located in the mandible and on the maxillae, which form the upper jaw and a portion of the hard palate. The zygomatic bones are the cheekbone prominences, and the nasal bones form the bridge of the nose. Other bones make up the nasal septum, which divides the nose cavity into two regions.

Whereas the ears are formed only by cartilage and not by bone, the nose is a mixture of bones, cartilage, and connective tissues. The lips and cheeks have a core of skeletal muscle.

The Vertebral Column and Rib Cage

The **vertebral column** (L. *vertebra,* "bones of backbone") supports the head and trunk and protects the spinal cord and the roots of the spinal nerves. It is a longitudinal axis that serves either directly or indirectly as an anchor for all the other bones of the skeleton.

MP3 The Vertebral Column and Thoracic Cage

Twenty-four vertebrae make up the vertebral column. Seven cervical vertebrae are located in the neck; 12 thoracic vertebrae are in the thorax; 5 lumbar vertebrae are in the lower back; 5 fused sacral vertebrae form a single sacrum; and several fused vertebrae are in the coccyx, or tailbone. Normally, the vertebral column has four curvatures, which provide more resilience and strength for an upright posture than could a straight column.

Intervertebral disks, composed of fibrocartilage, between the vertebrae provide padding. They prevent the vertebrae from grinding against one another and absorb shock caused by movements such as running, jumping, and even walking. The presence of the disks allows the vertebrae to move as we bend forward, backward, and from side to side. Unfortunately, these disks become weakened with age and can herniate and rupture. Pain results if a disk presses against the spinal cord and/or spinal nerves. The body may heal itself, or the disk can be removed surgically. If the latter, the vertebrae can be fused together, but this limits the flexibility of the body.

The thoracic vertebrae are a part of the *rib cage,* sometimes called the thoracic cage. The rib cage also contains the ribs, the costal cartilages, and the sternum, or breastbone (Fig. 39.7).

There are 12 pairs of ribs. The upper 7 pairs are "true ribs," because they attach directly to the sternum. The lower 5 pairs do not connect directly to the sternum and are called the "false ribs." Three pairs of false ribs attach by means of a common cartilage, and 2 pairs are "floating ribs," because they do not attach to the sternum at all.

The rib cage demonstrates how the skeleton is protective yet flexible. The rib cage protects the heart and lungs, yet it swings outward and upward on inspiration and then downward and inward on expiration.

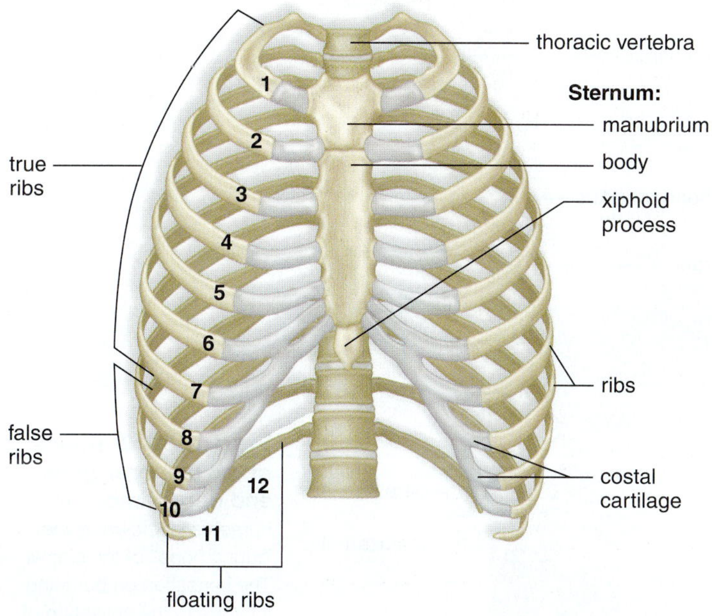

Figure 39.7 **The rib cage.** The rib cage consists of the thoracic vertebrae, the 12 pairs of ribs, the costal cartilages, and the sternum, or breastbone.

The Appendicular Skeleton

The **appendicular skeleton** (L. *appendicula,* dim. of *appendix,* "appendage") consists of the bones within the pectoral and pelvic girdles and the attached limbs (see Fig. 39.5). The pectoral (shoulder) girdle and upper limbs are specialized for flexibility, but the pelvic girdle (hipbones) and lower limbs are specialized for strength. A total of 126 bones make up the appendicular skeleton.

The Pectoral Girdle and Upper Limb

The components of the **pectoral girdle** are only loosely linked together by ligaments (Fig. 39.8). Each clavicle (collarbone) connects with the sternum and the scapula (shoulder blade), but the scapula is held in place only by muscles. This allows it to glide and rotate on the clavicle.

The single long bone in the arm, the humerus, has a smoothly rounded head that fits into a socket of the scapula. The socket, however, is very shallow and much smaller than the head. Although this means that the arm can move in almost any direction, the joint lacks stability. Therefore, this is the joint that is most apt to dislocate. The opposite end of the humerus meets the two bones of the lower arm, the ulna and the radius, at the elbow. (The prominent bone in the elbow is the topmost part of the ulna.) When the upper limb is held so that the palm is turned frontward, the radius and ulna are about parallel to one another. When the upper limb is turned so that the palm is next to the body, the radius crosses in front of the ulna, a feature that contributes to the easy twisting motion of the forearm.

The many bones of the hand increase its flexibility. The wrist has eight carpal bones, which look like small pebbles. From these, five metacarpal bones fan out to form a framework for the palm. The metacarpal bone that leads to the thumb is placed in such a way that the thumb can reach out and touch the other digits (*digit* is a term that refers to either fingers or toes). Beyond the metacarpals are the phalanges, the bones of the fingers and the thumb. The phalanges of the hand are long, slender, and lightweight.

The Pelvic Girdle and Lower Limb

The **pelvic girdle** (L. *pelvis,* "basin") (Fig. 39.9) consists of two heavy, large coxal bones (hip bones). The coxal bones are anchored to the sacrum, and together these bones form a hollow cavity called the pelvic cavity. The wider pelvic cavity in females compared to that of males accommodates pregnancy and childbirth. The weight of the body is transmitted through the pelvis to

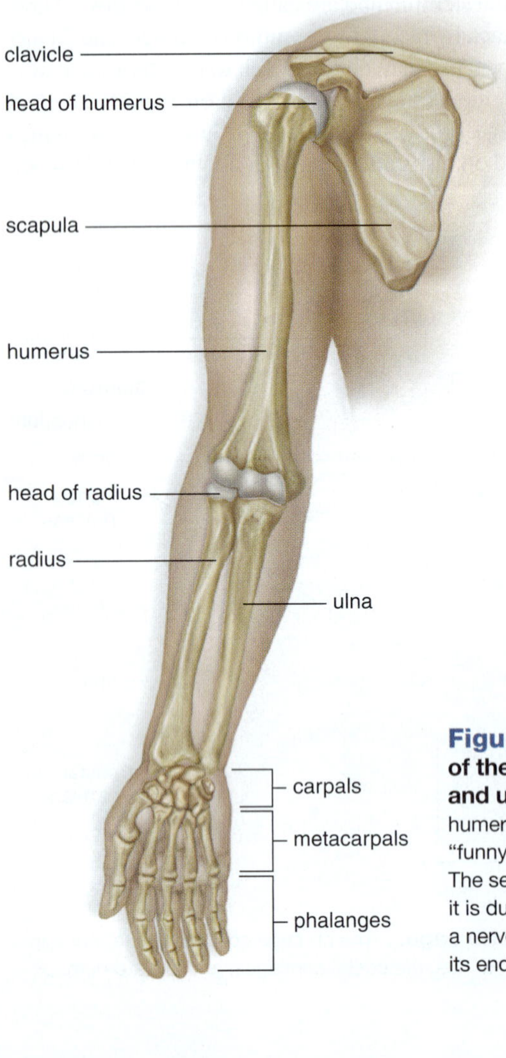

Figure 39.8 Bones of the pectoral girdle and upper limb. The humerus is known as the "funny bone" of the elbow. The sensation on bumping it is due to the activation of a nerve that passes across its end.

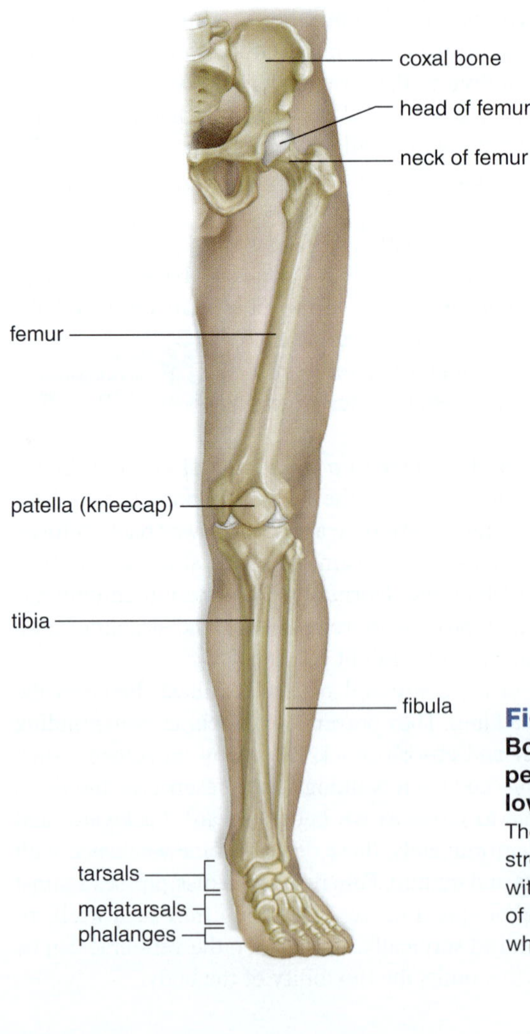

Figure 39.9 Bones of the pelvic girdle and lower limb. The femur is our strongest bone—it withstands a pressure of 540 kg per 2.5 cm^3 when we walk.

the lower limbs and then onto the ground. The largest bone in the body is the femur, or thighbone.

In the leg, the larger of the two bones, the tibia, has a ridge we call the shin. Both of the bones of the leg have a prominence that contributes to the ankle—the tibia on the inside of the ankle and the fibula on the outside of the ankle.

Although there are seven tarsal bones in the ankle, only one receives the weight and passes it on to the heel and the ball of the foot. If you wear high-heeled shoes, the weight is thrown toward the front of your foot. The metatarsal bones participate in forming the arches of the foot. There is a longitudinal arch from the heel to the toes and a transverse arch across the foot. These provide a stable, springy base for the body. If the tissues that bind the metatarsals together become weakened, "flat feet" are apt to result. The bones of the toes are called phalanges, just as are those of the fingers, but in the foot the phalanges are stout and extremely sturdy.

Classification of Joints

Bones are connected at the **joints,** which are classified as fibrous, cartilaginous, or synovial. Most fibrous joints, such as the sutures between the cranial bones, are immovable. Cartilaginous joints, such as those between the vertebrae, are slightly movable. The vertebrae are also separated by disks, which increase their flexibility. The two hip bones are slightly movable, because they are ventrally joined by cartilage at the pubic symphysis. Owing to hormonal changes, this joint becomes more flexible during late pregnancy, allowing the pelvis to expand during childbirth.

In freely movable **synovial joints,** the two bones are separated by a cavity. **Ligaments,** composed of fibrous connective tissue, bind the two bones to each other, holding them in place as they form a capsule. In a "double-jointed" individual, the ligaments are unusually loose. The joint capsule is lined by synovial membrane, which produces synovial fluid, a lubricant for the joint.

The shoulders, elbows, hips, and knees are examples of synovial joints (Fig. 39.10). In the knee, as in other freely movable joints, the bones are capped by a layer of articular cartilage. In addition, crescent-shaped pieces of cartilage called menisci (Gk. *meniscus,* "crescent") lie between the bones. These give added stability, helping support the weight placed on the knee joint. Unfortunately, athletes often suffer injury of the meniscus, known as torn cartilage. Thirteen fluid-filled sacs called bursae (sing., bursa) (L. *bursa,* "purse") occur around the knee joint. Bursae ease the friction between tendons and ligaments and between tendons and bones. Inflammation of the bursae is called bursitis. Tennis elbow is a form of bursitis.

Different types of synovial joints can be distinguished. The knee and elbow joints are *hinge joints* because, like a hinged door, they largely permit movement in one direction only. The joint between the first two cervical vertebrae, which permits side-to-side movement of the head, is an example of a *pivot joint,* which allows only rotation. More movable are the *ball-and-socket joints;* for example, the ball of the femur fits into a socket on the hip bone. Ball-and-socket joints allow movement in all planes and even rotational movement.

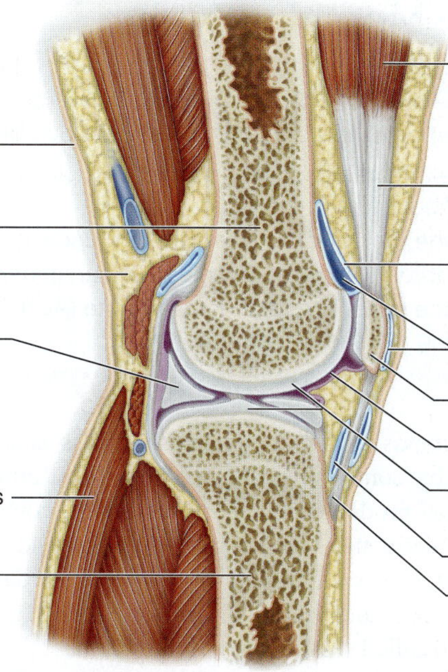

Figure 39.10 Knee joint. The knee is an example of a synovial joint. The cavity between the bones is encased by ligaments and lined by synovial membrane. The patella (kneecap) guides the quadriceps tendon over the joint when flexion or extension occurs.

All types of joints are subject to **arthritis,** or inflammation of the joints. The Arthritis Foundation estimates that nearly one in three adult Americans has some degree of chronic joint pain, and arthritis is secondary only to heart disease as a cause of work disability in the United States. The most common type of arthritis is osteoarthritis, which results from the deterioration of the cartilage in one or more synovial joints. The hands, hips, knees, lower back, and neck are most commonly affected. Rheumatoid arthritis is considered to be an autoimmune disease, in which the immune system attacks the joints for reasons that are not well understood.

Regardless of the cause, arthritis is most often treated by over-the-counter anti-inflammatory drugs, such as aspirin and ibuprofen, or more powerful prescription drugs, such as the corticosteroids. In severe cases, certain joints can be replaced with artificial versions made of ceramic, metal, and/or plastic. About 720,000 artificial knees and 330,000 artificial hips are installed in U.S. patients each year. About 4.5 million Americans, including about 1 in 20 over age 50, and 1 in 10 over age 80, have at least one artificial knee. These joint replacements generally last 15 to 20 years, after which they may need to be replaced, at an average cost of around $40,000.

Check Your Progress 39.2

1. Describe the functions of osteoblasts, osteoclasts, and osteocytes.
2. Distinguish between the structure and function of spongy bone and those of compact bone.
3. Determine whether each of the following bones belongs to the axial or appendicular skeleton: sacrum, frontal bone, humerus, tibia, vertebra, coxal bone, temporal bone, scapula, and sternum.

39.3 The Muscular System

Learning Outcomes

Upon completion of this section, you should be able to

1. Describe the macroscopic and microscopic structure of a muscle fiber.
2. Explain the molecular mechanism of muscle contraction.
3. Indicate three ways that muscle cells can generate ATP.
4. Explain the specific role of acetylcholine (ACh) in stimulating a muscle fiber to contract.

Muscles are composed of contractile tissue that is capable of changing its length by contracting and relaxing. Most animals rely on muscle tissue to produce movement—to swim, crawl, walk, run, jump, or fly. Clearly, there are strong evolutionary advantages to be able to move into new environments, to flee from danger, to seek and/or capture food, and to find new mates. Only a few animals are nonmotile (also called sessile), and most of these live in water, where currents can bring a supply of food to them.

MP3 Muscle Tissue

As discussed in Chapter 31, humans and other vertebrates have three distinct types of muscle tissue: smooth, cardiac, and skeletal. Most of the focus in this chapter is on skeletal muscle, or striated voluntary muscle, which is important in maintaining posture, providing support, and allowing for movement. The processes responsible for skeletal muscle contraction also release heat, which is distributed throughout the body, helping maintain a constant body temperature.

Macroscopic Anatomy and Physiology

The nearly 700 skeletal muscles and their associated tissues make up approximately 40% of the weight of an average human. Muscle tissue is approximately 15% more dense than fat tissue, so a pound of muscle takes up less space than does a pound of fat. However, even at rest, muscle tissue consumes about three times more energy than adipose tissue.

Several of the major human superficial muscles are illustrated in Figure 39.11. Skeletal muscles are attached to the skeleton by bands of fibrous connective tissue called **tendons** (L. *tendo,* "stretch"). When muscles contract, they shorten. Therefore, muscles can only pull; they cannot push. Because of this, skeletal muscles must work in antagonistic pairs. One muscle of an antagonistic pair flexes the joint and bends the limb; the other one extends the joint and straightens the limb. Figure 39.12 illustrates this principle.

In the laboratory, if a muscle is given a rapid series of threshold stimuli—that is, stimuli strong enough to bring about action potentials, as described in section 37.2.—it can respond to the next stimulus without relaxing completely. In this way, muscle contraction builds, or summates, until maximal sustained contraction, called **tetany,** is achieved. Tetanic contractions ordinarily occur in the body's muscles whenever skeletal muscles are actively used.

Even when muscles appear to be at rest, they exhibit **tone,** in which some of their fibers are contracting. As you saw in Chapter 38, sensory receptors called muscle spindles and Golgi tendon organs are partly responsible for maintaining tone. Muscle tone is particularly important in maintaining posture. If all the fibers in the muscles of the neck, trunk, and legs were to suddenly relax, the body would collapse.

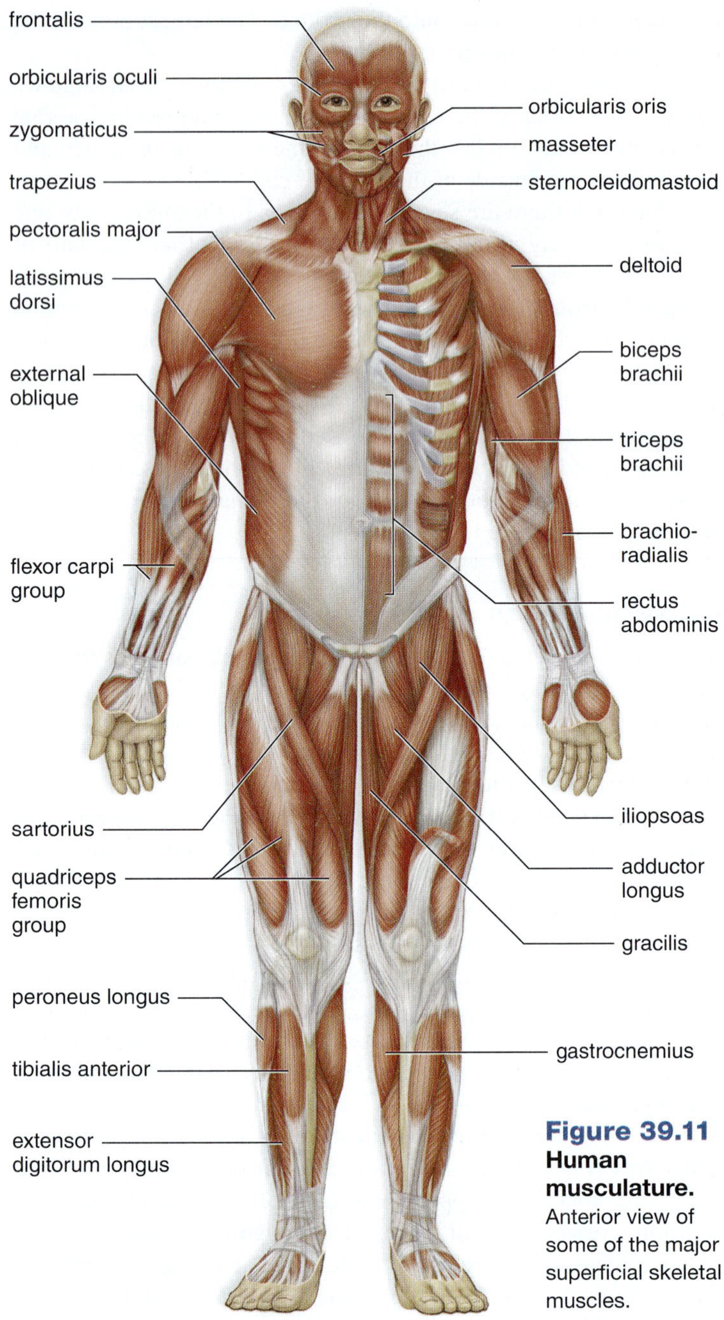

Figure 39.11 Human musculature. Anterior view of some of the major superficial skeletal muscles.

Muscle tone has also been implicated in the formation of facial wrinkles. As described in the Nature of Science feature, "The Accidental Discovery of Botox®," on page 744, medical injections of Botox® interfere with muscle contraction, smoothing wrinkles.

Microscopic Anatomy and Physiology

A vertebrate skeletal muscle is composed of a number of muscle fibers in bundles. Each muscle fiber is a cell containing the usual cellular components, but some components have special features (Fig. 39.13).

MP3 Muscle Structure

The **sarcolemma,** or plasma membrane, forms a transverse system, or T system. The T tubules penetrate, or dip down, into

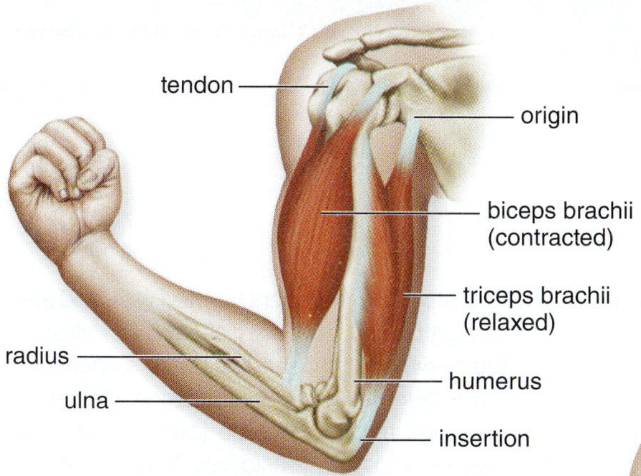

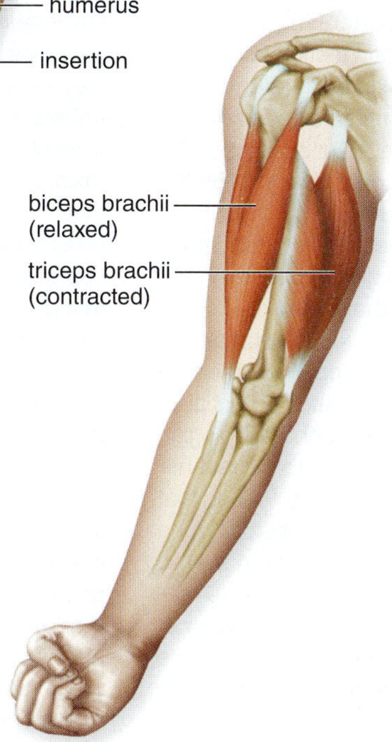

Figure 39.12
Antagonistic muscles. Muscles can exert force only by shortening; therefore, they often work as antagonistic pairs. The biceps and triceps brachii exemplify an antagonistic pair of muscles that act opposite to one another. The biceps brachii flexes the elbow joint, and the triceps brachii extends the elbow joint.

the cell, so that they come in contact—but do not fuse—with the **sarcoplasmic reticulum,** which consists of expanded portions of modified endoplasmic reticulum (ER). These expanded portions serve as storage sites for calcium ions (Ca^{2+}), which are essential for muscle contraction. The sarcoplasmic reticulum encases hundreds and sometimes even thousands of **myofibrils** (Gk. *myos,* "muscle"; L. *fibra,* "thread"), which are the contractile portions of a muscle fiber.

Myofibrils are cylindrical and run the length of the muscle fiber. The light microscope shows that a myofibril has light and dark bands, termed striations. These bands are responsible for skeletal muscle's striated appearance. The electron microscope reveals that the striations of myofibrils are formed by the placement of protein filaments within contractile units called **sarcomeres.**

Examining sarcomeres when they are relaxed shows that a sarcomere extends between two dark lines called Z lines (Fig. 39.13). There are two types of protein filaments: thick filaments, made up of **myosin,** and thin filaments, made up of **actin.** The I band is light-colored, because it contains only actin filaments attached to a Z line. The dark regions of the A band contain overlapping actin and myosin filaments, and its H zone has only myosin filaments.

Sliding Filament Model

Examining muscle fibers when they are contracted reveals that the sarcomeres within the myofibrils have shortened. When a sarcomere shortens, the actin (thin) filaments slide past the myosin (thick) filaments and approach one another. This causes the I band to shorten and the H zone to nearly or completely disappear.

Animation Sarcomere Contraction

Figure 39.13 Skeletal muscle fiber structure and function. A muscle fiber contains many myofibrils, divided into sarcomeres, which are contractile. When the myofibrils of a muscle fiber contract, the sarcomeres shorten: The actin (thin) filaments slide past the myosin (thick) filaments toward the center, so that the H zone gets smaller, to the point of disappearing.

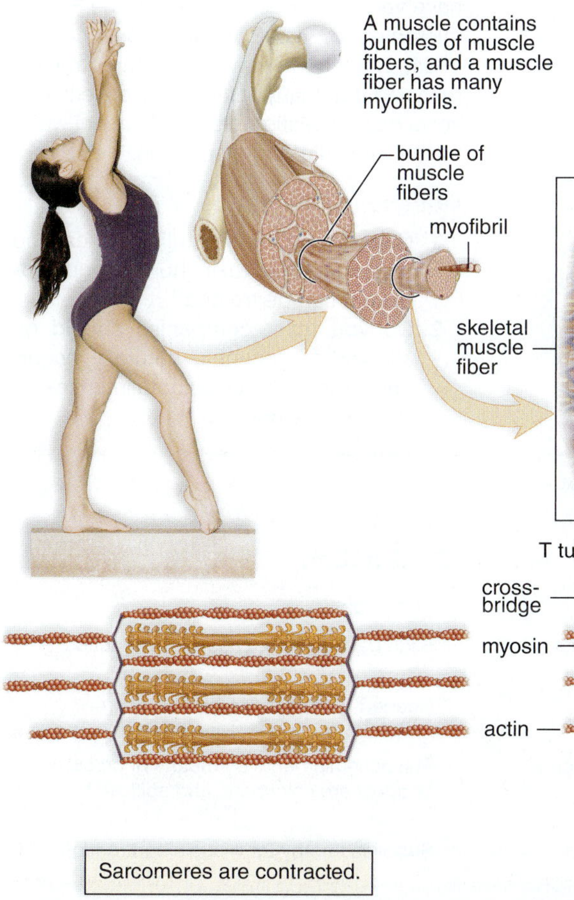

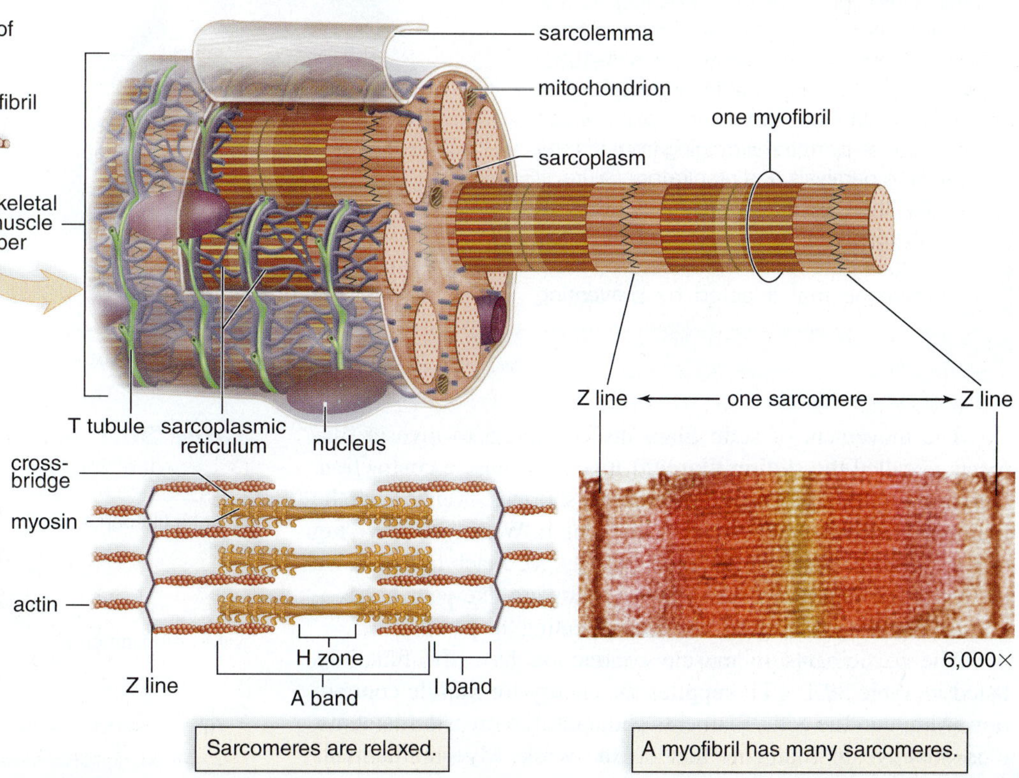

THEME Nature of Science

The Accidental Discovery of Botox®

Several of the most important bacterial pathogens that cause human diseases—including cholera, diphtheria, tetanus, and botulism—do so by secreting potent toxins capable of sickening or killing their victims. The botulinum toxin, produced by the bacterium *Clostridium botulinum,* is one of the most lethal substances known. Less than a microgram of the purified toxin can kill an average-size person, and 4 kilograms (8.8 pounds) would be enough to kill all the humans on Earth! Given this scary fact, it seems that the scientists who discovered the lethal activity of this bacterial toxin nearly 200 years ago could never have anticipated that the intentional injection of a very dilute form of botulinum toxin (now known as Botox®) would become the most common nonsurgical cosmetic procedure performed by many physicians.

As with many breakthroughs in science and medicine, the pathway from thinking about botulism as a deadly disease to using botulinum toxin as a beneficial treatment involved the hard work of many scientists, mixed with a considerable amount of luck.

In the 1820s, a German scientist, Justinus Kerner (1786–1862), was able to prove that the deaths of several people had been caused by their consumption of spoiled sausage (in fact, botulism is named for the Latin word for "sausage," *botulus*). A few decades later, a Belgian researcher named Emile Pierre van Ermengem (1851–1932) identified the specific bacterium responsible for producing the botulinum toxin, which could cause symptoms ranging from droopy eyelids to paralysis and respiratory failure.

By the 1920s, medical scientists at the University of California had obtained the toxin in pure form, which allowed them to determine that it acted by preventing nerves from communicating with muscles, specifically by interfering with the release of acetylcholine from the axon terminals of motor nerves (see section 39.3).

Scientists soon began testing very dilute concentrations of the toxin as a treatment for conditions in which the muscles contract too much, such as crossed eyes or spasms of the facial muscles or vocal cords, and in 1989 the FDA first approved diluted botulinum toxin (Botox®) for treating specific eye conditions called blepharospasm (eyelid spasms) and strabismus (crossing of the eyes).

Right around this time, a lucky break occurred that eventually would open the medical community's eyes to the greater potential of the diluted toxin. A Canadian ophthalmologist, Jean Carruthers, had been using it to treat her patients' eye conditions, when she noticed that some of their wrinkles had also subsided. One night at a family dinner, Dr. Carruthers shared this information with her husband, a dermatologist, who decided to investigate whether he could reduce the deep wrinkles of some of his patients by injecting the dilute toxin into their skin. The treatment worked well, and after trying it on several more patients (as well as on themselves!), the Canadian doctors spent several years presenting their findings at scientific meetings and in research journals. Although they were initially considered "crazy," the Carrutherses eventually were able to convince the scientific community that diluted botulinum toxin is effective in treating wrinkles; however, they never patented it for that use, so they missed out on much of the $1.3 billion in annual sales the drug now earns for the company that did patent it.

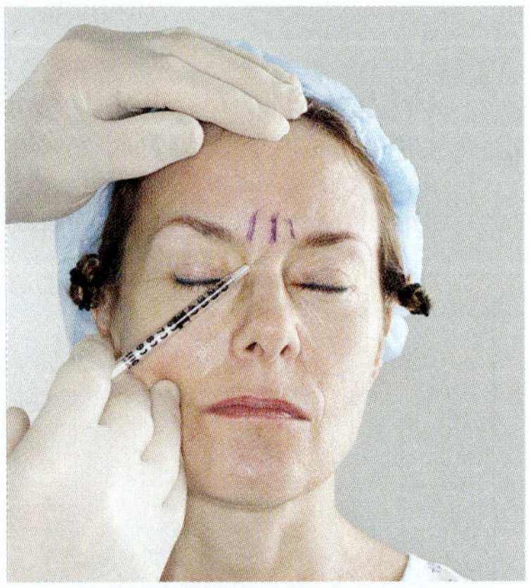

Figure 39A Treating wrinkles with diluted botulinum toxin.

The uses of diluted botulinum toxin have been growing since it was first FDA-approved for the treatment of frown lines in 2002 (Fig. 39A). It has now been approved for the treatment of chronic migraine headaches, excessive underarm sweating, and facial wrinkles known as "crow's-feet." The company is seeking approval for many other uses of diluted botulinum toxin. The annual market for Botox® is predicted to reach about $3 billion by 2018.

Perhaps all of this would have eventually happened even without the observations of an alert eye doctor, but progress would have very likely been slower. As the French microbiologist Louis Pasteur observed in 1854, "Chance favors the prepared mind," meaning that many scientific discoveries involve many investigators, and years of work, mixed with a flash of inspiration.

Questions to Consider

1. Considering that botulism is caused by a preformed toxin, how do you suppose it can be treated?
2. Do you think companies should be allowed to patent a naturally occurring molecule such as botulinum toxin? Why or why not?

The movement of actin filaments in relation to myosin filaments is called the **sliding filament model** of muscle contraction. During the sliding process, the sarcomere shortens, even though the filaments themselves remain the same length. When you play "tug of war," your hands grasp the rope, pull, let go, attach farther down the rope, and pull again. The myosin heads are like your hands—grasping, pulling, letting go, and then repeating the process.

The participants in muscle contraction have the functions listed in Table 39.1. ATP supplies the energy for muscle contraction. Although the actin filaments slide past the myosin filaments, it is the myosin filaments that do the work. Myosin filaments

Table 39.1 Muscle Contraction

Name	Function
Actin filaments	Slide past myosin, causing contraction
Ca^{2+}	Needed for myosin to bind to actin
Myosin filaments	Pull actin filaments by means of cross-bridges; are enzymatic and split ATP
ATP	Supplies energy for muscle contraction

break down ATP and form cross-bridges that attach to and pull the actin filaments toward the center of the sarcomere.

Animation Breakdown of ATP and Cross-Bridge Movement

Use of ATP in Contraction

ATP provides the energy for muscle contraction. Although muscle cells contain *myoglobin,* a molecule that stores oxygen, cellular respiration does not immediately supply all the ATP that is needed. In the meantime, muscle fibers rely on *creatine phosphate* (phosphocreatine), a storage form of high-energy phosphate. Creatine phosphate cannot directly participate in muscle contraction. Instead, it anaerobically regenerates ATP by the following reaction:

$$\text{creatine—P} + \text{ADP} \longrightarrow \text{ATP} + \text{creatine}$$

This reaction occurs in the midst of sliding filaments, and therefore this method of supplying ATP is the speediest energy source available to muscles.

When all of the creatine phosphate is depleted, mitochondria may by then be producing enough ATP for muscle contraction to continue. If not, fermentation is a second way for muscles to supply ATP without consuming oxygen. Fermentation, which is apt to occur when strenuous exercise first begins, supplies ATP for only a short time, and lactate builds up. Whether lactate causes muscle aches and fatigue on exercising is now being questioned.

We all have had the experience of needing to continue deep breathing following strenuous exercise. This continued intake of oxygen, which is required to complete the metabolism of lactate and restore cells to their original energy state, offsets what is known as **oxygen debt.** The lactate is transported to the liver, where 20% of it is completely broken down to carbon dioxide (CO_2) and water (H_2O). The ATP gained by this respiration is then used to reconvert 80% of the lactate to glucose.

In persons who regularly exercise, such as athletes in training, the number of mitochondria increases, and muscles rely on them rather than on fermentation to produce ATP. Less lactate is produced, and there is less oxygen debt.

Muscle Innervation

Muscles are stimulated to contract by motor nerve fibers. Nerve fibers have several branches, each of which ends at an axon terminal in close proximity to the sarcolemma of a muscle fiber. A small gap, called a synaptic cleft, separates the axon terminal from the sarcolemma. This entire region is called a **neuromuscular junction** (Fig. 39.14).

Animation Function of the Neuromuscular Junction

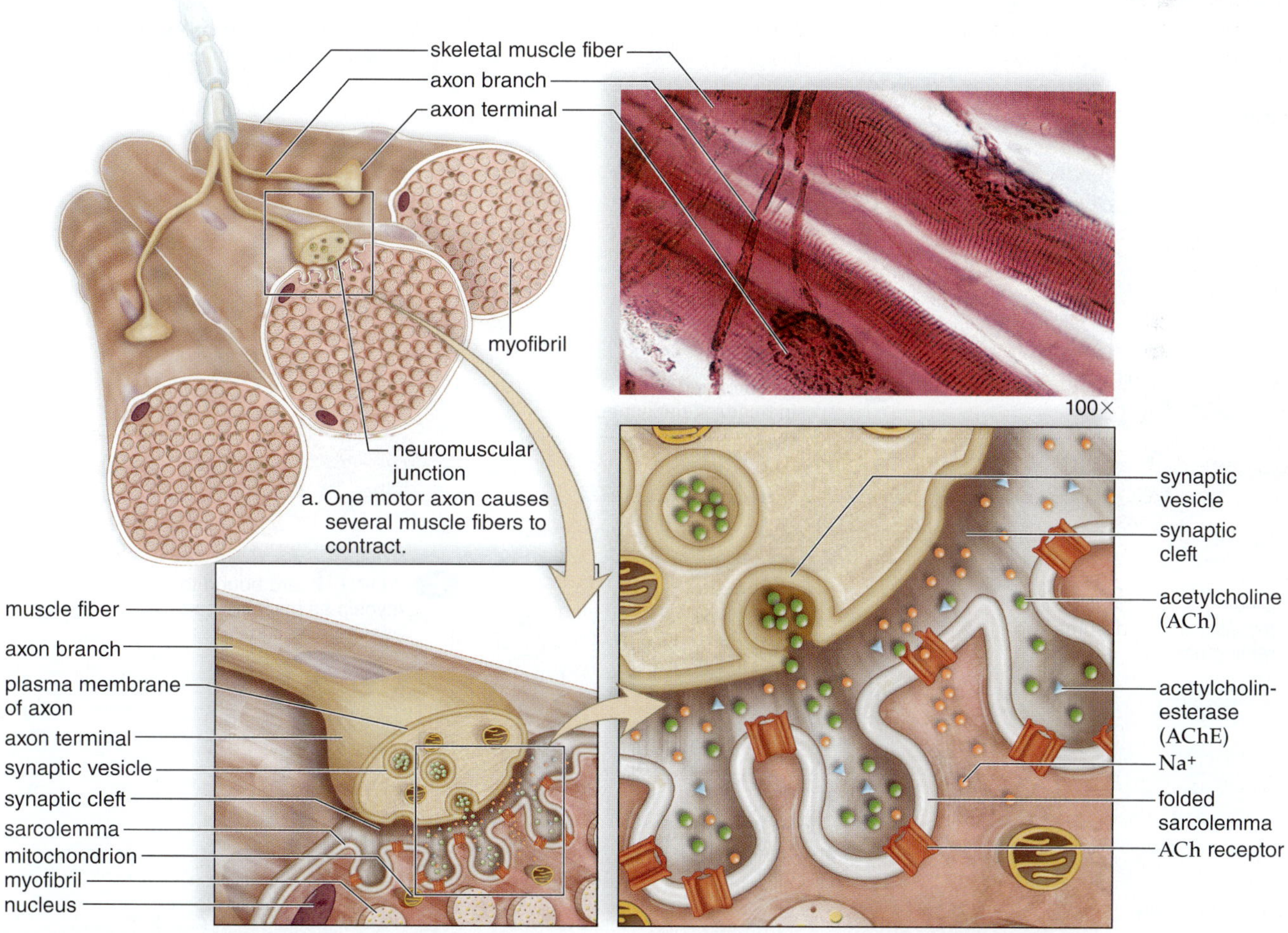

Figure 39.14 Neuromuscular junction. The branch of a motor nerve fiber (**a**) ends in an axon terminal (**b**) that meets but does not touch a muscle fiber. A synaptic cleft separates the axon terminal from the sarcolemma of the muscle fiber. Nerve impulses traveling down a motor fiber cause synaptic vesicles (**c**) to discharge a neurotransmitter that diffuses across the synaptic cleft. When the neurotransmitter is received by the sarcolemma of a muscle fiber, impulses begin, leading to muscle fiber contraction.

Axon terminals contain synaptic vesicles filled with the neurotransmitter acetylcholine (ACh). When nerve impulses traveling down a motor neuron arrive at an axon terminal, the synaptic vesicles release ACh into the synaptic cleft. ACh quickly diffuses across the cleft and binds to receptors in the sarcolemma. Now, the sarcolemma generates impulses that spread over the sarcolemma and down T tubules to the sarcoplasmic reticulum. The release of calcium from the sarcoplasmic reticulum causes the filaments in sarcomeres to slide past one another. Sarcomere contraction results in myofibril contraction, which in turn results in muscle fiber and finally muscle contraction.

Once a neurotransmitter has been released into a neuromuscular junction and has initiated a response, it is removed from the junction. When the enzyme acetylcholinesterase (AChE) breaks down acetylcholine, muscle contraction ceases due to reasons we will discuss next.

Role of Calcium in Muscle Contraction

Figure 39.15 illustrates the placement of two other proteins associated with a thin filament, which is composed of a double row of twisted actin molecules. Threads of tropomyosin wind about an actin filament, and troponin occurs at intervals along the threads. Calcium ions (Ca^{2+}) that have been released from the sarcoplasmic reticulum combine with troponin. After binding occurs, the tropomyosin threads shift their position, and myosin binding sites are exposed.

Thick filaments are bundles of myosin molecules with double globular heads. Myosin heads function as ATPase enzymes, splitting ATP into ADP and Ⓟ. This reaction activates the heads, so

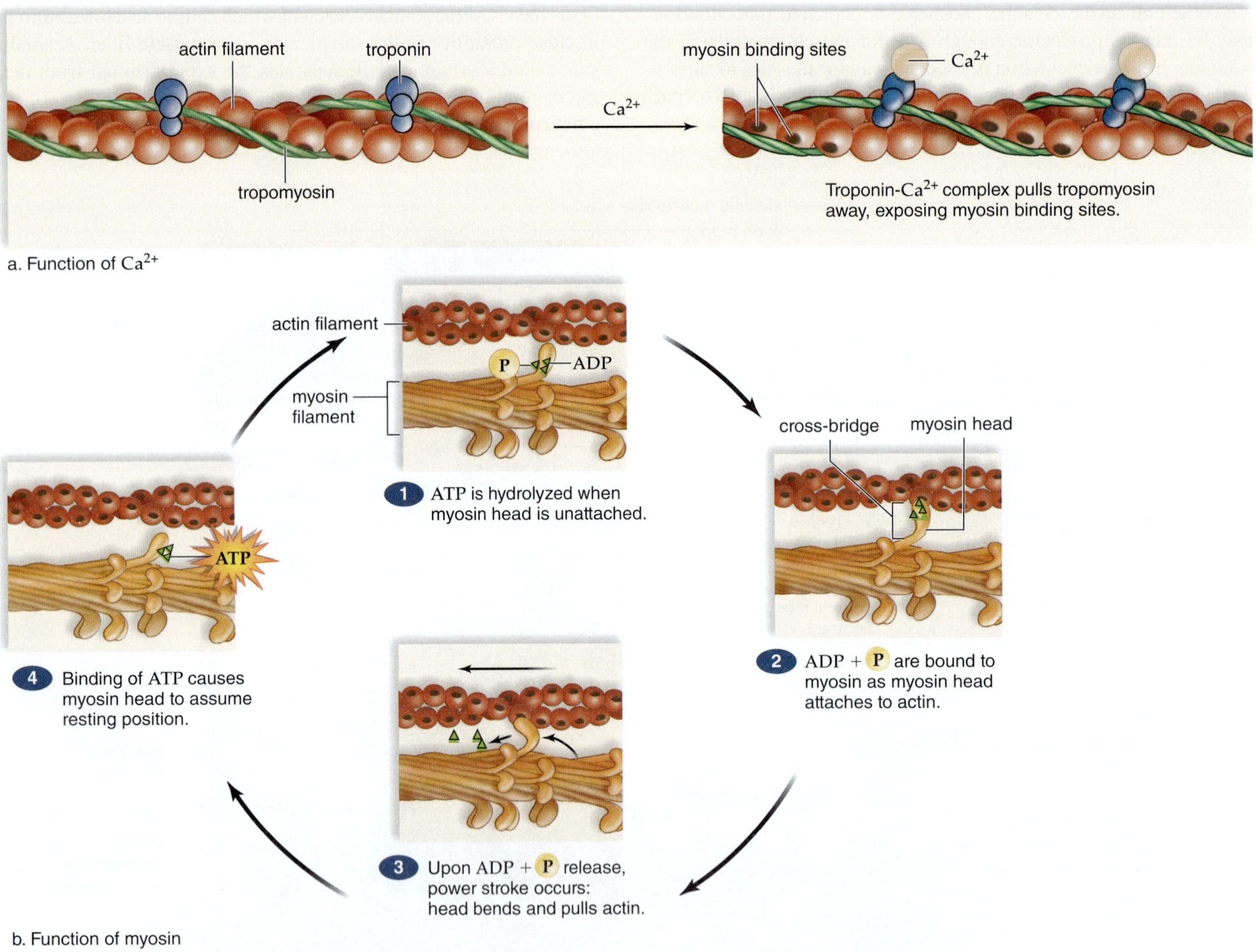

Figure 39.15 The role of calcium and myosin in muscle contraction. **a.** Upon release, calcium binds to troponin, exposing myosin binding sites. **b.** After breaking down ATP (1), myosin heads bind to an actin filament (2), and later, a power stroke causes the actin filament to move (3). When another ATP binds to myosin, the head detaches from actin (4), and the cycle begins again. Although only one myosin head is featured, many heads are active at the same time.

Tutorial
Skeletal Muscle Contraction

that they can bind to actin. The ADP and Ⓟ remain on the myosin heads until the heads attach to actin, forming cross-bridges. Now, ADP and Ⓟ are released, and this causes the cross-bridges to change their positions. This is the power stroke that pulls the thin filaments toward the middle of the sarcomere. When more ATP molecules bind to myosin heads, the cross-bridges are broken as the heads detach from actin. The cycle begins again; the actin filaments move nearer the center of the sarcomere each time the cycle is repeated.

Contraction continues until nerve impulses cease and calcium ions are returned to their storage sites. The membranes of the sarcoplasmic reticulum contain active transport proteins that pump calcium ions back into the calcium storage sites, and muscle relaxation occurs. When a person or an animal dies, ATP production ceases. Without ATP, the myosin heads cannot detach from actin, nor can calcium be pumped back into the sarcoplasmic reticulum. As a result, the muscles remain contracted, a phenomenon called *rigor mortis*.

Check Your Progress 39.3

1. Define an *antagonistic pair* of muscles.
2. Describe the microscopic levels of structure in a skeletal muscle.
3. Discuss the specific role of ATP in muscle contraction.

CONNECTING *the* CONCEPTS *with the* THEMES

Evolution

- Almost all animals have evolved some type of skeletal system to support the soft tissues of the body, provide protection, and provide surfaces for muscle attachment.
- The exoskeleton of arthropods and molluscs provides protection against enemies and prevents desiccation. It must be shed during molting, which can leave the animal vulnerable.
- Vertebrates, which have endoskeletons, have evolved limbs with many different types of joints, allowing for more complex movements.
- Muscles developed to facilitate locomotion, which can be critical to procuring food, escaping danger, finding mates, and/or moving into new environments.

Nature of Science

- Knowledge of the skeletal and muscular systems has come from studying a large variety of animals that have adapted to many environments on Earth.
- To understand the skeleton and muscles fully, we must be familiar with the nervous and sensory systems.
- The use of a very dilute form of botulinum toxin for treating many diseases, as well as for cosmetic purposes, provides an example of how basic scientific research can have unexpected practical applications.

Biological Systems

- Most animals have one of three types of skeletons: a hydrostatic skeleton, an exoskeleton, or an endoskeleton.
- The cartilaginous skeleton of the human fetus is converted into bones, which contain two types of bone tissue, compact bone and spongy bone.
- The human skeleton can be divided into an axial skeleton and an appendicular skeleton.
- Axon terminals of motor nerve fibers release acetylcholine at a neuromuscular junction, triggering muscle fiber contraction.
- Muscle fiber contraction is dependent on actin and myosin filaments, as well as a ready supply of ATP and calcium ions.

MEDIA STUDY TOOLS http://connect.mheducation.com

SMARTBOOK® Study smart. SmartBook helps you maximize your study time by identifying what you know and don't know. The recharge feature in SmartBook helps you prepare for the test by identifying the material that you are most likely to forget.

connect® BIOLOGY Find even more resources to learn the chapter concepts including animations, tutorial videos, and interactive practice questions.

MP3 Files	Animations	Tutorial
39.2 Calcium Homeostasis • Bone Structure • The Skull • The Vertebral Column and Thoracic Cage • The Appendicular Skeleton **39.3** Muscle Tissue • Muscle Structure	**39.2** Bone Growth • Osteoporosis **39.3** Sarcomere Contraction • Breakdown of ATP and Cross-Bridge Movement • Function of the Neuromuscular Junction	**39.3** Skeletal Muscle Contraction

SUMMARIZE

39.1 Diversity of Skeletons

Three types of skeletons are found in the animal kingdom: a **hydrostatic skeleton** (cnidarians, flatworms, and segmented worms); an **exoskeleton** (certain molluscs and arthropods); and an **endoskeleton** (sponges, echinoderms, and vertebrates). The rigid but jointed skeleton of arthropods and vertebrates helped them colonize the terrestrial environment. The overall shape of an animal's skeleton is adapted to its environment and the type(s) of locomotion it uses.

39.2 The Human Skeletal System

The human skeleton gives support to the body, helps protect internal organs, provides sites for muscle attachment, and is a storage area for calcium and phosphorus salts, as well as a site for blood cell formation.

Most bones are cartilaginous in the fetus but are converted to bone during development. A long bone undergoes endochondral ossification, in which a cartilaginous growth plate remains between the primary ossification center in the middle and the secondary centers at the ends of the bones. Growth of the bone is possible as long as the growth plates are present, but eventually they, too, are converted to bone.

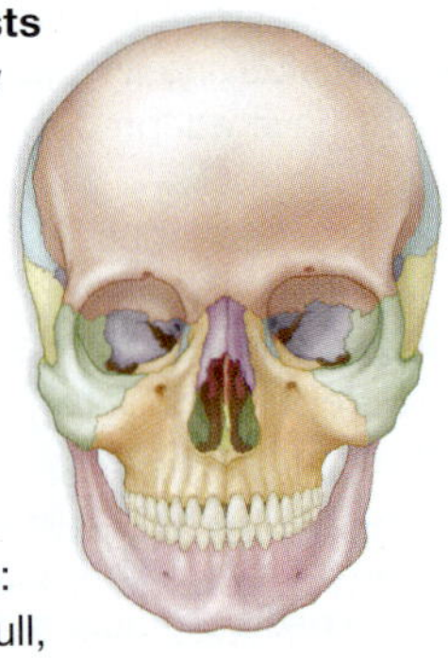

Bone is constantly being renewed; **osteoclasts** break down bone, and **osteoblasts** build new bone. **Osteocytes** are in the lacunae of osteons; a long bone has a shaft of **compact bone** and two ends that contain **spongy bone.** The shaft contains a medullary cavity with yellow marrow, and the ends contain **red bone marrow,** which produces blood cells. **Osteoporosis,** or loss of bone density, is a common disease in older adults.

The human skeleton is divided into two parts: (1) the **axial skeleton,** which is made up of the skull, the **vertebral column,** the sternum, and the ribs; and (2) the **appendicular skeleton,** which is composed of the **pectoral girdle,** the **pelvic girdle,** and their appendages. The skull of a newborn has membranous fontanels which usually close and become **sutures** by two years of age.

Joints are classified as immovable, as are those of the cranium; slightly movable, as are those between the vertebrae; and freely movable (synovial joints), as are those in the knee and hip. In **synovial joints, ligaments** bind the two bones together, forming a capsule containing synovial fluid. **Arthritis** is an inflammation of a joint.

39.3 The Muscular System

Muscle tissue can change its length by contracting and relaxing. Skeletal muscles attach to bones via **tendons.** Whole skeletal muscles can shorten only when they contract; therefore, they work in antagonistic pairs. For example, if one muscle flexes the joint and brings the limb toward the body, the other muscle of the antagonistic pair extends the joint and straightens the limb. A muscle at rest exhibits **tone,** meaning some fibers are contracting. **Tetany** refers to maximum sustained muscle contractions.

A whole skeletal muscle is composed of muscle fibers. Each muscle fiber is a cell that contains **myofibrils** in addition to the usual cellular components. The plasma membrane (**sarcolemma**) forms a transverse (T) system. T tubules penetrate the cell and contact the **sarcoplasmic reticulum,** which stores calcium ions. Longitudinally, myofibrils are divided into **sarcomeres,** which display the arrangement of actin and myosin filaments.

The **sliding filament model** of muscle contraction states that **myosin** filaments have cross-bridges, which attach to and detach from **actin** filaments, causing actin filaments to slide and the sarcomere to shorten. (The H zone disappears as actin filaments approach one another.) Myosin breaks down ATP, and this supplies the energy for muscle contraction. During **oxygen debt,** anaerobic creatine phosphate breakdown and fermentation can quickly generate ATP. Sustained exercise requires cellular respiration for the generation of ATP.

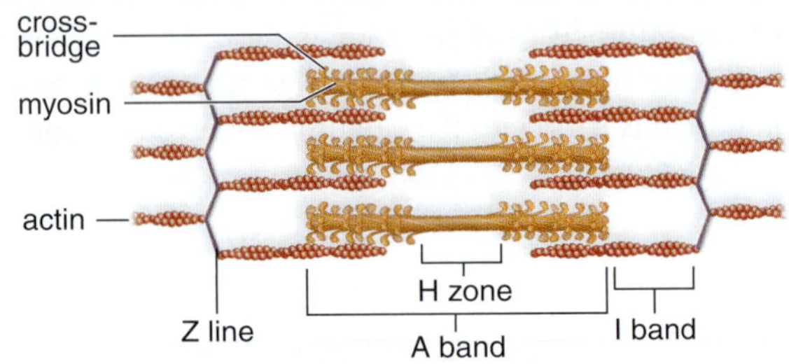

Nerves innervate muscles. A nerve impulse travels down a motor neuron to a **neuromuscular junction,** causing the release of ACh, which binds to receptors on the sarcolemma. Impulses begin and move down T tubules that approach the sarcoplasmic reticulum, where calcium is stored. Thereafter, calcium ions are released and bind to troponin. The troponin-Ca^{2+} complex causes tropomyosin threads winding around actin filaments to shift their position, revealing myosin binding sites. Myosin filaments are composed of many myosin molecules with double globular heads. When myosin heads break down ATP, they are ready to attach to actin. The release of ADP + Ⓟ causes myosin heads to change their position. This is the power stroke that causes the actin filament to slide toward the center of a sarcomere. When more ATP molecules bind to myosin, the heads detach from actin, and the cycle begins again.

ASSESS

Choose the best answer for each question.

39.1 Diversity of Skeletons

1. The coelom of an earthworm, the muscular foot of a clam, and the trunk of an elephant can all be considered what type of skeleton?
   a. exoskeleton
   b. hydrostatic
   c. jointed
   d. longitudinal
2. Which type of animal has an exoskeleton?
   a. arthropod
   b. cnidarian
   c. earthworm
   d. primate
3. Which of the following characteristics is not an advantage of a jointed vertebrate endoskeleton?
   a. allows flexible movement
   b. grows with the animal
   c. guards against dessication
   d. protects vital internal organs

39.2 The Human Skeletal System

4. The human skeletal system does not
   a. produce blood cells.
   b. store minerals.
   c. help produce movement.
   d. store fat.
   e. produce body heat.
5. Spongy bone
   a. is a storage area for fat.
   b. contains red bone marrow, where blood cells are formed.
   c. lends strength to bones.
   d. Both b and c are correct.
6. Which of the following is not a bone of the appendicular skeleton?
   a. the scapula
   b. a rib
   c. a metatarsal bone
   d. the patella
   e. the ulna
7. The vertebrae that articulate with the ribs are the
   a. lumbar vertebrae.
   b. sacral vertebrae.
   c. thoracic vertebrae.
   d. cervical vertebrae.
   e. coccyx.

For questions 8–11, match each bone to the location in the key.

Key:

a. upper arm
b. forearm
c. pectoral girdle
d. pelvic girdle
e. thigh
f. lower leg

8. ulna
9. tibia
10. clavicle
11. femur

39.3 The Muscular System

12. In a muscle fiber,
    a. the sarcolemma is connective tissue holding the myofibrils together.
    b. the sarcoplasmic reticulum stores calcium.
    c. both myosin and actin filaments have cross-bridges.
    d. there is a T system but no endoplasmic reticulum.
    e. All of these are correct.
13. According to the sliding filament model, when muscles contract,
    a. sarcomeres shorten.
    b. myosin heads break down ATP.
    c. actin slides past myosin.
    d. the H zone disappears.
    e. All of these are correct.
14. Nervous stimulation of muscles
    a. occurs at a neuromuscular junction.
    b. involves the release of ACh.
    c. results in impulses that travel down the T system.
    d. causes calcium to be released from the sarcoplasmic reticulum.
    e. All of these are correct.
15. Acetylcholine
    a. is active at somatic synapses but not at neuromuscular junctions.
    b. binds to receptors in the sarcolemma.
    c. precedes the buildup of ATP in mitochondria.
    d. is stored in the sarcoplasmic reticulum.
    e. Both b and d are correct.
16. Label this diagram of a muscle fiber, using these terms: myofibril, Z line, T tubule, sarcomere, sarcolemma, sarcoplasmic reticulum.

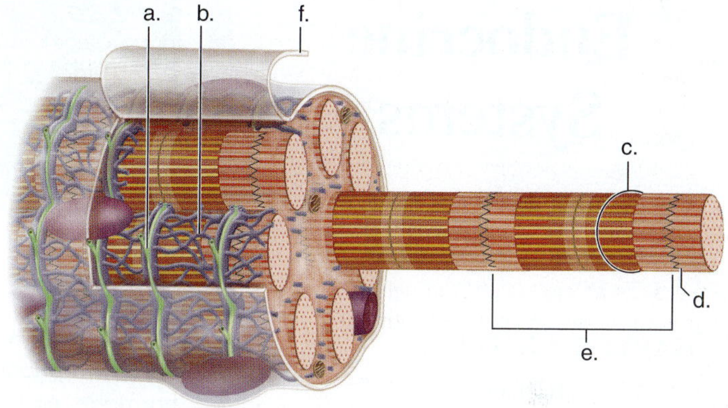

ENGAGE

1. It is observed that some motor neurons innervate only a few muscle fibers in the biceps brachii. Other motor neurons each innervate many muscle fibers. How might this observation correlate with our ability to pick up a pencil or a 2-liter soda bottle? On what basis would the brain bring about the correct level of contraction?
2. Some athletes believe that taking oral creatine will increase their endurance because it will increase the amount of phosphate available to their muscles for ATP synthesis. This statement can be regarded as two hypotheses: (a) oral creatine increases endurance, and (b) oral creatine increases the amount of creatine available in muscles for ATP synthesis. How could these two hypotheses be tested?
3. Similar to the activity in our muscle cells, certain bacteria can utilize sugars by aerobic cellular respiration (which requires oxygen) or by fermentation (which does not). However, like muscle cells, these bacteria only use fermentation when oxygen is not available. Why is aerobic respiration preferable? What is the specific role of oxygen in this process?
4. Most authorities emphasize the importance of calcium consumption for the prevention of osteoporosis. However, surveys show that osteoporosis is rare in some countries, such as China, where calcium intake is lower than in the United States. These studies often point to the higher level of animal protein consumed in the United States as having a detrimental effect on bone mass. If this is true, what physiological mechanism might explain it? Can you think of any other differences between the lifestyles in China and the United States that might be factors?

40

Hormones and Endocrine Systems

Chapter Outline

40.1 Animal Hormones 751

40.2 Hypothalamus and Pituitary Gland 755

40.3 Other Endocrine Glands and Hormones 758

Before You Begin

Before beginning this chapter, take a few moments to review the following discussions.

Section 3.3 What are the general structure and function of steroids?

Chapter 5 What are the general classifications of chemical signaling molecules?

Section 37.3 What are the location and function of the hypothalamus?

The sphinx moth (*Manduca sexta*) begins life as a caterpillar. The caterpillar molts and undergoes metamorphosis, as orchestrated by hormones.

Hormones, chemical messengers of the endocrine system, regulate the metamorphosis of many insects from wormlike larval stages to their adult forms. One hormone, ecdysone, initiates shedding of the exoskeleton as the larva passes through a series of growth stages. A decline in the production of another hormone triggers the final metamorphosis into an adult, as shown in the inset above for the sphinx moth (adult form), also referred to as the tobacco hornworm (caterpillar form).

Along with the nervous system, the endocrine system coordinates the activities of the body's other organ systems and helps maintain homeostasis. In contrast to the nervous system, the endocrine system is not centralized; instead, it consists of several organs scattered throughout the body. The hormones secreted by endocrine glands travel through the bloodstream and interstitial fluid to reach their target tissues. The metabolism of a cell changes when it has a plasma membrane or nuclear receptor for that hormone. In this chapter, you'll learn how hormones exert their slow but powerful influences on the body. You'll see how the endocrine system maintains homeostasis when working properly, as well as some consequences of endocrine malfunction.

As you read through this chapter, think about the following questions:

1. Why do more complex animals, such as mammals, tend to use some of the same hormones that are present in more primitive invertebrates, instead of evolving completely new hormones?
2. What are some specific examples in which the nervous system works with the endocrine system to control body functions?
3. What are some specific examples in which the endocrine system works independently?

Following *the* Themes

CHAPTER 40 HORMONES AND ENDOCRINE SYSTEMS

UNIT 7 COMPARATIVE ANIMAL BIOLOGY

Evolution	Animals have evolved two major organ systems that respond to changes in their internal and external environments: the nervous system, which is capable of rapid responses, and the endocrine system, which carries out slower, lasting responses.
Nature of Science	Advances in the understanding of how hormones function have led to effective treatments for many common disorders of the human endocrine system.
Biological Systems	A relatively limited number of peptide and steroid hormones are able to control a wide range of homeostatic processes in animals.

40.1 Animal Hormones

Learning Outcomes

Upon completion of this section, you should be able to

1. Distinguish between the mode of action of a neurotransmitter and that of a hormone.
2. Identify the major endocrine glands of the human body.
3. Compare the mechanisms of action of peptide and steroid hormones.

The nervous and endocrine systems work together to regulate the activities of the other organs. Both systems use chemical signals when they respond to changes that might threaten homeostasis. However, they have different means of delivering these signals (Fig. 40.1). As discussed in Chapters 37 and 38, sensory receptors detect changes in the internal and external environments and transmit that information to the CNS, which responds by stimulating muscles and glands. Communication depends on nerve signals, conducted in axons, and neurotransmitters, which cross synapses. Axon conduction occurs rapidly, as does the diffusion of a neurotransmitter across the short distance of a synapse. In other words, the nervous system is organized to respond rapidly to stimuli. This is particularly useful if the stimulus is an external event that endangers our safety—we can move quickly to avoid being hurt.

The **endocrine system** functions differently. The endocrine system is largely composed of glands (Fig. 40.2). These glands secrete **hormones,** such as insulin, which are carried by the bloodstream to target cells throughout the body. It takes time to deliver hormones, and it takes time for cells to respond, but the effect is longer-lasting. In other words, the endocrine system is organized for a slower but prolonged response.

Endocrine glands can be contrasted with exocrine glands. **Exocrine glands** secrete their products into ducts, which take them to the lumens of other organs or outside the body. For example, the salivary glands send saliva into the mouth by way of the salivary ducts. **Endocrine glands,** as stated, secrete their products into the bloodstream, which delivers them throughout the body.

Hormones influence almost every basic homeostatic function of an organism, including metabolism, growth, reproduction, osmoregulation, and digestion. Therefore, it is not surprising that hormones are produced by invertebrates as well as vertebrates. For example, the hormone insulin is a key regulator of metabolism in vertebrates, and insulin-related peptides have been identified in insects and molluscs, suggesting an early evolutionary origin of this hormone.

Hormones also control some processes that are unique to invertebrates. As mentioned in the chapter-opening story, hormones control insect *metamorphosis,* the dramatic transformation that some insects undergo while hatching from an egg as a wormlike larva, going through several molts in which the exoskeleton is shed, and maturing into adults. Several hormones control this process.

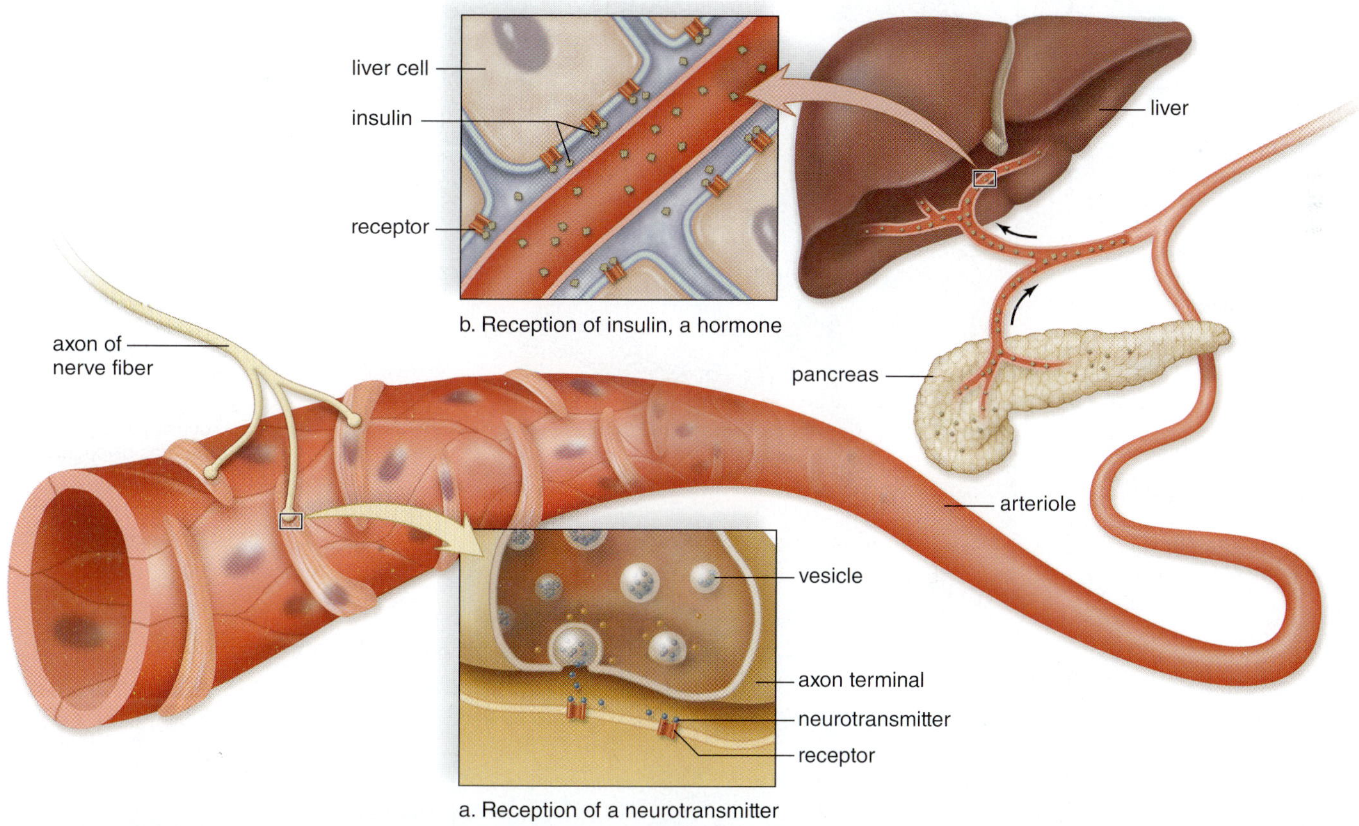

Figure 40.1 Modes of action of the nervous and endocrine systems. **a.** Nerve impulses passing along an axon cause the release of a neurotransmitter. The neurotransmitter, a chemical signal, binds to a receptor and causes the wall of an arteriole to constrict. **b.** The hormone insulin, a chemical signal, travels in the cardiovascular system from the pancreas to the liver, where it binds to a receptor and causes the liver cells to store glucose as glycogen.

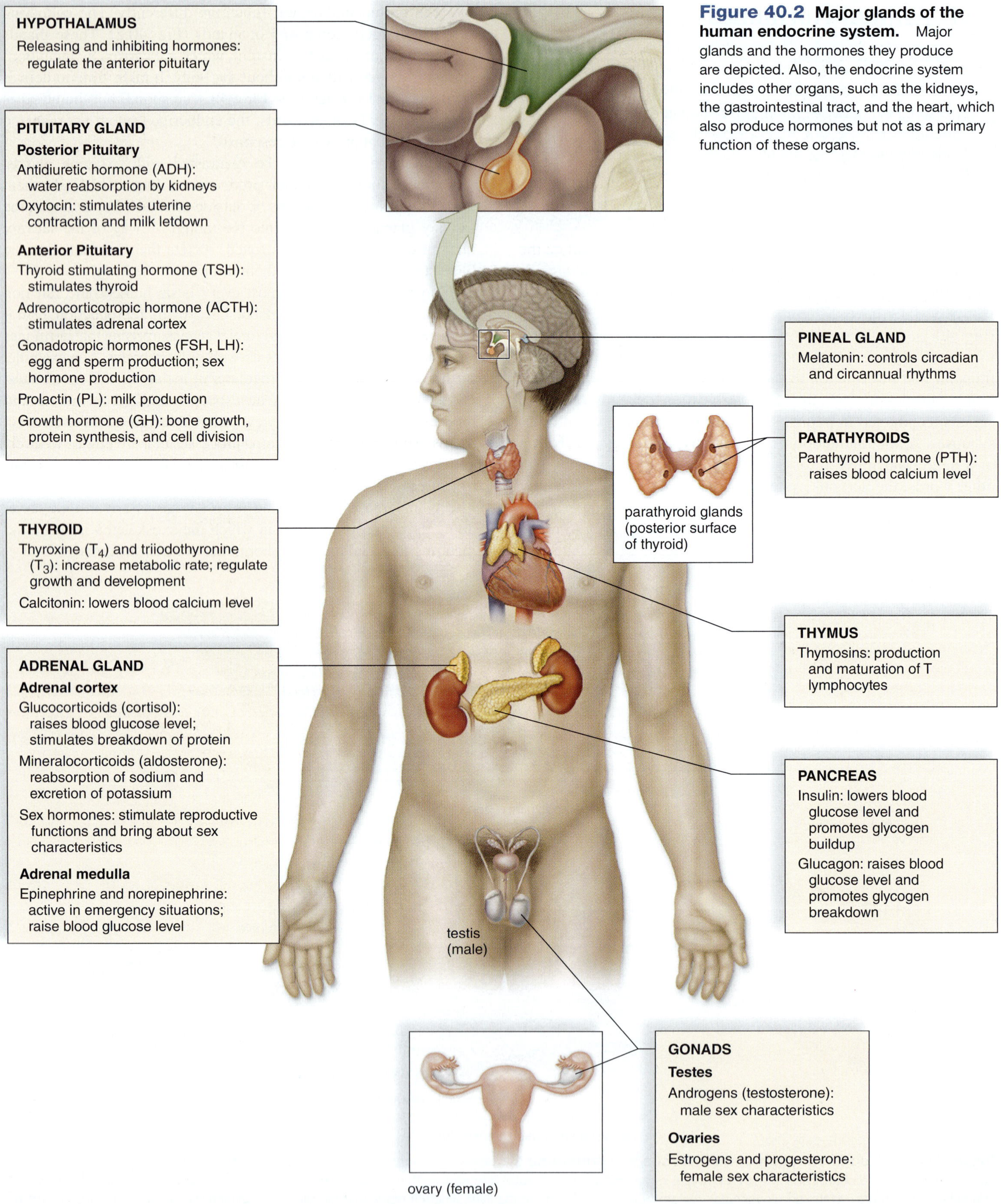

Figure 40.2 Major glands of the human endocrine system. Major glands and the hormones they produce are depicted. Also, the endocrine system includes other organs, such as the kidneys, the gastrointestinal tract, and the heart, which also produce hormones but not as a primary function of these organs.

Sometimes evolution produces new uses for the same hormones. In the freshwater snail *Lymnaea,* a peptide related to insulin is involved in body and shell growth, as well as in energy metabolism. Variable hormone functions are seen in vertebrates as well. All vertebrates synthesize thyroid hormones, which generally increase metabolism, as you'll see later in this chapter. In amphibians, a surge of thyroid hormones also seems to promote metamorphosis from a tadpole into an adult. In contrast, the hormone prolactin inhibits metamorphosis in amphibians, stimulates skin pigmentation in reptiles, initiates incubation of eggs in birds, and stimulates milk production in mammals.

Hormones Are Chemical Signals

Like other **chemical signals,** hormones are a means of communication between cells, between body parts, and even between individuals. However, only certain cells, called target cells, can respond to a specific hormone. A target cell for a particular hormone carries a receptor protein for that hormone (Fig. 40.3). The hormone and receptor protein bind together the way a key fits a lock. The target cell then responds to that hormone. For example, in a condition called *androgen insensitivity,* an individual has X and Y sex chromosomes, and the testes, which remain in the abdominal cavity, produce the sex hormone testosterone. However, the body cells lack receptors that are able to combine with testosterone, and the individual appears to be a normal female.

Chemical signals that influence the behavior of other individuals are called **pheromones.** Pheromones have been well documented in several animal species, although their influence has been more difficult to prove in humans. Women who live in the same household tend to have synchronized menstrual cycles, perhaps because pheromones released by a woman who is menstruating affect the menstrual cycle of other women in the household. Studies also suggest that women prefer the smell of t-shirts worn by men who have a different MHC type from themselves. As noted in Chapter 33, MHC molecules are involved in immunity, and choosing a mate of a different MHC type might improve the health of offspring.

In a small study reported in 2011, men had lower testosterone levels in their saliva after they smelled a jar containing tears from women who were sad, as compared to saline droplets that were trickled down the women's cheeks. While the significance of these studies is unclear, they suggest that humans may, in fact, be influenced by pheromones.

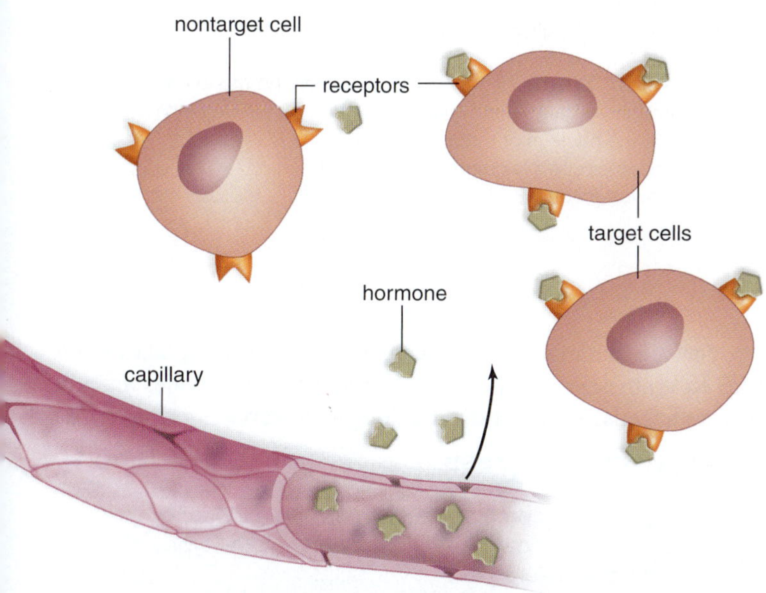

Figure 40.3 Target cell concept. Most hormones are distributed by the bloodstream to target cells. Target cells have receptors for the hormone, and the hormone combines with the receptor, as a key fits a lock.

The Action of Hormones

Hormones exert a wide range of effects on cells. Some hormones induce target cells to increase their uptake of particular molecules, such as glucose, or ions, such as calcium. Others bring about an alteration of the target cell's structure in some way.

Most endocrine glands secrete **peptide hormones.** These hormones are peptides, proteins, glycoproteins, and modified amino acids. **Steroid hormones,** in contrast, all have the same molecular complex of four carbon rings, because they are all derived from cholesterol (see Fig. 3.12).

The Action of Peptide Hormones. The actions of peptide hormones can vary, and as an example in this section, we concentrate on what happens in muscle cells after the hormone epinephrine binds to a receptor in the plasma membrane (Fig. 40.4). In muscle cells, the reception of epinephrine leads to the breakdown of glycogen to glucose, which provides energy for ATP production.

The immediate result of epinephrine binding is the formation of **cyclic adenosine monophosphate (cAMP)**, which contains one phosphate group attached to adenosine at two locations. Therefore, the molecule is cyclic. cAMP activates a protein kinase enzyme in the cell, and this enzyme in turn activates another enzyme, and so forth. The series of enzymatic reactions that follows cAMP formation is called an enzyme cascade or signaling cascade. Because each enzyme can be used over and over again, more enzymes become involved at every step of the cascade. Finally, many molecules of glycogen are broken down to glucose, which enters the bloodstream.

Typical of a peptide hormone, epinephrine never enters the cell. Therefore, the hormone is called the **first messenger,** whereas cAMP, which sets the metabolic machinery in motion, is called the **second messenger.** For example, imagine that the adrenal medulla, which produces epinephrine, is like a company's home office, which sends out a courier (the hormone epinephrine—the first messenger) to its factory (the cell). The courier doesn't have a pass to enter the factory but tells a supervisor through a screen door that the home office wants the factory to produce a particular product. The supervisor (cAMP—the second messenger) walks over and flips a switch that starts the machinery (the enzymatic pathway), and a product is made.

Animation Second Messengers

The Action of Steroid Hormones. Only the adrenal cortex, the ovaries, and the testes produce steroid hormones. Steroid

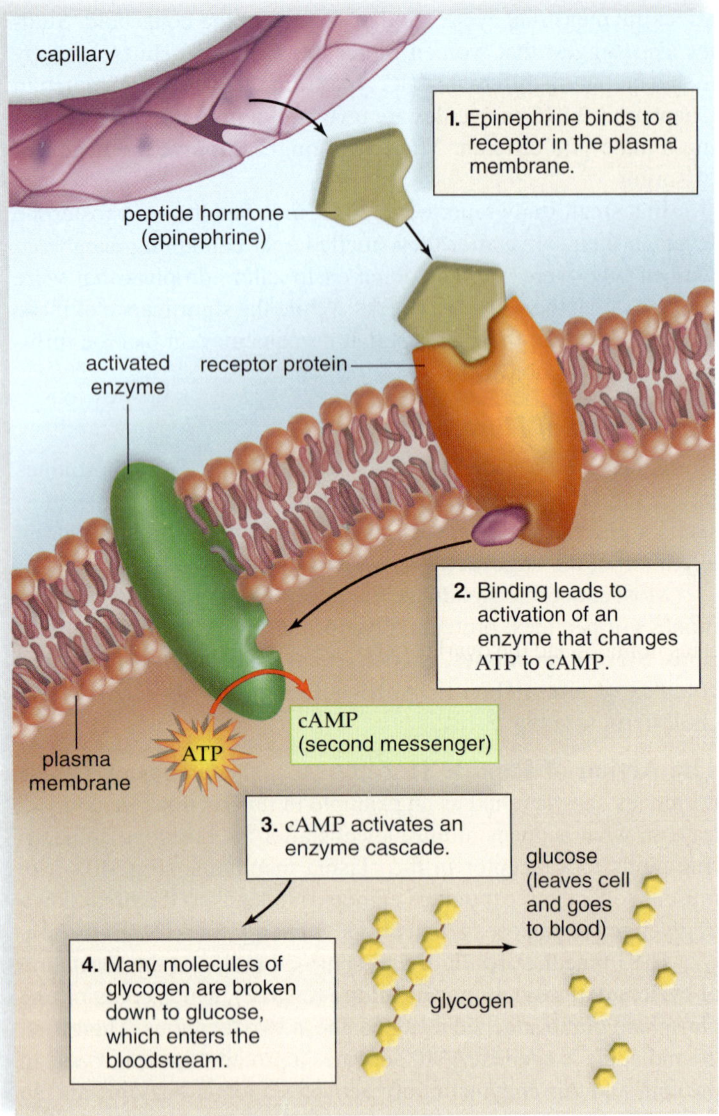

Figure 40.4 Epinephrine, a peptide hormone. Peptide hormones (epinephrine, in this example) act as first messengers, binding to specific receptors in the plasma membrane. First messengers activate second messengers (cAMP, in this case), which influence various cellular processes.

Tutorial Action of a Peptide Hormone

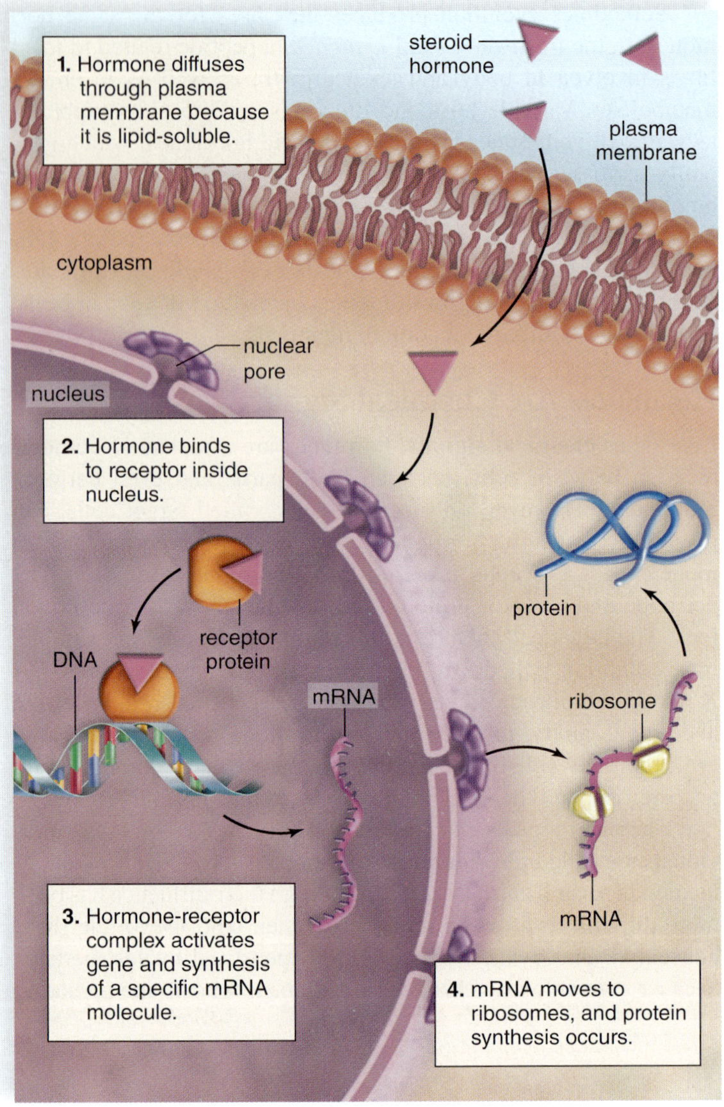

Figure 40.5 Steroid hormone. A steroid hormone passes directly through the target cell's plasma membrane before binding to a receptor in the nucleus or cytoplasm. The hormone-receptor complex binds to DNA, and gene expression follows.

Tutorial Action of a Steroid Hormone

hormones do not bind to plasma membrane receptors; instead, they are able to enter the cell because they are lipids (Fig. 40.5). Once inside, a steroid hormone binds to an internal receptor, usually in the nucleus but sometimes in the cytoplasm. Inside the nucleus, the hormone-receptor complex binds with DNA and activates certain genes. Messenger RNA (mRNA) moves to the ribosomes in the cytoplasm, and protein synthesis (e.g., an enzyme) follows. To continue the analogy, a steroid hormone is like a courier who has a pass to enter the factory (the cell). Once inside, the courier makes contact with the plant manager (DNA), who sees to it that the factory (cell) is ready to produce a product.

Animation Mechanism of Steroid Hormone Action

Steroids act more slowly than peptides, because it takes more time to synthesize new proteins than to activate enzymes already present in cells. Their action lasts longer, however.

Check Your Progress 40.1

1. Compare and contrast the nervous and endocrine systems with regard to their function and the types of signals they use.
2. Compare the location of the receptors for peptide and steroid hormones.
3. Explain why second messengers are needed for most peptide hormones.

40.2 Hypothalamus and Pituitary Gland

Learning Outcomes

Upon completion of this section, you should be able to

1. Describe the relationship between the hypothalamus and the pituitary gland.
2. List and describe the functions of the hormones released by the anterior and posterior pituitary gland.
3. Explain how some hormones are regulated by negative feedback, and some by positive feedback, and give an example of each.

The **hypothalamus** helps regulate the body's internal environment in two ways. Through the autonomic nervous system, it influences the heartbeat, blood pressure, appetite, body temperature, and water balance. It also controls the glandular secretions of the **pituitary gland** (hypophysis), a small gland about 1 cm in diameter connected to the hypothalamus by a stalklike structure. The pituitary has two portions: the posterior pituitary and the anterior pituitary.

MP3 Endocrine System

Animation Hormonal Communication

Posterior Pituitary

Neurons in the hypothalamus, called neurosecretory cells, produce the hormones **antidiuretic hormone** (**ADH**) (Gk. *anti,* "against"; *ouresis,* "urination") and oxytocin (Fig. 40.6, *left*). These hormones pass through axons into the **posterior pituitary,** where they are stored in axon terminals. Certain neurons in the hypothalamus are sensitive to the water-salt balance of the blood. When these cells determine that the blood is too concentrated, ADH is released from the posterior pituitary. Upon reaching the kidneys, ADH causes water to be reabsorbed. As the blood becomes dilute, ADH is no longer released. This is an example of control by **negative feedback**—the *effect* of the hormone (to dilute blood) shuts down the *release* of the hormone. Negative feedback maintains stable conditions and homeostasis.

If too little ADH is secreted, or if the kidneys become unresponsive to ADH, a condition known as *diabetes insipidus* results. Patients with this condition are usually very thirsty; they produce copious amounts of urine and can become severely dehydrated if the condition is untreated.

The consumption of alcohol inhibits ADH release. This effect helps explain the frequent urination associated with drinking alcohol.

Oxytocin (Gk. *oxys,* "quick"; *tokos,* "birth"), the other hormone made in the hypothalamus, causes uterine contractions during childbirth and milk letdown when a baby is nursing. The more the uterus contracts during labor, the more nerve impulses reach the hypothalamus, causing oxytocin to be released. Similarly, the more a baby suckles, the more oxytocin is released. In both instances, the release of oxytocin from the posterior pituitary is controlled by **positive feedback**—that is, the stimulus continues to bring about an effect that ever increases in intensity. Oxytocin may also play a role in the propulsion of semen through the male reproductive tract and may affect feelings of sexual satisfaction and emotional bonding.

Anterior Pituitary

A portal system, which consists of two capillary networks connected by a vein, lies between the hypothalamus and the anterior pituitary (Fig. 40.6, *right*). The hypothalamus controls the anterior pituitary by producing **hypothalamic-releasing hormones** and in some instances **hypothalamic-inhibiting hormones.** For example, one hypothalamic-releasing hormone stimulates the anterior pituitary to secrete a thyroid-stimulating hormone, and a particular hypothalamic-inhibiting hormone prevents the anterior pituitary from secreting prolactin.

Anterior Pituitary Hormones Affecting Other Glands

Some of the hormones produced by the **anterior pituitary** affect other glands. **Gonadotropic hormones** stimulate the gonads—the testes in males and the ovaries in females—to produce gametes and sex hormones. **Adrenocorticotropic hormone** (**ACTH**) stimulates the adrenal cortex to produce cortisol. **Thyroid-stimulating hormone** (**TSH**) stimulates the thyroid to produce thyroxine (T_4) and triiodothyronine (T_3). In each instance, the blood level of the last hormone in the sequence exerts negative feedback control over the secretion of the first two hormones. This is how it works for TSH:

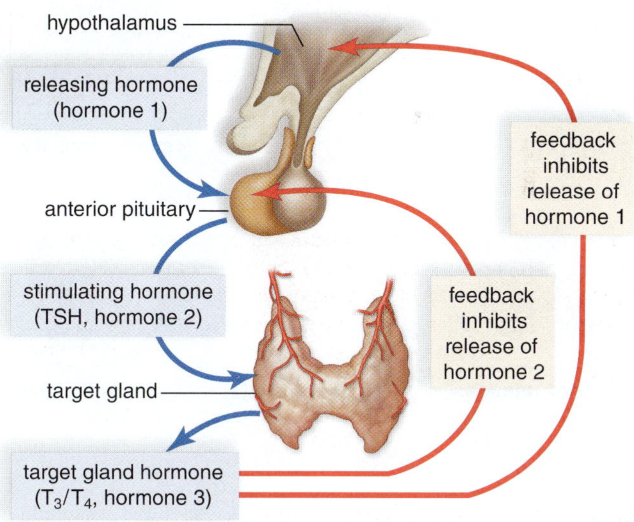

Anterior Pituitary Hormones Not Affecting Other Glands

Three hormones produced by the anterior pituitary do not affect other endocrine glands. **Prolactin** (**PRL**) (L. *pro,* "before"; *lactis,* "milk") is produced in quantity only after childbirth. It causes the mammary glands in the breasts to develop and produce milk. It also plays a role in carbohydrate and fat metabolism.

Melanocyte-stimulating hormone (**MSH**) (Gk. *melanos,* "black"; *kytos,* "cell") causes skin-color changes in many fishes, amphibians, and reptiles having melanophores, skin cells that

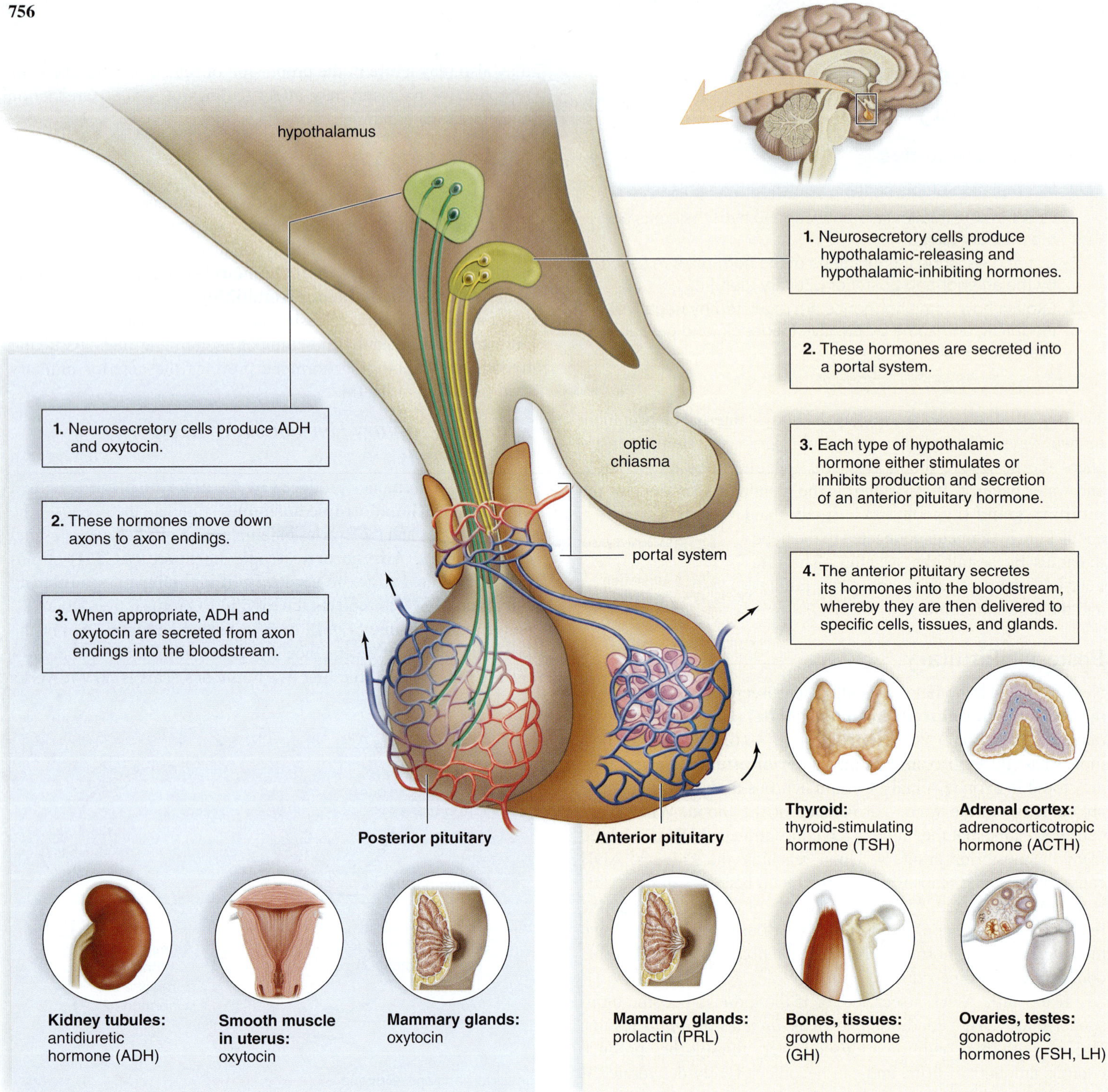

Figure 40.6 Hypothalamus and the pituitary. In this diagram, the name of the hormone is given below its target organ, which is depicted in the circle. *Left:* The hypothalamus produces two hormones, ADH and oxytocin, which are stored and secreted by the posterior pituitary. *Right:* The hypothalamus controls the secretions of the anterior pituitary, and the anterior pituitary controls the secretions of the thyroid, adrenal cortex, and gonads, which are also endocrine glands.

produce color variations. The concentration of this hormone in humans is very low.

Growth hormone (**GH**), or somatotropic hormone, promotes skeletal and muscular growth (Fig. 40.6, *right*). It increases the rate at which amino acids enter cells and protein synthesis occurs. It also promotes fat metabolism, as opposed to glucose metabolism. The amount of GH produced is greatest during childhood and adolescence.

If too little GH is produced during childhood, the individual has *pituitary dwarfism,* characterized by normal proportions but small stature. Such children also have problems with low blood sugar (hypoglycemia), because GH normally helps oppose the effect of insulin on glucose uptake. Through the administration of GH, growth patterns can be restored and blood sugar problems alleviated. If too much GH is secreted during childhood, the person may become a giant (Fig. 40.7*b*). Giants

a. b.

Figure 40.7 Effect of growth hormone. **a.** The amount of growth hormone produced by the anterior pituitary during childhood affects the height of an individual. Plentiful growth hormone produces very tall basketball players. **b.** Too much growth hormone can lead to gigantism, whereas an insufficient amount results in limited stature and even pituitary dwarfism.

usually have poor health, primarily because elevated GH cancels out the effects of insulin, promoting diabetes mellitus (see section 40.3).

On occasion, GH is overproduced in the adult, and a condition called *acromegaly* results. Because long bone growth is no longer possible in adults, only the feet, hands, and face (particularly the chin, nose, and eyebrow ridges) can respond, and these portions of the body become overly large (Fig. 40.8).

A quick Internet search reveals that many websites offer human growth hormone for sale as a "fountain of youth" that can help adults lose weight, add muscle, and reduce the effects of aging. Indeed, several professional actors and athletes have admitted to using human GH to build muscle and reduce body fat. Note that GH is not a *steroid,* which refers to testosterone or related steroid hormones. However, using human GH in this manner can have many undesired side effects, such as joint and muscle pain, high blood pressure, and diabetes mellitus.

MP3 Hormonal Secretion Action

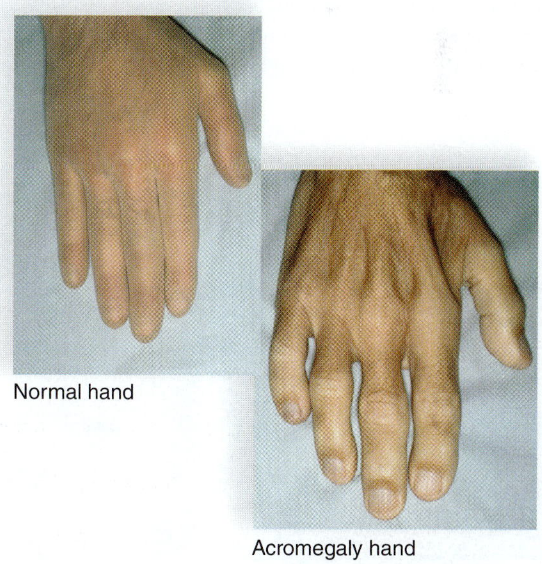

Figure 40.8 Acromegaly. Acromegaly is caused by overproduction of GH in the adult. It is characterized by enlargement of the bones in the face, fingers (as shown here), and toes as a person ages.

Check Your Progress 40.2

1. Explain how the hypothalamus communicates with the endocrine system.
2. List the hormones produced by the posterior pituitary, and provide a function for each.
3. List the hormones produced by the anterior pituitary, and provide a function for each.

40.3 Other Endocrine Glands and Hormones

Learning Outcomes

Upon completion of this section, you should be able to

1. Distinguish among the functions of T_3, T_4, calcitonin, and parathyroid hormone.
2. Compare and contrast the mineralocorticoids and glucocorticoids.
3. Identify the causes and major symptoms of the major conditions associated with the endocrine system.

In this section, we discuss the thyroid and parathyroid glands, adrenal glands, pancreas, pineal gland, thymus, and other tissues that produce hormones secondarily. All the hormone products of these glands and tissues play a role in health and homeostasis.

Thyroid and Parathyroid Glands

The **thyroid gland** is attached to the trachea just below the larynx (see Fig. 40.2). Weighing approximately 20 grams, the thyroid gland is composed of a large number of follicles, each a small, spherical structure made of thyroid cells that produce the hormones triiodothyronine (T_3), which contains three iodine atoms, and **thyroxine** (**T_4**), which contains four iodine atoms. Cells that reside outside the follicles of the thyroid gland produce the hormone calcitonin. The parathyroid glands, which produce parathyroid hormone, are embedded in the posterior surface of the thyroid gland.

Effects of T_3 and T_4

Both of these thyroid hormones increase the overall metabolic rate. They do not have a single target organ; instead, they stimulate most of the cells of the body to metabolize at a faster rate. More glucose is broken down, and more energy is used. Interestingly, even though T_3 and T_4 are peptide hormones because they are derived from the amino acid tyrosine, their receptor is actually located inside cells, more like a steroid hormone receptor.

Animation Mechanism of Thyroxine Action

To produce T_3 and T_4, the thyroid gland actively acquires iodine. The concentration of iodine in the thyroid gland is approximately 25 times that found in the blood. If a person consumes an insufficient amount of iodine, the thyroid gland is unable to produce the required amount of T_3 and T_4. This results in constant stimulation of the thyroid by the TSH released by the anterior pituitary. The thyroid gland enlarges, resulting in a *simple goiter* (Fig. 40.9*a*). In the 1920s, scientists discovered that the use of iodized salt allows the thyroid to produce thyroid hormones and therefore helps prevent simple goiter. However, iodine deficiency is still extremely common in some parts of the world, with an estimated 2 billion people (one-third of the world's population) still suffering from some degree of iodine deficiency.

An insufficiency of T_3 and T_4 in the newborn is called *congenital hypothyroidism* (cretinism) (Fig. 40.9*b*). Babies with this condition are short and stocky, and many are intellectually disabled. The causes vary from iodine deficiency in the pregnant mother to genetic defects affecting the production of TSH, T_3, T_4, or the receptor for any of these hormones. Once detected, iodine deficiency is easily treated by ensuring appropriate levels of iodine consumption. However, according to the American Thyroid Association, congenital hypothyroidism due to iodine deficiency remains the most common preventable cause of intellectual disability in the world.

Even mild iodine deficiency during pregnancy, which may be present in some women in the United States, may be associated with low intelligence in children. In cases due to problems with the thyroid gland itself, thyroxine therapy is curative, but it must begin as early as possible to avoid permanently stunted growth and intellectual disability.

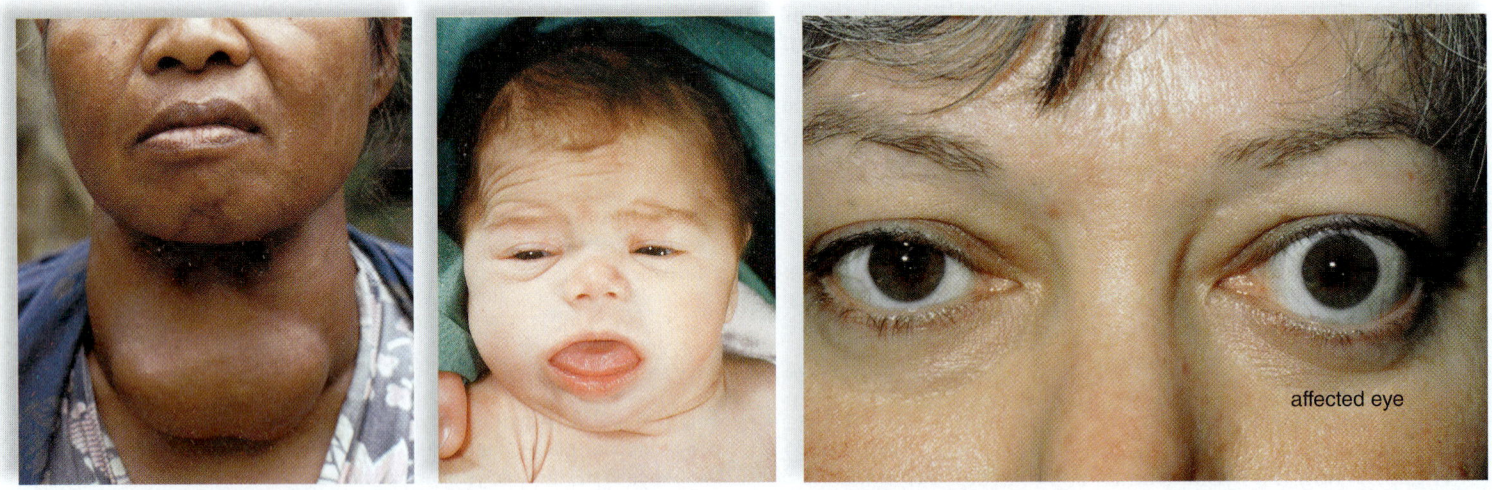

a. Simple goiter b. Congenital hypothyroidism c. Exophthalmic goiter

Figure 40.9 Abnormalities of the thyroid. **a.** An enlarged thyroid gland is often caused by a lack of iodine in the diet. Without iodine, the thyroid is unable to produce its hormones, and continued anterior pituitary stimulation causes the gland to enlarge. **b.** Individuals who develop hypothyroidism during infancy or childhood do not grow and develop as others do. Unless medical treatment is begun, the body is short and stocky; intellectual disability is also likely. **c.** In exophthalmic goiter, a goiter is due to an overactive thyroid, and the eye protrudes because of edema in eye socket tissue.

Hypothyroidism can also occur in adults, most often when the immune system produces antibodies that destroy the thyroid gland. Untreated hypothyroidism in adults results in *myxedema,* which is characterized by lethargy, weight gain, loss of hair, slower heart rate, lowered body temperature, and thickness and puffiness of the skin. The administration of thyroxine usually restores normal body functions and appearance.

Hyperthyroidism results from the oversecretion of T_3 or T_4. In Graves disease, antibodies are produced and react with the TSH receptor on thyroid follicular cells, mimicking the effect of TSH and causing the overproduction of T_3 and T_4. One typical sign of Graves disease is **exophthalmos** (exophthalmia), or excessive protrusion of the eyes due to edema in eye socket tissues and inflammation of the muscles that move the eyes (Fig. 40.9*c*). The patient usually becomes hyperactive, nervous, and irritable and suffers from insomnia. Graves disease is the most common cause of hyperthyroidism in children and adolescents, and it is five to ten times more common in females. Available treatments include drugs that block iodine uptake by the thyroid gland, surgical removal of part or all of the gland, and the administration of radioactive iodine to destroy the overactive tissue.

Hyperthyroidism can also be caused by thyroid cancer, the most common cancer of the endocrine system. Thyroid cancer is usually detected as a lump during physical examination. Again, the treatment is surgery in combination with the administration of radioactive iodine. The prognosis for most patients is excellent.

Effects of Calcitonin

Calcium (Ca^{2+}) plays a significant role in both nervous conduction and muscle contraction. It is also necessary for blood clotting and the maintenance of healthy bones and teeth. The blood calcium level is regulated in part by **calcitonin,** a hormone secreted by the thyroid gland when the blood calcium level rises. The primary effect of calcitonin is to bring about the deposit of calcium in the bones (Fig. 40.10, *top*). It does this by temporarily reducing the activity and number of osteoclasts. When the blood calcium lowers to normal, the thyroid's release of calcitonin is inhibited.

Although calcitonin appears to play a very important role in regulating calcium homeostasis in fish and a few other animals, it appears to be less significant in humans. As evidence for this, a deficiency of calcitonin (as occurs when the thyroid glands are removed) is not linked with any specific disorder. However, because of its bone-building effects, calcitonin is an FDA-approved drug for reducing bone loss in osteoporosis.

Parathyroid Glands

Parathyroid hormone (**PTH**), produced by the **parathyroid glands,** causes the blood calcium level to increase and the blood phosphate (HPO_4^{2-}) level to decrease. Low blood calcium stimulates the release of PTH, which promotes the activity of osteoclasts, releasing calcium from the bones. PTH also promotes the kidneys' reabsorption of calcium, lessening its excretion. In the kidneys, PTH also brings about activation of vitamin D. Vitamin D, in turn, stimulates the absorption of calcium from the small intestine (Fig. 40.10, *bottom*). These effects bring the blood calcium level back to the normal range, and PTH secretion stops. Calcitonin

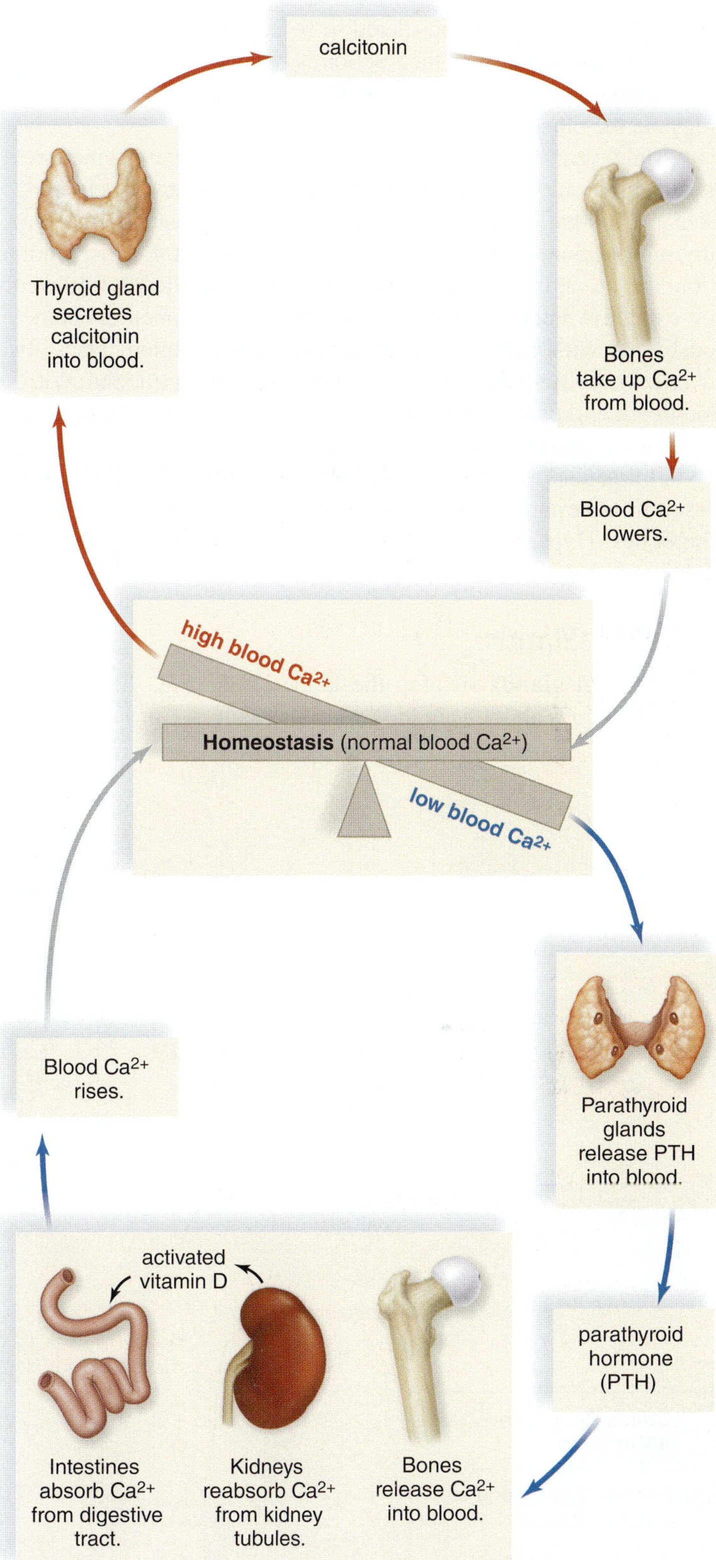

Figure 40.10 Regulation of blood calcium level. *Top:* When the blood calcium (Ca^{2+}) level is high, the thyroid gland secretes calcitonin. Calcitonin promotes the uptake of Ca^{2+} by the bones, and therefore the blood Ca^{2+} level returns to normal. *Bottom:* When the blood Ca^{2+} level is low, the parathyroid glands release parathyroid hormone (PTH). PTH causes the bones to release Ca^{2+} and the kidneys to reabsorb Ca^{2+} and activate vitamin D; thereafter, the intestines absorb Ca^{2+}. Therefore, the blood Ca^{2+} level returns to normal.

and PTH are therefore considered to be *antagonistic hormones,* because their action is opposite one another and both hormones work together to regulate the blood calcium level.

MP3 Calcium Homeostasis

Many years ago, the four parathyroid glands were sometimes mistakenly removed during thyroid surgery because of their small size and hidden location. Gland removal caused insufficient parathyroid hormone production, or *hypoparathyroidism.* This condition leads to a dramatic drop in the blood calcium level, followed by excessive nerve excitability. Nerve signals happen spontaneously and without rest, causing a phenomenon called tetany. In *tetany,* the body shakes from continuous muscle contraction. Without treatment (usually with intravenous calcium), severe hypoparathyroidism causes seizures, heart failure, and death.

Untreated *hyperparathyroidism* (oversecretion of PTH) can result in osteoporosis because of continuous calcium release from the bones. Hyperparathyroidism can also cause the formation of calcium kidney stones.

Adrenal Glands

The **adrenal glands** sit atop the kidneys (see Fig. 40.2). Each adrenal gland is about 5 cm long and 3 cm wide, weighs about 5 g, and consists of an inner portion called the **adrenal medulla** and an outer portion called the **adrenal cortex.** These portions, like the anterior and posterior pituitary, are two functionally distinct endocrine glands. Stress of all types, including both emotional and physical trauma, prompts the hypothalamus to stimulate both portions of the adrenal glands. The adrenal cortex is also involved in regulating the salt and water balance, as well as secreting a small amount of male and female sex hormones.

Adrenal Medulla

As noted in Chapter 37, during emergency situations that call for a "fight-or-flight" reaction, the hypothalamus sends nerve impulses by way of sympathetic nerve fibers to many organs, including the adrenal medulla (see Fig. 37.14). This neurological response to danger quickly dilates the pupils, speeds the heart, dilates the air passages, and inhibits many nonessential bodily functions. Meanwhile, the adrenal medulla secretes **epinephrine** (adrenaline) and **norepinephrine** (noradrenaline) into the bloodstream (Fig. 40.11). These hormones continue the response to stress throughout the body—for example, by accelerating the breakdown of glucose to form ATP, triggering the mobilization of glycogen reserves in skeletal muscle, and increasing the cardiac rate and force of

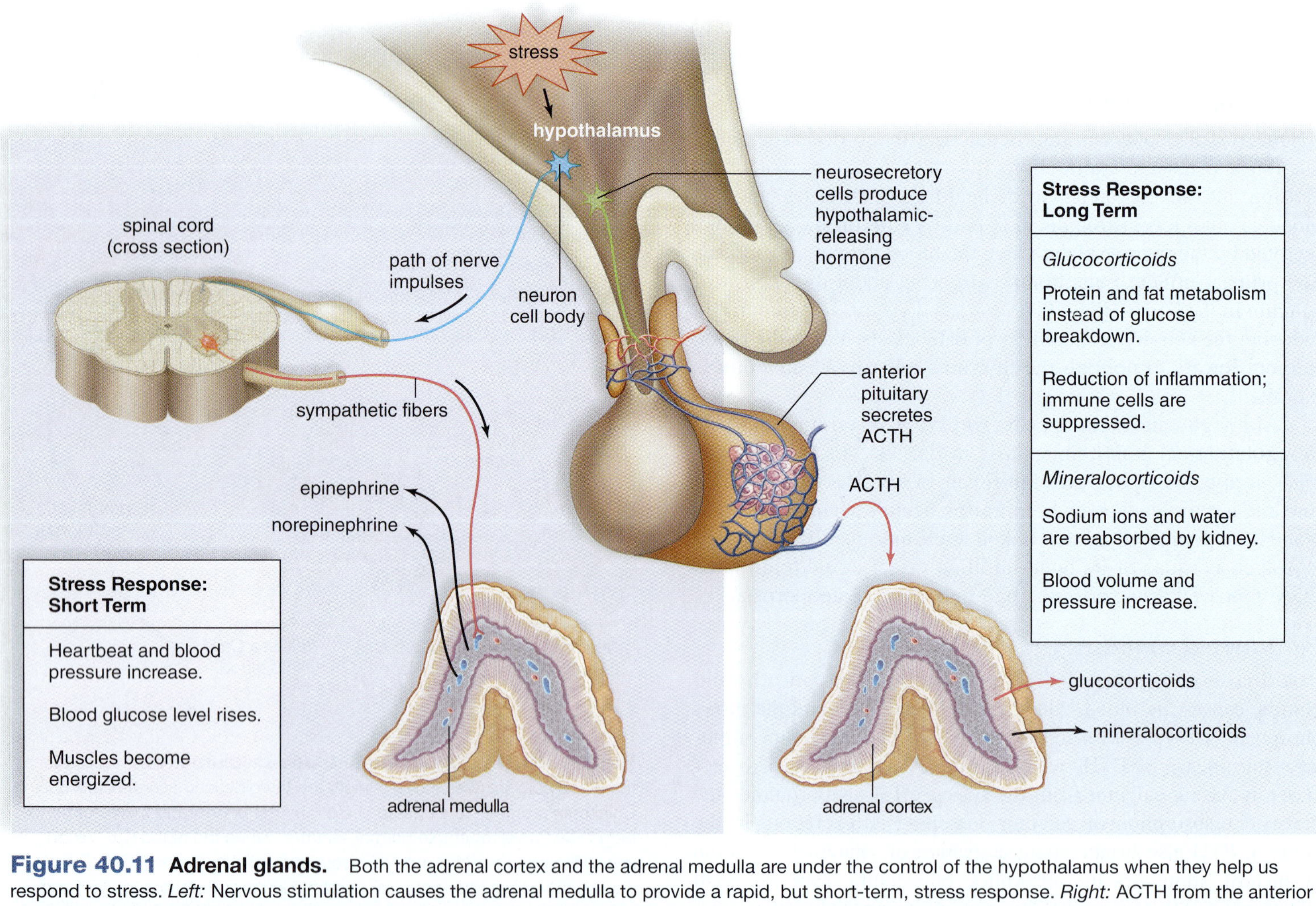

Figure 40.11 Adrenal glands. Both the adrenal cortex and the adrenal medulla are under the control of the hypothalamus when they help us respond to stress. *Left:* Nervous stimulation causes the adrenal medulla to provide a rapid, but short-term, stress response. *Right:* ACTH from the anterior pituitary causes the adrenal cortex to release glucocorticoids. Independently, the adrenal cortex releases mineralocorticoids. The adrenal cortex provides a slower, but long-term, stress response.

contraction. These effects are usually short-lived however, as these two hormones are rapidly metabolized by the liver.

Adrenal Cortex

In contrast to the rapid response of the sympathetic nervous system and adrenal medulla, the hypothalamus produces a longer-term response to stress by stimulating the anterior pituitary to produce ACTH, which in turn causes the adrenal cortex to secrete glucocorticoids.

Glucocorticoids. **Cortisol** is the most important **glucocorticoid** produced by the human adrenal cortex. Cortisol raises the blood glucose level in at least two ways:

1. It promotes the breakdown of muscle proteins to amino acids, which are taken up by the liver from the bloodstream and converted into glucose.
2. It promotes the catabolism of fatty acids rather than carbohydrates, and this spares glucose. The rise in blood glucose is beneficial to an animal under stress, because glucose is the preferred energy source for neurons.

Glucocorticoids also counteract the inflammatory response, including the type of reaction that leads to the pain and swelling of joints in arthritis and bursitis. Cortisone, a glucocorticoid, is often used to treat these conditions, because it reduces inflammation. However, very high levels of glucocorticoids in the blood can suppress the body's defense system, rendering an individual more susceptible to injury and infection.

Animation Glucocorticoid Hormones

Mineralocorticoids. **Mineralocorticoids** produced by the adrenal cortex regulate salt and water balance, leading to increases in blood volume and blood pressure. **Aldosterone** is the most important of the mineralocorticoids. Aldosterone primarily targets the kidneys, where it promotes the renal absorption of sodium (Na^+) and renal excretion of potassium (K^+), thereby helping regulate blood volume and blood pressure.

The secretion of mineralocorticoids is not controlled by the anterior pituitary. In Chapter 36, you saw that when the atria of the heart are stretched due to increased blood volume, cardiac cells release a hormone called **atrial natriuretic hormone (ANH)**, also called *atrial natriuretic peptide,* which inhibits the secretion of aldosterone from the adrenal cortex. (Note that the heart is one of several organs in the body that release a hormone but are not considered among the major endocrine organs.) The effect of ANH is to cause *natriuresis,* the excretion of sodium ions (Na^+). When sodium is excreted, so is water, and therefore blood pressure lowers to normal (Fig. 40.12, *top*). ANH can further lower blood pressure by dilating smooth muscle in blood vessels.

We also noted in Chapter 36 that when the blood sodium (Na^+) level, and therefore blood pressure, is low, the kidneys secrete **renin** (Fig. 40.12, *bottom*). Renin is an enzyme that converts the plasma protein angiotensinogen to angiotensin I, which is changed to angiotensin II by an enzyme in lung capillaries. Angiotensin II stimulates the adrenal cortex to release aldosterone. The effect of this renin-angiotensin-aldosterone system is to raise blood pressure in two ways: (1) Angiotensin II constricts the arterioles, and (2) aldosterone causes the kidneys to reabsorb sodium. When the blood sodium level rises, water is reabsorbed, in part because the

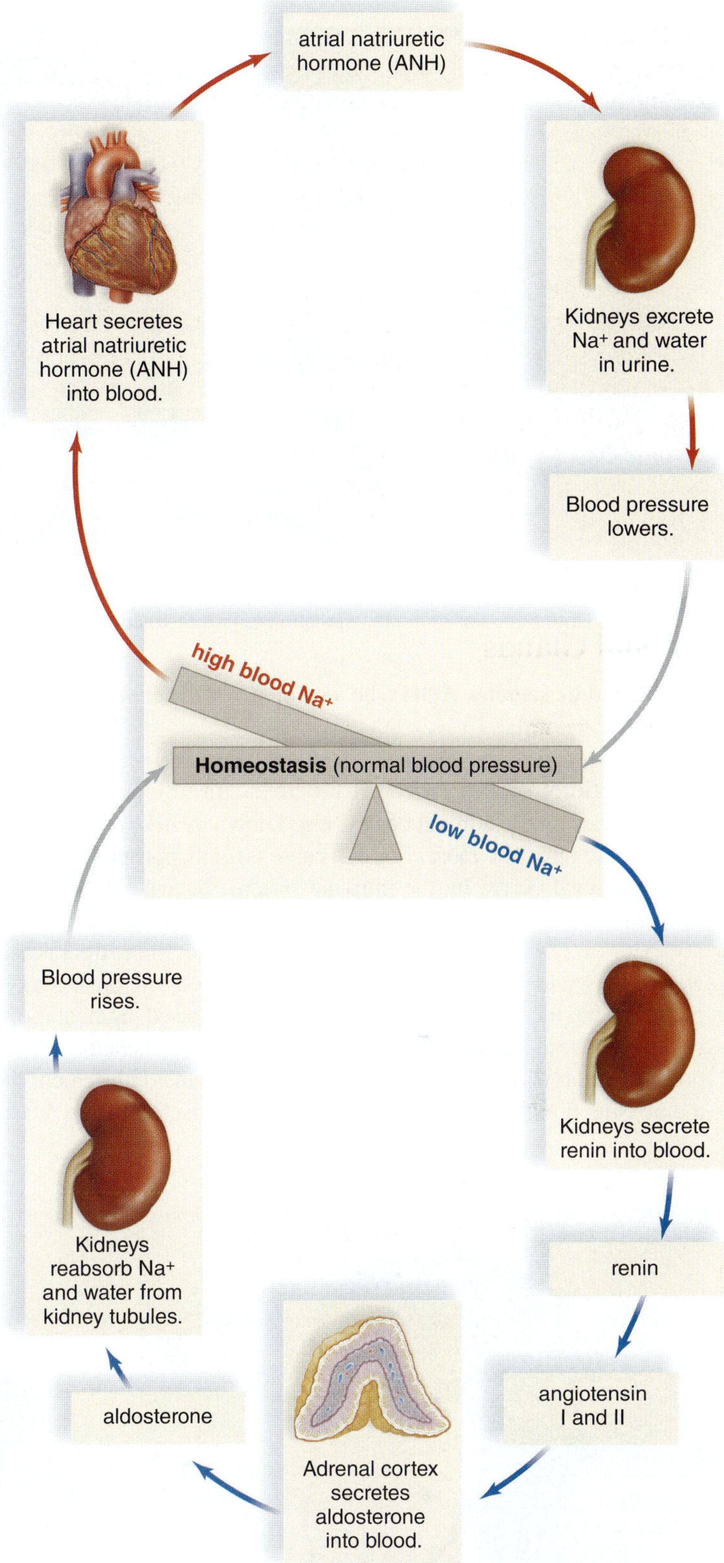

Figure 40.12 Regulation of blood pressure and volume. *Top:* When the blood Na^+ is high, a high blood volume causes the heart to secrete atrial natriuretic hormone (ANH). ANH causes the kidneys to excrete Na^+, and water follows. The blood volume and pressure return to normal. *Bottom:* When the blood sodium (Na^+) level is low, a low blood pressure causes the kidneys to secrete renin. Renin leads to the secretion of aldosterone from the adrenal cortex. Aldosterone causes the kidneys to reabsorb Na^+, and water follows, so that blood volume and pressure return to normal.

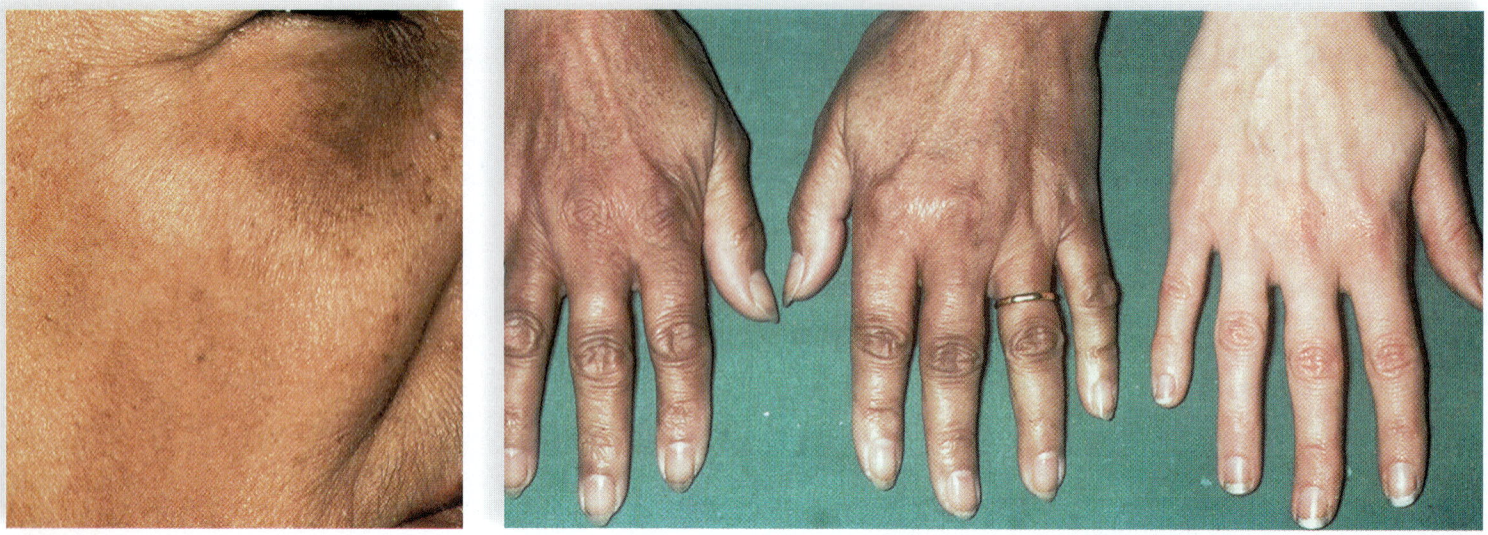

a. b.

Figure 40.13 Addison disease. Addison disease is characterized by a peculiar bronzing of the skin, particularly noticeable in light-skinned individuals. Note the color of the face (**a**) and the hands (**b**) compared to the hand of an individual without the disease.

hypothalamus secretes ADH (see section 40.2). Then, blood pressure rises to normal.

Malfunctions of the Adrenal Cortex. Insufficient secretion of hormones by the adrenal cortex, also known as *Addison disease,* is relatively rare. The most common cause is an inappropriate attack on the adrenal cortex by the immune system. Because the disease affects the secretion of both glucocorticoids and mineralocorticoids, a variety of symptoms may occur, such as dehydration, weakness, weight loss, and hypotension (low blood pressure). The presence of excessive but ineffective ACTH often causes increased pigmentation of the skin, because ACTH, like MSH, can stimulate melanocytes to produce melanin (Fig. 40.13). Treatment involves the replacement of the missing hormones. Left untreated, Addison disease can be fatal.

Excessive levels of glucocorticoids result in *Cushing syndrome*. This disorder can be caused by tumors that affect either the pituitary gland, resulting in excess ACTH secretion, or the adrenal cortex itself. The most common cause, however, is the administration of glucocorticoids to treat other conditions (e.g., to suppress chronic inflammation). Regardless of the source, excess glucocorticoids cause muscle protein to be metabolized and subcutaneous fat to be deposited in the midsection (Fig. 40.14). Excess production of adrenal male sex hormones in women may result in masculinization, including an increase in body hair, deepening of the voice, and beard growth. Depending on the cause, the treatment of Cushing syndrome may involve a careful reduction in the amount of cortisone being taken, the use of cortisol-inhibiting drugs, or surgery to remove any existing pituitary or adrenal tumor.

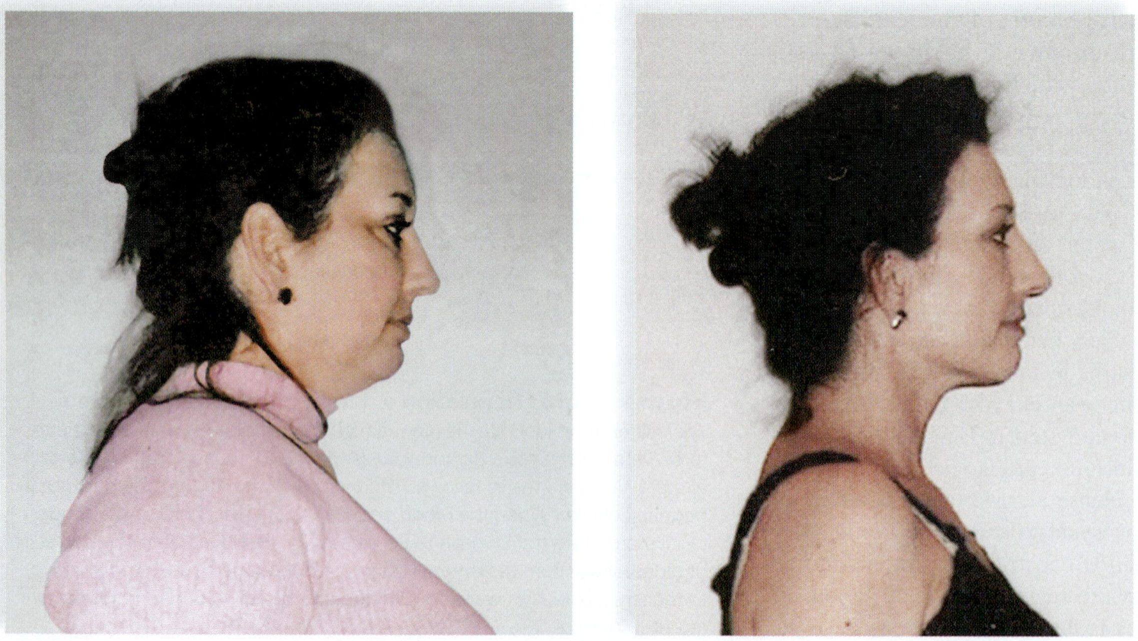

Figure 40.14 Cushing syndrome. This 40-year-old woman was diagnosed with a small tumor in her pituitary gland, which secreted large amounts of ACTH. The high ACTH levels stimulated the adrenal glands to produce excessive amounts of cortisol. *Left:* The patient at the time of surgery to remove her pituitary tumor. *Right:* The patient's appearance 1 year later.

Pancreas

The **pancreas** (see Fig. 40.2) is a slender, fish-shaped organ that stretches across the abdomen behind the stomach and near the duodenum of the small intestine. Approximately 6 inches long and weighing about 80 grams, the pancreas is composed of two types of tissue: Exocrine tissue produces and secretes digestive juices that go by way of ducts to the small intestine; endocrine tissue, called the **pancreatic islets** (islets of Langerhans), produces and secretes the hormones insulin and glucagon directly into the blood. The Nature of Science feature, "Identifying Insulin as a Chemical Messenger," on page 764 describes the discovery of insulin.

The pancreas is not under pituitary control. **Insulin** is secreted when there is a high blood glucose level, which usually occurs just after eating (Fig. 40.15, *top*). Insulin stimulates the uptake of glucose by cells, especially liver cells, muscle cells, and adipose tissue cells. In liver and muscle cells, glucose is then stored as glycogen, and in fat cells the breakdown of glucose supplies glycerol for the formation of fat. In these ways, insulin lowers the blood glucose level.

Glucagon is secreted from the pancreas, usually between meals, when blood glucose is low (Fig. 40.15, *bottom*). The major target tissues of glucagon are the liver and adipose tissue. Glucagon stimulates the liver to break down glycogen to glucose and to use fat and protein in preference to glucose as energy sources. Adipose tissue cells break down fat to glycerol and fatty acids. The liver takes these up and uses them as substrates for glucose formation. In these ways, glucagon raises the blood glucose level. Insulin and glucagon are another example of antagonistic hormones, which work together to maintain the blood glucose level.

Diabetes Mellitus

In 2013, the American Diabetes Association reported that nearly 26 million Americans (8.3% of the population) have *diabetes mellitus,* often referred to simply as diabetes. Of these, an estimated 7 million are undiagnosed. In people under age 20, about 215,000 (1 in 400) have diabetes. Diabetes is characterized by an inability of the body's cells, especially liver and muscle cells, to take up glucose as they should. This causes blood glucose to be higher than normal, and cells rely on other fuels, such as fatty acids, for energy. Therefore, cellular famine exists in the midst of plenty. As the blood glucose level rises, glucose, along with water, is excreted in the urine (Gr. *mellitus,* "honey, sweetness"). This results in frequent urination and causes the diabetic to be extremely thirsty.

Other symptoms of diabetes include fatigue, constant hunger, and weight loss. Diabetics often experience vision problems due to diabetic retinopathy (see section 38.2) and swelling in the lens of the eye due to the high blood sugar levels. If untreated, diabetics often develop serious and even fatal complications. Sores that don't heal develop into severe infections. Blood vessel damage causes kidney failure, nerve destruction, heart attack, or stroke.

Two types of diabetes have been identified: type 1 diabetes, sometimes called juvenile diabetes, and type 2 diabetes, or adult-onset diabetes. Despite their names, however, both diseases may occur in children or adults.

Type 1 Diabetes. In *type 1 diabetes,* the pancreas is not producing enough insulin. This condition is believed to be brought on by exposure to an environmental agent, most likely a virus, whose

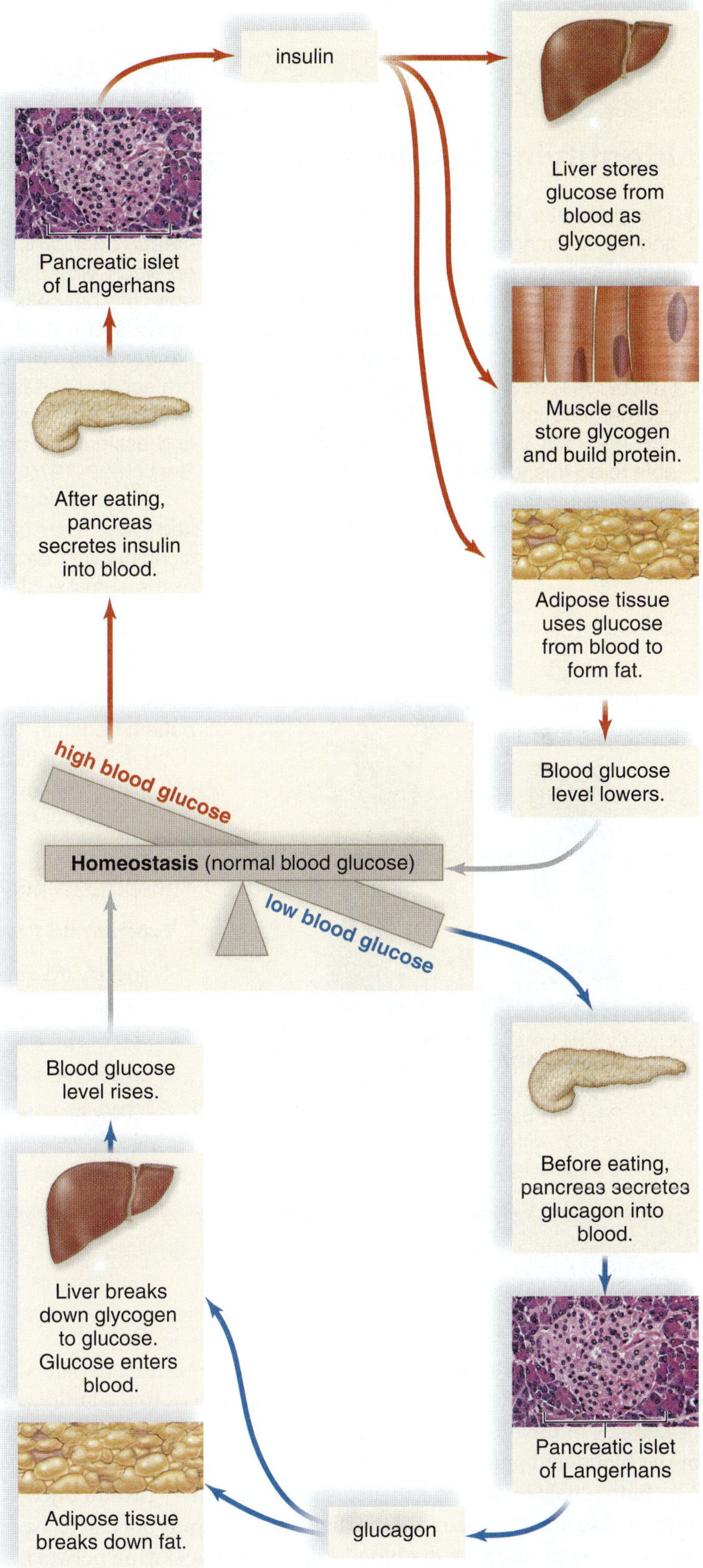

Figure 40.15 Regulation of blood glucose level.
Top: When the blood glucose level is high, the pancreas secretes insulin. Insulin promotes the storage of glucose as glycogen and the synthesis of proteins and fats (as opposed to their use as energy sources). Therefore, insulin lowers the blood glucose level to normal. *Bottom:* When the blood glucose level is low, the pancreas secretes glucagon. Glucagon acts opposite to insulin; therefore, glucagon raises the blood glucose level to normal.

THEME Nature of Science

Identifying Insulin as a Chemical Messenger

The pancreas is both an exocrine gland and an endocrine gland. It sends digestive juices to the duodenum by way of the pancreatic duct, and it secretes the hormones insulin and glucagon into the bloodstream.

In 1920, physician Frederick Banting (1891–1941) decided to try to isolate insulin in order to identify it as a chemical messenger. Previous investigators had been unable to do this, because the enzymes in the digestive juices destroyed insulin (a protein) during the isolation procedure. Banting hit upon the idea of ligating (tying off) the pancreatic duct, which he knew from previous research would lead to the degeneration only of the cells that produce digestive juices and not of the pancreatic islets (of Langerhans), where insulin is made. His professor, J. J. Macleod, made a laboratory available to him at the University of Toronto and assigned a graduate student, Charles Best (1899–1978), to assist him.

Figure 40A Early insulin experiments. Charles H. Best and Sir Frederick Banting in 1921 with the first dog to be kept alive by insulin.

Banting and Best (Fig. 40A) had limited funds and spent that summer working, sleeping, and eating in the lab. By the end of the summer, they had obtained pancreatic extracts that lowered the blood glucose level in diabetic dogs. Macleod then brought in biochemists, who purified the extract. Insulin therapy for the first human patient began in 1922, and large-scale production of purified insulin from pigs and cattle followed. Banting and Macleod received a Nobel Prize for their work in 1923.

The amino acid sequence of insulin was determined in 1953. Insulin is now synthesized using recombinant DNA technology, using the bacterium *E. coli* to produce the hormone. Banting and Best performed the required steps (given in the accompanying table) to identify a chemical messenger.

Questions to Consider

1. What type of disease or symptoms would you expect to occur after ligating the pancreatic ducts of dogs?
2. What are some advantages, and potential disadvantages, of producing a medicine destined to be injected into humans (such as insulin) in a bacterium such as *E. coli?*
3. Some people oppose the use of animals for medical research. Do you think that insulin would have eventually been discovered without animal experimentation? Why or why not?

Steps to Identify a Chemical Messenger as Used by Banting and Best

Experimental Procedure	Results from Banting and Best Experiment
1. Identify the source of the chemical	Pancreatic islets are source
2. Identify the effect to be studied	Presence of pancreas in body lowers blood glucose
3. Isolate the chemical	Insulin isolated from pancreatic secretions
4. Show that the chemical has the desired effect	Insulin lowers blood glucose

presence causes cytotoxic T cells to destroy the pancreatic islets. The body turns to the metabolism of fat, which leads to the buildup of ketones in the blood, called ketoacidosis, increasing the acidity of the blood and potentially leading to coma and death.

Individuals with type 1 diabetes must have daily insulin injections. These injections control the diabetic symptoms but can cause inconveniences, because the blood sugar level may swing between hypoglycemia (low blood glucose) and hyperglycemia (high blood glucose). Without testing the blood glucose level, it is difficult to be certain which of these is present, because the symptoms can be similar. These symptoms include perspiration, pale skin, shallow breathing, and anxiety. Whenever these symptoms appear, immediate attention is required to bring the blood glucose back within the normal range. If the problem is hypoglycemia, the treatment is one or two glucose tablets, hard candy, or orange juice. If the problem is hyperglycemia, the treatment is insulin. Better control of blood glucose levels can often be achieved with an insulin pump, a small device worn outside the body that is connected to a plastic catheter inserted under the skin.

Because diabetes is such a common problem, many researchers are working to develop more effective methods for treating it. The most desirable would be an artificial pancreas, defined as an automated system that would provide insulin based on real-time changes in blood sugar levels. It is possible to transplant a working pancreas, or even fetal pancreatic islet cells, into patients with type 1 diabetes. Another possibility is *xenotransplantation,* in which insulin-producing islet cells of another species, such as pigs, are placed inside a capsule that allows insulin to exit but prevents the immune system from attacking the foreign cells. Finally, researchers are now testing a "reverse vaccine," which, instead of stimulating an immune response, seems to block the immune system's attack on the islet cells, perhaps by inducing T cells capable of suppressing these responses.

Type 2 Diabetes. Most adult diabetics have *type 2 diabetes.* Often, the patient is overweight or obese, and adipose tissue produces a substance that impairs insulin receptor function. However, complex genetic factors can be involved, as shown by the tendency for type 2 diabetes to occur more often in certain families, or even ethnic groups. For example, the condition is 77% more common in African Americans than in non-Hispanic whites.

Normally, the binding of insulin to its plasma membrane receptor causes the number of protein carriers for glucose to increase, causing more glucose to enter the cell. In the type 2 diabetic, insulin still binds to its receptor, but the number of glucose carriers does not increase. Therefore, the cell is said to be insulin-resistant.

It is possible to prevent or at least control type 2 diabetes by adhering to a low-fat, low-sugar diet and exercising regularly. If this fails, oral drugs are available that stimulate the pancreas to secrete more insulin and enhance the metabolism of glucose in the liver and muscle cells. Millions of Americans may have type 2 diabetes without being aware of it; however, the effects of untreated type 2 diabetes are as serious as those of type 1 diabetes.

Testes and Ovaries

The activity of the testes and ovaries is controlled by the hypothalamus and pituitary. The **testes** are located in the scrotum, and the **ovaries** are located in the pelvic cavity. The testes produce **androgens** (e.g., **testosterone**), the male sex hormones. The ovaries produce **estrogens** and **progesterone,** the female sex hormones. These hormones provide feedback that controls the hypothalamic secretion of gonadotropin-releasing hormone (GnRH). The pituitary gland secretion of follicle-stimulating hormone (FSH) and luteinizing hormone (LH), the gonadotropic hormones, is controlled by feedback from the sex hormones, too. The activities of FSH and LH are discussed in Chapter 41.

Under the influence of the gonadotropic hormones, the testes release an increased amount of testosterone at the time of puberty, which stimulates growth of the penis and testes. Testosterone also brings about and maintains the male secondary sex characteristics that develop during puberty. These include the growth of facial, axillary (underarm), and pubic hair. It prompts the larynx and vocal cords to enlarge, causing the voice to lower. Testosterone also stimulates the activity of oil and sweat glands in the skin. Another side effect of testosterone is baldness. Genes for baldness are inherited by both sexes, but baldness is seen more often in males because of the presence of testosterone.

Testosterone is partially responsible for the muscular strength of males, and this is the reason some athletes take supplemental amounts of **anabolic steroids,** which are either testosterone or related chemicals. The dangerous side effects of taking anabolic steroids are listed in Figure 40.16.

The female sex hormones, estrogens (often referred to in the singular) and progesterone, have many effects on the body. In particular, estrogen secreted at the time of puberty stimulates the growth of the uterus and vagina. Estrogen is necessary for egg maturation and is largely responsible for the secondary sex characteristics in females, including female body hair and fat distribution.

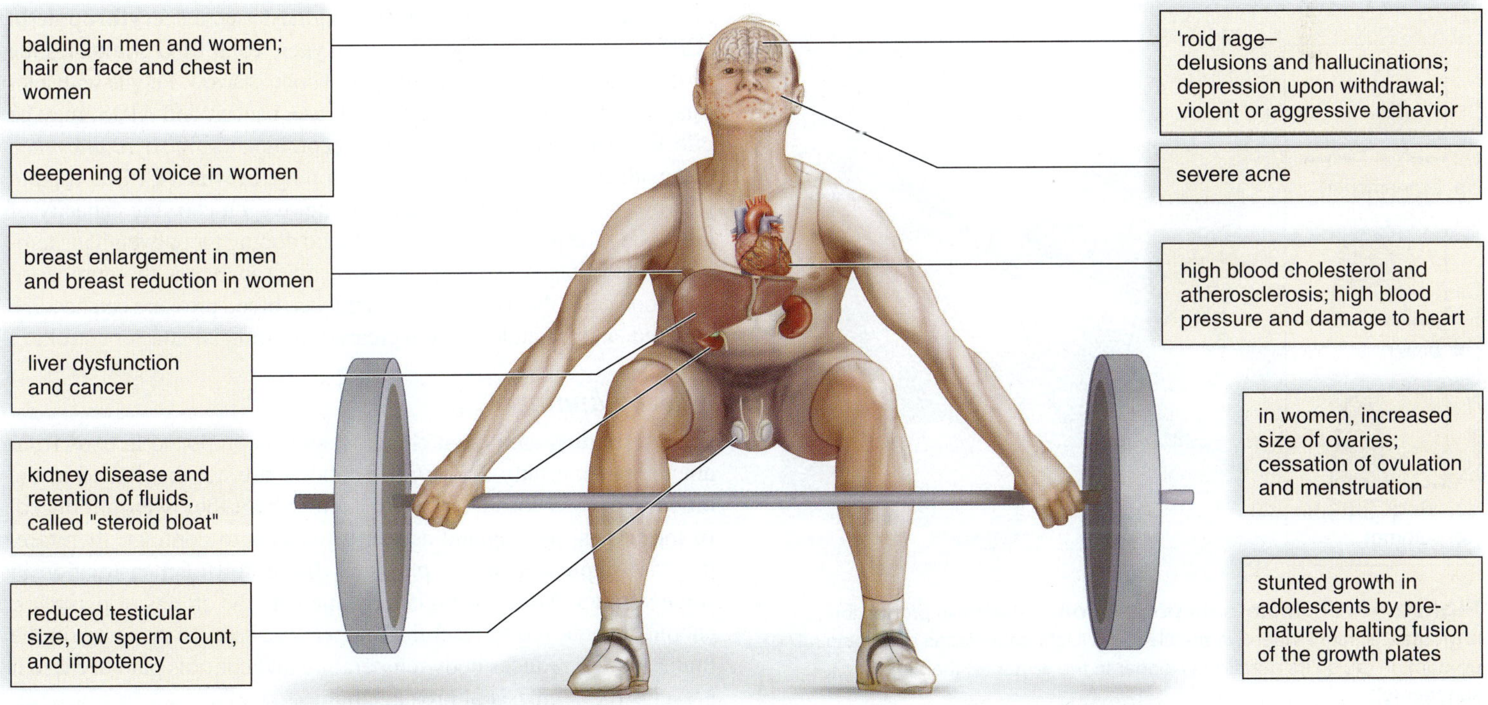

Figure 40.16 The effects of anabolic steroid use.

In general, females have a more rounded appearance than males because of a greater accumulation of fat beneath the skin. Also, the pelvic girdle is wider in females, resulting in a larger pelvic cavity. Both estrogen and progesterone are required for breast development and regulation of the uterine cycle. This includes monthly menstruation (discharge of blood and mucosal tissues from the uterus). The ovaries and adrenal glands of women also normally produce a small amount of testosterone, which plays a role in the development of muscle and bone strength, overall energy level, sex drive (libido), and sexual pleasure.

Pineal Gland

The **pineal gland** (epiphysis), located deep in the human brain (see Fig. 40.2), produces the hormone **melatonin,** primarily at night. Melatonin is involved in our daily sleep-wake cycle; normally, we grow sleepy at night when melatonin levels increase and awaken once daylight returns and melatonin levels are low (Fig. 40.17). Daily 24-hour cycles such as this are called **circadian rhythms** (L. *circum,* "around"; *dies,* "day"), and circadian rhythms are controlled by an internal timing mechanism called a biological clock.

Instead of being buried deep in the brain, the pineal gland of some vertebrates is on top of the brain, and in certain fossilized reptiles and even some primitive extant reptiles and amphibians, an additional opening in the skull is present, covered only by a thin layer of skin. This, along with the presence of light-sensing cells in the pineal gland, has led some investigators to conclude that this gland functioned as a "third eye" at some point in evolution. The exact functions of this structure are not completely understood, although it may have aided in determining the position of the sun or in establishing circadian rhythms.

Animal research suggests that melatonin also regulates sexual development. In keeping with these findings, it has been noted that children whose pineal glands have been destroyed due to a brain tumor experience early puberty.

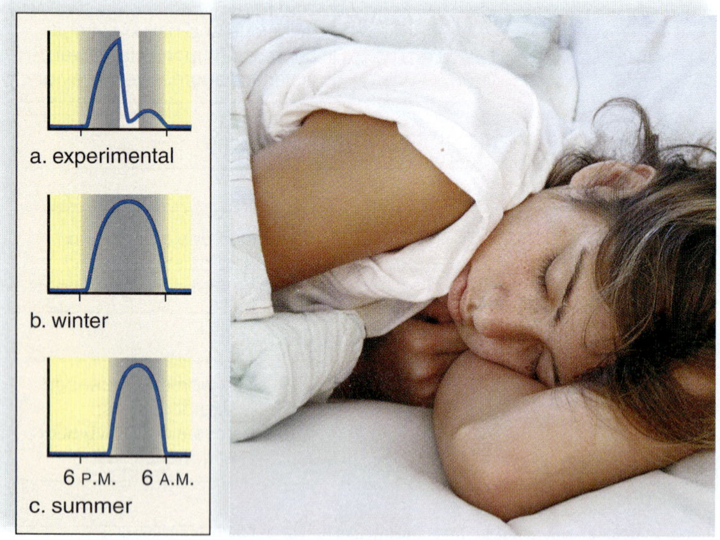

Figure 40.17 Melatonin production. Melatonin production is greatest at night when we are sleeping. Light suppresses melatonin production (**a**), so its duration is longer in the winter (**b**) than in the summer (**c**).

Thymus

The lobular **thymus** lies just beneath the sternum (see Fig. 40.2). This organ reaches its largest size and is most active during childhood. With aging, the thymus gets smaller and becomes fatty. Lymphocytes that originate in the bone marrow and then pass through the thymus become T lymphocytes. The lobules of the thymus are lined by epithelial cells that secrete hormones called thymosins. These hormones aid in the differentiation of T lymphocytes packed inside the lobules.

Hormones from Other Organs or Tissues

Some organs not usually considered endocrine glands do secrete hormones. Two examples already mentioned are renin excreted by the kidneys and atrial natriuretic hormone excreted by the heart. A number of other tissues produce hormones.

Leptin

Leptin is a peptide hormone produced by adipose tissue throughout the body. Leptin acts on the hypothalamus, where it signals satiety, or fullness. After leptin was discovered in the 1990s, researchers hoped that it could be used to control obesity in humans. Unfortunately, the trials have not yielded satisfactory results. In fact, the blood of obese individuals may be rich in leptin. It is possible that the leptin they produce is ineffective because of a genetic mutation or because their hypothalamic cells lack a suitable number of receptors for leptin.

Erythropoietin

As mentioned in Chapter 36, the kidneys secrete **erythropoietin (EPO)** in response to a low blood oxygen level. EPO stimulates the production of red blood cells in the red bone marrow. People with anemia, which is common in kidney disease, cancer, and AIDS, may be effectively treated with injections of recombinant EPO. In recent years, some athletes have practiced blood doping, in which EPO is used to improve performance by increasing the oxygen-carrying capacity of the blood. The potential dangers of blood doping far outweigh the temporary advantages, however. Because EPO increases the number of red blood cells, the blood becomes thicker, blood pressure can become elevated, and the athlete is at increased risk of heart attack or stroke.

Prostaglandins

Prostaglandins are potent chemical signals produced in cells from arachidonate, a fatty acid. Prostaglandins are not distributed in the blood. They act locally, quite close to where they were produced. In the uterus, prostaglandins cause muscles to contract; therefore, they are implicated in the pain and discomfort of menstruation in some women. Also, prostaglandins mediate the effects of *pyrogens,* chemicals believed to reset the temperature regulatory center in the brain. Aspirin reduces body temperature and controls pain because it prevents the synthesis of prostaglandins.

Certain prostaglandins increase the secretion of protective mucus in the stomach and thus are used to prevent gastric ulcers. Others lower blood pressure and have been used to treat hypertension. Still others inhibit platelet aggregation and have been used to prevent thrombosis. Because prostaglandins can affect various tissues, however, unwanted side effects can be a problem. For example, Misoprostol, a prostaglandin commonly used to prevent stomach ulcers, should not be taken by pregnant women, as it may cause uterine contractions, resulting in miscarriage or premature labor.

Check Your Progress 40.3

1. Explain how the renin-angiotensin-aldosterone system raises blood pressure.
2. List the endocrine gland that secretes each of the following hormones: aldosterone, melatonin, epinephrine, EPO, leptin, glucagon, ANH, cortisol, and calcitonin.
3. Name one hormone that stimulates the activity of osteoclasts and one that inhibits them.

CONNECTING *the* CONCEPTS *with the* THEMES

Evolution

- The cells of even the simplest multicellular animals have evolved the ability to communicate with each other using chemical signals.
- The nervous and endocrine systems have evolved together, so the two systems are structurally and functionally related.
- The nervous system has developed to be able to bring about a rapid response to stimuli, while the endocrine system provides a slower, longer-lasting response.

Nature of Science

- Through scientific research, we now have a better understanding of the structures and functions of the endocrine system.
- A relatively large number of disorders affect the human endocrine system, but many of these, such as diabetes and thyroid disorders, are treatable because of our understanding of how hormones work.

Biological Systems

- Invertebrates use hormones to control some unique processes, such as the metamorphosis of larvae to form adults.
- In vertebrates, the hypothalamus controls the function of the pituitary gland, which in turn controls several other endocrine glands.
- Hormones exhibit control and influence over a wide range of body systems and functions, including salt balance, growth, metabolism, sexual development, and responses to stimuli.
- Some tissues and organs having other main functions also produce hormones, and certain cells produce locally acting hormones.

MEDIA STUDY TOOLS http://connect.mheducation.com

SMARTBOOK® Study smart. SmartBook helps you maximize your study time by identifying what you know and don't know. The recharge feature in SmartBook helps you prepare for the test by identifying the material that you are most likely to forget.

connect BIOLOGY Find even more resources to learn the chapter concepts including animations, tutorial videos, and interactive practice questions.

MP3 Files	Animations	Tutorials
40.2 Endocrine System • Hormonal Secretion Action **40.3** Calcium Homeostasis	**40.1** Second Messengers • Mechanism of Steroid Hormone Action **40.2** Hormonal Communication **40.3** Mechanism of Thyroxine Action • Glucocorticoid Hormones	**40.1** Action of a Peptide Hormone • Action of a Steroid Hormone

SUMMARIZE

40.1 Animal Hormones

The nervous system and **endocrine system** both use chemical signals to maintain homeostasis. **Endocrine glands** secrete **hormones** into the bloodstream, and from there they are distributed to target organs or tissues. In contrast, **exocrine glands** secrete their products via ducts.

Hormones are **chemical signals** that usually act at a distance between body parts. **Pheromones** are chemical signals that influence the behavior of other individuals. Hormones are either peptides or steroids. Reception of **peptide hormones** at the plasma membrane activates an enzyme cascade inside the cell. The hormone is the **first messenger,** and key intermediates that are formed inside the cell, such as **cyclic adenosine monophosphate (cAMP),** are the **second messenger. Steroid hormones** combine with a receptor inside the cell, and the complex attaches to and activates DNA. Protein synthesis follows.

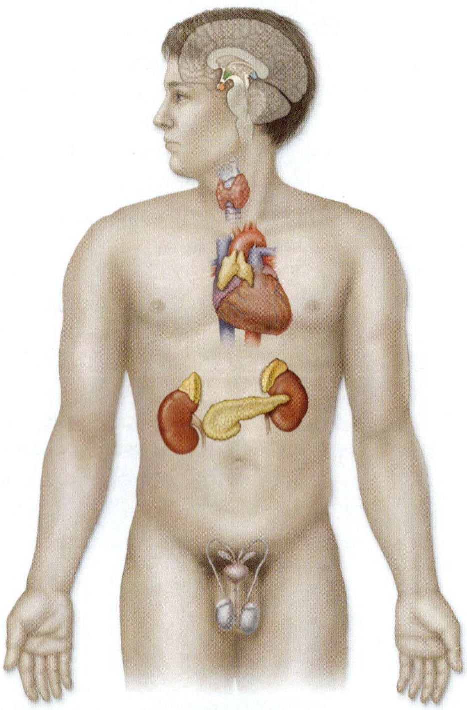

40.2 Hypothalamus and Pituitary Gland

The **hypothalamus** and **pituitary gland** exert control over the endocrine system. Neurosecretory cells in the hypothalamus produce **antidiuretic hormone** (**ADH**) and **oxytocin,** which are stored in axon endings in the **posterior pituitary** until they are released. As with most hormones, secretion of ADH is regulated by **negative feedback;** the effect of the hormone shuts down its release. In contrast, during childbirth and milk letdown, oxytocin secretion is regulated by **positive feedback;** a stimulus brings about ever-increasing hormone levels.

The hypothalamus produces **hypothalamic-releasing hormones** and **hypothalamic-inhibiting hormones,** which pass to the anterior pituitary by way of a portal system. The **anterior pituitary** produces several types of hormones. Some of these stimulate other endocrine glands: **gonadotropic hormones** influence the gonads; **adrenocorticotropic hormone** (**ACTH**) stimulates the adrenal glands; and **thyroid-stimulating hormone** (**TSH**) acts on the thyroid gland.

Three anterior pituitary hormones act directly on tissues: **prolactin** (**PRL**) causes mammary gland development and milk production; **melanocyte-stimulating hormone** (**MSH**) affects skin color; and **growth hormone** (**GH**) causes skeletal and muscular growth.

40.3 Other Endocrine Glands and Hormones

The **thyroid gland,** controlled by TSH, requires iodine to produce **thyroxine** (**T_4**) and triiodothyronine (T_3), which increase the metabolic rate. Depending on the age of an individual, a deficiency of T_3 and T_4 may result in congenital **hypothyroidism,** simple goiter, or myxedema. **Hyperthyroidism** may be the result of Graves disease (**exophthalmos** is a common sign) or thyroid cancer. The thyroid gland also produces **calcitonin,** which lowers blood calcium levels by increasing calcium deposition in bones.

The **parathyroid glands** secrete **parathyroid hormone** (**PTH**), which raises the blood calcium and decreases the blood phosphate levels.

The **adrenal glands** respond to stress: Immediately, the **adrenal medulla** secretes **epinephrine** and **norepinephrine,** which bring about responses we associate with emergency situations. On a long-term basis, the **adrenal cortex,** controlled by ACTH, produces the **glucocorticoids** (e.g., cortisol) and the **mineralocorticoids** (e.g., aldosterone). **Cortisol** stimulates hydrolysis of proteins to amino acids that are converted to glucose; in this way, it raises the blood glucose level. **Aldosterone** causes the kidneys to reabsorb sodium ions (Na^+) and to excrete potassium ions (K^+). Addison disease develops when the adrenal cortex is underactive, and Cushing syndrome occurs when the adrenal cortex is overactive.

Atrial natriuretic hormone (**ANH**) is secreted by the heart when its atria are stretched due to high blood pressure. ANH increases renal excretion of sodium ions and water. When blood pressure is too low, the kidneys release renin, resulting in the formation of angiotensin II, which causes arterioles to constrict, and aldosterone to be released.

The **pancreas** has **pancreatic islets** that secrete **insulin,** which lowers the blood glucose level, and **glucagon,** which has the opposite effect. The most common illness caused by hormonal imbalance is diabetes mellitus, which is due to the failure of the pancreas to produce insulin or the failure of the cells to take it up.

The testes and ovaries, controlled by gonadotropic hormones, produce the sex hormones. The major male sex hormone (**androgen**) is **testosterone,** and the major female sex hormones are **estrogen** and **progesterone.** Some athletes abuse **anabolic steroids** to increase strength and athletic performance.

Tissue and organs having other functions also produce hormones. The **pineal gland** produces **melatonin,** which may be involved in **circadian rhythms** and the development of the reproductive organs. The thymus produces hormones that aid in T lymphocyte development, **leptin** from adipose tissue regulates appetite, and **erythropoietin** from the kidneys stimulates the production of red blood cells. **Prostaglandins** are produced and act locally, with a variety of effects on different tissues.

ASSESS

Choose the best answer for each question.

40.1 Animal Hormones

For questions 1–5, match each hormone to a gland in the key.

Key:

a. pancreas
b. anterior pituitary
c. posterior pituitary
d. adrenal medulla
e. adrenal cortex

1. cortisol
2. growth hormone (GH)
3. oxytocin storage
4. insulin
5. epinephrine
6. Which of the following is not associated with the action of a peptide hormone?
   a. Hormone is released into the blood by an endocrine gland.
   b. Hormone binds to a plasma membrane receptor on target cells.
   c. Hormone enters cells and binds to internal receptors.
   d. Hormone triggers the activity of second messengers.
   e. All of these typically occur with peptide hormones.
7. Steroid hormones are secreted by
   a. the adrenal cortex.
   b. the gonads.
   c. the thyroid.
   d. Both a and b are correct.
   e. Both b and c are correct.

40.2 Hypothalamus and Pituitary Gland

8. Which one of the following statements about the pituitary gland is incorrect?
   a. The pituitary lies inferior to the hypothalamus.
   b. Growth hormone and prolactin are secreted by the anterior pituitary.
   c. The anterior pituitary and posterior pituitary communicate with each other.
   d. Axons run between the hypothalamus and the posterior pituitary.
9. The anterior pituitary controls the secretion(s) of both
   a. the adrenal medulla and the adrenal cortex.
   b. the thyroid and the adrenal cortex.
   c. the ovaries and the testes.
   d. Both b and c are correct.
10. Which hormone and condition are mismatched?
    a. insufficient ADH—diabetes insipidis
    b. too much ADH—diabetes mellitus
    c. insufficient growth hormone—pituitary dwarfism
    d. too much growth hormone in adults—acromegaly
    e. too much growth hormone in children—gigantism

40.3 Other Endocrine Glands and Hormones

11. Which association related to thyroid hormones is mismatched?
    a. iodine deficiency in pregnancy—congenital hypothyroidism
    b. hypothyroidism in adults—myxedema
    c. hyperthyroidism in adults—Graves disease
    d. calcitonin deficiency in adults—goiter
12. Parathyroid hormone causes
    a. the kidneys to excrete more calcium ions.
    b. bone tissue to break down and release calcium into the bloodstream.
    c. fewer calcium ions to be absorbed by the intestines.
    d. more calcium ions to be deposited in bone tissue.
13. Diabetes mellitus is associated with
    a. too much insulin in the blood.
    b. too much glucose in the blood.
    c. blood that is too dilute.
    d. Both b and c are correct.
14. The difference between type 1 and type 2 diabetes is that
    a. for type 2 diabetes, insulin is produced but not used; type 1 results from lack of insulin production.
    b. treatment for type 2 involves insulin injections, whereas type 1 can be controlled, usually by diet.
    c. only type 1 can result in complications such as kidney disease, reduced circulation, or stroke.
    d. type 1 can be a result of lifestyle, and type 2 is thought to be caused by a virus or other agent.
15. Which hormone is found in females?
    a. estrogen
    b. testosterone
    c. follicle-stimulating hormone
    d. Both a and c are correct.
    e. All of these are correct.

ENGAGE

THINKING SCIENTIFICALLY

1. Even though some of their functions overlap, why is it advantageous for animals to have both a nervous system and an endocrine system?
2. Caffeine inhibits the breakdown of cAMP in the cell. According to Figure 40.4, how would this influence a stress response brought about by epinephrine?
3. Certain endocrine disorders, such as Cushing syndrome, can be caused by excessive secretion of a hormone (in this case, ACTH) by the pituitary gland or by a problem with the endocrine gland itself (in this case, the adrenals). If you determined the ACTH levels of a Cushing patient, how could you tell the difference between a pituitary problem and a primary adrenal problem?
4. In animals, pheromones can influence many different behaviors. Because humans produce a number of airborne hormones, what human behaviors might be influenced by these? How would these hormones be received by others, and would we necessarily be aware of their effects on us?

41

Reproductive Systems

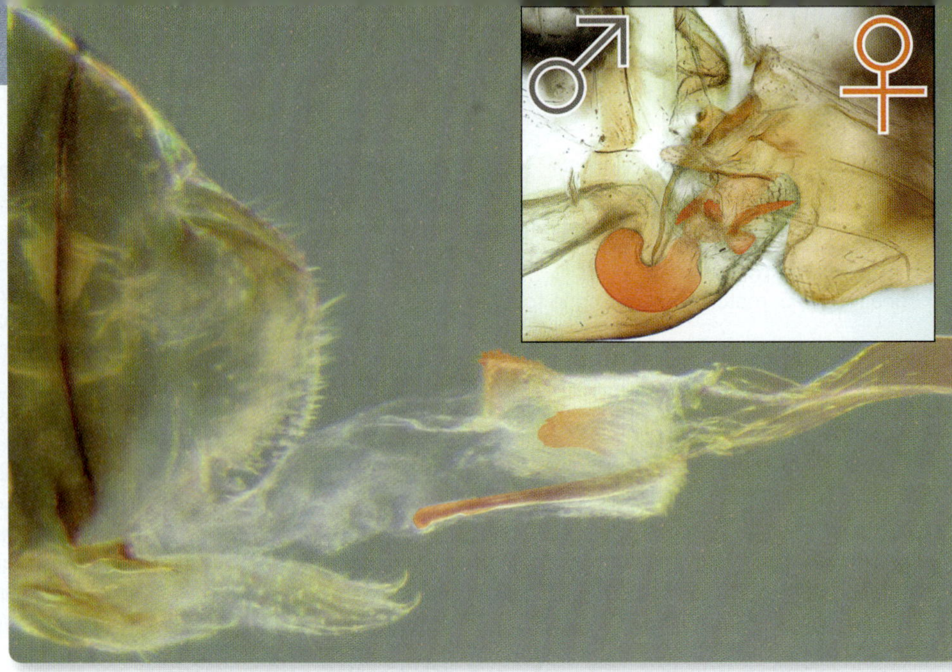

Some female Brazilian cave insects have an elaborate penislike organ which is used to obtain sperm from a vagina-like organ of males.

Chapter Outline

41.1 How Animals Reproduce 771

41.2 Human Male Reproductive System 773

41.3 Human Female Reproductive System 777

41.4 Control of Human Reproduction 781

41.5 Sexually Transmitted Diseases 785

Before You Begin

Before beginning this chapter, take a few moments to review the following discussions.

Chapter 10 How does meiosis halve the chromosome number to produce haploid cells?

Section 16.2 What influences does sexual selection have on mating patterns in animals?

Figure 40.6 What is the relationship between the hypothalamus/pituitary and the reproductive systems of males and females?

How do animals have sex? It seems like a simple question. In biology, however, simple questions can have complicated answers. For example, the penis of a male duck is actually an outpocketing of his cloaca, which remains inside the abdomen until just before copulation, when it becomes erect due to an influx of lymphatic fluid, not blood. Some male octopuses have a penis that detaches and swims to females, attaching to their vagina. Male seahorses even become pregnant, after a female uses a tiny structure called an ovipositor to insert her unfertilized eggs into a brood pouch on the male's abdomen, where they are fertilized by his sperm.

Until recently, though, at least most male animals could lay exclusive claim to possessing a true penis. In a recently discovered Brazilian cave insect of the genus *Neotrogla,* researchers were surprised to find that females have an erectable penis, or gynosome, which enters the vagina-like structure of a male, followed by sperm transfer from the male to the female. The scientists speculate that this arrangement may have developed as a strategy for the females to obtain sperm as a source of nutrition in their dark, nutrient-poor environment.

In this chapter, we focus on the more common strategies of asexual and sexual reproduction. Most of the chapter is devoted to the human male and female reproductive systems, methods that are available to help prevent unwanted pregnancies, and the sexually transmitted diseases that can have a detrimental effect on human reproduction.

As you read through this chapter, think about the following questions:

1. Considering this description of unusual animal reproduction, define the terms *male, female, sex,* and *pregnant.*
2. List some common examples of secondary sex characteristics that develop in male animals in order to attract female mates.

Following *the* Themes

CHAPTER 41 REPRODUCTIVE SYSTEMS

UNIT 7 COMPARATIVE ANIMAL BIOLOGY

Theme	
Evolution	Animals have evolved a large number of strategies, both sexual and asexual, for producing offspring.
Nature of Science	Knowledge about human reproduction has led to the development of efficient methods of birth control, techniques for helping infertile couples, and better prevention and treatment of sexually transmitted diseases.
Biological Systems	Animals reproducing asexually produce identical copies of themselves, whereas animals reproducing sexually have gonads that produce gametes and accessory organs for the storage and/or passage of gametes into or out of the body.

41.1 How Animals Reproduce

Learning Outcomes

Upon completion of this section, you should be able to

1. Compare and contrast asexual and sexual reproduction and list several animals that undergo each.
2. Describe some advantages of each of the following life histories: oviparous, ovoviviparous, and viviparous.

For an animal species to survive, individuals must reproduce. As discussed in Chapter 10, **sexual reproduction** involves a reshuffling of genetic material during *meiosis*. In animals, this process forms the sex cells, or **gametes,** which subsequently unite to form genetically unique offspring. The tremendous amount of genetic variation that is created during meiosis can be advantageous to the survival of a species, especially when the environment is changing.

In **asexual reproduction,** a single parent gives rise to offspring that are identical to the parent, unless mutations have occurred. The adaptive advantage of asexual reproduction is that organisms can reproduce rapidly and colonize favorable environments quickly. Although the majority of animals reproduce sexually, a few groups of animals are also capable of asexual reproduction.

Asexual Reproduction

Several types of invertebrates, such as sponges, cnidarians, flatworms, annelids, and echinoderms, can reproduce asexually. In cnidarians, such as hydras, new individuals may arise asexually as an outgrowth (bud) of the parent (Fig. 41.1). Some species of hydras can also reproduce sexually.

Many flatworms can reproduce asexually by splitting in half, generating two identical individuals. In the laboratory, a planarian can be cut into as many as ten pieces, each of which will grow into a new planarian. To varying degrees, many other animals, including sponges, annelids, and echinoderms, also have the ability to regenerate from fragments. Scientists are studying the ability of some animals to regenerate, with the hope of learning how to regenerate certain human tissues (see the Nature of Science feature, "Regenerative Medicine," in Chapter 31).

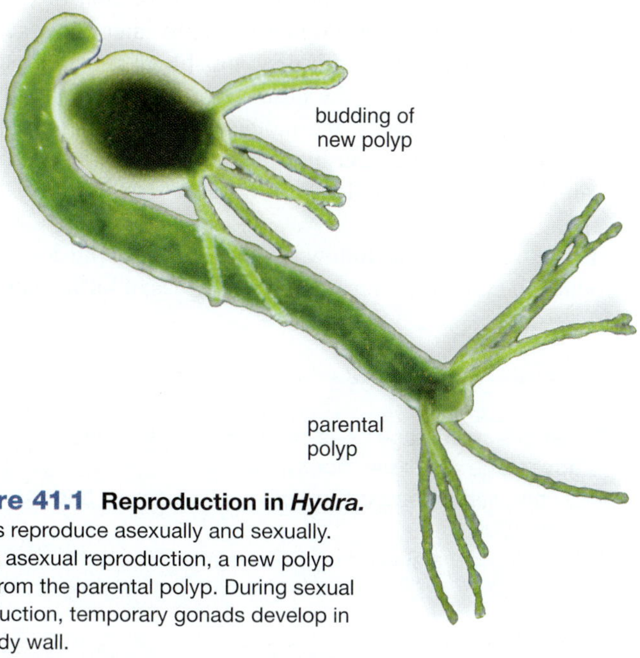

Figure 41.1 Reproduction in *Hydra.* Hydras reproduce asexually and sexually. During asexual reproduction, a new polyp buds from the parental polyp. During sexual reproduction, temporary gonads develop in the body wall.

Sexual Reproduction

The majority of animals are *dioecious,* which means having separate sexes. Usually during sexual reproduction, the egg of one parent is fertilized by the sperm of another. *Monoecious,* or **hermaphroditic,** animals have both male and female sex organs in a single body. Some hermaphroditic organisms, such as tapeworms, are capable of self-fertilization, but the majority, such as earthworms, practice cross-fertilization with other individuals. Sequential hermaphroditism, or sex reversal, also occurs. In coral reef fishes called wrasses, a male has a harem of several females. If the male dies, the largest female becomes a male.

Animals usually produce gametes in specialized organs called **gonads.** Sponges are an exception to this rule, because the collar cells lining the central cavity of a sponge give rise to sperm and eggs. Hydras and other cnidarians produce only temporary gonads in the fall, when sexual reproduction occurs. Animals in other phyla have permanent reproductive organs.

The gonads are **testes,** which produce sperm, and **ovaries,** which produce eggs. Eggs or sperm are derived from **germ cells,** which become specialized for this purpose during early development. Other cells in a gonad support and nourish the developing gametes or produce hormones necessary to the reproductive process. The reproductive system also usually has a number of accessory structures, such as ducts and storage areas, that aid in bringing the gametes together.

Several types of flatworms, roundworms, crustaceans, annelids, insects, fishes, lizards, and even some turkeys have the ability to reproduce parthenogenetically. **Parthenogenesis** is a modification of sexual reproduction in which an unfertilized egg develops into a complete individual. In honeybees, the queen bee makes and stores sperm, which she uses to selectively fertilize eggs. Any unfertilized eggs become haploid males.

Many aquatic animals practice *external fertilization,* in which the gametes are released and unite outside the bodies of the reproducing animals. Among fishes, methods of fertilization and care of the developing embryos vary greatly. Some species simply scatter large numbers of eggs and sperm in the environment, and they may even eat some of their eggs or offspring. Mouthbrooders scoop up their eggs or newly hatched young into their mouths, protecting them from predators. Other fish species build elaborate nests for their eggs, made from plant debris and saliva-coated bubbles.

Video Sea Urchin Reproduction

Video Coral Reef Spawning

An invertebrate aquatic species called palolo worms, which inhabit coral reefs in the South Pacific, exhibit a unique fertilization strategy. Always at the same time of year and during a particular phase of the moon, the worms break in half. During a period of just a few hours, hundreds of thousands of tail sections, containing the gonads, swim to the surface, releasing eggs and sperm. Meanwhile, local human populations take to the sea, using nets to scoop up large numbers of the half-worms, which are eaten raw or saved in buckets to be baked or fried and consumed at an annual feast.

Copulation is sexual union to facilitate the reception of sperm, resulting in *internal fertilization*. In terrestrial vertebrates, males typically have a penis for depositing sperm into the vagina of females. But not so in birds, which lack a penis and vagina. They have a cloaca, a chamber that receives products from the digestive, urinary, and reproductive tracts. A male transfers sperm to a female after placing his cloacal opening against hers. In damselflies, the female curls her abdomen forward to receive sperm previously deposited in a pouch by the male, but copulation in the usual sense doesn't occur (Fig. 41.2).

Video Frog Reproduction

Life History Strategies

Any animal that deposits an egg in the external environment is **oviparous.** Many oviparous animals have a larval stage, an immature form capable of feeding. Because the larva has a different lifestyle, it is able to use a different food source than the adult. Other animals, such as crayfish, do not have a larval stage; the egg hatches into a tiny juvenile with the same form as the adult.

Some invertebrates, fish, amphibians, and reptiles are **ovoviviparous,** meaning they retain their eggs in some way and release young that are able to fend for themselves. Notice that while developing, the young are still receiving nourishment from a yolk, as opposed to directly from the parent's body. Examples include certain molluscs, such as oysters, which retain their eggs in the mantle cavity, and male seahorses, which have a brood pouch for development of the fertilized eggs (as mentioned in the chapter-opening story), from which live offspring are released.

Most reptiles (e.g., turtles and crocodiles) lay a leathery-shelled egg that contains **extraembryonic membranes** to serve the needs of the embryo and prevent it from drying out. One membrane surrounds an abundant supply of **yolk,** which is a nutrient-rich material. The shelled egg with its internal extraembryonic membranes is a significant adaptation to the terrestrial environment. Birds lay and care for hard-shelled eggs; the newly hatched birds usually have to be fed before they are able to fly away and seek food for themselves (Fig. 41.3). Complex hormonal and neural regulation is involved in the reproductive behavior of parental birds.

Figure 41.2 Damselflies mating on land. On land, animals have various means to ensure that the gametes do not dry out. Here, a female damselfly receives sperm from the male.

Figure 41.3 Parenting in birds. Birds, such as the American goldfinch, *Carduelis tristis,* are oviparous and lay hard-shelled eggs. They are well known for incubating their eggs and caring for their offspring after they hatch.

Among mammals, the duckbill platypus and spiny anteater lay shelled eggs. Marsupials and placental mammals do not lay eggs. In marsupials, the yolk sac membrane briefly supplies the unborn embryo with nutrients acquired internally from the mother. Immature young finish their development in a pouch, where they are nourished on milk.

The placental mammals are termed **viviparous,** because they do not lay eggs and development occurs inside the female's body until offspring can live independently. Their **placenta** is a complex structure derived, in part, from the chorion, another of the reptilian extraembryonic membranes. The evolution of this type of placenta allowed the developing young to internally exchange materials with the mother until they are able to function on their own. Viviparity represents the ultimate in caring for the unborn, and in placental mammals, the mother continues to supply milk to nourish her offspring after birth. It should be noted that certain fish, amphibians, and reptiles are viviparous, though they lack a placenta.

In the sections that follow, we take an in-depth look at the reproductive systems of humans and the issues of birth control and health.

Check Your Progress 41.1

1. List one advantage of asexual reproduction and one advantage of sexual reproduction.
2. Distinguish among oviparous, ovoviviparous, and viviparous.
3. Explain how a shelled egg allows reproduction on land.

41.2 Human Male Reproductive System

Learning Outcomes

Upon completion of this section, you should be able to

1. Identify the structures of the human male reproductive system and provide a function for each.
2. Describe the location and stages of spermatogenesis.
3. Summarize how hormones regulate the male reproductive system.

The human male reproductive system includes the organs pictured in Figure 41.4 and listed in Table 41.1. The primary sex organs, or gonads, of males are the paired **testes** (sing., testis), which are suspended within the sacs of the scrotum.

MP3 Male Reproductive Anatomy and Physiology

The Testes

The testes produce **sperm** and the male sex hormones. They begin their development inside the abdominal cavity but descend into the scrotal sacs during the last 2 months of fetal development. If the testes fail to descend—and the male does not receive hormone therapy or undergo surgery to place the testes in the scrotum—male infertility may result. The reason is that undescended testes developing in the body cavity are subject to higher body temperatures than testes that have descended into the scrotum. The scrotum helps regulate testicular temperature by holding them closer to or farther away from the body. In fact, any activity that increases testicular temperature, such as taking hot baths, can decrease sperm production. However, despite the frequently held belief that wearing boxer-style underwear favors higher sperm production compared to briefs, this assumption is unsupported by research.

Sperm produced in the testes mature within the **epididymis** (pl., epididymides), a tightly coiled duct just outside each testis. Maturation seems to be required for sperm to swim to the egg. Once the sperm have matured, they enter the **vas deferens** (pl., vasa deferentia), also called the ductus deferens. Sperm may be stored

Table 41.1 Male Reproductive System

Organ	Function
Testes	Produce sperm and sex hormones
Epididymides	Sites of maturation and some storage of sperm
Vasa deferentia	Conduct and store sperm
Seminal vesicles	Contribute fluid to semen
Prostate gland	Contributes fluid to semen
Urethra	Conducts sperm (and urine)
Bulbourethral glands	Contribute fluid to semen
Penis	Organ of copulation

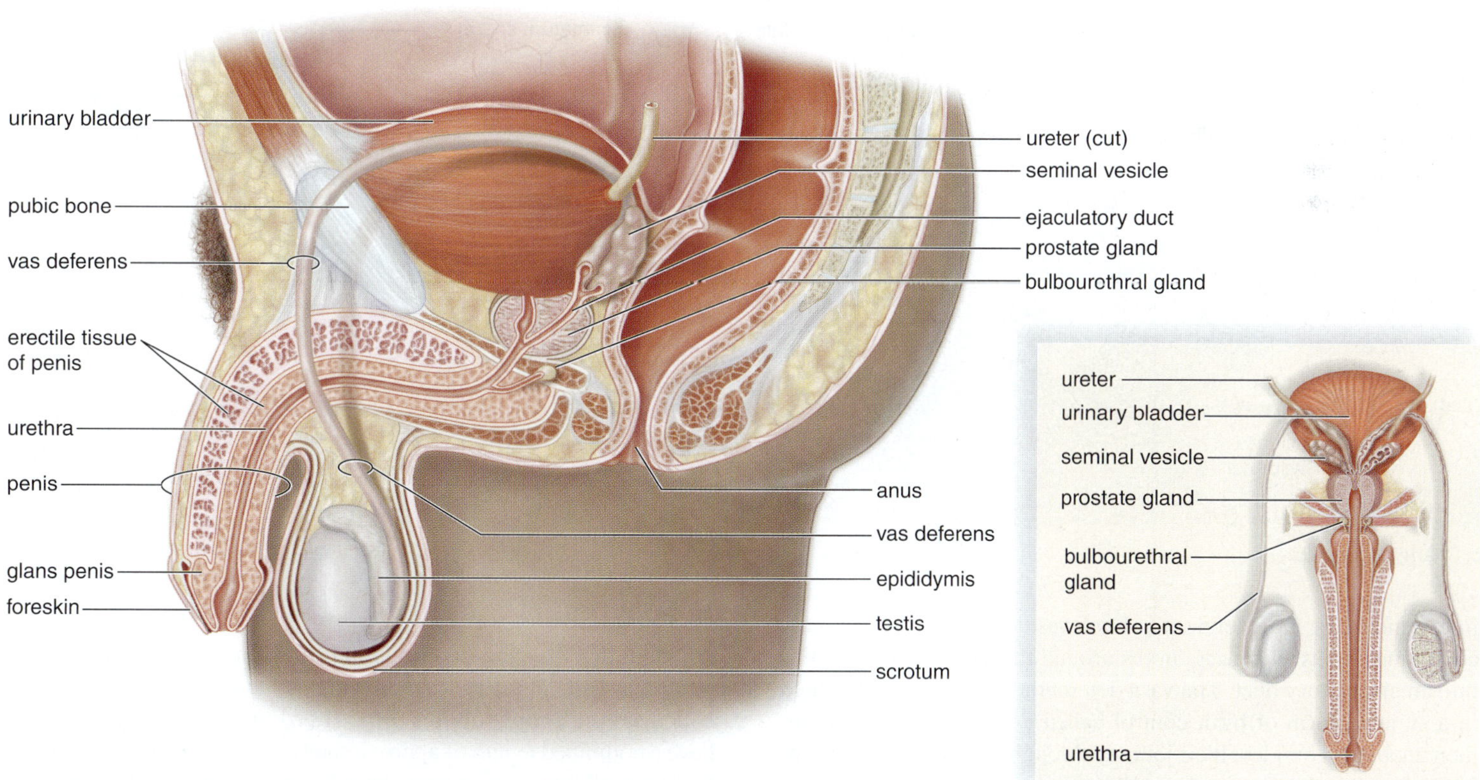

Figure 41.4 The male reproductive system. The testes produce sperm. The seminal vesicles, the prostate gland, and the bulbourethral glands provide a fluid medium for the sperm, which move from the vas deferens through the ejaculatory duct to the urethra in the penis. The foreskin (prepuce) is removed when a penis is circumcised.

Figure 41.5 Testis and sperm. **a.** The lobules of a testis contain seminiferous tubules. **b.** Light micrograph of a cross section of the seminiferous tubules, where spermatogenesis occurs. Note the location of interstitial cells in clumps among the seminiferous tubules. **c.** Diagrammatic representation of spermatogenesis, which occurs in the wall of tubules. **d.** A sperm has a head, a middle piece, and a tail. The nucleus is in the head, which is capped by the enzyme-containing acrosome.

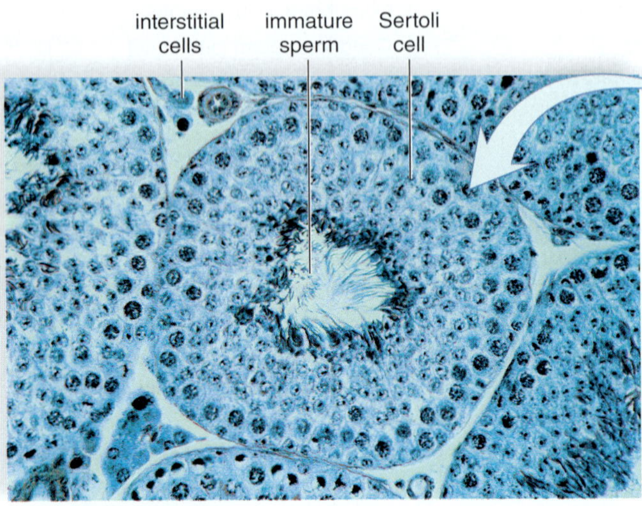

b. Seminiferous tubules

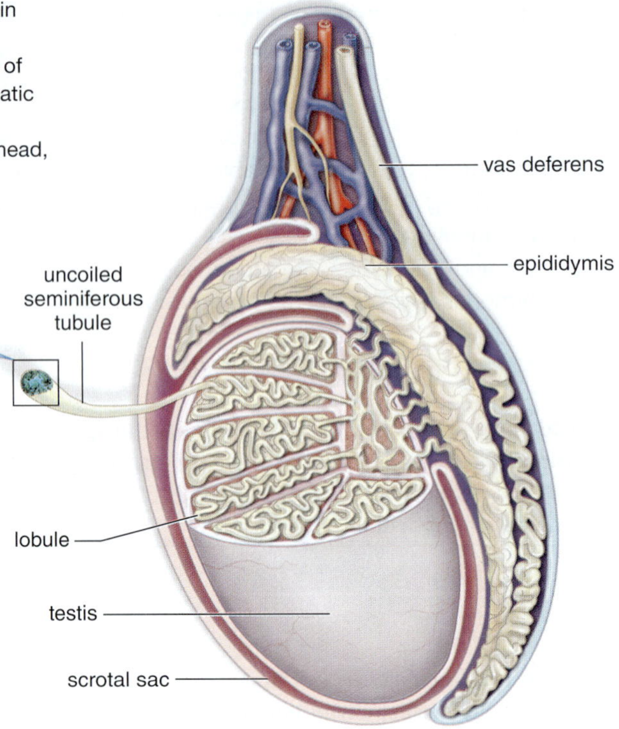

a. Testis, cut to show lobules

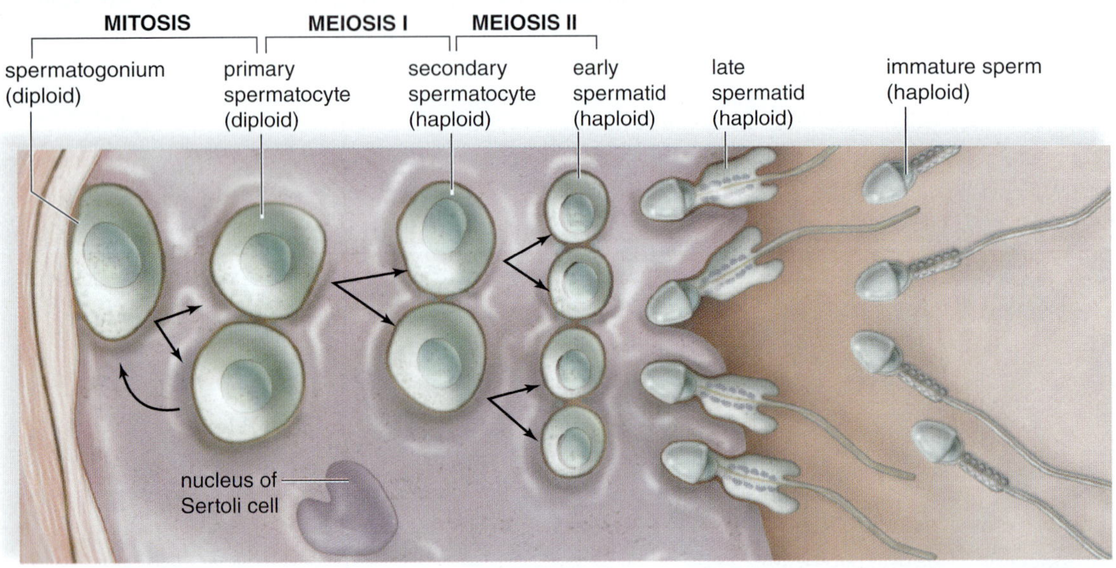

c. Spermatogenesis

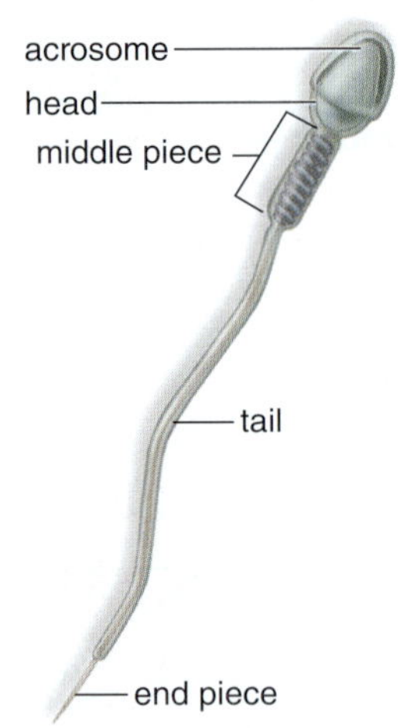

d. Sperm

for a time in the vasa deferentia. Each vas deferens passes into the abdominal cavity, where it curves around the bladder and empties into an ejaculatory duct. The vasa deferentia are severed or blocked in a surgical form of birth control called a *vasectomy*. Sperm are then unable to complete their journey through the male reproductive tract; they are phagocytized and destroyed by macrophages.

Testicular cancer is the most common type of cancer in young men between the ages of 15 and 34. The cure rate is high if early treatment is received. All men should perform regular self-exams to detect unusual lumps or swellings of the testes and report any changes to their doctor.

Production of Sperm

A longitudinal section of a testis shows that it is composed of compartments called lobules, each of which contains one to three tightly coiled **seminiferous tubules** (Fig. 41.5*a*). Altogether, these tubules have a combined length of approximately 250 meters (over two and a half football fields). This provides a large amount of surface area for sperm development. A microscopic cross section of a seminiferous tubule shows that it is packed with cells undergoing *spermatogenesis* (Fig. 41.5*b*), the production of sperm.

During spermatogenesis, spermatogonia divide to produce primary spermatocytes (which have the 2n number of chromosomes).

Primary spermatocytes move away from the outer wall, increase in size, and undergo meiosis I to produce secondary spermatocytes. Each secondary spermatocyte has only 23 chromosomes (Fig. 41.5*c*). Secondary spermatocytes (n) undergo meiosis II to produce four spermatids, each of which also has 23 chromosomes. Spermatids then differentiate into sperm (Fig. 41.5*d*). Note the presence of *Sertoli cells* (purple in Fig. 41.5*c*), which support and nourish the developing sperm and regulate spermatogenesis. It takes approximately 74 days for a spermatogonium to develop into sperm.

Animation Spermatogenesis

Mature sperm, or *spermatozoa,* have three distinct parts: a head, a middle piece, and a tail (Fig. 41.5*d*). Mitochondria in the middle piece provide energy for the movement of the tail, which is a flagellum. Like other eukaryotic flagella and cilia, microtubules in the sperm tail form the characteristic 9 + 2 pattern (see Fig. 4.20). The head contains a nucleus covered by a cap called the *acrosome* (Gk. *akros,* "at the tip"; *soma*, "body"), which stores enzymes needed to penetrate the thick membrane surrounding the egg.

Video Human Sperm

The Penis and Male Orgasm

The **penis** (Fig. 41.6) is the male organ of sexual intercourse. The penis has a long shaft and an enlarged tip called the glans penis. The glans penis is normally covered by a layer of skin called the *foreskin*. Male *circumcision,* the surgical removal of the foreskin, is performed on more than 50% of newborn males in the United States, often for cultural, religious, or health reasons. Research shows that circumcised males are 25–35% less likely to acquire certain STDs, including AIDS.

Three cylindrical columns of spongy, erectile tissue containing distensible blood spaces extend through the shaft of the penis. During sexual arousal, autonomic nerves release nitric oxide (NO). This stimulus leads to the production of cGMP (cycline guanosine monophosphate), a type of second messenger similar to cAMP (see section 40.1). The cGMP causes the smooth muscle of the incoming arterial walls to relax and the erectile tissue to fill with blood. The veins that normally take blood away from the penis are compressed, and the penis becomes erect.

Semen (seminal fluid) is a thick, whitish fluid that contains sperm and secretions from three glands (see Fig. 41.4 and Table 41.1). The paired **seminal vesicles** lie at the base of the bladder, and each has a duct that joins with a vas deferens. As sperm pass from the vasa deferentia into the ejaculatory ducts, the seminal vesicles secrete a thick, viscous fluid containing nutrients (fructose) for possible use by the sperm. The fluid also contains prostaglandins that stimulate smooth muscle contraction along the male and female reproductive tracts.

The **prostate gland** is a single, doughnut-shaped gland that surrounds the upper portion of the urethra just below the bladder. The prostate gland secretes a milky, alkaline fluid believed to activate or increase the motility of sperm. By age 50, over half of all men experience a non-cancerous enlargement of the prostate gland, or benign prostatic hypertrophy (BPH). Because of its location, BPH can result in an increased frequency of urination and a weak urinary stream. Also, prostate cancer is the second most common type of cancer among men in the United States (skin cancer is number one). Based on current rates, 13–16% of men born today can expect to be diagnosed with prostate cancer during their lifetime.

Bulbourethral glands are pea-sized organs that lie posterior to the prostate on each side of the urethra. They produce a clear, viscous secretion known as *pre-ejaculate*. Secretions from the bulbourethral glands are the first to enter the urethra, where they may cleanse the urethra of acidic residue from urine. Because this fluid can also pick up sperm that remain in the urethra from previous ejaculations, it is possible for sperm to enter the genital tract of a female during sexual intercourse even when ejaculation has not yet occurred.

As sexual stimulation intensifies, *ejaculation* may occur. During the first phase of ejaculation, called emission, the spinal cord sends sympathetic nerve impulses to the epididymides and vasa deferentia. Their muscular walls contract, causing sperm to enter the ejaculatory duct, whereupon the seminal vesicles, prostate gland, and bulbourethral glands release their secretions.

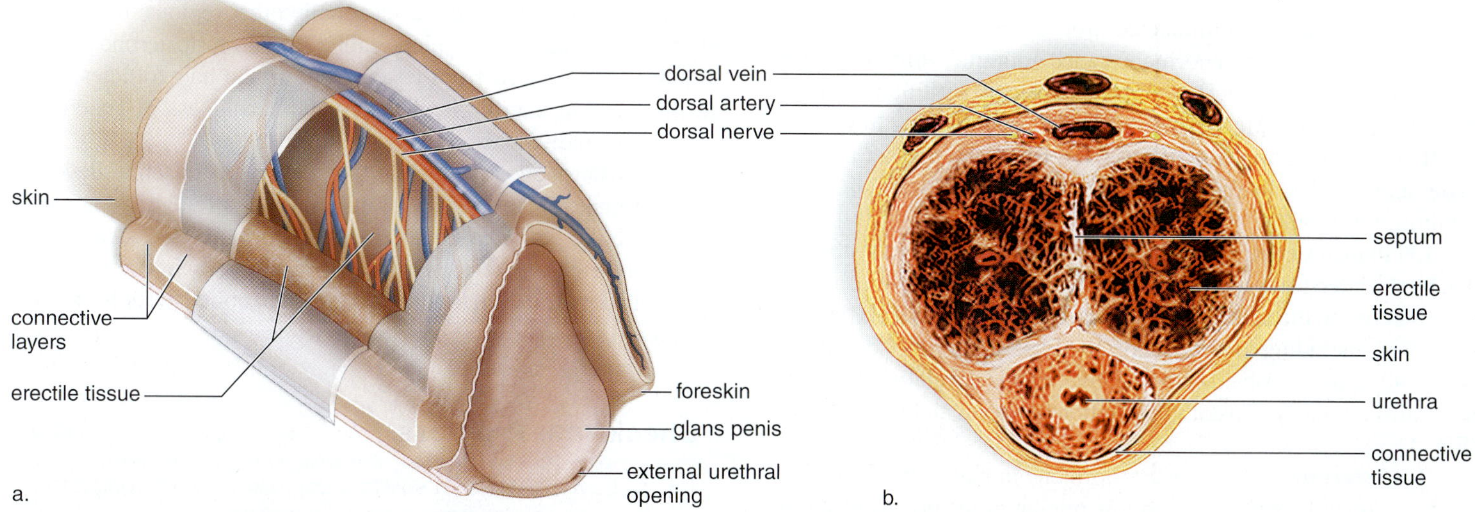

Figure 41.6 Penis anatomy. **a.** Beneath the skin and the connective tissue lies the urethra, surrounded by erectile tissue. This tissue expands to form the glans penis, which in uncircumcised males is partially covered by the foreskin (prepuce). **b.** Micrograph of shaft in cross section showing the location of erectile tissue. One column surrounds the urethra.

During the second phase of ejaculation, called expulsion, rhythmical contractions of muscles at the base of the penis and within the urethral wall expel semen in spurts from the opening of the urethra. During ejaculation, the urinary bladder sphincter normally contracts, so that no semen can enter the bladder. (The urethra carries either urine or semen at different times.)

The contractions that expel semen from the penis are a part of male *orgasm,* the physiological and psychological sensations that occur at the climax of sexual stimulation. The physiological reactions of both male and female orgasm involve the genital (reproductive) organs and assocated muscles, as well as the entire body. Marked muscular tension is followed by contraction and relaxation. In the male, following ejaculation and/or loss of sexual arousal, the penis returns to its flaccid state. After ejaculating, males typically experience a period of time, called the *refractory period,* during which stimulation does not bring about an erection. The length of the refractory period increases with age.

An average male releases 2–5 ml of semen during orgasm. The ejaculated semen of a normal human male contains between 50 and 150 million sperm per ml, meaning there may be more than 400 million sperm released in one ejaculation. According to the World Health Organization, a low sperm concentration, or *oligozoospermia,* is defined as less than 15 million sperm per ml of ejaculate. Because only one sperm typically enters an egg, however, men with low sperm counts can still be fertile.

Erectile dysfunction (*ED*), formerly called impotency, is the inability to achieve or maintain an erection suitable for sexual intercourse. ED may have a number of causes, including any condition that affects blood flow, certain medications or illicit drugs, and psychological factors. Since 1998, medications have been available for the treatment of ED. All these drugs—such as Viagra, Levitra, and Cialis—work by inhibiting an enzyme called PDE-5, found mainly in the penis, which normally breaks down cGMP. This extends the activity of cGMP, increasing the likelihood of achieving a full erection. Because the PDE-5 enzyme is also found in the retinas and certain other tissues, ED medications can cause vision problems and other undesirable side effects.

Hormonal Regulation in Males

The hypothalamus has ultimate control of the testes' sexual function, because it secretes a hormone called **gonadotropin-releasing hormone** (**GnRH**). GnRH stimulates the anterior pituitary to produce the **gonadotropic hormones, follicle-stimulating hormone** (**FSH**) and **luteinizing hormone** (**LH**), which are present in both males and females. In males, FSH promotes spermatogenesis in the seminiferous tubules.

LH in males controls the production of the androgen testosterone by the interstitial cells (Leydig cells), which are scattered in the spaces between the seminiferous tubules (see Fig. 41.5*b*). All these hormones, including *inhibin,* a hormone released by the seminiferous tubules, are involved in a negative feedback relationship that maintains the fairly constant production of sperm and testosterone (Fig. 41.7).

Testosterone is the main sex hormone in males. It is essential for the normal development and functioning of the organs listed in Table 41.1. Testosterone also brings about and maintains the male *secondary sex characteristics* that develop at the time of puberty. Males are generally taller than females and have broader shoulders

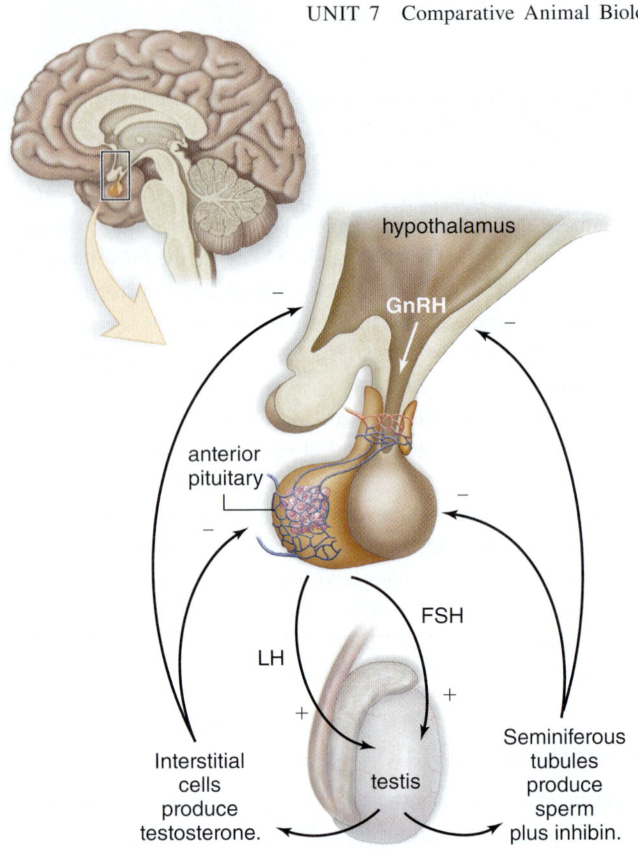

Figure 41.7 Hormonal control of the testes. GnRH (gonadotropin-releasing hormone) stimulates the anterior pituitary to produce FSH and LH. FSH stimulates the testes to produce sperm, and LH stimulates the testes to produce testosterone. Testosterone from interstitial cells and inhibin from the seminiferous tubules exert negative feedback control over the hypothalamus and the anterior pituitary, and this ultimately regulates the level of testosterone in the blood.

and longer legs relative to trunk length. The deeper voice of males compared to females is due to the development in males of a larger larynx with longer vocal cords. The "Adam's apple," which is a protrusion at the front of the larynx, is usually more prominent in males than in females. Testosterone causes males to develop noticeable hair on the face, chest, and occasionally other regions of the body, such as the back. A chemical called dihydrotestosterone (DHT), which is synthesized from testosterone in the hair follicles, leads to the receding hairline and male-pattern baldness in genetically susceptible individuals.

Testosterone is responsible for the greater muscular development in males. Knowing this, both males and females sometimes take anabolic steroids, either testosterone or related steroid hormones resembling testosterone, to build up their muscles. For more information about the risks of anabolic steroids, review Figure 40.16.

Check Your Progress 41.2

1. Trace the pathway a sperm must follow from its origin to its exit from the male reproductive tract.
2. List the glands that contribute fluids to the semen.
3. Describe the effects of FSH and LH in males.

41.3 Human Female Reproductive System

Learning Outcomes

Upon completion of this section, you should be able to

1. Identify the structures of the human female reproductive system and list a function for each.
2. Describe the process of oogenesis.
3. List the stages of the ovarian and uterine cycles and explain what is occurring in each stage.
4. Discuss the changes that occur in the uterus during menstruation, pregnancy, and menopause.

The human female reproductive system includes the organs depicted in Figure 41.8 and listed in Table 41.2. The female gonads, or primary sex organs, are paired **ovaries** (sing., ovary), located on each side of the upper pelvic cavity.

MP3 Female Reproductive Anatomy and Physiology

The Ovaries, Uterus, and Vagina

The ovaries produce the female sex hormones and a secondary **oocyte** each month. The uterine tubes, also called fallopian tubes, extend from the ovaries to the uterus; however, these tubes are not attached to the ovaries. Instead, they have fingerlike projections called fimbriae (sing., fimbria) that sweep over the ovaries. When an oocyte bursts from an ovary during ovulation, it usually is swept into a uterine tube by the combined action of the fimbriae and the beating of cilia that line the uterine tubes. Fertilization, if it occurs, normally takes place in a uterine tube. The developing embryo is propelled slowly by ciliary movement and tubular muscle contraction to the uterus. Some women opt to permanently prevent pregnancy by having a surgery called *tubal ligation,* in which the uterine tubes are closed off, so that sperm cannot reach the oocyte.

Normally, the ovaries alternate in producing one oocyte each month. For the sake of convenience, we will refer to the released oocyte as an **egg.** The ovaries also produce the female sex hormones—the **estrogens,** collectively called estrogen, and **progesterone**—during the ovarian cycle.

The **uterus** is a thick-walled, muscular organ about the size and shape of an inverted pear. The narrow end of the uterus is called the **cervix.** The embryo completes its development after embedding itself in the uterine lining, called the **endometrium** (Gk. *endon,* "within"; *metra,* "womb"). If, by chance, the embryo should embed itself in another location, such as a uterine tube, an *ectopic pregnancy* results.

A small opening in the cervix leads to the vaginal canal. The **vagina** is a tube at a 45-degree angle to the small of the back. The mucosal lining of the vagina lies in folds, and the vagina can be distended. This is especially important when the vagina serves as the birth canal, and it can facilitate intercourse, when the vagina receives the penis during copulation. Several different types of bacteria normally reside in the vagina and create an acidic environment. This environment is protective against the possible growth

Table 41.2 Female Reproductive Organs

Organ	Function
Ovaries	Produce egg and sex hormones
Uterine tubes (fallopian tubes)	Conduct egg; location of fertilization
Uterus (womb)	Houses developing embryo and fetus
Vagina	Receives penis during copulation and serves as birth canal

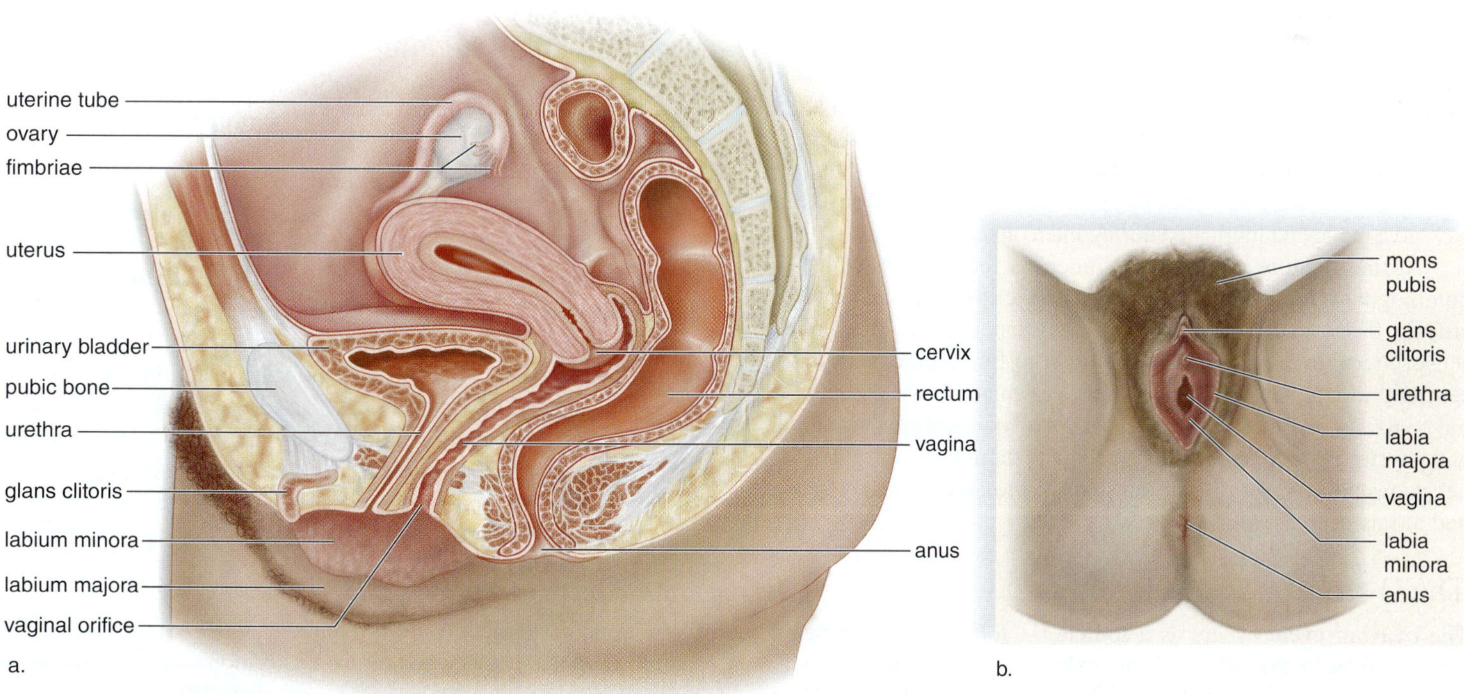

Figure 41.8 Female reproductive system. **a.** The ovaries produce one oocyte per month. Fertilization occurs in the uterine tube, and development occurs in the uterus. The vagina is the birth canal and organ of sexual intercourse. **b.** Vulva. At birth, the opening of the vagina is partially occluded by a membrane called the hymen. Physical activities and sexual intercourse disrupt the hymen.

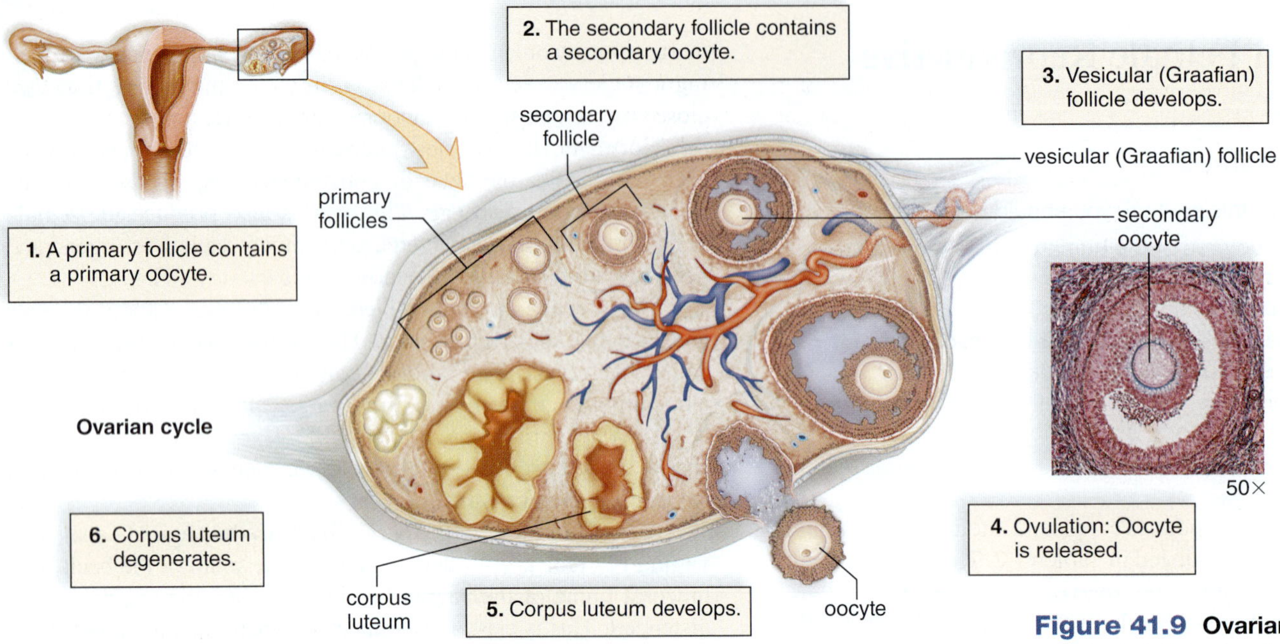

a.

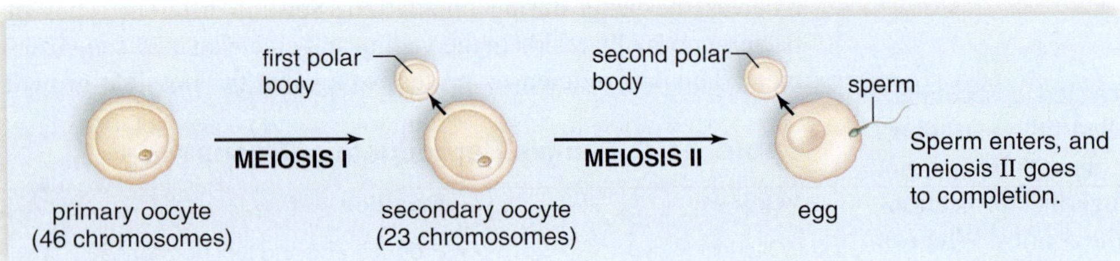

b.

Figure 41.9 Ovarian cycle.
a. A single follicle actually goes through all stages (1–6) in one place within the ovary. As a follicle matures, layers of follicle cells surround a secondary oocyte. Eventually, the mature follicle ruptures, and the secondary oocyte is released. The follicle then becomes the corpus luteum, which eventually disintegrates. **b.** During oogenesis, the chromosome number is reduced from 46 to 23. Fertilization restores the full number of chromosomes.

Tutorial Ovarian Cycle

of pathogenic bacteria, but sperm prefer the basic environment provided by semen.

The Female Orgasm

The external genital organs of a female are known collectively as the *vulva* (Fig. 41.8*b*). The *mons pubis* and two folds of skin called *labia minora* and *labia majora* are on each side of the urethral and vaginal openings. Beneath the labia majora, pea-sized greater *vestibular glands* (Bartholin glands) open on each side of the vagina. They keep the vulva moist and lubricated during intercourse.

At the juncture of the labia minora is the *clitoris,* which is homologous to the penis in males. The clitoris has a shaft of erectile tissue and is capped by a pea-shaped glans. The many sensory receptors of the clitoris allow it to function as a sexually sensitive organ. The clitoris has twice as many nerve endings as the penis. As in males, female orgasm involves both psychological and physiological reactions. In the female, there is also a release of neuromuscular tension in the muscles of the genital area, vagina, and uterus.

The Ovarian Cycle

The **ovarian cycle** occurs as a **follicle** (L. *folliculus,* "little bag") changes from a primary to a secondary and finally to a vesicular (Graafian) follicle (Fig. 41.9*a*) under the influence of follicle-stimulating hormone (FSH) and luteinizing hormone (LH) from the anterior pituitary (Fig. 41.10). Epithelial cells of a primary follicle surround a primary oocyte. Pools of follicular fluid surround the oocyte in a secondary follicle. In a vesicular follicle, a fluid-filled cavity increases to the point that the follicle wall balloons out on the surface of the ovary.

Oogenesis and Ovulation

As a follicle matures, *oogenesis,* a form of meiosis depicted in Figure 41.9*b,* is initiated and continues. The primary oocyte divides, producing two haploid cells. One cell is a secondary oocyte, and the other is a polar body. The vesicular follicle bursts, releasing the secondary oocyte (egg) surrounded by a clear membrane. This process is referred to as **ovulation.** Once a vesicular follicle has lost the secondary oocyte, it develops into a **corpus luteum** (L. *corpus,* "body"; *luteus,* "yellow"), a glandlike structure.

The secondary oocyte enters a uterine tube. If fertilization occurs, a sperm enters the secondary oocyte; then, the oocyte completes meiosis and becomes an egg. An egg with 23 chromosomes and a second polar body results. When the sperm nucleus unites with the egg nucleus, a zygote with 46 chromosomes is produced. If zygote formation and pregnancy do not occur, the corpus luteum begins to degenerate after about 10 days.

Phases of the Ovarian Cycle

The ovarian cycle is controlled by the gonadotropic hormones, FSH and LH (Fig. 41.11). The gonadotropic hormones are not present in constant amounts and instead are secreted at different rates during the cycle. For simplicity's sake, it is convenient to emphasize that during the first half, or **follicular phase,** of the cycle, FSH promotes the development of a follicle that primarily secretes estrogen.

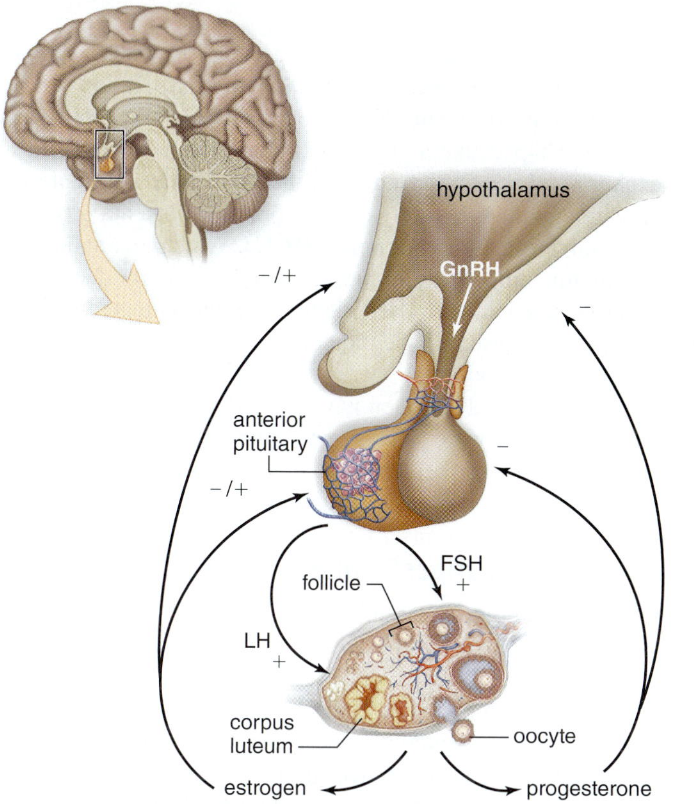

Figure 41.10 Hormonal control of ovaries. The hypothalamus produces GnRH (gonadotropic-releasing hormone). GnRH stimulates the anterior pituitary to produce FSH (follicle-stimulating hormone) and LH (luteinizing hormone). FSH stimulates the follicle to produce primarily estrogen, and LH stimulates the corpus luteum to produce primarily progesterone. Estrogen and progesterone maintain the sex organs (e.g., uterus) and the secondary sex characteristics, and they exert feedback control over the hypothalamus and the anterior pituitary. Feedback control regulates the relative amounts of estrogen and progesterone in the blood.

As the estrogen level in the blood rises, it exerts negative feedback control over the anterior pituitary secretion of FSH, so that the follicular phase comes to an end. At the same time that FSH along with LH release is being dampened by moderate amounts of estrogen, the synthesis of the gonadotropic hormones continues, and they build up in the anterior pituitary.

When the level of estrogen in the blood becomes very high, it exerts positive feedback on the hypothalamus and anterior pituitary. The hypothalamus is stimulated to suddenly secrete a large amount of GnRH. This leads to a surge of LH (and to a lesser degree, FSH) by the anterior pituitary and to ovulation at about the fourteenth day of a 28-day cycle (Fig. 41.11, *top*).

During the second half, or **luteal phase,** of the ovarian cycle, LH promotes the development of the corpus luteum, which primarily secretes progesterone. As the blood level of progesterone rises, it exerts negative feedback control over anterior pituitary secretion of LH, so that the corpus luteum begins to degenerate. As the luteal phase comes to an end, the low levels of progesterone and estrogen in the body cause menstruation to begin, as discussed next.

Animation Maturation of the Oocyte

The Uterine Cycle

The female sex hormones produced in the ovarian cycle (estrogen and progesterone) affect the endometrium of the uterus, causing the cyclical series of events known as the **uterine cycle** (Fig. 41.11, *bottom*). The 28-day cycle is divided into the three phases described next.

Phases of the Uterine Cycle

During *days 1–5,* there is a low level of female sex hormones in the body, causing the endometrium to disintegrate and its blood vessels to rupture. A flow of blood passes out of the vagina during **menstruation** (L. *menstrualis,* "happening monthly"), also known as the menstrual period. This is the *menstrual phase* of the uterine cycle.

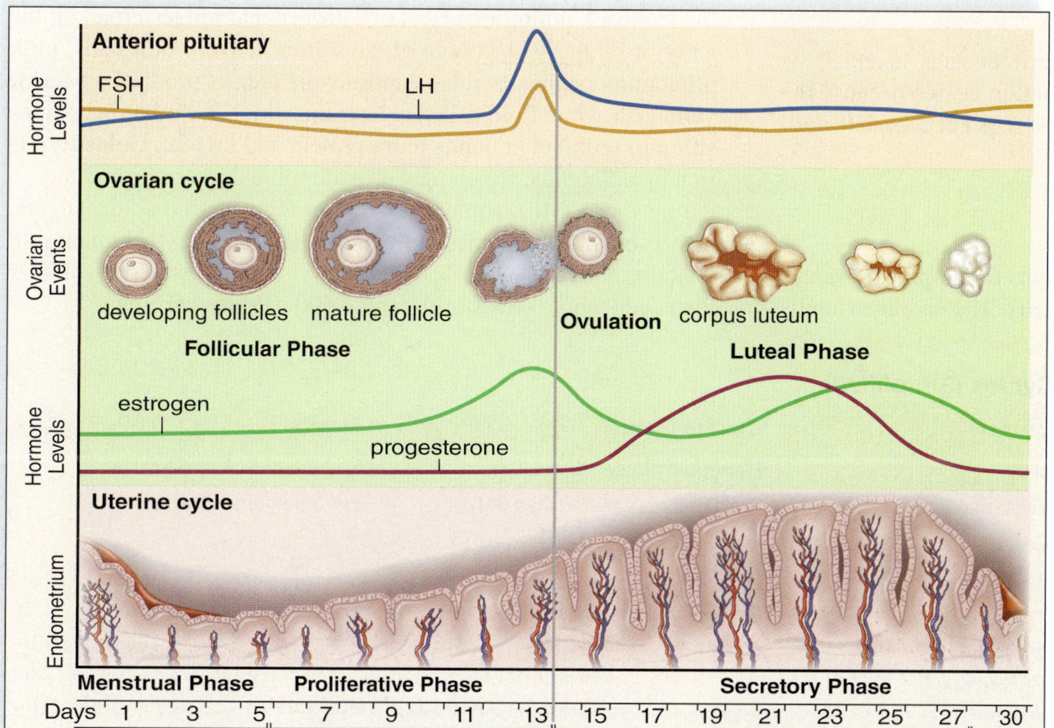

Figure 41.11 Female hormone levels during the ovarian and uterine cycles. During the follicular phase of the ovarian cycle (*top*), FSH released by the anterior pituitary promotes the maturation of a follicle in the ovary. The ovarian follicle produces increasing levels of estrogen, which causes the endometrium to thicken during the proliferative phase of the uterine cycle (*bottom*). After ovulation and during the luteal phase of the ovarian cycle, LH promotes the development of the corpus luteum. This structure produces increasing levels of progesterone, which causes the endometrium to become secretory. Menstruation and the proliferative phase begin when progesterone production declines to a low level.

During *days 6–13,* increased production of estrogen by an ovarian follicle causes the endometrium to thicken and to become vascular and glandular. This is called the *proliferative phase* of the uterine cycle.

Ovulation usually occurs on *day 14* of the 28-day cycle. During *days 15–28,* increased production of progesterone by the corpus luteum causes the endometrium to double in thickness and the uterine glands to mature, producing a thick, mucoid secretion. This is called the *secretory phase* of the uterine cycle. The endometrium now is prepared to receive the developing embryo. If pregnancy does not occur, the corpus luteum degenerates, and the low level of sex hormones in the female body causes the endometrium to break down as menstruation occurs. Table 41.3 compares the stages of the uterine cycle with those of the ovarian cycle.

Menstruation

Seven to ten days before the start of menstruation, some women suffer from premenstrual syndrome (PMS). During this time, a woman may exhibit breast enlargement and tenderness, achiness, headache, and irritability. The exact cause for PMS has yet to be discovered.

During menstruation, arteries that supply the lining constrict and the capillaries weaken. Blood spilling from the damaged vessels detaches layers of the lining, not all at once but in random patches. Mucus, blood, and degenerating endometrium descend from the uterus, through the vagina, creating menstrual flow. *Fibrinolysin,* an enzyme released by dying cells, prevents the blood from clotting. Menstruation lasts from 3 to 5 days, as the uterus sloughs off the thick lining that was 3 weeks in the making.

The first menstrual period, called *menarche,* typically occurs between the ages of 11 and 13. Menarche signifies that the ovarian and uterine cycles have begun. If menarche does not occur by age 16, or if normal uterine cycles are interrupted for 6 months or more without pregnancy being the cause, the condition is termed amenorrhea. Primary amenorrhea is usually caused by nonfunctional ovaries or developmental abnormalities. Secondary amenorrhea may be caused by weight loss and/or excessive exercise.

Menopause, which usually occurs between ages 45 and 55, is the time in a woman's life when menstruation ceases because the ovaries are no longer functioning. Menopause is not complete until menstruation is absent for a year.

Fertilization and Pregnancy

If fertilization does occur, an embryo begins development even as it travels down the uterine tube to the uterus. The endometrium is now prepared to receive the developing embryo, which becomes embedded in the lining several days following fertilization.

The placenta originates from both maternal and embryonic tissues. It is shaped like a large, thick pancake and is the site of the exchange of gases and nutrients between fetal and maternal blood, although the two rarely mix. At first, the placenta produces **human chorionic gonadotropin (HCG)**, which maintains the corpus luteum until the placenta begins its own production of progesterone and estrogen. HCG is the hormone detected in pregnancy tests; it is present in the mother's blood and urine as soon as 10 days after conception.

Progesterone and estrogen have two effects. They shut down the anterior pituitary, so that no new follicles mature, and they maintain the lining of the uterus, so that the corpus luteum is not needed. No menstruation occurs during pregnancy.

Estrogen and Progesterone

Estrogen, in particular, is essential for the normal development and functioning of the female reproductive organs listed in Table 41.2. Estrogen is also largely responsible for the secondary sex characteristics in females, including body hair and fat distribution. In general, females have a more rounded appearance than males because of a greater accumulation of fat beneath their skin. Also, the pelvic girdle enlarges, so that females have wider hips than males, and the thighs converge at a greater angle toward the knees. Both estrogen and progesterone are required for breast development as well.

The Female Breast

A female breast contains between 15 and 24 lobules, each with its own mammary duct (Fig. 41.12). A duct begins at the nipple and divides into numerous other ducts that end in blind sacs called alveoli.

Lactation, the production of milk by the cells of the alveoli, is stimulated by the hormone prolactin. Milk is not produced during pregnancy, because the production of prolactin is suppressed by the feedback inhibition effect of estrogen and progesterone on the anterior pituitary. A couple of days after delivery of a baby, milk production begins. In the meantime, the breasts produce a watery, yellowish-white fluid called colostrum, which is similar in composition to milk but contains more protein and less fat. Colostrum is rich in antibodies, which may provide some degree of immunity to the newborn. Milk contains water, proteins, amino acids, sugars, and lysozymes (enzymes with antibiotic properties). Human milk contains about 750 calories per liter, which is considerably higher than even whole cow's milk (about 500 calories per liter).

Table 41.3 Ovarian and Uterine Cycles (Simplified)

Ovarian Cycle	Events	Uterine Cycle	Events
Follicular phase—Days 1–13	FSH	Menstrual phase—Days 1–5	Endometrium breaks down
	Follicle maturation	Proliferative phase—Days 6–13	Endometrium rebuilds
	Estrogen		
Ovulation—Day 14*	LH spike		
Luteal phase—Days 15–28	LH	Secretory phase—Days 15–28	Endometrium thickens and glands are secretory
	Corpus luteum		
	Progesterone		

*Assuming a 28-day cycle

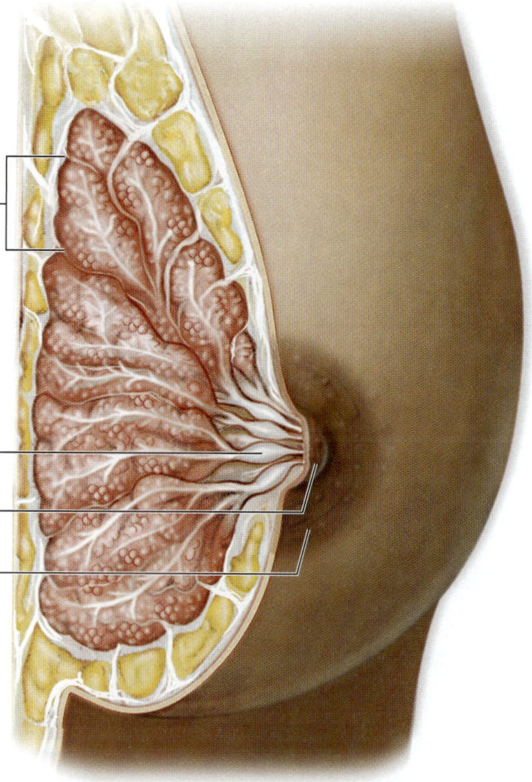

Figure 41.12 Anatomy of the breast. The female breast contains lobules consisting of ducts and alveoli. In the lactating breast, cells lining the alveoli have been stimulated to produce milk by the hormone prolactin.

Breast cancer is the second most common type of cancer among women in the United States; skin cancer is the most common. Based on current rates, about 12% of women born today can expect to be diagnosed with cancer of the breast during their lifetime. Women should regularly check their breasts for lumps and other irregularities and have mammograms as recommended by their physician. Although breast cancer genes have been described, most forms of breast cancer are nonhereditary. Breast cancer treatment options have expanded in recent years, resulting in increased survival rates. Currently, about 86% of women survive at least 5 years after their initial diagnosis.

Besides more traditional surgery, radiation, and chemotherapy, two newer types of breast cancer treatments are hormonal therapy and targeted therapy. Some breast cancers have receptors for estrogen, and they grow in response to increased estrogen levels. To treat these tumors, various drugs are available to either decrease estrogen synthesis or block the interaction of estrogen with its receptor. Targeted therapies usually involve the administration of a monoclonal antibody (see Chapter 33) that can react very specifically with antigens on tumor cells. The best known example, called Herceptin, blocks a protein called HER2, expressed on 20–30% of breast cancer cells, where it is involved in stimulating uncontrolled cell growth.

Check Your Progress 41.3

1. Name the structures of the female reproductive system that (a) produce the egg, (b) transport the egg, (c) house a developing embryo, and (d) serve as the birth canal.
2. Explain the effects of FSH and LH on the ovarian cycle.
3. Summarize the roles of estrogen and progesterone in the ovarian and uterine cycles.

41.4 Control of Human Reproduction

Learning Outcomes

Upon completion of this section, you should be able to

1. List five commonly available methods of birth control that are designed to prevent conception.
2. Note which birth control methods are effective at preventing the spread of sexually transmitted diseases.
3. Describe several reproductive technologies that can help infertile couples have children.

Several means are available to dampen or enhance the human reproductive potential. Contraceptives are medications and devices that reduce the chance of pregnancy.

Traditional Birth Control Methods

The most reliable method of birth control is abstinence—that is, not engaging in sexual intercourse. This form of birth control has the added advantage of preventing the transmission of sexually transmitted diseases. Male and female condoms also offer some protection against sexually transmitted diseases in addition to helping prevent pregnancy. Some of the common means of birth control used in the United States are shown in Figure 41.13.

Morning-After Pills

The term *morning-after pill,* or *emergency contraception,* refers to medications that can prevent pregnancy after unprotected intercourse. The expression "morning-after" is a misnomer, in that some treatments can be started up to 5 days after unprotected intercourse.

The first FDA-approved medication produced for emergency contraception is a kit called Preven. Preven includes four synthetic progesterone pills; two are taken up to 72 hours after unprotected intercourse, and two more are taken 12 hours later. The hormone upsets the normal uterine cycle, making it difficult for an embryo to implant in the endometrium. A recent study estimated that Preven is 85% effective in preventing unintended pregnancies. The Preven kit also includes a pregnancy test; women are instructed to take the test first before using the hormone, because the medication is not effective on an established pregnancy.

Another FDA-approved drug is called Plan B, which is up to 89% effective in preventing pregnancy if taken within 72 hours after unprotected sex. It is available without a prescription to women age 17 and older. In August 2010, ulipristal acetate (also known as ella) was also approved for emergency contraception. It can be taken up to 5 days after unprotected sex, and studies indicate it is somewhat more effective than Plan B. Unlike Plan B, however, a prescription is required.

Mifepristone, also known as RU-486 or the "abortion pill," can cause the loss of an implanted embryo by blocking the progesterone receptors of endometrial cells. This causes the endometrium to slough off, carrying the embryo with it. When taken in conjunction with a prostaglandin to induce uterine contractions, RU-486 is 95% effective at inducing an abortion up to the forty-ninth day of gestation. Because of its mechanism of action, the use of RU-486

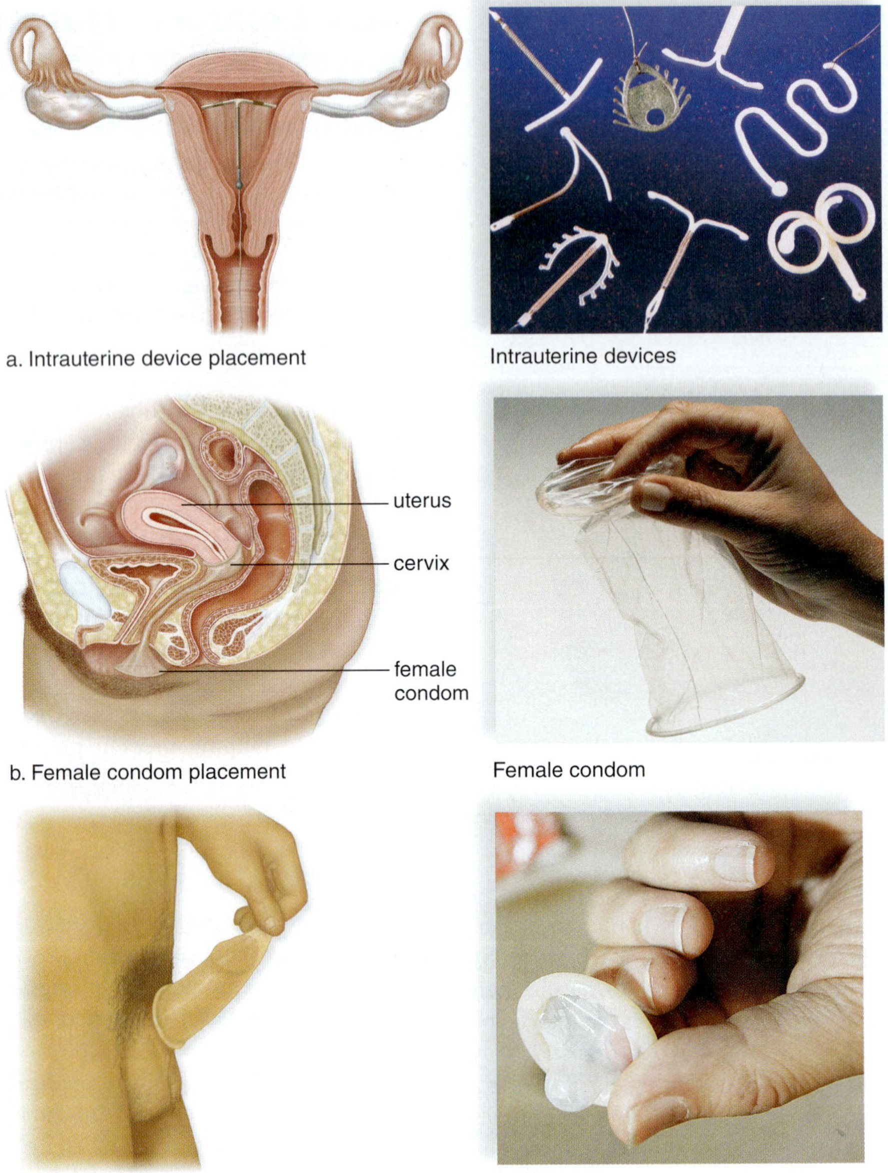

Figure 41.13 Various birth control methods. **a.** Intrauterine devices mechanically prevent implantation and can contain progesterone to prevent ovulation, prevent implantation, and thicken cervical mucus. **b.** A female condom fitted inside the vagina prevents sperm entry and protects against STDs. **c.** A male condom fits over the penis, preventing sperm from entering the vagina and protecting against STDs.

is more controversial compared to other medications, and while it is currently available by prescription in the United States for early medical abortion, it is not approved for emergency contraception.

Contraceptive Vaccines

Contraceptive vaccines are now being developed. For example, a vaccine intended to immunize women against HCG, the hormone so necessary to maintaining the implantation of the embryo, was successful in a limited clinical trial. Because HCG is not normally present in the body, no autoimmune reaction is expected, but the immunization does wear off with time. Others believe that it would also be possible to develop a safe antisperm vaccine that could be used in women. In some parts of the world, however, there is suspicion that such vaccines could be used to forcibly sterilize certain populations, and thus some of these vaccine trials have been met with resistance.

Male Birth Control

At noted in section 41.2, vasectomy is a surgical form of birth control for males. Over 500,000 American men have a vasectomy each year. Because this procedure is not easily reversed, a great deal of research is being devoted to developing safe and effective hormonal birth control for men. Implants, pills, patches, and injections are being explored as ways to deliver testosterone and/or progesterone at adequate levels to suppress sperm production. Even the most successful formulations are still in the experimental stage and are unlikely to be available outside of clinical trials for at least a few more years.

Reproductive Technologies

Infertility is the inability of a couple to achieve pregnancy after 1 year of regular, unprotected intercourse. The American Medical Association estimates that 15% of all couples are infertile. The cause of infertility can be attributed to the man (40%), the woman (40%), or both (20%).

Sometimes, the causes of infertility can be corrected by medical intervention, so that couples can have children. If no obstruction is apparent and body weight is normal, it is possible for women to take fertility drugs, which are gonadotropic hormones that stimulate the ovaries and bring about ovulation. Such hormone treatments may cause multiple ovulations and multiple births.

When reproduction does not occur in the usual manner, many couples adopt a child. Others sometimes try one of the assisted reproductive technologies (ARTs) developed to increase the chances of pregnancy. In these cases, sperm and/or oocytes are often retrieved from the testes and ovaries, and fertilization takes place in a clinical or laboratory setting.

Artificial Insemination by Donor (AID)

During artificial insemination, sperm from a donor are introduced into the woman's vagina by a physician. This technique is especially helpful if the male partner has a low sperm count, because the sperm can be collected over a period of time and concentrated, so that the sperm count is sufficient to result in fertilization. Often, however, a woman is inseminated by sperm acquired from a different donor. At times, a combination of partner and donor sperm is used.

A variation of AID is *intrauterine insemination* (*IUI*). In IUI, fertility drugs are given to stimulate the ovaries, and then the donor's sperm is placed into the uterus, rather than into the vagina.

If the prospective parents wish, sperm can be sorted into those believed to be X-bearing or Y-bearing, to increase the chances of having a child of the desired sex. First, the sperm are treated with a DNA-staining chemical. Because the X chromosome has slightly more DNA than the Y chromosome, it takes up more dye. When a laser beam shines on the sperm, the X-bearing sperm shine a little more brightly than the Y-bearing sperm. A machine sorts the sperm into two groups on this basis. Parents can expect about a 65% success rate when selecting for males and about 85% when selecting for females.

In Vitro Fertilization (IVF)

During IVF, fertilization occurs in laboratory glassware. Ultrasound machines can spot follicles in the ovaries that hold immature oocytes; therefore, the latest method is to forgo the administration of fertility drugs and retrieve immature oocytes by using a needle. The immature oocytes are then brought to maturity in glassware before concentrated sperm are added. After about 2–4 days, the embryos are ready to be transferred to the uterus of the woman, who is now in the secretory phase of her uterine cycle. If desired, the embryos can be tested for a genetic disease, as discussed in the Nature of Science feature, "Preimplantation Genetic Diagnosis," on page 784. If implantation is successful, development continues to term.

Video
In Vitro Fertilization

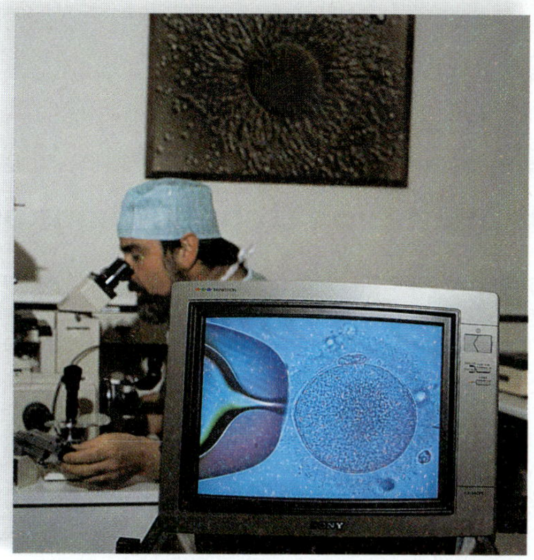

Figure 41.14 Intracytoplasmic sperm injection (ICSI). A microscope connected to a television screen is used to carry out in vitro fertilization. A pipette holds the oocyte steady while a needle (not visible) introduces the sperm into the oocyte.

Gamete Intrafallopian Transfer (GIFT)

Recall that the term **gamete** refers to a sex cell, either a sperm or an oocyte. Gamete intrafallopian transfer was devised to overcome the low success rate (15–20%) of in vitro fertilization. The method is the same as for in vitro fertilization, except the oocytes and sperm are placed in the uterine tubes immediately after they have been brought together. GIFT has the advantage of being a one-step procedure for the woman—the oocytes are removed and reintroduced at the same time. A variation on this procedure is to fertilize the eggs in the laboratory and then place the zygotes into the uterine tube.

Surrogate Mothers

In some instances, women are contracted and paid to carry babies as *surrogate mothers*. The sperm and even the oocyte can be contributed by the contracting parents.

Intracytoplasmic Sperm Injection (ICSI)

In this highly sophisticated procedure, a single sperm is injected into an oocyte (Fig. 41.14). It is used effectively when a man has severe infertility problems.

If all the assisted reproductive technologies discussed were employed simultaneously, it would be possible for a baby to have five parents: (1) sperm donor, (2) oocyte donor, (3) surrogate mother, (4) contracting mother, and (5) contracting father. The legal definitions of "parent" can become a matter of contention in some cases.

Check Your Progress 41.4

1. Identify which birth control methods discussed in this section physically block sperm from entering the uterus.
2. Rank these birth control methods regarding their effectiveness at preventing pregnancy: abstinence, female birth control pill, condoms, natural family planning, and vasectomy.
3. Briefly describe the following processes: artificial insemination by donor (AID), in vitro fertilization (IVF), gamete intrafallopian transfer (GIFT), and intracytoplasmic sperm injection (ICSI).

THEME **Nature of Science**

Preimplantation Genetic Diagnosis

If prospective parents are heterozygous for a genetic disorder, they may want the assurance that their offspring will be free of the disorder. Determining the genotype of an embryo can provide this assurance. For example, if both parents are *Aa* for a recessive disorder, the embryo will develop normally if it has the genotype *AA* or *Aa*. In contrast, if one of the parents is *Aa* for a dominant disorder, the embryo will develop normally only if it has the genotype *aa*.

Following in vitro fertilization (IVF), the zygote (fertilized egg) divides. When the embryo has six to eight cells (Fig. 41A*a*), the removal of one of these cells for testing purposes has no effect on normal development. Only embryos with a cell that tests negative for the genetic disorder of interest are placed into the uterus to continue developing.

So far, about 1,000 children with normal genotypes for genetic disorders that run in their families have been born worldwide following embryo testing. In the future, it's possible that embryos who test positive for a disorder will be treated by gene therapy, so that they, too, would be allowed to continue to term.

Testing the oocyte is possible if the condition of concern is recessive. Recall that meiosis in females results in a single egg and at least two polar bodies.[1] Polar bodies later disintegrate, and they receive very little cytoplasm, but they do receive a haploid number of chromosomes. When a woman is heterozygous for a recessive genetic disorder, about half the first polar bodies have received the genetic defect, and in these instances, the egg has received the normal allele. Therefore, if a polar body tests positive for a recessive defect, then the egg has received the normal dominant allele. Only normal eggs are then used for IVF. Even if the sperm should happen to carry the same mutant allele, the zygote will, at worst, be heterozygous. But the phenotype will appear normal (Fig. 41A*b*).

If, in the future, gene therapy becomes routine, an egg could be given genes that control traits desired by the parents, such as musical or athletic ability, prior to IVF.

[1]Once meiosis is complete, the oocyte is called an egg.

Questions to Consider

1. Some biomedical ethicists are concerned that embryo testing might lead to a form of eugenics, in which wealthier parents are able to afford to improve their children's physical traits. Others argue that this technology should be developed as much as possible to reduce the occurrence of terrible diseases. Which argument is more compelling to you, and why?
2. Some experts feel that preimplantation screening should be required for all embryos that are conceived by in vitro fertilization, to reduce the occurrence of disease. Do you agree or disagree?

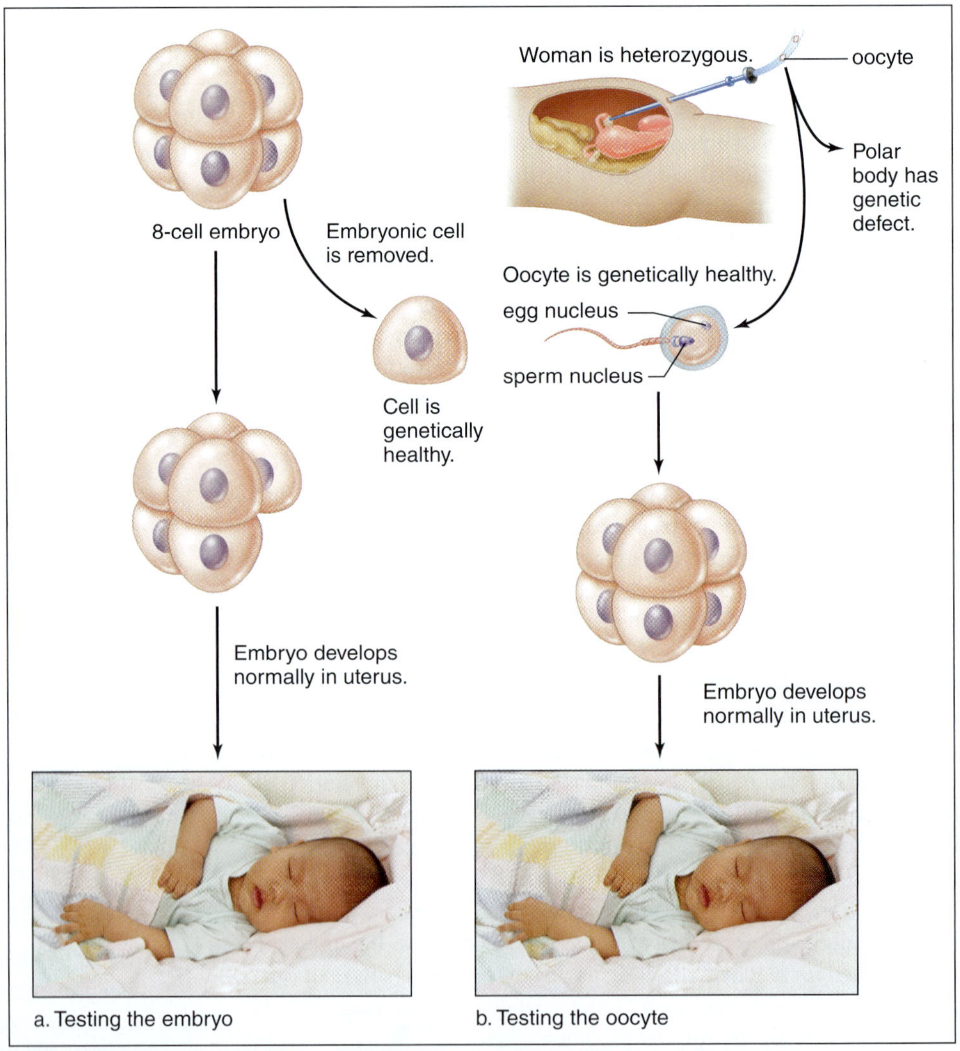

Figure 41A Preimplantation genetic diagnosis. **a.** Following IVF and cleavage, genetic analysis is performed on one cell removed from an eight-celled embryo. If it is found to be free of the genetic defect of concern, the seven-celled embryo is implanted in the uterus and develops into a fetus. **b.** Chromosome and genetic analysis is performed on a polar body attached to an oocyte. If the oocyte is free of a genetic defect, the egg is used for IVF, and the embryo is implanted in the uterus for further development.

41.5 Sexually Transmitted Diseases

Learning Outcomes

Upon completion of this section, you should be able to

1. List three STDs that are caused by viruses and three that are caused by bacteria.
2. Discuss how HIV affects the immune system and how this is related to the three categories of HIV infection.
3. Describe three STDs that are preventable by vaccination.

According to a 2013 report, about 20 million new cases of sexually transmitted diseases (STDs) occur each year in the United States. About half of these new cases occur in young people, ages 15–24. At any given time, about 110 million Americans, including about one in four college students, are infected with one or more of the 30 different STD-causing organisms ranging from viruses to arthropods; however, this discussion centers on the most prevalent and serious ones. In addition, the Biological Systems feature, "Preventing Transmission of STDs," on page 787 describes some methods to avoid the transmission of STDs.

Viral STDs

AIDS, genital herpes, genital warts, and hepatitis A and B are STDs caused by viruses. Therefore, they do not respond to treatment with traditional antibiotics. Antiviral drugs have been developed to treat some of these viral infections, but in many cases they persist for life.

AIDS

Acquired immunodeficiency syndrome (AIDS) is caused by the **human immunodeficiency virus (HIV)**. An introduction to HIV and AIDS was provided in Chapter 33. At the end of 2012, about 35.3 million people worldwide were infected with HIV, and about 36 million people have died of AIDS since 1981. The great majority of these people live in underdeveloped countries; an estimated 71% of HIV-infected individuals in the world live in sub-Saharan Africa.

In the United States, approximately 1.1 million people are infected with HIV, and about 636,000 have died from AIDS. Notably, about one in six HIV-infected Americans are unaware of their infection. Homosexual men make up the largest proportion of people living with HIV in the United States, but the fastest rate of increase is now seen in heterosexuals. Over half of new HIV infections are now occurring in people under the age of 25.

Transmission. HIV is transmitted by sexual contact with an infected person, including vaginal or anal intercourse and oral-genital contact. Also, needle-sharing among intravenous drug users is a very efficient way to transmit HIV. Some of the earliest cases of AIDS resulted from the transfusion of HIV-infected blood, but this is now extremely rare, thanks to effective screening methods. Babies born to HIV-infected mothers may become infected before or during birth or through breast-feeding. Transmission through tears, saliva, or sweat is thought to be highly unlikely.

Cultural factors may also play an important role in the transmission of HIV. For example, polygamy is common in many African countries, as are misconceptions about the effectiveness of condoms in preventing HIV transmission, and even about the fact that HIV causes AIDS. Such practices and beliefs may significantly increase the risk of HIV infection.

Stages and Symptoms of an HIV Infection. HIV infects helper T cells, macrophages, and other cells that have a molecule called CD4 on their surface (Fig. 41.15). As you may recall from Chapter 33, helper T cells control the activities of many other cells in the immune system. After HIV enters a host cell, it converts its viral RNA to DNA, which is then integrated into the host cell DNA.

Animation How the HIV Infection Cycle Works

In the early stages after an HIV infection, the person's blood contains a high amount of virus, even though the individual may experience only mild, flulike symptoms, which soon disappear. Even if an HIV-infected person undergoes an HIV test at this time, it would probably be negative, because this tests for the presence of anti-HIV antibodies, which are usually not detectable for an average of 25 days. However, the person can still be highly infectious at this stage.

After the initial surge of HIV replication, the immune system temporarily reduces the virus levels in the blood. Over an average period of 8–10 years, the HIV-infected cells gradually produce increasing amounts of the virus, which infect and destroy more and more helper T cells. As the virus reproduces, it also mutates, so that it is no longer recognized by the immune system.

Three categories of HIV infection are distinguished, based mainly on the type of disease seen in the patient. Within each category, the number of $CD4^+$ T cells is also monitored. Patients in category A usually have greater than 500 $CD4^+$ T cells per microliter of blood, and they either lack symptoms completely or have persistently enlarged lymph nodes. Patients in category B have a $CD4^+$ T-cell count between 200 and 499 per microliter, and they begin to suffer from opportunistic infections (OIs)—infections that would be unable to take hold in a person with a healthy immune system. Once the $CD4^+$ T-cell count falls below 200 per microliter, the patient is in category C, or true AIDS. The deficiency of helper T-cell functions in these patients results in their contracting one or more AIDS-defining OIs (see Chapter 33).

Treating or Preventing HIV Infection. Before the development of antiretroviral drugs, most HIV-infected individuals

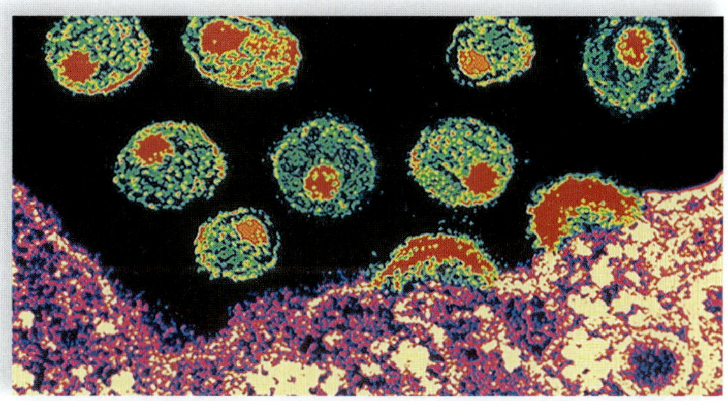

Figure 41.15 HIV, the AIDS virus. False-colored micrograph showing HIV particles budding from an infected helper T cell. These viruses can infect other helper T cells and macrophages, which work with helper T cells to stem infection.

eventually progressed to category C AIDS, which was almost invariably fatal. Although there is still no cure for HIV infection, total AIDS-related deaths have been decreasing worldwide since about 2004, mainly due to increased availability of antiretroviral therapy (ART).

According to the National Institute of Allergy and Infectious Diseases (NIAID), six major types of drugs have been developed for ART: (1) Entry inhibitors stop HIV from entering a cell by, for example, preventing the virus from binding to CD4; (2) fusion inhibitiors prevent the virus from fusing with a cellular membrane, which also interferes with viral entry; (3) reverse transcriptase inhibitors, such as AZT, interfere with the conversion of viral RNA into DNA; (4) integrase inhibitors prevent HIV from inserting its DNA into the DNA of the host cells; (5) protease inhibitors prevent a viral enzyme from processing newly created polypeptides; and (6) multiclass combination products combine more than one of these five types of ART into a single product. Treating patients with a combination of drug types means that the virus is less likely to become resistant to therapy, because several viral mutations would need to occur at once. This approach is sometimes called highly active antiretroviral therapy (HAART).

As of 2012, more than 9 million people worldwide were receiving ART. These numbers have increased greatly since 2003, with the implementation of the President's Emergency Plan for AIDS Relief, or PEPFAR, program, which began as a commitment from the United States to provide $15 billion in antiretroviral drugs to resource-poor countries from 2003 to 2008. The program was renewed and expanded in 2008, including additional funding to fight tuberculosis and malaria. A 2009 study showed that the PEPFAR program had cut AIDS death rates by more than 10% (1.1 million people) in targeted countries in Africa, although it had had no appreciable effect on the prevalence of the disease in those nations.

In 2013, the World Health Organization (WHO) recommended the use of ART for the prevention of HIV infection in pregnant women, young children, and other high-risk groups in countries with high rates of HIV infection. If an HIV-positive pregnant woman takes no ART during her pregnancy, there is a 20–25% chance that the baby will become infected. If drug therapy commences during pregnancy and delivery is by cesarean section, the chance of transmission from mother to infant is very slim (about 1%).

In recent years, claims that people have been "cured" of HIV/AIDS have surfaced in the media. In one example, a Berlin patient received a bone marrow transplant that apparently cured both his leukemia and his AIDS. Additional examples are two American babies who were born HIV-positive but who seemed to have had the virus eliminated from their systems after receiving high doses of ART. In both of the latter cases, doctors described the babies as being in remission rather than being cured, because the virus could return.

Despite the success of treatment, most experts believe that an end to the AIDS epidemic will not occur until a vaccine is available. Many different approaches have thus far been tried with limited success. A large 2009 study in Thailand showed that an experimental HIV vaccine reduced the risk of contracting the virus by nearly a third (31.2%). Researchers hope to build on these results to develop a more effective HIV vaccine.

Animation Treatment of HIV Infection

Human Papillomavirus

An estimated 20 million Americans are infected with the *human papillomavirus* (*HPV*). At least half of all Americans will be infected with HPV during their lifetime, although about 90% will recover from the infection without any consequences, or even knowledge that they were infected.

There are over 100 types of HPV. Many of these cause warts, and about 30–40 types specifically cause genital warts, which are an STD. Genital warts may appear as flat or raised lesions on the penis and foreskin of males (Fig. 41.16*a*) and on the vulva, vagina, and cervix in females. Note that if warts are present only on the cervix, there may be no outward signs of the disease. Newborns can also be infected with HPV during passage through the birth canal.

Warts can usually be treated effectively by surgery, freezing, application of an acid, or laser burning. Even after treatment,

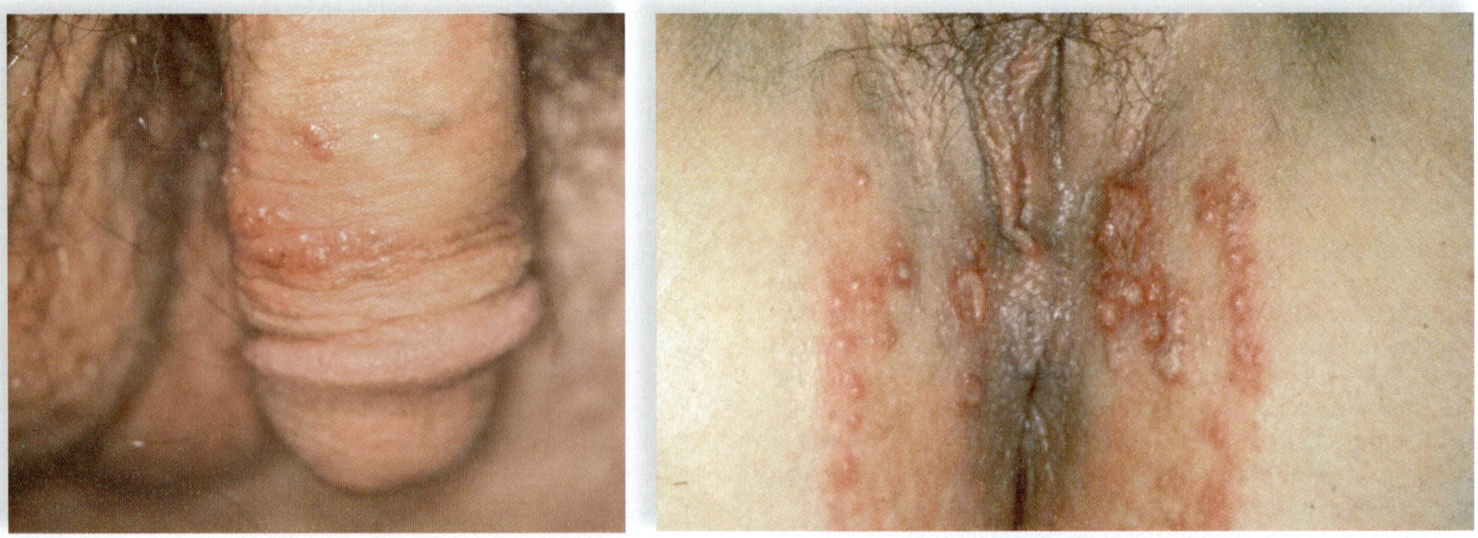

a. b.

Figure 41.16 Two STDs caused by viruses. **a.** Genital warts, caused by human papillomavirus and seen here on the penis, can also affect the anus, vulva, vagina, or cervix. **b.** Genital herpes, caused by the herpes simplex virus, can cause blisters on the labia (seen here), vagina, rectum, or penis.

THEME Biological Systems

Preventing Transmission of STDs

Being aware of how STDs are spread and then observing the following guidelines will greatly help prevent the transmission of STDs.

Sexual Activities Can Transmit STDs

Abstain from sexual intercourse or develop a long-term monogamous relationship (having only one sexual partner) with someone who is free of STDs (Fig. 41B).

Refrain from multiple sex partners or having relations with someone who has multiple sex partners. If you have sex with two other people, each of them has sex with two people, and so forth, the number of people who can be exposed to disease is quite large.

Remember that anyone can be at risk for HIV. The highest rate of increase in HIV infection is now occurring among heterosexuals.

Be aware that having sex with an intravenous drug user is risky, because needle-sharing can lead to infection with hepatitis and HIV. Also, anyone who already has another STD is more susceptible to becoming infected with HIV.

Uncircumcised males are more likely to become infected with an STD than circumcised males, at least partly because viruses and bacteria can remain under the foreskin for a long period of time.

Avoid anal-rectal intercourse (in which the penis is inserted into the rectum), because the lining of the rectum is thin and HIV can easily enter the body there.

Safer Sex Practices Can Help Prevent STD Transmission

Always use a latex condom during sexual intercourse if you do not know for certain that your partner is free of STDs. Be sure to follow the directions supplied by the manufacturer.

Avoid fellatio (kissing and insertion of the penis into a partner's mouth) *and cunnilingus* (kissing and insertion of the tongue into the vagina), because they may be a means of transmission. The mouth and gums often have cuts and sores that facilitate the entrance of infectious agents.

Be cautious about the use of alcohol or any drug that may prevent you from being able to control your behavior. Females have to be particularly aware of "date-rape" drugs, such as GHB (gamma-hydroxy-butyramine), which put a person in an uninhibited state with no memory of what has transpired. These drugs can easily be slipped into a beverage without the victim's knowledge.

Drug Use Transmits Hepatitis and HIV

Do not inject drugs. Be aware that hepatitis and HIV can be spread by blood-to-blood contact. If you are currently an IV drug user, stop immediately.

Always use a new, sterile needle for injection or one that has been cleaned in a bleach solution and then well rinsed if you are a drug user and cannot stop your behavior (Fig. 41C).

Questions to Consider

1. How can you be certain that a potential sexual partner doesn't have any STDs?
2. What does it mean if a person tells you he or she just had an HIV test and it was negative?
3. Which methods of birth control are most effective at preventing the transmission of STDs? Which are least effective or provide no protection?

Figure 41B Sexual activities can transmit STDs. People of all sexual orientations need to be aware of STD dangers and use precautions.

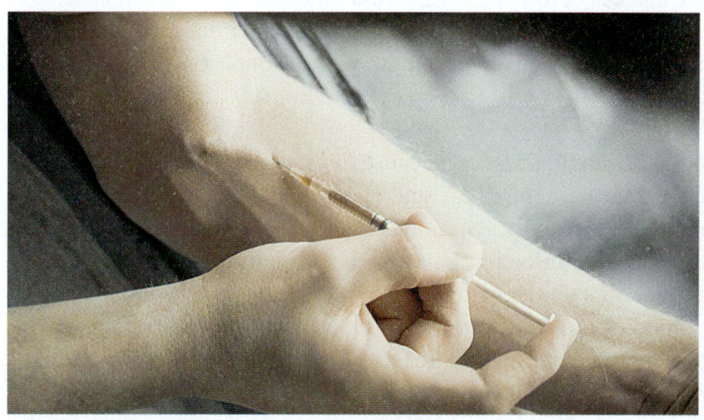

Figure 41C Sharing needles transmits STDs. Intravenous drug use with unclean, shared needles can transmit viral STDs.

however, the virus can sometimes be transmitted. Therefore, once someone has been diagnosed with genital warts, abstinence or the use of a condom is recommended.

About ten types of HPV are associated with cancer, which may affect the cervix, vulva, vagina, anus, penis, or oral cavity. Decades ago, cervical cancer was the leading cause of cancer death in women in the United States, but those rates have decreased significantly due to improved methods for diagnosis and treatment. However, cervical cancer is still a significant cause of death in developing countries. It can often be detected in its early stages by a *Pap smear,* in which a few cells are removed from the cervix for microscopic examination. If the cells are cancerous, a hysterectomy (removal of the uterus) may be recommended. Additionally, HPV now causes more cancers of the oropharynx than does tobacco, perhaps due to an increase in oral sex in combination with a decline in tobacco use.

Two vaccines are available to prevent females from infection with the types of HPV that cause the majority of cervical cancers. One of these (Gardasil) also protects women against genital warts and cancers of the vulva, vagina, and anus. Gardasil is also available for use in males. Both vaccines are most effective if administered at age 11 or 12, so that a protective immune response can develop before sexual activity begins.

Genital Herpes

Genital herpes is caused by the herpes simplex virus, of which there are two types. Type 1 usually causes cold sores and fever blisters, while type 2 more often causes genital herpes. In the United States, about one out of every six people ages 14–49 currently have genital herpes, but most infected people don't have symptoms. Even without symptoms, however, the virus can still be transmitted.

After becoming infected, most people experience a tingling or itching sensation before blisters appear on the genitals within 2–20 days (Fig. 41.16*b*). Once the blisters rupture, they leave painful ulcers, which take from 5 days to 3 weeks to heal. These symptoms may be accompanied by fever, pain on urination, swollen lymph nodes in the groin, and in women a watery vaginal discharge. At this time, the individual also has an increased risk of acquiring an HIV infection. Exposure to herpes during a vaginal delivery can cause an infection in the newborn, which can lead to neurological disorders and even death. Birth by cesarean section prevents this possibility.

After the blisters heal, the virus becomes latent (dormant). Blisters can recur repeatedly, at variable intervals and with milder symptoms. Sunlight, sexual intercourse, menstruation, and stress are associated with these recurrences. While the virus is latent, it resides in the ganglia of the sensory nerves associated with the infected skin. The ability of the virus to "hide" in this manner has made it difficult to develop an effective vaccine for herpes simplex viruses.

Viral Hepatitis

Several different viruses can cause *hepatitis,* or inflammation of the liver. Of these, hepatitis A and B viruses are the most important. Hepatitis A is usually acquired from sewage-contaminated drinking water, but this infection can also be sexually transmitted through oral-anal contact. Hepatitis B is spread through sexual contact and by blood-borne transmission (e.g., accidental needle-stick on the job, a contaminated blood transfusion, needles shared while injecting drugs, or transfer from mother to fetus). Simultaneous infection with hepatitis B virus and HIV is common, because both share the same routes of transmission. Fortunately, a combined vaccine is available for hepatitis A and B. It is recommended that all children receive the vaccine to prevent these infections.

Bacterial STDs

Chlamydia, gonorrhea, and syphilis are major bacterial STDs that are usually curable with antibiotic therapy, if diagnosed early enough. If properly used, latex condoms can help prevent the spread of most STDs.

Chlamydia

Infection with the tiny bacterium *Chlamydia trachomatis,* simply known as *chlamydia,* is the most commonly reported bacterial STD in the United States, with the highest rates of infection occurring in 15- to 19-year-olds.

Chlamydial infections of the lower reproductive tract are usually mild or asymptomatic, especially in women. About 8–21 days after infection, men may experience a mild burning sensation on urination and a mucoid discharge. Women may have a vaginal discharge along with symptoms resembling a urinary tract infection. Chlamydia also causes cervical ulcerations, which increase the risk of aquiring HIV. If the infection is misdiagnosed or if a woman does not seek medical help, the infection may spread to the uterine tubes, and pelvic inflammatory disease (PID) results.

PID is characterized by inflamed uterine tubes that become partially or completely blocked by scar tissue. As a result, the woman can become infertile or at increased risk of ectopic pregnancy, which can be a medical emergency. If a baby comes in contact with chlamydia during delivery, pneumonia or eye infections can result (Fig. 41.17).

Gonorrhea

Gonorrhea is caused by the bacterium *Neisseria gonorrhoeae* (Fig. 41.18). Most cases occur in people between 15 and 29 years of age, and rates among African Americans are about 20 times greater than among Caucasians. Although as many as 20% of infected males are asymptomatic, usually they experience pain on urination and a thick, greenish-yellow urethral discharge.

Untreated gonorrhea can cause infertility in both males and females by causing inflammation and scarring in either the vas deferens or the uterine tubes, respectively. If a baby is exposed during birth, an eye infection can lead to blindness. All vaginally delivered newborns are treated with antibiotic eyedrops to prevent this possibility.

Oral-genital contact can also cause *N. gonorrhoeae* infection of the mouth, throat, and tonsils. Anal intercourse can lead to gonorrhea proctitis, an inflammation of the anus characterized by

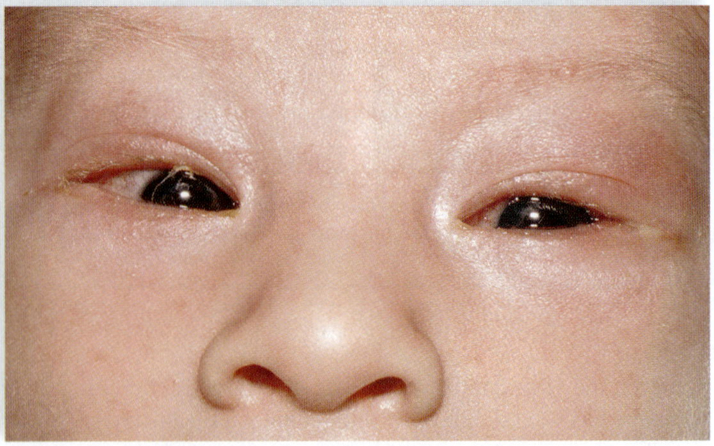

Figure 41.17 Chlamydial eye infection. This newborn's eyes were infected after passing through the birth canal of an infected mother.

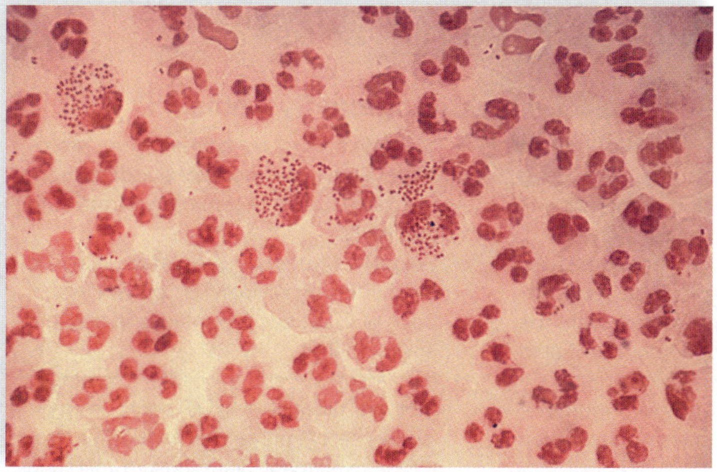

Figure 41.18 Gonorrhea. Photomicrograph of a urethral discharge from an infected male. In this image, the gonorrheal bacteria (*Neisseria gonorrhoeae*) have been phagocytosed by some of the neutrophils present in the discharge.

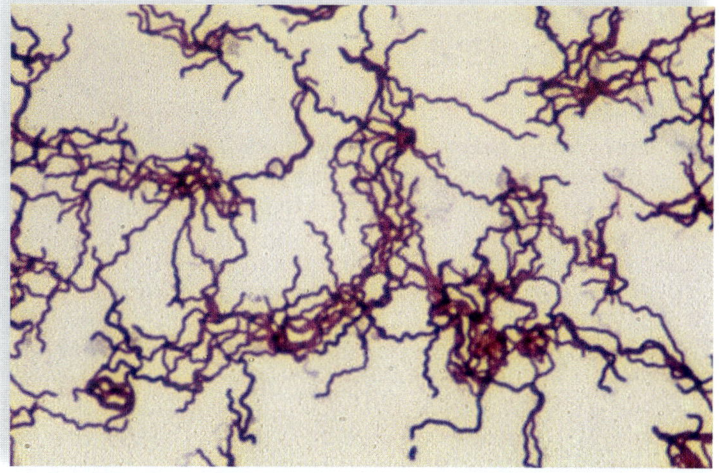

Figure 41.19 Syphilis. *Treponema pallidum,* the species of bacteria that cause syphilis, are spirochetes, as indicated by their corkscrewlike shape.

anal pain and blood or pus in the feces. Gonorrhea can spread to internal parts of the body, causing heart damage or arthritis. If, by chance, the person touches infected genitals and then touches his or her eyes, a severe eye infection can result.

Until recently, gonorrhea was considered to be curable by antibiotics. However, antibiotic-resistant strains of *N. gonorrhoeae* are becoming an increasingly urgent problem.

Syphilis

Syphilis is an STD that was first described hundreds of years ago. It is caused by the bacterium *Treponema pallidum* (Fig. 41.19). Untreated syphilis progresses through three stages. In the primary stage, a hard chancre (ulcerated sore with hard edges) indicates the site of infection. During the secondary stage, the individual breaks out in a rash that does not itch, coinciding with the replication and spread of the bacteria all over the body.

Not all cases of syphilis progress to the tertiary stage, during which syphilis may affect the cardiovascular or nervous system. Affected people may become blind or intellectually disabled, walk with a shuffle, or show signs of insanity. Gummas, which are large, destructive ulcers, may develop on the skin or in the internal organs.

T. pallidum bacteria can cross the placenta, causing birth defects or a stillbirth. If diagnosed and treated in its early stages, syphilis is easy to cure with antibiotics.

Vaginal Infections

Vaginitis, meaning vaginal inflammation from any cause, is the most commonly diagnosed gynecologic condition. Bacterial vaginosis (BV), or overgrowth of certain bacteria inhabiting the vagina, accounts for 40–50% of vaginitis cases in American women. A common culprit is the bacterium *Gardnerella vaginalis.* How women acquire this infection is not well understood, but it is more common in women who are sexually active.

In addition to BV, the yeast *Candida albicans* can also cause vaginitis. Yeast is normally found living in the vagina; under certain circumstances, its growth increases above normal, causing vaginitis. For example, women taking birth control pills or antibiotics may be prone to yeast infections. Both can alter the normal balance of vaginal organisms, causing a yeast infection. Typical signs of a yeast infection include a thickish, white vaginal discharge and itching of the vulva and/or vagina. Antifungal medications inserted into the vagina are used to treat yeast infections.

The protozoan *Trichomonas vaginalis* can also infect the urethra of males and the vagina of females. Infected males are usually asymptomatic, but they can pass the parasite to their partner through sexual intercourse. Symptoms of trichomoniases in females are a foul-smelling, yellow-green exudate and itching of the vulva/vagina. Effective antiprotozoal drugs are available by prescription, but if one partner remains infected, reinfection will occur. For this reason, both people should be treated simultaneously.

Check Your Progress 41.5

1. Describe four classes of antiretroviral drugs and the specific stage of the HIV reproduction cycle targeted by each class.
2. Describe the medical complications in women that are associated with HPV infection.
3. Identify a serious condition that can occur in women due to both chlamydia and gonorrhea.

CONNECTING *the* CONCEPTS *with the* THEMES

Evolution

- Most animals have two separate sexes and undergo sexual reproduction, which generates genetic diversity that can be advantageous to species survival.
- The evolution of a hard-shelled egg with extraembryonic membranes to protect and nourish the developing embryo was a significant adaptation to living on land.

Nature of Science

- By understanding the details of human reproduction, scientists are developing more methods of birth control that will be more efficient and easier to use.
- The invention of new technologies for enhancing human replication has provided many infertile couples with the opportunity to have children, but it also has raised many ethical concerns.
- Although many STDs are incurable, some effective treatments, and even vaccines, are becoming available.

Biological Systems

- Strategies used by animals that reproduce asexually include budding from the parent and regeneration from animal parts.
- Sexually reproducing animals may be oviparous, ovoviviparous, or viviparous. Each of these has advantages in different environments.
- In marsupials and placental mammals, mothers produce milk to continue nourishing the offspring after birth. This maternal investment helps ensure offspring survival.

MEDIA STUDY TOOLS http://connect.mheducation.com

SMARTBOOK® Study smart. SmartBook helps you maximize your study time by identifying what you know and don't know. The recharge feature in SmartBook helps you prepare for the test by identifying the material that you are most likely to forget.

connect BIOLOGY Find even more resources to learn the chapter concepts including animations, tutorial videos, and interactive practice questions.

MP3 Files	Animations	Tutorial	Videos
41.2 Male Reproductive Anatomy and Physiology **41.3** Female Reproductive Anatomy and Physiology	**41.2** Spermatogenesis **41.3** Maturation of the Oocyte **41.5** How the HIV Infection Cycle Works • Treatment of HIV Infection	**41.3** Ovarian Cycle	**41.1** Sea Urchin Reproduction • Coral Reef Spawning • Frog Reproduction **41.2** Human Sperm **41.4** In Vitro Fertilization

SUMMARIZE

41.1 How Animals Reproduce

Ordinarily, **asexual reproduction** may quickly produce a large number of offspring genetically identical to the parent. Sexual reproduction involves **gametes** and produces offspring that are genetically slightly different from the parents. The **gonads** are the primary sex organs, which produce **germ cells.** The **testes** produce sperm, and the **ovaries** produce eggs. The accessory sex organs include storage areas for sperm and ducts that conduct the gametes. Fertilization can be external or internal. Internal fertilization is facilitated by **copulation,** or sexual union.

Exceptions to the usual male-female patterns exist. In **hermaphroditic** animals, single individuals have both male and female gonads. In **parthenogenesis,** an unfertilized egg develops into an adult.

The egg of **oviparous** animals contains **yolk,** and in terrestrial animals a shelled egg prevents it from drying out. **Ovoviviparous** animals retain their eggs until the offspring have hatched; **viviparous** animals retain the embryo and give birth. Eggs of reptiles have **extraembryonic membranes** that allow them to develop on land; the same membranes are modified into a **placenta** for internal development in mammals. Placental mammals exemplify viviparous animals.

41.2 Human Male Reproductive System

In human males, **sperm** are produced in the **testes,** mature in each **epididymis,** and may be stored in each **vas deferens** before entering the urethra, along with **semen** (**seminal fluid**) produced by the **seminal vesicles, prostate gland,** and **bulbourethral glands.** Sperm and semen is ejaculated during male orgasm, when the **penis** is erect.

Spermatogenesis occurs in the **seminiferous tubules** of the testes, which also produce testosterone in interstitial cells. **Testosterone** brings about the maturation of the primary sex organs during puberty and promotes the secondary sex characteristics of males, such as low voice, facial hair, and increased muscle strength.

Follicle-stimulating hormone (FSH) from the anterior pituitary stimulates spermatogenesis, and luteinizing hormone (LH) stimulates testosterone production.

Gonadotropin-releasing hormone (**GnRH**) released from the hypothalamus controls anterior pituitary production of two **gonadotropic hormones, follicle-stimulating hormone** (**FSH**) **and luteinizing hormone** (**LH**). FSH stimulates spermatogenesis, and LH stimulates testosterone production. In males, the level of testosterone in the blood controls the secretion of GnRH and the gonadotropic hormones by a negative feedback system.

41.3 Human Female Reproductive System

In females, an **oocyte** (also called an **egg**) produced by an **ovary** enters a **uterine tube,** which leads to the **uterus.** At the lower end of the uterus, the **cervix** connects with the **vagina.** The external genital area of women includes the vaginal opening, the clitoris, the labia minora, and the labia majora.

During the **ovarian cycle,** one **follicle** a month (in either ovary) matures and produces first a primary, then a secondary oocyte. After **ovulation** (release of a secondary oocyte), the vesicular follicle develops into a **corpus luteum.** The follicle and the corpus luteum produce **estrogen** and **progesterone,** the female sex hormones. Ovulation marks the end of the **follicular phase** of the ovarian cycle, and the beginning of the **luteal phase.**

The **uterine cycle** occurs concurrently with the ovarian cycle. In the first half of these cycles (days 1–13, before ovulation), the anterior pituitary produces FSH and the follicle produces estrogen. Estrogen causes the **endometrium** to increase in thickness. In the second half of these cycles (days 15–28, after ovulation), the anterior pituitary produces LH and the follicle produces progesterone. Progesterone causes the endometrium to become secretory. Feedback control of the hypothalamus and anterior pituitary causes the levels of estrogen and progesterone to fluctuate. When they are at a low level, **menstruation** begins.

If fertilization occurs, a zygote is formed, and development begins. The resulting embryo travels down the oviduct and implants itself in the endometrium. A placenta, which is the region of exchange between the fetal blood and the mother's blood, forms. At first, the placenta produces **human chorionic gonadotropin** (**HCG**), which maintains the corpus luteum; later, it produces progesterone and estrogen.

The female sex hormones, estrogen and progesterone, also affect other traits of the body. Primarily, estrogen brings about the maturation of the primary sex organs during puberty and promotes the secondary sex characteristics of females, including body hair (usually less than in males), a wider pelvic girdle, a more rounded appearance, and the development of breasts. **Lactation** is the production of milk. Usually between the ages of 45 to 55, **menopause** occurs as a woman's ovarian function ceases.

41.4 Control of Human Reproduction

Numerous birth control methods and devices, such as the birth control pill, diaphragm, and condom, are available for those who wish to prevent pregnancy. A "morning-after pill," RU-486, is now available. **Contraceptive vaccines** are being developed that could prevent pregnancy.

If couples have **infertility,** or trouble conceiving a child, they may use assisted reproductive technologies. Artificial insemination and in vitro fertilization have been followed by more sophisticated techniques, such as intracytoplasmic sperm injection.

41.5 Sexually Transmitted Diseases

Important sexually transmitted diseases (STDs) caused by viruses include **acquired immunodeficiency syndrome** (**AIDS**), caused by the **human immodeficiency virus** (**HIV**). Other viral STDs include human papillomavirus (HPV), which causes genital warts and several types of cancer, as well as genital herpes and hepatitis A and B. Major bacterial STDs include chlamydia and gonorrhea, both of which may cause pelvic inflammatory disease (PID), and syphilis, which has cardiovascular and neurological complications if untreated. Vaginal infections have many causes; the most common are bacterial vaginosis (BV), yeast infection, and *Trichomonas* infection.

ASSESS

Choose the best answer for each question.

41.1 How Animals Reproduce

1. Which of these is a requirement for reproduction to be defined as sexual?
   a. separate male and female parents
   b. the production of gametes
   c. optimal environmental conditions
   d. an aquatic habitat
   e. All of these are correct.
2. Internal fertilization
   a. can prevent the drying out of gametes and zygotes.
   b. must take place on land.
   c. is practiced by humans.
   d. requires that males have a penis.
   e. Both a and c are correct.
3. Which term is incorrectly defined?
   a. oviparous—deposits egg(s) in the external environment
   b. ovoviviparous—retains eggs prior to releasing the young
   c. viviparous—lacks eggs and always has a placenta
   d. placenta—allows some mammals to be nourished inside the mother

41.2 Human Male Reproductive System

4. In tracing the path of sperm, you would mention vasa deferentia before
   a. testes.
   b. epididymides.
   c. urethra.
   d. seminiferous tubules.
   e. All of these are correct.
5. Which of these pairs is mismatched?
   a. interstitial cells—testosterone
   b. seminiferous tubules—sperm production
   c. vasa deferentia—seminal fluid production
   d. urethra—conducts sperm
   e. Both c and d are mismatched.
6. Testosterone is produced by
   a. interstitial cells.
   b. seminiferous tubules.
   c. the hypothalamus.
   d. the prostate.
   e. All of these are correct.

41.3 Human Female Reproductive System

7. Which of these is the correct path of an oocyte?
   a. uterine tube, fimbriae, uterus, vagina
   b. ovary, fimbriae, uterine tube, uterine cavity
   c. uterine tube, fimbriae, abdominal cavity
   d. fimbriae, uterine tube, uterine cavity, ovary
8. Fertilization of the oocyte usually occurs in
   a. the vagina.
   b. the uterus.
   c. the uterine tube.
   d. the ovary.
   e. All of these are correct.

9. The secretory phase of the uterine cycle occurs during which ovarian phase?
   a. follicular phase
   b. ovulation
   c. luteal phase
   d. menstrual phase
10. Following ovulation, the corpus luteum develops under the influence of
    a. progesterone.
    b. FSH.
    c. LH.
    d. estradiol.
11. During pregnancy,
    a. the ovarian and uterine cycles occur more quickly than before.
    b. GnRH is produced at a higher level than before.
    c. the ovarian and uterine cycles do not occur.
    d. the female secondary sex characteristics are not maintained.
    e. Both b and c are correct.

41.4 Control of Human Reproduction

12. Which means of birth control is most effective in preventing sexually transmitted diseases?
    a. condom
    b. pill
    c. diaphragm
    d. spermicidal jelly
    e. vasectomy
13. Using currently available technology, which is the most reliable method to achieve male sterilization?
    a. condom
    b. contraceptive vaccine
    c. hormonal implant
    d. vasectomy
14. In which process are oocytes and sperm placed directly into the uterine tubes?
    a. artificial insemination by donor
    b. in vitro fertilization
    c. intracytoplasmic sperm injection
    d. gamete intrafallopian transfer

41.5 Sexually Transmitted Diseases

15. Which of these is a sexually transmitted disease caused by a bacterium?
    a. gonorrhea
    b. hepatitis B
    c. genital warts
    d. genital herpes
    e. HIV
16. The human immunodeficiency virus has a preference for infecting
    a. B lymphocytes.
    b. cytotoxic T lymphocytes.
    c. helper T lymphocytes.
    d. All of these are correct.

For questions 17–19, match the descriptions with the sexually transmitted diseases in the key (use three from choices a–g).

Key:

a. AIDS
b. hepatitis B
c. genital herpes
d. genital warts
e. gonorrhea
f. chlamydia
g. syphilis

17. genital blisters and painful ulcers, which heal and then recur
18. flulike symptoms, jaundice, eventual liver failure possible
19. males have a thick, greenish-yellow urethral discharge; can lead to PID and infertility, can also infect the eyes, throat, and anus

ENGAGE

THINKING SCIENTIFICALLY

1. Compare the anatomy of the male and female reproductive tracts. Which organs serve similar functions? Which organs serve a unique function, with no analogous function in the other gender?
2. The average sperm count in males is now lower than it was several decades ago. The reasons for the lower sperm count usually seen today are not known. What data might be helpful in order to formulate a testable hypothesis?
3. Female athletes who train intensively often stop menstruating. The important factor appears to be the reduction of body fat below a certain level. Give a possible evolutionary explanation for a relationship between body fat in females and reproductive cycles.
4. In the animal kingdom, only primates menstruate. Other mammals come into season, or “heat,” during certain times of the year, while still others ovulate only after having sex. What would be some possible advantages of a monthly menstrual cycle?

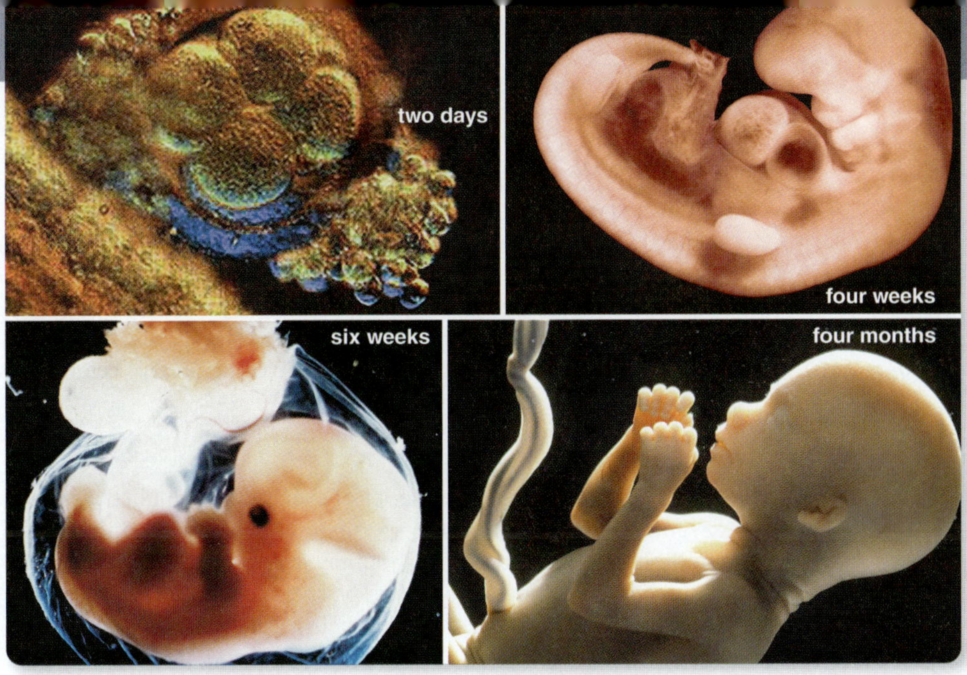

As development occurs, complexity increases.

42

Animal Development and Aging

Chapter Outline

42.1 Early Developmental Stages 794

42.2 Developmental Processes 798

42.3 Human Embryonic and Fetal Development 802

42.4 The Aging Process 809

Many of the fundamental processes that guide development, from fertilization onward, are shared by most of the animal kingdom. Whether the embryo will develop into a worm, an insect, an amphibian, a bird, or a mammal, it must first establish an anterior end and a posterior end. As the embryo continues to grow, the body segments form and give rise to structures such as appendages. The coordination of these developmental events is quite intricate and is easily disrupted by harmful environmental factors, such as chemicals and radiation. Interestingly, some of these environmental factors are thought to cause cellular damage, which over time contributes to the aging process.

Although the main focus of this chapter is on events that occur either at the beginning or toward the end of an animal's lifespan, we should note that once an animal has reached maturity, development does not cease. If an animal's body is injured, the wound must be repaired. Not surprisingly, many of the same genes involved in wound repair are those that were active during earlier stages in development. In this chapter, we examine the basic processes of early animal development, as well as human fetal development, and the consequences of aging.

As you read through this chapter, think about the following questions:

1. Why are so many similarities present in the early embryonic structures of animals as diverse as the lancelet, the frog, the chicken, and the human?
2. Why is the developing embryo of any species more susceptible to harmful environmental factors, compared to the mature animal?
3. What are some of the fundamental mechanisms by which cells can become specialized to form organs and tissues?

Before You Begin

Before beginning this chapter, take a few moments to review the following discussions.

Figure 13.7 How do transcription factors control the expression of eukaryotic genes?

Figure 41.8 In the human female reproductive system, where are the uterine tubes located relative to the ovaries? To the uterus?

Section 41.3 What are the roles of estrogen and progesterone in the ovarian and uterine cycles?

Following *the* Themes

CHAPTER 42 ANIMAL DEVELOPMENT AND AGING

UNIT 7 COMPARATIVE ANIMAL BIOLOGY

Evolution	Similarities in the developmental stages of embryos from animals as diverse as the lancelet, frog, chicken, and human provide evidence for the evolutionary relatedness of all animals.
Nature of Science	Research into the development of two model organisms, *Caenorhabditis elegans* (a roundworm) and *Drosophila melanogaster* (a fruit fly), has advanced the science of developmental biology and knowledge of the mechanisms of aging.
Biological Systems	Animal development begins when egg and sperm nuclei unite, and it progresses through complex stages controlled by the differential expression of a common set of genes found in all cells of an animal's body.

42.1 Early Developmental Stages

Learning Outcomes

Upon completion of this section, you should be able to

1. Identify the structures of an egg and a sperm that are directly involved in fertilization.
2. Compare and contrast the cellular, tissue, and organ stages of embryonic development.
3. Describe the major steps in the development of the central nervous system of vertebrates.

As noted in Chapter 41, animals that reproduce sexually use a wide variety of strategies to achieve **fertilization,** the union of a sperm and an egg to form a zygote. Figure 42.1 illustrates how fertilization occurs in mammals, including humans.

Details of Fertilization in Humans

Human sperm have three distinct parts: a tail, a middle piece, and a head. The head contains the sperm nucleus and is capped by a membrane-bound *acrosome*. The egg of a mammal (actually, the secondary oocyte) is surrounded by a few layers of adhering follicular cells, collectively called the *corona radiata.* These cells nourished the oocyte when it was in a follicle of the ovary. The oocyte also has an extracellular matrix termed the *zona pellucida* just outside the plasma membrane, but beneath the corona radiata.

Video Human Sperm

Fertilization requires a series of events that result in the diploid zygote. Although only one sperm actually fertilizes the oocyte, millions of sperm begin this journey, and perhaps a few hundred succeed in reaching the oocyte. The sperm cover the surface of the oocyte and secrete enzymes that help weaken the corona radiata. They squeeze through the corona radiata and bind to the zona pellucida. After a sperm head binds tightly to the zona pellucida,

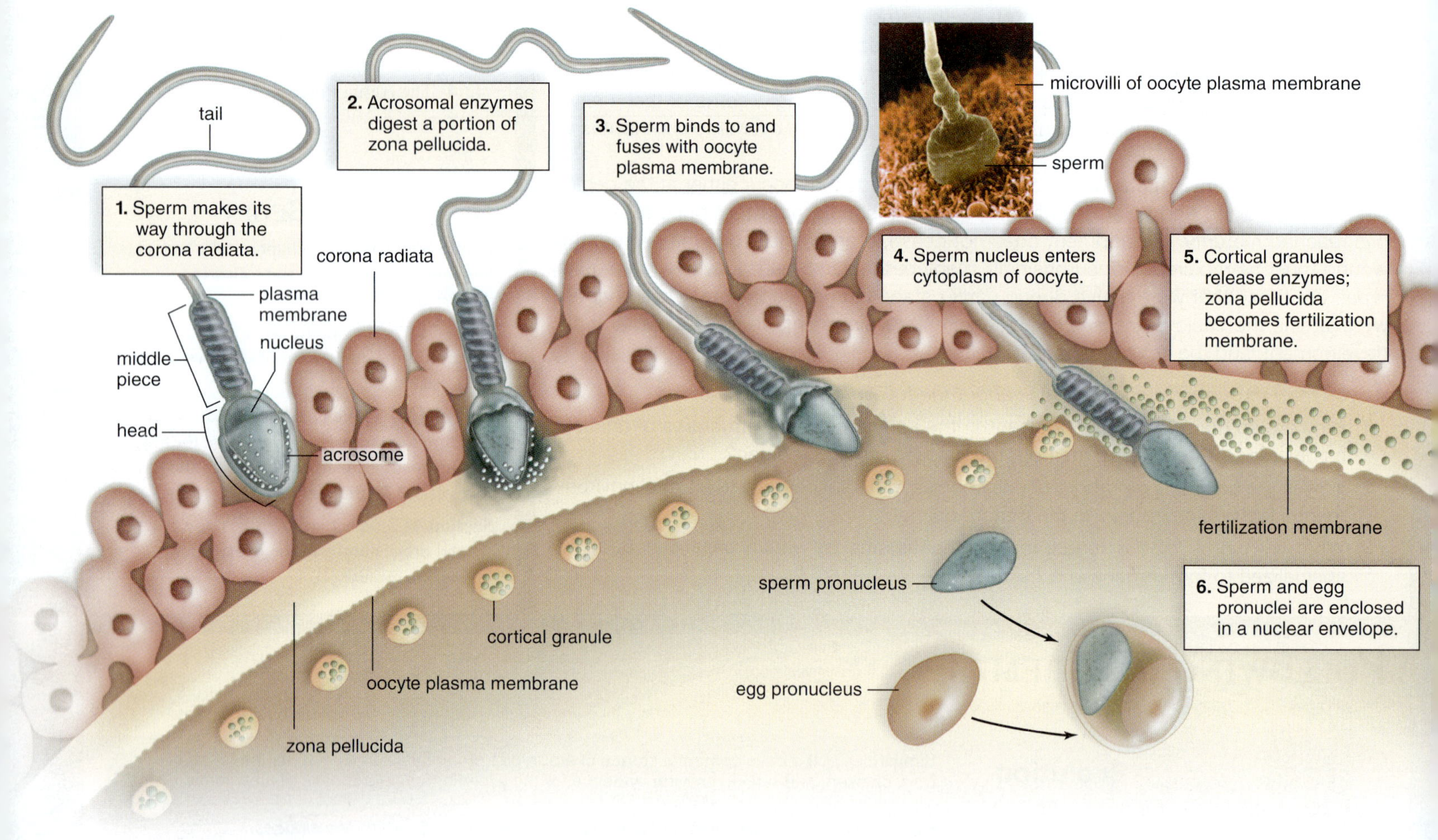

Figure 42.1 Fertilization. During fertilization, a single sperm is drawn into the oocyte by microvilli of its plasma membrane (micrograph). The head of a sperm has a membrane-bound acrosome filled with enzymes. When released, these enzymes digest a pathway for the sperm through the zona pellucida. After a sperm binds to the plasma membrane of the oocyte, changes occur that prevent other sperm from entering the oocyte. The oocyte finishes the second meiotic division and is now an egg. Fertilization is complete when the sperm pronucleus and the egg pronucleus contribute chromosomes to the zygote.

the acrosome releases digestive enzymes that forge a pathway for the sperm through the zona pellucida to the oocyte plasma membrane.

In April 2014, British scientists reported that they had discovered a key molecule, present under the zona pellucida, to which the sperm specifically bind. Named Juno, after an ancient Roman goddess of fertility, the molecule is required for sperm-egg fusion.

When the first sperm binds to the oocyte plasma membrane, the next few events prevent *polyspermy* (entrance of more than one sperm). As soon as a sperm touches the plasma membrane of an oocyte, the oocyte's plasma membrane depolarizes, and this change in charge, known as the "fast block," repels sperm only for a few seconds. Then, vesicles in the oocyte called *cortical granules* secrete enzymes that turn the zona pellucida into an impenetrable *fertilization membrane.* This longer-lasting cortical reaction is known as the "slow block." During this phase, Juno molecules are also released from the oocyte, preventing further specific interactions.

The last events of fertilization include the formation of a diploid zygote. Microvilli extending from the plasma membrane of the oocyte (Fig. 42.1) bring the entire sperm into the oocyte. The sperm nucleus releases its chromatin, which re-forms into chromosomes enclosed within the sperm *pronucleus.* Meanwhile, the secondary oocyte completes meiosis, becoming an egg whose chromosomes are also enclosed in a pronucleus. A single nuclear envelope soon surrounds both sperm and egg pronuclei. Cell division is imminent, and the centrosomes that give rise to a spindle apparatus are derived from the basal body of the sperm's flagellum. The two haploid sets of chromosomes share the first spindle apparatus of the newly formed zygote.

Embryonic Development

Development includes all the changes that occur during the life cycle of an organism. The basics of animal development were described in Chapter 28. Although development is a continuous, complex process, it can be divided into three major stages: (1) cellular stages, (2) tissue stages, and (3) organ stages.

Cellular Stages of Development

During the early stages of development, an organism is called an **embryo.** The cellular stages of development are:

1. Cleavage, resulting in a multicellular embryo;
2. Formation of the blastula.

Cleavage is cell division without growth; DNA replication and mitotic cell division occur repeatedly, and the cells get smaller with each division.

As shown in Figure 42.2, cleavage in a primitive animal called a lancelet results in uniform cells that form a **morula** (L. *morula,* "little mulberry"), which is a solid ball of cells. The morula continues to divide, forming a **blastula** (Gk. *blastos,* "bud"; L. *-ula,* "small"), a hollow ball of cells having a fluid-filled cavity called a **blastocoel.** The blastocoel forms when the cells of the morula extrude Na^+ into extracellular spaces, and water follows by osmosis, collecting in the center.

Video Blastocyst Formation

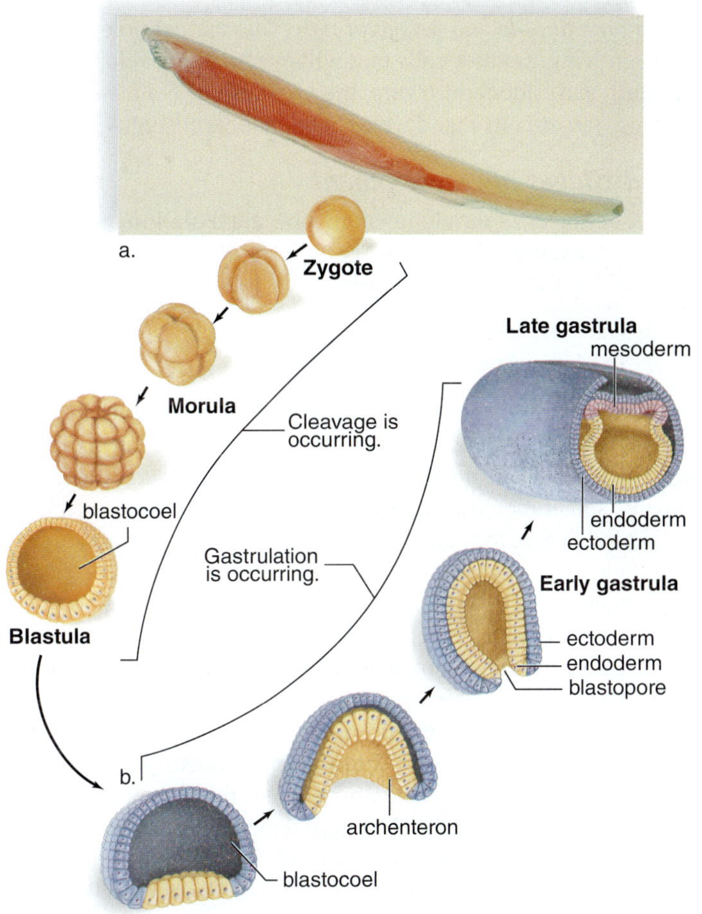

Figure 42.2 Embryonic stages of development. **a.** The early stages of development are exemplified in the lancelet. **b.** Cleavage produces a number of cells that form a cavity, the blastocoel. Invagination during gastrulation produces the germ layers ectoderm and endoderm. Then the mesoderm arises.

Tutorial Embryonic Stages of Development

All vertebrates have a blastula stage, but the appearance of the blastula can be different from that of a lancelet. Chickens lay a hard-shelled egg containing plentiful **yolk,** a dense nutrient material. Because yolk-filled cells do not participate in cleavage, the blastula is a layer of cells that spreads out over the yolk. The blastocoel then forms as a space that separates these cells from the yolk:

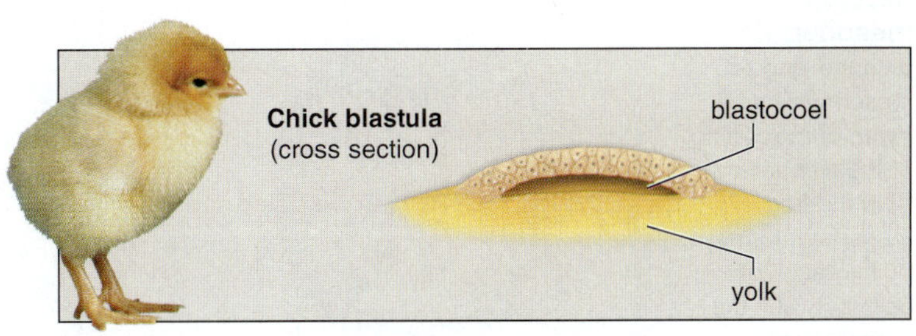

The zygotes of vertebrates, such as frogs, chickens, and humans, also undergo cleavage and form a morula. In frogs, cleavage is not equal because of the presence of the yolk. When yolk is present, the zygote and embryo exhibit polarity, and the embryo has an *animal pole* and a *vegetal pole*. The animal pole contains faster-growing, smaller cells that will eventually develop into the ectoderm and endoderm layers; the vegetal pole contains slower-growing, larger cells that develop into endoderm (Fig. 42.3).

Tissue Stages of Development

The tissue stages of development involve **gastrulation,** the formation of a **gastrula** (Gk. *gaster,* "belly"; L. *-ula,* "small") from the blastula. The two tissue stages are:

1. The early gastrula;
2. The late gastrula.

The early gastrula stage begins when certain cells begin to push, or invaginate, into the blastocoel, creating a double layer of cells (see Fig. 42.2). Cells migrate during this and other stages of development, sometimes traveling quite a distance before reaching a destination, where they continue developing. As cells migrate, they "feel their way" by changing their pattern of adhering to extracellular proteins.

An early gastrula has two layers of cells. The outer layer of cells is called the **ectoderm,** and the inner layer is called the **endoderm.** The endoderm borders the gut, but at this point, it is termed either the archenteron or the primitive gut. The pore, or hole, created by invagination (inward folding) is the **blastopore,** and in a lancelet, the blastopore eventually becomes the anus.

Gastrulation is not complete until three layers of cells that will develop into adult organs are produced. In addition to ectoderm and endoderm, the late gastrula has a middle layer of cells called the **mesoderm.**

Figure 42.2 illustrates gastrulation in a lancelet, and Figure 42.3 compares the lancelet, frog, and chicken late gastrula stages. In the lancelet, mesoderm formation begins as outpocketings from the primitive gut (Fig. 42.3). These outpocketings grow in size until they meet and fuse, forming two layers of mesoderm. The space between them is the coelom (see section 28.1). The coelom is a body cavity lined by mesoderm that contains internal organs. In humans, the coelom becomes the thoracic and abdominal cavities of the body.

Table 42.1 Embryonic Germ Layers

Embryonic Germ Layer	Vertebrate Adult Structures
Ectoderm (outer layer)	Nervous system; epidermis of skin and derivatives of the epidermis (hair, nails, glands); tooth enamel, dentin, and pulp; epithelial lining of oral cavity and rectum
Mesoderm (middle layer)	Musculoskeletal system; dermis of skin; cardiovascular system; urinary system; lymphatic system; reproductive system—including most epithelial linings; outer layers of respiratory and digestive systems
Endoderm (inner layer)	Epithelial lining of digestive tract and respiratory tract, associated glands of these systems; epithelial lining of urinary bladder; thyroid and parathyroid glands

In the frog, cells containing yolk do not participate in gastrulation and, therefore, they do not invaginate. Instead, a slitlike blastopore is formed when the animal pole cells begin to invaginate from above, forming endoderm. Animal pole cells also move down over the yolk, to invaginate from below. Some yolk cells, which remain temporarily in the region of the blastopore, are called the yolk plug. Mesoderm forms when cells migrate between the ectoderm and endoderm. Later, the mesoderm splits, creating the coelom.

A chicken egg contains so much yolk that endoderm formation does not occur by invagination. Instead, an upper layer of cells becomes ectoderm, and a lower layer becomes endoderm. Mesoderm arises by an invagination of cells along the edges of a longitudinal furrow in the midline of the embryo. Because of its appearance, this furrow is called the *primitive streak.* Later, the newly formed mesoderm splits to produce a coelomic cavity.

Ectoderm, mesoderm, and endoderm are called the embryonic **germ layers.** No matter how gastrulation takes place, the result is the same: Three germ layers are formed. It is possible to relate the development of future organs to these germ layers (Table 42.1).

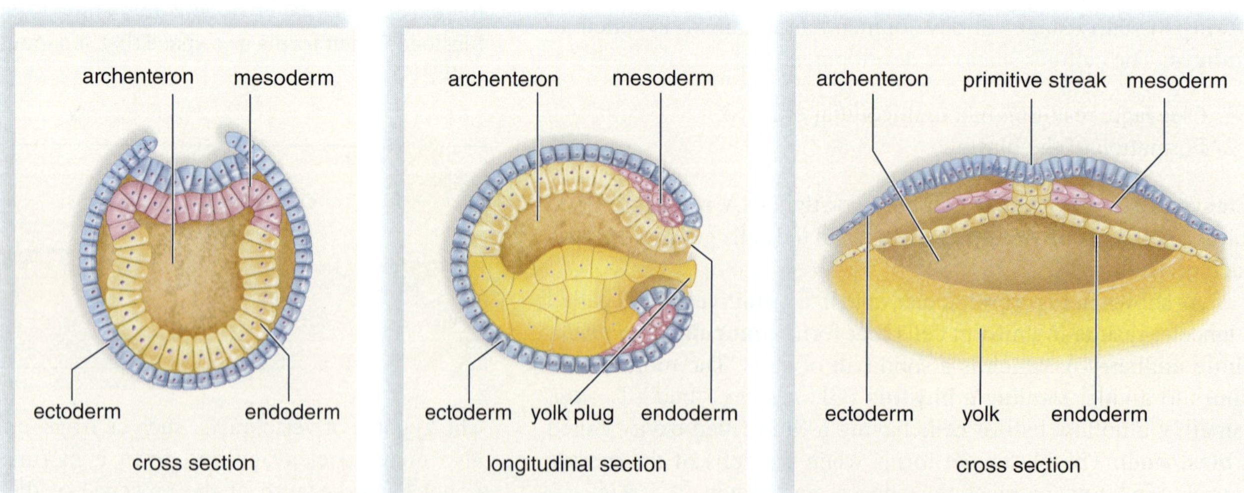

a. Lancelet late gastrula b. Frog late gastrula c. Chick late gastrula

Figure 42.3 Comparative development of mesoderm. **a.** In the lancelet, mesoderm forms by an outpocketing of the archenteron. **b.** In the frog, mesoderm forms by migration of cells between the ectoderm and endoderm. **c.** In the chick, mesoderm also forms by invagination of cells.

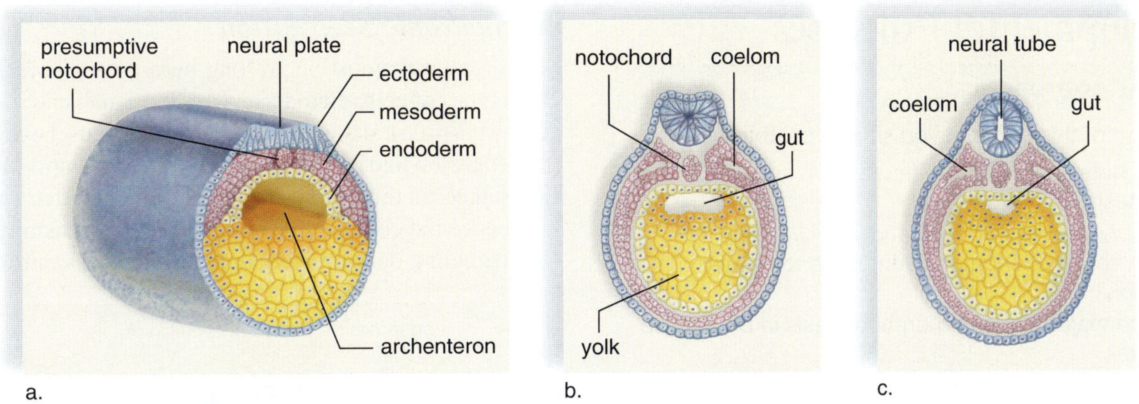

Figure 42.4 Development of neural tube and coelom in a frog embryo. **a.** Ectodermal cells that lie above the future notochord (called the presumptive notochord) thicken to form a neural plate. **b.** A splitting of the mesoderm produces a coelom, which is completely lined by mesoderm. **c.** A neural tube and a coelom have now developed.

Organ Stages of Development

The organs of an animal's body develop from the three embryonic germ layers. Much study has been devoted to how the nervous system develops in chordates, so we summarize that process here.

The newly formed mesoderm cells lie along the main longitudinal axis of the animal and coalesce to form a dorsal supporting rod called the **notochord.** The notochord persists in lancelets, but in frogs, chickens, and humans, it is replaced later in development by the vertebral column, giving these animals the name vertebrates.

The nervous system develops from midline ectoderm located just above the notochord. At first, a thickening of cells, called the **neural plate,** is seen along the dorsal surface of the embryo. Then, neural folds develop on each side of a neural groove, which becomes the **neural tube** when these folds fuse. Figure 42.4 shows cross sections of frog development to illustrate the formation of the neural tube. At this point, the embryo is called a *neurula*. Later, the anterior end of the neural tube develops into the brain, and the rest becomes the spinal cord.

The *neural crest* is a band of cells that develops where the neural tube pinches off from the ectoderm. These cells migrate to various locations, where they help form skin, muscles, the adrenal medulla, and the ganglia of the peripheral nervous system.

Midline mesoderm cells that did not contribute to the formation of the notochord now become two longitudinal masses of tissue. These two masses become blocked off into *somites,* which are serially arranged along both sides along the length of the notochord. Somites give rise to muscles associated with the axial skeleton and to the vertebrae. The serial origin of axial muscles and the vertebrae illustrates that vertebrates (including humans) are segmented animals. Lateral to the somites, the mesoderm splits, forming the mesodermal lining of the coelom.

In the development of other organs, a primitive gut tube is formed by endoderm as the body itself folds into a tube. The heart, too, begins as a simple tubular pump. Organ formation continues until the germ layers have given rise to the specific organs listed in Table 42.1. Figure 42.5 helps you relate the formation of vertebrate structures and organs to the three embryonic layers of cells: the ectoderm, the mesoderm, and the endoderm.

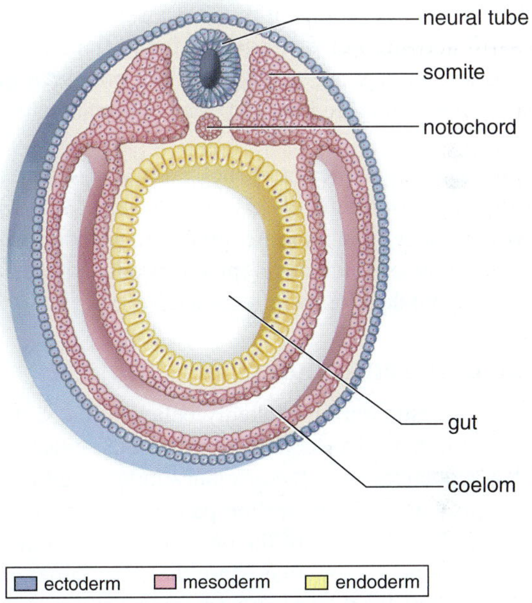

Figure 42.5 Vertebrate embryo, cross section. At the neurula stage, each of the germ layers, indicated by color (see key), can be associated with the later development of particular parts. The somites give rise to the muscles of each segment and to the vertebrae, which replace the notochord in vertebrates.

Check Your Progress 42.1

1. Describe the fast and slow blocks to polyspermy.
2. Name the germ layer of the gastrula that gives rise to the notochord, thyroid and parathyroid glands, nervous system, epidermis, skeletal muscles, kidneys, bones, and pancreas.
3. Identify the stage(s) of embryonic development in which cross sections of all chordate embryos closely resemble one another.

42.2 Developmental Processes

Learning Outcomes

Upon completion of this section, you should be able to

1. Distinguish between cellular differentiation and morphogenesis.
2. Define *cytoplasmic segregation* and *induction* and explain how these phenomena contribute to cellular differentiation.
3. Describe the major steps of morphogenesis in *Drosophila melanogaster.*

Development requires three interconnected processes: growth, cellular differentiation, and morphogenesis. **Cellular differentiation** occurs when cells become specialized in structure and function; that is, a muscle cell looks different and acts differently than a nerve cell. **Morphogenesis** produces the shape and form of the body. One of the earliest indications of morphogenesis is cell movement. Later, morphogenesis includes **pattern formation,** which means how tissues and organs are arranged in the body. **Apoptosis,** or programmed cell death (see Fig. 9.2), is an important part of pattern formation.

Developmental genetics has benefited from research using the roundworm *Caenorhabditis elegans* and the fruit fly *Drosophila melanogaster.* These organisms are referred to as model organisms, because the study of their development produced concepts that help us understand development in general.

Cellular Differentiation

At one time, investigators mistakenly believed that different cell types, such as human liver cells and brain cells, must inherit different DNA sequences from the original single-celled zygote. Perhaps, they speculated, the genes are parceled out as development occurs, and that is why cells of the body have a different structure and function. We now know that is not the case; rather, every cell in the body (except for the sperm and unfertilized egg) has a full complement of genes.

The zygote is **totipotent;** it has the ability to generate the entire organism and, therefore, must contain all the instructions needed by any other specialized cell in the body. For the first few days of cell division, all the embryonic cells are totipotent. When the embryonic cells begin to specialize and lose their totipotency, they do not lose genetic information. In fact, our ability to clone mammals such as sheep, mice, and cats from specialized adult cells shows that every cell in an organism's body has the same collection of genes.

The answer to this puzzle becomes clear when we consider that only muscle cells produce the proteins myosin and actin; only red blood cells produce hemoglobin; and only skin cells produce keratin. In other words, we now know that specialization is not due to a parceling out of genes; rather, it is due to differential gene expression. Certain genes, but not others, are turned on (transcribed) in differentiated cells. In recent years, investigators have turned their attention to discovering the mechanisms that lead to differential gene expression. Two mechanisms—cytoplasmic segregation and induction—seem to be especially important.

Cytoplasmic Segregation

Differentiation must begin long before we can recognize specialized types of cells. Ectodermal, endodermal, and mesodermal cells in the gastrula look quite similar, but they must be different, because they develop into different organs. The egg is now known to contain substances in the cytoplasm called **maternal determinants,** which influence the course of development. **Cytoplasmic segregation** is the parceling out of maternal determinants as mitosis occurs:

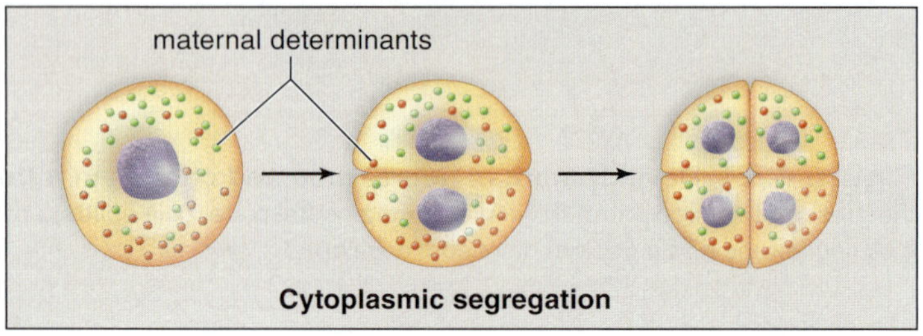

An experiment conducted in 1935 showed that the cytoplasm of a frog's egg is not uniform. It is polar, having both an anterior/posterior axis and a dorsal/ventral axis, which can be correlated with the *gray crescent,* a gray area that appears after the sperm fertilizes the egg (Fig. 42.6*a*). If the gray crescent is divided equally by the first cleavage, each experimentally separated daughter cell develops into a complete embryo (Fig. 42.6*b*). However, if the zygote divides so that only one daughter cell receives the gray crescent, only that cell becomes a complete embryo (Fig. 42.6*c*). This experiment allowed scientists to speculate that the gray crescent must contain particular chemical signals that are needed for development to proceed normally.

Induction and Frog Experiments

As development proceeds, the specialization of cells and formation of organs are influenced not only by maternal determinants but also by signals given off by neighboring cells. **Induction** (L. *in,* "into"; *duco,* "lead") is the ability of one embryonic tissue to influence the development of another tissue.

A frog embryo's gray crescent becomes the dorsal lip of the blastopore, where gastrulation begins. Because this region is necessary for complete development, the dorsal lip of the blastopore is called the *primary organizer.* The cells closest to the primary organizer become endoderm, those farther away become mesoderm, and those farthest away become ectoderm. This suggests that a molecular concentration gradient may act as a chemical signal to induce germ layer differentiation.

The gray crescent in the zygote of a frog marks the dorsal side of the embryo where the mesoderm becomes notochord and ectoderm becomes nervous system. In a classic experiment, researchers showed that presumptive (potential) notochord tissue induces the formation of the nervous system (Fig. 42.7). If presumptive nervous system tissue, located just above the presumptive notochord, is cut out and transplanted to the belly region of the embryo, it does not form a neural tube. But in contrast, if presumptive notochord tissue is cut out and transplanted beneath what would be belly ectoderm, this ectoderm does differentiate into neural tissue. Many other examples of induction are now known.

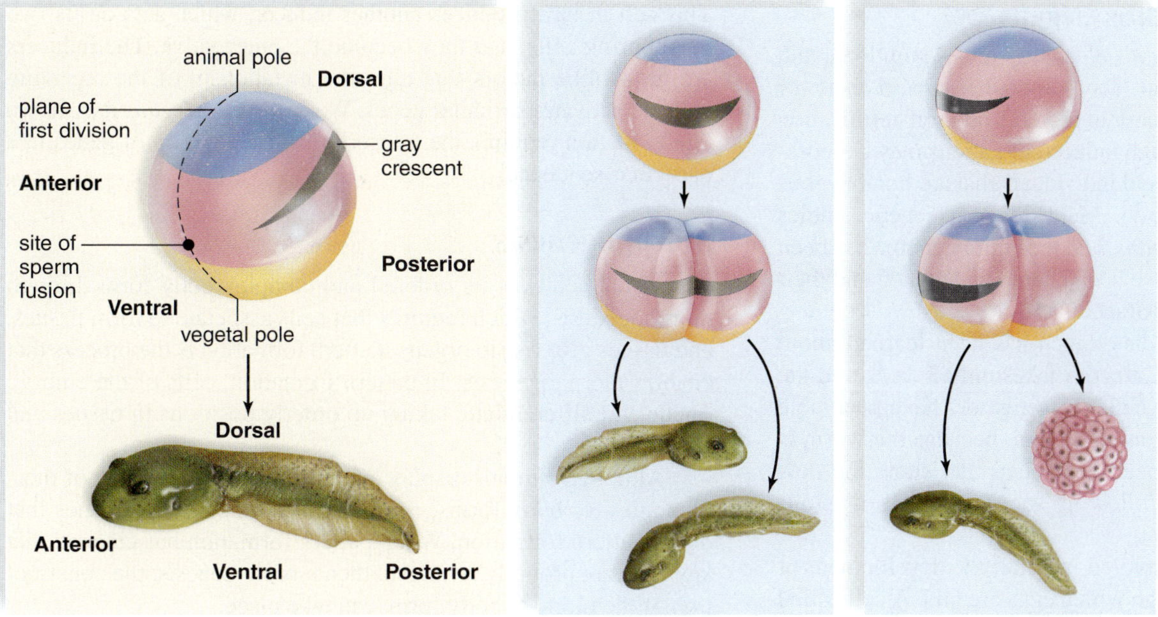

a. Zygote of a frog is polar and has axes.

b. Each cell receives a part of the gray crescent.

c. Only the cell on the left receives the gray crescent.

Figure 42.6 Cytoplasmic influence on development. **a.** The zygote of a frog has anterior/posterior and dorsal/ventral axes that correlate with the position of the gray crescent. **b.** The first cleavage normally divides the gray crescent in half, and each daughter cell is capable of developing into a complete tadpole. **c.** But if only one daughter cell receives the gray crescent, then only that cell can become a complete embryo. This shows that maternal determinants are present in the cytoplasm of a frog's egg.

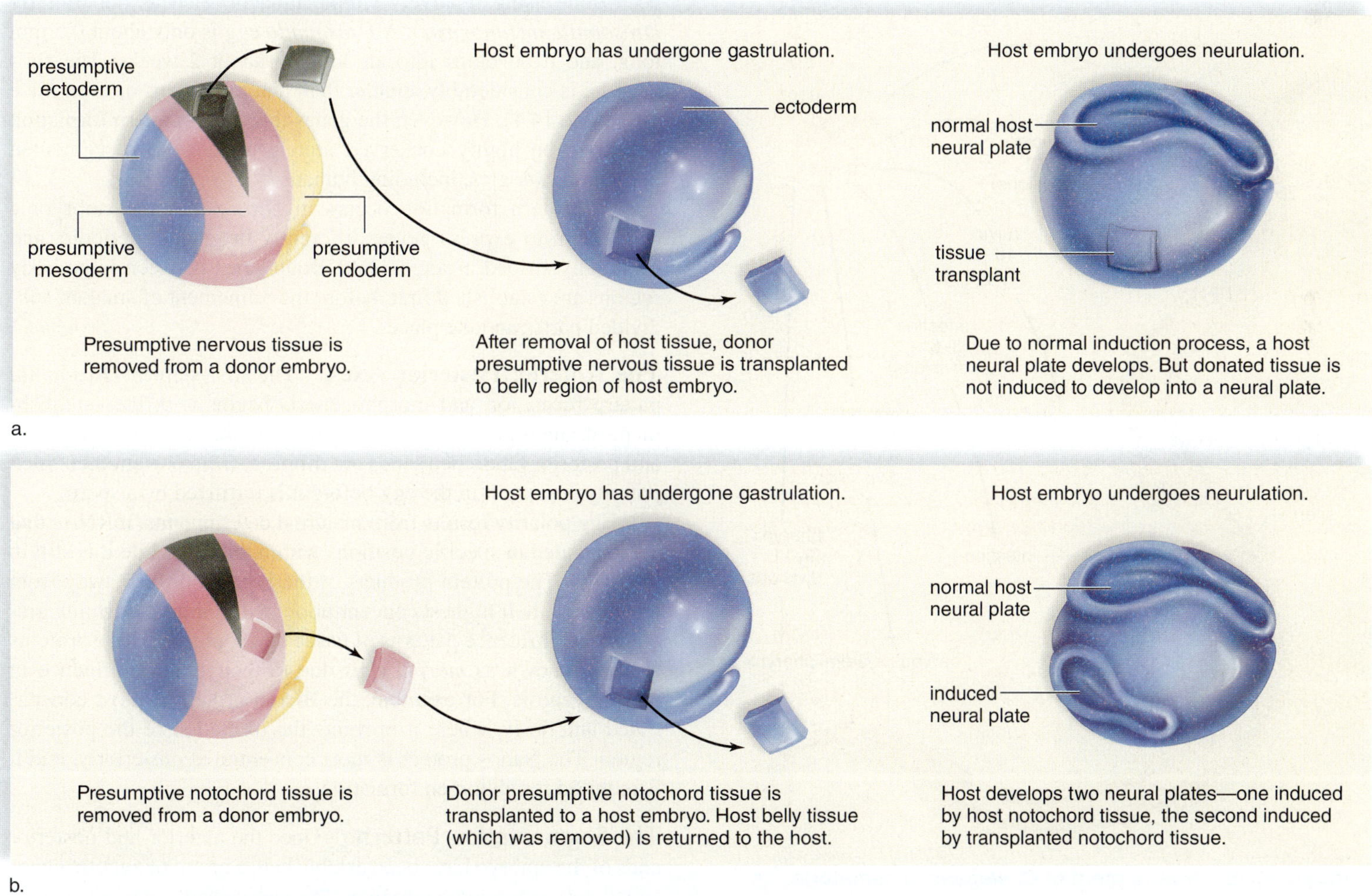

Figure 42.7 Control of nervous system development. **a.** In this experiment, the presumptive nervous system (*blue*) does not develop into the neural plate if moved from its normal location. **b.** In this experiment, the presumptive notochord (*pink*) can cause even belly ectoderm to develop into the neural plate (*blue*). This shows that the notochord induces ectoderm to become a neural plate, most likely by sending out chemical signals.

Induction in Caenorhabditis elegans

The tiny nematode *Caenorhabditis elegans* is only 1 mm long, and vast numbers can be raised in the laboratory either in petri dishes or a liquid medium. The worm is hermaphroditic, and self-fertilization is the rule. Therefore, even though induced mutations may be recessive, the next generation will yield individuals that are homozygous recessive and show the mutation. Many modern genetic studies have been performed on *C. elegans,* and the entire genome has been sequenced. Individual genes have been altered and cloned and their products injected into cells or extracellular fluid.

As a result of genetic studies, much has been learned about *C. elegans.* Development of *C. elegans* takes only 3 days, and the adult worm contains only 959 cells. Investigators have been able to watch the process from beginning to end, because the worm is transparent. *Fate maps* have been developed that show the destiny of each cell as it arises following successive cell divisions (Fig. 42.8).

Some investigators have studied in detail the development of the worm's vulva, a pore through which eggs are laid. A cell called the anchor cell induces the vulva to form. The cell closest to the anchor cell receives the most inducer and becomes the inner vulva.

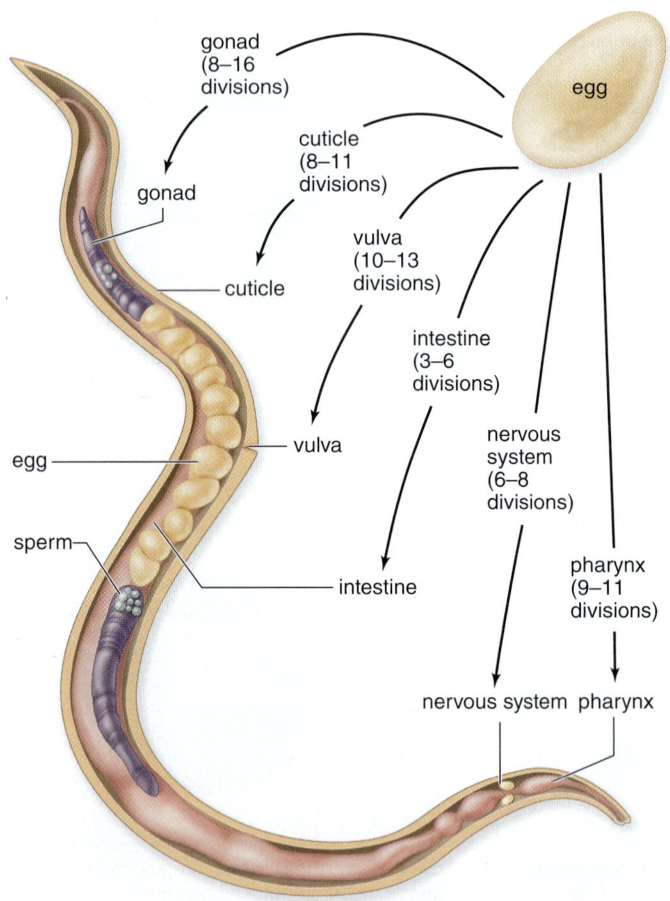

Figure 42.8 Development of *C. elegans,* a nematode. A fate map of the worm showing that as cells arise by cell division, they are destined to become particular structures.

This cell in turn produces another inducer, which acts on its two neighboring cells, and they become the outer vulva. The inducers are growthlike factors that alter the metabolism of the receiving cell and activate particular genes. Work with *C. elegans* has shown that induction requires the transcriptional regulation of genes in a particular sequence.

Morphogenesis

An animal achieves its ordered and complex body form through morphogenesis, which requires that cells associate to form tissues, and tissues give rise to organs. Pattern formation is the process that enables morphogenesis. In pattern formation, cells of the embryo divide and differentiate, taking up orderly positions in tissues and organs.

Although animals display an amazingly diverse array of morphologies, or body forms, most share common sets of genes that direct pattern formation. When pattern formation has ensured that key cells are properly arranged, then morphogenesis, the construction of the ultimate body form, can take place.

Morphogenesis in Drosophila melanogaster

Much of the study of pattern formation and morphogenesis has been done using relatively simple models, such as the fruit fly *Drosophila melanogaster.* A *Drosophila* egg is only about 0.5 mm long, and it develops into an adult in about 2 weeks. The fly's genome is considerably smaller than that of humans or even mice (see Table 14.1). However, the genes that direct pattern formation appear to be highly conserved among animals with segmented body morphologies, including humans.

As pattern formation occurs in *Drosophila,* the embryonic cells begin to express genes differently in graded, periodic, and eventually striped arrangements. Boundaries between large body regions are established first, before the refinement of smaller, subdivided parts can take place.

The Anterior/Posterior Axes. The first step in *Drosophila* pattern formation and morphogenesis begins with the establishment of anteroposterior polarity, meaning that the anterior (head) and posterior (abdomen) ends are different from one another. Such polarity is present in the egg before it is fertilized by a sperm.

Egg polarity results from maternal determinants, mRNAs that are deposited in specific positions within the egg while it is still in the ovary. The protein products of these genes diffuse away from the areas of their highest concentration in the embryo, forming gradients that influence patterns of tissue development. These proteins are also known as *morphogens* due to their crucial influence in morphogenesis. For example, the Bicoid protein is most concentrated anteriorly, where it prevents the formation of the posterior region. The Nanos protein is most concentrated posteriorly, and it is required for abdomen formation.

The Segmentation Pattern. Once the anterior and posterior ends of the embryo have been established, a group of zygotic genes called *gap* genes come into play. The task of gap genes is to divide the anteroposterior axis into broad regions (Fig. 42.9*a*). They are

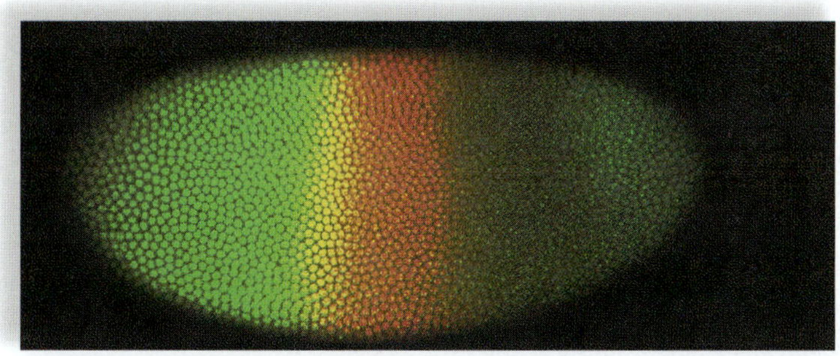

a. Protein products of gap genes

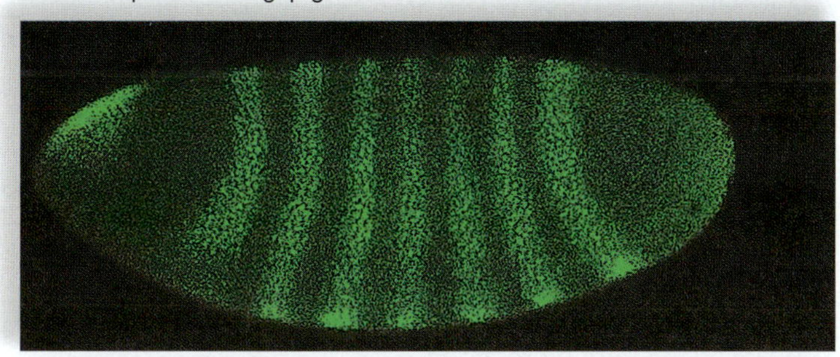

b. Protein products of pair-rule genes

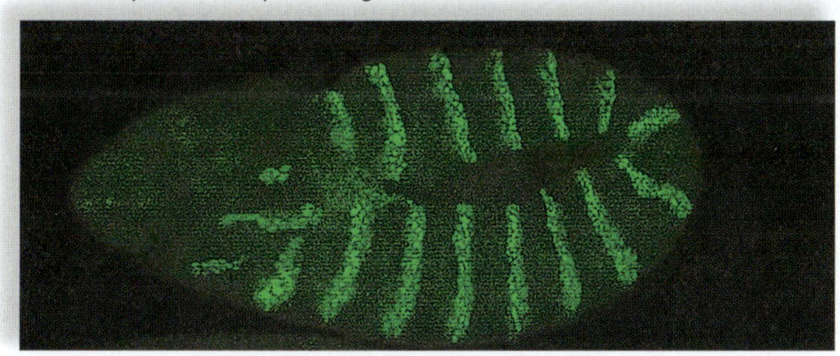

c. Protein products of segment-polarity genes

Figure 42.9 Development in *Drosophila,* a fruit fly. **a.** The different colors show that two different gap gene proteins are present from the anterior to the posterior end of an embryo. **b.** The green stripes show that a pair-rule gene is being expressed as segmentation of the fly occurs. **c.** Now segment-polarity genes help bring about the division of each segment into an anterior end and a posterior end.

a. SEM 50×

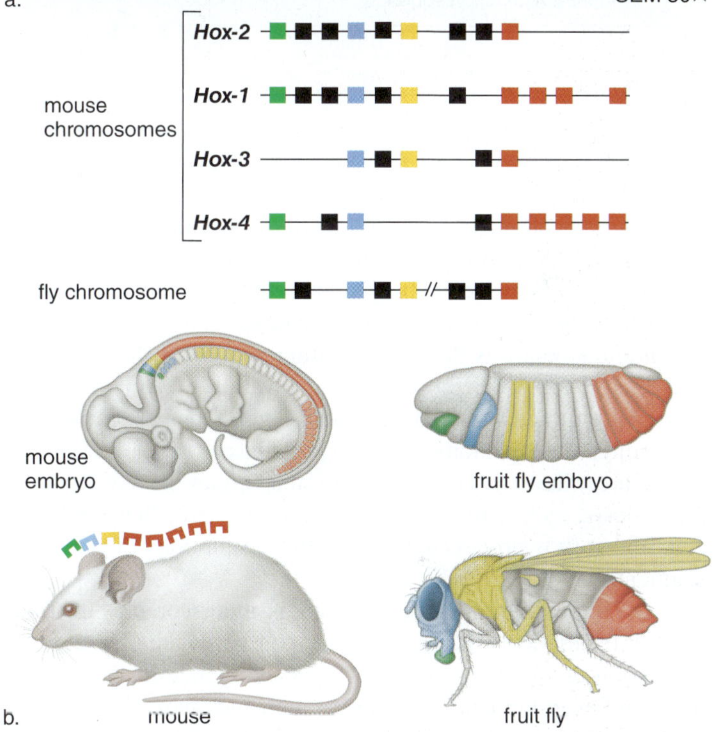

b.

Figure 42.10 Pattern formation in *Drosophila.* Homeotic genes control pattern formation, an aspect of morphogenesis. **a.** If homeotic genes are activated at inappropriate times, abnormalities such as a fly with four wings occur. **b.** The green, blue, yellow, and red colors show that homologous homeotic genes occur on four mouse chromosomes and on a fly chromosome in the same order. These genes are color coded to the region of the embryo, and therefore the adult, where they regulate pattern formation. The black boxes are homeotic genes that are not identical between the two animals. In mammals, homeotic genes are called *HOX* genes.

called gap genes because mutations in these genes result in gaps in the embryo, where large blocks of segments are missing. Gap genes are temporarily activated by the gradients of anterior and posterior morphogens and in turn activate the *pair-rule* genes.

The pair-rule genes are expressed periodically, in alternating stripes (Fig. 42.9*b*). They "rough out" a preliminary segmentation pattern along the anteroposterior axis. The products of pair-rule genes may stimulate or suppress the expression of other genes, particularly the *segment-polarity* genes. These genes ensure that each segment has boundaries, with distinct anterior and posterior halves. The segment-polarity genes are also expressed in a striped fashion, but with twice as many stripes as the pair-rule genes (Fig. 42.9*c*). Mutations in segment-polarity genes result in the loss of one part of each segment, as well as the duplication of another portion of the same segment.

Homeotic Genes. The **homeotic genes** are often referred to as *selector genes,* because they select for segmental identity—in other words, they dictate which body parts arise from the segments. Mutations in homeotic genes may result in the development of body parts in inappropriate areas, such as legs instead of antennae, or wings instead of tiny, balancing organs called halteres (Fig. 42.10*a*). Such alterations in morphology are known as *homeotic transformations.*

Interestingly, homeotic genes in *Drosophila* and other organisms have all been found to share a structural feature called a **homeobox.** (*HOX,* the term used for mammalian homeotic genes, is a shortened form of *homeobox.*) A homeobox is a sequence of nucleotides that encodes a 60-amino-acid sequence called a **homeodomain:**

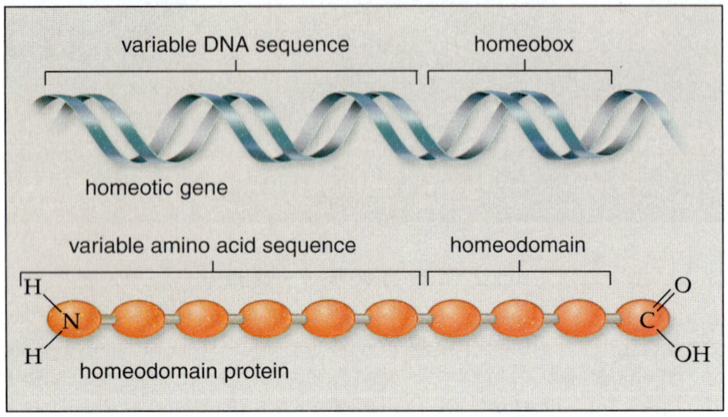

The homeodomain is a functionally important part of the protein encoded by a homeotic gene. Homeotic genes code for transcription factors, proteins that bind to regulatory regions of DNA and determine whether or not specific target genes are turned on. The homeodomain is the DNA-binding portion of the transcription factor, although the other, more variable sequences of a transcription factor determine which target genes are turned on. It is thought that the homeodomain proteins encoded by homeotic genes direct the activities of various target genes involved in morphogenesis, such as those involved in cell-to-cell adhesion. In the end, this orderly process determines the morphology of particular segments.

The importance of homeotic genes is underscored by the finding that the homeotic genes are highly conserved, being present in the genomes of many organisms, including mammals such as mice and even humans. Most homeotic genes have their loci on the same chromosome in *Drosophila,* while in mammals there are four clusters that reside on different chromosomes.

Notice that, in both flies and mammals, the position of the homeotic genes on the chromosome matches their anterior-to-posterior expression pattern in the body (Fig. 42.10*b*). The first gene clusters determine the final development of anterior segments, whereas those later in the sequence determine the final development of posterior segments of the animal's body.

Mutations in homeotic genes have effects in the mammalian body similar to those of the homeotic transformations observed in *Drosophila.* For instance, mutations in two adjacent *HOX* genes in the mouse result in shortened forelimbs that are missing the radius and ulna bones. In humans, mutations in a different *HOX* gene cause synpolydactyly, a rare condition in which there are extra digits (fingers and toes), some of which are fused to their neighbors.

Apoptosis. We have already discussed the importance of apoptosis (programmed cell death) in the normal, day-to-day operation of the immune system and in preventing the occurrence of cancer. Apoptosis is also an important part of morphogenesis. In tadpoles, for example, apoptosis is largely responsible for the disappearance of the tail. During human development, apoptosis is needed to remove the webbing between fetal fingers and toes.

Fate maps of *C. elegans* (see Fig. 42.8) indicate that apoptosis occurs in 131 cells as development takes place. When a cell-death signal is received, an inhibiting protein becomes inactive, triggering an apoptotic cascade that ends in enzymes destroying the cell.

Check Your Progress 42.2

1. Name and define two mechanisms of cellular differentiation.
2. Define the term *morphogen.*
3. Describe the function of the homeobox sequence in a homeotic gene.

42.3 Human Embryonic and Fetal Development

Learning Outcomes

Upon completion of this section, you should be able to

1. Name the membranes surrounding the human embryo and list their functions.
2. Chronologically list the major events that occur during embryonic and fetal development.
3. Describe the structure and functions of the placenta.

In humans, the length of time from conception (fertilization followed by implantation in the endometrium) to birth (parturition) is approximately 9 months. It is customary to calculate the time of birth by adding 280 days to the start of the last menstruation, because this date is usually known, whereas the day of fertilization is usually unknown. Because the time of birth is influenced by so many variables, only about 5% of babies actually arrive on the forecasted date.

Human development is often divided into embryonic development (months 1 and 2) and fetal development (months 3 through 9). During **embryonic development,** the major organs are formed, and during fetal development, these structures become larger and are refined.

Development can also be divided into *trimesters*. Each trimester can be characterized by specific developmental accomplishments. During the first trimester, embryonic and early fetal development occur. The second trimester is characterized by the development of organs and organ systems. By the end of the second trimester, the fetus appears distinctly human. In the third trimester, the fetus grows rapidly and the major organ systems become functional.

Extraembryonic Membranes

Before we consider human development chronologically, we must understand the placement of **extraembryonic membranes** (L. *extra,* "on the outside"). Extraembryonic membranes are best understood by considering their function in reptiles and birds. In reptiles, these membranes made development on land first possible. If an embryo develops in the water, the water supplies oxygen for the embryo and takes away waste products. The surrounding water prevents desiccation, or drying out, and provides a protective cushion. For an embryo that develops on land, all these functions are performed by the extraembryonic membranes.

In the chick, the extraembryonic membranes develop from extensions of the germ layers, which spread out over the yolk. Figure 42.11 shows the chick surrounded by the membranes. The **chorion** (Gk. *chorion,* "membrane") lies next to the shell and carries on gas exchange. The **amnion** (Gk. *amnion,* "membrane around fetus") contains the protective amniotic fluid, which bathes the developing embryo. The **allantois** (Gk. *allantos,* "sausage") collects nitrogenous wastes, and the **yolk sac** surrounds the remaining yolk, which provides nourishment.

The function of the extraembryonic membranes in humans has been modified to suit internal development. Their presence, however, shows that we are related to the reptiles.

The chorion develops into the fetal half of the placenta, the organ that provides the embryo/fetus with nourishment and oxygen and takes away its wastes. Blood vessels within the chorionic villi are continuous with the umbilical blood vessels. The blood vessels of the allantois become the umbilical blood vessels, and the allantois accumulates the small amount of urine produced by the fetal kidneys and later gives rise to the urinary bladder. The yolk sac, which lacks yolk, is the first site of blood cell formation. The amnion contains fluid to cushion and protect the embryo, which develops into a fetus.

It is interesting to note that all chordate animals develop in water—either in bodies of water or surrounded by amniotic fluid within a shell or uterus.

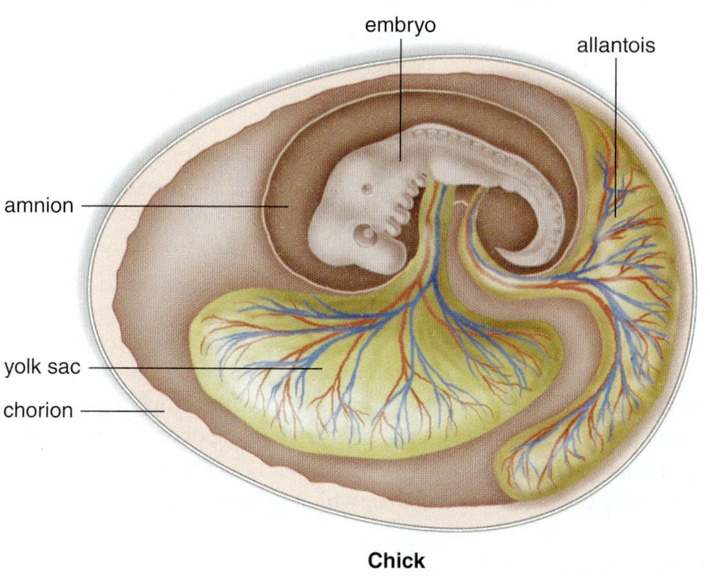

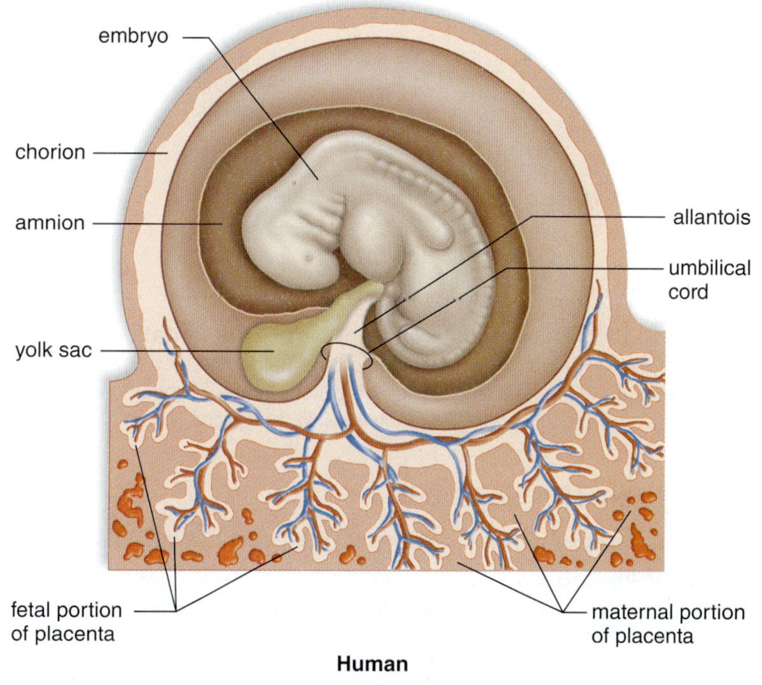

Figure 42.11 Extraembryonic membranes. Extraembryonic membranes, which are not part of the embryo, are found during the development of chicks and humans. Each has a specific function.

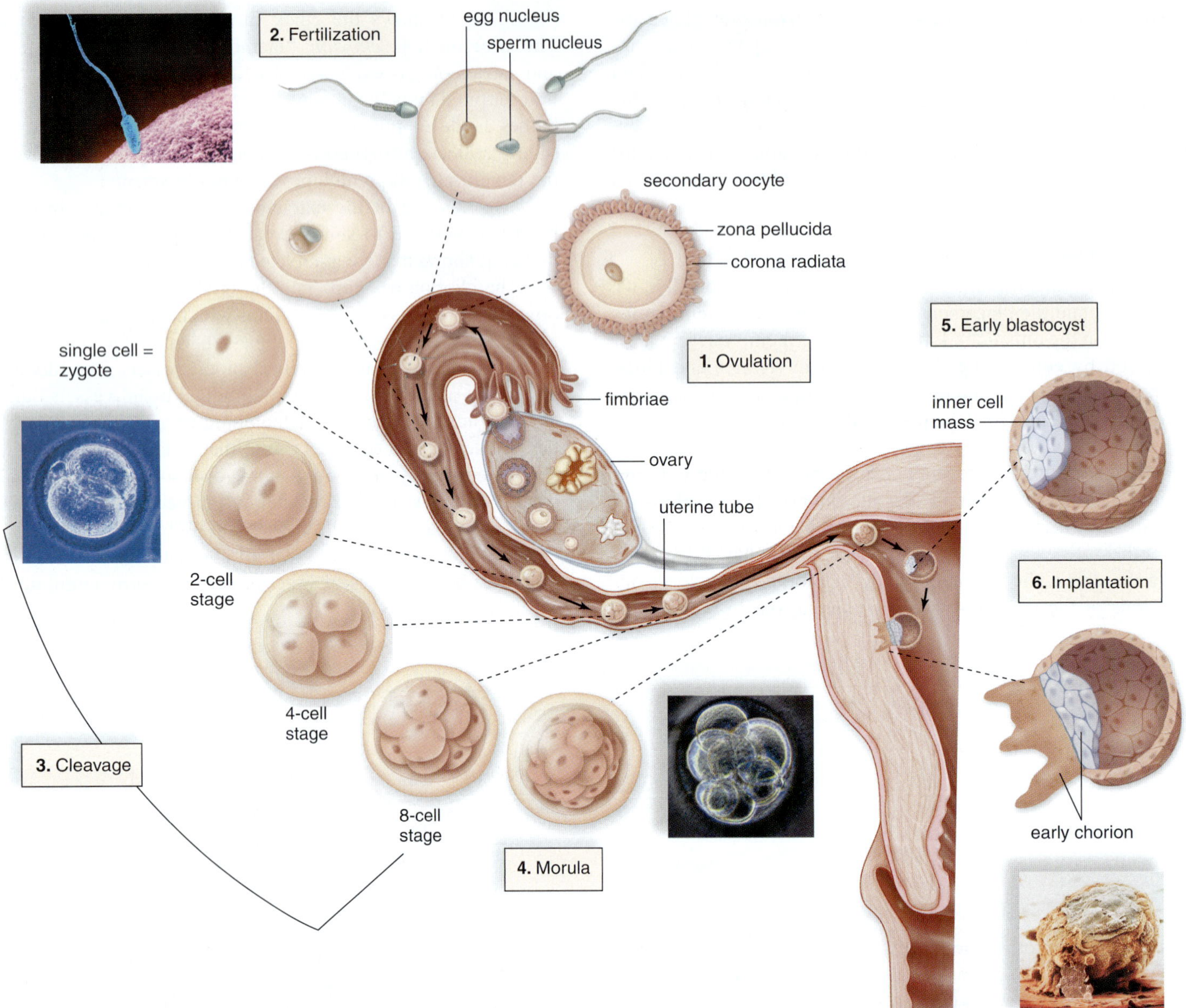

Figure 42.12 Human development before implantation. (1) At ovulation, the secondary oocyte leaves the ovary. A single sperm nucleus enters the oocyte, and (2) fertilization of the egg occurs in the uterine tube. As the zygote moves along the uterine tube, it undergoes (3) cleavage to produce (4) a morula. (5) The blastocyst forms and (6) implants itself in the uterine lining.

Embryonic Development

Embryonic development comprises the first 2 months of development.

The First Week

Fertilization usually occurs in the upper third of the uterine tube (Fig. 42.12). Cleavage begins 30 hours after fertilization and continues as the embryo passes through the uterine tube to the uterus. By the time the embryo reaches the uterus on the third day, it is a morula. By about the fifth day, the morula has transformed into a blastula, and in mammals, a blastocyst then forms.

Video Blastocyst Formation

The **blastocyst** has a fluid-filled cavity, a single layer of outer cells called the **trophoblast** (Gk. *trophe,* "food"; *blastos,* "bud"), and an inner cell mass. The early function of the trophoblast is to provide nourishment for the embryo. Later, the trophoblast, reinforced by a layer of mesoderm, gives rise to the chorion, one of the extraembryonic membranes (see Fig. 42.11). The inner cell mass eventually becomes the embryo, which develops into a fetus.

The Second Week

At the end of the first week, the embryo begins the process of **implantation** in the wall of the uterus. The trophoblast secretes enzymes to digest away some of the tissue and blood vessels of the endometrium of the uterus (Fig. 42.12). The embryo is now about the size of the period at the end of this sentence. The trophoblast begins to secrete **human chorionic gonadotropin** (**HCG**), the hormone that is the basis for the pregnancy test and that maintains the corpus luteum past the time it normally

THEME Biological Systems

Preventing and Testing for Birth Defects

Birth defects, or congenital disorders, are abnormal conditions that are present at birth. According to the Centers for Disease Control and Prevention (CDC), about 1 in 33 babies born in the United States has a birth defect. Those due to genetics can sometimes be detected before birth by one of the methods shown in Figure 42A; however, these methods are not without risk.

Some birth defects are not serious, and not all can be prevented. But women can take steps to increase their chances of delivering a healthy baby.

Eat a Healthy Diet

Certain birth defects occur because the developing embryo does not receive sufficient nutrition. For example, women of childbearing age are urged to make sure they consume adequate amounts of the vitamin folic acid in order to prevent neural tube defects, such as spina bifida and anencephaly. In spina bifida, part of the vertebral column fails to develop properly and cannot adequately protect the delicate spinal cord. With anencephaly, most of the fetal brain fails to develop. Anencephalic infants are stillborn or survive for only a few days after birth.

Fortunately, folic acid is plentiful in leafy green vegetables, nuts, and citrus fruits. It is also present in multivitamins, and many breads and cereals are fortified with it. The CDC recommends that all women of childbearing age get at least 400 micrograms of folic acid every day through supplements, in addition to eating a healthy, folic acid–rich diet. Unfortunately, neural tube birth defects can occur just a few weeks after conception, when many women are still unaware that they are pregnant—especially if the pregnancy is unplanned.

Avoid Alcohol, Smoking, and Drug Abuse

Alcohol consumption during pregnancy is a leading cause of birth defects. In severe instances, the baby is born with fetal alcohol syndrome (FAS), estimated to occur in 0.2 to 1.5 of every 1,000 live births in the United States. Children with FAS are frequently underweight, have an abnormally small head, abnormal facial development, and intellectual disability. As they mature, children with FAS often exhibit short attention span, impulsiveness, and poor judgment, as well as serious difficulties with learning and memory. Heavy alcohol use also reduces a woman's folic acid level, increasing the risk of the neural tube defects just described.

Cigarette smoking causes many birth defects. Babies born to smoking mothers typically have low birth weight, and are more likely to have defects of the face, heart, and brain than those born to nonsmokers.

Illegal drugs should also be avoided. For example, cocaine causes blood pressure fluctuations that deprive the fetus of oxygen. Cocaine-exposed babies may have problems with vision and coordination, and they may be intellectually disabled.

Alert Medical Personnel If You Are or May Be Pregnant

Several medications that are safe for healthy adults may pose a risk to a developing fetus. If pregnant women need to be immunized, they are usually given killed or inactivated forms of the vaccine, because live forms often present a danger to the fetus.

Because the rapidly dividing cells of a developing embryo or fetus are very susceptible to damage from radiation, pregnant women should avoid unnecessary X-rays. If a medical X-ray is unavoidable, the woman should notify the X-ray technician, so that her fetus can be protected as much as possible—for example, by draping a lead apron over her abdomen if this area is not targeted for X-ray examination.

Avoid Infections That Cause Birth Defects

Certain pathogens, such as the virus that causes rubella (German measles), can result in birth defects. In the past, this virus caused many birth defects, specifically intellectual disability, deafness, blindness, and heart defects. Rubella is much less of a danger in developed countries today, because most women have been vaccinated against rubella as children. Other infections that can cause birth defects include toxoplasmosis, herpes simplex, and cytomegalovirus.

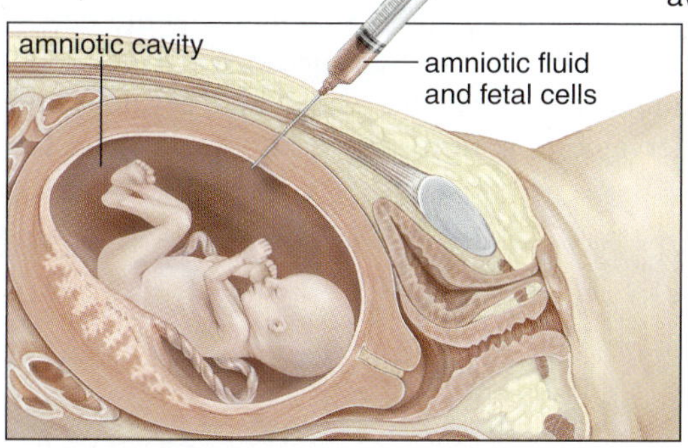

a. Amniocentesis

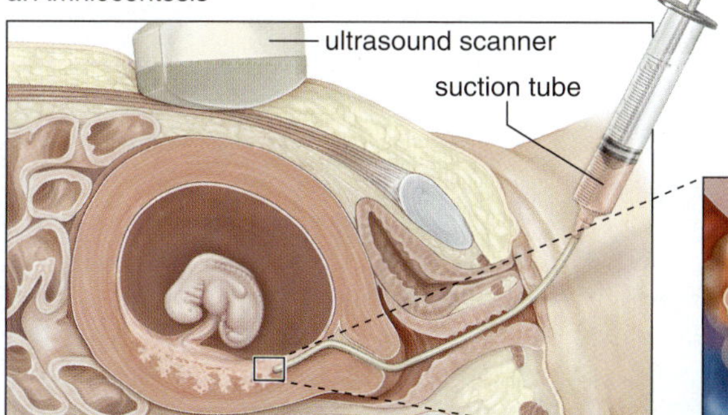

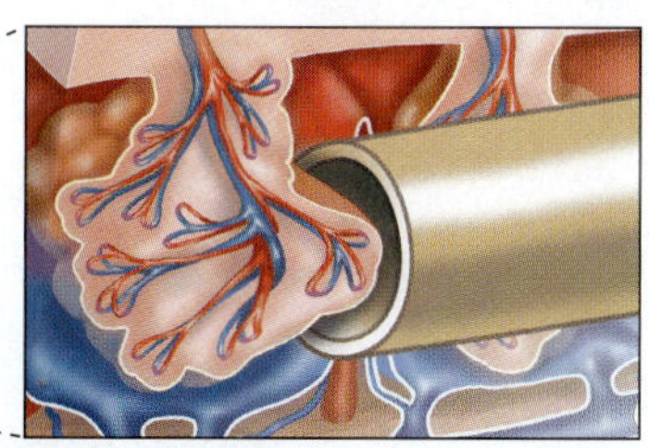

b. Chorionic villi sampling

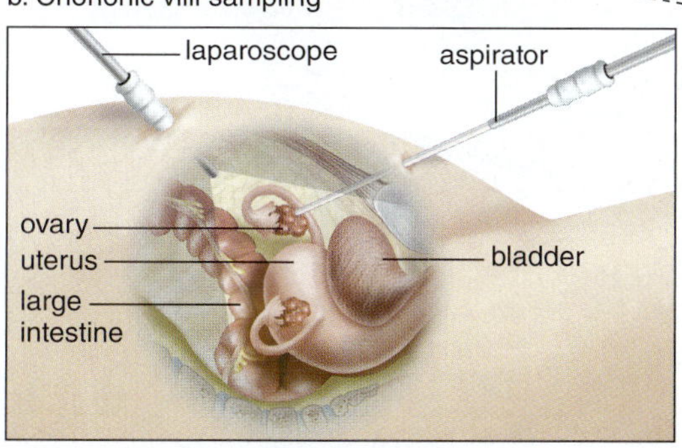

c. Preimplantation genetic diagnosis

Figure 42A Three methods for genetic defect testing before birth. About 20% of all birth defects are due to genetic or chromosomal abnormalities, which may be detected before birth. **a.** Amniocentesis is usually performed from the fifteenth to the seventeenth weeks of pregnancy. **b.** Chorionic villi sampling is usually performed from the eighth to the twelfth weeks of pregnancy. **c.** Preimplantation genetic diagnosis can be performed prior to in vitro fertilization, either on oocytes that have been collected from the woman or on the early embryo.

Questions to Consider

1. Why is it likely that the actual rate of birth defects is higher than what is reported?
2. Besides tobacco, alcohol, and illegal drugs, what other potential risks should a pregnant woman avoid? Why?

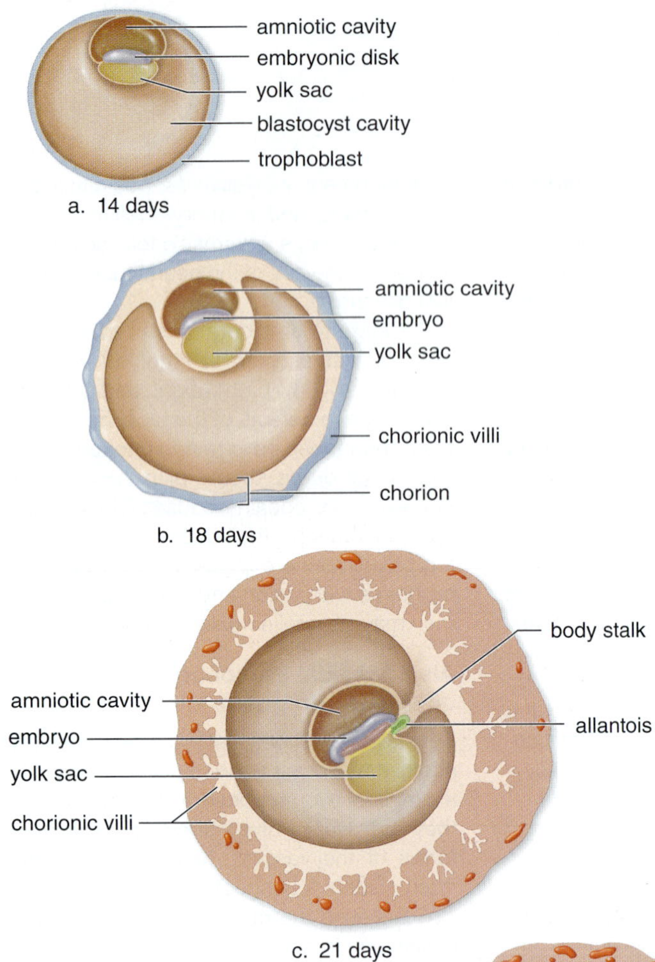

disintegrates. (Recall that the corpus luteum is a yellow body formed in the ovary from a follicle that has discharged its secondary oocyte.) Because of this, the endometrium is maintained, and menstruation does not occur.

As the week progresses, the inner cell mass detaches itself from the trophoblast, and two more extraembryonic membranes form (Fig. 42.13*a*). The yolk sac, which forms below the embryonic disk, has no nutritive function as it does in chickens, but it is the first site of blood cell formation. However, the amnion and its cavity are where the embryo (and then the fetus) develops. In humans, amniotic fluid acts as an insulator against cold and heat and absorbs shock, such as that caused by the mother exercising.

Gastrulation occurs during the second week. The inner cell mass now has flattened into the **embryonic disk,** composed of two layers of cells: ectoderm above and endoderm below. Once the embryonic disk elongates to form the primitive streak, the third germ layer, mesoderm, forms by invagination of cells along the streak. The trophoblast is reinforced by mesoderm and becomes the chorion (Fig. 42.13*b*). It is possible to relate the development of future organs to these germ layers (see Table 42.1).

The Third Week

Two important organ systems make their appearance during the third week. The nervous system is the first organ system to be visually evident. At first, a thickening appears along the entire dorsal length of the embryo; then the neural folds appear. When the neural folds meet at the midline, the neural tube, which later develops into the brain and the nerve cord, is formed (see Fig. 42.4). After the

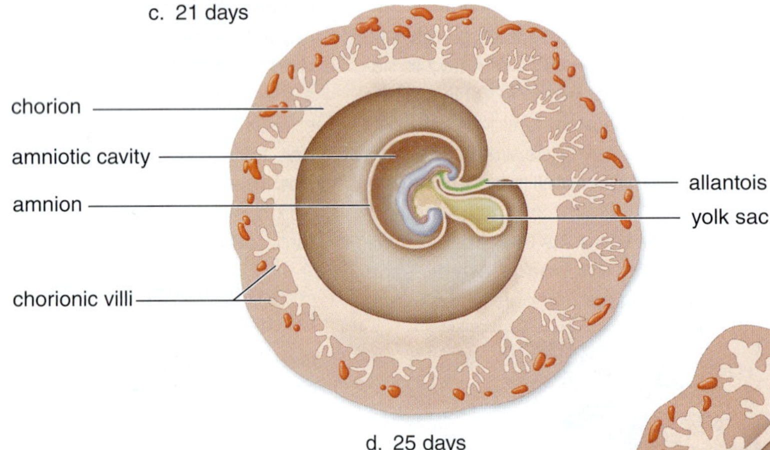

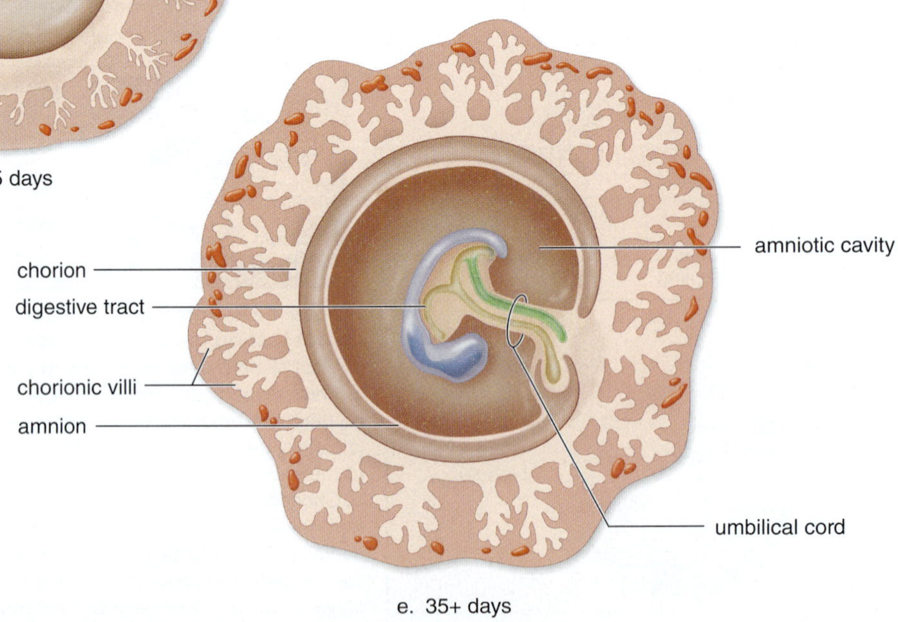

Figure 42.13 Human embryonic development. **a.** At first, the embryo contains no organs, only tissues. The amniotic cavity is above the embryo, and the yolk sac is below. **b.** The chorion develops villi, the structures so important to the exchange between mother and child. **c., d.** The allantois and yolk sac, two more extraembryonic membranes, are positioned inside the body stalk as it becomes the umbilical cord. **e.** At 35+ days, the embryo has a head region and a tail region. The umbilical cord takes blood vessels between the embryo and the chorion (placenta).

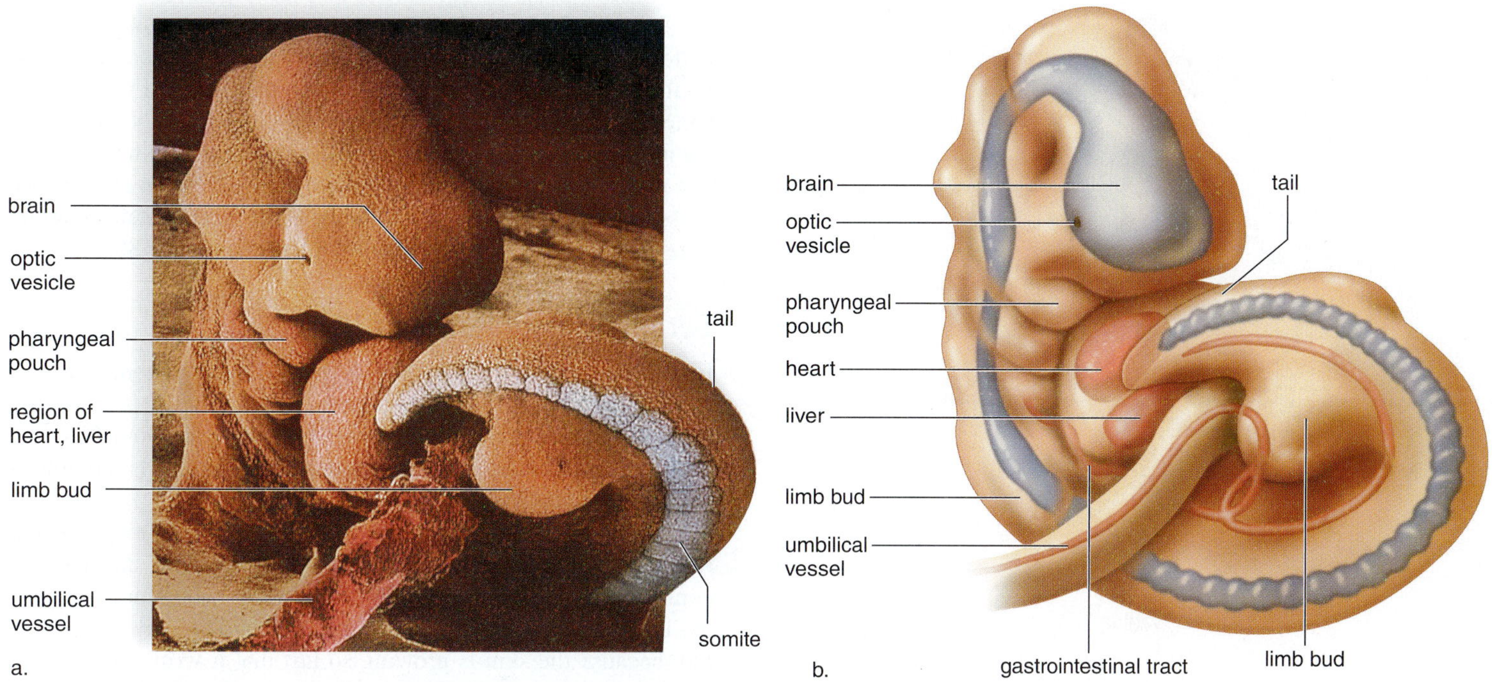

Figure 42.14 Human embryo at beginning of fifth week. **a.** Scanning electron micrograph. **b.** The embryo is curled, so that the head touches the heart, the two organs whose development is farther along than the rest of the body. The organs of the gastrointestinal tract are forming, and the arms and the legs develop from the bulges called limb buds. The tail is an evolutionary remnant; its bones regress and become those of the coccyx (tailbone). The pharyngeal arches become functioning gills only in fishes and amphibian larvae; in humans, the first pair of pharyngeal pouches becomes the auditory tubes. The second pair becomes the tonsils, while the third and fourth become the thymus and the parathyroid glands.

notochord is replaced by the vertebral column, the nerve cord is called the spinal cord.

Development of the heart begins in the third week and continues into the fourth week. At first, there are right and left heart tubes; when these fuse, the heart begins pumping blood, even though the chambers of the heart are not fully formed. The veins enter posteriorly, and the arteries exit anteriorly from this largely tubular heart, but later the heart twists, so that all major blood vessels are located anteriorly.

The Fourth and Fifth Weeks

At 4 weeks, the embryo is barely larger than the height of this print. A bridge of mesoderm called the body stalk connects the caudal (tail) end of the embryo with the chorion, which has treelike projections called **chorionic villi** (Gk. *chorion,* "membrane"; L. *villus,* "shaggy hair") that increase surface area for contact with the maternal portion of the placenta (Fig. 42.13*c, d, e*). The fourth extraembryonic membrane, the allantois, is contained within this stalk, and its blood vessels become the umbilical blood vessels. These structures form the **umbilical cord** (L. *umbilicus,* "navel"), which connects the developing embryo to the placenta (Fig. 42.13*e*).

Little protrusions called *limb buds* appear (Fig. 42.14); later, the arms and legs develop from the limb buds, and even the hands and feet become apparent. At the same time—during the fifth week—the head enlarges, and the developing eyes, ears, and nose are discernable.

The Sixth Through Eighth Weeks

During the sixth through eighth weeks of development, the embryo becomes easily recognizable as human. Concurrent with brain development, the head achieves its normal relationship with the body as a neck region develops. The nervous system is developed well enough to permit reflex actions, such as a startle response to touch. At the end of this period, the embryo is about 38 mm (1.5 in) long and weighs less than 1 g, even though all organ systems are established.

The Structure and Function of the Placenta

The **placenta** is a mammalian structure that functions in gas, nutrient, and waste exchange between the embryonic (later fetal) and maternal cardiovascular systems. The placenta begins formation once the embryo is fully implanted. At first, the entire chorion has chorionic villi that project into endometrium. Later, these disappear in all areas except where the placenta develops. By the tenth week, the placenta (Fig. 42.15) is fully formed and is producing progesterone and estrogen. These hormones have two effects:

- They prevent any new follicles from maturing because of negative feedback control of the hypothalamus and anterior pituitary.
- They maintain the lining of the uterus, so now the corpus luteum is not needed. No menstruation occurs during pregnancy.

The placenta has a fetal side contributed by the chorion and a maternal side consisting of uterine tissues. Notice in Figure 42.15 how the chorionic villi are surrounded by maternal blood, yet maternal and fetal blood do not mix under normal conditions, because exchange always takes place across plasma membranes.

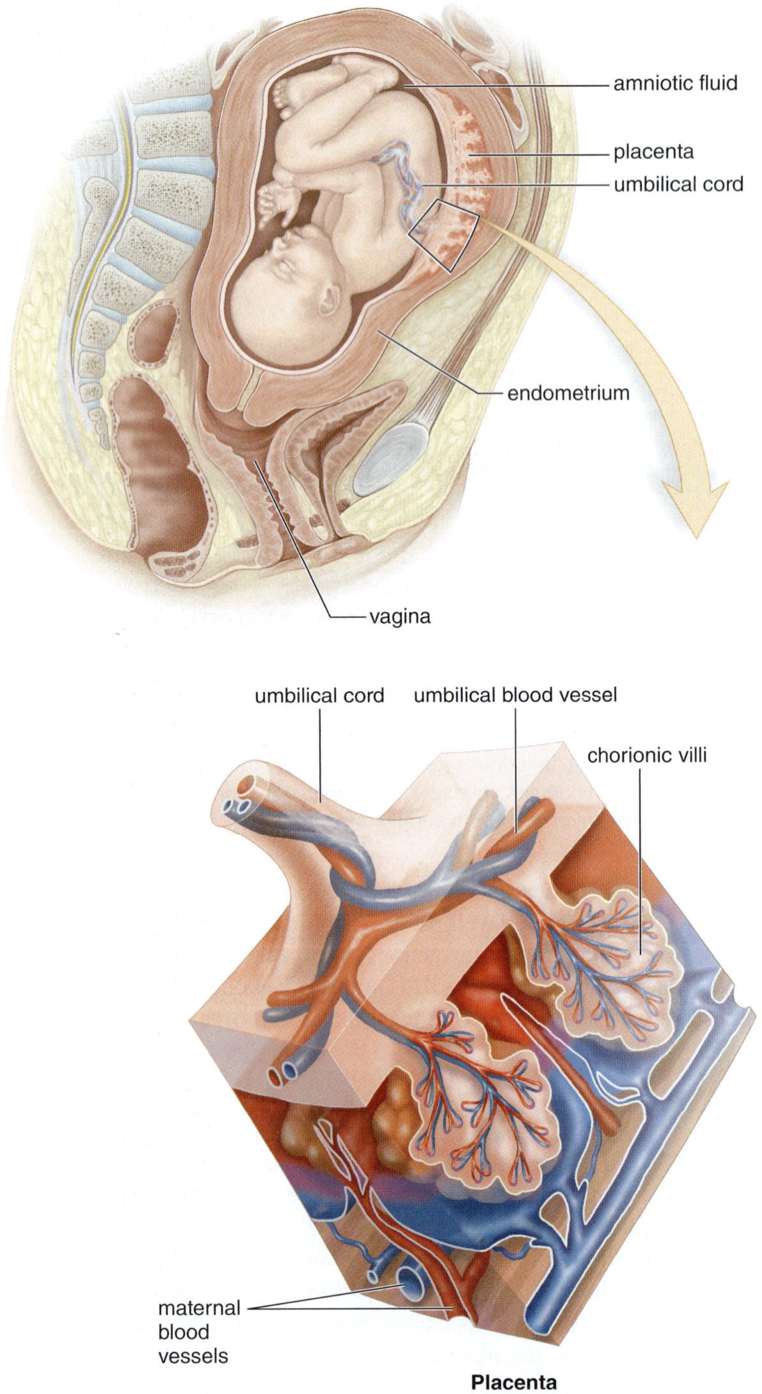

Figure 42.15 Anatomy of the placenta in a fetus at 6–7 months. The placenta is composed of both fetal and maternal tissues. Chorionic villi penetrate the uterine lining and are surrounded by maternal blood. Exchange of molecules between fetal and maternal blood takes place across the walls of the chorionic villi.

The umbilical cord is the lifeline of the fetus, because it contains the umbilical arteries and vein, which transport waste molecules (carbon dioxide and urea) to the placenta for disposal into the maternal blood and take oxygen and nutrient molecules from the placenta to the rest of the fetal circulatory system. If the placenta prematurely tears from the uterine wall, the lives of the fetus and the mother are endangered.

Late in fetal development, and peaking just before birth, maternal antibodies are transported across the placenta, into the fetal circulation. This helps ensure that the newborn will be protected against common pathogens and other microbes to which the mother has been exposed, until the newborn's immune system can mature and respond on its own. Harmful chemicals can also cross the placenta, as discussed in the Biological Systems feature, "Preventing and Testing for Birth Defects," on page 805.

Fetal Development and Birth

Fetal development (months 3–9) is marked by an extreme increase in size. Weight multiplies 600 times, going from less than 28 g to 3 kg. During this time, too, the fetus grows to about 50 cm in length. The genitalia appear in the third month, so it is possible to tell if the fetus is male or female.

Soon, hair, eyebrows, and eyelashes add finishing touches to the face and head. In the same way, fingernails and toenails complete the hands and feet. A fine, downy hair (lanugo) covers the limbs and trunk, only to disappear later. The fetus looks very old, because the skin is growing so fast that it wrinkles. A waxy, almost cheeselike substance (vernix caseosa) (L. *vernix,* "varnish"; *caseus,* "cheese") protects the wrinkly skin from the watery amniotic fluid.

From about the fourth month on, the mother can feel movements of the fetal limbs. At about the same time, the fetal heartbeat can be heard through a stethoscope. Conversely, the fetus is able to hear and respond to sounds by about 18 weeks. A fetus born at 24 weeks has a chance of surviving, although the lungs are still immature and often cannot capture oxygen adequately. As a fetus rapidly grows during the third trimester, its chances of surviving being born a month or two prematurely increase dramatically.

The Stages of Birth

When the fetal brain is sufficiently mature, the fetal hypothalamus stimulates the pituitary to release ACTH, causing the adrenal cortex to release androgens into the bloodstream. These diffuse into the placenta, which uses androgens as a precursor for estrogens, hormones that stimulate the production of prostaglandins and oxytocin. All three of these molecules cause the uterus to contract and expel the fetus.

The process of birth (parturition) includes three stages. During the first stage, the cervix dilates to allow passage of the baby's head and body. The amnion usually bursts about this time, an event termed the mother's water breaking. During the second stage, the baby is born and the umbilical cord is cut. During the third stage, the placenta is delivered (Fig. 42.16).

Check Your Progress 42.3

1. Identify the location where fertilization usually occurs in the human female.
2. Name the extraembryonic membrane that gives rise to each of the following: umbilical blood vessels, the first blood cells, and the fetal half of the placenta.
3. Describe the major changes that occur in the human embryo between the third and fifth weeks of development.

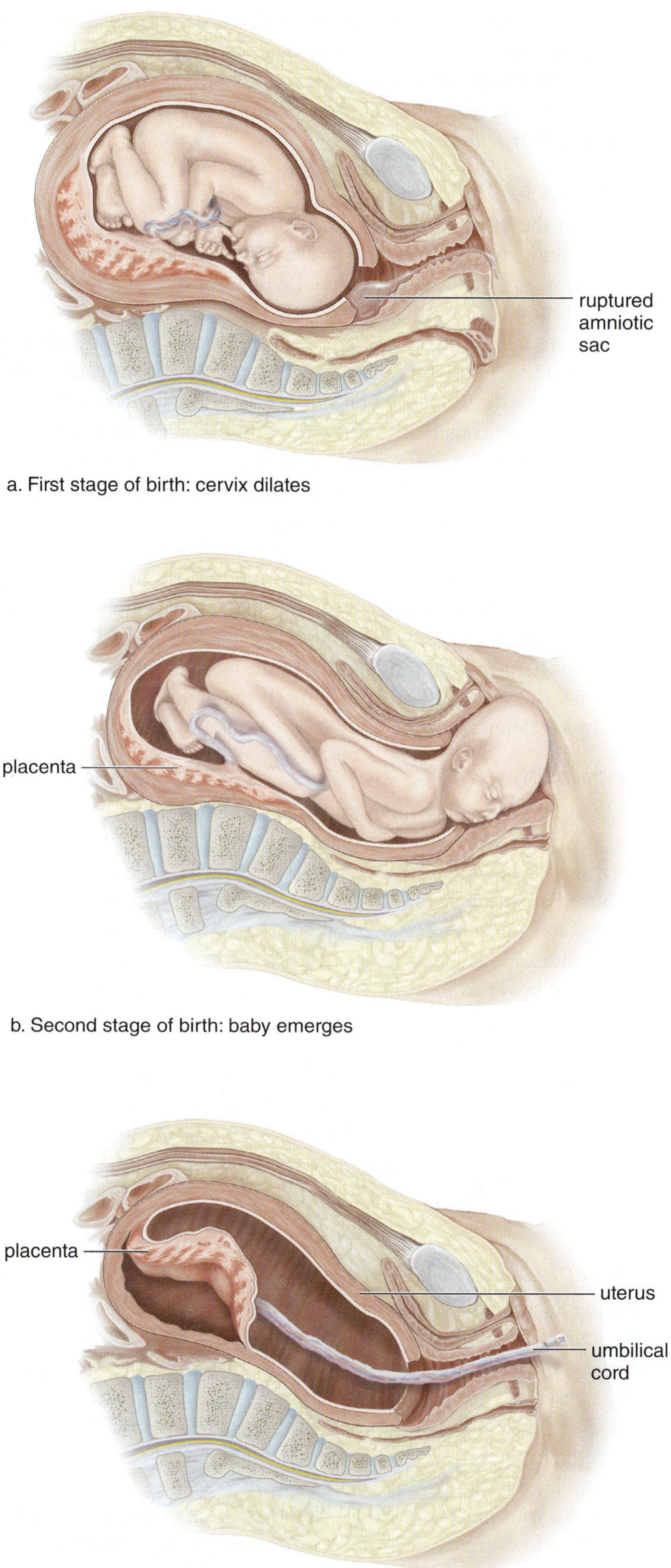

a. First stage of birth: cervix dilates

b. Second stage of birth: baby emerges

c. Third stage of birth: expelling afterbirth

Figure 42.16 Three stages of parturition. **a.** Position of fetus just before birth begins. **b.** Dilation of cervix and birth of baby. **c.** Expulsion of afterbirth.

42.4 The Aging Process

Learning Outcomes

Upon completion of this section, you should be able to

1. Discuss the effects of aging on various body systems.
2. Compare and contrast the preprogrammed theories with the damage accumulation theories of aging.

Aging and death are as much a part of biology as are development and birth. If animals—and indeed, all forms of life—did not age and eventually die, their exponentially increasing numbers would presumably overwhelm the ability of the Earth to support life. Moreover, aging and death are a critical part of the evolutionary process, because animals that age and die are continuously replaced by individuals with new genetic combinations that can adapt to environmental changes that occur.

The Effects of Aging on Organ Systems

Today's human population is older on average than ever before. In the next half-century, the number of people over age 65 will increase by 147%. Before discussing some possible mechanisms behind the aging process, it seems appropriate to first consider the effects of aging on the various body systems.

Integumentary System

As aging occurs, the skin becomes thinner and less elastic, because the number of elastic fibers decreases and the collagen fibers become increasingly cross-linked to each other, reducing their flexibility. There is also less adipose tissue in the subcutaneous layer; therefore, older people are more likely to feel cold. Together, these changes typically result in sagging and wrinkling of the skin.

As people age, the sweat glands also become less active, resulting in decreased tolerance to high temperatures. There are fewer hair follicles, so the hair on the scalp and the limbs thins out. Older people also experience a decrease in the number of melanocytes, making their hair gray and their skin pale. In contrast, some of the remaining pigment cells are larger, and pigmented blotches ("age spots") appear on the skin.

Cardiovascular System

Common problems with cardiovascular function are usually related to diseases, especially atherosclerosis (see Chapter 32). However, even with normal aging, the heart muscle does weaken somewhat, and it may increase slightly in size as it compensates for its decreasing strength. The maximum heart rate decreases even among the most fit older athlete, and it takes longer for the heart rate and blood pressure to return to normal resting levels following stress. Some part of this decrease in heart function may also be due to blood leaking back through heart valves that have become less flexible.

Aging also affects the blood vessels. The middle layer of arteries contains elastic fibers, which, like collagen fibers in the skin, become more cross-linked and rigid with time. These changes, plus a frequent decrease in the internal diameter of arteries due to atherosclerosis, contribute to a gradual increase in blood pressure with age. Indeed, nearly 50% of older adults have chronic hypertension.

Such changes are common in individuals living in Western industrialized countries but not in agricultural societies. This indicates that a diet low in cholesterol and saturated fatty acids, along with a sensible exercise program, may help prevent age-related cardiovascular disease.

Immune System

As people age, many of their immune system functions become compromised. Because a healthy immune system normally protects the entire body from infections, toxins, and at least some types of cancer, some investigators believe that losses in immune function can play a major role in the aging process.

As noted in Chapter 33, the thymus is an important site for T-cell maturation. Beginning in adolescence, the thymus begins to *involute,* gradually decreasing in size and eventually becoming replaced by fat and connective tissue. The thymus of a 60-year-old adult is about 5% of the size of the thymus of a newborn, resulting in a decrease in the ability of older people to generate T-cell responses to new antigens.

The evolutionary rationale for this may be that the thymus is energetically expensive for an organism to maintain, and that compared to younger animals that must respond to a high number of new infections and other antigens, older animals have already responded to most of the antigens to which they will be exposed in their life.

Aging also affects other immune functions. Because most B-cell responses are dependent on T cells, antibody responses also begin to decline. This, in turn, may explain why the elderly do not respond as well to vaccinations as young people do. This presents challenges in protecting older people against diseases such as influenza and pneumonia, which can otherwise be prevented by annual vaccination.

Not all immune functions decrease in older animals, though. The activity of natural killer cells, which are a part of the innate immune system, seems to change very little with age. Perhaps by investigating how these cells remain active throughout a normal human lifespan, researchers can learn to preserve other aspects of immunity in the elderly.

Digestive System

The digestive system is perhaps less affected by the aging process than other systems. Because the secretion of saliva decreases, more bacteria tend to adhere to the teeth, causing more decay and periodontal disease. Blood flow to the liver is reduced, resulting in less efficient metabolism of drugs or toxins. This means that, as a person gets older, less medication is needed to maintain the same level in the bloodstream.

Respiratory System

Cardiovascular problems are often accompanied by respiratory disorders, and vice versa. Decreasing elasticity of lung tissues means that ventilation is reduced. Because we rarely use the entire vital capacity, these effects may not be noticed unless the demand for oxygen increases, such as during exercise.

Excretory System

Blood supply to the kidneys is also reduced. The kidneys become smaller and less efficient at filtering wastes. Salt and water balance are difficult to maintain, and the elderly dehydrate faster than young people. *Urinary incontinence* (lack of bladder control) increases with age, especially in women. In men, an enlarged prostate gland may reduce the diameter of the urethra, causing frequent or difficult urination.

Nervous System

Between the ages of 20 and 90, the brain loses about 20% of its weight and volume. Neurons are extremely sensitive to oxygen deficiency, and neuron death may be due not to aging itself but to reduced blood flow in narrowed blood vessels. However, contrary to previous scientific opinion, recent studies using advanced imaging techniques show that most age-related loss in brain function is not due to whatever loss of neurons is occurring. Instead, decreased function may occur due to alterations in complex chemical reactions, or increased inflammation in the brain. For example, an age-associated decline in levels of dopamine can affect the brain regions involved in complex thinking.

Perhaps more important than the molecular details, recent studies have confirmed that lifestyle factors can affect the aging brain. For example, animals on restricted-calorie diets developed fewer Alzheimer-like changes in their brains. Other positive factors that may help maintain a healthy brain include attending college (the "use it or lose it" idea), regularly engaging in exercise, and getting sufficient sleep.

Sensory Systems

In general, with aging more stimulation is needed for taste, smell, and hearing receptors to function as before. A majority of people over age 80 have a significant decline in their sense of smell, and about 15% suffer from *anosmia,* or a total inability to smell. The latter condition can be a serious health hazard, due to the inability to detect smoke, gas leaks, or spoiled food. After age 50, most people gradually begin to lose the ability to hear tones at higher frequencies, and this can make it difficult to identify individual voices and to understand conversation in a group.

Starting at about age 40, the lens of the eye does not accommodate as well, resulting in *presbyopia,* or difficulty focusing on near objects, which causes many people to require reading glasses as they reach middle age. Finally, as noted in Chapter 38, cataracts and other eye disorders become much more common in the aged.

Musculoskeletal System

For the average person, muscle mass peaks between ages 16 and 19 for females and 18 and 24 for males. Beginning in the twenties or thirties, but accelerating with increasing age, muscle mass generally decreases, due to decreases in both the size and the number of muscle fibers. Most people who reach age 90 have 50% less muscle mass than when they were 20. Although some of this loss may be inevitable, regular exercise can slow this decline.

Like muscles, bones tend to shrink in size and density with age. Due to compression of the vertebrae, along with changes in posture, most of us lose height as we age. Those who reach age 80 will be about 2 inches shorter than they were in their twenties. Women lose bone mass more rapidly than men do, especially after menopause. As noted in Chapter 39, osteoporosis is a common disease in the elderly.

Although some decline in bone mass is a normal result of aging, certain extrinsic factors are also important. A proper diet and a moderate exercise program have been found to slow the progressive loss of bone mass.

Endocrine System

As with the immune system, aging of the hormonal system can affect many different organs of the body. These changes are complex, however, with some hormone levels tending to decrease with age and others increasing. The activity of the thyroid gland generally declines, resulting in a lower basal metabolic rate. The production of insulin by the pancreas may remain stable, but cells become less sensitive to its effects, resulting in a rise in fasting glucose levels of about 10 mg/dl each decade after age 50.

Human growth hormone (HGH) levels also decline with age, but it is very unlikely that taking HGH injections will "cure" aging, despite Internet claims. In fact, one study found that people with lower levels of HGH actually lived longer than those with higher levels.

Reproductive System

Testosterone levels are highest in men in their twenties. After age 30, testosterone levels decrease by about 1% per year. Extremely low testosterone levels have been linked to a decreased sex drive, excessive weight gain, a loss of muscle mass, osteoporosis, general fatigue, and depression. However, the levels below which testosterone treatment should be initiated remain controversial. Testosterone replacement therapy, whether through injection, patches, or gels, is associated with side effects such as enlargement of the prostate, acne or other skin reactions, and the production of too many red blood cells.

Menopause, the period in a woman's life during which the ovarian and uterine cycles cease, usually occurs between ages 45 and 55. The ovaries become unresponsive to the gonadotropic hormones produced by the anterior pituitary, and they no longer secrete estrogen or progesterone. At the onset of menopause, the uterine cycle becomes irregular, but as long as menstruation occurs, it is still possible for a woman to conceive. Therefore, a woman is usually not considered to have completed menopause (and thus be infertile) until menstruation has been absent for a year.

The hormonal changes during menopause often produce physical symptoms such as "hot flashes" (caused by circulatory irregularities), dizziness, headaches, insomnia, sleepiness, and depression. To ease these symptoms, female hormone replacement therapy (HRT) was routinely prescribed until 2002, when a large clinical study showed that in the long term, HRT caused more health problems than it prevented. For this reason, most doctors no longer recommend long-term HRT for the prevention of postmenopausal conditions.

It is also of interest that, as a group, females live longer than males. It is likely that estrogen offers women some protection against cardiovascular disorders when they are younger. Males suffer a marked increase in heart disease in their forties, but an increase is not noted in females until after menopause, when women lead men in the incidence of stroke. Men remain more likely than women to have a heart attack at any age, however.

Hypotheses About Why We Age

Aging is a complex process, and multiple factors can affect the aging process. The many hypotheses about why aging occurs can be grouped into two major categories: (1) theories suggesting that aging is mainly due to preprogrammed genetic events and (2) theories that aging is mainly due to the accumulation of cellular damage.

Preprogrammed Theories

Most scientists who study *gerontology,* or the science of aging, believe that aging is partly genetically preprogrammed. This idea is supported by the observation that longevity runs in families—that is, the children of long-lived parents tend to live longer than those of short-lived parents. As would also be expected, studies show that identical twins have a more similar lifespan than nonidentical twins.

Many laboratory studies of aging have been performed in the nematode *C. elegans,* in which single-gene mutations have been shown to influence the lifespan. For example, mutations that decrease the activity of a hormone receptor similar to the insulin receptor more than double the lifespan of the worms, which also behave and look like much younger worms throughout their prolonged lives. Interestingly, small-breed dogs, such as poodles or terriers, which may live 15–20 years, have lower levels of an analogous receptor, compared to large dog breeds that live 6–8 years.

Studies of the behavior of cells grown in the lab also suggest a genetic influence on aging. Most types of differentiated cells can divide only a limited number of times. One factor that may control the number of cell divisions is the length of the *telomeres,* sequences of DNA at the ends of chromosomes. As you learned in section 12.2, telomeres protect the ends of chromosomes from deteriorating or fusing with other chromosomes. Each time a cell divides, the telomeres normally shorten, and cells with shorter telomeres tend to undergo fewer divisions. Perhaps as we grow older, more and more cells are unable to divide, and instead, they undergo degenerative changes and die.

However, if aging were mainly a function of genes, we would expect much less variation in lifespan to be observed among individuals of a given species than is actually seen. For this reason, experts have estimated that in most cases genes account only for about 25% of what determines the length of life.

Damage Accumulation Theories

A second group of hypotheses postulate that aging involves the accumulation of damage over time. In 1900, the average human life expectancy was around 45 years. A baby born in the United States in 2009 is expected to live an average of about 78 years. Since human genes have presumably not changed much in such a short time, most of this gain in lifespan is due to better medical care, along with the application of scientific knowledge about how to prolong our lives.

Two basic types of cellular damage can accumulate over time. The first type can be thought of as agents that are unavoidable—for example, the accumulation of harmful DNA mutations or the buildup of harmful metabolites. For some time, medical researchers have known that proteins—such as the collagen fibers present in many support tissues—become increasingly cross-linked as people age.

This cross-linking may account for the inability of such organs as the blood vessels, the heart, and the lungs to function as they once did. Some researchers have now found that glucose has the tendency to attach to any type of protein, which is the first step in a cross-linking process. They are currently experimenting with drugs that can prevent cross-linking.

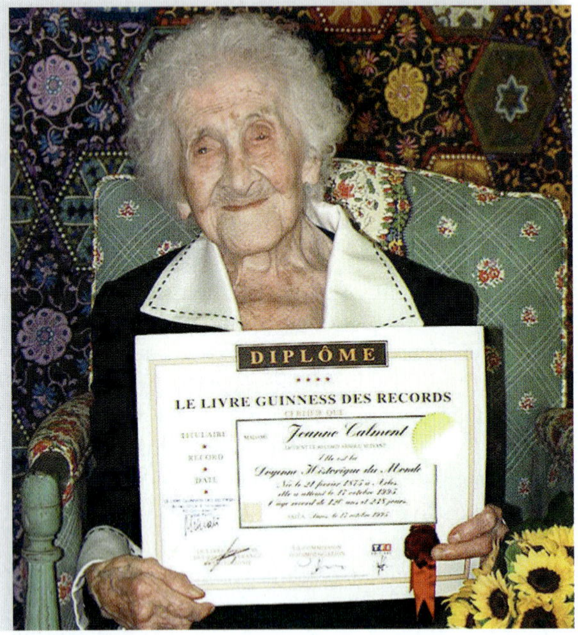

Figure 42.17 Supercentenarians. The oldest verified human life span was that of Jeanne Calment (1875–1997), who lived 122 years and was documented by the Guinness Book of Records as the "World's Oldest Living Person."

Other sources of cellular damage, however, may be avoidable, such as a poor diet or exposure to the sun. Recent work suggests that when an animal produces fewer free radicals, it lives longer. Free radicals are unstable molecules that have an unpaired electron. In order to stabilize themselves, free radicals react with another molecule, such as DNA or proteins (e.g., enzymes) or lipids found in plasma membranes. Eventually, these molecules are unable to function, and the cell is destroyed. By increasing one's consumption of natural antioxidants, such as those present in brightly colored and dark green vegetables, citrus fruits, nuts, fish, shellfish, and red wine, we can reduce our exposure to free radicals and slow the aging process.

The longest-lived human whose age has been verified by modern methods was Jeanne Calment (Fig. 42.17), who lived in France her entire life and died at the age of 122 years, 164 days (she had lived on her own until age 110, when she entered a nursing home). This is currently considered to be the longest potential human lifespan. However, our understanding of the aging process could be considered, ironically, to be in its infancy, and animal models and other studies suggest that it may be possible to extend human life far beyond that.

Check Your Progress 42.4

1. Describe how involution of the thymus can lead to decreased responses to vaccines in the elderly.
2. Define *menopause,* and explain why it occurs.
3. Distinguish between the two hypotheses regarding aging, and give examples of each.

CONNECTING *the* CONCEPTS *with the* THEMES

Evolution

- Many animals go through the same early embryonic stages, such as morula, blastula, and gastrula.
- Many of the genes involved in embryonic development—gap genes, pair-rule genes, and homeotic genes—are highly conserved throughout animal species, suggesting they arose very early in evolutionary history.
- Aging and the eventual death of individuals may provide an evolutionary advantage; an infinite population of immortal animals would quickly use up all resources, leading to extinction.

Nature of Science

- The study of early developmental forms of diverse animals, such as the lancelet, frog, and chick, has greatly advanced our understanding of development.
- Modern genetic techniques have allowed the identification of key genes and molecules involved in developmental processes.
- Scientists are just beginning to understand the complex array of processes that result in the phenomenon of aging.

Biological Systems

- Animal development begins with fertilization. Early development can be divided into cellular, tissue, and organ stages.
- All cells of the body possess the same DNA sequences. Cellular differentiation involves the processes of cytoplasmic segregation and induction, by which transcription and translation of these sequences are controlled.
- The aging process likely involves a combination of preprogrammed genetic events plus the accumulated damage from unavoidable events, such as DNA mutations and exposure to environmental agents.

MEDIA STUDY TOOLS http://connect.mheducation.com

SMARTBOOK® Study smart. SmartBook helps you maximize your study time by identifying what you know and don't know. The recharge feature in SmartBook helps you prepare for the test by identifying the material that you are most likely to forget.

connect BIOLOGY Find even more resources to learn the chapter concepts including animations, tutorial videos, and interactive practice questions.

Tutorial	Videos
42.1 Embryonic Stages of Development	**42.1** Human Sperm • Blastocyst Formation **42.3** Blastocyst Formation

SUMMARIZE

42.1 Early Developmental Stages

Development begins at **fertilization.** Only one sperm actually enters the oocyte. Both sperm and egg contribute chromosomes to the diploid zygote. **Development** includes all changes that occur during the life cycle of an organism.

The early developmental stages in animals proceed from cellular stages to tissue stages to organ stages. During the cellular stage, **cleavage** (cell division) occurs, but there is no overall growth. The result is a **morula,** which becomes the **blastula** when an internal cavity (the **blastocoel**) appears. During these early stages the organism is also called an **embryo.**

During the tissue stage, a **gastrula** forms from the blastula. **Gastrulation** (invagination of cells into the blastocoel) results in formation of the **germ layers: ectoderm, mesoderm,** and **endoderm.** Inward folding of the early gastrula creates the **blastopore.** Both the cellular and tissue stages can be affected by the amount of **yolk.**

Organ formation can be related to germ layers. For example, the nervous system develops from midline ectoderm, just above the **notochord.** During this process, a layer of cells called the **neural plate** contributes to formation of a **neural tube,** which gives rise to the brain and spinal cord.

42.2 Developmental Processes

The single-celled zygote is **totipotent,** meaning it can produce all types of speciallized cells in an organism. **Cellular differentiation** begins with **cytoplasmic segregation** in the egg. After the first cleavage of a frog embryo, only a daughter cell that receives certain **maternal determinants** (i.e., the gray crescent) is able to develop into a complete embryo.

Induction is also part of cellular differentiation. For example, the notochord induces the formation of the neural tube in frog embryos. In *C. elegans,* investigators have shown that induction is an ongoing process in which one type of tissue sequentially regulates the development of other types, through chemical signals coded for by particular genes.

Morphogenesis involves cell movement, which defines body shape and form. Some morphogen genes determine the axes of the body, and others regulate the development of segments. An important concept has emerged: During development, sequential sets of master genes code for morphogen gradients that activate the next set of master genes in turn. **Pattern formation,** the arrangement of tissues and organs, often involves **apoptosis** (programmed cell death).

Homeotic genes control pattern formation, such as the presence of antennae, wings, and limbs on the segments of *Drosophila.* Homeotic genes all contain a **homeobox,** a shared nucleotide sequence that codes for a sequence of 60 amino acids called a **homeodomain.** These proteins are also transcription factors, and the homeodomain is the portion of the protein that binds to DNA. Homologous homeotic genes have been found in a wide variety of organisms, and therefore they must have arisen early in the history of life and been conserved.

42.3 Human Embryonic and Fetal Development

Human development can be divided into **embryonic development** (months 1 and 2) and fetal development (months 3 through 9). Fertilization occurs in a uterine tube, and cleavage occurs as the embryo moves toward the uterus. The morula becomes the blastocyst before implanting in the endometrium of the uterus. During the second week, the **embryonic disk** forms from ectoderm and endoderm, and becomes the primitive streak that gives rise to the mesoderm layer of cells.

The **extraembryonic membranes** appear early in human development. The **trophoblast** of the **blastocyst** is the first sign of the **chorion,** which goes on to become the fetal part of the **placenta.** After **implantation,** the trophoblast secretes **human chorionic gonadotropin** (**HCG**), which maintains the corpus luteum and is also the basis of pregnancy testing.

Exchange occurs between fetal and maternal blood at the placenta. The chorion has **chorionic villi** that increase contact with maternal blood vessels in the placenta. The **amnion** contains amniotic fluid, which cushions and protects the embryo. The **yolk sac** and **allantois** form the **umbilical cord,** which is a lifeline between the mother and the developing embryo and fetus.

Organ development begins with neural tube and heart formation. There follows a steady progression of organ formation during embryonic development. During fetal development, refinement of organ systems occurs, and the fetus adds weight.

42.4 The Aging Process

The aging process is an essential part of the biology and evolution of animals. Aging affects every organ system of the human body, in mostly predictable ways. One example is **menopause,** during which a woman's ovarian and uterine cycles cease. Various hypotheses have been advanced to explain the aging process, which can be grouped into those suggesting that aging is genetically preprogrammed and those emphasizing the accumulation of various types of cellular damage over time.

ASSESS

Choose the best answer for each question.

42.1 Early Developmental Stages

1. Only one sperm enters and fertilizes a human ovum because
   a. sperm have an acrosome.
   b. the corona radiata gets larger.
   c. the zona pellucida lifts up.
   d. a fertilization membrane forms.
   e. All of these are correct.
2. Which of these stages is the first one out of sequence?
   a. cleavage
   b. blastula
   c. morula
   d. gastrula
   e. neurula
3. Which of these stages is mismatched?
   a. cleavage—cell division
   b. blastula—gut formation
   c. gastrula—three germ layers
   d. neurula—nervous system
   e. Both b and c are mismatched.

42.2 Developmental Processes

4. Developmental changes
   a. require growth, differentiation, and morphogenesis.
   b. stop occurring when one is grown.
   c. are dependent on a parceling out of genes into daughter cells.
   d. are dependent on activation of master genes in an orderly sequence.
   e. Both a and d are correct.
5. The ability of one embryonic tissue to influence the growth and development of another tissue is termed
   a. morphogenesis.
   b. pattern formation.
   c. apoptosis.
   d. cellular differentiation.
   e. induction.

For questions 6–9, match the statement with the terms in the key.

Key:

a. apoptosis
b. homeotic genes
c. fate maps
d. morphogen gradients
e. segment-polarity genes

6. Have been developed that show the destiny of each cell that arises through cell division during the development of *C. elegans*
7. Occurs when a cell-death cascade is activated
8. Can have a range of effects, depending on its concentration in a particular portion of the embryo
9. Genes that control pattern formation, the organization of differentiated cells into specific three-dimensional structures

42.3 Human Embryonic and Fetal Development

10. Label this diagram illustrating the placement of the extraembryonic membranes, and give a function for each membrane in humans.

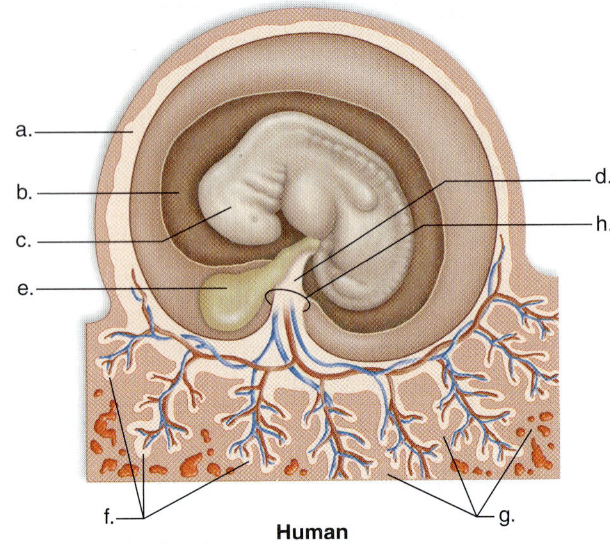

Human

11. Which developmental stage is mismatched?
    a. second week—implantation
    b. third week—heart development begins
    c. fourth and fifth week—limb buds appear
    d. sixth through eighth week—all organ systems present
    e. None of these stages are mismatched.
12. Near the time of parturition, which hormone stimulates the fetal adrenal cortex to release androgens?
    a. fetal ACTH
    b. fetal estrogen
    c. maternal ACTH
    d. maternal estrogen

42.4 The Aging Process

13. Which of the following is *not* an explanation for why the skin of older people tends to become wrinkled?
    a. increased subcutaneous fat deposition
    b. increased cross-linking of collagen fibers
    c. decreased sebaceous gland activity
    d. decreased number of elastic fibers
14. The onset of menopause is mainly due to
    a. a lack of ovarian follicles that are able to ovulate.
    b. an involution of the uterus that begins around age 45.
    c. an unresponsiveness of the ovaries to FSH and LH from the pituitary.
    d. an increased production of male hormones.
15. Possible causes of the cellular damage that can accumulate and contribute to aging include
    a. free radicals.
    b. spontaneous DNA mutations.
    c. cross-linked proteins.
    d. UV radiation.
    e. All of these are correct.

ENGAGE

THINKING SCIENTIFICALLY

1. Mitochondria contain their own genetic material. Mitochondrial mutations are inherited from mother to child, without any contributions from the father. Why?
2. *Drosophila* belongs to the insect order Diptera, in which one pair of wings is the norm. In *Drosophila, a* mutation in the *UBX* gene converts the halteres into an extra pair of wings located posterior to the normal pair (see Fig. 42.10*a*). Butterflies belong to the order Lepidoptera, in which two pairs of wings is normal. The ancestor of both flies and butterflies is thought to have possessed two pairs of wings. However, in butterflies, *UBX* mutations have been found to alter pigmentation, spot pattern, and scale morphology in the hindwings, but not the number of wing pairs. What does this tell us about the role of *UBX* in insect wing development?
3. A variety of fertility tests are available for women to purchase over the counter—for example, at drug stores. Especially as more women delay their first pregnancy until they are older, these tests are becoming more popular. Many of these tests incorporate antibodies that specifically react with hormones found in a woman's urine. What specific hormones could be measured in these tests, and how might these tests actually work?
4. The Biological Systems feature, "Preventing and Testing for Birth Defects," on page 805 suggests that women of childbearing age begin taking precautions to prevent birth defects (good nutrition, rubella vaccination, etc.). Why should these precautions begin before a woman becomes pregnant?

UNIT 8

Behavior and Ecology

Ecology is the science that describes the interaction between the living and nonliving components of the environment, the role that species play within the environment, and the impact that humans have upon the Earth.

Ecologists have two statements that summarize these interactions: "Everything is affected by everything else" and "There is no free lunch." For example, when tall smokestacks were constructed in the Midwest, the pollutants didn't go away; instead, they caused acid rain in the Northeast. As is so often the case, the only way to solve an ecological problem is to stop the behavior or activity that caused the problem. It must be dealt with head-on.

Ecology describes the biosphere (the portions of the sea, land, and air that contain living organisms), but today ecology is also an experimental science. Ecologists use the scientific process to determine how ecosystems function and how they will likely respond to human impact. One of the main goals of conservation biology is to help preserve species and manage ecosystems for sustainable human welfare.

UNIT OUTLINE

43 Behavioral Ecology 817

44 Population Ecology 836

45 Community and Ecosystem Ecology 855

46 Major Ecosystems of the Biosphere 881

47 Conservation of Biodiversity 905

Unit Learning Outcomes

The learning outcomes for this unit focus on three major themes in the life sciences.

Evolution	Discuss how natural selection and coevolution have played a significant role in shaping the diversity within ecosystems.
Nature of Science	Identify the aspects of biology that conservation biologists study in order to determine how best to manage natural resources.
Biological Systems	Explain the complex web of interactions among individuals, species, communities, and their environment.

Among honeybees, *Apis mellifera,* workers take care of the young.

43

Behavioral Ecology

Chapter Outline

43.1 Inheritance Influences Behavior 818

43.2 The Environment Influences Behavior 820

43.3 Animal Communication 824

43.4 Behaviors That Increase Fitness 828

Before You Begin

Before beginning this chapter, take a few moments to review the following discussions.

Section 13.3 How can the environment influence gene expression?

Section 16.2 What role does sexual selection and male competition play in the evolution of the species?

Section 40.2 What role does the endocrine system play in controlling behavior?

Honeybees (*Apis mellifera*) live in hives and are known to have a complex social structure. Roles within honeybee society include communication, food collection, nest building, production of offspring, and care of the young. These duties are conducted by the various social classes within the hive. Each hive contains one queen, which can live for several years. To establish her position, the queen secretes a pheromone that will attract drones, allowing her to mate. The pheromone will also inhibit the sexual development of the workers and promote tranquility within the hive. Offspring born from the queen's fertilized eggs will become female workers, which are unable to mate. These workers are responsible for collecting food, caring for the larvae, and guarding the hive.

Drones are male bees that are born from eggs that went unfertilized, resulting in individuals that have only one set of chromosomes. Drones lack a stinger, a pollen basket, or wax glands. Their main function is to mate with the queen, since this is the group that consists of the males. The result is a colony of bees that all share a degree of genetic relatedness. Members of the hive see to the propagation of their own genes when they help the queen reproduce. Behavioral ecology, as discussed in this chapter, is dedicated to the principle that natural selection shapes behavior, just as it does the anatomy and physiology of an animal.

As you read through this chapter, think about the following questions:

1. What are the benefits and drawbacks of living in a social group?
2. How do genetics and the environment work together to influence behavior?
3. Why are there so few mammal species that exhibit altruism?

Following *the* Themes

CHAPTER 43 BEHAVIORAL ECOLOGY

UNIT 8 BEHAVIOR AND ECOLOGY

Evolution	Natural selection has helped shape behavior by selecting for the individuals with the most well-adapted traits.
Nature of Science	Scientists have conducted experiments to show that both genetics and the environment help shape behavior.
Biological Systems	Animal communication requires a sender and a receiver that can interpret chemical, auditory, visual, and tactile signals.

43.1 Inheritance Influences Behavior

Learning Outcomes

Upon completion of this section, you should be able to

1. Explain the key aspects of studies that suggest behavior has a genetic basis.
2. Describe the body systems that play a role in influencing behavior.

Behavior encompasses any action that can be observed and described. For example, in Chapter 1, Figure 1.15 described the aggressive behavior of male mountain bluebirds toward other males during mating. An investigator was able to center in on this behavior, observe it, and record it objectively. In the same manner, scientists pose the question of whether genetics can determine the behavior an animal is capable of performing.

The "nature versus nurture" question asks to what extent our genes (nature) and the environment (nurture) influence behavior. We would expect that genes, which control the development of neural and hormonal mechanisms, also influence the behavior of an animal. The results of experiments done to discover the degree to which genetics controls behavior support the hypothesis that most behaviors have, at least in part, a genetic basis.

Experiments That Suggest Behavior Has a Genetic Basis

Among the many animal behavior studies, studies of lovebirds and garter snakes suggest that behavior has a genetic basis, and one type of study in humans attempts to evaluate nature versus nurture.

Behavior with a Genetic Basis: Nest Building in Lovebirds

Lovebirds are small, green-and-pink African parrots that nest in tree hollows. Several closely related species of lovebirds in the genus *Agapornis* build their nests differently. Fischer lovebirds cut large leaves (or in the laboratory, pieces of paper) into long strips with their bills. They use their bills to carry the strips (Fig. 43.1*a*) to the nest, where they weave them in with others to make a deep cup. In contrast, peach-faced lovebirds cut somewhat shorter strips and carry them to the nest in a very unusual manner. They pick up the strips in their bills and then insert them into their feathers (Fig. 43.1*b*). In this way, they can carry several of these short strips with each trip to the nest, whereas Fischer lovebirds can carry only one of the longer strips at a time.

Researchers hypothesized that if the behavior for obtaining and carrying nesting material is inherited, then hybrids might show intermediate behavior. When the two species of birds were mated, the hybrid birds were observed to have difficulty carrying nesting materials. They cut strips of intermediate length and then attempted to tuck the strips into their rump feathers. They did not push the strips far enough into the feathers, however, and when they walked or flew, the strips always came out. Hybrid birds eventually learned, after about 3 years, to carry the cut strips in

a. Fischer lovebird with nesting material in its beak.

b. Peach-faced lovebird with nesting material in its rump feathers.

Figure 43.1 Nest-building behavior in lovebirds. **a.** Fischer lovebirds, *Agapornis fischeri,* carry strips of nesting material in their bills, as do most other birds. **b.** Peach-faced lovebirds, *Agapornis roseicollis,* tuck strips of nesting material into their rump feathers before flying back to the nest.

Figure 43.2 Feeding behavior in garter snakes. The number of tongue flicks by inland and coastal garter snakes, *Thamnophis elegans,* is measured in terms of their response to slug extract on cotton swabs. Coastal snakes tongue-flicked more than inland snakes.

their beaks, but they still briefly turned their heads toward their rump before flying off. These studies support the hypothesis that behavior has a genetic basis.[1]

Bahavior with a Genetic Basis: Food Choice in Garter Snakes

A variety of experiments have been conducted to determine if food preference in garter snakes has a genetic basis. There are two types of garter snake populations in California. Inland populations are aquatic and commonly feed under water on frogs and fish. Coastal populations are terrestrial and feed mainly on slugs. In one study, inland adult snakes refused to eat slugs in the laboratory, whereas coastal snakes readily did so. The experimental results of matings between snakes from the two populations (inland and coastal) show that their newborns have a partial preference for slugs when feeding.

Differences between slug acceptors and slug rejecters appear to be inherited. Experiments were able to determine the physiological difference between the two populations. When snakes eat, their tongues carry chemicals to an odor receptor in the roof of the mouth. They use tongue flicks to recognize their prey. Even newborns will flick their tongues at cotton swabs dipped in fluids of their prey. Swabs were dipped in slug extract, and the number of tongue flicks were counted for newborn inland snakes and newborn coastal snakes. Coastal snakes had a higher number of tongue flicks than did inland snakes (Fig. 43.2).

[1]Dilger, W. "The Behavior of Lovebirds." *Scientific American* 206:89–98 (1962).

Apparently, inland snakes do not eat slugs because they cannot detect their smell as easily as coastal snakes can. A genetic difference between the two populations of snakes results in a physiological difference in their nervous systems. Although hybrids showed a great deal of variation in the number of tongue flicks, they tended to average between the number performed by coastal and inland snakes, as predicted by the genetic hypothesis.[2]

Twin Studies in Humans

On occasion, human twins have been separated at birth and raised under different environmental conditions. Studies of these twins have shown that they have similar food preferences and activity patterns, and they even select mates with similar characteristics. This type of study lends support to the hypothesis that at least certain behaviors are primarily influenced by nature (genes).

Animal Studies That Demonstrate Behavior Has a Genetic Basis

The nervous and endocrine systems are both responsible for the overall coordination of body systems. Studies have shown that the endocrine system plays a role in determining behavior.

Egg-Laying Behavior in Marine Snails

The egg-laying behavior in the marine snail *Aplysia* involves a specific sequence of movements. Following copulation, the animal

[2]Arnold, S. J. "Behavioral Variation in Natural Populations. I." *Evolution* 35:489–509 (1981).

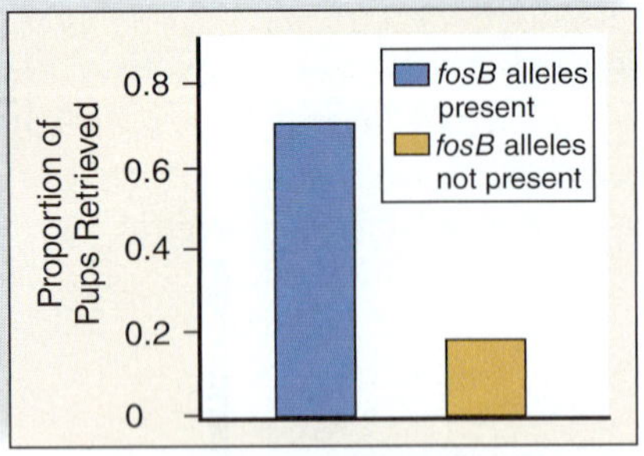

a.

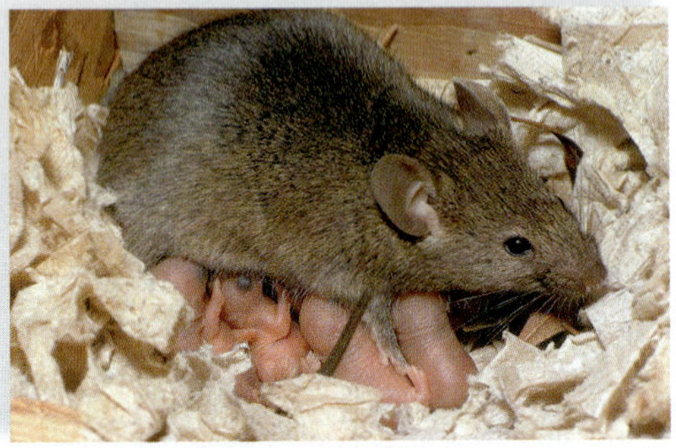

b. *fosB* alleles present.

c. *fosB* alleles not present.

Figure 43.3 Maternal care in mice. **a.** A mother mouse with *fosB* alleles spends time retrieving and crouching over its young, whereas mice that lack these alleles do not display these maternal behaviors. **b.** This typical mother mouse retrieves her young and crouches over them. **c.** This mouse, lacking *fosB,* does not retrieve her young and does not crouch over them.

extrudes long strings of more than a million egg cases. The animal takes the egg case string in its mouth, covers it with mucus, waves its head back and forth to wind the string into an irregular mass, and attaches the mass to a solid object, such as a rock.

Several years ago, scientists isolated and analyzed an egg-laying hormone (ELH) that causes the snail to lay eggs, even if it has not mated. ELH was found to be a small protein of 36 amino acids that diffuses into the circulatory system and excites the smooth muscle cells of the reproductive duct, causing them to contract and expel the egg string. Using recombinant DNA techniques, the investigators isolated the ELH gene. The gene's product turned out to be a protein with 271 amino acids. The protein can be cleaved into as many as 11 products, and ELH is one of these. ELH alone, or in conjunction with the other products, is thought to control all the components of egg-laying behavior in *Aplysia.*[3]

Nurturing Behavior in Mice

In another study, investigators found that maternal behavior in mice was dependent on the presence of a gene called *fosB*. Normally, when mothers first inspect their newborns, various sensory information from their eyes, ears, nose, and touch receptors travel to the hypothalamus. This incoming information causes *fosB* alleles to be activated, and a particular protein is produced. The protein begins a process during which cellular enzymes and other genes are activated. The end result is a change in the neural circuitry within the hypothalamus, which manifests itself in maternal nurturing behavior toward the young. Mice that do not engage in nurturing behavior were found to lack *fosB* alleles, and the hypothalamus failed to make any of the products or to activate any of the enzymes and other genes that lead to maternal nurturing behavior. Female mice with *fosB* alleles tended to retrieve their young and bring them back to them after they became separated. (Fig. 43.3).[4]

Check Your Progress 43.1

1. Compare the studies that show how behavior has a genetic basis.
2. Identify the body systems that influence behavior.

43.2 The Environment Influences Behavior

Learning Outcomes

Upon completion of this section, you should be able to

1. Choose the correct sequence of events required for classical and operant conditioning.
2. Identify examples of how the environment influences behavior.

Even though genetic inheritance serves as a basis for behavior, it is possible that environmental influences (nurture) also affect behavior. For example, behaviorists originally believed that some behaviors were unchanging behavioral responses known as **fixed action patterns (FAP)**. An FAP is elicited by a **sign stimulus**—a particular trigger in the environment. For example, male stickleback fish aggressively defend a territory

[3]Scheller, R. H., Jackson, J. F., et al. 1982. A "Family of Genes That Codes for ELH, a Neuropeptide Eliciting a Stereotyped Pattern of Behavior in *Aplysia.*" *Cell* 28:707–719 (1982).

[4]Brown, J. R., Ye, H., et al. "A Defect in Nurturing in Mice Lacking the Immediate Early Gene *fosB.*" *Cell* 86:297–309 (1996).

against other males. In laboratory studies, the male reacts more aggressively to any model that has a red belly like he has, rather than to a model that looks like a female stickleback fish. In this instance, the color red is a sign stimulus that triggers the aggressive behavior.

Investigators discovered that many behaviors that were originally thought to be FAPs could improve with practice due to the organism's learning. In this context, **learning** is defined as a durable change in behavior brought about by experience. Learning can be brought about by a wide variety of experiences. Habituation is a form of learning in which an animal no longer responds to a particular stimulus due to experience. Deer grazing on the side of a busy highway, ignoring traffic, is an example of habituation.

Instinct and Learning

Laughing gull chicks' begging behavior appears to be an FAP, because it is always performed the same way in response to the parent's red bill (the sign stimulus). A chick directs a pecking motion toward the parent's bill, grasps it, and strokes it downward (Fig. 43.4). Parents can bring about the begging behavior by swinging their bill gently from side to side. After the chick responds, the parent regurgitates food onto the floor of the nest. If need be, the parent then encourages the chick to eat. This interaction between the chicks and their parents suggests that the begging behavior involves learning.

To test this hypothesis, diagrammatic pictures of gull heads were painted on small cards, and then eggs were collected from nests in the field. The eggs were hatched in a dark incubator to eliminate visual stimuli before the test. On the day of hatching, each chick was allowed to make about a dozen pecks at the model. The chicks were returned to the nest, and then each was retested.

The tests showed that, on average, only one-third of the pecks by a newly hatched chick strike the model. But 1 day after hatching, more than half the pecks are accurate, and 2 days after hatching, the accuracy reaches a level of more than 75%. Investigators concluded that improvement in motor skills and learning can help improve the development of instinctive behaviors.[5]

Imprinting

Imprinting, a form of learning, occurs when a young animal forms an association with the first moving object it sees. Konrad Lorenz was one of the first individuals to study imprinting in birds. He observed chicks, ducklings, and goslings following the first moving object they saw after hatching, typically their mother. Imprinting in the wild has survival value and leads to reproductive success. This behavior enables an individual to recognize its own species and thus eventually to find an appropriate mate.

In the laboratory, however, investigators found that birds can seemingly be imprinted on any object—even a human or a red ball—if it is the first moving object they see during a sensitive period of 2 to 3 days after hatching. The term *sensitive period* means the period of time in which a particular behavior develops. A chick imprinted on

[5]Hailman, Jack P. "The Ontogeny of an Instinct." *Behavior* (Suppl. 13) (1967). (E. J. Brill Academic Publishers, Leiden, Germany.)

Figure 43.4 Pecking behavior in laughing gulls. **a.** At about 3 days, a laughing gull chick grasps the red bill of a parent, stroking it downward, and the parent then regurgitates food. **b.** The accuracy of a chick when striking a test probe, painted red. **c.** The chick-pecking accuracy graphically. Note from these diagrams that a chick markedly improves its ability (within only 2 days) to peck a blll, a behavior that normally causes a parent to regurgitate food.

a. Laughing gull adult and chick, *Leucophaeus atricilla*

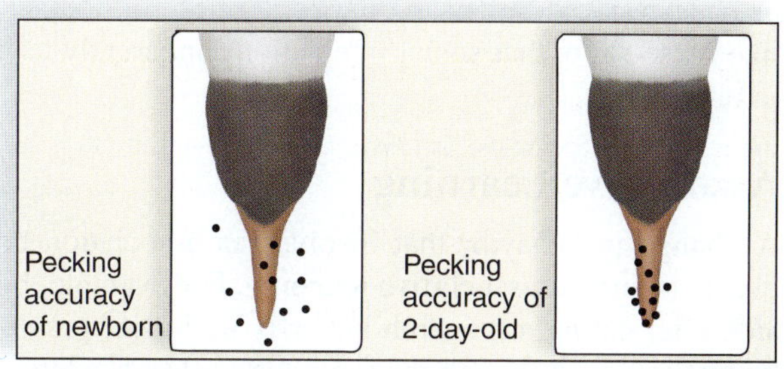

b.

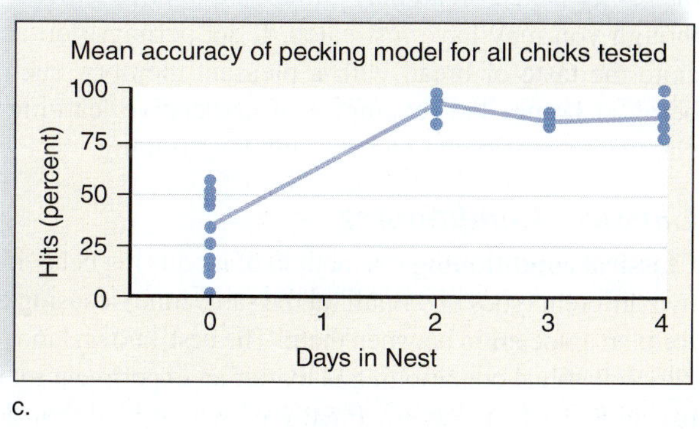

c.

a red ball follows it around and chirps whenever the ball is moved out of sight.

Social interactions between parent and offspring during the sensitive period seem to be a key to normal imprinting. For example, female mallards cluck the entire time that imprinting is occurring. It might be that vocalization before and after hatching is necessary to normal imprinting.

Goslings imprinted on an investigator

Social Interactions and Learning

White-crowned sparrows sing a species-specific song, but males of a particular region have their own dialect. Birds were caged in order to test the hypothesis that young white-crowned sparrows learn how to sing from older members of their species.

Three groups of birds were tested. Birds in the first group heard no songs at all. When grown, these birds sang a song that had a slight resemblance to the adult song. Birds in the second group heard tapes of white-crowns singing. When grown, they sang in that dialect, as long as the tapes had been played during a sensitive period, about age 10–50 days. White-crowned sparrows' dialects (or other species' songs) played before or after this sensitive period had no effect on the birds. Birds in a third group did not hear tapes and instead were given an adult tutor. No matter when the tutoring began, these birds sang the song of the tutor species. Results like these show that social interactions apparently assist learning in birds.

Associative Learning

A change in behavior that involves an association between two events is termed **associative learning.** For example, birds that get sick after eating a monarch butterfly no longer prey on monarch butterflies, even though they may be readily available. The smell of freshly baked bread may entice you to have a piece, even though you may have just eaten. If so, perhaps you associate the taste of bread with a pleasant memory, such as being at home. Two examples of associative learning are classical conditioning and operant conditioning.

Classical Conditioning

Classical conditioning is a method of modifying behavior by pairing two different types of stimuli (at the same time), causing an animal to form an association between them. The best-known laboratory example of classical conditioning is that of an experiment by the Russian psychologist Ivan Pavlov. First, Pavlov observed that dogs salivate when presented with food, so he began to ring a bell whenever the dogs were fed. Eventually, the dogs salivated excessively whenever the bell was rung, regardless of whether food was present (Fig. 43.5). The dogs had come to associate the ringing of the bell with being fed.

Classical conditioning suggests that an organism can be trained—that is, conditioned—to associate a response with a specific stimulus. Unconditioned responses occur naturally, as when salivation follows the presentation of food. Conditioned responses are learned, as when a dog learns to salivate when it hears a bell.

Advertisements attempt to use classical conditioning to sell products. Commercials often pair attractive people with a product being advertised in the hope that viewers will associate attractiveness with the product. This pleasant association may cause them to buy the product.

Some types of classical conditioning can be helpful when trying to increase beneficial behaviors. For example, it has been suggested that you hold a child on your lap when reading to him or her. The hope is that the child will associate a pleasant feeling with reading.

Operant Conditioning

Operant conditioning is a method of modifying behavior in which a stimulus-response connection is strengthened. Most people know that it is helpful to give an animal a reward, such as food or affection, when teaching it a trick. It is quite obvious that animal trainers use operant conditioning. They present a stimulus to the

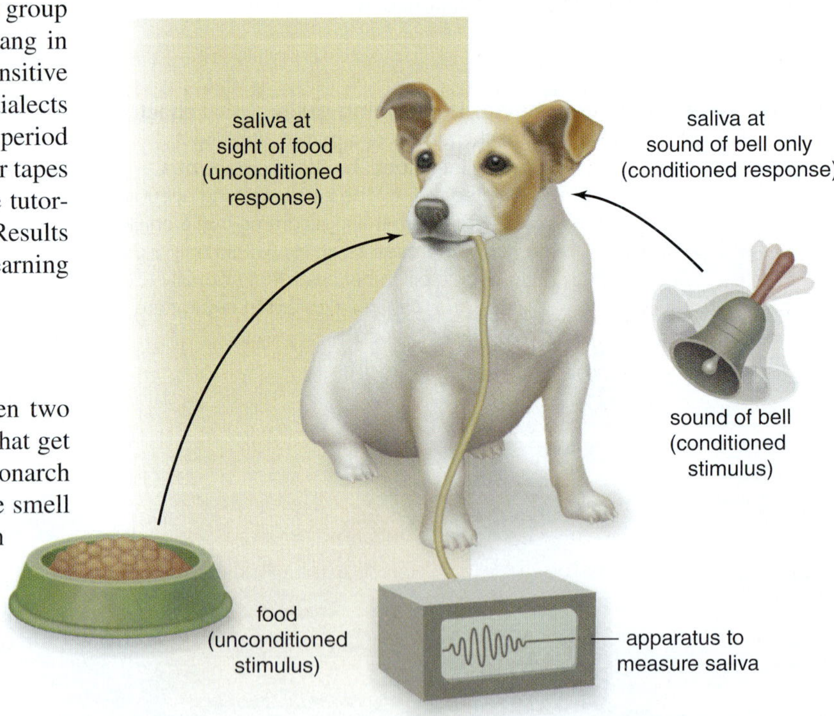

Figure 43.5 Classical conditioning. Ivan Pavlov discovered classical conditioning by performing this experiment with dogs. A bell is rung when a dog is given food. Salivation is measured. Eventually, the dog salivates when the bell is rung, even though no food is presented. Food is an unconditioned stimulus, and the sound of the bell is a conditioned stimulus that brings about the response—that is, salivation.

animal—say, a hoop—and then give a reward (food) for the proper response (jumping through the hoop). Sometimes, the reward does not need to be immediate; in latent operant conditioning, an animal makes an association without the immediate reward, as when squirrels make a mental map of where they have hidden nuts.

B. F. Skinner was well known for studying this type of learning in the laboratory. In the simplest type of experiment he performed, a caged rat inadvertently pressed a lever and was rewarded with sugar pellets, which it avidly consumed. Thereafter, the rat regularly pressed the lever whenever it wanted a sugar pellet. In more sophisticated experiments, Skinner even taught pigeons to play Ping-Pong by reinforcing the desired responses to stimuli.

In child rearing, it has been suggested that parents who give a positive reinforcement for good behavior will be more successful than parents who punish behaviors they believe are undesirable.

Orientation and Migratory Behavior

Migration is long-distance travel from one location to another. Loggerhead sea turtles hatch on a Florida beach and then travel across the Atlantic Ocean to the Mediterranean Sea, which offers an abundance of food. After several years, pregnant females return to the same beaches to lay their eggs. Every year, monarch butterflies fly from North America to Mexico, where they overwinter, and then return in the spring to breed.

At the very least, migration requires **orientation,** the ability to travel in a particular direction, such as south in the winter and north in the spring. Most of the research studying orientation has been done in birds. Many birds can use the sun during the day or the stars at night to orient themselves. The sun moves across the sky during the day, but the birds are able to compensate for this, because they have a sense of time. They are presumed to have a biological clock that allows them to know where the sun will be in relation to the direction they should be going anytime of the day.

Experienced birds can also **navigate,** which is the ability to change direction in response to environmental clues. These clues are apt to come from the Earth's magnetic field. Figure 43.6 shows a study that was done with starlings, which typically migrate from the Baltics to Great Britain and return. Test starlings were captured in Holland and transported to Switzerland. Experienced birds corrected their flight pattern and still got to Great Britain. Young, inexperienced birds ended up in Spain instead of Great Britain.

Migratory behavior has a proximate cause and an ultimate cause. The proximate cause consists of environmental stimuli that tell the birds it is time to travel. The ultimate cause is the possibility of reaching a more favorable environment for survival and reproduction. Is the benefit worth the cost—the dangers of the journey? It must be, or the behavior would not persist.

Cognitive Learning

In addition to the modes of learning already discussed, animals may learn through observation, imitation, and insight. For example, Japanese macaques learn to wash sweet potatoes before eating them by imitating others.

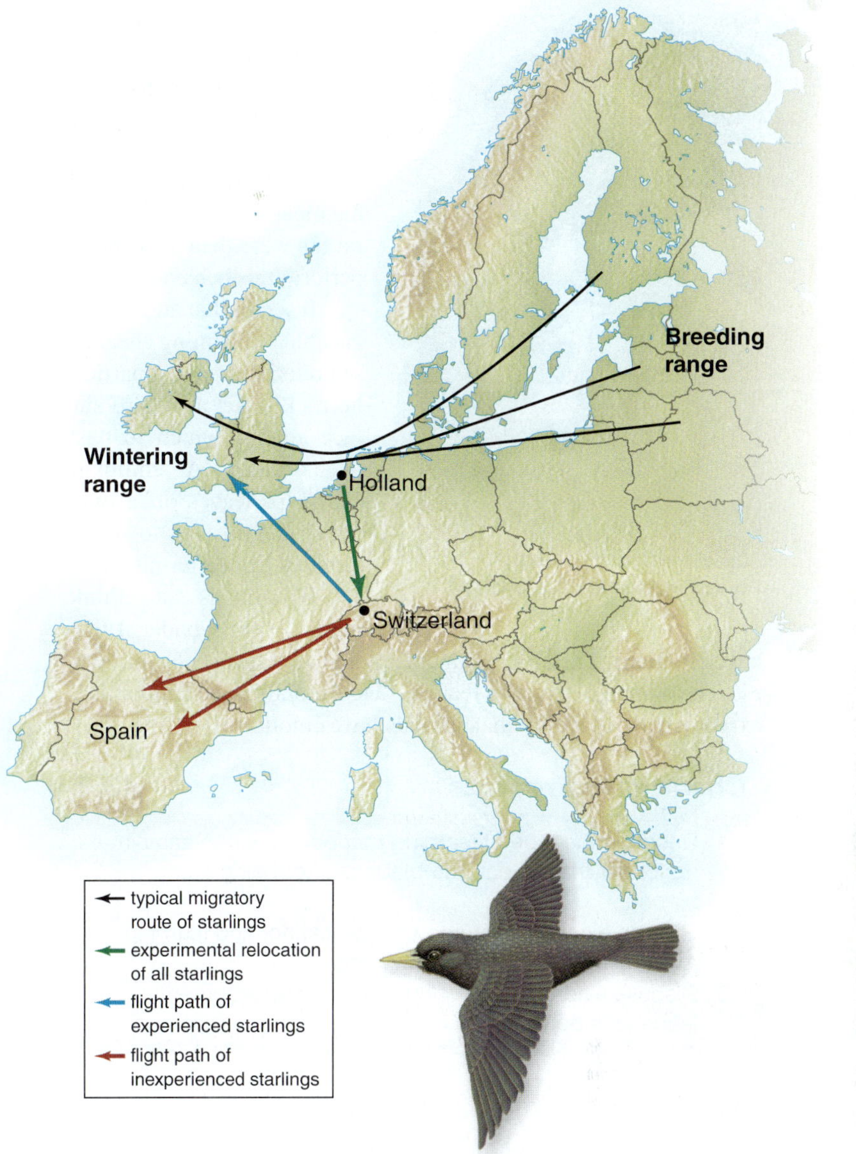

Figure 43.6 Starling migratory experiment. Starlings, *Sturnus vulgaris,* on their way from the Baltics to Great Britain were captured and released in Switzerland. Inexperienced birds kept flying in the same direction and ended up in Spain. Experienced birds had learned to navigate, as witnessed by the fact that they still arrived in Great Britain.

Insight learning occurs when an animal solves a problem without having any prior experience with the situation. The animal appears to call upon prior experience with other circumstances to solve the problem. For example, chimpanzees have been observed stacking boxes to reach bananas in laboratory settings.

Video Tool Use by Chimps

Other animals, too, aside from primates, seem to be able to reason things out. In one experiment, ravens were offered meat that was attached to string hanging from a branch in a confined aviary. The ravens were accustomed to eating meat, but they had no knowledge of how strings work. It took several hours, but eventually one raven flew to the branch, reached down, grabbed the string

Ravens learn to retrieve food.

with its beak, and pulled the string up over and over again, each time securing the string with its foot. Eventually, the meat was within reach, and the raven was able to grab the meat with its beak. Other ravens were then also able to perform this behavior.

It seems that animals are capable of planning ahead. A sea otter will save a particular rock to act as a hard surface against which to bash open clams. A chimpanzee strips leaves off a twig, which it then uses to secure termites from a termite nest.

If animals can think, many people wonder if they have emotions. This, too, is an unexplored area that is now of interest. The Nature of Science feature, "Do Animals Have Emotions?," explores the possibility that animals have emotions.

Check Your Progress 43.2

1. Describe the type of learning that occurs when an animal no longer tries to eat bumblebees after being stung by one.
2. Give an example that shows how instinct and learning may interact as behavior develops.
3. Discuss evidence that shows how the environment influences behavior.

Figure 43.7 Use of a pheromone. This male cheetah is spraying urine onto a tree to mark its territory.

43.3 Animal Communication

Learning Outcomes

Upon completion of this section, you should be able to

1. Recognize the various ways that animals can communicate.
2. Describe the advantages and disadvantages of chemical, auditory, visual, and tactile communication.

Animals exhibit a wide diversity of social behaviors. Some animals are largely solitary and join with a member of the opposite sex only for the purpose of reproduction. Others find mates and cooperate in raising offspring. Some form **societies** in which members of a species are organized in a cooperative manner, extending beyond sexual and parental behavior. We have already mentioned the social groups of honeybees and red deer (see Chapter 16). Social behavior in these and other animals requires that they communicate with one another.

Communicative Behavior

Communication is an action by a sender that may influence the behavior of a receiver. The communication can be purposeful, but it does not have to be. Bats send out a series of sound pulses and listen for the corresponding echoes to find their way through dark caves and locate food at night. Some moths have an ability to hear these sound pulses, and they begin evasive tactics when they sense that a bat is near. Bats are not purposefully communicating with the moths. The bat sounds are simply a cue to the moths that danger is near.

Chemical Communication

Chemical signals have the communicative advantage of being effective both night and day. A **pheromone** (Gk. *phero,* "bear, carry"; *monos,* "alone") is a chemical signal in low concentration that is passed between members of the same species. Some animals are capable of secreting different pheromones, each with a different meaning. Female moths secrete chemicals from abdominal glands, which are detected downwind by receptors on male antennae. The antennae are especially sensitive, and this ensures that only male moths of the correct species (and not predators) will be able to detect them.

Ants and termites mark their trails with pheromones. Cheetahs and other cats mark their territories by depositing urine, feces, and anal gland secretions at the boundaries (Fig. 43.7). Klipspringers (small antelope) use secretions from a gland below the eye to mark the twigs and grasses of their territory. Pheromones are known to control the behavior of social insects, as when worker bees slavishly care for the offspring produced by a queen.

Researchers are trying to determine to what degree pheromones, along with hormones, affect the behavior of animals. Researchers are trying to determine if pheromones are responsible for whether the animal will carry out parental care, become aggressive, or engage in courtship behavior. Some researchers

THEME Nature of Science

Do Animals Have Emotions?

In the light of recent YouTube video postings portraying the obvious expression of joy by dogs when their owners have returned from long absences, investigators have become more interested in determining to what extent animals have emotions. The body language of animals can be interpreted to suggest that they have feelings. When wolves reunite, they wag their tails to and fro, whine, and jump up and down. Many young animals play with one another or even by themselves, as when dogs chase their own tails. On the other end of the spectrum, on the death of a friend or parent, chimps are apt to sulk, stop eating, and even die.

It seems reasonable to hypothesize that animals are "happy" when they reunite, "enjoy" themselves when they play, and are "depressed" over the loss of a close friend or relative. Most people would agree that an animal is feeling some type of emotion when it exhibits certain behaviors (Fig. 43A).

In the past, scientists found it expedient to collect data only about observable behavior and to ignore the possible mental state of an animal. B. F. Skinner, whose research method is described in this chapter, was able to condition animals to respond automatically to a particular stimulus. He and others never considered that animals might have feelings. But now, some scientists believe they have sufficient data to suggest that some animals express emotions such as fear, joy, embarrassment, jealousy, anger, love, sadness, and grief. One possible definition describes emotion as a psychological phenomenon that helps animals direct and manage their behavior.

Multiple instances of soldiers returning from combat have shown how dogs react to seeing their owners after they have been absent for extended periods of time. On many occasions, the dogs run in circles, jump up and down, and excitedly lick their owner's face. Cries and squeals of excitement can be heard coming from them continuously as they greet their owners.

The Gabriela Cowperthwaite documentary *Blackfish,* about captive killer whales at SeaWorld, shows a scene in which keepers remove a young killer whale from its mother in order to relocate the young whale to another facility. When the mother cannot locate her offspring within the enclosure, she begins to shake violently and scream a long-range-frequency call in an attempt to find her child.

Perhaps it would be reasonable to consider the suggestion of Charles Darwin, who said that animals are different in degree rather than in kind. This means that animals can feel love but perhaps not to the degree that humans can. When you think about it, it is unlikely that emotions first appeared in humans with no evolutionary homologies in animals.

Dopamine controls the reward and pleasure center in the brain, as well as regulating emotional responses. High levels of dopamine are found in the brain of rats when they play, with the levels increasing when the rats anticipate the opportunity to play. Field research is providing information that is helping researchers learn how animal emotions correlate with their behavior, just as emotions influence human behavior.

Questions to Consider

1. If it can be shown that animals have emotions, should animals still be kept in captivity?
2. Do you think invertebrate animals have emotions? If so, what emotions might they experience?

Is the dog happy to see his owner?

Does the mother feel love for her calf?

Figure 43A Emotions in animals.

a.

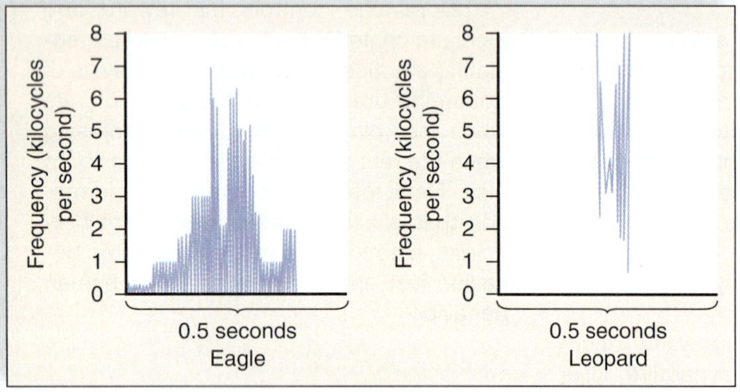

b.

Figure 43.8 Auditory communication. **a.** Vervet monkeys, *Cercopithecus aethiops,* are responding to an alarm call. Vervet monkeys can give different alarm calls according to whether a troop member sights an eagle or a leopard, for example. **b.** The frequency per second of the sound differs for each type of call.

Figure 43.9 Male baboon displaying full threat.

also maintain that human behavior is influenced by pheromones wafting through the air—pheromones that are not perceived consciously. They have discovered that like mice, humans have an organ in the nose, called the vomeronasal organ (VNO), that can detect not only odors but also pheromones. The neurons from this organ lead to the hypothalamus, the part of the brain that controls the release of many hormones in the body.

Auditory Communication

Auditory communication is communication through sound. It has some advantages over other kinds of communication. It is faster than chemical communication, and unlike visual communication, it is effective both night and day. Further, auditory communication can be modified not only by loudness but also by pattern, duration, and repetition. In an experiment with rats, a researcher discovered that an intruder can avoid attack by increasing the frequency with which it makes an appeasement sound.

Male crickets have calls, and male birds have songs for a variety of occasions. For example, birds may have one song for distress, another for courting, and still another for marking territories. Sailors have long heard the songs of humpback whales transmitted through the hulls of ships. But only recently has it been shown that the song has six basic themes. Each theme is composed of its own phrases that can vary in length and be interspersed with sundry cries and chirps. The purpose of the song is probably sexual, advertising the singer's availability. Bottlenose dolphins have one of the most complex languages in the animal kingdom.

Language is the ultimate auditory communication. Humans can produce a large number of different sounds and put them together in many ways. Nonhuman primates have different vocalizations, each having a definite meaning, as when vervet monkeys give alarm calls (Fig. 43.8). Although chimpanzees can be taught to use an artificial language, they do not progress beyond the capability level of a 2-year-old human child. It has also been difficult to prove that chimps understand the concept of grammar or can use their language to reason. It still seems as though humans possess a communication ability greater than that of other animals.

Video Meerkat Warning Calls

Visual Communication

Visual signals are most often used by species that are active during the day. Contests between males often make use of threat postures that possibly prevent outright fighting. A male baboon displaying full threat is an awesome sight that establishes his dominance and keeps peace within the baboon troop (Fig. 43.9). Hippopotamuses perform territorial displays that include mouth opening.

Many animals use complex courtship behaviors and displays. The plumage of a male Raggiana Bird of Paradise allows him to put on a spectacular courtship dance to attract a female, giving her a basis on which to select a mate. Defense and courtship displays are exaggerated and always performed in the same way, so that their meaning is clear. Fireflies use a flash pattern to signal females of the same species (Fig. 43.10).

Visual communication allows animals to signal others of their intentions without the need to provide any auditory or chemical messages. The body language of students during a lecture provides an example. Students who are leaning forward in their seats and making eye contact with the instructor appear interested and engaged with the material. Students who are leaning back in their chairs and gazing around the room or doodling are indicating that they have lost interest. Instructors can use students' body language to determine whether they are effectively presenting the material and make changes accordingly.

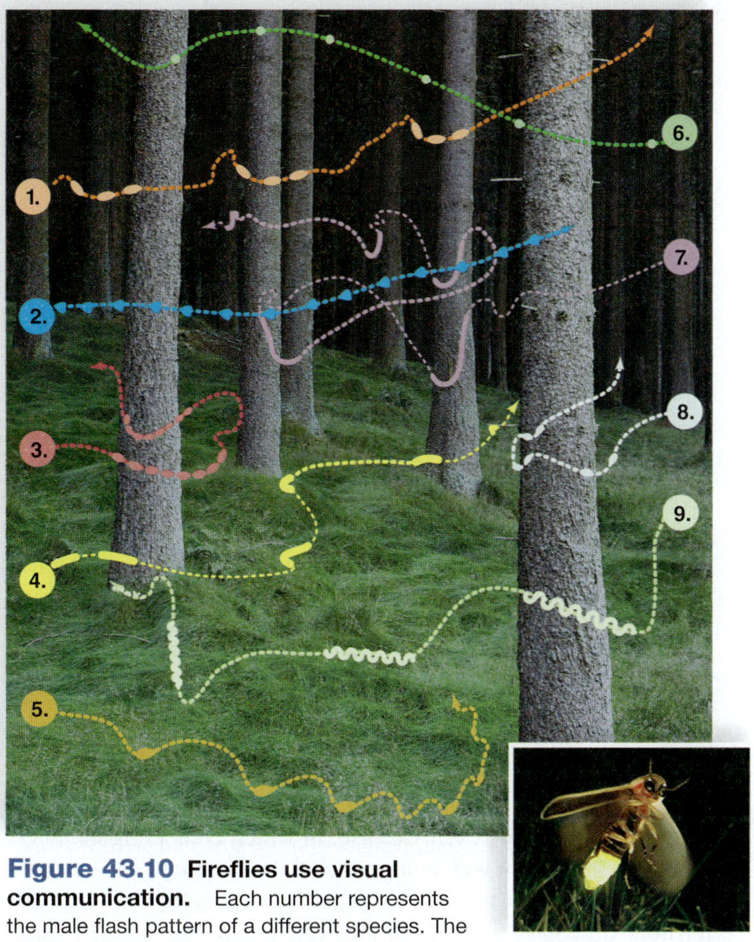

Figure 43.10 Fireflies use visual communication. Each number represents the male flash pattern of a different species. The patterns are a behavioral reproductive isolation mechanism.

The hairstyle and dress of a person, or the way he or she walks and talks, are also ways to send messages. Some studies have suggested that women tend to dress in a more sexually appealing manner when they are ovulating.[6] People who dress in black, move slowly, fail to make eye contact, and sit alone may be telling others that they are unhappy, or simply that they do not want to be socially engaged. Psychologists have long tried to understand how visual clues can be used to better understand human emotions and behavior. Similarly, researchers are evaluating body language in animals to determine whether they also have emotions.

Tactile Communication

Tactile communication occurs when information is passed from one animal to another by touch. For example, recall that laughing gull chicks peck at the parent's bill to induce the parent to feed them (see Fig. 43.4). A male leopard nuzzles the female's neck to calm her and to stimulate her willingness to mate. In primates, grooming—one animal cleaning the coat and skin of another—helps cement social bonds within a group.

Honeybees use a combination of communication methods, but especially tactile ones, to impart information about the environment. When a foraging bee returns to the hive, it performs a "waggle dance," which indicates the distance and direction of a food source (Fig. 43.11). As the bee moves between the two loops of a figure 8, it buzzes noisily and shakes its entire body in so-called waggles. Outside the hive, the dance is done on a horizontal surface, and the straight run indicates the direction of the food. Inside the hive, the angle of the straight run to that of the direction of gravity is the same as the angle of the food source to the sun. In other words, a 40° angle to the left of vertical means that food is 40° to the left of the sun.

a.

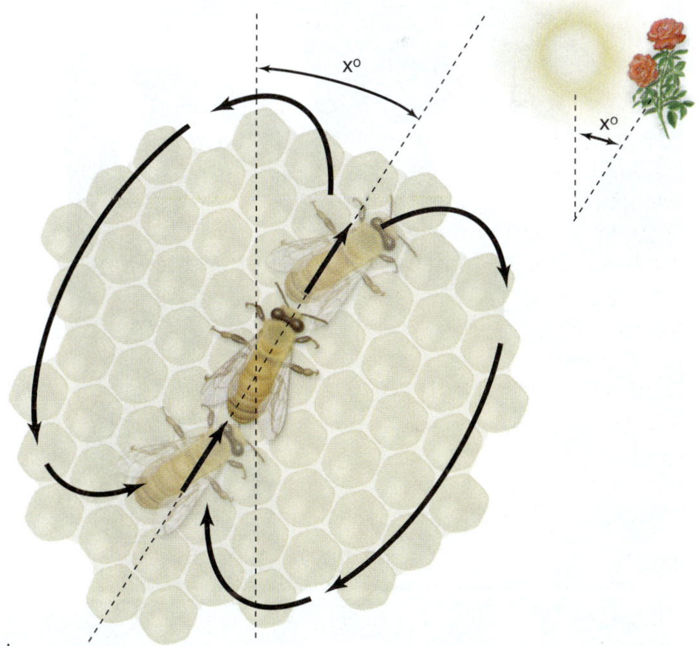

b.

Figure 43.11 Communication among bees. **a.** Honeybees do a waggle dance to indicate the direction of food. **b.** If the dance is done outside the hive on a horizontal surface, the straight run of the dance will point to the food source. If the dance is done inside the hive on a vertical surface, the angle of the straightaway to that of the direction of gravity is the same as the angle of the food source to the sun.

---

[6]Haselton, M. G., et al. "Ovulation and Human Female Ornamentation: Near Ovulation, Women Dress to Impress." *Hormones and Behavior* 51:41–45 (2007).

Bees can use the sun as a compass to locate food because they have a biological clock, an internal means of telling time, which allows them to compensate for the sun's movement in the sky. Today, we know that the timing of the clock, in both insects and mammals (including humans), requires alterations in the expression of a gene called *period.*

Check Your Progress 43.3

1. Describe examples of how communication is meant to affect the behavior of the receiver.
2. Give an advantage and disadvantage of each type of communication.
3. Identify the human receptor for each type of communication.

Figure 43.12 Male and female gibbons. Siamang gibbons, *Hylobates syndactylus,* are monogamous, and they share the task of raising offspring. They also share the task of marking their territory by singing. As is often the case in monogamous relationships, the sexes are similar in appearance. Male is above and female is below.

43.4 Behaviors That Increase Fitness

Learning Outcomes

Upon completion of this section, you should be able to

1. Describe how territoriality and the foraging techniques associated with it increase fitness.
2. Recognize various strategies that increase an individual's fitness.

Behavioral ecology is the study of how natural selection shapes behavior. It has been shown that behavior has a genetic basis, and investigators believe that certain behaviors lead to increased survival and production of offspring. Therefore, the behavior we can observe today in a species is still present because it has an adaptive value. The following types of behaviors, in particular, have been studied for their adaptive value: territoriality, reproductive strategies, social behaviors, and altruistic behaviors.

Territoriality and Fitness

In order to gather food, many animals have a home range where they carry out daily activities. The portion of the animal's range that it defends for its exclusive use, in which competing members of its species are not welcome, is called its **territory.** The behavior of defending one's territory is called **territoriality.** An animal's territory may have a good food source, and it may be the area in which the animal will reproduce.

As an example, gibbons live in the tropical rain forest of South and Southeast Asia. Normally, they can travel their home range in about 3–4 days. Gibbons are also monogamous and territorial. Territories are maintained by loud singing (Fig. 43.12). Males sing just before sunrise, and mated pairs sing duets during the morning. Males, but not females, show evidence of fighting to defend their territory in the form of broken teeth and scars. Defense of a territory has a certain cost; it takes energy to sing and fight off others.

In order for territoriality to continue, it must have an adaptive value. Chief among the benefits of territoriality are access to a source of food, breeding opportunities, and a place to rear young. The territory has to be the right size for the animal: Too large a territory cannot be defended, and too small a territory may not contain enough resources.

Cheetahs require a large territory in order to hunt for their prey. As a result, they need to mark their territory in a fashion that will last for a while. As shown in Figure 43.7, cheetahs, like many dogs, use urine to mark their territory. Hummingbirds are also known to defend a very small territory, because they depend on only a small patch of flowers as their food source.

Territoriality is more likely to occur during reproductive periods. Seabirds have very large home ranges consisting of hundreds of kilometers of open ocean, but when they reproduce they become fiercely territorial. Each bird has a very small territory, consisting of only a small patch of beach where they place their nest.

Video Cichlid Territoriality

Foraging for Food

Foraging for food is when animals are actively looking for nutrition. An animal needs to acquire a food source that will provide more energy than is expended in the effort of acquiring the food. In one study, it was shown that shore crabs prefer to eat intermediate-sized mussels, if they are given equal numbers of a variety of sizes. The net energy gained from medium mussels is greater than if the crabs eat larger mussels.The large mussels take too much energy to open per the amount of energy they provide (Fig. 43.13). The **optimal foraging model** states that it is adaptive for foraging behavior to be as energetically efficient as possible.

Even though it can be demonstrated that animals that take in more energy during foraging are more likely to produce a greater number of offspring, other factors also come into play in survival. If an animal forages in a way that puts it in danger and it is killed and eaten, it has no chance of producing offspring. Animals often face trade-offs that lead to modification of their behavior toward maximizing their success.

Reproductive Strategies and Fitness

Many primate species are **polygamous,** meaning that a single male mates with multiple females. Because of gestation and lactation, females tend to invest more energy in their offspring than do males. Under these circumstances, it is adaptive for females to be concerned about a good food source. When food sources are clumped, females congregate in small groups around the food. Because only a few females are expected to be receptive at a time, males will likely be able to defend these few from other males. Males are expected to compete with other males for the limited number of receptive females available (Fig. 43.14).

A limited number of primates are **polyandrous,** meaning that a female mates with multiple males. Tamarins are squirrel-sized New World monkeys that live in Central and South America. Tamarins live together in groups of one or more families in which one female mates with more than one male. The female normally gives birth to twins of such a large size that the fathers, not the mother, carry them about. This may be the reason these animals are polyandrous. Polyandry also occurs when the environment does not have sufficient resources to support several young at a time.

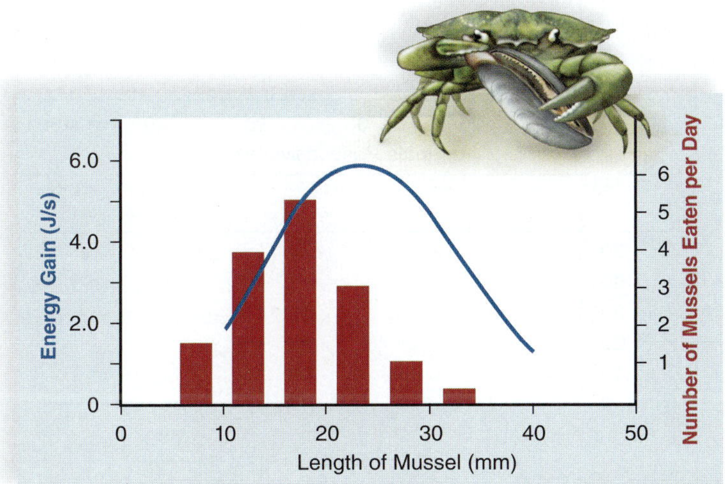

Figure 43.13 Foraging for food. When offered a choice of an equal number of each size of mussel, the shore crab, *Carcinus maenas,* prefers the intermediate size. This size provides the highest rate of net energy return measured in Joules per second (J/s). Net energy is determined by the energetic yield of the crab minus the energetic costs of breaking open the shell and digesting the crab.

Figure 43.14 Hamadryas baboons. Among Hamadryas baboons, *Papio hamadryas,* a male, which is silver-white and twice the size of a female, keeps and guards a harem of females with which he mates exclusively.

We have already mentioned the reproductive strategy of gibbons. They are **monogamous,** which means that they pair bond, and both males and females help with the rearing of the young. Males are active fathers, frequently grooming and handling infants. Monogamy is relatively rare in primates, which includes prosimians, monkeys, and apes (only about 18% are monogamous). In primates, monogamy occurs when males have limited mating opportunities, territoriality exists, and the male is fairly certain the offspring are his. In gibbons, females are evenly distributed in the environment, which can lead to an increase in aggression between females. Investigators note that females will attack an electronic speaker when it plays female sounds in their territory.

Sexual Selection

Sexual selection is a form of natural selection that favors features that increase an animal's chances of mating. In other words, these features are adaptive in the sense that they lead to increased fitness.

Sexual selection most often occurs due to female choice, which leads to male competition. Because most females produce a limited number of eggs relative to the number of sperm produced by the male, it is adaptive for females to be choosy about their mate. If they choose a mate that passes on features to a male offspring that will cause him to be chosen by other females, the parents' fitness has increased.

Whether females actually choose features that are adaptive to the environment is in question. For example, peahens are likely to choose peacocks that have the most elaborate tails. Such a fancy tail could otherwise be detrimental to the male and make him more likely to be captured by a predator. In one study, an extra ornament was attached to a father zebra finch, and the daughters of this bird underwent the process of imprinting (see section 43.2). Then, those females were more likely to choose a mate that also had the same artificial ornament.

THEME Evolution

Sexual Selection in Male Bowerbirds

At the start of the breeding season, male bowerbirds use small sticks and twigs to build elaborate display areas called bowers. They clear the space around the bower and decorate the area with items used to attract females. After the bower is complete, the male spends most of his time near his bower, calling to females, renewing his decorations, and guarding his work against possible raids by other males.

A female may approach, and then the male begins his display. He faces her, fluffs up his feathers, and flaps his wings to the beat of a call. If the female enters the bower, the two will mate. Some males mate with up to 25 females per year. Biologists discovered that females often chose males that had well-built bowers with well-decorated platforms.

Male bowerbirds are not gaudy in appearance, but their displays are highly intense and aggressive. Their courtship displays are similar to those used by males during aggressive encounters with other males. Males must display intensely to be attractive, but males that are too intense too soon can startle females. Females may benefit from mating with the most intensely displaying males, but if they are startled, they may not be able to efficiently assess male traits.

Communication between the sexes might maximize the potential benefits of intense male courtship displays while minimizing the potential costs. Females make crouching motions, and the degree of crouching reflects the level of display intensity she will tolerate without being startled. By giving higher-intensity displays when females increase their crouching, males could increase their courtship success by displaying intensely enough to be attractive without threatening the females.

The hypothesis tested was that males respond to female crouching signals by adjusting their intensity, and that a particular male's ability to respond to female signals is related to his success in courtship. A male's ability to modify his courtship display was difficult to measure in natural courtships, because it was not clear whether males were responding to females, or vice versa. To solve this problem, a robotic female bowerbird was used.

Using these "fembots," researchers were able to control female signals and measure male responses in experimental courtships. In general, male satin bowerbirds modulated their displays in response to robotic female crouching (Fig. 43B*a*). Video cameras monitored their behaviors at bowers, allowing each male's courtship/mating success to be measured. It was found that males who modulated their displays more effectively in response to robotic female signals startled real females less often in natural courtships and were thus more successful in courting females (Fig. 43B*b*).

The results suggest that females prefer intensely displaying males as mates and that successful males modulate their intensity in response to female signals, thus producing displays attractive to females without threatening them.[1]

Male responsiveness to female signals may be an important part of successful courtship in many species. Sexual selection may favor the ability of males to read female signals and adjust courtship displays accordingly.

Questions to Consider

1. Why would it be more adaptable for a male bowerbird to be aggressive and have an intense courtship display instead of being gaudy?
2. How would an extremely intense male bowerbird mate?

[1]Patricelli, G. L., Uy, J. A. C., Walsh, G., and Borgia, G. "Sexual Selection: Male Displays Adjusted to Female's Response." *Nature* 415:279–280 (2002).

Figure 43B Courtship success of males. **a.** Among satin bowerbirds, *Ptilonorhynchus violaceus,* the male bowerbird prepares the bower and typically decorates its platform with blue objects. **b.** Some males are better able to vary the intensity of their courtship display depending on the crouch rate of a robotic female, and these males startle real females less often. **c.** Experimenters found that males that respond best under experimental conditions do startle live females less and do have better courtship success.

Figure 43.15 Competition. During the mating season, bull elk, *Cervus elaphus,* males may find it necessary to engage in antler wrestling in order to have sole access to females in a territory.

While females can always be sure an offspring is theirs, males do not have this certainty. However, males produce a plentiful supply of sperm. The best strategy for males to increase their fitness, therefore, is to have as many offspring as possible. Competition may be required for them to gain access to females, and ornaments such as antlers can enhance a male's ability to fight (Fig. 43.15). When bull elk compete, they issue a number of loud screams that give way to a series of grunts. If a clear winner hasn't emerged, the two bulls walk in parallel to show each other their physique. If this doesn't convince one of them to back off, the pair resorts to ramming each other with their antlers. The intent of the fight is for one male to gain dominance over the other, not to kill his rival—although these encounters may have fatal consequences due to injury or mishap.

Societies and Fitness

The principles of evolutionary biology can be applied to the study of social behavior in animals. Sociobiologists hypothesize that living in a society has a greater reproductive benefit than reproductive cost. A cost-benefit analysis can help determine whether this hypothesis is supported.

Group living has its benefits. It can help an animal avoid predators, rear offspring, and find food. For instance, a group of impalas is more likely than a single impala to hear an approaching predator. Many fish moving rapidly in multiple directions might distract a would-be predator. Weaver birds form giant colonies, which help protect them from predators, and the birds may share information about food sources. Members of a baboon troop signal to one another when they have found an especially bountiful fruit tree. Lions working together are able to capture large prey, such as zebra and buffalo.

Video
Harris Hawks

Group living also has disadvantages. When animals are crowded together into a small area, disputes can arise over access to the best feeding places and sleeping sites. Dominance hierarchies are one way to apportion resources, but this puts subordinates at a disadvantage. Among red deer, sons are preferable, because, sons, as harem masters, will result in a greater number of grandchildren. However, sons, being larger than daughters, need to be nursed more frequently and for a longer period of time. Subordinate females do not have access to enough food resources to adequately nurse sons, and therefore they tend to rear only daughters. Still, like the subordinate males in a baboon troop, subordinate females in a red deer harem may be better off in terms of fitness if they stay with a group, despite the cost involved.

Living in close quarters exposes individuals to illness and parasites that can easily pass from one animal to another. Social behavior helps offset some of the proximity disadvantages. For example, baboons and other types of social primates invest a significant amount of time grooming one another. This most likely decreases their chances of contracting parasites. Humans use extensive medical care to help offset the health problems that arise from living in the densely populated cities of the world.

Altruism Versus Self-Interest

Altruism (L. *alter,* "the other") is a behavior that has the potential to decrease the lifetime reproductive success of the altruist while benefiting the reproductive success of another member of the society. In some insect societies, reproduction is limited to only one pair, the queen and her mate. For example, among army ants, the queen is inseminated only during her nuptial flight. The rest of her life is spent producing her offspring (Fig. 43.16).

Figure 43.16 Queen ant. A queen ant, *Solenopsis geminata,* has a large abdomen for egg production and is cared for by small ants, called nurses. The idea of inclusive fitness suggests that relatives, in addition to offspring, increase an individual's reproductive success. Therefore, sterile nurses are being altruistic when they help the queen produce offspring to which they are closely related.

The army-ant society has three sizes of sterile female workers. The smallest workers (3 mm), called the nurses, take care of the queen and larvae, feeding them and keeping them clean. The intermediate-sized workers, constituting most of the population, go out on raids to collect food. The soldiers (14 mm), with huge heads and powerful jaws, run along the sides and rear of raiding parties and protect the column of ants from attack by intruders.

The altruistic behavior of sterile workers can be explained in terms of fitness, which is judged by reproductive success. Genes are passed from one generation to the next in two, quite different ways. The first way is direct: A parent can pass a gene directly to an offspring. The second way is indirect: A relative that reproduces can pass a shared gene to the next generation. *Direct selection* is adaptation to the environment due to the reproductive success of an individual. *Indirect selection,* called **kin selection,** is adaptation to the environment due to the reproductive success of the individual's relatives. The **inclusive fitness** of an individual includes personal reproductive success as well as the reproductive success of relatives.

Examples of Inclusive Fitness

Among social bees, social wasps, and ants, the queen is diploid (2n), but her mate is haploid (n). If the queen has had only one mate, sister workers will be closely related to each other. They will share, on average, 75% of their genes, because they inherit the same allele from their father. Their potential offspring would share, on average, only 50% of their genes with the queen. Therefore, a worker can achieve a greater inclusive fitness by helping her mother (the queen) produce additional sisters than by directly reproducing. Under these circumstances, a behavior that appears altruistic is more likely to evolve.

Indirect selection can also occur among animals whose offspring receive only a half set of genes from both parents. Consider that your brother or sister shares 50% of your genes, your niece or nephew shares 25% of your genes, and so on. Therefore, the survival of two nieces or nephews is worth the survival of one sibling, assuming they both go on to reproduce.

Among chimpanzees in Africa, a female in estrus frequently copulates with several members of the same group, and the males make no attempt to interfere with each other's matings. Genetic relatedness appears to underlie their apparent altruism. Members of a group share more than 50% of their genes, because members never leave the territory in which they are born.

Reciprocal Altruism

In some bird species, offspring from a previous clutch of eggs may stay at the nest to help parents rear the next batch of offspring. In a study of Florida scrub jays, the number of fledglings produced by an adult pair doubled when they had helpers. In certain mammalian groups, like meerkats, the offspring have been observed to help their parents (Fig. 43.17). Among jackals in Africa, solitary pairs managed to rear an average of 1.4 pups, whereas pairs with helpers reared 3.6 pups.

There are several benefits of staying behind to help raise young. First, a helper is contributing to the survival of its own kin. Therefore, the helper actually gains a fitness benefit. Second, a helper is more likely than a nonhelper to inherit a parental territory, which may include other helpers. Helping, then, involves making a short-term reproductive sacrifice in order to increase future reproductive potential. Therefore, helpers at the nest are also practicing a behavior called **reciprocal altruism.**

Figure 43.17 Inclusive fitness. A meerkat is acting as a babysitter for its young sisters and brothers while their mother is away. Researchers point out that the helpful behavior of the older meerkat can lead to increased inclusive fitness.

Reciprocal altruism also occurs in animals that are not closely related. In this event, an animal helps or cooperates with another animal, gaining no immediate benefit. However, the animal that was helped repays the debt at some later time. Reciprocal altruism usually occurs in groups of animals that are mutually dependent. Cheaters in reciprocal altruism are recognized and are not reciprocated in future events.

An example of this type of reciprocal altruism occurs in vampire bats that live in the tropics. Bats returning to the roost after a feeding activity share their blood meal with other bats in the roost. If a bat fails to share blood with one that had previously shared blood with it, the cheater bat will be excluded from future blood sharing.

Check Your Progress 43.4

1. Explain how territoriality is related to foraging for food.
2. Compare and contrast reproductive strategies and forms of sexual selection.
3. Describe examples of how altruistic behavior increases an individual's fitness.

Connecting the Concepts with the Themes

Evolution

- Animal behavior, like other characteristics, can be acted upon by natural selection.
- For a behavior to persist for long periods of time, it cannot have negative consequences on the population.
- Behaviors that increase fitness have a high probability of persisting in a population.

Nature of Science

- Research has shown that both nature and nurture—that is, genes and the environment—determine behavior.
- Behavior can be modified by experience (learning).

Biological Systems

- Animals possess a number of systems that enable them to respond to stimuli—primarily the nervous system and the endocrine system, which have far-reaching effects.
- Sexual selection has helped shape mating behaviors by helping ensure that those with highest fitness reproduce the most.

MEDIA STUDY TOOLS http://connect.mheducation.com

SMARTBOOK® Study smart. SmartBook helps you maximize your study time by identifying what you know and don't know. The recharge feature in SmartBook helps you prepare for the test by identifying the material that you are most likely to forget.

connect BIOLOGY Find even more resources to learn the chapter concepts including animations, tutorial videos, and interactive practice questions.

Videos

43.2 Tool Use by Chimps
43.3 Meerkat Warning Calls
43.4 Cichlid Territoriality • Harris Hawks

SUMMARIZE

43.1 Inheritance Influences Behavior

Investigators have long been interested in the degree to which nature (genetics) or nurture (environment) influences **behavior.** Hybrid studies with lovebirds produce results consistent with the hypothesis that behavior has a genetic basis. Garter snake experiments indicate that the nervous system controls behavior. Studies of human twins have shown that certain types of behavior are apparently inherited. DNA studies of marine snails indicate that the endocrine system also controls behavior.

43.2 The Environment Influences Behavior

Fixed action patterns (**FAPs**) are behavioral responses elicited by a **sign stimulus.** Research has shown that FAPs can be modified by **learning.** The red bill of laughing gulls initiates chick begging behavior. However, with experience, the chick's begging behavior improves as the chick learns. It demonstrates an increased ability to recognize its parents.

Imprinting is a form of learning in which an animal develops an association with the first moving object it sees during a sensitive period. Song learning in birds involves various elements—including the existence of a sensitive period, during which an animal is primed to learn—and the positive benefit of social interactions.

Associative learning is a change in behavior that involves an association between two events. **Classical conditioning** and **operant conditioning** are two forms of associative learning. In classical conditioning, the pairing of two different types of stimuli causes an animal to form an association between them. In this way, dogs can be conditioned to salivate at the sound of a bell. In operant conditioning, animals learn behaviors because they are rewarded when they perform them.

Migration occurs when an animal travels from one location to another. In order to do this, **orientation,** or the ability to travel in a particular direction, is required. **Navigation** is the ability to change direction in response to environmental cues. Animals often use the sun, the stars, and the Earth's magnetic field during migration.

Insight learning occurs when an animal solves a problem without having prior experience with the situation.

43.3 Animal Communication

Members of various species of animals are organized in a cooperative manner, forming **societies.** Social behavior requires the passing of information from one individual to another. **Communication** is an action by a sender that affects the behavior of a receiver. Chemical, auditory, visual, and tactile signals are forms of communication that foster cooperation benefiting both the sender and the receiver. **Pheromones** are chemical signals passed between members of the same species. **Auditory communication** is the passing of information through sound, which can include language. **Visual communication** is the passing of information through the use of visual cues. This allows animals to signal others without the need for auditory or chemical messages. **Tactile communication** is the passing of information through the use of touch; it is often associated with sexual behavior.

43.4 Behaviors That Increase Fitness

Behavioral ecology is the study of how natural selection shapes behavior. Traits that promote reproductive success are expected to provide an overall advantage to the individual. Animals often have a **territory** in which they can obtain food, acquire living space, and reproduce. The act of defending one's territory is called **territoriality.** When animals choose foods that return the most net energy, they have more energy left over for reproduction. The **optimal foraging model** is an adaptive behavior in which foraging is as energetically efficient as possible.

Depending on the environment and social structure, animals use a specific type of reproductive strategy. Many primates are **polygamous,** meaning one male will mate with multiple females. **Polyandrous** means one female mating with multiple males. Gibbons are **monogamous,** meaning that a male and a female form a pair bond and mate only with each other. **Sexual selection** is a form of natural selection for traits that increase an animal's fitness. Males produce many sperm and are expected to compete to inseminate females. Females produce few eggs and are expected to be selective about their mates.

Living in a social group can have its advantages (e.g., ability to avoid predators, raise young, and find food). It also has disadvantages (e.g., tension between members, spread of illness and parasites, and reduced reproductive potential). When animals live in groups, the benefits must outweigh the costs, or the behavior would not exist.

In some animal societies, **altruism** is present. This is a behavior in which an individual benefits the reproductive success of another member of society while limiting its own. **Kin selection** occurs when individuals increase the reproductive success of their relatives. In this context, it is necessary to consider **inclusive fitness,** which includes personal reproductive success as well as the reproductive success of relatives. Sometimes, individuals help their parents rear siblings. Social insects help increase their mother's chance of reproductive success, but this behavior seems reasonable when we consider that siblings share 75% of their genes. Among mammals, a parental helper may be likely to inherit the parent's territory. In **reciprocal altruism,** animals aid one another for future benefits.

ASSESS

Choose the best answer for each question.

43.1 Inheritance Influences Behavior

1. Behavior is
   a. any action that is learned.
   b. all responses to the environment.
   c. any action that can be observed and described.
   d. all activity that is controlled by hormones.
   e. unique to birds and humans.
2. Which of the following are considered, by behaviorists, to control (in part or in whole) animal behavior?
   a. circulatory and respiratory systems
   b. respiratory and digestive systems
   c. digestive and nervous systems
   d. nervous and endocrine systems
   e. All systems of the body control behavior.
3. Which of the following is not an example of a genetically based behavior?
   a. Inland garter snakes do not eat slugs, whereas coastal populations do.
   b. One species of lovebird carries nesting strips one at a time, whereas another carries several.
   c. One species of warbler migrates, whereas another does not.
   d. Snails lay eggs in response to an egg-laying hormone.
   e. Wild foxes raised in captivity are not capable of hunting for food.

43.2 The Environment Infuences Behavior

4. How would the following graph differ if pecking behavior in laughing gulls were a fixed action pattern?
   a. It would be a diagonal line with an upward incline.
   b. It would be a diagonal line with a downward incline.
   c. It would be a horizontal line.
   d. It would be a vertical line.
   e. None of these are correct.

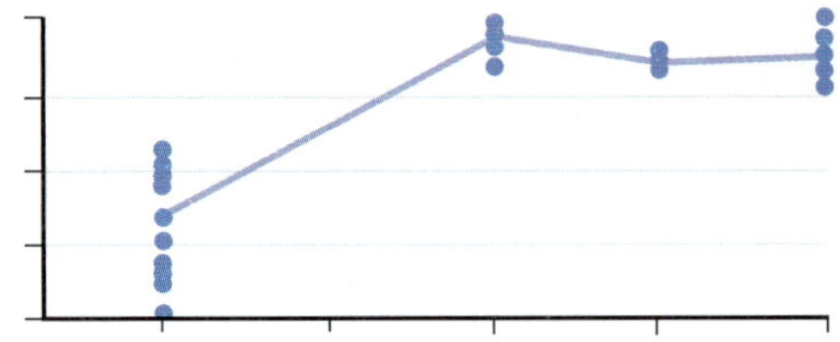

5. In white-crowned sparrows, social experience exhibits a very strong influence over the development of singing patterns. What observation led to this conclusion?
   a. Birds learned to sing only when they were trained by other birds.
   b. The sensitive period in which birds learned from other birds was wider than that when birds learned from tape recordings.
   c. Birds could learn different dialects only from other birds.
   d. Birds that learned to sing from a tape recorder could change their song when they listened to another bird.
6. Which of the following best describes classical conditioning?
   a. the gradual strengthening of stimulus-response connections that seemingly are unrelated
   b. a type of associative learning in which there is no contingency between response and reinforcer
   c. learning behavior in which an organism follows the first moving object it encounters
   d. learning behavior in which an organism exhibits a fixed action pattern from the time of birth

43.3 Animal Communication

For questions 7–11, match the type of communication in the key with its description. Answers may be used more than once.

Key:

a. chemical communication
b. auditory communication
c. visual communication
d. tactile communication

7. Aphids (insects) release an alarm pheromone when they sense they are in danger.
8. Male peacocks exhibit an elaborate display of feathers to attract females.
9. Ground squirrels give an alarm call to warn others of the approach of a predator.
10. Male silk moths are attracted to females by a sex attractant released by the female moth.
11. Sage grouses perform an elaborate courtship dance.

43.4 Behaviors That Increase Fitness

12. All of the following are benefits obtained through territoriality except
    a. access to mates.
    b. access to more food.
    c. access to more places to hide.
    d. access to more predators.
    e. increased space to raise more offspring.
13. The observation that male bowerbirds decorate their nests with blue objects favored by females can best be associated with
    a. insight learning.
    b. imprinting.
    c. sexual selection.
    d. altruism.
14. Which reproductive strategy is characterized by a female mating with multiple males?
    a. polygamy
    b. polyandry
    c. monogamy
    d. kin selection
    e. altruism

ENGAGE

THINKING SCIENTIFICALLY

1. Meerkats are said to exhibit altruistic behavior, because certain members of a population act as sentries. How would you test the hypothesis that sentries are engaged in altruistic behavior?
2. You are testing the hypothesis that human infants instinctively respond to higher-pitched voices. Your design is to record head turns toward a speaker that will play voices in different pitches. When you do the experiment with several different infants, your data support your hypothesis. However, infant exposure to different voices, prior to the experiment, can produce flawed results. How can you design an experiment that produces the greatest validity of results?
3. The 2013 film *Blackfish* attempts to expose the darker sider of keeping killer whales in captivity. It chronicles the 2010 death of a SeaWorld trainer along with multiple other incidences in which 5-ton orca whales appeared to intentionally attack their trainers. The movie has many people questioning the ethical nature of keeping animals in captivity.

   If we keep animals in captivity, are we depriving them of their freedom? Some point out that freedom is never absolute. Even an animal in the wild is restricted in various ways by its abiotic and biotic environments. Many modern zoos and aquariums keep animals in habitats that nearly match their natural ones, so that they have some freedom to roam and behave naturally—but is this really possible with an animal such as an orca?

   Today, reputable zoos and aquariums rarely go out and capture animals in the wild—they usually get their animals from other facilities. Most people feel it is not a good idea to take animals from the wild. Certainly, zoos and aquariums should not be involved in the commercial and often illegal trade of wild animals that still goes on today. Many zoos and aquariums today are involved in the conservation of animals, rather than just their exploitation for exhibit. They provide the best home possible while animals are recovering from injury or are increasing their numbers until they can be released back into the wild. In your opinion, which animals are suited to a life in captivity, and which are not?

44

Population Ecology

Asian carp in the Mississippi River system.

Chapter Outline

44.1 Scope of Ecology 837
44.2 Demographics of Populations 838
44.3 Population Growth Models 841
44.4 Regulation of Population Size 844
44.5 Life History Patterns 846
44.6 Human Population Growth 849

Before You Begin

Before beginning this chapter, take a few moments to review the following discussions.

Section 6.1 How does energy flow from one organism to another impact population growth?

Figure 10.7 Explain what role the human life cycle plays in determining population growth.

Section 16.1 Describe how microevolution is measured within a population.

In the early 1970s, several species of Asian carp were imported into Arkansas for the biocontrol of algal blooms in aquaculture facilities. Unfortunately, they escaped into the middle and lower Mississippi drainage. Over time, they have spread throughout the majority of the Mississippi River system.

Asian carp mainly eat the microscopic algae and zooplankton in freshwater ecosystems. They can achieve weights up to a hundred pounds and grow to a length of more than 4 feet. Because of their voracious appetite, they have the potential to reduce the populations of native fish and mussels due to competition for the same food sources.

The main fear of biologists is that the Asian carp will put extreme pressure on the zooplankton populations, which can then result in a dense planktonic algal bloom. This has the potential to completely disrupt the ecology of the ecosystems they invade. These fish also pose an economic threat by fouling the nets of commercial fishers. The other major concern is the impact these fish may have on the Great Lakes' 7-billion-dollar fishing industry.

This chapter previews the principles of population ecology, a field that is critical to the preservation and management of species, as well as the maintenance of the diversity of life on Earth.

As you read through this chapter, think about the following questions:

1. Does the old saying "90% of the fish are found in 10% of the lake" have any truth to it?
2. Do population growth models apply to humans the same way they apply to other species?

Following the Themes

CHAPTER 44 POPULATION ECOLOGY

UNIT 8 BEHAVIOR AND ECOLOGY

Evolution	A change in one population or in the environment will have an impact on the other populations in the community.
Nature of Science	Ecologists study the interactions between a variety of living organisms and their environment.
Biological Systems	Population growth is based on birthrates, mortality rates, the age structure of the population, and a variety of other factors.

44.1 Scope of Ecology

Learning Outcomes

Upon completion of this section, you should be able to

1. Identify what aspects of biology the study of ecology encompasses.
2. Identify the ecological levels that exist within the field of ecology.

In 1866, the German zoologist Ernst Haeckel (1834–1919) coined the word **ecology** (Gk. *oikos,* "home"; *-logy,* "study of"), which he defined as the study of the interactions among all organisms and with their physical environment. Haeckel also pointed out that ecology and evolution are intertwined. Ecological interactions act as selection pressures that result in evolutionary change, which in turn affects ecological interactions.

Ecology, like so many biological disciplines, is wide-ranging. At one of its lowest levels, ecologists study how an individual organism is adapted to its environment. For example, they study how a fish is adapted to and survives in its **habitat** (the place where the organism lives). Most organisms do not exist singly; rather, they are part of a population, the functional unit that interacts with the environment and on which natural selection operates (Fig. 44.1). A **population** is defined as all the organisms belonging to the same species within an area at the same time. At this level of study, ecologists are interested in factors that affect the growth and regulation of population size.

A **community** consists of all the populations of multiple species interacting at a locale. In a coral reef, there are numerous populations of algae, corals, crustaceans, fishes, and so forth. At this level, ecologists want to know how various populations interact with each other. An **ecosystem** is composed of the community of populations along with the abiotic variables (e.g., the availability of sunlight for plants). Energy flow and chemical cycling are significant aspects of how an ecosystem functions. Ecosystems rarely have rigid boundaries. Usually, a transition zone called an ecotone, which has a mixture of organisms from adjacent ecosystems, exists between ecosystems. The **biosphere** encompasses the zones of the Earth's soil, water, and air where living organisms are found.

Video Coral Reef Ecosystem

Modern ecology is not just descriptive; it is predictive. It analyzes levels of organization and develops models and hypotheses that can be tested. A central goal of modern ecology is to develop models that explain and predict the distribution and abundance of organisms. Ultimately, ecology considers not one particular area but the distribution and abundance of populations in the biosphere. For instance, ecologists try to determine which factors have selected for the mix of plants and animals in a tropical rain forest at one latitude and in a desert at another.

Although modern ecology is useful in and of itself, it also has unlimited application possibilities, including the proper management of plants and wildlife, the identification of and efficient use of renewable and nonrenewable resources, the preservation of habitats and natural cycles, the maintenance of food resources, and the ability to predict the impact and course of diseases, such as malaria or AIDS.

Check Your Progress 44.1

1. Distinguish between a population and a community.
2. Explain what is meant by *abiotic variables.*
3. Describe the central goal of modern ecological studies.

Figure 44.1 Ecological levels. The study of ecology encompasses levels of organization that proceed from the individual organism to the population, to the community, and finally to an ecosystem.

44.2 Demographics of Populations

Learning Outcomes

Upon completion of this section, you should be able to

1. Recognize how environmental conditions affect the density and distribution patterns of a population.
2. Interpret survivorship curves and life tables.
3. Recognize how the proportion of individuals at varying reproductive stages determines a population's age distribution.

Demography is the statistical study of a population, including information about its density, distribution, and rate of growth (which are dependent on the population's mortality pattern and age distribution).

Density and Distribution

Population density is the number of individuals per unit area; for example, there are 86.8 persons per square mile in the United States. Population density figures make it seem that individuals are uniformly distributed, but this often is not the case. For example, most people in the United States live in cities, where the number of people per unit area is dramatically higher than in the country. And even within a city, more people live in particular neighborhoods than in others, and such distributions can change over time. Therefore, basing ecological models solely on population density can lead to inaccurate results.

Population distribution is the pattern of dispersal of individuals across a given area. The availability of resources can affect where populations live. **Resources** are nonliving (abiotic) and living (biotic) components of an environment that support living organisms. Light, water, space, mates, and food are some important resources for populations. **Limiting factors** are environmental aspects that determine where an organism lives. For example, trout live only in cool mountain streams, where the oxygen content is high, but carp and catfish are found in rivers near the coast, because they can tolerate warm waters with a lower concentration of oxygen. The timberline is the limit of tree growth in mountainous regions or in high latitudes. Trees cannot grow above the high timberline because of low temperatures and the fact that water remains frozen most of the year. Biotic factors can also play a role in the distribution of organisms. In Australia, the red kangaroo does not live outside arid inland areas because it is adapted to feeding on the grasses that grow there.

Three descriptions—*clumped, random,* and *uniform*—are often used to characterize patterns of distribution. Suppose you considered the distribution of a species across its full range. A range is that portion of the globe where the species can be found; for example, red kangaroos live in Australia. On that scale, you would expect to find a clumped distribution. However, organisms are located in areas suitable to their adaptations; as mentioned, red kangaroos live in grasslands, and catfish live in warm river water near the coast.

Within a smaller area, such as a single body of water or a single forest, the availability of resources influences which pattern of distribution is exhibited for a particular population. For example, a study of the distribution of hard clams in a bay on the south shore of Long Island, New York, showed that clam abundance is associated with sediment shell content. Investigators hope to use this information to increase the density and number of clams in areas that have low clam populations.

Distribution patterns need not be constant. In a study of desert shrubs, investigators found that the distribution changed from clumped to random to a uniform distribution pattern as the plants matured. As time passed, it was found that competition for belowground resources caused the distribution pattern to change over time (Fig. 44.2).

Figure 44.2 Distribution patterns of the creosote bush. **a.** Photograph of mature shrub distribution. **b.** Young, small desert shrubs are clumped. **c.** Medium shrubs are randomly distributed. **d.** Mature shrubs are uniformly distributed.

a.

b.

Figure 44.3 Biotic potential. A population's maximum growth rate under ideal conditions—that is, its biotic potential—is greatly influenced by the number of offspring produced in each reproductive event. **a.** Mice, which produce many offspring that quickly mature to produce more offspring, have a much higher biotic potential than the rhinoceros (**b**), which produces only one or two offspring per infrequent reproductive event.

Other factors besides resource availability can influence distribution patterns. Breeding golden eagles, like many other birds, exhibit territoriality, and this behavioral characteristic discourages a clumped distribution. In contrast, cedar trees tend to be clumped near the parent plant, because seeds are not widely dispersed.

Population Growth

The **rate of natural increase** (***r***), or growth rate, is determined by the number of individuals born each year minus the number of individuals that die each year. It is assumed that immigration and emigration are equal and need not be considered in the calculation of the growth rate. Populations grow when the number of births exceeds the number of deaths. If the number of births is 30 per year and the number of deaths is 10 per year per 1,000 individuals, the growth rate is

$$(30 - 10)/1{,}000 = 0.02 = 2.0\%$$

The highest possible rate of natural increase for a population is called **biotic potential** (Fig. 44.3). Whether the biotic potential is high or low depends on the number of limiting factors that reduce or slow the population's potential reproduction, such as the following:

- Number of offspring per reproductive event that survive until the age of reproduction
- Amount of competition within the population
- Age and number of reproductive opportunities
- Presence of disease and predators

Mortality Patterns

Population growth patterns assume that populations are made up of identical individuals. Actually, the individuals of a population are in different stages of their lifespan. A **cohort** is all the members of a population born at the same time. Some investigators study population dynamics and construct life tables that show how many members of a cohort are still alive after certain intervals of time.

For example, Table 44.1 is a life table for a bluegrass cohort. The cohort contains 843 individuals. The table tells us that after 3 months, 121 individuals have died, and therefore the mortality rate is 0.143 per capita. Another way to express the same statistic, however, is to consider that 722 individuals are still alive—have survived—after 3 months. **Survivorship** is the probability of newborn individuals of a cohort surviving to particular ages. If we plot the number surviving at each age, a survivorship curve is produced (Fig. 44.4*b*).

The results of investigations like this have revealed that each species tends to fit one of three typical survivorship curves: I, II, and III (Fig. 44.4*a*). The type I curve is characteristic of a

Table 44.1 Life Table for a Bluegrass Cohort

Age (months)	Number Observed Alive	Number Dying	Mortality Rate per Capita	Avg. Number of Seeds per Individual
0–3	843	121	0.143	0
3–6	722	195	0.271	300
6–9	527	211	0.400	620
9–12	316	172	0.544	430
12–15	144	95	0.626	210
15–18	54	39	0.722	60
18–21	15	12	0.800	30
21–24	3	3	1.000	10
24	0	—	—	—

Figure 44.4 Survivorship curves. Survivorship curves show the number of individuals of a cohort that are still living over time. **a.** Three generalized survivorship curves. **b.** The survivorship curve for bluegrass represents a combination of the type I and type II curves. **c.** The survivorship curve for lizards generally fits a type II curve. **d.** The survivorship curve for mosquitoes is a type III curve.

population in which most individuals survive well past the midpoint of the lifespan and death does not come until near the end of the lifespan. Animals that have this type of survivorship curve include large mammals and humans in more-developed countries. In contrast, the type III curve is typical of a population in which most individuals die very young. This type of survivorship curve occurs in many invertebrates, fishes, and most plants. In the type II curve, survivorship decreases at a constant rate throughout the lifespan. In many songbirds and small mammals, death is usually unrelated to age; thus, they represent a type II survivorship curve.

The survivorship curves of natural populations do not always fit these three idealized curves. In a bluegrass cohort, as shown in Table 44.1, for example, most individuals survive until 6–9 months, and then the chances of survivorship diminish at an increasing rate (Fig. 44.4*b*). Statistics for a lizard cohort are close enough to classify the survivorship curve in the type II category (Fig. 44.4*c*), while a mosquito cohort has a type III curve (Fig. 44.4*d*).

Much can be learned about the life history of a species by studying its life table and the survivorship curve that can be constructed based on that table. In a population with a type III survivorship curve, it would be predicted that natural selection would favor those individuals that had more offspring. Because death comes early for most members, only a few would live long enough to reproduce, and therefore natural selection would favor those that produced more offspring.

Other types of information are also available from studying life tables. Looking again at Table 44.1, we can see that per-capita seed production increases as plants mature, and then seed production drops off.

Age Distribution

When the individuals in a population reproduce repeatedly, several generations may be alive at the same time. From the perspective of population growth, a population contains three major age groups: prereproductive, reproductive, and postreproductive. Populations differ according to what proportion of the population falls into each age group. At least three **age structure diagrams** are possible (Fig. 44.5).

Having a birthrate that is higher than the death rate results in the prereproductive group being the largest of the three groups. This produces a pyramid-shaped diagram. Under such conditions, even if the growth for that year were matched by the deaths for that year, the population would continue to grow in the following years. More individuals would be entering than leaving the reproductive group. Eventually, as the size of the reproductive group equals the size of the prereproductive group, a bell-shaped diagram results.

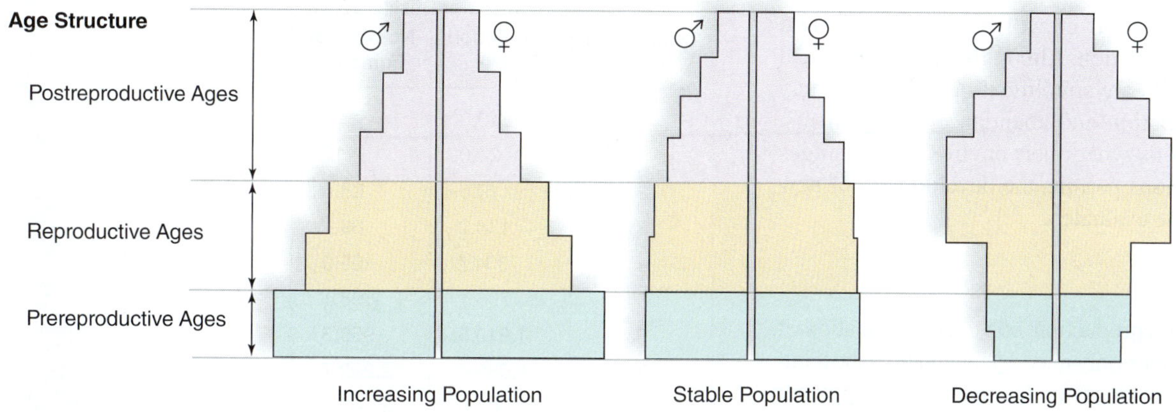

Figure 44.5 Age structure diagrams. Typical age structure diagrams for hypothetical populations that are increasing, stable, or decreasing. Different numbers of individuals in each age class create these distinctive shapes. In each diagram, the left half represents males, whereas the right half represents females.

The postreproductive group is still the smallest, however, because of mortality. If the birthrate falls below the death rate, the prereproductive group becomes smaller than the reproductive group. The age structure diagram is urn-shaped, because the postreproductive group is now the largest.

The age distribution reflects the past and future history of a population. Because a postwar baby boom occurred in the United States between 1946 and 1964, the postreproductive group will soon be the largest group.

Check Your Progress 44.2

1. Distinguish between population density and population distribution.
2. Describe the differences among type I, II, and III survivorship curves.
3. Explain why a bell-shaped age pyramid indicates a growing population.

44.3 Population Growth Models

Learning Outcomes

Upon completion of this section, you should be able to

1. Describe exponential population growth and the circumstances that encourage it.
2. Identify the features of logistic growth and the carrying capacity of a population.

Based on observation and natural selection principles, ecologists have developed two working models for population growth. In the pattern called **semelparity** (Gk. *seme,* "once"; L. *parous,* "to bear or bring forth"), the members of the population have only a single reproductive event in their lifetime. When the time for reproduction draws near, the mature adults cease to grow and expend all their energy in reproduction, and then die. Many insects, such as winter moths, and annual plants, such as zinnias, follow this pattern of reproduction growth. They produce a resting stage of development, such as eggs or seeds that can survive unfavorable conditions and resume growth the next favorable season. In other words, semelparity is an adaptation to an unstable environment.

In the pattern called **iteroparity** (Gk. *itero,* "repeat"), members of the population experience many reproductive events throughout their lifetime. They continue to invest energy in their future survival, which increases their chances of reproducing again. Iteroparity is an adaptation to a stable environment in which chances of survival for offspring are relatively high. Most vertebrates, shrubs, and trees have this pattern of reproduction.

Reproduction does not always fit these two patterns, though (Fig. 44.6). Even so, ecologists have found it useful to develop

a.

b.

Figure 44.6 Patterns of reproduction. In general, organisms tend to reproduce continuously or have a single reproductive event. **a.** Aphids use asexual reproduction during the summer months and then shift to sexual reproduction right before the onset of winter. **b.** Annual plants tend to produce a large number of seeds per reproductive event. The number that germinate often depends on environmental factors.

mathematical models of population growth based on these two very different patterns of reproduction. Although the mathematical models described in this section are simplifications, they still may be used to predict the distribution and abundance of organisms, or the responses of populations when their environment changes in some way. Testing predictions permits the development of new hypotheses, which can then be evaluated.

Exponential Growth

As an example of semelparous reproduction, consider a population of insects in which females reproduce only once a year and then the adult population dies. Each female produces on the average 4.8 eggs per generation, half of which will develop into female offspring, which reproduce the following year. In the next generation, the females produce an average of 4.8 eggs. In this case of discrete breeding, R = net reproductive rate.[1] Net reproductive rate is used because it is the observed rate of natural increase after deaths have occurred.

Figure 44.7*a* shows how the population would grow year after year for 10 years, assuming that R stayed constant from generation to generation. This growth is equal to the size of the population, because all members of the previous generation have died. Mayflies, featured in Figure 44.7, have one reproductive event, which occurs in the spring. The development of the next generation requires as many as 50 molts during the winter. Figure 44.7*b* shows the growth curve for such a population. This growth curve, which is roughly J shaped, depicts exponential growth. With **exponential growth,** the number of individuals added each generation increases rapidly due to the total number of reproductive females increasing in the population.

3D Animation Population Ecology: Exponential Population Growth

Video Exponential Growth

Notice that the growth curve in Figure 44.7*b* has two phases:

Lag phase: During this phase, growth is slow, because the population is small.

Exponential growth phase: During this phase, growth is accelerating.

Figure 44.7*c* gives the mathematical equation that allows you to calculate growth and size for any population that has discrete (nonoverlapping) generations. In other words, all members of the previous generation die off before the new generation appears. To use this equation to determine future population size, it is necessary to know R, which is the net reproductive rate determined after gathering mathematical data regarding past population increases. Notice that even though R remains constant, growth is exponential, because the number of individuals added each year is increasing. Therefore, the growth of the population is accelerating.

For exponential growth to continue unchecked, plenty of habitat, food, shelter, and any other requirements to sustain growth must be available. But in reality, environmental conditions prevent exponential growth. Eventually, any further growth is impossible because of limiting factors—the food supply runs out, and waste products begin to accumulate. Also, as the population increases in size, so do the effects of predation, parasites, disease, and competition among members.

Generation	Population Size	Number of Females
0	10.0	5
1	24.0	12
2	57.6	28.8
3	138.2	69.1
4	331.7	165.9
5	796.1	398.1
6	1,910.6	955.3
7	4,585.4	2,292.7
8	11,005.0	5,502.5
9	26,412.0	13,206.1
10	63,388.8	31,694.5

a.

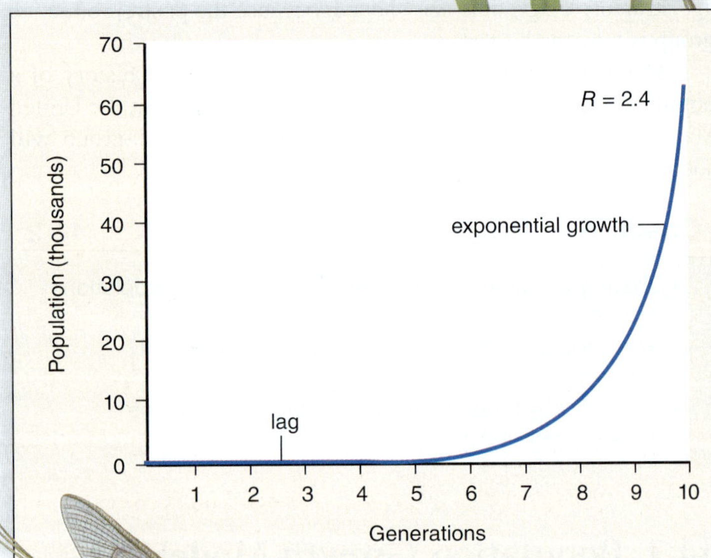

b.

To calculate population size from year to year, use this formula:

$$N_{t+1} = RN_t$$

N_t = number of females already present

R = net reproductive rate

N_{t+1} = population size the following year

c.

Figure 44.7 Model for exponential growth. When the data for discrete reproduction in (**a**) are plotted, the exponential growth curve in (**b**) results. **c.** This formula produces the same results as (**a**) and generates the same graph.

Tutorial Patterns of Population Growth

Logistic Growth

A second type of growth curve results when limiting environmental factors that oppose growth come into play. In 1930, Raymond Pearl developed a method for estimating the number of yeast cells

[1]Changing r to R is simply customary in discrete (nonoverlapping) breeding calculations; both coefficients deal with the same thing (birth minus death).

Growth of Yeast Cells in Laboratory Culture		
Time (t) (hours)	Number of individuals (N)	Number of individuals added per 2-hour period $\left(\frac{\Delta N}{\Delta t}\right)$
0	9.6	0
2	29.0	19.4
4	71.1	42.1
6	174.6	103.5
8	350.7	176.1
10	513.3	162.6
12	594.4	81.1
14	640.8	46.4
16	655.9	15.1
18	661.8	5.9

a.

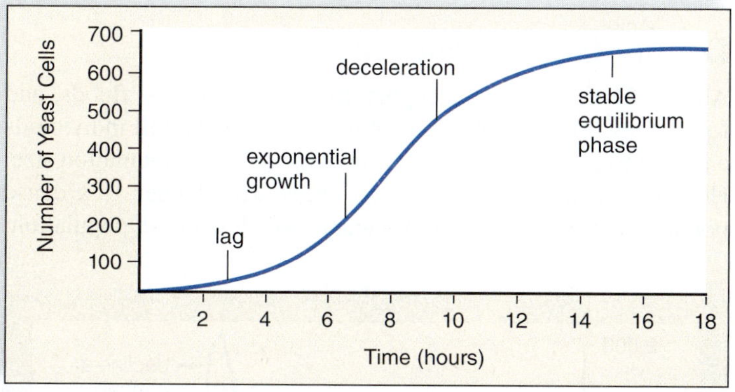

b.

To calculate population growth as time passes, use this formula:

$$\frac{N}{t} = rN\left(\frac{K-N}{K}\right)$$

N = population size

N/t = change in population size

r = rate of natural increase

K = carrying capacity

$\frac{K-N}{K}$ = effect of carrying capacity on population growth

c.

Figure 44.8 Model for logistic growth. When the data for repeated reproduction (**a**) are plotted, the logistic growth curve in (**b**) results. **c.** This formula produces the same results as (**a**) and generates the same graph.

Tutorial Patterns of Population Growth

accruing every 2 hours in a laboratory culture vessel. His data are shown in Figure 44.8*a*. When the data are plotted, the growth curve has the appearance shown in Figure 44.8*b*. This type of growth curve is a sigmoidal (S) or S-shaped curve.

3D Animation Population Ecology: Logistic Population Growth

Notice that this **logistic growth** has four phases:

Lag phase: During this phase, growth is slow, because the population is small.

Exponential growth phase: During this phase, growth is accelerating.

Deceleration phase: During this phase, growth slows down.

Stable equilibrium phase: During this phase, there is little if any growth, because births and deaths are about equal.

Figure 44.8*c* gives the mathematical equation that allows us to calculate logistic growth. (The curve is termed *logistic* because the exponential portion of the curve produces a straight line when the log of N is plotted.) The entire equation for logistic growth is

$$\frac{N}{t} = rN\frac{(K-N)}{K}$$

but let's consider each portion of the equation separately.

Because the population has repeated reproductive events, we need to consider growth as a function of change in time (Δ):

$$\frac{\Delta N}{\Delta t} = rN$$

If the change in time is very small, then we can use differential calculus, and the instantaneous population growth (d) is given by

$$\frac{dN}{dt} = rN$$

This portion of the equation applies to the first two phases of growth—the lag phase and the exponential growth phase. Here, also, we do not expect exponential growth to continue. Charles Darwin calculated that a single pair of elephants could have over 19 million live descendants after 750 years. Others have calculated that a single female housefly could produce over 5 trillion flies in 1 year! Such explosive growth does not occur, however, because environmental conditions, both abiotic and biotic, cause population growth to slow. The yeast population mentioned earlier was grown in a vessel in which food would run short and waste products would accumulate. Such environmental conditions prevent exponential growth from continuing.

Look again at Figure 44.8. Following exponential growth, a population is expected to enter a deceleration phase and then a stable equilibrium phase of the logistic growth curve. Now the population is at the carrying capacity of the environment.

Carrying Capacity

The environmental **carrying capacity (*K*)** is the maximum number of individuals of a given species the community can support. The closer a population size nears the carrying capacity of the community, the more likely it is that resources will become scarce and that biotic effects, such as competition and predation, will become evident. The birthrate is expected to decline, and the death rate is expected to increase. This results in a decrease in population growth; eventually, the population stops growing and its size remains stable. Carrying capacity in any community can vary throughout time, depending on fluctuating conditions—for example, the amount of rainfall from one year to the next.

How does the mathematical model for logistic growth take this process into account? To our equation for growth under conditions of exponential growth, we add the following term:

$$\frac{(K - N)}{K}$$

In this expression, K is the carrying capacity of the environment. The easiest way to understand the effects of this term is to consider two extreme possibilities. First, consider a time at which the population size is well below carrying capacity. Resources are relatively unlimited, and we expect rapid, nearly exponential growth to take place. The model predicts this. When N is very small relative to K, the term $(K-N)/K$ is very nearly $(K-0)/K$, or approximately 1. Therefore, $(dN)/(dt)$ is approximately equal to rN.

Similarly, consider what happens when the population reaches carrying capacity. Here, we predict that growth will stop and the population will stabilize. When N is equal to K, the term $(K-N)/K$ declines from nearly 1 to 0, and the population growth slows to zero.

As mentioned, the model predicts that exponential growth occurs only when population size is much lower than the carrying capacity. So as a practical matter, if humans are using a fish population as a continuous food source, it is best to maintain the size of the fish population in the exponential phase of the logistic growth curve. Biotic potential can have its full effect, and the birthrate is the highest it can be during this phase. If we overfish, the population will sink into the lag phase, and it will be years before exponential growth recurs.

In contrast, if we are trying to limit the growth of a pest, it is best, if possible, to reduce the carrying capacity rather than reduce the population size. Reducing the population size only encourages exponential growth to begin once again. Farmers can reduce the carrying capacity for a pest by alternating rows of different crops, rather than growing one type of crop throughout an entire field.

Check Your Progress **44.3**

1. Explain how carrying capacity (K) limits exponential growth.
2. Explain the conditions that would cause a population to undergo logistic growth.

44.4 Regulation of Population Size

Learning Outcomes

Upon completion of this section, you should be able to

1. Compare the density-independent and density-dependent factors that affect population size.
2. Describe the intrinsic factors that can impact population size and growth.

In a study of winter moth population dynamics, researchers discovered that a large proportion of eggs did not survive the winter and exponential growth never occurred. Perhaps a low number of individuals at the start of each season helps prevent the occurrence of exponential growth.

It is possible that exponential growth causes population size to rise above the carrying capacity of the environment, and as a consequence a population crash may occur. As one example, in 1911, 4 male and 21 female reindeer were released on St. Paul Island in the Bering Sea off Alaska. St. Paul Island had an undisturbed environment, and there was little hunting pressure and no predators. The herd grew exponentially to about 2,000 reindeer by 1938, overgrazed the habitat, and then abruptly declined to only 8 animals by 1950 (Fig. 44.9).

This pattern of a population explosion, eventually followed by a population crash, is called irruptive, or Malthusian, growth. It is named in honor of the eighteenth-century economist Thomas Robert Malthus, who had a great influence on Charles Darwin. Populations do not ordinarily undergo Malthusian growth because of factors that regulate population growth.

Ecologists have long recognized that both biotic and abiotic conditions play an important role in regulating population size in natural environments.

Density-Independent Factors

Abiotic factors include droughts, freezes, hurricanes, floods, and forest fires. Any one of these natural disasters can kill individuals and lead to a sudden and catastrophic reduction in population size. However, such an event does not necessarily happen to a dense population more often than it happens to a less dense population.

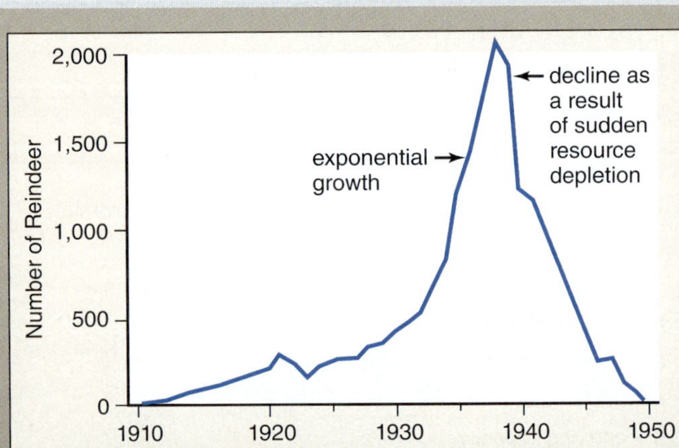

Figure 44.9 Density-dependent effect. On St. Paul Island, Alaska, reindeer, *Rangifer,* grew exponentially for several seasons and then underwent a sharp decline as a result of overgrazing the available range.

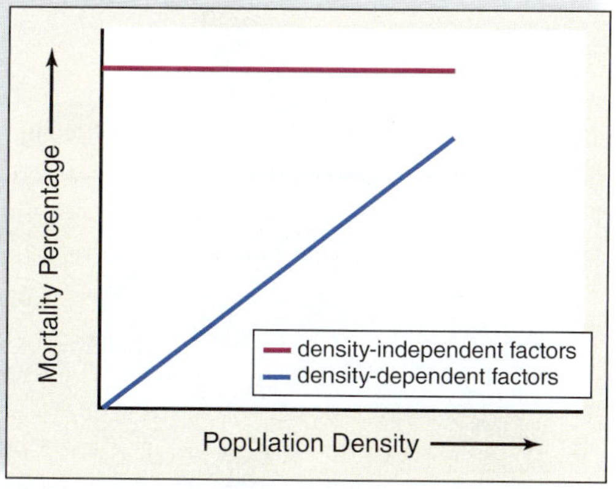

Figure 44.10 **Percentage of individuals that die per density of population.**

Therefore, an abiotic factor is usually a **density-independent factor,** meaning that the intensity of the effect does not increase with increased population density (Fig. 44.10).

3D Animation Population Ecology: Density-Independent Factors

For example, the percentage of individuals killed in a flash flood event is independent of density—floods do not necessarily kill a larger percentage of a dense population than of a less dense population. Nevertheless, the larger the population, the greater the number of individuals probably affected. In Figure 44.11, the impact of a flash flood on a low-density population of mice living in a field was 3 out of 5 (60% mortality), whereas the impact on a high-density population was 12 out of 20 (also 60% mortality).

Density-Dependent Factors

Biotic factors are considered **density-dependent factors,** because the percentage of the population affected does increase as the density of the population increases. Competition, predation, disease, and parasitism are all biotic factors that increase in intensity as the density increases.

3D Animation Population Ecology: Density-Dependent Factors

Competition can occur when members of a species attempt to use the same resources (such as light, food, space) that are in limited supply. As a result, not all members of the population can have access to the resources to the degree necessary to ensure survival or reproduction. As an example, let's consider a western bluebird (*Sialia mexicana*) population in which members have to compete for nesting sites. Each pair of birds requires a tree hole to raise offspring. If there are more holes than breeding pairs, each pair can have a hole in which to lay eggs and rear young birds. But if there are fewer holes than there are breeding pairs, then each pair must compete to acquire a nesting site. Pairs that fail to gain access to holes will be unable to contribute new members to the population.

Video Booby Chick Competition

Competition for food also controls population growth. However, resource partitioning among age groups is a way to reduce competition for food. As mentioned earlier in this text, the life cycle of butterflies includes caterpillars, which require a different food from the adults. The caterpillars graze on leaves, while the adults feed on nectar produced by flowers. Therefore, parents do not compete with their offspring for food.

Predation occurs when one living organism, the predator, eats another, the prey. In the broadest sense, predation impacts every species within a community at some level. The effect of predation on a prey population generally increases as the prey population grows denser, because prey are easier to find. Consider a field inhabited by a population of mice (Fig. 44.12). Each mouse must have a hole in which to hide to avoid being eaten by a hawk. If there are 100 holes, and a low density of 102 mice, then only 2 mice will be left out in the open. It might be hard for the hawk to find only 2 mice in the field. If neither mouse is caught, then the predation rate is 0/2 = 0%. However, if there are 100 holes, and a high density of 200 mice, then the chance is greater that the hawk will be able to find some of these 100 mice without holes. If half of the exposed mice are caught, the predation rate is 50/100 = 50%. Therefore, increasing the density of the available prey has increased the proportion of the population preyed upon.

Video Harris Hawk

Parasites, such as blood-sucking ticks, are generally much smaller than their hosts. Although parasites do not always kill their hosts, they do usually weaken them over time. A highly parasitized individual is less apt to produce as many offspring than it would if it were healthy. In this way, parasitism also plays a role in regulating population size.

Disease is much more likely to spread in a dense population due to the amount of contact between healthy individuals and

a. Low density of mice

b. High density of mice

Figure 44.11 **Density-independent effects.** The impact of a density-independent factor, such as weather or natural disasters, is not influenced by population density. The impact of a flash flood on (**a**) a low-density population (mortality rate of 3/5, or 60%) is similar to the impact on (**b**) a high-density population (mortality rate of 12/20, also 60%).

a. Low density of mice

b. High density of mice

Figure 44.12 Density-dependent effects—predation. The impact of predation on a population is directly proportional to the density of the population. In a low-density population (**a**), the chances of a predator finding the prey are low, resulting in little predation. But in the higher-density population (**b**), there is a greater likelihood of the predator locating potential prey, resulting in a greater predation rate.

infected individuals. Consider the spread of the *Yersinia pestis* plague, known as the Black Death, in Europe during the fourteenth century; this outbreak of bacterial disease reduced the population of Europe by 30–60%.

Other Considerations

Density-independent and density-dependent factors are extrinsic to the organism. Intrinsic factors—those based on the anatomy, physiology, or behavior of the organism—can also affect population size and growth rates. Territoriality and dominance hierarchies are behaviors that affect population size and growth rates. Recruitment and migration are other intrinsic social means by which the population sizes of more complex organisms are regulated.

Outside of any regulating factors, it could be that some populations have an innate instability. Ecologists have developed models that predict complex, erratic changes in even simple systems. For example, a computer model of Dungeness crab populations assumed that adults produce many larvae and then die. Most of the larvae do not survive, and those that do stay close to home. Under these circumstances, the model predicted wild fluctuations in population size without a recurring pattern. This type of complex, nonrandom generation is called *deterministic chaos,* or simply *chaos.*

Population growth–regulating factors can serve as selective agents. Some members of a population may possess traits that make it more likely that they, rather than other members of the population, will survive and reproduce when these particular density-independent or density-dependent factors are present in the environment. Therefore, these traits will be more prevalent in the next generation whenever these factors are a part of the environment (see Fig. 15.11).

Check Your Progress 44.4

1. Describe the effect that population density can have on competition and predation.
2. Provide examples that show how a density-independent factor can act as a selective agent.
3. Identify various intrinsic factors that influence population size and growth.

44.5 Life History Patterns

Learning Outcomes

Upon completion of this section, you should be able to

1. Compare the two life history patterns of species.
2. List organisms that exemplify *r*-selection and those that exhibit *K*-selection.

Populations vary on such particulars as the number of births per reproductive event, the age at reproduction, the lifespan, and the probability of living the entire lifespan. These particulars are part of a species' life history. Life histories contain characteristics that can be thought of as trade-offs. Each population is able to capture only so much of the available energy, and how this energy is distributed among its lifespan (short versus long), reproduction events (few versus many), care of offspring (little versus much), and so forth has evolved over time. Natural selection shapes the final life history of individual species. Related species, such as frogs and toads, may have different life history patterns, if they occupy different environmental roles (Fig. 44.13).

The logistic population growth model has been used to suggest that members of some populations are subject to *r*-selection and members of other populations are subject to *K*-selection.

Animation Life History Patterns

r-Selected Populations

In fluctuating or unpredictable environments, density-independent factors keep populations in the lag or exponential phase of population growth. Population size is low relative to *K,* and ***r*-selection** favors *r*-strategists, which tend to be small individuals that mature early and have a short lifespan. Most energy goes into producing many relatively small offspring, and minimal energy goes into parental care. The more offspring, the more likely it is that some of them will survive to reproductive age.

Because of low population densities, density-dependent mechanisms, such as predation and intraspecific competition, are unlikely to play a major role in regulating population size and growth rates most of the time. Such organisms are often very good dispersers and colonizers of new habitats. Classic examples of such *opportunistic species* are bacteria, some fungi, many insects, rodents, and annual plants (Fig. 44.14, *left*).

a. Mouth-brooding frog, *Rhinoderma darwinii*

b. Strawberry poison arrow frog, *Dendrobates pumilio*

c. Midwife toad, *Alyces obstetricans*

Figure 44.13 Parental care among frogs and toads. **a.** In mouth-brooding frogs of South America, the male carries the larvae in a vocal pouch (*brown area*), which elongates the full length of his body, before the froglets are released. **b.** In poison arrow frogs of Costa Rica, after the eggs hatch, the tadpoles wiggle onto the parent's back (*at white arrow*) and are then carried to water. **c.** The midwife toad of Europe carries strings of eggs entwined around his hind legs and takes them to water when they are ready to hatch.

K-Selected Populations

Some environments are relatively stable and predictable, and in these environments populations tend to be near *K,* with minimal fluctuations in size. Resources such as food and shelter are relatively scarce for these individuals, and those that are best able to compete have the greatest reproductive success. ***K*-selection** favors *K*-strategists, species that allocate energy to their own growth and survival, as well as to the growth and survival of their offspring. Therefore, they are usually fairly large, are late to mature, and have relatively few offspring (fecundity). They also have a limited number of reproductive events (parity) and have a fairly long lifespan.

Because these organisms, termed *equilibrium species,* are strong competitors, they can become established and exclude opportunistic species. They are specialists rather than colonizers and tend to become extinct when their normal way of life is destroyed. The best possible examples of *K*-strategists include long-lived plants (saguaro cacti, oaks, cypresses, and pines), birds of prey (hawks and eagles), and large mammals (whales, elephants, bears, and humans) (Fig. 44.14, *right*). Another example of a *K*-strategist is the Florida panther, the largest mammal in the Florida Everglades. It requires a very large range and produces few offspring, which require parental care. Currently, the Florida panther is unable to compensate for a reduction in its range and is therefore on the verge of extinction.

The Two Strategies in Nature

Nature is actually more complex than the two possible life history patterns suggest. It now appears that *r*-strategist and *K*-strategist populations are at the ends of a continuum, and most populations lie somewhere between these two extremes. For example, recall that plants

Figure 44.14 Life history strategies. Are dandelions *r*-strategists with the characteristics noted, and are bears *K*-strategists with the characteristics noted? Most often, the distinctions between these two possible life strategies are not clear-cut.

THEME Biological Systems

When a Population Grows Too Large

White-tailed deer (*Odocoileus virginianus*) (Fig. 44A*a*), which live from southern Canada to below the equator in South America, are prolific breeders. In one study, investigators found that 2 male and 4 female deer were able to produce 160 offspring in 6 years. Theoretically, the number of offspring could have reached 300, because the majority of does (female deer) tend to produce 2 young each time they reproduce.

A century ago, the white-tailed deer population across the eastern United States was less than half a million. Today, it is well over 200 million—even more than existed when Europeans first colonized America. This dramatic increase in population size can be attributed to a reduction in natural predators and a readily available supply of food. The natural predators of deer, such as wolves and mountain lions, are now absent from most regions, due mostly to humans' killing of these animals as well as a significant loss of available habitat. In addition, deer hunting is tightly controlled by government agencies and even banned in some urban areas.

The sad reality is that in areas where deer populations have become too large, the deer suffer from starvation as they deplete their own food supply. For example, after deer hunting was banned on Long Island, New York, the deer population quickly outgrew available food resources. The animals became sickly and weak and weighed so little that their ribs, vertebrae, and pelvic bones were visible through their skin.

A very large deer population can also cause humans many problems. A homeowner is dismayed to see new plants decimated and evergreen trees damaged because of munching deer. The economic damage that large deer populations cause to agriculture, landscaping, and forestry exceeds a billion dollars per year. More alarming, over a million deer-vehicle collisions take place in the United States each year (Fig. 44A*b*), resulting in over a billion dollars in insurance claims, thousands of human injuries, and hundreds of human deaths—not to mention the injuries and deaths of deer. Lyme disease, transmitted by deer ticks to humans, infects over 3,000 people annually and can lead to debilitating arthritic symptoms.

Deer overpopulation hurts not only deer and humans but other species as well. The forested areas that are overpopulated by deer have fewer understory plants. Furthermore, the deer selectively eat certain species of plants, while leaving others alone. This selection can cause long-lasting changes in the number and diversity of trees in forests, leading to a negative economic impact on logging and forestry. The number of songbirds, insects, squirrels, mice, and other animals declines with an increasing deer population. It is important, therefore, that we learn to properly manage deer populations.

One example of a successful strategy is that in some states, such as Texas, large landowners now set aside a portion of their property for a deer herd. They improve the nutrition of the herd and restrict the hunting of young bucks, but they allow hunting of does. The result is a self-sustaining herd that brings the landowner economic benefits through outfitting services.

Questions to Consider

1. If a deer herd becomes too large, should the government pay hunters to reduce its size?
2. What is the ideal size of a deer herd for a given area?
3. Is trophy deer hunting an ideal way to control herd size?

a.

b.

Figure 44A White-tailed deer. **a.** Buck (male deer) in rut during mating season. **b.** Deer killed in a car accident.

have an alternation-of-generations life cycle—the sporophyte generation and the gametophyte generation. Ferns, which could be classified as *r*-strategists, distribute many spores and leave the gametophyte to fend for itself. In contrast, gymnosperms (e.g., pine trees) and angiosperms (e.g., oak trees), which could be classified as *K*-strategists, retain and protect the gametophyte. They produce seeds that contain the next sporophyte generation plus stored food. The added investment is significant, but these plants still release large numbers of seeds.

Also, adult size is not always a determining factor for the life history pattern. For example, a cod is a rather large fish weighing up to 12 kg and measuring nearly 2 m in length—but cod release gametes in vast numbers, the zygotes form in the sea, and the parents make no further investment in the developing offspring. Of the 6–7 million eggs released by a single female cod, only a few will become adult fish. Cod are generally considered *r*-strategists.

The Biological Systems feature, "When a Population Grows Too Large," describes the consequences of unchecked population growth in white-tailed deer, a *K*-selected species for which a limiting factor, predation, has been reduced.

Check Your Progress 44.5

1. Compare and contrast the general characteristics of a *K*-strategist and an *r*-strategist.
2. Explain why a population may vary between *K* and *r* strategies.

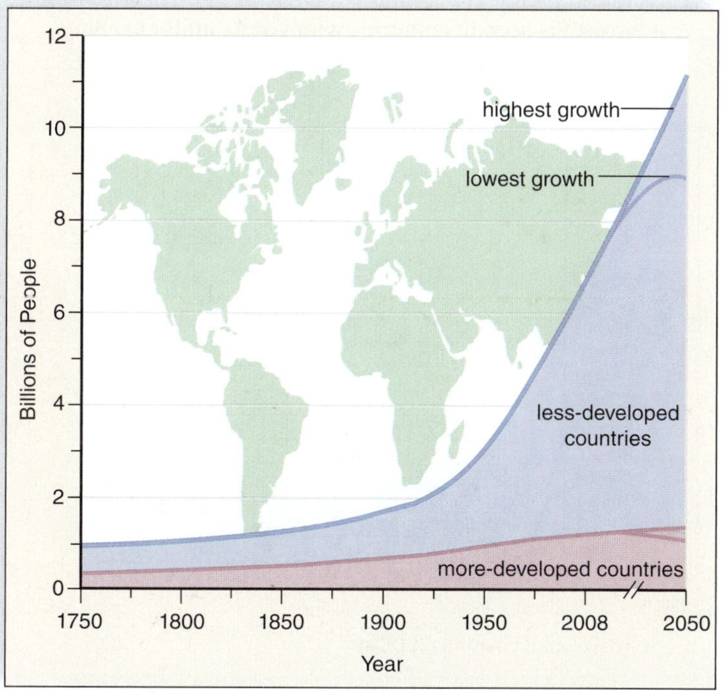

a.

Figure 44.15 World population growth. **a.** The world's population size is now slightly over 7 billion. It is predicted that it may level off between 9 and 11 billion by 2050, depending on the speed with which the growth rate declines. The lifestyle of most individuals in the (**b**) MDCs is contrasted with that of most individuals in the (**c**) LDCs.

44.6 Human Population Growth

Learning Outcomes

Upon completion of this section, you should be able to

1. Describe the past and present growth patterns of the human population.
2. Identify the pressures the human population places on the Earth's resources.

The world's population has risen steadily to its present size of over 7.1 billion people (Fig. 44.15). Prior to 1750, the growth of the human population was relatively slow, but as more reproducing individuals were added, growth increased, until the curve began to slope steeply upward, indicating that the population was undergoing exponential growth. The number of people added annually to the world population peaked at about 87 million around 1990, and currently it is a little over 79 million per year. This is roughly equal to the current populations of Argentina, Ecuador, and Peru combined.

The potential for future population growth can be appreciated by considering the **doubling time,** the length of time it takes for the population size to double. Currently, the doubling time is estimated to be between 50 and 60 years. Such an increase in population size would put extreme demands on our ability to produce and distribute resources. In 50 to 60 years, the world would need double the

b. Lifestyle in a city of a MDC

c. Lifestyle in a LDC

amount of food, jobs, water, energy, and so on just to maintain the present standard of living.

Many people are gravely concerned about the increased amount of resources that will be needed to support the additional growth in the human population. The first billion didn't occur until 1800; the second billion was attained in 1930; the fourth billion in 1974; and today there are over 7.2 billion. The world's population may level off around 8, 10.5, or 14.2 billion, depending on the speed at which the growth rate declines. Zero population growth cannot be achieved until the birthrate is equal to the death rate.

More-Developed Versus Less-Developed Countries

The countries of the world can be divided into two general groups. In the **more-developed countries (MDCs)**, typified by the United States, Germany, Japan, and Australia, population growth is low, and the people enjoy a high standard of living. In the **less-developed countries (LDCs)**, such as certain countries in Latin America, Africa, and Asia, population growth is expanding rapidly, and the majority of people live in poverty.

The MDCs doubled their populations between 1850 and 1950. This change was largely due to a decline in the death rate, the development of modern medicine, and improved socioeconomic conditions. The decline in the death rate was followed shortly thereafter by a decline in the birthrate, so populations in the MDCs experienced only modest growth between 1950 and 1975. This sequence of events (decreased death rate followed by decreased birthrate) is termed a **demographic transition.** Yearly growth of the MDCs has now stabilized at about 0.3%.

In contrast to the other MDCs, there is no end in sight for U.S. population growth. Although yearly growth of the U.S. population is only 0.8%, many people immigrate to the United States each year. In addition, an unusually large number of births occurred between 1946 and 1964 (called a baby boom), resulting in a large number of women still of reproductive age.

Most of the world's population (approximately 80%) now lives in LDCs. Although the death rate began to decline steeply in the LDCs following World War II because of the importation of modern medicine from the MDCs, the birthrate remained high. The yearly growth of the LDCs peaked at 2.5% between 1960 and 1965. Since that time, a demographic transition has occurred, and the death rate and the birthrate have fallen. The collective growth rate for the LDCs is now 1.5%, but many countries in sub-Saharan Africa have not participated in this decline. In some of these, women average more than five children each.

The population of the LDCs may explode from 5.9 billion today to over 8 billion by 2050. Some of this increase will occur in Africa, but most will occur in Asia, because the Asian population is now about four times the size of the African population. Asia already has 60% of the world's population, living on 31% of the world's arable (farmable) land. Therefore, Asia is expected to experience acute water scarcity, a significant loss of biodiversity, and increased urban pollution. Twelve of the world's most polluted cities are in Asia.

The following are suggestions for reducing the expected population increase in the LDCs:

1. Establish or strengthen family planning programs. A decline in growth is seen in countries with good family planning programs supported by community leaders.

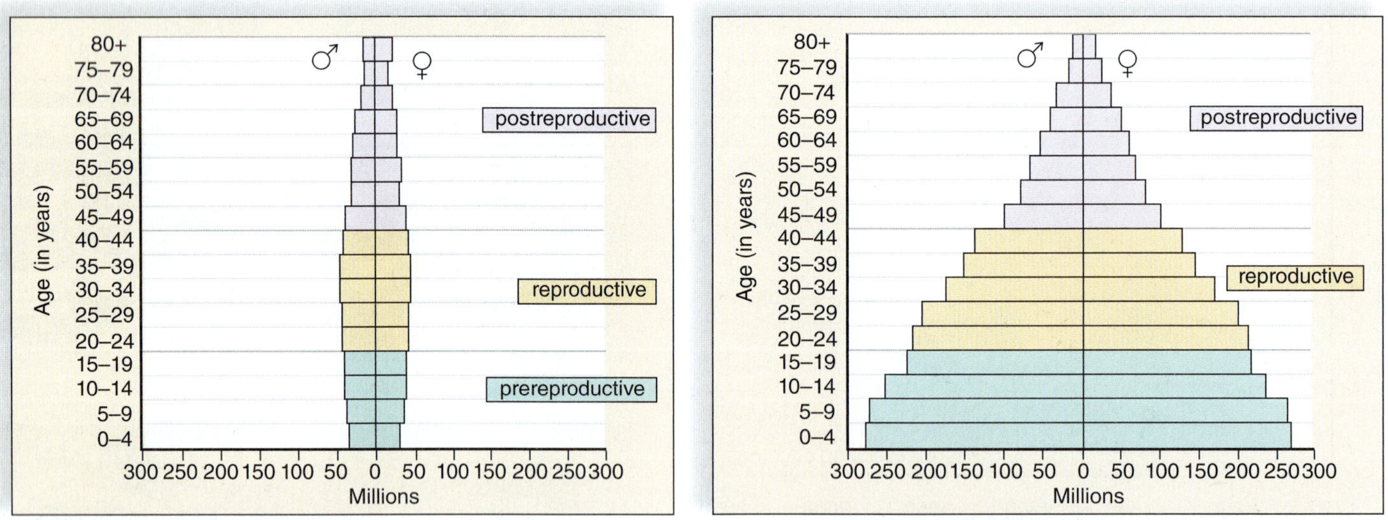

a. More-developed countries (MDCs)

b. Less-developed countries (LDCs)

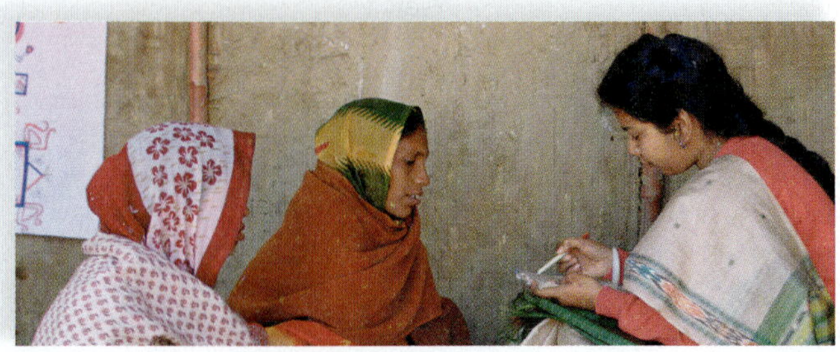

c.

Figure 44.16 Age-structure diagrams. The shape of these age-structure diagrams allows us to predict that (**a**) the populations of MDCs are approaching stabilization and (**b**) the populations of LDCs will continue to increase for some time. **c.** Improved women's rights and increasing contraceptive use could change this scenario. Here, a community health worker is instructing women in India about the use of contraceptives.

2. Use social progress to reduce the desire for large families. Providing education, raising the status of women, and reducing child mortality are social improvements that could reduce this desire.
3. Delay the onset of childbearing. This, along with wider spacing of births, could help birthrate decline.

Age Distributions

The populations of the MDCs and LDCs can be divided into the three age groups introduced earlier: prereproductive, reproductive, and postreproductive (Fig. 44.16). Currently, the LDCs are experiencing a population momentum, because they have more young women entering the reproductive years than older women leaving them.

Most people are under the impression that if each couple has two children, **zero population growth** (no increase in population size) will take place immediately. However, most countries will continue growing due to the age structure of the population and the human lifespan. If there are more young women entering the reproductive years than there are older women leaving them, **replacement reproduction** will still result in population growth.

Many MDCs have a stable age structure, but most LDCs have a youthful profile—a large proportion of the population is younger than 15 years old. For example, in Nigeria, 43.8% of the population is under age 15. On average, Nigerian women have a fertility rate of 5.25 per woman. As a result of these numbers, the population of Nigeria is expected to increase from 177 million to over 391 million by 2050. The population of the continent of Africa is projected to increase from slightly over 1.1 billion to around 2.2 billion between 2014 and 2050. This means that the LDC populations will still expand, even after replacement reproduction is attained. The more quickly replacement reproduction is achieved, however, the sooner zero population growth can also be achieved.

Population Growth and Environmental Impact

Population growth is putting extreme pressure on each country's social organization, the Earth's resources, and the biosphere. Because the population of the LDCs is still growing at a significant rate, it might seem that their population increase will be the greater cause of future environmental degradation. But this is not necessarily the case, because the MDCs consume a much larger proportion of the Earth's resources than do the LDCs. This consumption leads to environmental degradation, which is a major concern.

Environmental Impact

The environmental impact (EI) of a population is measured not only in terms of population size but also in terms of resource consumption per capita and the pollution produced by this consumption. In other words,

$$\begin{aligned} \text{EI} &= \text{population size} \times \text{resource consumption per capita} \\ &= \text{pollution per unit of resource used} \end{aligned}$$

Therefore, there are two possible types of overpopulation: The first is simply due to population growth, and the second is due to increased resource consumption caused by population growth. The first type of overpopulation is more obvious in LDCs, and the second type is more obvious in MDCs, because the per-capita consumption is so much higher in the MDCs. For example, an average family in the United States, in terms of per-capita resource consumption and waste production, is the equivalent of 30 people in India. We need to realize, therefore, that only a limited number of people can be sustained at anywhere near the standard of living in the MDCs.

The current environmental impact caused by MDCs and LDCs is shown in Figure 44.17. The MDCs account for only about one-fourth of the world population (Fig. 44.17*a*). But the MDCs have a significantly greater per person consumption of oil (Fig. 44.17*b*) and coal (Fig. 44.17*c*). While China leads the

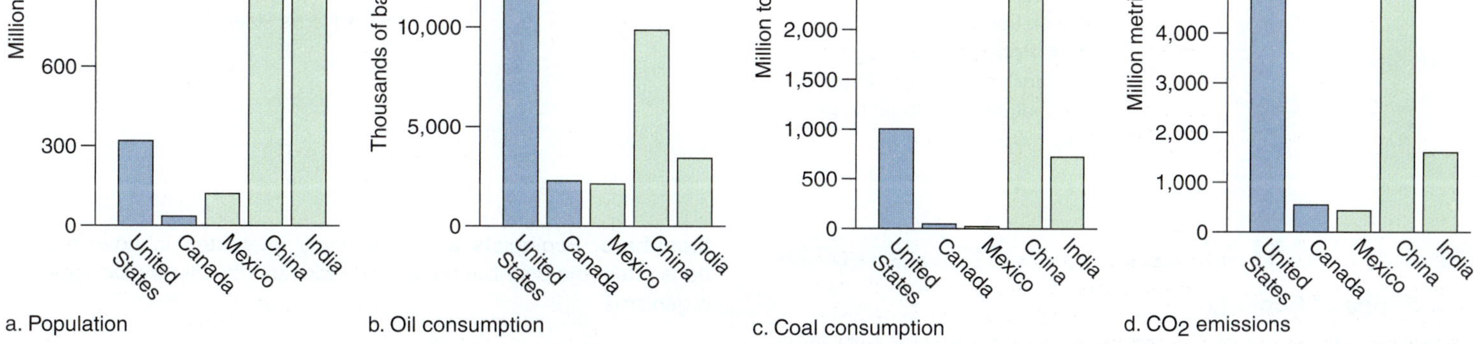

Figure 44.17 Environmental impact caused by MDCs and LDCs. **a.** MDCs have significantly smaller populations than LDCs. **b.** In the United States, per-person oil consumption is 8.1 times greater than that in China. **c.** In the United States, per-person oil consumption is 5.4 times greater than that in India. **d.** In the United States, per-person CO_2 production is 4.9 times greater than that in Mexico.

world in CO_2 emissions (Fig. 44.17*d*), they also have the largest percentage of the human population, making their impact smaller than that of MDCs.

As the LDCs become more industrialized, their per-capita consumption will also rise and, in some LDCs, it may nearly equal that of a more-developed country. For example, China's economy is growing rapidly and, because China has such a large population (1.3 billion), it is already competing with the United States for oil and metals on the world markets. It could be that as developing countries consume more, people in the United States will have no choice but to consume less.

Consumption of resources has a negative effect on the environment. In Chapter 45, you'll learn how resource consumption leads to various pollution problems, and in Chapter 47, you'll see how our increasing environmental impact may cause a mass extinction of wildlife that will significantly impact our entire planet.

Check Your Progress 44.6

1. Compare the population growth of the LDCs with that of the MDCs.
2. Describe why replacement reproduction still leads to continued population growth.
3. Explain how an increase in the consumption of resources by LDCs would affect consumption by MDCs.

CONNECTING *the* CONCEPTS *with the* THEMES

Evolution

- Ecology is integral to the evolution of a species, because interactions among members of a community along with the abiotic environment act as agents of natural selection.
- The evolution of lifespan, reproductive pattern, and number of offspring per reproductive event is shaped by the dynamics of the environment.

Nature of Science

- Modern ecological studies are focusing on how population change in one species impacts other species within a community.
- Principles of population growth and decline revealed by ecological research can apply to human populations and their environments.

Biological Systems

- An ecosystem can be seen as a large-scale biological system in which the many species and abiotic factors interact dynamically.
- Exponential population growth occurs when populations do not face limits on resources. When resources become limited, population growth exhibits a logistic curve.

MEDIA STUDY TOOLS http://connect.mheducation.com

SMARTBOOK® Study smart. SmartBook helps you maximize your study time by identifying what you know and don't know. The recharge feature in SmartBook helps you prepare for the test by identifying the material that you are most likely to forget.

connect BIOLOGY Find even more resources to learn the chapter concepts including animations, tutorial videos, and interactive practice questions.

Animation	3D Animations	Tutorial	Videos
44.5 Life History Patterns	**44.3** Population Ecology: Exponential Population Growth • Population Ecology: Logistic Population Growth **44.4** Population Ecology: Density-Independent Factors • Population Ecology: Density-Dependent Factors	**44.3** Patterns of Population Growth	**44.1** Coral Reef Ecosystem **44.3** Exponential Growth **44.4** Booby Chick Competition • Harris Hawk

SUMMARIZE

44.1 Scope of Ecology

Ecology is the study of the interactions of organisms with other organisms and with the physical environment. Some ecologists study how an organism is adapted to its individual **habitat,** whereas others look at larger levels of ecology. The additional levels of study include the organism, **population, community, ecosystem,** and **biosphere.** Ecologists are particularly interested in how interactions affect the distribution, abundance, and life history strategies of organisms.

44.2 Demographics of Populations

Demography studies provide statistical data about a population. For example, **population density** is the number of individuals per unit area or volume. Distribution of individuals can be uniform, random, or clumped. A **population's distribution** is often determined by abiotic factors, which include **resources** such as water and sunlight, temperature, and availability of nutrients. **Limiting factors** determine where an organism will live.

Population growth is dependent on number of births and number of deaths, as well as immigration and emigration. The number of births minus the number of deaths results in the **rate of natural increase** (r) (growth rate). **Biotic potential** is the highest possible rate of natural increase for a population. Population growth is measured by tracking a **cohort,** or members of a population that were born at the same time. Mortality (deaths per capita) within a population can be recorded in a life table and illustrated by a **survivorship** curve. The pattern of population growth is reflected in **age structure diagrams,** which are a reflection of the age distribution of a population, which consists of prereproductive, reproductive, and postreproductive segments.

44.3 Population Growth Models

Two working models are used to describe population growth. **Semelparity** describes a population that has one reproductive event before it dies. **Iteroparity** describes a population that has multiple reproductive events throughout its lifetime. If the environment offers unlimited resources and the number of reproductive females in the population reaches an optimum level, **exponential growth** results and produces a J-shaped curve.

Most environments restrict growth, and exponential growth cannot continue indefinitely. Under these circumstances, an S-shaped, or **logistic, growth** curve results. The growth of the population is given by the equation $N/t = rN(K-N)/K$ for populations in which individuals have repeated reproductive events. The term $(K-N)/K$ represents the unused portion of the **carrying capacity** (**K**). When the population reaches carrying capacity, the population stops growing, because environmental conditions oppose biotic potential.

44.4 Regulation of Population Size

Population growth can be limited by **density-independent factors** (e.g., abiotic factors, such as weather events) and **density-dependent factors** (biotic factors such as **competition** and **predation**). Other means of regulating population growth exist in some populations. For example, territoriality is possibly a means of population regulation. Other populations do not seem to be regulated, and their population size fluctuates widely.

44.5 Life History Patterns

The logistic growth model has been used to suggest that the environment promotes either ***r*-selection** or ***K*-selection.** In unpredictable environments, *r*-selection occurs due to density-independent factors. Energy is allocated to producing as many small offspring as possible. Adults remain small and do not invest in parental care of offspring. In relatively stable environments, *K*-selection occurs and density-dependent factors affect population size. Energy is allocated to survival and repeated reproductive events. The adults are large and invest in parental care of offspring. Actual life histories contain trade-offs between these two patterns, and such trade-offs are subject to natural selection.

44.6 Human Population Growth

The potential for human population growth can be predicted by determining the **doubling time.** Currently, the human population is still expanding, but deceleration has begun. **More-developed countries** (**MDCs**) experience a high standard of living and low birthrates. **Less-developed countries** (**LDCs**), on the other hand, experience a low standard of living and high birthrates. A **demographic transition** occurs when a country experiences a decrease in the death rate followed by a decrease in the birthrate. This helps slow the growth rate of a country. Encouraging people to follow a **zero population growth** pattern or **replacement reproduction** will also help slow population growth. It is unknown when the world population size will level off, but it may occur by 2050. The majority of population increase will come from the LDCs, but we also need to consider the significant environmental impact that MDCs have on the planet as we look to address future concerns.

ASSESS

Choose the best answer for each question.

44.1 Scope of Ecology

1. Which of these levels of ecological study involves an interaction between abiotic and biotic components?
   a. organisms
   b. populations
   c. communities
   d. ecosystem
   e. All of these are correct.
2. In which pair is the first included in the second?
   a. habitat population
   b. population community
   c. community ecosystem
   d. ecosystem biosphere
   e. All of these except a are correct.

44.2 Demographics of Populations

3. When phosphorus is made available to an aquatic community, the algal populations suddenly bloom. This indicates that phosphorus is
   a. a density-dependent regulating factor.
   b. a reproductive factor.
   c. a limiting factor.
   d. an *r*-selection factor.
   e. All of these are correct.
4. If a population has a type I survivorship curve (most have a long lifespan), which of these would you also expect?
   a. a single reproductive event per adult
   b. overlapping generations
   c. reproduction occurring near the end of the lifespan
   d. a very low birthrate
   e. None of these are correct.
5. A pyramid-shaped age distribution means that
   a. the prereproductive group is the largest group.
   b. the population will grow for some time in the future.
   c. the country is more likely an LDC than an MDC.
   d. fewer women are leaving the reproductive years than entering them.
   e. All of these are correct.

44.3 Population Growth Models

6. A J-shaped growth curve can be associated with
   a. exponential growth.
   b. biotic potential.
   c. unlimited resources.
   d. rapid population growth.
   e. All of these are correct.

7. Which type of growth pattern is represented by the graph below?
   a. exponential
   b. logistic
   c. carrying capacity
   d. biotic potential
   e. lag growth

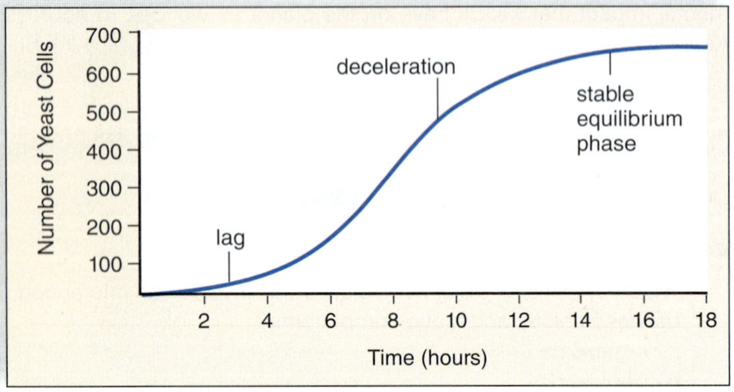

44.4 Regulation of Population Size

8. Which of these is a density-independent regulating factor?
   a. competition
   b. predation
   c. weather
   d. resource availability
   e. average age at which childbearing begins
9. Fluctuations in population growth can correlate to changes in
   a. predation.
   b. weather.
   c. resource availability.
   d. parasitism.
   e. All of these are correct.

44.5 Life History Patterns

10. A species that has many relatively small offspring, lives for a short period of time, and provides minimal parental care is exemplifying
    a. *r*-selection.
    b. *K*-selection.
    c. a mixture of *r*-selection and *K*-selection.
    d. density-dependent and density-independent regulation.
    e. a pyramid-shaped age-structure diagram.
11. In a tropical rain forest, which group of organisms is most likely to exhibit *K*-selection?
    a. frogs
    b. grasshoppers
    c. monkeys
    d. algae
    e. bacteria

44.6 Human Population Growth

12. Because of the postwar baby boom, the U.S. population
    a. can never level off.
    b. presently has a pyramid-type age distribution.
    c. is subject to regulation by density-dependent factors.
    d. is more conservation-minded than before.
    e. None of these are correct.
13. Which country leads the world in coal consumption per person?
    a. China
    b. Canada
    c. United States
    d. India
    e. Mexico

ENGAGE

THINKING SCIENTIFICALLY

1. In the winter moth life cycle, parasites are a less important cause of mortality than cold winter weather and predators. Give an evolutionary explanation for the inefficiency of parasites in controlling population size.
2. You are a river manager charged with maintaining the flow through the use of dams, so that trees, which have equilibrium life histories, can continue to grow along the river. What would you do?
3. The answer to how to curb the expected increase in the world's population lies in discovering how to curb the rapid population growth of the less-developed countries. In these countries, population experts have discovered what they call the "virtuous cycle." Family planning leads to healthier women, healthier women have healthier children, and the cycle continues. Women no longer need to have many babies for only a few to survive. More education is also helpful, because better-educated people are more interested in postponing childbearing and promoting women's rights. Women who have equal rights with men tend to have fewer children.

   "There isn't any place where women have had the choice that they haven't chosen to have fewer children," says Beverly Winikoff at the Population Council in New York City. "Governments don't need to resort to force." Bangladesh is a case in point. When women were provided the opportunity to make decisions about their lives, they made decisions about having smaller families. In 1990, the birthrate was 4.9 children per woman; in 2013, it was 2.5.

   Recently, some of the less-developed countries, faced with economic crises, have cut back on their family planning programs, and the more-developed countries have not taken up the slack. Indeed, some foreign donors have also cut back on aid—the United States by one-third. Are you in favor of foreign aid to help countries develop family planning programs? Why or why not?

A reintroduced wolf in Yellowstone National Park.

45

Community and Ecosystem Ecology

Chapter Outline

45.1 Ecology of Communities 856

45.2 Community Development 866

45.3 Dynamics of an Ecosystem 868

Before You Begin

Before beginning this chapter, take a few moments to review the following discussions.

Section 6.1 What role does thermodynamics play in the flow of energy through an ecosystem?

Section 43.2 Does a species' behavior determine the type of relationship it has with other species?

Section 44.4 How do density-dependent and density-independent factors influence community dynamics?

By 1926, the last known wolf pack had been eliminated from Yellowstone National Park. The extinction of the wolves produced a multitude of ecological effects that would ripple through the entire park for the next 70 years. Fewer wolves meant less predation upon the elk and other large herbivores. An increase in herbivores meant more pressure upon their food sources. Cottonwoods, willows, and streamside vegetation soon became depleted due to the increasing elk population.

Overbrowsing of the streamside vegetation produced a ripple effect throughout the stream ecosystem. This decreased the stability of the stream bank, which in turn caused increased erosion and sediment load in the streams. Fish, insects, birds, and all the wildlife associated with the streams were negatively impacted.

The reintroduction of wolves into Yellowstone began in 1995. Since then, biologists have seen signs of the stream bank stabilizing due to the decreased browsing by elk and other herbivores. This decrease in browsing is partly due to the wolves' preying upon the elk and other large herbivores in the park. Many of the problems that were caused by the eradication of wolves from Yellowstone have begun to be repaired as a direct result of their reintroduction. This chapter examines various types of interactions among the populations of a community. It also looks at the interactions with the nonliving component of an ecosystem.

As you read through this chapter, think about the following questions:

1. In a community, are some species interactions more common than others?
2. Do communities ever truly reach a climax status?
3. What would happen to the trophic levels in the community if the producer base were changed?

Following *the* Themes

CHAPTER 45 COMMUNITY AND ECOSYSTEM ECOLOGY

UNIT 8 BEHAVIOR AND ECOLOGY

Evolution	Changes to a species' niche can cause the niche of other species to change as well. This is the basis of coevolution.
Nature of Science	By studying the biogeochemical cycles of ecosystems, humans can better understand how to maintain ecosystem health.
Biological Systems	Communities are assemblages of populations that interact with each other as well as the abiotic environment. Populations interact with each other to form communities, which then interact with the abiotic environment as part of the greater ecosystem.

45.1 Ecology of Communities

Learning Outcomes

Upon completion of this section, you should be able to

1. Recognize the features of a biological community and species richness.
2. Describe the factors that define an organism's ecological niche within its community.
3. Identify how the interactions among species organize a community.

Populations do not occur as single entities; they are part of a community. A **community** is an assemblage of populations of different species interacting with one another in the same environment. Communities come in different sizes, and it is sometimes difficult to decide where one community ends and another begins.

A fallen log can be considered a community, because the many different populations living on and in it interact with one another. The fungi break down the log and provide food for the earthworms and insects living in and on the log. Those insects may feed on one another, too. If birds flying throughout the forest feed on the insects and worms living in and on the log, then they are also part of the larger forest community. In this section, we consider the characteristics of communities and the interactions that take place in them.

Community Structure

Two characteristics—species composition and diversity—allow us to compare communities. The species composition of a community, also known as **species richness,** is simply a listing of the various species found in that community. **Species diversity** is the measure of species richness and the relative abundance of the different species within the community.

Species Composition

It is apparent, by comparing the photographs in Figure 45.1, that a coniferous forest has a composition different from that of a tropical rain forest. The narrow-leaved evergreen trees are prominent in a coniferous forest, and broad-leaved evergreen trees are numerous in a tropical rain forest. As the list of mammals demonstrates, a coniferous community and a tropical rain forest community contain different types of mammals. Ecologists comparing these two communities would also find differences in the other plants and animals as well. In the end, ecologists would conclude that not only are the species compositions of these two communities different but the tropical rain forest has more species, and therefore, higher species richness.

a. b.

Figure 45.1 Community structure. Communities differ in their composition, as witnessed by their predominant plants and animals. The diversity of communities is described by the richness of species and their relative abundance. **a.** These are mammals commonly found in a coniferous forest. **b.** These are mammals commonly found in a tropical rain forest.

Species Diversity

The *diversity* of a community goes beyond composition, because it includes not only a listing of all the species in a community but also the relative abundance of each species. To take an extreme example, a forest sampled in West Virginia has 76 yellow poplar trees but only 1 American elm (among other species). If we were simply walking through this forest, we could miss seeing the American elm. If, instead, the forest had 36 poplar trees and 41 American elms, the forest would seem more diverse to us, and indeed it would be more diverse. The greater the species richness and the more even the distribution of the species in a community, the greater the species diversity.

Video Tallgrass Prairie Ecology

The Nature of Science feature, "Island Biogeography Pertains to Biodiversity," on page 860, discusses one of the landmark studies in community structure, which led to the **island biogeography model** of species diversity. This model postulates that species diversity on an island depends on the distance from the mainland (closer islands have more diversity than islands farther away) and the total area of the island (large islands have more diversity than small islands). Island biogeography studies play a role in determining the ideal size for conservation areas, among other questions.

Community Interactions

This chapter examines the various types of community interactions and their importance to the structure of a community. Such interactions illustrate some of the most important evolutionary selection pressures acting on individuals. They also help us develop an understanding of how biodiversity can be preserved.

Habitat and Ecological Niche

Each species occupies a particular position in the community, both in a spatial sense (where it lives) and in a functional sense (what role it plays). A particular place where a species lives and reproduces is its **habitat.** The habitat might be the forest floor, a swift stream, or the sea shore. The **ecological niche** is the role a species plays in its community. The niche includes the methods the species uses to acquire the resources it needs to meet energy, nutrient, and survival demands.

Video Finches Natural Selection

For a dragonfly larva, its habitat is a pond or lake, where it eats other insects. The pond must contain vegetation where the dragonfly larva can hide from its predators, such as fish and birds. In addition, the water must be clear enough for it to see its prey and warm enough for it to be in active pursuit.

Because it is difficult to study all aspects of a niche, some observations focus only on one aspect, such as study of the Galápagos finches (see section 15.2). Each of these bird species is a specialist species, because each has a limited diet, tolerates only small changes in environmental conditions, and lives in a specific habitat. Some species, such as raccoons, roaches, and house sparrows, are generalist with a broad range of niches. These organisms have a diversified diet, tolerate a wide range of environmental conditions, and can live in a variety of places.

Because a species' niche is determined by both abiotic factors (such as climate and habitat) and biotic factors (such as competitors, parasites, and predators), ecologists like to distinguish between the fundamental and the realized niche. The *fundamental niche* comprises all the abiotic conditions under which a species can survive when adverse biotic conditions are absent. The *realized niche* comprises those conditions under which a species does survive when adverse biotic interactions, such as competition and predation, are present. Therefore, a species' fundamental niche tends to be larger than its realized niche.

Video Cichlid Specialization

Competition Between Populations

Competition occurs when members of different species try to use a resource (such as light, space, or nutrients) that is in limited supply. If the resource is not in limited supply, there is no competition. In the 1930s, Russian ecologist G. F. Gause grew two species of *Paramecium* in one test tube containing a fixed amount of food. Although each population survived when grown separately, only one survived when they were grown together (Fig. 45.2). The successful *Paramecium* population had a higher biotic potential than the unsuccessful population. After observing the outcome of other, similar experiments in the laboratory, ecologists formulated the **competitive exclusion principle,** which states that no two species can indefinitely occupy the same niche at the same time.

It is possible for two species to have different ecological niches, so that extinction of one species is avoided. In another

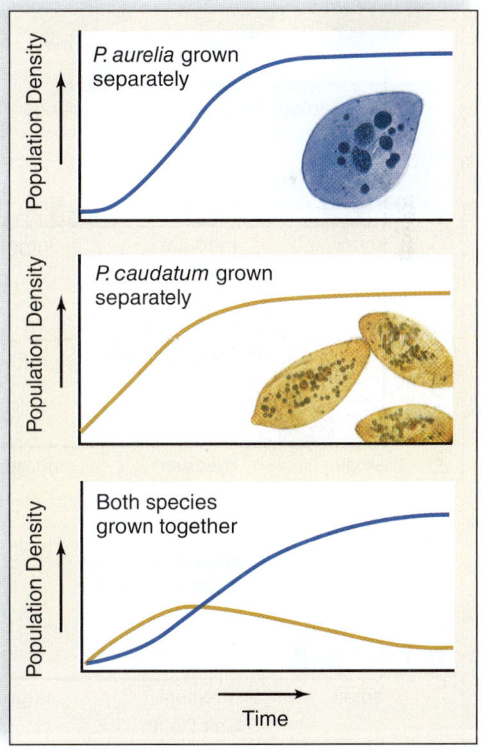

Figure 45.2 Competition between two laboratory populations of *Paramecium*. When grown alone in pure culture, *Paramecium aurelia* and *Paramecium caudatum* exhibit sigmoidal growth. When the two species are grown together in mixed culture, *P. aurelia* is the better competitor, and *P. caudatum* dies out. Both attempted to exploit the same resources, which led to competitive exclusion.
Source: Data from G. F. Gause, *The Struggle for Existence,* 1934, Williams & Wilkins Company, Baltimore, MD.

laboratory experiment, Gause found that two species of *Paramecium* did continue to occupy the same tube when one species fed on food at the bottom of the tube and the other fed on food suspended in solution. Individuals of each species that can avoid competition have a reproductive advantage. **Resource partitioning** is the apportioning of resources in order to decrease competition between two species. This often leads to increased niche specialization and less niche overlap. An example of resource partitioning involves owl and hawk populations. Owls and hawks feed on similar prey (small rodents), but owls are nocturnal hunters and hawks are diurnal hunters. What could have been one niche became two, more specialized niches because of a divergence of behavior.

It is possible to observe the process of niche specialization in nature. When three species of ground finches of the Galápagos Islands occur on separate islands, their beaks tend to be the same intermediate size, enabling each to feed on a wider range of seeds (Fig. 45.3; see also Fig. 15.9). Where the three species co-occur, selection has favored divergence in beak size, because the size of the beak affects the kinds of seeds that can be eaten. In other words, competition has led to resource partitioning and, therefore, niche specialization.

Video Finches Adaptive Radiation

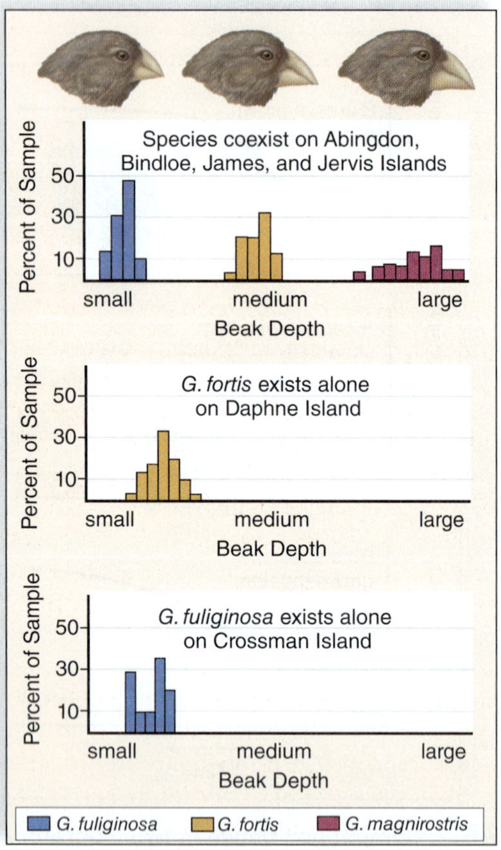

Figure 45.3 Character displacement in finches on the Galápagos Islands. When *Geospiza fuliginosa, G. fortis,* and *G. magnirostris* are on the same island, their beak sizes are appropriate to eating small-, medium-, and large-sized seeds respectively. When *G. fortis* and *G. fuliginosa* are on separate islands, their beaks have the same intermediate size, which allows them to eat seeds that vary in size.

Character displacement is the tendency for characteristics to be more divergent when populations belong to the same community than when they are isolated. Character displacement is often used as evidence that competition and resource partitioning have taken place.

Niche specialization can be subtle. Five different species of warblers that occur in North American forests are all nearly the same size, and all feed on a type of spruce tree caterpillar. For all these species to exist in the same community, they must be avoiding direct competition. In a famous study, ecologist Robert MacArthur recorded the amount of time each of five warbler species spent in different regions of spruce trees to determine where each species did most of its feeding (Fig. 45.4). He discovered that each species primarily used different parts of the tree and, in that way, had a more specialized niche.

As another example, consider that swallows, swifts, and martins all eat flying insects and parachuting spiders. These birds even frequently fly together in mixed flocks. But each type of bird has a different nesting site and migrates at a slightly different time of year. In doing so, they are not competing for the same food source at the same time.

In all these cases of niche partitioning, we have merely assumed that what we observe today is due to competition in the past. Joseph Connell has studied the distribution of barnacles on the Scottish coast. Small barnacles live on the high part of the intertidal zone, and large barnacles live on the lower part (Fig. 45.5). Free-swimming larvae of both species attach themselves to rocks randomly throughout the intertidal zone; however, in the lower zone, the large *Balanus* barnacles seem to either force the smaller *Chthamalus* individuals off the rocks or grow over them.

To test his observation, Connell removed the larger barnacles and found that the smaller barnacles grew equally well on all parts of the rock. The entire intertidal zone is the fundamental niche for *Chthamalus,* but competition is restricting the range of *Chthamalus* on the rocks. *Chthamalus* is more resistant to drying out than is *Balanus;* therefore, it has an advantage that permits it to grow in the upper intertidal zone, where it is exposed more often to the air. In other words, the upper intertidal zone becomes the realized niche for *Chthamalus.*[1]

Predator-Prey Interactions

Predation occurs when one living organism, called a **predator,** feeds on another, called **prey.** In its broadest sense, predaceous consumers include not only animals such as lions, which kill zebras, but also filter-feeding blue whales, which strain krill from ocean waters, and herbivorous deer, which browse on trees and shrubs.

Parasitism can be considered a type of predation, because one individual obtains nutrients from another (see page 863). Parasitoids are organisms that lay their eggs inside a host. The developing larvae feed on the host and sometimes cause the host to die.

Predation and parasitism tend to increase the abundance of the predator and parasite, but at the expense of the prey or host population.

[1]Connell, J. H. "The Influence of Interspecific Competition . . . ." *Ecology* 42:710–723 (1961).

Figure 45.4 Niche specialization among five species of coexisting warblers. The diagrams represent spruce trees. The time each species spent in various portions of the trees was determined; each species spent more than half its time in the blue regions.

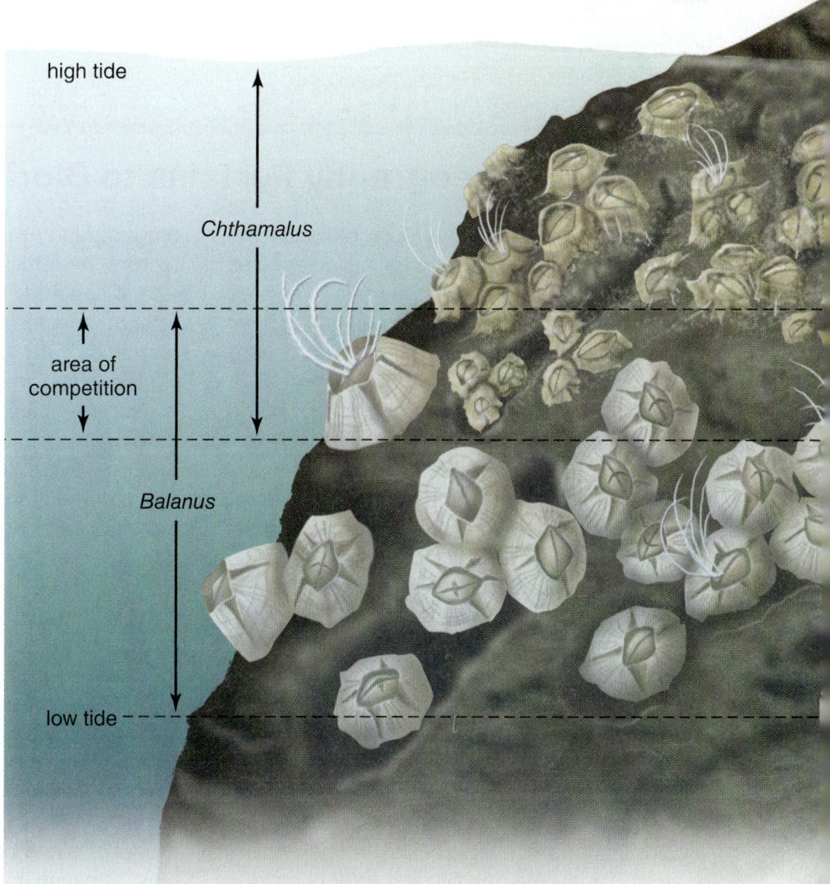

Figure 45.5 Competition between two species of barnacles. Competition prevents two species of barnacles from occupying as much of the intertidal zone as possible. Both exist in the area of competition between *Chthamalus* and *Balanus.* Above this area only *Chthamalus,* a small barnacle, survives, and below it only *Balanus,* a large barnacle, survives.

Predator-Prey Population Dynamics

Predators play a role in determining the population density of prey. In another classic experiment, G. F. Gause reared the ciliated protozoans *Paramecium caudatum* (prey) and *Didinium nasutum* (predator) together in a culture medium. He observed that *Didinium* ate all the *Paramecium* and then died of starvation.

In nature, we can find a similar example. When a gardener took prickly-pear cactus to Australia from South America, the cactus spread out of control until millions of acres were covered with nothing but cacti. The cacti were brought under control when a moth from South America, whose caterpillar feeds only on the cactus, was introduced. The caterpillar was a voracious predator on the cactus, efficiently reducing the cactus population. Now both cactus and moth are found at greatly reduced densities in Australia.

This raises an interesting point: The population density of the predator can be affected by the prevalence of the prey. In other words, the predator-prey relationship is actually a two-way street. In that context, consider that at first the biotic potential (maximum reproductive rate) of the prickly-pear cactus was maximized, but factors that oppose biotic potential came into play after the moth was introduced. And the biotic potential of the moth was maximized when it was first introduced, but the carrying capacity decreased after its food supply was diminished.

Video
Harris Hawk

THEME **Nature of Science**

Island Biogeography Pertains to Biodiversity

Would you expect larger coral reefs to have a greater number of species, called species richness, than smaller coral reefs? The area (space) occupied by a community can have a profound effect on its biodiversity. American ecologists Robert MacArthur and E. O. Wilson developed a general model of island biogeography to explain and predict the effects of (1) distance from the mainland and (2) size of an island on community diversity.

Imagine two new islands that contain no species at all. One of these islands is near the mainland, and one is far from the mainland. Which island will receive more immigrants from the mainland? Most likely, the near one, because it's easier for immigrants to get there. Similarly, imagine two islands that differ in size. Which island will be able to support a greater number of species? The large one, because its greater amount of resources can support more populations, while species on the smaller island may eventually face extinction due to scarce resources.

MacArthur and Wilson studied the biodiversity on many island chains, including the West Indies, and discovered that species richness does correlate positively with island distance from mainland and island size. They developed a model of island biogeography that takes into account both factors. An equilibrium is reached when the rate of species immigration matches the rate of species extinction due to limited resources (Fig. 45A*a*). Notice that the equilibrium point is highest for a large island that is near the mainland. The equilibrium could be dynamic (new species keep on arriving, and new extinctions keep on occurring), or the composition of the community could remain steady unless disturbed.

Biodiversity

Conservationists note that the trends graphed in Figure 42A*a* apply to their studies. Humans often create preserved islands of habitat surrounded by farms, towns, cities, or even water. For example, in Panama, Barro Colorado Island (BCI) (Fig. 45A*b*) was created in the 1920s when a river was dammed to form a lake. As predicted by the model of island biogeography, BCI lost species, because it was a small island that had been cut off from the mainland. Among the species that became extinct were the top predators on the island—the jaguar, puma, and ocelot. Thereafter, medium-sized terrestrial mammals, such as the coatimundi, increased in number. Because the coatimundi is an avid predator of bird eggs and nestlings, soon there were fewer bird species on BCI, even though the island is large enough to support them.

The model of island biogeography suggests that the larger the conserved area, the better the chance of preserving more species. Is it possible to increase the amount of habitat without using more area? If the environment has patches, it has a greater number of habitats—and thus greater diversity. As gardeners, we are urged to create patches in our yards if we wish to attract more butterflies and birds. One way to introduce patchiness is through stratification, the use of layers. Just as a high-rise apartment building allows more human families to live in an area, so can stratification within a community provide more and different types of living space for different species. Stratification could be accomplished by having low plants, medium shrubs, and taller trees.

Questions to Consider

1. Why is the mammalian and reptilian diversity so low on islands?
2. How can city planners apply island biogeography to urban development?

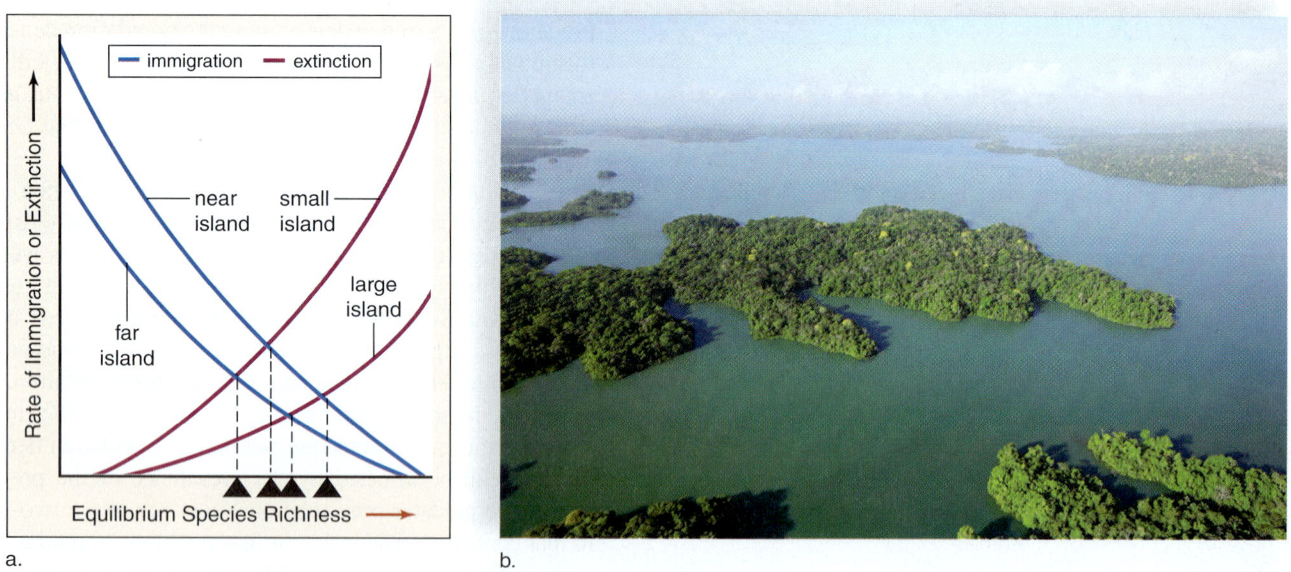

Figure 45A Species richness. **a.** The balance between immigration and extinction for four possible types of islands. A large island that is near the mainland will have a higher equilibrium species richness than the other types of islands. **b.** Barro Colorado Island, Panama.

Sometimes, instead of remaining in a steady state, predator and prey populations fluctuate in size. We can appreciate that an increase in predator population size is dependent on an increase in prey population size. A decrease in population size instead of the establishment of a steady population size can also occur. At least two possibilities account for the reduction:

1. Perhaps the biotic potential (reproductive rate) of the predator is so great that its increased numbers overconsume the prey, and then as the prey population declines, so does the predator population.
2. Perhaps the biotic potential of the predator is unable to keep pace with the prey, and the prey population overshoots the carrying capacity and suffers a crash. Now the predator population follows suit because of a lack of food.

In either case, the result is a series of peaks and valleys, with the predator population size lagging slightly behind that of the prey population.

A famous example of predator-prey cycles occurs between the snowshoe hare (*Lepus americanus*) and the Canada lynx (*Lynx canadensis*), a type of small, predatory cat (Fig. 45.6). The snowshoe hare is a common herbivore in the coniferous forests of North America, where it feeds on terminal twigs of various shrubs and small trees. The Canada lynx feeds on snowshoe hares but also on ruffed and spruce grouse, two types of birds. Studies have revealed that the hare and lynx populations cycle regularly, as graphed in Figure 45.6.

Investigators at first assumed that the lynx was the only factor that would bring about a decline in the hare population, and that this accounts for the cycling. But others have noted that the decline in snowshoe hare abundance is accompanied by low growth and reproductive rates, which could be signs of a food shortage. Experiments were done to test whether predation or lack of food causes the decline in the hare population. The results indicate that the hare cycle is produced by interactions among three trophic levels, predators, hares, and their food supplies. Its biological impacts produce a ripple effect across multiple species of predator and prey species in the taiga.

Prey Defenses

Prey defenses are mechanisms that thwart the possibility of being eaten by a predator. Prey species have evolved a variety of mechanisms that enable them to avoid predators, including heightened senses, speed, protective armor, protective spines or thorns, tails and appendages that break off, and chemical defenses.

Video Forest Blue Butterfly

One common strategy to avoid capture by a predator is **camouflage,** or the ability to blend into the background, which helps the prey avoid being detected by the predator. Some animals

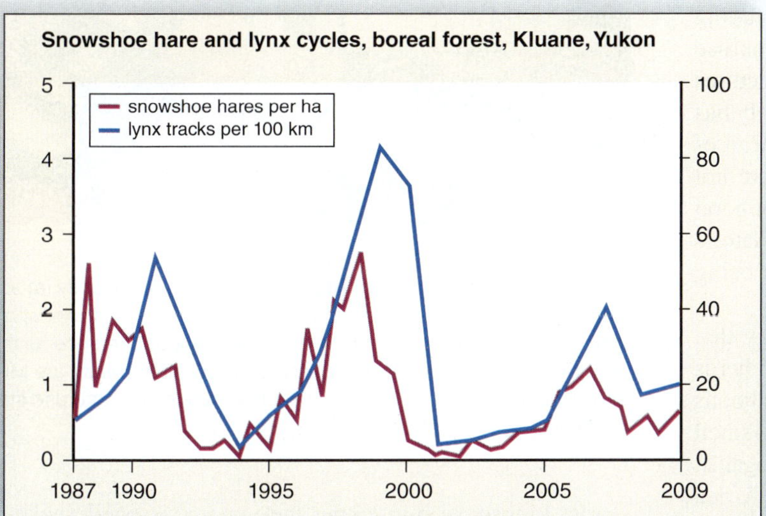

b.

Source: data from Krebs, 2010

a.

Figure 45.6 Predator-prey interaction between a lynx and a snowshoe hare. **a.** A Canada lynx, *Lynx canadensis,* is a solitary predator. A long, strong forelimb with sharp claws grabs its main prey, the snowshoe hare, *Lepus americanus*. **b.** Research indicates that the snowshoe hare population reaches a peak abundance before that of the lynx by a year or more. A study conducted from 1987 to 2009 shows a cycling of the lynx and snowshoe hare populations.

a. Camouflage b. Warning colorization c. Fright

Figure 45.7 Antipredator defenses. **a.** Flounder is a fish that blends in with its background. **b.** The skin secretions of poison arrow frogs make them dangerous to touch. **c.** The large false head of the South American lantern fly may startle a potential predator.

have cryptic coloration that blends them into their surroundings. For example, flounders can take on the same coloration as their background (Fig. 45.7*a*). Many examples of protective camouflage are known: Walking sticks look like twigs; katydids look like sprouting green leaves; some caterpillars resemble bird droppings; and some insects, moths, and lizards blend into the bark of trees.

Video Flounder Camouflage

Video Leaf-Tailed Gecko

Another common antipredator defense among animals is *warning coloration,* which tells the predator that the prey is potentially dangerous. As a warning to possible predators, poison arrow frogs are brightly colored (Fig. 45.7*b*). Also, many animals, including caterpillars, moths, and fishes, possess false eyespots that confuse or startle another animal. Other animals have elaborate structures that cause the *startle response*. The South American lantern fly has a large false head with false eyes, making it resemble the head of an alligator (Fig. 45.7*c*). However, antipredator defenses are not always false. A porcupine certainly looks formidable, and for good reason. Its arrowlike quills have barbs that dig into the predator's flesh and penetrate even deeper as the enemy struggles after being impaled. In the meantime, the porcupine runs away.

Association with other prey is another common strategy that may help avoid capture. Flocks of birds, schools of fish, and herds of mammals stick together as protection against predators. Baboons detect predators visually, and antelopes detect predators by smell and sometimes forage together, gaining double protection against stealthy predators. The gazellelike springboks of southern Africa jump stiff-legged 2–4 m into the air when alarmed. Such a jumble of shapes and motions might confuse an attacking lion, allowing the herd to escape.

Mimicry

Mimicry is the resemblance of one species to another that possesses an overt antipredator defense. A mimic that lacks the defense of the organism it resembles is called a Batesian mimic (named for Henry Bates [1825–92], who described the phenomenon). Once an animal experiences the defense of the model organism, it remembers the coloration and avoids all animals that look similar. Figure 45.8*a, b* shows two insects (flower fly and longhorn beetle) that resemble the yellow jacket wasp shown in Figure 45.8*c* but lack the wasp's ability to sting. Classic examples of Batesian mimicry include the scarlet kingsnake mimicking the venomous coral snake and the *Papilio memnon* butterfly mimicking foul-tasting butterflies.

a. Flower fly b. Longhorn beetle c. Yellow jacket wasp d. Bumblebee

Figure 45.8 Mimicry among insects. A flower fly (**a**) and a longhorn beetle (**b**) are Batesian mimics, because they are incapable of stinging another animal, yet they have the same appearance as the yellow jacket wasp (**c**). A bumblebee (**d**) and a yellow jacket wasp are Müllerian mimics, because they have a similar appearance and both use stinging as a defense.

Some species that actually have the same defense resemble each other. Many stinging insects—bees, wasps, hornets, and bumblebees—have the familiar alternating black and yellow bands. Once a predator has been stung by a black and yellow insect, it is wary of that color pattern in the future. Mimics that share the same protective defense are called Müllerian mimics, after Fritz Müller (1821–97), who discovered this type of mimicry. The bumblebee in Figure 45.8*d* is a Müllerian mimic of the yellow jacket wasp, because both of them can sting.

Symbiotic Relationships

Symbiosis is the close association between two different species over long periods of time. Symbiotic relationships are of three

Table 45.1 Symbiotic Relationships

Interaction	Species 1	Species 2
Parasitism*	Benefited	Harmed
Commensalism	Benefited	No effect
Mutualism	Benefited	Benefited

*Can be considered a type of predation.

types, as shown in Table 45.1. Some biologists argue that the amount of harm or good two species do to one another is dependent on which variable the investigator is measuring.

Parasitism is similar to predation in that an organism, called a **parasite,** derives nourishment from another, called a **host.** Parasitism is a type of symbiosis in which one of the species causes some harm to the other but does not attempt to kill it (Table 45.1). Viruses (such as HIV, which reproduces inside human lymphocytes) are always parasitic, and parasites occur in all the kingdoms of life. Bacteria (e.g., strep infection), protists (e.g., malaria), fungi (e.g., rusts and smuts), plants (e.g., indian pipe), and animals (e.g., tapeworms and fleas) all have parasitic members. Although small parasites can be endoparasites (e.g, heartworms) (Fig. 45.9), larger ones are more likely to be ectoparasites (e.g., leeches).

The effects of parasites on the health of the host can range from slightly weakening them to actually killing them over time. When host populations are at a high density, parasites readily spread from one host to the next, causing intense infestations and a subsequent decline in host density. Parasites that do not kill their host can still play a role in reducing the host's population density, because an infected host is less fertile and becomes more susceptible to other causes of death.

In addition to nourishment, host organisms also provide their parasites with a place to live and reproduce, as well as a mechanism for dispersing offspring to new hosts. Many parasites have both a primary and a secondary host. The secondary host may be a vector that transmits the parasite to the next primary host. Usually, both hosts are required in order to complete the life cycle, as in malaria (see Fig. 21.15). The association between parasite and host is so intimate that parasites will often parasitize only a few select species.

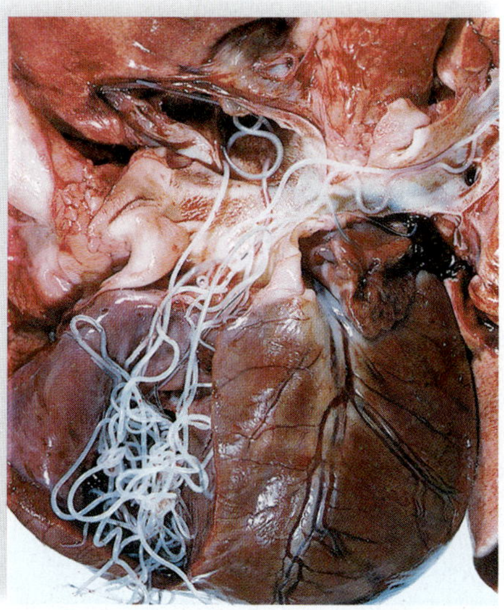

Figure 45.9 Heartworm. *Dirofilaria immitis* is a parasitic nematode spread by mosquitoes. The worms, which live in the heart and pulmonary blood artery, can cause death of the host.

Commensalism

Commensalism is a symbiotic relationship between two species in which one species is benefited and the other is neither benefited nor harmed.

Instances are known in which one species provides a home and/or transportation for the other species. Barnacles that attach themselves to the backs of whales or the shells of horseshoe crabs are provided with both a home and transportation. It is possible, though, that the movement of the host is impeded by the presence of the attached animals, and therefore some are reluctant to use these as instances of commensalism.

Epiphytes, such as Spanish moss and some species of orchids and ferns, grow in the branches of trees, where they receive light, but they take no nourishment from the trees. Instead, their roots obtain nutrients and water from the air. Clownfishes live within the waving mass of tentacles of sea anemones. Because most fishes avoid the stinging tentacles of the anemones, clownfishes are protected from predators. Perhaps this relationship borders on mutualism, because the clownfishes actually may attract other fishes on which the anemones can feed, or they may provide some cleaning services for the anemones.

Commensalism often turns out, on closer examination, to be an instance of either mutualism or parasitism. Cattle egrets are so named because these birds stay near cattle, which flush out their prey—insects and other animals—from vegetation. The relationship becomes mutualistic when egrets remove ectoparasites from the cattle (Fig. 45.10). Remoras are fishes that attach themselves

Figure 45.10 Egret symbiosis. Cattle egrets eat insects off and around various animals, such as this African cape buffalo.

to the bellies of sharks by means of a modified dorsal fin acting as a suction cup. Remoras benefit by getting a free ride and feeding on a shark's leftovers. However, the shark benefits when remoras remove its ectoparasites.

Mutualism

Mutualism is a symbiotic relationship in which both members benefit. As with other symbiotic relationships, it is possible to find numerous examples among all organisms. Bacteria that reside in the human intestinal tract acquire food, but they also provide us with vitamins that we cannot synthesize for ourselves. Termites would not be able to digest wood if not for the protozoans that inhabit their intestinal tracts and digest cellulose. Mycorrhizae are mutualistic associations between the roots of plants and fungal hyphae. The hyphae improve the uptake of nutrients for the plant, protect the plant's roots against pathogens, and produce plant growth hormones. In return, the plant provides the fungus with carbohydrates.

Video Fruit Bat Seed Dispersal

Video Ant-Caterpillar Mutualism

Some sea anemones make their homes on the backs of crabs. The crab uses the stinging tentacles of the sea anemone to gather food and to protect itself, while the sea anemone gets a free ride that allows it greater access to food. Lichens can grow on rocks, because their fungal member conserves water and leaches minerals, which are provided to the algal partner. The algae photosynthesize and provide organic food for both organisms. However, it's been suggested that the fungus is parasitic, at least to a degree, on the algae.

In Mexico and Central America, the bullhorn acacia tree is adapted to provide a home for ants of the species *Pseudomyrmex ferruginea*. Unlike other acacias, this species has swollen thorns with a hollow interior, where ant larvae can grow and develop. In addition to housing the ants, acacias provide them with food. The ants feed from nectaries at the base of the leaves and eat fat- and protein-containing nodules called Beltian bodies, found at the tips of the leaves. The ants constantly protect the plant from herbivores and other plants that might shade it.

Video Thorn Tree Ants

Figure 45.11 Clark's nutcrackers. Mutualism can take many forms, as when birds such as Clark's nutcrackers, *Nucifraga columbiana,* feed on but also disperse the seeds of whitebark pine trees.

Figure 45.12 Cleaning symbiosis. A cleaner wrasse, *Labroides dimidiatus,* in the mouth of a spotted sweetlip, *Plectorhincus chaetodontoides,* is feeding off parasites. Does this association improve the health of the sweetlip, or is the sweetlip being exploited? Investigation is under way.

The relationship between plants and their pollinators, mentioned previously, is a good example of mutualism. Perhaps the relationship began when certain herbivores feasted on certain plants. The plant's provision of nectar may have spared the pollen while allowing the animal to become an agent of pollination. Eventually, pollinator mouthparts became very specific to gathering the nectar of a particular plant species.The plant species, in turn, often becomes dependent on the pollinator for dispersing pollen. The mutualistic relationships between plants and their pollinators are examples of **coevolution.** The Evolution feature, "Interactions and Coevolution," discusses some examples of coevolution.

Video Pollinators

The outcome of mutualism is an intricate web of species interdependencies critical to the community. For example, in areas of the western United States, the branches and cones of whitebark pine are turned upward, meaning that the seeds do not fall to the ground when the cones open. Birds called Clark's nutcrackers eat the seeds of the whitebark pine trees and store them in the ground (Fig. 45.11), thus acting as critical seed dispersers for the trees. Also, grizzly bears find the stored seeds and consume them.

Whitebark pine seeds do not germinate unless their seed coats are exposed to fire. When natural forest fires in the area are suppressed, whitebark pine trees decline in number, and so do Clark's nutcrackers and grizzly bears. When lightning-ignited fires are allowed to burn, or prescribed burning is used in the area, the whitebark pine populations increase, as do the populations of Clark's nutcrackers and grizzly bears.

Cleaning symbiosis is a symbiotic relationship in which crustaceans, fish, and birds act as cleaners for a variety of vertebrates. Large fish in coral reefs line up at cleaning stations and wait their turn to be cleaned by small fish, which even enter the mouths of the large fish (Fig. 45.12). Whether cleaning symbiosis is an example of mutualism is still being questioned.

Check Your Progress 45.1

1. Explain the difference between species richness and diversity.
2. Identify the difference between an organism's habitat and its niche.
3. Describe the two factors that can cause predator and prey populations to cycle in a predictable manner.

THEME Evolution

Interactions and Coevolution

Coevolution is present when two species adapt in response to selective pressure imposed by the other. Symbiosis (close association between two species), which includes parasitism, commensalism, and mutualism, is especially prone to the process of coevolution.

Flowering plants often use brightly colored petals or enticing scents to attract animal pollinators (see the Evolution feature, "Plants and Their Pollinators," in Chapter 27). Butterfly-pollinated flowers often contain a platform for the butterfly to land on, and the butterfly has a proboscis that allows it to feed on the flower.

Video Pollinators

Coevolution also occurs between predators and prey. For example, a cheetah sprints forward to catch its prey, which selects for the gazelles that are fast enough to avoid capture. Over generations, the adaptation of the prey may put selective pressure on the predator for an adaptation to the prey's defense mechanism. In this way, an evolutionary "arms race" can develop.

The process of coevolution has been studied in the brown-headed cowbird, a social parasite that reproduces at the expense of other birds by laying its eggs in their nests. It is a strange sight to see a small bird feeding a cowbird nestling several times its size. Investigators discovered that in order to "trick" a host bird, the female cowbird has to quickly lay an egg that mimics the host's egg while the host is away from the nest. The cowbird will leave most of the host's eggs in the nest to prevent the host from deserting a nest with only one egg. (The cowbird chick hatches first and is behaviorally adapted to pushing other eggs out of the nest [Fig. 45B*a*].)

At this stage in the "arms race," the cowbird appears to have the upper hand; however, selection may favor host birds that are able to distinguish the cowbird eggs from their own. In the case of the yellow warbler, the adults have evolved the mechanism of building a new nest on top of the cowbird eggs in order to avoid being brood parasitized.

Coevolution can take many forms. The sexual portion of the life cycle of *Plasmodium,* the cause of malaria, occurs within mosquitoes (the vector), while the asexual portion occurs in humans. The human immune system uses surface proteins to detect pathogens. *Plasmodium* stays one step ahead of the host's immune system by changing the surface proteins of its cells to avoid detection. A similar capability of HIV has added to the difficulty of producing an AIDS vaccine.

The relationship between parasite and host can even include the ability of parasites to seemingly manipulate the behavior of their hosts in self-serving ways. Ants infected with the lance fluke mysteriously cling to blades of grass with their mouthparts. There, the infected ants are eaten by grazing sheep, transmitting the flukes to the next host in their life cycle. Similarly, snails of the genus *Succinea* are parasitized by worms of the genus *Leucochloridium.* As the worms mature, they invade the snail's eyestalks, making them resemble edible caterpillars. Birds eat the infected snails, allowing the parasites to complete their development inside the urinary tracts of birds.

The traditional view was that as host and parasite coevolved, each would become more tolerant of the other. Eventually, parasites could become commensal, or harmless to the host. Then over time, the parasite and host might even become mutualists. In fact, the evolution of the eukaryotic cell by endosymbiosis is based on the supposition that some early bacteria took up residence inside a larger cell, and then the parasite and cell became mutualists.

However, this argument is too teleological for some; after all, no organism is capable of "looking ahead" at its evolutionary fate. Rather, if an aggressive parasite could transmit more of itself in less time than a benign one, aggressiveness would be favored by natural selection, because the most aggressive would reproduce. Other factors, such as the life cycle of the host, can determine whether aggressiveness is beneficial or not. For example, a benign parasite of newts would do better than an aggressive one. Why? Because newts take up solitary residence in the forest, outside of ponds, for 6 years, and parasites have to wait that long before they are likely to meet up with another potential host. If a parasite kills its host before it can reach another, it loses not only its food source but also its home.

Questions to Consider

1. What happens to a species if it cannot coevolve along with the species it is interacting with?
2. Why is it necessary for a parasite to avoid killing its host?

a. b.

Figure 45B Social parasitism. The brown-headed cowbird, *Molothrus ater,* is a social parasite of more than 220 species of birds. **a.** The blue eggs of the eastern bluebird and the speckled egg of the cowbird in a nest. **b.** A cowbird chick that is outcompeting its nestmates for food.

45.2 Community Development

Learning Outcomes

Upon completion of this section, you should be able to

1. Choose the correct sequence of events that occur during ecological succession.
2. Compare the two types of ecological succession.

Each community has a history that can be surveyed over time. We know that the distribution of life has been influenced by dynamic changes occurring during the history of Earth. We have previously discussed how continental drift contributed to various mass extinctions during the past. For example, as the continents slowly came together, forming the supercontinent Pangaea, many forms of marine life became extinct. And when the continents drifted toward the poles, immense glaciers drew water from the oceans and even chilled once-tropical lands. During ice ages, glaciers moved southward; then between ice ages, glaciers retreated, changing the environment and allowing life to colonize the land once again. Over time, complex communities evolved. Many ecologists, however, try to observe changes as they occur during much shorter timescales.

Ecological Succession

Communities are subject to disturbances that can range in severity from a storm blowing down a patch of trees to a beaver damming a pond to a volcanic eruption. We know from observation that after disturbances, changes occur in the community over time.

Ecological succession is a change within a community involving a series of species replacements. *Primary succession* is the formation of soil from exposed rock due to wind, water, and other abiotic factors. This type of succession occurs in areas where there is no base soil, such as following a volcanic eruption or a glacial retreat (Fig. 45.13). *Secondary succession* is a disturbance-based succession in which there is a progressive change from grasses to shrubs to a mixture of shrubs and trees. This type of succession occurs in areas where soil is already present, as when a cultivated field returns to a natural state.

Pioneer species are the first producers to inhabit a community after a disturbance. These are the first species that begin secondary succession. The area then progresses through the series of stages over time (Fig. 45.14). The initial stage begins with small, short-lived species and proceeds through stages of species of mixed sizes and lifespans, until finally large, long-lived species of trees predominate. Ecologists have tried to determine the processes and

a. Rock/lichen stage
b. Shrub stage
c. Low tree stage
d. High tree stage

Figure 45.13 Primary succession. This is an example of primary succession as glaciers retreated from an area called Glacier Bay, Alaska. **a.** Lichens and mosses invade the area left vacant by the retreating glacier. **b.** Small bushes grow in the newly abundant soil. **c.** Fast-growing alder trees help build the soil nutrients and water content. **d.** Improved soil conditions allow the white spruce–Western hemlock community to develop, eventually moving toward a more mature and stable community.

mechanisms by which the changes described in Figures 45.13 and 45.14 take place—and whether these processes always have the same end point of community composition and diversity.

Models About Succession

In 1916, F. E. Clements (1874–1945) proposed that succession in a particular area will always lead to a mature and stable community, which he called a **climax community.** He hypothesized that climate, in particular, determined whether succession resulted in a desert, a type of grassland, or a particular type of forest. This is the reason, he said, that coniferous forests occur in northern latitudes, deciduous forests in temperate zones, and tropical rain forests in the tropics. Secondarily, he hypothesized that soil conditions might also affect the results. Shallow, dry soil might produce a grassland where otherwise a forest might be expected, or the rich soil of a riverbank might produce a woodland where a prairie would be expected.

Further, Clements hypothesized that each stage facilitated the invasion and replacement by organisms of the next stage. Shrubs can't grow on dunes until dune grass has caused soil to develop. Similarly, in the example in Figure 45.14, shrubs can't arrive until grasses have made the soil suitable for them. Each group of species prepares the way for the next, so that grass–shrub–forest development occurs in a sequential way. Therefore, in what is sometimes called "climax theory," this is known as the *facilitation model* of succession.

Aside from this facilitation model, an *inhibition model* also has been proposed to help explain succession. That model predicts that colonists hold on to their space and inhibit the growth of other plants until the colonists die or are damaged. Still another model, called the *tolerance model,* predicts that different types of plants can colonize an area at the same time. Sheer chance determines which seeds arrive first, and successional stages may simply reflect the length of time it takes species to mature. This alone could account for the grass–shrub–forest development that is often seen (Fig. 45.14).

The length of time it takes for trees to develop might give the impression that there is a recognizable series of plant communities, from the simple to the complex. In reality, succession occurs gradually, and the models mentioned here are probably not mutually exclusive but a mixture of multiple, complex processes.

Although the dynamic nature of natural communities may not have been apparent to early ecologists, we now recognize it as a community's most outstanding characteristic. Also, it seems obvious now that the most complex communities are ones that contain various stages of succession. Each successional stage has its own mix of plants and animals, and if a sample of all stages is present, community diversity is greatest. Furthermore, we do not know whether succession continues to specific end points, because the process may not be complete anywhere on Earth.

Check Your Progress 45.2

1. Identify the events that occur during succession.
2. Compare and contrast the various models used to explain succession.

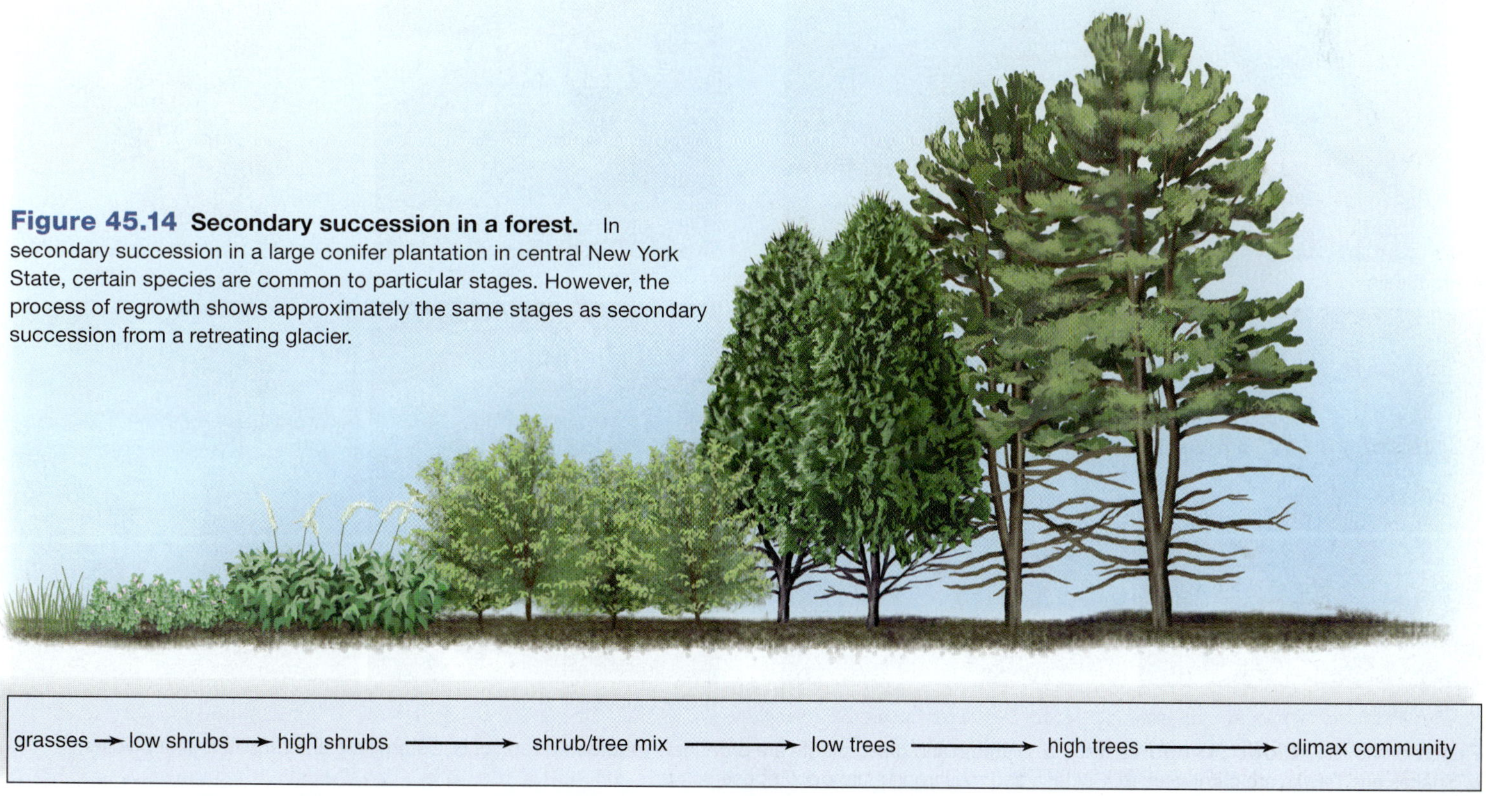

Figure 45.14 Secondary succession in a forest. In secondary succession in a large conifer plantation in central New York State, certain species are common to particular stages. However, the process of regrowth shows approximately the same stages as secondary succession from a retreating glacier.

45.3 Dynamics of an Ecosystem

Learning Outcomes

Upon completion of this section, you should be able to

1. Describe the interactions of organisms with their environment that comprise an ecosystem.
2. Identify the ways autotrophs, photoautotrophs, and heterotrophs obtain nutrients.
3. Interpret the energy flow and biogeochemical cycling within and among ecosystems.
4. Explain the energy flow among populations through food webs and ecological pyramids.

An **ecosystem** is composed of the interactions between populations as well as their physical environment. The abiotic (nonliving) components of an ecosystem include the atmosphere, water, and soil. The biotic (living) components can be categorized according to their food source as autotrophs or heterotrophs (Fig. 45.15).

Autotrophs

Autotrophs are organisms that require only inorganic nutrients and an outside energy source to produce organic nutrients for their own use and for all the other members of a community. **Producers** are a type of autotroph, because they produce food from inorganic nutrients. Autotrophs include photosynthetic organisms, such as land plants and algae (photoautotrophs). They possess chlorophyll and carry on photosynthesis in freshwater and marine habitats. Algae are aquatic autotrophs that make up the phytoplankton load in most bodies of water. Green plants are the dominant photosynthesizers on land.

Video Plants

Some autotrophic bacteria are chemosynthetic. They obtain energy by oxidizing inorganic compounds, such as ammonia, nitrites, and sulfides, and they use this energy to synthesize organic compounds. Chemoautotrophs have been found to support communities in some caves as well as at hydrothermal vents along deep-sea oceanic ridges where sunlight is unavailable.

Heterotrophs

Heterotrophs are organisms that need preformed organic nutrients they can use as an energy source. They are called **consumers** because they consume food that was generated by a producer.

Herbivores are animals that graze directly on plants or algae. In terrestrial habitats, insects are small herbivores; antelopes and bison are large herbivores. In aquatic habitats, zooplankton are small herbivores; fishes and manatees are large herbivores. **Carnivores** feed on other animals; birds that feed on insects are carnivores, and so are hawks that feed on birds and small mammals. **Omnivores** are animals that feed on both plants and animals. Chickens, raccoons, and humans are omnivores.

a. Producers

c. Carnivores

b. Herbivores

d. Decomposers

Figure 45.15 Biotic components. **a.** Diatoms and green plants are producers (photoautotrophs). **b.** Caterpillars and rabbits are herbivores. **c.** Snakes and hawks are carnivores. **d.** Bacteria and mushrooms are decomposers.

Some animals are scavengers, such as vultures and jackals, which eat the carcasses of dead animals.

Video
Snake Eating

Detritivores are organisms that feed on detritus, which consists of decomposing particles of organic matter. Marine fan worms filter detritus from the water, whereas burrowing clams take it from the substratum. Earthworms, some beetles, termites, and ants are all terrestrial detritivores.

Bacteria and fungi, including mushrooms, are **decomposers;** they acquire nutrients by breaking down dead organic matter, including animal wastes. Decomposers perform a valuable service, because they release inorganic substances, which are taken up by plants once more. Otherwise, plants would be completely dependent only on physical processes, such as the release of minerals from rocks, to supply them with inorganic nutrients.

Video
Decomposers

A diagram of all the biotic components of an ecosystem illustrates that every ecosystem is characterized by two fundamental phenomena: energy flow and chemical cycling (Fig. 45.16). Energy flow begins when producers absorb solar energy, and chemical cycling begins when producers take in inorganic nutrients from the physical environment. Thereafter, via photosynthesis, producers make organic nutrients (food) directly for themselves and indirectly for the other populations within the ecosystem.

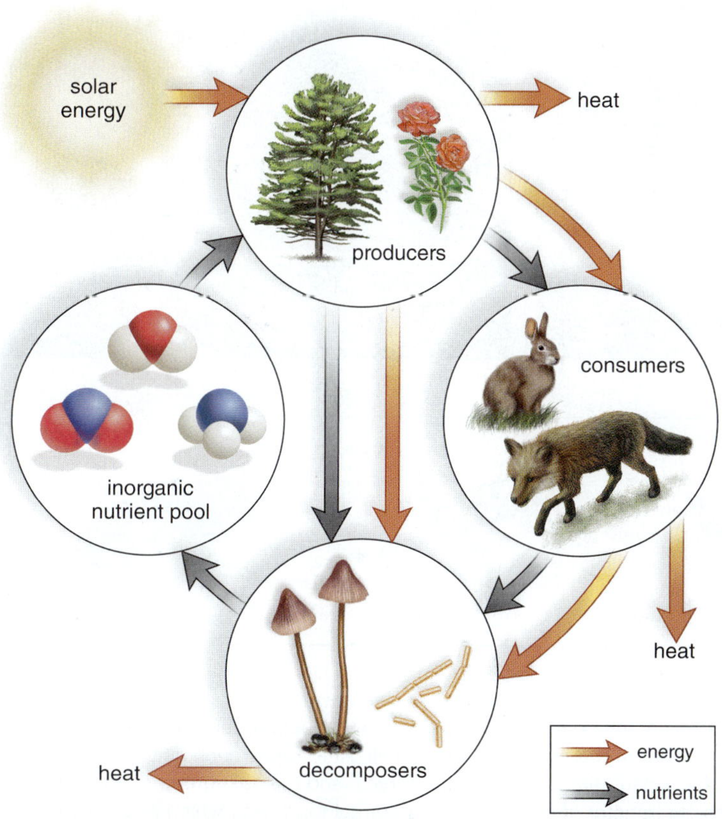

Figure 45.16 Nature of an ecosystem. Chemicals cycle, but energy flows through an ecosystem. As energy transformations repeatedly occur, all the energy derived from the sun eventually dissipates as heat.

Tutorial
Cycling of Energy and Nutrients in an Ecosystem

Figure 45.17 Energy balances. Only about 10% of the food energy taken in by an herbivore is passed on to carnivores. A large portion goes to detritus feeders via defecation, excretion, and death, and another large portion is used for cellular respiration.

Energy Flow

Energy flows through an ecosystem following various pathways. As organic nutrients pass from one component of the ecosystem to another, as when an herbivore eats a plant or a carnivore eats an herbivore, only a portion of the original amount of energy is transferred. Eventually, the energy dissipates into the environment as heat. Therefore, the vast majority of ecosystems cannot exist without a continual input of solar energy.

Only a portion of the organic nutrients made by producers is passed on to consumers, because plants use organic molecules to fuel their own cellular respiration. Similarly, only a small percentage of nutrients consumed by lower-level consumers, such as herbivores, is available to higher-level consumers, or carnivores. As Figure 45.17 demonstrates, a certain amount of the food eaten by an herbivore is not digested and is eliminated as feces. Metabolic nitrogenous wastes are excreted as urine. Of the assimilated energy, a large portion is used during cellular respiration for the production of ATP and thereafter becomes heat. Only the remaining energy, which is converted into increased body weight or additional offspring, becomes available to carnivores.

The elimination of feces and urine by a heterotroph, and the death of an organism, does not mean that organic nutrients are lost to an ecosystem. Instead, they represent the organic nutrients made available to decomposers. Decomposers convert the organic nutrients, such as glucose, back into inorganic chemicals, such as carbon dioxide and water, and release them to the soil or atmosphere. When producers absorb these inorganic chemicals, they have completed their cycle within an ecosystem.

The laws of thermodynamics, which were first described in section 6.1, support the concept that energy flows through an

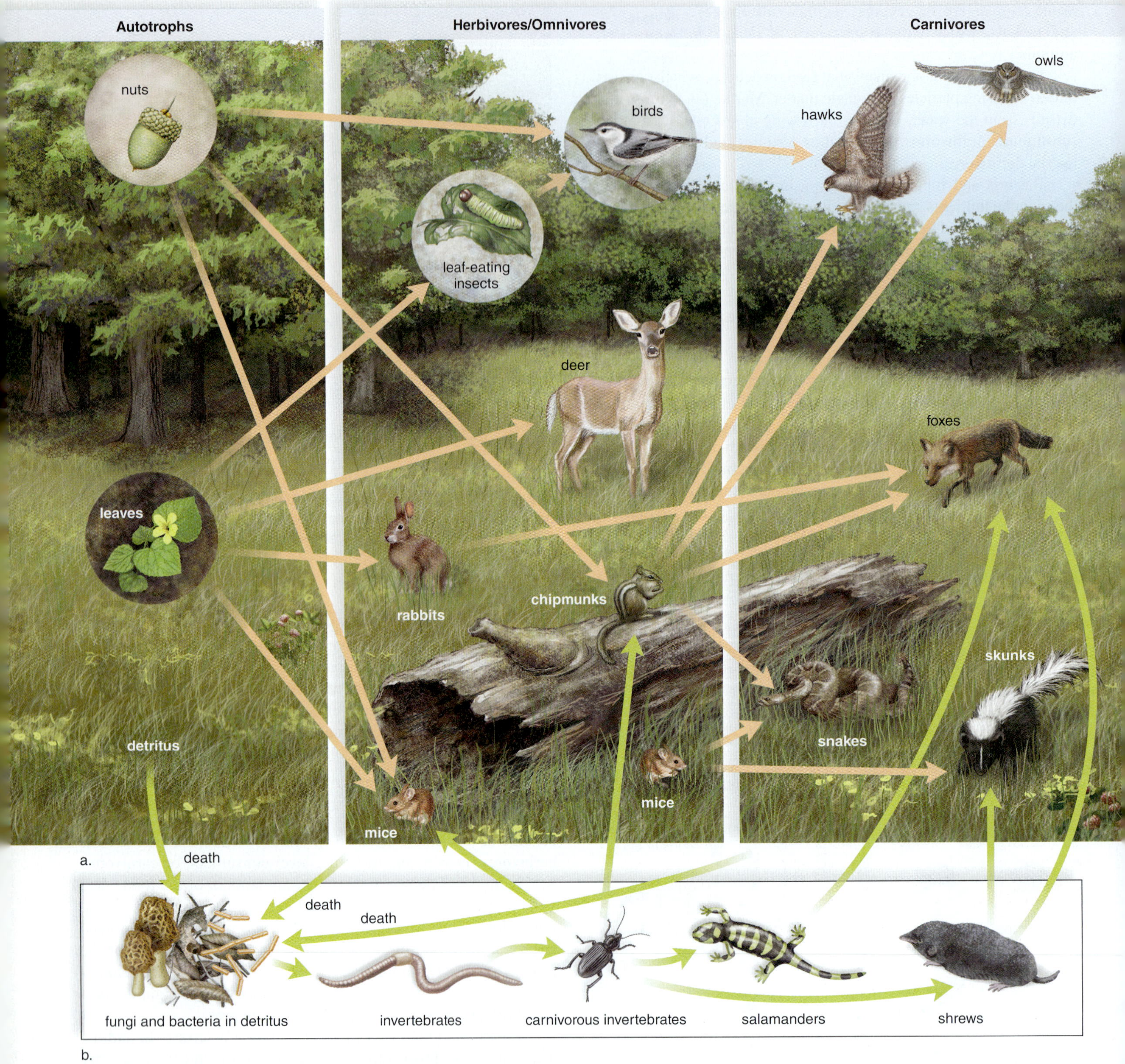

Figure 45.18 Grazing and detrital food web. Food webs are descriptions of who eats whom. **a.** Tan arrows illustrate the possible transfer of energy within a grazing food web. For example, nuts provide energy to birds, which are then fed upon by a hawk. Autotrophs such as the tree are producers (first trophic, or feeding, level), the first series of animals are primary consumers (second trophic level), and the next group of animals are secondary consumers (third trophic level). **b.** Green arrows illustrate the possible transfer of energy within a detrital food web, which begin with detritus—the bacteria and fungi of decay and the remains of dead organisms. A large portion of these remains are from the grazing food web illustrated in (**a**). The organisms in the detrital food web are sometimes fed on by animals in the grazing food web, as when robins feed on earthworms. Thus, the grazing food web and the detrital food web are connected to one another.

ecosystem. The first law states that energy can be neither created nor destroyed. This explains why ecosystems are dependent on a continual outside source of energy, usually solar energy, which photosynthesizers use to produce organic nutrients. The second law states that, with every transformation, some energy is degraded into a less available form, such as heat. Because plants carry on cellular respiration, for example, only about 55% of the original energy absorbed by plants is available to an ecosystem.

An Example of Energy Flow

Let's apply the principles presented so far to an actual ecosystem—a forest of 132,000 m^2 in New Hampshire. The various interconnecting paths of energy flow are represented by a **food web,** a diagram that describes trophic (feeding) relationships. Figure 45.18*a* is a grazing food web that begins with a producer and moves through succesive trophic levels. A portion of the energy available in the oak tree will be transferred to the organisms that feed on the acorns and leaf tissue. A portion of the energy available in the form of herbivores such as caterpillars, mice, rabbits, and deer will be transferred to the carnivores, such as the hawks and owls. The organisms that feed on both the producers and the herbivores are classified as omnivores.

Figure 45.18*b* is a detrital food web, which begins with detritus. Detritus provides energy to soil organisms such as earthworms. Earthworms in turn provide energy to carnivorous invertebrates, which then provide energy to carnivores, such as shrews or salamanders. Because the members of a detrital food web may become food for aboveground carnivores, the detrital and grazing food webs are connected.

We naturally tend to think that aboveground plants, such as trees, are the largest storage form of organic matter and energy, but this is not necessarily the case. In a deciduous forest, the organic matter lying on the forest floor and mixed into the soil contains over twice as much energy as the leaf matter of living trees. Therefore, more energy in a forest may be funneling through the detrital food web than through the grazing food web.

Trophic Levels

The arrangement of the species in Figure 45.18 suggests that organisms are linked to one another in a straight line, according to feeding relationships, or who eats whom. Diagrams that show a single path of energy flow in an ecosystem are called **food chains.** For example, in the grazing food web, we might find this grazing food chain:

leaves ⟶ caterpillars ⟶ birds ⟶ hawks

And in the detrital food web, we might find this detrital food chain:

detritus ⟶ earthworms ⟶ salamanders

A **trophic level** is a level of nutrients within a food web or chain. In the grazing food web in Figure 45.18*a,* the green plants are the producers in the first trophic level, the animals in the center are the primary consumers in the second trophic level, and the last group of animals are the secondary consumers in the third trophic level.

Ecological Pyramids

The shortness of food chains can be attributed to the loss of energy between trophic levels. In general, only about 10% of the energy of one trophic level is available to the next trophic level. Therefore, if an herbivore population consumes 1,000 kg of plant material, only about 100 kg is converted into the body tissue of an herbivore, 10 kg is converted into the body tissue of the first-level carnivores, and 1 kg into the body tissue of the second-level carnivores. This "10% rule" explains why few top-level carnivores can be supported in a food web. The flow of energy with large losses between successive trophic levels is sometimes depicted as an **ecological pyramid** (Fig. 45.19).

Energy flow from one trophic level to the next produces a pyramid based on the number of organisms or the amount of biomass at each trophic level. When constructing such pyramids, problems arise, however. For example, in Figure 45.18, each tree would contain numerous caterpillars; therefore, there would be more herbivores than autotrophs. The explanation, of course, has to do with size. An autotroph can be as tiny as a microscopic alga or as big as a beech tree; similarly, an herbivore can be as small as a caterpillar or as large as an elephant.

Pyramids of biomass eliminate size as a factor because **biomass** is the number of organisms multiplied by the dry weight of the organic matter within one organism. The biomass of the producers is expected to be greater than the biomass of the herbivores,

Figure 45.19 Ecological pyramid. The biomass, or dry weight (g/m^2), for trophic levels in a grazing food web in a bog at Silver Springs, Florida. There is a sharp drop in biomass between the producer level and herbivore level, which is consistent with the common knowledge that the detrital food web plays a significant role in bogs.

and that of the herbivores is expected to be greater than that of the carnivores. In aquatic ecosystems, such as some lakes and open seas where algae are the only producers, the herbivores may have a greater biomass than the producers, because the algae are consumed at such a high rate. Such pyramids, which have more herbivores than producers, are called inverted pyramids:

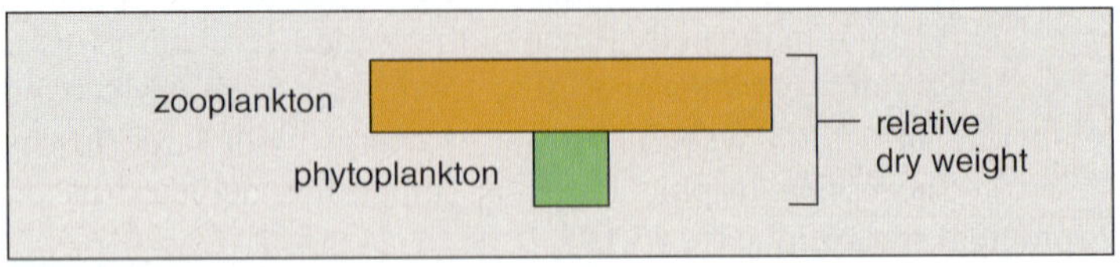

These kinds of drawbacks are making some ecologists hesitant about using pyramids to describe ecological relationships. One more problem is what to do with the decomposers, which are rarely included in pyramids, even though a large portion of energy becomes detritus in many ecosystems.

Chemical Cycling

The pathways by which chemicals circulate through ecosystems involve both living (biotic) and nonliving (geologic) components known as the **biogeochemical cycles.** Here we focus on the four main biogeochemical cycles:

- Water cycle
- Carbon cycle
- Phosphorus cycle
- Nitrogen cycle

A biogeochemical cycle may be sedimentary or gaseous. The phosphorus cycle is a sedimentary cycle; the chemical is absorbed from the soil by plant roots, passed to heterotrophs, and eventually returned to the soil by decomposers. The carbon and nitrogen cycles are gaseous, meaning that the chemical returns to and is withdrawn from the atmosphere as a gas.

Chemical cycling involves the components of ecosystems shown in Figure 45.20. A *reservoir* is a source normally unavailable to producers, such as the carbon present in calcium carbonate shells on ocean bottoms. An *exchange pool* is a source from which organisms can take chemicals, such as the atmosphere or soil.

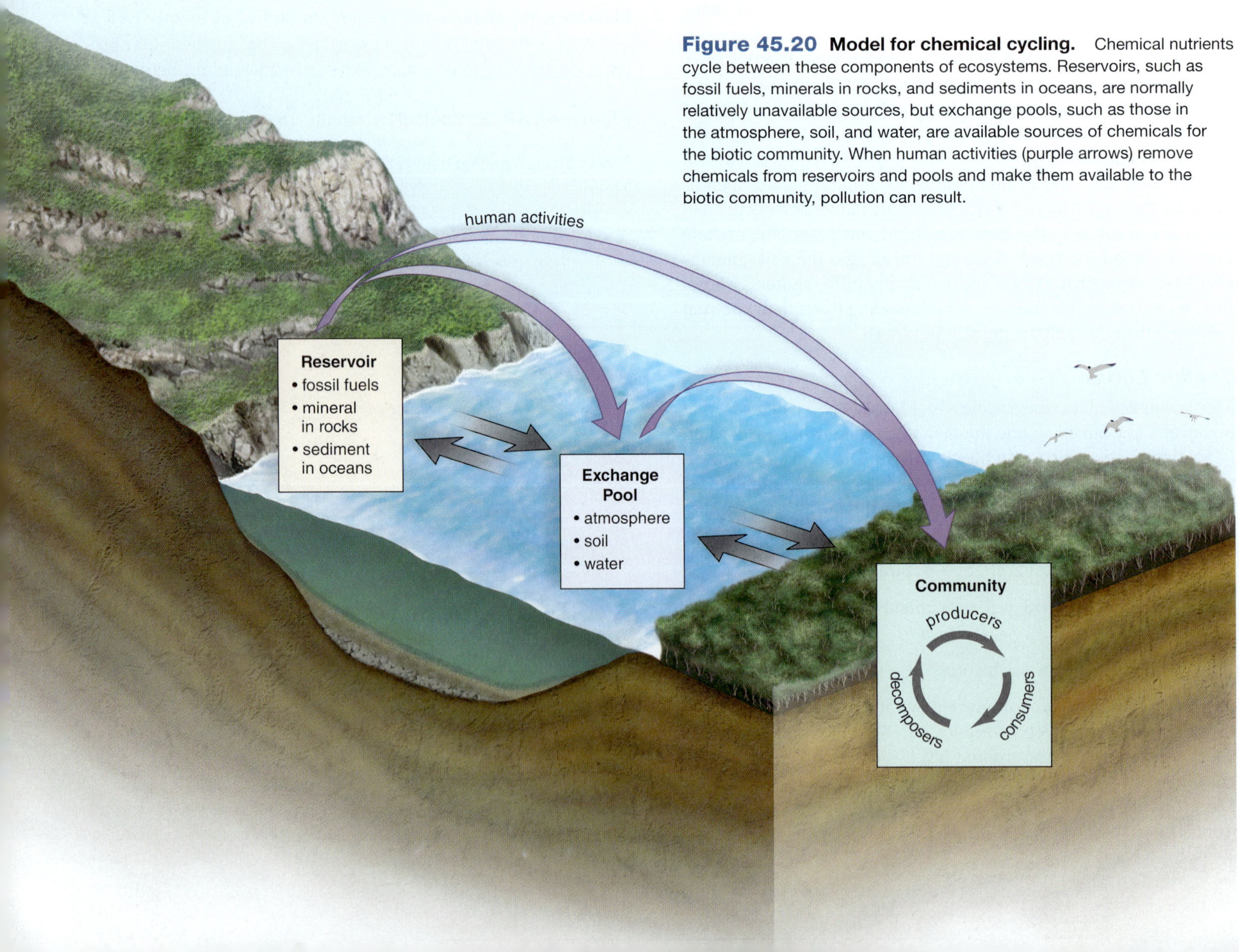

Figure 45.20 Model for chemical cycling. Chemical nutrients cycle between these components of ecosystems. Reservoirs, such as fossil fuels, minerals in rocks, and sediments in oceans, are normally relatively unavailable sources, but exchange pools, such as those in the atmosphere, soil, and water, are available sources of chemicals for the biotic community. When human activities (purple arrows) remove chemicals from reservoirs and pools and make them available to the biotic community, pollution can result.

Certain chemicals can move along food chains in a *biotic community* and never enter an exchange pool.

Human activities (purple arrows in Fig. 45.20) remove chemicals from reservoirs and exchange pools and make them available to the biotic community. In this way, human activities result in pollution, because they upset the normal balance of nutrients for producers in the environment.

The Water Cycle

The **water cycle,** also called the *hydrological cycle,* is described in Figure 45.21. A **transfer rate** is defined by the amount of a substance that moves from one component of the environment to another within a specified period of time. The width of the arrows in Figure 45.21 indicates the transfer rate of water.

During the water cycle, fresh water is first distilled from salt water through evaporation. During evaporation, a liquid, in this case water, changes to a gaseous state. The sun's rays cause fresh water to evaporate from the seawater, and the salts are left behind. Next, condensation occurs. During condensation, a gas is converted into a liquid. For example, vaporized fresh water rises into the atmosphere, collects in the form of a cloud, cools, and falls as rain over the oceans and the land.

Water evaporates from land and from plants (evaporation from plants is called transpiration) and from bodies of fresh water. Because land lies above sea level, gravity eventually returns all fresh water to the sea. In the meantime, water is contained within standing bodies (lakes and ponds), flowing bodies (streams and rivers), and groundwater.

Some of the water from precipitation (e.g., rain, snow, sleet, hail, and fog) makes its way into the ground and saturates the earth to a certain level. The top of the saturation zone is called the groundwater table, or simply, the water table. Because water infiltrates through the soil and rock layers, groundwater can be located in rock layers called aquifers. Water is usually released in appreciable quantities to wells or springs. Aquifers are recharged when rainfall and melted snow percolate into the soil.

Human Activities and the Water Cycle. In some parts of the United States, especially the arid West and southern Florida, withdrawals from aquifers exceed the level of recharge. This is called "groundwater mining." In these locations, the groundwater level is dropping, and residents may run out of groundwater, at least for irrigation purposes, within a few years.

Fresh water makes up about 3% of the world's supply of water. Water is considered a renewable resource, because a new supply is always being produced as a result of the water cycle. It is possible to run out of fresh water, however, when the rate of consumption exceeds the rate of production or when the water has become so polluted that it is not usable.

Video Thames River

Figure 45.21 The water (hydrological) cycle. Evaporation from the ocean exceeds precipitation, so there is a net movement of water vapor onto land, where precipitation results in surface water and groundwater, which flow back to the sea. On land, transpiration by plants contributes to evaporation.

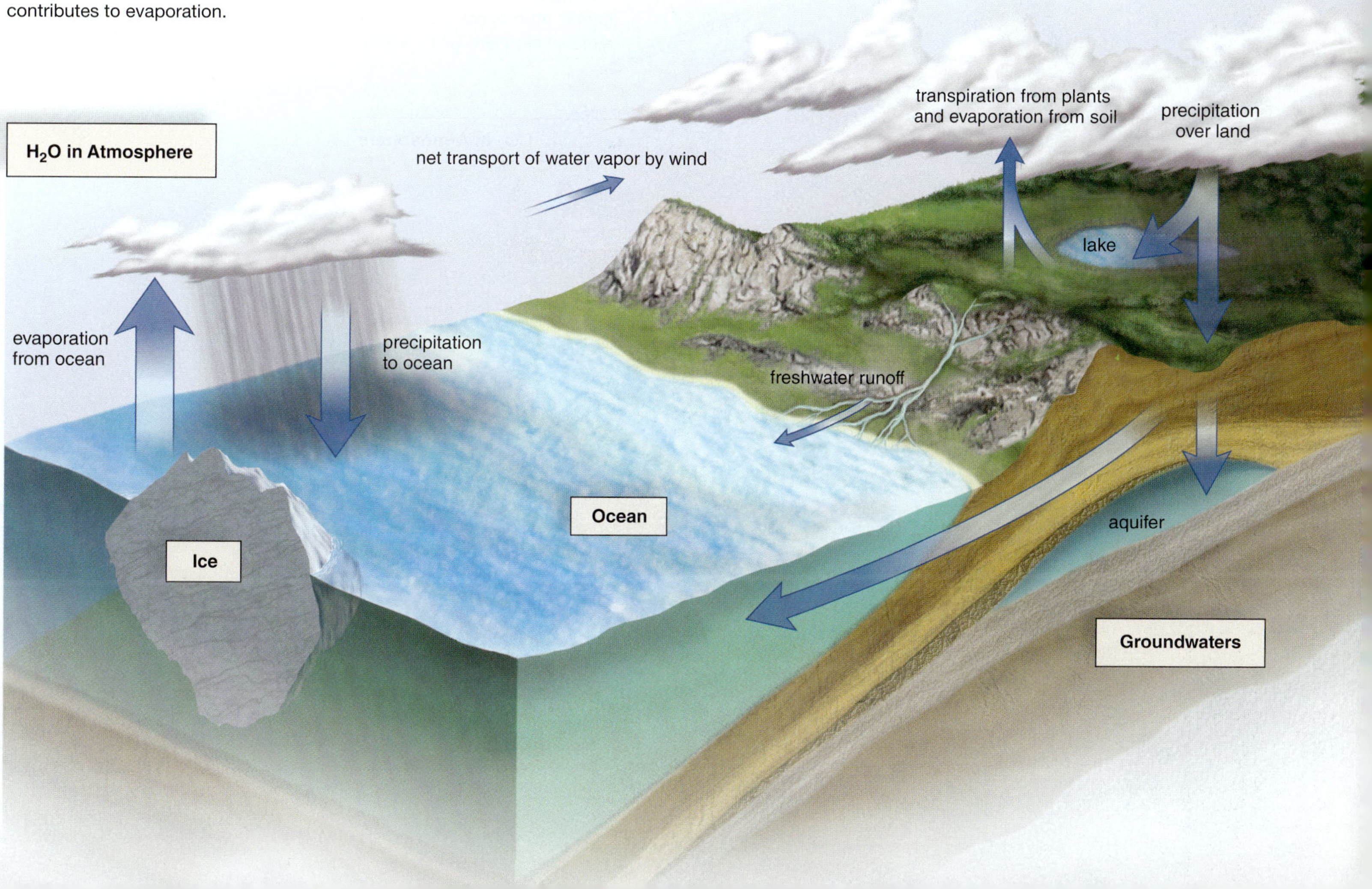

The Carbon Cycle

In the carbon cycle, organisms in both terrestrial and aquatic ecosystems exchange carbon dioxide (CO_2) with the atmosphere (Fig. 45.22). Therefore, the CO_2 in the atmosphere is the exchange pool for the carbon cycle. On land, plants take up CO_2 from the air through photosynthesis. The CO_2 is incorporated into nutrients, which are used by both autotrophs and heterotrophs alike. When organisms, including plants, respire, carbon is returned to the atmosphere as CO_2. Carbon dioxide then recycles to plants by way of the atmosphere.

In aquatic ecosystems, the exchange of CO_2 with the atmosphere is indirect. Carbon dioxide from the air combines with water to produce bicarbonate ion (HCO_3^-). This is the main source of carbon for algae. Similarly, when aquatic organisms respire, the CO_2 they give off becomes HCO_3^-. The amount of bicarbonate in the water is in equilibrium with the amount of CO_2 in the air.

Reservoirs Hold Carbon. Living and dead organisms contain organic carbon and serve as a reservoir for the carbon cycle. The world's biotic components, particularly trees, contain 800 billion tons of organic carbon. An additional 1,000–3,000 billion tons are estimated to be held in the remains of plants and animals in the soil. Ordinarily, decomposition of organisms returns CO_2 to the atmosphere.

Some 300 MYA, plant and animal remains were transformed into coal, oil, and natural gas, the materials we call fossil fuels. Another reservoir for carbon is the inorganic carbonate that accumulates in limestone and in calcium carbonate shells. Many marine organisms have calcium carbonate shells that remain in sediments long after the organisms have died. Geologic forces change these sediments into limestone.

Human Activities and the Carbon Cycle. More CO_2 is being released into the atmosphere than is being removed, largely due to the burning of fossil fuels and the destruction of forests. When humans do away with forests, we reduce a reservoir and the very organisms that take up excess carbon dioxide. Today, the amount of CO_2 released into the atmosphere is about twice the amount that can be absorbed by the producers on the planet. Much of the excess CO_2 dissolves into the ocean.

Several other gases are emitted into the atmosphere due to human activities. These include nitrous oxide (N_2O) from fertilizers and animal wastes and methane (CH_4) from bacterial decomposition that occurs in anaerobic environments. These gases are known as **greenhouse gases,** because, just like the panes of a greenhouse, they allow solar radiation to pass through but hinder the escape of infrared rays (heat) back into space. This phenomenon has come to be known as the **greenhouse effect.**

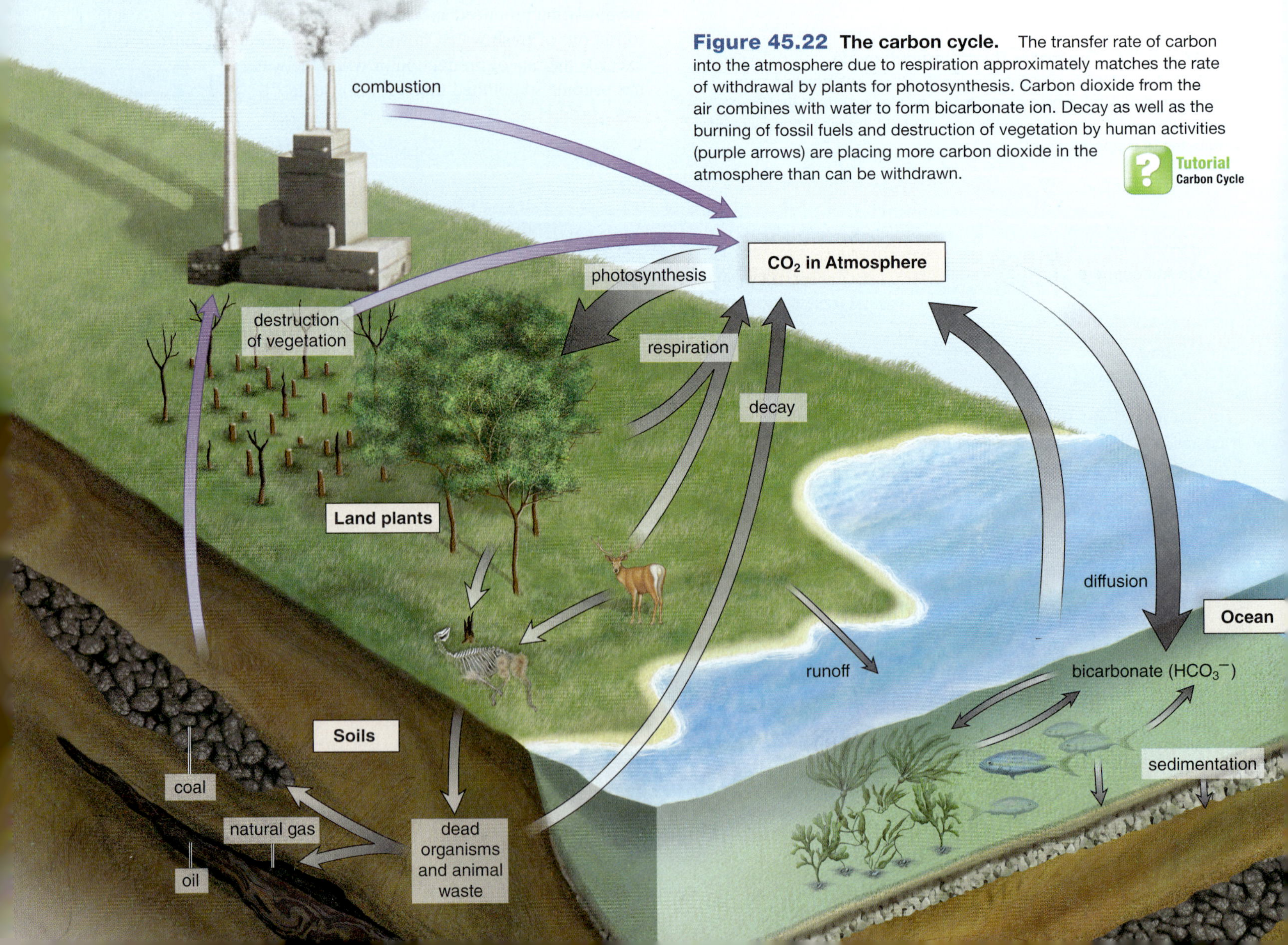

Figure 45.22 The carbon cycle. The transfer rate of carbon into the atmosphere due to respiration approximately matches the rate of withdrawal by plants for photosynthesis. Carbon dioxide from the air combines with water to form bicarbonate ion. Decay as well as the burning of fossil fuels and destruction of vegetation by human activities (purple arrows) are placing more carbon dioxide in the atmosphere than can be withdrawn.

Tutorial Carbon Cycle

The greenhouse gases are contributing significantly to an overall rise in the Earth's ambient temperature, a trend called **global warming,** which is leading to climate change. The expression **climate change** refers to the recent changes in the Earth's climate. The global climate has already warmed about 0.6°C since the Industrial Revolution. Computer models are unable to consider all possible variables, but the Earth's temperature may rise 1.5–4.5°C by 2100 if greenhouse emissions continue at the current rates. The Nature of Science feature, "Global Climate Change," on page 877 discusses these effects further.

Video Karoo Global Warming

Video Warming Hurts Rice

It is predicted that, as the oceans warm, temperatures in the polar regions will rise to a greater degree than in other regions. If so, glaciers will melt and sea level will rise, not only due to this melting but also because water expands as it warms. Increased rainfall is likely along the coasts, while drier conditions are expected inland. Coastal agricultural lands, such as the deltas of Bangladesh and China, will be inundated with seawater, and billions of dollars will have to be spent to keep coastal cities such as New Orleans, New York, Boston, Miami, and Galveston from disappearing into the sea.

The Phosphorus Cycle

Figure 45.23 depicts the phosphorus cycle. Phosphorus from oceanic sediments moves onto land due to a geologic uplift. On land, the very slow weathering of rocks places phosphate ions (PO_3^- and HPO_4^+) into the soil. Some of these become available to plants, which use phosphate in a variety of molecules, including phospholipids, ATP, and the nucleotides that become a part of DNA and RNA. Animals eat producers and incorporate some of the phosphate into their teeth, bones, and shells. However, eventually the death and decay of all organisms and the decomposition of animal wastes make phosphate ions available to producers once again. Because the available amount of phosphate is already being used within food chains, phosphate is usually a limiting inorganic nutrient for plants. Phosphate levels influence the size of populations within an ecosystem.

Some phosphate naturally runs off into aquatic ecosystems, where algae acquire it from the water before it can become trapped in sediments. Phosphate in marine sediments does not become available to producers on land until a geologic upheaval exposes sedimentary rocks on land. The cycle then begins again.

Human Activities and the Phosphorus Cycle. Human beings boost the supply of phosphate by mining phosphate ores used in the production of fertilizer and detergents. Runoff of phosphate and nitrogen from fertilizer use, animal wastes from livestock feedlots, and discharge from sewage treatment plants result in **eutrophication** (overenrichment) of waterways.

The Nitrogen Cycle

Nitrogen gas (N_2) makes up about 78% of the gases in the atmosphere, but plants cannot make use of nitrogen in its gaseous form.

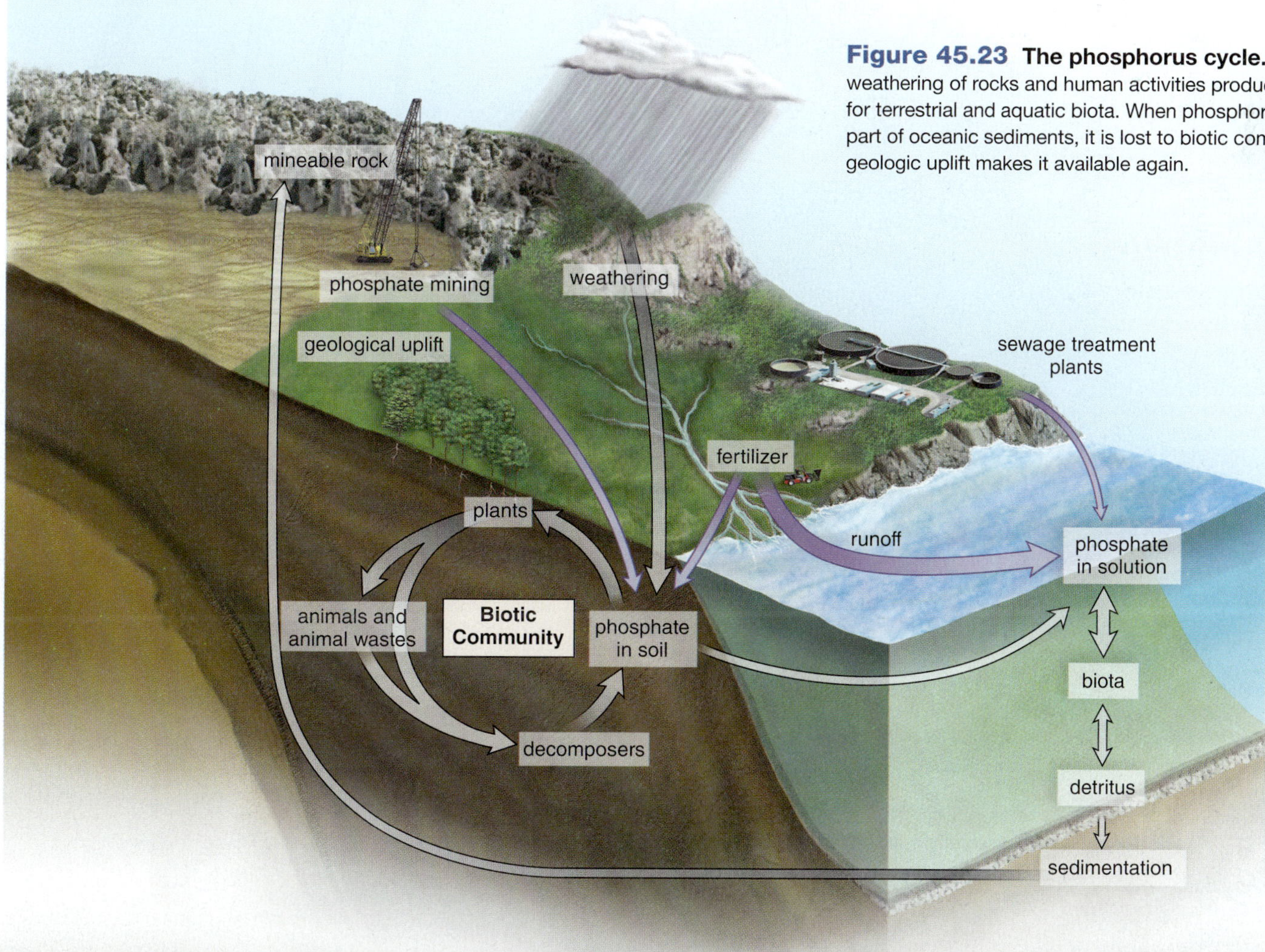

Figure 45.23 The phosphorus cycle. Both the weathering of rocks and human activities produce phosphorus for terrestrial and aquatic biota. When phosphorus becomes a part of oceanic sediments, it is lost to biotic communities until geologic uplift makes it available again.

The availability of nitrogen in an ecosystem can limit the size of the producer populations. First, let's consider that **N_2 (nitrogen) fixation** occurs when nitrogen gas (N_2) is converted to ammonium (NH_4^+), a form plants can use (Fig. 45.24). Some cyanobacteria in aquatic ecosystems and some free-living bacteria in soil are able to fix atmospheric nitrogen in this way. Other nitrogen-fixing bacteria live in nodules on the roots of legumes, such as beans, peas, and clover. They make organic compounds containing nitrogen available to the host plants, so that the plant can form proteins and nucleic acids.

Plants can also use nitrates (NO_3^-) as a source of nitrogen. The production of nitrates during the nitrogen cycle is called **nitrification.** Nitrification can occur in two ways:

1. Nitrogen gas (N_2) is converted to NO_3^- in the atmosphere when cosmic radiation, meteor trails, and lightning provide the energy needed for nitrogen to react with oxygen.
2. Ammonium (NH_4^+) in the soil from various sources, including the decomposition of organisms and animal wastes, is converted to NO_3^- by nitrifying bacteria in soil. Specifically, NH_4^+ (ammonium) is converted to NO_2^- (nitrite), and then NO_2^- is converted to NO_3^- (nitrate).

During the process of **assimilation**, plants take up NH_4^+ and NO_3^- from the soil and use these ions to produce proteins and nucleic acids. Notice in Figure 45.24 that the subcycle involving the biotic community, which occurs on land and in the ocean, need not depend on the presence of nitrogen gas at all.

Finally, **denitrification** is the conversion of nitrate back to nitrogen gas, which then enters the atmosphere. Denitrifying bacteria living in the anaerobic mud of lakes, bogs, and estuaries carry out this process as a part of their own metabolism. In the nitrogen cycle, denitrification would counterbalance nitrogen fixation if not for human activities.

Video Dung Beetles

Figure 45.24 The nitrogen cycle. Nitrogen is primarily made available to biotic communities by internal cycling of the element. Without human activities, the amount of nitrogen returned to the atmosphere (denitrification in terrestrial and aquatic communities) exceeds withdrawal from the atmosphere (N_2 fixation and nitrification). Human activities (purple arrows) result in an increased amount of NO_3^- in terrestrial communities, with resultant runoff to aquatic biotic communities.

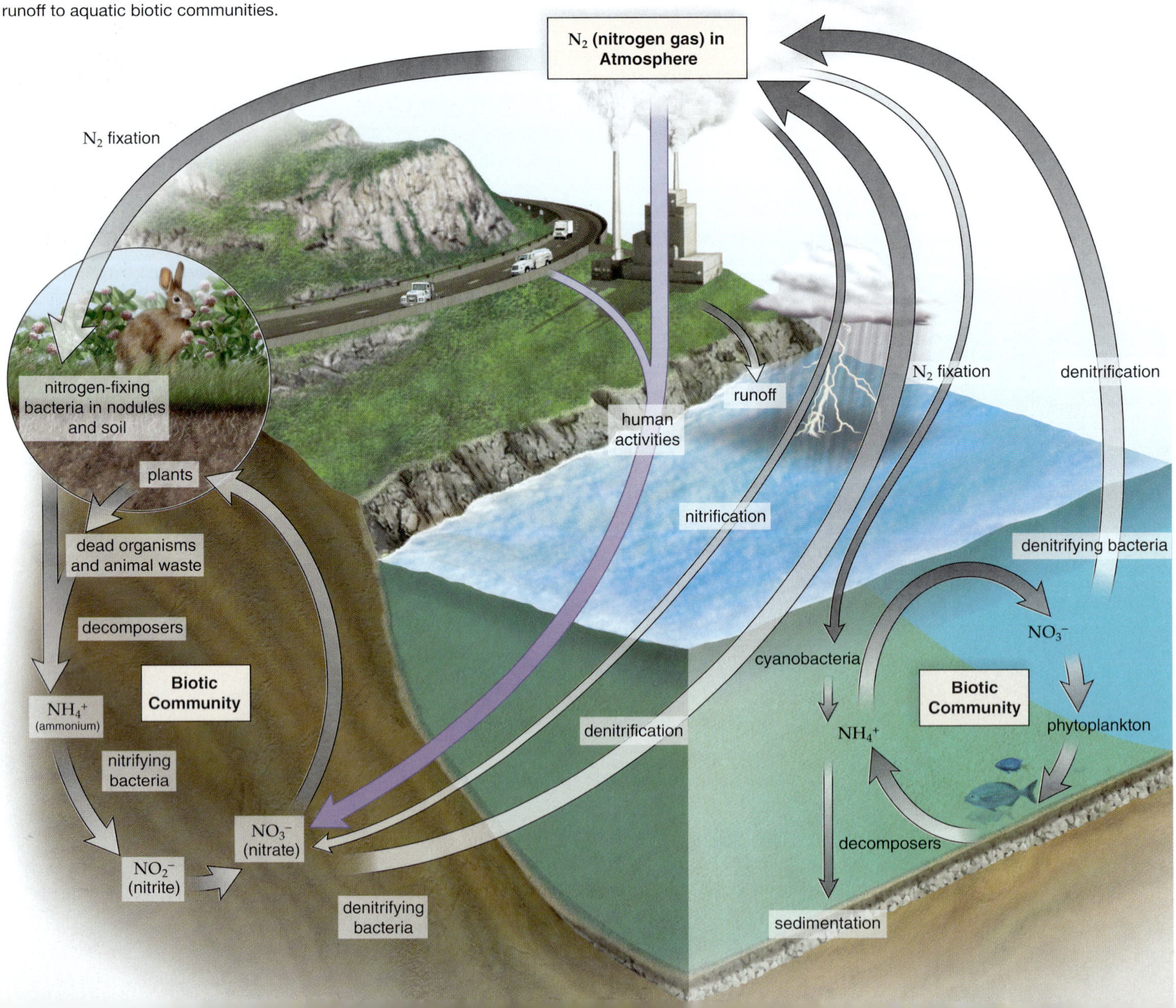

THEME **Nature of Science**

Global Climate Change

Scientists around the world are working on collecting and interpreting environmental data that will help us understand how and why the Earth's climate is changing. Changes in the average temperature of a region, precipitation patterns, sea levels, and greenhouse gas concentrations are all indicators that our climate is changing.

Since 1901 the average temperature across the United States has risen, with 2000–2009 being the warmest decade on record worldwide with 30–60% of the United States experiencing drought conditions.

Average precipitation rates have also increased by 6% over the past century. Since 1990, the United States has experienced 8 of the top 10 years of extreme precipitation events. Increases in sea surface temperatures produce a more active hurricane season. Since the mid-1990s, the Atlantic Ocean, the Caribbean, and the Gulf of Mexico have seen 6 of the 10 most active hurricane seasons.

Sea levels worldwide have risen an average of 1 inch per decade due to the overall increase in the surface temperature of the world's oceans. Over half of the human population lives within 60 miles of the coast. Climate models suggest that we will see a rise in sea levels of 3 to 4 feet over the next century. New York City ranges from 5 feet to 16 feet above sea level, while the Florida Keys are an average of 3 to 4 feet above sea level. Even if the rising waters don't produce flooding, many coastal areas will be exposed to increasingly severe storms and storm surges that could lead to significant economic losses.

Between 1990 and 2000, global carbon dioxide emissions increased by 1.5 times (Fig. 45C*a*). Production of electricity is the largest producer of greenhouse gas emissions in the United States, followed by transportation. We are facing a global problem that will require global cooperation in order to be solved.

The Kyoto Protocol was initially adopted on December 11, 1997, and entered into force on February 16, 2005. The goal of the protocol was to achieve stabilization and reduction of the greenhouse gas concentrations in the atmosphere. As of November 2009, 187 countries had ratified the protocol with the goal of reducing emissions by an average of 5.2% by the year 2012. As of 2008, the United States was responsible for approximately 19% of the global CO_2 emissions from fossil fuel combustion and is the largest per-capita emission producer in the world. Unfortunately, the United States still has not ratified the agreement (Fig. 45C*b*).

The Copenhagen conference in 2009 ended without any type of binding agreement for long-term action against climate change. It did produce a collective commitment by many developed nations to raise 30 billion dollars to be used to help poor nations cope with the effects of and combat climate change.

The 2010 Cancun summit on climate change helped solidify this agreement. Because deforestation produces about 15% of the global carbon emissions, many developing countries will be able to receive incentives to prevent the destruction of their rain forests.

The 2013 UN Climate Change Conference in Warsaw was successful in establishing ways to help developing nations reduce greenhouse emissions as a result of deforestation as well as establishing finance commitments to assist developing nations. The conference also helped keep governments on track toward a universal climate agreement that will be presented in 2015, which will be implemented in 2020.

The main concern of the conferences is that the hard decisions of making significant changes to our greenhouse emissions continues to be pushed off into the future. The longer we delay in making changes, the higher the risks become.

Questions to Consider

1. Should the United States and other developed nations pay developing nations to preserve their forests?
2. Do individuals have a personal responsibility to help prevent climate change, or is it a governmental responsibility?

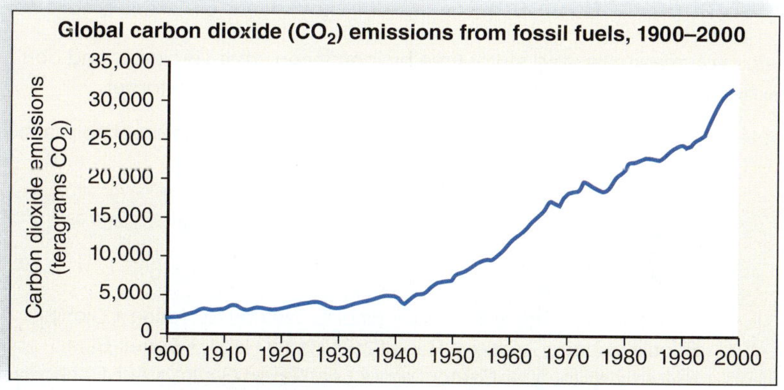

a.

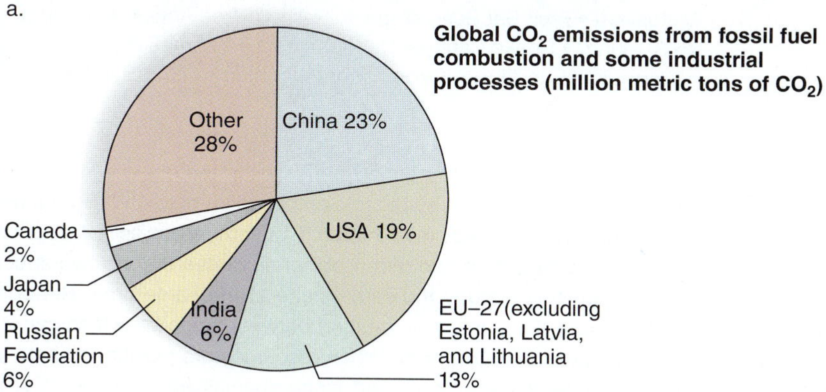

b.

Figure 45C Global warming and climate change. **a.** CO_2 emissions have increased significantly in the past hundred years, causing changes in global climate patterns, including more severe droughts and erratic patterns of precipitation. **b.** Although China is the largest producer of CO_2, the United States produces more CO_2 per person (or capita).

Tutorial Global Warming and Climate Change

Human Activities and the Nitrogen Cycle. Humans significantly alter the transfer rates in the nitrogen cycle by producing fertilizers from N_2. Fertilizer runs off into lakes and rivers and results in an overgrowth of algae and rooted aquatic plants. When the algae die off, enlarged populations of decomposers use up all the oxygen in the water, and the result is a massive fish kill.

Acid deposition occurs because nitrogen oxides (NO_x) and sulfur dioxide (SO_2) enter the atmosphere from the burning of fossil fuels. Both these gases combine with water vapor to form acids, which eventually return to the earth. Acid deposition has drastically affected forests and lakes in northern Europe, Canada, and the northeastern United States by altering the pH of their soils and surface waters. Acid deposition also reduces agricultural yields and corrodes marble, metal, and stonework.

Check Your Progress 45.3

1. Identify the various types of populations that are at the base of an ecological pyramid and the start of a food chain.
2. Compare the flow of energy to the flow of chemicals through an ecosystem.
3. Provide examples of how human activities can alter the biogeochemical cycles.

CONNECTING *the* CONCEPTS *with the* THEMES

Evolution

- Changes in one species in a community will lead to corresponding changes in other species.
- As succession occurs in a community, species must adapt or they will become extinct.

Nature of Science

- Understanding ecological succession can help us determine the age of a community.
- Ecologists study the movement of energy and nutrients through communities in order to determine community health.

Biological Systems

- Interspecific interactions help determine population sizes and community diversity.
- Ecosystems have biotic and abiotic components that determine species richness.
- Disturbances and climate variability can produce changes within the biodiversity of a community.

MEDIA STUDY TOOLS http://connect.mheducation.com

SMARTBOOK® Study smart. SmartBook helps you maximize your study time by identifying what you know and don't know. The recharge feature in SmartBook helps you prepare for the test by identifying the material that you are most likely to forget.

connect® BIOLOGY Find even more resources to learn the chapter concepts including animations, tutorial videos, and interactive practice questions.

Tutorials

45.3 Cycling of Energy and Nutrients in an Ecosystem • Carbon Cycle • Global Warming and Climate Change

Videos

45.1 Tallgrass Prairie Ecology • Finches Natural Selection • Cichlid Specialization • Adaptive Radiation • Harris Hawk • Forest Blue Butterfly • Flounder Camouflage • Leaf-Tailed Gecko • Ant-Caterpillar Mutualism • Fruit Bat Seed Dispersal • Thorn Tree Ants • Pollinators
45.3 Plants • Snake Eating • Decomposers • Thames River • Warming Hurts Rice • Karoo Global Warming • Dung Beetles

SUMMARIZE

45.1 Ecology of Communities

A **community** is an assemblage of populations interacting with one another in the same environment. Communities differ in their composition, also known as **species richness.** The **species diversity** includes both species richness and relative abundance. The **island biogeography model** of diversity helps us determine the ideal size of conservation areas.

An organism's **habitat** is where it lives in the community. An **ecological niche** is defined by the role an organism plays in its community, including its habitat and how it interacts with other species in the community. The **competitive exclusion principle** states that no two species can indefinitely occupy the same niche at the same time. **Resource partitioning** decreases competition between species, which can lead to **character displacement.** When resources are partitioned between two or more species, increased niche specialization occurs. But the difference between species can be more subtle, as when warblers feed at different parts of the tree canopy. Barnacles competing on the Scottish coast may be an example of ongoing competition.

Predation occurs when a **predator** feeds upon **prey.** This type of interaction is especially influenced by the size of the predator population and the amount of resources available for the prey. A cycling of population densities may occur. Prey defenses take many forms: **Camouflage,** warning coloration, and **mimicry** are a few mechanisms. Batesian mimicry occurs when one species has the warning coloration but lacks the defense. Müllerian mimicry occurs when two species with the same warning coloration have the same defense.

Symbiosis, the close association of two species over a long period of time, comes in several forms. **Parasitism** occurs when a **parasite** takes nourishment from its **host.** Whether parasites are aggressive (kill their host) or benign probably depends on which results in the highest fitness. **Commensalism** occurs when one species gains a benefit from the relationship and there is no impact on the other. Mutualism occurs when both species involved in a relationship receive a positive benefit. The continuous interaction between two species results in **coevolution,** a change in one species in response to a change in the other.

45.2 Community Development

Ecological succession involves a series of species replacements in a community. Primary succession occurs where there is no soil present. Secondary succession occurs where soil is present and certain plant species can begin to grow. **Pioneer species** are the first producers to inhabit an area after a disturbance. A **climax community** forms when stages of succession lead to a particular type of community.

45.3 Dynamics of an Ecosystem

Populations interact with each other and the abiotic environment to form an **ecosystem.** The biotic components of an ecosystem consist of **autotrophs,** which are the producers of the community, and **heterotrophs,** which are the consumers of the community. Heterotrophs can be classified as **herbivores, carnivores, omnivores, detritivores,** and **decomposers** based on how they obtain their nutrients. Abiotic components are resources such as nutrients and conditions such as type of soil and temperature.

Ecosystems are characterized by energy flow and chemical cycling. Energy flows in one direction, because as food passes from one population to the next, each population makes energy conversions that result in a loss of usable energy. Chemicals cycle because they pass from one population to the next until decomposers return them once more to the producers.

Ecosystems contain **food webs** in which the various organisms are connected by trophic relationships. In grazing food webs, **food chains** begin with a producer. In a detrital food web, food chains begin with detritus. **Trophic levels** depict the levels of nutrients available in a food web or chain. As energy flows from one level to the next, it is depicted as an **ecological pyramid.** Ecological pyramids are graphic representations of the number of organisms multiplied by their **biomass,** representing the energy content of trophic levels.

Biogeochemical cycles are the pathways in which chemicals circulate through ecosystems. Chemical cycling involves a reservoir, an exchange pool, and a biotic community.

In the **water cycle** (hydrological cycle), evaporation over the ocean is not compensated for by precipitation. The **transfer rate** is the amount of substance that moves from one component of the environment to another. Precipitation over land results in bodies of fresh water plus groundwater, including aquifers. Eventually, all water returns to the oceans. In the carbon cycle, carbon dioxide in the atmosphere is an exchange pool; both terrestrial and aquatic plants and animals exchange carbon dioxide with the atmosphere. Living and dead organisms serve as reservoirs for the carbon cycle, because they contain organic carbon. Human activities increase the level of CO_2 and other **greenhouse gases,** which help cause a phenomenon known as the **greenhouse effect. Global warming** is leading to **climate change** as the ambient temperature of the Earth continues to increase.

In the phosphorus cycle, geologic upheavals move phosphorus from the ocean to land. Slow weathering of rocks returns phosphorus to the soil. Most phosphorus is recycled within a community. Excessive accumulation of phosphorus in an aquatic community can lead to **eutrophication.**

In the nitrogen cycle, plants cannot use nitrogen gas from the atmosphere. During **nitrogen fixation,** N_2 converts to ammonium, making nitrogen available to plants. **Nitrification** is the production of nitrates, whereas **denitrification** is the conversion of nitrate back to N_2, which enters the atmosphere. Human activities increase transfer rates in the nitrogen cycle. **Acid deposition** occurs when nitrogen oxides enter the atmosphere, combine with water vapor, and return to Earth in precipitation.

ASSESS

Choose the best answer for each question.

45.1 Ecology of Communities

1. The species composition of a community is called
   a. species richness.
   b. species diversity.
   c. climax community.
   d. pioneer species.
   e. ecological niche.
2. The ecological niche of an organism
   a. is the same as its habitat.
   b. includes how it competes for and acquires food.
   c. is specific to the organism.
   d. is usually occupied by another species.
   e. Both b and c are correct.
3. What type of interaction is it when an alfalfa plant gains fixed nitrogen from the bacterial species *Rhizobium* in its root system and the *Rhizobium* gains carbohydrates from the plant?
   a. mutualism
   b. parasitism
   c. commensalism
   d. competition
   e. predation
4. What type of interaction is it when both foxes and coyotes in an area feed primarily on a limited supply of rabbits?
   a. mutualism
   b. parasitism
   c. commensalism
   d. competition
   e. predation

45.2 Community Development

5. Which is the correct sequence of events during primary succession?
   a. low shrub–high shrub–grass–shrub-tree–low tree–high tree
   b. high shrub–low shrub–grass–low tree–high tree–shrub-tree
   c. grass–shrub-tree–high tree–low tree–low shrub–high shrub
   d. grass–low shrub–low tree–shrub-tree–high shrub–high tree
   e. grass–low shrub–high shrub–shrub-tree–low tree–high tree

6. Mosses growing on bare rock will eventually help create soil. These mosses are involved in ______________ succession.
   a. primary
   b. secondary
   c. tertiary
   d. complete
   e. incomplete

45.3 Dynamics of an Ecosystem

7. In what way are decomposers like producers?
   a. Either may be the first member of a grazing or a detrital food chain.
   b. Both produce oxygen for other forms of life.
   c. Both require nutrient molecules and energy.
   d. Both are present only on land.
   e. Both produce organic nutrients for other members of ecosystems.
8. When a heterotroph takes in food, only a small percentage of the energy in that food is used for growth, because
   a. some food is not digested and is eliminated as feces.
   b. some metabolites are excreted as urine.
   c. some energy is given off as heat.
   d. All of these are correct.
   e. None of these are correct.
9. During chemical cycling, inorganic nutrients are typically returned to the soil by
   a. autotrophs.
   b. detritivores.
   c. decomposers.
   d. tertiary consumers.
10. How do plants contribute to the carbon cycle?
    a. When plants respire, they release CO_2 into the atmosphere.
    b. When plants photosynthesize, they consume CO_2 from the atmosphere.
    c. When plants photosynthesize, they provide oxygen to heterotrophs.
    d. Both a and b are correct.
11. Choose the statement that is true concerning this food chain:

    grass → rabbits → snakes → hawks

    a. Each predator population has a greater biomass than its prey population.
    b. Each prey population has a greater biomass than its predator population.
    c. Each population is omnivorous.
    d. Each population returns inorganic nutrients and energy to the producer.
    e. Both a and c are correct.
    f. Both a and b are correct.
12. Label this diagram of an ecosystem:

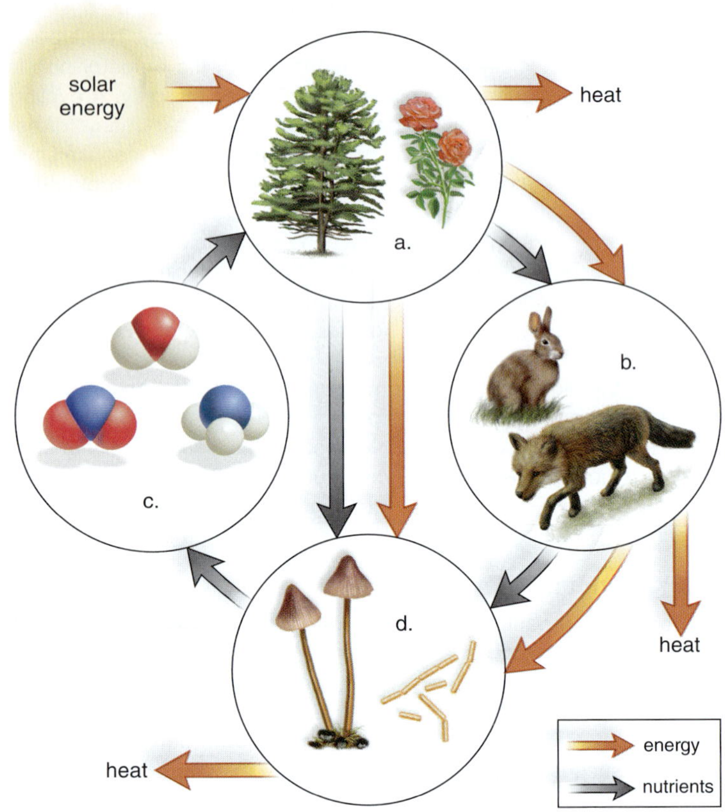

ENGAGE

THINKING SCIENTIFICALLY

1. Residents that live 25 miles downstream from a new coal-burning power plant are complaining about a strange taste that has started to appear in their drinking water. How would you explain the connection between the new power plant and the strange taste in the drinking water?
2. Upland game hunters in Illinois have been noticing a decrease in the rabbit and pheasant populations over the past 5 years. They have lobbied the Department of Natural Resources and the state legislature to allow them to shoot red-tailed hawks, one of the rabbit's and pheasant's main predators. Explain why shooting hawks will not ultimately solve the problem of a decreased rabbit and pheasant population.
3. If you observe three species of frogs in the same general area, how would you test a hypothesis that they occupy different niches?
4. In order to improve species richness, you decide to add phosphate to a pond. How might you determine how much phosphate to add in order to avoid eutrophication?

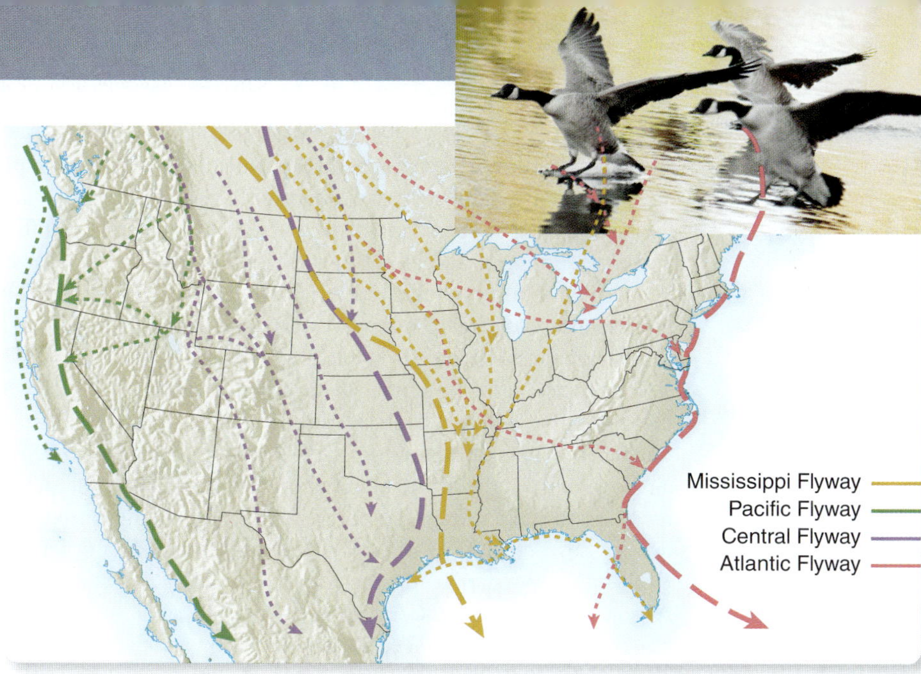

Migration route of Canada geese.

46 Major Ecosystems of the Biosphere

Chapter Outline

46.1 Climate and the Biosphere 882
46.2 Terrestrial Ecosystems 885
46.3 Aquatic Ecosystems 895

Before You Begin

Before beginning this chapter, take a few moments to review the following discussions.

Section 7.5 How does the climate determine the producer base of a biome?

Section 45.1 How do symbiotic relationships enable species to survive in harsh environments?

Figure 45.21 How does the water cycle connect terrestrial and aquatic habitats?

Toward fall, as temperatures drop and food becomes scarce, Canada geese (*Brantas canadensis*) begin their annual migration. Their primary migration corridor extends from north of the Arctic Circle to south of the Tropic of Cancer. Millions of geese will follow the Mississippi, Pacific, Central, or Atlantic flyways to reach specific wintering grounds. Birds can travel at speeds of 30 to 55 miles per hour and have been known to travel up to 650 miles in a single day. Some migrations will cover 2,000–3,000 miles.

During the winter migration they stop at key staging points to rest and feed. Wetlands, marshes, grasslands, and chaparral regions serve as vital habitat for their migration. The high concentration of birds in these areas makes them more susceptible to disease outbreaks and predation. Natural predators range from foxes and owls to bald eagles and humans. Waterfowl hunting in Illinois contributes over 82 million dollars a year to the state's economy.

After the birds have wintered in the southern portion of their range they begin the spring migration toward their breeding grounds. This migration will take them through the same flyways and staging points that the winter migration does. Nesting occurs from the middle of March to the middle of May. By late August the new birds have grown enough to begin the cycle over again. A variety of biomes play critical roles in the survival of this species.

As you read through this chapter, think about the following questions:

1. What advantage do migratory birds have over their counterparts that do not migrate?
2. How would changes in one land biome impact the other land biomes?

Following *the* Themes

CHAPTER 46 MAJOR ECOSYSTEMS OF THE BIOSPHERE

UNIT 8 BEHAVIOR AND ECOLOGY

Theme	
Evolution	The biodiversity of the Earth's biomes has evolved over billions of years.
Nature of Science	New species are being discovered as scientists explore previously unexplored biomes.
Biological Systems	Over time, changes in one biome influence changes in the other biomes.

46.1 Climate and the Biosphere

Learning Outcomes

Upon completion of this section, you should be able to

1. Describe how solar radiation produces variations in Earth's climate.
2. Explain how global air circulation patterns and physical geographic features are associated with the Earth's temperature and rainfall patterns.

Climate is the prevailing weather conditions in a particular region. Climate is dictated by temperature and rainfall, which are influenced by the following factors:

- Variations in solar radiation distribution due to the tilt of the Earth as it orbits about the sun
- Other effects caused by topography (surface features) and whether a body of water is nearby

Effect of Solar Radiation

Because the Earth is a sphere, it receives direct sunlight at the equator but indirect sunlight at the poles (Fig. 46.1*a*). The region between latitudes approximately 26.5° north and south of the equator is considered the tropics. The tropics are warmer than the areas north of 23.5°N and south of 23.5°S, known as the temperate regions.

The tilt of the Earth as it orbits around the sun causes one pole or the other to be angled toward the sun (except at the spring and fall equinoxes, when sunlight aims directly at the equator). This accounts for the seasons that occur in all parts of the Earth except at the equator (Fig. 46.1*b*). When the Northern Hemisphere is having winter, the Southern Hemisphere is having summer, and vice versa.

If the Earth were standing still and were a solid, uniform ball, all air movements would be in two directions. Air at the equator warmed by the sun would rise and move toward the poles where it would cool and sink. Rising air creates zones of lower air pressure. However, because the Earth rotates on its axis daily and its surface consists of continents and oceans, the flow of warm and cold air is modified into three large circulation cells in each hemisphere (Fig. 46.2).

At the equator, the sun heats the air and evaporates water. Warm, moist air rises, and as it cools it loses most of its moisture as rain. The greatest amounts of rainfall on Earth are near the equator. The rising air flows toward the poles, but at about 30° north and south latitude, it cools and sinks toward the Earth's surface before reheating. As the dry air descends and warms, areas of high pressure are generated. High-pressure regions are zones of low rainfall. The great deserts of Africa, Australia, and the Americas occur at these latitudes.

At the Earth's surface, the air flows toward the poles and the equator. As dry air moves across the Earth, moisture from both land and water gets absorbed. At about 60° north and south latitude, the warmed air rises and cools, producing another low-pressure area with high rainfall. This moisture supports the great forests of the temperate zone. Part of this rising air flows toward the equator, and part continues toward the poles, where it will descend. The poles are high-pressure areas that have low amounts of precipitation.

Besides affecting precipitation, the spinning of the Earth also affects the winds (Fig. 46.2). In the Northern Hemisphere, large-scale winds generally bend clockwise, and in the Southern Hemisphere, they bend counterclockwise. The curving pattern of the winds, ocean currents, and cyclones is the result of the fact that the Earth rotates in an eastward direction. At about 30° north latitude and 30° south latitude, the winds blow from the east-southeast in the Southern Hemisphere and from the east-northeast in the Northern Hemisphere (the east coasts of continents at these latitudes are wet). The doldrums, regions of calm, occur at the equator. The winds blowing from the

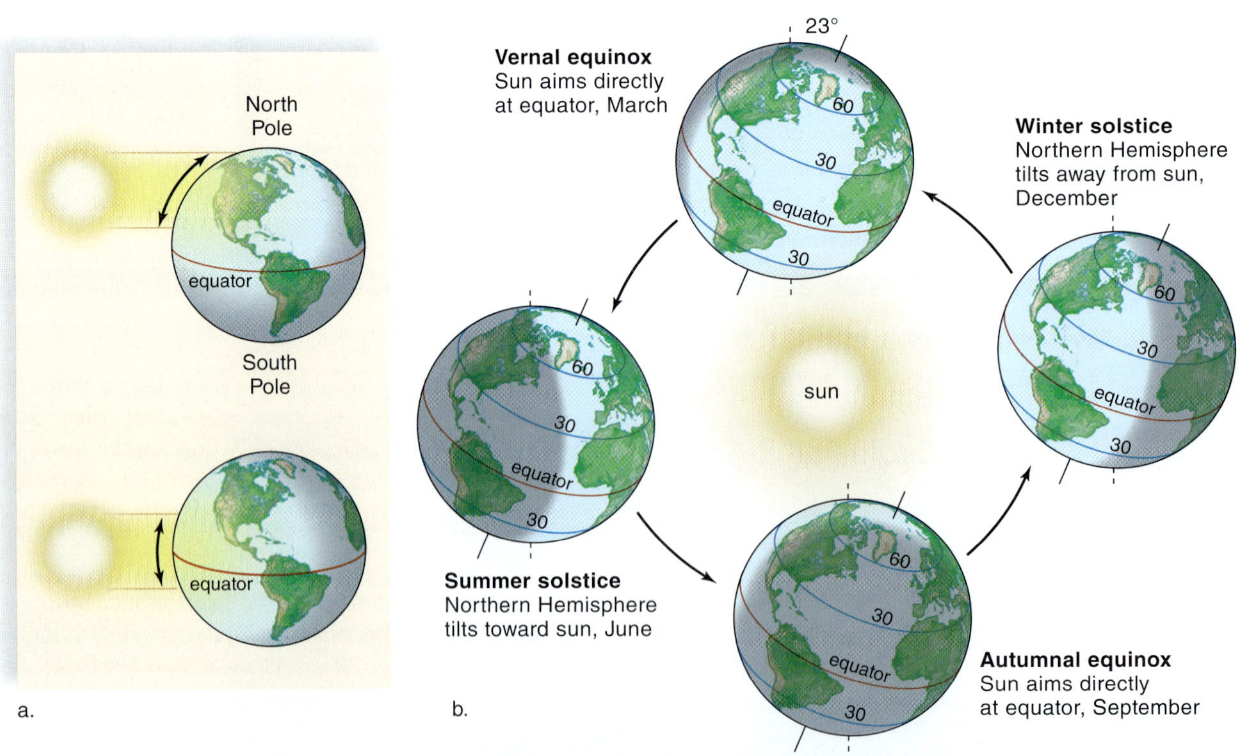

Figure 46.1 Distribution of solar energy. **a.** Because the Earth is a sphere, beams of solar energy striking the Earth near one of the poles are spread over a wider area than similar beams striking the Earth at the equator. **b.** The seasons of the Northern and Southern Hemispheres are due to the tilt of the Earth on its axis as it rotates about the sun.

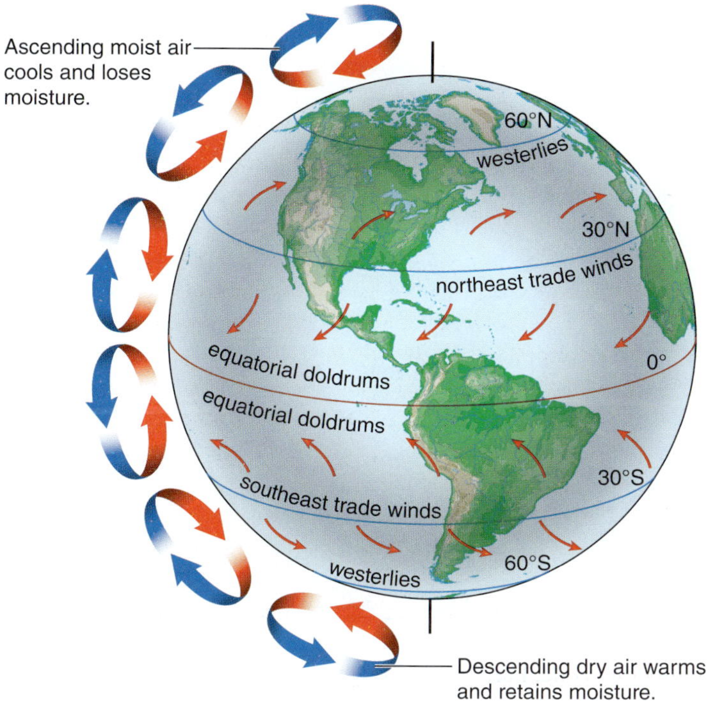

Figure 46.2 Global wind circulation. Air ascends and descends as shown because the Earth rotates on its axis. Also, the trade winds move from the northeast to the west in the Northern Hemisphere, and from the southeast to the west in the Southern Hemisphere. The westerlies move toward the east.

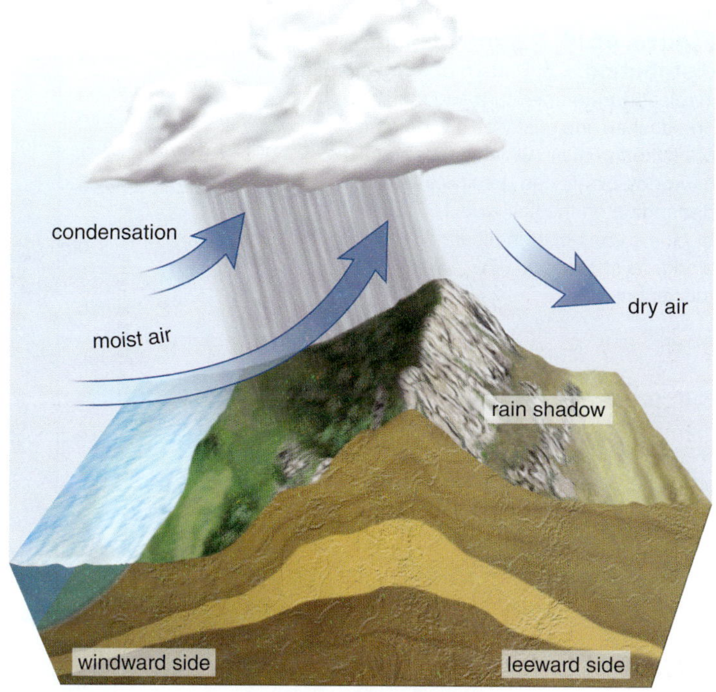

Figure 46.3 Formation of a rain shadow. When winds from the sea cross a coastal mountain range, they rise and release their moisture as they cool this side of a mountain, called the windward side. The leeward side of a mountain receives relatively little rain and is therefore said to lie in a "rain shadow."

doldrums toward the poles are called trade winds because sailors depended on them to fill the sails of their trading ships.

Between 30° and 60° north and south latitudes, strong winds called the prevailing westerlies blow from west to east. The west coasts of the continents at these latitudes are wet, as is the Pacific Northwest, where a massive evergreen forest is located. Weaker winds, called the polar easterlies, blow from east to west at still higher latitudes of their respective hemispheres.

Other Effccts

Topography means the physical features of the land. One physical feature that affects climate is the presence of mountains. As air blows up and over a coastal mountain range, it rises and cools. The coastal side of the mountain, called the windward side, receives more rainfall than the other side. On the interior side of the mountain or leeward side, the air descends, absorbs moisture from the ground, and produces dry air and clear weather (Fig. 46.3). The leeward side of the mountain or **rain shadow,** tends to be much drier than the windward side.

In the Hawaiian Islands, for example, the windward side of the mountains receives more than 750 cm of rain a year, while the leeward side gets on average only 50 cm of rain and is generally sunny. In the United States, the western side of the Sierra Nevada Mountains is lush, while the eastern side is a semidesert.

The temperature of the oceans is more stable than that of landmasses. Oceanic water gains or loses heat more slowly than do terrestrial environments. This difference causes coasts to have a unique weather pattern that is not observed inland. During the day, the land warms more quickly than the ocean, and the air above the land rises, pulling a cool sea breeze in from the ocean. At night, the reverse happens; the breeze blows from the land toward the sea.

India and some other countries in southern Asia have a **monsoon** climate, in which wet ocean winds blow onshore for almost half the year. The land heats more rapidly than the waters of the Indian Ocean during spring. The difference in temperature between the land and the ocean causes an enormous circulation of air: Warm air rises over the land, and cooler air comes in off the ocean to replace it. As the warm air rises, it loses its moisture, and the monsoon season begins. As just discussed, rainfall is particularly heavy on the windward side of hills. Cherrapunji, a city in northern India, receives an annual average of 1,090 cm of rain a year because of its high altitude. This weather pattern has reversed by November. The land is now cooler than the ocean; therefore, dry winds blow from the Asian continent across the Indian Ocean. In the winter, the air over the land is dry, the skies cloudless, and temperatures pleasant. The chief crop of India is rice, which starts to grow when the monsoon rains begin.

In the United States, people often speak of the "lake effect," meaning that in the winter, arctic winds blowing over the Great Lakes become warm and moisture-laden. When these winds rise and lose their moisture, snow begins to fall. Places such as Buffalo, New York, receive heavy snowfalls due to the lake effect, and snow is on the ground there for an average of 90–140 days every year.

Check Your Progress 46.1

1. Identify the conditions that account for a warm climate at the equator.
2. Name two physical features that can affect rainfall.

Figure 46.4 Pattern of biome distribution. **a.** Pattern of world biomes in relation to temperature and moisture. The dashed line encloses a wide range of environments in which either grasses or woody plants can dominate the area, depending on the soil type. **b.** The same type of biome can occur in different regions of the world, as shown on this global map.

Tutorial
Factors Influencing the Distribution of Biomes

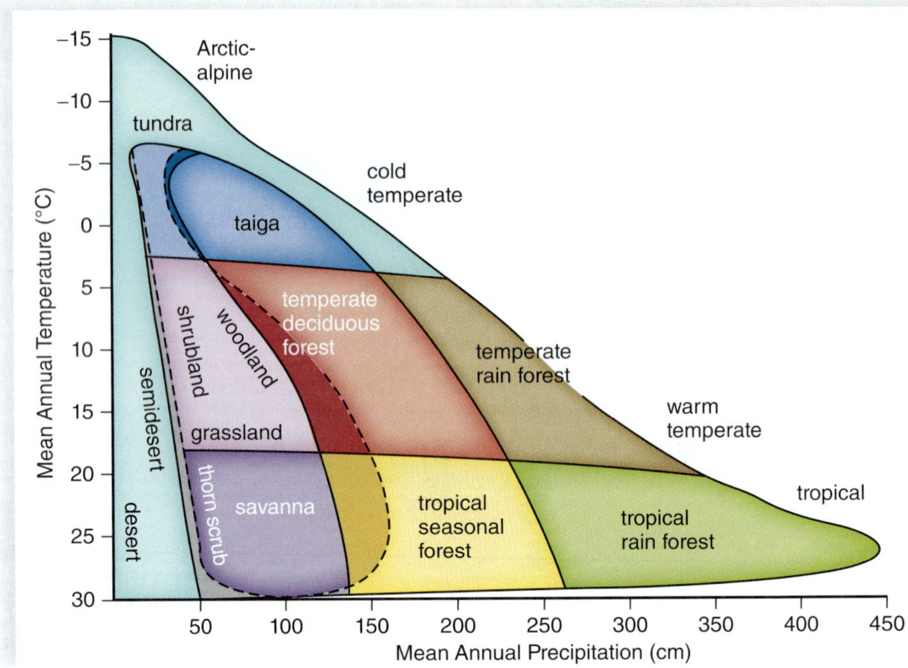

a. Biome pattern of temperature and precipitation

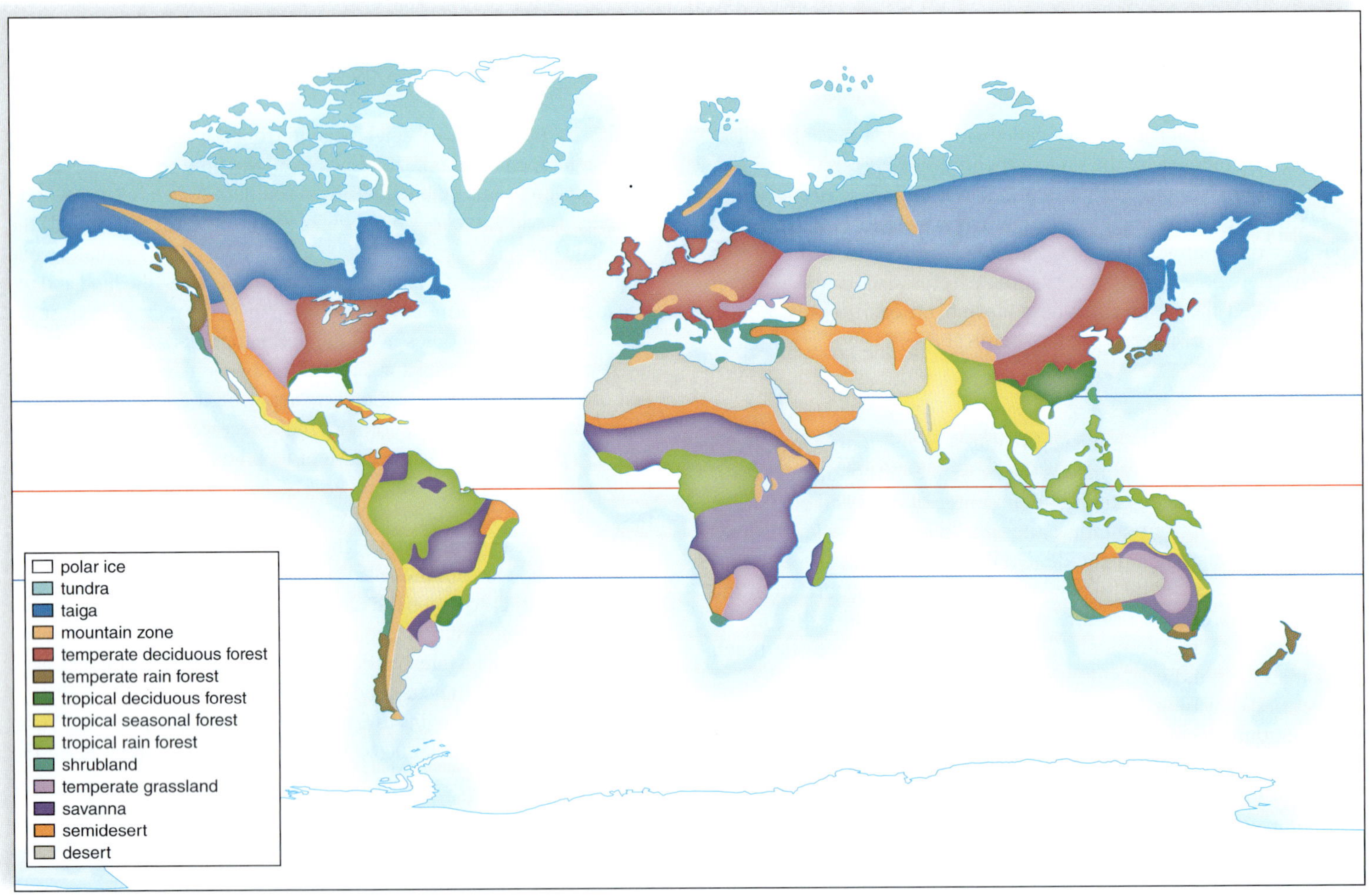

b. Distribution of biomes

46.2 Terrestrial Ecosystems

Learning Outcomes

Upon completion of this section, you should be able to

1. Recognize the geographic distribution of the major terrestrial biomes.
2. Identify the key characteristics of the major terrestrial biomes.

Major terrestrial ecosystems, called biomes, are characterized by their climate and geography. A **biome** has a particular mix of plants and animals that are adapted to living within a specific range of environmental conditions in a given region.

When terrestrial biomes are plotted according to their mean annual temperature and mean annual precipitation, particular patterns result (Fig. 46.4*a*). The distribution of biomes is shown in Figure 46.4*b*. Even though Figure 46.4 shows definite dividing lines, keep in mind that the change from one biome to another is gradual. Each biome has a connection to all the other terrestrial and aquatic ecosystems of the biosphere.

Animation Biomes

The distribution of the biomes and their corresponding populations are determined principally by differences in climate. Latitude and altitude both play a role in determining temperature gradients. If you travel from the equator to the North Pole, you can observe tropical rain forests, followed by a temperate deciduous forest, a coniferous forest, and tundra. A similar sequence is seen when ascending a mountain (Fig. 46.5). The coniferous forest of a mountain is called a **montane coniferous forest,** and the tundra near the peak of a mountain is called an **alpine tundra.** When going from the equator to the South Pole, you would not reach a region corresponding to a coniferous forest or tundra because of the lack of landmasses in the Southern Hemisphere.

Each biome supports characteristic types of animals; however, as described in the opening essay, many animals are able to migrate from one region to another. Animals may breed in one biome, for example, but spend the nonbreeding period in another biome. The Nature of Science feature, "Wildlife Conservation and DNA," on page 887 describes how DNA analysis is helping to clarify not only species relationships but also migration patterns.

We now examine the features of the major biomes, beginning with the tundra.

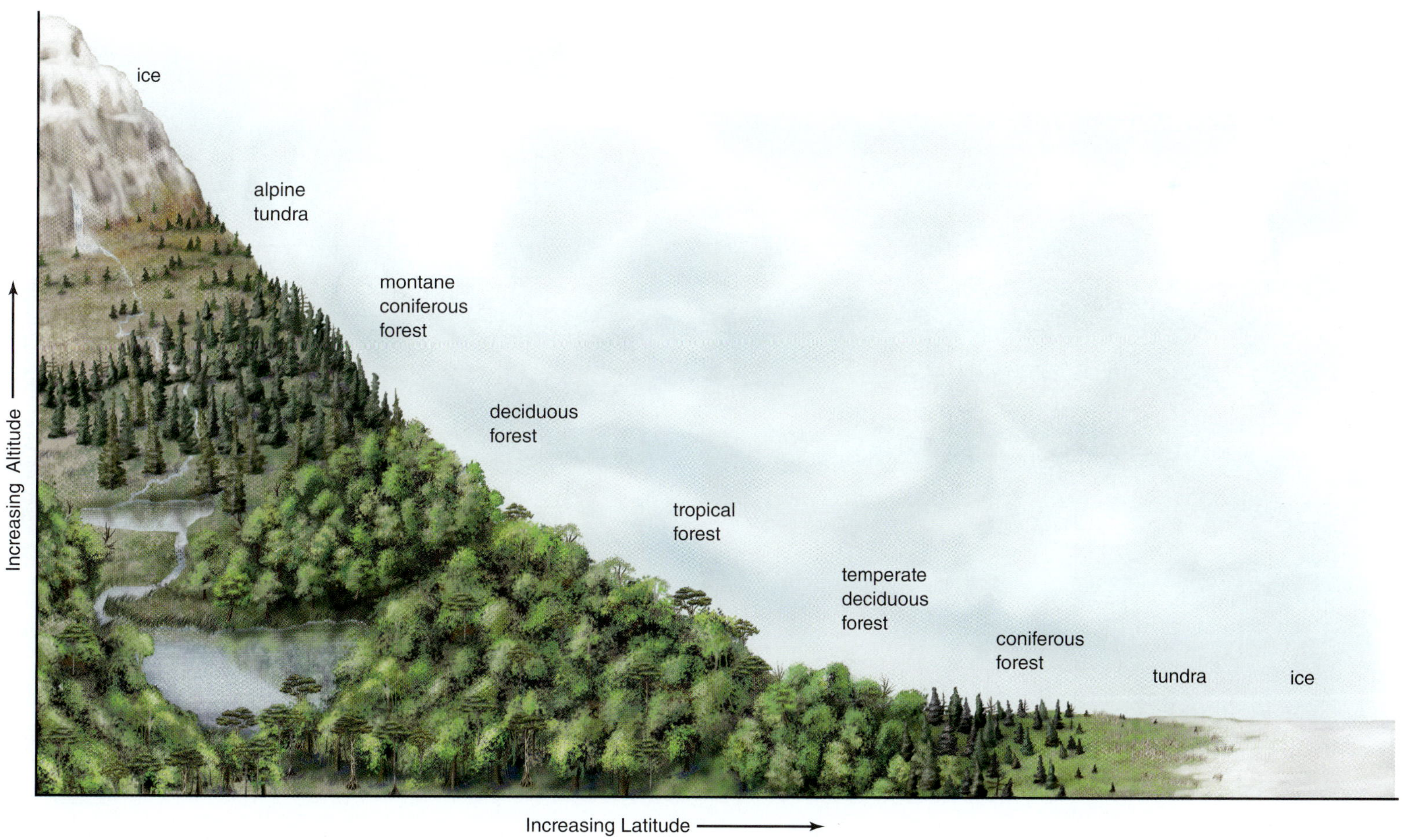

Figure 46.5 Climate and biomes. Biomes change with altitude just as they do with latitude because vegetation is partly determined by temperature. Precipitation also plays a significant role, which is one reason grasslands, instead of tropical or deciduous forests, are sometimes found at the base of mountains.

Tundra

The **Arctic tundra** biome is characterized as being cold and dark much of the year. It has extremely long, cold, harsh winters and short summers (6–8 weeks). Because rainfall amounts to only about 20 cm a year, the tundra could be considered a desert, but melting snow creates a landscape of pools and bogs in the summer. Only the topmost layer of soil thaws; **permafrost** is the layer that remains permanently frozen resulting in minimal drainage. The available soil in the tundra is nutrient poor. Arctic tundra, which encircles the Earth just south of ice-covered polar seas in the Northern Hemisphere, covers about 20% of the Earth's land surface (Fig. 46.6). (As mentioned, a similar ecosystem, the alpine tundra, occurs above the timberline on mountain ranges.)

Trees are not found in the tundra because the growing season is too short. The roots cannot penetrate the permafrost and they cannot become anchored in the shallow boggy soil during the summer. In the summer, the ground is covered with short grasses and sedges, as well as numerous patches of lichens and mosses. Dwarf woody shrubs, such as dwarf birch, flower and seed during the short growing season.

A few animals live in the tundra year-round. For example, the mouselike lemming stays beneath the snow; the ptarmigan, a grouse, burrows in the snow during storms; and the musk ox conserves heat because of its thick coat and short, squat body. Other animals that live in the tundra include snowy owls, lynxes, voles, Arctic foxes, and snowshoe hares. In the summer, the tundra is alive with numerous insects and birds, particularly shorebirds and waterfowl that migrate inland. Caribou in North America and reindeer in Asia and Europe also migrate to and from the tundra, as do the wolves that prey upon them. Polar bears are common near the coastal regions. All species have adaptations for living in extreme cold with a short growing season.

Figure 46.6 The tundra. **a.** In this biome, which is nearest the polar regions, the vegetation consists principally of lichens, mosses, grasses, and low-growing shrubs. The melting snow forms pools of water on the permanently frozen ground, attracting many birds. **b.** Caribou will feed on lichens, grasses, and shrubs during the summer then migrate south during the winter.

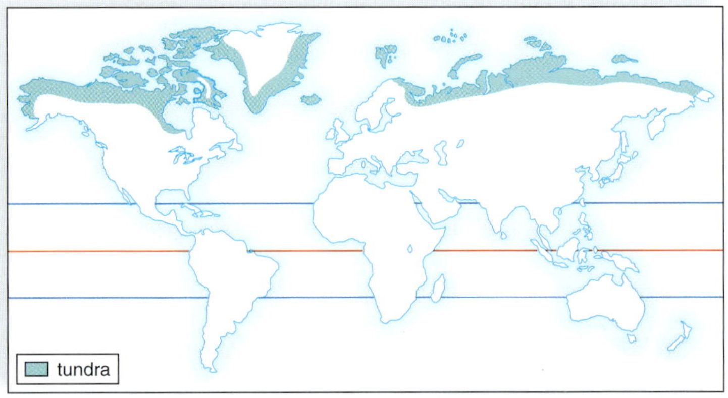

a. Tundra

b. Bull caribou

THEME Nature of Science

Wildlife Conservation and DNA

After DNA analysis, scientists were amazed to find that some 60% of loggerhead turtles drowning in the nets and hooks of fisheries in the Mediterranean Sea were from beaches in the southeastern United States. Because the unlucky creatures were a good representative sample of the turtles in the area, that meant more than half of the young turtles living in the Mediterranean Sea had hatched from nests on beaches in Florida, Georgia, and South Carolina (Fig. 46A*a*). Some 20,000–50,000 loggerheads die each year due to the Mediterranean fisheries, which may partly explain the decline in loggerheads nesting on southeastern U.S. beaches for the last 25 years.

a.

b.

Figure 46A DNA studies. **a.** Many loggerhead turtles found in the Mediterranean Sea are from the southeastern United States. **b.** No matter where they are found in their range, all jaguars belong to the same species.

Jaguars (*Panthera onca*) (Fig. 46A*b*) are the third largest cat in the world behind lions and tigers. They are the largest cat in the Western Hemisphere. Their natural range extends from as far north as Mexico to as far south as Argentina. Currently they are listed as "Near Threatened" by the International Union for the Conservation of Nature (IUCN). In order to conserve a top-level predator that has an extensive range, it requires support from all of the various countries that are home to jaguars.

Detailed genetic analysis of jaguar DNA has indicated that, whether they live in Mexico, Argentina, or anywhere in between, they are all the same species. They are the only wide-ranging carnivore in the world that shows genetic continuity across their entire range. This genetic information led to the formation of the Jaguar Corridor Initiative (JCI). The goal of the initiative is to create a genetic corridor that links jaguar populations in all of the 18 countries in Latin America, from Mexico to Argentina, and hopefully ensuring the survival of this species.

In a classic example of how DNA analysis might be used to protect endangered species from future ruin, scientists from the United States and New Zealand carried out discreet experiments in a Japanese hotel room on whale sushi bought in local markets. Sushi, a staple of the Japanese diet, is a rice and fish mixture wrapped in seaweed. Armed with a miniature DNA sampling machine, the scientists found that, of the 16 pieces of whale sushi they examined, many were from whales that are endangered or protected under an international moratorium on whaling. "Their findings demonstrated the true power of DNA studies," says David Woodruff, a conservation biologist at the University of California, San Diego.

One sample was from an endangered humpback, four were from fin whales, one was from a northern minke, and another from a beaked whale. Stephen Palumbi of the University of Hawaii says the technique could be used for monitoring and verifying catches. Until then, he says, "no species of whale can be considered safe."

Meanwhile, Ken Goddard, director of the unique U.S. Fish and Wildlife Service Forensics Laboratory in Ashland, Oregon, is already on the watch for wildlife crimes in the United States and 122 other countries that send samples to him for analysis. "DNA is one of the most powerful tools we've got," says Goddard, a former California police crime-lab director.

The lab has blood samples, for example, for all of the wolves being released into Yellowstone National Park—"for the obvious reason that we can match those samples to a crime scene," says Goddard. The lab has many cases currently pending in court that he cannot discuss. But he likes to tell the story of the lab's first DNA-matching case. Shortly after the lab opened in 1989, California wildlife authorities contacted Goddard. They had seized the carcass of a trophy-sized deer from a hunter. They believed the deer had been shot illegally on a 3,000-acre preserve owned by actor Clint Eastwood. The agents found a gut pile on the property but had no way to match it to the carcass. The hunter had two witnesses to deny the deer had been shot on the preserve.

Goddard's lab analysis made a perfect match between tissue from the gut pile and tissue from the carcass. Says Goddard: "We now have a cardboard cutout of Clint Eastwood at the lab saying 'Go ahead: Make my DNA.'"

Questions to Consider

1. Would DNA evidence from a crime scene be enough to convict someone of a crime involving a protected species?
2. What degree of DNA difference is necessary to qualify two organisms as different species?

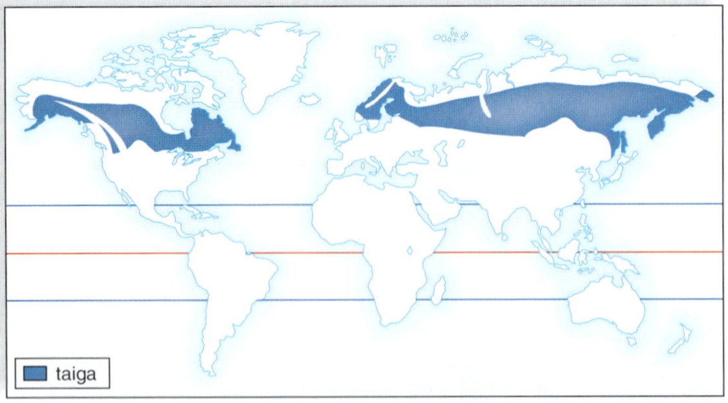

a. Spruce trees in the taiga biome

b. Bull moose, *Alces americanus*, a large mammal

Figure 46.7 The taiga. The taiga, which means "swampland," spans northern Europe, Asia, and North America. The appellation "spruce-moose" refers to the (**a**) dominant presence of spruce trees and (**b**) moose, which frequent the ponds.

Coniferous Forests

Coniferous forests have long, cold, snowy winters with warm and humid summers. They tend to be dominated by cone-bearing trees and can be found in three primary locations: in the **taiga,** which extends around the world in the northern part of North America and Eurasia; near mountaintops (where it is called a montane coniferous forest); and along the Pacific coast of North America, as far south as northern California.

The taiga typifies the coniferous forest with its cone-bearing trees, such as spruce, fir, and pine. These trees are well adapted to the cold because both the leaves and bark have thick coverings. Also, the needlelike leaves can withstand the weight of heavy snow. There is a limited understory of plants, but the forest floor is covered by low-lying mosses and lichens beneath a layer of needles. Birds harvest the seeds of the conifers, while bears, deer, moose, beavers, and muskrats live around the lakes and streams. Wolves prey on these larger mammals. A montane coniferous forest also harbors the wolverine and the mountain lion. The taiga, or boreal forest, exists south of the tundra and covers approximately 11% of the Earth's landmasses (Fig. 46.7). There are no comparable biomes in the Southern Hemisphere because no large landmasses exist at that latitude.

The coniferous forest that runs along the west coast of Canada and the United States is sometimes called a **temperate rain forest** due to the plentiful rainfall and rich soil. As the prevailing winds move in off the Pacific Ocean, they drop their moisture when they meet the coastal mountain range. This environment has produced some of the tallest conifer trees ever in existence, including the coastal redwoods. This forest is also called an old-growth forest because some trees are as old as 800 years. It truly is an evergreen forest because of the abundance of green plants year-round.

Squirrels, lynxes, and numerous species of amphibians, reptiles, and birds inhabit the temperate rain forest. The northern spotted owl, *Strix occidentalis caurina,* is an endangered species found in this ecosystem that has received recent conservation attention and efforts.

Temperate Deciduous Forests

Temperate deciduous forests have a moderate climate with relatively high rainfall (75–150 cm per year). The seasons are well defined, and the growing season ranges between 140 and 300 days. The trees, such as oak, beech, sycamore, and maple, have broad leaves. These are termed deciduous trees because they lose their leaves in the fall and regrow them in the spring. Deciduous forests can be found south of the taiga in eastern North America, eastern Asia, and much of Europe (Fig. 46.8).

The tallest trees form a canopy, an upper layer of leaves that are the first to receive sunlight. Enough sunlight penetrates to provide energy for another layer of trees, called understory trees. Beneath these trees are shrubs and herbaceous plants that may flower in the spring before the trees have put forth their leaves. Still another layer of plant growth—mosses, lichens, and ferns—resides beneath the shrub layer. This stratification provides a variety of habitats for insects and birds. Ground animals are also plentiful. Squirrels, rabbits, woodchucks, and chipmunks are small herbivores. These and ground birds such as turkeys, pheasants, and grouse are preyed on by various carnivores. In contrast to the taiga, amphibians and reptiles occur in this biome because the winters are not as cold. Frogs and turtles prefer an aquatic existence, as do beavers and muskrats. Autumn fruits, nuts, and berries provide a supply of food for the winter, and the leaves, after turning brilliant colors and falling to the ground, contribute to the rich layer of humus. The minerals within the rich soil are washed far into the ground by spring rains, but the deep tree roots capture them and take them back up into the forest system again.

Figure 46.8 Temperate deciduous forest. **a.** The Shawnee National Forest in Illinois is home to many varied plants and animals. **b.** Marsh marigolds may be found in wetland areas, chipmunks feed on acorns, and bobcats prey on these and other small mammals.

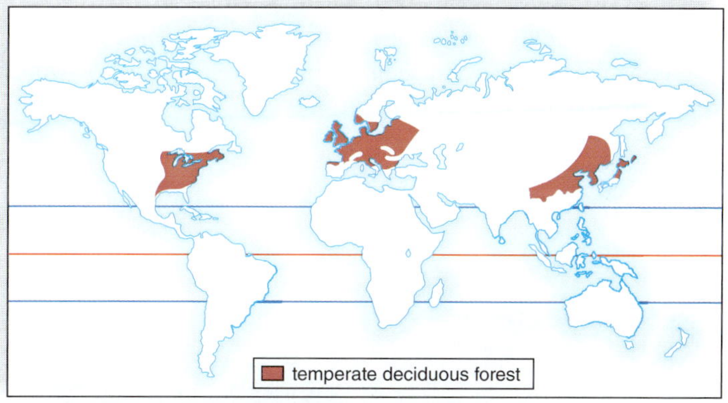

a.

marsh marigolds

eastern chipmunk

bobcat

b.

Tropical Forests

Tropical rain forests are characterized by weather that is always warm (between 20° and 25°C), rainfall that is plentiful (with a minimum of 190 cm per year), and approximately 12 hours of sunlight every day. These are the richest land biomes on Earth. It is estimated that approximately 50% of the species found on Earth live in the tropical rain forests. South America, Africa, and the Indo-Malayan region near the equator all contain tropical rainforests.

A tropical rain forest has a complex structure, with many levels of life, including the forest floor, understory, and canopy. The vegetation of the forest floor tends to be very sparse because most of the sunlight is filtered out by the canopy. The understory consists of smaller plants that are specialized for life in the shade. The canopy, topped by the crowns of tall trees, is the most productive level of the tropical rain forest (Fig. 46.9).

Some of the broadleaf evergreen trees can grow from 15–50 m or more in height. These tall trees often have trunks buttressed at ground level to prevent their toppling over. Lianas, or woody vines, encircle the tree as it grows and help strengthen the trunk. The diversity of species is enormous—a 10-km^2 area of tropical rain forest may contain 750 species of trees and 1,500 species of flowering plants.

Although some animals live on the forest floor (e.g., pacas, agoutis, peccaries, and armadillos), most live in the trees (Fig. 46.10). Insect life is so abundant that the majority of species have not been identified yet. Termites play a vital role in the decomposition of woody plant material, and ants are found everywhere, particularly in the trees. The various birds, such as hummingbirds, parakeets, parrots, and toucans, are often beautifully colored. Amphibians and reptiles are well represented by many types of frogs, snakes, and lizards. Lemurs, sloths, and monkeys are well-known primates that feed on the fruits of the trees. The largest carnivores are the big cats—the jaguars in South America and the leopards in Africa and Asia.

Many animals spend their entire lives in the canopy, as do some plants. **Epiphytes** are plants that grow on other plants but usually have roots of their own, which absorb moisture and minerals from the air. Epiphytes such as bromeliads catch rain and debris by forming vases of overlapping leaves. The most common epiphytes are related to pineapples, orchids, and ferns.

Figure 46.9 Levels of life in a tropical rain forest. The primary levels in a tropical rain forest are the canopy, the understory, and the forest floor. But the canopy (solid layer of leaves) contains levels as well, and some organisms spend their entire lives in one particular level. Long lianas (hanging vines) climb into the canopy, where they produce leaves. Epiphytes are air plants that grow on the trees but do not parasitize them.

We usually think of tropical forests as being nonseasonal rain forests, but tropical forests that have wet and dry seasons are found in India, Southeast Asia, West Africa, South and Central America, the West Indies, and northern Australia. Here, there are deciduous trees, with many layers of growth beneath them.

Contrary to popular belief, the soil of a tropical rain forest biome is nutrient poor. Due to the high amount of herbivory, only a small amount of leaf litter makes it to the forest floor. Productivity is high because of high temperatures, a year-long growing season, and nearly 12 hours of sunlight year-round. Slash-and-burn agriculture can be somewhat successful if it is done on a small scale, but becomes destructive and unsustainable on a large scale. Trees are felled and burned, and the ashes provide enough nutrients for several harvests. Thereafter, the forest must be allowed to regrow, and a new section must be cut and burned. In addition, in humid tropical forests, iron and aluminum oxides occur at the surface, causing a reddish residue known as laterite. When the trees are cleared, laterite bakes in the hot sunlight to a bricklike consistency, which will not support crops.

It is estimated that 2.4 acres of rain forest are destroyed per second. This rate equates to nearly 78 million acres annually. Unless conservation strategies are employed soon, rain forests will be destroyed beyond recovery, taking with them unique and interesting life-forms. Ecologists estimate that approximately 137 species are driven to extinction every day in rain forests.

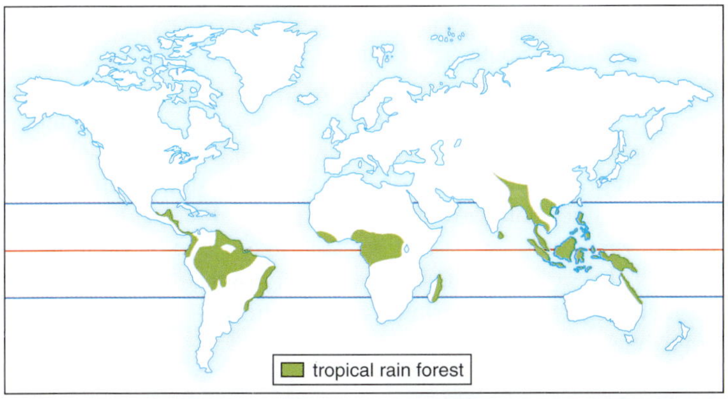

Figure 46.10 Tropical rain forests. Some of the representative animals found in the tropical rain forests.

poison-dart frog

spike-headed katydid

panther

blue and gold macaw

blue morpho butterfly

orchid

chameleon

Shrublands

Shrublands are regions that tend to have dry summers and receive most of their rainfall in the winter. In general, shrubs are shorter than trees (4.5–6 m) with woody, persistent stems but no large, central trunk. Shrubs have small but thick evergreen leaves, which are often coated with a waxy material that prevents loss of moisture. Their thick underground roots can survive dry summers and frequent fires and take moisture from the deep soil. Shrubs are adapted to withstand arid conditions and can quickly sprout new growth after a fire. You may recall from Chapter 45 that a shrub stage is part of the process of both primary and secondary succession.

Shrublands occur along the cape of South Africa, the western coast of North America, and the southwestern and southern shores of Australia, as well as around the Mediterranean Sea and in central Chile. **Chaparral** (Fig. 46.11) is a type of shrubland that lacks an understory and ground litter and is highly flammable. The seeds of many species require the heat and scarring action of fire to induce germination. Other shrubs sprout from the roots after a fire. Typical animals of the chaparral include mule deer, rodents, lizards, and greater roadrunners. Chaparral regions occur in California.

A northern shrub area that lies to the west of the Rocky Mountains is sometimes classified as a cold desert.This region is dominated by sagebrush, which provides the resources to support various species of birds.

Grasslands

Grasslands occur where annual rainfall is greater than 25 cm but generally insufficient to support trees, despite the fertile soil. For example, in temperate areas, where rainfall is between 25 and 75 cm, it is too dry for forests and too wet for deserts to form.

Grasses are well adapted to a changing environment and can tolerate a high degree of grazing, flooding, drought, and sometimes fire. Where rainfall is high, tall grasses that reach more than 2 m in height (e.g., pampas grass) can flourish. In drier areas, shorter grasses (between 5 and 10 cm) are dominant. Low-growing bunch grasses (e.g., grama grass) grow in the United States near deserts.

The growth of grasses is seasonal. As a result, grassland animals (such as bison) migrate and others (such as ground squirrels) hibernate when there is little grass for them to eat.

Temperate Grasslands

The **temperate grasslands** are characterized as having winters that are bitterly cold and summers that are hot and dry. When traveling across the United States from east to west, the line between the temperate deciduous forest and a tall-grass prairie is roughly along the border between Illinois and Indiana. The tall-grass prairie receives more rainfall than does the short-grass prairie, which occurs near deserts. Temperate grasslands include the Russian steppes, the South American pampas, and the North American prairies (Fig. 46.12).

Video Tallgrass Prairie Ecology

Large herds of bison—estimated at hundreds of thousands—once roamed the prairies, as did herds of pronghorn antelope. Now, small mammals, such as mice, prairie dogs, and rabbits, typically live belowground, but usually feed aboveground. Hawks, snakes, badgers, coyotes, and foxes feed on these mammals. Virtually all of these grasslands, however, have been converted to agricultural lands because of their fertile soils.

Video Dung Beetles

Savannas

Savannas occur in regions where a relatively cool dry season is followed by a hot rainy season (Fig. 46.13). The savanna is

a. Shrubland overview

b. Wildlife of the chaparral

Figure 46.11 Shrubland. **a.** Shrublands, such as chaparral in California, are subject to raging fires, but the shrubs are adapted to quickly regrow. **b.** Greater roadrunners find a home in the chaparral.

characterized by large expanses of grasses with relatively few trees. The plants of the savanna have extensive and deep root systems, which enable them to survive drought and fire. One tree that can survive the severe dry season is the thorny flat-topped *Acacia,* which sheds its leaves during a drought. The largest savannas are in central and southern Africa. Other savannas exist in Australia, Southeast Asia, and South America.

The African savanna supports the greatest variety and number of large herbivores of all the biomes. Elephants and giraffes are browsers that feed on tree vegetation. Antelopes, zebras, wildebeests, water buffalo, and rhinoceroses are grazers that feed on the grasses. Any plant litter that is not consumed by grazers is attacked by a variety of small organisms, among them termites. Termites build towering nests in which they tend fungal gardens, their source of food. The herbivores support a large population of carnivores. Lions, hyenas, cheetahs, and leopards all prey upon the abundant herbivore populations.

Video
Thorn Tree Ant

Deserts

Deserts are characterized by days that are hot because a lack of cloud cover allows the sun's rays to penetrate easily, but nights are cold because heat escapes easily into the atmosphere. The winds

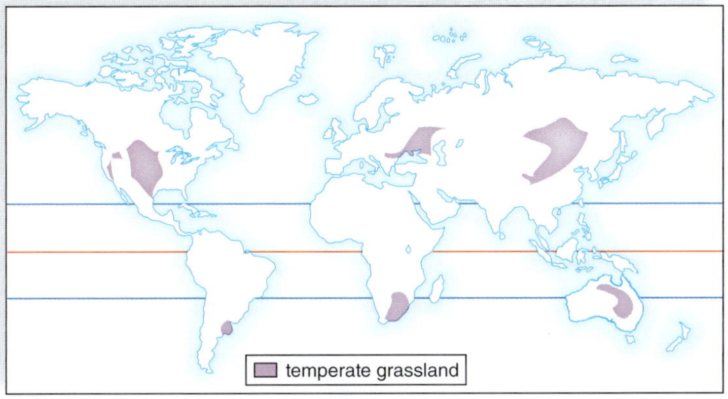

a. Tall-grass prairie

b. American bison

Figure 46.12 Temperate grassland. **a.** Tall-grass prairies are seas of grasses dotted by pines and junipers. **b.** Bison, once abundant, are now being reintroduced into certain areas.

a. Herbivores of the savanna biome

b. A carnivore of the savanna biome

Figure 46.13 The savanna. The African savanna varies from grassland to widely spaced shrubs and trees. **a.** This biome supports a large assemblage of herbivores (e.g., zebras, wildebeests, and giraffes). **b.** Carnivores (e.g., cheetahs) prey on these.

that descend in these regions lack moisture, resulting in an annual rainfall of less than 25 cm. Deserts are usually found at latitudes of about 30°, in both the Northern and Southern Hemispheres. Deserts cover nearly 30% of the Earth's land surface.

The Sahara, which stretches all the way from the Atlantic coast of Africa to the Arabian peninsula, along with a few other deserts, has little or no vegetation. However, most deserts contain a variety of plants that are highly adapted to survive long droughts, extreme heat, and extreme cold (Fig. 46.14*a*). Adaptations to these conditions include thick epidermal layers, water-storing succulent stems and leaves, and the ability to set seeds quickly in the spring. The best-known desert perennials in North America are the spiny cacti, which have stems that store water and carry on photosynthesis. Also common are nonsucculent shrubs, such as the many-branched sagebrush with silvery gray leaves and the spiny-branched ocotillo, which produces leaves during wet periods and sheds them during dry periods.

Some animals are adapted to the desert environment. To conserve water, many desert animals such as reptiles and insects are nocturnal or burrowing and have a protective outer body covering. A desert has numerous insects, which pass through the stages of development in synchrony with the periods of rain. Reptiles, especially lizards and snakes, are perhaps the most characteristic group of vertebrates found in deserts, but burrowing birds (e.g., burrowing owls) and rodents (e.g., the kangaroo rat) are also well known (Fig. 46.14*b, c*). Larger mammals, such as the kit fox (Fig. 46.14*d*), prey on the rodents, as do hawks.

Check Your Progress 46.2

1. Identify the progression of biomes, starting at the equator and moving toward the North Pole.
2. Contrast the vegetation of the tropical rain forest with that of a temperate deciduous forest.

Figure 46.14 The desert. Plants and animals that live in a desert are adapted to arid conditions. **a.** The plants are either succulents, which retain moisture, or shrubs with woody stems and small leaves, which lose little moisture. **b.** Among the animal life, the kangaroo rat feeds on seeds and other vegetation. **c.** Burrowing owls feed on rodents, reptiles, and insects. **d.** The kit fox is a desert carnivore.

a.

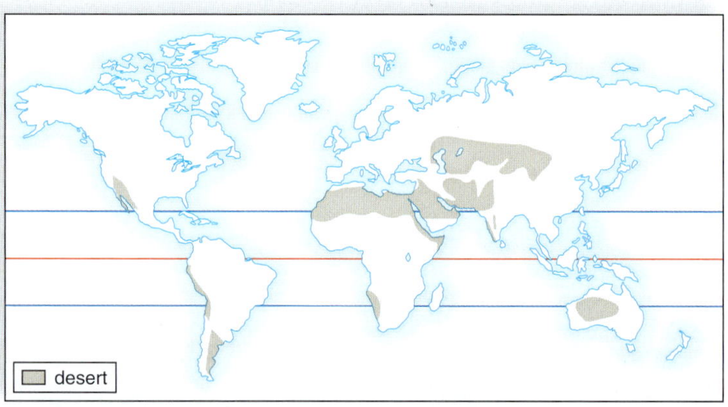

b. Bannertail kangaroo rat

c. Burrowing owls

d. Kit fox

46.3 Aquatic Ecosystems

Learning Outcomes

Upon completion of this section, you should be able to

1. Compare the characteristics of freshwater and saltwater ecosystems.
2. Interpret the manner in which ocean currents affect the climate and weather over the continents.

Aquatic ecosystems are classified in two categories: freshwater (inland) systems or saltwater systems. Brackish water, however, is a mixture of fresh and salt water. Figure 46.15 shows how these ecosystems are joined physically as well as some of the organisms that are adapted to live in them.

In the water cycle, the sun's rays cause seawater to evaporate, and the salts are left behind. As discussed in Chapter 45, the evaporated fresh water rises into the atmosphere, cools, and falls as rain. When rain falls, some of the water sinks, or percolates, into the ground and saturates it. The layer of water in the ground is called the groundwater table, or simply the water table.

Because land lies above sea level, gravity eventually returns all fresh water to the sea. Along the route to the sea it is contained in lakes and ponds, or streams and rivers. Sometimes groundwater is also located in underground rivers called aquifers. Whenever the Earth contains basins or channels, water will rise to the level of the water table.

Wetlands

Wetlands are areas that are wet for at least part of the year. Generally, wetlands are classified by their vegetation. **Marshes** are wetlands that are frequently or continually inundated by water. They are characterized by the presence of rushes, reeds, and other grasses, which provide excellent habitat for waterfowl and small mammals. Marshes are one of the most productive ecosystems on Earth. **Swamps** are wetlands that are dominated by either woody plants or shrubs. Common swamp trees include cypress, red maple, and tupelo. The American alligator is a top predator in many swamp ecosystems. **Bogs** are wetlands characterized by acidic waters, peat deposits, and sphagnum moss. Bogs receive most of their water from precipitation and are nutrient poor. Several species of plants thrive in bogs, including cranberries, orchids, and insectivorous plants such as Venus flytraps and pitcher plants. Moose and a number of other animals are inhabitants of bogs in the northern United States and Canada.

Humans have historically channeled and diverted rivers and filled in wetlands with the idea that "useless land" was being improved. However, these activities degrade ecosystems, can cause seasonal flooding, and eliminate food and habitats for many unique fishes, waterfowl, and other wildlife. Wetlands also purify waters by filtering them and by diluting and breaking down toxic wastes and excess nutrients. Wetlands directly absorb storm waters and overflow from lakes and rivers. In this way, they protect farms, cities, and towns from the devastating effects of floods. Federal and local laws have been enacted for the protection of wetlands as more people recognize their value.

Video Thames River

Figure 46.15 Freshwater and saltwater ecosystems. *Center:* Mountain streams have cold, clear water that flows over waterfalls and rapids. As streams merge, a river forms and gets increasingly wider and deeper until it meanders across broad, flat valleys. At its mouth, a river may divide into many channels, where wetlands and estuaries are located, before flowing into the sea. *To Sides:* Mayfly larvae are found in clean water with a high oxygen content. Trout are a major predator of mayflies. Carp are adapted to water that contains little oxygen and much sediment. Blue crabs are found in estuary regions.

Lakes

Lakes are bodies of fresh water that can be classified by their nutrient status. Oligotrophic (nutrient-poor) lakes are characterized by a small amount of organic matter and low productivity (Fig. 46.16*a*). Eutrophic (nutrient-rich) lakes are characterized by plentiful organic matter and high productivity (Fig. 46.16*b*). These latter lakes are usually situated in naturally nutrient-rich regions or are enriched by agricultural or urban and suburban runoff. **Eutrophication** is the process in which a body of water receives a large input of nutrients.

Lake Overturn

In the temperate zone, deep lakes are stratified in the summer and winter and have distinct vertical zones. In summer, lakes in the temperate zone have three layers of water that differ in temperature (Fig. 46.17). The surface layer, the epilimnion, is warm from solar radiation; the middle layer, the thermocline, experiences an abrupt drop in temperature; and the lowest layer, the hypolimnion, is cold. These differences in temperature prevent mixing. The warmer, less dense water of the epilimnion "floats" on top of the colder, more dense water of the thermocline, which floats on top of the hypolimnion.

Phytoplankton found in the sunlit epilimnion use up various nutrients as they photosynthesize. Photosynthesis releases oxygen, giving this layer a ready supply. Detritus accumulates at the bottom of the lake, and there oxygen is used up as decomposition occurs. Decomposition releases nutrients, however. As the season progresses, the epilimnion becomes nutrient poor, while the hypolimnion begins to be depleted of oxygen.

In the fall, as the epilimnion cools, and in the spring, as it warms, an overturn occurs. In the fall, the upper epilimnion waters become cooler than the hypolimnion waters. This causes the surface water to sink and the deep water to rise. The **fall overturn** is the mixing of the layers until the temperature is uniform throughout the lake. At this point, wind aids in the circulation of water so that mixing occurs. Eventually, oxygen and nutrients become evenly distributed.

As winter approaches, the water cools and ice forms. Ice has an insulating effect, preventing further cooling of the water below. This permits aquatic organisms to survive the winter in the water beneath the ice.

a. Oligotrophic lake

b. Eutrophic lake

Figure 46.16 Types of lakes. Lakes can be classified according to whether they are (**a**) oligotrophic (nutrient-poor) or (**b**) eutrophic (nutrient-rich). Eutrophic lakes tend to have large populations of algae and rooted plants, resulting in a large population of decomposers, which use up much of the oxygen and leave little oxygen for fishes.

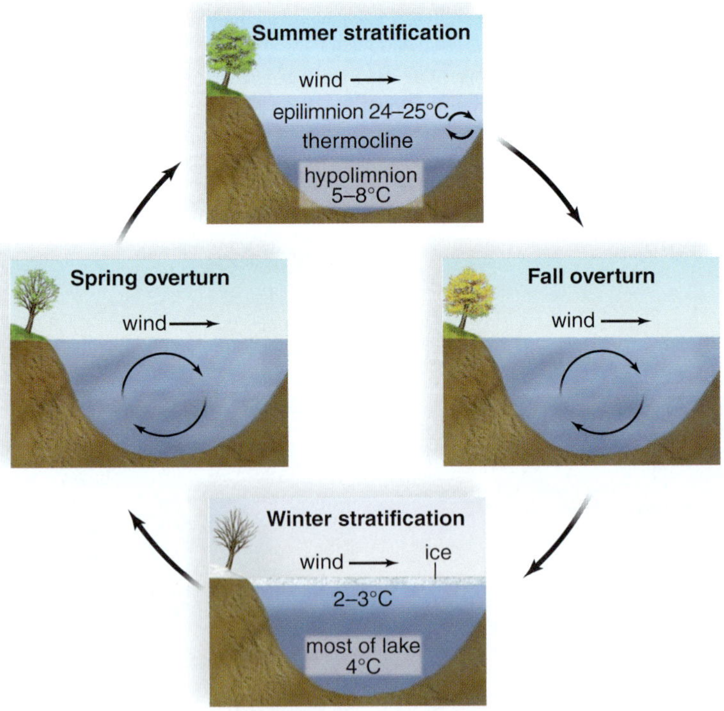

Figure 46.17 Lake stratification in a temperate region. Temperature profiles of a large oligotrophic lake in a temperate region vary with the season. During the spring and fall overturns, the deep waters receive oxygen from surface waters, and surface waters receive inorganic nutrients from deep waters.

In the spring, as the ice melts, the cooler water on top sinks below the warmer water on the bottom. The **spring overturn** is the mixing of the layers until the temperature is uniform throughout the lake. At this point, wind again aids in the circulation of the water. When the surface waters absorb solar radiation, thermal stratification occurs once more.

The vertical stratification and seasonal change of temperatures in a lake influence the seasonal distribution of fish and other aquatic life. For example, coldwater fish move to the deeper water in summer and inhabit the upper water in winter. In the fall and spring just after mixing occurs, phytoplankton growth at the surface is most abundant.

Life Zones

In both fresh and salt water, free-drifting microscopic organisms called *plankton* (Gk. *planktos,* wandering) are important components of the ecosystem. **Phytoplankton** (Gk. *phyton,* plant; *planktos,* wandering) are photosynthesizing algae that act as the producer base of the lake ecosystem. They become noticeable when a green scum or red tide appears on the water. **Zooplankton** (Gk. *zoon,* animal; *planktos,* wandering) are tiny animals that feed on the phytoplankton.

Lakes and ponds can be divided into several life zones (Fig. 46.18). The *littoral zone* is closest to the shore, the *limnetic zone* forms the sunlit body of the lake, and the *profundal zone* is below the level of light penetration. The *benthic zone* includes the sediment at the soil-water interface.

Aquatic plants are rooted in the shallow littoral zone of a lake, providing habitat for numerous protozoans, invertebrates, fishes, and some reptiles. Largemouth bass are a type of ambush predator that waits among vegetation around the margins of lakes and surges out to capture passing prey. Wading birds are commonly seen feeding in the littoral zone. Some organisms, such as the water strider, live at the water-air interface and can literally walk on water. In the limnetic zone, small fishes, such as minnows and killifish, feed on plankton and serve as food for larger fish, such as bass. In the profundal zone, zooplankton, invertebrates, and fishes

Video
Plankton Diversity

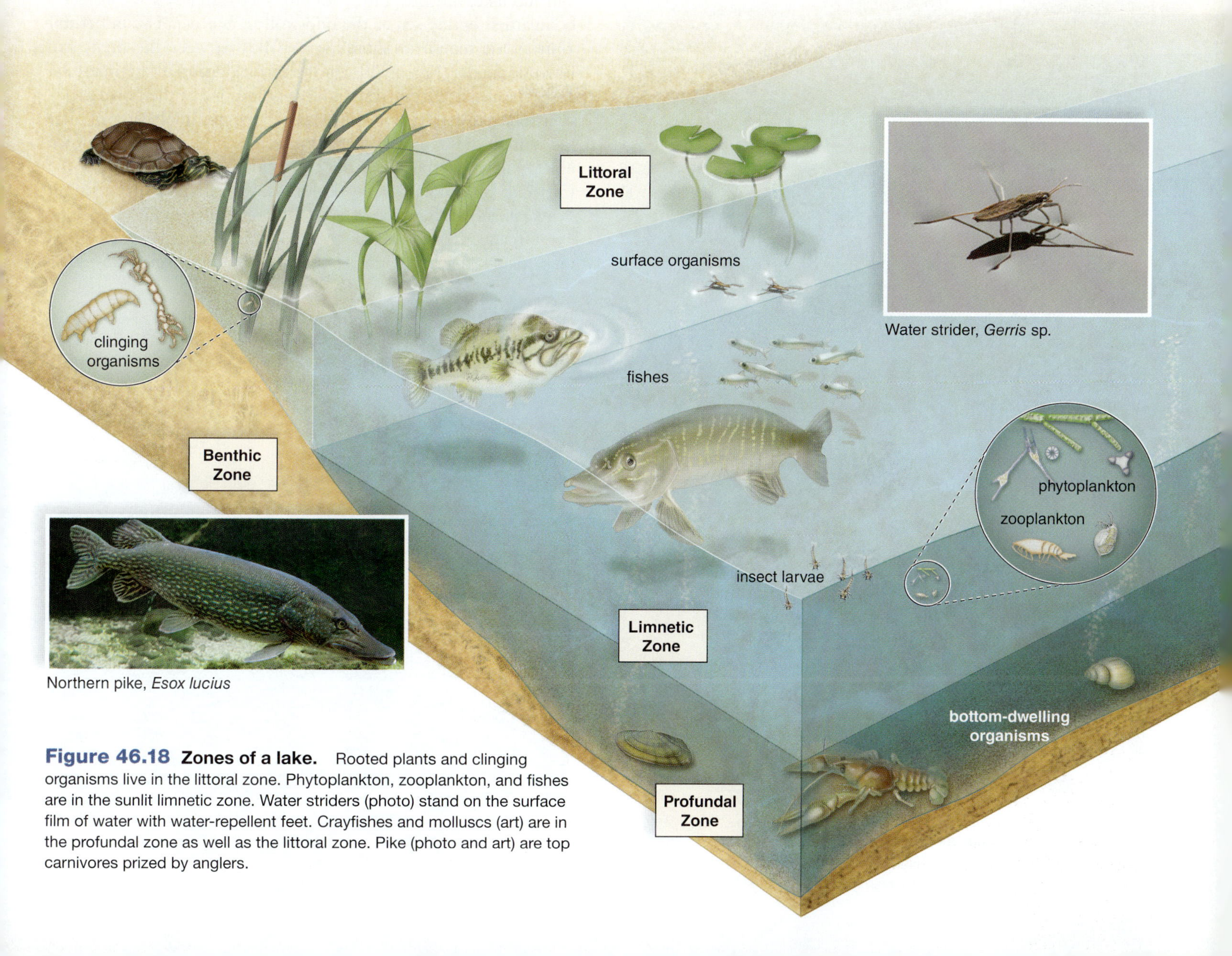

Figure 46.18 Zones of a lake. Rooted plants and clinging organisms live in the littoral zone. Phytoplankton, zooplankton, and fishes are in the sunlit limnetic zone. Water striders (photo) stand on the surface film of water with water-repellent feet. Crayfishes and molluscs (art) are in the profundal zone as well as the littoral zone. Pike (photo and art) are top carnivores prized by anglers.

such as catfish and whitefish feed on debris that falls from higher zones.

The bottom of the lake, the benthic zone, is composed of mostly silt, sand, inorganic sediment, and dead organic material (detritus). Bottom-dwelling organisms are known as benthic species and include worms, snails, clams, crayfishes, and some insect larvae. Decomposers, such as bacteria, are also found in the benthic zone and break down wastes and dead organisms into nutrients that are eventually used by the producers.

Coastal Ecosystems

An **estuary** is a portion of the ocean where fresh water and salt water meet and mix. Mudflats, mangrove swamps, and rocky shores, featured in Figure 46.19, are examples of estuaries. Mangrove swamps develop in subtropical and tropical zones, while marshes and mudflats occur in temperate zones. Coastal bays, fjords (inlets of water between high cliffs), and some lagoons (bodies of water separated from the sea by a narrow strip of land) are also classified as estuaries. Therefore, the term estuary has a very broad definition.

Organisms living in an estuary must be able to withstand constant mixing of waters and rapid changes in salinity. Organisms adapted to the estuarine environment benefit from its abundance of nutrients. An estuary acts as a nutrient trap because the sea prevents the rapid escape of nutrients brought by a river. Estuaries are biologically diverse and highly productive communities.

Phytoplankton and shore plants thrive in the nutrient-rich estuaries, providing an abundance of food and habitat for animals. It is estimated that nearly two-thirds of marine fishes and shellfish spawn and develop in the protective and rich environment of estuaries, making the estuarine environment the nursery of the sea. An abundance of larval, juvenile, and mature fish and shellfish attract a wide variety of predators.

Rocky shores (Fig. 46.19*c*) and sandy shores are constantly bombarded by the sea as the tides roll in and out. The **intertidal zone** is the region of shoreline that lies between the high- and low-tide marks (Fig. 46.20). In the upper portion of the intertidal zone, barnacles are glued so tightly to the stone by their own secretions that their calcareous outer plates remain in place even after the animal dies. In the midportion of the intertidal zone, brown algae, known as rockweed, may overlie the barnacles. Below the intertidal zone, macroscopic seaweeds, which are the main photosynthesizers, anchor themselves to the rocks by holdfasts.

Organisms cannot attach themselves to shifting, unstable sands on a sandy beach; therefore, nearly all the permanent residents

a. Mudflat

b. Mangrove swamp

c. Rocky shore

Figure 46.19 Coastal ecosystems. **a.** Mudflats are frequented by migrant birds. **b.** Mangrove swamps skirt the coastlines of many tropical and subtropical lands. **c.** Some organisms of a rocky coast live in tidal pools.

Figure 46.20 Ocean ecosystems. Organisms live in the well-lit waters of the euphotic zone and in the increasing darkness of the deep-sea waters of the pelagic zones (see Fig. 46.21).

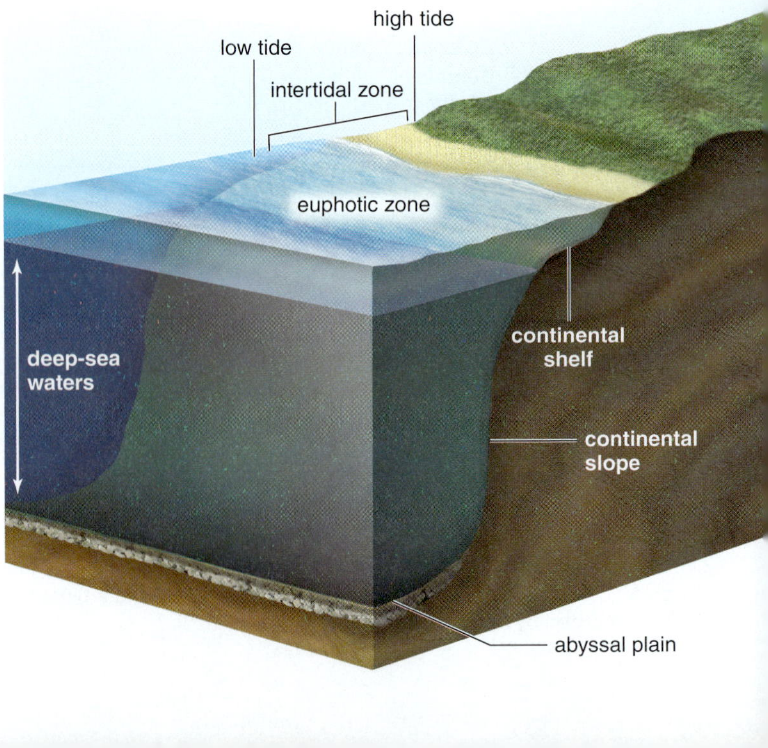

dwell underground. Either they burrow during the day and surface to feed at night or they remain permanently within their burrows and tubes. Ghost crabs and sandhoppers (amphipods) burrow themselves above the high-tide mark and feed at night when the tide is out. Sandworms and sand (ghost) shrimp remain within their burrows in the intertidal zone and feed on detritus whenever possible. Still lower in the sand, clams, cockles, and sand dollars are found. A variety of shorebirds visit the beaches and feed on various invertebrates and fishes.

Oceans

Shallow ocean waters (called the *euphotic zone*) contain the greatest concentration of organisms in the sea (Fig. 46.20). Here, phytoplankton is food not only for zooplankton but also for small fishes. These attract a number of predatory and commercially valuable fishes. On the continental shelf, seaweed can be found growing, even on outcroppings as the water gets deeper. Clams, worms, and sea urchins are preyed upon by sea stars, lobsters, crabs, and brittle stars.

Coral reefs are areas of biological abundance that are primarily found in shallow, warm, tropical waters. Their chief constituents are stony corals, animals that have a calcium carbonate (limestone) exoskeleton, and calcareous red and green algae. Corals provide a home for microscopic algae called *zooxanthellae*. The corals feed at night, and the algae photosynthesize during the day, forming a mutualistic relationship. The algae need sunlight for photosynthesis, which is why coral reefs typically develop in shallow, sunlit waters.

Video
Coral Reef Ecosystems

A reef is densely populated with life. The large number of crevices and caves provide shelter for filter feeders (sponges, sea squirts, and fanworms) and for scavengers (crabs and sea urchins). The barracuda, moray eel, and shark are top predators in coral reefs. Many types of small, beautifully colored fishes live here. These become food for larger fishes, including snappers caught for human consumption.

Video
Reef Balls

Most of the ocean's volume lies within the **pelagic zones,** as noted in Figure 46.21. The *epipelagic zone* lacks the inorganic nutrients of shallow waters, which means it does not have as high a concentration of phytoplankton as the shallow areas. Still, the photosynthesizers are food for a large assembly of zooplankton, which then become food for schools of various fishes. A number of porpoise and dolphin species visit and feed in the epipelagic zone. A variety of whales can be found in this zone. Baleen whales strain krill (small crustaceans) from the water, and toothed sperm whales feed primarily on the common squid.

Animals in the deeper waters of the *mesopelagic zone* are carnivores and are adapted to the absence of light. Many of these organisms tend to be translucent, red colored, or even luminescent; among these are some species of shrimps, squids, and fishes.

Figure 46.21 Ocean inhabitants Different organisms are characteristic of the epipelagic, mesopelagic, and bathypelagic zones of the pelagic division compared to the abyssal zone of the benthic division.

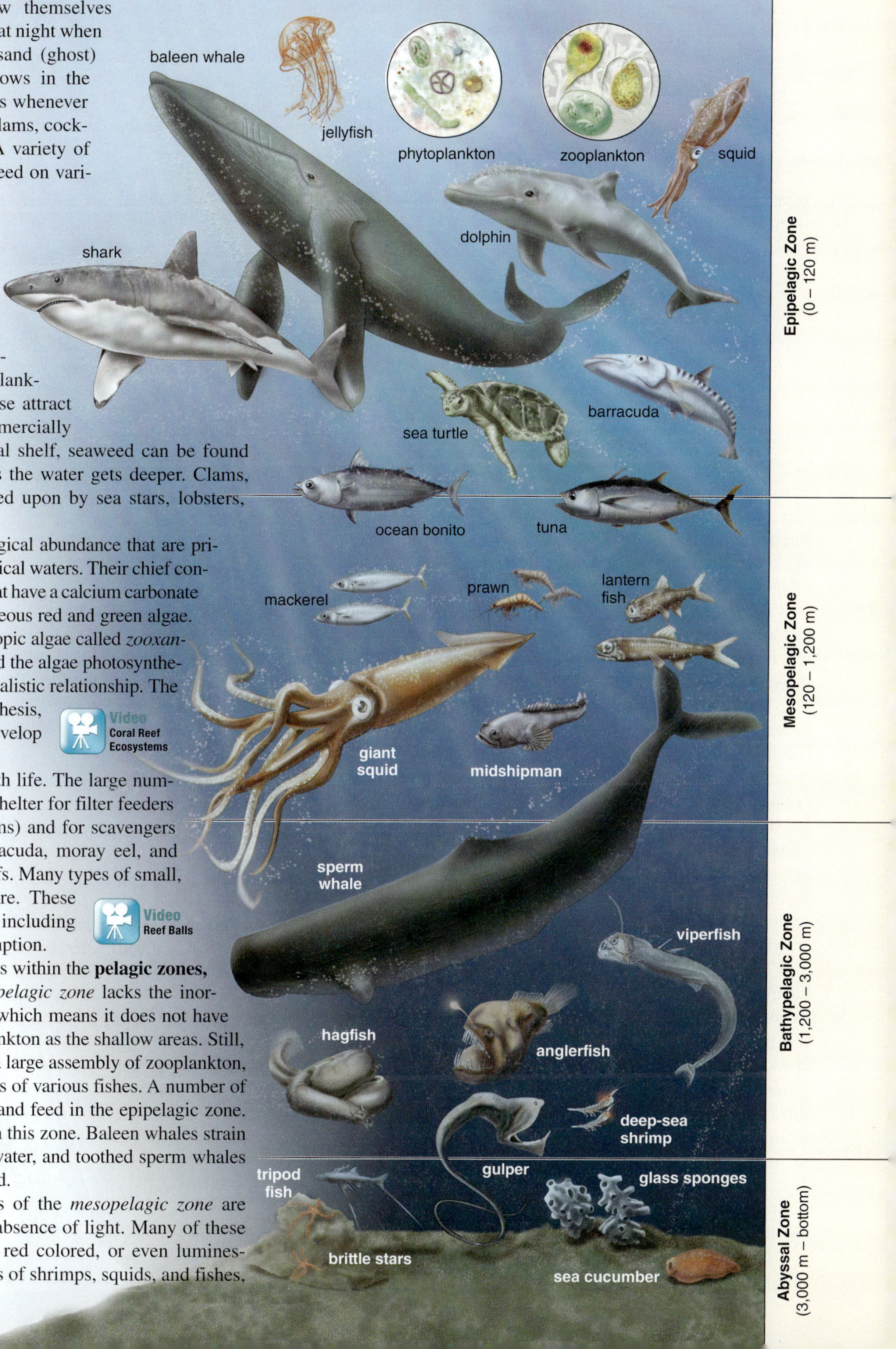

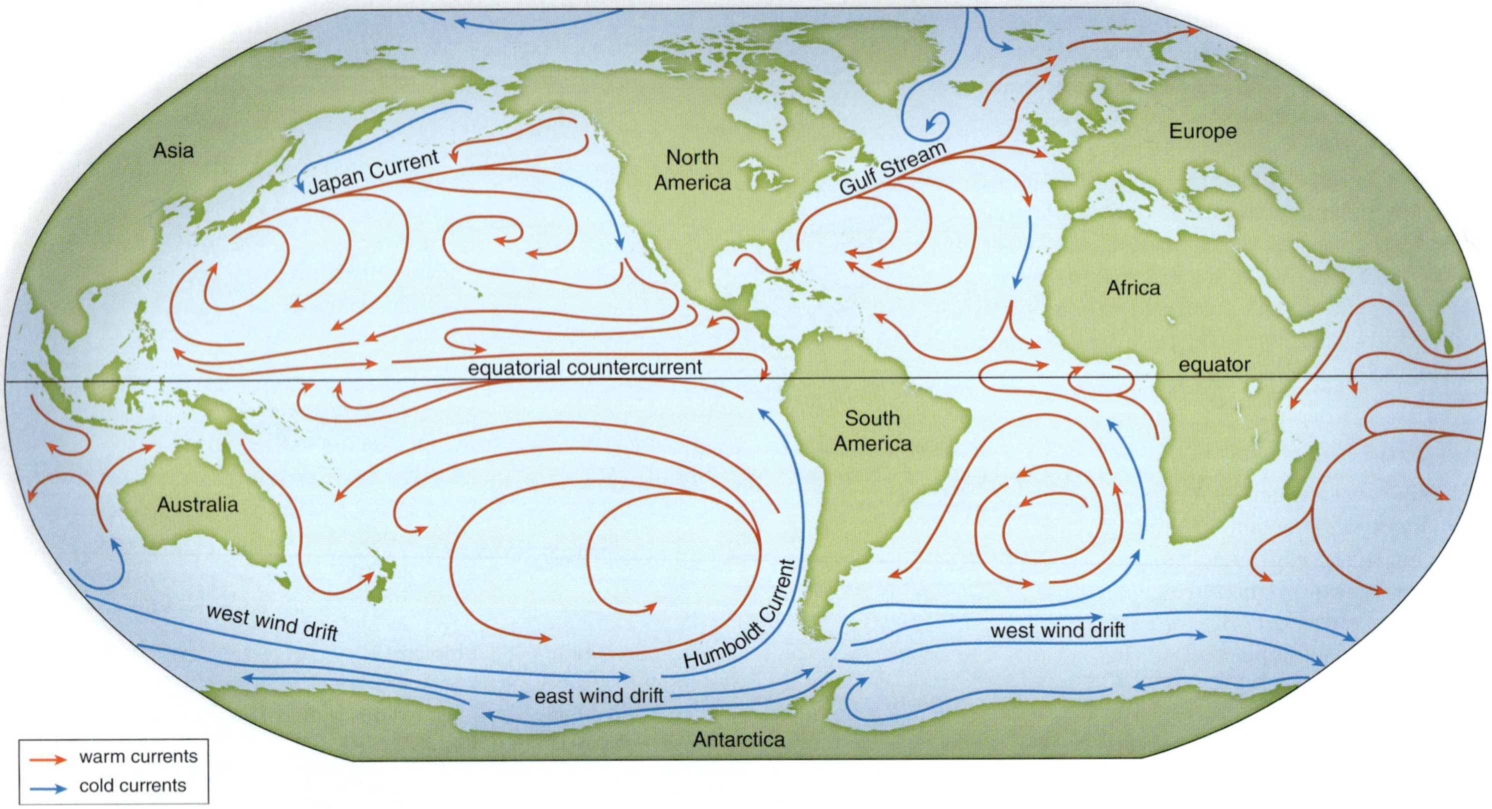

Figure 46.22 Ocean currents. The arrows on this map indicate the locations and directions of the major ocean currents set in motion by the global wind circulation. By carrying warm water to cool latitudes (e.g., the Gulf Stream) and cool water to warm latitudes (e.g., the Humboldt Current), these currents have a major effect on the world's climates.

including lantern and hatchet fishes. Various species of zooplankton, invertebrates, and fishes migrate from the mesopelagic zone to the surface to feed at night.

The deepest waters of the *bathypelagic zone* are in complete darkness except for an occasional flash of bioluminescent light. Carnivores and scavengers are found in this zone. Strange-looking fishes with distensible mouths and abdomens and small, tubular eyes feed on infrequent prey.

It once was thought that minimal life existed on the *abyssal plain* beneath the bathypelagic zone due to the intense pressure and extreme cold. However, many invertebrates survive there by feeding on debris floating down from the mesopelagic zone. Sea lilies (crinoids) rise above the seafloor; sea cucumbers and sea urchins crawl around on the sea bottom; and tube worms burrow in the mud.

The flat abyssal plain is interrupted by enormous underwater mountain chains called oceanic ridges. Along the axes of the ridges, crustal plates spread apart, and molten magma rises to fill the gap. At **hydrothermal vents,** seawater percolates through cracks and is heated to about 350°C, causing sulfate to react with water and form hydrogen sulfide (H_2S). Chemoautotrophic bacteria that obtain energy from oxidizing hydrogen sulfide exist freely or live mutualistically with organisms at the vents. They are the start of food chains for an ecosystem that includes huge tube worms, clams, crustaceans, echinoderms, and fishes. This ecosystem can exist where light never penetrates because, unlike photosynthesis, chemosynthesis does not require light energy.

In section 45.3, we discussed food webs and energy flow through ecosystems. Although energy is lost with transfer to each level, from producers to primary and then secondary consumers, pollutants can become concentrated as they move up the web—an effect termed **biomagnification.** The Biological Systems feature "Biomagnification of Mercury," discusses biomagnification as it occurs with mercury pollution, particularly in aquatic ecosystems.

Ocean Currents

Climate is driven by the sun, but the oceans play a major role in redistributing heat in the biosphere. Water tends to be warm at the equator and much cooler at the poles because of the distribution of the sun's rays (see Fig. 46.1*a*). Air takes on the temperature of the water below, and warm air moves from the equator to the poles. In other words, the oceans make the winds blow. Landmasses also play a role, but the oceans hold heat longer and remain cool longer during periods of changing temperature than do continents.

When wind blows strongly and steadily across a great expanse of ocean for a long time, friction from the moving air begins to drag the water along with it. Once the water has been set in motion, its momentum, aided by the wind, keeps it moving in a steady flow called a current. Because the ocean currents eventually strike land, they move in a circular path—clockwise in the Northern Hemisphere and counterclockwise in the Southern Hemisphere (Fig. 46.22). As the currents flow, they take warm water from the equator to the poles. One such current, called the Gulf Stream, brings tropical Caribbean water to the east coast of North America and the higher latitudes of western Europe. Without the Gulf Stream, Great Britain, which has a relatively warm temperature, would be as cold as Greenland. In the Southern

THEME Biological Systems

Biomagnification of Mercury

Scientists have known since the 1950s that the emissions of mercury into the environment can lead to serious health effects for humans. Studies show that fish and wildlife exposed to mercury emissions are negatively impacted. Humans are impacted if they come into contact with affected fish and wildlife. Recent fish studies have shown a widespread contamination of mercury in streams, wetlands, reservoirs, and lakes throughout the majority of the United States.

Mercury becomes a serious environmental risk when it undergoes bioaccumulation in an organism's body. Bioaccumulation occurs when an organism accumulates a contaminant faster than it can eliminate it. Most organisms can eliminate about half the mercury in their bodies every 70 days, if they can avoid ingesting any additional mercury during this time. Problems arise when organisms cannot eliminate the mercury before they ingest more.

Mercury tends to enter ecosystems at the base of the food chain and increase in concentration as it moves up each successive trophic level. Top-level predators and organisms that are long lived are the most susceptible to high levels of mercury accumulation in their body tissues.

Mercury exposure for humans generally occurs due to eating contaminated fish or breathing mercury vapor. Methylmercury is the form that leads to health problems such as sterility in men, damage to the central nervous system, and in severe cases, birth defects in infants. Developing fetuses and children can have health consequences from intake levels 5–10 times lower than adults.

Studies have shown such elevated levels of mercury in sharks, tuna (Fig. 46B*a*), and swordfish that the EPA has advisories against eating these fish for women who may become pregnant, are pregnant, or are nursing mothers and for young children. Every state in the United States, in conjunction with federal agencies, has developed fish advisories for certain bodies of water in that state. Currently, 45 states warn pregnant women to limit their fish consumption from their waters.

Mercury poisoning isn't limited to just aquatic species. Research conducted in the northeastern United States and Canada showed the presence of mercury in a variety of birds ranging from thrushes to bald eagles. It is no surprise that loons and bald eagles can have high levels of mercury accumulation due to their consumption of contaminated fish. It was the presence of mercury in the songbirds that raised serious concerns among ecologists. Some speculate that songbirds in the northeast are ingesting mercury when they feed upon insects that have picked up the toxin from eating smaller insects, which ingested it from vegetation. This raises concerns about mercury's ability to enter food webs and bioaccumulate in previously unknown ways.

Ultimately the blame for mercury pollution falls squarely on the shoulders of humans. Every ecosystem on the planet has some degree of exposure to this pollutant, and with this exposure comes the risk of mercury contamination (Fig. 46B*b*). Studies on species ranging from polar bears to sharks show that there are no limits to where mercury can be found. With coal-burning power plants being the largest human-caused source of mercury emissions, it will be up to us to find a solution to this global problem.

Questions to Consider

1. Would you support higher regulations on coal-fired power plants to reduce their mercury emissions even if it meant an increase in energy costs?
2. What is the easiest way to prevent the bioaccumulation of mercury in an organism?
3. Are there any species in an ecosystem that are not impacted by exposure to mercury?
4. Is there a way to limit the movement of mercury from one ecosystem to the next?

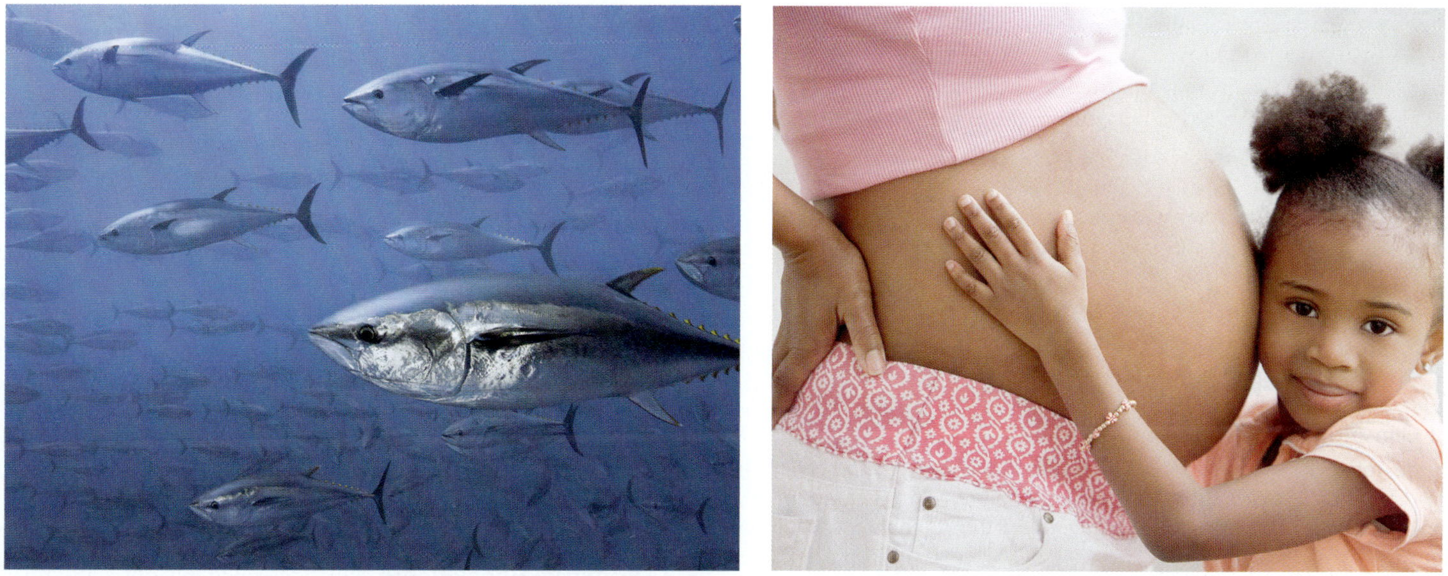

a. b.

Figure 46B Biomagnification of mercury. **a.** Tuna may contain high levels of mercury due to biomagnification. **b.** Eating tuna or other fish that contains high levels of mercury can lead to problems with fetal development.

Hemisphere, another major ocean current warms the eastern coast of South America.

Also in the Southern Hemisphere, a current called the Humboldt Current flows toward the equator. The Humboldt Current carries phosphorus-rich cold water northward along the west coast of South America. During a process called **upwelling,** cold offshore winds cause cold nutrient-rich waters to rise and take the place of warm nutrient-poor waters. In South America, the enriched waters cause an abundance of marine life, which supports the fisheries of Peru and northern Chile. Birds feeding on these organisms deposit their droppings on land, where they are mined as guano, a commercial source of phosphorus. When the Humboldt Current is not as cool as usual, upwelling does not occur, stagnation results, the fisheries decline, and climate patterns change globally. This phenomenon is called an **El Niño–Southern Oscillation** (**ENSO**).

Check Your Progress 46.3

1. Identify the abiotic features of freshwater and saltwater ecosystems.
2. Describe what occurs during an upwelling.

CONNECTING the CONCEPTS with the THEMES

Evolution

- The biosphere is the product of interactions between living organisms over billions of years.
- The organisms found in biomes have evolved, partially, in response to changes in abiotic conditions, such as rainfall and temperature in terrestrial biomes, and temperature and nutrients in aquatic biomes.

Nature of Science

- Humans have transformed many terrestrial biomes to suit their needs, such as conversion of grasslands to agricultural use.
- Humans depend on the biodiversity of the Earth's biomes for health and well-being; altering a biome can have far-reaching effects.
- Climate change brought about by human activities can alter biomes and lead to extinctions.

Biological Systems

- The interactions within and between biomes affect every living organism on Earth.
- Runoff from terrestrial ecosystems has an impact on the aquatic ecosystems.
- The quality of aquatic ecosystems influences terrestrial ecosystems through the hydrologic cycle as well as through the services provided for land protection and diversity.

MEDIA STUDY TOOLS http://connect.mheducation.com

SMARTBOOK® Study smart. SmartBook helps you maximize your study time by identifying what you know and don't know. The recharge feature in SmartBook helps you prepare for the test by identifying the material that you are most likely to forget.

connect | BIOLOGY Find even more resources to learn the chapter concepts including animations, tutorial videos, and interactive practice questions.

Animation	Tutorial	Videos
46.2 Biomes	**46.2** Factors Influencing the Distribution of Biomes	**46.2** Tallgrass Prairie Ecology • Dung Beetles • Thorn Tree Ants **46.3** Thames River • Plankton Diversity • Coral Reef Ecosystems • Reef Balls

SUMMARIZE

46.1 Climate and the Biosphere

Climate refers to the prevailing weather conditions in a particular region. The position of a region on the Earth, along with the amount of solar radiation, plays a role in determing the region's weather patterns.

Warm air rises near the equator, loses its moisture, and then descends at about 30° north and south latitude to the poles. When the air descends, it absorbs moisture from the land, and therefore the great deserts of the world are formed at 30° latitudes. Because the Earth rotates on its axis daily, the winds blow in opposite directions above and below the equator. **Topography** also plays a role in the distribution of moisture. Air rising over coastal ranges loses its moisture on the windward side. The **rain shadow** is on the leeward side, which tends to be arid. A **monsoon** climate occurs in southern Asia when wet ocean winds blow onshore, carrying heavy rains.

46.2 Terrestrial Ecosystems

A **biome** is a major type of terrestrial community. Biomes are distributed according to climate—that is, temperature and rainfall influence the pattern of biomes about the world. The effect of temperature causes the same sequence of biomes when traveling to northern latitudes as when traveling up a mountain. A **montane coniferous forest** is located on the side of mountain while the tundra near the peak is the **alpine tundra.**

The **Arctic tundra** is the northernmost biome and consists largely of short grasses, sedges, and dwarf woody plants. Because of cold winters and short summers, most of the water in the soil is frozen year-round. This is called **permafrost.**

The **taiga,** a coniferous forest, has less rainfall than other types of forests. The **temperate rain forest** is a coniferous forest on the west coast of Canada and the United States. The **temperate deciduous forest** has trees that gain and lose their leaves because of the alternating seasons of summer and winter. **Tropical rain forests** are continually warm and wet. These are the most complex and productive of all biomes. **Epiphytes** are plants in the rain forest that grow on other plants.

Shrublands usually occur along coasts that have dry summers and receive most of their rainfall in the winter. **Chaparral** regions are a type of shrubland that is highly flammable. **Grasslands** receive more rainfall than deserts but not enough to be a forest. **Temperate grasslands,** such as that found in the central United States, have a limited variety of vegetation and animal life. **Savanna,** a tropical grassland, supports the greatest number of different types of large herbivores.

Deserts are characterized by a lack of water—they are usually found in places with less than 25 cm of precipitation per year. Some desert plants, such as cacti, are succulents with thick stems and leaves, and others are shrubs that are deciduous during dry periods.

46.3 Aquatic Ecosystems

Aquatic ecosystems are divided into freshwater and saltwater systems. **Wetlands** are areas that are wet for part of the year. **Marshes, swamps,** and **bogs** are different types of wetlands. **Lakes** are classified by their level of nutrients. **Eutrophication** is a rapid increase in the nutrient load of an aquatic ecosystem. In deep lakes of the temperate zone, the temperature and the concentration of nutrients and gases in the water vary with depth. The **fall overturn** and **spring overturn** cycle the nutrients and oxygen throughout the lake. **Phytoplankton** and **zooplankton** are the microscopic organisms that live in aquatic ecosystems. Lakes and ponds have three life zones. Rooted plants and clinging organisms live in the littoral zone, plankton and fishes live in the sunlit limnetic zone, and bottom-dwelling organisms such as crayfishes and molluscs live in the profundal zone.

Marine ecosystems are divided into coastal ecosystems and the oceans. The **estuary** is where fresh water meets salt water. Estuaries (and associated salt marshes, mudflats, and mangrove forests) are near the mouth of a river. Estuaries are considered the nurseries of the sea. The **intertidal zone** is the region between the high- and low-tide marks.

An ocean is divided into the pelagic zone and the ocean floor. **Coral reefs** are areas of high biological abundance in shallow, warm, tropical waters. The **pelagic zone** (open waters) has three zones. The epipelagic zone receives adequate sunlight and supports the most life. The mesopelagic zone contains organisms adapted to minimal or no light. The bathypelagic zone is in complete darkness. The ocean floor includes the continental shelf, the continental slope, and the abyssal plain. **Hydrothermal vents** form where seawater percolates through cracks in the floor of the ocean. Chemoautotrophic bacteria form the base of these communities. As pollutants move through aquatic ecosystems they undergo an increase in concentration called **biomagnification.** Ocean currents move in a circular pattern. During an **upwelling,** cold, nutrient-rich waters rise to the surface. The **El Niño–Southern Oscillation** (**ENSO**) is a reversal of the normal circulation pattern that prevents the upwelling from occurring.

ASSESS

Choose the best answer for each question.

46.1 Climate and the Biosphere

1. The seasons are best explained by
   a. the distribution of temperature and rainfall in biomes.
   b. the tilt of the Earth as it orbits about the sun.
   c. the daily rotation of the Earth on its axis.
   d. the fact that the equator is warm and the poles are cold.
2. Which region on Earth receives the largest amount of direct solar radiation throughout the year?
   a. Northern Hemisphere above 60°
   b. Northern Nemisphere between 30° and 60°
   c. Southern Nemisphere below 60°
   d. Southern Nemisphere between 30° and 60°
   e. the equator, between 30° north and 30° south
3. Why are lush evergreen forests present in the Pacific Northwest of the United States?
   a. The rotation of the Earth causes this.
   b. Winds blow from the ocean, bringing moisture.
   c. They are located on the leeward side of a mountain range.
   d. Both b and c are correct.
   e. All of these are correct.
4. Which biome is associated with the rain shadow of a mountain?
   a. desert
   b. tropical rain forest
   c. taiga
   d. coniferous forest
   e. chaparral

46.2 Terrestrial Ecosystems

5. Which of these influences the location of a particular biome?
   a. latitude
   b. average annual rainfall
   c. average annual temperature
   d. altitude
   e. All of these are correct.
6. What is the geographic distribution of a temperate deciduous forest?
   a. the northern part of North America and Eurasia
   b. just south of the polar caps in the Northern Hemisphere
   c. eastern North America and eastern Asia
   d. near the equator in South America and Africa
   e. central United States and central Asia
7. Which of these pairs is mismatched?
   a. tundra—permafrost
   b. savanna—*Acacia* trees
   c. prairie—epiphytes
   d. coniferous forest—evergreen trees

8. All of these phrases describe the tundra except
   a. low-lying vegetation.
   b. northernmost biome.
   c. short growing season.
   d. many different types of species.

46.3 Aquatic Ecosystems

9. The forest with a multilevel canopy is the
   a. tropical rain forest.
   b. coniferous forest.
   c. tundra.
   d. temperate deciduous forest.
10. An estuary acts as a nutrient trap because of the
    a. action of rivers and tides.
    b. depth at which photosynthesis can occur.
    c. amount of rainfall received.
    d. height of the water table.
11. Phytoplankton are more likely to be found in which life zone of a lake?
    a. limnetic zone
    b. profundal zone
    c. benthic zone
    d. All of these are correct.
12. The mild climate of Great Britain is best explained by
    a. the winds called the westerlies.
    b. the spinning of the Earth on its axis.
    c. Great Britain being a mountainous country.
    d. the flow of ocean currents.
13. El Niño
    a. prevents the ocean current from upwelling.
    b. causes stagnation of the nutrients.
    c. changes global climate patterns.
    d. decreases the local fisheries.
    e. All of these are correct.

ENGAGE

THINKING SCIENTIFICALLY

1. Pharmaceutical companies are "bioprospecting" the tropical rain forests. These companies are looking for naturally occurring compounds in plants or animals that can be used as drugs for a variety of diseases. The most promising compounds act as antibacterial or antifungal agents. Even discounting the fact that the higher density of species in tropical rain forests would produce a wider array of compounds than another biome, why would antibacterial and antifungal compounds be more likely to evolve in this biome?
2. When hurricanes come ashore, the ocean tide may surge dozens of feet above normal into coastal communities. Recently, Hurricane Sandy had devastating effects on New York City, causing 19 billion dollars in damages, making it one of the costliest natural disasters in U.S. history. Sandy was the second most powerful Atlantic tropical cyclone on record. How would a storm like Sandy impact the biomes located in the eastern part of the United States?
3. Runoff from agricultural fertilizer, golf courses, ski resorts, and homeowners are all contributing to water contamination. Some farmers are using irrigation methods that deliver water directly to plant roots, no-till agriculture that reduces the loss of topsoil and decreases the need for herbicides, and integrated pest management, which relies heavily on good bugs to kill bad bugs. Encouraged—in some cases, compelled—by state and federal agents, dairy farmers have built sheds, concrete containments, and underground liquid storage tanks to hold wastes when it rains. Later, the manure is trucked to fields and spread as fertilizer.

   Builders in Traverse County, Minnesota, are required to control soil erosion with filter fences, to steer rainwater away from exposed soil, to build sediment basins, and to plant protective buffers. Presently, homeowners must have a 25-foot setback from wetlands and a 50-foot setback from lakes and creeks. They are also encouraged to pump out their septic systems every 2 years. Do you approve of legislation that requires farmers and homeowners to protect freshwater supplies? Why or why not?

Lionfish are an invasive species in the Caribbean.

47

Conservation of Biodiversity

The lionfish, *Pterois volitans* and *Pterois miles,* are natives of the coral reefs around Southeast Asia and Indonesia. In their natural environment, lionfish play an important role as one of the top predators, relying on their rapid reflexes and camouflage to capture small fish. While they are a venomous fish, the chemicals they produce are designed mostly for defense. Their warning coloration acts as a hazard sign to potential predators.

However, their beautiful coloration also made them attractive for aquarium owners. And, either intentionally or unintentionally, starting around 1985, some of these fish were dumped into the Atlantic Ocean, probably off the coast of Miami. DNA analyses suggest that the initial population of lionfish may have been as few as six individuals.

That is where the problem began. Lionfish lack natural predators in the Atlantic, and by 2000 their populations had increased to the point that they had become one of the most predominant predators in the Caribbean. Lionfish are known to eat over 50 types of fish in the Caribbean, many of which are commercially important, such as snapper and grouper. Even a small population of lionfish can remove up to 80% of the small feeder fish on a reef. This upsets the natural ecology of the reef and adversely affects the populations of larger reef predators. The introduction of lionfish into this area has drastically reduced coral reef biodiversity.

In this chapter we will explore the science of conservation biology and investigate some of the practices by which scientists are attempting to conserve the biodiversity on the planet. We will also explore some of the threats to biodiversity, including those presented by invasive alien species, such as lionfish.

As you read through this chapter, think about the following questions:

1. In what ways do scientists measure biodiversity?
2. What are some of the other threats to biodiversity?

Chapter Outline

47.1 Conservation Biology and Biodiversity 906

47.2 Value of Biodiversity 908

47.3 Causes of Extinction 911

47.4 Conservation Techniques 916

Before You Begin

Before beginning this chapter, take a few moments to review the following discussions.

Section 18.3 Why are the current extinctions of such a great concern to scientists?

Section 44.6 How is human population growth contributing to extinction rates?

Section 45.3 What role does biodiversity play in ecosystem health?

Following *the* Themes

CHAPTER 47 CONSERVATION OF BIODIVERSITY	
UNIT 8 BEHAVIOR AND ECOLOGY	
Evolution	For most species, evolutionary change does not occur rapidly enough to allow for adaptations to the human-related changes in the Earth's environment. The result is an increase in the rate of extinction.
Nature of Science	Conservation biology is supported by multiple fields of science ranging from ecology and systematics to evolution and genetics.
Biological Systems	The identification of both biodiversity hotspots and keystone species is critical for successful conservation.

47.1 Conservation Biology and Biodiversity

Learning Outcomes

Upon completion of this section, you should be able to

1. Identify the role of conservation biology with regard to biodiversity.
2. Describe how conservation biology is an applied, goal-oriented, multidisciplinary field.
3. Recognize the spectrum of biodiversity.

Conservation biology (L. *conservatio,* "keep, save") is a relatively new discipline of biology that studies all aspects of biodiversity with the goal of conserving natural resources for this generation and all future generations. Conservation biology is unique in that it is concerned with both the development of scientific concepts and the application of these concepts to the everyday world. The primary goal of conservation biology is the management of **biodiversity,** the variety of life on Earth. To achieve this goal, conservation biologists come from many subfields of biology, which only recently have been brought together into a cohesive whole.

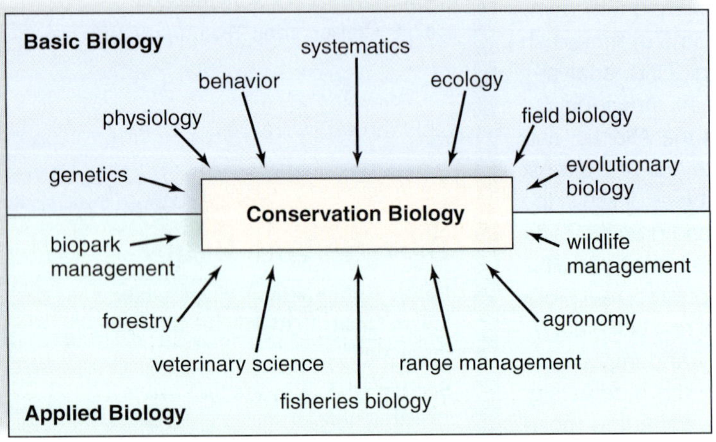

Conservation biologists must be aware of the latest theoretical and practical findings. They need to use this knowledge to identify the source of problems and suggest suitable courses of action. Often, conservation biologists work with government officials from the local to the federal level. Public education is another important component of this profession.

For conservation biology to be an effective field of study, scientists must look at the larger connections within the biosphere. Overall, a high level of biodiversity is desirable, because it positively influences the stability of ecosystems. In turn, stable ecosystems contribute to the overall health of the human population. The reverse also tends to be true—if biodiversity decreases, then ecosystem stability decreases, often resulting in negative consequences for the human population.

Conservation biology has emerged in response to the extinction crisis the Earth is experiencing. Estimates vary, but at least 10–20% of all species now living most likely will become extinct in the next 20 to 50 years unless planned, coordinated actions are taken. It is urgent that all citizens understand the concept and importance of biodiversity, the causes of present-day extinctions, how to prevent future extinctions from occurring, and the potential consequences of decreased biodiversity.

To protect biodiversity, scientists can apply the science of **bioinformatics:** the collecting of, analyzing of, and making readily available biological information, using modern computer technology. Throughout the world, molecular, descriptive, and biogeographical information on organisms is being collected for study purposes.

Biodiversity—Three Levels of Organization

It is common practice to describe biodiversity in terms of the number of species present among various groups of organisms. Figure 47.1 accounts for the species that have so far been described. It has been estimated that there are between 10 and 50 million species living on Earth.

According to the U.S. Fish and Wildlife Service (FWS), as of 2014, there are over 470 animal species and 720 plant species in the United States that are in danger of extinction. Worldwide there are nearly 30,000 species in danger of extinction. An **endangered species** is one that is in peril of immediate extinction throughout all or most of its range. Examples of endangered species include the giant panda, hawksbill turtle, California condor, and snow leopard. **Threatened species** are organisms that are likely to become endangered in the near future. Examples of threatened species include the Navajo sedge, northern spotted owl, and coho salmon.

To develop a meaningful understanding of life on Earth, we need to know more about species than just their total number. Ecologists describe biodiversity as a combination of three levels of biological organization:

- Genetic diversity
- Community diversity
- Landscape diversity

Genetic Diversity

Genetic diversity refers to genetic variations among the members of a population. Populations with high genetic diversity are more likely to

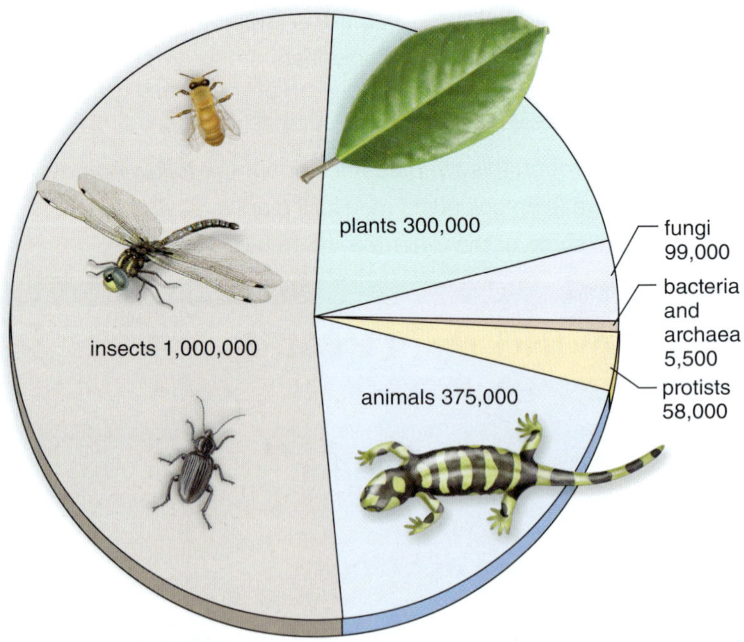

Figure 47.1 Number of described species. There are about 1.8 million described species, with insects making up over half of them. Estimates range between 10 and 50 million total species on Earth.

have some individuals that can survive a change in the structure of their ecosystem. These populations tend to be more adaptable than those with limited biodiversity. The 1846 potato blight in Ireland, the 1922 wheat failure in the Soviet Union, and the 1984 outbreak of citrus canker in Florida were all made worse by limited genetic variation among these crops. If a species' population is quite small and isolated, it is more likely to become extinct because of a limited genetic diversity. As a population's size decreases, its genetic diversity decreases as well.

Community Diversity

Community diversity is dependent on the interactions between species in a community. The species composition of one community can be completely different from that of another community. Diverse community compositions increase the levels of biodiversity in the biosphere.

Past conservation efforts frequently concentrated on saving a single species with human appeal, termed a charismatic species—such as the the black-footed ferret or spotted owl. This approach, however, is short-sighted. A more effective approach is to conserve species that play a critical role in an ecosystem. Saving an entire community can save many species, whereas disrupting a community threatens the existence of all species.

One example of a short-sighted approach was the introduction of opossum shrimp, *Mysis relicta,* into Flathead Lake in Montana and its tributaries in the early 1980s as food for salmon. The shrimp ate so much zooplankton that, in the end, far less food was available for the fish. This then resulted in a decrease in the available food for the grizzly bears and bald eagles (Fig. 47.2).

Landscape Diversity

Landscape diversity involves studying **landscape** interactions. A landscape is a group of interacting ecosystems, such as mountains, rivers, and grasslands. Sometimes, ecosystems are so fragmented that they are connected only by small patches or strips of undeveloped land that allow organisms to move from one ecosystem to the next. Fragmentation of the landscape reduces reproductive capacity and food availability and can disrupt seasonal behaviors.

Distribution of Biodiversity

Biodiversity is not evenly distributed throughout the biosphere. If we are forced to protect only small areas, we should focus our efforts on regions of the greatest biodiversity. Biodiversity is typically higher in the tropics, and it declines toward the poles. The coral reefs of the Indonesian archipelago contain the greatest biodiversity of all aquatic ecosystems.

Video
Coral Reef Ecosystems

Some regions of the world are called **biodiversity hotspots,** because they contain a large concentration of species. The biodiversity found in the hotspots accounts for about 44% of all known higher plant species and 35% of all terrestrial vertebrate species. Hotspots cover only about 1.4% of the Earth's land area. The island of Madagascar, the Cape region of South Africa, Indonesia, the coast of California, and the Great Barrier Reef of Australia are locations of biodiversity hotspots.

Check Your Progress 47.1

1. Explain the role of conservation biology.
2. Describe how conservation biology is supported by a variety of disciplines.
3. Define *biodiversity,* and explain what is meant by a biodiversity hotspot.

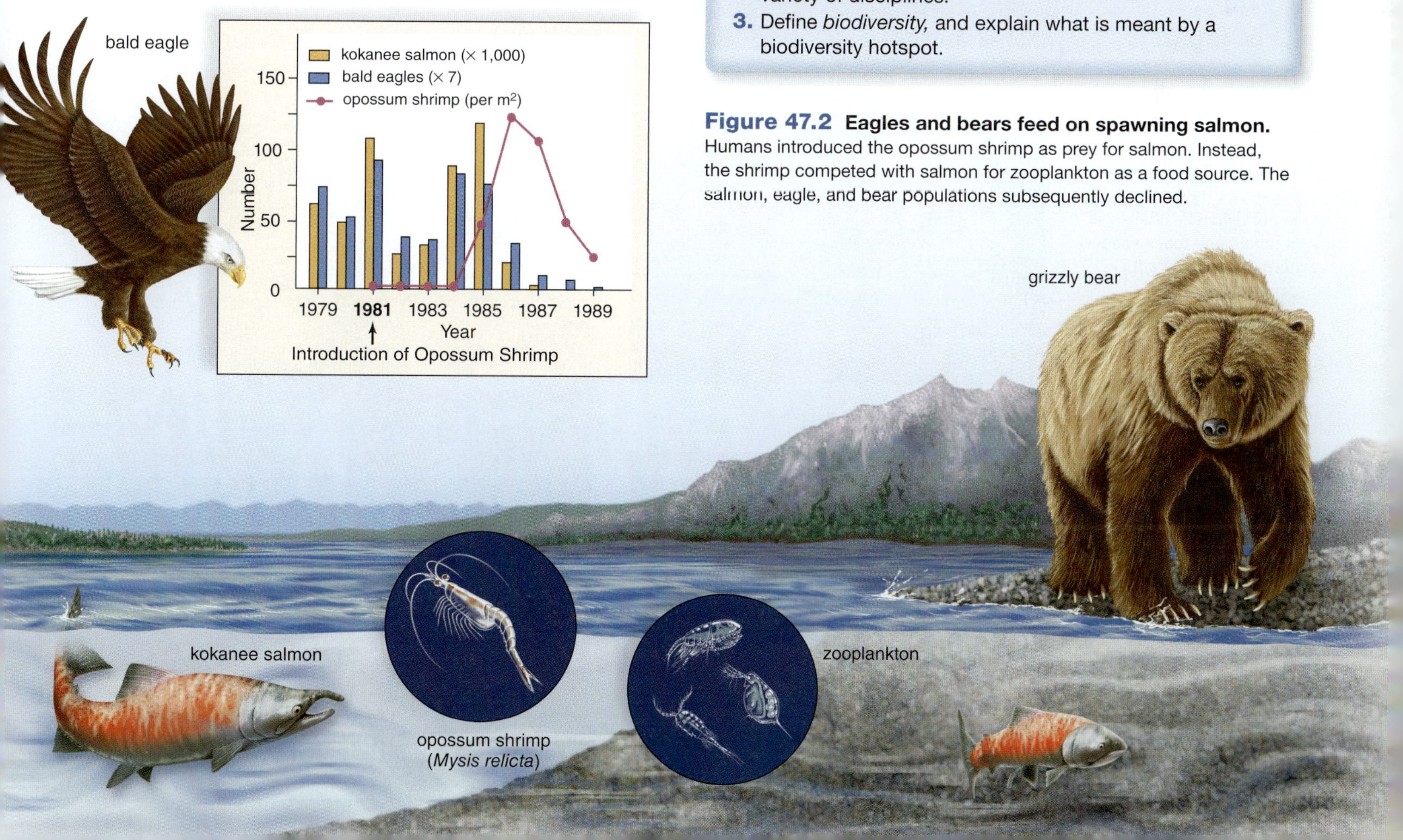

Figure 47.2 Eagles and bears feed on spawning salmon. Humans introduced the opossum shrimp as prey for salmon. Instead, the shrimp competed with salmon for zooplankton as a food source. The salmon, eagle, and bear populations subsequently declined.

47.2 Value of Biodiversity

Learning Outcomes

Upon completion of this section, you should be able to

1. Compare the direct and indirect values of biodiversity.
2. Describe the role biodiversity plays in a natural ecosystem.

Conservation biology strives to reverse the trend toward possible species extinction. Educating people about the value of biodiversity is a key step in making conservation possible.

Direct Value

Species that perform services from which humans can derive an economic value are considered to have a direct value to humans. Here we discuss a few of the key direct values provided by various species (Fig. 47.3).

Medicinal Value

Video Indigenous Medicine

Humans often associate the worth of a resource with its value expressed in monetary terms. Medicines derived from living organisms are a good example.

Most of the prescription drugs used in the United States, valued at over 200 billion dollars, were originally derived from living organisms. The rosy periwinkle from Madagascar is an excellent example of a tropical plant that has provided us with useful medicines. Potent chemicals from this plant are now used to treat two forms of cancer: leukemia and Hodgkin disease. Because of these drugs, the survival rate for childhood leukemia has gone from 10% to 90%, and Hodgkin disease is now usually curable. Researchers estimate that hundreds of new drugs are yet to be found in tropical rain forests, and that they may have an economic value of over 100 billion dollars.

The antibiotic penicillin is derived from a fungus, and certain species of bacteria produce the antibiotics tetracycline and streptomycin. These drugs have proved to be indispensable in the treatment of bacterial infections.

Leprosy is among those diseases for which no cure is currently available. The bacterium that causes leprosy will not grow in the laboratory, but scientists discovered that it grows naturally in the nine-banded armadillo. Having a source for the bacterium may make it possible to find a cure for leprosy. The blood of horseshoe crabs, *Limulus,* contains a substance called limulus amoebocyte lysate, which is used to ensure that medical devices, such as pacemakers, surgical implants, and prosthetic devices, are free of bacteria (see the Nature of Science feature, "How Horseshoe Crabs Save Human Lives," in Chapter 32).

Agricultural Value

Crops such as wheat, corn, and rice have been derived from wild plants that were modified to increase their yields. The same high-yield, genetically similar strains tend to be grown worldwide.

Figure 47.3 Direct value of wildlife. Many wild species have a significant monetary value to humans. The challenge is to manage these species correctly to ensure their continued survival.

Wild species, like the rosy periwinkle, *Catharanthus roseus,* are sources of many medicines.

Wild species, like these fish, provide us with food.

Wild species, like the nine-banded armadillo, *Dasypus novemcinctus,* play a role in medical research.

When rice crops in Africa were being devastated by a virus, researchers grew wild rice plants from thousands of seed samples until they found one that contained a gene for resistance to the virus. These wild plants were then used in a breeding program to transfer the gene into high-yield rice plants. If this variety of wild rice had become extinct before it could be discovered, rice cultivation in Africa could have collapsed.

Video Warming Hurts Rice

Using natural pest controls is preferable to using chemical pesticides. When a rice pest called the brown planthopper became resistant to pesticides through natural selection, farmers began to use the natural predator of the brown planthopper. The economic savings were calculated at well over a billion dollars. Similarly, cotton growers in Cañete Valley, Peru, found that pesticides were no longer working against the cotton aphid because of the resistance that had evolved in the aphid population. Researchers have identified the aphids' natural predators and are using them with great success.

Animals are the main pollinators for flowering plants. The domesticated honeybee, *Apis mellifera,* pollinates almost 10 billion dollars worth of food crops annually in the United States. This dependency on a single species has inherent dangers: Colony collapse disorder has wiped out more than 30% of the commercial honeybee population in the United States. The USDA estimates that honeybee pollination supports 15 billion dollars worth of agricultural production, including more than 130 fruits and vegetables, per year. In order to find honeybees resistant to the mites, researchers must look to the wild populations.

Video Pollinators

Consumptive Use Value

Humans have had much success cultivating crops, domesticating animals, growing trees in plantations, and so forth. But so far, aquaculture, the growing of fish and shellfish for human consumption, has contributed only minimally to human welfare. Most freshwater and marine harvests depend on catching wild animals, such as fishes (e.g., trout, cod, tuna, and flounder), crustaceans (e.g., lobsters, shrimps, and crabs), and mammals (e.g., whales). These aquatic organisms are an invaluable source of biodiversity.

The environment provides a variety of other products that are sold commercially worldwide, including wild fruits and vegetables, skins, fibers, beeswax, and seaweed. Hunting and fishing are the primary methods of obtaining meat for many human populations. In one study, researchers calculated that the economic value of wild pigs in the diet of native hunters in Sarawak, East Malaysia, was approximately 40 million dollars per year.

Similarly, many trees are still felled in the natural environment for their wood. Researchers have calculated, however, that a species-rich forest in the Peruvian Amazon is worth far more if the forest is used for fruit and rubber production than for timber production. Fruit and the latex needed to produce rubber can be brought to market for an unlimited number of years. Unfortunately, once the trees are gone, none of these products, including timber, can be harvested.

Wild species, like the lesser long-nosed bat, are pollinators of agricultural and other plants.

Wild species, like ladybugs, play a role in biological control of agricultural pests.

Wild species, like rubber trees, can provide a product indefinitely if the forest is not destroyed.

Figure 47.4 Indirect value of ecosystems. Natural ecosystems provide many physiological and psychological benefits to humans.

Indirect Value

The wild species we have been discussing all play a role in their respective ecosystems. If we want to preserve them, it is more beneficial to save large portions of their ecosystems. The indirect value of these ecosystems is based on services they provide that do not have a measurable economic value. Biogeochemical cycles, waste recycling, and provision of fresh water all have indirect values to human society. In addition, these natural regions have intangible value; a walk in the woods or time spent at the local park can have a calming and restorative effect on many people. Reconnecting with nature has proven physiological and psychological benefits to human health (Fig. 47.4). Our very survival depends on the functions that ecosystems perform for us.

Biogeochemical Cycles

Ecosystems are characterized by energy flow and chemical cycling (see Chapter 45). The biodiversity within ecosystems contributes to the workings of the water, carbon, nitrogen, phosphorus, and other biogeochemical cycles. We are dependent on these cycles for fresh water, the removal of carbon dioxide from the atmosphere, the uptake of excess soil nitrogen, and the provision of phosphate. When human activities upset the natural workings of biogeochemical cycles, the environmental consequences often result in negative consequences for humans. Technology is limited in its ability to mimic the biogeochemical cycles.

Video Dung Beetles

Waste Recycling

Decomposers break down dead organic matter and other types of wastes to inorganic nutrients, which are used by the producers in ecosystems. This function aids humans immensely, because we dump millions of tons of waste material into natural ecosystems each year. If it were not for decomposition, waste would soon cover the entire surface of our planet. We can build sewage treatment plants, but they are expensive, and few of them break down solid wastes completely to inorganic nutrients. It is less expensive and more efficient to provide plants and trees with partially treated wastewater and let soil bacteria cleanse it completely.

Video Decomposers

Biological communities are also capable of breaking down and immobilizing pollutants, such as heavy metals and pesticides. A review of wetland functions in Canada assigned a value of 50,000 dollars per hectare (100 acres, or 10,000 m^2) per year to the ability of natural areas to purify water and take up pollutants.

Provision of Fresh Water

Few terrestrial organisms are adapted to living in a salty environment—they need fresh water. The water cycle continually supplies fresh water to terrestrial ecosystems. Humans use fresh water in innumerable ways, from drinking water to irrigating crops. Freshwater ecosystems, such as rivers and lakes, provide us with a large diversity of species we can use as a source of food.

Video Thames River

Unlike other commodities, there is no substitute for fresh water. We can remove salt from seawater to obtain fresh water, but the cost of desalination is about four to eight times the average cost of fresh water acquired via the water cycle.

Forests and other natural ecosystems exert a "sponge effect." They soak up water and then release it at a regular rate. When rain falls in a natural area, plant foliage and dead leaves lessen its impact, and the soil slowly absorbs it, especially if it has been aerated by soil organisms. The water-holding capacity of forests reduces the possibility of flooding. The value of a marshland outside Boston, Massachusetts, has been estimated at 72,000 dollars per hectare per year, solely for its ability to reduce floods. Forests release water slowly for days or weeks after the rains have ceased. Rivers flowing through forests in West Africa release twice as much water halfway through the dry season, and between three and five times as much at the end of the dry season, as do rivers from agricultural lands.

Prevention of Soil Erosion

Intact ecosystems naturally retain soil and prevent soil erosion. The importance of this ecosystem attribute is especially observed following deforestation. In Pakistan, the world's largest dam, the Tarbela Dam, is losing its storage capacity of 12 billion m^3 many years sooner than expected, because sediment is building up behind the dam due to deforestation. At one time, the Philippines exported 100 million dollars worth of oysters, mussels, clams, and cockles each year. Now, silt carried down rivers following deforestation is smothering the mangrove ecosystem that serves as a nursery for the sea. Most coastal ecosystems are not as bountiful as they once were because of deforestation and a myriad of other problems.

Regulation of Climate

At the local level, trees provide shade and reduce the need for fans and air conditioners during the summer. Proper placement of shade trees near a home can reduce energy bills 10–20%.

Globally, forests restore the climate, because they take up carbon dioxide. The leaves of trees use carbon dioxide when they photosynthesize, the bodies of the trees store carbon, and oxygen is released as a by-product. When trees are cut and burned, carbon dioxide is released into the atmosphere. The reduction in forests reduces the carbon dioxide uptake and the oxygen output. This change in the atmospheric gases, especially greenhouse gases such as CO_2, affects the amount of solar radiation retained on the Earth's surface. Large-scale deforestation is affecting the global atmosphere, and in turn changing the Earth's climate.

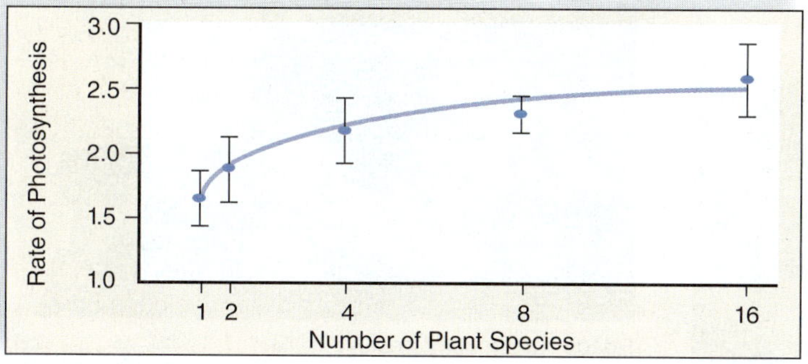

Figure 47.5 The influence of biodiversity on community productivity. Research results show that the higher the biodiversity (measured by the number of plant species), the greater the rate of photosynthesis in an experimental community.

Ecotourism

In the United States, millions of people enjoy vacationing in a natural setting. To do so, they spend billions of dollars each year on fees, travel, lodging, and food. Many tourists want to go sport fishing, whale watching, boating, hiking, birdwatching, and the like. With 1,200 miles of sand beaches and over 86 million tourists a year, tourism generates 67 billion dollars a year for Florida's economy. Others simply want to immerse themselves in the beauty and serenity of a natural environment. Many underdeveloped countries in tropical regions are taking advantage of this by offering "ecotours" of the local biodiversity. Providing guided tours of forests is more profitable than destroying them.

Biodiversity and Natural Ecosystems

Massive changes in biodiversity, such as deforestation, have a significant impact on ecosystems. Researchers are interested in determining whether a high degree of biodiversity also helps ecosystems function more efficiently. To test the benefits of biodiversity in a Minnesota grassland habitat, researchers sowed plots with seven levels of plant diversity. Their study found that ecosystem performance improves with increasing diversity. A similar study in California also showed greater overall resource use in more diverse plots because of resource partitioning among the plants.

Another group of experimenters tested the effects of an increase in diversity at four levels: producers, herbivores, parasites, and decomposers. They found that the rate of photosynthesis increased as diversity increased (Fig. 47.5). A computer simulation has shown that the response of a deciduous forest to elevated carbon dioxide is a function of species diversity. A more complex community, composed of nine tree species, exhibited a 30% greater amount of photosynthesis than a community composed of a single species.

More studies are needed to test whether biodiversity maximizes resource acquisition and retention in an ecosystem. Diverse ecosystems are better able to withstand environmental changes and invasions by other species. Fragmentation affects the functioning of and distribution of organisms in an ecosystem. These topics and many more are the focus of current conservation biology research.

Check Your Progress 47.2

1. Explain the difference between a direct value and an indirect value of biodiversity.
2. Recognize the benefit of biodiversity to natural ecosystems.

47.3 Causes of Extinction

Learning Outcomes

Upon completion of this section, you should be able to

1. Classify the causes of extinction.
2. Compare natural and human-influenced causes of extinction.

To stem the tide of extinction due to human activities, it is first necessary to identify its causes. Researchers examined the records of 1,880 threatened and endangered wild species in the United States and found that habitat loss was involved in 85% of the cases (Fig. 47.6*a*). Exotic species had a hand in nearly 50%, pollution was a factor in 24%, overexploitation in 17%, and disease in 3%. (The percentages add up to more than 100%, because most of these species are imperiled for more than one reason.) Macaws are a good example of how a combination of factors can lead to a species decline (Fig. 47.6*b*). Not only has their habitat been reduced by encroaching timber and mining companies, but macaws are also hunted for food and collected for the pet trade.

Habitat Loss

Habitat loss is occurring in all ecosystems, but concern has now centered on tropical rain forests and coral reefs, because they are particularly rich in species.

A sequence of events in Brazil offers a fairly typical example of the manner in which rain forest is converted to land uninhabitable for wildlife. The construction of a major highway into the forest first provided a way to reach the interior of the forest (Fig. 47.6*c*). Small towns and industries sprang up along the highway, and roads branching off the main highway gave rise to even more roads. The result was fragmentation of the once immense forest.

The government offered subsidies to anyone willing to take up residence in the forest, and the people who came cut and burned trees in patches (Fig. 47.6*c*). Tropical soils contain limited nutrients, but when the trees are burned, nutrients are released that support a lush growth for the grazing of cattle for about 3 years. However, once the land was degraded (Fig. 47.6*c*), the farmers moved on to another portion of the forest to start over again.

Loss of habitat also affects freshwater and marine biodiversity. Coastal degradation is mainly due to the large concentration of people living on or near the coast. Already, 60% of coral reefs have been destroyed or are on the verge of destruction; it is possible that all coral reefs will disappear during the next 40 years unless our behaviors drastically change. Mangrove forest destruction is also a problem; Indonesia, with the most mangrove acreage, has lost 45% of its mangroves, and the percentage is even higher for other tropical countries. Wetland areas, estuaries, and seagrass beds are also being rapidly destroyed by human actions.

Exotic Species

Exotic species, sometimes called invasive species, are nonnative members of an ecosystem. Ecosystems around the globe are characterized by unique assemblages of organisms that have evolved together in one location. Migration of some species to a new

b. Macaws

Roads cut through forest

Forest occurs in patches

Destroyed areas

c. Wildlife habitat is reduced.

Figure 47.6 Habitat loss. **a.** In a study that examined records of imperiled U.S. plants and animals, habitat loss emerged as the greatest threat to wildlife. **b.** Macaws that reside in South American tropical rain forests are endangered for many of the same reasons listed in the graph in (**a**). **c.** Habitat loss due to road construction in Brazil. *Top:* Road construction opened up the rain forest and subjected it to fragmentation. *Middle:* The result was patches of forest and degraded land. *Bottom:* Wildlife could not live in destroyed portions of the forest.

location is not usually possible because of barriers such as oceans, deserts, mountains, and rivers. Humans, however, have introduced exotic species into new ecosystems in a variety of ways.

Colonization

Europeans, in particular, took a number of familiar species with them when they colonized new places. For example, the pilgrims brought the dandelion to the United States as a familiar salad green. In addition, they introduced pigs to North America that have since become feral, reverting to their wild state. In some parts of the United States, feral pigs have become quite destructive.

Horticulture and Agriculture

Some exotics, which have escaped from cultivated areas, are now taking over vast tracts of land. Kudzu is a vine from Japan that the U.S. Department of Agriculture thought would help prevent soil erosion. The plant now covers much landscape in the South, including smothering walnut, magnolia, and sweet gum trees (Fig. 47.7*a*). The water hyacinth was introduced to the United States from South America because of its beautiful flowers. Today, it clogs up waterways and diminishes natural diversity.

Accidental Transport

Global trade and travel accidentally bring new species from one region to another. Researchers found that the ballast water released from ships into Coos Bay, Oregon, contained 367 marine species from Japan. The zebra mussel from the Caspian Sea was accidentally introduced into the Great Lakes in 1988. It now forms dense beds that squeeze out native mussels. Other organisms accidentally introduced into the United States include the Formosan termite, the Argentine fire ant, and the nutria, a type of large rodent.

Video Alien Invasion

Exotic species can disrupt food webs. As mentioned earlier, opossum shrimp introduced into a lake in Montana added a trophic level that, in the end, meant less food for bald eagles and grizzly bears (see Fig. 47.2).

The Impact of Exotics on Islands

Islands are particularly susceptible to environmental discord caused by the introduction of exotic species. Islands have unique assemblages of native species that are closely adapted to one another and cannot compete well against exotics. Myrtle trees, *Myrica faya,* introduced into the Hawaiian Islands from the Canary Islands, are symbiotic with a type of bacterium that is capable of nitrogen fixation. This feature allows the species to establish itself on nutrient-poor volcanic soil, a distinct advantage in Hawaii. Once established, myrtle trees disrupt the normal succession of native plants on volcanic soil.

The brown tree snake has been introduced onto a number of islands in the Pacific Ocean. The snake eats adult birds, their eggs, and nestlings. On Guam, it has reduced ten native bird species to the point of extinction. On the Galápagos Islands, black rats have reduced populations of giant tortoise, while goats and feral pigs have changed the vegetation from highland forest to pampaslike grasslands and destroyed stands of cacti. In Australia, mice and rabbits have stressed native marsupial populations. Mongooses introduced into the Hawaiian Islands to control rats also prey on native birds (Fig. 47.7*b*).

a. b.

Figure 47.7 Exotic species. **a.** Kudzu, a vine from Japan, was introduced into several southern states to control erosion. Today, kudzu has taken over and displaced many native plants. Here it has engulfed an abandoned house. **b.** Mongooses were introduced into Hawaii to control rats, but they also prey on native birds.

Pollution

In the present context, **pollution** can be defined as any contaminant introduced into the environment that adversely affects the lives and health of living organisms. Pollution has been identified as the third main cause of extinction. Pollution can also weaken organisms and lead to disease. Biodiversity is particularly threatened by acid deposition, eutrophication, ozone depletion, and synthetic organic chemicals.

Acid Deposition

Both sulfur dioxide from power plants and nitrogen oxides in automobile exhaust are converted to acids when they combine with water vapor in the atmosphere. These acids return to Earth as either wet deposition (acid rain or snow) or dry deposition (sulfate and nitrate salts). Sulfur dioxide and nitrogen oxides are not always deposited in the same location where they are emitted but may be carried far downwind. Acid deposition causes trees to weaken and increases their susceptibility to disease and insects. It also kills small invertebrates and decomposers, so that entire ecosystems are disrupted. Many lakes in the northern United States are now lifeless because of the effects of acid deposition.

Eutrophication

Lakes are also under stress due to overenrichment. When lakes receive excess nutrients due to runoff from agricultural fields and fertilized lawns, as well as wastewater from sewage treatment, algae begin to grow in abundance. An algal bloom is apparent as a thick, green layer of algae on the surface of the water. Upon death, the decomposers break down the algae, but in so doing, they use up oxygen. A decreased amount of oxygen is available to fish, leading sometimes to a massive fish kill.

Ozone Depletion

The ozone shield is a layer of ozone (O_3) in the stratosphere, some 50 km above the Earth. The ozone shield absorbs most of the wavelengths of harmful ultraviolet (UV) radiation, so that they do not strike the Earth. The cause of ozone depletion can be traced to chlorine atoms (Cl^-) that come from the breakdown of chlorofluorocarbons (CFCs). The best-known CFC is Freon, a heat-transfer agent still found in older refrigerators and air conditioners. Severe ozone shield depletion can impair crop and tree growth and kill plankton (microscopic plant and animal life) that sustain oceanic life. The immune system and the ability of all organisms to resist infectious diseases can also become weakened.

Organic Chemicals

Our modern society uses synthetic organic chemicals in all sorts of ways. Organic chemicals called nonylphenols are used in products ranging from pesticides to dishwashing detergents, cosmetics, plastics, and spermicides. These chemicals mimic the effects of hormones, which can harm wildlife.

As one example, salmon are born in fresh water but mature in salt water. After investigators exposed young fish to nonylphenol, they found that 20–30% were unable to make the transition between fresh and salt water. Nonylphenols cause the pituitary to produce prolactin, a hormone that may prevent saltwater adaptation in these fish.

Climate Change

As mentioned in Chapter 45, **climate change** refers to recent changes in the Earth's climate. Our planet is experiencing erratic temperature patterns, more severe storms, and melting glaciers as a result of an increase in the Earth's temperature. You may also recall from Chapter 45 that the majority of carbon dioxide and methane produced today is the result of human-based activities. These gases are known as greenhouse gases, because they allow solar radiation to reach the Earth but hinder the escape of its heat back into space. Data collected around the world show a steady rise in CO_2 concentration. These data are used to generate computer models that predict how the Earth will continue to warm in the near future (Fig. 47.8*a*). An upward shift in temperatures could influence everything from growing seasons in plants to migratory patterns of animals.

As temperatures rise, regions of suitable climate for human needs may shift toward the poles and higher elevations. Extinctions are expected to increase as species attempt to migrate to more suitable climates. (Plants migrate when seeds disperse and growth occurs in a new locale.) For example, for beech trees to remain in a

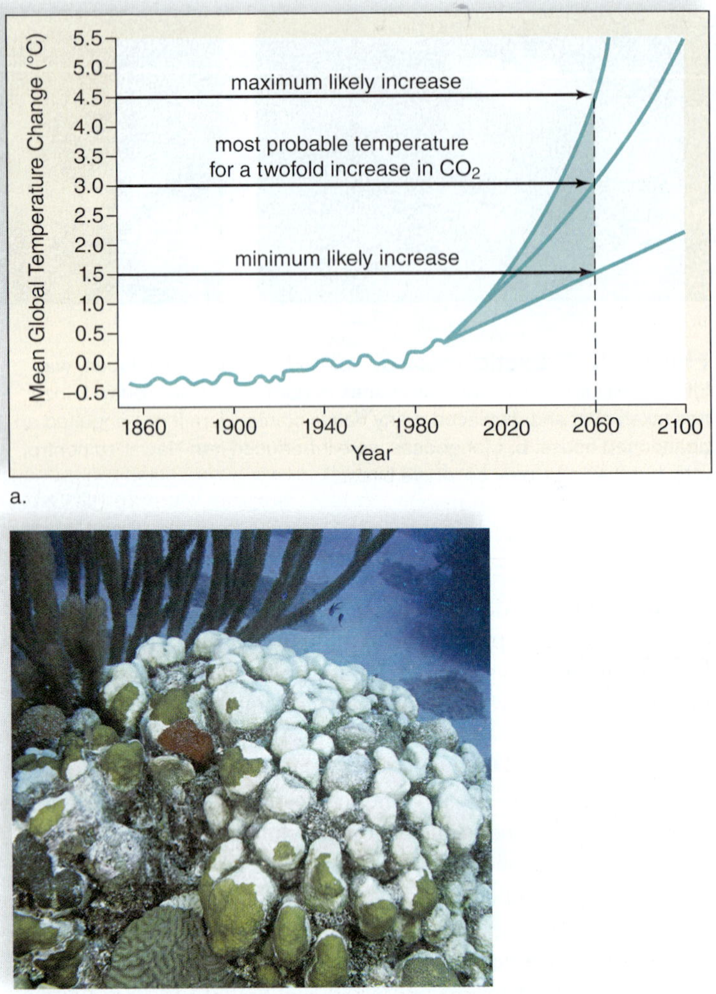

a.

b.

Figure 47.8 Climate change. **a.** Mean global temperature is expected to rise due to the introduction of greenhouse gases into the atmosphere. **b.** The resulting changes in climate have the potential to significantly affect the world's biodiversity. A temperature rise of only a few degrees causes coral reefs to "bleach" and become lifeless. If, in the meantime, migration occurs, coral reefs could move northward.

favorable habitat, it's been calculated that the rate of beech migration would have to be 40 times faster than has ever been observed. It seems unlikely that beech or any other type of tree would be able to meet the pace required.

Many species are confined to relatively small patches of habitat surrounded by agricultural or urban areas that prevent natural migrations. Even if species have the capacity to disperse to new sites, suitable habitats may not be available. If the global climate changes faster than organisms can migrate, extinction of many species is likely to occur, while other species may experience population increases. For example, parasites and pests that are usually killed by cold winters would be able to survive in greater numbers. The tropics may expand, and whether present-day temperate-zone agriculture will survive is questionable.

Video Karoo Global Warming

Overexploitation

Overexploitation is the process of taking more individuals from a wild population than can be naturally replaced, resulting in a decrease in the population. A positive feedback cycle explains overexploitation: The smaller the population, the more valuable its members, and the greater the incentive to capture the few remaining organisms. Poachers are very active in the collecting and sale of endangered and threatened species, because it has become so lucrative. The overall international value of trading wildlife species is approximately 20 billion dollars, of which 8 billion dollars is attributed to the illegal sale of rare species. The Nature of Science feature, "Overexploitation of Asian Turtles," provides a case in point.

Markets for rare plants and exotic pets support both legal and illegal trade in wild species. Rustlers dig up rare cacti, such as the crested saguaros, and sell them to gardeners for as much as \$15,000 each. Parrots are among the birds taken from the wild for sale to pet owners. For every bird delivered alive, many more have died in the collection process. The same holds true for tropical fish, which often come from the coral reefs of Indonesia and the Philippines. Divers dynamite reefs or use plastic squeeze-bottles of cyanide to stun the fish; in the process, many fish and valuable corals are killed.

The Convention of International Trade of Endangered Species (CITES) was an agreement established in 1973 to ensure that international trade of species does not threaten their survival. Today, over 35,600 species of plants and animals receive some level of protection from over 172 countries worldwide.

Poachers still hunt for hides, claws, tusks, horns, and bones of many endangered mammals. Because of its rarity, a single Siberian tiger is now worth more than \$500,000—its bones are pulverized and used as a medicinal powder. The horns of rhinoceroses become ornate carved daggers, and their bones are ground up to sell as a medicine. The ivory of an elephant's tusk is used to make art objects, jewelry, and piano keys. The fur of a Bengal tiger sells for as much as \$100,000 in Tokyo.

The U.N. Food and Agricultural organization tells us that we have now overexploited 11 of 15 major oceanic fishing areas. Fish are a renewable resource if harvesting does not exceed the ability of the fish to reproduce. Modern society uses larger and more efficient fishing fleets to decimate fishing stocks. Pelagic species, such as tuna, are captured by purse-seine fishing, in which a very large net surrounds a school of fish. The net is then closed in the same manner as a drawstring purse. Thousands of dolphins that swim above schools of tuna are often captured and then die in this type of net. Many tuna suppliers now advertise their product as "dolphin safe"; however, this label's meaning varies, depending on the agency or organization that licenses the label.

Other fishing boats drag huge trawling nets, large enough to accommodate 12 jumbo jets, along the seafloor to capture bottom-dwelling fish (Fig. 47.9*a*). Only large fish are kept, while the undesirable small fish and sea turtles are discarded back into the ocean to die. Trawling has been called the marine equivalent of clear-cutting forest, because after the net goes by, the sea bottom is devastated (Fig. 47.9*b*).

Today's fishing practices don't allow fisheries to recover. Cod and haddock, once the most abundant bottom-dwelling fish along the northeast coast of the United States, are now often outnumbered by dogfish and skate.

A marine ecosystem can be disrupted by overfishing, as exemplified on the U.S. west coast. When sea otters began to decline in

THEME **Nature of Science**

Overexploitation of Asian Turtles

Conservation Alert

Collection and trade of terrestrial tortoises and freshwater turtles for human consumption and other uses has surged in Asia over the past two decades and is now spreading to areas around the globe (Fig. 47A). With 40–60% of all types of turtles already endangered, these practices have virtually wiped out many tortoise and freshwater turtle populations from wide areas of Asia and have brought others to the brink of extinction in a matter of years.

The wild-collection trade started in Bangladesh in order to supply consumption demands in South China, and then it quickly spread across tropical Asia as one area after another became depleted. Currently, the practice has spread to the United States.

Tortoises and turtles mature late (on the order of 10 to 20+ years), experience great longevity (measured in decades), and have low annual reproductive rates. Traders prefer to collect larger, and therefore breeding-age, individuals. This means that wild populations are not likely to recover after they have been plundered. Presently, the stocking of thousands of new turtle farms (ranging from backyard to industrial-scale operations) is also causing a further run on the last remaining wild stocks of many endangered species. Then, too, there is a very active illegal pet trade in rare tortoises and freshwater turtles for wealthy patrons. This illegal pet trade is now vexing enforcement authorities in Asia, Europe, and the United States.

Major Challenges Today and in the Future

Basic scientific knowledge about the range, natural history, and conservation needs of individual species of tortoises and turtles is lacking. In fact, some traded species are so poorly known, and so endangered, that specimens have sometimes only been found by researchers in wildlife markets. Wildlife inspectors and enforcement officials are often unaware of conservation concerns and are unable to identify turtle species and to associate them with pertinent regulations.

Because the wild-collection trade is now worldwide, it is essential that all nations and states with (remaining) wild turtle populations ensure that their domestic legislation is adequate to secure the future of their turtle populations. Too often, trade starts up and occurs faster than regulatory measures can be put in place to prevent local populations from being decimated.

Figure 47B Aquaculture farms. Red-eared slider turtles (*Trachemys scripta elegans*) are now raised in aquaculture farms. Although this takes pressure off wild turtle populations, such turtles reinforce the habit of turtle consumption and can become invasive species when they escape into the wild.

Figure 47A Turtles for sale. In 2011, during a religious holiday in Dhaka, Bangladesh, over 100,000 turtles were butchered for consumption. Large numbers of wild-caught tortoises and freshwater turtles are offered for sale at markets in East Asia. Shown here are steppe tortoises (*Testudo horsefieldii*) and elongated tortoises (*Indotestudo elongata*)

Another issue that needs attention is the threat of invasive species and the spread of diseases from aquaculture facilities to wild populations. It is known that Chinese soft-shell turtles and red-eared slider turtles are highly adaptable to new environments (Fig. 47B). Reports abound from around the globe that turtles have either escaped or have been deliberately released from aquaculture facilities. Ecologists can only speculate about how they will impact wild populations when they become competitors, predators, hybridizers, and disease vectors. Impact studies are desirable, as is a regulatory framework governing the transport and handling of nonlocal turtles.

International conservation groups, such as World Conservation Society and the World Wildlife Fund, have been working on educating the public about the eminent threat to turtle species everywhere. Without more awareness and action by everyday citizens, many species of turtles will be extinct within the foreseable future.

Questions to Consider

1. Does the buyer or the seller hold more responsibility for the overexploitation of the Asian turtles?
2. What type of ecological impact could an invasive turtle population have on a native turtle population?

Written by Peter Paul van Dijk with the assistance of Bruce J. Weissgold.

a. Fishing by use of a drag net

b. Result of drag net fishing

Figure 47.9 Trawling. **a.** These Alaskan pollock were caught by dragging a net along the seafloor. **b.** Appearance of the seafloor after the net has passed.

numbers, investigators found that they were being eaten by orcas (killer whales). Usually, orcas prefer seals and sea lions over sea otters. The orcas began eating sea otters when the seal and sea lion populations decreased due to the overfishing of the perch and herring. Ordinarily, sea otters keep the population of sea urchins under control. With fewer sea otters around, the sea urchin population exploded and decimated the kelp beds. Overfishing set in motion a chain of events that detrimentally altered the food web of this ecosystem.

Check Your Progress **47.3**

1. Identify the five main causes of extinction.
2. Explain why the introduction of exotic species can be detrimental to biodiversity.

47.4 Conservation Techniques

Learning Outcomes

Upon completion of this section, you should be able to

1. Describe the value of preserving biodiversity hotspots.
2. Distinguish between keystone species and flagship species.
3. Identify and explain the most useful procedures for habitat restoration.

Despite the value of biodiversity to our very survival, human activities are causing the extinction of thousands of species a year. Clearly, we need to reverse this trend and preserve as many species as possible. Habitat preservation and restoration are important in preserving biodiversity.

Habitat Preservation

Preservation of a species' habitat is of primary concern, but first we must prioritize which species to preserve. As mentioned previously, the biosphere contains biodiversity hotspots, relatively small areas having a concentration of endemic (native) species not found anyplace else. In the tropical rain forests of Madagascar, 93% of the primate species, 99% of the frog species, and over 80% of the plant species are endemic, found only in Madagascar. Preserving these forests and other hotspots would save a wide variety of organisms.

Keystone Species

Keystone species are species that influence the viability of a community, although their numbers may not be excessively high. The extinction of a keystone species can lead to other extinctions and a loss of biodiversity. For example, bats are designated a keystone species in tropical forests of the Old World. They are pollinators that also disperse the seeds of trees. When bats are killed off and their roosts destroyed, the trees fail to reproduce. The grizzly bear is a keystone species in the northwestern United States and Canada (Fig. 47.10*a*). Bears disperse the seeds of berries; as many as 7,000 seeds may be in one dung pile. Grizzly bears kill and eat the young of many hoofed animals and thereby keep their populations under control. Grizzly bears are also a principal mover of soil when they dig up roots and prey upon hibernating ground squirrels and marmots. Other keystone species are beavers in wetlands, bison in grasslands, alligators in swamps, and elephants in grasslands and forests.

Keystone species should not be confused with **flagship species,** which evoke a strong emotional response in humans. Flagship species are considered charismatic and are treasured for their beauty, "cuteness," strength, and/or regal nature. These species can help motivate the public to preserve biodiversity. Flagship species include Monarch butterflies, lions, tigers, dolphins, and giant pandas. Some keystone species are also flagship species—for example, the grizzly bear—but a keystone species is valued because of its role in an ecosystem.

Metapopulations

The grizzly bear population is actually a **metapopulation** (Gk. *meta,* "higher-order"), a large population that has been subdivided

a. Grizzly bear, *Ursus arctos horribilis*

b. Old-growth forest; northern spotted owl, *Strix occidentalis caurina* (inset)

Figure 47.10 Habitat preservation. When particular species are protected, other wildlife benefit. **a.** The Greater Yellowstone Ecosystem has been delineated in an effort to save grizzly bears, which need a very large habitat. **b.** Currently, the remaining portions of old-growth forests in the Pacific Northwest are not being logged in order to save the northern spotted owl (*inset*).

into several small, isolated populations due to habitat fragmentation. Originally, there were an estimated 50,000–100,000 grizzly bears south of Canada. This number has been reduced as the result of human housing communities encroaching on their home range. Bears are often killed when they come in contact with people. Currently, there are six virtually isolated subpopulations totaling about 1,000 individuals. The Yellowstone National Park population numbers 200, but the others are even smaller.

Saving metapopulations sometimes requires determining which of the populations is a source and which are sinks. A **source population** is one that is stable or growing in size, producing an excess of individuals. Often, its birthrate is higher than its death rate. The excess individuals from source populations will move into **sink populations,** where the environment is not as favorable and where the birthrate equals or is often lower than the death rate. When trying to save the northern spotted owl, conservationists determined that it was best to avoid having owls move into sink habitats, since they could not sustain the owls. The northern spotted owl reproduces successfully in old-growth rain forests of the Pacific Northwest (Fig. 47.10*b*), but not in nearby immature forests that are in the process of recovering from logging. Scientists have also recognized that competition from the barred owl is also a major threat to the recovery of the northern spotted owl.

Landscape Preservation

Grizzly bears inhabit a number of different types of ecosystems, including plains, mountains, and rivers. Saving any one of these types of ecosystems alone would not be sufficient to preserve the grizzly bears, however. Instead, it is necessary to save diverse ecosystems that are connected. As mentioned earlier, a landscape encompasses different types of ecosystems. An area called the Greater Yellowstone Ecosystem, where bears are free to roam, has now been defined. It contains millions of acres in Yellowstone National Park; state lands in Montana, Idaho, and Wyoming; five national forests; various wildlife refuges; and even private lands.

Landscape protection for one species is often beneficial for other wildlife that share the same space. The last of the contiguous 48 states' harlequin ducks, bull trout, westslope cutthroat trout, lynx, pine martens, wolverines, mountain caribou, and great gray owls are found in areas occupied by grizzly bears. Gray wolves have also returned to this territory. Grizzly bear range also overlaps with 40% of Montana's vascular plants of special conservation status.

The Edge Effect. An *edge* is a transition from one habitat to another. In the case of land development, an edge may be very defined, as when a forest abuts farm fields. The edge reduces the amount of habitat typical of an ecosystem, because the edge habitat is slightly different from that of the interior of the patch—this is termed the **edge effect.** For example, forest edges are brighter, warmer, drier, and windier, with more vines, shrubs, and weeds than the forest interior. Also, Figure 47.11*a* shows that a small and a large patch of habitat both have the same depth of edge; therefore, the effective habitat shrinks as a patch gets smaller.

Many popular game animals, such as turkeys and white-tailed deer, are more plentiful in the edge region of a particular area. However, today it is known that creating edges can be detrimental to wildlife because of habitat fragmentation.

The edge effect can also have a serious impact on population size. Songbird populations west of the Mississippi have been declining, and ornithologists have noticed that the nesting success of songbirds is quite low at the edge of a forest. The cause turns out to be the brown-headed cowbird, a social parasite of songbirds. Adult cowbirds prefer to feed in open agricultural areas, and they

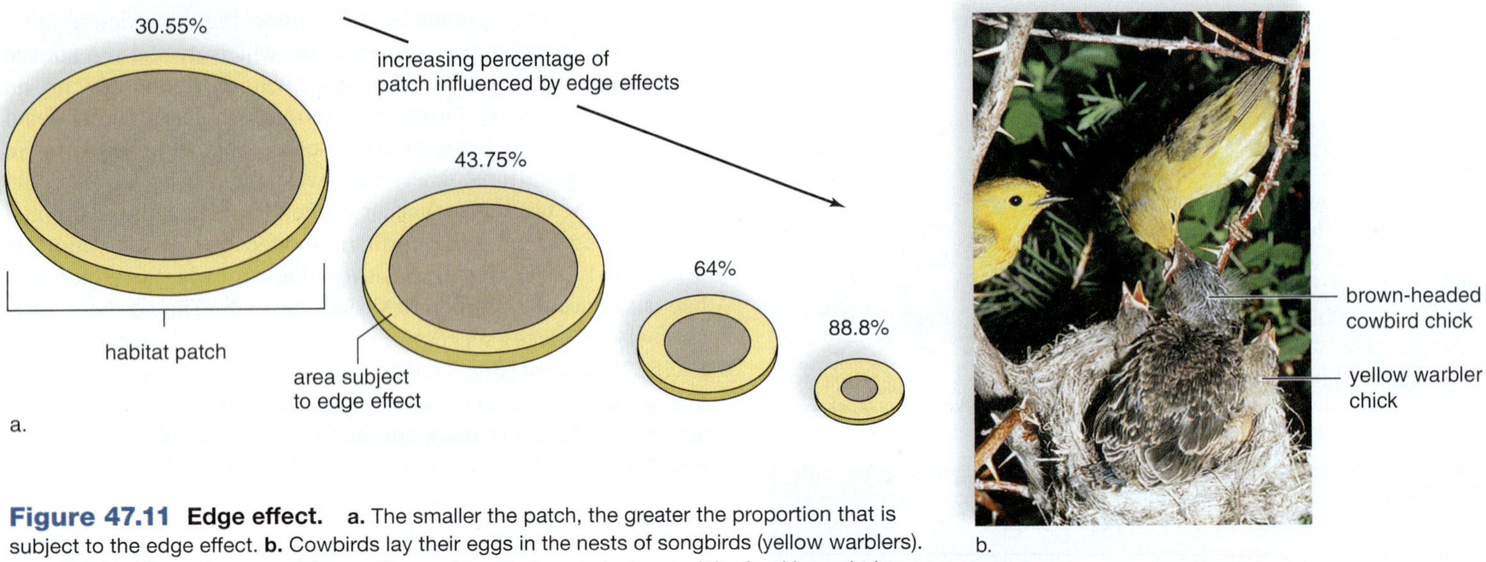

Figure 47.11 Edge effect. **a.** The smaller the patch, the greater the proportion that is subject to the edge effect. **b.** Cowbirds lay their eggs in the nests of songbirds (yellow warblers). A cowbird is bigger than a warbler nestling and is able to acquire most of the food brought by the warbler parent.

only briefly enter the forest when searching for a host nest in which to lay their eggs (Fig. 47.11*b*). Cowbirds are therefore benefited, while songbirds are disadvantaged, by the edge effect.

Habitat Restoration

Restoration ecology is a subdiscipline of conservation biology that looks for scientific ways to return ecosystems to their natural state. Restoration ecology has three key principles. First, it should start as quickly as possible to avoid losing the remaining fragments of the original habitat.

Second, once the natural histories of the species in the habitat are understood, it is best to mimic natural processes to bring about restoration. Natural processes include using controlled burns to bring back grassland habitats, biological pest controls to rid the area of exotic species, and bioremediation techniques to clean up pollutants.

Third, the goal is sustainable development, the ability of an ecosystem to be self-sustaining while still providing services to humans. The Nature of Science feature, "Emiquon Floodplain Restoration," presents an example of a successful project in west-central Illinois.

Check Your Progress 47.4

1. Explain why it is more important to preserve hotspots than other areas.
2. Identify examples of keystone species.
3. List and explain the three principles of habitat restoration.

CONNECTING the CONCEPTS with the THEMES

Evolution

- The Earth's biodiversity has taken millions of years to establish itself, yet many regions have seen a decline in biodiversity only in the past few hundred years.
- Ecosystem stability is dependent on the coevolution of the species that inhabit it. Therefore, the removal or decline of one species may have an impact beyond what we expect.

Nature of Science

- Conservation biology seeks to balance the needs of humans with the rest of the species on Earth.
- Educating the public about the many values of biodiversity is critical to gaining support for species and ecosystem preservation.

Biological Systems

- Stable ecosystems tend to have a high level of biodiversity within them; this diversity is vulnerable to sudden changes that do not allow adaptation.
- A keystone species is a species that has the greatest influence on community diversity.
- Exotic species tend to disrupt the stability of the ecosystems into which they are introduced. They often outcompete native species for resources due to the lack of natural predators that would keep their population in check.

THEME Nature of Science

Emiquon Floodplain Restoration

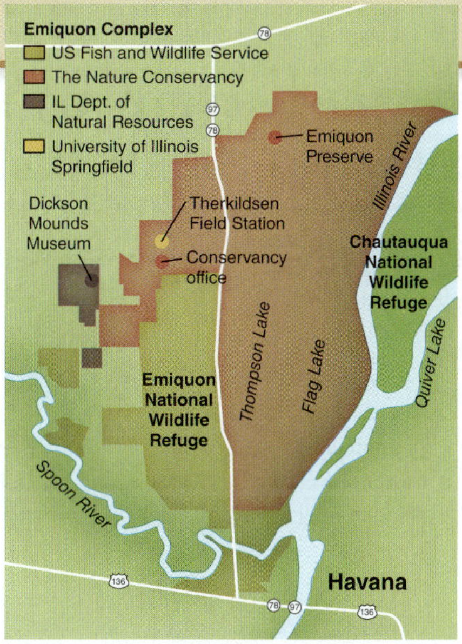

Figure 47C The Emiquon Restoration Project. Restoration plans call for allowing the river to return to some of its former range.

Emiquon was once a vital component of the Illinois River system, which is a major tributary of the Mississippi River. It included a complex system of backwater wetlands and lakes that originally covered an area of approximately 14,000 acres (Fig. 47C). The vast diversity of native plants and animals found within Emiquon have supported human civilizations for over 10,000 years. Early cultures accessed the abundant inland fisheries and a variety of mussel species that could be found there. River otters, minks, beavers, and migrating birds as well as a variety of prairie grasses on the floodplains would have been used by native peoples (Fig. 47D).

In the early 1900s, the Illinois River was one of the most economically significant river ecosystems in the United States. It boasted the highest mussel abundance and productive commercial fisheries per mile of any stream in the United States. Each year during the seasonal flooding, the river deposited nutrient-laden sediment onto the floodplain. This created a highly productive floodplain ecosystem. With the abundance of nutrients, a wide diversity of plant life would have been supported. This broad producer base would have been able to support a number of trophic levels, as well as a wide diversity of food webs.

Throughout the twentieth century, modern cultures altered the natural ebb and flow of the Illinois River to better suit their needs. Levees were constructed, resulting in the isolation of nearly half the original floodplain from the river. Native habitats were converted to agricultural land. Corn and soybeans soon replaced the variety of prairie and wetland vegetation that were once a hallmark of the Emiquon floodplain. The river itself was also altered to better facilitate the export and import of a variety of commercial goods by boat. With the isolation of these areas from the seasonal flooding, a decrease in the amount of nutrients being deposited also occurred. This resulted in a continuous decline in the natural productivity of the floodplain.

Restoration Plan

The National Research Council has identified Emiquon as one of three large-floodplain river ecosystems in the United States that can be restored to some semblance of its original diversity and function. The floodplain restoration work at Emiquon is a key part of The Nature Conservancy's efforts to conserve the Illinois River. The Emiquon Complex was designated as the United States' 32nd Wetland of International Importance alongside prestigious wetlands such as Chesapeake Bay and San Francisco Bay.

The Emiquon Science Advisory Council is composed of over 40 scientists and managers who provide guidance for the restoration and managed connection between Emiquon's restored floodplain and the Illinois River. This connection allows for the seasonal flooding that once existed there. This seasonal flooding is helping naturalize the flow of the river and restore the cyclical process of flooding and drying out. Flooding provides more nutrients for wetland plants as well as helping improve the water quality. Over 100 wetland species of plants have been documented since the restoration began. Access between the river and the floodplain allows access for various aquatic species, such as paddlefish and gar. Both species need a variety of habitats for reproduction and survival. The wetlands in turn provide vital nutrients to the species that reside in the river. The Conservancy is continuing to work with the Illinois Natural History Survey, the University of Illinois, Springfield, and others to survey the community diversity. Since 2007, they have documented over 250 species of birds, including over 90% of Illinois endangered and threatened wetland bird species.

In 2009, the normal operation of Emiquon was estimated to have contributed 1.1 million dollars to the local economy. As usage continues to increase, so does the economic impact Emiquon is having on local economies. To date, scientists, students, and decision makers from five continents have visited Emiquon. The knowledge gained from the research at Emiquon and The Nature Conservancy's work on the Illinois River and within the Upper Mississippi River system is helping influence floodplain restoration projects around the world.

Questions to Consider

1. In what ways might scientists assess whether the restoration plan was successful in its goal?
2. How might the Emiquon Floodplain Restoration plan be used as a model for other restoration projects?

Figure 47D Forty-nine black-crowned night-heron nests have been recorded at Emiquon since the restoration project began. Research at Emiquon is providing critical information about wetland restoration techniques.

MEDIA STUDY TOOLS http://connect.mheducation.com

SMARTBOOK® Study smart. SmartBook helps you maximize your study time by identifying what you know and don't know. The recharge feature in SmartBook helps you prepare for the test by identifying the material that you are most likely to forget.

connect® BIOLOGY Find even more resources to learn the chapter concepts including animations, tutorial videos, and interactive practice questions.

Videos

47.1 Coral Reef Ecosystems
47.2 Indigenous Medicine • Warming Hurts Rice • Pollinators • Dung Beetles • Decomposers • Thames River
47.3 Alien Invasion • Karoo Global Warming

SUMMARIZE

47.1 Conservation Biology and Biodiversity

Conservation biology is the scientific study of **biodiversity** and its management for sustainable human welfare. The unequaled present rate of extinctions has drawn together scientists and environmentalists in basic and applied fields to address the problem by using the science of **bioinformatics.**

Biodiversity is described as the number of species in a given habitat. Conservation biology looks to identify which species are **endangered** or **threatened,** making them the primary focus of conservation. Biodiversity is a combination of three levels of organization: **genetic diversity, community diversity,** and **landscape diversity,** in which the interacting ecosystems form a **landscape.**

Conservationists have discovered that biodiversity is not evenly distributed in the biosphere, and therefore saving particular areas may protect more species than saving other areas. Areas with exceptionally high concentrations of biodiversity are termed **biodiversity hotspots.**

47.2 Value of Biodiversity

The direct value of biodiversity is seen in the economic value that a species provides to humans. Wild species are our best source of new medicines to treat human ills, and they help meet other medical needs as well.

Wild species have agricultural value. Domesticated plants and animals are derived from wild species, and wild species can be used as a source of genes for the improvement of their genetics. Instead of synthetic pesticides, wild species can be used as biological controls, and most flowering plants benefit from animal pollinators. Much of our food, particularly fish and shellfish, is still caught in the wild. Hardwood trees from natural forests supply us with lumber for many purposes.

The indirect services provided by ecosystems are largely unseen and difficult to quantify, but they are absolutely necessary to our well-being. These services include the workings of biogeochemical cycles, waste recycling, the provision of fresh water, the prevention of soil erosion, and the regulation of climate. Many people enjoy living and vacationing in natural settings. Studies show that more diverse ecosystems function better than less diverse systems.

47.3 Causes of Extinction

Researchers have identified the major causes of extinction. Habitat loss is the most frequent cause, followed by the introduction of **exotic species, pollution** and resulting diseases, **climate change,** and **overexploitation.**

Habitat loss has occurred in all parts of the biosphere, but concern has now centered on tropical rain forests and coral reefs, where biodiversity is especially high. Exotic species have been introduced into foreign ecosystems through colonization, horticulture and agriculture, and accidental transport. Various causes of pollution include fertilizer runoff, industrial emissions, and improper disposal of wastes. Climate change is contributing to species extinction, but at this point the full impact is still being explored. Overexploitation is exemplified by commercial fishing, which is so efficient that the fisheries of the world are collapsing.

47.4 Conservation Techniques

To preserve species, it is necessary to preserve key habitats that contain the most species. **Keystone species** are ones that influence the viability of the community and are of significant importance. **Flagship species** are those that evoke an emotional response by humans and can be used to help promote conservation. Some emphasize the need to preserve biodiversity hotspots because of their richness. Often today, it is necessary to save **metapopulations** because of past habitat fragmentation. If so, it is best to determine the **source populations** and prioritize those over the **sink populations.**

When preserving landscapes, it is best to preserve large tracts of land due to the decreased biodiversity caused by habitat fragmentation and the **edge effect.**

Because many ecosystems have been degraded, habitat restoration may be necessary before sustainable development is possible. Three principles of restoration are (1) start before sources of wildlife and seeds are lost; (2) use simple biological techniques that mimic natural processes; and (3) aim for sustainable development, so that the ecosystem can persist while fulfilling the needs of humans.

ASSESS

Choose the best answer for each question.

47.1 Conservation Biology and Biodiversity

1. Which of these is not within the realm of conservation biology?
   a. helping manage a national park
   b. restoring an ecosystem
   c. writing textbooks and/or popular books on the value of biodiversity
   d. introducing endangered species back into the wild
   e. All of these are concerns of conservation biology.
2. Which statement accepted by conservation biologists best shows that they support ethical principles?
   a. Biodiversity is the variety of life observed at various levels of biological organization.
   b. Wild species directly provide us with all sorts of goods and services.
   c. New technologies can help determine conservation plans.
   d. There are three principles of restoration biology that need to be adhered to in order to restore ecosystems.
   e. Biodiversity is desirable and has value in and of itself, regardless of any practical benefit.
3. Which group of organisms represents the largest percentage of biodiversity on Earth?
   a. plants
   b. protists
   c. fungi
   d. insects
   e. bacteria and archaea

47.2 Value of Biodiversity

4. The services provided to us by ecosystems are unseen. This means
   a. they are not valuable.
   b. they are noticed particularly when the service is disrupted.
   c. biodiversity is not needed for ecosystems to keep functioning as before.
   d. we should be knowledgeable about them and protect them.
   e. Both b and d are correct.
5. Consumptive value
   a. means we should think of conservation in terms of the long run.
   b. means we are placing too much emphasis on living organisms that are useful to us.
   c. means some organisms, other than domestic crops and farm animals, are valuable as sources of food.
   d. is a type of direct value.
   e. Both c and d are correct.
6. Which of these is not an indirect value of species?
   a. participate in biogeochemical cycles
   b. participate in waste disposal
   c. help provide fresh water
   d. prevent soil erosion
   e. All of these are indirect values.

47.3 Causes of Extinction

7. Which of these is a true statement?
   a. Habitat loss is the most frequent cause of extinctions today.
   b. Exotic species are often introduced into ecosystems by accidental transport.
   c. Climate change may cause many extinctions but also expands the ranges of other species.
   d. Overexploitation of fisheries could lead to a complete collapse of the fishing industry.
   e. All of these statements are true.
8. Which of these is expected if the average global temperature increases?
   a. the inability of some species to migrate to cooler climates as environmental temperatures rise
   b. the possible destruction of coral reefs
   c. an increase in the number of parasites in the temperate zone
   d. a population decline for some species but an increase for others
   e. All of these are correct.
9. Complete the following graph by labeling each bar (a–e) with a cause of extinction, from the most influential to the least influential.

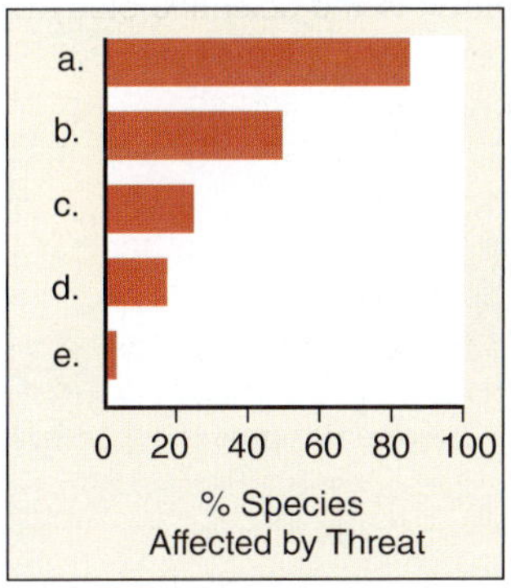

47.4 Conservation Techniques

10. Most likely, ecosystem performance improves
    a. the more diverse the ecosystem.
    b. as long as selected species are maintained.
    c. as long as species have both direct and indirect value.
    d. if extinctions are diverse.
    e. Both a and b are correct.
11. Why is the grizzly bear considered a keystone species that exists within a metapopulation?
    a. Grizzly bears require many thousands of miles of preserved land, because they are large animals.
    b. Grizzly bears have functions that increase biodiversity, but presently the population is subdivided into isolated subpopulations.
    c. When grizzly bears are present, so are many other species within a diverse landscape.
    d. Grizzly bears are a source population for many other types of organisms across several population types.
    e. All of these are correct.

12. Which habitat restoration method is not part of a natural restoration plan?
    a. using controlled burns to bring back natural grasslands
    b. using biological pest controls to eliminate invasive species
    c. applying bioremediation techniques to clean up oil spills
    d. using ladybugs to prey upon cotton aphids
    e. introducing new prey species into a community to increase the producer base

ENGAGE

THINKING SCIENTIFICALLY

1. The scale at which conservation biologists work often makes direct experimentation difficult. But computer models can assist in predicting the future of populations or ecosystems. Some scientists feel that these models are inadequate, because they cannot reproduce all the variables found in the real world. If you were trying to predict the impact on songbirds of clear-cutting a portion of a forest, what information would you need to develop a good model?
2. Bioprospecting is the search for medically useful molecules derived from living organisms. The desire for monetary and medicinal gains from such discoveries can lead to the overexploitation of endangered species. However, bioprospecting can protect ecosystems, and in this way, saves many species rather than individual species. What types of living organisms would bioprospectors be most interested in?
3. It has been close to 20 years since the reintroduction of wolves into Yellowstone National Park. Many ranchers and hunters want to be allowed to hunt wolves that leave the boundaries of the park. Do you think it should be legal to hunt and kill wolves if they are outside the park boundaries? What type of information would you need in order to make this decision?

Appendix A

Answer Key

Chapter 1

Check Your Progress

1.1: 1. Populations consist of all of the organisms of a species in a given area, while an ecosystem contains groups of populations (a community) plus the physical environment. 2. Organization, acquisition and use of energy, homeostasis, response to the environment, reproduction and development, and adaptation. 3. Adaptations allow organisms to survive in changing environments; the changes as a result of adaptations represent evolution. **1.2:** 1. Members of a population that have new adaptive traits reproduce more than other members, and therefore pass on these traits to the next generation in a greater proportion. 2. Domain, Kingdom, Phylum, Class, Order, Family, Genus, Species. 3. Protists are generally unicellular organisms; fungi are multicellular filamentous organisms that absorb food; plants are multicellular organisms that are usually photosynthetic; animals are multicellular organisms that must ingest and process their food. **1.3:** 1. Experiments test the contribution of the experimental variable to a given observation. 2. Test groups are exposed to the experimental variables, while control groups are not. 3. Hypotheses are tested by designing experiments that examine the contribution of a single experimental variable to the observation, and by statistically comparing the response of a test group to that of a control group in an experiment. **1.4:** 1. Technology applies scientific knowledge to the interests of humans. Discoveries made through scientific investigations may or may not be applied to a new technology. 2. Humans benefit from biodiversity because we depend on many different organisms for food, medicines, and raw materials. Well-functioning ecosystems that have high biodiversity ensure stable conditions suitable for the survival of humans and other organisms.

Assess

1. c; **2.** b; **3.** b; **4.** c; **5.** c; **6.** a; **7.** b; **8.** d; **9.** c; **10.** d; **11.** c; **12.** d

Chapter 2

Check Your Progress

2.1: 1. Atomic number is the number of protons in an atom. Atomic mass is the approximate sum of the masses of the protons and neutrons. 2. Groups indicate the number of electrons in the outer, or valence, shell of an atom. In Group VII, all the atoms including chlorine have seven electrons in the valence shell. Periods indicate the number of shells present in an atom. All atoms in Period 3, including chlorine, have three shells. 3. Two different isotopes of an element have the same number of protons but differ in the number of neutrons. Their atomic number is the same but they have different atomic masses. **2.2:** 1. Ionic and covalent bonds are similar in that both involve electrons, but differ as to whether the electrons are shared or not. Ionic bonds are created when one atom gives up electron(s) and another gains electron(s) so that both atoms have outer shells filled. Covalent bonds are formed when two atoms share electrons to fill their outer shells. 2. Ions are formed when an electron is transferred away from an atom, creating a positively charged particle, or when an electron is accepted by an atom, creating a negatively charged particle. 3. Methane forms covalent bonds with each of four hydrogen atoms. The electrons of the hydrogens and the carbon are shared equally and are nonpolar. The oxygen in water attracts the electrons it shares with hydrogen more strongly, resulting in greater electronegativity and a polar covalent bond. **2.3:** 1. Water forms hydrogen bonds because the molecule is asymmetrical and the hydrogens, which are slightly positive, are attracted to the slightly negatively charged oxygen atoms on other molecules. 2. Hydrogen bonding results in the molecules of water clinging together and demonstrating the property of cohesion. This results in water's ability to absorb a lot of heat energy, to resist evaporation, and to demonstrate high surface tension. 3. When water evaporates off the surface of an animal it absorbs large amounts of body heat because of its high heat of evaporation. **2.4:** 1. An acid dissociates in water to release hydrogen ions. A base either takes up hydrogen ions or releases hydroxide ions. 2. Moving down the pH scale, each unit is 10 times as acidic as the previous unit. Therefore a solution of pH 2.0 is 100 times more acidic than a pH 4.0 solution. 3. Buffers maintain the pH of body fluids within the narrow limits required for life.

Assess

1. e; **2.** c; **3.** a; **4.** b; **5.** c; **6.** a; **7.** b; **8.** c; **9.** 3; **10.** d; **11.** a; **12.** c; **13.** b; **14.** c; **15.** b

Chapter 3

Check Your Progress

3.1: 1. A carbon atom bonds with up to four different elements. Carbon–carbon bonds are stable, so long chains can be built, chains can have branching patterns, and chains can form rings. Isomers are possible. 2. Both molecules with hydroxyl groups (alcohols) and those with carboxyl groups (acids) are polar and soluble in water however the carboxyl group more easily donates its hydrogen and functions as an acid. 3. The degradation, or hydrolysis, of a biomolecule requires water. If no water is present the molecule won't be cut into pieces without some other factor, such as heat. **3.2:** 1. Carbohydrates have a carbon to hydrogen to oxygen ratio of 1:2:1 and are chains of varying lengths of single sugar molecules. Carbohydrates are energy sources for organisms and function as structural elements. 2. Disaccharides are formed when a bond is created and a molecule of water is removed by a dehydration reaction between two monosaccharides. 3. Humans have no digestive enzymes capable of breaking the bonds in cellulose. **3.3:** 1. Triglycereides provide insulation, arc long tcrm energy stores, are a food source, are used in cooking, and are alternative fuel sources. Phospholipids are a component of cell membranes and are essential to cell function. Steroids such as cholesterol are a component of cell membranes and are precursors to steroid hormones. Waxes are involved in skin and fur maintenance, providing protection. Bees use wax to store honey in their combs. 2. A saturated fatty acid contains no double bonds between carbon atoms, while an unsaturated fatty acid contains one or more double bonds. Saturated fatty acids are flat and tend to stack densely in specific locations such as blood vessels. Unsaturated fatty acids are kinked, and do not stack as densely. 3. Phospholipids, which have both hydrophilic and hydrophobic domains, arrange themselves so that their hydrophilic heads are adjacent to water and the hydrophobic tails point inward toward each other. **3.4:** 1. Proteins function in organisms in metabolism, structure, transportation, defense, hormone regulation, and contraction producing motion. 2. Two amino acids are linked together to form a dipeptide by a dehydration reaction. The –OH of the carboxyl group of one amino acid and the –H of the amine group of the other are removed. 3. The primary structure of a protein is the linear sequence of amino acids. The secondary structure occurs when the polypeptide coils or folds in a particular way. The tertiary structure is the folding that results in the final three-dimensional shape of the polypeptide. Quaternary structure occurs when more than one polypeptide interact to form the functional protein. 4. Incorrect protein folding changes the shape of the protein and affects its function. This can result in diseases such as cystic fibrosis, TSEs, and Alzheimer's. **3.5:** 1. A nucleic acid stores information in the sequences of bases within a gene. Gene sequences specify amino acid sequences, thus assembling the proteins that carry out cellular functions necessary for life. 2. Phosphate, 5-carbon sugar, and nitrogen-containing base. 3. ATP contains 3 phosphate molecules that are strongly negatively charged and repel each other. The covalent bonds attaching them contain considerable potential energy, which can be captured when the bonds are broken and used to do cellular work.

Assess

1. e; **2.** c; **3.** a; **4.** c; **5.** c; **6.** b; **7.** b; **8.** d; **9.** d; **10.** c; **11.** d; **12.** a; **13.** b; **14.** c; **15.** d

Chapter 4

Check Your Progress

4.1: 1. Unlike macromolecules, cells have the ability to reproduce. All cells come from preexisting cells. 2. Living organisms are made up of one or more cells, which form the basic unit of life. New cells are produced from old cells. Energy flows within cells. 3. The movement of sufficient materials into and out of a cell requires an adequately large surface area relative to cell volume. Cell size is limited by the amount of surface area. **4.2:** 1. Prokaryotic cells lack a membrane-bound nucleus, which is present in a eukaryotic cell. 2. Cell envelope: mesosome, plasma membrane, cell wall, glycocalyx. Cytoplasm: inclusion bodies, nucleoid, ribosomes. Appendages: conjugation pilus, fimbriae, flagellum. **4.3:** 1. Separates various metabolic processes; localizes enzymes, substrates, and products; and allows cells to become specialized. 2. Organelles effectively divide cellular work, so that incompatible processes can occur simultaneously. This division of labor enables the cell to more quickly respond to changing environments. 3. The nucleus likely evolved from the invagination of the plasma membrane. Mitochondria and chloroplasts are thought to have evolved when a larger eukaryotic cell engulfed a smaller prokaryotic cell and eventually became dependent on the smaller organism. **4.4:** 1. Chromatin is made up of strands of DNA combined with proteins. When these strands condense and coil up, rod like structures called chromosomes are formed. 2. Cells tightly regulate what enters and exits the nucleus by the use of nuclear pores. 3. Hereditary information encoded in genes located in the nucleus is transcribed into RNA molecules, some of which specify amino acid sequence in proteins. Amino acid sequence is determined by RNA sequence through translation. Polypeptides produced from RNA are then folded into a three-dimensional structure that has biological function. **4.5:** 1. Rough ER contains ribosomes, while smooth ER does not. Rough ER synthesizes proteins and modifies them, while smooth ER synthesizes lipids, among other activities. 2. Membrane-bound organelles work together as part of a larger endomembrane system within the cell. Transport vesicles from the ER proceed to the Golgi apparatus, which modifies their contents and repackages them in new vesicles to be delivered to specific locations within the cell based on a molecular address. Incoming food and other particles are packaged inside vesicles and delivered to lysosomes, which help to break down the food for energy or for building cellular structures. 3. Golgi malfunction would prevent materials from reaching their necessary destinations, thus disrupting cellular function and likely leading to cell death. **4.6:** 1. Both peroxisomes and lysosomes are membrane-bound vesicles that enclose enzymes. Lysozymes have hydrolytic-digestive enzymes at a low pH which digest material taken into the cell and destroy nonfunctional organelles and parts of the cytoplasm. Peroxisomes have enzymes that break down fatty acids. 2. The enzymes of the lysosomes are made in the endoplasmic reticulum and loaded into the vesicle by the Golgi apparatus. Enzymes in the peroxisomes are made in the cytoplasm and transported into the vesicle. **4.7:** 1. Both mitochondria and chloroplasts have roles in energy-related reactions in the cell. Chloroplasts carry out photosynthesis in which solar energy is used to synthesize carbohydrates. Mitochondria carry out cellular respiration in which carbohydrates are broken down to produce ATP which is used in the energy requiring reactions of the cell. 2. Chloroplasts and mitochondria are similar in size to bacteria, also have multiple membranes, replicate via binary fission, and have circular genomes. 3. The complex internal membranes of the mitochondria provide a large surface area that houses enzymes involved in cellular respiration. The complex membranes of the thylakoid in chloroplasts provide a large area to contain chlorophyll and other pigments involved in photosynthesis. **4.8:** 1. Microtubules help maintain cellular shape and provide an internal road system used to transport intracellular materials. Intermediate filaments anchor cells to each other and components in the external environment. Actin filaments are used in some forms of cellular movement. 2. Cilia and flagella are both composed of microtubules arranged in a particular pattern and enclosed by the plasma membrane. Cilia are shorter than flagella. ATP is used to produce cellular movement via microtubules. 3. Cellular movement from microtubules is accomplished via motor molecules that use ATP.

Assess

1. c; **2.** d; **3.** d; **4.** a; **5.** a; **6.** c; **7.** e; **8.** a; **9.** c; **10.** a; **11.** c; **12.** c; **13.** b; **14.** d; **15.** c; **16.** a; **17.** c; **18.** c; **19.** d; **20.** e

Chapter 5

Check Your Progress

5.1: 1. Phospholipids, because they have both a hydrophobic and a hydrophilic end, form a bilayer that can be used by the cell to create compartments. The cell is separated from the external environment and compartments are created within the cell which allows many different chemical processes to function at the same time. 2. Proteins are embedded or associated with membranes, creating a mosaic. These proteins influence membrane function. Membranes are considered a fluid mosaic because the phospholipids and membrane proteins can freely move within the membrane. 3. Membrane pores are hydrophobic due to the fatty acid tails of phospholipids. Hydrophobic substances can freely move across the membrane. In contrast, hydrophilic substances require transport mediated by membrane proteins or other mechanisms. **5.2:** 1. Diffusion is the movement of molecules from an area of high concentration to a low concentration along a concentration gradient. Osmosis is specific to the movement of water along a concentration gradient across a semipermeable membrane. Both involve diffusion which requires no energy. 2. Water inside a cell in a hypertonic solution would move from the higher concentration within the cell to the lower concentration outside of the cell. The cell would shrink. A cell in a hypotonic solution would swell up as water moved from the higher concentration outside the cell to the lower concentration within the cell. 3. Both move molecules from high to low concentrations. Diffusion does not require a membrane or transport proteins, while facilitated transport does. **5.3:** 1. Both require a carrier protein, but active transport also requires energy. 2. Active transport requires energy because molecules are moved against a concentration gradient. 3. Bulk transport is used to transport large molecules across plasma membranes. The molecules are contained in vesicles and do not mix with the surroundings. **5.4:** 1. The extracellular matrix is a network of polysaccharides and proteins that have many different functions. 2. Tight junctions between cells connect the plasma membranes to form a rigid barrier that prevents the passage of molecules. Adhesion junctions mechanically connect two cells in a flexible attachment. Gap junctions join similar plasma membrane channels between two cells which allow small molecules and ions to pass from one cell to another. 3. Plamodesmata allow water and small solutes to pass from one cell to another in a plant.

Assess

1. b; **2.** e; **3.** d; **4.** c; **5.** c; **6.** c; **7.** b; **8.** e; **9.** d; **10.** c; **11.** e; **12.** c

Chapter 6

Check Your Progress

6.1: 1. The potential energy stored in a breakfast meal is converted into the kinetic energy of the morning run. 2. Cells can convert one form of energy to another according to the first law of thermodynamics. However with each change of one form of energy to another, some energy is lost, usually as heat. 3. By the second law of thermodynamics each energy transformation in a cell results in more entropy. Less energy is available to do work. As a result all organisms depend on the constant supply of energy from the sun. **6.2:** 1. ATP holds energy but gives it up because it is unstable and the last phosphate group can be easily removed, releasing energy. 2. In a cell ATP is constantly regenerated by combining ADP and inorganic phosphate using the energy from the breakdown of glucose and other organic molecules. 3. ATP can donate a phosphate to energize a compound for a reaction. The addition of a high-energy negatively charged phosphate can change the molecule's shape and in that way bring about a structural change that in turn changes its function. **6.3:** 1. Enzymes bring reactants together at their active site or they position a substrate so it is ready to react. Because they interact very specifically with their reactants, the products produced from enzymes can be regulated as is seen in metabolic pathways. 2. Enzymes have an active site where the substrates fit specifically. This reduces the energy required for the reaction to occur. 3. Environmental factors that affect the speed of enzymatic reactions include substrate concentration, pH, temperature, and the availability of cofactors and inhibitors. **6.4:** 1. During photosynthesis carbon dioxide from the air is fixed to create carbohydrate molecules used for energy and plant structure. During respiration carbohydrates are broken down to derive energy for life processes, and carbon dioxide is produced as a by-product. There is a cyclical relationship between photosynthesis and cellular respiration. 2. Energized electrons are used to create ATP via proton gradients across membranes in plants and animals.

Assess

1. c; **2.** e; **3.** b; **4.** d; **5.** b; **6.** c; **7.** d; **8.** d; **9.** b; **10.** a; **11.** d; **12.** c

Chapter 7

Check Your Progress

7.1: 1. Plants, algae, and cyanobacteria. 2. Thylakoid membrane absorbs solar energy, and carbohydrate forms in stroma. **7.2:** 1. In photosynthesis carbon dioxide is reduced to form carbohydrates and water is oxidized to form oxygen. 2. Enzymes are used to regulate chemical reactions during photosynthesis. **7.3:** 1. Visible light is a small part of the overall electromagnetic spectrum. It ranges from purple to red light. 2. Low energy electrons from water within chlorophyll in Photosystem II are excited to a higher energy state by sunlight. The high potential energy of these excited electrons is siphoned off during redox reactions in the electron transport chain, creating proton gradients across the thylakoid membrane. These electrons are excited again to a higher energy state by Photosystem I, which harnesses the energy to create NADPH. **7.4:** 1. Step 1: A molecule of carbon dioxide is attached to RuBP which then splits to form 2 molecules of 3PG. Step 2: Each molecule of 3PG is reduced to form G3P. Step 3: G3P molecules are converted to RuBP. 2. For every 3 turns of the Calvin cycle 5 molecules of G3P are used, to re-form 3 molecules of RuBP. **7.5:** 1. C_4 plants include many grasses, sugarcane, and corn; CAM plants include cacti, stonecrops, orchids, and bromeliads. 2. C_4 plants reduce photorespiration by preventing oxygen from competing with carbon dioxide for an active site on the enzyme rubisco.

Assess

1. d; **2.** a; **3.** c; **4.** e; **5.** d; **6.** a; **7.** e; **8.** T; **9.** F; **10.** F; **11.** e; **12.** c; **13.** e

Chapter 8

Check Your Progress

8.1: 1. In respiration, hydrogen atoms are removed from glucose along with electrons so glucose is oxidized to form carbon dioxide. Also, hydrogen atoms along with electrons are added to oxygen, reducing it to form water. 2. NAD^+ and FAD are coenzymes that can accept high energy electrons and a hydrogen ion(s) and transport them to the electron transport chain. There the electrons are released to produce ATP. 3. Phase 1: Glucose is broken down to 2 molecules of pyruvate, producing NADH. Phase 2: Pyruvate is broken down to an acetyl group and one molecule of CO_2. Phase 3: The acetyl group enters the citric acid cycle and NADH and $FADH_2$ are formed. Phase 4: NADH and $FADH_2$ give up their electrons in the electron transport chain and ATP is produced. **8.2:** 1. During the energy investment steps, ATP breakdown provides the phosphate groups to activate substrates. During the energy-harvesting steps, NADH and ATP are produced. 2. ADP and phosphate from the cytoplasm are connected together during substrate-level ATP synthesis via enzymes from glycolysis. 3. Two ATPs are used to add a phosphate to glucose in the energy-investment step. In the energy harvesting steps NAD is reduced, ATP is produced, and water generated. **8.3:** 1. During fermentation the NADH produced in glycolysis is oxidized to NAD^+ as pyruvate is reduced to lactate. 2. Fermentation in animals and bacteria results in the formation of lactate from pyruvate. In plants or yeast, alcohol is formed. 3. Advantages: production of certain foods, alcoholic beverages, and industrial chemicals; source of rapid bursts of cellular energy in the absence of oxygen. Disadvantage: low yield of ATP. **8.4:** 1. For each glucose molecule, the first 2 CO_2 molecules are produced during the preparatory reaction, and the remaining 4 CO_2 molecules are produced during the citric acid cycle. 2. The electron transport chain produces the most ATP during glycolysis breakdown. 3. A dam holds back water, just as the inner membrane holds back hydrogen ions. As water flows over a dam, electricity is produced. As hydrogen ions flow down their concentration gradient through an ATP synthase complex, ATP is produced. **8.5:** 1. Catabolic and anabolic processes are regulated by enzymes, which allow a cell to respond to environmental changes. 2. Both have an inner membrane (thylakoids in chloroplasts; cristae in mitochondria) where complexes form an ETC, and ATP is produced by chemiosmosis. Both have a fluid-filled interior. In the stroma of chloroplasts, NADPH helps reduce CO_2 to a carbohydrate and in the matrix of mitochondria, NAD^+ helps oxidize glucose products with the release of CO_2.

Assess

1. c; **2.** d; **3.** c; **4.** c; **5.** d; **6.** d; **7.** c; **8.** b; **9.** c; **10.** a; **11.** b; **12.** b; **13.** b; **14.** c

Chapter 9

Check Your Progress

9.1: 1. G_1, S, G_2, and M stage. The cell increases in size and grows organelles in G_1. DNA is replicated during S stage. Cell division occurs during M stage. During G_2 proteins are synthesized that help in cell division. 2. DNA damage stops cell division at G_1 checkpoint. G_2 checkpoint stops cell division if there has been a failure to properly replicate the DNA. If there has been a failure of chromosomes to attach properly to the spindle, cell division will be stopped at the M checkpoint. 3. Apoptosis controls the overall density of cells by targeting excess cells, or cells with damage, for cell death. **9.2:** 1. Euchromatin is the loosely coiled chromatin that contains genes that are being actively transcribed. Heterochromatin is a part of the chromosome that is highly compacted and stains darkly. 2. A single DNA strand wraps around histone proteins. Three-dimensional zigzag structures with binding proteins are formed. Radial loops are formed as coiling occurs. The radial loops compact to form heterochromatin. A metaphase chromosome forms. **9.3:** 1. Prophase: The nuclear membrane fragments, the nucleus disappears, the chromosomes condense, and the spindle begins to form. Metaphase: The chromosomes align along a central line in the cell. Anaphase: The sister chromatids separate and head to opposite poles. Telophase: The nuclear envelope re-forms, chromosomes de-condense; the spindle disassociates. 2. Animal cells furrow and plant cells develop a cell plate. Because plant cells have rigid cell walls, they cannot furrow. 3. Because the number of times most cells can divide is controlled, stem cells provide source material for tissues that need to continuously replace cells. **9.4**: 1. Cancer cells lack differentiation, have abnormal nuclei, fail to undergo apoptosis, may form tumors that may undergo angiogenesis, and may also metastasize throughout the body. 2. Malignant tumor cells enter circulation and lodge in other tissues of the body. Benign tumors remain in the tissue in which they form. 3. a. Cell commits to cell division even in the absence of proper stimuli. b. Cell fails to stop dividing because the proper stimuli to stop are absent. **9.5:** 1. Binary fission involves inward growth of plasma membrane and cell wall concomitant with the separation of the duplicated chromosome attached to the plasma membrane. Mitosis always involves a mitotic spindle to distribute the daughter chromosomes. 2. Prokaryotes usually have a single, small circular chromosome with a few genes and only a few associated proteins. Eukaryotes have many long, linear chromosomes, which contain many thousands of genes and many more proteins.

Assess

1. b; **2.** d; **3.** a; **4.** c; **5.** d; **6.** e; **7.** d; **8.** c; **9.** b; **10.** c; **11.** e; **12.** c; **13.** c; **14.** b; **15.** c

Chapter 10

Check Your Progress

10.1: 1. Homologous chromosomes represent two copies of the same kind of chromosome, each containing the same genes for traits in the same order. They are also similar in length and location of the centromere. 2. Homologous chromosomes replicate prior to meiosis I, producing sister chromatids. When they separate each daughter cell receives one pair of sister chromatids, resulting in a reduction from 2n to n chromosomes. Meiosis II splits the sister chromatids, producing n chromosomes in each daughter cell. 3. A bivalent is formed between homologous chromosomes to enable crossing-over and strand exchange. This increases the potential genetic variability of the offspring. **10.2:** 1. Independent assortment of chromosomes increases the number of possible combinations of chromosomes in each gamete. Crossing-over shuffles the alleles between homologous chromosomes to create even more variation. 2. There are 2^4 or 16 possible combinations. 3. Genetic variability ensures that at least some individuals have traits that may allow a species to survive new adverse conditions. **10.3:** 1. In metaphase I homologous chromosome pairs align at the metaphase plate. In metaphase II single chromosomes align at the metaphase plate. 2. Two daughter cells that share the same parent cell from meiosis I are identical unless crossing-over has occurred. 3. Pairs of homologous chromosomes would not split, resulting in cells with incorrect chromosome combinations. Without proper reduction in chromosome number, offspring would have incorrect chromosome combinations and would likely not survive. **10.4:** 1. In metaphase I of meiosis, homologous chromosomes are paired at the metaphase plate with each homologue facing opposite spindle poles. In metaphase II and mitotic metaphase, homologous chromosomes are not paired, and sister chromatids are attached to spindle fibers from opposite spindle poles. 2. Meiosis II resembles mitosis because sister chromatids are separated during both processes. Meiosis II differs from mitosis because the cells are haploid and not diploid. **10.5:** 1. In males, the primary spermatocytes are located within the testes. In females, the primary oocytes are located within the ovaries. 2. Oogenesis produces a single cell, containing the bulk of the cytoplasm and other cellular contents that will undergo embryonic development. Spermatogenesis produces four haploid cells, each of which could potentially merge with the haploid egg to create a diploid organism. **10.6:** 1. Nondisjunction in meiosis may cause aneuploidy, an extra or missing chromosome. 2. Sex chromosome aneuploidy is more common because only one of the X chromosomes is active. Any extra X chromosomes become Barr bodies. 3. An inversion involves the reversal of a piece of a chromosome from within, and normally does not cause symptoms. A translocation is swapping of two chromosome fragments from one to the other, and while not usually troublesome, may cause severe problems in offspring if the two chromosomes go into separate cells.

Assess

1. d; **2.** b; **3.** a; **4.** d; **5.** a; **6.** e; **7.** b; 8. c; **9.** d; **10.** b; **11.** a; **12.** d; **13.** a; **14.** d; **15.** d; **16.** d; **17.** d; **18.** c

Chapter 11

Check Your Progress

11.1: 1. The blending theory of inheritance assumed that the characteristics of the parents were blended together in the offspring to produce an intermediate appearance. With Mendel's particulate theory of inheritance hereditary units of the parents were shuffled to produce new combinations in the offspring. 2. The garden pea has easily observed traits, a relatively short generation time, each plant produces many offspring (peas), and cross pollination is only possible by hand. **11.2:** 1. The law of independent assortment states that during meiosis pairs of factors segregate independently and that all combinations are possible in the offspring. 2. Both the *TT* and *Tt* individual express fully functional proteins for the trait. 3. The *Aabb* individual occurs when one parent contributes an *Ab* gamete (probability = $\frac{1}{4}$) and another parent contributes an *ab* gamete (probability = $\frac{1}{4}$). For both gametes to combine has a probability of $\frac{1}{4} \times \frac{1}{4} = \frac{1}{16}$. Since there are two possibilities of how these combine the probability of an *Aabb* individual is $\frac{1}{16} + \frac{1}{16} = \frac{1}{8}$. **11.3** 1. Autosomal recessive children have unaffected parents, heterozygotes have normal phenotype, and two affected parents always have affected children. Autosomal dominant affected children have an affected parent, the heterozygotes are affected, two affected parents can produce an unaffected child, and two unaffected parents will not have affected children. 2. Ivar's father carried the dominant allele and passed it to Ivar who had to have been homozygous recessive. Ivar's wife was homozygous recessive because both parents were normal.

11.4 1. When the individual exhibiting incomplete dominance of a trait is crossed with itself the offspring show a phenotypic ratio of 1:2:1. 2. X-linked traits have a different pattern of inheritance than alleles that are autosomal because the Y chromosome lacks the allele for the trait. 3. Multifactorial traits are controlled by polygenes that are subject to environmental influences.

Assess

1. b; **2.** d; **3.** d; **4.** a; **5.** b; **6.** b; **7.** a; **8.** b; **9.** c; **10.** c; **11.** b; **12.** a; **13.** d; **14.** b

Chapter 12

Check Your Progress

12.1: 1. DNA is a right-handed double helix with two strands that run in opposite directions. The backbone is composed of alternating sugar phosphate groups, and the molecule is held together in the center by hydrogen bonds between interacting bases. A always hydrogen-bonds to T, and G to C. 2. Chargoff determined that the amount of adenine in a cell equals the amount of thymine and guanine equals cytosine. The amount of the four bases varies with each species. This indicated that adenine bonds to thymine and guanine binds to cytosine. Franklin used X-ray diffraction to determine that DNA was a helix with some feature repeated many times. **12.2:** 1. (1) The DNA strands are separated by DNA helicase (2) new nucleotides are positioned by complementary base pairing (3) new nucleotides are joined together by DNA polymerase to form a new DNA strand. 2. The strands of DNA are antiparallel and synthesis can only proceed in a 5′ to 3′ direction. Synthesis on the lagging strand occurs in short 5′ to 3′ segments that must be joined. 3. Prokaryotic DNA replication begins at a single origin of replication and usually proceeds in both directions toward a termination region on the opposite side of the chromosome. Eukaryotic DNA replication begins at multiple origins of replication and continues until the replication forks meet. **12.3:** 1. Information from a gene coded within the DNA is copied into mRNA. mRNA sequence dictates the sequence of amino acids in a protein, which produces the observable traits for an organism. The flow of genetic information is DNA to RNA to protein to trait. 2. mRNA carries information from DNA to direct the synthesis of a protein. rRNA makes up part of the ribosomes that are used to translate messenger RNAs. tRNA transfers amino acids to the ribosome during protein synthesis. 3. Several different codons may specify the same amino acid. **12.4:** 1. When mRNA is synthesized, RNA polymerase joins the nucleotides together in a 5′ to 3′ direction. 2. During mRNA processing, introns are spliced out and the exons joined together, and a 5′ guanosine cap and a 3′ poly-A tail are added. 3. A pre-mRNA transcript contains noncoding sequences called introns that are spliced out, leaving coding sequences, called exons, in the mature mRNA. 4. The ability to create multiple mRNA molecules from a single gene through alternative splicing enables a cell to create a wider variety of proteins whose diversity may positively affect organism survival. **12.5:** 1. Transfer RNA delivers amino acids to the ribosome by binding to the appropriate codon on the mRNA being translated. 2. A ribosome consists of a small and large subunit. Each subunit is composed of a mixture of protein and rRNA. Large and small ribosomal subunits work to connect amino acids together in a sequence dictated by the mRNA. 3. Initiation: All components of the translational complex, including the first tRNA carrying methionine, are assembled. Elongation: Amino acids are delivered one by one as tRNA molecules pair with the codons on the mRNA. Termination: A stop codon is reached, a release factor binds to it, and the completed protein is cleaved from the last tRNA as the ribosomal subunits dissociate.

Assess

1. d; **2.** a; **3.** e; **4.** b; **5.** c; **6.** a; **7.** a; **8.** d; **9.** a; **10.** b; **11.** c; **12. a.** GGA GGA CUU ACG UUU; **b.** CCU CCU GAA UGC AAA; **c.** glycine-glycine-leucine-threonine-phenylalanine

Chapter 13

Check Your Progress

13.1: 1. An operon is a group of genes that are regulated in a coordinated manner. A promotor is a regulatory region that controls the initiation of transcription. 2. In a repressible operon, the operon is normally on and is turned off by the action of a repressor. In an inducible operon, the operon is normally off and is turned on by an environmental condition. 3. A gene under positive control is transcribed when it is regulated by a protein that is an activator and not a repressor, whereas one under negative control is not transcribed when it is regulated by a protein that is a repressor. 4. The *trp* operon is anabolic because it is involved with the synthesis of tryptophan and the *lac* operon is catabolic because its gene products are involved with the metabolism of lactose. **13.2:** 1. Chromatin, transcriptional, posttranscriptional, translational, and posttranslational 2. Packing genes into heterochromatin inactivates a gene, whereas it is genetically active when held in loosely packed euchromatin. 3. siRNA molecules participate in posttranscriptional regulation while proteosomes are active in posttranslational regulation. **13.3:** 1. Spontaneous: errors in DNA replication and natural chemical changes in the bases in DNA; Induced: organic chemicals and physical mutagens like X-rays and UV radiation. 2. A frameshift mutation affects all the codons downstream from the mutation, resulting in a completely altered sequence that yields nonfunctional proteins. 3. Mutations in these genes disrupt the normal operation of the checkpoints in the cell cycle, allowing cell division to proceed without regulation.

Assess

1. e; **2.** a; **3.** b; **4. a.** DNA; **b.** regulator gene; **c.** promoter; **d.** operator; **e.** active repressor; **5.** e; **6.** b; **7.** b; **8.** b; **9.** d; **10.** b; **11.** e; **12.** e

Chapter 14

Check Your Progress

14.1: 1. Restriction enzymes are used to cut the foreign DNA and the plasmid in identical ways such that the ends of the DNA are sticky and can bind to each other, thus allowing the insertion of the foreign DNA into the plasmid. 2. With recombinant DNA technology a piece of DNA of interest is inserted into a plasmid, which is then introduced into a bacteria that makes many copies of the plasmid. PCR makes many copies of the DNA of interest without having to create a vector. 3. DNA molecules amplified by PCR can be used to create DNA profiles, or fingerprints, which can be analyzed to identify the source of the DNA sample. **14.2:** 1. Bacteria: production of medicine and drugs, bioengineering, production of organic chemicals; Animals: gene pharming, studying human disease; Plants: increased food production, production of human medicines. 2. A transgenic animal contains recombinant DNA molecules in addition to its genome, whereas a cloned animal is genetically identical to the one from which it was created and does not contain rDNA. **14.3:** 1. Liposomes, nasal sprays, and adenoviruses are currently being used to deliver genes to cells for gene therapy. 2. Ex vivo gene therapy is being used to treat SCID and familial hypercholesterolemia. In vivo gene therapy is being used to treat cystic fibrosis. **14.4:** 1. The genome is the sum of all of the genes in a cell whereas the proteome represents the sum of all of the proteins encoded by the genome. The proteome is typically bigger than the genome. 2. A tandem repeat consists of repeated sequences that are one next to the other, whereas transposons are short sequences of DNA that may move within the genome. 3. Microarrays allow scientists to analyze the expression of genes under different environmental conditions. Bioinformatics allows scientists to rapidly analyze experimental data and generate large databases of information.

Assess

1. c; **2.** c; **3.** c; **4.** d; **5.** d; **6.** d; **7.** b; **8.** d; **9.** e; **10.** e; **11.** c; **12.** e; **13.** e; **14.** e

Chapter 15

Check Your Progress

15.1: 1. Catastrophism, put forth by Cuvier, proposed that changes in the types of fossils in strata are explained by local mass extinctions followed by new species repopulating the area. 2. Lamarck's inheritance of acquired characteristics is flawed as an explanation of biological diversity because features acquired by an individual during their lifetime are not heritable to offspring. Evolution occurs only in traits that are heritable. 3. 1707-1778 Linnaeus (classification); 1731-1802; Darwin (first theory of evolution); 1769-1832 Cuvier (catastrophism); 1744-1829 Lamarck (acquired characteristics); 1797-1875 Lyell (uniformitarianism); 1831-1836 voyage of H.M.S *Beagle;* 1859 Darwin publishes *Origin of Species*. **15.2:** 1. Fossils, biogeography, change over time (geology). 2. Individuals have variation that is heritable; organisms compete for resources; differential reproductive success; adaptation to environmental change. 3. Dogs, *Brassica* and wheat have all evolved through artificial selection. Finches, when separated into different islands with different food sources, evolved different beaks. Changes in the DNA code regulate genes producing new phenotypes that may have increased reproductive success. **15.3:** 1. Transitional fossils represent intermediate evolutionary forms of life in transition from one type to another, or a common ancestor of these types. *Tiktaalik* and *Basilosaurus* are examples of intermediate fossils. 2. Homologous structures are those that are shared because of common ancestry. Analogous structures are shared because of convergent evolution and do not inform us about common ancestry. 3. Biomolecules may change over time as a result of natural selection. Examples are the cytochrome *c* protein in cellular metabolism and the homeobox, or *Hox,* genes that have been shaped by natural selection to orchestrate the development of the body plan in animals. 4. Evolution proposes that life-forms changed as a result of random events. New variation arises due to random processes, but natural selection is not a random process; it can only shape preexisting variation.

Assess

1. d; **2.** e; **3.** e; **4.** b; **5.** a; **6.** e; **7.** e; **8.** a; **9.** e; **10.** a; **11.** b, c, d, e; **12.** b, d, e; **13.** c; **14.** e, e; **15.** b

Chapter 16

Check Your Progress

16.1: 1. a. Random mating - allele frequencies do not change. b. No selection - allele frequencies change. c. No mutation - allele frequencies change. d. No migration - allele frequencies change. e. Large population (no genetic drift) - allele frequencies change. 2. $AA = p^2 = 0.01$, $Aa = 2pq = 0.18$, $aa = q^2 = 0.81$. **16.2:** 1. Stabilizing selection – intermediate phenotype increases in frequency. Directional selection – an extreme phenotype is favored and the average phenotype value is changed. Disruptive selection – two extreme phenotypes are favored and two new average phenotype values result. 2. See Figure 16.10 on page 287. 3. Sexual selection is a form of natural selection because traits associated with mating determine whether or not traits are passed on to the next generation. **16.3:** 1. Mutation, migration, sexual reproduction, genetic drift. 2. In Africa where malaria is common, heterozygotes for the sickle-cell trait have an advantage because it provides some resistance to malaria. Those who are homozygous for normal trait lack natural resistance; those homozygous for the sickle-cell trait die young from sickle-cell disease. Heterozygotes for the cystic fibrosis trait have resistance to typhoid fever. As with sickle-cell trait, heterozygotes are maintained because of stabilizing selection that selects against homozygotes for the cystic fibrosis trait (who have the disease) and homozygous normal trait (who get typhoid fever).

Assess

1. c; **2.** c; **3.** c; **4.** b; **5.** e; **6.** e; **7.** b; **8.** c; **9.** e; **10.** a; **11.** c; **12.** c; **13.** b

Chapter 17

Check Your Progress

17.1: 1. Microevolution results from genetic drift, natural selection, mutation, and migration. Macroevolution involves the same processes but on a larger scale. It results from accumulation of microevolution changes that split one species into two. 2. (Three of the following) Biological species concept: Individuals of a population cannot interbreed. Phylogenetic species concept: Two populations are on different evolutionary lineages in a phylogeny that are monophyletic. Morphological species concept: A species is defined by a diagnostic trait (or traits). Evolutionary species concept: Relies on identification of certain morphological diagnostic traits that define a different evolutionary pathway. 3. If the frogs cannot recognize each other as mates, they will be reproductively isolated. **17.2:** 1. Allopatric speciation. 2. Show that each of the cats is adapted to a different environment. 3. In a newly discovered African Rift Valley lake, you would expect to see in the fishes the same shape of head and teeth for feeding in similar habitats as those observed in other lakes. Similar forms would evolve independently in this new lake—each of which would match the form of a fish that feeds in the same habitat in a different lake. **17.3:** 1. Punctuated equilibrium model proposes periods of rapid evolution and periods of little evolution (stasis). Cuvier proposed that this pattern in the fossil record was because new groups of organisms moved into an area after a mass extinction (catastrophe). 2. *Hox* genes have a powerful effect on development and offer a mechanism by which evolution could occur rapidly. 3. *Pax6* – development of eyes. *Tbx5* – development of limbs. *Hox* – overall shape. *Pitx1* – development of pelvic-fin in stickleback fish.

Assess

1. e; **2.** c; **3.** e; **4.** c; **5.** e; **6. a.** species 1; **b.** geographic barrier; **c.** genetic changes; **d.** species 2; **e.** genetic changes; **f.** species 3; **7.** b; **8.** b; **9.** b; **10.** a; **11.** e; **12.** a; **13.** b; **14.** c; **15.** e

Chapter 18

Check Your Progress

18.1: 1. Stage 1 – organic monomers evolved from inorganic compounds. Stage 2 – organic monomers are joined to form polymers. Stage 3 – protocells (protobionts) are formed as organic polymers are enclosed in a membrane. Stage 4 – Probionts aquire the ability to replicate and conduct cellular processes. 2. a. Iron-sulfur world hypothesis – iron-nickel sulfides at thermal vents act as catalysts to join organic molecules. b. Protein-first hypothesis – amino acids collected in small pools and the heat of the sun formed proteinoids, small polypeptides with catalytic properties. c. RNA-first hypothesis – RNA carried out the processes commonly associated with DNA and proteins before either of these evolved. 3. Protocell – single fatty acid hydrophilic tail and membrane of a single layer of fatty acids. Modern cell – double fatty acid hydrophilic tail and a bilayer of phospholipids. **18.2:** 1. Two half-lives of C^{14} = 2(5730 years) = 11,460 years. 2. DNA content and double-membrane structure of the mitochondria and chloroplasts 3. Chemical evolution, evolution of first cells, evolution of eukaryotic cells by endosymbiosis, and first heterotrophic protists before photosynthetic protists. 4. See Table 18.1 page 326. **18.3:** 1. The theory of plate tectonics says that the Earth's crust is fragmented into slab-like plates that float on a lower hot mantle layer. Growth of mountains, earthquakes, and the layout of the continents are all evidence. 2. Distribution of animals would be similar on continents that were once joined together. Marsupials in South America and Australia are an example.

3. Ordovician – continental drift and cooling of the Earth. Devonian – impact of a meteorite with the Earth. Permian – excess carbon dioxide. Triassic – meteorite collision with Earth. Cretaceous – impact of an asteroid with Earth.

Assess

1. c; **2.** d; **3.** a; **4.** b; **5.** b; **6.** b; **7.** a; **8.** b; **9.** c; **10.** d; **11.** a; **12.** c; **13.** e; **14.** a; **15.** e

Chapter 19

Check Your Progress

19.1: 1. Classification is the naming of organisms; taxonomy is the assignment of groups to a taxon; systematic biology is the study of classification and taxonomy, or more generally, the study of biodiversity. 2. Domain: Eukarya; Kingdom: Animalia; Phylum: Chordata; Class: Mammalia; Order: Primates; Family: Hominidae; Genus: *Homo;* Species: *Homo sapiens*. 3. Wings in birds and bats are analogous because they have independent evolutionary origins. Thus, birds and bats are not a natural group. **19.2:** 1. Archaea have differences in their rRNA, their cell wall structure, and the lipids in their plasma membrane. 2. Molecular data (RNA/DNA sequence data) indicate that fungi are related to animals. 3. Eukarya. **19.3:** 1. Ancestral traits are found in the common ancestor of a lineage and may be present in the descendants. Derived traits are found in the descendants but not in the common ancestor. 2. A phylogeny that requires the fewest number of evolutionary steps is the most parsimonious explanation of evolutionary history, and is thus the best hypothesis. 3. Traits include fossil, morphological, behavioral, and molecular.

Assess

1. e; **2.** e; **3.** d; **4.** a, b, c; **5.** b, c, d, e; **6.** a; **7.** c; **8.** d; **9.** a; **10.** d; **11.** e

Chapter 20

Check Your Progress

20.1: 1. All viruses have a nucleic acid (which can be either RNA or DNA and double stranded or single stranded) and a capsid. All have enzymes and some have an outer envelope. They are the size of a macromolecule. 2. Viruses are host specific because the molecules of the capsid or spikes on an enveloped virus specifically bind to receptors on the host cell. 3. A virus in a lytic cycle reproduces inside a cell many times and then lyses the cell, releasing the virus particles. In the lysogenic cycle, the viral DNA is inserted into the cell's DNA and is passed on as the cell divides. At some future point that cell can enter the lytic cycle. 4. If a virus's host survives, many more copies of the virus will be produced and spread to other hosts than if the host dies. **20.2:** 1. Bacteria exist in the air and sterilization kills bacteria. 2. The bacterial nucleoid is not surrounded by a membrane; the eukaryote nucleus has its own membrane 3. Transduction: movement of genetic material from one bacterium to another via a virus; transformation: pick up and incorporate genetic material from the environment; conjugation: exchange genetic material between bacteria. **20.3:** 1. The peptidoglycan layer is much thicker in Gram-positive cells. In Gram-negative cells this layer is thinner and is located between layers of the plasma membrane. 2. Autotrophs use the energy from the sun or from inorganic compounds to produce organic compounds. Heterotrophs obtain carbon and energy from organic nutrients produced by other living organisms. 3. Cyanobacteria are considered responsible for the production of an oxygen-rich atmosphere. **20.4:** 1. Archaea and bacteria differ in rRNA base sequences. The plasma membrane of archaea can survive elevated temperature because they contain different lipids. The cell walls of archaea do not contain peptidoglycans like bacteria. 2. Methanogens, halophiles, thermoacidophiles. 3. Archaea and eukaryotes share some of the same ribosomal proteins, initiate transcription in the same way, and have similar tRNA.

Assess

1. b; **2.** c; **3.** b; **4.** c; **5.** c; **6.** c; **7.** b; **8.** b; **9.** d; **10.** c; **11.** c; **12.** d; **13.** d; **14.** e; **15.** a

Chapter 21

Check Your Progress

21.1: 1. Endosymbiosis: Mitochondria were free-living bacteria and chloroplasts were derived from cyanobacteria that were engulfed and incorporated into a eukaryotic protocell. 2. Algae are photoautotrophs; they get their food from photosynthesis; protozoans are heterotrophs and ingest their food from the environment. 3. a. Archaeplastida; b. Opisthokonta. **21.2:** 1. Green algae contain chlorophyll *a* and *b,* have a cell wall that contains cellulose, store food as starch, can be symbiotic with plants, animals, and fungi, and can be single-celled, filamentous, or colonial. 2. Archaeplastids reproduce through an alteration-of-generation life cycle. **21.3:** 1. Alveolates are all single celled and have small sacs (alveoli) below their plasma membrane that are used for support and membrane transport. Stamenopiles do not have alveoli and can exist in colonial forms. 2. Brown algae are used as human food, fertilizer, and a source of algin. Diatoms produce diatomaceous earth which is used in filters, construction, and as polishing agents. 3. See Figure 21.15 on page 384. **21.4:** 1. Excavates have atypical or absent mitochondria and distinctive flagella and/or deep oral grooves. 2. Trypanosome: African sleeping sickness and Chagas disease; *Giardia:* giardiasiss. **21.5:** 1. Amoebozoans use pseudopods to move and feed. 2. Animals and fungi are classified as opisthokonts. 3. Radiolarian tests in layers of sediment are used as an indicator of oil deposits and as a way to date sedimentary rock.

Assess

1. d; **2.** b; **3.** c; **4.** a; **5.** a; **6.** b; **7.** b; **8.** c; **9.** b; **10.** c; **11.** e; **12.** a; **13.** b; **14.** d; **15.** a

Chapter 22

Check Your Progress

22.1: 1. Animals are heterotrophs: They get their nutrition by consuming outside nutrients. Fungi are saprotrophs: They externally digest organic material and absorb nutrients. 2. Plant cell walls are made of cellulose; fungal cell walls are made of chitin. 3. Reproductive structure that can form a new organism without fusing with another reproductive cell. **22.2:** 1. Microsporidia are obligate intracellular animal parasites that lack mitochondria. They have the smallest known eukaryotic genome. Chytrids are aquatic with flagellated gametes and spores. They are the simplest fungi known and appear as single cells or as branched aseptate hyphae. 2. Fruiting bodies raise spores upward, exposing them to wind, insects, and rain, each of which can increase the dispersal of the spores. See Table 22.1 on page 398 for a description of the spores present in each phylum. 3. Athletes' foot: *Trichophyton* (tinea); candidiasis: *Candida;* fungal flu: *Histoplasma*. **22.3:** 1. Lichens have a sac fungus and either a cyanobacterium or green alga. 2. Lichens reproduce asexually by releasing fragments that contain hyphae and an algal cell. In fruticose lichens, the sac fungus reproduces sexually. 3. Fungus enters the cortex of the roots of plants, giving the plants a greater absorptive surface for nutrients.

Assess

1. b; **2.** e; **3.** b; **4.** c; **5.** c; **6. a.** spores; **b.** sporangium; **c.** sporangiophore; **d.** stolon; **e.** rhizoids; **7.** b; **8.** b; **9.** b; **10.** c; **11.** a; **12.** e

Chapter 23

Check Your Progress

23.1: 1. A cellulose cell wall produced in same way; apical cells that produce new tissue; plasmodesmata between cells; transfer of nutrients from haploid cells of previous generation to zygote of new generation. 2. Development of a cuticle to reduce water loss; tracheids to transport water and minerals upward; three-dimensional tissues; diploid genome. 3. The diploid sporophyte produces haploid spores by meiosis. The haploid gametophyte produces gametes. **23.2:** 1. Bryophytes have a dominant gametophyte generation, produce flagellated sperm that swim to the egg, and can also reproduce asexually. 2. Asexual and sexual reproduction, ability to survive in harsh environments, sporophyte is protected from drying out. 3. The gametophyte portion is most dominant. The sporophyte is dependent upon the gametophyte. **23.3:** 1. In lycophytes, the dominant sporophyte has vascular tissue, and therefore roots, stems, and leaves. 2. The walls of xylem contain lignin, a strengthening agent. 3. Homosporous—having spores that grow into one type of gamete; heterosporous—having microspores that grow into a male gametophyte and megaspores that grow into a female gametophyte. **23.4:** 1. In ferns, but not mosses, the sporophyte is dominant and separate from the gametophyte. 2. Megaphylls have broad leaves with several branches of vascular tissue. 3. See Figure 23.16 on page 422. **23.5:** 1. (1) Water is not required for fertilization because pollen grains (male gametophytes) are windblown, and (2) ovules protect female gametophytes and become seeds that disperse the sporophyte, the generation that has vascular tissue. 2. Conifers: cone bearing and evergreen; cycads: cone bearing, evergreen, and wind pollinated; ginkgoes: cone bearing, deciduous; gnetophytes: cone bearing and insect pollinated. 3. The stamen contains the anther and the filament. Pollen forms in the pollen sac of the anther. The carpel contains the stigma, style, and ovary. An ovule in the ovary becomes a seed, and the ovary becomes the fruit.

Assess

1. e; **2. a.** sporophyte; **b.** meiosis; **c.** gametophyte; **d.** fertilization; **3.** c; **4.** a; **5.** b; **6.** e; **7.** e; **8.** b; **9.** e; **10.** e; **11.** b; **12.** b

Chapter 24

Check Your Progress

24.1: 1. Protoderm, ground, and procambium meristem 2. Epidermal tissue made up of epidermal cells; ground tissue made up of parenchyma, collenchyma, and sclerenchyma cells; vascular tissue made up of tracheids and vessel elements of the xylem and sieve-tube members and companion cells of the phloem. 3. Xylem transports water and minerals throughout the plant. Phloem transports sucrose and other organic molecules usually from the leaves to the roots. **24.2:** 1. See Figure 24.6 on page 440. 2. Vegetative organs are the leaves (photosynthesis), the stem (support, new growth, transport), and the root (absorb water and minerals). 3. Monocots: embryo with single cotyledon; xylem and phloem in a ring in the root; scattered vascular bundles in the stem; parallel leaf veins; flower parts in multiples of three. Eudicots: embryo with two cotyledons; phloem located between arms of xylem in the root; vascular bundles in a ring in the stem; netted leaf veins; flower parts in multiples of fours or fives. **24.3:** 1. The root apical meristem is located at the tip of the root and is covered by the root cap. 2. Endodermis forms a boundary between the cortex and inner vascular cylinder that contains the Casparian strip made up of lignin which prevents water and minerals from moving between the cells. 3. See Figure 24.12 on page 444. **24.4:** 1. A vascular bundle contains xylem and phloem. 2. Vascular bundles are scattered in monocot stems and form a ring in eudicot stems. 3. Primary growth is growth in length and is nonwoody; secondary growth is growth in girth and is woody. 4. Bark is composed of cork, cork cambium, cortex, and phloem. 5. An annual ring is composed of one year's growth of wood—one layer of spring wood made up of wide vessel elements with thin walls followed by one layer of summer wood which has fewer vessels. **24.5:** 1. Photosynthesis, which produces organic food for a plant, occurs in the mesophyll. 2. The palisade mesophyll contains elongated cells that carry out photosynthesis; the spongy mesophyll contains irregular cells that increase the surface area for gas exchange. 3. Leaves can be simple with a single blade or compound, which is divided into many leaflets. Pinnate leaves have leaflets occurring in pairs while palmately compound leaves have all the leaflets attached at a single point. Leaves are adapted for variations in environmental conditions of light and moisture.

Assess

1. e; **2.** d; **3.** e; **4.** d; **5.** c; **6.** d; **7.** c; **8.** c; **9.** c; **10.** b; **11.** b; **12.** b; **13.** a; **14. a.** upper epidermis; **b.** palisade mesophyll; **c.** vascular bundle; **d.** spongy mesophyll; **e.** lower epidermis

Chapter 25

Check Your Progress

25.1: 1. Nitrogen, phosphorus, and potassium. 2. Humus improves soil aeration, soil texture, increases water holding capacity, decomposes to release nutrients for plant growth, and helps retain positively charged minerals and make them available for plant uptake. 3. Plants extract Ca^{2+} and K^{+} from negatively charged clay and humus in soil by exchanging H^{+} for them. 4. (1) Helps prevent soil erosion; (2) helps retain moisture; and (3) as the remains decompose, nutrients are returned to the soil. **25.2:** 1. Minerals must pass through the endodermal cell membrane by passive transport or active transport requiring ATP from cellular respiration. The Casparian strip prevents the backflow of minerals to the soil. 2. The bacteria convert atmospheric nitrogen to nitrate or ammonium, which can be taken up by plant roots. The host plant provides food and a space to live for the bacteria. 3. The fungus obtains sugars and amino acids from the plant. The plant obtains inorganic nutrients and water from the fungus. **25.3:** 1. Evaporation of water from leaf surfaces causes water to be under tension in stems due to the cohesion of the water molecules. 2. When water molecules are pulled upward during transpiration, their cohesiveness creates a continuous water column. Adhesion allows water molecules to cling to the sides of xylem vessels, so the column of water does not slip down. 3. Sugars enter sieve tubes at sources and water flows in as well, creating pressure. The pressure is relieved at the other end when sugars and water are removed at the sink.

Assess

1. b; **2.** a; **3.** c; **4.** e; **5.** a; **6.** e; **7.** e; **8.** c; **9.** e; **10.** d; **11.** b; **12.** c

Chapter 26

Check Your Progress

26.1: 1. Hormones coordinate the responses of plants to stimuli. 2. Auxins concentrate on the shady side of a stem where they cause the wall to weaken, stretch, and eventually be rebuilt as elongated on one side. This bends the stem. 3. ABA maintains seed and bud dormancy and closes stomata. **26.2:** 1. Growth toward a stimulus is a positive tropism and growth away from a stimulus is a negative tropism. 2. Sedimentation of statoliths (starch granules) is the mechanism by which root cells perceive gravity. 3. Internal stimuli: action potentials, hormones, turgor pressure. External stimuli: light, touch, gravity. 4. The loss of water from cells of a leaf results in reduced turgor pressure and the leaf goes limp. **26.3:** 1. Phytochrome is composed of two proteins that contain light-sensitive regions. When triggered by red light phytochrome causes various genes to become active or inactive leading to seed germination, shoot elongation, and flowering responses. 2. The plant is responding to a short night, not to the length of the day. 3. The conversion between phytochrome red and phytochrome far-red promotes seed germination, inhibits shoot elongation, promotes

flowering, and affects plant spacing and the accumulation of chlorophyll.
4. (1) occur every 24 hours (2) occur with or without day/night lighting (3) can be reset if external cues are provided.

Assess

1. c; **2.** d; **3.** b; **4.** e; **5.** a; **6.** c; **7.** d; **8.** a; **9.** b; **10.** c; **11.** a; **12.** d; **13.** e; **14.** b; **15.** b

Chapter 27

Check Your Progress

27.1: 1. Male gametophytes (sperm-bearing pollen grains) are produced in the anther of the stamen. The female gametophyte (egg-bearing embryo sac) is produced in an ovule within the ovary of the carpel. 2. Each microspore produces a two-celled pollen grain. The generative cell produces two sperm, and the tube cell produces a pollen tube. One of the four megaspores resulting from meiosis of the mother cell in the ovule produces a seven-celled female gametophyte, called the embryo sac. 3. In order to accommodate the larger size of the pollinator, natural selection in the flowering plant may lead to an increase in the structural support of the part that the pollinator lands upon. It may also lead to an increase in the size of the flower to accommodate a larger pollinator. **27.2:** 1. The embryo is derived from the zygote; the stored food is derived from the endosperm; and the seed coat is derived from the ovule wall. 2. The ovule is a sporophyte structure produced by the female parent. Therefore, the wall (which becomes seed coat) is 2n. The embryo inside the ovule is the product of fertilization and is, therefore, 2n. 3. Cotyledons are embryonic leaves that are present in seeds. Cotyledons store nutrients derived from endosperm (in eudicots). **27.3:** 1. Dry fruits, with a dull, thin, and dry covering derived from the ovary, are more apt to be windblown. Fleshy fruits, with a juicy covering derived from the ovary and possibly other parts of the flower, are more apt to be eaten by animals. 2. Eudicot seedlings have a hook shape, and monocot seedlings have a sheath to protect the first true leaves. **27.4:** 1. (1) the newly formed plant is often supported nutritionally by the parent plant until it is established; (2) if the parent is ideally suited for the environment, the offspring will be as well; (3) if distance between individuals make cross pollination unlikely, asexual reproduction is a good alternative. 2. Stolons and rhizomes produce new shoots and roots; fruit trees produce suckers; stem cuttings grow new roots and become a shoot system. 3. Tissue from leaves, meristem, and anthers can become whole plants in tissue culture.

Assess

1. d; **2.** a; **3.** b; **4.** e; **5.** c; **6.** a; **7.** a; **8.** e; **9.** b; **10.** c; **11.** e; **12.** d; **13.** e; **14.** a

Chapter 28

Check Your Progress

28.1: 1. Multicellular, usually with specialized tissues, ingest food, diploid life cycle. 2. Animals are descended from an ancestor that resembled a hollow spherical colony of flagellated cells. Individual cells became specialized for reproduction. Two tissue layers arose by invagination. 3. Radial symmetry: body organized circularly; examples: cnidarians and ctenophores. Bilateral symmetry: have right and left halves; examples: most other animals. 4. Deuterostomes: blastopore becomes anus, radial cleavage, coelum forms from gut; Protostomes: blastopore becomes mouth, spiral cleavage, coelom forms from mesoderm. **28.2:** 1. Sponges are multicellular with no symmetry and no digestive cavity. Cnidarians have true tissues, are radially symmetrical and have a gastrovascular cavity. 2. Interior of sponges has canals lined with flagellated cells called choanocytes. Flagella produce water current that carries food particles that are filtered out. 3. Polyps have mouths directed upward. Medusae are bell-shaped with tentacles around the opening of the bell and mouth directed downward.
4. Stinging cells called cnidocytes have a fluid-filled capsule called a nematocyst in which a hollow threadlike structure is coiled and is discharged when stimulated. Some trap prey; others contain paralyzing toxins. **28.3:** 1. They all have bilateral symmetry, three tissue layers, and protostome development. 2. Annelids and molluscs have a complete digestive tract, a true coelom, and a circulatory system (closed in annelids and open in molluscs). Flatworms have a gastrovascular cavity with only one opening, no coelom, and no circulatory system. 3. See Figure 28.13 page 526 for the life cycle of *Schistosoma,* a blood fluke. See Figure 28.14 page 527 for the life cycle of *Taenia,* a tapworm.
4. Bryozoa, Phoronida, Brachiopoda. **28.4:** 1. Roundworms and arthropods are protostomes that molt. 2. Crustaceans breathe by gills and have swimmerets. Insects breathe by tracheae, and they may have wings. 3. The first pair of appendages is the chelicerae (modified fangs), and the second pair is the pedipalps (hold, taste, chew food). They have a cephalothorax and abdomen. **28.5:** 1. Both echinoderms and chordates follow a deuterostome pattern of development and molecular data indicates they are closely related. 2. The larval stage is bilaterally symmetrical.
3. Skin gills are tiny, fingerlike extensions of the skin that project through the body wall that are used for respiration. Tube feet are a part of the water vascular system on the oral surface and are used in locomotion, feeding, gas exchange, and sensory reception. The stomach is located in the central disc and has two parts. The cardiac stomach can be inverted and extended into bivalves where it secretes digestive enzymes. Partly digested food is taken into the pyloric stomach inside the sea star where digestion continues. 4. The water vascular system functions in locomotion, feeding, gas exchange, and sensory reception.

Assess

1. d; **2.** e; **3.** b; **4.** d; **5.** a; **6.** e; **7.** b; **8.** b; **9.** e; **10.** c; **11. a.** head; **b.** antenna; **c.** simple eye; **d.** compound eye; **e.** thorax; **f.** tympanum; **g.** abdomen; **h.** forewing; **i.** hindwing; **j.** ovipositor; **k.** spiracles; **l.** air sac; **m.** spiracle; **n.** tracheae; **12.** e; **13.** e; **14.** d; **15.** e

Chapter 29

Check Your Progress

29.1: 1. Humans are chordates that have the four chordate characteristics during the embryonic period of their life cycle. Notochord is replaced by vertebral column during development. Dorsal tubular nerve cord becomes the spinal cord. Pharyngeal pouches (the first pair of pouches) develop into auditory tubes. Post-anal tail is present in developing embryo, but lost during development. 2. A sea squirt larva has the four characteristics as a larva, and then undergoes metamorphosis to become an adult, which has gill slits but none of the other characteristics. **29.2**: 1. Endoskeleton protects internal organs, provides a place of attachment for muscles, and permits rapid, efficient movement. 2. Four legs are useful for locomotion on land, where the body is not supported by water. **29.3:** 1. All fishes are aquatic vertebrates and ectothermic. They all live in water, breathe by gills, and have a single circulatory loop (Fig. 29.11*a*, page 551). 2. Ray-finned bony fishes have fan-shaped fins supported by thin, bony rays. Lobe-finned bony fishes have fleshy fins supported by bones. **29.4:** 1. a. Two pairs of limbs; smooth, nonscaly skin that stays moist; lungs; a three-chambered heart with a double-loop circulatory pathway; sense organs adapted for a land environment; ectothermic; and have aquatic reproduction. b. Lobe-finned fishes and amphibians both have lungs and internal nares that allow them to breathe air. The same bones are present in the front fins of the lobe-finned fishes as in the forelimbs of early amphibians. 2. Usually, amphibians carry out external fertilization in the water. The embryos develop in the eggs until the tadpoles emerge. They then undergo metamorphosis, growing legs and reabsorbing the tail, to become adults. **29.5:** 1. Alligators and crocodiles live in fresh water, have a thick skin, two pairs of legs, powerful jaws, and a long muscular tail that allows them to capture and eat other animals in or near the water. Snakes have no limbs and have relatively thin skin. They live close to or in the ground and can escape detection. They use smell (Jacobson's organ) and vibrations to detect prey. Some use venom to subdue prey, which they eat whole because their jaws are distensible. 2. Birds are reptiles: feathers are modified scales; birds have clawed feet and a tail that contains vertebrae. **29.6:** 1. Mammals have hair or fur and mammary glands. 2. Monotremes (have a cloaca and lay eggs), marsupials (young are born immature and finish development in a pouch), and placental (eutherians) mammals (development occurs internally and the fetus is nourished by placenta).

Assess

1. e; **2.** b; **3. a.** pharyngeal pouches; **b.** dorsal tubular nerve cord; **c.** notochord; **d.** postanal tail; **4.** c; **5.** c; **6.** a; **7.** c; **8.** c; **9.** e; **10.** a; **11.** d; **12.** e; **13.** d; **14.** e; **15.** a, e; **16.** c; **17.** b

Chapter 30

Check Your Progress

30.1: 1. Mobile limbs, grasping hands, flattened face and stereoscopic vision, large complex brain, reduced reproductive rate. 2. Refer to Figure 30.4 on page 567, focusing on the bars on the right of the figure. 3. Both humans and chimpanzees are hominines and they share similar structure of their genomes, plus the general characteristics of all primates. **30.2:** 1. Ardipithecines lived primarily in trees, whereas the australopiths lived both in and out of trees. The brain size of the australopiths was larger, and this group was better adapted for bipedalism.
2. Larger brain size and the appearance of bipedalism, which freed the hands to use tools and hold babies, are evolutionary developments that occurred with hominins. **30.3:** 1. increased size of brain cavity, increased height, eventual use of tools and fire. 2. Bipedalism allowed for organisms to move young more easily; increased brain size allowed for higher intellect and thus adaptation to nonforest environments. **30.4:** 1. The replacement model suggests that humans evolved from one group in Africa, and then migrated to other locations. This means that different groups of Cro-Magnon humans could adapt to different locations, eventually forming the major human ethnic groups.
2. Increased brain size and reliance on tool use and a hunter-gatherer lifestyle required development of communication.

Assess

1. c; **2.** d; **3.** a; **4.** b; **5.** b; **6.** d; **7.** a; **8.** c; **9.** d; **10.** b; **11.** b; **12.** d; **13.** d; **14. a.** modern humans; **b.** archaic humans; **c.** *Homo erectus;* **d.** *Homo erectus;* **e.** *Homo ergaster*

Chapter 31

Check Your Progress

31.1: 1. Squamous epithelium: flat cells that line the blood vessels and air sacs of lungs; cuboidal epithelium: cube shaped cells that line the kidney tubules and various glands; columnar epithelium: rectangular cells that line the digestive tract; stratified epithelium: layers of cells that protect various body surfaces; glandular epithelium: modified to secrete products, like the goblet cells of the digestive tract. 2. Fibrous connective tissue has collagen and elastic fibers in a jellylike matrix between fibroblasts. Supportive connective tissue has protein fibers in a solid matrix between collagen or bone cells. Fluid connective tissue lacks fibers and has a fluid matrix that contains blood cells or lymphatic cells. 3. Skeletal muscle, which is striated with multiple nuclei, causes bones to move when it contracts. Smooth muscle, which is spindle-shaped with a single nucleus, causes the walls of internal organs to constrict. Cardiac muscle, which has branching, striated cells, each with a single nucleus, causes the heart to beat.
4. Dendrites conduct signals toward the cell body; the cell body contains most of the cytoplasm and the nucleus and it carries on the usual functions of the cell; the axon conducts nerve impulses. **31.2:** 1. An organ is two or more types of tissues working together to perform a function; an organ system is many organs working to perform a process. 2. The lymphatic and immune systems work together to protect the body from disease (other answers are possible). 3. The dorsal cavity contains the cranial cavity and the vertebral cavity; the ventral cavity contains the thoracic, abdominal, and pelvic cavities. **31.3:** 1. The skin protects the deeper tissues from pathogens, trauma, and dehydration in all animals. Only birds have feathers, a derivative of skin that is involved in flying. 2. The epidermis is stratified squamous epithelium, and it protects and prevents water loss. The dermis is dense fibrous connective tissue, and it helps regulate body temperature and provides sensory reception. 3. A dark skinned person in a low sunlight area might develop a deficiency of vitamin D, which is required for normal bone growth.
4. Nails are keratinized cells that grow from the nail root; hair is made of dead epithelial cells that have been keratinized; sweat glands are tubules originating from the dermis; oil glands are usually associated with hair follicles. **31.4:** 1. Homeostasis, the dynamic equilibrium of the internal environment, maintains body conditions within a range appropriate for cells to continue living. 2. Circulatory system brings nutrients and removes waste from interstitial fluid; respiratory system carries out gas exchange; urinary system excretes metabolic wastes and maintains water-salt balance and pH of blood.
3. When conditions go beyond or below a set point, a correction is made to bring conditions back to normality again.

Assess

1. b; **2.** e; **3.** e; **4.** c; **5.** d; **6.** a; **7.** b; **8.** b; **9.** a, c, g; **10.** b, d, e; **11.** b, c, f; **12.** intestinal lining is columnar epithelium that protects and absorbs; heart has cardiac muscle that pumps blood; compact bone supports body and protects organs; **13.** c; **14.** d; **15.** c; **16.** a; **17.** b; **18.** e; **19.** c; **20.** c; **21.** e; **22.** d; **23.** e

Chapter 32

Check Your Progress

32.1: 1. Circulatory systems carry nutrients and oxygen to cells and remove their wastes. 2. Hemolymph is present in animals with open circulatory systems and carries nutrients. Blood is found in closed circulatory systems and contains respiratory pigments that allow it to carry oxygen to cells and carbon dioxide away from them as well as transport nutrients. 3. With a closed circulatory system, oxygen must diffuse across walls of capillaries. **32.2:** 1. Arteries carry blood away from the heart, capillaries exchange their contents with interstitial fluid, and veins return blood back to the heart. 2. Veins contain valves, because they are thin-walled and there is insufficient blood pressure in veins to return blood to the heart. Valves plus skeletal muscle contractions are needed. Arteries have thick walls and higher blood pressure so they don't need valves to move blood to the tissues. 3. One-circuit pathways are less efficient, because blood supplying the tissues (after leaving the gills) is under low pressure. Two-circuit pathways supply oxygen to tissues more efficiently, which helps land animals meet the increased demands of locomotion on land vs. in water. **32.3:** 1. From the body: vena cava, right atrium, tricuspid valve, right ventricle, pulmonary semilunar valve, pulmonary trunk and arteries (carrying oxygen-poor blood). From the lungs: pulmonary veins, left atrium, bicuspid valve, left ventricle, aortic semilunar valve, aorta. 2. First the atria contract, then the ventricles contract, and then they both rest. The *lub* sound occurs when the atrioventricular valves close, and the *dub* sound occurs when the semilunar valves close. 3. Systolic pressure results from when the ventricles contract, creating high pressure. Diastolic pressure is that lower pressure present when the ventricles relax. 4. Thromboembolism, stroke, heart attack. **32.4:** 1. Blood contains plasma and formed elements. Plasma transports many substances to and from the capillaries, helps defend against pathogen invasion, helps regulate body temperature, and helps with clotting to prevent excessive blood loss. The formed elements include red blood cells that carry oxygen to tissues and return some carbon dioxide, white blood cells that help protect the body from infections, and platelets that are involved in blood clotting. 2. Platelets accumulate at the site of injury and release a clotting factor that results in the synthesis of thrombin. Thrombin synthesizes fibrin threads that provide a framework for the clot. 3. Any fetal Rh-positive blood cells that enter the circulation of an Rh-negative mother are recognized as foreign antigens, which the mother makes antibodies against. An Rh-positive woman does not make antibodies against the Rh-positive cells of the fetus. 4. Capillary exchange is affected by osmotic pressure (tends to cause water to enter capillaries) and blood pressure (tends to cause water to leave capillaries).

Assess

1. d; **2.** b; **3.** a; **4.** a; **5.** b; **6.** b; **7.** d; **8.** b; **9.** c; **10.** e; **11. a.** superior vena cava; **b.** aorta; **c.** left pulmonary artery; **d.** pulmonary trunk; **e.** left pulmonary veins; **f.** right pulmonary artery; **g.** right pulmonary veins; **h.** semilunar valve; **i.** left atrium; **j.** right atrium; **k.** left atrioventricular (bicuspid) valve; **l.** right atrioventricular (tricuspid) valve; **m.** right ventricle; **n.** chordae tendinae; **o.** septum; **p.** left ventricle; **q.** inferior vena cava; **12.** c; **13.** b; **14.** b

Chapter 33

Check Your Progress

33.1: 1. Under certain conditions, slime molds develop specialized sentinel cells that engulf pathogens and toxins. 2. PAMPs are highly conserved microbial molecules such as double-stranded viral RNA as well as carbohydrates, lipids, and proteins found only in bacterial cell walls. 3. Immune cells expressing receptors for specific antigens can multiply and differentiate, leading to an increased response to an antigen. Immunological memory and the ability to respond to continuously evolving pathogens become possible. **33.2:** 1. The lymphatic capillaries absorb excess interstitial fluid and lipids which are returned to the bloodstream in a one-way system powered by the contraction of skeletal muscles. The production and distribution of lymphocytes occurs in the organs and vessels of the lymphatic system. The circulatory system is involved with the transportation of oxygen and CO_2, nutrients, erythrocytes, and lymphocytes. 2. The lymphatic system absorbs fats, returns excess interstitial fluid to the bloodstream, produces lymphocytes, and helps defend the body against pathogens. 3. Red bone marrow is a spongy, semisolid red tissue located in certain bones (e.g., ribs, clavicle, vertebral column, heads of femur, and humerus), which produces all the blood cells of the body. The thymus is a soft, bilobed gland located in the thoracic cavity between the trachea and the sternum where T lymphocytes mature. Lymph nodes are small ovoid structures located along lymphatic vessels through which the lymph travels. The spleen is an oval organ with a dull purplish color that has both immune as well as red blood cell maintenance functions. **33.3:** 1. Physical barriers include the skin, mucus membranes, and ciliated epithelia; chemical barriers include lysozyme, stomach acid, and protective proteins (e.g., complement). 2. Four cardinal signs of inflammation are heat, swelling, redness, and pain. Inflammation generally directs other components of the immune system (molecules and cells) to the inflamed area. 3. Mast cells initiate inflammation; phagocytes (dendritic cells, macrophages, neutrophils) devour pathogens; natural killer cells kill virus-infected cells and cancer cells by cell-to-cell contact. 4. Complement proteins assist other immune defenses by initiating inflammation, enhancing phagocytosis of pathogens, and forming a membrane attack complex. **33.4:** 1. With antibody-mediated immunity, B-cells bind to specific antigens. T cells secrete cytokines that stimulate these B cells to differentiate into many plasma cells and memory cells. The plasma cells secrete antibodies against a pathogen. During cell-mediated immunity, cytotoxic T cells bind to antigens presented by the MHC on the surface of a cell that is infected by a virus or is cancerous, marking it for destruction. Helper T cells regulate adaptive immunity. 2. Antibodies have a constant region and two variable regions that bind to a specific antigen in a lock-and key-manner. Many different antigen-binding sites can be generated. 3. Active immunity is induced in an individual by natural infection, vaccination, or exposure to a toxin; passive immunity is produced by one individual and transferred into someone else, such as maternal antibodies, administration of antibodies to treat diseases, and bone marrow transplants. **33.5:** 1. Most primary immunodeficiencies are due to a genetic defect in a component of the immune system. 2. Autoimmune diseases (e.g., rheumatoid arthritis) are generally treated with drugs that suppress the immune system. 3. The transplanting of animal tissues or organs into humans.

Assess

1. a; **2.** d; **3.** c; **4.** b; **5.** c; **6.** d; **7.** c; **8.** e; **9.** b; **10.** c; **11.** e; **12.** b; **13.** c; **14.** IgG; **15.** IgM; **16.** IgA; **17.** IgE; **18.** e; **19.** d; **20.** e; **21.** b; **22.** c; **23.** d

Chapter 34

Check Your Progress

34.1: 1. Planarians have an incomplete digestive tract, called the gastrovascular cavity; the complete tract of earthworms has a pharynx, crop, gizzard, and intestine. 2. An incomplete tract has only one opening, with parts that are not very specialized, because they serve multiple functions. 3. Carnivores tend to have pointed incisors and enlarged canine teeth to tear off pieces of food small enough to quickly swallow. The molars are jagged for efficient chewing of meat. Herbivores have reduced canines but sharp and even incisors to clip grasses. The large flat molars grind and crush tough grasses. **34.2:** 1. Mouth, pharynx, esophagus, stomach, duodenum, jejunum, ileum, large intestine. 2. Taste buds enable an animal to discern the difference between nutritious and non-nutritious foods that could contain potentially dangerous substances that should not be consumed. 3. The stomach mechanically churns food, while acid and pepsin begin protein digestion. The small intestine finishes the digestion of proteins, fats, carbohydrates, and nucleic acids. The villi and microvilli of the intestinal wall greatly increase the surface area, thereby enhancing absorption of the final products of digestion. The large intestine absorbs excess water, and its diameter is larger to permit storage of undigested food prior to defecation. 4. Bile from the liver (stored in the gallbladder) emulsifies fat. The pancreas secretes sodium bicarbonate which neutralizes acid chyme from the stomach. It also secretes amylase, trypsin, and lipase which digest carbohydrates, proteins, and lipids. The liver has many other functions, such as the storage of glucose as glycogen. **34.3:** 1. Starch digestion begins in the mouth where salivary amylase digests starch to maltose. Pancreatic amylase continues this same process in the small intestine. In the small intestine maltase and brush-border enzymes digest maltose to glucose, which enters a blood capillary. Protein digestion starts in the stomach where pepsin digests protein to peptides and continues in the small intestine where trypsin carries out this same process. 2. Carbohydrates are digested to simple sugars (monosaccharides), proteins are digested to amino acids, and fats are digested to glycerol plus fatty acids. **34.4:** 1. Vegetables, if properly chosen, can supply limited calories and all necessary amino acids and vitamins. Much urea results when excess amino acids from proteins are metabolized. The loss of water needed to excrete urea can result in dehydration and loss of calcium ions. 2. A diet high in saturated fats tends to raise LDL cholesterol, which is associated with atherosclerosis. 3. A vitamin is an organic molecule that is required in the diet because it cannot be synthesized.

Assess

1. e; **2.** d; **3.** a; **4.** d; **5.** e; **6.** b; **7.** c; **8.** a; **9.** a; **10.** c; **11.** Test tube 1: no digestion (as a control); Test tube 2: some digestion (enzyme but no acid); Test tube 3: no digestion (no enzyme); Test tube 4: digestion (both enzyme and acid are present); **12.** d; **13.** b; **14.** c

Chapter 35

Check Your Progress

35.1: 1. Air has a drying effect, and respiratory surfaces have to be moist. Hydras are aquatic, while earthworms live in moist earth, and salamanders have skin glands that provide this moisture. These animals also have a large surface area compared to their size, or they have many capillaries close to the skin to facilitate gas exchange. 2. When blood containing a low O_2 level flows in an opposite direction than the O_2-rich water passing over the gills, a higher percentage of oxygen is transferred than if both flowed in the same direction. The highest value a concurrent mechanism could achieve would be 50% of the O_2 content in the water, because equilibration would occur. 3. Hemolymph distributes some O_2 in the hemocoel, but it is inefficient. Tracheae are air tubes that branch into ever smaller tracheoles, which deliver oxygen to most cells. Tracheae open at spiracles, and some larger insects have air sacs that can expand and contract. Some aquatic insects have tracheal gills, expansions of the body wall to provide more oxygen-absorbing surface area. **35.2:** 1. During inspiration, the rib cage moves up and out, and the diaphragm contracts and moves down. As the thoracic cavity expands like a flexible container, air flows into the lungs due to decreased air pressure in the lungs. During expiration, the rib cage moves down and the diaphragm relaxes and moves up to its former position. Air flows out as a result of increased pressure in the lungs. 2. The carotid bodies and aortic bodies contain chemoreceptors that send stimulatory messages to the respiratory center if the pH or O_2 levels are too low. 3. In the lungs, oxygen entering pulmonary capillaries combines with hemoglobin (Hb) in red blood cells to form oxyhemoglobin (HbO_2). In the tissues, Hb gives up O_2, while CO_2 enters the blood and the red blood cells. Some CO_2 combines with Hb to form carbaminohemoglobin ($HbCO_2$). Most CO_2 combines with water to form carbonic acid, which dissociates into H^+ and HCO_3^-. The H^+ is absorbed by the globin portions of hemoglobin to form reduced hemoglobin HbH^+. This helps stabilize the pH of the blood. The HCO_3^- is carried in the plasma. **35.3:** 1. Penicillin is an antibiotic that kills bacteria, but most colds are caused by viruses. 2. Narrowing of the airways is seen in bronchitis and asthma; reduced lung expansion is seen in pulmonary fibrosis and emphysema. 3. Smoking causes or contributes to acute and chronic bronchitis, asthma, pulmonary fibrosis, emphysema, and lung cancer (along with many cardiovascular disorders).

Assess

1. a. external respiration; **b.** CO_2; **c.** CO_2; **d.** tissue cells; **e.** internal respiration; **f.** O_2; **g.** O_2; **2.** b; **3.** c; **4.** c; **5.** f; **6.** c; **7.** a; **8.** d; **9.** a; **10.** b; **11.** b; **12.** b; **13.** c; **14.** c; **15.** a; **16.** a; **17.** c

Chapter 36

Check Your Progress

36.1: 1. Osmoregulation involves the balance between salts and water; metabolic wastes are removed by excretion. 2. Urea is not as toxic as ammonia, and it does not require as much water to excrete; uric acid takes more energy to prepare than urea. 3. Kangaroo rats are active at night, have convoluted nasal passages to capture moisture in exhaled air, fur to prevent water loss from skin, secrete a hypertonic urine, and eliminate very dry feces. **36.2:** 1. The kidneys alone secrete nitrogenous wastes, but share responsibility for regulating blood pressure, pH, and water-salt balance. Certain hormones secreted by the kidneys, such as erythropoietin and ADH, have unique functions. 2. In response to low blood pressure, the kidneys secrete renin, which converts angiotensinogen to angiotensin I, which is converted to angiotensin II. Angiotensin II causes blood vessel constriction and stimulates the adrenal glands to release aldosterone, which acts on kidney tubules to increase salt reabsorption. Blood pressure rises in response. 3. The kidneys maintain blood pH by secreting H^+ ions and reabsorbing bicarbonate ions, as needed.

Assess

1. d; **2.** a; **3.** d; **4.** c; **5.** e; **6.** c; **7.** c; **8.** e; **9.** d; **10. a.** glomerular capsule; **b.** proximal convoluted tubule; **c.** loop of the nephron; **d.** descending limb; **e.** ascending limb; **f.** distal convoluted tubule; **g.** collecting duct; **h.** renal artery; **i.** afferent arteriole; **j.** glomerulus; **k.** efferent arteriole; **l.** peritubular capillary network; **m.** renal vein

Chapter 37

Check Your Progress

37.1: 1. A nerve net is a simple nervous system consisting of interconnected neurons, with no CNS. A ganglion is a cluster of neuron (nerve cell) bodies. In animals with a CNS and a PNS, it is a cluster of neurons located outside the CNS. A centrally located brain controls the ganglia and associated nerves. 2. The hindbrain controls essential functions like breathing, heart functions, and basic motor activity; the midbrain is a relay station connecting the hindbrain with the forebrain; the forebrain, which receives sensory input, includes the hypothalamus (involved with homeostasis) as well as the cerebrum (involved with higher functions). 3. The more recently evolved parts of the brain are the outer portions, such as the cerebral cortex, and (in mammals) the neocortex. **37.2:** 1. Nerve impulses travel more quickly down myelinated axons due to saltatory conduction ("jumping") of action potentials from one node of Ranvier to the next. 2. Na^+ moves from the outside of the axon membrane to the inside; K^+ moves from the inside of the axon membrane to the outside. 3. Inhibition of AChE, the enzyme that normally breaks down acetylcholine (ACh), would result in increased activity of nerves that use ACh as a neurotransmitter. **37.3:** 1. Sensory information from internal organs travels via spinal nerves, through the dorsal root ganglia to synapses on spinal cord interneurons whose axons travel in tracts to the brain. These then send motor impulses through tracts, to the ventral root ganglia, and then to the effector organ (smooth muscle in the intestine). 2. The four major lobes of the human brain are the frontal, parietal, temporal, and occipital. 3. Parkinson disease, with symptoms such as tremors, difficult speech, and trouble walking or standing, is associated with a loss of dopamine-producing cells in the basal nuclei of the forebrain; multiple sclerosis is an autoimmune disease that damages the myelin sheaths of neurons in the CNS, resulting in fatigue and in problems with visual and muscular function. **37.4:** 1. A reflex arc can travel from the sensory receptor, synapse on interneurons in the spinal cord, then travel back to the effector (muscle) without being perceived by the brain first. 2. Eating a big meal mainly stimulates the parasympathetic branch of the autonomic nervous system, diverting blood supply to the digestive tract and away from the muscles. Exercising will engage the sympathetic system, reducing blood flow to the stomach and inhibiting its function, causing it to ache. 3. The parasympathetic ("rest and digest") division dominates as you sleep, but being startled causes a sudden increase in sympathetic ("fight or flight") activity.

Assess

1. a; **2.** c; **3.** c; **4.** a; **5.** a; **6.** c; **7.** c; **8.** c; **9.** c; **10.** c; **11.** b; **12.** d; **13.** b; **14.** c; **15.** d; **16.** d; **17.** c; **18. a.** central canal; **b.** gray matter; **c.** white matter; **d.** dorsal root; **e.** cell body of sensory neuron in dorsal root ganglion; **f.** spinal nerves; **g.** cell body of motor neuron; **h.** interneuron

Chapter 38

Check Your Progress

38.1: 1. Sensory transduction is the conversion of some type of environmental stimulus into a nerve impulse. 2. Some snakes can perceive infrared radiation; bats, dolphins, and whales perceive very high- or low-frequency sound waves (echolocation); a dog's sense of smell is far more sensitive than a human's. **38.2:** 1. Both are chemical senses that use chemoreceptors to detect molecules in the air (smell) and moist food (taste). 2. Sweet, sour, salty, bitter, umami. 3. Neurons transmit signals from different sensory organs to different areas of the brain, where they are interpreted as different types of information. **38.3:** 1. Rods are located in the peripheral region of the retina and are very sensitive to light. They function in night vision, peripheral vision and the perception of motion. Cones are concentrated in the fovea centralis and are for color perception and fine detail. They are best-suited for bright light. Many rods may excite a single ganglion cell, but much smaller numbers of cones excite individual ganglion cells. 2. Sclera, choroid, retina. Light must pass through the ganglion and bipolar cell layers before reaching the photoreceptor cells. 3. An eyeball that is too long results in nearsightedness; an eyeball that is too short results in farsightedness; an uneven cornea results in astigmatism. **38.4:** 1. a. middle; b. outer; c. inner; d. inner; e. inner; f. outer. 2. Auditory canal, tympanic membrane (eardrum), ossicles (malleus, incus, and stapes), oval window, cochlea. 3. The utricle and saccule are responsible for gravitational equilibrium; the semicircular canals for rotational equilibrium. **38.5:** 1. A person lacking muscle spindles would have trouble walking, sitting, or doing other activities due to a lack of muscle tone. A person lacking nociceptors would be prone to injury due to an absence of warning signs associated with pain. 2. Pain is generally associated with potential harm: i.e., something to be avoided. An animal that quickly adapted to pain would have an increased chance of being injured or killed by a potentially dangerous stimulus.

Assess

1. c; **2.** c; **3.** b; **4.** c; **5.** b; **6.** a; **7.** d; **8.** c; **9.** a. retina—contains photoreceptors; **b.** choroid—absorbs stray light; **c.** sclera—protects and supports eye; **d.** optic nerve—transmits impulses to brain; **e.** fovea centralis—makes acute vision possible; **f.** muscle in ciliary body—holds lens in place, also accommodation; **g.** lens—refracts and focuses light rays; **h.** iris—regulates light entrance; **i.** pupil—admits light; **j.** cornea—refracts light rays; **10.** b; **11.** e; **12.** b; **13.** c; **14.** c; **15.** a; **16.** c; **17.** c; **18.** c

Chapter 39

Check Your Progress

39.1: 1. Hydrostatic skeleton, exoskeleton, endoskeleton 2. The tongue is a muscular hydrostat. 3. Because the muscle layers surrounding the coelom no longer contract, the hydrostatic skeleton cannot provide support for the body. **39.2:** 1. Osteoblasts build bone and osteoclasts break it down. Osteocytes live within the lacunae where they affect the timing and location of bone remodeling. 2. Compact bone, which serves mainly to support the body, contains many osteons, in which central canals are surrounded by a hard matrix with lacunae. Spongy bone is lighter, with numerous bars and plates as well as spaces filled with red bone marrow, which produces the blood cells. 3. Sacrum - axial, frontal - axial, humerus - appendicular, tibia - appendicular, vertebra - axial, coxal - axial, temporal - axial, scalpula - appendicular, sternum - axial. **39.3:** 1. A pair of muscles that work opposite to one another; for example, if one muscle flexes (bends) the joint the other extends (straightens) it. 2. Myofibrils are tubular contractile units that are divided into sarcomeres. Each sarcomere contains actin (thin filaments) and myosin (thick filaments). 3. Cleavage of ATP allows myosin heads to bind to actin filaments, pulling them toward the center of the sarcomere.

Assess

1. b; **2.** a; **3.** c; **4.** e; **5.** d; **6.** b; **7.** c; **8.** b; **9.** f; **10.** c; **11.** e; **12.** b; **13.** e; **14.** e; **15.** b; **16. a.** T tubule; **b.** sarcoplasmic reticulum; **c.** myofibril; **d.** Z line; **e.** sarcomere; **f.** sarcolemma of muscle fiber (see Figure 39.13, p. 743)

Chapter 40

Check Your Progress

40.1: 1. The nervous system responds rapidly to both external and internal stimuli, while the endocrine system responds more slowly, but often has longer-lasting effects. Both systems use chemicals to communicate with other body systems; the nervous system at synapses, the endocrine system via hormones secreted mainly into the bloodstream. 2. Peptide hormones have receptors in the plasma membrane. Steroid hormones have receptors that are generally in the nucleus, sometimes in the cytoplasm. 3. Since peptide hormones usually bind to receptors on the outside of the cell, they must communicate with the inside of the cell via second messengers. **40.2:** 1. The hypothalamus communicates with the endocrine system via the pituitary gland. Two hormones produced by the hypothalamus are stored in the posterior pituitary. Several others are produced by the anterior pituitary in response to hypothalamic-releasing factors that reach the anterior pituitary via a portal system. 2. ADH conserves body water by causing reabsorption of water by the kidneys; oxytocin causes uterine contractions during labor and milk letdown during nursing. 3. Thyroid-stimulating hormone (TSH) stimulates release of thyroid hormones; adrenocorticotropic hormone (ACTH) stimulates the adrenal glands to produce glucocorticoids, prolactin (PRL) causes breast development and milk production; growth hormone (GH) promotes bone and muscle growth; the gonadotropic hormones FSH and LH stimulate the testes or ovaries to produce gametes and sex hormones; melanocyte-stimulating hormone (MSH) causes skin color changes in some animals. **40.3:** 1. Angiotensin II causes arterioles to constrict; aldosterone causes reabsorption of Na^+, accompanied by water, in the kidneys. 2. Aldosterone - adrenal cortex, melatonin - pineal gland, epinephrine - adrenal medulla, EPO - kidneys, leptin - adipose tissue, glucagon - pancreas, ANH - heart, cortisol - adrenal cortex, calcitonin - thyroid gland. 3. PTH stimulates osteoclasts and calcitonin inhibits them.

Assess

1. e; **2.** b; **3.** c; **4.** a; **5.** d; **6.** c; **7.** d; **8.** c; **9.** d; **10.** b; **11.** d; **12.** b; **13.** b; **14.** a; **15.** e

Chapter 41

Check Your Progress

41.1: 1. Asexual reproduction allows organisms to reproduce rapidly and colonize favorable environments quickly. Sexual reproduction produces offspring with a new combination of genes that may be more adaptive to a changed environment. 2. An oviparous animal lays eggs that hatch outside the body. A viviparous animal gives birth after the offspring have developed within the mother's body. Ovoviviparous animals retain fertilized eggs within a parent's body until they hatch; the parent then gives birth to the young. 3. A shelled egg contains extraembryonic membranes which keep the embryo moist, carry out gas exchange, collect wastes, and provide yolk as food. **41.2:** 1. Seminiferous tubule, epididymis, vas deferens, ejaculatory duct, urethra 2. Seminal vesicles, prostate gland, and bulbourethral glands. 3. In males, FSH stimulates spermatogenesis and LH stimulates testosterone production. **41.3:** 1. (a)Ovary, (b)oviduct, (c)uterus, (d)cervix, vagina. 2. During the follicular phase, FSH stimulates ovarian follicles to produce primarily estrogen. A surge of LH (and FSH) also triggers ovulation. During the luteal phase, LH stimulates the corpus luteum to produce primarily progesterone. 3. Estrogen secreted by the developing follicle inhibits FSH secretion by the anterior pituitary, ending the follicular phase. Progesterone secreted by the developing corpus luteum inhibits LH secretion by the anterior pituitary, ending the luteal phase. Rising estrogen levels cause the endometrium to thicken (proliferative phase), and progesterone causes uterine glands to mature (secretory phase). If no pregnancy occurs, low levels of estrogen and progesterone initiate menstruation. **41.4:** 1. Male and female condoms and the diaphragm prevent sperm from coming in contact with the egg. 2. Abstinence (100%) and vasectomy (nearly 100%) are the most effective. The birth control pill and condoms are the next most effective. Natural family planning is least effective. 3. In AID, sperm are placed in the vagina or sometimes the uterus. In IVF, fertilization takes place outside of the body and embryos are transferred to the woman's uterus. In GIFT, eggs and sperm are brought together in laboratory glassware, and placed in the uterine tubes immediately afterward. In ICSI, one sperm is injected directly into an egg. **41.5:** 1. Antiretroviral drug categories include entry inhibitors (viral attachment to host receptor), reverse transcriptase inhibitors (production of DNA from viral RNA), integrase inhibitors (insertion of viral DNA into host DNA), and protease inhibitors (processing of viral proteins). 2. HPV induced genital warts can be painful and disfiguring, and some strains cause cancer of the cervix, vagina, vulva, anus, and mouth and throat. 3. Chlamydia and gonorrhea are associated with pelvic inflammatory disease and infertility.

Assess

1. b; **2.** e; **3.** d; **4.** c; **5.** c; **6.** a; **7.** b; **8.** c; **9.** c; **10.** c; **11.** c; **12.** a; **13.** d; **14.** d; **15.** a; **16.** c; **17.** c; **18.** b; **19.** e

Chapter 42

Check Your Progress

42.1: 1. The fast block is the depolarization of the egg's plasma membrane that occurs upon initial contact with a sperm. The slow block occurs when the secretion of cortical granules converts the zona pellucida into the fertilization membrane. 2. Notochord - mesoderm, thyroid and parathyroid glands - endoderm, nervous system - ectoderm, epidermis - ectoderm, skeletal muscle - mesoderm, kidneys - mesoderm, bones - mesoderm, pancreas - endoderm. 3. Neurula stage. **42.2:** 1. Cytoplasmic segregation is the parceling out of maternal determinants as mitosis occurs. Induction is the influence of one embryonic tissue on the development of another. 2. Morphogens are proteins that diffuse away from the areas of high concentration in the embryo, forming gradients that influence patterns of tissue development. 3. The homeobox encodes the homeodomain region of the protein product of the gene. The homeodomain is the DNA-binding region of the protein, which is a transcription factor. **42.3:** 1. Oviduct 2. Umbilical blood vessels - allantois, first blood cells - yolk sac, fetal half of the placenta - chorion. 3. Structures that begin to develop between the third and fifth weeks include the nervous system, heart, chorionic villi, umbilical cord, and limb buds. **42.4:** 1. The thymus is the site where T cells finish their development. T cells are needed to stimulate other types of immune cells, including B cells that produce antibodies. 2. Menopause is the time when ovarian and uterine cycles cease. The ovaries become unresponsive to the gonadotropic hormones from the anterior pituitary and estrogen and progesterone secretion stops. 3. Preprogrammed theories suggest aging is partly genetically programmed; single gene mutations influence aging in *C. elegans*. Damage accumulation theories suggest aging is due to an accumulation of cellular damage: e.g., DNA mutations, cross linking of proteins, or oxidation by free radicals.

Assess

1. d; **2.** b; **3.** b; **4.** e; **5.** e; **6.** c; **7.** a; **8.** d; **9.** b; **10. a.** chorion: contributes to forming placenta where wastes are exchanged for nutrients and oxygen; **b.** amnion: protects embryo and prevents desiccation; **c.** embryo; **d.** allantois: blood vessels become umbilical blood vessels; **e.** yolk sac: first site of blood cell formation; **f.** chorionic villi: embryonic portion of placenta; **g.** maternal portion of placenta; **h.** umbilical cord: connects developing embryo to placenta (see Figure 42.11, page 803); **11.** e; **12.** a; **13.** a; **14.** c; **15.** e

Chapter 43

Check Your Progress

43.1: 1. Gene for egg-laying hormone in *Aplysia* was isolated and its protein product controls egg-laying behavior. The gene *fosB* has been found to control maternal behavior in mice. 2. Nervous and endocrine **43.2:** 1. Associative learning 2. Just hatched, laughing gull chicks instinctively peck at parent's bill to be fed but their accuracy improves after a few days. 3. Parents of chicks shape their begging behavior. Animals or even objects can imprint behaviors in birds during sensitive periods. Social interactions can influence the songs white-crowned sparrows learn. Macaques learn to wash potatoes by imitating others. **43.3:** 1. Pheromones are used to mark a territory so other individuals of that species will stay away; honeybees do a waggle dance to guide other bees to a food source; vervet monkeys have calls that make other vervets run away. 2. Chemical (effective all the time, not as fast as auditory); auditory (can be modified but the recipient has to be present when message is sent); visual (need not be accompanied by chemical or auditory, needs light in order to receive); tactile (permits bonding; recipient must be close). 3. Chemical: taste buds and olfactory receptors; auditory: ears; visual: eyes; tactile: touch receptors in skin. **43.4:** 1. A benefit of territoriality is to ensure a source of food. 2. Both an animal's reproductive strategy (polygamous, polyandrous, or monogamous) and form of sexual selection favors features that increase an animal's chance of leaving offspring. Most often sexual selection is due to female choice, which forces males to compete. 3. Altruistic behavior is supposed to be selfless but when, for example, an offspring helps its parents raise siblings, the helper may be increasing the presence of some of its own genes in the next generation.

Assess

1. c; **2.** d; **3.** e; **4.** c; **5.** c; **6.** a; **7.** a; **8.** c; **9.** b; **10.** a; **11.** c; **12.** d; **13.** c; **14.** b

Chapter 44

Check Your Progress

44.1: 1. A population is all the members of one species that inhabit a particular area. A community is all the populations that interact within that area. 2. Abiotic variables in an ecosystem are those that are nonliving such as sunlight, temperature, wind, and terrain. 3. To develop models that explain and predict the distribution and abundance of organisms. **44.2:** 1. Population density is the number of individuals per unit area. Population distribution is the pattern of dispersal of individuals across an area of interest. 2. In a type I survivorship curve, most individuals survive well past the midpoint of the life span and death does not come until near the end of the life span. In type II, survivorship decreases at a constant rate throughout the life span. In type III, most individuals die young. 3. In a bell-shaped age pyramid, the pre-reproductive members represent the largest portion of the population. **44.3:** 1. Exponential growth ceases when the environment cannot support a larger population size; that is, the point at which the size of the population has reached the environment's carrying capacity. 2. Logistic growth occurs when some environmental factors such as food and space, become limited. The scarce resources lead to increased competition and predation. **44.4:** 1. As population density increases, competition and predation become more intense. 2. If a flash flood occurs, mice that can stay afloat will survive and reproduce whereas those that quickly sink will not survive and will not reproduce. In this way the ability to stay afloat will be more prevalent in the next generation. 3. Intrinsic factors: anatomy, physiology, behaviors such as territoriality, dominance hierarchies, recruitment, and migration. **44.5:** 1. A *K*-strategist species: allocate energy to their own growth and survival and to the growth and survival of their limited number of offspring. *r*-strategist species: allocate energy to producing a large number of offspring and little or no energy goes into parental care. 2. Populations may vary between *K* and *r* strategies based upon environmental conditions. **44.6:** 1. Less-developed countries have a high rate of population growth while more-developed countries have a low rate of population growth. 2. If the age structure of the population is such that there are more young women entering their reproductive years, then the population will continue to grow even if it is experiencing replacement reproduction. 3. Since resources are in limited supply, an increase in consumption by the LDC will cause more competition for resources in all countries.

Assess

1. d; **2.** e; **3.** c; **4.** b; **5.** e; **6.** e; **7.** b; **8.** c; **9.** e; **10.** a; **11.** c; **12.** e; **13.** c

Chapter 45

Check Your Progress

45.1: 1. Species richness is a list of all species found in the community. Species diversity considers species richness and the relative abundance of each species. 2. An organism's habitat is the place where it lives and reproduces. The niche is the role it plays in its community, such as whether it is a producer or consumer. 3. The two factors are (1) the predator causes the prey population to decline which leads to a decline in the predator population; later when the prey population recovers so does the predator population; (2) lack of food causes the prey population to decline followed by the predator population; later when food is available to the prey population they both recover. **45.2:** 1. Primary succession, during which soil is formed, occurs first. Secondary succession occurs as one species is replaced by another, usually progressing from grasses to shrubs to trees. 2. The facilitation model predicts that a community will grow toward becoming a climax community. The inhibition model predicts that colonists will inhibit the growth of a community. The tolerance model predicts that various plants can colonize an area at the same time. **45.3:** 1. Producers of food (photosynthesizers) such as cyanobacteria and algae are at the base of an ecological pyramid. 2. Energy passes from one population to the next, and at each step more is converted to heat until all of the original input has become heat. Therefore energy flows through an ecosystem. Chemicals pass from one population to the next and then recycle back to the producer populations again. 3. Return of CO_2 to the atmosphere because humans burn fossil fuels and destroy forests that take up CO_2; increase phosphate in surface water by runoff of applied fertilizer to croplands.

Assess

1. a; **2.** e; **3.** a; **4.** d; **5.** e; **6.** a; **7.** c; **8.** d; **9.** c; **10.** d; **11.** b; **12. a.** producers; **b.** consumers; **c.** inorganic nutrient pool; **d.** decomposers

Chapter 46

Check Your Progress

46.1: 1. Because the Earth is a sphere, the sun's rays hit the equator straight on but are angled as they reach the poles. 2. The windward side of the mountain receives more rainfall than the other side. Winds blowing over bodies of water collect moisture that they lose when they reach land. **46.2:** 1. Tropical rain forest, temperate grassland, semidesert, tropical seasonal forest, desert, shrubland, tropical deciduous forest, temperate deciduous forest, mountain zone, taiga, tundra, polar ice. 2. A tropical rain forest has a canopy (tops of a great variety of tall evergreen hardwood trees) with buttressed trunks at ground level. Long lianas (hanging vines) climb into the canopy. Epiphytes grow on the trees. The understory consists of smaller plants and the forest floor is very sparse. A temperate deciduous forest contains trees (oak, beech, sycamore, and maple) that lose their leaves in the fall. Enough light penetrates the canopy to allow a layer of understory trees. Shrubs, mosses, and ferns grow at ground level. **46.3:** 1. In freshwater and saltwater ecosystems the heat of the sun causes evaporation that eventually condenses and falls back as rain which eventually moves to the sea. In both systems wind moves surface water. This can cause overturns in lakes and currents as well as upwellings in oceans. 2. An upwelling occurs when cold offshore winds cause cold nutrient-rich water to rise up, taking the place of nutrient-poor warm waters.

Assess

1. b; **2.** e; **3.** b; **4.** a; **5.** e; **6.** c; **7.** c; **8.** d; **9.** a; **10.** a; **11.** a; **12.** d; **13.** e

Chapter 47

Check Your Progress

47.1: 1. Conservation biology is focused on studying biodiversity and on conserving natural resources necessary for preserving all species now and in the future. 2. Many disciplines in basic and applied biology provide data that support conservation biology and its goal of preserving biodiversity. 3. Biodiversity includes the number of species on Earth; genetic diversity (variations in a species); ecosystem diversity (interactions of species); and landscape diversity (interactions of ecosystems). A hotspot is a specific region with a large amount of biodiversity. **47.2:** 1. Direct value is a service that is immediately recognizable, while indirect value may not be as noticeable immediately but none-the-less eventually contributes to biodiversity. 2. Research is ongoing to measure how high biodiversity helps ecosystems function more efficiently. Rates of photosynthesis, response to elevated carbon dioxide levels, degree of resource acquisition and retention within an ecosystem, and the ability to withstand environmental changes and invasion of pathogens are thought to be correlated with levels of biodiversity. **47.3:** 1. Habitat loss, introduction of exotic species, pollution, climate change, overexploitation 2. Exotic plants displace native plants; predators introduced to kill pests also kill native animals; escaped animals may compete with, prey on, hybridize with, or introduce diseases into native populations. **47.4:** 1. Hotspots contain high levels of endemic species not found elsewhere. Preserving areas that are hotspots would be most effective at preserving biodiversity 2. Keystone species include bats in tropical forest of the Old World, grizzly bears in the northwestern United States and Canada, beavers in wetlands, bison in grasslands, alligators in swamps, and elephants in grasslands and forests. 3. Habitat restoration should start as quickly as possible, mimic natural processes of the habitats involved, and support ecosystems to be self-sustaining while still serving humans.

Assess

1. e; **2.** e; **3.** d; **4.** e; **5.** e; **6.** e; **7.** e; **8.** e; **9. a.** habitat loss; **b.** exotic species; **c.** pollution; **d.** overexploitation; **e.** disease; **10.** e; **11.** b; **12.** c

Appendix B

Tree of Life

The Tree of Life depicted in this appendix is based on the phylogenetic (evolutionary) trees presented in the text. Figure 1.5 showed how the three domains of life—Bacteria, Archaea, and Eukarya—are related. This relationship is also apparent in the Tree of Life, which combines the individual trees given in the text for eukaryotic protists, plants, fungi, and animals. In combining these trees, we show how all organisms may be related to one another through the evolutionary process.

The text also described the organisms that are included in the tree. These descriptions are repeated here.

Prokaryotes

Domains Bacteria and Archaea (Chapter 20) constitute the prokaryotic organisms that are characterized by their simple structure but a complex metabolism. The chromosome of a prokaryote is not bounded by a nuclear envelope, and therefore, these organisms do not have a nucleus. Prokaryotes carry out all the metabolic processes performed by eukaryotes and many others besides. However, they do not have organelles, except for plentiful ribosomes.

Domain Bacteria

Bacteria are the most plentiful of all organisms, capable of living in most habitats, and carry out many different metabolic processes. While most bacteria are aerobic heterotrophs, some are photosynthetic, and some are chemosynthetic. Motile forms move by flagella consisting of a single filament. Their cell wall contains peptidoglycan and they have distinctive RNA sequences.

Domain Archaea

Their cell walls lack peptidoglycan, their lipids have a unique branched structure, and their ribosomal RNA sequences are distinctive.

Methanogens. Obtain energy by using hydrogen gas to reduce carbon dioxide to methane gas. They live in swamps, marshes and intestines of mammals.

Extremophiles. Able to grow under conditions that are too hot, too cold, and too acidic for most forms of life to survive.

Nonextreme archaea. Grow in wide variety of environments that are considered within the normal range for living organisms.

Domain Eukarya

Eukarya have a complex cell structure with a nucleus and several types of organelles that compartmentalize the cell. Mitochondria that produce ATP and chloroplasts that produce carbohydrate are derived from prokaryotes that took up residence in a larger nucleated cell. Protists tend to be unicellular, while plants, fungi, and animals are multicellular with specialized cells. Each multicellular group is characterized by a particular mode of nutrition. Flagella, if present, have a 9 + 2 organization.

Protists

The protists (Chapter 21) are a catchall group for any eukaryote that is not a plant, fungus, or animal. Division into six supergroups is a working hypothesis that is subject to change as more is known about the evolutionary relationships of the protists. A supergroup is a major eukaryotic group and six supergroups encompass all members of the domain Eukarya including protists, plants, fungi, and animals.

Archaeplastids. A supergroup of photosynthesizers with plastids derived from endosymbiotic cyanobacteria. Includes land plants and other photosynthetic organisms, such as green and red algae and charophytes, exemplified by the stoneworts, which share a common ancestor with land plants.

Chromalveolates. A supergroup that includes the Stramenopiles, which have a unique flagella, and the Alveolates, which have small sacs under plasma membrane.

Stramenopiles. **Includes brown algae, such as** *Laminaria* **and** *Fucus,* **diatoms, golden brown algae, and water molds.**

Alveolates. **Includes dinoflagellates, ciliates such as** *Paramecium,* **and apicomplexans, such as** *Plasmodium vivax.*

Excavates. Have an excavated oral grove and form a supergroup that includes zooflagellates, such as euglenids (e.g., *Euglena*); diplomonads, such as *Giardia lambia;* and kinetoplastids (have a DNA granule called a kinetoplast), such as trypanosomes.

Amoebozoans. Supergroup of amoeboid cells that move by pseudopodia. Includes amoeboids, such as *Amoeba proteus,* and slime molds.

Rhizarians. Supergroup of amoeboid cells with tests. They form a supergroup that includes the foraminiferans and the radiolarians.

Opisthokonts. Supergroup named for members that have a single posterior flagellum (Gk., *opistho,* rear and *kontos,* pole). Includes animals and choanoflagellates that may be related to the common ancestor of animals and the fungi.

Plants

Plants (Chapter 23) are photosynthetic eukaryotes that became adapted to living on land. Includes aquatic green algae called charophytes, which have a haploid life cycle and share certain traits with the land plants.

Land Plants (embryophytes)

Have an alternation of generation life cycle; protect a multicellular sporophyte embryo; produce gametes in gametangia; possess apical tissue that produces complex tissues; and a waxy cuticle that prevents water loss.

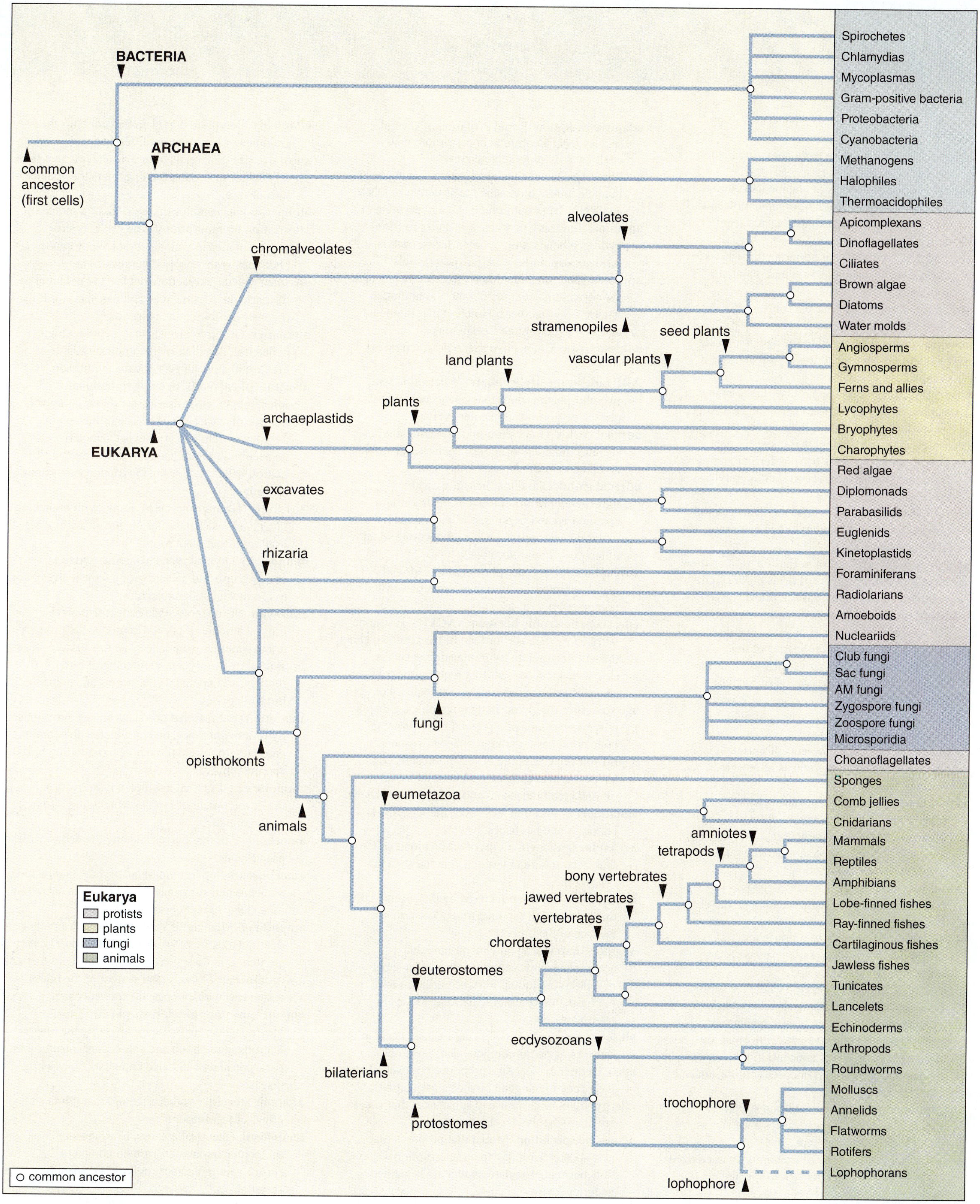

BACTERIA
ARCHAEA
EUKARYA
common ancestor (first cells)
Spirochetes
Chlamydias
Mycoplasmas
Gram-positive bacteria
Proteobacteria
Cyanobacteria
Methanogens
Halophiles
Thermoacidophiles
alveolates
chromalveolates
stramenopiles
Apicomplexans
Dinoflagellates
Ciliates
Brown algae
Diatoms
Water molds
seed plants
vascular plants
land plants
plants
archaeplastids
Angiosperms
Gymnosperms
Ferns and allies
Lycophytes
Bryophytes
Charophytes
Red algae
excavates
Diplomonads
Parabasilids
Euglenids
Kinetoplastids
rhizaria
Foraminiferans
Radiolarians
Amoeboids
Nucleariids
fungi
opisthokonts
Club fungi
Sac fungi
AM fungi
Zygospore fungi
Zoospore fungi
Microsporidia
Choanoflagellates
animals
eumetazoa
Sponges
Comb jellies
Cnidarians
amniotes
tetrapods
bony vertebrates
jawed vertebrates
vertebrates
chordates
deuterostomes
Mammals
Reptiles
Amphibians
Lobe-finned fishes
Ray-finned fishes
Cartilaginous fishes
Jawless fishes
Tunicates
Lancelets
Echinoderms
ecdysozoans
bilaterians
protostomes
trochophore
lophophore
Arthropods
Roundworms
Molluscs
Annelids
Flatworms
Rotifers
Lophophorans
Eukarya
protists
plants
fungi
animals
common ancestor

Glossary

A

abiogenesis Origin of life from nonliving matter, such as occurred on the early Earth.

abiotic synthesis Process of chemical evolution that resulted in the formation of organic molecules (amino acids, monosaccharides, etc.) from inorganic material.

abscisic acid (ABA) Plant hormone that causes stomata to close and initiates and maintains dormancy.

abscission Dropping of leaves, fruits, or flowers from a land plant.

absolute dating Determining the age of a fossil by direct measurement, usually involving radioisotope decay.

absorption spectrum For photosynthetic pigments, a graph of how much solar radiation is absorbed versus the wavelength of light.

accessory pigment Protein that assists in the photosynthetic process by transferring energy from the photons to the central chlorophyll molecules.

acid Molecules tending to raise the hydrogen ion concentration in a solution and thus to lower its pH numerically.

acid deposition The return to Earth in rain or snow of sulfate or nitrate salts of acids produced by commercial and industrial activities.

acquired immunodeficiency syndrome (AIDS) Disease caused by the HIV virus that destroys helper T cells and macrophages of the immune system, thus preventing an immune response to pathogens; caused by sexual contact with an infected person, intravenous drug use, and transfusions of contaminated blood (rare).

actin One of two major proteins of muscle; makes up thin filaments in myofibrils of muscle fibers. *See also* myosin.

actin filament Component of the cytoskeleton; plays a role in the movement of the cell and its organelles; a protein filament in a sarcomere of a muscle, its movement shortens the sarcomere, yielding muscle contraction.

action potential Electrochemical changes that take place across the axon membrane; the nerve impulse.

active immunity Ability to produce antibodies due to the immune system's response to a microorganism or a vaccine.

active site Region of an enzyme where the substrate binds and where the chemical reaction occurs.

active transport Use of a plasma membrane carrier protein to move a molecule or an ion from a region of lower concentration to one of higher concentration; it opposes equilibrium and requires energy.

adaptation Species modification in structure, function, or behavior that makes a species more suitable to its environment.

adaptive immunity Type of immunity characterized by the action of lymphocytes to specific antigens.

adaptive radiation Rapid evolution of several species from a common ancestor into new ecological or geographical zones.

adenine (A) One of four nitrogen-containing bases in nucleotides composing the structure of DNA and RNA. Pairs with uracil (U) and thymine (T).

adhesion Tendency of water molecules to form hydrogen bonds with polar surfaces, such as the inside of capillaries and transport vessels.

adhesion junction Junction between cells in which the adjacent plasma membranes do not touch but are held together by intercellular filaments attached to buttonlike thickenings.

adipose tissue Connective tissue in which fat is stored.

ADP (adenosine diphosphate) Nucleotide with two phosphate groups that can accept another phosphate group and become ATP.

adrenal cortex Outer portion of the adrenal gland; secretes mineralocorticoids, such as aldosterone, and glucocorticoids, such as cortisol.

adrenal gland Gland that lies atop a kidney; the *adrenal medulla* produces the hormones epinephrine and norepinephrine, and the *adrenal cortex* produces the glucocorticoid and mineralocorticoid hormones.

adrenal medulla Inner portion of the adrenal gland; secretes the hormones epinephrine and norepinephrine.

adrenocorticotropic hormone (ACTH) Hormone secreted by the anterior lobe of the pituitary gland that stimulates activity in the adrenal cortex.

aerobic Chemical process that requires air (oxygen); phase of cellular respiration that requires oxygen.

age structure diagram In demographics, a display of the age groups of a population; a growing population has a pyramid-shaped diagram.

agglutination Clumping of red blood cells due to a reaction between antigens on red blood cell plasma membranes and antibodies in the plasma.

agnathans Fishes that lack jaws; namely, the lampreys and hagfishes.

agranular leukocyte Form of white blood cell that lacks spherical vesicles (granules) in the cytoplasm.

aldosterone Hormone secreted by the adrenal cortex that regulates the sodium and potassium ion balance of the blood.

allantois Extraembryonic membrane that accumulates nitrogenous wastes in the eggs of reptiles, including birds; contributes to the formation of umbilical blood vessels in mammals.

allele Alternative form of a gene; alleles occur at the same locus on homologous chromosomes.

allele frequency Relative proportion of each allele for a gene in the gene pool of a population.

allergy Immune response to substances that usually are not recognized as foreign.

allopatric speciation Model that proposes that new species arise due to an interruption of gene flow between populations that are separated geographically.

alloploidy Polyploid organism that contains the genomes of two or more different species.

allosteric site Site on an allosteric enzyme that binds an effector molecule; binding alters the activity of the enzyme.

alpine tundra Tundra near the peak of a mountain.

alternation of generations Life cycle, typical of land plants, in which a diploid sporophyte alternates with a haploid gametophyte.

altruism Social interaction that has the potential to decrease the lifetime reproductive success of the member exhibiting the behavior.

alveolates Group of protists that includes single-celled dinoflagellates, apicomplexans, and ciliates; alveoli support plasma membrane.

alveolus (pl., alveoli) In humans, terminal, microscopic, grapelike air sac found in lungs.

Alzheimer disease (AD) Disease of the central nervous system (brain) that is characterized by an accumulation of beta amyloid protein and neurofibrillary tangles in the hippocampus and amygdala.

AM fungi (Glomeromycota) Fungi with branching invaginations (arbuscular mycorrhizae, or AM) used to invade plant roots.

amino acid Organic molecule composed of an amino group and an acid group; covalently bonds to produce peptide molecules.

ammonia Nitrogenous end product that takes a limited amount of energy to produce but requires much water to excrete because it is toxic.

amnion Extraembryonic membrane of birds, reptiles, and mammals that forms an enclosing, fluid-filled sac.

amniote Vertebrate that produces an egg surrounded by four membranes, one of which is the amnion; amniote groups are the reptiles, (including birds), and mammals.

amniotic egg Egg that has an amnion, as seen during the development of reptiles, (including birds), and mammals.

amoeboid Cell that moves and engulfs debris with pseudopods.

amoebozoan Supergroup of eukaryotes that includes amoebas and slime molds and is characterized by lobe-shaped pseudopodia.

amphibian Member of vertebrate class Amphibia that includes frogs, toads, and salamanders; they are tied to a watery environment for reproduction.

amygdala Part of the limbic system of the brain; associated with emotional experiences.

amyotrophic lateral sclerosis (ALS) Neurodegenerative disease affecting the motor neurons in the brain and spinal cord, resulting in paralysis and death; also known as Lou Gehrig disease.

anabolic steroid Synthetic steroid that mimics the effect of testosterone.

anabolism Chemical reaction in which smaller molecules (monomers) are combined to form larger molecules (polymers); anabolic metabolism.

anaerobic Chemical reaction that occurs in the absence of oxygen; an example is the fermentation reactions.

analogous, analogous structure Structure that has a similar function in separate lineages but differs in anatomy and ancestry.

analogy Similarity of function but not of origin.

anaphase Fourth phase of mitosis; chromosomes move toward the poles of the spindle.

anaphylactic shock Severe systemic form of anaphylaxis involving bronchiolar constriction, impaired breathing, vasodilation, and a rapid drop in blood pressure with a threat of circulatory failure.

anapsid Characteristic of a vertebrate skull in which there is no opening in the skull behind the eye socket (orbit).

ancestral traits Traits that are found in a common ancestor and its descendants.

androgen Male sex hormone (e.g., testosterone).

aneuploidy Condition in which a cell does not contain the correct number, or combinations, of chromosomes.

angina pectoris Condition characterized by thoracic pain resulting from occluded coronary arteries; may precede a heart attack.

angiogenesis Formation of new blood vessels; rapid angiogenesis is a characteristic of cancer cells.

angiosperm Flowering land plant; the seeds are borne within a fruit.

angiotensin II Hormone produced from angiotensinogen (a plasma protein) by the kidneys and lungs; raises blood pressure.

animals Multicellular, heterotrophic eukaryote that undergoes development to achieve its final form. In general, animals are mobile organisms, characterized by the presence of muscular and nervous tissue.

annelid The segmented worms, such as the earthworm and the clam worm.

annual ring Layer of wood (secondary xylem) usually produced during one growing season.

anterior pituitary Portion of the pituitary gland that is controlled by the hypothalamus and produces six types of hormones, some of which control other endocrine glands.

antheridium (pl., antheridia) Sperm-producing structures, as in the moss life cycle.

anthropoid Group of primates that includes monkeys, apes, and humans.

antibody Protein produced in response to the presence of an antigen; each antibody combines with a specific antigen.

antibody-mediated immunity Specific mechanism of defense in which plasma cells derived from B cells produce antibodies that combine with antigens.

anticodon Three-base sequence in a transfer RNA molecule base that pairs with a complementary codon in mRNA.

antidiuretic hormone (ADH) Hormone secreted by the posterior pituitary that increases the permeability of the collecting ducts in a kidney.

antigen Foreign substance, usually a protein or a polysaccharide, that stimulates the immune system to react, such as to produce antibodies.

antigen-presenting cell (APC) Cell that displays an antigen to certain cells of the immune system so they can defend the body against that particular antigen.

antigen receptor Receptor protein in the plasma membrane of immune system cells whose shape allows them to combine with a specific antigen.

anus Outlet of the digestive tube.

aorta In humans, the major systemic artery that takes blood from the heart to the tissues.

aortic body Sensory receptor in the aortic arch sensitive to the O_2, CO_2, and H^+ content of the blood.

apical dominance Influence of a terminal bud in suppressing the growth of axillary buds.

apical meristem In vascular land plants, masses of cells in the root and shoot that reproduce and elongate as primary growth occurs.

apicomplexan Parasitic protozoans, formerly called sporozoans, that lack mobility and form spores; now named for a unique collection of organelles.

apoptosis Programmed cell death; involves a cascade of specific cellular events leading to death and destruction of the cell.

appendicular skeleton Part of the vertebrate skeleton comprising the appendages, shoulder girdle, and hip girdle.

appendix In humans, small, tubular appendage that extends outward from the cecum of the large intestine.

aquaporin Channel protein through which water can diffuse across a membrane.

arboreal Living in trees.

archaeans (archaea) Prokaryotic organisms that are members of the domain Archaea.

archaeplastid Supergroup of eukaryotes that includes land plants and red and green algae. Developed from endosymbiotic cyanobacteria.

archegonium Egg-producing structures, as in the moss life cycle.

Arctic tundra Biome that encircles the Earth just south of ice-covered polar seas in the Northern Hemisphere.

arteriole Vessel that takes blood from an artery to capillaries.

artery Blood vessel that transports blood away from the heart.

arthritis Condition characterized by an inflammation of the joints; two common forms are osteoarthritis and rheumatoid arthritis.

arthropod Invertebrates, with an exoskeleton and jointed appendages, such as crustaceans and insects.

artificial selection Intentional breeding of certain traits, or combinations of traits, over others to produce a desirable outcome.

ascus Fingerlike sac in which nuclear fusion, meiosis, and ascospore production occur during sexual reproduction of sac fungi.

asexual reproduction Reproduction that requires only one parent and does not involve gametes.

associative learning Acquired ability to associate two stimuli or between a stimulus and a response.

assortative mating Mating of individuals with similar phenotypes.

asthma Condition in which bronchioles constrict and cause difficulty in breathing.

astigmatism Uneven shape of the cornea or lens of the eye; causes a distortion in the light reaching the retina of the eye.

astrocyte Nervous system cell that provides metabolic and structural support to the neurons.

asymmetry Lack of any symmetrical relationship in the morphology of an organism.

atherosclerosis Form of cardiovascular disease characterized by the accumulation of fatty materials (usually cholesterol) in the arteries.

atom Smallest particle of an element that displays the properties of the element.

atomic mass Average of atom mass units for all the isotopes of an atom.

atomic number Number of protons within the nucleus of an atom.

atomic symbol One or two letters that represent the name of an element—e.g., H stands for a hydrogen atom, and Na stands for a sodium atom.

ATP (adenosine triphosphate) Nucleotide with three phosphate groups. The breakdown of ATP into ADP + P makes energy available for energy-requiring processes in cells.

ATP synthase Complex of proteins in the cristae of mitochondria and thylakoid membrane of chloroplasts that produces ATP from the diffusion of hydrogen ions across a membrane.

atrial natriuretic hormone (ANH) Hormone secreted by the heart that increases sodium excretion.

atrioventricular valve Heart valve located between an atrium and a ventricle.

atrium Chamber; particularly an upper chamber of the heart lying above a ventricle.

auditory communication Sound that an animal makes for the purpose of sending a message to another individual.

auditory tube Also called the eustachian tube; connects the middle ear to the nasopharynx for the equalization of pressure.

australopithecine (australopith) One of several species of *Australopithecus,* a genus that contains the first generally recognized humanlike hominins.

autoimmune disease Disease that results when the immune system mistakenly attacks the body's own tissues.

autonomic system Portion of the peripheral nervous system that regulates internal organs.

autoploidy Polyploid organism with multiple chromosome sets all from the same species.

autosome Chromosome pairs that are the same between the sexes; in humans, all but the X and Y chromosomes.

autotroph Organism that can capture energy and synthesize organic molecules from inorganic nutrients.

auxin Plant hormone regulating growth, particularly cell elongation; also called indoleacetic acid (IAA).

axial skeleton Part of the vertebrate skeleton forming the vertical support or axis, including the skull, the rib cage, and the vertebral column.

axon Elongated portion of a neuron that conducts nerve impulses, typically from the cell body to the synapse.

B

bacillus A rod-shaped bacterium; also a genus of bacteria, *Bacillus*.

bacteriophage Virus that infects bacteria.

bacterium (pl., bacteria) Member of the domain Bacteria.

bark External part of a tree, containing cork, cork cambium, and phloem.

Barr body Dark-staining body in the cell nuclei of female mammals that contains a condensed, inactive X chromosome; named after its discoverer, Murray Barr.

base Molecules tending to lower the hydrogen ion concentration in a solution and thus raise the pH numerically.

basidium Clublike structure in which nuclear fusion, meiosis, and basidiospore production occur during sexual reproduction of club fungi.

basophil White blood cell with a granular cytoplasm; able to be stained with a basic dye.

B cell Lymphocyte that matures in the bone marrow and, when stimulated by the presence of a specific antigen, gives rise to antibody-producing plasma cells.

B-cell receptor (BCR) Molecule on the surface of a B cell that binds to a specific antigen.

behavior Observable, coordinated responses to environmental stimuli.

behavioral ecology Study of how natural selection shapes behavior.

benign Mass of cells derived from a single mutated cell that has repeatedly undergone cell division but has remained at the site of origin.

bicarbonate ion Ion that participates in buffering the blood, and the form in which carbon dioxide is transported in the bloodstream.

bilateral symmetry Body plan having two corresponding or complementary halves.

bile Secretion of the liver that is temporarily stored and concentrated in the gallbladder before being released into the small intestine, where it emulsifies fat.

binary fission Splitting of a parent cell into two daughter cells; serves as an asexual form of reproduction in bacteria.

binomial nomenclature Scientific name of an organism, the first part of which designates the genus and the second part of which designates the specific epithet.

biocultural evolution Phase of human evolution in which cultural events affect natural selection.

biodiversity Total number of species, the variability of their genes, and the communities in which they live.

biodiversity hotspot Region of the world that contains unusually large concentrations of species.

biogeochemical cycle Circulating pathway of elements such as carbon and nitrogen involving exchange pools, storage areas, and biotic communities.

biogeography Study of the geographical distribution of organisms.

bioinformatics Area of scientific study that utilizes computer technologies to analyze large sets of data, typically in the study of genomics and proteomics.

biological clock Internal mechanism that maintains a biological rhythm in the absence of environmental stimuli.

biological species concept The concept that defines species as groups of populations that have the potential to interbreed and that are reproductively isolated from other groups.

biology The branch of science that is concerned with the study of life and living organisms.

biomagnification The accumulation of pollutants as they move up the food web.

biomass The number of organisms multiplied by their weight.

biome One of the biosphere's major communities, characterized in particular by certain climatic conditions and particular types of plants.

biomolecule Organic molecules such as proteins, nucleic acids, carbohydrates, and fats.

biosphere Zone of air, land, and water at the surface of the Earth in which living organisms are found.

biotechnology products Commercial or agricultural products that are made with or derived from transgenic organisms.

biotic potential Maximum population growth rate under ideal conditions.

bird Endothermic reptile that has feathers and wings, is often adapted for flight, and lays hard-shelled eggs.

bivalent Homologous chromosomes, each having sister chromatids that are joined by a nucleoprotein lattice during meiosis; also called a tetrad.

bivalve Type of mollusc with a shell composed of two valves; includes clams, oysters, and scallops.

blade Broad, expanded portion of a land plant leaf that may be single or compound leaflets.

blastocoel Fluid-filled cavity of a blastula.

blastocyst Early stage of human embryonic development that consists of a hollow, fluid-filled ball of cells.

blastopore Opening into the primitive gut formed at gastrulation.

blastula Hollow, fluid-filled ball of cells occurring during animal development prior to gastrula formation.

blind spot Region of the retina, lacking rods or cones, where the optic nerve leaves the eye.

blood Fluid circulated by the heart through a closed system of vessels; type of connective tissue.

blood pressure Force of blood pushing against the inside wall of blood vessels.

body cavity In vertebrates, defined regions of the body in which organs reside.

bog Wet, spongy ground in a low-lying area, usually acidic and low in organic nutrients.

bolide Term for a large, crater-forming meteorite that strikes the Earth's surface.

bone Connective tissue having protein fibers and a hard matrix of inorganic salts, notably calcium salts.

bony fishes Vertebrates belonging to the class of fish called Osteichthyes that have a bony, rather than cartilaginous, skeleton.

bottleneck effect Type of genetic drift; occurs when a majority of genotypes are prevented from participating in the production of the next generation as a result of a natural disaster or human interference.

brain Ganglionic mass at the anterior end of the nerve cord; in vertebrates, the brain is located in the cranial cavity of the skull.

brain stem In mammals; portion of the brain consisting of the medulla oblongata, pons, and midbrain.

bronchiole In terrestrial vertebrates, small tube that conducts air from a bronchus to the alveoli.

bronchus (pl., bronchi) In terrestrial vertebrates, branch of the trachea that leads to the lungs.

brown algae Marine photosynthetic protists with a notable abundance of xanthophyll pigments; this group includes well-known seaweeds of northern rocky shores.

bryophyte A nonvascular land plant—including the mosses, liverworts, and hornworts—in which the gametophyte is dominant.

budding Asexual form of reproduction whereby a new organism develops as an outgrowth of the body of the parent.

buffer Substance or group of substances that tend to resist pH changes of a solution, thus stabilizing its relative acidity and basicity.

bulbourethral glands Male sex glands that produce pre-ejaculate fluid that neutralizes acid in the urethra.

bulk transport Movement of substances, usually large particles, across the plasma membrane using vesicles.

C

C_3 plant Plant that fixes carbon dioxide via the Calvin cycle; the first stable product of C_3 photosynthesis is a 3-carbon compound.

C_4 plant Plant that fixes carbon dioxide to produce a C_4 molecule that releases carbon dioxide to the Calvin cycle.

calcitonin Hormone secreted by the thyroid gland that increases the blood calcium level.

calorie Amount of heat energy required to raise the temperature of 1 gram of water 1°C.

Calvin cycle reaction Portion of photosynthesis that takes place in the stroma of chloroplasts and can occur in the dark; it uses the products of the light reactions to reduce CO_2 to a carbohydrate.

calyx The sepals collectively; the outermost flower whorl.

CAM Crassulacean-acid metabolism; a form of photosynthesis in succulent plants that separates the light-dependent and Calvin reactions by time.

camera-type eye Type of eye found in vertebrates and certain molluscs; a single lens focuses an image on closely packed photoreceptors.

camouflage Process of hiding from predators in which an organism's behavior, form, and pattern of coloration allow it to blend into the background and prevent detection.

cancer Malignant tumor whose nondifferentiated cells exhibit loss of contact inhibition, uncontrolled growth, and the ability to invade tissue and metastasize.

capillary Microscopic blood vessel; gases and other substances are exchanged across the walls of a capillary between blood and interstitial fluid.

capsid Protective protein containing the genetic material of a virus.

capsule A form of glycocalyx that consists of a gelatinous layer; found in blue-green algae and certain bacteria.

carbaminohemoglobin Hemoglobin carrying carbon dioxide.

carbohydrate Class of organic compounds that typically contain carbon, hydrogen, and oxygen in a 1:2:1 ratio; includes the monosaccharides, disaccharides, and polysaccharides.

carbon dioxide fixation Process by which carbon dioxide gas is attached to an organic compound; in photosynthesis, this occurs in the Calvin cycle reactions.

carbonic anhydrase Enzyme in red blood cells that speeds the formation of carbonic acid from water and carbon dioxide.

carcinogen Environmental agent that causes mutations leading to the development of cancer.

cardiac cycle One complete cycle of systole and diastole for all heart chambers.

cardiac muscle Striated, involuntary muscle tissue found only in the heart.

cardiac output Blood volume pumped by each ventricle per minute (not total output pumped by both ventricles).

carnivore Consumer in a food chain that eats other animals.

carotenoid An accessory photosynthetic pigment of plants and algae that is often yellow or orange in color; consists of two classes—the xanthophylls and the carotenes.

carotid body Structure located at the branching of the carotid arteries; contains chemoreceptors sensitive to the O_2, CO_2, and H^+ content in blood.

carpel Ovule-bearing unit that is a part of a pistil.

carrier Heterozygous individual who has no apparent abnormality but can pass on an allele for a recessively inherited genetic disorder.

carrier protein Protein in the plasma membrane that combines with and transports a molecule or an ion across the plasma membrane.

carrying capacity (K) Largest number of organisms of a particular species that can be maintained indefinitely by a given environment.

cartilage Connective tissue in which the cells lie within lacunae embedded in a flexible, proteinaceous matrix.

cartilaginous fish Vertebrates that belong to the class of fish called the Chondrichthyes: possess a cartilaginous, rather than bony, skeleton; include sharks, rays, and skates.

Casparian strip Layer of impermeable lignin and suberin bordering four sides of root endodermal cells; prevents water and solute transport between adjacent cells.

catabolism Metabolic process that breaks down large molecules into smaller ones; catabolic metabolism.

cataracts Condition where the lens of the eye becomes opaque, preventing the transmission of light to the retina.

catastrophism Belief, proposed by Georges Cuvier, that periods of catastrophic extinctions occurred, after which repopulation of surviving species took place, giving the appearance of change through time.

cell The smallest unit of life that displays all the properties of life; composed of cytoplasm surrounded by a plasma membrane.

cell body Portion of a neuron that contains a nucleus and from which dendrites and an axon extend.

cell cycle An ordered sequence of events in eukaryotes that involves cell growth and nuclear division; consists of the stages G_1, S, G_2, and M.

cell envelope In a prokaryotic cell, the portion composed of the plasma membrane, the cell wall, and the glycocalyx.

cell-mediated immunity Specific mechanism of defense in which T cells destroy antigen-bearing cells.

cell plate Structure across a dividing plant cell that signals the location of new plasma membranes and cell walls.

cell recognition protein Glycoproteins in the plasma membrane that identify self and help the body defend itself against pathogens.

cell suspension culture Small clumps of naked plant cells grown in tissue culture that produce drugs, cosmetics, or agricultural chemicals, among others.

cell theory One of the major theories of biology, which states that all organisms are made up of cells; cells are capable of self-reproduction and come only from preexisting cells.

cell wall Cellular structure that surrounds a plant, protistan, fungal, or bacterial cell and maintains the cell's shape and rigidity; composed of polysaccharides.

cellular differentiation Process and developmental stages by which a cell becomes specialized for a particular function.

cellular respiration Metabolic reactions that use the energy from carbohydrate, fatty acid, or amino acid breakdown to produce ATP molecules.

cellular response Response to the transduction pathway in which proteins or enzymes change a signal to a format that the cell can understand, resulting in the appropriate response.

cellular slime mold Free-living amoeboid cells that feed on bacteria and yeasts by phagocytosis and aggregate to form a plasmodium that produces spores.

cellulose Polysaccharide that is the major complex carbohydrate in plant cell walls.

centipede Elongated arthropod characterized by having one pair of legs to each body segment; they may have 15 to 173 pairs of legs.

central dogma Processes that dictate the flow of information from the DNA to RNA to protein in a cell.

central nervous system (CNS) Portion of the nervous system consisting of the brain and spinal cord.

central vacuole In a plant cell, a large, fluid-filled sac that stores metabolites. During growth, it enlarges, forcing the primary cell wall to expand and the cell surface-area-to-volume ratio to increase.

centriole Cell structure, existing in pairs, that occurs in the centrosome and may help organize a mitotic spindle for chromosome movement during animal cell division.

centromere Constriction where sister chromatids of a chromosome are held together.

centrosome Central microtubule organizing center of cells. In animal cells, it contains two centrioles.

cephalization Having a well-recognized anterior head with a brain and sensory receptors.

cephalochordate Small, fishlike invertebrate that is a member of the phylum Chordata. Probably the closest living relative to vertebrates.

cephalopod Type of mollusc in which the head is prominent and the foot is modified to form two arms and several tentacles; includes squids, cuttlefish, octopuses, and nautiluses.

cerebellum In terrestrial vertebrates, portion of the brain that coordinates skeletal muscles to produce smooth, graceful motions.

cerebral cortex Outer layer of cerebral hemispheres; receives sensory information and controls motor activities.

cerebral hemisphere Either of the two lobes of the cerebrum in vertebrates.

cerebrospinal fluid Fluid found in the ventricles of the brain, in the central canal of the spinal cord, and in association with the meninges.

cerebrum Largest part of the brain in mammals.

cervix Narrow end of the uterus, which leads into the vagina.

channel protein Protein that forms a channel to allow a particular molecule or ion to cross the plasma membrane.

chaparral Biome characterized by broad-leafed evergreen shrubs forming dense thickets.

character displacement Tendency for character-istics to be more divergent when similar species belong to the same community than when they are isolated from one another.

charophyte Type of living green algae that on the basis of nucleotide sequencing and cellular features is most closely related to land plants.

chelicerate Arthropods (e.g., horseshoe crabs, sea spiders, arachnids) that exhibit a pair of pointed appendages used to manipulate food.

chemical energy Energy associated with the interaction of atoms in a molecule.

chemical signal Molecule that brings about a change in a cell, a tissue, an organ, or an individual when it binds to a specific receptor.

chemiosmosis Process by which mitochondria and chloroplasts use the energy of an electron transport chain to create a hydrogen ion gradient that drives ATP formation.

chemoautotroph Organism able to synthesize organic molecules by using carbon dioxide as the carbon source and the oxidation of an inorganic substance (such as hydrogen sulfide) as the energy source.

chemoheterotroph Organism that is unable to produce its own organic molecules and therefore requires organic nutrients in its diet.

chemoreceptor Sensory receptor that is sensitive to chemical stimulation—for example, receptors for taste and smell.

chitin Strong but flexible nitrogenous polysaccharide found in the exoskeleton of arthropods and in the cell walls of fungi.

chlorophyll Green photosynthetic pigment of algae and plants that absorbs solar energy; occurs as chlorophyll *a* and chlorophyll *b*.

chlorophyte Most abundant and diverse group of green algae, including freshwater, marine, and terrestrial forms that synthesize. Chlorophytes share chemical and anatomical characteristics with land plants.

chloroplast Membrane-bound organelle in algae and plants with chlorophyll-containing membranous thylakoids; where photosynthesis takes place.

choanoflagellate Single-celled choanoflagellates have one and colonial forms have many collar cells like those of sponges; choanoflagellates are the protists most closely related to animals.

cholesterol A steroid found in the plasma membrane of animal cells and from which other types of steroids are derived.

chordate Animal that has a dorsal tubular nerve cord, a notochord, pharyngeal gill pouches, and a postanal tail at some point in its life cycle; includes a few types of invertebrates (e.g., sea squirts and lancelets) and the vertebrates.

chorion Extraembryonic membrane functioning for respiratory exchange in birds and reptiles; contributes to placenta formation in mammals.

chorionic villus In placental mammals treelike extension of the chorion, projecting into the maternal tissues at the placenta.

choroid Vascular, pigmented middle layer of the eyeball.

chromalveolate Supergroup of eukaryotes that includes alveolates and stramenopiles.

chromatid Following replication, a chromosome consists of a pair of sister chromatids, held together at the centromere; each chromatid is comprised of a single DNA helix.

chromatin Network of DNA strands and associated proteins observed within a nucleus of a cell.

chromosome The structure that transmits the genetic material from one generation to the next; composed of condensed chromatin; each species has a particular number of chromosomes that is passed on to the next generation.

chyme Thick, semiliquid food material that passes from the stomach to the small intestine.

chytrid (Chytridiomycota) Mostly aquatic fungi with flagellated spores that may represent the most ancestral fungal lineage.

cilia (sing., cilium) Short, hairlike projections from the plasma membrane, occurring usually in larger numbers.

ciliary muscle Within the ciliary body of the vertebrate eye, the ciliary muscle that controls the shape of the lens.

ciliate Complex single-celled protist that moves by means of cilia and digests food in food vacuoles.

circadian rhythm Biological rhythm with a 24-hour cycle.

circulatory system In animals, an organ system that moves substances to and from cells, usually via a heart, blood, and blood vessels.

cirrhosis Chronic, irreversible injury to liver tissue; commonly caused by frequent alcohol consumption.

citric acid cycle Cycle of reactions in mitochondria that begins with citric acid. This cycle breaks down an acetyl group and produces CO_2, ATP, NADH, and $FADH_2$; also called the Krebs cycle.

clade Evolutionary lineage consisting of an ancestral species and all of its descendants, forming a distinct branch on a cladogram.

cladistics Method of systematics that uses derived characters to determine monophyletic groups and construct cladograms.

cladogram In cladistics, a branching diagram that shows the relationship among species in regard to their shared derived characters.

class One of the categories, or taxa, used by taxonomists to group species; the taxon above the order level.

classification Process of naming organisms and assigning them to taxonomic groups (taxa).

classical conditioning Type of learning whereby an unconditioned stimulus that elicits a specific response is paired with a neutral stimulus, so that the response becomes conditioned.

cleavage Cell division without cytoplasmic addition or enlargement; occurs during the first stage of animal development.

cleavage furrow Indentation in the plasma membrane of animal cells during cell division; formation marks the start of cytokinesis.

climate Generalized weather patterns of an area, primarily determined by temperature and average rainfall.

climate change Recent changes in the Earth's climate; evidence suggests that this is primarily due to human influence, including the increased release of greenhouse gases.

climax community In ecology, community that results when succession has come to an end.

clonal selection theory States that the antigen selects which lymphocyte will undergo clonal expansion and produce more lymphocytes bearing the same type of receptor.

cloning Production of identical copies. In organisms, the production of organisms with the same genes; in genetic engineering, the production of many identical copies of a gene.

closed circulatory system A type of circulatory system where blood is confined to vessels and is kept separate from the interstitial fluid.

clotting Also called coagulation, the response of the body to an injury in the vessels of the circulatory system; involves platelets and clotting proteins.

club fungi (Basidiomycota) Fungi that produce spores in club-shaped basidia within a fruiting body; includes mushrooms, shelf fungi, and puffballs.

cnidarian Invertebrate existing as either a polyp or medusa with two tissue layers and radial symmetry.

coacervate droplet An aggregate of colloidal droplets held together by electrostatic forces.

coccus A spherical-shaped bacterium.

cochlea Spiral-shaped structure of the vertebrate inner ear containing the sensory receptors for hearing.

codominance Inheritance pattern in which both alleles of a gene are equally expressed in a heterozygote.

codon Three-base sequence in messenger RNA that during translation directs the addition of a particular amino acid into a protein or directs termination of the process.

coelom Body cavity of an animal; the method by which the coelom is formed (or lack of formation) is an identifying characteristic in animal classification.

coenzyme Nonprotein organic molecule that aids the action of the enzyme to which it is loosely bound.

coevolution Mutual evolution in which two species exert selective pressures on the other species.

cofactor Nonprotein assistant required by an enzyme in order to function; many cofactors are metal ions, others are coenzymes.

cohesion Tendency of water molecules to form hydrogen bonds with other water molecules.

cohesion-tension model Explanation for upward transport of water in xylem based upon transpiration-created tension and the cohesive properties of water molecules.

cohort Group of individuals having a statistical factor in common, such as year of birth, in a population study.

coleoptile Protective sheath that covers the young leaves of a seedling.

collagen fiber White fiber in the matrix of connective tissue, giving flexibility and strength.

collecting duct Duct within the kidney that receives fluid from several nephrons; the reabsorption of water occurs here.

collenchyma Plant tissue composed of cells with unevenly thickened walls; supports growth of stems and petioles.

colonial flagellate hypothesis Hypothesis that all animals are descended from an ancestor that resembled a hollow colony of flagellated cells, most likely resembling modern choanoflagellates.

colony Loose association of cells, each remaining independent for most functions.

columnar epithelium Type of epithelial tissue with cylindrical cells.

comb jelly Invertebrate that resembles a jelly fish and is the largest animal to be propelled by beating cilia.

commensalism Symbiotic relationship in which one species is benefited, and the other is neither harmed nor benefited.

common ancestor Ancestor common to at least two lines of descent.

communication Signal by a sender that influences the behavior of a receiver.

community Assemblage of species interacting with one another within the same environment.

community diversity Variety of species in a particular locale, dependent on the species interactions.

compact bone Type of bone that contains osteons consisting of concentric layers of matrix and osteocytes in lacunae.

comparative genomics Study of genomes through the direct comparison of their genes and DNA sequences from multiple species.

competition Results when members of a species attempt to use a resource that is in limited supply.

competitive exclusion principle Theory that no two species can occupy the same niche in the same place and at the same time.

competitive inhibition Form of enzyme inhibition where the substrate and inhibitor are both able to bind to the enzyme's active site. Only when the substrate is at the active site will product form.

complement Collective name for a series of enzymes and activators in the blood, some of which may bind to an antibody and may lead to rupture of a foreign cell.

complementary base pairing Hydrogen bonding between particular purines and pyrimidines; responsible for the structure of DNA, and some RNA, molecules.

complementary DNA (cDNA) DNA that has been synthesized from mRNA by the action of reverse transcriptase.

complete digestive tract Digestive tract that has both a mouth and an anus.

compound Substance having two or more different elements in a fixed ratio.

compound eye Type of eye found in arthropods; it is composed of many independent visual units.

concentration gradient Gradual change in chemical concentration between two areas of differing concentrations.

conclusion Statement made following an experiment as to whether or not the results support the hypothesis.

cone Reproductive structure in conifers made up of scales bearing sporangia; pollen cones bear microsporangia, and seed cones bear megasporangia.

cone cell Photoreceptor in vertebrate eyes that responds to bright light and makes color vision possible.

conidiospore Spore produced by sac and club fungi during asexual reproduction.

conifer Member of a group of cone-bearing gymnosperm land plants that includes pine, cedar, and spruce trees.
conjugation Transfer of genetic material from one cell to another.
conjugation pilus (pl., conjugation pili) In a bacterium, elongated, hollow appendage used to transfer DNA to other cells.
conjunctiva Delicate membrane that lines the eyelid, protecting the sclera.
connective tissue Type of animal tissue that binds structures together, provides support and protection, fills spaces, stores fat, and forms blood cells; adipose tissue, cartilage, bone, and blood are types of connective tissue; living cells in a nonliving matrix.
conservation biology Discipline that seeks to understand the effects of human activities on species, communities, and ecosystems and to develop practical approaches to preventing the extinction of species and the destruction of ecosystems.
consumer Organism that feeds on another organism in a food chain generally; primary consumers eat plants, and secondary consumers eat animals.
continental drift The movement of the Earth's crust by plate tectonics, resulting in the movement of continents with respect to one another.
contraceptive vaccine Under development, this birth control method immunizes against the hormone HCG, crucial to maintaining implantation of the embryo.
control Sample that goes through all the steps of an experiment but does not contain the variable being tested; a standard against which the results of an experiment are checked.
convergent evolution Similarity in structure in distantly related groups generally due to similiar selective pressures in like environments.
copulation Sexual union between a male and a female.
coral reef Coral formations in shallow tropical waters that support an abundance of diversity.
corepressor Molecule that binds to a repressor, allowing the repressor to bind to an operator in a repressible operon.
cork Outer covering of the bark of trees; made of dead cells that may be sloughed off.
cork cambium Lateral meristem that produces cork.
cornea Transparent, anterior portion of the outer layer of the eyeball.
corolla The petals, collectively; usually, the conspicuously colored flower whorl.
corpus luteum Follicle that has released an egg and increases its secretion of progesterone.
cortex In plants, ground tissue bounded by the epidermis and vascular tissue in stems and roots; in animals, outer layer of an organ, such as the cortex of the kidney or adrenal gland.
cortisol Glucocorticoid secreted by the adrenal cortex that responds to stress on a long-term basis; reduces inflammation and promotes protein and fat metabolism.
cost-benefit analysis A weighing-out of the costs and benefits (in terms of contributions to reproductive success) of a particular strategy or behavior.
cotyledon Seed leaf for embryo of a flowering plant; provides nutrient molecules for the developing plant before photosynthesis begins.
countercurrent exchange Fluids flow side-by-side in opposite directions, as in the exchange of fluids in the kidneys.
coupled reactions Reactions that occur simultaneously; one is an exergonic reaction that releases energy, and the other is an endergonic reaction that requires an input of energy in order to occur.
covalent bond Chemical bond in which atoms share one pair of electrons.
cranial nerve Nerve that arises from the brain.
crenation In animal cells, shriveling of the cell due to water leaving the cell when the environment is hypertonic.
cristae (sing., crista) Short, fingerlike projections formed by the folding of the inner membrane of mitochondria.
Cro-Magnon Common name for the first fossils to be designated *Homo sapiens*.
crossing-over Exchange of segments between nonsister chromatids of a bivalent during meiosis.
crustacean Member of a group of aquatic arthropods that contains, among others, shrimps, crabs, crayfish, and lobsters.
cryptic species Species that are very similar in appearance but are considered separate species based on other characteristics, such as behavior or genetics.
cuboidal epithelium Type of epithelial tissue with cube-shaped cells.
cutaneous receptor Sensory receptors of the dermis that are activated by touch, pain, pressure, and temperature.
cuticle Waxy layer covering the epidermis of plants that protects the plant against water loss and disease-causing organisms.
cyanobacterium (pl., cyanobacteria) Photosynthetic bacterium that contains chlorophyll and releases oxygen; formerly called a blue-green alga.
cycad Type of gymnosperm with palmate leaves and massive cones; cycads are most often found in the tropics and subtropics.
cyclic adenosine monophosphate (cAMP) ATP-related compound that acts as the second messenger in peptide hormone transduction; it initiates activity of the metabolic machinery.
cyclin Protein that cycles in quantity as the cell cycle progresses; combines with and activates the kinases that promote the events of the cycle.
cyst In protists and invertebrates, resting structure that contains reproductive bodies or embryos.
cystic fibrosis (CF) Genetic disease caused by a defect in the *CFTR* gene, which is responsible for the formation of a transmembrane chloride ion transporter; causes the mucus of the body to be viscous.
cytochrome Any of several iron-containing protein molecules that are members of the electron transport chain in photosynthesis and cellular respiration.
cytokine Type of protein secreted by a T lymphocyte that attacks viruses, virally infected cells, and cancer cells.
cytokinesis Division of the cytoplasm following mitosis or meiosis.
cytokinin Plant hormone that promotes cell division; often works in combination with auxin during organ development in plant embryos.
cytoplasm Region of a cell between the nucleus, or the nucleoid region of a bacterium, and the plasma membrane; contains the organelles of the cell.
cytoplasmic segregation Process that parcels out the maternal determinants, which play a role in development, during mitosis.
cytosine (C) One of four nitrogen-containing bases in the nucleotides composing the structure of DNA and RNA; pairs with guanine.
cytoskeleton Internal framework of the cell, consisting of microtubules, actin filaments, and intermediate filaments.
cytotoxic T cell T lymphocyte that attacks and kills antigen-bearing cells.

D

data (sing., datum) Facts or information collected through observation and/or experimentation.
day-neutral plant Plant whose flowering is not dependent on day length—e.g., tomato and cucumber.
deamination Removal of an amino group (—NH_2) from an amino acid or other organic compound.
deciduous Land plant that sheds its leaves annually.
decomposer Organism, usually a bacterium or fungus, that breaks down organic matter into inorganic nutrients that can be recycled in the environment.
deductive reasoning The use of general principles to predict specific outcomes. Often uses "if . . . then" statements.
dehydration reaction Chemical reaction in which a water molecule is released during the formation of a covalent bond.
delayed allergic response Allergic response initiated at the site of the allergen by sensitized T cells, involving macrophages and regulated by cytokines.
deletion Change in chromosome structure in which the end of a chromosome breaks off or two simultaneous breaks lead to the loss of an internal segment; often causes abnormalities—e.g., cri du chat syndrome.
demographic transition Due to industrialization, a decline in the birthrate following a reduction in the death rate, so that the population growth rate is lowered.
demography Properties of the rate of growth and the age structure of populations.
denatured Loss of a protein's or an enzyme's normal shape, so that it no longer functions; usually caused by a less than optimal pH and temperature.
dendrite Part of a neuron that sends signals toward the cell body.
dendritic cell Antigen-presenting cell of the epidermis and mucous membranes.
denitrification Conversion of nitrate or nitrite to nitrogen gas by bacteria in soil.
dense fibrous connective tissue Type of connective tissue containing many collagen fibers packed together; found in tendons and ligaments, for example.
density-dependent factor Biotic factor, such as disease or competition, that affects population size in a direct relationship to the population's density.
density-independent factor Abiotic factor, such as fire or flood, that affects population size independent of the population's density.
deoxyribose Pentose sugar found in DNA.

derived trait Structural, physiological, or behavioral trait that is present in a specific lineage and is not present in the common ancestor for several related lineages.

dermis In mammals, thick layer of the skin underlying the epidermis.

desert Ecological biome characterized by a limited amount of rainfall; deserts have hot days and cool nights.

desmosome(s) Intercellular junctions that connect cytoskeletons of adjacent cells.

detritivore Any organism that obtains most of its nutrients from the detritus in an ecosystem.

deuterostome Group of coelomate animals in which the second embryonic opening is associated with the mouth; the first embryonic opening, the blastopore, is associated with the anus.

development Process of regulated growth and differentiation of cells and tissues.

diabetic retinopathy Complication of diabetes that causes the capillaries in the retina to become damaged, potentially causing blindness.

diagnostic traits Characteristics that distinguish species from one another.

diaphragm In mammals, dome-shaped, muscularized sheet separating the thoracic cavity from the abdominal cavity; contraceptive device that prevents sperm from reaching the egg.

diapsid Characteristic of a vertebrate skull in which there are two openings in the skull behind the eye socket (orbit).

diarrhea Excessively frequent and watery bowel movements.

diastole Relaxation period of a heart chamber during the cardiac cycle.

diatom Golden-brown alga with a cell wall in two parts, or valves; significant component of phytoplankton.

differentiation Specialization in the structure or function of a cell typically caused by the activation of specific genes.

diffusion Movement of molecules or ions from a region of higher to lower concentration; it requires no energy and tends to lead to an equal distribution (equilibrium).

dihybrid cross Cross between parents that differ in two traits.

dikaryotic Having two haploid nuclei that stem from different parent cells; during sexual reproduction, sac and club fungi have dikaryotic cells.

dinoflagellate Photosynthetic single-celled protist with two flagella, one whiplash and the other located within a groove between protective cellulose plates; significant part of phytoplankton.

dinosaur General term used to describe the large reptiles that existed prior to the start of the Cretaceous period.

dioecious Having unisexual flowers or cones, with the male flowers or cones confined to certain land plants and the female flowers or cones of the same species confined to other different plants.

diploid (2n) Cell condition in which two of each type of chromosome are present.

diplomonad Protist that has modified mitochondria, two equal-sized nuclei, and multiple flagella.

directional selection Outcome of natural selection in which an extreme phenotype is favored, usually in a changing environment.

disaccharide Sugar that contains two monosaccharide units; e.g., maltose.

disruptive selection Outcome of natural selection in which the two extreme phenotypes are favored over the average phenotype, leading to more than one distinct form.

distal convoluted tubule Final portion of a nephron that joins with a collecting duct; associated with tubular secretion.

diverge Process by which a new evolutionary path begins; on a phylogenetic tree, this is indicated by branching lines.

DNA (deoxyribonucleic acid) Nucleic acid polymer produced from covalent bonding of nucleotide monomers that contain the sugar deoxyribose; the genetic material of nearly all organisms.

DNA ligase Enzyme that links DNA fragments; used during production of recombinant DNA to join foreign DNA to vector DNA.

DNA microarray Glass or plastic slide containing thousands of single-stranded DNA fragments arranged in an array (grid); used to detect and measure gene expression; also called gene chips.

DNA polymerase During replication, an enzyme that joins the nucleotides complementary to a DNA template.

DNA repair enzyme One of several enzymes that restore the original base sequence in an altered DNA strand.

DNA replication Synthesis of a new DNA double helix prior to mitosis and meiosis in eukaryotic cells and during prokaryotic fission in prokaryotic cells.

domain Largest of the categories, or taxa, used by taxonomists to group species; the three domains are Archaea, Bacteria, and Eukarya.

domain Archaea One of the three domains of life; contains prokaryotic cells that often live in extreme habitats and have unique genetic, biochemical, and physiological characteristics; its members are sometimes referred to as *archaea*.

domain Bacteria One of the three domains of life; contains prokaryotic cells that differ from archaea because they have their own unique genetic, biochemical, and physiological characteristics.

domain Eukarya One of the three domains of life, consisting of organisms with eukaryotic cells; includes protists, fungi, plants, and animals.

dominance hierarchy Organization of animals in a group that determines the order in which the animals have access to resources.

dominant allele Allele that exerts its phenotypic effect in the heterozygote; it masks the expression of the recessive allele.

dormancy In plants, a cessation of growth under conditions that seem appropriate for growth.

dorsal root ganglion Mass of sensory neuron cell bodies located in the dorsal root of a spinal nerve.

double fertilization In flowering plants, one sperm nucleus unites with the egg nucleus, and a second sperm nucleus unites with the polar nuclei of an embryo sac.

double helix Double spiral; describes the three-dimensional shape of DNA.

doubling time Number of years it takes for a population to double in size.

dryopithecine Tree-dwelling primate existing 12–9 MYA; ancestral to apes.

duodenum First part of the small intestine, where chyme enters from the stomach.

duplication Change in chromosome structure in which a particular segment is present more than once in the same chromosome.

E

ecdysozoan Protostome characterized by periodic molting of its exoskeleton. Includes the roundworms and arthropods.

echinoderm Invertebrates such as sea stars, sea urchins, and sand dollars; characterized by radial symmetry and a water vascular system.

ecological niche Role an organism plays in its community, including its habitat and its interactions with other organisms.

ecological pyramid Visual depiction of the biomass, number of organisms, or energy content of various trophic levels in a food web—from the producer to the final consumer populations.

ecological release In an ecosystem, the freedom of a species to expand its use of available resources due to an elimination of competition.

ecological succession The gradual replacement of communities in an area following a disturbance (secondary succession) or the creation of new soil (primary succession).

ecology Study of the interactions of organisms with other organisms and with the physical and chemical environment.

ecosystem Biological community together with the associated abiotic environment; characterized by a flow of energy and a cycling of inorganic nutrients.

ectoderm Outermost primary tissue layer of an animal embryo; gives rise to the nervous system and the outer layer of the integument.

ectotherm Organism having a body temperature that varies according to the environmental temperature.

edge effect Phenomenon in which the edges around a landscape patch provide a slightly different habitat than the favorable habitat in the interior of the patch.

effector Muscle or gland that receives signals from motor fibers and thereby allows an organism to respond to environmental stimuli.

egg Also called an ovum; haploid cell that is usually fertilized by a sperm to form a diploid zygote.

elastic cartilage Type of cartilage composed of elastic fibers, allowing greater flexibility.

elastic fiber Yellow fiber in the matrix of connective tissue, providing flexibility.

electrocardiogram (ECG) Recording of the electrical activity associated with the heartbeat.

electron Negative subatomic particle, moving about in an energy level around the nucleus of an atom.

electronegativity The ability of an atom to attract electrons toward itself in a chemical bond.

electron shell The average location, or energy level, of an electron in an atom. Often drawn as concentric circles around the nucleus.

electron transport chain (ETC) Process in a cell that involves the passage of electrons along a series of membrane-bound electron carrier molecules from a higher to lower energy level; the energy released is used for the synthesis of ATP.

element Substance that cannot be broken down into substances with different properties; composed of only one type of atom.

El Niño–Southern Oscillation (ENSO) Warming of water in the eastern Pacific equatorial region such that the Humboldt Current is displaced, with possible negative results, such as reduction in marine life.

elongation Middle stage of translation in which additional amino acids specified by the mRNA are added to the growing polypeptide.

embryo Stage of a multicellular organism that develops from a zygote before it becomes free-living; in seed plants, the embryo is part of the seed.

embryo sac Female gametophyte (megagametophyte) of flowering plants.

embryonic development In humans, the first 2 months of development following fertilization, during which the major organs are formed.

embryonic disk During human development, flattened area during gastrulation from which the embryo arises.

embryophyta Bryophytes and vascular plants; both of which produce embryos.

emergent property A function or trait that appears as biological complexity increases.

emerging viruses Newly identified viruses that are becoming more prominent, usually because they cause serious disease.

emphysema Disorder of the respiratory system, specifically the lungs, that is characterized by damage to the alveoli, thus reducing the ability to exchange gases with the external environment.

endangered A species that is in peril of immediate extinction throughout all or most of its range (e.g., California condor, snow leopard).

endergonic reaction Chemical reaction that requires an input of energy; opposite of exergonic reaction.

endocrine gland Ductless organ that secretes hormone(s) into the bloodstream.

endocrine system Organ system involved in the coordination of body activities; uses hormones as chemical signals secreted into the bloodstream.

endocytosis Process by which substances are moved into the cell from the environment; includes phagocytosis, pinocytosis, and receptor-mediated endocytosis.

endoderm Innermost primary tissue layer of an animal embryo that gives rise to the linings of the digestive tract and associated structures.

endodermis Internal plant root tissue forming a boundary between the cortex and the vascular cylinder.

endomembrane system Cellular system that consists of the nuclear envelope, endoplasmic reticulum, Golgi apparatus, and vesicles.

endometrium Mucous membrane lining the interior surface of the uterus.

endoplasmic reticulum (ER) System of membranous saccules and channels in the cytoplasm, often with attached ribosomes.

endoskeleton Protective internal skeleton, as in vertebrates.

endosperm In flowering plants, nutritive storage tissue that is derived from the union of a sperm nucleus and polar nuclei in the embryo sac.

endospore Spore formed within a cell; certain bacteria form endospores.

endosymbiosis *See* endosymbiotic theory.

endosymbiotic theory Explanation of the evolution of eukaryotic organelles by phagocytosis of prokaryotes.

endotherm Organism in which maintenance of a constant body temperature is independent of the environmental temperature.

energy Capacity to do work and bring about change; occurs in a variety of forms.

energy of activation (E_a) Energy that must be added in order for molecules to react with one another.

enhancer DNA sequence that acts as a regulatory element to increase the level of transcription when regulatory proteins, such as transcription activators, bind to it.

entropy Measure of disorder or randomness in a system.

enzymatic protein Protein that catalyzes a specific reaction; may be found in the plasma membrane or the cytoplasm of the cell.

enzyme Organic catalyst, usually a protein, that speeds a reaction in cells due to its particular shape.

enzyme inhibition Means by which cells regulate enzyme activity; may be competitive or noncompetitive inhibition.

eosinophil White blood cell containing cytoplasmic granules that stain with acidic dye.

epidermal tissue Exterior tissue, usually one cell thick, of leaves, young stems, roots, and other parts of plants.

epidermis In mammals, the outer, protective layer of the skin; in plants, tissue that covers roots, leaves, and stems of nonwoody organisms.

epididymis Location of sperm maturation in an adult human; located in the testis.

epigenetic inheritance An inheritance pattern in which a nuclear gene has been modified but the changed expression of the gene is not permanent over many generations; the transmission of genetic information by means that are not based on the coding sequences of a gene.

epiglottis Structure that covers the glottis, the air-tract opening, during the process of swallowing.

epinephrine Hormone secreted by the adrenal medulla in times of stress; adrenaline.

epiphyte Plant that takes its nourishment from the air because its placement in other plants gives it an aerial position.

epithelial tissue Tissue that lines hollow organs and covers surfaces.

erythropoietin (EPO) Hormone produced by the kidneys that speeds red blood cell formation.

esophagus Muscular tube for moving swallowed food from the pharynx to the stomach.

essential nutrient In plants, substance required for normal growth, development, or reproduction; in humans, a nutrient that cannot be produced in sufficient quantities by the body and must be obtained from the diet.

estrogen Female sex hormone that helps maintain sexual organs and secondary sex characteristics.

estuary Portion of the ocean located where a river enters and fresh water mixes with salt water.

ethylene Plant hormone that causes ripening of fruit and is involved in abscission.

etiolate The yellowing of leaves and increase in shoot length that occur when a plant is grown in the dark.

euchromatin Chromatin with a lower level of compaction and therefore accessible for transcription.

eudicot (Eudicotyledone) Flowering plant group; members have two embryonic leaves (cotyledons), net-veined leaves, vascular bundles in a ring, flower parts in fours or fives and their multiples, and other characteristics.

euglenid Flagellated and flexible freshwater single-celled protist that usually contains chloroplasts and has a semirigid cell wall.

eukaryotic cell (eukaryote) Type of cell that has a membrane-bound nucleus and membranous organelles; found in organisms within the domain Eukarya.

euploidy Condition in which a cell contains the correct number, and combinations, of chromosomes.

eutrophication Enrichment of water by inorganic nutrients used by phytoplankton. Often, overenrichment caused by human activities leads to excessive bacterial growth and oxygen depletion.

evergreen Land plant that sheds leaves over a long period, so some leaves are always present.

evolution Genetic change in a species over time, resulting in the development of genetic and phenotypic differences that are the basis of natural selection; descent of organisms from a common ancestor.

evolutionary species concept Every species has its own evolutionary history, which is partly documented in the fossil record.

excavate Supergroup of eukaryotes that includes euglenids, kinetoplastids, parabasalids, and diplomonads.

excretion Elimination of metabolic wastes by an organism at exchange boundaries such as the plasma membrane of single-celled organisms and excretory tubules of multicellular animals.

exergonic reaction Chemical reaction that releases energy; opposite of endergonic reaction.

exocrine gland Gland that secretes its product to an epithelial surface directly or through ducts.

exocytosis Process in which an intracellular vesicle fuses with the plasma membrane, so that the vesicle's contents are released outside the cell.

exon Segment of mRNA containing the protein-coding portion of a gene that remains within the mRNA after splicing has occurred.

exophthalmos Condition associated with an enlargement of the thyroid gland accompanied by an abnormal protrusion of the eyes.

exoskeleton Protective external skeleton, as in arthropods.

exotic species Nonnative species that migrate or are introduced by humans into a new ecosystem; also called alien species.

experiment A test of a hypothesis that examines the influence of a single variable. Often involves both control and test groups.

experimental design Methodology by which an experiment will seek to support the hypothesis.

experimental variable Factor of the experiment being tested.

expiration Act of expelling air from the lungs; exhalation.

exponential growth Growth, particularly of a population, in which the increase occurs in the same manner as compound interest.

extant Species, or other levels of taxa, that are still living.

external respiration Exchange of oxygen and carbon dioxide between alveoli of the lungs and blood.

extinction Total disappearance of a species or higher group.

extracellular matrix (ECM) Nonliving substance secreted by some animal cells; is composed of protein and polysaccharides.

extraembryonic membrane Membrane that is not a part of the embryo but is necessary to the continued existence and health of the embryo.

ex vivo gene therapy Gene therapy in which cells are removed from an organism, and DNA is injected to correct a genetic defect; the cells are returned to the organism to treat a disease or disorder.

F

facilitated transport Passive transfer of a substance into or out of a cell along a concentration gradient by a process that requires a protein carrier.

facultative anaerobe Prokaryote that is able to grow in either the presence or the absence of gaseous oxygen.

FAD Flavin adenine dinucleotide; a coenzyme of oxidation-reduction that becomes $FADH_2$ as oxidation of substrates occurs in the mitochondria during cellular respiration; FAD then delivers electrons to the electron transport chain.

FAD+ (flavin adenine dinucleotide) A coenzyme that delivers electrons to the electron transport chain; when reduced forms $FADH_2$.

fall overturn Mixing process that occurs in fall in stratified lakes, whereby oxygen-rich top waters mix with nutrient-rich bottom waters.

family One of the categories, or taxa, used by taxonomists to group species; the taxon located above the genus level.

farsighted Condition in which an individual cannot focus on objects closely; caused by the focusing of the image behind the retina of the eye.

fat Organic molecule that contains glycerol and three fatty acids; energy storage molecule.

fatty acid Molecule that contains a hydrocarbon chain and ends with an acid group.

fermentation Anaerobic breakdown of glucose that results in a gain of two ATP and end products, such as alcohol and lactate; occurs in the cytoplasm of cells.

fern Member of a group of land plants that have large fronds; in the sexual life cycle, the independent gametophyte produces flagellated sperm, and the vascular sporophyte produces windblown spores.

fertilization Fusion of sperm and egg nuclei, producing a zygote that develops into a new individual.

fibroblast Cell found in loose connective tissue that synthesizes collagen and elastic fibers in the matrix.

fibrocartilage Cartilage with a matrix of strong collagenous fibers.

fibrous root system In most monocots, a mass of similarly sized roots that cling to the soil.

filament End-to-end chains of cells that form as cell division occurs in only one plane; in plants, the elongated stalk of a stamen.

fimbria (pl., fimbriae) Small, bristlelike fiber on the surface of a bacterial cell, which attaches bacteria to a surface; also fingerlike extension from the oviduct near the ovary.

fin In fishes and other aquatic animals, membranous, winglike, or paddlelike process used to propel, balance, or guide the body.

first messenger Chemical signal such as a peptide hormone that binds to a plasma membrane receptor protein and alters the metabolism of a cell because a second messenger is activated.

fishes Aquatic, gill-breathing vertebrates that usually have fins and skin covered with scales; fishes were among the earliest vertebrates that evolved.

fitness Ability of an organism to reproduce and pass its genes to the next fertile generation; measured against the ability of other organisms to reproduce in the same environment.

five-kingdom system System of classification that contains the kingdoms Monera, Protista, Plantae, Animalia, and Fungi.

fixed action pattern (FAP) Innate behavior pattern that is stereotyped, spontaneous, independent of immediate control, genetically encoded, and independent of individual learning.

flagellum (pl., flagella) Long, slender extension used for locomotion by some bacteria, protozoans, and sperm.

flagship species Species that evoke a strong emotional response in humans; charismatic, cute, regal (e.g., lions, tigers, dolphins, pandas).

flatworm Invertebrates such as planarians and tapeworms with extremely thin bodies, a three-branched gastrovascular cavity, and a ladder-type nervous system.

flower Reproductive organ of a flowering plant, consisting of several kinds of modified leaves arranged in concentric rings and attached to a modified stem called the receptacle.

fluid-mosaic model Model for the plasma membrane based on the changing location and pattern of protein molecules in a fluid phospholipid bilayer.

follicle Structure in the ovary of animals that contains an oocyte; site of oocyte production.

follicular phase First half of the ovarian cycle, during which the follicle matures and much estrogen (and some progesterone) is produced.

food chain The order in which one population feeds on another in an ecosystem, thereby showing the flow of energy from a detritivore (detrital food chain) or a producer (grazing food chain) to the final consumer.

food web In ecosystems, a complex pattern of interlocking and crisscrossing food chains.

foraminiferan A protist bearing a calcium carbonate test with many openings through which pseudopods extend.

formed elements Portion of the blood that consists of erythrocytes, leukocytes, and platelets (thrombocytes).

formula A group of symbols and numbers used to express the composition of a compound.

fossil Any past evidence of an organism that has been preserved in the Earth's crust.

founder effect Cause of genetic drift due to colonization by a limited number of individuals who, by chance, have different genotype and allele frequencies than the parent population.

fovea centralis Region of the retina consisting of densely packed cones; responsible for the greatest visual acuity.

frameshift mutation Insertion or deletion of at least one base, so that the reading frame of the corresponding mRNA changes.

free energy Energy in a system that is capable of performing work.

frond Leaf of a fern, palm, or cycad.

fruit Flowering plant structure consisting of one or more ripened ovaries that usually contain seeds.

fruiting body Spore-producing and spore-disseminating structure found in sac and club fungi.

functional genomics Study of gene function at the genome level. It involves the study of many genes simultaneously and the use of DNA microarrays.

functional group Specific cluster of atoms attached to the carbon skeleton of organic molecules that enters into reactions and behaves in a predictable way.

fungus (pl., fungi) Eukaryotic saprotrophic decomposer; the body is made up of filaments called hyphae that form a mass called a mycelium.

G

gallbladder Organ attached to the liver that stores and concentrates bile.

gamete Haploid sex cell; e.g., egg or sperm.

gametogenesis Development of the male and female sex gametes.

gametophyte Haploid generation of the alternation-of-generations life cycle of a plant; produces gametes that unite to form a diploid zygote.

ganglion (pl., ganglia) Collection or bundle of neuron cell bodies usually outside the central nervous system.

gap junction Junction between cells formed by the joining of two adjacent plasma membranes; it lends strength and allows ions, sugars, and small molecules to pass between cells.

gastropod Mollusc with a broad, flat foot for crawling (e.g., snails and slugs).

gastrovascular cavity Blind digestive cavity in animals that have a sac body plan.

gastrula Stage of animal development during which the germ layers form, at least in part, by invagination.

gastrulation Formation of a gastrula from a blastula; characterized by an invagination to form cell layers of a caplike structure.

gel electrophoresis Process that separates molecules, such as proteins and DNA, based on their size and electrical charge by passing them through a matrix.

gene Unit of heredity existing as alleles on the chromosomes; in diploid organisms, typically two alleles are inherited—one from each parent.

gene cloning DNA cloning to produce many identical copies of the same gene.

gene flow Sharing of genes between two populations through interbreeding.

gene mutation Altered gene whose sequence of bases differs from the original sequence.

gene pharming Production of pharmaceuticals using transgenic organisms, usually agricultural animals.

gene pool Total of the alleles of all the individuals in a population.

gene therapy Correction of a detrimental mutation by the insertion of DNA sequences into the genome of a cell.

genetically modified organism (GMO) Organism whose genetic material has been altered or enhanced using DNA technology.

genetic code Universal code that has existed for eons and allows for conversion of DNA and RNA's chemical code to a sequence of amino acids in a protein. Each codon consists of three bases that stand for one of the 20 amino acids found in proteins or directs the termination of translation.

genetic diversity Variety among members of a population.

genetic drift Mechanism of evolution due to random changes in the allelic frequencies of a population; more likely to occur in small populations or when only a few individuals of a large population reproduce.

genetic profile An individual's genome, including any possible mutations.

genetic recombination Process in which chromosomes are broken and rejoined to form novel combinations; in this way, offspring receive alleles in combinations different from their parents.

genomics Area of study that examines the genome of a species or group of species.

genotype Genes of an organism for a particular trait or traits; often designated by letters—for example, *BB* or *Aa*.

genus One of the categories, or taxa, used by taxonomists to group species; contains those species that are most closely related through evolution.

geologic timescale History of the Earth based on the fossil record and divided into eras, periods, and epochs.

germ cell During zygote development, cells that are set aside from the somatic cells and that will eventually undergo meiosis to produce gametes.

germinate Beginning of growth of a seed, spore, or zygote, especially after a period of dormancy.

germ layer Primary tissue layer of a vertebrate embryo—namely, ectoderm, mesoderm, or endoderm.

gibberellin Plant hormone promoting increased stem growth; also involved in flowering and seed germination.

gills Respiratory organ in most aquatic animals; in fish, an outward extension of the pharynx.

ginkgo Member of phylum Ginkgophyta; maidenhair tree.

girdling Removing a strip of bark from around a tree.

gland Epithelial cell or group of epithelial cells that are specialized to secrete a substance.

glaucoma Condition in which the fluid in the eye (aqueous humor) accumulates, increasing pressure in the eye and damaging nerve fibers.

global warming Predicted increase in the Earth's temperature due to human activities that promote the greenhouse effect.

glomerular capsule Cuplike structure that is the initial portion of a nephron.

glomerular filtration Movement of small molecules from the glomerulus into the glomerular capsule due to the action of blood pressure.

glomerulus Capillary network within the glomerular capsule of a nephron.

glottis Opening for airflow in the larynx.

glucagon Hormone produced by the pancreas that stimulates the liver to break down glycogen, thus raising blood glucose levels.

glucocorticoid Type of hormone secreted by the adrenal cortex that influences carbohydrate, fat, and protein metabolism; *see also* cortisol.

glucose Six-carbon monosaccharide; used as an energy source during cellular respiration and as a monomer of the structural polysaccharides.

glycerol Three-carbon carbohydrate with three hydroxyl groups attached; a component of fats and oils.

glycocalyx Gel-like coating outside the cell wall of a bacterium. If compact, it is called a capsule; if diffuse, it is called a slime layer.

glycogen Storage polysaccharide found in animals; composed of glucose molecules joined in a linear fashion but having numerous branches.

glycolipid Lipid in plasma membranes that contains an attached carbohydrate chain; assembled in the Golgi apparatus.

glycolysis Anaerobic breakdown of glucose that results in a gain of two ATP and the production of pyruvate; occurs in the cytoplasm of cells.

glycoprotein Protein in plasma membranes that has an attached carbohydrate chain; assembled in the Golgi apparatus.

gnathostome Term used to describe any vertebrates with jaws.

gnetophyte Member of one of the four phyla of gymnosperms; Gnetophyta has only three living genera, which differ greatly from one another—e.g., *Welwitschia* and *Ephedra*.

golden brown algae Single-celled organism that contains pigments, including chlorophyll *a* and *c* and carotenoids, that produce its color.

Golgi apparatus Organelle consisting of sacs and vesicles that processes, packages, and distributes molecules about or from the cell.

gonad Organ that produces gametes; the ovary produces eggs, and the testis produces sperm.

gonadotropic hormone Substance secreted by the anterior pituitary that regulates the activity of the ovaries and testes; principally, follicle-stimulating hormone (FSH) and luteinizing hormone (LH).

granular leukocyte Form of white blood cell that contains spherical vesicles (granules) in its cytoplasm.

granum (pl., grana) Stack of chlorophyll-containing thylakoids in a chloroplast.

grassland Biome characterized by rainfall greater than 25 cm/yr, grazing animals, and warm summers; includes the prairie of the U.S. Midwest and the African savanna.

gravitational equilibrium Maintenance of balance when the head and body are motionless.

gravitropism Growth response of roots and stems of plants to the Earth's gravity; roots demonstrate positive gravitropism, and stems demonstrate negative gravitropism.

gray matter Nonmyelinated axons and cell bodies in the central nervous system.

green algae Members of a diverse group of photosynthetic protists; contain chlorophylls *a* and *b* and have other biochemical characteristics like those of plants.

greenhouse effect Reradiation of solar heat toward the Earth, caused by an atmosphere that allows the sun's rays to pass through but traps the heat in the same manner as the glass of a greenhouse.

greenhouse gases Gases in the atmosphere, such as carbon dioxide, methane, water vapor, ozone, and nitrous oxide, that are involved in the greenhouse effect.

ground tissue Tissue that constitutes most of the body of a plant; consists of parenchyma, collenchyma, and sclerenchyma cells that function in storage, basic metabolism, and support.

growth factor A hormone or chemical, secreted by one cell, that may stimulate or inhibit growth of another cell or cells.

growth hormone (GH) Substance secreted by the anterior pituitary; controls size of an individual by promoting cell division, protein synthesis, and bone growth.

guanine (G) One of four nitrogen-containing bases in nucleotides composing the structure of DNA and RNA; pairs with cytosine.

guard cell One of two cells that surround a leaf stoma; changes in the turgor pressure of these cells cause the stoma to open or close.

guttation Liberation of water droplets from the edges and tips of leaves.

gymnosperm Type of woody seed plant in which the seeds are not enclosed by fruit and are usually borne in cones, such as those of the conifers.

H

habitat Place where an organism lives and is able to survive and reproduce.

hair follicle Tubelike depression in the skin in which a hair develops.

halophile Type of archaean that lives in extremely salty habitats.

haploid (n) Cell condition in which only one of each type of chromosome is present.

Hardy-Weinberg principle (equilibrium) Mathematical law stating that the gene frequencies in a population remain stable if evolution does not occur due to nonrandom mating, selection, migration, and genetic drift.

heart Muscular organ whose contraction causes blood to circulate in the body of an animal.

heart attack Damage to the myocardium due to blocked circulation in the coronary arteries; myocardial infarction.

heat Type of kinetic energy associated with the random motion of molecules.

helper T cell Secretes lymphokines, which stimulate all kinds of immune cells.

heme Iron-containing group found in hemoglobin.

hemizygous Possessing only one allele for a gene in a diploid organism; males are hemizygous for genes on the X chromosome.

hemocoel Residual coelom found in arthropods, which is filled with hemolymph.

hemoglobin (Hb) Iron-containing respiratory pigment occurring in vertebrate red blood cells and in the blood plasma of some invertebrates.

hemolymph Circulatory fluid that is a mixture of blood and interstitial fluid; seen in animals that have an open circulatory system, such as molluscs and arthropods.

hepatitis Inflammation of the liver. Viral hepatitis occurs in several forms.

herbaceous stem Nonwoody stem; herbaceous plants tend to die back to ground level at the end of the growing season.

herbivore Primary consumer in a grazing food chain; a plant eater.

hermaphroditic Type of animal that has both male and female sex organs.

heterochromatin Highly compacted chromatin that is not accessible for transcription.

heterosporous Seed plant that produces two types of spores—microspores and megaspores. A plant that produces only one type of spore is *homosporous*.

heterotroph Organism that cannot synthesize needed organic compounds from inorganic substances and therefore must take in organic food.

heterozygote advantage Situation in which individuals heterozygous for a trait have a selective advantage over those who are homozygous dominant or recessive; an example is sickle-cell disease.

heterozygous Possessing unlike alleles for a particular trait.

hexose Any monosaccharide that contains six carbons; examples are glucose and galactose.

hippocampus Region of the central nervous system associated with learning and memory; part of the limbic system.

histamine Substance, produced by basophils in blood and mast cells in connective tissue, that causes capillaries to dilate.

histone A group of proteins involved in forming the nucleosome structure of eukaryote chromatin.

homeobox 180-nucleotide sequence located in all homeotic genes.

homeodomain Conserved DNA-binding region of transcription factors encoded by the homeobox of homeotic genes.

homeostasis Maintenance of normal internal conditions in a cell or an organism by means of self-regulating mechanisms.

homeothermic Ability to regulate the internal body temperature around an optimum value.

homeotic genes Genes that control the overall body plan by controlling the fate of groups of cells during development.

hominid Taxon that includes humans, chimpanzees, gorillas, and orangutans.

hominin Taxon that includes humans and species very closely related to humans plus chimpanzees.

hominine Taxon that includes humans, chimpanzees, and gorillas.

hominoid Taxon that includes the hominids plus the gibbons.

homologous chromosome (homologue) Member of a pair of chromosomes that are alike and come together in synapsis during prophase of the first meiotic division; a *homologue*.

homologous gene Gene that codes for the same protein, even if the base sequence may be different.

homologous, homologous structure A structure that is similar in different types of organisms because these organisms descended from a common ancestor.

homology Similarity of parts or organs of different organisms caused by evolutionary derivation from a corresponding part or organ in a remote ancestor, usually having a similar embryonic origin.

homosporous A plant that produces only one type of asexual spore.

homozygous Possessing two identical alleles for a particular trait.

hormone Chemical messenger produced in one part of the body that controls the activity of other parts.

hornwort A bryophyte of the phylum Anthocerophyta with a thin gametophyte and tiny sporophyte that resembles a broom handle.

horsetail A seedless vascular plant having only one genus (*Equisetum*) in existence today; characterized by rhizomes, scalelike leaves, strobili, and tough, rigid stems.

host Organism that provides nourishment and/or shelter for a parasite.

human chorionic gonadotropin (HCG) Gonadotropic hormone produced by the chorion that maintains the uterine lining.

Human Genome Project (HGP) Initiative to determine the complete sequence of the human genome and to analyze this information.

human immunodeficiency virus (HIV) Virus responsible for the disease AIDS.

humus Decomposing organic matter in the soil.

hunter-gatherer A hominin that hunted animals and gathered plants for food.

hyaline cartilage Cartilage whose cells lie in lacunae separated by a white, translucent matrix containing very fine collagen fibers.

hybridization Interbreeding between two different species; typically prevented by prezygotic isolation mechanisms.

hydrogen bond Weak bond that arises between a slightly positive hydrogen atom of one molecule and a slightly negative atom of another molecule or between parts of the same molecule.

hydrogen ion (H^+) Hydrogen atom that has lost its electron and therefore bears a positive charge.

hydrolysis reaction Splitting of a chemical bond by the addition of water, with the H^+ going to one molecule and the OH^- going to the other.

hydrophilic Type of molecule, often polar, that interacts with water by dissolving in water and/or by forming hydrogen bonds with water molecules.

hydrophobic Type of molecule that is typically nonpolar and therefore does not interact easily with water.

hydroponics Technique for growing plants by suspending them with their roots in a nutrient solution.

hydrostatic skeleton Fluid-filled body compartment that provides support for muscle contraction resulting in movement; seen in cnidarians, flatworms, roundworms, and segmented worms.

hydrothermal vent Hot springs in the seafloor along ocean ridges where heated seawater and sulfate react to produce hydrogen sulfide; here, chemosynthetic bacteria support a community of varied organisms.

hydroxide ion (OH^-) One of two ions that result when a water molecule dissociates; it has gained an electron and therefore bears a negative charge.

hyperthyroidism Caused by the oversecretion of hormones from the thyroid gland; symptoms include hyperactivity, nervousness, and insomnia.

hypertonic solution Higher-solute concentration (less water) than the cytoplasm of a cell; causes cell to lose water by osmosis.

hypha Filament of the vegetative body of a fungus.

hypertension Form of cardiovascular disease characterized by blood pressure over 140/95 (over 45 years of age) or 130/90 (under 45 years of age).

hypothalamic-inhibiting hormone One of many hormones produced by the hypothalamus that inhibits the secretion of an anterior pituitary hormone.

hypothalamic-releasing hormone One of many hormones produced by the hypothalamus that stimulates the secretion of an anterior pituitary hormone.

hypothalamus In vertebrates, part of the brain that helps regulate the internal environment of the body—for example, heart rate, body temperature, and water balance.

hypothesis Supposition established by reasoning after consideration of available evidence; it can be tested by obtaining more data, often by experimentation.

hypothyroidism Caused by the undersecretion of hormones from the thyroid gland; symptoms include weight gain, lethargic behavior, and depression.

hypotonic solution Solution that contains a lower-solute (more water) concentration than the cytoplasm of a cell; causes cell to gain water by osmosis.

I

immediate allergic response Allergic response that occurs within seconds of contact with an allergen; caused by the attachment of the allergen to IgE antibodies.

immune system System associated with protection against pathogens, toxins, and some cancerous cells; in humans, this is an organ system.

immunity Ability of the body to protect itself from foreign substances and cells, including disease-causing agents.

immunization Strategy for achieving immunity to the effects of specific disease-causing agents.

immunoglobulin (Ig) Globular plasma protein that functions as an antibody.

implantation In placental mammals, the embedding of an embryo at the blastocyst stage into the endometrium of the uterus.

imprinting Learning to make a particular response to only one type of animal or object.

inbreeding Mating between closely related individuals; influences the genotype ratios of the gene pool.

inclusive fitness Fitness that results from personal reproduction and from helping nondescendant relatives reproduce.

incomplete digestive tract Digestive tract that has a single opening, usually called a mouth.

incomplete dominance Inheritance pattern in which an offspring has an intermediate phenotype, as when a red-flowered plant and a white-flowered plant produce pink-flowered offspring.

incomplete penetrance Dominant alleles that are either not always or partially expressed.

independent assortment Alleles of unlinked genes segregate independently of each other during meiosis, so that the gametes can contain all possible combinations of alleles.

index fossil Deposits found in certain layers of strata; similar fossils can be found in the same strata around the world.

induced fit model Change in the shape of an enzyme's active site that enhances the fit between the active site and its substrate(s).

induced mutation Mutation that is caused by an outside influence, such as organic chemicals or ionizing radiation.

inducer Molecule that brings about activity of an operon by joining with a repressor and preventing it from binding to the operator.

induction Ability of a chemical or a tissue to influence the development of another tissue.

inductive reasoning Using specific observations and the process of logic and reasoning to arrive at general scientific principles.

infertility Inability to have as many children as desired.

inflammatory response Tissue response to injury that is characterized by redness, swelling, pain, and heat.

ingroup In a cladistic study of evolutionary relationships among organisms, the group that is being analyzed.

inheritance of acquired characteristics Lamarckian belief that characteristics acquired during the lifetime of an organism can be passed on to offspring.

initiation First stage of translation in which the translational machinery binds an mRNA and assembles.

innate immunity An immune response that does not require a previous exposure to the pathogen.

inner ear Portion of the ear consisting of a vestibule, semicircular canals, and the cochlea; where equilibrium is maintained and sound is transmitted.

inorganic chemistry Branch of science that studies the chemical reactions and properties of all of the elements, except hydrogen and carbon.

insect Type of arthropod. The head has antennae, compound eyes, and simple eyes; the thorax has three pairs of legs and often wings; and the abdomen has internal organs.

insight learning Ability to apply prior learning to a new situation without trial-and-error activity.

inspiration Act of taking air into the lungs; inhalation.

insulin Hormone released by the pancreas that lowers blood glucose levels by stimulating the uptake of glucose by cells, especially muscle and liver cells.

integration Summing up of excitatory and inhibitory signals by a neuron or by some part of the brain.

interferon Antiviral agent produced by an infected cell that blocks the infection of another cell.

intergenic sequence Region of DNA that lies between genes on a chromosome.

interkinesis Period of time between meiosis I and meiosis II during which no DNA replication takes place.

intermediate filament Ropelike assemblies of fibrous polypeptides in the cytoskeleton that provide support and strength to cells; so called because they are intermediate in size between actin filaments and microtubules.

internal respiration Exchange of oxygen and carbon dioxide between blood and interstitial fluid.

interneuron Neuron located within the central nervous system that conveys messages between parts of the central nervous system.

internode In vascular plants, the region of a stem between two successive nodes.

interphase Stages of the cell cycle (G_1, S, G_2) during which growth and DNA synthesis occur when the nucleus is not actively dividing.

interspersed repeat Repeated DNA sequence that is spread across several regions of a chromosome or across multiple chromosomes.

interstitial fluid Fluid that surrounds the body's cells; consists of dissolved substances that leave the blood capillaries by filtration and diffusion.

intertidal zone Region along a coastline where the tide recedes and returns.

intron Intervening sequence found between exons in mRNA; removed by RNA processing before translation.

inversion Change in chromosome structure in which a segment of a chromosome is turned around 180°; this reversed sequence of genes can lead to altered gene activity and abnormalities.

invertebrate Animal without a vertebral column or backbone.

ion Charged particle that carries a negative or positive charge.

ionic bond Chemical bond in which ions are attracted to one another by opposite charges.

iris Muscular ring that surrounds the pupil and regulates the passage of light through this opening.

iron-sulfur world Hypothesis that ocean thermal vents provided all of the materials needed for abiotic synthesis of the first molecules.

island biogeography model Proposes that the biodiversity on an island is dependent on its distance from the mainland, with islands located a greater distance having a lower level of diversity.

isomer Molecules with the same molecular formula but a different structure, and therefore a different shape.

isotonic solution Solution that is equal in solute concentration to that of the cytoplasm of a cell; causes cell to neither lose nor gain water by osmosis.

isotope Atoms of the same element having the same atomic number but a different mass number due to a variation in the number of neutrons.

iteroparity Repeated production of offspring at intervals throughout the life cycle of an organism.

J

jaundice Yellowish tint to the skin caused by an abnormal amount of bilirubin (bile pigment) in the blood, indicating liver malfunction.

jawless fishes (agnathans) Type of fishes that lack jaws; includes hagfishes and lampreys.

joint Articulation between two bones of a skeleton.

junction protein Protein in the cell membrane that assists in cell-to-cell communication.

K

karyotype Chromosomes arranged by pairs according to their size, shape, and general appearance in mitotic metaphase.

keystone species Species whose activities significantly affect community structure.

kidneys Paired organs of the vertebrate urinary system that regulate the chemical composition of the blood and produce a waste product called urine.

kinetic energy Energy associated with motion.

kinetochore An assembly of proteins that attaches to the centromere of a chromosome during mitosis.

kinetoplastid Single-celled, flagellate protist characterized by the presence in their single mitochondrion of a kinetoplast (a structure containing a large mass of DNA).

kingdom One of the categories, or taxa, used by taxonomists to group species; the taxon above phylum.

kin selection Indirect selection; adaptation to the environment due to the reproductive success of an individual's relatives.

***K*-selection** Favorable life-history strategy under stable environmental conditions characterized by the production of a few offspring with much attention given to offspring survival.

L

lactation Secretion of milk by mammary glands, usually for the nourishment of an infant.

lacteal Lymphatic vessel in an intestinal villus; aids in the absorption of fats.

lacuna Small pit or hollow cavity, as in bone or cartilage, where a cell or cells are located.

lake Body of fresh water, often classified by nutrient status, such as oligotrophic (nutrient- poor) or eutrophic (nutrient-rich).

landscape A number of interacting ecosystems.

landscape diversity Variety of habitat elements within an ecosystem (e.g., plains, mountains, and rivers).

large intestine In vertebrates, portion of the digestive tract that follows the small intestine; in humans, consists of the cecum, colon, rectum, and anal canal.

larynx Cartilaginous organ located between the pharynx and the trachea; in humans, contains the vocal cords; sometimes called the voice box.

last universal common ancestor (LUCA) The first living organism on the planet, from which all life evolved.

lateral line Canal system containing sensory receptors that allow fishes and amphibians to detect water currents and pressure waves from nearby objects.

law Universal principle that describes the basic functions of the natural world.

law of independent assortment Mendelian principle that explains how combinations of traits appear in gametes; *see also* independent assortment.

law of segregation Mendelian principle that explains how, in a diploid organism, alleles separate during the formation of the gametes.

laws of thermodynamics Two laws explaining energy and its relationships and exchanges. The first, also called the "law of conservation," says that energy cannot be created or destroyed but can only be changed from one form to another; the second says that energy cannot be changed from one form to another without a loss of usable energy.

leaf Lateral appendage of a stem, highly variable in structure, often containing cells that carry out photosynthesis.

leaf vein Vascular tissue within a leaf.

learning Relatively permanent change in an animal's behavior that results from practice and experience.
lens Transparent, disclike structure found in the vertebrate eye behind the iris; brings objects into focus on the retina.
leptin Hormone produced by adipose tissue that acts on the hypothalamus to signal satiety (fullness).
less-developed country (LDC) Country that is becoming industrialized; typically, population growth is expanding rapidly, and the majority of people live in poverty.
leukocyte Another name for a white blood cell (WBC).
lichen Symbiotic relationship between certain fungi and either cyanobacteria or algae, in which the fungi possibly provide inorganic food or water and the algae or cyanobacteria provide organic food.
life cycle Recurring pattern of genetically programmed events by which individuals grow, develop, maintain themselves, and reproduce.
ligament Tough cord or band of dense, fibrous tissue that binds bone to bone at a joint.
light reaction Portion of photosynthesis that captures solar energy and takes place in thylakoid membranes of chloroplasts; it produces ATP and NADPH.
lignin Chemical that hardens the cell walls of land plants.
limbic system In humans, functional association of various brain centers, including the amygdala and hippocampus; governs learning and memory and various emotions, such as pleasure, fear, and happiness.
limiting factor Resource or environmental condition that restricts the abundance and distribution of an organism.
lineage Line of descent represented by a branch in a phylogenetic tree.
lipase Fat-digesting enzyme secreted by the pancreas.
lipid Class of organic compounds that tend to be soluble in nonpolar solvents; includes fats and oils.
liposome Droplet of phospholipid molecules formed in a liquid environment.
liver Large, dark red internal organ that produces urea and bile, detoxifies the blood, stores glycogen, and produces the plasma proteins, among other functions.
liverwort Bryophyte with a lobed or leafy gametophyte and a sporophyte composed of a stalk and capsule.
lobe-finned fishes Fishes with limblike fins; also called the sarcopterygians.
locus Physical location of a trait (or gene) on a chromosome.
logistic growth Population increase that results in an S-shaped curve; growth is slow at first, steepens, and then levels off due to environmental resistance.
long-day plant Plant that flowers when day length is longer than a critical length; e.g., wheat, barley, clover, and spinach.
loop of the nephron Portion of a nephron between the proximal and distal convoluted tubules; functions in water reabsorption.
loose fibrous connective tissue Tissue composed mainly of fibroblasts widely separated by a matrix containing collagen and elastic fibers.
lophophoran A general term to describe several groups of lophotrochozoans that have a feeding structure called a lophophore.
lophotrochozoa Main group of protostomes; widely diverse. Includes the flatworms, rotifers, annelids, and molluscs.
lumen Cavity inside any tubular structure, such as the lumen of the digestive tract.
lung cancer Uncontrolled cell growth that affects any component of the respiratory system.
lungs Internal respiratory organ containing moist surfaces for gas exchange.
lungfishes Type of lobe-finned fishes that utilize lungs in addition to gills for gas exchange.
luteal phase Second half of the ovarian cycle, during which the corpus luteum develops and much progesterone (and some estrogen) is produced.
lycophyte Club mosses, among the first vascular plants to evolve and to have leaves. The leaves of the lycophytes are microphylls.
lymph Fluid, derived from interstitial fluid, that is carried in lymphatic vessels.
lymphatic capillary Smallest vessel of the lymphatic system; closed-ended; responsible for the uptake of fluids from the surrounding tissues.
lymphatic system Organ system consisting of lymphatic vessels and lymphatic organs; transports lymph and lipids, and aids the immune system.
lymphatic vessel Vessel that carries lymph.
lymph node Mass of lymphatic tissue located along the course of a lymphatic vessel.
lymphocyte Specialized white blood cell that functions in specific defense; occurs in two forms—T lymphocytes and B lymphocytes.
lymphoid (lymphatic) organ Organ other than a lymphatic vessel that is part of the lymphatic system; the lymphatic organs are the lymph nodes, tonsils, spleen, thymus, and bone marrow.
lysogenic cell Cell that contains a prophage (virus incorporated into DNA), which is replicated when the cell divides.
lysogenic cycle Bacteriophage life cycle in which the virus incorporates its DNA into that of a bacterium; occurs preliminary to the lytic cycle.
lysosome Membrane-bound vesicle that contains hydrolytic enzymes for digesting macromolecules and bacteria; used to recycle worn-out cellular organelles.
lytic cycle Bacteriophage life cycle in which the virus takes over the operation of the bacterium immediately upon entering it and subsequently destroys the bacterium.

M

macroevolution Large-scale evolutionary change, such as the formation of new species.
macronutrient Essential element needed in large amounts for plant growth, such as nitrogen, calcium, or sulfur.
macrophage In vertebrates, large phagocytic cell derived from a monocyte that ingests microbes and debris.
macular degeneration Condition in which the capillaries supplying the retina of the eye become damaged, resulting in reduced vision and blindness.
malignant The power to threaten life; cancerous.
maltase Enzyme produced in the small intestine that breaks down maltose to two glucose molecules.
mammal Endothermic vertebrate characterized especially by the presence of hair and mammary glands.
mantle In molluscs, an extension of the body wall that covers the visceral mass and may secrete a shell.
marsh Soft wetland that is treeless.
marsupial Member of a group of mammals bearing immature young nursed in a marsupium, or pouch—for example, kangaroo and opossum.
mass extinction Episode of large-scale extinction in which large numbers of species disappear in a few million years or less.
mass number Mass of an atom equal to the number of protons plus the number of neutrons within the nucleus.
mast cell Connective tissue cell that releases histamine in allergic reactions.
maternal determinant One of many substances present in the egg that influences the course of development.
matrix Unstructured, semifluid substance that fills the space between cells in connective tissues or inside organelles.
matter Anything that takes up space and has mass.
mechanoreceptor Sensory receptor that responds to mechanical stimuli, such as pressure, sound waves, or gravity.
medulla oblongata In vertebrates, part of the brain stem that is continuous with the spinal cord; controls heartbeat, blood pressure, breathing, and other vital functions.
medusa Among cnidarians, bell-shaped body form that is directed downward and contains much mesoglea.
megafauna Large animals, such as humans, bears, and deer. Often defined as those over 100 pounds (44 kg) in adult size.
megaphyll Large leaf with several to many veins.
megaspore One of the two types of spores produced by seed plants; develops into a female gametophyte (embryo sac).
meiosis Type of nuclear division that reduces the chromosome number from 2n to n; daughter cells receive the haploid number of chromosomes in varied combinations.
melanocyte Specialized cell in the epidermis that produces melanin, the pigment responsible for skin color.
melanocyte-stimulating hormone (MSH) Substance that causes melanocytes to secrete melanin in most vertebrates.
melatonin Hormone, secreted by the pineal gland, that is involved in biorhythms.
membrane-first hypothesis Proposes that the plasma membrane was the first component of the early cells to evolve.
memory Capacity of the brain to store and retrieve information about past sensations and perceptions; essential to learning.
memory B cell Forms during a primary immune response but enters a resting phase until a secondary immune response occurs.
memory T cell T cell that differentiated during an initial infection and responds rapidly during subsequent exposure to the same antigen.
meninges Protective membranous coverings around the central nervous system.

meningitis Inflammation of the brain or spinal cord meninges (membranes).

menopause Termination of the ovarian and uterine cycles in older women.

menstruation Periodic shedding of tissue and blood from the inner lining of the uterus in primates.

meristem Undifferentiated embryonic tissue in the active growth regions of plants.

mesoderm Middle primary tissue layer of an animal embryo that gives rise to muscle, several internal organs, and connective tissue layers.

mesoglea Transparent, jellylike substance located between the endoderm and ectoderm of some sponges and cnidarians.

mesophyll Inner, thickest layer of a leaf consisting of palisade and spongy mesophyll; the site of most of photosynthesis.

mesosome In a bacterium, plasma membrane that folds into the cytoplasm and increases surface area.

messenger RNA (mRNA) Type of RNA formed from a DNA template and bearing coded information for the amino acid sequence of a polypeptide.

metabolic pathway Series of linked reactions, beginning with a particular reactant and terminating with an end product.

metabolic pool Metabolites that are the products of and/or the substrates for key reactions in cells, allowing one type of molecule to be changed into another type, such as carbohydrates converted to fats.

metabolism The sum of the chemical reactions that occur in a cell.

metamorphosis Change in shape and form that some animals, such as insects, undergo during development.

metaphase Third phase of mitosis; chromosomes are aligned at the metaphase plate.

metaphase plate A disk formed during metaphase in which all of a cell's chromosomes lie in a single plane at right angles to the spindle fibers.

metapopulation Population subdivided into several small and isolated populations due to habitat fragmentation.

metastasis Spread of cancer from the place of origin throughout the body; caused by the ability of cancer cells to migrate and invade tissues.

methanogen Type of archaean that lives in oxygen-free habitats, such as swamps, and releases methane gas.

MHC (major histocompatibility complex) protein Protein marker that is a part of cell-surface markers anchored in the plasma membrane, which the immune system uses to identify "self."

micelle Single layer of fatty acids (or phospholipids) that orientate themselves in an aqueous environment.

microevolution Change in gene frequencies between populations of a species over time.

microglia Supportive cells of the nervous system that help remove bacteria and debris, thus supporting the activity of the neurons.

micronutrient Essential element needed in small amounts for plant growth, such as boron, copper, and zinc.

microphyll Small leaf with one vein.

microRNA (miRNA) Short sequences of RNA, usually less than 22 nucleotides, that are involved in posttranscriptional regulation of gene expression. These molecules either inhibit, or reduce, the expression of specific genes.

microsphere Formed from proteinoids exposed to water; has properties similar to those of today's cells.

microspore One of the two types of spores produced by seed plants; develops into a male gametophyte (pollen grain).

microtubule Small, cylindrical organelle composed of tubulin protein around an empty central core; present in the cytoplasm, centrioles, cilia, and flagella.

midbrain In mammals, the part of the brain located below the thalamus and above the pons.

middle ear Portion of the ear consisting of the tympanic membrane, the oval and round windows, and the ossicles, where sound is amplified.

migration Regular back-and-forth movement of animals between two geographic areas at particular times of the year.

millipede More-or-less cylindrical arthropod characterized by having two pairs of short legs on most of its body segments; may have 13 to almost 200 pairs of legs.

mimicry Superficial resemblance of two or more species; a survival mechanism that avoids predation by appearing to be noxious.

mineral Naturally occurring inorganic substance containing two or more elements; certain minerals are needed in the diet.

mineralocorticoid Hormones secreted by the adrenal cortex that regulate salt and water balance, leading to increases in blood volume and blood pressure.

mitochondria (sing., mitochondrion) Membrane-bound organelle in which ATP molecules are produced during the process of cellular respiration.

mitosis The stage of cellular reproduction in which nuclear division occurs; process in which a parent nucleus produces two daughter nuclei, each having the same number and kinds of chromosomes as the parent nucleus.

mitotic spindle A complex of microtubules and associated proteins that assist in separating the chromatids during cell division.

mixotrophic Organism that can use autotrophic and heterotrophic means of gaining nutrients.

model Simulation of a process that aids conceptual understanding until the process can be studied firsthand; a hypothesis that describes how a particular process could be carried out.

mold Various fungi whose body consists of a mass of hyphae (filaments) that grow on and receive nourishment from organic matter, such as human food and clothing.

mole The molecular weight of a molecule expressed in grams; contains 6.023×10^{23} molecules.

molecular clock Idea that the rate at which mutational changes accumulate in certain genes is constant over time and is not involved in adaptation to the environment.

molecule Union of two or more atoms of the same element; also, the smallest part of a compound that retains the properties of the compound.

mollusc Invertebrates including squids, clams, snails, and chitons; characterized by a visceral mass, a mantle, and a foot.

monoclonal antibody One of many antibodies produced by a clone of hybridoma cells that all bind to the same antigen.

monocot (Monocotyledone) Flowering plant group; members have one embryonic leaf (cotyledon), parallel-veined leaves, scattered vascular bundles, flower parts in threes or multiples of three, and other characteristics.

monocyte Type of agranular leukocyte that functions as a phagocyte, particularly after it becomes a macrophage, which is also an antigen-presenting cell.

monoecious Having male flowers or cones and female flowers or cones on a single plant.

monogamous Breeding pair of organisms that reproduce only with each other through their lifetime.

monohybrid cross Cross between parents that differ in only one trait.

monomer Small molecule that is a subunit of a polymer—e.g., glucose is a monomer of starch.

monophyletic Group of species including the most recent common ancestor and all its descendants.

monosaccharide Simple sugar; a carbohydrate that cannot be broken down by hydrolysis—e.g., glucose; also, any monomer of the polysaccharides.

monosomy Chromosome condition in which a diploid cell has one less chromosome than normal; designated as 2n-1.

monotreme Egg-laying mammal—e.g., duckbill platypus or spiny anteater.

monsoon Climate in India and southern Asia caused by wet ocean winds that blow onshore for almost half the year.

montane coniferous forest Coniferous forest of a mountain.

more-developed country (MDC) Country that is industrialized; typically, population growth is low, and the people enjoy a good standard of living overall.

morphogenesis Emergence of shape in tissues, organs, or entire embryo during development.

morphological species concept Definition of a species that defines species by specific diagnostic traits.

morphology Physical characteristics that contribute to the appearance of an organism.

morula Spherical mass of cells resulting from cleavage during animal development prior to the blastula stage.

mosaic evolution Concept that human characteristics did not evolve at the same rate; for example, some body parts are more humanlike than others in early hominins.

moss Bryophyte that is typically found in moist habitats.

motor (efferent) neuron Nerve cell that conducts nerve impulses away from the central nervous system and innervates effectors (muscle and glands).

mouth In humans, organ of the digestive tract where food is chewed and mixed with saliva.

mRNA transcript mRNA molecule formed during transcription that has a sequence of bases complementary to a gene.

mucosa Epithelial membrane containing cells that secrete mucus; found in the inner cell layers of the digestive (first layer) and respiratory tracts.

multicellular Organism composed of many cells; usually has organized tissues, organs, and organ systems.

multiple alleles Inheritance pattern in which there are more than two alleles for a particular trait; each individual has only two of all possible alleles.

multiple sclerosis (MS) Disease of the central nervous system characterized by the breakdown of myelin in the neurons; considered to be an autoimmune disease.

muscularis Smooth muscle layer found in the digestive tract.

muscular (contractile) tissue Type of animal tissue composed of fibers that shorten and lengthen to produce movements.

mutagen Chemical or physical agent that increases the chance of mutation.

mutations Any change made in the nucleotide sequence of DNA; source of new variation for a species.

mutualism Symbiotic relationship in which both species benefit in terms of growth and reproduction.

mycelium Tangled mass of hyphal filaments composing the vegetative body of a fungus.

mycorrhizae (sing., mycorrhiza) Mutualistic relationship between fungal hyphae and roots of vascular plants.

myelin sheath White, fatty material—derived from the membrane of neurolemmocytes—that forms a covering for nerve fibers.

myofibril Specific muscle cell organelle containing a linear arrangement of sarcomeres, which shorten to produce muscle contraction.

myosin Muscle protein making up the thick filaments in a sarcomere; it pulls actin to shorten the sarcomere, yielding muscle contraction.

N

NAD$^+$ (nicotinamide adenine dinucleotide) Coenzyme in oxidation-reduction reactions that accepts electrons and hydrogen ions to become NADH + H^+ as oxidation of substrates occurs. During cellular respiration, NADH carries electrons to the electron transport chain in mitochondria.

NADP$^+$ (nicotinamide adenine dinucleotide phosphate) Coenzyme in oxidation-reduction reactions that accepts electrons and hydrogen ions to become NADPH + H^+. During photosynthesis, NADPH participates in the reduction of carbon dioxide to a carbohydrate.

nail Flattened epithelial tissue from the stratum lucidum of the skin; located on the tips of fingers and toes.

natural killer (NK) cell Lymphocyte that causes an infected or cancerous cell to burst.

natural group In systematics, groups of organisms that possess a shared evolutionary history.

natural selection Mechanism of evolutionary change caused by environmental selection of organisms most fit to reproduce; results in adaptation to the environment.

navigate To steer or manage a course by adjusting one's bearings and following the result of the adjustment.

Neandertal Hominin with a sturdy build that lived during the last Ice Age in Europe and the Middle East; hunted large game and left evidence of being culturally advanced.

nearsighted (myopic) Condition in which an individual cannot focus on objects at a distance; caused by the focusing of the image in front of the retina of the eye.

negative feedback Mechanism of homeostatic response by which the output of a system suppresses or inhibits activity of the system.

nematocyst In cnidarians, a capsule that contains a threadlike fiber, the release of which aids in the capture of prey.

nephridium (pl., nephridia) Segmentally arranged, paired excretory tubules of many invertebrates, as in the earthworm.

nephron Microscopic kidney unit that regulates blood composition by glomerular filtration, tubular reabsorption, and tubular secretion.

nerve Bundle of long axons outside the central nervous system.

nerve fiber Axon; conducts nerve impulses away from the cell. Axons are classified as either myelinated or unmyelinated based on the presence or absence of a myelin sheath.

nerve net Diffuse, noncentralized arrangement of nerve cells in cnidarians.

nervous tissue Tissue that contains nerve cells (neurons), which conduct impulses, and neuroglia, which support, protect, and provide nutrients to neurons.

neural plate Region of the dorsal surface of the chordate embryo that marks the future location of the neural tube.

neural tube Tube formed by closure of the neural groove during development. In vertebrates, the neural tube develops into the spinal cord and brain.

neurodegenerative disease Disease, usually caused by a prion, virus, or bacterium, that damages or impairs the function of nervous tissue.

neuroglia Nonconducting nerve cells that are intimately associated with neurons and function in a supportive capacity.

neuromuscular junction Region where an axon bulb approaches a muscle fiber; contains a presynaptic membrane, a synaptic cleft, and a postsynaptic membrane.

neuron Nerve cell that characteristically has three parts: dendrites, cell body, and an axon.

neurotransmitter Chemical stored at the ends of axons that is responsible for transmission across a synapse.

neutron Neutral subatomic particle, located in the nucleus and assigned one atomic mass unit.

neutrophil Granular leukocyte that is the most abundant of the white blood cells; first to respond to infection.

nitrification Process by which nitrogen in ammonia and organic compounds is oxidized to nitrites and nitrates by soil bacteria.

nitrogen fixation Process whereby free atmospheric nitrogen is converted into compounds, such as ammonium and nitrates, usually by bacteria.

node In plants, the place where one or more leaves attach to a stem.

nodes of Ranvier Gaps in the myelin sheath around a nerve fiber.

noncompetitive inhibition Form of enzyme inhibition where the inhibitor binds to an enzyme at a location other than the active site; while at this site, the enzyme shape changes, the inhibitor is unable to bind to its substrate, and no product forms.

noncyclic pathway Light-dependent photosynthetic pathway that is used to generate ATP and NADPH; because the pathway is noncyclic, the electrons must be replaced by the splitting of water (photolysis).

nondisjunction Failure of the homologous chromosomes or sister chromatids to separate during either mitosis or meiosis; produces cells with abnormal chromosome numbers.

nomenclature In systematics, the process of assigning names to taxonomic groups; usually determined by governing organizations.

nonpolar covalent bond Bond in which the sharing of electrons between atoms is fairly equal.

nonrandom mating Mating among individuals on the basis of their phenotypic similarities or differences, rather than mating on a random basis.

nonseptate Lacking cell walls; some fungal species have hyphae that are nonseptate.

nonvascular plants Bryophytes, such as mosses and liverworts, that have no vascular tissue and either occur in moist locations or have special adaptations for living in dry locations.

norepinephrine Neurotransmitter of the postganglionic fibers in the sympathetic division of the autonomic system; also, a hormone produced by the adrenal medulla.

notochord Cartilage-like supportive dorsal rod in all chordates at some time in their life cycle; replaced by vertebrae in vertebrates.

nuclear envelope Double membrane that surrounds the nucleus in eukaryotic cells and is connected to the endoplasmic reticulum; has pores that allow substances to pass between the nucleus and the cytoplasm.

nucleariid Protist that may be related to fungi although nucleariids lack the same type of cell wall and have threadlike pseudopods.

nuclear pore Opening in the nuclear envelope that permits the passage of proteins into the nucleus and ribosomal subunits out of the nucleus.

nucleic acid Polymer of nucleotides; both DNA and RNA are nucleic acids.

nucleoid Region of prokaryotic cells where DNA is located; it is not bound by a nuclear envelope.

nucleolus Dark-staining, spherical body in the nucleus that produces ribosomal subunits.

nucleoplasm Semifluid medium of the nucleus containing chromatin.

nucleosome In the nucleus of a eukaryotic cell, a unit composed of DNA wound around a core of eight histone proteins, giving the appearance of a string of beads.

nucleotide Monomer of DNA and RNA consisting of a 5-carbon sugar bonded to a nitrogenous base and a phosphate group.

nucleus Membrane-bound organelle within a eukaryotic cell that contains chromosomes and controls the structure and function of the cell.

O

obligate anaerobe Prokaryote unable to grow in the presence of free oxygen.

observation Initial step in the scientific method that often involves the recording of data from an experiment or natural event.

octet rule The observation that an atom is most stable when its outer shell is complete and contains eight electrons; an exception is hydrogen, which requires only two electrons in its outer shell to have a completed shell.

oil Triglyceride, usually of plant origin, that is composed of glycerol and three fatty acids and is liquid in consistency due to many unsaturated bonds in the hydrocarbon chains of the fatty acids.

oil gland Gland of the skin, associated with a hair follicle, that secretes sebum; sebaceous gland.

olfactory cell Modified neuron that is a sensory receptor for the sense of smell.

oligodendrocyte Type of glial cell that forms myelin sheaths around neurons in the CNS.

omnivore Organism in a food chain that feeds on both plants and animals.

oncogene Cancer-causing gene formed by a mutation in a proto-oncogene; codes for proteins that stimulate the cell cycle and inhibit apoptosis.

oocyte (egg) Immature egg that is undergoing meiosis; upon completion of meiosis, the oocyte becomes an egg.

oogenesis Production of eggs in females by the process of meiosis and maturation.

open circulatory system Arrangement of internal transport in which blood bathes the organs directly, and there is no distinction between blood and interstitial fluid.

operant conditioning Learning that results from rewarding or reinforcing a particular behavior.

operator In an operon, the sequence of DNA that serves as a binding site for a repressor, and thereby regulates the expression of structural genes.

operon Group of structural and regulating genes that function as a single unit.

opisthokont Supergroup of eukaryotes that includes choanoflagellates, animals, nucleariids, and fungi.

opposable thumb Fingers arranged in such a way that the thumb can touch the ventral surface of the fingertips of all four fingers.

optimal foraging model Analysis of behavior as a compromise of feeding costs versus feeding benefits.

order One of the categories, or taxa, used by taxonomists to group species; the taxon located above the family level.

organ Combination of two or more different tissues performing a common function.

organelle Small, membranous structures in the cytoplasm having a specific structure and function.

organic chemistry Branch of science that deals with organic molecules, including those that are unique to living organisms.

organic molecule Molecule that always contains carbon and hydrogen, and often contains oxygen as well; organic molecules are associated with living organisms.

organ of Corti Structure in the vertebrate inner ear that contains auditory receptors (also called spiral organ).

organ system Group of related organs working together; examples are the digestive and endocrine systems.

orientation In birds, the ability to know present location by tracking stimuli in the environment.

osmoregulation Regulation of the water-salt balance to maintain a normal balance within internal fluids.

osmosis Diffusion of water through a selectively permeable membrane.

osmotic pressure Measure of the tendency of water to move across a selectively permeable membrane; visible as an increase in liquid on the side of the membrane with higher solute concentration.

ossicle One of the small bones of the vertebrate middle ear—malleus, incus, and stapes.

osteoblast Bone-forming cell.

osteoclast Cell that is responsible for bone resorption.

osteocyte Mature bone cell located within the lacunae of bone.

osteoporosis Condition characterized by a loss of bone density; associated with levels of sex hormones and diet.

ostracoderm Earliest vertebrate fossils of the Cambrian and Devonian periods; these fishes were small, jawless, and finless.

otolith Calcium carbonate granule associated with sensory receptors for detecting movement of the head; in vertebrates, located in the utricle and saccule.

outer ear Portion of the ear consisting of the pinna and the auditory canal.

outgroup In a cladistic study of evolutionary relationships among organisms, a group that has a known relationship to, but is not a member of, the taxa being analyzed.

ovarian cycle Monthly changes occurring in the ovary that determine the level of sex hormones in the blood.

ovary In flowering plants, the enlarged, ovule-bearing portion of the carpel that develops into a fruit; female gonad in animals that produces an egg and female sex hormones.

overexploitation When the number of individuals taken from a wild population is so great that the population becomes severely reduced in numbers.

oviparous Type of reproduction in which development occurs in an egg, laid by mother, in reptiles.

ovoviviparous Animals that produce eggs that develop internally and hatch at around the same time as they are released to the environment; mostly aquatic.

ovulation Bursting of a follicle when a secondary oocyte is released from the ovary; if fertilization occurs, the secondary oocyte becomes an egg.

ovule In seed plants, a structure that contains the female gametophyte and has the potential to develop into a seed.

oxidation Loss of one or more electrons from an atom or molecule; in biological systems, generally the loss of hydrogen atoms.

oxidation-reduction reaction A paired set of chemical reactions in which one molecule gives up electrons (oxidized) while another molecule accepts electrons (reduced); commonly called a redox reaction.

oxygen debt Amount of oxygen required to oxidize lactic acid produced anaerobically during strenuous muscle activity.

oxyhemoglobin Compound formed when oxygen combines with hemoglobin.

oxytocin Hormone released by the posterior pituitary that causes contraction of the uterus and milk letdown.

ozone shield Accumulation of O_3, formed from oxygen in the upper atmosphere; a filtering layer that protects the Earth from ultraviolet radiation.

P

p53 The protein produced from a tumor suppressor gene that (1) attempts to repair DNA damage, or (2) stops the cell cycle, or (3) initiates apoptosis.

pacemaker Cells of the sinoatrial node of the heart; electrical device designed to mimic the normal electrical patterns of the heart.

paleontology Study of fossils that results in knowledge about the history of life.

palisade mesophyll Layer of tissue in a plant leaf containing elongated cells with many chloroplasts.

pancreas Internal organ that produces digestive enzymes and the hormones insulin and glucagon.

pancreatic amylase Enzyme that digests starch to maltose.

pancreatic islet Masses of cells that constitute the endocrine portion of the pancreas.

panoramic vision Vision characterized by having a wide field of vision; found in animals with eyes to the side.

parabasalid Single-celled protist that lacks mitochondria; possesses flagella in clusters near the anterior of the cell.

parasite Species that is dependent on a host species for survival, usually to the detriment of the host species.

parasitism Symbiotic relationship in which one species (the *parasite*) benefits in terms of growth and reproduction to the detriment of the other species (the *host*).

parasympathetic division Division of the autonomic system that is active under normal conditions; uses acetylcholine as a neurotransmitter.

parathyroid gland Gland embedded in the posterior surface of the thyroid gland; it produces parathyroid hormone.

parathyroid hormone (PTH) Hormone secreted by the four parathyroid glands that increases the blood calcium level and decreases the phosphate level.

parenchyma Plant tissue composed of the least specialized of all plant cells; found in all organs of a plant.

Parkinson disease (PD) Progressive deterioration of the central nervous system due to a deficiency in the neurotransmitter dopamine.

parsimony In systematics, the simplest solution in the analysis of evolutionary relationships.

partial pressure Pressure exerted by each gas in a mixture of gases.

parthenogenesis Development of an egg cell into a whole organism without fertilization.

passive immunity Protection against infection acquired by transfer of antibodies to a susceptible individual.

pathogen Disease-causing agent, such as viruses, parasitic bacteria, fungi, and animals.

pattern formation Positioning of cells during development that determines the final shape of an organism.

pectoral girdle Portion of the vertebrate skeleton that provides support and attachment for the upper (fore) limbs; consists of the scapula and clavicle on each side of the body.

pelagic zone Open portion of the ocean.

pelvic girdle Portion of the vertebrate skeleton to which the lower (hind) limbs are attached; consists of the coxal bones.
penis Male copulatory organ; in humans, the male organ of sexual intercourse.
pentose Five-carbon monosaccharide. Examples are deoxyribose found in DNA and ribose found in RNA.
pepsin Enzyme secreted by gastric glands that digests proteins to peptides.
peptidase Intestinal enzyme that breaks down short chains of amino acids to individual amino acids that are absorbed across the intestinal wall.
peptide Two or more amino acids joined together by covalent bonding.
peptide bond Type of covalent bond that joins two amino acids.
peptide hormone Type of hormone that is a protein, a peptide, or derived from an amino acid.
peptidoglycan Polysaccharide that contains short chains of amino acids; found in bacterial cell walls.
perennial Flowering plant that lives more than one growing season because the underground parts regrow each season.
pericycle Layer of cells surrounding the vascular tissue of roots; produces branch roots.
periderm Protective tissue that replaces epidermis; includes cork, cork cambium.
peripheral nervous system (PNS) Nerves and ganglia that lie outside the central nervous system.
peristalsis Wavelike contractions that propel substances along a tubular structure, such as the esophagus.
permafrost Permanently frozen ground, usually occurring in the tundra, a biome of Arctic regions.
peroxisome Enzyme-filled vesicle in which fatty acids and amino acids are metabolized to hydrogen peroxide that is broken down to harmless products.
petal A flower part that occurs just inside the sepals; often conspicuously colored to attract pollinators.
petiole The part of a plant leaf that connects the blade to the stem.
phagocytosis Process by which cells engulf large substances, forming an intracellular vacuole.
pharyngitis Inflammation of the pharynx; often caused by viruses or bacteria.
pharynx In vertebrates, common passageway for both food intake and air movement; located between the mouth and the esophagus.
phenomenon Observable natural event or fact.
phenotype Visible expression of a genotype—e.g., brown eyes or attached earlobes.
pheromone Chemical messenger that works at a distance and alters the behavior of another member of the same species.
phloem Vascular tissue that conducts organic solutes in plants; contains sieve-tube members and companion cells.
phoronids Lophophorans of the phylum Phoronida that are characterized by a long protective tube formed of chitin.
phospholipid Molecule that forms the bilayer of the cell's membranes; has a polar, hydrophilic head bonded to two nonpolar, hydrophobic tails.
photoautotroph Organism able to synthesize organic molecules by using carbon dioxide as the carbon source and sunlight as the energy source.
photoperiod (photoperiodism) Relative lengths of daylight and darkness that affect the physiology and behavior of an organism.
photoreceptor Sensory receptor that responds to light stimuli.
photorespiration Series of reactions that occurs in plants when carbon dioxide levels are depleted but oxygen continues to accumulate, and the enzyme RuBP carboxylase fixes oxygen instead of carbon dioxide.
photosynthesis Process, usually occurring within chloroplasts, that uses solar energy to reduce carbon dioxide to carbohydrate.
photosystem Photosynthetic unit where solar energy is absorbed and high-energy electrons are generated; contains a pigment complex and an electron acceptor; occurs as PS (photosystem) I and PS II.
phototropism Growth response of plant stems to light; stems demonstrate positive phototropism.
pH scale Measurement scale for hydrogen ion concentration. Based on the formula $-\log[H^+]$.
phylogenetic species concept Definition of a species that is determined by analysis of a phylogenetic tree to determine a common ancestor.
phylogenetic tree Diagram that indicates common ancestors and lines of descent among a group of organisms.
phylogeny Evolutionary history of a group of organisms.
phylum One of the categories, or taxa, used by taxonomists to group species; the taxon located above the class level.
phytochrome Photoreversible plant pigment that is involved in photoperiodism and other responses of plants, such as etiolation.
phytoplankton Part of plankton containing organisms that photosynthesize, releasing oxygen to the atmosphere and serving as food producers in aquatic ecosystems.
phytoremediation The use of plants to restore a natural area to its original condition.
pineal gland Gland—either at the skin surface (fish, amphibians) or in the third ventricle of the brain (mammals)—that produces melatonin.
pinocytosis Process by which vesicle formation brings macromolecules into the cell.
pioneer species Early colonizer of barren or disturbed habitats that usually has rapid growth and a high dispersal rate.
pith Parenchyma tissue in the center of some stems and roots.
pituitary gland Small gland that lies just inferior to the hypothalamus; consists of the anterior and posterior pituitary, both of which produce hormones.
placenta Organ formed during the development of placental mammals from the chorion and the uterine wall; allows the embryo, and then the fetus, to acquire nutrients and rid itself of wastes; produces hormones that regulate pregnancy.
placental mammal Also called eutherian; species that rely on internal development whereby the fetus exchanges nutrients and wastes with its mother via a placenta.
placoderms First jawed vertebrates; heavily armored fishes of the Devonian period.
plankton Freshwater and marine organisms that are suspended on or near the surface of the water; include phytoplankton and zooplankton.
plants Multicellular, photosynthetic eukaryotes that increasingly became adapted to live on land.
plasma In vertebrates, the liquid portion of blood; contains nutrients, wastes, salts, and proteins.
plasma cell Mature B cell that mass-produces antibodies.
plasma membrane Membrane surrounding the cytoplasm that consists of a phospholipid bilayer with embedded proteins; functions to regulate the entrance and exit of molecules from cell.
plasmid Extrachromosomal ring of accessory DNA in the cytoplasm of prokaryotes.
plasmodesmata In plants, cytoplasmic connections in the cell wall that connect two adjacent cells.
plasmodial slime mold Free-living mass of cytoplasm that moves by pseudopods on a forest floor or in a field, feeding on decaying plant material by phagocytosis; reproduces by spore formation.
plasmolysis Contraction of the cell contents due to the loss of water.
plastid Organelle of plants and algae that is bound by a double membrane and contains internal membranes and/or vesicles (i.e., chloroplasts, chromoplasts, leucoplasts).
plate tectonics Concept that the Earth's crust is divided into a number of fairly rigid plates whose movements account for continental drift.
platelet Component of blood that is necessary to blood clotting.
pleiotropy Inheritance pattern in which one gene affects many phenotypic characteristics of the individual.
pneumonia Condition of the respiratory system characterized by the filling of the bronchi and alveoli with fluid; caused by a viral, fungal, or bacterial pathogen.
poikilothermic Body temperature that varies depending on environmental conditions; informally termed "cold-blooded."
point mutation Change of only one base in the sequence of bases in a gene.
polar body Nonfunctional product of oogenesis produced by the unequal division of cytoplasm in females during meiosis; in humans three of the four cells produced by meiosis are polar bodies.
polar covalent bond Bond in which the sharing of electrons between atoms is unequal.
pollen grain In seed plants, structure that is derived from a microspore and develops into a male gametophyte.
pollen tube In seed plants, a tube that forms when a pollen grain lands on the stigma and germinates. The tube grows, passing between the cells of the stigma and the style to reach the egg inside an ovule, where fertilization occurs.
pollination In gymnosperms, the transfer of pollen from pollen cone to seed cone; in angiosperms, the transfer of pollen from anther to stigma.
pollution Any environmental change that adversely affects the lives and health of living organisms.
polyandrous Practice of female animals having several male mates; found in the New World monkeys where the males help in rearing the offspring.
polygamous Practice of males having several female mates.

polygenic inheritance Pattern of inheritance in which a trait is controlled by several allelic pairs.

polygenic trait Traits that are under the control of multiple genes as opposed to monogenic (single-gene) traits.

polymer Macromolecule consisting of covalently bonded monomers; for example, a polypeptide is a polymer of monomers called amino acids.

polymerase chain reaction (PCR) Technique that uses the enzyme DNA polymerase to produce millions of copies of a particular piece of DNA.

polyp Among cnidarians, body form that is directed upward and contains much mesoglea; in anatomy, small, abnormal growth that arises from the epithelial lining.

polypeptide Polymer of many amino acids linked by peptide bonds.

polyploidy Having a chromosome number that is a multiple greater than twice that of the monoploid number.

polyribosome String of ribosomes simultaneously translating regions of the same mRNA strand during protein synthesis.

polysaccharide Polymer made from carbohydrate monomers; the polysaccharides starch and glycogen are polymers of glucose monomers.

pons Portion of the brain stem above the medulla oblongata and below the midbrain; assists the medulla oblongata in regulating the breathing rate.

population Group of organisms of the same species occupying a certain area and sharing a common gene pool.

population density The number of individuals per unit area or volume living in a particular habitat.

population distribution The pattern of dispersal of individuals living within a certain area.

population genetics The study of gene frequencies and their changes within a population.

portal system Pathway of blood flow that begins and ends in capillaries, such as the portal system located between the small intestine and liver.

positive feedback Mechanism of homeostatic response in which the output of the system intensifies and increases the activity of the system.

posterior pituitary Portion of the pituitary gland that stores and secretes oxytocin and antidiuretic hormone produced by the hypothalamus.

posttranscriptional control Gene expression following transcription that regulates the way mRNA transcripts are processed.

posttranslational control Alternation of gene expression by changing a protein's activity after it is translated.

postzygotic isolating mechanism Anatomical or physiological difference between two species that prevents successful reproduction after mating has taken place.

potential energy Stored energy in a potentially usable form, as a result of location or spatial arrangement.

predation Interaction in which one organism (the predator) uses another (the prey) as a food source.

predator Organism that practices predation.

prediction Step of the scientific process that follows the formulation of a hypothesis and assists in creating the experimental design.

preparatory (prep) reaction Reaction that oxidizes pyruvate with the release of carbon dioxide; results in acetyl CoA and connects glycolysis to the citric acid cycle.

pressure-flow model Explanation for phloem transport; osmotic pressure following active transport of sugar into phloem produces a flow of sap from a source to a sink.

prey Organism that provides nourishment for a predator.

prezygotic isolating mechanism Anatomical, physiological, or behavioral difference between two species that prevents the possibility of mating.

primary lymphoid organ Location in the lymphatic system where lymphocytes develop and mature; e.g., bone marrow and thymus.

primate Member of the order Primates; includes prosimians, monkeys, apes, and hominins, all of whom have adaptations for living in trees.

primordial soup hypothesis Another name for the hypothesis of abiogenesis proposed by Alexander Oparin and J.B.S. Haldane that the first cells evolved in the oceans from chemicals present in the early atmosphere.

principle Theory that is generally accepted by an overwhelming number of scientists; also called a law.

prion Infectious particle consisting of protein only and no nucleic acid.

producer Photosynthetic organism at the start of a grazing food chain that makes its own food—e.g., green plants on land and algae in water.

product Substance that forms as a result of a reaction.

progesterone Female sex hormone that helps maintain sexual organs and secondary sex characteristics.

proglottid Segment of a tapeworm that contains both male and female sex organs and becomes a bag of eggs.

prokaryote Organism that lacks the membrane-bound nucleus and the membranous organelles typical of eukaryotes.

prokaryotic cell Cells that generally lack a membrane-bound nucleus and organelles; the cell type within the domains Bacteria and Archaea.

prolactin (PRL) Hormone secreted by the anterior pituitary that stimulates the production of milk from the mammary glands.

prometaphase Second phase of mitosis; chromosomes are condensed but not fully aligned at the metaphase plate.

promoter In an operon, a sequence of DNA where RNA polymerase binds prior to transcription.

prophase First phase of mitosis; characterized by the condensation of the chromatin; chromosomes are visible, but scattered in the nucleus.

proprioceptor Class of mechanoreceptors responsible for maintaining the body's equilibrium and posture; involved in reflex actions.

prosimian Group of primates that includes lemurs and tarsiers, and may resemble the first primates to have evolved.

prostaglandin Hormone that has various and powerful local effects.

prostate gland Gland in male humans that secretes an alkaline, cloudy, fluid that increases the motility of sperm.

proteasome Cellular structure containing proteases that is involved in the destruction of tagged proteins; used by cells for posttranslational control of gene expression.

protease Enzyme that breaks the peptide bonds between amino acids in proteins, polypeptides, and peptides.

protein Polymer of amino acids; often consisting of one or more polypeptides and having a complex three-dimensional shape.

protein-first hypothesis In chemical evolution, the proposal that protein originated before other macromolecules and made possible the formation of protocells.

proteinoid Abiotically polymerized amino acids that, when exposed to water, becomes a microsphere having cellular characteristics.

proteome Sum of the expressed proteins in a cell.

proteomics Study of the complete collection of proteins that a cell or organism expresses.

protists The group of eukaryotic organisms that are not a plant, fungus, or animal. Protists are generally a microscopic complex single cell; they evolved before other types of eukaryotes in the history of Earth.

protocell (protobiont) In biological evolution, a possible cell forerunner that became a cell once it acquired genes.

proton Positive subatomic particle located in the nucleus and assigned one atomic mass unit.

proto-oncogene Gene that promotes the cell cycle and prevents apoptosis; may become an oncogene through mutation.

protostome Group of coelomate animals in which the first embryonic opening (the blastopore) is associated with the mouth.

protozoan Heterotrophic, single-celled protist that moves by flagella, cilia, or pseudopodia.

proximal convoluted tubule Portion of a nephron following the glomerular capsule where tubular reabsorption of filtrate occurs.

pseudocoelom Body cavity lying between the digestive tract and body wall that is incompletely lined by mesoderm.

pseudopod Cytoplasmic extension of amoeboid protists; used for locomotion and engulfing food.

pteridophyte Ferns and their allies (horsetail and whisk fern).

pulmonary circuit Circulatory pathway between the lungs and the heart.

pulmonary fibrosis Respiratory condition characterized by the buildup of connective tissue in the lungs; typically caused by inhalation of coal dust, silica, or asbestos.

pulmonary tuberculosis Respiratory infection caused by the bacterium *Mycobacterium tuberculosis*.

pulse Vibration felt in arterial walls due to expansion of the aorta following ventricle contraction.

Punnett square Visual representation developed by Reginald Punnett that is used to calculate the expected results of simple genetic crosses.

pupil Opening in the center of the iris of the vertebrate eye.

R

radial symmetry Body plan in which similar parts are arranged around a central axis, like spokes of a wheel.

radiocarbon dating Process of radiometric dating that measures the decay of ^{14}C to ^{14}N.

radiolarian Protist that has a glassy silicon test, usually with a radial arrangement of spines; pseudopods are external to the test.

rain shadow Leeward side (side sheltered from the wind) of a mountainous barrier, which receives much less precipitation than the windward side.

rate of natural increase (r) Growth rate dependent on the number of individuals that are born each year and the number of individuals that die each year.

ray-finned bony fishes Group of bony fishes with fins supported by parallel bony rays connected by webs of thin tissue.

reactant Substance that participates in a reaction.

receptor Type of membrane protein that binds to specific molecules in the environment, providing a mechanism for the cell to sense and adjust to its surroundings.

receptor-mediated endocytosis Selective uptake of molecules into a cell by vacuole formation after they bind to specific receptor proteins in the plasma membrane.

receptor protein Proteins located in the plasma membrane or within the cell; bind to a substance that alters some metabolic aspect of the cell.

recessive allele Allele that exerts its phenotypic effect only in the homozygote; its expression is masked by a dominant allele.

reciprocal altruism The trading of helpful or cooperative acts, such as helping at the nest, by individuals—the animal that was helped will repay the debt at some later time.

recombinant DNA (rDNA) DNA that contains genes from more than one source.

red algae Marine photosynthetic protists with a notable abundance of phycobilin pigments; includes coralline algae of coral reefs.

red blood cell Erythrocyte; contains hemoglobin and carries oxygen from the lungs or gills to the tissues in vertebrates.

red bone marrow Vascularized, modified connective tissue that is sometimes found in the cavities of spongy bone; site of blood cell formation.

red tide A population bloom of dinoflagellates that causes coastal waters to turn red. Releases a toxin that can lead to paralytic shellfish poisoning.

redox reaction A paired set of chemical reactions in which one molecule gives up electrons (oxidized) while another molecule accepts electrons (reduced); also called an oxidation-reduction reaction.

reduction Gain of electrons by an atom or molecule with a concurrent storage of energy; in biological systems, the electrons are accompanied by hydrogen ions.

reflex action Automatic, involuntary response of an organism to a stimulus.

refractory period Time following an action potential when a neuron is unable to conduct another nerve impulse.

regulator gene Gene that controls the expression of another gene or genes; in an operon, regulator genes code for repressor proteins.

reinforcement Connection between natural selection and reproductive isolation that occurs when two closely related species come back into contact after a period of isolation.

relative dating Determining the age of fossils by noting their sequential relationships in strata; *absolute dating* relies on radioactive dating techniques to assign an actual date.

renin Enzyme released by the kidneys that leads to the secretion of aldosterone and a rise in blood pressure.

repetitive DNA element Sequence of DNA on a chromosome that is repeated several times.

replacement reproduction Population in which each person is replaced by only one child.

replication fork In eukaryotic DNA replication, the location where the two parental DNA strands separate.

repressor In an operon, protein molecule that binds to an operator, preventing transcription of structural genes.

reproduce To produce a new individual of the same kind.

reproductive cloning Used to create an organism that is genetically identical to the original individual.

reproductive isolation Model by which new species arise when gene flow is disrupted between two populations, genetic changes accumulate, and the populations are subsequently unable to mate and produce viable offspring.

reproductively isolated Descriptive term that indicates that a population is incapable of interbreeding with another population.

reptile Terrestrial vertebrate with internal fertilization, scaly skin, and an egg with a leathery shell; includes snakes, lizards, turtles, crocodiles, and birds.

resource Abiotic and biotic components of an environment that support or are needed by living organisms.

resource partitioning Mechanism that increases the number of niches by apportioning the supply of a resource, such as food or living space, between species.

respiration Sequence of events that results in gas exchange between the cells of the body and the environment.

respiratory center Group of nerve cells in the medulla oblongata that sends out nerve impulses on a rhythmic basis, resulting in involuntary inspiration on an ongoing basis.

responding variable Result or change that occurs when an experimental variable is utilized in an experiment.

resting potential Membrane potential of an inactive neuron.

restriction enzyme Bacterial enzyme that stops viral reproduction by cleaving viral DNA; used to cut DNA at specific points during production of recombinant DNA.

reticular activating system (RAS) Area of the brain that contains the reticular formation; acts as a relay for information to and from the peripheral nervous system and higher processing centers of the brain.

reticular fiber Very thin collagen fiber in the matrix of connective tissue, highly branched and forming delicate supporting networks.

retina Innermost layer of the vertebrate eyeball containing the photoreceptors—rod cells and cone cells.

retinal detachment Condition characterized by the separation of the retina from the choroid layer of the eye.

retrovirus RNA virus containing the enzyme reverse transcriptase that carries out RNA/DNA transcription.

reverse transcriptase Viral enzyme found in retroviruses that is capable of converting their RNA genome into a DNA copy.

rhizarian Supergroup of eukaryotes that includes foraminiferans and radiolarians.

rhizoid Rootlike hair that anchors a plant and absorbs minerals and water from the soil.

rhizome Rootlike underground stem.

rhodopsin Light-absorbing molecule in rod cells and cone cells that contains a pigment and the protein opsin.

ribose Pentose sugar found in RNA.

ribosomal RNA (rRNA) Structural form of RNA found in the ribosomes.

ribosome Site of protein synthesis in a cell; composed of proteins and ribosomal RNA (rRNA).

ribozyme RNA molecule that functions as an enzyme that can catalyze chemical reactions.

RNA (ribonucleic acid) Nucleic acid produced from covalent bonding of nucleotide monomers that contain the sugar ribose; occurs in many forms, including messenger RNA, ribosomal RNA, and transfer RNA.

RNA-first hypothesis In chemical evolution, the proposal that RNA originated before other macromolecules and allowed the formation of the first cell(s).

RNA interference Cellular process that utilizes miRNA and siRNA molecules to reduce, or inhibit, the expression of specific genes.

RNA polymerase During transcription, an enzyme that creates an mRNA transcript by joining nucleotides complementary to a DNA template.

rod cell Photoreceptor in vertebrate eyes that responds to dim light.

root apical meristem Meristem tissue located under the root cap; the site of the majority of cell division in the root.

root cap Protective cover of the root tip, whose cells are constantly replaced as they are ground off when the root pushes through rough soil particles.

root hair Extension of a root epidermal cell that collectively increases the surface area for the absorption of water and minerals.

root nodule Structure on plant root that contains nitrogen-fixing bacteria.

root pressure Osmotic pressure caused by active movement of mineral into root cells; elevates water in xylem for a short distance.

root system Includes the main root and any and all of its lateral (side) branches.

rotational equilibrium Maintenance of balance when the head and body are suddenly moved or rotated.

rotifer Microscopic invertebrates characterized by ciliated corona that when beating looks like a rotating wheel.

rough ER (endoplasmic reticulum) Membranous system of tubules, vesicles, and sacs in cells; has attached ribosomes.

roundworm Invertebrates with nonsegmented cylindrical body covered by a cuticle that molts;

some forms are free-living in water and soil, and many are parasitic.

***r*-selection** Favorable life history strategy under certain environmental conditions; characterized by a high reproductive rate with little or no attention given to offspring survival.

RuBP carboxylase An enzyme that starts the Calvin cycle reactions by catalyzing attachment of the carbon atom from CO_2 to RuBP.

rumen The first chamber in the digestive tract of a ruminant mammal; microorganisms here break down complex carbohydrates.

S

saccule Saclike cavity in the vestibule of the vertebrate inner ear; contains sensory receptors for gravitational equilibrium.

sac fungi (Ascomycota) Fungi that produce spores in fingerlike sacs called asci within a fuiting body; includes morels, truffles, yeasts, and molds.

salivary amylase In humans, enzyme in saliva that digests starch to maltose.

salivary gland In humans, gland associated with the mouth that secretes saliva.

salt Solid substance formed by ionic bonds that usually dissociates into individual ions in water.

saltatory conduction Movement of nerve impulses from one node to another along a myelinated axon.

saprotroph Organism that secretes digestive enzymes and absorbs the resulting nutrients back across the plasma membrane.

sarcolemma Plasma membrane of a muscle fiber; also forms the tubules of the T system involved in muscular contraction.

sarcomere One of many units, arranged linearly within a myofibril, whose contraction produces muscle contraction.

sarcoplasmic reticulum Smooth endoplasmic reticulum of skeletal muscle cells; surrounds the myofibrils and stores calcium ions.

saturated fatty acid Fatty acid molecule that lacks double bonds between the carbons of its hydrocarbon chain. The chain bears the maximum number of hydrogens possible.

savanna Terrestrial biome that is a grassland in Africa, characterized by few trees and a severe dry season.

Schwann cell Cell that surrounds a fiber of a peripheral nerve and forms the myelin sheath.

scientific method Process by which scientists formulate a hypothesis, gather data by observation and experimentation, and come to a conclusion.

scientific theory Concept, or a collection of concepts, widely supported by a broad range of observations, experiments, and data.

sclera White, fibrous, outer layer of the eyeball.

sclerenchyma Plant tissue composed of cells with heavily lignified cell walls; functions in support.

scolex Tapeworm head region; contains hooks and suckers for attachment to host.

sea star An echinoderm with noticeable 5-pointed radial symmetry; found along rocky coasts where they feed on bivalves.

secondary lymphoid organ Location in the lymphatic system where lymphocytes are activated by antigens; example is the lymph nodes.

secondary metabolite Molecule not directly involved in growth, development, or reproduction of an organism; in plants, these molecules, which include nicotine, caffeine, tannins, and menthols, can discourage herbivores.

secondary oocyte In oogenesis, the functional product of meiosis I; becomes the egg.

second messenger Chemical signal such as cyclic AMP that causes the cell to respond to the first messenger—a hormone bound to plasma membrane receptor protein.

secretion Release of a substance by exocytosis from a cell.

sediment Particulate material (sand, clay, etc.) that is carried by streams and rivers and deposited in areas of slow water movement.

sedimentation Process by which particulate material accumulates and forms a stratum.

seed Mature ovule that contains an embryo, with stored food enclosed in a protective coat.

seed plant Vascular plants that disperse seeds; the gymnosperms and angiosperms.

seedless vascular plant Collective name for club mosses (lycophytes) and ferns (pteridophytes). Characterized by windblown spores.

segmentation Repetition of body units, as seen in the earthworm.

selectively permeable Property of the plasma membrane that allows some substances to pass but prohibits the movement of others.

semelparity Condition of having a single reproductive effort in a lifetime.

semen (seminal fluid) Thick, whitish fluid consisting of sperm and secretions from several glands of the male reproductive tract.

semicircular canal One of three half-circle-shaped canals of the vertebrate inner ear; contains sensory receptors for rotational equilibrium.

semiconservative replication Process of DNA replication that results in two double helix molecules, each having one parental and one new strand.

semilunar valve Valve resembling a half moon located between the ventricles and their attached vessels.

seminal vesicles Glands in male humans that secrete a viscous fluid that provides nutrition to the sperm cells.

seminiferous tubule Long, coiled structure contained within chambers of the testis where sperm are produced.

senescence Sum of the processes involving aging, decline, and eventual death of a plant or plant part.

sensory (afferent) neuron Nerve cell that transmits nerve impulses to the central nervous system after a sensory receptor has been stimulated.

sensory receptor Structure that receives either external or internal environmental stimuli and is a part of a sensory neuron or transmits signals to a sensory neuron.

sensory transduction Process by which a sensory receptor converts an input to a nerve impulse.

sepal Outermost, leaflike covering of the flower; usually green in color.

septate Having cell walls; some fungal species have hyphae that are septate.

septum Partition or wall that divides two areas; the septum in the heart separates the right half from the left half.

serosa Outer embryonic membrane of birds and reptiles; chorion.

seta (pl., setae) A needlelike, chitinous bristle in annelids, arthropods, and others.

sexual dimorphism Species that have distinct differences between the sexes, resulting in male and female forms.

sexual reproduction Reproduction involving meiosis, gamete formation, and fertilization; produces offspring with chromosomes inherited from each parent with a unique combination of genes.

sexual selection Changes in males and females, often due to male competition and female selectivity, leading to increased fitness.

shoot apical meristem Group of actively dividing embryonic cells at the tips of plant shoots.

shoot system Aboveground portion of a plant consisting of the stem, leaves, and flowers.

short-day plant Plant that flowers when day length is shorter than a critical length—e.g., cocklebur, poinsettia, and chrysanthemum.

short tandem repeat (STR) sequence Procedure of analyzing DNA in which PCR and gel electrophoresis are used to create a banding pattern; these are usually unique for each individual; process used in DNA barcoding.

shrubland Arid terrestrial biome characterized by shrubs and tending to occur along coasts that have dry summers and receive most of their rainfall in the winter.

sieve-tube member Member that joins with others in the phloem tissue of plants as a means of transport for nutrient sap.

sign stimulus The environmental trigger that causes a fixed action pattern or unchanging behavioral response.

signal Molecule that stimulates or inhibits an event in the cell.

signal peptide Sequence of amino acids that binds with a signal recognition particle (SRP), causing a ribosome to bind to the endoplasmic reticulum (ER).

signal transduction Process that occurs within a cell when a molecular signal (protein, hormone, etc.) initiates a response within the interior of the cell.

sink In the pressure-flow model of phloem transport, the location (roots) from which sugar is constantly being removed. Sugar will flow to the roots from the source.

sink population Population that is found in an unfavorable area where at best the birthrate equals the death rate; sink populations receive new members from source populations.

sister chromatid One of two genetically identical chromosomal units that are the result of DNA replication and are attached to each other at the centromere.

skeletal muscle Striated, voluntary muscle tissue that comprises skeletal muscles; also called striated muscle.

skin Outer covering of the body; can be called the integumentary system because it contains organs such as sense organs.

sliding filament model An explanation for muscle contraction based on the movement of actin filaments in relation to myosin filaments.

small intestine In vertebrates, the portion of the digestive tract that precedes the large intestine;

in humans, consists of the duodenum, jejunum, and ileum.

smooth (visceral) muscle Nonstriated, involuntary muscles found in the walls of internal organs.

smooth ER (endoplasmic reticulum) Membranous system of tubules, vesicles, and sacs in eukaryotic cells; site of lipid synthesis; lacks attached ribosomes.

society Group in which members of species are organized in a cooperative manner, extending beyond sexual and parental behavior.

sodium-potassium pump Carrier protein in the plasma membrane that moves sodium ions out of and potassium ions into cells; important in the function of nerve and muscle cells in animals.

soil Accumulation of inorganic rock material and organic matter that is capable of supporting the growth of vegetation.

soil erosion Movement of topsoil to a new location due to the action of wind or running water.

soil horizon Major layer of soil visible in vertical profile; for example, topsoil is the A horizon.

soil profile Vertical section of soil from the ground surface to the unaltered rock below.

solute Substance that is dissolved in a solvent, forming a solution.

solution Fluid (the solvent) that contains a dissolved solid (the solute).

solvent Liquid portion of a solution that dissolves a solute.

somatic cell Body cell; excludes cells that undergo meiosis and become sperm or eggs.

somatic system Portion of the peripheral nervous system containing motor neurons that control skeletal muscles.

source In the pressure-flow model of phloem transport, the location (leaves) of sugar production. Sugar will flow from the leaves to the sink.

source population Population that can provide members to other populations of the species because it lives in a favorable area, and the birthrate is most likely higher than the death rate.

speciation Origin of new species due to the evolutionary process of descent with modification.

species Group of similarly constructed organisms capable of interbreeding and producing fertile offspring; organisms that share a common gene pool; the taxon at the lowest level of classification.

species diversity Variety of species that make up a community.

species richness Number of species in a community.

specific epithet In the binomial system of taxonomy, the second part of an organism's name; it may be descriptive.

sperm Male gamete having a haploid number of chromosomes and the ability to fertilize an egg, the female gamete.

spermatogenesis Production of sperm in males by the process of meiosis and maturation.

spicule Skeletal structure of sponges composed of calcium carbonate or silicate.

spinal cord In vertebrates, the nerve cord that is continuous with the base of the brain and housed within the vertebral column.

spinal nerve Nerve that arises from the spinal cord.

spirillum (pl., spirilla) Long, rod-shaped bacterium that is twisted into a rigid spiral; if the spiral is flexible rather than rigid, it is called a spirochete.

spirochete Long, rod-shaped bacterium that is twisted into a flexible spiral; if the spiral is rigid rather than flexible, it is called a spirillum.

spleen Large, glandular organ located in the upper left region of the abdomen; stores and filters blood.

sponge Invertebrates that are pore-bearing filter feeders whose inner body wall is lined by collar cells that resemble a single-celled choanoflagellate.

spongy bone Type of bone that has an irregular, meshlike arrangement of thin plates of bone.

spongy mesophyll Layer of tissue in a plant leaf containing loosely packed cells, increasing the amount of surface area for gas exchange.

spontaneous mutation Mutation that arises as a result of anomalies in normal biological processes, such as mistakes made during DNA replication.

sporangium (pl., sporangia) Structure that produces spores.

spore Asexual reproductive or resting cell capable of developing into a new organism without fusion with another cell, in contrast to a gamete.

sporophyll Modified leaf that bears a sporangium or many sporangia.

sporophyte Diploid generation of the alternation-of-generations life cycle of a plant; produces haploid spores that develop into the haploid generation.

spring overturn Mixing process that occurs in spring in stratified lakes whereby oxygen-rich top waters mix with nutrient-rich bottom waters.

squamous epithelium Type of epithelial tissue that contains flat cells.

stabilizing selection Outcome of natural selection in which extreme phenotypes are eliminated and the average phenotype is conserved.

standard deviation A statistical analysis of data from an observation or experiment; measures how much the data varies.

stamen In flowering plants, the portion of the flower that consists of a filament and an anther containing pollen sacs where pollen is produced.

starch Storage polysaccharide found in plants that is composed of glucose molecules joined in a linear fashion with few side chains.

statolith Sensors found in root cap cells that cause a plant to demonstrate gravitropism.

stem Usually the upright, vertical portion of a plant that transports substances to and from the leaves.

stereoscopic vision Vision characterized by depth perception and three-dimensionality.

steroid Type of lipid molecule having a complex of four carbon rings—e.g., cholesterol, estrogen, progesterone, and testosterone.

steroid hormone Type of hormone that has the same complex of four carbon rings, but each one has different side chains.

stigma In flowering plants, portion of the carpel where pollen grains adhere and germinate before fertilization can occur.

stolon Stem that grows horizontally along the ground and may give rise to new plants where it contacts the soil—e.g., the runners of a strawberry plant.

stomach In vertebrates, muscular sac that mixes food with gastric juices to form chyme, which enters the small intestine.

stomata (sing., stoma) Small openings between two guard cells on the underside of leaf epidermis through which gases pass.

stramenopile Group of protists that includes water molds, diatoms, and golden brown algae and is characterized by a "hairy" flagellum.

strata (sing. stratum) Ancient layers of sedimentary rock; result from slow deposition of silt, volcanic ash, and other materials.

striated Having bands; in cardiac and skeletal muscle, alternating light and dark bands produced by the distribution of contractile proteins.

strobilus In club mosses, terminal clusters of leaves that bear sporangia.

stroke Condition resulting when an arteriole in the brain bursts or becomes blocked by an embolism; cerebrovascular accident.

stroma Region within a chloroplast that surrounds the grana; contains enzymes involved in the synthesis of carbohydrates during the Calvin cycle of photosynthesis.

stromatolite Domed structure found in shallow seas consisting of cyanobacteria bound to calcium carbonate.

structural gene Gene that codes for the amino acid sequence of a peptide or protein.

structural genomics Study of the sequence of DNA bases and the amount of genes in organisms.

style Elongated, central portion of the carpel between the ovary and stigma.

submucosa Tissue layer just under the epithelial lining of the lumen of the digestive tract (second layer).

substrate Reactant in an enzyme-controlled reaction.

substrate-level ATP synthesis Process in which ATP is formed by transferring a phosphate from a metabolic substrate to ADP.

supergroup Systematic term that refers to the major groups of eukaryotes.

surface-area-to-volume ratio Ratio of a cell's outside area to its internal volume; the relationship limits the maximum size of a cell.

surface tension Force that holds moist membranes together due to the attraction of water molecules through hydrogen bonds.

survivorship Probability of newborn individuals of a cohort surviving to particular ages.

suture Line of union between two nonarticulating bones, as in the skull.

swamp Wet, spongy land that is saturated and sometimes partially or intermittently covered with water.

sweat gland Skin gland that secretes a fluid substance for evaporative cooling; sudoriferous gland.

swim bladder In fishes, a gas-filled sac whose pressure can be altered to change buoyancy.

symbiosis Relationship that occurs when two different species live together in a unique way; it may be beneficial, neutral, or detrimental to one or both species.

symbiotic relationship *See* symbiosis.

symmetry Pattern of similarity in an object.

sympathetic division Division of the autonomic system that is active when an organism is under stress; uses norepinephrine as a neurotransmitter.

sympatric speciation Origin of new species in populations that overlap geographically.

synapse Junction between neurons consisting of the presynaptic (axon) membrane, the synaptic cleft, and the postsynaptic (usually dendrite) membrane.

synapsid Characteristic of a vertebrate skull in which there is a single opening in the skull behind the eye socket (orbit).

synapsis Pairing of homologous chromosomes during meiosis I.

synaptic cleft Small gap between presynaptic and postsynaptic cells of a synapse.

synaptonemal complex Protein structure that forms between the homologous chromosomes of prophase I of meiosis; promotes the process of crossing-over.

synovial joint Freely moving joint in which two bones are separated by a cavity.

systematics Study of the diversity of life for the purpose of understanding the evolutionary relationships between species.

systemic circuit Circulatory pathway of blood flow between the tissues and the heart.

systole Contraction period of the heart during the cardiac cycle.

T

tactile communication Communication through touch; for example, when a chick pecks its mother for food, chimpanzees groom each other, and honeybees "dance."

taiga Terrestrial biome that is a coniferous forest extending in a broad belt across northern Eurasia and North America.

tandem repeat Repetitive DNA sequence in which the repeats occur one after another in the same region of a chromosome.

taproot Main axis of a root that penetrates deeply and is used by certain plants (such as carrots) for food storage.

taste bud Structure in the vertebrate mouth containing sensory receptors for taste; in humans, most taste buds are on the tongue.

taxon (pl., taxa) Group of organisms that fills a particular classification category.

taxonomist Scientist that investigates the identification and naming of new organisms.

taxonomy Branch of biology concerned with identifying, describing, and naming organisms.

T cell Lymphocyte that matures in the thymus and exists in four varieties, one of which kills antigen-bearing cells outright.

T-cell receptor (TCR) Molecule on the surface of a T cell that can bind to a specific antigen fragment in combination with an MHC molecule.

technology Application of scientific knowledge for a practical purpose.

telomere Tip of the end of a chromosome that shortens with each cell division and may thereby regulate the number of times a cell can divide.

telophase Final phase of mitosis; daughter cells are located at each pole.

temperate deciduous forest Forest found south of the taiga; characterized by deciduous trees such as oak, beech, and maple, moderate climate, relatively high rainfall, stratified plant growth, and plentiful ground life.

temperate grassland Grasslands characterized by very cold winters and dry, hot summers; examples are the North American prairies and Russian steppes.

temperate rain forest Coniferous forest—e.g., that running along the west coast of Canada and the United States—characterized by plentiful rainfall and rich soil.

template Parental strand of DNA that serves as a guide for the complementary daughter strand produced during DNA replication.

tendon Strap of fibrous connective tissue that connects skeletal muscle to bone.

terminal bud Bud that develops at the apex of a shoot.

termination End of translation that occurs when a ribosome reaches a stop codon on the mRNA that it is translating, causing release of the completed protein.

territoriality Marking and/or defending a particular area against invasion by another species member; area often used for the purpose of feeding, mating, and caring for young.

territory Area occupied and defended exclusively by an animal or a group of animals.

testcross Cross between an individual with a dominant phenotype and an individual with a recessive phenotype to determine whether the dominant individual is homozygous or heterozygous.

testes Male gonads that produce sperm and the male sex hormones.

testosterone Male sex hormone that helps maintain sexual organs and secondary sex characteristics.

tetany Severe spasm caused by involuntary contraction of the skeletal muscles due to a calcium imbalance.

tetrapod Four-footed vertebrate; includes amphibians, reptiles, birds, and mammals.

thalamus In vertebrates, the portion of the diencephalon that passes on selected sensory information to the cerebrum.

therapeutic cloning Used to create mature cells of various cell types. Facilitates study of specialization of cells and provides cells and tissue to treat human illnesses.

thermoacidophile Type of archaean that lives in hot, acidic, aquatic habitats, such as hot springs or near hydrothermal vents.

thermoreceptor Sensory receptor that detects heat.

thigmotropism In plants, unequal growth due to contact with solid objects, as the coiling of tendrils around a pole.

threatened Species that is likely to become an endangered species in the foreseeable future (e.g., gray wolf, Louisiana black bear).

thylakoid Flattened sac within a granum of a chloroplast; membrane contains chlorophyll; location where the light reactions of photosynthesis occur.

thymine (T) One of four nitrogen-containing bases in nucleotides composing the structure of DNA; pairs with adenine.

thymus Lymphoid organ involved in the development and functioning of the immune system; T lymphocytes mature in the thymus.

thyroid gland Large gland in the neck that produces several important hormones, including thyroxine, triiodothyronine, and calcitonin.

thyroid-stimulating hormone (TSH) Substance produced by the anterior pituitary that causes the thyroid to secrete thyroxine and triiodothyronine.

thyroxine Hormone secreted from the thyroid gland that promotes growth and development; in general, it increases the metabolic rate in cells.

tight junction Junction between cells when adjacent plasma membrane proteins join to form an impermeable barrier.

tissue Group of similar cells combined to perform a common function.

tissue culture Process of growing tissue artificially, usually in a liquid medium in laboratory glassware.

tone Continuous, partial contraction of muscle.

tonicity The solute concentration (osmolarity) of a solution compared to that of a cell. If the solution is isotonic to the cell, there is no net movement of water; if the solution is hypotonic, the cell gains water; and if the solution is hypertonic, the cell loses water.

topography Surface features of the Earth.

totipotent Cell that has the full genetic potential of the organism, including the potential to develop into a complete organism.

toxin Poisonous substance produced by living cells or organisms. Toxins are often proteins that are capable of causing disease on contact with or absorption by body tissues.

trachea (pl., tracheae) In insects, air tubes located between the spiracles and the tracheoles. In tetrapod vertebrates, air tube (windpipe) that runs between the larynx and the bronchi.

tracheid In vascular plants, type of cell in xylem that has tapered ends and pits through which water and minerals flow.

tract Bundle of myelinated axons in the central nervous system.

trait A characteristic of an organism; may be based on the physiology, morphology, or the genetics of the organism.

transcription First stage of gene expression; process whereby a DNA strand serves as a template for the formation of mRNA.

transcription activator Protein that participates in the initiation of transcription by binding to the enhancer regulatory regions.

transcriptional control Control of gene expression by the use of transcription factors, and other proteins, that regulate either the initiation of transcription or the rate at which it occurs.

transcription factor In eukaryotes, protein required for the initiation of transcription by RNA polymerase.

transduction Exchange of DNA between bacteria by means of a bacteriophage.

transduction pathway Series of proteins or enzymes that change a signal to one understood by the cell.

trans fat Unsaturated fatty acid chain in which the configuration of the carbon-carbon double bonds is such that the hydrogen atoms are across from each other, as opposed to being on the same side (cis).

transfer rate Amount of a substance that moves from one component of the environment to another within a specified period of time.

transfer RNA (tRNA) Type of RNA that transfers a particular amino acid to a ribosome during protein synthesis; at one end it binds to the amino acid, and at the other end it has an anticodon that binds to an mRNA codon.

transformation Taking up of extraneous genetic material from the environment by bacteria.

transgenic organism An organism whose genome has been altered by the insertion of genes from another species.

transitional fossil Fossil that bears a resemblance to two groups that in the present day are classified separately.

translation During gene expression, the process whereby ribosomes use the sequence of codons in mRNA to produce a polypeptide with a particular sequence of amino acids.

translational control Gene expression regulated by influencing the interaction of the mRNA transcripts with the ribosome.

translocation Movement of a chromosomal segment from one chromosome to another nonhomologous chromosome, leading to abnormalities—e.g., Down syndrome.

transpiration Plant's loss of water to the atmosphere, mainly through evaporation at leaf stomata.

transposon DNA sequence capable of randomly moving from one site to another in the genome.

trichocyst Found in ciliates; contains long, barbed threads useful for defense and capturing prey.

trichomes In plants, specialized outgrowths of the epidermis (e.g., root hairs).

triglyceride Neutral fat composed of glycerol and three fatty acids; typically involved in energy storage.

triplet code During gene expression, each sequence of three nucleotide bases stands for a particular amino acid.

trisomy Chromosome condition in which a diploid cell has one more chromosome than normal; designated as $2n + 1$.

trochozoan Type of protostome that produces a trochophore larva; also has two bands of cilia around its middle.

trophic level Feeding level of one or more populations in a food web.

trophoblast Outer membrane surrounding the embryo in mammals; when thickened by a layer of mesoderm, it becomes the chorion, an extraembryonic membrane.

tropical rain forest Biome near the equator in South America, Africa, and the Indo-Malay regions; characterized by warm weather, plentiful rainfall, a diversity of species, and mainly tree-living animal life.

tropism In plants, a growth response toward or away from a directional stimulus.

true coelom Body cavity completely lined with mesoderm; found in certain protostomes and all deuterostomes.

trypsin Protein-digesting enzyme secreted by the pancreas.

tubular reabsorption Movement of primarily nutrient molecules and water from the contents of the nephron into blood at the proximal convoluted tubule.

tubular secretion Movement of certain molecules from blood into the distal convoluted tubule of a nephron, so that they are added to urine.

tumor Cells derived from a single mutated cell that has repeatedly undergone cell division; benign tumors remain at the site of origin, while malignant tumors metastasize.

tumor suppressor gene Gene that codes for a protein that ordinarily suppresses the cell cycle; inactivity due to a mutation can lead to a tumor.

turgor movement In plant cells, pressure of the cell contents against the cell wall when the central vacuole is full.

turgor pressure Pressure of the cell contents against the cell wall; in plant cells, determined by the water content of the vacuole; provides internal support.

tympanic membrane Membranous region that receives air vibrations in an auditory organ; in humans, the eardrum.

U

umbilical cord Cord connecting the fetus to the placenta through which blood vessels pass.

uniformitarianism Belief, supported by James Hutton, that geologic forces act at a continuous, uniform rate.

unsaturated fatty acid Fatty acid molecule that contains double bonds between some carbons of its hydrocarbon chain; contains fewer hydrogens than a saturated hydrocarbon chain.

upwelling Upward movement of deep, nutrient-rich water along coasts; it replaces surface waters that move away from shore when the direction of prevailing wind shifts.

uracil (U) Pyrimidine base that occurs in RNA, replacing thymine.

urea Main nitrogenous waste of terrestrial amphibians and most mammals.

ureter Tubular structure conducting urine from the kidney to the urinary bladder.

urethra Tubular structure that receives urine from the bladder and carries it to the outside of the body.

uric acid Main nitrogenous waste of insects, reptiles, and birds.

urinary bladder Organ where urine is stored.

urine Liquid waste product made by the nephrons of the vertebrate kidney through the processes of glomerular filtration, tubular reabsorption, and tubular secretion.

urochordate Group of aquatic invertebrate chordates that consists of the tunicates (sea squirts).

uterine cycle Cycle that runs concurrently with the ovarian cycle; it prepares the uterus to receive a developing zygote.

uterus In mammals, expanded portion of the female reproductive tract through which eggs pass to the environment or in which an embryo develops and is nourished before birth.

utricle Cavity in the vestibule of the vertebrate inner ear; contains sensory receptors for gravitational equilibrium.

V

vacuole Membrane-bound sac, larger than a vesicle; usually functions in storage and can contain a variety of substances. In plants, the central vacuole fills much of the interior of the cell.

vagina Component of the female reproductive system that serves as the birth canal; receives the penis during copulation.

valence shell The outer electron shell of an atom. Contains the valence electrons, which determine the chemical reactivity of the atom.

vas deferens Also called the ductus deferens; storage location for mature sperm before they pass into the ejaculatory duct.

vascular bundle In plants, primary phloem and primary xylem enclosed by a bundle sheath.

vascular cambium In plants, lateral meristem that produces secondary phloem and secondary xylem.

vascular cylinder In eudicots, the tissues in the middle of a root, consisting of the pericycle and vascular tissues.

vascular plant Plant that has xylem and phloem.

vascular tissue Transport tissue in plants, consisting of xylem and phloem.

vector In genetic engineering, a means to transfer foreign genetic material into a cell—e.g., a plasmid.

vein Blood vessel that arises from venules and transports blood toward the heart.

vena cava Large systemic vein that returns blood to the right atrium of the heart in tetrapods; either the superior or inferior vena cava.

ventilation Process of moving air into and out of the lungs; breathing.

ventricle Cavity in an organ, such as a lower chamber of the heart or the ventricles of the brain.

venule Vessel that takes blood from capillaries to a vein.

vertebral column Portion of the vertebrate endoskeleton that houses the spinal cord; consists of many vertebrae separated by intervertebral disks.

vertebrate Chordate in which the notochord is replaced by a vertebral column.

vertigo Equilibrium disorder that is often associated with problems in the receptors of the semicircular canals in the ear.

vesicle Small, membrane-bound sac that stores substances within a cell.

vessel element Cell that joins with others to form a major conducting tube found in xylem.

vestibule Space or cavity at the entrance to a canal, such as the cavity that lies between the semicircular canals and the cochlea.

vestigial structure Remnant of a structure that was functional in some ancestor but is no longer functional in the organism in question.

villus Small, fingerlike projection of the inner small intestinal wall.

viroid Infectious strand of RNA devoid of a capsid and much smaller than a virus.

virus Noncellular parasitic agent consisting of an outer capsid and an inner core of nucleic acid.

visual accommodation Ability of the eye to focus at different distances by changing the curvature of the lens.

visual communication Form of communication between animals using their bodies, including various forms of display.

vitamin Organic nutrient that is required in small amounts for metabolic functions. Vitamins are often part of coenzymes.

viviparous Animal that gives birth after partial development of offspring within mother.

vocal cord In humans, fold of tissue within the larynx; creates vocal sounds when it vibrates.

W

water column In plants, water molecules joined together in xylem from the leaves to the roots.

water (hydrological) cycle Interdependent and continuous circulation of water from the ocean, to the atmosphere, to the land, and back to the ocean.

water mold Filamentous organisms having cell walls made of cellulose; typically decomposers of dead freshwater organisms, but some are parasites of aquatic or terrestrial organisms.

water potential Potential energy of water; a measure of the capability to release or take up water relative to another substance.

water vascular system Series of canals that takes water to the tube feet of an echinoderm, allowing them to expand.

wax Sticky, solid, water-repellent lipid consisting of many long-chain fatty acids usually linked to long-chain alcohols.

wetland Area that is covered by water at some point in the year. (*See also* bog, marsh, swamp.)

whisk fern Common name for seedless vascular plant that consists only of stems and has no leaves or roots.

white blood cell Leukocyte, of which there are several types, each having a specific function in protecting the body from invasion by foreign substances and organisms.

white matter Myelinated axons in the central nervous system.

whorl Cluster of branches, or other plant structures, that occurs in a circular pattern.

wood Secondary xylem that builds up year after year in woody plants and becomes the annual rings.

X

X-linked Allele that is located on an X chromosome; not all X-linked genes code for sexual characteristics.

xylem Vascular tissue that transports water and mineral solutes upward through the plant body; it contains vessel elements and tracheids.

Y

yeast Single-celled fungus that has a single nucleus and reproduces asexually by budding or fission, or sexually through spore formation.

yolk Dense nutrient material in the egg of a bird or reptile.

yolk sac One of the extraembryonic membranes that, in shelled vertebrates, contains yolk for the nourishment of the embryo, and in placental mammals is the first site for blood cell formation.

Z

zero population growth No growth in population size.

zooflagellate Nonphotosynthetic protist that moves by flagella; typically, zooflagellates enter into symbiotic relationships, and some are parasitic.

zooplankton Part of plankton containing protozoans and other types of microscopic animals.

zoospore Spore that is motile by means of one or more flagella.

zygospore Thick-walled resting cell formed during sexual reproduction of zygospore fungi.

zygospore fungi (Zygomycota) Fungi, such as black bread mold, which reproduce by forming windblown spores in sporangia; sexual reproduction involves a thick-walled zygospore.

zygote Diploid cell formed by the union of two gametes; the product of fertilization.

Credits

Photographs

CHAPTER 1 Opener: (tortoise): © Cleveland Metroparks Zoo. McGraw-Hill Education; opener: (researcher on ship): © NOAA; opener:(scientist in lab): © Blend Images/Alamy RF; 1.1(bacteria): © Eye of Science/Science Source; 1.1(paramecium): © Michael Abbey/Science Source; 1.1(morel): © Carol Wolfe, photographer; 1.1(sunflower): © Photodisc Green/Getty RF; 1.1(octopus): © Erica S. Leeds; 1.3: © David Tipling/Alamy RF; 1.5: © Photodisc/Getty RF; 1.8: © Eye of Science/Science Source; 1.9: © A.B. Dowsett/SPL/Science Source; 1.10(protist): © M. I. Walker/Science Source; 1.10(fungi): © Pixtal/agefotostock RF; 1.10(plant): © Tinke Hamming/Ingram Publishing RF; 1.10(animal): © Fuse/Getty RF; 1.13(Science magazines): © McGraw-Hill Education. Mark Dierker, photographer; 1.14(students): © image100 Ltd RF; 1.14(surgeon): © Phanie/Science Source; 1.15a: © Frank & Joyce Burek/Getty RF; 1.15b: © Dr. Phillip Dustan; p. 16: Science magazines: © McGraw-Hill Education. Mark Dierker, photographer.

CHAPTER 2 Opener: © NASA/JPL-Caltech/MSSS; 2.1: © Design Pics/Don Hammond RF; 2.4a: © Southern Illinois University/Science Source; 2.4b(left): © Mazzlota et al./Science Source; 2.4b(right): © National Institutes of Health; 2.5a(both): © Tony Freeman/PhotoEdit; 2.5b: © Mark Kostich/Getty RF; 2.7b(both): © Evelyn Jo Johnson; 2.10b: © Grant Taylor/Stone/Getty Images; 2.12b: © E+/Stockcam/Getty RF.

CHAPTER 3 Opener: © 5Blue Media; 3.1a: © Peter Austin/E+/Getty RF; 3.1b: © Design Pics/Don Hammond RF; 3.1c: © Science Photo Library RF/Getty RF; 3.1d: © Creatas/PunchStock RF; 3.4a: © Brand X Pictures/PunchStock RF; 3.4b: © Ingram Publishing/Alamy RF; 3.4c: © H. Pol/CNRI/SPL/Science Source; 3.7a: © Jeremy Burgess/SPL/Science Source; 3.7b: © Don W. Fawcett/Science Source; 3.8: © Scimat/Science Source; 3.12: © Ernest A. Janes/Bruce Coleman/Photoshot; 3.13a: © Harald Theissen/Getty RF; 3.13b: © IT Stock Free/Alamy RF; 3.17a: © vgajic/E+/Getty RF; 3.17b: © Image Source/Alamy RF; 3.17c: © francois dion/Flickr/Getty RF; 3.20a: © Photodisk Red/Getty Images.

CHAPTER 4 Opener: © Oliver Meckes/Nicole Ottawa/Science Source; 4.1a: © Geoff Bryant/Science Source; 4.1b: © Ray F. Evert/University of Wisconsin, Madison; 4.1c: © U. S. Fish & Wildlife Service/Lewis Gorman; 4.1d: © Gary Delong/Science Source; 4Aa: © Eric V. Grave/Science Source; 4Ab: © Biophoto Associates/Science Source; 4Ac: © Andrew Syred/Science Source; 4B(bright-field): © Ed Reschke; 4B(bright-field stained): © Biophoto Associates/Science Source; 4B(differential): © M. I. Walker/Science Source; 4B(phase contrast): © Walter Dawn/Science Source; 4B(dark-field): © Stephen Durr; 4.4: © Sercomi/Science Source; 4.6: © Alfred Pasieka/Science Source; 4.7–4.8(top right): © Biophoto Associates/Science Source; 4.8(bottom left): © Don W. Fawcett/Science Source; 4.10: © Martin M. Rotker/Science Source; 4.11: © Biophoto Associates/Science Source; 4.12: © Don W. Fawcett/Science Source; 4.14: © EM Research Services, Newcastle University; 4.15: © Biophoto Associates/Science Source; 4.17a: © Science Source; 4.18a: © Keith R. Porter/Science Source; 4.19a(actin filaments): © Thomas Deerinck/Science Source; 4.19a(*Chara*): © Bob Gibbons/Alamy; 4.19b(intermediate): © Cultura Science/Alvin Telser, PhD/Getty Images; 4.19b(peacock): © Vol. 86/Corbis RF; 4.19c(microtubules): © Dr. Gopal Murti/Science Source; 4.19c(chameleon): © Photodisc/Vol. 6/Getty RF; 4.20b: © Don W. Fawcett/Science Source; 4.21(sperm): © David M. Phillips/Science Source; 4.21(basal body): © Biophoto Associates/Science Source; 4.21(flagellum): © Steve Gschmeissner/Science Source.

CHAPTER 5 Opener: © Axel Fassio/Photographer's Choice/Getty RF; 5Aa(human egg): © Anatomical Travelogue/Science Source; 5Aa(embryo): © Neil Harding/Stone/Getty Images; 5Aa(baby): © Photodisc Collection/Getty RF; 5.11d: © Don W. Fawcett/Science Source; 5.13a: © SPL/Science Source; 5.13b–c: © David M. Phillips/Science Source; 5.14: © Biophoto Associates/Science Source.

CHAPTER 6 Opener: © PhotoAlto/PunchStock RF; 6A: © Chikumo Chiaki/AP Images; 6.8b: © MedioImages/SuperStock RF; 6.8c: © Creatas/PunchStock RF; 6.9: © G.K. & Vikki Hart/The Image Bank/Getty Images; 6.12(leaves): © Comstock/PunchStock RF; 6.12(runner): © Photodisc/Getty RF.

CHAPTER 7 Opener: © aylinstock/E+/Getty RF; 7.1a(*Oscillatoria*): © Sinclair Stammers/SPL/Science Source; 7.1a(kelp): © Chuck Davis/The Image Bank/Getty Images; 7.1a(sequoias): © Dynamic Graphics Group/Creatas/Alamy RF; 7.1b(bamboo trees): © Corbis RF; 7.1b(panda): © Keren Su/Getty RF; 7.2: © Science Source; 7.3: © Nigel Cattlin/Science Source; 7.4: © Georg Gerster/Science Source; 7.11: © Herman Eisenbeiss/Science Source; 7.12a: © Brand X Pictures/PunchStock RF; 7.12b: © USDA/Doug Wilson, photographer; 7.13: © S. Alden/PhotoLink/Getty RF; p. 128 #1–green alga: © Stephen Durr.

CHAPTER 8 Opener: © Image Source/Javider Perini CM/RF; 8.1: © E. & P. Bauer/zefa/Corbis; 8A(bread, wine, cheese): © McGraw Hill Education./John Thoeming, photographer; 8A(yogurt, pickles/vinegar): © Bruce M. Johnson; 8.6: © Keith R. Porter/Science Source; 8.10: © C Squared Studios/Getty RF.

CHAPTER 9 Opener: © SPL/Science Source; 9.2: © Steve Gschmeissner/Science Source; 9.3a: © Dr. Gopal Murti/Phototake; 9.3b: © Dr. Jerome Rattner, *Cell Biology and Anatomy*, University of Calgary; 9.3c: © Ulrich Laemmli and J.R. Paulson, Dept. of Molecular Biology, University of Geneva, Switzerland; 9.3d–e: © Peter Engelhardt/Department of Pathology and Virology, Haartman Institute/Centre of Excellence in Computational Complex Systems Research, Biomedical Engineering and Computational Science, Faculty of Information and Natural Sciences, Helsinki University of Technology, Helsinki, Finland; 9.4a: © Andrew Syred/Science Source; 9.5 animal cell(early prophase, prophase, metaphase, anaphase, telophase): © Ed Reschke; 9.5 animal cell(prometaphase): © Michael Abbey/Science Source; 9.5 plant cell(early prophase, prometaphase): © Ed Reschke; 9.5 plant cell(prophase, metaphase, anaphase, telophase): © Kent Wood/Science Source; 9.6(top): © National Institutes of Health(NIH)/USHHS; 9.6(bottom): © Steve Gschmeissner/SPL/Getty RF; 9.7, 9.9d: © Biophoto Associates/Science Source; 9.10: © Scimat/Science Source.

CHAPTER 10 Opener: © George Doyle/Getty RF; 10.1a: © L. Willatt/Science Source; 10.3a: © Dr. D. Von Wettstein; 10.10a–10.11(both): © CNRI/SPL/Science Source; 10A: © Suzanne L. Collins/Science Source; 10.13b: © The Williams Syndrome Association; 10.14b: © Fred, Herbert, and Hendrik van Dijk. "Images of Memorable Cases, 50 Years at the Bedside: Case 133."

CHAPTER 11 Opener: © McGraw-Hill Education./Jill Braaten, photographer; 11.1a: © Ned M. Seidler/National Geographic Stock; 11.1b: © Michael Mahovlich/Radius Images/Getty RF; 11.12: © Division of Medical Toxicology, University of Virginia; 11.15(tissue): © Ed Reschke; p. 199(sickle cell): © Phototake, Inc./Alamy.

CHAPTER 12 Opener: © Design Pics/Don Hammond RF; 12.2b: © Lee D. Simon/Science Source; 12.4a–c: © Science Source; 12.5a(DNA model): © Photodisk Red/Getty RF; 12.5d: © A. Barrington Brown/Science Source; 12.13a: © Oscar L. Miller/Science Source; 12.13b: © UIG via Getty Images; 12.13c: © Science Photo Library/Alamy RF; 12Aa: © Frans Lanting/Mint Images/Getty Images; 12Ab: © CB2/ZOB/WENN/Newscom; 12Ac: © Dr. Gopal Murti/Science Source; 12.16d: © Science Source.

CHAPTER 13 Opener: © UIG via Getty Images; 13.5a: © Alfred Pasieka/Science Source; 13.6: © cgbaldauf/E+/Getty RF; 13A: © Hero/Corbis/Glow Images RF; 13.12a–b(both): © Eye of Science/Science Source.

CHAPTER 14 Opener: © Tim Ridley/Dorling Kindersley/Getty Images; 14.5(bacteria): © Medical-on-Line/Alamy; 14.5(researcher): © CEK/AP Images; 14.10c: © Mondae Leigh Baker; Table14.3(human): © Image Source/JupiterImages RF; Table14.3(plant): © USDA/Peggy Greb, photographer; Table14.3(mouse): © Redmond Durrell/Alamy RF; Table14.3(fruit fly): © Herman Eisenbeiss/Science Source; Table14.3(roundworm): © CMSP/J.L. Carson/Getty Images; Table14.3(yeast): © G. Wanner/Getty Images; 14A: © Deco/Alamy RF; 14.11: © Geno Zhou/FeatureChina/Newscom; 14B: © Bart Coenders/E+/Getty RF.

CHAPTER 15 Opener: © Tyler Keillor, University of Chicago; 15.1a: © National Library of Medicine; 15.1c: © Nicole Duplaix/National Geographic/Getty Images; 15.1d: © Christian Heinrich/Getty RF; 15.1e: © Nick Green/Photolibrary/Getty Images; 15.1f: © Chad Ehlers/Photographer's Choice/Getty Images; 15.1g: © Galen Rowell/Corbis; 15.4(*Lepus*): © Daniel Trim Photography/Moment/Getty RF; 15.4(*Dolichotis*): © Juan & Carmecita Munoz/Science Source; 15.5a: © Kevin Schafer/Corbis; 15.5b: © Tim Laman/National Geographic/Getty Images; 15.6: © Lisette Le Bon/SuperStock; 15.7(wolf): © Nicholas James McCollum/Moment Open/Getty RF; 15.7(Irish wolfhound): © Ralph Reinhold/Index Stock Imagery/Photolibrary RF; 15.7(Boston terrier): © Robert Dowling/Corbis; 15.8(cabbage): © Image Source/PunchStock RF; 15.8(Brussel sprouts): © Anne Murphy/Moment/Getty RF; 15.8(kohlrabi): © Foodcollection RF; 15.8(mustard): © Medioimages/PunchStock RF; 15.9a: © Miguel Castro/Science Source; 15.9b: © Celia Mannings/Alamy; 15.9c: © Michael Stubblefield/Alamy RF; 15.11(all): © Nicolas Gompel and Benjamin Prud'homme; 15A(bacteria,top): © Eye of Science/Science Source; 15A(bacteria,left): © CDC/Janice Haney Carr; 15A(circular Tree of Life): © EMBL/Figure by Christian von Mering; 15A(mushrooms): © IT Stock/agefotostock RF; 15.14(sugar glider): © ANT Photo Library/Science Source; 15.14(wombat): © Photodisc Blue/Getty RF; 15.14(kangaroo): © George Holton/Science Source.

CHAPTER 16 Opener: © Paul Gunning/Science Source; 16.11b(both): © John Cancalosi/Getty Images; 16.13: © Panoramic Images/Getty Images; 16.14a: © Westend61/SuperStock RF; 16.14b: © Neil McIntre/Getty Images; 16A(both): © Ingram Publishing+M496/SuperStock RF; 16.15(*P. o. obsoleta*): © Robert Hamilton/Alamy RF; 16.15(*P. o. quadrivittata*): © Millard H. Sharp/Science Source; 16.15(*P. o. rossalleni*): © R. Andrew Odum/Photolibrary/Getty Images; 16.15(*P. o. spiloides*): © F. Teigler/Blickwinkel/agefotostock; 16.15(*P. o. lindheimeri*): © Michelle Gilders/agefotostock/SuperStock.

CHAPTER 17 Opener: (frogs on pencil): © Indraneil Das/Alamy RF; Opener: (frog on leaf): © Sam Yue/Alamy; 17.1(Rio Grande leopard frog): © Danita Delimont/Alamy; 17.1(Southern leopard frog): © Robin Chittenden/Alamy; 17.1(Northern leopard frog): © Michelle Gilders/Alamy; 17.3a: © Danita Delimont/Getty Images; 17.6: © Henri Leduc/Moment Open/Getty RF; 17.7(horse): © Creatas/PunchStock RF; 17.7(donkey): © Photodisc Collection/Getty RF; 17.7(mule): © Radius Images/Alamy RF; p. 304(lizard/dewlap): © Jonathan Losos; 17.10(*C. concinna*): © Steffen Hauser/botanikfoto/Alamy; 17.10(*C. virgata*): © Dr. Dean Wm. Taylor/Jepson Herbarium, UC Berkeley; 17.10(*C. pulchella*): © agefotostock/Alamy; 17B(top): © Ralph Lee Hopkins/Getty RF; 17B(center): © Minden Pictures/SuperStock; 17B(bottom): © Kevin Schafer/Corbis;

17C(both): © Dr. Arkhat Abzhanov, Harvard University, Dept. of OEB; 17.14(left): © Carolina Biological Supply/Science Source; 17.14(center): © Vol. OS02/Photodisc/Getty RF; 17.14(right): © Elmer Frederick Fischer/Corbis RF; 17.15(both): © Prof. Dr. Walter Gehring, University of Basel, Switzerland; p. 314(horse): © Creatas/PunchStock RF; p. 314(donkey): © Photodisc Collection/Getty RF; p. 314(mule): © Radius Images/Alamy RF.

CHAPTER 18 Opener: © Joe Tucciarone/Science Source; 18.3: © Ralph White/Eureka Premium/Corbis; 18.6a: © Kenneth Murray/Science Source; 18.7(trilobite): © Francois Gohier/Science Source; 18.7(ichthyosaur): © R. Koenig/Blickwinkel/age Fotostock; 18.7(fossil fish): © Alan Morgan; 18.7(fern): © George Bernard/NHPA/Photoshot; 18.7(ammonites): © Sinclair Stammers/SPL/Science Source; 18.7(dinosaur footprint): © fotocelia/iStock/360/Getty RF; 18.9a: © Francois Gohier/Science Source; 18.9b: © Dr. J. William Schopf; 18.10a: © Sinclair Stammers/Science Source; 18.10b: © DeAgostini/SuperStock; 18.11*(Marrella)*: © O. Louis Mazzatenta/National Geographic/Getty Images; 18.11*(Thaumaptilon)*: © Simon Conway Morris, University of Cambridge; 18.11*(Vauxia)*: © Alan Siruinikoff/Science Source; 18.11*(Wiwaxia)*: © O. Louis Mazzatenta/National Geographic/Getty Images; 18.12a: © Richard Bizley/Science Source; 18.12b: © powerofforever/iStock/360/Getty RF; 18.12c: © Ablestock.com RF; 18.13–18.14: © Chase Studio/Science Source; 18.15: © Gianni Dagli Orti/Corbis.

CHAPTER 19 Opener: © NIBSC/Science Source; 19.1(all): © Sylvia S. Mader; 19.2a: © Pixtal/agefotostock RF; 19.2b: © Neil Fletcher/Dorling Kindersley/Getty Images; 19.2c: © Falk Kienas/iStock/360/Getty RF; 19A(bee, both): © Stefan Sollfors/Alamy RF; 19B: © Lars Klove/The New York Times/Redux; 19C(tuna): © Miscellaneoustock/Alamy RF; 19C(tilapia): © Ammit/iStock/360/Getty RF; 19.8: © David Dilcher and Ge Sun.

CHAPTER 20 Opener: © BSIP SA/Alamy RF; 20.1a: © Biophoto Associates/Science Source; 20.1b: © Eye of Science/Science Source; 20.1c: © Science Source; 20.1d: © CDC/Cynthia Goldsmith; 20.2b: © Eye of Science/Science Source; 20.7: © CNRI/SPL/Science Source; 20.8b: © CDC/Dr. William A. Clark; 20.9a: © Ed Reschke/Getty Images; 20.9b: © Eye of Science/Science Source; 20.9c: © Scimat/Science Source; 20.10: © Nigel Cattlin/Alamy; 20.11: © Alfred Pasieka/SPL/Science Source; 20.12a: © Michael Abbey/Science Source; 20.12b: © Sinclair Stammers/SPL/Science Source; 20.13a(main): © Marco Regalia Sell/Alamy RF; 20.13a(inset): © Eye of Science/Science Source; 20.13b(main): © Jeff Lepore/Science Source; 20.13b(inset): © Eye of Science/Science Source; 20.13c(main): © Susan Rosenthal/Corbis; 20.13c(inset): © Dr. M. Rohde, GBF/Science Source.

CHAPTER 21 Opener: © FLPA/SuperStock; Table 21.1(volvox): © Stephen Durr; Table 21.1(diatoms): © M.I. Walker/Science Source; Table 21.1*(Giardia)*: © CDC/Dr. Stan Erlandsen and Dr. Dennis Feely; Table 21.1(amoeba): © Melba Photo Agency/PunchStock RF; Table 21.1(radiolarians): © Eye of Science/Science Source; Table 21.1(choanoflagellate): © D. J. Patterson/provided courtesy microscope.mbl.edu; 21.2: © Biophoto Associates/Science Source; 21.4(both): © Manfred Kage/Science Source; 21.5a: © Minden Pictures/SuperStock; 21.6b: © M.I. Walker/Science Source; 21.7a: © Bob Gibbons/Alamy; 21.7b: © Kingsley Stern; 21.8: © Steven P. Lynch; 21.9: © D.P Wilson/Eric & David Hosking/Science Source; 21.10a: © Dr. Ann Smith/Science Source; 21.10b: © Biophoto Associates/Science Source; 21.12: © Noble Proctor/Science Source; 21.13a: © Jeff Foott/Discovery Channel Images/Getty Images; 21.13b: © Corpus Christi Caller-Times, Todd Yates/AP Images; 21.14a: © Carolina Biological Supply/Phototake; 21.14b: © Ed Reschke/Photolibrary/Getty Images; 21.14c: © Eric Grave/Science Source; 21.16b: © Kage Mikrofotografie/Phototake; 21.17: © Kallista Images/Getty Images; 21A: © AP Images; 21D(woman): © Wavebreakmedia Ltd/Lightwavemedia 360/Getty RF; 21D(Earth): © NASA image by Reto Stockli, based on data from NASA and NOAA; 21E(boy): © Author's Image/PunchStock RF; 21E(tsetse fly): © Frank Greenaway/Getty Images; 21E(pig): © G.K. & Vikki Hart/Getty RF; 21E(goat): © Eureka/Alamy RF; 21.18a: © Eye of Science/Science Source; 21.20(left): © NHPA/SuperStock; 21.20(right): © NaturePL/SuperStock; 21.22a(cliffs): © Stockbyte/Getty RF; 21.22a*(Globigerina)*: © NHPA/SuperStock; 21.22b(tests): © Eye of Science/Science Source; p. 391(nuclearia): © Dr. Sc. Yuuji Tsukii/Protist Information Server, URL: http://protist.I.hosei.ac.jp.

CHAPTER 22 Opener: © Ecovative Design, LLC/www.ecovativedesign.com; 22.2a: © Matt Meadows/Photolibrary/Getty Images; 22.2b: © Andrew Syred/Science Source; 22.3a: © Biophoto Associates/Science Source; 22.3b: © Dr. Hilda Canter-Lund; 22.4(zygospore): © Biophoto Associates/Science Source; 22.4(sporangium): © Ed Reschke/Photolibrary/Getty Images; 22.5a: © Science Photo Library RF/Getty RF; 22.5b: © Dennis Kunkel Microscopy, Inc./Phototake; 22.6a(cup fungi): © Carol Wolfe; 22.6b(morel): © Corbis RF; 22.6b(ascospores): © Carolina Biological Supply/Phototake; 22.7a: © Alex Wild Photography; 22.7b: © New York State Department of Environmental Conservation, Al Hicks/AP Images; 22.8a: © Dr. M. A. Ansary/Science Source; 22.8b: © John Hadfield/SPL/Science Source; 22.9(basidium): © Biophoto Associates/Science Source; 22.10a: © imagebroker/Alamy RF; 22.10b: © De Agostini/Getty Images; 22.10c: © Dr. Robert Siegel/Stanford University; 22.10d: © Ed Reschke/Getty Images; 22.10e: © David Plummer/Alamy; 22A: © De Agostini/Getty Images; 22B: © Wildlife GmbH/Alamy; 22.11a: © Marvin Dembinsky Photo Associates/Alamy; 22.11b: © Patrick Lynch/Science Source; 22.12a: © Digital Vision/Getty RF; 22.12b: © Steven P. Lynch; 22.12c: © yogesh more/Alamy RF; 22.14(truffle): © bonzami emmanuelle/Alamy RF; 22.14(oak trees): © Biosphoto/SuperStock.

CHAPTER 23 Opener: © Blend Images/Getty RF; 23.2a: © Biology Pics/Science Source; 23.2b: © Garry DeLong/Science Source; 23.3*(Chara)*: © Heather Angel/Natural Visions; 23.3*(Coleochaete)*: © Dr. Charles F. Delwiche, University of Maryland; 23.4a: © Ingram Publishing/agefotostock RF; 23.4b: © Dr. Jeremy Burgess/Science Source; 23.7a: © Ed Reschke/Getty Images; 23.7b: © J.M. Conrarder/National Audubon Society/Science Source; 23.7c: © Ed Reschke/Oxford Scientific/Getty Images; 23.8: © Steven P. Lynch; 23.9(sporophyte): © Peter Arnold, Inc./Alamy; 23.9(gametophyte): © Steven P. Lynch; 23A(both): © Catherine La Farge; 23.10: © Hans Steur, the Netherlands; 23.13(whole plant): © Robert P. Carr/Bruce Coleman/Photoshot; 23.13(whorl): © blickwinkel/Alamy; 23.14(whisk fern): © Biophoto Associates/Science Source; 23.14(sporangia): © Steven P. Lynch; 23C(fossil ferns): © Sinclair Stammers/SPL/Science Source; 23.15(leatherleaf fern): © Gregory Preest/Alamy; 23.15(tree fern): © Danita Delimont/Getty Images; 23.15(hart's tongue fern): © Organics image library/Alamy RF; 23.16: © Steven P. Lynch; 23.17a(seed cones): © Steven P. Lynch; 23.17a(pollen cones): © Maria Mosolova/Photolibrary/Getty RF; 23.17a(forest): © Steven P. Lynch; 23.17b: © Ed Reschke/Peter Arnold/Getty Images; 23.17c(tree): © Steffen Hauser/botanikfoto/Alamy; 23.18(pollen): © Ed Reschke/Photolibrary/Getty Images; 23.19a(tree): © Hoberman Collection/Alamy; 23.19a(cone): © PlazaCameraman/Getty RF; 23.19b(Gingko branch): © blickwinkel/Alamy; 23.19b(Gingko ovule): © Ken Robertson, University of Illinois/INHS; 23.19c(ephedra plant): © Evelyn Jo Johnson; 23.19c(ephedra inset): © Karl J. Niklas; 23.19d: © TravelMuse/Alamy RF; 23.20a: © UZH; 23.20b: © Stephen McCabe/Arboretum at University of California Santa Cruz; 23D: © Axel Fassio/Photographer's Choice/Getty RF; 23.21a: © Ed Reschke/Photolibrary/Getty Images; 23.21b: © Steven P. Lynch.

CHAPTER 24 Opener(neem tree): © Dinodia Photos/Alamy; CO24(neem products): © bdspn/iStock/360/Getty RF; 24.2a: © Nigel Cattlin/Alamy; 24.2b: © Anthony Pleva/Alamy; 24.2c: © Biophoto Associates/Science Source; 24.2d: © Kingsley Stern; 24.3(all): © Biophoto Associates/Science Source; 24Aa: © Niels Poulsen/Alamy; 24Ab: © George Grall/National Geographic/Getty Images; 24.4a: © Garry DeLong/Science Source; 24.5: © Robert Knauft/Biology Pics/Science Source; 24.7: © Steven P. Lynch; 24.9a(root tip): © Ray F. Evert/University of Wisconsin, Madison; 24.9b: © Carolina Biological Supply/Phototake; 24.10: © Lee Wilcox; 24.11a: © McGraw Hill Education./Al Telser, photographer; 24.11b: © George Ellmore, Tufts University; 24.12a: © Martin Harvey/Photodisc/Getty RF; 24.12b: © NokHoOkNoi/iStock/360/Getty RF; 24.12c: © FLPA/Mark Newman/agefotostock; 24.12d: © DEA/S Montanari/agefotostock; 24.14a: © Steven P. Lynch; 24B(flooring): © Roman Borodaev/Alamy RF; 24B(shoots): © Koki Iino/Getty RF; 24B(bamboo grove): © Michele Westmorland/Getty Images; 24.15(left): © Ed Reschke; 24.15(right): © Ray F. Evert/University of Wisconsin, Madison; 24.16(top): © Carolina Biological Supply/Phototake; 24.16(bottom): © Kingsley Stern; 24.17(circular cross section): © Ed Reschke/Getty Images; 24.19a: © Ardea London/Ardea.com; 24.20a: © Evelyn Jo Johnson; 24.20b: © Science Pictures Limited/Science Source; 24.20c–d: © McGraw-Hill Education./Carlyn Iverson, photographer; 24.22: © Ray F. Evert, University of Wisconsin, Madison; 24.24a: © Nature Picture Library RF; 24.24b–c: © Steven P. Lynch.

CHAPTER 25 Opener: © Maurizio Migliorato/Alamy RF; 25.3b: © Ricochet Creative Productions LLC/Leah Moore, photographer; 25.6a: © Carolina Biological Supply/Phototake; 25A: © Yann Arthus-Bertrand/Corbis; 25.7(left): © Dr. Jeremy Burgess/Science Source; 25.7(right): © Garry DeLong/Science Source; 25.8: © Dr. Keith Wheeler/Science Source; 25.9a: © Kevin Schafer/Corbis; 25.9b(sundew plant): © blickwinkel/Alamy; 25.9b(sundew leaf): © Dr. Jeremy Burgess/Science Source; 25C(both): © McGraw Hill Education. Ken Cavanagh, photographer; 25.12a: © Martyn F. Chillmaid/Science Source; 25.12b: © H. Schmidbauer/Blickwinkel/agefotostock; 25.14: © Adalberto Rios Szalay/Sexto Sol/Stockbyte/Getty RF; 25.15(both): © Jeremy Burgess/SPL/Science Source; 25.16: © Kolling et al. *Plant Methods* 2013 9:45; 25.19a: © M. H. Zimmermann/Harvard Forest, Harvard University; 25.19b: © Steven P. Lynch.

CHAPTER 26 Opener: © USDA/Bruce Fritz, photographer; 26.4a: © Science Source; 26.4b: © Amnon Lichter, The Volcani Center; 26.5a: © Biophoto Associates/Science Source; 26.5b-d: © Alan Darvill and Stefan Eberhard, Complex Carbohydrate Research Center, University of Georgia; 26.6: © Oxford/iStock/360/Getty RF; 26.7: © Dr. Donald R. McCarty, University of Florida; 26.9(both): © Kingsley Stern; 26B: © Steven P. Lynch; 26.12a: © Cathlyn Melloan/Stone/Getty Images; 26.12b: © Alison Thompson/Alamy; 26.13a: © Martin Shields/Alamy; 26.13b: © Randy Moore/BioPhot; 26D: © Nigel Cattlin/Alamy; 26.14(both): © John Kaprielian/Science Source; 26.17: © Nigel Cattlin/Alamy; 26.18(both): © Gretel Guest; 26.18b(left): © Michael Jenner/Alamy; 26.18b(right): © JupiterImages/Photos.com/360/Getty RF.

CHAPTER 27 Opener: © Jaap Hart/E+/Getty RF; 27.3a: © Marg Cousens/Getty Images; 27.3b: © Pat Pendarvis; 27.4a: © Steven P. Lynch; 27.4b: © Garden World Images/agefotostock; 27.5(pollen grain): © Graham Kent; 27.5(embryo sac): © Ed Reschke; 27.6a: © Tim Gainey/Alamy; 27.6b: © Medical-on-line/Alamy; 27.6c: © Steve Gschmeissner/Science Photo Library/Getty RF; 27Aa: © Steven P. Lynch; 27Ab: © Creatas/PunchStock RF; 27Ba: © Anthony Mercieca/Science Source; 27Bb: © Dr. Merlin D. Tuttle/Bat Conservation International; 27.7b–d: © Dr. Chun-Ming Liu; 27.7e(torpedo): © Biology Media/Science Source; 27.7f(mature embryo): © Steven P. Lynch; 27.8a(bean): © Dwight Kuhn; 27.8b(corn): © Ray F. Evert/University of Wisconsin, Madison; 27.9a: © Peter Fakler/Alamy RF; 27.9b: © D. Hurst/Alamy RF; 27.9c: © Foodcollection RF; 27.9d: © Robert Llewellyn/Corbis RF; 27.9e: © Ricochet Creative Productions LLC/Kim Scott, photographer; 27.9f: © Ingram Publishing/Alamy RF; 27.10a: © Dimitri Vervitsiotis/Getty RF; 27.10b: © Stockbyte RF; 27.10c: © Scott Camazine/Science Source; 27.10d: © M. Brodie/Alamy; 27.11a(bean): © Ed Reschke; 27.11b(corn): © Ed Reschke/Photolibrary/Getty Images; 27.12: © NHPA/SuperStock; 27.13a: © Biophoto Associates/Science Source; 27.13b–c: © Prof. Dr. Hans-Ulrich Koop, from *Plant Cell Reports,* 17:601-604; 27.13d: © Louise Murray/Visuals Unlimited/Corbis; 27.14: © Hillside Nursery, Shelburne Falls, MA.

CHAPTER 28 Opener: © Peter Arnold/Digital Vision/Getty RF; 28.1b: © Dwight Kuhn; 28.1c: © Steve Bloom/Taxi/Getty; 28.1d: © Carolina Biological Supply/Phototake; 28A(phase I): © Yoav Levy/Phototake; 28A(phase II, all): © Jim Langeland, Steve Paddock, and Sean Carroll/University of Wisconsin, Madison; 28B(chicken): © Getty

Images/Photodisc RF; 28B(mouse): © Redmond Durrell/Alamy RF; 28B(snake): © IT Stock/PunchStock RF; 28.6a: © Andrew J. Martinez/Science Source; 28.7a: © Jeff Rotman; 28.7b: © Mark Conlin/Alamy; 28.8a: © Comstock Images/PictureQuest RF; 28.8b: © Ron Taylor/Bruce Coleman/Photoshot; 28.8c: © Islands in the Sea 2002, NOAA/OER; 28.8d: © Amos Nachoum/Corbis; 28.9: © NHPA/M. I. Walker/Photoshot RF; 28.10a: © The Natural History Museum/Alamy; 28.11: © Diane R. Nelson; 28.12e: © NHPA/M. I. Walker/Photoshot RF; 28.13a: © SPL/Science Source; 28.14(proglottid): © CDC; 28.14(scolex): © C. James Webb/Phototake; 28.16b: © Clouds Hill Imaging, Ltd./Science Source; 28.17a: © ANT Photo Library/Science Source; 28.17b: © Michael Lustbader/Science Source; 28.18a: © Wolfgang Polzer/Alamy; 28.18b: © Farley Bridges; 28.18c: © Douglas Faulkner/Science Source; 28.18d: © Georgette Douwma/Science Source; 28.19c: © MikeLane45/iStock360/Getty RF; 28.20b: © Diane R. Nelson; 28.20c: © St. Bartholomew's Hospital/SPL/Science Source; 28.21a: © P&R Fotos/agefotostock/SuperStock; 28.21b: © CDC; 28.21c: © Vanessa Vick/The New York Times/Redux; 28.22c: © Tommy IX/iStock/360/Getty RF; 28.23a: © Michael Lustbader/Science Source; 28.23b: © DEA/C. Galasso/De Agostini Picture Library/Getty Images; 28.23c: © L. Newman & A. Flowers/Science Source; 28.25a: © Larry Miller/Science Source; 28.25b: © David Aubrey/Corbis; 28.26(mealybug, leafhopper, dragonfly): © Farley Bridges; 28.26(beetle): © George Grall/Getty Images; 28.26(louse): © Alastair Macewen/Getty Images; 28.26(wasp): © James H. Robinson/Science Source; 28.28a: © U.S. Fish & Wildlife Service/Robert Pos; 28.28b: © Tom McHugh/Science Source; 28.28c: © Mark Kostich/Getty RF; 28D: © BSIP/Newscom; 28.29b: © Randy Morse, GoldenStateImages.com; 28.29c: © Roberto Nistri/Alamy; 28.29d: © Fuse/Getty RF.

CHAPTER 29 Opener: © FLPA/SuperStock; 29.2: © Heather Angel/Natural Visions; 29.3: © Corbis RF; 29.6: © Heather Angel/Natural Visions; 29.8a: © Alastair Pollock Photography/Getty RF; 29.8b: © Comstock Images/PictureQuest; 29.9b: © powerofforever/iStock 360/Getty RF; 29.9c: © Jane Burton/Bruce Coleman/Photoshot; 29.9d: © F616Team Hymas/Getty RF; 29.9e: © Franco Banfi/WaterFrame/Getty Images; 29.10: © Peter Scoones/SPL/Science Source; 29.13a: © Suzanne L. Collins & Joseph T. Collins/Science Source; 29.13b: © NHPA/T. Kitchin & V. Hurst; 29.13c: © FLPA/Alamy; 29Aa: © MedioImages/SuperStock RF; 29Ab: © Allan Friedlander/SuperStock; 29Ac: © PTH/Phototake; 29.16a: © Michael Patrick O'Neill/Alamy; 29.16b: © Design Pics/SuperStock RF; 29.16c: © Joel Sartorie/National Geographic/Getty Images; 29.16d: © Tim Cuff/Alamy; 29.16e: © Martin Harvey/Gallo Images/Corbis; 29.18a: © Creatas/PunchStock RF; 29.18b: © Joe McDonald/Corbis; 29.18c: © IT Stock/PunchStock RF; 29.18d: © Jeremy Woodhouse/Getty RF; 29.20a: © D. Parer & E. Parer-Cook/Ardea; 29.20b: © John White Photos/Getty RF; 29.20c: © John Macgregor/Photolibrary/Getty Images.

CHAPTER 30 Opener: © Frank Vinken/Max Planck Institute for Evolutionary Anthropology, Leipzig, Germany; 30.1(tarsier): © Magdalena Biskup Travel Photography/Getty RF; 30.1(monkey): © Paul Souders/Corbis; 30.1(baboon): © St. Meyers/Okapia/Science Source; 30.1(orangutan): © Tim Davis/Science Source; 30.1(chimpanzee): © Martin Harvey/Photolibrary/Getty Images; 30.1(humans): © Comstock Images/JupiterImages RF; 30.5b: © Carolina Biological/Visuals Unlimited/Corbis; 30.6(*A. ramidus*): © Richard T. Nowitz/Science Source; 30.6(*A. afarensis*): © Scott Camazine/Alamy; 30.6(*A. africanus*): © Philippe Plailly/Science Source; 30.6(*H. habilis*): © Kike Calvo VWPics/SuperStock; 30.6(*H. sapiens*): © Kenneth Garrett/Getty Images; 30.8a: © Dan Dreyfus and Associates; 30.8b: © John Reader/Science Source; 30.9: © AP Images; 30.11: © Tomas Abad/agefotostock/SuperStock; 30.12a: © Photodisc/Getty RF; 30.12b: © Sylvia S. Mader; 30.12c: © B & C Alexander/Science Source; 30Cb: © Mark Thiessen/National Geographic Stock.

CHAPTER 31 Opener: © Science Photo Library/Science Source; 31.1(all)– 31.3a–b, d–e: © Ed Reschke; 31.3c: © McGraw-Hill Education./Dennis Strete, Photographer; 31.5a, c: © Ed Reschke; 31.5b: © McGraw-Hill Education./Dennis Strete, Photographer; 31.6b: © Ed Reschke; 31A(both): © Justin C. Rosenberg; 31.9: © Eye of Science/Science Source; 31Ca–b: © Dr. P. Marazzi/Science Source; 31Cc: © James Stevenson/SPL/Science Source.

CHAPTER 32 Opener: © AP Images; 32.1a: © NHPA/M. I. Walker/Photoshot RF; 32.1b: © Lester V. Bergman/Corbis; 32.1c: © Randy Morse, GoldenStateImages.com; 32.6b: © McGraw-Hill Education./Professor David F. Cox; 32.7b: © Thomas Deerinck, NCMIR/Science Source; 32.8d: © Biophoto Associates/Science Source; 32.9b–c: © Ed Reschke; 32.9d: © David Joel/MacNeal Hospital/Getty Images; 32A: © Biophoto Associates/Science Source; 32.15: © J. C. Revy/Phototake; 32.16: © J. C. Revy/Phototake; 32.17: © Eye of Science/Science Source; 32B: © Andrew J. Martinez/Science Source.

CHAPTER 33 Opener: © Ricochet Creative Productions LLC/Leah Moore, photographer; 33.1a: © Stephen Durr; 33.1b: © Carolina Biological/Visuals Unlimited/Corbis; 33.3a–b, d: © Ed Reschke/Photolibrary/Getty Images; 33.3c: © McGraw-Hill Education./Al Telser, photographer; 33.6: © Dennis Kunkel Microscopy, Inc./Phototake; 33.12b: © Steve Gschmeissner/Science Source; p. 633(AIDS patient): © A. Ramey/PhotoEdit; p. 633(shingles): © CDC; p. 633(candidiasis): © Dr. M.A. Ansary/Science Source; p. 633(pneumonia): © CDC; 33.13: © McGraw-Hill Education./Jill Braaten, photographer; 33.14: © Digital Vision/Getty RF; 33.15(girl): © Damien Lovegrove/SPL/Science Source; 33.15(allergens): © David Scharf/SPL/Science Source; 33.16: © Southern Illinois University/Science Source; 33.17: © Scott Camazine/Alamy.

CHAPTER 34 Opener: © Goodshoot/PunchStock RF; 34.9b: © Ed Reschke/Photolibrary/Getty Images; 34.10(villi): © Kage-mikrofotografie/Phototake; 34.10(microvilli): © Science Photo Library RF/Getty RF; 34.14: © Jeffrey Coolidge/Stone/Getty Images.

CHAPTER 35 Opener: © Timothy A. Clary/AFP/Getty Images; 35.4a(fish): © Dr. Jeffrey Isaacson, Nebraska Wesleyan University; 35.4c(gills): © David M. Phillips/Science Source; 35.5–35.6b: © Ed Reschke; 35.12: © Andrew Syred/Science Source; 35.13: © Dr. P. Marazzi/Science Source; 35.15a–b: © Matt Meadows/Photolibrary/Getty Images; 35.15c: © Biophoto Associates/Science Source; 35B: © Martina Paraninfi/Moment/Getty RF.

CHAPTER 36 Opener: © Georgette Douwma/Science Source; 36.4: © Bob Calhoun/Bruce Coleman/Photoshot; 36.5: © Eric Hosking/Science Source; 36.6a: © James Cavallini/Science Source; 36.8b: © Science Photo Library RF/Getty RF; 36.8c–d: © Ed Reschke/Photolibrary/Getty Images; 36.9b: © Joseph F. Gennaro, Jr./Science Source; 36A: © AFP/Getty Images.

CHAPTER 37 Opener: © Slaven Vlasic/Getty Images Entertainment; 37.4a: © McGraw Hill Education. Dr. Dennis Emery, Dept of Zoology and Genetics, Iowa State University; 37.4c: © David M. Phillips/Science Source; 37.7a, 37A: © Science Source; 37.8b: © Colin Chumbley/Science Source; 37.12c: © Karl E. Deckart/Phototake.

CHAPTER 38 Opener(snake): © Nutscode/T Service/Science Source; opener(mouse): © Ted Kinsman/Science Source; 38.1a: © Robert Marien/Corbis RF; 38.1b: © James R.D. Scott/Moment/Getty RF; 38.1c: © Jim Parkin/iStock/360/Getty RF; 38.2b–c: © Science Source; 38.4: © Farley Bridges; 38.5(both): © Heather Angel/Natural Visions; 38.8: © Science Source; 38.9b: © Biophoto Associates/Science Source; 38.14: © P. Motta/SPL/Science Source; 38.15(both): © Dr. Yeohash Raphael, the University of Michigan, Ann Arbor.

CHAPTER 39 Opener: © Jae C. Hong/AP Images; 39.2: © Martin Shields/Alamy; 39.3: © E. R. Degginger/Science Source; 39.4(hyaline cartilage, compact bone): © Ed Reschke; 39.4(osteocyte): © Biophoto Associates/Science Source; 39.13(gymnast): © Corbis RF; 39.13(myofibril): © Biology Media/Science Source; 39A: © Thinkstock/Getty RF; 39.14a(right): © Ed Reschke/Oxford Scientific/Getty Images.

CHAPTER 40 Opener(caterpillar): © Daniel Borzynski/Alamy; opener(moth): © James Urbach/SuperStock; 40.7a: © Wide World Photos/AP Images; 40.7b: © General Photographic Agency/Getty Images; 40.8: © Bart's Medical Library/Phototake; 40.9a: © Bruce Coleman, Inc./Alamy; 40.9b: © Medical-on-Line/Alamy; 40.9c: © Dr. P. Marazzi/Science Source; 40.13a: © Custom Medical Stock Photo; 40.13b: © NMSB/Custom Medical Stock Photo; 40.14(both): © Shannon Halverson; 40.15(pancreatic islet, top and bottom): © Peter Arnold, Inc./Alamy; 40A: © Bettmann/Corbis; 40.17: © Evelyn Jo Johnson.

CHAPTER 41 Opener(both): © Dr. Kazunori Yoshizawa; 41.1: © Natural Visions/Alamy; 41.2: © Herbert Kehrer/zefa/Corbis; 41.3: © Anthony Mercieca/Science Source; 41.5b: © Ed Reschke; 41.6b: © Anatomical Travelogue/Science Source; 41.9a: © Ed Reschke/Photolibrary/Getty Images; 41.13a: © Saturn Stills/Science Source; 41.13b: © Keith Brofsky/Getty RF; 41.13c: © McGraw-Hill Education. Lars A. Niki, photographer; 41.14: © CC Studio/SPL/Science Source; 41Aa–b(both): © Brand X/SuperStock RF; 41.15: © Scott Camazine/Science Source; 41.16a(left): © CDC/Dr. M. F. Rein; 41.16b(right): © Bart's Medical Library/Phototake; 41B(left): © Corbis CD Vol 161/Corbis RF; 41B(right): © Corbis CD Vol 178/Corbis RF; 41C: © Brand X Pictures RF; 41.17: © Dr. Allan Harris/Phototake; 41.18: © CDC/Joe Miller; 41.19: © Melba Photo Agency/Alamy RF.

CHAPTER 42 Opener(2 days): Courtesy of the film, Building Babies, © ICAM/Mona Lisa/Phototake; opener(4 weeks): © Science Source; opener(6 weeks): © Claude Edelmann/Science Source; opener(4 months): © Tissuepix/Science Source; p. 795(chick): © Photodisc/Getty RF; 42.1: © David M. Phillips/Science Source; 42.2a: © Patrick J. Lynch/Science Source; 42.9(all): © Steve Paddock, Howard Hughes Medical Research Institute; 42.10a: © David Scharf/Science Source; 42.12(fertilization): © Don W. Fawcett/Science Source; 42.12(2-cell): © Rawlins-CMSP/Getty Images; 42.12(morula): © RBM Online/epa/Corbis; 42.12(implantation): © Bettmann/Corbis; 42.14a: © Lennart Nilsson/Scanpix; 42.17: © Georges Gobet/AFP/Getty Images.

CHAPTER 43 Opener: © Juniors Bildarchiv GmbH/Alamy; 43.1a: © Juniors Bildarchiv/agefotostock; 43.1b: © Refuge for Saving the Wildlife, Inc.; 43.2(inland): © John Serrao/Science Source; 43.2(coastal): © Chris Mattison/Alamy; 43.3b: © Tom McHugh/Science Source; 43.3c: © Biosphoto/SuperStock; p. 820(fish): © fotolincs/Alamy; p. 822(imprinting): © Nina Leen/Time Life Pictures/Getty Images; p. 824(raven): © Dr. Bernd Heinrich; 43.7: © Gregory G. Dimijian/Science Source; 43A(left): © The Forum, Jay Pickthorn/AP Images; 43A(right): © Bryce Flynn/Moment/Getty RF; 43.8a(main): © Arco Images/GmbH/Alamy; 43.8a(inset): © Flavio Vallenari/E+/Getty RF; 43.9: © Raul Arboleda/AFP/Getty Images; 43.10(firefly): © Phil Degginger/Alamy; 43.10(trees): © PhotoLink RF; 43.11a: © Scott Camazine/Alamy; 43.12: © Nicole Duplaix/Photolibrary/Getty Images; 43.14: © Thomas Dobner 2006/Alamy RF; 43Ba: © T & P Gardener/Bruce Coleman/Photoshot; 43.15: © D. Robert & Lorri Franz/Corbis; 43.16: © Mark Moffett/Minden Pictures/National Geographic Creative; 43.17: © Biosphoto/SuperStock.

CHAPTER 44 Opener: © Illinois River Biological Station via the Detroit Free Press, Nerissa Michaels/AP Images; 44.1: © David Hall/Science Source; 44.2a: © Evelyn Jo Johnson; 44.3a: © E. R. Degginger/Science Source; 44.3b: © Corbis RF; 44.4b: © Holt Studios/Science Source; 44.4c: © Bruce M. Johnson; 44.4d: © Digital Vison/Getty RF; 44.6a: © Hombil Images/Alamy; 44.6b: © Lyn Holly Coorg/Photographer's Choice RF/Getty RF; 44.9: © SuperStock/Alamy; 44.13a: © Minden Pictures/SuperStock; 44.13b: © Richard La Val/Animals Animals/agefotostock; 44.13c: © Tom McHugh/Science Source; 44.14(dandelions): © Elena Elisseeva/Alamy RF; 44.14(bears): © Winfried Wisniewski/The Image Bank/Getty Images; 44Aa: © Minden Pictures/SuperStock; 44Ab: © Altrendo Images/Getty Images; 44.15b(top): © iStockphoto.com/YinYang RF; 44.15b(bottom): © Michael Coyne/Lonely Planet Images/Getty Images; 44.16c: © Tim Graham/Getty Images.

CHAPTER 45 Opener: © NPS photo by Barry O'Neill; 45.1a(forest): © Charlie Ott/Science Source; 45.1a(squirrel): © Stephen Dalton/Science Source; 45.1a(wolf): © Renee Lynn/Science Source; 45.1b(forest): © Digital Vision/PunchStock RF; 45.1b(kinkajou): © Alan & Sandy Carey/Science Source; 45.1b(sloth): © McGraw-Hill Education; 45Ab: © Minden Pictures/SuperStock; 45.6a: © Alan & Sandy Carey/Science Source; 45.7a: © Richard Thompson/Alamy; 45.7b: © MedioImages/SuperStock RF; 45.7c: © A. Cosmos Blank/Science Source; 45.8a: © Pauline S. Mills/

Getty RF; 45.8b: © blickwinkel/Alamy; 45.8c: © Rami Aapasuo/Alamy; 45.8d: © James H. Robinson/Science Source; 45.9: © Sharon Patton/University of Tennessee College of Veterinary Medicine; 45.10: © James Hager/Getty Images; 45.11: © All Canada Photos/Alamy; 45.12: © Bill Wood/Bruce Coleman/Photoshot; 45Ba: © Daniel Dempster Photography/Alamy; 45Bb: © Rolf Nussbaumer/Alamy; 45.13a(glacial rock): © Alaska Stock/Alamy; 45.13a(lichen): © Background Abstracts/Getty RF; 45.13b(shrub): © Don Paulson/Alamy RF; 45.13b(*Dryas* flower): © Science Photo Library/Alamy; 45.13c: © Bruce Heinemann/Getty RF; 45.13d: © Don Paulson/Alamy RF; 45.15a(diatom): © Ed Reschke; 45.15a(tree): © Herman Eisenbeiss/Science Source; 45.15b(caterpillar): © Corbis RF; 45.15b(rabbit): © Gerald C. Kelley/Science Source; 45.15c(snake/rabbit): © Derrick Hamrick/imagebroker/Corbis; 45.15c(hawk/snake): © Tze-hsin Woo/Getty RF; 45.15d(bacteria): © Image Source/Getty RF; 45.15d(mushroom): © Denise McCullough; 45.17: © George D. Lepp/Science Source.

CHAPTER 46 Opener: © Andrew Michael/Alamy; 46.6a: © Johnny Johnson/Getty Images; 46.6b: © John Shaw/Photoshot; 46Aa: © Porterfield/Chickering/Science Source; 46Ab: © Mint Images-Frans Lanting/Getty Images; 46.7a: © Creatas/Jupiterimages RF; 46.7b: © Jupiterimages/liquidlibrary/360/Getty RF; 46.8a: © David S. Rice/Science Source; 46.8b(marigolds): © 4FR/iStock/360/Getty RF; 46.8b(chipmunk): © Elena Elisseeva/Tetra Images/Getty RF; 46.8b(bobcat): © Tom McHugh/Science Source; 46.10(frog): © National Geographic+M676/Getty Images; 46.10(katydid): © G. Fischer/Blickwinkel/agefotostock; 46.10(jaguar): © Professor David F. Cox, Lincoln Land Community College; 46.10(macaw): © PhotoLink/Getty RF; 46.10(butterfly): © Kevin Schafer/Getty Images; 46.10(orchids): © Professor David F. Cox, Lincoln Land Community College; 46.10(lizard): © J.H. Pete Carmichael/Getty Images; 46.11a(chaparral): © Doug Wechsler/Animals Animals/agefotostock; 46.11b(roadrunner): © John Cancalosi/Photolibrary/Getty Images; 46.12a: © Jim Steinberg/Science Source; 46.12b: © Fuse/Getty RF; 46.13a: © Danita Delimont/Alamy; 46.13b: © Digital Vision/Getty RF; 46.14a: © Corbis RF; 46.14b: © Bob Calhoun/Bruce Coleman/Photoshot; 46.14c: © Lee Karney/USFWS; 46.14d: © Kevin Schafer/Getty Images; 46.15(mayfly): © Michael Windelspecht; 46.15(trout): © William H. Mullins/Science Source; 46.15(carp): © blickwinkel/Alamy; 46.15(crab): © Sandy Franz, iStock/360/Getty RF; 46.16a: © Roger Evans/Science Source; 46.16b: © McGraw-Hill Education. Pat Watson, photographer; 46.18(pike): © abadonian/iStock/360/Getty RF; 46.18(water strider): © Matti Suopajarvi/Getty RF; 46.19a: © Marc Lester/Getty Images; 46.19b: © shakzu/iStock/360/Getty RF; 46.19c: © NOAA; 46Ba: © Brian J. Skerry/Getty Images; 46Bb: © Jose Luis Pelaez Inc./Blend Images LLC RF.

CHAPTER 47 Opener: © Dirk-Jan Mattaar/iStock/360/Getty RF; 47.3(periwinkle): © Steven P. Lynch; 47.3(armadillo): © Photodisc/Getty RF; 47.3(fishing): © Herve Donnezan/Science Source; 47.3(bat): © Dr. Merlin D. Tuttle/Bat Conservation International; 47.3(ladybug): © Martin Ruegner/Masterfile RF; 47.3(rubber tree): © Bryn Campbell/Stone/Getty Images; 47.4: © David Buffington/Blend Images/Getty RF; 47.6b: © IT Stock/PunchStock RF; 47.6c(road cuts): © Universal Images Group Limited/Alamy; 47.6c(forest patches): © Eco Images/UIG/Getty Images; 47.6c(destroyed areas): © L. Hobbs/PhotoLink/Getty RF; 47.7a: © Chuck Pratt/Bruce Coleman/Photoshot; 47.7b: © Chris Johns/National Geographic Stock; 47.8b: © Secret Sea Visions/Photolibrary/Getty Images; 47A–B: © Peter Paul van Dijk; 47.9a: © StrahilDimitrov/iStock/360/Getty RF; 47.9b: © Peter Auster/University of Connecticut; 47.10a: © Design Pics/Deb Garside RF; 47.10b(forest): © Buddy Mays/Corbis RF; 47.10b(owl): © Pat & Tom Leeson/Science Source; 47.11b: © Jeff Foott/Discovery Channel Images/Getty Images; 47D(bird): © NHPA/SuperStock; 47D(researcher): © Marilyn Kok, University of Illinois Springfield.

Line Art and Text

CHAPTER 1 Figure 1.4: Sylvia Mader and Michael Windelspecht, *Essentials of Biology*, 4e. New York, NY: McGraw-Hill Education. Copyright © 2015 McGraw-Hill Education. All rights reserved. Used with permission; 1.11: Sylvia Mader and Michael Windelspecht, *Essentials of Biology*, 4e. New York, NY: McGraw-Hill Education. Copyright © 2015 McGraw-Hill Education. All rights reserved. Used with permission; 1.12: Sylvia Mader and Michael Windelspecht, *Human Biology*, 13e. New York, NY: McGraw-Hill Education. Copyright © 2014 McGraw-Hill Education. All rights reserved. Used with permission; 1.14: Sylvia Mader and Michael Windelspecht, *Essentials of Biology*, 4e. New York, NY: McGraw-Hill Education. Copyright © 2015 McGraw-Hill Education. All rights reserved. Used with permission.

CHAPTER 2 Figure 2A: Source: EPA.gov.

CHAPTER 4 Figure 4.14: Sylvia Mader, *Lab Manual for Inquiry into Life*, 14/e. New York, NY: McGraw-Hill Education. Copyright © 2014 McGraw-Hill Education. All rights reserved. Used with permission; 4.16: Sylvia Mader, *Lab Manual for Inquiry into Life*, 14/e. New York, NY: McGraw-Hill Education. Copyright © 2014 McGraw-Hill Education. All rights reserved. Used with permission; pg. 79 (right and left): Sylvia Mader, *Lab Manual for Inquiry into Life*, 14/e. New York, NY: McGraw-Hill Education. Copyright © 2014 McGraw-Hill Education. All rights reserved. Used with permission.

CHAPTER 5 Figure 5.4: Sylvia Mader, *Lab Manual for Inquiry into Life*, 14/e. New York, NY: McGraw-Hill Education. Copyright © 2014 McGraw-Hill Education. All rights reserved. Used with permission.

CHAPTER 6 Figure 6.2: Sylvia Mader, *Lab Manual for Inquiry into Life*, 14/e. New York, NY: McGraw-Hill Education. Copyright © 2014 McGraw-Hill Education. All rights reserved. Used with permission.

CHAPTER 7 Figure 7A: "Nature Climate Change," Nature.com, October 2012.

CHAPTER 9 Figure 9.7: Sylvia Mader, *Lab Manual for Inquiry into Life*, 14/e. New York, NY: McGraw-Hill Education. Copyright © 2014 McGraw-Hill Education. All rights reserved. Used with permission.

CHAPTER 10 Figure 10.2: Sylvia Mader, *Lab Manual for Inquiry into Life*, 14/e. New York, NY: McGraw-Hill Education. Copyright © 2014 McGraw-Hill Education. All rights reserved. Used with permission.

CHAPTER 11 Figure 11A: Sylvia Mader and Michael Windelspecht, *Human Biology*, 13e. New York, NY: McGraw-Hill Education. Copyright © 2014 McGraw-Hill Education. All rights reserved. Used with permission; 11.17: Sylvia Mader and Michael Windelspecht, *Essentials of Biology*, 4e. New York, NY: McGraw-Hill Education. Copyright © 2015 McGraw-Hill Education. All rights reserved. Used with permission.

CHAPTER 12 Figure 12.6: Robert Brooker, et al, *Biology*, 2e. New York, NY: McGraw-Hill Education. Copyright © 2011 McGraw-Hill Education. All rights reserved. Used with permission.

CHAPTER 14 Figure 14.1: Sylvia Mader and Michael Windelspecht, *Human Biology*, 13e. New York, NY: McGraw-Hill Education. Copyright © 2014 McGraw-Hill Education. All rights reserved. Used with permission.

CHAPTER 15 Figure 15.2: Sylvia Mader, *Lab Manual for Inquiry into Life*, 14/e. New York, NY: McGraw-Hill Education. Copyright © 2014 McGraw-Hill Education. All rights reserved. Used with permission.

CHAPTER 16 Figure 16.16: Sylvia Mader, *Lab Manual for Inquiry into Life*, 14/e. New York, NY: McGraw-Hill Education. Copyright © 2014 McGraw-Hill Education. All rights reserved. Used with permission.

CHAPTER 17 Figure 17.8: Sylvia Mader and Michael Windelspecht, *Essentials of Biology*, 4e. New York, NY: McGraw-Hill Education. Copyright © 2015 McGraw-Hill Education. All rights reserved. Used with permission.

CHAPTER 20 Figure 20.8: Peter Raven, et al, *Biology*, 10e. New York, NY: McGraw-Hill Education. Copyright © 2014 McGraw-Hill Education. All rights reserved. Used with permission.

CHAPTER 21 Figure 21B: "Pathogen & Environment," November 14, 2013. www.cdc.gov; 21C: "Number of Case-Reports of Primary Amebic Meningoencephalitis by State of Exposure," April 14, 2014. www.cdc.gov.

CHAPTER 28 Page 512: Charles Darwin, *On the Origin of Species*. London, England: John Murray, 1859.

CHAPTER 30 Figure 30.6: Sylvia Mader, *Concepts of Biology*, 3e. New York, NY: McGraw-Hill Education. Copyright © 2014 McGraw-Hill Education. All rights reserved. Used with permission.

CHAPTER 31 Figure 31.6a: Sylvia Mader, *Lab Manual for Inquiry into Life*, 14/e. New York, NY: McGraw-Hill Education. Copyright © 2014 McGraw-Hill Education. All rights reserved. Used with permission.

CHAPTER 32 Figure 32.9a: Sylvia Mader, *Lab Manual for Inquiry into Life*, 14/e. New York, NY: McGraw-Hill Education. Copyright © 2014 McGraw-Hill Education. All rights reserved. Used with permission.

CHAPTER 38 Figures 38.11a-c: Sylvia Mader and Michael Windelspecht, *Human Biology*, 13e. New York, NY: McGraw-Hill Education. Copyright © 2014 McGraw-Hill Education. All rights reserved. Used with permission.

CHAPTER 43 Figure 43.2: Sylvia Mader, *Lab Manual for Inquiry into Life*, 14/e. New York, NY: McGraw-Hill Education. Copyright © 2014 McGraw-Hill Education. All rights reserved. Used with permission.

CHAPTER 44 Figure 44.10: Sylvia Mader, *Concepts of Biology*, 3e. New York, NY: McGraw-Hill Education. Copyright © 2014 McGraw-Hill Education. All rights reserved. Used with permission.

CHAPTER 45 Figure 45.2: Data from G. F. Gause, *The Struggle for Existence*. Baltimore, MD: William & Wilkins Company, 1934; 45.6b: Sylvia Mader, *Lab Manual for Inquiry into Life*, 14/e. New York, NY: McGraw-Hill Education. Copyright © 2014 McGraw-Hill Education. All rights reserved. Used with permission; 45.6c: "Trends in Population Cycles," July 3, 2014. www.biodivcanada.ca; 45.20-45.21: Sylvia Mader, *Lab Manual for Inquiry into Life*, 14/e. New York, NY: McGraw-Hill Education. Copyright © 2014 McGraw-Hill Education. All rights reserved. Used with permission; 45Ca: National CO2 Emissions from *Fossil-Fuel Burning, Cement Manufacture, and Gas Flaring*, pp. 1751-2008; 45Cc: T.A. Boden, G. Marland, and R.J. Andres, "Global, Regional, and National Fossil-Fuel CO2 Emissions" (2010). Oak Ridge, TN: U.S. Department of Energy.

CHAPTER 46 Figure 46.1-46.6: Sylvia Mader, *Lab Manual for Inquiry into Life*, 14/e. New York, NY: McGraw-Hill Education. Copyright © 2014 McGraw-Hill Education. All rights reserved. Used with permission; 46.8: Sylvia Mader, *Lab Manual for Inquiry into Life*, 14/e. New York, NY: McGraw-Hill Education. Copyright © 2014 McGraw Hill Education. All rights reserved. Used with permission; 46.10: Sylvia Mader, *Lab Manual for Inquiry into Life*, 14/e. New York, NY: McGraw-Hill Education. Copyright © 2014 McGraw-Hill Education. All rights reserved. Used with permission; 46.12: Sylvia Mader, *Lab Manual for Inquiry into Life*, 14/e. New York, NY: McGraw-Hill Education. Copyright © 2014 McGraw-Hill Education. All rights reserved. Used with permission; 46.14: Sylvia Mader, *Lab Manual for Inquiry into Life*, 14/e. New York, NY: McGraw-Hill Education. Copyright © 2014 McGraw-Hill Education. All rights reserved. Used with permission; 46.17: Sylvia Mader, *Lab Manual for Inquiry into Life*, 14/e. New York, NY: McGraw-Hill Education. Copyright © 2014 McGraw-Hill Education. All rights reserved. Used with permission; 46.21: Sylvia Mader, *Lab Manual for Inquiry into Life*, 14/e. New York, NY: McGraw-Hill Education. Copyright © 2014 McGraw-Hill Education. All rights reserved. Used with permission.

CHAPTER 47 Figure 47.2: Sylvia Mader, *Lab Manual for Inquiry into Life*, 14/e. New York, NY: McGraw-Hill Education. Copyright © 2014 McGraw-Hill Education. All rights reserved. Used with permission; 47.8a: Sylvia Mader, *Lab Manual for Inquiry into Life*, 14/e. New York, NY: McGraw-Hill Education. Copyright © 2014 McGraw-Hill Education. All rights reserved. Used with permission.

Index

This text contains a substantial number of supporting graphics. To ease index use, regular page numbers indicate both discussion and figure on the same page; page numbers with *f* indicate additional figures on a different page; *t* indicates a table.

A

ABA (abscisic acid)
growth response control of flowering plants, 478t, 482–83
stomata opening and closing, 470, 483
Abdominal cavity, 590
Abiogenesis, 319
Abiotic resource and factor
ecosystem dynamics, 837, 857, 868
population density and distribution, 838, 843, 844–45
Abiotic synthesis, 319
ABO system, 614
Abortion pill. *See* RU-486
Abscisic acid (ABA)
growth response control of flowering plants, 478t, 482–83
stomata opening and closing, 470, 483
Abscission, 482, 483
Absolute dating of fossils, 324
Absorption spectrum, 119
Abstinence, 781, 787
Abyssal plain, 898*f*, 899*f*, 900
Acacia, bullhorn, 864
Accessory organ, 649–51
gallbladder, 645f, 649, 651
liver, 645f, 649–51, 654
pancreas, 645f, 647, 649, 752f, 763–65
Accessory pigment, 119, 379, 380, 381
Accessory structure, skin, 592–94
Accidental transport of exotic species, 912
Acetobacter aceti, 135
Acetylcholine (ACh), 698
botulinum toxin, 744
muscle contraction, 746
parasympathetic system, 709, 710f
somatic system, 708
Acetylcholinesterase (AChE), 698, 746
ACh (acetylcholine), 698
botulinum toxin, 744
muscle contraction, 746
parasympathetic system, 709, 710f
AChE (acetylcholinesterase), 698, 746
Achromatopsia, 290–91
Acid, 30, 31
Acid deposition, 31, 878, 913
Acid rain, 31, 813
Acid reflux, 646
Acid-base balance, 685, 687–88
Acne, 594
Acquired immunity. *See* Adaptive immune defense
Acquired immunodeficiency syndrome (AIDS). *See* AIDS (acquired immunodeficiency syndrome)
Acromegaly, 757
Acrosome, 774*f*, 775, 794–95
Actacus fluviatilis, 534*f*
ACTH (adrenocorticotropic hormone), 755, 760*f*, 761, 762
Actin filament
cytoskeleton, 75
muscle fibers and contraction, 586, 743, 744–45, 746–47
stomata opening and closing, 483
Action potential, 696–97
propagation, 697
transmission across a synapse, 698
Active immunity, 632–34
Active site, 105, 108*f*
Acute bronchitis, 669–70
Acute phase protein, 626–27
Acyclovir, 358
AD (Alzheimer disease), 705
Adalimumab, 630
Adam's apple, 664, 776
Adaptation
adaptive radiation, 306–7
characteristic of life, 5–6
definition, 5
desert ecosystem, 894
diet
carnivores, 643, 644
herbivores, 643, 644
omnivores, 643–44
natural selection, 267–68
seed plant evolution and full adaption to land, 423–31
angiosperms, 426–31
gymnosperms, 329, 421, 423–26
Adaptive immune defense, 628–34
active vs. passive immunity, 632–34
B cells and antibody-mediated immunity, 628–30
evolution, 622–23
T cells and cell-mediated immunity, 628f, 630–32
Adaptive radiation, 306–7
Addison disease, 762
Adenosine diphosphate (ADP), 52, 103, 746–47
Adenosine triphosphate (ATP), 52
cycle, 103f
definition, 50, 52
function, 104
metabolic reactions, 103–4
mineral uptake in plants, 461–62
muscle contraction, 104f, 744–45, 746–47
origin of life, 322, 325
production, 123, 132f, 138, 139–40
structure, 53f, 103
Adenovirus, 354*f*, 356
ADH (antidiuretic hormone), 685–87, 755, 756*f*
Adhesion, 28–29, 468
Adhesion junction, 95, 582
Adipocyte, 584
Adipose tissue
carbohydrates, 652
definition, 584
glucagon, 763
glucose, 763
leptin, 766
subcutaneous layer of skin, 592
Adler tree, 866*f*
ADP (adenosine diphosphate), 52, 103, 746–47
Adrenal cortex, 761–62
anatomy, 752f, 760
anterior pituitary, 756f
cortisol, 755, 761
functions, 760, 761–62
malfunctions, 762
Addison disease, 762
Cushing syndrome, 762
steroid hormones, 753–54
Adrenal gland, 752*f*, 761–62
adrenal cortex, 761–62
anatomy, 752f, 760
anterior pituitary, 756f
cortisol, 755, 761
functions, 760, 761–62
malfunctions, 762
steroid hormones, 753–54
adrenal medulla, 752f, 753, 760–61
Adrenal medulla, 752*f*, 753, 760–61
Adrenaline. *See* Epinephrine
Adrenocorticotropic hormone (ACTH), 755, 760*f*, 761, 762
Adrenoleukodystrophy, 204
Adult stem cell, 157
Adult-onset diabetes. *See* Type 2 diabetes
AED (automatic external defibrillator), 609
Aequorea victoria, 219
Aeration, soil, 459, 460
Aerial root, 444
Aerobic reaction, 131
Aerotropism, 486
Afferent arteriole, 682
Afferent neuron, 696
Aflatoxin, 402
African ape, 565*f*
African cape buffalo, 863*f*
African human trypanosomiasis, 388, 389*f*
African sleeping sickness, 388, 389*f*
Afterbirth. *See* Placenta
Agapornis, 818–19
Agar, 379
Age distribution of population, 840–41, 850*f*, 851
Age of Amphibians. *See* Carboniferous period
Age of Fishes. *See* Devonian period
Age spot, 809
Age structure diagram, 840–41, 850*f*
Agent Orange, 481
Agglutination, 614
Aggregate fruit, 504, 505*f*
Aging process, 809–12
cardiovascular system, 809–10
digestive system, 810
endocrine system, 811
excretory system, 810
hypotheses of aging, 811–12
damage accumulation theories, 811–12
preprogrammed theories, 811
immune system, 810
integumentary system, 809
musculoskeletal system, 810–11
nervous system, 810
reproductive system, 811
respiratory system, 810
sensory system, 810
summary and assessment, 813–15
Agnathan. *See* Jawless fish evolution
Agranular leukocyte, 615. *See also* Lymphocyte; Monocyte
Agriculture
animal pharming, 555
biodiversity direct value, 908–9
Cro-Magnons, 575
exotic species, 912
Agrobacterium tumefaciens, 362
Agrobacterium-mediated plant transformation, 362
AID (artificial insemination by donor), 783
AIDS (acquired immunodeficiency syndrome), 784–86. *See also* HIV
fungal infections, 402, 672
opportunistic infections, 384, 632, 633, 672, 785
pregnancy, 786
stages and symptoms, 784
toxoplasmosis, 384
transmission, 784, 787
treatment and prevention, 784–85, 786
Air pressure (breathing), 665–66
Air pressure (climate), 883
Air sac
birds, 557–58, 666
insects, 660, 663
Alagille syndrome, 182
Alarm call, 826
Albumin, 613
Alces americanus, 888
Alcohol
abuse, 700
birth defects, 805
cardiovascular health, 612
sexually transmitted diseases, 787
urination, 755
Aldosterone, 687, 761
Alexandrium catanella, 382–83
Algae. *See also* Protist evolution and diversity
algal bloom, 382–83, 913
Archaeplastida supergroup, 374, 375t, 376t, 377–78, 405, 412, 413f
charophytes, 378, 412, 413f
chlorophytes, 377–78
Chromalveolata supergroup, 374, 375t, 380–84
coral reef, 379, 899
development of plants, 412, 413f
ecology dynamics, 868
green algae, 374, 375t, 376t, 377–78, 405, 412, 413f
lichen, 405
red algae, 374, 375t, 376t, 379
Stramenopiles, 380–82
supergroups, 374
symbiotic relationships, 377, 864
Algae, blue-green. *See* Cyanobacterium
Algae, brown. *See* Chromalveolata supergroup
Algae, golden brown. *See* Chromalveolata supergroup
Algae, green. *See* Archaeplastida supergroup
Algae, red. *See* Archaeplastida supergroup
Algal bloom, 382–83, 913
Algin, 381
Allantois, 803, 806*f*, 807
Allele, 167, 190, 280
Alleiic trait, multiple, 197–98
Allen's rule, 577
Allergy, 636–37
anaphylactic shock, 621, 636–37
delayed allergic response, 637
immediate allergic response, 636–37
Alligator, 553*f*, 556–57
Allopatric speciation, 303–4, 305*f*
Alloploidy, 305
Allosteric site, 108
Alopecia, 594
Alpha helix, 48
Alpha-linolenic fatty acid, 653
Alpine tundra, 884, 885*f*, 886
ALS (amyotrophic lateral sclerosis), 702
Alternation of generation
brown algae, 380
chytrids, 399
definition, 172, 176
green algae, 378
land plants, 414, 415, 416, 847–49
Alternative mRNA splicing, 219
Altman, Sidney, 320

Altruism *vs.* self-interest, 831–32
Alu element, 568
Aluminum, 458
Alvarez, Luis, 333
Alvarez, Walter, 333
Alveolate evolution and diversity, 382–84
  apicomplexans, 384
  ciliates, 383
  dinoflagellates, 381f, 382–83
Alveoli
  lungs, 664
  mammary duct, 780, 781f
Alyces obstetricans, 847f
Alzheimer disease (AD), 705
AM fungus
  evolution and diversity of fungi, 398, 400
  mycorrhizae, 406–7
Amanita mushroom, 218, 404
Amanitin, 404
Amborella trichopoda, 426–27
Ambulocetus natans, 271
Ambystoma tigrinum, 552f
Amebocyte, 617
Amenorrhea, 780
American bison, 893, 894f
American crocodile, 556f
American goldfinch, 772f
Ames test, 239f
Amino acid, 47–48
  conversion to glucose, 649
  essential, 653
  tracing phylogeny, 348–49
Ammonia, 678, 688
Ammonium, 876
Amniocentesis, 805f
Amnion, 803, 808
Amniote
  anapsid skull, 553
  diapsid skull, 553
  evolution, 546t, 548, 553–54
  synapsid skull, 553
Amniotic egg
  monotremes, 560
  reptiles, 547–48, 553, 556f
Amoeba, social, 622
Amoeba proteus, 389
Amoebic dysentery, 389
Amoeboid
  evolution and diversity, 389
  Naegleria fowleri, 373, 386–87
  organelles, 374
Amoebozoa supergroup, 389–90
  amoeboids, 373, 374, 386–87, 389
  evolution and diversity, 374, 375t, 389–90
  slime molds, 389–90
Amphetamine psychosis, 700
Amphibian
  atria and ventricles, 604
  characteristics, 520t, 551
    aquatic reproduction, 551
    double-loop circulatory pathway, 551
    ectothermy, 551
    limbs, 551
    lungs, 551, 661
    sense organs, 551
    skin, 551, 591
  diversity, 551–52
    caecilians, 551, 552
    frogs and toads, 551, 552
    salamanders and newts, 551–52
  evolution, 546t, 547, 551–52
  history of life, 329
  hormones and endocrine system, 753
  metamorphosis, 551, 753
  respiration, 551, 661, 665
  skin color, 755–56
Amphipathic molecule, 83–84
Amygdala, 704, 705
Amylase
  pancreatic, 649, 651
  salivary, 646, 651
Amyloplast, 487
Amyotrophic lateral sclerosis (ALS), 702
Anabolic steroid, 765, 776
Anabolism, 141–42
Anaerobe, obligate
  bacteria, 364, 365
  methanogens, 368
Anaerobic reaction, 131
Analgesic, 730
Analogous structure, 273, 348
Analogous trait, 307
Analogy, 347
Anaphase, 153
Anaphylactic shock, 621, 636–37
Anapsid skull, 553–54
Anatomical evidence for evolution, 273–74
Ancestry. *See also* Common ancestor
  ancestral traits, 344–45
  colonial flagellate hypothesis, 515
  LUCA (last universal common ancestor), 318, 341
  phylogenetic species concept, 298
  plant, 412–15
Anchor cell, 800
Androgen, 765
Androgen insensitivity, 753
Androgenetic alopecia, 594
Anencephaly, 805
Aneuploidy, 177–80
Angina pectoris, 611
Angiogenesis, 159
Angioplasty, 611
Angiosperm evolution and full adaption to land, 426–31. *See also* Flowering plant
  definition, 428
  features, 413t
  flower structure, 428, 429t
  flowering plant life cycle, 428–31
  flowers and diversification, 429
  life history strategy, 849
  monocots and eudicots, 428
  origination and radiation, 426–28
  ovules, 423, 426
  sporophytes, 415f
Angiotensin I, 687, 761
Angiotensin II, 687, 761
ANH (atrial natriuretic hormone), 687, 761
Animal. *See also* Human; Invertebrate; Vertebrate
  classification, 344, 514
  communication, 824–28
    auditory communication, 826
    chemical communication, 824–26
    tactile communication, 827–28
    visual communication, 826–27
  comparison of animal and plant cells, 436t
  definition, 8
  development and aging, 793–815
    aging process, 809–12
    developmental processes, 798–802
    early developmental stages, 794–97
    human embryonic and fetal development, 777, 793f, 794–97, 802–9
    summary and assessment, 813–15
  embryonic development, 516–17
    blastula formation, 517, 795, 804
    cleavage, 517, 795, 804
    coelom development, 517, 796
    developmental processes, 798–800
    organ stages of development, 797
  emotions, 825
  evolution
    body plan, 515, 518–19
    colonial flagellate hypothesis, 515
    features, 514, 515–17
    invertebrate, 514–40
    phylogeny, 515–16
  excretory systems, 678–81
    invertebrate excretory organs, 679
    nitrogenous waste products, 678
    osmoregulation by aquatic vertebrates, 677, 679–80
    osmoregulation by terrestrial vertebrates, 680–81
  fruit dispersal, 504, 506f
  genetically modified, 248–49, 555
  history of life, 325f, 328, 329
  hormone and endocrine system, 751–54
    action, 753–54
    chemical signals and pheromones, 753
  organization and homeostasis, 580–99
    homeostasis, 581, 594–96
    integumentary system, 591–94
    organs, organ systems, and body cavities, 589–90
    summary and assessment, 597–99
    tissue types, 582–89
  pollination, 429, 500–501, 864, 865, 909
  reproductive systems, 514, 771–72
    asexual reproduction, 771
    life history strategies, 772
    sexual reproduction, 771–72
Animal cell
  anatomy, 65f
  comparison of animal and plant cells, 436t
  cytokinesis, 154–55, 156f
  extracellular matrix, 95f
  mitosis, 154f, 155f
  surface modification, 95–96
Animal pharming, 555
Animal pole, 796
Animalia kingdom, 341
Ankle, 741
Ankyrin-1 gene, 197
Annelid. *See also* Earthworm
  anatomy, 530–32
  characteristics, 520t
  closed circulatory system, 602
  diversity, 531–32
  hydrostatic skeleton, 530, 734
  leeches, 531f, 532
  oligochaete annelid, 530
  polychaete annelid, 530, 531–32
  sessile tube worms, 531–32
Annual plant, 441, 841
Annual ring, 448–49
Anodonta, 528, 529f. *See also* Clam
Anolis, 304
Anopheles
  malaria, 384
  survivorship, 840
Anosmia, 810
Anoxygenic photosynthesis, 365
Ant
  altruism vs. self-interest, 831–32
  army, 831–32
  coevolution, 865
  fungal garden, 401
  leaf cutter, 401
  mutualism, 864
  pheromones, 824
  soil dwelling, 460
Antagonistic hormone, 760, 763
Antagonistic pairing of muscles, 742, 743f
Antelope, 862
Antenna
  chemical communication, 824
  crustaceans, 534, 535
Antennule, 534
Anterior pituitary, 755–57
  hormones affecting other glands
    adrenocorticotropic hormone (ACTH), 755, 760f, 761, 762
    gonadotropic hormones, 755, 765, 776, 778–79
    thyroid-stimulating hormone (TSH), 755, 758
  hormones not affecting other glands, 755–57
    growth hormone (GH), 756–57
    melanocyte-stimulating hormone (MSH), 755–56
    prolactin (PRL), 755
  hypothalamic-inhibiting hormones, 755
  hypothalamic-releasing hormones, 755
  major endocrine glands, 752f
Anteroposterior polarity, 800
Anther, 428, 429f, 497
Antheridium, 414
  bryophytes, 415, 416
  ferns, 423
  water molds, 382
Anthoceros, 416
Anthrax, 366
Anthropoid, 566, 567f
Antibiotic
  resistance, 279, 366, 789
  sac fungi, 402
  use, 366
Antibody
  aging process, 810
  development, 629
  monoclonal antibodies, 630, 781
  plasma, 613, 614, 629
  structure, 629, 630f
Antibody titer, 634
Antibody-mediated immunity, 628–30
Anticodon, 220
Antidepressant drug, 698
Antidiuresis, 687
Antidiuretic hormone (ADH), 685–87, 755, 756f
Antigen
  antigen-presenting cells (APCs), 630–31, 635
  binding site, 630f
  blood typing, 614
  molecular mimicry, 637
  receptors, 622, 628, 629
  T cells, 631
  transplant rejection, 638
Antigen-presenting cells (APCs), 630–31, 635
Anti-hCG monoclonal antibody, 630
Antihistamine, 637
Anti-inflammatory medication, 730, 741
Antioxidant, 812
Antiparallelism, 211
Antiretroviral drug (ART), 358, 785, 786
Antithrombin, 555
Antler, 345
Anubis baboon, 565f
Anus
  blastospore, 517, 796
  complete digestive tract, 642, 643
  earthworms, 642f, 643
  human, 645f, 648, 788–89
Anvil. *See* Incus
Aorta, 605f, 609
Aortic body, 667
Aortic valve abnormality, 600
Apatosaurus, 330
APCs (antigen-presenting cells), 630–31, 635
Ape. *See also* Primate order
  brachiation, 570
  history of life, 331
  phylogeny, 345
  primate evolution, 565f, 566
Ape, African, 565f
Ape, Asian, 565f
Apetala 1 gene, 441
Aphid, 472, 841f
Apical dominance, 479
Apical meristem
  differentiation, 436
  plant life cycle, 441

root system, 440, 443, 502
shoot system, 440, 445
stem system, 502
Apicomplexan, 384
Apis mellifera, 817, 827–28, 909. *See also* Bee
Aplysia, 819–20
Apoptosis, 149
B cells, 629
cancer cells, 158
G_1 checkpoint, 150
morphogenesis, 798, 802
NK cells, 627
T cells, 631
Appendicitis, 648
Appendicular skeleton, 740–41
in axial skeleton, 738f
pectoral girdle and upper limb, 740
pelvic girdle and lower limb, 740–41, 766, 780
Appendix, 625, 648
Appetite suppressant, 655
Aquaporin, 86, 685
Aquatic ecosystem, 895–902
coastal ecosystems, 895f, 898–99
lakes, 895f, 896–98
oceans, 895f, 899–902
overexploitation, 914–16
wetlands, 895
Aquatic reproduction, 551
Aqueous humor, 720
Aquifer, 873, 895
Ara ararauna, 558f, 891f
Arabidopsis, 471f, 486, 488
Arachnid
anatomy, 537
carnivorous diet, 644
excretory organs, 679
Arboreal life of primates, 565
Arbuscular mycorrhizal fungus. *See* AM fungus
Archaea
domain classification, 7–8, 341, 343–44, 368
history of life, 325
structure, 361, 368
cell wall, 343–44, 368
flagella, 361
plasma membrane, 343, 361
S-layer, 361
symbiotic relationships, 365–66
commensalism, 365, 369
definition, 365
mutualism, 365, 369
parasitism, 365
types, 368–69
halophiles, 344, 368f, 369
methanogens, 344, 368–69
thermoacidophiles, 368f, 369
Archaefructus liaoningensis, 347f, 426f
Archaeocyte, 642
Archaeopteryx, 271
Archaeplastida supergroup, 377–79
evolution and diversity, 374, 375t, 377–79, 412
green algae, 374, 375t, 376t
charophytes, 378, 412, 413f
chlorophytes, 377–78
development of plants, 412, 413f
lichen, 405
red algae, 374, 375t, 376t, 379
Archaic humans, 573–75
Cro-Magnons, 574–75
Denisovans, 574, 576
Neandertals, 564, 574, 576
Archegonium, 414
bryophytes, 415, 416, 417f
ferns, 423
Archenteron, 796
Archosaur, 553–54
Arctic tundra, 884, 885f, 886
Ardipithecus kadabba, 570
Ardipithecus ramidus, 570–71
Ardipithicine, 570–71
Argus II artificial retina, 723
Aristotle, 338
Arm
bones, 740
muscles, 743f
Armadillo, nine-banded, 908
Armstrong, Lance, 686
Army ant, 831–32
Arrector pili muscle, 594
Arrow cichlid, 304
ART (antiretroviral drug), 358, 785, 786
Art, Cro-Magnon, 575
Arteriole, 603, 604f, 610, 682
Artery
aging process, 809
anatomy, 603
coronary, 609, 611, 612f
hepatic, 649
Arthritis, 637, 741, 761
Arthritis, rheumatoid, 637
Arthropod, 533–39
centipedes and millipedes, 535–36
characteristics, 520t, 533–34
exoskeleton, 533–34
metamorphosis, 534
nervous system, 534
respiratory organs, 534
segmentation, 534
chelicerates, 537
arachnids, 537, 644, 679
horseshoe crab, 537, 617, 908
crustacean, 534–35
barnacles, 534–35, 858, 859f, 863, 898
chemical senses, 716
copepods, 534
crayfish, 534f, 535, 897f
decapods, 534
krill, 534, 643
respiratory system, 661–62
excretory organs, 679
exoskeleton, 533–34, 734
insects, 536–37
anatomy, 536
characteristics, 536–37
chemical senses, 716
entomophagy, 536
excretory organs, 679
grasshopper, 536–37, 602f
history of life, 329
metamorphosis, 536, 537, 750, 751
omnivorous diet, 644
pollination, 429, 500–501, 909
respiratory system, 536–37, 600, 602f, 661f, 662–63
sound wave detection, 724
vision, 718
metamorphosis, 534, 536, 537, 750, 751
nervous system
crab, 692–93f
crayfish, 535
open circulatory system, 601, 602f
respiratory system
crustacean, 661–62
insects, 536–37, 600, 602f, 660, 661f, 662–63
sensory organs
balance, 727
chemical senses, 716
compound eye, 533f, 534, 535, 718
Artificial insemination by donor (AID), 783
Artificial joint, 741
Artificial lung, 671
Artificial pancreas, 764
Artificial selection, 268, 427
Artiodactyl, 344–45
Artiodactyla order, 345
Asbestos, 672
Ascaris, 532f, 533
Ascocarp, 400, 401
Ascomycota
evolution and diversity of fungi, 398, 400–402
food industry, 402
lichen, 405
mycorrhizae, 406–7
poisonous, 404
reproduction, 400–401
structure, 396
symbiotic relationships, 401
Ascospore, 401
Ascus, 400f, 401
Aseptate fungus, 397
Asexual reproduction
definition, 507, 771
flowering plant, 507–9
cell suspension culture, 509
stems, roots, and cuttings, 507–9
tissue culture, 508f, 509
fungi, 397, 399, 400–401
genetic variation, 169, 170
invertebrates
cnidarians, 522, 523, 771
sponges, 521
lichen, 405
parthenogenesis, 171, 771
protists, 374
apicomplexans, 384
ciliates, 383
diatoms, 381
dinoflagellates, 383
euglenids, 385
green algae, 377, 378
water molds, 382
single-celled organisms, 161, 162
Asian ape, 565f
Asian carp, 836
Asian turtle, 915
Asparagus
pollination, 499f
tissue culture propagation, 509
Aspergillus, 135, 402
Assimilation (nitrogen cycle), 876
Association (prey defense), 862
Associative learning, 822–23
classical conditioning, 822
definition, 822
operant conditioning, 822–23
Assortive mating, 285
Aster, 153
Asteroidea, 539
Asthma, 636, 670
Astigmatism, 724
Astrocyte, 587, 695
Astronaut homeostasis, 581
Asymmetry, 515–16, 518
Atherosclerosis, 611, 612f, 809
Athlete's foot, 402, 592
Atom, 2, 20–21
Atomic mass, 21
Atomic mass unit, 21
Atomic number, 21
Atomic symbol, 20
Atomic theory, 20
ATP (adenosine triphosphate). *See* Adenosine triphosphate (ATP)
ATP synthase complex, 138
Atrial natriuretic hormone (ANH), 687, 761
Atrial natriuretic peptide. *See* Atrial natriuretic hormone (ANH)
Atriopore, 546
Atrioventricular (AV) node, 608
Atrioventricular valve, 606, 607f
Atrium
amphibians, 604
fish, 603
human, 606, 607
lancelet gills, 546
Attachment
lock-and-key, 629
viral, 356, 357f
Auditory canal, 725
Auditory communication, 826
Auditory nerve. *See* Cochlear nerve
Auditory tube, 725, 807f
Aurelia, 523f
Australopith. *See* Australopithecine
Australopithecine, 570f, 571–72
Australopithecus aethiopicus, 571
Australopithecus afarensis, 571, 572f
Australopithecus africanus, 572
Australopithecus boisei, 571
Australopithecus robustus, 572
Australopithecus sediba, 572
Autoimmune disease, 637
monoclonal antibodies, 630
multiple sclerosis (MS), 704
Automatic external defibrillator (AED), 609
Autonomic system, 708–10
functions, 707, 708–9
nervous system organization, 694f
parasympathetic division, 707, 709, 710f
sympathetic division, 707, 709, 710f
Autoploidy, 305
Autosomal dominant disorder, 195f, 196–97
Autosomal pattern of inheritance, 194–97
Autosomal recessive disorder, 195–96
Autosome, 194
Autotroph
autotrophic bacteria, 364–65
chemoautotrophic bacteria, 365, 368, 900
photoautotrophic bacteria, 364–65
definition, 115
ecosystem dynamics, 868, 870f, 871
Autumnal equinox, 882
Auxin
growth response control of flowering plants, 477f, 478–81
apical dominance, 479
fruit retention, 479–81
organ development, 481, 482f
uses, 481
movement response of plants
gravitropism, 479, 487
phototropism, 479, 480f, 486
AV (atrioventricular) node, 608
Avery, Oswald, 209
Axial skeleton, 738–39
skull, 738–39
vertebral column and rib cage, 739
Axil, 450
Axillary bud
apical dominance, 479
leaf, 451f, 452f
Axillary hair, 765
Axillary lymph node, 623f
Axolotl salamander, 588–89
Axon
neuron component, 587, 695, 751
spinal reflex, 708
synaptic integration, 699
terminal and muscle innervation, 745–46
Axon hillock, 699
Axonal membrane
acting potential, 696–97
resting potential, 696
Azadirachta indica, 435
Azalea, 497f
AZT, 358

B

B cell
aging process, 810
allergic responses, 636–37
antibody-mediated immunity, 628–30
autoimmune disorders, 637

B-cell receptor (BCR), 628–29
development, 624, 628
functions, 615
memory B cell, 629, 634
B cell receptor (BCR), 628–29
Baboon
polygamy, 829f
prey defense, 862
primate diversity, 565f
visual communication, 826
Baboon, Anubis, 565*f*
Baboon, Hamadryas, 829*f*
Baby boom, 841, 850
Bacillus, 62, 363, 364*f*
Bacteria, domain, 7–8, 341–43. *See also* Bacterium
Bacterial vaginosis (BV), 789
Bacteriophage, 354*f*, 356–57, 363
Bacteriorhodopsin, 369
Bacterium (bacteria). *See also* Cyanobacterium; *specific bacteria name*
antibiotics, 279, 366, 402, 789
cell characteristics, 363
Gram-negative bacteria, 363, 364f
Gram-positive bacteria, 363, 364f
peptidoglycan, 363
ecosystem dynamics, 868f, 869
fermentation, 135
genetically modified, 247–48
history of life, 325–27
human diseases, 365–66
intestinal, 648
metabolism, 363–65
autotrophic bacteria, 364–65
heterotrophic bacteria, 365
oxygen, 364–65
nitrogen fixation, 365, 367, 460, 462, 463f, 481, 876
reproduction, 363
spontaneous generation, 360, 361f
structure
flagella, 361
glycocalyx, 361
S-layer, 361
summary and assessment, 370–72
symbiotic relationships, 365–66
commensalism, 365
definition, 365
mutualism, 365, 367, 864
parasitism, 365–66, 863
transformation, 208–9
DIY biology, 362
prokaryote reproduction, 363
Balance and hearing, 724–28
disorders, 728
hearing loss, 728
vertigo, 728
human ear, 725–27
anatomy, 725–26
pitch, 726
volume, 726–27
sense of balance, 727–28
gravitational equilibrium, 727, 728
invertebrates, 726f
rotational equilibrium, 727–28
sound wave detection, 724–25
Balanced translocation, 181
Balanus, 858, 859*f*
Bald eagle, 557*f*, 558*f*, 901, 907, 912
Baldness, 765, 776, 809
Baleen whale, 643
Ball-and-socket joint, 741
Bamboo, 446
Banana, 509
Bang, Frederick, 617
Bangham, Alec, 321
Banting, Frederick, 764
Barb, 557
Barbule, 557
Barcoding. *See* DNA barcoding
Bariatric surgery, 655
Bark, 447–48, 449*f*
Barnacle, 534–35, 858, 859*f*, 863, 898
Barr, Murray, 179
Barr body, 179, 234
Barred tiger salamander, 552*f*
Barrett's esophagus, 647
Barriers, immune defense, 625
Barro Colorado Island, 860
Bartholin gland. *See* Vestibular gland
Basal body, 77, 361, 362*f*
Basal cell carcinoma, 593
Basal ganglia. *See* Basal nuclei
Basal nuclei, 704
Base
acid-base balance, 685, 687–88
definition, 30
Base substitution, 239
Basement membrane, 582
Basidiocarp, 402
Basidiomycota
biology, 402–3
evolution and diversity of fungi, 398, 402–5
smuts and rusts, 402, 405
structure, 396, 402
Basidium, 402
Basilar membrane, 726
Basilosaurus, 271
Basophil granule, 615
Bat
communication, 824
echolocation, 716
keystone species, 916
pollination, 501, 909f
reciprocal altruism, 832
white nose syndrome, 401
Bat, lesser long-nosed, 909*f*
Bat, vampire, 832
Batesian mimic, 862
Bath salts, 700
Bathypelagic zone, 899*f*, 900
Batrachochytrium dendrobatidis, 399
BCR (B cell receptor), 628–29
Beadle, George, 215
Beagle (ship), 262, 263*f*, 265, 267
Beak
bird characteristic, 557
bird classification, 558
morphology of finches, 269f, 308–9f, 858
Bean
seed development, 503
seed germination, 506, 507f
thigmotropism in runner bean, 486
Bear
community diversity, 907, 912
keystone species, 916
K-selected populations, 847f
landscape preservation, 917
metapopulation, 916–17
Bear, grizzly, 907, 912, 916–17
Beaumont, William, 647
Becquerel, Antoine-Henri, 22
Bee
behavioral ecology, 817
bumblebee, 862
colony collapse disorder (CCD), 513, 909
honey, 537
honeybee, 817, 827–28, 909
inclusive fitness, 832
mimicry, 862
parthenogenesis, 771
pheromones, 824
pollination, 429, 500, 501, 909
tactile communication, 827–28
vision, 718
Beetle
anatomy, 536f
mimicry, 862
Beetle, longhorn, 862
Begging behavior, 821
Behavior and behavioral ecology, 817–32
animal communication, 824–28
auditory communication, 826
chemical communication, 824–26
tactile communication, 827–28
visual communication, 826–27
behaviors that increase fitness, 828–32
altruism vs. self-interest, 831–32
reproductive strategies and fitness, 829–31
societies and fitness, 831
territoriality and fitness, 828–29
definition, 5, 818, 828
environment influences behavior, 820–24
associative learning, 822–23
cognitive learning, 823–24
imprinting, 821–22, 829
instinct and learning, 821
orientation and migratory behavior, 823
social interactions and learning, 822
inheritance influences behavior, 818–20
garter snakes, 819
lovebirds, 818–19
marine snails, 819–20
mice, 820
twin studies, 819
summary and assessment, 833–35
Behavioral isolation, 301*f*, 302
Behavioral trait, 348
Beltian body, 864
Beneficial nutrient, 457–58
Benign cell growth, 158
Benign positional vertigo (BPV), 728
Benign prostatic hypertrophy (BPH), 775
Benthic species, 898
Benthic zone, 897, 898
Bergmann's rule, 577
Berry, true, 505*f*
Best, Charles, 764
Beta sheet, 48
Bicarbonate ion
buffer system, 32, 687
carbon cycle, 874
respiration, 668
Bicep, 743*f*
Bicoid gene, 518
Bicuspid tooth. *See* Premolar
Bicuspid valve, 607*f*
Biennial plant, 441
Bigfin reef squid, 530*f*
Bilateral symmetry
animal evolution, 516, 518–19, 692
nervous system, 692
Bile, 647, 649, 651
Bile duct, 649
Bile salt, 649
Bilirubin, 649
Binary fission, 161–62, 363
Binomial nomenclature, 8–9, 339
Bioaccumulation, 901
Biochemical evidence for evolution, 274, 275*f*
Biocultural evolution, 575
Biodiversity. *See also* Diversity; Systematic biology
causes of extinction, 911–16
climate change, 913–14
disease, 911, 913
exotic species, 905, 911–13
habitat loss, 13–14, 911, 912f
overexploitation, 911, 914–16
pollution, 911, 913
challenges facing science, 13–14
conservation, 905–22
causes of extinction, 911–16
conservation biology and biodiversity, 906–7
techniques, 916–18
value of biodiversity, 908–11
conservation biology and biodiversity, 906–7
biodiversity organization, 906–7
definition, 906
distribution of biodiversity, 907
definition, 13, 906
described species, 906f
distribution, 907
hotspot, 907, 916
island biogeography, 860
organization, 906–7
community diversity, 906, 907
genetic diversity, 906–7
landscape diversity, 906, 907
value, 908–11
biodiversity and natural ecosystems, 911
direct value, 908–9
indirect value, 910–11
Biogeochemical absorption, 463
Biogeochemical cycle, 872–78
carbon cycle, 874–75
indirect value of biodiversity, 910
nitrogen cycle, 873, 875–78
phosphorus cycle, 873, 875
water cycle, 873, 895, 910
Biogeography, 265–67, 271–73, 857, 860
Bioinformatics, 255, 906
Biological clock
bee communication, 828
circadian rhythms, 492–93, 766
orientation and migration, 823
Biological species concept, 299
Biological system, 31
Biology, 2
Bioluminescence
bathypelagic zone, 900
comb jellies, 522
dinoflagellates, 382
Biomagnification, 900, 901
Biomass, 871–72
Biome
climate, 885
coniferous forest, 857f, 884, 885f, 888
definition, 885
deserts, 893–94
global distribution, 884f
grassland, 892–93
shrublands, 892
temperate deciduous forest, 885f, 888–89
tropical forest, 885f, 890–91
tundra, 884, 885f, 886
Biomolecule, 36, 38, 318
Bioremediation, 248
Biosphere, 2, 837. *See also* Ecosystem
Biosynthesis, viral, 356, 357*f*, 358
Biotechnology, 244–59
biotechnology products, 247–49
genetically modified animals, 248–49, 555
genetically modified bacteria, 247–48
genetically modified plant, 248
DNA cloning, 245–47
polymerase chain reaction, 246–47
recombinant DNA technology, 245
reproductive, 156, 157
therapeutic, 156, 157
transgenic animal, 248–49
gene therapy, 245, 250
blindness, 723
Parkinson disease, 706
preimplantation genetic diagnosis, 784, 805f
genomics, 251–55
bioinformatics, 255, 906
eukaryote genome, 251–52
functional and comparative genomics, 253–55
gene definition, 5, 252–53
genome sequencing, 251
intergenic sequences, 252
proteomics, 225, 255

summary and assessment, 257–59
Biotic community, 874
Biotic potential, 839, 859, 861
Biotic resource and factor
ecosystem dynamics, 857, 868
population density and distribution, 838, 843, 844–45
Bipedalism, 331, 569–71, 735
Bird
characteristics, 520t, 557–58
endothermy, 558
feathers, 557, 591
respiration, 557–58
sense organs and nervous system, 558
skeleton, 557
classification, 558
communication, 826
diapsid skull, 553
dinosaurs, 330
diversity, 556, 558
evolution, 557–58
hormones and endocrine system, 753
learning, 821–22, 829
lung, 661
migration, 558, 823, 881
orientation, 823
osmoregulation, 680–81
poaching, 914
pollination, 429, 501
reciprocal altruism, 832
reproduction, 773
respiratory system, 557–58, 661, 666
tool use, 576
Bird flu, 359
Bird's-nest fungus, 402, 403*f*
Birth
homeostasis with positive feedback, 595–96, 755
stages, 808, 809f
Birth control, 781–83
abstinence, 781
condoms, 781, 782f
contraceptive vaccines, 782
intrauterine devices, 782f
male, 774, 782–83
morning-after pills, 781–82
population control, 850–51
traditional methods, 781, 782f
tubal ligation, 777
vasectomy, 774, 782
Birth defect, 805
Birth weight, 287*f*
Birthrate
age distribution, 840–41, 850f, 851
population growth
human, 850
models, 842, 843
Bison, American, 893, 894*f*
Biston betularia, 270, 280–83
Bitter taste receptor, 717
Bivalent, 168
Bivalve anatomy, 528–29
Black mamba, 555
Black rat, 914
Black walnut tree, 485
Black widow spider, 537*f*
Black-and-white spitting cobra, 555
Blackberry, 505*f*
Blackburnian warbler, 859*f*
Black-crowned night-heron, 919*f*
Blackfish (2013), 825
Black-throated green warbler, 859*f*
Blade, leaf, 450, 451*f*
Blaine, David, 669
Blastocoel, 795
Blastocyst, 804
Blastopore, 517, 796, 798
Blastula, 517, 795, 804
Blending concept of inheritance, 187
Blind spot, 720
Blindness
color blindness, 202, 290, 720–21
diabetic retinopathy, 722, 763
eye disorders, 722–24
sexually transmitted diseases, 788
Blood, 613–18
calcium, 736
capillary exchange, 616–18
circulatory circuits, 609–10
pulmonary, 609
systemic, 609–10
clotting, 615–17
concentration, 755
formed elements, 585–86, 613–16
platelets, 585f, 586, 613, 615–16
red blood cells, 585, 613–15
white blood cells, 585f, 586, 613, 615
invertebrates, 601
liver functions, 649
path through body, 609–10
path through the heart, 606–7
plasma, 585, 613, 614, 649
spleen, 624
types, 614–15
ABO system, 614
Rh system, 614–15
Blood clotting, 615–17
Blood doping, 686, 766
Blood fluke, 526
Blood glucose level
liver, 649, 763
pancreas, 763
Blood pressure
aging process, 809
anaphylactic shock, 636–37
capillary exchange, 616
hypertension, 611, 767, 809
mechanisms, 607, 610–11
water-salt balance in humans, 687, 761–62
Blood type
ABO system, 614
Rh system, 614–15
Blood vessel
arteries
aging process, 809
anatomy, 603
coronary, 609, 611, 612f
hepatic, 649
capillaries
anatomy, 603, 604f
blood pressure, 610
exchange, 616–18
inflammatory response, 626
small intestines, 648
veins
anatomy, 603
blood pressure, 610
hepatic portal vein, 649
hepatic vein, 649
varicose, 610
Bloom
algal, 382–83, 913
phosphate, 367
Blue crab, 895*f*
"Blue Fugate" family, 194
Blue light
phototropism, 486
signal transduction in plants, 477f
stomata opening and closing, 470
Blue morpho butterfly, 891*f*
Blue-and-yellow macaw, 558*f*, 891*f*
Bluebird, western, 845
Blue-footed booby, 302*f*
Bluegrass, 839, 840
Blue-green algae. *See* Cyanobacterium
Blue-spotted stingray, 549*f*
BMI (body mass index), 654
BMP4 gene (bone morphogenetic protein 4), 309*f*
Bobcat, 889*f*
Body cavity, 582, 590
Body fluid regulation, 677–90
animal excretory systems, 678–81
invertebrate excretory organs, 679
nitrogenous waste products, 678, 679, 685, 688
osmoregulation by aquatic vertebrates, 677, 679–80
osmoregulation by terrestrial vertebrates, 680–81
reptiles, 553
human urinary system, 681–88
aging process, 810
homeostasis and kidneys, 685–88
kidneys, 681–82
urine formation, 682–85
summary and assessment, 688–90
Body language, 826
Body mass index (BMI), 654
Body plan, 515, 518–19
Bog, 895
Bohr, Niels, 23
Bohr model, 23
Boletus, 403*f*
Bolide, 333–34
Bolus, 646
Bone
calcium deposits, 759
compact bone, 585, 736–38
growth and renewal, 736
human
compact bone, 736–38
dermal bone, 736
long bone anatomy, 736–38
osteoporosis, 736
spongy bone, 585, 736, 738
supportive connective tissue, 584, 585
Bone marrow
compact bone, 585
red, 585, 614, 615, 623f, 624, 737
spongy bone, 585, 738
transplant, 634, 636
yellow, 585, 737
Bone mass, 736, 810–11
Bone morphogenetic protein 4 (BMP4 gene), 309*f*
Bony fish
evolution, 549–50
lobe-finned, 550, 551, 552f
ray-finned, 549–50
gills, 550, 662f
osmoregulation, 680
Booby, blue-footed, 302*f*
Book lung, 534
Booster vaccine, 634
Boreal forest, 888
Botox, 744
Bottleneck effect, 285
Bottlenose dolphin, 826
Botulinum toxin, 744
Botulism, 366, 744
Boulengerula fischeri, 552*f*
Bovidae family, 345
Bovine spongiform encephalopathy, 360
Bowerbird, 830
Bowman's capsule. *See* Glomerular capsule
BPH (benign prostatic hypertrophy), 775
BPV (benign positional vertigo), 728
Brachiation, 570
Brachiopod, 524
Brachiosaurus, 554
Bradykinesis, 706
Brain
aging process, 810
anatomy, 702f
brain stem, 704, 708
central nervous system, 702–5
brain stem, 704, 708
cerebellum, 693, 699, 704
cerebral cortex, 694, 703–4
cerebrum, 693, 694, 703–4
hypothalamus, 693, 704
limbic system, 704–5
medulla oblongata, 693, 704
reticular activating system (RAS), 704, 705f
cerebellum, 693, 704
cerebrospinal fluid, 702
cerebrum, 703–4
basal nuclei, 704
cerebral cortex, 694, 703–4
cerebral hemispheres, 703
nervous system organization, 693
cocaine use, 701f
cranial nerve, 707
deep brain stimulation, 706
embryonic development, 806–7
hypothalamus, 693, 704
learning and memory, 705
limbic system, 704–5
medulla oblongata, 693, 704
meninge, 702
pineal gland, 704, 752f, 766
primate characteristic and evolution, 565, 566, 568
reticular activating system (RAS), 704, 705f
thalamus, 693, 704, 722
tumor, 587
ventricles, 702, 703
vertebrate nervous system organization, 693, 694
Brain stem, 704, 738
Branch
phylogeny, 300, 344
shoot system, 440
Branching
root system, 443f
vascular plants, 418, 420f
Branchiostoma, 545–46
cladogram, 346, 347f
cleavage, 795
gastrula, 796
Brassica napa, 463
Brassica rapus, 463
Brazilian cave insect, 770
BRCA I gene, 159
Bread
Neurospora crassa mold, 402
Rhizopus stolonifer mold, 399
yeast, 135, 402
Breast
cancer, 781
development, 766, 780
feeding, 634, 755, 780
Breastbone. *See* Sternum
Breathing and transport of gases, 665–69
breathing, 660, 665–67
amphibians, 665
fish, 662
mammals, 665–66
modifications in human, 666–67
reptiles, 665–66
gas exchange and transport
partial pressure, 668
surfaces, 660–64
transport of oxygen and carbon dioxide, 668–69
oxygen debt, 745
Brewing, 481
Bristlecone pine, 423
Brittle star, 540
Broca's area, 703
Bronchiole, 664
Bronchitis, 669–70
Bronchus
definition, 664
respiratory tract disorders, 669–70
asthma, 636, 670
bronchitis, 669–70
choking, 669
Brown, Robert, 58
Brown algae. *See* Chromalveolata supergroup

Brown tree snake, 913
Brown-headed cowbird, 865, 917–18
Brush-border enzyme, 648
Bryophyte
evolution and colonization of land, 412, 415–17
features, 413t, 415
hornworts, 416
liverworts, 415–16
mosses, 416–17
glaciers, 418
Bryozoan, 524
Bud
leaf, 451f, 452f
shoot system, 440, 441f, 445
Budding
fungi, 397, 400–401
hydras, 523, 771
sponges, 521
viral, 357–58
Buffer system, 32, 687
Buffon, Count, 262
Bulbourethral gland, 773f, 773t, 775
Bulk transport, 86, 93–94
Bullfrog
Linnaean classification hierarchy, 340
specialization, 552f
Bullfrog, North American, 340
Bullhorn acacia, 864
Bumblebee, 862
Burdock, 504, 506f
Burkitt lymphoma, 182
Burrowing owl, 894f
Bursa, 741
Bursitis, 741, 761
Butterfly
competition for food, 845
migration, 823
mimicry, 862
pollination, 500–501, 865
rain forest diversity, 891f
Butterfly, blue morpho, 891f
Butterfly, monarch, 823
BV (bacterial vaginosis), 789

C

C_3 photosynthesis, 125f
C_4 photosynthesis, 125–26
Cactus, 452, 894
Cactus, prickly-pear, 859
Caecilian, 551, 552
Caenorhabditis elegans, 532, 800, 802, 811
Caffeine, 700
Calcitonin, 736, 758, 759
Calcium
bones, 736
calcitonin, 736, 758, 759
parathyroid, 759–60
Calcium carbonate, 874
Calcium ion
binding of myosin and actin, 744
blood, 736
role in muscle contraction, 746–47
storage in muscles, 743
Calcium phosphate, 736
Callinectes sapidus, 895f
Callus, 481, 508f, 509
Calmodulin (CAM) gene, 309f
Calorie, 27
Calvin, Melvin, 118
Calvin cycle reaction, 118, 123f, 124
Calyx, 428, 429f, 497
CAM (calmodulin) gene, 309f
CAM photosynthesis, 126
Cambarus, 535f. *See also* Crayfish
Cambrian period, 326t, 328
Camelina, 114
Camera-type eye, 719
Camouflage, 861–62
CAMP (cyclic adenosine monophosphate), 753
Canada lynx, 861
Canadian goldenrod, 499f
Canadian goose, 881
Canaliculi, 585
Cancer
breast, 781
cell characteristics, 158–59
cell cycle, 158–59
cervical, 787–88
esophagus, 647
etiology, 240–41
Hodgkin disease, 908
human papillomavirus (HPV), 787–88
leukemia, 908
lung, 672
lymph nodes, 624
microorganism treatments, 353
monoclonal antibodies, 630, 781
origin, 159, 160f
progression, 158f
prostate, 635, 775
skin, 592, 593
teste, 774
testicular, 774
thyroid, 759
vaccines, 635
Canda, 524f
Candida albicans, 402, 789
Candidiasis, 633
Canine tooth, 644
Cannabis. *See* Marijuana
Cannibalism
invertebrates, 644
kuru, 360
Canola, 463
Canopy, forest, 889, 890
CAP (catabolite activator protein), 231
Cape buffalo, African, 863f
Cape May warbler, 859f
Cape sundew, 464
Capillary
anatomy, 603, 604f
blood pressure, 610
exchange, 616–18
inflammatory response, 626
small intestines, 648
Capillary bed, 603, 604f, 609, 617
Capillary exchange, 616–18
Capsaicin, 82
Capsella, 413f
Capsid, viral, 356
Capsule, prokaryotic cell, 62, 361, 362f
Carbaminohemoglobin, 668
Carbohydrate, 39–42
disaccharide, 39–41
fiber, 652–53
human digestion and nutrition, 651, 652–53, 652f
monosaccharide, 39
polysaccharide, 41–42
storage in roots, 442, 444
Carbon, 36–38
Carbon cycle, 874–75
Carbon dating, 324
Carbon dioxide
carbon cycle, 874–75
citric acid cycle, 138
climate change, 122, 874, 877
deforestation, 910
emissions, 851–52, 874, 877, 910
fixation, 123–24, 874
human respiratory system
acid-base balance, 687
breathing rate, 667
transport, 668–69
Permian mass extinction, 334
reduction, 124, 877
stomata opening and closing, 470
Carbon monoxide, 668
Carbonic anhydrase, 668, 669
Carboniferous period, 326t, 329f, 421
Carcinogen, 239
Carcinus maenas, 829
Cardiac cycle, 607–8
Cardiac muscle, 586, 587, 608
Cardiac output, 607
Cardinal, 558f
Cardinalis cardinalis, 558f
Cardiovascular disease (CVD), 611–13
aging process, 809–10
atherosclerosis, 611, 809
cholesterol, 653, 654
death, 611
hypertension, 611
obesity, 654
recent findings, 612
stroke and heart attack, 611, 654
Cardiovascular system, 600–620
blood, 613–18
capillary exchange, 616–18
formed elements, 613–16
plasma, 613
blood vessels
arteries, 603, 609, 611, 612f, 649, 809
capillaries, 603, 604f, 610, 616–18, 626, 648
veins, 603, 605f, 606, 609, 610, 649
definition, 603
human, 605–12
aging process, 809–10
blood pressure, 607, 610–11, 616, 636–37, 687, 761–62, 809
cardiovascular disease, 611–13, 653, 654, 809–10
heart, 605–9
pulmonary circuit, 609
systemic circuit, 609–10
invertebrate transport, 601–2
closed circulatory system, 602
open circulatory system, 601, 602f
vertebrate transport, 603–4
circulatory pathways comparison, 603–4
Carduelis tristis, 772f
Caribou, 886
Carnivore
diet adaptation, 643, 644
ecosystem dynamics, 868, 870f, 871, 872
Carnivorous plant, 452–53, 464
Carotenoid, 720
Carotenoid accessory pigment, 119, 381
Carotid body, 667
Carp
ecosystem dynamics, 895f
exotic species, 836
Carp, Asian, 836
Carpal bone, 740
Carpel, 428, 429f, 430f, 497
Carrageenan, 379
Carrier, genetic, 195
Carrier protein, 85
Carroll, Sean, 269
Carruthers, Jean, 744
Carrying capacity, 843–44
K-selected population, 847–49
predator-prey interactions, 859, 861
Cartilage, 584–85
costal cartilage, 739
elastic cartilage, 585
endochondral ossification, 736
fibrocartilage, 585
hyaline cartilage, 584–85, 737
knee, 741
osteoarthritis, 741
Cartilaginous fish, 549, 680
Cartilaginous joint, 741
Casparian strip, 442f, 443, 461
Caspase, 149
Cat
chemical communication, 824
nervous system, 692f, 693
toxoplasmosis, 384
Catabolism, 141, 761
Catabolite activator protein (CAP), 231
Cataract, 722
Catastrophism, 263
Caterpillar, 485, 868f
Caterpillar, hawkmoth, 485
Catharanthus roseus, 908
Cation exchange, 459
Cation exchange capacity (CEC), 459
Cattle
African sleeping sickness, 388
ancestral and derived traits, 344–45
bovine spongiform encephalopathy, 360
classification, 345
commensalism, 863
digestive system, 641, 644
Cattle egret, 863
Cavy, Patagonian, 266f
CBOL (Consortium for the Barcode of Life), 342–43
CBP protein, 197
CCD (colony collapse disorder), 513, 909
CDKN2A gene, 147
CDNA, 358
Cebidae, 345
Cebus capucinus, 565f
CEC (cation exchange capacity), 459
Cech, Thomas, 320
Cecum, 643, 648
Celiac disease, 650
Cell. *See also* Cell cycle and cellular reproduction; Cellular respiration
comparison of animal and plant, 436t
definition, 2, 58
flow of energy
cells and entropy, 102
thermodynamic laws, 101–2
flowering plant cells and tissues, 436–39
epidermal tissue, 436–37
ground tissue, 436
vascular tissue, 436
structure and function, 57–81
cell size, 59
cellular level of organization, 58–61
cytoskeleton, 75–77
endomembrane system, 69–71
energy-related organelles, 73–75
eukaryotic cells, 64–67
microbodies and vacuoles, 72–73
nucleus and ribosomes, 67–69
prokaryotic cells, 62–63
Cell body (neuron component), 587, 695
Cell cycle and cellular reproduction, 147–65
cell cycle, 148–50
cell cycle control, 149
interphase, 148
cell cycle and cancer, 158–59
cancer cell characteristics, 158–59
origin of cancer, 159, 160f
eukaryotic chromosome, 151–52
mitosis and cytokinesis, 152–56
chromosome number, 152
cytokinesis in animal and plant cells, 154–55, 156f
functions of mitosis, 156
mitosis preparation, 152–53
phases of mitosis, 153–54, 155f
stem cells, 156
prokaryotic cell division, 161–62
binary fission, 161–62
comparing prokaryotes and eukaryotes, 162
prokaryotic chromosome, 161
summary and assessment, 163–65
Cell death. *See* Apoptosis
Cell division
binary fission, 161–62
comparing prokaryotes and eukaryotes, 162
cytokinin, 481

definition, 149
prokaryotic chromosome, 161
Cell envelope, 62, 361
Cell junction, 95–96
Cell plate, 155
Cell recognition protein, 85
Cell signaling, 87, 241*f*
Cell size, 59
Cell structure and function, 57–81
cell size, 59
cellular level of organization, 58–61
cytoskeleton, 75–77
actin filaments, 75
centrioles, 77
cilia and flagella, 77, 78f
intermediate filaments, 75
microtubules, 75–77
endomembrane system, 69–71
endoplasmic reticulum, 69
Golgi apparatus, 69–70
lysosome, 70–71
energy-related organelles, 73–75
chloroplast, 73, 74f
mitochondrium, 73–75
eukaryotic cells, 64–67
microbodies and vacuoles, 72–73
peroxisome, 72
vacuole, 72–73
nucleus, 67–68
prokaryotic cells, 62–63
ribosomes, 68–69
summary and assessment, 79–81
Cell surface modification, 95–96
Cell suspension culture, 509
Cell theory, 58
Cell wall
domain Archaea, 343–44, 368
plant, 96
prokaryotic cell, 62
synthesis and antibiotics, 366
Cell-mediated immunity, 630–32
Cellular damage theory of aging, 811–12
Cellular differentiation
animal development, 798–800
cytoplasmic segregation, 798
induction, 798–800
seed development, 502
Cellular organization, 58–61
Cellular respiration, 129–45
ATP and muscle contractions, 747
definition, 73, 74
efficiency, 140
fermentation, 134
glycolysis, 132, 133f
metabolism, 141–42
anabolism, 141–42
catabolism, 141
energy flow, 142
energy organelles, 142
mitochondria, 110, 136–40
citric acid cycle, 137–38
electron transport chain, 138–40
preparatory reaction, 136–37
overview, 130–31
NAD+ and FAD, 130–31
phases, 131
root (plant), 457
summary and assessment, 143–45
Cellular response and signal transduction, 477
Cellular slime mold, 390, 622
Cellular stages of development, 795–96
Cellulose, 41, 641, 644
Cenozoic era, 326*t*, 330–31, 332*f*
Centipede and millipede, 535–36
Central dogma of molecular biology, 69, 216
Central nervous system, 699–706. *See also* Brain; Spinal cord
definition, 694
functions, 699
Central vacuole, 72–73
Centriole, 77, 153
Centromere, 152
Centrosome, 76, 152–53
Cephalization
animal evolution, 516, 692
invertebrate nervous systems, 692, 693
vertebrate characteristic, 547
Cephalochordate, 545–46
Cephalopod
anatomy, 529–30
nervous system, 693
Cephalothorax, 535, 537
Cercopithecus aethiops, 826
Cerebellum, 693, 699, 704
Cerebral cortex, 694, 703–4
Cerebral hemisphere, 703
Cerebrospinal fluid, 702
Cerebrum, 703–4
basal nuclei, 704
cerebral cortex, 694, 703–4
cerebral hemispheres, 703
nervous system organization, 693
Cerumen, 725
Cervical cancer, 787–88
Cervical vertebra, 739
Cervidae, 345
Cervix
anatomy, 777
birth, 808, 809f
cervical cancer, 787–88
Cervus elaphus, 831
Cestode. *See* Tapeworm
CF (cystic fibrosis), 196, 293, 294, 672–73
CFC (chlorofluorocarbon), 913
CFTR gene, 196, 249, 293, 672–73
CGMP (cycline guanosine monophosphate), 775, 776
Chagas disease, 388
Chalfie, Martin, 219
Chambered nautilus, 530*f*
Chameleon, 556, 891*f*
Chancre, 789
Channel protein, 85
Channelrhodopsin, 723
Chaos, 846
Chaparral, 892
Chaperone protein, 50
Chara, 379, 412, 413*f*
Character displacement, 858
Chargaff, Erwin, 209
Charged tRNA, 221
Charismatic species, 907
Charophyte, 378, 412, 413*t*
Chase, Martha, 209
Checkpoint, cell cycle, 149, 150
Cheekbone, 739
Cheese, 402
Cheetah, 824, 828, 894*f*
Chelicera, 537
Chelicerate
anatomy and characteristics, 537
arachnids, 537, 644, 679
horseshoe crab, 537, 617, 908
Chelonia mydas, 556*f*
Chemical communication, 753, 824–26
Chemical cycling
ecosystem dynamics, 869, 872–78
carbon cycle, 874–75
nitrogen cycle, 873, 875–78
phosphorus cycle, 873, 875
water cycle, 873, 895, 910
life requirement, 4f
Chemical digestion, human, 645
Chemical ecology of plants, 484–85
Chemical element. *See* Element
Chemical energy, 101
Chemical sense organ, 716–18
sense of smell in humans, 717–18
sense of taste in humans, 716f, 717
vomeronasal organ (VNO), 826
Chemical weathering, 459
Chemiosmosis, 138, 140, 462
Chemistry, 19–34. *See also* Organic chemistry
acids and bases, 30–32
acid, 30, 31
base, 30
buffers and pH, 32, 687
pH scale, 30
elements, 20–23
atomic number and mass number, 21
atoms, 20–21
electrons and energy, 23
elements, 20
periodic table, 21–22
radioactive isotopes, 22–23
inorganic chemistry, 36
molecules and compounds, 24–25
covalent bonding, 25
ionic bonding, 24
summary and assessment, 32–34
water chemistry, 26–29
hydrogen bonding, 26–27
properties of water, 27–29
Chemoautotroph
archaea, 368, 369
bacteria, 365, 900
ecology dynamics, 868, 900
Chemoheterotroph, 365, 369
Chemoreception, 716–18
sense of smell in humans, 717–18
sense of taste in humans, 716f, 717
Chemoreceptor, 715, 716, 717
Chemosynthesis, 365
Chemotropism, 486
Chicken
development
extraembryonic membrane, 803
gastrula, 796
yolk, 795, 796, 803
vertebrae and body plan, 518–19
Childbirth. *See* Birth
Chimaera (ratfish), 549
Chimpanzee
auditory communication, 826
evolution, 565f
genome comparison with Homo sapiens, 568
inclusive fitness, 832
phylogeny and molecular traits, 348
Chinese liver fluke, 526–27
Chipmunk, Eastern, 889*f*
Chitin
arthropod exoskeleton, 533–34, 734
fungi, 397
structure, 41
Chlamydia, 788
Chlamydia trachomatis, 788
Chlamydomonas, 377
Chlorine, 24, 913
Chlorofluorocarbon (CFC), 913
Chlorophyll, 116, 377, 380
Chlorophyll *a*, 119, 377, 380
Chlorophyll *b*, 119, 377
Chlorophyll *c*, 380
Chlorophyte, 377–78
Chloroplast
euglenids, 385
origin of life, 327
photosynthesis, 110, 116
structure, 73, 74f
Choanocyte, 521
Choanoflagellate
animal ancestry, 515
evolution and diversity, 374, 390
Choking, 669
Cholesterol
alcohol, 612
atherosclerosis, 611, 612f
cardiovascular disease, 653, 654
functions, 653
omega-3 fatty acids, 612
sources, 653
structure, 45
Chordae tendineae, 606
Chordate. *See also* Vertebrate
characteristics, 520t, 545
dorsal tubular nerve cord, 545
notochord, 545
pharyngeal pouches, 545
postanal tail, 545
cladistics, 346
evolution, 545–46
nonvertebrate, 539, 545–46
cephalochordates, 545–46
urochordates, 546
Chorion, 803, 804, 806
Chorionic villi sampling, 805*f*
Chorionic villus, 807, 808*f*
Choroid, 719
Christmas tree worm, 531–32
Chromalveolata supergroup, 380–84
Alveolates, 382–84
evolution and diversity, 374, 375t, 380–84
Stramenopiles, 380–82
Chromatid, 148, 152
Chromatin, 67–68, 152, 232–34
Chromatin remodeling complex, 233
Chromatin structure, 232
Chromoplast, 73
Chromosomal DNA, 251*f*
Chromosome
comparison of human and chimpanzee, 568
definition, 68
duplication, 152, 153f, 181
eukaryotic, 151–52
homologous, 167–68, 170, 190f
mutation, 181f
number, 152, 177–80, 181f, 182
Philadelphia, 182
polyploidy speciation, 305
prokaryotic, 161
structure, 180–82
Chronic bronchitis, 670
Chronic inflammation, 626
Chronic obstructive pulmonary disease (COPD), 670
Chthamalus, 858, 859*f*
Chylomicron, 652
Chyme, 647, 649
Chytrid, 396, 398, 399
Chytridiomycota, 396, 398, 399
Cicada, 735*f*
Cichlid
convergent evolution, 307
speciation, 304
Cichlid, arrow, 304
Cichlid, midas, 304
Cigarette smoking, 673, 700
Ciliary muscle, 720
Ciliate, 383
Cilium
ciliates, 383
contractile vacuole, 374
definition, 77
endosymbiosis, 327
human lungs, 663, 664f
trochophores, 524
Cinchona ledgeriana, 509
Circadian rhythm
biological clock, 492–93, 766
phytochrome, 492–93
stomata opening and closing, 470, 492
Circulation and cardiovascular system, 600–620. *See also* Blood
blood, 613–18
capillary exchange, 616–18
formed elements, 613–16
plasma, 613
blood vessels
arteries, 603, 609, 611, 612f, 649, 809
capillaries, 603, 604f, 610, 616–18, 626, 648

veins, 603, 605f, 606, 609, 610, 649
human cardiovascular system, 605–12
blood pressure, 607, 610–11, 616, 636–37, 687, 761–62, 809
cardiovascular disease, 611–13, 653, 654, 809–10
heart, 605–9
pulmonary circuit, 609
systemic circuit, 609–10
invertebrate transport, 601–2
closed circulatory system, 602
open circulatory system, 601, 602f
summaries and assessment, 618–20
vertebrate transport, 603–4
circulatory pathways, 551f, 603–4
Circulation cell, 882
Circulatory pathway
amphibians, 551
double-loop, 551, 559, 604
mammals, 551f, 559
single-loop, 551f, 603
verebrates, 551f, 603–4
Circulatory system
blood vessels
veins, 603, 605f, 606, 609, 610, 649
closed
invertebrates, 530, 602
vertebrates, 603–4
definition, 601
open circulatory system, 601, 602f
reptile characteristic, 553
Circumcision, 775, 787
Cirrhosis, 651
Cis configuration, 42
Citric acid, 402
Citric acid cycle, 131, 137–38
Citrus canker, 907
Clade, 346, 347*f*
Cladistics, 346, 347*f*
Cladogram, 346, 347*f*
Clam
anatomy, 528–29
filter feeders, 643–44
foot, 734
population distribution, 838
Clam worm, 531, 532
Clarkia, 305
Clark's nutcracker, 864
Class, 7, 339. *See also* Taxonomy, systematics, and phylogeny
Classical conditioning, 822
Classification of life, 7–9. *See also* Taxonomy, systematics, and phylogeny
Claviceps purpurea. See Ergot fungus
Clavicle, 740
Clay, soil, 459
Cleaning symbiosis, 864
Cleavage
definition, 517, 795
embryonic development in animals, 517
gray matter, 798, 799f
human embryo, 804
Cleavage furrow, 154
Clements, F. E., 867
Climate
biomes, 885
coniferous forests, 888
desert, 893–94
grassland, 893
shrubland, 892
temperate deciduous forests, 885f, 888
tropical rain forests, 890
tundra, 886
definition, 882
ecosystem, 882–84
bodies of water, 882, 883, 900–902
solar radiation, 882–84
topography, 882
ice ages, 331
ocean currents, 900–902
regulation and biodiversity, 910
transpiration, 468
Climate change
bryophytes and glaciers, 418
carbon cycle, 874–75
definition, 14, 122, 875, 913
extinction threat, 913–14
global, 877
greenhouse gases, 874–75, 877, 913
pathogenic protists, 386–87
reactions, 877
Climax community, 867
Clinostat, 487
Clitellum, 531
Clitoris, 778
Cloaca
birds, 773
monotremes, 560
salamanders, 551
uric acid, 678
Clonal selection theory, 628, 629*f*, 631
Cloning, 245–47
plants, 507
polymerase chain reaction, 246–47
recombinant DNA technology, 245
reproductive, 156, 157
therapeutic, 156, 157
transgenic animal, 248–49
Clonorchis sinensis, 526–27
Closed circulatory system. *See also* Cardiovascular system
annelids, 530, 602
invertebrate transport, 602
vertebrates, 603–4
Clostridium botulinum, 744
Clostridium for cancer treatment, 353
Clostridium tetani, 365
Clothing, bamboo, 446
Clotting, 615–17
Clover
fruit dispersal, 504
photoperiodism, 490, 491f
Clownfish, 863
Club fungus. *See* Basidiomycota
Club moss. *See* Lycophyte
Clumped distribution, 838, 839
Cnidarian, 522–23
characteristics, 520t
diversity, 522–23
nerve net, 693
radial symmetry, 515
reproduction, 522, 523, 771
skeleton, 734
transport, 601
Cnidocyte, 522
Coacervate droplet, 321, 322
Coal, 421
Coastal ecosystem, 895*f*, 898–99
Coastal erosion, 460
Coastal redwood, 423
Cobra, black-and-white spitting, 555
Cocaine and crack, 701, 805
Coccus, 62, 363, 364*f*
Coccyx, 738, 739, 807*f*
Cochlea, 725, 726
Cochlear canal, 726
Cochlear implant, 728
Cochlear nerve, 726
Cocklebur
fruit dispersal, 504
photoperiodism, 490, 491f
Cod, 849
Codominance, 197
Codon, 216, 217*f*, 223
Codonosiga, 390
Coelacanth, 550
Coelom
embryonic development in animals, 517, 796, 797f
human body cavities, 590
true coelom, 517
vertebrates, 547
Coenzyme, 50, 654
Coevolution, 500–501, 864, 865
Cognitive learning, 823–24
imitation, 823
insight, 823–24
observation, 823
Cohesion, 28–29, 468
Cohesion-tension model of xylem transport, 468, 469*f*
Cohort, 839
Cold, common, 669
Cold-bloodedness. *See* Poikilothermism
Coleochaete, 412, 413*f*
Coleoptile
gravitropism, 487
phototropism, 478–79
Collagen fiber
connective tissue, 583, 584f
fibrocartilage, 585
hyaline cartilage, 584, 737
dermis, 592
Collar cell, 521
Collarbone
birds, 557
human, 740
Collecting duct, 682, 683*f*, 685*f*
Collenchyma cell, 437, 438
Colon, 648, 649
Colonial flagellate hypothesis, 515, 516
Colonization and exotic species, 912
Colony
bryozoans, 524
protists, 374
algae, 374, 377, 378
choanoflagellates, 390
ciliates, 383
Colony collapse disorder (CCD), 513, 909
Color blindness, 202, 290, 720–21
Color vision, 566, 718–19, 720
Coloration
camouflage, 861–62
leaf color change, 482
mimicry, 862
warning, 862
Colostrum, 780
Columnar epithelium, 583
Comb jelly, 520*t*, 522
Comet extraterrestrial origins hypothesis, 320
Commensalism
bacteria and archaea, 365, 369
definition, 863
ecology of communities, 863–64
Common ancestor
cladistics, 346, 347f
fungi, 396
homology, 347
human ethnicity, 577
LUCA (last universal common ancestor), 318, 341
phylogeny, 344, 345, 346, 348
primate evolution, 566, 567f, 569, 577
protists, 375t
Common cold, 669
Common name, 339
Communication
animal, 824–28
auditory communication, 826
chemical communication, 824–26
tactile communication, 827–28
visual communication, 826–27
cell signaling, 87
definition, 824
Community. *See also* Community and ecosystem ecology
definition, 2, 837, 856
diversity, 906, 907
Community, climax, 867
Community and ecosystem ecology, 855–80
community development, 866–67
ecological succession, 866–67
models about succession, 867
dynamics of an ecosystem, 868–78
autotrophs, 868
chemical cycling, 869, 872–78
energy flow, 869–72
heterotrophs, 868–69
ecology of communities, 856–65
community interactions, 857–58
community structure, 856–57
predator-prey interactions, 858–62, 865
symbiotic relationships, 862–64
summary and assessment, 878–80
Compact bone, 585, 736
Companion cell, 439, 465
Comparative genomics, 253–55
Competition
male competition, 288–89
phytochrome, 491
between populations, 857–58
regulation of population size, 845
resource, 267, 857–58
sexual selection, 831
Competitive exclusion principle, 857
Competitive inhibition, 109
Complement protein, 625, 627, 637
Complementary base pairing, 51, 211, 213–14
Complete digestive tract
earthworms, 642–43
flatworms, 525
humans, 645
Complete flower, 428, 429*t*, 497
Complete metamorphosis, 536
Composite flower, 429*t*, 501
Compound, 24–25
Compound eye, 533*f*, 534, 535, 718
Compound fruit, 504
Compound leaf, 452
Compound light microscope, 60
Concentration gradient, 86
Conclusion, 12–13
Concurrent flow, 662
Condensation, 873
Conditioned response, 822
Condom, 781, 782*f*, 787, 788
Conduction, heart, 608
Condylactis, 523*f*
Cone
conifers, 423, 424f, 425f
cycads, 424
Cone cell
color vision, 566
compound eye, 718
human eye, 720, 721
Confocal microscopy, 61
Confuciusornis, 271
Congenital disorder. *See* Birth defect
Congenital heart defect, 600
Congenital hypothyroidism, 758
Conidiophore, 401
Conidiospore, 401
Conifer
cypress, 424f
evolution and adaption to land, 423–24
pine tree, 423, 424f, 425f
spruce, 424f
Coniferous forest, 857*f*, 884, 885*f*, 888
Conjugation
ciliates, 383
prokaryotes, 363
Spirogyra, 378, 379f
Conjugation pilus, 63, 362*f*, 363
Conjunctiva, 719
Connective tissue
animal organization, 582, 583–86
categories
fibrous connective tissues, 584
fluid connective tissues, 585–86
supportive connective tissues, 584–85
components, 583–84

ground substance, 583–84
protein fibers, 583–84
specialized cells, 583
Connell, Joseph, 858
Conservation biologist, 906
Conservation biology, 906–7
Conservation of biodiversity, 905–22
causes of extinction, 911–16
climate change, 913–14
diseases, 911, 913
exotic species, 911–13
habitat loss, 911, 912f
overexploitation, 911, 914–16
pollution, 911, 913
conservation biology and biodiversity, 906–7
biodiversity organization, 906–7
definition, 906
distribution of biodiversity, 907
conservation techniques, 916–18
habitat preservation, 916–18
habitat restoration, 918–19
and DNA, 887
plant tissue culture, 509
summary and assessment, 920–22
value of biodiversity, 908–11
biodiversity and natural ecosystems, 911
direct value, 908–9
indirect value, 910–11
Consortium for the Barcode of Life (CBOL), 342–43
Consumer, 868, 869, 871
Consumptive use value of biodiversity, 909
Contact dermatitis, 637
Continental drift, 332–33, 334
Continental shelf, 898*f*, 899
Continental slope, 898*f*
Continuous feeder, 643–44
Contraception. *See* Birth control
Contractile tissue. *See* Muscular tissue
Contractile vacuole, 374, 389
Contraction, muscular
adenosine triphosphate (ATP), 104f, 744–45, 746–47
role of calcium, 746–47
sliding filament model, 743–45
Contrast, 61
Control center, negative feedback, 595, 596*f*
Control group, 10
Controlled study, 12*f*
Convergent evolution, 307, 339, 347
Cooksonia, 418, 419*f*
COPD (chronic obstructive pulmonary disease), 670
Copepod, 534
Copulation, 772
Copy number variation, 255
Coral
cup coral, 523f
poaching, 914
reef community, 837, 899
structure, 522, 523f
zooxanthellae, 382
Coral reef
algae, 379, 899
bleaching, 914f
cnidarians, 522
community, 837
habitat loss, 13f, 911
ocean ecosystems, 899
Coral snake, 862
Corepressor, 230
Corey, Robert, 48
Cork, 437, 448, 449
Cork cambium, 437, 446, 447, 448, 449
Corm, 449, 450*f*
Corn
dormancy and abscisic acid (ABA), 482
evolution of maize, 427
pollen, 500
seed development, 503
seed germination, 506, 507f
Corn smut, 405
Cornea, 718, 719, 724
Corolla, 428, 429*f*, 497
Corona, 527
Corona radiata, 794
Coronary artery, 609, 611, 612*f*
Coronary bypass, 611
Coronary vascular disease (CVD), 44
Corpus callosum, 703
Corpus luteum, 778, 779, 780, 804–6
Cortex
bark, 448
roots, 443
stem system, 445, 446
Cortex, cerebral. *See* Cerebral cortex
Cortical granule, 795
Cortisol, 755, 761
Cortisone, 761
Costal cartilage, 739
Cost-benefit analysis, 288
Cotyledons
monocots and eudicots, 428, 440, 502, 503
seed development, 502, 503
seed germination, 506, 507f
Countercurrent exchange, 662
Coupled reaction, 104
Courtship
birds, 558, 826, 830
displays, 826, 830
fireflies, 826, 827f
Covalent bonding, 25
Cowbird, brown-headed, 865, 917–18
Coxal bone, 740, 741
Coxal gland, 679
Crab
aquatic ecosystem, 895f
blue, 895f
crustacean diversity, 534f
deterministic chaos, 846
Dungeness, 846
mutualism, 864
nervous system, 692–93f
optimal foraging model, 829
shore, 829
Crack cocaine, 701
Cranial cavity, 590
Cranial nerve, 707
Cranium, 738, 739*f*
Crank. *See* Methamphetamine
Crayfish
anatomy, 534f, 535
European, 534f
profundal and littoral zone, 897f
C-reactive protein, 626–27
Creatine phosphate, 745
Crenation, 90
Creosote bush, 838
Cretaceous period, 326*t*, 330, 333, 334
Cretinism. *See* Congenital hypothyroidism
Cri du chat syndrome, 182
Crick, Francis H.C., 211
Cricket, 826
Crinoidea, 540
Crista, 74
Crocodile, 554, 556–57
Crocodilian, 556–57, 604
Crocodylus acutus, 556*f*
Cro-Magnon, 574–75
agriculture, 575
language use, 575
tool use, 575
Crop, 642, 643
Crossing-over, 169–70
Cross-pollination, 499–501
Crotalus atrox, 267–68, 556
Crustacean, 534–35. *See also* Crab
barnacles, 534–35, 858, 859f, 863, 898
chemical senses, 716
copepods, 534
crayfish, 534f, 535, 897f
decapods, 534
excretory organs, 679
krill, 534, 643
respiratory system, 661–62
Crustose lichen, 405, 406*f*
Cryptic species, 297, 298
Crystal jellyfish, 219
Ctenophore, 522
Cuboidal epithelium, 582*f*, 583
Cucumber, 452*f*
Cup coral, 523*f*
Curie, Marie, 22
Curiosity rover, 19
Current, ocean, 900–902
Cuscuta pentagona, 464, 485
Cushing syndrome, 762
Cutaneous receptor, 729
Krause end bulbs, 729
Meissner corpuscles, 729
Merkel disks, 729
Pacinian corpuscles, 729
root hair plexus, 729
Ruffini endings, 729
Cutaneous respiration, 551, 661
Cuticle
bryophytes, 415
ecdysozoans, 532
human nails, 593
leaf, 450, 451f
plant adaptation, 414
plant epidermal cells, 436, 446
water transport, 469–70
Cutoff point, homeostasis, 595
Cutting, asexual reproduction, 507–9
Cuvier, Georges, 262
CVD. *See* Cardiovascular disease; Coronary vascular disease
Cyanide gas, 108
Cyanobacterium
definition, 63
domain Bacteria, 341–43
hornworts, 416
lichen, 367, 405
mutualism, 367
nitrogen fixation, 367, 876
photosynthesis, 365, 367
stromatolite, 325
traits, 367
Cycads, 329–30, 424, 425*f*
Cycas revoluta, 424
Cyclic adenosine monophosphate (cAMP), 753
Cyclic pathway, 121*f*
Cyclin, 149
Cycline guanosine monophosphate (cGMP), 775, 776
Cyclospora cayetanensis, 384
Cyclosporiasis, 384
Cyclosporine, 402
Cyclotella, 381*f*
Cynognathus, 332
Cypress, 424*f*
Cyprinus carpio, 895*f*
Cyst
Giardia lamblia, 385
protist reproduction, 374
tapeworm, 527
Cystic fibrosis (CF), 196, 293, 294, 672–73
Cystic fibrosis transmembrane regulator (CFTR), 196, 249, 293, 672–73
Cysticercosis, 527
Cytochrome, 138
Cytochrome c, 349
Cytokine
cell-mediated immunity, 631
inflammatory response, 626
interferons, 628
T cells, 631
therapeutic agents, 632
Cytokinesis, 148, 154–55, 156*f*
Cytokinin, 478t, 481–82
Cytomegalovirus, 633, 805
Cytoplasm, 361, 362*f*
Cytoplasm, prokaryotic cell, 62–63, 361
Cytoplasmic segregation, 798
Cytoskeleton, 66, 75–77
Cytotoxic T cell, 628*f*, 630, 631

D

Dalton, John, 20
Damage accumulation theory of aging, 811–12
Damselfly, 773
Dandelion, 504, 506*f*, 847*f*
Darwin, Charles, 262, 264*f*. *See also* Darwin and evolution
Darwin, Erasmus, 262
Darwin, Francis, 478
Darwin and evolution, 261–78. *See also* Evolution
auxins, 478
Darwin's theory of evolution, 265–70
biogeographical observations, 265–67
natural selection and adaptation, 267–68
observations of change over time, 265
evidence for evolution, 270–75
anatomical evidence, 273–74
biochemical evidence, 274, 275f
biogeographical evidence, 271–73
criticism of evolution, 274–75
evidence from developmental biology, 274
fossil evidence, 270–71
finches, 306, 308–9f
history of evolutionary thought, 262–64
late eighteenth/early nineteenth century influences, 262–64
mid-eighteenth century influences, 262
last universal common ancestor (LUCA), 318
Dasypus novemcinctus, 908
Data presentation and analysis, 10–11
Date rape drug, 700–701, 787
Daughter cell, 152, 169
Daylily, 497*f*
Day-neutral plants, 490
De la Chapelle syndrome, 179
De Mairan, Jean, 492
Deafness, 728, 810
Deamer, David, 321
Deamination, 141, 649, 654
Death
deadly fungi, 404
enzyme inhibitors and, 108
Death cap mushroom, 404
Death rate
definition, 840–41
human population growth, 850
population growth models, 842, 843
Decapod, 534
Deceleration phase, 843
Deciduous forest
temperate, 885f, 888–89
tropical, 891
Deciduous plant, 450, 889
Decomposition
benthic zone, 898
ecosystem dynamics, 868f, 869
indirect value of biodiversity, 910
saprotrophs, 365, 396, 402, 868f, 869
slime molds, 389
soil nutrition, 460
Deductive reasoning, 10
Deep brain stimulation, 706
Deer
ancestral and derived traits, 344–45
classification, 345

dominance, 831
overpopulation, 848
Deer, red, 831
Deer, white-tailed, 848
Defense display, 826
Defense strategy
defense display, 826
plants, 438
hormones, 484–85
leaves, 450
secondary metabolites, 484
trichomes, 437
prey defenses, 861–62
association, 862
camouflage, 861–62
mimicry, 862
startle response, 862
warning coloration, 862
territoriality, 828–29
Deforestation
biodiversity, 891, 909, 910, 911
climate change, 874, 877
Degeneracy, 216
Dehiscent fruit, 504, 505*f*
Dehydration reaction, 38
Delayed allergic response, 637
Deletion, chromosome, 181, 182
Dementia, 238, 705
Demographic transition, 850
Demographics of populations, 838–41. *See also* Population ecology
density and distribution, 838–39
clumped distribution, 838, 839
random distribution, 838
regulation of population size, 844–46
uniform distribution, 838
population growth, 839–41
age distribution, 840–41, 850f, 851
mortality patterns, 839–40
Demography, 839
Denaturation of protein, 50, 106
Dendrite, 587, 695
Dendritic cell
function, 627
inflammatory response, 626
T cell, 631
Dendrobates pumilio, 847*f*
Denisova hominin, 574, 576
Denisovan evolution, 574, 576
Denitrification, 876
Dense fibrous connective tissue, 584, 740
Density-dependent factor, 845–46
Density-independent factor, 844–45, 846
Dentition. *See* Tooth
Deoxyribonucleic acid. *See* DNA
Deoxyribose, 39
Dependent variable, 10
Depolarization, 697
Depressant, 700
Depression, 698
Derived trait, 344–45, 346, 347*f*
Dermal bone, 736
Dermal denticle, 549
Dermatophye, 592
Dermis, 591, 592, 729
Described species, 906*f*
Desert
cold, 892
ecosystem, 838, 893–94
Desert shrub, 838
Designer drug, 700
bath salts, 700
K2, 701
MDMA (methylenedioxymethamphetamine), 700
methamphetamine, 700
MPTP, 706
Desmosome, 606*f*
Deterministic chaos, 846
Detritivores, 869, 870*f*
Deuterostome, 539–40
characteristics, 520t, 539
chordates, 539, 545–46
echinoderms, 520t, 539–40
embryonic development, 516–17
Development and aging, 793–815
aging process, 809–12
cardiovascular system, 809–10
digestive system, 810
endocrine system, 811
excretory system, 810
hypotheses of aging, 811–12
immune system, 810
integumentary system, 809
musculoskeletal system, 810–11
nervous system, 810
reproductive system, 811
respiratory system, 810
sensory system, 810
cellular stages of development, 795–96
characteristic of life, 5
definition, 502, 795
developmental processes, 798–802
cellular differentiation, 502, 798–800
morphogenesis, 798, 800–802
early developmental stages, 794–97
developmental processes, 798–802
embryonic development, 515–17, 795–97, 802–7
fertilization, 794–95, 804
human
aging process, 809–12
embryonic and fetal development, 777, 793f, 794–97, 802–9
organ stages of development, 797
tissue stages of development, 796
Developmental biology, 274*f*
Developmental genes and macroevolution, 310–12
eye development, 310–11
human evolution, 312
limb development, 311, 312
pelvic fin genes, 312
shape development, 311–12
Developmental similarity, 274*f*
Devonian period, 326*t*, 329, 333, 334
Dew, 468
DHT (dihydrotestosterone), 776
Diabetes insipidus, 755
Diabetes mellitus, 763–65
diabetic retinopathy, 722, 763
obesity, 654
type 1 diabetes, 654, 763–65
type 2 diabetes, 654, 765
urination, 685
Diabetic retinopathy, 722, 763
Diagnostic traits
evolutionary species concept, 297, 299f
morphological species concept, 297
Diamondback rattlesnake, 267–68, 556*f*
Diaphragm
breathing, 665
vomiting, 646
Diapsid skull, 553–54
Diarrhea, 648
Diastole, 607, 610
Diastolic pressure, 610
Diatom, 381, 868*f*
Diatomaceous earth, 381
DiCaprio, Leonardo, 122
Dicot. *See* Eudicot
Dictyostelium discoideum, 622
Didelphis virginianus, 560*f*
Didinium nasutum, 859
Diencephalon, 704, 705*f*
Diet
adaptation
carnivores, 643, 644
herbivores, 643, 644
omnivores, 643–44
genetic basis for behavior, 819
human
birth defect prevention, 805
gluten-free, 650
low-carbohydrate, 653
obesity, 653, 654–55
vegan and vegetarian, 653
vitamins and minerals, 654–56
Differential reproductive success, 267
Differentiation
cancer cell, 158
cellular development, 502, 798–800
seed development, 502
Diffusion
definition, 88
gas-exchange surfaces and respiration, 660, 663
Digestive system, 641–58
digestive enzymes, 651–52
lipase, 649, 652
maltase, 651
pancreatic amylase, 649, 651
pepsin, 647, 651
peptidase, 651, 652
salivary amylase, 646, 651
trypsin, 651
digestive tracts, 642–44
adaptations to diet, 643–44
complete, 642–43
complete digestive tract, 525, 642–43, 645
continuous feeders, 643
discontinuous feeders, 643
human, 645–51
incomplete, 525, 642
functions, 642
human digestive system, 645–51
accessory organs, 645f, 649–51
aging process, 810
chemical digestion, 645
large intestine, 645f, 646, 648–49
mechanical digestion, 645
mouth, 645–46
pharynx and esophagus, 645f, 646, 647
small intestine, 645f, 646, 647–48
stomach, 645f, 646, 647
human nutrition and health, 652–56
carbohydrates, 652–53
diet and obesity, 654–55
lipids, 653
proteins, 653–54
vitamins and minerals, 654–56
summary and assessment, 657–58
Digestive tract, 642–44
adaptations to diet
carnivores, 643, 644
herbivores, 643, 644
omnivores, 643–44
complete digestive tract, 525, 642–43, 645
continuous feeders, 643
discontinuous feeders, 643
filter feeders, 643–44
human, 645–51
accessory organs, 645f, 649–51
large intestine, 645f, 648–49
mouth, 645–46
pharynx and esophagus, 645f, 646, 647
small intestine, 645f, 647–48
stomach, 645f, 646, 647
incomplete digestive tract, 525, 642
organ stages of development, 797
Digitalis, 509
Dihybrid cross, 191
Dihydrotestosterone (DHT), 776
Dikaryotic hypha, 397
Dikaryotic mycelium, 402
Dimorphism, 288*f*
Dinoflagellate, 381*f*, 382–83
Dinosaur
history of life, 330
nests, 317
reptile evolution, 554
Dioecious animal, 771
Dioecious plant
flowers, 497, 498f
gymnosperms, 424
Dionaea, 452–53, 489
Diplo arrangement, 363
Diploblastic animal, 516
Diploid genome and plant adaptation to land, 414
Diploid number, 152, 167
Diplomonad, 385
Direct selection, 832
Direct value of biodiversity, 908–9
agricultural value, 908–9
consumptive use value, 909
medicinal value, 908
Directional selection, 286, 287*f*
Dirofilaria immitis, 863
Disaccharide, 39–41
Discontinuous feeder, 643
Disease
emerging, 13–14, 358–60
extinction cause, 911, 913
neglected tropical diseases (NTD), 388
regulation of population size, 845–46
sexually transmitted disease (STD), 785–89
Disease reservoirs, 388
Disruptive selection, 287, 288*f*
Distal convoluted tubule, 682, 683*f*
Distribution, population, 838–39
Diuresis, 685–87
Divergence (phylogeny), 344
Diversity. *See also* Biodiversity
artificial selection, 427
community diversity, 906, 907
fungi evolution and, 395–409
landscape, 906, 907
life, 2f
maintenance, 290–93
heterozygote advantage, 292–93
natural selection, 290–92
organizing, 7–9
plant evolution and, 410–34
protist evolution and, 373–94
species, 856, 857, 890, 911
Diving response, 669
DIY biology, 362
DNA (deoxyribonucleic acid)
chromosomal, 251f
cloning, 245–47
polymerase chain reaction, 246–47
recombinant DNA technology, 245
reproductive, 156, 157
therapeutic, 156, 157
transgenic animal, 248–49
comparison of human and chimpanzee, 568
conservation of wildlife, 887
conversion of RNA in viruses, 358
definition, 5
discovery, 209
DNA fingerprinting, 246, 247f
function, 50
G_1 stage checkpoint, 150
"junk," 236, 252
methylation, 237f
Neandertals, 564
origin of life, 322, 327
phylogenetic species concept, 298–99
repetitive elements, 252
replacement model of human evolution, 574, 577
replication, 213–15
complementary base pairing, 213–14
prokaryote vs. eukaryote, 214–15
research, 208
steroid hormones, 754
structure, 50–52, 209–12, 215t
tracing phylogeny, 348–49

transcription, 217–20
unique noncoding, 252
unwinding, 213
wildlife conservation, 887
x-ray diffraction, 210–11, 212f
DNA barcoding, 340, 342–43
DNA cloning, 245–47
polymerase chain reaction, 246–47
recombinant DNA technology, 245
reproductive, 156, 157
therapeutic, 156, 157
transgenic animal, 248–49
DNA fingerprinting, 246, 247*f*
DNA helicase, 213
DNA ligase, 214, 245
DNA microarray, 253, 254
DNA repair enzyme, 239
Dodder, 464, 485
Dog
chemoreceptors, 715f, 716
classical conditioning, 822
emotions, 825
evolution, 544
heartworm disease, 533
Dog, New Guinea singing, 544
Dogfish shark, 555
Doldrum, 883
Dolichotis patagonuma, 266*f*
Dolphin
auditory communication, 826
convergent evolution, 307
echolocation, 716
fishing industry, 914
teeth, 644
Dolphin, bottlenose, 826
Domain, 7–8, 339
Domestication of cattle, 641
Dominance, incomplete, 197–98
Dominance hierarchy, 288, 831
Dominant allele, 190
Dominant generation, 415
Dopamine
animal emotions, 825
cocaine, 701
functions, 698
Parkinson disease, 706
tobacco use, 700
Doping, 686, 766
Dormancy
bryophytes, 418
plant hormones, 481, 482–83
plant survival mechanism, 438
seed germination, 504–6
Dorsal cavity, 590
Dorsal root, 707
Dorsal root ganglion, 707
Dorsal tubular nerve cord, 545
Double fertilization, 429, 430*f*, 499
Double helix, 211
Double-loop circulatory pathway
amphibian characteristic, 551
mammals, 559
vertebrates, 604
Doubling time, 849–50
Douglas, Gabrielle “Gabby,” 742
Down syndrome, 178–79
Downy mildew, 382
Dragonfly
anatomy, 533f, 536f
habitat, 857
vision, 718
Drone, 817
Drosera capensis, 464
Drosophila biarmipes, 269–70
Drosophila melanogaster
immune system evolution, 622
model organism, 10, 798
morphogenesis, 800–802
Parkinson disease, 255
segmentation, 518, 800–802
wing color mutations, 269f
Drug abuse, 700–701
alcohol, 700
birth defects, 805
cocaine and crack, 701
date rape drugs, 700–701, 787
designer drugs, 700
growth hormone, 757
heroin, 701
marijuana and K2, 701
nicotine, 700
sexually transmitted diseases, 785, 787
Drug testing, 686
Drupe, 505*f*
Dry fruit, 504, 505*f*
Dryocopus pileatus, 558*f*
Dryopithecine, 568–69
Duck, 770
Duckbill platypus, 560, 772
Due date, 802
Dugesia, 525
Dungeness crab, 846
Duodenum, 647
Duplication, chromosome, 152, 153*f*, 181
Dutch elm disease, 401
Dwarfism, pituitary, 756, 757*f*
Dye, 219
Dysentery
amoebic, 389
Shigella dysenteriae, 366

E

Eagle, bald, 557*f*, 558*f*, 901, 907, 912
Ear
disorders, 728
aging process, 810
hearing loss, 728, 810
vertigo, 728
human, 725–27
functions, 725
inner ear, 725, 726–27
middle ear, 725
outer ear, 725
mammal characteristic, 725
snakes, 556
vertebrate nervous system organization, 693
Eardrum. *See* Tympanic membrane
Earth
as a system, 122
temperature, 875, 877, 884f, 885, 913–14
tilt, 882
Earthquake, 333
Earthworm
anatomy, 530–31
closed circulatory system, 602
complete digestive tract, 642–43
excretory organs, 679
hydrostatic skeleton, 734
nervous system, 692
respiration, 660, 661f
soil, 460
Earwax, 725
East African australopith, 571, 572*f*
Easterly, polar, 883
Eastern chipmunk, 889*f*
Ebola virus, 357
Ecdysone, 750
Ecdysozoan, 532–39
arthropods, 533–39
centipedes and millipedes, 535–36
characteristics, 520t, 533–34
chelicerates, 537, 617, 908
crustacean, 534–35
insects, 536–37
characteristics, 520t, 532
roundworms, 532–33
anatomy, 532
characteristics, 520t
parasitic, 532f, 533
ECG (electrocardiogram), 608–9
Echidna, 560, 772
Echinoderm, 539–40
brittle stars, 540
characteristics, 520t, 539
feather star, 540
radial symmetry, 539
sand dollar, 540
sea cucumber, 539–40
sea stars, 539, 540, 601, 644
sea urchin, 539f, 540, 916
skeleton, 735
transport, 601
Echinoidea, 540
Echolocation, 715*f*, 716
E-cigarette, 673, 700
Ecological niche, 857–59
character displacement, 858
competitive exclusion principle, 857
niche specialization, 858, 859f
realized niche, 857
resource partitioning, 858
Ecological pyramid, 871–72
Ecological release, 306
Ecological succession, 866–67
Ecology
behavioral ecology, 817–32
animal communication, 824–28
behaviors that increase fitness, 828–32
definition, 828
environment influences behavior, 820–24
inheritance influences behavior, 818–20
community and ecosystem ecology, 855–80
community development, 866–67
dynamics of an ecosystem, 868–78
ecology of communities, 856–65
definition, 813, 837
ecological succession, 866–67
population ecology, 836–54
demographics of populations, 838–41
human population growth, 849–52
life history patterns, 846–49
population growth models, 841–44
regulation of population size, 844–46
scope of ecology, 837
restoration ecology, 918–19
scope, 837
Ecology, restoration, 918–19
Ecosystem
aquatic, 895–902
coastal ecosystems, 895f, 898–99
lakes, 895f, 896–98
oceans, 895f, 899–902
overexploitation, 914–16
wetlands, 895
biodiversity and performance, 911
biosphere, 881–904
community and ecosystem ecology, 855–80
community development, 866–67
dynamics of an ecosystem, 868–78
ecology of communities, 856–65
definition, 2, 837, 868
dynamics of an ecosystem, 868–78
autotrophs, 868
chemical cycling, 4f, 869, 872–78
energy flow, 869–72
heterotrophs, 115, 868–69
ecology of communities, 856–65
community interactions, 857–58
community structure, 856–57
predator-prey interactions, 858–62, 865
symbiotic relationships, 862–64
landscape diversity, 906, 907
landscape preservation, 917–18
major, 881–904
aquatic ecosystems, 895–902
climate and the biosphere, 882–84
terrestrial ecosystems, 885–94
major ecosystems of the biosphere, 881–904
aquatic ecosystems, 895–902
climate and the biosphere, 882–84
terrestrial ecosystems, 885–94
summary and assessment, 902–4
terrestrial, 885–94
biome distribution, 884f
coniferous forest, 857f, 884, 885f, 888
deserts, 893–94
grasslands, 892–93
shrublands, 892
temperate deciduous forest, 885f, 888–89
tropical rain forest, 885f, 890–91
tundra, 884, 885f, 886
Ecotone, 837
Ecotourism, 911
Ecstasy. *See* MDMA (methylenedioxy-methamphetamine)
Ectoderm, 795*f*, 796, 797, 798, 799*f*
Ectoparasite, 863
Ectopic pregnancy, 777
Ectothermy
amphibian characteristic, 551
fish, 548
reptile characteristic, 553
ED (erectile dysfunction), 776
Edema, 624, 758*f*, 759
Edge, 917
Edge effect, 917–18
Ediacaran invertebrate, 327, 328*f*
Effector, 708
Efferent neuron. *See* Motor neuron
Egg
amniotic egg, 547–48, 553, 556f, 560
birds, 772
extraembryonic membrane, 772, 803
human oocyte, 176, 777, 778–79, 784, 794–95, 804
mammal, 772
monotremes, 559t, 560
oviparous animals, 772
ovoviviparous animals, 772
reptiles, 547–48, 553, 556f, 772, 803
yolk, 772, 795, 796, 803
Egg-laying hormone (ELH), 820
Egret, cattle, 863
Ejaculation, 775–76
Ejaculatory duct, 773*f*, 774, 775
Ejection, dispersal by, 504, 506*f*
El Niño–Southern Oscillation (ENSO), 902
Elastic cartilage, 585
Elastic fiber
connective tissue, 583–84, 585
dermis, 592
Elastin, 583–84
Elbow, 740, 741
Electrocardiogram (ECG), 608–9
Electromagnetic receptor, 714, 715
Electron
definition, 20–21
energy, 23
flow in light reactions, 119–20
Electron shell, 21
Electron transport chain, 120, 131, 138–40, 142
Element, 20–23
atomic number and mass number, 21
atoms, 20–21
electrons and energy, 23
periodic table, 21–22
radioactive isotopes, 22–23
Elephant, 914
Elephant seal, 669
Elephantiasis, 532*f*, 533
ELH (egg-laying hormone), 820
Elk, 831
Ella, 781

Elm, Dutch, 401
Elongated tortoise, 915*f*
Elongation, 223–24
Embolus, 611
Embryo
definition, 795
development in animals, 516–17
blastula formation, 517, 795, 804
cellular stages of development, 795–96
cleavage, 517, 795, 804
coelom development, 517, 796
developmental processes, 798–800
organ stages of development, 797
human
development, 777, 795–97, 802–7
fertilization, 780, 794–95, 804
preimplantation genetic diagnosis, 784
seed development, 502–3
Embryo sac, 499. *See also* Gametophyte
Embryonic disk, 806
Embryonic stem cell, 157
Emergency contraception, 781–82
Emergent property, 2
Emerging disease, 13–14, 358–60
Emerging viruses
influenza, 359
origin map, 359f
reproduction, 358–60
Emiquon floodplain restoration, 919
Emission, 775
Emotion, animal, 825
Emphysema, 670*f*, 672
Encephalitis
toxoplasmic, 633
West Nile, 358
Encephalitozoon cuniculi, 398
Endangered species, 906
Endergonic reaction, 103
Endocardium, 605
Endocarp, 504, 505*f*
Endochondral ossification, 736
Endocrine gland
definition, 583, 751
glandular epithelia, 583
pancreas, 649, 763, 764
Endocrine system, 750–69
aging process, 811
animal hormones, 751–54
action, 753–54
chemical signals and pheromones, 753
antagonistic hormones, 760, 763
functions, 750, 751
hypothalamus and pituitary gland, 755–57
anterior pituitary, 755–57
posterior pituitary, 755, 756f
kidney secretion, 685, 687
major glands, 752f
mode of action, 751, 753–54
peptide hormones, 753, 754f
steroid hormones, 753–54
other endocrine glands and hormones, 758–67
adrenal glands, 752f, 753–54, 755, 756f, 760, 761–62
erythropoietin (EPO), 614, 685, 686, 766
leptin, 655, 766
pancreas, 645f, 647, 649, 752f, 763–65
parathyroid, 752f, 758, 759–60
pineal gland, 704, 752f, 766
prostaglandins, 766–67, 775
testes and ovaries, 752f, 753–54, 755, 765–66, 773–75, 777, 811
thymus, 623f, 624, 752f, 766
thyroid, 752f, 755, 756f, 758–59
summary and assessment, 768–69
Endocuticle, 533*f*
Endocytosis, 93–94
Endoderm, 795*f*, 796, 797, 798, 799*f*
Endodermis, 443, 461
Endolymph, 726
Endomembrane system, 69–71
endoplasmic reticulum, 69
Golgi apparatus, 69–70
Endometrium, 777, 779–80
Endoparasite, 863
Endoplasmic reticulum (ER), 69, 743
Endorphin, 698, 730
Endoskeleton, 735
advantages, 735
composition, 734, 735
echinoderms, 735
jawed fishes, 548
vertebrate characteristic, 547, 735
Endosperm, 429, 499, 502*f*
Endospore, 366
Endosymbiosis
cilia, 327
parabasalids and diplomonads, 385
protists, 374
zooxanthellae, 382
Endosymbiotic theory, 64, 325–27, 374
Endothermy, 558
Energy
activation, 106
cellular respiration, 142
definition, 4, 101
electron, 23
energy flow and cells, 101–2
cells and entropy, 102
energy forms, 101
thermodynamic laws, 101–2
energy flow and ecosystem dynamics, 4f, 869–72
ecological pyramids, 871–72
food web, 870f, 871
thermodynamic laws, 869–71
forms, 101
life requirement, 4
metabolic reactions and energy transformation, 103–4
storage, 41
Energy drinks, 700
Energy flow
cells and the flow of energy, 101–2
cells and entropy, 102
energy forms, 101
thermodynamic laws, 101–2
ecosystem dynamics, 4f, 869–72
ecological pyramids, 871–72
food web, 870f, 871
thermodynamic laws, 869–71
trophic levels, 871
Energy of activation, 106
Energy organelles, 142
Energy-harvesting phase, 132
Energy-investment phase, 132
Enhancer, 235
Ensatina salamander, 303–4
ENSO (El Niño–Southern Oscillation), 902
Entamoeba histolytica, 389
Entomophagy, 536, 537
Entrainment, 492
Entropy, 102
Entry inhibitor, 786
Envelope
prokaryotic cell, 361
virus, 356, 357, 358
Environment
influence on behavior, 820–24
associative learning, 822–23
cognitive learning, 823–24
imprinting, 821–22
instinct and learning, 821
orientation and migratory behavior, 823
social interactions and learning, 822
influence on inheritance, 200–201
interaction with, 5
Environment, global. *See* Climate change; Ecosystem
Enzymatic protein, 85
Enzyme
actions, 105f
brush-border enzymes, 648
Calvin cycle, 142
carbonic anhydrase, 668, 669
cofactors, 107, 108f
digestive, 651–52
lipase, 649, 652
maltase, 651
pancreatic amylase, 649, 651
pepsin, 647, 651
peptidase, 651, 652
salivary amylase, 646, 651
trypsin, 651
function, 38
inhibition, 108–9
metabolic pathway, 105–9
energy of activation, 106
enzymatic speed, 106–7
enzyme inhibition, 108–9
enzyme-substrate complex, 105–6
origin of life, 320, 322
restriction, 245, 246f
speed, 106–7, 108f
Enzyme cascade, 753
Eocene epoch, 331
Eosinophil, 615, 627
Ependymal cell, 695
Ephedra, 424*f*, 425*f*, 426
Epibatidine, 555
Epicotyl, 503
Epidermal tissue, 436–37
Epidermis
human skin, 591, 592
plants, 436
herbaceous stems, 446
roots, 443
Epididymis, 773, 775
Epigenetic inheritance, 234, 237
Epiglottis, 646, 664
Epilimnion, 896
Epinephrine, 637, 753, 760
Epipelagic zone, 899
Epiphysis. *See* Pineal gland
Epiphyte, 444, 863, 890
Episodic memory, 705
Epithelial tissue, 582–83
glandular epithelia, 583
junctions, 582
simple epithelia, 582–83
columnar epithelium, 583
cuboidal epithelium, 582f, 583
squamous epithelium, 582–83
stratified epithelia, 583
Epitheliomuscular cell, 523
Epithelium. *See* Epithelial tissue
EPO (erythropoietin), 614, 685, 686, 766
Eptifibatide, 555
Equilibrium (balance)
gravitational, 727, 728
rotational, 727–28
Equilibrium (population)
island biogeography, 860
species, 847
Equinox, 882
Equisetum, 420, 421
Equus
evolution, 312–13
herbivore diet, 644
ER (endoplasmic reticulum), 69, 743
Erectile dysfunction (ED), 776
Ergot fungus, 404
Ergotism, 404
Erosion, soil, 460, 910
Erythrocyte. *See* Red blood cell
Erythropoietin (EPO), 614, 685, 686, 766
Escherichia coli
attachment, 357f
bacilli shape, 364f
division, 162
plasmids, 245
Eskimo, 575*f*, 577
Esophagus
cancer, 647
gastroesophageal reflux disease (GERD), 647
human digestion, 645f, 646
Esox lucius, 897*f*
An Essay on the Principle of Population (Malthus), 264
Essential nutrient
human, 653
plants, 457–58
Estrogen
breast cancer, 781
function, 765–66, 780
ovarian cycle, 778–79
placenta, 807
production, 765, 777
structure, 45–46
uterine cycle, 766, 779–80
Estuary, 898
Ethnicity, human, 575–77
Ethylene, 483–84
growth response control of flowering plants
abscission, 482, 483
ripening, 483–84
structure and use, 478t
Etiolation, 491
Eucalyptus, 463
Euchromatin, 152, 233
Eudicot
evolution and full adaption to land, 428
features, 428t, 441t
flowers, 497
herbaceous stems, 446
organs and organ systems, 440–41
root system, 442f, 443, 444
seed development, 502–3
seed germination, 506–7
woody plants, 447
Eudicotyledone. *See* Eudicot
Euglena deses, 385
Euglenid, 385
Eukarya, domain, 7–8, 341, 343*t*, 344
Eukaryote. *See also* Animal; Fungus; Plant; Protist evolution and diversity
cell division, 162
classification, 8, 341
DNA replication, 214–15
five-kingdom system, 341
gene expression regulation, 232–38
chromatin structure, 232–34
posttranscriptional control, 235–36
posttranslational control, 238
transcriptional control, 234–35
translational control, 238
genome, 251–52
reproduction, 363
supergroup evolution, 374, 376t
three-domain system, 341, 343t, 344
Eukaryotic cell, 62, 64–67
Eukaryotic chromosome, 151–52
Euphotic zone, 898*f*, 899
Euploidy, 177
European crayfish, 534*f*
European hare, 266*f*
European royal families and hemophilia, 203
Eustachian tube. *See* Auditory tube
Eutherian. *See* Placental mammal
Eutrophic lake, 896
Eutrophication, 875, 896, 913
Evaporation (water cycle), 873, 895
Evaporation, heat of, 27
Evening primrose, 719*f*
Evergreen, 450
Evolution, 261–78

amniotes, 546t, 548, 553–54
animal, 514–20
  body plans, 515, 518–19
  colonial flagellate hypothesis, 515
  features, 514, 515–17
  phylogeny, 515–16
biocultural, 575
classification, 7–9
Darwin's theory of, 265–70
  biogeographical observations, 265–67
  evidence for evolution, 270–75
  finches, 306, 308–9f
  history of evolutionary thought, 262–64
  natural selection and adaptation, 267–68
  observations of change over time, 265
definition, 6
dog, 544
evidence for, 270–75
  anatomical evidence, 273–74
  biochemical evidence, 274, 275f
  biogeographical evidence, 271–73
  criticism of evolution, 274–75
  evidence from developmental biology, 274
  fossil evidence, 270–71
fungi, 395–409
  diversity of fungi, 398–405
  evolution and characteristics, 396–98
  symbiotic relationships, 405–7
geological factors, 332–34
history of evolutionary thought, 262–64
history of life, 323–32
  Cenozoic era, 330–31
  clock timetable, 323
  fossils tell a story, 323–24
  geologic timescale, 324–25, 326t
  Mesozoic era, 329–30
  Paleozoic era, 327–29
  Precambrian time, 324–27
human, 564–79
  early Homo species, 572–73
  humanlike hominins, 569–72
  later Homo species, 573–77
  primate evolution, 565–69, 577
immune systems, 622–23
  cellular slime molds, 622
  Drosophila, 622
  rise of adaptive immunity, 622–23
invertebrate, 513–40
  animal evolution, 514–20
  deuterostomes, 539–40
  ecdysozoan diversity, 532–39
  lophotrochozoan diversity, 524–32
  nervous system, 692–93
  simplest invertebrates, 521–23
macroevolution and speciation, 296–316
  anatomy of speciation, 300f
  definition, 297, 303
  how new species evolve, 297–303
  modes of speciation, 303–8
  phylogenetic species concept, 300
  principles of macroevolution, 308–12
of maize, 427
mammal, 546t, 559–61
  marsupials, 272–73, 332–33, 559t, 560
  monotremes, 559t, 560
  nervous system, 694
  placental mammals, 559t, 560–61, 772
meiosis and sexual reproduction, 183
microevolution, 280–85, 300
mosaic, 571
natural selection, 6–7, 267–68
nervous system, 692–94
  invertebrates, 692–93
  mammals, 694
  vertebrates, 692f, 693–94
plant, 411–34
  ancestry and features of land plants, 412–15
  bryophytes and colonization of land, 412, 415–17
  lycophytes evolution and vascular tissue, 412, 417–19, 421
  pteridophytes and megaphylls, 419–23
  seed plant evolution and adaption to land, 423–31
  tree, 7f
population, 279–95
  genes, populations, and evolution, 280–85
  maintenance of diversity, 290–93
  natural selection, 286–89
principle, 11
protists, 373–94
  Amoebozoa, Opisthokonta, and Rhizaria supergroups, 374, 375t, 389–91
  Archaeplastida supergroup, 374, 375t, 377–79
  biology of protists, 374–76
  Chromalveolata supergroup, 374, 375t, 380–84
  Excavata supergroup, 374, 375t, 385–89
summary and assessment, 276–78
tree of life, 7f, 272
vertebrate, 544–63
  amphibians, 546t, 547, 551–52
  chordates, 545–46
  fishes, 546t, 547, 548–50, 552t
  mammals, 546t, 559–61
  nervous system, 693–94
  reptiles, 546t, 547–48, 553–58
  vertebrates, 547–48
Evolutionary species concept, 297, 299f
Evolutionary tree. See Phylogeny
Ex vivo gene therapy, 250
Excavata supergroup, 374, 375t, 385–89
Exchange pool, 872f, 873–74
Excitation, 698, 699
Excretion, 678
Excretory system, 677–90
  animal excretory systems, 678–81
    invertebrate excretory organs, 679
    nitrogenous waste products, 678, 679, 685, 688
    osmoregulation by aquatic vertebrates, 677, 679–80
    osmoregulation by terrestrial vertebrates, 680–81
  reptiles, 553
  definition, 678
  human urinary system, 681–88
    aging process, 810
    homeostasis and kidneys, 685–88
    kidneys, 681–82
    urine formation, 682–85
  summary and assessment, 688–90
Excurrent siphon, 529
Exergonic reaction, 103
Exhalation. See Expiration
Exocarp, 504, 505f
Exocoetus volitans, 550f
Exocrine gland
  definition, 583, 751
  pancreas, 649, 763, 764
Exocuticle, 533f
Exocytosis, 93
Exon shuffling, 220
Exophthalmia. See Exophthalmos
Exophthalmic goiter, 758f
Exophthalmos, 758f, 759
Exoskeleton, 734–35
  arthropods, 533–34, 734
  echinoderms, 539
  molluscs, 734
  zooids, 524
Exotic species
  extinction, 911–13
    accidental transport, 912
    colonization, 912
    horticulture and agriculture, 912
    impact on islands, 912–13
  wild-collection trade, 905, 915
Experiment, 10
Experimental design, 10, 12
Experimental variable, 10
Expiration, 660, 665–66
Exponential growth, 842
Exponential growth phase, 842, 843
Expulsion, 776
Extant, 262
External fertilization, 771–72
External respiration, 660, 667, 668
Exteroceptor, 715
Extinction
  causes, 911–16
    climate change, 913–14
    diseases, 911, 913
    exotic species, 911–13
    habitat loss, 911, 912f
    overexploitation, 911, 914–16
    pollution, 911, 913
  definition, 13, 327
  dinosaurs, 330, 554
  endangered species, 906
  fossil record, 265f, 297
  geologic timescale, 326t
  habitat preservation, 916–18
    keystone species, 916
    landscape preservation, 917–18
    metapopulations, 916–17
  habitat restoration, 918–19
  mass extinction, 327–28, 329, 330, 333–34, 554
  megafauna, 331
  Neandertals, 576
  phylogeny, 300
  Pleistocene overkill, 575
  threatened species, 906
Extracellular matrix, 95
Extraembryonic membrane, 772, 803
Extraterrestrial origins hypothesis, 320
Eye
  anatomy, 719–20
  artificial retinas, 722, 723
  camera-type, 719
  compound, 533f, 534, 535, 718
  development of, 310–11
  disorders, 722–24
    aging process, 810
    glaucoma and cataracts, 722
    retinal diseases, 722
    visual focus disorders, 722–24
  focusing, 720, 722–24
  human, 719–24
    aging process, 810
    anatomy, 719–20
    disorders, 722–24, 810
    focusing, 720, 722–24
    integration of visual signals, 721–22
    photoreceptors, 719, 720–21
    from the retina to the visual cortex, 722
  pineal, 556
  primary visual area of occipital lobes, 703
  simple, 534, 537
  vertebrate nervous system organization, 693
Eyelid
  amphibians, 551
  human variations, 577

F

F_1 generation, 188–89
F_2 generation, 189
Face
  facial bones, 738–39
  flat face trait in primates, 565
Facial hair, 765, 776
Facilitated transport, 91
Facilitation model of succession, 867
Facultative anaerobe, 364
FAD+ (flavin adenine dinucleotide), 107, 130–31
Fall overturn, 896
Fallopian tube. See Uterine tube
False rib, 739
Familial hypercholesterolemia, 198
Family, 7, 339
Family planning. See Birth control
Family tree. See Phylogeny
Fan worm, 531–32
FAP (fixed action patterns), 820–21
Far-red wavelength, 490, 491
Farsightedness, 724
FAS (fetal alcohol syndrome), 805
Fat
  adipose tissues, 584, 592, 652
  human digestion and nutrition, 35, 652, 653, 6
  structure and function, 42–46
Fat absorption inhibitor, 655
Fate map, 800, 802
Fatty acid
  omega-3 fatty acid, 612, 653, 654
  protocell, 321
  saturated, 42, 43f, 44
  structure, 42
  unsaturated, 42, 43f
FBNI gene, 198
Feather
  bird characteristic, 557, 591
  dinosaurs, 554, 557
Feather star, 540
Featherduster worm, 531–32
Feces
  energy flow, 869
  human, 648–49
Female choice, 287–88
Female gametophyte, 499
Female reproductive system, human, 777–81. See also Pregnancy
  aging process, 811
  anatomy, 777f, 778
  estrogen and progesterone
    breast cancer, 781
    function, 765–66, 780
    ovarian cycle, 778–79
    placenta, 807
    production, 765, 777
    structure, 45–46
    uterine cycle, 766, 779–80
  female breast, 766, 780–81
  female orgasm, 778
  fertilization, 777, 778, 780, 794–95, 804
  ovaries
    gonadotropic hormones, 778–79
    ovarian cycle, 777, 778–79, 780t
    ovaries, 752f, 765–66, 777, 811
      gonadotropic hormones, 755
      steroid hormones, 753–54
  uterus, 777, 779–80
  vagina, 777–78
Femur, 740f, 741
Fermentation, 131, 134, 135, 745
Fern
  evolution, 420, 421, 422–23
  fossil, 332
  gametophytes, 415f
  life cycle, 422f
  life history strategy, 849
  traits, 413t, 422
Fern, hart's tongue, 422
Fern, leatherleaf, 422
Fern, whisk, 420
Fertilization
  double, 429, 430f, 499
  external fertilization, 771–72

genetic variation, 170
human, 777, 778, 780, 783, 794–95, 804
internal fertilization, 772
IVF (in vitro fertilization), 783, 784
Fertilization membrane, 795
Fertilizer
fish kill, 878, 913
greenhouse gases, 874
nitrogen cycle, 877
plant nutrition, 458
Fetal alcohol syndrome (FAS), 805
Fetus, human
embryonic development, 793f, 802–9
early, 794–97
extraembryonic membranes, 803
fetal development, 802, 808
placenta structure and function, 807–8, 809f
preimplantation genetic diagnosis, 784
Fever, 627
FGF8 gene, 518
Fiber
connective tissue, 583–85
collagen fibers, 583, 584, 585
elastic fibers, 583–84, 585, 592
reticular fibers, 583
muscle, 586, 743, 744–45, 746–47
muscular tissue, 583
nervous tissue, 583, 587
sclerenchyma cells, 438, 439
skin, 592
Fiber (carbohydrate), 652–53
Fibrin, 615f, 616
Fibrinogen, 616
Fibrinolysin, 780
Fibroblast, 584
Fibrocartilage, 585
Fibrous connective tissue, 584
dense fibrous connective tissues, 584, 740
loose fibrous connective tissues, 584
Fibrous joint, 741
Fibrous protein, 48, 49f
Fibrous root system, 444
"Fight or flight" response, 709, 760–61
Filament
flowers, 428, 429f, 497
fungi, 397
protists, 378, 379
Filament model, sliding, 743–45
Filarial worm, 532f, 533
Filter feeder
baleen whale, 643
clam, 643–44
Fimbria
cell structure, 63
human ovaries, 777
pathogens, 366
prokaryotes, 361, 362f
Fin
convergent evolution, 307
fish characteristic, 548
pectoral-fin, 548, 549
pelvic fin, 312, 548
Finch
American goldfinch, 772f
ecological niche, 857, 858
Galápagos, 269f, 306, 857, 858
sexual selection, 829
speciation, 269f, 306, 308–9f
zebra, 829
Fingerprint, 592
Fingerprinting, DNA, 246, 247f
Firefly, 826, 827f
First messenger, 753
Fischer lovebird, 818–19
Fish
breathing and transport of gases, 661–62
characteristics, 520t, 548
ectothermy, 548
endoskeleton, 548
gills, 548
scales, 548
chemical senses, 716
circulatory pathway, 551f
cleaning symbiosis, 864
convergent evolution, 307
DNA barcoding, 343
evolution, 546t, 547, 548–50
bony fishes, 549–50, 552f
cartilaginous fishes, 549
jawed fish, 546t, 547, 548–49
jawless fish, 546t, 547, 548
lobe-finned fishes, 550, 551, 552f
fertilization and reproduction, 772
history of life, 329
lateral line system, 724–25
limiting factors, 838
omega-3 fatty acids, 612
osmoregulation, 677, 680
overexploitation, 914–15
poaching, 914
single-loop circulatory pathway, 603
skin color, 755–56
Fish kill
dinoflagellates, 382–83
fertilizers, 878, 913
"Fishapod," 261
Fishing industry, 908f, 909, 914–15
Fitness
behaviors that increase fitness, 828–32
altruism vs. self-interest, 831–32
inclusive fitness, 831f, 832
reproductive strategies and fitness, 829–31
societies and fitness, 831
territoriality and fitness, 828–29
definition, 267
sexual selection, 287
Five-kingdom system, 341
Fixed action patterns (FAP), 820–21
Flagellum
Chara, 379
Chlamydomonas, 377
choanoflagellates, 390
chytrids, 399
colonial flagellate hypothesis, 515
dinoflagellates, 382
endosymbiosis, 327
euglenids, 385
prokaryotes, 63, 361, 362f
structure, 77, 78f
Volvox colony, 378
Flagship species, 916
Flame cell, 679
Flamingo tongue shell, 530f
Flat foot, 741
Flatworm, 524–26. *See also* Planarian
asexual reproduction, 771
characteristics, 520t
incomplete digestive tract, 525, 642
parasitic, 526–27
flukes, 526–27
tapeworms, 527, 642
transport, 601
Flavin adenine dinucleotide (FAD+), 107, 130–31
Flavin pigment, 470
Fleshy fruit, 504
Floodplain restoration, 919
Floral industry
flower color and conducting system, 456
global, 495
photoperiodism, 490
Florida panther, 847
Florida scrub jay, 832
Flounder, 862
Flow
energy
cells and the flow of energy, 101–2
ecosystem dynamics, 4f, 869–72
pressure-flow model of phloem transport, 471–72
Flower. *See also* Flowering plant
color and conducting system, 456
diversification, 429
life cycle of flowering plants, 428–31, 496
phytochrome and photoperiodism, 490, 491f
structure, 428, 429f, 496–97
calyx, 428, 429f, 497
carpels, 428, 429f, 430f, 497
monocots and eudicots, 440–41
petals, 428, 429f, 497
sepals, 428, 429f, 497
stamen, 428, 430f, 497
terminology, 429t
Flower fly, 862
Flowering
phytochrome and photoperiodism, 490, 491f
timing, 497
Flowering plant
control of growth responses, 476–94
growth and movement responses, 485–89
phytochrome responses, 489–93
plant hormones, 477–85
summary and assessment, 493–94
definition, 428
evolution and full adaption to land, 426–31
flower structure, 428, 429t
flowering plant life cycle, 428–31
flowers and diversification, 429
monocots and eudicots, 428
origination and radiation, 426–28
ovules, 423, 426
sporophytes, 415f
features, 413t
life history strategy, 849
nutrition and transport, 456–75
plant nutrition and soil, 457–60
summary and assessment, 473–75
transport mechanisms, 465–73
water and mineral uptake, 461–64
photosynthesis, 116
reproduction, 495–511
asexual reproductive strategies, 507–9
fruit types and seed dispersal, 504–7
life cycle overview, 496
seed development, 502–3
sexual reproductive strategies, 496–501
summary and assessment, 510–11
structure and organization, 435–55
cells and tissues, 436–39
leaf organization and diversity, 450–53
organs, 440–41
roots organization and diversity, 442–44
stem organization and diversity, 445–50
summary and assessment, 453–55
Flu. *See* Influenza virus
Fluid connective tissue, 585–86. *See also* Blood; Lymph
Fluid regulation. *See* Body fluid regulation
Fluid regulation and excretory system, 677–90
animal excretory systems, 678–81
invertebrate excretory organs, 679
nitrogenous waste products, 678, 679, 685, 688
osmoregulation by aquatic vertebrates, 677, 679–80
osmoregulation by terrestrial vertebrates, 680–81
reptiles, 553
human urinary system, 681–88
aging process, 810
homeostasis and kidneys, 685–88
kidneys, 681–82
urine formation, 682–85
Fluid-mosaic model, 83f, 84
Fluke
Chinese liver fluke, 526–27
coevolution, 865
schistosomiasis, 526
Fluke, Chinese liver, 526–27
Fluke, lance, 865
Fly
compound eye, 718f
flower, 862
mimicry, 862
South American lantern, 862
startle response, 862
Fly agaric, 404
Flycatcher, 304
Flying fish, 550f
Focusing the human eye, 720
disorders, 722–24
astigmatism, 724
farsightedness, 724
nearsightedness, 724
presbyopia, 810
Folic acid, 805
Foliose lichen, 405, 406f
Follicle (ovarian cycle), 778
Follicle, hair, 592f, 594, 809
Follicle-stimulating hormone (FSH), 765, 776, 778–79
Follicular phase, 778
Fontanel, 738
Food
bamboo industry, 446
biodiversity direct value, 908–9
fat in, saturated and trans fats, 44
fermentation and food production, 135
foraging, 828–29
fungi and food industry, 402
insects as, 538
Food chain, 871, 874
Food vacuole, 389
Food web, 870f, 871, 913, 916
Foot
bird classification, 558
human
athlete's foot, 402, 592
bones, 740f, 741
dexterity, 735
molluscs, 528
muscular hydrostats, 734
sea star, 540, 644
Foraging for food, 828–29
Foramen magnum, 738
Foraminiferan, 374, 391
Forebrain, 693, 694
Forelimb mobility in primates, 565
Foreskin, 773f, 775
Forest
carboniferous, 421
climate regulation, 910
coniferous forest, 857f, 884, 885f, 888
deforestation, 874, 877, 891, 909, 910, 911
destruction of tropical rain forest, 122
ecological succession, 866, 867
soil profile, 460
temperate deciduous forest, 885f, 888–89
tropical forest, 885f, 890–91
water-holding capacity, 910
Forest floor
temperate deciduous forests, 889
tropical rain forest, 890
Formula, 24
FosB gene, 820
Fossil
continental drift, 332–33
dating, 324, 346–47, 391
definition, 270–71, 323

evolutionary species concept, 297
foraminiferans and radiolarians, 391
history of life timetable, 323–24
horse, 312
living, 544
morphological species concept, 297
prokaryotes, 360
trait, 346–47
Fossil fuel
acid deposition, 878, 913
carbon reservoir, 874
Carboniferous forests, 421
climate change, 877
term, 115
Founder effect, 285
Fovea centralis, 719f, 720, 722
Fox, George, 368
Fox, John, 600
Fox, kit, 894f
Fox, Michael J., 691
Fox, Sidney, 320
Fracture from osteoporosis, 736
Fragaria, 449, 450f, 507, 508f
Fragmentation
biodiversity, 907, 911
fungi, 397
habitat, 911, 912f, 917
hornworts, 416
sponges, 521
Frameshift mutation, 240
Free energy, 103
Free nerve ending. See Pain receptor
Free radical, 812
French paradox, 612
Frog
behavioral traits, 348
characteristics, 551, 552
coloration, 862
development
cleavage, 795
cytoplasmic segregation, 798
gastrula, 796
induction, 798–99
neural tube, 797
leopard, 297, 298f, 348
Linnaean classification hierarchy, 340
medicinal uses, 555
North American bullfrog, 340
parasitism, 399
parental care, 847f
poison-dart, 552, 555, 847f, 862, 891f
speciation, 296, 297, 298f
temporal isolation, 299–302
tree, 552
Frond, fern, 422
Frontal bone, 738, 739f
Frontal lobe, 694, 703, 705
Fructose, 652
Fruit
auxins and fruit retention, 479–81
definition, 504
ethylene and ripening, 483–84
flower life cycle, 496
flower structure, 429
fruit dispersal, 504, 506f
seed dispersal, 429, 504–7
seed germination, 504–7
types, 504, 505f
Fruit fly
immune system evolution, 622
model organism, 10, 798
morphogenesis, 800–802
Parkinson disease, 255
segmentation, 518, 800–802
wing color mutations, 269f
Fruiting body
basidiomycetes, 402, 403f
sac fungi, 401
Fruticose lichen, 405, 406f
FSH (follicle-stimulating hormone), 765, 776, 778–79
Fucus, 380
Functional genomics, 253–55
Functional group, 37–38
Fundamental niche, 857
Fungal infections, 402, 672
Fungicide, 402
Fungus (fungi). See also Water mold
dermatophyes, 592
ecosystem dynamics, 869
evolution and diversity, 395–409
deadly, 404
diversity of fungi, 398–405
evolution and characteristics, 396–98
summary and assessment, 407–9
symbiotic relationships, 400, 401, 405–7
five-kingdom system, 341
food industry, 402
lichen, 367, 405–6
mycorrhizae, 406–7, 444, 462, 464, 864
nucleariids, 391
phylogeny, 8, 398
reproduction, 397, 399, 400–401, 402–3
slime molds, 389
structure, 396, 397
three-domain system, 344
types, 397
Fusion inhibitor, 786

G

G_1 stage, 148, 150
G_2 stage, 148
G3P (glyceraldehyde-3-phosphate), 124
GABA, 700
Galápagos finch, 269f, 306, 857, 858
Galápagos tortoise, 266f
Galeocerdo cuvier, 549
Gall (trees), 481
Gallbladder, 645f, 649, 651
Gallic acid, 402
Gallstones, 651
Gametangium, 400, 414
Gamete, 167, 172, 771, 783
Gamete intrafallopian transfer (GIFT), 783
Gamete isolation, 302
Gametophyte
alternation of generations, 176, 414, 415, 849
angiosperms, 428–29, 430f, 496, 497–99
bryophytes, 415, 416–17
development, 497–99
gymnosperms, 425f, 426
pteridophytes, 423
Gamma-hydroxybutyric acid (GHB), 700, 787
Ganglion, 525, 692
Gap gene, 518, 800–801
Gap junction, 95–96
epithelial tissues, 582
human heart, 606f
Gardasil, 788
Garden pea, 187, 188f
Gardnerella vaginalis, 789
Garrdo, Archibald, 215
Gart gene, 179
Garter snake, 819
Gas, noble, 21
Gas exchange, animal
breathing and transport of gases, 667–69
amphibians, 660, 665–67
birds, 666
fish, 662
mammals, 665–66
modifications in human, 666–67
reptiles, 665–66
earthworms, 602, 660, 661f
gas exchange and transport
partial pressure, 668
surfaces, 89f, 660–64
transport of oxygen and carbon dioxide, 668–69
skin, 591, 661
surfaces, 660–64
gills of fish, 661–62
human lungs, 89f, 663–64
overview, 660–61
respiration steps, 660
trachea system of insects, 662–63
Gas exchange, plant
leaf stomata, 450
lenticles, 437
Gaseous biogeochemical cycle, 872
Gastric banding, 655
Gastric gland, 647
Gastric juice, 647
Gastric mill, 535
Gastric pit, 647
Gastric ulcer, 647, 767
Gastroesophageal reflux disease (GERD), 647
Gastroesophageal sphincter, 646
Gastrointestinal tract, 807f. See also Digestive tract
Gastropod, 529, 530f
Gastrovascular cavity, 522
Gastrula, 796
Gastrulation, 795f, 796, 799f, 806
Gause, C. F., 857, 859
GBS (Guillain-Barré syndrome), 709
Gecko, 556
Gehring, Walter, 310
Gel electrophoresis, 246
Gelidium, 379
Gemma, 416
Gemmae cup, 416
Gene, 207–27
ankyrin-1, 197
Apetala 1, 441
bicoid, 518
BMP4, 309f
BRCA I, 159
CAM, 309f
CDKN2A, 147
CFTR, 196, 249, 293, 672–73
definition, 5, 68, 252–53
FBNI, 198
FGF8, 518
fosB, 820
Gart, 179
gene expression regulation, 225, 228–43
eukaryotic regulation, 232–38
gene mutation, 238–41
prokaryotic regulation, 229–31
homologous, 255
Hox, 274, 311–12, 515, 518–19, 801–2
LEAFY, 441
MC1R, 207
molecular biology of the gene, 207–27
genetic code of life, 215–17
genetic material, 208–12
replication of DNA, 213–15
transcription, 217–20
translation, 220–25
moving between species, 219
mutation, 238–41
organization, 68
p16, 241
p53, 159
Pitx1 gene, 312
ras, 159
regulator, 229
SRY, 179, 249
structural, 229, 230
summary and assessment, 226–27
Tbx5, 311
tumor suppressor, 159
Gene cloning, 245. See also DNA cloning
Gene expression regulation, 228–43
definition, 225
eukaryotic regulation, 232–38
chromatin structure, 232–34
posttranscriptional control, 235–36
posttranslational control, 238
transcriptional control, 234–35
translational control, 238
gene mutation, 238–41
effect of mutations on protein activity, 239–40
mutations can cause cancer, 240–41
prokaryotic regulation, 229–31
summary and assessment, 242–43
Gene flow, 284
Gene mutation, 238–41
effect of mutations on protein activity, 239–40
evolution, 312
mutations can cause cancer, 240–41
Gene pharming, 248
Gene pool, 280
Gene therapy, 250
blindness, 723
cloning, 245
Parkinson disease, 706
preimplantation genetic diagnosis, 784, 805f
Genetic diagnosis, 784, 805f
Genetic diversity, 906–7
Genetic drift, 284
Genetic engineering. See under Genetically modified
Genetic profile, 253
Genetic recombination, 169–70
Genetic testing, 254, 784, 805f
Genetic variation, 169–70
genetic recombination, 169–70
independent assortment of homologous chromosomes, 170
significance of genetic variation, 170
Genetically modified animal, 248–49, 555
Genetically modified bacterium, 247–48
Genetically modified organism (GMO), 245, 247
Genetically modified plant, 248, 427
Genital herpes, 786f, 788, 805
Genital wart, 592, 786–87
Genome sequencing, 251
Genomic imprinting, 234
Genomics, 251–55
bioinformatics, 255, 906
comparison of Homo sapiens and chimpanzees, 568
eukaryote genome, 251–52
functional and comparative, 253, 255
functional and comparative genomics, 253–55
gene definition, 252–53
genome sequencing, 251
intergenic sequences, 252–53
proteomics, 225, 255
structure of eukaryotic genome, 251–52
summary and assessment, 257–59
Genotype, 190–91
Genus, 7, 339
Geographic isolation, 303–4, 306–7
Geology
factors that influence evolution, 332–34
continental drift, 332–33
mass extinctions, 333–34
timescale, 324–25, 326t
Geomyces destructans, 401
Geospiza fortis, 269, 857, 858
GERD (gastroesophageal reflux disease), 647
Germ cell, 771
Germ layer
embryonic development, 516, 795f, 796, 798, 806
humans, 806
organ development, 797
German measles, 805
Germination, 490–91, 504–7
Gerontology, 811

GFP (green fluorescent protein), 219, 362, 588
GH (growth hormone), 736, 756–57, 811
GHB (gamma-hydroxybutyric acid), 700, 787
Ghrelin, 655
Giantism, 756–57
Giardia lamblia, 57, 385
Gibberellic acid, 481
Gibberellin, 478*t*, 481, 482
Gibbon, 828, 829
GIFT (gamete intrafallopian transfer), 783
Gila monster, 556
Gill
arthropods, 534, 535
chordates, 546
definition, 660
fish, 661–62
bony fishes, 550, 662f
jawed fish characteristics, 548
pharyngeal pouches, 545
sea star, 540
single-loop circulatory pathway, 603
tracheal, 663
Gill (mushroom), 402
Gill arch and evolution of jaw, 548
Ginkgo, 424–26
Ginkgo biloba, 424–26
Giraffa camelopardalis, 298–99, 301*f*
Giraffe, 298–99, 301*f*
Girdling, 448, 471
Gizzard, earthworm, 642, 643
Glaciation, 331
Glaciers and bryophytes, 418
Gladiolus, 449, 450f
Gland. *See also* Endocrine system
definition, 583
glandular epithelium, 583
major human glands, 752f
Glandular epithelium, 583
Glans penis, 775
Glaucoma, 722
Glial cell. *See* Neuroglia
Glial cell–derived growth factor, 587
Global warming, 14, 875. *See also* Climate change
Globigerina, 391*f*
Globular protein, 48, 50
Globular stage of seed development, 502, 503*f*
Gloeocapsa, 367*f*
Glomeromycota
evolution and diversity of fungi, 398, 400
mycorrhizae, 406–7
reproduction, 400–401
Glomerular capsule, 681, 682, 683*f*, 684*f*
Glomerular filtrate, 682–83
Glomerular filtration, 682–83, 684*f*
Glomerulus, 682
Glossina, 388, 389*f*
Glossopteris, 332
Glottis, 646, 664
Glucagon, 763, 764
Glucocorticoid, 760*f*, 761, 762
Glucose
breakdown, 131f
cortisol, 761
diabetes, 654, 763, 764, 765
digestive enzymes, 651
liver, 649, 652, 751f, 763
metabolism, 139–40
structure, 39
tubular reabsorption, 683–85
Gluten, 650
Gluten sensitivity, 650
Gluten-free diet, 650
Glyceraldehyde-3-phosphate (G3P), 124*f*
Glycerol, 42, 649
Glycocalyx, 62, 85, 361
Glycogen
function, 41
fungi, 397
human liver, 649, 652, 751f, 763
structure, 40f
Glycolipid, 84–85
Glycolysis, 131, 133*f*, 322
Glycoprotein, 84–85, 361
Glyptodon, 265
GMO (genetically modified organism), 247
Gnathostome, 546*t*
Gnetophyte, 424*f*, 425*f*, 426
Gnetum, 426
GnRH (gonadotropin-releasing hormone), 765, 776, 779
Goiter, 758
Gold, 463
Golden brown algae. *See* Chromalveolata supergroup
Goldenrod, 499*f*
Goldfinch, American, 772*f*
Golgi, Camillo, 69–70
Golgi apparatus, 69–70
Golgi tendon organ, 729, 742
Gonad
hormone and endocrine system, 752f, 755, 756f
sexual reproduction, 771, 773–74, 776
Gonadotropic hormone
follicle-stimulating hormone (FSH), 765, 776, 778–79
function, 755, 776
gonadotropin-releasing hormone (GnRH), 765, 776, 779
luteinizing hormone (LH), 765, 776, 778–79
Gonadotropin-releasing hormone (GnRH), 765, 776, 779
Gondwana, 332, 334
Gonorrhea, 788–89
Gonorrhea proctitis, 788–89
Gonyaulax, 381*f*
Good genes hypothesis, 287
Goose, Canadian, 881
Goose bumps, 594
Gooseneck barnacle, 534*f*
Gout, 678
Graafian follicle. *See* Vesicular follicle
Gracilaria, 379
Gracility of australopithecine, 571, 572
Gradualistic model of macroevolution, 308–9, 310*f*
Grain
indehiscence, 504
refinement, 653
Gram stain, 363, 364*f*
Gram-negative bacterium, 363, 364*f*, 365, 367
Gram-positive bacterium, 363, 364*f*
Granite moss, 416
Grant, Peter, 268
Grant, Rosemary, 268
Granular leukocyte, 615
basophils, 615
eosinophils, 615, 627
neutrophils, 615, 627
Granum, 73, 116
Granzyme, 631, 632*f*
Grapes
downy mildew, 382
gibberellins, 481
Grapsus grapsus, 534*f*
Grass
bluegrass life table, 839, 840
ecological succession, 866, 867
pollen, 499f
Grasshopper, 536–37, 602*f*
Grassland
ecosystem, 892–93
savannas, 892–93
soil profile, 460
Graves disease, 759
Gravitational equilibrium, 727, 728
Gravitropism, 479, 485, 487
Gray crescent, 798, 799*f*
Gray matter
basal nuclei, 704
cerebral cortex, 703
spinal cord, 702
Grazer, 644
Grazing food web, 870*f*, 871
Great white shark, 549
Greater roadrunner, 893*f*
Greater Yellowstone Ecosystem, 917
Green algae. *See* Archaeplastida supergroup
Green fluorescent protein (GFP), 219, 362, 588
Green gland, 679
Green sea turtle, 556*f*
Green sulfur bacterium, 365
Greenhouse effect, 874, 877. *See also* Climate change
Greenhouse gas, 874–75, 877, 913
Griffith, Frederick, 208
Grizzly bear, 907, 912, 916–17
Grooming, 827, 831
Ground meristem, 445, 502
Ground pine, 419
Ground substance, 583–84
Ground tissue, 436, 437–38
Groundwater mining, 873
Groundwater table, 873, 895
Growth factor, 149
Growth hormone (GH), 736, 756–57, 811
Growth plate, 736
Growth rate. *See* Rate of natural increase
Growth response control of flowering plants, 476–94
growth and movement responses, 485–89
external stimuli, 485–87
internal stimuli, 487–89
phytochrome responses, 489–93
circadian rhythms, 492–93
phytochrome, 490–91
plant hormones, 477–85
abscisic acid (ABA), 478t, 482–83
auxins, 477f, 478–81, 482f
cytokinins, 478t, 481–82
ethylene, 478t, 483–84
gibberellins, 478t, 481, 482
signal transduction, 477
Guard cell
leaf, 451
water transport, 437, 469, 470
Guillain-Barré syndrome (GBS), 709
Gulf Stream, 900
Gull, 821
Gumma, 789
Gut. *See* Digestive tract
Guttation, 468
Gymnosperm
evolution and full adaption to land, 329, 421, 423–26
conifers, 423–24
cycads, 424, 425f
ginkgoes, 424–26
gnetophytes, 424f, 425f, 426
features, 413t
life history strategy, 849
ovules, 423
sporophytes, 415f
Gynosome, 770

H

H (hemagglutinin) spike, 359
H5N1 flu, 359
H7N9 flu, 359
HAART (highly active antiretroviral therapy), 786
HAB (harmful algal bloom), 382–83
Habitat
community interactions, 857
definition, 837, 857
island biogeography model, 860
isolation, 299, 301f
loss, 13–14, 911, 912f
preservation, 916–18
keystone species, 916
landscape preservation, 917–18
metapopulations, 916–17
restoration, 918–19
Habituation, 821
Haeckel, Ernst, 341, 837
Hagfish, 548
Hair
human
axillary and pubic, 765
baldness, 765, 776, 809
facial hair, 765, 776
hair follicle, 592f, 594, 809
variations, 577, 594
loss, 594
mammal characteristic, 559, 591
Hair cell, inner ear, 725
Hair follicle, human, 592*f*, 594, 809
Haliaetus, 557*f*, 558*f*, 901, 907, 912
Hallucinogenic mushroom, 404
Halocynthia, 546
Halophile, 344, 368*f*, 369
Hamadryas baboon, 829*f*
Hamilton, Tyler, 686
Hammer. *See* Malleus
Hand
bones, 740
dexterity, 735
primate evolution, 565f, 566f
styloid process, 576
Haploid generation, 414
Haploid number, 152, 167
Hardy, Godfrey H., 280
Hardy-Weinberg equilibrium, 280–85
migration, 284
mutation, 283–84
natural selection, 285
nonrandom mating, 285
small population size, 284–85
Hardy-Weinberg principle, 281
Hare, European, 266*f*
Hare, snowshoe, 861
Harmful algal bloom (HAB), 382–83
Hart's tongue fern, 422
Haustorium, 464, 485
Haversian system. *See* Osteon
Hawaiian honeycreeper, 306
Hawk, 868*f*
Hawkmoth caterpillar, 485
Hay fever, 636
Hb. *See* Hemoglobin
HCG (human chorionic gonadotropin), 630, 780, 804
HCO3– buffer system, 32, 687
HDL (high-density lipoprotein), 654
HDN (hemolytic disease of the newborn), 615
Head louse, 536*f*
Hearing and balance, 724–28
disorders, 728
aging process, 810
hearing loss, 728, 810
vertigo, 728
human ear, 725–27
aging process, 810
anatomy, 725–26
pitch, 726
volume, 726–27
sense of balance, 727–28
gravitational equilibrium, 727, 728
invertebrates, 726f
rotational equilibrium, 727–28
sound wave detection, 724–25
Hearing loss, 728, 810
Heart. *See also* Atrium; Ventricle (heart)
cardiac muscle, 586, 587, 608
human, 605–9
aging process, 809

conduction, 608
congenital defects, 600
embryonic development, 807
heartbeat, 607–9
path of blood through the heart, 606–7
septum, 604, 605–6
structure, 604–6
Heart attack, 611, 654
Heart murmur, 608
Heart stage of seed development, 502, 503*f*
Heartbeat, human, 607–9
cardiac cycle, 607–8
electrocardiogram (ECG), 608–9
pacemaker, 608
pulse, 608
Heartburn, 646
Heartwood, 448–49
Heartworm, 533, 863
Heat, 101
Heat capacity, 27
Heat of evaporation, 27
Heimlich maneuver, 669
Helicobacter pylori, 11–13, 647
Helium model, 21*f*
Helix aspersa, 530*f,* 644
Heloderma suspectum, 556
Helper T cell, 628*f,* 629, 630
Hemagglutinin spike, 359
Hematocrit, 686
Heme, 668
Hemerocallis, 497*f*
Hemizygosity, 202
Hemlock, Western, 866*f*
Hemocoel
arthropods, 535, 537
molluscs, 528, 529
Hemocyanin, 528, 535, 617
Hemoglobin
earthworms, 602
oxygen transport, 668
red blood cells, 585, 613–14, 668
Hemolymph
arthropods
crayfish, 535
grasshopper, 537, 602f
horseshoe crabs, 617
respiration, 662
invertebrates, 601, 602f
molluscs, 528
Hemolysis, 90
Hemolytic disease of the newborn (HDN), 615
Hemophilia, 203
Hepatic artery, 649
Hepatic portal vein, 649
Hepatic vein, 649
Hepatitis, 649–51, 785, 787, 788
Herbaceous stem, 445–46, 448*f*
Herbicide, 481
Herbivore
diet adaptation, 643, 644
ecosystem dynamics, 868, 870f, 871, 872, 893, 894f
Herceptin, 630, 781
Hereditary spherocytosis, 197
Heritable variation, 267
Hermaphroditism. *See also under* Monoecious
animals, 771
earthworms, 530–31
planarians, 525
sequential, 771
Heroin, 701
Herpes, genital, 786*f,* 788, 805
Herpes simplex virus, 786*f,* 788, 805
Hershey, Alfred, 209
Heterochromatin, 152, 233, 234
Heterocyst, 367
Heterosporous vascular plants
definition, 419
flowering plants, 428, 497
seed plants, 423
Heterotroph
animals, 514, 868–69
bacteria, 343, 365
ecosystem dynamics, 115, 868–69
protists, 374
Heterotrophic bacterium, 343, 365
Heterozygosity, 190, 292–93, 784
Hexose, 39
HFCS (high-fructose corn syrup), 653
HGH (human growth hormone). *See* Growth hormone (GH)
High blood pressure (hypertension), 611, 767, 809
High-density lipoprotein (HDL), 654
High-fructose corn syrup (HFCS), 653
Highly active antiretroviral therapy (HAART), 786
Hindbrain, 693, 704
Hindlimb
mobility in primates, 565
reduction, 312
Hinge joint, 741
Hip
artificial, 741
joint, 741
Hipbone. *See* Coxal bone; Pelvic girdle
Hippocampus, 704, 705
Hippocampus kuda, 550*f,* 770, 772
Hippopotamus, 826
Hirudo medicinalis, 531*f,* 532
Histamine, 626, 636, 637
History of evolutionary thought, 262–64
History of life, 323–32
Cenozoic era, 326t, 330–31, 332f
fossils tell a story, 323–24
clock timetable, 323
dating fossils, 324
geologic timescale, 324–25, 326t
Mesozoic era, 326t, 329–30, 332f
origin of life, 318–23
evolution of monomers, 318, 319–20
evolution of polymers, 318, 320
evolution of protocells, 318, 320–22
evolution of self-replication system, 318, 322
Paleozoic era, 326t, 327–29, 332f
Precambrian time, 324–27
eukaryotic cells arise, 325–26
multicellularity, 327
HIV (human immunodeficiency virus), 784–86. *See also* AIDS
categories, 633, 785
evolution, 360
host specificity, 356
opportunistic infections, 384, 632, 633, 672, 785
phylogeny, 337, 349
pregnancy, 786
reproduction, 358
stages and symptoms, 784
T cells, 631–32, 633, 785
transmission, 784, 787, 788
treatment and prevention, 784–85, 786
virulence and lysogenic cycle, 357
HMS *Beagle,* 262, 263*f,* 265, 267
Hobbit. *See Homo floresiensis*
Hodgkin disease, 908
Holdfast structure, 380
Holly, 483, 497, 498*f*
Holocene epoch, 331
Holothuroidea, 539–40
Home range, 828
Homeobox gene, 274, 311–12, 515, 518–19, 801–2
Homeodomain, 802
Homeostasis, 594–99
animal organization, 581, 594–96
negative feedback, 595, 596f, 755
positive feedback, 595–96, 755
blood, 613
definition, 4–5, 594
endocrine system, 751
human kidneys, 685–88
acid-base balance, 685, 687–88
water-salt balance, 685–87, 760, 761
hypothalamus, 693
lymphatic system, 623
summary and assessment, 597–99
Homeothermism, 594
Homeotic gene. *See Hox* gene
Homeotic transformation, 801
Hominid
classification, 569t
definition, 566
evolution, 331, 567f, 570f
Hominidae, 345
Hominin
definition, 566
Homo species, early, 569t, 570f, 572–73
Homo ergaster and Homo erectus, 572, 573, 575
Homo floresiensis, 573
Homo habilis and Homo rudolfensis, 572
Homo species, later, 569t, 570f, 573–77
common ancestry, 577
Cro-Magnons, 574–75
Denisovans, 574, 576
human variations, 575–77
hypotheses, 573–74
Neandertals, 564, 574
humanlike hominins, 569–72
ardipithicines, 570–71
australopithecines, 570f, 571–72
bipedalism, 569–71
South African australopiths, 572
Hominine
definition, 566
evolution, 567f
Hominoid
definition, 566
evolution, 567f, 568
Homo erectus, 572, 575
Homo ergaster, 572, 573
Homo floresiensis, 573
Homo habilis, 572
Homo heidelbergensis, 573, 574
Homo neandertalensis
characteristics, 574
extinction, 576
genome, 564
tool use, 576
Homo rudolfensis, 572
Homo sapiens. See *Homo* species, later; Human
Homo species, early
classification, 569t
evolution, 570f, 572–73
Homo ergaster and Homo erectus, 572, 573, 575
Homo floresiensis, 573
Homo habilis and Homo rudolfensis, 572
Homo species, later
and chimpanzees, genomic comparison, 568
classification, 569t
evolution, 570f, 573–77
common ancestry, 577
Cro-Magnons, 574–75
Denisovans, 574, 576
human variations, 575–77
hypotheses, 573–74
Neandertals, 564, 574
multiregional continuity hypothesis, 574
replacement model hypothesis, 573–74, 577
Homologous chromosome, 167–68, 170, 190*f*
Homologous gene, 255
Homologous structure, 273, 347
Homologous trait, 307
Homologue, 167
Homology, 347
Homosporous vascular plant, 419
Honey, 537
Honeybee, 817, 827–28, 909. *See also* Bee
Honeydew, 472
Hook, 361, 362*f*
Horizon, soil, 460
Hormonal therapy
breast cancer, 781
hormone replacement therapy (HRT), 811
Hormone, 750–69. *See also* Endocrine system; *specific hormone*
animal hormones, 751–54
action, 753–54
chemical signals and pheromones, 753
antagonistic hormones, 760, 763
flowering plant growth response control, 477–85
functions, 750, 751
hypothalamus and pituitary gland, 755–57
anterior pituitary, 755–57
male hormone regulation, 776
posterior pituitary, 755, 756f
kidney secretion, 685, 687
major glands, 752f
mode of action, 751, 753–54
peptide hormones, 753, 754f
steroid hormones, 753–54
other endocrine glands and hormones, 758–67
adrenal glands, 752f, 753–54, 755, 756f, 760, 761–62
erythropoietin (EPO), 614, 685, 686, 766
leptin, 655, 766
pancreas, 645f, 647, 649, 752f, 763–65
parathyroid, 752f, 758, 759–60
pineal gland, 704, 752f, 766
prostaglandins, 766–67, 775
testes and ovaries, 752f, 753–54, 755, 765–66, 773–75, 777, 811
thymus, 623f, 624, 752f, 766
thyroid, 752f, 755, 756f, 758–59
plant, 477–85
regulation in human males, 776
summary and assessment, 768–69
Hormone replacement therapy (HRT), 811
Hornet, 862
Hornwort, 412, 415, 416
Horse
evolution, 312–13
herbivore diet, 644
Horseshoe crab, 537, 617, 908
Horsetail, 420, 421
Horticulture and exotic species, 912. *See also* Agriculture
Host (parasitism)
alternate, 405
coevolution, 865
definition, 863
secondary, 526, 527, 863
Host specificity (viruses), 356
Hot springs, 368–69
Hotspot, biodiversity, 907, 916
House mouse, 340
Hox gene, 274, 311–12, 515, 518–19, 801–2
HOXC8 gene, 518–19
HPV (human papillomavirus), 592, 786–88
HRT (hormone replacement therapy), 811
Human
bipedalism, 571f
cardiovascular system, 605–12

aging process, 809–10
blood pressure, 607, 610–11, 616, 636–37, 687, 761–62, 809
cardiovascular disease, 611–13, 653, 654, 809–10
heart, 605–9
pulmonary circuit, 609
systemic circuit, 609–10
classification, 569t
digestive system and nutrition, 645–51
accessory organs, 645f, 649–51
aging process, 810
chemical digestion, 645
large intestine, 645f, 648–49
mechanical digestion, 645
mouth, 645–46
pharynx and esophagus, 645f, 646, 647
small intestine, 645f, 647–48
stomach, 645f, 646, 647
ear, 725–27
aging process, 810
anatomy, 725–27
functions, 725
inner ear, 725, 726–27
middle ear, 725
outer ear, 725
evolution, 564–79
early Homo species, 570f, 572–73
humanlike hominins, 566, 569–72
later Homo species, 570f, 573–77
primate evolution, 565–69, 570f, 577
summary and assessment, 578–79
eye, 719–24
aging process, 810
anatomy, 719–20
disorders, 722–24, 810
focusing, 720, 722–24, 810
integration of visual signals, 721–22
photoreceptors, 719, 720–21
from the retina to the visual cortex, 722
genome comparison to chimpanzees, 568
hormone and endocrine system, 750–69
hypothalamus and pituitary gland, 755–57
other endocrine glands and hormones, 758–67
immune and lymphatic system, 621–40
adaptive immune defenses, 628–34
aging process, 810
immune system disorders and adverse reactions, 636–38
innate immune defenses, 622
lymphatic system, 623–25
muscular system, 742–47
macroscopic anatomy and physiology, 742
microscopic anatomy and physiology, 742–47
muscular tissue, 586–87
neuron and nervous system. See also Brain; Spinal cord
aging process, 810
central nervous system, 694, 699–706
embryonic development, 799f
nervous tissue, 582, 587–88, 695–99
peripheral nervous system, 694, 707–10
phylogeny and molecular traits, 348
population growth, 849–52
age distributions, 850f, 851
environmental impact, 851–52
more-developed vs. less-developed countries, 849f, 850–51
reproductive system
aging process, 811
control of reproduction, 781–84
female reproductive system, 777–81, 811
male reproductive system, 773–76, 811
sexually transmitted diseases, 385, 784–89
respiratory system
aging process, 810
lungs, 89f, 661, 663–64
modifications in breathing, 666–67
oxygen debt, 745
skeletal system, 736–41
aging process, 810–11
appendicular skeleton, 738f, 740–41, 766, 780
axial skeleton, 738–39
bone growth and renewal, 736
cartilaginous fetal, 585
compared to Ardipithecus ramidus, 571f
functions, 736
joint classification, 741
long bone anatomy, 736–38
somatic senses, 729–30
teeth, 644
urinary system, 681–88
aging process, 810
homeostasis and kidneys, 685–88
kidneys, 681–82
urine formation, 682–85
vision, 719–24
aging process, 810
anatomy, 719–20
disorders, 722–24, 810
focusing, 720, 722–24
integration of visual signals, 721–22
photoreceptors, 719, 720–21
from the retina to the visual cortex, 722
Human chorionic gonadotropin (HCG), 630, 780, 804
Human Genome Project, 251
Human growth hormone (HGH). *See* Growth hormone (GH)
Human immunodeficiency virus. *See* HIV
Human papillomavirus (HPV), 592, 786–88
Humboldt Current, 902
Humerus, 740
Humira, 630
Hummingbird
pollination, 501
territoriality, 828
Humor, 629
Humoral immunity. *See* Antibody-mediated immunity
Humpback whale, 826
Humus, 459
Hunter-gatherer, 575
Hunting, subsistence, 909
Huntington disease, 197
Hurricane, 877
Hutton, James, 264
Hyaline cartilage, 584–85, 737
Hybridization
alloploidy, 305
postzygotic isolating mechanisms, 301f, 302–3
prezygotic isolating mechanisms, 299
Hybridoma, 630
Hydra
anatomy, 522
asexual reproduction, 523, 771
nerve net, 692
respiration, 660, 661f
sexual reproduction, 771
skeleton, 734
transport, 601
Hydrochloric acid, 647
Hydrogen bonding, 26–27
Hydrogen ion
acids and bases, 30
cellular respiration, 138
human respiratory system, 667
mineral uptake in plants, 461–62
opening and closing of stomata, 470
phototropism, 479, 480f
pressure-flow model of phloem transport, 471
Hydrogenation, 35
Hydrolysis origin reaction, 38
Hydrolytic cycle, 873
Hydrophilic molecule, 28
Hydrophobic molecule, 28
Hydroponics, 458
Hydrostatic skeleton
annelids, 530, 734
cnidarian, 522
Hydrothermal vents
ecology and ecosystems, 868, 900
origin of life, 319, 320f, 322, 365
Hydroxide ion, 30
Hydrozoan, 522
Hylobates syndactylus, 828*f*
Hyman, Flo, 198
Hymen, 777*f*
Hypercholesterolemia, familial, 198
Hyperglycemia, 764
Hyperopic vision, 724
Hyperparathyroidism, 760
Hypertension, 611, 767, 809
Hyperthyroidism, 759
Hypertonic solution, 90–91
Hypha
fungi, 397
sac fungi, 401
zygospore fungi, 399–400
lichen, 405
Hypocotyl, 503
Hypodermis. *See* Subcutaneous layer
Hypoglycemia, 764
Hypolimnion, 896
Hypoparathyroidism, 760
Hypophysis. *See* Pituitary gland and hypothalamus
Hypothalamus and pituitary gland, 755–57
anterior pituitary, 755–57
hormones affecting other glands, 755, 765
hormones not affecting other glands, 755–57
hypothalamic-inhibiting hormones, 755
hypothalamic-releasing hormones, 755
major endocrine glands, 752f
hypothalamus
anatomy, 704
vertebrate organization, 693
ovarian cycle, 779
posterior pituitary, 752f, 755, 756f
Hypothesis, 9–10
Hypothyroidism, 758, 759
Hypotonic solution, 90
Hyracotherium, 312
Hysterectomy, 787–88

I

IAA (indoleacetic acid), 479. *See also* Auxin
Iberomesornis, 271
Ice
freezing of water, 29
lake ecosystems, 896
Ice age, 331
Ichthyophthirius, 383
ICSI (intracytoplasmic sperm injection), 783
ICTV (International Committee on Taxonomy of Viruses), 355
ICZN (International Code of Zoological Nomenclature), 340
Identical twin, 237*f*
Ig (immunoglobulin). *See* Immunoglobulin (Ig)
IgA, 629
IgE, 629, 636, 637
IgG, 629, 634, 637
IgM, 629
Iguana
characteristics, 556
speciation, 304
Illinois River, 919
Illumination, 61
Imitation, learning by, 823
Immediate allergic response, 636–37
Immune and lymphatic system, 621–40. *See also* White blood cell
adaptive immune defenses, 628–34
active vs. passive immunity, 632–34
B cells and antibody-mediated immunity, 628–30
evolution, 622–23
T cells and cell-mediated immunity, 630–32
aging process, 810
evolution of immune systems, 622–23
cellular slime molds, 622
Drosophila, 622
rise of adaptive immunity, 622–23
immune system disorders and adverse reactions, 636–38
allergies, 636–37
autoimmune diseases, 637
Guillain-Barré syndrome (GBS), 709
immunodeficiencies, 636
transplant rejection, 638
innate immune defenses, 622, 625–28
inflammatory response, 625–27
phagocytes and natural killer cells, 625, 627
physical and chemical barriers, 625
protective proteins, 625, 627–28
lymphatic system, 623–25
homeostatic functions, 623
lymphatic vessels, 623–24
lymphoid organs, 624–25
Immune complex, 629
Immunization and active immunity, 632–34
Immunodeficiency disorder, 636
Immunoglobulin (Ig)
aging process, 810
development, 629
monoclonal antibodies, 630, 781
plasma, 613, 614, 629
structure, 629, 630f
Immunokine, 555
Imperfect flower, 429*t*, 497
Implantation, 777, 804
Impotence. *See* Erectile dysfunction (ED)
Imprinting, 821–22, 829
In vitro fertilization (IVF), 783, 784, 805*f*
In vivo gene therapy, 250
Inbreeding, 285, 290–91
Incisor, 644
Inclusive fitness, 831*f*, 832
Incomplete digestive tract, 525, 642
Incomplete dominance, 197–98
Incomplete flower, 428, 429*t*, 497
Incomplete metamorphosis, 537
Incomplete penetrance, 197–98
Incontinence, urinary, 810
Incus, 725
Indehiscent fruit, 504, 505*f*
Independent assortment, law of, 170, 191–92
Index fossil, 324
Indirect selection. *See* Kin selection
Indirect value of biodiversity, 910–11
biogeochemical cycles, 910
climate regulation, 910
ecotourism, 911
fresh water provision, 910
soil erosion protection, 910
waste recycling, 910
Indoleacetic acid (IAA), 479. *See also* Auxin

Indotestudo elongata, 915*f*
Induced fit model, 106
Induced mutation, 239
Inducer, 231
Inducible operon, 231
Induction, 798–800
Inductive reasoning, 9
Indusium, 422
Infection. *See also* Bacterium; Virus
fungal, 402, 672
opportunistic, 384, 632, 633, 672, 785
Inferior vena cava, 606
Infertility, 783
Inflammatory response, 625–27
analgesics, 730
anti-inflammatory medication, 730, 741
anti-TNF treatments, 632
capillaries, 626
chronic inflammation, 626–27
complement, 627
fever, 627
glucocorticoids, 761
white blood cells, 626
Inflorescence, 429*t*
Influenza virus
classification, 355
emerging viruses, 358–60
Infrared spectrum, 714, 715, 716
Ingroup (cladistics), 346
Inguinal lymph nodes, 623*f*
Inhalation. *See* Inspiration
Inheritance
acquired characteristics, 264
behavioral ecology, 818–20
garter snakes, 819
lovebirds, 818–19
marine snails, 819–20
mice, 820
twin studies, 819
beyond Mendelian inheritance, 197–204
incomplete dominance and incomplete penetrance, 197–98
multiple allelic traits, 197–98
pleiotropic effects, 198–99
polygenic inheritance, 199–201
blending concept of inheritance, 187
epigenetic, 234
Mendel's laws, 188–94
law of segregation, 188–90, 192f
Mendel and the laws of probability, 192–93
Mendel's cross, 190–91
Mendel's law of independent assortment, 170, 191–92
testcrosses, 193–94
Mendel's patterns of inheritance and human disease, 194–97
autosomal dominant disorder, 195f, 196–97
autosomal recessive disorder, 195–96, 195f
particulate theory of inheritance, 187
Inhibin, 776
Inhibition, 698, 699
Inhibition model of succession, 867
Initiation, 222–23
Innate immune defense, 625–28
cellular slime molds and fruit flies, 622
inflammatory response, 625–27
capillaries, 626
chronic inflammation, 626–27
complement, 627
fever, 627
white blood cells, 626
phagocytes and natural killer cells, 625, 627
eosinophils, 627
macrophages, 615, 626, 627
natural killer (NK) cells, 627
neutrophils, 627
physical and chemical barriers, 625
protective proteins, 625, 627–28
complement, 627
interferons, 628
Innate immunity, 622
Inner ear, 725, 726–27
Innervation, muscle, 745–46
Inorganic chemistry, 36
Inorganic molecule and origin of life, 318, 319
Insect
anatomy, 536
characteristics, 536–37
chemical senses, 716
entomophagy, 536, 537
excretory organs, 679
as food, 538
history of life, 329
metamorphosis, 536, 537, 750, 751
omnivorous diet, 644
pollination, 429, 500–501, 909
respiratory system, 536–37, 600, 602f, 661f, 662–63
sound wave detection, 724
vision, 718
Insight learning, 823–24
Insoluble fiber, 653
Inspiration, 660, 665–66
Instinct and learning, 821
Insulin, 654, 751*f*, 763, 764, 765
Insulin pump, 764
Integral protein, 84
Integrase inhibitor, 786
Integration
central nervous system, 699
synaptic, 699
viral, 356, 357f
visual signals, 721–22
Integumentary system, 591–94. *See also* Skin
accessory structures, 592–94
hair follicles, 592f, 594
nails, 592
oil glands, 593, 625
sweat glands, 593
aging process, 809
Interaction with environment, 5
Intercalary meristem, 436
Intercalated disk
cardiac muscle, 586f, 587
heart anatomy, 606f
path of blood through the heart, 606
Intercostal muscle, 665*f*, 666, 667*f*
Interferon, 625, 628
Intergenic sequence, 252
Interkinesis, 172
Intermediate filament, 75
Intermediate host. *See* Secondary host
Internal development (mammal characteristic), 559
Internal fertilization, 772
Internal organ
loose fibrous connective tissues, 584
mammal characteristic, 559
vertebrate characteristic, 547
Internal respiration, 660, 667, 668–69
International Code of Botanical Nomenclature, 340
International Code of Zoological Nomenclature (ICZN), 340
International Committee on Taxonomy of Viruses (ICTV), 355
Interneuron, 695*f*, 696, 702, 708
Internode
leaf, 451f
shoot system, 440, 445
Interoceptor, 715
Interphase, 148
Interspersed repeat, 252
Interstitial cell, 774*f*, 776
Interstitial fluid
capillaries, 603
capillary exchange, 616
lymphatic vessels, 586, 623–24
Intertidal zone, 898
Intervertebral disk, 739
Intestine
adaptations to diet, 644
earthworms, 642, 643
human
large, 645f, 646, 648–49
small intestine, 645f, 646, 647–48, 650
Intracellular digestion, 643
Intracytoplasmic sperm injection (ICSI), 783
Intramembranous ossification, 736
Intrauterine device (IUD), 782*f*
Intrauterine insemination (IUI), 783
Intron, 219–20
Invagination, 795*f*, 796
Invasive species. *See* Exotic species
Inversion, chromosome, 181
Invertebrate
balance, 726f
cannibalism, 644
characteristics, 520t
definition, 514
evolution, 512–43
animal evolution, 514–20
deuterostomes, 539–40
ecdysozoan diversity, 532–39
lophotrochozoan diversity, 524–32
nervous system, 692–93
simplest invertebrates, 521–23
summary and assessment, 541–43
excretory organs, 679
coxal glands, 679
flame cells, 679
green glands, 679
Malpighian tubules, 679
maxillary glands, 679
nephridia, 679
history of life, 327, 328f, 329
hormones and endocrine system, 751
nervous system, 692–93
cephalization, 692
nerve net, 692
reproduction
cnidarians, 522, 523, 771
sponges, 521, 771
transport, 601–2
closed circulatory system, 602
open circulatory system, 601, 602f
Involution of thymus, 810
Iodine, 758, 759
Ion, 24, 30
Ionic bonding, 24
Ipomoea leptophylla, 492
Iris, 450*f*, 507, 719
Iron-sulfur world hypothesis, 319, 320
Irruptive growth, 844
Ischemia, 600
Ischemic stroke, 703
Island
biogeography and biodiversity, 860
exotic species, 912–13
island biogeography model, 857, 860
Islet of Langerhans. *See* Pancreatic islet
Isomer, 38
Isotonic solution, 89–90
Isotope
definition, 21
radioactive, 22–23
Iteroparity, 841
IUD (intrauterine device), 782*f*
IUI (intrauterine insemination), 783
Ivanowsky, Dimitri, 355
IVF (in vitro fertilization), 783, 784, 805*f*

J

Jackal, 832
Jacob, François, 229
Jacobs syndrome, 180
Jacobson's organ, 556, 716
Jaguar, 887
Jasmonic acid, 484
Jatropha, 498*f*
Jaundice, 649
Jawed fish
characteristics, 547
ectothermy, 548
endoskeleton, 548
gills, 548
evolution, 546t, 547, 548–49
Jawless fish evolution, 546*t*, 547, 548
Jeholodens, 330
Jellyfish
green fluorescent protein, 219
nerve net, 692
radial symmetry, 515
structure, 522, 523f
Jellyfish, crystal, 219
Jet lag, 492
Jock itch, 402
Joining, DNA, 214
Joint, human, 740, 741
Juglans nigra, 485
Junction protein, 85, 582
"Junk" DNA, 236, 252
Juno molecule, 795
Jurassic period, 326*t*, 330
Juvenile diabetes. *See* Type 1 diabetes
Juxtaglomerular apparatus, 687

K

K. *See* Carrying capacity
K2, 701
Kangaroo, red, 838
Kangaroo rat, 680, 681*f*, 894
Kaposi sarcoma, 633
K-Ar dating, 324
Karyotype, 179
Katydid, spike-headed, 891*f*
Keel, 557
Kelp, 380
Kenyan giant scorpion, 537*f*
Keratin, 592, 593, 594
Kerner, Justinus, 744
Ketamine, 700–701
Ketoacidosis, 764
Keystone species, 916
Kidney
human urinary system, 681–88
homeostasis and kidneys, 685–88
kidneys, 681–82
urine formation, 682–85
osmoregulation, 679, 685–87, 760, 761
parathyroid hormone (PTH), 759–60
Kidney stone, 760
Kidston, Barry, 706
Killer whale. *See* Orca
Kin selection, 832
Kinetic energy, 101
Kinetochore, 152
Kinetochore spindle fiber, 153
Kinetoplast, 388
Kinetoplastid, 388
Kingdom
five-kingdom system, 341
Linnaean classification hierarchy, 7, 339
Kingsnake, scarlet, 862
Kinocilium, 728
Kit fox, 894*f*
Klinefelter syndrome, 180
Klipspringer, 824
Knee, human, 740, 741
Knockout mouse, 249
Koala, 560, 644
Kokanee salmon, 907
Krause end bulb, 729
Krill, 534, 643
K-selection, 847–49
Kubiak, Gary, 600
Kudzu, 912, 913*f*
Kuru, 360

Kyoto Protocol, 877

L

Labia majora, 778
Labia minora, 778
Labor, 595–96, 755
Labroides dimidiatus, 864
Lac operon, 230–31
Lactate, 745
Lactation
definition, 780
passive immunity, 634, 780
prolactin (PRL), 755, 766
Lacteal, 586, 648
Lactobacillus bulgaricus, 135
Lactose intolerance, 41, 652
Lacuna
bone, 585, 737
cartilage, 584
Ladder-type nervous system, 525
Ladybug, 909*f*
Lag phase, 842, 843
Lagging strand, 214
Lake
acid deposition, 913
ecosystem, 895f, 896–98
lake effect, 883
life zones, 897–98
benthic zone, 897, 898
limnetic zone, 897
littoral zone, 897
profundal zone, 897–98
overturn, 896–97
LAL (limulus amoebocyte lysate), 617, 908
De Lamarck, Jean-Baptiste, 263–64
Lamellae, 662
Laminaria, 380
Laminarin, 380
Lamprey, 548
Lance fluke, 865
Lancelet, 545–46
cladogram, 346, 347f
cleavage, 795
gastrula, 796
Land
development of life on, 328–29
landscape diversity and preservation, 906, 907, 917–18
plant adaptation, 413–14
bryophytes and colonization of land, 415–17
seed plant evolution and full adaption to land, 423–31
Land snail
omnivorous diet, 644
shell, 530f
Landscape, 907
Landscape diversity, 906, 907
Landscape preservation, 917–18
Language use
chimpanzees, 826
Cro-Magnons, 575
dolphins, 826
humans, 826
Lap-Band, 655
Larch, 449*f*
Large intestine, 645*f,* 646, 648–49
Larix decidua, 449*f*
Larval stage
clams, 529
reproduction and life history strategy, 772
Larynx
amphibians, 551
human, 664, 776
LASIK (laser-assisted in situ keratomileusis), 724
Last universal common ancestor (LUCA), 318, 341
Late prophase, 153
Lateral line system, 549, 724–25
Laterite, 891
Latimeria chalumnae, 550
Latin and scientific name, 8–9, 339
Latrodectus, 537f
Laughing gull, 821
Laurasia, 332
Law of segregation, 188–90. *See also* Mendel's law; Thermodynamic law
LDC (less-developed country), 849*f,* 850–52
LDL (low-density lipoprotein), 654
Leading strand, 214
Leaf
adaptation to land, 414
anatomy, 116f
arrangement, 452
color change and cytokinins, 482
definition, 450
microphylls, 419, 420f
shoot system, 440
structure and organization, 450–53
classification, 452
diversity, 452–53
morphology, 450–51
veins and vascular tissue, 439, 440
Leaf cutter ant, 401
Leaf primordia, 445
Leaf vein, 439, 440
Leafhopper, 536*f*
LEAFY gene, 441
Leafy liverwort, 415–16
Learning
associative learning, 822–23
classical conditioning, 822
operant conditioning, 822–23
cognitive learning, 823–24
imitation, 823
insight, 823–24
observation, 823
definition, 821
habituation, 821
imprinting, 821–22, 829
instinct and learning, 821
memory in human brain, 705
orientation and migratory behavior, 823
social interactions and learning, 822
Leatherleaf fern, 422
Leclerc, Georges-Louis, 262
Leech, 531*f,* 532
Leeuwenhoek, Antonie van, 360
Leg bones, human, 740–41
Legume
fruit type, 505f
nitrogen fixation, 365, 876
Lens
anatomy, 719–20
cataracts, 722
focusing disorders, 724, 810
Lenticel, 437, 448
Leopard frog, 297, 298*f,* 348
Leprosy, 908
Leptin, 655, 766
Lepus americanus, 861
Lepus europaeus, 266*f*
Less-developed country (LDC), 849*f,* 850–52
Lesser long-nosed bat, 909*f*
Leucochloridium, 865
Leucophaeus atricilla, 821
Leucoplast, 73
Leukemia, 908
Leukocyte. *See* White blood cell
Lewontin, Richard, 577
LH (luteinizing hormone), 765, 776, 778–79
Liana, 890
Lichen
cyanobacteria, 367
ecological succession, 866
symbiotic relationships, 367, 405–6, 864
weathering and soil creation, 459
Lidy Hot Springs (ID), 368–69
Life, 1–17
challenges facing science, 13–14
biodiversity and habitat loss, 13–14
climate change, 14, 877
emerging disease, 14
characteristics, 2–6
adaptation, 5–6
homeostasis, 4–5
material and energy requirements, 4
organization, 2, 3f
reproduction and development, 5
response to environment, 5
DNA barcoding of, 342–43
evolution and classification, 6–9. See also Evolution; Taxonomy, systematics, and phylogeny
classification, 7–9
evolution, 6–7
origin and history, 317–36
Cenozoic era, 330–31
clock timetable, 323
fossils tell a story, 323–24
geologic timescale, 324–25, 326t
geological factors that influence evolution, 332–34
Mesozoic era, 329–30
monomer evolution, 318, 319–20
Paleozoic era, 327–29
polymer evolution, 318, 320
Precambrian time, 324–27
protocell evolution, 318, 320–22
self-replication system evolution, 318, 322
summary and assessment, 335–36
scientific method, 9–13
data presentation and analysis, 10–11
hypothesis, 9–10
observation, 9
predictions and experiments, 10
scientific theory, 11
summary and assessment, 15–17
Life cycle
algae, 377, 378
conifer, 425f
flowering plant, 428–31
human, 176–77, 178f
Life expectancy, 811
Life history patterns, 846–49
K-selected populations, 847–49
r-selected populations, 846, 847
Life history strategy, 772
Life process and organ system, 590*t*
Life table, 839–40
Life zone
lake, 897–98
benthic zone, 897, 898
limnetic zone, 897
littoral zone, 897
profundal zone, 897–98
ocean
abyssal plain, 898f, 899f
euphotic zone, 898f, 899
pelagic zone, 899–900
Lifespan, 811–12
Ligament
dense fibrous connective tissue, 584, 740
synovial joints, 741
Light
animal detection and vision, 718–19
blue light, 470, 477f, 486
far-red wavelength, 490, 491
human eye, 720
light reaction, 117, 119–20
sense of vision, 714, 715
ultraviolet
nectar guides, 501, 718, 719f
ozone depletion, 913
wavelengths, 714
Light reaction, 117, 119–20
Lignin
Casparian strip, 443
sclerenchyma cells, 438
vascular plants, 418
Lilium, 339
Lily, 339, 497*f*
Limb
amphibian characteristic, 551
development, 311, 312, 807
paired limbs in reptiles, 553
primate characteristic, 565
Limb bud, 807
Limb development, 311, 312, 807
Limbic system, 704–5
amygdala, 704, 705
hippocampus, 704, 705
Limestone, 874
Limiting factor, 838, 842
Limnetic zone, 897
Limulus, 537, 617, 908
Limulus amoebocyte lysate (LAL), 617, 908
Lineage
phylogeny, 300, 344
skull amniote lineage, 553–54
Linnaean classification hierarchy, 339–40
DNA barcoding, 340, 342–43
International Code of Botanical Nomenclature, 340
International Code of Zoological Nomenclature (ICZN), 340
Linnaean taxonomy, 339
binomial nomenclature, 339
specific epithet, 339
Linnaeus, Carolus, 262, 297, 339
Linoleic acid. *See* Omega-3 fatty acid
Lion, 644
Lionfish, 550*f,* 905
Lipase, 649, 652
Lipid. *See also* Cholesterol
human diet and nutrition, 35, 652, 653, 654
structure and function, 42–46
wax, 46
Liposome, 321
Littoral zone, 897
Liver
anatomy, 645f, 649–51
deamination, 654
disorders, 649–51
cirrhosis, 651
hepatitis, 649–51
jaundice, 649
glycogen storage, 751f, 763
inflammatory response, 626–27
insulin, 763
transplants, 651
Liver fluke, Chinese, 526–27
Liverwort
evolution, 412
reproduction, 416
traits, 415
Liverwort, leafy, 415–16
Living fossil, 544
Lizard
characteristics, 556
parthenogenesis, 171
reproductive isolation, 304
survivorship, 840
Lizard, monitor, 556
Lizard, whiptail, 171
Loading in phloem transport, 471
Loam, 459
Lobaria, 406*f*
Lobe-finned fish, 550, 551, 552*f*
Lobule, 649
Lock-and-key attachment
antigens, 629
viral, 356
Locomotion and support system, 733–49
diversity of skeletons, 734–35

endoskeletons, 734, 735
exoskeletons, 734–35
hydrostatic skeleton, 734
human skeletal system, 736–41
appendicular skeleton, 738f, 740–41, 766, 780
axial skeleton, 738–39
bone growth and renewal, 736
functions, 736
joint classification, 741
long bone anatomy, 736–38
muscular system, 742–47
macroscopic anatomy and physiology, 742
microscopic anatomy and physiology, 742–47
summary and assessment, 748–49
Locus, 190
Loggerhead sea turtle, 823, 887
Logistic growth, 842–44, 846
Long bone
anatomy, 736–38
compact bone, 736–38
spongy bone, 736, 738
endochondral ossification, 736
Long-day plant, 490
Longevity, 811–12
Longhorn beetle, 862
Longitudinal fissure, 703
Long-term memory, 705
Long-term stress response, 760f, 761
Loop of the Henle. *See* Loop of the nephron
Loop of the nephron, 682, 683f, 685
Loose fibrous connective tissue, 584
Lophophoran, 524
Lophophore, 524
Lophotrochozoa
characteristics, 520t, 524
diversity, 524–32
lophophorans, 524
trochozoans, 524–32
phylogeny, 516t
Lorica, 381
Loriciferan, 660
Lou Gehrig disease. *See* ALS (amyotrophic lateral sclerosis)
Louse, head, 536f
Lovebird, 818–19
Low-density lipoprotein (LDL), 654
Lower respiratory tract disorder, 669–73
lungs, 670–73
cystic fibrosis (CF), 196, 293, 294, 672–73
emphysema, 670f, 672
lung cancer, 672
pneumonia, 208, 633, 670–71, 672
pulmonary fibrosis, 670f, 672
pulmonary tuberculosis, 670f, 672
trachea and bronchi, 669–70
asthma, 636, 670
bronchitis, 669–70
choking, 669
chronic obstructive pulmonary disease (COPD), 670
Lub-dub sound, 608
LUCA (last universal common ancestor), 318, 341
Lucy *(Australopithecus afarensis)*, 571, 572f
Lucy's baby *(Australopithecus afarensis)*, 571
Lumbar vertebra, 739
Lumbricus terrestris
anatomy, 530–31
closed circulatory system, 602
complete digestive tract, 642–43
excretory organs, 679
hydrostatic skeleton, 734
nervous system, 692
respiration, 660, 661f
soil, 460
Lumen, 646
Lung. *See also* Pulmonary circuit
amphibians, 551, 661
arthropods, 534
artificial, 671
bird, 661
chemoreceptors, 717
human, 661, 663–64, 670–73, 687
lower respiratory tract disorders
cystic fibrosis (CF), 196, 293, 294, 672–73
pneumonia, 633, 670–71, 672
mammals
human, 89f
reptiles, 553
respiratory tract disorders, 670–73
cystic fibrosis (CF), 672–73
emphysema, 670f, 672
lung cancer, 672
pneumonia, 670–71
pulmonary fibrosis, 670f, 672
pulmonary tuberculosis, 670f, 672
Lung, book, 534
Lung cancer, 672
Lungfish, 550
Lunula, 593
Lupus, 637
Luteal phase, 779
Luteinizing hormone (LH), 765, 776, 778–79
Lycophyte
evolution and colonization of land, 413t
evolution and vascular tissue, 412, 417–19, 421
Lycopodium, 419
Lyell, Charles, 264
Lyme disease, 848
Lymnaea, 753. *See also* Snail
Lymph
capillary exchange, 616
definition, 586, 624
Lymph gland. *See* Lymph node
Lymph node, 586, 623f, 624
Lymph nodule. *See* Peyer patch
Lymphatic and immune system, 621–40
adaptive immune defenses, 628–34
active vs. passive immunity, 632–34
B cells and antibody-mediated immunity, 628–30
evolution, 622–23
T cells and cell-mediated immunity, 630–32
evolution of immune systems, 622–23
cellular slime molds, 622
Drosophila, 622
rise of adaptive immunity, 622–23
immune system disorders and adverse reactions, 636–38
allergies, 636–37
autoimmune diseases, 637
immunodeficiencies, 636
transplant rejection, 638
innate immune defenses, 622, 625–28
inflammatory response, 625–27
phagocytes and natural killer cells, 625, 627
physical and chemical barriers, 625
protective proteins, 625, 627–28
lymphatic system, 623–25
homeostatic functions, 623
lymphatic vessels, 623–24
lymphoid organs, 624–25
summary and assessment, 639–40
Lymphatic capillary
anatomy, 623–24
function, 617f
small intestines, 648
Lymphatic duct, 623f, 624
Lymphatic vessel, 623–24
Lymphocyte, 615. *See also* B cell; T cell
development, 624
function, 586
natural killer (NK) cells, 627
primary lymphoid organs, 624
Lymphoid organ
primary
red bone marrow, 585, 614, 615, 623f, 624, 737
thymus, 623f, 624, 752f, 766
secondary, 624–25
lymph nodes, 623f, 624
Peyer patches, 625, 646
spleen, 623f, 624, 625
tonsils, 623f, 625
vermiform appendix, 625
Lymphoma, Burkitt, 182
Lynx, Canada, 861
Lynx canadensis, 861
Lysergic acid, 404
Lysogenic cell, 356–57
Lysogenic cycle, 356–57
Lysosome, 70–71, 642
Lysozyme, 356
Lystrosaurus group, 332
Lytic cycle, 356, 357, 357f

M

M (miotic) stage, 148–49
Maasai, 575f, 577
MAC (*Mycobacterium avium* complex), 633
Macaque, 228
MacArthur, Robert, 858, 860
Macaw, 558f, 891f, 911, 912f
Macaw, blue-and-yellow, 558f, 891f
MacLeod, Colin, 209
Macrocystis, 380, 381
Macroevolution and speciation, 296–316
anatomy of speciation, 300f
definition, 297, 303
how new species evolve, 297–303
reproductive isolating mechanisms, 299–303
what is a species, 297–99
modes of speciation, 303–8
adaptive radiation, 306–7
allopatric speciation, 303–4, 305f
convergent evolution, 307
sympatric speciation, 304–5
phylogenetic species concept, 300
principles of macroevolution, 308–12
developmental genes and macroevolution, 310–12
macroevolution is not goal-oriented, 312
Macronutrient, 457
Macrophage
functions, 627
inflammatory response, 626
monocytes, 615
Macrospore, 419, 431f
Macular degeneration, 555, 722
Mad cow disease, 360
Madreporite, 539f, 540
Magnification, 61
Maiasaura, 317
Maize. *See* Corn
Major histocompatibility complex (MHC) protein. *See* MHC protein
Malaria, 384, 865
Male competition, 288–89
Male gametophyte, 499
Male reproductive system, human, 773–76
aging process, 811
hormonal regulation, 776
penis and male orgasm, 773f, 775–76
production of sperm, 773, 774–75
teste
cancer, 774
functions, 765–66, 771, 773–74
gonadotropic hormones, 755
major glands, 752f
steroid hormones, 753–54
Malignant cell growth, 158
Malleus, 725
Malpighian tubule, 536, 537f, 679
Maltase, 651
Malthus, Thomas, 264, 844
Malthusian growth, 844
Maltose, 40f, 651
Mambalgin, 555
Mammal
body cavities, 590
Cenozoic era, 331
characteristics, 520t, 559
ear, 725
hair, 559, 591
internal development, 559
internal organs, 559
mammary glands, 559
skeleton, 559
skull, 549
teeth, 549
circulatory pathway, 551f, 559
diapsid skull, 553
diet adaptations, 644
evolution, 546t, 559–61
marsupials, 272–73, 332–33, 559t, 560
monotremes, 559t, 560
nervous system, 694
placental mammals, 559t, 560–61, 772
hormones and endocrine system, 753
Mesozoic era, 330
middle ear, 725
nervous system organization, 694
osmoregulation in marine mammals, 680–81
respiration, 665–66
Mammal, pouched. *See* Marsupial evolution
Mammary duct, 780, 781f
Mammary gland
mammal characteristic, 559
prolactin (PRL), 755
Mandible
crustaceans, 534
human, 739
Manduca sexta, 750
Mangrove swamp, 898, 911
Manta ray, 549
Mantle, 528, 530f
Mantle cavity, 528
Maple, 504, 505f
Marchantia, 416
Marfan syndrome, 198, 199f
Marigold, marsh, 889f
Marijuana, 701
Marine mammal osmoregulation, 680–81
Marine snail, 819–20
Marrella, 328f
Marsh, 895
Marsh marigold, 889f
Marshall, Barry, 647
Marsupial evolution, 272–73, 332–33, 559t, 560
Masculinization, 762
Mass extinction, 327–28, 329, 330, 333–34, 554
Mass number, 21
Mast cell, 626, 636f
Mastoid sinus, 738
Mastoiditis, 738
Material and energy requirements, 4
Maternal care behavior, 820
Maternal determinant, 798
Mating and genetic diversity, 285. *See also* Courtship; Reproductive strategies and fitness
Matrix
connective tissue
blood, 586
bone, 585
cartilage, 584–85
mitochondria, 74

Matter, 20
Maturation, viral, 356, 357*f*, 358
Maxilla
  crustaceans, 534
  human, 739
Maxillary gland, 679
Mayfly, 842, 895*f*
Mayr, Ernst, 303
MC1R gene, 207
McCarty, Maclyn, 209
McClintock, Barbara, 252
MDC (more-developed country), 849*f*, 850–52
MDMA (methylenedioxymethamphetamine), 700
Mealybug, 536*f*
Measles, German, 805
Mechanical digestion, 645
Mechanical energy, 101
Mechanical isolation, 302
Mechanical weathering, 459
Mechanoreception, 724
Mechanoreceptor, 715, 725, 729
Mediastar, 539*f*
Medicine
  horseshoe crabs, 617
  medicinal value of biodiversity, 908
  regenerative, 588–89
  from vertebrates, 555
  vertebrates and human medicine, 555
Medulla oblongata, 693, 704
Medullary cavity, 736
Medusa, 522
Meerkat, 832
Megafauna, 331
Megakaryocyte, 615
Megaphyll, 419–20, 422
Megaspore
  angiosperms, 428, 496, 497, 499
  gymnosperms, 425f
Meiosis and sexual reproduction, 166–85
  changes in chromosome number and structure, 177–82
    aneuploidy, 177–80
    changes in chromosome structure, 180–82
  cycle of life, 176–77, 178f
  genetic variation, 169–70
    genetic recombination, 169–70
    independent assortment of homologous chromosomes, 170
    significance of genetic variation, 170
  meiosis compared to mitosis, 174–75
  meiosis overview, 167–69
    fate of daughter cells, 169
    homologous chromosomes, 167–68
    meiosis is reduction division, 168–69
  meiosis phases, 172–73
  overview, 167–69
  parthenogenesis, 171, 771
  spermatogenesis and oogenesis, 176–77, 178f, 774f, 775, 778, 778f
  summary and assessment, 183–85
Meiosis I, 173*f*, 174*f*, 175, 774*f*, 778*f*
Meiosis II, 172, 173*f*, 175, 774*f*, 775, 778*f*
Meissner corpuscle, 729
Melanin
  ACTH (adrenocorticotropic hormone), 762
  hair color, 594
  skin cancer, 593
  skin color, 207, 592
Melanocyte, 199, 592, 762, 809
Melanocyte-stimulating hormone (MSH), 755–56
Melanoma, 147, 593
Melanophore, 755–56
Melatonin, 704, 766
Membrane attack complex, 627
Membrane potential, 696
Membrane structure and function, 82–99. *See also* Plasma membrane
  active transport across a membrane, 91–94
    bulk transport, 93–94
  cell surface modification, 95–96
    animal cell, 95–96
    plant cell, 96
  passive transport across a membrane, 88–91
    facilitated transport, 91
    osmosis, 88–91
  plasma membrane structure and function, 62, 83–86
    fluid-mosaic model, 83f, 84
    glycoproteins and glycolipids, 84–85
    permeability, 86
    plasma membrane components, 83–84, 320–21
    protein function, 85
  summary and assessment, 97–99
Membrane-first hypothesis, 321
Memory
  brain, 705
  sense of smell, 717–18
Memory B cell, 629, 634
Memory T cell, 631
Menarche, 780
Mendel, Gregor, 187. *See also* Mendelian patterns of inheritance
Mendeleev, Dmitri, 21
Mendelian patterns of inheritance, 186–206
  beyond Mendelian inheritance, 197–204
    incomplete dominance and incomplete penetrance, 197–98
    multiple allelic traits, 197–98
    pleiotropic effects, 198–99
    polygenic inheritance, 199–201
  blending concept of inheritance, 187
  Mendel's laws, 188–94
    law of segregation, 188–90, 192f
    Mendel and the laws of probability, 192–93
    Mendel's cross, 190–91
    Mendel's law of independent assortment, 170, 191–92
    testcrosses, 193–94
  Mendel's patterns of inheritance and human disease, 194–97
    autosomal dominant disorder, 195f, 196–97
    autosomal recessive disorder, 195–96, 195f
  particulate theory of inheritance, 187
  summary and assessment, 204–6
Mendel's cross, 190–91
Mendel's law, 188–94
  law of segregation, 188–90
  Mendel and the laws of probability, 192–93
  Mendel's law of independent assortment, 191–92
Meniere disease, 728
Meninge, 702
Meningitis, 702
Meniscus, 741
Menkes syndrome, 202
Menopause, 780, 811
Menstrual cycle. *See* Uterine cycle
Menstrual phase, 779, 780
Menstruation, 753, 779, 780
Mercury, 612, 901
Meristem cell, 436, 508*f*, 509. *See also* Apical meristem
Merkel disk, 729
Merychippus, 312
Mesocarp, 504, 505*f*
Mesoderm
  coelom development, 517, 796
  embryonic development, 795f, 796, 799f
  organ development, 797
Mesoglea, 522
Mesopelagic zone, 899–900
Mesophyll tissue, 451
Mesosome, 62
Mesozoic era, 326*t*, 329–30, 332*f*
Messenger RNA (mRNA)
  codons, 217f
  functions, 50, 68
  pre-messenger RNA, 235f
  processing, 218–20
  production, 217–18
  splicing, 219
  steroid hormones, 754
  transcript, 217
Metabolic pathway, 105
  energy of activation, 106
  enzymatic speed, 106–7
  enzyme inhibition, 108–9
  enzyme-substrate complex, 105–6
Metabolic pool, 141
Metabolic rate, human, 758
Metabolic reactions
  energy transformation, 103–4
  life requirement, 100
Metabolic waste, 685. *See also* Body fluid regulation; Excretory system
Metabolism
  anabolism, 141–42
  bacteria, 363–65
    autotrophic, 365
    heterotrophic, 365
    oxygen, 364–65
  catabolism, 141
  definition, 4, 103
  energy and enzymes, 100–113
    cells and the flow of energy, 101–2
    metabolic pathways and enzymes, 105–9
    metabolic reactions and energy transformation, 100, 103–4
    oxidation-reduction reactions and metabolism, 109–10
    summary and assessment, 111–13
  energy organelles, 142
Metacarpal bone, 740
Metagenomics, 256
Metamorphosis
  amphibians, 551, 753
  arthropods, 534, 536, 537, 750, 751
Metaphase, 153
Metaphase I, 172
Metaphase plate, 153
Metapopulations, 916–17
Metastasis, 159
Metatarsal bone, 741
Meteorite
  extraterrestrial origins hypothesis, 320
  mass extinction, 333–34
Methadone, 701
Methamphetamine, 700
Methane
  cattle, 641
  greenhouse effect, 874
  methanogens, 368
Methanogen, 344, 368–69
Methanogenesis, 368
Methemoglobinemia, 195–96
Methicillin-resistant *Staphylococcus aureus* (MRSA), 279, 366
Methylenedioxymethamphetamine (MDMA), 700
Methylmercury, 901
MG (myasthenia gravis), 637, 709
MHC protein
  natural killer (NK) cells, 627
  pheromones, 753
  T cells, 630, 631f
  transplant rejection, 638
Micelle, 321
Microbody, 72–73
Microevolution, 280–85, 300
Microglium, 587
Microhyla nepenthicola, 296
Micronutrient, 457
Microorganism
  importance, 352
  phylogenetic species concept, 298
Microphyll, 419, 420*f*
Microraptor, 317
MicroRNA, 219, 236
Microscopy, 60–61
Microsphere, 320
Microsporangium, 425*f*, 431*f*
Microspore
  angiosperms, 428–29, 431f, 496, 497–99
  gymnosperms, 425f
  vascular plants, 419
Microsporidia, 396, 398–99
Microtubule, 75–77
Microvillus, 648
Midas cichlid, 304
Midbrain, 693, 704
Middle ear, 725
Midwife toad, 847*f*
Mifepristone, 781–82
Migration
  birds, 558, 823, 881
  definition, 823
  gene flow, 284
  hominins, 572
  orientation and migratory behavior, 823
  population size, 846
Milk. *See* Lactation
Milkweed, 504
Miller, Stanley, 319
Miller-Urey experiment, 319
Millipede and centipede, 535–36
Mimicry
  molecular mimicry and autoimmune diseases, 637
  orchids, 429, 500
  prey defense, 862
Mimosa pudica, 488–89
Mineral
  cohesion-tension model of xylem transport, 468
  definition, 457
  discovery with plants, 463
  human diet and nutrition, 656
  phytoremediation, 462–63
  plant nutrition and uptake, 457–58, 461–64
Mineralocorticoid, 760*f*, 761, 762
Miocene epoch, 331
Miotic (M) stage, 148–49
Mississippi River, 836
Mistletoe, 8–9
Mite, 537
Mitochondrial disease, 75
Mitochondrial DNA (mtDNA)
  replacement model of human evolution, 574, 577
  tracing phylogeny, 349
Mitochondrial Eve hypothesis, 574
Mitochondrium
  cellular respiration, 110, 136–40
    citric acid cycle, 137–38
    electron transport chain, 138–40
    preparatory reaction, 136–37
  cholorplasts, 110
  citric acid cycle, 137–38
  definition, 73–75
  electron transport chain, 138–40
  endosymbiosis, 374
  functions, 136–40
  origin of life, 322, 325, 327
  preparatory reaction, 136–37
Mitosis
  function, 156
  human sperm production, 774f
  meiosis compared, 174–75

phases, 153–54, 155f
preparation, 152–53
protists, 374, 377
Mitotic spindle, 149
Mitotic stage, 148–49
Mitral atrioventricular valve, 607*f*
Mixotroph, 374
euglenids, 385
golden brown algae, 381
Mode of speciation, 303–8
adaptive radiation, 306–7
allopatric speciation, 303–4, 305f
convergent evolution, 307
sympatric speciation, 304–5
alloploidy, 305
autoploidy, 305
polyploidy, 305
Model organism
Arabidopsis, 488
definition, 10
fruit fly, 10, 798
roundworms, 532, 798, 800
sac fungi, 402
Model system, 10
Molar, 644
Mold
bread
Neurospora crassa mold, 402
Rhizopus stolonifer mold, 399
food industry, 402
sac fungi, 400–401
Mold, slime. *See* Slime mold
Mole (energy release), 103
Mole (skin), 592, 593
Molecular biology of the gene, 207–27
genetic code of life, 215–17
the genetic code, 216–17
RNA carries the information, 215–16
genetic material, 208–12
DNA structure, 50–52, 209–12, 215t
transformation of bacteria, 208–9
replication of DNA, 213–15
complementary base pairing, 213–14
joining, 214
prokaryote vs. eukaryote, 214–15
unwinding, 213
summary and assessment, 226–27
transcription, 217–20
messenger RNA production, 217–18
RNA processing, 218–20
translation, 220–25
gene expression, 225
ribosomal RNA, 221–22
steps, 222–25
transfer RNA, 220–21
Molecular clock, 328, 349
"Molecular fossil," 129
Molecular mimicry, 637
Molecular trait, 348–49
Molecule
amphipathic, 83–84
biomolecule, 36, 38, 318
chemistry, 24–25
definition, 2, 24
energy storage, 41
hydrophilic, 28
hydrophobic, 28
motor, 75
organic, 36–38
biomolecules of cells, 38
carbon atom, 36–37
carbon skeleton and functional group, 37–38
origin of life, 318, 319
structural, 41
water, 26f
Mollusc, 527–30
anatomy, 528
bivalves, 528–29
cephalopods, 529–30
characteristics, 520t
digestive systems, 643
exoskeleton, 734
gastropods, 529, 530f
nervous system, 692–93f
omnivorous diet, 644
profundal and littoral zone, 897f
respiratory system, 661
Molly. *See* MDMA (methylenedioxymethamphetamine)
Molothrus ater, 865
Molting
arthropods, 533f, 534, 734, 735f, 751
birds, 557
reptiles, 553
Monarch butterfly, 823
Monera, 341
Mongoose, 914
Monitor lizard, 556
Monkey
auditory communication, 826
evolution, 566f, 567f
New World, 565f, 566
Old World, 565f, 566
phylogeny, 345, 348f
Monkey, vervet, 826
Monkey, white-faced, 565*f*
Monoclonal antibodies, 630, 781
Monocot
evolution and full adaption to land, 428
features, 428t, 441t
flowers, 497
herbaceous stems, 446
organs and organ systems, 440–41
root system, 443–44
seed germination, 506–7
woody plants, 447
Monocotyledone. *See* Monocot
Monocyte, 615, 626
Monod, Jacques, 229
Monoecious animal, 525, 771
Monoecious plant
flower, 497, 498f
gymnosperms, 423
Monogamy, 828, 829
Monohybrid cross, 188*f*
Monomer
cell structure, 38
origin of life, 318, 319–20
Monophyly
phylogenetic species concept, 298, 300
protists, 374
reptiles, 553
Monosaccharide, 39
Monosomy, 178
Monotreme, 559*t*, 560
Mons pubis, 778
Monsoon climate, 883
Montane coniferous forest, 884, 885*f*, 888
Moose, 888
More-developed country (MDC), 849*f*, 850–52
Morel mushroom, 400*f*, 401*f*
Morgan, Thomas Hunt, 201–2
Morning glory, 492
Morning-after pill, 781–82
Morphogen, 800
Morphogenesis, 798, 800–802
Morphological species concept, 297
Morphological trait, 347–48
Morphology
definition, 297
evolutionary species concept, 297
morphological species concept, 297
morphological trait, 347–48
Mortality patterns, 839–40
Morula, 795, 804
Mosaic evolution, 571
Mosquito
malaria, 384
survivorship, 840
Moss
ecological succession, 866f
evolution, 412
gametophytes, 414, 415f, 416–17
glaciers, 418
reproduction, 416–17
traits, 415, 416
Moth
communication, 824
genetic basis of body color, 270, 280–83
metamorphosis, 750
pollination, 500–501
Moth, peppered, 270, 280–83
Moth, sphinx, 750
Mother of pearl, 528
Motility, 514
Motor molecule, 75
Motor neuron, 695–96
Motor output and control, 699, 703
Mountain
alpine tundra, 884, 885f, 886
climate, 883
montane coniferous forest, 884, 885f, 888
Mouse
biotic potential, 839f
Linnaean classification hierarchy, 340
model organism, 10
nurturing behavior, 820
pattern formation, 801f
predation and population size, 845, 846f
transgenic, 249
vertebrae and body plan, 518
Mouse, house, 340
Mouse, knockout, 249
Mouth
blastospore, 517
digestive tracts, 642
earthworms, 642, 643
human, 645–46
incomplete digestive tracts, 642
human, 645–46
palate, 645–46
salivary glands, 645f, 646
uvula, 646
lophophore, 524
Mouthbrooder, 772, 847*f*
MPTP, 706
MRNA (messenger RNA). *See* Messenger RNA (mRNA)
MRNA transcript, 217
MRSA (methicillin-resistant *Staphylococcus aureus*), 279, 366
MS (multiple sclerosis), 637, 704
MSH (melanocyte-stimulating hormone), 755–56
MtDNA (mitochondrial DNA)
replacement model of human evolution, 574, 577
tracing phylogeny, 349
Mucosa, 646
Mucous gland, 591
Mucus
digestive tract, 646
innate immune defense, 625
respiratory system, 663, 664f
Mudflat, 898
Mulberry, 462
Mule, 302*f*, 303
Müllerian mimic, 862
Mullis, Kary, 246
Multicellular organism, 2
Multifactorial trait, 200–201
Multiple fruit, 504, 505*f*
Multiple sclerosis (MS), 637, 704
Multipolar shape, 695, 696
Multiregional continuity hypothesis, 574
Mus musculus
biotic potential, 839f
Linnaean classification hierarchy, 340
model organism, 10
nurturing behavior, 820
pattern formation, 801f
predation and population size, 845, 846f
transgenic, 249
vertebrae and body plan, 518
Muscaridine, 404
Muscarine, 404
Muscle contraction
adenosine triphosphate (ATP), 104f, 744–45, 746–47
sliding filament model, 743–45
Muscle fiber, 586, 743, 744–45, 746–47
Muscle mass, 810
Muscle relaxation, 747
Muscle spindle, 729, 742
Muscle tone, 729, 742
Muscular dystrophy, 202–4
Muscular hydrostat, 734
Muscular system, human, 742–47
aging process, 810–11
antagonistic pairing of muscles, 742, 743f
macroscopic anatomy and physiology, 742
microscopic anatomy and physiology, 742–47
ATP use in contraction, 104f, 744–45, 746–47
muscle innervation, 745–46
role of calcium in muscle contraction, 746–47
sliding filament model, 743–45
muscular tissue, 586–87
cardiac muscle, 586, 587, 608
skeletal muscle, 586, 742–47
smooth muscle, 586
tissue types, 582
superficial musculature, 742f
Muscular tissue, 586–87
cardiac muscle, 586, 587, 608
skeletal muscle, 586, 742–47
smooth muscle, 586
tissue types, 582
Muscularis, 646
Mushroom
Amanita, 218, 404
Basidiomycota, 402–4
Boletus, 403f
death cap, 404
ecosystem dynamics, 868f, 869
hallucinogenic, 404
morel, 400f, 401f
poisonous, 218, 402, 404
puffball, 402, 403f
Mussel
anatomy, 528, 529f
exotic species, 912
Mussel, zebra, 912
Mutation
causes, 238–41
emerging viruses, 358–60
evolution, 312
genetic diversity, 5, 283–84
human chromosomal, 181f
prokaryotes, 363
Mutualism
bacteria, 365, 367, 864
cross-pollination, 500–501
definition, 864
ecology of communities, 863t, 864
fungi, 401, 406–7, 444, 462, 464, 864
lichen, 367, 405, 864
mycorrhizae, 406–7, 444, 462, 464, 864
Myasthenia gravis (MG), 637, 709
Mycelium, 397, 402
Mycobacterium avium complex (MAC), 633
Mycobacterium tuberculosis, 670*f*, 672
MycoBond, 395
Mycorrhiza, 406–7, 444, 462, 464, 864
Myelin
axons, 587, 695
Guillain-Barré syndrome (GBS), 709

multiple sclerosis, 637
nerve fibers, 587
oligodendrocytes, 587, 695
Mylodon darwinii, 265
Myocardial infarction. *See* Heart attack
Myocardium, 605
Myofibril, 743, 746
Myoglobin, 745
Myopic vision, 724
Myosin filament, 586, 743, 744–45, 746–47
Myrica faya, 913–14
Myrtle tree, 913–14
Mysis relicta, 907, 912
Mytilus edulis, 528, 529*f*
Myxedema, 759

N

NAD+ (nicotinamide adenine dinucleotide), 107, 130–31
NADP+ (nicotinamide adenine dinucleotide phosphate), 107, 117
Naegleria fowleri, 373, 386–87
Nail, human, 592
Naked virus, 354*f,* 355*f,* 356, 357, 358
Name, common, 339
Name, scientific, 339
Nasal bone, 738, 739, 739*f*
Nasal septum, 739
Nasopharynx, 646
Nastic movement. *See* Turgor movement
Natriuresis, 761
Natural group, 338
Natural killer (NK) cell, 627
Natural selection
adaption, 267–68
characteristic of life, 6–7
populations, 285, 286–92
Tree of Life, 272
"Nature *vs.* nurture" question, 818, 820
Nautilus, chambered, 530*f*
Nautilus belauensis, 530*f*
Navigation, 823
NE (norepinephrine), 698, 709, 710*f,* 760
Neandertal
characteristics, 574
extinction, 576
genome, 564
tool use, 576
Nearsightedness, 724
Nectar, 429, 500, 501
Nectar guide, 501, 718, 719*f*
Neem, 435
Negative feedback
blood concentration, 755
homeostasis, 595, 596f, 755
ovarian cycle, 779
production of sperm and testosterone, 776
Negative tropism, 485
Neglected tropical disease (NTD), 388
Neisseria gonorrhoeae, 788–89
Nematocyst, 522, 523
Nematode. *See* Roundworm
Neocortex, 694
Neogene period, 330–31
Neotrogla, 770
Nepenthes ampullaria, 296
Nephridiopore, 679*f*
Nephridiostome, 679*f*
Nephridium, 530, 679
Nephron, 681–82, 683*f*
Nereis, 531, 532
Nereocystis, 380
Nerve, 587, 707. *See also* Neuron
Nerve cell. *See* Nervous tissue; Neuron
Nerve cord, 545, 806–7
Nerve fiber. *See* Axon
Nerve impulse transmission, 696–98
action potential, 696–97
propagation of action potentials, 697
resting potential, 696
transmission across a synapse, 697–98
Nerve net, 523, 692
Nervous system, 691–713. *See also specific components*
aging process, 810
bird characteristic, 558
breathing, 667
central nervous system, 694, 699–706. See also Brain; Spinal cord
embryonic development, 799f, 806–7
evolution of nervous system, 692–94
invertebrates, 692–93
mammals, 694
vertebrates, 693–94
functions, 587
ladder-type, 525
mode of action, 696–98, 751
nervous tissue, 695–99
anatomy, 587–89, 695–96
functions, 587
neuroglia, 587–88, 695
neurons, 587, 695–96
synaptic integration, 699
tissue types, 582
transmission of nerve impulses, 696–98
neurons
anatomy, 695
components, 587
definition, 587, 695
types, 695–96
peripheral nervous system, 694, 707–10
anatomy, 707
autonomic system, 694f, 708–10
somatic system, 694f, 707, 708, 709t
summary and assessment, 711–13
Nervous tissue, 695–99
anatomy, 587–89, 695–96
functions, 587
neuroglia, 587–88, 695
astrocytes, 587, 695
microglia, 587, 695
oligodendrocytes, 587, 695
neurons, 587, 695–96
axon, 587, 695
cell body, 587, 695
definition, 587, 695
dendrites, 587, 695
synaptic integration, 699
tissue types, 582
transmission of nerve impulses, 696–98
action potential, 696–97
propagation of action potentials, 697
resting potential, 696
transmission across a synapse, 697–98
Nest
fish, 772
lovebirds, 818–19
Net reproductive rate, 842
Neural crest, 797
Neural plate, 797, 799*f*
Neural tube
development, 797
human, 805, 806–7
Neuraminidase spike, 359
Neurodegenerative disease, 360
Neuroglia, 587–88
astrocytes, 587, 695
ependymal cell, 695
functions, 695
microglia, 587, 695
node of Ranvier, 695
oligodendrocytes, 587, 695
satellite cell, 695
Schwann cell, 695
Neuromuscular junction, 745
Neuron, 691–713. *See also* Nervous system
anatomy, 695
central nervous system, 694, 699–706. See also Brain; Spinal cord
components, 587
definition, 587, 695
embryonic development, 799f, 806–7
evolution of nervous system, 692–94
functions, 587
ladder-type, 525
mode of action, 696–98, 751
nervous tissue, 695–99
anatomy, 587–89, 695–96
functions, 587
neuroglia, 587–88, 695
neurons, 587, 695–96
synaptic integration, 699
tissue types, 582
transmission of nerve impulses, 696–98
summary and assessment, 711–13
types, 695–96
Neurospora crassa, 402
Neurotransmitter, 697–98, 751
Neurula, 797
Neurulation, 799*f*
Neutron, 20–21
Neutrophil, 615, 627
New Guinea singing dog, 544
New World monkey, 565*f,* 566
Newt, 551–52
Niche, ecological, 857–59
character displacement, 858
competitive exclusion principle, 857
niche specialization, 858, 859f
realized niche, 857
Niche specialization, 858, 859*f*
Nickel, 458
Nicotiana attenuate, 485
Nicotinamide adenine dinucleotide (NAD+), 107, 130–31
Nicotinamide adenine dinucleotide phosphate (NADP+), 107, 117
Nicotine
abuse, 700
defense mechanism, 485
vaping, 673
Night-heron, black-crowned, 919*f*
Nine-banded armadillo, 908
Nitrate
nitrification, 876
NPK ratio of fertilizers, 458
Nitrification, 876
Nitrite, 876
Nitrogen
cycle, 873, 875–78
NPK ratio of fertilizers, 458
Nitrogen cycle, 873, 875–78
Nitrogen fixation
archaea, 369
bacteria, 365, 367, 460, 462, 463f, 481, 876
cyanobacteria, 367, 876
cytokinins, 481
root nodules, 444, 462, 463f
Nitrogen oxide, 878, 913
Nitrogenous waste product
ammonia, 678, 688
urea, 678, 685
uric acid, 678, 679
Nitrous oxide, 874
NK (natural killer) cell, 627
Noble gas, 21
Nociceptor. *See* Pain receptor
Nodal tissue, 608
Node (phylogeny), 300
Node (plant)
leaf, 451f
rhizomes, 507–9
shoot system, 440, 445
stolons, 507
stoneworts, 412
Node of Ranvier, 695, 697
Nomenclature
binomial, 8–9, 339
standardization, 340
Noncompetitive inhibition, 108, 109*f*
Noncyclic pathway, 119, 120*f*
Nondisjunction, 177
Nonfunctional protein, 240
Nonliving thing, 2
Nonpolar covalent bond, 25
Nonrandom mating, 285
Nonsteroidal anti-inflammatory drugs (NSAIDS), 730
Nonvascular plant, 415, 421. *See also* Bryophyte
Nonylphenol, 913
Noradrenaline. *See* Norepinephrine (NE)
Norepinephrine (NE), 698, 709, 710*f,* 760
Northern pike, 897*f*
Northern spotted owl, 888, 917
Nosema ceranae, 513
Notochord
chordates, 545
development, 797, 798, 799f
replacement with vertebrae, 547, 797
NPK ratio, 458
NSAIDS (nonsteroidal anti-inflammatory drugs), 730
NTD (neglected tropical disease), 388
Nucifraga columbiana, 864
Nuclear envelope, 68
Nuclear pore, 68
Nucleariid, 391
Nucleic acid, 50–52
ATP, 52
DNA and RNA structure, 50–52
virus categorization, 355–56
Nucleoid, 62, 161, 361, 362*f*
Nucleolus, 68
Nucleoplasm, 67
Nucleosome, 233
Nucleotide, 50, 51*f,* 210*f*
Nucleotide sequence, 348–49
Nucleus
cancer cell, 158
cell structure, 67–68
Nummulites, 391
Nurse (army ant), 831*f,* 832
Nurturing behavior, 820
Nutrient
availability, cell division, 150
cycling, 369. See also Nitrogen fixation
essential
human, 653
plant, 457–58
Nutrition, 652–58
human health and, 652–56
carbohydrates, 652–53
diet and obesity, 654–55
lipids, 653
proteins, 653–54
vitamins and minerals, 654–56
protocell evolution, 322
summary and assessment, 657–58
Nutrition, flowering plant, 456–75
plant nutrition and soil, 457–60
essential inorganic nutrients, 457–58
soil, 457, 459–60
transport mechanisms, 465–73
organic nutrient transport, 470–73
reviewing xylem and phloem structure, 465
stomata opening and closing, 469–70
water potential, 465–68
water transport, 468–69
water and mineral uptake, 461–64
mineral uptake, 461–64
water, 461

O

Obesity
adipose tissues, 584
carbohydrates, 653
definition, 654
high-fructose corn syrup (HFCS), 653

leptin, 766
treatment approaches, 655
Obligate anaerobe
bacteria, 364, 365
methanogens, 368
Obligate intracellular parasite, 354, 356
Observation
learning by, 823
scientific method, 9
Occipital bone, 738, 739*f*
Occipital lobe, 703
Ocean
currents, 900–902
ecosystems, 895f, 899–902
abyssal plain, 898f, 899f
euphotic zone, 898f, 899
pelagic zone, 899–900
temperature, 883
water cycle, 873, 895
Ochromonas, 381
Octet rule, 23
Octopus, 770
Odocoileus virginianus, 848
Oedogonium, 413*f*
Oenothera, 719*f*
Oil (lipid), 42
Oil consumption, 851–52
Oil gland, 593, 625
Okazaki fragment, 214
Old World monkey, 565*f,* 566
Olfactory bulb, 693, 704, 717–18
Olfactory cell, 717
Olfactory organ
Jacobson's organ, 556
vertebrate nervous system organization, 693
Oligocene epoch, 331
Oligochaete annelid, 530
Oligodendrocyte, 587, 695
Oligotrophic lake, 896
Oligozoospermia, 776
Omega-3 fatty acid, 612, 653, 654
Omnivore
diet adaptation, 643–44
ecosystem dynamics, 868–69, 870f, 871
On the Origin of Species (Darwin), 272
Oncogene, 159, 241*f*
One-circuit circulatory pathway. *See* Single-loop circulatory pathway
One-trait testcross, 193, 194*f*
One-way ventilation mechanism, 666
Oocyte, 176, 777, 778–79, 784, 794–95, 804
Oogenesis, 176–77, 178*f,* 778
Oogonium, 382
Oomycete. *See* Water mold
Oparin, Alexander, 321
Oparin-Haldane hypothesis, 319
Open circulatory system, 601, 602*f*
Operant conditioning, 822–23
Operator, 229
Operculum, 550, 662
Operon, 229, 230–31
Ophiuroidea, 540
Ophrys
aerial roots, 444
mycorrhizae, 464
pollination, 429, 500
rain forest biodiversity, 891f
sexual mimicry, 2, 429, 500
Opiate, 701
Opioid medications, 730
Opisthokonta supergroup, 374, 375*t,* 390–91
Opossum, Virginia, 560
Opossum shrimp, 907, 912
Opportunistic infection, 384, 632, 633, 672, 785
Opportunistic species, 846, 847*f*
Opposable thumb
derived trait, 344–45
human, 740
primate evolution, 565, 566f
Opsin, 720, 721*f*
Optic chiasma, 722
Optic lobe, 693
Optic nerve, 720, 722
Optic tract, 722
Optimal foraging model, 829
Opuntia, 452*f*
Oral groove, 374
Oral sex and sexually transmitted disease, 787, 788
Oral thrush, 402
Orangutan, 565*f*
Orca, 825, 916
Order, 7, 339
Ordovician period, 326*t,* 329, 333, 334
Organ
aging processes, 809–11
animal
organization, 2, 589–90
vertebrate characteristic, 547
xenotransplantation, 555, 638, 764–65
body cavities, 590
definition, 590
formation
cytokinin, 481, 482f
embryonic development, 806–7
stages of development, 797
plants
definition, 436
flowering plant life cycle, 441
flowering plant structure, 440–41
Organ of Corti, 726, 728
Organ system
animal organization, 2, 589–90
body cavities, 590
definition, 590
homeostasis, 594
life processes, 590t
Organelle
amoeboid, 374
cell structure, 64, 73, 74f
energy organelle, 142
Organic chemical as extinction cause, 913
Organic chemistry, 35–56
carbohydrate, 39–42
disaccharide, 39–41
monosaccharide, 39
polysaccharide, 41–42
definition, 36
lipid, 42–46
phospholipid, 44–45
steroid, 45–46
triglyceride, 42–44
wax, 46
nucleic acid, 50–52
ATP, 52
DNA and RNA structure, 50–52
organic molecule, 36–38
biomolecules of cells, 38
carbon atom, 36–37
carbon skeleton and functional group, 37–38
protein, 46–50
amino acid, 47–48
folding, 50
structure, 48–50
summary and assessment, 54–56
Organic molecule, 36–38
carbon atom, 36–37
carbon skeleton and functional group, 37–38
origin of life, 318
Organism, 5
Organization
animal, 2, 580–99
homeostasis, 581, 594–96
integumentary system, 591–94
organs, organ systems, and body cavities, 589–90
summary and assessment, 597–99
tissue types, 582–89
of life, 2, 3f
Organogenesis, 509
Orgasm
female, 778
male, 776
Orientation and migratory behavior, 823
Origin and history of life, 317–36
geological factors that influence evolution, 332–34
continental drift, 332–33
mass extinctions, 333–34
history of life, 323–32
Cenozoic era, 330–31
clock timetable, 323
fossils tell a story, 323–24
geologic timescale, 324–25, 326t
Mesozoic era, 329–30
Paleozoic era, 327–29
Precambrian time, 324–27
origin of life, 318–23
evolution of monomers, 318, 319–20
evolution of polymers, 318, 320
evolution of protocells, 318, 320–22
evolution of self-replication system, 318, 322
Origin of replication, 215
Ornament and sexual selection, 829–31
Ornithorhynchus anatinus, 560*f,* 772
Oscillatoria, 367*f*
Osmolarity, 683
Osmoregulation
aquatic vertebrates, 677, 679–80
cartilaginous fishes, 680
freshwater bony fishes, 680
kidneys, 679
marine bony fishes, 680
definition, 678
human kidneys, 679, 685–87, 760, 761
terrestrial vertebrates, 680–81
kangaroo rat, 680, 681f
marine mammals and sea birds, 680–81
Osmosis, 88–91
definition, 88, 683
human urinary system, 683
pressure-flow model of phloem transport, 471
Osmotic gradient, 685
Osmotic potential, 467
Osmotic pressure
capillary exchange, 616
definition, 89
water potential, 465–67
water uptake, 461
Ossicle (echinoderm skeleton), 539
Ossicle (middle ear), 725
Ossification, 736
Osteoarthritis, 741
Osteoblast, 736
Osteoclast, 736, 759
Osteocyte, 736, 737*f*
Osteogenesis imperfecta, 196–97
Osteon, 585, 737
Osteoporosis, 736, 759, 810
Ostium, 602*f*
Ostracoderm, 548
Otolith, 728
Otolithic membrane, 728
Otter, sea, 914–16
Outer ear, 725
Outgroup (cladistics), 346
Out-of-Africa hypothesis, 573–74
Oval window, 725, 726
Ovarian cycle, 777, 778–79, 780*t*
Ovary
functions, 771
human, 752f, 765–66, 777
aging process, 811
gonadotropic hormones, 755, 778–79
ovarian cycle, 777, 778–79, 780t
steroid hormones, 753–54
plants
flower life cycle, 496
flower structure, 428, 429f, 497
flowering plant reproduction, 499
fruits, 504, 505f
Overexploitation and extinction, 911, 914–16
Overturn, lake, 896–97
Oviparous animal, 772
Ovoviviparous animal, 772
Ovulation, 778, 779, 780, 804*f,* 827
Ovule
angiosperm, 423, 426
flower structure, 428, 429f, 497
flowering plant life cycle, 430f
pine life cycle, 425f
seed plants, 423
Owl
burrowing, 894f
northern spotted, 888, 917
Oxalis, 492
Oxidation-reduction reaction, 109
chloroplasts and photosynthesis, 110
mitochondria and cellular respiration, 110
Oxygen. *See also* Respiratory system
bacteria metabolism, 364–65
transport in human respiratory system, 668–69
Oxygen debt, 745
Oxygenic photosynthesis, 365
Oxyhemoglobin, 668
Oxytocin, 755, 756*f*
Oyster, 772
Ozone shield
depletion, 913
history of life, 325

P

P generation, 188
P muscle contraction, 746–47
P wave, 608
P16 gene, 241
P53 gene, 159
Pacemaker, 608
Pacinian corpuscle, 729
Pain medication, 730
Pain receptor, 729–30
Pair-rule gene, 518, 801
Pakicetus, 271
Palate, 645–46
Paleogene period, 330–31
Paleontology, 262, 323
Paleozoic era, 326*t,* 327–29, 332*f*
Palisade mesophyll, 451
Palmately compound leaf, 452*f*
Palolo worm, 772–73
PAM (primary amoebic meningoencephalitis), 386
PAMP (pathogen-associated molecular pattern), 622–23
Pan troglodyte. See Chimpanzee
Pancreas, 645*f,* 647, 649, 752*f,* 763–65
Pancreatic amylase, 649
Pancreatic exocrine gland, 583*f*
Pancreatic islet, 763, 764, 765
Pancreatic juice, 647, 649
Pandinus, 537*f*
Pangaea, 332
Panoramic vision, 719
Panther
Florida, 847
K-selection, 847
rain forest diversity, 891f
Panthera onca, 887
Pap smear, 787
Papilio memnon, 862
Papillae, 716*f,* 717
Papio anubis, 565*f*
Papio hamadryas, 829*f*
Parabasalid and diplomonad, 385

Paralysis, 702
Paramecium
cilia, 383
competition, 857, 858
diversity of life, 2
predator-prey interactions, 859
Paramylon, 385
Parapodium, 531
Parasaurolophus, 330
Parasite and parasitism
alternate host, 405
bacteria, 365–66, 863
coevolution, 865
definition, 863
dodder vine, 464, 485
ecology of communities, 863
ectoparasite, 863
endoparasite, 863
flatworm, 526–27
flukes, 526–27
tapeworms, 527, 642
fungi
chytrids, 399
Microsporidia, 398–99
sac fungi, 400, 401
lampreys, 548
lichens, 405
mineral uptake in plants, 464
pathogens, 365–66
predator-prey interactions, 858
protists
African sleeping sickness, 388
amoeboids, 389
apicomplexans, 384
regulation of population size, 845
roundworms, 532f, 533
secondary host, 526, 527, 863
viruses, 354, 356, 863
Parasitic flatworm, 526–27
flukes, 526–27
tapeworms, 527, 642
Parasympathetic division, 707, 709, 710*f*
Parathyroid, 752*f,* 758, 759–60, 807*f*
Parathyroid hormone (PTH), 736, 759–60
Parenchyma cell
ground tissue, 437–38
ovary, 499
root cortex, 443
vascular tissue, 439
xylem and phloem rays, 447
Parent cell, 152
Parietal bone, 738, 739*f*
Parietal lobe, 703
Parkinson disease (PD), 691, 704, 706
Parmelia, 406*f*
Parrot, 558
Parsimony, 346
Parthenogenesis, 171, 771
Partial pressure, 668
Particle, subatomic, 20–21
Particulate theory of inheritance, 187
Parturition. *See* Birth
Passive immunity
active immunity vs., 632, 634
from mother, 780, 808
Passive transport, 88–91
Pasteur, Louis, 135, 355, 360, 361*f*
Patagonian cavy, 266*f*
Patella, 741*f*
Pathogen
endospore, 366
parasitism, 365–66
pathogen-associated molecular patterns (PAMPs), 622–23
pathogenic protists and climate change, 386–87
white blood cells, 586
Pathogen-associated molecular pattern (PAMP), 622–23
Pattern formation, 798, 800, 801*f*
Pauling, Linus, 48
Pavlov, Ivan, 822
Pax6 gene, 310–11
PD (Parkinson disease), 691, 704, 706
PDE-5 enzyme, 776
Pea, garden, 187, 188*f*
Peach, 505*f*
Peach-faced lovebird, 818–19
Peacock, 829
Peahen, 829
Pearl, 528
Pearl, Raymond, 842–43
Peat moss, 416
Pecten, 528, 529*f*
Pectoral fin, 548, 549
Pectoral girdle, 740
Pedipalp, 537
Pelagic zone, 899–900
Pellicle, 385
Pelvic cavity, 590, 740–41, 766
Pelvic fin, 312, 548
Pelvic girdle, 740–41, 766, 780
Pelvic inflammatory disease (PID), 788
Penetrance, incomplete, 197–98
Penetration, virus, 356, 357*f*
Penguin, 558
Penicillin
discovery, 9–10
food industry, 402
fungal derivation, 908
Penicillium
discovery, 9–10
food industry, 402
fungal derivation, 908
Penis
Brazilian cave insects, 770
duck, 770
human, 773f, 775–76
octopus, 770
Pentose, 39
Peppered moth, 270, 280–83
Pepsin, 647, 651
Peptidase, 651, 652
Peptide (cell structure), 48
Peptide (human digestion), 651, 652
Peptide bond, 48
Peptide hormone, 753, 754*f*
Peptidoglycan, 41, 363, 364*f,* 366
Perception, 715
Perennial plant, 441
Perfect flower, 429*t,* 497
Perforin, 631, 632*f*
Pericardium, 605
Pericarp, 504, 505*f*
Pericycle, 443
Periderm, 437
Perilymph, 726
Period. *See* Menstruation
Period gene, 828
Periodic table, 21–22
Peripheral nervous system, 707–10
anatomy, 707
autonomic system, 694f, 707, 708–10
parasympathetic division, 707, 709, 710f
sympathetic division, 707, 709, 710f
definition, 694, 707
somatic system, 694f, 707, 708, 709t
Peripheral protein, 84
Peristalsis, 646
Peritoneum, 646
Peritonitis, 648
Peritubular capillary network, 683, 684*f*
Periwinkle, rosy, 908
Permafrost, 886
Permeability, selective, 86
Permian period, 326*t,* 329, 333, 334
Peroxisome, 72
Pest control, 909
PET (positron-emission tomography), 22
Petal, 428, 429*f,* 497
Petiole, 450, 451*f,* 452
Petromyzon, 548
Peyer patch, 625, 646
pH
acid deposition, 878
buffer, 32, 687
human acid-base balance, 685, 687–88
optimal, 106
scale, 30
thermoacidophiles, 369
Phagocyte and natural killer cell, 627
dendritic cell, 626, 627
eosinophils, 615, 627
macrophages
functions, 627
inflammatory response, 626
monocytes, 615, 621
natural killer (NK) cells, 627
neutrophils, 615, 627
Phagocytosis
action, 93
amoeboids, 389
complement, 627
evolution of immune systems, 622
planarians, 642
white blood cells, 586
Phalange, 740, 741
Pharming, animal, 555
Pharyngeal arch, 807*f*
Pharyngeal pouch
chordates, 545
humans, 807f
Pharyngitis, 669
Pharynx
earthworms, 642, 643
fish gills, 662
human
digestion, 645f, 646
respiration, 664
planarians, 642
Phascolarctos cinereus, 560*f,* 644
Phaseolus, 486
Phenomenon, 9
Phenotype, 189*f,* 190–91
Phenylalanine, 186
Phenylketonuria, 186, 196
Pheomelanin, 594
Pheromone
chemical communication, 824–26
chemoreception, 716
humans, 753, 826
Philadelphia chromosome, 182
Phloem
bark, 448
features, 439
herbaceous stems, 446
leaves, 451f
plant transport systems
organic nutrient transport, 470–73
structure, 465–66
procambium, 445
root system, 443
sap
nutrient transport, 470–72
water transport, 465
structure, 439, 440, 465
vascular plant, 418
woody plants, 447
Phloem rays, 447
Phoradendron tomentosum, 8–9
Phoronid, 524
Phosphate
bloom, 367
NPK ratio of fertilizers, 458
phosphorus cycle, 875
Phosphocreatine. *See* Creatine phosphate
Phospholipid, 44–45
Phosphorus
cycle, 873, 875
guano, 902
NPK ratio of fertilizers, 458
Phosphorylation, 104
Phot, 486
Photoautotroph
bacteria, 364–65
dinoflagellates, 382
ecosystem dynamics, 868
protists, 374
Photomorphogenesis, 491
Photoperiodism, 490, 491*f,* 492
Photoreceptor
animals, 715, 718
human eye, 719, 720–21
phototropism, 486
Photorespiration, 125
Photosynthesis, 114–28
accessory pigments, 119, 379, 380, 381
archaea, 369
cellular respiration vs., 142f
chloroplasts, 110, 116
cyanobacteria, 365, 367
definition, 4, 115
ecology dynamics, 868
mesophyll tissue, 451
organs, 73
other types, 125–26
C_4 photosynthesis, 125–26
CAM photosynthesis, 126
photosynthesis and adaptation to the environment, 126
photoautotrophic bacteria, 364–65
photosynthetic organisms, 115–16
flowering plants, 116
plants convert solar energy, 119–23
ATP production, 123
electron flow in light reactions, 119–20
organization of the thylakoid membrane, 120, 121f
pigments and photosystems, 119
plants fix carbon dioxide, 123–24, 874
Calvin cycle importance, 124
process, 117–18
NADP+/NADPH, 117
reactions, 117–18
protists, 374, 379
summary and assessment, 127–28
Photosynthetic pigment, 119, 379, 380, 381
Photosystem, plant, 119
Phototropism, 485–86
auxins, 478, 480f, 486
coleoptile, 478
sunflower, 476
Phrenic nerve, 666, 667*f*
Phylogenetic species concept, 298–99, 300, 301*f*
Phylogeny, 337–51
anatomy of speciation, 300
animal evolution, 515–16
chordate, 546t
cladistics, 346, 347f
common ancestor, 344, 345, 346, 348
definition, 338
five-kingdom system, 341
fungi, 398
interpreting a phylogeny, 344–45
Linnaean classification hierarchy, 339–40
monophyly, 298
phylogenetic species concept, 298–99, 300, 301f
phylogeny, 344–49
cladistics, 346, 347f
common ancestor, 344, 345, 346, 348
interpreting a phylogeny, 344–45
tracing phylogeny, 346–49
plant, 412
systematic biology, 338–40
definition, 338
field of taxonomy, 338–39
Linnaean classification hierarchy, 339–40
Linnaean taxonomy, 339
three-domain system, 341–44
defining the domains, 341
domain Archaea, 341, 343–44

domain Bacteria, 341–43
tracing phylogeny, 346–49
Phylum, 7, 339
Physalia, 522, 523*f*
Phytochrome and phytochrome response, 489–93
circadian rhythms, 492–93
biological clock, 492–93
competition, 491
flowering and photoperiodism, 490, 491f
germination, 490–91
Phytophthora infestans, 381
Phytoplankton
lake ecosystems, 896, 897
ocean ecosystems, 899
Phytoremediation, 462–63
PID (pelvic inflammatory disease), 788
Pig
African sleeping sickness, 388
xenotransplantation, 555, 638
Pigment
photosynthetic, 119, 379, 380, 381
skin pigmentation, 207
visual, 720, 721f
Pigmy rattlesnake, 555
Pike, Northern, 897*f*
Pileated woodpecker, 558*f*
Pine
evolution, 419
life cycle, 423, 425f
mutualism, 864
Pine, bristlecone, 423
Pine, ground, 419
Pine, whitebark, 864
Pineal eye, 556
Pineal gland, 704, 752*f*, 766
Pineapple, 505*f*
Pingelap, 290–91
Pinna, 725
Pinnately compound leaf, 452*f*
Pinocytosis, 93–94
Pinus longaeva, 423
Pinworm, 533
Pioneer species, 866
Pisum sativum, 187, 188*f*
Pit, 439
Pit of Bones, 576
Pit viper
infrared sensory organ, 714
medicine from, 555
Pitch (hearing), 726
Pitcher plant, 296
Pith
root system, 443
stem system, 445, 446
wood, 449f
Pituitary dwarfism, 756, 757*f*
Pituitary gland and hypothalamus, 755–57
anterior pituitary, 755–57
hormones affecting other glands, 755, 765
hormones not affecting other glands, 755–57
hypothalamic-inhibiting hormones, 755
hypothalamic-releasing hormones, 755
major endocrine glands, 752f
definition, 755
posterior pituitary, 752f, 755, 756f
vertebrate nervous system organization, 693
Pitx1 gene, 312
Pivot joint, 741
Placenta
human, 780, 807–8, 809f
marsupials, 560
placental mammals, 560, 772
Placental mammal
continental drift, 333
evolution, 559t, 560–61, 772
life history strategy, 772
orders, 561t
Placoderm, 548–49
Plague, 846
Plan B, 781
Planarian
anatomy, 525–26
asexual reproduction, 771
chemical senses, 716
excretory organs, 679
incomplete digestive tract, 642
light detection, 718
nerve net, 692
respiration, 660
skeleton, 734
transport, 601
Plankton
diatoms, 381
golden brown algae, 381
lake ecosystems, 896, 897
ocean ecosystems, 899
protists, 374
Plant. *See also* Flowering plant; Photosynthesis
chemical ecology of, 484–85
classification, 341, 344
comparison of animal and plant cells, 436t
definition, 8, 412
ecology dynamics, 868
evolution and diversity, 411–34
ancestry and features of land plants, 412–15
bryophytes and colonization of land, 412, 415–17
lycophytes evolution and vascular tissue, 412, 417–19, 421
pteridophytes and megaphylls, 419–23
seed plant evolution and adaption to land, 423–31
summary and assessment, 432–34
evolution and full adaption to land, 426–31
flower structure, 428, 429t
flowering plant life cycle, 428–31
flowers and diversification, 429
monocots and eudicots, 428
origination and radiation, 426–28
ovules, 423, 426
sporophytes, 415f
genetically modified, 248
growth response control of flowering plants, 476–94
growth and movement responses, 485–89
phytochrome responses, 489–93
plant hormones, 477–85
history of life, 325f, 329–30
importance of, 410
mycorrhizae, 406–7, 444, 462, 464, 864
nutrition and transport, 456–75
plant nutrition and soil, 457–60
transport mechanisms, 465–73
water and mineral uptake, 461–64
phylogeny, 412
polyploidy, 305
reproduction, 495–511
asexual reproductive strategies, 507–9
fruit types and seed dispersal, 504–7
life cycle overview, 496
seed development, 502–3
sexual reproductive strategies, 496–501
structure and organization, 435–55
cells and tissues, 436–39
leaf organization and diversity, 450–53
organs, 440–41
roots organization and diversity, 442–44
stem organization and diversity, 445–50
survival mechanisms, 438
Plant cell
anatomy, 66f
central vacuole, 72–73
cytokinesis, 155, 156f
mitosis, 154f, 155f
walls, 96
Plantae
five-kingdom system, 341
key groups of plants and their features, 413t
Plantlet, 508*f*, 509
Plaque (atherosclerosis), 611, 612*f*
Plasma, 585, 613, 614, 629, 649. *See also* Complement protein
Plasma cell, 629
Plasma membrane
archaea, 343, 361
origin of life, 320–21
prokaryotic cells, 361, 362f
sensory organs, 715
sodium-potassium pump, 93, 696
structure and function, 62, 83–86
fluid-mosaic model, 83f, 84
glycoproteins and glycolipids, 84–85
permeability, 86
plasma membrane components, 83–84, 320–21
viruses, 357
water and mineral uptake in plants, 461
water potential in plants, 465–67
Plasmid
DIY biology, 362
DNA cloning, 245
prokaryotes, 361, 362
Plasmin, 616
Plasmodesma, 96, 96*f*
Plasmodesmatum, 461
Plasmodial slime mold, 389, 390*f*
Plasmodium, 389, 390*f*
Plasmodium, 384, 865
Plasmodium vivax, 384
Plasmolysis, 90
Plasmopara viticola, 382
Plastid
archaeplastids, 377
chromoplasts, 73
endosymbiosis, 374
leucoplasts, 73
protists, 374
Plate tectonics, 334
Platelet
blood clotting, 615–16
blood elements, 585f, 586, 613, 615–16
Platypus, duckbill, 560, 772
Plectorhincus chaetodontoides, 864
Pleiotropic effect, 198–99
Pleistocene epoch, 331
Pleistocene overkill, 575
Pleurobrachia pileus, 522*f*
Pliocene epoch, 331
Plumule, 503
PMS (premenstrual syndrome), 780
Pneumatophore, 444
Pneumocystis pneumonia, 633, 672
Pneumonia, 670–71
HIV opportunistic infection, 633, 672
vaccine experiments, 208
Poaching, 914
Poikilothermism, 594
Point mutation, 239–40
Pointed molar, 644
Poison ivy, 438
Poison-dart frog, 552, 555, 847*f*, 862, 891*f*
Polar body, 176, 778, 784
Polar covalent bond, 25
Polar easterly, 883
Polar spindle fiber, 153
Polarity development, 796, 800
Pollen cone, 424, 425*f*
Pollen grain. *See also* Pollination
angiosperms, 428, 431f, 496, 499
conifers, 425f
definition, 423
fossil, 426f
monocots and eudicots, 440–41
Pollen tube
angiosperms, 429, 430f, 496, 499
gymnosperms, 423, 424, 425f, 426
Pollination
animal, 429, 500–501, 864, 865, 909
bee, 429, 500, 501, 909
bird, 429, 501
coevolution, 500–501, 864, 865
cross-pollination, 499–501
definition, 423, 499
flower diversity, 429
self-pollination, 499
wind, 429, 499, 501
Pollution
acid deposition, 31, 813, 878, 913
carbon dioxide emissions, 851–52, 874, 877, 910
definition, 914
extinction, 911, 913
acid deposition, 913
eutrophication, 913
organic chemicals, 913
ozone depletion, 913
human population growth, 851–52
waste recycling, 910
water, 460, 910
Poly-A tail, 218
Polyandry, 829
Polychaete annelid, 530, 531–32
Polydactyly, 198
Polygamy, 829
Polygenic inheritance, 199–201
Polygenic trait, 286
Polymer
definition, 38
origin of life, 318, 320
Polymerase chain reaction, 246–47
Polyorchis penicillatus, 522*f*
Polyp
cnidarian body form, 522
colon, 649
coral, 522, 523f
Hydra asexual reproduction, 771f
sea anemone, 522, 523f
Polypeptide, 48
Polyploidy, 305
Polyribosome, 68, 222
Polysaccharide, 41–42
Polyspermy, 795
Polytrichum, 417*f*
Poly-x female, 180
Pongo pygmaeus, 565*f*
Pons, 704
Poplar, 462–63
Population. *See also* Population ecology; Population evolution
definition, 2, 280, 837
sink population, 917
source population, 917
Population density, 838, 848
Population distribution, 838–39
Population ecology, 836–54
demographics of populations, 838–41
density and distribution, 838–39
population growth, 839–41
human population growth, 849–52
age distributions, 850f, 851
doubling time, 849–50
environmental impact, 851–52
more-developed vs. less-developed countries, 849f, 850–51
life history patterns, 846–49
K-selected populations, 847–49
r-selected populations, 846, 847
population growth models, 841–44
exponential growth, 842

    iteroparity, 841
    logistic growth, 842–44, 846
    semelparity, 841, 842
  regulation of population size, 844–46
    density-dependent factors, 845–46
    density-independent factors, 844–45, 846
    other considerations, 846
    predator-prey interactions, 845, 846f, 859–61
  scope of ecology, 837
  summary and assessment, 852–54
Population evolution, 279–95
  genes, populations, and evolution, 280–85
  Hardy-Weinberg equilibrium and microevolution, 270, 280–85
    migration, 284
    mutation, 283–84
    natural selection, 285
    nonrandom mating, 285
    small population size, 284–85
  maintenance of diversity, 290–93
    heterozygote advantage, 292–93
    natural selection, 290–92
  natural selection, 286–89
    sexual selection, 287–89
    types, 286–87
  summary and assessment, 294–95
Population growth, 839–41
  age distribution, 840–41, 850f, 851
  biotic potential, 839
  human, 849–52
    age distributions, 850f, 851
    environmental impact, 851–52
    more-developed vs. less-developed countries, 849f, 850–51
  models, 841–44
    exponential growth, 842
    iteroparity, 841
    logistic growth, 842–44, 846
    semelparity, 841, 842
  mortality patterns, 839–40
Porcupine, 862
Pore mushroom, 402
Porphyra, 379
Porphyria, 198–99
Portal system, 609, 755
Portuguese man-of-war, 522, 523*f*
Positive feedback
  homeostasis, 595–96, 755
  ovarian cycle, 779
Positive tropism, 485
Positron-emission tomography (PET), 22
Possum, 560
Postanal tail, 545
Posterior pituitary, 752*f*, 755, 756*f*
Posterior/anterior axis, 800
Postganglionic fiber, 709
Postreproductive age group, 840–41, 850*f*, 851
Posttranscriptional control, 232, 235–36
Posttranslational control, 232, 238
Postzygotic isolating mechanism, 301*f*, 302–3, 304
Potash, 458
Potassium
  NPK ratio of fertilizers, 458
  opening and closing of stomata, 470
Potassium-argon (K-Ar) dating, 324
Potato
  famine, 381, 907
  tuber, 449, 450f, 507–9
Potential energy
  definition, 101
  water potential, 467
Pouched mammal. *See* Marsupial evolution
Power stroke, 746*f*, 747
Prairie, 893, 894*f*
Praying mantis, 644
Precambrian time, 324–27
  eukaryotic cells arise, 325–26
  multicellularity, 327
Precapillary sphincter, 617
Precipitation
  biome
    coniferous forests, 888
    desert, 893–94
    distribution, 884f, 885
    grassland, 893
    shrubland, 892
    temperate deciduous forests, 888
    tropical rain forests, 890
    tundra, 886
  climate change, 877
  lake effect, 883
  monsoon, 883
  rainshadow, 883
  spinning of the Earth, 882
  water cycle, 873, 895
Predation
  definition, 845, 858
  predator-prey interactions, 858–62
    coevolution, 865
    parasitism, 858
    population dynamics, 845, 846f, 859–61
    prey defenses, 861–62
Predator
  definition, 845, 858
  predator-prey interactions, 858–62
    coevolution, 865
    population dynamics, 845, 846f, 859–61
    prey defenses, 861–62
Prediction, 10
Pre-ejaculate, 775
Preganglionic fiber, 709
Pregnancy
  development, 793f, 802–9
    due date, 802
    early, 794–97
    extraembryonic membranes, 803
    fetal, 802, 808
    placenta structure and function, 807–8
    stages of birth, 808, 809f
    trimester, 803
  ectopic, 777
  fertilization, 777, 778, 780, 783, 794–95, 804
  HIV/AIDS, 786
  implantation, 777, 804
  placenta structure and function, 807–8, 809f
  Rh incompatibility, 615
  seahorses, 770, 772
  tests, 630, 780
  toxoplasmosis, 384, 805
Prehypertension, 611
Premenstrual syndrome (PMS), 780
Pre-messenger RNA, 235*f*
Premolar, 644
Preparatory reaction, 131, 136–37
Preprogrammed theories of aging, 811
Prepuce. *See* Foreskin
Prereproductive age group, 840–41
Presbyopia, 810
Preservation, habitat, 916–18
  keystone species, 916
  landscape preservation, 917–18
  metapopulations, 916–17
Pressure. *See also* Blood pressure
  partial, 668
  potential, 467
Pressure-flow model of phloem transport, 471–72
Prevailing westerly, 883
Preven, 781
Prey
  defenses, 861–62
    association, 862
    camouflage, 861–62
    mimicry, 862
    startle response, 862
    warning coloration, 862
  definition, 845, 858
  predator-prey interactions, 858–62
    coevolution, 865
    defenses, 861–62
    population dynamics, 845, 846f, 859–61
Prey birds, 558
Prezygotic isolating mechanism, 299–302
  behavioral isolation, 301f, 302, 827f
  gamete isolation, 302
  habitat isolation, 299, 301f
  mechanical isolation, 302
  reinforcement, 304
  temporal isolation, 299–302
Prickly-pear cactus, 859
Primaevifilum, 327*f*
Primary amenorrhea, 780
Primary amoebic meningoencephalitis (PAM), 386
Primary asomatosensory area, 729
Primary follicle, 778
Primary growth
  plants, 440
  woody plants, 446–47
Primary host, 526, 527, 863
Primary immunodeficiency disorder, 636
Primary lymphoid organ
  definition, 624
  red bone marrow, 585, 614, 615, 623f, 624, 737
  thymus, 623f, 624, 752f, 766
Primary meristem, 445
Primary motor area, 703
Primary nondisjunction, 178
Primary olfactory area, 717*f*
Primary oocyte, 176, 778
Primary organizer, 798
Primary ossification center, 736
Primary protein structure, 48, 49*f*
Primary somatosensory area, 703
Primary spermatocyte, 774–75
Primary succession, 866
Primary taste area, 703
Primary tissue (woody plants), 446
Primary visual area, 703
Primate order. *See also* Human
  ancestral and derived traits, 344–45
  characteristics, 565–66
    complex brain, 566
    mobile limbs, 565
    reduced reproductive rate, 566
    stereoscopic vision, 565–66
  classification, 568, 569t
  diversity, 565f
  evolution, 565–69
    characteristics, 565–66
    common ancestor, 566, 567f, 569, 577
    history of life, 331
    sequence, 566–69
  molecular traits, 348–49
  monogamy, 829
  phylogeny, 345
  tool use, 576
Primitive gut, 517
Primitive streak, 796
Primordial soup hypothesis, 319
Principle, 11
Principles of Geology (Hutton), 264
Prion, 50, 360
PRL (prolactin), 755, 780, 781*f*
Probability, law of, 192–93
Probiont. *See* Protocell evolution
Problem solving, 823–24
Procambium, 445, 503
Proconsul, 568–69
Producer, 868, 869, 871
Product, 103
Product rule of probability, 192–93
Production, 765–66
Proembryo stage of seed development, 502
Profiling, DNA, 246
Profundal zone, 897–98
Progesterone, 765–66, 777, 779, 780, 807
Proglottid, 527
Programmed cell death. *See* Apoptosis
Progymnosperm
  Carboniferous period, 421
  Devonian period, 423
Prokaryote, 360–63. *See also* Archaea; Bacterium
  cell division, 161–62
  definition, 7–8, 62–63, 360
  DNA replication, 214–15
  five-kingdom system, 341
  gene expression regulation, 229–31
  history of life, 325
  reproduction, 363
  structure, 62, 360–61, 362f
  summary and assessment, 370–72
  three-domain system, 341–44
Prokaryotic cell, 62–63
Prolactin (PRL), 755, 780, 781*f*
Proliferative phase, 779*f*, 780
Prometaphase, 153
Promoter, 217, 229
Pronucleus, 795
Prop roots, 444
Propagation of action potentials, 697
Prophage, 356–57
Prophase, 153, 172
Proprioceptor, 729
Prosimian
  brain, 566
  evolution, 567f, 569
  primate diversity, 565f
Prostaglandin, 766–67, 775
Prostate cancer, 635, 775
Prostate gland, 773*f*, 773*t*, 775, 810
Protease, 238
Protease inhibitor, 786
Proteasome, 238
Protective protein, 627–28
  complement, 625, 627
  interferons, 625, 628
Protein, 46–50
  amino acid, 47–48
  chaperone, 50
  denaturation, 50
  fibrous, 48, 49f
  folding, 50
  function, 46–47
  globular, 48, 50
  integral, 84
  nonfunctional, 240
  origin of life, 320, 322
  peripheral, 84
  phylogeny tracing, 348–49
  plasma membrane, 85
  point mutation, 239–40
  protective proteins, 625, 627–28
  structure, 48–50
  synthesis and antibiotics, 366
  synthesis in prokaryotes, 361
Protein (dietary)
  human diet and nutrition, 653–54
  human digestion, 651–52
Protein fiber, 583–85
  collagen fibers, 583, 585
  elastic fibers, 583–84
  reticular fibers, 583
Proteinaceous infectious particles. *See* Prion
Protein-first hypothesis, 320, 322
Proteinoid, 320
Proteome, 255
Proteomics, 225, 255
Proterospongia, 390
Prothrombin activator, 616
Protist evolution and diversity, 373–94

Amoebozoa, Opisthokonta, and Rhizaria supergroups, 374, 375t, 389–91
Archaeplastida supergroup, 374, 375t, 377–79
biology of protists, 374–76
Chromalveolata supergroup, 374, 375t, 380–84
classification, 8, 341, 344
definition, 374
Excavata supergroup, 374, 375t, 385–89
history of life, 325f
pathogenic protists and climate change, 386–87
summary and assessment, 392–94
Protista kingdom, 341
Protocell evolution, 318, 320–22
Protoderm, 445
Proton, 20–21
Proton pump
mineral uptake in plants, 461–62
water transport in plants, 470
Protonema, 416
Proto-oncogene, 159
Protoplast, 248
Protostome. *See also* Lophotrochozoa
characteristics, 520t, 524
embryonic development, 516–17
Protozoan. *See* Ciliate; Protist evolution and diversity
Provenge, 635
Proximal convoluted tubule, 681, 683
Pseudocoelom
roundworm, 532
transport, 601
Pseudocolochirus, 539*f*
Pseudogene, 568
Pseudomyr mexferruginea, 864
Pseudoplasmodium, 390
Pseudopod
Amoebozoa supergroup, 374, 389
foraminiferans and radiolarians, 391
Pseudostratification, 583
Psilocybe *mexicana*, 404
Psilocybin, 404
Psilotum, 420
Psychosis, amphetamine, 700
Pteridophyte
definition, 419
evolution, 419–23
ferns, 421, 422–23
horsetails, 420, 421
whisk ferns, 420
features, 413t
Pterois miles, 905
Pterois volitans, 550*f*, 905
PTH (parathyroid hormone), 736, 759–60
Ptilonorhynchus violaceus, 830
Puberty, 765, 766, 776
Pubic hair, 765
Pubic symphysis, 741
Publication, scientific, 11
Puffball mushroom, 402, 403*f*
Pulmonary circuit
double-loop circulatory pathway, 604
human, 609
Pulmonary fibrosis, 670*f*, 672
Pulmonary trunk, 605*f*, 606, 609
Pulmonary tuberculosis, 670*f*, 672
Pulse, 608
Pulvini, 492
Punctuated equilibrium model of macroevolution, 309, 310*f*
Punnett square, 192, 193*f*
Pupil, 719
Purkinje fiber, 608
Purple bacterium, 365
Pus, 615, 627
Pyramid, ecological, 871–72
Pyrenoid, 377
Pyrogen, 766
Pyruvate, 131

Q

QRS complex, 608
Quadriceps tendon, 741*f*
Quaternary period, 326*t*, 330–31
Quaternary protein structure, 49*f*, 50
Quillwort, 419

R

R. *See* Rate of natural increase
Rabbit, 868*f*
Radial cleavage, 517
Radial symmetry
animal evolution, 515, 518–19
cnidarians and comb jellies, 515
echinoderms, 539
Radiata, 520*t*
Radiation level, 22, 23*f*
Radicle, 503
Radioactive isotope, 22–23
Radioactivity, 22
Radiocarbon dating of fossils, 324
Radiolarian, 374, 391
Radiole, 531–32
Radius, 740
Radula, 528, 643
Raggiana Bird of Paradise, 826
Ragnarsson, Ivar, 197
Rain forest
temperate rain forest, 888
tropical rain forest
biome, 885f, 890–91
destruction, 122, 877
habitat loss, 911, 912f
soil profile, 460
species richness, 856, 890
transpiration, 468, 469f
Rainbow trout, 895*f*
Rainfall. *See* Precipitation
Rainforest, destruction of tropical, 122
Rainshadow, 883
Rana catesbeiana
Linnaean classification hierarchy, 340
specialization, 552f
Random distribution, 838
Rangifer, 844
RAS (reticular activating system), 704, 705*f*
Ras gene, 159
Rat
black, 914
kangaroo, 680, 681f, 894
Rate of natural increase
calculation, 839
r-selected population, 846, 847
Ratfish, 549
Rattlesnake, diamondback, 267–68, 556*f*
Rattlesnake, pigmy, 555
Raven, 823–24
Ray, 549
Ray-finned bony fish, 549–50
Reactant, 103
Realized niche, 857
Reasoning. *See also* Learning
deductive, 10
inductive, 9
Reception during plant movement, 485
Receptor
antigen, 622, 628, 629
B-cell receptor (BCR), 628–29
chemoreceptors, 715, 716, 717
cutaneous receptors, 729
pain receptors, 729–30
sensory receptors, 715–16
signal transduction, 477
target cells, 753, 754
Receptor protein, 85
Receptor-mediated endocytosis, 94
Recessive allele, 190
Reciprocal altruism, 832
Recombinant DNA technology, 245
Recombinant human EPO (rHuEPO), 686, 766
Recombination, genetic, 170
Recruitment and population size, 846
Rectum, 645*f*, 648
Red algae. *See* Archaeplastida supergroup
Red blood cell
blood element, 585, 613–15
blood types, 614–15
Red deer, 831
Red kangaroo, 838
Red pulp, 625
Red tide, 382–83
Red wavelength, 490, 491
Red wine effect, 612
Red-eared slider turtle, 915
Redox reaction, 109
Reduction, carbon dioxide, 124, 877
Reduction division, 168–69
Refinement, 653
Reflex action, 702, 708, 709
Reflex arc, 708
Refractory period, 697, 776
Regeneration
hydras, 523
planarians, 525
regenerative medicine, 588–89
salamander, 588–89
sea stars, 540
sponges, 522, 771
Regulation of population size, 844–46
density-dependent factors, 845–46
density-independent factors, 844–45, 846
other considerations, 846
Regulator gene, 229
Reindeer, 844
Reinforcement of reproductive isolation, 304
Release, viral, 356, 357–58
Remora, 863–64
Renal cortex, 681, 682*f*, 683*f*
Renal medulla, 681, 682*f*, 683*f*, 685
Renal pelvis, 681, 682
Renin, 687, 761
Repetitive DNA element, 252
Replacement model of human evolution, 573–74, 577
Replacement reproduction, 851
Replication, DNA, 213–15
complementary base pairing, 213–14
prokaryote vs. eukaryote, 214–15
Replication fork, 215
Repolarization, 697
Repressible operon, 230
Repressor, 229
Reproduction. *See* Asexual reproduction; Reproductive system; Sexual reproduction
Reproductive age group, 840–41, 850*f*, 851
Reproductive cloning, 156, 157
Reproductive isolating mechanisms, 299–303
definition, 284, 299
postzygotic isolating mechanisms, 301f, 302–3
hybrid inviability, 303
hybrid sterility, 303
prezygotic isolating mechanisms, 299–302
behavioral isolation, 301f, 302, 827f
gamete isolation, 302
habitat isolation, 299, 301f
mechanical isolation, 302
temporal isolation, 299–302
reinforcement, 304
Reproductive strategies and fitness, 829–31
monogamy, 829
polyandry, 829
polygamy, 829
sexual selection, 829–31
Reproductive success, 267–68
Reproductive system, 770–92. *See also* Asexual reproduction; Sexual reproduction
amphibians, 551
animals, 514, 771–72
asexual reproduction, 771
life history strategies, 772
sexual reproduction, 771–72
bacteria, 363
characteristic of life, 5
eukaryote, 363
flowering plant, 495–511
asexual reproductive strategies, 507–9
fruit types and seed dispersal, 504–7
seed development, 502–3
sexual reproductive strategies, 496–501
human
aging process, 811
control of reproduction, 774, 777, 781–84
female reproductive system, 777–81
male reproductive system, 773–76
reproductive technology, 783–84
sexually transmitted diseases, 385, 785–89
iteroparity, 841
net reproductive rate, 842
prokaryote, 363
binary fission, 363
conjugation, 363
transduction, 363
transformation, 363
protists, 374
reduced reproductive rate, 566
semelparity, 842
summary and assessment, 790–92
viruses, viroids, and prions, 356–60
animal viruses, 356–59
bacteriophages, 356–57
emerging viruses, 358–60
retroviruses, 358
steps, 356
Reproductive technology, 783–84
artificial insemination by donor (AID), 783
gamete intrafallopian transfer (GIFT), 783
intracytoplasmic sperm injection (ICSI), 783
surrogate mothers, 783
in vitro fertilization (IVF), 783
Reptile
characteristics, 520t, 553
amniotic egg, 547–48, 553, 803
circulation, 553
ectothermy, 553
excretion, 553
lungs, 553
paired limbs, 553
skin, 553, 591
circulatory pathway, 551f
diversity, 556–57
birds, 556, 558
crocodilians, 556–57
lizards, 556
snakes, 556
tuataras, 556
turtles and tortoises, 556
evolution, 546t, 547–48, 553–57
amniotes, 553–54
dinosaur, 554
turtle, 553–54
history of life, 330
hormones and endocrine system, 753
poaching, 914
reproduction and life history strategy, 772
respiration, 665–66

skin color, 755–56
ventricle, 604
Reservoir (chemical cycling), 872*f,* 873, 874
Reservoir (disease), 388
Resolution, 61
Resource and population, 838, 851–52
Resource competition, 267, 857–58
Resource partitioning, 858
Respiration. *See also* Respiratory system
birds, 557–58, 666
cellular, 660
cutaneous, 551, 661
external respiration, 660, 667, 668
internal respiration, 660, 667, 668–69
oxygen debt, 745
reptiles, 665
steps, 660
external respiration, 660, 667, 668
internal respiration, 660, 667, 668–69
ventilation, 660
ventilation, 660
Respiratory center, 666–67
Respiratory pump, 611
Respiratory syncytial virus (RSV), 630
Respiratory system, 659–76
aging process, 810
arthropods, 534
breathing and transport of gases, 665–69
breathing, 665–67
gas exchange and transport, 667–69
oxygen debt, 745
disorders, 669–73
lower, 669–73
upper, 669
gas-exchange surfaces, 660–64
gills of fish, 661–62
human lungs, 663–64
overview, 660–61
respiration steps, 660
trachael system of insects, 662–63
respiration and human health, 669–73
lower respiratory tract disorders, 669–73
upper respiratory tract disorders, 669
respiration steps, 660
external respiration, 660, 667, 668
internal respiration, 660, 667, 668–69
ventilation, 660
summary and assessment, 674–76
Respiratory tract disorder, 669–73
lower, 669–73
lungs, 670–73
trachea and bronchi, 669–70
upper, 669
common cold, 669
strep throat, 669
Responding variable, 10
Response to environment, 5
Response to signal transduction, 477
Resting potential, 696
Restoration ecology, 918–19
Restriction enzyme, 245, 246*f*
Results, 12–13
Resveratrol, 612
Reticular activating system (RAS), 704, 705*f*
Reticular fiber, 583, 584*f*
Retina
artificial, 722, 723
diseases, 722
eye anatomy, 719, 720
focusing eye, 720, 724
integration of visual signals, 721–22
from retina to visual cortex, 722
structure and function, 721f
Retinal, 720, 721*f*
Retinal detachment, 722
Retroviruses, 358
Reverse transcriptase, 322, 358
Reverse transcriptase inhibitor, 786
Reverse vaccine for diabetes mellitus, 765
Rh system, 614–15
Rheumatic fever, 637
Rheumatoid arthritis, 637, 741
Rhinoceros
biotic potential, 839f
poaching, 914
Rhinoderma darwinii, 847f
Rhizaria supergroup, 391
evolution and diversity, 374, 375t, 389–91
foraminiferans, 374, 391
radiolarians, 374, 391
Rhizobium, 462
Rhizoid
charophytes, 379
liverworts, 416
molds, 399
mosses, 416, 417f
Rhizome
flowering plants, 449, 450f, 507–9
lycophytes, 419
pteridophytes, 420, 423
Rhizopus stolonifer, 399
Rhododendron, 497*f*
Rhodoglossum affine, 379*f*
Rhodopsin, 720, 721*f*
RHuEPO (recombinant human EPO), 686, 766
Rhyniophyte, 418, 419*f*
Rib, 739
Rib cage, 738, 739
Ribose, 39
Ribosomal RNA (rRNA)
archaea, 368
function, 50, 68, 221–22
production, 68
structure, 221
three-domain system, 341
Ribosome, 68–69
prokaryotic cell, 63, 361, 362f
structure and function, 221–22
Ribozyme, 105, 322
Rice, 908–9
Rigor mortis, 747
Ringworm, 402, 492
River ecosystem, 895
RNA (ribonucleic acid)
carrier of genetic code of life, 215–16
conversion to DNA in retroviruses, 358
function, 50, 68, 215–16
messenger RNA (mRNA)
codons, 217f
functions, 50, 68
pre-messenger RNA, 235f
processing, 218–20
production, 217–18
splicing, 219
steroid hormones, 754
transcript, 217
origin of life, 320, 322
processing, 218–20
ribosomal RNA (rRNA)
archaea, 368
function, 50, 68, 221–22
production, 68
structure, 221
three-domain system, 341
small RNA, 236
small-interfering RNA (siRNA), 236
structure, 50–52, 216f
transcription
messenger RNA production, 217–18
RNA processing, 218–20
transfer RNA (tRNA), 50, 68, 220–21
translation
ribosomal RNA, 221–22
transfer RNA, 220–21
viroids, 360
RNA polymerase, 217–18, 404
RNA-first hypothesis, 320, 322
Roadrunner, greater, 893*f*
Robustness of australopithecine, 571, 572
Rock
dating, 324
ecological succession, 866
lichen, 367, 864
phosphorus cycle, 875
Rockweed. *See Fucus*
Rocky shore ecosystem, 898
Rod cell
color vision, 566
human eye, 720, 721–22
Rodhocetus, 271
Rohypnol, 700
Romalea, 536–37
Roofie, 700
Root (phylogeny), 300
Root (plant)
asexual reproduction, 507–9
cellular respiration, 457
features, 442
lycophytes, 419
organization and diversity, 440, 442–44
diversity, 444
eudicots, 442f, 443
monocots, 443–44
pressure, 468
vascular cylinder, 439
water and mineral uptake, 461–64
Root apical meristem, 443
Root cap, 443, 487
Root hair, 443
function, 442
plant epidermal cells, 436–37
water uptake, 461
zone of maturation, 443
Root hair plexus, 729
Root nodule
cytokinins, 481
nitrogen fixation, 444, 462, 463f, 876
Root pressure, 468
Rosy periwinkle, 908
Rotational equilibrium, 727–28
Rotifer, 520*t,* 527, 528*f*
Rough endoplasmic reticulum, 69
Round window, 725, 726
Roundworm, 532–33
aging process, 802, 811
anatomy, 532
characteristics, 520t
hydrostatic skeleton, 734
model organism, 532, 798, 800
parasitic
Ascaris, 532f, 533
filarial worm, 532f, 533
pinworms, 533
Trichinella, 532f, 533
Royal families in Europe and hemophilia, 203
RRNA (ribosomal RNA). *See* Ribosomal RNA (rRNA)
R-selection, 846, 847
RSV (respiratory syncytial virus), 630
RU-486, 781–82
Rubber tree, 909
Rubella, 805
RuBP carboxylase, 124
Ruffini ending, 729
Rumen, 641, 644
Runaway hypothesis, 287
Runner. *See* Stolon
Runner bean, 486
Rust, 402, 405
Rye, 404

S

S stage, 148
SA (sinoatrial) node, 608, 609
Sac fungus
evolution and diversity of fungi, 398, 400–402
food industry, 402
lichen, 405
mycorrhizae, 406–7
poisonous, 404
reproduction, 400–401
structure, 396
symbiotic relationships, 401
Saccharomyces cerevisiae
fermentation, 135
logistic growth, 842–43
sac fungi, 400–402
Saccule, 728
Sacral vertebra, 739
Sacrum, 738, 739, 740
Sago palm, 424
Sahelanthropus tchadensis, 570
Salamander
characteristics, 551–52
regeneration, 588–89
respiration, 661
speciation, 303–4
Salamander, axolotl, 588–89
Salamander, barred tiger, 552*f*
Salem witch hysteria, 404
Salicylic acid, 484
Saliva, 646, 651
Salivary amylase, 646, 651
Salivary gland, 645*f*
Salix, 443*f*
Sally lightfoot crab, 534*f*
Salmon
community diversity, 907
organic chemical extinction threat, 913
speciation, 304, 305f
Salmon, kokanee, 907
Salmon, sockeye, 304, 305*f*
Salmonella, 366
Salt
halophiles, 368f, 369
table, 24f
water-salt balance, human, 685–87, 760, 761
Salt gland, 677, 680–81
Saltatory conduction, 697
Salty taste receptor, 717
Samara, 505*f*
Sand dollar, 540
Sand particle size in soil, 459
Sandy shore ecosystem, 898–99
Sap
nutrient transport, 470–72
water transport, 465
Saprolegnia, 382
Saprotroph
basidiomycetes, 402
chemoheterotrophic bacteria, 365
ecosystem dynamics, 868f, 869
fungi, 396
Sapwood, 448
Sarcolemma, 742–43, 745, 746
Sarcomere, 743, 744–45, 746–47
Sarcoplasmic reticulum, 743, 746
Sarcoscypha, 400*f*
Sarin gas, 108
Satellite cell, 695
Satin bowerbird, 830
Saturated fat, 653, 654
Saturated fatty acid, 42, 43*f,* 44
Savanna, 892–93
Sawfish ray, 549
Saxitoxin, 382–83
Scale
bird feathers, 557
fish, 548, 549, 591
reptiles, 553, 591
Scallop, 528, 529*f*
Scanning electron microscope (SEM), 60
Scapula, 740
Scarlet kingsnake, 862
Schistosoma, 526
Schistosomiasis, 526
Schleiden, Matthias, 58

Schwann, Theodore, 58
Schwann cell, 695
SCID (severe combined immunodeficiency), 636
Science
challenges facing, 13–14
biodiversity and habitat loss, 13–14
climate change, 14, 877
emerging disease, 14
process (scientific method), 9–13
Scientific method, 9–13
Scientific name, 8–9, 339
Scientific publication, 11
Scientific theory, 11
Sclereid, 438
Sclerenchyma cell, 437, 438, 439
Sclerosis, 704
Scolex, 527
Scorpion, 537
Scrotum, 773
Scrub jay, Florida, 832
Sea anemone
cnidarian diversity, 522, 523f
mutualism, 864
nerve net, 692
Sea apple, 539*f*
Sea bird
osmoregulation, 680–81
territoriality, 828
Sea cucumber, 539–40
Sea lettuce, 378
Sea level, 875, 877
Sea otter, 914–16
Sea squirt, 546
Sea star
anatomy, 539, 540
carnivorous diet, 644
circulatory system, 601
Sea turtle
conservation, 877
green, 556f
loggerhead, 823, 887
migration, 823
reptile diversity, 556f
salt gland, 677, 681
Sea urchin, 539*f*, 540, 916
Seahorse, 550*f*, 770, 772
Seal, elephant, 669
Season, 882
Seaweed
brown algae, 380
green algae, 377
intertidal zone, 898
red algae, 379
Sebaceous gland, 594
Second messenger, 753
Secondary amenorrhea, 780
Secondary follicle, 778
Secondary growth
herbaceous stems, 448f
vascular cambium, 436
woody plants, 447, 448
Secondary host, 526, 527, 863
Secondary immunodeficiency disorder, 636. *See also* Opportunistic infection
Secondary lymphoid organ, 624–25
lymph nodes, 623f, 624
Peyer patches, 625, 646
spleen, 623f, 624, 625
tonsils, 623f, 625
vermiform appendix, 625
Secondary metabolite, 484
Secondary nondisjunction, 178
Secondary oocyte, 176, 778, 794, 804
Secondary protein structure, 48, 49*f*
Secondary sex characteristic, 765, 776, 780
Secondary spermatocyte, 775
Secondary succession, 866, 867
Secondary tissue of woody plants, 446
Secondary xylem, 448
Secretory phase, 779*f*, 780
Sedgewick, Adam, 265
Sediment
foraminiferans and radiolarians tests, 391
fossils, 323
Sedimentary biogeochemical cycle, 872
Sedimentation, 323
Seed
definition, 423
development, 502–3
eudicots, 502–3
globular stage, 502, 503f
heart stage and torpedo stage embryos, 502, 503f
mature embryo, 502f, 503
zygote and proembryo stages, 502
dispersal, 429, 504–7
dormancy, 438, 504–6
flowering plant life cycle, 430f, 496, 498f, 499
germination, 490–91, 504–7
Seed coat, 423, 502*f*, 503
Seed cone, 424, 424*f*, 425*f*
Seed fern
Carboniferous period, 421
continental drift, 332
Devonian period, 423
Seed plant
definition, 423
evolution and full adaption to land, 423–31
angiosperms, 426–31
gymnosperms, 329, 421, 423–26
features, 413t
Seedless vascular plant
origins, 418
pteridophytes, 419–23
ferns, 422–23
horsetails, 420
whisk ferns, 420
Segmentation
annelids, 530
arthropods, 534, 800–802
bilateral symmetry, 518
chordates, 546
vertebrates, 797
Segmented worms. *See* Annelid
Segment-polarity gene, 801
Segregation, law of, 188–90, 192*f*
Selam *(Australopithecus afarensis),* 571
Selective permeability, 86
Selenium
beneficial plant nutrient, 458
phytoremediation, 463
Self-interest *vs.* altruism, 831–32
Self-pollination, 499
Self-replication system and origin of life, 318, 322
"Self-splicing," 218–19
SEM (scanning electron microscope), 60
Semantic memory, 705
Semelparity, 841, 842
Semen, 775, 776
Semicircular canal, 725, 727–28
Semiconservative replication, 213
Semilunar valve, 606, 607*f*
Seminal fluid, 775, 776
Seminal vesicle, 773*f*, 773*t*, 775
Seminiferous tubule, 774, 776
Senescence, 481–82
Sense organs, 714–32
aging process, 810
amphibian characteristic, 551
bird characteristic, 558
chemical senses, 716–18
crustaceans, 716
fish, 716
insects, 716
planarians, 716
sense of smell in humans, 717–18
sense of taste in humans, 716f, 717, 718
snakes, 716
hearing and balance, 724–28
disorders, 728
human ear, 725–27
sense of balance, 727–28
sound wave detection, 724–25
sensory receptors, 715–16
chemoreceptors, 715, 716, 717
electromagnetic receptors, 715
mechanoreceptors, 715, 725, 729
photoreceptors, 715, 718, 719, 720–21
thermoreceptors, 714, 715
somatic senses, 729–30
cutaneous receptors, 729
pain receptors, 729–30
proprioceptors, 729
summary and assessment, 730–32
vertebrate nervous system organization, 693
vision, 718–24
animal detection of light, 718–19
human eye, 719–24
vomeronasal organ (VNO), 826
Sensitive period, 821
Sensitive plant, 488–89
Sensor in negative feedback, 595, 596*f*
Sensory input, 699
Sensory neuron, 696
Sensory receptor, 715–16
exteroceptors
chemoreceptors, 715, 716, 717
electromagnetic receptors, 714, 715
mechanoreceptors, 715, 725, 729
photoreceptors, 715, 718, 719, 720–21
reflex action, 708
thermoreceptors, 715
interoceptors, 715
Sensory transduction, 715
Sentinel cell, 622
Sepal, 428, 429*f*, 497
Sepioteuthis lessoniana, 530*f*
Septate fungus, 397, 402
Septum
fungi, 397
nasal, 739
ventricle, 604, 605–6
Sequential hermaphroditism, 771
Sequoia sempervirens, 423
Serosa, 646
Serotonin, 698
Sertoli cell, 775
Sessile ciliate, 383
Sessile tube worm, 531–32
Set point, 595
Seta
annelids, 530
crayfish, 535
Severe combined immunodeficiency (SCID), 636
Sex chromosome number, abnormal, 179–80
Sex hormone
adrenal glands, 760
bone growth and bone mass, 736
testes and ovaries, 765–66, 776, 777
Sex reversal, 771
Sex selection, 783
Sexual development
pineal gland, 766
testes and ovaries, 765–66, 773
Sexual dimorphism, 287
Sexual mimicry, 2, 429, 500
Sexual reproduction
animal, 514, 771–72
conjugation, 378, 379f, 383
definition, 167, 771
flowering plant, 496–501
cross-pollination, 499–501
flower structure, 496–97
gametophyte development, 497–99
life cycle in detail, 497–501
life cycle overview, 496
sporophyte development, 499
fungi, 397
Basidiomycota, 402–3
Zygomycota, 399–400
human
aging process, 811
control of reproduction, 781–84
female reproductive system, 777–81, 811
male reproductive system, 773–76, 811
sexually transmitted diseases, 385, 784–89
invertebrates
cnidarians, 523, 771
sponges, 521, 771
lichen, 405
meiosis, 166–85
changes in chromosome number and structure, 177–82
compared to mitosis, 174–75
cycle of life, 176–77, 178f
genetic variation, 169–70
overview, 167–69
parthenogenesis, 771
phases, 172–73
spermatogenesis and oogenesis, 176–77, 178f, 774f, 775, 778, 778f
parthenogenesis, 771
protists, 374
apicomplexans, 384
ciliates, 383
diatoms, 381
green algae, 377–78, 379
water molds, 382
Sexual selection, 287–89, 829–31
Sexual union. *See* Copulation
Sexually transmitted disease (STD), 785–89. *See also* AIDS; HIV
bacterial, 788–89
chlamydia, 788
gonorrhea, 788–89
syphilis, 789
prevention, 787, 788
trichomoniasis, 385, 789
vaginal infections, 789
viral, 785–88
genital herpes, 786f, 788, 805
hepatitis, 785, 787, 788
HIV/AIDS, 785–86
human papillomavirus (HPV), 592, 786–88
Shape
animal development, 311–12, 515
bacteria, 62, 363, 364f
human variations, 577
virus, 355–56
Shark
characteristics, 549
commensalism, 863–64
dogfish, 555
great white, 549
medicinal uses, 555
mercury, 901
tiger, 549
Shattering, 427
Shelf fungus, 402, 403*f*
Shell, 529, 530*f*
Shepherd's purse plant, 413*f*
Shiga toxin, 366
Shigella dysenteriae, 366
Shimomura, Osamu, 219
Shingles, 633
Shivering, 595
Shoot apical meristem, 445, 502
Shoot system, 440, 441, 502
Shore crab, 829
Shorebird, 558
Short tandem repeat sequence (STR), 247

Short-day plants, 490
Short-term memory, 705
Short-term stress response, 760*f*
Shotgun sequencing, 256
Shoulder. *See* Pectoral girdle
Shoulder blade. *See* Scapula
Shrimp
  ecosystem dynamics, 907, 912
  excretory organs, 679
  opossum, 907, 912
Shrub
  chaparral, 892
  desert ecosystems, 838, 894
  ecological succession, 866, 867
  layer of temperate deciduous forest, 889
  shrubland ecosystem, 892
Shrubland ecosystem, 892
Sialia mexicana, 845
Siamang gibbon, 828*f*
Sickle-cell disease, 199, 292–93
Sieve plate, 539*f,* 540
Sieve tube, 465, 471
Sieve-tube member, 439, 445, 465
Sign stimulus, 820–21
Signal, 149
Signal peptide, 69
Signal transduction pathway, 477
Signaling cascade, 753
Silt particle size and soil, 459
Silurian period, 326*t,* 329
Silversword alliance, 306
Sima de los Huesos (Pit of Bones), 576
Simple epithelial tissue, 582–83
  columnar epithelium, 583
  cuboidal epithelium, 582f, 583
  squamous epithelium, 582–83
Simple eye, 534, 537
Simple fruit, 504
Simple goiter, 758
Simple leaf, 452
Single nucleotide polymorphism, 251
Single-loop circulatory pathway, 551*f*
Single-stranded binding protein, 213
Sink (phloem), 471
Sink population, 917
Sinoatrial (SA) node, 608, 609
Sinornis, 271
Sinus, 738
Sinus venosus, 603
Siphon
  molluscs, 529, 643
  sea squirts, 546
Sipuleucel-T, 635
SiRNA (small-interfering RNA), 236
Sister chromatid, 148, 152
Skate, 549
Skeletal muscle, 742–47
  anatomy, 586, 742
  macroscopic anatomy and physiology, 742
  microscopic anatomy and physiology, 742–47
    ATP use in contraction, 104f, 744–45, 746–47
    muscle innervation, 745–46
    role of calcium in muscle contraction, 746–47
  sliding filament model, 743–45
Skeleton
  bird characteristic, 557
  cartilaginous fetal, 585
  cartilaginous fishes, 549
  chordates, 545
  cnidarian, 522, 734
  diversity, 734–35
    endoskeletons, 734, 735
    exoskeletons, 734–35
    hydrostatic skeleton, 734
  endoskeleton, 735
    advantages, 735
    composition, 734, 735
    echinoderms, 735
    jawed fishes, 548
    vertebrate characteristic, 547, 735
  exoskeleton, 734–35
    arthropods, 533–34, 734
    echinoderms, 539
    molluscs, 734
    zooids, 524
  human, 736–41
    aging process, 810–11
    appendicular skeleton, 738f, 740–41, 766, 780
    axial skeleton, 738–39
    bone growth and renewal, 736
    cartilaginous fetal, 585
    compared to Ardipithecus ramidus, 571f
    functions, 736
    joint classification, 741
    long bone anatomy, 736–38
  hydrostatic skeleton
    annelids, 530, 734
    cnidarian, 522
  mammal characteristic, 559
  monkey compared to Proconsul, 569f
  summary and assessment, 748–49
  vertebrate characteristic, 547, 735
Skill memory, 705
Skin
  adhesion junctions, 582
  adipose tissue, 584, 592
  amphibians, 551, 591, 755–56
  cutaneous receptor, 729
  fish, 755–56
  functions, 591
  gas exchange, 591, 661
  human
    accessory structures, 592–94
    aging process, 809
    anatomy, 591–92
    color, 199–200, 207, 577, 592
    dermis, 591, 592, 729
    epidermis, 591, 592
    hair follicles, 592f, 594, 809
    nails, 592
    oil glands, 593, 625
    regions, 591–92
    skin cancer, 592, 593
    subcutaneous layer, 591, 592
    sweat glands, 583f, 594, 595, 809
    thick, 592
    thin, 592
    wrinkles, 742, 744
  innate immune defense, 625
  reptiles, 553, 591, 755–56
Skin cancer, 592, 593
Skin color
  human, 199–200, 207, 577, 592
  melanocyte-stimulating hormone (MSH), 755–56
Skin gill, 540
Skink, 556
Skinner, B. F., 823
Skototropism, 486
Skull
  amniote lineage
    anapsid, 553
    diapsid, 553–54
    synapsid, 553
  human, 738–39
  vertebrate characteristic, 547, 549
S-layer of prokaryotic cell, 361
Sleep
  circadian rhythms, 492
  melatonin, 704, 766
  obesity, 655
Sleeping sickness, African, 388, 389*f*
Sliding filament model, 743–45
Slime layer, 361
Slime mold
  cellular slime molds, 390
  evolution and diversity, 389–90
  immune system evolution, 622
  plasmodial slime molds, 389, 390f
Slip, 509
Slug (social amoeba), 622
Small intestine
  anatomy, 645f, 647–48
  celiac disease, 650
  lumen, 646
Small RNA (sRNA), 236
Small-interfering RNA (siRNA), 236
Smell
  anosmia, 810
  human sense of smell, 717–18, 810
Smoking
  birth defects, 805
  cigarettes, 673, 700
  nicotine abuse, 700
  "vaping" vs., 673
Smooth endoplasmic reticulum, 69
Smooth muscle, 586
Smut, 402, 405
Snail
  anatomy, 529
  coevolution, 865
  egg-laying, 819–20
  hormone function, 753
  land, 530f, 644
  marine, 819–20
  omnivorous diet, 644
  shell, 530f
Snake
  brown tree, 913
  carnivore, 868f
  characteristics, 556
  chemical senses, 716
  coloration, 267–68, 862
  exotic species impact, 913
  food choice in garter snakes, 819
  garter, 819
  infrared sensory organ, 714, 715, 716
  medicinal uses, 555
  vertebrae and body plan, 518–19
Snowshoe hare, 861
Social amoeba, 622
Social behavior
  animal communication, 824–28
    auditory communication, 826
    chemical communication, 824–26
    tactile communication, 827–28
    visual communication, 826–27
  learning and social interactions, 822
Social interactions and learning, 822
Social parasite, 865*f*
Social status, 228
Social unit, 228
Society
  definition, 824
  fitness and societies, 831
Sockeye salmon, 304, 305*f*
Sodium, 24, 458. *See also* Salt
Sodium bicarbonate, 649
Sodium chloride (table salt), 24*f*
Sodium-potassium pump, 92*f,* 93, 696
Soft palate, 646
Soft spot. *See* Fontanel
Soil
  acid deposition, 878
  definition, 459
  erosion, 460, 910
  formation, 459
  permafrost, 886
  plant nutrition, 457, 459–60
    erosion, 460
    humus, 459, 460
    living organisms, 460
    particles, 459–60
    profile, 460
  tropical rain forest, 891
Soil erosion, 460, 910
Soil horizon, 460
Soil profile, 460
Solar radiation effect on climate, 882–84
Soldier (army ant), 832
Solidago canadensis, 499*f*
Soluble fiber, 653
Solute, 28, 88
Solute concentration
  water potential, 467
  water transport, 468
Solution
  acid, 30, 31
  base, 30
  definition, 28, 88
  hypertonic, 90–91
  hypotonic, 90
  isotonic, 89–90
Solvent, 28, 88
Somatic cell, 149
Somatic sense organ, 729–30
  cutaneous receptors, 729
  pain receptors, 729–30
  proprioceptors, 729
Somatic system, 708
  functions, 707
  nervous system organization, 694f
  pathways comparison, 709t
  somatic sense organ, 729–30
    cutaneous receptors, 729
    pain receptors, 729–30
    proprioceptors, 729
Somatosensory area, primary, 703, 729
Somatotropic hormone. *See* Growth hormone (GH)
Somite, 797
Songbird
  auditory communication, 826
  characteristics, 558
  edge effect, 917–18
  mercury, 901
  social interactions and learning, 822
  tracing phylogeny, 349
Sorus, 422
Sound wave detection, 724–25
Sour taste receptor, 717
Source (phloem), 471
Source population, 917
South African australopiths, 572
South American lantern fly, 862
Soy sauce, 135
Sparrow, white-crowned, 822
Special K. *See* Ketamine
Specialization. *See* Cellular differentiation
Speciation and macroevolution, 296–316
  anatomy of speciation, 300f
  definition, 297, 303
  how new species evolve, 297–303
    reproductive isolating mechanisms, 299–303
    what is a species, 297–99
  modes of speciation, 303–8
    adaptive radiation, 306–7
    allopatric speciation, 303–4, 305f
    convergent evolution, 307
    sympatric speciation, 304–5
  phylogenetic species concept, 300
  principles of macroevolution, 308–12
    developmental genes and macroevolution, 310–12
    macroevolution is not goal-oriented, 312
  summary and assessment, 314–16
Species. *See also* Speciation and macroevolution
  community structure, 856–57
  cryptic species, 297, 298
  definition, 297
  described, 906f
  keystone, 916
  Linnaean classification hierarchy, 7, 339–40
  pioneer species, 866
Species composition, 856
Species diversity, 856, 857, 890, 911
Species richness, 856, 860, 890

Sperm, human
  concentration, 776
  fertilization, 794–95, 804
  production, 773, 774–75
  spermatogenesis, 176–77, 774–75, 776
Spermatid, 775
Spermatocyte, 774–75
Spermatogenesis, 176–77, 774–75, 776
Spermatogonium, 775
Spermatozoa, 774*f*, 775
Sphagnum, 416
Sphenodon punctatus, 556
Sphenoid bone, 738
Spherocytosis, hereditary, 197
Sphincter, gastroesophageal, 646
Sphinx moth, 750
Sphygmomanometer, 610
Spice. *See* K2
Spicule, 521
Spider
  black widow, 537f
  carnivorous diet, 644
  characteristics, 537
Spike (virus), 355, 356, 357, 358, 359
Spike moss, 419
Spike-headed katydid, 891*f*
Spina bifida, 805
Spinal cord
  functions, 702
  human embryonic development, 806–7
  meninge, 702
  parasympathetic division, 709, 710f
  reflex action, 708
  spinal nerve, 707
  sympathetic division, 709, 710f
  vertebrate nervous system organization, 692f, 693f, 694
Spinal nerve, 707
Spiny anteater. *See* Echidna
Spiracle
  cartilaginous fishes, 549
  insects, 663
Spiral cleavage, 517
Spirillum, 62, 363, 364*f*
Spirillum volutans, 364*f*
Spirobranchus giganteus, 531–32
Spirochete, 62
Spirogyra, 378, 379*f*
Spleen, 623*f*, 624, 625
Sponge, 521–22
  asexual reproduction, 521
  digestion, 642
  regeneration, 522, 771
  transport, 601
Spongin, 521
Spongy bone, 585, 736, 738
Spongy mesophyll, 451
Spontaneous generation, 360, 361*f*
Spontaneous mutation, 238–39
Sporangiophore, 399, 400
Sporangium
  bryophytes, 417
  fungi, 399, 400
  lycophytes, 419
  plants, 414
  pteridophytes
    ferns, 422, 423
    whisk ferns, 420
  rhyniophytes, 418
  slime molds, 389–90
Spore
  anaphase II, 172
  bryophytes
    liverwort, 416
    mosses, 417
    plants, 414
    protist reproduction, 374
    slime molds, 389–90
  fungi, 397
    club fungi, 402
    sac fungi, 400f, 401
    smuts and rusts, 405
    zygospore fungi, 400
  lycophytes, 419
  pteridophytes
    ferns, 422, 423
    horsetails, 420
  vascular plants, 418
Sporophyte
  alternation of generation, 176, 414, 415, 849
  angiosperms
    conifers, 430f
    flowering plant reproduction, 496, 498f, 499
  bryophytes
    hornworts, 416
    liverworts, 416
    mosses, 416–17
  dominant generation, 415, 418
  gymnosperms, 425f
  lycophytes, 418
  pteridophytes, 422, 423
Sporozoan. *See* Apicomplexan
Sports, EPO misuse in, 686
Spotted owl, 888, 917
Spotted sweetlip, 864
Spring overturn, 897
Spring wood, 448
Springbok, 862
Spruce
  conifer diversity, 424f
  coniferous forest, 888
  ecological succession, 866f
  white, 866f
Squalamine, 555
Squamous cell carcinoma, 593
Squamous epithelium, 582–83
Squid
  cephalopod diversity, 530f
  discontinuous feeding, 643
  nervous system, 692–93f
Squid, bigfin reef, 530*f*
SRNA (small RNA), 236
SRY gene, 179, 249
St. Anthony's Fire, 404
Stabilizing selection, 286
Stable equilibrium phase, 843
Stamen
  flower structure, 428, 429f, 497
  flowering plant life cycle, 430f
Stapes, 725, 726
Staph arrangement, 363
Staphylococcus aureus, methicillin-resistant (MRSA), 279, 366
Starch, 40*f*, 41
Starfish. *See* Sea star
Starling, 823
Start and stop signal, 216
Start codon, 223
Startle response, 862
Statistical data, 10
Statistical significance, 11
Statocyst, 535, 726*f*, 727
Statolith, 487, 726*f*, 727
STD. *See* Sexually transmitted disease
Stegosaurus, 330
Stem
  asexual reproduction, 507–9
  elongation with gibberellins, 481
  lenticels, 437
  organization and diversity, 445–50
    diversity, 449–50
    herbaceous stems, 445–46, 448f
    structure, 445
    woody stems, 445f, 446–49
  shoot system, 440, 445
  underground, 449
  vascular bundle, 439, 440
Stem cell, 156
  adult stem cell, 157
  cloning, 156, 157
  embryonic stem cell, 157
  in fibrous connective tissue, 584f
  Parkinson disease, 706
  salamander regeneration, 588
Stent, 611
Stentor, 383
Steppe tortoise, 915*f*
Stereoscopic vision, 565–66, 719
Sterility
  postzygotic isolating mechanisms, 301f, 302–3
  sympatric speciation, 305
Sternum, 739
Steroid hormone
  actions, 753–54
  anabolic, 765, 776
  structure, 45–46
Stickelback fish, 820–21
Stigma, 428, 429*f*, 497
Stimulant, 701
Stimulatory pathway, 159
Stimulus
  classical conditioning, 822
  fixed action patterns (FAP), 820–21
  operant conditioning, 822–23
  plant movement, 485
    external stimuli, 485–87
    internal stimuli, 487–89
  plant transport, 477
  sign stimulus, 820–21
  vision, 715
Stingray, 549
Stinkhorn fungus, 402, 403*f*
Stirrup. *See* Stapes
Stolon, 399, 449, 450*f*, 507, 508*f*
Stoma
  bryophytes, 415
  functions, 116, 450–51
  guard cells, 437, 469
  opening and closing
    abscisic acid (ABA), 470, 483
    circadian rhythms, 492
  plant adaptation, 414
  water transport, 469–70
Stomach
  discontinuous feeding, 643
  human
    anatomy, 645f, 647
    bariatric surgery, 655
    gastroesophageal sphincter, 646
    Lap-Band, 655
    lumen, 646
  squid, 643
Stone Age dog. *See* New Guinea singing dog
Stone cell. *See* Sclereid
Stonewort, 379, 412, 413*f*
Storage (root), 442, 444
STR (short tandem repeat sequence), 247
Stramenopile, 380–82
  brown algae, 380–81
  diatoms, 381
  golden brown algae, 381
  water molds, 381–82
Stratification
  ecological, 860
  fossil record, 263, 323
  lake overturn, 896–97
  temperate deciduous forest, 889
Stratified epithelium, 583
Stratum, 263, 323
Strawberry, 449, 450*f*, 507, 508*f*
Stream ecosystem, 895
Strep arrangement, 363
Strep throat, 669
Streptococcus mutans, 244
Streptococcus pneumoniae, 208
Streptococcus pyogenes, 637, 669
Streptococcus thermophilus, 135, 364*f*
Stress and adrenal glands, 760–61
Striation
  cardiac muscle, 586f, 587
  skeletal muscle, 586, 743
Strix occidentalis caurina, 888, 917
Strobilus, 419, 420
Stroke, 611, 654, 703–4
Stroma, 73
Stromatolite, 325, 327*f*
Strongylocentrotus, 539*f*
Structural gene, 229, 230
Structural genomics, 251
Structural molecule, 41
Sturnus vulgaris, 823
Style (flower structure), 497
Stylet, 472
Styloid process, 576
Subatomic particle, 20–21
Subcutaneous layer, 591, 592
Subduction zone, 333
Suberin, 443, 448
Submucosa, 646
Subordination in societies, 831
Subspecies, 292*f*
Substance abuse, 700–701
Substantia nigra, 706
Substrate, 105, 106
Substrate-level ATP synthesis, 132
Substrate-level phosphorylation, 132
Subthalamic nucleus, 706
Succession, ecological, 866–67
Succession, soil, 460
Succinea, 865
Sucrose, 471, 652
Suctoria, 383
Sulcus, 703
Sulfur dioxide, 878, 913
Sum rule of probability, 193
Summer wood, 448
Sun and solar radiation, 882–83
Sundew, Cape, 464
Sunflower, 476
Sunscreen, 593
Superantigen, 365
"Superbug," 279. *See also* Antibiotic
Supergroup, protist, 374
Superior vena cava, 606
Superposition, law of, 323
Support systems. *See* Locomotion and support system; Muscular system; Skeleton
Supportive connective tissue, 584–85
  bone, 584, 585
    compact bone, 585
    spongy bone, 585
  cartilage, 584–85
    elastic cartilage, 585
    fibrocartilage, 585
    hyaline cartilage, 584–85, 737
    knee, 741
Supraorbital ridge, 738
Surface tension, 29
Surface-area-to-volume ratio, 59
Surrogate mother, 783
Survival mechanism, plant, 437, 438, 484
Survivorship, 839–40
Sushi, 887
Suspensor, 502
Suture (skull), 738, 739*f*
Swallowing, 646
Swamp
  ecosystem, 895
  mangrove swamp, 898
  methanogens, 368–69
Sweat, 594, 595
Sweat gland
  aging process, 809
  anatomy, 583f
  functions, 594, 595
Sweet potato
  propagation, 509
  root system, 444
Sweet taste receptor, 717
Sweetlip, spotted, 864
Swim bladder, 549–50
Swimmeret, 535
Swordfish

anatomy, 550f
mercury, 901
Swyer syndrome, 179
Symbiosis. *See also* Symbiotic relationship
cleaning, 864
definition, 365, 862
Symbiotic relationship
algae, 377, 864
archaea, 365–66, 369
bacteria, 365–66, 367, 416, 863, 864
commensalism
bacteria, 365
ecology of communities, 863–64
ecology of communities, 862–64
commensalism, 863–64
mutualism, 863t, 864
parasitism, 863
fungi, 400, 401, 405–7, 444, 462, 464, 864
hornworts and cyanobacteria, 416
lichens, 405–6, 864
mutualism
bacteria, 365, 367, 864
cross-pollination, 500–501, 864
definition, 864
ecology of communities, 863t, 864
fungi, 401, 406–7, 444, 462, 464, 864
lichen, 367, 405, 864
mycorrhizae, 406–7, 444, 462, 464, 864
mycorrhizae, 406–7, 444, 462, 464, 864
parasitism
alternate host, 405
bacteria, 365–66, 863
coevolution, 865
definition, 863
dodder vine, 464, 485
ecology of communities, 863
ectoparasite, 863
endoparasite, 863
flatworm, 526–27
fungi, 398–99, 400, 401
lampreys, 548
lichens, 405
mineral uptake in plants, 464
pathogens, 365–66
predator-prey interactions, 858
protists, 384, 388, 389
regulation of population size, 845
roundworms, 532f, 533
secondary host, 526, 527, 863
viruses, 354, 356, 863
Symmetry
animal evolution, 515–16, 692
bilateral symmetry, 516, 518–19, 692
body plans, 518–19
definition, 515
radial symmetry, 515, 518–19, 539
Sympathetic division, 707, 709, 710f
Sympatric speciation, 304–5
alloploidy, 305
autoploidy, 305
polyploidy, 305
Synapse
definition, 697
integration, 699
transmission across, 697–98
Synapsid skull, 553, 560
Synapsis, 168
Synaptic cleft, 697, 698, 745
Synaptic vesicle, 745, 746
Synaptonemal complex, 168
Synovial joint, 741
Synovial membrane, 741
Synpolydactyly, 802
Synthetic biology, 362
Syphilis, 789
System, Earth as a, 122
Systematic biologist, 344
Systematic biology, 338–40
definition, 338
field of taxonomy, 338–39
Linnaean classification hierarchy, 339–40, 342–43
Linnaean taxonomy, 339
Systematics, 7, 338
Systemic circuit
double-loop circulatory pathway, 604
human, 609–10
Systemic lupus erythematosus, 637
Systemin, 484
Systole, 607, 610
Systolic pressure, 610

T

T cell
aging process, 810
allergic responses, 637
autoimmune disorders, 637
cell-mediated immunity, 630–32
cytotoxic T cell, 628f, 630, 631
delayed allergic response, 637
development, 624, 628
functions, 615
helper T cell, 628f, 629, 630
HIV/AIDS, 631–32, 633, 785
thymus, 766
T cell receptor (TCR), 630
T system, 742
T tubule, 742, 746
T wave, 608
T_3. *See* Triiodothyronine
T_4. *See* Thyroxine
Tactile communication, 827–28
Taenia solium, 527
Taeniura lymmagill, 549f
Taiga, 888
Tail
histone molecule, 233
human, 807f
postanal, 545
Tailbone. *See* Coccyx
Tall-grass prairie, 894f
Tamarin, 829
Tandem repeat, 252
Tanning, 592, 593
Tapetum lucidum, 719
Tapeworm, 527, 642
Taproot, 444
Target cell, 753, 754
Targeted therapy for breast cancer, 781
Tarsal bone, 741
Tarsier, 565f, 566f
Taste
parietal lobes, 703
sense of taste in humans, 716f, 717, 718
taste bud, 646, 716f, 717
Taste bud, 646, 716f, 717
Taste pore, 717
Tatum, Edward, 215
Taxon
definition, 7, 338
hierarchy, 339–40
Taxonomist, 297, 338
Taxonomy, systematics, and phylogeny, 337–51
definition, 7, 338
five-kingdom system, 341
phylogeny, 344–49
cladistics, 346, 347f
common ancestor, 344
interpreting a phylogeny, 344–45
tracing phylogeny, 346–49
summary and assessment, 350–51
supergroups, 374
systematic biology, 338–40
definition, 338
field of taxonomy, 338–39
Linnaean classification hierarchy, 339–40
Linnaean taxonomy, 339
three-domain system, 341–44
defining the domains, 341
domain Archaea, 341, 343–44
domain Bacteria, 341–43
domain Eukarya, 341, 343t, 344
Tay-Sachs disease, 70–71
TB (tuberculosis), 637, 670f, 672
Tbx5 gene, 311
TCR (T cell receptor), 630
Teardrop Glacier, 418
Technology, 13
Tectorial membrane, 726
Tegument, 526
Telomere, 159, 811
Telophase, 153–54, 172
Telophase I, 172
TEM (transmission electron microscope), 60
Temperate deciduous forest, 885f, 888–89
Temperate grassland, 893, 894f
Temperate rain forest, 888
Temperature
Earth
biome distribution, 884f, 885
climate change, 875, 877, 913–14
enzyme speed impact, 107, 108f
fever, 627
lake, 896–97
skeletal muscle, 742
water, 27f
Temperature regulation
capillary exchange, 618
homeostasis with negative feedback, 595, 596f
prostaglandins, 766
skeletal muscle, 742
sweat glands, 594, 595
Template, 213
Template strand, 217
Temporal bone, 738, 739f
Temporal isolation, 299–302
Temporal lobe, 703
Tendon
dense fibrous connective tissue, 584
skeletal muscle, 586, 742
Tendril, 449, 452, 486–87
Teosinte, 427
Terminal bud
apical dominance, 479
leaf, 451f
shoot system, 440, 441f, 445
Termination, 224–25
Termite
mutualism, 864
pheromones, 824
Terrestrial ecosystem, 885–94
biome distribution, 884f
coniferous forest, 857f, 884, 885f, 888
deserts, 893–94
grasslands, 892–93
shrublands, 892
temperate deciduous forest, 885f, 888–89
tropical rain forest, 885f, 890–91
tundra, 884, 885f, 886
Territoriality, 289, 828–29
Territory
definition, 289, 828
territoriality, 289, 828–29
Tertiary period, 326t, 330–31, 334
Tertiary protein structure, 48, 49f, 50
Test (foraminiferan and radiolarian), 391
Testcross, 193–94
Testis
cancer, 774
functions, 765–66, 771, 773–74
gonadotropic hormones, 755
major glands, 752f
steroid hormones, 753–54
Testosterone
aging process, 811
functions, 45
pheromones, 753
production, 765, 766, 776
women, 766
Testudo horsefieldii, 915f
Tetanus, 365
Tetany, 742, 760
Tetrad, 168
Tetrahydrocannabinol (THC), 701
Tetrapod, 261, 546t, 547, 551
T-even bacteriophage, 354f
Thalamus, 693, 704, 722
Thallose liverwort, 415–16
Thallus, 415, 416
Thamnophis elegans, 819
Thaumaptilon, 328f
THC (tetrahydrocannabinol), 701
Thecodont, 554
Theory, 11
Therapeutic cloning, 156, 157
Thermoacidophile, 368f, 369
Thermocline, 896
Thermodynamic law
cells and the flow of energy, 101–2
ecosystem dynamics, 869–71
Thermoreceptor, 715, 729
Thermus aquaticus, 246
Theropod, 330
Thick skin, 592
Thighbone. *See* Femur
Thigmomorphogenesis, 486
Thigmotropism, 485, 486–87
Thin skin, 592
Thoracic cage. *See* Rib cage
Thoracic cavity, 590, 665
Thoracic duct, 623f, 624
Thoracic vertebra, 739
Thorn, 438
Threatened species, 906
3-phosphoglycerate (3PG), 124
Three-domain system, 341–44
Archaea, 341, 343–44
Bacteria, 341–43
defining the domains, 341
Eukarya, 341, 343t, 344
Threshold, 697
Thrombin, 616
Thrombocyte. *See* Platelet
Thromboembolism, 611
Thrombosis, 767
Thrombus, 611
Thrush, 633
Thumb, opposable
derived trait, 344–45
human, 740
primate evolution, 565, 566f
Thylakoid, 63, 73, 116
Thylakoid membrane, 120, 121f
Thymine dimer, 239
Thymosin, 766
Thymus
aging process, 810
development, 807f
endocrine system, 752f
functions, 624, 766
lymphatic system organization, 623f
Thyroid, 752f, 755, 756f, 758–59, 811
Thyroid-stimulating hormone (TSH), 755, 758
Thyroxine, 755, 758, 759
Tibia, 741
Tick, 537
Tidal pool, 898
Tidal ventilation mechanism, 666
Tide, red, 382–83
Tiger, 914
Tiger shark, 549
Tight junction
cells, 95
epithelial tissues, 582
Tiktaalik roseae, 261, 270–71, 551
Timberline, 838
Tineas, 402
Tinsel flagellum, 385

Tissue, animal
definition, 2, 582
tissue engineering, 638, 671
tissue types, 582–89
connective tissue, 582, 583–86
epithelial tissue, 582–83
muscular tissue, 582, 586–87
nervous tissue, 582, 587–88
Tissue, plant
asexual reproduction with tissue culture, 508f, 509
definition, 2, 436
flowering plant cells and tissues, 436–39
epidermal tissue, 436–37
ground tissue, 436, 437–38
vascular tissue, 436, 439
plant adaptation to land, 414
Tissue culture of plants, 508*f*, 509
Tissue engineering of organs, 638, 671
Tissue stages of development, 796
Titer, antibody, 634
Tmesipteris, 420
TNF (tumor necrosis factor), 630, 632
Toad
characteristics, 551, 552
parental care, 847f
Toad, midwife, 847*f*
Tobacco
birth defects, 805
defense mechanism, 485
nicotine abuse, 700
vaping, 673
Tobacco hornworm, 750
Tobacco mosaic virus, 355, 356
Tolerance model of succession, 867
Tomato
defense response, 484–85
ethylene ripening, 483–84
nutrient deficiencies, 458f
true berry, 505f
Tone, muscle, 729, 742
Tongue
amphibians, 551
human, 646
sense of taste, 716f, 717
Tonicity, 89
Tonsil, 623*f*, 625, 807*f*
Tool use
birds, 576
Cro-Magnons, 575
human evolution, 576
primates, 576
Tooth
adaptations to diet, 644
canines, 644
human, 644, 739
incisors, 644
mammal characteristics, 549
molars, 644
premolars, 644
Tooth socket, 739
Topography and climate, 883
Topsoil, 460, 910
Torpedo stage of seed development, 502, 503*f*
Torpor, 551
Tortoise
characteristics, 556
elongated, 915f
Galápagos, 266f
overexploitation, 915
reptile diversity, 556
steppe, 915f
Totipotency
bryophytes, 418
flowering plants, 509
zygote, 798
Toxin
botulinum, 744
saxitoxin, 382–83
Shiga toxin, 366
tetanus, 365
Toxoplasma gondii, 384, 633, 805
Toxoplasmic encephalitis, 633
Toxoplasmosis, 384, 633, 805
Trachea
arthropods, 534
insects, 536–37, 602f, 661f, 662–63
mammals
dog, 661f
human, 636, 664, 669–70
respiratory tract disorders, 669–70
asthma, 636, 670
bronchitis, 669–70
choking, 669
Tracheal gill, 663
Tracheid
plant adaptation, 414
plant survival mechanisms, 438
structure, 439, 465
Trachemys scripta elegans, 915*f*
Tracheole, 663
Tracheostomy, 669
Tracheotomy, 669
Tract (interneuron bundle), 702
Trade wind, 883
Trait
analogous trait, 307
ancestral trait, 344–45
behavioral trait, 348
cladistics, 346, 347f
derived trait, 344–45, 346, 347f
fossil, 346–47
molecular trait, 348–49
morphological trait, 347–48
multifactorial trait, 200–201
multiple alleliic trait, 197–98
polygenic trait, 286
systemic biology, 338
X-linked trait, 203
Trans configuration, 42
Trans fat
nutrition and health, 35, 44, 653, 654
structure, 42
Transcription, 217–20
definition, 216
homeotic genes, 802
messenger RNA production, 217–18
posttranscriptional control, 232, 235–36
RNA processing, 218–20
transcriptional control, 234–35
Transcription activator, 235
Transcription factor, 235*f*
Transcriptional control, 232, 234–35
Transduction
plant hormones, 477
plant movement, 485
prokaryote reproduction, 363
Transduction pathway, 477
Transfer rate of carbon cycle, 873, 874*f*
Transfer RNA (tRNA), 50, 68, 220–21
Transform boundary, 333
Transformation of bacteria, 208–9
DIY biology, 362
prokaryote reproduction, 363
Transfusion, blood, 614
Transgenic animal, 248–49
Transgenic organism, 245. *See also under* Genetically modified
Transient ischemic attack, 600
Transitional fossil, 270, 271*f*
Translation, 220–25
control, 232
definition, 216
eukaryotic regulation, 235–36
ribosomal RNA, 221–22
steps, 222–25
transfer RNA, 220–21
Translocation, chromosome, 181, 182
Transmissible spongiform encephalopathy (TSE), 50, 360
Transmission across a synapse, 697–98
Transmission electron microscope (TEM), 60
Transmission of nerve impulses, 696–98
action potential, 696–97
propagation of action potentials, 697
resting potential, 696
transmission across a synapse, 697–98
Transpiration
water cycle, 873
water transport in plants, 468, 469f
Transplant
artificial lungs, 671
artificial retina, 722, 723
bone marrow, 634, 636
liver, 651
pancreas, 764–65
rejection, 638
xenotransplantation, 555, 638, 764–65
Transport, animal. *See also* Circulation and cardiovascular system
breathing and transport of gases, 667–69
amphibians, 660, 665–67
birds, 666
fish, 662
human, 666–69
mammals, 665–66
reptiles, 665–66
gas exchange and transport
partial pressure, 668
surfaces, 89f, 660–64
transport of oxygen and carbon dioxide, 668–69
human respiratory system, 668–69
invertebrate transport, 601–2
closed circulatory system, 602
open circulatory system, 601, 602f
vertebrate transport, 603–4
Transport, plant, 465–73
organic nutrient transport, 470–73
reviewing xylem and phloem structure, 465
stomata opening and closing, 469–70
water potential, 465–68
water transport
cohesion-tension model of xylem, 468, 469f
transpiration, 468, 469f
Transport across a membrane
active, 91–94
bulk transport, 93–94
passive, 88–91
facilitated transport, 91
osmosis, 88–91
Transposon, 252, 253*f*, 568
Trastuzumab, 630
Traumotropism, 486
Trawling, 914, 916*f*
Tree
annual rings, 448–49
coastal erosion, 460
ecological succession, 866, 867
timberline, 838
trunk, 449f
Tree fern, 421, 422
Tree frog, 552
The Tree of Life Project, 272
Trematode. *See* Fluke
Treponema pallidum, 789
Triad, 649
Triassic period, 326*t*, 330, 333, 334
Tribe (taxonomy), 569*t*
Tricep, 743*f*
Triceratops, 330
Trichinella, 532*f*, 533
Trichinosis, 533
Trichocyst, 383
Trichome
defense mechanisms, 485
leaf, 450
plant epidermal cells, 437
Trichomonas vaginalis, 385, 789
Trichomoniasis, 385, 789
Tricuspid valve, 606
Trigger hair, 489
Triglyceride, 42–44, 653
Triiodothyronine, 755, 758, 759
Trilobite, 328
Trimester, 803
Triplet code, 216
Triploblastic animal, 516–17
deuterostomes, 516–17
protostomes, 516–17
Triploid plant, 305
Trisomy, 178
Trisomy 21, 178–79
TRNA (transfer RNA). *See* Transfer RNA (tRNA)
Trochozoan, 524–32
annelids, 530–32. See also Earthworm
flatworms, 524–26
flukes, 526–27
parasitic, 526–27, 642
planarians, 525–26
flukes, 526–27
molluscs, 527–30
anatomy, 528
bivalves, 528–29
cephalopods, 529–30
gastropods, 529, 530f
rotifers, 527, 528f
Trophic level, 870*f*, 871
Trophoblast, 804, 806
Tropical forest, 890–91
Tropical rain forest
biome, 885f, 890–91
destruction, 122, 877
habitat loss, 911, 912f
soil profile, 460
species richness, 856, 890
transpiration, 468, 469f
Tropism, 485, 486–87
Tropomyosin, 746
Troponin, 746
Trout, rainbow, 895*f*
Trp operon, 229*f*, 230
True berry, 505*f*
True coelom, 517
True moss, 416
True rib, 739
True-breeding, 187
Truffle, 406–7
Trumpet vine, 504
Trypanosoma brucei, 388, 389*f*
Trypanosoma cruzi, 388
Trypanosome, 388
Trypsin, 649, 651
TSE (transmissible spongiform encephalopathy), 50, 360
Tsetse fly, 388
TSH (thyroid-stimulating hormone), 755, 758
Tsien, Roger, 219
Tuatara, 556
Tubal ligation, 777
Tubastrea, 523*f*
Tube foot, 540, 644, 734
Tube worm, 643–44
Tuber, 449, 450*f*, 507–9
Tuber melanosporum, 407
Tuberculosis (TB), 637, 670*f*, 672
Tubular reabsorption, 683–85
Tubular secretion, 684*f*, 685
Tulip, 495
Tumor, 158–59, 587
Tumor necrosis factor (TNF), 630, 632
Tumor suppressor gene, 159
Tuna, 901, 914
Tundra, 884, 885*f*, 886
Tunic, 546
Tunicate, 546
Turgor movement, 485, 488–89
Turgor pressure
opening and closing of stomata, 469

plant movement, 479, 480f, 488
wilting, 90, 467
Turner syndrome, 179–80
Turtle
anapsid skull, 553
Asian, 915
characteristics, 556
conservation, 877
evolution, 553–54
migration, 823
overexploitation, 914, 915
phylogeny, 347
red-eared slider, 915
reptile diversity, 556, 556f
salt gland, 677, 681
Twin
identical, 237f
studies, 819
Two-circuit circulatory pathway. See Double-loop circulatory pathway
Two-trait testcross, 194
Tympanic canal, 726
Tympanic membrane, 725
Tympanum, 536, 537f, 724
Type 1 diabetes, 654, 763–65
Type 2 diabetes, 654, 765
Typhlosole, 642
Tyrannosaurus rex, 330, 554

U

Ulcer, 647, 767
Ulipristal acetate, 781
Ulna, 740
Ultraviolet (UV) light
nectar guides, 501, 718, 719f
ozone depletion, 913
tanning, 592, 593
Ulva, 378
Umami taste receptor, 717
Umbilical cord, 807, 808
Unambiguousness, 216
Unbalanced translocation, 182
Unconditioned response, 822
Understory, 889, 890
Undescended testis, 773
Unidirectional tropism, 485
Uniform distribution, 838
Uniformitarianism, 264
Unipolar shape, 696
Unique noncoding DNA, 252
Unpacking, 233
Unsaturated fat, 653, 654
Unsaturated fatty acid, 42, 43f
Unwinding, DNA, 213
Upper respiratory tract disorder, 669
common cold, 669
strep throat, 669
Upwelling, 902
Uracil, 215
Urea, 678
production, 649, 654
tubular reabsorption, 685
Ureter, 681, 682f
Urethra, 681, 682f
human male, 773f, 773t, 775–76
Urey, Harold, 319
Uric acid
definition, 678
gout, 678
insects, 679
Urinary bladder, 681, 682f, 776
Urinary incontinence, 810
Urinary system, human, 681–88
antidiuretic hormone (ADH), 755
homeostasis and kidneys, 685–88
acid-base balance, 685, 687–88
water-salt balance, 685–87, 760, 761
kidneys, 681–82
nephrons, 681–82, 683f
urine formation, 682–85
glomerular filtration, 682–83, 684f
tubular reabsorption, 683–85
tubular secretion, 684f, 685
Urine
definition, 679
earthworms, 679
energy flow, 869
fish, 680
human, 682–85
acid-base balance, 688
formation, 682–85
glomerular filtration, 682–83, 684f
tubular reabsorption, 683–85
tubular secretion, 684f, 685
kangaroo rat, 680
Urochordate, 546
Ursus arctos horribilis, 907, 912, 916–17
Ustilago, 405
Uterine cycle, 779–80
estrogen and progesterone, 766, 779–80
menstrual phase, 779, 780
pheromones, 753
proliferative phase, 779f, 780
prostaglandins, 766
secretory phase, 779f, 780
Uterine tube, 777, 788
Uterus
human, 777
hysterectomy, 787–88
uterine cycle, 779–80
marsupials, 560
placental mammals, 560t
uterine cycle, 766, 779–80
Utricle, 728
UV radiation. See Ultraviolet light
UVA ray, 593
UVB ray, 593
Uvula, 646

V

Vaccine
active immunity, 632–34
aging process, 810
booster, 634
cancer, 635
contraceptive vaccines, 782
diabetes mellitus, 765
hepatitis, 788
HIV, 786
human papillomavirus (HPV), 788
pneumonia experiments, 208
smoking, 700
Vacuole
amoeboids, 374
planarians, 642
plant, 72–73
Vagina, 385, 777–78, 789
Vaginitis, 385, 789
Vaginosis, bacterial, 789
Vagus nerve, 707
Valence shell, 23
Valve
heart
atrioventricular valves, 606, 607f
mitral atrioventricular valve, 607f
semilunar valves, 607f
lymphatic system, 624
veins, 603, 610
Vampire bat, 832
Van Ermengem, Emile Pierre, 744
Van Niel, C. B., 117
"Vaping," 673, 700
Variation, heritable, 267
Varicose vein, 610
Vas deferens, 773–74, 775
Vascular bundle
definition, 439
stem structure, 440, 445
herbaceous stems, 446
woody plants, 447
Vascular cambium
bark, 448
definition, 436, 445
wood, 448, 449f
woody stems, 446, 447
Vascular cylinder
definition, 439
root system, 442f, 443
Vascular plant
dominance, 417
features, 413t
megaphylls, 420f
plant evolution, 412, 413t, 417–19
Vascular ray, 439
Vascular tissue, plant
definition, 415
evolution of lycophytes, 417–18
features, 436, 439
monocots and eudicots, 440–41
root system, 443
structure, 439
Vasectomy, 774, 782
Vauxia, 328f
Vector
African sleeping sickness, 388
malaria, 384
plasmid, 245
Veganism, 653
Vegetal pole, 796
Vegetarianism, 653
Vegetative cell, 377
Vegetative organ, 440–41
Vein
human
anatomy, 603
blood pressure, 610
hepatic portal vein, 649
hepatic vein, 649
varicose, 610
vena cava, 605f, 606, 609
leaf (plant), 439, 440
Velocity, blood, 610
Vena cava, 605f, 606, 609
Venation, 440
Venom, 555
Ventilation. See also Breathing and transport of gases
definition, 660
fish, 662
human rate, 666
methods, 665
Ventral cavity, 590
Ventral nerve cord, 535
Ventral root, 707
Ventricle (brain), 702, 703
Ventricle (heart)
amphibians, 604
fish, 603
human, 606, 607
reptiles, 604
septum, 604
Ventricular fibrillation, 608f, 609
Venule, 603, 604f, 609, 610
Venus flytrap, 452–53, 489
Vermiform appendix. See Appendix
Vernal equinox, 882
Vertebra, 739, 740, 797
Vertebral cavity, 590
Vertebral column
human, 738f, 739, 797
vertebrate characteristic, 547, 549
Vertebrate. See also Reproductive system
characteristics, 520t, 547, 549
definition, 514
evolution, 544–63
amniotes, 546t, 548, 553–54
amphibians, 546t, 547, 551–52
chordates, 545–46
fishes, 546t, 547, 548–50, 552t
gnathostomes, 546t
mammals, 546t, 559–61
nervous system, 692f, 693–94
reptiles, 546t, 547–48, 553–58
summary and assessment, 562–63
tetrapods, 546t, 547
vertebrates, 547–48
history of life, 329
medicines from, 555
nervous system, 692f, 693–94
central nervous system, 694
mammals, 694
organization, 693–94
peripheral nervous system, 694
osmoregulation
aquatic vertebrates, 677, 679–80
terrestrial vertebrates, 680–81
kangaroo rat, 680, 681f
marine mammals and sea birds, 680–81
transport, 551f
Vertigo, 728
Vervet monkey, 826
Vesicle, 66, 321
Vesicular follicle, 778
Vessel element
structure, 465
vascular tissue of plants, 439
Vestibular canal, 726
Vestibular gland, 778
Vestibular nerve, 728
Vestibule, 725, 728
Vestigial structure, 262, 273
Video-enhanced contrast microscopy, 61
Villus, 647–48
Virchow, Rudolph, 58
Virginia opossum, 560
Viroid, 360
Virus, 354–60. See also Archaea; Bacterium
classification, 355
definition, 354
discovery, 355
emerging viruses, 358–60
host specificity, 356
human viral diseases, 354t
origins, 354–55
parasitic nature, 354, 356, 863
prions, 360
prokaryotes, 360–63
reproduction, 363
structure, 360–61, 362f
reproduction, 356–60
retroviruses, 358
structure, 355–56
summary and assessment, 370–72
viroids, 360
Visceral mass, 528
Visceral muscle. See Smooth muscle
Visceroceptor, 715
Vision, 718–24
animal detection of light, 718–19
artificial retinas, 722, 723
camera-type eye, 719
compound eye, 533f, 534, 535, 718
human eye, 719–24
aging process, 810
anatomy, 719–20
disorders, 722–24, 810
focusing, 720, 722–24
integration of visual signals, 721–22
photoreceptors, 719, 720–21
from the retina to the visual cortex, 722
panoramic, 719
stereoscopic, 565–66, 719
Visual accommodation, 720
Visual communication, 826–27
Visual cortex, 722
Vitamin A, 720
Vitamin D, 593, 759
Vitamins, 654–56
coenzymes, 107
definition, 654
folic acid, 805
minerals, 656
Viviparity, 772
VNO (vomeronasal organ), 826

Vocal cord, 664, 776
Vocal pouch, 847*f*
Voice box. *See* Larynx
Voltage, 696
Volume, 726–27
Volume and breathing, 665–66
Voluntary muscle. *See* Skeletal muscle
Volvox, 2, 378
Vomeronasal organ (VNO), 826
Vomiting, 646
Vulva, 777*f,* 778, 800

W

Wächtershäuser, Günter, 319
Waggle dance, 827
Wallace, Alfred Russell, 267
Warbler
blackburnian, 859f
black-throated green, 859f
Cape May, 859f
coevolution, 865
edge effect, 918f
speciation, 858, 859f
yellow, 865, 918f
Warfarin, 108
Warm-bloodedness. *See* Homeothermism
Warning coloration, 862
Warren, Robin, 647
Wart
genital, 592, 786–87
skin, 592
Wasp
anatomy, 536f
inclusive fitness, 832
mimicry, 862
orchid pollination, 429
pollination, 500
yellow jacket, 862
Waste recycling and indirect value of biodiversity, 910
Water. *See also* Aquatic ecosystem
chemistry, 26–29
hydrogen bonding, 26–27
properties of water, 27–29
cycle, 873, 895, 910
fresh water provision and biodiversity, 910
groundwater mining, 873
ice
freezing of water, 29
lake ecosystems, 896
molecule, 26f
plant nutrition
soil, 457
water and mineral uptake, 461–64
plant transport mechanisms
water potential, 465–68
water transport, 468–69
pollution, 460, 910
properties, 27–29
temperature, 27f
water-salt balance, human, 685–87, 760, 761
xylem and phloem sap, 465
Water column, 468
Water cycle, 873, 895, 910
Water hyacinth, 912
Water mold, 374, 381–82
Water potential, 465–68
Water pressure
auxin, 479
water potential, 465, 467
Water strider, 897*f*
Water table, 873, 895
Water transport, 468–69
cohesion-tension model of xylem, 468, 469f
transpiration, 468, 469f
water potential, 465–68
Water vascular system, 539, 540
Waterfowl characteristic, 558
Water-salt balance. *See also* Osmoregulation
adrenal glands, 760, 761
human kidneys, 685–87
Watson, James, 211
Watson and Crick model of DNA, 211–12
Wax, 46
Weather
biomes
coniferous forests, 888
desert, 893–94
grassland, 893
shrubland, 892
temperate deciduous forests, 888
tropical rain forests, 890
tundra, 886
bodies of water, 882, 883, 900–902
climate change, 877
ocean currents, 900–902
solar radiation, 882–83
topography, 883
Weathering
phosphorus cycle, 875
soil creation, 459
Wegener, Alfred, 332
Weight, human birth, 287*f*
Weinberg, Wilhelm, 280
Welwitschia, 425*f,* 426
Went, Frits W., 478–79
West Nile encephalitis, 358
Westerly, prevailing, 883
Western bluebird, 845
Western hemlock, 866*f*
Wetland
ecosystem, 895
habitat loss, 911
restoration, 919
waste recycling, 910
Whale
auditory communication, 826
echolocation, 715f, 716
filter feeding, 643
sushi, 887
teeth, 644
Whale, baleen, 643
Whale, humpback, 826
Wheat rust, 405
Whiptail lizard, 171
Whisk fern, 420
White blood cell. *See also* B cell; T cell
agranular leukocytes, 615
antigen binding, 622
blood element, 585f, 586, 613, 615
development, 624
in fibrous connective tissue, 584f
function, 586
granular leukocytes, 615
basophils, 615
eosinophils, 615, 627
neutrophils, 615, 627
inflammatory response, 626
lymphocyte, 615
development, 624
function, 586
natural killer (NK) cells, 627
primary lymphoid organs, 624
monocytes, 615
natural killer (NK) cells, 627
White matter, 702
White nose syndrome, 401
White pulp, 625
White spruce, 866*f*
Whitebark pine, 864
White-crowned sparrow, 822
White-faced monkey, 565*f*
White-tailed deer, 848
Whittaker, R. H., 341
Wild-collection trade, 914, 915
Wildlife conservation. *See* Conservation of biodiversity
Wilkins, Maurice H. F., 211
Williams syndrome, 181*f,* 182
Willow, 443*f*
Wilson, E. O., 860
Wilting, 90, 125, 467
Wind
circulation, 882–83
fruit dispersal, 504
lake overturn, 896–97
ocean currents, 900–902
-pollination, 429, 499, 501
Windpipe. *See* Trachea
Wine, 382
Wing, 557–58
Wiwaxia, 328*f*
Wobble hypothesis, 220
Woese, Carl, 341, 368
Wolf, 855, 887, 917
Womb. *See* Uterus
Wood, 448–49
Woodpecker, 558
Woody stem, 446–49
bark, 447–48
organization, 445f
stem diversity, 449
wood, 448–49
Worker (army ant), 832
Wrasse, 771, 864
Wrinkle, 742, 744
Wrist, 740
Wuchereria bancrofti, 532*f,* 533

X

X. *See* MDMA (methylenedioxymethamphetamine)
Xanthoria, 406*f*
Xenical, 655
Xenotransplantation, 555, 638, 764–65
X-inactivation, 234*f*
Xiphias gladius, 550*f,* 901
X-linked agammaglobulinemia (XLA), 636
X-linked disorder, 202–4
X-linked inheritance, 201–4
X-linked trait, 203
X-ray
birth defects, 805
diffraction of DNA, 210–11
"XX male," 179
"XY female," 179
Xylem
features, 439
herbaceous stems, 446
leaves, 451f
mineral uptake, 462
plant survival mechanisms, 438
plant transport systems, 465–68
procambium, 445
root system, 443
structure, 439, 440, 445, 465
vascular plant, 418
water transport
cohesion-tension model, 468, 469f
transpiration, 468
woody plants, 447
bark, 447
wood, 448, 449f
Xylem ray, 447
Xylem sap, 465, 468

Y

Yeast
fermentation, 135
logistic growth, 842–43
sac fungi, 400–402
Yeast infection, 402, 789
Yellow jacket wasp, 862
Yellow tube sponge, 521*f*
Yellow warbler, 865, 918*f*
Yellow-rumped warbler, 859*f*
Yellowstone National Park, 855, 887, 917
Yersinia pestis, 846
Yolk, 772, 795, 796, 803
Yolk sac, 806

Z

Zamia pumila, 424
Zeatin, 481. *See also* Cytokinin
Zebra finch, 829
Zebra mussel, 912
Zero population growth, 850, 851
Zona pellucida, 794
Zone of cell division, 442*f,* 443
Zone of elongation, 442*f,* 443
Zone of maturation, 442*f,* 443
Zooflagellate, 374
Zooid, 524
Zooplankton, 897, 899
Zoospore
chytrids, 399
definition, 378
water molds, 382
Zooxanthella, 382, 899
Zygomatic bone, 739
Zygomycota, 396, 398, 399–400
Zygospore, 377–78, 400
Zygospore fungus. *See* Zygomycota
Zygote
angiosperms, 429, 430f, 496
animal development, 582, 795, 798
conifer life cycle, 425f
definition, 167, 299
differentiation, 582
fern, 422f
human fertilization, 777, 778, 780, 794–95, 804
land plants, 412, 413f, 414
mosses, 417f
seed development stage, 502
totipotency, 798

A Brief History of Biology

Year	Name	Contribution
1628	William Harvey	Demonstrates that the blood circulates and the heart is a pump.
1665	Robert Hooke	Uses the word cell to describe compartments he sees in cork under the microscope.
1668	Francesco Redi	Shows that decaying meat protected from flies does not spontaneously produce maggots.
1673	Antonie van Leeuwenhoek	Uses microscope to view living microorganisms.
1735	Carolus Linnaeus	Initiates the binomial system of naming organisms.
1809	Jean B. Lamarck	Supports the idea of evolution by the inheritance of acquired characteristics.
1825	Georges Cuvier	Founds the science of paleontology and shows that fossils are related to living forms.
1828	Karl E. von Baer	Establishes the germ layer theory of development.
1838	Matthias Schleiden	States that plants are multicellular organisms.
1839	Theodor Schwann	States that animals are multicellular organisms.
1851	Claude Bernard	Concludes that a relatively constant internal environment allows organisms to survive under varying conditions.
1858	Rudolf Virchow	States that cells come only from preexisting cells.
1858	Charles Darwin, Alfred Wallace	Independently present evidence that natural selection guides the evolutionary process.
1865	Louis Pasteur	Disproves the theory of spontaneous generation for bacteria; shows that infections are caused by bacteria.
1866	Gregor Mendel	Proposes basic laws of genetics based on his experiments with garden peas.
1882	Robert Koch	Establishes the germ theory of disease and develops many techniques used in bacteriology.
1902	Walter S. Sutton, Theodor Boveri	Suggest that genes are on the chromosomes, after noting the similar behavior of genes and chromosomes.
1904	Ivan Pavlov	Shows that conditioned reflexes affect behavior, based on experiments with dogs.
1910	Thomas H. Morgan	States that each gene has a locus on a particular chromosome, based on experiments with Drosophila.
1924	Hans Spemann, Hilde Mangold	Show that induction occurs during development, based on experiments with frog embryos.
1929	Sir Alexander Fleming	Discovers the toxic effect of a mold product he called penicillin on certain bacteria.
1937	Konrad Z. Lorenz	Founds the study of ethology and shows the importance of imprinting in learning.
1937	Sir Hans A. Krebs	Discovers the reactions of a cycle that produces carbon dioxide during cellular respiration.
1940	George Beadle, Edward Tatum	Develop the one gene—one enzyme theory, based on red bread mold studies.
1944	O. T. Avery, Colin MacLeod, Maclyn McCarty	Demonstrate that DNA alone from virulent bacteria can transform nonvirulent bacteria.
1945	Melvin Calvin, Andrew A. Benson	Discover the individual reactions of a cycle that reduces carbon dioxide during photosynthesis.
1950	Barbara McClintock	Discovers transposons (jumping genes) while doing experiments with corn.
1952	Alfred D. Hershey, Martha Chase	Find that only DNA from viruses enters cells and directs the reproduction of new viruses.
1953	James Watson, Francis Crick, Rosalind Franklin	Establish that the molecular structure for DNA is a double helix.
1953	Harold Urey, Stanley Miller	Demonstrate that the first organic molecules may have arisen from the gases of the primitive atmosphere.
1954	Linus Pauling	States that disease-causing abnormal hemoglobins are due to mutations.
1954	Jonas Salk	Develops a vaccine that protects against polio.
1958	Matthew S. Meselson, Franklin W. Stahl	Demonstrate that DNA replication is semiconservative.
1961	Francois Jacob, Jacques Monod	Discover that genetic expression is controlled by regulatory genes.
1964	Marshall W. Nirenberg, Philip Leder	Produce synthetic RNA, enabling them to break the DNA code.
1973	Stanley Cohen	Uses recombinant DNA technique (genetic engineering) to place plant and animal genes in Escherichia coli.
1977	Carl Woese	Based on differences in ribosomal RNA sequences, proposes the three domain system of classifications.
1978	Peter Mitchell	Determines chemiosmotic mechanism by which ATP is produced in chloroplasts and mitochondria.
1989	Sidney Altman, Thomas R. Check	Independently discover that some RNA molecules can act as enzymes.
1990	R. Michael Blaese, W. French Anderson, Kenneth W. Culver	Develop procedure to infuse genetically engineered blood cells for treatment of immune system disorder—first gene therapy used in a human.
1997	Ian Wilmut	Clones an adult mammal for the first time.
2003	Human Genome Project	Complete human genome sequenced, creating push to discover genomic links to health and disease.
2010	Craig Venter	Leads team that develops the first synthetic life form.

Metric System

Unit and Abbreviation	Metric Equivalent	Approximate English-to-Metric Equivalents
Length		
nanometer (nm)	$= 10^{-9}$ m (10^{-3} µm)	
micrometer (µm)	$= 10^{-6}$ m (10^{-3} mm)	
millimeter (mm)	= 0.001 (10^{-3}) m	
centimeter (cm)	= 0.01 (10^{-2}) m	1 inch = 2.54 cm 1 foot = 30.5 cm
meter (m)	= 100 (10^{2}) cm = 1,000 mm	1 foot = 0.30 m 1 yard = 0.91 m
kilometer (km)	= 1,000 (10^{3}) m	1 mi = 1.6 km
Weight (mass)		
nanogram (ng)	$= 10^{-9}$ g	
microgram (µg)	$= 10^{-6}$ g	
milligram (mg)	$= 10^{-3}$ g	
gram (g)	= 1,000 mg	1 ounce = 28.3 g 1 pound = 454 g
kilogram (kg)	= 1,000 (10^{3}) g	= 0.45 kg
metric ton (t)	= 1,000 kg	1 ton = 0.91 t
Volume		
microliter (µl)	$= 10^{-6}$ l (10^{-3} ml)	
milliliter (ml)	$= 10^{-3}$ l = 1 cm^3 (cc) = 1,000 mm^3	1 tsp = 5 ml 1 fl oz = 30 ml
liter (l)	= 1,000 ml	1 pint = 0.47 l 1 quart = 0.95 l 1 gallon = 3.79 l
kiloliter (kl)	= 1,000 l	

°C	°F	
100	212	Water boils at standard temperature and pressure.
71	160	Flash pasteurization of milk
57	134	Highest recorded temperature in the United States, Death Valley, July 10, 1913
41	105.8	Average body temperature of a marathon runner in hot weather
37	98.6	Human body temperature
13.7	56.66	Human survival is still possible at this temperature.
0	32.0	Water freezes at standard temperature and pressure.

Units of Temperature

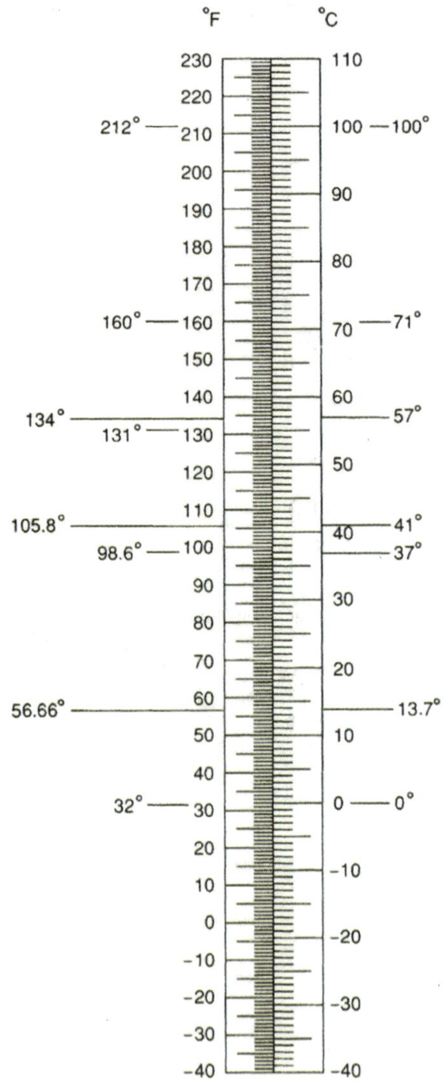

To convert temperature scales:

$$°C = \frac{(°F - 32)}{1.8}$$

$$°F = 1.8(°C) + 32$$